Introduction to Circuit Analysis and Design

Tildon H. Glisson, Jr.

Introduction to Circuit Analysis and Design

Prof. T. H. Glisson, Jr.
3072 EB-II, Box 7911
N.C. State University
Raleigh, NC 27695
glisson@ncsu.edu

The solution manual and other supplementary material are provided to those who are using or considering using the book as the textbook for a related course. To gain access to the solution manual and other supplementary material, go to http://www.springer.com/978-90-481-9442-1, click on the appropriate link under Additional Information on the right side of that page, and follow the instructions given there.

ISBN: 978-94-017-8074-2 ISBN: 978-90-481-9443-8 (eBook)

DOI: 10.1007/978-90-481-9443-8

Softcover re-print of the Hardcover 1st edition 2011

Printed on acid-free paper

springer.com

To Andy Sage, Ben O'Neal, and Sy Matthews. Each knows why.

Preface

This book is intended as the textbook for a first course in electric circuits. It has evolved from notes that I developed over several years, while I taught the introductory circuits course to sophomore electrical- and computer-engineering students at North Carolina State University.

As the chapter titles indicate, the major topics are those covered in almost all books of this kind. But selection and treatment of topics was guided not only by what is traditional for the target course, but also by the meaning and purpose of circuit analysis and (ultimately) design. This book reflects the views that circuit analysis is concerned primarily with finding and interpreting the *relation* a circuit establishes between excitation and response, that circuit design is concerned primarily with creating a circuit that enforces a specified *relation* between excitation and response, and that in design in particular, the devil is in the details. These perspectives led to somewhat different approaches to some aspects of the subject and to including some topics not treated in outwardly similar books. As a bonus, these topics give a practical flavor to and stimulate interest in some otherwise dry subjects.

Compared with other books for the same audience, this book gives more attention to transfer functions, input and output impedance, and two-port models; to power dissipation and power transfer, topics whose importance in practice is far greater than their treatment in most books would suggest; to the physical origins of capacitance and inductance, so students can gain some understanding of the origins of stray, residual, and parasitic capacitance and inductance, and how consideration of such effects influences component selection and placement; and to other ways in which physical components differ from idealized components, such as variation of parameters with temperature and frequency. Treatment of op amps is not limited to the ideal model, but covers topics such as output swing, output current limit, slew rate, bias-current compensation, power dissipation, stability, and gain-bandwidth product, all of which are important in any practical design. One unfamiliar with these topics cannot understand an op-amp data sheet or intelligently select an op amp for any particular application.

The book contains more than 1,100 problems and more than 800 examples. Many of the problems are assemblages of similar problems; for example, "find the input impedance and output impedance of each circuit in Fig. ___." So there are really many more than 1,100 problems. There is an adequate number of problems calling for numerical answers, through which students can gain confidence. But consistent with the focus on excitation-response relations, most problems ask the student to *obtain an expression for* a quantity or a *relation* among quantities. Some problems go further, and ask for an analysis or discussion of the influence of one or more parameters on a relation of interest. There is an adequate number of design problems in which there are fewer specifications than free parameters, requiring students to exercise some judgment.

Different instructors approach this subject in different but equally valid ways, so I will not presume to suggest a syllabus. I note only that the book contains ample material for a two-semester or three-quarter course, but can be used selectively in a one-semester or two-quarter course. Most chapters and the book as a whole are organized such that topics generally included in a one-semester course are covered first.

The only prerequisite for successful study of this book is facility with differential and integral calculus. Some exposure to electricity and magnetism, such as provided by a freshman physics course, might be helpful, but is not required. Some problems require use of a mathematics package such as *Matlab*© or *Mathcad*©, and some require use of a simulation package such as *Pspice*© or *Electronics Workbench*©. Problems requiring computer assistance are identified by the symbol 💻. There is no preference for or attempt to teach any particular software package, and it is possible to avoid these requirements altogether by simply not assigning such problems.

A companion website provides solutions to all exercises and errata and (for faculty members that adopt the book) solutions to all problems in the book, selected supplementary material, and a mechanism for submitting corrections, suggestions, and comments.

I have many to thank for whatever is good about the book, and only myself to blame for everything else. Thanks to James Kang and Bruno Osorno for their helpful comments on an early draft; to Art Davis, John Hauser, S.C. Dutta Roy, Joel Trussell, and Gary Ybarra for their thorough and helpful reviews; to Chris Lunsford for his contributions to several chapters; to Hari Chandrasekar, Vinodh Kotipalli, Misha Kumar, Christina Lee, and Satish Naidu, all excellent graduate students who spent many hours proofreading the text and checking my solutions to examples, exercises, and problems; and to all the companies, publishers, and individuals that permitted me to use or adapt parts of their intellectual property. Thanks to Jeff Kahler of Nuhertz Technologies and Siegfried Linkwitz for allowing me to use several of their designs in examples and problems. Special thanks to Art Davis and John Hauser for many correspondences and discussions that helped clarify or otherwise refine my presentations of various topics. Special thanks also, to those at Springer: Mark de Jongh, Senior Publishing Editor for Electrical Engineering; Cindy Zitter, his Senior Assistant; and Project Manager R. Samuel Devanand, for being so capable and helpful.

Raleigh, North Carolina Tildon H. Glisson Jr.

Contents

Chapter 1
Introduction

Electrical engineering is among the largest and most diverse professions. The *Institute of Electrical and Electronic Engineering* (IEEE), which is the principal technical and professional organization for electrical engineers, is the largest professional organization in the world. Within the IEEE alone, there are more than 40 technical societies, each focused on a sub-area of electrical engineering. These are listed in Table 1.1.

Electrical engineers are employed in every major industry, working not only with other electrical engineers, but also with people from all walks of life. Their work ranges from very applied to highly theoretical. They work indoors and out, in government and in private industry, alone and in large groups. They contribute to the design and manufacture of virtually every product you can name. In short, almost anyone inclined toward engineering can find an exciting, fulfilling, and rewarding career in electrical engineering. You have chosen well.

1.1 Electric Circuits

The one thing that almost all electrical engineers have in common is the fact that they design, manufacture, maintain, teach about, or sell devices, equipment, or systems whose operation depends *primarily* upon *manipulation of electricity*. In such systems, electricity is the primary means for conveying or converting energy, transmitting or processing data, monitoring or controlling other equipment or processes, or measuring or observing various phenomena. Manipulating electricity toward useful ends is what circuits are all about. Thus it is no accident that almost all introductory EE courses and textbooks focus on circuits.

In the broadest sense, a circuit is something that makes electricity do something useful. Circuits in computers use electricity to store, retrieve, transmit, and process data at unimaginable speed. Circuits deliver tremendous quantities of electrical energy to our homes and industries at bargain-basement prices. Circuits allow us to transmit and receive hundreds of radio and TV programs. Circuits make our automobiles run cleaner and more efficiently. They help us store and prepare our food, wash our clothes, and cool our homes. They help aircraft, ships, and space shuttles navigate. They help us defend our borders and warn us of impending danger. They help monitor and predict the weather and help physicians monitor our health and perhaps save our lives. It is difficult to think of anything we do in which we are not assisted by one or more electric circuits. Certainly there are other important subjects in EE, but without circuits, most of those subjects would either not exist or have no practical application. We might be able to imagine a world without circuits, but few of us would want to live there. All in all, study of circuits is an excellent way to begin study of electrical engineering.

The principal quantities of interest in circuit analysis and design are current, voltage, and power, none of which is visible.[1] For the most part, not even the effects of currents and voltages in the innards of a circuit are visible. Indeed, without a powerful microscope, we cannot even see the innards of many modern circuits. We must rely on measurements and mathematics to tell us whether a circuit is working as it should. Thus it might appear that circuit analysis and

[1]Current in an arc is visible, but unless we are designing an electric welder, we usually try to avoid arcs.

T.H. Glisson, *Introduction to Circuit Analysis and Design*,
DOI 10.1007/978-90-481-9443-8_1, © Springer Science+Business Media B.V. 2011

Table 1.1 IEEE technical societies

IEEE Aerospace and Electronic Systems Society	IEEE Instrumentation and Measurement Society
IEEE Antennas and Propagation Society	IEEE Lasers & Electro-Optics Society
IEEE Broadcast Technology Society	IEEE Magnetics Society
IEEE Circuits and Systems Society	IEEE Microwave Theory and Techniques Society
IEEE Communications Society	IEEE Nanotechnology Council
IEEE Components Packaging, and Manufacturing Technology Society	IEEE Nuclear and Plasma Sciences Society
IEEE Computational Intelligence Society	IEEE Oceanic Engineering Society
IEEE Computer Society	IEEE Photonics Society
IEEE Consumer Electronics Society	IEEE Power Electronics Society
IEEE Control Systems Society	IEEE Power and Energy Society
IEEE Council on Superconductivity	IEEE Product Safety Engineering Society
IEEE Dielectrics and Electrical Insulation Society	IEEE Professional Communication Society
IEEE Education Society	IEEE Reliability Society
IEEE Electromagnetic Compatibility Society	IEEE Robotics & Automation Society
IEEE Electron Devices Society	IEEE Sensors Council
IEEE Engineering Management Society	IEEE Signal Processing Society
IEEE Engineering in Medicine and Biology Society	IEEE Society on Social Implications of Technology
IEEE Geoscience & Remote Sensing Society	IEEE Solid-State Circuits Society
IEEE Industrial Electronics Society	IEEE Systems, Man, and Cybernetics Society
IEEE Industry Applications Society	IEEE Ultrasonics, Ferroelectrics, and Frequency Control Society
IEEE Information Theory Society	IEEE Vehicular Technology Society
IEEE Intelligent Transportation Systems Society	

design is inherently more abstract than, for example, mechanism analysis and design. We can see mechanisms operate, whereas most electric circuits look the same whether operating or not. Unless something is seriously wrong, the components of a circuit do not jump about, make clanking noises, emit smoke, or squirt messy fluids here and there.

But in truth, the level of abstraction of circuit analysis and design relative to that of other subjects is overstated. It is true that we cannot see current or voltage, but we cannot see force, pressure, or temperature, either. Nor can we see a mechanism we have not yet designed. Any extraordinary difficulties with circuits probably arise more from lack of familiarity than level of abstraction. Almost everyone has some intuitive understanding of fluid flow and pressure and the functions of pipes, pumps, and valves. But few laypersons have an equally well-developed intuition regarding current and voltage and the functions of circuit components such as resistors, transistors, and capacitors.

In the beginning, analogies can be helpful. Current is flow of electric charge and is analogous to fluid flow (flow of molecules) in a hydraulic system. Voltage is analogous to pressure, in that voltage can cause current in a wire, much like pressure can cause fluid flow in a pipe. Circuit components and circuits themselves are described in terms of the current-voltage relationships they establish at their terminals, just as hydraulic components are described in terms of flow and pressure at their terminals (inlets and outlets); for example, a pump can be described by the flow it can provide at one or more pressures (e.g., 3 gal/min at 100 lb/in^2) and a generator can be described by the current it can provide at the specified terminal voltage (e.g., 50 A at 120 V). If we know the flow-pressure characteristics of each component in a hydraulic system, we can calculate flow and pressure anywhere in the system. Likewise, if we know the current-voltage characteristics of each component in a circuit, we can calculate current and voltage anywhere in the circuit.

Although analogies can be helpful, they also can be misleading if carried too far. There are circuits and devices in which current does not behave like fluid flow. Also, circuits can do many things that hydraulic systems cannot (and vice versa), in which case there is no analogy to fall back on. Ultimately, if we are to design circuits, we must come to understand current, voltage, and circuits on their own terms.

When you have mastered the material presented in this book, you will have a basic understanding of what current and voltage are and how they behave. You will be able to analyze and design a number of basic (but important) kinds of circuits. You will have accumulated a repertoire of physical laws and mathematical procedures that are essential in study of more advanced circuits and systems for communication, control, computing, and other applications. You will have begun to speak the language of electrical engineering, which is the first step toward becoming an electrical engineer and enjoying a challenging and rewarding career.

1.2 How to Study This Book

Because this is presumably one of your first engineering courses, you might be receptive to some advice about how to study an engineering subject – and this book, in particular. Most engineering subjects have pretty much the same anatomy, illustrated by Fig. 1.1:

- At the core are (usually very few) **undefined quantities**; for example, time and charge, and essential properties of those quantities.[2]
- Next are **definitions**. These include definitions of important quantities, such as current and voltage, in terms of the core undefined quantities. Also included here are important symbols, terminology, and notation associated with the subject.

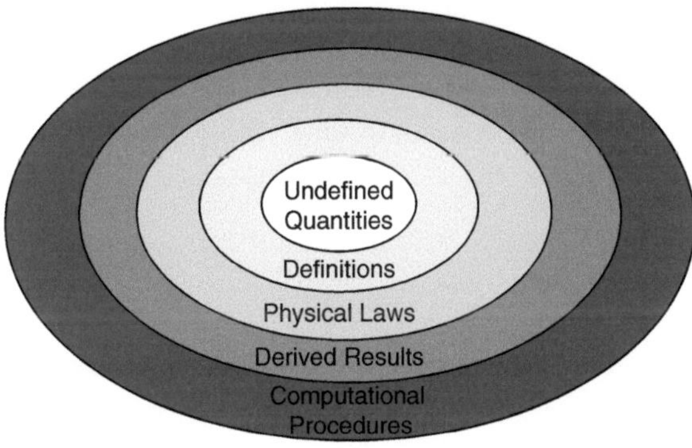

Fig. 1.1 Anatomy of a technical subject

- **Physical laws** prescribe relations among defined quantities; for example, in circuits, important physical laws are Ohm's law and Kirchhoff's laws.
- **Derived results** are important relations among the defined quantities that are obtained using physical laws, definitions, and mathematics; Thévenin's theorem (Chapter 4) is an example of an important derived result.
- **Computational procedures** systematically use physical laws, derived results, and mathematics to solve problems (e.g., to analyze or design circuits). Examples are the node-voltage and mesh-current methods for finding voltages and currents in circuits.

As you make your way through this book, note whether each new thing you encounter as a definition, a law, a derived result, or a procedure. Establish in your mind the relation of the new thing to things learned previously.

- *You should memorize each definition, when it is first encountered, before proceeding.* If you do not, chances are you will have difficulty understanding subsequent text, derivations, and examples. Definitions are given for a reason, the reason being frequent subsequent use of the thing defined. Special symbols comprise an important class of defined terms and also should be memorized; for example, in this book the letters i and v always stand for current and voltage, respectively. Certain dimensions and units also are defined quantities and must be committed to memory.
- *You must also memorize physical laws*, because they do not necessarily follow from anything else presented in this book.[3] Fortunately, there are relatively few laws to be memorized.
- Derived results are obtained using definitions, laws, limiting assumptions, and other derived results. You should be sure you understand important limiting assumptions before proceeding. In almost all subjects, there are many more derived results than physical laws. Your teacher can guide you regarding which derived results you should memorize and which you should be able to derive, yourself, possibly under slightly different conditions.

[2]The essential property of time is perhaps best captured by Yogi Berra's definition: "Time is what keeps everything from happening all at once." The essential properties of charge are given in Chapter 2.

[3]One subject's laws can be another subject's derived results; for example, Ohm's law (a law in circuits courses) is a derived result in some more fundamental courses.

- A procedure is essentially a recipe for solving a class of problems, based upon certain laws and derived results. The class of problems to which a procedure can be applied is defined by the conditions under which the underlying laws and derived results apply. For example, mesh analysis, which is a procedure for finding voltages and currents in a circuit, is based upon Kirchhoff's voltage law, which in turn is applicable only under certain restrictions on the natures (e.g., wavelengths) of the voltages and currents applied to the circuit.

Exercise 1.1. Recall from your study of physics that electric charge is conserved. In absence of any further information, would you regard this claim as a definition, a physical law, or a derived result?

Viewing the content of a subject in the manner illustrated by Fig. 1.1 can help you organize your notes and thinking, especially at the end, when you are preparing for a final examination. From the outset, it can help you plan your outline and identify what is truly important or general and what is less important or specific as you proceed through the text. Before each exam, for example, you might be sure that you can answer the following questions:

- What quantities have been defined or described and what are the definitions or defining properties? Include symbols, terms, and basic units.
- What fundamental laws have been presented and what relations do they establish among defined quantities?
- What derived results have been obtained and under what conditions do they apply? What are the dimensions and units of derived quantities and how are they related to the dimensions and units of quantities defined or derived previously? What laws and mathematical operations are needed to derive the results?
- What procedures have been presented and under what conditions are they useful? What laws, derived results, and mathematical operations do they use?
- What is a paradigm example[4] for each law, derived result, and procedure?

[4]A *paradigm example* is one that illustrates application of a law, derived result, or procedure in the simplest possible setting, uncluttered by need for other laws, results, or procedures.

If you can do all that, you can be confident that you have a good grasp of the material. The only thing left to do is to hone your skills through practice; e.g., by reworking homework problems and other problems that are like those your teacher assigned or recommended. As you work through examples and problems, attempt to discern what definitions, laws, derived results, and procedures the example or problem is intended to reinforce.

1.3 Dimensions and Units

A **quantity** is something that is *quantifiable*; i.e., something that can be counted, measured, or calculated. In everyday life and especially in engineering and science, we deal with many different quantities; for example, time, length, weight, temperature, pressure, speed, area, and cost. A quantity has a numerical value and a *dimension* and is expressed by a number and a *unit*. Both the numerical value and the unit are necessary. One dollar and 1 yen have the same *numerical* value (one) but have quite different values in the marketplace.

A quantity can be expressed in various units; for example, length can be expressed in inches, meters, miles, furlongs, yards, angstroms, rods, light-years, and other units. No matter how it is expressed, length remains length. Length is *what* is being measured or specified. The unit describes *how* (on what scale) length is expressed. Length is a dimension and the meter is a unit of length. Similarly, time is a dimension and the second is a unit of time.

Exercise 1.2. Which of the following are quantities (dimensioned or dimensionless), which are units, and which are neither? (a) Time, (b) centrifugal force, (c) electron-volt, (d) Newton-meter, (e) light-year, (f) age, (g) checking account balance, (h) purple, (i) longitude, (j) relative humidity, (k) north pole, (l) virtue, (m) valor, (n) micron.

There are many different quantities, but only seven **basic dimensions** (basic quantities). The dimensions of all other known quantities can be expressed in terms of these seven and are called **derived dimensions**.

Derived dimensions and units also are called *compound dimensions* and *compound units* because they are products and quotients of basic dimensions and units; for example, the dimension of area is length squared and the dimension of speed is length/time. The dimensions of area and speed are derived (compound) dimensions.

Choosing the seven basic dimensions, the associated basic units, and names of units for important derived quantities establishes a *system of units*. The official system of units for engineering and science is the SI system,[5] which was adopted by a number of very important people at a long meeting in Paris in 1960. In the SI system, the seven basic dimensions are length, time, mass, electric current, temperature, amount of substance, and luminous intensity.[6] So far as we know, none of the seven can be expressed in terms of the other six. Of the seven basic quantities, only five are used in this book. These five and the associated SI units are given in Table 1.2.

Note that electric current is the basic electrical quantity in the SI system. Electric charge would serve as well, and because current in a circuit is flow of charge (charge per unit time), charge would seem to be the more logical choice. But there is a good reason for choosing current, rather than charge, as a fundamental quantity. It turns out that there are two kinds of current: *Conduction current*, which is flow of charge, and *displacement current*, which can exist where no charge is present, as in a time-varying electromagnetic field (e.g., a radio wave).

Table 1.2 Basic SI quantities and units

Quantity (basic dimension)	Name of SI unit	SI abbreviation (symbol) for unit
Length	Meter	m
Time	Second	s
Mass	Kilogram	kg
Electric current	Ampere	A
Temperature	Kelvin	K

[5]*SI System* is redundant, because SI stands for Systeme Internationale, which (obviously) is French for international system. Nonetheless, "the SI system" is what everyone calls it.

[6]Actually, which seven dimensions are defined as basic is a matter of choice. Any seven independent dimensions would do, but history and common sense have ruled in favor of the seven given here. There is much wider agreement on which seven quantities are basic than on the standard units for those seven.

Thus, from an engineering perspective, current appears to be a more basic quantity than charge.

In any particular discipline, many quantities of interest are derived quantities; for example, in EE, we deal not only with current and time, but also with voltage, work (or energy), power, and many other quantities. Repeated use of compound units for all these quantities, such as kg m^2 s^{-2} for energy, is cumbersome. Compound units for important derived quantities are given their own names, most of which are chosen in honor of great engineers or scientists somehow associated with the quantity. Table 1.3 gives the names, symbols, and SI units for quantities (both basic and derived) used in this book, in alphabetical order. Table 1.3 also gives compound equivalents for derived units. Compound equivalents are useful in checking expressions for dimensional correctness (see below). Quantities used only occasionally in this book are not included in the table but are defined and discussed appropriately where they are introduced.

There are two kinds of **dimensionless quantities**: (1) Those whose definitions do not involve any dimensioned quantities, such as atomic number and valence; and (2) those defined in terms of dimensioned quantities, but in such a way that all of the basic units involved cancel, such as the fine-structure constant and an angle. We use angles extensively in this book, so a little discussion of their nature is in order.

Fundamentally, an angle is the quotient of two lengths (arc length/radius). The dimensions of length cancel from the quotient, so an angle is dimensionless (the compound unit of an angle is unity). Most dimensionless quantities are expressed without units. Angles are exceptions because an angle can be specified in either radians (rad) or degrees (deg), where $\pi\ rad = 180$ deg. Although dimensionless, radians and degrees are different scales of measurement. In absence of any prior agreement, we must attach the correct one of those units to the numerical specification of an angle. *Except in Chapter 14, we specify angles in radians*, so we do not attach the dimensionless unit *rad* to any such specification. In this book, specifying an angle as (e.g.) $\theta = 1.47$ means $\theta = 1.47$ radians.

Exercise 1.3. Give a simpler unit for each of the following: (a) J s^{-1}, (b) V A s, (c) C s^{-1}, (d) A^2 Ω, (e) Ω H F^2 s^{-2}, (f) kg m^2 s^{-2}.

Table 1.3 Symbols and SI units. Names of basic dimensions are in **bold** type

Quantity	Symbol(s)	SI unit (symbol)	Equivalent unit(s)
Admittance	Y	siemen (S)	Ω^{-1}, A V^{-1}
Angle	$\theta, \phi, \alpha, \beta, \Phi, \Theta$	radian (rad)	None
Angular frequency	ω	radians per second (rad/s)	s^{-1}
Capacitance	C	farad (F)	AsV^{-1}. CV^{-1}
Charge	q	coulomb (C)	A·s
Complex frequency	s	radians per second (rad/s)	s^{-1}
Complex power	S	volt·ampere (VA)	J s^{-1}, W
Conductance	G, g	siemens (S)	Ω^{-1}, A V^{-1}
Conductivity	σ	siemens per meter (S m^{-1})	$(\Omega m)^{-1}$
Current	i, I	ampere (A)	Basic dimension; Cs^{-1}, $V\Omega^{-1}$
Distance or length	x, y, z, h, d, l	meter (m)	Basic dimension
Electric potential	Φ	volt (V)	J C^{-1}
Electric field strength	E	newtons per coulomb or volts per meter	NC^{-1} or Vm^{-1}
Energy	w	joule (J)	V C, N m, W s
Force	f, F	newton (N)	kg m s^{-2}, V C m^{-1}
Frequency	f	hertz (Hz)	s^{-1}
Impedance	Z	ohm (Ω)	V A^{-1}
Inductance	L	henry (H)	V s A^{-1}, WbA^{-1}
Magnetic flux	ϕ	weber (Wb)	Vs
Mass	m	kilogram (kg)	Basic dimension
Period or interval	T	second (s)	S
Permeability	μ	henries per meter (H m^{-1})	V s A^{-1} m^{-1}
Permittivity	ε	farads per meter (F m^{-1})	A s V^{-1} m^{-1}
Pole	p	radians per second	s^{-1}
Power	p, P	watt (W)	J s^{-1}, V A
Reactance	X	ohm (Ω)	V A^{-1}
Reactive power	Q	volt ampere reactive (VAR)	J s^{-1}, V A, W
Resistance	R, r	ohm (Ω)	V A^{-1}
Resistivity	ρ	ohm meter (Ω m)	V A^{-1} m
Speed or velocity	u	meters per second (m s^{-1})	m s^{-1}
Susceptance	B	siemen (S)	Ω^{-1}, A V^{-1}
Temperature	T	kelvin (K)	Basic dimension
Time	t	second (s)	Basic dimension
Time constant	τ	second (s)	S
Voltage	v, V	volt (V)	J C^{-1}, A Ω
Work	w	joule (J)	V C, N m, W s
Zero	z	radians per second	s^{-}

There are several important conventions and rules regarding dimensioned quantities:

- *The symbol for a quantity stands for both the numerical value and the associated unit*, as in $x = 10$ m and $t = 5$ s. Constructions such as "x meters" and "t seconds" are redundant.
- Reserved symbols for quantities (e.g., t for time) carry subscripts where more than one such quantity is of interest. For example, times at which various events occur in a circuit might be designated t_0, t_1, t_2, and so on. The presence of a subscript does not alter the dimension or unit of a quantity.
- Terms that are equated, compared (using $<$ or $>$), added, or subtracted must have the same dimension and, if numerical, must be expressed in the same unit.
- In particular, *all equations must be dimensionally homogeneous*; that is, all terms in an equation must have the same dimension.
- The dimension of the product of quantities is the product of the dimensions of the quantities.
- The dimension of the quotient of quantities is the quotient of the dimensions of the quantities. The quotient of like quantities is dimensionless.

- The dimension of a differential is that of the associated variable; e.g., the dimension of dt is that of t.
- Dimensionally, a derivative is a quotient and an integral is a product; e.g., the dimension of dq/dt is that of q/t and the dimension of $\int i\,dt$ is that of $i{\cdot}t$.
- Definite limits of integration must have the dimension of the differential; for example, in $\int_a^b f(x)\,dx$, both a and b must have the dimension of x. A similar condition applies to limits in general.
- The units of infinite limits of integration ($\pm\infty$) are understood to be those of the differential; e.g., in $\int_{-\infty}^{t_1} i(t)\,dt$ it is understood that $-\infty$ stands for an exceedingly large and negative *time*. A similar condition applies to limits in general.
- With very few exceptions, arguments of functions must be dimensionless; for example, we cannot take the logarithm of 50 kg. The reason why is that $\ln(x)$ can be expressed as an infinite power series, and we cannot (e.g.) add kg to kg^2 to kg^3 . . . .[7]
- Finally, voltage, current, power, and time, if zero or infinite, are zero or infinite in any system of units. We do not attach units to those quantities when they are zero or infinite; for example, we say $t\to\infty$, not $t\to\infty$ s, because if time approaches infinity, it does so whether measured in seconds, minutes, or weeks.[8] Similarly, we say $v = 0$, not $v =$ 0 V, because if a voltage equals zero, it is zero in any system of units.

It is poor practice to write statements such as "$v = v_0 e^{-t}$, where time t is expressed in milliseconds." Such practice makes it impossible to check relations for dimensional consistency and thereby robs us of an opportunity to catch careless errors. Further, such practice robs us of many opportunities for developing insight; e.g., by hiding the all-important time constant in the exponential relation above. It is far better to adopt the convention that *t always stands for time in seconds* and to write the statement above as "$v = v_0 e^{-t/\tau}$, where $\tau = 1$ ms."

[7]Of course, the coefficients of a Taylor series can be dimensioned, and one could define different series for (e.g.) the logarithm of different physical quantities. Such an approach would be cumbersome, at best.

[8]Some mathematical software (e.g., Mathcad) that allow units to be used insist that they be attached to all quantities, even those having zero magnitude. For example, if you ask Mathcad to compare 10 s to 0, you will get an error. You must compare 10 s to 0 s.

Units and dimensions are not only essential parts of the values of physical quantities, but also provide insight into the meaning and reasonableness of derived results. Sometimes they even guide derivations of new relations. They also provide a powerful means of checking correctness of equations and answers. An equation that is dimensionally inconsistent is incorrect. There is no reason to proceed with a solution until the error is found and corrected.

Make it a habit to check all derived equations and relations for dimensional consistency. An equation is dimensionally consistent if all terms have the same unit. We use SI units exclusively, so you may check units instead of dimensions, which is somewhat easier, because it is easier to remember the units of quantities than it is to remember the (usually compound) dimensions for the quantities. We use $\mathrm{SI}(\beta)$ to denote the SI unit of a quantity β, where

$$\mathrm{SI}(\beta) = \begin{cases} \text{the SI unit of } \beta, \text{ if } \beta \text{ is dimensioned,} \\ 1, \text{ if } \beta \text{ is dimensionless,} \end{cases} \tag{1.1}$$

and

$$\mathrm{SI}(\alpha\beta) = \mathrm{SI}(\alpha)\ \mathrm{SI}(\beta), \qquad \mathrm{SI}\left(\frac{\alpha}{\beta}\right) = \frac{\mathrm{SI}(\alpha)}{\mathrm{SI}(\beta)}. \tag{1.2}$$

Example 1.1. We wish to show that the dimension of the quantity RC (resistance times capacitance) is that of time. From Table 1.3,

$$\mathrm{SI}(RC) = \mathrm{SI}(R)\mathrm{SI}(C) = \left(\mathrm{VA}^{-1}\right)\left(\mathrm{A}\cdot\mathrm{sV}^{-1}\right) = \mathrm{s}.$$

Thus the unit of the quantity RC is seconds (s) and the dimension of RC is time.

Example 1.2. We wish to show that the expression

$$i = C\frac{dv}{dt}$$

is dimensionally consistent, where i is current, C is capacitance, v is voltage, and t is time.

Current i stands alone on the left, so we seek to reduce the unit of the term on the right to the ampere. From Table 1.3

$$\mathrm{SI}\left(C\frac{dv}{dt}\right)=\mathrm{SI}(C)\frac{\mathrm{SI}(v)}{\mathrm{SI}(t)}=\left(\mathrm{AsV}^{-1}\right)\left(\frac{\mathrm{V}}{\mathrm{s}}\right)=\mathrm{A}.$$

Thus the equation is dimensionally consistent.

Exercise 1.4. Show that the relation

$$i=\frac{1}{L}\int_{-\infty}^{t}v(t')dt'$$

where i is current, L is inductance, v is voltage, and t is time, is dimensionally consistent.

Exercise 1.5. Check each relation for dimensional consistency.

$$\text{(a)}\,L\frac{di}{dt}+Ri-\frac{1}{RC}\int_{-\infty}^{t}i(t')dt'=5v_1,$$

$$\text{(b)}\,\frac{V}{I}=\frac{\sqrt{R^2+(\omega L)^2}}{1+\omega RC}.$$

As an aside, the SI system includes no unit for weight. Weight is force due to gravity and is approximately a fixed multiple of mass everywhere on the earth's surface, given by $f=mg$, where m is mass and g is acceleration due to gravity. Thus, on the earth's surface, mass serves as well as weight to express how heavy one thing is relative to others; that is, if one thing has twice the mass of another, it also has twice the weight of the other (in the same location). Consequently, there is really no need for a quantity called weight. Mass serves just as well. Conversions from mass in kilograms to weight in pounds build in the acceleration due to gravity (approximately 9.8 m s^{-2}) and are really conversions from mass to force on the mass due to gravity. The conversion factor works out to be about 0.454 kg lb^{-1}, so 1 kg (mass) of butter weighs about 2.2 lb.[9]

The size (scale) of a unit often is inappropriate to a particular problem or even to an entire discipline; for example, the meter is much too large a unit to use in atomic physics and much too small a unit to use in astrophysics. Even though the SI system is the official system of units, other units that have proved useful remain in use (e.g., the angstrom and the light-year). Also, engineers must communicate effectively with those in other professions and trades, not all of whom have embraced the SI system so warmly, and it often is necessary to convert one unit of measure to another (e.g., meters to inches). We use the SI system almost exclusively in this book so we omit a table of conversion factors. Such tables can be found in most standard handbooks.[10]

In the SI system, very large multiples and very small fractions of units are made more manageable by using named prefixes that denote multiplication by a power of ten. The power of ten denoted by most prefixes is a positive or negative whole multiple of three. Those that are not are not widely used in electrical engineering, except for the centimeter (0.01 m) and angstrom (10^{-10}m). Table 1.4 gives selected SI prefixes.

Table 1.4 SI prefixes

Prefix	Abbreviation	Value
Femto	f	10^{-15}
Pico	p	10^{-12}
Nano	n	10^{-9}
Micro	μ	10^{-6}
Milli	m	10^{-3}
Centi	c	10^{-2}
Kilo	k	10^{3}
Mega	M	10^{6}
Giga	G	10^{9}
Tera	T	10^{12}
Peta	P	10^{15}

[9]The abbreviation lb stands for *libra*, an ancient Roman unit of weight which was presumably the predecessor to the English pound. See Yunus A. Cengel and Michael Boles, *Thermodynamics: An Engineering Approach* (3rd Ed.), McGraw-Hill, 1998.

[10]An especially complete table of conversion factors is given in the *Handbook of Chemistry and Physics* (76th Ed.), edited by David R. Lide, CRC Press, New York, 1995.

There are three basic rules regarding use of prefixes:

- Prefixes are not to be repeated or used with other prefixes; e.g., use pF, not μμF and ns, not μms.
- Where a prefix is used with a symbol, the prefix and the symbol become one quantity that can be raised to a power or used in a fraction without parentheses; for example, $1\ \text{cm}^2 = 10^{-4}\text{m}^2$ and 4 m/ms = $4000\ \text{m s}^{-1}$.
- Prefixes are never used alone, even for purely numerical quantities.

These rules are designed to eliminate ambiguity; for example, although the symbol m for meter and the prefix m for milli are identical, there is no possibility for confusion because m for milli is never used alone or repeated (mm always stands for millimeter, never milli-milli).

Nonetheless, there remain possibilities for confusion. For example, does mV mean millivolt or meter-volt? In this case, we can avoid confusion by writing mV for millivolt and Vm for volt-meter. But what about Vms? Does it mean volt-millisecond or volt-meter-second? In such cases we can use spaces, hyphens, dots, or parentheses to avoid ambiguity. For example, Vms means volt-millisecond and Vm s, Vm-s, Vm s, and (Vm)s all mean volt-meter-second.

There is no need (at present) to memorize any entries in the tables above. You will learn many of them through repeated use in subsequent sections. But you should learn the three rules above and become proficient at checking dimensions (units) using the procedure illustrated in the examples above. Again, you should form the habit of checking units with reasonable frequency as you work through a problem. Doing so will help you catch many careless errors and help you avoid the fruitless labor of solving incorrect equations.

Exercise 1.6. Express each of the following without prefixes: (a) $100\ \mu\text{As}^{-1}$ (b) $10\ \text{V ms}^{-1}$, (c) 100 kV A, (d) $10\ \text{mm}\ \mu\text{s}^{-2}$.

Exercise 1.7. Using prefixes, express each of the following in at least two other ways: (a) $5\ \text{mVs}^{-1}$, (b) 25 kAμs, (c) $100\ \text{mJms}^{-1}$.

Exercise 1.8. Specify whether each m stands for meter or milli in each of the following: (a) Vms^{-1}, (b) m Ams^{-1}, (c) Nm·s^{-1}, (d) m^2s^{-1}, (e) mms^{-1}.

1.4 Symbols and Notation

Symbols for quantities and units used in this book are those in Table 1.3. Because the alphabet is finite, some symbols have more than one meaning. To the extent possible, we avoid conflicts as follows:

- Symbols for quantities are always in *italics* and symbols for units are always in roman; e.g., C (italic) is the symbol for capacitance (a quantity) and C (roman) is the symbol for the coulomb (the SI unit of charge). This convention is not entirely satisfactory because it is difficult to distinguish italic and roman letters in handwritten notes and assignments. Where conflicts arise (and confusion is possible), we suggest using an underscore to denote roman (e.g., $\underline{\text{C}}$ for coulomb).
- Symbols for complex representations of currents and voltages are printed under a tilde; for example, $\tilde{I}$ and $\tilde{V}$ denote complex quantities called *phasors*, defined in Chapter 12, and used to represent sinusoidal currents and voltages. We use a tilde in these cases because V and I (no tilde) denote associated real quantities (the magnitudes of the corresponding complex quantities).
- We rely on context when there is no tolerable alternative; for example, the symbols for period and temperature are identical, and we must rely on context to tell us what T stands for.
- The principal quantities of interest in circuit analysis and design are time, current, voltage, work, and power. In this book, t always stands for time and i and v and their italic capitals always stand for current and voltage, respectively.
- To avoid clutter, explicit time dependence often is omitted; for example, v and i alone denote $v(t)$ and $i(t)$, respectively and dv/dt stands for $dv(t)/dt$.

The mathematical notation used in the text is generally that used in standard prerequisite mathematics

Table 1.5 Notation used in the text

Symbol	Meaning
$\to$	Approaches
$\cong$	Is approximately equal to
$\equiv$	Is identically equal to
$\Rightarrow$	Implies or yields
$\Leftrightarrow$	Implies and is implied by
iff	if and only if
e.g.	For example
i.e.	Namely, or that is, or in other words

courses. Table 1.5 gives symbols and abbreviations used throughout the text. For example, we might write that $3x + 4 = 10 \Rightarrow x = 2$, which reads as "three x plus four equals ten *implies* that (or *yields*) x equals two."

Equations are numbered only for cross-reference. A numbered equation is not necessarily an important equation.

Boldface type is used when an important quantity or term is introduced or defined. *Italic* type is used to highlight secondary definitions, identify important terms, and emphasize important comments.

As is standard in technical works, we use a number of Greek letters. Table 1.6 gives the Greek alphabet and the usual meaning (if any) assigned to each Greek letter in this book. For the most part, these choices reflect the meanings assigned to symbols in practice. Some (such as δ) have multiple meanings, whereas others are not used. Also, the usual meaning is not necessarily the only meaning used in this book. We occasionally assign a meaning not given in Table 1.6 for temporary use in a particular development or section.

1.5 Symbols Versus Numbers

In general, *symbolic relations* among quantities are much more valuable than particular numerical values. For example, Newton's law $f = ma$ is much more valuable and informative than would be a table of particular values for force, mass, and acceleration. The symbolic relation $f = ma$ tells us that acceleration is proportional to force, whereas knowing only that a 1 kg mass subjected to a 2 N force experiences an acceleration of 2 ms^{-2} is of little general value. In engineering and especially in conceptual design, *relations* among quantities are of far more interest than particular numerical values for those quantities. Indeed, much of electrical engineering focuses on designing things (such as circuits) that create desirable *relations* among physical quantities, such as voltages.

Resist the urge to prematurely replace quantities by their numerical values, even when those values are known and fixed. Instead, form the habit of working with symbols until the desired relation is obtained, checking units along the way. Only when you are certain you have a correct relation should you seek a particular numerical answer (with correct units), if that is what is called for in the problem. Jumping to numbers too quickly eliminates the possibility of checking dimensions, greatly limits the implications that can be drawn from the solution, and multiplies the number of different problems that must be solved by a large factor. To encourage you to use symbols and symbolic relations and to help you become proficient in that regard, many of the problems in this book ask you to *obtain relationships among* or (symbolic) *expressions for* quantities.

Also, put your years of mathematics study to good use. Assign symbols to quantities that occur frequently in a development. For example, if you find the quantity $\omega L/R$ occurs repeatedly in a particular development, you might define $Q = \omega L/R$ to simplify further manipulations. Where possible, use a symbol that suggests the dimension of the quantity; for example, if the quantity RC occurs repeatedly in a particular analysis, you might let t_0 or τ denote RC, because the dimension of RC is time.

1.6 Presentation of Calculations

Intermediate steps in the calculation of a value from a symbolic expression can be presented in basically two ways: The first (and generally best for students) is to exhibit both the value and the unit of each quantity in the expression. The second is to simply exhibit values, attaching the correct unit to only the final expression. For example, suppose we must calculate the value of

$$|Z| = \sqrt{R^2 + (2\pi f L)^2},$$

where $R = 1\,\mathrm{k\Omega}$, $f = 10$ kHz, and $L = 10$ mH. Any of the following presentations would be considered

Table 1.6 Greek alphabet and common uses for characters

Lower-case	Name	Usual meaning	Upper-case	Usual meaning
α	Alpha	Peaking factor	A	Not used
β	Beta	Tolerance	B	Not used
χ	Chi	Not used	X	Not used
δ	Delta	skin depth; damping factor; loss angle	Δ	Increment prefix; e.g., Δx means a small increase in a quantity x
ε	Epsilon	Error, permittivity, energy	E	Not used
ϕ	Phi	Angle	Φ	Angle, electric potential
γ	Gamma	Ripple factor	Γ	Not used
η	Eta	Efficiency	H	Not used
ι	Iota	Not used	I	Not used
κ	Kappa	Not used	K	Not used
λ	Lambda	Wavelength	Λ	Not used
μ	Mu	Permeability; voltage gain	M	Not used
ν	Nu	Not used	N	Not used
o	Omicron	Not used	O	Not used
π	Pi	3.14159...	Π	Product
θ	Theta	Angle	Θ	Angle
ρ	Rho	Resistivity	P	Not used
σ	Sigma	Conductivity; damping ratio	Σ	Sum
τ	Tau	Time constant	T	Not used
υ	Upsilon	Not used	Υ	Not used
ω	Omega	Angular frequency	Ω	Ohm (unit)
ξ	Xi	Not used	Ξ	Not used
ψ	Psi	Angle	Ψ	Angle
ζ	Zeta	Not used	Z	Not used

acceptable by most publishers of technical journals and textbooks:

$$|Z| = \sqrt{[1{,}000\Omega]^2 + [2\pi(10^4\text{Hz})(10^{-2}\text{H})]^2} = 1{,}181\Omega,$$

$$|Z| = \sqrt{[1\,\text{k}\Omega]^2 + [2\pi(10\,\text{kHz})(10\,\text{mH})]^2} = 1.18\,\text{k}\Omega,$$

$$|Z| = \sqrt{[1,000]^2 + [2\pi(10^4)(10^{-2})]^2} = 1,181\,\Omega = 1.18\,\text{k}\Omega.$$

In the first two of the three presentations above, the units of the quantities involved are shown throughout. In the third, the unit is attached to only the ultimate result, and the unit (Ω) transcends the equal sign. All three presentations are considered dimensionally correct. Which of these presentations you use will depend upon your instructor's preference, your confidence in the dimensional correctness of the symbolic expression, and your familiarity with units and prefixes (e.g., you might recognize that the k in kHz and the m in mH in the second of the two presentations cancel). Generally, it is good to use the first form in the beginning, progressing to the third as you gain familiarity with various expressions and the quantities involved.

1.7 Approximations

Approximations are valuable and even inherent in engineering analysis and design. They are valuable because they can simplify computations and aid insight. In many cases, engineers employ without hesitation approximations that can introduce errors as large as 10%. Approximations are inherent because circuit parameters are imprecise. For example, the actual capacitance of a 10 nF capacitor might be anywhere in the range 10 nF $\pm$ 20%. As a reminder, we use the symbol $\cong$ to denote *is approximately equal to*; e.g., $\pi \cong 3.14$.

Opportunities for approximation arise repeatedly in circuit analysis and design. Almost all approximations are based in some way upon the **basic approximation**:

$$|x_1| >> |x_2| \Rightarrow x_1 + x_2 \cong x_1. \tag{1.3}$$

In many cases, the basic approximation (1.3) is acceptable if $|x_1| \geq 10|x_2|$ (error $\cong 10\%$) because, as it turns out, values of circuit parameters often are even more uncertain than that.

Example 1.3. Assume $|R_1| \gg |R_2|$ and find the approximate value of

$$G = \frac{1}{R_1} + \frac{1}{R_2}.$$

Solution:

$$G = \frac{1}{R_1} + \frac{1}{R_2} = \frac{R_1 + R_2}{R_1 R_2} \cong \frac{R_1}{R_1 R_2} = \frac{1}{R_2}.$$

Or

$$|R_1| \gg |R_2| \Rightarrow \frac{1}{R_1} \ll \frac{1}{R_2} \Rightarrow \frac{1}{R_1} + \frac{1}{R_2} \cong \frac{1}{R_2}.$$

Example 1.4. Assume $|x| \ll 1$ and find the approximate value of $y = \sin(x)$.

Solution: Using Taylor's expansion gives

$$y = \sin(x) = x - \frac{x^3}{3!} + \frac{x^5}{5!} - \cdots,$$

which can be written

$$y = \sin(x) = x\left[1 - \frac{x^2}{3!} + \frac{x^4}{5!} - \cdots\right].$$

Because $|x| \ll 1$,

$$\left[1 - \frac{x^2}{3!} + \frac{x^4}{5!} - \cdots\right] \cong 1,$$

and so

$$y = \sin(x) \cong x; \ |x| \ll 1.$$

Example 1.5. Certain irrational numbers such as π and $\sqrt{2}$ arise often in expressions for quantities of interest in circuit analysis. Faced with such an expression, and needing to make a quick mental or pencil-and-paper calculation, we might use

$$\pi = 3.0000 + 0.1416\cdots \cong 3 \text{ (error} < 5\%),$$
$$\sqrt{2} = 1.000 + 0.4142\cdots \cong 1.5 \text{ (error} \cong 6\%).$$

Certainly the wide availability of pocket calculators and personal computers has greatly reduced need for hand computation and therefore the need for approximations in numerical computations. On the other hand, ability to make approximations that simplify *symbolic* expressions is valuable, because such simplifications often facilitate insight; e.g., in identifying which components of a circuit are critical to performance. Judicious use of approximations can foster insights that even the most careful computer calculation cannot reveal.

Example 1.6. It is found that a performance measure called *voltage gain* for a certain electronic circuit is given by

$$A_v = \frac{\beta R_C R_L}{(R_B + r_b + R_S)(R_C + R_L)},$$

where all quantities on the right are circuit parameters. It is known that $R_B \gg r_b$, $R_B \gg R_S$, and $R_L \gg R_C$. Using reasonable approximations, simplify the expression on the right and identify the circuit parameters that are critical to obtaining a specified voltage gain.

Solution: Using (1.3), we have

$$R_B \gg r_b \Rightarrow R_B + r_b + R_S \cong R_B + R_S,$$
$$R_B \gg R_S \Rightarrow R_B + R_S \cong R_B,$$
$$R_L \gg R_C \Rightarrow R_L + R_C \cong R_L.$$

Thus

$$A_v = \frac{\beta R_C R_L}{(R_B + r_b + R_S)(R_C + R_L)} \cong \frac{\beta R_C R_L}{R_B(R_C + R_L)}$$
$$\cong \frac{\beta R_C R_L}{R_B R_L} = \frac{\beta R_C}{R_B}.$$

Under the conditions given, the critical parameters are β, R_C, and R_B. The parameters R_S, R_L and r_b have relatively little influence on voltage gain.

Exercise 1.9. The voltage gain of a certain circuit is given by

$$A_v = \frac{a}{1 + a\,b},$$

where a and b are circuit parameters. Assume $|a\,b| \gg 1$ and give an approximate expression for A_v. If this assumption holds, which circuit parameter determines the voltage gain of the circuit?

Some care is required in applying (1.3) to quantities raised to powers. For example, even if $|x| \ll 1$, it is not necessarily true that $(1 + x)^n \cong 1$, especially if $|n|$ is large. In such cases, the binomial series expansion

$$(1+x)^n = 1 + nx + \frac{n(n-1)}{2!}x^2 + \frac{n(n-1)(n-2)}{3!}x^3 + \cdots$$

often justifies the approximation

$$(1 + x)^n \cong 1 + nx. \tag{1.4}$$

The approximation (1.4) is justified if the quadratic term is much smaller (in magnitude) than the linear term; i.e., if

$$|(n-1)\,x| \ll 2. \tag{1.5}$$

For example, the true value of $(1.05)^4$ is 1.21550625. Equation (1.4) gives

$$(1.05)^4 = (1 + 0.05)^4 \cong 1 + (4)(0.05) = 1.2,$$

which is within 1.3% of the true value.

An **asymptotic approximation** to a function is the limiting form of the function for either exceedingly large or exceedingly small values of the independent variable. As an example, functions of the form

$$f(x) = 1 \pm e^x$$

are encountered frequently in electrical engineering. Such functions arise in analysis of circuits containing capacitors and inductors, where (usually) $x < 0$ and in analysis of semiconductor devices, such as diodes and transistors where (often) $x > 0$. Thus two asymptotic approximations of this function are important:

$$x < 0, |x| \gg 1 \Rightarrow 1 \pm e^x \cong 1,$$
$$x \gg 1 \Rightarrow 1 \pm e^x \cong \pm\, e^x. \tag{1.6}$$

In some cases, these approximations are deemed valid if $|x| \geq 3$, because $e^3 \cong 20$ is 20 times larger than one (5% error) and $e^{-3} \cong 0.05$ is only about 5% of one. In others, we might require $|x| \geq 5$ or more, in which case the error in (1.6) is less than 1%.

Exercise 1.10. An asymptotic approximation used extensively in a subsequent chapter is

$$\log(x + 1) \cong \log(x); \quad x \gg 1.$$

For what values of x is the error in this approximation less than 10%?

Approximations for elementary functions can be obtained from Taylor series expansions. For example,

$$\exp(x) = 1 + x + \frac{x^2}{2!} + \cdots \cong 1 + x, \quad |x| \ll 1$$
$$\cos(x) = 1 - \frac{x^2}{2!} + \frac{x^4}{4!} - \cdots \cong 1, \quad |x| \ll 1$$
$$\sin(x) = x - \frac{x^3}{3!} + \frac{x^5}{5!} - \cdots \cong x, \quad |x| \ll 1$$
$$\ln(x + 1) = x - \frac{x^2}{2} + \frac{x^3}{3} - \cdots \cong \left\{ \begin{array}{c} 0, \ |x| \ll 1 \\ \ln(x),\ x \gg 1 \end{array} \right\} \tag{1.7}$$

Exercise 1.11. Without using a calculator or computer, give approximate values for (a) $\frac{1}{2.02}$, (b) $\ln(1+e^5)$, (c) $\sin(0.501\pi)$, (d) $2+e^{0.02}$, (e) $\frac{1}{5}+\frac{1}{0.02}$.

Exercise 1.12. A particular current in an electric circuit is found to be

$$i_1 = \frac{R_2}{R_1+R_2} i_0.$$

Find approximations for i_1 (a) if $R_2 \gg R_1$ and (b) as R_1 becomes arbitrarily large (i.e., as $R_1 \to \infty$).

You should cultivate your ability to make reasonable approximations, partly because approximations will help you determine whether a particular answer or expression is reasonable, but mainly because approximations often foster useful insight; e.g., as to which components of a circuit are critical to acceptable performance.

1.8 Precision and Tolerance

Values of circuit components are never known with infinite precision. Rather, there is a precision associated with every such value; for example, the value of a resistance might be specified as $1\ \text{k}\Omega \pm 5\%$, meaning that the actual value is somewhere between 950.0 and 1,050.0 Ω. Thus, there also are precisions associated with calculated values of functions of circuit parameters.

Example 1.7. In the expression $I=V/R$, the value of V is known to be in the range $V_0 \pm 2\%$ and the value of R is known to be in the range $R_0 \pm 10\%$. What is the precision of the calculated value of I?
Solution: Let V_0, R_0 denote the nominal (zero-error) values for V and V, and let $I_0 = V_0/R_0$ denote the zero-error value for I. The maximum error occurs if V has its maximum value and R has its smallest value. The minimum error occurs if V has its minimum value and R has its maximum value. Thus

$$I_{\max} = \frac{V_{\max}}{R_{\min}} = \frac{1.02V_0}{0.90R_0} = 1.13I_0,$$
$$I_{\min} = \frac{V_{\min}}{R_{\max}} = \frac{0.98V_0}{1.10R_0} = 0.89I_0.$$

It follows that $0.89I_0 \le I \le 1.13I_0$, or $I = I_0 + 13\%$ or -11%, which might be written $I = I_0 + 13\%/{-11\%}$.

In electrical engineering, limits on precision (e.g., the +13% and −11% in Example 1.7) are called **tolerances**. For example, if the resistance of a resistor is known with precision ±5%, we say that the tolerance of the resistor is ±5%. Tolerances can be symmetric, as in ±5%, but often are not. Capacitances often have asymmetric tolerances; e.g., $C = C_0 + 20\%/-10\%$.

1.9 Engineering Notation

In **engineering notation** (a special kind of scientific notation), values are typically written using the appropriate number of significant digits and a multiplier of ten raised to a power that is a whole multiple of three; for example, 1.56×10^6 and 67.0×10^{-3}. Usually, the factor 10^n is then replaced by a prefix on whatever unit is associated with the value written (see Table 1.4). For example, we would write 62.63×10^3 V(volts) as 62.63 kV (kilovolts). Using engineering notation can avoid ambiguity associated with trailing zeros on a whole number. For example, whereas the significance of the trailing zero in 320 is ambiguous, it is clear that 3.2×10^2 has two significant digits. Similarly, 3200 is ambiguous, but 3.20×10^3 has three significant digits.

Exercise 1.13. Write each number using engineering notation. Assume trailing zeros are not significant. (a) 2015, (b) 16,380,000, (c) 0.759, (d) 0.000462, (e) 47.92×10^2.

Circuit components are available only in certain standard sizes and tolerances; for example, standard carbon-film resistors[11] are readily available in tolerances of $\pm10\%$ and $\pm5\%$, meaning that the actual resistance is within those percentages of the specified value. One can obtain high-precision components, such as metal-film resistors having tolerances of 0.1% or less; however, normal variations in ambient temperature and humidity or heating during soldering can cause the actual values to drift by 1% or 2% or more from the specified values. In general, actual values of circuit parameters are seldom within even $\pm1\%$ of specified values. Thus, when analyzing or designing circuits, it is almost always unnecessary and sometimes misleading to specify more than three significant figures in the numerical value of any quantity. Using too many significant figures can also cause you to miss opportunities for laborsaving approximations; for example, by leading you to believe that two resistors having resistances of 1.01256 kΩ and 1.01274 kΩ are different.

It would be troublesome and distracting to give the precision of every computed value throughout our introduction to linear circuits, so unless otherwise stated in a problem or example, we shall for the most part treat values as if they are known exactly. Again, this is a relatively harmless practice in instructional material, so long as you are forewarned that numerical results are often specified more precisely than is truly justified. In actual practice, however, it is usually essential to specify the precision of circuit parameters, based upon the accuracy required of various currents and voltages.

1.10 Problems

Section numbers in shaded boxes indicate prerequisite sections for problems that follow.

Section 1.2 is prerequisite for the following problems.

P 1.1 Pick one of the IEEE technical societies from the list on page 1.1 that sounds interesting to you. Visit the IEEE website (www.ieee.org) to find out more about the society. Briefly summarize the interests and purpose of the society, and give an example of an item that is related to the society's work.

P 1.2 Use the procedure described in Section 1.2 to outline the content of this chapter; i.e., list (or give a reference to) the defined quantities (e.g., symbols, special notation), physical laws, derived results, and procedures (if any).

Section 1.3 is prerequisite for the following problems.

P 1.3 The symbols in each quantity below are as defined in Table 1.2. Determine the SI unit of the quantity.

(a) $\frac{1}{C}\int_{-\infty}^{t} i(t')dt'$, (b) $VI\cos(\theta)$, (c) $\frac{dq}{dt}$, (d) Ri^2,

(e) $\frac{v^2}{R}$, (f) $\frac{Rv}{\tilde{Z}}$, (g) $\frac{L}{R}\frac{di}{dt}$, (h) Cv.

P 1.4 Check each equation for dimensional consistency.

(a) $\frac{R_1R_2}{R_1+R_2}i+R_1C\frac{dv}{dt}=0$,

(b) $\frac{1}{C}\int_{-\infty}^{t} i(t')dt'+L\frac{di}{dt}+Ri=v$,

(c) $C\frac{dv}{dt}+\frac{v}{R}+\frac{1}{L}\int_{-\infty}^{t} v(t')dt'=i$,

(d) $R_1R_2C_1C_2\frac{d^2v}{dt^2}+\frac{R_1}{L}\frac{dv}{dt}+R_2i=I$,

(e) $\frac{\tilde{Z}_1\tilde{Z}_2}{R+\tilde{Z}_1}\tilde{I}+\frac{\tilde{V}}{\tilde{Y}}=\tilde{V}_0$,

(f) $LC\frac{d^2v}{dt^2}+RC\frac{dv}{dt}+v=RC\frac{di}{dt}$.

P 1.5 Each quantity below is dimensionless. Express the SI unit of the variable a in terms of the SI units for current (A), voltage (V), and time (s).

(a) $\frac{aV}{RL}$ (b) $\frac{IV}{aLC}$ (c) $aV^2\frac{dv}{dt}$ (d) $\frac{aRI}{L\left(\frac{di}{dt}\right)}$

(e) $\frac{1}{a}\int_0^{t_0} v(t)\,dt$ (f) $\frac{a\sqrt{LC}}{RC}$

(g) $\frac{apt}{Cv^2}$ (h) $a^2\left[\frac{C\left(\frac{dv}{dt}\right)}{L\left(\frac{di}{dt}\right)}\right]^2$

[11]Resistance is defined in Chapter 2.

P 1.6 Express the speed of light in furlongs per fortnight. (Refer to your dictionary for definitions of furlong and fortnight.)

P 1.7 What is the SI equivalent of one light-year?

P 1.8 Express 1 kV/MHz in V/Hz.

P 1.9 Express the speed of light in m/μs.

P 1.10 Obtain the dimension and SI unit of $\sqrt{LC}$, where L is inductance and C is capacitance.

P 1.11 Obtain the dimension and SI unit of L/R, where L is inductance and R is resistance.

P 1.12 Obtain the dimension and unit of $\tilde{I}\,Z$, where $\tilde{I}$ is current and Z is impedance.

P 1.13 Obtain the dimension and unit of $\sqrt{\mu\varepsilon}$, where μ is permeability and ε is permittivity.

P 1.14 Express the charge of an electron in μ C, n C, and p C.

P 1.15 Electrical utilities (and power-system engineers) often express energy in kWh (kilowatt-hours). Express 1 kWh in J (joules).

P 1.16 Find the mass, in kilograms, of a man that weighs 200 lb.

P 1.17 List as many units as you can for each of the following quantities and name disciplines in which such units would be used: (a) length, (b) area, (c) volume, (d) energy.

P 1.18 The *fine-structure constant* is $\alpha = \frac{1}{2}\mu_0\, c\, q_e^{\,2}\, h^{-1}$, where μ_0 is the permeability of a vacuum, c is the speed of light in a vacuum, q_e is the charge of an electron, and h is Planck's constant. Show that the fine-structure constant is dimensionless.

P 1.19 The *Stefan-Boltzmann constant* is $\sigma = (\pi^2/60)\, k^4\, h^{-3}\, c^{-2}$, where k is the Boltzmann constant, h is Planck's constant, and c is the speed of light in a vacuum. (a) Express the unit of the Stefan-Boltzmann constant in terms of basic SI units. (b) Show that the unit of the Stefan-Boltzmann constant can be expressed as W m^{-2} K^{-4}.

P 1.20 The SI unit for resistivity is the Ω m, but this unit is inconvenient (too large) in many cases, so resistivity is often expressed in μΩ cm. Find the factor that converts Ω m to μΩ cm.

P 1.21 Show that

$$i(t) = \frac{1}{L}\int_{-\infty}^{t} v(t')dt'$$

is dimensionally consistent.

P 1.22 Show that $1\text{Vm}^{-1} = 1\text{N C}^{-1}$.

Section 1.7 is prerequisite for the following problems.

P 1.23 For each function given below: (i) Find the small-x and large-x linear asymptotes. (ii) If the asymptotes intersect, calculate the values of the function and either asymptote at that point. (iii) Sketch a graph of the function, using the asymptotes as guides.

$$\text{(a)}\, f(x) = \tan^{-1}(x), \quad \text{(b)}\, f(x) = \frac{1}{1+x^2},$$

$$\text{(c)}\, f(x) = \frac{1}{x}, \quad \text{(d)}\, f(x) = \begin{cases} 0, & x < 0, \\ e^{-x}, & x \geq 0, \end{cases}$$

$$\text{(e)}\, f(x) = \begin{cases} 0, & x < 0, \\ 1 - e^{-x}, & x \geq 0. \end{cases}$$

P 1.24 A certain voltage V_{out} is expressed in terms of another voltage V_n by

$$V_{out} = \frac{R_{out} - \mu R_1}{R_{out} + R_1} V_n,$$

where R_1, R_{out} are resistances and μ is dimensionless. It is known that $R_1 \gg R_{out}$ and $\mu \gg 1$. Find a reasonable approximation for V_{out}.

P 1.25 A certain voltage V_L in an electric circuit is found to be

$$V_L = \frac{R_L}{R_T + R_L} V_T,$$

(a) Find a reasonable approximation for V_L if $R_L \gg R_T$.
(b) Find a reasonable approximation for V_L if $R_T \gg R_L$.
(c) Find V_L as R_T becomes arbitrarily large (i.e., as $R_T \to \infty$).

P 1.26 A performance measure called *current gain* for a particular circuit is found to be

$$A_i = \frac{R_{in} R_{out} R_S\, g}{(R_{out} + R_L)(R_{in} + R_S)},$$

where all quantities on the right are circuit parameters.

(a) Find a reasonable approximation for A_i if $R_{out} \gg R_L$.
(b) Find a reasonable approximation for A_i if $R_{in} \gg R_S$.
(c) Find a reasonable approximation for A_i if both $R_{out} \gg R_L$ and $R_{in} \gg R_S$.

(d) The current gain A_i is dimensionless. What is the SI unit of g?

P 1.27 Give asymptotic approximations to $\log\sqrt{1+(f/f_0)^2}$ for (a) $0<f\ll f_0$ and (b) $f\gg f_0>0$.

P 1.28 Given that $R_1\gg R_2\gg R_3$, obtain an approximation to $\frac{1}{R_2}+\frac{R_1+R_3}{R_1R_3}$.

P 1.29 Given that $R_1\gg R_2\gg R_3$, obtain an approximation to $\frac{R_1R_2R_3}{R_1R_2+R_1R_3+R_2R_3}$.

P 1.30 Let

$$G(f)=-\log\left[1+(f/f_0)^2\right]+\log\left[1+(f/f_1)^2\right]-\log\left[1+(f/f_2)^2\right],$$

where $0\ll f_0\ll f_1\ll f_2$. Give asymptotic approximations for $G(f)$ for (a) $f_0\ll f\ll f_1$, (b) $f_1\ll f\ll f_2$, and (c) $f\gg f_2$

P 1.31 Let

$$H(f)=\frac{K\sqrt{1+(f/f_1)^2}}{\sqrt{1+(f/f_0)^2}\sqrt{1+(f/f_2)^2}};\ f_0\ll f_1\ll f.$$

Give asymptotic approximations for $H(f)$ for (a) $f_0\ll f\ll f_1$, (b) $f_1\ll f\ll f_2$, and (c) $f\gg f_2$

P 1.32 Show that if x and y are both positive, then

$$\frac{1}{x}+\frac{1}{y}>\frac{1}{x+y}.$$

Section 1.8 is prerequisite for the following problems.

P 1.33 Estimate the precision of the indicated quantity:

(a) $\frac{R_1}{R_2}$, where the precision of each R is $\pm 5\%$

(b) $\frac{R_1R_2}{R_1+R_2}$, where the precision of each R is $\pm 5\%$

(c) $(f/f_0)^2$, where the precision of f_0 is $\pm 1\%$. Does the magnitude of the error depend upon the variable f?

(d) $20\log[1+(f/f_0)^2]$, where f is known exactly, $f\gg f_0$, and the precision of f_0 is $\pm 5\%$. Does the magnitude of the error depend upon the variable f?

(e) $\exp(-t/RC)$ for $t=20\,\mu s$, where $R=1\,k\Omega\pm 5\%$ and $C=20\,nF\pm 20\%$.

(f) $\frac{R_1R_2R_3}{R_1R_2+R_1R_3+R_2R_3}$, where the precision of each R is $\pm 1\%$

Section 1.9 is prerequisite for the following problems.

P 1.34 Express each of the following using engineering notation. Omit ambiguous zeros.

(a) 14,600 V, (b) 9,870,000 W, (c) 48.3×10^8, (d) Five million people, (e) 0.0056, (f) 0.0001, (g) 0.000708.

P 1.35 Express each of the following using unit prefixes; for example, as in $47{,}000\,\Omega=47\,k\Omega$.

(a) 12 million volts, (b) 4.2×10^{-6} A, (c) 56×10^5 W, (d) 0.050 V, (e) 0.0002 A, (f) $4.8\times 63.4\times 109.0$m, (g) $0.0506\,g\div 2048$, (h) 10×10^9 Hz, (i) $34.7\times 10^{13}\,\Omega$.

P 1.36 Express each of the following using engineering notation, without using unit prefixes; for example, $123\,kW=123\times 10^3$ W.

(a) 55GHz, (b) 12 ns, (c) 100 MW, (d) $2.2\,k\Omega$s, (e) 100nF, (f) $14.6\,\mu s$, (g) 12mm, (h) $47\,\mu F$, (i) 12pF (j) $108\,\mu\Omega$ cm, (k) 75 kV A, (l) $100\,T\Omega$, (m) 1 fA.

Chapter 2
Current, Voltage, and Resistance

In this chapter, we define current and voltage, which are among the principal quantities of interest in circuit analysis. We present Ohm's law, which is one of the fundamental laws of electrical engineering, and which is the defining relation for resistance and conductance.

2.1 Charge and Current

Charge is a property of electrons and protons, which are building blocks of atoms. The SI unit of electric charge is the **coulomb** (C).[1] A quantity of charge is represented by the symbol q and is in general a function of time.

Physicists are still hard at work trying to determine what charge is made of, but we do not need to await the results. We already know enough about how charge behaves to make it do many useful things. We accept charge as an undefined quantity whose properties are well understood. For electrical engineers, the important properties of charge are the following:

- There are two kinds of charge: The negative charge of an electron and the positive charge of a proton. The charge of an electron is $q_e \cong -1.602 \times 10^{-19}$C. The charge of a proton is $q_p = -q_e$.
- Like charges (charges having the same sign) repel each other and unlike charges attract each other. The force on a point charge[2] q_1 due to the presence of another point charge q_2 is given (in a vacuum) by *Coulomb's law*

$$f(x) = \frac{q_1 q_2}{4\,\pi\,\varepsilon_r\,\varepsilon_0\,x^2} = k\,\frac{q_1\,q_2}{x^2}, \tag{2.1}$$

where x is the distance between the charges and $k = (4\,\pi\,\varepsilon_r\,\varepsilon_0)^{-1}$ is a constant whose value depends upon the medium. In free space, k is approximately $9 \times 10^9\,\mathrm{N\ m^2\ C^{-2}}$. The force on q_1 given by (2.1) is attractive (directed toward q_2) if q_1 and q_2 have opposite signs and repulsive (directed away from q_2) if q_1 and q_2 have the same sign. The fact that charges exert forces on one another allows us to establish conditions under which some charges make other charges do useful things.
- The charge of an electron (or proton) is the smallest unit of charge and all charges are whole multiples of that fundamental unit. This fact is unimportant in ordinary circuit analysis, where so many electrons and/or ions are involved in the processes of interest that we can treat quantity of charge as a real variable–one that can assume any value. But it might be important in future ultra-small devices, where relatively few electrons will be involved in processes of interest.
- Charge can be neither created nor destroyed; that is, total charge is conserved.[3] If we enclose a volume

[1]After the French physicist Charles Augustin de Coulomb (1736–1806).

[2]In practical applications, two charges can be considered point charges if their radii are much smaller than their separation.

[3]Total charge is the algebraic sum of charge. Thus, the net charge of a hydrogen atom, which consists of one electron and one proton, is zero.

T.H. Glisson, *Introduction to Circuit Analysis and Design*,
DOI 10.1007/978-90-481-9443-8_2, © Springer Science+Business Media B.V. 2011

with a surface and monitor the charge passing through the surface, any increase or decrease in the net charge enclosed in the volume will be accounted for by passage of charge through the surface. It is good that charge is conserved, because if charge could suddenly appear (or disappear) here and there, it would be difficult to keep track of and control.

- A moving charge creates a magnetic field and a magnetic field exerts a force on a moving charge. These facts make electric generators, motors, and a host of other devices possible.

Because of the properties above, charge can be made to do useful work. We can use electric charge to move enormous amounts of energy easily from one place to another, transmit information from one place to another, perform computations at unimaginable speed, convert mechanical energy to electrical form (and vice versa), and do many other useful things.

To do work, charge must move under the influence of a force. Consequently, moving charge usually is of more interest to electrical engineers than charge itself. Motion or flow of charge is called **current**, denoted by i. By convention, *the positive direction of current is opposite to the direction of electron flow.*[4] Current i through a plane (e.g., a cross section of a wire) is defined by[5]

$$i = \frac{dq}{dt}, \tag{2.2}$$

where dq is a quantity of charge passing through the plane in time dt and i is the current through the plane in the direction of the flow of positive charge. Alternatively, if q is a quantity of charge in an enclosed space, then (2.2) gives the current passing *into* the space through the enclosing surface. The SI unit of current is the **ampere** (A),[6] where 1 A equals 1 C s^{-1}.

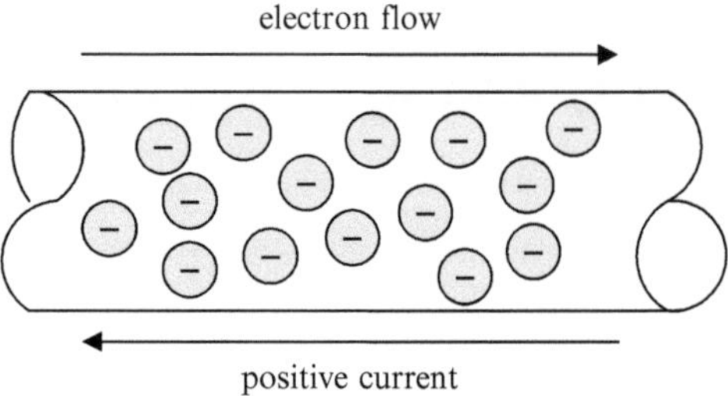

Fig. 2.1 The direction of positive current is opposite to the direction of electron flow. See Example 2.1

Example 2.1. A steady flow of 10^{20} electrons per second is directed from left to right through a cross section of a wire, as illustrated in Fig. 2.1. The current from left to right through the cross section is constant and equals

$$i = \left(10^{20}\,\frac{\text{electrons}}{\text{second}}\right)\left(-1.602\times10^{-19}\,\frac{\text{coulombs}}{\text{electron}}\right).$$
$$= -16.02\,\text{C}\,\text{s}^{-1} = -16.02\,\text{A}$$

The current from left to right is negative because negative charge is moving from left to right. The current from right to left is positive and equals +16.02 A.

Exercise 2.1. An electric current is established in a tank of salt water, as shown in Fig. 2.2. The positively charged sodium ions are drawn left to the negative electrode and the negatively charged chlorine ions are drawn right to the positive electrode. Each ion is singly charged and N ions reach each electrode during each interval T. Give an expression for the current I. Show that the expression is dimensionally consistent.

[4] We have Ben Franklin to thank for this unfortunate accident of history. He decided to call the static charge induced on glass by rubbing the glass with silk *positive* and stated that *the direction of current is from positive to negative* (from glass to silk). Had he made either of these choices the other way, the positive direction of current would be in the direction of electron flow.

[5] Explicit time-dependence often is omitted for the sake of brevity. Thus, in (2.2) q stands for $q(t)$ and i for $i(t)$. Also, current defined by (2.2) is called *conduction current*. When you study electromagnetism, you will learn about another kind of current called *displacement current*, that does not involve flow of charge. Conduction current is the current of interest in ordinary circuit analysis, where we are relatively unconcerned with the inner workings of circuit elements.

[6] After the French Mathematician and Physicist André Marie Ampére (1775–1836), who discovered basic laws of electromagnetism.

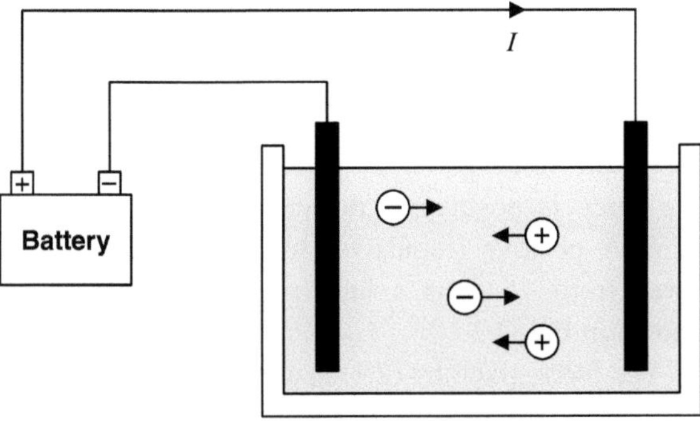

Fig. 2.2 See Exercise 2.1

2.2 Electric Field

In this section, we review the definition of work and define the *electric field* established by a charge, drawing on an analogy with gravity and the gravitational field. This section is preparation for the next, where voltage is defined in terms of work done on a charge by an electric field.

Recall from physics that potential energy is capacity for doing work and that the work w done by a force $f(x)$ on an object as the object moves from x_1 to x_2 is given by[7]

$$w = \int_{x_1}^{x_2} f(x)\, dx. \tag{2.3}$$

The work done by the force is positive if the force is in the direction of motion and negative if the force is opposite to the direction of motion. Another way to say this is that the force does work on the object if the force and motion are in the same direction and work is done on whatever provides the force if the force and motion are in opposite directions. The SI unit of work is the *joule* (J),[8] where (in mechanical units) 1 J equals 1 N-m.

Work and energy are expressed in the same unit (J) and are physically the same thing. The distinction between work and energy is semantic, but useful. **Energy** describes the *state* of something; for example, we might speak of the potential energy of a mass that has been raised to some height or the kinetic energy of a moving mass. **Work** is a *quantity of energy transferred from one thing to another or converted from one form to another*; for example, w in (2.3) is work because it is a quantity of energy transferred from whatever supplies the force f to an object acted on by the force. Similarly, a quantity of electrical energy converted to mechanical form by an electric motor is called work.

An electric field and its interaction with charge are analogous to a gravitational field and its interaction with mass. On or near the earth's surface, the gravitational force f on a mass m is nearly independent of position, is directed toward the center of the earth, and is given by $f = mg$, where g is the acceleration due to gravity (approximately 9.8 m s^{-2} at sea level). For purposes of ordinary (non-relativistic) mechanics, we can define gravitational field strength F (a vector) near the earth's surface as the *force per unit mass due to gravity*; that is, as the gravitational force on a mass divided by the mass. Thus, the magnitude of the field strength F is $F = f/m = (mg)/m = g$ and its direction is toward the center of the earth.

A gravitational field possesses potential energy because it does work on a mass that is allowed to fall. The work done *on* the mass *by* the gravitational field as the mass falls from height h to the ground is $w =$ *force x distance* $= mgh$. To put this another way, the work $w = mgh$ is the amount of energy transferred *from the field to the mass* as the mass falls from height h. The work the gravitational field can do on the mass (the energy that can be transferred from the field to the mass) increases with the height of the mass. Thus, energy is stored in the gravitational field when a mass is raised and energy is released by the field (transferred to the mass) when the mass falls.[9] The *gravitational potential* at a distance h above (but near) the earth's surface is defined by $\varphi(h) = g\,h$ and equals the work done per unit mass as a mass falls from or is

[7]To avoid using vector notation, we assume the force $f(x)$ is directed along the x-axis, in either the same direction as the motion or opposite to the direction of motion.

[8]After the English physicist James Prescott Joule (1818–1889).

[9]In this discussion, we are concerned only with energy exchanged between the gravitational field and a single mass. There may be other forces acting on the mass in addition to that due to gravity, but they do not affect the energy exchange between the mass and the field; for example, we can throw a mass toward the ground, in which case additional energy is transferred to the mass, but the energy transferred to the mass *from the field* is unaffected by the throwing.

raised to a height h. Thus the work w done is given by $w = m\,\varphi(h)$. If the mass falls, work is done by the field (energy is transferred from the field to the mass). If the mass is raised, work is done on the field (energy is transferred from the mass to the field).

Example 2.2. An object having mass $m = 2$ kg is raised from the ground to a height $h = 10$ m and released. The potential energy stored in the field when the object is raised is

$$w = mgh \cong (2\text{kg})(9.8\,\text{m}\cdot\text{s}^{-2})(10\text{m}) = 196\,\text{J}.$$

When the object is released, this energy is given back to the object as kinetic energy. The kinetic energy of the object when it hits the ground is 196 J. When the object is raised, 196 J are stored in the field. When the object is released, the field gives up the stored energy by doing work $w = 196$ J on the object.

Exercise 2.2. An object having mass m is tossed up. When it strikes the ground, its velocity (downward) is u. Give an expression for the height attained by the object. Show that the expression is dimensionally consistent.

Below, by analogy with gravitation, we define electric field strength and electric potential. We illustrate these definitions using the field and potential due to a point charge. *The stated definitions are general, but the expressions given apply to a point charge only.*

Consider the situation shown in Fig. 2.3, where positive point charges q_1 and q_2 are separated by a distance x. We assume the charge q_2 is fixed and we can somehow move the charge q_1 about.

By Coulomb's law (2.1), the force $f(x)$ on q_1 due to the presence of q_2 is given by

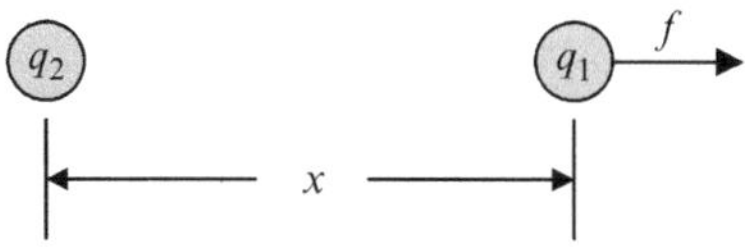

Fig. 2.3 Coulomb force

$$f(x) = k\,\frac{q_1 q_2}{x^2},$$

where $k = 9 \times 10^9\,\text{Nm}^2\text{C}^{-2}$. Because like charges repel, the force given by (2.1) is repulsive; that is, the force is positive if the charges have the same sign. A positive (repulsive) force on q_1 is directed away from q_2 along a line from q_2 through q_1, as shown in Fig. 2.3.

The force given by (2.1) is attributed to an *electric field* produced by the charge q_2. The electric field strength $\boldsymbol{E}$ (vector) at a point is defined as the force *per unit charge* on a *positive* charge at that point. Thus, the electric field strength at a distance x from an isolated point charge q_2 is given by

$$E(x) = k\,\frac{q_2}{x^2}. \tag{2.4}$$

The coulomb force given by (2.1) and the electric field strength given by (2.4) for a point charge depend only upon distance from the charge; that is, electric field strength is directed radially outward from a positive charge, as illustrated in Fig. 2.4. If q_2 were negative, the electric field strength given by (2.4) would be negative, meaning the force on a positive charge is directed radially inward toward q_2. Note that (2.4) gives electric field strength a distance x from a *point charge*. It is not a general result. Other charge distributions lead to different expressions for electric field strength.

If the charge q_1 in Fig. 2.3 is free to move, and if both charges have the same sign, then the charge q_1 will accelerate away from q_2, taking energy from the electric field as it accelerates. On the other hand, if we somehow force the charge q_1 toward the charge q_2, we

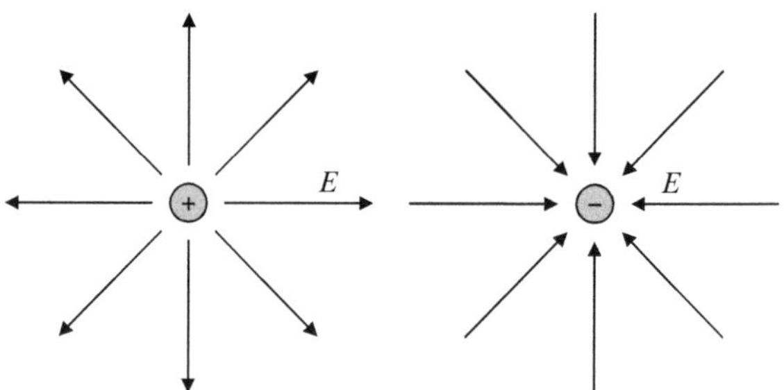

Fig. 2.4 Electric field due to a point charge. Electric field lines of force emanate from positive charge and terminate on negative charge

store potential energy in the field because we increase the capacity of the field to do work on the charge q_1. If a positive charge moves in the direction of electric field strength, then the motion and the force are in the same direction and energy is transferred *from the field* to the charge. If a positive charge moves opposite to the field, energy is transferred *to the field* from whatever moves the charge.

Recall that a mass moving toward the earth's surface (in the direction of the gravitational field) takes energy from the gravitational field and a mass moving away from the earth's surface (opposite to the direction of the gravitational field) gives energy to the gravitational field. Similarly, a positive charge moving in the direction of an electric field takes energy from the field and a positive charge moving opposite to the direction of the field gives energy to the field. Currents in electric circuits are due primarily to electron flow, but we may treat a flow of electrons in one direction as a flow of positively charged particles in the opposite direction. Thus *energy is taken from an electric field by a current moving in the direction of the field and energy is given to the field by current moving opposite to the direction of the field.*

Continuing the point-charge example, we can calculate the work done *on* the charge q_1 *by* the field using (2.3), where the force *f(x)* is the product of the charge q_1 and the electric field strength (the force per unit charge). From (2.4), the force on the charge is given by

$$f(x) = E(x)\,q_1 = k\,\frac{q_1\,q_2}{x^2}.$$

Suppose both charges have the same sign, q_2 is stationary, and the charge q_1 moves from point a to point b on a line drawn from point a to q_2. From (2.3), the work done *on* the charge *by* the field is given by

$$\begin{aligned} w &= \int_a^b f(x)\,dx = k\,q_1\,q_2 \int_a^b \frac{dx}{x^2} \\ &= k\,q_1\,q_2\left[-\frac{1}{x}\right]_a^b = -k\,q_1\,q_2\left(\frac{1}{b}-\frac{1}{a}\right). \end{aligned} \quad (2.5)$$

If $a < b$, the charge q_1 moved away from q_2 (in the direction of the field), the work w is positive, and the field has done work on the charge. If $a > b$, the charge q_1 moved toward q_2 (opposite to the direction of the field), w is negative, and work has been done on the field. In the first case, energy is removed from the field (transferred to the charge). In the second, energy is stored in the field.

Example 2.3. In Fig. 2.3, the movable charge $q_1 = 1$ mC and the fixed charge $q_2 = 5$ mC. If q_1 moves from $a = 10$ m to $b = 20$ m (away from q_2), the work done by the field on q_1 is given by (2.5) and equals

$$\begin{aligned} w &= -k\,q_1\,q_2\left(\frac{1}{b}-\frac{1}{a}\right) \\ &\cong -(45\times 10^3\,\mathrm{N\cdot m^2})\left(\frac{1}{20\mathrm{m}}-\frac{1}{10\mathrm{m}}\right) = 2,250\,\mathrm{J} \end{aligned}$$

which is approximately the work done in lifting a 225 kg mass to a height of 1 m (a 225 kg mass weighs about 496 lb).

Exercise 2.3. Point charges q, $-q$ are a distance d apart. Obtain an expression for the charge q in terms of the separation d and the force f each charge exerts on the other. Show that the expression is dimensionally consistent.

2.3 Electric Potential and Voltage

We continue with the point-charge example. From the discussion above, the potential energy of the field due to q_2 depends upon the position of q_1 relative to that of q_2. The closer is q_1 to q_2, the greater is the potential energy of the field and the farther is q_1 from q_2, the smaller is the potential energy of the field.

The **electric potential** at a point is denoted by Φ and is defined as the work done *on an electric field* (or by an external force) *per unit charge* when a *positive* charge is brought from infinity (where the electric field strength is assumed to be zero) to the point.[10] Thus the electric potential at a point is numerically (but not

[10] Note that electric potential is not the same as potential energy. The unit of electric potential is that of energy divided by charge.

dimensionally) equal to the work required of an external force to bring a one-coulomb positive charge from infinity to the point. The work done and thus the potential can be positive or negative, depending upon whether the charge is forced to move against the electric field or is allowed to drift in the direction of the field. The SI unit of electric potential is the **volt** (V),[11] where 1 V equals 1 J C^{-1}.

Equation (2.5) gives the work done *by the field* on a point charge when the charge is moved from a to b in the neighborhood of another point charge. The work done *on the field* is just the negative of the work done *by the field*. Changing the sign of the right side of (2.5), dividing by q_1, and taking the limit as $a \longrightarrow \infty$, we find that the electric potential (the work done on the field per unit charge) in bringing a charge from $x \longrightarrow \infty$ to $x = b$ is given by

$$\Phi(b) = \frac{k q_2}{b}. \tag{2.6}$$

Point b can be any point (any distance from q_2), so the electric potential at distance x from a point charge q_2 is given by

$$\Phi(x) = \frac{k q_2}{x}. \tag{2.7}$$

For a point charge, the electric field strength given by (2.4) and the electric potential given by (2.7) are related as

$$E(x) = \frac{\Phi(x)}{x}. \tag{2.8}$$

Electric field strength is defined as force per unit charge, so electric field strength can be expressed in units of newtons per coulomb (N C^{-1}). Equation (2.8) for electric field strength is specific to a point charge, but illustrates the fact that electric field strength also can be expressed in **volts per meter** (V m^{-1}), which is the preferred unit in electrical engineering (both are SI units).

[11]After Count Alessandro Volta (1745–1827), an Italian inventor who discovered hydrolysis and invented the battery and the electric condenser (now called a capacitor).

Example 2.4. From (2.4) the electric field strength at a point 10 m from a 5 μC point charge is

$$E = \frac{kq}{x^2} \cong \frac{(9\times 10^9 \mathrm{N\cdot m^2\cdot C^{-2}})(5\times 10^{-6}\mathrm{C})}{(10\,\mathrm{m})^2}$$
$$= 450\,\mathrm{NC^{-1}} = 450\,\mathrm{Vm^{-1}}.$$

From (2.7), the electric potential at the point is

$$\Phi = \frac{kq}{x} \cong \frac{(9\times 10^9 \mathrm{N\cdot m^2\cdot C^{-2}})(5\times 10^{-6}\mathrm{C})}{(10\,\mathrm{m})}$$
$$= 4.50\,\mathrm{kV}.$$

Exercise 2.4. Equal point charges are fixed in space as shown in Fig. 2.5 and exert a force f on one another. Find an expression for the electric potential due to the charges at the point $x_1 > x_0 > 0$ in terms of the distances x_0, x_1. and the force f Show that the expression is dimensionally consistent. *Hint*: The total electric potential due to two point charges is the sum of the potentials due to each charge individually.

What causes water to flow through a pipe is a pressure difference from one end of the pipe to the other. It is pressure *difference* that causes flow, not absolute pressure at either end. Similarly, what causes charge to flow is a potential difference. In electric circuits, we are interested in *potential differences*, not absolute potentials.

The **voltage** at point a with respect to point b, denoted by v_{ab}, is the potential difference

$$v_{ab} = \Phi(a) - \Phi(b), \tag{2.9}$$

q

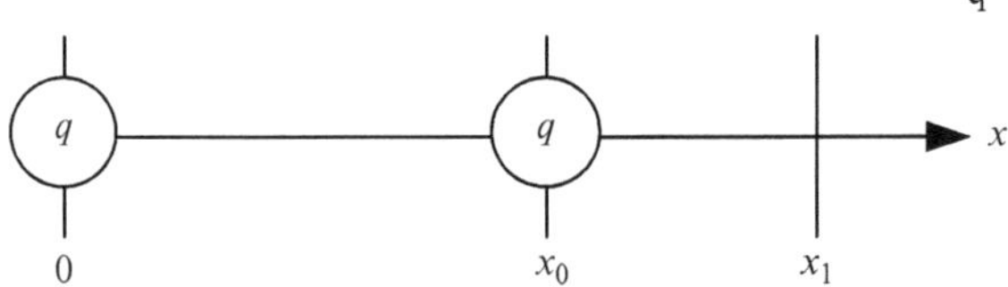

Fig. 2.5 See Exercise 2.4

where $\Phi(a)$ is the potential at point a and $\Phi(b)$ is the potential at point b The SI unit for electric potential is the volt, so the SI unit of voltage is also the volt (V). Note that the voltage v_{ab} is the *work done on the field per unit charge* in moving a charge from point b to point a.

The difference of two potentials can be formed in two ways and the order of the subscripts a and b is significant. Two points are required to specify (or calculate, or measure) a voltage and one of the points must be identified as the reference. By convention, the second subscript on the voltage defined by (2.9) identifies the reference point; e.g., in (2.9) the potential $\Phi(b)$ is the **reference potential.** To measure v_{ab} with a voltmeter, we would touch the black (–) probe to the reference point b and the red (+) probe to point a. If we made the measurement the other way around, we would be measuring the negative of v_{ab} because, by comparison with (2.9),

$$v_{ba} = \Phi(b) - \Phi(a) = -v_{ab}. \tag{2.10}$$

Example 2.5. From (2.9) and (2.7), the voltage at a point a with respect to another point b in the neighborhood of an isolated point charge q_2, is given by

$$v_{ab} = \frac{k\,q_2}{x_a} - \frac{k\,q_2}{x_b} = k\,q_2\left(\frac{1}{x_a} - \frac{1}{x_b}\right),$$

where x_a is the distance from point a to q_2 and x_b is the distance from point b to q_2. If q_2 is positive, the voltage v_{ab} is positive if a is nearer q_2 than b (if $x_a < x_b$) and negative if a is farther from q_2 than b (if $x_a > x_b$).

Exercise 2.5. The point a is at a distance x_a from an isolated point charge q, as shown in Fig. 2.6. Let v_{ab} denote the voltage between a and another point b. Express the distance d in terms of q, x_a, and v_{ab}. Show that the expression is dimensionally consistent.

If a voltage v_{ab} is positive, the potential energy of a positive charge is greater at the point a than it is at point b. In that case, a positive charge experiences a

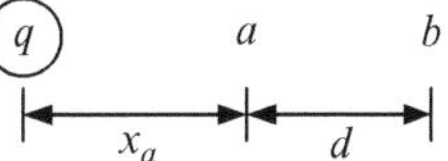

Fig. 2.6 See Exercise 2.5

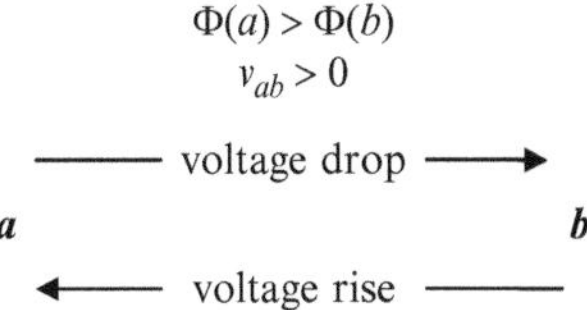

Fig. 2.7 Definitions of voltage rise and voltage drop

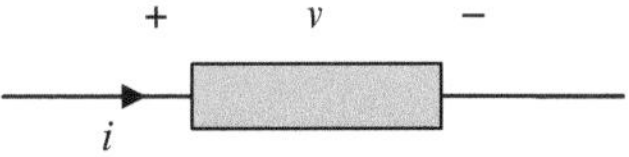

Fig. 2.8 Voltage polarity and current direction in Ohm's law

voltage drop in moving from a to b and a **voltage rise** in moving from b to a.[12] If v_{ab} is negative, the reverse is true. Thus current from a to b goes through a voltage drop if v_{ab} is positive and a voltage rise if v_{ab} is negative. Figure 2.7 illustrates these definitions.

2.4 Ohm's Law and Resistance

Ohm's law states that *the voltage across a piece of material is proportional to the current through the material*; for example, if the object in Fig. 2.8 represents a sample of conducting material, the voltage v across the sample is related to the current i through the sample according to[13]

$$v = R\,i, \tag{2.11}$$

where R is called the **resistance** of the sample. The SI unit of resistance is the **ohm** (Ω),[14] where 1 Ω equals 1 V/A (1 VA^{-1}). Despite its simplicity, Ohm's law is one of the most important laws in all of electrical engineering.

[12] The terms *rise* and *drop* arise from analogy with gravity, where a mass gains potential energy if we raise it and loses potential energy if we lower (or drop) it.

[13] In general (and usually), both voltage and current vary with time.

[14] After the German physicist Georg Simon Ohm (1787–1854).

In Ohm's law (2.11) *the positive direction of current is in the direction of a voltage drop* (from a higher to a lower potential), as shown in the figure. If either the assumed voltage polarity or the assumed current direction is opposite to that shown in Fig. 2.8, the right side of (2.11) must be negated.

Example 2.6. In Fig. 2.8, the current i is equal to 5 A in the direction shown and the resistance of the sample is 100 Ω. From (2.11), the voltage v is 500 V.

Exercise 2.6. Refer to Fig. 2.9. Find the current i_2 if the voltage v is 10 V and the resistance of the sample is 1 kΩ.

2.5 Resistivity

The resistance of a sample of material depends upon the kind and state of the material (e.g., copper at 20°C) and upon the size and shape of the sample. The effects of the kind and state of the material can be separated from those of size and shape. A material can be characterized by its **resistivity**, denoted by ρ, which is independent of size and shape. The resistance of a homogeneous sample of material having resistivity ρ, length l, and *uniform* cross-sectional area A is given by

$$R = \frac{\rho l}{A}. \tag{2.12}$$

The SI unit of resistivity is the **ohm meter** (Ω m).[15] Depending upon the material, resistivity can depend strongly or weakly upon the state of the material (e.g., temperature and pressure).

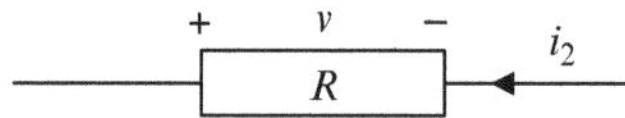

Fig. 2.9 See Exercise 2.6

[15]In reference works, resistivity often is specified in μΩ cm (10^{-8} Ω m).

Example 2.7. The resistivity of pure copper at 20°C is $1.72 \times 10^{-8}\,\Omega\text{m}$. At 20°C, the resistance of a pure copper cylindrical wire having diameter $d = 0.2$ cm and length $l = 5$km is

$$R = \frac{\rho l}{A} \cong \frac{(1.72 \times 10^{-8}\,\Omega\text{m})(5000\,\text{m})}{(\pi)(0.001\text{m})^2} \cong 27.4\ \Omega.$$

If the current in the wire is 15 A, the voltage drop from end to end is

$$v = Ri \cong (27.4\,\Omega)(15\text{A}) \cong 411\text{V}.$$

Exercise 2.7. A certain conductor on a printed circuit board has a rectangular cross section. The conductor has width w_0 and is built from a material having resistivity ρ_0. Suppose we want to build the conductor from a different material having resistivity ρ_1. Find an expression for the new width w_1 if we want to maintain the length, height, and resistance of the original conductor

Many materials are classified according to their resistivity as *insulators* (very high resistivity) or *conductors* (very low resistivity). Materials having resistivities between these extremes are called *semiconductors*. Glass and air are good insulators. Most metals are good conductors.[16] Silicon containing certain impurities is a semiconductor.

2.6 Conductance and Conductivity

The reciprocal of resistance is called **conductance** and is denoted by G:

$$G = \frac{1}{R}. \tag{2.13}$$

[16]Metals are good conductors because the outer electrons of metal atoms are only loosely held and are relatively free to "drift" through the solid under the influence of an applied voltage.

The unit of conductance is the **siemen** (S), where $1\ \mathrm{S} = 1\ \Omega^{-1}$. The reciprocal of resistivity is called **conductivity**, denoted by σ:

$$\sigma = \frac{1}{\rho}. \tag{2.14}$$

The unit of conductivity is **siemens per meter** ($\mathrm{S\ m^{-1}}$). From (2.12)–(2.14), the conductance of a homogeneous piece of material having conductivity σ, length l and uniform cross-sectional area A is given by

$$G = \frac{\sigma A}{l}. \tag{2.15}$$

Using conductance, Ohm's law is written

$$i = G v. \tag{2.16}$$

Conductance often is used symbolically in place of resistance when writing circuit equations that would otherwise contain many terms of the form v/R; however, in specifications of values, resistance is almost always the quantity specified.

2.7 Resistors

Resistance is introduced (intentionally) in circuits using components called **resistors**. Resistors are made of various materials and in various configurations. A *composition resistor* is essentially a cylindrical core of material consisting of very fine carbon or metallic granules imbedded in a non-conducting or semi-conducting material and incased in plastic. The resistance is determined by the concentration of carbon or metal in the core material. There are two main types of *film resistors*. One kind is made by depositing a film of metal or carbon on a non-conducting cylindrical core and then cutting away some of the material to leave a spiral ribbon of the material. The other kind consists of a thin film of (usually) metal or certain metal oxides on a planar surface. Many chip resistors and surface-mount resistors are of this kind, as are resistors in integrated circuits. A *wirewound resistor* is just what the name suggests: A wire wound on a non-conducting core. The resistance of a wirewound resistor is determined by the resistivity, cross-sectional area, and length of the wire. If adjacent windings are in contact, the wire is insulated.

Resistors are fixed (have fixed values) or variable. Variable resistors can be continuously variable, like potentiometers and rheostats, or have contacts at only certain angular or linear positions. A continuously variable resistor incorporates a tap or a slide that can be positioned and repositioned along a strip of conducting material or wirewound core. Those having only a limited set of possible values sometimes are called encoders because they provide a precise relation between position and resistance. Figures 2.10 and 2.11 show assorted fixed and variable resistors.

Years ago, carbon-composition resistors were the most common (numerous) kinds of resistors. Nowadays, the most numerous resistors are film resistors, and carbon-composition resistors have almost disappeared. Chip resistors, most small axial-lead resistors, and resistors in integrated circuits are constructed of thin carbon or metallic films (often, nichrome or tantalum nitride). In chip and integrated resistors, the film is deposited on an insulating substrate and joined to contact pads or conductors at each end, as illustrated in Fig. 2.12. The resistance of a thin-film resistor is given by the usual relation

$$R = \frac{\rho l}{A} = \frac{\rho l}{w h}. \tag{2.17}$$

where w and h are the width and thickness, respectively, of the strip, ρ is the resistivity of the film, and l is the length of the strip.

It is conventional and convenient to define the **sheet resistance** of a film resistor as

$$\frac{\rho}{h} = \text{sheet resistance}. \tag{2.18}$$

Sheet resistance has the dimension of resistance and usually is expressed in either ohms or ohms per square. With this definition (2.17) becomes

$$R = \left(\frac{\rho}{h}\right)\left(\frac{l}{w}\right) = (\text{sheet resistance}) \times (\text{number of squares}). \tag{2.19}$$

The number of squares in a film resistor is dimensionless and is obtained by dividing the length of the film into squares having sides equal to the width of the film, as illustrated by Fig. 2.13.

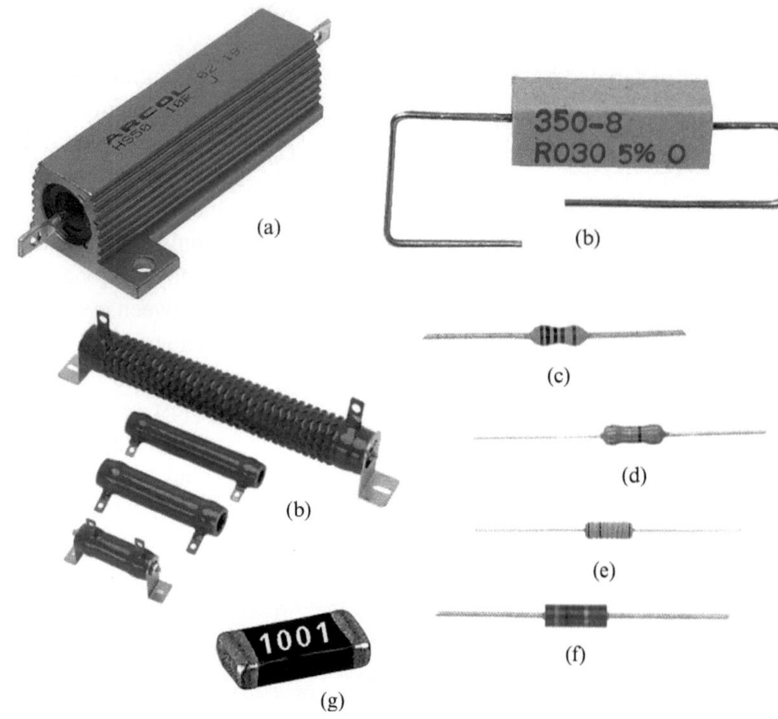

Fig. 2.10 Fixed resistors (not to scale): (**a**) Aluminum-encased wirewound, (**b**) power wirewound, (**c**) metal-oxide, (**d**) carbon-film, (**e**) metal-film, (**f**) carbon composition, (**g**) surface-mount chip (Photographs courtesy Rapid Electronics, Ltd.)

Fig. 2.11 Variable Resistors (not to scale) (Photos courtesy Rapid Electronics, Ltd.)

The resistance of a thin-film resistor is determined by the sheet resistance, which is a function of the material and film thickness, and the number of squares. The resistance is independent of the size of a square. For example, a film 2 μm wide by 10 μm long has the same resistance as one 200 nm wide and 1μm long, as illustrated by Fig. 2.14, if both have the same sheet resistance.

Tantalum nitride (TaN) is widely used for thin-film resistors, partly because the resistivity of TaN can be varied over a wide range by varying the composition. A common composition has a resistivity of about 250 μΩ cm at 25°C. A common thickness for thin-film resistors is 50 nm. A 50 nm film of 250 μΩ cm TaN has a sheet resistance of $50\,\Omega/\text{sq}$, which proves to be a convenient value.

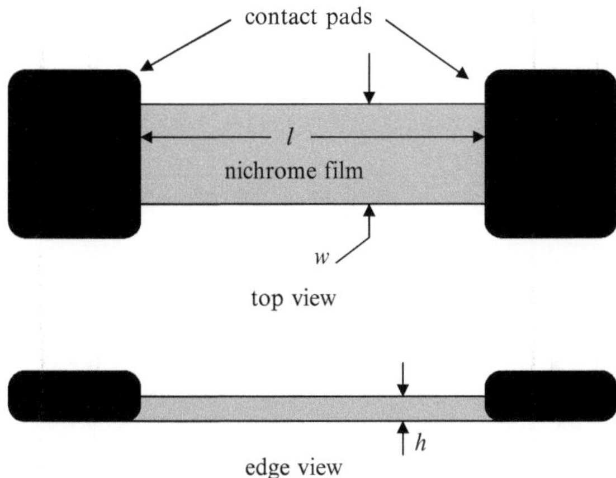

Fig. 2.12 Thin-film resistor

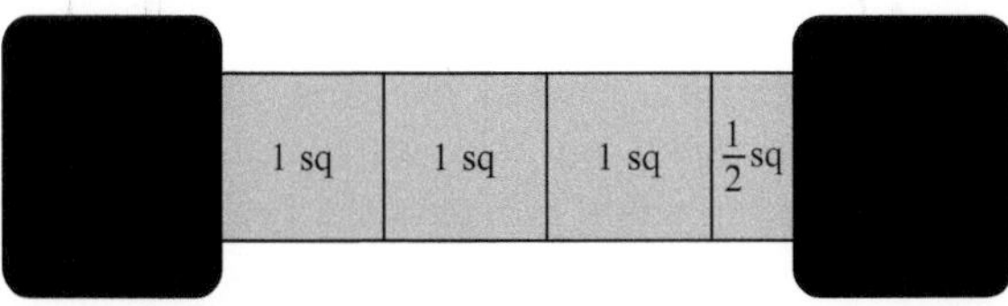

Fig. 2.13 Thin-film resistor squares

Exercise 2.8. The resistivity of a certain composition of nichrome is108 μΩ cm at 25°C. What is the thickness of a layer of this material that will have a sheet resistance of 75 Ω/sq at 25°C?

Exercise 2.9. The length of the film in a 500 Ω thin film resistor is fixed while its width is halved. What is the new resistance?

Exercise 2.10. The resistivity of a certain material at 25°C is 95 μΩ cm. (a) What is the sheet resistance of a 100 nm thick film of the material? (b) What is the resistance at 25°C of a thin film resistor made of 100 nm film that is 75 nm wide and 1μm long?

The natural tolerances of thin film resistors are at best about ± 10%; however, a thin film resistor can be trimmed to a very precise value by monitoring the resistance while using a laser to cut one or more notches in the film, as illustrated by Fig. 2.15. It is difficult to specify the width and depth of the notch because of variations in the thickness and even the composition of the film (which is why the natural tolerances are relatively poor). Ability to trim thin film resistors to such precise values is important to design and construction of high-precision electronic circuits.

2.8 E Series, Tolerance, and Standard Resistance Values

It is impossible for parts suppliers to stock resistors in every possible value, so a convention for defining a reasonable set of *standard values* is necessary. The convention used is based on what are called **E series**, which are approximately logarithmic divisions of a decade. Table 2.1 gives the values of the E12, E24, E48, E96, and E192 series, which are the most commonly used series for resistors. The E12 series divides 1 decade into 12 values, the E24 series divides 1 decade into 24 values, and so on. The E3 and E6 series are seldom used for resistors and are not shown here. The E3 series contains every other value from the E6 series, which contains every other value from the E12 series, which contains every other value from the E24 series. Similarly, the E48 series contains every other value from the E96 series, which contains every other value from the E192 series. The E24 series does not contain every other value from the E48 series, because E48 values have three significant digits, whereas E24 values have only two. But one can say that the E24 series contains rounded-off values from the E48 series.

The resistance of a resistor from any particular E series consists of a number from the series, multiplied by an integer power of ten. For example, the value 133 appears in the E48 series, so one can readily obtain resistors having nominal resistances of (e.g.) 133 Ω, 13.3 kΩ, and 1.33 MΩ. Every standard (off-the-shelf) resistor value equals the product of a number from an E series and an integer power of ten, but not all such values are necessarily available from a higher or lower series. For example, one can buy ± 10% and ± 5% (E12 and E24) resistors having values of 33, 330 Ω, 3.3 kΩ, and so on, but none of these values are available

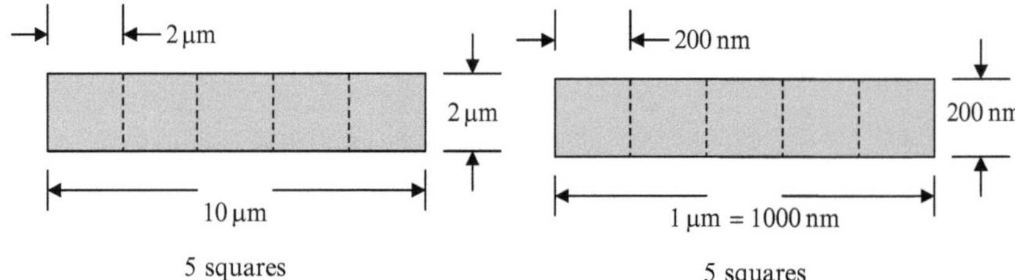

Fig. 2.14 Illustrating calculation of sheet resistance

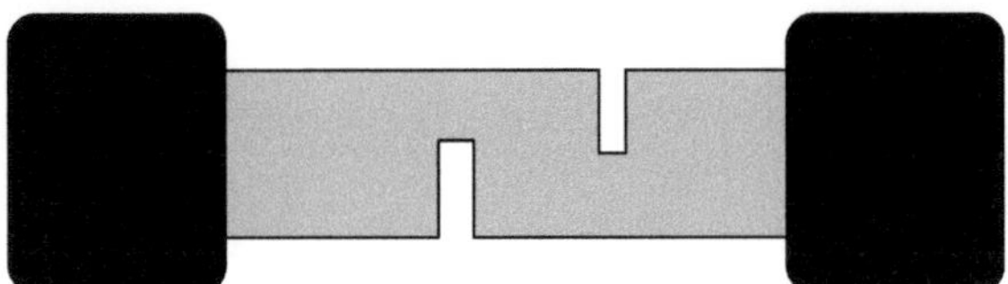

Fig. 2.15 Trimmed thin-film resistor

in the higher-precision (E48, E96, and E192) series. As another example, resistors having values $383 \times 10^n\,\Omega$, with $n = -1, 0, 1, 2, \cdots$ are available in the E48, E96, and E192 series, but not in the E12 or E24 series.

A standard tolerance is associated with each value from an E series. Standard tolerances for off-the-shelf resistors are $\pm 20\%$ (E6 – rarely used, nowadays), $\pm 10\%$ (E12), $\pm 5\%$ (E24), and $\pm 1\%$ (E96). Standard tolerances for high-precision resistors (obtainable from some large suppliers and manufacturers) are $\pm 0.5\%$, $\pm 0.1\%$, and $\pm 0.05\%$. Resistors having tolerances as tight as $\pm 0.001\%$ are available in selected resistances from certain manufacturers. However, high-precision resistors are not bulk-produced in every value from any particular E series, but typically are produced in only certain values having wide application (e.g., in precision instrumentation and measurement systems).

Some manufacturers can produce special-order resistors having virtually any nominal value, with tolerances of better than 0.05%. Ordering a batch of such resistors might be justified if a particular value and tight tolerance are required for a circuit being produced in quantity.

Example 2.8. A calculation indicates that a resistor having resistance $R = 61.7\,\text{k}\Omega$ is needed in a certain circuit. Refer to Table 2.1 and give the nearest standard value from each series.
Solution: The E12 value nearest to the specified value is 56 kΩ. The E24 value nearest to the specified value is 62 kΩ. The value from the E48, E96, and E192 series nearest to the specified value is 61.9 kΩ.

Example 2.9. One kind of ammeter measures the current in a wire by measuring the voltage across a precision resistor called a *shunt* connected in series with the wire. In a particular case, the shunt has resistance $R = 0.05\,\Omega \pm 0.05\%$ and the voltage across the shunt is $V = 1.872\,\text{mV}$. What is the current I through the shunt?
Solution: The current is in the range $I_{\min} \le I \le I_{\max}$, where

$$I_{\min} = \frac{V}{R_{\max}} = \frac{1.872\,\text{mV}}{(0.05\,\Omega)(1.0005)} = 37.421\,\text{mA},$$
$$I_{\max} = \frac{V}{R_{\min}} = \frac{1.872\,\text{mV}}{(0.05\,\Omega)(0.9995)} = 37.459\,\text{mA}.$$

Exercise 2.11. A calculation indicates that a resistor having resistance $R = 50\,\text{k}\Omega \pm 1\%$ is needed in a certain application. What value would you specify and from which series?

2.9 Resistor Marking

Axial-lead composition, coated wire-wound, carbon-film, and metal-film resistors are color-coded for resistance and tolerance by either 4-band or 5-band codes

Table 2.1 E series for standard resistor values

E192				E96		E48	E24	E12
± 0.5%				± 1%		± 2%	± 5%	± 10%
100	178	316	562	100	316	100	10	10
101	180	320	569	102	324	105	11	12
102	182	324	576	105	332	110	12	15
104	184	328	583	107	340	115	13	18
105	187	332	590	110	348	121	15	22
106	189	336	597	113	357	127	16	27
107	191	340	604	115	365	133	18	33
109	193	344	612	118	374	140	20	39
110	196	348	619	121	383	147	22	47
111	198	352	626	124	392	154	24	56
113	200	357	634	127	402	162	27	68
114	203	361	642	130	412	169	30	82
115	205	365	649	133	422	178	33	
117	208	370	657	137	432	187	36	
118	210	374	665	140	442	196	39	
120	213	379	673	143	453	205	43	
121	215	383	681	147	464	215	47	
123	218	388	690	150	475	226	51	
124	221	392	698	154	487	237	56	
126	223	397	706	158	499	249	62	
127	226	402	715	162	511	261	68	
129	229	407	723	165	523	274	75	
130	232	412	732	169	536	287	82	
132	234	417	741	174	549	301	91	
133	237	422	750	178	562	316		
135	240	427	759	182	576	332		
137	243	432	768	187	590	348		
138	246	437	777	191	604	365		
140	249	442	787	196	619	383		
142	252	448	796	200	634	402		
143	255	453	806	205	649	422		
145	258	459	816	210	665	442		
147	261	464	825	215	681	464		
149	264	470	835	221	698	487		
150	267	475	845	226	715	511		
152	271	481	856	232	732	536		
154	274	487	866	237	750	562		
156	277	493	876	243	768	590		
158	280	499	887	249	787	619		
160	284	505	898	255	806	649		
162	287	511	909	261	825	681		
164	291	517	919	267	845	715		
165	294	523	931	274	866	750		
167	298	530	942	280	887	787		
169	301	536	953	287	909	825		
172	305	542	965	294	931	866		
174	309	549	976	301	953	909		
176	312	556	988	309	976	953		

Table 2.2 Resistor color codes

Color	Value	Multiplier	Precision
Black	0	10^0	
Brown	1	10^1	±1%
Red	2	10^2	±2%
Orange	3	10^3	
Yellow	4	10^4	
Green	5	10^5	±0.5%
Blue	6	10^6	±0.25%
Voilet	7	10^7	±0.1%
Grey	8	10^8	±0.05%
White	9	10^9	
Gold		10^{-1}	±5%
Silver		10^{-2}	±10%
None			±20%

as defined in Table 2.2. A **4-band code** is used for resistors from the E24 and lower series. A **5-band code** is used for resistors from the E48 and higher series. The first band is closer to one end of the resistor than the last band is to the other. The first two bands denote the first two significant digits of the resistance. In a 4-band code, the third band is the multiplier, which is an integer power of ten, and the fourth band denotes the tolerance. In a 5-band code, the third band denotes the third significant digit, the fourth band denotes the multiplier, and the fifth band is the tolerance.

Example 2.10. Give the resistance of resistors marked as follows:

(a) Red, violet, yellow, green; (b) red, violet, brown, orange, violet.

Answers: (a) $270\,\text{k}\Omega \pm 0.5\%$, (b) $271\,\text{k}\Omega$ $\pm 0.1\%$.

Exercise 2.12. Give the resistance of resistors marked as follows:

(a) Yellow, orange, red, gold; (b) yellow, green, white, orange, violet.

In addition to resistance and tolerance, properies of resistors include power-dissipation rating, self-heating coefficient (described in Chapter 5), maximum operating temperature, and temperature coefficient of resistance (described in the next section). Most manufacturers and vendors provide that information in datasheets for their offerings.

Exercise 2.13. Using a search engine of your choosing, do a search on "resistors" and locate several manufacturers' web sites. Visit each site and examine the technical data they provide on their offerings. Identify the technologies that provide (a) the lowest temperature coefficient of resistance, (b) the highest precision, and (c) the highest power-dissipation ratings.

2.10 Variation of Resistivity and Resistance with Temperature

Resistivities of materials and thus resistances of conductors and resistors vary with temperature.[17] Temperature dependence of resistivities of certain metals is of interest because conductors in electric circuits are metallic. Figure 2.16 shows resistivity as a function of temperature for aluminum (Al), copper (Cu), and gold (Au).[18] The scales on the axes are logarithmic. The vertical (resistivity) scale is $\log(\rho/\rho_0)$, where $\rho_0 = 10^{-8}\ \Omega\,\text{m}$ (1 μΩ cm). The horizontal (temperature) scale is $\log(T/T_0)$, where $T_0 = 1\,\text{K}$ (one kelvin). For example, for $\log(T/T_0) = 2$, we have $T/T_0 = 100$ and the temperature is $T = 100\,T_0 = 100\,\text{K}$. At that temperature, the resistivity of copper is given by $\log(\rho/\rho_0) = -0.5$, so $\rho = \rho_0 \times 10^{-0.5} = 0.316\,\mu\Omega\text{cm}$. The vertical reference line indicates $T = 298$ K $=$ 25°C (77°F), where the values are $\rho_{Al}/\rho_0 = 2.709$, $\rho_{Cu}/\rho_0 = 1.712$, and $\rho_{Au}/\rho_0 = 2.255$.

Figure 2.16 suggests that resistivity is very nearly a linear function of temperature for temperatures above about $10^{2.25}\,\text{K} \cong 178\,\text{K}$, which is approximately -95°C. Operating temperatures at or above room temperature are common for linear circuits, so we are particularly interested in temperatures above about 298 K = 20°C or so. Figure 2.17 shows graphs of resistivity versus temperature (both K and °C) for silver (Ag), aluminum (Al), gold (Au), copper (Cu), and tungsten (W). For $200\,\text{K} \le T \le 900\,\text{K}$ ($25°\text{C} \le T \le 673°\text{C}$), resistivity is very nearly a linear function of temperature. Actually, the linear range for these metals extends from about -100°C to almost 1000°C.

For a sufficiently small change ΔT in temperature, the derivative of resistivity with respect to temperature can be approximated by

$$\frac{d\rho(T)}{dT} \cong \frac{\rho(T+\Delta T) - \rho(T)}{\Delta T},$$

which gives, for the variation of resistivity with temperature,

$$\begin{aligned}\rho(T+\Delta T) &\cong \rho(T) + \frac{d\rho(T)}{dT}\Delta T \\ &= \rho(T)[1 + \alpha(T)\,\Delta T], \qquad (2.20)\end{aligned}$$

where

$$\alpha(T) = \frac{1}{\rho(T)}\frac{d\rho(T)}{dT} \qquad (2.21)$$

is the **temperature coefficient of resistivity** at temperature T. In engineering applications of (2.20), temperature T often is expressed in degree Celsius and resistivities of materials and the associated temperature coefficients are commonly specified at $T_{25} = 25°\text{C} = 298\,\text{K}$. In such applications it is usual to specialize (2.20) and write

$$\hat{\rho}_T = \rho_{25}\,[1 + \alpha_{25}(T - T_{25})], \qquad (2.22)$$

where $T_{25} = 25°\text{C} = 298\,\text{K}$; the parameters ρ_{25}, α_{25} are the resistivity and temperature coefficient, respectively; and $\hat{\rho}_T$ is an approximation to the true (measured) resistivity at temperature T. For many metals (2.22) can be used to estimate resistivity over a wide range – from about -100°C to almost 1000°C.

The numerical value of a temperature coefficient is the same whether expressed in $°\text{C}^{-1}$ or K^{-1}, because the quantity $T - T_{25}$ in (2.22) is the same, whether both temperatures are expressed in degree Celsius or Kelvin. In practice (e.g., on component data sheets), temperature coefficients are commonly expressed in **parts per million** (ppm), which is 10^6 K times the coefficient expressed in K^{-1} (or $°\text{C}^{-1}$):

[17] Resistivity can also vary with pressure, strain, strength and direction of an applied magnetic field, and other things. The operation of various transducers (e.g., strain gauges) is based upon such effects.

[18] *Handbook of Chemistry and Physics* (76th Ed.), edited by David R. Lide, CRC Press, 1995.

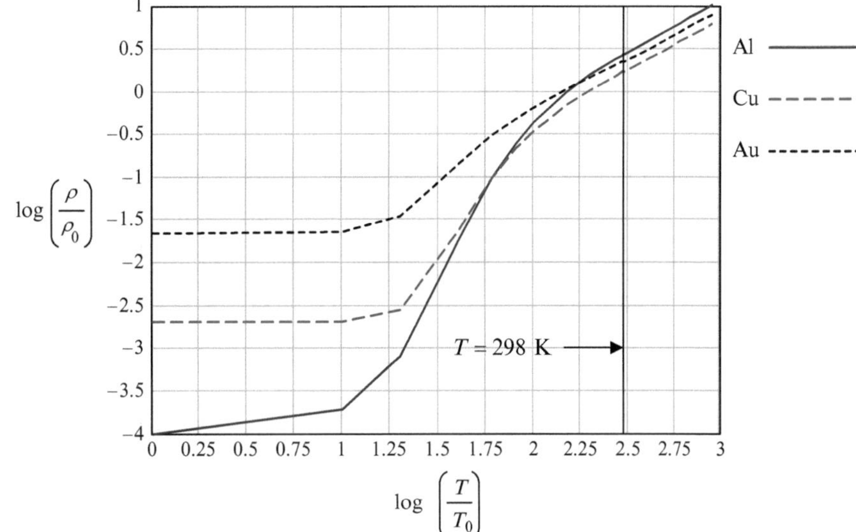

Fig. 2.16 Graphs of resistivity versus temperature (log-log scale) for pure aluminum (Al), copper (Cu), and gold (Au), where $\rho_0 = 10^{-8}\ \Omega\text{m} = 1\ \mu\Omega\text{cm}$ and $T_0 = 1\,\text{K}$. (See Footnote 18)

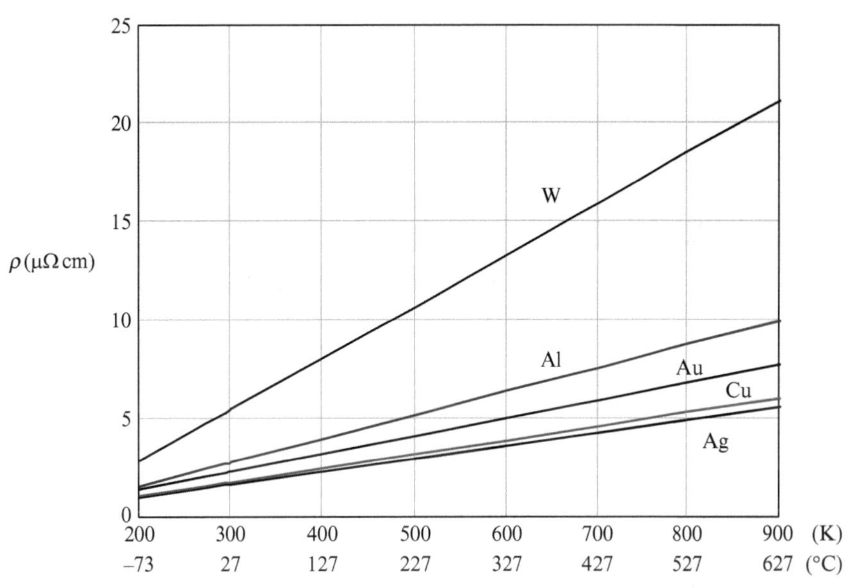

Fig. 2.17 Resistivities of selected metals (See Footnote 18)

$$\alpha\,(\text{ppm}) = \alpha\left(\text{K}^{-1}\right) \times 10^6\,\text{K}. \tag{2.23}$$

Table 2.3 gives resistivities and temperature coefficients at 25°C for selected metals. Be aware that such data are available from many sources, and temperature coefficients in particular can vary significantly from one source to the next. Also, the values in Table 2.3 are for pure metals, whereas wires and cables used in electrical components and apparatus often are alloys whose resistivities and temperature coefficients can differ significantly from those for the dominant pure metal. The data given in Table 2.3 are intended for use only in examples and problems given in this book, not for actual engineering design.

Figure 2.18 shows graphs of the percent error versus temperature for resistivities calculated using (2.22), calculated as

$$\text{error}(T) = 100\frac{\hat{\rho}_T - \rho_T}{\rho_T} \tag{2.24}$$

Table 2.3 Resistivities and temperature coefficients for selected metals[19]

Metal	Resistivity at 25°C (μΩ cm)	Temperature coefficient at 25°C (K^{-1})	Temperature coefficient at 25°C (ppm)
Silver (Ag)	1.62	4.053×10^{-3}	4053
Aluminum (Al)	2.71	4.438×10^{-3}	4438
Gold (Au)	2.26	4.022×10^{-3}	4022
Copper (Cu)	1.71	4.138×10^{-3}	4138
Tungsten (W)	5.39	4.849×10^{-3}	4849

For temperatures between 298 K (25°C) and 898 K (625°C), the maximum error (magnitude) is less than 5%. For that range of temperatures, the maximum error (magnitude) for copper (a popular material for wires) is approximately 2%.

Exercise 2.14. Refer to Fig. 2.18. Why is the error at 298 K equal to zero for all four linear approximations?

Example 2.11. A copper wire is heated from 25°C to 100°C. What is the percent increase in resistivity?

Solution: From Table 2.3, the resistivity and temperature coefficient for copper at 25°C are $\rho_{25} = 1.71\ \mu\Omega$ cm and $\alpha_{25} = 4{,}138$ppm, respectively. From (2.22), the resistivity at 100°C is

$$\begin{aligned}\rho_{100} &= \rho_{25}[1 + \alpha_{25}(T - T_{25})] \\ &= (1.71\ \mu\Omega\ \text{cm})[1 + (4{,}138 \times 10^{-6\circ}\text{C}^{-1}) \\ &\quad \times (100^\circ\text{C} - 25^\circ\text{C})] \\ &= 2.24\ \mu\Omega\ \text{cm}.\end{aligned}$$

The percent increase is

$$100 \times \frac{\rho_{100} - \rho_{25}}{\rho_{25}} \cong 31\%.$$

[19]Ibid. The temperature coefficients were obtained from the slopes of linear-least-squares fits to the resistivity data given there.

Exercise 2.15. The resistivity of a certain metal is $1.6 \times 10^{-8}\,\Omega$ m at 10°C and $2.0 \times 10^{-8}\,\Omega$ m at 70°C. Estimate the temperature coefficient of resistivity for the metal at 10°C.

The resistivities of materials used to make resistors also vary with temperature. Such variation is approximately linear over the range of operating temperatures for any particular resistor. In practice, it usually is more convenient to deal directly with resistance than resistivity. For any particular resistor, resistance is proportional to resistivity, so (2.22) gives

$$R_T = R_{25}\,[1 + \alpha_{25}(T - T_{25})]. \tag{2.25}$$

In view of (2.25), the temperature coefficient of resistivity is also called the **temperature coefficient of resistance**, usually called just the **temperature coefficient** or abbreviated as **TCR**.

Temperature coefficients for resistors vary considerably. Generally, carbon composition and carbon-film resistors have temperature coefficients on the order of 10^{-3}K^{-1}, whereas wirewound and some metal-film and metal-oxide resistors have temperature coefficients on the order of 10^{-5}K^{-1}. In general, resistances of wirewound, metal-film, and metal-oxide resistors are much less sensitive to temperature than carbon-film and carbon-composition resistors.

The resistivities of many materials approach zero (the materials become *superconductors*) at temperatures approaching 0 K. The resistivity of a superconductor is less than $10^{-22}\,\Omega$cm, whereas the lowest (room temperature) resistivity found in metals is on the order of $10^{-6}\,\Omega$cm. Thus resistivities for superconductors are about a factor of 10^{-16} smaller than those of materials ordinarily used for conductors in electric circuits. Some materials become superconductors at temperatures near 77 K (the boiling point of nitrogen), which are much easier to attain and maintain than those near absolute zero. As of this writing, use of superconducting materials is limited to a few specialized applications.

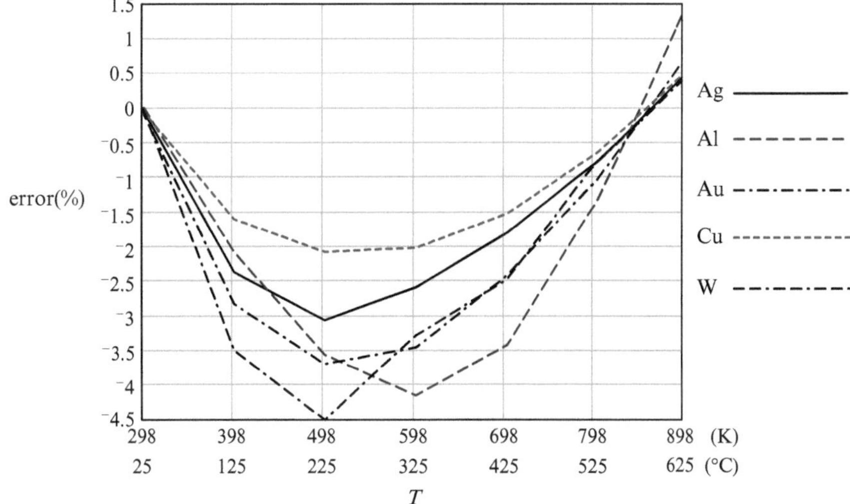

Fig. 2.18 Percent error in a linear approximation to resistivity as a function of temperature for five metals. See (2.24). The curves are not smooth because each contains only 11 data points

Table 2.4 American wire gauge and wire diameters

AWG	6	8	10	12	14	16	18	20	22
d (mm)	4.115	3.264	2.588	2.052	1.628	1.290	1.024	0.813	0.643

2.11 American Wire Gauge (AWG) and Metric Wire Gauge (MWG)

In the United States, the size of wire is specified by a whole number called *gauge* (AWG, or *American wire gauge*). Table 2.4 gives the diameters of selected AWG wires. The diameter of any AWG wire is given by

$$D_{AWG} = 8.251e^{-(0.1159)AWG}\,\text{mm}. \tag{2.26}$$

For AWG 00, 000, and 0000 sizes, use $AWG = -1, -2, -3$, respectively, in (2.26).

Example 2.12. The diameter of AWG 18 wire is

$$D_{18} = 8.251\,\text{mm}\,e^{-(0.1159)(18)} = 1.024\,\text{mm}$$

and the diameter of AWG 000 wire is

$$D_{000} = 8.251\,\text{mm}\,e^{-(0.1159)(-2)} = 10.404\,\text{mm}.$$

The resistance of a length L (m) of AWG copper wire is given by

$$R_{\text{Cu}} \cong \left(320 \times 10^{-6}\,\Omega\,\text{m}^{-1}\right) L\; e^{(0.232)\,(AWG)}. \tag{2.27}$$

For other materials, use

$$R = \left(\frac{\rho}{\rho_{\text{Cu}}}\right) R_{Cu}. \tag{2.28}$$

Example 2.13. The resistance of 1 km of AWG 000 copper wire is

$$R_{\text{Cu}} \cong \left(320 \times 10^{-6}\,\Omega\,\text{m}^{-1}\right)(1\text{km})\; e^{(0.232)(-2)} \cong 0.201\,\Omega.$$

and the resistance of 1 km of AWG 000 aluminum wire is

$$R_{\text{Al}} = \frac{\rho_{\text{Al}}}{\rho_{\text{Cu}}} R_{\text{Cu}} = \left(\frac{2.709}{1.712}\right)(0.202\,\Omega) = 0.320\,\Omega.$$

Exercise 2.16. The diameter of a certain wire must be at least 5 mm. What is the largest possible AWG designation of a wire meeting this requirement?

In Europe, wire size is specified by diameter in mm or less often by metric wire gauge (MWG), which is ten times the diameter in mm. For example, the diameter of MWG 30 wire is 3 mm. Simply specifying wire diameter in mm seems to make a great deal of sense. However, an advantage of the AWG convention is that equal steps in AWG designation correspond to equal resistance ratios; for example, if decreasing the AWG designation of a certain kind of wire from 22 to 20 decreases the resistance of the wire by a factor of 0.6, then so would a decrease from 20 to 18, or 18 to 16, and so on.

Exercise 2.17. What is the MWG designation for AWG 18 wire?

2.12 DC and AC

Historically, dc and ac arose as abbreviations for *direct current* and *alternating current*, respectively, where *direct* was synonymous with *constant* and *alternating* was synonymous with *sinusoidal*. Nowadays, dc and ac are applied to either current or voltage. Thus folks speak of *dc voltage*, despite the seeming contradiction, or *ac current*, despite the seeming redundancy. Furthermore, ac is commonly used to mean a time-varying (not just sinusoidal) current or voltage. But for the present, we use *ac* as a synonym for *sinusoidal*.

Sinusoidal currents and voltages are especially important in electrical engineering, for reasons you will come to understand as you progress through this book and the rest of your curriculum. We may regard a dc current or voltage as a sinusoidal current or voltage having frequency $f = 0$; e.g., if $f = 0$, then $V\cos(2\pi f\,t) = V\cos(0) = V$.

Conventionally, constant currents and voltages and constant parameters having the dimensions of current or voltage are denoted by capital letters; e.g., $I = 20\,\text{mA}$ or $V = 5\,\text{V}$. Time-varying currents and voltages are denoted by lower-case letters; e.g., $i(t) = I\cos(2\pi f\,t)$ and $v(t) = V\cos(2\pi f\,t)$, where the current i and the voltage v vary with time and the parameters I and V are constant.

Most circuit components respond differently to ac than to dc. In particular, the resistance of a conductor (such as a wire) is larger for a rapidly varying current than for one that varies more slowly or is constant. The resistance of a conductor carrying a sinusoidal current increases with the frequency of the current, as described in the next section.

2.13 Skin Effect and Proximity Effect

In this section, we use $\mathcal{R}$ (script) to denote radius of a cylindrical conductor and R (italic) to denote resistance.

Refer to Fig. 2.19, which shows (qualitatively) a cross section of a conductor carrying a sinusoidal current $i = I\cos(2\pi f\,t)$ and above that, a graph illustrating (again qualitatively) how the current is distributed in the conductor. The abscissa is radial distance from the center of the conductor and the ordinate is current density J, which is current per unit

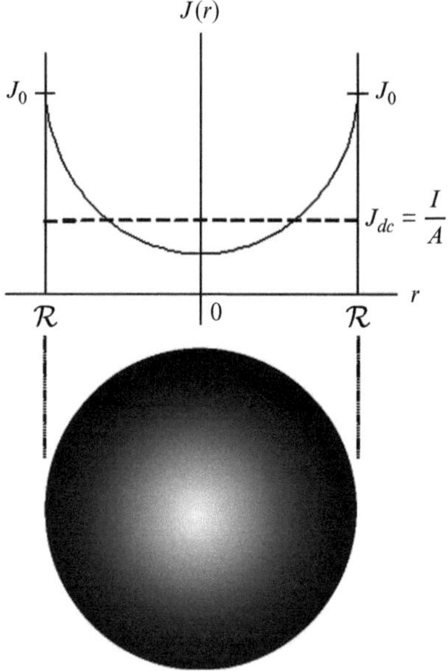

Fig. 2.19 An illustration of skin effect

cross-sectional area. At dc (for $f = 0$), the total current I is uniformly distributed across any diameter, as indicated by the dashed line at $J = I/A$, where $A = \pi \mathcal{R}^2$ is the cross-sectional area of the conductor. As the frequency of the current is increased, the current density becomes non-uniform, being greater at the surface of the conductor than at the center. At sufficiently high frequencies, most of the current is concentrated in a relatively thin concentric layer of material near the surface. This phenomenon is called **skin effect**. Under conditions described below, the current density J at a distance r from the center of a cylindrical conductor is given by

$$J(r) = J_0 e^{-(\mathcal{R}-r)/\delta(f)};\ 0 \le r \le \mathcal{R}, \tag{2.29}$$

where $\mathcal{R}$ is the radius of the conductor, J_0 is the current density at the surface of the conductor, and f is the frequency of the time-varying current. The parameter $\delta(f)$ in (2.29) is a function of frequency called **skin depth**, given by

$$\delta(f) = \sqrt{\frac{\rho}{\pi f \mu_r \mu_0}}, \tag{2.30}$$

where ρ is the resistivity of the material (Ω m), $\mu_0 = 4\pi \times 10^{-7}\,\mathrm{Hm}^{-1}$ is the *permeability* of a vacuum, and μ_r is the *relative permeability* of the material (dimensionless). For non-magnetic materials, such as copper and aluminum, $\mu_r = 1$. From (2.29), skin depth is the distance from the surface of a conductor at which the current density in the conductor is $e^{-1} \cong 0.37$ times the value at the surface.

Skin effect concentrates current near the surface of a conductor, reducing the effective cross-sectional area A in the relation $R = \rho l/A$ and thereby makes the resistance of a conductor larger for a sinusoidal current (**ac resistance**) than that for a constant current (**dc resistance**). In what follows, we obtain an expression for the ratio of ac resistance to dc resistance.

The (total) current in a conductor having cross-sectional area A is given by

$$i = \int_A J dA, \tag{2.31}$$

where J (A m^{-2}) is the current density over the cross section. In words, the total current i in a conductor is obtained by integrating the current density J over a cross-sectional area A of the conductor. For a cylindrical conductor (a wire) having radius $\mathcal{R}$, we may express (2.31) in polar coordinates as

$$i = \int_0^{2\pi} \int_0^{\mathcal{R}} J(r, \theta) r\, dr\, d\theta, \tag{2.32}$$

where i is the current in the conductor at any particular time and $J(r, \theta)$ is the current density over a cross section of the conductor at the same time. From (2.29) and (2.32),

$$\begin{aligned} i &= \int_0^{2\pi} \int_0^{\mathcal{R}} J_0 e^{-(\mathcal{R}-r)/\delta} r\, dr\, d\theta \\ &= 2\pi J_0 e^{-\mathcal{R}/\delta} \int_0^{\mathcal{R}} e^{r/\delta} r\, dr \\ &= 2\pi J_0 e^{-\mathcal{R}/\delta} \left[\delta^2 + \mathcal{R}\delta e^{\mathcal{R}/\delta} - \delta^2 e^{\mathcal{R}/\delta}\right] \\ &= 2\pi J_0 \left[\delta^2 e^{-\mathcal{R}/\delta} + \mathcal{R}\delta - \delta^2\right] \\ &= 2\pi J_0 \mathcal{R}\delta \left[1 + \frac{\delta}{\mathcal{R}}\left(e^{-\mathcal{R}/\delta} - 1\right)\right]. \end{aligned} \tag{2.33}$$

For $\mathcal{R} \gg \delta$, (2.33) reduces to

$$i \cong 2\pi J_0 \mathcal{R}\delta, \tag{2.34}$$

which is exactly what we would obtain if the current density were uniform and equal to J_0 in a hollow tube having outer radius $\mathcal{R}$ and inner radius $\mathcal{R} - \delta$ (wall thickness δ). We can show this as follows. If the current density were given by

$$J(r) = \begin{cases} J_0, & \mathcal{R} - \delta < r \le \mathcal{R}, \\ 0, & r \le \mathcal{R} - \delta, \end{cases} \tag{2.35}$$

then the current would be given by

$$\begin{aligned} i &= \int_0^{2\pi} \int_{\mathcal{R}-\delta}^{\mathcal{R}} J_0 r\, dr\, d\theta = 2\pi J_0 \left[\frac{r^2}{2}\right]_{\mathcal{R}-\delta}^{\mathcal{R}} \\ &= \pi J_0 \left[\mathcal{R}^2 - (\mathcal{R} - \delta)^2\right] = \pi J_0 \left(2\delta\mathcal{R} - \delta^2\right) \\ &= 2\pi J_0 \mathcal{R}\delta \left(1 - \frac{\delta}{2\mathcal{R}}\right) \cong 2\pi J_0 \mathcal{R}\delta;\ \mathcal{R} \gg \delta, \end{aligned}$$

which agrees with (2.34), as claimed above. It follows that the *effective cross-sectional area* seen by a current in a cylindrical conductor is the cross-sectional area of the tube described above, given by

$$A = \pi\left[\mathcal{R}^2 - (\mathcal{R}-\delta)^2\right] = 2\pi\mathcal{R}\delta - \pi\delta^2 = \pi\delta(2\mathcal{R}-\delta). \tag{2.36}$$

For $\mathcal{R} \gg \delta$, (2.36) gives

$$A \cong 2\pi\mathcal{R}\delta. \tag{2.37}$$

If the radius of a cylindrical conductor is much larger than the skin depth, skin effect appears to reduce a solid cylindrical conductor having radius $\mathcal{R}$ to a tubular conductor having outer radius $\mathcal{R}$ and inner radius $\mathcal{R}-\delta$. Whereas the actual cross-sectional area of a conductor having radius $\mathcal{R}$ – that seen by a dc current – is $A_{dc} = \pi\mathcal{R}^2$, the *effective* cross-sectional area seen by an ac current is given by (2.36). Consequently, the ratio of ac resistance to dc resistance for the conductor is given by

$$\frac{R_{ac}}{R_{dc}} \cong \frac{\rho l}{A_{ac}}\left(\frac{\rho l}{A_{dc}}\right)^{-1} = \frac{A_{dc}}{A_{ac}} = \frac{\pi\mathcal{R}^2}{\pi\delta(2\mathcal{R}-\delta)} = \frac{\mathcal{R}^2}{\delta(2\mathcal{R}-\delta)}. \tag{2.38}$$

It is conventional to describe wires in terms of diameter D, rather than radius $\mathcal{R}$. From (2.38), the ratio of ac resistance R_{ac} to dc resistance R_{dc} for a cylindrical conductor having diameter $D = 2\mathcal{R}$ is given by

$$\frac{R_{ac}}{R_{dc}} \cong \frac{D^2}{4\delta(D-\delta)}, \tag{2.39}$$

where $\delta = \delta(f)$ is skin depth, given by (2.30). Figure 2.20 shows a graph of this ratio versus frequency for five American wire gauge (AWG) sizes of copper wire at $T = 25°\text{C}$ (77°F).

The fact that current density decreases exponentially with distance from the surface as $e^{-r/\delta}$ suggests that (2.39) provides a good approximation to R_{ac}/R_{dc} if the radius of the conductor exceeds three skin depths. Taking $D = 6\delta$ ($\mathcal{R} = 3\delta$) in (2.39) gives

$$\frac{R_{ac}}{R_{dc}} \geq \frac{(6\delta)^2}{4\delta(6\delta-\delta)} = \frac{36}{20} = 1.8, \tag{2.40}$$

which means that the curves in Fig. 2.20 are valid above the horizontal line at $R_{ac}/R_{dc} = 1.8$. Also, the ac resistance must approach the dc resistance at low frequencies, so the curves must approach unity, somewhat as shown, for frequencies approaching zero. Finally, there is no reason for the curves to make other than a smooth transition from their behavior above 1.8 to their behavior for frequencies approaching zero. In other words, the curves in

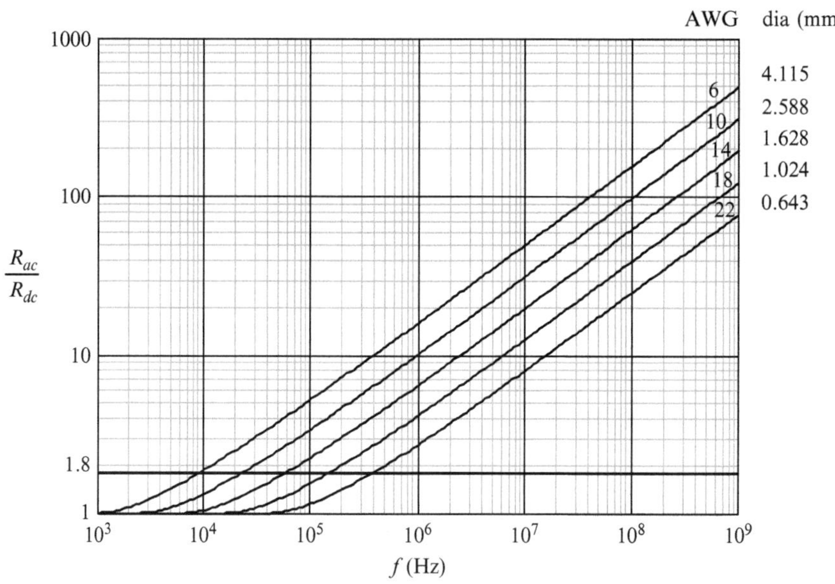

Fig. 2.20 Ratio of ac to dc resistance versus frequency for a 1-m length of each of five wire sizes (copper at 25°C)

Fig. 2.20 cannot be too far wrong, even for skin depths larger than one-third of a radius (below the line at $R_{ac}/R_{dc} = 1.8$).

Figure 2.20 illustrates the fact that the ratio of ac to dc resistance increases with wire diameter. You might conclude from this that countering skin effect in any particular case requires using smaller wire. However, this is not the case, because dc and ac resistance (individually) decrease with increasing wire diameter (decreasing AWG designation).

The ratio R_{ac}/R_{dc} can be expressed in terms of AWG wire sizes, as follows. From (2.39) and (2.26),

$$\left(\frac{R_{ac}}{R_{dc}}\right)_{AWG} \cong \frac{D_{AWG}{}^2}{4\delta(D_{AWG} - \delta)},$$

$$D_{AWG} \cong 8.251e^{-(0.1159)AWG}\,\text{mm}. \tag{2.41}$$

Exercise 2.18. The dc resistance of a 1-m length of AWG 22 wire at 25°C is 0.0526 Ω. (a) What is the ac resistance of the wire at (i) 10 MHz and 25°C? (ii) At 1 GHz and 25°C? (b) Repeat for a 1-m length of AWG 20 wire at the same temperature and made from the same material.

For a non-magnetic material, $\mu_r = 1$. In that case, a useful expression for skin depth is obtained by expressing resistivity ρ in μΩ cm and frequency f in MHz. With these adaptations (2.30) can be written

$$\delta(f) = \sqrt{\frac{\rho}{\pi f \mu_0}}$$
$$= \sqrt{\frac{1\,\mu\Omega\text{ cm}}{(\pi)(4\pi 10^{-7}\text{Hm}^{-1})(1\text{ MHz})}}\sqrt{\frac{\rho/(1\,\mu\Omega\text{ cm})}{f/(1\text{ MHz})}}$$

and finally as

$$\delta(f) = d\sqrt{\frac{\rho/(1\,\mu\Omega\text{ cm})}{f/(1\text{ MHz})}};$$
$$d = \sqrt{\frac{1\,\mu\Omega\text{ cm}}{(\pi)(4\pi 10^{-7}\text{Hm}^{-1})(1\text{ MHZ})}} = 50.33\,\mu\text{m}. \tag{2.42}$$

For example, the resistivity of aluminum at 25°C is $\rho = 2.71\,\mu\Omega$ cm, so the skin depth of aluminum at 25°C and 100 MHz is

$$\delta = d\sqrt{\frac{\rho/(1\,\mu\Omega\text{ cm})}{f/(1\text{ MHz})}}$$
$$= (50.33\,\mu\text{m}) \times \sqrt{\left(\frac{2.71}{100}\right)} = 8.29\,\mu\text{m}$$

and the skin depth of copper under the same conditions is

$$\delta = d\sqrt{\frac{\rho/(1\,\mu\Omega\text{ cm})}{f/(1\text{ MHz})}}$$
$$= (50.33\,\mu\text{m}) \times \sqrt{\left(\frac{1.71}{100}\right)} = 6.58\,\mu\text{m}$$

Figure 2.21 shows graphs of skin depth versus frequency for copper and aluminum at 25°C. At 25°C, skin depth in an aluminum conductor is about 25% larger than that in a copper conductor.

Example 2.14. Assume the relations derived in this section involving ac and dc resistance require $D \geq 6\,\delta(f)$. Derive an expression for the minimum frequency at which the formulas are valid for copper wire at 25°C. Construct a graph of the minimum frequency versus AWG wire size for $0 \leq \text{AWG} \leq 22$. Compare your results with Fig. 2.20.

Solution: From (2.42), the skin depth of copper at 25°C and frequency f is given by

$$\delta(f) = 50.33\sqrt{\frac{1.71}{f/(1\text{ MHz})}}\,\mu\text{m}$$
$$= \frac{65.82}{\sqrt{f/(1\text{ MHz})}}\,\mu\text{m}.$$

If we require $D \geq 6\delta(f)$, then

$$\delta(f) = \frac{65.82}{\sqrt{f/(1\text{ MHz})}}\,\mu\text{m} \leq \frac{D}{6}$$
$$\Rightarrow \frac{f}{1\text{ MHz}} \geq \left(\frac{395\,\mu\text{m}}{D}\right)^2$$

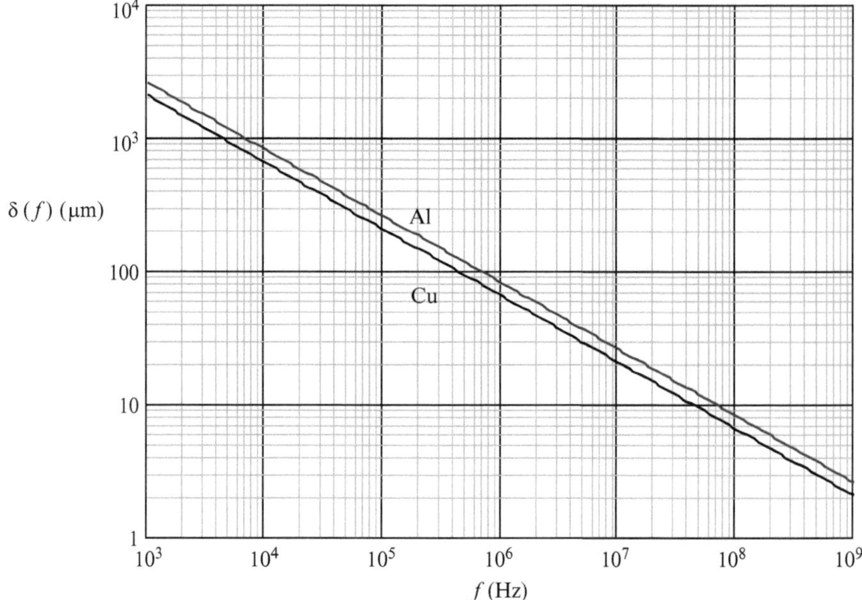

Fig. 2.21 Skin depth versus frequency for aluminum and copper at 25°C (77°F)

From (2.41), the diameter of an AWG-designated wire is given by

$$D_{AWG} = 8.251e^{-(0.1159)AWG}\ \text{mm}$$
$$= 8251e^{-(0.1159)AWG}\ \mu\text{m},$$

and so we require

$$\frac{f}{1\,\text{MHz}} \geq \left(\frac{395}{8251e^{-(0.1159)AWG}}\right)^2$$
$$\Rightarrow f \geq 2.292e^{(0.2318)AWG}\ \text{kHz}.$$

Figure 2.22 shows a graph of the minimum frequency given above versus AWG wire size.

Computed values below are in good agreement with the frequencies at which the lines in Fig. 2.20 intersect the line $R_{ac}/R_{dc} = 1.8$

AWG	6	10	14	18	22
Minimum Frequency (kHz)	9.21	23.3	58.9	149	376

Exercise 2.19. Construct graphs (on the same axes) of skin depth versus frequency for copper for $T = 0°\text{C}$ and $T = 100°\text{C}$ and for $1\,\text{kHz} \leq f \leq 1\,\text{GHz}$. Use logarithmic scales on the axes, as in Fig. 2.21.

Not all conductors are cylindrical. For example, conductors in integrated circuits are more nearly rectangular (in cross section). Skin-effect calculations for conductors in integrated circuits (IC's) can be complicated. For one thing, IC's are built in layers, so conditions above and below a conductor can be different, and skin depths can differ significantly on different surfaces of such a conductor. Typically, about all one would attempt is to determine the theoretical skin depth in relation to conductor cross-section dimensions. As of this writing, the frequencies of operation of IC's and the dimensions of conductors in IC's are such that skin effect is negligible.

Example 2.15. A copper conductor in a certain high-speed integrated circuit has an approximately rectangular cross section with height $h = 50\,\text{nm}$ and width $w = 25$ nm. The highest frequency of interest is $f = 30\,\text{GHz}$ and the operating temperature is $T = 100°\text{C}$. Is skin effect significant? If not, at what frequency would skin effect become significant?

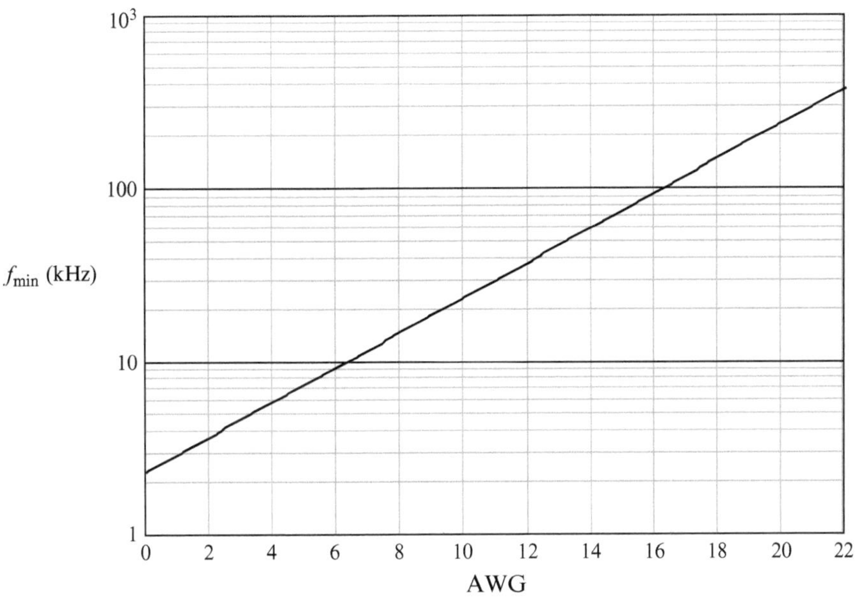

Fig. 2.22 Graph of Example 2.14

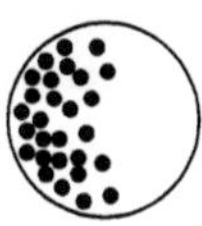
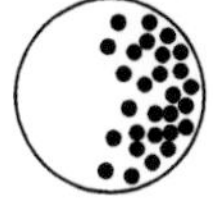

(a) currents in opposite directions

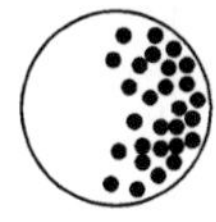
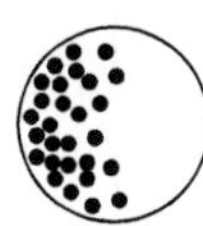

(b) currents in the same direction

Fig. 2.23 Illustrating proximity effect

Solution: The resistivity of copper at 100°C is

$$\begin{aligned}\rho_{100} &\cong \rho_{25}[1 + \alpha_{25}(100^\circ\mathrm{C} - 25^\circ\mathrm{C})] \\ &\cong (1.71\,\mu\Omega\,\mathrm{cm})[1 + (4138 \times 10^{-6})(75)] \\ &\cong 2.24\,\mu\Omega\,\mathrm{cm}.\end{aligned}$$

The skin depth at 30 GHz is

$$\begin{aligned}\delta &= d\sqrt{\frac{\rho/(1\,\mu\Omega\,\mathrm{cm})}{f/(1\,\mathrm{MHz})}} \\ &\cong (50.33\,\mu\mathrm{m})\sqrt{\frac{2.24}{30 \times 10^3}} \cong 435\,\mathrm{nm},\end{aligned}$$

which is much larger than the largest dimension of the conductor, so skin effect is insignificant. For skin effect to be significant, the skin depth must be on the order of the largest dimension. The corresponding frequency is

$$(50.33\,\mu\mathrm{m})\sqrt{\frac{2.24}{f/1\,\mathrm{MHz}}} \cong 50\,\mathrm{nm}$$

$$\Rightarrow f \cong 2.27\,\mathrm{THz},$$

which lies in the infrared region of the electromagnetic spectrum.

The *proximity effect*, like the skin effect, is a essentially a manifestation of the facts that (1) a moving charge creates a magnetic field and (2) a magnetic field exerts a force on a moving charge. As a result, currents parallel to and in proximity to one another exert forces on each other. The proximity effect distorts the current density in conductors running side-by-side. If the currents in a pair of parallel conductors are in opposite directions, the current will be concentrated in the halves of the conductors farthest from each other, as illustrated by Fig. 2.23(a). If the currents in a pair of parallel conductors are in the same direction, the current will be concentrated in the half of the each conductor nearest the other, as illustrated by Fig. 2.23(b). The proximity-effect force on a current is also felt by the conductor carrying the current. The force can be quite large, and in some industrial plants where conductors carrying large currents are side by side, it is necessary to strap the conductors down to keep them in place.

The proximity effect, like the skin effect, increases with the frequency of the current, causing an increase in effective resistance with frequency because the conducting electrons are crowded into smaller and smaller cross-sections of each conductor. In some tightly wound multilayer coils, such as found in transformers and inductors,[20] current crowding caused by the proximity effect can cause an increase in resistance with increasing frequency that is as great or even greater than that caused by skin effect. Expressions for the combined influence of proximity effect and skin effect are complicated and derivations of such expressions are beyond the scope of this text. Our purpose in introducing these phenomena is primarily to describe why the resistance of a conductor increases with the frequency of a sinusoidal current through the conductor. Problems at the end of the chapter that call for quantitative analyses of ac resistance consider only skin effect.

2.14 Concluding Remark

Practicing circuit designers must be aware of E-series standard values, temperature-dependence of resistance, skin effect, and many other things. But including all of these details in every example and problem in this introductory book would obscure the main point of the example or problem. Unless standard value and tolerance are an essential part of an example or problem, we treat resistors as if they can have any value whatever. Unless variation of resistance with temperature or frequency is an essential part of an example or problem, we ignore those effects, as well.

[20]We treat transformers and inductors in Chapter 9.

2.15 Problems

Section 2.1 is prerequisite for the following problems.

P 2.1 The peak current in a strong lightning bolt is about 250 kA and persists for about 150 μs. Assuming the cloud is positively charged and that only electrons are involved in the current, approximately how many electrons are transferred and in which direction?

P 2.2 In a certain medium, electrons pass from left to right through an imaginary plane at the rate 10^{20} electrons per second and singly-charged positive ions pass from right to left at the rate 10^{20} ions per second. What is the current (magnitude and direction) through the plane?

P 2.3 A 1.5 V alkaline D cell (flashlight battery) can provide a current of 100 mA for about 80 h. How much charge exits the positive terminal during that time?

P 2.4 The table below shows ampere-hour (Ah) ratings of three types of 9 V batteries commonly advertised as alkaline, heavy-duty, and general-purpose. Alkaline batteries cost about 75% more than heavy-duty batteries and 130% more than general-purpose batteries. Which is the cheapest per ampere-hour delivered? By what factor?

Battery type (all are 9 V)	mAh at 10 mA
Alkaline	470
Heavy duty	180
General purpose	160

P 2.5 The table below shows ampere-hour rating, mass, and approximate cost (circa 2009) for four sizes of alkaline batteries. Which is the cheapest per ampere-hour? Which is the most effective in terms of mA h g^{-1}? If a remote control uses 2 AA batteries and has a mass of 100 g with the AA batteries in, what would the mass be if it used D cells? What would it weigh in both cases?

Size	mAh	Mass (g)	Cost ($)
AAA	400	12	0.94
AA	1000	22	0.94
C	3200	64	1.29
D	8000	125	1.29

P 2.6 A 3 V lithium button battery is rated at 150 mAh and will last approximately 1 year when used as a memory backup battery in a desktop computer, depending upon how much the computer is used (the battery is used only while the computer is off). If the computer is on 8 h/day 7 days/week and the battery lasts 1 year, what is the idle current drawn by the memory?

P 2.7 For $t > 0$, the charge to the right of an infinite plane is given by

$$q(t) = \begin{cases} 0, & t < 0, \\ q_0\left(1 - e^{-t/\tau}\right), & t > 0, \end{cases}$$

where $q_0 = 100\,\mu\text{C}$ and $\tau = 2\,\text{ms}$. (a) Obtain an expression for the current through the plane from right to left. (b) Compute the values of the current for $t = 0$, $t = 1\text{ms}$, and $t = 10\,\text{ms}$.

P 2.8 For $t > 0$, the current inward through the surface of a sphere is given by

$$i(t) = \begin{cases} 0, & t < 0, \\ i_0\left(1 - e^{-t/\tau}\right), & t > 0, \end{cases}$$

where $i_0 = 2$ mA and $\tau = 1$ ms. The charge inside the sphere is zero for $t = 0$. Find the charge inside the sphere for $t = 1$ ms and $t = 5$ ms.

P 2.9 The automatic cannon on a certain fighter aircraft fires N rounds per second. If each bullet carries a charge q, what is the current at the muzzle of the gun?

P 2.10 The current in a wire is $I = 5$ A. If the current is due entirely to electron flow, how many electrons pass through a fixed cross section of the wire in 1 s?

P 2.11 An electric current is established in an electrolyte, as shown in Fig. P 2.1. Positive ions drift to the left and the negative ions drift to the right. During each interval T, n_1 singly charged ions and n_2 doubly charged ions reach each electrode. Obtain an expression for the current I.

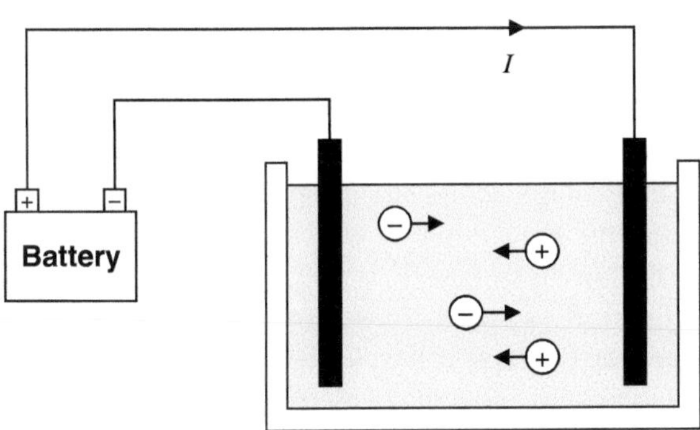

Fig. P 2.1 See Problem P 2.11

P 2.12 The net positive charge on the left side of a plane surface is given at any time t by

$$q = q_0 \cos(2\pi f t),$$

where $q_0 = 5\,\mu\text{C}$ and $f = 1{,}000\,\text{s}^{-1}$. What is the current from right to left through the surface?

P 2.13 Repeat Problem P 2.12 for

$$q = \begin{cases} 0, & t < 0, \\ q_0\, e^{-t/\tau}, & t \geq 0, \end{cases}$$

with $q_0 = 5\ \mu\text{C}$ and $\tau = 1$ ms.

P 2.14 The current required to change the state of one bit in a certain memory chip is on the order of 1 μA. The time required to change the state is 15 ns. How much charge is involved? How many electrons?

P 2.15 Suppose the current through the starter on an automobile is about 75 A for 1.2 s and is due entirely to electron flow. About how many electrons pass through the starter in that time? If the voltage drop across the starter is 12 V, what is the work done by the electrons on the starter?

P 2.16 The charge that passes through a surface in an interval $t_1 \leq t < t_2$ is given by

$$q(t_2) - q(t_1) = \Delta q = \int_{t_1}^{t_2} i(t)\, dt$$

Calculate the charge that passes through a surface in an interval $0 \leq t < 1$ ms if the current is given by

(a) $i = I_0 \cos(2\pi f t)$ with $I_0 = 10$ mA and $f = 1\,\text{kHz}$
(b) $i = I_0\, e^{-t/\tau}$ with $I_0 = 5$ mA and $\tau = 2$ ms.

Section 2.2 is prerequisite for the following problems.

P 2.17 Assuming the proton (nucleus) and electron comprising an atom of hydrogen can be treated as point charges, calculate the force each exerts on the other when the atom is in its ground state. The (ground-state) mean distance from the nucleus (proton) to the electron is approximately 5.3×10^{-11}m.

P 2.18 Refer to Problem P 2.17. What is the electric field strength due to the nucleus at the ground-state distance of the electron from the nucleus?

P 2.19 Refer to Problem P 2.17. How much work is required to move the electron from its ground state to an infinite distance from the nucleus? Express the result in joules and in electron-volts.

P 2.20 Imagine that a positive point charge Q_1 is attached to a 30,000 kg truck and a negative point charge $-Q_1$ is attached to a crane 3 m above the truck. (a) What is the minimum magnitude of the charge that would lift the truck? (b) How much work is done in lifting the truck 2 m? (c) How fast is the truck moving when it reaches that height?

P 2.21 A particle having charge q moves from point a to point b under the influence of an electric field E, directed from a to b. When the particle reaches point b, the electric field is reversed and the particle moves back to point a. What is the energy lost by the particle?

Section 2.3 is prerequisite for the following problems.

P 2.22 Using the fact that a voltage V_{ab} equals work done *per unit charge* on charge moving from a to b, and assuming that the terminal voltage of a battery remains constant throughout the life of the battery (it doesn't), calculate the total energy stored in each battery of Problem P 2.5. (The terminal voltage of each battery is 1.5 V.)

P 2.23 A certain 12 V marine battery is advertised as an 800 A h battery. Calculate the total energy stored in the battery. If the battery weighs 30 lb, what is the energy density in SI units?

P 2.24 The electric field strength at a point 5 m from a point charge is 200 V m^{-1}. What is the magnitude of the charge? If the charge is positive, what is the electric potential at the point?

P 2.25 An electron at rest is released at point a and travels in a straight line to point b 1 cm away, where the velocity of the electron is 10^6 m s^{-1}. What is the voltage at point b with respect to point a?

P 2.26 In a cathode-ray video tube, electrons are released at a heated cathode and accelerated toward the face of the tube. The potential at the face is about 25 kV greater than that at the cathode. If the initial velocity of an electron equals zero, what is the velocity of the electron when it strikes the face of the tube?

P 2.27 Dry air at atmospheric pressure will ionize and conduct electricity (break down) in an electric field whose strength is on the order of 1 MVm^{-1}. After scuffing across a carpet under these conditions, you can draw a ¼ in. spark from your finger to a metal doorframe. Approximately what was the voltage from your fingertip to the doorframe just before the arc was struck?

P 2.28 See Problem P 2.27. If a cloud from which a lightning-bolt originates is 4000 ft from the ground, what is the magnitude of the voltage from the cloud to the ground?

P 2.29 The electric field strength 1.5 m from a point charge is 1.5 Vm^{-1}, directed away from the charge. What is the electric potential at that point?

P 2.30 The electric field strength between uniformly and oppositely charged infinite parallel plates 0.5 cm apart is 1 kVm^{-1}. (a) What is the voltage on the positive plate with respect to the negative plate? (b) How much work is done if a 10 nC charge is moved from the negative to the positive plate? Is the work done on the field or by the field?

P 2.31 The electric potential 2 cm from a point charge is 100 V. What is the magnitude of the charge?

P 2.32 The electric potential a distance x from a 0.05 nC point charge is 50 V. What is the distance x?

Section 2.4 is prerequisite for the following problems.

P 2.33 A certain copper wire has resistance per unit length $r = 1\text{m}\Omega\,\text{m}^{-1}$. A voltage $v = 1$ mV is applied across a 10 m length of the wire. The resulting current is due entirely to electrons. How many electrons have passed through a cross section of the wire 1 ms after the voltage is applied?

P 2.34 If 10^{19} electrons enter the left end of a 1-cm long piece of material each second and exit the right end having experienced a voltage rise of 5 V, what is the resistance of the piece of material?

P 2.35 The resistance of a sample of material is 1.5 kΩ. (a) If the current through the sample is 3 mA, what is the voltage across the sample? (b) If the voltage across the sample is 6 V, what is the current?

P 2.36 A voltage V is applied (end-to-end) to a cylindrical sample of material having length L and diameter D. Express the current through the cylindrical sample in terms of the electric field strength in the sample and the sample diameter. Show that the expression is dimensionally consistent.

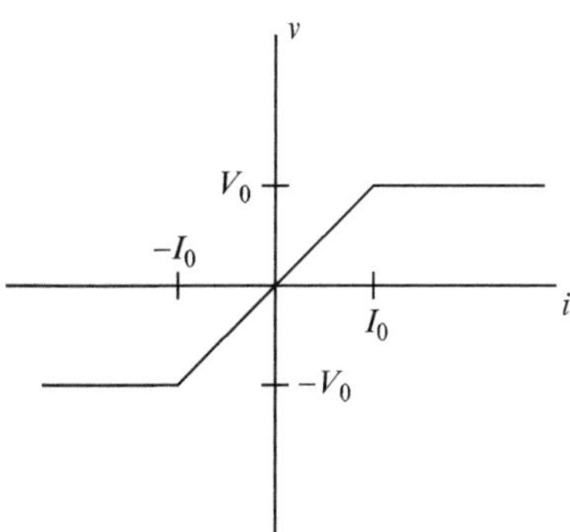

Fig. P 2.2 See Problem P 2.37

P 2.37 Resistance as defined and discussed in this chapter is called *static resistance*, by which is meant that it is the ratio of a *constant* voltage to the resulting *constant* current. *Dynamic resistance*, on the other hand, is defined as the rate of change of voltage with respect to current; that is, if we denote dynamic resistance by r, then the dynamic resistance of a device is given by $r = dv/di$, where v is the voltage across the device and i is the current through the device. With this in mind, refer to Fig. P 2.2, which shows a graph of voltage versus current for a certain hypothetical device. Draw graphs of the static resistance R and the dynamic resistance r of the device versus current.

Section 2.5 is prerequisite for the following problems.

P 2.38 The resistivity of a certain aluminum alloy used for electrical wire is 2.83×10^{-8} Ωm. What is the diameter of an aluminum wire that has the same resistance per meter as a copper wire whose diameter is 0.2 cm?

P 2.39 The resistance of a cylindrical sample of material is R. Express the resistivity of the material in terms of the resistance, diameter, and length of the sample.

P 2.40 Express the lengthwise current through a rectangular sample of material in terms of the resistivity of the material, the voltage applied to the sample, and the sample dimensions (length, height, width). Show that your expression is dimensionally consistent. Assuming that all other parameters are held constant, what happens to the current through the sample as the length of the sample is increased?

P 2.41 Refer to Problem P 2.36. Express the current through the sample in terms of the resistivity of the sample material, the cross-sectional area of the sample, and the electric field strength in the sample. Show that the expression is dimensionally consistent.

P 2.42 The cost per unit weight of copper is about twice that of aluminum. At 20°C, the resistivity of copper is 1.71 μΩ cm, the resistivity of aluminum is 2.71 μΩ cm, the density of copper is 8.96 g cm^{-3}, and that of aluminum is 2.70 g cm^{-3}. An electric power company needs to construct a transmission line from a power plant to a substation 10 km from the plant. The number and therefore the cost of the line supports is directly proportional to the weight of the wire they must support. The voltage drop from end to end of the line must not exceed 400 V when the current is 25 A. Considering all these factors, which of the two materials should be used for the transmission line?

P 2.43 The cold resistance of the tungsten filament in a 60 W incandescent light bulb is about 20 Ω. The diameter of the filament wire is about 50μm and the resistivity of tungsten (at 25°C) is about 5.39 μΩ cm. How long is the wire from which the filament is made? How is such a length of wire made to fit in such a small space?

P 2.44 A metal pipe has inside diameter r_i and outside diameter r_o. Let ρ denote the resistivity of the metal and obtain an expression for the resistance per unit length of the pipe.

P 2.45 A sample of a certain material is a disk having radius r and thickness d. Let ρ denote the resistivity of the material and obtain an expression for the resistance of the sample from one edge to another, where the contacts are at the endpoints of a diagonal. *Hint*: As you probably recall from physics (and as shown in Chapter 4), resistances in series add.

Section 2.6 is prerequisite for the following problems.

P 2.46 The resistance of a certain resistor is in the range $R_0(1 \pm 0.01\beta)$, where R_O is the nominal resistance and β is the precision (or tolerance), expressed in %. (a) Express the corresponding range of values of the conductance in terms of R_O and β. (b) For $R_0 = 1\,\text{k}\Omega$, express the tolerances of the conductance in percent for $\beta = 1\%, 5\%, 10\%$ and 20%. Under what condition is the tolerance on the conductance approximately equal to $\pm\beta$? (c) The voltage across a resistor having resistance $1\,\text{k}\Omega \pm 20\%$ is $V_0 = 5\,\text{V}$. What is the range of possible values of the resulting current? If the current must be $5\,\text{mA} \pm 2\%$, what precision is required of the resistor?

P 2.47 A voltage $v(t)$ is applied to a sample of material for $0 \le t < t_0$. Obtain an expression for the conductance G of the sample in terms of the net charge Q that exits the sample during this time, for (a) $v(t) = V_0$, (b) $v(t) = V_0 \exp(-t/\tau)$, and (c) $v(t) = V_0 \sin(\omega t)$.

P 2.48 The cross-section of a sample of material is an equilateral triangle. Denote the base of the triangular cross-section by b, the length of the sample by L and the conductivity of the material by σ. The voltage V across the sample and the resulting current I through the sample are known. Obtain an expression for the height of the sample in terms of the length, conductivity, applied voltage, and resulting current. Show that the expression is dimensionally consistent.

Section 2.7 is prerequisite for the following problems.

P 2.49 The resistivity of a certain tantalum nitride film resistor is $\rho = 250\,\mu\Omega$ cm at 25°C. The film is of uniform but uncertain thickness and width. The actual (measured) resistance at 25°C is 1.1 kΩ. The desired resistance is 1.2 kΩ, and is to be attained by trimming one edge of the film along its length and parallel to the opposite edge. By what percent must the width of the film be reduced?

P 2.50 A certain thin film resistor is designed and fabricated assuming the resistivity of the film is $250\,\mu\Omega$ cm at the operating temperature. The true value is $232\,\mu\Omega$ cm at that temperature. If the fabricated resistor has precisely the correct dimensions, by what percent must the width of the film be reduced (by trimming) to achieve the desired resistance?

P 2.51 A certain thin film resistor having uniform thickness and width is trimmed as shown in Fig. P 2.3. The film thickness is h and the resistivity of the material is ρ. (a) Express the new resistance in terms of the old and the dimensions shown on the figure. (b) Show that if $w < l$, $\Delta w \ll w$, and $\Delta l \ll l$, then a good approximation to the resistance after trimming is

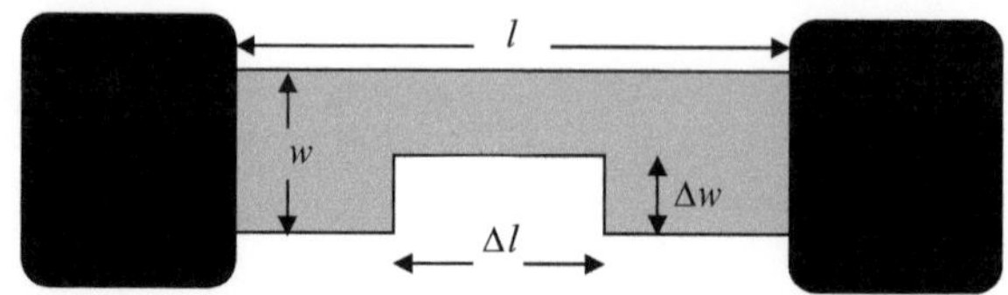

Fig. P 2.3 See Problem P 2.51

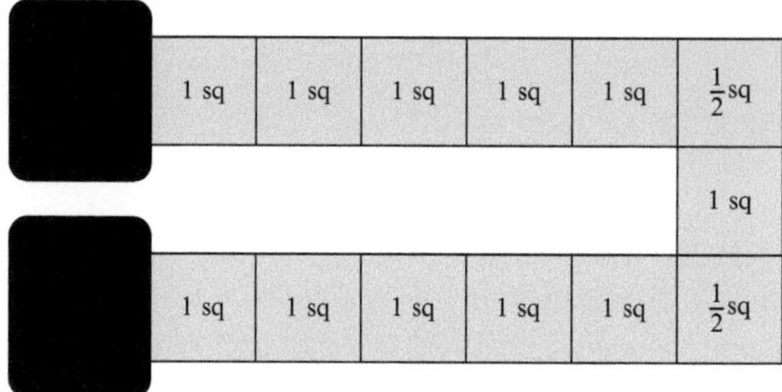

Fig. P 2.4 See Problem P 2.52

$$R \cong R_{sheet}\left(\frac{l}{w} + \frac{\Delta l \Delta w}{w^2}\right).$$

P 2.52 For large resistance values, thin film resistors often are built in zigzag fashion. In such geometries, corner squares count approximately as half squares, as illustrated by Fig. P 2.4, when computing resistance. Justify this approximation. (The correct value, obtained by modeling and analysis, is 0.558 square.)

Section 2.8 is prerequisite for the following problems.

P 2.53 If two resistors R_1, R_2 have the same precision $\pm\beta$, what is the precision of $R_1R_2/(R_1 + R_2)$? Give both exact and first-order approximate relations. What are the actual and approximate precisions for $\beta = 0.05$ (5%)?

P 2.54 If two resistors R_1, R_2 have the same precision $\pm\beta$, what is the precision of $R_1/(R_1 + R_2)$? Give both exact and first-order approximate relations. What are the actual and approximate precisions for $\beta = 0.05$ (5%)?

P 2.55 Suppose your company is considering a standard offering of resistors whose values are separated by 0.2%; i.e., such that $R_{n+1} = 1.002R_n$ for any two consecutive values. Approximately how many such values would be required to cover the range $1\,\Omega \leq R \leq 1\,\mathrm{M}\Omega$?

Section 2.9 is prerequisite for the following problems.

P 2.56 Give the nominal resistance and tolerance indicated by the following color codes:

(a) Gray, red, green, silver
(b) Green, blue, yellow, gold
(c) Brown, black, orange, gold
(d) Yellow, violet, green, red, violet
(e) White, orange, brown, orange, violet
(f) Green, brown, brown, orange, green

P 2.57 Give the nominal resistance and tolerance indicated by the following color codes:

(a) Red, yellow, orange, gold
(b) Brown, green, brown, white
(c) Red, violet, orange, white
(d) Orange, brown, blue, black, white
(e) Brown, orange, orange, red, brown
(f) Red, brown, gray, orange, violet

Section 2.10 is prerequisite for the following problems.

P 2.58 The temperature coefficient of a metal film resistor is typically in the neighborhood of 25 ppm. What is the percent increase in the resistance of a metal-film resistor if the temperature of the resistor is raised from 25°C to 140°C?

P 2.59 Show that if R_1 and R_2 have the same temperature coefficient, then the quantity R_1/R_2 is independent of temperature.

P 2.60 Show that if R_1 and R_2 have the same temperature coefficient, then the quantity $R_1/(R_1 + R_2)$ is independent of temperature.

P 2.61 Show that if R_1 and R_2 have the same temperature coefficient α, then the temperature coefficient of the quantity $R_1R_2/(R_1 + R_2)$ is also α.

P 2.62 Suppose a particular material at temperature T_0 has temperature coefficient of resistivity α. A sample of the material at the same temperature has resistance R_0. A voltage v is applied to the sample, and the temperature is allowed to vary a small amount ΔT from T_0. Obtain a *linear* approximation for the current through the material in terms of the temperature change ΔT, the temperature coefficient α and the applied voltage.

P 2.63 The resistance of a certain resistor is given by $R = R_0[1 + \alpha(T - T_0)]$, where T is the temperature of the resistor in degrees Celsius and $T_0 = 25°\mathrm{C}$. A constant voltage V_0 is applied to the resistor. After a short time, the current stabilizes at I_0. Obtain an expression for the temperature of the resistor as a function of the current I_0.

P 2.64 Suppose a constant current I_0 is forced through the resistor described in Problem P 2.63. After a short time, the voltage across the resistor stabilizes at V_0. Obtain an expression for the temperature of the resistor as a function of the voltage V_0.

Section 2.11 is prerequisite for the following problems.

P 2.65 The conductance of a 1-m sample of 24 AWG wire is measured at 25°C, and is found to be 9.1 S. If the wire is made of either copper, aluminum, or gold, which is most likely?

P 2.66 For a fixed length and kind of wire, by what factor does the resistance of the wire increase if the gauge (AWG) is increased from (a) 6 to 8? (b) from 8 to 10? (c) from 10 to 12?

P 2.67 Let R_N denote the resistance of a 1 m length of wire having AWG designation N. Show that the ratio R_N/R_{N+k}, where k is any whole number, is a fixed function of k, independent of N.

P 2.68 If the length of a certain wire must be doubled, but the resistance must remain the same, by what number must the AWG designation be decreased?

P 2.69 In a certain application, the required size for a copper wire is AWG 12. For reasons of cost, it is desired to use an aluminum wire, instead. What should be the size (AWG) of the aluminum wire?

Sections 2.12 and 2.13 are prerequisite for the following problems.

P 2.70 (a) Show that if the diameter of a cylindrical conductor is much larger than the skin depth, the ratio of ac to dc resistance is approximately $(D + \delta)/(4\delta)$.

(b) Show that the approximation of part (a) is more accurate than the (perhaps more obvious) approximation $D/(4\delta)$.

P 2.71 Obtain an expression for the skin depth δ_1 of a cylindrical conductor at frequency f_1 in terms of the skin depth δ_0 at frequency f_0. Assume the temperature of the conductor is fixed.

P 2.72 A certain web site gives the following formula for the ac resistance of a wire:

$$R_{ac} = R_{dc} k\sqrt{f},$$

where (quoting the web page) "f is frequency of ac in MHz" and k is called the "wire gauge factor," whose values for various AWG wire sizes are given in an accompanying table. (a) What is the SI unit of k? (b) The formula assumes that $D \gg \delta$, where D is the wire diameter and δ is the skin depth. Derive the formula from (2.39) in the text and obtain an expression for k. Show that the formula has the correct dimension. (c) What are the magnitudes of k for AWG 22 copper wire at 25°C and 100°C? (d) For what frequencies is the formula approximately valid for AWG 22 copper wire at 25°C?

P 2.73 A certain coil is wound with 50 m of AWG 32 copper wire. What is the dc resistance of the coil at 25°C? If the increase in resistance with frequency is due entirely to skin effect, what is the ac resistance of the coil at 10 MHz and 25°C?

P 2.74 In a certain application, the temperature of a 22 AWG copper wire varies from 0°C to 100°C and the frequency of the current through the wire ranges from 3 to 30 MHz. What is the maximum percent increase in the ac resistance of the wire from the value at 0°C and 3 MHz?

P 2.75 In integrated circuits (ICs), conductors are strips of metal, often aluminum or copper, having approximately rectangular cross section. At this writing (circa 2009), the smallest conductors are approximately 20 nm wide and 50 nm thick and the highest frequency of interest in digital IC's is about 30 GHz. Assume an operating temperature of 100°C and calculate the skin depth for copper at 30 GHz. Is skin effect significant in such devices?

P 2.76 As IC manufacturing technology advances, the smallest lines (minimum conductor widths) and thicknesses decrease and the achievable operating frequencies increase. Do these advances make skin effect more significant or less so?

P 2.77 Electric utilities use large rectangular-cross-section copper conductors called *bus bars* to connect adjacent equipment in distribution substations. A 2500 A bus bar is about 10 mm thick and 200 mm wide. Calculate the skin depth of copper at 60 Hz and determine whether the bus bar could be thinner. Assume an operating temperature of 100°C. Ignore corners and the short edges.

Chapter 3
Circuit Elements, Circuit Diagrams, and Kirchhoff's Laws

In this chapter, we define a few important circuit elements and introduce circuit diagrams. We describe how circuit diagrams are annotated; that is, how voltages and currents are denoted in a manner that permits and facilitates analysis. We present Kirchhoff's laws, which are the fundamental laws of circuit analysis, and the principle of superposition, which is a powerful tool for analysis of linear circuits.

3.1 Schematics and Circuit Diagrams

Components of a physical circuit are connected by wires or other physical conductors, such as foils on a printed-circuit board or metallic strips in integrated circuits. A **schematic** or **wiring diagram** is a picture showing how a physical circuit is constructed, where conductors are represented by lines and other components, such as transistors, are represented by standardized graphic symbols.

A **circuit diagram**, on the other hand, depicts the *mathematical relations* that exist among various *idealized circuit elements*, which individually or in groups, comprise mathematical models for physical circuit components. For the present, we may regard a circuit diagram as an abstraction of a *schematic* (wiring diagram). But keep in mind that schematics and circuit diagrams are different things. The various parts of a schematic are in one-to-one correspondence with the components they represent; that is, each symbol in a schematic stands for a physical component in the associated physical circuit. On the other hand, a circuit diagram is composed of *mathematical models* for physical components. A model for a single physical component (such as a transistor) may consist of more than one circuit element, or a few interconnected circuit elements can represent a larger number of physical components. Consequently, *the individual circuit elements shown in a circuit diagram do not necessarily represent physical components.* In some cases, the circuit diagram representing a physical circuit may look like the associated schematic, but in general, a circuit diagram represents a physical circuit only to the extent that results of mathematical analysis of the circuit diagram correctly describe the observed behavior of the associated physical circuit. *A schematic is a guide for construction. A circuit diagram is a guide for analysis and design.*

To illustrate points made above, we refer to Fig. 3.1, which shows both a schematic and a circuit diagram for a simple transistor amplifier. In the circuit diagram of Fig. 3.1(b), the transistor (a component) has been replaced by an equivalent circuit model (a collection of idealized elements). At this point, do not be concerned with what the symbols mean, how the model is obtained, or how the circuit works. Instead, focus on these points:

- The circuit model for the transistor does not look anything like a real transistor. Even if we dissect a transistor and examine the innards under a microscope, we will not find either of the two elements comprising the circuit model. *The elements of the model are non-physical*; nonetheless, the model describes the transistor, in the sense that mathematical analysis of the circuit diagram yields results in good agreement with the observed behavior of the actual amplifier (under normal operating conditions).

T.H. Glisson, *Introduction to Circuit Analysis and Design*,
DOI 10.1007/978-90-481-9443-8_3, © Springer Science+Business Media B.V. 2011

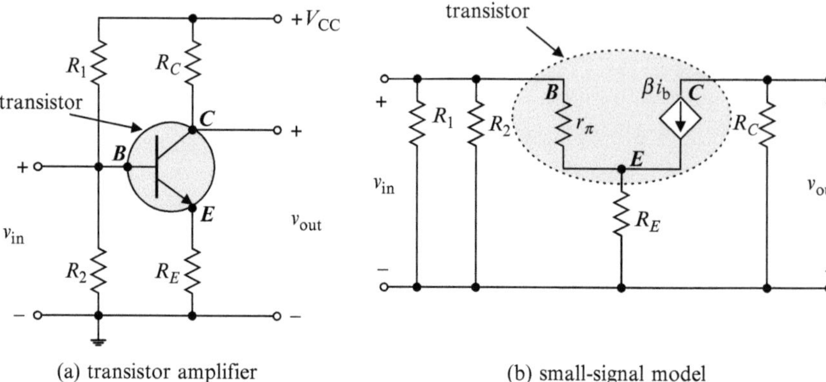

Fig. 3.1 (**a**) A transistor amplifier (schematic, or wiring diagram) and (**b**) an associated model (circuit diagram)

- The number of elements in the circuit diagram exceeds the number of (physical) components in the schematic. The transistor is represented by one component (symbol) in the schematic and by two elements in the circuit diagram. Often, a single transistor is modeled by four or more elements. The elements in a circuit diagram are rarely in one-to-one correspondence with those in the associated physical circuit.
- To the untrained eye, the circuit diagram (model) does not look like the physical circuit it represents. The elements in the circuit diagram are not interconnected in the same way as in the physical circuit. But again, a correct mathematical analysis of the circuit diagram would yield results in agreement with observed behavior of the physical circuit (under conditions for which the model is applicable).

Where we refer to a *circuit*, unless otherwise noted, we are referring to a *circuit diagram* (mathematical model), not to a physical circuit or schematic. Likewise, where we refer to a *circuit element*, we are referring to an *idealized mathematical model*, not a physical circuit component. The distinction between the elements of a circuit diagram and the components of a (physical) circuit will become clearer as we progress.

Because circuit analysis deals exclusively with analysis of idealized models, and for the sake of economy, we omit the adjective *ideal* when referring to circuit elements and circuit diagrams. It is understood, for example, that an element called a *conductor* is an *ideal* conductor, not a physical conductor (wire).

a ——— b

$$v_{ab} = \Phi(a) - \Phi(b) = 0$$

Fig. 3.2 The symbol for a conductor is a line. The voltage v_{ab} across the conductor equals zero, regardless of the current through the conductor

3.2 Conductors and Connections

A **conductor** is an idealization of a physical conductor (e.g., a wire) and is represented by a line, as illustrated by Fig. 3.2. The defining property of a conductor is that *the electric potential is the same everywhere on a conductor* (and on all other conductors to which it is connected). In Fig. 3.2 the potential Φ (a) at the left end (point a) equals the potential Φ (b) at the right end (point b), or anywhere in between. *The voltage v_{ab} across the conductor equals zero, regardless of the current through the conductor*. In other words, *the resistance of a conductor equals zero.*

In circuit diagrams, a **connection** of intersecting conductors is denoted by a heavy dot, reminiscent of a lump of solder. There is no such dot at the intersection of conductors that cross (viewed from above) but are not connected. Refer to the circuit diagram in Fig. 3.3, where lines represent conductors and boxes represent (unspecified) circuit elements. The conductor connecting elements 1 and 3 is connected at the point a to the conductor connecting elements 2 and 4. The electric potential (or the voltage with respect to some other point) is the same everywhere on both of those conductors. The conductor connecting elements 3 and 6 is not connected to the conductor connecting

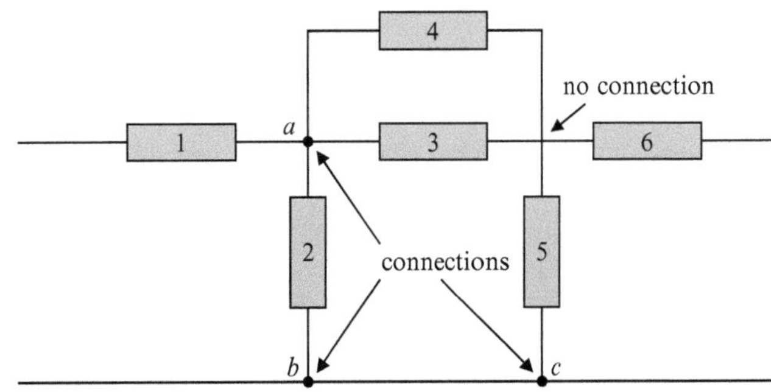

Fig. 3.3 A dot is used to indicate a connection, where two or more conductors (or other elements) are connected

elements 4 and 5. The potentials on those conductors may be different.

3.3 Annotating Circuit Diagrams

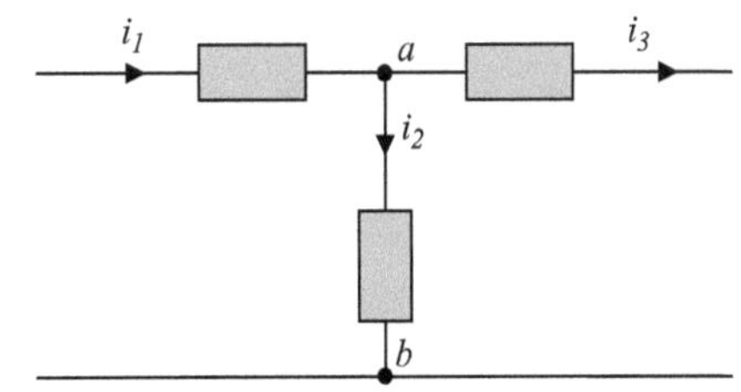

Fig. 3.4 Representing currents in a circuit diagram

Currents are denoted in a circuit diagram by labeled arrows, as illustrated in Fig. 3.4. Subscripts are used to distinguish various currents from each other. An arrow indicates the *positive direction* of the associated current, which we are free to define and which therefore is not necessarily the actual direction of the current. In fact, in most interesting circuits, the magnitudes and directions of most currents change with time. A current whose direction (sign) changes with time cannot possibly be always in the direction of the associated arrow. Again, the arrow indicates the direction defined as positive, not necessarily the actual direction of the associated current. It's like the arrow pointing north on a roadmap. The arrow defines what is meant by north, but does not imply that everyone always drives north.

Voltages are denoted in circuit diagrams in three ways, as illustrated by Fig. 3.5 and explained below.

In Fig. 3.5(a), points of interest (nodes) are labeled *a, b, c* and voltages can be specified unambiguously using **double-subscript notation**; e.g., as v_{ab}, v_{bc}, and v_{ac}.

If a particular point (node) is identified as *the* reference node, then some other voltages can be specified using **single-subscript notation**, where it is understood that the omitted second subscript is the reference point. For example, if the node *c* in Fig. 3.5(a) is specified as the reference node, then $v_a = v_{ac}$ and $v_b = v_{bc}$. However, if point *c* is the reference, we cannot use single-subscript notation to denote v_{ab} because v_{ab} is not referred to point *c* (see (3.1), below).

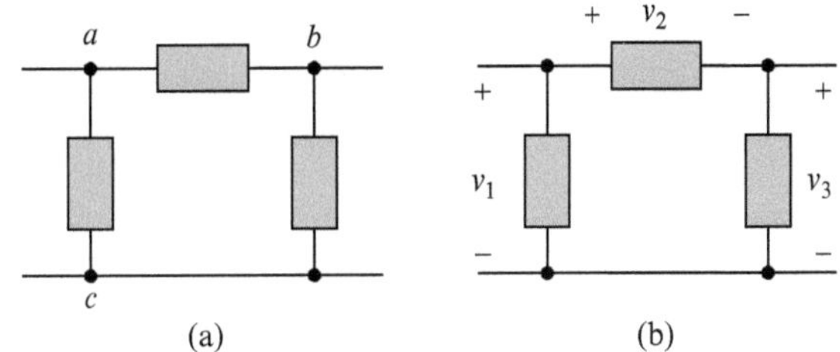

Fig. 3.5 Conventions for specifying voltages on a circuit diagram (see text)

In Fig. 3.5(b), voltages are specified using **plus-minus notation**, where the – sign identifies the reference point and the + sign identifies the point at which *v* is specified with respect to the reference. The +/– marks are reference marks that define what is meant by a positive value of *v*, just as *a* and *b* in Fig. 3.5(a) are reference marks that define what is meant by a positive value of v_{ab}. If the point marked (+) is at a higher potential than the point marked (−), the associated voltage *v* is positive. If the reverse is true, *v* is negative. For example, v_2 in Fig. 3.5(b) corresponds to v_{ab} in Fig. 3.5(a).

All three conventions (double-subscript, single-subscript, and +/–) are widely used. Two or even all three of these notations might used in a single circuit diagram; for example, one voltage of interest might be specified using single-subscript notation and another using +/– notation. Understanding all three notations

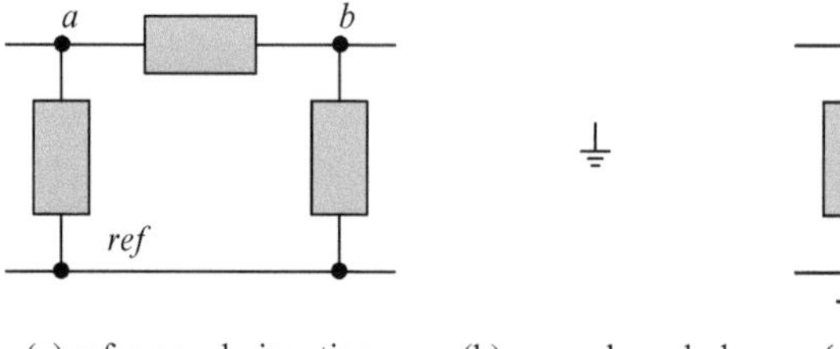

Fig. 3.6 Conventions for designating a voltage reference point

is essential. Otherwise it is unnecessarily difficult to obtain and interpret equations describing circuits.

A voltage symbol having a subscript is not necessarily an instance of what is defined above as single-subscript or double-subscript notation. For example, the subscripts on the voltage symbols in Fig. 3.5(b) serve simply to distinguish each voltage from the others. The notation in use is +/– notation, not single-subscript notation, because the subscripts do not refer to labeled points in the circuit (and because the voltage symbols appear between polarity marks).

When single-subscript notation is used, that is, when one point (or conductor) in a circuit is chosen as a reference and other voltages in the circuit are specified with respect to that point, we are in effect treating the reference point as if its potential were zero, but in fact we usually do not know that – nor do we care. Again, we are concerned with potential differences (voltages), not absolute potentials.

The point chosen as the reference in a circuit diagram often is indicated (in this book) by the label *ref*, as illustrated in Fig. 3.6(a). Alternatively, the **ground**[1] symbol shown in Fig. 3.6(b) may be used to denote the reference point, as illustrated in Fig. 3.6(c).

We refer again to Fig. 3.5(a) to make another important point. If $\Phi(a)$, $\Phi(b)$, and $\Phi(c)$ are the electric potentials at points, a, b, and c, then by definition

$$v_{ac} = \Phi(a) - \Phi(c), \quad v_{bc} = \Phi(b) - \Phi(c),$$
$$v_{ab} = \Phi(a) - \Phi(b).$$

It follows that

$$v_{ab} = v_{ac} - v_{bc}, \qquad v_{ba} = v_{bc} - v_{ac} = -v_{ab}. \quad (3.1)$$

[1]So called because earth ground (the grounded part of an electrical receptacle) is often the reference point.

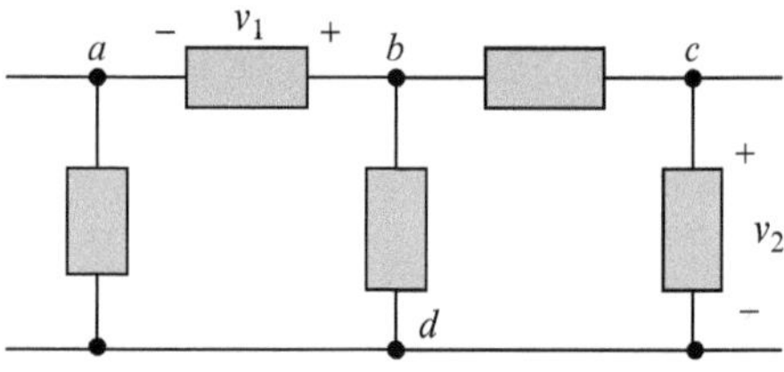

Fig. 3.7 See Example 3.1

Such relations are needed often in writing equations that describe circuits.

Example 3.1. In Fig. 3.7, the electric potentials at the labeled points are $\Phi(a) = 520$ V, $\Phi(b) = 515$ V, $\Phi(c) = 505$ V, and $\Phi(d) = 500$ V. Find the voltages v_{ad}, v_{ab}, v_{ac} and v_2. Using point d as the reference, find v_a and v_c.

Solution: By definition and from (3.1)

$$v_{ad} = \Phi(a) - \Phi(d) = 520 - 500 = 20\,\text{V},$$
$$v_{ab} = \Phi(a) - \Phi(b) = 520 - 515 = 5\,\text{V},$$
$$v_{ac} = \Phi(a) - \Phi(c) = 520 - 505 = 15\,\text{V},$$
$$v_2 = v_{cd} = \Phi(c) - \Phi(d) = 5\,\text{V}$$

If we select the point d as the reference point for all voltages and use single-subscript notation, we have

$$v_a = v_{ad} = \Phi(a) - \Phi(d) = 20\,\text{V},$$
$$v_c = v_{cd} = \Phi(c) - \Phi(d) = 5\,\text{V}$$

Exercise 3.1. With reference to Example 3.1, find the voltages v_{bd}, v_{cd}, v_{da}, v_1, v_b (point d is the reference for the last).

3.4 Series and Parallel Connections

A point on a circuit element or in a circuit to which we connect or envision connecting something (e.g., another element or a measuring instrument) is called a **terminal**. A circuit element or circuit having only two accessible terminals is called a **two-terminal element** or a **two-terminal circuit**.

Two or more two-terminal elements or circuits can be connected in **series**, as shown in Fig. 3.8(a), or in **parallel**, as shown in Fig. 3.8(b). *The same current passes through elements in series and the same voltage exists across elements in parallel.* Series and parallel connections are undefined for elements having more than two terminals.

Example 3.2. In Fig. 3.9(a), elements 1 and 2 are not in series because of the intervening connection to element 3, which allows some of the current passing through element 1 to bypass element 2; however, element 1 is in series with the parallel connection of elements 2 and 3. In Fig. 3.9(b), elements 1 and 2 are not in parallel because of the intervening element 3; however, element 1 is in parallel with the series connection of elements 2 and 3.

Exercise 3.2. Use the terms *series* and *parallel* to describe (in words) the connections shown in Fig. 3.10.

3.5 Open Circuits and Short Circuits

A pair of terminals brought out from a circuit but not connected to each other, either by a conductor or a circuit element, is said to be **open** or to form an **open circuit**. If the terminals are connected by a conductor, the terminals are said to be **shorted** or to form a **short circuit**. Figure 3.11 illustrates this terminology. The *rest of the circuit* might represent an audio amplifier and the small circles at the ends of the lines might represent the terminals to which you would normally connect a loudspeaker. The lines themselves represent conductors, indicating that the terminals are connected to points inside the circuit. The terms open and short are used with reference to the terminals alone, regardless of how or to what the associated conductors are connected inside the rest of the circuit. *The current through an open circuit equals zero, regardless of the voltage across the open circuit, and the voltage across a short circuit equals zero, regardless of the current through the short circuit.* If an element connecting two terminals is removed, it is said that the terminal pair

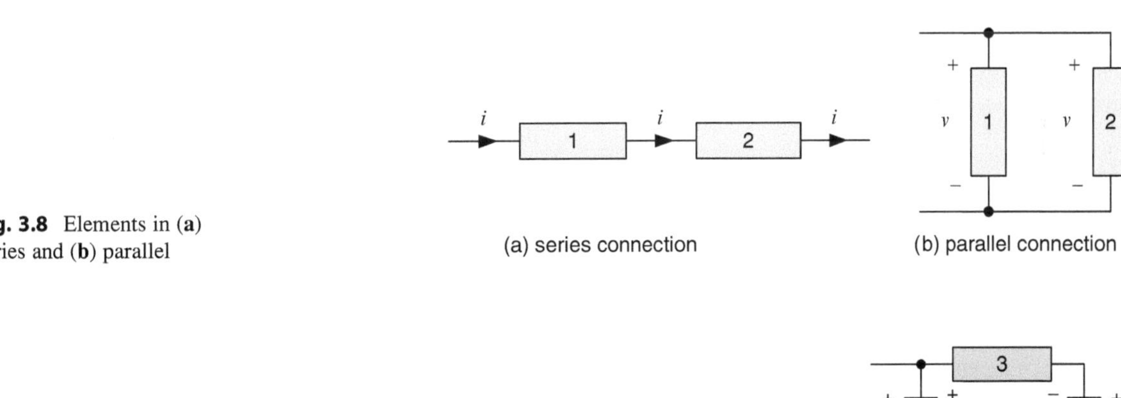

Fig. 3.8 Elements in (**a**) series and (**b**) parallel

Fig. 3.9 See Example 3.2

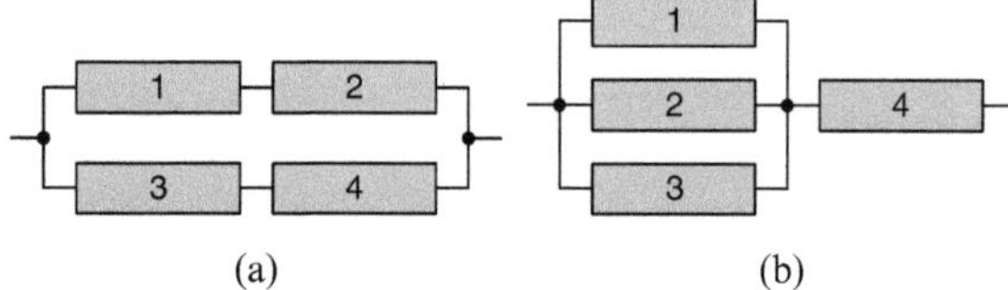

Fig. 3.10 See Exercise 3.2

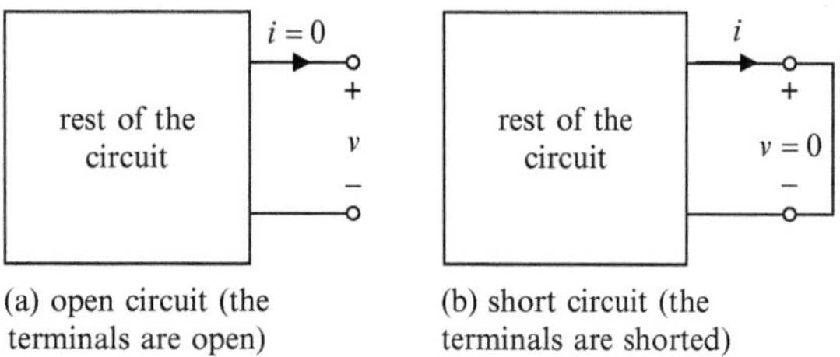

Fig. 3.11 Open and shorted terminals

is *opened*, or that the terminal pair is made an open circuit. If a conductor is connected between two terminals, it is said that the terminals are *shorted*, or that the terminal pair is made a short circuit.

The terms *open circuit* and *short circuit* (or open and short) are most correctly used when neither is the normal condition; i.e., when the opened or shorted condition is temporary, whether intentional or accidental. For example, if you disconnect a loudspeaker from the output terminals of your home audio amplifier, you are opening those terminals. If you accidentally allow your screwdriver to contact both terminals at once, you have shorted the terminals. Nonetheless, both open and shorted are sometimes used to describe normal conditions; for example, a conductor might be called a short circuit, even if the conductor occupies its intended place in a circuit.

3.6 Basic Circuit Elements: Resistors and Independent Sources

In this section, we define three two-terminal circuit elements: A resistor, an independent voltage source, and an independent current source.[2] A number of useful circuits can be described using only these elements, and one or more of these elements appears in virtually every circuit model of practical importance.

[2]The meaning of *independent* in this context is described in the sequel.

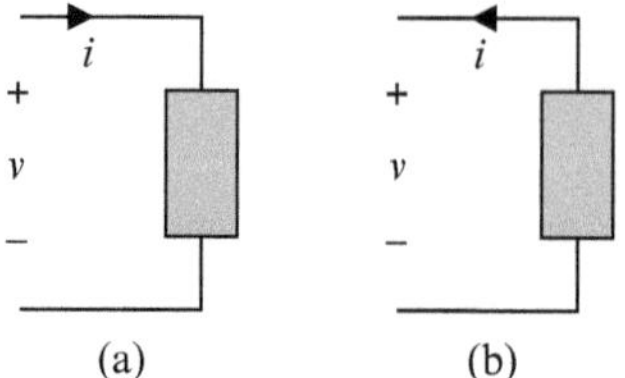

Fig. 3.12 Pertaining to the definition of the terminal characteristic of a two-terminal element

Table 3.1 Symbols[3] and terminal characteristics for three basic circuit elements

Element	Symbol	Terminal characteristic
resistor	i, $+\ v\ -$, R	$v = iR$
voltage source	i, $+\ v\ -$, v_0	$v = v_0$ (independent of i)
current source	i, $+\ v\ -$, i_0	$i = i_0$ (independent of v)

The **terminal characteristic** of a two-terminal circuit element is the relation the element establishes between the voltage across and current through its terminals. We may take the positive direction of current as either *into* the terminal called positive (in the direction of a voltage drop) as in Fig. 3.12(a) or *out of* the terminal called positive (in the direction of a voltage rise), as in Fig. 3.12(b), and we may take either current or voltage as the independent variable. That is, we may express the terminal characteristic of an element as either $i = f(v)$ or $v = f^{-1}(i)$, where the positive direction of current is either into or out of the positive terminal of the element. In general, we may use whichever of these four formulations is most convenient.

Table 3.1 gives the names, graphic symbols, and terminal characteristics of the three elements defined in this section. *The terminal characteristic of an element is independent of what is connected to the element.* For example, the current through and voltage across a resistor are related by $v = i\,R$, regardless of

[3]The polarity marks and arrows and the letters v and i (without subscripts) define the voltage and current that appear in the terminal characteristic. They are not part of the symbol for the element.

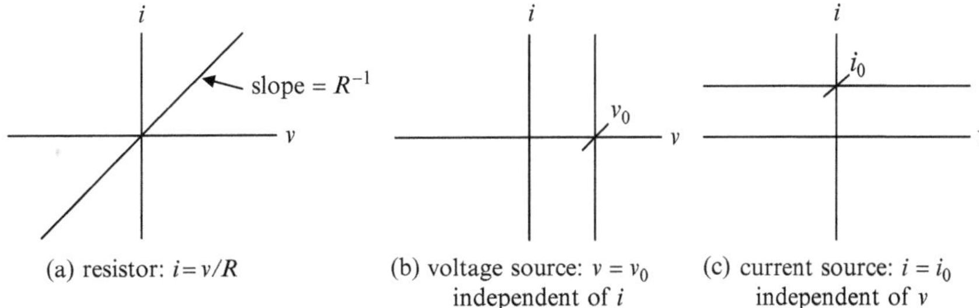

Fig. 3.13 Graphical representations of the terminal characteristics of the elements defined in Table 3.1

what else is connected to the resistor and regardless of the nature of the current or voltage. Of course, this is an idealization. The terminal characteristic $v = i\,R$ describes a physical resistor only under certain operating conditions. For example, a resistor can be destroyed by a sufficiently large current, after which it is no longer described by $v = R\,i$.

Figure 3.13 shows graphical representations of the terminal characteristics of the elements defined in Table 3.1. The current i_0 and voltage v_0 can be time-varying, so you should regard Fig. 3.13(b) and (c) as snapshots (still photos), unless the current i_0 and voltage v_0 are constant.

A label usually is written by the symbol for an element. For example, the label R is attached to the resistor in Table 3.1. The label serves double duty: It represents both the name and the resistance of the resistor. For example, when written next to the symbol for a resistor, R_1 is the name of the resistor (distinguishes that resistor from others labeled $R_2, R_3, \cdots$) and also stands for the resistance of the resistor. For example, we might say that resistor R_1 has resistance $R_1 = 1\ \text{k}\Omega$.

When writing equations describing circuits, it often is convenient to use conductance, rather than resistance. The **conductance** of a resistor having resistance R is denoted by G and is defined by

$$G = R^{-1}. \tag{3.2}$$

Using conductance, Ohm's law is written

$$i = G\,v. \tag{3.3}$$

We regard resistance as the fundamental quantity and conductance mainly as a convenient way to express reciprocals of resistances in circuit equations. Thus, in this book, *a label or a numerical value appearing beside the symbol for a resistor always denotes resistance*, not conductance, as illustrated by Fig. 3.14.

Fig. 3.14 Labeling a resistor using conductance

The voltage across a **voltage source** is called the **source voltage** (denoted by v_0 in Table 3.1) and is *independent of the current through the voltage source.* In other words, the voltage across the voltage source shown in Table 3.1 equals v_0, no matter what is connected to the source. The source voltage may be constant but in general varies with time. A conductor is equivalent to a voltage source whose source voltage equals zero. Conversely, a voltage source whose source voltage equals zero is equivalent to a conductor (a short circuit). Thus we say that *the internal resistance of a voltage source equals zero.*

The current through a **current source** is called the **source current** (denoted by i_0 in Table 3.1) and is *independent of the voltage across the current source.* In other words, the current through the current source in Table 3.1 equals i_0, no matter what is connected to the source. The source current may be constant but in general varies with time. An open circuit is equivalent to a current source whose source current equals zero. Conversely, a current source whose source current equals zero is equivalent to an open circuit. Thus we say that *the internal conductance of a current source equals zero* (or that the internal resistance is infinite).

Circuit elements cannot be connected in ways that are self-contradictory; i.e., in ways that cause conflicts among defined quantities. For example, the circuit in Fig. 3.15(a) is self-contradictory unless $v_1 = v_2$, in which case one of the sources is superfluous. Similarly, the connection in Fig. 3.15(b) is self-contradictory unless $i_1 = i_2$, in which case one of the sources is superfluous. Ideal voltage sources cannot be connected in parallel and Ideal current sources cannot be connected in series.

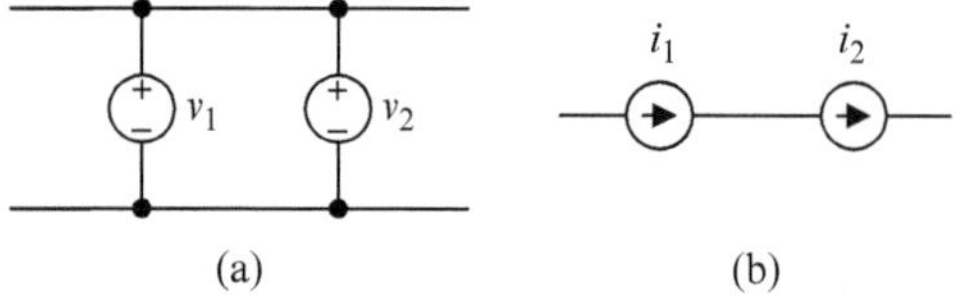

Fig. 3.15 Self-contradictory connections

Labels for current and voltage sources serve both as names for the sources and as symbols for the source current or voltage. For example, we might say that the source voltage for the source v_0 is $v_0 = 5$ V, or that the source current for the current source i_0 is $i_0 = 25$ mA.

The sources defined in Table 3.1 are called **independent sources**, as in *independent voltage source* and *independent current source*, because the source voltage or source current depends only upon time (e.g., is independent of any other currents or voltages). In a subsequent chapter, we describe *dependent sources*, whose source voltages or source currents depend upon other quantities; e.g., currents or voltages in other branches of the circuit.

Example 3.3. In the circuit of Fig. 3.16, the source voltages are $v_1 = 5$ V and $v_2 = 15$ V. Find the voltage v_x across the (unspecified) circuit element.

Solution: We label three points, as shown in the figure, where, by the definition of an ideal voltage source,

$$v_{ac} = v_1, \qquad v_{bc} = v_2.$$

It follows from (3.1) that

$$v_{ab} = v_x = v_{ac} - v_{bc} = v_1 - v_2 = -10 \text{ V}$$

Exercise 3.3. Refer to Fig. 3.17. Find the voltages v_{ab}, v_{ac}, v_{bd}

Figure 3.18(a) shows an alternate symbol for a *constant-voltage source* alongside the conventional symbol in Fig. 3.18(b). The symbol in Fig. 3.18(a) was originally inspired by the architecture of a battery, in which metallic plates are separated by an electrolyte. In electronic circuits, the symbol in Fig. 3.18(a) is used primarily to denote a constant-voltage power supply; e.g., the battery in a cell phone.

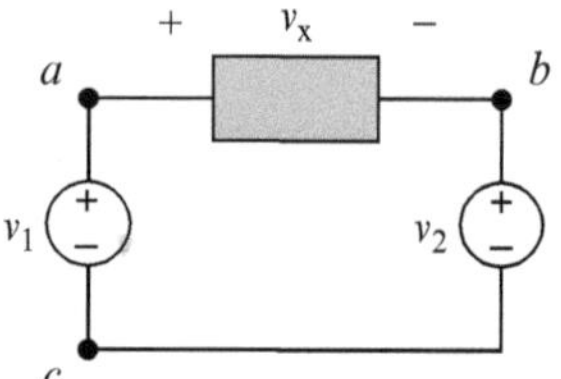

Fig. 3.16 See Example 3.3

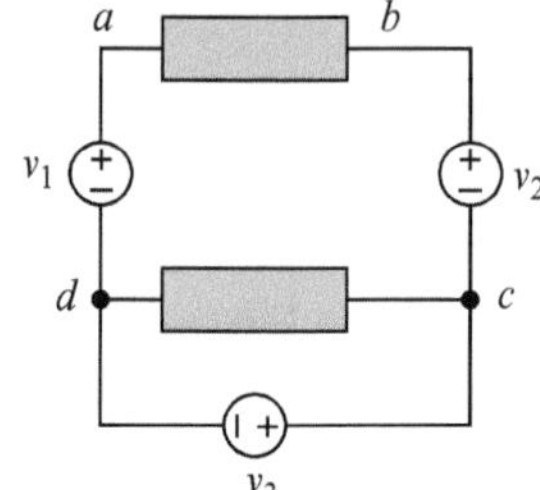

Fig. 3.17 See Exercise 3.3

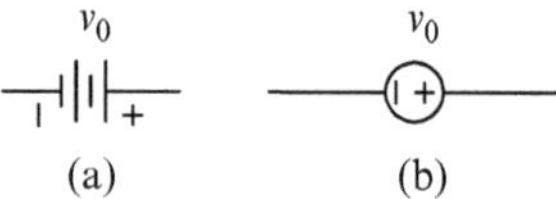

Fig. 3.18 Equivalent symbols for a constant-voltage source

The (ideal) sources defined above do not exist. For example, no physical voltage source, such as a battery, can maintain a constant voltage across its terminals regardless of the current through the terminals. Nonetheless, ideal sources either alone or in combination with other elements can be good models for physical sources.

Because of the nature of known methods for generating electricity, most everyday physical sources (e.g., batteries, generators, solar cells, and fuel cells) behave more like voltage sources than current sources. However, electronic circuits approximating current sources can be constructed and are quite useful.

3.7 Kirchhoff's Current Law and Node Analysis

Kirchhoff's current law[4] states that *the algebraic sum of the currents leaving any enclosed volume in a*

[4]After the German physicist Gustav Robert Kirchhoff (1824–1887).

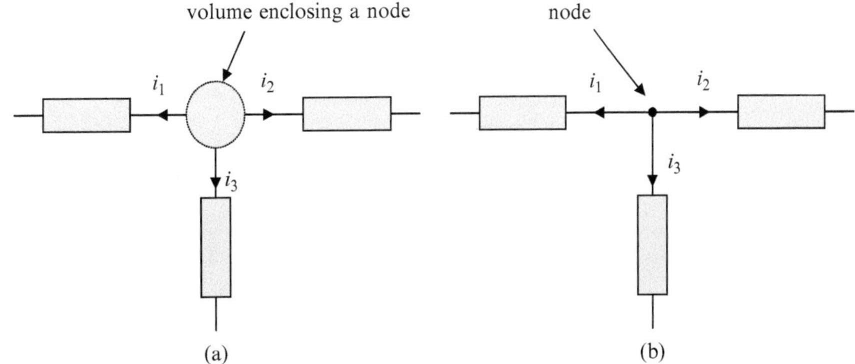

Fig. 3.19 An illustration of Kirchhoff's current law

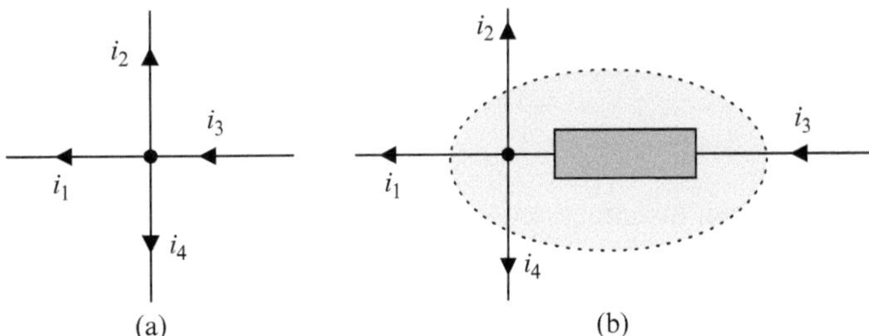

Fig. 3.20 Another Illustration of Kirchhoff's current law

circuit equals zero.[5] We may express Kirchhoff's current law as

$$\sum_n i_n = 0. \tag{3.4}$$

where the i_n are currents leaving an enclosed volume and the sum is over all possible current paths entering or exiting the volume. In (3.4), currents entering the volume are negative.

In applications of Kirchhoff's current law to electric circuits, volumes of interest enclose connections joining two or more elements and conductors provide current paths into and out of the volume, as illustrated in Fig. 3.19(a), where

$$i_1 + i_2 + i_3 = 0.$$

In such cases, we may think of the currents as leaving the *point* at which the elements are connected, as illustrated in Fig. 3.19(b). A point at which two or more elements are connected is called a **node**, and Kirchhoff's current law often is stated as *the algebraic sum of the currents leaving any node equals zero.*

In the algebraic sum of currents *leaving* a volume or node, currents *entering* the volume or node are negated.[6] For example, in both Fig. 3.20(a) and (b),

$$i_1 + i_2 - i_3 + i_4 = 0.$$

Exercise 3.4. Is the circuit in Fig. 3.21 self-consistent? Use Kirchhoff's current law to justify your answer

Kirchhoff's current law is derived from more fundamental principles in textbooks on electromagnetics. We shall accept Kirchhoff's current law as an empirical law, noting that it amounts to requiring that current through any plane intersecting the path of the current be continuous; that is, the currents on each side of the plane are the same.[7] From another point of view,

[5] Kirchhoff's current law can also be stated as "the sum of currents *entering* a volume equals zero" or as "the sum of the currents entering a volume equals the sum of the currents leaving the volume." For the present, it is best to stick with one version of the law.

[6] In "leaving" and "entering," we are referring to the direction of the arrow associated with the current.

[7] Here, "current" includes both conduction current and displacement current.

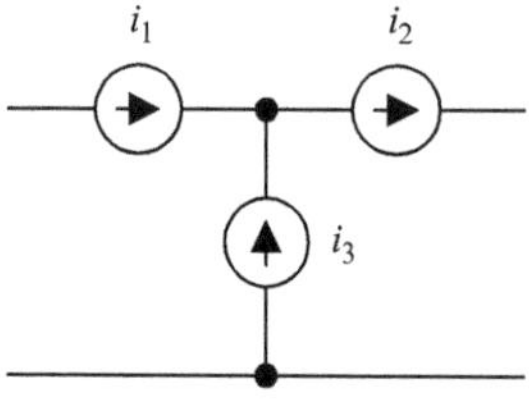

Fig. 3.21 See Exercise 3.4

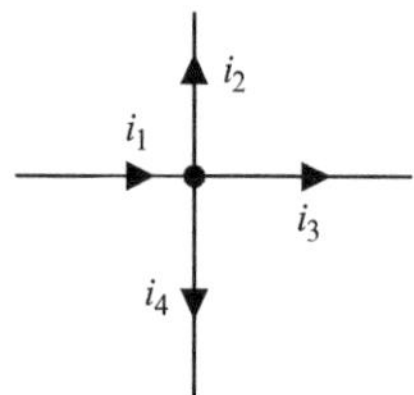

Fig. 3.22 See Example 3.4

charge is not manufactured at a node, nor does charge accumulate at a node, so (because charge is conserved) the net charge leaving a node or volume must be zero.

Example 3.4. In Fig. 3.22, $i_1 = 5$ A, $i_2 = 1$ A, $i_3 = 2$ A. Find the current i_4.
Solution: Kirchhoff's current law gives

$$(-i_1) + i_2 + i_3 + i_4 = 0,$$

It follows that

$$i_4 = i_1 - i_2 - i_3 = 5 - 1 - 2 = 2\text{A}$$

Applying Kirchhoff's current law to circuit analysis leads to what is called **node analysis** (or *nodal analysis*), in which *node voltages are taken as the unknown quantities*. If one or more currents are the quantities of ultimate interest, they may be expressed via Ohm's law in terms of the voltages and resistances.

Node analysis proceeds as follows:

- Identify and label a reference node for other voltages in the circuit.
- Identify and label each other node for which the voltage at that node, relative to the reference node, is unknown.
- Write Kirchhoff's current law at each such node, using Ohm's law to express the currents in terms of the node voltage and the resistance of the resistors attached to the node.

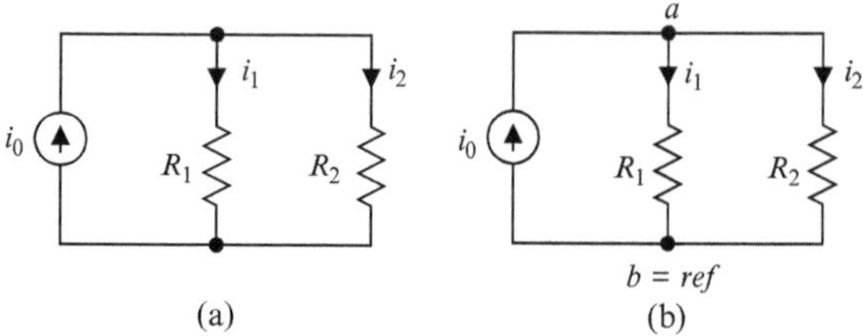

Fig. 3.23 See Example 3.5

- Solve the resulting equations for the node voltages.
- Use the node voltages and Ohm's law to find any currents of interest.

Example 3.5. Obtain expressions for the currents i_1, i_2 in the circuit of Fig. 3.23(a).
Solution: We identify and label the nodes in the circuit as shown in Fig. 3.23(b), choosing node b as the reference node. The only other node in the circuit is node a, and the (unknown) voltage at that node relative to the reference is denoted in single-subscript notation by v_a. The net current leaving node a must be zero, so by Kirchhoff's current law,

$$-i_0 + i_1 + i_2 = 0 \Rightarrow i_1 + i_2 = i_0.$$

By Ohm's law,

$$i_1 = \frac{v_a}{R_1}, \quad i_2 = \frac{v_a}{R_2},$$

and Kirchhoff's current law is expressed in terms of the unknown node voltage v_a by

$$\frac{v_a}{R_1} + \frac{v_a}{R_2} = v_a\left(\frac{1}{R_1} + \frac{1}{R_2}\right) = i_0 \Rightarrow v_a = \frac{R_1 R_2}{R_1 + R_2} i_0.$$

Thus

$$i_1 = \frac{v_a}{R_1} = \frac{R_2}{R_1 + R_2} i_0, \quad i_2 = \frac{v_a}{R_2} = \frac{R_1}{R_1 + R_2} i_0$$

The circuit in Fig. 3.23 is called a **current divider** because the current provided by the source is divided between the two resistors in the manner specified by the relations above.

Exercise 3.5. Refer to Fig. 3.24. Obtain expressions for the currents through the three resistors.

Example 3.6. Obtain an expression for the voltage v_2 across the resistor R_2 in the circuit of Fig. 3.25(a).

Solution: We identify and label the nodes and choose a reference node as shown in Fig. 3.25 (b). By the definition of an independent voltage source, $v_a = v_0$. Thus the only unknown node voltage is $v_b = v_2$. By Kirchhoff's current law and Ohm's law,

$$i_1 + i_2 = \frac{v_{ba}}{R_1} + \frac{v_b}{R_2} = \frac{v_b - v_a}{R_1} + \frac{v_b}{R_2}$$
$$= \frac{v_b - v_0}{R_1} + \frac{v_b}{R_2} = 0,$$

which yields

$$v_2 = v_b = \frac{R_2}{R_1 + R_2} v_0$$

The circuit of Fig. 3.25 is called a **voltage divider** because the voltage provided by the source is divided between the two resistors in agreement with the relation above

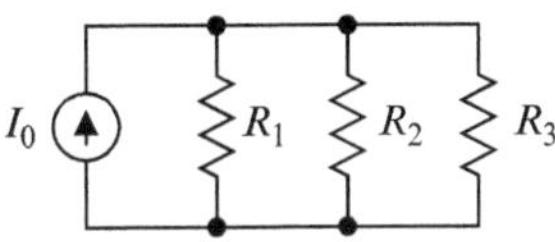

Fig. 3.24 See Exercise 3.5

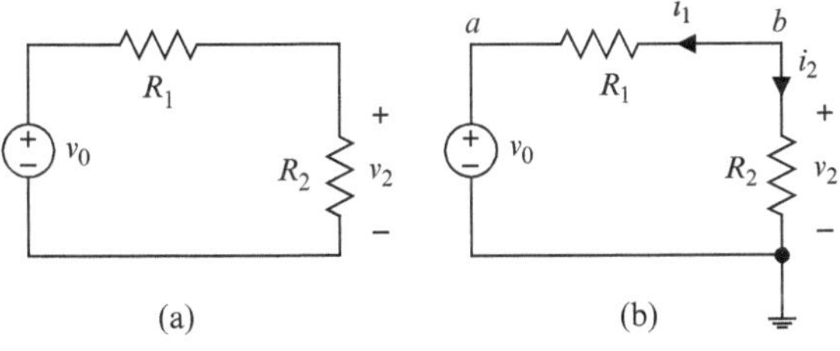

Fig. 3.25 See Example 3.6

Exercise 3.6. Refer to Fig. 3.25. Express the voltage v_{ab} across the resistor R_1 in terms of R_1, R_2, and v_0

Example 3.7. Obtain an expression for the current i_2 in the circuit of Fig. 3.26(a). The resistances and the source voltages are presumed to be known.

Solution: We identify and label the nodes in the circuit as shown in Fig. 3.26(b), choosing node d as the reference node. By the definition of an ideal voltage source,

$$v_a = v_1, \quad v_c = v_2,$$

so v_b is the only unknown node voltage. We apply Kirchhoff's current law at node b and obtain

$$i_1 + i_2 + i_3 = \frac{v_b - v_a}{R_1} + \frac{v_b}{R_2} + \frac{v_b - v_c}{R_3}$$
$$= \frac{v_b - v_1}{R_1} + \frac{v_b}{R_2} + \frac{v_b - v_2}{R_3} = 0,$$

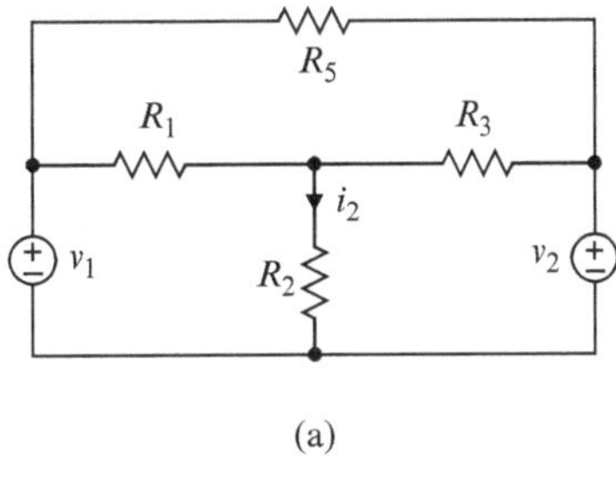

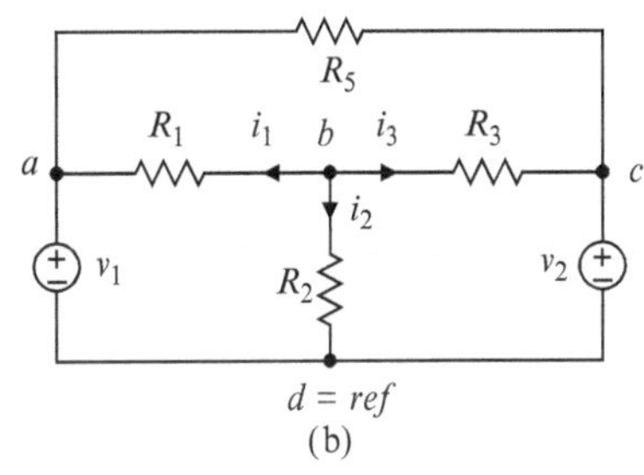

Fig. 3.26 See Example 3.7

in which the only unknown is v_b. We find

$$v_b = \left(\frac{1}{R_1}+\frac{1}{R_2}+\frac{1}{R_3}\right)^{-1}\left(\frac{v_1}{R_1}+\frac{v_2}{R_2}\right).$$

Finally, by Ohm's law

$$i_2 = \frac{v_b}{R_2} = \frac{R}{R_2}\left(\frac{v_1}{R_1}+\frac{v_2}{R_2}\right),$$

$$R = \left(\frac{1}{R_1}+\frac{1}{R_2}+\frac{1}{R_3}\right)^{-1}$$

Exercise 3.7. Refer to Fig. 3.26. Obtain expressions for the currents i_1, i_3

Example 3.8. Express the voltages v_1, v_2 in the circuit of Fig. 3.27(a) in terms of the source voltage v_0, where $R_1 = 1$ kΩ, $R_2 = 2.2$ kΩ, and $R_3 = 10$ kΩ.

Solution: We identify and label the nodes as shown in Fig. 3.27(b), choosing node d as the reference node. By the definition of an ideal voltage source, $v_a = v_0$. The node voltages $v_b = v_{bd}$ and $v_c = v_{cd}$ are unknown, so we apply Kirchhoff's current law to those nodes. Thus

$$i_1+i_2+i_3 = \frac{v_b - v_0}{R_1}+\frac{v_b}{R_2}+\frac{v_b - v_c}{R_1} = 0$$

$$\Rightarrow \left(\frac{2}{R_1}+\frac{1}{R_2}\right)v_b - \frac{v_c}{R_1} = \frac{v_0}{R_1},$$

$$-i_3+i_4+i_5 = \frac{v_c - v_b}{R_1}+\frac{v_c}{R_2}+\frac{v_c - v_0}{R_3} = 0$$

$$\Rightarrow -\frac{v_b}{R_1}+\left(\frac{1}{R_1}+\frac{1}{R_2}+\frac{1}{R_3}\right)v_c = \frac{v_0}{R_3}.$$

To simplify the notation, we use conductances in place of resistances; e.g., $G_1 = 1/R_1$.

Thus

$$\left(\frac{2}{R_1}+\frac{1}{R_2}\right)v_b - \frac{v_c}{R_1} = \frac{v_0}{R_1}$$

$$\Rightarrow (2G_1+G_2)v_b - G_1 v_c = G_1 v_0,$$

$$-\frac{v_b}{R_1}+\left(\frac{1}{R_1}+\frac{1}{R_2}+\frac{1}{R_3}\right)v_c = \frac{v_0}{R_3}$$

$$\Rightarrow -G_1 v_b + (G_1+G_2+G_3)v_c = G_3 v_0.$$

To simplify further, let

$$G_b = \frac{2}{R_1}+\frac{1}{R_2} \cong 2.455 \text{ mS},$$

$$G_c = \frac{1}{R_1}+\frac{1}{R_2}+\frac{1}{R_3} \cong 1.555 \text{ mS}.$$

whence

$$G_b v_b - G_1 v_c = G_1 v_0, \quad -G_1 v_b + G_c v_c = G_3 v_0,$$

which yield

$$v_b = v_1 = \frac{G_1(G_c+G_3)}{G_b G_c - G_1{}^2} v_0 \cong 0.588 v_0,$$

$$v_c = v_2 = \frac{G_b G_3 + G_1{}^2}{G_b G_c - G_1{}^2} v_0 \cong 0.442 v_0.$$

Exercise 3.8. Refer to Fig. 3.27. Express the voltages v_{ac}, v_{bc} in terms of the source voltage v_0.

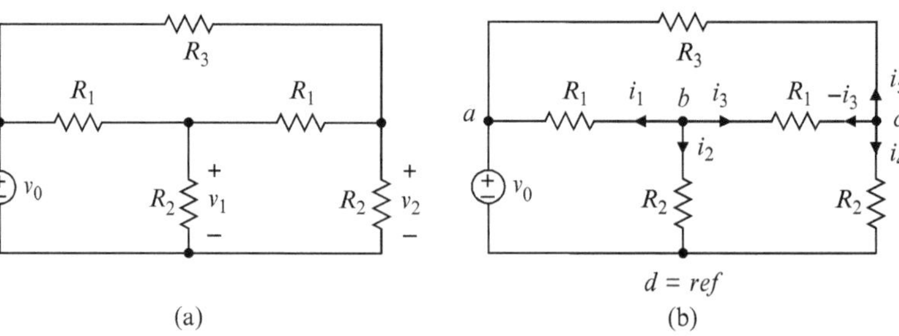

Fig. 3.27 See Example 3.8

The next example illustrates application of Kirchhoff's current law to a circuit in which two nodes, neither of which is the reference node, are connected by an independent voltage source. Each independent voltage source that appears in a circuit reduces the number of unknown node voltages by one.

Example 3.9. Refer to Fig. 3.28(a). Express the voltage v_2 in terms of the circuit parameters.
Solution: We choose the reference node and label the remaining nodes, as shown in Fig. 3.28(b), and denote the current through the source v_0 by i_x. By the definition of an ideal voltage source,

$$v_c = v_1, \quad v_2 = v_b = v_a + v_0.$$

If v_a were known, the voltage v_b would be determined. Writing Kirchhoff's current law at nodes a and b gives

$$-i_0 + \frac{v_a}{R_1} + i_x = 0, \quad -i_x + \frac{v_b - v_1}{R_3} + \frac{v_b}{R_2},$$

which yield

$$\frac{v_a + v_0 - v_1}{R_3} + \frac{v_a + v_0}{R_2} = i_0 - \frac{v_a}{R_1}$$
$$\Rightarrow v_a = \frac{R_1\left[R_2 R_3\, i_0 + R_2\, v_1 - (R_2 + R_3)\, v_0\right]}{R_1 R_2 + R_1 R_3 + R_2 R_3}.$$

Thus,

$$\begin{aligned} v_2 = v_b &= v_a + v_0 \\ &= \frac{R_1\left[R_2 R_3\, i_0 + R_2\, v_1 - (R_2 + R_3)\, v_0\right]}{R_1 R_2 + R_1 R_3 + R_2 R_3} + v_0 \\ &= \frac{R_2(R_1 R_3 i_0 + R_1 v_1 + R_3 v_0)}{R_1 R_2 + R_1 R_3 + R_2 R_3}. \end{aligned}$$

Once you have solved two or three problems such as Example 3.9, where two non-reference nodes are connected by an independent voltage source, you will realize that it is unnecessary to introduce an extra unknown, such as i_x. You can simply look ahead and express the current passing through the source in terms of the dependent node voltage at the other end of the source (node b, in Example 3.9).

Example 3.10. Refer to Fig. 3.29(a). Express the voltage v_0 in terms of the circuit parameters.
Solution: We choose the reference node and label the remaining nodes, as shown in Fig. 3.30(b). By the definition of an ideal voltage source, $v_c = v_b + v_1$, $v_d = v_c + v_2 = v_b + v_1 + v_2$.

Applying Kirchhoff's current law to node a gives

$$-i_1 + \frac{v_a}{R_1} + \frac{v_a - v_b}{R_2} = 0. \tag{3.5}$$

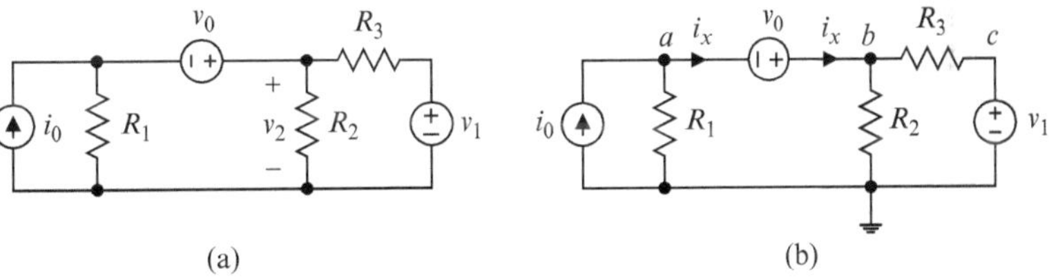

Fig. 3.28 See Example 3.9

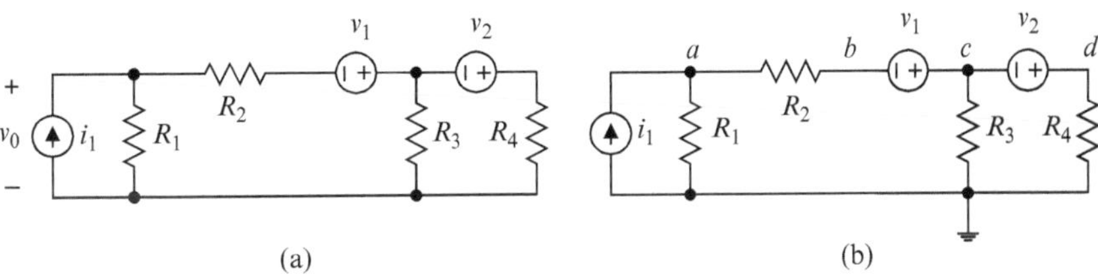

Fig. 3.29 See Example 3.10

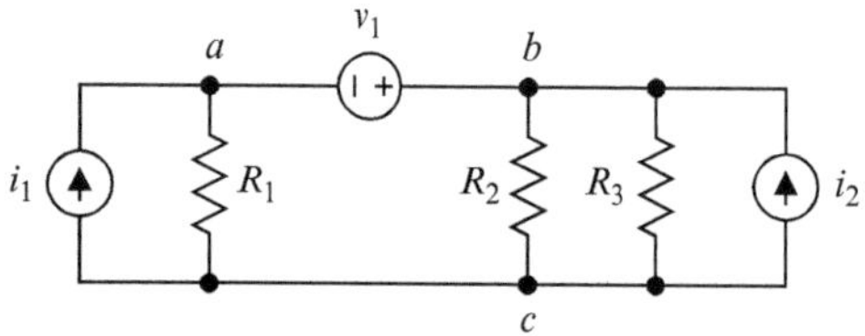

Fig. 3.30 See Exercise 3.11

The current leaving node b through the source v_1 equals $v_c/R_3 + v_d/R_4$. Applying Kirchhoff's current law to node b gives

$$\frac{v_b - v_a}{R_2} + \frac{v_b + v_1}{R_3} + \frac{v_b + v_1 + v_2}{R_4} = 0. \quad (3.6)$$

Using conductance notation and collecting terms in (3.5) and (3.6) gives

$$(G_1+G_2)v_a - G_2v_b = i_1,$$
$$-G_2v_a + (G_2+G_3+G_4)v_b$$
$$= -(G_3+G_4)v_1 - G_4v_2. \quad (3.7)$$

Eliminating v_b yields

$$v_a = v_0 = \frac{(G_2+G_3+G_4)i_1 - G_2[(G_3+G_4)v_1 + G_4v_2]}{(G_2+G_3+G_4)(G_1+G_2) - G_2^2}.$$

Exercise 3.9. Refer to Fig. 3.29(b) in Example 3.10 above. Choose nodes a and d as the independent nodes and write node equations at those nodes. The voltages v_a and v_d must be the only unknown quantities that appear in the equations.

Exercise 3.10. Refer to Fig. 3.29(b) in Example 3.10 above. Choose nodes b and c as the independent nodes and write node equations at those nodes. The voltages v_b and v_c must be the only unknown quantities that appear in the equations.

Exercise 3.11. Refer to Fig. 3.30. Obtain an expression for the voltage v_{bc}.

3.8 Kirchhoff's Voltage Law and Mesh Analysis

Kirchhoff's voltage law states that *the algebraic sum of the voltage drops*[8] *around any closed path equals zero*. In the algebraic sum of voltage drops around a closed path, voltage rises are negated. With reference to Fig. 3.31, Kirchhoff's voltage law around the dotted path can be expressed as

$$\sum_n v_n = 0. \quad (3.8)$$

Kirchhoff's voltage law follows from the definitions of electric potential and voltage. The electric potential at any particular point and time has a single, specific value, which does not depend upon how we arrive at the point. Thus, if we imagine that we instantaneously follow a closed path that begins and ends at the same point (a) but is otherwise arbitrary, the potential at the end of the path is the same as that at the beginning; that is, the net potential difference (voltage drop) around the closed path is

$$\Phi(a) - \Phi(a) = 0.$$

Kirchhoff's voltage law follows, because the point can be any point and the path can be any path through the point.

Example 3.11. In Fig. 3.32, $v_1 = 3$ V, $v_2 = 10$ V, $v_4 = 15$ V Find the voltage v_3.
Solution: Kirchhoff's voltage law gives

$$v_1 + v_2 + v_3 - v_4 = 0.$$

[8]Kirchhoff's voltage law can also be stated as "the sum of voltage rises around any closed path equals zero" or as "the sum of the rises equals the sum of the drops." Again, for the present, it is best to stick with one version of the law.

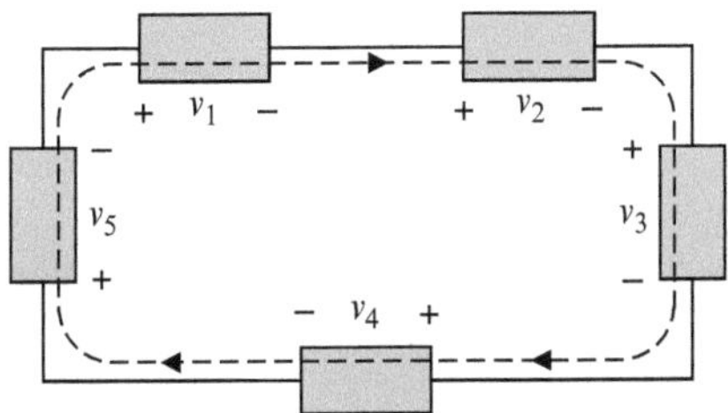

Fig. 3.31 Illustration of Kirchhoff's Voltage Law – see (3.8)

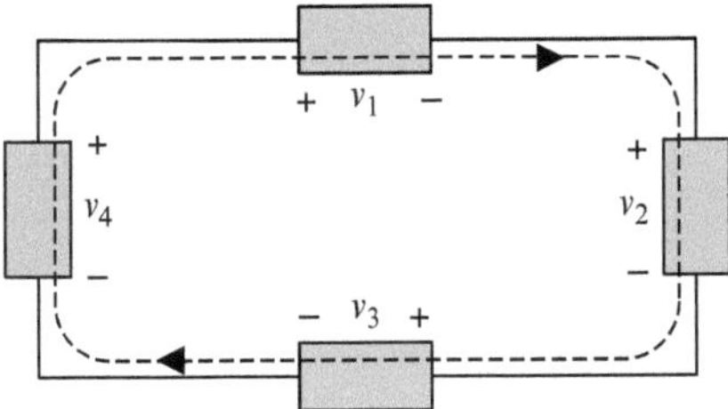

Fig. 3.32 See Example 3.11

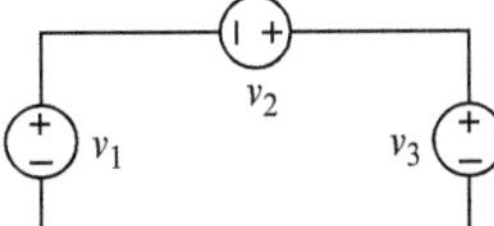

Fig. 3.33 See Exercise 3.12

where v_4 is negated because it is a voltage rise (in the direction of the path). Solving for v_3 gives

$$v_3 = v_4 - v_1 - v_2 = 15 - 3 - 10 = 2 \text{ V}.$$

Exercise 3.12. Is the circuit in Fig. 3.33 self-consistent? Use Kirchhoff's voltage law to justify your answer.

A **loop** is a closed path in or around a circuit. All of the dashed lines in Fig. 3.34 are loops. A **mesh** is a loop that does not enclose (encircle) any other loops. Paths 1, 2, 3 are meshes. Paths *b* and *c* are not meshes. Path *b* encloses two meshes. Path *c* encloses three.

Kirchhoff's voltage law can be applied to any loop or mesh, but in the beginning, it is best to use only meshes, as that is the surer way to obtain the required number of independent equations. Applying Kirchhoff's voltage law to each mesh in a circuit and solving for the mesh currents is called **mesh analysis**. In mesh analysis, we regard each mesh as the path taken by a hypothetical current, called a **mesh current**. If one or more voltages are the quantities of ultimate interest, they may be expressed via Ohm's law in terms of the currents and resistances.

Mesh analysis proceeds as follows:

- Identify each mesh and associate a mesh current with each.
- Write Kirchhoff's voltage law around each mesh, using Ohm's law to express the voltages across resistors in the mesh in terms of the mesh current and the resistances of the resistors.
- Solve the resulting equations for the mesh currents.
- Use the mesh currents and Ohm's law to find any voltages of interest.

A few examples should clarify the procedure.

Example 3.12. (Compare with Example 3.6): Obtain an expression for the voltage across the resistor R_2 in the circuit of Fig. 3.35. The resistances and source voltage are presumed to be known.

Solution: There is only one mesh. We define a path and associated mesh current, as shown in Fig. 3.35. We write Kirchhoff's voltage law clockwise around the path and obtain

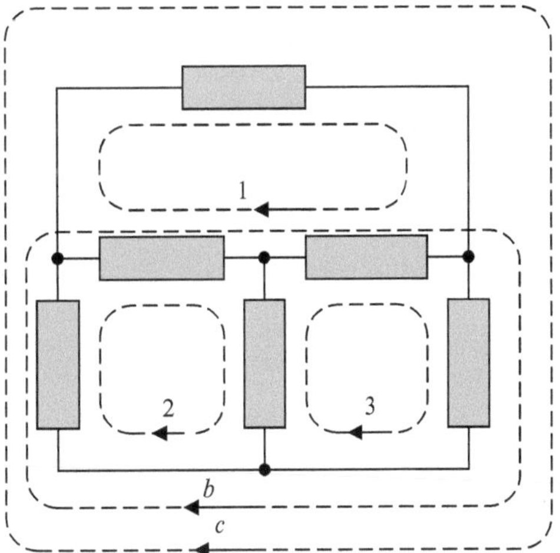

Fig. 3.34 Illustrating the definitions of mesh and loop

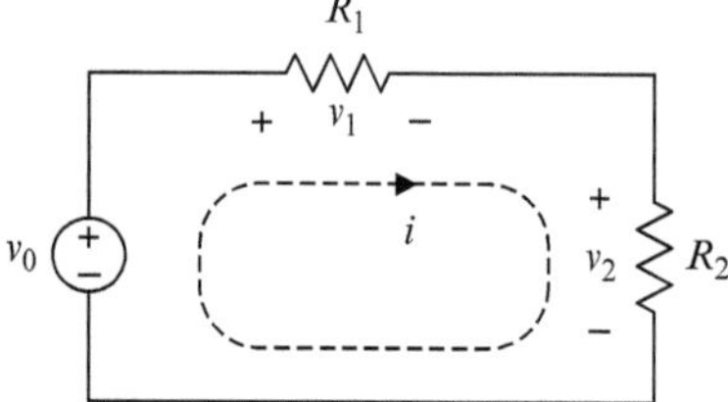

Fig. 3.35 See Example 3.12

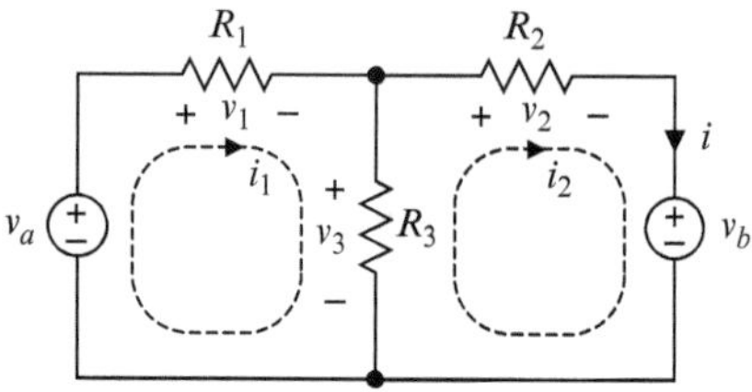

Fig. 3.36 See Example 3.13

$$-v_0 + v_1 + v_2 = 0,$$

where, by Ohm's law

$$v_1 = R_1\, i, \quad v_2 = R_2\, i,$$

so Kirchhoff's voltage law may be written

$$R_1\, i + R_2\, i = v_0,$$

which yields

$$i = \frac{v_0}{R_1 + R_2}.$$

By Ohm's law

$$v_2 = R_2\, i = \frac{R_2}{R_1 + R_2} v_0.$$

Example 3.13. Obtain an expression for the current i in the circuit of Fig. 3.36. The resistances and source voltages are presumed to be known.
Solution: There are two meshes and associated mesh currents, as shown in Fig. 3.36, noting that the current of interest is $i = i_2$. We write Kirchhoff's voltage law around each path to obtain

$$v_1 + v_3 - v_a = 0,$$
$$v_2 + v_b - v_3 = 0.$$

By Ohm's law

$$v_1 = R_1\, i_1, \quad v_2 = R_2\, i_2, \quad v_3 = R_3\,(i_1 - i_2),$$

where i_1–i_2 is the total current down through the resistor R_3. Using these relations in the Kirchhoff-voltage-law equations above gives two equations in the two unknown currents i_1, i_2:

$$R_1 i_1 + R_3(i_1 - i_2) = v_a \Rightarrow (R_1 + R_3) i_1 - R_3 i_2 = v_a,$$
$$R_2 i_2 - R_3(i_1 - i_2) = -v_b$$
$$\Rightarrow -R_3 i_1 + (R_2 + R_3) i_2 = -v_b.$$

To eliminate i_1, we add R_3 times the first equation to $(R_1 + R_3)$ times the second. This gives

$$-{R_3}^2\, i_2 + (R_1 + R_3)(R_2 + R_3)\, i_2$$
$$= R_3\, v_a - (R_1 + R_3)\, v_b,$$

which yields

$$i = i_2 = \frac{R_3\, v_a - (R_1 + R_3)\, v_b}{R_1 R_2 + R_1 R_3 + R_2 R_3}.$$

Exercise 3.13. Obtain an expression for the current i_1 in Fig. 3.36.

The next example illustrates application of Kirchhoff's voltage law to a circuit containing a current source. Each independent current source in a circuit reduces the number of unknown mesh currents by one.

Example 3.14. Obtain an expression for the current i_1 through the resistor R_1 in the circuit of Fig. 3.37.

Solution: Although there are two meshes, there is only one unknown current, because, by Kirchhoff's current law and the definition of an independent current source, the current through the resistor R_2 is

$$i_2 = i_1 + i_0,$$

and i_0 is known. Thus we need only one equation. Writing Kirchhoff's voltage law clockwise around the indicated path gives

$$-v_a + v_1 + v_2 + v_b = 0.$$

By Ohm's law,

$$v_1 = R_1 i_1 , \quad v_2 = R_2 i_2 = R_2(i_1 + i_0).$$

Thus Kirchhoff's voltage law yields

$$-v_a + R_1 i_1 + R_2(i_1 + i_0) + v_b = 0 \Rightarrow i_1 = \frac{v_a - v_b - R_2 i_0}{R_1 + R_2}.$$

When applying Kirchhoff's voltage law to a circuit containing one or more independent current sources, *it is never necessary to choose a path that passes through an independent current source* because, by definition, the current through an independent current source is known.

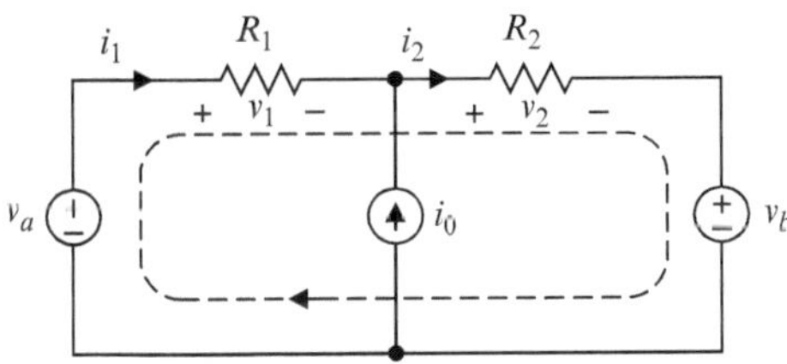

Fig. 3.37 See Example 3.14

Example 3.15. Refer to Fig. 3.38. Use mesh analysis to obtain an expression for the voltages v_1, v_2, v_3.

Solution: There are four meshes, as shown in Fig. 3.38, but two of the mesh currents (i_0 and $3i_0$) are known. Thus we need only the two equations for the unknown mesh currents i_1, i_2.

$$-R_1 i_0 + (R_1 + R_2 + R_3)i_1 - R_3 i_2 = 0,$$
$$-3R_4 i_0 - R_3 i_1 + (R_3 + R_4 + R_5)i_2 = 0.$$

To simplify notation, let

$$R_A = R_1 + R_2 + R_3, \quad R_B = R_3 + R_4 + R_5.$$

Then

$$R_A\, i_1 - R_3\, i_2 = R_1\, i_0, \quad R_3\, i_1 + R_B\, i_2 = 3R_4\, i_0,$$

which yield

$$i_1 = \frac{R_1 R_B + 3R_3 R_4}{R_A R_B - R_3{}^2} i_0, \quad i_2 = \frac{R_1 R_3 + 3R_A R_4}{R_A R_B - R_3{}^2} i_0,$$

and thus

$$v_1 = (i_0 - i_1)R_1 = \frac{R_A R_B - R_3{}^2 - R_B R_1 - 3R_4 R_3}{R_A R_B - R_3{}^2} R_1 i_0,$$
$$v_2 = (i_1 - i_2)R_3 = \frac{R_B R_1 + 3R_3 R_4 - R_1 R_3 - 3R_A R_4}{R_A R_B - R_3{}^2} R_3 i_0,$$
$$v_3 = i_2 R_5 = \frac{R_1 R_3 + 3R_A R_4}{R_A R_B - R_3{}^2} R_5 i_0.$$

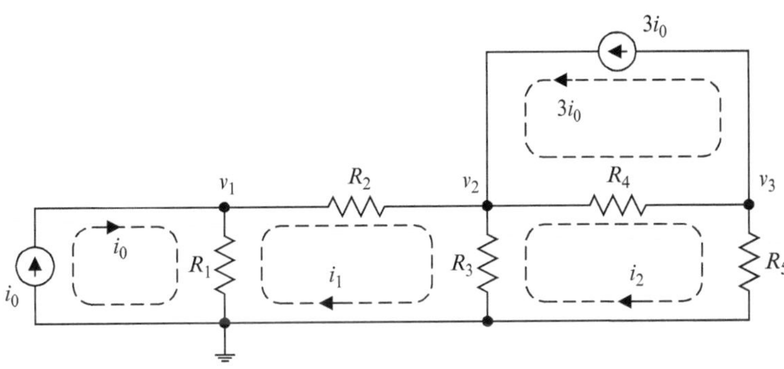

Fig. 3.38 See Example 3.15

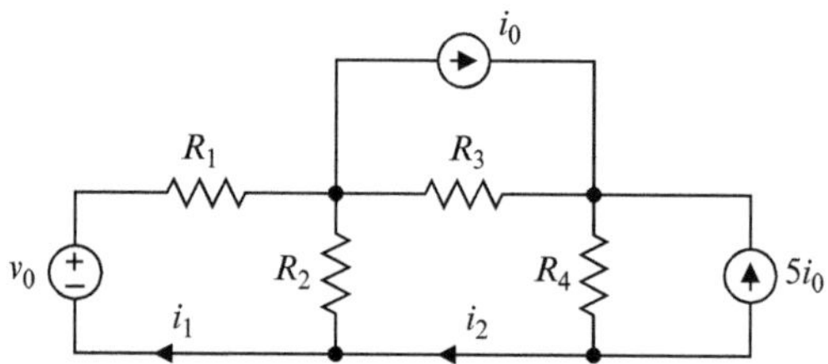

Fig. 3.39 See Exercise 3.14

Exercise 3.14. Refer to Fig. 3.39. Obtain expressions for the currents i_1, i_2.

3.9 Voltage and Current Dividers

Figure 3.40 shows the simplest form of a **voltage divider**, so named because the source voltage v_0 is divided between the resistors R_1, R_2. The voltage across the resistor R_2 is given by (see Example 3.6)

$$v_2 = \frac{R_2}{R_1 + R_2} v_0. \tag{3.9}$$

Equation (3.9) applies only if all of the current that passes through the resistor R_1 also passes through the resistor R_2. A common error is applying (3.9) to a circuit such as the one shown in Fig. 3.41, where some of the current passing through R_1 is diverted to a load R_3 (not all passes through R_2). In Fig. 3.41, the voltage v_a is *not* given by (3.9), unless the resistance R_3 is so much larger than R_2 that the current i_3 through R_3 is a negligible fraction of i_1.

For a specific example, consider the circuit of Fig. 3.42(a), in which the load on the voltage divider is expressed as a multiple of the resistance R. The voltage v_2 is given by

$$v_2 = \frac{k}{2k+1} v_0 = \frac{v_0}{2 + k^{-1}}, \tag{3.10}$$

as you are asked to show in Exercise 3.15. The load voltage v_2 is approximately half the source voltage if k is sufficiently large, as shown by the graph in Fig. 3.42 (b). For example, for $k = 100$, $v = 0.498v_0$.

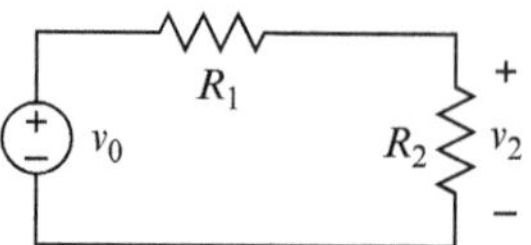

Fig. 3.40 Voltage divider

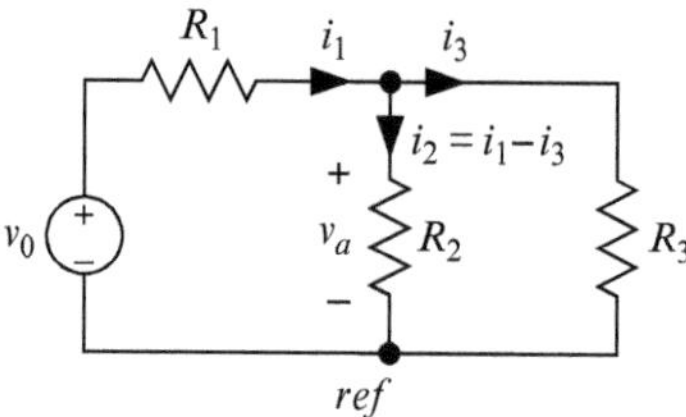

Fig. 3.41 Circuit used to illustrate erroneous application of voltage division. See text

Exercise 3.15. Show that the voltage v_2 in Fig. 3.42(a) is given by (3.10).

Figure 3.43 shows the simplest form of a **current divider**, so named because the source current is divided between the resistors R_1, R_2. The current through the resistor R_2 is given by (see Example 3.5)

$$i_2 = \frac{R_1}{R_1 + R_2} i_0. \tag{3.11}$$

Voltage and current dividers are quite common, and you should memorize the circuits in Figs. 3.40 and 3.43 and the associated relations (3.9) and (3.11). Voltage and current dividers can contain more than two resistors (can divide voltages or currents into more than two parts), but two-part dividers are by far the most common.

3.10 Superposition

As it applies to resistive circuits, the **principle of superposition** can be stated as follows:

Suppose a circuit contains two or more *independent* current or voltage sources. Any particular current or voltage in the circuit can be obtained as the sum of the corresponding currents or voltages found by applying the sources one at a time, all others being set to zero.

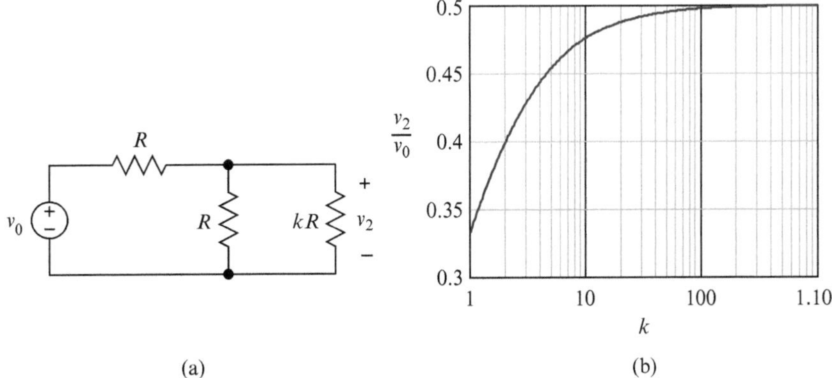

Fig. 3.42 The voltage v_2 is approximately half the source voltage for large k

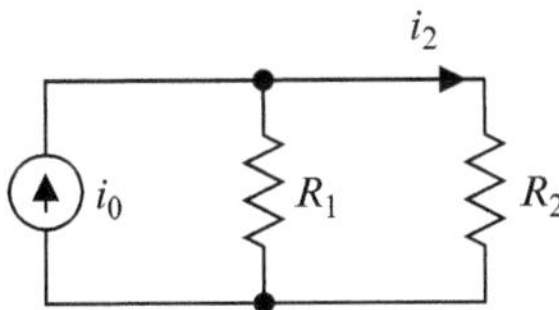

Fig. 3.43 Current divider

Note that setting a source voltage to zero is equivalent to replacing the associated source with a short circuit and setting a source current to zero is equivalent to replacing the associated source with an open circuit.[9]

Example 3.16. Using superposition, obtain an expression for the current i_x in the circuit shown in Fig. 3.44(a).

Solution: We first find the current i due to the voltage source v_0 acting alone; that is, with the source current i_0 set to zero (replaced by an open circuit), as illustrated by Fig. 3.44(b). This gives

$$i|_{i_0=0} = \frac{v_0}{R_1 + R_2 + R_3}.$$

Next we find the current i due to the current source i_0 acting alone; that is, with the source voltage v_0 set to zero (replaced by a short circuit), as illustrated by Fig. 3.44(c). We obtain

$$i|_{v_0=0} = \frac{R_2 i_0}{R_1 + R_2 + R_3}.$$

There are no other sources, so the (total) current i_x is given by

$$i = i|_{i_0=0} + i|_{v_0=0} = \frac{v_0 + R_2 i_0}{R_1 + R_2 + R_3}.$$

In applications of superposition, the various sources may be applied in any order.

The principle of superposition is a consequence of Kirchhoff's laws and the fact that the terminal characteristic of a resistor is *linear*. We can show this as follows: Applying either or both of Kirchhoff's laws to a resistive circuit leads to a set of equations of the form

$$\begin{aligned}
&a_{11}\,x_1 + a_{12}\,x_2 + \cdots + a_{1N}\,x_N \\
&\quad = b_{11}\,y_1 + b_{12}\,y_2 + \cdots + b_{1M}\,y_M, \\
&a_{21}\,x_1 + a_{22}\,x_2 + \cdots + a_{2N}\,x_N \\
&\quad = b_{21}\,y_1 + b_{22}\,y_2 + \cdots + b_{2M}\,y_M, \\
&\vdots \\
&a_{N1}\,x_1 + a_{N2}\,x_2 + \cdots + a_{NN}\,x_N \\
&\quad = b_{N1}\,y_1 + b_{N2}\,y_2 + \cdots + b_{NM}\,y_M,
\end{aligned} \tag{3.12}$$

where the y's are known source currents or voltages, the x's are unknown currents or voltages, and the a's and b's are (in general) functions of circuit parameters (resistances and conductances) and are independent of the known and unknown currents and voltages. The solution of such a set of equations has the form

[9] Review the definitions of independent current and voltage sources in Section 3.6.

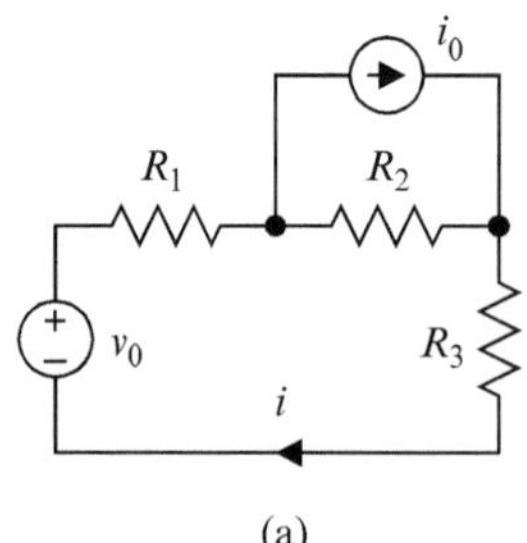

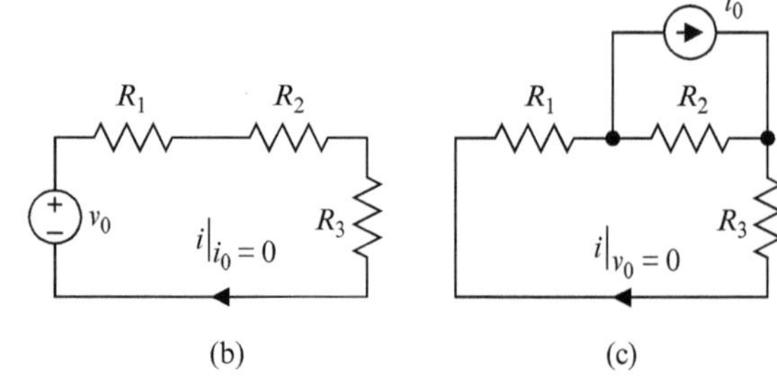

Fig. 3.44 See Example 3.16

$$
\begin{aligned}
x_1 &= c_{11}y_1 + c_{12}y_2 + \cdots + c_{1M}y_M,\\
x_2 &= c_{21}y_1 + c_{22}y_2 + \cdots + c_{2M}y_M,\\
&\vdots\\
x_N &= c_{N1}y_1 + c_{N2}y_2 + \cdots + c_{NM}y_M,
\end{aligned}
\tag{3.13}
$$

where the c's are functions of the circuit parameters (the a's and b's) but do not depend upon the source currents and voltages (the y's). For example, in Example 3.13 we obtain the equation

$$
\begin{aligned}
(R_1 + R_3)\, i_1 - R_3\, i_2 &= v_a,\\
-R_3\, i_1 + (R_2 + R_3)\, i_2 &= -v_b,
\end{aligned}
$$

which have the form given in (3.12), with

$$
\begin{aligned}
&N = 2,\ x_1 = i_1,\ x_2 = i_2,\\
&a_{11} = R_1 + R_3,\ a_{12} = -R_3,\\
&a_{21} = -R_3,\ a_{22} = R_2 + R_3,\\
&M = 2,\ y_1 = v_a,\ y_2 = v_b,\\
&b_{11} = 1,\ b_{12} = 0,\ b_{21} = 0,\ b_{22} = -1.
\end{aligned}
$$

The solution (see Example 3.13 and Exercise 3.13) is

$$
\begin{aligned}
i_1 &= \frac{(R_2+R_3)v_a - R_3 v_b}{R_1R_2+R_1R_3+R_2R_3} = \left[\frac{(R_2+R_3)}{R_1R_2+R_1R_3+R_2R_3}\right] v_a\\
&\quad + \left[\frac{-R_3}{R_1R_2+R_1R_3+R_2R_3}\right] v_b,\\
i_2 &= \frac{R_3 v_a - (R_1+R_3)v_b}{R_1R_2+R_1R_3+R_2R_3} = \left[\frac{R_3}{R_1R_2+R_1R_3+R_2R_3}\right] v_a\\
&\quad + \left[\frac{-(R_1+R_3)}{R_1R_2+R_1R_3+R_2R_3}\right] v_b.
\end{aligned}
\tag{3.14}
$$

which has the form given in (3.13), with

$$
\begin{aligned}
c_{11} &= \frac{(R_2+R_3)}{R_1R_2+R_1R_3+R_2R_3},\ c_{12} = \frac{-R_3}{R_1R_2+R_1R_3+R_2R_3},\\
c_{21} &= \frac{R_3}{R_1R_2+R_1R_3+R_2R_3},\ c_{22} = \frac{-(R_1+R_3)}{R_1R_2+R_1R_3+R_2R_3}.
\end{aligned}
$$

Equations obtained by applying Kirchhoff's laws to *any resistive circuit* can be put in the form (3.12) and the solution for such a set can *always* be written in the form (3.13). Equation (3.13) (or, e.g., (3.14)) expresses each current and voltage in a resistive circuit as a *linear combination* of source currents and voltages. To find a particular current or voltage due to a particular source current or voltage, we set all source currents and voltages except the one of interest to zero. For example, with reference to (3.14), suppose we wish to obtain an expression for that part of the current i_2 that can be attributed to the source v_b. To do so, we set $v_a = 0$ in the expression for i_2. This gives

$$
i_2|_{v_a=0} = \left[\frac{-(R_1+R_3)}{R_1R_2+R_1R_3+R_2R_3}\right] v_b. \tag{3.15}
$$

Similarly, if we wish to obtain an expression for that part of the current i_2 that can be attributed to the source v_a, we set $v_b = 0$ in the expression for i_2. This gives

$$
i_2|_{v_b=0} = \left[\frac{R_3}{R_1R_2+R_1R_3+R_2R_3}\right] v_a. \tag{3.16}
$$

An expression for the (total) current i_2, i.e., the second of Equations (3.14), can be obtained by adding the right sides of (3.15) and (3.16), in agreement with the principle of superposition.

An important by-product of the proof of superposition given above is the following: *If a linear resistive circuit contains exactly one independent source, then every current and voltage in the circuit is proportional to the source current or voltage*. This conclusion is reached by setting all but one source equal to zero in (3.13).

Superposition can be generalized to more complex linear circuits, as shown in a subsequent chapter, and is one of the most important principles in circuit analysis. Although it usually does not greatly simplify analysis of resistive circuits excited by simple (e.g., constant or sinusoidal) sources, it is of tremendous importance in analysis of more general linear circuits subjected to more complex excitations. But superposition is most easily understood and applied in the context of resistive circuits and simple excitations, and mastering the concept in that setting will pay dividends in more realistic and practical applications.

3.11 Problems

Section 3.1 is prerequisite for the following problems.

P 3.1 Under what conditions does an ideal conductor accurately model a physical conductor in a circuit diagram? How else might you model a physical conductor?

P 3.2 Refer to Fig. P 3.1. Express the potentials at points e, f, g, h in terms of the potentials at points a, b, c, d.

P 3.3 Figure P 3.2 shows a circuit diagram and an associated table of potentials. Complete the table.

P 3.4 Figure P 3.3 shows a circuit diagram and an associated table of potentials. Complete the table.

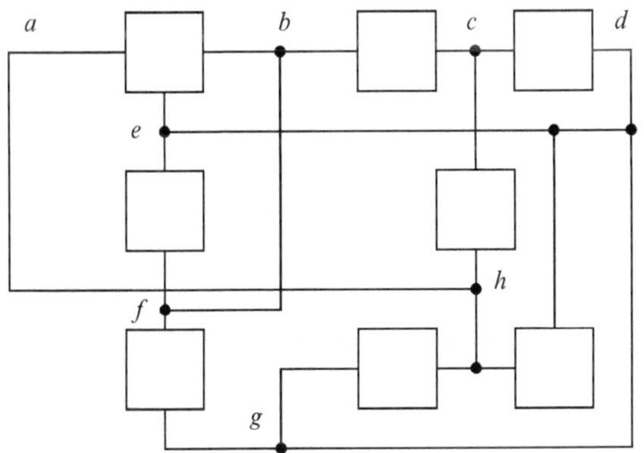

Fig. P 3.1 See Problem P 3.2

Section 3.2 is prerequisite for the following problems.

P 3.5 In Problem P 3.2, g is the reference node and $v_a = 15$ V, $v_b = 10$ V, $v_c = -10$ V, $v_d = 12$ V. (a) Find v_e, v_f, v_{bh}, v_{dh}, v_{fh}, v_{ca}. (b) Label the diagram with the voltage across each element using plus–minus notation.

P 3.6 Refer to Problem P 3.3. (a) Take point i as reference and find the voltages v_{ab}, v_{bg}, v_{cf}, v_{cd}, v_{fe}, v_a, v_b, v_c, v_f, v_h. (b) Label the diagram with the voltage across each element using plus–minus notation.

P 3.7 Refer to Problem P 3.4. (a) Take point k as reference and find the voltages v_a, v_d, v_h, v_g, v_{bd}, v_{dg}, v_{ih}, v_{bf}. (b) Label the diagram with the voltage across each element using plus–minus notation.

P 3.8 In a certain circuit, the potentials at points a, b, c, and d are $\Phi(a) = 50$ V, $\Phi(b) = 25$ V, $\Phi(c) = 15$ V, and $\Phi(d) = 100$ V. Take point d as the reference and find the voltages v_a, v_b, v_c, v_{ab}, v_{bc}, v_{ac}.

P 3.9 In a certain circuit, the potentials at points a, b, c, and d are $\Phi(a) = 5$ V, $\Phi(b) = 15$ V, $\Phi(c) = 25$ V, and $\Phi(d) = 10$ V. Take point d as the reference and find the voltages v_a, v_b, v_c, v_{ab}, v_{bc}, v_{ac}.

Section 3.3 is prerequisite for the following problems.

P 3.10 Use the terms *series* and *parallel* to describe (in words) the connections shown in Fig. P 3.4. (Note: It is possible that neither of the terms apply to a particular arrangement.)

P 3.11 Draw diagrams (e.g., see Fig. P 3.4) depicting the connections described below:

(a) The parallel connection of elements 1 and 2 is in series with the parallel connection of elements 3 and 4.
(b) Elements 1, 2 and 3 are in parallel with the series connection of elements 4 and 5.
(c) Elements 1, 2, and 3 are in series with the parallel connection of elements 4 and 5.

Section 3.4 is prerequisite for the following problems.

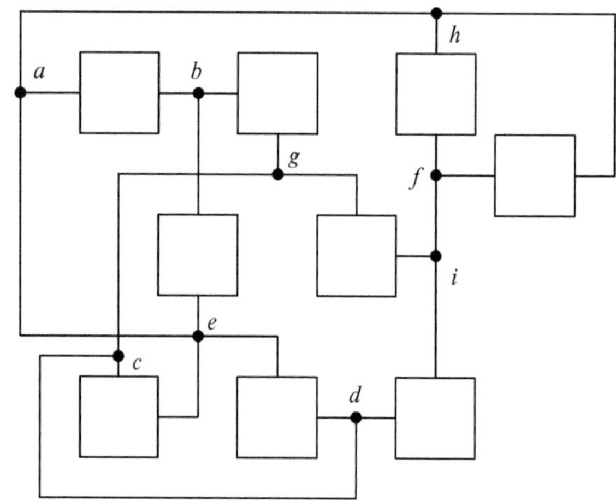

point	potential (V)
a	10
b	6
c	
d	4
e	
f	1
g	
h	
i	

Fig. P 3.2 See Problem P 3.3

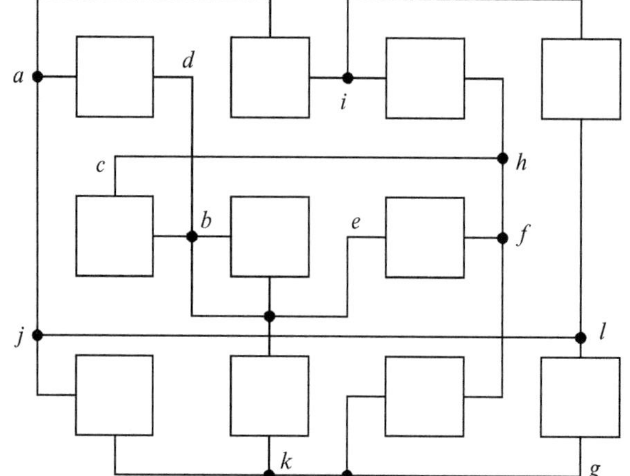

point	potential (V)
a	2
b	
c	
d	
e	5
f	
g	
h	3
i	6
j	
k	10
l	

Fig. P 3.3 See Problem P 3.4

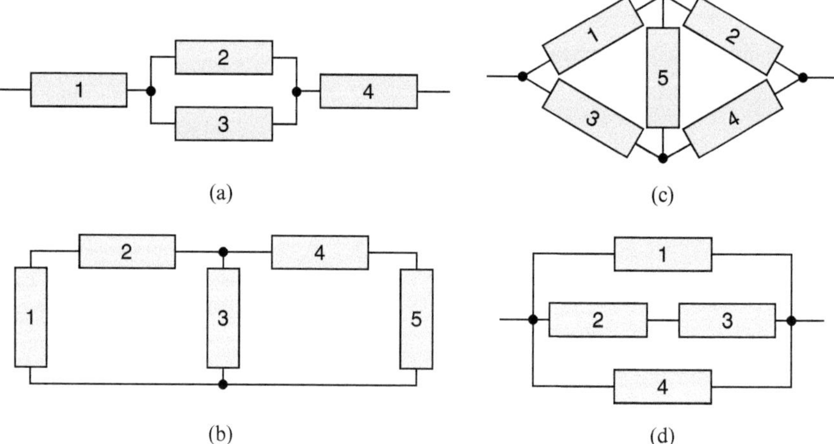

Fig. P 3.4 See Problem P 3.10

P 3.12 What resistance is equivalent to (a) a short circuit and (b) an open circuit?

P 3.13 What conductance is equivalent to (a) a short circuit and (b) an open circuit?

P 3.14 A switch can be idealized as a short circuit in the *on* position and as an open circuit in the *off* position. Suppose a battery, switch, and lamp are connected as shown in Fig. P 3.5. If the terminal voltage of the battery is v_0 and the resistance of the lamp is R, what are the voltage across and current through the lamp when the switch is in (a) the *off* position and (b) the *on* position?

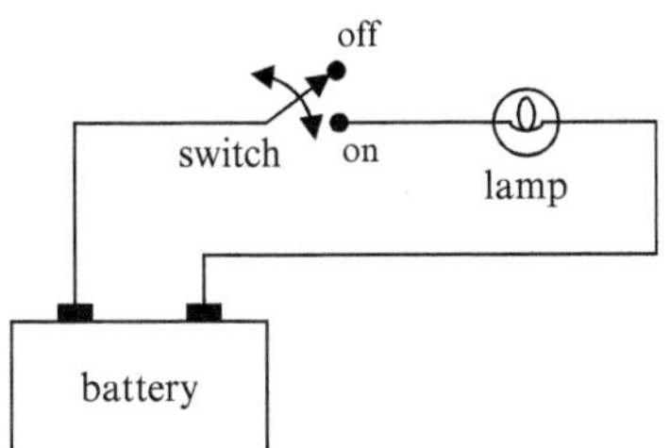

Fig. P 3.5 See Problem P 3.14

Section 3.5 is prerequisite for the following problems.

P 3.15 For each circuit shown in Fig. P 3.6, express the voltages v_{ab}, v_{ac}, v_{ad}, v_{ae} in terms of the source voltages.

P 3.16 Refer to Fig. P 3.7. Employ the definitions and techniques described in Sections 3.1 through 3.6 to express each labeled voltage and current (e.g., v_x, i_x) in terms of the resistances and source currents and voltages.

P 3.17 There is something wrong (self-contradictory) with each circuit in Fig. P 3.8 where the source currents and voltages are arbitrary. What is it?

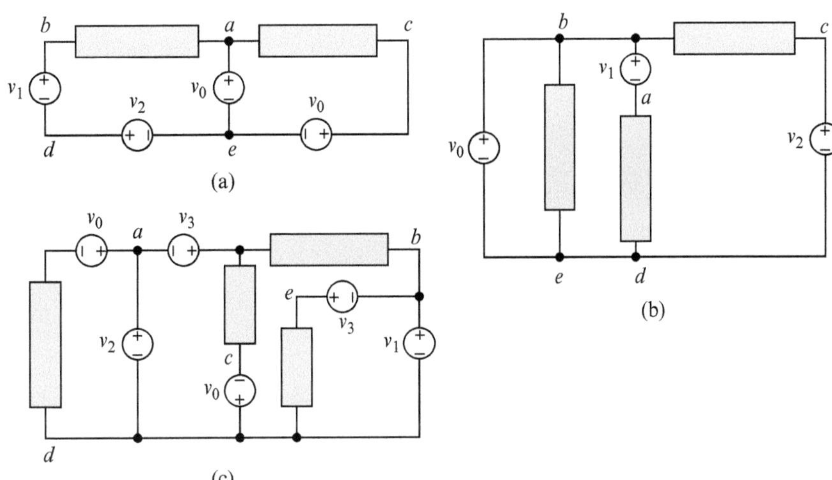

Fig. P 3.6 See Problem P 3.15

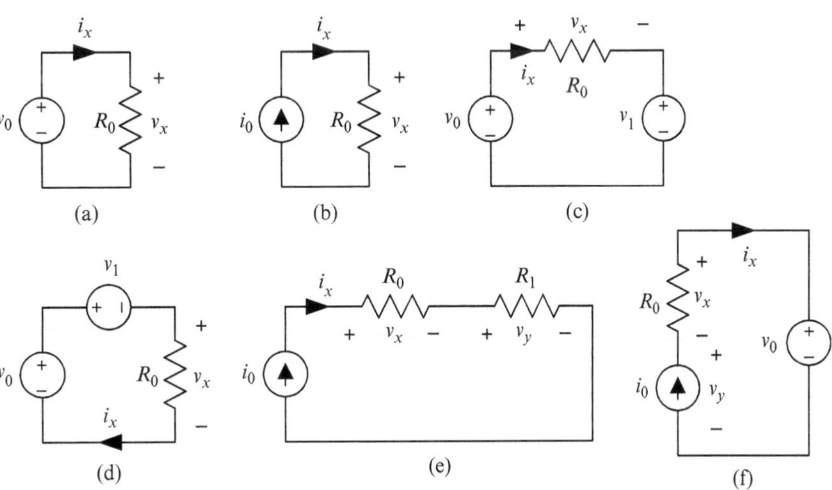

Fig. P 3.7 See Problem P 3.16

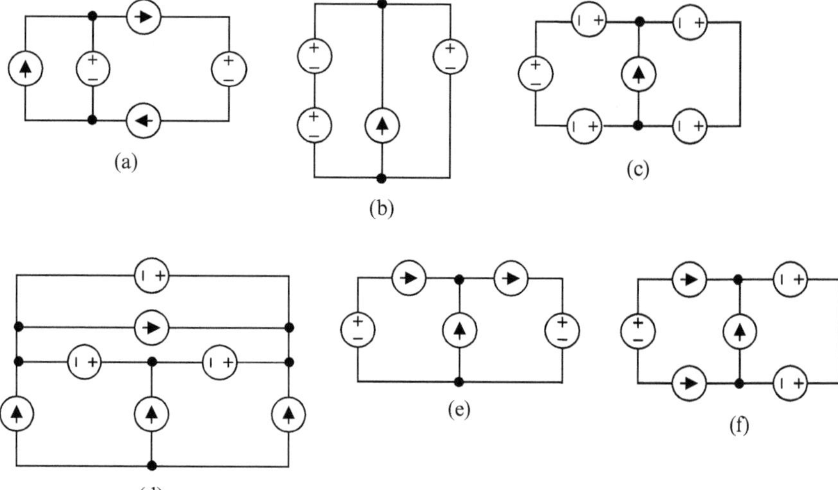

Fig. P 3.8 See Problem P 3.17

Section 3.6 is prerequisite for the following problems.

Note: In subsequent problems, the symbols in Fig. P 3.9 denote a voltmeter (V) having infinite resistance and an ammeter (A) having zero resistance. The polarity marks indicate the voltage polarity that gives a positive voltmeter reading and the arrow indicates the current direction that gives a positive ammeter reading.

P 3.18 Refer to Fig. P 3.10, where $V_S = 13$ V. Ammeter 1 reads 5.04 mA, ammeter 2 reads 1.42 mA, and the voltmeter reads 7.96 V. Find the resistances R_S, R_1, R_2. What are the nearest standard values for R_1 and R_2 from the lowest possible E series?

P 3.19 Refer to Fig. P 3.11, where $V_0 = 12$ V. Ammeter 1 reads 2.14 mA, ammeter 2 reads 4.87 mA, and the voltmeter reads 5.46 V. Find the resistances R_0, R_1, R_2. What are the nearest standard values for R_0, R_1, and R_2 from the lowest possible E series?

P 3.20 Find the current i_x in the circuit of Fig. P 3.12, where $i_1 = 5$ mA, $i_2 = 10$ mA, and $v_0 = 5$ V. To what extent does the current i_x depend upon the voltage v_0?

P 3.21 In the circuit of Fig. P 3.13, $v_0 = 15$ V, $v_{ad} = 10$ V, $v_{bd} = 5$ V, and $R = 1$ kΩ. Find the indicated currents and the voltages v_{ab}, v_{bc}, v_{ac}.

P 3.22 In the circuit of Fig. P 3.14, $I_S = 100$ mA and $R_S = 10$ kΩ. It is required that the current through R_1 equal 25 mA and the current through R_2 equal 50 mA. Find the resistances R_1, R_2. Then specify the nearest

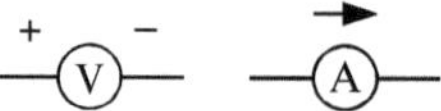

Fig. P 3.9 Symbols for voltmeter and ammeter

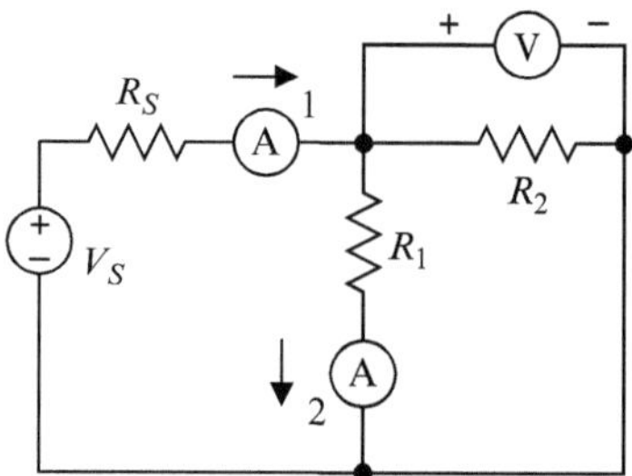

Fig. P 3.10 See Problem P 3.18

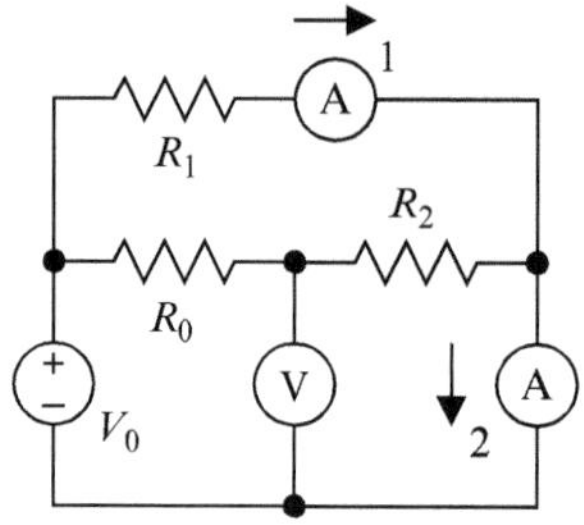

Fig. P 3.11 See Problem P 3.19

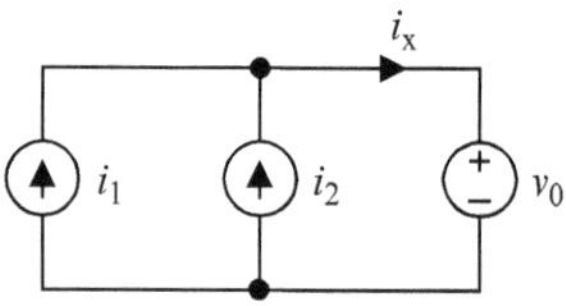

Fig. P 3.12 See Problem P 3.20

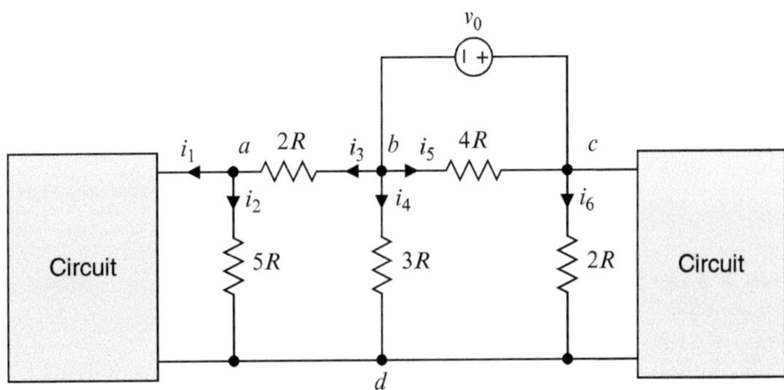

Fig. P 3.13 See Problem P 3.21

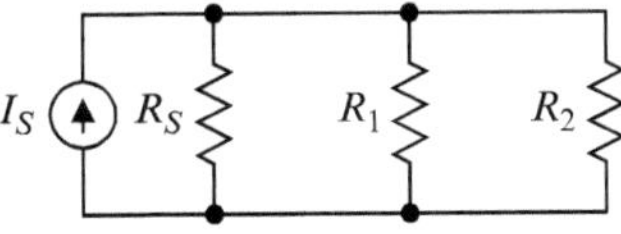

Fig. P 3.14 See Problem P 3.22

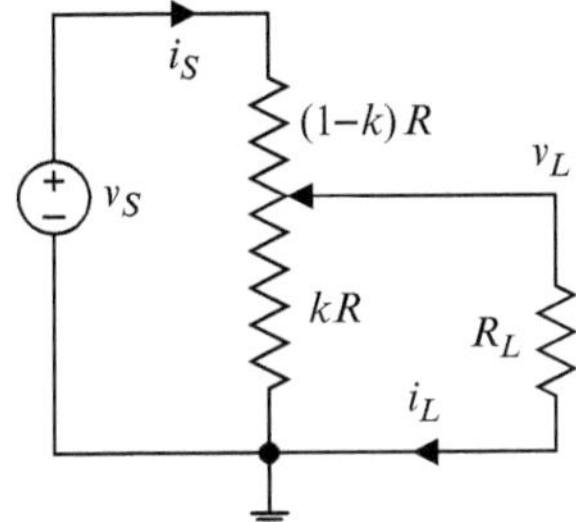

Fig. P 3.15 See Problem P 3.23

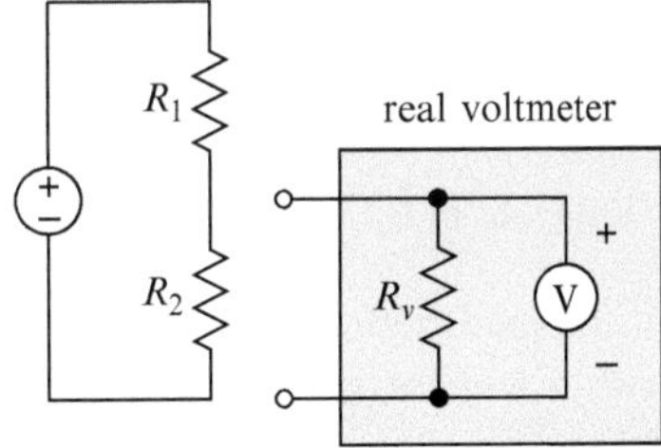

Fig. P 3.16 See Problem P 3.25

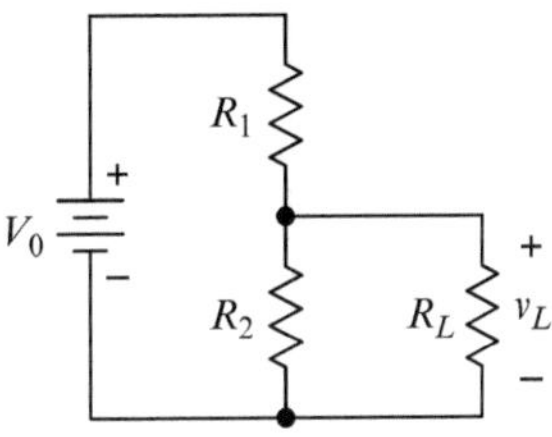

Fig. P 3.17 See Problem P 3.26

values from the E192 series and obtain an expression for the percent error in each current. (Assume the values given for I_S, R_S are exact.)

P 3.23 In Fig. P 3.15, the resistor on the left is a variable resistor, called a **rheostat**. The resistance from the movable tap (the arrow) to one end of the rheostat is kR and the total resistance of the rheostat is R. The voltage v_L across the load resistor R_L must equal one-half of the source voltage v_S. Express the required rheostat setting k in terms of the resistances R, R_L.

P 3.24 Refer to Problem P 3.23 and Fig. P 3.15. The current i_L through the load resistor R_L must equal one-third of the current i_S. Express the required rheostat setting k in terms of the resistances R, R_L.

P 3.25 An ideal voltmeter has infinite internal resistance. A real voltmeter has a large but finite internal resistance and can be modeled as an ideal voltmeter in parallel with a resistor. In Fig. P 3.16, $R_1 = 10\,\text{k}\Omega$, $R_2 = 53\,\text{k}\Omega$, and $R_v = 500\,\text{k}\Omega$. The voltage across the series connection of R_1 and R_2 is fixed. When the (real) voltmeter is connected to R_2, it reads 5.6 V. What is the true voltage across R_2 (before the meter is connected)?

P 3.26 In Fig. P 3.17, $V_0 = 12$ V and $900\,\Omega \leq R_L \leq 1{,}200\,\Omega$. The voltage v_L must be $9\,\text{V} \pm 0.1\,\text{V}$. Specify the resistances R_1, R_2. What are the minimum and maximum currents drawn from the source? Use the nearest E192-series values for R_1, R_2 and calculate the resulting minimum and maximum values of the voltage v_L. Does the circuit still meet the specifications?

P 3.27 A simple model for a battery consists of an ideal voltage source in series with a resistance, called the *internal resistance* of the battery. The open-circuit (no-load) terminal voltage of the battery equals the terminal voltage of the ideal source. A battery having

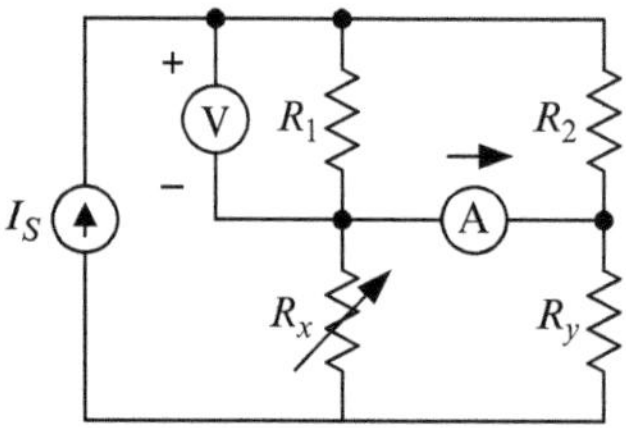

Fig. P 3.18 See Problem P 3.28

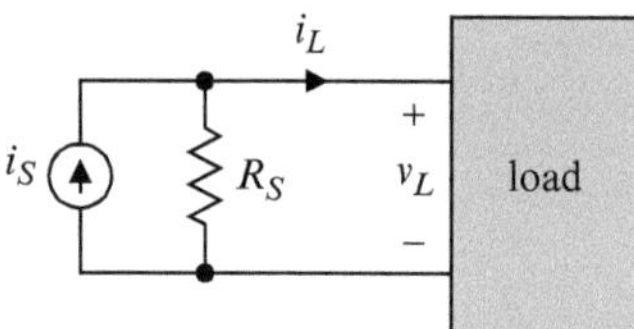

Fig. P 3.19 See Problem P 3.29

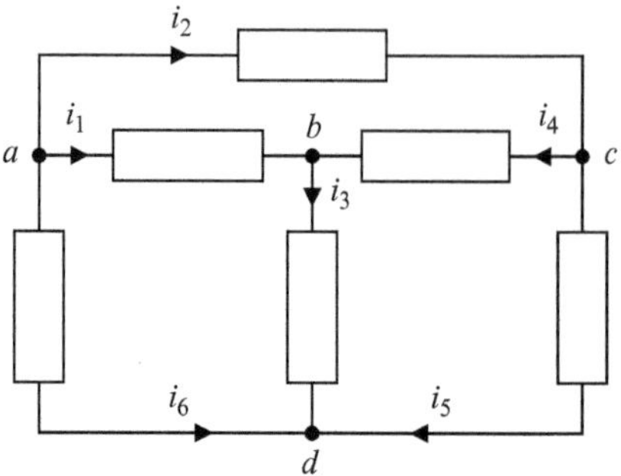

Fig. P 3.20 See Problem P 3.30

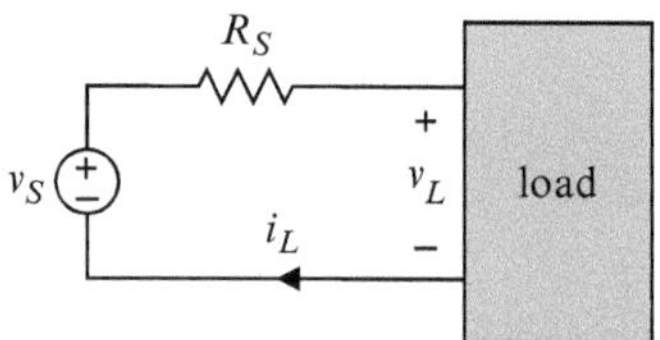

Fig. P 3.21 See Problem P 3.31

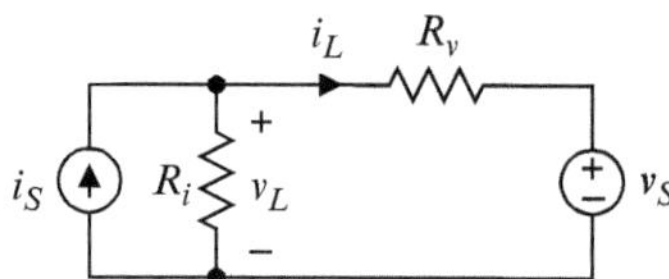

Fig. P 3.22 See Problem P 3.32

open-circuit (no-load) terminal voltage V_0 is connected in series with a resistor having resistance R_L. Using such a model, obtain expressions for the battery internal resistance R_i in terms of (a) the current I_R through the resistor R_L and (b) the voltage V_R across the resistor.

P 3.28 Refer to Fig. P 3.9. In Fig. P 3.18, $R_1 = 10\,\text{k}\Omega$ and $R_2 = 28\,\text{k}\Omega$. The variable resistor R_x is adjusted until the ammeter reads zero, at which point the voltmeter reads 7.3 V and the resistance $R_x = 15.4\,\text{k}\Omega$. Find the resistance R_y and the source current I_S.

P 3.29 Refer to Fig. P 3.19. Use Kirchhoff's current law to show that the relation between the voltage v_L and the current i_L is independent of the load (of whatever is in the box).

Section 3.7 is prerequisite for the following problems.

P 3.30 Refer to Fig. P 3.20. Make d the reference node and write every relation you can think of among the node voltages and the voltages across the various elements; for example, $v_{ab} = v_a - v_b$. Then write every relation you can think of among the branch currents; for example, $i_1 + i_2 + i_6 = 0$.

P 3.31 Refer to Fig. P 3.21. Use Kirchhoff's voltage law to express the voltage v_L in terms of the current i_L, the source voltage v_S and the resistance R_S. Note that the relation thus obtained is independent of the load (of whatever is in the box).

P 3.32 Refer to Fig. P 3.22. According to Problem P 3.29, the relation between the voltage v_L and the current i_L is independent of the current i_S and the resistance R_i. According to Problem P 3.31, the relation between the voltage v_L and the current i_L is independent of the voltage v_S and the resistance R_v. Does this mean that the relation between the voltage v_L and the current i_L is independent of all four quantities v_S, R_v, i_S, R_i? From another point of view, the relation between the voltage v_L and the current i_L depends only upon v_S, R_v and (simultaneously) only upon v_S, R_i. How can this be? Explain.

P 3.33 Refer to Fig. P 3.23. (a) Obtain symbolic expressions for the voltage v_x and the current i_x. (b) Let $I_0 = 1\text{mA}$, $I_1 = 2\text{mA}$, $V_0 = 5\text{V}$, $V_1 = 2\text{V}$, $R_0 = 1\ \text{k}\Omega$, $R_1 = 2\ \text{k}\Omega$. Calculate the values of the voltage v_x and the current i_x.

P 3.34 💻 Use simulation to verify your answers to Problem P 3.33.

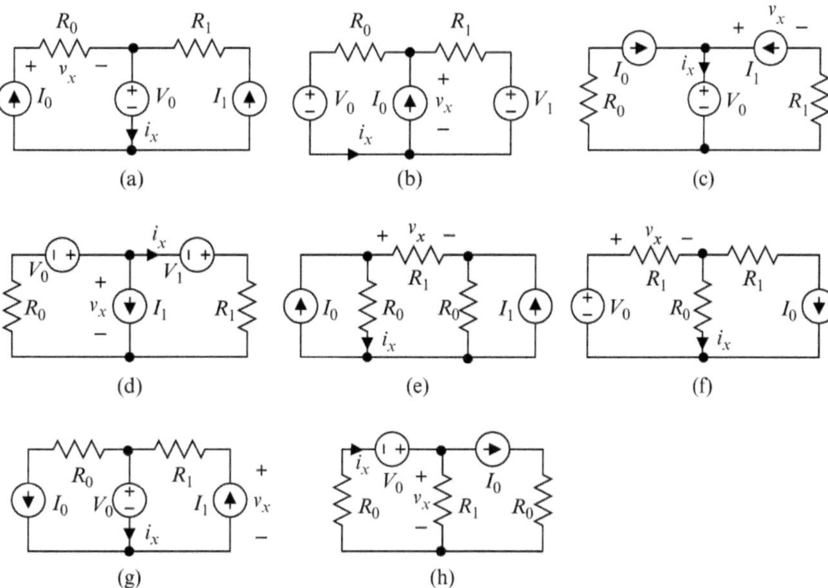

Fig. P 3.23 See Problem P 3.33

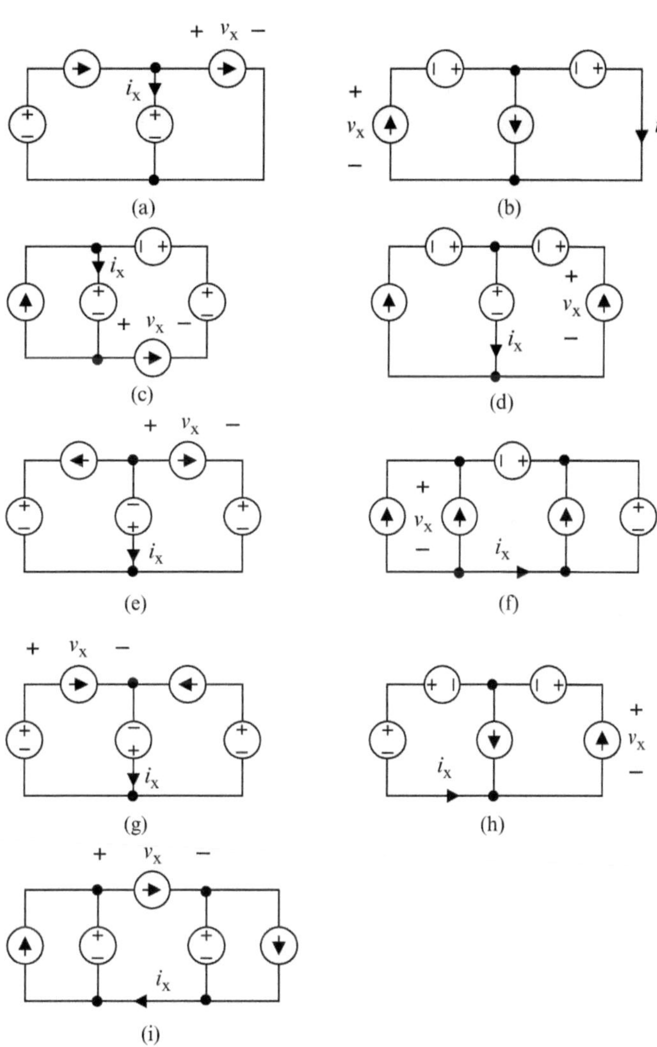

Fig. P 3.24 See Problem P 3.35

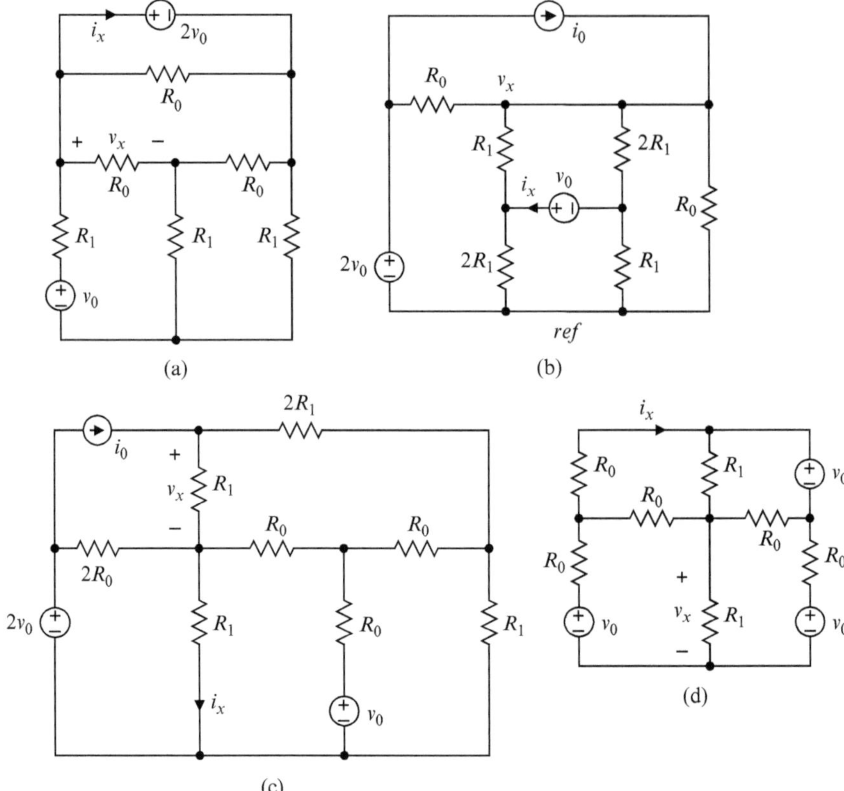

Fig. P 3.25 See Problem P 3.36

P 3.35 Refer to Fig. P 3.24. (a) Obtain symbolic expressions for the current i_x and the voltage v_x. (b) Let each source voltage be 5 V and each source current be 10 mA. Calculate the values of the current i_x and the voltage v_x.

P 3.36 🖳 For each circuit in Fig. P 3.25: (i) Write the appropriate number of mesh (Kirchhoff's voltage law) equations. (ii) Let $v_0 = 1.5$ V, $i_0 = 1$ mA, $R_0 = 1$ kΩ, $R_1 = 16$ kΩ. Calculate the mesh currents, the voltage v_x, and the current i_x. (iii) Use simulation to verify the results.

P 3.37 🖳 Repeat Problem P 3.36 using node analysis (Kirchhoff's current law). Omit the simulations if you did them when working Problem P 3.36.

Section 3.8 is prerequisite for the following problems.

P 3.38 It is required that the output of a voltage divider be between 74.5% and 75.5% of the input and that the total resistance be at least 400 kΩ. Assume that both resistors have the same precision (tolerance) and

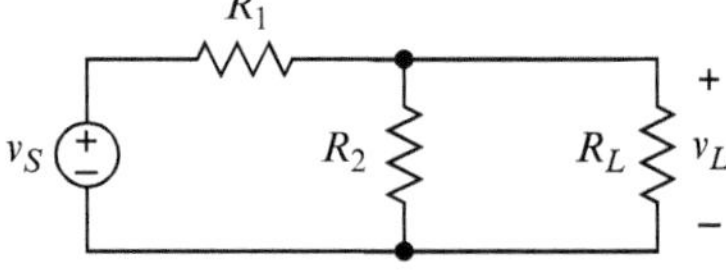

Fig. P 3.26 See Problem P 3.39

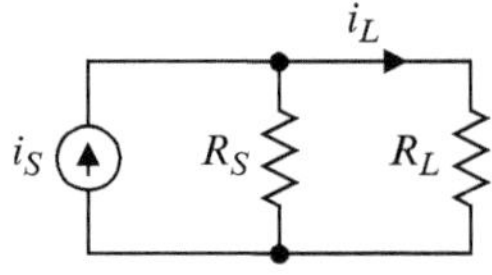

Fig. P 3.27 See Problem P 3.40

specify the values and precision of the least precise resistors that can be used.

P 3.39 In Fig. P 3.26, $R_1 = 100\,\text{k}\Omega \pm 5\%$, $R_2 = 300\,\text{k}\Omega \pm 5\%$, and $v_S = 5\,\text{V}$. (a) What is the smallest permissible value of the load resistance R_L if the load voltage v_L must be at least 70% of the source voltage v_S? (b) If R_L has the value determined in part (a), what is the maximum possible value of the load voltage?

P 3.40 In Fig. P 3.27, $R_S = 500\,\text{k}\Omega$. What is the largest permissible value of the load resistance R_L if the load current i_L must be at least 90% of the source current i_S?

P 3.41 In Fig. P 3.28, the source voltage V_0 is constant and the voltmeter reads 5 V before the load R_L is connected to the circuit. What does the voltmeter read after the load is attached? What is the source voltage V_0?

P 3.42 Refer to Fig. P 3.29, where the current I is fixed. Before the switch is closed, $I_1 = 4\,\text{mA}$. After the switch is closed, the current I is unchanged. What is the new value of I_1? What is the load current I_L? What kind of source is driving this circuit?

P 3.43 Refer to Fig. P 3.30, where $V_2 = 5\,\text{V}$ before the switch is closed. Find the value of k before the switch is closed. Find the value of k such that $V_2 = 5\,\text{V}$ after the switch is closed.

P 3.44 Refer to Fig. P 3.31, where the switches are *ganged* (tied together), such that they open and close together. (a) Find R_2 such that $V_2 = 9\,\text{V}$ when the switches are open. (b) Then find R_0 such that V_2 remains equal to 9 V when the switches are closed.

P 3.45 Refer to Fig. P 3.32, where the load resistance R_L is nominally equal to 1 kΩ, but can vary by ±30%. Other circuitry (not shown) must keep the load current I_L equal to 80% of the source current I_0 by adjusting the variable resistance R. What must be the range of R?

P 3.46 Refer to Fig. P 3.33, where the load resistance R_L is nominally equal to 5 kΩ, but can vary by ±40%. Other circuitry (not shown) must keep the load voltage V_L equal to 90% of the source voltage V_0 by adjusting the variable resistance R. What must be the range of R?

P 3.47 Refer to Fig. P 3.34. The load resistance R_L varies randomly within the range 15 – 25 Ω. Other circuitry (not shown) monitors the load current I_L and varies the source current I_0 to keep the load current

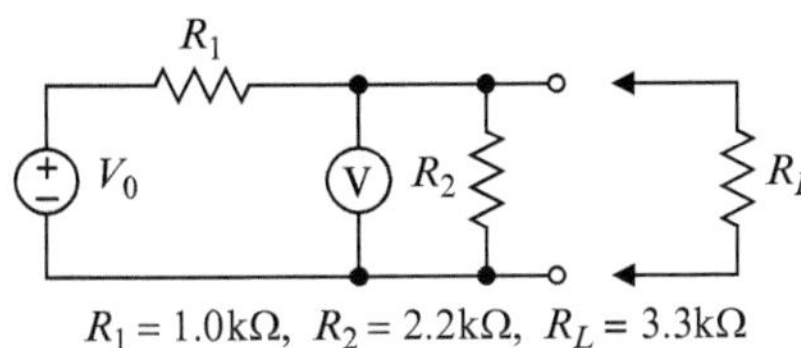

Fig. P 3.28 See Problem P 3.41

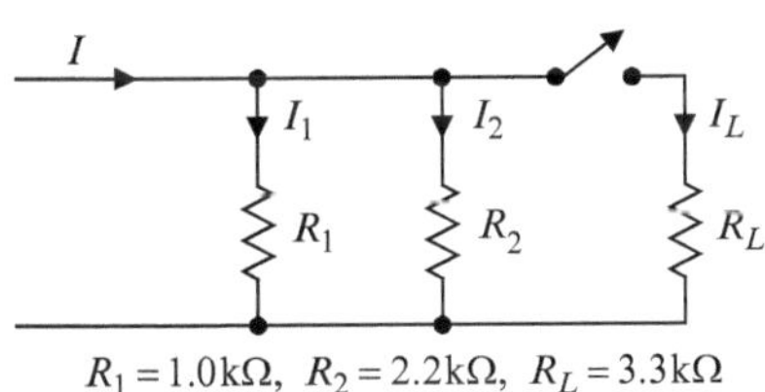

Fig. P 3.29 See Problem P 3.42

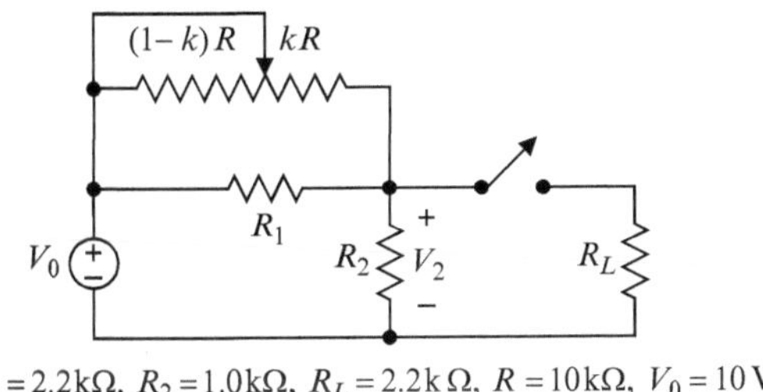

Fig. P 3.30 See Problem P 3.43

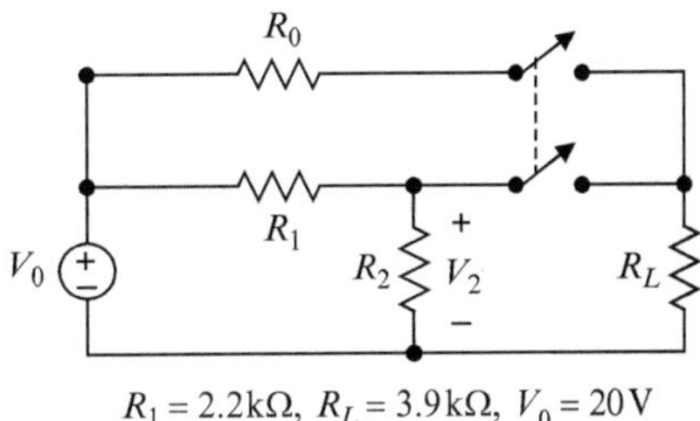

Fig. P 3.31 See Problem P 3.44

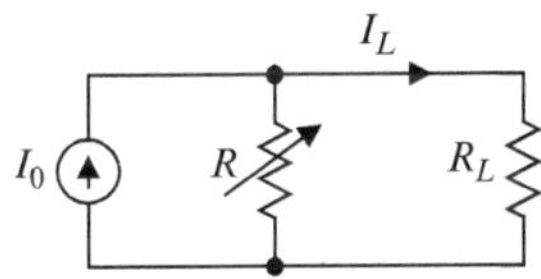

Fig. P 3.32 See Problem P 3.45

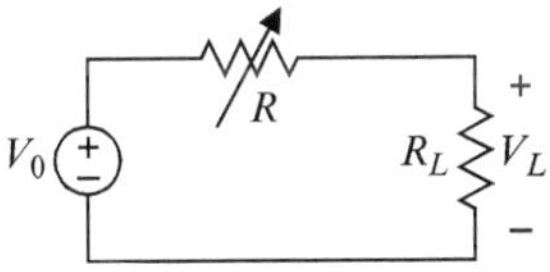

Fig. P 3.33 See Problem P 3.46

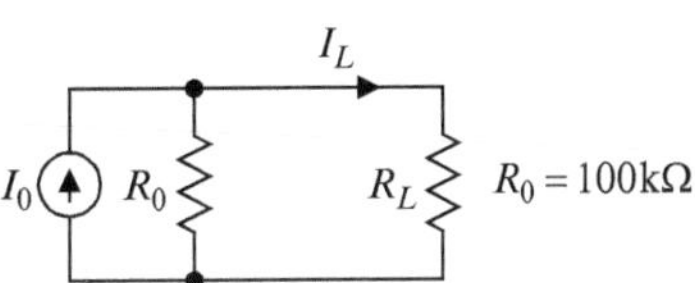

Fig. P 3.34 See Problem P 3.47

within 1% or 10 A. What must be the range of the source current?

P 3.48 Refer to Fig. P 3.35. The load resistance R_L varies randomly within the range 15 – 25Ω. Other circuitry (not shown) monitors the load voltage V_L and varies the source voltage V_0 to keep the load voltage within 1% of 20 V. What must be the range of the source voltage?

P 3.49 Refer to Fig. P 3.36, where the temperature coefficients at 25° for R_1 and R_2 are 0.005 $K^{-1}$and 0.008 K^{-1}, respectively. What is the maximum range of V_L if the temperature of the circuit ranges from 25°C to 50°C?

P 3.50 Refer to Fig. P 3.37. The resistor R_2 is switched in and out, randomly. What is the percent variation of the voltage V_1?

P 3.51 Refer to Fig. P 3.38. The load current I_L must not exceed 15 A, but can have any value less than 15 A without harm. The current I is nominally 10 A, but can occasionally jump to 50 A for a short time. Other circuitry (not shown) monitors the current I and closes the switch instantly if I exceeds 12 A, diverting some current through the resistor R. What is the maximum permissible value for R?

P 3.52 Explain why it is best to use resistors of the same kind (e.g., metal film, wirewound) in any particular voltage divider or current divider. *Hint*: Temperature is not necessarily constant.

P 3.53 It is proposed that a length of wire be used as the probe (sensor) in a temperature-measurement system, as illustrated in Fig. P 3.39, where R_W is the resistance of the wire and the voltage v is to be measured as an indication of temperature. Assume the resistivity of the wire is given by $\rho(T) = \rho_0[1 + \alpha_0(T - T_0)]$, where T is temperature in °C and $T_0 = 25°C$.

(a) Obtain an expression for the voltage v as a function of temperature in °C. The source voltage V_S, the source resistance R_S, the parameters ρ_0 and α_0, and the length L and diameter d of the wire are presumed to be known.
(b) Describe conditions under which the relation between temperature T and the measured voltage v is approximately linear (a straight-line relation).
(c) Let $V_s = 12$ V, $R_s = 100$ Ω, $\rho_0 = 1.543$ μΩ cm, $\alpha_0 = 0.0044°C^{-1}$, $L = 1$ m, and $d = 0.643$ mm. Use a computer mathematics program to plot the voltage v versus probe temperature for $0°C \leq T \leq 20°C$.

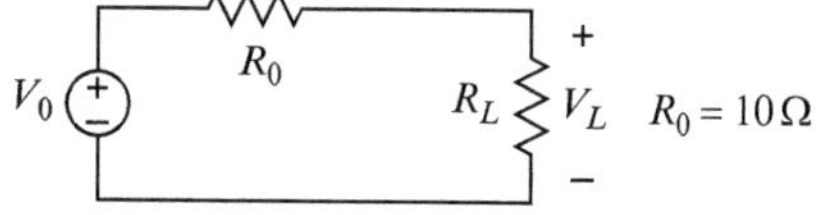

Fig. P 3.35 See Problem P 3.48

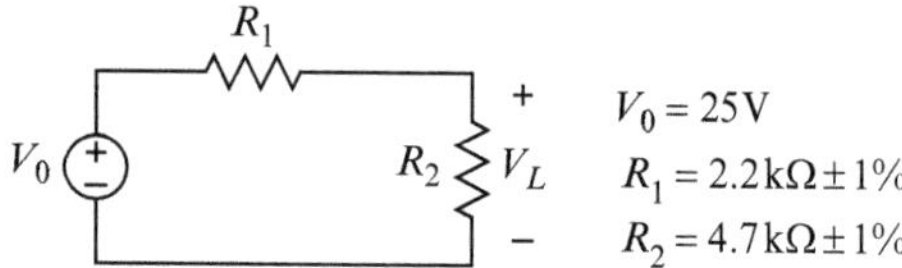

Fig. P 3.36 See Problem P 3.49

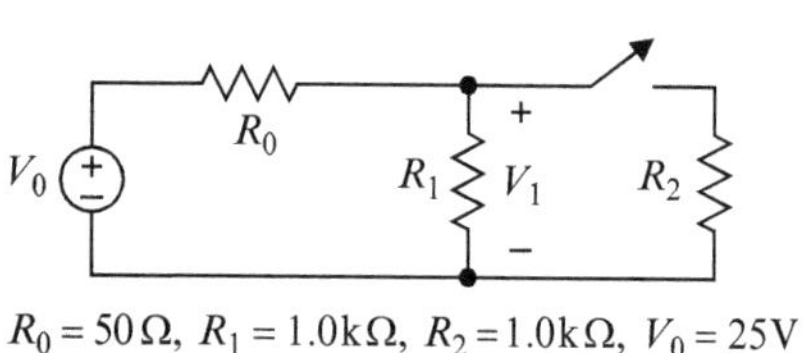

Fig. P 3.37 See Problem P 3.50

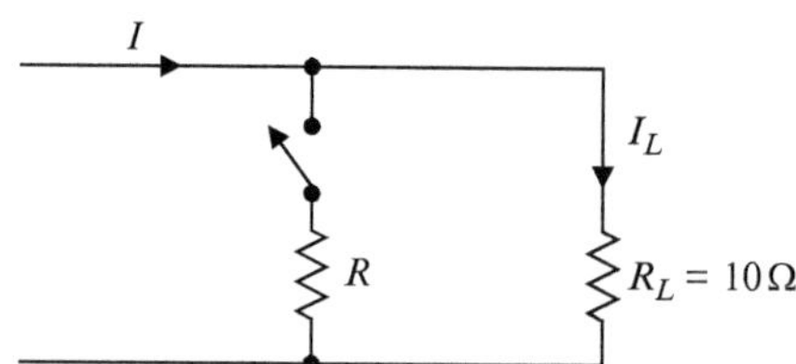

Fig. P 3.38 See Problem P 3.51

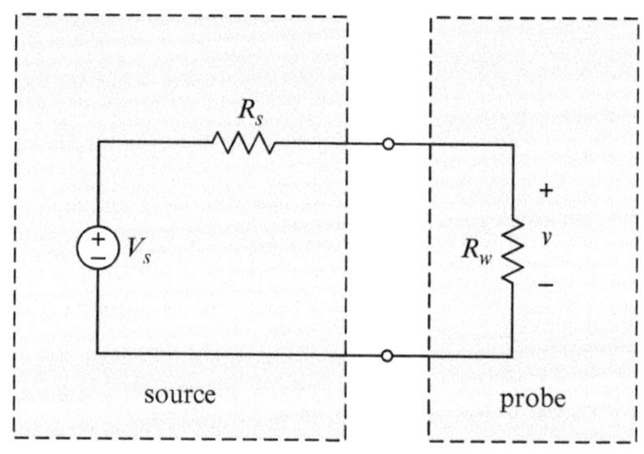

Fig. P 3.39 See Problem P 3.53

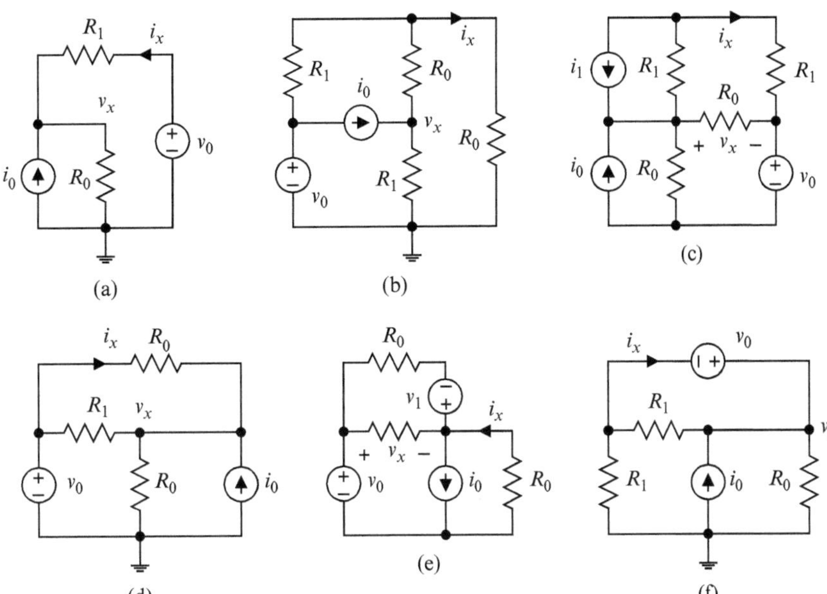

Fig. P 3.40 See Problem P 3.54

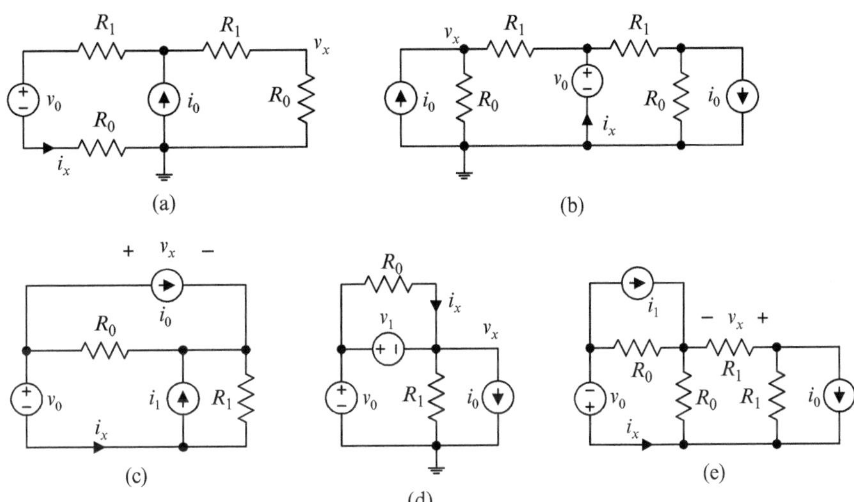

Fig. P 3.41 See Problem P 3.55

P 3.54 Refer to Fig. P 3.40.

(i) Obtain expressions for the voltage v_x and the current i_x in terms of circuit parameters. Show that the expressions are dimensionally consistent.

(ii) Let $i_0 = 5\,\text{mA}$, $i_1 = 10\,\text{mA}$, $v_0 = 5\,\text{V}$, $v_1 = 10\,\text{V}$, $R_0 = 1\,\text{k}\Omega$, $R_1 = 2.2\,\text{k}\Omega$. Calculate the values of v_x and i_x.

(iii) If you used node analysis in Part (ii), check your work using mesh analysis. If you used mesh analysis, check using node analysis.

(iv) Simulate the circuit and verify the calculated values.

P 3.55 Refer to Fig. P 3.41.

(i) Obtain expressions for the voltage v_x and the current i_x in terms of circuit parameters. Show that the expressions are dimensionally consistent.

(ii) Let $i_0 = 5\,\text{mA}$, $i_1 = 10\,\text{mA}$, $v_0 = 5\,\text{V}$, $v_1 = 10\,\text{V}$, $R_0 = 1\,\text{k}\Omega$, $R_1 = 2.2\,\text{k}\Omega$. Calculate the values of v_x and i_x.

(iii) If you used node analysis in Part (ii), check your work using mesh analysis. If you used mesh analysis, check using node analysis.

(iv) Simulate the circuit and verify the calculated values.

Section 3.9 is prerequisite for the following problems.

P 3.56 Refer to Fig. P 3.42, where $V_0 = 5\,\text{V}$, $V_1 = 10\,\text{V}$, $I_0 = 2\,\text{mA}$, $I_1 = 5\,\text{mA}$, $R_0 = 1\,\text{k}\Omega$, and $R_1 = 2.2\,\text{k}\Omega$. Use superposition to find the voltage v_x and the current i_x. Verify your results with straightforward application of Kirchhoff's laws.

P 3.57 Use simulation to verify your answers to Problem P 3.56.

P 3.58 Refer to Problem P 3.23 for a description of a rheostat. In Fig. P 3.43, $V_1 = 10\,\text{V}$, $V_2 = 15\,\text{V}$, $R_1 = 3.3\,\text{k}\Omega, R_2 = 2.2\,\text{k}\Omega$, and $R = 5\,\text{k}\Omega$. Find a value of k for which the current i equals zero. Is it possible to find such a value if $V_2 < 0$? Show how or explain why not.

P 3.59 Refer to Problem P 3.23 for a description of a rheostat. In Fig. P 3.44, $I_1 = 10\,\text{mA}$, $I_2 = 15\,\text{mA}$, $R_1 = 3.3\,\text{k}\Omega, R_2 = 2.2\,\text{k}\Omega$, and $R = 5\,\text{k}\Omega$. Find a value of k for which the current i equals zero. Is it possible to find such a value if $I_2 < 0$? Show how or explain why not.

P 3.60 Refer to Fig. P 3.45, where $V_0 = 120\,\text{V}$, $R_0 = 0.1\,\Omega$, and the load resistance is nominally $R_L = 10\,\Omega$, but can vary by $\pm 50\%$. Additional monitoring circuitry (not shown) attempts to keep the load voltage between 100 and 125 V by switching sources in or out as the load resistance varies. (a) Construct graphs of load voltage versus number of switches in position b, for (i) $R_L = 5\,\Omega$, (ii) $R_L = 10\,\Omega$, and (iii) $R_L = 15\,\Omega$ (on the same axes). (b) How many switches must be in position b in each case? (c) If the number of sources is unlimited, what is the maximum possible load voltage in each case? (d) Are there values of the load resistance for which it would be impossible to keep the load voltage between 100 and 125 V, and if so, what are they? (e) What is the minimum number of sources required for any load?

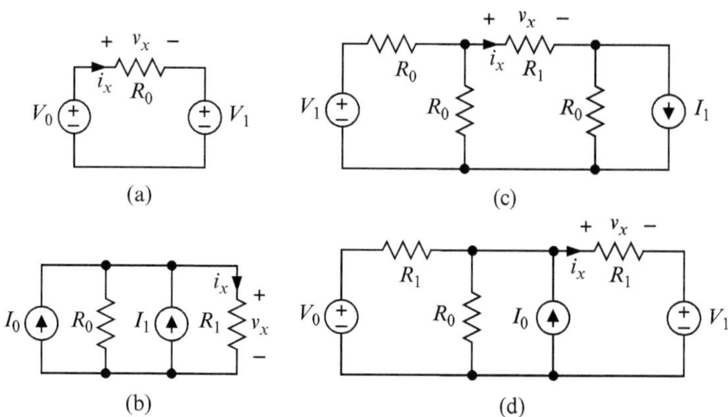

Fig. P 3.42 See Problem P 3.56

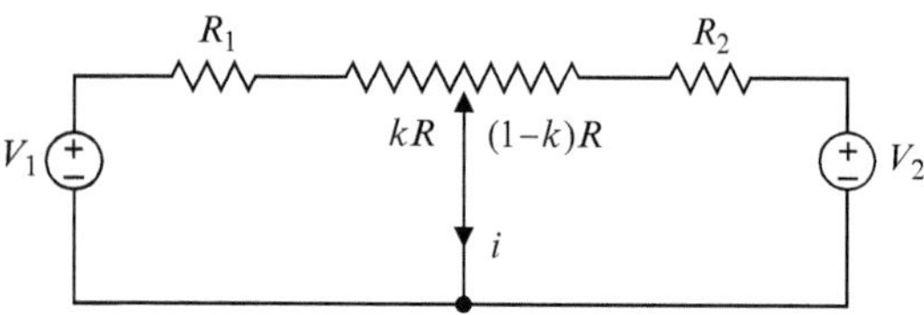

Fig. P 3.43 See Problem P 3.58

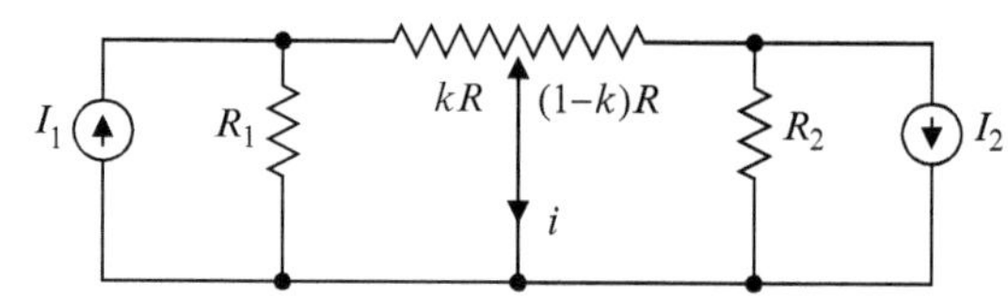

Fig. P 3.44 See Problem P 3.59

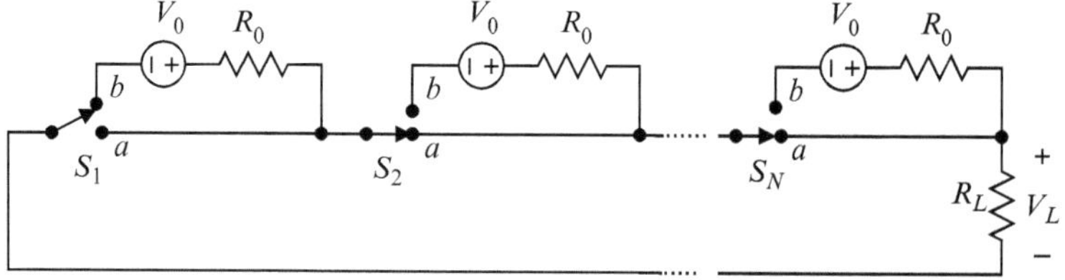

Fig. P 3.45 See Problem P 3.60

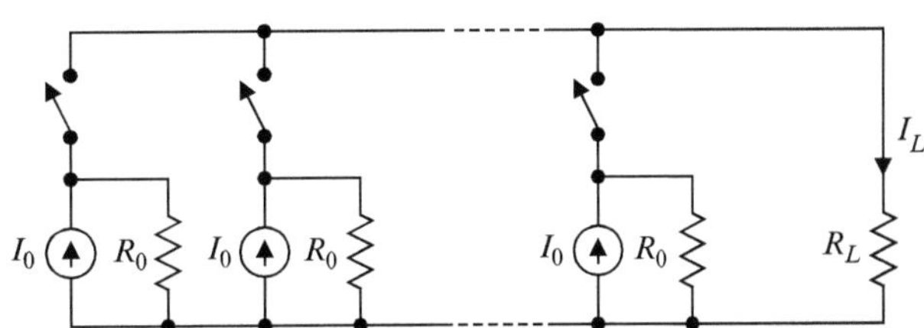

Fig. P 3.46 See Problem P 3.61

P 3.61 Refer to Fig. P 3.46, where $I_0 = 1\,\text{A}$, $R_0 = 100\,\text{k}\Omega$, and load resistance is nominally $R_L = 500\,\Omega$, but can vary by $\pm 30\%$. Monitoring circuitry (not shown) attempts to keep the load current between 3.75 and 4.5 A by closing or opening switches as the load resistance varies. (a) Construct graphs of load current versus number of switches closed for (i) $R_L = 350\,\Omega$, (ii) $R_L = 500\,\Omega$, and (iii) $R_L = 650\,\Omega$ (on the same axes). (b) How many switches must be closed in each case, and what is the actual load current in each case? (c) If the number of sources is unlimited, what is the maximum possible load current in each case? (d) Are there values of the load resistance for which it would be impossible to keep the load current between 3.75 and 4 A, and if so, what are they? (e) What is the minimum number of sources required for any load? (f) What load resistance would draw the minimum possible current within the range $3.5\,\text{A} \le I_L \le 4.5\,\text{A}$?

Chapter 4
Equivalent Circuits

Two-terminal elements or circuits having identical terminal characteristics are *equivalent* at their terminals. For example, any two-terminal element or circuit having the terminal characteristic $v = Ri$, where v and i are voltage across and current through the terminals of the element and R is independent of v and i, is equivalent at those terminals to a resistor having resistance R. Equivalent elements or circuits cannot be distinguished by any measurements made at their terminals, and replacing one by the other in a larger circuit causes no change in any voltage or current in the rest of the circuit.

In this chapter, we show how using equivalent circuits can simplify analysis. We introduce Thévenin's theorem, which permits reduction of any resistive circuit (at a terminal pair) to a voltage source in series with a resistor, and Norton's theorem, which permits reduction of any resistive circuit (at a terminal pair) to a current source in parallel with a resistor. We begin with a discussion of terminal characteristics.

4.1 Terminal Characteristics

Recall that the **terminal characteristic** for a two-terminal element is the relation the element establishes between voltage across and current into (or out of) the terminals. Recall also that *the terminal characteristic for an element or circuit is independent of whatever is attached to the element or circuit at the terminals in question*. For example, a resistor (idealized) obeys Ohm's law, regardless of what is connected to the resistor. Similarly, the source voltage for an independent voltage source is independent of whatever is connected to the source.

To specify or determine a terminal characteristic, we must specify a direction for positive current in relation to the polarity assigned to terminal voltage. For a resistor, for example, the relation $v = iR$ holds if the positive direction of current is into the positive terminal of the resistor. If the positive direction for current is out of the positive terminal, then the terminal characteristic is $v = -iR$. For a resistor and some other circuit elements, the positive direction for current is conventionally into the positive terminal. But in general, choice of the positive direction for current in defining a terminal characteristic often depends upon whether we regard the element or circuit in question as a *source* or a *load*.[1] Generally, if the element is regarded as a source, the positive direction of the terminal current is out of the positive terminal. If the element is regarded as a load, the positive direction of the terminal current is into the positive terminal.

To obtain the terminal characteristic for a two-terminal element or circuit, we assign a polarity (+/−) to the terminal pair, assign a positive direction to current, and find the relation between the current and voltage thus defined (e.g., using Kirchhoff's laws).

Example 4.1. Obtain the terminal characteristic at the terminals *a–b* for the circuit shown in Fig. 4.1(a), regarding the circuit as a source.

[1]Loosely, an element or circuit is regarded as a source if the element or circuit delivers energy to a circuit and is regarded as a load if it removes energy from the circuit (converts electrical energy to another form); e.g., a generator is ordinarily regarded as a source and a motor is ordinarily regarded as a load. As another example, from the perspective of an amplifier in a home audio system, a CD player is a source and a loudspeaker is a load.

T.H. Glisson, *Introduction to Circuit Analysis and Design*,
DOI 10.1007/978-90-481-9443-8_4, © Springer Science+Business Media B.V. 2011

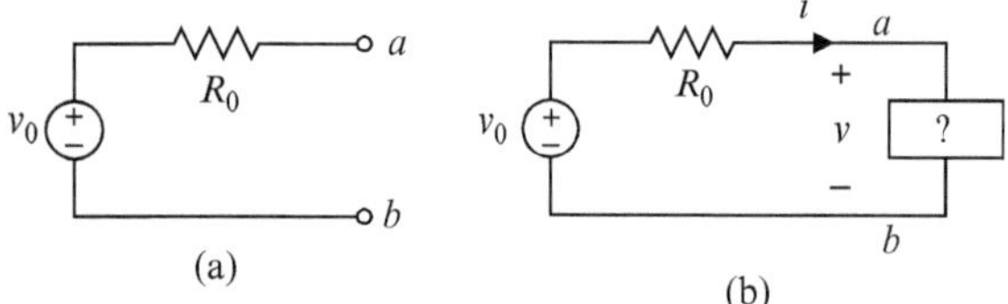

Fig. 4.1 See Example 4.1

Solution: Refer to Fig. 4.1(b). We identify terminal a as the positive terminal. Because we regard the circuit as a source, we take the positive direction of current out of the positive terminal. We assume an unspecified (arbitrary) load is attached to the terminals (to provide a closed path for current). From Kirchhoff's voltage law,

$$-v_0 + R_0\,i + v = 0 \Rightarrow i = \frac{v_0 - v}{R_0}.$$

Thus the terminal characteristic is expressed by

$$i = \frac{v_0 - v}{R_0}$$

and displayed graphically as in Fig. 4.2.

Example 4.2. Obtain the terminal characteristic at the terminals a–b for the circuit shown in Fig. 4.3, regarding the circuit as a load.

Solution: We choose terminal a as the positive terminal and b as the reference. Because we regard the circuit as a load, we take the positive direction for current into terminal a. Writing Kirchhoff's voltage law around the path defined by the terminals and the two resistors gives

$$-v + i R_1 + v_c = 0 \Rightarrow v_c = v - R_1\,i.$$

To eliminate the node voltage v_c, we write Kirchhoff's current law at the node joining the two resistors. This gives

$$\frac{v_c - v}{R_1} + \frac{v_c}{R_0} = i_0$$

$$\Rightarrow v_c = \left(\frac{R_0 R_1}{R_0 + R_1}\right)\left(\frac{v}{R_1} + i_0\right).$$

Eliminating v_c yields

$$v - R_1\,i = \left(\frac{1}{R_1} + \frac{1}{R_0}\right)^{-1}\left(\frac{v}{R_1} + i_0\right)$$

$$\Rightarrow i = \frac{v - R_0\,i_0}{R_1 + R_0}.$$

As in the previous example, a graphical representation of the terminal characteristic is a straight line, as shown in Fig. 4.4 (the graph is drawn for $i_0 > 0$).

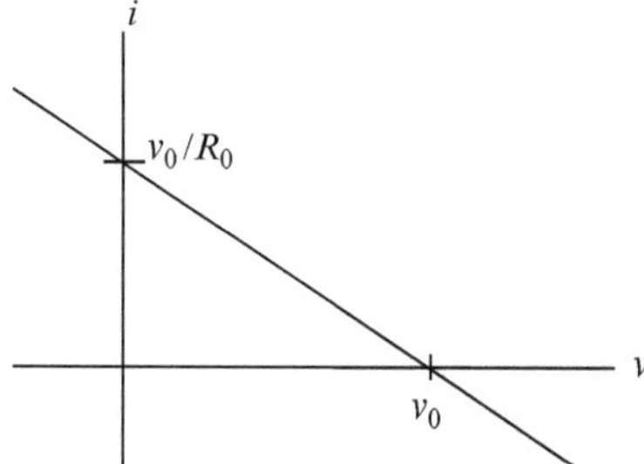

Fig. 4.2 See Example 4.1

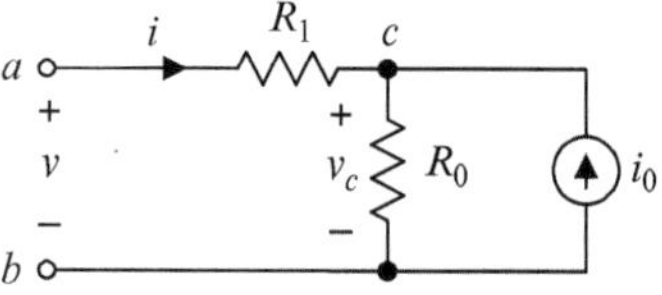

Fig. 4.3 See Example 4.2

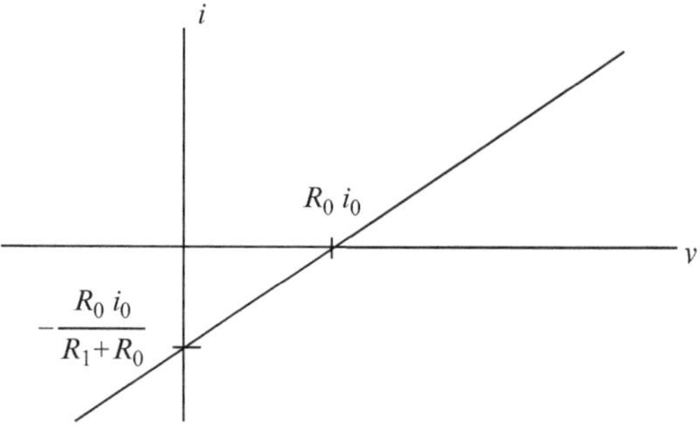

Fig. 4.4 See Example 4.2

Exercise 4.1. Obtain the terminal characteristic at the terminals *a–b* for the circuit shown in Fig. 4.5, regarding the circuit as a source.

4.2 Equivalent Circuits

Consider two circuits and designated terminal pairs, as shown in Fig. 4.6. The circuits are **equivalent** at designated terminal pairs if their terminal characteristics at those terminal pairs are identical; i.e., if the positive direction for current is the same (either into or out of the positive terminal) in both cases and the functions $f_1(v)$, $f_2(v)$ are identical. Two circuits can be equivalent at some specified terminal pair(s) but not others. If the terminal pairs in question are clear from context, or if each circuit has only one terminal pair of interest (to the outside world), then we say simply that the circuits are equivalent.

If we enclose two equivalent circuits in boxes, such that we have access to only the terminal pairs at which the circuits are equivalent, no electrical measurements at the terminals can distinguish one circuit from the other. Consequently, a collection of elements (a circuit) accessible by only a single terminal pair can be replaced by an equivalent collection of elements without altering the behavior of the circuit as a whole. Often, a large number of elements can be replaced (mathematically, at least) by a much smaller number of elements, facilitating further analysis and interpretation of results. In this section, we show how various two-terminal resistive circuits can be reduced to simpler, equivalent two-terminal circuits. Such reductions often simplify circuit analysis and design.

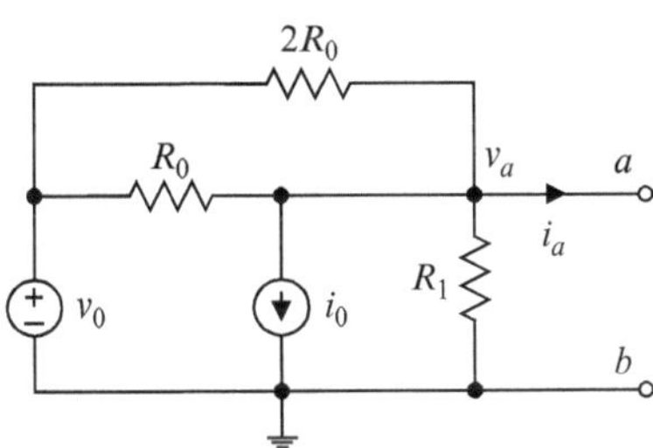

Fig. 4.5 See Exercise 4.1

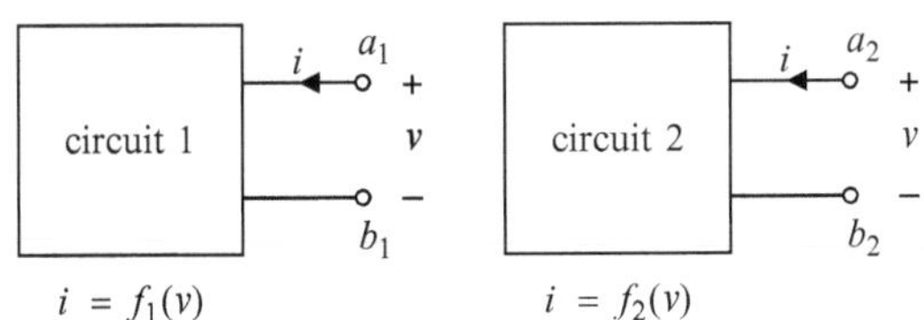

Fig. 4.6 Two two-terminal circuits having terminal characteristics $f_1(v), f_2(v)$

4.2.1 Resistors in Series Are Additive

Refer to Fig. 4.7, which shows a circuit consisting of a series connection of N resistors. We seek the terminal characteristic of the circuit at the terminals *a–b*. We regard the circuit as a load, so we take the positive direction for current into the positive terminal of the circuit. Because the same current passes through each element in a series connection, it follows from Ohm's law and Kirchhoff's voltage law that the voltage across the series connection of resistors is

$$v = R_1\, i + R_2\, i + \cdots R_N\, i = (R_1 + R_2 + \cdots + R_N)i.$$

Thus the terminal characteristic and equivalent resistance for the series connection are given by

$$i = \frac{v}{R_{eq}}; \qquad R_{eq} = R_1 + R_2 + \cdots + R_N. \tag{4.1}$$

This result shows that a series connection of N resistors is equivalent to a single resistor whose resistance is the sum of the resistances in the series connection. That is, for R_{eq} given by (4.1), the two circuits in Fig. 4.7 have the same terminal characteristic at the terminals *a–b*. Thus, wherever we encounter a series connection of resistors, we can replace the series connection by a single equivalent resistor, simplifying the circuit.

4.2.2 Conductances in Parallel Are Additive

Refer to Fig. 4.8, which shows a parallel connection of N resistors. Because the same voltage exists across

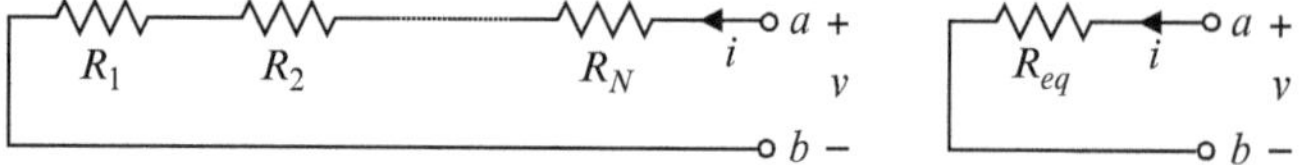

Fig. 4.7 Resistors in series are additive. See (4.1)

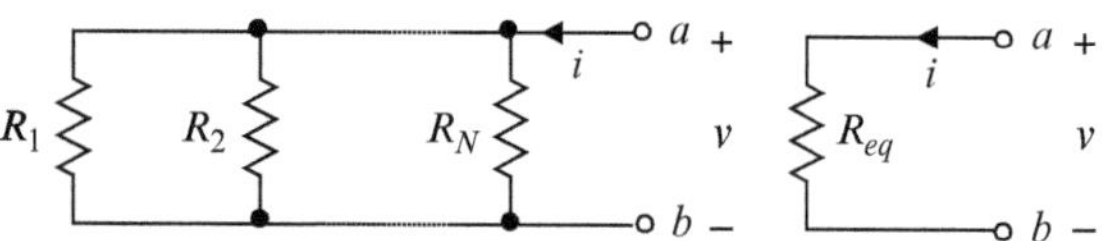

Fig. 4.8 Conductances in parallel are additive. See (4.4)

elements in parallel, it follows from Ohm's law and Kirchhoff's current law that

$$\begin{aligned} i &= G_1 v + G_2 v + \cdots + G_N v \\ &= (G_1 + G_2 + \cdots + G_N)v, \end{aligned} \tag{4.2}$$

where $G_n = R_n{}^{-1}$. Equation (4.2) may be written as

$$i = G_{eq}\, v; \quad G_{eq} = G_1 + G_2 + \cdots + G_N. \tag{4.3}$$

This result shows that the *conductances of resistors in parallel are additive*. Thus the equivalent resistance of resistors in parallel is given by

$$\frac{1}{R_{eq}} = \frac{1}{R_1} + \frac{1}{R_2} + \cdots + \frac{1}{R_N}. \tag{4.4}$$

For R_{eq} given by (4.4), the two circuits in Fig. 4.8 have the same terminal characteristic at the terminals *a–b*.

Resistors in parallel are encountered often, especially in analysis of electronic circuits, so it is convenient to have a shorthand notation for the cumbersome relation (4.4). The notation

$$\begin{aligned} R_{eq} &= R_1 \| R_2 \| R_3 \| \cdots \| R_N \\ \Rightarrow \frac{1}{R_{eq}} &= \frac{1}{R_1} + \frac{1}{R_2} + \cdots + \frac{1}{R_N} \end{aligned} \tag{4.5}$$

is widely used, at least in practice. This notation does not speed computation, but often facilitates interpretation of circuit equations.

For the special (but important) case $N = 2$, (4.4) gives

$$R_{eq} = R_1 \| R_2 = \frac{R_1 R_2}{R_1 + R_2}. \tag{4.6}$$

The expression $R_1 \| R_2$ is read or said as "R_1 in parallel with R_2".

Example 4.3. Obtain a single, equivalent resistance at the terminals *a–b* for the circuit of Fig. 4.9.

Solution: From (4.6), an equivalent, single resistance for the parallel connection of R_2 and R_3 is

$$R_a = \frac{R_2 R_3}{R_2 + R_3}.$$

This equivalent resistance is in series with R_1 and R_4. Thus the equivalent resistance at terminals *a–b* is

$$R_{eq} = R_1 + R_4 + R_a.$$

Example 4.4. Obtain an expression for the equivalent resistance at the terminals *a–b* of the circuit shown in Fig. 4.10.

Solution: Figure 4.11 illustrates steps in the solution. The resistors R_2, R_3 are connected in parallel. The equivalent resistance for the parallel connection is given by (see (4.6))

$$R_a = \frac{R_2 R_3}{R_2 + R_3}.$$

The equivalent resistance for the series connection of R_a, R_4 is given by

$$R_b = R_a + R_4 = \frac{R_2 R_3}{R_2 + R_3} + R_4.$$

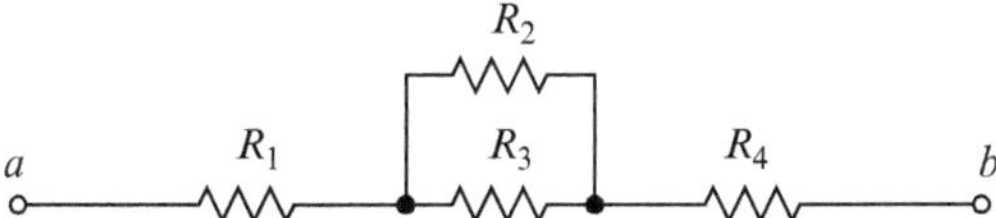

Fig. 4.9 See Example 4.3

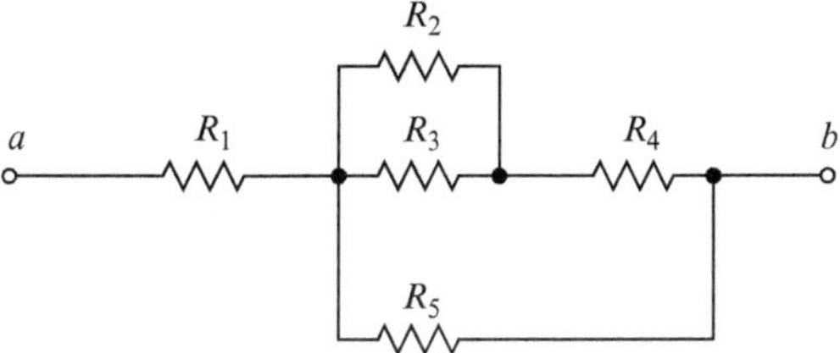

Fig. 4.10 See Example 4.4

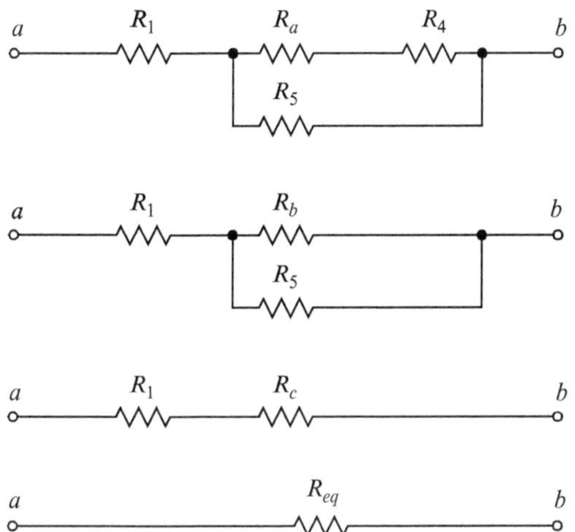

Fig. 4.11 See Example 4.4

The equivalent resistance of the parallel connection of R_b, R_5 is given by

$$R_c = \frac{R_b R_5}{R_b + R_5} = \frac{\left(\frac{R_2 R_3}{R_2 + R_3} + R_4\right) R_5}{\frac{R_2 R_3}{R_2 + R_3} + R_4 + R_5} = \frac{[R_2 R_3 + R_4 (R_2 + R_3)] R_5}{R_2 R_3 + (R_2 + R_3)(R_4 + R_5)}.$$

Finally, the equivalent resistance of the series connection of R_1, R_c is given by

$$R_{eq} = R_1 + R_c = R_1 + \frac{[R_2 R_3 + R_4 (R_2 + R_3)] R_5}{R_2 R_3 + (R_2 + R_3)(R_4 + R_5)}.$$

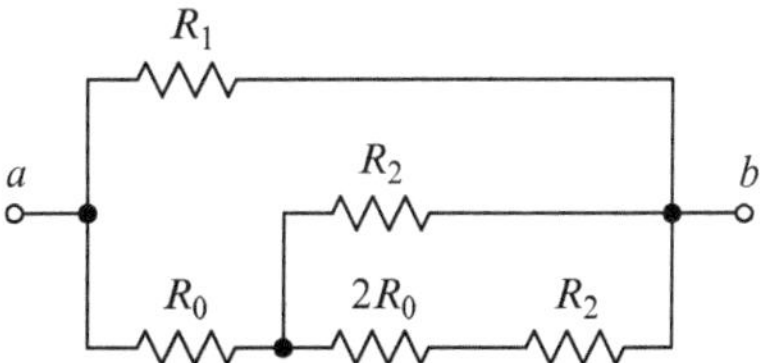

Fig. 4.12 See Exercise 4.2

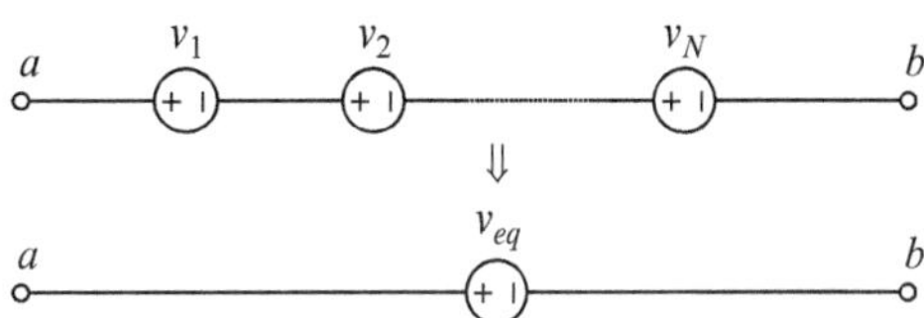

Fig. 4.13 Voltage sources in series are additive

Exercise 4.2. Obtain an expression for the equivalent resistance at the terminals *a*–*b* of the circuit shown in Fig. 4.12.

4.2.3 Voltage Sources in Series Are Additive

Applying Kirchhoff's voltage law to the top circuit in Fig. 4.13 yields

$$v_{ab} = v_1 + v_2 + \cdots + v_N. \tag{4.7}$$

A series connection of voltage sources is equivalent to a single voltage source whose source voltage equals the sum of the original source voltages. Whenever ideal voltage sources appear in series in a circuit, they can be replaced by a single voltage source whose source voltage is the algebraic sum of the original source voltages. We do not consider voltage sources in parallel because such an arrangement contradicts the definitions of both a parallel connection and an ideal voltage source (unless all the sources are identical, in which case any one of them would do).

4.2.4 Current Sources in Parallel Are Additive

Applying Kirchhoff's current law to the circuit on the left in Fig. 4.14 yields

$$i_{eq} = i_1 + i_2 + \cdots + i_N, \tag{4.8}$$

which shows that any number of current sources in parallel can be reduced to a single current source, where the source current equals the sum of the source currents for the original (constituent) sources. In any circuit where ideal current sources appear in parallel, they can be replaced by a single current source whose source current is the algebraic sum of the original source currents. We do not consider current sources in series because such an arrangement contradicts the definition of a series connection (unless all the sources are identical, in which case any one of them would do).

The sums in (4.7) and (4.8) are algebraic. If the polarity of a voltage source in Fig. 4.13 or the direction of a current source in Fig. 4.14 is reversed, the corresponding term in the associated sum must be negated.

4.2.5 Elements in Series Commute, as do Elements in Parallel

Commuting two elements means swapping their places in a circuit. If the resulting circuit is equivalent to the original, we say the swapped elements **commute**. Resistors and sources in series commute, as do resistors and sources in parallel; that is, the terminal characteristic of a series (or parallel) connection of such elements is independent of the order in which the elements are connected. Consequently, resistors in series add, as do voltage sources in series, no matter how ordered in the series connection. Conductances in parallel add, as do current sources in parallel, no matter how ordered in the parallel connection. For example, the series circuits in Fig. 4.15 are equivalent at the terminals *a–b*, as are the parallel circuits in Fig. 4.16.

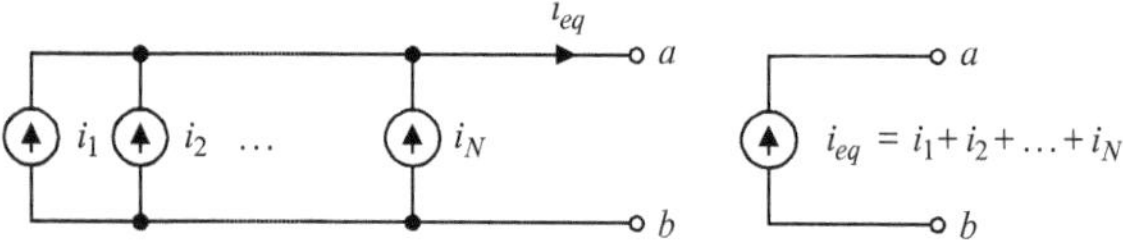

Fig. 4.14 Current sources in parallel are additive

Example 4.5. Simplify the circuit shown in Fig. 4.17; that is, find a simpler circuit that is equivalent at the terminals *a–b* to the circuit shown.

Solution: See Fig. 4.18. We first combine the parallel elements to obtain the circuit of Fig. 4.18(a), where

$$R_a = \frac{(3R)(5R)}{3R+5R} = \frac{15R}{8},$$
$$R_b = \frac{(4R)(5R)}{4R+5R} = \frac{20R}{9}.$$

The two rightmost current sources add to give a net (downward) current $5\,i_0 - 3\,i_0 = 2\,i_0$.

We then combine the series elements to obtain the circuit of Fig. 4.18(b), where

$$R_c = 2R + R_a = \frac{31R}{8}.$$

The two voltage sources add to give a net voltage (positive on the left). $2\,v_0 - v_0 = v_0$.

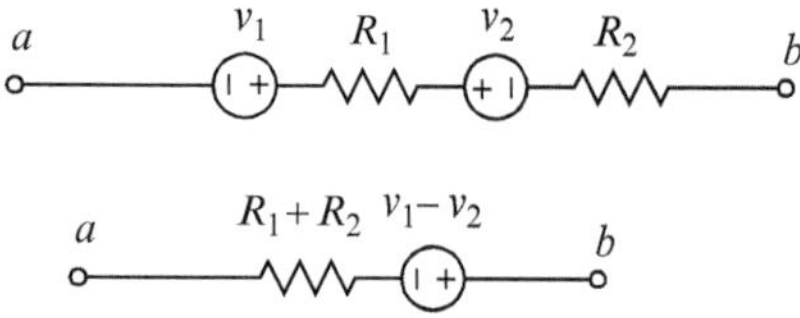

Fig. 4.15 Equivalent series circuits

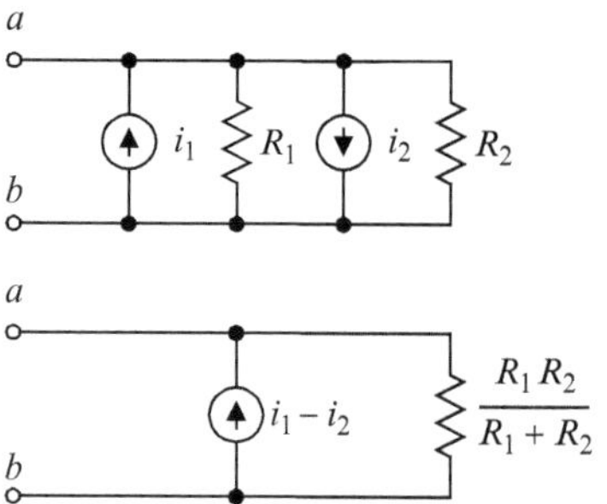

Fig. 4.16 Equivalent parallel circuits

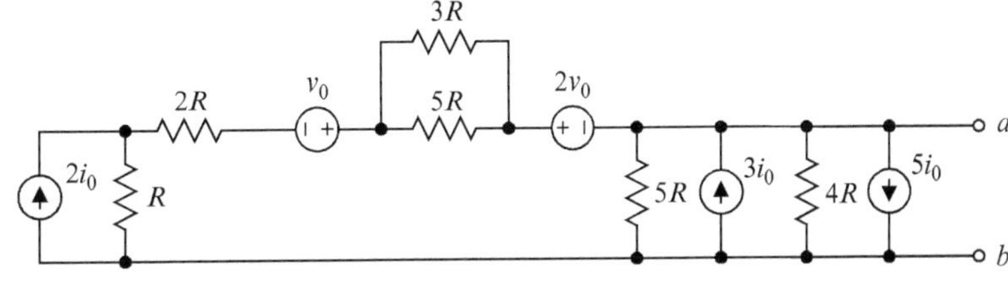

Fig. 4.17 See Example 4.5

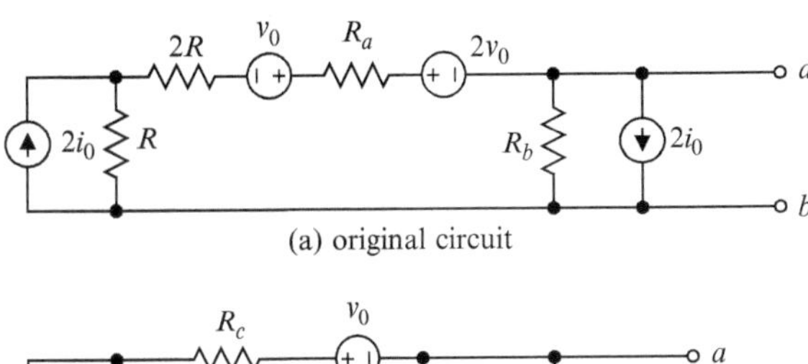

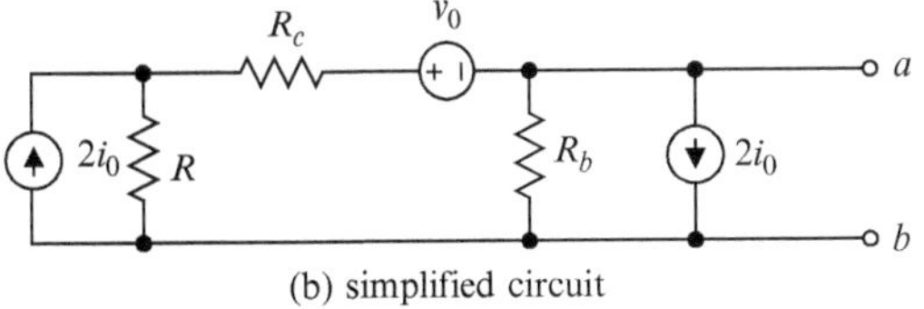

Fig. 4.18 See Example 4.5

Using only methods described above, we cannot simplify the circuit further; e.g., the two remaining current sources are not in parallel because the same voltage does not appear across both. In Section 4.3, we show how such circuits can be simplified further.

Exercise 4.3. Find values of α, β, γ, δ such that the circuits shown in Fig. 4.19(a) and Fig. 4.19(b) are equivalent at the terminals *a-b*.

4.2.6 Finding Equivalent Resistance Using a Known Source

If the current into the positive terminal and voltage across a terminal pair of a device or circuit obey a relation of the form $i = v/R$, then by definition the device or circuit is equivalent to a resistor having resistance R at the terminal pair in question. If it is known in advance that a particular device or circuit is equivalent to a resistor at a particular terminal pair, then we can find the equivalent resistance, either experimentally or mathematically, by attaching a known voltage source v to the terminal pair and measuring or calculating the resulting current i into the terminal pair. Alternatively, we may attach a known current source i and measure or calculate the resulting voltage v across the terminal pair. In either case, the equivalent resistance is given by $R_{eq} = v/i$ as illustrated by Fig. 4.20. This approach is useful when a circuit cannot be simplified further by reducing series and parallel connections of resistors to single, equivalent resistors, as illustrated by the next example.

Example 4.6. Find the equivalent resistance at the terminals *a–b* of the circuit shown in Fig. 4.21(a).

Solution: We attach a known voltage source v to the terminals *a–b*, as shown in Fig. 4.21(b), obtain an expression for the current i, and use $R_{eq} = v/i$. With reference to Fig. 4.21(b), Kirchhoff's current law gives

$$i = i_x + i_y = \frac{v - v_x}{R} + \frac{v - v_y}{2R},$$

where b is the reference node. Writing Kirchhoff's current law for nodes x and y gives

$$\frac{v_x - v}{R} + \frac{v_x - v_y}{5R} + \frac{v_x}{4R} = 0,$$
$$\frac{v_y - v}{2R} + \frac{v_y - v_x}{5R} + \frac{v_y}{3R} = 0,$$

which yield

$$v_x \cong 0.777\, v, \quad v_y \cong 0.634\, v.$$

Thus

$$i \cong \frac{v - 0.777\, v}{R} + \frac{v - 0.634\, v}{2R} \cong 0.406\, \frac{v}{R}$$

which gives

$$R_{eq} = \frac{v}{i} \cong \frac{R}{0.406} \cong 2.46 R.$$

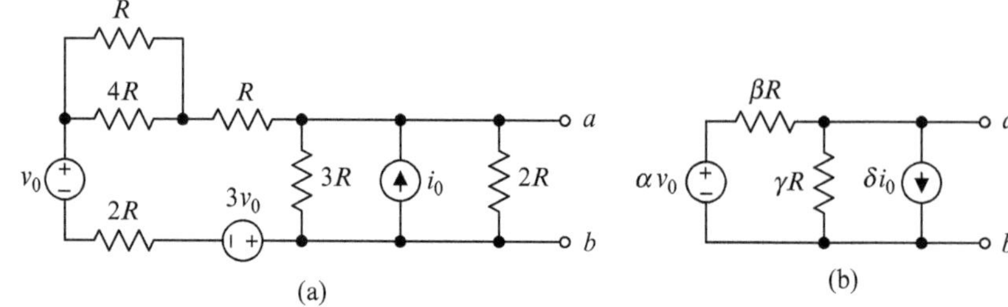

Fig. 4.19 See Exercise 4.3

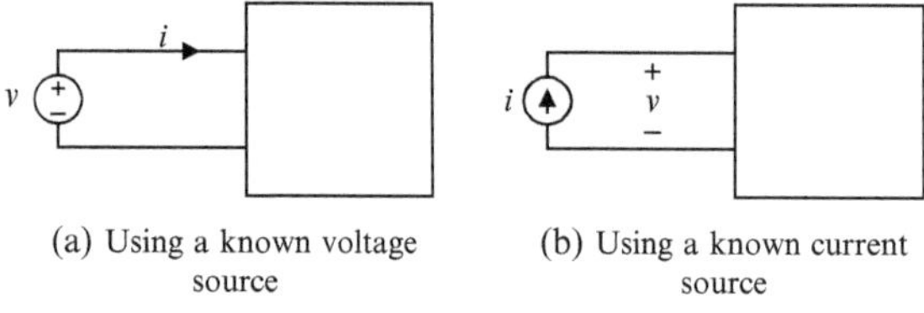

Fig. 4.20 Finding the equivalent resistance $R_{\mathrm{eq}} = v/i$ at a terminal pair

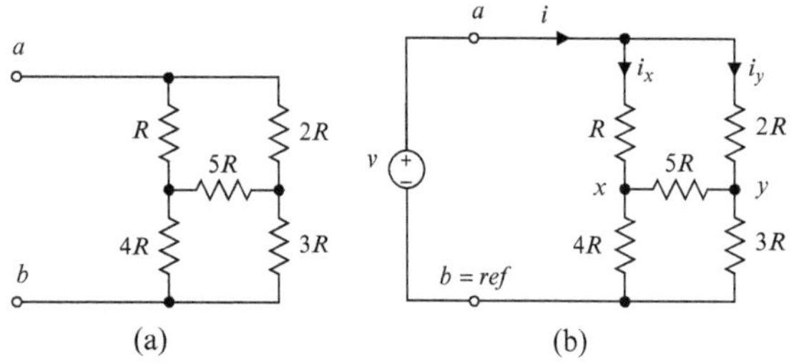

Fig. 4.21 See Example 4.6

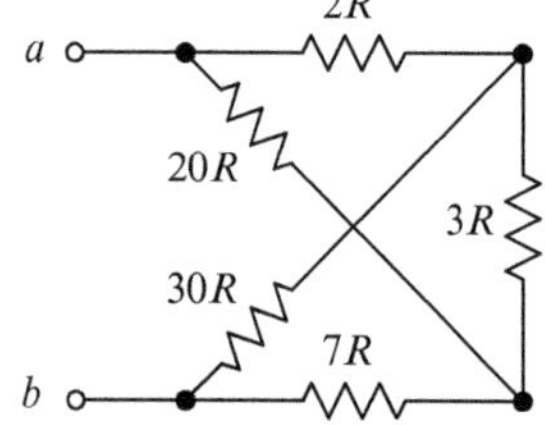

Fig. 4.22 See Exercise 4.4

Exercise 4.4. Refer to Fig. 4.22, where $R = 1\Omega$. Find the equivalent resistance at the terminals *a–b*.

4.2.7 Approximations

The resistance of a series connection of resistors is larger than the largest single resistance in the series connection. If the resistance of one resistor in a series connection is much larger than the total resistance of the others, the equivalent resistance of the series connection often can be approximated by that of the larger resistor. Similarly, the resistance of a parallel connection of resistors is smaller than the smallest single resistance in the connection. If the resistance of one resistor in a parallel connection is much smaller than that of the equivalent resistance of the others, the equivalent resistance of the parallel connection often can be approximated by that of the smaller resistor. In making these approximations, the meanings of "much larger than" and "much smaller than" depend upon the application at hand. Sometimes, a factor of 10 is sufficient; for example, if $R_2 > 10R_1$, then $R_1 + R_2$ (series connection) often can be approximated by R_2 and $R_1R_2/(R_1 + R_2)$ (parallel connection) by R_1. More often, a factor of 100 is sufficient, and almost always, a factor of 1000 is more than sufficient. As R_2 ranges from 10 R_1 to 1000 R_1, the error in these approximations ranges from about 10% to about 0.1%. Such approximations often provide useful checks on the results of more exact calculations.

Example 4.7. (a) Find the equivalent resistance at the terminals *a–b* for the circuit shown in Fig. 4.23. (b) Check the result using reasonable approximations.

Solution: (a) The equivalent resistance is

$$R_{eq} = R_1 + R_4 + \frac{R_2 R_3}{R_2 + R_3} \cong 10.6\,\mathrm{k}\Omega.$$

(b) Because $R_3 \ll R_2$ we approximate the parallel connection of those resistors by R_3 alone, which then appears in series with R_1 and R_4 Because $R_1 \gg R_3 + R_4$ the equivalent

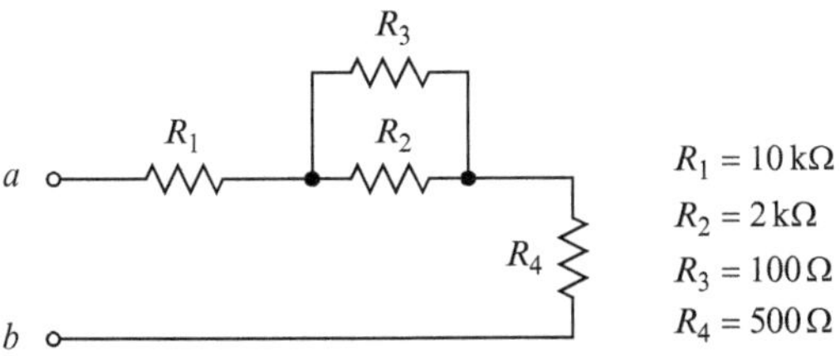

Fig. 4.23 See Example 4.7

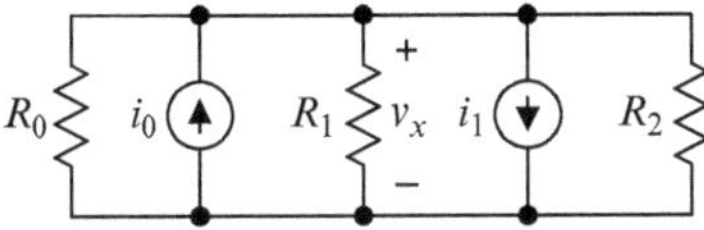

Fig. 4.24 See Exercise 4.5

resistance at the terminals *a–b* is approximately $R_1 = 10\,\text{k}\Omega$. The result obtained in (a) above is reasonable. (Also, the error in the approximation is only about 6%).

Similar approximations can be made for voltage sources in series, where the magnitude of one source voltage is much larger than the sum of the others, and for two current sources in parallel, where the magnitude of one source current is much larger than the sum of the others.

Exercise 4.5. Refer to Fig. 4.24, where $R_0 \gg R_2$, $i_1 \gg i_0$. (a) Obtain an exact expression for v_x. (b) Find a reasonable approximation to v_x. (c) Let $R_0 = 10\,\text{k}\Omega$, $R_1 = 5\,\text{k}\Omega$, $R_2 = 4.7\,\Omega$, $i_0 = 1\,\text{mA}$, $i_1 = 100\,\text{mA}$. What is the percentage error in the approximation?

4.3 Source Transformations

In this section, we show that a voltage source in series with a resistor is equivalent to a current source in parallel with an identical resistor. This equivalence allows what are called **source transformations**, which, in conjunction with methods described above, often can be used to reduce a complex circuit to a much simpler one.

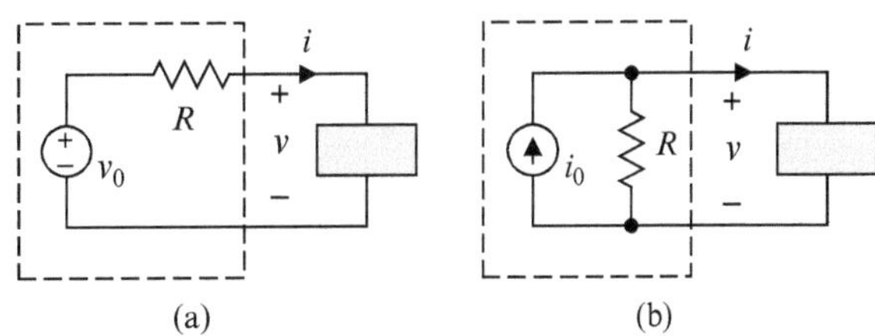

Fig. 4.25 Equivalent source models: $i_0 = v_0/R$

We wish to show that the circuits inside the dotted boxes in Fig. 4.25 are equivalent at their terminals if $i_0 = v_0/R$ and the two resistors are identical. The gray boxes represent arbitrary (unspecified) elements or circuits.

Writing Kirchhoff's voltage law for the circuit in Fig. 4.25(a) gives

$$-v_0 + iR + v = 0,$$

which yields

$$i = -\frac{v}{R} + \frac{v_0}{R}. \tag{4.9}$$

Writing Kirchhoff's current law for the circuit in Fig. 4.25(b) gives

$$-i_0 + \frac{v}{R} + i = 0,$$

which yields

$$i = -\frac{v}{R} + i_0. \tag{4.10}$$

The terminal characteristics (4.9) and (4.10) are identical (the circuits in Fig. 4.25 are equivalent at the terminals *a–b*) if $i_0 = v_0/R$.

The significance of the development above is the following: Whenever we wish, we may replace a voltage source in series with a resistor with a current source in parallel with an identical resistor, or vice-versa, as illustrated by Fig. 4.26. Such source transformations, in conjunction with reduction of series and parallel connections as described above, can be useful in circuit analysis. But note that an isolated voltage source (no series resistance) cannot be transformed to a current source. Similarly, an isolated current source (no parallel resistance) cannot be transformed to a voltage source. This is not a serious limitation, because all realistic source models possess internal resistance.

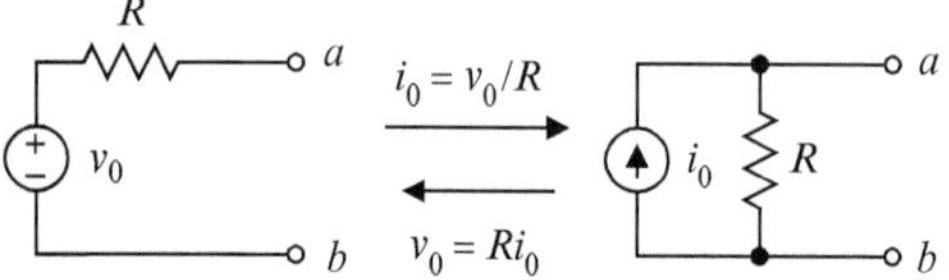

Fig. 4.26 Source transformations. Note the polarity of the voltage source and the direction of the current source. If either is reversed, the other is also

Example 4.8. Obtain an expression for the voltage v in the circuit of Fig. 4.27(a).

Solution: We proceed as shown in Fig. 4.27 (b)–(e): We transform the voltage source v_1 and series resistor R_1 to an equivalent current source and parallel resistor R_1, as shown in Fig. 4.27(b), where

$$i_3 = v_1/R_1. \tag{4.11}$$

We combine the parallel current sources i_1, i_3 and resistors R_1, R_2, as shown in Fig. 4.27(c), where

$$i_4 = i_1 + i_3, \quad R_a = \frac{R_1 R_2}{R_1 + R_2}. \tag{4.12}$$

We transform the current source i_4 and parallel resistor R_a to a voltage source and series resistor as shown in Fig. 4.27(d), where

$$v_3 = R_a i_4. \tag{4.13}$$

We add the series voltage sources and add the series resistors to obtain the circuit shown in Fig. 4.27(e), where

$$v_4 = v_3 + v_2, \; R_b = R_a + R_3. \tag{4.14}$$

The circuit shown in Fig. 4.27(e) is a voltage divider. Thus we have

$$v = \frac{R_4 v_4}{R_4 + R_b}, \tag{4.15}$$

which, using (4.11) through (4.14), yields

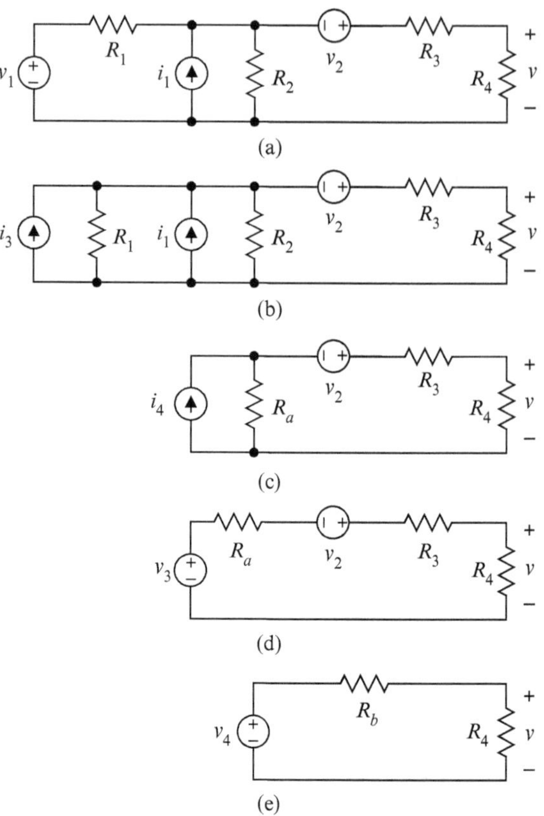

Fig. 4.27 See Example 4.8

$$v = \frac{R_4[R_1R_2i_1 + R_2v_1 + (R_1 + R_2)v_2]}{(R_4 + R_3)(R_1 + R_2) + R_1R_2}.$$

Exercise 4.6. Simplify the circuit of Fig. 4.19(a) at the terminals *a*–*b* to (a) a resistance R_T in series with a voltage source v_T (positive polarity nearest terminal *a*) and (b) a resistance R_N in parallel with a current source i_N (arrowhead pointing to the conductor attached to terminal *a*).

Example 4.9. Refer to Fig. 4.28. (a) Find the voltage across the load resistor R_L. (b) Repeat, using reasonable approximations, and compare with the result obtained in (a).

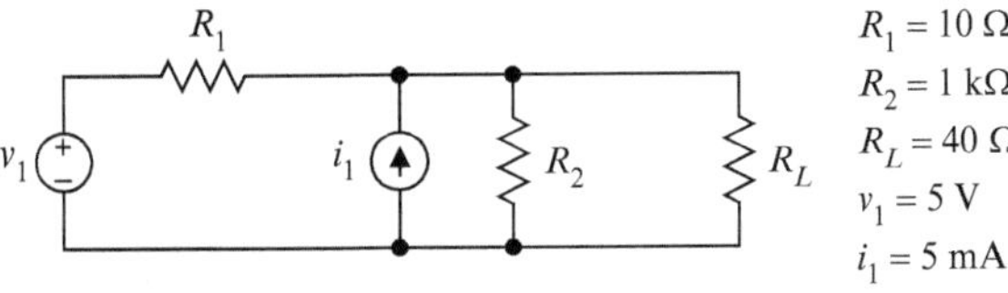

Fig. 4.28 See Example 4.9

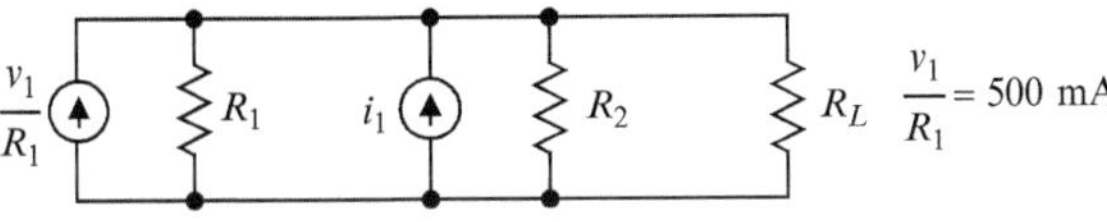

Fig. 4.29 See Example 4.9

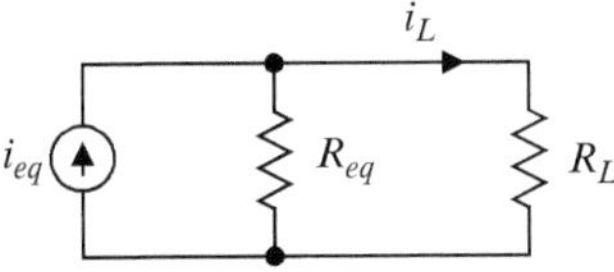

Fig. 4.30 See Example 4.9

Solution: Refer to Fig. 4.29. We transform the voltage source in series with resistor R_1 to an equivalent current source in parallel with resistor R_1, as shown in Fig. 4.29.

(a) The circuit shown in Fig. 4.29 reduces to the current divider shown in Fig. 4.30, where

$$i_{eq} = \frac{v_1}{R_1} + i_1 \cong 505\,\text{mA};$$
$$R_{eq} = \frac{R_1 R_2}{R_1 + R_2} \cong 9.90\,\Omega.$$

The load current and load voltage are

$$i_L = \frac{R_{eq}}{R_{eq} + R_L} i_{eq} \cong 100.2\ \text{mA},$$
$$v_L = i_L R_L \cong (100.2\ \text{mA})(40\,\Omega) \cong 4.01\text{V}.$$

(b) In the circuit shown in Fig. 4.29,

$$\frac{v_1}{R_1} = 500\,\text{mA} \gg i_1, \quad R_1 \ll R_2.$$

Thus we ignore the current source i_1 and the resistor R_2. Then with reference to Fig. 4.30, we have that

$$i_{eq} \cong \frac{v_1}{R_1} = 100\,\text{mA}, \quad R_{eq} \cong R_1 = 10\,\Omega.$$

Thus the load current and load voltage are (approximately)

$$i_L \cong \frac{R_{eq}}{R_{eq} + R_L} \frac{v_1}{R_{eq}} = \frac{R_1}{R_1 + R_L} \frac{v_1}{R_1} = 100\,\text{mA},$$
$$v_L = i_L R_L = 4.00\,\text{V}.$$

The error in the approximation is about 0.2% for the load current and about 0.25% for the load voltage.

Again, even in cases where errors introduced by approximations are intolerable, finding an approximate solution often provides a good check on the reasonableness of a more exact result. You should make it a habit to perform such checks, in addition to dimension checks and limit checks.

4.4 Thévenin and Norton Equivalent Circuits

A **resistive circuit** is a circuit containing only resistors and sources.

Figure 4.31 shows an arbitrary resistive circuit. We choose any terminal pair *a–b* that satisfies Kirchhoff's current law, meaning that the current exiting one terminal equals the current entering the other. **Thévenin's theorem** states that the terminal characteristic at such a terminal pair is

$$\frac{i}{i_{sc}} = 1 - \frac{v}{v_{oc}}, \tag{4.16}$$

where i and v are the terminal current and voltage, respectively, and i_{sc} and v_{oc} are the **short-circuit current** and the **open-circuit voltage** at the terminals *a–b*, whose definitions are illustrated by Fig. 4.32. It does not matter what is in the two-terminal gray box

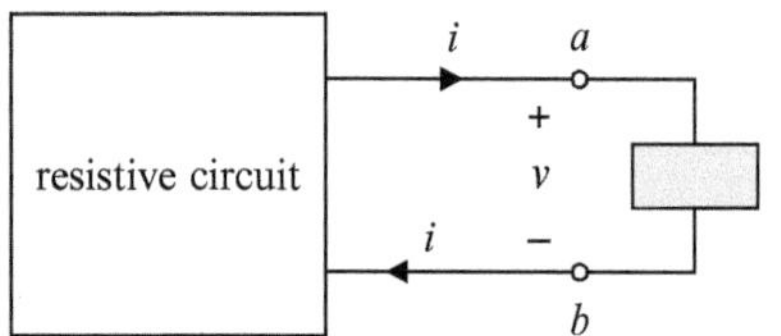

Fig. 4.31 Pertaining to Thévenin's theorem (see (4.16))

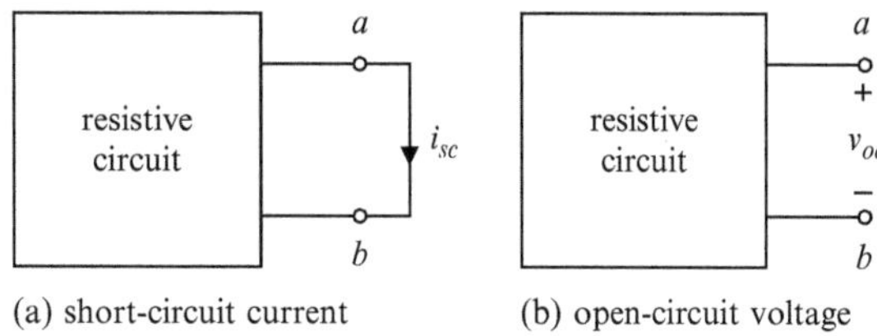

Fig. 4.32 Definitions of short-circuit current and open-circuit voltage

which, for present purposes, simply provides a path for the terminal current i. Thévenin's theorem is among the most important derived results in electrical engineering.

Essentially, Thévenin's theorem states only that the relation between terminal voltage and terminal current is given by (4.16). The theorem says nothing about the internal structure of the circuit, except that it contains only resistors and sources. Nonetheless, it is useful to interpret (4.16) as the terminal characteristic of what is called the **Thévenin equivalent** for the actual circuit in question, as follows: We write (4.16) as

$$v = v_{oc} - \left(\frac{v_{oc}}{i_{sc}}\right) i, \tag{4.17}$$

and then as

$$v = v_T - R_T\, i, \tag{4.18}$$

or as

$$i = -\frac{v}{R_T} + \frac{v_T}{R_T}, \tag{4.19}$$

where

$$\begin{aligned} v_T &= v_{oc} = \textit{Thevenin equivalent voltage}, \\ R_T &= \frac{v_{oc}}{i_{sc}} = \textit{Thevenin equivalent resistance}. \end{aligned} \tag{4.20}$$

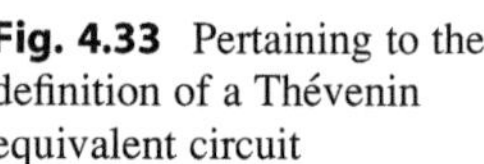
Fig. 4.33 Pertaining to the definition of a Thévenin equivalent circuit

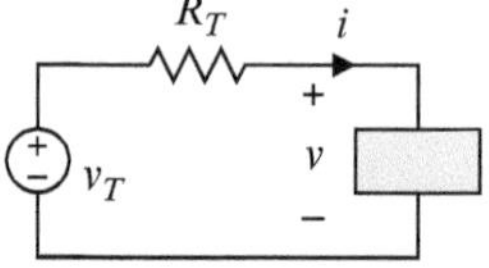

We may interpret (4.18) as Kirchhoff's voltage law written for the circuit shown in Fig. 4.33. From this perspective, Thévenin's theorem and (4.20) state that a two-terminal resistive circuit is equivalent at the terminals to a voltage source v_T in series with a resistor R_T. Again, it does not matter what is in the two-terminal gray box which, for present purposes, simply provides a path for the terminal current i. This development leads to the following restatement of Thévenin's theorem: *Any resistive circuit is equivalent at any particular terminal pair satisfying Kirchhoff's current law to an independent source in series with a resistor, where the source voltage and the resistance of the resistor are given by* (4.20). The source voltage R_T is called the **Thévenin equivalent voltage** and the resistance R_T of the resistor is called the **Thévenin equivalent resistance**. The combination of the source and resistor is called the **Thévenin equivalent circuit** at the terminal pair under consideration. A Thévenin equivalent voltage is called the **Thévenin voltage**, a Thévenin equivalent resistance is called the **Thévenin resistance**, and a Thévenin equivalent circuit is called the **Thévenin equivalent**. A circuit can have more than one terminal pair. A Thévenin equivalent for such a circuit is specific to the terminal pair chosen.

Neither the Thévenin equivalent voltage source nor the Thévenin equivalent resistance is necessarily physical. That is, neither necessarily corresponds to any particular element in the associated physical circuit. A Thévenin equivalent should be viewed simply as a means of depicting the fundamental relation (4.16) by a circuit diagram.

The usual purpose of representing a circuit by its Thévenin equivalent is to treat the circuit as a source to which a load is to be attached; e.g., to study current and voltage delivered by the circuit to various loads. Consequently, *the positive direction for current in the terminal characteristic* (4.19) *for a Thévenin equivalent is conventionally out of the positive terminal, as shown in* Fig. 4.33.

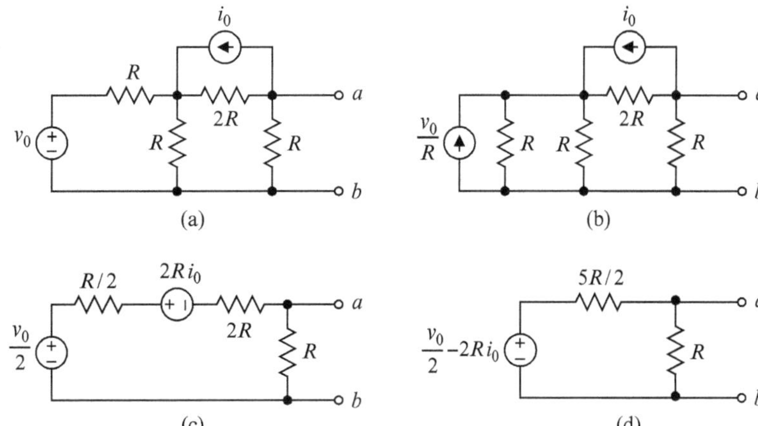

Fig. 4.34 See Example 4.10

The Thévenin equivalent for a circuit is unique. We may use any workable method to find the Thévenin equivalent circuit parameters R_T, including straightforward application of Kirchhoff's laws, reducing series and parallel connections to simpler equivalents, and source transformations.

Example 4.10. Obtain the Thévenin equivalent for the circuit of Fig. 4.34(a) at the terminals *a–b*.

Solution: We use source transformations and series-parallel reductions to simplify the circuit as illustrated by Fig. 4.34(b)–(d). The circuit of Fig. 4.34(d) is a voltage divider and the open-circuit voltage is given by

$$v_{oc}=\frac{R}{R+5R/2}\left(\frac{v_0}{2}-2R\,i_0\right)=\frac{2}{7}\left(\frac{v_0}{2}-2R\,i_0\right).$$

The short-circuit current is given by

$$i_{sc}=\frac{\left(\frac{v_0}{2}-2R\,i_0\right)}{5R/2}=\frac{(v_0-4R\,i_0)}{5R}.$$

Thus the Thévenin voltage and Thévenin resistance are given by

$$v_T=v_{oc}=\frac{2}{7}\left(\frac{v_0}{2}-2R\,i_0\right)=\frac{v_0}{7}-\frac{4R\,i_0}{7},$$

$$R_T=\frac{v_{oc}}{i_{sc}}=\frac{5R}{7}.$$

Exercise 4.7. Obtain the Thévenin equivalent for the circuit of Fig. 4.35 at the terminals *a–b*.

The terminal characteristic of a two-terminal circuit is a property of the circuit and is independent of any attachment to the terminals. A heuristic proof of Thévenin's theorem based upon this fact follows:

We attach an independent voltage source v to the terminals *a–b* of a resistive circuit, as illustrated by Fig. 4.36. We assume that the terminals *a–b* are neither shorted nor open nor connected directly to another source inside the box, in which case the source voltage v is the terminal voltage. Together, the circuit in the box and the source v comprise a resistive circuit, so the terminal current i is a linear function of all the sources in the box and of the source v; that is, the terminal current is given by[2]

$$i=a_1v_1+a_2v_2+\cdots+a_Nv_N+b_1i_1+b_2i_2+\cdots + b_Mi_M+m\,v,$$

where $v_1, v_2, \cdots, v_N$ are all of the voltage sources inside the box and $i_1, i_2, \cdots, i_M$ are all of the current sources inside the box. All these sources are fixed, and we may write the relation above more simply as

$$i=mv+b, \tag{4.21}$$

where $b=a_1v_1+a_2v_2+\cdots+a_Nv_N+b_1i_1+b_2i_2+\cdots+b_Mi_M$. The relation above describes a straight

[2]See Section 3.10 in Chapter 3.

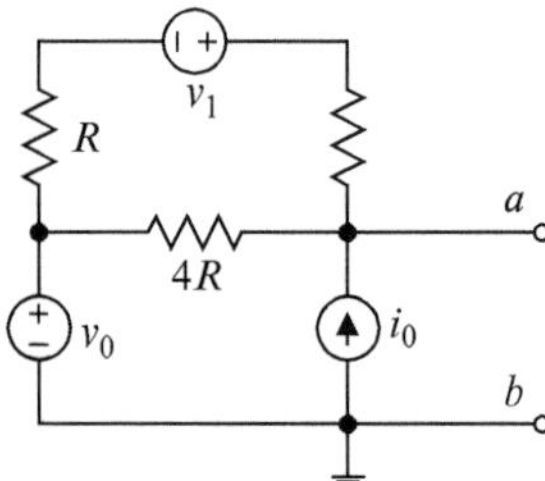

Fig. 4.35 See Exercise 4.7

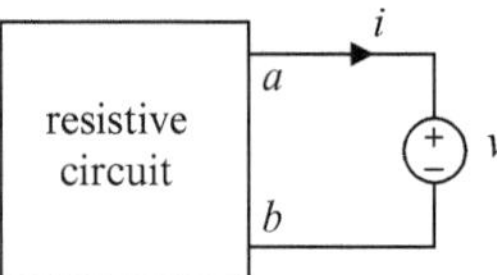

Fig. 4.36 Pertaining to the development of Thévenin's theorem

line whose slope m and intercept b can be determined using any two points on the line. Setting $v = 0$ in (4.21) is equivalent to shorting the terminals a–b and gives $i = b$, so the parameter b is the short-circuit current; i.e.,

$$b = i_{sc}. \tag{4.22}$$

Setting v to the open-circuit voltage v_{oc} is equivalent to opening the terminals, in which case $i = 0$ and (4.21) becomes $v_{oc} = -b/m = -i_{sc}/m$, which yields

$$m = -\frac{i_{sc}}{v_{oc}}. \tag{4.23}$$

From (4.21–4.23),

$$i = -\frac{i_{sc}}{v_{oc}} v + i_{sc}.$$

Dividing both sides of (4.23) by i_{sc} and rearranging terms gives (4.16).

Using a source transformation, the Thévenin equivalent for a circuit can be transformed to a current source in parallel with a resistor, as shown in Fig. 4.37. The resulting circuit is called the **Norton equivalent circuit** or simply the **Norton equivalent.** The source current i_N is called **the Norton current** and the resistance R_N is called the **Norton resistance.** Writing Kirchhoff's current law for the circuit on the right in Fig. 4.37 yields the terminal characteristic for a Norton equivalent circuit:

$$i = -\frac{v}{R_N} + i_N. \tag{4.24}$$

From (4.24), the Norton source current i_N is the short-circuit current:

$$i_N = i_{sc}. \tag{4.25}$$

From Fig. 4.26, the transformation relations are

$$R_N = R_T, \qquad i_N = \frac{v_T}{R_T}, \tag{4.26}$$

which are just the usual source-transformation relations made specific to the notation used here. Figure 4.37 illustrates relations between the Thévenin and Norton equivalents for the same circuit. *In general, elements comprising the Thévenin or Norton equivalent for a circuit are non-physical and do not correspond to components in an associated physical circuit.*

As noted above, a circuit regarded as a source often is represented by the Thévenin or Norton equivalent for the circuit. In this context, the Thévenin (or Norton) equivalent resistance for the circuit also is called the **internal resistance**, the **output resistance**, or the **source resistance** of the circuit (the source). Generally, the term *internal resistance* is used in reference to the Thévenin or Norton equivalent for a physical source, such as a battery, a power supply, or an electric generator, and the term *output resistance* is used when the source in question is a circuit. For example, we might refer to the *internal resistance* of an alkaline battery and the *output resistance* of an audio amplifier (at the speaker terminals). The term *source resistance* often is used in either case or when the nature of the source is unspecified. But the terminology police do not vigorously enforce these conventions, and you may treat internal resistance, output resistance, and source resistance as synonyms without fear of arrest and imprisonment.

Example 4.11. Obtain the Norton equivalent for the circuit of Example 4.10 at the terminals *a–b*.

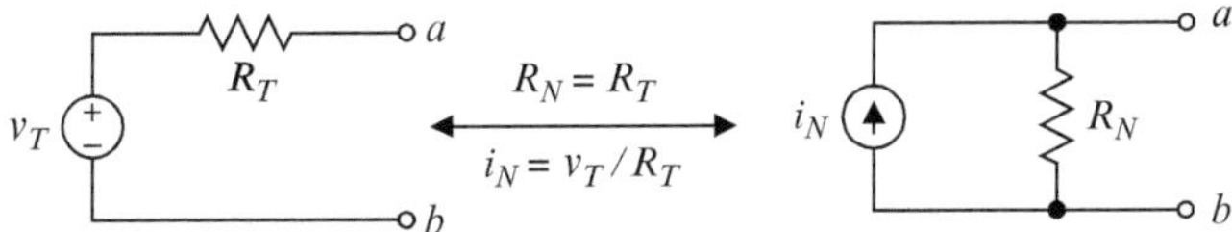

Fig. 4.37 Thévenin and Norton equivalent circuits

Solution: The Norton current is the short-circuit current. The Norton resistance is the Thévenin resistance. From Example 4.10

$$i_N = i_{sc} = \frac{(v_0 - 4R\,i_0)}{5R}, \quad R_N = R_T = \frac{5R}{7}.$$

Exercise 4.8. Obtain the Norton equivalent for the circuit of Exercise 4.7 at the terminals *a–b*.

Example 4.12. Obtain the Thévenin and Norton equivalents at the terminals *x–y* for the circuit of Fig. 4.38, where $R = 10\ \text{k}\Omega$ and $v_0 = 25$ V.

Solution: We find the Thévenin equivalent by finding the open-circuit voltage and the short-circuit current. First, we find the open-circuit voltage. We label the nodes as shown in the figure and select node c as the reference node. Writing Kirchhoff's current law at nodes *a* and *b* gives

$$G(v_a - v_0) + \frac{G}{2} v_a + G(v_a - v_b) = 0,$$
$$G(v_b - v_a) + \frac{G}{2}(v_b - v_0) = 0,$$

where $G = R^{-1}$ Upon dividing both equations by *G*/2 and collecting terms, we obtain

$$5\,v_a - 2\,v_b = 2v_0,$$
$$-2\,v_a + 3\,v_b = v_0.$$

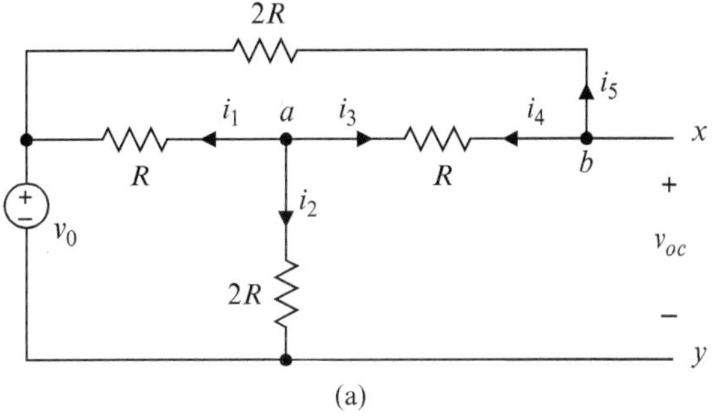

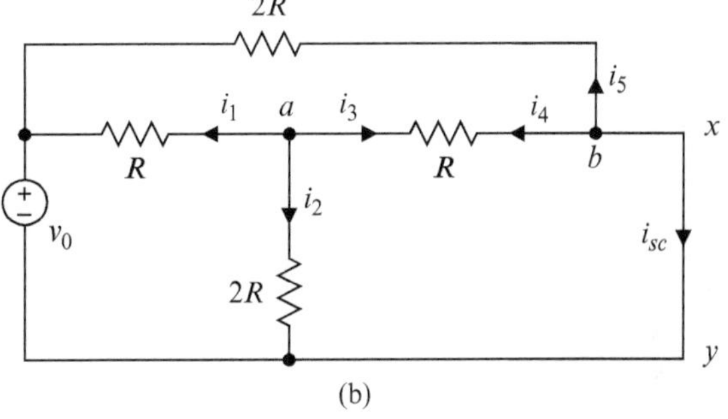

Fig. 4.38 Circuit of Example 4.12: (**a**) With the terminals *x–y* open for calculating the open-circuit voltage v_{oc} and (**b**) with the terminals *x–y* shorted for calculating the short-circuit current i_{sc}

The open circuit voltage is $v_{oc} = v_{bc} = v_b$. We eliminate v_a and obtain

$$v_{oc} = v_b = \frac{9\,v_0}{11}.$$

To determine the short-circuit current, we short the terminals *x–y* as shown in the figure. With the terminals shorted, $v_b = 0$. Writing Kirchhoff's current law at node a leads to

$$\frac{5}{2} v_a = v_0 \;\Rightarrow\; v_a = \frac{2\,v_0}{5}.$$

Writing Kirchhoff's current law at node *b* (with $v_b = 0$) gives

$$i_4 + i_5 + i_{sc} = G(-v_a) + \frac{G}{2}(-v_0) + i_{sc} = 0,$$

which yields

$$i_{sc} = G v_a + \frac{G}{2} v_0 = \frac{2 v_0}{5R} + \frac{v_0}{2R} = \frac{9 v_0}{10R},$$

where we have used the value for v_a obtained above. Thus we have

$$v_T = v_{oc} = \frac{9 v_0}{11} \cong 20.5 \text{ V},$$

$$i_N = i_{sc} = \frac{9 v_0}{10R} \cong 2.25 \text{ mA},$$

$$R_T = R_N = \frac{v_{oc}}{i_{sc}} = \frac{10R}{11} \cong 9.09 \text{ k}\Omega.$$

Figure 4.39 shows a graph of the terminal characteristic for a Thévenin or Norton equivalent circuit. You will find it helpful to memorize Fig. 4.39. From (4.19) or from (4.24), the Thévenin or Norton equivalent resistance for a circuit is the negative reciprocal of the slope of the terminal characteristic:

$$R_T = R_N = -\left(\frac{di}{dv}\right)^{-1} = \frac{v_{oc}}{i_{sc}}. \quad (4.27)$$

This relation helps us remember that the internal resistance of an independent voltage source equals zero (because the terminal characteristic is a vertical line and $di/dv \to \infty$) and that the internal resistance of an independent current source is infinite (because the terminal characteristic is a horizontal line and $di/dv = 0$). The general form of the terminal characteristic of a resistive circuit, given by (4.16), is easily deduced from the graph in Fig. 4.39.

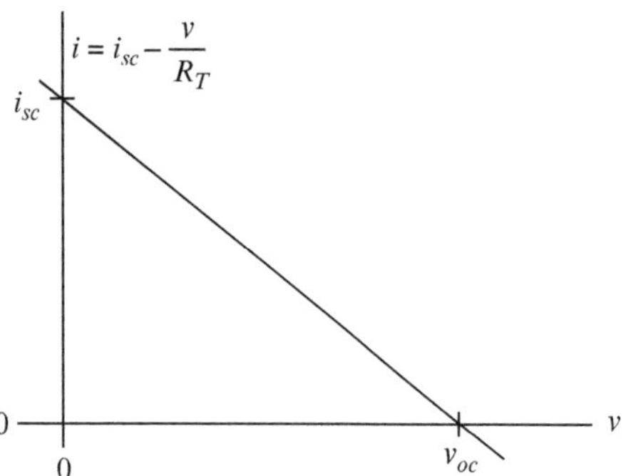

Fig. 4.39 Terminal characteristic for a Thévenin or Norton equivalent (for a two-terminal resistive circuit)

Although the graph in Fig. 4.39 suggests that the short-circuit current and the open-circuit voltage are both positive, that is not necessarily the case. Both can be negative, in which case we would draw the equivalent circuits and a graph of the terminal characteristic as shown in Fig. 4.40. *For a resistive circuit, the open-circuit voltage and the short-circuit current must have the same sign.* (Why?)

The Thévenin equivalent for a circuit can be obtained experimentally, by measuring two pairs of i–v values. Any two pairs will do, because the terminal characteristic is a straight line, as shown by Fig. 4.39. Although open-circuit voltage and short-circuit current are convenient values for mathematical analysis, it may be unwise to attempt to measure short-circuit current at the terminals of a physical circuit. The next example shows how one obtains the Thévenin equivalent for a circuit from an arbitrary pair of i–v values.

Example 4.13. The two pairs of values v_i, i_1 and v_2, i_2 are measured at a terminal pair of a resistive circuit. Obtain an expression for the Thévenin equivalent at the terminal pair.

Solution: Figure 4.41 illustrates the problem. The points (v_i, i_1) and (v_2, i_2) are sufficient to define the straight-line iv (terminal) characteristic, which has the form

$$i = -\frac{v}{R_T} + \frac{v_T}{R_T}, \quad (4.28)$$

where $-1/R_T$ is the slope and v_T/R_T is the (current) intercept. From Fig. 4.41,

$$-\frac{1}{R_T} = \frac{i_2 - i_1}{v_2 - v_1} \Rightarrow R_T = -\frac{v_2 - v_1}{i_2 - i_1}. \quad (4.29)$$

Using this result and the pair (v_1, i_1) in (4.28) gives

$$\frac{v_T}{R_T} = i_1 - m v_1 \Rightarrow v_T = i_1 R_T + v_1. \quad (4.30)$$

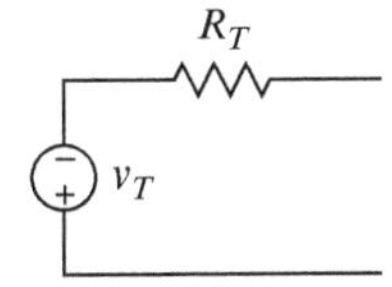

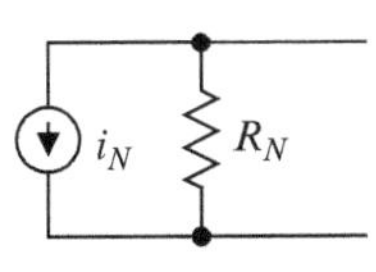

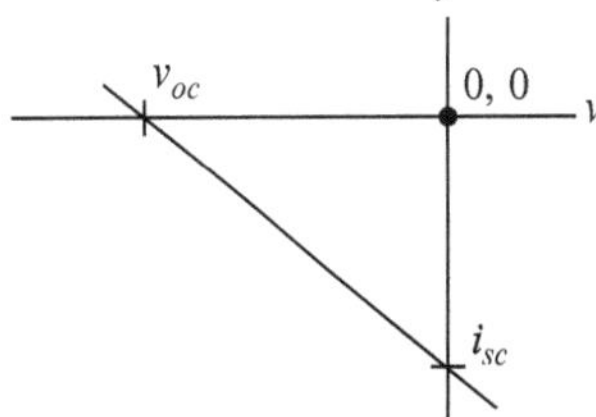

Fig. 4.40 Thévenin and Norton equivalent circuits if the open-circuit voltage and short-circuit current are negative

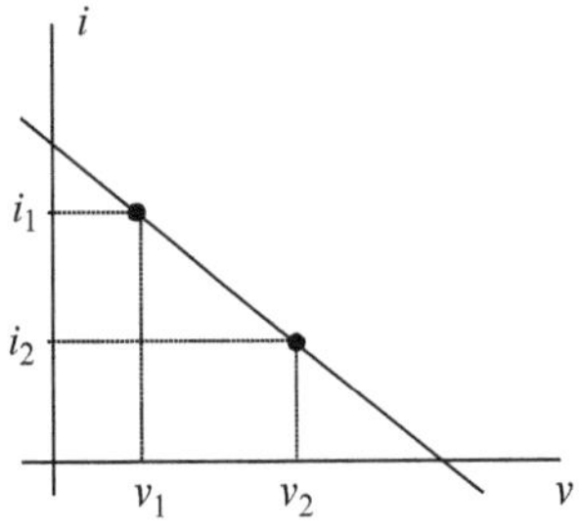

Fig. 4.41 See Example 4.13

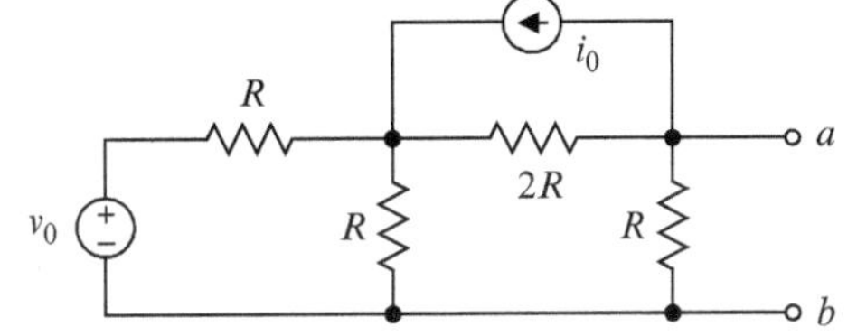

Fig. 4.42 See Example 4.14

Exercise 4.9. A resistor R_1 is connected to terminals *a–b* of a resistive circuit, and the resulting current out of terminal *a* is found to be i_1. Then R_1 is removed and another resistor R_2 is connected to the terminals. The voltage v_{ab} across the second resistor is found to be v_2. Obtain the Thévenin equivalent for the resistive circuit at terminals *a–b* in terms of R_1, R_2, i_1, and v_2.

We can find the Thévenin or Norton equivalent resistance of a *resistive circuit* using the so-called **look-back method**, as follows:[3]

- Replace all independent voltage sources by short circuits (because the internal resistance of an independent voltage source equals zero).
- Replace all independent current sources by open circuits (because the internal conductance of an independent current source equals zero).
- Find the equivalent resistance at the terminals of interest.

[3]In Chapter 6, we introduce sources called dependent sources. In general, the look-back method fails for circuits containing such sources.

Example 4.14. (Compare with Example 4.10.) Obtain an expression for the Thévenin equivalent resistance at the terminals *a–b* of the circuit shown in Fig. 4.42.

Solution: We replace the voltage source with a short circuit (a conductor) and replace the current source with an open circuit (remove the current source) to obtain the circuit shown in Fig. 4.43(a). We then use series/parallel reduction to obtain the Thévenin (Norton) equivalent resistance, as shown in Fig. 4.43(b)–(d).

Exercise 4.10. Use the look-back method to determine the Norton equivalent resistance at the terminals *a–b* of the circuit shown in Fig. 4.44.

Both the open-circuit voltage v_{oc} and the short-circuit current i_{sc} equal zero for a circuit containing no sources. It follows that both the Thévenin equivalent voltage $v_T = v_{oc}$ and the Norton equivalent current $i_N = i_{sc}$ for such a circuit equal zero. Consequently, we cannot use (4.20) to obtain the Thévenin or Norton equivalent resistance for such a circuit because $R_T = v_{oc}/i_{sc}$ is indeterminate. We point out this special case only for sake of completeness.

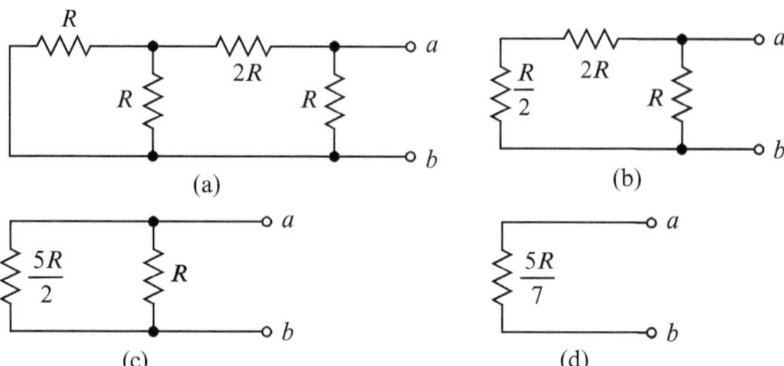

Fig. 4.43 See Example 4.14

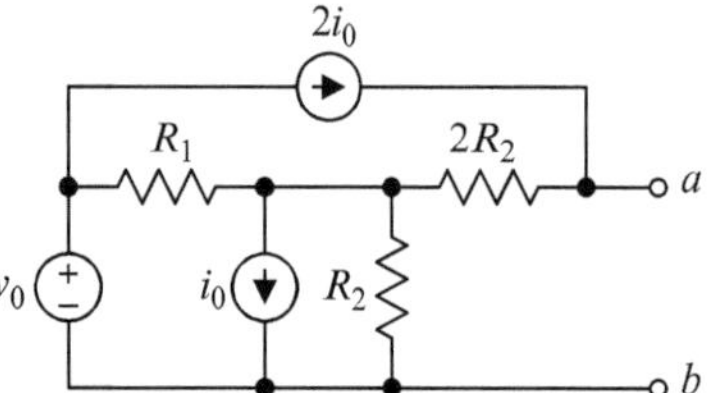

Fig. 4.44 See Exercise 4.10

In practice, Thévenin and Norton equivalents are not used to represent two-terminal circuits containing no sources. Such a circuit would instead be represented as an equivalent resistance, which can be found using methods described in Section 4.2 above.

Thévenin's theorem is both plausible and remarkable. It is plausible because circuit-reduction and transformation methods described above allow us to reduce virtually any planar resistive circuit[4] to a voltage source in series with a resistor. It is remarkable because it allows us to use a very simple model to describe even a very complex circuit (at a single terminal pair). Even more remarkable, Thévenin's theorem can be generalized to any circuit composed of any combination of sources, resistors, capacitors, and inductors, as described in subsequent chapters.

We conclude this section with four remarks:

- *The terminal (i–v) characteristic for a Thévenin or Norton equivalent circuit is independent of whatever is connected to the terminals of the circuit.* The voltage across and current through the terminals are related by (4.16) no matter what.
- *The Thévenin and Norton equivalents for a particular circuit at a particular terminal pair are unique.* The Thévenin equivalent source voltage, the Norton equivalent source current, and Thévenin/Norton equivalent resistance can be found by any legitimate method. Nonetheless, you should regard the method based upon open-circuit voltage and short-circuit current as the fundamental (analytical) method for obtaining a Thévenin or Norton equivalent.
- A Thévenin or Norton equivalent is only *externally* equivalent to the associated circuit. *In general, the components of a Thévenin or Norton equivalent are non-physical.* They do not necessarily correspond to any physical components in an associated physical circuit. Thus, for example, the voltage across a Thévenin resistance does not necessarily have any *physical* significance (is not necessarily measurable).
- *An independent voltage source does not possess a Norton equivalent* (because the short-circuit current is undefined). Similarly, *an independent current source does not possess a Thévenin equivalent* (because the open-circuit voltage is undefined). However, these limitations are of no practical significance.

4.5 Notation: Constant and Time-Varying Current and Voltage

As we introduce more and increasingly complex devices and circuits, we need more suggestive notation to help us keep track of quantities and parameters

[4]A planar circuit is one that can be drawn in two dimensions without crossing conductors.

involved in analysis and design. In this section, we describe notational conventions for current and voltage and for parameters having dimensions of current or voltage. Notation introduced here is used throughout the remaining chapters of this book.

During a time interval of interest, a current or voltage can be constant or time-varying. The almost universal convention is to use upper-case letters to denote constant currents or voltages and lower-case letters to denote time-varying currents or voltages. For example, $V_0 = 15\text{V}$ and $i_1 = 5\cos(\omega t)\,\text{mA}$.

We also use upper-case letters to denote a constant parameter having the dimension of a current or voltage, but which is not necessarily a current or voltage in a circuit under study (or in any other circuit). For example, we express a sinusoidally time-varying voltage as

$$v_1(t) = V_1 \cos(\omega t).$$

The peak amplitude V_1 is a constant parameter of the expression above for a sinusoidal voltage.

If a current or voltage might be either constant or time-varying, e.g., where an analysis is valid for either case, the current or voltage is represented by a lower-case i or v. In this sense, i and v are more general than I and V; that is, a constant current or voltage is a special case of a time-dependent current or voltage.

4.6 Significance of Terminal Characteristics and Equivalence

There are three important points to be made regarding terminal characteristics and equivalent circuits:

- A circuit element is defined by its terminal characteristic. We do not need to know what is inside the element or how the element works to apply Kirchhoff's laws to a circuit containing the element.
- If two two-terminal elements or circuits have identical terminal characteristics, then they are equivalent at those terminals; that is, one can be replaced by the other *at the terminals in question*. This is the basis for the circuit-reduction methods described in Sections 4.2 and 4.3 and is a reason why Thévenin's theorem is so important. Keep in mind, however, that the equivalence is only at the terminals. The internal structures of equivalent circuits are not necessarily (and usually are not) the same. In particular, we may use the Thévenin or Norton equivalent for a circuit to study current through, voltage across, and power dissipated in a load attached to the circuit, but *not* to study any of those quantities internal to the circuit.
- The task of studying the response of a circuit for various possible loads is greatly simplified by finding the Thévenin (or Norton) equivalent for the circuit at the terminals to which the load is to be attached. If the load is resistive, the circuit and load are reduced to a voltage divider (or current divider), and further analysis is trivial.

4.7 Problems

Section 4.1 is prerequisite for the following problems.

P 4.1 In Fig. P 4.1, $v_1 = 1.5$ V, $v_2 = 5$ V, $i_1 = 5$ mA, $i_2 = 10$ mA, $R_1 = 1\,\text{k}\Omega$, and $R_2 = 3.3\,\text{k}\Omega$. (i) Obtain the terminal characteristic at the terminals a–b for each circuit, regarding the circuit as a source. (ii) Draw and label fully a graph of the terminal characteristic (current as a function of voltage). (iii) Assume a voltage source $v_0 = 2$ V is connected to the terminals a–b of each circuit, with the positive terminal at a. Find the resulting current exiting terminal a and use the graph obtained in part (ii) to illustrate the solution. (iv) Simulate the circuit and verify your answers.

P 4.2 Figure P 4.2 shows a generalized graph of the terminal characteristic at terminals a–b of each circuit in Fig. P 4.3, where each is regarded as a source. (i) Obtain expressions for the intercepts i_{SC}, v_{OC} in terms of circuit parameters. (ii) Let $i_0 = 5\,\text{mA}$, $v_0 = 1.5\,\text{V}$, $R_1 = 1\,\text{k}\Omega$, and $R_2 = 3.3\,\text{k}\Omega$. Calculate the values of the intercepts and simulate the circuit to verify the calculated values.

P 4.3 In Fig. P 4.4, $v_1 = 1.5$ V, $i_1 = 1$ mA, $R_1 = 1.5\,\text{k}\Omega$, and $R_2 = 4\,\text{k}\Omega$. (a) Obtain symbolic expressions for the terminal characteristics for the circuits at the labeled terminals. Treat the terminals a

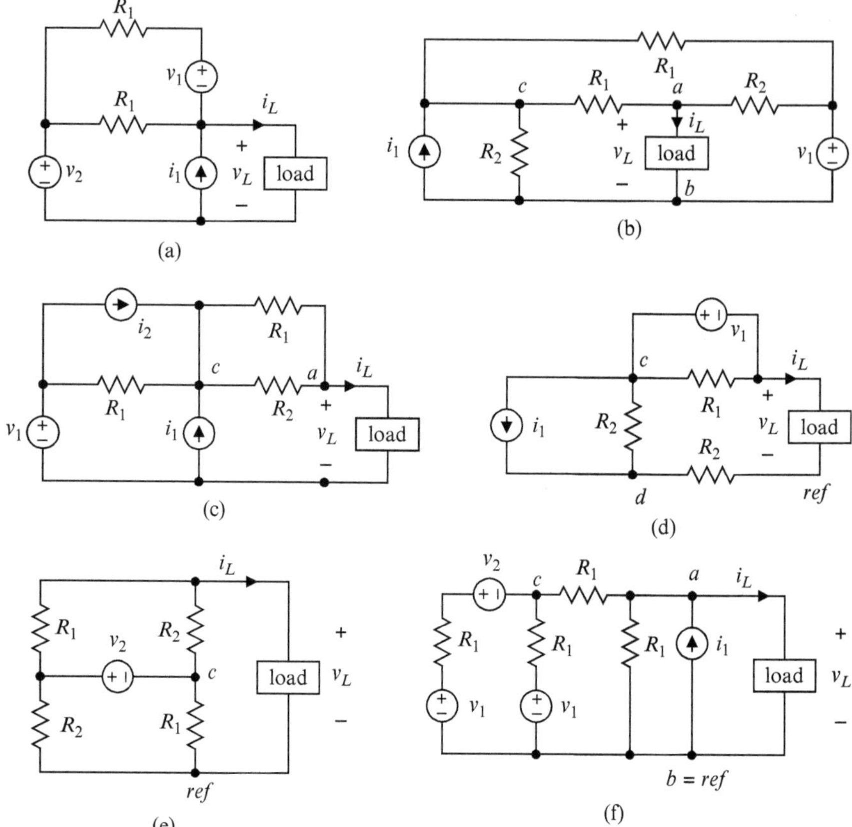

Fig. P 4.1 See Problem P 4.1

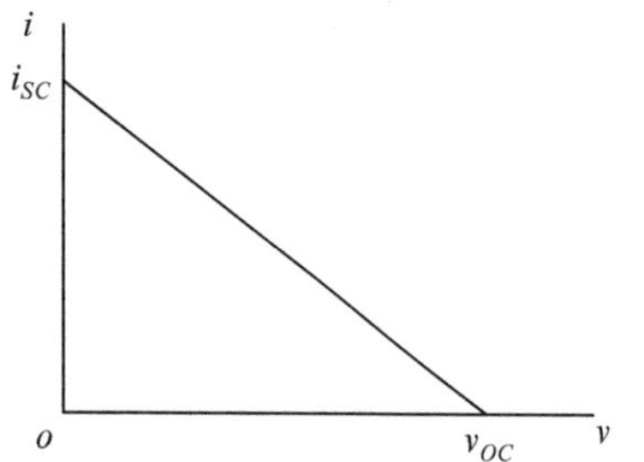

Fig. P 4.2 See Problem P 4.2

and c as positive. Treat the circuit on the left as a source and the circuit on the right as a load. (b) The circuits are connected: a to c and b to d. What is the resulting voltage v_{ab}? What is the resulting current i entering terminal c? (c) Plot the terminal characteristics obtained in Problem P 4.3 for each circuit on the same axes. (d) At what point (v, i) do the terminal characteristics intersect? Would this point be different if you treat the circuit on the left as a load and the circuit on the right as a source? (e) Simulate the circuit and verify the calculated values of the terminal voltage and current.

P 4.4 Imagine you are presented with a resistive circuit sealed in a box, but with access to a terminal pair. You have only a voltmeter and a few different resistors. Assume it is safe to connect any of the resistors to the terminal pair. What are the simplest (and fewest) measurements you would make to determine the terminal characteristic of the circuit at the accessible terminal pair? Justify your answer.

P 4.5 💻 (a) Obtain an expression for the terminal characteristic at the terminals a–b of the circuit shown in Fig. P 4.5, regarding the circuit as a load. (b) The circuit is connected to the terminals of a source whose terminal characteristic is shown in Fig. P 4.6, with terminal a connected to terminal c. Obtain an expression for $v_{ab} = v_{cd}$ in terms of i_1, v_1, and circuit parameters. (c) The connections are reversed, so that terminal a is connected to terminal d. Obtain an expression for $v_{ab} = v_{cd}$ in terms of i_1, v_1, and circuit parameters. (d) Let $v_1 = 10\,\text{V}$, $i_1 = 8\,\text{mA}$, $R_1 = 1\,\text{k}\Omega$,

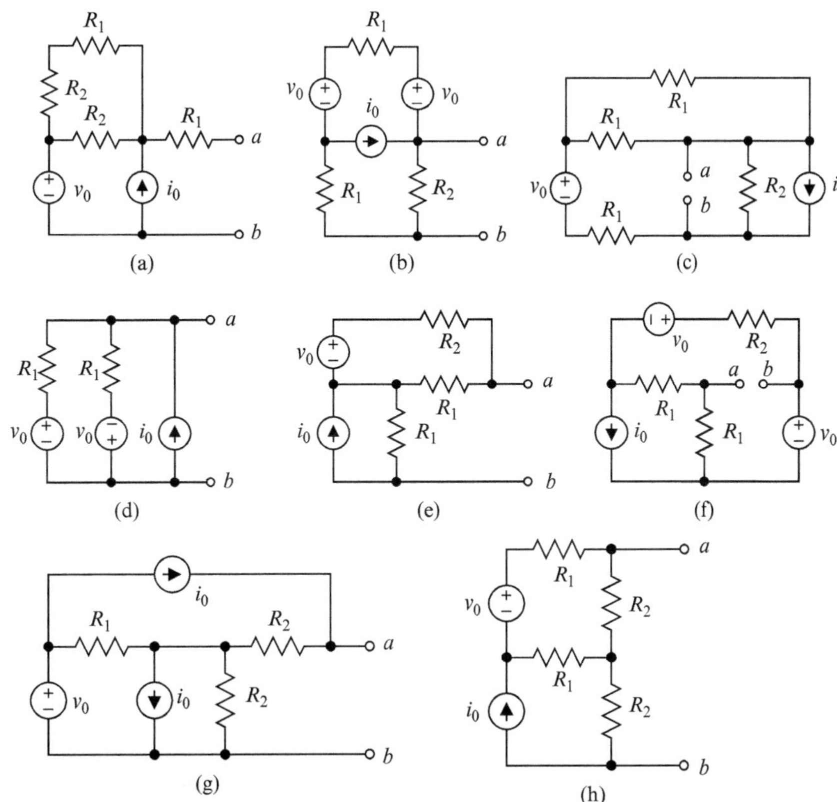

Fig. P 4.3 See Problem P 4.2, 47

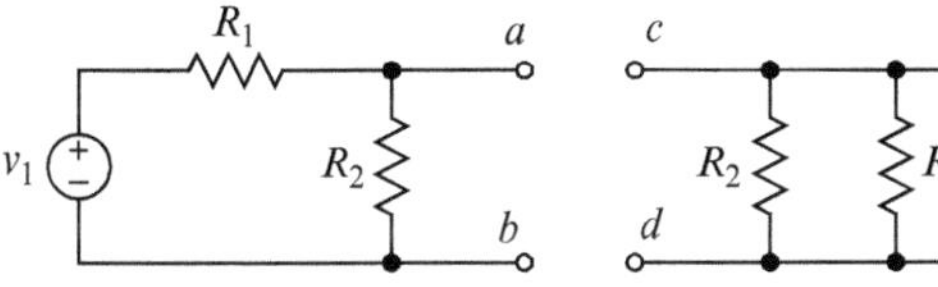

Fig. P 4.4 See Problem P 4.3

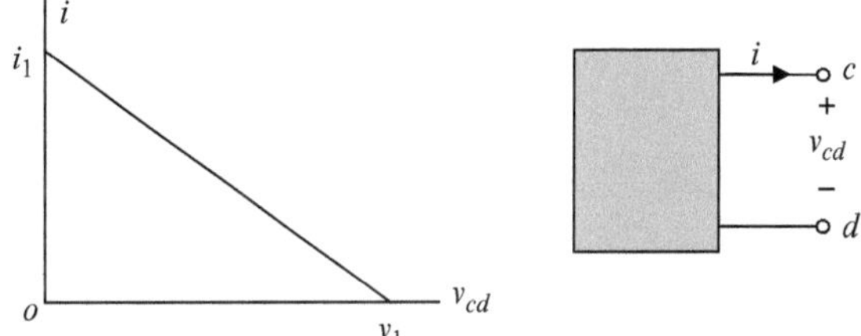

Fig. P 4.6 See Problem P 4.5

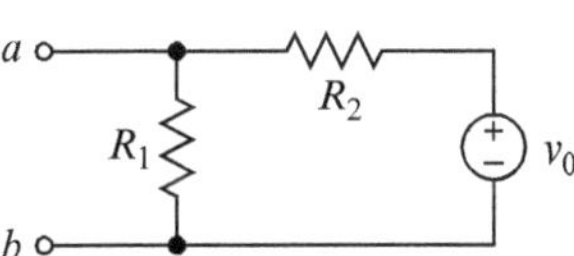

Fig. P 4.5 See Problem P 4.4

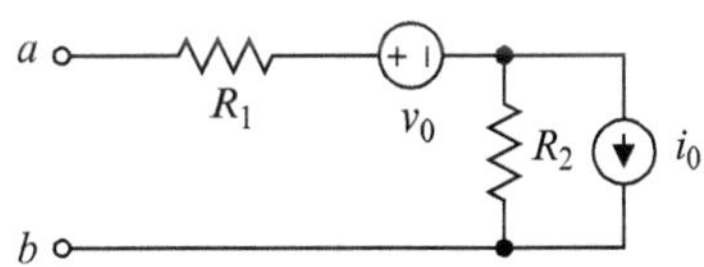

Fig. P 4.7 See Problem P 4.6

$R_2 = 2.2\,\text{k}\Omega$ and $v_0 = 15\,\text{V}$. Calculate the values of the terminal voltages obtained in parts (b) and (c). (d) Use simulation to verify the results obtained in parts (b) and (c).

P 4.6 Repeat Problem P 4.5 for the circuit shown in Fig. P 4.7, with $i_0 = 5\,\text{mA}$.

P 4.7 Repeat Problem P 4.5 for the circuit shown in Fig. P 4.8, with $i_0 = 5\,\text{mA}$, $R_0 = 470\,\Omega$, and $v_2 = 12\,\text{V}$.

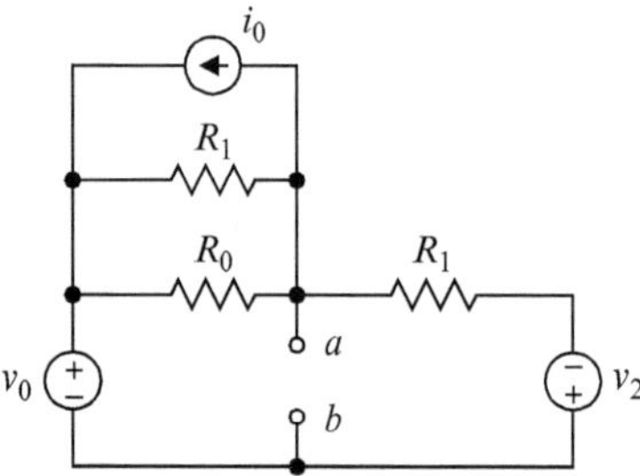

Fig. P 4.8 See Problem P 4.7

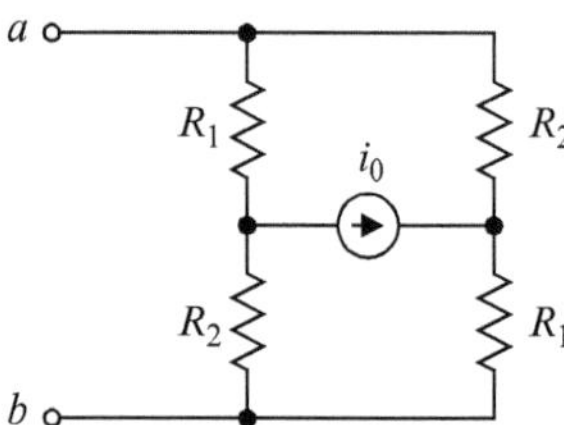

Fig. P 4.9 See Problem P 4.8

P 4.8 Repeat Problem P 4.5 for the circuit shown in Fig. P 4.9.

Section 4.2 is prerequisite for the following problems.

P 4.9 Show that the tolerance of a series connection of two resistors is *in general* bounded by the tolerance of the least precise resistor. Do not assume the (nominal) resistances are equal.

P 4.10 Repeat Problem P 4.9 for a parallel connection of two resistors.

P 4.11 A 470 $\Omega \pm 5\%$ resistor is connected in series with a 121 k$\Omega \pm 1\%$ resistor. Give the nominal resistance and the tolerance of the series connection.

P 4.12 A 470 $\Omega \pm 5\%$ resistor is connected in parallel with a 121 k$\Omega \pm 1\%$. Give the nominal resistance and the tolerance of the parallel connection.

P 4.13 The tolerance of a parallel connection of two resistors R_1, R_2 must be $\pm 2\%$ or better. What tolerance is required of each resistor?

P 4.14 A trimmer resistor is in series with a 91 k$\Omega \pm 5\%$ fixed resistor. The range of the trimmer must be such that the total (series) resistance can be set to 100 kΩ. What is the required resistance of the trimmer?

P 4.15 Two resistors having 25°C resistances R_1, R_2 and 25°C temperature coefficients a_1, a_2 are connected in series. (a) Obtain an expression for the 25°C temperature coefficient of the equivalent resistance. (It is a function of the individual nominal resistances and temperature coefficients.) (b) What is the temperature coefficient if the two resistors have equal nominal resistances and temperature coefficients? (c) If they have equal temperature coefficients but different nominal resistances? (d) Equal nominal resistances but different temperature coefficients?

P 4.16 Two resistors having 25°C resistances $R_1 = 1\,\text{k}\Omega$, $R_2 = 2.2\,\text{k}\Omega$ and 25°C temperature coefficients $\alpha_1 = 0.001\,\text{K}^{-1}$, and $\alpha_2 = 0.0005\,\text{K}^{-1}$ are connected in series. (a) Obtain a symbolic expression for the equivalent resistance as a function of the temperature difference $\Delta T = T - T_0$, where $T_0 = 25°\text{C}$. (b) Calculate the equivalent resistance for $T = -50°\text{C}$ and $T = 100°\text{C}$, and use the two data points to obtain a linear (straight-line) approximation to the equivalent resistance as a function of temperature. (c) What is the theoretical error in the linear expression obtained in Part (b)? Calculate the actual error at the endpoints of the range $-50°\text{C} \le T \le 100°\text{C}$.

P 4.17 Two resistors having 25°C resistances $R_1 = 100\,\text{k}\Omega$, $R_2 = 220\,\text{k}\Omega$ and 25°C temperature coefficients $\alpha_1 = 0.003\,\text{K}^{-1}$, $\alpha_2 = 0.004\,\text{K}^{-1}$ are connected in parallel. (a) Obtain a symbolic expression for the equivalent resistance as a function of the temperature difference $\Delta T = T - T_0$, where $T_0 = 25°\text{C}$. (b) Calculate the equivalent resistance for $T = -50°\text{C}$ and $T = 100°\text{C}$, and use the two data points to obtain a linear (straight-line) approximation to the equivalent resistance as a function of temperature. (c) Construct graphs of the true equivalent resistance and the linear approximation on the same axes, for $-50°\text{C} \le T \le 100°\text{C}$. (d) Calculate and plot (on another pair of axes) the percent error in the approximation over the same range.

P 4.18 Obtain an expression for the equivalent resistance at the terminals *a–b* for each circuit shown in Fig. P 4.10.

P 4.19 Obtain expressions for the resistors R_1, R_2, R_3 in terms of R_a, R_b, R_c such that the circuits in Fig. P 4.11 are equivalent at each pair of terminals (*w–x*, *y–z*, *w–y*).

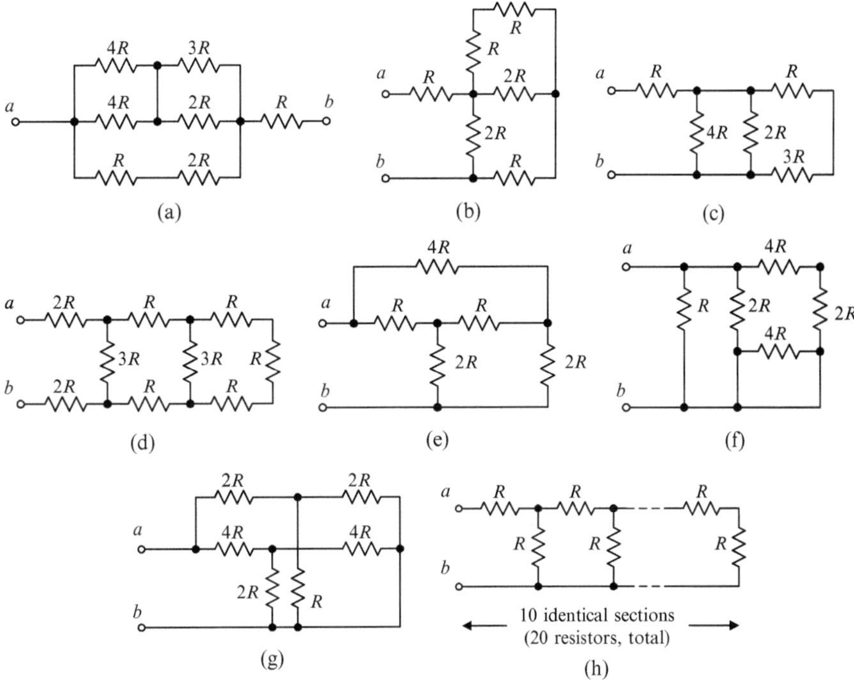

Fig. P 4.10 See Problem P 4.18

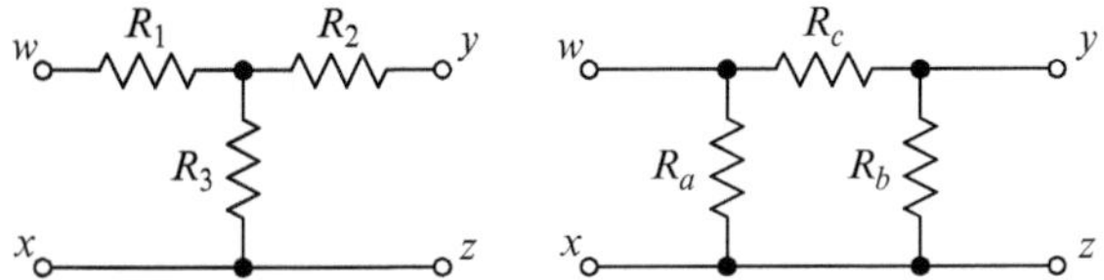

Fig. P 4.11 See Problem P 4.19, 20

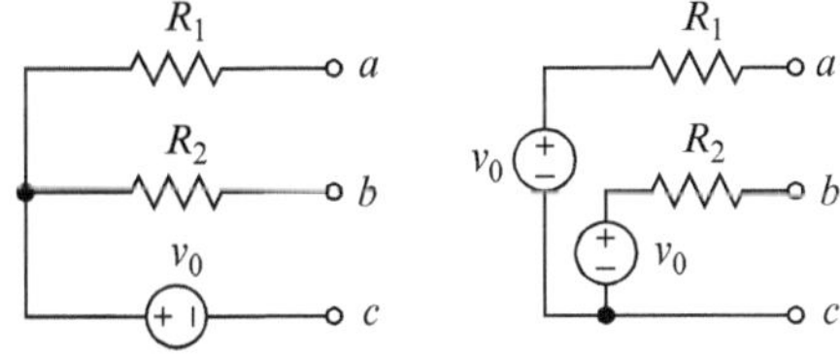

Fig. P 4.12 See Problem P 4.21

P 4.20 Obtain expressions for the resistors R_a, R_b, R_c in terms of R_1, R_2, R_3, such that the circuits of Fig. P 4.11 are equivalent at each pair of terminals (w–x, y–z, w–y).

P 4.21 Show that the circuits shown in Fig. P 4.12 are equivalent at each of the three possible terminal pairs.

P 4.22. Show that the circuits shown in Fig. P 4.13 are equivalent at each of the three possible terminal pairs.

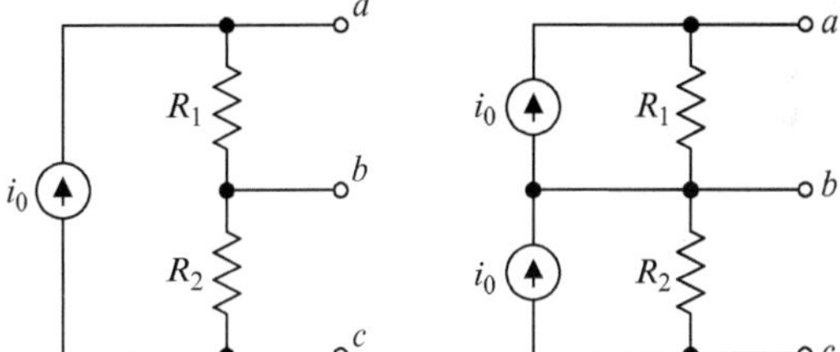

Fig. P 4.13 See Problem P 4.22

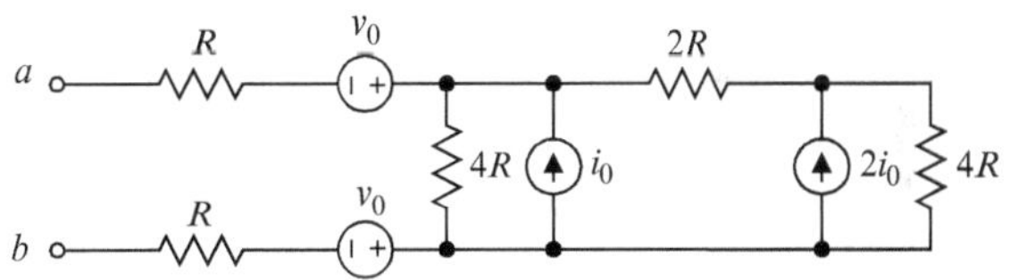

Fig. P 4.14 See Problem P 4.23

Section 4.3 is prerequisite for the following problems.

P 4.23 See Fig. P 4.14. (a) Use source transformations to obtain a circuit consisting of a voltage source in series with a resistance that is equivalent to the original circuit at the terminals *a–b*. Express the source voltage and series resistance as functions of the original circuit parameters. (b) Let $i_0 = 5\,\text{mA}$, $v_0 = 10\,\text{V}$, and $R = 1\,\text{k}\Omega$. Construct a graph of the terminal characteristic for the reduced equivalent circuit, regarding the circuit as a source. (c) Simulate the equivalent circuit with a load $R_L = 2.2\,\text{k}\Omega$ attached to the terminals *a–b* and verify that the load current and load voltage are consistent with the terminal characteristic obtained above. (d) Simulate the original circuit using the same load and verify that the load current and load voltage equal those obtained in part (c) (from the equivalent circuit).

P 4.24 🖳 Repeat Problem P 4.23 for the circuit shown in Fig. P 4.15.

P 4.25 🖳 Repeat Problem P 4.23 for the circuit shown in Fig. P 4.16.

P 4.26 🖳 Repeat Problem P 4.23 for the circuit shown in Fig. P 4.17.

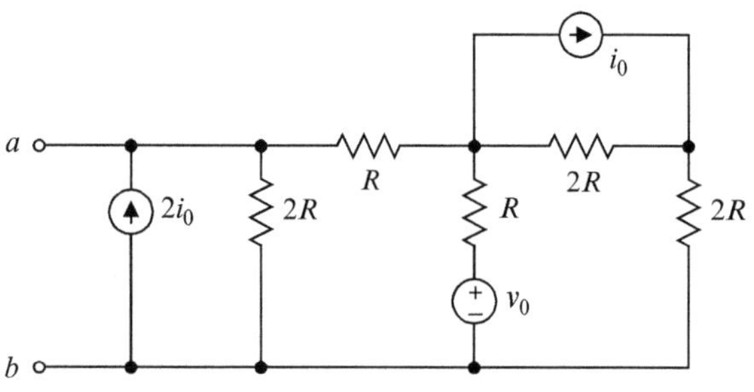

Fig. P 4.15 See Problem P 4.24

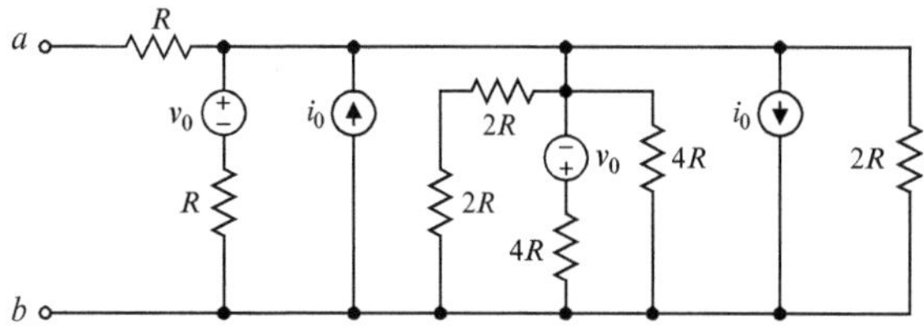

Fig. P 4.16 See Problem P 4.25

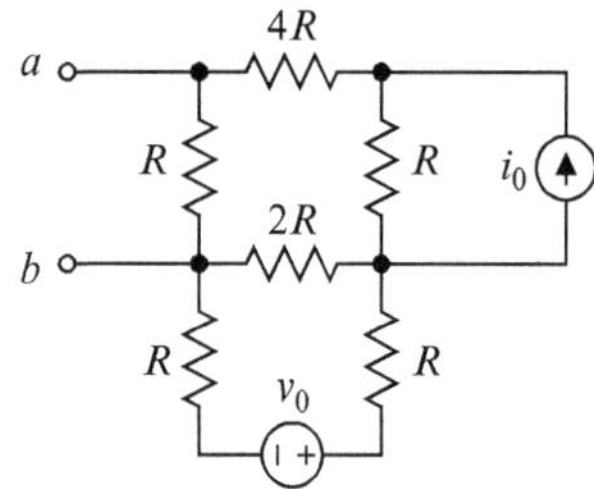

Fig. P 4.17 See Problem P 4.26

P 4.27 Repeat Problem P 4.23 for the circuit shown in Fig. P 4.18.

P 4.28 Repeat Problem P 4.23 for the circuit shown in Fig. P 4.19.

P 4.29 For each circuit of Fig. P 4.20, use source transformations to obtain an expression for the voltage v_x. Check your answers using node or mesh analysis.

Section 4.4 is prerequisite for the following problems.

P 4.30 🖳 In Fig. P 4.21, $v_0 = 5\text{V}$, $v_1 = 10\text{V}$, $i_0 = 5\text{mA}$, $i_1 = 10\text{ mA}$, $R_0 = 1\,\text{k}\Omega$, and $R_1 = 2.2\,\text{k}\Omega$. (a) Use simulation to obtain the open-circuit voltage and short-circuit current at the terminals *a–b*. (b) Simulate the original circuit, the Thévenin equivalent, and the Norton equivalent with a load $R_L = 2.7\,\text{k}\Omega$ attached to the terminals *a–b* and show that the load current (or voltage) is the same for all three cases (allowing for small computational errors).

P 4.31 🖳 See Fig. P 4.22, where $i_0 = 5\text{mA}$, $v_0 = 10\text{V}$, $R_0 = 1\,\text{k}\Omega$, and $R_2 = 1.5\,\text{k}\Omega$. (a) Use simulation to obtain the Thévenin equivalent for the source at the terminals to which the load is connected. (b) Draw a graph of the terminal characteristic for the source. (c) Use the Thévenin equivalent with the load attached and plot the load current i_L as a function of the load resistance R_L, for $0\,\Omega \leq R_L \leq 10\,\text{k}\Omega$.

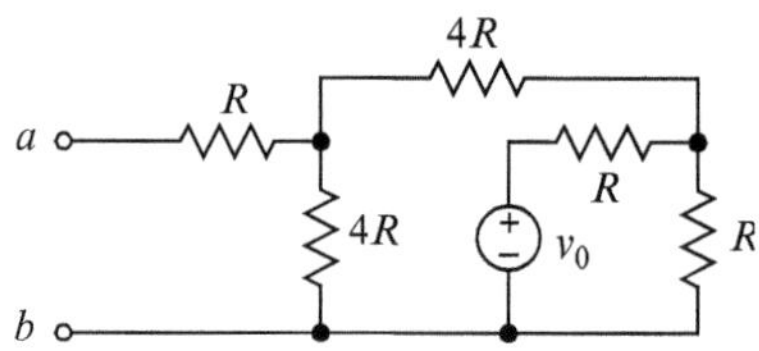

Fig. P 4.18 See Problem P 4.27

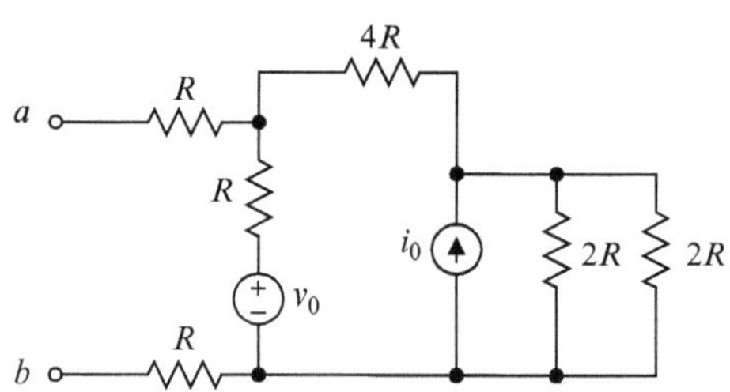

Fig. P 4.19 See Problem P 4.28

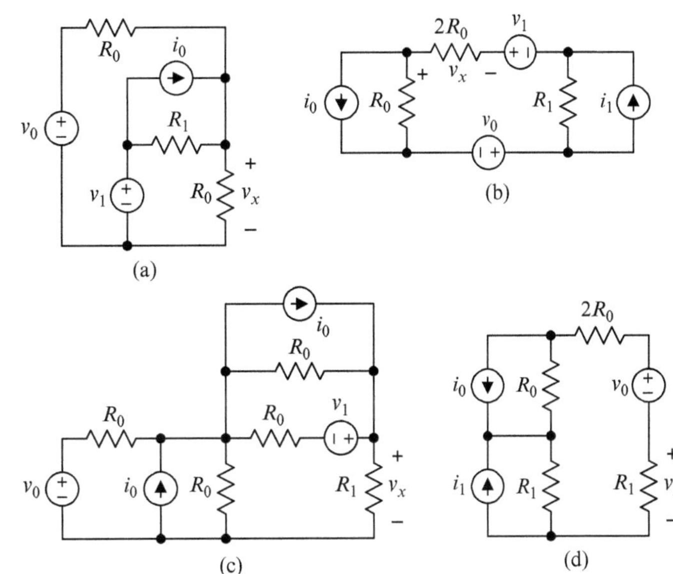

Fig. P 4.20 See Problem P 4.29

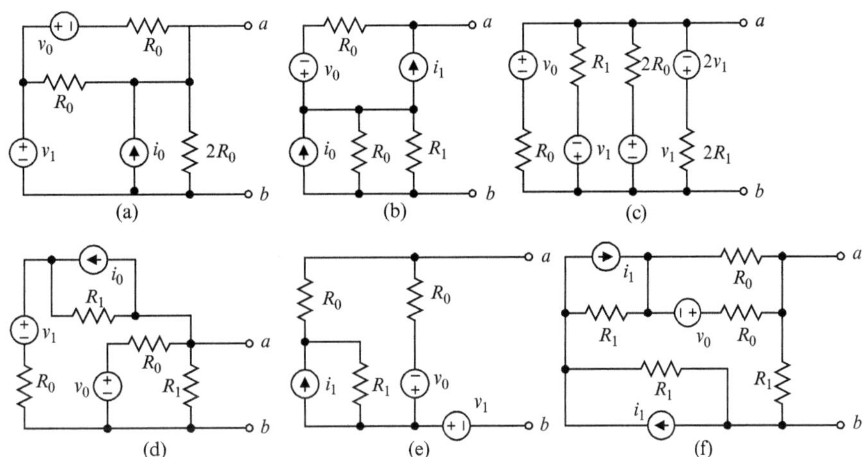

Fig. P 4.21 See Problem P 4.30

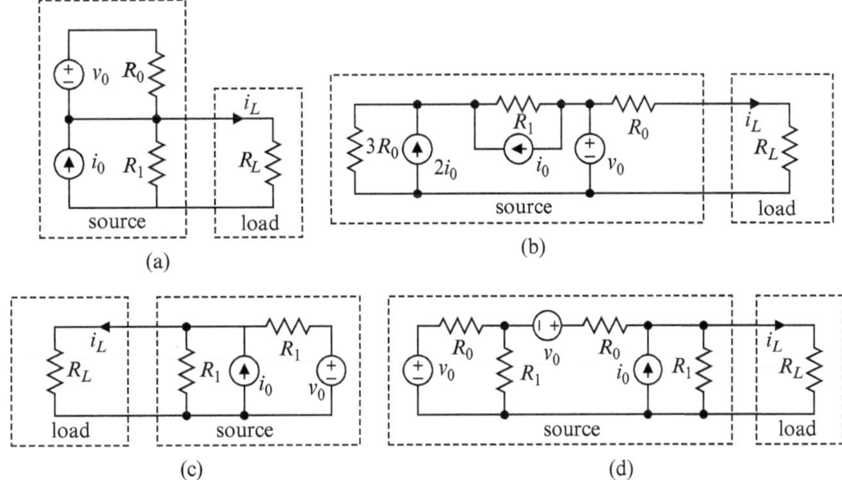

Fig. P 4.22 See Problem P 4.31

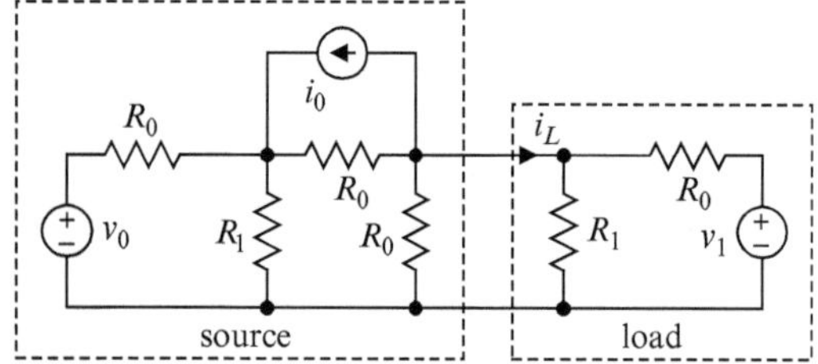

Fig. P 4.23 See Problem P 4.32

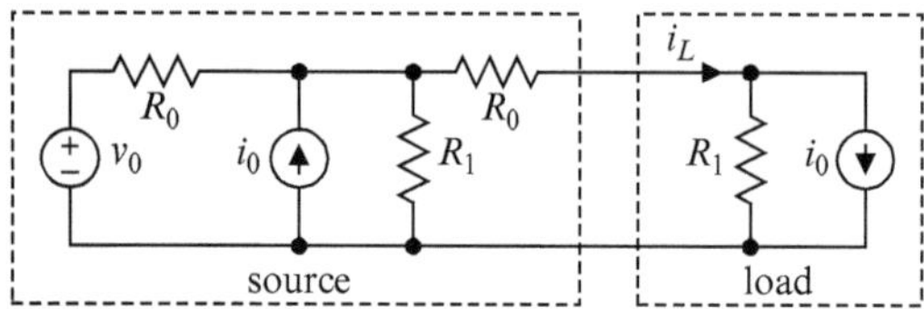

Fig. P 4.24 See Problem P 4.33

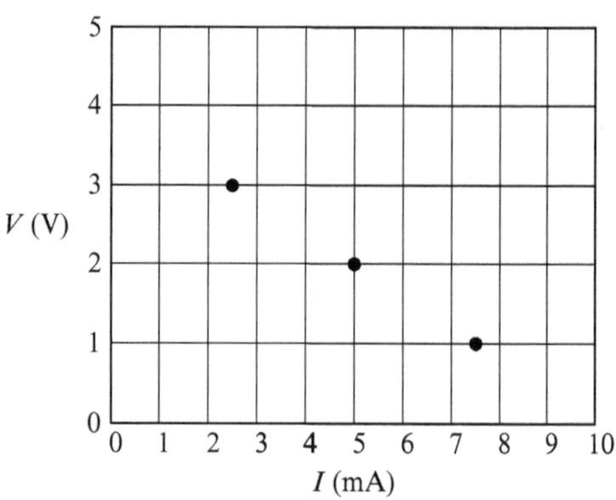

Fig. P 4.25 See Problem P 4.34

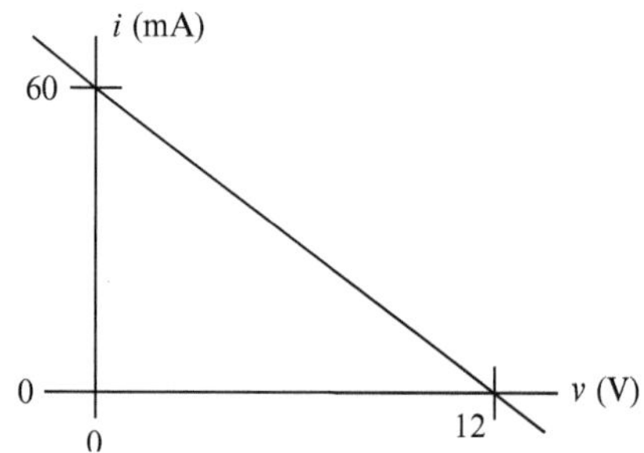

Fig. P 4.26 See Problem P 4.35

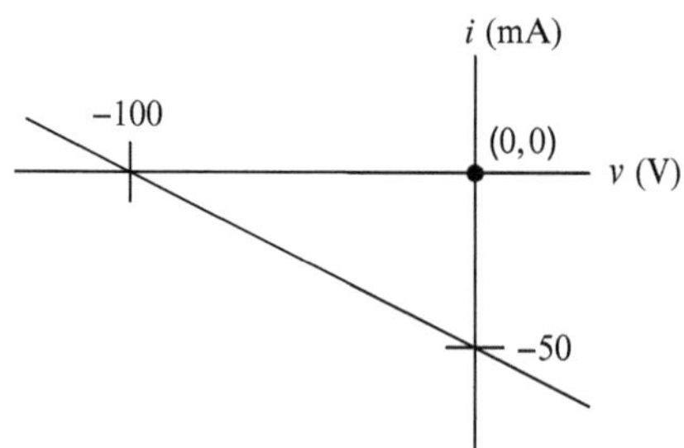

Fig. P 4.27 See Problem P 4.36

P 4.32 In Fig. P 4.23, $i_0 = 5\text{mA}$, $v_0 = 10\text{V}$, $v_1 = 12\text{V}$, $R_0 = 1\,\text{k}\Omega$, and $R_1 = 2.2\,\text{k}\Omega$. (a) Use a method of your choice to obtain the terminal characteristics of the source and load. (b) Plot the terminal characteristics of the source and load on the same axes, and find (graphically) the load current i_L that results when the source and load are connected. (c) Using a method of your choice, but different from that used in part (a), find the load current i_L when the load is connected to the source. Compare the result with that obtained in part (b) and explain any differences.

P 4.33 Repeat Problem P 4.32 for the source and load shown in Fig. P 4.24.

P 4.34. Measurements of voltage and current are made at terminals *a–b* of a resistive circuit for three different resistive loads. Figure P 4.25 shows a graph of the measured values. Find the Thévenin and Norton equivalents for the circuit at the terminals *a–b*.

P 4.35. Figure P 4.26 shows the terminal characteristic for a certain two-terminal circuit. The positive direction for the terminal current is out of the positive terminal of the circuit.

(a) Find the Thévenin and Norton equivalents for the circuit.
(b) A 400 Ω resistor is connected to the terminals of the circuit. Find the voltage across and current through the resistor.
(c) The 400 Ω resistor is removed and a 15V independent source is connected to the terminals of the circuit, with the positive terminal of the source connected to the positive terminal of the circuit. Find the current out of the positive terminal of the 15 V source.

P 4.36 Repeat Problem P 4.35 for the graph in Fig. P 4.27.

P 4.37 Use a method specified by your instructor to find the Thévenin and Norton equivalent circuits at the terminals *a–b* for each circuit in Fig. P 4.28, where $i_0 = 5\text{mA}$, $v_0 = 5\text{V}$, and $R = 1\text{k}\Omega$.

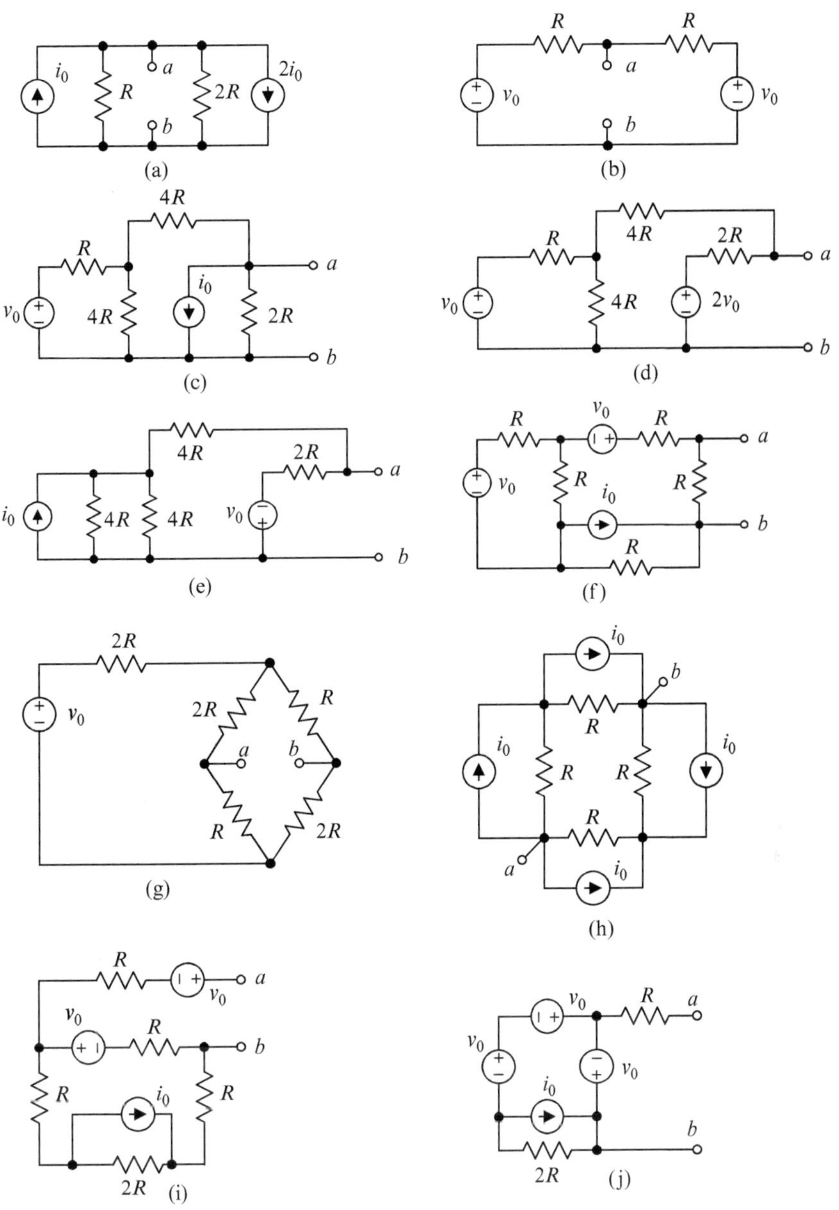

Fig. P 4.28 See Problem P 4.37

P 4.38 N identical Norton source models are connected in series, as in Fig. P 4.29. Find the (single) Norton equivalent at the terminals a–b.

P 4.39 N identical Thévenin source models are connected in parallel, as in Fig. P 4.30. Find the (single) Thévenin equivalent at the terminals a–b.

P 4.40 Refer to Fig. P 4.31, where the resistance R is known and fixed and the voltage v_0 is variable. It is found that for $v_0 = v_1$, $v_R = v_a$ and that for $v_0 = v_2$, $v_R = v_b$. Obtain an expression for the voltage v_R as a function of the voltage v_0 and in terms of the known quantities v_1, v_2, v_a, v_b, R.

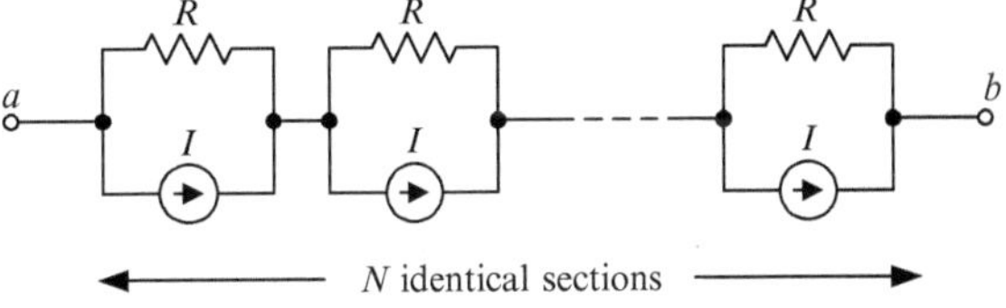

Fig. P 4.29 See Problem P 4.38

P 4.41 Refer to Fig. P 4.31, where the voltage v_0 is known and fixed and the resistance R is variable. It is found that for $R = R_1$, $v_R = v_a$ and that for $R = R_2$, $v_R = v_b$. Obtain an expression for the voltage v_R as a

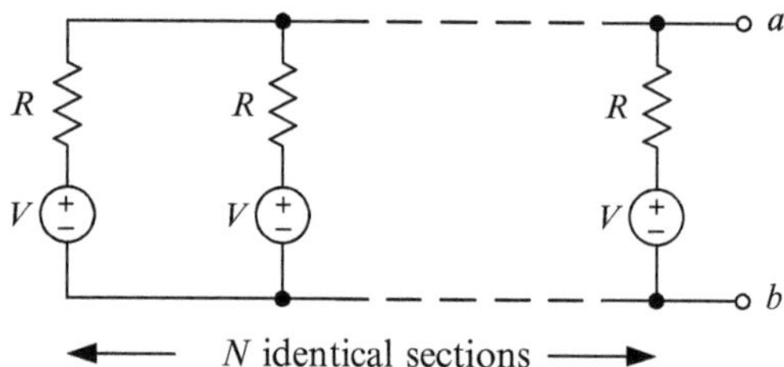

Fig. P 4.30 See Problem P 4.39

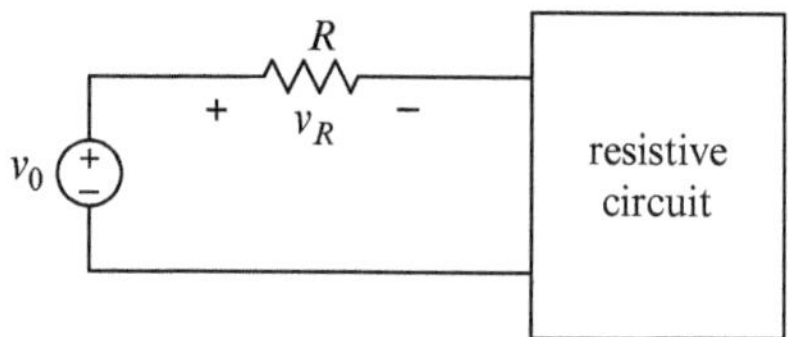

Fig. P 4.31 See Problem P 4.40, 41

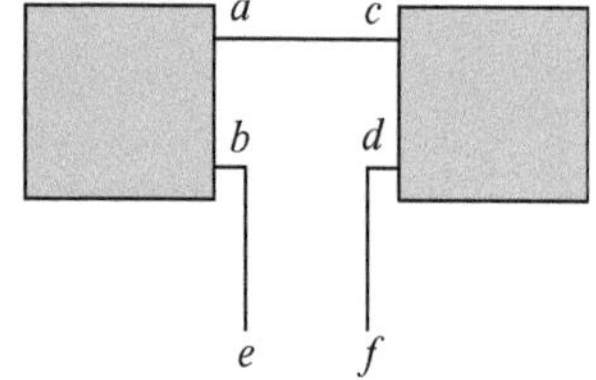

Fig. P 4.32 See Problem P 4.42

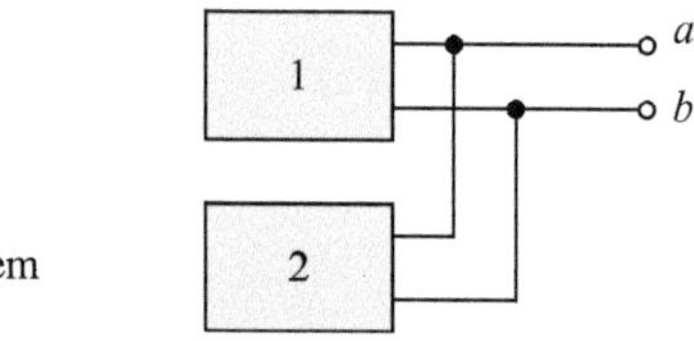

Fig. P 4.33 See Problem P 4.43, 44

function of the resistance R and in terms of the known quantities R_1, R_2, v_a, v_b, v_0.

P 4.42 Refer to Fig. P 4.32. Let v_{T1}, R_{T1} and v_{T2}, R_{T2} denote the Thévenin voltage and resistance at the terminals *a–b* and *c–d*, respectively, where *a* and *c* are the positive terminals. Find the Thévenin equivalent at the terminals *e–f*.

P 4.43 Refer to Fig. P 4.33, where the boxes enclose resistive circuits. Let v_{T1}, R_{T1} and v_{T2}, R_{T2} denote the Thévenin voltage and resistance at the terminals *a–b* for circuits 1 and 2, respectively, where *a* is the positive terminal. Obtain expressions for the Thévenin equivalent voltage and resistance at the terminals *a–b* in terms of v_{T1}, R_{T1} and v_{T2}, R_{T2}.

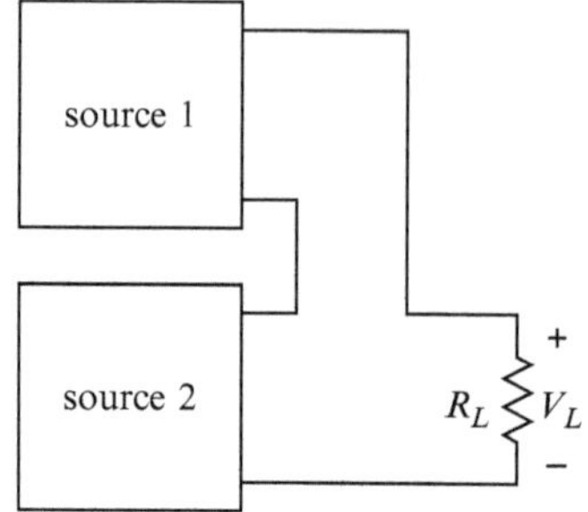

Fig. P 4.34 See Problem P 4.45, 46

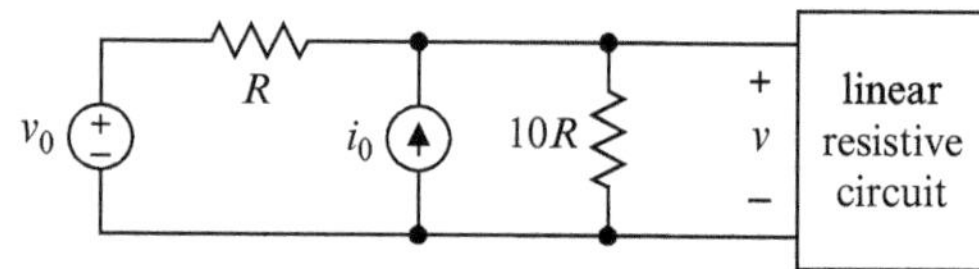

Fig. P 4.35 See Problem P 4.48

P 4.44 Refer to Fig. P 4.33, where the boxes enclose resistive circuits. Let v_{T1}, R_{T1} denote the Thévenin voltage and resistance for circuit 1. Let i_N, R_N denote the Norton current and resistance at the terminals *a–b*. Obtain expressions for the Thévenin equivalent voltage and resistance for circuit 2 in terms of v_{T1}, R_{T1} and i_N, R_N.

P 4.45 Two sources having the same open-circuit voltage $V_{OC} = 10\,\text{V}$ are connected in series, and then to a load $R_L = 200\,\Omega$, as shown in Fig. P 4.34. It is found that $V_L = 5\,\text{V}$. Obtain the Thévenin equivalent for the series connection of the sources.

P 4.46 Two sources having the same short-circuit current $I_{SC} = 100\,\text{mA}$ are connected in series, and then to a load $R_L = 500\,\Omega$, as shown in Fig. P 4.34. It is found that $V_L = 15\,\text{V}$. Obtain the Thévenin equivalent for the *parallel* connection of the sources.

P 4.47 In essence, Problem P 4.2 asks for the terminal characteristics of the circuits in Fig. P 4.3, treating the circuits as sources. Using results of that problem, obtain the terminal characteristic at *a–b* for each circuit, treating each as a load.

P 4.48 Refer to Fig. P 4.35, where $R = 2\,\Omega$ and the sources v_0, i_0 are independently variable. It is found that for $v_0 = 10\text{V}$ and $i_0 = 2\text{A}$, the voltage $v = 12\text{V}$, and that for $v_0 = 20\text{V}$ and $i_0 = 6\text{A}$, the voltage $v = 24\text{V}$. Find the voltage v if $v_0 = 30\text{V}$ and $i_0 = 10\text{A}$.

P 4.49 💻 Refer to Fig. P 4.36, where the boxes enclose resistive circuits. The Thévenin voltage and resistance, respectively, at the terminals *a–b* for the circuit in box 1 are $v_{T1} = 5\text{V}$, $R_{T1} = 2\text{k}\Omega$, where *a* is the positive terminal. A known resistance *R* is connected as shown, and the voltage *v* is measured. It is found that for $R = 1\,\text{k}\Omega$, $v = 1.143\text{V}$ and that for $R = 2\,\text{k}\Omega$, $v = 1.600\,\text{V}$. Find the Thévenin equivalent for the circuit in box 2. Use simulation to verify your answer.

P 4.50 A current source $i_0 = 5\,\text{mA}$ is connected to terminals *a–b* of a resistive circuit, with the positive direction of current into terminal *a*. It is found that $v_{ab} = 10\text{V}$. Then the current source is removed and it is found that $v_{ab} = 5\text{V}$. Find the Thévenin equivalent for the circuit at the terminals *a–b*.

P 4.51 A voltage source $v_0 = 5\text{V}$ is connected to terminals *a–b* of a resistive circuit, with the positive terminal connected to circuit terminal *a*. It is found that the current into the positive terminal of the source is $i = 10\,\text{mA}$. Then the voltage source is removed, a 2 kΩ resistor is connected to the terminals *a–b*, and the voltage across the resistor is found to be 2 V. Find the Thévenin equivalent for the circuit at the terminals *a–b*.

P 4.52 A current source $i_0 = 5\text{mA}$ is connected to terminals *a–b* of a resistive circuit, with the positive direction of current into terminal *a*. It is found that $v_{ab} = 10\text{V}$. Then a 2 kΩ resistor is connected in parallel with the current source and it is found that $v_{ab} = 5\text{V}$. Find the Thévenin equivalent for the circuit at the terminals *a–b*.

P 4.53 A voltage source $v_0 = 5\text{V}$ is connected to terminals *a–b* of a resistive circuit, with the positive terminal connected to circuit terminal *a*. It is found that the current into the positive terminal of the source is $i = 10\,\text{mA}$. Then a 2 kΩ resistor is connected in series with the voltage source, the series connection is connected to the terminals *a–b*, and it is found that $v_{ab} = 2\,\text{V}$. Find the Thévenin equivalent for the circuit at the terminals *a–b*. Use simulation to verify your answer.

P 4.54 A variable resistor is connected to terminals *a–b* of a resistive circuit. Neither terminal is directly connected (inside the circuit) to an independent source. Does the current through the resistor increase or decrease as the resistance is decreased? Use Thévenin's theorem to justify your answer.

P 4.55 A variable resistor is connected to a terminal pair of a resistive circuit, as shown in Fig. P 4.37. The current through the resistor is measured and plotted versus the resistance *R*. The resulting plot also is shown in Fig. P 4.37. Find the Norton equivalent for the circuit at the terminals *a–b*.

P 4.56 A voltage source is connected to the-terminals of a resistive circuit, as shown in Fig. P 4.38. The Thévenin voltage and resistance for the circuit are $v_T = 20\text{V}$ and $R_T = 10\,\text{k}\Omega$, respectively. Construct a graph of the current *i* through the source versus the source voltage v_0 for $-10\text{V} \le v_0 \le 30\text{V}$. Show how you would obtain the Thévenin equivalent for the circuit from the graph if v_T and R_T were unknown.

P 4.57 Refer to Fig. P 4.39. The circuits enclosed in the boxes are resistive and are equivalent at the

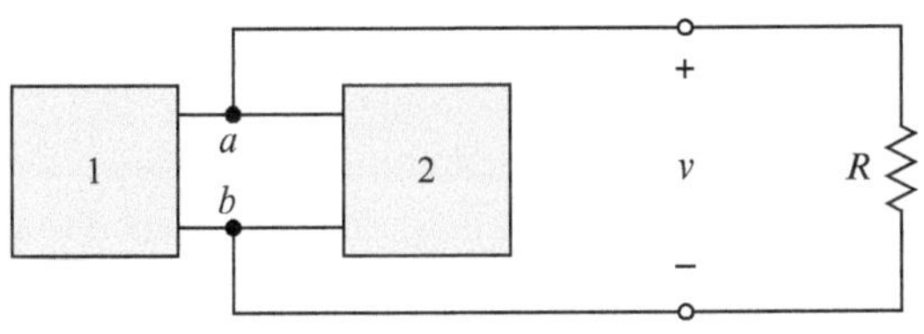

Fig. P 4.36 See Problem P 4.49

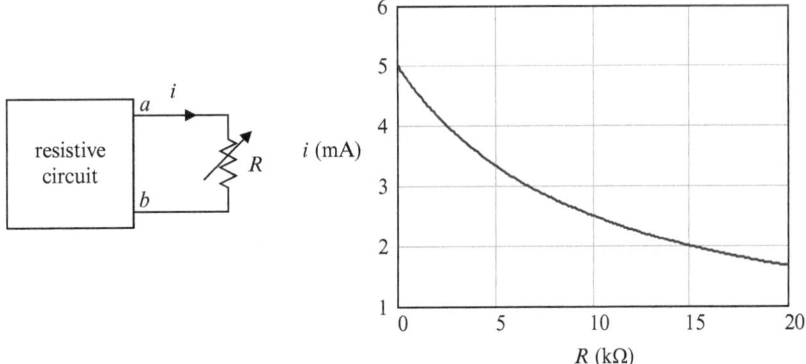

Fig. P 4.37 See Problem P 4.55

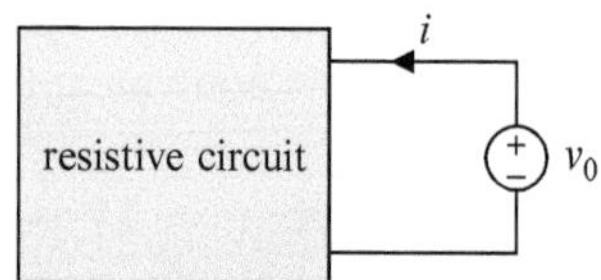

Fig. P 4.38 See Problem P 4.56

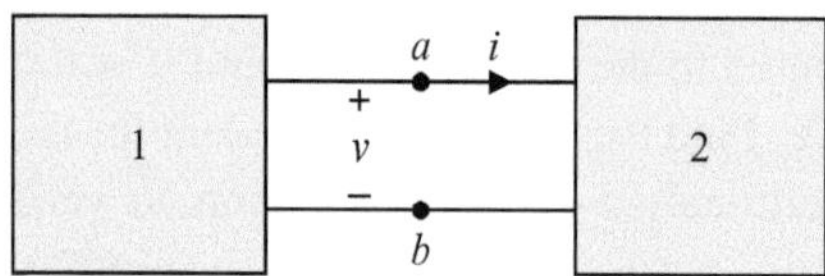

Fig. P 4.39 See Problem P 4.57

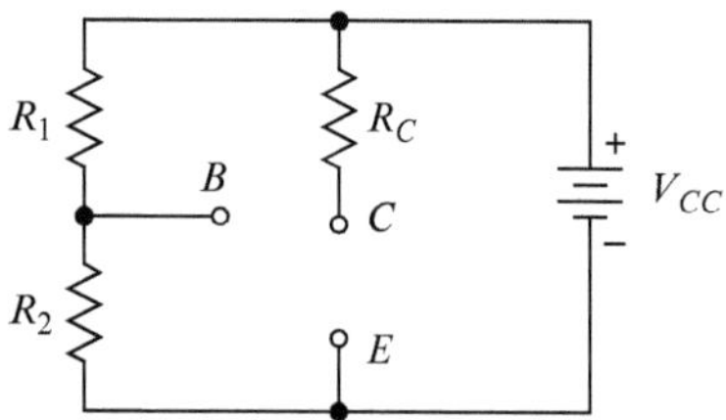

Fig. P 4.40 See Problem P 4.58

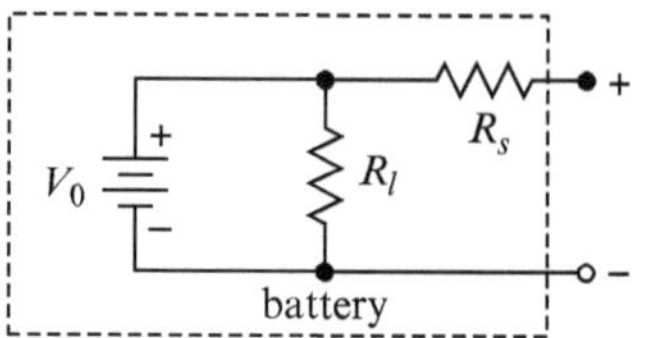

Fig. P 4.41 See Problem P 4.62

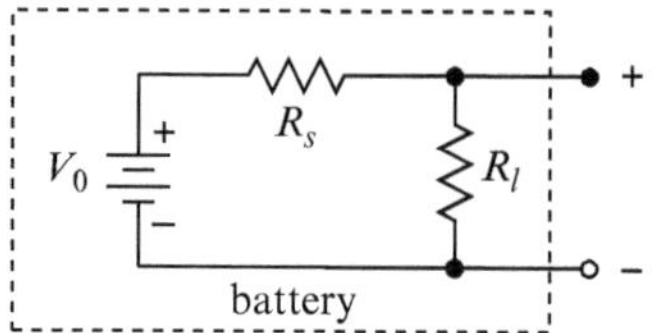

Fig. P 4.42 See Problem P 4.64

terminals *a–b*, each having open-circuit voltage $v_{OC} = 50\,\text{mV}$. Find the voltage v and the current i.

P 4.58 Refer to Fig. P 4.40. Find the Thévenin equivalent circuit at (a) the terminals *C–E* and (b) the terminals *B–E*. (c) Draw graphs of the two terminal characteristics.

P 4.59 Suppose we agree that a source represented by a Thévenin equivalent can be considered a good voltage source if the load voltage varies by no more than 1% (of the Thévenin voltage) over the range of expected values of load resistance. If a certain source model has Thévenin resistance $R_T = 1\Omega$, for what values of load resistance can the source be considered a good voltage source?

P 4.60 Suppose we agree that a source represented by a Norton equivalent can be considered a good current source if the load current varies by no more than 1% (of the Norton current) over the range of expected values of load resistance. If a certain source model has Norton resistance $R_N = 10\,\text{k}\Omega$, for what values of load resistance can the source be considered a good current source?

P 4.61 Imagine you are presented with a resistive circuit sealed in a box, but with access to a terminal pair. You also have access to a full complement of laboratory instruments. Describe fully all of the methods you can think of for finding the Thévenin equivalent for the circuit at the exposed terminals.

P 4.62 An automotive lead-acid battery can be modeled as shown in Fig. P 4.41. The *leakage resistance* R_ℓ accounts for the fact that the battery will eventually *self-discharge*, even if not connected to a load. The series resistance R_s accounts (approximately) for the fact that the terminal voltage decreases as load current increases. Obtain the Norton equivalent for the battery at the terminals. What happened to the leakage resistance? In what sense is the Norton model equivalent to the battery?

P 4.63 Repeat Problem P 4.62 using a Thévenin equivalent for the battery.

P 4.64 Figure P 4.42 shows another model for the battery described in Problem P 4.62. (a) Obtain the Norton equivalent for this model. (b) Obtain the Thévenin equivalent for this model. (c) What kinds of studies *cannot* be performed using the Thévenin or Norton models for the battery? What kinds can be performed? (d) Under what condition are the models in Figs. P 4.42 and P 4.41 equivalent at their terminals?

Chapter 5
Work and Power

Work is an amount of energy converted from one form to another and **power** is the rate of such conversion. The **instantaneous power** $p(t)$ corresponding to *instantaneous work* $w(t)$ is given by

$$p = \frac{dw}{dt}. \tag{5.1}$$

Brightness of a light bulb or computer display, loudness of music from a loudspeaker, and time required for a pump to drain a flooded basement all depend upon *rate* of energy conversion. Consequently, electrical engineers usually are more interested in the rate at which work is done (power) than work itself. Exceptions are design of storage batteries, where energy stored per unit volume and per unit mass are prime concerns, and in management of electric-utility companies, which sell energy. But even in those cases, engineers are interested in the rate at which stored energy can be drawn from a battery and the rate at which energy can be delivered to a customer.

Power is of concern in circuit analysis and design for various reasons, including:

- An electric circuit must do work *at a specified rate* on a load, such as a light bulb, a loudspeaker, a transmitting antenna, or an electric motor. Each component of the circuit must be capable of doing the work required of the component at the specified rate.
- The power supply for a circuit must be able to provide energy to the circuit at a rate sufficient to support operation of the circuit. That is, the power supply must be able to provide power equal to or greater than the total power dissipated (converted to another form) by the circuit.
- Work done on circuit components generates heat, which must be removed quickly enough to keep the temperatures of the components from rising above acceptable values. The temperature of a dissipative device will rise until the rate at which heat is removed equals the rate at which the heat is produced. This equilibrium condition must be reached at a temperature the device can endure. As a rule, the lifetime of a device decreases as the equilibrium operating temperature increases. Also, because resistivity generally increases with increasing temperature, performance usually decreases with increasing temperature. Consequently, it usually is desirable to keep the equilibrium temperature of a device below the maximum permissible value.
- Most circuit components, as a by-product of normal operation, generate electrical noise (e.g., thermal noise in resistors) that contaminates other voltages and currents of interest. Electrical noise produced by most components increases with increasing temperature, which, in turn, usually increases with the rate at which work is done on or by the component.
- Many circuits are described at least in part by *power gain* (or loss), which is the ratio of the power delivered by a circuit to a load to the power provided to the circuit by a source. Power gain (or loss) is an especially important parameter in communication systems.

In electrical engineering, there is a rich vocabulary associated with work and power. Practicing engineers refer to power being absorbed, consumed, dissipated, generated, produced, delivered, transmitted or distributed. Strictly speaking, it is energy, not power, that is absorbed, generated, distributed, or the like. Instantaneous power is the *rate* at which *energy* is being

T.H. Glisson, *Introduction to Circuit Analysis and Design*,
DOI 10.1007/978-90-481-9443-8_5, © Springer Science+Business Media B.V. 2011

converted or transmitted *at a particular time and place*. Power does not go anywhere and power is not generated or dissipated. Nonetheless, practicing engineers understand what is meant by such terminology, and the terminology is useful because it makes descriptions and discussions much more compact than they would be otherwise.

In this chapter, we obtain expressions for electrical work and power as functions of voltage, current, and circuit parameters.

5.1 Instantaneous Power and the Passive Sign Convention

Instantaneous power, denoted by p, is the instantaneous rate at which energy is converted from one form to another. Equivalently, instantaneous power is the rate at which the associated work is done. Instantaneous power corresponding to instantaneous work w is given by (5.1), repeated below for ready reference.

$$p = \frac{dw}{dt}. \tag{5.2}$$

The SI unit of power is the *watt* (W), where $1\,\mathrm{W} = 1\,\mathrm{J\,s^{-1}}$. Instantaneous power is (in general) a function of time.

We are interested in work and power in the context of circuit analysis, so we wish to express instantaneous power in terms of voltage and current. Because voltage is work done (on an electric field) by a charge per unit charge, we may write

$$v = \frac{dw}{dq}, \tag{5.3}$$

where dq is a small quantity of positive charge and dw is the work done *by* the charge. It follows that

$$dw = v\,dq. \tag{5.4}$$

Using the right side of (5.4) for dw in (5.2) gives

$$p = v\frac{dq}{dt} = v\,i. \tag{5.5}$$

In (5.5), v and i denote voltage across and current through a terminal pair of a device or circuit. The power p given by (5.5) is called the **instantaneous power dissipated** by the device or circuit (at the terminal pair in question).

Above, we define positive work as work done *by* a charge dq. Because energy is conserved, we could as well define positive work as work done *on* the charge, but defining positive work as work done *by* the charge is consistent with the **passive sign convention** used in almost all circuits textbooks, according to which power is positive if the positive direction of current in (5.5) is in the direction of a voltage drop; i.e., into the positive terminal of a device or circuit element under consideration. Thus positive power means that positive work is being done *by* the current (the device or circuit is consuming energy) and negative power means that work is being done *on* the current (the device or circuit is producing energy). Figure 5.1 illustrates the definition of power dissipated according to the passive sign convention.

Generally, we follow the passive sign convention, to the extent that power *dissipated* by an element, device, or circuit means power dissipated in accord with the passive sign convention. But in many cases, it is more convenient (and meaningful) to refer to power *delivered* by an element, device, or circuit, where power delivered is the negative of that determined using the passive sign convention. Specifically, for a resistor, we almost always compute power dissipated (consumed), but for a source, we frequently compute power delivered to a load. For example, for the circuit shown in Fig. 5.2, the instantaneous power *dissipated*

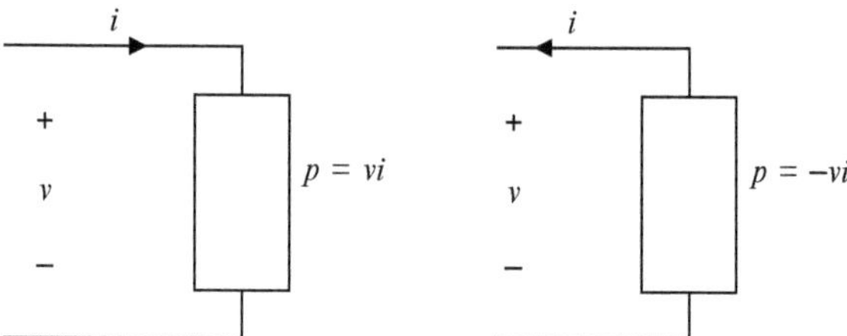

Fig. 5.1 Passive sign convention: The instantaneous power dissipated at a terminal pair is the product of the instantaneous voltage across the terminals and the instantaneous current *into* the positive terminal

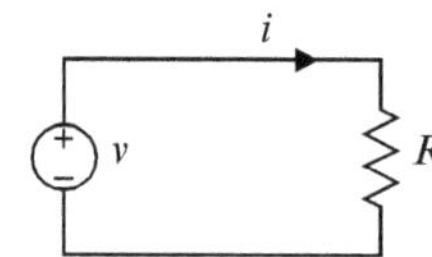

Fig. 5.2 Power *delivered* by the source is *dissipated* by the resistor (see text)

by the resistor and the power *delivered* by the source are both given by $p = v\,i$, where the positive direction for current is out of the positive terminal of the source and into the positive terminal of the resistor.

Example 5.1. The voltage across a certain electrical device and the current entering the positive terminal of the device are given by

$$v(t) = V_0 \cos(\omega t), \quad i(t) = I_0 \cos(\omega t),$$

respectively. Obtain an expression for the instantaneous power dissipated by the device.
Solution: The positive direction for current is consistent with the passive sign convention. From (5.5), the instantaneous power dissipated is given by

$$p = v\,i = V_0 I_0 \cos^2(\omega t) = \frac{V_0 I_0}{2}\,[1 + \cos(2\,\omega t)],$$

where we have used the identity

$$\cos^2(\alpha) \equiv \frac{1}{2}\,[1 + \cos(2\,\alpha)].$$

Exercise 5.1. The voltage across a certain electrical device and the current entering the positive terminal of the device are given by

$$v(t) = V_0 \cos(\omega t), \quad i(t) = I_0 \cos(\omega t + \theta),$$

respectively. Show that the instantaneous power dissipated by the device is given by

$$p(t) = \frac{V_0 I_0}{2}[\cos(\theta) + \cos(2\omega t + \theta)].$$

Example 5.2. Obtain expressions for the instantaneous power dissipated by each component in the circuit shown in Fig. 5.3, where $I_0 > 0$ and $V_0 > 0$. Is the current source producing energy or consuming energy? Is the voltage source producing energy or consuming energy?

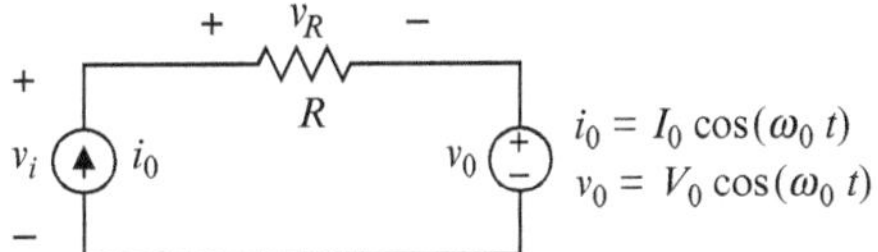

Fig. 5.3 See Example 5.2

Solution: From (5.5) and Ohm's law, the instantaneous power dissipated by the resistor is given by

$$p_R = v_R\, i_0 = (i_0 R)\, i_0 = R\, {i_0}^2 = R\, {I_0}^2 \cos^2(\omega_0\, t)$$

and is independent of the source voltage v_0. The power *dissipated* by the voltage source is given by

$$p_v = i_0 v_0 = I_0\, V_0 \cos^2(\omega t).$$

Because the positive direction for the current i_0 is into the positive terminal of the voltage source. The power *dissipated* by the current source is given by $p_i = -i_0\, v_i$ because the positive direction for the current i_0 is out of the positive terminal of the current source. Kirchhoff's voltage law gives

$$-v_i + v_R + v_0 = 0 \Rightarrow v_i = v_R + v_0 = i_0 R + v_0$$
$$= (I_0 R + V_0)\cos(\omega t).$$

Thus the power *dissipated* by the current source is given by

$$p_i = -i_0\, v_i = -I_0 (I_0 R + V_0) \cos^2(\omega t).$$

Because the quantities R, I_0, V_0, and $\cos^2(\omega t)$ are all positive, the power dissipated by the current source is negative and the power dissipated by the voltage source is positive. It follows that the current source is producing energy, whereas the voltage source is consuming energy.

Exercise 5.2. Obtain expressions for the instantaneous power dissipated by each component in the circuit shown in Fig. 5.4, where

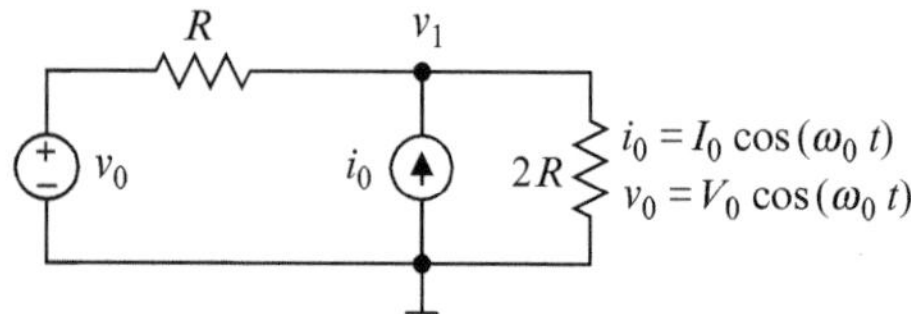

Fig. 5.4 See Exercise 5.2

$I_0 > 0$ and $V_0 > 0$. Is the current source producing energy or consuming energy? Show that the powers dissipated by the resistors are positive. Is the voltage source producing energy or consuming energy?

Examples and exercises above illustrate the fact that a current or voltage source in a circuit containing two or more sources can either dissipate or deliver power, depending upon the configuration of the circuit.[1] As a physical example, a battery connected to a light bulb *delivers* power to the bulb (the battery produces energy), whereas a battery connected to a battery charger *receives* energy from the charger (the battery consumes energy). The examples and exercises also illustrate the fact that the power dissipated by a physical resistor is always positive. If you obtain a negative value for the power dissipated by a physical resistor, you have made an error.[2]

According to the passive sign convention, power at the terminals of an electric heater, for example, is positive, whereas power at the terminals of an electric generator is negative. In practice, despite preferences of textbook authors, what is meant by positive power usually depends upon circumstances. In describing an electric heater, which converts electrical energy to heat energy, one might define positive power as the rate at which work is done on the heating element, which is consistent with the passive sign convention. On the other hand, in describing an electric generator, we are interested in the rate at which the generator can convert mechanical energy to electrical energy, and we might define positive power in opposition to the passive sign convention. No manufacturer of electric generators would advertise a "−5 kW generator."

Nonetheless, if applied consistently and interpreted correctly, the passive sign convention facilitates analysis. We do not need to determine whether an element is consuming or generating energy before choosing reference directions for voltage and current. We simply assign directions to currents and polarities to voltages and then use the passive sign convention (Fig. 5.1) for power calculations; that is, we use $p = v\,i$ if the positive direction for current is into the positive terminal of a device, and $p = -v\,i$ if the positive direction for current is out of the positive terminal of a device. If the power thus obtained is positive, then electrical energy is being converted to some other form (the device dissipates power). If the power turns out to be negative, then some other form of energy is being converted to electrical energy (the device generates power).

5.2 Instantaneous Power Dissipated by a Resistor: Joule's Law

Using Ohm's law $v = R\,i$ in (5.5) shows that instantaneous power dissipated by a resistor is given by

$$p = v\,i = R\,i^2 = \frac{v^2}{R}. \tag{5.6}$$

Because (physical) resistance is non-negative, the instantaneous power dissipated by a physical resistor is non-negative and equals zero only if the current through or voltage across the resistor equals zero.[3] Equation (5.6) is essentially an expression of *Joule's law*.

Integrating (5.6) over an interval $t_1 \le t < t_2$ gives the total work done on a resistor in the interval:

$$w = \int_{t_1}^{t_2} p(t)\,dt = R\int_{t_1}^{t_2} i^2(t)dt = \frac{1}{R}\int_{t_1}^{t_2} v^2(t)\,dt. \tag{5.7}$$

Electrical energy delivered to a resistor is converted to heat or other radiant energy (as by a light bulb).

[1]Also, a source can dissipate power during some time intervals and deliver power during others.

[2]Using electronic devices, it is possible to construct a two-terminal circuit that has the properties of a negative resistance; however, such a circuit is not a physical resistor.

[3]Certain two-terminal circuits are equivalent to a negative resistance at their terminals.

The radiant energy produced in an interval $t_1 \leq t < t_2$ is given by (5.7). The instantaneous *rate* at which the radiant energy is produced is given by (5.6).

Example 5.3. The voltage across a resistor having resistance $R = 1\,\mathrm{k\Omega}$ is $V = 10\,\mathrm{V}$.

Calculate the instantaneous power dissipated by the resistor, the current through the resistor, and the total work done on the resistor in 1 s.

Solution: From (5.6), the power dissipated is

$$p = \frac{V^2}{R} = \frac{(10\,\mathrm{V})^2}{10^3\,\Omega} = 0.100\,\mathrm{W} = 100\,\mathrm{mW}.$$

Because $p = i^2 R$, the current is

$$I = \sqrt{\frac{p}{R}} = \sqrt{\frac{0.100\,\mathrm{W}}{10^3\,\Omega}} = 0.010\,\mathrm{A} = 10\,\mathrm{mA}.$$

Check:

$$p = VI = (10\,\mathrm{V})(0.01\,\mathrm{A}) = 0.10\,\mathrm{W} = 100\,\mathrm{mW}.$$

Because the applied voltage and the resistance are independent of time, the instantaneous power and current also are independent of time.

From (5.7), the total work done in 1 s is

$$w = \int_{t_1}^{t_2} p(t)\,dt = \int_{0}^{1\,\mathrm{s}} (100\,\mathrm{mW})\,dt$$
$$= (100\,\mathrm{mW})(1\,\mathrm{s}) = 100\,\mathrm{mWs} = 100\,\mathrm{mJ}.$$

Exercise 5.3. The current through a resistor having resistance $R = 1\mathrm{k\Omega}$ is $I = 5\mathrm{mA}$.

Calculate the instantaneous power dissipated by the resistor, the voltage across the resistor, and the total work done on the resistor in 2 s.

Because a resistor converts electrical energy to radiant energy which is then lost forever from the circuit, *a resistor always dissipates power*. You might think *dissipates* carries a negative connotation, and sometimes it does; for example, the cost to an electric power company of energy lost in the resistance of power-transmission lines can be significant. But dissipation of electric power as heat or other radiant energy also can be useful. Everyday examples include light bulbs, electric ovens, electric heaters, electric ranges, and electric water heaters. Moreover, even a power company finds the resistance of its lines useful during winter, when heat caused by current in the lines can melt ice that otherwise might accumulate and break the lines.

Virtually all circuit components, and resistors in particular, generate heat while operating, which is why a television receiver, for example, is warmer when on than when off. Some devices become hot enough to burn skin when operating normally. But there are limits to the temperatures that resistors, transistors, and other electrical devices can endure and still operate as intended; for example, a typical off-the-shelf discrete bipolar junction transistor will operate normally at temperatures below about 75°C = 167°F. Bipolar transistors built to military specifications (*mil-spec* devices) will operate normally at somewhat higher temperatures, up to about 125°C = 257°F. Typically, circuits are designed such that components operate below the maximum permissible temperatures because operation at higher temperatures shortens component lifetime, affects cooling requirements, increases electrical noise produced by components, and increases the resistance of (and power losses in) conductors and resistors. Ability of a component to operate at a safe temperature in ambient air depends upon the size, composition, and surface area of the component; e.g., a ½ W resistor is larger than a ¼ W resistor of the same kind.

Barring catastrophic failure, the temperature of a device rises until an equilibrium temperature is reached, at which point the power dissipated as heat equals the rate at which heat is removed from the device. A sustained temperature above the maximum temperature a device can endure will damage the device or at least cause permanent changes to properties of the device. Most devices are rated, based upon maximum permissible temperature, in terms of the average power that can be dissipated in *ambient air* (in absence of heat sinks and forced cooling)[4] without damage; for example, a ½ W resistor can dissipate ½ W in ambient air without harm, but if called upon

[4]The ambient-air power-dissipation rating of a component can be exceeded if heat sinks and/or forced cooling are used.

to dissipate more than that would either exhibit increased resistance (due to overheating) or fail, depending upon the actual power delivered to the resistor and the time for which the power dissipated exceeds the rated dissipation.[5] Specifying an appropriate power-dissipation rating for each component in a circuit is as important as specifying any other circuit parameter, such as resistance. Specifying an unnecessarily large power-dissipation rating needlessly increases the size, weight, and cost of a circuit. Specifying a too-small power-dissipation rating for one or more components can cause a circuit to function poorly and fail prematurely because of overheating.

In undemanding, low-power applications, radiation, conduction through the leads to a circuit board, and convection by the surrounding air might provide sufficient cooling to keep device temperatures below maximum permissible values. But in some applications and in many integrated circuits in particular, it is impossible or impractical to make components large enough to provide sufficient power dissipation in ambient air. In such cases heat sinks, fans, or even liquid cooling might be required to maintain acceptable operating temperatures.[6] Some large mainframe computers require liquid cooling. Even small desktop computers use fans to remove heat from their enclosures and high-performance microprocessors come packaged with an integral heat sink and cooling fan.[7]

Example 5.4. In steady state, the power radiated from a heated body is given (approximately) by the Stefan-Boltzmann law

$$p = \varepsilon\sigma A T^4, \tag{5.8}$$

where ε is the *emissivity* of the body (dimensionless), $\sigma = 5.67 \times 10^{-8}\,\mathrm{W\,m^{-2}\,K^{-4}}$ is the *Stefan-Boltzmann constant*, A is the surface area of the body, and T is the temperature of the body (K). A certain ¼ W cylindrical composition resistor must operate at or below 150°C to maintain its resistance within ±5% of the nominal (specified) value. The emissivity of the surface of the resistor is $\varepsilon = 0.8$. What should the surface area of the resistor be (assuming radiation is the only heat-loss mechanism)?

Solution: At 150°C and in steady state, the power radiated by the resistor must equal ¼ W, denoted by p_0. Thus

$$\begin{aligned} p_0 = \varepsilon\sigma A T^4 \Rightarrow A &= \frac{p_0}{\varepsilon\sigma T^4} \\ &= \frac{0.25\,\mathrm{W}}{(0.8)(5.67 \times 10^{-8}\,\mathrm{W\,m^{-2}\,K^{-4}})(423\,\mathrm{K})^4} \\ &= 1.72 \times 10^{-4}\,\mathrm{m}^2. \end{aligned}$$

The surface area of a cylindrical resistor having diameter d and length l is given by $A = \pi d l$. If the resistor is 1.25 cm long, then the diameter must be

$$\begin{aligned} d = \frac{A}{\pi l} = \frac{1.72 \times 10^{-4}\,\mathrm{m}^2}{\pi(1.25 \times 10^{-2}\,\mathrm{m})} &= 4.38 \times 10^{-3}\,\mathrm{m} \\ &= 4.38\,\mathrm{mm}. \end{aligned}$$

The Stefan-Boltzmann law (5.8) gives a pessimistic value for the power-dissipation rating of a resistor because heat also is lost through convection to the surrounding air and conduction through the leads. In absence of heat sinks or forced cooling, radiation accounts for only about 20% of heat loss. Thus the surface area of a resistor need not be as large as that implied by the Stefan-Boltzmann law.

Exercise 5.4. A certain cylindrical resistor is 2 cm long and 0.6 cm in diameter. The operating temperature of the resistor must not exceed 150°C. Assume that radiation is the

[5] Common ratings for ordinary composition resistors are 1/8, 1/4, 1/2, and 1 W. Composition resistors having ratings up to 5 W are easily obtained. If a much higher power-dissipation rating is warranted, one can use metallic-film or wire-wound resistors, or use forced cooling, or both.

[6] Many books give guidelines for sizing heat sinks with and without forced-air cooling. See, for example, Paul Horowitz and Winfield Hill, *The Art of Electronics* (2nd Ed.), Cambridge University Press, pp 312 *ff*.

[7] The power dissipated by a modern high-performance microprocessor is on the order of 100 W cm^{-2}.

principal heat-removal mechanism and use the Stefan-Boltzmann law to find the power-dissipation rating of the resistor if the emissivity of the surface is 0.8.

5.3 Conservation of Power

Energy is conserved. If all energy sources (e.g., current and voltage sources) and circuit components are accounted for, the total energy dissipated (work done) in a circuit equals zero (recall that, according to the passive sign convention, energy generated is negative). We may express the total instantaneous work done in a circuit comprising N elements as

$$w = w_1 + w_2 + \cdots + w_N = 0, \quad (5.9)$$

where w_k is the work done in the kth element. It follows that total power dissipated by the circuit also equals zero; that is

$$\frac{dw}{dt} = \frac{dw_1}{dt} + \frac{dw_2}{dt} + \cdots + \frac{dw_N}{dt} = 0 \quad (5.10)$$

or

$$p = p_1 + p_2 + \cdots + p_N = 0. \quad (5.11)$$

The most frequent uses of (5.11) are (1) calculating the power requirements of a circuit (sizing a power supply) and (2) specifying heat-removal methods, such as heat sinks, fans, and liquid cooling. Also, the fact that power is conserved can provide a useful check on the correctness of an analysis.[8]

Example 5.5. Find the power dissipated by each element in the circuit of Fig. 5.5. Show that the total power dissipated equals zero.

[8]Equation (5.11) states that the total power dissipated equals zero. It is also true that the total power delivered equals zero, or that the total power dissipated equals the total power delivered (in a circuit).

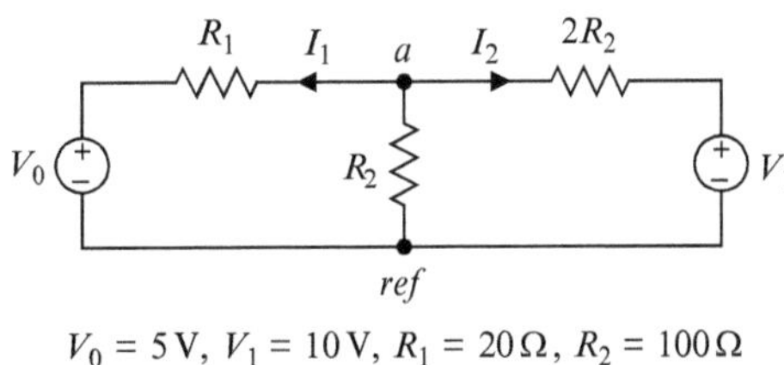

$V_0 = 5\,\text{V},\ V_1 = 10\,\text{V},\ R_1 = 20\,\Omega,\ R_2 = 100\,\Omega$

Fig. 5.5 See Example 5.5

Solution: Writing Kirchhoff's current law at the node a gives

$$\frac{V_a - V_0}{R_1} + \frac{V_a}{R_2} + \frac{V_a - V_1}{2R_2} = 0$$
$$\Rightarrow V_a = \frac{2R_2V_0 + R_1V_1}{3R_1 + 2R_2},$$

whence

$$I_1 = \frac{V_a - V_0}{R_1} = \frac{V_1 - 3V_0}{3R_1 + 2R_2},$$

and

$$I_2 = \frac{V_a - V_1}{2R_2} = \frac{R_2V_0 - (R_1 + R_2)V_1}{R_2(3R_1 + 2R_2)}.$$

From (5.5) and (5.6), the powers dissipated by the circuit elements are

$$p_{R_1} = R_1I_1^2 = \frac{R_1(V_1 - 3V_0)^2}{(3R_1 + 2R_2)^2} \cong 7.40 \text{ mW},$$

$$p_{R_2} = \frac{V_a^2}{R_2} = \frac{(2R_2V_0 + R_1V_1)^2}{R_2(3R_1 + 2R_2)^2} \cong 213.02 \text{ mW},$$

$$p_{2R_2} = 2R_2I_2^2 = \frac{2[R_2V_0 - (R_1 + R_2)V_1]^2}{R_2(3R_1 + 2R_2)^2}$$
$$\cong 144.97 \text{ mW},$$

$$p_{V_0} = V_0I_1 = \frac{V_0(V_1 - 3V_0)}{3R_1 + 2R_2}$$
$$\cong -96.15 \text{ mW},$$

$$p_{V_1} = V_1I_2 = \frac{V_1[R_2V_0 - (R_1 + R_2)V_1]}{R_2(3R_1 + 2R_2)}$$
$$\cong -269.23 \text{ mW}.$$

The total power dissipated by the circuit is

$$p_{R_1} + p_{R_2} + p_{2R_2} + p_{v_0} + p_{v_1} = 0.$$

Note the use of the passive sign convention; e.g., to compute the power dissipated by the voltage source, we take the positive direction for current into the positive terminal of the source.

Exercise 5.5. Show that the total power dissipated in the circuit of Exercise 5.2 equals zero.

Electrical energy provided by a source can be derived from an electrochemical reaction (as in a battery), from electromechanical devices (such as a generator), from radiation (as in a solar cell), or other means. The electrical energy provided to a circuit component might be converted to rotational mechanical energy (as in an electric motor), to heat (as in a resistor or an electric heater) or other electromagnetic radiation (as in a radio transmitting antenna), or other form. In all cases, total energy and power are conserved. Also in all cases, at least some electrical energy is converted to heat, not all of which can be recovered for good use. No energy-conversion device is 100% efficient.

5.4 Peak Power

The **peak power**, denoted by $\hat{p}$, associated with a current and voltage is the maximum value of the *magnitude* of the instantaneous power caused to be dissipated by the current and voltage. In general,

$$\hat{p} = \max|p(t)| = \max|v(t)\, i(t)|. \tag{5.12}$$

From (5.12) and Ohm's law, the peak power dissipated by a resistor having resistance R is given by

$$\hat{p} = R \max\left|i^2(t)\right| = \frac{1}{R}\max\left|v^2(t)\right|, \tag{5.13}$$

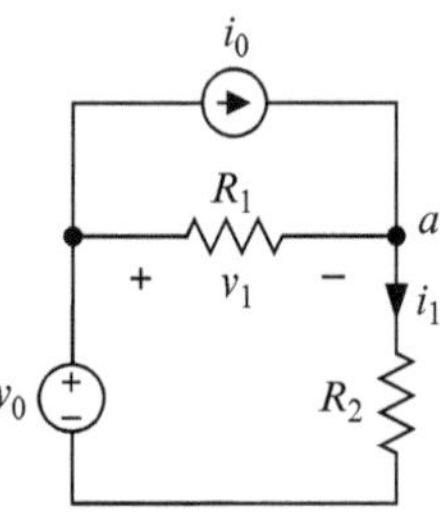

Fig. 5.6 See Example 5.6

where i and v are the current through and voltage across the resistor, respectively. Unlike instantaneous power, *peak power is not conserved* (peak power is non-negative).

Example 5.6. See Fig. 5.6, where $v_0 = V_0\cos(\omega t), i_0 = I_0\cos(\omega t)$. Obtain expressions for the instantaneous and peak powers dissipated by each element.
Solution: Writing Kirchhoff's current law at node a gives

$$\frac{v_a}{R_2} + \frac{v_a - v_0}{R_1} - i_0 = 0 \Rightarrow v_a = \frac{R_2(v_0 + R_1 i_0)}{R_1 + R_2},$$

whence

$$i_1 = \frac{v_a}{R_2} = \frac{v_0 + R_1 i_0}{R_1 + R_2}, \quad v_1 = v_0 - v_a$$
$$= \frac{R_1(v_0 - R_2 i_0)}{R_1 + R_2}.$$

From (5.5) and (5.6), the instantaneous powers dissipated by the circuit elements are

$$p_{R_1} = \frac{v_1^2}{R_1} = \frac{R_1(V_0 - R_2 I_0)^2}{(R_1 + R_2)^2}\cos^2(\omega t),$$
$$p_{R_2} = R_2 i_1^2 = \frac{R_2(R_1 I_0 + V_0)^2}{(R_1 + R_2)^2}\cos^2(\omega t),$$
$$p_{v_0} = -v_0 i_1 = -\frac{V_0(R_1 I_0 + V_0)}{R_1 + R_2}\cos^2(\omega t),$$
$$p_{i_0} = v_1 i_0 = \frac{R_1 I_0(V_0 - R_2 I_0)}{R_1 + R_2}\cos^2(\omega t).$$

As a check, we confirm that $p_{R_1} + p_{R_2} + p_{v_0} + p_{i_0} = 0$. Because $\max[\cos^2(\omega t)] = 1$, the peak powers dissipated by the elements are given by

$$\hat{p}_{R_1} = \frac{R_1(V_0 - R_2 I_0)^2}{(R_1 + R_2)^2},$$
$$\hat{p}_{R_2} = \frac{R_2(R_1 I_0 + V_0)^2}{(R_1 + R_2)^2},$$
$$\hat{p}_{v_0} = \frac{|V_0(R_1 I_0 + V_0)|}{R_1 + R_2},$$
$$\hat{p}_{i_0} = \frac{|R_1 I_0(V_0 - R_2 I_0)|}{R_1 + R_2}.$$

Exercise 5.6. The instantaneous power dissipated by a certain device is given by

$$p = P_0 \cos(\omega t)\cos(\omega t + \theta), P_0 > 0.$$

Obtain an expression for the peak power dissipated.

Peak power is of interest partly because there can occur relatively short-duration (transient) current or voltage *spikes* in a circuit, some of which can have very large amplitudes. A large spike can transfer a large amount of energy to a device in a very short time, which means that the instantaneous power dissipated by the device is very large during that time. That energy might be sufficient to destroy the device, even if the device were capable of absorbing the same amount of energy over a longer time. As an analogy, the total energy transferred to a tree by an average lightning strike might be only a fraction of the energy received by the tree from the sun in the course of an average month. But such a strike could destroy the tree, whereas the sunlight would not.

Peak power is of interest also because all physical sources (and other components) are peak-power-limited; that is, there is a limit to the magnitude of the instantaneous power that can be dissipated by any physical source or component.

A peak-power limit on an ideal voltage source limits the current the source can provide (the current through the source). Similarly, a peak-power limit on an ideal current source limits the voltage the source can produce. For example, if a current source produces current I and the maximum possible voltage across the source is V_{max}, then the source is peak-power limited to $p_{max} = V_{max} I$.

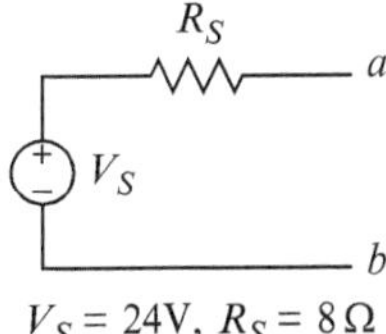

Fig. 5.7 See Example 5.7

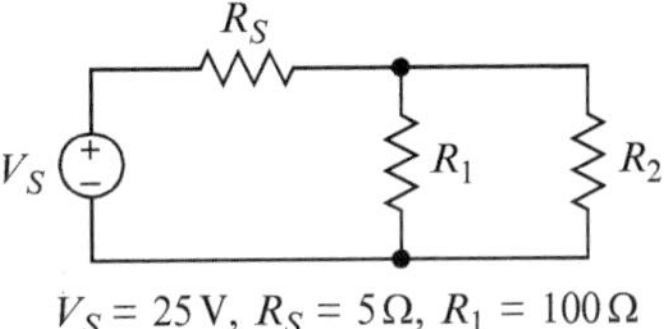

Fig. 5.8 See Example 5.8

Example 5.7. See Fig. 5.7. The source V_S is peak-power limited to 48 W. What is the minimum resistance that one should connect to the terminals *a*–*b*?

Solution: The maximum current the source can deliver is $I_{max} = p_{max}/V_S = 2$ A. (a) The minimum resistance one should connect to the terminals is obtained as follows:

$$\frac{V_S}{R_S + R_{min}} = I_{max} \Rightarrow R_{min} = \frac{V_S}{I_{max}} - R_S = 4\,\Omega.$$

Example 5.8. In Fig. 5.8, the maximum instantaneous power the source V_S can deliver is 50 W. What is the maximum possible current through the source? What is the minimum allowable value of the resistance R_2?

Solution: The maximum possible current through the source is

$$I_{max} = \frac{p_{max}}{V_S} = \frac{50\,\text{W}}{25\,\text{V}} = 2\,\text{A}.$$

The actual current through the source is given by

$$I = \frac{V_S}{R_S + R_1 \| R_2}.$$

It follows that

$$(R_S + R_1\|R_2)_{min} = \frac{V_S}{I_{max}} = \frac{25\,\text{V}}{2\,\text{A}} = 12.5\,\Omega$$

and

$$(R_1\|R_2)_{min} = \frac{R_1 (R_2)_{min}}{R_1 + (R_2)_{min}} = 12.5\,\Omega - R_S = 7.5\,\Omega = R,$$

which yields

$$(R_2)_{min} = \frac{R_1 R}{R_1 - R} = \frac{(100\,\Omega)(7.5\,\Omega)}{100\,\Omega - 7.5\,\Omega} \cong 8.12\,\Omega.$$

5.5 Available Power

A physical source can be modeled using either the Thévenin or Norton equivalent for the source at the source terminals, as illustrated by Fig. 5.9. If V_{OC}, I_{SC} denote the open-circuit voltage and short-circuit current, respectively, at the terminals of the source, then

$$V_S = V_{OC},\ R_S = \frac{V_{OC}}{I_{SC}},\ I_S = I_{SC}. \tag{5.14}$$

A Thévenin or Norton source model is by its nature peak-power limited. To show this, we consider the circuit in Fig. 5.10, where the power delivered to the load R_L by the Thévenin source model is given by

$$P_L = \frac{{V_L}^2}{R_L} = \frac{1}{R_L}\left(\frac{V_S R_L}{R_S + R_L}\right)^2 = \frac{{V_S}^2 R_L}{(R_S + R_L)^2}. \tag{5.15}$$

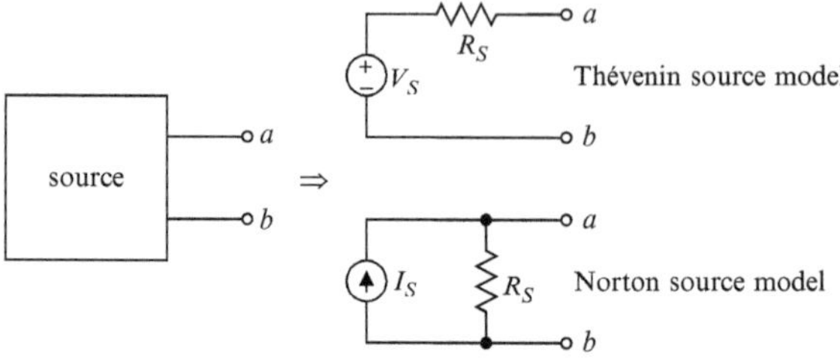

Fig. 5.9 Equivalent source models, where $V_S = R_S I_S$

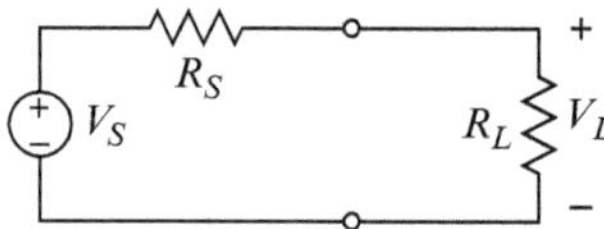

Fig. 5.10 Pertaining to the definition of available power (see (5.17))

The power delivered to the load equals zero for both $R_L = 0$ and $R_L \to \infty$, so must have a maximum value, which we can find by solving

$$\frac{dP_L}{dR_L} = 0 \Rightarrow \frac{d}{dR_L}\left[\frac{{V_S}^2 R_L}{(R_S + R_L)^2}\right] = \frac{{V_S}^2}{(R_S + R_L)^2} - \frac{2{V_S}^2 R_L}{(R_S + R_L)^3} = 0. \tag{5.16}$$

Thus the load resistance that draws maximum power from the source is $R_L = R_S$, and the maximum power that can be drawn from the source is given by

$$P_{L\,max} = \frac{{V_S}^2 R_S}{(R_S + R_S)^2} = \frac{{V_S}^2}{4R_S}. \tag{5.17}$$

In other words, the Thévenin source model is peak power limited. The maximum power that can be drawn from the source, given by (5.17), is called the **available power** for the source. The resistance R_S is called the **internal resistance** of the source. These important quantities are discussed at length in Section 5.16.

Exercise 5.7. Show without using (5.17) that the available power for a Norton source model is given by

$$P_{L\max} = \frac{{I_S}^2 R_S}{4}. \quad (5.18)$$

Exercise 5.8. A certain fully-charged battery has an open-circuit terminal voltage of 12.6V and an internal resistance of approximately 0.1Ω. Draw circuit diagrams for the Thévenin and Norton models for the battery. What is the maximum power that can be drawn from the battery? What is the maximum current that can be drawn from the battery?

Exercise 5.9. The power available from a certain source is 50W. A 10 Ω load draws 1A from the source, where positive current exits the positive terminal of the source. Obtain the Thévenin and Norton models for the source.

5.6 Time Averages

In most circuit applications, *average* power dissipated (or delivered) is of more interest than instantaneous power. Household light bulbs, electric heaters, electric water heaters, electric generators, electric motors, audio amplifiers, radio transmitters, and a host of other devices, circuits, and systems are rated according to *average* power dissipated or delivered, not instantaneous power.

Intuitively, average power dissipated over an interval is the total work done in the interval, divided by the duration of the interval. For average power to be meaningful, the averaging interval must be long enough to obtain a true average; that is, the duration of the interval must be such that increasing the duration yields no significant change in the calculated average power. Physical instruments that measure average power must use a finite interval. Otherwise, a measurement could never be completed. But using mathematical models for physical currents and voltages, we can calculate average power over an infinite interval, which is easier than attempting to decide how long to make the interval based upon knowledge of each particular instantaneous power of interest.

Average current and voltage, such as the average voltage produced by a power supply and the average current provided by a battery charger, also are of interest. Average currents and voltages also are of considerable importance in electronic circuits, as you will find when you study that subject.

In this section, we introduce time averages in general. We describe how one computes the average value of a function of time, regardless of whether the function represents power, current, voltage, or other physical quantity. The relations developed in this section are used repeatedly in subsequent sections and chapters, and most likely in subsequent courses and later, in practice.

The **time average** of a function of time $x(t)$ is denoted by $\overline{x}$ and is defined by

$$\overline{x} = \lim_{T\to\infty} \frac{1}{T} \int_0^T x(t)\,dt. \quad (5.19)$$

Note: In some texts, and probably in later work, you will see the time average of a function defined as

$$\overline{x} = \lim_{T\to\infty} \frac{1}{T} \int_{-T/2}^{T/2} x(t)\,dt. \quad (5.20)$$

It is easy to cook up functions for which the two definitions give different results; for example,

$$x(t) = \begin{cases} 5\,\text{V}, & t \le 0, \\ 10\,\text{V}, & t > 0. \end{cases}$$

For functions that are either constant or periodic, the two definitions give identical results, in which cases (5.19) is often easier to apply than (5.20). Also, (5.19) is more consistent with measurements, which must begin at some finite time.

The dimension of an average is that of the quantity averaged; e.g., if $x(t)$ in (5.19) is a voltage, the dimension of the average $\overline{x}$ is voltage. Because time averages are the only averages we use, we refer to the right side of (5.19) as simply the **average** of $x(t)$.

Two specific averages are especially useful. First, *the average of a constant is the constant itself*:

$$\overline{a} = a. \tag{5.21}$$

Second, *the average of a sinusoid equals zero*:

$$\overline{\cos(\omega t + \theta)} = 0; \quad \omega \neq 0. \tag{5.22}$$

Equation (5.21) is obtained from (5.19) as follows: Because a is independent of time,

$$\begin{aligned}\lim_{T\to\infty} \frac{1}{T}\int_0^T a\,dt &= \lim_{T\to\infty}\frac{1}{T}[a\,t]_{t=0}^{t=T} = \lim_{T\to\infty}\frac{1}{T}[a\,T]\\ &= \lim_{T\to\infty}(a) = a.\end{aligned}$$

Equation (5.22) is obtained from (5.19) as follows:

$$\begin{aligned}\lim_{T\to\infty}\frac{1}{T}\int_0^T \cos(\omega t+\theta)\,dt &= \lim_{T\to\infty}\frac{1}{T}\left[\frac{1}{\omega}\sin(\omega t+\theta)\right]_{t=0}^{t=T}\\ &= \lim_{T\to\infty}\frac{1}{T}\left[\frac{1}{\omega}\sin(\omega T+\theta) - \frac{1}{\omega}\sin(\theta)\right] = 0.\end{aligned}$$

because (for $\omega \neq 0$) the quantity in the brackets is finite.

The time average of a function of time exists if the function is bounded; i.e., if $|x(t)| < \infty$ for all values of time t.[9] We are concerned only with averages of functions representing *physical* currents, voltages, and powers, all of which are bounded. Furthermore, in this introductory treatment, we are concerned only with averages of constant and periodic functions. You might learn more about time averages in general in subsequent courses.

Two useful properties of averages in general follow immediately from (5.19):

First, if a is a constant

$$\overline{ax} = a\overline{x}. \tag{5.23}$$

In words, the average of the product of a constant and a function of time is the product of the constant and the average of the function of time. Equation (5.23) follows from the fact that the constant a can be taken outside both the integral and the limiting operation in (5.19). Second, if $x_1, x_2, \cdots, x_N$ denote functions of time, then

$$\overline{x_1 + x_2 + \cdots + x_N} = \overline{x_1} + \overline{x_2} + \cdots + \overline{x_N}. \tag{5.24}$$

In words, the time average of a sum of functions equals the sum of the time averages of the functions. Equation (5.24) follows from the fact that both integration and limiting distribute over a sum.

In addition to the relations given above, a few trigonometric identities are useful in finding averages of sinusoidal functions of time, as illustrated by the following examples.

Example 5.9. Obtain an expression for the average value of $x(t) = \cos^2(\omega t + \theta)$, where $\omega > 0$ and θ is a constant but otherwise arbitrary angle.

Solution: We use the identity

$$\cos^2(\alpha) \equiv \frac{1}{2} + \frac{1}{2}\cos(2\alpha)$$

to obtain

$$x(t) = \cos^2(\omega t + \theta) = \frac{1}{2} + \frac{1}{2}\cos(2\omega t + 2\theta).$$

From (5.24),

$$\overline{x} = \overline{\frac{1}{2} + \frac{1}{2}\cos(2\omega t + 2\theta)} = \overline{\frac{1}{2}} + \overline{\frac{1}{2}\cos(2\omega t + 2\theta)}.$$

From (5.21),

$$\overline{\frac{1}{2}} = \frac{1}{2}.$$

From (5.23) and (5.22),

[9] This condition is more stringent than necessary, but in this book, we do not need to compute time averages of functions not satisfying the condition.

$$\overline{\frac{1}{2}\cos(2\,\omega\,t+2\,\theta)}=\frac{1}{2}\overline{\cos(2\,\omega\,t+2\,\theta)}=0.$$

Thus

$$\bar{x}=\overline{\cos^2(\omega\,t+\theta)}=\frac{1}{2}.$$

Exercise 5.10. Obtain an expression for the average of $x(t)=\sin^2(\omega\,t+\theta)$, where $\omega>0$ and θ is a constant but otherwise arbitrary angle. You may use the result obtained in Example 5.9.

Example 5.10. Obtain an expression for the average of $x(t)=\cos(\omega\,t+\theta_1)\cos(\omega\,t+\theta_2)$, where $\omega>0$ and θ_1, θ_2 are constant but otherwise arbitrary angles.
Solution: We use the identity

$$\cos(\alpha)\,\cos(\beta)\equiv\frac{1}{2}\cos(\alpha+\beta)+\frac{1}{2}\cos(\alpha-\beta)$$

to obtain

$$\begin{aligned}x(t)&=\cos(\omega\,t+\theta_1)\,\cos(\omega\,t+\theta_2)\\&\equiv\frac{1}{2}\cos(2\,\omega\,t+\theta_1+\theta_2)+\frac{1}{2}\cos(\theta_1-\theta_2).\end{aligned}$$

From (5.24),

$$\begin{aligned}\bar{x}&=\overline{\frac{1}{2}\cos(2\,\omega\,t+\theta_1+\theta_2)+\frac{1}{2}\cos(\theta_1-\theta_2)}\\&=\overline{\frac{1}{2}\cos(2\,\omega\,t+\theta_1+\theta_2)}+\overline{\frac{1}{2}\cos(\theta_1-\theta_2)}.\end{aligned}$$

From (5.23) and (5.22),

$$\begin{aligned}&\overline{\frac{1}{2}\cos(2\,\omega\,t+\theta_1+\theta_2)}\\&=\frac{1}{2}\overline{\cos(2\,\omega\,t+\theta_1+\theta_2)}=0.\end{aligned}$$

From (5.21),

$$\overline{\frac{1}{2}\cos(\theta_1-\theta_2)}=\frac{1}{2}\cos(\theta_1-\theta_2),$$

because $\theta_1-\theta_2$ is a constant. Thus

$$\bar{x}=\overline{\cos(\omega t+\theta_1)\cos(\omega t+\theta_2)}=\frac{1}{2}\cos(\theta_1-\theta_2).$$

Exercise 5.11. Obtain an expression for the average of $x(t)=[\cos(\omega\,t+\theta_1)+\cos(\omega\,t+\theta_2)]^2$, where $\omega>0$ and θ_1, θ_2 are constant but otherwise arbitrary angles.

Exercise 5.12. Show by example that if x_1, x_2 are both functions of time, then

$$\overline{x_1x_2}\neq\overline{x_1}\,\overline{x_2}.$$

That is, show that the average of the product of two functions is in general *not* equal to the product of the average values.

Example 5.11. Obtain an expression for the average of $x(t)=\cos(\omega_1t+\theta_1)+\cos(\omega_2t+\theta_2)$, where $\omega_2>\omega_1>0$ and θ_1,θ_2 are constant but otherwise arbitrary angles.
Solution: We use the identity

$$\cos(\alpha)\,\cos(\beta)\equiv\frac{1}{2}\cos(\alpha+\beta)+\frac{1}{2}\cos(\alpha-\beta)$$

to obtain

$$\begin{aligned}\bar{x}&=\overline{\cos(\omega_1\,t+\theta_1)\cos(\omega_2\,t+\theta_2)}\\&\equiv\overline{\frac{1}{2}\cos[(\omega_1+\omega_2)\,t+\theta_1+\theta_2]}\\&\quad+\overline{\frac{1}{2}\cos[(\omega_1-\omega_2)\,t+\theta_1-\theta_2]}.\end{aligned}$$

From (5.24) and (5.23)

$$\bar{x} = \frac{1}{2}\overline{\cos[(\omega_1+\omega_2)t+\theta_1+\theta_2]} + \frac{1}{2}\overline{\cos[(\omega_1-\omega_2)t+\theta_1-\theta_2]}.$$

From (5.22), the last two terms equal zero because $\omega_1 \neq \omega_2$. Thus

$$\bar{x} = \overline{\cos(\omega_2 t+\theta_1)\cos(\omega_1 t+\theta_2)} = 0, \ \omega_1 \neq \omega_2.$$

Exercise 5.13. Obtain an expression for the average of $x(t)=[\cos(\omega_1 t+\theta_1)+\cos(\omega t+\theta_2)]^2$, where $\omega_2>\omega_1>0$ and θ_1,θ_2 are constant but otherwise arbitrary angles.

Results obtained in Example 5.9, Example 5.10 and Example 5.11 are collected below for future reference.

$$\begin{aligned}&\overline{\cos^2(\omega t+\theta)} = \frac{1}{2}, \ \omega \neq 0,\\ &\overline{\cos(\omega t+\theta_1)\cos(\omega t+\theta_2)} = \frac{1}{2}\cos(\theta_1-\theta_2), \ \omega \neq 0,\\ &\overline{\cos(\omega_2 t+\theta_1)\cos(\omega_1 t+\theta_2)} = 0, \ \omega_1 \neq \omega_2.\end{aligned} \tag{5.25}$$

5.7 Average Power

The **average power** dissipated by a two-terminal element is conventionally denoted by P rather than $\bar{p}$, and is defined by

$$P = \bar{p} = \overline{vi}. \tag{5.26}$$

In (5.26), p is the instantaneous power dissipated by the element and v and i are the instantaneous voltage across and current through the element. *Average power is conserved*, as can be shown by applying (5.24) to (5.11).

Equation (5.26) applies to any two-terminal element. From (5.26) and Ohm's law, the **average power dissipated by a resistor** having resistance R is given by

$$P = \overline{vi} = \overline{R i^2} = R\overline{i^2} \quad \text{or} \quad P = \overline{vi} = \overline{\frac{1}{R}v^2} = \frac{1}{R}\overline{v^2}, \tag{5.27}$$

where i and v are the current through and voltage across the resistor. The quantities $\overline{i^2}$ and $\overline{v^2}$ are called the **mean squared amplitudes** of the current i and the voltage v, respectively.[10]

If the current through a resistor is a constant given by $i = I_0$, then from (5.6), the instantaneous power dissipated by the resistor is a constant given by $p = {I_0}^2 R$, where R is the resistance of the resistor. It follows that if a resistor having resistance R is subjected to a *constant current* $i = I_0$ or a *constant voltage* $v = V_0$, the instantaneous power dissipated, the peak power dissipated, and the average power dissipated are equal and given by

$$p = \hat{p} = P = {I_0}^2 R \quad \text{or} \quad p = \hat{p} = P = \frac{{V_0}^2}{R}. \tag{5.28}$$

Example 5.12. The voltage across a 100 Ω resistor is 24 V. From (5.6), the instantaneous power dissipated is

$$\begin{aligned}p &= VI = V\frac{V}{R} = \frac{V^2}{R}\\ &= \frac{(24\,\text{V})^2}{100\,\Omega} \cong 5.76\text{W}.\end{aligned} \tag{5.29}$$

From (5.28), the average power dissipated and peak power dissipated are

$$P = \hat{p} \cong 5.76\text{W} \tag{5.30}$$

From (5.27) and (5.25), the average power dissipated by a resistor having resistance R subjected to a sinusoidal current or voltage is given by

[10] In this context, *mean* is a synonym for *average*.

$$i(t) = I_0 \cos(\omega t + \theta) \Rightarrow P = R\overline{I_0{}^2\cos^2(\omega t + \theta)}$$
$$= \frac{I_0{}^2 R}{2}, \omega \neq 0,$$
$$v(t) = V_0 \cos(\omega t + \theta) \Rightarrow P = \frac{1}{R}\overline{V_0{}^2\cos^2(\omega t + \theta)}$$
$$= \frac{V_0{}^2}{2R}, \quad \omega \neq 0. \tag{5.31}$$

Example 5.13. The voltage at the terminals of a certain 1500 W electric heater when the heater is operating is

$$v(t) = V_0 \cos(2\pi f t); \quad V_0 = 165\text{V}.$$

Find the hot resistance of the heater element.
Solution: From (5.31),

$$P = \frac{V_0{}^2}{2R} = 1500\text{W} \Rightarrow R = \frac{V_0{}^2}{2P} = \frac{(165\text{ V})^2}{(2)(1500\text{ W})}$$
$$\cong 9.08\,\Omega.$$

Exercise 5.14. The current through a resistor having resistance $R = 15\,\Omega$ is given by

$$i = I_0 \cos(2\pi f t),$$

with $I_0 = 10$ A and $f = 60$ Hz. Find the instantaneous power, the peak power, and the average power dissipated by the resistor.

Example 5.14. Obtain expressions for the average powers dissipated by the elements in the circuit shown in Fig. 5.11.

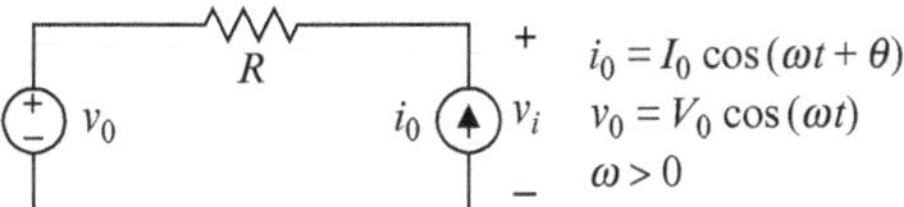

Fig. 5.11 See Example 5.14

Solution: By the definition of an ideal current source, the current through the resistor equals i_0. From (5.31), the average power dissipated by the resistor is given by

$$P_R = \frac{I_0{}^2 R}{2}.$$

The current entering the positive terminal of the voltage source is i_0. Thus (passive sign convention) the average power dissipated by the voltage source is given by

$$P_v = \overline{v_0\, i_0} = \overline{V_0 I_0 \cos(\omega t) \cos(\omega t + \theta)}.$$

From (5.23) and (5.25),

$$P_v = V_0 I_0 \overline{\cos(\omega t) \cos(\omega t + \theta)} = \frac{V_0 I_0}{2} \cos(\theta).$$

By Kirchhoff's voltage law, the voltage v_i across the current source is given by

$$-v_0 - R\, i_0 + v_i = 0 \Rightarrow v_i = v_0 + R\, i_0.$$

The current i_0 exits the terminal designated as positive in the definition of v_i, so the average power dissipated by the current source (passive sign convention) is given by

$$P_i = \overline{-v_i\, i_0} = -\overline{(v_0 + R\, i_0)\, i_0} = -\overline{v_0\, i_0} - \overline{R\, i_0{}^2}.$$

From (5.24) and results already obtained,

$$P_i = -\overline{v_0\, i_0} - \overline{R\, i_0{}^2} = -P_v - P_R$$
$$= -\frac{V_0 I_0}{2} \cos(\theta) - \frac{I_0{}^2 R}{2},$$

as expected, because total average power is conserved.

Exercise 5.15. Obtain expressions for the average powers dissipated by the elements in the circuit shown in Fig. 5.12.

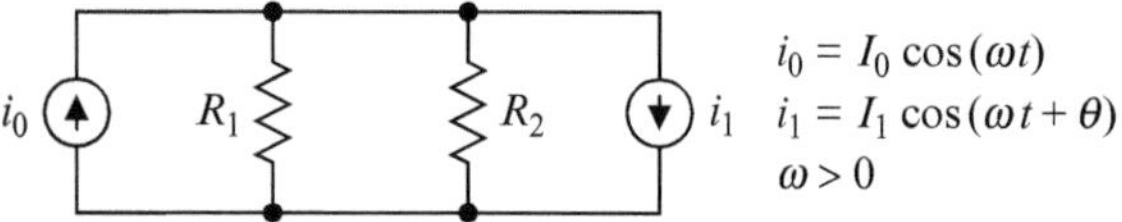

Fig. 5.12 See Exercise 5.15

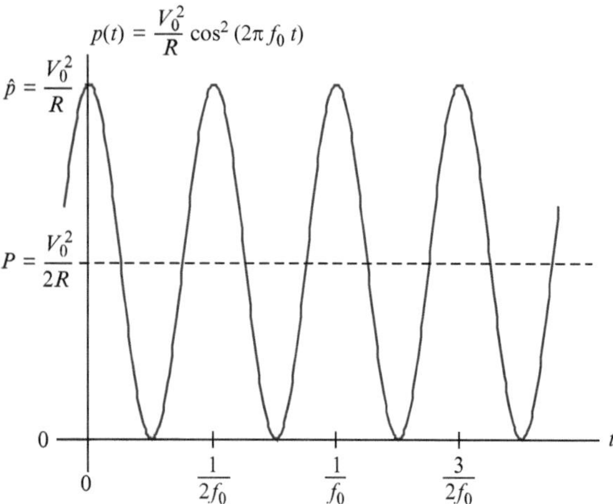

Fig. 5.13 An illustration of the definitions of peak power and average power

Figure 5.13 shows a graph of instantaneous power for a sinusoidal voltage applied to a resistor. Indicated on the figure are the peak power $\hat{p} = V_0{}^2/R$ and the average power $P = V_0{}^2/(2R)$. For a *sinusoidal* current or voltage applied to a *resistor*, the average power equals one-half of the peak power.

In summary, the instantaneous power p, the peak power $\hat{p}$, and the average power P dissipated by a *resistor* having resistance R and subjected to a *sinusoidal* current $i(t) = I_0 \cos(\omega t + \theta)$ or *sinusoidal* voltage $v(t) = V_0 \cos(\omega t + \theta)$ are given by

$$\left.\begin{array}{l} p = \dfrac{I_0{}^2 R}{2}[1 + \cos(2\omega t)], \hat{p} = I_0{}^2 R, P = \dfrac{\hat{p}}{2} = \dfrac{I_0{}^2 R}{2} \\ \text{or} \\ p = \dfrac{V_0{}^2}{2R}[1 + \cos(2\omega t)], \hat{p} = \dfrac{V_0{}^2}{R}, P = \dfrac{\hat{p}}{2} = \dfrac{V_0{}^2}{2R} \end{array}\right\} \omega > 0. \tag{5.32}$$

Equation (5.32) applies only when *the applied current or voltage is sinusoidal and the resistance R is independent of both time and applied current or voltage*. It is not difficult to err in this regard. The most common error is applying (5.32) to a resistor subjected to a non-sinusoidal current or voltage. As an example of a more subtle error, consider an ordinary incandescent light bulb driven by a fixed-amplitude sinusoidal voltage source. The resistance of the light-bulb filament is larger when the filament is hot than when cold. If the frequency of the applied voltage is sufficiently large, the filament does not cool significantly between peaks of the applied voltage, the resistance of the filament remains nearly constant (at the hot value), and the average power delivered to the bulb is given by (5.32). But if the frequency of the applied voltage is sufficiently low, the bulb varies between bright and dark, the resistance of the filament varies between hot and cold (or warm) values, and the average power applied to the bulb cannot be computed using (5.32). In such a case, we might express the resistance as a function of the applied current, which is a function of time, and calculate the average value of the instantaneous power $p = i^2(t) R[i(t)]$, but such problems are beyond the scope of this book.

Example 5.15.

(a) Obtain expressions for the instantaneous power dissipated (passive sign convention) by each element in the circuit shown in Fig. 5.14.

(b) Let $v = V_0 = 5\,\text{V}, i = I_0 = 2\,\text{mA}$ and calculate the peak and average powers dissipated by each element.

(c) Let $v_0 = V_0 \cos(\omega t)$, $i_0 = I_0 \cos(\omega t)$, with $V_0 = 5\,\text{V}$, $I_0 = 2\,\text{mA}$, and $\omega > 0$. Calculate the peak and average powers dissipated by each element.

Solution:

(a) We first obtain expressions for the current through and voltage across each element. Kirchhoff's current law yields

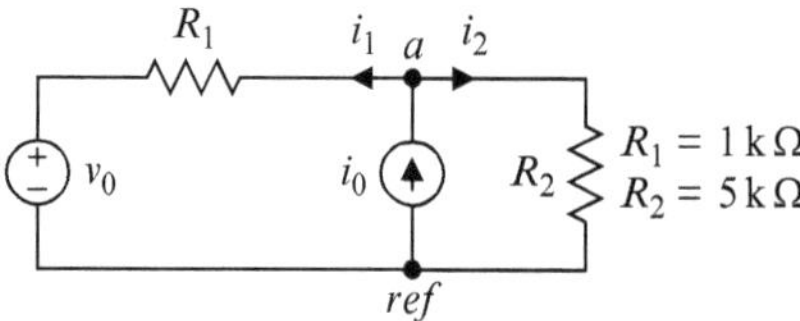

Fig. 5.14 See Example 5.15

$$\frac{v_a - v_0}{R_1} - i_0 + \frac{v_a}{R_2} = 0$$
$$\Rightarrow v_a = \frac{R_2\,(v_0 + R_1\,i_0)}{R_1 + R_2},$$

whence

$$i_1 = \frac{v_a - v_0}{R_1} = \frac{R_2(v_0 + R_1\,i_0)}{R_1(R_1 + R_2)} - \frac{v_0}{R_1}$$
$$= \frac{R_2\,i_0 - v_0}{R_1 + R_2},\, i_2 = \frac{v_a}{R_2} = \frac{v_0 + R_1\,i_0}{R_1 + R_2}.$$

Let p_v denote the power dissipated by the voltage source, p_i denote the power dissipated by the current source, p_1 the power dissipated by resistor R_1, and p_2 the power dissipated by resistor R_2. The instantaneous powers dissipated are given by

$$p_v = v_0\,i_1 = v_0\left(\frac{R_2\,i_0 - v_0}{R_1 + R_2}\right),$$
$$p_i = -i_0\,v_a = -i_0\left[\frac{R_2\,(v_0 + i_0\,R_1)}{R_1 + R_2}\right],$$
$$p_1 = {i_1}^2 R_1 = \left(\frac{R_2\,i_0 - v_0}{R_1 + R_2}\right)^2 R_1,$$
$$p_2 = {i_2}^2 R_2 = \left(\frac{v_0 + R_1\,i_0}{R_1 + R_2}\right)^2 R_2.$$

(b) Because both sources are constant, all currents or voltages in the circuit are constant. Thus the instantaneous, peak, and average powers dissipated by each component are the same. From part (a) above,

$$p_v = \hat{p}_v = P_v = V_0\left(\frac{R_2\,I_0 - V_0}{R_1 + R_2}\right)$$
$$= (5\,\mathrm{V})\left[\frac{(5\,\mathrm{k\Omega})(2\,\mathrm{mA}) - 5\,\mathrm{V}}{1\,\mathrm{k\Omega} + 5\,\mathrm{k\Omega}}\right]$$
$$\cong 4.17\ \mathrm{mW},$$
$$p_i = \hat{p}_i = P_i = -I_0\left[\frac{R_2\,(V_0 + I_0\,R_1)}{R_1 + R_2}\right]$$
$$= (-2\,\mathrm{mA})\left\{\frac{[5\,\mathrm{k\Omega}][5\,\mathrm{V} + (2\,\mathrm{mA})(1\,\mathrm{k\Omega})]}{1\,\mathrm{k\Omega} + 5\,\mathrm{k\Omega}}\right\}$$
$$\cong -11.67\mathrm{mW},$$

$$p_1 = \hat{p}_1 = P_1 = {i_1}^2 R_1 = \left(\frac{R_2\,I_0 - V_0}{R_1 + R_2}\right)^2 R_1$$
$$= \left[\frac{(5\,\mathrm{k\Omega})(2\,\mathrm{mA}) - 5\,\mathrm{V}}{1\,\mathrm{k\Omega} + 5\,\mathrm{k\Omega}}\right]^2 (1\,\mathrm{k\Omega})$$
$$\cong 694\,\mu W,$$
$$p_2 = \hat{p}_2 = P_2 = {i_2}^2 R_2 = \left(\frac{V_0 + R_1\,I_0}{R_1 + R_2}\right)^2 R_2$$
$$= \left[\frac{5\,\mathrm{V} + (1\,\mathrm{k\Omega})(2\,\mathrm{mA})}{1\,\mathrm{k\Omega} + 5\,\mathrm{k\Omega}}\right]^2 (5\,\mathrm{k\Omega})$$
$$\cong 6.81\,\mathrm{mW}.$$

(c) From part (a), the instantaneous powers dissipated are

$$p_v = V_0\,\cos(\omega\,t)$$
$$\times\left(\frac{R_2\,I_0\,\cos(\omega\,t) - V_0\cos(\omega\,t)}{R_1 + R_2}\right)$$
$$= V_0\left(\frac{R_2\,I_0 - V_0}{R_1 + R_2}\right)\cos^2(\omega\,t)$$
$$= \frac{V_0}{2}\left(\frac{R_2\,I_0 - V_0}{R_1 + R_2}\right)[1 + \cos(2\,\omega\,t)]$$
$$\cong 2.08\,[1 + \cos(2\,\omega\,t)]\ \mathrm{mW},$$

$$p_i = -I_0\cos(\omega\,t)$$
$$\times\left[\frac{R_2\,(V_0\cos(\omega\,t) + I_0\cos(\omega\,t)\,R_1)}{R_1 + R_2}\right]$$
$$= -I_0\left[\frac{R_2\,(V_0 + R_1 I_0)}{R_1 + R_2}\right]\cos^2(\omega\,t)$$
$$= -\frac{I_0}{2}\left[\frac{R_2\,(V_0 + R_1 I_0)}{R_1 + R_2}\right]$$
$$\times [1 + \cos(2\,\omega\,t)]$$
$$= -5.83[1 + \cos(2\,\omega\,t)]\mathrm{mW},$$

$$p_1 = \left(\frac{R_2\,I_0\cos(\omega\,t) - V_0\cos(\omega\,t)}{R_1 + R_2}\right)^2 R_1$$
$$= \left(\frac{R_2\,I_0 - V_0}{R_1 + R_2}\right)^2 R_1\cos^2(\omega\,t)$$
$$= \frac{1}{2}\left(\frac{R_2\,I_0 - V_0}{R_1 + R_2}\right)^2 R_1\,[1 + \cos(2\,\omega\,t)]$$
$$\cong 347[1 + \cos(2\,\omega\,t)]\mu W,$$

and

$$p_2 = \left(\frac{V_0\cos(\omega t) + R_1 I_0\cos(\omega t)}{R_1 + R_2}\right)^2 R_2$$
$$= \left(\frac{V_0 + R_1 I_0}{R_1 + R_2}\right)^2 R_2\cos^2(\omega t)$$
$$= \frac{1}{2}\left(\frac{V_0 + R_1 I_0}{R_1 + R_2}\right)^2 R_2[1 + \cos(2\omega t)]$$
$$\cong 3.40\,[1 + \cos(2\omega t)]\text{mW}.$$

The peak and average powers are [see (5.32)]

$$\hat{p}_v \cong 4.17\,\text{mW},\ \hat{p}_i \cong -11.67\,\text{mW},$$
$$\hat{p}_1 \cong 694\,\mu\text{W}, \hat{p}_2 \cong 6.81\,\text{mW},$$

and

$$P_v \cong 2.08\,\text{mW},\ P_i \cong -5.83\,\text{mW},$$
$$P_1 \cong 347\,\mu\text{W},\ P_2 \cong 3.40\,\text{mW},$$

respectively.

All physical circuit components are peak-power-limited and average-power limited. Typically, the average power a component can dissipate is less than the (short-term) peak power the component can dissipate. For example, an audio amplifier rated at 100 W might be able to provide as much as 150 W for a short time (a few milliseconds), but cannot provide more than 100 W on average (over a long time).

5.8 Root Mean Squared (RMS) Amplitude of a Current or Voltage

The **root mean squared amplitudes** (rms amplitude) of a current and voltage are denoted by the subscript *rms* and are defined by

$$I_{rms} = \sqrt{\overline{i^2(t)}}, \qquad V_{rms} = \sqrt{\overline{v^2(t)}}. \tag{5.33}$$

We use upper-case letters to denote rms amplitudes because an rms amplitude is independent of time. Thus I_{rms} denotes the rms amplitude of $i(t)$, V_{rms} denotes the rms amplitude of $v(t)$, $I_{1\,rms}$ denotes the rms amplitude of $i_1(t)$, and so on. The dimension of the rms amplitude of a current or voltage is current or voltage, respectively.

The *rms* amplitude of a current or voltage is the *positive* square *root* of the *mean* (average) of the *squared* current or voltage. The order of the operations is important: First square the current or voltage, then compute the average, then take the positive square root. In general, these operations are not interchangeable.

Example 5.16. Obtain an expression for the rms amplitude of a constant current $i = I_0$, where I_0 might be positive or negative.
Solution: From (5.21), the mean squared amplitude is

$$\overline{i^2} = \overline{I_0{}^2} = I_0{}^2.$$

From (5.33), the rms amplitude is

$$I_{rms} = \sqrt{I_0{}^2} = |I_0|.$$

Example 5.17. Obtain an expression for the rms amplitude of a sinusoidal voltage $v = V_0\cos(\omega t + \theta)$.

Solution: From (5.25), the mean squared amplitude of a sinusoidal voltage having peak amplitude $|V_0|$ is given by

$$\overline{v^2} = \overline{V_0{}^2\cos^2(\omega t + \theta)} = \frac{V_0{}^2}{2}.$$

From (5.33), the rms amplitude of a sinusoidal voltage having peak amplitude $|V_0|$ is given by

$$V_{rms} = \sqrt{\overline{v^2(t)}} = \sqrt{\frac{V_0{}^2}{2}} = \frac{|V_0|}{\sqrt{2}}.$$

Example 5.17 shows that the rms amplitudes of a *sinusoidal* current and a *sinusoidal* voltage are given by

$$\left.\begin{aligned} i(t) = I_0\cos(\omega t+\theta) \Rightarrow I_{rms} = \frac{|I_0|}{\sqrt{2}} \\ v(t) = V_0\cos(\omega t+\theta) \Rightarrow V_{rms} = \frac{|V_0|}{\sqrt{2}} \end{aligned}\right\}\omega>0. \quad (5.34)$$

The mean squared and rms amplitudes of a sinusoidal current or voltage having frequency $\omega > 0$ are independent of both the frequency ω and the initial phase θ of the current or voltage. Equation (5.33) applies to any current or voltage. Equation (5.34) applies only to *sinusoidal* currents and voltages. Because the rms amplitude of a sinusoidal current or voltage is independent of the initial phase of the current or voltage (of the angle θ in (5.34)), the rms amplitude of a sinusoidal current or voltage is independent of the algebraic sign of the current or voltage and of whether the current or voltage is expressed as a cosine or sine; for example,

$$v(t) = V_0\sin(\omega t+\theta) = V_0\cos\left(\omega t + \underbrace{\theta - \pi/2}_{\theta'}\right)$$

$$\Rightarrow V_{rms} = \frac{|V_0|}{\sqrt{2}}.$$

Exercise 5.16. Give the rms amplitude of each of the following: (a) $i(t) = -25\,\text{mA}$, (b) $v(t) = 5\text{V}$, (c) $v(t) = -10\cos(2\pi f t + \pi/4)\text{V}$, (d) $i(t) = 10\sin(2\pi f t - \pi/4)\text{mA}$.

In general, *rms amplitudes are not additive*, because (in general)

$$\sqrt{\overline{(v_1+v_2)^2}} \neq \sqrt{\overline{v_1{}^2}} + \sqrt{\overline{v_2{}^2}}. \quad (5.35)$$

For example, if $V = V_1 + V_2$, where $V_1 = 5\text{V}$ and $V_2 = -5\text{V}$, then $V = V_1 + V_2 = 0$ and $V_{rms} = 0$ whereas $V_{1\,rms} + V_{2\,rms} = 10\text{V}$.

5.9 Average Power Dissipated in a Resistive Load

From (5.33), $I_{rms}{}^2 = \overline{i^2}$ and $V_{rms}{}^2 = \overline{v^2}$, so we may express the power dissipated by a resistive load as

$$P = I_{rms}{}^2 R \quad \text{or} \quad P = \frac{V_{rms}{}^2}{R}. \quad (5.36)$$

Under the passive sign convention, the power dissipated by a resistor is positive. Thus, from (5.36), the average power dissipated by a resistor having resistance R also can be expressed as

$$P = \sqrt{P^2} = \sqrt{\left(\frac{V_{rms}{}^2}{R}\right)\left(I_{rms}{}^2 R\right)} = V_{rms} I_{rms}, \quad (5.37)$$

where I_{rms} and V_{rms} are the rms amplitudes of the current through and voltage across the resistor. *Equations* (5.36) *and* (5.37) *apply (in general) only to resistive loads and to power delivered by a source.*

Example 5.18. Refer to Fig. 5.15, where $i_0 = I_0\cos(\omega t)$ and $v_0 = V_0\cos(\omega t)$, with $\omega > 0$. Obtain expressions for the rms amplitudes of v_R, v_1, i_R. Then obtain expressions for the average powers dissipated in the resistors.
Solution: Kirchhoff's current law gives

$$\frac{v_1}{R} + \frac{v_1 - v_0}{R} - i_0 = 0$$

$$\Rightarrow v_1 = \frac{Ri_0 + v_0}{2} = \frac{RI_0 + V_0}{2}\cos(\omega t),$$

whence

$$V_{1\,rms} = \frac{|RI_0 + V_0|}{2\sqrt{2}}.$$

It follows that

$$i_R = \frac{v_1}{R} \Rightarrow I_{R\,rms} = \frac{V_{1\,rms}}{R} = \frac{|RI_0 + V_0|}{2\sqrt{2}R}$$

and

$$v_R = v_0 - v_1 = \frac{V_0 - RI_0}{2}\cos(\omega t)$$
$$\Rightarrow V_{R\,rms} = \frac{|V_0 - RI_0|}{2\sqrt{2}}.$$

The average power dissipated in the resistor on the left is given by

$$P_1 = \frac{V_{R\,rms}{}^2}{R} = \frac{(V_0 - RI_0)^2}{8R}.$$

The average power dissipated in the resistor on the right is given by-

$$P_2 = \frac{V_{1\,rms}{}^2}{R} = \frac{(RI_0 + V_0)^2}{8R}$$

and by

$$P_2 = I_{R\,rms}{}^2 R = \left(\frac{RI_0 + V_0}{2\sqrt{2}R}\right)^2 R = \frac{(RI_0 + V_0)^2}{8R}.$$

Exercise 5.17. Refer to Fig. 5.15 and Example 5.17. Assume $V_0 > 0$, $I_0 > 0$. Obtain expressions for the average powers dissipated by the sources. Show that the total power dissipated equals zero.

In the electric utility industry, where currents and voltages are sinusoidal and where power is a primary concern, it is convenient and conventional to specify currents and voltages in terms of their rms amplitudes, rather than peak amplitudes. For example, the voltage available at a residential wall receptacle in North America is commonly said to be 120 V, which actually is the rms amplitude of a 60-Hz sinusoidal voltage whose peak amplitude is approximately 170 V. Using rms amplitudes in power calculations involving sinusoidal currents and voltages avoids cluttering expressions with the factors $1/2$ and $1/\sqrt{2}$.

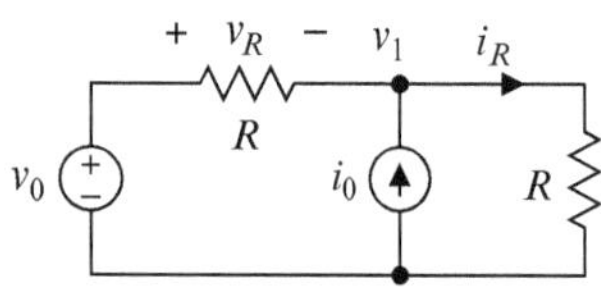

Fig. 5.15 See Example 5.18

It follows from (5.36) that any two currents or voltages having the same rms amplitudes deliver the same average power to a resistive load, even if the current or voltage waveforms are different. In other words, two currents or two voltages having equal rms amplitude are *equally effective* in delivering power to a resistive load.[11]

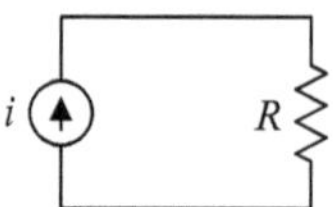

Fig. 5.16 See Example 5.19

Example 5.19. In the circuit shown in Fig. 5.16, $R = 1.5\ \text{k}\Omega$ and

$$i(t) = I_0 \cos(2\pi f t);\ \ I_0 = 3\text{mA}.$$

Find the average powers dissipated by the resistor and the source.

Solution: The rms amplitude of the current is

$$I_{rms} = \frac{|I_0|}{\sqrt{2}} = 2.12\ \text{mA}.$$

Thus the average power dissipated by the resistor is

$$P_R = I_{rms}{}^2 R = \left(2.12 \times 10^{-3}\ \text{A}\right)^2 (1500\ \Omega)$$
$$= 6.75\ \text{mW}.$$

Because power is conserved, the average power dissipated by the source (passive sign convention) is

$$P_i = -P_R = -6.75\ \text{mW}.$$

[11]For this reason, the rms amplitude of a current or voltage also is called the *effective value* of the current or voltage; however, this terminology is falling into disuse.

5.10 Summary: Power Relations

Table 5.1 summarizes definitions and relations given above.

5.11 Notation

Conventionally, rms is appended to the unit of a current or voltage to denote rms amplitude; for example, $V = 10$ Vrms. However, keep in mind that Vrms and Arms are not true units. For example, the relation between rms amplitude and peak amplitude of a current or voltage depends upon the waveform of the current or voltage. Alternatively, we often use expressions such as $V_{rms} = 10$ V, which means that the rms amplitude of *v*(*t*) is 10 V. We do not express time-dependent currents or voltages using rms amplitudes because rms amplitude is independent of time. For example, we do not write expressions such as $i_{rms}(t) = 10\cos(\omega t)$ V. Moreover, we avoid expressions such as $v(t) = \sqrt{2}V_{rms}\cos(\omega t)$, which is invalid for $\omega = 0$.

Other pseudo units you will encounter are Vac or VAC, which are synonymous with Vrms, and Vdc or VDC, which indicate a constant voltage. For example, a nameplate rating of 120 VAC @ 60 Hz in and 5 VDC out on small plug-in power supply means the supply should be plugged into an outlet providing a 60 Hz sinusoidal voltage having rms amplitude 120 V, and that under that condition the output is a constant voltage having amplitude 5 V. Such terminology is common in practice, and you should understand it. But in analysis and design, stick with precise terminology.

5.12 Measurement of RMS Amplitude

No instrument can take the limits specified by the *mathematical* definitions

$$I_{rms} = \sqrt{\lim_{T\to\infty} \frac{1}{T_0}\int_0^{T_0} i^2(t)dt}\,, \qquad V_{rms} = \sqrt{\lim_{T\to\infty} \frac{1}{T_0}\int_0^{T_0} v^2(t)\,dt}, \tag{5.38}$$

because such measurements would take forever. Fortunately, we do not need to average forever to measure the rms amplitude of a physical current or voltage. The rms amplitude of a current or voltage can be measured with any necessary precision by an instrument that computes rms amplitude using an average over a *finite* but sufficiently long interval. The required averaging time T_0 depends upon the nature of the current or voltage and upon the required precision of the result. For example, let $i = I_0\cos(\omega t)$. Then, for finite T_0 a measurement of rms amplitude yields

Table 5.1 Expressions for power dissipation in a resistor

Quantity	General (Definition)	Current through or voltage across a resistor (any waveform)	*Constant* current I_0 through or voltage V_0 across a resistor	*Sinusoidal* current $I_0\cos(\omega t)$ through or voltage $V_0\cos(\omega t)$ across a resistor
Instantaneous power	$p = vi$	$p = i^2 R = \frac{v^2}{R}$	$p = {I_0}^2 R = \frac{{V_0}^2}{R} = V_0 I_0$	$p = {I_0}^2 R\cos^2(\omega t) = \frac{{V_0}^2}{R}\cos^2(\omega t) = I_0 V_0\cos^2(\omega t)$
Peak power	$\hat{p} = \max\lvert p\rvert$	$\hat{p} = R\max\lvert i^2\rvert = \frac{\max\lvert v^2\rvert}{R}$	$\hat{p} = {I_0}^2 R = \frac{{V_0}^2}{R} = V_0 I_0$	$\hat{p} = {I_0}^2 R = \frac{{V_0}^2}{R} = V_0 I_0$
Average power	$P = \lim_{T\to\infty}\frac{1}{T}\int_0^T p\,dt$	$P = {I_{rms}}^2 R = \frac{{V_{rms}}^2}{R} = V_{rms} I_{rms}$	$P = {I_0}^2 R = \frac{{V_0}^2}{R} = V_0 I_0$	$P = \frac{{I_0}^2 R}{2} = \frac{{V_0}^2}{2R} = \frac{V_0 I_0}{2}$

$$
\begin{aligned}
I_{rms} &\cong \sqrt{\frac{1}{T_0}\int_0^{T_0} i^2(t)\,dt} = \sqrt{\frac{1}{T_0}\int_0^{T_0} {I_0}^2\cos^2(\omega\,t)\,dt} \\
&= \sqrt{\frac{{I_0}^2}{2\,T_0}\int_0^{T_0} [1+\cos(2\,\omega\,t)]\,dt} \\
&= \sqrt{\frac{{I_0}^2}{2\,T_0}\left[T_0 + \frac{1}{2\,\omega}\sin(2\,\omega\,T_0)\right]} \\
&= \sqrt{\frac{{I_0}^2}{2} + \frac{{I_0}^2}{4\,\omega\,T_0}\sin(2\,\omega\,T_0)} \\
&= \frac{|I_0|}{\sqrt{2}}\sqrt{1+\frac{\sin(2\,\omega\,T_0)}{2\,\omega\,T_0}}. \qquad (5.39)
\end{aligned}
$$

The factor $|I_0|/\sqrt{2}$ is the mean squared amplitude (the desired result). Because $|\sin(2\,\omega\,T_0)| \le 1$, the error (magnitude) is bounded as

$$|\varepsilon| \le \frac{|I_0|}{\sqrt{2}}\sqrt{1+\frac{1}{2\,\omega\,T_0}} - \frac{|I_0|}{\sqrt{2}}.$$

We assume $2\,\omega\,T_0 \gg 1$ and use the approximation $\sqrt{1+x} \cong 1 + x/2$ to obtain

$$|\varepsilon| \le \frac{|I_0|}{\sqrt{2}}\left(1+\frac{1}{4\,\omega\,T_0}\right) - \frac{|I_0|}{\sqrt{2}} = \frac{|I_0|/\sqrt{2}}{4\,\omega\,T_0}.$$

Thus the error expressed as a percent of the true value is bounded as

$$|\varepsilon|(\%) \le \frac{100}{4\,\omega\,T_0} \cong \frac{4}{f\,T_0}, \qquad (5.40)$$

where f is frequency in hertz ($\omega = 2\pi f$). For example, if the magnitude of the error must be smaller than 1% of the actual rms amplitude, we would require

$$\frac{4}{f\,T_0} < 1 \Rightarrow T_0 > \frac{4}{f}.$$

For any specified percent error, the required averaging time decreases as the frequency is increased. A longer averaging time is needed for a low-frequency sinusoid than for a higher frequency one. Conversely, if the averaging time T_0 is fixed, increasing (decreasing) the frequency of an applied current or voltage increases (decreases) the accuracy of the measurement. Again, the required averaging time depends upon both the required precision of the result and one or more properties of the current or voltage (frequency, in this example).

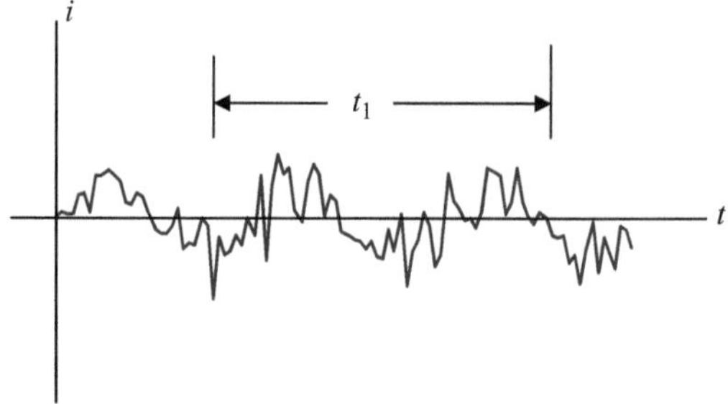

Fig. 5.17 Pertaining to the time required to measure the rms amplitude of a current or voltage (see text)

More generally, to measure the rms amplitude of a non-sinusoidal, time-varying current or voltage, we need only average over a time much longer than the duration of the most slowly varying component of the current or voltage. For example, inspection of the current waveform shown in Fig. 5.17 suggests that the period of the most slowly varying component of the waveform is (approximately) equal to the time t_1 shown on the graph. To measure the rms amplitude of the current shown in Fig. 5.17, we would average for a time much longer than t_1.

Example 5.20. The averaging time employed by a certain rms voltmeter is fixed and is such that the maximum percent error in a measurement of the rms amplitude of a 100 Hz sinusoidal voltage is 0.1%. What is the maximum percent error in a measurement of the rms amplitude of a 10 Hz sinusoidal voltage?
Solution: We use (5.40) to find the averaging time:

$$\frac{4}{f\,T_0} = 0.1 \Rightarrow T_0 = \frac{40}{f} = \frac{40}{100\,\mathrm{Hz}} = 0.4\,\mathrm{s}.$$

We use (5.40) with $T_0 = 0.4\,\mathrm{s}$ and $f = 10\,\mathrm{Hz}$ to obtain

$$\max[|\varepsilon|(\%)] \cong \frac{4}{f\,T_0} = \frac{4}{(10\,\mathrm{Hz})(0.4\,\mathrm{s})} = 1\%.$$

Exercise 5.18. A certain rms voltmeter is advertised as providing ±0.1% accuracy for sinusoidal voltages having frequencies from 10 Hz to 10 MHz. (a) What is the minimum averaging time used by the voltmeter? (b) If the error is ±0.1% at 10 Hz, what is the error at 10 MHz? Assume that finite averaging time is the only source of error.

Exercise 5.19. The rms amplitude of a certain voltage is to be measured. If it is known that the voltage is sinusoidal, is there a fast way to perform the measurement that does not rely on averages and provides high accuracy?

Finite averaging time is a significant source of rms measurement error for low-frequency signals, but diminishes in effect as frequency is increased (see (5.40)). This does not mean that rms measurements are arbitrarily precise for high-frequency signals, however, because finite averaging time is only one of several sources of error. Other sources of error dominate at high frequencies, where the error due to finite averaging time is minimal. In fact, measurement accuracy generally increases with increasing frequency up to a point, but then begins to fall as frequency is increased further. Manufacturers of measurement devices supply data sheets for their products which specify the cumulative effects of the most significant error sources at various frequencies. In practice, engineers rely on these data sheets to determine the accuracy of a particular measurement.

5.13 Dissipation Derating

The ability of a resistor to safely dissipate power generally decreases with increasing temperature. For example, a ½ W resistor that can safely dissipate ½ W when its temperature is 25°C might be able to safely dissipate only ¼ W when its temperature is 120°C. In practice, the power-dissipation ratings of resistors (and other components) are reduced (*derated*) by a percentage depending upon the expected maximum operating temperature. Derating is based upon information given on manufacturers' data sheets. For example, Fig. 5.18 shows part of the data sheet (highlights added) for a particular family of thick-film resistors. The derating specification means that a resistor from this family can dissipate at least 35 W at case temperatures up to 25°C, but the power dissipation rating decreases linearly above 25°C, reaching zero at 150°C. Derating specifications also are often presented graphically. For example, the derating specification given in Fig. 5.18 can be presented graphically, as in Fig. 5.19.

Example 5.21. Refer to Fig. 5.19. What is the safe dissipation rating for a resistor of that family if the operating (case) temperature is 75°C?

Solution: From the graph, the safe dissipation rating is 21 W, or 60% of rated dissipation (at 25°C).

A resistor carrying current exhibits **self heating** due to Joule's law ($P = I_{rms}{}^2R$). The increase in temperature due to self heating is calculated using the **self-heating coefficient** φ defined by

$$\varphi = \frac{\Delta T}{\Delta P}, \tag{5.41}$$

where ΔP is a change in power dissipated and ΔT is the associated change in temperature. The self-heating coefficient for a resistor is the reciprocal of the slope of the derating characteristic. For temperatures above 25°C, the self-heating coefficient for a resistor having the derating characteristic in Fig. 5.19 is

$$\varphi = \frac{150°\text{C} - 25°\text{C}}{35\,\text{W}} = 3.57°\text{C}\,\text{W}^{-1}. \tag{5.42}$$

The operating temperature of a resistor is determined by adding the ambient (environmental) temperature to the temperature increase due to self heating. Once the operating temperature has been determined, the resistor can be derated using the manufacturer's specifications.

FEATURES

- 35 Watt power rating at 25°C
- SMD - DPAK package configuration
- Heat resistance to cooling plate: R_{th} < 4.28°C/W
- A molded case for environmental protection.
- Resistor element is electrically insulated from the metal sink tab.

SPECIFICATIONS

Material

Terminal: Copper

Terminal Plating: Lead Free Solder (97% Tin, 3% Silver)

Electrical

Resistance Range: 0.05 Ω to 10 KΩ other values on request

Tolerance: ±1% to ±10% (0.5% on request)

Max. Operating Voltage: 350 V

Insulation Resistance: 10 GΩ min.

Power Rating: Depends upon case temperature. See derating curve.

Working Temperature Range: –55°C to +175°C

Solder Process: The TDH35P cannot exceed 220°C (260°C for the TDH35H) for more than 10 seconds during soldering process.

Derating: 100% @ 25°C to 0% @ 150°C curve referenced to case temperature

Dielectric Strength: 1,800VAC

Operating Temperature Range: –55°C to +150°C

Temperature Coefficient: 10 Ω and above, ±50ppm/°C, referenced to 25°C, ΔR taken at +105°C. Between 1 and 10 Ω, ±(100ppm+0.002 Ω) / °C, referenced to 25°C, ΔR taken at +105°C. For under 10 Ω:
0R6 - 9R9 ..... 100PPM
0R4 - 0R59 ... 150PPM
0R2 - 0R39 ... 250PPM
0R1 - 0R19 ... 500PPM
0R05 - 0R09 = 1000PPM

Inductance: less than 20 nanohenries

Flatness: less than 0.1mm tolerance

Fig. 5.18 Extracted from the data sheet for Ohmite's TDH series of 35 W thick-film power surface mount resistors (highlights added) (Courtesy of Ohmite Manufacturing Corporation)

Example 5.22. A 13.9 Ω resistor having the derating characteristic in Fig. 5.19 will carry rms current $I_{rms} = 1.00\,\text{A}$. What is the maximum safe ambient temperature?

Solution: The power that must be safely dissipated by the resistor is

$$P = I_{rms}{}^2R = (1.00\,\text{A})^2(13.9\,\Omega) = 13.9\text{W}.$$

From (5.42), the self-heating coefficient is $3.57°\text{C}\,\text{W}^{-1}$. Thus the temperature increase due to self heating is

$$\Delta T = \left(3.57°\text{C}\,\text{W}^{-1}\right)(10.0\,\text{W}) = 35.7°\text{C}.$$

From Fig. 5.19, the maximum operating temperature for which the resistor can safely dissipate 13.9W is approximately $T = 100°\text{C}$. Thus the maximum safe ambient temperature is

$$T_A = T - \Delta T = 100°\text{C} - 35.7°\text{C} = 76.3°\text{C}$$

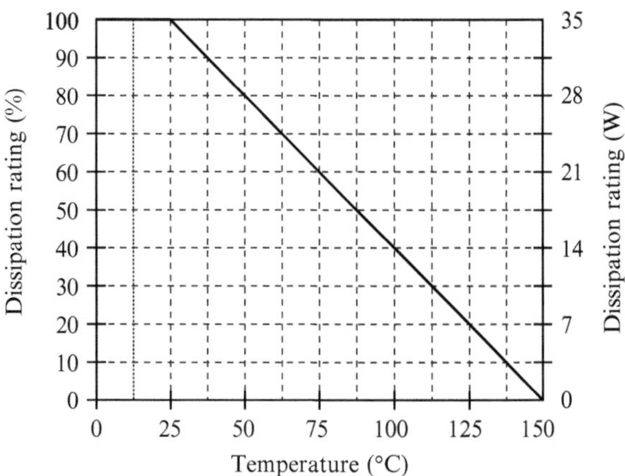

Fig. 5.19 Power dissipation derating specification for Ohmite's TDH series of 35 W thick-film power surface mount resistors, expressed graphically

Example 5.23. A particular application requires a 5Ω resistor to carry rms current $I_{rms} = 2\,\text{A}$ in an environment where the ambient temperature is $T_a = 50°\text{C}$. Can a 35W

resistor having the derating characteristic in Fig. 5.19 be used in this application?
Solution: The power to be dissipated is

$$P = I_{rms}{}^2 R = (2\,\text{A})^2 (5\,\Omega) = 20\,\text{W}.$$

From (5.42), the self-heating coefficient is $\varphi = 3.57°\text{C}\,\text{W}^{-1}$. Thus the temperature increase due to self heating is

$$\Delta T = \varphi P = \left(3.57°\text{C}\,\text{W}^{-1}\right)(20\,\text{W}) = 71.4°\text{C}.$$

The operating temperature is

$$T = T_a + \Delta T = 50°\text{C} + 71.4°\text{C} = 121.4°\text{C}.$$

From Fig. 5.19, the maximum safe dissipation at this temperature is only slightly more than 7W, whereas the required dissipation is 20W. We must look for a resistor having a larger power rating or a smaller self-heating coefficient (or both). Alternatively, we might use some kind of forced cooling, which can be expensive but is often necessary.

Example 5.24. The power-dissipation rating for a certain resistor is 250 mW at 25°C, derated linearly to zero at 125°C. (a) What is the self-heating coefficient for this resistor? (b) What is the power-dissipation rating at 100°C? (c) Draw a graph of permissible power dissipation versus temperature for this resistor.
Solution: (a) From (5.41), the self-heating coefficient is

$$\varphi = \frac{\Delta T}{\Delta P} = \frac{(125 - 25)°\text{C}}{(0.25 - 0)\,\text{W}} = 400°\text{CW}^{-1}. \quad (5.43)$$

(b) For an operating temperature of 100°C, the temperature increase above the rated operating temperature is $\Delta T = 100°\text{C} - 25°\text{C} = 75°\text{C}$ and the corresponding decrease in permissible power dissipation is given by:

$$\Delta P = \frac{\Delta T}{\varphi} = \frac{75°\text{C}}{400°\text{CW}^{-1}} \cong 188\,\text{mW}.$$

Thus the permissible power dissipation at 100°C is 250 − 188 mW = 62 mW.

(c) Figure 5.20 shows a graph of power-dissipation rating versus temperature.

Resistance varies with temperature, and operating temperature is one of the two main sources of uncertainty regarding the resistance of a resistor. The other is the **initial tolerance**, which is the precision of the nominal resistance at the temperature corresponding to rated power dissipation, often 25°C. Initial tolerance can range from ±20% for some carbon-composition resistors down to ±0.05% or less for high-precision resistors.

From Chapter 2, resistance R as a function of temperature T is given by

$$R = R_0\left[1 + \alpha_0(T - T_0)\right], \quad (5.44)$$

where R_0 is the resistance at temperature T_0 and α_0 is the **temperature coefficient of resistance** (**TCR**) at that temperature. For present purposes, the temperature of a resistor is given by $T = T_A + \Delta T$, where T_A is ambient temperature and ΔT is the temperature increase due to self heating, given by $\Delta T = \varphi P = \varphi I_{rms}{}^2 R_0$, where φ is the self-heating coefficient and I_{rms} is the rms current through the resistor. Using these relations in (5.44) gives

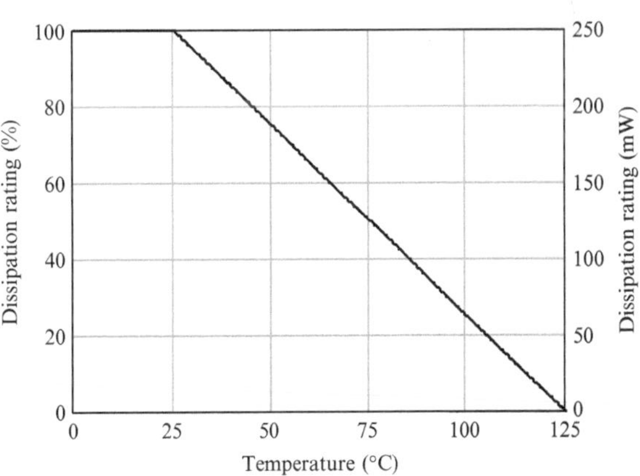

Fig. 5.20 See Example 5.24

$$R = R_0\left[1 + \alpha_0\left(T_A + \varphi I_{rms}{}^2 R_0 - T_0\right)\right]. \tag{5.45}$$

It follows that the error due to operating temperature is given by

$$\Delta R = R_0\alpha_0\left(T_A + \varphi I_{rms}{}^2 R_0 - T_0\right). \tag{5.46}$$

Thus the effective tolerance (%) is given by

$$\begin{aligned}\beta &= \beta_0 + 100\frac{\Delta R}{R_0}\\ &= \beta_0 + 100\alpha_0\left(T_A + \varphi I_{rms}{}^2 R_0 - T_0\right),\end{aligned} \tag{5.47}$$

where β_0 is the initial tolerance (%).

Example 5.25. A certain resistor has resistance $R_0 = 100\,\mathrm{k\Omega} \pm 5\%$ and temperature coefficient of resistance $\alpha_0 = 0.001°\mathrm{C}^{-1}$, both at 25°C. The power-dissipation rating is 250 mW at 25°C, derated linearly to zero at 100°C. The rms current through the resistor is $I_{rms} = 1\,\mathrm{mA}$ and the ambient temperature is $T_A = 40°\mathrm{C}$. (a) What is the operating temperature of the resistor? (b) What are the resistance and effective tolerance at that temperature? (c) What is the power-dissipation rating at the operating temperature?
Solution: (a) The self-heating coefficient is

$$\varphi = \frac{\Delta T}{\Delta P} = \frac{(100-25)°\mathrm{C}}{(0.250)\,\mathrm{W}} = 300°\mathrm{CW}^{-1}.$$

The temperature increase ΔT due to self heating is given by

$$\begin{aligned}\Delta T &= \varphi\,\Delta P = \varphi I_{rms}{}^2 R_0\\ &= \left(300°\mathrm{C\,W}^{-1}\right)(1\,\mathrm{mA})^2(100\,\mathrm{k\Omega}) = 30°\mathrm{C}.\end{aligned}$$

The operating temperature is $T = T_A + \Delta T = 40°\mathrm{C} + 30°\mathrm{C} = 70°\mathrm{C}$.
(b) The resistance at that temperature is

$$\begin{aligned}R &= R_0(1 + \alpha_0 T)\\ &= (100\,\mathrm{k\Omega})\left[1 + \left(0.001°\mathrm{C}^{-1}\right)(70°\mathrm{C})\right]\\ &= 107\,\mathrm{k\Omega},\end{aligned}$$

so the percent error due to heating alone is 7%. The effective tolerance is the sum of the error due to heating and the initial tolerance.

$$\text{effective tolerance} = 5\% + 7\% = 12\%.$$

(c) For temperatures above 25°C, the power dissipation must be derated. The permissible dissipation is expressed as a percentage of the 25°C rating by

$$D = 100\left(1 - \frac{T - 25°\mathrm{C}}{T_{\max} - 25°\mathrm{C}}\right);\ T_{\max} = 100°\mathrm{C}.$$

Thus

$$D = 100\left(1 - \frac{70°\mathrm{C} - 25°\mathrm{C}}{75°\mathrm{C}}\right) = 40\%.$$

So the permissible dissipation is $(0.40)(250\,\mathrm{mW}) \cong 100$ mW.

Example 5.26. A certain 1 W resistor has resistance $R = 1\,\mathrm{k\Omega} \pm 1\%$ at 25°C. The rms amplitude of the current through the resistor is $I_{rms} = 8\,\mathrm{mA}$. The self-heating coefficient is $\varphi = 100°\mathrm{CW}^{-1}$. The temperature coefficient of resistance at 25°C is $\alpha = 5 \times 10^{-4}\ °\mathrm{C}^{-1}$ (5000 ppm) and the environmental temperature due to heat from other components ranges from 15°C to 70°C. What is the maximum value of the actual resistance?
Solution: The maximum resistance at 25°C is $1\mathrm{k\Omega}+1\% = 1.010\mathrm{k\Omega}$. The corresponding power dissipated by the resistor is

$$P = (8\,\mathrm{mA})^2(1.010\,\mathrm{k\Omega}) \cong 64.6\,\mathrm{mW}.$$

The temperature increase due to self heating is

$$\Delta T_{SH} = \varphi P = \left(100°\mathrm{CW}^{-1}\right)(64.6\mathrm{mW}) = 6.46°\mathrm{C}$$

and the corresponding increase in resistance due to self heating is

$$\Delta R_{SH} = \alpha \Delta T_{SH} R$$
$$= (5 \times 10^{-4}\,°\mathrm{C}^{-1})(6.46°\mathrm{C})(1.010\,\mathrm{k}\Omega)$$
$$\cong 3.26\,\Omega.$$

The maximum increase in resistance due to ambient temperature is

$$\Delta R_A = \alpha T_{\max} = (5 \times 10^{-4}\,°\mathrm{C}^{-1})(70°\mathrm{C})(1.010\mathrm{k}\Omega)$$
$$\cong 35.4\Omega.$$

Thus the maximum possible resistance is

$$R_{\max} = (1.010\mathrm{k}\Omega) + 3.26\Omega + 35.4\Omega \cong 1.049\mathrm{k}\Omega.$$

So the actual upper limit on the resistance is 4.9%, not 1%.

Exercise 5.20. What is the minimum resistance of the resistor considered in Example 5.26 above?

Anticipating and allowing for resistance variation with temperature is an important part of circuit design. In demanding applications, specifying small temperature coefficients and small self-heating coefficients is at least as important as specifying a small tolerance. If sufficiently small temperature coefficients and self-heating coefficients cannot be achieved, some sort of forced cooling is indicated.

5.14 Power Dissipation in Physical Components and Circuits

Elements in a circuit diagram (model) do not necessarily correspond to physical components in an associated physical circuit.[12] *Power dissipated by a non-physical element is itself non-physical and often meaningless.* For example, the average powers dissipated (individually) by a Thévenin equivalent resistance and Thévenin equivalent source are in general meaningless, because the Thévenin resistance and the Thévenin source are non-physical elements. An expression or value for the power dissipated by an element in a circuit diagram (circuit model) is meaningful only if the element corresponds to a physical component in an associated physical circuit. Alternatively, the power dissipated by a two-terminal circuit is meaningful only if the voltage across and current through the circuit equal the voltage across and current through a physical component modeled by the circuit.

Example 5.27. Consider the circuit shown in Fig. 5.21(a), where each resistor corresponds to a physical resistor and the source corresponds to a physical source (e.g., a battery). The average power *delivered* by the source is given by

$$P_S = V_0\left(\frac{V_0}{4R}\right) = \frac{{V_0}^2}{4R}.$$

This is a meaningful calculation, because the source represents a physical component. In a design context, the calculation tells us that the voltage source must be capable of delivering the power thus calculated.

Now consider Fig. 5.21(b), where the circuit has been replaced by its Thévenin equivalent. In this case, the power delivered by the Thévenin source is given by

$$P'_S = \left(\frac{2V_0/3}{4R + 4R/3}\right)\frac{2V_0}{3} = \frac{{V_0}^2}{12R},$$

which is only one-third of the value obtained above. The power delivered by the Thévenin

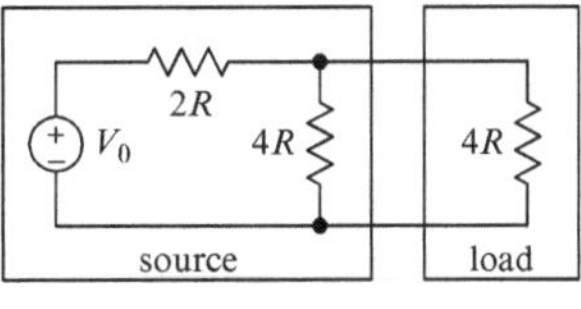

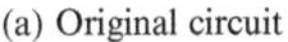

(a) Original circuit

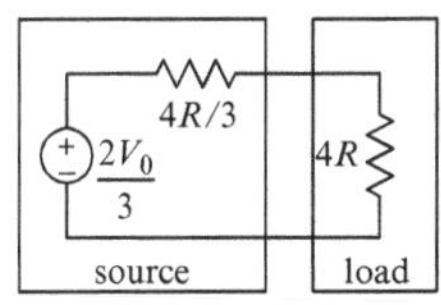

(b) Source replaced by Theveinin equivalent

Fig. 5.21 See Example 5.27

[12]Review Section 3.1 of Chapter 3.

equivalent source is non-physical, and cannot be interpreted as the power required of a physical source. In general, meaningful power calculations must be based upon physical circuit models; i.e., models in which circuit elements correspond to physical components.

Exercise 5.21. Show that the power dissipated by the load in Fig. 5.21(b) equals the power dissipated by the load in Fig. 5.21(a).

When calculating the power dissipated by a particular physical component, it is permissible to replace the rest of the circuit by the (non-physical) Thévenin or Norton equivalent, because doing so does not change the terminal characteristics (the $i - v$ relationship) at the terminals of the component in question.

Example 5.28. In the circuit shown in Fig. 5.22(a),

$$I_0 = 10\,\text{mA},\ V_0 = 500\,\text{mV},\ R_1 = 50\,\Omega,\ R_2 = 100\,\Omega.$$

(1) Calculate the power dissipated by each resistor. (2) Repeat for the "equivalent" circuit shown in Fig. 5.22(b), where the circuit to the left of the resistor R_2 is replaced by its Norton equivalent. Compare the computed values and explain.
Solution: (1) Applying Kirchhoff's current law to the circuit in Fig. 5.22(a) gives

$$\frac{V_a - V_0}{R_1} + \frac{V_a}{R_2} - I_0 = 0$$

$$\Rightarrow V_a = \frac{(V_0 + R_1 I_0) R_2}{R_1 + R_2} = 667\,\text{mV}.$$

The powers P_1, P_2 dissipated by R_1, R_2, respectively, are

$$P_1 = \frac{(V_a - V_0)^2}{R_1} = 556\,\mu\text{W},$$

$$P_2 = \frac{V_a^{\,2}}{R_2} = 4.44\,\text{mW}.$$

For the "equivalent" circuit in Fig. 5.22(b), we obtain (current division)

$$I_1 = \frac{R_2 (I_0 + V_0/R_1)}{R_1 + R_2},\quad I_2 = \frac{R_1 (I_0 + V_0/R_1)}{R_1 + R_2},$$

and so

$$P_1 = I_1^{\,2} R_1 = 8.89\text{mW}, P_2 = I_2^{\,2} R_2 = 4.44\text{mW}.$$

The power dissipated by R_2 is the same for both circuits because the circuit seen by R_2 in Fig. 5.22(b) is the Norton equivalent of that seen by R_2 in Fig. 5.22(a); that is, the terminal characteristic of the circuit to the left of R_2 is the same in both cases. The power dissipated by R_1 is different for the two circuits because the circuit seen by R_1 in Fig. 5.22(b) is *not* the Norton equivalent of that seen by R_1 in Fig. 5.22(a). In other words, if the resistor R_2 in Fig. 5.22(a) represents a physical resistor, then so does the resistor R_2 in Fig. 5.22(b), because only the rest of the circuit has been transformed. On the other hand, if the voltage source and the resistor R_1 in Fig. 5.22(a) represent a physical source and resistor, then the current source and resistor R_1 in Fig. 5.22(b) do not. Do not be misled by the fact that the original resistor and the Norton equivalent resistor are both labeled R_1, which means only that the two resistors have the same resistance, *not* that they are the same physical resistor.

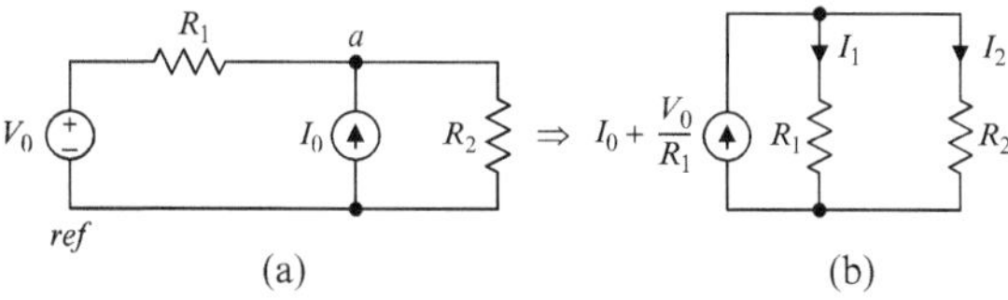

Fig. 5.22 See Example 5.28

To calculate average and peak power for each component in a circuit, we must know (in general) the instantaneous voltage across and current into the terminals of each component under all operating conditions. In many cases, we must also account for resistance in wires that otherwise might be represented by ideal conductors in a circuit model, and we must know the efficiency with which some components convert electrical energy to the desired form (some is converted to heat); for example, much of the energy delivered to a loudspeaker is converted to acoustic energy, not to heat (the efficiency of a typical high-quality loudspeaker is about 70% or smaller).

5.15 Active and Passive Devices, Loads, and Circuits

In the discussion that follows, *device* means *physical component*, not idealized circuit element. For example, a battery is a physical component, whereas an independent voltage source is an idealized circuit element (which often is a useful model for a battery).

Devices often are described as being either active or passive. In common usage, an **active device** is a device that allows one current or voltage to control another. Transistors are important active devices, as are most integrated circuits. Without active devices, there would be no electronic circuits (and little need for electrical engineers).

A device that is not active is a **passive device**. Most passive devices have only two terminals, whereas active devices have three or more terminals. A resistor is a passive device.

An **active circuit** is a circuit incorporating one or more active devices and which allows some currents or voltages to control others. An audio amplifier is an active circuit because it incorporates active devices and because it allows the current or voltage from (e.g.) a microphone to control the current or voltage delivered to a loudspeaker. The power delivered to the loudspeaker is provided by a power supply, not by the microphone, but is controlled by the current or voltage produced at the terminals of the microphone.

A **passive circuit** is a circuit containing only passive devices. Passive circuits are important components of many larger, active circuits; for example, one or more passive circuits may be needed to establish the conditions under which a transistor can perform as an active device. Thus, although virtually all electrical systems for communication, control, and other applications require active devices and circuits, they also contain a great many passive circuits.

The terms *load* and *source* are used in various ways in various contexts. Generally, a two-terminal element, device, or circuit is regarded as a **load** if the average power dissipated by the element, device, or circuit is positive and as a **source** if the average power dissipated by the element, device, or circuit is negative. When we say that one element, device, or circuit *loads* a second element, device, or circuit, we mean that the first draws average power from the second; i.e., that the total average power dissipated by the first (the load) is positive and that the total average power dissipated by the second (the source) is negative. When we say that a source is *lightly loaded*, we mean that the power delivered by the source is a small fraction of the power available from the source. Conversely, the average power delivered by a *heavily loaded* source is a large fraction of the available power. As an analogy, the engine of a car idling at a traffic light is lightly loaded, whereas the engine of a car pulling a heavy trailer up a steep grade might be heavily loaded. For an electrical example, refer to Fig. 5.23, which shows a source driving a load. For simplicity, the voltage V_S is constant. Thus the current I is constant and the average power delivered by the source V_S to the load R_L is given by

$$P_L = I_S{}^2 R_L = \frac{V_S{}^2 R_L}{(R_S + R_L)^2}.$$

From (5.17), the available power is given by

$$P_{L\,\max} = \frac{V_S{}^2}{4R_S},$$

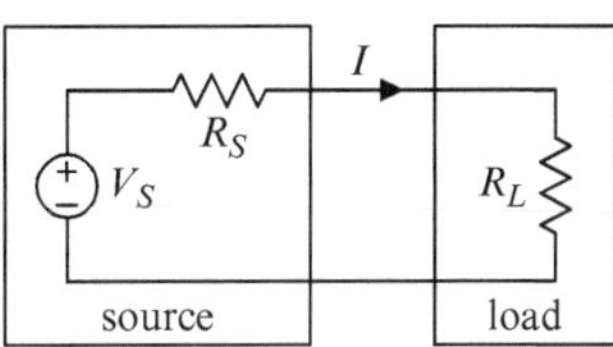

Fig. 5.23 Source driving a load

so the fraction of the available power delivered to the load is given by

$$\alpha = \frac{P_L}{P_{L\,\max}} = \frac{{V_S}^2 R_L}{(R_S + R_L)^2} \frac{4R_S}{{V_S}^2} = \frac{4R_L R_S}{(R_S + R_L)^2}.$$

If $R_L \cong R_S$, then $\alpha \cong 1$ and the source is heavily loaded. If either $R_L \gg R_S$ or $R_L \ll R_S$, then $\alpha \ll 1$ and the source is lightly loaded.

Loads (two-terminal circuits) also are called **active** or **passive**, depending upon whether they require a separate source of power to be effective. An active load is essentially an active device, in the sense that an active load controls (or dissipates) power provided by a supply that is separate from the current or voltage driving the load. In some applications, a dc electric motor is wired such that the armature current is provided by a fixed source and the relatively small field current produced by an electronic circuit controls the speed of the motor. A dc motor operated in this manner is an active load. A passive load requires no separate power source. Whereas the power dissipated by an active load comes partly (or mostly) from a separate power supply, the power dissipated by a passive load comes entirely from the circuit driving the load.

Do not attach too much significance to the definitions above of the terms *active* and *passive*. The terms are useful and much-used by electrical engineers, but a rigorous definition that would allow us to determine whether any particular circuit is active or passive would be of little value in circuit analysis and design. Also be aware that not all authors are in agreement regarding the definitions.

5.16 Power Transfer and Power Transfer Efficiency

Power transfer refers to *average power transferred to a load from a source*, where the source, the load, or both might be circuits. The power transferred to a load is the power dissipated in the load.

Refer to the model shown in Fig. 5.24, where a source drives a load. We wish to express the power transferred to the load in terms of the model parameters v_S, R_S, R_L. The power transferred to the load is given by

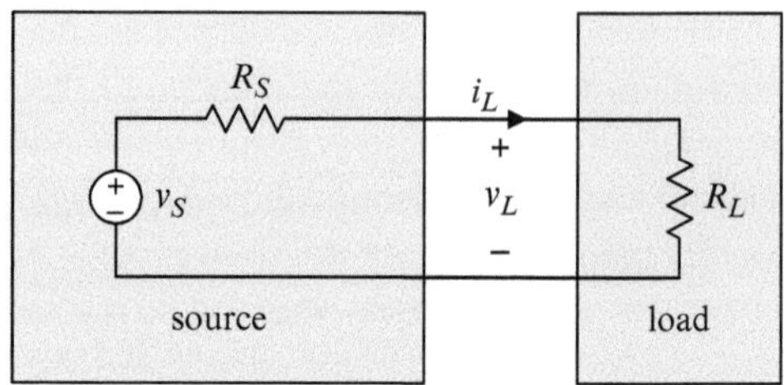

Fig. 5.24 Model used to study power transfer

$$P_L = {I_{L\,rms}}^2 R_L = \left(\frac{V_{S\,rms}}{R_S + R_L}\right)^2 R_L = \frac{R_L\, {V_{S\,rms}}^2}{(R_S + R_L)^2}. \tag{5.48}$$

We consider power transfer from two perspectives: First, from the perspective of someone who is designing the circuit that will serve as the load and second from the perspective of one who is designing the circuit that will serve as the source.

If the source is fixed but the load resistance is variable, the load resistance that maximizes power transfer is that for which

$$\frac{dP_L}{dR_L} = 0 \Rightarrow \frac{d}{dR_L}\left[\frac{R_L\, {V_{S\,rms}}^2}{(R_S + R_L)^2}\right] = \frac{(R_S + R_L)^2 {V_{S\,rms}}^2 - 2R_L\, {V_{S\,rms}}^2 (R_S + R_L)}{(R_S + R_L)^4} = 0,$$

which yields

$$R_L = R_S. \tag{5.49}$$

If the source is fixed, then the load resistance that maximizes power transfer is equal to the source resistance, provided the source resistance is not zero. Recall from Section 5.5 that the maximum power that can be drawn from a source is called the *available power* for the source. With reference to Fig. 5.24, the load that draws the available (maximum) power from a source equals the source resistance R_S.

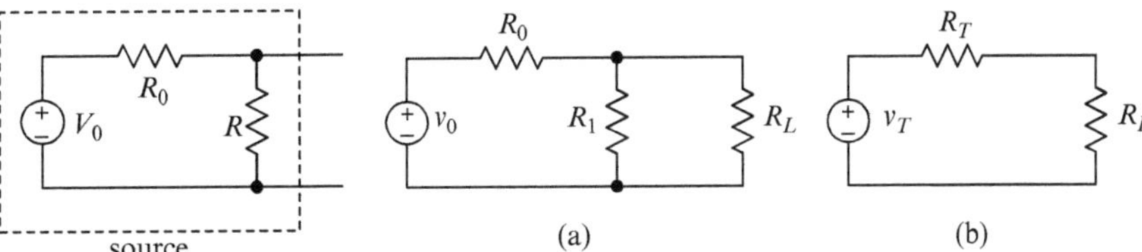

Fig. 5.25 See Exercise 5.22

Fig. 5.26 A circuit (**a**) and its Thévenin equivalent (**b**)

Exercise 5.22. Obtain an expression for the maximum power that can be transferred to a load by the source (voltage divider) in Fig. 5.25. If the maximum power is delivered to a load (not shown), what is the power *delivered* by the source V_0? What is the power *dissipated* in each of R_0 and R?

Inspection of (5.48) reveals that if the load resistance R_L is fixed and *the voltage* v_S *is independent of the source resistance*, then minimizing the source resistance maximizes the power transferred to the load. From the perspective of one designing a circuit regarded as a source, minimizing the source resistance is desirable, whether or not maximum power transfer is a design goal, because power dissipated in the source is wasted and generates unwanted heat. For example, we may regard an audio amplifier as a source and a loudspeaker as a load. One designing such an amplifier would strive to achieve the lowest possible output resistance to avoid wasting power and generating unwanted heat. But there is a limit to the current an amplifier can provide without suffering damage, so the designer would also call for fuses or circuit breakers to protect the amplifier in case someone attaches a load that would otherwise draw too much current.

In general, to maximize the power transferred to a load, we would minimize the source resistance and then make the load resistance equal to the source resistance (assuming the source can provide the theoretical maximum power). But there is a subtle trap one can fall into regarding maximum power transfer. Inspired by the derivation above, one might be tempted to try to achieve maximum power transfer by first finding the Thévenin equivalent for a source and then finding the values of the variable circuit parameters for which the Thévenin equivalent resistance is minimum. This approach can fail if, as is usual, both the Thévenin resistance and the Thévenin voltage depend upon some of the same parameters. In that case, decreasing the Thévenin resistance might also decrease the Thévenin voltage and decrease, rather than increase, the power delivered to the load. For example, suppose we wish to maximize the power transferred to the fixed load R_L by the circuit shown in Fig. 5.26(a). Suppose we proceed by first obtaining the Thévenin equivalent shown in Fig. 5.26(b), where

$$v_T = \frac{R_1 v_0}{R_0 + R_1}, \quad R_T = R_0 \| R_1 = \frac{R_0 R_1}{R_0 + R_1}. \qquad (5.50)$$

Suppose further that R_0 and R_1 represent physical resistors, where R_0 is fixed and R_1 is somewhat variable. With reference to the Thévenin equivalent in Fig. 5.26(b), the analysis above tells us to minimize the Thévenin equivalent resistance R_T. Because R_0 is fixed, the only way to do that is to minimize the resistance R_1. However, minimizing R_1 *minimizes* the power transferred to the load. To achieve maximum power transfer in this case, we must maximize R_1 which *maximizes* the Thévenin equivalent resistance. Again, keep in mind that the elements of the Thévenin (or Norton) equivalent for a source are non-physical, and consequently that the powers dissipated by those elements (individually) might have little or no relation to the power dissipated by elements in the associated physical circuit.

In general, finding conditions for maximum power transfer if source parameters are variable requires obtaining a circuit model whose elements correspond to physical components, writing an expression for the power transferred to the load in terms of the variable parameters, and maximizing the power transferred to the load, subject to whatever constraints are imposed on those parameters (e.g., the permissible values of a certain variable resistance might be $10\,\Omega \leq R \leq 5\text{k}\Omega$).

Problems of this kind generally require numerical methods or techniques based on the calculus of variations, and are beyond the scope of this book.

The **power transfer efficiency** of the circuit in Fig. 5.24 is denoted by η and is defined by

$$\eta = \frac{P_L}{P_T}, \tag{5.51}$$

where P_L is the power delivered to the load and P_T is the total power produced by the voltage source. The difference $P_T - P_L$ is dissipated (wasted) in the source resistance, usually as heat. From (5.48), the power transferred to the load is given by

$$P_L = \frac{R_L {V_{S\,rms}}^2}{(R_S + R_L)^2}. \tag{5.52}$$

The total power produced by the voltage source is given by

$$\begin{aligned} P_T &= {I_{L\,rms}}^2 (R_L + R_S) \\ &= \left(\frac{V_{S\,rms}}{R_S + R_L}\right)^2 (R_L + R_S) = \frac{{V_{S\,rms}}^2}{R_L + R_S}. \end{aligned} \tag{5.53}$$

Therefore the power transfer efficiency is given by

$$\eta = \frac{P_L}{P_T} = \frac{R_L {V_{S\,rms}}^2}{(R_S + R_L)^2} \frac{R_L + R_S}{{V_{S\,rms}}^2} = \frac{R_L}{R_L + R_S}. \tag{5.54}$$

To maximize power transfer *efficiency*, we would minimize the source resistance and *maximize* the load resistance, possibly subject to a constraint on the minimum power that must be transferred to the load. Here again, elements in the source must correspond to physical components. If v_S, R_S are the (non-physical) parameters of a Thévenin source model, then minimizing R_S by varying the physical parameters upon which R_S depends usually will not maximize power transfer efficiency.

Exercise 5.23. If the circuit in Fig. 5.24 achieves maximum power transfer, what is the power transfer efficiency of the circuit?

Refer again to Fig. 5.25. Equations (5.48) and (5.54) show that reducing the source resistance R_S increases both power transfer and power transfer efficiency. Consequently, if either power transfer or power transfer efficiency is desirable, minimizing the source impedance is a design goal.

Achieving maximum power transfer is of interest in a few applications, mainly in the output stages of circuits driving energy-conversion devices, such as motors and loudspeakers, or in the output stages of circuits driving certain signal-transmission components, such as long cables and transmitting antennas. But for most other electronic systems, achieving maximum power transfer *efficiency* is more important than maximizing power transfer. Power dissipated in a source resistance usually is dissipated as heat that can shorten the lives of nearby components, increase electrical noise, and require extra measures for cooling. Also, power dissipated in the source resistance must be provided by the power supply, even though it is wasted power that contributes nothing to the purpose of the circuit. Moreover, even where actual power transfer is of interest, it is usually the case that a source is being designed to drive a *fixed load*. Consequently, one designing such a source would try to minimize the source output resistance, thus maximizing both power transfer efficiency and actual power transfer.

The development above can be summarized as follows:

- If the load is variable and the source is fixed, maximum power transfer is achieved by making the load resistance equal to the (Thévenin equivalent) source resistance.
- If one or more physical resistors in the source (and, possibly, the load) are variable, we must express the power transferred to the load or power transfer efficiency as a function of the (physical) resistances and find the values of those resistances that maximize that expression. *We cannot (in general) simply minimize the Thévenin equivalent resistance of the source.*

In practice, there arise cases where specifications on power transfer or power transfer efficiency cannot be met because of insufficient variability of the source or load parameters; i.e., where the source and load cannot be made equal. In such cases, it often is possible to achieve maximum (or near maximum) power transfer or power transfer efficiency by inserting a transformer or other matching network between the source and the load, as described in Chapter 13.

However, in many applications, power transfer as a function of load resistance has a broad maximum (is not sharply peaked), so that moderate departure from the ideal (5.49) is acceptable. We can illustrate this remark and other conclusions drawn above as follows. Refer to Fig. 5.27, where a source drives a load R_L. We assume *the source is fixed* and the load is variable and we examine power transfer and power transfer efficiency as functions of load resistance. In this analysis, the source model consisting of v_S and R_S can be the Thévenin equivalent for a more complicated circuit, because we are interested in the power dissipated in the load and not in the powers dissipated in the source model.

In Fig. 5.27, let $R_L = \alpha R_S$. From (5.48), the power delivered to the load can be expressed as

$$P_L = \frac{R_L {V_{S\,rms}}^2}{(R_S + R_L)^2} = \frac{\alpha {V_{S\,rms}}^2}{(1+\alpha)^2 R_S}. \tag{5.55}$$

The power delivered to the load is maximum for $R_L = R_S$ and is given by

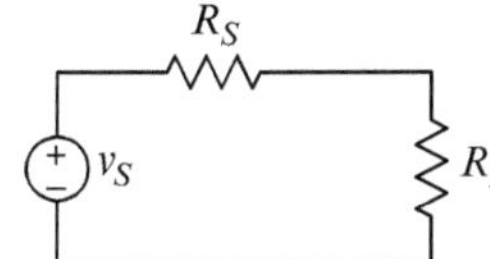

Fig. 5.27 Circuit used for illustrative maximum power calculation (see text)

$$\max(P_L) = \frac{R_S {V_{S\,rms}}^2}{(R_S + R_S)^2} = \frac{{V_{S\,rms}}^2}{4R_S}.$$

The actual power delivered to the load, expressed as a fraction of the maximum, is given by

$$\frac{P_L}{\max(P_L)} = \left(\frac{\alpha {V_{S\,rms}}^2}{(1+\alpha)^2 R_S}\right)\left(\frac{4R_S}{{V_{S\,rms}}^2}\right) = \frac{4\alpha}{(1+\alpha)^2}. \tag{5.56}$$

From (5.54), the power transfer efficiency is given by

$$\eta = \frac{R_L}{R_L + R_S} = \frac{\alpha}{1+\alpha}. \tag{5.57}$$

Figure 5.28 shows graphs of the functions defined by (5.56) and (5.57) versus the variable $\alpha = R_L/R_S$. The graph of $P_L/\max(P_L)$ can be misleading if not interpreted correctly. Equation (5.49) is derived under the assumption that the source resistance R_S is fixed and the load resistance R_L is variable, so those assumptions apply also to (5.56) and the graph of $P_L/\max(P_L)$. That is, although $\alpha = R_L/R_S$ depends upon both R_L and R_S, we must regard R_S as fixed when interpreting the graph.

As expected, the graphs show that maximum power transfer and 50% efficiency occur for $\alpha = 1$ (for $R_L = R_S$). The peak in power transfer is broad,

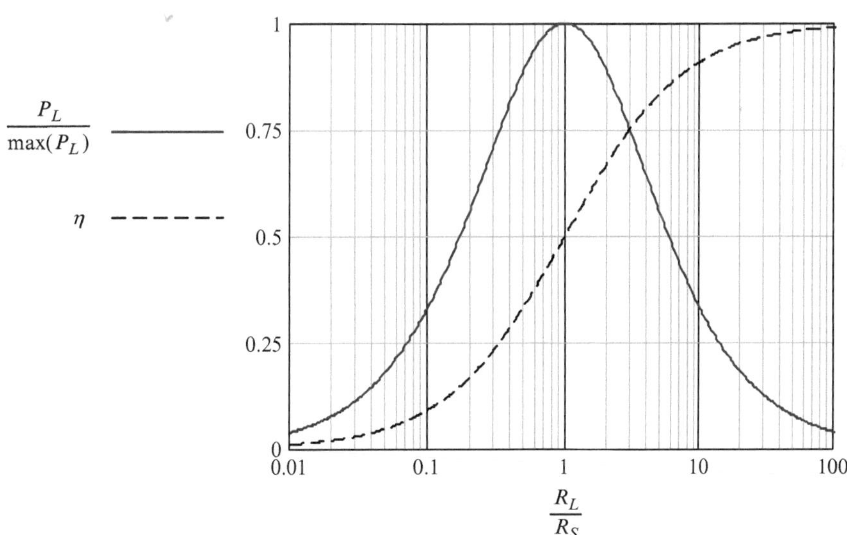

Fig. 5.28 Power transfer and power transfer efficiency as functions of load resistance for the circuit shown in Fig. 5.27

and the actual power delivered to the load exceeds 90% of the maximum possible for (approximately) $0.5\,R_S < R_L < 2\,R_S$. Thus, if 90% of maximum power transfer is acceptable, then it is acceptable to match the load to the source resistance to within a factor of two.

Example 5.29. A certain fixed source has source resistance $R_S = 1\text{k}\Omega$. The source must deliver 95% or more of the maximum possible power to a load R_L. Find the range of acceptable load resistance.

Solution: Let $R_L = \alpha R_S$. With reference to (5.56), we require

$$\frac{4\,\alpha}{(1+\alpha)^2} \geq 0.95,$$

which leads to

$$\alpha^2 - 2.21\,\alpha + 1 \leq 0.$$

The roots of the quadratic (parabolic) left side are

$$\alpha_1 = 0.635, \quad \alpha_2 = 1.576,$$

as illustrated by Fig. 5.29.

Therefore, 95% or more of the maximum possible power is delivered to a load R_L constrained by

$$0.635\,R_S \leq R_L \leq 1.575\,R_S$$

or, for $R_S = 1\text{k}\Omega$,

$$635\,\Omega \leq R_L \leq 1575\,\Omega.$$

It is advisable to make a calculation such as this before embarking on a (possibly expensive) load-matching process. It might well be that the marginal gain in power delivered is not worth the cost of matching.

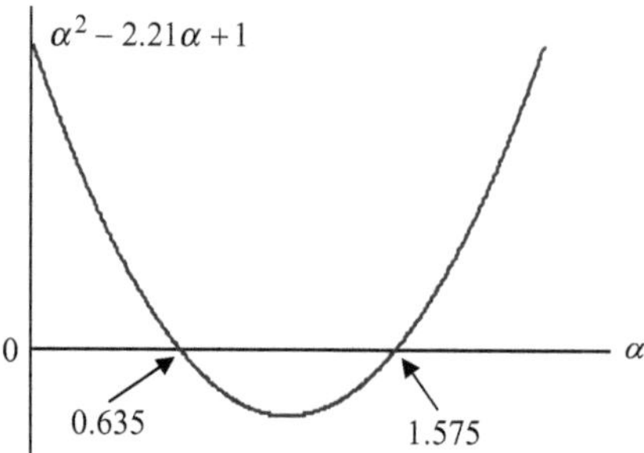

Fig. 5.29 See Example 5.29

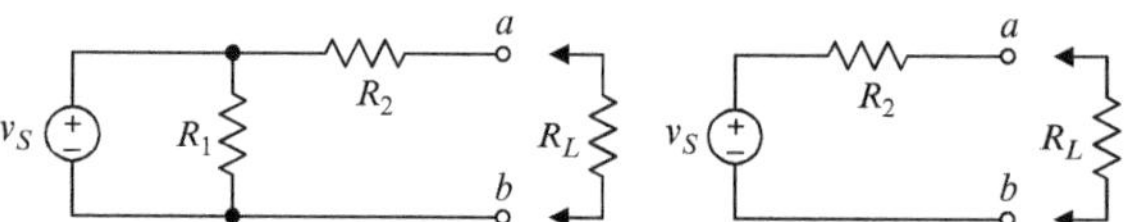

Fig. 5.30 Internal losses reduce power transfer efficiency (see text)

The introductory treatment above of power transfer efficiency ignores certain internal losses that occur in any physical source. Some such losses are incurred even if no load is attached to a source, but are not reflected in a Thévenin model for the source. For example, consider the circuits (sources) shown in Fig. 5.30. The two circuits are equivalent at the terminals *a–b*, but are not internally equivalent. In the circuit on the left, power is dissipated internally (in R_1) when the terminals *a-b* are open (when there is no load), whereas no power is dissipated in the circuit on the right under the same condition. If a load R_L is attached to each circuit, the power delivered by the source in the circuit on the left is given by

$$P_{S1} = \frac{{V_{S\,rms}}^2}{R_1 \| (R_2 + R_L)}$$

and for the circuit on the right is given by

$$P_{S2} = \frac{{V_{S\,rms}}^2}{R_2 + R_L}.$$

The power delivered to the load for each circuit is the same, and is given by

$$P_L = \frac{{V_{S\,rms}}^2 R_L}{(R_2 + R_L)^2}.$$

Thus the power transfer efficiency for the circuit on the left is

$$\eta_1 = \frac{P_L}{P_{S1}} = \frac{R_L[R_1 \| (R_2 + R_L)]}{(R_2 + R_L)^2}$$

and for the circuit on the right by

$$\eta_2 = \frac{P_L}{P_{S2}} = \frac{R_L}{R_2 + R_L},$$

which can be written

$$\eta_2 = \frac{R_2 + R_L}{R_1 \| (R_2 + R_L)} \eta_1.$$

Because $R_2 + R_L \geq R_1 \| (R_2 + R_L)$, the power transfer efficiency for the circuit on the right exceeds that for the circuit on the left. The point of the discussion and example above is that the Thévenin model for a source can conceal internal losses that can be important to power transfer efficiency. Designing a circuit (source) for maximum power transfer efficiency often involves considerations other than simply minimizing the Thévenin equivalent resistance (as seen at the output terminals).

Thévenin and Norton source models can be good models for physical sources, but are not necessarily so. Limits on the current and power that can be drawn from any particular physical source are not necessarily related as indicated by a linear Thévenin model for the source, often because the physical source becomes nonlinear for large terminal current. The Thévenin model for a source implies that if we short the terminals of the source, the resulting current is given by $i_{SC} = v_S / R_S$. But the physical source so modeled might be unable to deliver that current. As another example, the current any electronic amplifier can deliver is limited, and many such amplifiers (and other circuits), regarded as sources, have very small source resistances. Although a linear model for the amplifier might be valid under or near normal operating conditions, shorting the amplifier terminals can ask for more current (and power) than the amplifier can deliver without damage. For this reason, most modern high-power home audio amplifiers have circuit breakers in the output stage that open if the output current exceeds a pre-set limit.

5.17 Superposition of Power

In certain cases, mean squared amplitudes are additive. The only such case we need consider is that of a current or voltage that is expressed as a sum of sinusoids, no two of which have the same frequency, and one of which may have zero frequency (may be dc). Using (5.25), it can be shown that

$$\begin{aligned} &\overline{[A_0 + A_1 \cos(\omega_1 t) + A_2 \cos(\omega_2 t + \theta_2) + \cdots]^2} \\ &= A_0^2 + \frac{A_1^2}{2} + \frac{A_2^2}{2} + \cdots ;\ 0 < \omega_1 < \omega 2 < \cdots . \end{aligned} \tag{5.58}$$

It follows that if the current through a resistor is given by

$$\begin{aligned} i = I_0 + I_1 \cos(\omega_1 t) + I_2 \cos(\omega_2 t + \theta_2) \\ + \cdots ;\ 0 < \omega_1 < \omega_2 < \cdots , \end{aligned}$$

then the average power dissipated by the resistor is given by

$$\begin{aligned} P &= I_0^2 R + \frac{I_1^2 R}{2} + \frac{I_2^2 R}{2} + \cdots \\ &= \left(I_{0rms}^2 + I_{1rms}^2 + I_{2rms}^2 + \cdots\right) R;\ 0 < \omega_1 < \omega_2 < \cdots \end{aligned} \tag{5.59}$$

Likewise, if the voltage across a resistor is given by

$$\begin{aligned} v = V_0 + V_1 \cos(\omega_1 t) + V_2 \cos(\omega_2 t + \theta_2) \\ + \cdots ;\ 0 < \omega_1 < \omega_2 < \cdots , \end{aligned} \tag{5.60}$$

then the average power dissipated by the resistor is given by

$$\begin{aligned} P &= \frac{V_0^2}{R} + \frac{V_1^2}{2R} + \frac{V_2^2}{2R} + \cdots \\ &= \left(V_{0\,rms}^2 + V_{1\,rms}^2 + V_{2\,rms}^2 + \cdots\right) \frac{1}{R}; \\ &\quad 0 < \omega_1 < \omega_2 < \cdots . \end{aligned} \tag{5.61}$$

Equations (5.59) and (5.61) comprise a special case of the principle of **superposition of power**. Superposition of power applies where the voltages or currents involved are sinusoids having *different, non-zero frequencies*. Superposition of power also applies where the currents or voltages involved are derived from *physically independent sources and have no dc components*, and where a current or voltage is represented by a sum of orthogonal functions.

Example 5.30. Calculate the average power dissipated by the resistor R_2 in the circuit of Fig. 5.31.

Solution: The circuit is a current divider. The current through the resistor R_2 is given by

$$i_x = \frac{R_1}{R_1 + R_2}(I_1 + i_2 + i_3).$$

The currents I_1, i_2, i_3 have different frequencies, so we may use superposition of power. The average power dissipated by the resistor R_2 is

$$\begin{aligned} P &= \left(\frac{R_1}{R_1 + R_2}\right)^2 \left(I_{1\,rms}{}^2 + I_{2\,rms}{}^2 + I_{3\,rms}{}^2\right) R_2 \\ &= \left(\frac{R_1}{R_1 + R_2}\right)^2 \left(I_0{}^2 + \frac{1}{2}(2I_0)^2 + \frac{1}{2}I_0{}^2\right) R_2 \\ &= \frac{1}{9}\left(1 + 2 + \frac{1}{2}\right) I_0{}^2 R_2 \\ &= \frac{7}{18}(5\text{mA})^2(2\text{k}\Omega) = 19.44\,\text{mW} \end{aligned}$$

Exercise 5.24. Refer to Fig. 5.32, where $R_0 = 10\,\text{k}\Omega$, $R_1 = 100\,\Omega$, $R_L = 5\,\text{k}\Omega$, $i_0 = I_0 \cos(\omega_0 t)$, and $v_0 = V_0 \cos(2\omega_0 t)$, with $I_0 = 2\,\text{mA}$ and $V_0 = 25\,\text{V}$. Obtain an expression for the power dissipated in the load R_L. Then calculate the value of the power dissipated in the load. Assume $\omega_0 > 0$.

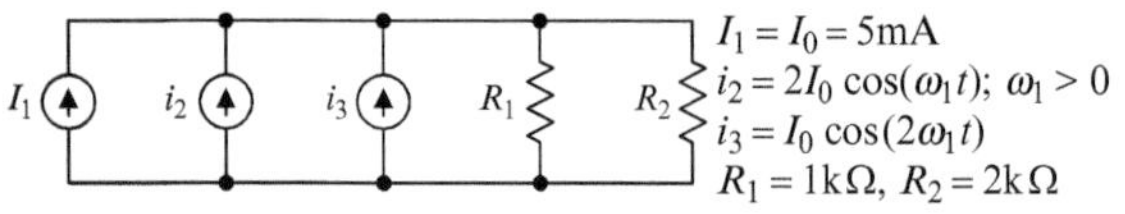

Fig. 5.31 See Example 5.30

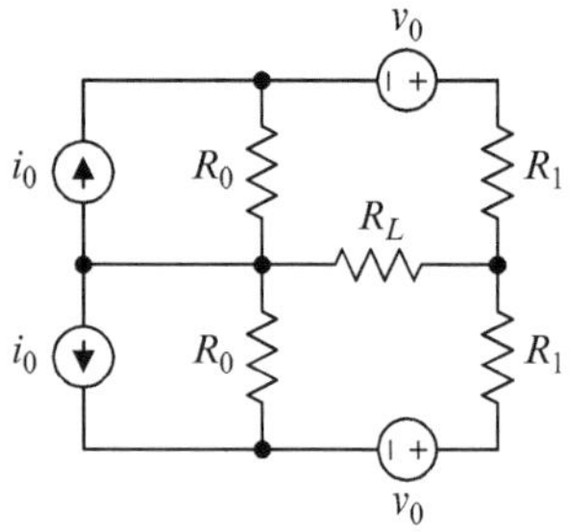

Fig. 5.32 See Exercise 5.24

5.18 Problems

Section 5.1 is prerequisite for the following problems.

P 5.1 A 12-V automobile battery causes a current $I = 15\text{A}$ through the automobile headlights for total time $t = 1\text{h}$. Calculate the power dissipated by the headlights, the total charge that passes through the headlights, and the total work done on the headlights during that time.

P 5.2 One calorie is the energy required to raise the temperature of 1cm^3 of water 1°C. In winter in the mid-Atlantic states, the average temperature of city water is about 50°F. If electric power is \$0.10 per kW h, how much does it cost to heat 40 gal of water to 140°F?

P 5.3 An ordinary incandescent lamp converts only about 8% of applied electrical power to visible light. Most of the remaining 92% is infrared radiation. If a 25 W fluorescent lamp provides as much visible light as a 100 W incandescent lamp, what fraction of electrical power applied to a fluorescent lamp is converted to visible light?

P 5.4 A 100 W incandescent lamp delivers about 8 W of visible light. A compact fluorescent lamp (CFL) delivers about the same light while consuming 25 W. If a certain light in your home is on 8 h each day, 365 days per year, and if electric power costs \$0.10 per kW h, how much would you save per year if you replace the 100 W incandescent lamp with a 25 W CFL?

P 5.5 A certain car operating at maximum efficiency can travel about 30 miles on 1 gal of gasoline. If the car were electric (all else equal), how far could it travel on one 12 V, 100 A h battery?

P 5.6 The maximum *instantaneous* electrical power *available* to many single-family homes built since about 1990 in the US is approximately 190 kW (limited by one or more circuit breakers). If the voltage is sinusoidal and has peak amplitude $V_0 = 340\,\text{V}$, what is the peak amplitude of the (sinusoidal) current when maximum instantaneous power is being delivered?

P 5.7 Express the instantaneous power dissipated by each element in the circuit shown in Fig. P 5.1 in terms of the quantities i_0, v_0, R_1, R_2.

Section 5.2 is prerequisite for the following problems.

P 5.8 The voltage across a certain resistor having resistance R is given by

$$v(t) = V_0 \cos(2\pi f t).$$

(a) Obtain expressions for the current through the resistor, the instantaneous power dissipated by the resistor, and the total work done on the resistor during an interval $0 \leq t \leq t_1$. (b) Let $V_0 = 155\,\text{V}, f = 60\,\text{Hz}$, and $R = 10\,\Omega$. Calculate the total work done on the resistor during the interval $0 \leq t \leq 5\text{s}$.

P 5.9 What is the largest constant current to which a 1kΩ, ½ W resistor should be subjected in ambient air? What is the largest constant voltage?

P 5.10 Refer to Fig. P 5.2. (a) Obtain an expression for the power dissipated by the resistor R_2. (b) Assume the resistor R_1 is fixed and resistor R_2 is variable, and obtain an expression for the value of R_2 that maximizes the power dissipated by R_2. (c) Assume the resistor R_2 is fixed and resistor R_1 is variable, and obtain an expression for the value of R_1 that maximizes the power dissipated by R_2.

P 5.11 Which resistor in Fig. P 5.3 must have the largest power-dissipation rating?

P 5.12 Which resistor in Fig. P 5.4 must have the largest power-dissipation rating?

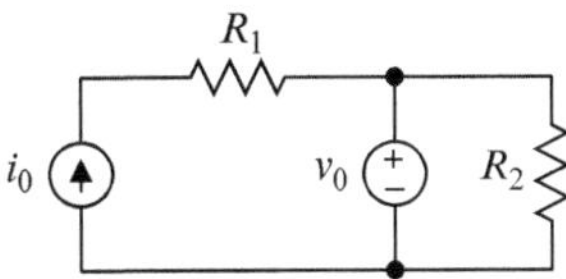

Fig. P 5.1 See Problem P 5.7

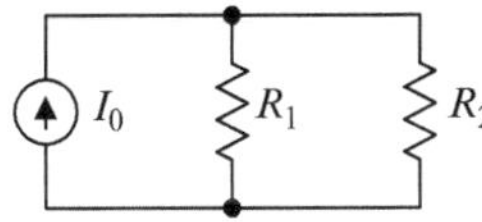

Fig. P 5.2 See Problem P 5.10

P 5.13 Refer to Fig. P 5.5, where $V_0 = 50\,\text{V}$. The power delivered by the source is 90.2 mW, the voltage V_A equals 32 V, and $R_3 = 91\,\text{k}\Omega$. Find R_1 and R_2.

P 5.14 A resistor having resistance R is connected first to the terminals of one resistive linear circuit and then to the terminals of another. In each case, the power dissipated by the resistor is the same. Does this mean the circuits are equivalent? Give at least two examples that justify your answer.

P 5.15 Two identical resistors are connected in series. The power-dissipation rating of each is denoted by P. What is the power-dissipation rating of the series connection?

P 5.16 Two identical resistors are connected in parallel. The power-dissipation rating of each is denoted by P. What is the power-dissipation rating of the parallel connection?

P 5.17 A ½ W, 1 kΩ resistor is connected in series with a 1 W, 2.2 kΩ resistor. What are the maximum allowable constant current through and voltage across the series connection?

P 5.18 A ½ W, 1 kΩ resistor is connected in parallel with a 1 W, 2.2 kΩ resistor. What are the maximum allowable constant current through and voltage across the parallel connection?

P 5.19 Refer to Fig. P 5.6, where the circuit parameters V_0 and R_0 are known and R_1 is unknown, but it

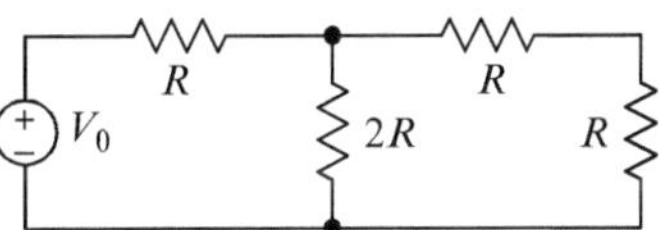

Fig. P 5.3 See Problem P 5.11

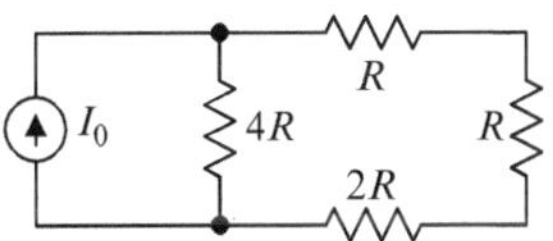

Fig. P 5.4 See Problem P 5.12

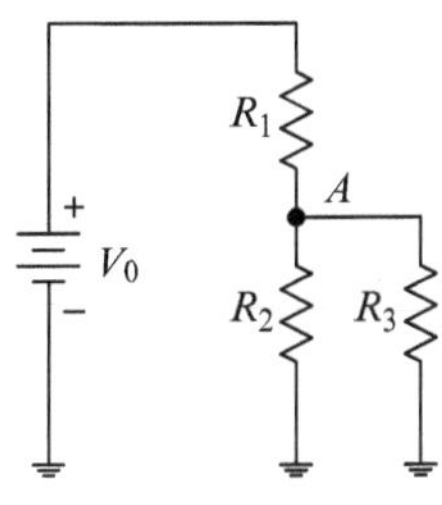

Fig. P 5.5 See Problem P 5.13

is known that resistor R_1 dissipates power P_1. Under what condition(s) is the value of R_1 determined uniquely by this information?

Section 5.3 is prerequisite for the following problems.

P 5.20 In Fig. P 5.7, the power delivered by the source V_S is $P_S = 200\,\text{mW}$. Use conservation of power to find the resistance R_2.

P 5.21 Obtain an expression for the total power delivered by the source in the circuit of Fig. P 5.8.

P 5.22 In Fig. P 5.9, the power delivered by the source is $p_s = 2\text{W}$. Use conservation of power to find the resistance R_3.

P 5.23 In Fig. P 5.10, the power delivered by the source is 1 W. Use conservation of power to find the resistance R_1.

P 5.24 Refer to Fig. P 5.11. The power delivered by the source is 3.16 W and the power dissipated in R_1 and R_2 is 500 mW. Find the source resistance R_S. Assume R_S is non-zero.

P 5.25 Refer to Fig. P 5.12. A certain 90 V dc motor is 83% efficient, meaning that 83% of the electrical power provided to the motor is delivered to a mechanical load and that 90 V is the maximum voltage that should be applied continuously to the motor terminals. The motor can deliver at most 1.5 hp to a mechanical load without damage. If the motor operates at rated load (1.5 hp) and voltage ($V_M = 90\,\text{V}$) from a source having open-circuit voltage $V_S = 100\,\text{V}$, what are the losses (in W) in the source resistance and motor (individually)?

P 5.26 Refer to Fig. P 5.13. The total power *dissipated* in the three resistors is 250 mW and the

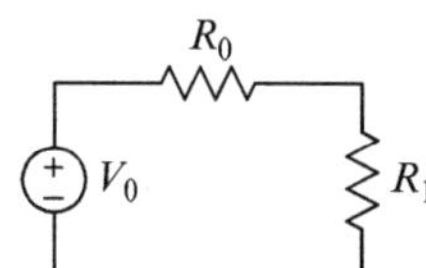

Fig. P 5.6 See Problem P 5.19

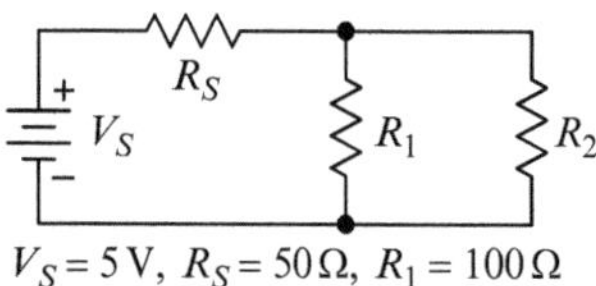

Fig. P 5.7 See Problem P 5.20

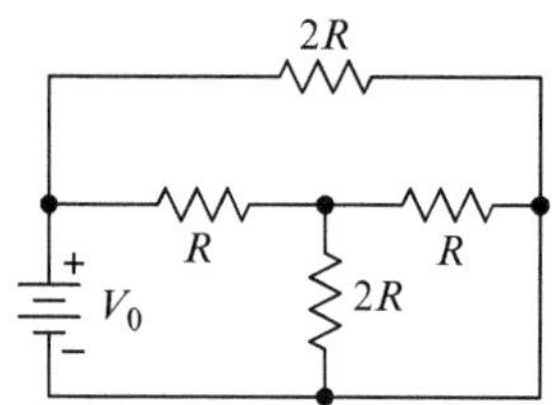

Fig. P 5.8 See Problem P 5.21

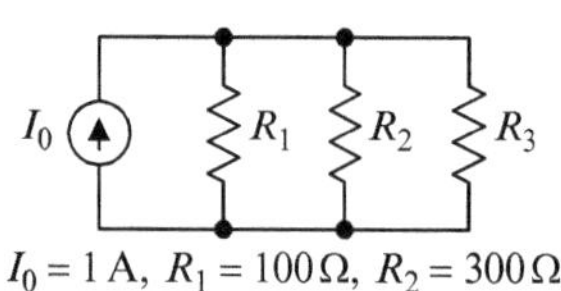

Fig. P 5.9 See Problem P 5.22

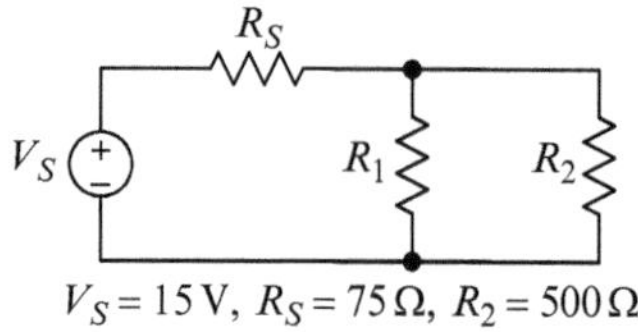

Fig. P 5.10 See Problem P 5.23

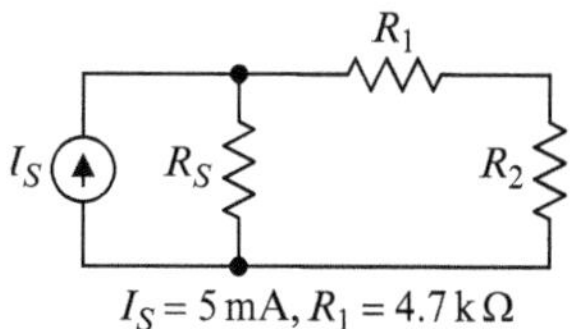

Fig. P 5.11 See Problem P 5.24

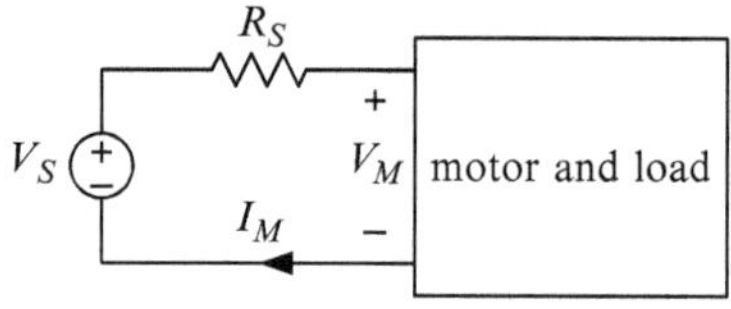

Fig. P 5.12 See Problem P 5.25

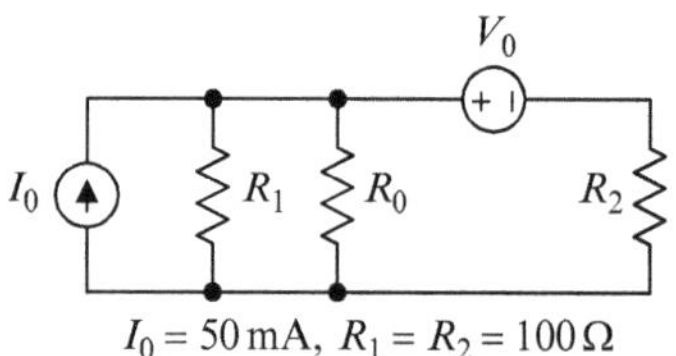

Fig. P 5.13 See Problem P 5.26

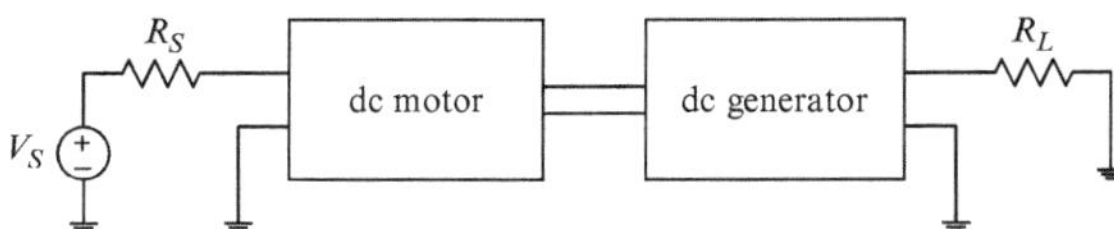

Fig. P 5.14 See Problem P 5.27

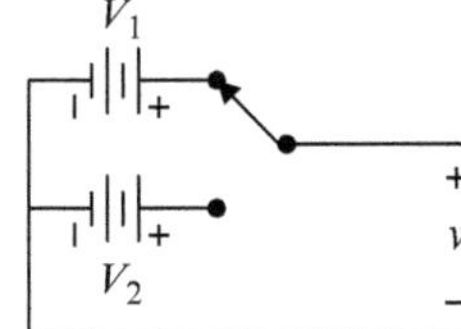

Fig. P 5.15 See Problem P 5.30

power *delivered* by the current source is 187.5 mW. Find the resistance R_0 and the source voltage V_0.

P 5.27 A motor drives a generator, as shown in Fig. P 5.14. The efficiency of the motor is 80% and the efficiency of the generator is 90%. This means that the mechanical power delivered to the generator equals 80% of the electrical power delivered to the motor, and that the electrical power delivered to the load R_L equals 90% of the mechanical power delivered to the generator. The motor is rated at 90 V and 1 hp, meaning that the mechanical output is 1 hp when the voltage applied to the *motor terminals* equals 90 V (assuming the source can deliver the required current). Assume that the motor is operating at rated mechanical output, and that the generator is producing 200 V at its output terminals. (a) What is the load resistance R_L? (b) What is the electrical power delivered to the motor? (c) If the source voltage is $V_S = 100\,\text{V}$, what is the electrical loss in the source resistance R_S? (d) What is the source resistance?

Section 5.6 is prerequisite for the following problems.

P 5.28 The open-circuit voltage for a certain source is 10V. The source delivers 8W to a 2Ω load. Obtain Thévenin and Norton models for the source.

P 5.29 As a battery discharges, its terminal voltage under load decreases. A certain 24V battery, when fully charged, delivers 20A to a 1Ω load, but when 50% discharged, delivers 19A to the same load. What is the internal resistance of the battery in each case?

P 5.30. Refer to Fig. P 5.15, where the switch changes position at times $t_n = n\,\Delta T$. Obtain expressions for the average and mean-squared values of $v(t)$.

P 5.31 Obtain the time average of each of the following functions:

(a) $\sin^2(\omega t);\ \omega>0$,

(b) $x(t) = \left\{ \begin{array}{ll} [1-\exp(-\sigma t)], & t>0 \\ 0 & t \le 0 \end{array} \right\}$,

(c) $y(t) = [a\cos(\omega t) + b\sin(\omega t)]^2$,

(d) $z(t) = \left\{ \begin{array}{ll} \left(\frac{t}{\tau}\right)^2 \exp\left(-\frac{t}{\tau}\right), & t>0 \\ 0 & t \le 0 \end{array} \right\}$.

P 5.32 By definition, the *dc component* of a current or voltage is the average of the current or voltage. Thus, the average of $v(t)$ is $\bar{v} = V_{dc}$. The *ac component* of a current or voltage is what remains if the dc component is removed; i.e. $v_{ac}(t) = v(t) - V_{dc}$. Study the definitions just given and then answer the following questions. (a) Does the ac component of a current or voltage have a dc component? (b) Does the dc component of a current or voltage have an ac component? (c) Is a current or voltage the sum of the ac and dc components of the current or voltage? (d) Is the mean-squared amplitude of a current or voltage equal to the sum of the mean-squared amplitudes of the ac and dc components of the current or voltage? (e) Does Ohm's law hold for the ac and dc components (individually) of the current through and voltage across a resistor? Justify all of your answers.

Section 5.7 is prerequisite for the following problems.

P 5.33 A sinusoidal current passes through a sample of material having resistance R. Express the peak amplitude of the current in terms of the average power dissipated by the sample.

P 5.34 The voltage across a certain 10 Ω resistor is 500 V for 10 s, then zero for 20 s, then 600 V for 5 s, then off for 60 s. This cycle is repeated indefinitely. What is the average power dissipated by the resistor?

P 5.35 The current through a sample of material having resistance $R = 1.5\text{k}\Omega$ is given by

$$i = I_0[3 + 2\cos(\omega_1 t) + \cos(\omega_2 t)];\ \ 0 < \omega_1 < \omega_2,$$
$$I_0 = 25\,\text{mA}.$$

Find the average power dissipated by the sample.

P 5.36 When a 12 V battery (a dc source having negligible internal resistance) is connected across a sample of material, the resulting current is $I = 2$A. When the battery is disconnected and a sinusoidal voltage is applied to the same sample, the average power dissipated is 12 W. What is the peak amplitude of the sinusoidal voltage?

P 5.37 The full-load current rating of a dc motor is the average current the motor can draw without overheating. A certain motor has a full-load current rating of 12 A. If the motor is on for 20 s then off for 20 s, then on again for 20 s, then off for 20 s, and so on indefinitely, what is the maximum current the motor can draw during the on times?

P 5.38 In Fig. P 5.16, all resistors have the same power-dissipation rating $P_{\max}$. (a) Express the maximum permissible value of the source voltage in terms of R and $P_{\max}$ if the source voltage is a constant given by $v_0 = V_0$. (b) Repeat for $v_0 = V_0 \cos(\omega t)$.

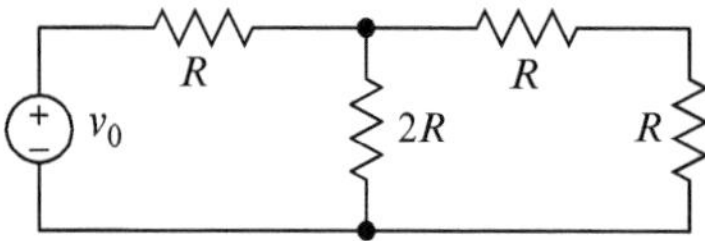

Fig. P 5.16 See Problem P 5.38

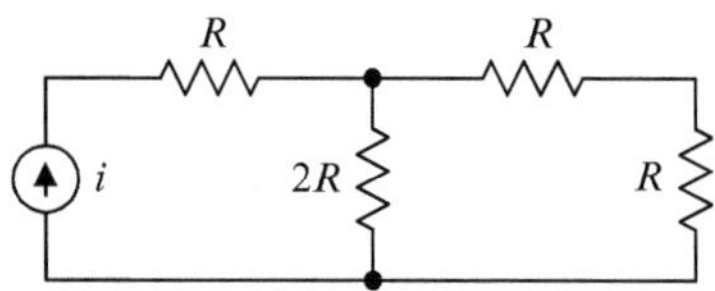

Fig. P 5.17 See Problem P 5.39

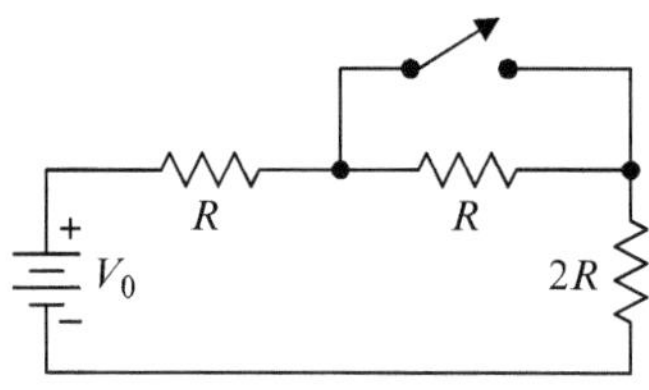

Fig. P 5.18 See Problem P 5.40

P 5.39 In Fig. P 5.17, all resistors have the same power-dissipation rating $P_{\max}$. (a) Express the maximum permissible value of the source current in terms of R and $P_{\max}$ if the source current is a constant given by $i = I_0$. (b) Repeat for $i = I_0 \cos(\omega t)$.

P 5.40 Refer to Fig. P 5.18. The switch is closed at times $t = 0, T, 2T, 3T, \cdots$, and opened at times $t = T/2, 3T/2, 5T/2, \cdots$. Obtain an expression for the average power dissipated in each element.

P 5.41 Figure P 5.19 shows one cycle of a periodic voltage that is applied to a resistor. (a) If the resistance of the resistor is 100 Ω, what must the power-dissipation rating of the resistor be? (b) If the power-dissipation rating of the resistor is 1 W, what is the minimum allowable value of the resistance?

Section 5.8 is prerequisite for the following problems.

P 5.42 Show that Ohm's law applies to the rms current through a resistor and the rms voltage across the resistor; that is, show that if v is the voltage across and i is the current through a resistor having resistance R, then $V_{rms} = R I_{rms}$. Show this in general, not just for constant or sinusoidal current and voltage. Is the relation $V_{rms} = R I_{rms}$ valid if the resistance R is a function of time? Is the relation $V_{rms} = R I_{rms}$ valid if the resistance R is a function of the current through the resistor? Justify your answers.

P 5.43 Does Kirchhoff's current law hold for rms currents? Explain.

P 5.44 Does Kirchhoff's voltage law hold for rms voltages? Explain.

P 5.45 Can you express the rms amplitude of $v = v_1 + v_2$ in terms of $V_{1\,rms}$ and $V_{2\,rms}$? Do so or explain why not.

P 5.46 The ratio of the peak amplitude to the rms amplitude of the *ac component* of a *periodic* current or voltage is called the **crest factor** of the current or voltage. For example, the crest factor for a sinusoidal waveform is $\sqrt{2}$. Find the crest factor for each waveform shown in Fig. P 5.20.

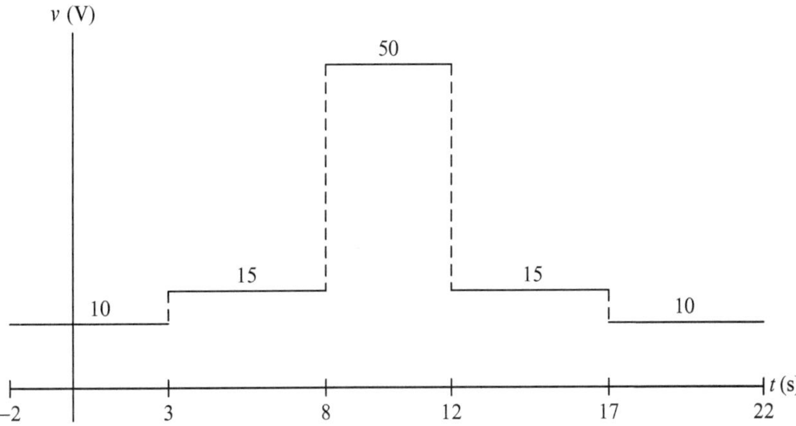

Fig. P 5.19 See Problem P 5.41

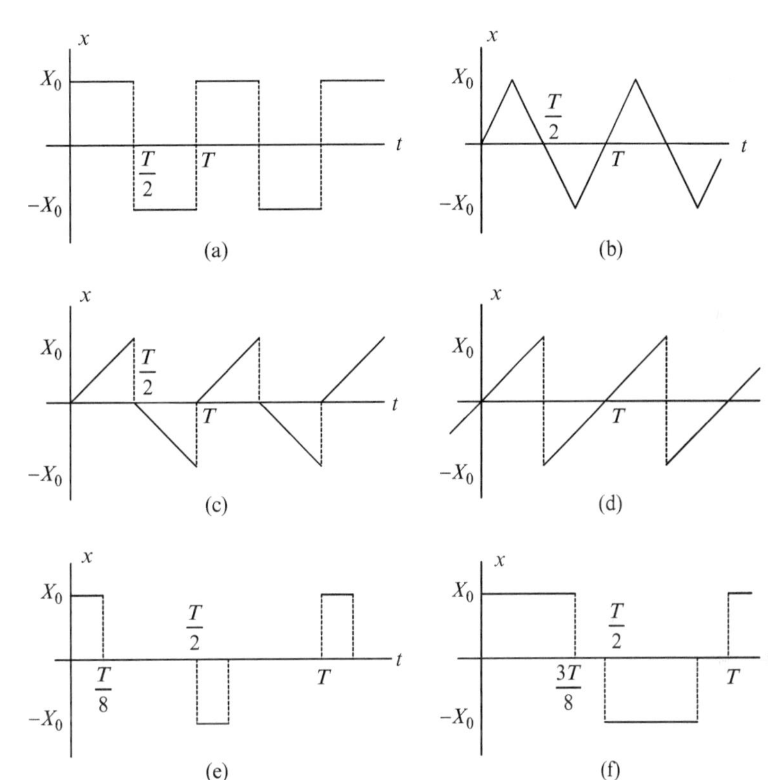

Fig. P 5.20 See Problem P 5.46, 48

P 5.47 Refer to Problem P 5.46. Can the crest factor of a waveform be less than unity? Explain why or why not. Give two examples of waveforms having unity crest factors.

P 5.48 Refer to Problem P 5.46. Assume each periodic waveform shown in Fig. P 5.20 is a voltage across a resistor having resistance R. Express the peak power dissipated in terms of the average power dissipated and the crest factor for the waveform.

P 5.49 Refer to Fig. P 5.21. Obtain an expression for the average power dissipated by the unspecified element in terms of the rms amplitudes of the terminal current and voltage.

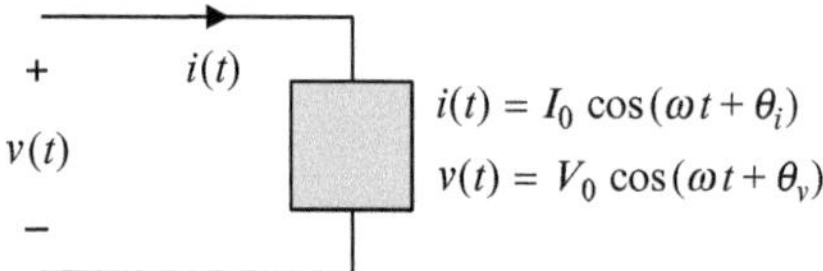

Fig. P 5.21 See Problem P 5.49

Fig. P 5.22 See Problem P 5.52

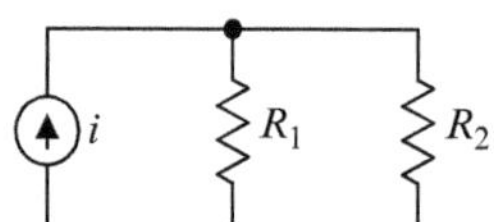

Fig. P 5.23 See Problem P 5.54

P 5.50 For what fraction of a period does the magnitude of a sinusoidal current or voltage exceed the rms amplitude of the current or voltage?

P 5.51 A certain 100 Ω resistor is rated at ½ W. (a) What is the peak amplitude of the largest sinusoidal current the resistor can safely carry? (b) The resistor will be damaged if the *instantaneous* power dissipated exceeds ½ W for more than 200 ms. If the amplitude of a sinusoidal current through the resistor equals 100 mA, what is the minimum safe frequency of the current?

P 5.52 In Fig. P 5.22, $R = 1\,\text{k}\Omega$, $V_0 = 5\,\text{V}$, and $v(t) = v_{ac}(t) + V_{dc}$, where $V_{dc} = 2\,\text{V}$ and $v_{ac}(t)$ has peak amplitude 25 V and crest factor 2.2. Find the average power dissipated in the resistor.

Section 5.10 is prerequisite for the following problems.

P 5.53 The resistance of a certain transmission line is $0.6\,\Omega\,\text{km}^{-1}$. The voltage applied to the line is 65 kV (rms). The line delivers 88% of the input power to a $5\,\text{k}\Omega$ load. If the line resistance is the only loss mechanism, what is the length of the line?

P 5.54 In the circuit shown in Fig. P 5.23, $R_2 > R_1$. (a) Which resistor dissipates more power? (b) Does the answer change if the current source is replaced with a voltage source?

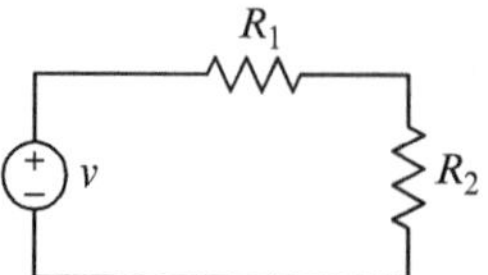

Fig. P 5.24 See Problem P 5.55

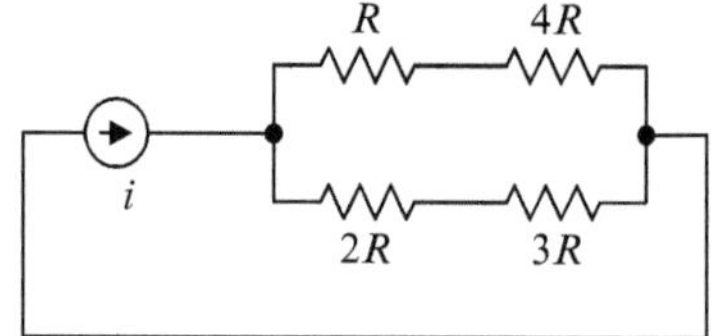

Fig. P 5.25 See Problem P 5.56

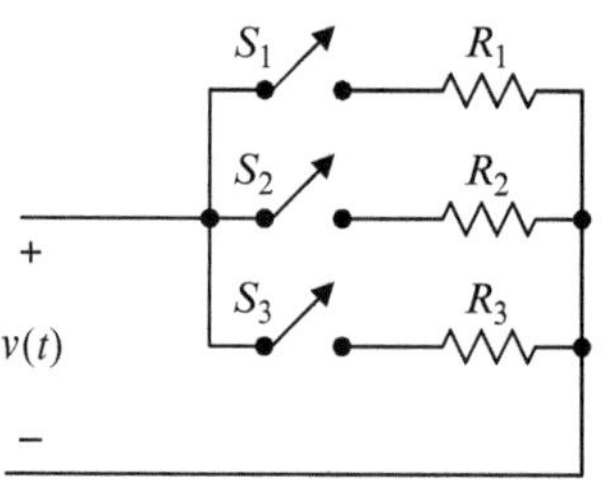

Fig. P 5.26 See Problem P 5.58

P 5.55 In the circuit shown in Fig. P 5.24, $R_2 > R_1$. Which resistor dissipates more power? Does the answer change if the voltage source is replaced with a current source?

P 5.56 Refer to Fig. P 5.25. (a) Which resistor dissipates more power than any of the others? (b) What fraction of the total power dissipated is dissipated by the resistor having resistance 4 *R*? (c) Do the answers to any of the questions above change if the current source is replaced with a voltage source?

P 5.57 What is the smallest permissible resistance for a 1 W resistor that must endure a terminal voltage $V_{rms} = 120\,\text{V}$ in ambient air?

P 5.58 The voltage *v*(*t*) at a residential wall outlet (in the US) is a sinusoidal voltage having rms amplitude $V_{rms} \cong 120\,\text{V}$. Refer to Fig. P 5.26, which represents a small space heater having low (S_1 closed, S_2, S_3 open), medium (S_1, S_2 closed, S_3 open), and high (S_1, S_3 closed, S_2 open) settings, which, when the heater is connected to a wall outlet, provide 750, 1000, and 1500 W, respectively. Find the values of R_1, R_2, R_3.

P 5.59 The standard power-dissipation ratings of ordinary composition resistors are 1/8, 1/4, 1/2, and 1 W. Assume these ratings apply if the resistance

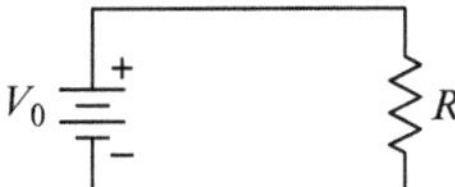

Fig. P 5.27 See Problem P 5.60

equals the specified value and that the rated dissipation must not be exceeded. The standard precisions for such resistors are $\pm 1\%$, $\pm 5\%$, $\pm 10\%$, and $\pm 20\%$. Express the maximum permissible rms current through and the maximum permissible rms voltage across a composition resistor in terms of the precision, the resistance, and the power-dissipation rating.

P 5.60 See Problem P 5.59, above. A resistor having nominal resistance R must be able to dissipate the rated power P for any resistance within the specified tolerance; e.g., a $1\ \mathrm{k\Omega} \pm 10\%$, ½ W resistor must be able to dissipate ½ W if the actual resistance is anywhere in the range $1\ \mathrm{k\Omega} \pm 10\%$. In Fig. P 5.27, the resistor has tolerance $\pm 100\alpha\ (\%)$. Let P denote the power dissipation rating for the resistor and express the actual power dissipated as a fraction of the power dissipation rating if the resistor happens to have its nominal resistance.

P 5.61 The temperature coefficient for a certain $1\ \mathrm{k\Omega} \pm 1\%$ resistor is $0.001\ \mathrm{K}^{-1}$ at 25°C. The resistor will be used in an environment where the temperature ranges from 0°C to 80°C. If the voltage across the resistor will be given by $v(t) = V_0 \cos(2\pi f\, t)$, where the maximum value of v_0 is 15 V, what should be the power-dissipation rating for the resistor? Does the frequency of the voltage matter?

P 5.62 Assume that the lifetime of the filament in an incandescent lamp is inversely proportional to the mean squared current through the filament, provided the rms current does not exceed 1.5 A. A certain manufacturer's 100 W lamp for residential use is advertised to have a lifetime of 1000 h under normal operating conditions ($V_{rms} = 120$ V). Approximately how long would the lamp last if the rms amplitude of the applied voltage were 130 V? 110 V?

P 5.63 Figure P 5.28 shows a voltage divider driving a resistive load. It is known that the rms amplitude of the source voltage $v(t)$ will not exceed a known value V_{rms}, but R_L is unknown. Specify safe power-dissipation requirements for the resistors R_1, R_2 in terms of R_1, R_2, and V_{rms}.

P 5.64 The labels on the resistors in the circuit in Fig. P 5.29 denote the resistance and power-dissipation ratings of the resistors. Obtain expressions for safe values of the power-dissipation ratings in terms of R_1, R_2, and V_a, where V_a is the rms amplitude of the voltage v_{ab}. By *safe values*, we mean dissipation ratings that are safe regardless of the values or relative values of R_1 and R_2.

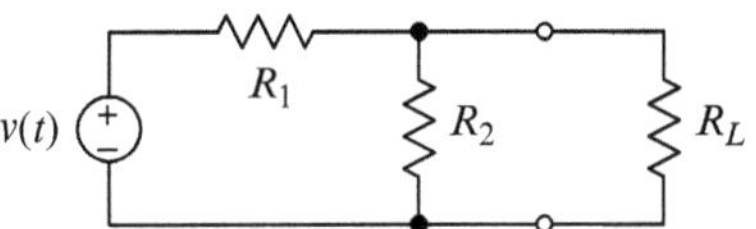

Fig. P 5.28 See Problem P 5.63

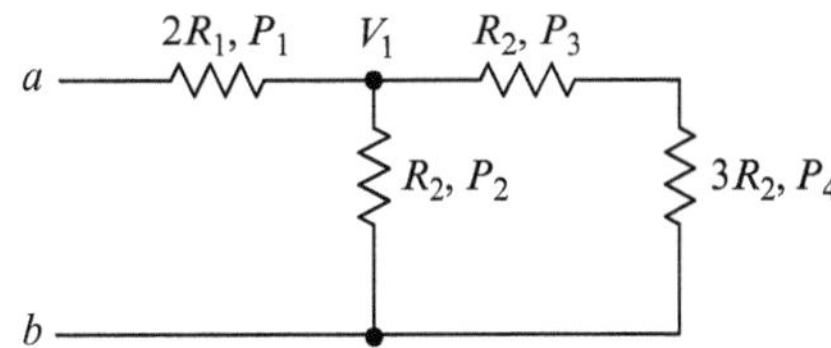

Fig. P 5.29 See Problem P 5.64

P 5.65 A certain amplifier drives four 16 Ω speakers and one 8 Ω speaker, all in parallel. The Thévenin equivalent for the amplifier at the speaker terminals has open-circuit voltage 30 V (rms) and output resistance 1.5 Ω. (a) What is the power delivered to each speaker? What is the total power delivered by the Thévenin source? (b) The owner wishes to add more speakers. What is the maximum number of 16 Ω speakers that can be added if the amplifier would be damaged by a power load exceeding 250 W? (d) What is the maximum number of 8 Ω speakers that can be added if the amplifier would be damaged by a power load exceeding 250 W?

P 5.66 (Review skin effect in Chapter 2) Wire is rated according to its current-carrying capacity, called **ampacity**, which depends upon the size (e.g., AWG), material (e.g., copper), insulation, and environment (e.g., force cooled or ambient, in the open or in conduit, etc.). Ordinary plastic-coated AWG 24 copper wire, similar to what you might buy at your local electronics store, has an ampacity of about 3.5 A in the open (in ambient air) and can safely dissipate about $1.25\ \mathrm{Wm}^{-1}$ in open air when carrying 3.5 A at dc. If the current is sinusoidal, having rms amplitude 3.5 A, at what frequency would skin effect cause the power dissipation to exceed $1.25\ \mathrm{Wm}^{-1}$ by 15% in an AWG 24 copper wire carrying 3.5 A (rms)?

P 5.67 Currents encountered in electric-utility substations and in high-power radio and television transmitters can be quite large, and large conductors are needed to keep losses to acceptable levels. Because of skin depth, conductors for large ac currents do not need to be solid cylinders, but can often be hollow tubes. Such construction can save a significant amount of expensive copper. Suppose a certain copper conductor must carry a sinusoidal current having rms amplitude I_{rms} and frequency f, and let φ denote the maximum permissible power loss per unit length. (a) Obtain an expression for the required cross-sectional area of the copper conductor. (b) Calculate the required cross-sectional area if the rms amplitude of the current and the allowable loss are 1 kA and $1\,\mathrm{Wm}^{-1}$ respectively. What is the diameter of a solid cylindrical wire having that cross-sectional area? (c) If the frequency of the current is 60 Hz, what is the skin depth? What fraction of the cross-sectional area of the solid conductor referred to above is utilized by the current? What is the loss in that case? (d) 💻 What are the inner and outer radii of a hollow copper tube that would meet the loss specification for the specified current and frequency? (e) What is the diameter of a solid copper conductor that meets the loss specification? (f) What is the volume of copper saved per meter by using the tube instead of a solid conductor? What is the percent of the copper saved by using a hollow-tube conductor?

P 5.68 Household incandescent lamps are designed to operate at 60 Hz, with an applied voltage of 120 V (rms). The tungsten filament in a 100 W lamp operates at about 3000 K and will burn out almost instantly if the power dissipated exceeds 175 W. The diameter of the filament is approximately 30 μm and the resistivity of tungsten at 3000 K is about 92 μΩ cm. If the current through the filament is held constant at the design value and the frequency is increased, at what frequency will the lamp burn out? (Note: The answer will be inexact because the filament in an incandescent lamp is tightly and doubly coiled, and the expression given in Chapter 2 for skin depth does not apply to such a geometry. Regard this as a drill problem, not as a real-world design problem.)

Section 5.11 is prerequisite for the following problems.

P 5.69 If $v(t) = V_0$, where V_0 is constant, how would you write an expression for the rms amplitude of $v(t)$?

P 5.70 If a current is expressed in standard form by $i(t) = I_0 \cos(\omega_0 t + \theta)$ would you write $i(t) = \sqrt{2} I_{rms} \cos(\omega_0 t + \theta)$?

P 5.71 A certain plug-in power supply (adapter) is labeled

input: 120 VAC 60 Hz 12 W,
output: 12 VDC 500 mA.

(a) What are the rms and peak amplitudes of the input voltage? (b) What are the allowable effective loads (Ω) on the adapter? (c) What is the maximum power output? (d) To what does "12 W" refer?

P 5.72 The voltage across and current through a certain resistor having resistance R are denoted by $v(t)$ and $i(t)$, respectively, where the positive direction of the current is into the positive terminal. (a) Is it correct to say that $R = v(t)/i(t)$? (b) If not, why not, and how would you express the relation between $v(t)$, $i(t)$, and R?

Section 5.12 is prerequisite for the following problems.

P 5.73 What is the minimum averaging time necessary to measure the rms value of a 1 Hz sinusoidal voltage with ±0.1% accuracy?

P 5.74 A certain voltmeter measures rms amplitude by averaging, and is advertised not to exceed 0.5% error (magnitude) for a 10 Hz sinusoidal input. Assuming the averaging time is fixed, what is the error at 1 kHz? (Neglect sources of error other than finite averaging time.)

P 5.75 💻 The rms amplitude of a voltage or current is measured by averaging for a finite time. If we start averaging at time zero and stop averaging at time t, the measured rms voltage is given by

$$\hat{v}_{rms}(t) = \sqrt{\frac{1}{t}\int_0^t v^2(t')dt'}.$$

Given a voltage $v(t) = V_0[\cos(2\pi f_1 t) + \cos(2\pi f_2 t)]$, with $V_0 = 5\,\mathrm{V}$, $f_1 = 100\,\mathrm{Hz}$, and $f_2 = 1\,\mathrm{kHz}$: (a) What is the true rms amplitude of $v(t)$? (b) Obtain an expression for the measured rms amplitude $\hat{v}_{rms}(t)$ defined above. Plot the measured rms amplitude versus time for $0 \le t \le 5/f_1$. Show the true rms amplitude on the same graph. (c) Calculate and plot the percent error in the measured rms voltage over the same interval.

P 5.76 Some inexpensive multimeters measure the rms amplitude of the *ac component* of a voltage as $v_{rms} = v_{peak}/\sqrt{2}$, where v_{peak} is the maximum amplitude of the ac component of the voltage. This measurement is correct only for a sinusoidal voltage. Find the measurement error (%) for each of the voltages defined in Fig. P 5.30. Show how to use the crest factors to correct the measured rms amplitude.

P 5.77 The accuracy with which a "true rms" multimeter can measure the rms amplitude of a current or voltage depends upon the crest factor of the current or voltage. The error generally increases with increasing crest factor (see Problem P 5.46). Table P 5.1 shows the rated accuracy of the Agilent 3458 multimeter when used to measure the rms amplitude of a *sinusoidal* voltage and additional errors incurred for various crest factors, where the upper limit on the CF is not included in the range of CF's; e.g., the additional error for CF = 3 is 0.25%, not 0.15%. What is the accuracy with which this voltmeter can measure the rms value of each waveform shown in Fig. P 5.30? For this problem, assume the frequency of a periodic signal equals the reciprocal of the period; i.e., $f = T^{-1}$.

P 5.78 Discuss the following remark:

The rms amplitude of a sinusoid is independent of the frequency of the sinusoid.

P 5.79 How long would it take to measure to within 0.1% the rms amplitude of a sinusoid whose frequency is 10^{-4} Hz?

Section 5.13 is prerequisite for the following problems.

P 5.80 The power-dissipation rating for a certain resistor is 500 mW at 25°C, derated linearly to zero at 175°C. (a) Draw a graph of permissible power dissipation versus temperature for this resistor. (b) What is the self-heating coefficient for this resistor?

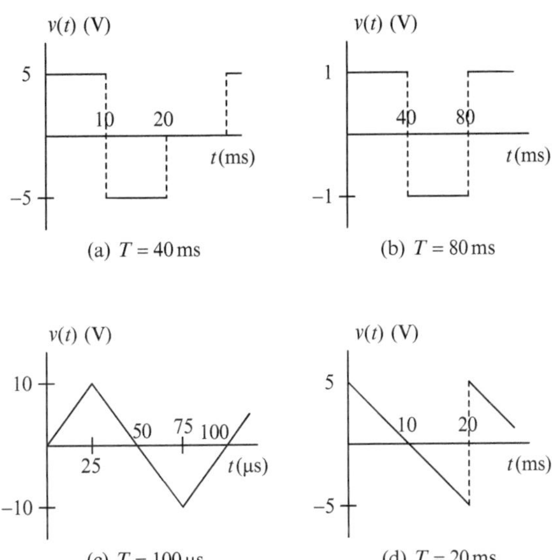

Fig. P 5.30 See Problem P 5.76

Table P 5.1 Selected accuracy specifications of the Agilent 3458 multimeter (Data Courtesy of Agilent Corporation)

(a) Rated Accuracy					
Frequency Range	10 Hz ≤ f < 20 Hz	20 Hz ≤ f < 40 Hz	40 Hz ≤ f < 100 Hz	100 Hz ≤ f < 20 kHz	20 kHz ≤ f < 50 Hz
Maximun Error (%)	0.40	0.15	0.06	0.03	0.15
(b) Additional Errors for Non-Sinusoidel Inputs					
Crest Factor Range	1–2	2–3	3–4	4–5	
Additional Error (%)	0.00	0.15	0.25	0.40	

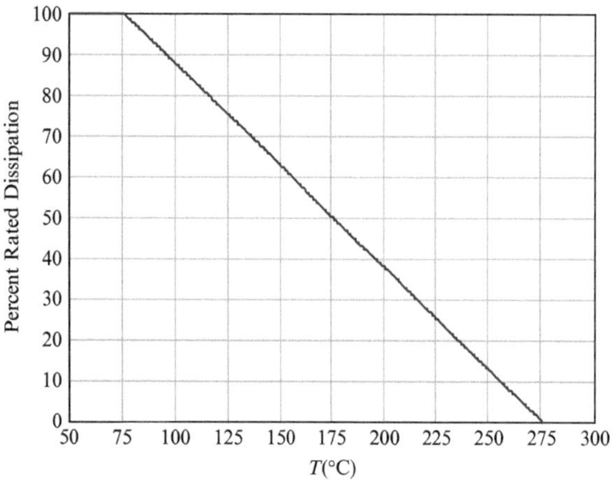

Fig. P 5.31 See Problem P 5.81

P 5.81 Figure P 5.31 shows the dissipation derating characteristic for a certain family of high-power, wirewound resistors. Estimate the self-heating coefficient for a 20 W resistor from this family.

P 5.82 The power-dissipation rating for a certain resistor is 500mW at 25°C. The self-heating coefficient for the resistor is $300°\mathrm{CW}^{-1}$ for temperatures above 25°C. (a) Draw a graph of percent rated dissipation versus temperature for this resistor. (b) What is the power-dissipation rating for this resistor if the temperature of the resistor is 75°C?

P 5.83 The temperature coefficient of resistance at 25°C for the resistor described in Problem P 5.80 is 4500 ppm. The resistance of the resistor at 25°C is $1\,\mathrm{k}\Omega \pm 10\%$. What is the maximum resistance of this resistor if it carries an rms current of 5 mA in an enclosure where the ambient temperature is 75°C?

P 5.84 The temperature coefficient of resistance at 75°C for the family of resistors described in Problem P 5-81 is 450 ppm. A $10\Omega \pm 5\%$ resistor from this family is rated at 5 W 75°C and carries an rms current of 500 mA in an enclosure where the ambient temperature is 100°C. What is the range of operating resistances for the resistor? What is the maximum power dissipation? Is the 5 W rating adequate?

P 5.85 The temperature coefficient of resistance at 25°C for the resistor described in Problem P 5.82 is 2500 ppm. The resistance of the resistor at 25°C is $1\,\mathrm{k}\Omega \pm 10\%$ and the resistor must carry an rms current of 15 mA in an enclosure where the ambient temperature is 70°C. What is the derated power-dissipation rating for the resistor?

P 5.86 The resistance of a certain resistor is specified as $2.2\,\mathrm{k}\Omega \pm 5\%$ at 25°C. The permissible power dissipation at 25°C is 5 W, derated linearly to zero at 225°C. The temperature coefficient of resistance at 25°C equals 450 ppm. (a) Draw a graph of permissible power dissipation versus temperature for this resistor. (b) What is the maximum resistance of this resistor if it carries current 5 mA rms in an enclosure where the ambient temperature is 75°C?

P 5.87 The rated power dissipation for a certain $2.2\,\mathrm{k}\Omega \pm 5\%$ resistor is 5 W at 25°C, derated linearly to zero at 150°C. The temperature coefficient of resistance at 25°C is 400 ppm. (a) What are the nominal, minimum, and maximum resistances of this resistor if it carries an rms current of 5 mA in an enclosure where the ambient temperature is 80°C? (b) What are the high, low, and nominal derated dissipation ratings?

P 5.88 A certain resistor has nominal resistance $R = 10\,\mathrm{k}\Omega$ at 25°C, temperature coefficient of resistance $\alpha = 0.002°\mathrm{C}^{-1}$ (2 ppm) at 25°C, and a power-dissipation rating of 250 mW at 25°C, derated linearly to zero at 125°C. The rms current through the resistor is 1.5 mA. If the ambient temperature is 25°C, what is the nominal steady-state temperature of the resistor and what is the nominal resistance at that temperature?

P 5.89 The power-dissipation rating for a certain $1\mathrm{k}\Omega$ (nominal) resistor is 500 mW at 25°C, derated linearly to zero at 135°C. (a) What is the power-dissipation rating when the operating temperature is 70°C? If the rms current through the resistor in that case is 10 mA, what is the ambient temperature? (b) If the operating temperature is 85°C and the ambient temperature is 30°C, what is the current through the resistor?

P 5.90 A certain resistor has nominal resistance 2.2 kΩ at 25°C and self-heating coefficient $\varphi = 180°\mathrm{CW}^{-1}$. The dissipation derating is linear above 25°C and the temperature of the resistor is 70°C in ambient (25°C) air when the power dissipated equals 250 mW. (a) What is the power-dissipation rating at 25°C? (b) At what temperature does the power-dissipation rating go to zero? (c) What is the operating temperature of the resistor the rms voltage across the resistor is 15V, the ambient temperature is 25°C, and the TCR is 500 ppm?

P 5.91 Figure P 5.32 shows a voltage divider, where the resistors R_1, R_2 have identical temperature coefficients and identical self-heating coefficients.

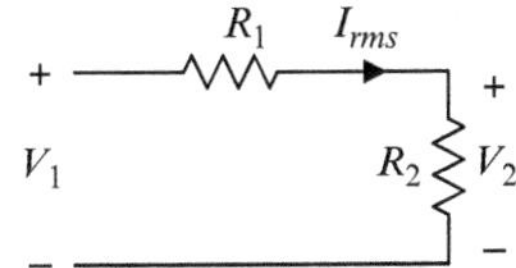

Fig. P 5.32 See Problem P 5.91

v_S R_S R_L i $\frac{v_S}{R_S}$ R_S R_L v_L

Fig. P 5.34 See Problem P 5.93

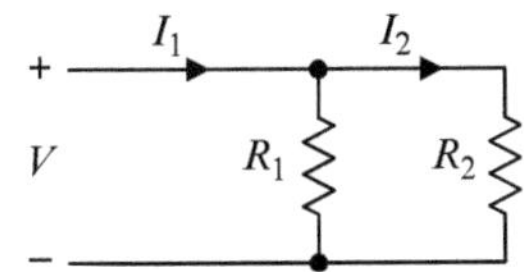

Fig. P 5.33 See Problem P 5.92

$V_0 = 10\,\text{V},\ R_1 = 10\,\Omega,\ R_2 = 30\,\Omega,\ R_L = 100\,\Omega$

Fig. P 5.35 See Problem P 5.94

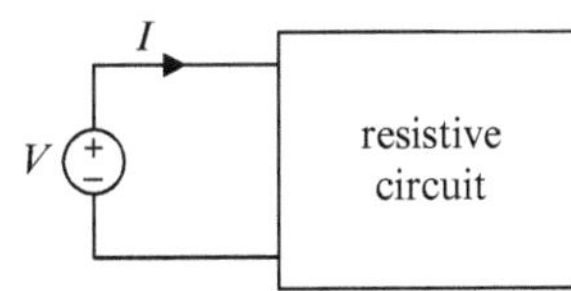

Fig. P 5.36 See Problem P 5.95

The ambient temperature for both is 25°C. (a) Obtain an expression for the voltage ratio $H = V_2/V_1$ as a function of the rms current I_{rms} through the resistors. (b) Let $\alpha = TCR = 4500$ ppm, φ = self − heating *coefficient* = 200°C W^{-1}, $R_1 = 1\,\text{k}\Omega$, $R_2 = 2.2\,\text{k}\Omega$ (at 25°C) and construct a graph of the voltage ratio for $0 \le I_{rms} \le 20\,\text{mA}$. (c) If you design the voltage divider assuming the resistances have their 25°C values (are independent of the current I_{rms}), what is the percent error in the voltage ratio for $I_{rms} = 20\,\text{mA}$?

P 5.92 Figure P 5.33 shows a current divider, where the resistors R_1, R_2 have identical temperature coefficients and identical self-heating coefficients. The ambient temperature for both is 25°C. (a) Obtain an expression for the current $H = I_2/I_1$ as a function of the rms voltage V_{rms} across the resistors. (b) Let $\alpha = TCR = 4500$ ppm, φ = self − heating *coefficient* = 200°C W^{-1}, $R_1 = 1\,\text{k}\Omega$, $R_2 = 2.2\,\text{k}\Omega$ (at 25°C) and construct a graph of the current ratio for $0 \le V_{rms} \le 10\,\text{V}$. (c) If you design the current divider assuming the resistances have their 25°C values (are independent of the current I), what is the percent error in the current ratio for $V_{rms} = 10\,\text{V}$?

Section 5.14 is prerequisite for the following problems.

P 5.93 Figure P 5.34 shows a load driven by (the Thévenin-equivalent for) a source and the same load driven by an equivalent Norton model for the source. Obtain expressions for the powers dissipated by all elements in each circuit and discuss the results.

P 5.94 Refer to Fig. P 5.35. (a) Calculate the power dissipated in each element. Check your result by showing that the total power dissipated (passive sign convention) equals zero. (b) Obtain the Thévenin equivalent for the circuit at the terminals *a–b* (with the load R_L removed). Then connect the load to the Thévenin equivalent and calculate the power dissipated by each element. Check your result by showing that the total power dissipated (passive sign convention) equals zero. Is the power dissipated by the Thévenin resistance equal to the total power dissipated by the resistors R_1, R_2 in the original circuit? Is the power dissipated by the Thévenin source equal to the power dissipated by the source in the original circuit? Is the power dissipated by the load the same in each case? (c) Repeat part (b) for the Norton equivalent. (d) Is the power dissipated by the Thévenin source equal to that dissipated by the Norton source? Is the power dissipated by the Thévenin resistance equal to that dissipated by the Norton resistance?

P 5.95 In Fig. P 5.36, the resistive circuit contains only resistors. The rms voltage V and rms current I are known. Your lab partner claims that the maximum power rating required for any resistor in the network does not exceed $P = V_{rms}I_{rms}$. (a) Is he correct? (b) Is he still correct if the circuit also contains unknown current or voltage sources?

Section 5.15 is prerequisite for the following problems.

P 5.96 Give three everyday examples of devices containing active circuits. You probably carry at least two such devices.

P 5.97 Discuss the following statements:

(a) All active devices have three or more terminals.
(b) An active circuit must have at least two sources of power, one of which is the input.
(c) All energy-conversion devices are active.

P 5.98 (Web research) Which of the following devices can be active?

(a) Silicon-controlled rectifier (SCR)
(b) Thyristor
(c) MOSFET
(d) JFET
(e) BJT
(f) Selenium rectifier
(g) Solar cell
(h) DC electric motor
(i) DC generator

P 5.99 Classify each of the following as active (contains active components) or passive.

(a) Flashlight
(b) Microprocessor
(c) Portable radio
(d) Pocket calculator
(e) Digital wristwatch
(f) Cell phone
(g) Automobile battery

P 5.100 Refer to Fig. P 5.37. For each of the following conditions, describe the source as lightly loaded or heavily loaded.

(a) $v_L(t) \cong 0.9\, v_S(t)$
(b) $|i_L(t)| \ll |v_S(t)/(2R_S)|$ for all times t
(c) $R_L \gg R_S$
(d) $R_L \ll R_S$
(e) $I_{L\,rms} V_{L\,rms} \gg {I_{L\,rms}}^2 R_S$
(f) $v_S(t) \cong 1.01\, i_L(t) R_S$
(g) $V_{L\,rms}/I_{L\,rms} \cong 0.01\, R_S$

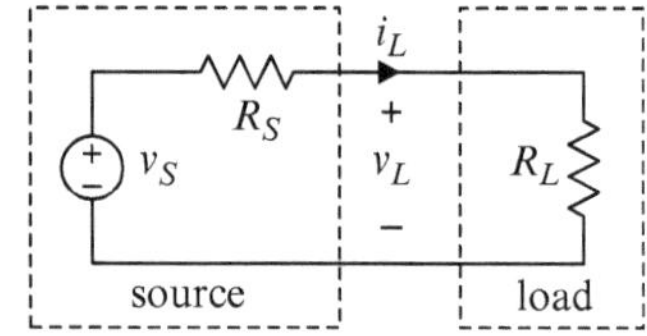

Fig. P 5.37 See Problem P 5.100

Section 5.16 is prerequisite for the following problems.

P 5.101 The internal resistance of a 9 V alkaline battery is about 2 Ω. What is the maximum power the battery can deliver to a resistive load?

P 5.102 A resistor having resistance R is connected to the terminals of a linear resistive circuit. It is found that the power dissipated by the resistor is P_1. An identical resistor is then connected in parallel with the first, after which it is found that the total power dissipated by the two resistors is very nearly $2P_1$. What can you deduce about the value of the Thévenin equivalent resistance, relative to R, for the circuit from these measurements?

P 5.103 Repeat Problem P 5.102 for a case where the two powers are approximately equal.

P 5.104 A certain linear resistive circuit having a dc output delivers 80 W to an 8 Ω resistive load, under which condition the power-transfer efficiency is 0.8 (80%). Can you obtain the Thévenin equivalent for the circuit from these data? If so, show how. If not, explain why not.

P 5.105 A certain linear resistive voltage source having a sinusoidal output can deliver at most 80 W (average) to an 8 Ω resistive load, under which condition the power-transfer efficiency is 0.8 (80%). Can you obtain the Thévenin equivalent for the circuit from these data? If so, show how. If not, explain why not.

P 5.106 It is required to maximize the power delivered by two identical batteries to a fixed but unknown resistive load. Is it better to connect the batteries in series or parallel, or does it matter? Justify your answer.

P 5.107 Refer to Fig. P 5.38, where the load R_L is fixed. (a) Obtain an expression for the power delivered

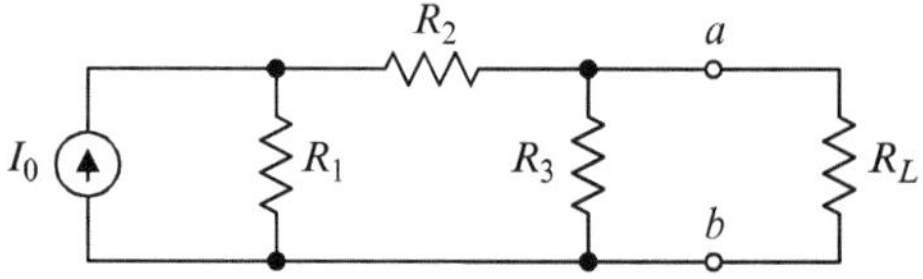

Fig. P 5.38 See Problem P 5.107

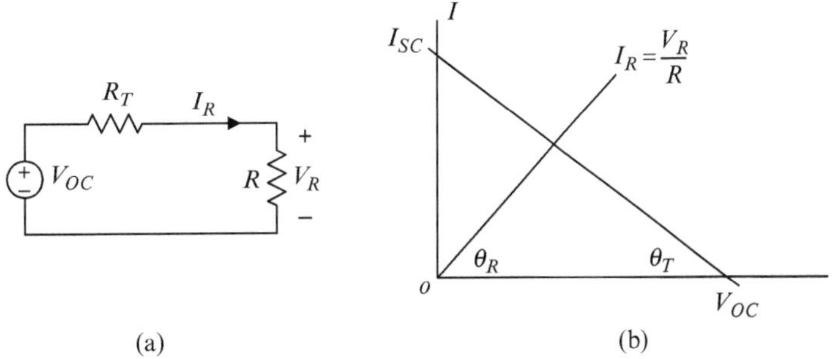

Fig. P 5.39 See Problem P 5.108

to the load. (b) For what values of the resistances R_1, R_2, R_3 is the power delivered to the load maximum? (c) Obtain an expression for the Thévenin equivalent resistance of the circuit to the left of the terminals a–b. (d) For what value of the Thévenin equivalent resistance is maximum power delivered to the load? Discuss.

P 5.108 Figure P 5.39 shows a resistor R connected to a Thévenin source model and graphs of the terminal characteristics of the source and resistor on the same axes. The Thévenin resistance is fixed. What is the relationship between the angles θ_R, θ_T when the power dissipated in the resistor has its maximum value?

P 5.109 Show that the power that can be drawn from an independent current source having infinite output resistance is unlimited.

P 5.110 Suppose you must design an amplifier that will drive a load whose nominal resistance is 8 Ω. If maximum power transfer is essential, what value would you attempt to achieve for the output resistance of the amplifier?

P 5.111 Even if power transfer from an amplifier to a load is important, why might it be unwise to minimize the output resistance of the amplifier? *Hint*: What limits the current drawn from the amplifier?

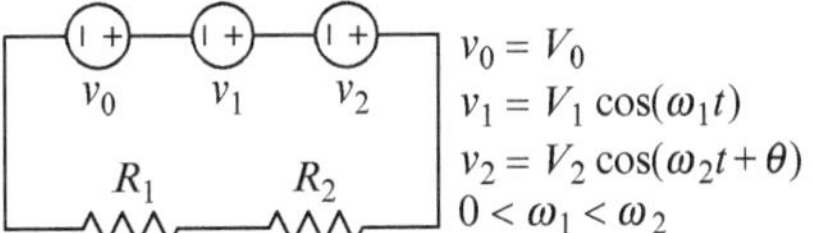

Fig. P 5.40 See Problem P 5.112

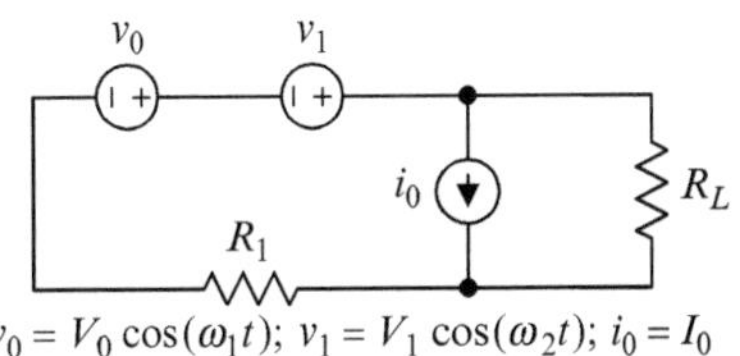

Fig. P 5.41 See Problem P 5.114

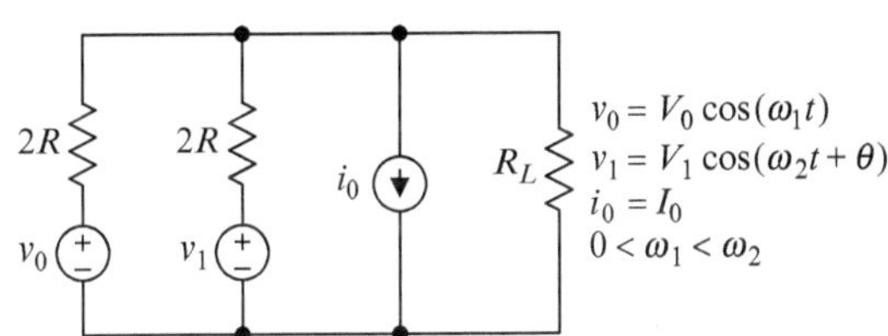

Fig. P 5.42 See Problem P 5.115

Section 5.17 is prerequisite for the following problems.

P 5.112. Refer to Fig. P 5.40. Obtain expressions for the average power dissipated by each resistor.

P 5.113 (a) Obtain an expression for the rms amplitude of

$$i(t) = I_0 + \sum_{k=1}^{N} I_k \cos(\omega_k t + \theta_k),$$

where $0 < \omega_1 < \omega_2 < \cdots < \omega_N$. (b) Suppose $N = 3$, $I_0 = 0.5\,\text{mA}$, $I_1 = 1\,\text{mA}$, $I_2 = 1\,\text{mA}$, $I_3 = 1\,\text{mA}$. Calculate the rms amplitude of $i(t)$. What is the average power dissipated if $i(t)$ is applied to a 4.7 kΩ resistor?

P 5.114 Refer to Fig. P 5.41 where $0 < \omega_1 < \omega_2$. Obtain an expression for the average power dissipated in the resistor R_L.

P 5.115 Refer to Fig. P 5.42 where $0 < \omega_1 < \omega_2$. Obtain an expression for the average power dissipated in the resistor R_L.

Chapter 6
Dependent Sources and Unilateral Two-Port Circuits

Circuits are designed and used to establish specified relations among certain currents and voltages, where some currents and voltages are identified as **inputs** (causes) and others as **outputs** (effects). Inputs also are called **excitations**, and outputs also are called **responses**.

Inputs are applied to and outputs are extracted from terminal pairs. A terminal pair to which an input is applied or from which an output is extracted is called a **port**. To be considered a port, a terminal pair must satisfy Kirchhoff's current law (current entering one terminal equals current out exiting the other).

A **two-port circuit** is a *physical* circuit having one input port and one output port. Figure 6.1 shows an abstract representation of a two-port circuit, with v_1 or i_1 as an input (excitation) and v_2 or i_2 as the corresponding output (response). A **two-port model** is a particular, simplified representation of a two-port circuit or device. Whereas a two-port *circuit* might contain a great many physical components, a two-port *model* for the circuit contains no more than four elements, and often fewer than four elements; for example, some operational amplifiers,[1] available as integrated circuits, contain a dozen or more transistors, but a particular and useful two-port *model* for an operational amplifier contains only one element. It follows that the circuit diagram for a two-port model bears little or no resemblance to the physical structure of the associated physical circuit or device. Moreover, few, if any components of a two-port model correspond to physical components. Rather, a two-port model for a physical circuit uses idealized circuit elements to depict *relations* established by the circuit among the input and output currents and voltages. In other words, a two-port model expresses the terminal characteristics of a two-port circuit, but says nothing about how those terminal characteristics are achieved. The circuit diagram depicting a two-port model is a far cry from a wiring diagram; nonetheless, specifying parameters of a two-port model often is a first step in circuit design.

For simplicity, we often refer to a two-port circuit or the associated model as simply a **two-port**. Also, we usually draw circuit diagrams such that the input port is on the left and the output port is on the right.

The two-ports treated in this chapter are composed of only resistors and elements called *dependent sources*, defined in Section 6.2. Few useful physical circuits are purely resistive, so it might seem that models containing only resistors and sources are of limited value. However, *the principles and methods described here are applicable to virtually any linear circuit - even those containing capacitors and inductors.*[2] In fact, although we have to this point treated only resistive circuits, everything in previous chapters and everything in this chapter are readily extended to circuits that are not entirely resistive. Subsequent chapters describe how.

Topics covered in this chapter are prerequisite to fruitful study of subsequent chapters that treat amplifiers and other signal-processing circuits.

[1]Operational amplifiers are treated in Chapter 7.

[2]Capacitors and inductors are treated in subsequent chapters.

T.H. Glisson, *Introduction to Circuit Analysis and Design*,
DOI 10.1007/978-90-481-9443-8_6,

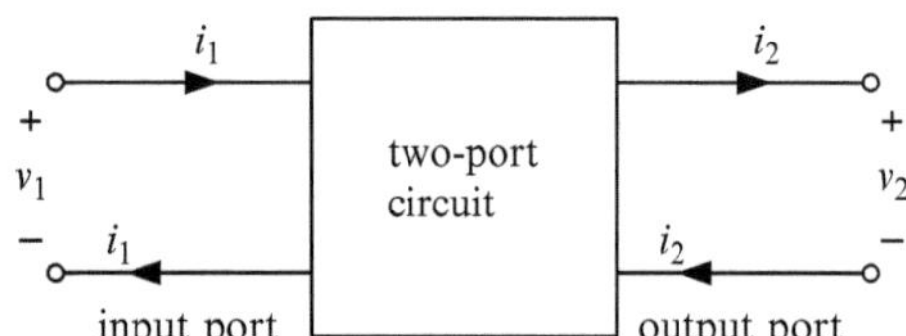

Fig. 6.1 A representation of a two-port circuit

6.1 Input Resistance and Output Resistance

The **input resistance** R_{in} of a two-port circuit can be obtained *mathematically* (from a circuit model) using either an ideal voltage source or an ideal current source, as illustrated by Fig. 6.2, where

$$R_{in} = \frac{V_1}{I_1}. \tag{6.1}$$

In general, input resistance depends upon load resistance, so an appropriate load must be attached to the output terminals when making such a calculation.

The **output resistance** R_{out} of a two-port can be obtained *mathematically* (from a circuit model) using either an ideal voltage source or an ideal current source, as illustrated by Fig. 6.3, where

$$R_{out} = \frac{V_2}{I_2}. \tag{6.2}$$

In general, output resistance depends upon source resistance, so an appropriate source resistance must be attached to the input terminals when making such a calculation.

The input and output resistances of a two-port are *measured* essentially the same way they are calculated, except we might need to account for the internal resistance R_S of the (non-ideal) source used to drive the measurement. In Fig. 6.4(a) the input resistance is given in terms of the applied voltage V_1 and the measured current I_1 by

$$R_{in} = \frac{V_1}{I_1} - R_S, \tag{6.3}$$

where the internal resistance R_S of the source is subtracted because it is not part of the input resistance. Similarly, in Fig. 6.4(b),

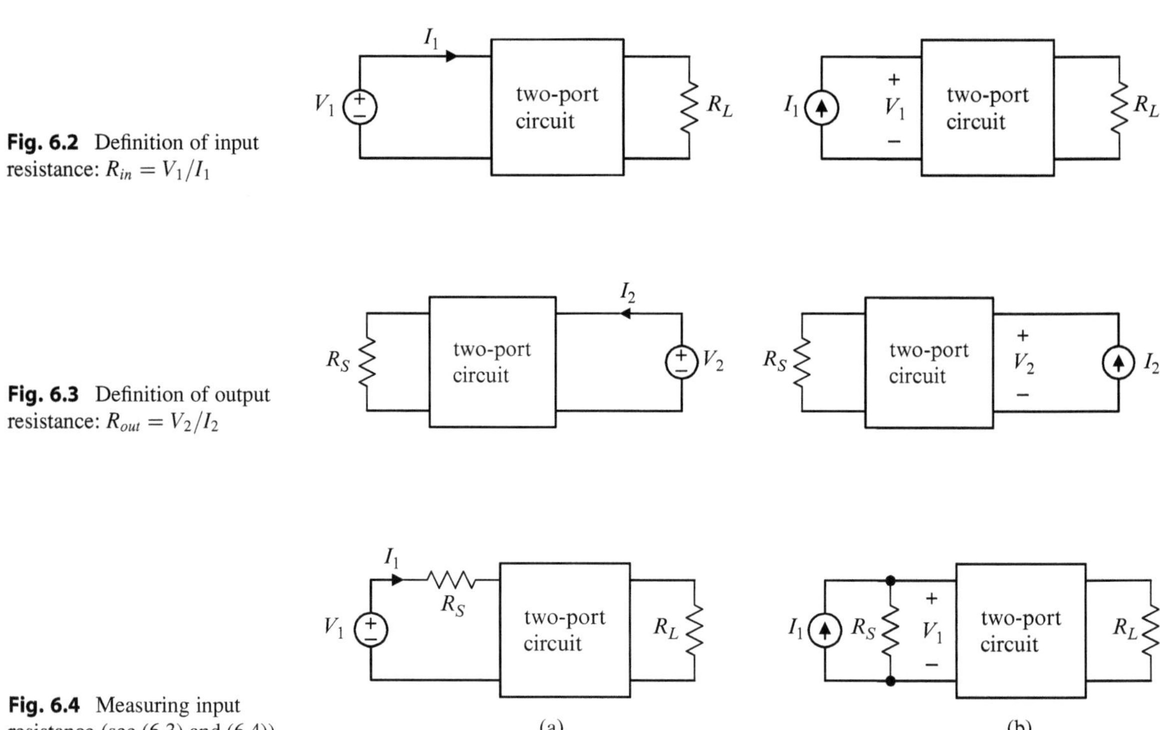

Fig. 6.2 Definition of input resistance: $R_{in} = V_1/I_1$

Fig. 6.3 Definition of output resistance: $R_{out} = V_2/I_2$

Fig. 6.4 Measuring input resistance (see (6.3) and (6.4))

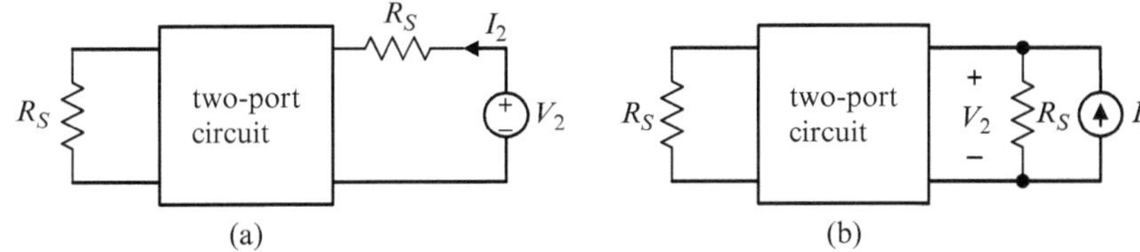

Fig. 6.5 Measuring output resistance (see (6.5) and (6.6))

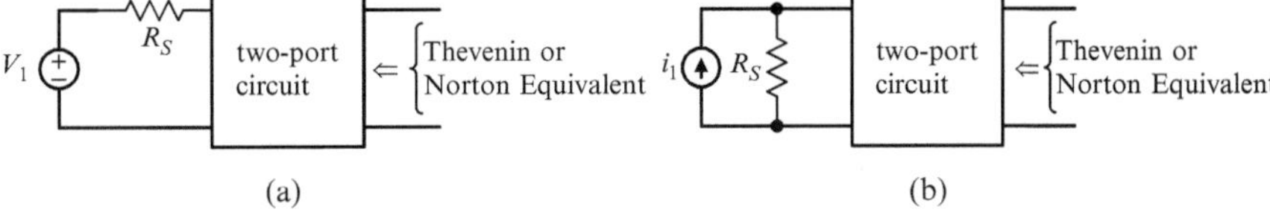

Fig. 6.6 Output resistance is the Thévenin or Norton equivalent resistance

$$G_{in} = \frac{1}{R_{in}} = \frac{I_1}{V_1} - \frac{1}{R_S}. \quad (6.4)$$

Likewise, in Fig. 6.5(a),

$$R_{out} = \frac{V_2}{I_2} - R_S \quad (6.5)$$

and, in Fig. 6.5(b),

$$G_{out} = \frac{1}{R_{out}} = \frac{I_2}{V_2} - \frac{1}{R_S}. \quad (6.6)$$

Note that V_2,I_2 in Fig. 6.5(a) are not the same as V_2, I_2 in Fig. 6.5(b).

We can also determine the output resistance of a two-port by finding the Thévenin equivalent of the two-port at the output terminals with either a voltage source or current source attached to the input, as illustrated by Fig. 6.6. The output resistance is the Thévenin (or Norton) equivalent resistance.

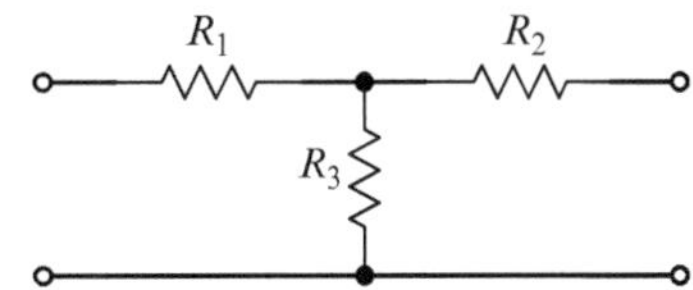

Fig. 6.7 See Example 6.1

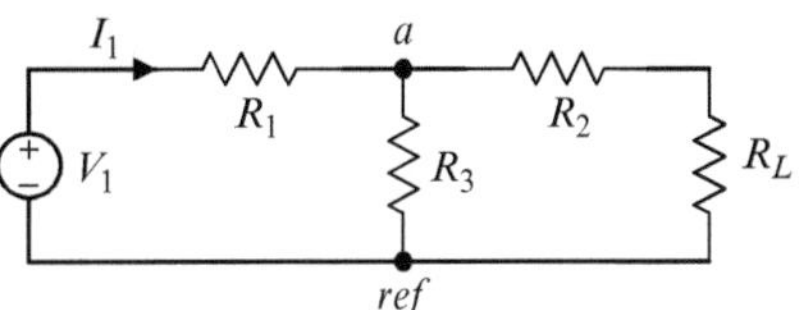

Fig. 6.8 See Example 6.1

Example 6.1. Obtain an expression for the input resistance of the circuit in Fig. 6.7.

Solution: Because we are obtaining input resistance from a circuit diagram (as opposed to a laboratory measurement), we may assume an ideal test source as shown in Fig. 6.8, where we have connected a load R_L, as required. The input resistance of the circuit is given by

$$R_{in} = \frac{V_1}{I_1} = \left(\frac{V_1}{V_1 - V_a}\right) R_1. \quad (6.7)$$

Applying Kirchhoff's current law to the circuit in Fig. 6.8 gives

$$\frac{V_a - V_1}{R_1} + \frac{V_a}{R_3} + \frac{V_a}{R_2 + R_L} = 0. \quad (6.8)$$

which yields

$$V_a = \frac{R_3(R_2 + R_L)V_1}{R_3(R_2 + R_L) + R_1(R_2 + R_3 + R_L)}. \quad (6.9)$$

Using (6.9) in (6.7) yields

$$R_{in} = \frac{R_3(R_2+R_L)+R_1(R_2+R_3+R_L)}{R_2+R_3+R_L}. \quad (6.10)$$

We could as well have used series/parallel reduction to find the equivalent resistance at the input terminals, which also is the input resistance. Thus

$$\begin{aligned} R_{in} &= R_1+R_3\|(R_2+R_L) \\ &= R_1+\frac{R_3(R_2+R_L)}{R_3+R_2+R_L} \\ &= \frac{R_3(R_2+R_L)+R_1(R_2+R_3+R_L)}{R_2+R_3+R_L}. \end{aligned} \quad (6.11)$$

as above. If $R_2 \gg R_L$, the input resistance is only weakly dependent upon the load resistance as is evident from the circuit diagram in Fig. 6.7.

Example 6.2. Use the procedure illustrated by Fig. 6.3(b) to obtain an expression for the output resistance of the circuit shown in Fig. 6.7.

Solution: Because we are obtaining output resistance from a circuit diagram (as opposed to a laboratory measurement), we may use an ideal dc test source as shown in Fig. 6.9, where

$$R_{out} = \frac{V_2}{I_2} = \left(\frac{V_2}{V_2 - V_a}\right)R_2. \quad (6.12)$$

Kirchhoff's current law gives

$$\frac{V_a}{R_1+R_S} + \frac{V_a}{R_3} + \frac{V_a - V_2}{R_2} = 0, \quad (6.13)$$

which yields

$$V_a = \frac{R_3(R_1+R_S)V_2}{R_3(R_1+R_S)+R_2(R_1+R_3+R_S)}. \quad (6.14)$$

Using this result in (6.12) gives

$$R_{out} = \frac{R_3(R_1+R_S)+R_2(R_1+R_3+R_S)}{R_1+R_3+R_S}. \quad (6.15)$$

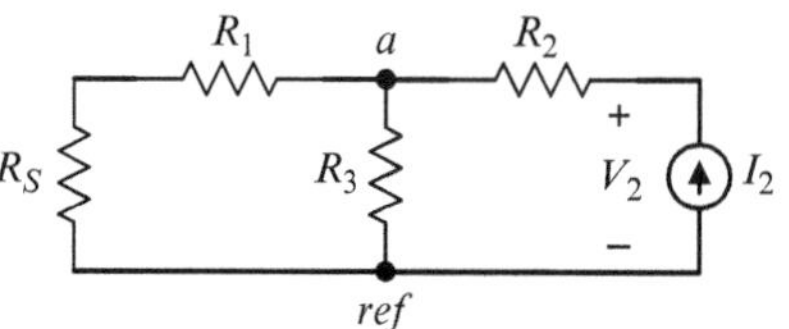

Fig. 6.9 See Example 6.2

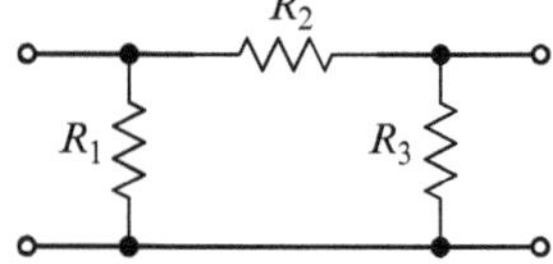

Fig. 6.10 See Exercise 6.1

Here also, we could have used series/parallel reduction to obtain the equivalent resistance at the output terminals. If $R_1 \gg R_S$, the output resistance is only weakly dependent upon the source resistance.

Exercise 6.1. (a) Use the procedures illustrated by Figs. 6.2(a) and 6.3(a) to obtain expressions for the input resistance and output resistance of the circuit shown in Fig. 6.10. (b) Repeat, using series/parallel reduction. (c) Obtain an expression for the output resistance by attaching a test source to the input and finding the Thévenin equivalent at the output terminals.

6.2 Dependent Sources

A **dependent source** is a source whose terminal voltage or terminal current *depends upon* or is *controlled by* another voltage or current called the **control voltage** or **control current**. The control current or voltage for a dependent source may be in the same circuit as the dependent source or in another circuit remote from the dependent source. It is *not* necessary that a circuit containing a dependent source be connected to the circuit containing the control current

Table 6.1 Dependent (controlled) sources

Name	Symbol	Control	Terminal characteristic
Voltage-controlled voltage source (VCVS)	μv_c; i; + v −	v_c	$v = \mu v_c$
Voltage controlled current source (VCCS)	$g v_c$; i; + v −	v_c	$i = g v_c$
Current controlled voltage source (CCVS)	$r i_c$; i; + v −	i_c	$v = r i_c$
Current controlled current source (CCCS)	βi_c; i; + v −	i_c	$i = \beta i_c$

or voltage by a wire, resistor, or other element. For example, you might think of a radio receiver as containing a dependent source controlled by a current or voltage in a distant transmitter.

Dependent sources also are called **controlled sources**, which in many applications is a more descriptive term. We limit our discussion to the four *linear* dependent sources whose terminal currents or voltages are proportional to one other current or voltage.[3] Table 6.1 gives the circuit-diagram symbols and terminal characteristics for these four sources.

By convention, and *for the present, we assume the parameters μ,g,r,β are positive*. The sign of the terminal characteristic for a source can be reversed by reversing either the direction of i or the polarity of v on the associated symbol given in Table 6.1. The dimensionless parameters μ and β are the **intrinsic voltage gain** and the **intrinsic current gain**, respectively, of the associated sources. The parameters r and g are the **intrinsic transresistance** and **intrinsic transconductance**, respectively, of the associated sources.[4] Transresistance has the dimension of resistance and is expressed in ohms (Ω). Transconductance has the dimension of conductance and is expressed in siemens (S).

Unless confusion is otherwise possible, we omit the adjective *intrinsic* in references to parameters of dependent sources; e.g., if no confusion is possible, we refer to the intrinsic voltage gain of a VCVS as the voltage gain of the source. But where other like-named quantities are present, we are specific. For example, we might need to refer to the (*overall*) voltage gain of a circuit that contains a VCVS having *intrinsic* voltage gain μ.

A dependent source is a *non-physical* but useful element for exhibiting a relation between two currents, two voltages, or a current and a voltage. A dependent source exhibits only the relation, not the physical mechanism that establishes the relation. This is similar to using the symbol for a resistor to depict a relation (Ohm's law) between current through and voltage across the terminals of a resistor. Like the symbol for a dependent source, the symbol for a resistor says nothing about how that relation is established (physically - in a solid). Like a dependent source, a resistor in a circuit

[3]In general, the source current or voltage can be a linear or nonlinear *function* of one or more other currents and voltages. More generally, the input or output can be a function of a non-electrical quantity, such as temperature, pressure, or torque. For example, an electric motor can be regarded as a controlled source whose input is current and whose output is torque.

[4]These names are contractions of *transfer resistance* and *transfer conductance*, respectively.

diagram might be non-physical; e.g., the Thévenin equivalent resistance in a model for a two-terminal circuit is in general non-physical. Still, you might think, a resistor is more real than a dependent source, because you can buy resistors, but you cannot find dependent sources listed in electronics catalogs. However, you can buy devices called transistors and operational amplifiers (introduced in the next chapter) that behave like dependent sources under certain operating conditions. Thus it can be argued that dependent sources are as real as resistors, capacitors, and inductors, the main difference being that devices that behave like dependent sources are sold under other names.

By and large, analysis of circuits containing dependent sources proceeds in the same manner as analysis of circuits not containing such sources, as illustrated by the following examples.

Example 6.3. Transform the circuit shown in Fig. 6.11(a) to the equivalent circuit shown in Fig. 6.11(b).

Solution: The transformation proceeds exactly as if the dependent sources were independent sources, following the procedure for converting a Norton source model to a Thévenin source model. The Norton and Thévenin equivalent resistances are equal and the controlled-source voltage $R\beta i_b$ is obtained by multiplying the controlled-source current βi_b by the Norton equivalent resistance.

Exercise 6.2. Transform the circuit shown in Fig. 6.12(a) to the equivalent circuit shown in Fig. 6.12(b) (express the transconductance g in terms of the circuit parameters).

Example 6.4. Refer to Fig. 6.13. Obtain an expression for the voltage v_L.

Solution: By voltage division,

$$v_L = \left(\frac{R_L}{R_o + R_L}\right)\left(\frac{R_i}{R_S + R_i}\right)\mu v_S.$$

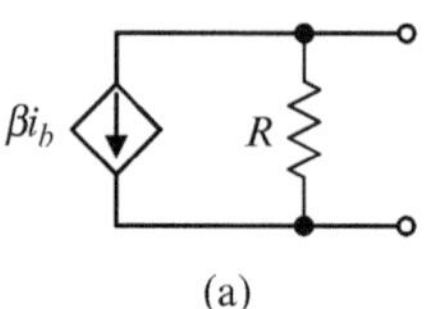

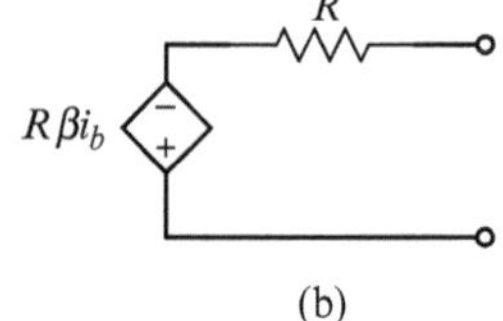

Fig. 6.11 See Example 6.3

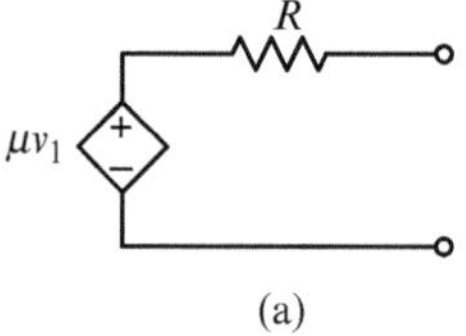

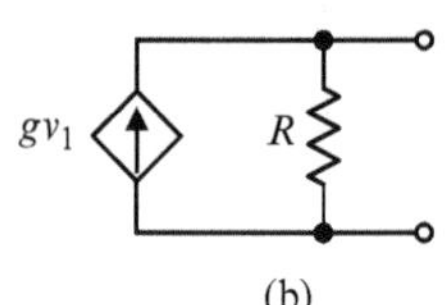

Fig. 6.12 See Exercise 6.2

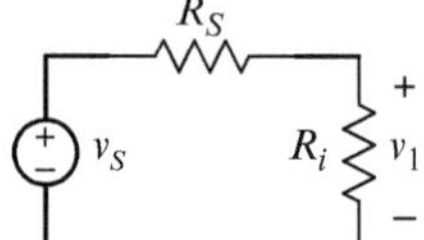

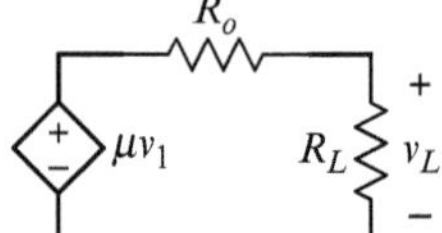

Fig. 6.13 See Example 6.4

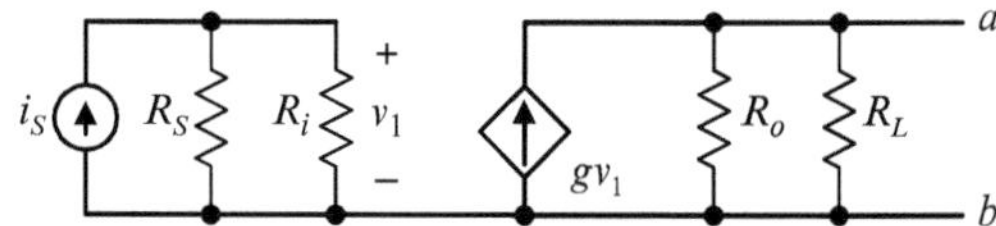

Fig. 6.14 See Exercise 6.3

Exercise 6.3. Refer to Fig. 6.14. Obtain an expression for the voltage v_{ab}.

Example 6.5. Find the equivalent resistance R_{eq} one would measure at the terminals a-b of the circuit shown in Fig. 6.15.

Solution: To find the equivalent resistance, we use Ohm's law. We apply (mathematically) a

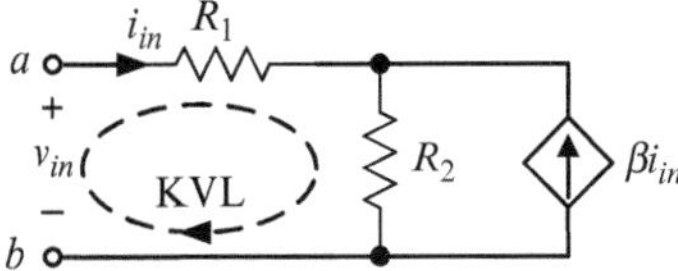

Fig. 6.15 See Example 6.5

voltage v_{in} at the terminals a–b, find the resulting current i_{in}, and obtain

$$R_{eq} = \frac{v_{in}}{i_{in}}.$$

Kirchhoff's voltage law gives

$$-v_{in} + R_1 i_{in} + (1+\beta)R_2 i_{in} = 0,$$

which yields

$$R_{eq} = \frac{v_{in}}{i_{in}} = R_1 + (1+\beta)R_2.$$

Note that if the internal resistance of the dependent current source were infinite, as it is for an independent current source, the equivalent resistance would be $R_1 + R_2$, as one might be misled to expect from an inspection of the circuit.

Exercise 6.4. The equivalent or *apparent resistance* at a terminal pair is given by $R_{eq} = V_S/I$, where V_S is a voltage applied to the pair and I is the resulting current entering the positive terminal. Find the apparent resistance at the terminals a–b of the circuit shown in Fig. 6.16.

Example 6.6. Refer to Fig. 6.17. Obtain the Thévenin equivalent for the circuit at the terminals a–b.

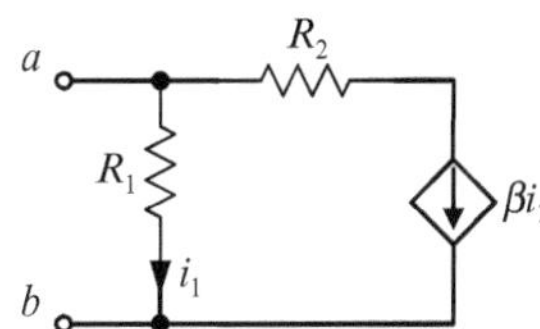

Fig. 6.16 See Exercise 6.4

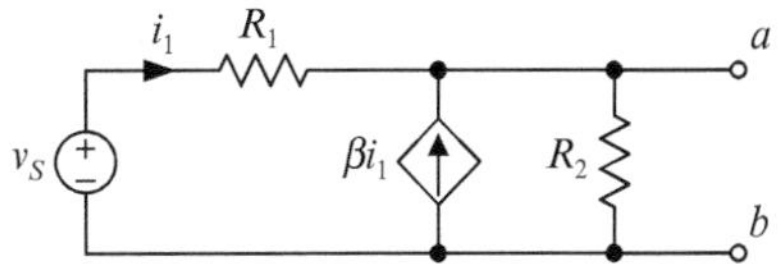

Fig. 6.17 See Example 6.6

Solution: We use Kirchhoff's current law to obtain an expression for the open-circuit voltage:

$$-i_1 - \beta i_1 + \frac{v_{oc}}{R_2} = 0 \Rightarrow v_{oc} = (\beta+1)R_2 i_1.$$

But

$$i_1 = \frac{v_S - v_{oc}}{R_1}.$$

It follows that

$$v_{oc} = (\beta+1)R_2 i_1 = (\beta+1)R_2\left(\frac{v_S - v_{oc}}{R_1}\right).$$

Thus the Thévenin-equivalent voltage is given by

$$v_T = v_{oc} = \frac{(\beta+1)R_2 v_S}{R_1 + (\beta+1)R_2}.$$

The short-circuit current is given by

$$\begin{aligned} -i_1 - \beta i_1 + i_{sc} = 0 \Rightarrow i_{sc} &= (\beta+1)i_1 \\ &= (\beta+1)\frac{v_S}{R_1}. \end{aligned}$$

Thus the Thévenin-equivalent resistance is given by

$$\begin{aligned} R_T = \frac{v_{oc}}{i_{sc}} &= \frac{(\beta+1)R_2 v_S}{R_1 + (\beta+1)R_2}\left[(\beta+1)\frac{v_S}{R_1}\right]^{-1} \\ &= \frac{R_1 R_2}{R_1 + (\beta+1)R_2}. \end{aligned}$$

Exercise 6.5. Refer to Fig. 6.14 (Exercise 6.3). Obtain the Thévenin equivalent for the circuit at the terminals *a–b*.

As examples above illustrate, analysis of a circuit containing one or more dependent sources is carried out in the same way as that of a circuit containing no such sources. But there is one caveat: When applying *superposition* to a circuit containing dependent sources, proceed as follows:

1. Select an independent source. Set all other *independent* sources to zero (replace independent voltage sources by short circuits and independent current sources by open circuits).
2. Examine all dependent sources. Set to zero each dependent source whose terminal current or voltage is proportional to the terminal current or voltage of an independent source that was set to zero in step (1) above, but leave all other dependent sources in place. In other words, do not zero a dependent source whose terminal current or voltage depends upon a current or voltage produced by the non-zero independent source under consideration.
3. Repeat steps (1) and (2) until all independent sources have been accounted for, then add the individual responses obtained, as usual.

The next example is contrived to illustrate the procedure described above with as little distracting mathematics as possible.

Example 6.7. Use superposition to obtain an expression for the voltage v_x in the circuit shown in Fig. 6.18(a).

Solution: Figure 6.18(b)–(d) illustrate the procedure.

Response to v_1: We set the other two independent sources (v_2 and i_1) to zero (replace v_2 by a short circuit and i_1 by an open circuit). Because $v_2 = 0$, then $gv_2 = 0$ as well, and we replace the dependent current source with an open circuit. The control current i_{c1} for the dependent voltage source is not necessarily

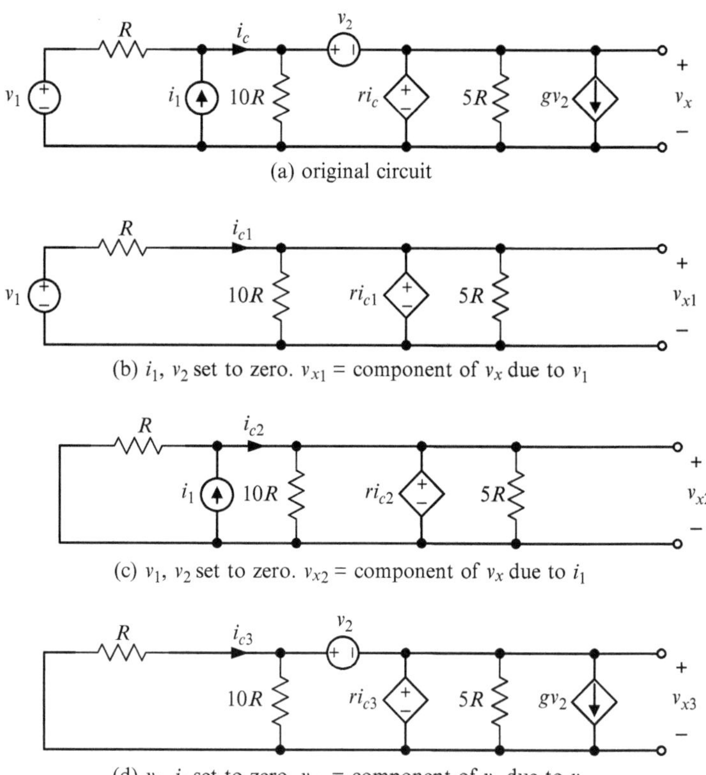

Fig. 6.18 See Example 6.7

zero, so we do not set the dependent voltage source to zero. Figure 6.18(b) shows the resulting circuit. By inspection,

$$v_{x3} = ri_{c3}$$

and

$$i_{c1} = \left(\frac{v_1 - ri_{c1}}{R}\right) \Rightarrow i_{c1} = \frac{v_1}{R+r}.$$

Thus

$$v_{x1} = \frac{rv_1}{R+r}. \tag{6.16}$$

Response to i_1: Beginning again with the circuit in Fig. 6.18(a), we set v_1 and v_2 to zero, which makes gv_2 zero as well. Thus we replace the independent voltage sources with short circuits and the dependent current source with an open circuit. Here again, the control current i_{c2} for the dependent voltage source is not necessarily zero, so we do not set the dependent voltage source to zero. Figure 6.18(c) shows the resulting circuit. By inspection,

$$v_{x2} = ri_{c2}.$$

Kirchhoff's current law gives

$$i_{c2} + \frac{ri_{c2}}{R} - i_1 = 0 \Rightarrow i_{c2} = \frac{Ri_1}{R+r}.$$

Thus

$$v_{x2} = \frac{rRi_1}{R+r}. \tag{6.17}$$

Response to v_2: Beginning again with the circuit in Fig. 6.18(a), we set v_1 and i_1 to zero. In this case, $gv_2 \neq 0$, so we do not set the dependent current source to zero. Figure 6.18(d) shows the resulting circuit. By inspection,

$$v_{x1} = ri_{c1}$$

and

$$i_{c3} = -\frac{v_2 + v_{x3}}{R}$$

Thus

$$v_{x3} = -\frac{r(v_2 + v_{x3})}{R} \Rightarrow v_{x3} = -\frac{rv_2}{R+r}$$

All independent sources have been accounted for, and the complete response is given by

$$v_x = v_{x1} + v_{x2} + v_{x3} = \frac{r(v_1 + Ri_1 - v_2)}{R+r}.$$

Exercise 6.6. Use superposition to obtain an expression for the voltage v_{out} in the circuit shown in Fig. 6.19.

We present the procedure and example above for completeness, but such problems rarely arise in practice. In analysis of electronic systems, superposition is useful mainly where an input (the current or voltage at a single input port) can be modeled using a number of

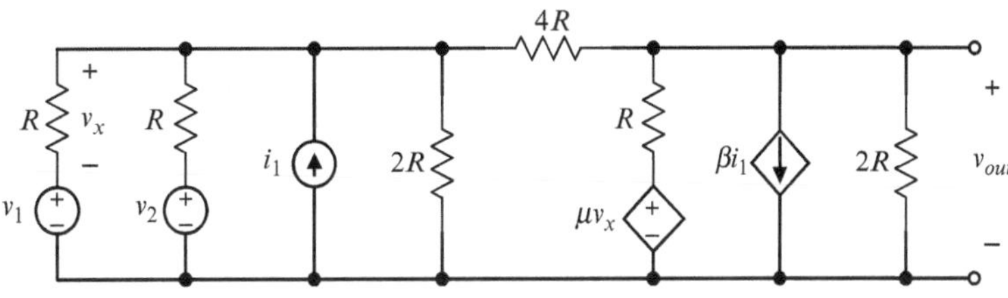

Fig. 6.19 See Exercise 6.6

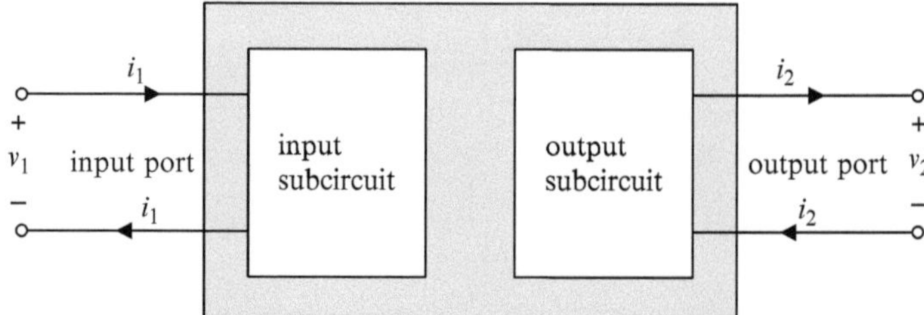

Fig. 6.20 Two-port model

independent voltage sources in series or a number of independent current sources in parallel. In most realistic circuit models, internal sources are dependent sources, all of which are controlled by currents or voltages produced by the input. Circuits having independent sources scattered about internally or in which different controlled sources depend upon different independent sources are found mainly in textbooks.

6.3 Linear Two-Port Models

Review the introductory discussion at the beginning of this chapter.

Because the currents entering and exiting the input port of a two-port circuit are equal and the currents entering and exiting the output port of a two-port circuit are equal, a two-port circuit can be *modeled* as if it were two separate circuits, called the *input subcircuit* and the *output subcircuit*, as shown in Fig. 6.20.[5]

A two-port model is **bilateral** if a current or voltage in either subcircuit can give rise to a current or voltage in the other subcircuit. A two-port model is **unilateral** if currents and voltages in the output subcircuit cannot produce currents or voltages in the input subcircuit. Think of *unilateral* as meaning *one-way*, from input to output, but not vice-versa. In a unilateral two-port model, the **controlling current or voltage** is a current or voltage in the input subcircuit and the **controlled current or voltage** is a dependent source in the output subcircuit.

Exercise 6.7. Under what condition(s) is the circuit shown in Fig. 6.21 unilateral?

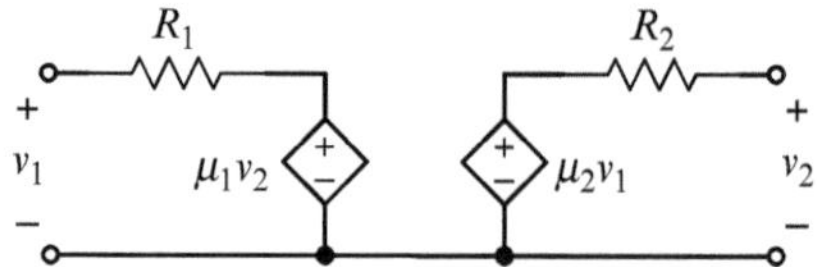

Fig. 6.21 See Exercise 6.7

For present purposes and for many important applications, unilateral models are sufficient, and *we limit our development in this chapter to unilateral models.* In addition, we assume that a model contains only dependent sources and resistors (no independent sources).[6] Consequently, the model is linear, and *every current and voltage in the model is proportional to a current or voltage applied at the input port.*

From the viewpoint of a source connected to the input port, the input subcircuit in Fig. 6.20 is a load. Because the model is resistive, the load is resistive, and is the equivalent resistance seen at the input port of the model. This resistance is the **input resistance** of the unilateral two-port model.

From the viewpoint of a load attached to the output port, the output subcircuit in Fig. 6.20 is a source. Because the circuit is linear (contains no independent sources), the output (current i_2 or voltage v_2) is proportional to the input (current i_1 or voltage v_1). It follows that the output subcircuit is a Thévenin or Norton model in which the source is dependent upon the input current or voltage. The Thévenin or Norton equivalent resistance at the output port is the **output resistance** of the unilateral two-port circuit.

The input resistance of a unilateral two-port is independent of the load attached to the output terminals, because currents and voltages in the output subcircuit do not give rise to currents or voltages in the input subcircuit. Likewise, because a source attached

[5]Remember that Fig. 6.20 is the structure of a two-port *model*, not of the associated physical circuit.

[6]Two-port models can be extended to cases where a circuit in question contains independent sources, but we have no need of that generalization in this book.

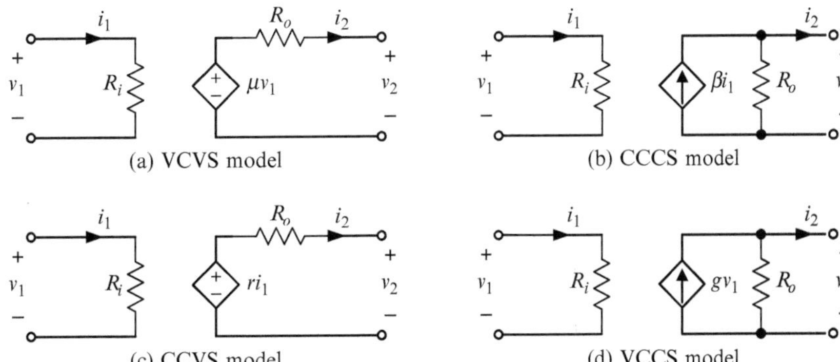

Fig. 6.22 Linear models for a unilateral two-port circuit

to the output cannot produce a current or voltage in the input subcircuit, the output resistance of a unilateral two-port is independent of what is attached to the input terminals. In other words, *the output resistance of a unilateral two-port is independent of the source, and the input resistance of a unilateral two-port is independent of the load.*

The form of a **unilateral** two-port model is determined by which of input current or voltage is the controlling quantity and by which of a Thévenin or Norton model is used to represent the output subcircuit. Figure 6.22 shows four possible unilateral linear two-port models. Any one of the four *models* can be made equivalent *at its terminals* to any resistive unilateral linear two-port *circuit*. Furthermore, any one of the four models in Fig. 6.22 can be transformed to any other using a source transformation and the relation $v_T = R_T i_N$ *provided both the input resistance* R_i *and the output resistance* R_o *are non-zero and finite* (as they are for any physical circuit). The four models in Fig. 6.22 are equivalent, so choice of a particular model in any application may be based entirely upon ease and objectives of subsequent analysis and interpretation.

Example 6.8. Refer to Fig. 6.22. Use source transformation to obtain a CCCS two-port model from the equivalent CCVS model.

Solution: The input resistance is unchanged. The output resistance is also unchanged, because the Thévenin and Norton equivalent resistances are the same for equivalent circuits. By the usual rules for source transformation, and with reference to Fig. 6.22(b) and (c), we have

$$\beta i_1 = \frac{ri_1}{R_o} \Rightarrow \beta = \frac{r}{R_o}.$$

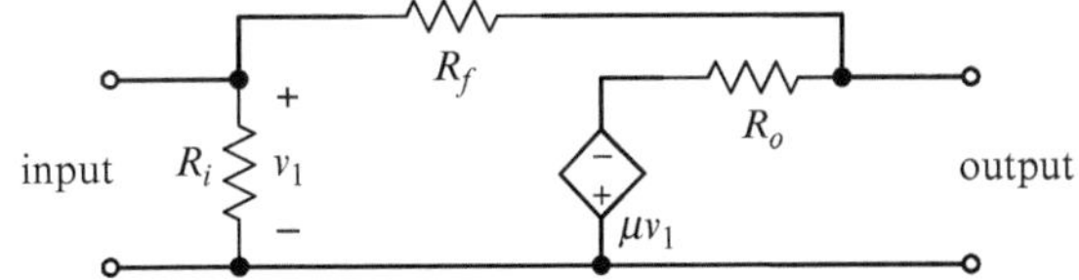

Fig. 6.23 See Exercise 6.9

Exercise 6.8. Refer to Fig. 6.22. Use source transformation to obtain a VCVS two-port model from the equivalent VCCS model.

The components of a two-port model are non-physical. Neither the input resistance nor the output resistance is necessarily that of a single resistor or even a combination of resistors. Both the input resistance and the output resistance of an electronic circuit, for example, might be determined primarily by properties of a transistor or integrated circuit. Similarly, the dependent source in the output subcircuit is not necessarily associated with any particular physical component.

Exercise 6.9. Obtain expressions for the input resistance and the output resistance of the circuit shown in Fig. 6.23. Use two different methods (or test sources) for each.

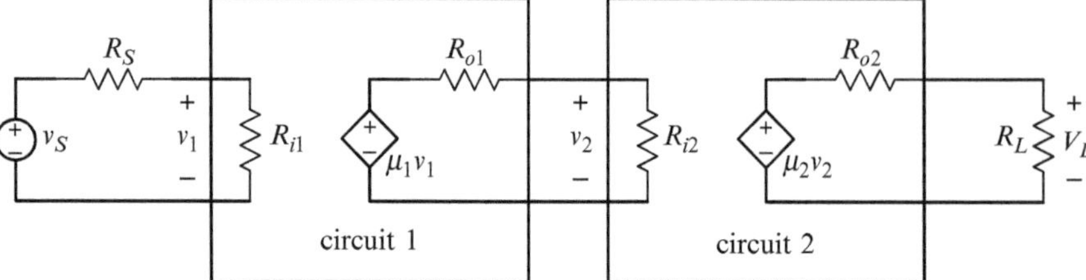

Fig. 6.24 Two-ports in cascade

6.4 Two-Ports in Cascade

As noted above, many complex circuits can be designed and analyzed as interconnections of two-ports. **Cascade connections**, where one two-port drives or is driven by another, are common in electronic systems. Figure 6.24 shows a cascade connection of two two-ports. Analysis of a cascade connection of *unilateral* two-ports is straightforward, as illustrated by the following example.

Example 6.9. Refer to Fig. 6.24. (a) Express the output voltage v_L in terms of the source voltage v_S and circuit parameters. (b) Reduce the cascade connection to a single equivalent VCVS two-port.

Solution: (a) Using voltage division and working from right to left gives

$$\begin{aligned} v_L &= \frac{R_L}{R_L+R_{o2}}\mu_2 v_2 = \frac{R_L}{R_L+R_{o2}}\mu_2 \frac{R_{i2}}{R_{i2}+R_{o1}}\mu_1 v_1 \\ &= \frac{R_L}{R_L+R_{o2}}\mu_2 \frac{R_{i2}}{R_{i2}+R_{o1}}\mu_1 \frac{R_{i1}}{R_{i1}+R_S} v_S \\ &= \left(\frac{R_L}{R_L+R_{o2}}\right)\left(\frac{R_{i2}}{R_{i2}+R_{o1}}\right)\left(\frac{R_{i1}}{R_{i1}+R_S}\right)\mu_2\mu_1 v_S. \end{aligned} \tag{6.18}$$

(b) By inspection, the input resistance and the output resistance of the cascade connection are R_{i1} and R_{o2}, respectively. The open-circuit output voltage is given by (6.18) for $R_L \to \infty$:

$$\begin{aligned} v_{oc} &= \lim_{R_L\to\infty}\left(\frac{R_L}{R_L+R_{o2}}\right)\left(\frac{R_{i2}}{R_{i2}+R_{o1}}\right) \\ &\quad \times \left(\frac{R_{i1}}{R_{i1}+R_S}\right)\mu_2\mu_1 v_S \\ &= \left(\frac{R_{i2}}{R_{i2}+R_{o1}}\right)\left(\frac{R_{i1}}{R_{i1}+R_S}\right)\mu_2\mu_1 v_S \\ &= \left(\frac{R_{i2}}{R_{i2}+R_{o1}}\right)\mu_2\mu_1 v_1. \end{aligned}$$

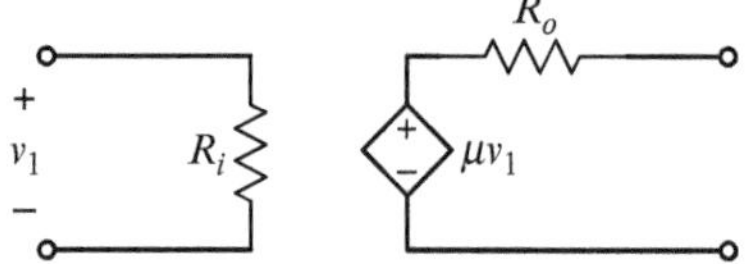

Fig. 6.25 A single VCVS two-port that is equivalent to the cascade connection of two-ports in Fig. 6.24. See Example 6.9

Thus the circuit in Fig. 6.24 can be reduced to the circuit shown in Fig. 6.25, where

$$\mu = \left(\frac{R_{i2}}{R_{i2}+R_{o1}}\right)\mu_2\mu_1,$$
$$R_o = R_{o2},\ R_i = R_{i1}.$$

Exercise 6.10. (a) Obtain a single VCVS two-port that is equivalent to the cascade connection of two-ports in Fig. 6.26. (b) From the single VCVS model, obtain equivalent VCCS, CCVS, and CCCS models.

6.5 Voltage, Current, and Power Transfer

This section describes conditions for optimum transfer of voltage, current, and power from a linear source to a linear load. The source and load can be circuits, in which case the source resistance is the output resistance of the first circuit and the load resistance is the input resistance of the second circuit.

We approach our study first from the perspective of one designing a load; e.g., a circuit that is to be driven by another circuit. The source is considered to be fixed, and can be represented by its Thévenin or

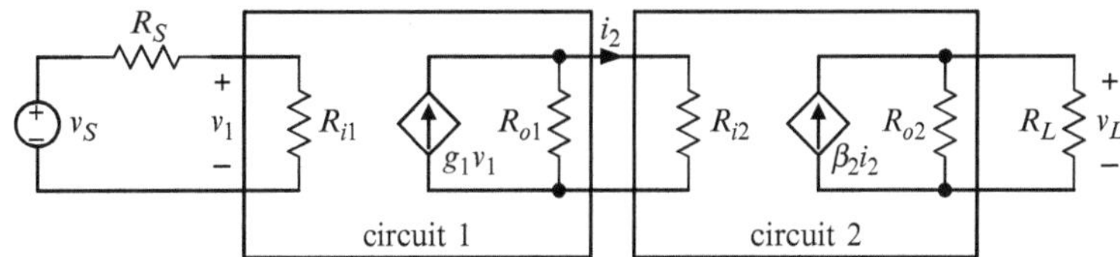

Fig. 6.26 See Exercise 6.10

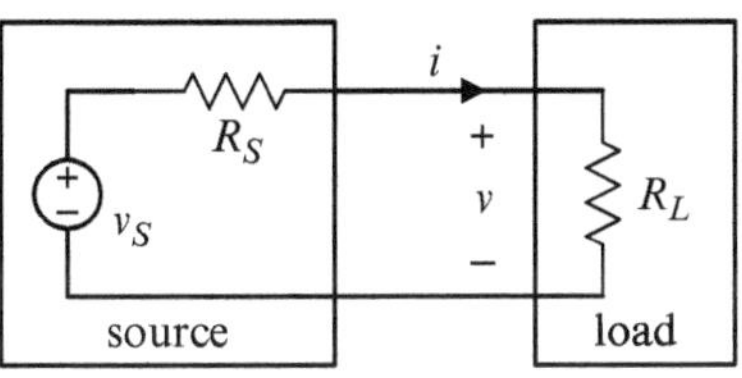

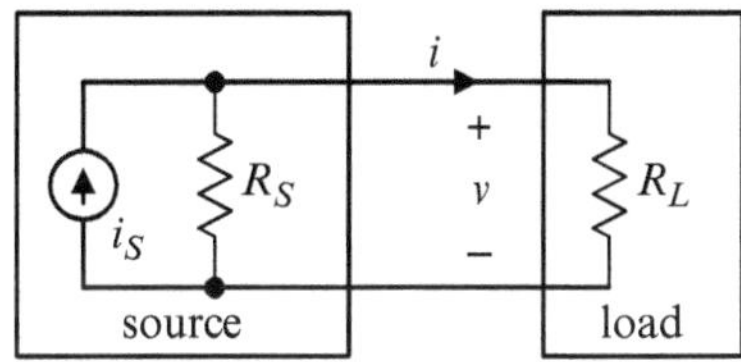

Fig. 6.27 Circuit models for analysis of voltage, current, and power transfer

Norton equivalent, as shown in Fig. 6.27, where the voltage and current transferred from the source to the load are v and i, respectively. *We presume the source models represent the same physical source, and therefore are equivalent at their terminals*, with $v_S = R_S i_S$. The resistance R_S is the output resistance of the source and is called the **source resistance**. The resistance R_L is the input resistance of the load and is called the **load resistance**. Although Fig. 6.27 shows the sources as independent sources, they could be dependent.

Keep in mind that the elements of the Thévenin and Norton equivalent circuits and the input resistance of a load are in general *non-physical*. We cannot in general change those elements directly. We can only change certain physical parameters upon which the Thévenin (or Norton) equivalent for the source and the input resistance of the load depend.

The open-circuit source voltage v_S is called the **available voltage** for that source. The short-circuit source current i_S is called the **available current** for the source. This terminology reflects the fact that *at any particular time*, the available voltage (available current) is the largest (in magnitude) voltage (current) that can be transferred by the source to a resistive load.

Exercise 6.11. The current and voltage available from a certain source are $I = 2\text{A}$ and $V = 15\text{V}$, respectively. Draw circuit diagrams for the Thévenin and Norton models for the source.

The **available power** P_A is the maximum average power that can be transferred from the source to a load. From Chapter 5, we know that the available power from a source is the power delivered by the source to a load whose input resistance equals the output resistance of the source.

Exercise 6.12. The power available from a certain source having internal (source) resistance $R_S = 10\ \Omega$ is $P = 500$ mW. Can you determine both the magnitudes and signs of the available voltage and current?

Exercise 6.13. The current and voltage available from a certain source are $I = 2\text{A}$ and $V = 15\text{V}$, respectively. Find the power available from the source.

Exercise 6.14. If the source resistance is halved and the available voltage is unchanged, by what factor is the available power multiplied?

We examine voltage, current, and power transfer from a fixed source to a load in terms of the available voltage v_S, the available current i_S, and the available power P_A. As we show below, efficacy of voltage, current, and power transfer is determined by the relation of the load resistance to the source resistance.

The source-load model in Fig. 6.27(a) is a voltage divider. The voltage transferred to the load is given by

$$v = \frac{R_L v_S}{R_L + R_S} = \frac{R_L}{R_L + R_S} \times (\textit{available voltage}). \tag{6.19}$$

The voltage transferred to the load is less than the source voltage by the *voltage loading factor*

$$\frac{R_L}{R_L + R_S} \leq 1.$$

If the load resistance R_L approaches infinity, the voltage loading factor approaches unity, and the voltage transferred to the load equals the available voltage. As a practical matter, almost all of the available voltage is transferred to the load if $R_L \gg R_S$.

To maximize voltage transfer, make the load resistance as large as possible. Note that if the voltage transferred to the load equals the available (open-circuit) voltage, both the current and the power transferred to the load equal zero.

Exercise 6.15. If it is required that at least 95% of the voltage available from a source be transferred to a load R_L, what are the permissible values of R_L/R_S, where R_S is the source resistance?

The source-load model in Fig. 6.27(b) is a current divider. The current transferred to the load is given by

$$i = \frac{v_S}{R_S + R_L} = \frac{R_S i_S}{R_S + R_L} = \frac{R_S}{R_S + R_L} \times (\textit{available current}). \tag{6.20}$$

The current transferred to the load by the source is less than the available current by the *current loading factor*

$$\frac{R_S}{R_S + R_L} \leq 1.$$

If the load resistance R_L approaches zero, the current loading factor approaches unity, and the current transferred to the load equals the available current. As a practical matter, almost all of the available current is transferred to the load if $R_L \ll R_S$.

To maximize current transfer, make the load resistance as small as possible. But keep in mind that all sources are current limited. For each source, there is a maximum load (minimum load resistance) the source can support. If a load demands more current than a source can supply, the source might be damaged or destroyed. For example, if you accidentally lay a wrench across the terminals of an automotive battery, the battery might explode. Note also that if the current transferred to the load equals the available (short-circuit) current, both the voltage and the power transferred to the load equal zero.

Exercise 6.16. If it is required that at least 95% of the current available from a source be transferred to a load R_L, what are the permissible values of R_L/R_S, where R_S is the source resistance?

The power available from a source depends upon the source resistance. If the source resistance equals zero, then the available power is unlimited (mathematically). For example, imagine an independent source v_S (having zero source resistance) driving a resistive load R_L. The average power dissipated by the load, given by $P = {V_{Srms}}^2/R_L$, is unbounded for $R_L \rightarrow 0$.

No physical source can provide unlimited power. Moreover, every physical source exhibits internal losses, so the source resistance of any realistic model for a physical source is non-zero. A realistic model for a physical source must contain non-zero source resistance and cannot deliver unlimited power to a load. In the discussion that follows, *we assume the source resistance is non-zero*. Because no power can be transferred to a load having zero or infinite resistance, *we also assume the load resistance is non-zero and finite*.

With reference to Fig. 6.27, the power transferred to the load is given by

$$P_L = V_{rms}I_{rms} = \left(\frac{R_L V_{S\,rms}}{R_L + R_S}\right)\left(\frac{V_{S\,rms}}{R_L + R_S}\right) = \frac{R_L {V_{S\,rms}}^2}{(R_L + R_S)^2}. \quad (6.21)$$

From Chapter 5, maximum power is transferred from a *fixed source* to a load if the load resistance R_L equals the source resistance R_S. Therefore the **available power** from a source is given by (6.21) with $R_L = R_S$:

$$\textit{available power} = P_A = \frac{R_S {V_{S\,rms}}^2}{(2R_S)^2} = \frac{{V_{S\,rms}}^2}{4R_S}. \quad (6.22)$$

The available power given by (6.22) is the maximum average power that can be drawn from a source by a load. Again, the available power is unbounded if $R_S = 0$, but that condition is not achieved by any physical source. From (6.22),

$${V_{S\,rms}}^2 = 4R_S P_A. \quad (6.23)$$

Using this relation in (6.21) gives

$$P_L = \frac{4R_L R_S P_A}{(R_L + R_S)^2} = \frac{4R_L R_S}{(R_L + R_S)^2} \times (\textit{available power}). \quad (6.24)$$

If the load resistance is not equal to the source resistance, the power transferred to the load is less than the available power by the factor

$$\frac{4R_L R_S}{(R_L + R_S)^2} \leq 1, \quad R_S \text{ fixed}. \quad (6.25)$$

To maximize power transfer, make the load resistance equal to the source resistance. As a practical matter (for a resistive source and load), the power transferred to a load has a broad maximum, so this requirement is not very strict. For example, if $R_L = 2R_S$ (twice the optimum value), then from (6.24)

$$P_L = \frac{8{R_S}^2}{(3R_S)^2} \times (\textit{available power}) \cong 0.889 \times (\textit{available power}).$$

If $R_L = 2R_S$, approximately 89% of the available power is transferred to the load.

Exercise 6.17. What fraction of the power available from a source is transferred to a load whose resistance is half the source resistance?

Exercise 6.18. If it is required that at least 95% of the power available from a source be transferred to a load R_L, what are the permissible values of R_L/R_S, where R_S is the source resistance?

Equations (6.19), (6.20), and (6.24) quantify penalties for non-optimum transfer of voltage, current, and power from a source to a load. These equations also imply desirable values for or relations between the input resistance of a load and the output resistance of a source, depending upon which of voltage transfer, current transfer, or power transfer is most important. Although the input resistance and the output resistance of a circuit are generally non-physical,[7] they are functions of circuit parameters or device parameters that are to some extent dependent upon design and device selection. Knowing desirable values for input resistance and output resistance helps a designer specify components and the configuration of a circuit.

The development above assumes a fixed source and variable load. Study of voltage, current, and power transfer from a variable source to a fixed load is more difficult, because such study cannot be based upon only the parameters of the Thévenin equivalent for the source. It might appear from (6.19), for example, that voltage transfer is maximized by minimizing the Thévenin-equivalent source resistance R_S, but that is not necessarily the case because a particular physical parameter (e.g., a physical resistance) might affect both the Thévenin-equivalent resistance and the Thévenin-equivalent source voltage. For example, consider the circuit shown in Fig. 6.28(a) and the associated Thévenin equivalent shown in

[7]That is, neither is necessarily associated with any particular physical resistor.

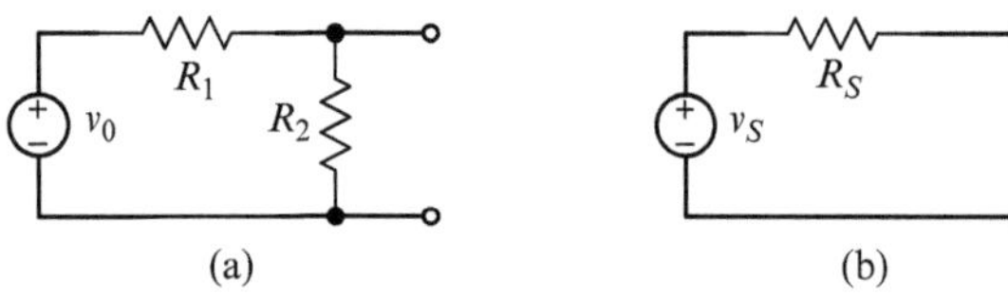

Fig. 6.28 A case where minimizing the Thévenin-equivalent resistance of a source also minimizes the voltage transferred to a load. See (6.26), (6.27), and related discussion

Fig. 6.28(b). The output (Thévenin-equivalent) resistance of the circuit is given by

$$R_S = R_1 \| R_2. \tag{6.26}$$

We presume the elements of the circuit in Fig. 6.28(a) are physical. Suppose that R_1 is fixed but that R_2 is somewhat under our control. We can decrease the output resistance by decreasing R_2. However, because the Thévenin-equivalent source voltage is given by

$$v_S = \frac{R_2 v_0}{R_1 + R_2}, \tag{6.27}$$

decreasing R_2 actually decreases the available voltage, in spite of the fact that decreasing R_2 decreases the output (Thévenin-equivalent) resistance of the source. If R_1 is fixed and we wish to maximize any or all of current, voltage, or power transfer, we would make R_2 as large as possible even though doing so *maximizes* the source resistance.

To maximize voltage, current, or power transfer from a variable source to a load, we must express the quantity of interest (e.g., load voltage) in terms of actual (physical) circuit parameters and find the values of those parameters for which the quantity is maximized. This is in general a difficult problem, because such expressions can be complicated and there are always constraints on the values of the variable parameters. However (and fortunately), it often is possible to minimize the output resistance of an electronic circuit without reducing the available voltage. In such cases, one almost always attempts to minimize the output (source) resistance, because so doing maximizes all three of voltage transfer, current transfer, and power transfer to a fixed load.

The factors that multiply available voltage in (6.19) and available current in (6.20) often are called **loading factors or coupling factors**. For example, in an electronics textbook, factors appearing in (6.18) (Example 6.9) might be identified as follows:

$$v_L = \underbrace{\left(\frac{R_{i1}}{R_{i1}+R_s}\right)}_{\text{input loading}} \underbrace{\left(\frac{R_{i2}}{R_{i2}+R_{o1}}\right)}_{\text{interstage loading}} \underbrace{\left(\frac{R_L}{R_L+R_{02}}\right)}_{\text{output loading}} \mu_1 \mu_2 v_s. \tag{6.28}$$

There are three generally undesirable consequences of loading:

- The most obvious consequence is attenuation; e.g., as introduced by each loading factor in (6.28). Attenuation is often the least troublesome consequence of loading because attenuation can be made up by amplification.
- A more serious consequence of loading is that *loading makes the overall relation between the source voltage and the load voltage dependent upon properties of the source and load.* If the relation between an input and output depends strongly upon properties of the source or load, a nonlinear source or load can wreak havoc with the intended function of the circuit.
- Finally, loading increases the number of parameters that affect the relation between input and output and therefore increases the number of parameters that must be considered by a designer. As a rule, it is desirable to make performance dependent upon as few parameters as possible, and to then specify those parameters as tightly as possible.

For these reasons, circuit designers often strive to eliminate loading. We discuss this point further below.

6.6 Transfer Characteristics, Transfer Ratios, and Gain

The primary purpose of any particular two-port circuit is to establish a specified relation between the current or voltage *available* to the circuit from a source and the associated current or voltage transferred from the circuit to a load. For a resistive circuit, the relation is a proportional one. With reference to Fig. 6.29, *where the Thévenin and Norton source models are equivalent*, we may write

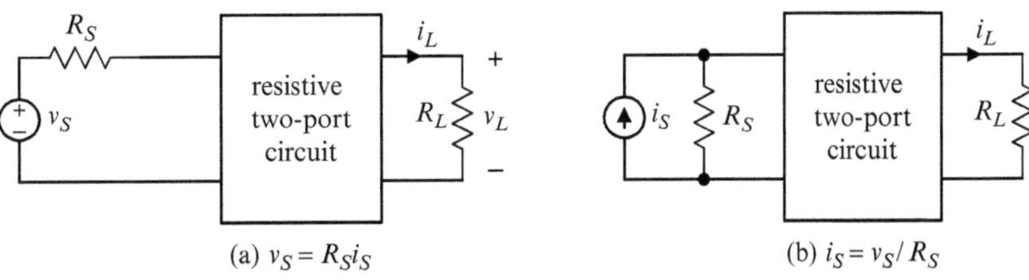

Fig. 6.29 Pertaining to the definitions of transfer characteristics for equivalent sources

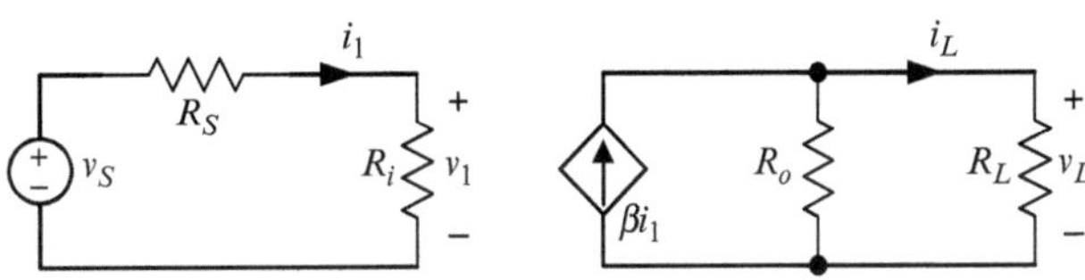

Fig. 6.30 See Example 6.10

$$v_L = H_v v_S, \tag{6.29}$$

$$v_L = H_r i_S, \tag{6.30}$$

$$i_L = H_i i_S, \tag{6.31}$$

and

$$i_L = H_g v_S. \tag{6.32}$$

The relations (6.29–6.32) are called **transfer characteristics**. The parameters H_v, H_r, H_i, H_g are called **transfer ratios**. For example (6.29) is a voltage transfer characteristic and H_4 is the associated transfer ratio. Conversely,

$$\begin{aligned} H_v &= \frac{v_L}{v_S}, \ H_g = \frac{i_L}{v_S}, \ v_S \neq 0; \\ H_i &= \frac{i_L}{i_S}, \ H_z = \frac{v_L}{i_S}, \ i_S \neq 0. \end{aligned} \tag{6.33}$$

In general, the source current i_S and source voltage v_S in (6.33) are functions of time, and can equal zero at times. The quantities v_L/v_S, v_L/i_S, i_L/v_S, and v_L/i_S are undefined at those times. But currents and voltages in resistive circuits are proportional, so whenever any one of v_L, v_S, i_L, and i_S is non-zero, all are. Thus the definitions (6.33), where v_S and i_S must be non-zero, are acceptable.

For a resistive two-port model, the transfer ratios H_v, H_r, H_i, H_g are functions of only the model parameters (R_i, R_o and one of μ, β, r, and g), the source resistance R_S, and the load resistance R_L. In a subsequent chapter, we generalize these definitions to circuits containing other elements (e.g., capacitance and inductance).

The transfer characteristics and the transfer ratios for a circuit are always specified or determined with a source and load attached to the circuit. To repeat, a transfer ratio is in general a function not only of the circuit parameters but also of the load resistance R_S and the source resistance R_S.

Example 6.10. Refer to Fig. 6.30. Find each of the transfer characteristics and transfer ratios defined above.

Solution: For the output subcircuit, we have

$$v_L = \beta i_1 (R_o \| R_L) \tag{6.34}$$

and

$$i_L = \frac{v_L}{i_L} = \frac{R_o \beta i_1}{R_o + R_L}. \tag{6.35}$$

For the input subcircuit,

$$i_1 = \frac{v_S}{R_i + R_S} \tag{6.36}$$

From the equivalence of Thévenin and Norton source models,

$$v_S = R_S i_S. \tag{6.37}$$

From (6.34) and (6.36),

$$v_L = \frac{\beta (R_o \| R_L)}{R_i + R_S} v_S \Rightarrow H_v = \frac{\beta (R_o \| R_L)}{R_i + R_S}. \tag{6.38}$$

From (6.37) and the left half of (6.38),

$$v_L = \frac{\beta (R_o \| R_L)}{R_i + R_S} R_S i_S \Rightarrow H_r = \frac{\beta (R_o \| R_L) R_S}{R_i + R_S}. \tag{6.39}$$

From (6.35) and (6.36),

$$i_L = \frac{R_o \beta v_S}{(R_o + R_L)(R_i + R_S)} \Rightarrow H_g = \frac{R_o \beta}{(R_o + R_L)(R_i + R_S)}. \tag{6.40}$$

Finally, from (6.37) and the left half of (6.40)

$$i_L = \frac{R_o \beta R_S i_S}{(R_o + R_L)(R_i + R_S)} \Rightarrow H_i = \frac{R_o \beta R_S}{(R_o + R_L)(R_i + R_S)}. \tag{6.41}$$

Exercise 6.19. Refer to Fig. 6.31. Find each of the transfer characteristics and transfer ratios defined above.

One or more gains are of interest in various applications. These are voltage gain, current gain, transresistance, and transconductance, defined with reference to Fig. 6.29 as follows:

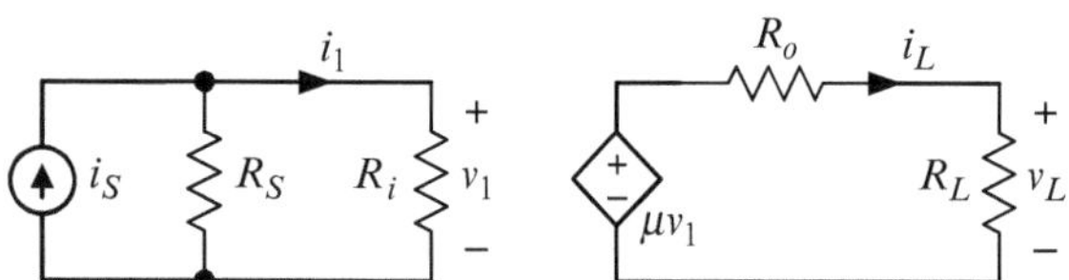

Fig. 6.31 See Exercise 6.19

Voltage Gain:

$$A_v = |H_v| = \frac{V_{L\,rms}}{V_{S\,rms}}, \tag{6.42}$$

Current Gain:

$$A_i = |H_i| = \frac{I_{L\,rms}}{I_{S\,rms}}, \tag{6.43}$$

Transresistance:

$$A_r = |H_r| = \frac{V_{L\,rms}}{I_{S\,rms}}, \tag{6.44}$$

Transconductance:

$$A_g = |H_g| = \frac{I_{L\,rms}}{V_{S\,rms}}. \tag{6.45}$$

Transconductance has the dimension of conductance (S) and transresistance has the dimension of resistance (Ω). Voltage gain and current gain are dimensionless. Although (for convenience), we refer to all of the quantities A_v, A_i, A_r, A_g as gains, transresistance and transconductance are not gains in the usual sense because current and voltage have different dimensions. It makes no sense to say a current is larger or smaller than some voltage, just as it makes no sense to say a dollar is larger or smaller than a kilogram.

Defining the gains in terms of rms amplitudes avoids the minor difficulty that arises when the instantaneous source current or voltage passes through zero, discussed in relation to (6.33). Also, in circuits containing capacitance or inductance,[8] the currents and voltages appearing in each definition are not necessarily instantaneously proportional, nor do they necessarily even have the same waveform. Thus gain must be defined (in general) in terms of rms amplitudes.[9] To save writing, some omit the rms subscripts when obtaining or presenting expressions for gain, but the rms subscript is implied in such expressions. In the context of this chapter, we could as well use the definitions

[8]These elements are treated in subsequent chapters.

[9]Peak amplitudes can be (and often are) used if the voltages or currents involved have the same waveform.

$$\begin{aligned} A_v &= \left|\frac{v_L}{v_S}\right|, \quad A_g = \left|\frac{i_L}{v_S}\right|, \quad v_S \neq 0; \\ A_i &= \left|\frac{i_L}{i_S}\right|, \quad A_z = \left|\frac{v_L}{i_S}\right|, \quad i_S \neq 0, \end{aligned} \tag{6.46}$$

which are equivalent to (6.42–6.45) for resistive circuits.

Example 6.11. Obtain expressions for (a) the voltage gain, (b) the current gain, (c) the transresistance, and (d) the transconductance of the two-port circuit in Fig. 6.32(a).

Solution: We attach a source and a load to the circuit, as shown in Fig. 6.32(b), before seeking the transfer characteristics.

(a) By inspection,

$$i_1 = \frac{vLs}{R_S + R_i} \tag{6.47}$$

and

$$v_L = -\beta i_1 (R_o \| R_L). \tag{6.48}$$

Thus

$$\begin{aligned} v_L &= -\beta\left(\frac{v_S}{R_S + R_i}\right)(R_o \| R_L) \\ &= -\beta\left(\frac{R_o \| R_L}{R_S + R_i}\right) v_S \end{aligned} \tag{6.49}$$

and the voltage gain of the circuit is given by

$$A_v = \left|-\beta \frac{(R_o \| R_L)}{R_S + R_i}\right| = \beta \frac{(R_o \| R_L)}{R_S + R_i}. \tag{6.50}$$

(b) By current division,

$$\begin{aligned} i_L &= -\beta i_1 \frac{R_o}{R_o + R_L} \\ &= -\beta\left(\frac{v_S}{R_S + R_i}\right)\frac{R_o}{R_o + R_L} \\ &= \frac{-\beta R_o R_S i_S}{(R_S + R_i)(R_o + R_L)}, \end{aligned} \tag{6.51}$$

where we have made use of $v_s = R_S i_S$. The current gain is given by

$$\begin{aligned} A_v &= \left|\frac{-\beta R_o R_S}{(R_S + R_i)(R_o + R_L)}\right| \\ &= \frac{\beta R_o R_S}{(R_S + R_i)(R_o + R_L)}. \end{aligned} \tag{6.52}$$

(c) From (6.49)

$$\begin{aligned} v_L &= -\beta\left(\frac{R_o \| R_L}{R_S + R_i}\right) v_S \\ &= -\beta\left(\frac{R_o \| R_L}{R_S + R_i}\right) R_S i_S, \end{aligned} \tag{6.53}$$

and the transresistance is given by

$$\begin{aligned} A_v &= \left|-\beta\left(\frac{R_o \| R_L}{R_S + R_i}\right) R_S\right| \\ &= \beta\left(\frac{R_o \| R_L}{R_S + R_i}\right) R_S. \end{aligned} \tag{6.54}$$

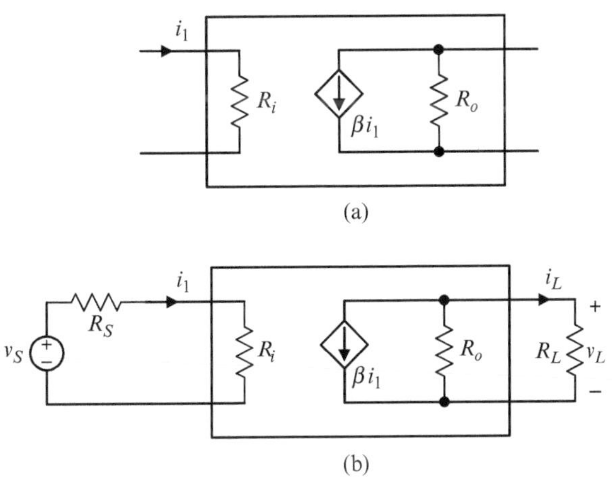

Fig. 6.32 See Example 6.11

(d) From (6.51)

$$i_L = -\beta\left(\frac{v_S}{R_S + R_i}\right)\frac{R_o}{R_o + R_L}, \tag{6.55}$$

and the transconductance is given by

$$A_g = \left|-\beta\left(\frac{1}{R_S + R_i}\right)\frac{R_o}{R_o + R_L}\right| = \frac{\beta R_o}{(R_o + R_L)(R_S + R_i)}. \tag{6.56}$$

Where necessary to avoid confusion, we refer to the gains A_v, A_i, A_g, A_r defined by (6.46) as *overall* gains to distinguish them from the *intrinsic* gains defined in Section 6.2. For example, when we refer to the current gain of a two-port or circuit, we mean the overall current gain. Thus, the circuit considered in Example 6.11 has *intrinsic* current gain β and *overall* current gain A_i given by (6.52).

Exercise 6.20. Obtain expressions for the voltage gain, current gain, transconductance, and transresistance of the two-port in Fig. 6.33.

Exercise 6.21. Obtain expressions for the voltage gain, current gain, transconductance, and transresistance of the circuit shown in Fig. 6.34.

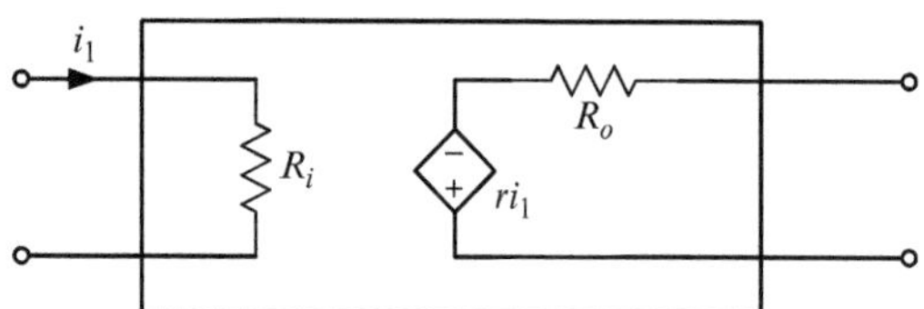

Fig. 6.33 See Exercise 6.20

In Fig. 6.29, because the source models are equivalent, we have $v_S = R_S i_S$. Also, by Ohm's law, $v_L = R_L i_L$. These relations allow us to express any gain in terms of any other gain.

Example 6.12. Express the current gain, transconductance, and transresistance of a circuit in terms of the voltage gain.
Solution: From (6.43) and (6.42)

$$A_i = \frac{I_{L\,rms}}{I_{S\,rms}} = \frac{V_{Lrms}/R_L}{V_{Srms}/R_S} = \frac{R_S}{R_L}A_v. \tag{6.57}$$

From (6.44)

$$A_r = \frac{V_{L\,rms}}{I_{S\,rms}} = \frac{V_{L\,rms}}{V_{Srms}/R_S} = R_S A_v. \tag{6.58}$$

From (6.45)

$$A_g = \frac{I_{L\,rms}}{V_{S\,rms}} = \frac{V_{Lrms}/R_L}{V_{S\,rms}} = \frac{1}{R_L}A_v. \tag{6.59}$$

Exercise 6.22. Express the voltage gain, transconductance, and transresistance of a circuit in terms of the current gain.

Exercise 6.23. In each of the following, express the parameter α in terms of the source resistance and load resistance. (a) $A_i = \alpha A_r$, (b) $A_g = \alpha A_i$, (c) $A_v = \alpha A_r$, (d) $A_i A_v = \alpha A_r A_g$.

Equations (6.57)–(6.59) show that *at most one of the four gains A_v, A_i, A_g, A_r can be independent of the source and load resistances*. For example, if the voltage gain of a circuit is independent of the source and

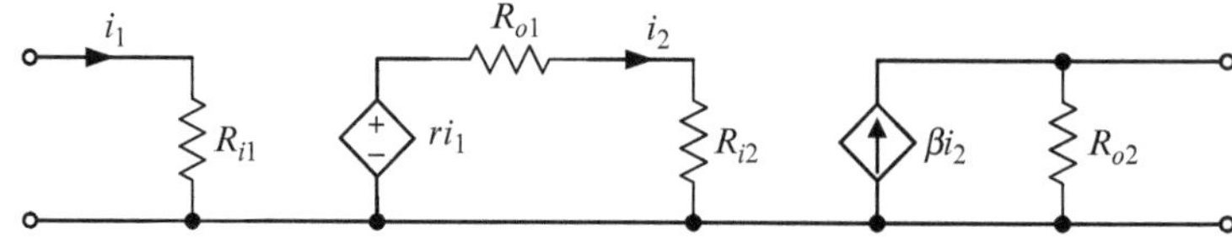

Fig. 6.34 See Exercise 6.21

load resistances, then (6.57)–(6.59) show that A_i, A_g, and A_r depend upon the source and load resistances. To pursue this further, we consider the circuit treated in Example 6.11, for which the voltage gain is given by

$$A_v = \beta \frac{(R_o \| R_L)}{R_S + R_i}. \tag{6.60}$$

If the input resistance is much larger than the source resistance (if $R_i \gg R_S$) and the output resistance is much smaller than the load resistance (if $R_o \ll R_r$), then $R_S + R_i \cong R_i, R_o \| R_L \cong R_o$, and the voltage gain is given to a good approximation by

$$A_v \cong \beta \frac{R_o}{R_i}, \tag{6.61}$$

which is independent of the source and load resistances. But under those same conditions, the current gain, given by (6.52), becomes

$$A_i \cong \frac{\beta R_o R_S}{R_i R_L},$$

and is strongly dependent upon both the source resistance and the load resistance.

The point of the discussion above is the following: The purpose of a two-port circuit is to establish some specified relation between an *available current or voltage* and the associated *load current or voltage*. For reasons given at the end of Section 6.5, it usually is desirable that the relation be independent of properties of the source and load. The development leading to (6.61) shows that the voltage gain of the circuit in Fig. 6.32 is independent of the source and load if the input resistance is much larger than the source resistance and the output resistance is much smaller than the load resistance. Although this conclusion is drawn for a specific two-port circuit, it is true in general. The implication for design is that *if the voltage gain of a circuit is to be independent of source and load, then make the input resistance of the circuit as large as possible and the output resistance as small as possible.*

Example 6.13. A certain two-port circuit has input resistance $R_i = 100\ \text{k}\Omega$ and output resistance $R_o = 1\ \text{k}\Omega$. For what values of source and load resistance is the voltage gain of the circuit approximately independent of the source and load resistances?

Solution: Of course the answer depends upon the quantitative meaning of *approximately*. If we want the effect of source and load resistances on voltage gain to be about 1% or smaller, we might require

$$R_S < 0.01 R_i \Rightarrow R_S < 1\ \text{k}\Omega,$$
$$R_L > 100 R_o \Rightarrow R_L > 100\ \text{k}\Omega.$$

At first glance, it might seem odd that the gains of a two-port circuit are defined such that they are dependent upon properties of the source and load. After all, the source and load are not parts of the circuit under consideration. But a gain of a circuit is essentially a performance measure, and it does make sense for a gain to depend upon how well the circuit makes use of available current, voltage, or power, which depends upon the input resistance *relative to the source resistance*, and upon how effectively the circuit transfers current, voltage, or power to a load, which depends upon the output resistance *relative to the load resistance*.

6.7 Power Gain

Refer to Fig. 6.29. The **power gain** of a two-port circuit is defined by

$$A_p = \frac{\textit{power delivered to the load}}{\textit{available power}}$$
$$= \frac{V_{Lrms}{}^2 / R_L}{V_{Srms}{}^2 / (4R_S)} = \frac{4R_S}{R_L} A_v{}^2. \tag{6.62}$$

From the definition (6.62), it is clear that the power gain of a circuit must be calculated (or measured) with a load attached to the output terminals.

Other, equivalent expressions can be obtained from (6.62) and relations among the various gains, such as (6.57–6.59). For example, from (6.57),

$$A_v = \frac{R_L}{R_S} A_i. \tag{6.63}$$

Using this expression for A_v in (6.62) gives

$$A_P = \frac{4R_L}{R_S} A_i^2. \quad (6.64)$$

Exercise 6.24. Show that the power gain of a two-port circuit can be expressed in each of the following ways:

$$A_p = \frac{4}{R_L R_S} A_r^2 = 4R_L R_S A_g^2 = 4A_v A_i$$
$$= 4A_r A_g. \quad (6.65)$$

Exercise 6.25. From (6.62), it appears that increasing the output resistance of a source driving a circuit increases the power gain of the circuit. Discuss whether this is a sensible way to increase power gain.

Exercise 6.26. From (6.64), increasing the output resistance of the source driving a circuit appears to *decrease* the power gain of the circuit. Are (6.64) and (6.62) contradictory in this regard? Discuss.

6.8 Gains and Relative Values in Decibels (dB)

Virtually all early electronic systems (telephone, radio, television) were intended to produce outputs (sound and pictures) directly perceptible to humans. Human sensory perception is approximately logarithmic. For example, doubling the apparent loudness of an audio system requires squaring the power delivered to the loudspeaker. An analogous statement holds for light incident on the retina.

Human sensory perception also has an enormous dynamic range. For example, the dynamic range of human hearing spans about twelve orders of magnitude, corresponding to acoustic power incident on the eardrum ranging from about 10^{-12} W m^{-2}(the threshold of hearing) to about 1 W m^{-2} (the threshold of pain). This means, loosely, that the powers of interest at the output of an audio system span 12 orders of magnitude and that the associated currents or voltages of interest span about six orders of magnitude (because power is proportional to the squares of current and voltage). It is impossible to display (graphically) such ranges of power, current, or voltage on a linear scale.

Because human hearing, in particular, is approximately logarithmic and has such a large dynamic range, the Bell Telephone Company many years ago adopted a logarithmic measure of **relative power** called the **decibel** (**dB**) defined by

$$P_{dB} = 10 \log\left(\frac{P}{P_0}\right), \quad (6.66)$$

where the logarithm is base ten, P is a power of interest, and P_0 is a reference power. The inverse relation is

$$P = P_0\, 10^{P_{\text{dB}}/10}. \quad (6.67)$$

For example, the standard reference for sound intensity is a power density of $P_0 = 1$ pWm^{-2}. Thus, if the sound intensity on stage with a rock band is 120 dB, the power density incident on a band member's eardrum is

$$P = P_0\, 10^{P_{dB}/10} = P_0 \times 10^{12} = 1\text{Wm}^{-2}.$$

Sustained intensity levels above about 90 dB can cause permanent hearing loss. Intensities above about 120 dB can rupture an eardrum instantly.

If the power of interest and the reference power P_0 are dissipated in the same or equivalent loads, then (6.66) gives

$$P_{dB} = 10 \log\left(\frac{V_{rms}{}^2/R}{V_{0rms}{}^2/R}\right)$$
$$= 20 \log\left(\frac{V_{rms}}{V_{0\,rms}}\right) \quad (6.68)$$

or

$$P_{dB} = 10 \log\left(\frac{I_{rms}{}^2 R}{I_{0\,rms}{}^2 R}\right) = 20 \log\left(\frac{I_{rms}}{I_{0\,rms}}\right). \quad (6.69)$$

In general, it is incorrect to express the logarithm of a quotient of dimensioned quantities as the difference of logarithms; for example, it is **incorrect** to write

$$P_{dB} = 20 \log\left(\frac{I_{rms}}{I_{0\,rms}}\right)$$
$$\stackrel{?}{=} 20 \log(I_{rms}) - 20 \log(I_{0\,rms}),$$

because the argument of the logarithm function must be dimensionless. It is true that if both currents are expressed in the same unit (e.g., μA), the result will be correct. Nonetheless, such practice is discouraged because it is prone to error. It also is unnecessary in this age of pocket computers.

Since its creation and adoption by the Bell Telephone Company, the decibel has become a standard dimensionless unit for relative power in virtually all electronic systems. In particular, the **power gain of a circuit in decibels (dB)** is defined by

$$A_{P\,dB} = 10 \log(A_P)\,\text{dB}. \tag{6.70}$$

In this case, the reference power is the available power.

Example 6.14. The power gain of a certain two-port circuit is 85 dB. What is the actual power gain (not dB)?
Solution: From (6.70),

$$10 \log(A_P) = 85\text{ dB} \Rightarrow A_p = 10^{85/10}$$
$$= 10^{0.5} \times 10^8 = 3.16 \times 10^8.$$

Exercise 6.27. The power gain of a certain two-port circuit is 108 dB. What is the actual power gain (not dB)?

Exercise 6.28. The power gain of a certain two-port circuit is 70 dB. If the power delivered to the load is 5 W, what is the available power (from the source)?

Power gain in dB can be expressed in terms of voltage gain and current gain. From (6.62) and (6.64), we have

$$A_{P\,dB} = 10 \log\left(\frac{4R_S}{R_L}A_v^{\,2}\right)$$
$$= 10 \log\left(\frac{4R_S}{R_L}\right) + 10 \log\left(A_v^{\,2}\right) \tag{6.71}$$

and

$$A_{P\,dB} = 10 \log\left(\frac{4R_L}{R_S}A_i^{\,2}\right)$$
$$= 10 \log\left(\frac{4R_L}{R_S}\right) + 10 \log\left(A_i^{\,2}\right). \tag{6.72}$$

Exercise 6.29. Show that power gain in dB can be expressed as

$$A_{P\,dB} = 6.02 + 10 \log(A_r A_g)$$
$$= 6.02 + 10 \log(A_v A_i). \tag{6.73}$$

Equations (6.71) and (6.72) inspire the following definitions of **voltage gain in dB** and **current gain in dB**:

$$A_{v\,dB} = 10 \log\left(A_v^{\,2}\right) = 20 \log(A_v) \tag{6.74}$$

and

$$A_{i\,dB} = 10 \log\left(A_i^{\,2}\right) = 20 \log(A_i). \tag{6.75}$$

Exercise 6.30. Obtain expressions for the voltage gain, current gain, and power gain, all in dB, for the circuit shown in Fig. 6.35.

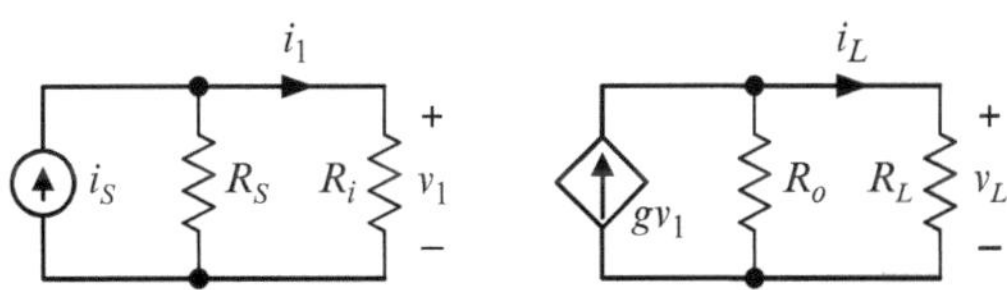

Fig. 6.35 See Exercise 6.30

Exercise 6.31. The power gain and voltage gain of a certain two-port are $A_{p\,dB} = 105\text{dB}$, $A_{v\,dB} = 84\text{dB}$. Find the current gain (dB).

Example 6.15. The power gain of a certain two-port circuit equals 50 dB. The source resistance and the load resistance are 15 and 25 Ω, respectively. Find the voltage gain and the current gain, both in dB and actual value.

Solution: From (6.71) and (6.74)

$$A_{v\,dB} = A_{P\,dB} - 10\,\log\left(\frac{4R_S}{R_L}\right)$$
$$= 50 - 10\,\log\left(\frac{4\times 15}{25}\right) = 46.2\text{dB}.$$

From (6.72) and (6.75),

$$A_{i\,dB} = A_{P\,dB} - 10\,\log\left(\frac{4R_L}{R_S}\right)$$
$$= 50 - 10\,\log\left(\frac{4\times 25}{15}\right) = 41.8\text{dB}.$$

It follows that the voltage and current gains (not dB) are

$$A_v = 10^{2.31} = 204, \quad A_i = 10^{2.09} = 122.$$

Transresistance and transconductance are dimensioned and cannot be expressed in dB, because we cannot take the logarithm of a dimensioned quantity. However, transresistance and transconductance can be normalized to dimensionless quantities, using some appropriate resistance or conductance as the normalizing factor, and the normalized quantities can then be expressed in dB.

In practice, current gain, voltage gain, and power gain are almost always expressed in dB (as are other ratios, such as an important power ratio called signal-to-noise ratio). You should commit the definitions (6.70), (6.74), and (6.75) to memory.

6.9 Design Considerations

The primary purpose of a circuit is to establish a specified relation between input current or voltage and output current or voltage. The circuit design is guided by the nature of the source, the nature of the load, and what the circuit is intended to achieve. Developments above suggest the following guidelines:

- *It is almost always desirable to minimize output resistance.* Minimizing output resistance maximizes all four of voltage transfer, current transfer, power transfer, and power transfer efficiency to any load having a finite non-zero resistance. Do not exceed current- and power-delivery capabilities of the circuit.
- To maximize voltage transfer from a source to a circuit input, maximize the input resistance of the circuit.
- To maximize current transfer from a source to a circuit input, minimize the input resistance of the circuit, but do not allow the source current to exceed the maximum current the source can provide.
- To maximize power transfer from a source to a circuit input, make the circuit input resistance equal to the source output resistance, but do not exceed the maximum power the source can provide.
- To maximize power transfer efficiency from a source to a circuit input, maximize the input resistance of the circuit.

Specifying the input resistance of a circuit hinges on which of voltage transfer, current transfer, or power transfer is most important, and that often is determined by properties of the source. Although current and voltage are the primary quantities of interest in circuit analysis and design, they are ultimately only intermediaries that convey other quantities, such as audio, video, and data. Where electric circuits are used to process or transmit such signals, it is often the case that one or the other of current and voltage best represents the quantity of ultimate interest. For example, a mechanical tachometer produces a terminal *voltage* proportional to the angular velocity of the armature. Consequently, if we are designing an amplifier to scale the output of a tachometer, we might specify a large input resistance to maximize voltage transfer. Here again, *large* and *small* are relative terms that must be quantified to be of use in design.

6.10 Problems

In the problems that follow, two-port circuits are drawn with inputs on the left and outputs on the right, unless otherwise labeled.

Section 6.1 is prerequisite for the following problems.

P 6.1 Refer to Fig. P 6.1, where $R = 1\text{k}\Omega$, the source resistance $R_S = 10\Omega$, and the load resistance $R_L = 5\text{k}\Omega$. Find the input resistance and the output resistance of each two-port shown.

P 6.2 In Fig. P 6.2, the source has open-circuit voltage V_{oc} and short-circuit current I_{sc}. Express the input resistance of the two-port in terms of V_{oc}, I_{sc}, I.

P 6.3 In Fig. P 6.3, the source has open-circuit voltage V_{oc} and short-circuit current I_{sc}. Express the input resistance of the two-port in terms of V_{oc}, I_{sc}, V.

P 6.4 In Fig. P 6.4, the source has open-circuit voltage V_{oc} and short-circuit current I_{sc}. For $R = R_1$, $I = I_1$. For $R = R_2 \neq R_1$, $I = I_2 \neq I_1$. Obtain an expression for the input resistance of the two-port.

P 6.5 In Fig. P 6.5, the source has open-circuit voltage V_{oc} and short-circuit current I_{sc}. For $R = R_1$, $V = V_1$. For $R = R_2 \neq R_1$, $V = V_2 \neq V_1$. Obtain an expression for the input resistance of the two-port.

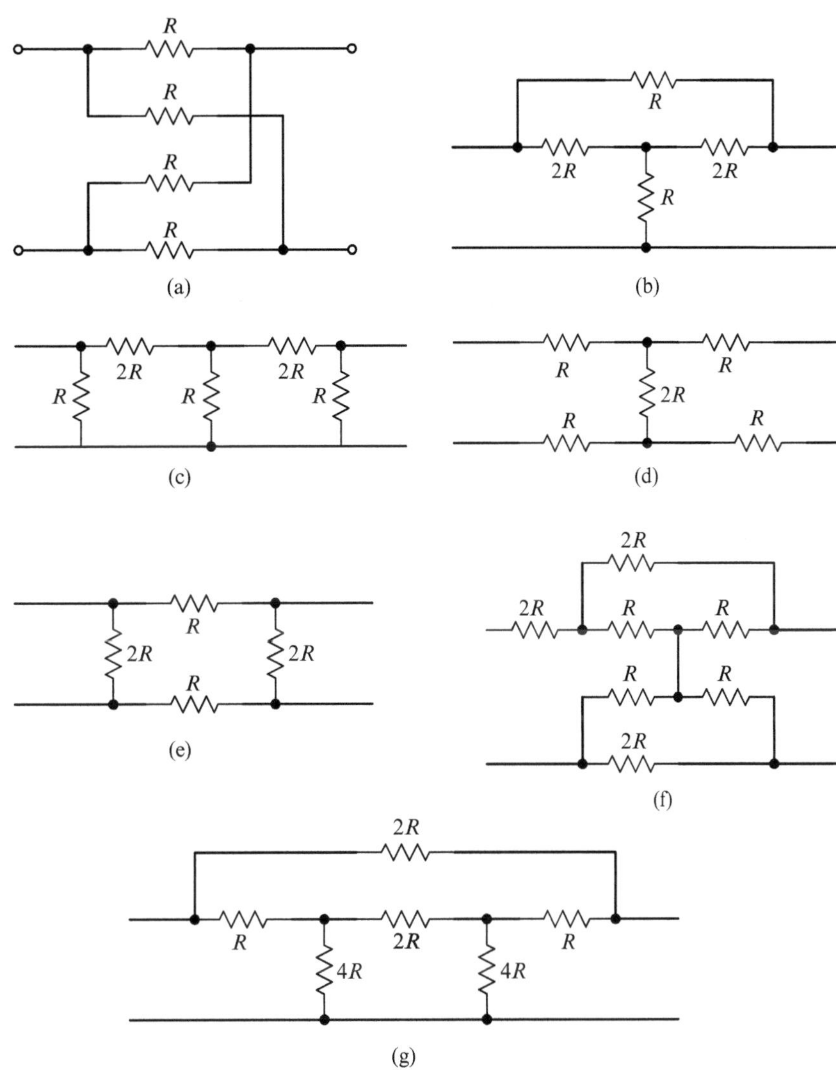

Fig. P 6.1 See Problem P 6.1

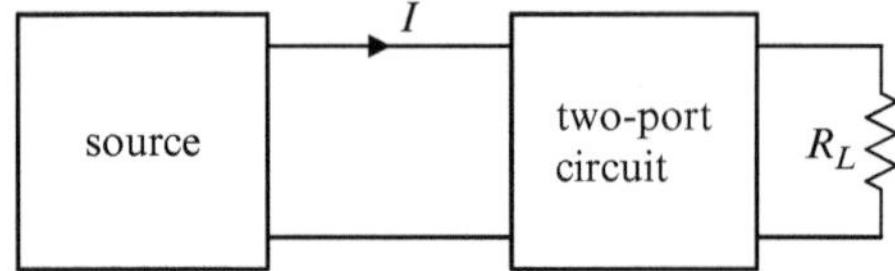

Fig. P 6.2 See Problem P 6.2

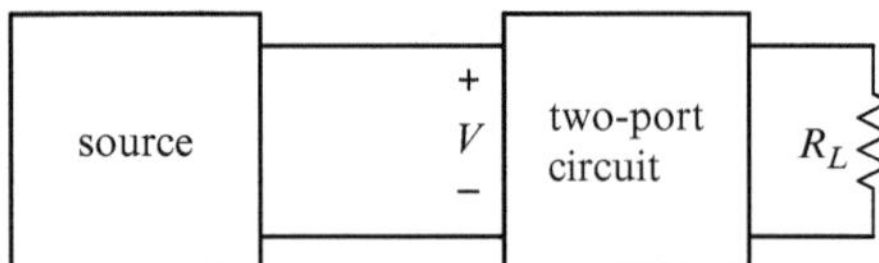

Fig. P 6.3 See Problem P 6.3

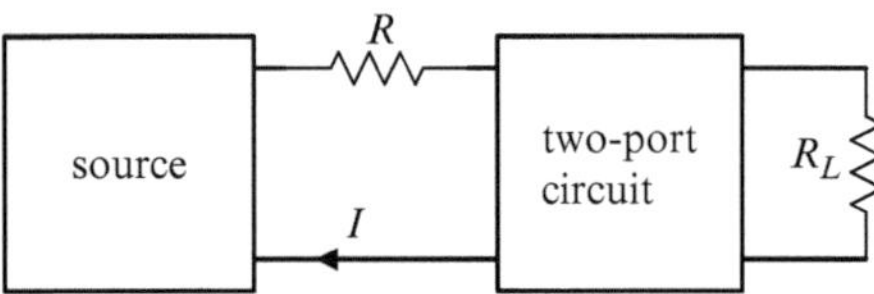

Fig. P 6.4 See Problem P 6.4

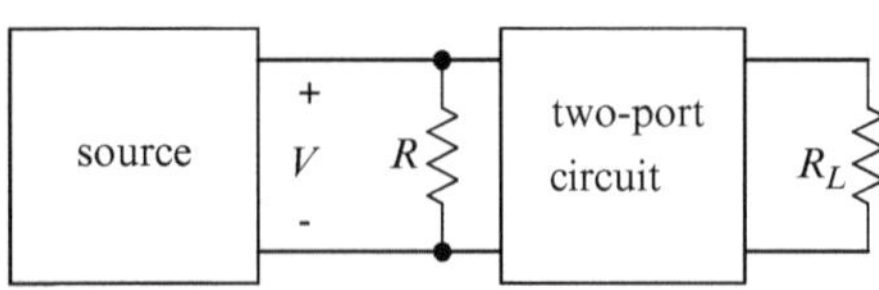

Fig. P 6.5 See Problem P 6.5

Section 6.2 is prerequisite for the following problems.

P 6.6 Show that a resistor can be represented as a dependent voltage source.

P 6.7 Show that a resistor can be represented as a dependent current source.

P 6.8 See Fig. P 6.6. (a) Obtain an expression for the voltage V_a. (b) Let $V_s = 10\text{V}$, $R = 1\text{k}\Omega$ and use a simulation to check your expression (numerically).

P 6.9 Refer to Fig. P 6.7. (a) Express the load voltage v_L in terms of the input voltage V_S and the circuit parameters. (b) Perform a simulation using $R_e = 100\ \Omega$, $R_b = 1\text{k}\Omega$, $R_c = 5\text{k}\Omega$, $R_L = 2\ \text{k}\Omega$, and $r = 2\text{k}\Omega$. Show that the voltage v_L thus obtained agrees with that computed using the relation found in part (a).

P 6.10 Refer to Fig. P 6.8. Express the load voltage v_L and the load current i_L in terms of the available source voltage v_S and the circuit parameters R_S, μ, R_L.

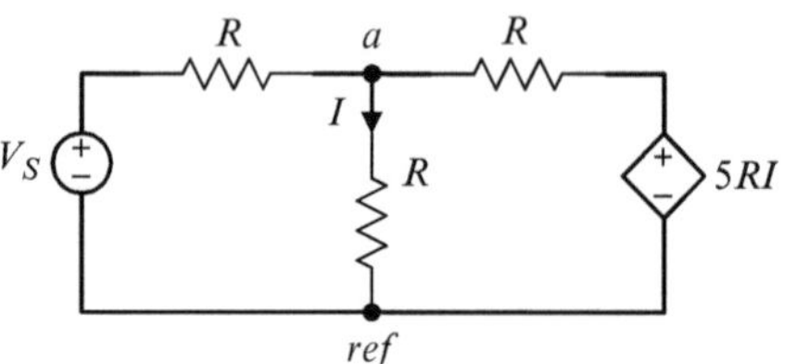

Fig. P 6.6 See Problem P 6.8

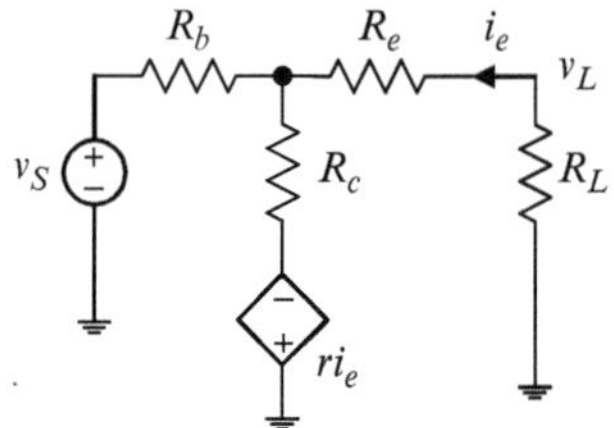

Fig. P 6.7 See Problem P 6.9

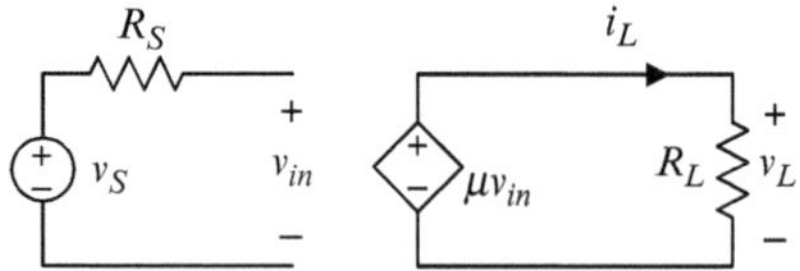

Fig. P 6.8 See Problem P 6.10

P 6.11 Refer to Fig. P 6.9. Obtain the Thévenin equivalent circuit at the terminals *a*–*b*. Express the Thévenin source voltage in terms of the independent source voltage v_S.

P 6.12 Refer to Fig. P 6.10. Express the voltage v_L in terms of the voltage v_S and the circuit parameters. No other currents or voltages may appear in the expression.

Section 6.3 is prerequisite for the following problems.

In problems below, the four models referred to are those defined in Fig. 6.22 in the text.

P 6.13 Obtain all four two-port models for the circuit shown in Fig. P 6.11. Express all model parameters in terms of the original circuit parameters. Is the circuit unilateral or bilateral?

P 6.14 Figure P 6.12(a) shows the circuit diagram symbol for an npn bipolar junction transistor (BJT).

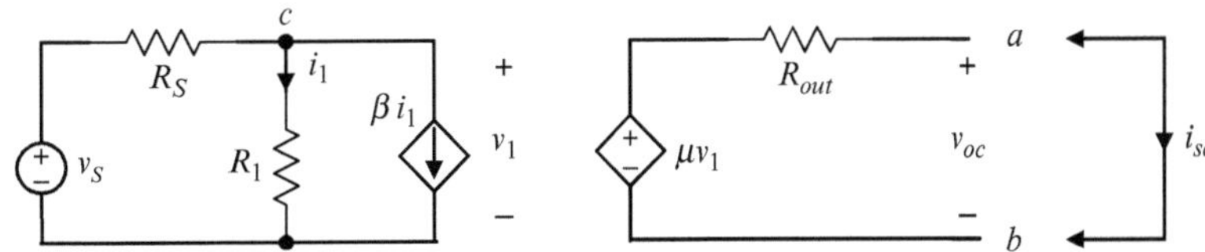

Fig. P 6.9 See Problem P 6.11

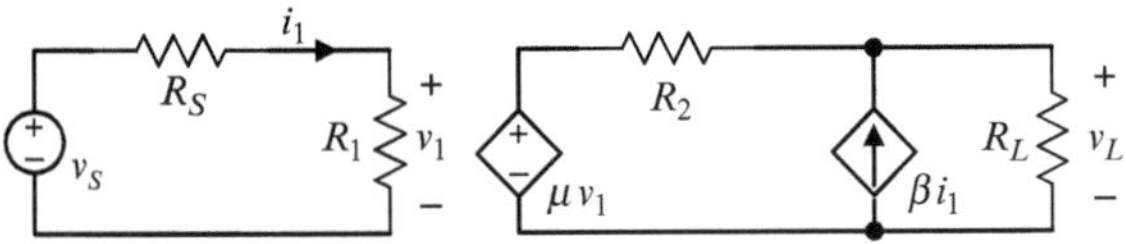

Fig. P 6.10 See Problem P 6.12

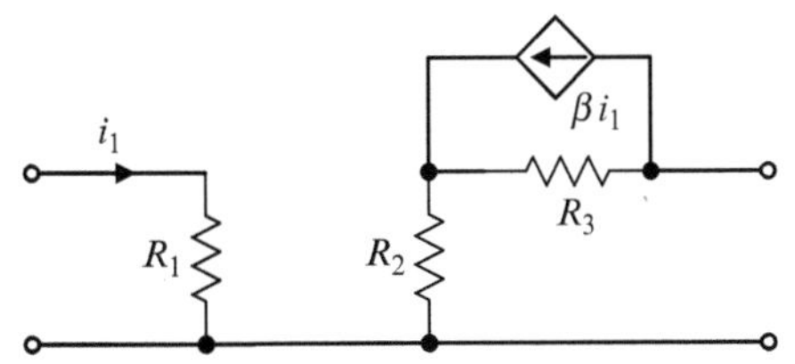

Fig. P 6.11 See Problem P 6.13

The terminals are designated *b* (for *base*), *c* (for *collector*), and *e* (for *emitter*). Figure P 6.12(b) shows a simplified linear ac model for an npn BJT. Such transistors are used as linear amplifiers in three basic configurations, as shown by the circuit diagrams in Fig. P 6.12(c), (d), and (e). For each of the three configurations, replace the transistor with the linear ac model in Fig. P 6.12(b) and express the output in terms of the voltage V_s and the circuit parameters.

P 6.15 Refer to Problem P 6.14. Figure P 6.13 shows an alternative equivalent model to that in Fig. P 6.12(a) for a BJT. Repeat problem P 6.14 using the model in Fig. P 6.13.

P 6.16 Figure P 6.14 shows a transistor circuit called an *emitter-coupled pair*. The inputs are the voltages v_{b1},v_b and the output is the voltage $v_{c_1} - v_{c_2}$. Replace each transistor with the CCCS model in Fig. P 6.12(b). The voltage V_{CC} is the supply voltage for the circuit and the terminal labeled V_{CC} is connected to ground (*ref*) in the small-signal linear model. (The dc supply voltage appears as a short circuit for other currents and voltages.) (a) Obtain an expression for the output in terms of the two input voltages.

P 6.17 Repeat Problem P 6.16 using the VCCS model in Fig. P 6.13.

P 6.18 Find a single equivalent VCVS two-port model for the circuit shown in Fig. P 6.15. Express the parameters of the resultant two-port in terms of the original circuit parameters.

P 6.19 Find a single equivalent VCVS two-port model for the circuit shown in Fig. P 6.16. Express the parameters of the resultant two-port in terms of the original circuit parameters.

P 6.20 Known independent sources are attached to the terminals of a two-port, as shown in Fig. P 6.17. It is found that $i_1 = 1$mA when $v_1 = 5$V, that $v_2 = 100$ V when $i_2 = 0$, and that $v_2 = 125$V when $i_2 = 100$ mA. It is known that the two-port is unilateral. Find the parameters of each of the four models for the two-port.

Section 6.4 is prerequisite for the following problems.

P 6.21 For each circuit in Fig. P 6.18, obtain expressions for the input resistance and the output resistance. Express the load voltage v_L in terms of the voltage v_S and the circuit parameters.

P 6.22 For each circuit in Fig. P 6.19, obtain expressions for the input resistance and the output resistance. Express the load voltage v_L in terms of the voltage v_S and the circuit parameters.

Section 6.5 is prerequisite for the following problems.

In each problem below, all sources are linear and all two-ports are unilateral, unless the problem implies otherwise.

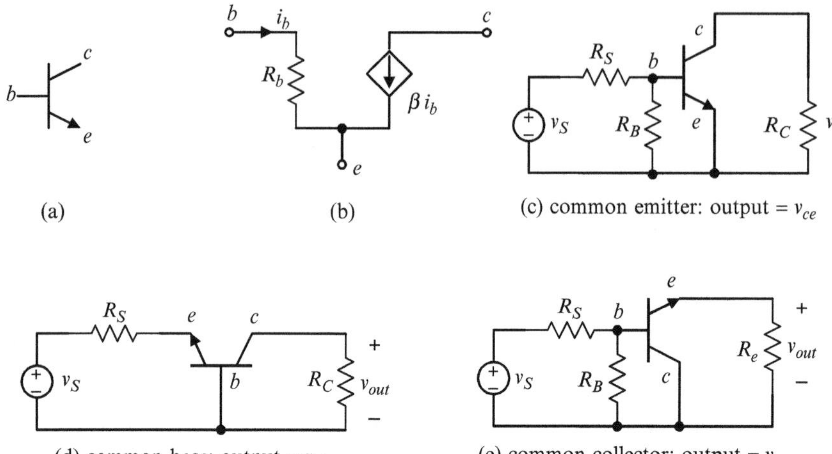

Fig. P 6.12 See Problem P 6.14, 15, 16

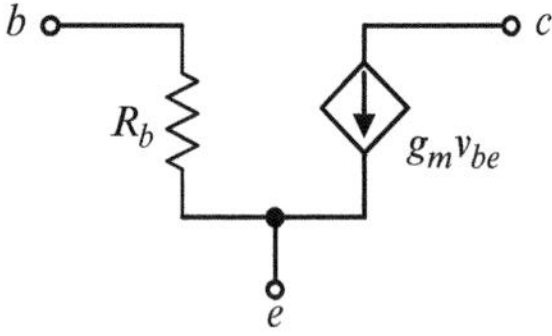

Fig. P 6.13 See Problem P 6.15, 17

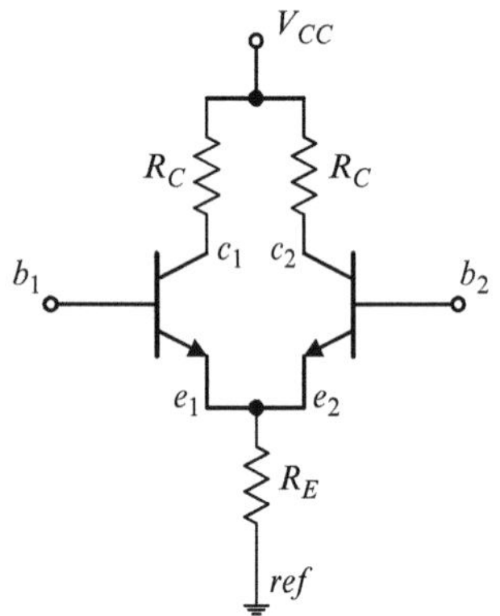

Fig. P 6.14 See Problem P 6.16

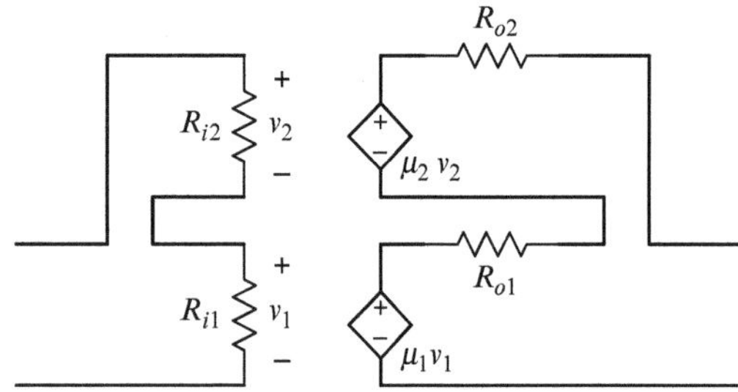

Fig. P 6.15 See Problem P 6.18

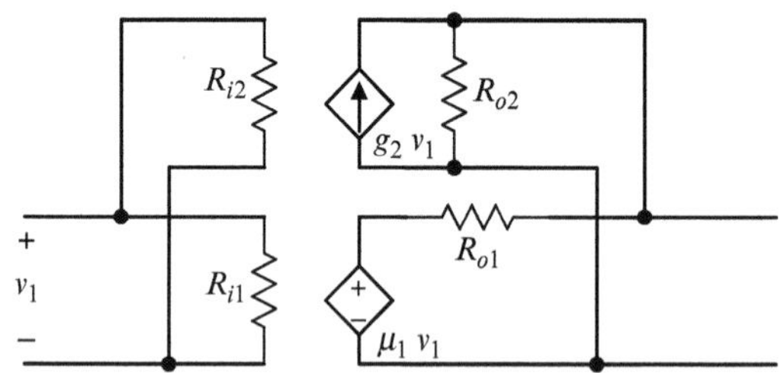

Fig. P 6.16 See Problem P 6.19

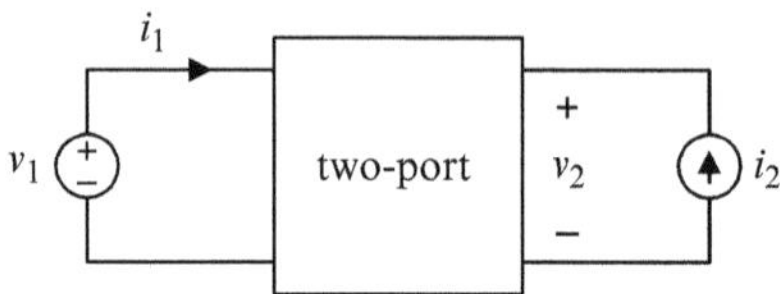

Fig. P 6.17 See Problem P 6.20

P 6.23 The output resistance of certain source is 1.2 Ω. The current available from the source is 1 A. Draw circuit diagrams for Thévenin and Norton models for the source and give the values of the model parameters.

P 6.24 If the current available from a source is doubled, with all other parameters unchanged, by what factors are the available voltage and power multiplied?

P 6.25 If the voltage available from a source is doubled, with all other parameters unchanged, by what factors are the available current and power multiplied?

P 6.26 If the output resistance of a source is halved, with all other parameters unchanged, by what

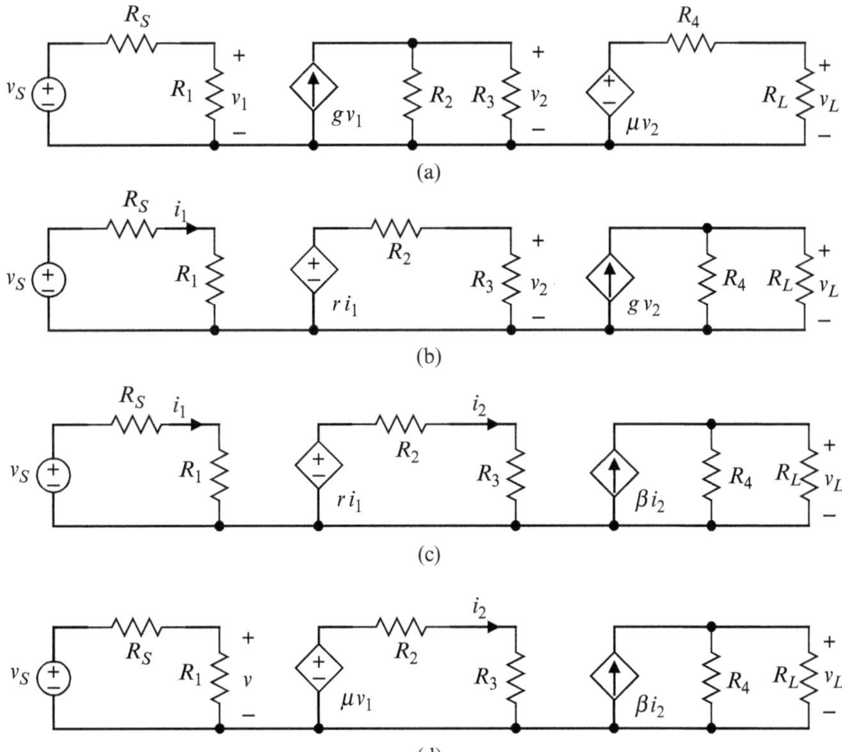

Fig. P 6.18 See Problem P 6.21

factors are the available voltage, current, and power multiplied?

P 6.27 A resistive load $R_L = 100\ \Omega$ is attached to the terminals of a source whose available current and available power are 5 A and 100 W, respectively. What is the power delivered to the load? What resistive load would draw the available power from the source?

P 6.28 The open-circuit voltage and short-circuit current for a certain source are 15 V and 15 mA, respectively. Obtain a graph of the power delivered by the source to a load R_L versus the resistance of the load, for $100\ \Omega \le R_L \le 10\ \text{k}\Omega$. Use a logarithmic scale for the load resistance.

Section 6.6 is prerequisite for the following problems.

P 6.29 Construct a table that gives the voltage gain for each of the four unilateral two-port models (VCVS, VCCS, CCVS, CCCS).

P 6.30 Find the input resistance and voltage transfer ratio of the circuit shown in Fig. P 6.20. Is the two-port circuit unilateral? Justify your answer.

P 6.31 Assume the four two-port models in Fig. P 6.21 are equivalent. Complete the table below, which shows how to convert any model to any other.

Convert To → From ↓	VCVS	CCVS	VCCS	CCCS
VCVS	$\mu = \mu$		$g = \mu/R_o$	$\beta = \mu R_i/R_o$
CCVS		$r = r$		
VCCS	$\mu = g R_o$		$g = g$	$\beta = g R_i$
CCCS		$r = \beta R_o$		$\beta = \beta$

P 6.32 The input resistance and output resistance of a certain unilateral two-port circuit are 10 kΩ and 100 Ω, respectively. When driven by a source having source resistance $R_s = 200\ \Omega$ and driving a resistive load $R_L = 2\ \text{k}\Omega$, the voltage transfer ratio is $H_v = -400$. Obtain all four unilateral two-port models for the circuit.

P 6.33 Figure P 6.22 illustrates measurements made on a certain two-port circuit. With both switches open,

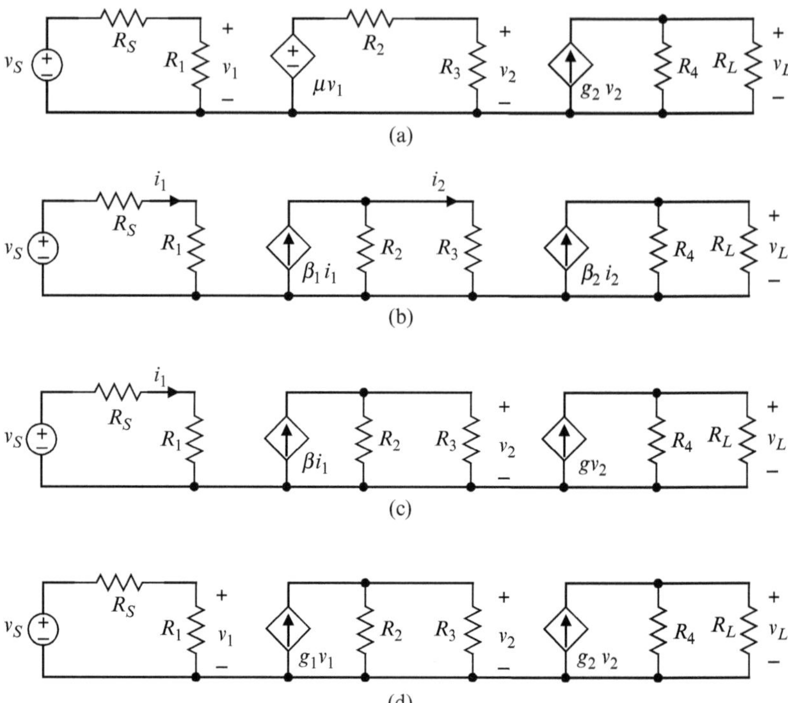

Fig. P 6.19 See Problem P 6.22

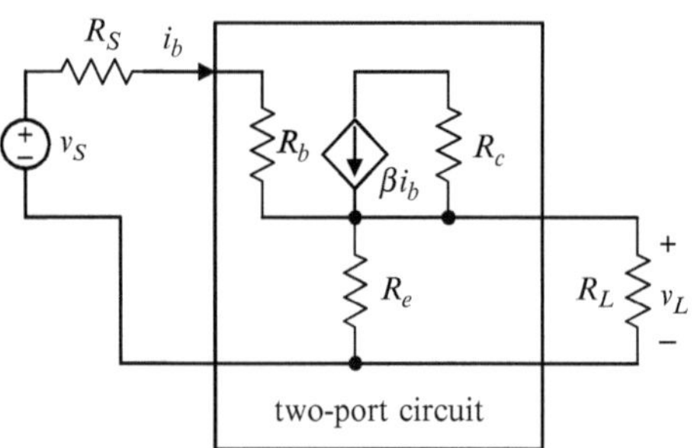

Fig. P 6.20 See Problem P 6.30

the voltage V_1 is set to 10 mV. When switch S_1 is closed, the voltage V_1 drops to 9.5 mV, the current I_1 becomes 1 μA, and the voltage V_2 becomes 1.5 V. When switch S_2 is closed, the voltage V_2 drops to 1.3 V. Find the output resistance of the source and the three parameters of a VCVS model for the circuit.

P 6.34 In a certain application, a two-port circuit will be driven by a source from which the available current and voltage are $I_A = 25$ mA and $V_A = 500$ mV, respectively. The circuit must provide a voltage transfer ratio of at least H_v while driving a resistive load $R_L = 4$ kΩ. Specify the two-port parameters to the extent possible.

P 6.35 Obtain expressions for the input resistance, output resistance, voltage gain, current gain, transresistance, and transadmittance for each of the three transistor amplifier circuits in Fig. P 6.12 (Problem P 6.14).

Section 6.8 is prerequisite for the following problems.

In problems below, *circuit* means *two-port circuit* or a cascade of two or more two-port circuits.

P 6.36 Obtain expressions for the power gains of VCVS and CCVS two-port models. Then show that if the models are equivalent, the expressions for power gain are identical.

P 6.37 Obtain expressions for the power gains of VCVS and VCCS two-port models. Then show that if the models are equivalent, the expressions for power gain are identical.

P 6.38 Obtain expressions for the power gains of VCVS and CCCS two-port models. Then show that if the models are equivalent, the expressions for power gain are identical.

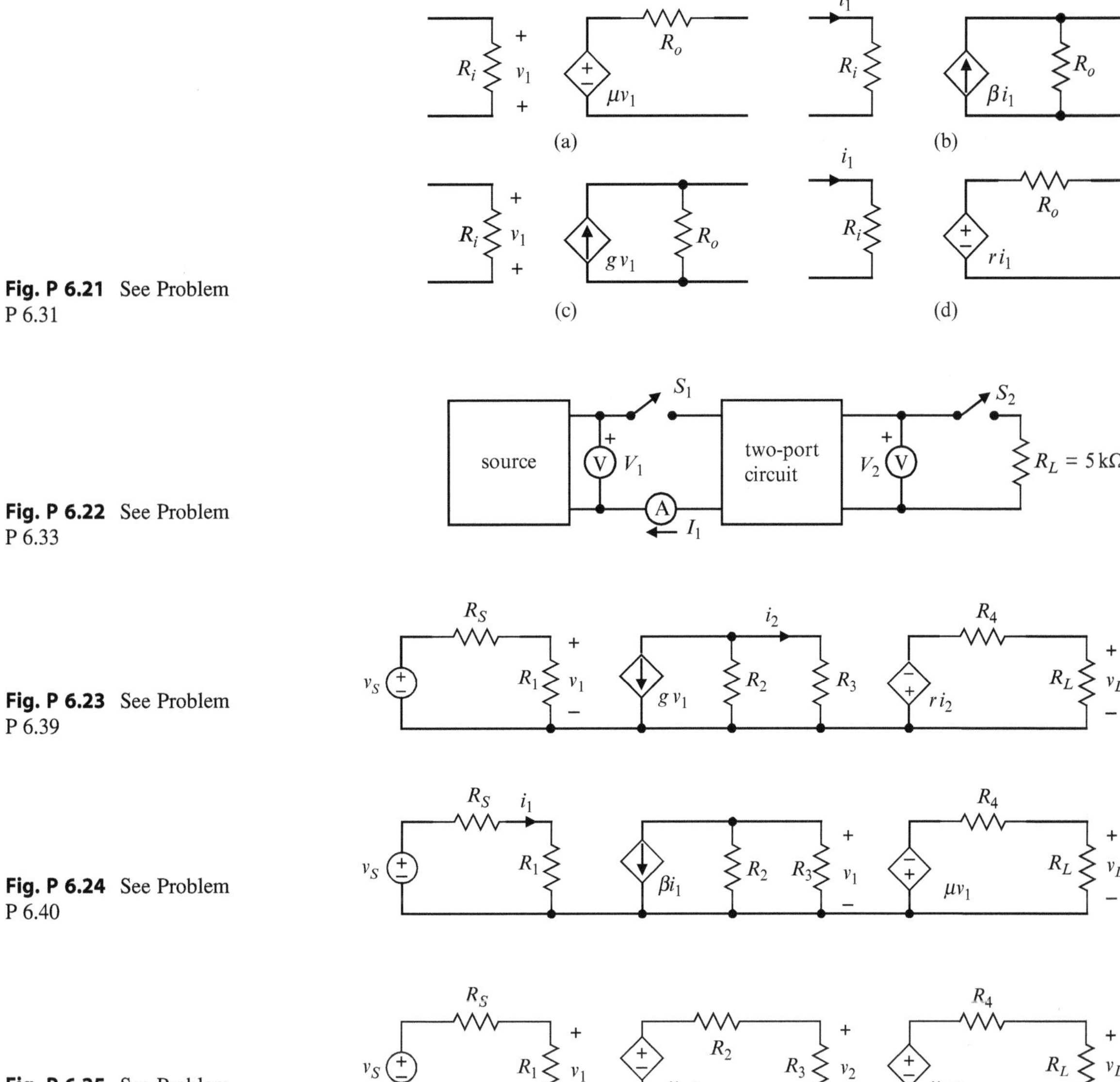

Fig. P 6.21 See Problem P 6.31

Fig. P 6.22 See Problem P 6.33

Fig. P 6.23 See Problem P 6.39

Fig. P 6.24 See Problem P 6.40

Fig. P 6.25 See Problem P 6.41

P 6.39 (a) Obtain an expression for the power gain of the circuit in Fig. P 6.23. (b) Let $R_S = 50\,\Omega$, $R_1 = R_2 = R_3 = 10\,\text{k}\Omega$, $R_4 = 100\,\Omega$, $R_L = 5\,\text{k}\Omega$, $g = 10\,\text{mS}$, $r = 8\,\text{k}\Omega$ and express the power gain in dB.

P 6.40 (a) Obtain an expression for the power gain of the circuit in Fig. P 6.24. (b) Let $R_S = 50\,\Omega$, $R_1 = R_2 = R_3 = 10\,\text{k}\Omega$, $R_4 = 100\,\Omega$, $R_L = 5\,\text{k}\Omega$, $\beta = 100$, $\mu = 50$ and express the power gain in dB.

P 6.41 (a) Obtain an expression for the power gain of the circuit in Fig. P 6.25. (b) Let $R_S = 50\,\Omega$, $R_1 = R_2 = R_3 = 10\,\text{k}\Omega$, $R_4 = 100\,\Omega$, $R_L = 5\,\text{k}\Omega$, $\mu_1 = 50$, $\mu_2 = 25$ and express the power gain in dB.

P 6.42 (a) Obtain an expression for the power gain of the circuit in Fig. P 6.26. (b) Let $R_S = 50\Omega$, $R_1 = R_2 = R_3 = 10\text{k}\Omega$, $R_4 = 100\Omega$, $R_L = 5\text{k}\Omega$, $\mu = 75$, $r = 20\text{k}\Omega$ and express the power gain in dB.

P 6.43 A certain circuit has input resistance $R_{in} = 20\,\text{k}\Omega$, output resistance $R_{out} = 100\,\Omega$, and no-load voltage gain $\mu = 200$. The circuit is driven by a source having output resistance $R_S = 50\ \Omega$ and available power $P = 15$ W. The circuit is to drive a resistive load R_L. (a) Obtain symbolic expressions for the current gain, voltage gain, transresistance, transadmittance, and power gain of the circuit. (b) Find the

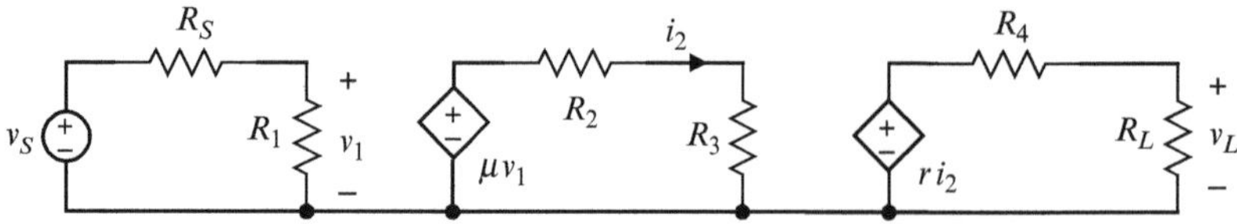

Fig. P 6.26 See Problem P 6.42

value of the load resistance R_L that maximizes the current gain and express the current gain in dB for that load. (c) Find the value of the load resistance R_L that maximizes the voltage gain and express the voltage gain in dB for that load. (d) Find the value of the load resistance R_L that maximizes the normalized transresistance A_r/R_L and express the normalized transresistance in dB for that load. (e) Find the value of the load resistance R_L that maximizes the normalized transconductance A_gR_L and express the normalized transconductance in dB for that load. (f) Find the value of the load resistance R_L that maximizes the power gain and express the power gain in dB for that load.

P 6.44 Repeat Problem P 6.43 for a circuit having input resistance $R_{in} = 1\,\text{k}\Omega$, output resistance $R_{out} = 50\,\Omega$, and no-load voltage gain $\mu = 1{,}000$. The circuit is driven by a source having output resistance $R_S = 20\,\text{k}\Omega$ and available current $I_S = 100\,\text{mA}$. The circuit is to drive a resistive load R_L.

P 6.45 Repeat Problem P 6.43 for a circuit having input resistance $R_{in} = 1\,\text{M}\Omega$, output resistance $R_{out} = 30\,\Omega$, and no-load voltage gain $\mu = 500$. The circuit is driven by a source having output resistance $R_S = 1\,\text{k}\Omega$ and available voltage $V_S = 25\,\text{mV}$. The circuit is to drive a resistive load R_L.

P 6.46 A certain circuit has input resistance $R_{in} = 20\,\text{k}\Omega$, output resistance $R_{out} = 100\,\Omega$, and no-load voltage gain $\mu = 200$. The circuit is driven by a source having output resistance $R_S = 50\,\Omega$ and available power $p = 15\,\text{W}$. The circuit is to drive a load having unspecified resistance R_L. Construct a graph of the power gain of the circuit in dB versus load resistance, for $1\,\Omega \le R_L \le 10\,\text{k}\Omega$. Use a logarithmic scale for R_L.

P 6.47 The voltage gain of a certain circuit is given by

$$A_v = \left(\frac{R_{in}}{R_{in} + R_S}\right)\left(\frac{R_L}{R_L + R_{out}}\right)\mu,$$

where R_S is the output resistance of a source, R_L is the resistance of a load, R_{in} is the input resistance of the circuit, R_{out} is the output resistance of the circuit, and μ is the no-load voltage gain of the circuit. (1) Draw a circuit diagram for a VCVS two-port model for the circuit. (2) Assuming all other parameters are fixed, specify the values of R_{in} and R_{out} that maximize (a) current gain, (b) voltage gain, and (c) power gain.

P 6.48 The power gain of a certain circuit is 75 dB. The power delivered to the load is 5 W. What is the power available from the source?

P 6.49 The power gain of a certain circuit is 20 dB. The power available from the source is 10 mW. What is the power delivered to the load?

P 6.50 The voltage gain of a certain circuit is 55 dB. The voltage available from the source is 5 mV. What is the voltage transferred to the load?

P 6.51 The current gain of a certain circuit is 75 dB. The output resistance of the source is $R_S = 10\,\text{k}\Omega$ and the load resistance is $R_L = 5\,\text{k}\Omega$. (a) What is the voltage gain in dB? (b) If the voltage across the load is 50 V, what is the current available from the source? (c) If the input resistance is 100 kΩ and the output resistance of the circuit is 100 Ω, what is the no-load voltage gain of the circuit?

P 6.52 For a particular source and load, the voltage gain and power gain of a certain circuit are 120 and 150 dB, respectively. What is the current gain in dB?

P 6.53 For a particular source and load, the voltage gain and current gain of a certain circuit are $A_v = 125$ dB and $A_i = 30$ dB, respectively. What is the power gain in dB?

P 6.54 For a particular source and load, the power gain and voltage gain of a certain circuit are $A_{p\,dB} = 150$dB, $A_{v\,dB} = 50$ dB. What is the current gain in dB?

P 6.55 For a particular source and load, the power gain and current gain of a certain circuit are $A_{p\,dB} = 150$ dB, $A_{v\,dB} = 50$ dB. What is the voltage gain in dB?

Section 6.9 is prerequisite for the following problems.

P 6.56 Refer to Fig. P 6.27, which is proposed as a temperature-measuring circuit, where $V_0 = 15\,\text{V}$,

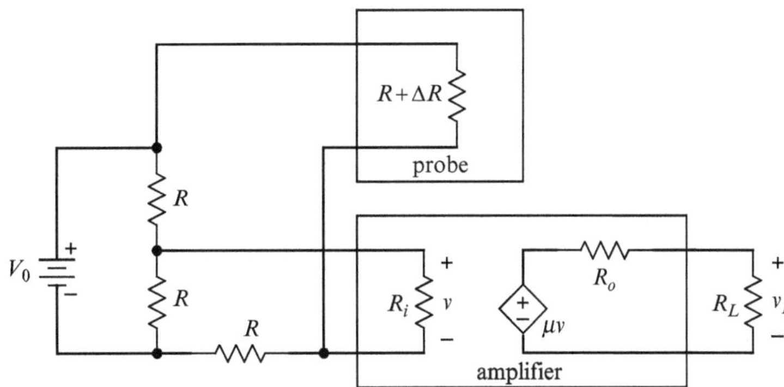

Fig. P 6.27 See Problem P 6.56

$R_L = 1\,\text{M}\Omega, R = 500\,\text{k}\Omega$, and $\alpha = 50\,\Omega(°\text{C})^{-1}$. The incremental resistance ΔR is given by $\Delta R = \alpha T$, where T°C is temperature in °C, and the temperature range of interest is $0°\text{C} \le T \le 100°\text{C}$.

(a) Express the voltage v as a function of vR. Assume that $\Delta R \ll R$ and simplify the expression. Under what condition is the voltage v proportional to the incremental resistance ΔR (and thus to temperature)? Specify a reasonable value for the amplifier input resistance R_i.
(b) Express the voltage v_L as a function of temperature (in °C). Specify a reasonable value for the amplifier output resistance R_o.
(c) The voltage v_L must change by 100 mV for each 1°C change in temperature T. Specify the amplifier no-load voltage gain μ.
(d) If the amplifier no-load voltage gain is 100 and cannot be changed, what other parameter would you adjust to achieve the specified 100 mV °C^{-1}, and what value would you specify for that parameter?

P 6.57 Refer to Fig. P 6.28. It is required that $v_1 = v_0/2$, and the power drawn from the source must be as small as possible. (a) It is known that $R_i \ge 5\,\text{M}\Omega$. What values would you choose for R_1, R_2, and why? (b) It is known that $R_L = 5\,\text{k}\Omega$ and $R_o < 10\,\Omega$. It is required that $v_L = 25v_0$. What value would you specify for μ?

P 6.58 You are asked to design a circuit that will accept as input a voltage from a photocell and produce as output a voltage related in some known way to the intensity of light incident on the photocell. Based upon your current knowledge, write down each question you must address before you begin designing the circuit.

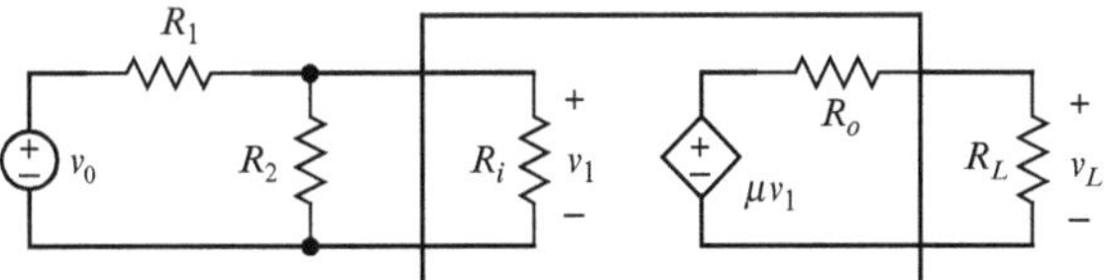

Fig. P 6.28 See Problem P 6.57

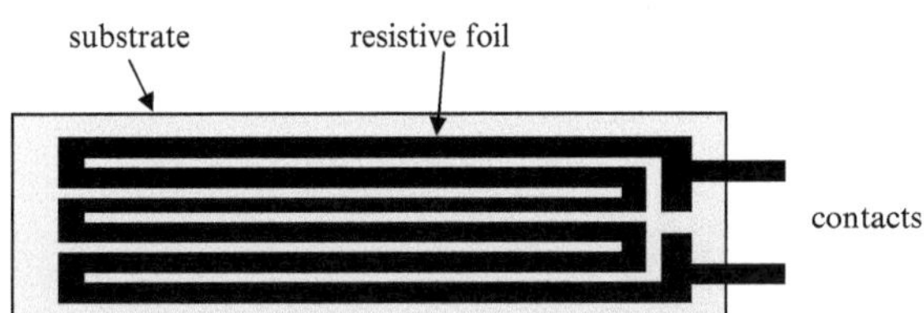

Fig. P 6.29 See Problem P 6.59

P 6.59 Strain is elongation of a body per unit length, denoted by $\Delta L/L$, which results from application of force to the body. For example, when a beam is subjected to a downward load in the center, the top of the beam is compressed ($\Delta L < 0$) and the bottom of the beam is stretched ($\Delta L > 0$). For most structural materials, strain (below the fracture point) is on the order of 1 mm/m. Strain is measured using devices called *strain gages*. The simplest and most commonly used strain gage consists of a foil resistor attached to a flexible substrate, as illustrated by Fig. P 6.29. When the strain gage is stretched or compressed, the resistance of the foil changes in approximate proportion to strain. The relation is

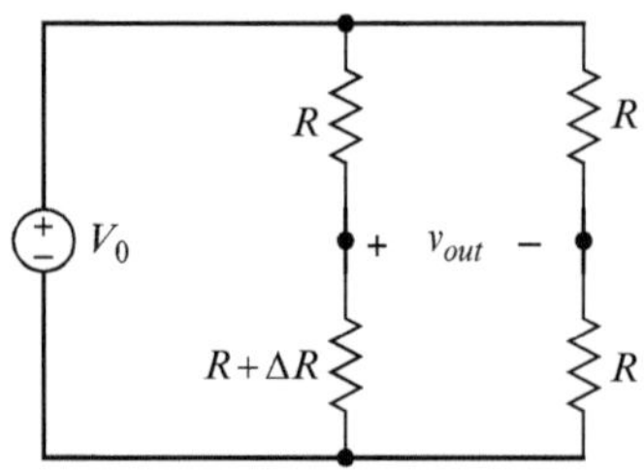

Fig. P 6.30 See Problem P 6.59, 60

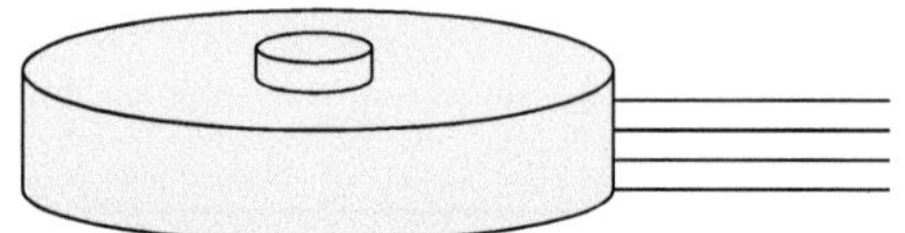

Fig. P 6.31 See Problem P 6.60

$$\frac{\Delta L}{L} = \frac{\Delta R}{RF},$$

where L and R are the nominal length and resistance, respectively, of the gage, ΔL and ΔR are the changes in length and resistance, respectively, and F is a constant called the *gage factor*. The nominal resistance of resistive strain gages is typically about 100 Ω to 1 kΩ and the gage factor is typically on the order of 1.

Strain gages are often used in four-arm bridge configurations, as illustrated by Fig. P 6.30, where R denotes a fixed resistor and $R+\Delta R$ denotes a strain gage. Let $V_0 = 1$V.

(a) Obtain the Thévenin equivalent at the output terminals of the circuit in Fig. P 6.30.

(b) Show that for $\Delta R \ll R$, the open-circuit output is given (approximately) by

$$v_{out} \cong \frac{V_0}{4}\left(\frac{\Delta R}{R}\right).$$

(c) The gage factor of the strain gage is $F = 1.8$ and the nominal resistance of the gage is $R = 350\ \Omega$. The voltage v_{out} is to be applied to the input of an amplifier whose output is a voltage proportional to strain; that is,

$$v = k\left(\frac{\Delta L}{L}\right),$$

with $k = 1$ V. The output of the amplifier is to be sensed by a voltmeter having input resistance $R_{VM} = 10\,\text{M}\Omega$. Specify the parameters of a unilateral two-port model for the amplifier. Which of the three parameters has the greatest effect on the overall accuracy of a strain measurement? Obtain an expression for the amplifier output voltage without making any approximations. Construct a plot of the percent error in the approximate relation for $10^{-6} \leq \Delta L/L \leq 10^{-3}$. Explore effects of varying the amplifier input and output resistances.

(d) Why might one expect the gage factor F for a resistive strain gage to be on the order of 1?

P 6.60 Strain gages (see Problem P 6.59) are used in devices called *load cells* that are used to measure weight (or force). There are various kinds of load cells having different configurations and purposes. Figure P 6.31 shows a sketch of a *button load cell*. The four wires connect internally to a bridge circuit, such as the one shown in Fig. P 6.30 – two for the power supply and two for the output. When the button is depressed by a force, one or more strain gages inside the device are deformed, and an output voltage proportional to the force is produced. Small button gages are used in small scales (such as postal scales), and can sense weights ranging from zero to a few hundred pounds with accuracies of about 1%. Much larger load cells are used in the scales at truck weigh stations along the interstates.

Do a little research on the web and write a one-page paper (not counting figures) on the structure and operation (circuit model) for any particular kind of load cell. Include a web-address list of your sources.

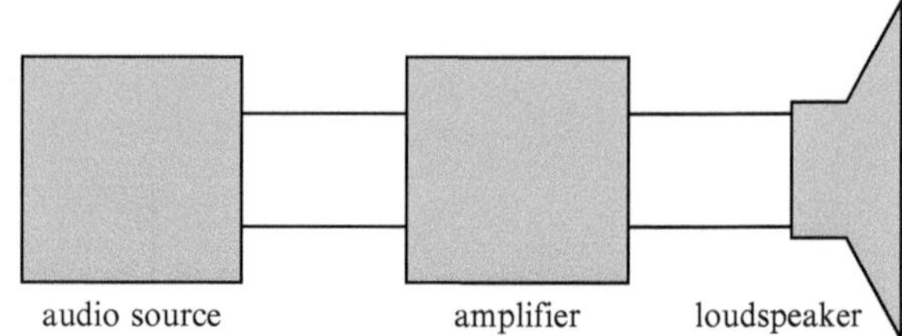

Fig. P 6.32 See Problem P 6.61

P 6.61 Refer to Fig. P 6.32. The audio source has output resistance $R_S = 250\,\Omega$ and open-circuit voltage $V_{S\,rms} = 25$ mV. The loudspeaker has input resistance $R_L = 8\,\Omega$. Specify the parameters of a unilateral VCVS two-port model for the amplifier, such that the power delivered to the loudspeaker is 100 W. You are free to make each parameter as large or small as you wish, but justify each specification.

Chapter 7
Operational Amplifiers I

Essentially, an **amplifier** is a circuit that for some inputs delivers more power to a load than is received from an input. Amplification is achieved by using a small current or voltage (an input) to control a larger current or voltage provided by a power supply and delivered to a load, as illustrated by Fig. 7.1. For example, in a home audio amplifier, the voltage from a CD player (the source) controls the power delivered to the loudspeaker (the load). Nearly all of the power delivered to the speaker terminals is provided by the amplifier's power supply, which draws its power from a wall outlet. In almost all applications of amplifiers, the power provided by an input is negligible in comparison to that provided by the power supply.

Amplifiers are core components of a great many circuits and systems, such as active filters, oscillators, radio transmitters, control systems, and laboratory instruments. Amplifiers can be constructed using discrete components (transistors), or can be purchased (as integrated or modular circuits). It is not a stretch to claim that amplifiers are the most important linear circuits in existence, largely because of what can be achieved using amplifiers in conjunction with feedback.

An amplifier is a two-port circuit and, as such, can be partly described by one or more of voltage gain, current gain, transconductance, and transresistance. Most amplifiers are designed such that one of the four gains is approximately independent of source and load resistance. The desired independence is achieved in part by specifying the input resistance and output resistance relative to the source resistance and the load resistance, respectively. Amplifiers also are classified according to which of the gains is approximately independent of source and load resistance. For example, an amplifier whose current gain is approximately independent of source and load resistances is called a *current amplifier*.

Exercise 7.1. In the table below, R_i, R_o are the input resistance and output resistance, respectively, of an amplifier. The amplifier is driven by a source having output resistance R_S and drives a load having input resistance R_L. Match each amplifier type on the left with one of the constraints on the right.

Amplifier type	Constraint
(a) Current amplifier	(i) $R_i \gg R_S$, $R_O \ll R_L$
(b) Transresistance amplifier	(ii) $R_i \ll R_S$, $R_O \gg R_L$
(c) Voltage amplifier	(iii) $R_i \gg R_S$, $R_O \gg R_L$
(d) Transconductance amplifier	(iv) $R_i \ll R_S$, $R_O \ll R_L$

An amplifier can be represented by any of the two-port circuit models presented in Chapter 6. The kind of amplifier being represented often suggests which of the four two-port models is most appropriate. For example, a transconductance amplifier (voltage in, current out) is most naturally represented by a VCCS model.

Exercise 7.2. Match each amplifier type on the left with the most natural model on the right.

Amplifier type	Two-port model
(a) Current amplifier	(i) VCVS
(b) Transresistance amplifier	(ii) VCCS
(c) Voltage amplifier	(iii) CCCS
(d) Transconductance amplifier	(iv) CCVS

T.H. Glisson, *Introduction to Circuit Analysis and Design*,
DOI 10.1007/978-90-481-9443-8_7, © Springer Science+Business Media B.V. 2011

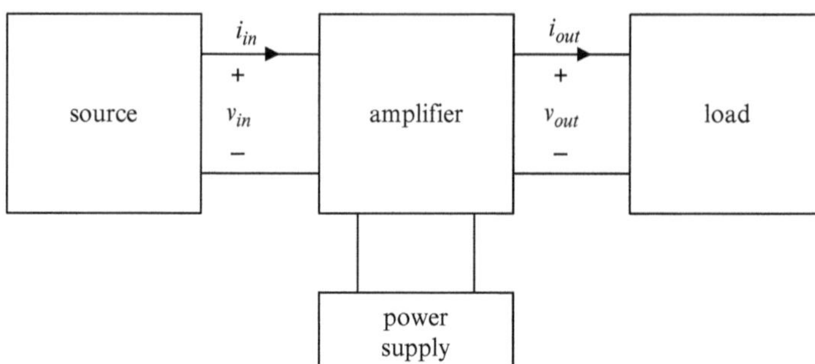

Fig. 7.1 An amplifier driven by a source and driving a load

Amplifiers are used throughout many electronic systems. Amplifiers often appear as the first (input) stage, which receives an input (e.g., a current or voltage from a sensor), as the last (output) stage, which delivers a current or voltage to a load, and as interior circuits, where they help perform a variety of signal-processing functions. Amplifiers used as the first or last stages can be any one of the four kinds named above, depending upon the nature of the source and load. Amplifiers used as interior stages for filtering and other signal-processing operations are almost invariably voltage amplifiers. This is because a voltage amplifier typically wastes less power internally (in the output resistance) than does a current or transconductance amplifier and demands less power from a source (input or previous stage) than does a current or transresistance amplifier. Consequently, voltage amplifiers generally require less power from a power supply and generate less heat than do other kinds of amplifiers performing basically the same task.

An **operational amplifier** or **op amp** is essentially a voltage amplifier having a very large intrinsic dc voltage gain. Hundreds of different op amps are available in integrated-circuit (IC) form. Low-power integrated op amps dissipate a half watt or so, are smaller than a dime, and range in cost from less than 50¢ to a few dollars. High-power op amps dissipate from tens of watts up to a few hundred watts and can cost several hundred dollars. Figure 7.2 shows three integrated op amps. The single 741 and the quad 324 can dissipate about 500 mW and deliver up to approximately 25 mA to a load. The PA03© can dissipate 500 W and deliver up to approximately 30 A to a load.

Modern op amps have intrinsic dc voltage gains ranging from about 10^5 to about 10^7, input resistances ranging from about 1 MΩ to about 1 TΩ $\left(10^{12}\,\Omega\right)$, output resistances ranging from about 1 Ω to about

Fig. 7.2 From left to right: A single 741 op amp, a quad 324 op amp (four op amps in one IC package), and a high-power PA03© op amp (Photograph of the PA03© courtesy of Apex Precision Power Division of Cirrus Logic, Inc)

100 Ω, and power-dissipation ratings from a few milliwatts to a few hundred watts. Most manufacturers' web sites provide extensive data on their various offerings. Integrated op amps contain 15 or more transistors, about the same number of resistors, and a few diodes.[1] Most also contain one (intentional) capacitor, in addition to the junction capacitances inherent in the transistors themselves.[2]

Op amps are used in countless applications, including amplifiers, active filters, other linear signal-processing circuits (e.g., integrators and differentiators), and in a host of nonlinear applications, such as oscillators and detectors. This chapter describes operational amplifiers and treats selected circuits using operational amplifiers in *feedback configurations.* We describe op amps primarily in terms of their terminal characteristics. You will learn more about the internal structure and operation of op amps in a subsequent course on electronic devices and circuits.

[1] In integrated op amps, transistors with the base connected to the collector often serve as diodes.

[2] We treat capacitance in Chapter 8.

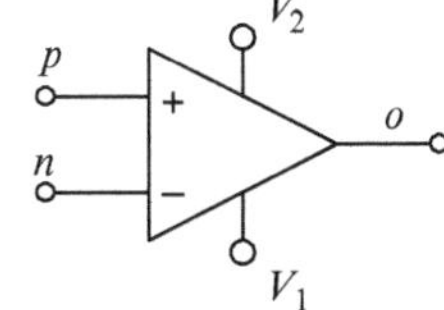

Fig. 7.3 Circuit-diagram symbol for an operational amplifier

7.1 Operational Amplifier Terminals and Voltage Reference

Figure 7.3 shows a circuit-diagram symbol for an op amp. The symbol in Fig. 7.3 has five terminals:[3] The **positive input terminal** or *p* terminal, the **negative input terminal** or *n* terminal, the **output terminal** or *o* terminal, and the **power supply terminals**, labeled V_2 and V_1.[4] There is no common or ground terminal on an op amp. The reference point for terminal voltages is established by the manner in which a power supply and load are connected to an op amp, as explained below.

As does any amplifier, an op amp requires an external source of power to achieve amplification. Power is provided to an op amp by dc sources (supply voltages) connected to the terminals labeled V_2 and V_1 in Fig. 7.3. For proper operation of an op amp, the voltage V_2 must exceed the voltage V_1 by at least an amount V_B, but must not exceed V_1 by more than an amount V_A, where V_A and V_B depend upon the internal structure of the op amp and vary from one op amp to another. Formally, the supply voltages must satisfy an inequality of the form

$$V_A \geq V_2 - V_1 \geq V_B. \tag{7.1}$$

For integrated general-purpose op amps, V_B is typically about 2–10 V and V_A is typically about 10–50 V. For example, for National Semiconductor's LF147 op amp, $V_A = 44$ V and $V_B = 5$ V, which means V_2 must be at least $V_1 + 5$ V, but cannot exceed $V_1 + 44$ V.

In linear circuits, op amps are often powered by symmetric bipolar supplies that provide equal-magnitude positive and negative dc voltages, as shown in Fig. 7.4a. Usually, the supply sources are not shown explicitly. Rather, the supply voltages are simply written next to the supply terminals, as shown in Fig. 7.4b. Typical values for V_{CC} are on the order of 2–15 V for small, low-power op amps, ranging up to about 200 V for high-power op amps.

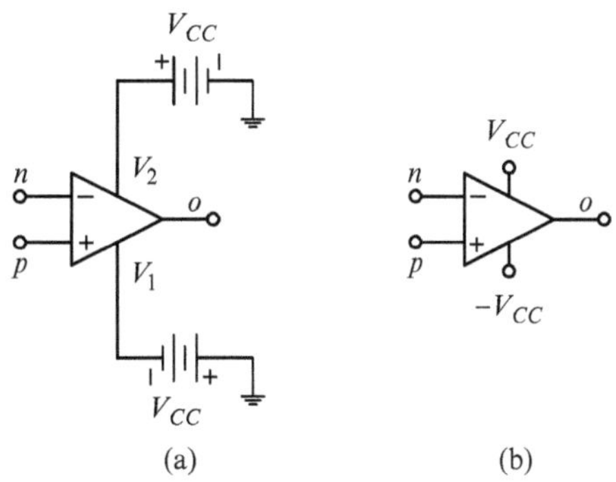

Fig. 7.4 Circuit-diagram symbol for an op amp, showing how the external supply is connected. The meaning of the ground symbol is explained below

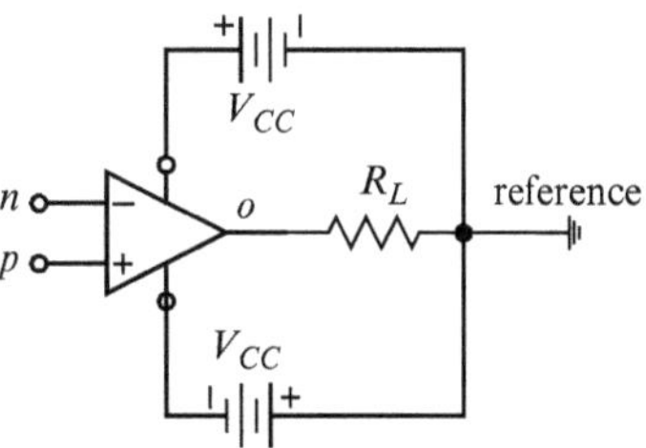

Fig. 7.5 Establishing the reference for op-amp terminal voltages

As noted above, the reference point for the terminal voltages of an op amp is established by the manner in which the supply and the load are connected to the op amp and to each other. Figure 7.5 shows an op amp driving a load R_L. In Fig. 7.5, the power supplies are shown explicitly and the load is connected between the op-amp *o* (signal-output) terminal and the neutral terminal of the supply, which establishes the neutral terminal of the supply as the reference for all op-amp terminal voltages, as indicated by the ground symbol. Thus, v_n means the voltage from the *n* terminal to the reference and v_p means the voltage from the *p* terminal to the reference.

Often, the supply voltage sources and the supply terminals themselves are omitted from circuit diagrams. In such cases, it usually is implied that the power supply is bipolar and symmetric and that the load is connected from the op-amp *o* terminal to the neutral terminal of the supply. The neutral terminal of the supply is then

[3] Real op amps have additional terminals whose functions you will learn when you study electronics.

[4] Because of the manner in which the internal circuitry is connected to the supplies, the supplies or supply terminals often are called *rails*.

the reference for all terminal voltages, and is denoted by the ground symbol in a circuit diagram, as in Fig. 7.5.

In practice, the labels n, p, and o shown in Fig. 7.5 also are usually omitted in circuit diagrams. It is understood that the terminal labeled "−" is the negative input (n) terminal, the terminal labeled "+" is the positive input (p) terminal, and the unlabeled terminal at the opposite vertex of the triangular symbol is the output (o) terminal. Nonetheless, for sake of clarity, we often include the labels n, p, and o on circuit diagrams. You might find it helpful to do this also, especially when preparing to replace the op-amp symbol by an equivalent circuit, as we do in the next section.

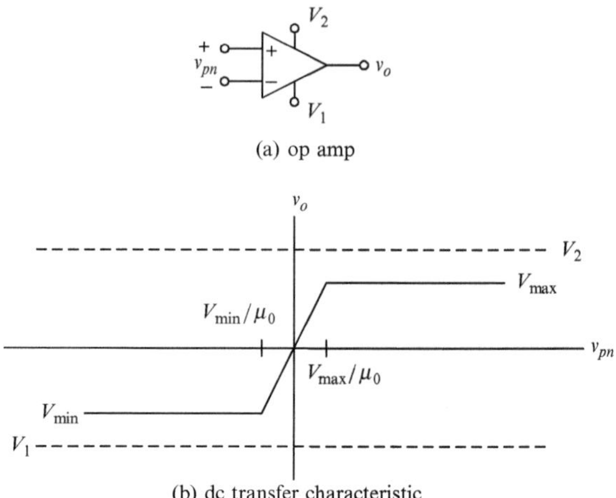

Fig. 7.6 DC transfer characteristic for an op amp (typical)

7.2 DC Circuit Model for an Op Amp

In this chapter, we consider the dc behavior of op amps in order to present some basic ideas in an uncluttered setting. The model and analyses presented in this chapter are strictly applicable only for dc or slowly varying excitation. In Chapter 17, we introduce frequency response and other refinements.

The **no-load dc transfer characteristic** for the dc (or low-frequency) model for an op amp has the form

$$v_o = \begin{cases} V_{\min}, & \mu_0 v_{pn} \le V_{\min}, \\ \mu_0 v_{pn}, & V_{\min} < \mu_0 v_{pn} < V_{\max}, \\ V_{\max}, & \mu_0 v_{pn} \ge V_{\max}, \end{cases} \tag{7.2}$$

where the parameter μ_0 is the **intrinsic dc voltage gain**. The minimum and maximum output voltages V_{min}, V_{max} are determined by the internal structure of the op amp and the supply voltages V_1, V_2. The output v_o cannot be less than V_{min} and cannot exceed V_{max}. Whenever the output of an op amp equals either V_{min} or V_{max}, i.e., whenever $v_{pn} \le V_{min}/\mu_0$ or $v_{pn} \ge V_{max}/\mu_0$, the op amp is said to be **saturated**. Figure 7.6 shows a graphical representation of the transfer characteristic (7.2), where $V_2 > V_{max} > V_{min} > V_1$. The voltage v_{pn} equals zero where the graph crosses the horizontal axis, but the voltage v_o at that point is not necessarily zero because the voltages V_2, V_{max}, V_{min}, V_1 might all have the same sign.[5]

With reference to Fig. 7.6, it is necessary that $V_2 > V_1$, but it is not necessary that the supply voltages have opposite signs. This fact is sometimes useful; e.g., we might wish to amplify an input voltage that has a dc offset (a non-zero average value), such that the input is always positive. But in many linear applications, signal voltages have no dc components and op amps are powered by symmetric bipolar supplies, meaning that $V_1 = -V_{CC}$, $V_2 = V_{CC}$.

For some op amps, V_{min} is about 1 V larger than V_1 and V_{max} is about 1 V smaller than V_2. The output of such an op amp powered by a symmetric bipolar supply $-V_1 = V_2 = 15\text{ V}$ (or $\pm V_{CC} = \pm 15\text{ V}$) is confined (approximately) to the range $\pm$ 14V. However, there are so-called **rail-to-rail** op amps for which $V_{min} = V_1$ and $V_{max} = V_2$.

To avoid cluttering our treatment with too many details, *we assume rail-to-rail operation and a symmetric bipolar supply*, which means that the no-load dc transfer characteristic is of the form

$$v_o = \begin{cases} -V_{CC}, & \mu_0 v_{pn} \le -V_{CC}, \\ \mu_0 v_{pn}, & -V_{CC} < \mu_0 v_{pn} < V_{CC}, \\ V_{CC}, & \mu_0 v_{pn} \ge V_{CC}, \end{cases} \tag{7.3}$$

as shown in Fig. 7.7. The range of values $-V_{CC} < \mu_0 v_{pn} < V_{CC}$ is called the **range of linear operation**, whether one is referring to the input or output; that is, the range of linear operation can mean either the

[5] Also, a small dc offset voltage afflicts virtually all op amps, as described in Section 7.8.

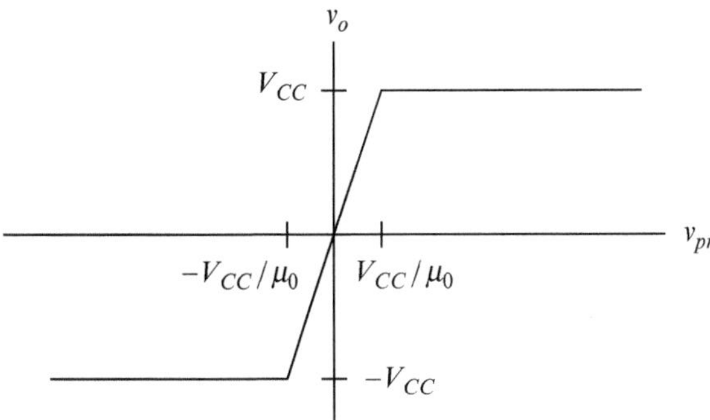

Fig. 7.7 Idealized DC Transfer characteristic for a rail-to-rail op amp having a symmetric bipolar supply

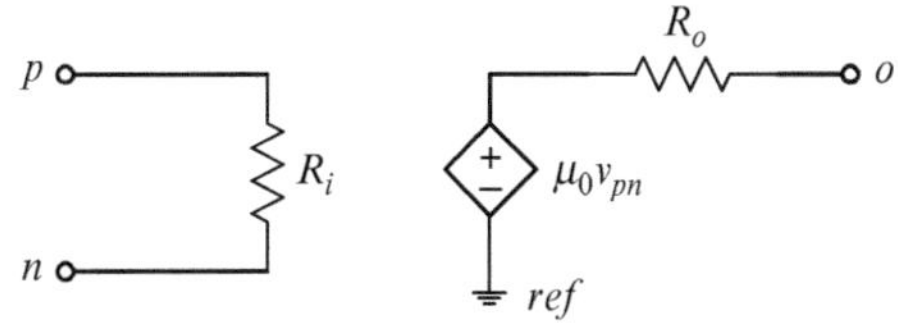

Fig. 7.8 Linear model for an op amp at dc (for $f = 0$)

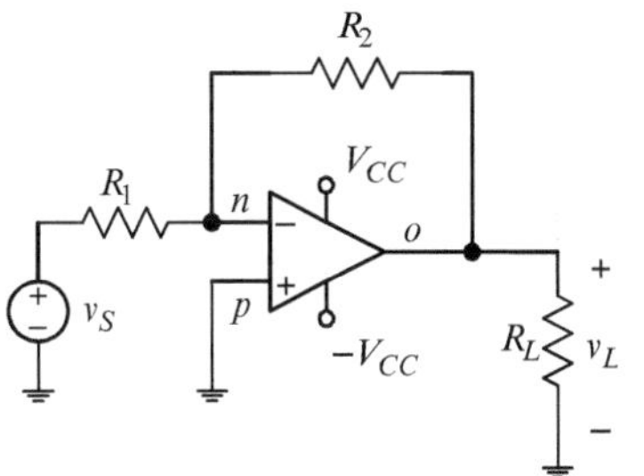

Fig. 7.9 See Example 7.1

values of an input v_{pn} that satisfy $-V_{CC}/\mu_0 < v_{pn} < V_{CC}/\mu_0$ or (equivalently) the values of an output v_o that satisfy $-V_{CC} < v_o < V_{CC}$.

It is questionable whether any real op amp is truly linear over the so-called range of linear operation. In fact, the voltage gain of an op amp is so large that it is difficult to even measure the voltage transfer characteristic. Nonetheless, the transfer characteristic (7.3) is helpful to explaining and understanding op amp behavior and circuit design, so we shall assume it is approximately true. There is no harm in this assumption because, as we show later, it does not matter in most applications whether the transfer characteristic of an op amp is actually linear over the so-called range of linear operation, so long as $|v_o|$ is a great deal larger than $|v_{pn}|$ for all values of v_{pn} in that range.

Figure 7.8 shows a linear model for an op amp at dc that is consistent with the linear portion of the no-load dc transfer characteristic (7.3). The parameters R_i, R_o are the **input resistance** and **output resistance**, respectively, and the parameter μ_0 is the **intrinsic dc voltage gain**. An op amp is approximately unilateral, so the input resistance is independent of a load attached to the output terminals and the output resistance is independent of a source attached to the input terminals. Typical values for these parameters are as follows:

- Input resistance R_i: from about 1 MΩ to more than 1 TΩ
- Output resistance R_o: from about 1 Ω to about 1 kΩ, but typically <100 Ω
- Intrinsic dc voltage gain μ_o: from about 10^5 to about 10^7

The model in Fig. 7.8 does not include effects such as saturation, limits on output current, and frequency dependence, all of which are important in all applications. In a subsequent chapter, we treat some of these other effects and refine the model in Fig. 7.8 to account for the fact that the intrinsic voltage gain of an op amp decreases as the frequency of an input increases. *Relations obtained in this chapter are applicable only for low-frequency inputs.* We cannot quantify *low-frequency* at this point, because its meaning depends upon the structure of a circuit at hand.

Example 7.1. In Fig. 7.9, $V_{CC} = 10\,\text{V}$, $R_1 = 10\,\text{k}\Omega$, $R_2 = 50\,\text{k}\Omega$, $R_L = 5\,\text{k}\Omega$, and the source voltage v_S is independent of time. The model parameters for the op amp (see Fig. 7.8) are $R_i = 1\,\text{M}\Omega$, $R_o = 50\,\Omega$, $\mu_0 = 10^6$. Using the linear model in Fig. 7.8 for the op amp, obtain an expression for the voltage transfer characteristic for the circuit.

Solution: We replace the op-amp symbol with the linear model shown in Fig. 7.8 to obtain the circuit diagram shown in Fig. 7.10. We have used the fact that $v_p = 0$ and thus $\mu_0 v_{pn} = \mu_0(v_p - v_n) = -\mu_0 v_n$.

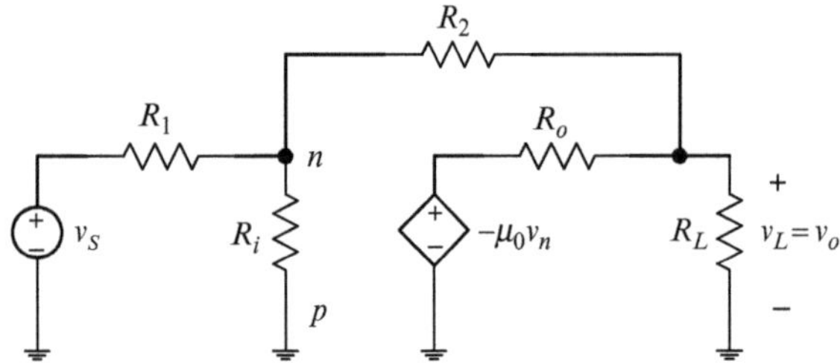

Fig. 7.10 See Example 7.1

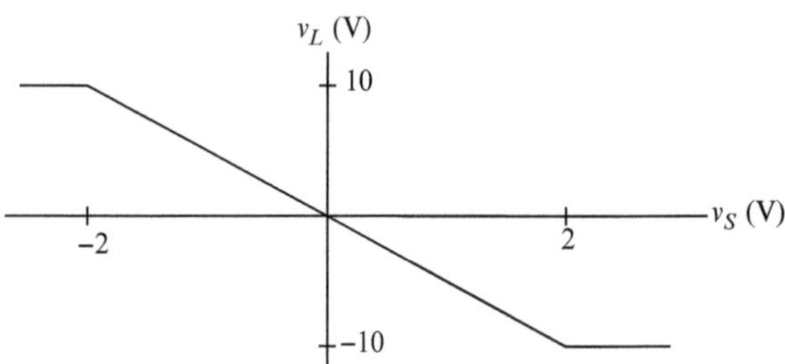

Fig. 7.11 See Example 7.1

For the circuit in Fig. 7.10, Kirchhoff's current law gives

$$\text{Node } n\text{: } G_1(v_n - v_S) + G_i v_n + G_2(v_n - v_L) = 0, \tag{7.4}$$

$$\text{Node } o\text{: } G_o(v_L + \mu_0 v_n) + G_2(v_L - v_n) + G_L v_L = 0. \tag{7.5}$$

where we use conductances to simplify the notation; e.g., $G_1 = R_1^{-1}$. From (7.4),

$$v_L - \frac{(G_1 + G_i + G_2)}{G_2} v_n = -\frac{G_1 v_S}{G_2}. \tag{7.6}$$

From (7.5),

$$v_n = \frac{(G_o + G_2 + G_L)}{(G_2 - G_o \mu_0)} v_L. \tag{7.7}$$

Note that for $\mu_0 \to \infty$, (7.7) gives $v_n \to 0$, in which case (7.6) reduces to

$$v_L \cong -\frac{G_1 v_S}{G_2} = -\frac{R_2}{R_1} v_S = -5\, v_S. \tag{7.8}$$

To obtain an exact solution, we use (7.7) to eliminate v_n in (7.6). This gives

$$(G_1 + G_i + G_2)\frac{(G_o + G_2 + G_L)}{(G_2 - G_o \mu_0)} v_L - G_2 v_L = G_1 v_S, \tag{7.9}$$

which yields

$$\frac{v_L}{v_s} = \frac{G_1(G_2 - \mu_0 G_o)}{(G_1 + G_i + G_2)(G_o + G_2 + G_L) - G_2(G_2 - \mu_0 G_o)} \cong -4.99997 = H_v. \tag{7.10}$$

Here again, if we let $\mu_0 \to \infty$, (7.10) reduces to (7.8).

The difference (0.0006%) between the exact solution (7.10) and the approximate solution (7.8) is insignificant because the difference is much smaller than typical uncertainties in parameter values; for example, even in the best of circumstances, the resistances R_1, R_2 would probably be known only to within ±0.05%. Uncertainties in values of the op-amp model parameters might be larger, perhaps as large as ±20%.

The accuracy of the approximation (7.8) is determined primarily by the magnitude of the intrinsic gain μ_0 and to a lesser degree by the values of the external resistances R_1, R_2, R_L relative to the internal (model) resistances R_i, R_0.

Equations (7.8) and (7.10) hold for $-10\,\text{V} < v_L < 10\,\text{V}$ or $-2\,\text{V} < v_S < 2\,\text{V}$. For inputs outside that range, the op amp is saturated. Thus a more complete statement of the transfer characteristic is

$$v_L \cong \begin{cases} 10\,\text{V}, & v_S \le -2\text{V}, \\ -5\, v_S, & -2\text{V} < v_S < 2\text{V}, \\ -10\text{V}, & v_S > 2\text{V}. \end{cases} \tag{7.11}$$

Figure 7.11 shows a graph of the transfer characteristic (7.11)

7.3 The Ideal Op Amp and Some Basic Op-Amp Circuits at DC

For most op amps and virtually all linear applications, the approximation $\mu_0 \to \infty$ is so good that it is usually incorporated in the op-amp model itself, at least for first-pass design and analysis of linear op-amp circuits. The approximation $\mu_0 \to \infty$ has two important consequences: First, because the output $v_o = \mu_0 (v_p - v_n)$ must be finite, the voltage $v_{pn} = v_p - v_n$ must be exceedingly small.[6] Second, because the voltage $v_{pn} = v_p - v_n$ is exceedingly small and the op-amp input resistance is large, the current entering either input terminal must also be exceedingly small. Thus the **ideal op amp** is defined by the following properties:

$$\mu_0 \to \infty, \tag{7.12}$$

$$v_p - v_n \to 0 \Rightarrow v_p \cong v_n, \tag{7.13}$$

$$i_p \cong 0, \; i_n \cong 0. \tag{7.14}$$

Equations (7.12), (7.13) and (7.14) are the **fundamental relations** for preliminary analysis and design of linear op-amp circuits.

In a physical circuit, the voltage difference $v_p - v_n$ and the currents $i_p = -i_n$ are not zero, because if either the current i_p or the voltage v_{pn} were zero, no power (no energy) would be delivered to the op amp input, and no physical system can respond to an excitation that conveys no energy to the system. But the output voltage v_{out} (limited by the supply) is typically on the order of a few volts and the intrinsic gain μ_0 is typically on the order of 10^5–10^6, so the voltage $v_{pn} = v_{out}/\mu_0$ is typically on the order of (at most) about 100 μV. Because the voltage v_{pn} is at most about 100 μV and the input resistance R_i is at least about 1 MΩ and can exceed 10^{12} Ω, the currents i_p, i_n are no larger than a few picoamperes (100 μV/1 MΩ = 100 pA) and usually are much smaller than that. *In linear applications, the voltage v_{pn} and the currents i_p, i_n are so small relative to other voltages and currents in the external circuit that they are practically zero.* Thus (7.14) and (7.13) are true for all practical purposes, *provided the op amp is not saturated*; i.e., provided $|v_{out}| < V_{CC}$. For an op amp in saturation, the approximation $v_p = v_n$ ($v_{pn} = 0$) might not hold, but the approximation $i_p = i_n = 0$ usually remains good, because even in saturation the voltage v_{pn} usually is no more than a few volts and the op-amp input resistance is quite large.

[6] Although v_{pm} is small, the individual voltages v_p, v_n are not necessarily small.

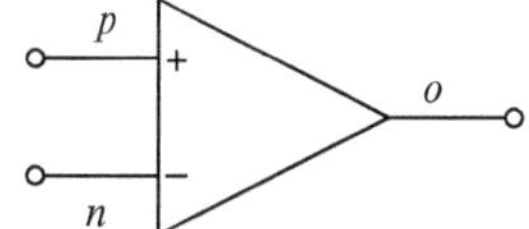

Fig. 7.12 Circuit-diagram symbol for an ideal op amp. See (7.13) and (7.14)

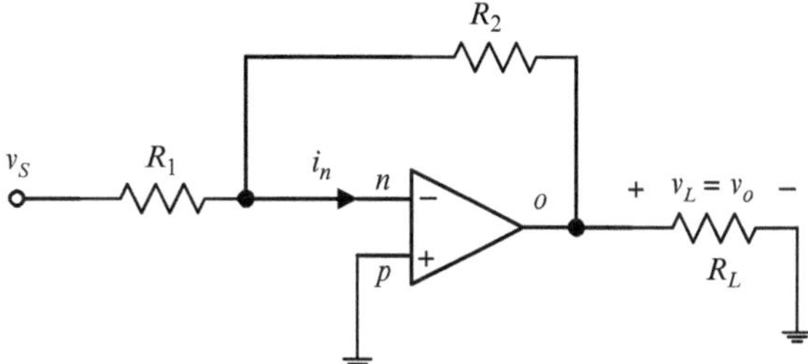

Fig. 7.13 See Example 7.2

Because the intrinsic dc gain of an ideal op amp is infinite and the input voltage v_{pn} is zero, we cannot construct a dependent-source circuit model for an *ideal* op amp. The output would be indeterminate, of the form $\infty \cdot 0$. An ideal op amp is simply denoted by the symbol shown in Fig. 7.12. The power-supply terminals often are omitted, the implication being that the supply is adequate for the anticipated ranges of input and output amplitudes.

Example 7.2. (Compare with Example 7.1.) Obtain an expression for the voltage transfer characteristic for the circuit shown in Fig. 7.13. Assume the op amp is ideal.

Solution: Kirchhoff's current law at node *n* gives

$$\frac{v_n - v_S}{R_1} + \frac{v_n - v_L}{R_2} + i_n = 0.$$

The positive input terminal is connected to ground (to the reference node), so $v_p = 0$. From the property (7.13) of an ideal op amp, $v_n = v_p$, so $v_n = 0$. Thus the equation above reduces to

$$-\frac{v_S}{R_1} - \frac{v_L}{R_2} + i_n = 0.$$

From (7.14), $i_n = 0$. It follows that the transfer characteristic is

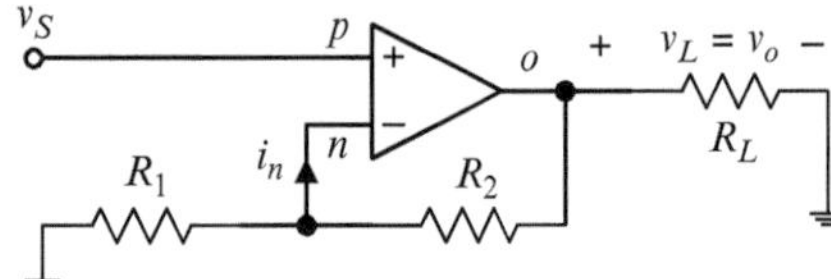

Fig. 7.14 See Example 7.3

$$v_L = -\frac{R_2}{R_1} v_S$$

and the gain is given by

$$A_{v\,dc} = \left|\frac{V_L}{V_S}\right| = \frac{R_2}{R_1}.$$

The circuit in Fig. 7.13 is called an **inverting amplifier** or an **inverter** because the input v_S and the output v_L have opposite signs. Note that *the voltage gain is independent of the load resistance* R_L but depends upon the resistance R_1. In many applications, the resistance R_1 equals the resistance of an external resistor in series with the source resistance. Usually, the resistance of the external resistance R_1 is much larger than the source resistance, in which case the voltage gain is essentially independent of both the source resistance and the load resistance. In such cases, the inverting amplifier is a good voltage amplifier.

Example 7.3. Obtain an expression for the voltage transfer characteristic for the circuit shown in Fig. 7.14. The op amp is ideal.
Solution: By Kirchhoff's current law,

$$\frac{v_n}{R_1} + \frac{v_n - v_L}{R_2} + i_n = 0.$$

From (7.13), $v_n = v_p = v_S$. From (7.14), $I_n = 0$. It follows that

$$\frac{v_S}{R_1} + \frac{v_S - v_L}{R_2} = 0 \Rightarrow v_L = \left(1 + \frac{R_2}{R_1}\right) v_S. \quad (7.15)$$

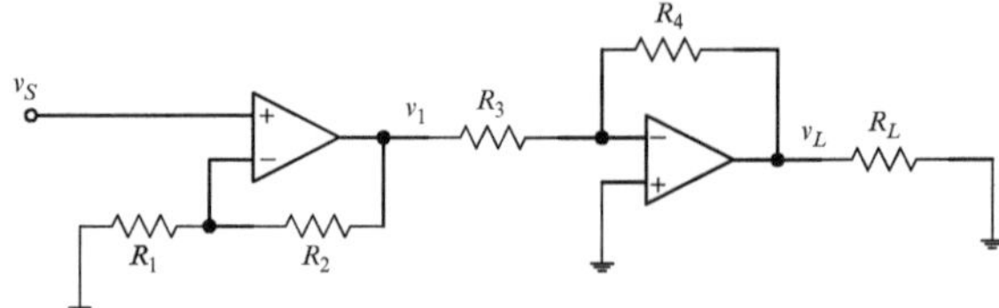

Fig. 7.15 See Exercise 7.3

The circuit in Fig. 7.14 is called a **non-inverting amplifier** because the input and the output have the same sign (the transfer characteristic is positive). Note that *the voltage transfer characteristic is independent of the load resistance* R_L. No source resistance is shown in the circuit diagram, but including same would not matter, because no current enters the positive terminal of the op amp; i.e., there would be no drop across the source resistance and the relation $v_p = v_S$ would remain true.

Because the voltage transfer characteristic for an inverting or non-inverting amplifier is independent of the load on the amplifier, the overall voltage transfer characteristic of two or more such amplifiers in cascade equals the product of the transfer characteristics of the individual amplifiers.

Exercise 7.3. Obtain an expression for the voltage transfer characteristic for the circuit shown in Fig. 7.15.

Example 7.4. Obtain an expression for the voltage transfer characteristic for the circuit shown in Fig. 7.16. The op amp is ideal.
Solution: By inspection, $v_L = v_n$. From (7.13), $v_n = v_p$. Again by inspection, $v_p = v_S$. Thus

$$v_L = v_S.$$

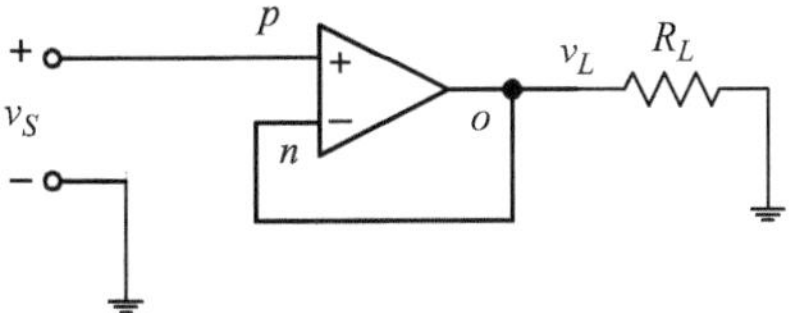

Fig. 7.16 See Example 7.4

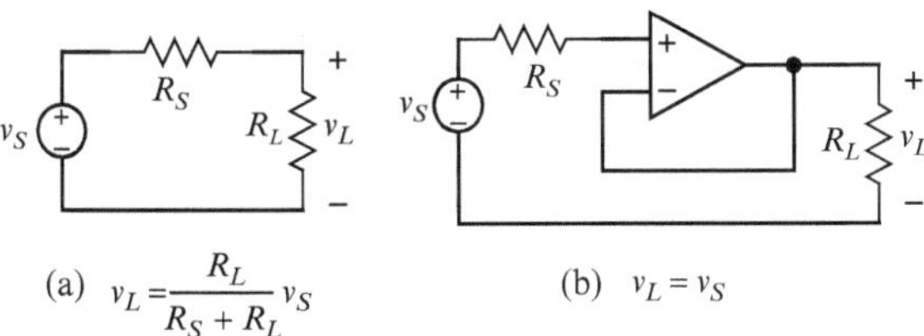

Fig. 7.17 Using a follower to buffer a load

The circuit in Fig. 7.16 is called a **voltage follower** because the output voltage equals (follows) the input voltage. Here again, the voltage transfer characteristic is independent of the load R_L. Would including a source resistance affect the result above?

A follower usually is used to provide a large input resistance and a small output resistance. For example, suppose a fixed load is to be driven by a fixed source, as illustrated by Fig. 7.17. If we drive the load directly with the source, as in Fig. 7.17(a), then the load voltage is given by

$$v_L = \frac{R_L}{R_L + R_S} v_S.$$

The voltage transferred to the load depends upon the load and is less than that available from the source. The reduction in available voltage can be significant if the source output resistance R_S and load input resistance R_L are comparable. But if we insert a follower between the source and the load, as in Fig. 7.17(b), the load voltage equals the source (available) voltage. A follower used as described above and illustrated by Fig. 7.17(b) often is called a **buffer**.

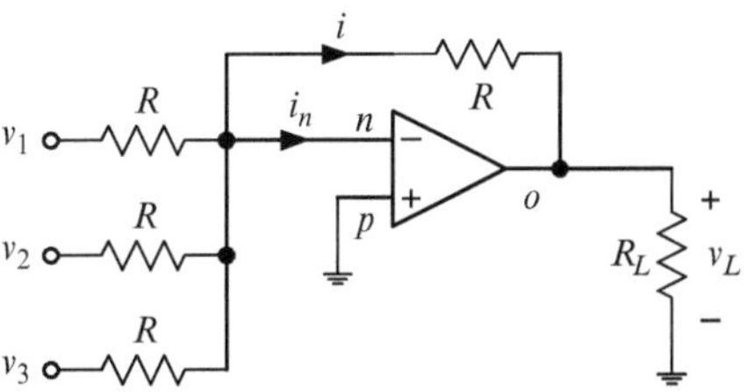

Fig. 7.18 See Example 7.5

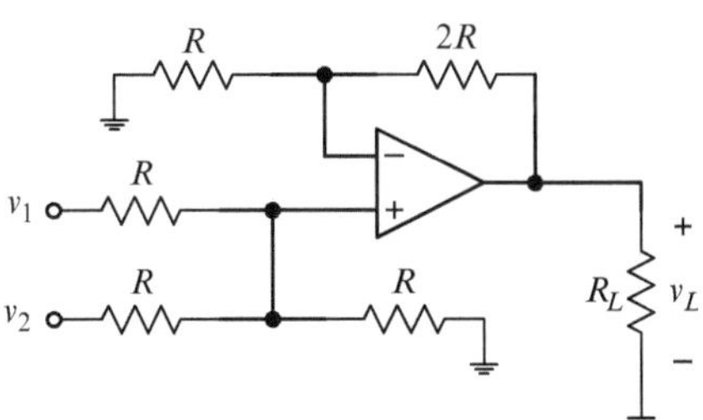

Fig. 7.19 See Exercise 7.4

Example 7.5. Refer to Fig. 7.18. Assume the op amp is ideal and express the load voltage v_L in terms of the inputs v_1, v_2, v_3 and the resistance R.

Solution: Because $v_n = v_p = 0$ and $i_n = 0$, we have from Kirchhoff's current law that

$$i = \frac{v_1}{R} + \frac{v_2}{R} + \frac{v_3}{R}.$$

It follows that

$$v_L = -R\,i = -(v_1 + v_2 + v_3).$$

The circuit is an **inverting adder**. The output is independent of the load. If the resistance R is much larger than the largest of the source resistance, the output also is independent of the source resistances.

Exercise 7.4. Refer to Fig. 7.19. Express the load voltage v_L in terms of the inputs v_1, v_2 and the resistance R.

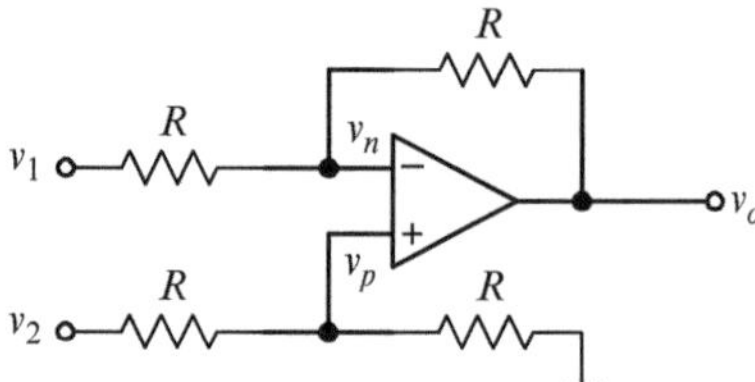

Fig. 7.20 See Example 7.6

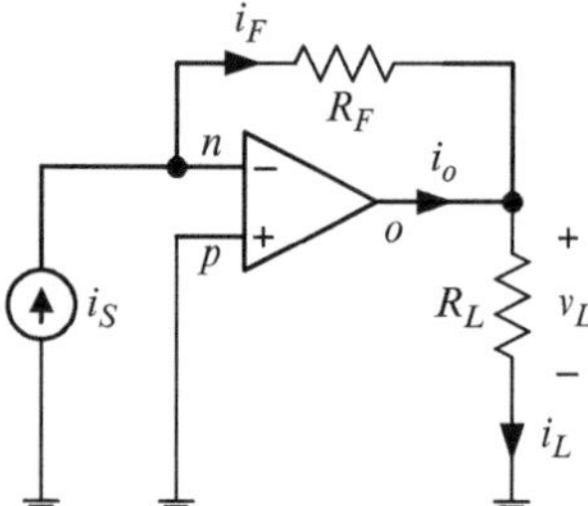

Fig. 7.21 See Example 7.7

Example 7.6. Refer to Fig. 7.20. Assume the op amp is ideal and express the output v_o as a function of the inputs v_1, v_2 and the resistance R.
Solution: By Kirchhoff's current law,

$$\frac{v_n - v_1}{R} + \frac{v_n - v_o}{R} = 0 \Rightarrow v_o = 2v_n - v_1. \quad (7.16)$$

By voltage division,

$$v_p = \frac{v_2}{2}.$$

The op amp is ideal, so $v_n = v_p$ and (7.16) gives

$$v_o = v_2 - v_1.$$

The circuit in Fig. 7.20 is an example of a **difference amplifier**. The output is independent of the load. If the resistance R is much larger than the largest of the source resistance, the output also is independent of the source resistances.

Example 7.7. Refer to Fig. 7.21. Assume the op amp is ideal. Express the load voltage v_L and the load current in terms of the source current i_S and the resistances R_F and R_L.
Solution: Because the op amp is ideal $v_n = v_p = 0$, and the current entering the negative input terminal equals zero. Thus $i_F = i_S$ and

$$v_L = v_o = -R_F i_S.$$

The circuit is a **current-to-voltage converter**; that is, the output voltage is proportional to the input current. The circuit also could be called a **transresistance amplifier**, because the transfer characteristic relating output voltage to input current is real and independent of the load resistance (so long as $R_L > 0$).

By Ohm's law,

$$i_L = \frac{v_L}{R_L} = -\frac{R_F}{R_L} i_S.$$

The current-to-current transfer characteristic is not independent of the load, so the circuit is not a good current amplifier. Can we obtain expressions for the transconductance and voltage gain of this circuit? Show how or explain why not.

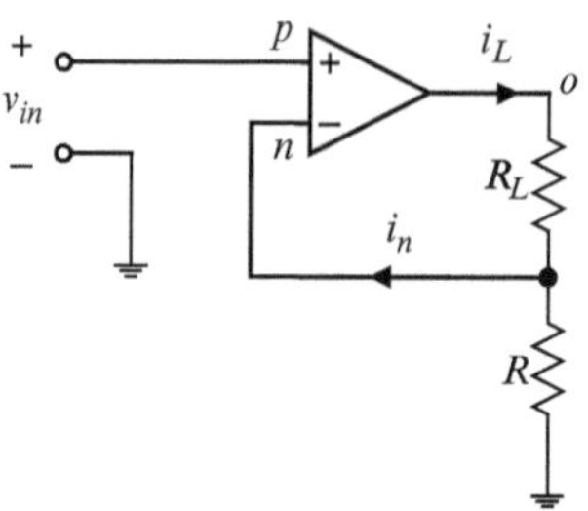

Fig. 7.22 See Example 7.8

Example 7.8. Refer to Fig. 7.22, where the op amp is ideal. Obtain an expression for the load current i_L in terms of the input voltage v_{in} and the circuit parameters. Thus show that the load current is independent of the load resistance R_L.

Solution: The op amp is ideal, so $v_n = v_p = v_{in}$. It follows that the voltage across the resistance R also equals the input voltage v_{in}. By Kirchhoff's current law,

$$i_L - i_n = \frac{v_{in}}{R}.$$

The op amp is ideal, so $i_n = 0$, and

$$i_L = \frac{v_{in}}{R}.$$

The load current is independent of the load resistance R_L. The circuit is a **voltage-to-current converter** or **transconductance amplifier**. A disadvantage of this particular circuit is the fact that the load floats; i.e., neither end of the load shares a ground with the input. This can be a fatal disadvantage if the load is another circuit sharing the same power supply. It might not be a problem if the load is a passive terminal load, such as a loudspeaker.

The ideal model for an op amp can be used to analyze and design a number of important circuits. Moreover, it is difficult to use other than the ideal model for initial design because more refined models (and the resulting equations) contain many parameters. In most cases, *initial (pencil-and paper) design of a linear circuit using op amps essentially requires use of the ideal model*. Subsequently, the design can be refined, if necessary, using one of several circuit simulation programs and more complete models.

7.4 Feedback and Stability of Op-Amp Circuits

Because the intrinsic dc gain μ_0 of an op amp is so large, the slightest electrical disturbance at the input terminals is sufficient to drive the output to saturation (to $\pm V_{CC}$). Consequently, op amps are almost always used in *feedback configurations*, where one or more current paths are provided from the output terminal to one or both input terminals. Feedback from the output to the positive (p) terminal is called **positive feedback** and feedback from the output to the negative *(n)* terminal is called **negative feedback**. Positive feedback drives the input voltage v_{pn} in the same direction as that taken by the output, and can cause the op amp to saturate. Negative feedback drives the input voltage V_{pn} in the opposite direction of that taken by the output; i.e., if the output increases, V_{pn} decreases, and vice versa. Negative feedback reduces the gain of an op amp to usable values by continuously forcing the input of the op amp toward zero. The intrinsic dc voltage gain of an op amp also is called the **open-loop dc voltage gain** because it is the voltage gain of the op amp in the absence of feedback; i.e., in the absence of a *closed loop* connecting the input to the output.

The feedback is negative in all of the examples in Section 7.3, which illustrate three important virtues of feedback:

- Within the accuracy of the approximations employed, and as a result of negative feedback, *the overall voltage gain is determined by parameters (resistances) external to the op amp*. Because the external resistances can be specified precisely, the transfer characteristic of the feedback amplifier can be specified precisely, even if the values of the op amp intrinsic dc gain μ_0, input resistance R_i, and output resistance R_o are not known precisely, so long as the intrinsic gain μ_0 is large enough.
- The circuit transfer characteristic is linear, *even if the op amp itself is nonlinear* in the range $|v_{out}| < V_{CC}$ (again, provided the intrinsic gain μ_0 is sufficiently large throughout the range of possible input amplitudes).
- Without feedback, an op amp is almost useless. Feedback makes an op amp a useful and widely used circuit element.

Some circuits use a combination of negative and positive feedback. Such feedback wields a two-edged sword, because positive feedback can make an op amp circuit unstable, in which case the output either oscillates uncontrollably or locks up (saturates) at one or the other of the supply voltages. A familiar example of unstable (oscillatory) operation of an amplifier is the loud squeal issued by a public-address system if the microphone gets too close to the loudspeaker, in which case the acoustic power delivered by the loudspeaker to the microphone is sufficient to initiate the oscillation. The frequency of the oscillation is determined in part by the distance (delay time) between the loudspeaker and microphone.

A purely resistive circuit cannot oscillate, so the output of an unstable resistive op-amp circuit will be either $+V_{CC}$ or $-V_{CC}$, depending upon the sign of the voltage V_{pn}. Below, we give a condition for stable operation of a resistive op amp circuit that incorporates some positive feedback.

To prevent saturation, feedback in an op amp circuit must continuously drive the voltage $v_{pn} = v_p - v_n$ toward zero.[7] Loosely, this means that the voltage fed back from the output terminal to the negative input terminal must exceed that fed back to the positive input terminal. For linear operation, the *net feedback must be negative*. Otherwise, the effect of feedback is to drive the voltage v_{pn} away from zero, and the amplifier will simply saturate (or *lock up*) at one or the other of the supply voltages. Such saturation in an unstable circuit is inevitable, even if no input is applied to the circuit, because of inherent imbalance between the input terminals of an op amp and also because of ever-present electrical noise.

In examples in Section 7.3 above, resistive feedback is presented only to the negative input terminal of the op amp, which ensures stable operation. If feedback also is presented to the positive input terminal, we must ensure that the *net* feedback is negative. If the feedback is negative and the circuit operates linearly, we say the circuit is **stable**. If the feedback is positive, then the output is not linearly dependent upon the input and the circuit is **unstable**. To illustrate conditions for stable operation, we consider the circuit shown in Fig. 7.23, which incorporates both negative and positive feedback. The output voltage is given by

$$v_{out} = \mu_0 (v_p - v_n) = \mu_0 \, v_{pn}. \tag{7.17}$$

According to (7.17), the output voltage responds instantly (in zero time) to the change in the voltage v_{pn}. In a physical circuit, there is some time delay Δt,

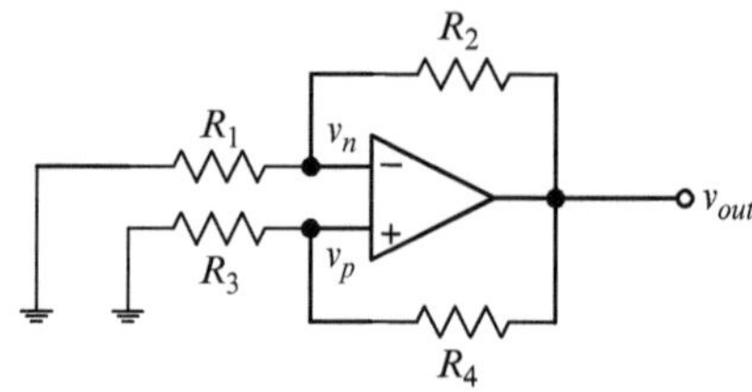

Fig. 7.23 Circuit referred to in obtaining the condition for stable operation

albeit small, between a change in the input v_{pn} and the resulting change in the output v_{out}. A more realistic input-output relation is

$$v_{out}(t + \Delta t) = \mu_0 \, v_{pn}(t). \tag{7.18}$$

In words, a change in the input v_{pn} at time t produces a change in the output v_{out} at a slightly later time $t + \Delta t$. To examine the stability of the circuit, we imagine that some small disturbance, such as electrical noise, gives rise to a momentary, non-zero value for the input v_{pn}. No current enters the op-amp input terminals, so we may use voltage division to obtain

$$\begin{aligned} v_p &= \frac{R_3}{R_3 + R_4} v_{out} = a_p \, v_{out}, \\ v_n &= \frac{R_1}{R_1 + R_2} v_{out} = a_n \, v_{out}, \end{aligned} \tag{7.19}$$

where the definitions of a_p, a_n are evident. Thus $v_{pn} = (a_p - a_n) \, v_{out}$ and (7.18) becomes

$$v_{out}(t + \Delta t) = \mu_0 \, (a_p - a_n) \, v_{out}(t). \tag{7.20}$$

Retaining only the first two terms of Taylor's expansion for the left side of (7.20) yields

$$v_{out} + \frac{dv_{out}}{dt} \Delta t = \mu_0 \, (a_p - a_n) \, v_{out},$$

which simplifies to

$$\frac{dv_{out}}{dt} = \alpha \, v_{out}; \quad \alpha = \frac{\mu_0 \, (a_p - a_n) - 1}{\Delta t}. \tag{7.21}$$

The solution of the differential equation (7.21) is

$$v_{out} = C \, e^{\alpha t}, \tag{7.22}$$

[7]Again, the gain of an op amp is so large that if the voltage $v_p - v_n$ is greater than μ_0/V_{CC}, the amplifier will saturate.

as can be verified by substituting the right side of (7.22) for v_{out} in (7.21). The value of the constant C depends upon the initial state of the op amp; i.e., from (7.22), $C = v_{out}(0)$, which is non-zero if for no other reason than ever-present electrical noise. The actual value of C is irrelevant. It is sufficient to note that if $\alpha > 0$, the magnitude of the output increases exponentially with time until the output reaches one of the saturation limits $\pm V_{\max}$. In other words, the circuit is unstable for $\alpha > 0$. But if $\alpha < 0$, the output goes exponentially toward zero, and the circuit is stable. Because μ_0 and Δt are both positive, the circuit in Fig. 7.23 is stable if

$$\alpha < 0 \Rightarrow \frac{\mu_0 (a_p - a_n) - 1}{\Delta t} < 0 \Rightarrow a_p < a_n + \frac{1}{\mu_0}. \quad (7.23)$$

It follows from (7.23) and the definitions of a_p, a_n in (7.19) that the requirement for stable operation of the circuit in Fig. 7.23 is

$$\frac{R_3}{R_3 + R_4} < \frac{R_1}{R_1 + R_2} + \frac{1}{\mu_0}. \quad (7.24)$$

The number $1/\mu_0$ is in the neighborhood of 10^{-5}–10^{-7}. As a practical matter, we may say that for stable, linear operation of an op-amp circuit, *the voltage fed back to the negative input terminal must* ***exceed*** *the voltage fed back to the positive input terminal.* Although demonstrated above for a specific circuit, this condition holds for linear resistive op-amp circuits in general.

It might appear that we could have based the development above on (7.17) and avoided using a differential equation to obtain the condition (7.24). Such an argument would be flawed because (7.17) is an *instantaneous* relation in which v_{pm} and v_{out} have always the same sign (because μ_0 is positive).

Example 7.9. Obtain the voltage transfer characteristic for the circuit shown in Fig. 7.24. The op amp is ideal.
Solution: Assuming $v_p = v_n = 0$ and proceeding exactly as in Example 7.2, we obtain

$$\frac{v_S}{R_1} + \frac{v_L}{R_2} = 0 \Rightarrow \frac{v_L}{v_S} = -\frac{R_2}{R_1} \;??$$

However, this result is meaningless (and incorrect) because the net feedback is to the positive input terminal of the op amp and the circuit is unstable. The output would immediately saturate at one or the other of the supply voltages.

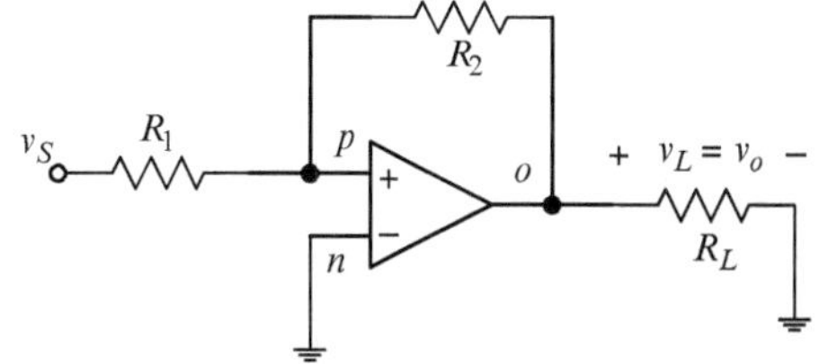

Fig. 7.24 See Example 7.9

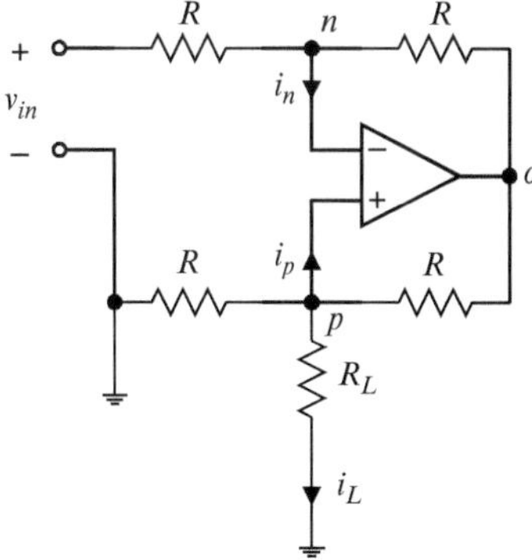

Fig. 7.25 See Example 7.10

Example 7.10. Refer to Fig. 7.25, where the op amp is ideal. (a) Obtain an expression for the load current i_L in terms of the input voltage v_{in} and circuit parameters. Thus show that the load current is independent of the load resistance. (b) Show that the circuit is stable.
Solution: (a) Because the op amp is ideal, $i_n = i_p = 0$ and $v_n = v_p$. Let $v_n = v_p = v$. By Kirchhoff's current law,

$$\frac{v - v_{in}}{R} + \frac{v - v_o}{R} = 0 \Rightarrow v_o = 2v - v_{in},$$

$$\frac{v}{R} + \frac{v}{R_L} + \frac{v - v_o}{R} = 0 \Rightarrow v_o = \left(2 + \frac{R}{R_L}\right) v.$$

Equate the expressions on the right above (eliminate v_0) to obtain

$$\left(2+\frac{R}{R_L}\right)v = 2v - v_{in} \Rightarrow v = -\frac{R_L}{R}v_{in}$$

$$\Rightarrow i_L = \frac{v}{R_L} = -\frac{v_{in}}{R}.$$

So the load current is independent of the load resistance. The circuit is a **voltage-to-current converter** or **transconductance amplifier**. Unlike the circuit in Example 7.8, the load does not float. But the relation $i_L = -v_{in}/R$ depends upon the four resistors being identical, which can be difficult to achieve in practice.

(b) Based on (7.23), we wish to show that $a_p < a_n$ or, equivalently, that $a_p - a_n < 0$. For $v_{in} = 0$, Kirchhoff's current law gives

$$\frac{v_n}{R} + \frac{v_n - v_o}{R} = 0 \Rightarrow v_n = \frac{1}{2}v_o,$$

$$\frac{v_p}{R} + \frac{v_p}{R_L} + \frac{v_p - v_o}{R} = 0 \Rightarrow v_p = \frac{R_L}{2R_L + R}v_o,$$

so

$$a_n = \frac{1}{2}; \quad a_p = \frac{R_L}{2R_L + R}$$

and

$$a_p - a_n = \frac{-R}{2(2R_L + R)} < 0, \quad R_L \text{ finite},$$

and the circuit is stable, provided R_L is finite. If $R_L \to \infty$ (open circuit), the circuit is unstable because $\lim_{R_L \to \infty}(a_p - a_n) = 0$.

7.5 Input Resistance and Output Resistance of Op-Amp Circuits

As discussed in Chapter 6, the input resistance and output resistance of a two-port circuit can be determined (mathematically) as shown in Fig. 7.26. Other approaches are possible; for example, we can use current sources instead of voltage sources as the test sources. Also, we can determine the output resistance by finding the Thévenin equivalent resistance at the output terminals.

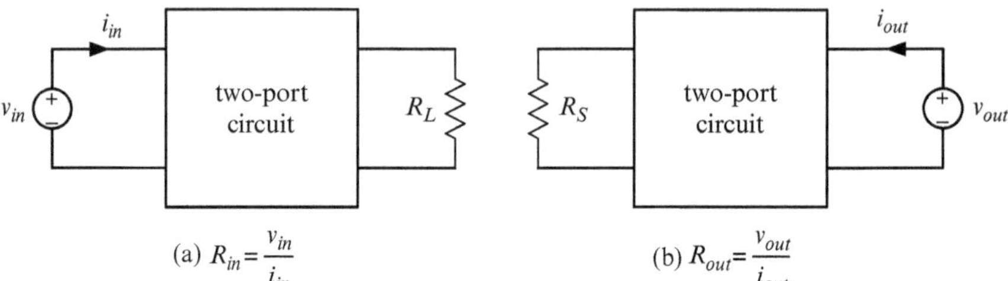

Fig. 7.26 Definitions of input resistance and output resistance of a two-port circuit

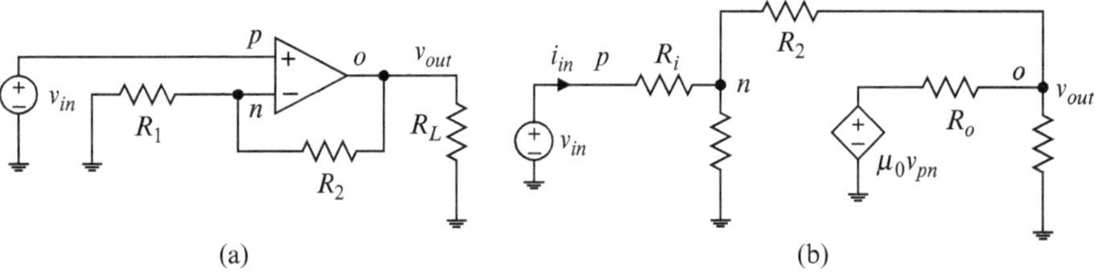

Fig. 7.27 See Example 7.11

Example 7.11. (a) Obtain an expression for the input resistance of the non-inverting amplifier shown in Fig. 7.27(a). Do not assume the op amp is ideal (use the dc linear model). Show that for $\mu_0 \gg R_2/R_1 \gg 1$, the input resistance is given to a good approximation by

$$R_{in} \cong \frac{\mu_0}{A_v} R_i,$$

where A_v is the voltage gain of the non-inverting amplifier.

(b) For a typical op amp and non-inverting amplifier circuit, $\mu_0 \gg 1, R_i \gg R_1, R_L \gg R_o, R_1 \gg R_o$, and $\mu_0 R_1 \gg R_2$. Assume these inequalities hold and simplify the expression obtained in part (a).

Solution: (a) We replace the op amp with the linear dc model to obtain the circuit diagram shown in Fig. 7.27(b), where R_i, R_o, μ_0 are the input resistance, output resistance and intrinsic dc gain, respectively, of the op amp.

Applying Kirchhoff's current law to the *n* and *o* nodes in Fig. 7.27(b) gives

$$\frac{v_n - v_{in}}{R_i} + \frac{v_n}{R_1} + \frac{v_n - v_o}{R_2} = 0 \Rightarrow \left(\frac{1}{R_i} + \frac{1}{R_1} + \frac{1}{R_2}\right) v_n - \frac{1}{R_2} v_o = \frac{1}{R_i} v_{in},$$

$$\frac{v_o - v_n}{R_2} + \frac{v_o - \mu_0(v_{in} - v_n)}{R_o} + \frac{v_o}{R_L} = 0 \Rightarrow \left(\frac{\mu_0}{R_o} - \frac{1}{R_2}\right) v_n + \left(\frac{1}{R_o} + \frac{1}{R_L} + \frac{1}{R_2}\right) v_o = \frac{\mu_0}{R_o} v_{in}.$$

Eliminating v_o from these equations yields

$$v_n = \frac{[R_L(\mu_0 R_i + R_o) + R_2(R_o + R_L)] R_1 v_{in}}{(\mu_0 + 1) R_L R_i R_1 + (R_1 + R_i)[R_2(R_L + R_o) + R_o R_L] + R_i R_1 R_o}.$$

The input resistance of the non-inverting amplifier is given by

$$R_{in} = \frac{v_{in}}{i_{in}} = \frac{v_{in}}{(v_{in} - v_n)/R_i} = \frac{v_{in} R_i}{v_{in} - v_n}, \tag{7.25}$$

which, using the expression above for v_n, becomes

$$R_{in} = \frac{(\mu_0 + 1) R_L R_i R_1 + (R_1 + R_i)[R_2(R_L + R_o) + R_o R_L] + R_i R_1 R_o}{R_2(R_o + R_L) + R_L(R_o + R_1) + R_1 R_o}.$$

(b) If $\mu_0 \gg 1, R_i \gg R_1, R_L \gg R_o, R_1 \gg R_o$ and $\mu_0 R_1 \gg R_2$, then the expression above for R_{in} reduces to

$$R_{in} \cong \frac{\mu_0 R_L R_1 + R_2 R_L + R_o R_L + R_1 R_o}{R_2 R_L + R_L R_1 + R_1 R_o} R_i \cong \frac{\mu_0 R_1 + R_2}{R_2 + R_1} R_i \cong \frac{\mu_0 R_1}{R_2 + R_1} R_i = \frac{\mu_0}{A_v} R_i, \tag{7.26}$$

where $A_v = 1 + R_2/R_1$ is the voltage gain of the non-inverting amplifier. Note that under the conditions imposed in this example, the input resistance of the amplifier is independent of the load resistance.

Exercise 7.5. Repeat Example 7.11, but using a current source as the test input.

Example 7.12. Obtain an expression for the output resistance of the non-inverting amplifier considered in Example 7.11. Do not assume the op amp is ideal (use the dc linear model). Show that under reasonable assumptions (see Example 7.11) the output resistance of the non-inverting amplifier is given to a good approximation by

$$R_{out} = \frac{A_v}{\mu_0 + A_v} R_o,$$

where $A_v = 1 + R_2/R_1$ is the voltage gain of the non-inverting amplifier.

Solution: We attach a source resistance R_S to the input and attach a voltage source to the output, as shown in Fig. 7.28. We seek an expression for $R_{out} = v_{out}/i_{out}$.

Applying Kirchhoff's current law and $v_{pn} = v_p - v_n$ gives

$$\frac{v_{out} - \mu_0(v_p - v_n)}{R_o} - i_{out} + \frac{v_{out} - v_n}{R_2} = 0,$$

$$\frac{v_n}{R_i'} + \frac{v_n}{R_1} + \frac{v_n - v_{out}}{R_2} = 0,$$

where $R_i' = R_i + R_S$. By voltage division,

$$v_p = \frac{R_S}{R_i'} v_n.$$

Using the latter two equations to eliminate v_n and v_p from the first equation yields

$$R_{out} = \frac{v_{out}}{i_{out}} = \frac{R_2 R_o R_a^2}{(R_2 + R_o)R_a^2 - \mu_0 R_1 R_2 R_S + R_1 R'_i(\mu_0 R_2 - R_o)},$$

where $R_a^2 = R_1 R_2 + R_1 R_i' + R_2 R_i'$. For a typical op amp and non-inverting amplifier circuit, $R_i \gg R_S \Rightarrow R_i' \cong R_i$, $R_i \gg R_1, R_i \gg R_2$, and $R_2 \gg R_o$. If these inequalities hold, then $R_a^2 \cong R_i(R_1 + R_2)$ and the expression above for R_{out} reduces to

$$R_{out} \cong \frac{R_2 R_o R_i (R_1 + R_2)}{R_2 R_i (R_1 + R_2) + \mu_0 R_1 R_2 (R_i - R_S)} \cong \frac{R_o(R_1 + R_2)}{R_1 + R_2 + \mu_0 R_1} = \frac{A_v}{\mu_0 + A_v} R_o.$$

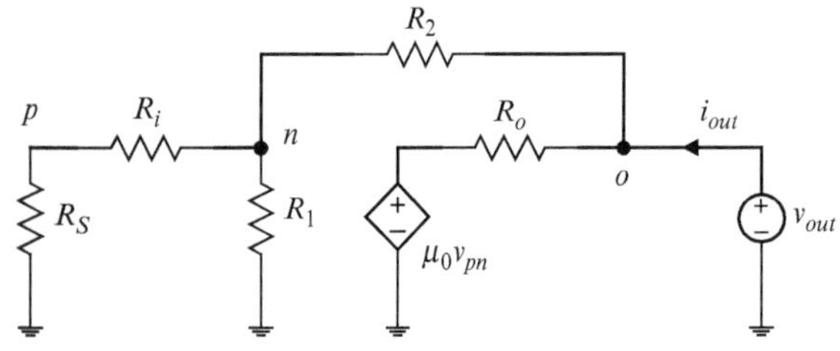

Fig. 7.28 See Example 7.12

Exercise 7.6. Repeat Example 7.12, but find the output resistance by finding the Thévenin equivalent at the output terminals (load removed) with an ideal independent voltage source attached to the input. You may assume the approximations used in Example 7.12 are valid.

Exercise 7.7. Refer to Fig. 7.27. The parameters of a certain non-inverting amplifier are $R_1 = 1\,\text{k}\Omega$, $R_2 = 100\,\text{k}\Omega$, $R_L = 5\,\text{k}\Omega, R_S = 100\,\Omega$. The op amp has input resistance $R_i = 10\,\text{M}\Omega$, output resistance $R_o = 80\,\Omega$, and intrinsic dc voltage gain $\mu_0 = 10^7$. Compute the output resistance and input resistance of the amplifier using both the exact and approximate expressions obtained in Example 7.11 and Example 7.12 above.

7.6 Properties of Common Op-Amp Circuits

Voltage amplification and buffering are common needs, so it is useful to tabulate and compare the essential properties of inverting amplifiers, non-inverting amplifiers, and voltage followers. Equations (7.28–7.36) below give exact (dc), approximate, and limiting ($\mu_0 \to \infty$) expressions for the transfer characteristic, input resistance, and output resistance for each circuit. We omit the derivations, which are straightforward but tedious. In the expressions that follow,

$$R_{1S} = R_1 + R_S; \quad R_{iS} = R_i + R_S. \tag{7.27}$$

Exercise 7.8. Show that the input resistance of the current-to-voltage converter treated in Example 7.7 is given (approximately) by $R_{in} \cong R_F/\mu_0$.

7.6.1 Inverting Amplifier

See Fig. 7.29.

7.6.1.1 Voltage Transfer Characteristic

$$v_{out} = \frac{(R_o - \mu_0 R_2) R_i R_L v_{in}}{(\mu_0 + 1) R_{1S} R_i R_L + (R_o R_L + R_o R_2 + R_L R_2)(R_i + R_{1S}) + R_o R_{1S} R_i} \underset{\mu_0 \to \infty}{\longrightarrow} -\frac{R_2}{R_{1S}} v_{in}. \tag{7.28}$$

7.6.1.2 Input Resistance

$$R_{in} = \frac{(\mu_0 + 1) R_L R_i R_1 + (R_1 + R_i)(R_L + R_o) R_2 + [(R_1 + R_i) R_L + R_i R_1] R_o}{(\mu_0 + 1) R_L R_i + (R_L + R_o) R_2 + (R_L + R_i) R_o} \underset{\mu_0 \to \infty}{\longrightarrow} R_1. \tag{7.29}$$

7.6.1.3 Output Resistance

$$\begin{aligned} R_{out} &= \frac{(R_i R_2 + R_{1S} R_2 + R_{1S} R_i) R_o}{(R_o + R_2)(R_i + R_{1S}) + (\mu_0 + 1) R_{1S} R_i} \\ &\cong \frac{(R_2 + R_{1S}) R_o}{R_2 + R_{1S} + \mu_0 R_{1S}} = \frac{1 + A_v}{1 + A_v + \mu_0} R_o \end{aligned} \tag{7.30}$$

If $\mu_0 \to \infty$, then

$$R_{out} \to 0. \tag{7.31}$$

If $\mu_0 \gg A_v \gg 1$, then

$$R_{out} \cong \frac{A_v}{\mu_0} R_o. \tag{7.32}$$

7.6.2 Non-inverting Amplifier

See Fig. 7.30.

7.6.2.1 Voltage Transfer Characteristic

$$v_{out} = \frac{[\mu_0 R_{iS}(R_1 + R_2) + R_o R_1] R_L v_{in}}{(\mu_0 + 1) R_1 R_{iS} R_L + (R_o R_L + R_o R_2 + R_L R_2)(R_{iS} + R_1) + R_o R_1 R_{iS}} \underset{\mu_0 \to \infty}{\longrightarrow} \left(1 + \frac{R_2}{R_1}\right) v_{in}. \tag{7.33}$$

7.6.2.2 Input Resistance

$$R_{in} = \frac{(\mu_0 + 1) R_L R_i R_1 + (R_1 + R_i)[R_2(R_L + R_o) + R_o R_L] + R_i R_1 R_o}{R_2(R_o + R_L) + R_L(R_o + R_1) + R_1 R_o} \cong \frac{\mu_0}{A_v} R_i \underset{\mu_0 \to \infty}{\longrightarrow} \infty;\ A_v = 1 + \frac{R_2}{R_1}. \tag{7.34}$$

7.6.2.3 Output Resistance

$$R_{out} = \frac{(R_1 R_2 + R_2 R_{iS} + R_1 R_{iS}) R_o}{(R_o + R_2)(R_{iS} + R_1) + (\mu_0 + 1) R_1 R_{iS}} \cong \frac{A_v}{\mu_0 + A_v} R_o \cong \frac{A_v}{\mu_0} R_o \underset{\mu_0 \to \infty}{\longrightarrow} 0;\ \mu_0 \gg A_v = 1 + \frac{R_2}{R_1} \gg 1. \tag{7.35}$$

7.6.3 Voltage Follower

See Fig. 7.31.

7.6.3.1 Voltage Transfer Characteristic

$$v_{out} = \frac{(R_o + \mu_0 R_{iS}) R_L v_{in}}{(\mu_0 + 1) R_{iS} R_L + R_{iS} R_o + R_L R_o} \underset{\mu_0 \to \infty}{\longrightarrow} v_{in}. \tag{7.36}$$

7.6.3.2 Input Resistance

$$R_{in} = \frac{(\mu_0 + 1) R_L R_i + R_i R_o + R_o R_L}{R_L + R_o} \cong \mu_0 R_i \underset{\mu_0 \to \infty}{\longrightarrow} \infty. \tag{7.37}$$

7.6.3.3 Output Resistance

$$R_{out} = \frac{R_{iS} R_o}{(\mu_0 + 1) R_{iS} + R_o} \cong \frac{R_o}{\mu_0} \underset{\mu_0 \to \infty}{\longrightarrow} 0. \tag{7.38}$$

Table 7.1 summarizes properties given above for circuits in which the op amp is assumed to be ideal.

Exercise 7.9. In Figs. 7.29 and 7.30, let $R_S = 100\Omega$, $R_1 = 10\text{k}\Omega$, $R_2 = 1\text{M}\Omega$, $R_L = 5\text{k}\Omega$, $R_i = 10\text{M}\Omega$, $R_o = 100\Omega$, and $\mu_0 = 10^6$ and calculate the voltage gain, input resistance, and output resistance of each circuit using (a) the exact expressions and (b) the approximate expressions given in Table 7.1.

Table 7.1 Properties of common op-amp circuits for $\mu_0 \to \infty$. See (7.28–7.38)

Circuit	Voltage Transfer Ratio H_v	Input Resistance	Output Resistance
Inverting amplifier	$-R_2/R_{1S}$	R_1	$A_v R_o/\mu_0 \to 0$
Non-inverting amplifier	$1 + R_2/R_1$	$R_i \mu_0/A_v \to \infty$	$A_v R_o/\mu_0 \to 0$
Follower	1	$\mu_0 R_i \to \infty$	$R_o/\mu_0 \to 0$

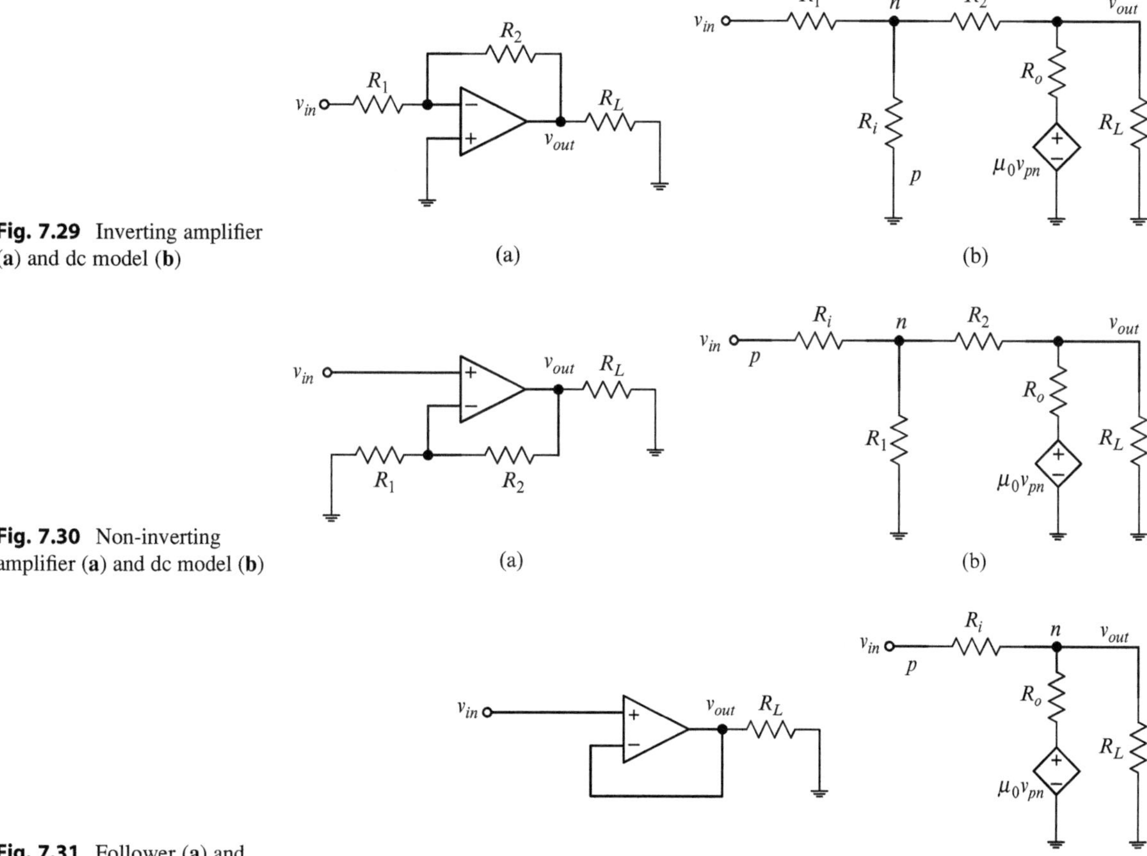

Fig. 7.29 Inverting amplifier (**a**) and dc model (**b**)

Fig. 7.30 Non-inverting amplifier (**a**) and dc model (**b**)

Fig. 7.31 Follower (**a**) and dc model (**b**)

A few of the differences between inverting and non-inverting amplifiers are:

- An inverting amplifier introduces a sign inversion. A non-inverting amplifier does not. This difference is seldom important.
- The input resistance of an inverting amplifier is approximately the resistance R_1, which is typically on the order of 10 kΩ or larger, depending upon the op amp in use. The input resistance of a non-inverting amplifier is larger than that of the op amp, which typically exceeds 1 MΩ and can be as large as 1 TΩ. However, in some applications, as discussed in Chapter 17, it is necessary to add a resistance R_X from the p input terminal to ground, in which case the input resistance is no larger than R_X.
- An inverting amplifier can provide a fractional gain (attenuation), whereas the gain of a non-inverting amplifier must exceed unity.[8]

[8]Achieving gains less than 10 with an inverting amplifier is attended by several problems, as discussed in Chapter 17.

Of these three differences, the difference in input resistance is most significant. However, even this advantage of a non-inverting amplifier disappears in many applications, as we show in a subsequent chapter. For various reasons, some of which we treat in subsequent chapters, inverting amplifiers are used more often than non-inverting amplifiers. If the relatively low input resistance of an inverting amplifier is a problem, the amplifier can be preceded by a voltage follower, which provides a very high input resistance.

For the same voltage gains (magnitudes), the inverting and non-inverting amplifiers have approximately the same output resistance. In a typical application, the voltage gain is on the order of 100. For a typical op amp, $\mu_0 \cong 10^6$ and $R_o \cong 100$, so the output resistances of typical inverting and non-inverting amplifiers are on the order of 10 mΩ. The output resistance of a voltage follower is on the order of 100 $\mu\Omega$. In most applications, the output resistances of inverting and non-inverting amplifiers are negligible compared to the load resistance.

The remarks above pertain only to the dc models. We treat ac models for op amps in Chapter 17, where we show that voltage gain, input resistance, and output resistance can all be strong functions of the frequency of a sinusoidal input. We also show how certain additional components required in practical op-amp circuits can alter the conclusions reached above.

7.7 Op Amp Structure and Properties

Op amps can be built using various kinds of transistors, such as bipolar junction transistors (BJT's), junction field-effect transistors (JFET's), and metal-oxide field-effect transistors (MOSFET's). The input resistances of op amps range from about 1 MΩ for BJT-based op amps to more than 1 TΩ for MOSFET-based op amps. Intrinsic gains range from about 10^5 to 10^6 for BJT-based op amps and from about 10^4 to 10^5 for MOSFET-based op amps.[9] Op amps using JFET's in the input stage have input resistances and intrinsic gains intermediate to those for BJT-based and MOSFET-based op amps.

As we show in Chapter 17, the input resistance of an op amp influences the choice of resistors external to an op amp in inverting or non-inverting configurations. With reference to Figs. 7.29 and 7.30, *a good rule of thumb is to make R_1 as large as possible but no more than than 1% of the op-amp input resistance.*

7.8 Output Current Limit

Almost all op amps incorporate internal circuitry that limits the output current to protect internal devices from damage in case the output terminals are shorted (or the op amp is overloaded). Typical limits range from about 1 mA to about 100 mA for small integrated general purpose op amps to more than 25 A for high-power op amps.

Because output current is limited, the voltage an op amp can impress on a load also is limited; e.g., any particular rail-to-rail op amp can provide the full supply voltage to a load only if the resulting load current does not exceed the limit imposed by the op amp. For example, National Semiconductor's LM8272 rail-to-rail op amp allows a maximum supply voltage of $\pm V_{CC\ \max} = \pm 12\,\text{V}$ and a maximum output current of $I_{out\ \max} = 100\,\text{mA}$. Consequently, the op amp cannot drive a resistive load having resistance less than $R_{L\ \min} = V_{CC\ \max}/I_{out\ \max} = 12/0.1 = 120\,\Omega$ and maintain full ($\pm$12V) output swing. If it is necessary to drive a load having resistance less than 120 Ω at full swing (± 12 V), we must use a different op amp. The voltage transfer characteristic shown in Fig. 7.7 assumes that the output current limit is not exceeded.

Example 7.13. Refer to Example 7.1. (a) What is the minimum acceptable output current limit for the op amp? (b) What is the minimum acceptable load resistance if the op-amp output current limit is 10 mA?
Solution: (a) Assume full-scale output (± 10 V) is possible. The maximum load current is approximately

$$i_L = \frac{10\,\text{V}}{5\,\text{k}\Omega} = 2\,\text{mA},$$

so the op amp need supply no more than 2 mA. Almost any op amp can satisfy this requirement. (b) If the op amp can provide 10 mA and if the supply voltage is 10 V, then (assuming rail-to-rail operation) the minimum load resistance is

$$\frac{10\,\text{V}}{10\,\text{mA}} = 1\,\text{k}\Omega.$$

The solutions above ignore the op amp output resistance, which is a negligible fraction of the load resistances considered.

7.9 Input Offset Voltage

Because of their internal structures, all op amps exhibit a slight imbalance such that an input must be

[9]This parameter is not specified on manufacturers' data sheets and is difficult to measure. The values given here are deduced from other performance data. Related topics are discussed in more detail in Chapter 17.

slightly different from zero to produce an output equal to zero. Thus the output of an op amp is given by

$$v_o = \mu_0 (v_{pn} - v_{pn0}),$$

where the (small) voltage v_{pn0} is called the **input offset voltage**. Input offset voltages range from about 100 μV to about 10 mV for general-purpose op amps.

The input offset voltage and the high dc gain of an op amp will cause a powered, open-loop op amp to saturate even if the input terminals are shorted (even if $v_{pn} = 0$). Most op amps provide terminals for nulling the input offset voltage; e.g., by connecting a variable resistor to the terminals with the wiper connected to one or the other of the supply voltages, as illustrated by Fig. 7.32. The terminals and recommended nulling circuitry vary from one op amp to another, and are given in manufacturers' data sheets.

High-precision applications may warrant the use of so-called precision op amps that have small offsets and low drift and provide for external nulling circuitry. Unfortunately, the input offset voltage varies with temperature and also tends to drift with time, even if the temperature is constant. Consequently, it is difficult to maintain a perfect null and we must rely on feedback to keep the offset from driving the output to saturation, which means that *any practical linear op-amp circuit must provide negative feedback at dc. For present purposes, this means that there must be a resistive path from the output (o) terminal to the negative (n) input terminal.* The resistance can be zero, as in a follower.

Even if the input offset voltage does not cause the output to saturate, it still causes a corresponding offset in the output voltage. For example, Fig. 7.33 shows a circuit model for the effect of the dc offset voltage in

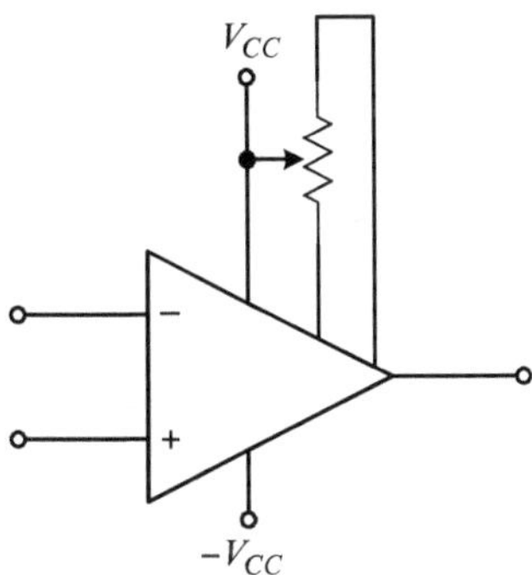

Fig. 7.32 Circuit for nulling the dc offset voltage (typical)

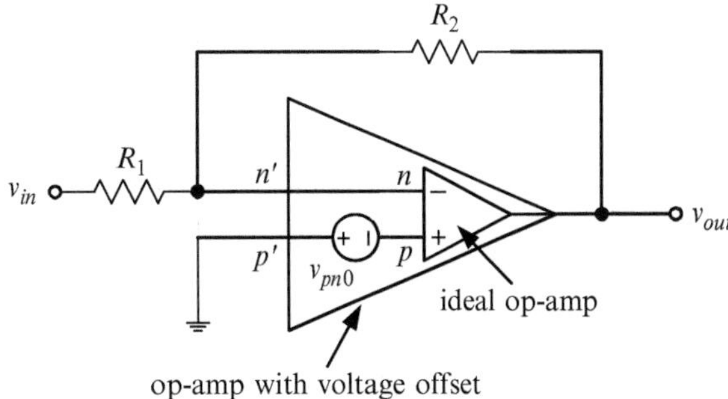

Fig. 7.33 Circuit model for input dc offset in an op amp, connected in an inverting configuration

an inverting amplifier. Straightforward application of Kirchhoff's current law shows that the output voltage is given by

$$v_{out} = -\frac{R_2}{R_1} v_{in} - \left(1 + \frac{R_2}{R_1}\right) v_{pn0},$$

where the second term on the right is the output offset. Whether the output offset is troublesome depends upon the magnitude of the offset relative to that of the desired output and upon the precision demanded in an application at hand.

7.10 Input Bias Currents

Input bias currents are small direct (constant) currents that enter (or exit) the input (*p* and *n*) terminals of a powered op amp. *Any circuit in which an op amp is imbedded must provide dc paths to ground from both the n and p input terminals.* Otherwise, one or both of the bias currents will be blocked, and the resulting input voltage will drive the op amp to saturation. Both input bias currents are the same direction and have approximately the same magnitude, which ranges from less than 1 nA to a few micro amperes. The direction and magnitude of the input bias currents depend upon the internal structure of the op amp.

Even if there is a dc path to ground from each input of an op-amp, the input bias currents will cause the voltage v_{pn} (and thus the output) to assume some nonzero value (offset) that is difficult to predict (because of the complexity of the input circuitry in a typical op amp), but which can be large enough to be troublesome, especially in high-precision applications. Below, we show how the offset can be reduced by a compensating resistor connected to one of the op-amp input terminals.

We first consider an inverting amplifier. Figure 7.34 shows an inverting amplifier, where the dc input bias currents are represented by currents I_n, I_p. We seek the output v_o due to the bias currents alone, so we set the dc offset voltage and the input v_S to zero (superposition). The purpose of the resistor R_X is to make $v_p = v_n$, such that the voltage v_o due to the bias currents equals zero.

Applying Kirchhoff's current law to the model shown in Fig. 7.34 gives

$$\frac{v_n}{R_1'} + I_n + \frac{v_n - v_o}{R_2} = 0, \quad \frac{v_p}{R_X} + I_p = 0 \Rightarrow v_p = -I_p R_X, \tag{7.39}$$

where

$$R_1' = R_1 + R_S. \tag{7.40}$$

We assume $I_p = I_n$ and require $v_p = v_n$. This gives

$$\begin{aligned} &\frac{-I_p R_X}{R_1'} + I_p + \frac{-I_p R_X - v_o}{R_2} = 0 \\ &\Rightarrow v_o = -\frac{R_X R_2 + R_X R_1' - R_1' R_2}{R_1'} I_p. \end{aligned} \tag{7.41}$$

Requiring $v_o = 0$ yields

$$R_X = \frac{R_1' R_2}{R_1' + R_2} = (R_1 + R_S) \| R_2 \tag{7.42}$$

Typically, the source resistance R_S is much smaller than the resistance R_1, in which case $R_1' \cong R_1$ and

$$R_X \cong R_1 \| R_2, \quad R_S = R_1. \tag{7.43}$$

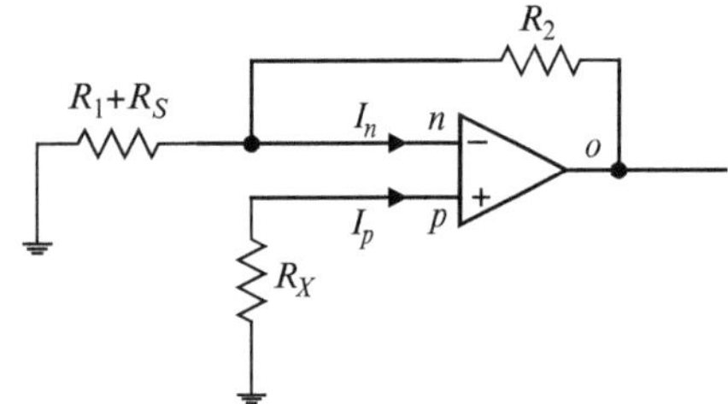

Fig. 7.34 Input dc bias currents in an inverting amplifier

Moreover, the dc gain is usually greater than ten, in which case

$$R_X \cong R_1 \| R_2 \cong R_1, \quad R_2 \gg R_1. \tag{7.44}$$

Exercise 7.10. By an analysis like that leading to (7.41), show that the input bias current compensating resistance R_X in a current-to-voltage converter shown in Fig. 7.35 is given by (7.42) with $R_1 = 0$ and $R_2 = R_F$, provided $R_S \gg R_F$.

In a direct-coupled non-inverting amplifier, the compensating resistor R_X is connected to the op-amp p terminal in series with the source, as shown in Fig. 7.36. If the compensating resistor were connected to ground (to the reference node), the compensating resistor would be shunted by the source and rendered ineffective. Here again, we seek an expression for the compensating resistance such that $v_S = 0$ implies $v_n = v_p$ and $v_o = 0$. From Fig. 7.36 and Kirchhoff's voltage law, $v_S = 0$ and $v_o = 0$ imply that

$$v_p = -I_p(R_S + R_X), \quad v_n = -I_n(R_1 \| R_2).$$

For $I_p = I_n$, it follows that

$$R_S + R_X = R_1 \| R_2 \Rightarrow R_X = (R_1 \| R_2) - R_S. \tag{7.45}$$

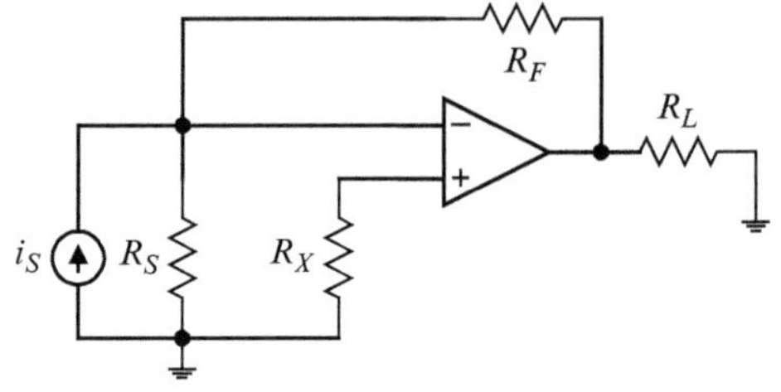

Fig. 7.35 See Exercise 7.10

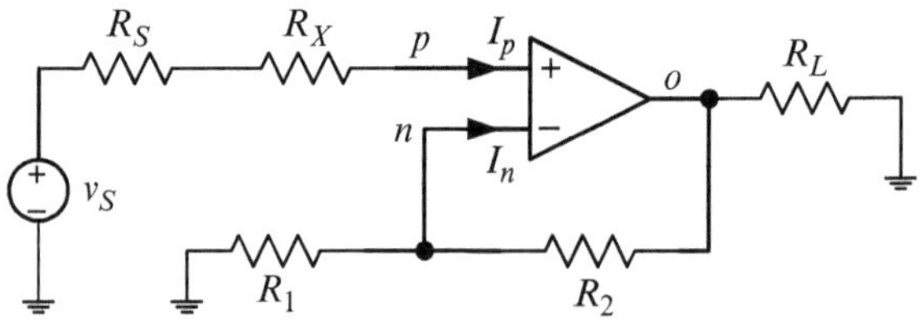

Fig. 7.36 DC bias compensation for a non-inverting amplifier

Simple resistive compensation is not possible if $(R_1\|R_2)<R_S$, but that would be unusual. In most applications, $(R_1\|R_2) \gg R_S$, in which case

$$R_X \cong (R_1\|R_2), \quad (R_1\|R_2) \gg R_S. \tag{7.46}$$

Here also, the dc gain usually is large enough that

$$R_X \cong (R_1\|R_2) \cong R_1, \quad R_2 \gg R_1 \gg R_S. \tag{7.47}$$

To obtain (7.42) and (7.45), we assumed that $I_p = I_n$. In a real op amp, the bias currents entering the n and p terminals might differ (perhaps by as much as 20%). Also, the relation $R_X \cong R_1$ is inexact, so the compensation described above is imperfect. In practice, the compensating resistance often is in part a variable resistor (trimmer) that allows in-situ adjustment. Even then, the dc bias currents can depend strongly on operating temperature, and perfect balance is difficult to achieve and maintain. Nonetheless, an amplifier is more precise with compensation than without, and compensation usually is advisable.

We conclude this section by showing that bias-current compensation is not required for a voltage follower. Figure 7.37 shows the relevant circuit diagram and corresponding model for analysis, where we set the source voltage v_S to zero because we are interested in only the output voltage due to the input bias currents. Kirchhoff's current law gives

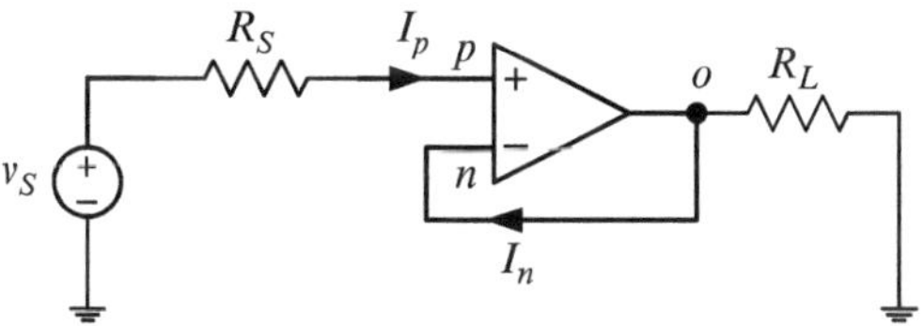

(a) circuit diagram

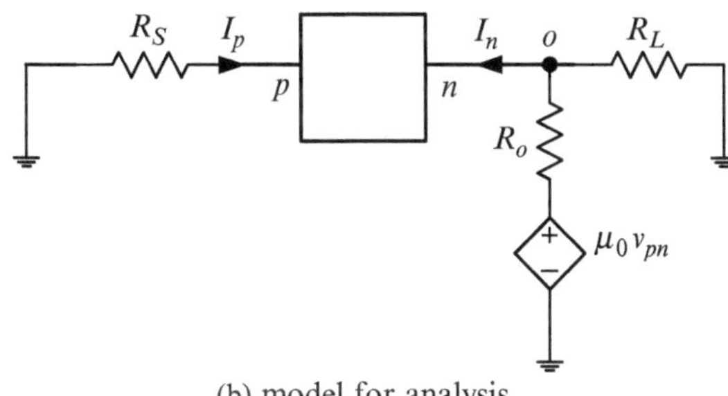

(b) model for analysis

Fig. 7.37 Voltage follower circuit

$$\frac{v_p}{R_S} + I_p = 0; \qquad I_n + \frac{v_n - \mu_0(v_p - v_n)}{R_o} + \frac{v_n}{R_L} = 0. \tag{7.48}$$

To obtain the output voltage $v_o = v_n$ due to the bias currents alone, we assume $I_p = I_n$ and eliminate v_p from (7.48). We obtain

$$v_n = -I_n \frac{(R_o + \mu_0 R_S)R_L}{R_o + R_L + \mu_0 R_L}. \tag{7.49}$$

For $\mu_0 \to \infty$, (7.49) reduces to $v_n \cong -I_n R_S$. From (7.48), we have $v_p = -I_p R_S$, which also is apparent from the circuit diagram. Because $I_p = I_n$, as assumed above, we have $v_p = v_n$, $\mu_0(v_p - v_n) = 0$. It follows that no compensating resistor is necessary for the voltage follower shown in Fig. 7.37a.

Although bias-current compensating resistors usually are included in real circuits, they do not enter into analyses of resistive op-amp amplifiers at dc and alter ac analyses in only a few cases. To avoid distracting clutter, we usually omit showing them in circuit diagrams unless they must be included to obtain a correct analysis.

7.11 Power Dissipation in Op Amps and Op-Amp Circuits

Removing heat from a circuit can be expensive, as can providing the wasted power that produces the heat. Circuits that operate hot generally produce more electrical noise than similar circuits that operate at lower temperatures. Also, parameter values change with increasing temperature, and component lifetime generally decreases with increasing operating temperature. For these and other reasons, specifying power-dissipation ratings for circuit components and planning for heat dissipation is an essential part of circuit design.

Typical small integrated op amps can dissipate from about 500 mW to a little more than 1 W. Much larger power op amps can dissipate 100 W or more.

The power dissipated by an op amp depends upon a number of things, including the supply voltage, the input voltage, the load (output) voltage, the load resistance, and properties of other surrounding circuitry (e.g., the resistors in a feedback network). Thus expressions for power dissipated can be complicated and difficult to interpret. A circuit designer needs a simple, approximate, and conservative relation that allows initial specification of the power dissipation rating required of an op amp. If necessary, the specification can be refined by subsequent more detailed analysis, by simulation, or by construction and testing. In this section, we derive a few simple relations that are useful in design of linear resistive op-amp circuits. Relations given here are strictly applicable to only op-amp circuits subjected to dc excitation, but can provide useful bounds for a circuit subjected to any excitation. More general relations for circuits subjected to ac excitation are given in Chapter 17.

The total power delivered to an op-amp circuit consists of the power P_S delivered by the supply and the power P_{in} delivered to the circuit by an input. The total power dissipated in an op-amp circuit consists of the power P_A dissipated in the op amp and the power P_R dissipated in the load and other circuit components external to the op amp. Power is conserved, so the power delivered equals the power dissipated; that is,

$$P_S + P_{in} = P_A + P_R. \tag{7.50}$$

In virtually all linear applications, the power delivered by an input is negligible ($P_{in} \ll P_S$). The power dissipated by an op amp in a linear circuit is given to a good approximation by

$$P_A \cong P_S - P_R. \tag{7.51}$$

In what follows, we obtain an expression for the power dissipated by an op amp in a resistive circuit, assuming *dc excitation and linear operation.*

The power dissipated in circuit components external to the op amp is given to a good approximation by

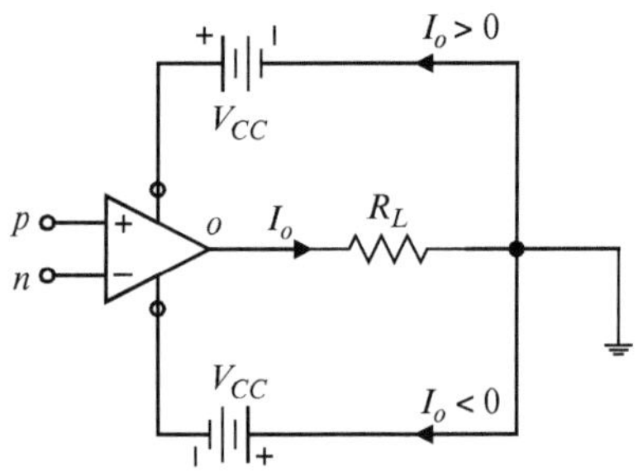

Fig. 7.38 Supply current in a symmetrically powered op amp operating linearly and having a positive output ($v_o > 0$)

$$P_R \cong \frac{{V_L}^2}{R'_L}, \tag{7.52}$$

where V_L is the load voltage and R'_L is the **effective load** on the op amp, which is the resistance seen by the current exiting the output (o) terminal of the op amp. To minimize waste, we would like most of the power P_R to be dissipated in the actual load R_L.

Figure 7.38 shows how we obtain an expression for the power P_S delivered by the supply to an op amp having a *dc input*. Because of the internal structure of an op amp, positive output current I_o passes into the positive supply terminal and out of the op-amp output terminal.[10] Negative output current passes into the negative supply terminal and out through the output terminal. In either case, the power *delivered* by the power supply is given approximately by[11]

$$P_S \cong V_{CC}\,|I_o| = V_{CC}\frac{|V_L|}{R'_L}, \tag{7.53}$$

where I_o is the current exiting the output (o) terminal of the op amp.

The effective load R'_L depends upon the actual load on the circuit and (in general) components of the feedback network. For an inverting amplifier and resistive load R_L, the effective load is $R'_L = R_L \parallel R_2$. For a non-inverting amplifier and resistive load R_L, the effective load is $R'_L = R_L \| (R_1 + R_2)$. Often, the specified gain is large enough that $R_2 \gg R_1$, in which case the effective load for a non-inverting amplifier is approximately $R'_L = R_L \| R_2$. To minimize the load on the op amp (to maximize the effective load resistance), the resistance R_2 should be as large as possible, relative to the load R_L. The maximum effective load resistance (the minimum load) is the actual load R_L.

From (7.51), (7.52), and (7.53), the power dissipated by the op amp is given by

[10] The output stage of an op amp is typically a so called *push-pull* configuration of two transistors, where one transistor amplifies the positive parts of an applied voltage and the other amplifies the negative parts.

[11] This is the power delivered by the supply to the output stage. A typical op amp is a three-stage amplifier, and even if no input is applied (even if $i_0 = 0$), some dc power is required to keep the op amp in an active (ready) state.

$$P_A = P_S - P_R = V_{CC}\frac{|V_L|}{R'_L} - \frac{{V_L}^2}{R'_L} = \frac{(V_{CC} - |V_L|)V_L}{R'_L}. \quad (7.54)$$

From (7.54), the power dissipated by the op amp for a *dc input* is maximum for $|V_L| = V_{CC}/2$ and is given by

$$P_{A\,\max} = \frac{{V_{CC}}^2}{4R'_L}. \quad (7.55)$$

We show in Chapter 17 that (7.55) gives the worst-case value for the power dissipated by an op amp in a linear resistive circuit. Consequently, (7.55) often is used for initial screening of op amps suitable for a particular application (assuming the supply voltage and the equivalent load resistance are known).

Example 7.14. (Follower) In Fig. 7.39, the supply voltage is $V_{CC} = 25\text{V}$ and the load resistance is $R_L = 1\,\text{k}\Omega$. Specify the worst-case power-dissipation rating for the op amp. *Solution*: The current i_o equals the load current i_L because the current entering the n terminal of the op amp is negligible. Thus the effective load is the actual load on the circuit ($R'_L = R_L$). The maximum power dissipated by the op amp (the specified power-dissipation rating) for constant (dc) excitation is

$$P_{A\max} = \frac{{V_{CC}}^2}{4R'_L} \cong 156\,\text{mW}.$$

Exercise 7.11. The supply voltage to a follower is ±25 V. The circuit is to drive a 2 kΩ resistive load. Specify the power-dissipation rating for the op amp.

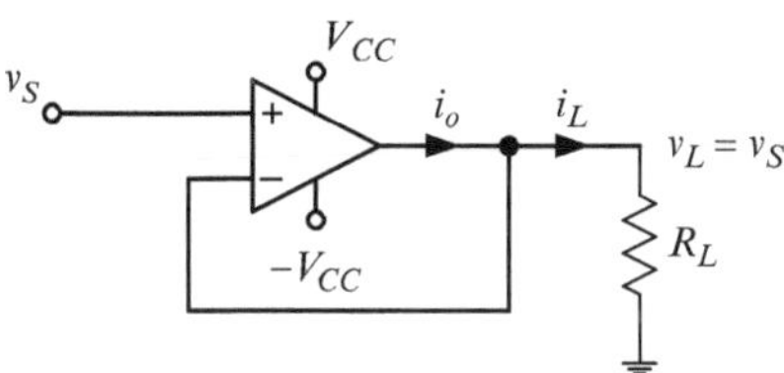

Fig. 7.39 See Example 7.14

Example 7.15. (Inverting amplifier) In Fig. 7.40, $V_{CC} = 25\,\text{V}$, $R_L = 100\,\Omega$, $R_1 = 10\,\text{k}\Omega$, and $R_2 = 500\,\text{k}\Omega$. Specify the worst-case power-dissipation rating for the op amp. *Solution*: Because $v_n = 0$, the effective resistance seen by the current exiting the op amp output terminal (the effective load) is

$$R'_L = R_L \| R_2 \cong \frac{(500\,\text{k}\Omega)(1\,\text{k}\Omega)}{510\,\text{k}\Omega} = 998\,\Omega$$

and the maximum power dissipated by the op amp (the specified power-dissipation rating) is

$$P_{A\max} = \frac{{V_{CC}}^2}{4R'_L} = \frac{(25\,\text{V})^2}{(4)(998\,\Omega)} \cong 157\,\text{mW}$$

Exercise 7.12. (Non-inverting amplifier) In Fig. 7.41, $V_{CC} = 25\,\text{V}$, $R_L = 1\,\text{k}\Omega$, $R_1 = 10\,\text{k}\Omega$, and $R_2 = 500\,\text{k}\Omega$. Specify the worst-case power-dissipation rating for the op amp.

Equation (7.54) implies that the average power dissipated by an op amp equals zero if the load voltage $v_L = 0$. Actually, an idle op amp draws a dc quiescent current, which usually is given on the manufacturer's data sheet. For a PA03© high-power op amp, for example, the quiescent supply current for a ±(75/2)

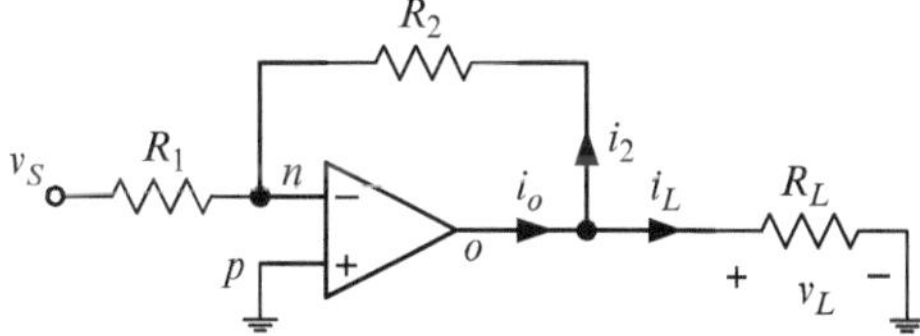

Fig. 7.40 See Example 7.15

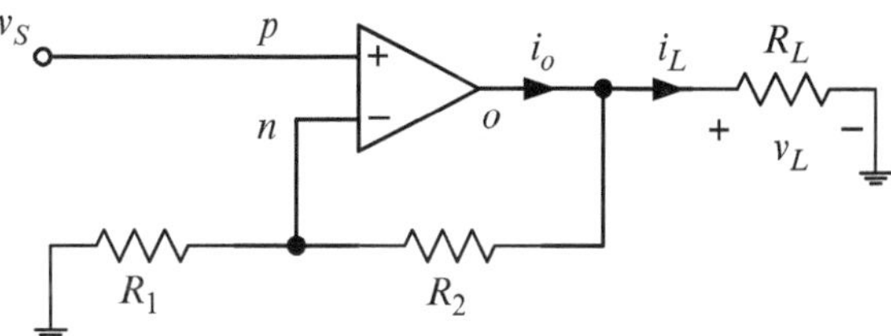

Fig. 7.41 See Exercise 7.12

$= \pm 37.5$ V supply is about 300 mA, which means that the average power dissipated by an idling PA03© can be as much as (2)(75/2)(0.3) = 22.5 W. In many cases, the idle power is a negligible fraction of the active power, but not always.

7.12 Design Considerations

Initial (conceptual) design of linear resistive op amp circuits is based upon the ideal model for an op amp:

- The voltage v_{pn} is zero (or the intrinsic dc voltage gain μ_0 is infinite).
- The currents entering the *n* and *p* terminals are zero.

In addition, **three fundamental rules** govern design of linear op amp circuits:

- The net voltage feedback must be negative, which means that the voltage fed back to the *n* terminal must exceed that fed back to the *p* terminal (if any).
- There must be negative voltage feedback at dc, which requires a dc path from the *o* terminal to the *n* terminal.
- There must be a dc path to ground from each of the *n* and *p* terminals (for each of the input bias currents).

All standard linear op-amp circuits obey these rules.

Once a conceptual design is complete, we must select physical components as follows:

- The power-dissipation rating of each component must exceed the expected power dissipation.
- The precision of each component must be such that any specification on overall precision (e.g., of gain) is met (but see *As an aside*, below).
- The output current required of each op amp must not exceed the maximum current the op amp can supply.
- Each op amp must be able to accept power supply voltages that can accommodate the required output swing.
- The effective load on each op amp should be minimized, which means the effective load resistance on each should be as large as is practical. This means that the feedback resistor R_2 should be as large as possible, which for any particular gain also makes R_1 as large as possible and maximizes the input resistance. A rule of thumb for amplifiers using general-purpose, integrated BJT-input op amps is that the resistor R_1 should be no larger than about 1% of the op amp input resistance.[12]

Other guidelines based upon frequency response and output slew rate are treated in Chapter 17.

As an aside, the precision of component parameters in an integrated circuit is limited; for example, the resistance of a resistor might differ by 10% from the design value. However, such errors are correlated, which means that if a resistor is 10% over- or under-valued, then another resistor in the same area of the circuit will probably be off by about the same amount and in the same direction. Thus, although it might be difficult to achieve a specific resistance with high precision, it is possible to achieve fairly precise resistance ratios, which means, for example, that the gains of integrated inverting and non-inverting amplifiers can be precisely established. For example, if both R_1 and R_2 in an inverting or non-inverting amplifier differ from their design values by the same fraction α, the voltage gain is nonetheless precise, because

$$\frac{(1+\alpha)R_2}{(1+\alpha)R_1} = \frac{R_2}{R_1}.$$

Thus, in such a case, the main consequence of imprecise resistance values is an imprecise input resistance, which usually is inconsequential.

Example 7.16. A certain photodiode produces a current given by

$$i_S = k_i \lambda, \tag{7.56}$$

where λ is the intensity of the incident light in watts per square meter and $k_i = 1\,\text{mA}\,\text{W}^{-1}\text{m}^2$. The output resistance of the photodiode is approximately 10 kΩ. We wish to produce a voltage whose magnitude is given by

$$|v_o| = k_v \lambda, \tag{7.57}$$

where $k_v = 5\,\text{V}\,\text{W}^{-1}\text{m}^2$. The sign of the voltage is immaterial. The maximum intensity

[12] We justify this rule in Chapter 17.

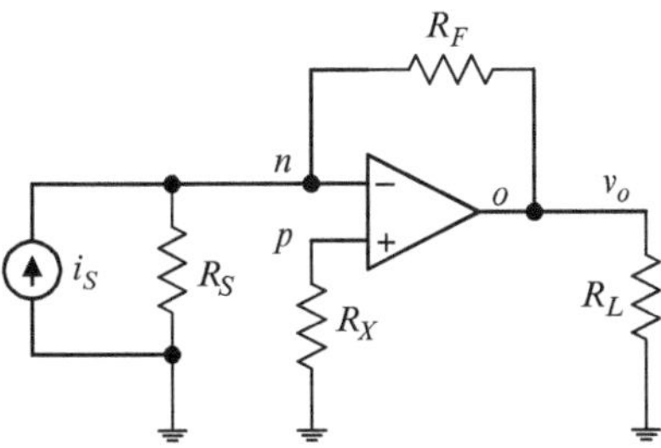

Fig. 7.42 See Example 7.16

of the incident light is 3.4 Wm^{-2}. The output voltage will be applied to another circuit having input resistance $R_L = 100\,\text{k}\Omega$.
Solution: We would use a current-to-voltage converter (see Example 7.7), as shown in Fig. 7.42, where

$$v_o = -R_F i_S = -R_F k_i \lambda. \tag{7.58}$$

Equations (7.57) and (7.58) imply

$$R_F = \frac{k_v}{k_i} = 5\,\text{k}\Omega. \tag{7.59}$$

Compensating for input bias current completes the conceptual design. From (7.42), we have for an inverting configuration that

$$R_X = (R_1 + R_S)\|R_2.$$

In this example, $R_1 = 0$, $R_S = 10\,\text{k}\Omega$, and $R_2 = R_F = 5\,\text{k}\Omega$.
Thus

$$R_X = \frac{(5\,\text{k}\Omega)(10\,\text{k}\Omega)}{5\,\text{k}\Omega + 10\,\text{k}\Omega} = 3.33\,\text{k}\Omega.$$

The feedback is entirely negative (there is no feedback to the p terminal), the feedback path is resistive, so there is feedback at dc, and there are dc paths from both inputs to ground.
From (7.57),

$$|v_o|_{\max} = k_v \lambda_{\max} = \left(5\,\text{V}\,\text{W}^{-1}\text{m}^2\right) \times \left(3.4\,\text{Wm}^{-2}\right) = 17\,\text{V}. \tag{7.60}$$

The op amp must be able to accommodate a ±17 V output swing. The effective load on the op amp is

$$R_L' = R_F\|R_L \cong R_F, \tag{7.61}$$

so the op amp must be able to provide an output current of

$$|i_o|_{\max} = \frac{|v_o|_{\max}}{R_L'} = \frac{17\,\text{V}}{5\,\text{k}\Omega} = 3.4\,\text{mA}. \tag{7.62}$$

Virtually any op amp can meet this requirement. Assuming rail-to-rail operation, the required power-supply voltage is $V_{CC} = 17\,\text{V}$. From (7.55), the power-dissipation rating for the op amp should be

$$P_A = \frac{V_{CC}{}^2}{4R_L'} = \frac{(17\,\text{V})^2}{(4)(5\,\text{k}\Omega)} = 14.5\,\text{mW}. \tag{7.63}$$

Virtually any op amp can meet this requirement. The current through the resistor R_x is negligible. From (7.56), the maximum current through the feedback resistor is

$$\begin{aligned} |i_S|_{\max} &= k_i \lambda_{\max} \\ &= \left(1\,\text{mA}\,\text{W}^{-1}\text{m}^2\right)\left(3.4\,\text{Wm}^{-2}\right) \\ &= 3.4\,\text{mA}, \end{aligned} \tag{7.64}$$

and the required power-dissipation rating for the resistor R_F is

$$P_F = (3.4\,\text{mA})^2(5\,\text{k}\Omega) = 57.8\,\text{mW} \tag{7.65}$$

so a 1/8 W (125 mW) resistor is more than adequate. The total power requirement for the circuit is the total power dissipated by the circuit (all components). In this circuit, the total power dissipated is approximately that dissipated by the op amp (14.5 mW) plus that dissipated by the feedback resistor (57.8 mW). The powers dissipated by the compensating resistor R_x, the op-amp input resistance, the source resistance, and the load resistance are negligible by comparison. Thus the circuit will demand approximately

$$P_{total} \cong 14.5\,\text{mW} + 57.8\,\text{mW} = 72.3\,\text{mW}$$

from the power supply. We might add 10% or 20% (to provide a safety margin) and specify 80 or 90 mW. This figure can be checked in the simulation phase of the design.

The overall accuracy of the system is approximately the tolerance of the resistor R_F; e.g., for an overall accuracy of $\pm 1\%$, we would specify that tolerance for R_F.

7.13 Problems

All op amps are ideal unless the problem statement indicates otherwise. In problems calling for simulation, use an op-amp model having the parameters given in Table P 7.1.

Section 7.2 is prerequisite for the following problems.

Table P 7.1 Op-amp simulation parameters

Parameter	Value
Open-loop gain (intrinsic dc voltage gain)	400 MV V^{-1}
Input resistance	10 GΩ
Output resistance	5 Ω
Output swing	±25 V
Input bias current	0
Input offset current[a]	0
Input offset voltage	0
Unity-gain bandwidth (gain-bandwidth product)[a]	1.5 GHz
Slew Rate[a]	500 kVs^{-1}
Compensation capacitance[a]	10 pF

[a]These parameters are treated in Chapter 17

P 7.1 Re-draw the circuits in Fig. P 7.1, replacing the op amp with a linear model. Label the *n*, *p*, and *o* nodes.

P 7.2 Refer to Fig. P 7.2. For a certain op amp to operate linearly, the voltage V_2 must exceed V_1 by at least 5 V, but by no more than 30 V. If a symmetric bipolar supply $\pm V_{CC}$ is to be used, what are the minimum and maximum permissible values of V_{CC}?

P 7.3 In Fig. P 7.3, $V_{CC} = 15\,\text{V}$, $R_1 = 10\,\text{k}\Omega$, $R_2 = 500\,\text{k}\Omega$, $R_L = 5\,\text{k}\Omega$, $R_S = 50\,\Omega$. The op amp has input resistance $R_i = 2\,\text{M}\Omega$, output resistance $R_o = 30\,\Omega$, and intrinsic (no-load) dc voltage gain $\mu_0 = 10^5$. (a) Replace the op amp by a linear model and obtain an expression for the voltage transfer characteristic for the circuit. (b) If the op amp is rail-to-rail, for what input amplitudes is the circuit linear? Use a graph of the voltage transfer characteristic to illustrate your answer. (c) Give the voltage gain, current gain, power gain, transresistance, and transconductance of the circuit within the range of linear operation. Express the current, voltage, and power gains in dB. (d) Obtain a simpler expression for the voltage gain by assuming $\mu_0 \to \infty$ in the expression you obtained in part (a). Then repeat part (c).

P 7.4 Repeat Problem P 7.3 for the circuit shown in Fig. P 7.4, where $R_S = 75\,\Omega$, $R_L = 1\text{k}\Omega$, $V_{CC} = 12\text{V}$, $R_i = 1\text{M}\Omega$, $R_o = 50\,\Omega$, $\mu_0 = 2 \times 10^5$.

P 7.5 Figure P 7.5 shows (in part) the pin-out diagram for a particular integrated op amp, where V_- and V_+ are the lower (smaller) and upper supply voltages. Ignore the pins that are not labeled. Re-draw Fig. P 7.3, using the diagram in Fig. P 7.5 for the op amp. Use battery symbols to show how the supply voltages are supplied. Make the diagram as neat and clear as possible, and try to minimize the number of points at which conductors cross but are not connected.

P 7.6 Repeat Problem P 7.5 for the circuit in Fig. P 7.4.

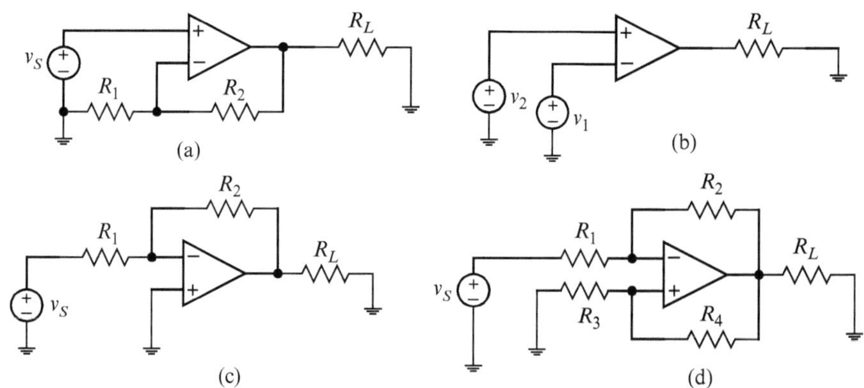

Fig. P 7.1 See Problem P 7.1

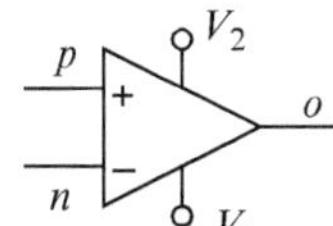

Fig. P 7.2 See Problem P 7.2

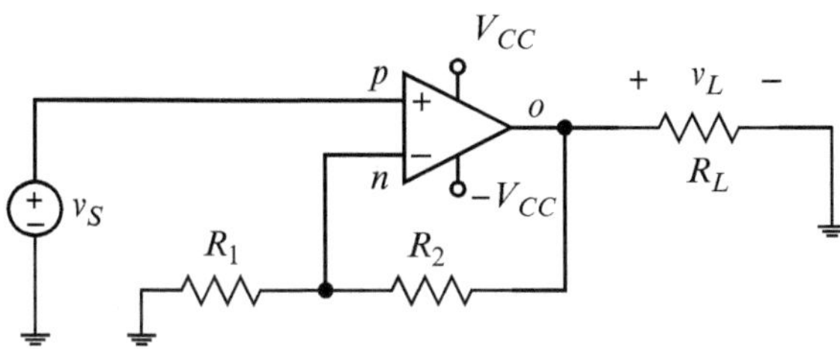

Fig. P 7.3 See Problem P 7.3, 5

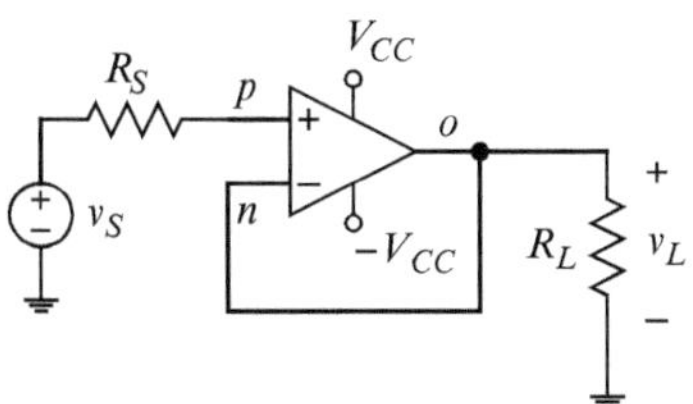

Fig. P 7.4 See Problem P 7.4, 6

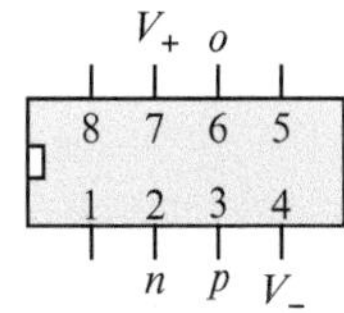

Fig. P 7.5 See Problem P 7.5

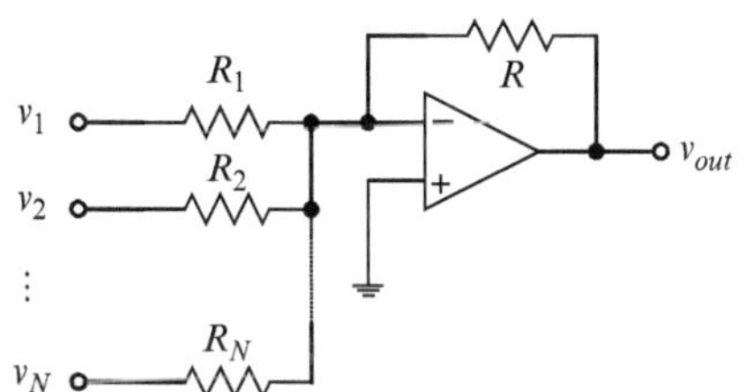

Fig. P 7.6 See Problem P 7.7

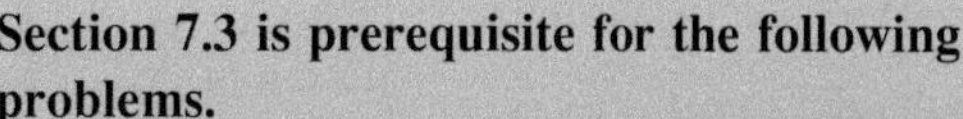

Section 7.3 is prerequisite for the following problems.

P 7.7 Refer to Fig. P 7.6. (a) Express the voltage v_{out} in terms of the voltages $v_1, v_2, \cdots, v_N$. (b) Let $N = 5$, $v_1 = v_2 = \cdots = v_5 = 1\,\text{V}$, $R_1 = R_2 = \cdots = R_3 = R = 10\,\text{k}\Omega$, and simulate the circuit. If the simulated result does not agree with the relation obtained in part (a), explain why.

P 7.8 Refer to Fig. P 7.7. (a) Express the voltage v_{out} in terms of the voltages v_1 and v_2. (b) Let $v_1 = 1\,\text{V}, v_2 = 2\,\text{V}$, $R = 10\,\text{k}\Omega$, and simulate the circuit. If the simulated result does not agree with the relation obtained in part (a), explain why.

P 7.9 Refer to Fig. P 7.8. (a) Obtain an expression for the voltage gain. (b) Let $v_{in} = 1\,\text{V}$, $R_1 = 20\,\text{k}\Omega$, $R_2 = R_4 = 10\,\text{k}\Omega$, $R_3 = 15\,\text{k}\Omega$, and $R_L = 5\,\text{k}\Omega$. Simulate the circuit and show that the simulated output agrees with the analysis in part (a).

P 7.10 (a) Obtain an expression for the voltage gain A_v for each circuit in Fig. P 7.9. (b) Let $R = 10\,\text{k}\Omega$, $a = 0.8$, $v_S = 1\text{V}$, $R_S = 20\,\Omega$, $R_L = 5\,\text{k}\Omega$, and use simulations to check your answers.

P 7.11 Obtain an expression for the dc voltage gain of each circuit shown in Fig. P 7.10.

P 7.12 Refer to Fig. P 7.11. (a) Obtain an expression for the current i_L. What kind of circuit is this?

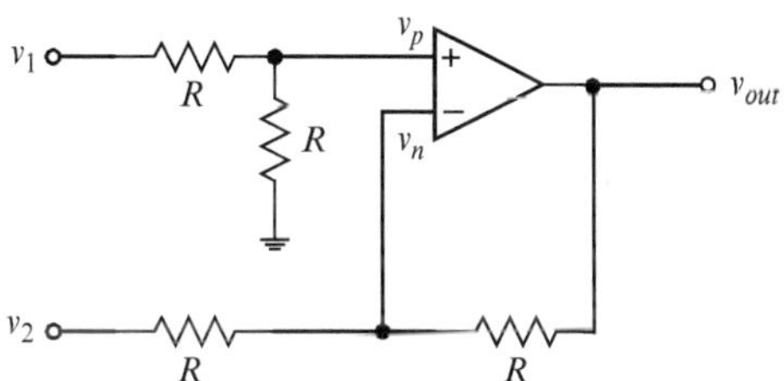

Fig. P 7.7 See Problem P 7.8

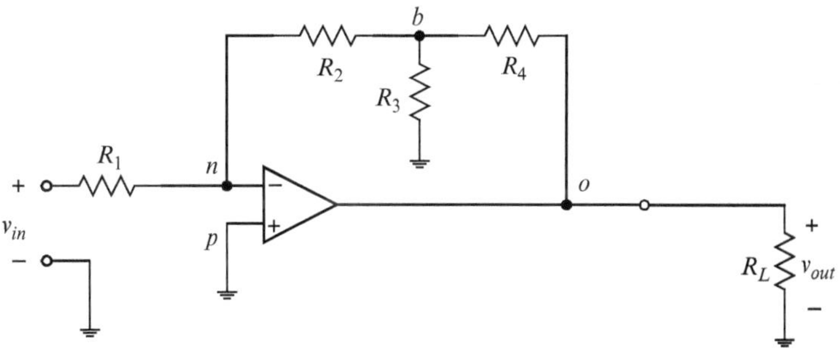

Fig. P 7.8 See Problem P 7.9

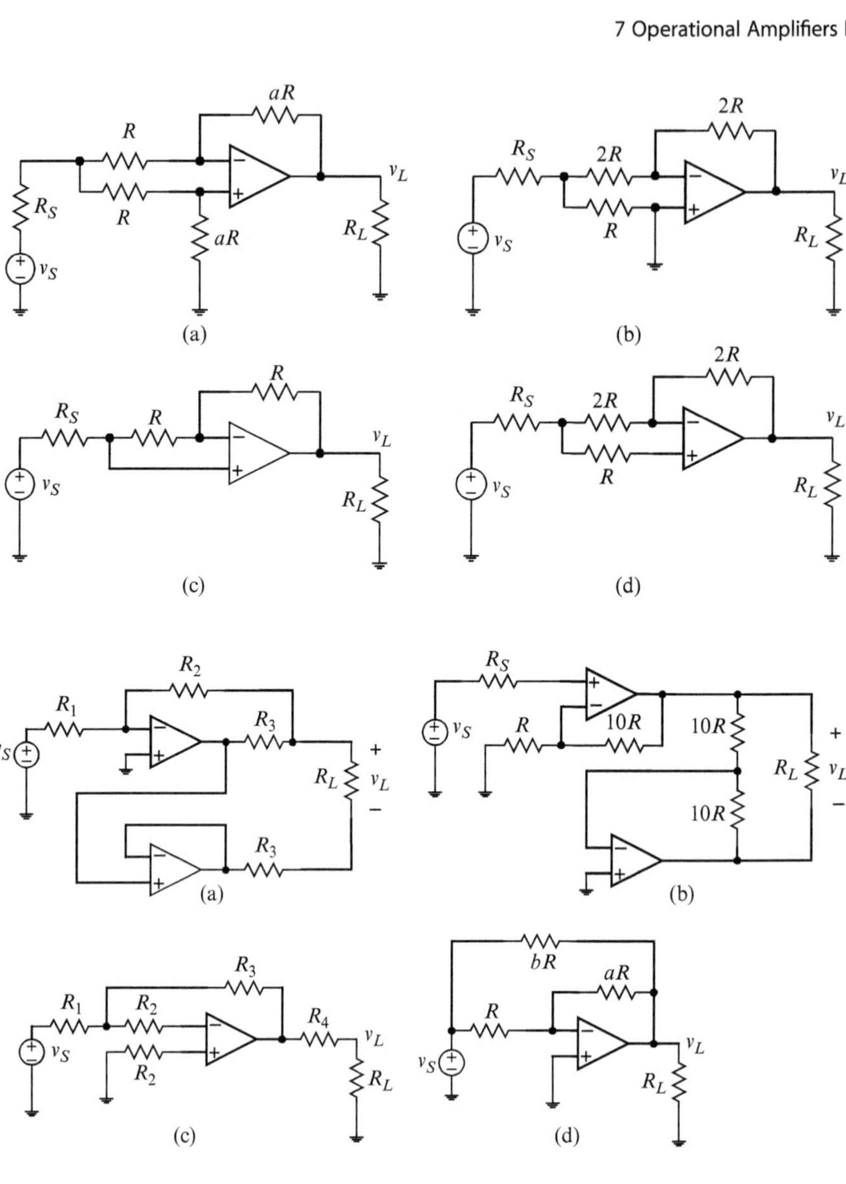

Fig. P 7.9 See Problem P 7.10

Fig. P 7.10 See Problem P 7.11

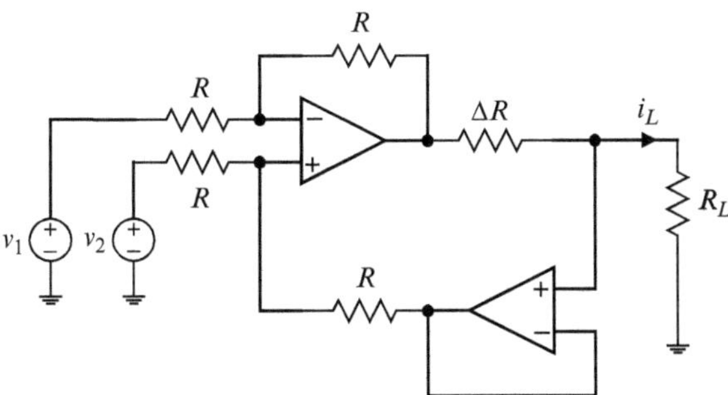

Fig. P 7.11 See Problem P 7.12

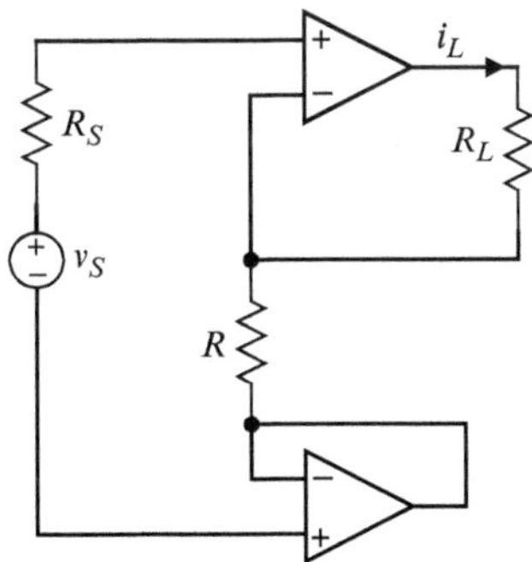

Fig. P 7.12 See Problem P 7.13

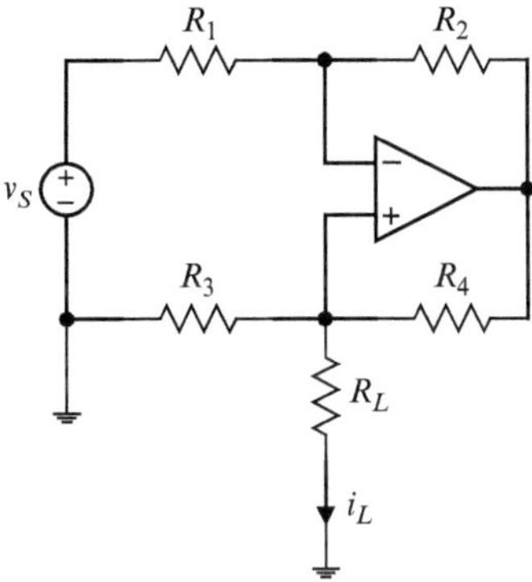

Fig. P 7.13 See Problem P 7.14

(b) Let $v_i = 100\,\text{mV}$, $v_2 = 200\,\text{mV}$, $R = 10\,\text{k}\Omega$, $\Delta R = 100\,\Omega$, $R_L = 5\,\text{k}\Omega$ and simulate the circuit to check your answer.

P 7.13 Refer to Fig. P 7.12. (a) Express the load current i_L in terms of the source voltage v_S. (b) Draw a circuit diagram showing how a power supply would be connected to the op amps consistent with your answer to part (a).

P 7.14 Refer to Fig. P 7.13, where $R_2/R_1 = R_4/R_3$. Obtain an expression for the load current i_L in terms of the input voltage v_S and thus show that the circuit is a voltage-to-current converter (a transconductance amplifier). *Hint*: Let $R_2 = aR_1$, $R_4 = aR_3$. (b) It is given that $R_L = 1\,\text{k}\Omega$, $v_S = V_{CC} = 20\,\text{V}$ and it is required that $|I_L| = 10\,\text{mA}$. Specify the circuit parameters, making sure that the circuit operates linearly (that the op amp does not saturate). (c) Simulate the circuit to verify your design. (d) Suppose the resistance ratios are not exactly matched, such that $R_2 = aR_1$, $R_4 = (a + \varepsilon)R_3$, where $\varepsilon = a$. Under what condition does the circuit remain a good transconductance amplifier? (e) Assume $a > 0$ and show that the circuit is stable (let $v_S = 0$ and show that $v_n > v_p$).

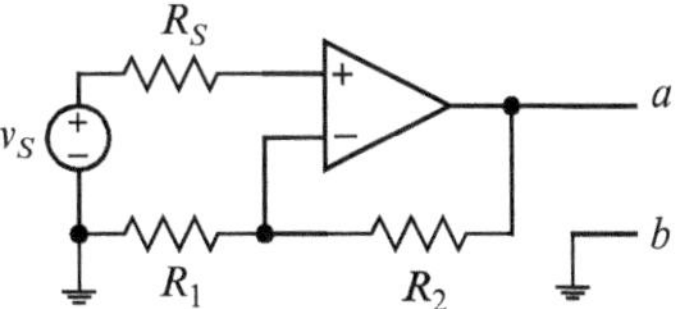

Fig. P 7.14 See Problem P 7.15, 16

P 7.15 Can you find the Thévenin equivalent at the terminals *a*–*b* for the circuit shown in Fig. P 7.14, where the op amp is ideal? Do so or explain why not.

P 7.16 Can you find the Norton equivalent at the terminals *a*–*b* for the circuit shown in Fig. P 7.14, where the op amp is ideal? Do so or explain why not.

P 7.17 Refer to Fig. P 7.15(a). What is the voltage v_{np}? Does this mean that the circuit in Fig. P 7.15(a) is equivalent to the circuit in Fig. P 7.15(b)? Justify your answer.

Section 7.4 is prerequisite for the following problems.

P 7.18 Refer to Fig. P 7.16. (a) Obtain an expression for the transconductance. (b) Show that the circuit is stable. (c) Let $v_S = 1\,\text{V}$, $R = 100\,\text{k}\Omega$, $R_L = 10\,\text{k}\Omega$ and perform a simulation to check the relation obtained in part (a).

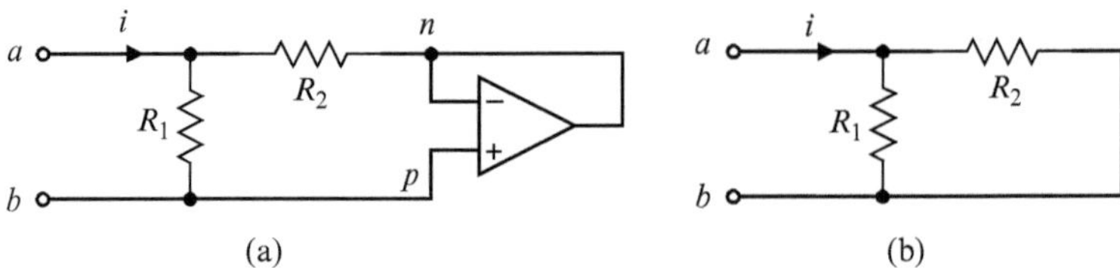

Fig. P 7.15 See Problem P 7.17

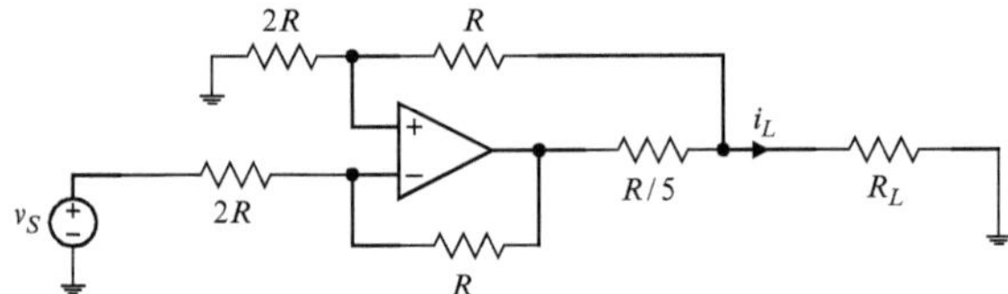

Fig. P 7.16 See Problem P 7.18

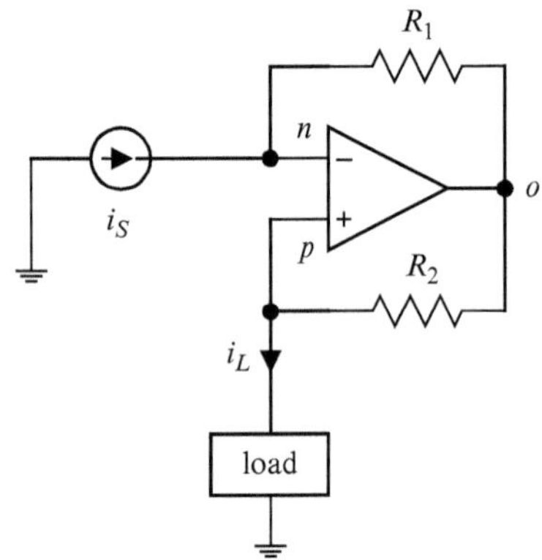

Fig. P 7.17 See Problem P 7.19

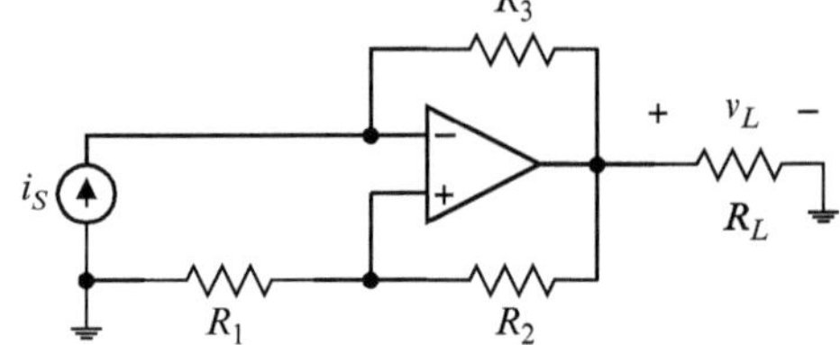

Fig. P 7.18 See Problem P 7.20

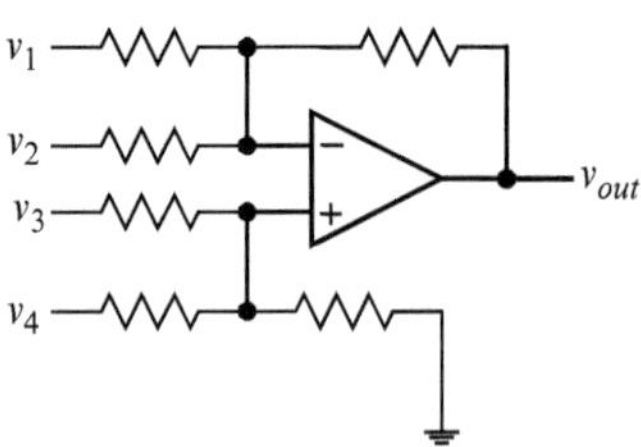

Fig. P 7.19 See Problem P 7.21

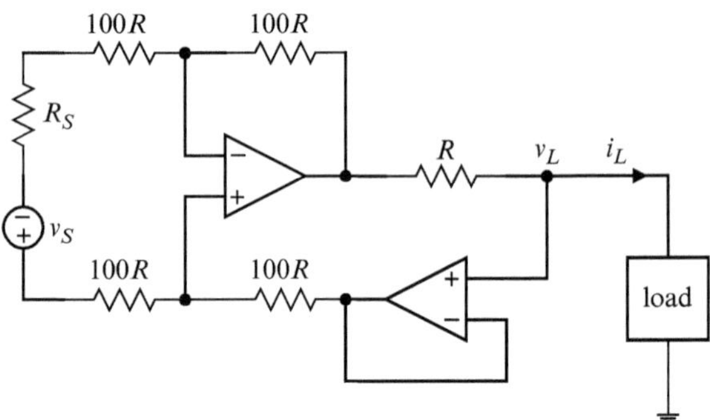

Fig. P 7.20 See Problem P 7.22

P 7.19 (a) Obtain an expression for the current transfer ratio H_i of the circuit in Fig. P 7.17. (b) Show that the circuit is stable. (c) Let $i_S = 1\,\text{mA}$, $R_1 = 20\,\text{k}\Omega$, and $R_2 = 2\,\text{k}\Omega$. If the load is resistive, and the supply voltage is 25 V, what is the maximum load resistance for which the circuit is linear? (d) 💻 Use $R_L = 200\,\Omega$ and the parameters given in part (c) and simulate the circuit to check the relation obtained in part (a).

P 7.20 (a) Obtain an expression for the transresistance H_r of the circuit in Fig. P 7.18. (b) Is the circuit stable? (c) 💻 Let $i_S = 50\,\mu\text{A}$, $R_1 = 2\text{k}\Omega$, $R_2 = 10\text{k}\Omega$, $R_3 = 20\text{k}\Omega$, $R_L = 100\text{k}\Omega$, and simulate the circuit as a check on your answers to parts (a) and (b).

P 7.21 Refer to Fig. P 7.19, where all resistors have the same value R. Obtain an expression for the output voltage v_{out}. Is the circuit stable?

P 7.22 Refer to Fig. P 7.20, where $R_S < R$ and the op amps are rail-to-rail. (a) Obtain an expression for the load current i_L. (b) Is the circuit stable? (c) Assume a symmetric supply $\pm V_{CC}$. As the load voltage v_L is increased (by increasing the input v_S), which op amp saturates first, and why? Obtain an expression for the load voltage corresponding to saturation of the left op amp. (d) Assume the load is resistive and obtain an expression for the maximum allowable load resistance. (e) Assume the load is resistive and obtain an expression for the maximum allowable input v_S.

P 7.23 Refer to Fig. P 7.21. (a) Obtain an expression for the output v_o in terms of the current i_S and

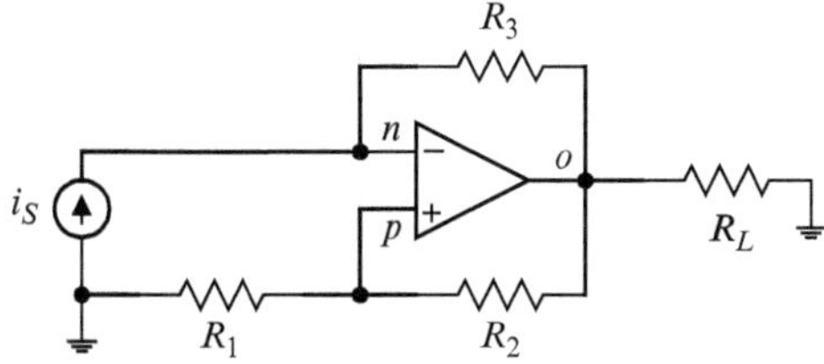

Fig. P 7.21 See Problem P 7.23

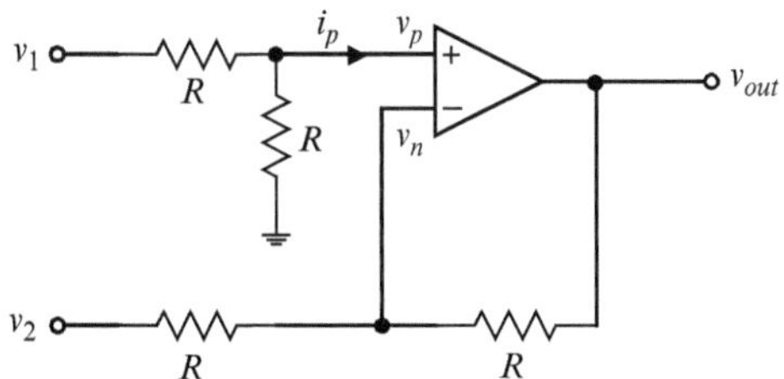

Fig. P 7.22 See Problem P 7.24

the circuit parameters. (b) Is the circuit stable? (c) If the circuit is stable, obtain an expression for the voltage across the current source. (d) Let $R_1 = 20\,\text{k}\Omega$, $R_2 = 50\,\text{k}\Omega$, $R_L = 5\,\text{k}\Omega$, $i_S = 100\,\mu\text{A}$. Simulate the circuit as a check on the answer you obtained in part (a).

Section 7.5 is prerequisite for the following problems.

P 7.24 Refer to Fig. P 7.22. (a) Obtain expressions for the input resistances seen at each input, assuming the other input is connected to an ideal voltage source. (b) Let $R = 22\,\text{k}\Omega$ and use a simulation to check your answers.

P 7.25 Refer to Fig. P 7.23. Replace the op amp by a dc model with $R_i = 5\text{M}\Omega$, $R_o = 75\,\Omega$, and $\mu_0 = 2 \times 10^5$, let $R = 1\,\text{k}\Omega$, and obtain an expression for the output resistance. Use a simulation to check your answer.

P 7.26 Refer to Fig. P 7.23. (a) Obtain an expression for the input resistance of the circuit. How do you interpret the result? (b) Let $R = 1\,\text{k}\Omega$ and simulate the circuit to check your answer.

P 7.27 Refer to Fig. P 7.24. Obtain an expression for the input resistance. (b) Obtain an expression for the output resistance. (c) Let $R_S = 50\Omega$, $R = 10\text{k}\Omega$,

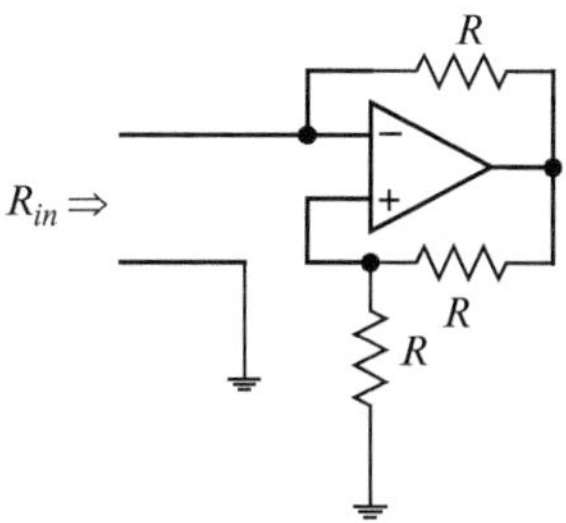

Fig. P 7.23 See Problem P 7.25, 26

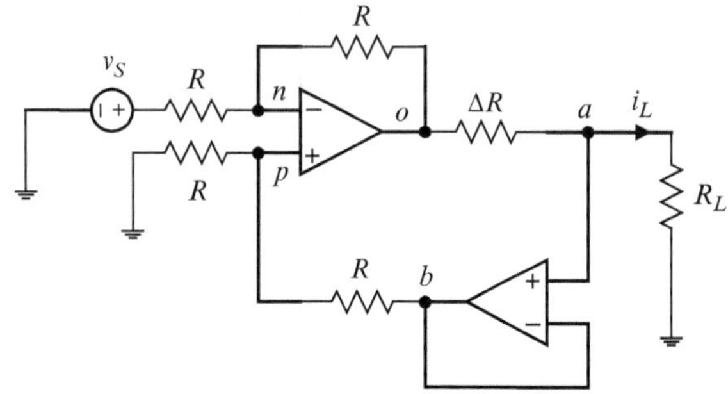

Fig. P 7.24 See Problem P 7.27

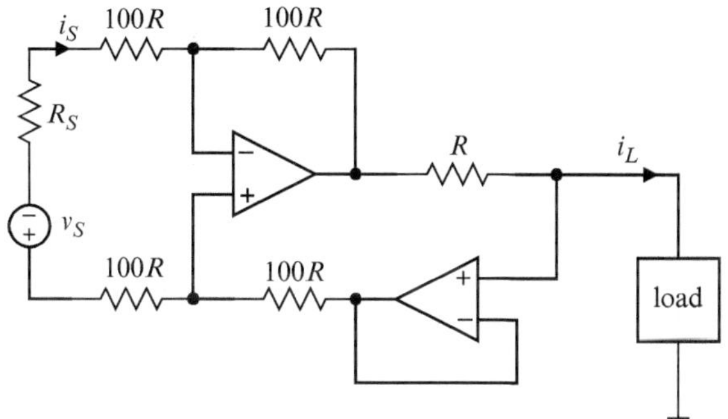

Fig. P 7.25 See Problem P 7.28

$\Delta R = 100\Omega$, $R_L = 3.4\text{k}\Omega$, $v_S = 10\text{mA}$, and use a simulation to check the expression obtained in part (a).

P 7.28 Refer to Fig. P 7.25. Obtain an expression for the input resistance. *Hint*: $R_{in} = -v_S/i_S - R_s$.

P 7.29 Refer to the current-to-voltage converter circuit in Fig. P 7.26. Assume that the input is an ideal current source, that the input resistance of the op amp is much larger than the feedback resistance R_f, and that the feedback resistance is much larger than the output resistance of the op amp. Show that for $\mu_0 \to \infty$, the output resistance of the circuit approaches R_o/μ_0, where R_o and μ_0 are the output

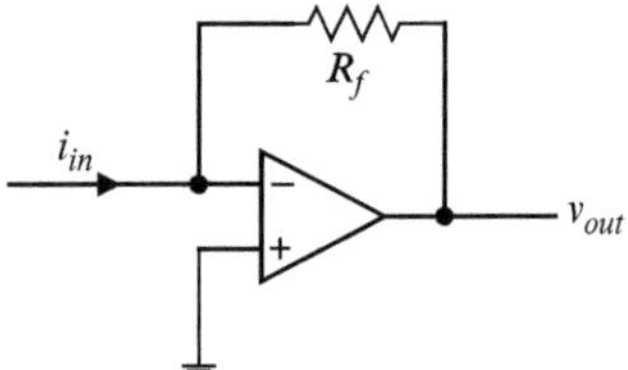

Fig. P 7.26 See Problem P 7.29

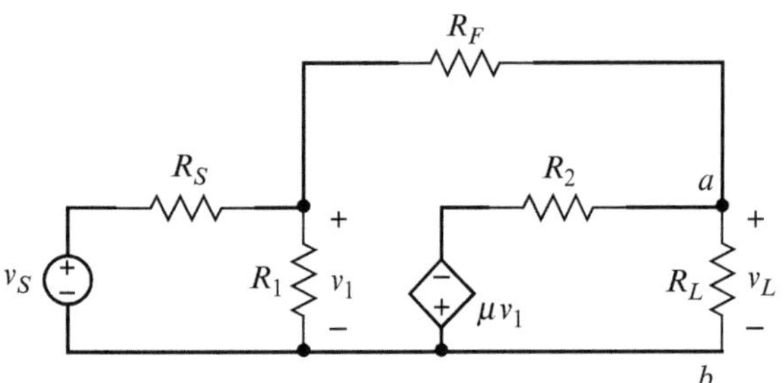

Fig. P 7.27 See Problem P 7.30

resistance and intrinsic dc gain, respectively, of the op amp.

P 7.30 Refer to Fig. P 7.27.

(a) Remove the load and obtain expressions for the voltage v_1, the open-circuit voltage, and the short-circuit current at the terminals *a–b*.
(b) Using the expressions obtained in part (a), obtain expressions for the input resistance, the output resistance, and voltage transfer ratio H_v.
(c) Obtain the limits of the expressions obtained in parts (a) and (b) as $\mu_0 \to \infty$.
(d) Let $R_S = 100\,\Omega$, $R_1 = 2\,\text{M}\Omega$, $R_F = 22\,\text{k}\Omega$, $R_2 = 75\,\Omega$, and $R_L = 5\,\text{k}\Omega$. Plot the output resistance and voltage gain (separate plots) versus μ_0 for $10 \leq \mu_0 \leq 10^6$.

Use a logarithmic scale for μ_0.

Section 7.6 is prerequisite for the following problems.

P 7.31 Figure P 7.28 shows circuit diagrams for a non-inverting amplifier and a VCVS model for same. The model parameters are the source resistance R_S, the *amplifier* (not op-amp) input resistance R_{in}, the *amplifier* output resistance R_{out}, and a no-load voltage-transfer parameter μ_v. *Note*: The parameter μ_v in Fig. P 7.28 is *not* the op-amp intrinsic voltage transfer ratio μ_0. (a) Obtain an expression for the parameter μ_v in terms of the *amplifier* model parameters and the voltage transfer ratio H_v for the amplifier. (b) Show that if $R_{in} \gg R_S$ and $R_{out} = R_L$, then $\mu_v \cong H_v$.

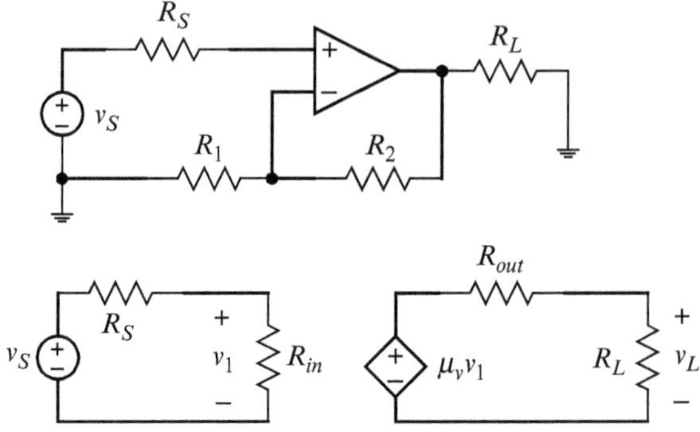

Fig. P 7.28 See Problem P 7.31

P 7.32 Figure P 7.29. shows circuit diagrams for an inverting amplifier and a VCVS model for same. The model parameters are the source resistance R_S, the *amplifier* (not op-amp) input resistance R_{in}, the *amplifier* output resistance R_{out}, and a voltage-transfer parameter μ_v. *Note*: The parameter μ_v in Fig. P 7.29 is *not* the op-amp intrinsic voltage transfer ratio μ_0. (a) Obtain an expression for the parameter μ_v in terms of the *amplifier* model parameters and the voltage transfer ratio H_v for the amplifier. (b) Show that if $R_i \gg R_S$ and $R_o = R_L$, then $\mu_v \cong H_v$.

In solving problems P 7.33 through P 7.38, use the relations given in Section 7.6.

P 7.33 The parameters of a certain inverting amplifier and associated op amp are $R_1 = 10\,\text{k}\Omega$, $R_2 = 500\,\text{k}\Omega$, $R_S = 50\,\Omega$, $R_o = 50\,\Omega$, $R_i = 2\,\text{M}\Omega$, $R_L = 5\,\text{k}\Omega$. Plot the approximate voltage gain R_2/R_1 and the actual gain given in Section 7.6 in the text versus the no-load voltage gain μ_0, for $1 \leq \mu_0 \leq 10^5$. Use a logarithmic scale for μ_0.

P 7.34 Repeat Problem P 7.33 for a non-inverting amplifier.

P 7.35 Ignore irrelevant parameter values given in Problem P 7.33 and repeat that problem for a voltage follower.

P 7.36 Using the parameter values given in Problem P 7.33, plot the actual and approximate input

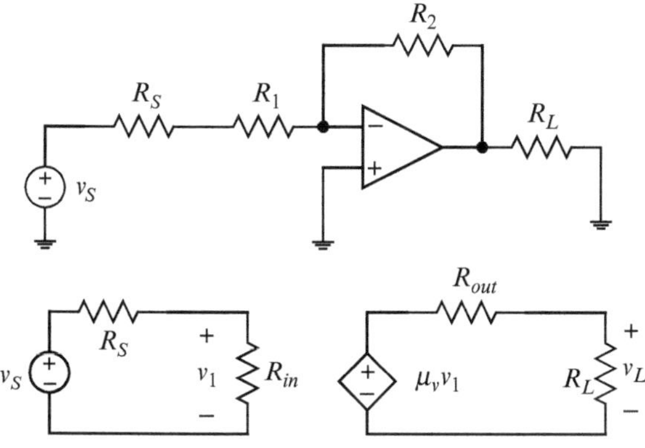

Fig. P 7.29 See Problem P 7.32

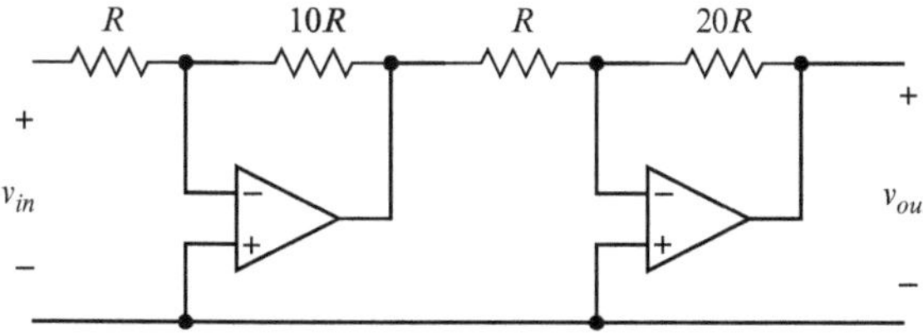

Fig. P 7.30 See Problem P 7.39

resistances and the actual and approximate output resistances for an inverting amplifier versus μ_0, for $1 \leq \mu_0 \leq 10^5$. Use logarithmic scales for both R_{in} and μ_0.

P 7.37 Repeat Problem P 7.36 for a non-inverting amplifier.

P 7.38 Repeat Problem P 7.36 for a voltage follower. (Ignore irrelevant parameters.)

Section 7.10 is prerequisite for the following problems.

P 7.39 Refer to Fig. P 7.30, where each op amp is powered by a symmetric ±15 V supply. (a) Obtain an expression for the voltage gain A_v. (b) Assume the op amps are rail-to-rail. What is the maximum input voltage for which the output voltage is given by $|v_{out}| = A_v|v_{in}|$? (c) If each op amp has an output current limit $i_{\max} = 750\,\text{mA}$ and if $|v_{out}| \leq V_{CC}$, what is the minimum load resistance for which the output voltage is given by $|v_{out}| = A_v|v_{in}|$?

P 7.40 Refer to Fig. P 7.31, where each op amp is powered by a symmetric ±12 V supply. (a) Obtain an expression for the voltage gain $A_v = v_{out}/v_{in}$. (b) Assume the op amps are rail-to-rail. What is the maximum input voltage for which the output voltage is given by $v_{out} = A_v v_{in}$? (c) If each op amp has an output current limit $i_{\max} = 1\,\text{A}$ and if $|v_{out}| \leq V_{CC}$, what is the minimum load resistance for which the output voltage is given by $v_{out} = A_v v_{in}$?

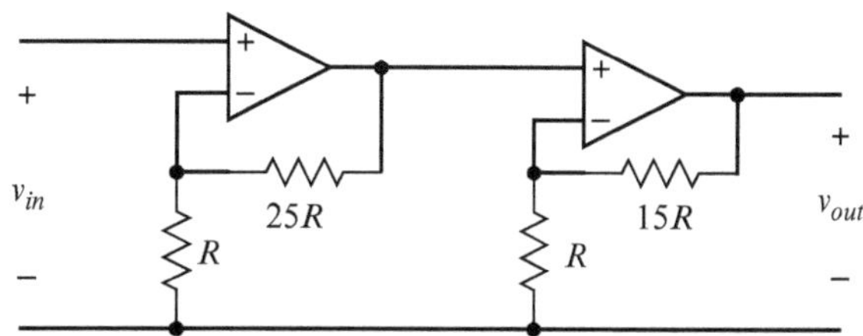

Fig. P 7.31 See Problem P 7.40

Note: Most op amps provide a means for nulling the input offset voltage, as described in the text. Manufacturer's data sheets specify how such nulling is achieved, usually via a variable resistor connected to the power supply and to additional terminals provided for that purpose. However, the simplest op amp models in simulators such as Pspice and Electronic Workbench do not provide special pins for nulling circuitry. The 1 mV offset is troublesome for inputs less than about 100 mV, and in such cases can be cancelled by connecting a dc source in series with an input, as illustrated for an inverting amplifier by Fig. P 7.32. (The 9.8 kΩ resistor compensates for the input bias currents). As you can see, the output for $V_s = 1$ mV is approximately $(-R_2/R_1)V_S = 50\,\text{mV}$, as it should be. That would not be the case if the −1 mV offset cancellation (at the *p* input) were not present.

Another way to handle this problem is to modify the virtual op-amp model by reducing the input offset voltage to zero, as suggested by Table P 7.1. Your instructor can specify how you should handle input offset cancellation in simulation problems.

Input offset voltage becomes less significant as the amplitude of the input increases (or for ac coupling, as described in Chapter 17). For example, if the input in Fig. P 7.32 were 100 mV instead of 1 mV, the output would be within about 1% of 5 V.

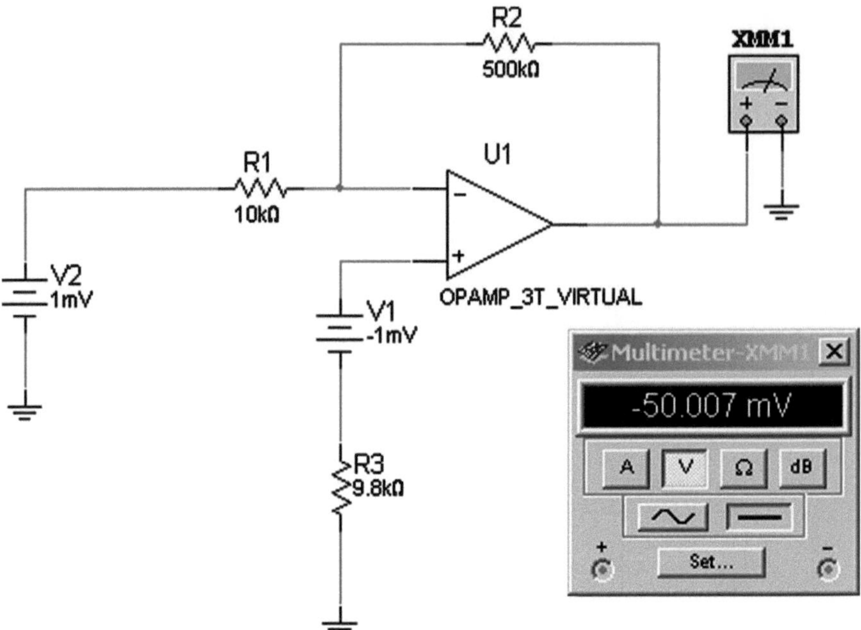

Fig. P 7.32 See note following Problem P 7.40

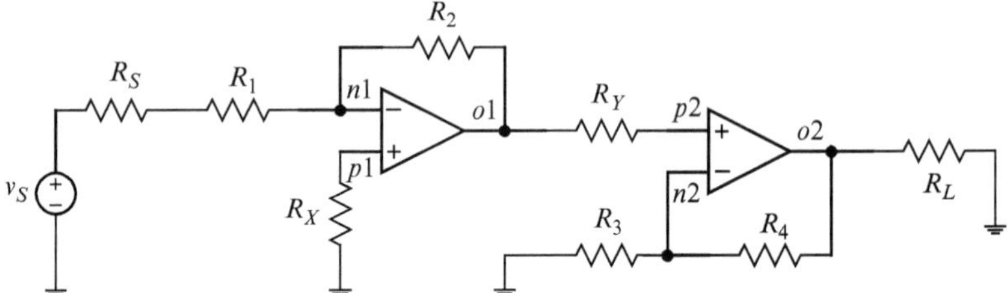

Fig. P 7.33 See Problem P 7.41, 48

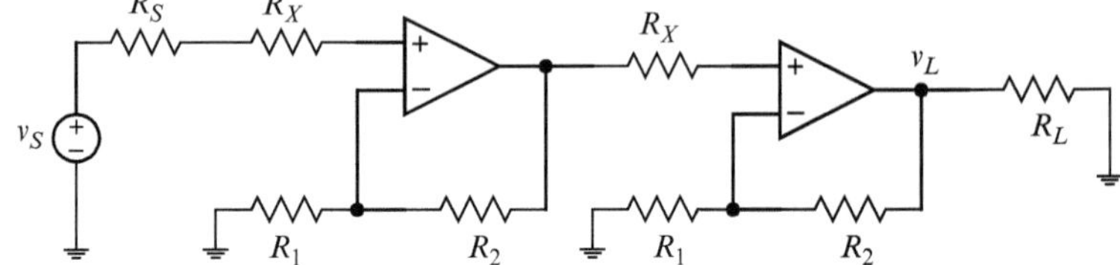

Fig. P 7.34 See Problem P 7.42, 49

P 7.41 Refer to Fig. P 7.33, where $R_S = 25\,\Omega$ and $R_L = 5\,\text{k}\Omega$. The overall voltage gain of the circuit is to be approximately 4000 and the input resistance is to be at least 20 kΩ. (a) Assume the op amps are ideal and specify reasonable values for all resistors shown. Justify each choice. (b) 💻 Use simulation to verify your design.

P 7.42 Refer to Fig. P 7.34, where $R_S = 100\,\Omega$ and $R_L = 15\,\text{k}\Omega$. The overall voltage gain of the circuit is to be approximately 10^4 and the input resistance is to be at least 10 kΩ. (a) Assume the op amps are ideal and specify reasonable values for all resistors shown. Justify each choice. (b) 💻 Use simulation to verify your design.

P 7.43 Show how to compensate for input bias currents in the circuit in (a) Fig. P 7.6. (b) Fig. P 7.7. (c) Fig. P 7.8. Assume the source resistance(s) are negligible.

Section 7.11 is prerequisite for the following problems.

P 7.44 Re-draw the circuit shown in Fig. P 7.3 to show how the reference point for pin voltages is established by the manner in which the supply and load are connected to the op amp. Use battery symbols to represent the positive and negative supplies. Show the effective load and the paths taken by the current exiting the op-amp output terminal for both positive and negative output voltages. Keep in mind that no (or negligible) current enters the op-amp input terminals.

P 7.45 Repeat Problem P 7.44 for the circuit in Fig. P 7.4.

P 7.46 Repeat Problem P 7.44 for an inverting amplifier.

P 7.47 What is the effective load on the op amp in the circuit of Fig. P 7.8?

P 7.48 Refer to the two-stage amplifier shown in Fig. P 7.33. (a) Obtain expressions for the power delivered by the supply to each stage, the power dissipated by the resistive components of each stage, and the power dissipated by each op amp. (b) Let $V_{CC} = 15\,\text{V}$, $v_S = 100\,\text{mV}$, $R_S = 75\,\Omega$, $R_1 = R_3 = 10\,\text{k}\Omega$, $R_2 = R_4 = 100\,\text{k}\Omega$, $R_L = 5\,\text{k}\Omega$ and calculate the total power required of the supply. Assume ideal op amps, a symmetric supply, and rail-to-rail operation.

P 7.49 Repeat Problem P 7.48 for the circuit in Fig. P 7.34.

P 7.50 Refer to Fig. P 7.35, which shows (a) an inverting amplifier and (b) a non-inverting amplifier. The op amps are ideal and the source resistance is negligible.

1. Show that the amplifiers in Fig. P 7.35 have equal voltage gains.
2. Show that if the inverting and non-inverting amplifiers have equal dc inputs V_S, equal symmetric supplies $\pm V_{CC}$, and equal resistive loads R_L, then (i) both amplifiers draw the same power from their respective power supplies and (ii) the powers dissipated by the op amps in the two amplifiers are equal.
3. Let I denote an inverting amplifier and N denote a non-inverting amplifier. A two-stage amplifier is to be built as one of the cascades II, IN, NI, or NN. The two-stage amplifier is to be powered by a symmetric supply $\pm V_{CC}$ and must drive a resistive load R_L. Which of the four two-stage cascades requires the least power from the supply?
4. Let $R_1 = 10\,\text{k}\Omega$, $R_2 = 100\,\text{k}\Omega$, $R_L = 5\,\text{k}\Omega$, $V_{CC} = 15\,\text{V}$, and $V_S = 100\,\text{mV}$. Calculate the total power required by each of the two-stage cascades.

P 7.51 Obtain an expression for the power required by the circuit in Fig. P 7.6, with $N = 3$. Assume source resistances are negligible, that the op amps are ideal and rail-to-rail, that the supply is symmetric (denoted by $\pm V_{CC}$), and that the load is resistive (denoted by R_L).

P 7.52 Obtain an expression for the power required by the circuit in Fig. P 7.8. Assume source resistances are negligible, that the op amps are ideal and rail-to-rail, that the supply is symmetric (denoted by $\pm V_{CC}$), and that the load is resistive (denoted by R_L).

P 7.53 Repeat Problem P 7.51 for the circuit in Fig. P 7.36.

P 7.54 Obtain an expression for the power dissipated by the op amp in the circuit in Fig. P 7.7. Assume source resistances are negligible, that the op amps are ideal and rail-to-rail, that the supply is symmetric (denoted by $\pm V_{CC}$), and that the load is resistive (denoted by R_L).

P 7.55 Repeat Problem P 7.54 for the circuit in Fig. P 7.8.

P 7.56 Repeat Problem for the circuit in Fig. P 7.36.

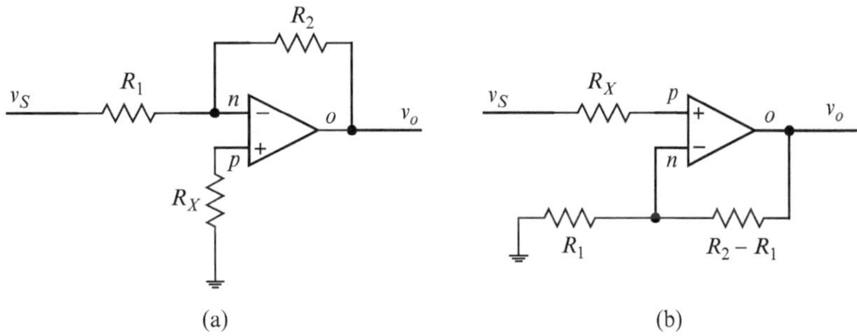

Fig. P 7.35 See Problem P 7.50

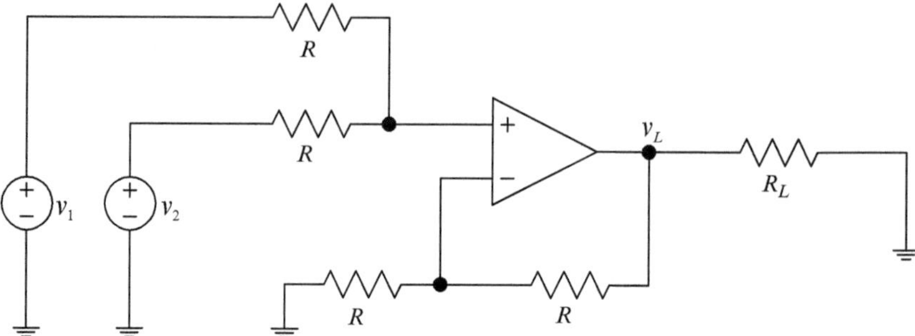

Fig. P 7.36 See Problem P 7.53, 56

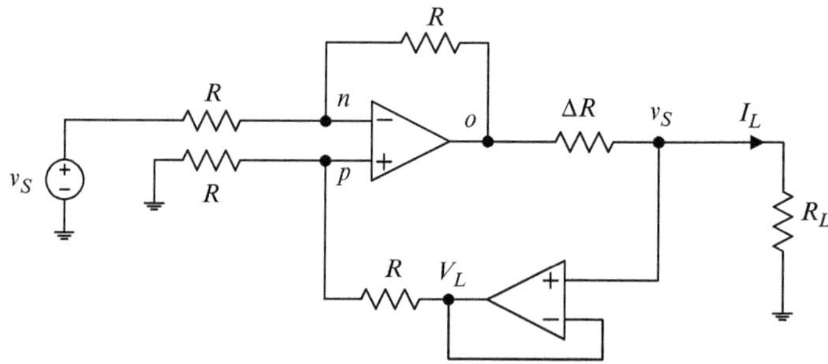

Fig. P 7.37 See Problem P 7.59

P 7.57 The power supply for a circuit must be able to provide the total power dissipated by the circuit. Usually, a power supply will be specified with a margin of 20% or more; for example, if the circuit to be powered dissipates 100 W, the designer might specify a 120 W supply. Use a 25% margin and specify the capacity of a power supply for the circuit in P 7.8, where $v_{in} = 700\,\text{mV}$, $V_{CC} = 20\,\text{V}$, $R_1 = 10\,\text{k}\Omega$, $R_2 = R_4 = 50\,\text{k}\Omega$, $R_3 = 22\,\text{k}\Omega$, and $R_L = 5\,\text{k}\Omega$.

Section 7.12 is prerequisite for the following problems. Note: Material presented in this chapter permits conceptual design of certain op amp circuits, based upon the ideal dc model for an op amp. Resulting circuits would work as specified if the op amps were ideal and if the currents and voltages involved are constant or slowly varying, but might not work with real op amps or for rapidly varying inputs. More realistic circuits typically require additional components whose roles we are at this point unprepared to discuss. Chapter 17 offers a more complete treatment of op amps and op amp circuits, where design can be discussed in more depth. The problems that follow are to be interpreted as conceptual design problems, with additional constraints imposed by supply voltage and maximum output current. In all cases, you should verify (and perhaps fine-tune) your design using simulation software.

P 7.58 Design an amplifier whose transresistance equals 1 kΩ and is independent of the load resistance. The source resistance is 500 kΩ. If the maximum output current (magnitude) is 35 mA, and the load resistance ranges from 2 to 4 kΩ, what is the maximum permissible supply voltage? Use a simulation to verify your design.

P 7.59 (a) Show that the circuit in Fig. P 7.37 is a transconductance amplifier. (b) If the supply voltage, the load resistance, and ΔR are fixed, what is the maximum permissible magnitude of the excitation v_S (for linear operation)? (c) Design an amplifier whose transconductance equals 2 mS and is independent of the load resistance. (d) Use a simulation to verify your design.

P 7.60 A certain tachometer produces a voltage given by $v = k_T n$, where n is the angular velocity of the rotor in rpm and $k_T = 100\,\text{mV}(\text{rpm})^{-1}$. (a) Design a circuit that will convert the voltage v produced by the tachometer to a current $i = k_A v$ that will drive a 50 Ω load. As the speed of the tachometer ranges from zero to 200 rpm, the magnitude of the current provided to the load must range from zero to 40 mA. (b) Will the circuit operate linearly over the specified range of the tachometer output? (c) Use a simulation to verify your design. *Hint*: See Problem P 7.59.

P 7.61 A small, low-power portable appliance requires a 9 V battery or supply and draws no more than 20 mA. (a) Design an op amp circuit that efficiently converts the 12 V available from an automobile battery to the 9 V required by the appliance. Show how to power the op amp from the automobile battery, given that the input is positive and that the output is positive and less than 12 V. (b) What dissipation rating is required for the op amp? Why would you use such a circuit instead of a simple resistive voltage divider? (c) How efficient is the circuit? (d) Use a simulation to verify your design.

P 7.62 The voltage produced by a certain source ranges from zero to 2 V. Design a circuit that effectively converts the voltage to a current that ranges from zero to −20 mA. Use a simulation to verify your design. *Hint*: See Problem P 7.59.

P 7.63 A photodiode produces a current that is approximately proportional to the intensity of incident light. The angular velocity of a small permanent-magnet dc motor is proportional to the voltage impressed on the motor armature. Design a circuit that makes the speed of the motor proportional to the light incident on the photodiode.

P 7.64 The resistance of a certain pressure sensor is given by

$$R(p) = R_0 + \alpha\,(p - p_0),$$

where p is atmospheric pressure, $p_0 = 101\,\text{kPa}$ is a reference pressure (approximately equal to average sea-level atmospheric pressure), $\alpha = 1\,\Omega\,\text{kPa}^{-1}$, and $R_0 = 100\,\Omega$ is the resistance for $p = p_0$. Design a circuit which, when driven by the pressure sensor, produces a voltage given by

$$v = k(p - p_0),\quad 0.8\,p_0 \le p \le 1.2\,p_0,$$

where $k = 100\,\text{mV}\,\text{kPa}^{-1}$. Use a Norton source model incorporating the pressure sensor, assuming you have available a good current source whose output resistance is much, much larger than $R\ (p)$ for any pressure of interest.

P 7.65 A certain photocell produces a current given by $i = k_0\lambda$, where λ is the intensity of the incident light in Wm^{-2} and $k_0 = 1\,\text{mA}\,\text{W}^{-1}\text{m}^2$. The internal resistance of the photocell exceeds 50 kΩ. In a certain application, an array of four such photocells is to be used to measure the average intensity of light incident on a surface. Design a circuit whose inputs are the currents from the photocells and whose output is a voltage having magnitude $|v_L| = k_V\Lambda$, where Λ is the average of the intensities incident on the four photocells and $k_V = 10\,\text{V}\,\text{W}^{-1}\text{m}^2$. The maximum intensity of the incident light is $\lambda_{max} = 1.7\,\text{Wm}^{-2}$ and the load resistance exceeds 10 kΩ.

Chapter 8
Capacitance

An ideal resistor is a purely dissipative element. A resistor does not generate electrical energy, nor does it store electrical energy for release at a later time. The terminal voltage v at any time is given by $v = i R$, where i is the current at the *same* time. Consequently, the response of a resistor to current through or voltage across its terminals is instantaneous, depending only on the present value of the voltage or current. Simply put, a resistor has no electrical memory.[1] Consequently, a resistive circuit as a whole has no electrical memory. The output voltage or current at any instant depends only upon the input voltage or current at the same instant. Some useful circuits are, for all practical purposes, resistive. But most useful signal-processing operations, such as integration and filtering, require memory and cannot be performed by a purely resistive circuit.

At any time t, the output of a circuit or circuit element having memory depends upon values of the input at times earlier than t.[2] There are two passive elements that exhibit memory: *capacitors and inductors*. These elements exhibit memory because they can store energy for later release, which allows the present output of a circuit containing these elements to depend upon previous inputs. In a capacitor, energy is stored in an electric field. In an inductor, energy is stored in a magnetic field. We treat capacitance in this chapter and inductance in the next.

8.1 Capacitance

Figure 8.1 shows two parallel conducting plates separated by an insulator (e.g., air) and connected by wires to a battery. When the battery is first connected to the plates, electrons are drawn away from the upper plate by the positive charge on the positive terminal of the battery and forced through the battery to the lower plate. As electrons migrate away from the positive plate and toward the negative plate[3] current exists in the circuit in the direction shown. As the current continues, uncovered positive charges left behind by departing electrons make it increasingly difficult for the battery to pull electrons away from the upper plate. Because charge cannot pass through the insulating medium, the current decreases with time until the voltage across the plates is equal in magnitude and opposite in sign of that across the battery terminals.[4] At that point, the battery can no longer draw electrons away from the upper plate and the current goes to zero. The time during which all this happens depends upon the magnitude of the current (the rate at which the charge is transferred), which in turn depends upon

[1]Resistors do possess thermal memory because resistance changes with temperature and the temperature of a resistor depends upon past current through the resistor. This particular form of memory has not been found useful in signal-processing circuits.

[2]So far as we know, the present output of a circuit (or any other physical system) cannot depend upon future values of an input. If it were otherwise, we could predict the future.

[3]Recall that the positive direction for current is opposite to the direction of electron flow.

[4]Electrons migrate until the work the battery can do on a charge equals the work necessary to move the charge from the positive (upper) plate through the circuit to the negative (lower) one, or, until Kirchhoff's voltage law is satisfied, whichever perspective you prefer.

T.H. Glisson, *Introduction to Circuit Analysis and Design*,
DOI 10.1007/978-90-481-9443-8_8,

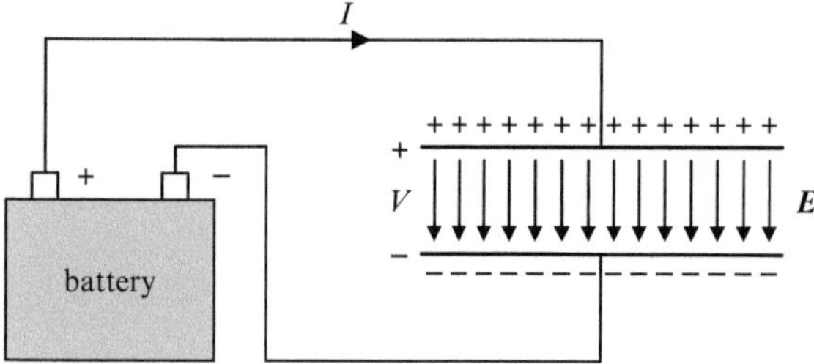

Fig. 8.1 Charged parallel conducting plates separated by an insulating medium

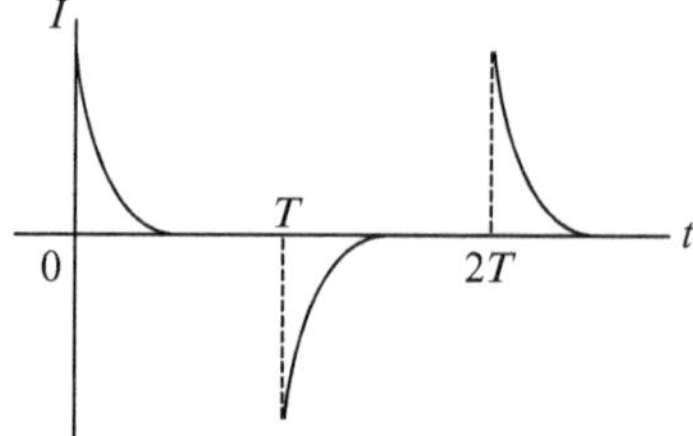

Fig. 8.2 Qualitative graph of current in the circuit of Fig. 8.1 when the polarity of the battery is periodically reversed (at the times 0, T, $2T$, ...)

the ability of the battery to supply current, the area and separation of the plates, the resistance of the wires and plates, and the internal resistance of the battery itself.

Exercise 8.1. Refer to Fig. 8.1. Draw a qualitative graph of the voltage V versus time, beginning at $t = 0$ when the battery is first connected to the plates and assuming the voltage equals zero at that time.

If the battery is disconnected from the plates, the plates remain charged, because there is no path by which the electrons can return to their original locations.[5] If the battery is reconnected, but with polarity reversed, a current will again exist until the plates are fully charged in the opposite manner. This process – disconnecting and then reconnecting the battery with opposite polarity – could be repeated indefinitely, in which case the current might be as shown (qualitatively) in Fig. 8.2.

As illustrated by Fig. 8.2, current can exist in the circuit of Fig. 8.1 even though no charges move through the insulator separating the plates. This phenomenon puzzled scientists for years because of their firm (and correct) belief that current is continuous on a closed path. The puzzle was solved by Maxwell,[6] who showed that there are two kinds of current: *Conduction current*, which is motion of charge, and *displacement current*, which can exist even where no charges are present – even in free space – if time-varying electric fields are present. Total current (conduction + displacement) is continuous. In the circuit of Fig. 8.1, the current in the wires is predominantly conduction current, the current in the insulating medium between the plates is predominantly displacement current, the total current everywhere in the circuit is the same, and the total current is nonzero only while the electric field between the plates is changing. Once the electric field is established and unchanging, the displacement current goes to zero, the space between the plates is an open circuit, and the total current goes to zero as well.[7]

Example 8.1. A 15-V battery is connected to a pair of plates separated by an insulating medium, as shown in Fig. 8.1. The initial charge on each plate equals zero. The battery is connected to the plates at $t = 0$, after which time the current through the battery is given by

$$i(t) = \begin{cases} 0, & t < 0, \\ I_0\, e^{-t/\tau}, & t \geq 0, \end{cases} \qquad (8.1)$$

where $I_0 = 50\,\text{mA}$ and $\tau = 20\,\text{ps}$. Find the charge on the positive plate after the battery has been connected for a very long time (for $t \to \infty$).

[5]The charge will gradually leak through the imperfect insulator separating the plates, but for a good insulator, the leakage will occur very slowly.

[6]James Clerk Maxwell (1831–1879).

[7]Further discussion of displacement current is beyond the scope of this book. Conduction current is the dominant current in elements and circuits considered in this book, except in insulating media *inside* capacitors, where the current is predominantly displacement current. Even in that case, we are concerned only with the conduction current into and out of the *terminals* of the capacitor.

Solution: Charge is the integral of current. Thus, for $t \geq 0$,

$$q(t) = \int_{-\infty}^{t} i(t')\,dt' = \int_{0}^{t} I_0\, e^{-t'/\tau}\,dt'$$
$$= -I_0\,\tau\left[e^{-t'/\tau}\right]_0^t = I_0\,\tau\left(1 - e^{-t/\tau}\right). \quad (8.2)$$

For $t \rightarrow \infty$, the exponential vanishes and

$$q(\rightarrow \infty) = I_0\,\tau = 1\,\text{pC}. \quad (8.3)$$

8.2 Capacitors

The parallel-plate arrangement in Fig. 8.1 constitutes a **capacitor**. Capacitors for use in circuits are constructed of parallel conductors, called **plates**, separated by an insulating medium, such as air, paper, glass, mica, and various ceramics and plastics. The insulating material separating the plates is called the **dielectric** for the capacitor. Often, the plates are long strips of metal foil or metalized plastic, separated by an insulator and wrapped into a tight cylinder, with a terminal (a wire) connected to each of the foil strips.

It is shown in introductory physics textbooks that the charge (magnitude) on either of the parallel plates is given by

$$q = Cv, \quad (8.4)$$

where v is the voltage across the plates (from positive to negative), q is the magnitude of the charge on either plate, and C is a constant called the **capacitance** of the parallel-plate arrangement. The unit of capacitance is the *farad* (F),[8] where 1 F equals 1 C V^{-1} or, equivalently, 1 A s V^{-1}. A two-terminal device so constructed that its dominant electrical property is capacitance is called a **capacitor**.

It is customary to refer to the charge on the positive plate of a capacitor as the *charge on the capacitor*, even though the net charge on the capacitor as a whole is zero. It also is customary to say the capacitor is *charged* to v (volts), even though charge and voltage are different things. Such language, though imprecise, is commonly used and understood by electrical engineers.

The capacitance of a parallel-plate capacitor[9] is given approximately by[10]

$$C = \frac{\varepsilon_0\,\varepsilon_r\,A}{d}, \quad (8.5)$$

where $\varepsilon_0 = 8.854 \times 10^{-12}\,\text{F m}^{-1}$ is the permittivity of a vacuum, $\varepsilon_r \geq 1$ (dimensionless) is the **relative permittivity** of the dielectric ($\varepsilon_r = 1$ for a vacuum or air), A is the area of one plate, and d is the distance between the plates. The approximation (8.5) is quite good if the separation d is very much smaller than both the width and the length of a plate.

Achieving a large capacitance requires a dielectric having large relative permittivity, a large plate area, and a small plate separation. However, the dielectric thickness (plate separation) must be large enough to withstand normal working voltages without breaking down (without allowing an arc between the plates).

Example 8.2. The capacitance of a parallel-plate capacitor having an air (free-space) dielectric is 1.00 F. The separation of the plates is 100 μm (about the thickness of a sheet of notebook paper). Find the area of either plate.

Solution: From (8.5),

$$A = \frac{C\,d}{\varepsilon_0} = \frac{(1\text{F})(100 \times 10^{-6}\text{m})}{8.85 \times 10^{-12}\,\text{Fm}^{-1}}$$
$$\cong 11.3 \times 10^6\,\text{m}^2. \quad (8.6)$$

If the plates were square, each would be about 3.4 km (2 miles) on a side.

As you can see from Example 8.2, a 1 F capacitor is quite large. Most capacitors used in electronic circuits have capacitances ranging from about 10 nF to about

[8] After the English scientist Michael Faraday (1791–1867).

[9] In a parallel-plate capacitor, the plates are of equal area and uniformly separated with their edges aligned.

[10] Eq. (8.5) is derived in virtually all university-level introductory physics textbooks.

1 μF. Those used in power supplies, often the largest capacitors in an electronic system, have capacitances on the order of 20 – 2000 μF.

Some off-the-shelf capacitors, like resistors, are commonly available in only certain sizes, again from E series. Unlike resistors, the E3 and E6 series of values are widely used for some common types of capacitors because of the difficulty of manufacturing those types to higher precision. The E6 and E3 (in bold) series are (**10**, 15, **22**, 33, **47**, 68). The corresponding capacitance values are the series numbers multiplied by an integer power of ten; e.g., 4.7 pF, 47 pF, 470 pF, and so on. Higher series values for certain types of capacitors can be obtained at higher cost. But also, improved manufacturing processes and better materials have made it possible to manufacture some kinds of capacitors to almost any value at relatively high precision. It is not difficult to obtain (in bulk) capacitors having capacitances of 1.0, 2.0, 3.0, 4.0, 5.0, ... pF, although none but the first of these values is from an E series.

In addition to capacitance and kind of dielectric, capacitors are characterized by the maximum voltage they can endure without breakdown (conduction through the dielectric). All else equal, a thin dielectric provides greater capacitance than a thicker one but offers less resistance to breakdown. Thus a capacitor might be rated at 120 pF and 50 VDC, meaning that 50 V is the maximum sustained voltage (magnitude) that should be applied to the capacitor.

Capacitance, like resistance, can vary with operating temperature. For example, the dielectric might expand or shrink with changes in temperature, which would alter the distance between the plates and thus the capacitance. Overheating due to careless soldering could change the capacitance of a capacitor (permanently) by 5% or more. Also, it is difficult in manufacture to maintain uniform dielectric thickness and various electrical properties (such as relative permittivity) over the required area. Consequently, capacitance of an off-the-shelf capacitor typically is specified only to within ±20% of the nominal value. Some have large asymmetric tolerances, such as +30% to −80%. More accurate capacitors are obtainable but, of course, at higher prices. As a rule, capacitors having mica, glass, or plastic (e.g., polystyrene or polycarbonate) dielectrics exhibit good accuracy, low leakage current through the dielectric, and good temperature stability, whereas those using an impregnated paper dielectric are poorer on those counts. The dielectric in an electrolytic capacitor is a very thin oxide layer, created and maintained by an electrochemical process. Thus an electrolytic capacitor is polarized, with one terminal marked positive, meaning that that terminal must always be at a higher potential than the other. Electrolytic capacitors are typically used to smooth voltages that are always positive, but which have ac components. Figures 8.3 and 8.4 show assorted fixed and variable capacitors.

Capacitors are labeled in various ways, as illustrated by Fig. 8.5. The capacitance of capacitor (a) is clearly marked as 470 μF. The marking on capacitor (b) is a common three-digit code for capacitance in picofarads. The first two digits are significant figures and the last is the number of following zeros. The capacitance is 100 pF. The marking on capacitor (c) is the numerical value of the capacitance in μF. The colored-dot code illustrated by capacitor (d) (example colors) is no longer used, probably because no one could ever figure it out. Suffice it to say that there are numerous schemes for indicating the capacitance, maximum voltage rating, and tolerance of a capacitor. You will find some that are not marked at all, probably because the manufacturer could not come up with a sufficiently obscure code. The best way to decipher a particular code on a particular capacitor is to visit the manufacturer's website.

Some capacitors are constructed by depositing metal layers on each side of a thin sheet of a plastic film, such as polyester or polypropylene. The metal layers are the plates and the film is the dielectric. Capacitors in integrated circuits are constructed by laying down a layer of metal, then a layer (a *film*) of dielectric material, then another layer of metal. Discrete and integrated capacitors fabricated in this manner are called **film capacitors** or **thin-film capacitors**. The thickness of the film (the distance between the plates) can be as small as about 1 μm, so sizable capacitances can be attained within reasonable surface areas.

A film capacitor, like a film resistor, is a planar device. By analogy with sheet resistance, capacitance of film capacitors often is described in terms of a quantity called **sheet capacitance**, which is the *capacitance per unit area* of the capacitor. Sheet capacitance is denoted by C_{sheet}. Thus the capacitance of a film capacitor can be expressed as

$$C = \frac{\varepsilon_r \varepsilon_0 A}{d} = A\, C_{\text{sheet}}, \tag{8.7}$$

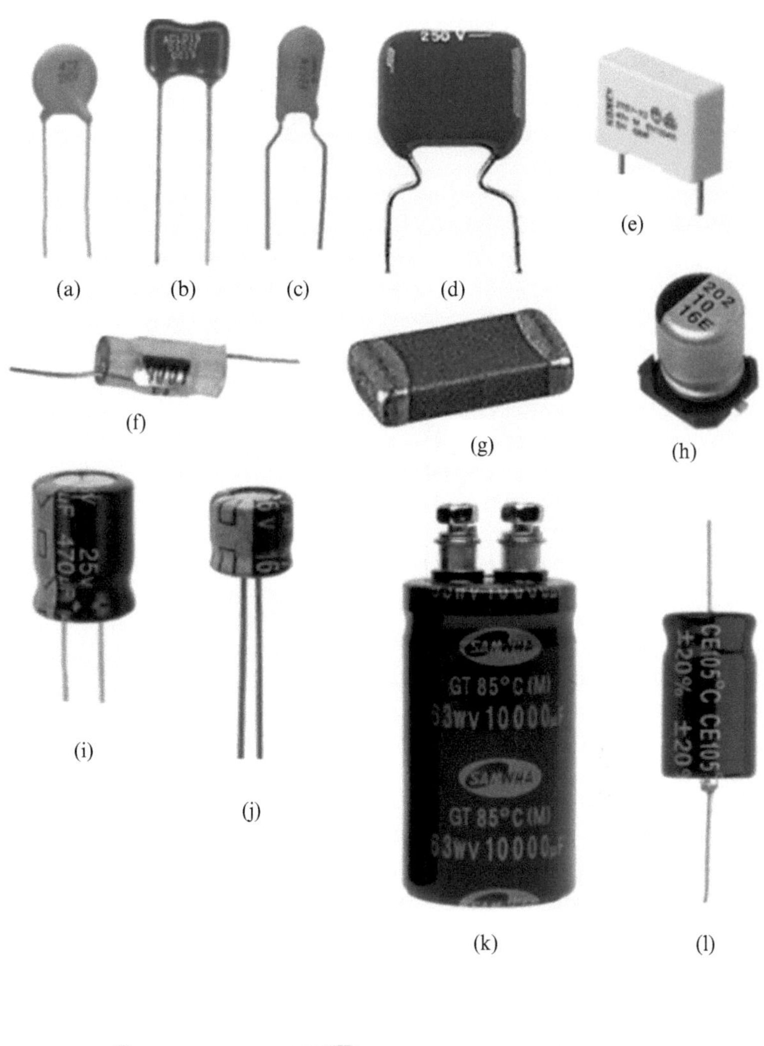

Fig. 8.3 Fixed capacitors (not to scale): (**a**) ceramic, (**b**) mica, (**c**) tantalum, (**d**), (**e**) polypropylene, (**f**) polystyrene, (**g**) ceramic chip, (**h–l**) electrolytic (Photographs courtesy of Rapid Electronics, Ltd)

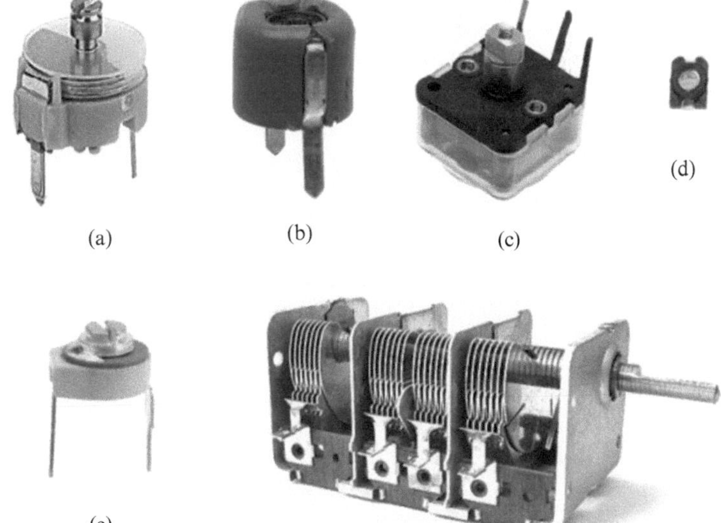

Fig. 8.4 Variable capacitors (not to scale). (**a–e**) small trimmer capacitors, (**f**) four-gang air-dielectric tuning capacitor (Photographs (**a–e**) courtesy of Rapid Electronics, Ltd. (**f**) courtesy of West Florida Components)

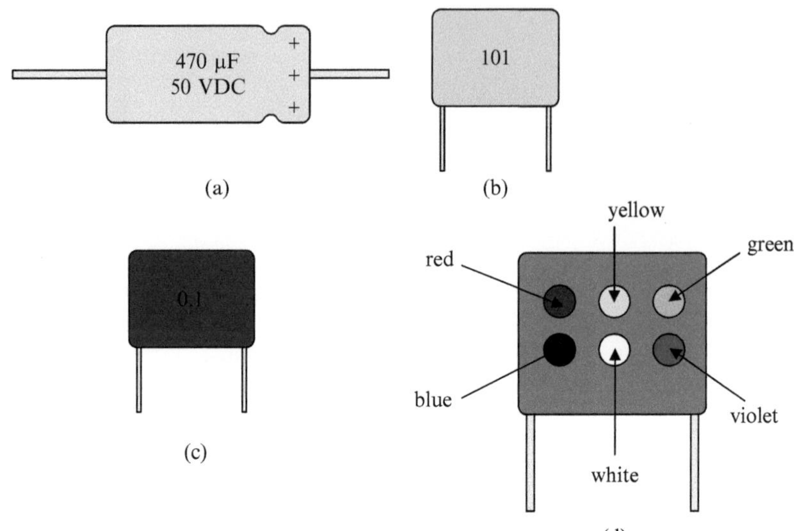

Fig. 8.5 Capacitance markings. In (d), the colors are for illustration only (have no special significance).

where A is the surface area. It follows that

$$C_{\text{sheet}} = \frac{\varepsilon_r \varepsilon_0}{d}. \tag{8.8}$$

For example, if an IC manufacturing facility (a *foundry*) can deposit layers as thin as 1 μm using a dielectric material having relative permittivity $\varepsilon_r = 7$, that foundry can provide a sheet capacitance of

$$C_{\text{sheet}} = \frac{\varepsilon_r \varepsilon_0}{d} = \frac{(7)(8.854 \times 10^{-12}\,\text{F m}^{-1})}{10^{-6}\,\text{m}}$$
$$\cong 62\,\mu\text{F m}^{-2} = 62\,\text{pF mm}^{-2}.$$

8.3 Terminal Characteristics of an Ideal Capacitor

Figure 8.6 shows circuit-diagram symbols for non-polarized and polarized (electrolytic) capacitors. To avoid distractions from our main purpose, we use the symbol for a non-polarized capacitor almost exclusively.

From (8.4),

$$i(t) = \frac{dq}{dt} = C\frac{dv}{dt}, \tag{8.9}$$

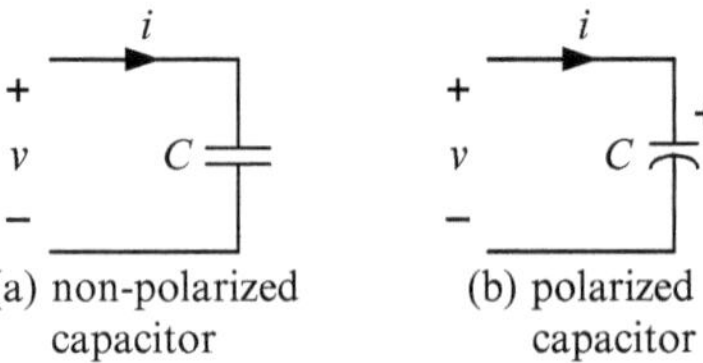

Fig. 8.6 Circuit-diagram symbols for capacitors

which implies

$$dv(t) = \frac{1}{C}i(t)dt.$$

It follows that

$$\int_{t_0}^{t} dv(t) = v(t) - v(t_0) = \frac{1}{C}\int_{t_0}^{t} i(t')dt' \tag{8.10}$$

Eq. (8.10) is usually written

$$v(t) = v(t_0) + \frac{1}{C}\int_{t_0}^{t} i(t')dt' \tag{8.11}$$

Equation (8.11) shows that a capacitor has memory: The present terminal voltage depends upon past values of the terminal current.

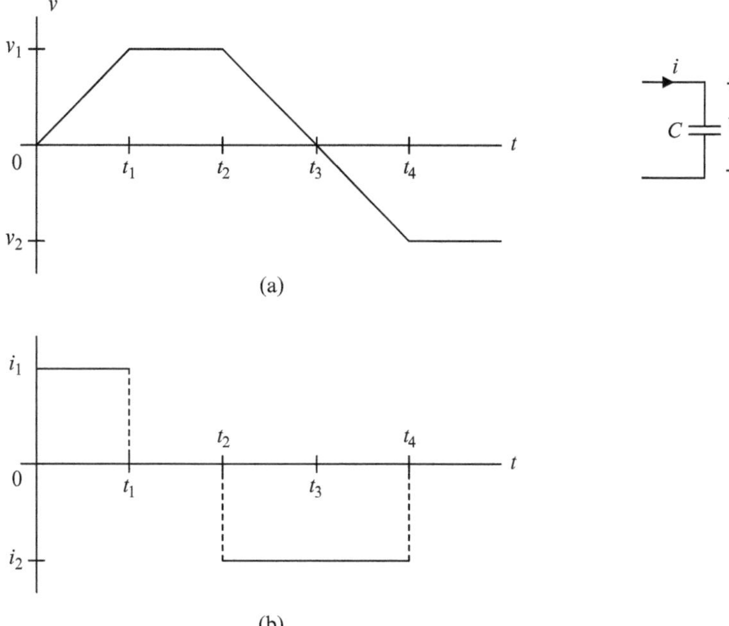

Fig. 8.7 See Example 8.3

Again, from (8.4), we have

$$i = \frac{dq}{dt} = C\frac{dv}{dt}. \tag{8.12}$$

Equations (8.11) and (8.12) are the terminal characteristics for an *ideal* capacitor. In this chapter we assume all capacitors are ideal. More realistic models for capacitors are described in subsequent chapters.

Equation (8.12) relates the voltage across a capacitor to the current entering the positive terminal of the capacitor. For economy, we say that (8.12) relates the *voltage across a capacitor to the current through the capacitor*. By implication, the positive direction of the current is into the positive terminal.

Equation (8.12) has several important implications:

- Current is nonzero whenever the terminal voltage is changing (when dv/dt is nonzero) and is zero otherwise. *A capacitor appears as an open circuit to a constant voltage.*
- The magnitude of the terminal current is proportional to the *rate of change* of the terminal voltage. *The more rapid the voltage change, the larger the required current.*
- Changing the terminal voltage instantaneously would require infinite current. Thus, in physical circuits, the voltage across a capacitor is a continuous function of time.[11] In other words, *the voltage across a capacitor cannot change instantaneously.*
- The current through a capacitor is proportional to the capacitance C. *More current (or more time) is required to charge a larger capacitor to a specified voltage than is required for a smaller one.*

A circuit in which all currents and voltages are constant is said to be in **dc steady state**. If the voltage across a capacitor is constant, the current through the capacitor is zero. Thus, *for any circuit in dc steady state, all capacitors in the circuit are effectively open circuits.*

Example 8.3. Figure 8.7(a) shows a graph of voltage across a capacitor as a function of time. Draw a graph of the current i through the capacitor versus time.

[11]Continuity of the voltage across a capacitor also follows from (8.10).

Solution: Figure 8.7(b) shows the current through the capacitor, given by (8.12). The direction of the current is into the terminal taken as positive in the definition of the voltage v, where, from (8.12), the currents i_1 and i_2 in Fig. 8.7(b) are the slopes of the lines directly above in Fig. 8.7(a), multiplied by the capacitance C. Thus,

$$i_1 = \frac{Cv_1}{t_1}, \quad i_2 = \frac{C(v_2 - v_1)}{t_4 - t_2}.$$

Exercise 8.2. Figure 8.8 shows a graph of the voltage across a 20 pF capacitor. Draw a graph of the current through the capacitor.

Example 8.4. The voltage across a 0.1 μF capacitor is given by

$$v(t) = V\cos(2\pi f\,t),$$

with $V = 10\text{V}$ and $f = 1$ MHz. Obtain an expression for the current through the capacitor.

Solution: From (8.12),

$$\begin{aligned} i(t) &= C\frac{dv}{dt} = -2\pi f\,V C\sin(2\pi f\,t) \\ &= -I\sin(2\pi f\,t), \end{aligned}$$

with

$$\begin{aligned} I &= 2\pi f\,V\,C = 2\pi\,(1\text{MHz})(10\text{V})(1\text{nF}) \\ &\cong 6.28\,\text{A}. \end{aligned}$$

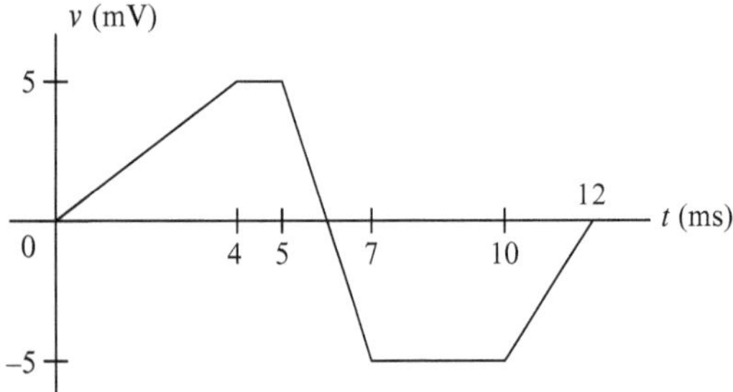

Fig. 8.8 See Exercise 8.2

Exercise 8.3. The voltage across a 0.1 μF capacitor is given by

$$v(t) = V\sin(2\pi f\,t),$$

with $V = 5\text{V}$ and $f = 10$ MHz. Obtain an expression for the current through the capacitor. Find the maximum magnitude of the current.

Example 8.5. The current through a 0.5 μF capacitor is given by

$$i(t) = \begin{cases} 0, & t < 0, \\ Ie^{-t/\tau}, & t \geq 0, \end{cases}$$

where $I = 5$ mA and $\tau = 10$ ms. Obtain an expression for the current through the capacitor.

Solution: From (8.10),

$$\begin{aligned} v(t) &= \frac{1}{C}\int_{-\infty}^{t} i(t')dt' = \frac{I}{C}\int_{0}^{t} e^{-t'/\tau}dt' \\ &= -\frac{I\tau}{C}e^{-t'/\tau}\bigg|_0^t \\ &= \frac{I\tau}{C}\left[1 - e^{-t/\tau}\right] = V\left[1 - e^{-t/\tau}\right], \end{aligned}$$

with

$$V = \frac{I\tau}{C} = \frac{(5\,\text{mA})(10\,\text{ms})}{500\,\text{nF}} = 100\,\text{V}.$$

Exercise 8.4. Figure 8.9 shows a graph of the current through a 5 nF capacitor. The current equals zero for times not shown. Draw a graph of the voltage across the capacitor.

Example 8.6. Refer to Fig. 8.10. What is the voltage v_C across the capacitor a very long time after the switch is moved from a to b?

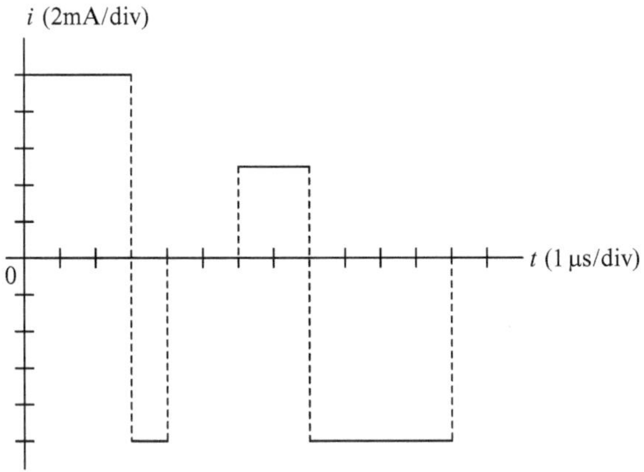

Fig. 8.9 See Exercise 8.4

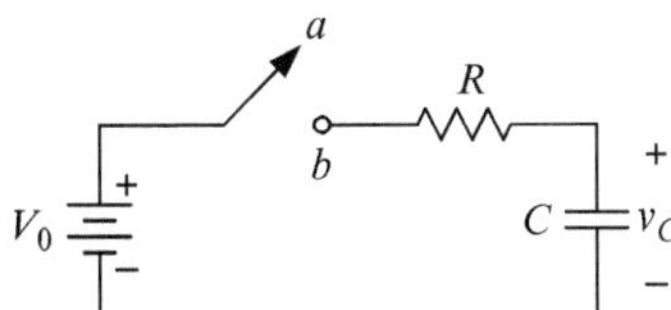

Fig. 8.10 See Example 8.6

Solution: After a sufficiently long time, the capacitor appears as an open circuit (the voltage source is dc), the current through the resistor equals zero, and, by Kirchhoff's voltage law, the voltage across the capacitor equals the source voltage V_0.

Exercise 8.5. Refer to Fig. 8.11, where the current source is constant. What is the voltage v_C across the capacitor a very long time after the switch is moved from *a* to *b*?

Because the voltage across a capacitor cannot change instantaneously, circuit models implying such changes are inconsistent with reality. Figure 8.12 shows two examples of such models, often posed as trick questions on examinations. In Fig. 8.12(a), a battery is suddenly connected (at $t = 0$) to a capacitor whose terminal voltage equals zero for $t < 0$. The model implies that the voltage across the capacitor

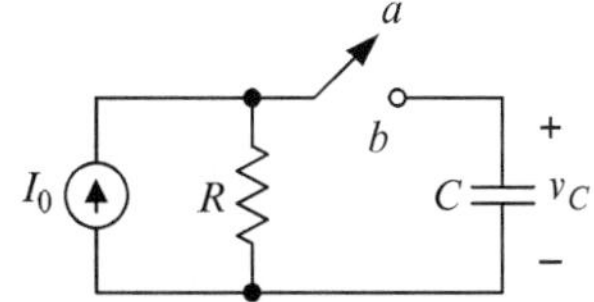

Fig. 8.11 See Exercise 8.5

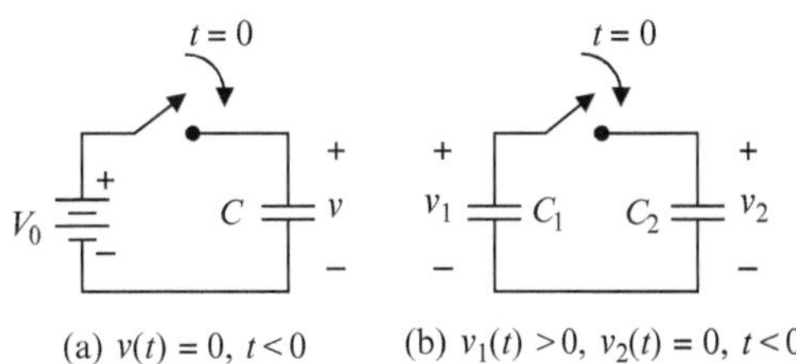

Fig. 8.12 Unrealistic circuit models implying instantaneous changes in capacitor terminal voltages

changes instantaneously, which would require infinite current. Omitted from the model is the resistance of the connecting wires (among other things). Including those parameters would make the model more realistic. In Fig. 8.12(b), a charged capacitor is suddenly connected (at $t = 0$) to an uncharged capacitor. This model also implies that the voltages across the capacitors must change instantaneously, which would require infinite current. Moreover, comparing the initial and final voltages across the capacitors shows that energy somehow disappears from the circuit, in spite of the fact that there are no energy-loss mechanisms (e.g., resistors) in the model.[12] Again, the resistance of the connecting wires is omitted from the model. Including that resistance (no matter how small, so long as it is non-zero) would make the model realistic and would account for the lost energy.

Often, it is possible and even helpful to use unrealistic models similar to those in Fig. 8.12 for *qualitative* analysis (for discerning or describing how a circuit works) and even for calculations that do not require exact expressions for terminal current or voltage,[13] but obtaining correct expressions for currents or voltages

[12] Section 8.9 treats energy stored in the electric field in a capacitor.

[13] See Section 8.10.8 for an example.

as functions of time requires using models consistent with physical reality.

From another point of view, the circuit in Fig. 8.12(a) is unrealistic because the internal resistance of the source (battery) appears to be zero. All physical sources possess internal resistance. Whenever you encounter a voltage source having no series resistance or a current source having no parallel resistance in a circuit, be aware that any quantitative analysis of the circuit might lead to incorrect results.

Exercise 8.6. What is wrong with the circuit (model) in Fig. 8.13, where $i_0 > 0$ for all time? Is it possible to find a value for v_C at any time?

8.4 Charge-Discharge Time Constant

A circuit composed of only resistors and one capacitor is called an ***RC* circuit**. In this section, we obtain expressions for currents and voltages in an *RC* circuit subjected to an abrupt change in excitation.

The following notation is helpful: Let t_0 denote a particular time. Then t_0^- and t_0^+ denote times infinitesimally before and after t_0. For example, if $t = 0$ denotes the time at which switch is closed, then $t = 0^-$ is the time just before closure and $t = 0^+$ is the time just after closure.

Refer to Fig. 8.14, where a capacitor that has been previously charged to a voltage V_0 is connected at $t = 0$ to a resistor having resistance R. By Kirchhoff's current law, we have for $t > 0$ that

$$C\frac{dv}{dt} + \frac{v}{R} = 0 \Rightarrow \frac{dv}{dt} = -\frac{v}{RC}. \tag{8.13}$$

According to (8.13), the voltage v and its derivative have the same mathematical form (are proportional). The function having that property is the exponential function. You can show by substitution that the solution to (8.13) is

$$v(t) = K\exp\left(-\frac{t}{RC}\right),\ t>0, \tag{8.14}$$

For $t = 0^+$, (8.14) gives $v(0^+) = K$, so we have

$$v(t) = v(0^+)\exp\left(-\frac{t}{RC}\right),\ t>0. \tag{8.15}$$

From the problem statement, we know that $v(0^-) = V_0$. We also know that the voltage across a capacitor cannot change instantaneously. It follows that $v(0^+) = V_0$ and thus

$$v(t) = V_0\exp\left(-\frac{t}{RC}\right),\ t>0, \tag{8.16}$$

as illustrated by Fig. 8.14.

Equation (8.16) usually is written

$$v(t) = V_0\exp\left(-\frac{t}{\tau}\right),\ t>0, \tag{8.17}$$

where

$$\tau = RC. \tag{8.18}$$

is called the **time constant** for the circuit consisting of the capacitor and resistor. *The time constant determines how fast the capacitor discharges through the resistor*. Because $e^{-1} \cong 0.37$, the voltage across the capacitor drops to 37% of its initial value in a time equal to one time constant. Similarly, $e^{-3} \cong 0.05$, so the voltage drops to 5% of its initial value in a time equal to three time constants. You can show that the voltage is less than 1% of its initial value when t equals five time constants.

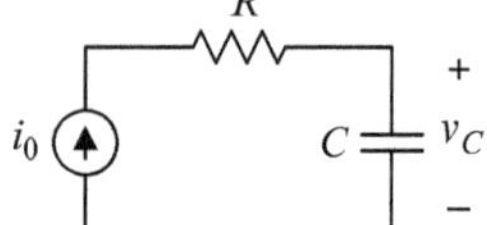

Fig. 8.13 See Exercise 8.6

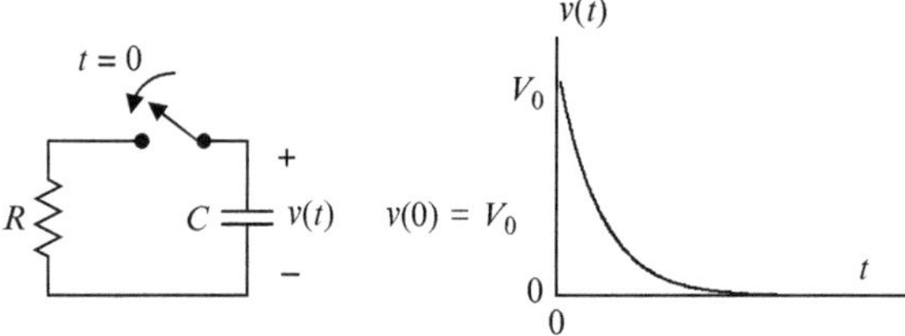

Fig. 8.14 The voltage across a charged capacitor decreases exponentially with time after the capacitor is connected to a resistor

Exercise 8.7. In the circuit shown in Fig. 8.14, $C = 100$ nF and the voltage $v(t)$ has dropped to half of its initial value 7 μs after the switch is closed. Find the resistance R.

Now refer to Fig. 8.15, where a capacitor that is initially uncharged is connected through a resistor to a constant voltage source at $t = 0$. Because the capacitor is initially uncharged, $v(0^-) = 0$. By Kirchhoff's voltage law

$$Ri + v(0^-) + \frac{1}{C}\int_{0^-}^{t} i(t')dt' = Ri + \frac{1}{C}\int_{0^-}^{t} i(t')dt' = V_0. \tag{8.19}$$

Differentiating once with respect to time and rearranging terms gives

$$\frac{di}{dt} = -\frac{i}{RC},$$

which we now know implies that

$$i = i(0^+)\exp\left(-\frac{t}{RC}\right), \quad t > 0. \tag{8.20}$$

From (8.19)

$$Ri(0^+) + \frac{1}{C}\int_{0^-}^{0^+} i(t')dt' = Ri(0^+) = V_0 \Rightarrow i(0^+) = \frac{V_0}{R}.$$

Thus,

$$i = \frac{V_0}{R}\exp\left(-\frac{t}{\tau}\right), \quad t > 0,$$

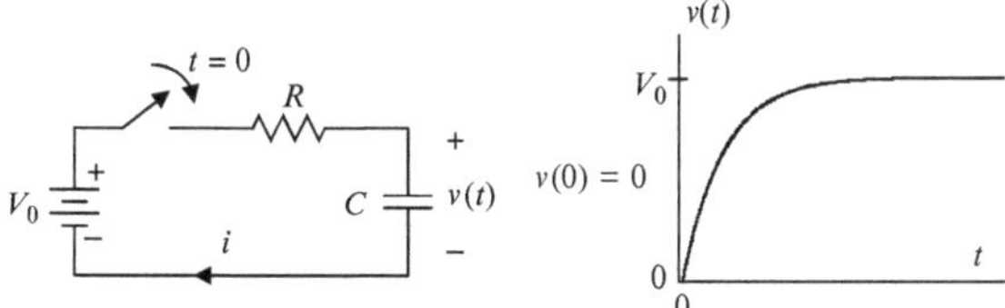

Fig. 8.15 When an uncharged capacitor is connected through a resistor to a constant voltage V_0, the voltage across the capacitor increases exponentially toward the voltage V_0

where $\tau = RC$ is the time constant for the circuit. The voltage $v(t)$ across the capacitor is given by

$$\begin{aligned} v(t) &= v(0) + \frac{1}{C}\int_0^t i(t')dt' \\ &= 0 + \frac{V_0}{\tau}\int_0^t \exp\left(-\frac{t'}{\tau}\right)dt', \end{aligned}$$

which yields

$$v(t) = V_0\left[1 - \exp\left(-\frac{t}{\tau}\right)\right], t \geq 0 \tag{8.21}$$

as illustrated in Fig. 8.15. Again, *the time constant determines how fast the capacitor charges when connected to a constant source through a resistor*. Because $1 - e^{-1} \cong 0.63$, the voltage equals about 63% of its final value V_0 for $t = \tau = RC$. You can show that for $t = 3\tau$ and $t = 5\tau$ the voltage equals about 95% and 99%, respectively, of V_0.

In general, *the time constant for a circuit consisting of any number of resistors and one capacitor is the product of the capacitance of the capacitor and the Thévenin equivalent resistance seen at the terminals of the capacitor.*

The time constant for charging and discharging a capacitor and the mathematical forms of the voltages during charging and discharging are important. You will encounter these concepts in many guises and countless times during your college days and most likely during your career as an electrical engineer. Study Fig. 8.14 and Equation (8.17) and Fig. 8.15 and Equation (8.21) until they are firmly fixed in your mind. Commit to memory that the time constant is a measure of how fast a capacitor can charge and discharge, and that the time required for a capacitor having capacitance C to fully discharge or fully charge through a resistor having resistance R is approximately $5RC$, where *fully* means a discharge to less than 1% of the initial value or a charge to more than 99% of the final value.

Exercise 8.8. In the circuit shown in Fig. 8.15, $R = 50$ kΩ, $V_0 = 5$ V, and the switch is closed at $t = 0$. At $t = 1$ μs the voltage $v(t)$ equals 2 V. Find the capacitance C.

Exercise 8.9. In Fig. 8.16, the switch is moved at $t = 0$ from position 0 to position 1 and left there until the capacitor is fully charged. The switch is then immediately moved to position 2 until the capacitor is fully discharged. (a) How long must the switch reside in each position? (b) Draw a graph (a neat, fully labeled sketch) of the voltage $v(t)$ across the capacitor for $t \geq 0$.

Example 8.7. In Fig. 8.17, the capacitor is initially charged such that $v(t) = V_0$ for $t < 0$. The switch is then closed at $t = 0$. (a) Obtain an expression for the voltage $v(t)$ for $t > 0$. (b) Draw graphs (neat, fully labeled sketches) of the voltage $v(t)$ for $t \geq 0$ and for (i) $V_0 < V_1$ and (ii) $V_0 > V_1$.

Solution: (a) Kirchhoff's voltage law gives

$$Ri + v(0) + \frac{1}{C}\int_0^t i(t')dt' = V_1.$$

Differentiating once with respect to time and rearranging terms yields

$$\frac{di}{dt} = -\frac{1}{RC}i,$$

which implies

$$i(t) = K\exp\left(-\frac{t}{RC}\right).$$

The voltage across the capacitor must be continuous, so the voltage across the resistor (left to right) at $t = 0$ must be $V_1 - V_0$. It follows that

$$i(0^+) = \frac{V_1 - V_0}{R} = K$$
$$\Rightarrow i(t) = \left(\frac{V_1 - V_0}{R}\right)\exp\left(-\frac{t}{RC}\right), \quad t > 0.$$

From (8.11),

$$\begin{aligned} v(t) &= V_0 + \left(\frac{V_1 - V_0}{RC}\right)\int_0^t \exp\left(-\frac{t'}{RC}\right)dt' \\ &= V_1 - (V_1 - V_0)\exp\left(-\frac{t}{RC}\right), \quad t > 0. \end{aligned} \tag{8.22}$$

(b) Figure 8.18 shows graphs of the voltage $v(t)$ given by (8.22), where the solid line is for $V_0 < V_1$ and the dashed line is for $V_0 > V_1$. In either case, the transition from V_0 to V_1 is essentially complete in about five time constants.

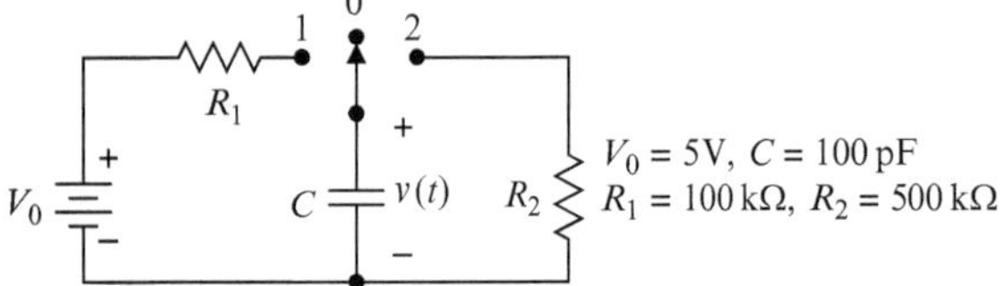

Fig. 8.16 See Exercise 8.9

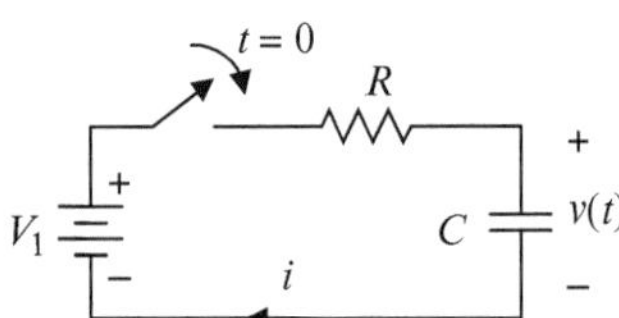

Fig. 8.17 See Example 8.7

The relation (8.22) obtained in Example 8.7 above can be written in a more general and useful form as

$$v(t) = v(\infty) - [v(\infty) - v(0^+)]\exp(-t/\tau), \quad t > 0. \tag{8.23}$$

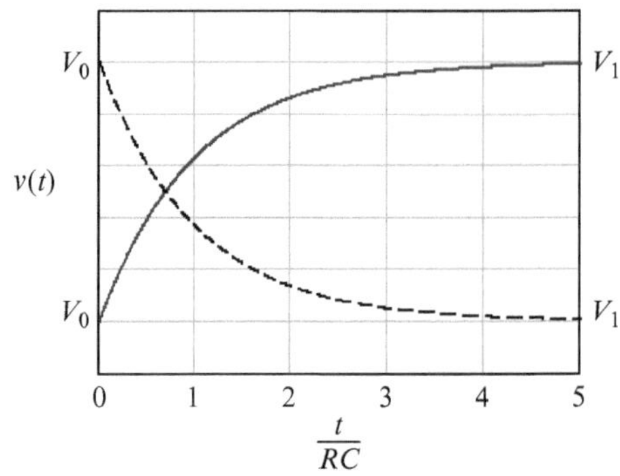

Fig. 8.18 See Example 8.7

The time constant τ in (8.23) is given by

$$\tau = R_T C, \tag{8.24}$$

where R_T *is the Thévenin equivalent resistance seen at the terminals of the capacitor*. Equation (8.23) can be applied to *any voltage* in any circuit consisting of any number of resistors and one capacitor, where the initial and final voltages are given or can be deduced.

Equation (8.23) describes the transition of the voltage across a capacitor from an initial value $v(0^+)$ *toward* a final value $v(\infty)$.

Example 8.8. Refer to Fig. 8.19. The switch is open for $t < 0$ and closed at $t = 0$. (a) Obtain an expression for the voltage $v_C(t)$. (b) Use (8.23) to obtain an expression for the voltage $v_1(t)$. (c) Show that your answers satisfy $v_1(t) + v_C(t) = V_0$ (Kirchhoff's voltage law) for $t > 0$.

Solution: (a) Because the switch was open for $t < 0$, $v_C(0^-) = 0$. Because the voltage across a capacitor cannot change instantaneously, $v_C(0^+) = 0$. After the switch has been closed for a long time, the capacitor appears as an open circuit, and

$$v_C(\infty) = \frac{R_2 V_0}{R_1 + R_2}.$$

The time constant for $t > 0$ is given by

$$\tau = R_T C,$$

where

$$R_T = R_1 \| R_2 = \frac{R_1 R_2}{R_1 + R_2}$$

is the Thévenin equivalent resistance at the terminals of the capacitor. Thus, from (8.23)

$$\begin{aligned} v_C(t) &= v_C(\infty) - [v_C(\infty) - v_C(0^+)]\exp\left(-\frac{t}{\tau}\right) \\ &= \frac{R_2 V_0}{R_1 + R_2}\left[1 - \exp\left(-\frac{t}{\tau}\right)\right], \quad t > 0. \end{aligned}$$

(b) From part (a), $v_C(0^+) = 0$. By Kirchhoff's voltage law,

$$v_1(0^+) = V_0 - v_C(0^+) = V_0.$$

For $t \to \infty$,

$$\begin{aligned} v_1(\infty) &= V_0 - v_C(\infty) \\ &= V_0 - \frac{R_2 V_0}{R_1 + R_2} = \frac{R_1 V_0}{R_1 + R_2} \end{aligned}$$

Thus

$$\begin{aligned} v_1(t) &= \frac{R_1 V_0}{R_1 + R_2} - \left(\frac{R_1 V_0}{R_1 + R_2} - V_0\right)\exp\left(-\frac{t}{\tau}\right) \\ &= \frac{R_1 V_0}{R_1 + R_2} + \frac{R_2 V_0}{R_1 + R_2}\exp\left(-\frac{t}{\tau}\right), \quad t > 0 \end{aligned}$$

Note that $v_1(0^-) = 0$, whereas $v_1(0^+) = V_0$, which illustrates the fact that a current or a voltage other than the voltage across a capacitor can change instantaneously.

(c)

$$\begin{aligned} v_1(t) + v_C(t) &= \frac{R_1 V_0}{R_1 + R_2} + \frac{R_2 V_0}{R_1 + R_2}\exp\left(-\frac{t}{\tau}\right) \\ &\quad + \frac{R_2 V_0}{R_1 + R_2}\left[1 - \exp\left(-\frac{t}{\tau}\right)\right] \\ &= \frac{R_1 V_0}{R_1 + R_2} + \frac{R_2 V_0}{R_1 + R_2} = V_0 \end{aligned}$$

Again, (8.23) is applicable not only to the voltage across a capacitor, but to *any node voltage or any branch current in any circuit consisting of any number of resistors and one capacitor*, where the initial and final voltages are given or can be deduced. If used to express a branch current, (8.23) becomes

$$i(t) = i(\infty) - [i(\infty) - i(0^+)]\exp\left(-\frac{t}{\tau}\right), \quad t > 0. \tag{8.25}$$

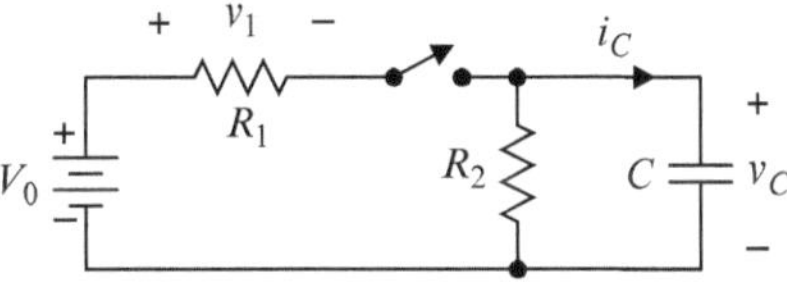

Fig. 8.19 See Example 8.8

Example 8.9. Refer to Fig. 8.19. Use (8.25) to obtain an expression for the current $i_C(t)$.

Solution: We deduce that $i_C(0^+) = V_0/R_1$ and $i_C(\infty) = 0$. From (8.25),

$$i_C(t) = \frac{V_0}{R_1}\exp\left(-\frac{t}{\tau}\right),\ \tau = (R_1 \| R_2)C,\ t > 0.$$

Exercise 8.10. Which of the currents and voltages labeled in Fig. 8.20 cannot change instantaneously?

The next example illustrates how (8.23) can be used to construct a solution to a problem involving multiple switching times and/or a piecewise-constant source.

Example 8.10. Refer to Fig. 8.21, where the source voltage v_S equals zero for times not shown. Construct a graph of the voltage $v_C(t)$ across the capacitor for $-2\text{ ms} \le t \le 8\text{ ms}$.

Solution: Let $\tau = RC = 500\,\mu\text{s}$, $V_0 = 100\,\text{mV}$, $t_n = n\,\text{ms}$. We use (8.23) to piece together the response $v_C(t)$.

The source voltage equals zero for $t \le 0$, so

$$v_C(t) = 0,\ t \le 0. \tag{8.26}$$

The first non-zero segment of v_S equals $200\,\text{mV} = 2V_0$ for $0 < t \le t_2$, where $t_2 = 2\,\text{ms}$. The initial value of the response is $v(0^+) = 0$ and the response approaches $v(\infty) = 2V_0$. Thus, from (8.23)

$$\begin{aligned} v_C(t) &= 2V_0 - (2V_0 - 0)\exp\left(-\frac{t}{\tau}\right) \\ &= 2V_0\left[1 - \exp\left(-\frac{t}{\tau}\right)\right], \\ &0 < t \le t_2. \end{aligned} \tag{8.27}$$

The next segment of v_S extends from $t = t_2{}^+$ to $t = t_3$. The response in that interval is initially equal to the final value $v_C(t_2)$ of the response in the previous interval and approaches $-200\,\text{mV} = -2V_0$. Thus, for $t_2 < t \le t_3$, we use (8.23) with $v_C(0^+) = v_C(t_2)$ and $v_C(\infty) = -2V_0$. We must also shift the response to the beginning time t_2 for the interval to which it applies. This gives

$$\begin{aligned} v_C(t) = -2V_0 - [-2V_0 - v_C(t_2)] \\ \exp\left(-\frac{t - t_2}{\tau}\right),\ t_2 < t \le t_3, \end{aligned} \tag{8.28}$$

where, from (8.27)

$$v_C(t_2) = 2V_0\left[1 - \exp\left(-\frac{t_2}{\tau}\right)\right].$$

The next segment of v_S extends from t_3 to t_4. The response in that interval is initially equal to the final value $v_C(t_3)$ of the response in the previous interval and approaches $300\,\text{mV} = 3V_0$. Thus, for $t_3 < t \le t_4$, we use (8.23) with $v_C(0^+) = v_C(t_3)$ and $v_C(\infty) = 3V_0$. We must also shift the response to the beginning time t_3 for the interval to which it applies. This gives

$$\begin{aligned} v_C(t) = 3V_0 - [3V_0 - v_C(t_3)] \\ \exp\left(-\frac{t - t_3}{\tau}\right),\ t_3 < t \le t_4, \end{aligned} \tag{8.29}$$

where, from (8.28)

$$\begin{aligned} v_C(t_3) = -2V_0 - [-2V_0 - v_C(t_2)] \\ \exp\left(-\frac{t_3 - t_2}{\tau}\right). \end{aligned}$$

Continuing in this way, we obtain

$$\begin{aligned} v_C(t) = -V_0 - [-V_0 - v_C(t_4)] \\ \exp\left(-\frac{t - t_4}{\tau}\right),\ t_4 < t \le t_6, \end{aligned} \tag{8.30}$$

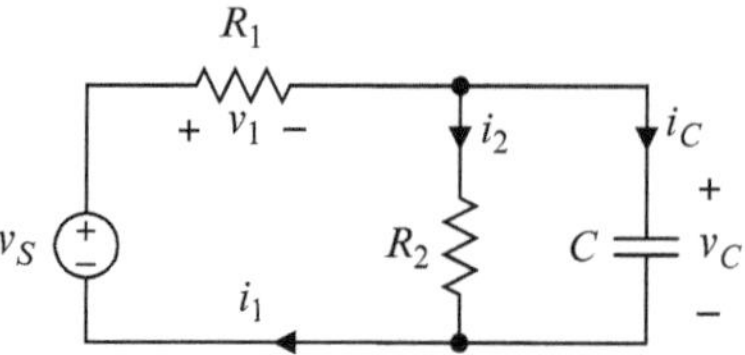

Fig. 8.20 See Exercise 8.10

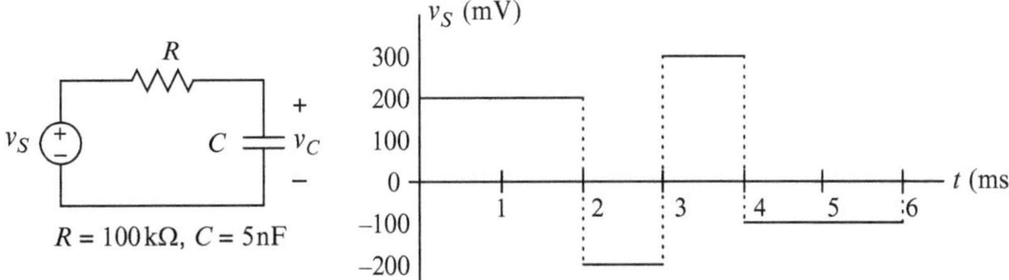

Fig. 8.21 See Example 8.10

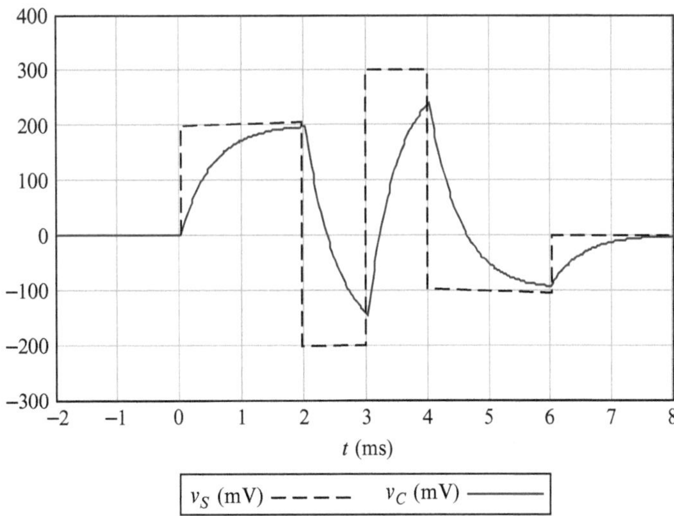

Fig. 8.22 See Example 8.10

where, from (8.29)

$$v_C(t_4) = 3V_0 - [3V_0 - v_C(t_3)]\exp\left(-\frac{t_4 - t_3}{\tau}\right),$$

and

$$v_C(t) = 0 - [0 - v_C(t_6)] \exp\left(-\frac{t - t_6}{\tau}\right),\ t_6 < t, \qquad (8.31)$$

where, from (8.30)

$$v_C(t_6) = -V_0 - [-V_0 - v_C(t_4)] \exp\left(-\frac{t_6 - t_4}{\tau}\right).$$

Figure 8.22 shows a graph of the response.

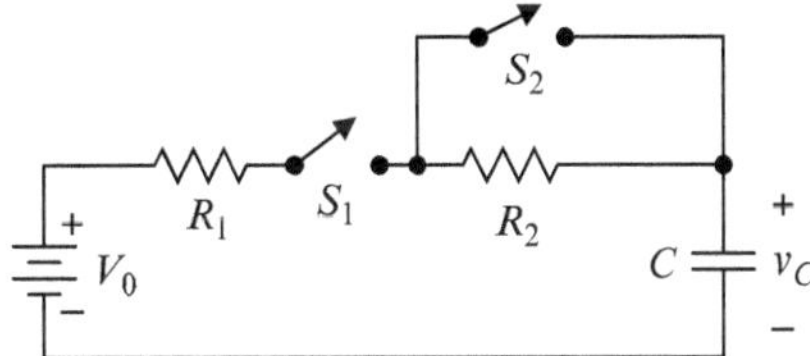

Fig. 8.23 See Example 8.11

When using (8.23) or (8.25) to express a voltage or current for multiple switching times, be sure to use the correct time constant for each interval.

Example 8.11. In Fig. 8.23, both switches are open for $t < 0$, and $v_C(0^-) = 0$. Switch S_1 is closed at $t = 0$ and switch S_2 is closed at $t = t_1$. Obtain an expression for the voltage $v_C(t)$ across the capacitor.

Solution: We use (8.23) (twice). We are given $v_C(0^-) = 0$, which implies $v_C(0^+) = 0$, because the voltage across a capacitor is continuous. For $0 < t \le t_1$, the voltage $v_C(t)$ rises toward V_0 with time constant given by

$$\tau_1 = (R_1 + R_2)C.$$

Thus for $0 < t \le t_1$, $v(\infty) = V_0$, and (8.23) becomes

$$v_C(t) = V_0\left[1 - \exp\left(-\frac{t}{\tau_1}\right)\right]; \quad \tau_1 = (R_1 + R_2)C, \ 0 < t \le t_1.$$

For $t_1 < t < \infty$, the resistor R_2 is bypassed by a short circuit, so the time constant is given by

$$\tau_2 = R_1 C.$$

If we take t_1 as the time origin for the second segment, the initial and final values of $v_C(t)$ are given by

$$v_C(0^+) = V_0\left[1 - \exp\left(-\frac{t_1}{\tau_1}\right)\right]; \quad v_C(\infty) = V_0.$$

We then reset the time origin to zero (delay the response by t_1) and obtain

$$v_C(t) = V_0 - \left\{V_0 - V_0\left[1 - \exp\left(-\frac{t_1}{\tau_1}\right)\right]\right\} \exp\left(-\frac{t-t_1}{\tau_2}\right) = V_0 - V_0 \exp\left(-\frac{t_1}{\tau_1}\right)\exp\left(-\frac{t-t_1}{\tau_2}\right), \quad \tau_2 = R_1 C, \ t > t_1.$$

Therefore

$$v_C(t) = V_0 \begin{cases} 0, & t < 0 \\ [1 - \exp(-t/\tau_1)], & 0 < t \le t_1 \\ 1 - \exp(-t/\tau_1)\exp[-(t - t_1)/\tau_2], & t > t_1 \end{cases}$$

where

$$\tau_1 = (R_1 + R_2)C, \ \tau_2 = R_1 C.$$

8.5 Capacitors in Series and Parallel

Figure 8.24a shows n capacitors in series. Because the same current passes through each capacitor, Kirchhoff's voltage law gives

$$\begin{aligned} v &= \frac{1}{C_1}\int_{-\infty}^{t} i(t')\,dt' + \frac{1}{C_2}\int_{-\infty}^{t} i(t')\,dt' + \cdots + \frac{1}{C_n}\int_{-\infty}^{t} i(t')\,dt' \\ &= \left(\frac{1}{C_1} + \frac{1}{C_2} + \cdots + \frac{1}{C_n}\right)\int_{-\infty}^{t} i(t')\,dt', \end{aligned} \tag{8.32}$$

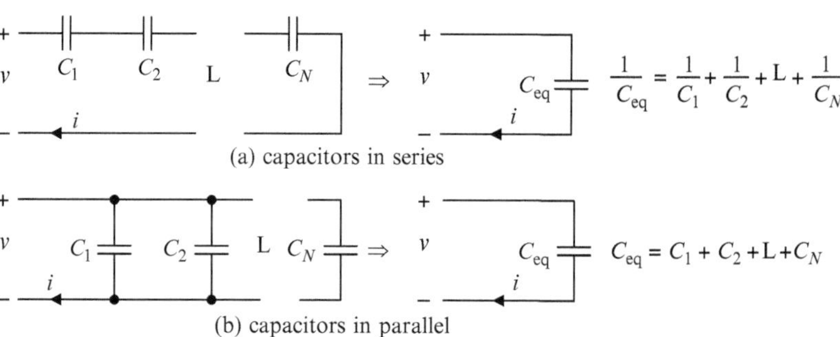

Fig. 8.24 Capacitors in (**a**) series and (**b**) parallel

which can be written

$$v = \frac{1}{C_{eq}} \int_{-\infty}^{t} i(t')\, dt',$$
$$\frac{1}{C_{eq}} = \frac{1}{C_1} + \frac{1}{C_2} + \cdots + \frac{1}{C_n}. \tag{8.33}$$

Thus *a series connection of capacitors is equivalent at the terminals to a single capacitor having capacitance C_{eq} given by* (8.33). Combining capacitors in series yields an equivalent capacitance that is smaller than the smallest of the individual capacitances.

Figure 8.24b shows n capacitors in parallel. Because the same voltage appears across each capacitor, Kirchhoff's current law gives

$$i = C_1 \frac{dv}{dt} + C_2 \frac{dv}{dt} + \cdots + C_n \frac{dv}{dt}$$
$$= (C_1 + C_2 + \cdots + C_n) \frac{dv}{dt}, \tag{8.34}$$

which can be written

$$i = C_{eq} \frac{dv}{dt}, \quad C_{eq} = C_1 + C_2 + \cdots + C_n. \tag{8.35}$$

Therefore *the equivalent capacitance of n capacitors in parallel is the sum of the individual capacitances*. Combining capacitors in parallel yields an equivalent capacitance that is larger than the largest of the individual capacitances.

Example 8.12. Find the equivalent capacitance at the terminals of the circuit shown in Fig. 8.25(a).

Solution: See Fig. 8.25(b). First combine the two rightmost capacitors using (8.35). Then combine the two series capacitors using (8.33). Finally, combine the remaining two parallel capacitors, again using (8.35).

Exercise 8.11. Refer to Fig. 8.26. Obtain an expression for the equivalent capacitance at the terminals *a–b*.

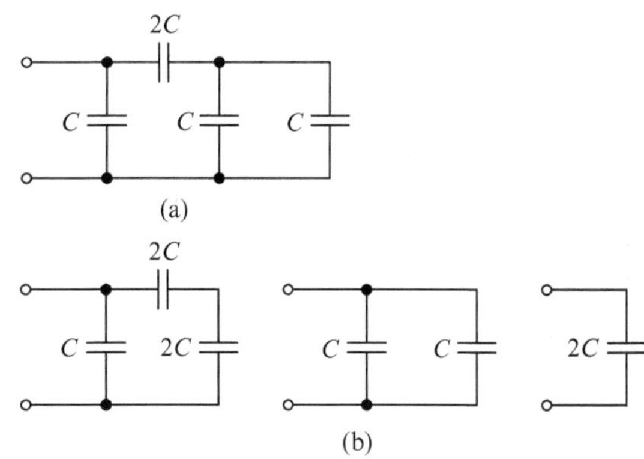

Fig. 8.25 See Example 8.12

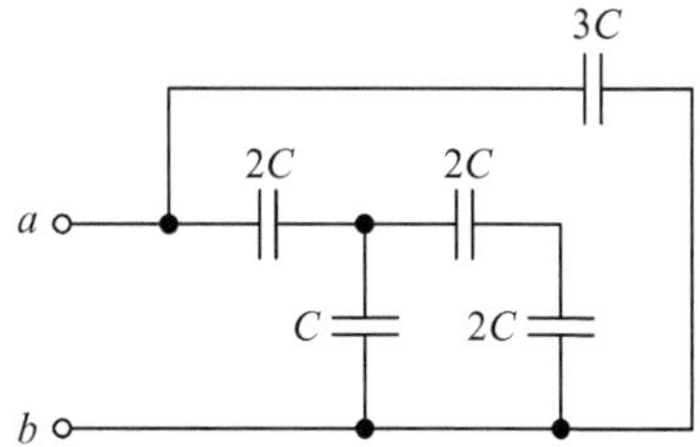

Fig. 8.26 See Exercise 8.11

In practical problems, we rarely need to find the equivalent capacitance of more than two capacitors in series or parallel. Even this need arises primarily in only two circumstances. One is in modeling, where various approximations occasionally reduce more complex arrangements to a point where remaining capacitances (or other elements) appear in series or parallel. The other is when a small variable capacitor is used in parallel with a larger one to allow trimming (fine adjustment) of the total capacitance. Otherwise, few if any circuits are designed using capacitors in series or parallel (where it is cheaper and easier to use one).

Note: The double-bar notation is used only for resistors in parallel. Do *not* use the notation $C_1 \| C_2$ to denote capacitors in parallel because the implied method for obtaining the equivalent capacitance of two capacitors in parallel is incorrect. Use $C_1 + C_2$ instead.

8.6 Leakage Resistance

Dielectrics used in capacitors have resistivities that are on the order of 1to100 MΩ cm. Thus a dielectric is not a perfect insulator and some conduction current passes between the plates. This current is called **leakage current** and the resistance of the dielectric is called **leakage resistance**. From Chapter 2, the resistance of a sample of homogeneous material having length l, cross-section area A, and resistivity ρ is given by

$$R = \frac{\rho\, l}{A}.$$

Thus the leakage resistance for a parallel-plate capacitor is given by

$$R_{leak} = \frac{\rho\, d}{A}, \tag{8.36}$$

where A is the area of one plate (the cross-sectional area of the dielectric), ρ is the resistivity of the dielectric, and d is the separation of the plates (the length of the dielectric). Leakage resistance in modern capacitors ranges from about 1 MΩ to almost 1 TΩ (10^{12} Ω). For capacitors with the same geometry and identical dielectric and plate materials, leakage resistance is inversely proportional to capacitance; that is, from (8.36) and (8.5),

$$R_{leak} = \frac{\rho\, \varepsilon_0\, \varepsilon_r}{C}. \tag{8.37}$$

Equation (8.37) shows that leakage resistance decreases with increasing capacitance. It follows that leakage current generally increases with capacitance, although a large capacitor having a very good (high-resistivity) dielectric might exhibit lower leakage than a smaller capacitor having a poor dielectric.

Exercise 8.12. Would including the leakage resistances of the capacitors in Fig. 8.12 make the circuit models realistic? Explain how or why not.

Example 8.13. Refer to Fig. 8.27(a), where V_0 is constant and the capacitors are initially uncharged, such that $v_1(0) = v_2(0) = 0$. Obtain expressions for the voltages v_1, v_2 a very long time after the switch is closed.

Solution: Because V_0 is constant, the current through the capacitors equals zero for $t \to \infty$. However, from Kirchhoff's voltage law,

$$v_1 + v_2 = V_0,$$

so the voltages across the capacitors are nonzero. For $t \to \infty$, the voltages are determined by the leakage resistances, as shown in Fig. 8.27(b), and not by the capacitances. The circuit in Fig. 8.27(b) is a voltage divider, and in dc steady state all of the current I passes through the resistors (none through the capacitors). It follows that

$$v_1 = \frac{R_1 V_0}{R_1 + R_2}, \quad v_2 = \frac{R_2 V_0}{R_1 + R_2}. \tag{8.38}$$

If we do not know the leakage resistance (the usual case), but know the dielectric materials, we can estimate the leakage resistances using (8.37):

$$R_1 = \frac{\rho_1\, \varepsilon_0\, \varepsilon_1}{C_1}, \quad R_2 = \frac{\rho_2\, \varepsilon_0\, \varepsilon_2}{C_2},$$

in which case (8.38) becomes

$$v_1 = \frac{V_0}{1 + C_1\rho_2\varepsilon_2/(C_2\rho_1\varepsilon_1)},$$

$$v_2 = \frac{V_0}{1 + C_2\rho_1\varepsilon_1/(C_1\rho_2\varepsilon_2)}.$$

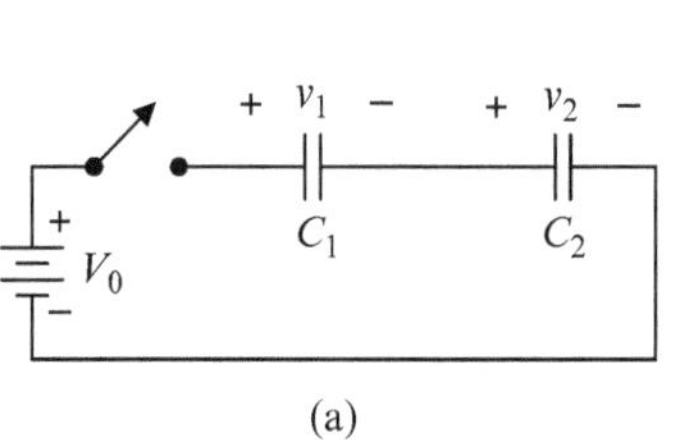

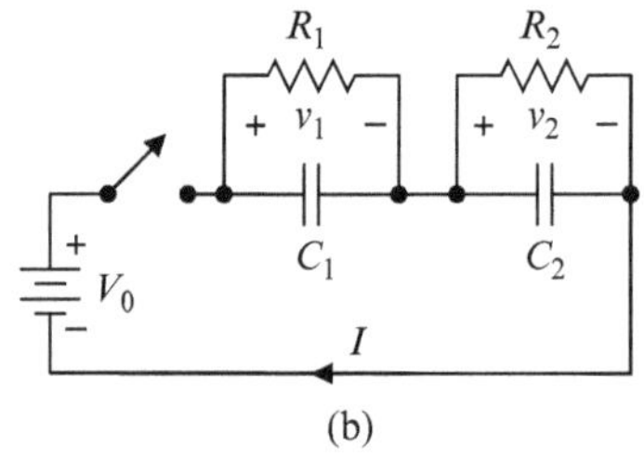

Fig. 8.27 See Example 8.13

If the capacitors are of the *same type* (same dielectric) and are *precise*, then $\rho_1 = \rho_2$, $\varepsilon_1 = \varepsilon_2$ and

$$v_1 = \frac{C_2 V_0}{C_1 + C_2}, \quad v_2 = \frac{C_1 V_0}{C_1 + C_2}.$$

In addition to shunt leakage resistance, physical capacitors exhibit series resistance due partly to the leads (if long), partly to conduction in the plates, and partly to loss mechanisms in the dielectric that are manifested externally as resistance. The total series resistance is important at high frequencies and high switching rates, as occur in modern digital computers, and also in large capacitors used to smooth the supply voltage in some power supplies, where the power dissipated by the series resistance generates heat and speeds aging of the capacitor.

8.7 Stray and Parasitic Capacitance; Capacitive Coupling

Like resistance, capacitance is intentionally introduced in most circuits to achieve certain ends. But whether we like it or not, capacitance is present wherever current-carrying conductors are separated by an insulating medium. For example, capacitance exists between wires on a circuit board and between wires strung on utility poles. Capacitance that is more or less *widely distributed* in a circuit, arising, for example, from parallel conductors on the same or opposite sides of a printed-circuit board or from conductors near ground planes, usually is called **stray capacitance**. Unintentional, more localized capacitance also exists in devices such as diodes and transistors because of the geometries of those devices. Such capacitance often is called *junction capacitance* because of its association with device structures called junctions. Capacitance in passive elements such as resistors and inductors often is called *residual capacitance*. *Localized* unintentional capacitance, such as residual capacitance and junction capacitance in electronic devices, often is called **parasitic capacitance**.

Stray and parasitic capacitance can be significant, depending upon the geometry involved and the rate at which the relevant voltage varies. Parasitic capacitance in transistors limits the speed with which a transistor can respond to rapidly changing voltages and thereby limits the frequency response of transistor amplifiers, although the limit can be quite high. Stray and parasitic capacitance can be especially troublesome in computers and in microwave circuits, where voltages change very quickly. Stray capacitance can be important between telephone lines that are parallel over great distances, even though the voltages might not change as quickly. If not prevented or properly accounted for, stray and parasitic capacitance can prevent a circuit from working as intended.

For example, suppose that a certain application calls for a resistive voltage divider, as shown in Fig. 8.28(a), where $v_S(t) = V_0 \cos(2\pi f_0 t)$. If the frequency f_0 of the source $v_S(t)$ is high enough, the residual (or parasitic) capacitances of the resistors come into play, as shown in Fig. 8.28(b). We are at this point unprepared for detailed analysis of the circuit in Fig. 8.28(b), but you can see that for a time-varying source voltage $v_S(t)$, the actual output voltage $v_b(t)$ will differ from the desired (or intended) output voltage $v_a(t)$. If it is known that that residual capacitance will be significant in a resistive voltage divider, a designer can compensate for that by intentionally introducing capacitances C_1, C_2 (as in Fig. 8.28(b) that are much larger than the stray capacitances, and then designing the resistive-capacitive

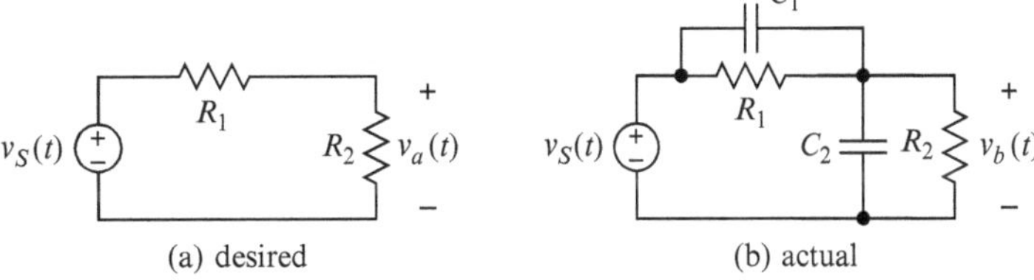

Fig. 8.28 Residual capacitance can affect resistive voltage dividers at high frequencies

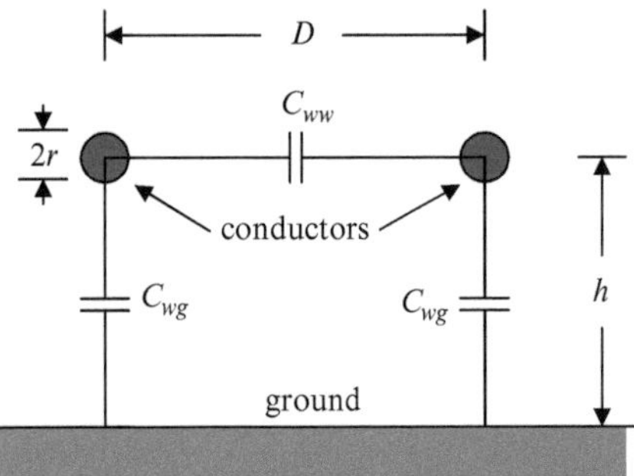

Fig. 8.29 Stray conductor-to-conductor and conductor-to-ground capacitance

voltage divider accordingly. Such a voltage divider is said to be *compensated.*

Parallel conductors and conductors parallel to ground planes can be significant sources of stray capacitance. Refer to Fig. 8.29, which represents parallel conductors in a vacuum (or air) above a ground plane; for example, a two-conductor power transmission line strung on utility poles above ground, or a smaller two-conductor wire running parallel to a grounded metal chassis. The wire-to-wire capacitance and the wire-to-ground capacitance per meter are given by[14]

$$\frac{\Delta C_{ww}}{\Delta l} = \pi\varepsilon_0\left[\ln\left[\frac{d}{r\sqrt{1+0.25(d/h)^2}}\right]\right]^{-1},$$
$$\frac{\Delta C_{wg}}{\Delta l} = 2\pi\varepsilon_0\left[\ln\left(\frac{2h-r}{r}\right)\right]^{-1}, \tag{8.39}$$

where d is the center-to-center distance between the conductors, h is the distance of the conductors from the ground plane, r is the radius of the conductors, and $\varepsilon_0 = 8.842 \times 10^{-12}\,\mathrm{F\,m^{-1}}$ is the permittivity of air.

Example 8.14. A pair of straight AWG 24 copper wires are parallel to each other and to a metal chassis. The distance between the wires is 1 mm and the distance of the pair from the chassis is 1 cm. Calculate the wire-to-wire and wire-to-ground capacitance per cm of length and the total capacitance if the length of the pair of wires is 4 cm.

Solution: From Chapter 2, the radius of AWG 24 wire is

$$r = \frac{1}{2}(8.251\,\mathrm{mm})\exp\left(-\frac{24}{8.628}\right)$$
$$= 256\,\mu\mathrm{m}.$$

From (8.39), with $r = 256\,\mu\mathrm{m}, d = 1\,\mathrm{mm}$, $h = 1\,\mathrm{cm}$,

$$\frac{\Delta C_{ww}}{\Delta l} = \pi\varepsilon_0\left[\ln\left[\frac{d}{r\sqrt{1+0.25(d/h)^2}}\right]\right]^{-1}$$
$$= 0.204\,\mathrm{pF\,cm^{-1}},$$
$$\frac{\Delta C_{wg}}{\Delta l} = 2\pi\varepsilon_0\left[\ln\left(\frac{2h-r}{r}\right)\right]^{-1}$$
$$= 0.128\,\mathrm{pF\,cm^{-1}}.$$

The total capacitance per unit length is given by

$$\frac{\Delta C}{\Delta l} = \frac{\Delta C_{ww}}{\Delta l} + \frac{\Delta C_{wg}}{\Delta l} \cong 0.332\,\mathrm{pF\,m^{-1}},$$

so the capacitance of a 4-cm length of wire is

$$C \cong 1.33\,\mathrm{pF}.$$

As you can see, wires a few or more cm in length can introduce a few picofarads of stray capacitance. Such capacitance limits the rate at which the voltage across a load can change. For example, if a 4 cm length of the conductors of this example drives a 10 kΩ load, the effective time constant due to the wire-to-ground capacitance alone is about 13 ns, which means that the time required for the load voltage to change from one value to another is approximately 65 ns. This means the load voltage can switch between two levels no faster than about

[14]Corcoran, George F., and Henry R. Reed, *Introductory Electrical Engineering*, Wiley, New York, 1957, pp 419–426.

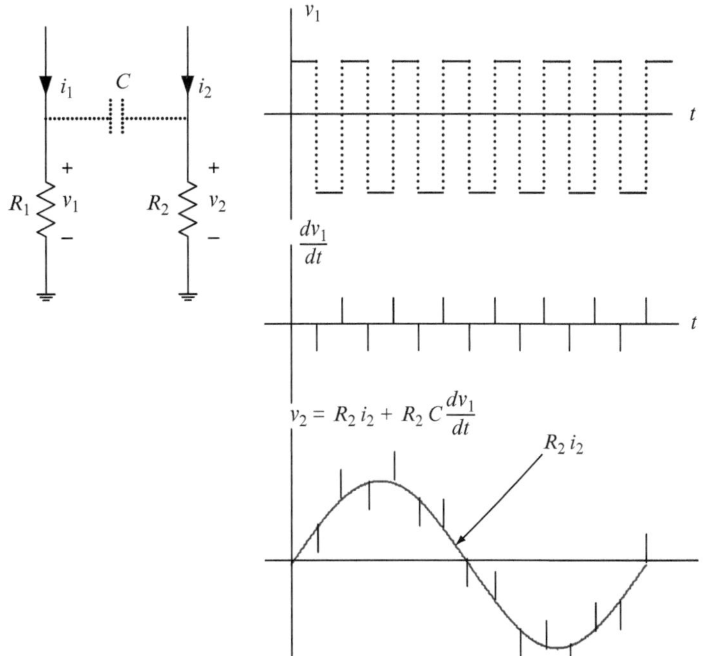

Fig. 8.30 Illustrating one possible effect of stray capacitive coupling. (Adapted from Horowitz, Paul et al. 1995.)

15 million times per second. If we are talking about a digital communication system, for example, we could transmit no more than a few million bits per second over the line to the load, which would be unacceptably slow in today's world.

Equations (8.39) ignore whatever structures support the wires, insulation surrounding the wires, and connections at the ends of the wires. They should not be regarded as accurate relations for real-world circuits, in which many wires of various sizes run in many directions, at various distances from each other and from ground planes. A circuit designer will not attempt to actually calculate stray capacitance using (8.39), but will be guided by relations embodied in those expressions. For example, increasing the distance d between wires, decreasing the radius r of wires, and increasing the distances from ground planes all act to decrease stray capacitance. The relative permittivity of an insulating material cannot be less than unity, so binding the wires to each other with an insulating material can increase wire-to-wire capacitance. Thus, for example, using the smallest wire necessary, using insulating material (if required) having a low relative permittivity, and avoiding long runs near ground planes and other wires can help limit stray wiring capacitance.

One of the great advantages of integrated circuits is their microscopic size. Conductors are extremely short and have extremely small cross-sectional area. Thus stray wiring capacitance in an integrated circuit can be quite small, even though the distances between conductors are also microscopic.

Sometimes, *stray capacitive coupling* in a circuit can induce the derivative of a signal in one conductor into a neighboring conductor, as illustrated by Fig. 8.30. In absence of capacitive coupling, the voltage v_2 across the resistor R_2 equals $R_2 i_2$. In presence of stray capacitive coupling, as indicated by the dashed capacitor, Kirchhoff's current law at the node joining the capacitor to the resistor R_2 gives

$$C\,\frac{d(v_2 - v_1)}{dt} + \frac{v_2}{R_2} - i_2 = 0$$

$$\Rightarrow v_2 = R_2 i_2 - R_2 C\,\frac{dv_2}{dt} + R_2 C\,\frac{dv_1}{dt}.$$

In presence of capacitive coupling, the voltage v_2 contains a term proportional to the derivative of the voltage v_1. If the voltage v_1 varies much more rapidly than the voltage v_2, then (on average)

$$\left|\frac{dv_1}{dt}\right| \gg \left|\frac{dv_2}{dt}\right|.$$

It follows that

$$v_2 \cong R_2\, i_2 + R_2\, C\,\frac{dv_1}{dt},$$

which shows that a voltage proportional to the derivative of v_1 is superimposed on the (desired) voltage $R_2\, i_2$, again as illustrated (qualitatively) by Fig. 8.30. Usually, such coupling is undesirable and must be minimized by shielding and careful conductor layout. If shielding and conductor layout do not eliminate the problem, and if v_1 varies much more rapidly than v_2, then a bypass capacitor in parallel with R_2 might reduce the induced interference to an acceptable level.

Stray capacitance and capacitive coupling are difficult to compute for complex geometries, such as exist in modern circuits, but a general knowledge of origins and effects of stray capacitance and capacitive coupling can help keep a circuit designer out of trouble.

8.8 Variation of Capacitance with Temperature

For most dielectric materials, relative permittivity varies with temperature, so the capacitance of capacitors made from those materials varies with temperature. Manufacturers describe the relation of capacitance to temperature in mainly one of two ways. For some materials, the variation of capacitance with temperature is approximately linear. In such cases, manufacturers data sheets give a **temperature coefficient of capacitance** (**TCC**) α, defined by

$$\alpha = \frac{C_T - C_0}{(T - T_0)C_0} = \frac{1}{C_0}\frac{\Delta C}{\Delta T}, \qquad (8.40)$$

where

$$\begin{aligned} C_T &= \text{capacitance at actual temperature } T, \\ C_0 &= \text{capacitance at reference temperature } T_0 \\ &\quad (\text{usually}, 25°\text{C}). \end{aligned}$$

Thus the capacitance at temperature T is approximated as

$$C_T = C_0(1 + \alpha\,\Delta T) = C_0[1 + \alpha\,(T - T_0)]. \quad (8.41)$$

Manufacturers' commonly specify the temperature coefficient in parts per million (ppm), in which case (8.41) becomes

$$C_T = C_0\left[1 + \left(\alpha_{ppm} \times 10^{-6}\right)(T - T_0)\right], \qquad (8.42)$$

where the unit of α_{ppm} is K^{-1} or $(°C)^{-1}$. It does not matter which unit is used, because the numerical value of the difference $T - T_0$ is the same, whether the temperatures are expressed in Kelvin (K) or degrees Celsius.

Example 8.15. The capacitance and temperature coefficient for a certain capacitor at 25°C are 50 nF and 700 ppm, respectively. Find the capacitance at 65°C.

Solution: From (8.42),

$$\begin{aligned} C_{35} &= 50\,\text{nF} \times \left[1 + \left(70 \times 10^{-6}\right)(65 - 25)\right] \\ &= 51.4\,\text{nF}. \end{aligned}$$

For many dielectric materials, the variation of capacitance with temperature is nonlinear, as illustrated by the hypothetical graph in Fig. 8.31. The horizontal axis is temperature and the vertical axis is the percent deviation of the capacitance from the capacitance at 25°C, given by

$$\frac{\Delta C}{C_{25}} \times 100 = \frac{C - C_{25}}{C_{25}} \times 100.$$

For $T = 25°\text{C}$, $C = C_{25}$ and the deviation equals zero. The deviation is within $\pm\,5\%$ for approximately $5°\text{C} \le T \le 160°\text{C}$. Thus a specification of $\pm\,5\%$

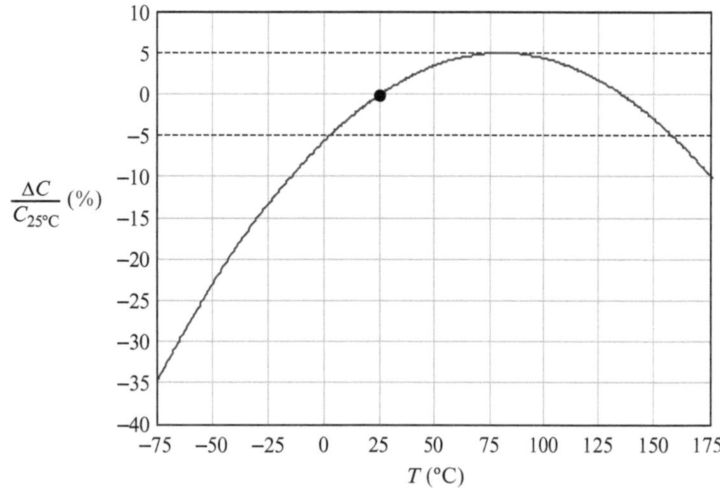

Fig. 8.31 Variation of capacitance with temperature (hypothetical example)

Table 8.1 Variation of capacitance with temperature for various dielectric materials (Courtesy of AVX Corporation)

Dielectric	Ceramic (NPO)	Mica	Polyester	Poly-carbonate	Poly-propylene	Polystyrene	Tantalum	Aluminum electrolytic
Capacitance range	1 pF – 100 nF	1 pF – 90 nF	1 nF – 10 μF	1 pF – 100 nF	47 pF – 47 nF	100 pF – 27 nF	10 nF – 1000 μF	500 nF – 10 μF
Temp. range (°C)	−55 + 85	−55 + 125	−55 + 125	−55 + 125	−55 + 85	−55 + 70	−55 + 125	−40 + 85
$\Delta C/C_{25°C}$ (%)	± 0.3	−0.4 + 1.8	±12	± 2	± 2.5	± 1	± 8	±10

Table 8.2 Definition of codes for variation of capacitance with temperature

Minimum temperature (°C)		Maximum temperature (°C)		Change in capacitance with temperature (%)	
X	−55	**2**	+45	**A**	± 1
Y	−30	**4**	+65	**B**	± 1.5
Z	+10	**5**	+85	**C**	± 2.2
		6	+10	**D**	± 3.3
		7	+125	**E**	± 4.7
				F	± 7.5
				P	±10
				R	± 15
				S	± 22
				T	+ 22, −33
				U	+ 22, −56
				V	+ 22, −82

would be accompanied by the temperature range $5°C \leq T \leq 160°C$.

Table 8.1 shows typical variations for various dielectric materials. For example, for polyester, the variation of capacitance with temperature can be as much as $\pm$ 12% for the temperature range $-55°C \leq T \leq 125°C$, whereas for polycarbonate, it is no more than $\pm$ 2% over the same range.

To standardize such specifications, the industry developed the code defined by Table 8.2. The code consists of three characters: A letter (X, Y, or Z), a number (2, 4.5, 6, or 7), and another letter (one of A–F, P, R, S–V); for example, X7R. The First letter and the number specify a temperature range and the last letter specifies the temperature variation within that range. For example, the X7 in X7R specifies the temperature range −55 to 125°C and the R means that the capacitance will vary by no more than ±15% from the nominal value over that range.

Exercise 8.13. Specify the tolerance and temperature range corresponding to each of the following designations: X5R, Y5V, Z5U.

Exercise 8.14. List the designations for the capacitors that could replace a Y5U capacitor based upon specifications of only capacitance and tolerance alone.

In circuit-simulation programs, dependence of capacitance on temperature often is approximated using a quadratic function of temperature of the form

$$C = C_0 + \alpha\,\Delta T + \beta(\Delta T)^2,\ \Delta T = T - T_0, \quad (8.43)$$

where T is temperature (°C), T_0 is a reference temperature (often, 25°C), and C_0 is the capacitance at the reference temperature. The coefficients α $(F\,K^{-1})$ and β $(F\,K^{-2})$ are called the *first temperature coefficient* and the *second temperature coefficient*, respectively. Pspice and other simulation programs allow a user to specify the temperature coefficients.

Variation of capacitance with temperature is an important consideration in circuit design. Indeed, capacitors are classified on that basis, according to whether they are sensitive or relatively insensitive to temperature variations. In some applications, such as coupling and bypass (described below), large capacitance is relatively more important than temperature variation. In others, such as timing, differentiating, integrating, and filtering, stability of capacitance is relatively more important than actual value, because the important quantity is the product of a capacitance and a resistance, and the latter can be precise.

To complicate matters further, capacitance can vary not only with temperature, but also with any applied dc bias voltage and with the frequency of an applied sinusoidal voltage. We discuss some of these details in a subsequent chapter. Here, we note only that some manufacturers provide fairly complete Pspice models for some of their offerings that take such variations into account.

8.9 Energy Storage and Power Dissipation in a Capacitor

The capacitor of Fig. 8.1 will remain charged when disconnected from the battery. That is, a nonzero voltage will exist across the terminals of the capacitor, which means that energy is stored in the electric field between the plates. If a charged capacitor is connected to a resistance, for example, current will pass through the resistor as the capacitor discharges and work will be done according to Joule's law. If the capacitor is allowed to discharge completely, the work done will equal the energy stored in the field. Ability to store energy for later release is what gives a capacitor electrical memory.

The instantaneous power dissipated in a capacitor having capacitance C is given by

$$p = v\,i = C\,v\frac{dv}{dt} = \frac{dw}{dt}, \quad (8.44)$$

where $v(t)$ is the instantaneous voltage across the capacitor and $w(t)$ is the instantaneous energy stored in the electric field produced by the voltage. Equation (8.44) implies that

$$dw = Cv\,dv \Rightarrow w(t) = C\int_{v(-\infty)}^{v(t)} v'\,dv'$$
$$= \frac{1}{2}C\left[v(t)^2 - v^2(-\infty)\right].$$

We assume the voltage equals zero for $t \to -\infty$ (that the capacitor is initially uncharged). Thus the instantaneous energy stored in the electric field produced by a capacitor is given by

$$w(t) = \frac{1}{2}C\,v^2(t), \quad (8.45)$$

where $v(t)$ is the instantaneous voltage across the capacitor. For economy, *the instantaneous energy stored in the electric field produced by a capacitor* usually is referred to as *the energy stored by (or in) the capacitor.*

The energy stored in the electric field produced by a voltage makes the voltage resistant to change, because it takes time to transfer energy to or from the electric field. In other words, the electric field created by a voltage gives the voltage a kind of inertia that resists changes in the voltage. A capacitor enhances this effect by concentrating and intensifying the electric field produced by a voltage across the capacitor.

Example 8.16. Continuing Example 8.4, the energy stored in the capacitor at any time t is given by

$$w = \frac{1}{2}Cv^2 = \frac{1}{2}CV^2\cos^2(2\pi f\,t)$$
$$= \frac{CV^2}{4}[1 + \cos(4\pi f\,t)]$$
$$= 2.5 \times 10^{-6}[1 + \cos(4\pi f\,t)]\ \text{J}.$$

Exercise 8.15. Refer to Fig. 8.8 and Exercise 8.2. Draw a graph of the energy stored in the electric field produced by the capacitor.

Exercise 8.16. A capacitor having capacitance $C = 250\,\text{pF}$ is initially charged such that the terminal voltage is V_0. The capacitor is then connected (at $t = 0$) across a resistor having resistance $R = 10\,\text{k}\Omega$. At what time is the energy stored in the capacitor equal to half of its initial value?

The following fact is often useful: *When a capacitor is fully charged by a constant source through a fixed resistance, the total energy delivered by the source is exactly twice the final energy stored by the capacitor.*

We can show this as follows. In Fig. 8.32, the switch is in position a for $t < 0$ and is moved to position b at $t = 0$. The capacitor charges to $v_C(\infty) = V_0$, at which point the energy stored on the capacitor is given by

$$w_C = \frac{1}{2}CV_0^2.$$

The charge stored by the capacitor is given by $Q = CV_0$. This charge must pass through the source, so the work done by the source on this charge (the energy delivered by the source) is given by

$$w_S = QV_0 = C{V_0}^2,$$

which is exactly twice the energy stored by the capacitor. Half of the energy delivered by the source is dissipated in the resistor, independent of the resistance of the resistor. We make use of this result in Section 8.10.9.

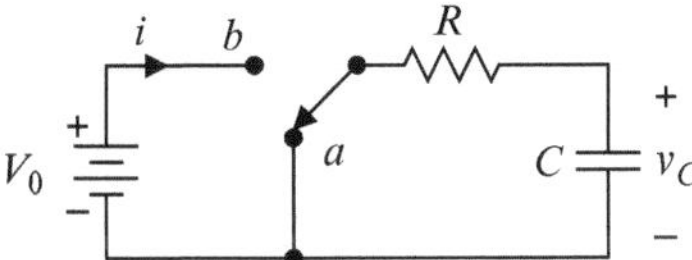

Fig. 8.32 Pertaining to energy dissipation and storage in an RC series circuit (see text)

The *average* power dissipated in an *ideal* capacitor equals zero, provided the applied current or voltage is bounded (as it always is in a physical circuit). We can show this as follows: The average power P_T dissipated in a capacitor over an interval $t_0 \le t < t_0 + T$ is the change in stored energy divided by the duration of the interval, given by

$$P_T = \frac{C}{2T}\left[v^2(t_0 + T) - v^2(t_0)\right]. \tag{8.46}$$

In any physical problem, voltages are finite, so the quantity in the brackets is finite and

$$P = \lim_{T\to\infty} P_T = 0. \tag{8.47}$$

The average power dissipated in a capacitor *over a finite time interval*, given by (8.46) is not necessarily zero. The *instantaneous* power dissipated in a capacitor is not zero, nor is the peak power.

A capacitor can absorb energy at some times and return energy to a circuit at other times. None of the energy absorbed (by an *ideal* capacitor) is lost to the circuit, but can be returned on demand. This is in contrast to a resistor, which converts electrical energy to heat energy that is then lost forever to the circuit. On an instantaneous basis, a capacitor behaves sometimes like a sink and other times like a source, but on average (over a sufficiently long time), is neither.

Example 8.17. Find the average powers dissipated in the resistor and the source in the circuit shown in Fig. 8.33, where $R = 2\,\text{k}\Omega$ and

$$v(t) = V_0\cos(2\pi f t);\ V_0 = 15\,\text{V}.$$

Solution: The average power dissipated in the resistor is

$$P_R = \frac{V_0^2}{2R} = \frac{225\text{V}^2}{4000} = 56.3\,\text{mW}.$$

Because power is conserved and because the average power dissipated in a capacitor

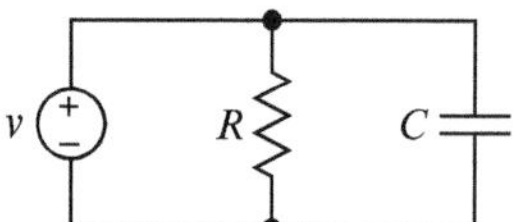

Fig. 8.33 See Example 8.17

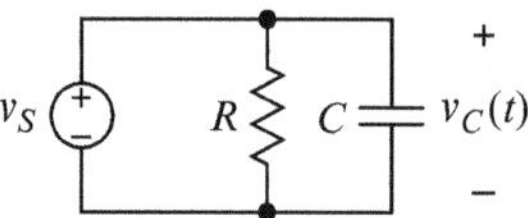

Fig. 8.34 See Exercise 8.17

equals zero, the power dissipated in the source is

$$P_v = -P_R = -56.3\,\text{mW}.$$

Exercise 8.17. In Fig. 8.34, $R = 10\,\text{k}\Omega$ and $C = 250\,\text{nF}$. The instantaneous power dissipated in the resistor is given by $p_R(t) = p_0 + p_1 \cos^2(2\omega_0 t)$, with $p_0 = 20\,\text{mW}$, $p_1 = 50\,\text{mW}$, and $f_0 = 20\,\text{kHz}$. The dc component of the source voltage v_S is positive and the source voltage is maximum at $t = 0$. Obtain an expression for the instantaneous voltage $v_C(t)$ across the capacitor. *Hint*: Superposition of power is applicable (Why?).

The discussion in this section pertains to an ideal capacitor. Real capacitors exhibit leakage resistance and what is called *equivalent series resistance* (ESR), so the average power dissipated by a real capacitor is not zero. Section 8.6 above describes leakage resistance. We describe ESR in a subsequent chapter.

8.10 Applications

Capacitors are used in basically six ways:

- To discriminate between sinusoids based on frequency (including dc), as in bypass, coupling, and filtering applications (e.g., Sections 8.10.3 and 8.10.6, below);
- To produce a current or voltage proportional to the derivative or integral of another current or voltage (see Sections 8.10.1 and 8.10.2, below);
- To store energy or charge for later release; e.g., as in the electronic flash unit on a camera and in switched-capacitor resistors (see Section 8.10.8, below);
- As a timing device; e.g., where some action is triggered when voltage across a capacitor reaches a certain threshold, as in pulse-detection circuits, timing circuits, and some oscillators (not treated in this book);
- To change the phase of one sinusoid relative to that of another, as in power-factor correction, some oscillators, and the starting and run capacitors on electric motors. (Power-factor correction is treated in a subsequent chapter);
- As components of circuit models for physical devices; e.g., parasitic (junction) capacitance in a transistor and parasitic (residual) capacitance in a resistor or an inductor. (Discussed briefly in a subsequent chapter).

At this point in our development, we are unprepared to illustrate all of the applications listed above or even to treat any particular one in great detail. This section describes a few applications and effects of capacitance at an introductory level. Additional applications and more detailed treatments are given in subsequent chapters, where we introduce useful circuits in which capacitance is essential or unavoidable.

8.10.1 Differentiating Circuits

Applying Kirchhoff's current law to the circuit shown in Fig. 8.35 gives

$$-C\frac{dv_{in}}{dt} - \frac{v_{out}}{R} = 0 \Rightarrow v_{out} = -RC\frac{dv_{in}}{dt}. \quad (8.48)$$

The output is proportional to the derivative with respect to time. The circuit is called a *differentiator*. The circuit shown works reasonably well for slowly varying inputs. Practical differentiators have additional components, the purposes of which are discussed in subsequent chapters.

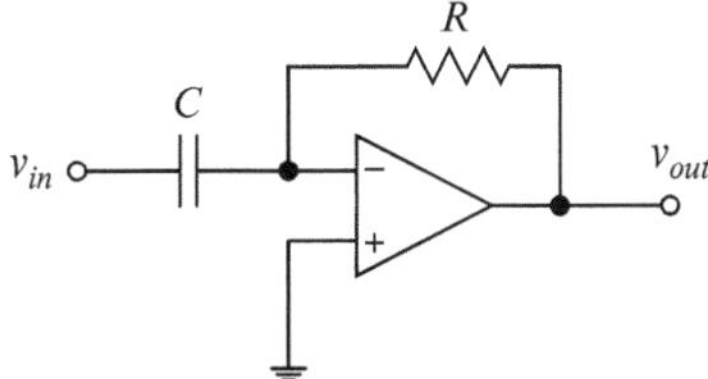

Fig. 8.35 Differentiating circuit

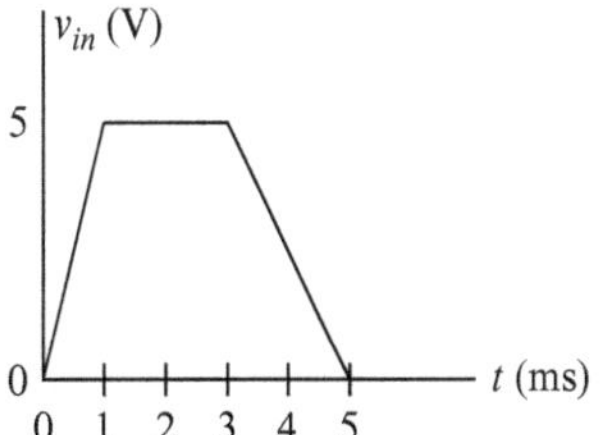

Fig. 8.36 See Example 8.19

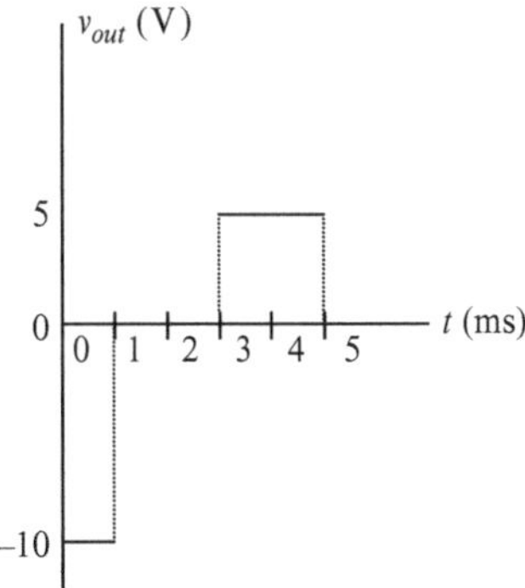

Fig. 8.37 See Example 8.19

Example 8.18. In Fig. 8.35, let

$$v_{in}(t) = V_0 \cos(2\pi f t).$$

The (ideal) output is

$$v_{out} = -RC\frac{dv_{in}}{dt} = 2\pi f RC\, V_0 \sin(2\pi f t).$$

Exercise 8.18. In Fig. 8.35, $R = 10\,\text{k}\Omega$, $C = 22\,\text{pF}$, and the supply voltage for the op amp (assumed to be rail-to-rail) is $\pm V_{CC} = \pm 15\,\text{V}$. The input v_{in} is given by

$$v_{in} = V_0 \cos(2\pi f\, t).$$

(a) If $V_0 = 1\,\text{V}$, what is the highest frequency for which the circuit operates linearly? (b) If $f = 1\,\text{MHz}$, what is the largest value of V_0 for which the circuit operates linearly?

Example 8.19. In Fig. 8.35, $R = 200\,\text{k}\Omega$, $C = 10\,\text{nF}$, and the input v_{in} is as shown in Fig. 8.36. Draw a graph of the output.

Solution: The output is given by

$$\begin{aligned} v_{out}(t) &= -RC\frac{dv_{in}(t)}{dt} \\ &= -RC \times [\text{instantaneous slope of } v_{in}(t)]. \end{aligned}$$

For the values given, $RC = 2 \times 10^{-3}\,\text{s}$. For $0 \le t < 1\,\text{ms}$, the derivative (slope) is $(5\,\text{V})/(1\,\text{ms}) = 5{,}000\,\text{Vs}^{-1}$. For $3\,\text{ms} \le t < 5\,\text{ms}$, the derivative (slope) is $-(5\,\text{V})/(2\text{ms}) = -2{,}500\,\text{V}\,\text{s}^{-1}$. For all other times, the slope equals zero. Thus

$$v_{out} = \begin{cases} (-2\times10^{-3})(5{,}000) = -10\text{V}, & 0 \le t < 1\text{ms}, \\ (-2\times10^{-3})(-2{,}500) = 5\text{V}, & 3\text{ms} \le t < 5\text{ms}, \\ 0, & \text{all other times}. \end{cases}$$

Figure 8.37 shows a graph of the output.

8.10.2 Integrating Circuits

Under certain conditions, the circuit shown in Fig. 8.38 can produce an output proportional to the integral of the input, in which case the circuit is called an *integrator*. Because $v_n = 0$, the input current is given by $i = v_{in}/R$. This is also the current through the capacitor, because the current entering the *n*

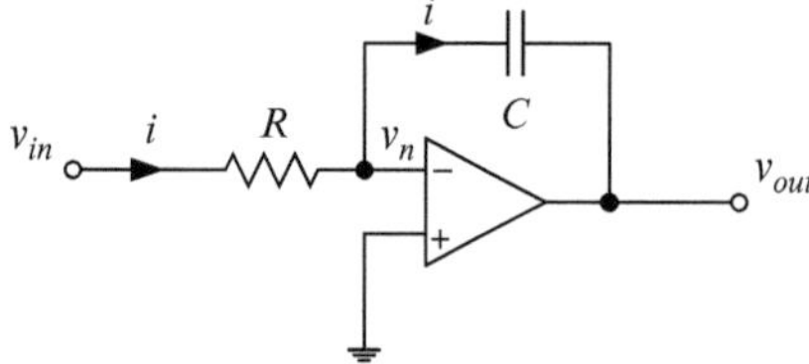

Fig. 8.38 Integrating circuit

terminal of the op amp equals zero. Thus the voltage across the capacitor (the output voltage) is given by

$$v_{out} = -\frac{1}{C}\int_{-\infty}^{t} i(t')\,dt' = -\frac{1}{RC}\int_{-\infty}^{t} v_{in}(t')\,dt'. \quad (8.49)$$

In practical op-amp circuits, there must be feedback at dc. If the capacitor in Fig. 8.38 is ideal, there is no such feedback. Physical capacitors exhibit leakage resistance, and in some applications, leakage resistance provides sufficient dc feedback. For present purposes, we assume that is the case.

Exercise 8.19. In Fig. 8.38, let

$$v_{in}(t) = \begin{cases} 0, & t \le 0, \\ V_0 \cos(2\pi f t), & t > 0. \end{cases}$$

Obtain an expression for the output v_{out}.

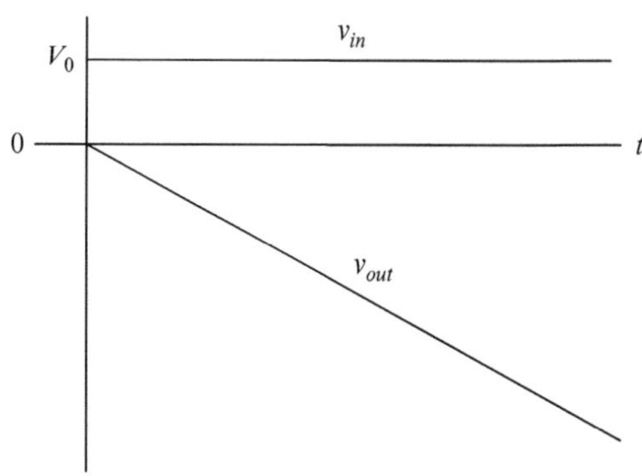

Fig. 8.39 See Example 8.20

Example 8.20. In Fig. 8.38, let

$$v_{in} = \begin{cases} 0, & t \le 0, \\ V_0, & t > 0. \end{cases}$$

Obtain an expression for the output $v_{out}(t)$.

Solution: From (8.49)

$$v_{out} = \begin{cases} -\dfrac{1}{RC}\displaystyle\int_{-\infty}^{t} 0\,dt' = 0, & t \le 0, \\ -\dfrac{1}{RC}\displaystyle\int_{0}^{t} V_0\,dt' = -\dfrac{V_0\,t}{RC}, & t > 0. \end{cases}$$

Figure 8.39 shows a graph of the output. What limits the output?

Example 8.21. In Fig. 8.38, $R = 200\,\text{k}\Omega$, $C = 10\,\text{nF}$, and the input v_{in} is as shown in Fig. 8.40. Draw a graph of the output.

Solution: The output is given by

$$v_{out}(t) = -\frac{1}{RC}\int_{-\infty}^{t} v_{in}(t')dt' = -\frac{1}{RC}\times\begin{bmatrix}\text{instantaneous net area bounded by} \\ v_{in}(t') \text{ and the } t' \text{ axis to the left of } t' = t\end{bmatrix}.$$

A graph of $v_{in}(t')$ versus the variable of integration t' is obtained by simply re-labeling the axes in Fig. 8.40. To perform the integration, imagine the time t (the upper limit of integration) advancing from left to right on the t' axis. For each value of t, compute the area bounded by $v(t')$ and the t' axis to the left of t.

For $t < 0$, $v_{in}(t') = 0$ and the area bounded by $v_{in}(t')$ to the left of t equals zero.

For $0 \le t < 1$ ms, the area bounded by $v_{in}(t')$ to the left of t equals $(-10\,\text{V}) \times t$.

For $1\,\text{ms} \le t < 3\,\text{ms}$, the area bounded by $v_{in}(t')$ to the left of t equals $(-10\,\text{V}) \times (1\,\text{ms}) = -10^{-3}\,\text{V s}$. For $3\,\text{ms} \le t < 5\,\text{ms}$, the area bounded by $v_{in}(t')$ to the left of t equals $-10^{-3}\,\text{V s} + (5\,\text{V}) \times (t - 3\,\text{ms})$. For $t > 5\,\text{ms}$, the area bounded by $v_{in}(t')$ to the left of t is constant and equal to $-10^{-3}\,\text{V s} + (5\,\text{V}) \times (5\,\text{ms} - 3\,\text{ms}) = 0$. Thus

$$v_{out} = -\frac{1}{RC}\begin{cases} 0, & t < 0, \\ (-10\,\text{V})t, & 0 \le t < 1\,\text{ms}, \\ -10^{-3}\,\text{V s}, & 1\,\text{ms} \le t < 3\,\text{ms}, \\ -10^{-3}\,\text{V s} + (5\,\text{V})(t - 3\,\text{ms}), & 3\,\text{ms} \le t < 5\,\text{ms}, \\ 0, & t \ge 5\,\text{ms} \end{cases}$$

where $RC = 2\,\text{ms}$. Figure 8.41 shows a graph of the output.

In Example 8.20, the magnitude of the output increases linearly with time, and will drive the op amp to saturation, at which point the circuit no longer acts as an integrator. Exercise 8.19 seems to illustrate that the circuit in Fig. 8.38 can function as an integrator for a long time if the input contains no dc component. However, in any real such circuit, a small but nonzero dc offset at the input is almost unavoidable, and will eventually drive the output to saturation, because there is no (or very little) feedback at dc (an ideal capacitor is an open circuit at dc). Practical integrators provide a means of resetting the output to zero after a predetermined time to avoid saturation. Figure 8.42 shows such a circuit, in which the switch is closed periodically to discharge the capacitor and reset the output to zero (the switch usually is electronic, not mechanical). The circuit in Fig. 8.42 is called a *finite-time integrator*, because the integration ends when the switch is closed. Finite-time integrators are used in a number of applications, including analog computers and various pulse-detection circuits.

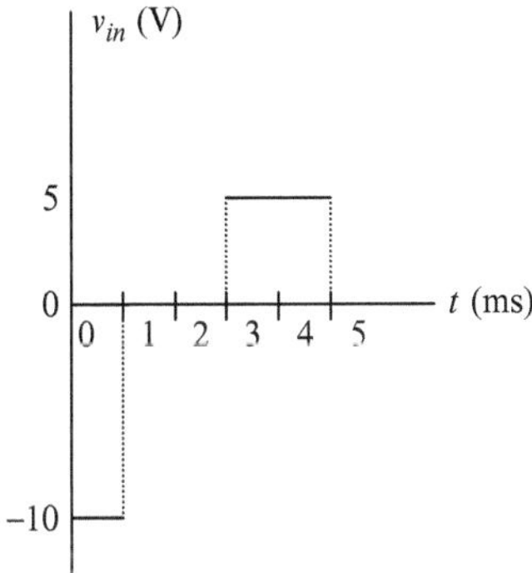

Fig. 8.40 See Example 8.20

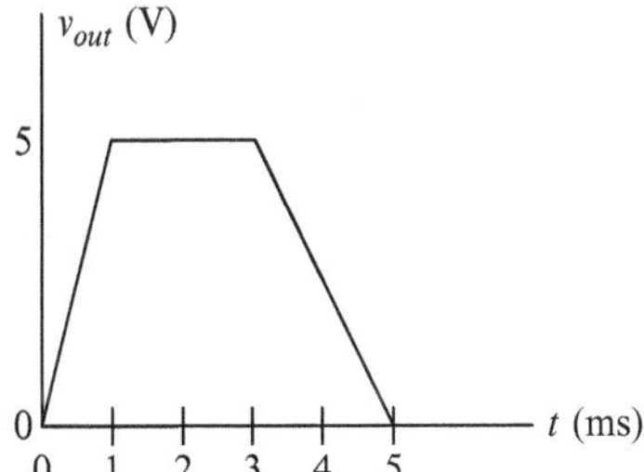

Fig. 8.41 See Example 8.20

Exercise 8.20. In Fig. 8.42, $R_1 = 2\,\text{k}\Omega$, $C = 10\,\text{pF}$, and $R_2 = 100\,\Omega$. The circuit must be able to integrate a constant 5 mV input for 100 µs (before the op amp saturates). (a) What is the required power supply voltage? (b) When the integrator is reset (when the switch is closed), how long does it take for the capacitor to discharge, before another integration can begin? (Give and justify a reasonable time).

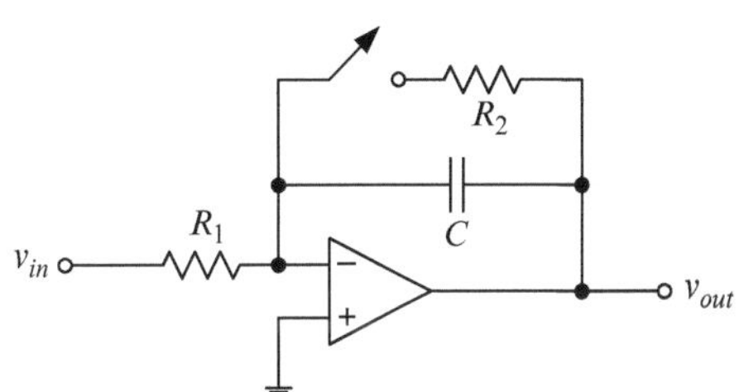

Fig. 8.42 A finite-time (resettable) integrator

8.10.3 Bypass Capacitors

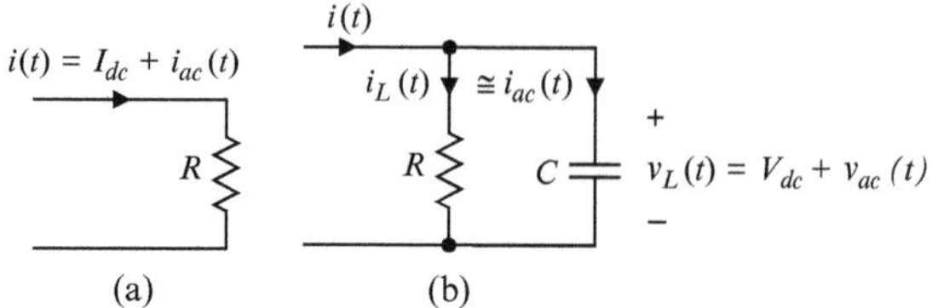

Fig. 8.43 Illustrating the function of a bypass capacitor

The purpose of a **bypass capacitor** is to shunt undesirable time-varying components of a current, called **ripple**, around a resistive load, while desirable dc components pass through the load, such that the load voltage is nearly constant. There are at least three major areas of application:

- Currents demanded by various components in linear systems fluctuate with varying demand. For example, in an audio amplifier, current delivered to the loudspeaker fluctuates with loudness, so the current demanded of the amplifier's output stage varies with the instantaneous loudness of the audio output. Such varying demand causes the current drawn from the supply to fluctuate, and the fluctuations can be coupled to signal lines; e.g., through stray capacitance or by causing the power supplied to earlier stages to fluctuate. Bypass capacitors can reduce such fluctuations. Plastic-dielectric capacitors are often used in such applications, but if high precision is required, silver-mica capacitors might be used.
- In power supplies, relatively large bypass capacitors (often, 1000 μF or more) are used to smooth the ripples in a rectified sinusoidal voltage. The capacitors most commonly used in larger power supplies are aluminum electrolytics because of the large capacitances achievable with such capacitors. Tantalum electrolytics might be used in smaller (low-voltage) power supplies. Electrolytic capacitors are typically imprecise. For example, the tolerance on an aluminum electrolytic might be +60%/−40%. Conservative sizing is necessary where such capacitors are used.
- In digital systems, bypass capacitors are attached from the power pin to ground at virtually every integrated circuit (IC) to keep the supply voltage at the pin within the operating range for the IC. Typical values range from about 10 to 100 nF, and occasionally to 500 nF. Ceramic capacitors are often used in this application, partly because of their large capacitance to volume ratios.

In this subsection, we describe bypassing in applications where the ripple can be approximated by a sinusoid. In the next, we describe bypassing for rectifier circuits, and in the next after that, bypassing for digital systems. We do not treat bypassing for very high-frequency applications because a useful treatment of that topic would take us too far afield.

Refer to Fig. 8.43(a). Let $i(t) = I_{dc} + i_{ac}(t)$ denote the current through a resistive load R, where the dc component I_{dc} is the desirable component and the ac component $i_{ac}(t)$ is undesirable. In such cases, an appropriate figure of merit is the **ripple factor**, defined as

$$\gamma = \frac{I_{ac\,rms}}{|I_{dc}|} = \frac{I_{ac\,rms}R}{|I_{dc}|R} = \frac{V_{ac\,rms}}{|V_{dc}|}. \tag{8.50}$$

To reduce the ripple in the load, we may connect a bypass capacitor in parallel with the load, as shown in Fig. 8.43(b). The purpose of the bypass capacitor is to shunt most of the ac component $i_{ac}(t)$, such that the load current $i_L(t)$ is largely the dc component I_{dc}.

A simplified approach to specifying a bypass capacitor is as follows. We assume the ripple is approximately sinusoidal. Thus let

$$i_{ac}(t) \cong I_{ac}\cos(\omega_0 t).$$

If the capacitor is absent, all of the current passes through the resistor, and the ripple factor is given by

$$\gamma_0 = \frac{I_{ac\,rms}}{|I_{dc}|} = \frac{|I_{ac}|}{\sqrt{2}|I_{dc}|}. \tag{8.51}$$

With the capacitor in place, all of the dc current passes through the resistor (a capacitor does not pass dc), so the dc component of the load voltage is given by

$$V_{dc} = I_{dc} R.$$

Ideally, almost all of the ripple current is diverted through the capacitor. In that case, the load voltage is given approximately by

$$v_L(t) \cong v_L(0) + \frac{1}{C}\int_0^t i_{ac}(t')dt'$$
$$= v_L(0) + \frac{I_{ac}}{\omega_0 C}\sin(\omega_0 t).$$

Thus the ac component of the load voltage is given by

$$v_{ac}(t) = \frac{I_{ac}}{\omega_0 C}\sin(\omega_0 t)$$

and the ripple factor is given (approximately) by

$$\gamma = \frac{V_{ac\,rms}}{|V_{dc}|} \cong \frac{|I_{ac}|}{\omega_0 C\sqrt{2}|I_{dc}|R} = \frac{\gamma_0}{\omega_0 RC}. \qquad (8.52)$$

Thus the bypass capacitance required to achieve a specified (small) ripple factor γ in a resistive load R is given by

$$C \cong \frac{1}{\omega_0 R}\left(\frac{\gamma_0}{\gamma}\right) = \frac{1}{2\pi f_0 R}\left(\frac{\gamma_0}{\gamma}\right), \qquad (8.53)$$

where γ_0 is the ripple factor in absence of bypassing and f_0 is the frequency of a sinusoidal approximation to the ripple.

A somewhat more careful analysis yields (compare with (8.52))

$$\gamma = \frac{V_{ac\,rms}}{|V_{dc}|} = \frac{\gamma_0}{\sqrt{1+D^2}}, \quad D = \omega_0 RC \qquad (8.54)$$

and (compare with (8.53))

$$C \cong \frac{1}{\omega_0 R}\sqrt{\left(\frac{\gamma_0}{\gamma}\right)^2 - 1}. \qquad (8.55)$$

The difference between (8.53) and (8.55) is less than 1% if $\gamma_0 > 7\gamma$, as is true in virtually all bypassing applications.

Example 8.22. In Fig. 8.43, $I_{dc} = 50\,\text{mA}$, $I_{ac} = 5\,\text{mA}$, $f_0 = 10\,\text{kHz}$, and $R = 2\,\Omega$. (a) Find the ripple factor in absence of the capacitor. (b) Find the capacitance that reduces the ripple factor to 0.1% (0.001).

Solution: (a) From (8.51),

$$\gamma_0 = \left|\frac{I_{ac}}{I_{dc}\sqrt{2}}\right| = \frac{5}{50\sqrt{2}} = 70.7 \times 10^{-3}.$$

From (8.53)

$$C \cong \frac{1}{\omega_0 R}\left(\frac{\gamma_0}{\gamma}\right) \cong \frac{1}{\omega_0 R}\left(\frac{70.7\times10^{-3}}{1.00\times10^{-3}}\right)$$
$$\cong \frac{70.7}{(2\pi\times10^4\,\text{Hz})(2\,\Omega)} \cong 563\,\mu\text{F}.$$

Exercise 8.21. Refer to Example 8.22. Calculate the required capacitance using (18.55).

If a bypass capacitor is effective for a sinusoidal current having frequency f_0 it is even more effective for sinusoids having frequencies greater than f_0.

Exercise 8.22. About 99% of any sinusoidal current having frequency greater than 100 Hz is to be diverted (bypassed) around a $1\,\text{k}\Omega$ resistive load by a capacitor in parallel with the load. What capacitance is required?

Equation (8.53) is not strictly applicable if the ripple is not sinusoidal. Nonetheless, we often may use that relation if the ripple is approximately periodic and if we can estimate the period T of the ripple. In such cases, it is advisable to incorporate a margin of 25% or so; i.e.,

$$C \cong \frac{1.25\,\gamma_0 T}{\gamma\, 2\pi R}. \qquad (8.56)$$

For example, if the period of the ripple is about 1 ns, the equivalent load is about 1Ω, and we require $\gamma_0/\gamma = 100$, the required capacitance is about 20 nF.

8.10.4 Bypass Capacitors (Filter Capacitors) in Rectifier Circuits

Equations (8.52) and (8.54) are strictly applicable only to *linear* circuits in which the load is resistive and the ripple is *sinusoidal*. They might be approximately correct if the circuit is approximately linear, the load is predominately resistive, and the ripple is approximately sinusoidal, but such applications must be considered on a case-by-case basis. An application in which the relations above are generally *inapplicable* is in power supplies where bypass capacitors (called *filter capacitors*) are used to suppress the ripple introduced by rectifier circuits. In such applications, the circuit is decidedly nonlinear and the ripple is decidedly non-sinusoidal.

Figure 8.44 shows a half-wave rectifier circuit driving a resistive load R_L. Figures 8.45 and 8.46 show results of an analysis of the circuit.[15] Figure 8.45 shows the load voltages for $C = 100\,\mu\text{F}$ and $C = 1{,}000\,\mu\text{F}$. The peak-to-peak load voltage ripple is approximately 102 V for $C = 100\,\mu\text{F}$ and approximately 13.4 V for $C = 1{,}000\,\mu\text{F}$. The rms ripple is approximately 37.6 V for $C = 100\,\mu\text{F}$ and approximately 4.92 V for $C = 1{,}000\,\mu\text{F}$. The dc load voltage is approximately 93.8 V for $C = 100\,\mu\text{F}$ and approximately 108 V for $C = 1{,}000\,\mu\text{F}$.[16] The measured (calculated) ripple factors are

$$\gamma_{100\,\mu\text{F}} \cong \frac{37.6\,\text{V}}{93.8\,\text{V}} \cong 0.401,$$

$$\gamma_{1000\,\mu\text{F}} \cong \frac{4.92\,\text{V}}{108\,\text{V}} \cong 0.046.$$

Increasing the capacitance by an order of magnitude decreases the ripple by almost an order of magnitude.

A price of increasing the filter capacitance is an increase in the diode current, which determines the size and cost of the diode. Figure 8.46 shows the initial current in the diode for two values of the filter (bypass) capacitor C. The maximum diode current is approximately 5 A for $C = 100\,\mu\text{F}$ and 13.5 A for $C = 1{,}000\,\mu\text{F}$. These currents would be even larger

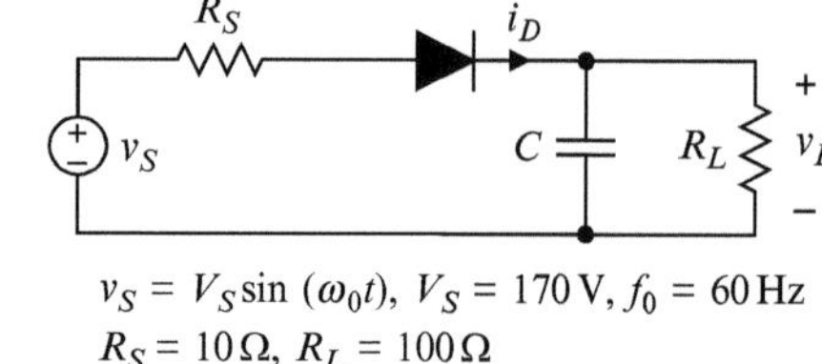

Fig. 8.44 Bypass (filter) capacitor in a half-wave rectifier circuit

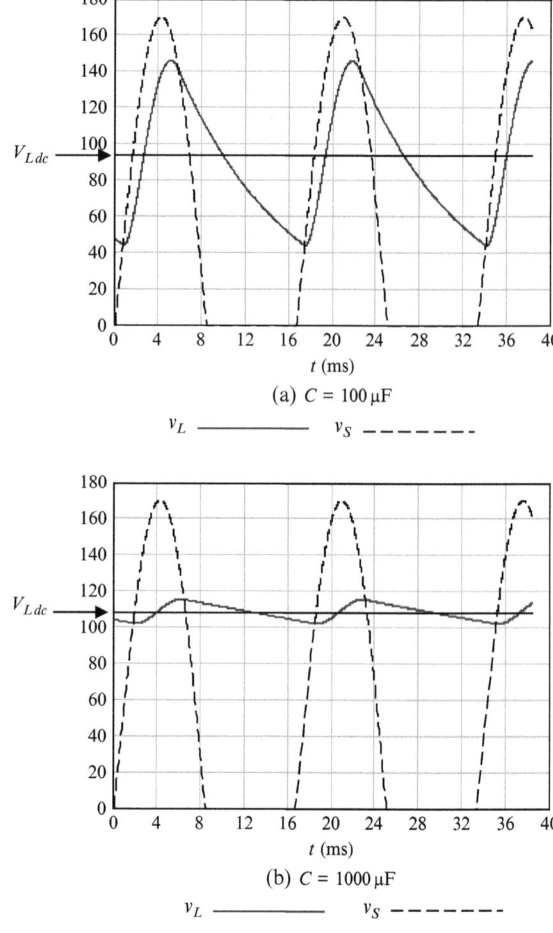

Fig. 8.45 Load-voltage waveforms in a half-wave rectifier circuit for two values of the filter capacitance

[15] The equations describing the circuit are nonlinear and cannot be solved in closed form. The analysis was performed numerically, using methods described in John R. Hauser, *Numerical Methods for Nonlinear Engineering Models*, Springer, 2009.

[16] All these values were obtained from the numerical analysis.

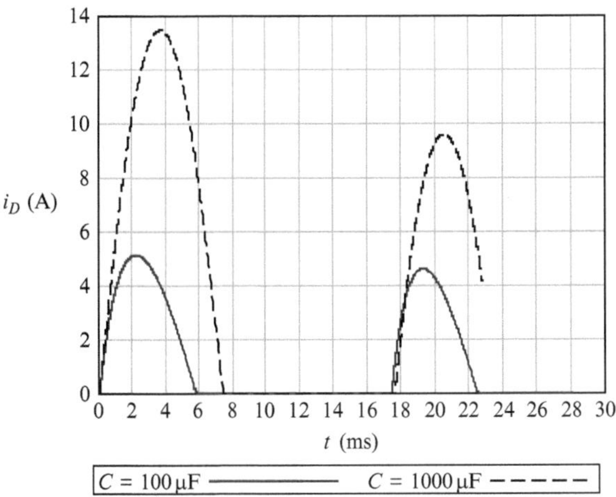

Fig. 8.46 Diode current in a half-wave rectifier circuit for two values of the filter capacitance

if the circuit were energized on a peak of the source voltage.

We can obtain approximate expressions for the rms ripple and ripple factor, as follows: When the diode in Fig. 8.45 is conducting, it is approximately a short circuit (because the diode drop is much smaller than the source voltage), and the parameters of the Théve-nin equivalent for the circuit seen by the capacitor are

$$v_T = \frac{R_L v_S}{R_L + R_S},\ R_T = R_S \| R_L.$$

When the diode is not conducting, it is approximately an open circuit, and the capacitor discharges through the load resistor R_L. Thus the charging and discharging time constants are different, and are given by

$$\text{Charging}: \tau_S - (R_S \,\|\, R_L)C,\ \text{Discharging}: \tau_L - R_L C.$$

In this example and often in applications, $R_S \ll R_L$, and the charging time constant is much smaller than the discharge time constant. Thus the load voltage almost follows the source voltage during charging, but not during discharge. If the discharge time constant is much larger than the period of the source voltage, as is usually the case, the discharge will occupy the lion's share of each period, and we may estimate the peak-to-peak ripple as

$$V_{RPP} = V_{\max} - V_{\min}, \tag{8.57}$$

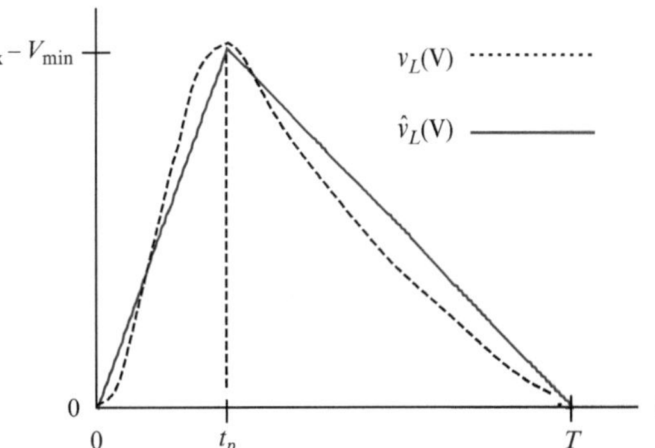

Fig. 8.47 Piecewise-linear (triangular) approximation to the load voltage in a half-wave rectifier circuit

where

$$V_{\min} \cong V_{\max} \exp\left(-\frac{T}{\tau_L}\right), \tag{8.58}$$

and thus

$$V_{\max} - V_{\min} \cong V_{\max}\left[1 - \exp\left(-\frac{T}{\tau_L}\right)\right]. \tag{8.59}$$

The dotted line in Fig. 8.47 shows one period of the ripple, translated to the origin. In the figure, $V_{\max}$ and $V_{\min}$ denote the maximum and minimum values of the load voltage, T is the period of the ripple, and t_p is the time at which the peak of the ripple occurs.

The solid line in Fig. 8.47 represents a piecewise-linear (triangular-wave) approximation to the ripple. For $t_p \leq t < T$ (approximately), the ripple follows an exponential decay, with time constant $\tau_L = R_L C$, as given above. For $\tau_L \gg T$, the exponential decay is approximately linear during $t_p \leq t < T$. It can be shown that the rms amplitude of the ripple, based upon the triangular-wave approximation, is given by

$$V_{L\,ac\,rms} = \frac{V_{\max} - V_{\min}}{\sqrt{12}}. \tag{8.60}$$

If the ripple is small, as is usual, then $V_{\max} \cong V_{DC}$, where V_{DC} is the dc component of the load voltage. With this approximation, (8.60) and (8.59) give

$$V_{L\,ac\,rms} = \frac{V_{\max} - V_{\min}}{\sqrt{12}} = \frac{V_{\max}}{\sqrt{12}}\left[1 - \exp\left(-\frac{T}{\tau_L}\right)\right]$$
$$\cong \frac{V_{DC}}{\sqrt{12}}\left[1 - \exp\left(-\frac{T}{\tau_L}\right)\right],$$

and the ripple factor is given by

$$\gamma = \frac{V_{L\,ac\,rms}}{V_{DC}} \cong \frac{1}{\sqrt{12}}\left[1 - \exp\left(-\frac{T}{\tau_L}\right)\right],$$
$$\tau_L = R_L C. \tag{8.61}$$

To achieve a small ripple, we want the discharge time constant τ_L to be large relative to the period T of the ripple, such that the load voltage does not change much during the discharge part of each cycle. If that is the case, the factor in the brackets in (8.61) is approximately T/τ_L and the ripple factor is approximately

$$\gamma = \frac{V_{L\,ac\,rms}}{V_{DC}} \cong \frac{T}{\tau_L\sqrt{12}} = \frac{T}{R_L C\sqrt{12}}. \tag{8.62}$$

For the circuit in Fig. 8.44, the ripple factors calculated using (8.62) are 0.481 for $C = 100\,\mu\text{F}$ and 0.048 for $C = 1{,}000\,\mu\text{F}$, compared to 0.401 and 0.046 obtained by a numerical analysis of the circuit.

From (8.62), the capacitance C required to achieve a specified *small* ripple factor γ is given by

$$C = \frac{T}{R_L\gamma\sqrt{12}}. \tag{8.63}$$

From (8.63), achieving a ripple factor of 0.046 in the circuit of Fig. 8.44 requires a capacitance of $1{,}046\,\mu\text{F}$, which is close to the $1{,}000\,\mu\text{F}$ used to achieve that ripple factor.

The approximate relation (8.63) improves as the required ripple factor decreases, and is a useful design relation for half-wave rectifier circuits. But keep in mind that increasing the filter capacitance increases the diode current and the required power-dissipation rating for the diode.

Note that T is the period of the rectified voltage, which equals the period of the source for half-wave rectification and equals half the period of the source for full-wave rectification. Keep in mind that several approximations were invoked in deriving (8.63). One assumed the ripple is triangular. Another assumed that the maximum amplitude of the load voltage is approximately equal to the dc component of the load voltage. This approximation is poor unless the rms ripple is no more than a few percent of the dc component. Nonetheless, using a filter capacitor prescribed by (8.63) achieves a slightly smaller ripple than the specified value (γ). A capacitance calculated using (8.63) can be quite large for low-frequency applications. For example, for a 60 Hz source and $100\,\Omega$ load, the capacitance required to achieve a 1% ripple factor exceeds $4{,}000\,\mu\text{F}$. Thus other components usually supplement the smoothing provided by filter capacitors in such applications.

8.10.5 Bypassing in Digital Systems

An important application of bypass capacitors is in smoothing power-supply voltages at the power-input terminals of each integrated circuit (IC) in a larger digital system, such as a computer or memory bank. Each IC can contain a great many individual switching circuits (components of logic gates). Whenever a logic gate changes state, it must charge or discharge stray and parasitic capacitances, and draws a current pulse (+ or −) from the supply to do so. Thus, as various circuits in the IC switch from one state to another, the current drawn by the IC from the power supply varies, and wiring and source resistance can cause the supply voltage to vary at the IC. The variation can be increased by operation of other IC's drawing from the same supply. If the variation exceeds a certain tolerance, the circuit can malfunction. The job of a bypass capacitor is to keep the power-supply voltage at the IC within the range acceptable to the IC. Supply-voltage fluctuations at an IC are decidedly non-periodic, and specifying a bypass capacitor for a digital integrated circuit is done differently from specifying bypass capacitors for linear and rectifier circuits, as described below.

On average, the voltage across a bypass capacitor equals the supply voltage. The capacitor reduces ripple by momentarily supplying current when the voltage tries to fall and momentarily sinking current (absorbing charge) when the voltage tries to rise. The bypass capacitor's terminal characteristic

$$i = C\frac{dv}{dt}$$

governs the relation between current and voltage change. Using the approximation $dv/dt \cong \Delta v/\Delta t$ and solving for the capacitance gives

$$C = i\frac{\Delta t}{\Delta v}, \tag{8.64}$$

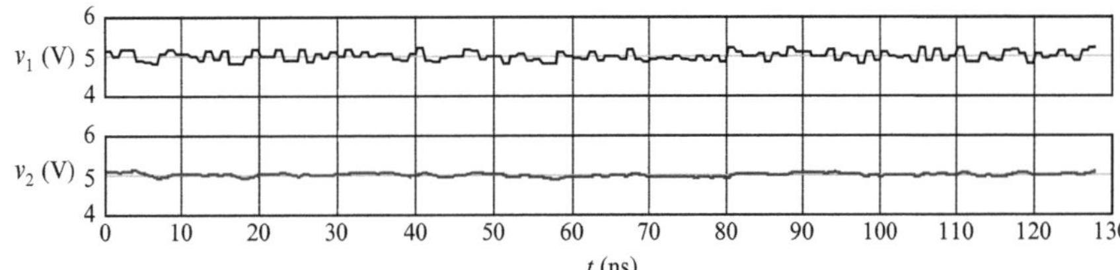

Fig. 8.48 Supply voltage at the terminals of an IC with and without bypass

which we interpret as follows: The current i is the current sourced or sunk by the capacitor in order to limit the change in voltage to Δv during an interval Δt. We may use (8.64) to estimate the bypass capacitance needed by a digital circuit, as illustrated by the next example.

Example 8.23. A certain digital integrated circuit (IC) is operating at a clock frequency of 1 GHz. As it operates, it can require a current of 50 mA above or below that provided (on average) by the supply. It is required that the supply voltage at the IC be kept within $\Delta v = 5\,\text{mV}$ of the specified value. Specify a bypass capacitor.

Solution: From (8.64)

$$C = (50\,\text{mA})\left(\frac{1\,\text{ns}}{5\,\text{mV}}\right) = 10\,\text{nF}.$$

In most applications (in digital systems), bypass capacitances fall in the range from about 10 to 500 nF. Such capacitors can be quite small, because the operating voltages are small (typically, 5 V or less).

Figure 8.48 illustrates the smoothing effect of a bypass capacitor. The upper trace is the voltage at the power pin of an un-bypassed integrated logic circuit in a larger logic system. The fluctuations due to switching in this and other circuits in the larger system are evident. The lower trace shows the voltage at the same pin after adding a bypass capacitor. While not perfectly smooth, the bypassed voltage is well within specifications on the IC.

Usually, one would add a little (10–25%) to the capacitance given by (8.64) to account for the occasional larger-than-expected current demand. On the theory that bigger is better, you might be tempted to choose a capacitance much larger than that. But that is not always wise. Parasitic effects (e.g., lead inductance and losses in the dielectric) and size (space occupied) increase with capacitance, as does the current required to charge or discharge the capacitor, so a larger-than-necessary capacitor can cause as many problems as it solves.

8.10.6 Coupling Capacitors

Figure 8.49 shows a source connected to a resistive load through a *coupling capacitor C*. The source and load are said to be **capacitively coupled.**[17] The usual purpose of capacitive coupling is to prevent a dc component of the source voltage from reaching the load while allowing ac components above a certain frequency to pass through to the load. Below, we describe how to specify a coupling capacitor.

We assume the source voltage is of the form

$$v_S(t) = V_{dc} + v_{ac}(t),$$

where $v_{ac}(t)$ has no dc component. The circuit is linear, so superposition is applicable. The capacitor will block the dc component V_{dc} by maintaining an average charge V_{dc}, so the load voltage has no dc component. We need concern ourselves with only the ac component. We assume a sinusoidal model

$$v_{ac}(t) = V_{ac}\cos(\omega_0 t), \tag{8.65}$$

which could be the lowest-frequency component of a sum of such voltages. The circuit is linear, so the voltage across the capacitor is also sinusoidal, of the form

$$v_C(t) = V_C\cos(\omega_0 t + \theta). \tag{8.66}$$

[17]Direct coupling also is called **dc coupling** and capacitive coupling also is called **ac coupling**.

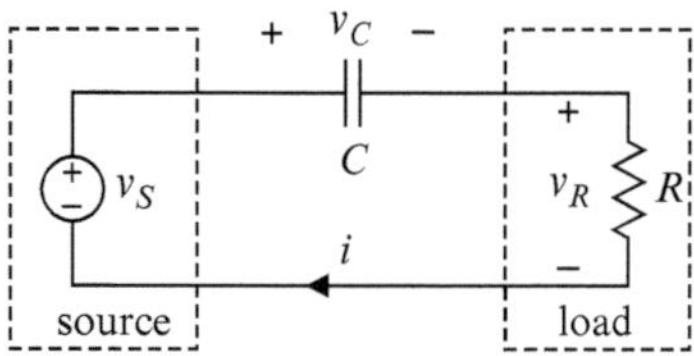

Fig. 8.49 Capacitive coupling of a source to a load

From (8.12), the current through the capacitor is given by

$$i = C\frac{dv_C}{dt} = -\omega_0 CV_C \sin(\omega_0 t + \theta).$$

It follows that the load voltage is given by

$$\begin{aligned} v_R(t) &= i(t)R = -\omega_0 RCV_C \sin(\omega_0 t + \theta) \\ &= V_R \sin(\omega_0 t + \theta), \end{aligned} \tag{8.67}$$

where

$$V_R = -\omega_0 RCV_C \Rightarrow V_C = -\frac{V_R}{\omega_0 RC}. \tag{8.68}$$

We wish to express the rms load voltage as a fraction of the rms ac source voltage. Applying Kirchhoff's voltage law to the circuit gives

$$\begin{aligned} v_{ac}(t) &= v_R(t) + v_C(t) \Rightarrow V_{ac} \cos(\omega_0 t) \\ &= V_R \sin(\omega_0 t + \theta) + V_C \cos(\omega_0 t + \theta), \end{aligned}$$

From trigonometry,

$$a \cos(\alpha) + b \sin(\alpha) \equiv \sqrt{a^2 + b^2} \cos(\alpha - \varphi),$$
$$\varphi = \tan^{-1}\left(\frac{b}{a}\right).$$

Thus

$$V_{ac} \cos(\omega_0 t) = \sqrt{{V_R}^2 + {V_C}^2} \cos(\omega_0 t + \theta - \varphi),$$

so the rms amplitudes are related as

$$\begin{aligned} \frac{|V_{ac}|}{\sqrt{2}} &= \frac{\sqrt{{V_R}^2 + {V_C}^2}}{\sqrt{2}} \Rightarrow {V_{ac}}^2 = {V_R}^2 + {V_C}^2 \\ &= {V_R}^2 + \left(\frac{V_R}{\omega_0 RC}\right)^2. \end{aligned}$$

If we specify that

$$|V_R| = k|V_{ac}|, \; 0 < k < 1,$$

meaning that the rms ac load voltage is a specified fraction k of the rms ac source voltage, then

$$\begin{aligned} {V_{ac}}^2 &= k^2 {V_{ac}}^2 \left[1 + \left(\frac{1}{\omega_0 RC}\right)^2\right] \\ &\Rightarrow \omega_0 RC = \frac{k}{\sqrt{1 - k^2}}. \end{aligned}$$

Therefore the coupling capacitance required to ensure that the rms ac load voltage equals at least a fraction k of the rms ac source voltage is given by

$$C \geq \frac{k}{\omega_0 R\sqrt{1 - k^2}}. \tag{8.69}$$

The analysis leading to (8.69) ignores the source resistance (not shown in Fig. 8.49). This is equivalent to assuming that the load resistance is much larger than the source resistance, which is true in the majority of applications where capacitive coupling is employed. A much more tedious analysis that does not ignore the source resistance leads to

$$C \geq \frac{k}{2\pi f_0 R_L \sqrt{1 - k^2}} \frac{1}{\sqrt{1 - \frac{k^2}{1 - k^2} \frac{R_S}{R_L}\left(2 + \frac{R_S}{R_L}\right)}}. \tag{8.70}$$

In applying either (8.69) or (8.70), keep in mind that tolerances on capacitors can be large – 20% or more, and consider incorporating a margin of 25% or so; e.g., by requiring

$$C \geq \frac{1.25k}{\omega_0 R\sqrt{1 - k^2}}.$$

Example 8.24. The first stage of a two-stage amplifier is capacitively coupled to the second stage, as shown in Fig. 8.50. The output resistance of the first stage (the source resistance) is negligible. The input resistance of the second stage is 20 kΩ. The voltage gain of each amplifier is 25. The overall voltage gain must be at least 500 for any sinusoid whose frequency exceeds 30 Hz. Specify the coupling capacitor.

Solution: If the amplifiers were direct coupled, the overall voltage gain would be $(25)^2 = 625$. The specified gain at 30 Hz is 500, so

$$k = \frac{500}{625} = 0.8$$

and we require

$$C \geq \frac{k}{2\pi f_0 R\sqrt{1-k^2}}$$
$$\Rightarrow C \geq \frac{0.8}{2\pi(30\,\text{Hz})(20\times 10^3\,\Omega)(0.6)}$$
$$\Rightarrow C \geq 354\,\text{nF}.$$

The simulation shown in Fig. 8.51 below illustrates this result. (The voltmeters indicate rms amplitude.) The voltage gain of each inverting amplifier equals 25, the coupling capacitance is that calculated above, and the overall voltage gain at 30 Hz (to three significant figures) is

$$A_v = \frac{5.00\,\text{V}}{10.0\,\text{mV}} = 500,$$

as specified.

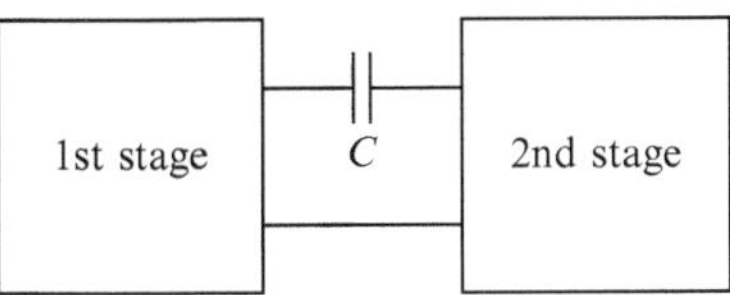

Fig. 8.50 See Example 8.24

Exercise 8.23. The two stages of an amplifier are coupled as shown in Fig. 8.50, where $C = 500\,\text{nF}$. The output resistance of the first stage is negligible and the input resistance of the second stage is 10 kΩ. For what frequencies of a sinusoidal output from the first stage is the overall gain *reduced* by a factor of 0.5 or more?

Effects of capacitive coupling in cascaded stages are cumulative. For example, refer to Fig. 8.52, where $\omega_0 > 0$. We may assume (without loss of generality)

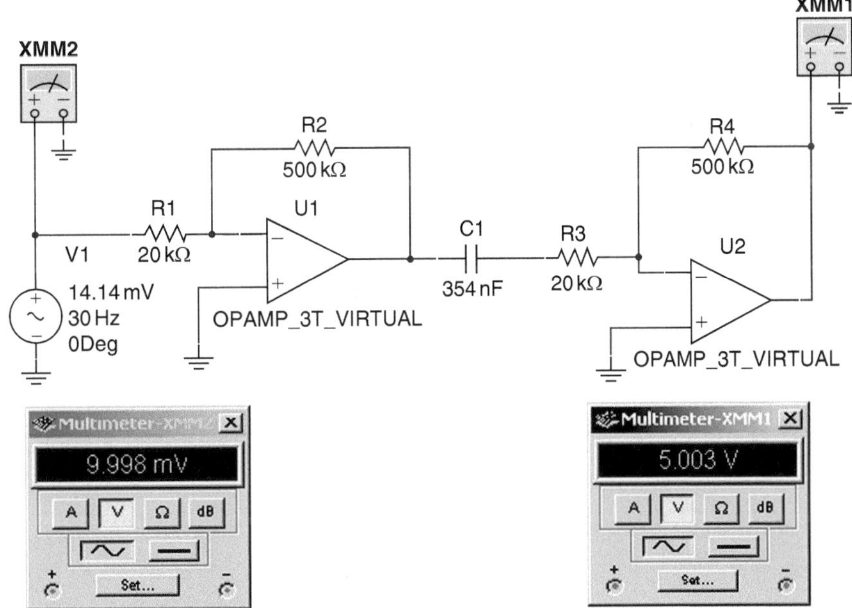

Fig. 8.51 See Example 8.24

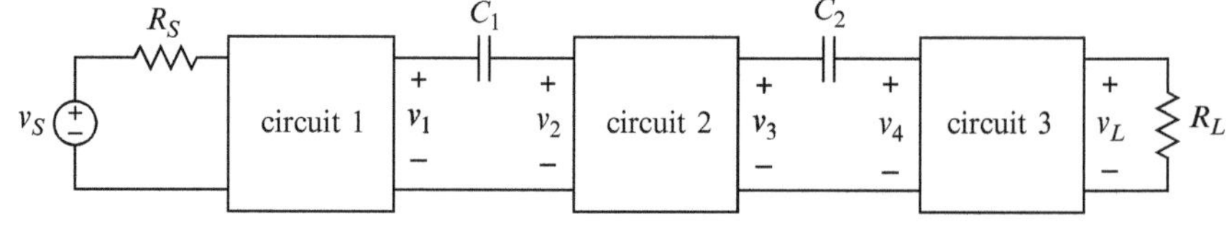

Fig. 8.52 Circuits in cascade, with capacitive coupling

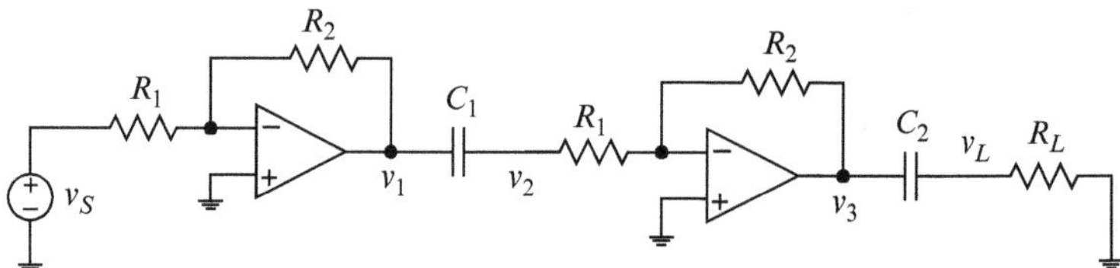

Fig. 8.53 See Example 8.25

that the initial phase angles θ_n are such that $V_1, V_2, \cdots V_L$ are positive. We also assume that the output resistance of each circuit (or the source) is much smaller than the input resistance of the succeeding circuit (or the load). Suppose the coupling capacitors are specified such that

$$V_2 = k_1 V_1; \; V_4 = k_2 V_3$$

for frequency f_0. Then

$$\begin{aligned} V_L &= A_{v3} V_4 = A_{v3} k_2 V_3 = A_{v3} k_2 A_{v2} V_2 \\ &= A_{v3} k_2 A_{v2} k_1 V_1 = A_{v3} k_2 A_{v2} k_1 A_{v1} V_S. \end{aligned}$$

The overall voltage gain is given by

$$A_v = \frac{V_L}{V_S} = A_{v3} A_{v2} A_{v1} k_1 k_2.$$

If we want the overall voltage gain to be a fraction k of the product of the individual gains, then we must specify the coupling capacitors such that $k_1 k_2 = k$. If the input resistances of circuit 2 and circuit 3 are R_2 and R_3, respectively, and if the output resistances of circuit 1 and circuit 2 are negligible, then we would require

$$C_1 \geq \frac{k_1}{2\pi f_0 R_2 \sqrt{1 - k_1^2}}, \; C_2 \geq \frac{k_2}{2\pi f_0 R_3 \sqrt{1 - k_2^2}}. \tag{8.71}$$

Example 8.25. Figure 8.53 shows a two-stage amplifier, where $v_S = V_S \cos(2\pi f_0 t)$, with $f_0 = 80\,\text{Hz}$ and $V_S = 5\,\text{mV}$. Assume the op amps are ideal. The source resistance and the output resistances of the individual amplifiers are negligible.

(a) Assume the coupling capacitors are conductors and specify the resistances R_1, R_2 such that the overall voltage gain is $A_v = 2,500$.
(b) Specify the coupling capacitors such that the overall voltage gain for $f \geq f_0$ equals or exceeds $0.8 A_v = 2,000$.

Solution:

(a) The overall voltage gain is given by

$$A_v = \frac{V_{L\,rms}}{V_{S\,rms}} = k_1 k_2 \left(\frac{R_2}{R_1}\right)^2,$$

where

$$k_1 = \frac{V_{2\,rms}}{V_{1\,rms}}, \; k_2 = \frac{V_{L\,rms}}{V_{3\,rms}}.$$

If the capacitors are conductors, then $k_1 \cong 1$, $k_2 \cong 1$, and

$$A_v = 2,500 \cong \left(\frac{R_2}{R_1}\right)^2 = (50)^2 \Rightarrow R_2 = 50 R_1.$$

We choose

$$R_2 = 1\text{M}\Omega \Rightarrow R_1 = 20\,\text{k}\Omega.$$

(b) Let $k_1 = k_2 = \sqrt{0.8}$ for $f = f_0$. From (8.71)

$$C_1 = C_2 = \frac{\sqrt{0.8}}{2\pi f_0 R_1 \sqrt{1-0.8}} = 199\,\text{nF} \cong 200\,\text{nF}.$$

Equations (8.69) and (8.70) are applicable to a sinusoidal current or voltage, or to a sum of sinusoidal currents or voltages, where the lowest frequency represented equals f_0. For piecewise-constant signals, we can approach the problem of specifying a coupling capacitor from another point of view. Refer to Fig. 8.54, where a rectangular pulse train source $v_S(t)$ is capacitively coupled to a resistive load R_L. The dc component of the pulse train is blocked by the capacitor, so the average value of the load voltage must be zero. Also, the voltage across the capacitor cannot change instantaneously, so when the source voltage changes by $\pm V_0$, the load voltage must also change by $\pm V_0$. The dc component of the source voltage v_S is given by

$$V_{S\,dc} = \frac{1}{T}\int_0^T v_S(t)dt = \frac{V_0 t_0}{T},$$

so the maximum and minimum amplitudes of the ac component of v_S are given by

$$V_{\max} = V_0\left(1 - \frac{t_0}{T}\right),\; V_{\min} = V_{\max} - V_0. \tag{8.72}$$

Figure 8.55 shows a qualitative graph of the load voltage v_L (exaggerated), superimposed on a graph (dotted) of the ac component of the source voltage. By inspection,

$$V_2 = V_1 \exp\left(-\frac{t_0}{\tau}\right), \tag{8.73}$$

where

$$\tau = R_L C. \tag{8.74}$$

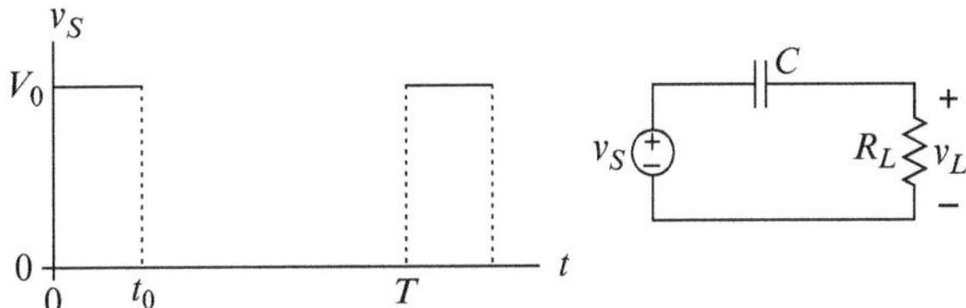

Fig. 8.54 Rectangular pulse-train source capacitively coupled to a load (see also Fig. 8.55)

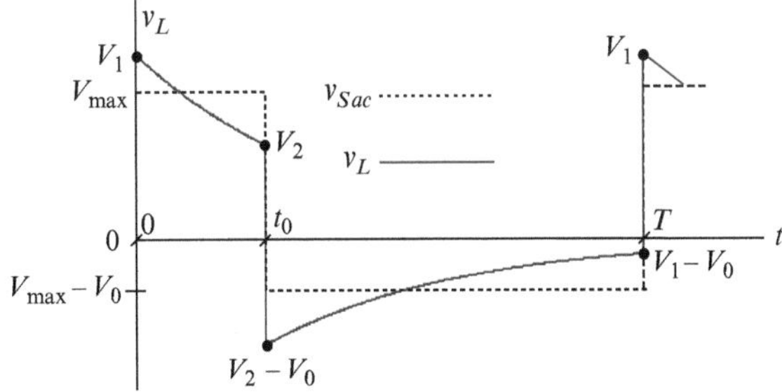

Fig. 8.55 Qualitative graph of the response of the circuit in Fig. 8.54

We know that

$$V_2 < V_{\max} < V_1. \tag{8.75}$$

Thus, a measure of the difference between the load voltage v_L and the ac component of v_S is given by

$$\begin{aligned} V_1 - V_2 &= V_1\left[1 - \exp\left(-\frac{t_0}{\tau}\right)\right] \\ \Rightarrow \frac{V_1 - V_2}{V_1} &= 1 - \exp\left(-\frac{t_0}{\tau}\right), \end{aligned} \tag{8.76}$$

where we have used (8.73). If the difference $V_1 - V_2$ is sufficiently small, as is usually the case in problems of this kind, then $V_1 - V_2$ is approximately twice the maximum error ΔV, and we may write

$$\frac{2\Delta V}{V_1} = 1 - \exp\left(-\frac{t_0}{\tau}\right). \tag{8.77}$$

Finally, if the error is sufficiently small, we may use $V_1 \cong V_{\max}$ in (8.77) to obtain

$$\begin{aligned} \frac{2\Delta V}{V_{\max}} &= 1 - \exp\left(-\frac{t_0}{\tau}\right) \\ \Rightarrow \tau &= -\frac{t_0}{\ln(1 - 2\Delta V/V_{\max})}. \end{aligned} \tag{8.78}$$

Equation (8.78) gives the time constant required to achieve a specified error. The relation is applicable to circuits equivalent to the circuit in Fig. 8.54 and is valid if $\Delta V \ll V_{\max}$. For ΔV small, the relation tends to give a time constant slightly larger than necessary (an error smaller than specified). The parameter t_0 is the duration of the positive pulse in each period of the input, $V_{S\,\max}$ is the maximum amplitude of the ac component of the input, and ΔV is (approximately) the maximum difference between the load voltage and the ac component of the input.

Example 8.26. In Fig. 8.54, $V_0 = 10\,\text{V}$, $T = 10\,\text{ms}$, $t_0 = 2\,\text{ms}$, and $R_L = 5\,\text{k}\Omega$. The load voltage v_L must be within 100 mV of the ac component of the source voltage v_S at all times. Specify the coupling capacitor C. Calculate the voltages V_1 and V_2 defined by Fig. 8.55 to verify your answer.

Solution: The dc component of the source voltage is

$$\frac{V_0 t_0}{T} = 2\,\text{V}.$$

Thus, in (8.78) $V_{\max} = 8\,\text{V}$ and $\Delta V = 100\,\text{mV}$. We find

$$\tau = -\frac{t_0}{\ln(1 - 2\Delta V/V_{\max})} = -\frac{2\,\text{ms}}{\ln(1 - 0.2/10)}$$
$$\cong 100\,\text{ms} \Rightarrow C = \frac{\tau}{R_L} = \frac{100\,\text{ms}}{5\,\text{k}\Omega} = 20\,\mu\text{F}.$$

From Fig. 8.55, we have

$$V_2 = V_1 \exp\left(-\frac{t_0}{\tau}\right),$$
$$V_1 - V_0 = (V_2 - V_0)\exp\left(-\frac{T - t_0}{\tau}\right). \quad (8.79)$$

Using the first relation to replace V_2 in the second gives

$$V_1 - V_0 = \left[V_1 \exp\left(-\frac{t_0}{\tau}\right) - V_0\right]\exp\left(-\frac{T - t_0}{\tau}\right)$$
$$\Rightarrow V_1 = V_0 \frac{\exp\left(-\frac{T - t_0}{\tau}\right) - 1}{\exp\left(-\frac{T}{\tau}\right) - 1} \cong 8.079\,\text{V},$$

whence

$$V_2 = V_1 \exp\left(-\frac{t_0}{\tau}\right) = 7.919\,\text{V}.$$

The errors are

$$V_1 - V_{\max} = 79\,\text{mV}, V_{\max} - V_2 = 81\,\text{mV}.$$

both of which are less than 100 mV.

Exercise 8.24. In practice, a non-negative rectangular pulse train usually is described in terms of its duty cycle d, period T, and peak amplitude V_0, where the *duty cycle* is the fraction of each cycle for which the pulse train is positive. Express the requirement (8.78) for faithful transmission of the ac component of a rectangular pulse train in terms of the duty cycle of the pulse train.

8.10.7 Input Bias Current Compensation in Capacitively Coupled Amplifiers

Recall from Chapter 7 that input bias currents in an op amp can cause a dc offset voltage at the output of an amplifier, and that the offset can be largely corrected by inserting a compensating resistor R_X as shown in Fig. 8.56 for direct-coupled inverting and non-inverting amplifiers. The approximations assume $R_1 \gg R_S$ and $R_2 \gg R_1$, which often hold for such amplifiers.

Input-bias-current compensation for capacitively coupled amplifiers is different from that for direct-coupled amplifiers. The compensating resistor for a capacitively coupled inverting amplifier is connected from the positive input (p) terminal to ground, as shown in Fig. 8.57. But in this case, there is no dc path back through the source, so no dc bias current passes through R_1. Requiring that the output v_o due to the input bias currents alone equals zero requires $v_p = v_n$, which, again assuming $I_p = I_n$, implies

$$I_n\left(R_2 \middle\| \frac{R_i}{2}\right) = I_p\left(R_X \middle\| \frac{R_i}{2}\right) \Rightarrow R_X = R_2, \quad (8.80)$$

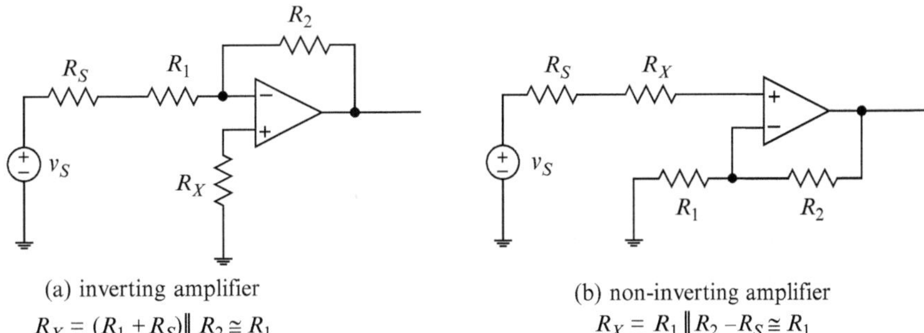

(a) inverting amplifier
$R_X = (R_1 + R_S) \| R_2 \cong R_1$

(b) non-inverting amplifier
$R_X = R_1 \| R_2 - R_S \cong R_1$

Fig. 8.56 Input bias-current compensation in direct-coupled inverting and non-inverting amplifiers

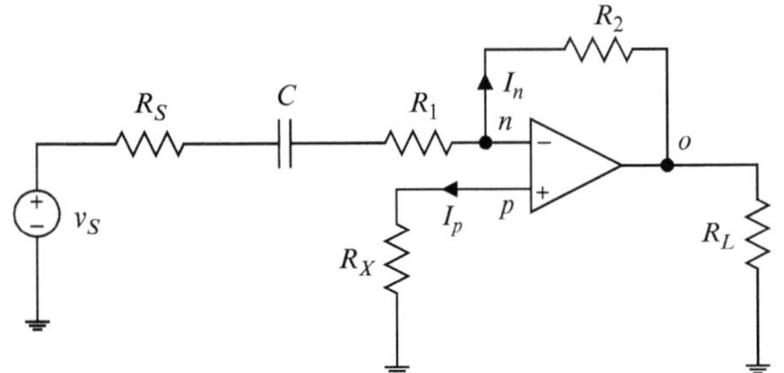

Fig. 8.57 DC bias compensation for a capacitively coupled inverting amplifier, where $R_X = R_2$

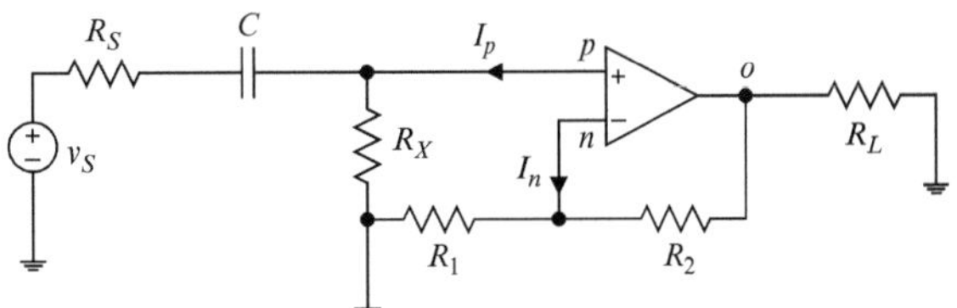

Fig. 8.58 Input bias current compensation for a capacitively coupled non-inverting amplifier, where $R_X \cong R_1$

where R_i is the input resistance of the op amp. Thus, for a capacitively coupled inverting amplifier, the input bias current compensating resistance equals the feedback resistance R_2, not the input resistance R_1, as in direct-coupled amplifiers.

In a capacitively coupled non-inverting amplifier, the compensating resistor must be connected from the p terminal to ground, as shown in Fig. 8.58, because there is no dc path back through the source. Requiring that the output v_o due to the input bias currents alone equals zero requires $v_p = v_n$, which (assuming $I_p = I_n$) leads to

$$I_p\left(R_X \left\| \frac{R_i}{2}\right.\right) = I_n(R_1 \| R_2) \left\| \frac{R_i}{2}\right.$$
$$\Rightarrow R_X = R_1 \| R_2, \tag{8.81}$$

and again,

$$R_X = R_1 \| R_2 \cong R_1, \; R_2 \gg R_1. \tag{8.82}$$

Input bias current compensation might be optional for direct-coupled amplifiers and for capacitively coupled *inverting* amplifiers, because the resulting dc offset in the output might be insignificant in some applications. But *input bias current compensation is required for a capacitively coupled non-inverting amplifier*, because otherwise the bias current I_p passes through only the resistance $R_i/2$ and the resulting voltage is large enough to cause the op amp to saturate.

One effect of bias-current compensation in a capacitively coupled non-inverting amplifier is a substantial reduction of the amplifier input resistance, typically from 1 MΩ or more to approximately $R_X = R_1 \| R_2$. In a typical application, $R_2 \gg R_1$ so the input resistance of a bias-current-compensated capacitively coupled non-inverting amplifier is about the same as that for an inverting amplifier having the same dc voltage gain. If direct coupling to a source is possible, a non-inverting amplifier presents a much larger resistance (magnitude) to the source than does a comparable inverting amplifier, but if capacitive coupling is necessary (to block a dc component), a non-inverting amplifier offers no particular advantage over an inverting amplifier in that regard. A capacitively coupled follower would require a dc path to ground from the positive terminal, which would reduce the input resistance and might eliminate the reason for using the follower.

Example 8.27. Figure 8.59(a) shows a preliminary design for a two-stage amplifier that will drive a resistive load $R_L = 2\,\text{k}\Omega$. The available input voltage is $v_S = V_S \cos(2\pi f_0 t)$, with $V_S \leq 10\,\text{mV}$ and $f_0 \geq 50\,\text{Hz}$. The source resistance R_S is no larger than $50\,\Omega$. For frequencies greater than or equal to f_0, the *minimum* overall voltage gain is to be $A_v = 800$, with about equal gain from each stage. The first stage is to be capacitively coupled to the second, as shown, in case imperfect cancellation of the input voltage offset in the first stage produces an undesirable dc component in the first-stage output v_1. The available power-supply voltages are $\pm V_{CC} = \pm 15\,\text{V}$. The power-supply output resistance and associated wiring resistance is less than $2\,\Omega$. Complete the design. Include bias-current compensation resistors.

Solution: Figure 8.59(b) shows a circuit diagram for the amplifier. The power-supply circuitry is omitted to avoid cluttering the diagram.

The voltage gain for frequencies much larger than 50 Hz is not specified, but we know it must be larger than 800. The larger we make the overall high-frequency voltage gain, the smaller will be the parameter k in (8.69) and (consequently) the smaller will be the required coupling capacitance. The maximum allowable voltage gain is determined by the maximum amplitude of the input (10 mV) and the supply voltage (15 V). Assuming rail-to-rail operation,

$$\max(A_v) = \frac{15\,\text{V}}{10\,\text{mV}} = 1,500.$$

To provide a little margin, we choose, somewhat arbitrarily,

$$A_v = 35^2 = 1,225.$$

It follows that

$$k = \frac{800}{1,225} = 0.653.$$

The feedback networks in the inverting and non-inverting amplifiers are the same, so that the voltage gains of the two stages are approximately the same and approximately equal to 35. We choose $R_2 = 1\text{M}\Omega$ to minimize the load on each op amp and to allow a large input resistance. Thus, for the inverting amplifier,

$$\frac{R_2}{R_1} = 35,\ R_2 = 1\text{M}\Omega \Rightarrow R_1 = \frac{R_2}{35} \cong 28.6\,\text{k}\Omega.$$

For the non-inverting amplifier,

$$1 + \frac{R_4}{R_3} = 35,\ R_4 = 1\text{M}\Omega \Rightarrow R_3 = \frac{R_4}{34} = 29.4\,\text{k}\Omega.$$

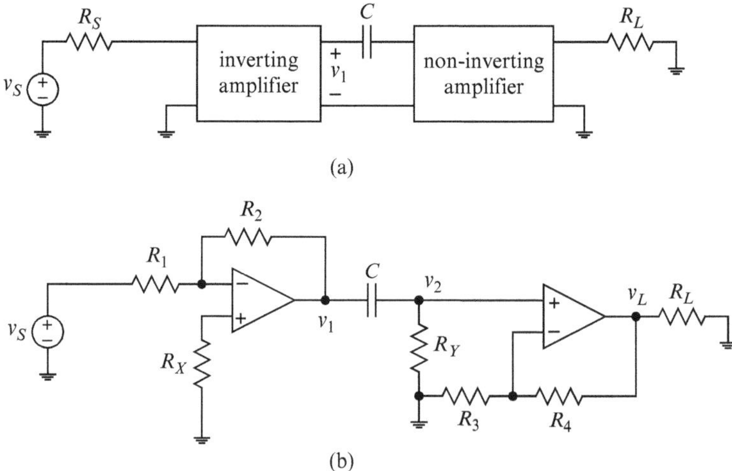

Fig. 8.59 See Example 8.27

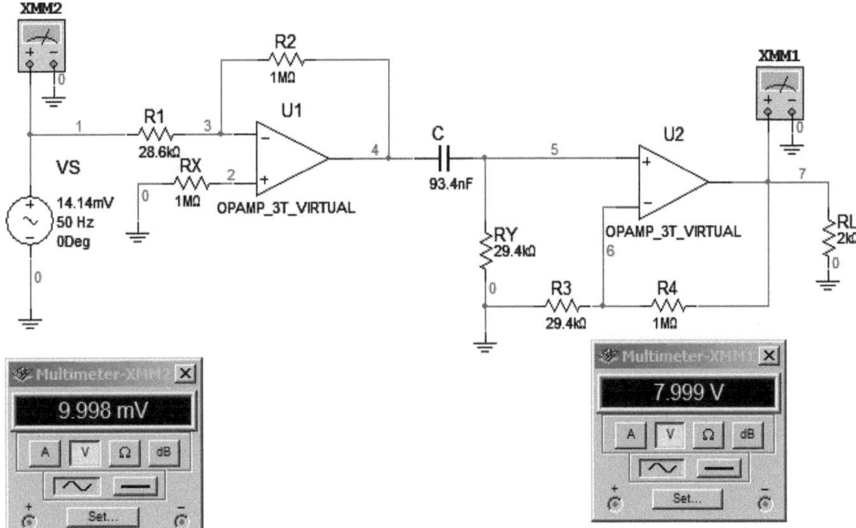

Fig. 8.60 Simulation of the circuit in Fig. 8.59(b)

From (8.80) and (8.82), the bias-current compensation resistances are

$$R_X = R_2 = 1\,\mathrm{M\Omega},\ R_Y = R_3 = 29.4\,\mathrm{k\Omega}.$$

The lowest frequency of interest is 50 Hz and the input resistance for the second stage is approximately R_Y. From (8.69), we require a coupling capacitance

$$C = \frac{k}{2\pi f_0 R_3 \sqrt{1-k^2}} \Rightarrow C = 93.4\,\mathrm{nF}.$$

The more precise relation (8.70) gives

$$C = \frac{k}{2\pi f_0 R_3 \sqrt{1-k^2}} \frac{1}{\sqrt{1 - \dfrac{k^2}{1-k^2}\dfrac{R_S}{R_3}\left(2 + \dfrac{R_S}{R_3}\right)}}$$

$$= 93.5\,\mathrm{nF}.$$

The difference, given typical tolerances on capacitors, is insignificant. Figure 8.60 below shows a simulation of the circuit. The measured voltage gain at 50 Hz is

$$A_v = \frac{7.999\,\mathrm{V}}{9.998\,\mathrm{mV}} \cong 800.$$

8.10.8 Switched Capacitor Circuits

Refer to Fig. 8.61. The circuit diagrams illustrate that the switches are alternately and oppositely open and closed; i.e., when switch 1 is closed, switch 2 is open, and vice-versa. The timing diagram on the right indicates for how long each switch is closed. We assume that the voltages v_1 and v_2 are approximately constant during any time interval having duration ΔT. Each time a switch is closed (or open), it remains so for a time $\Delta T/2$. When switch 1 is closed, the capacitor is discharged. When switch 2 is closed, the capacitor is charged to $\Delta v = v_1 - v_2$, which means that a quantity of charge given by $\Delta q = C\Delta v$ flows through a load attached to the port on the right. Each time switch 1 is closed, switch 2 is open and no charge flows through the load. The *average* current through the load over one complete switching cycle is given by

$$\bar{i} = C\frac{\Delta v}{\Delta T} = Cf_s \Delta v; \quad \Delta v = v_1 - v_2, \tag{8.83}$$

where $f_s = \Delta T^{-1}$ is the **switching frequency** (the number of complete switching cycles per second) and $v_1 - v_2$ is the voltage across the capacitor when switch 2 is closed.

From (8.83) the average current through the capacitor is proportional to the voltage across the capacitor, which means that the switched capacitor acts as a

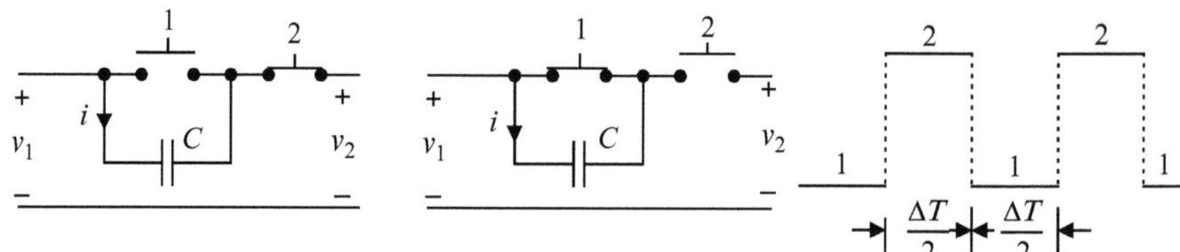

Fig. 8.61 Switched capacitor as a resistor

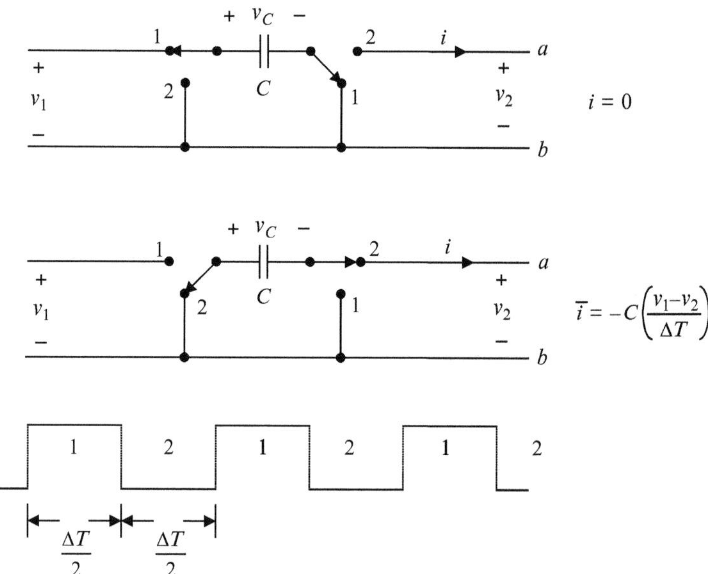

Fig. 8.62 A switched-capacitor, negative-resistance circuit

resistor. By Ohm's law, the effective resistance of the switched-capacitor arrangement is given by

$$R_{eq} = \frac{\Delta v}{\bar{i}} = \frac{\Delta v}{Cf_s\,\Delta v} = \frac{1}{Cf_s} = \frac{\Delta T}{C}. \tag{8.84}$$

For the analysis above and the resulting relation (8.84) to be valid, the voltages v_1 and v_2 must be slowly varying relative to the switching frequency f_s. We are unprepared for a detailed examination of this issue, but we note that in practice, if the voltages are sinusoidal, the switching rate might be 50–100 times the frequency of the sinusoid.

Figure 8.62 shows another switched-capacitor arrangement, where the load is connected from terminal a to terminal b. We assume that the voltages v_1 and v_2 are approximately constant during any interval of duration $\Delta T/2$. When the switches are in position 1, the voltage across the capacitor equals v_1. When the switches are in position 2, the voltage across the capacitor equals $-v_2$. Thus the voltage across the capacitor changes from v_1 to $-v_2$ during the time $\Delta T/2$ that the switch is in position 2. The total change in the voltage is $\Delta v = v_1 + v_2$. The corresponding change in the charge on the capacitor is $\Delta q = C\Delta v$, with the positive direction of charge flow is *from right to left*, whereas the positive polarity of the voltage v_C across the capacitor is from *left to right*. In other words, during the time $\Delta T/2$ that the switch is in position 2, the direction of charge flow is from the negative terminal to the positive terminal on the capacitor. The average current during one complete switching period is given by

$$i = \frac{\Delta q}{\Delta T} = -C\frac{\Delta v}{\Delta T}$$

and the apparent resistance is given by

$$R_{eq} = -\frac{\Delta T}{C} = -\frac{1}{Cf_s}. \tag{8.85}$$

Thus the circuit in Fig. 8.62 acts as a negative resistance.

In a circuit, a positive switched-capacitor resistor (Fig. 8.61) can *float*; that is it is unnecessary for the capacitor and load to have a common ground, because the capacitor discharges through a closed loop. A negative-resistance switched-capacitor filter cannot float (relative to the load) because the capacitor must discharge through the load.

In implementations of switched-capacitor resistors, the switches are transistors, which act as single-pole single-throw switches. Thus a capacitor and two transistors are required to implement the single switched-capacitor positive resistor illustrated by Fig. 8.61. Nonetheless, switched-capacitor resistors often are preferred to ohmic resistors in integrated circuits, for several reasons:

- A switched-capacitor resistor can occupy less area on a chip than an equivalent ohmic resistor.
- The equivalent resistance of a switched-capacitor resistor can be set precisely (e.g., by adjusting the switching rate), whereas it is difficult to achieve precise ohmic resistances in integrated circuits.
- The equivalent resistance of a switched-capacitor resistor can be varied (by varying the switching rate), whereas the resistance of an ohmic resistor is fixed.
- In many practical circuits, performance depends not upon the resistance of a single resistor, but instead upon ratios of one resistance to another; e.g., in inverting and non-inverting op-amp amplifier circuits. It is difficult to achieve a precise (single) capacitor in integrated form, but it is not as difficult to achieve precise capacitance ratios. A precise ratio of one switched-capacitor resistor to another in an integrated circuit can be achieved.

On the downside, switched capacitors can be noisy (switching transients can contaminate voltages of interest). Also, transistors are not perfect switches, in that they introduce additional voltage drops. The capacitor voltage does not quite reach the source voltage during the charging period and the maximum load voltage is slightly less than the initial capacitor voltage at the beginning of the discharge. Thus there is a minimum source voltage for which a switched-capacitor resistor will work. You will learn more about these issues when you study electronics.

8.10.9 Power Dissipation in Switched-Capacitor Circuits

The electronic switches used in switched-capacitor resistors have a small but non-zero **on-resistance** R_{on} when closed and a large but finite **off-resistance** R_{off} when open. The on-resistance of the switches ranges from a few microohms to a few ohms and the off-resistance from a few megaohms to hundreds of gigaohms, depending upon the devices used. Figure 8.63 shows a circuit model for a switched-capacitor resistor, including the on- and off-resistances of the switches. In the following analysis, we assume $R_{off} \gg R_{on}$, $\Delta T \gg R_{on}C$, and that the applied voltage v_1 is approximately constant during any interval of duration ΔT. These assumptions are generally valid for any practical implementation using a switched-capacitor as a resistor.

We assume $R_{off} \gg R_{on}$. For either switch position, the Thévenin equivalent resistance seen by the capacitor is

$$R_T = R_{on} \| (R_{on} + R_{off}) \cong R_{on}.$$

Thus the charge and discharge time constants are the same, both given by

$$\tau \cong R_{on}C.$$

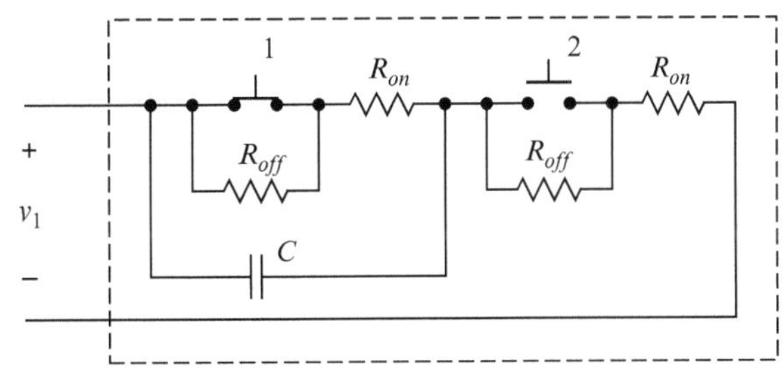

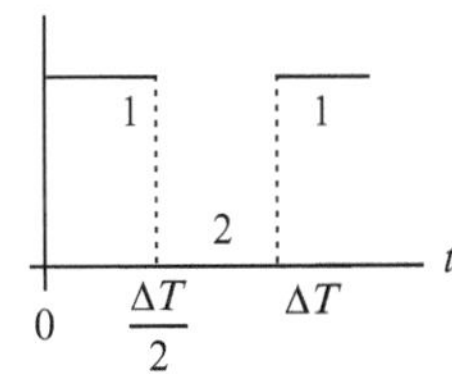

Fig. 8.63 Circuit model for a switched-capacitor resistor

Switched-capacitor resistors generally are designed such that the capacitor charges and discharges completely each cycle. Thus

$$\tau \ll \Delta T.$$

In what follows, we consider a single switching interval beginning when switch 2 is closed, and take the time origin as the beginning of that interval. Let v_1 denote the applied voltage (approximately constant) during the interval. While switch 2 is closed, the capacitor charges from zero to

$$\frac{(R_{off} + R_{on})\,v_1}{2R_{on} + R_{off}} \cong v_1.$$

While switch 1 is closed, the capacitor discharges from v_1 to

$$\frac{R_{on}\,v_1}{2R_{on} + R_{off}} \cong \frac{R_{on}}{R_{off}} v_1 \cong 0.$$

The charge placed on the capacitor while switch 2 is closed is given by

$$\Delta Q \cong C v_1.$$

Thus the energy stored by the capacitor while switch 2 is closed is given by

$$\Delta w_C = \frac{1}{2} C {v_1}^2.$$

Recall from Section 8.9 that when a capacitor is fully charged by a constant source through a resistance, the total energy provided by the source equals twice that finally stored by the capacitor. Thus the total energy provided by the source in Fig. 8.63 is given by

$$\Delta w_S = 2\Delta w_C = C {v_1}^2.$$

The average power dissipated by the switched-capacitor resistor is given by

$$P = \frac{\Delta w_S}{\Delta T} = \frac{C {v_1}^2}{\Delta T} = \frac{{v_1}^2}{R_{eq}},$$

where

$$R_{eq} = \frac{\Delta T}{C}$$

is the equivalent resistance of the switched-capacitor resistor. We may conclude that the power dissipated by a switched-capacitor resistor equals that dissipated by an ordinary resistor having the same resistance, provided that (1) the applied voltage is approximately constant during any switching interval, (2) the off-resistance of the switches is much larger than the on-resistance, and (3) the capacitor charges and then discharges completely during each switching cycle. For the model in Fig. 8.63, the power is dissipated by the switches, because the average power dissipated by an ideal capacitor equals zero. This might not be true in a physical circuit, because physical capacitors are not entirely lossless.

In addition to their intentional use as resistors, switched-capacitor configurations are found in memory and logic circuits (in digital computers and other digital systems). In memory and logic circuits, the switched capacitance is mainly device junction capacitance and stray (wiring) capacitance, and predominately the latter. Below, we obtain an expression for power dissipated in such circuits, which bears on power supply and cooling requirements.

Refer to Fig. 8.64, where the switch toggles between positions 1 and 2, being in each position for a time $\Delta T/2$. We assume the time constants R_1C and R_2C for the charge and discharge are both much smaller than the time $\Delta T/2$, so the capacitor charges fully

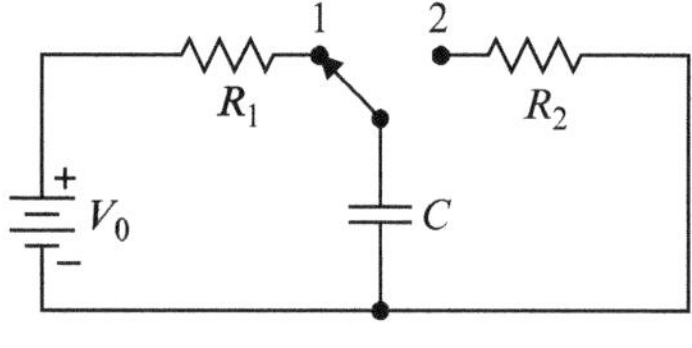

(a) switched-capacitor circuit

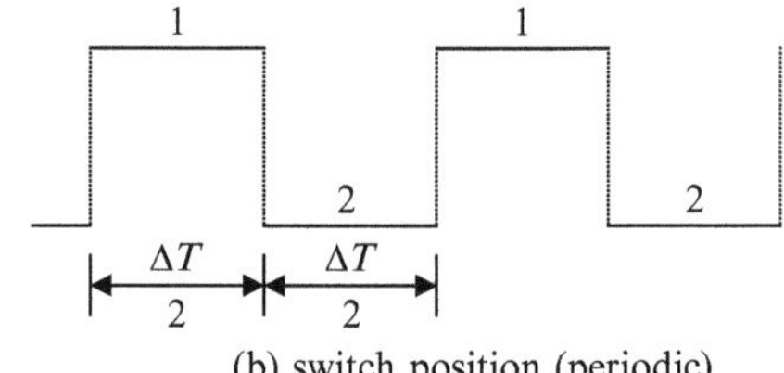

(b) switch position (periodic)

Fig. 8.64 Circuit used to calculate power dissipation in a switched capacitor

(to V_0) while the switch is in position 1 and discharges completely while the switch is in position 2. Thus the work done in each position is given by

$$W = \frac{1}{2} C V_0^2$$

and the average power dissipated during one complete switching cycle is given by

$$P = \frac{2W}{\Delta T} = \frac{C V_0^2}{\Delta T}. \tag{8.86}$$

The switching in Fig. 8.64 is periodic, so the average power (over all time) equals the average power dissipated during any one interval of duration ΔT.

Equation (8.86) can be written

$$P = C V_0^2 f_s,\ f_s = \frac{1}{\Delta T}, \tag{8.87}$$

where f_s is called the **switching frequency**. In a digital logic circuit, switching operations are controlled by a *clock* waveform, which is approximately a square wave. A logic gate can change state at most once during a complete clock cycle. Thus the frequency of the clock driving a switched capacitor is twice the switching frequency f_s appearing in (8.87). If ΔT is the duration of a complete switching cycle, the *clock frequency* would be $f_C = 2/\Delta T$ or $\Delta T = 2/f_C$ and (8.87) becomes

$$P = \frac{1}{2} C V_0^2 f_C. \tag{8.88}$$

Equation (8.88) is strictly applicable only to a circuit that switches periodically, at a rate f_C. However, in a typical memory or logic circuit, there are very many gates switching in what can be modeled as a random pattern. As a result, the total power dissipated can be estimated as

$$P_T = \frac{N}{2} C V_0^2 f_C, \tag{8.89}$$

where N is the average number of gates that switch during any one clock period.

Example 8.28. In a certain microprocessor, the clock rate is 2 GHz and the logic-level voltage is 1 V. The capacitance associated with each gate or memory cell is about 10 fF (10^{-14} F), which is a realistic value. (a) What is the power dissipation per gate or memory cell? (b) The microprocessor contains a total of 10^8 gates and memory cells, but on average, only 10% are switching at any given time. If (8.88) is valid, what is the average power dissipated by the microprocessor?

Solution: (a) From (8.88), the power dissipation per device is

$$P = \frac{1}{2} C V_0^2 f_C = \frac{1}{2} (10\,\text{fF})(1\,\text{V})^2 (2\,\text{GHz}) = 10\,\mu\text{W}. \tag{8.90}$$

(b) If only about 10% of the gates are switching at any given time, the total power dissipated is

$$P_T = \left(0.1 \times 10^8\right)(10\,\mu\text{W}) = 100\,\text{W}.$$

Without extra measures for cooling, the microprocessor would become quite hot. Modern microprocessors are packaged with an integral heat sink and fan, and dissipate on the order of 100 W cm^{-2} (chip area). Current projections predict an ultimate density of about 10^{12} devices per cm^2, so heat dissipation will become an even greater problem in the future. Even today, high-performance mainframe computers require elaborate liquid cooling systems.

Equation (8.87) has profound implications for high-speed logic design, because it shows that average power dissipated increases only linearly with capacitance, only linearly with switching rate (clock rate), but as the *square* of the voltage V. Generally, the minimum achievable capacitance is dictated by device geometry and chip layout, whereas demand for increased performance drives the clock rate f_c to ever-higher values. As a result, integrated logic-circuit designers are driven to make logic-level voltages as small as possible. For example, if the voltage V_0 in (8.87) is decreased from 5 to 2 V, the power dissipation is multiplied by the factor $(2/5)^2$=0.16, which is a decrease of about 84%. Such a reduction has significant implications for power-supply requirements (e.g., battery weight and lifetime) and for measures necessary to remove heat.

Example 8.29. The clock rate for a certain integrated logic circuit is 1.2 GHz. By what percentage must the supply voltage be decreased if the clock rate is to be increased by 50% to 1.8 GHZ and the power dissipation must remain the same?

Solution: The power dissipated is fixed. From (8.87)

$$P = \frac{1}{2}C{V_1}^2 f_{C1} = \frac{1}{2}C{V_2}^2 f_{C2} \Rightarrow V_2 = \sqrt{\frac{f_{C1}}{f_{C2}}}V_1$$

$$\cong 0.82V_1.$$

The supply voltage must be decreased by about 18%.

8.11 Problems

Section 8.2 is prerequisite for the following problems.

P 8.1 Two metal plates 4 m long and 100 mm wide are parallel and separated by a dielectric whose relative permittivity is 3.5. The capacitance of the arrangement is 35 nF. What is the thickness of the dielectric? Ignore edge effects.

P 8.2 Single-layer ceramic disc capacitors usually are made by plating each side of a thin ceramic disk (the dielectric) with a suitable metal, such as silver. The relative permittivities of some ceramics exceed 10^4, so ceramic disk capacitors can provide relatively large capacitances in a small volume. The relative permittivity of a particular ceramic is 12,000. What is the thickness of a single-layer disc capacitor utilizing that dielectric if the capacitance is 50 nF, the diameter is 1 cm, and the thickness of each plate is 25 μm? Neglect edge effects (fringing).

P 8.3 Two parallel-plate capacitors C_1, C_2 have the same plate area A. The relative permittivities of the dielectrics are $\varepsilon_1, \varepsilon_2$. What is the relation between the plate separations if the capacitances are the same?

P 8.4 The plates of one parallel-plate capacitor are square, 1 m on a side, and 100 μm apart. The capacitance is equal to that of a second parallel-plate air-dielectric capacitor whose plates are 2 × 4 m and 50 μm apart. What is the relative permittivity ε_r of the dielectric material between the plates of the first capacitor?

P 8.5 A parallel-plate air-dielectric capacitor is charged to 100 V and disconnected from the source. The distance between the plates is then halved. What is the new voltage across the capacitor?

P 8.6 The capacitance of a certain air-dielectric parallel-plate capacitor is 2 pF. The plate separation is halved and the space between the plates is filled with an oil whose relative permittivity is 400. What is the new capacitance?

P 8.7 The capacitance of a certain rectangular air-dielectric parallel-plate capacitor is 4 pF. Each dimension of each plate is halved and the space between the plates is filled with a glass whose relative permittivity is 5. What is the new capacitance?

P 8.8 Figure P 8.1 shows the circuit-diagram symbol for a *varactor diode*, which when reverse-biased (when $v > 0$) acts as a voltage-controlled capacitor having capacitance

$$C = \frac{C_0}{[1 + v/v_0]^{\alpha}}.$$

For a certain varactor diode, $C_0 = 67\,\text{pF}$, $v_0 = 1.18\,\text{V}$, and $\alpha = 0.44$. (a) Construct a graph of capacitance versus voltage for this diode for $100\,\text{mV} \le v \le 100\,\text{V}$. Use logarithmic scales for both axes. (b) Let $v = V_{dc} + V_{ac}\cos(2\pi f\,t)$, with $V_{dc} = 2\,\text{V}$, $V_{ac} = 1\,\text{V}$, and $f = 10\,\text{kHz}$. Construct a graph of the capacitance versus time (one period).

Fig. P 8.1 See Problem P 8.8

Section 8.3 is prerequisite for the following problems.

P 8.9 The charge on one plate of a certain capacitor is given by

$$q(t) = Q_0 \sin(2\pi f\, t),\; Q_0 > 0.$$

Let V_{rms} denote the rms voltage across the capacitor and obtain expressions for the capacitance, the rms current I_{rms} through the capacitor, and the ratio V_{rms}/I_{rms}.

P 8.10 Refer to Fig. P 8.2, where $C = 400\,\text{nF}$, $R = 10.\Omega$, and the voltage across the capacitor is as shown. Draw a graph of the current i versus time.

P 8.11 Refer to Fig. P 8.3. Draw a graph of the source voltage v_S versus time.

P 8.12 Figure P 8.4 shows graphs of the voltage $v(t)$ across and current $i(t)$ through a capacitor, where

$$v(t) = V_0 \cos(2\pi f\, t), \quad i(t) = -I_0 \sin(2\pi f\, t),$$

with $V_0 = 5\,\text{V}$ and $I_0 = 31\,\text{mA}$. The current enters the positive terminal of the capacitor. Find the capacitance of the capacitor.

P 8.13 The voltage across a capacitor cannot change instantaneously because such change would require infinite current. Can the capacitance of a variable capacitor change instantaneously?

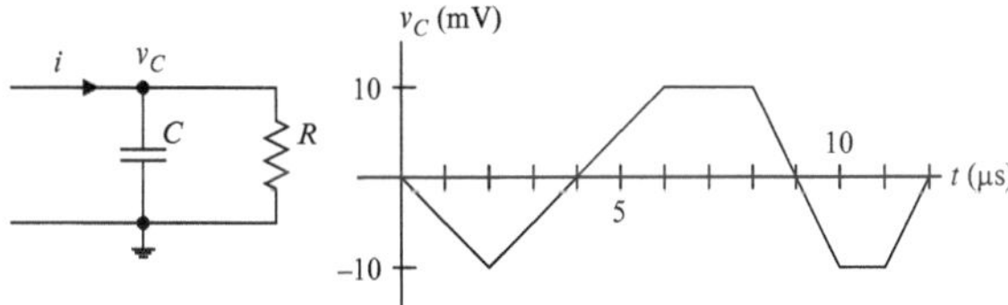

Fig. P 8.2 See Problem P 8.10

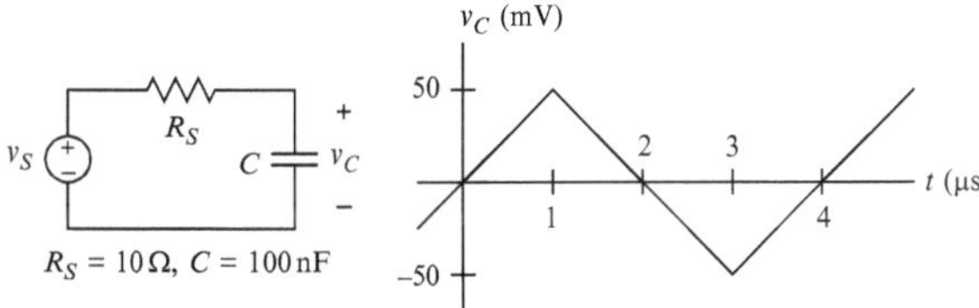

Fig. P 8.3 See Problem P 8.11

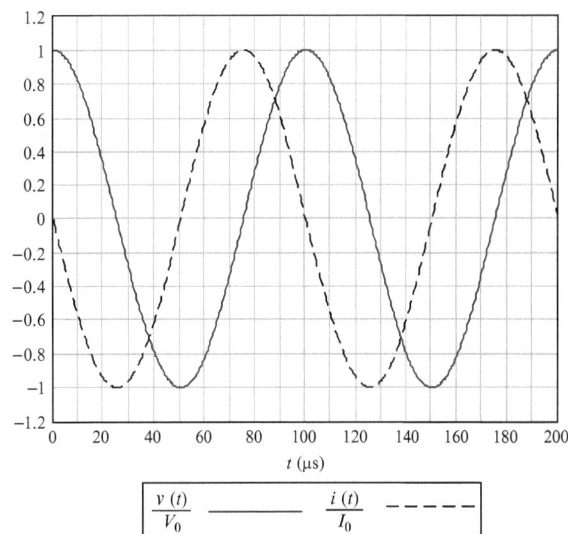

Fig. P 8.4 See Problem P 8.12

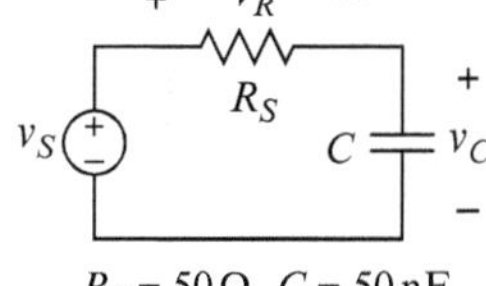

Fig. P 8.5 See Problem P 8.14, 15

P 8.14 In Fig. P 8.5, $v_C = V_C \cos(2\pi f_0\, t)$. Show that the source voltage can be expressed as a single sinusoidal function of time of the form $v_S(t) = V_S \cos(2\pi f_0\, t + \theta)$.

P 8.15 In Fig. P 8.5, the voltage across the resistor is given by $v_R = V_R \cos(2\pi f_0 t + 0.25\pi)$, with $V_R = 10.64\,\text{V}$. Find the voltage $v_C(t)$ across the capacitor.

P 8.16 Refer to Fig. P 8.6. Obtain an expression for the source voltage $v_S(t)$. Give the values of all parameters in the expression.

P 8.17 A variable capacitor is set to 200 pF and connected to a source having open-circuit (available) voltage $V_0 = 15\,\text{V}$. After it is fully charged, the capacitor is disconnected from the source and the capacitance is increased to 5 nF. What is the new voltage across the capacitor?

P 8.18 (i) Show that the circuit in Fig. P 8.7(a) is equivalent at the terminals a–b to the circuit in Fig. P 8.7(b). Thus show that the circuit in Fig. P 8.7(a) is a **capacitance multiplier**. (ii) What limits the charge (voltage) that can be placed on the capacitor?

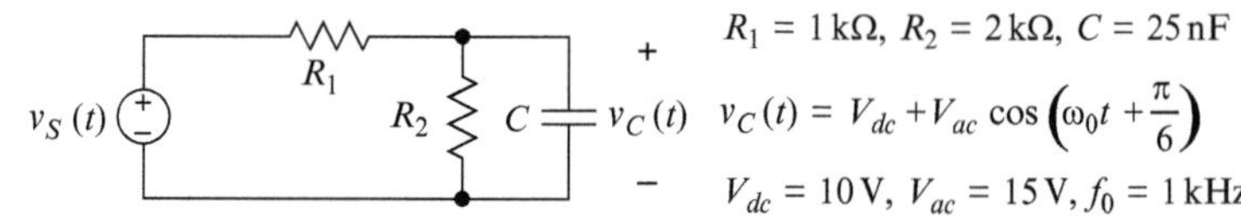

Fig. P 8.6 See Problem P 8.16

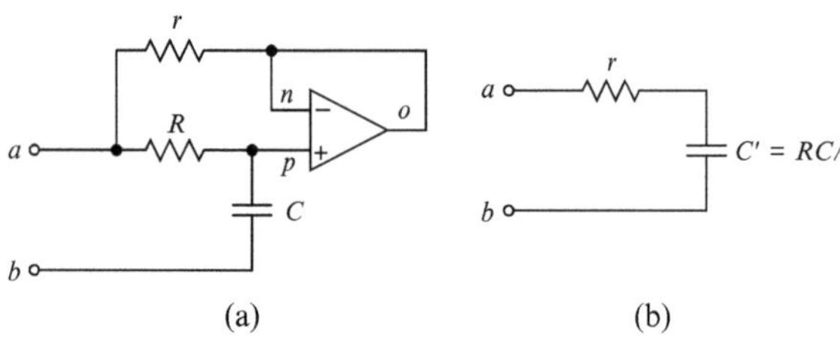

Fig. P 8.7 See Problem P 8.18

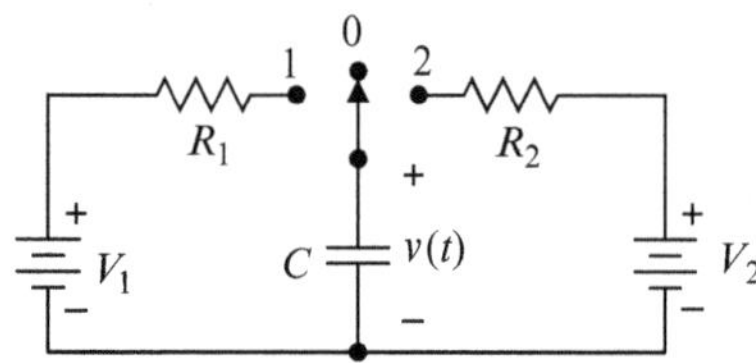

Fig. P 8.8 See Problem P 8.19

Section 8.4 is prerequisite for the following problems.

P 8.19 Refer to Fig. P 8.8, where $v(0) < V_1 < V_2$. The switch is moved from position 0 to position 1 at $t = 0$, then to position 2 at $t = t_1$, then back to position 0 at $t = t_2$. Sketch neatly and label fully a graph of the voltage $v(t)$ for $0 \le t \le t_2$. Then give an expression for the voltage $v(t)$ for $t > 0$.

P 8.20 In Fig. P 8.9, the switch is closed at $t = 0$. The graph shows the voltage across the capacitor for $t > 0$. Estimate the parameters of the Thévenin source.

P 8.21 In Fig. P 8.10, the switch is moved from a to b at $t = 0$, having been at a for a very long time. The graph shows the voltage across the resistor as a function of time. Estimate the source voltage V_0 and the capacitance C.

P 8.22 A capacitor $C = 10\,\text{nF}$ in series with a resistor $R = 10\,\Omega$ is suddenly connected (at $t = 0$) to a dc source $V_0 = 20\,\text{V}$. What is the final $(t \to \infty)$ value of the charge (C) on the capacitor? If the initial charge on the capacitor equals zero, what is the initial value of the current? At what time does the charge reach 99% of its final value?

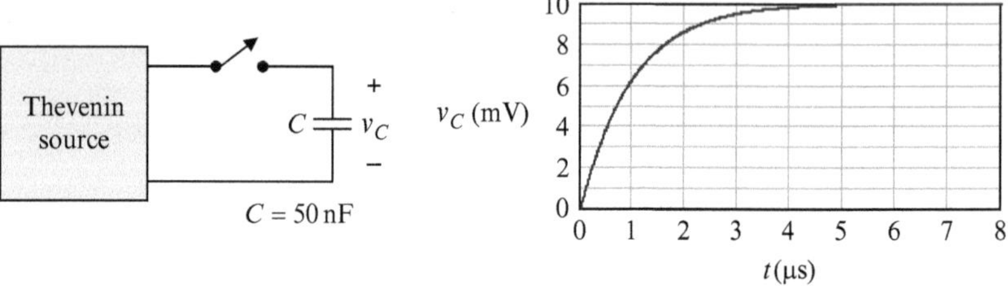

Fig. P 8.9 See Problem P 8.20

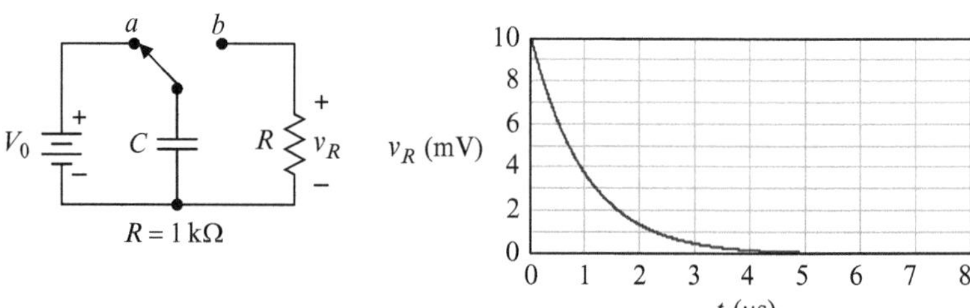

Fig. P 8.10 See Problem P 8.21

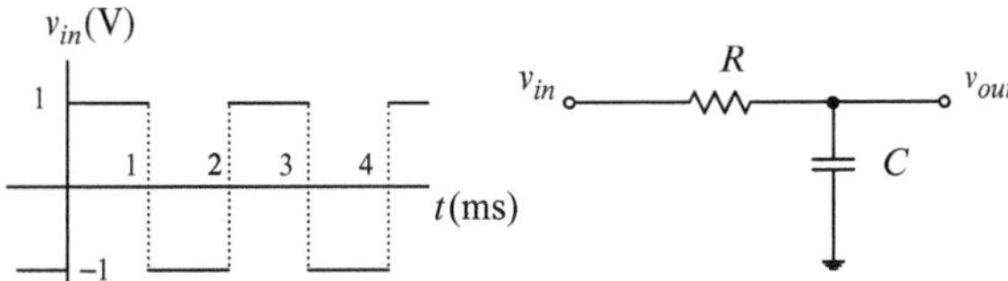

Fig. P 8.11 See Problem P 8.23

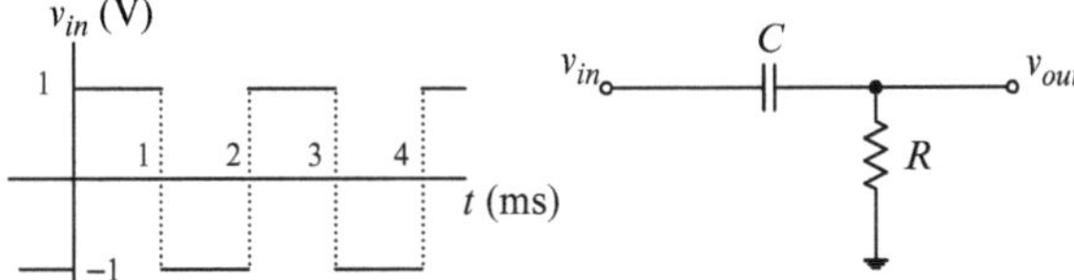

Fig. P 8.12 See Problem P 8.24

P 8.23 Refer to Fig. P 8.11, where the circuit is in steady state. (a) What is required of R and C if the output voltage is to be approximately equal to the input voltage? (b) Construct plots of the voltage $v_{out}(t)$ versus time for one complete period of the input for $C = 100\,\text{nF}$ and (a) $R = 10\,\text{k}\Omega$, (b) $R = 2\,\text{k}\Omega$. (c) Calculate the average power dissipated by the resistor for (i) $R = 10\,\text{k}\Omega$ and (ii) $R = 2\,\text{k}\Omega$.

P 8.24 Refer to Fig. P 8.12, where the circuit is in steady state. (a) What is required of R and C if the output voltage is to be approximately equal to the input voltage? (b) Construct plots of the voltage $v_{out}(t)$ versus time for one complete period of the input for $C = 100\,\text{nF}$ and (i) $R = 10\,\text{k}\Omega$, (ii) $R = 2\,\text{k}\Omega$. (c) Calculate the average power dissipated by the resistor for (i) $R = 10\,\text{k}\Omega$ and (ii) $R = 2\,\text{k}\Omega$.

P 8.25 Refer to Fig. P 8.13, where the output is in steady-state. Sketch a graph of the output voltage v_{out} versus time. You need not determine the voltages involved, but the graph should closely resemble the actual output.

P 8.26 In Fig. P 8.14, the switch is closed (at $t = 0$) after being open for a very long time. Obtain symbolic expressions for the initial and final values of the voltage v_C, the initial value of the current i, and the time at which the voltage v_C equals 99% of the final value.

P 8.27 In Fig. P 8.15, the switch has been at a for a very long time. It is moved from a to b at $t = 0$ and back to a at $t = RC$. Sketch neatly and label fully a graph of the voltage $v_C(t)$ for $0 \le t \le 5RC$.

P 8.28 A capacitor is charged to a voltage V_0 and then connected to two identical resistors through switches, as shown in Fig. P 8.16. Switch S_1 is closed at $t = 0$ and switch S_2 is closed at $t = RC$. Sketch neatly and label fully a graph of the voltage v_C (t) for $0 \le t \le 5RC$.

P 8.29 A capacitor is charged to a voltage V_0 and then connected to two identical resistors and two switches, as shown in Fig. P 8.17. Switch S_1 is closed

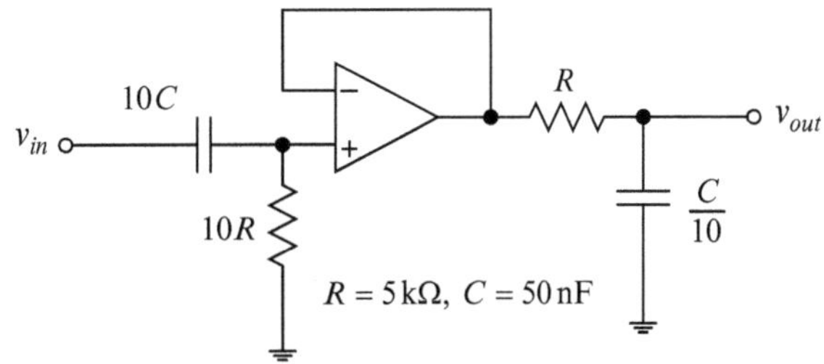

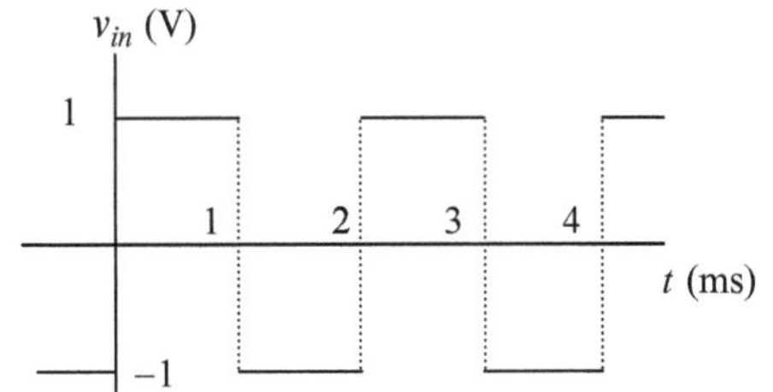

Fig. P 8.13 See Problem P 8.25

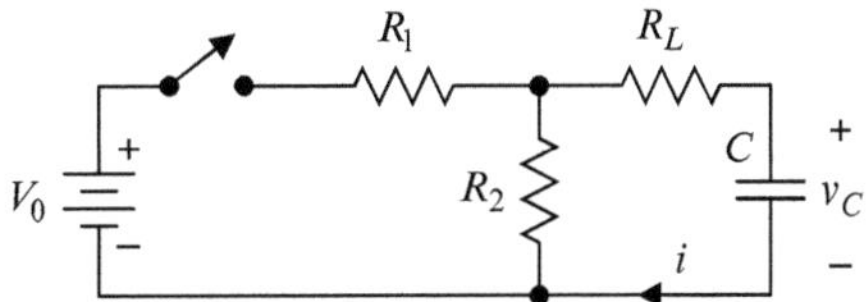

Fig. P 8.14 See Problem P 8.26

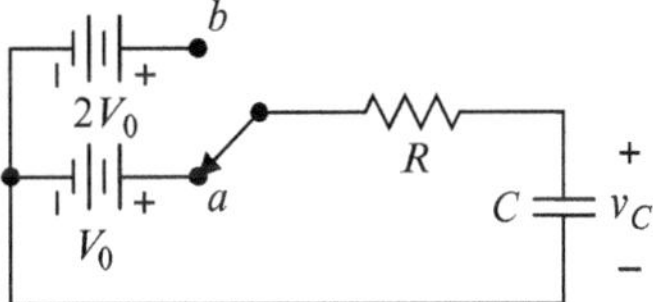

Fig. P 8.15 See Problem P 8.27

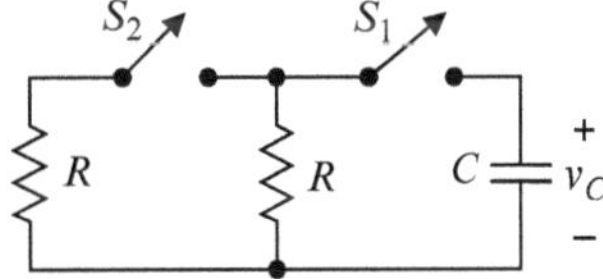

Fig. P 8.16 See Problem P 8.28

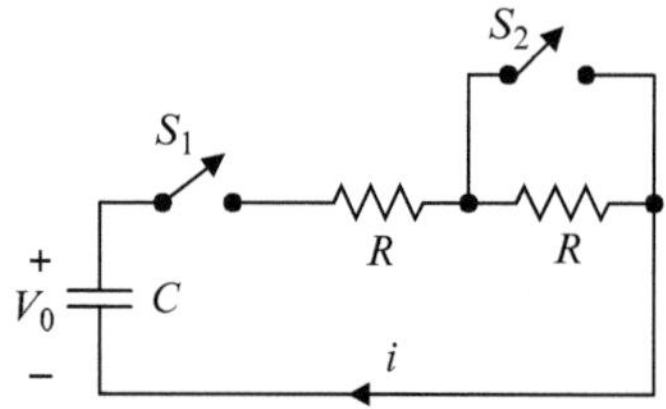

Fig. P 8.17 See Problem P 8.29

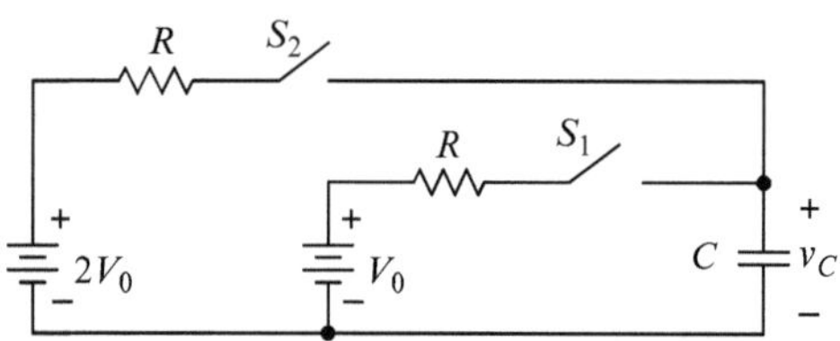

Fig. P 8.18 See Problem P 8.30

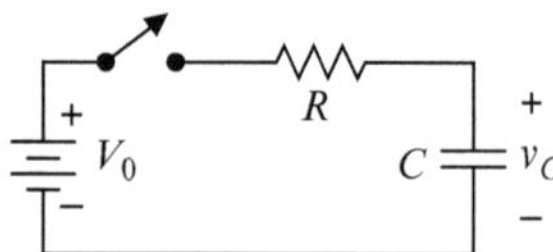

Fig. P 8.19 See Problem P 8.31

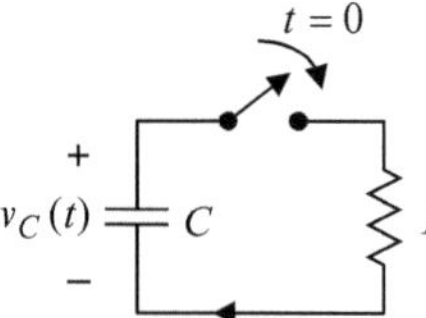

Fig. P 8.20 See Problem P 8.32

at $t = 0$ and switch S_2 is closed at $t = RC$. Sketch neatly and label fully a graph of the current $i(t)$ for $0 \leq t \leq 5RC$.

P 8.30 In Fig. P 8.18, the capacitor is initially uncharged. Switch S_1 is closed at $t = 0$ and switch S_2 is closed at $t = RC$. Sketch neatly and label fully a graph of the voltage $v_C(t)$ for $0 \leq t \leq 5RC$. Show the time constants associated with each exponential rise.

P 8.31 In Fig. P 8.19, $R = 2.2\,\text{k}\Omega \pm 10\%$, $C = 50\,\text{nF} \pm 20\%$, and $V_0 = 10\,\text{V}$. The switch is closed at $t = 0$. Construct graphs showing upper and lower bounds on the voltage $v_C(t)$ for $t > 0$.

P 8.32 If a capacitor is charged to a voltage $v_C(0^-) = V_0$ and subsequently discharged through a resistor, as depicted by Fig. P 8.20, the voltage $v_C(t)$ decreases exponentially, as $v_C(t) = V_0 \exp(-t/\tau)$, where $\tau = RC$. You might have heard it said that "*theoretically, neither the voltage nor the current ever actually reach zero, but only approach zero asymptotically*." In essence, this problem asks you to evaluate that claim. (a) Can the magnitude of the charge on the capacitor be non-zero and less than that of one electron? Obtain an expression for the time at which the charge on the capacitor becomes less than that of one electron (in magnitude). (b) Let $V_0 = 10\,\text{V}$, $C = 100$ nF, and $R = 10\,\text{k}\Omega$. Calculate the time defined in part (a). (c) The relation $v_C(t) = V_0 \exp(-t/\tau)$ is derived from a differential equation under the assumption that charge and current can be represented by real variables. In reality, charge is discrete (is a whole multiple of the charge on one electron). Are the differential equation and the exponential solution valid when only a few electrons participate in the current?

P 8.33 In the circuit shown in Fig. P 8.21, switch S_2 is closed and switch S_1 is open for all $t < 0$. At $t = 0$ switch S_2 is opened and switch S_1 is closed. At $t = t_1$ switch S_2 is closed. Finally, at $t = t_2$, switch S_1 is opened. (a) Sketch neatly and label fully a graph of the voltage $v(t)$ versus t. Do not assume $v(t_1) = 0$ or $v(t_2) = 0$. (b) Obtain an expression for the voltage $v(t)$.

P 8.34 In Fig. P 8.22, the switch is in position a for $t < 0$ and is moved to b at $t = 0$. Show that the total energy dissipated in the resistor is independent of the resistance R.

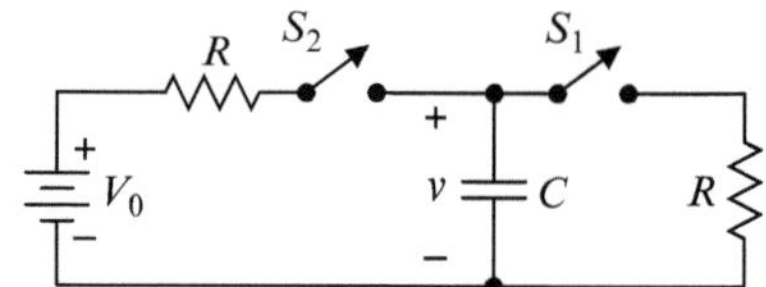

Fig. P 8.21 See Problem P 8.33

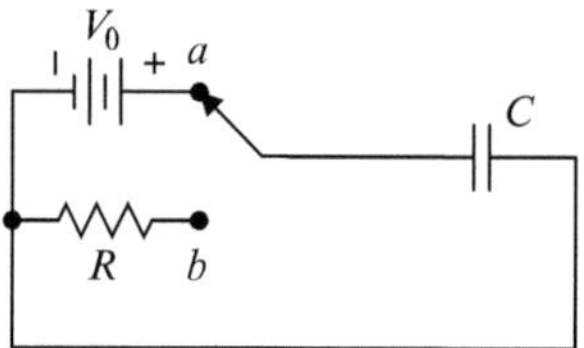

Fig. P 8.22 See Problem P 8.34

Section 8.5 is prerequisite for the following problems.

P 8.35 A large, parallel-plate, air-dielectric capacitor is charged to $V_{ab} = 50\,\text{V}$. Then a material having relative permittivity $\varepsilon_r = 4$ is inserted halfway between the plates, as illustrated by the shading in Fig. P 8.23. What is the new value of V_{ab}? Assume the plates are large enough that edge effects are negligible.

P 8.36 Refer to Problem P 8.35. One can vary the capacitance of a parallel-plate capacitor by occupying more or less of the space between the plates with a dielectric material, the remainder occupied by air. Describe such a capacitor whose capacitance can be varied from 10 to 50 pF. Assume square plates and a separation of 1 mm. What is the area of each plate? What relative permittivity is required of the inserted dielectric material?

P 8.37 Figure P 8.24 shows an air-dielectric variable capacitor having 16 rotating and 15 fixed leaves.[18] The distance between each fixed leaf and a neighboring rotating leaf is 500 μm. For simplicity, assume the leaves have the shapes shown, where the cross indicates the centerline of the shaft. Ignore edge effects and the small semicircular bumps on the leaves (for the shaft) and obtain an expression for the capacitance of the capacitor as a function of the angle θ. Give the range of the variable capacitance for $0 \leq \theta \leq \pi$. Is the expression least accurate for θ near zero or θ near π?

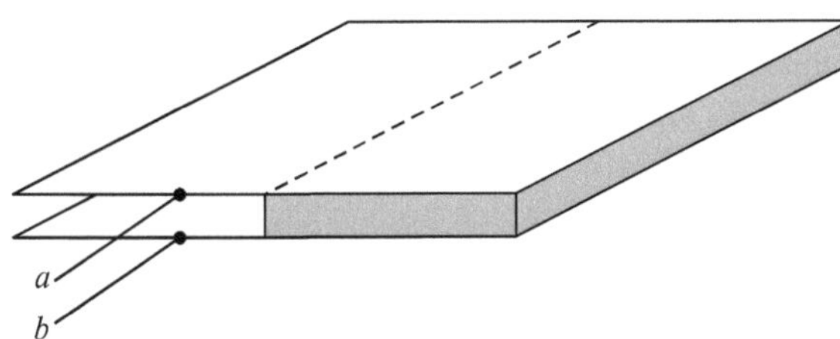

Fig. P 8.23 See Problem P 8.35

[18]Photograph courtesy MTM Scientific, Corp.

P 8.38 In Fig. P 8.25, $C_2 > C_1$ and the capacitors are initially uncharged. The switch is closed at $t = 0$. Which of the two currents i_1, i_2 decreases to $0.1\ V_0/R$ first?

P 8.39 In Fig. P 8.26, $C_2 > C_1$ and the capacitors are initially uncharged. The switch is closed at $t = 0$. Which of the two voltages v_1, v_2 reaches $0.9\ V_0$ first?

P 8.40 Refer to Fig. P 8.27. Capacitors C_1 and C_2 are individually charged to voltages V_1 and V_2, respectively, and then connected in series with a switch and resistor. The switch is closed at $t = 0$. (a) Obtain an expression for the voltage across the resistor for $t > 0$. (b) Obtain an expression for the time at which $|v(t)| = 0.1|v(0^+)|$.

P 8.41 An accurate, high-precision capacitor having capacitance $C = 51.5\,\text{nF}$ is needed in a certain application. The required accuracy and precision are to be obtained as illustrated by Fig. P 8.28, where a fixed capacitor having capacitance $C_0 = 47\,\text{nF} \pm 5\%$ is *trimmed* by a smaller variable capacitor ΔC while the total capacitance is monitored by a very accurate capacitance meter. (a) Specify the variable capacitor. (b) Could you achieve an equivalent result using a series connection of the same two capacitors?

P 8.42 Two identical capacitors having capacitance $C = C_0 \pm 10\%$ are connected in series. Obtain expressions for the capacitance and the tolerance of the series connection.

P 8.43 Two identical capacitors having capacitance $C = C_0 \pm 10\%$ are connected in parallel. Obtain expressions for the capacitance and the tolerance of the parallel connection.

Note. The circuit models in Problems P 8.44–P 8.47 below are unrealistic, because they presume that the voltage across a capacitor can change instantaneously. However, the models do allow correct calculation of total charge transfer

P 8.44 Refer to Fig. P 8.29. Obtain an expression for the voltage V_2 after the switch is moved from a to b if $V_2 = 0$ before the switch is moved. How much charge is transferred from C_1 to C_2?

P 8.45 Refer to Fig. P 8.30. Capacitor C_1 is charged to voltage V_1 and then connected to the parallel connection of C_2 and C_3, which are uncharged. Obtain

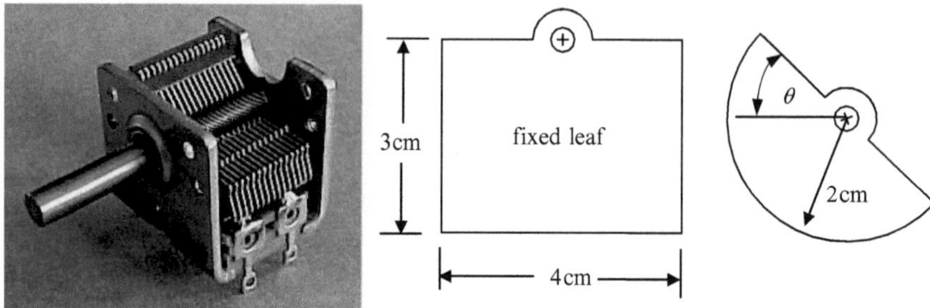

Fig. P 8.24 See Problem P 8.37

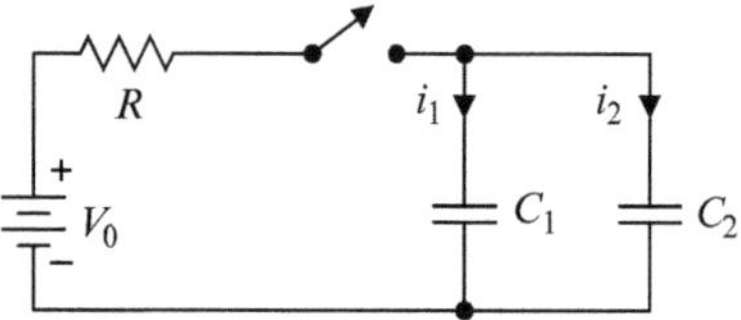

Fig. P 8.25 See Problem P 8.38

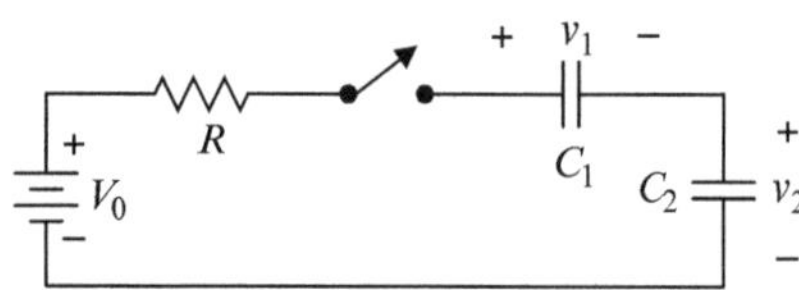

Fig. P 8.26 See Problem P 8.39

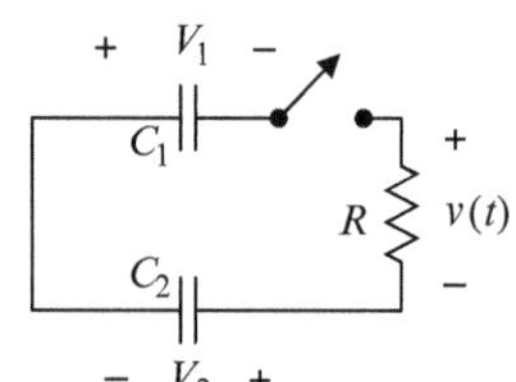

Fig. P 8.27 See Problem P 8.40

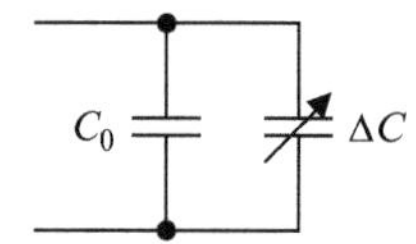

Fig. P 8.28 See Problem P 8.41

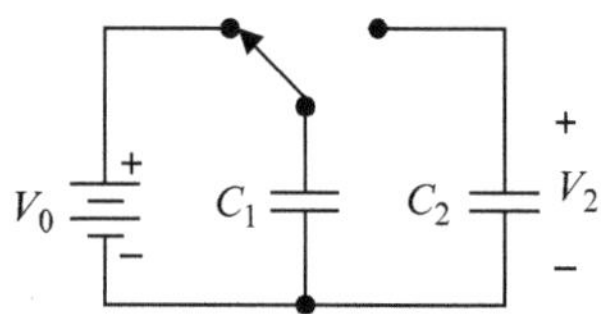

Fig. P 8.29 See Problem P 8.44

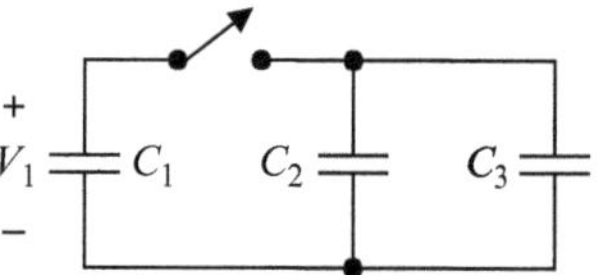

Fig. P 8.30 See Problem P 8.45

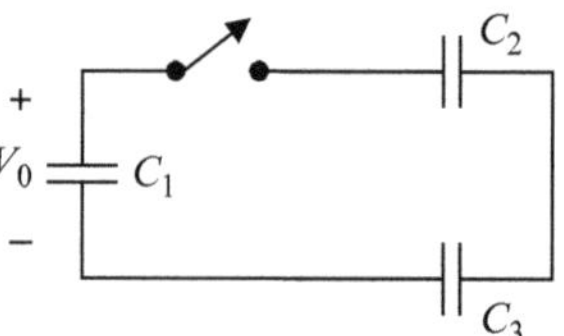

Fig. P 8.31 See Problem P 8.46

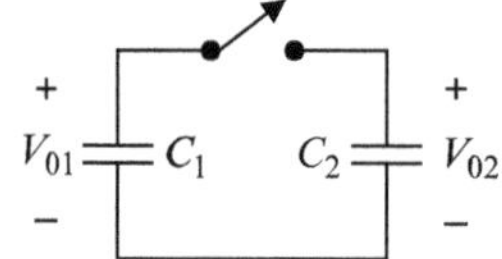

Fig. P 8.32 See Problem P 8.47

expressions for the charge on each capacitor (in coulombs) after the switch is closed.

P 8.46 Refer to Fig. P 8.31. Capacitor C_1 is charged to voltage V_0 and then connected to the series connection of C_2 and C_3, which are uncharged. Express the voltages across the capacitors after the switch is closed in terms of the initial voltage V_0 and the capacitances C_1, C_2, C_3.

P 8.47 Ideal capacitors C_1, C_2 are charged individually to voltages V_{01}, V_{02} and then connected as shown in Fig. P 8.32. Obtain expressions for the voltages across the capacitors after the switch is closed.

Section 8.6 is prerequisite for the following problems.

P 8.48 A capacitor is connected in series with a voltage source $v(t) = V_0 \cos(\omega_0 t)$ and an rms ammeter having zero internal resistance, as shown in Fig. P 8.33.

(a) Assume the leakage resistance of the capacitor is infinite and obtain an expression for the current $i(t)$. Denote the peak amplitude of the current by I_0.
(b) Assume the leakage resistance of the capacitor is infinite and express the capacitance in terms of the frequency of the source and the rms amplitudes of the current $i(t)$ and the voltage $v(t)$.
(c) Assume the leakage resistance is finite and obtain an expression for the current $i(t)$. Denote the peak amplitude of the current by I_0.
(d) Assume the leakage resistance of the capacitor is finite and express the capacitance in terms of the leakage resistance, the frequency of the source, and the rms amplitudes of the current $i(t)$ and the voltage $v(t)$.

P 8.49 A current source is used to charge a capacitor having capacitance C and leakage resistance R_l, as shown in Fig. P 8.34. What is the maximum voltage that can be impressed on the capacitor? If the capacitor is disconnected from the source after being fully charged, how long will it take for the capacitor to discharge?

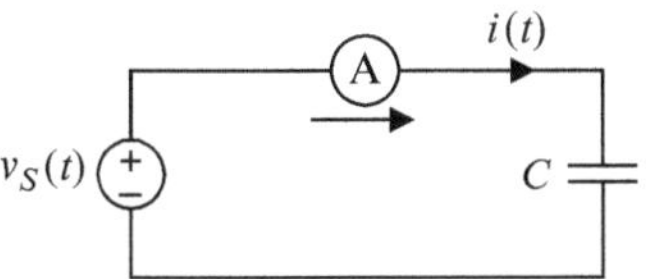

Fig. P 8.33 See Problem P 8.48

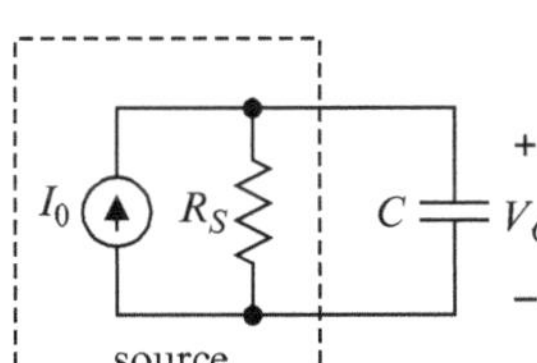

Fig. P 8.34 See Problem P 8.49, 50, 51

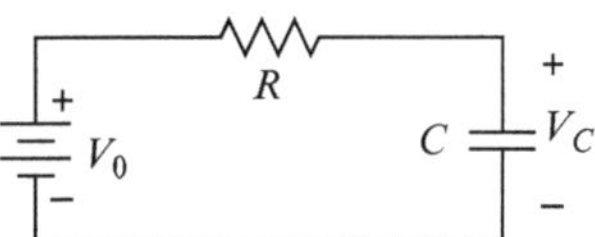

Fig. P 8.35 See Problem P 8.52, 53

P 8.50 Refer to Fig. P 8.34. Under what condition will the final capacitor voltage be within 1% of the voltage available from the source?

P 8.51 In Fig. P 8.34, $I_0 = 10\,\mu\text{A}$, $R_S = 1\,\text{M}\Omega$, and the voltage V_C is steady at 7 V. (a) What is the leakage resistance of the capacitor? (b) The charged capacitor is disconnected from the circuit and connected to a dc voltmeter having internal resistance $R_m = 10\,\text{M}\Omega$. The voltage across the capacitor drops to half its initial value in 8 s. What is the capacitance C?

P 8.52 Refer to Fig. P 8.35, where the components have been connected as shown for a very long time and the capacitor has leakage resistance R_l. Under what condition is V_C within 0.1% of V_0?

P 8.53 In Fig. P 8.35, $V_0 = 15\,\text{V}$ and $R = 100\,\text{k}\Omega$, and the voltage V_C is steady at 14.8 V. (a) What is the leakage resistance of the capacitor? (b) The charged capacitor is disconnected from the circuit and connected to a dc voltmeter having internal resistance $R_m = 10\,\text{M}\Omega$. The voltage across the capacitor drops to half its initial value in 4 s. What is the capacitance C?

Section 8.7 is prerequisite for the following problems.

P 8.54 Let C denote the end-to-end parasitic capacitance for an axial-lead, cylindrical resistor having resistance R. Assume the voltage across the resistor is given by $v = V_0 \cos(\omega_0 t)$. Obtain an expression for the frequency f_0 at which the rms current through the parasitic capacitance equals 1% of the rms current through the resistor. Calculate the frequency for $R = 100\,\text{k}\Omega$ and $C = 1\,\text{pF}$.

P 8.55 The wire-to-wire capacitance of a parallel pair of identical conductors in air far above the ground is given (approximately) by

$$C = \frac{\pi \varepsilon_0 L}{\ln[(d - r)/r]},$$

where $\varepsilon_0 = 8.854 \times 10^{-12}\,\mathrm{F\,m^{-1}}$ is the permittivity of free space (or air), d is the center-to-center separation of the wires, r is the radius of each wire, and L is the length of the pair. (a) Find the capacitance of a 100-mile long pair of AWG 00 copper wires 1.2 m apart and well above the ground. (b) A 440 V 60 Hz generator is attached to one end of the wires. Ignoring everything but the wire-to-wire capacitance, what is the rms amplitude of the current drawn from the generator if no load is attached to the far end?

P 8.56 Refer to Problem P 8.55. Suppose the AWG wire size used for a certain long-distance power transmission line is increased from AWG 0 to AWG 00. The wires are well above ground. By what factor must the distance between the wires be increased if the wire-to-wire capacitance must remain the same?

P 8.57 Proximity to ground increases the wire-to-wire capacitance of a pair of suspended conductors. The capacitance of such a pair is given by

$$C = \frac{2\pi\,\varepsilon_0 L}{\ln\left[\dfrac{2hd}{r\sqrt{4h^2+d^2}}\right]},$$

where $\varepsilon_0 = 8.854 \times 10^{-12}\,\mathrm{F\,m^{-1}}$ is the permittivity of free space (or air), d is the center-to-center separation of the wires, r is the radius of each wire, h is the height of each wire above the ground, and L is the length of the pair. (a) Find the capacitance of a 100-mile long pair of AWG 00 copper wires 1.2 m apart and 20 ft above the ground. (b) A 440 V 60 Hz generator is attached to one end of the wires. Ignoring everything but the wire-to-wire capacitance, what is the current drawn from the generator if no load is attached to the far end?

P 8.58 Refer to Problem P 8.57. Suppose the AWG wire size used for a certain long-distance power transmission line is increased from AWG 00 to AWG 000 and the distance between the wires remains the same. The height above ground is much larger than the separation between the wires. Is it necessary to increase the height of the wires above ground if the wire-to-wire capacitance must remain the same?

Section 8.9 is prerequisite for the following problems.

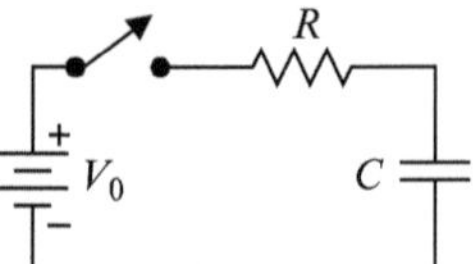

Fig. P 8.36 See Problem P 8.60

P 8.59 Show that the instantaneous energy $w(t)$ stored in the electric field in an ideal capacitor is given by

$$w(t) = \frac{1}{2C}Q^2(t),$$

where C is the capacitance and Q is the charge on either plate.

P 8.60 Refer to Fig. P 8.36, where the capacitor is initially uncharged and the switch is closed at $t = 0$. Show that for $t \to \infty$ the total energy dissipated in the resistor equals the total energy stored in the capacitor.

P 8.61 To increase the energy-storage capability of a capacitor, should one increase or decrease the thickness of the dielectric?

P 8.62 Show that the stored energy per unit volume (per m^3) for a capacitor is given by

$$\frac{w}{vol} = \frac{1}{2}\varepsilon_r\varepsilon_0 E^2,$$

where E is the electric field strength ($\mathrm{Vm^{-1}}$) in the dielectric.

P 8.63 Refer to Problem P 8.62. How does the maximum stored energy per unit volume (volumetric energy density) for a capacitor compare with that for gasoline?

P 8.64 Refer to Problem P 8.8. The parameters of the varactor diode are $C_0 = 50\,\mathrm{pF}$, $v_0 = 1\,\mathrm{V}$, and $\alpha = 0.5$. The voltage across the varactor diode is given by

$$v = V_{dc} + V_{ac}\cos(2\pi f\,t);\ V_{dc} = 500\,\mathrm{mV},$$
$$V_{ac} = 100\,\mathrm{mV}, f = 20\,\mathrm{kHz}.$$

Calculate the average power dissipated by the varactor diode.

Section 8.10 is prerequisite for the following problems.

P 8.65 Refer to Fig. P 8.37, where the input is sinusoidal, having frequency f, and the op amp is

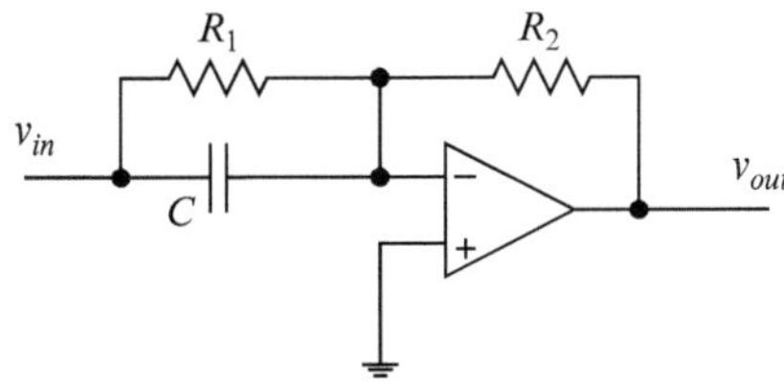

Fig. P 8.37 See Problem P 8.65

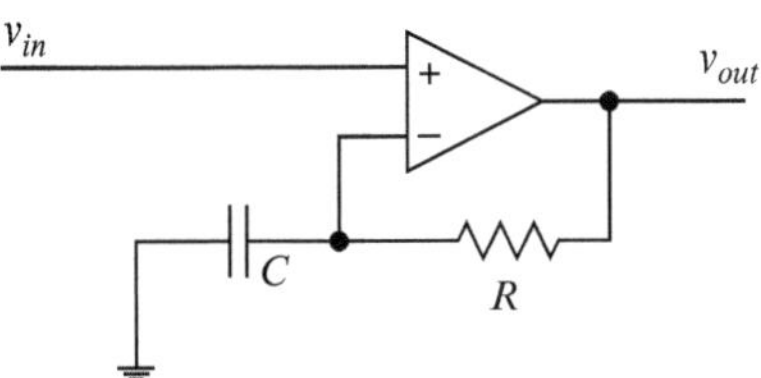

Fig. P 8.38 See Problem P 8.66

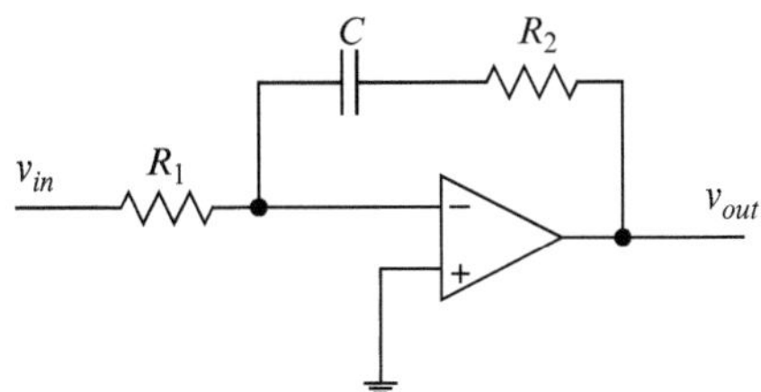

Fig. P 8.39 See Problem P 8.67

ideal. Under what conditions is the output approximately proportional to the derivative of the input?

P 8.66 Refer to Fig. P 8.38, where the input is sinusoidal, having frequency *f*, and the op amp is ideal. Under what conditions is the output approximately proportional to the derivative of the input?

P 8.67 Refer to Fig. P 8.39, where the input is sinusoidal, having frequency *f*, and the op amp is ideal. Under what conditions is the output approximately proportional to the integral of the input?

P 8.68 Refer to Fig. P 8.40, where the input is sinusoidal, having frequency *f*, and the op amp is ideal. Under what conditions is the output approximately proportional to the integral of the input?

P 8.69 Refer to Fig. P 8.41, where the input is sinusoidal, having frequency *f*, and the op amp is ideal. Under what conditions is the output approximately proportional to the integral of the input?

P 8.70 A 15 V supply having internal resistance $R_S = 1\,\Omega$ supplies power to ten circuits, each of which represents an average load of $10\,\Omega$ and each of which is bypassed at its power pin by a 50 nF capacitor.

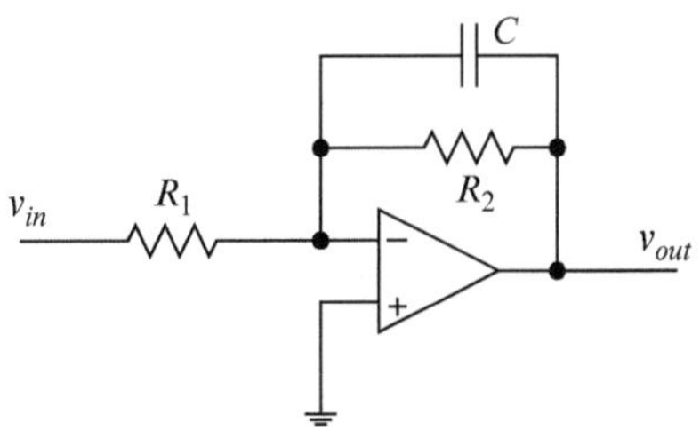

Fig. P 8.40 See Problem P 8.68

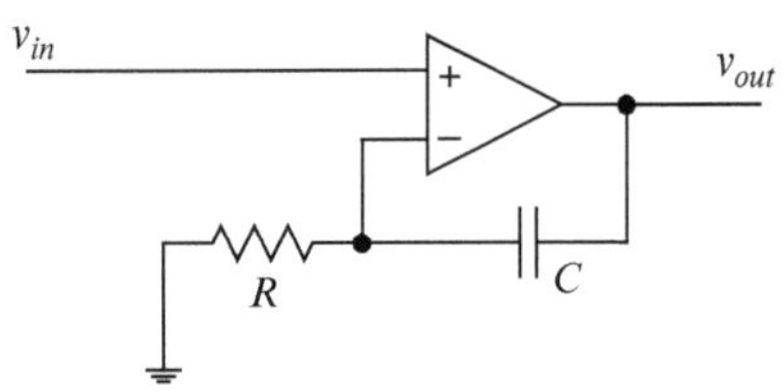

Fig. P 8.41 See Problem P 8.69

(a) What is the steady-state current that must be provided by the supply?
(b) What is the initial current (when the circuit is first energized and all capacitors are discharged)?
(c) What is the time constant for the initial surge current and what is the total additional energy that must be provided by the supply during the initial surge?
(d) By what factor is the additional energy multiplied if each bypass capacitance is doubled in an attempt to further reduce ripple?

P 8.71 In the absence of a bypass capacitor, the ripple factor at the power pin of a certain circuit is 0.015. A 10 nF bypass capacitor reduces the ripple factor to 0.002. What bypass capacitance is required to reduce the ripple factor to 0.0001?

P 8.72 Without bypass, the ripple factor at the power pin of a certain circuit is 0.02. The ripple is approximately sinusoidal, with a frequency of 180 Hz. A 75 μF bypass capacitor reduces the ripple factor to 0.002. What is the Thévenin-equivalent resistance at the power pin?

P 8.73 A certain digital IC has eight outputs. The load seen by each is equivalent to a 40 pF capacitor, and the logic-level voltage and the supply voltage are both 5 V. The outputs can change state in 4 ns. (a) What is the additional supply current (pulse) demanded by the IC if all eight outputs switch from off (0 V) to on (5 V) at once? (b) What bypass capacitance should be used at the IC's power pin if the IC is clocked at 100 MHz (the minimum duration of a state is 10 ns) and the supply voltage must be held to within 1% of the nominal value?

P 8.74 Figure P 8.42 shows a simulation of a half-wave rectifier circuit. Multimeter XMM2 indicates the rms amplitude of the ac (ripple) component of the load voltage and multimeter XMM1 indicates the dc component, where for $C = 200\,\mu\text{F}$, the rms amplitude of the ripple is approximately 19.1 V and the dc component of the load voltage is approximately 102.5 V

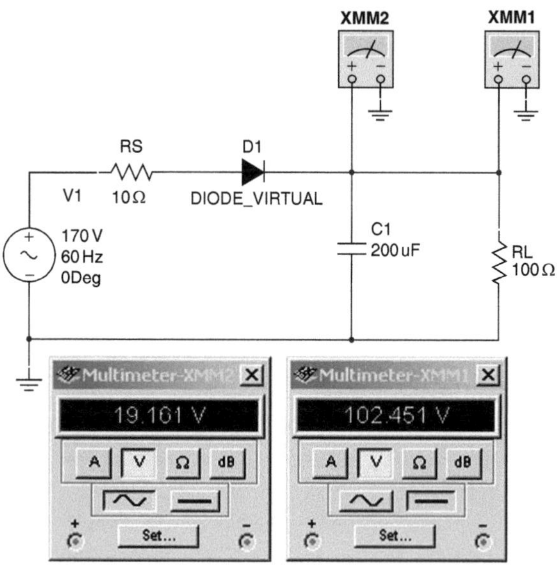

Fig. P 8.42 See Problem P 8.74

(a) What is the ripple factor for the circuit as shown? How good is the agreement of the measured ripple factor with the approximate formula developed in the text?

(b) Use the approximate relation developed in the text to find the capacitance needed to reduce the ripple factor to 0.015. Then use a simulation to determine the actual ripple for that value of capacitance. Be sure to give the simulation time to achieve steady state (wait until the voltage readings change very little with time).

(c) For the capacitance found in part (b), what is the worst-case (maximum) current through the diode when the circuit is first switched on?

P 8.75 Figure P 8.43 shows a simulation of a full-wave rectifier circuit, with an oscillograph of the ac (ripple) component load voltage and measurements

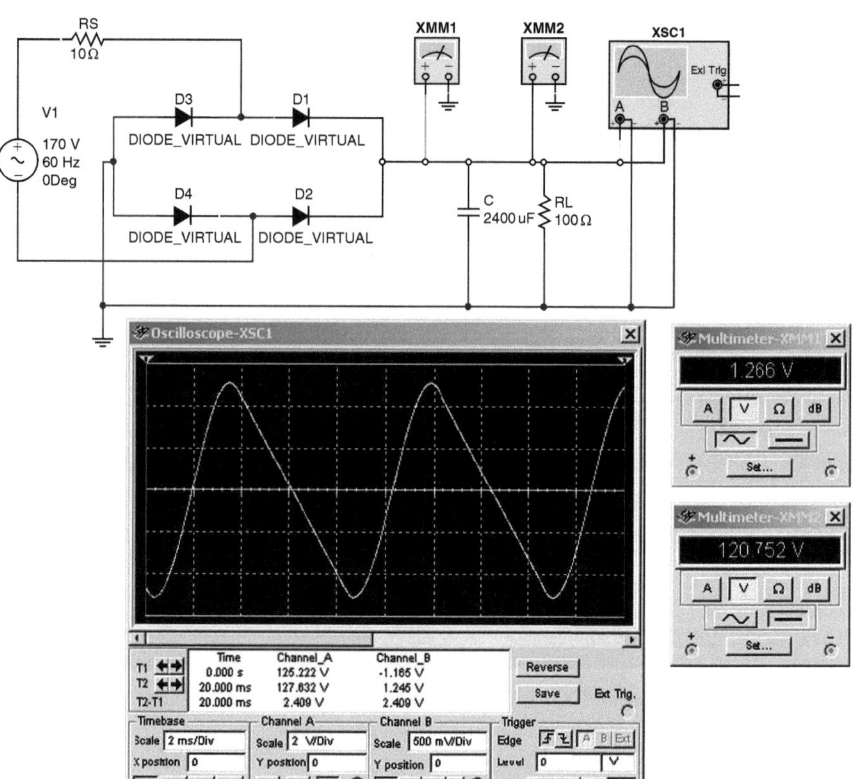

Fig. P 8.43 See Problem P 8.75

of the dc load voltage ($\cong 121$ V) and rms ripple ($\cong 1.27$ V). Note that the source floats.

(a) What is the load-voltage ripple factor?
(b) Trace the paths traveled by the current i for $v_S > 0$ and $v_S < 0$. If the capacitor is removed (if $i_L = i$), what is the waveform of the load voltage v_L?
(c) Use the relation

$$C = \frac{T}{R_L \gamma \sqrt{12}}$$

developed in the text to calculate the capacitance required to reduce the ripple factor to 0.005. Use a simulation to check the result. Be sure to give the simulation time to achieve steady state.

P 8.76 Figure P 8.44 shows a full-wave rectifier circuit using two identical sources.

(a) Trace the paths traveled by the current i for $v_S>0$ and $v_S<0$. If the capacitor is removed (if $i_L = i$), what is the waveform of the load voltage v_L?
(b) Simulate the circuit and plot the steady-state load voltage waveform for $C = 1{,}000\ \mu$F. Measure the rms amplitude of the ripple voltage and the dc component of the load voltage. Calculate the ripple from both the measured values and the formula developed in the text.
(c) What capacitance would be required to reduce the peak-to-peak ripple to 1 V?
(d) Simulate the circuit using the capacitance you found in part (c) and verify your calculation.
(e) If the source resistance for each source is 5 Ω, what is the maximum possible surge current through either diode?

P 8.77 Refer to Fig. P 8.45. Specify the coupling capacitors such that the overall voltage gain equals or exceeds 450 for all frequencies of interest. The source resistance is negligible, the op amps are ideal, and the output resistance of each stage is negligible. Use a simulation to check your work.

P 8.78 Refer to Fig. P 8.46. What coupling capacitance C is required if the voltage gain for $f_0 = 1$ kHz must be at least 95% of the voltage gain for $f_0 = 10$ kHz? Use a simulation to check your work.

P 8.79 (a) Specify the components of a cascade of two or more inverting amplifiers such that the overall voltage gain would be 10^4 if the amplifiers were direct coupled. The feedback resistances cannot exceed 1MΩ and the input resistance of each inverting amplifier must be at least 25 kΩ. Assume the op amps are ideal.

(b) In actual use, the source will be capacitively coupled to the first stage, each stage will be capacitively coupled to the next, and a 10 kΩ load resistance will be capacitively coupled to the last stage. The overall voltage gain must be at least 9800 for any sinusoidal input whose frequency exceeds 1 kHz. Specify the coupling capacitors.
(c) Use a simulation to verify your design.

P 8.80 Refer to Fig. P 8.47. Specify the coupling capacitors such that the overall voltage gain at 1.5 kHz

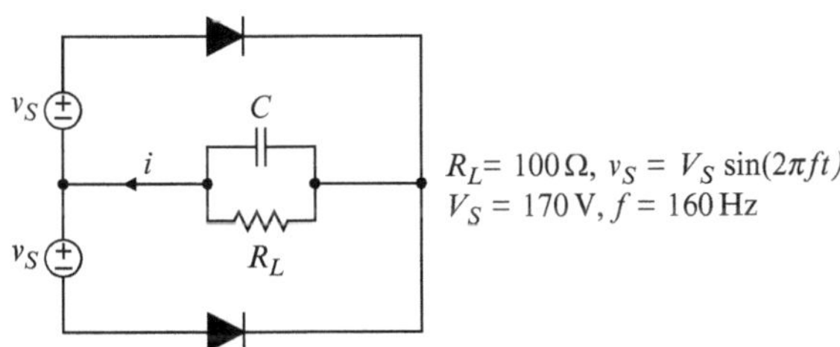

Fig. P 8.44 See Problem P 8.76

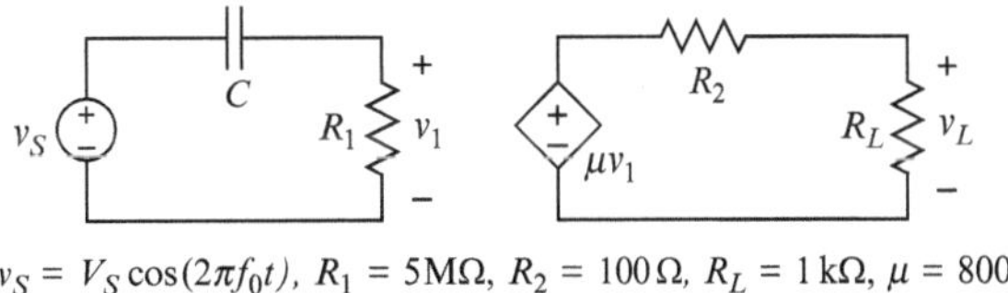

Fig. P 8.46 See Problem P 8.78

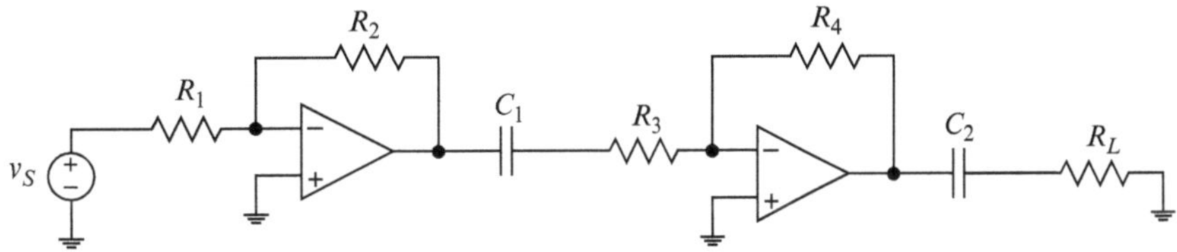

Fig. P 8.45 See Problem P 8.77

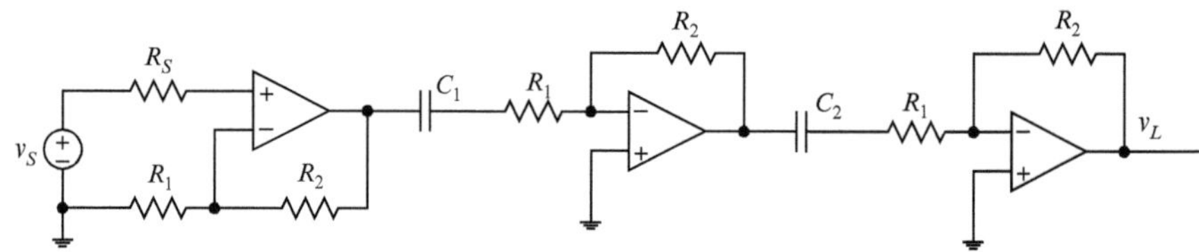

$v_S = V_0 + V_S\cos(2\pi f_0 t)$, $f_0 \geq 1.5\,\text{kHz}$, $V_0 = 500\,\text{mV}$, $V_S \leq 2\,\text{mV}$
$R_S = 100\,\Omega$, $R_1 = 15\,\text{k}\Omega$, $R_2 = 305\,\text{k}\Omega$, $V_{cc} = 15\,\text{V}$

Fig. P 8.47 See Problem P 8.80

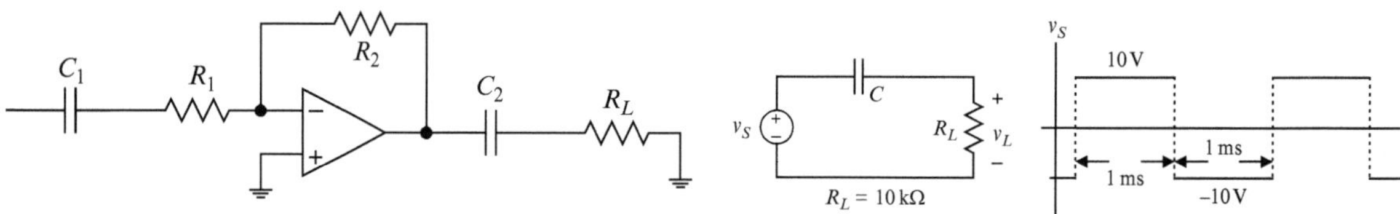

Fig. P 8.48 See Problem P 8.81

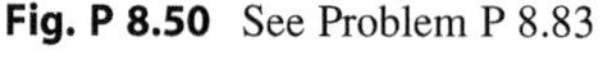

Fig. P 8.50 See Problem P 8.83

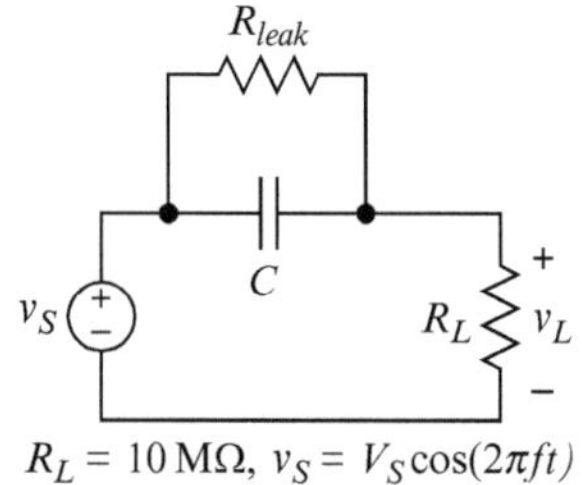

Fig. P 8.49 See Problem P 8.82

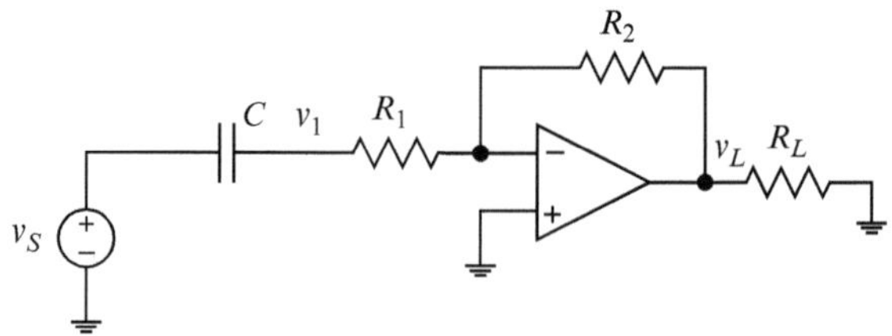

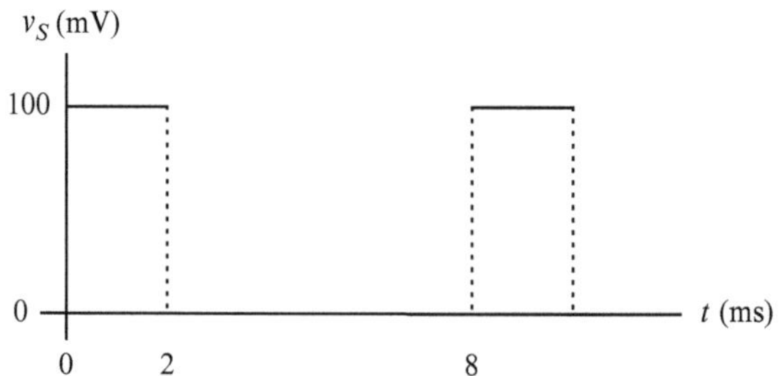

Fig. P 8.51 See Problem P 8.84

is at least 98% of the gain would be achieved if the stages were direct coupled. Use a simulation to verify your design.

P 8.81 Refer to Fig. P 8.48. The source resistance is negligible and the op amp is ideal. (a) The overall high frequency voltage gain must be 50 dB. The voltage gain at 30 Hz must be no less than 49 dB. Specify the components. (b) The maximum amplitude of the input is 100 mV. What is the minimum acceptable supply voltage? Assume a symmetric supply and rail-to-rail operation. (c) Perform a simulation to verify your design.

P 8.82 In Fig. P 8.49, the leakage resistance R_{leak} is ignored and the coupling capacitance C is specified such that $V_{L\,rms} = 0.8\,V_{S\,rms}$ for $f = 30\,\text{Hz}$. It is subsequently found that the capacitor leakage resistance is approximately equal to the load resistance R_L. What is the actual value of $V_{L\,rms}/V_{S\,rms}$ for $f = 30\,\text{Hz}$?

P 8.83 In Fig. P 8.50, the source is a square wave, as shown. In steady state, the load voltage v_L must be within 5% of the input v_S at all times. Specify the coupling capacitor C. Use a simulation to verify your answer.

P 8.84 Refer to Fig. P 8.51, The source resistance is negligible and the load resistance is $R_L = 5\,\text{k}\Omega$. The op amp is rail-to-rail and otherwise ideal. The input $v_S(t)$ is a rectangular pulse train, as shown. Let $v_{S\,ac}(t)$ denote the ac component of the input. The circuit is intended to amplify the ac component of the input, such that $v_L(t) \cong 100\,v_{S\,ac}$. The absolute error $|100 v_{S\,ac} - v_L(t)|$ must not exceed 100 mV. Specify the circuit components.

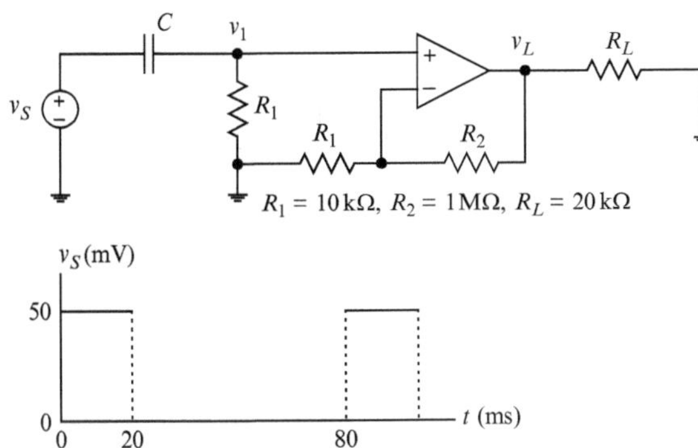

Fig. P 8.52 See Problem P 8.85

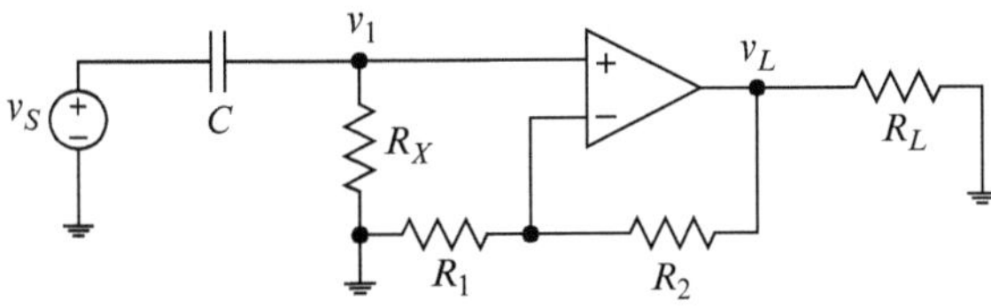

Fig. P 8.53 See Problem P 8.86

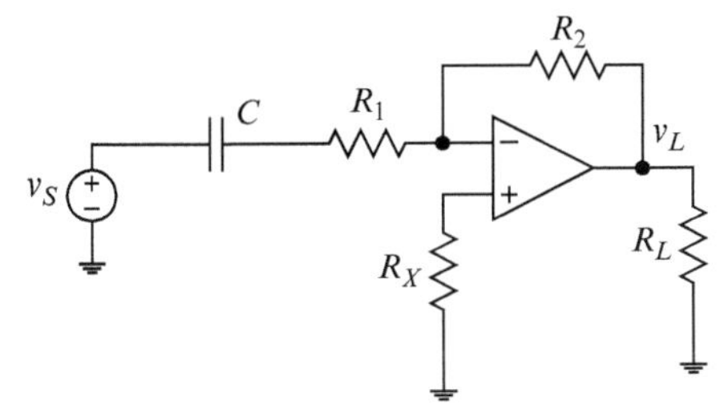

Fig. P 8.54 See Problem P 8.87

P 8.85 Refer to Fig. P 8.52. Specify the capacitor such that the voltage v_1 is always within 1% of the ac component of the source voltage v_S. Then determine the maximum percent difference between the load voltage v_L and a true rectangular pulse train. Use a simulation to check your work.

P 8.86 In Fig. P 8.53, $R_L = 5\,k\Omega$, $R_1 = 20\,k\Omega$, $R_2 = 1\,M\Omega$, $C = 50\,nF$, the op amp is ideal, and the supply voltages are $\pm 20\,V$. The source v_S has a constant (dc) component and one sinusoidal (ac) component; i.e.,

$$v_S(t) = V_{DC} + V_{AC}\cos(2\pi f_0\,t).$$

Specify the resistance R_X. What is the lowest frequency for which the voltage gain for the ac component of the source exceeds 30 dB?

P 8.87 In Fig. P 8.54, the source v_S has one constant and one sinusoidal component. Specify the resistors and the capacitor such that the voltage gain equals 30 dB at 50 Hz.

P 8.88 Figure P 8.55 shows two amplifiers in cascade. The op amps are ideal. At 1 kHz, the voltage gain and input resistance of each amplifier (not including the capacitors) are approximately 35 dB and $10\,k\Omega$. What are appropriate values for the input-bias-current compensating resistors R_X, R_Y?

P 8.89 Figure P 8.56 shows two amplifiers in cascade. The op amps are ideal. At 1 kHz, the voltage gain and input resistance of each amplifier (not including the capacitors) are approximately 40 dB and

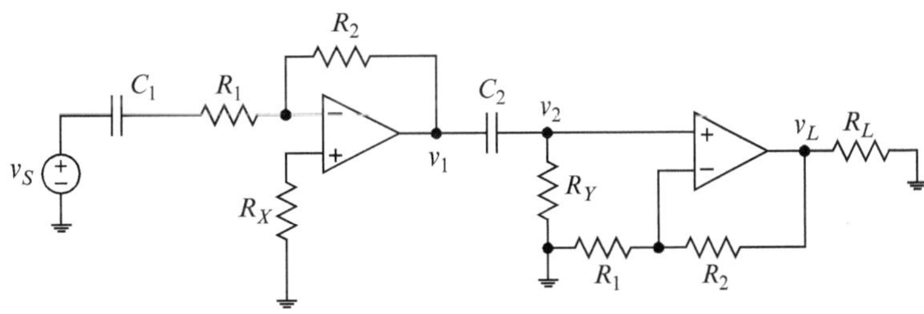

Fig. P 8.55 See Problem P 8.88

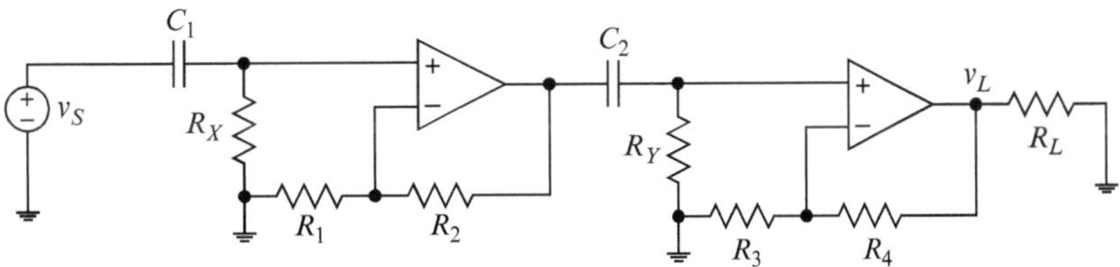

Fig. P 8.56 See Problem P 8.89

10 kΩ. What are appropriate values for the input-bias-current compensating resistors R_X, R_Y?

P 8.90 Figure P 8.57 shows two amplifiers in cascade. The op amps are ideal. At 1 kHz, the voltage gain and input resistance of each amplifier (not including the capacitors) are approximately 25 dB and 20 kΩ. What are appropriate values for the input-bias-current compensating resistors R_X, R_Y?

P 8.91 Refer to Fig. P 8.58, where the timing diagram indicates which switch is closed, as a function of time. Obtain a symbolic expression for the equivalent resistance of the switched capacitor. Then use the equivalent resistance to calculate the dc component (average value) of the load voltage v_L. Simulate the circuit to confirm your answer. Discuss the practicality of this particular switched-capacitor implementation.

P 8.92 Refer to Fig. P 8.59. The switches are in position 2 and the capacitors are uncharged at $t = 0^-$, at which time the switches begin operating as indicated by the timing diagram. (a) Plot the voltage $v(t)$ versus time for $0 \leq t \leq 10\,\mu$s. (b) Replace the switched capacitor by an equivalent resistor and plot the voltage $v(t)$ on the axes used in part (a). Do the two circuits exhibit the same (or very similar) transient response?

P 8.93 Refer to Fig. P 8.60. Switches 1 and 2 operate as indicated by the timing diagram. The voltages v_1 and v_2 are approximately constant during any interval

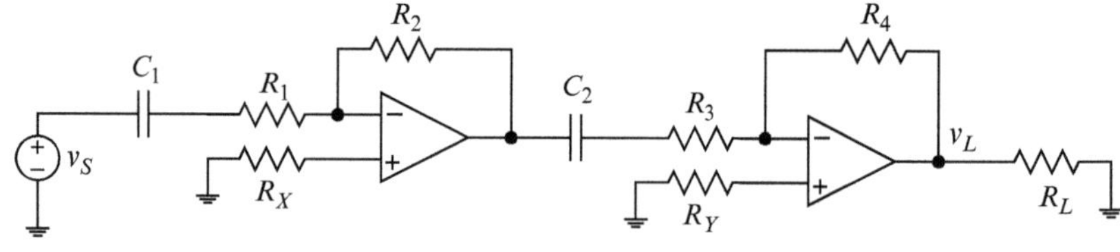

Fig. P 8.57 See Problem P 8.90

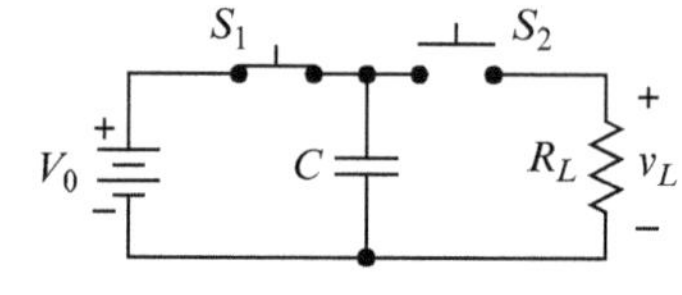

$V_0 = 10\,\text{V}$, $C = 100\,\text{pF}$, $R_L = 1\,\text{k}\Omega$, $\Delta T = 1\,\mu\text{s}$

Fig. P 8.58 See Problem P 8.91

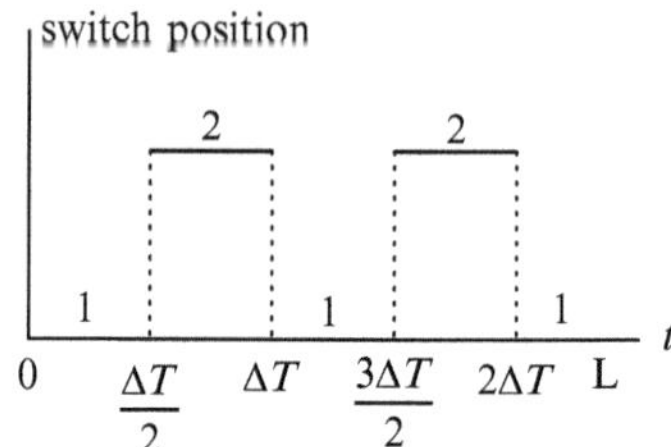

$V_0 = 10\,\text{V}$, $C_1 = 20\,\text{nF}$, $C_2 = 100\,\text{nF}$, $\Delta T = 2\,\text{ns}$, $R_S = 10\,\Omega$

Fig. P 8.59 See Problem P 8.92, 96

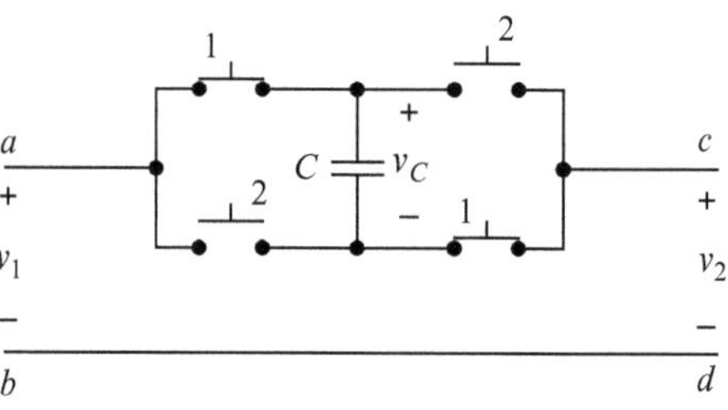

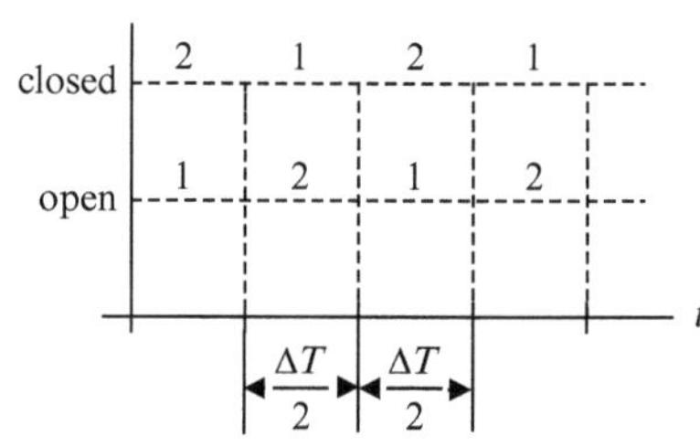

Fig. P 8.60 See Problem P 8.93

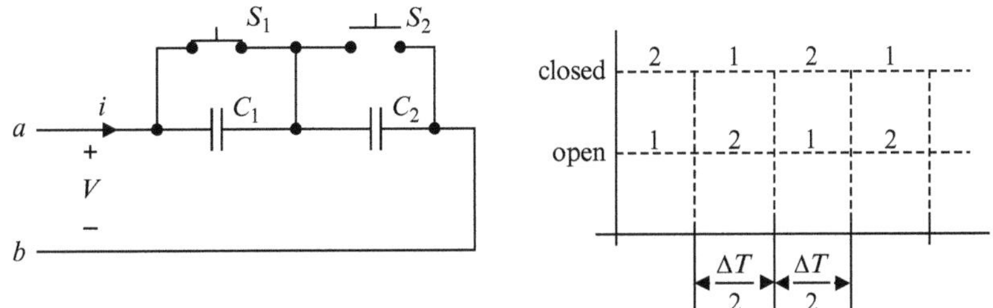

Fig. P 8.61 See Problem P 8.94

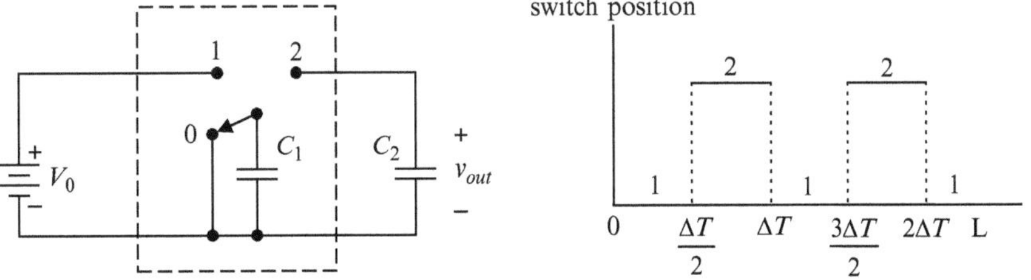

Fig. P 8.62 See Problem P 8.95

of duration $\Delta T/2$. (a) Show that the circuit is approximately equivalent to a resistor and obtain an expression for the resistance. (b) While terminals *a–b* are open and the switches are operating as described above and at a very fast rate, a resistor R_L and capacitor C_L are connected in parallel and then to terminals *c–d*. Then a voltage source $v_1(t) = V_0$ is connected to terminals *a–b*. Assume the switched capacitor can be represented by its equivalent resistance and obtain an expression for the output $v_2(t)$.

P 8.94 Refer to Fig. P 8.61, where the switch positions alternate, as indicated by the timing diagram. Obtain expressions for (a) the time average of the current *i*, (b) the equivalent resistance at the terminals *a–b*, (c) the average current through each switch.

P 8.95 In Fig. P 8.62, $V_0 = 10\,\text{V}$, $C_1 = 5\,\text{nF}$, $C_2 = 100\,\text{nF}$ and $\Delta T = 2\,\text{ns}$. The switch is at position 0 for $t < 0$ and begins toggling between positions 1 and 2 thereafter, as indicated by the timing diagram. The capacitor C_2 is initially uncharged, such that $v_{out}(0) = 0$. (a) Plot the voltage v_{out} versus time for $0 \le t \le 200\,\text{ns}$.

P 8.96 Refer to Fig. P 8.59 (Problem P 8.92). The capacitors are uncharged for $t = 0$, at which time the switches begin operating as indicated by the timing diagram.

(a) Plot the energy stored in C_2 versus time for $0 \le t \le 10\,\mu\text{s}$.
(b) On the same axes, plot the cumulative work done by the source for $0 \le t \le 10\,\mu\text{s}$.

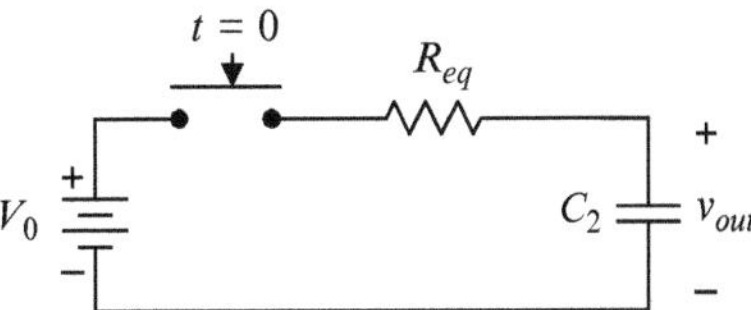

Fig. P 8.63 See Problem P 8.96

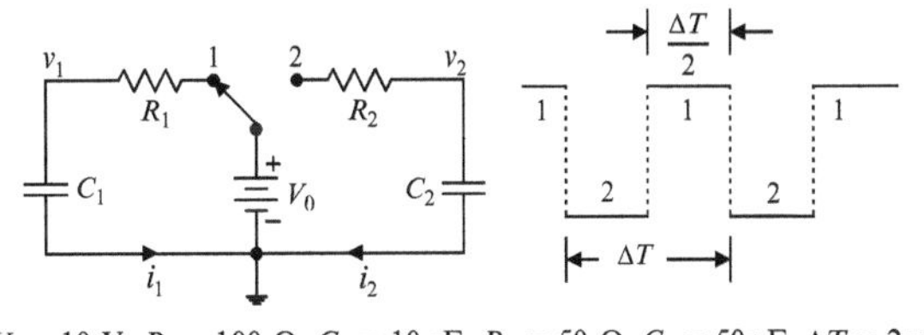

Fig. P 8.64 See Problem P 8.97

(c) Find the equivalent resistance R_{eq} of the switched capacitor. Use the equivalent *RC* circuit model shown in Fig. P 8.63 and plot the energy delivered by the source and the energy stored in C_2 for $0 \le t \le 5\tau$, where τ is the time constant for the equivalent *RC* circuit. Are the results consistent with those obtained in parts (a) and (b)?

P 8.97 Refer to Fig. P 8.64, Plot the voltages $v_1(t)$ and $v_2(t)$ and the instantaneous power delivered by the source on the same axes.

P 8.98 Refer to Fig. P 8.65, where the switch operates as indicated by the timing diagram. (a) Obtain an expression for the average power delivered by the

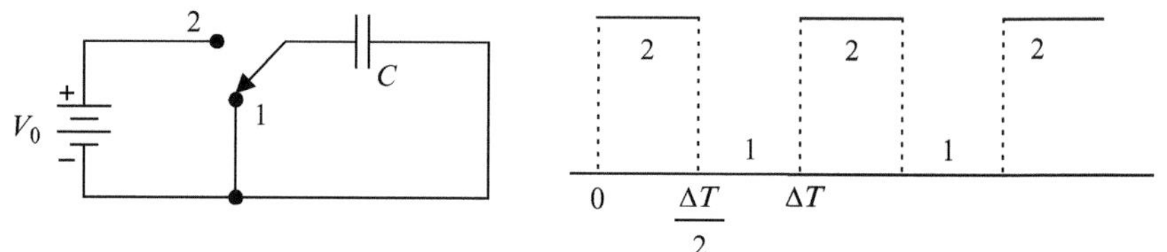

Fig. P 8.65 See Problem P 8.98, 99

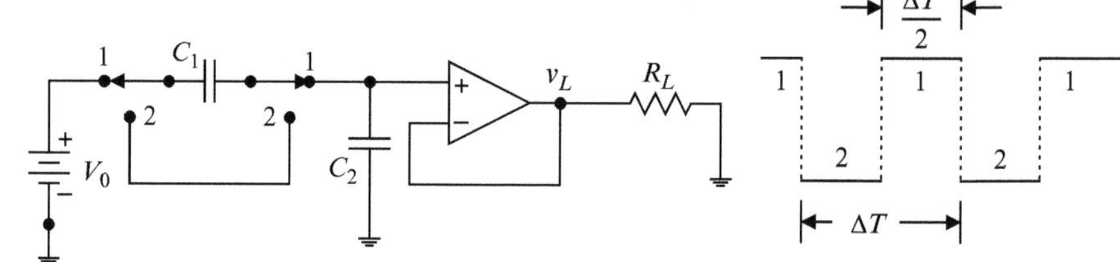

Fig. P 8.66 See Problem P 8.100

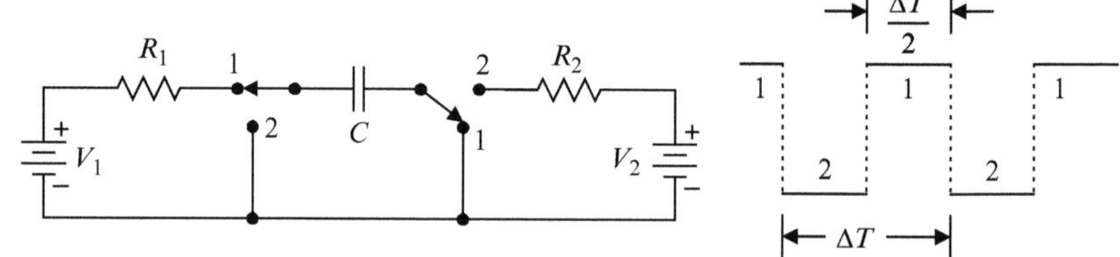

Fig. P 8.67 See Problem P 8.101

source. (b) It is shown in the text that the average power dissipated by a capacitor equals zero. If that is the case, where is the power delivered by the source dissipated?

P 8.99 Refer to Fig. P 8.65, where the switch operates as indicated by the timing diagram. (a) Obtain an expression for the energy stored in the capacitor while the switch is in position 2. (b) Obtain an expression for the work done by the source while the switch is in position 2. (c) Why are the expressions obtained in parts (a) and (b) different?

P 8.100 Refer to Fig. P 8.66, where the op amp is ideal and the switches have been operating as indicated for a very long time. The source resistance (not shown) is small but it is not zero. What is the average power delivered by the source V_0?

P 8.101 Refer to Fig. P 8.67, where the switches have been operating as indicated for a very long time. Assume that ΔT is much larger than each of $\tau_1 = R_1C$ and $\tau_2 = R_2C$. Obtain an expression for the power delivered by the sources (dissipated in the circuit).

Chapter 9
Inductance

A phenomenon called *inductance* is analogous to capacitance, but with the roles of current and voltage swapped. Whereas capacitance stores energy in an electric field produced by voltage, inductance stores energy in a magnetic field produced by current. Whereas capacitance resists changes in terminal voltage, inductance resists changes in terminal current. Inductance is relatively unimportant in low- to mid-frequency, low-power electronic circuits, where its desirable effects can be obtained by other means, but is important in high-frequency circuits and in electric power generation and distribution systems. Because inductance arises from a magnetic field, we begin with a discussion of magnetic fields.

9.1 Magnetic Field

If you sprinkle iron filings on a sheet of paper overlaying a magnet, the filings will arrange themselves along curved lines exiting (by convention) the north pole of the magnet and entering the south pole, as illustrated in Fig. 9.1. It is said that the magnet produces a **magnetic field** characterized by *lines of force* that align the filings. The stronger the magnet, the greater the density of lines passing through any plane perpendicular to the surface of the paper. Capitalize *Magnetic flux*, denoted by ϕ, is a measure of the density of the lines. The SI unit of magnetic flux is the **weber** (Wb).[1]

Energy is stored in a magnetic field because the field has the potential of doing work on magnetic material. For example, suppose we fix a strong magnet to a smooth surface and place a piece of iron against one of the poles of the magnet. Then we move the iron a short distance away from the magnet against the force of attraction. If we then release the iron, the field will do work on the iron by accelerating it toward the magnet. When we move the iron away from the magnet, we store energy in the field. When we release the iron, the iron gains kinetic energy from the field equal to the energy we stored in the field (ignoring friction).

A moving charge (current) produces a magnetic field. In introductory physics books, it is shown that the magnetic flux produced by a current in a long straight wire is concentric to the wire, as illustrated in Fig. 9.2. The direction of the magnetic flux produced by a current in a wire can be determined using the *right-hand rule*: If you wrap the fingers of your right hand around the wire with your thumb pointing in the direction of the current, your fingers point in the direction of the flux.[2]

The flux produced by the current is proportional to the current. So long as the current is steady, the field is steady, and there is no energy exchange between the field and the current producing the field. But energy is exchanged between the field and a *changing* current. If the current increases then the flux, being proportional to the current, increases as well and the energy stored in the field increases. If the current decreases, then the flux decreases and the field returns energy to the current. Because work must be done to store energy in or extract energy from the field, *the magnetic field*

[1]After the German physicist Wilhelm Eduard Weber (1804–1891).

[2]Of course, if you actually do this with a "live" wire, all of your appendages could point in different directions.

T.H. Glisson, *Introduction to Circuit Analysis and Design*,
DOI 10.1007/978-90-481-9443-8_9, © Springer Science+Business Media B.V. 2011

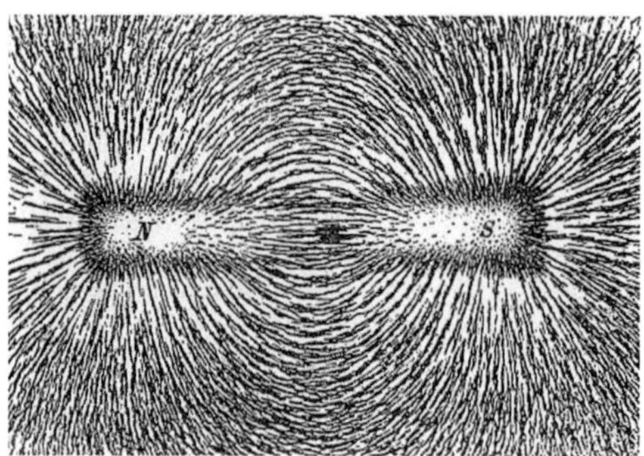

Fig. 9.1 Iron filings reveal magnetic lines of force from a permanent magnet

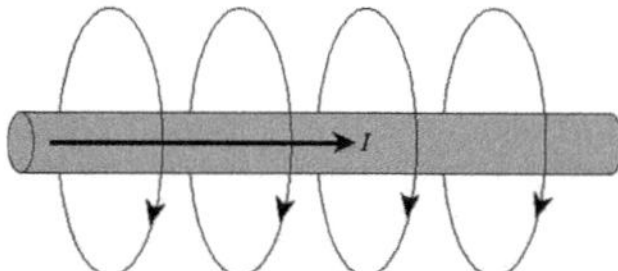

Fig. 9.2 Magnetic flux arising from current in a conductor

produced by a current always acts to oppose any change in the current. The magnetic field gives a kind of inertia to the moving charges comprising the current.

9.2 Self Inductance

The magnetic field produced by a current in a conductor opposes a change in that current by inducing a voltage that tends to keep the current from changing. The induced voltage v is given by *Faraday's law of induction*[3]

$$v = \frac{d\lambda}{dt}, \tag{9.1}$$

where λ is the magnetic **flux linkage** with the conductor. Flux linkage λ depends upon both the quantity ϕ of flux in the space containing the conductor and the geometry of the conductor; e.g., if a quantity ϕ of flux links every turn of an N-turn coiled conductor, the flux linkage is $\lambda = N\phi$. The polarity of the induced voltage is that which opposes a change of flux (Lenz's law). If the current is decreasing, the induced voltage tends to produce a current in the direction of the current already present. If the current is increasing, the induced voltage tends to produce a current in the direction opposite to that of the current already present.

Because magnetic flux is proportional to the current that produces the flux,[4] the flux linkage with a conductor also is proportional to the current producing the flux and we may write

$$\lambda = Li, \tag{9.2}$$

where L is a constant of proportionality called **inductance**. The SI unit of inductance is the **henry**[5] (H), where 1 H equals 1 Vs A^{-1}.

If the flux that induces a voltage in a conductor is due entirely to current in the *same* conductor, the associated inductance is called **self-inductance**. It also is possible for flux produced by current in one conductor to induce a voltage in another conductor, in which case the associated inductance is called **mutual inductance**. In practice and in this book, *inductance* alone means self-inductance. If mutual inductance is meant, the adjective *mutual* is included.

Inductance depends strongly upon the geometry of the conductor carrying the current and the medium in which the field is produced (e.g., air versus iron). The inductance of a conductor is increased by winding the conductor in a coil, as illustrated in Fig. 9.3. In this configuration, the flux produced is greater than that for a straight wire carrying the same current because the current enclosed by and creating *flux loops* is effectively N times the current in a single conductor, where N is the number of turns in the coil. Thus the total flux produced is N times that produced by a single conductor:

$$\phi = N\phi_1. \tag{9.3}$$

In (9.3), the flux ϕ_1 is proportional to the current producing the flux; i.e., $\phi_1 = ki$. Furthermore, flux linkage with an N-turn coil is N times that for a straight conductor because each loop links the coiled conductor N times. Thus

[3]After the English chemist and physicist Michael Faraday, (1791–1867). Faraday's law of induction is treated in virtually all introductory physics textbooks.

[4]This assertion must be modified for flux produced in a magnetic material by relatively large currents. We ignore that refinement here.

[5]After the American physicist Joseph Henry (1791–1878).

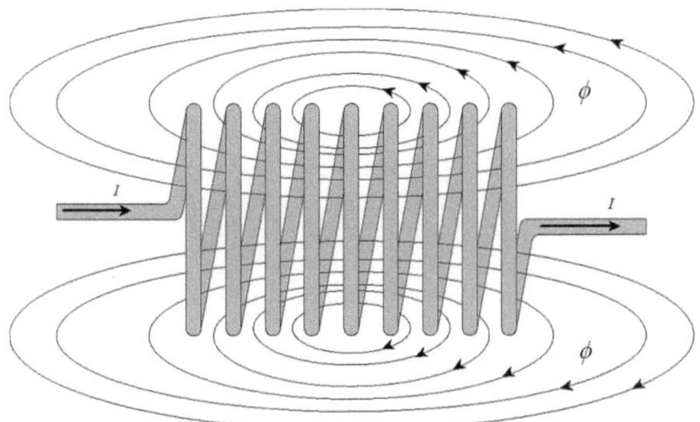

Fig. 9.3 Magnetic flux produced by current in a coil

$$\lambda = N\phi = N(N\phi_1) = kN^2 i = Li, \tag{9.4}$$

where, from (9.2), inductance L is given by

$$L = \frac{\lambda}{i} = kN^2 \tag{9.5}$$

The flux produced by current in a coil (and thus the inductance of the coil) can be further increased by winding the coil around a magnetic material, such as iron. The direction of the flux is again determined using a *right-hand rule*: Imagine wrapping the fingers of your right hand around the coil in the direction of the current; your thumb then points in the direction of the flux through the center of the coil.

9.3 Inductance of Air-Core Coils

The inductance of a *long, tightly wound, single-layer, helical air-core coil* having N turns, length l, and diameter d is given by[6]

$$L = \frac{\pi\mu_0 N^2 d^2}{4l} = \frac{\mu_0 N^2 A}{l}, \tag{9.6}$$

where $\mu_0 = 4\pi \times 10^{-7}\,\mathrm{H\,m^{-1}}$ is the permeability of a vacuum (also very nearly that of air), $A = \pi d^2/4$ is the cross-sectional area of the coil, and *long* means $l \gg d$. We may write (9.6) as

$$L = kN^2, \tag{9.7}$$

in agreement with (9.5).

[6]This relation is derived in all introductory, university-level physics textbooks.

Exercise 9.1. What is the SI unit of k in (9.7)?

Flux loops near each end of a coil do not link as many turns as those near the middle. The associated reduction in the total flux linkage can be neglected in long coils. But for short coils, it is necessary to correct for the reduced flux linkages. Various handbooks give various formulas for short coils, one being

$$L = \frac{\mu_0 N^2 d^2}{(0.579\,d + 1.273\,l)}, \tag{9.8}$$

where N is the number of turns, d is the diameter of the coil and l is the length of the coil. Equation (9.8) is accurate to about $\pm$ 1% for *tightly wound, single-layer, helical, air-core coils.*

Example 9.1. Find the number of turns needed on a paper tube 10 cm long and 1 cm in diameter to produce an inductor having inductance $L = 1$ mH. Assume the coil is helical and tightly wound in a single layer.

Solution: From (9.8),

$$\begin{aligned}
N &= \sqrt{\frac{(0.579\,d + 1.273\,l)L}{\mu_0 d^2}} \\
&= \sqrt{\frac{(0.579)(1\,\mathrm{cm}) + (1.273)(10\,\mathrm{cm})}{(4\pi \times 10^{-7}\,\mathrm{Hm^{-1}})(1\,\mathrm{cm})^2}(1\,\mathrm{mH})} \\
&= \sqrt{\frac{(13.31\,\mathrm{cm})(1\,\mathrm{mH})}{(4\pi \times 10^{-7}\,\mathrm{Hm^{-1}})(1\,\mathrm{cm})^2}} \\
&= \sqrt{\frac{13.31\,\mathrm{mH}}{(4\pi \times 10^{-7}\,\mathrm{Hm^{-1}})(1\,\mathrm{cm})}} \\
&= \sqrt{\frac{13.31\,\mathrm{mH}}{(4\pi \times 10^{-7}\,\mathrm{Hm^{-1}})(0.01\,\mathrm{m})}} \\
&= \sqrt{\frac{13.31 \times 10^{-3}\,\mathrm{H}}{(4\pi \times 10^{-9}\,\mathrm{H})}} = 1029.
\end{aligned}$$

Exercise 9.2. If the coil in Example 9.1 occupies the full length (10 cm) of the paper core and the thickness of the insulation is negligible, what is the maximum diameter of the wire? If the wire is copper, what is the resistance of the coil?

9.4 Inductors

Inductance can be introduced in circuits using components called **inductors**, to achieve certain desired effects. Figure 9.4 shows several inductors. Inductors are constructed by winding wire around a core, which might be air (a paper or plastic tube), iron, or other material. Inductors wound on a core of non-magnetic material, such as paper or plastic are called **air-core inductors** and inductors wound on a core of magnetic material such as iron, ferrite, or steel are called **iron-core inductors**.

Figure 9.5 shows circuit-diagram symbols for inductors. Figure 9.5(a) shows the conventional circuit-diagram symbol for an air-core or ideal inductor having inductance L. Figure 9.5(b) shows a variation sometimes used to denote an inductor wound on a core made of iron (or other magnetic material). Figure 9.5(c) shows the conventional symbol for a variable inductor, almost all of which have magnetic cores. A common type of variable inductor has a threaded cylindrical magnetic core. The inductance is varied by screwing the core farther into or out of the coils constituting the winding. As has become common practice, we use the symbol in Fig. 9.5(a) for an inductor, regardless of the core material. We also limit our treatment to inductors that are linear under normal operating conditions.

Because of the way they are constructed, inductors sometimes are called *coils*. Because of their purpose in certain applications, some iron-core inductors sometimes are called *chokes*. Often, iron-core inductors used in power supplies are called chokes, and iron-core and air-core inductors used in other applications

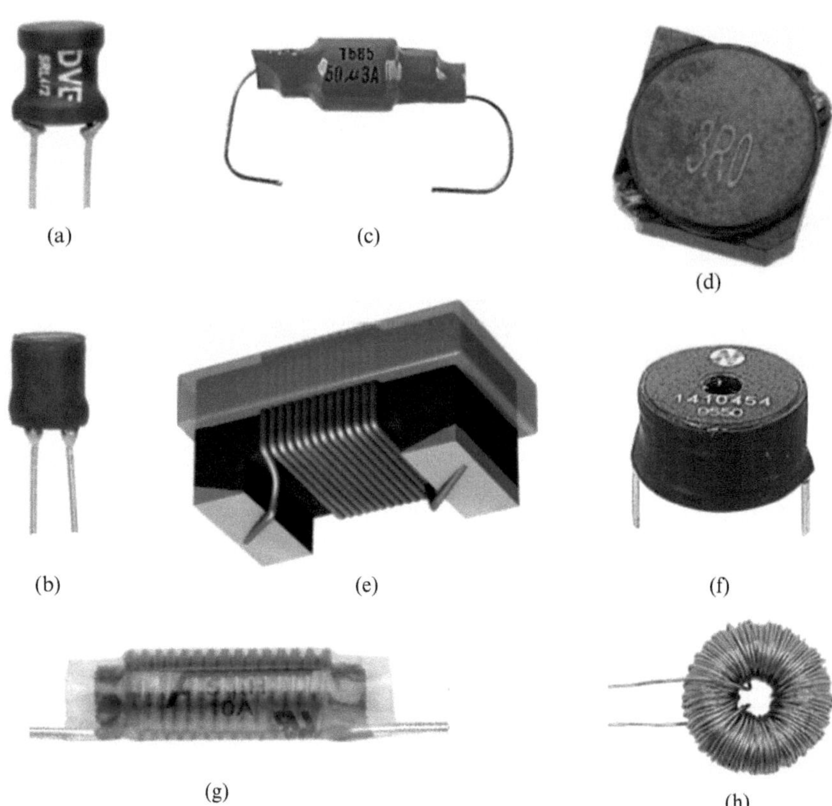

Fig. 9.4 Assorted inductors (not to scale): (**a**), (**b**) small radial-lead inductors, (**c**) small ferrite-core noise-suppression inductor, (**d**), (**e**) surface-mount inductors, (**f**) high-current PCB inductor, (**g**) axial-lead high-current inductor, (**h**) toroidal inductor (Photographs courtesy of Rapid Electronics, Ltd)

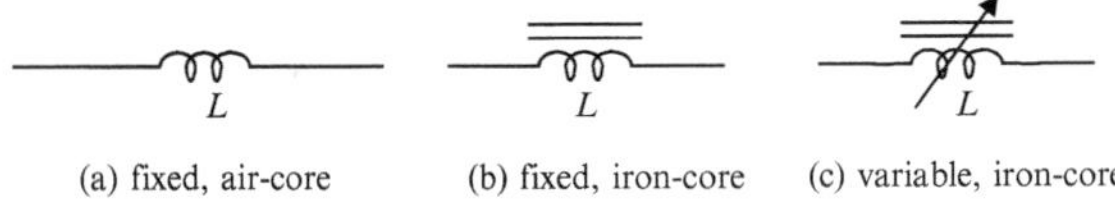

Fig. 9.5 Circuit-diagram symbols for inductors

are called coils, but this terminology is neither standard nor universal.

Inductance is introduced intentionally in circuits for various purposes, some of which are described in the sequel. In most applications where either inductance or capacitance could be used, such as in active filters and integrating and differentiating circuits, capacitance is preferred because of the smaller size and lower cost of capacitors, relative to inductors. Also, it is easier to produce a capacitor than an inductor of comparable precision in an integrated circuit, and it is easier to confine the electric field in a capacitor than it is to confine the magnetic field produced by an inductor, so capacitors produce less electrical interference than inductors.

Nonetheless, there are important applications for inductors, especially in some power supplies and in high-frequency circuits, where they are used in conjunction with capacitance to build very selective filters. Inductance is especially important in circuit models for electric power generation and distribution systems, because generators, motors, and transformers exhibit significant inductance. Inductance is only occasionally encountered in low- to moderate-frequency electronic circuits, where desirable effects of inductance can be introduced in other ways (avoiding need for bulky and heavy coils).

9.5 Terminal Characteristic of an Inductor

Using (9.2), we may rewrite Faraday's law (9.1) as

$$v = L\frac{di}{dt}, \tag{9.9}$$

where v is the voltage across the terminals of an inductor and i is the current entering the terminal designated positive, as shown in Fig. 9.6. Equation (9.9) is one expression of the terminal characteristic of an inductor.

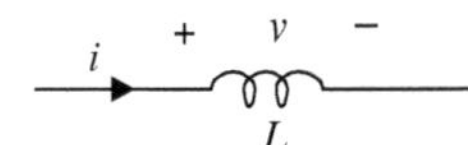

Fig. 9.6 Symbol for an inductor, annotated in agreement with (9.9)

Several implications of (9.9) are noteworthy:

- The voltage across an inductor is nonzero whenever the current through the inductor is changing (when di/dt is nonzero) and is zero otherwise. *An inductor acts as a conductor (a short circuit) for constant current.*
- The magnitude of the voltage across an inductor is proportional to the rate of change of the current through the inductor. The faster the current changes, the larger is the opposing voltage. *An inductor opposes changes in the current through the inductor.*
- Changing the current through an inductor instantaneously would require infinite voltage across the inductor. Thus, in physical circuits, the current through an inductor is a continuous function of time. In other words, *the current though an inductor cannot change instantaneously.*
- The voltage across an inductor is proportional to the inductance L. More voltage is required to change the current through an inductor at a specified rate in a larger inductor than a smaller one. Conversely, for a fixed rate of change in current, a large inductance produces a larger induced voltage than does a smaller inductance.

A circuit in which all currents and voltages are constant is said to be in **dc steady state**. If the current through an inductor is constant, the voltage across the inductor is zero. Thus, *for any circuit in dc steady state, all inductors in the circuit are effectively short circuits.*

Example 9.2. Figure 9.7(a) shows the current in an inductor having inductance L. From (9.9) the resulting (induced) voltage across the inductor is the product of the inductance and the derivative (instantaneous slope) of the current. Figure 9.7(b) shows the voltage across the inductor, where

$$v_1 = \frac{Li_1}{t_1}, \qquad v_2 = \frac{L(i_2 - i_1)}{t_4 - t_2}.$$

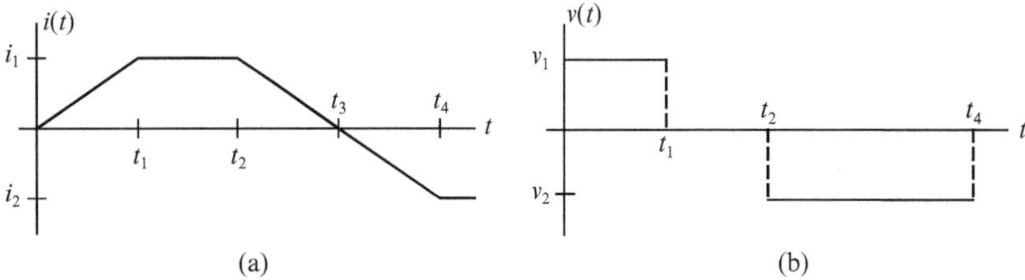

Fig. 9.7 Current through and voltage across the inductor of Example 9.2

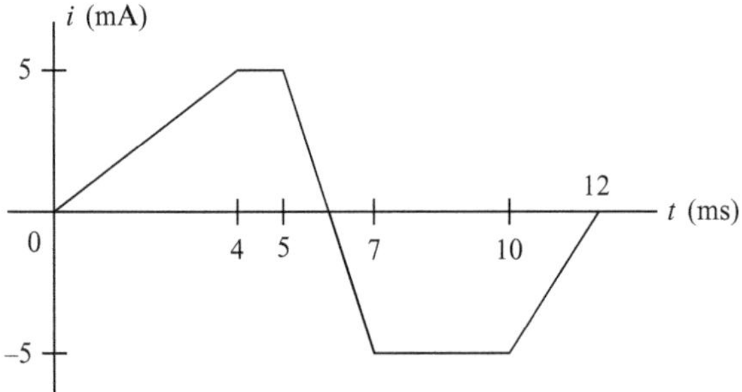

Fig. 9.8 See Exercise 9.3

Exercise 9.3. Figure 9.8 shows the current in an inductor having inductance $L = 10\,\text{mH}$. Draw a graph of the voltage across the inductor.

Example 9.3. The current in an inductor having inductance $L = 1$ mH (10^{-3} H) is given by $i(t) = I\cos(2\pi f t)$, with $I = 5$ mA and $f = 1\,\text{kHz}$. The voltage induced across the inductor is

$$v(t) = L\frac{d[I\cos(2\pi f t)]}{dt} = -2\pi f LI\sin(2\pi f t).$$

The maximum (peak) voltage is $2\pi f LI = 2\pi\,(1\,\text{kHz})(1\,\text{mH})(5\,\text{mA}) \cong 31.4\,\text{mV}$. If the frequency is increased to 100 kHz, the voltage increases to 3.14 V, illustrating that the voltage across an inductor is proportional to the rate of change of the current through the inductor.

Exercise 9.4. The current in an inductor having inductance $L = 10$ mH is given by $i(t) = I_0 + I_1\sin(2\pi f t + \pi/4)$. Obtain an expression for the voltage induced across the inductor.

Example 9.4. In the circuit shown in Fig. 9.9, the switch is moved from a to b at $t = 0$, having been in position a for $t < 0$. Obtain expressions for the voltage across the inductor just before the switch is moved (at $t = 0^-$) and just after the switch is moved (at $t = 0^+$).

Solution: Because the source is constant and because the switch is in position a for $t < 0$, the current i is constant and the inductor appears to be a short circuit for $t = 0^-$. The current through the inductor at $t = 0^-$ is given by $i = V_0/R$. For $t = 0^-$, Kirchhoff's voltage law gives

$$\begin{aligned} -V_0 + R\,i(0^-) + v(0^-) &= -V_0 + R\frac{V_0}{R} + v(0^-) \\ &= 0 \Rightarrow v(0^-) = 0. \end{aligned}$$

Because the current in an inductor cannot change instantaneously, the current is the same for $t = 0^-$ and $t = 0^+$. By Kirchhoff's voltage law, for $t = 0^+$,

$$\begin{aligned} R\,i(0^+) + v(0^+) &= R\frac{V_0}{R} + v(0^+) = 0 \\ &\Rightarrow v(0^+) = -V_0. \end{aligned}$$

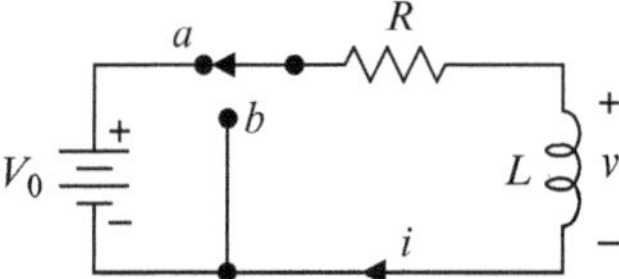

Fig. 9.9 See Example 9.4

Equation (9.9) expresses voltage across an inductor in terms of current through the inductor. From (9.9),

$$di(t) = \frac{1}{L}v(t)dt. \tag{9.10}$$

Integrating (9.10) from time t_0 to time t gives

$$\int_{t_0}^{t} di(t') = i(t) - i(t_0) = \frac{1}{L}\int_{t_0}^{t} v(t')dt'$$

or

$$i(t) = i(t_0) + \frac{1}{L}\int_{t_0}^{t} v(t')dt'. \tag{9.11}$$

Example 9.5. The voltage across a 100 mH inductor is given by

$$v(t) = \begin{cases} 0, & t \le 0, \\ V_0 \cos(2\pi f t), & t > 0, \end{cases}$$

with $V_0 = 20\,\text{V}$ and $f = 400$ Hz. The current through the inductor equals zero for $t = 0$. From (9.11) the current through the inductor is given by

$$\begin{aligned} i(t) &= i(0) + \frac{1}{L}\int_0^t v(t')dt' = \frac{1}{L}\int_0^t V_0 \cos(2\pi f t')dt' \\ &= \frac{V_0}{2\pi f L}\sin(2\pi f t) = I_0 \sin(2\pi f t), \end{aligned}$$

where

$$I_0 = \frac{V_0}{2\pi f L} = \frac{20\,\text{V}}{2\pi(400\,\text{Hz})(100\,\text{mH})} \cong 79.6\,\text{mA}.$$

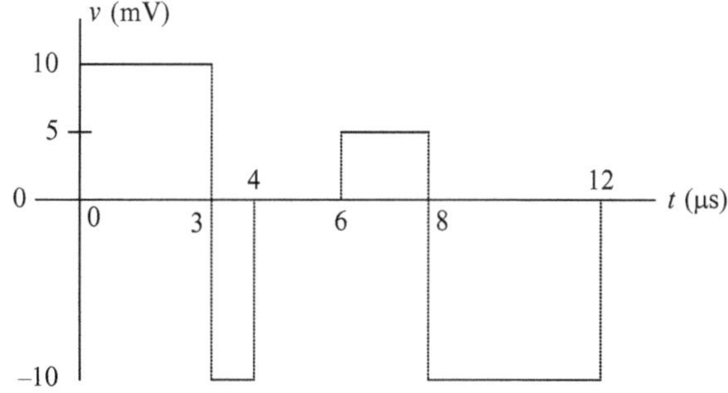

Fig. 9.10 See Exercise 9.6

Exercise 9.5. The voltage across a 100 mH inductor is given by

$$v(t) = \begin{cases} 0, & t \le 0, \\ V_0\, e^{-t/\tau}, & t > 0. \end{cases}$$

It also is known that the current through the inductor equals zero for $t < 0$. Obtain an expression for the current through the inductor.

Exercise 9.6. Figure 9.10 shows a graph of the voltage across a 20 mH inductor. The voltage equals zero for times not shown and the current through the inductor equals zero for $t < 0$. Draw a graph of the current through the inductor for $0 \le t \le 12\,\mu\text{s}$.

9.6 Time Constant

A circuit composed of only resistors and one inductor is called an ***RL circuit***. In this section, we obtain expressions for currents and voltages in an *RL* circuit subjected to an abrupt change in excitation.

Recall that ${t_0}^-$ and ${t_0}^+$ denote times infinitesimally before and after a time t_0. For example, if $t = 0$ denotes the time at which switch is closed, then $t = 0^-$ is the time just before the switch is closed and $t = 0^+$ is the time just after the switch is closed.

Refer to Fig. 9.11, where the switch is open for $t < 0$ and is closed at $t = 0$. The switch is open for $t = 0^-$, so $i(0^-) = 0$. Thus, from (9.11)

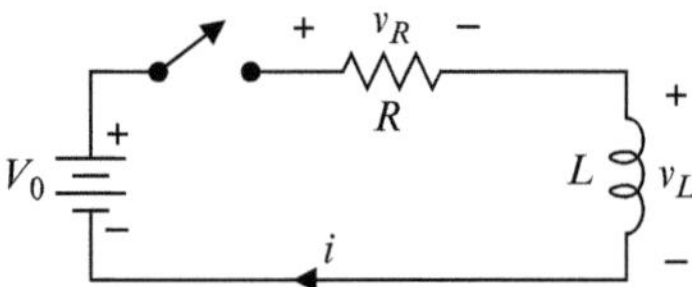

Fig. 9.11 A switched *RL* series circuit. See (9.12)

$$i(t) = i(0^-) + \frac{1}{L}\int_{0^-}^{t} v_L(t')dt' = \frac{1}{L}\int_{0^-}^{t} v_L(t')dt'$$

and Kirchhoff's current law gives, for $t > 0$

$$\frac{v_L(t) - V_0}{R} + \frac{1}{L}\int_{0^-}^{t} v_L(t')dt' = 0. \tag{9.12}$$

Differentiating (9.12) once with respect to time and rearranging terms gives, for $t > 0$

$$\frac{L}{R}\frac{dv_L(t)}{dt} = -v_L(t), \tag{9.13}$$

According to (9.14), the voltage v_L and its derivative have the same mathematical form (are proportional). The function having that property is the exponential function. You can show by substitution that the solution to (9.14) is

$$v_L(t) = K\exp\left(-\frac{Rt}{L}\right), \quad t > 0, \tag{9.14}$$

From (9.12),

$$\frac{v_L(0^+) - V_0}{R} + \frac{1}{L}\int_{0^-}^{0^+} v_L(t')dt' = \frac{v_L(0^+) - V_0}{R} = 0$$
$$\Rightarrow v_L(0^+) = V_0$$

Thus

$$v_L(t) = V_0\exp\left(-\frac{Rt}{L}\right), \quad t > 0 \tag{9.15}$$

Equation (9.15) usually is written

$$v_L(t) = V_0\exp\left(-\frac{t}{\tau}\right), \quad t > 0. \tag{9.16}$$

where

$$\tau = \frac{L}{R} \tag{9.17}$$

is the **time constant** for the series *RL* circuit. *The time constant determines how long it takes for the voltage across the inductor to vanish.* Because $e^{-1} \cong 0.37$, the voltage across the inductor drops to 37% of its initial value in a time equal to one time constant. Similarly, $e^{-3} \cong 0.05$, so the voltage drops to 5% of its initial value in a time equal to three time constants. You can show that the voltage is less than 1% of its initial value when t equals five time constants.

From Ohm's law,

$$i(t) = \frac{V_0 - v_L(t)}{R}. \tag{9.18}$$

Using (9.16) in (9.18) gives

$$i(t) = \frac{V_0}{R}\left[1 - \exp\left(-\frac{t}{\tau}\right)\right]; \quad t > 0, \quad \tau = \frac{L}{R}. \tag{9.19}$$

Figure 9.12 shows graphs of the voltage v_L and the current i, normalized to their maximum values, versus time as a multiple of the time constant. *The time constant determines how long it takes for the current through the inductor to reach the steady-state value* $i(\infty) = V_0/R$.

In Fig. 9.13, the source is piecewise-constant, switching from one value to another at $t = 0$:

$$v_S(t) = \begin{cases} V_1, & t < 0, \\ V_2, & t \geq 0. \end{cases} \tag{9.20}$$

The inductor presents zero resistance to dc. It follows that

$$i(0^+) = i(0^-) = \frac{V_1}{R}, \; i(\infty) = \frac{V_2}{R}. \tag{9.21}$$

By Kirchhoff's voltage law,

$$\begin{aligned} &Ri(t) + L\frac{di(t)}{dt} = V_2 \\ &\Rightarrow \frac{L}{R}\frac{di(t)}{dt} + i(t) = \frac{V_2}{R} = i(\infty), \; t > 0 \end{aligned} \tag{9.22}$$

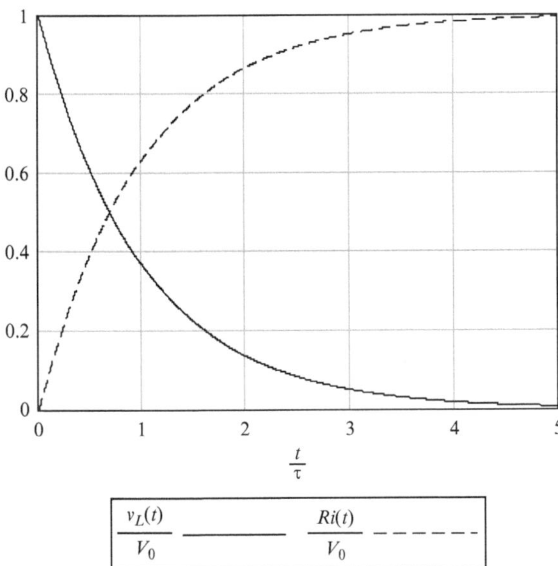

Fig. 9.12 Inductor voltage and current in the circuit shown in Fig. 9.11

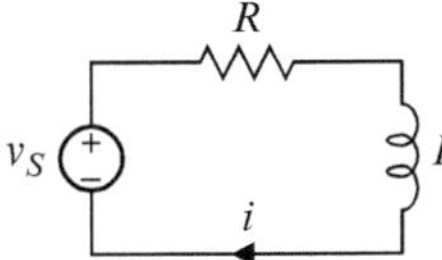

Fig. 9.13 An *RL* series circuit subjected to piecewise-constant excitation. See (9.20)

You can show by substitution that (9.22) is satisfied if

$$i(t) = i(\infty) - [i(\infty) - i(0^+)]\exp\left(-\frac{t}{\tau}\right), \quad \tau = \frac{L}{R}. \tag{9.23}$$

It can be shown that (9.23) is the complete solution to the differential equation (9.22). Equation (9.23) can be applied to any circuit that can be reduced to the form shown in Fig. 9.13, where $v_S(t)$ is piecewise constant. For example, refer again to the circuit in Fig. 9.11, where the switch is closed at $t = 0$. By inspection,

$$i(0^+) = 0,\ i(\infty) = \frac{V_0}{R},\ \tau = \frac{L}{R}.$$

From (9.23), for $t > 0$

$$i(t) = \frac{V_0}{R} - \left(\frac{V_0}{R} - 0\right)\exp\left(-\frac{t}{\tau}\right) = \frac{V_0}{R}\left[1 - \exp\left(-\frac{t}{\tau}\right)\right],$$

which we obtained by a more circuitous route above (see (9.19)).

Equation (9.23) is applicable not only to the current through an inductor, but to *any* branch current, in any circuit that is equivalent to the form shown in Fig. 9.13. Also, any node voltage in such a circuit can be expressed as

$$v(t) = v(\infty) - [v(\infty) - v(0^+)]\exp\left(-\frac{t}{\tau}\right), \quad \tau = \frac{L}{R}. \tag{9.24}$$

When applying (9.23) or (9.24), be careful to use the correct time constant for the interval of interest and to use $i(0^+)$ or $v(0^+)$, not $i(0^-)$ or $v(0^-)$, unless you are describing the current through an inductor, in which case $i(0^+) = i(0^-)$. Also keep in mind that $i(\infty)$ or $v(\infty)$ denote the eventual value of the current given by (9.23) or the voltage given by (9.24) if (9.23) or (9.24) remain valid for $t \to \infty$. But they are not necessarily the actual end-point values in any interval, because the validity of an instance of (9.23) or (9.24) might be terminated (e.g., by a switching) before the current or voltage reaches $i(\infty)$ or $v(\infty)$.

Example 9.6. Refer to Fig. 9.14, where the switch is open for $t < 0$ and is closed at $t = 0$. Obtain expressions for $i_1(t),\ i_2(t),\ i_L(t)$ for $t > 0$.

Solution: We need the circuit time constant and the initial and final values for each current. The Thévenin equivalent resistance at the terminals of the inductor is given by $R_T = R_1 \| R_2$, so the time constant is given by

$$\tau = \frac{L}{R_T} = \frac{(R_1 + R_2)L}{R_1 R_2}. \tag{9.25}$$

Because the switch is open for $t < 0$, and because the current through an inductor cannot change instantaneously, we know that

$$i_L(0^+) = 0 \Rightarrow i_1(0^+) = i_2(0^+) = \frac{V_0}{R_1 + R_2}.$$

Because an inductor presents a short circuit to a constant voltage or current, we know that

$$i_L(\infty) = \frac{V_0}{R_1},\ i_2(\infty) = 0,\ i_1(\infty) = i_L(\infty) = \frac{V_0}{R_1}.$$

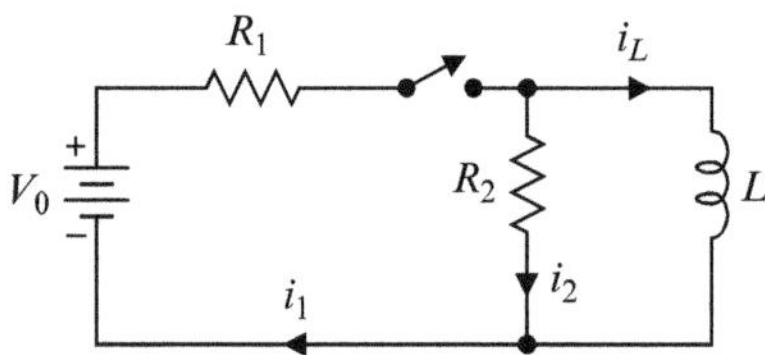

Fig. 9.14 See Example 9.6

We may now use (9.23) to obtain the requested expressions:

$$i_1(t) = i_1(\infty) - [i_1(\infty) - i_1(0^+)]\exp\left(-\frac{t}{\tau}\right)$$
$$= \frac{V_0}{R_1} - \left(\frac{V_0}{R_1} - \frac{V_0}{R_1+R_2}\right)\exp\left(-\frac{t}{\tau}\right),$$
$$i_2(t) = i_2(\infty) - [i_2(\infty) - i_2(0^+)]\exp\left(-\frac{t}{\tau}\right)$$
$$= \frac{V_0}{R_1+R_2}\exp\left(-\frac{t}{\tau}\right),$$
$$i_L(t) = i_L(\infty) - [i_L(\infty) - i_L(0^+)]\exp\left(-\frac{t}{\tau}\right)$$
$$= \frac{V_0}{R_1}\left[1 - \exp\left(-\frac{t}{\tau}\right)\right].$$

In all cases, the time constant τ is given by (9.25) and the expression is valid for $t > 0$.

Exercise 9.7. Show that the currents (expressions) obtained in Example 9.6 satisfy Kirchhoff's current law.

Exercise 9.8. Use the expression for $i_L(t)$ obtained in Example 9.6 to obtain an expression for the voltage across the inductor. Then use that expression and the source voltage to obtain expressions for the currents $i_1(t)$, $i_2(t)$. Show that these expressions for $i_1(t)$, $i_2(t)$ agree with those obtained in Example 9.6.

Exercise 9.9. Refer to Fig. 9.15. The switch is closed for $t < 0$ and opened at $t = 0$.

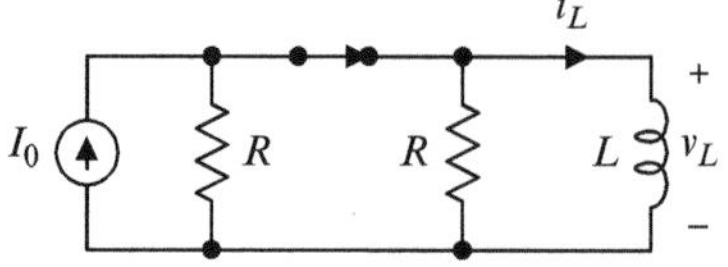

Fig. 9.15 See Exercise 9.9

Obtain expressions for the current i_L and the voltage v_L.

Example 9.7. Refer to Fig. 9.16. The source voltage $v_S(t)$ equals zero for times not shown. Obtain an expression for the voltage across the inductor for $t > 0$.

Solution: One approach is to obtain an expression for the current $i(t)$ and subsequently use

$$v_L(t) = v_S(t) - Ri(t).$$

The time constant is

$$\tau = \frac{L}{R} = 1\,\mu\text{s}.$$

For $t < 0$, the source appears as a short circuit, so $i(0^+) = i(0^-) = 0$. For $0 < t \leq t_1$, the current increases toward

$$i_1(\infty) = \frac{V_1}{R} = 10\,\text{mA},$$

where the subscript on i_1 indicates that the relation holds for the first segment of the response. From the relations above and (9.23), we have

$$i_1(t) = i_1(\infty) - [i_1(\infty) - i_1(0^+)]\exp\left(-\frac{t}{\tau}\right)$$
$$= \frac{V_1}{R} - \left[\frac{V_1}{R} - 0\right]\exp\left(-\frac{t}{\tau}\right)$$
$$= \frac{V_1}{R}\left[1 - \exp\left(-\frac{t}{\tau}\right)\right],$$
$$\tau = \frac{L}{R},\quad 0 < t \leq t_1.$$

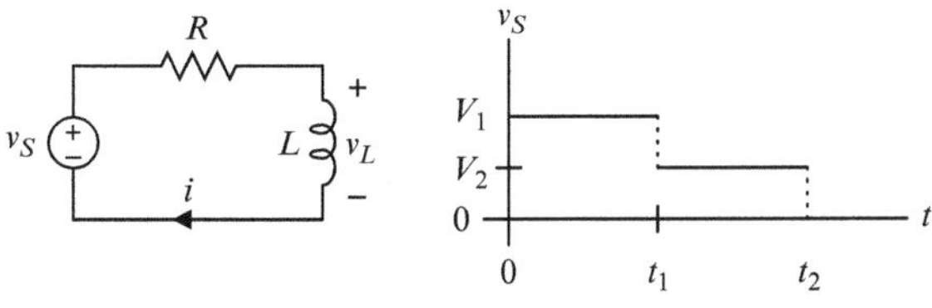

$R = 1\text{k}\Omega$, $L = 1\text{mH}$, $V_1 = 10\text{V}$, $V_2 = 5\text{V}$, $t_1 = 2\mu\text{s}$, $t_2 = 4\mu\text{s}$

Fig. 9.16 See Example 9.7

The final value of the first segment becomes the initial value for the second segment, because the current through an inductor is continuous. Thus, if we reset the time origin to t_1,

$$\begin{aligned}i_2(0^+) = i_1(t_1) &= \frac{V_1}{R}\left[1 - \exp\left(-\frac{t_1}{\tau}\right)\right]\\ &= (10\,\text{mA})[1 - \exp(-2)]\\ &\cong 8.65\,\text{mA}\end{aligned}$$

and

$$i_2(\infty) = \frac{V_2}{R} = 5\,\text{mA}.$$

We use these values in (9.23) and translate the resulting segment to t_1. Thus

$$\begin{aligned}i_2(t) &= i_2(\infty) - [i_2(\infty) - i_2(0^+)]\exp\left(-\frac{t-t_1}{\tau}\right)\\ &= \frac{V_2}{R} - \left[\frac{V_2}{R} - i_1(t_1)\right]\exp\left(-\frac{t-t_1}{\tau}\right),\\ &\quad t_1 < t \le t_2.\end{aligned}$$

The final value of the second segment becomes the initial value for the third segment, because the current through an inductor is continuous. Thus,

$$\begin{aligned}i_3(0^+) &= i_2(t_2)\\ &= \frac{V_2}{R} - \left[\frac{V_2}{R} - i_1(t_1)\right]\exp\left(-\frac{t_2-t_1}{\tau}\right)\\ &\cong 5\text{mA} - [5\text{mA} - 8.65\text{mA}]\exp(-2)\\ &\cong 4.73\text{mA}\end{aligned}$$

and

$$i_3(\infty) = 0.$$

We use these values in (9.23) and translate the resulting segment to t_2. Thus

$$\begin{aligned}i_3(t) &= i_3(\infty) - [i_3(\infty) - i_3(0^+)]\exp\left(-\frac{t-t_2}{\tau}\right)\\ &= 0 - [0 - i_2(t_2)]\exp\left(-\frac{t-t_1}{\tau}\right)\\ &= i_2(t_2)\exp\left(-\frac{t-t_2}{\tau}\right), \quad t > t_2.\end{aligned}$$

Thus

$$i(t) = \left\{\begin{array}{ll} 0, & t<0,\\ i_1(t), & 0<t\le t_1,\\ i_2(t), & t_1<t\le t_2,\\ i_3(t), & t>t_2.\end{array}\right\}$$

$$\Rightarrow v_L(t) = v_S(t) - Ri(t)$$

$$= \left\{\begin{array}{ll} 0, & t<0,\\ V_1 - Ri_1(t), & 0<t\le t_1,\\ V_2 - Ri_2(t), & t_1<t\le t_2,\\ -Ri_3(t), & t>t_2,\end{array}\right\}$$

$$= \left\{\begin{array}{ll} 0, & t<0,\\ V_1\exp\left(-\frac{t}{\tau}\right), & 0<t\le t_1,\\ [V_2 - Ri_1(t_1)]\exp\left(-\frac{t-t_1}{\tau}\right), & t_1<t\le t_2,\\ -Ri_2(t_2)\exp\left(-\frac{t-t_2}{\tau}\right), & t>t_2.\end{array}\right\}$$

Figure 9.17 shows a graph of the voltage $v_L(t)$ versus time.

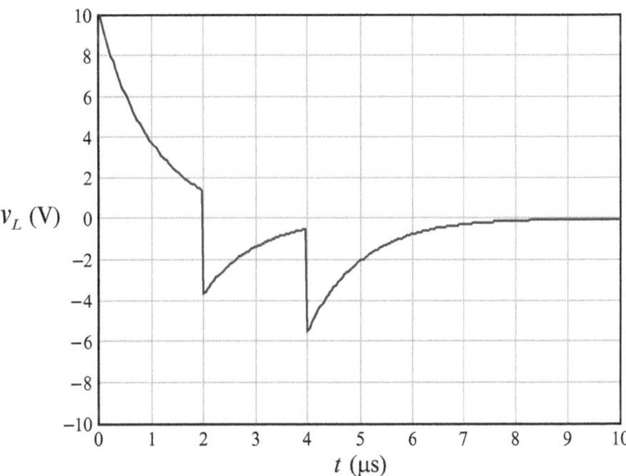

Fig. 9.17 Load voltage in the circuit in Fig. 9.16. See Example 9.7

9.7 Inductors in Series and Parallel

In this section, we assume that coils connected in series or in parallel are connected in such a way that there are no magnetic flux linkages between the coils. Section 9.12 treats magnetically coupled coils.

Refer to Fig. 9.18. By Kirchhoff's voltage law,

$$v = L_1 \frac{di}{dt} + L_2 \frac{di}{dt} + \cdots + L_N \frac{di}{dt}$$
$$= (L_1 + L_2 + \cdots + L_N) \frac{di}{dt}. \tag{9.26}$$

Therefore a series connection of inductors is equivalent (at the terminals) to one inductor having inductance L_{eq} given by

$$L_{eq} = L_1 + L_2 + \cdots + L_N. \tag{9.27}$$

Refer to Fig. 9.19. By Kirchhoff's current law,

$$i = \frac{1}{L_1}\int_{-\infty}^{t} v(t)\,dt + \frac{1}{L_2}\int_{-\infty}^{t} v(t)\,dt + \cdots + \frac{1}{L_N}\int_{-\infty}^{t} v(t)\,dt$$
$$= \left(\frac{1}{L_1} + \frac{1}{L_2} + \cdots + \frac{1}{L_N}\right)\int_{-\infty}^{t} v(t)\,dt. \tag{9.28}$$

Therefore a parallel connection of inductors is equivalent (at the terminals) to one inductor having inductance L_{eq} given by

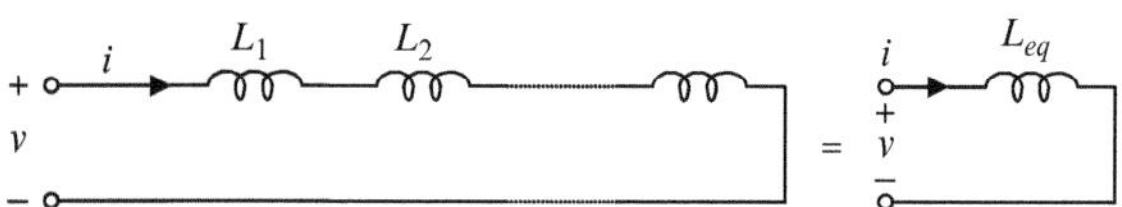

Fig. 9.18 Inductors in series

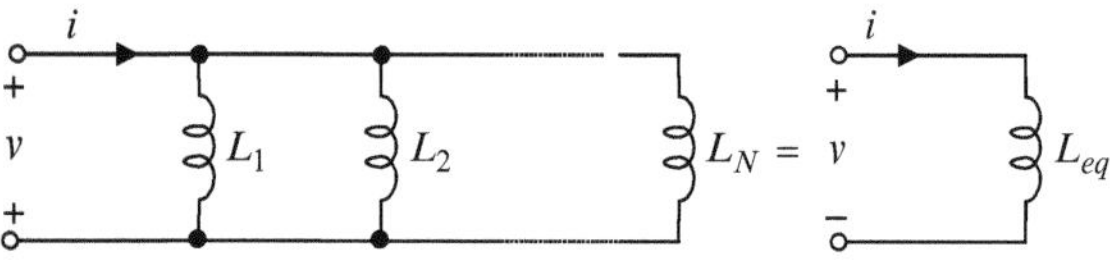

Fig. 9.19 Inductors in parallel

$$\frac{1}{L_{eq}} = \frac{1}{L_1} + \frac{1}{L_2} + \cdots + \frac{1}{L_N}. \tag{9.29}$$

Equations (9.27) and (9.29) assume that none of the flux produced by any one of the inductors links any of the other inductors; that is, that there is no mutual inductance between the coils. This assumption is invalid, as are (9.27) and (9.29), if the coils comprising the inductors are close enough together that magnetic coupling is significant.

Equation (9.29) ignores the winding resistances of the individual inductors, which might not be a bad approximation at low frequencies, for inductors having small dc resistances.

Equations (9.27) and (9.29) are rarely useful in practice. It might happen occasionally that certain approximations lead to a circuit model in which two inductors appear in series or parallel, in which case (9.27) or (9.29) allow further simplification. But (9.27) and (9.29) rarely arise in design because one would never specify two or more inductors where one would do. An exception might be where a very accurate and precise inductance is needed, in which case one might use a fixed inductance in series with a smaller, adjustable inductance.

Example 9.8. Obtain an expression for the equivalent inductance at the terminals *a*-*b* of the circuit shown in Fig. 9.20. Assume the magnetic fields produced by the inductors do not overlap.

Solution: From (9.29), the equivalent inductance of the parallel inductors L_2, L_3 is given by

$$\frac{1}{L_{23}} = \frac{1}{L_2} + \frac{1}{L_3} \Rightarrow L_{23} = \frac{L_2 L_3}{L_2 + L_3}.$$

From (9.27), the equivalent inductance of the circuit at the specified terminals is given by

$$L_{eq} = L_1 + L_{23} = L_1 + \frac{L_2 L_3}{L_2 + L_3}.$$

Exercise 9.10. Obtain an expression for the equivalent inductance at the terminals *a*–*b* of

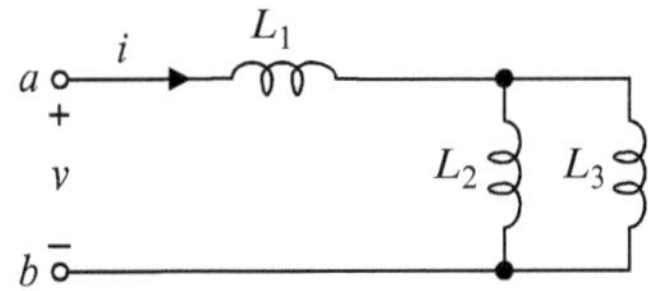

Fig. 9.20 See Example 9.8

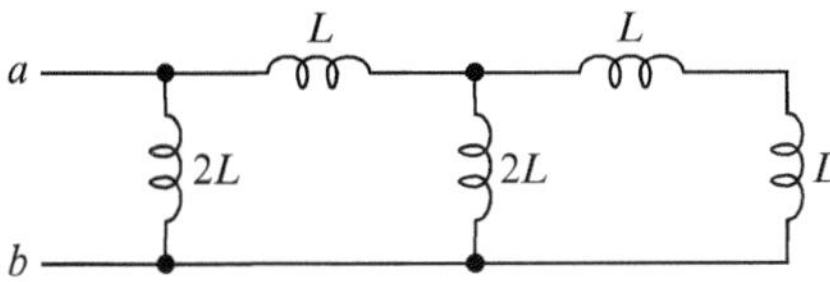

Fig. 9.21 See Exercise 9.10

the circuit shown in Fig. 9.21. Assume the magnetic fields produced by the inductors do not overlap.

9.8 Energy Storage and Power dissipation in an Inductor

The instantaneous power dissipated in an inductor having inductance L is given by

$$p = v\,i = L\frac{di}{dt}\,i = \frac{dw}{dt}, \tag{9.30}$$

where $i(t)$ is the instantaneous current through the inductor and $w(t)$ is the instantaneous energy stored in the magnetic field produced by the current. We have from (9.30) that $dw = L\,i\,di$, which implies

$$w(t) = L\int i\;di = \frac{1}{2}\,Li^2(t) + B$$

where B is a constant of integration. From (9.4), magnetic flux is proportional to current, so if $i(t) = 0$, the flux equals zero, the magnetic field strength equals zero, and the stored energy $w(t)$ equals zero. Thus $B = 0$ and the instantaneous energy stored in the magnetic field produced by an inductor is given by

$$w(t) = \frac{1}{2}L\,i^2(t), \tag{9.31}$$

where $i(t)$ is the instantaneous current through the inductor. For economy, *the instantaneous energy stored in the magnetic field produced by an inductor* usually is referred to as *the energy stored by (or in) the inductor*.

The energy stored in the magnetic field produced by a current makes the current resistant to change, because it takes time to transfer energy from the current to the magnetic field or vice-versa. In other words, the magnetic field created by a current gives the current a kind of inertia that resists changes in the current. An inductor enhances this effect by concentrating and intensifying the magnetic field produced by a current through the inductor.

Exercise 9.11. The current in an inductor having inductance $L = 1$ mH is given by $i(t) = I\,\cos(2\,\pi f\,t)$, with $I = 5$ mA and $f = 1$ kHz. Express the instantaneous energy stored in the inductor as a function of time.

The average power dissipated in an inductor over a *finite* time interval $t_0 \le t < t_0 + T$ is given by

$$\begin{aligned} P_T &= \frac{1}{T}\,[\varepsilon(t_0 + T) - \varepsilon(t_0)] \\ &= \frac{1}{T}\,\left[\frac{1}{2}\,L\,i^2(t_0 + T) - \frac{1}{2}\,L\,i^2(t_0)\right]. \end{aligned} \tag{9.32}$$

We assume the stored energy (or, equivalently, the current through the inductor) is finite always. Thus the terms in the brackets on the right side of (9.32) are finite. It follows that the average power dissipated in an inductor equals zero:

$$P = \lim_{T\to\infty}\,(P_T) = 0.$$

The *instantaneous* power dissipated in an inductor is not zero, nor is the peak power. An inductor can absorb energy at some times and return energy to a circuit at other times. None of the energy absorbed (by an *ideal* inductor) is lost to the circuit, but can be returned on demand. This is in contrast to a resistor, which converts electrical energy to heat energy that is then lost forever to the circuit. On an instantaneous basis, an inductor behaves sometimes like a sink and other times like a source, but on average (over a sufficiently long time), is neither.

Example 9.9. Find the average powers dissipated in the resistor and the source in the circuit shown in Fig. 9.22, where $R = 2\,\mathrm{k}\Omega$ and

$$i(t) = I_0 \cos(2\pi f t), \quad I_0 = 10\,\mathrm{mA}.$$

Solution: Because the average power dissipated in the inductor equals zero, and because power is conserved, the sum of the powers dissipated in the resistor and the source equals zero. The power dissipated in the resistor is

$$P_R = {i_{rms}}^2 R = \frac{{I_0}^2 R}{2} = 100\,\mathrm{mW}$$

and the power dissipated in the source is

$$P_i = -P_R = -100\,\mathrm{mW}.$$

Exercise 9.12. Refer to Fig. 9.23. What is the voltage v_L across the inductor a very long time after the switch is moved from a to b? What is the current i_L through the inductor at that time?

Exercise 9.13. Refer to Fig. 9.24. What is the voltage v_L across the inductor a very long time after the switch is moved from a to b? What is the current i_L through the inductor at that time?

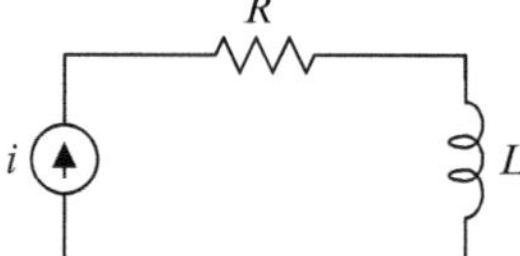

Fig. 9.22 See Example 9.9

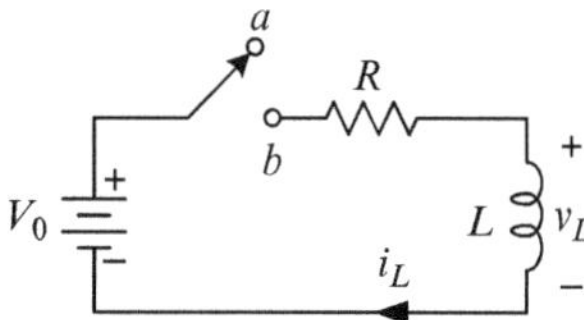

Fig. 9.23 See Exercise 9.12

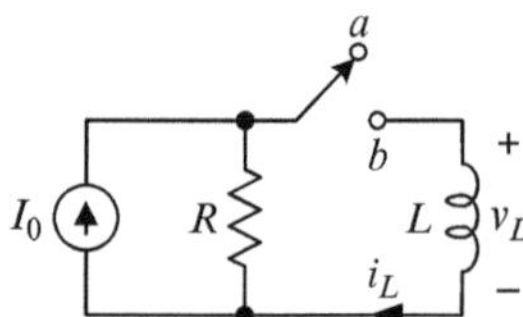

Fig. 9.24 See Exercise 9.13

9.9 Parasitic Self-Inductance

Self-inductance is present wherever current exists, whether we like it or not. **Parasitic self-inductance** refers to unintentional and generally undesirable inductance that is present in components other than inductors, such as conductors, resistors, and capacitors. In the simplest cases, parasitic self-inductance is due largely to current in the leads of such components and in conductors (wires) that interconnect devices.

Numerous formulas have been presented for the self inductance of a straight, isolated, non-magnetic wire.[7] One of these is

$$L = \frac{\mu_0 l}{2\pi}\left[\ln\left(\frac{4l}{d}\right) - 1\right], \quad \mu_0 = 4\pi \times 10^{-9}\,\mathrm{H\,cm^{-1}}, \tag{9.33}$$

where the length l of the wire is much larger than the diameter d. Figure 9.25 shows a graph of inductance in nH versus wire length in cm for $1\ \mathrm{cm} \leq l \leq 10\,\mathrm{cm}$, calculated from (9.33). Also shown (dotted line) is a linear approximation to (9.33) for that range of wire length. The slope of the linear approximation is approximately 12 $\mathrm{nH\,cm^{-1}}$, which implies that the self-inductance of an isolated straight conductor is about 12 $\mathrm{nH\,cm^{-1}}$. Inductances calculated using (9.33) might be a bit high, as other methods yield lower values. But in design, it usually is best to be conservative.

[7] Grover, F.W., *Inductance Calculations*, Dover, New York, 2004.

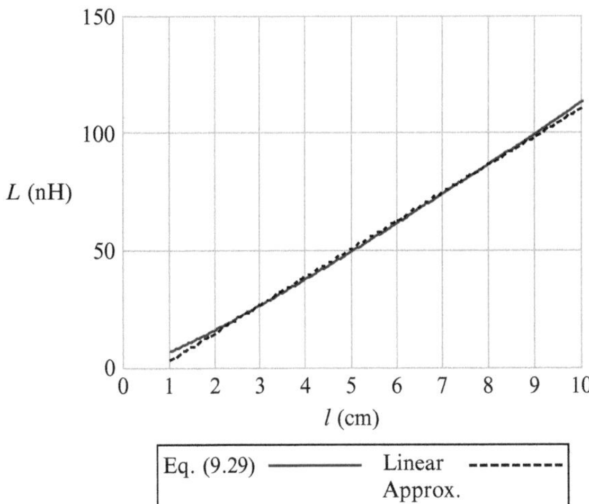

Fig. 9.25 Self-inductance of a straight wire as a function of wire length

In many cases, lumped-constant models fail before self-inductance of conductors becomes significant. Lumped-constant models are valid only for circuits whose dimensions are much larger than the wavelengths of currents and voltages involved. The wavelength λ of a sinusoidal current or voltage (in air) having frequency f is given by

$$\lambda = \frac{c}{f}, \tag{9.34}$$

where $c \cong 3 \times 10^8\,\mathrm{m\,s^{-1}}$ is the speed of light. For example, the wavelength of a 1 GHz sinusoid is 30 cm, so ordinary lumped-constant models can be applied to centimeter-size circuits operating at frequencies of 1 GHz or below, but probably cannot be applied to such circuits at frequencies of 10 GHz (wavelengths of 3 cm) or to circuits in which the lengths of conductors approach 30 cm. For example, although the graph in Fig. 9.25 indicates that a 5-cm conductor has a self-inductance of 50 nH, that fact is meaningless at frequencies higher than 1 GHz or so, because a wavelength of 5 cm corresponds to a frequency of 6 GHz.

Where a lumped-constant model is valid, self-inductances are significant if the associated currents and voltages are significant relative to currents and voltages of interest. For example, consider a circuit whose dimensions are on the order of 1 cm and in which currents and voltages of interest are on the order of 1 mA and 1 mV, respectively. From the development above, the self-inductance of a 1-cm wire is on the order of $(12\,\mathrm{nH\,cm^{-1}}) \times (1\,\mathrm{cm}) = 12\,\mathrm{nH}$. The associated self-induced voltage for a sinusoidal current through the wire is given by

$$\begin{aligned} v &= L\frac{di}{dt} = L\frac{d}{dt}[I_0 \cos(2\pi f t)] \\ &= -2\pi f L I_0 \sin(2\pi f t). \end{aligned} \tag{9.35}$$

The peak amplitude of the self-induced voltage equals 1 mV (is comparable to voltages of interest) if

$$2\pi f L I_0 = 1\,\mathrm{mV}, \tag{9.36}$$

which implies

$$f = \frac{1\,\mathrm{mV}}{(2\pi)\,(12\,\mathrm{nH})\,(1\,\mathrm{mA})} \cong 13.3\,\mathrm{MHz}. \tag{9.37}$$

The self-induced voltage along a 1-cm straight wire carrying a 1-mA sinusoidal current is less than 1 mV until the frequency of the current approaches 13 MHz. At that frequency and for that current, a lumped-constant (R, L, C) model remains valid for a centimeter-size circuit, because the wavelength of a 13 MHz current or voltage is

$$\lambda = \frac{c}{f} = \frac{3 \times 10^8}{13 \times 10^6} \cong 23\,\mathrm{m}, \tag{9.38}$$

which is very much larger than 1 cm. Thus in this case, self-inductance is both meaningful and significant.

Exercise 9.14. A sinusoidal current having frequency f and peak amplitude I passes through an axial-lead resistor having resistance R and series self-inductance L. Assuming a lumped-constant model is applicable, obtain an expression for the frequency at which the rms voltage drop across the self-inductance equals 1% of the rms voltage drop across the resistance.

Inductors are used in basically three ways:

- In series with a load, to suppress ripple or high-frequency noise;

- As components of circuit models for physical devices;. e.g., electric motors, generators, and transformers;
- In conjunction with capacitors, to construct frequency-selective circuits (filters).

We are as yet unprepared to treat circuits containing both inductors and capacitors. But we can describe how an inductor alone can be used to reduce ripple (or undesirable high-frequency noise) accompanying a desirable current or voltage.

9.10 Reducing Ripple

Recall that in some applications where the dc component V_{dc} of a voltage is the desirable component and the ac component v_{ac} is undesirable, the **ripple factor**, defined as

$$\gamma = \frac{V_{ac\,rms}}{|V_{dc}|} \tag{9.39}$$

is a useful figure of merit. Like capacitors, inductors can be used to reduce ac ripple in a resistive load. When used for this purpose, an inductor is called a **choke**. Whereas a bypass capacitor is connected in parallel with a load, a choke is connected in series with a load, as shown in Fig. 9.26, where the source v_S consists of a desirable dc component $V_{S\,dc}$ and an undesirable ac component (ripple) $v_{S\,ac}(t)$. If the choke is absent (is replaced by a conductor), the load voltage v_R equals the source voltage v_S and the ripple factor is given by

$$\gamma_0 = \frac{V_{S\,ac\,rms}}{|V_{S\,dc}|}. \tag{9.40}$$

Below, we obtain an expression for the ripple factor with the inductor present. We assume the ac component of the source voltage is sinusoidal. The required trigonometric and algebraic manipulations are simplest if we assume the initial phase of the ac component of the current i equals zero, such that

$$i = I_{dc} + I_{ac}\cos(\omega_0 t). \tag{9.41}$$

It follows that the load voltage is given by

$$v_R = RI_{dc} + RI_{ac}\cos(\omega_0 t) = V_{R\,dc} + V_{R\,ac}\cos(\omega_0 t)$$

and that the source voltage is given by an expression of the form

$$v_S = V_{S\,dc} + V_{S\,ac}\cos(\omega_0 t + \theta). \tag{9.42}$$

Superposition allows us to find the responses to the dc and ac component of the source individually. Because an inductor acts as a short circuit for dc, the dc component of the load voltage (the voltage across the resistor) equals the dc component of the source voltage. Thus

$$V_{R\,dc} = V_{S\,dc}. \tag{9.43}$$

The ac components of the voltages involved satisfy Kirchhoff's voltage law. Thus

$$v_{L\,ac} + v_{R\,ac} = v_{S\,ac} \Rightarrow L\frac{di_{ac}}{dt} + Ri_{ac} = V_{S\,ac}\cos(\omega_0 t + \theta).$$

From (9.41), $i_{ac} = I_{ac}\cos(\omega_0 t)$. Thus

$$\begin{aligned} &-L\omega_0 I_{ac}\sin(\omega_0 t) + RI_{ac}\cos(\omega_0 t) \\ &\quad = V_{S\,ac}\cos(\omega_0 t + \theta) \\ &\quad = V_{S\,ac}[\cos(\omega_0 t)\cos(\theta) - \sin(\omega_0 t)\sin(\theta)], \end{aligned}$$

which holds for all time only if the coefficients of $\cos(\omega_0 t)$ and $\sin(\omega_0 t)$ on the left side equal the corresponding coefficients on the right side. Thus we obtain two equations

$$RI_{ac} = V_{S\,ac}\cos(\theta), \quad L\omega_0 I_{ac} = V_{S\,ac}\sin(\theta),$$

which yield

$$(RI_{ac})^2 + (L\omega_0 I_{ac})^2 = (V_{S\,ac})^2. \tag{9.44}$$

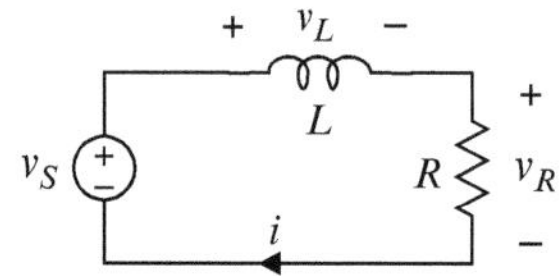

Fig. 9.26 Use of an inductive *choke* to reduce ripple

The ac component of the load voltage v_R is given by

$$V_{R\,ac} = RI_{ac},$$

so (9.44) becomes

$$(V_{R\,ac})^2 + \left(\frac{L\omega_0}{R} V_{R\,ac}\right)^2 = (V_{S\,ac})^2,$$

which reduces to

$$|V_{R\,ac}| = \frac{|V_{S\,ac}|}{\sqrt{1+Q^2}},$$

where we have defined

$$Q = \frac{\omega_0 L}{R}. \tag{9.45}$$

The ripple factor is given by

$$\gamma = \left|\frac{V_{Rac}}{V_{dc}\sqrt{2}}\right| = \frac{1}{|V_{dc}|\sqrt{2}} \frac{|V_{ac}|}{\sqrt{1+Q^2}} = \frac{\gamma_0}{\sqrt{1+Q^2}}, \tag{9.46}$$

where, from (9.40)

$$\gamma_0 = \left|\frac{V_{ac}}{V_{dc}\sqrt{2}}\right|. \tag{9.47}$$

Conversely, the inductance L required to attain a specified ripple factor γ is found as follows: From (9.46),

$$Q = \frac{\omega_0 L}{R} = \sqrt{\left(\frac{\gamma_0}{\gamma}\right)^2 - 1}, \tag{9.48}$$

whence

$$L = \frac{R}{\omega_0}\sqrt{\left(\frac{\gamma_0}{\gamma}\right)^2 - 1}; \quad \gamma_0 = \left|\frac{V_{ac}}{V_{dc}\sqrt{2}}\right|. \tag{9.49}$$

In many applications, $\gamma_0/\gamma \gg 1$, in which case

$$L \cong \frac{R\gamma_0}{\omega_0 \gamma}, \quad \gamma_0 \gg \gamma. \tag{9.50}$$

Example 9.10. In Fig. 9.26, $R = 2\,\Omega$ and $v_S = V_{dc} + V_{ac}\cos(\omega_0 t + \theta)$ with $V_{dc} = 100\,\text{V}$, $V_{ac} = 25\sqrt{2}\,\text{V}$, and $f_0 = 120\,\text{Hz}$. The required ripple factor is 0.1% (0.001). Specify the inductance L.

Solution: From (9.49)

$$\gamma_0 = \left|\frac{V_{ac}}{V_{dc}\sqrt{2}}\right| = \frac{25\sqrt{2}\,\text{V}}{100\sqrt{2}\,\text{V}} = 0.25,$$

$$L = \frac{R}{\omega_0}\sqrt{\left(\frac{\gamma_0}{\gamma}\right)^2 - 1}$$

$$= \left(\frac{2\,\Omega}{240\pi\,\text{s}^{-1}}\right)\sqrt{\left(\frac{0.25}{0.001}\right)^2 - 1} = 663\,\text{mH}.$$

We could have simplified the calculation somewhat by using (9.50), because

$$\frac{\gamma_0}{\gamma} = \frac{0.25}{0.001} = 250 \gg 1.$$

Thus

$$L \cong \frac{R\gamma_0}{\omega_0\gamma} = \frac{(2\,\Omega)(0.25)}{(240\pi\,\text{s}^{-1})(0.001)} = 663\,\text{mH}.$$

as before.

Exercise 9.15. The dc output of a certain power supply is 12 V. For a $4\,\Omega$ resistive load, the ripple factor of the load voltage is $\gamma = 10^{-4}$. (a) What is the rms amplitude of the ac component of the load voltage? (b) Assuming the ripple is a 120 Hz sinusoid, what is the ripple factor for the load voltage if a 200 mH inductor is connected in series with the load?

Another approach is to require that the ac component of the load voltage be a small fraction of the ac component of the source voltage, which is equivalent to requiring that the ac component of the inductor voltage be much larger than the ac component of the load voltage, which requires

$$\left|L\frac{di_{ac}}{dt}\right| \gg |Ri_{ac}|. \tag{9.51}$$

Using (9.41) in (9.51) gives

$$|\omega_0 LI_{ac}| \gg |RI_{ac}| \Rightarrow \frac{\omega_0 L}{R} = Q \gg 1, \tag{9.52}$$

which implies that requiring a large fraction of the total ac voltage to appear across the inductor is equivalent to requiring that

$$Q = \frac{\omega_0 L}{R} \gg 1. \tag{9.53}$$

Also, from (9.46), if Q is large,

$$\frac{\gamma_0}{\gamma} = \sqrt{1+Q^2} \cong Q;\ Q \gg 1,$$

or

$$\gamma = \frac{\gamma_0}{Q};\ Q \gg 1. \tag{9.54}$$

which yields

$$L = \frac{R\,\gamma_0}{\omega_0\,\gamma} \tag{9.55}$$

as before. Equations (9.50) and (9.55) are good approximations to (9.49) (less than about 1% error) if $\gamma_0/\gamma > 7$, which is usually true in applications.

Exercise 9.16. The voltage across a 2 Ω resistive load is given by

$$v = V_{dc} + V_{ac}\cos(2\pi f t),$$

with $f = 10\,\text{kHz}$, $V_{dc} = 5\,\text{V}$, and $V_{ac} = 500\,\text{mV}$. Specify a series choke such that the load ripple is reduced by a factor of 1000.

Using a series choke to reduce the ripple in a load voltage or current is not always an alternative to bypassing the load with a capacitor. A bypass capacitor reduces the voltage ripple across both the capacitor and the load, whereas the *total* ripple voltage across the series connection of a choke and load is not reduced by the choke. For example, suppose we wish to reduce the ac component of the voltage between two nodes a, b connected by a resistor. Connecting a capacitor in parallel with the resistor will reduce the ac component of v_{ab}, but connecting an inductor in series with the resistor will not.

Keep in mind that the analysis above assumes *sinusoidal* ripple, as given by (9.42). If the ac component of the source v_S in Fig. 9.26 is not sinusoidal, then the ac components of the source voltage and load voltage have different waveforms.[8] In that case, analysis of the load ripple is much more complicated and beyond the scope of this book.

9.11 Inductive Kick

Current through an inductor cannot change instantaneously. Because voltage across the terminals of an inductor is proportional to the time derivative of current through the inductor, the terminal voltage becomes as large as necessary to prevent an instantaneous change in current. For example, consider the situation illustrated by Fig. 9.27, where a switch attempts to interrupt the current through an inductor. When the switch is suddenly opened, a very large voltage can appear across the switch, leading to arcing and (sooner or later) failure of the switch. Such a high voltage in a switched inductive circuit is called **inductive kick.**[9] Inductive kick is a useful phenomenon in certain applications, such as automobile ignition systems, where the large voltage applied to the spark

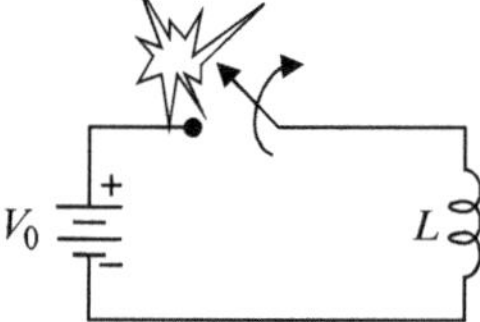

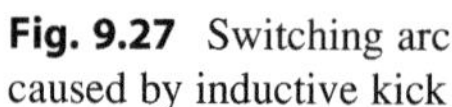

Fig. 9.27 Switching arc caused by inductive kick

[8]For example, if the current through an inductor has a triangular waveform, then the voltage across the inductor has a rectangular (piecewise-constant) waveform.

[9]This terminology is borrowed from Paul Horowitz and Winfield Hill, *The Art of Electronics*, (2nd Ed.), Cambridge University Press, New York, 1980.

plugs is obtained by interrupting the current through the primary of a transformer,[10] and in some high-voltage power supplies, where high voltages are produced in essentially the same manner. But in many cases, inductive kick is undesirable and potentially destructive.

Special measures often are necessary to protect switching apparatus and other components when switching large inductive loads such as transformers and electric motors. Such measures take various forms, ranging from none for small inductive loads carrying small currents to elaborate measures for large inductive loads carrying large currents. For example, if the inductance in Fig. 9.27 is very small, perhaps a sufficiently robust switch is all that is necessary. But if the inductance is large, additional provisions are necessary. One such provision for a dc motor is to shunt the switch with a diode, as illustrated by Fig. 9.28. The diode allows energy stored in the inductor to dissipate in the forward resistance of the diode and in the resistance of the motor windings. This method does not work for an ac motor, because in that case, the diode would shunt every other half cycle and useable power to the motor would be greatly reduced.[11]

Figure 9.29 shows a protection method sometimes employed for ac motors. The resistor is large enough that only a small fraction of the total current is diverted

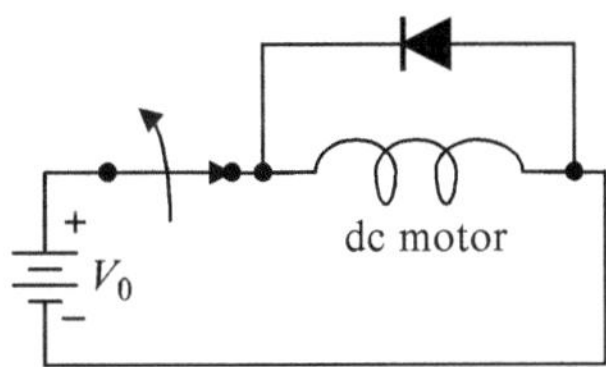

Fig. 9.28 A diode as an arc suppressor

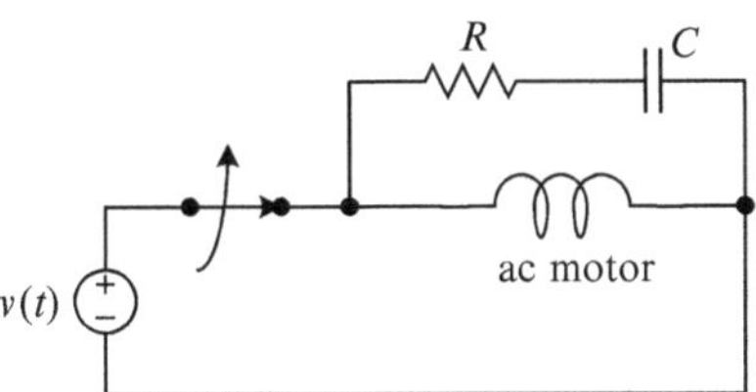

Fig. 9.29 An *RC* (snubber) circuit as an arc suppressor

[10]The ignition *coil* in an automobile is actually a transformer.

[11]In fact, the motor might not even run, unless it is a universal or ac/dc motor.

from the motor while it is running. But when the switch is opened, the current through the inductor is diverted to the *RC* branch, and energy is stored in the capacitor for subsequent dissipation in the resistor and the resistance of the motor windings. Thus the electrical energy stored in the magnetic field of the motor and the mechanical energy stored in the load, which otherwise would be dissipated in an arc at the switch, is dissipated instead as heat by the resistor. For small to medium (1/10–5 hp) motors, the resistance R is typically about 20–200 Ω and the capacitance C is typically about 0.01–0.1 μF. The *RC* protective circuit shown in Fig. 9.29 is called a *snubber*. More complex snubber circuits can more effectively convert the mechanical energy stored in a rotating machine to heat dissipation in a resistor, allowing faster decreases in motor speed. Analysis of such circuits is best done using methods treated in subsequent chapters.

9.12 Magnetically Coupled Coils and Mutual Inductance

Faraday's law of induction states that a change in magnetic flux linking a conductor induces a voltage in the conductor, regardless of the source of the flux. A magnetic field produced by a changing current in one conductor can induce a voltage in a neighboring conductor. That is, if changing flux arising from changing current in one conductor links a second conductor, then voltage will be induced in the second conductor. The polarity of the induced voltage is such that any current produced by the voltage will attempt to keep the flux from changing. Two conductors (or coils, or circuits) linked in this way are said to be **magnetically coupled** and induction of a voltage in one conductor by changing current in another is called **mutual inductance**.

We begin by obtaining a circuit model for magnetically coupled coils. We refer to Fig. 9.30, where the resistances R_1, R_2 represent the resistance of the wire forming the coils. The currents i_1, i_2 produce magnetic flux, where:

ϕ_{11} is the flux produced by the current i_1 that links only coil 1

ϕ_{12} is the flux produced by the current i_1 that links both coils

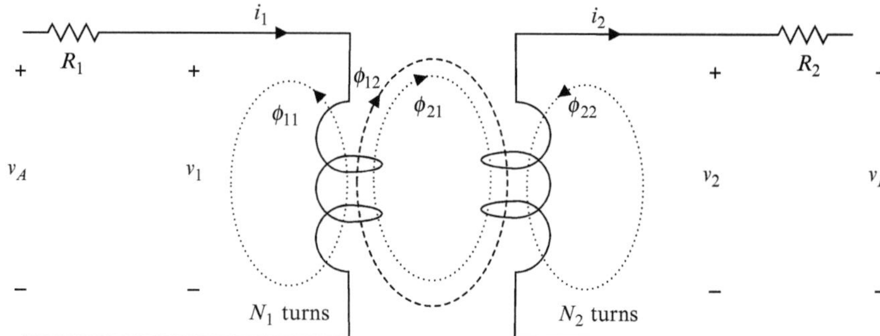

Fig. 9.30 Magnetically coupled coils

ϕ_{22} is the flux produced by the current i_2 that links only coil 2, and
ϕ_{21} is the flux produced by the current i_2 that links both coils.

The total flux ϕ_1 produced by the current through coil 1 and the total flux ϕ_2 produced by the current through coil 2 are given by

$$\phi_1 = \phi_{11} + \phi_{12}, \quad \phi_2 = \phi_{22} + \phi_{21}. \tag{9.56}$$

Actually, not all of the flux produced by the current through a coil links every turn of the coil. The fluxes defined above are *effective* fluxes that account for that fact. For example, the flux ϕ_{11} is an effective flux which, if linked with all the turns of coil 1, gives the correct total number of flux linkages.

We assume the coils are wound such that the total flux produced by a current is proportional to the current.[12] Thus the total flux produced by each coil is proportional to the current through the coil. A fraction k of the flux produced by coil 1 links with coil 2 and (by symmetry) the same fraction the flux produced by coil 2 links with coil 1. We have

$$\phi_1 = \alpha N_1 i_1, \quad \phi_{12} = k\phi_1 = k\alpha N_1 i_1 \tag{9.57}$$

and

$$\phi_2 = \alpha N_2 i_2, \quad \phi_{21} = k\phi_2 = k\alpha N_2 i_2, \tag{9.58}$$

where α depends upon the coil geometry and core material, but not on the current or flux.

[12]This is equivalent to assuming that the permeability of the core material is constant and independent of position.

From (9.57) and (9.58),

$$k = \frac{\phi_{12}}{\phi_1} = \frac{\phi_{21}}{\phi_2}. \tag{9.59}$$

The parameter k is the **coupling coefficient** for the coils because the fractions ϕ_{12}/ϕ_1 and ϕ_{21}/ϕ_2 indicate the degree of coupling between two coils. That is, how much of the flux produced by one coil links the other coil. The maximum possible value of each fraction is unity.

We are dealing with currents through and voltages across coils, so we seek an inductive model relating the currents and voltages in Fig. 9.30. With reference to (9.4), the **self-inductances** L_1, L_2 of the coils are given by

$$L_1 = \frac{N_1\phi_1}{i_1} \Rightarrow \phi_1 = \frac{L_1 i_1}{N_1}, \quad L_2 = \frac{N_2\phi_2}{i_2} \Rightarrow \phi_2 = \frac{L_2 i_2}{N_2}. \tag{9.60}$$

By analogy, we define the **mutual inductances** M_{21}, M_{12} as

$$\begin{aligned} M_{21} &= \frac{N_1\phi_{21}}{i_2} \Rightarrow \phi_{21} = \frac{M_{21} i_2}{N_1}, \\ M_{12} &= \frac{N_2\phi_{12}}{i_1} \Rightarrow \phi_{12} = \frac{M_{12} i_1}{N_2}. \end{aligned} \tag{9.61}$$

Using the right sides of (9.57) and (9.58) in (9.61) yields

$$\begin{aligned} &M_{21} = k\alpha N_1 N_2, \; M_{12} = k\alpha N_1 N_2 \\ &\Rightarrow M_{12} = M_{21} = M. \end{aligned} \tag{9.62}$$

From (9.60), (9.61), and (9.62), we obtain

$$L_2 L_2 = \frac{N_2\phi_2}{i_2}\frac{N_1\phi_1}{i_1},$$

$$M_{12}M_{21} = M^2 = \frac{N_2\phi_{12}}{i_1}\frac{N_1\phi_{21}}{i_2} \Rightarrow \frac{M^2}{L_2 L_2} = \frac{\phi_{12}\phi_{21}}{\phi_2\phi_1}.$$

It follows from this result and (9.59) that the **coupling coefficient** is given by

$$k = \sqrt{\frac{M^2}{L_2 L_2}} = \frac{|M|}{\sqrt{L_2 L_2}}. \quad (9.63)$$

Conventionally, as used in circuit equations, mutual inductance is positive, so the absolute-value sign in (9.63) is unnecessary. This remark is explained more fully in the sequel.

Using (9.62), the right side of (9.61) can be written

$$\phi_{21} = \frac{Mi_2}{N_1}, \; \phi_{12} = \frac{Mi_1}{N_2}. \quad (9.64)$$

The total flux linkages for the two coils are

$$\lambda_1 = N_1(\phi_1 \pm \phi_{21}), \; \lambda_2 = N_2(\phi_2 \pm \phi_{12}). \quad (9.65)$$

where the positive sign applies if the fluxes produced by each current are in the same direction (as shown in Fig. 9.30) and the negative sign applies otherwise. Using (9.60) and (9.64) in (9.65) gives

$$\lambda_1 = N_1\left(\frac{L_1 i_1}{N_1} \pm \frac{Mi_2}{N_1}\right) = L_1 i_1 \pm Mi_2,$$
$$\lambda_2 = N_2\left(\frac{L_2 i_2}{N_2} \pm \frac{Mi_1}{N_2}\right) = L_2 i_2 \pm Mi_1, \quad (9.66)$$

where $M \geq 0$. By Faraday's law, the voltages across the coils are given by

$$v_1 = \frac{d\lambda_1}{dt} = L_1\frac{di_1}{dt} \pm M\frac{di_2}{dt},$$
$$v_2 = \frac{d\lambda_2}{dt} = L_2\frac{di_2}{dt} \pm M\frac{di_1}{dt}. \quad (9.67)$$

To know whether to use the positive or negative signs on the mutual terms in (9.67), we must know whether the fluxes produced by i_1 and i_2 are in the same or

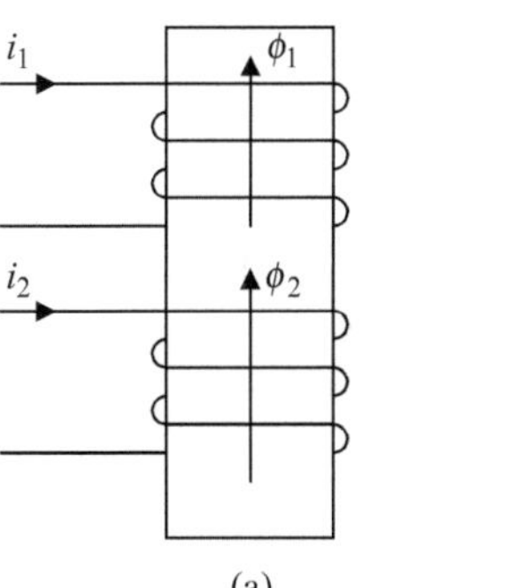

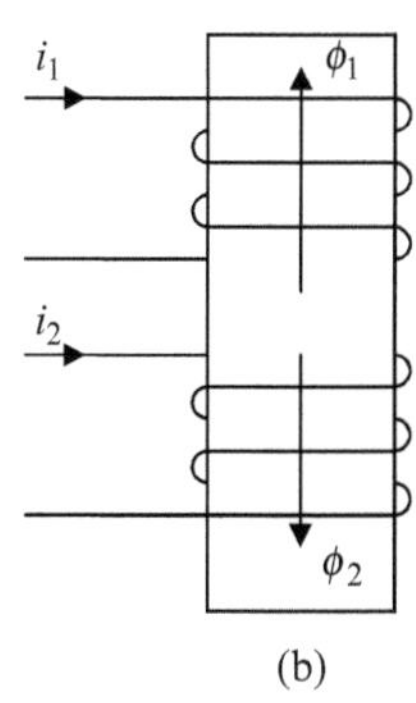

Fig. 9.31 The sign of the mutual terms in (9.67) is (**a**) positive or (**b**) negative, depending upon how the coils are wound and the directions of the currents in the coils

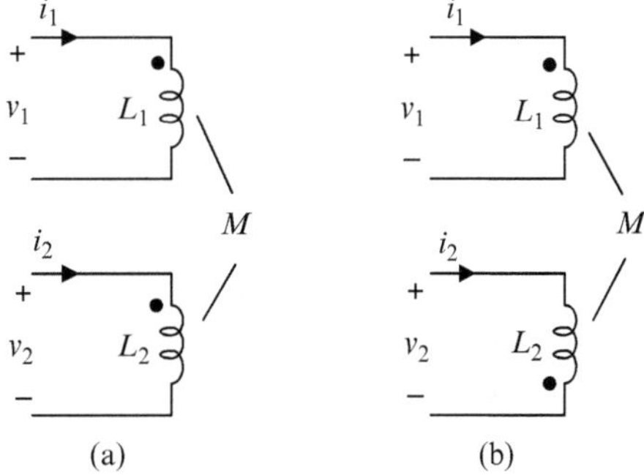

Fig. 9.32 Dot-convention labeling for the coils in Fig. 9.31

opposite directions. If the fluxes are in the same direction, we must use the positive sign. If the fluxes are in opposite directions, we must use the negative sign. If we know how the coils are wound – e.g., if we are presented with a picture like the one shown in Fig. 9.31 – we can determine which sign should be used. If the current directions and coil windings are as shown in Fig. 9.31(a), we would use the positive sign. If the current directions and coil windings are as shown in Fig. 9.31(b), we would use the negative sign.

But it is impractical to require such detailed representations of coils in circuit diagrams, so we use instead a convention called the **dot convention**, where dots are used to indicate corresponding ends of coupled coils. If the positive directions of the currents are either both into or both out of corresponding ends, we must use the positive sign. Otherwise, we must use the negative sign.

Figure 9.32 shows how the dot convention can be used to indicate the corresponding ends of the coils drawn in Fig. 9.31. For the arrangement in Fig. 9.32(a), which corresponds to Fig. 9.31(a), we would write

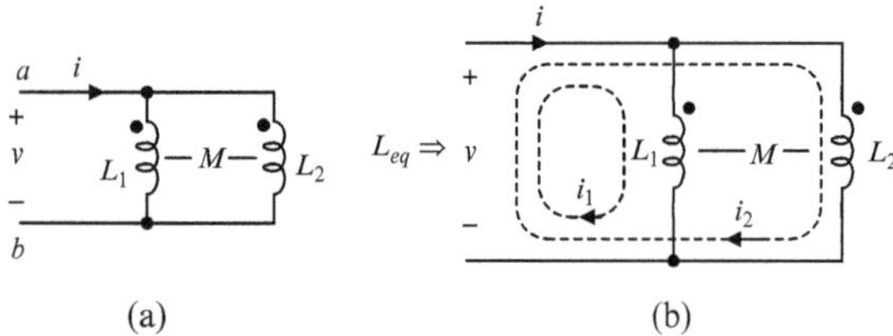

Fig. 9.33 See Example 9.11

$$v_1 = L_1\frac{di_1}{dt} + M\frac{di_2}{dt}, \qquad v_2 = L_2\frac{di_2}{dt} + M\frac{di_1}{dt},$$

whereas for the arrangement in Fig. 9.32(b), which corresponds to Fig. 9.31(b), we would write

$$v_1 = L_1\frac{di_1}{dt} - M\frac{di_2}{dt}, \qquad v_2 = L_2\frac{di_2}{dt} - M\frac{di_1}{dt}.$$

Example 9.11. Obtain the equivalent inductance at the terminals *a–b* of the circuit shown in Fig. 9.33(a).

Solution: We re-draw the circuit to obtain the circuit diagram in Fig. 9.33(b), which defines the loop currents such that each enters the dotted terminal of the associated coil. Applying Kirchhoff's voltage law to each of the indicated loops gives

$$\begin{aligned} v &= L_1\frac{di_1}{dt} + M\frac{di_2}{dt}, \\ v &= L_2\frac{di_2}{dt} + M\frac{di_1}{dt}, \end{aligned} \tag{9.68}$$

where we use the positive sign on the mutual term because both currents enter dotted terminals. We can write (9.68) in vector-matrix form as

$$\begin{bmatrix} L_1 & M \\ M & L_2 \end{bmatrix}\begin{bmatrix} i'_1 \\ i'_2 \end{bmatrix} = \begin{bmatrix} v \\ v \end{bmatrix}, \tag{9.69}$$

where $i'_1 = di_1/dt,\ i'_2 = di_2/dt$. The solution is

$$\begin{bmatrix} i'_1 \\ i'_2 \end{bmatrix} = \begin{bmatrix} L_1 & M \\ M & L_2 \end{bmatrix}^{-1}\begin{bmatrix} v \\ v \end{bmatrix} = \frac{v}{L_1L_2 - M^2}\begin{bmatrix} L_2 - M \\ L_1 - M \end{bmatrix},$$

which, when written out, becomes

$$\begin{aligned} i'_1 &= \frac{di_1}{dt} = \frac{L_2 - M}{L_1L_2 - M^2}v, \\ i'_2 &= \frac{di_2}{dt} = \frac{L_1 - M}{L_1L_2 - M^2}v, \end{aligned}$$

whence

$$\frac{di_1}{dt} + \frac{di_2}{dt} = \frac{d(i_1 + i_2)}{dt} = \frac{di}{dt} = \frac{L_2 + L_1 - 2M}{L_1L_2 - M^2}v,$$

which yields

$$v = \left(\frac{L_1L_2 - M^2}{L_2 + L_1 - 2M}\right)\frac{di}{dt}.$$

This last relation has the form

$$v = L_{eq}\frac{di}{dt},$$

where the equivalent inductance L_{eq} is given by

$$L_{eq} = \frac{L_1L_2 - M^2}{L_1 + L_2 - 2M}.$$

From (9.63),

$$M = \sqrt{L_1L_2} \Rightarrow k = 1. \tag{9.70}$$

Coils for which the coupling coefficient is near unity are said to be **tightly coupled**. Tight coupling between coils can only be achieved if the coils are wound on a common *magnetic* core.

Exercise 9.17. Refer to (9.68). Show that if the coupling coefficient equals unity, then

$$v_2 = -\sqrt{\frac{L_2}{L_1}}v_1; \quad k = 1. \tag{9.71}$$

Exercise 9.18. The solution given in Example 9.11 is invalid if the coupling coefficient is unity, because if $k = 1$, then $M = \sqrt{L_1L_2}$ and the matrix on the left in (9.69) is singular. (a) Show that if $M = \sqrt{L_1 L_2}$ then $v = 0$ and

$$\sqrt{L_2}\frac{di_2}{dt} = -\sqrt{L_1}\frac{di_1}{dt}.$$

(b) Does this mean that $i_2\sqrt{L_2} = -i_1\sqrt{L_1}$? Explain.

9.13 Parasitic Mutual Inductance

Parasitic mutual inductance between two long, straight, non-magnetic, identical wires depends upon wire spacing and pair geometry (e.g., parallel or twisted), but does not exceed the self inductance of either wire.

Refer to Fig. 9.34, which shows an end view of two neighboring conductors carrying current into the page. We identify four components of the total flux (per unit length of the conductors):

$$\begin{aligned}
\phi_{11} &= \text{flux produced by current } i_1 \\
&\quad \text{that links only conductor 1} \\
\phi_{12} &= \text{flux produced by current } i_1 \\
&\quad \text{that links both conductors} \\
\phi_{22} &= \text{flux produced by current } i_2 \\
&\quad \text{that links only conductor 2} \\
\phi_{21} &= \text{flux produced by current } i_2 \\
&\quad \text{that links both conductors}
\end{aligned} \tag{9.72}$$

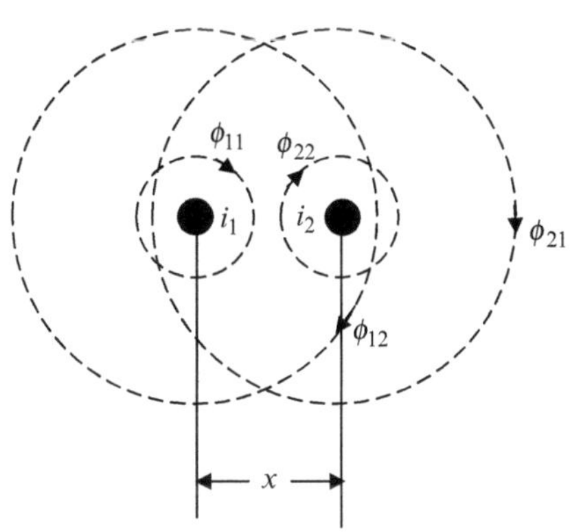

Fig. 9.34 Flux linkages between parallel current-carrying conductors

Thus the total flux produced by current i_1 equals $\phi_1 = \phi_{11} + \phi_{12}$, the total flux produced by current i_2 equals $\phi_2 = \phi_{22} + \phi_{21}$, and the self-inductances (per unit length) of the individual conductors are given by

$$\phi_1 = L_1\, i_1, \qquad \phi_2 = L_2\, i_2. \tag{9.73}$$

Similarly, the flux ϕ_{12} produced by current i_1 that links conductor 2 and the flux ϕ_{21} produced by current i_2 that links conductor 1 give rise to mutual inductances given by

$$\phi_{12} = M_{12}\, i_1, \quad \phi_{21} = M_{21}\, i_2. \tag{9.74}$$

By symmetry,

$$M_{12} = M_{21} = M. \tag{9.75}$$

The induced voltages in the conductors are given by

$$\begin{aligned}
v_1 &= L_1\frac{di_1}{dt} \pm M\frac{di_2}{dt}, \\
v_2 &= L_2\frac{di_2}{dt} \pm M\frac{di_1}{dt},
\end{aligned} \tag{9.76}$$

where the positive sign is used if the currents are in the same direction, as assumed in Fig. 9.34, and negative if the currents are in opposite directions. In other words, if the fluxes produced by the conductors are in the same direction (both clockwise or both counter-clockwise), then the self-induced and mutually-induced voltages have the same polarity. Otherwise, the self-induced and mutually-induced voltages have opposite polarity.

Parasitic mutual inductance can be significant in very large circuits, even if the frequencies of currents and voltages involved are relatively low; for example, where telephone lines run parallel to electric-power lines over a great distance, 60-cycle hum is induced in the telephone lines and telephone companies must employ special circuits (filters) to eliminate the hum.[13] Similarly, parasitic mutual inductance between

[13]Between-conductor capacitance also contributes to the coupling. Also, because of nonlinearities in power transformers, the third and fifth harmonics (180- and 300-Hz components) also are present in current on power lines. Telephone companies use filters to eliminate (block) components at and below 300 Hz.

parallel telephone lines can introduce what is called *crosstalk*, where one conversation is superimposed on another. Privacy considerations require telephone companies to take measures to avoid crosstalk. Both problems have been greatly alleviated by a shift to digital transmission.

Detailed treatment of such issues is beyond the scope of this book. A standard undergraduate course in electromagnetism will help prepare you to tackle such important problems in practice.

9.14 Transformers

Magnetic coupling is the basis of operation for devices called **transformers**, which use magnetically coupled coils to allow current in one or more coils to induce voltages in one or more other coils, according to Faraday's law of induction. One or more of the coils might be tapped at various points. The coils driven by sources are called **primary coils** and the coils driving loads are called **secondary coils**. For economy, practicing engineers omit *coil* and refer to a primary coil as a **primary** and to a secondary coil as a **secondary**. We limit our discussion to transformers having one primary and one secondary. The secondary might have a center tap.

The primary and secondary usually are wound on a common core. The core can be a magnetic material, such as iron, some steels, and ferrite, or a non-magnetic material, such as air (a hollow paper or plastic tube). A transformer having a magnetic core usually is called an *iron-core transformer*, regardless of the specific composition of the core. A transformer having a non-magnetic core usually is called an *air-core transformer*, although the core might be (at least in part) paper or plastic. Some transformers for radio-frequency applications have a magnetic core that can be screwed into (or out of) a hollow cylindrical paper or plastic tube on which both coils are wound, which effectively changes the coupling coefficient and the inductances. Adjustable power transformers have a sliding tap on the secondary which effectively changes the number of secondary turns (and thereby the secondary voltage. See (9.80)). Figure 9.35 shows circuit-diagram symbols for various kinds of transformers. Figure 9.36 shows a selection of transformers (not to scale).

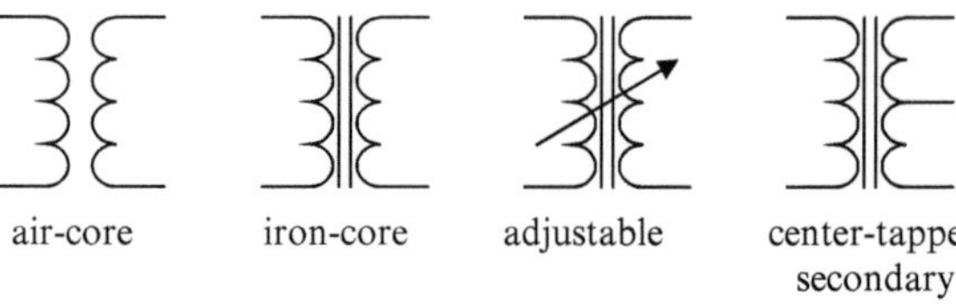

Fig. 9.35 Circuit-diagram symbols for transformers

Each term of the form $M(di/dt)$ in (9.67) is a voltage induced in one coil by current in the other. The voltage induced in one coil depends upon the mutual inductance M *and upon the rate of change of the current* in the other coil. Thus the voltage induced can be large, even if the coupling coefficient (or mutual inductance) is small. Also, air-core transformers do not exhibit various nonlinear effects and core losses inherent in iron-core transformers. For these and other reasons, most transformers for high-frequency applications (large di/dt) are air-core. Many for broadcast radio applications (medium di/dt) are either air-core or adjustable. Almost all transformers for audio ($\leq 20\,\text{kHz}$) and power (60 Hz) applications are iron-core. Coupling coefficients range from about 0.01 for some air-core transformers to about 0.99 for some iron-core power transformers. For most audio and high-frequency transformers, linearity is more important than efficiency. For power transformers, efficiency (tight coupling) is as important if not more important than linearity. Air-core transformers are linear, whereas iron-core transformers are at best only approximately so, more nearly linear for smaller than for larger currents and voltages. We limit our treatment to transformers for which linearity is an acceptable approximation, at least for preliminary analysis or design.

It is impractical to show the internal structure of a transformer (how the coils are wound) in a circuit diagram, so we use the dot convention described in Section 9.12, where dots are placed at corresponding ends of the primary and secondary windings, as illustrated by Fig. 9.37, where

$$v_1 = L_1\frac{di_1}{dt} + M\frac{di_2}{dt}; \quad v_2 = L_2\frac{di_2}{dt} + M\frac{di_1}{dt}. \quad (9.77)$$

When preparing to write circuit equations for a transformer, we are free to choose the positive directions for the currents. There is no reason to intentionally

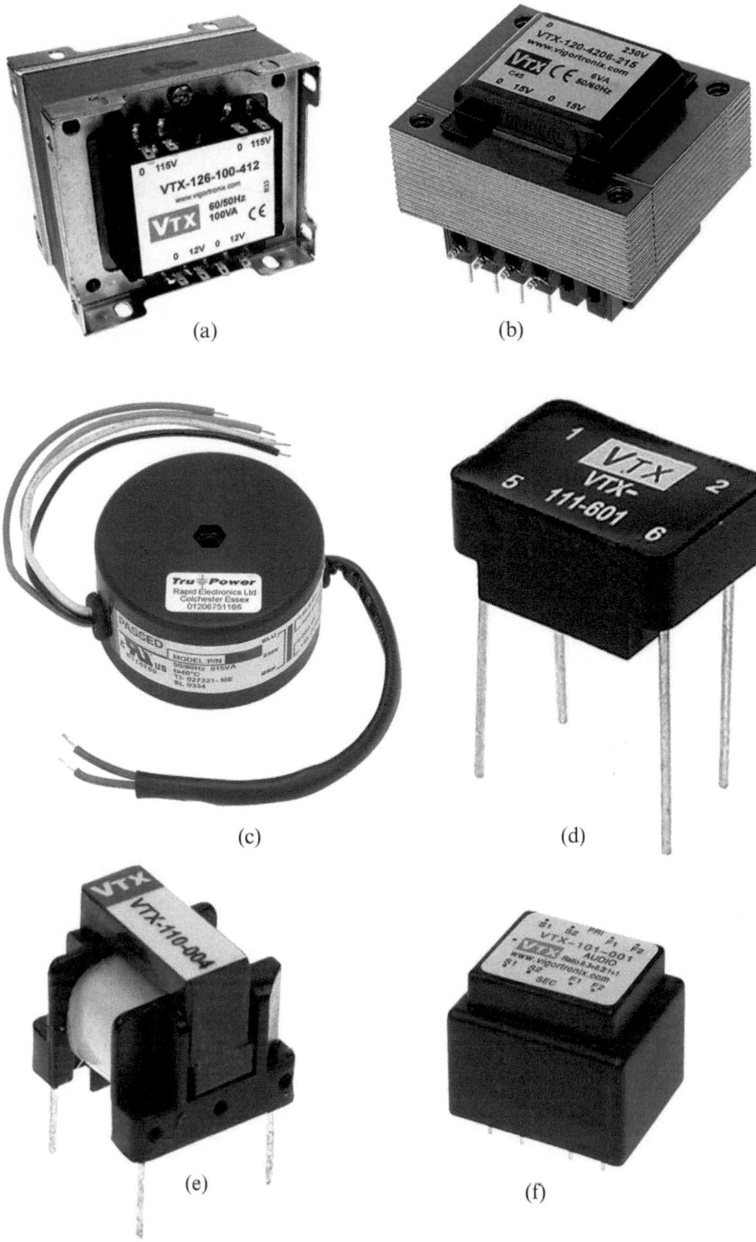

Fig. 9.36 Assorted transformers (not to scale): (**a**) chassis-mount power transformer, (**b**) PCB power transformer, (**c**) encapsulated power transformer, (**d**) (**e**), PCB-mount pulse transformers, (**f**) PCB-mount audio transformer (Photographs courtesy of Rapid Electronics, Ltd)

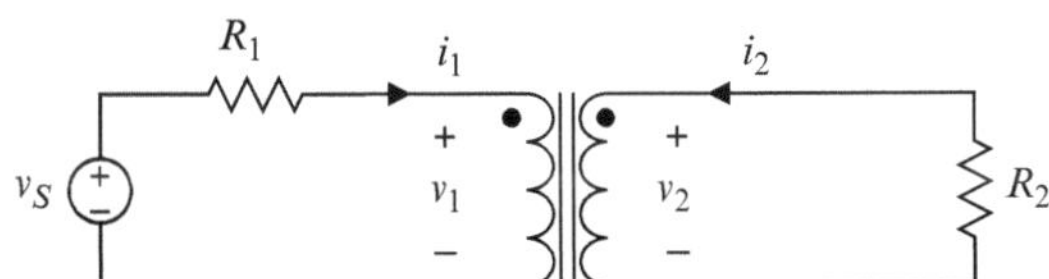

Fig. 9.37 Dot convention for a transformer

confuse ourselves, so we should always choose the directions and polarities such that both currents enter the dotted terminals, in which case the dotted terminals are positive. If we do that, we will always use the positive sign before the mutual inductance M in the circuit equations.

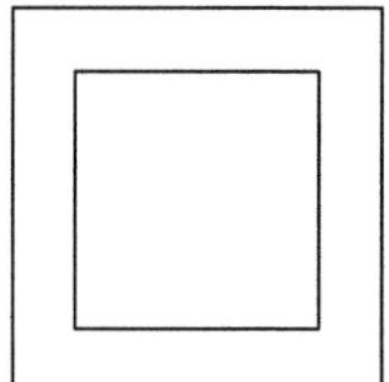

Fig. 9.38 See Exercise 9.19

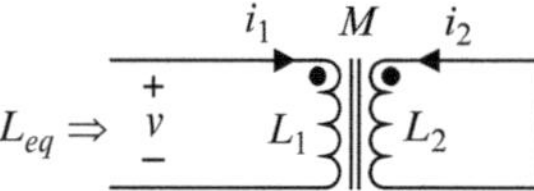

Fig. 9.39 See Example 9.12

The dot convention is extended to three and more coupled circuits by using various dot shapes (e.g., diamonds, squares, etc.), but we do not need that generalization.

Exercise 9.19. The shape in Fig. 9.38 represents a rectangular iron core. Draw coils on the left and right parts of the core such that (a) both coils are dotted at the top and (b) one coil is dotted at the top and the other at the bottom.

We are at this point unprepared to solve transformer problems, except in certain simple cases, one of which is the subject of the next example.

Example 9.12. Obtain an expression for the equivalent inductance seen at the primary winding of a transformer whose secondary is shorted.

Solution: Refer to Fig. 9.39. Kirchhoff's voltage law gives

$$L_1 \frac{di_1}{dt} + M \frac{di_2}{dt} = v; \qquad L_2 \frac{di_2}{dt} + M \frac{di_1}{dt} = 0.$$

Using the second equation to eliminate i_2 from the first gives

$$v = \left(L_1 - \frac{M^2}{L_2}\right) \frac{di_1}{dt} \Rightarrow L_{eq} = L_1 - \frac{M^2}{L_2}.$$

9.15 Ideal Transformers

An **ideal transformer** is a fictional *lossless* transformer whose *coupling coefficient is unity*. Because the coupling coefficient k equals unity, all of the flux produced by the primary links the secondary, and vice versa. That is, $\phi_{12} = \phi_1$, $\phi_{21} = \phi_2$, and (9.65) becomes

$$\begin{aligned} \lambda_1 &= N_1(\phi_1 \pm \phi_2), \\ \lambda_2 &= N_2(\phi_2 \pm \phi_1) = -N_2(\phi_1 \pm \phi_2). \end{aligned} \tag{9.78}$$

It follows (Faraday's law) that

$$\begin{aligned} v_1 &= \frac{d\lambda_1}{dt} = N_1 \frac{d(\phi_1 \pm \phi_2)}{dt}, \\ v_2 &= \frac{d\lambda_2}{dt} = -N_2 \frac{d(\phi_1 \pm \phi_2)}{dt}, \end{aligned} \tag{9.79}$$

and thus

$$v_2 = -\frac{N_2}{N_1} v_1, \; k = 1. \tag{9.80}$$

Because an ideal transformer is lossless, the winding resistances equal zero, and *all* of the power delivered to the primary is delivered to a load connected to the secondary. Refer to Fig. 9.40. The instantaneous power delivered to the primary is given by $p_{in} = v_1 \, i_1$. Consistent with the passive sign convention, the instantaneous power delivered to the load is given by $p_{out} = -v_2 \, i_2$. For a lossless transformer,

$$p_{in} = p_{out} \Rightarrow v_1 \, i_1 = -v_2 \, i_2 \Rightarrow \frac{v_2}{v_1} = -\frac{i_1}{i_2}. \tag{9.81}$$

Equation (9.80) holds if the coupling coefficient equals unity, even if losses are present, whereas (9.81) holds only if the transformer is lossless. However, if losses are present, the voltages v_1, v_2 are inaccessible (the resistance and inductance are distributed all along each winding), so the fact that (9.80) holds in presence of losses is of little interest unless the losses are small relative to the power delivered to the primary. Also, winding resistance is not the only loss mechanism in a transformer: Time-varying magnetic fields in an iron-core transformer produce eddy currents and attendant losses in the core material.

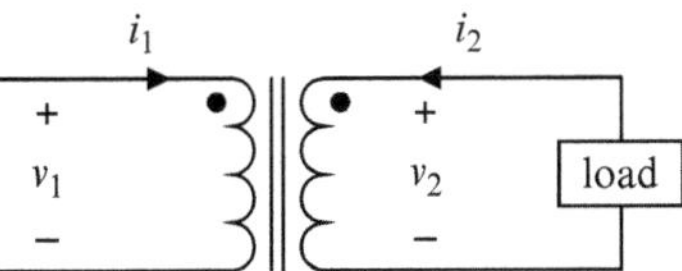

Fig. 9.40 Power transfer in an ideal transformer (see (9.81))

Equations (9.80) *and* (9.81) *hold only for a transformer having unity (perfect) coupling and no losses.* Nonetheless, (9.80) and (9.81) provide a starting point for understanding and designing circuits that incorporate real (not ideal) transformers.

The ratio of secondary to primary turns for a transformer is called the **turns ratio**[14] and is denoted by n.

$$n = \frac{N_2}{N_1} = \text{turns ratio}. \tag{9.82}$$

From (9.80), (9.81), and (9.82), the primary and secondary voltages and currents in an ideal transformer are related as

$$v_2 = \pm n\, v_1, \quad i_2 = \mp n^{-1}\, i_1. \tag{9.83}$$

Equations (9.83) are valid only for *a lossless transformer having unity coupling and only for time-varying (ac) voltages and currents.* A transformer (ideal or otherwise) does not pass dc.[15]

In (9.83), $v_2 = nv_1$ if the assumed polarities for v_1 and v_2 are the same with respect to the dotted terminals. Otherwise, $v_2 = -nv_1$. In (9.83), $i_2 = -n^{-1}i_1$ if the positive direction for both currents is either into or out of the dotted terminals. Otherwise, $i_2 = n^{-1}i_1$. To avoid confusion, it is best to *take the positive direction of primary current into the dotted terminal of the primary and the positive direction of secondary current out of the dotted terminal of the secondary, and to assign voltage polarities such that the dotted terminals are positive*, as shown in Fig. 9.41.[16] In that case, (9.83) becomes

[14] Some define the turns ratio as N_1/N_2.

[15] In certain configurations, a special kind of transformer, called an autotransformer, can pass dc. See Problems P 9.93 and P 9.94 at the end of the chapter.

[16] As a practical matter, there is little reason to do otherwise.

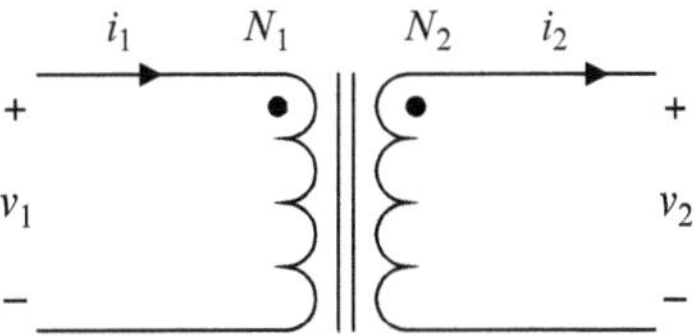

Fig. 9.41 Conventional current directions and voltage polarities for an ideal transformer. See (9.84)

$$\begin{aligned} v_2 &= \frac{N_2}{N_1} v_1 = n\, v_1, \\ i_2 &= \frac{N_1}{N_2} i_1 = n^{-1}\, i_1. \end{aligned} \tag{9.84}$$

For an ideal transformer, the least confusing positive directions and polarities of primary and secondary currents and voltages, relative to the dot positions, are those consistent with (9.84), as described by the *Italicized* statement above. For a non-ideal transformer, the least confusing positive directions of primary and secondary currents, relative to the dot positions, are those that make the sign before the mutual inductance positive.

With reference to Fig. 9.41, (9.84) specifies the terminal characteristics of an ideal transformer. Self and mutual inductance are irrelevant and do not appear in equations describing an ideal transformer. Thus, although the symbol for an ideal transformer (Fig. 9.41) suggests inductance in the primary and secondary, *an ideal transformer possesses neither self-inductance nor mutual inductance. For an ideal transformer, the only relevant relations are those given by* (9.84), *where the assumed current directions and voltage polarities are as shown in Fig. 9.41.*

Ideal transformers do not exist. However, iron-core transformers can achieve tight coupling (coupling coefficients approaching 0.99) and high efficiency (95% or more). *The ideal transformer is a somewhat realistic model only for nearly linear (small-signal) operation of a very efficient (low-loss) iron-core transformer.* An ideal transformer is a poor model for air-core transformers, whose coupling coefficients are much smaller than unity. Nonetheless, as noted above, the ideal transformer relations (9.84) provide a starting point for specifying transformers and a basis for understanding why and how transformers are used. Some applications follow.

9.16 Applications of Transformers

9.16.1 Source and Load Transformations: Matching Transformers

Recall from Chapter 6 that maximum power is transferred from a fixed source to a resistive load if the resistance of the load equals the source resistance. **Matching transformers** are used to alter the apparent resistance presented to a source by a load, or the apparent output resistance of a source, in order to achieve maximum power transfer in cases where the source and load resistance are different and cannot be changed. Load matching using transformers is based upon the fact that a transformer alters the apparent resistance of a load or source, as illustrated by the following examples and subsequent discussion. Throughout this section, we use only the *ideal model* for a transformer.

Example 9.13. Obtain an expression for the input resistance of the circuit shown in Fig. 9.42.

Solution: By definition,

$$R_{in} = \frac{v_1}{i_1}.$$

From (9.84)

$$R_{in} = \frac{n^{-1}\,v_2}{n\,i_2} = \frac{1}{n^2}\frac{v_2}{i_2}.$$

By Ohm's law, $v_2/i_2 = R$. It follows that

$$R_{in} = \frac{1}{n^2}R. \tag{9.85}$$

Thus the ideal transformer transforms the load seen by a source (v_1) from R to R/n^2, where n is the transformer turns ratio. Such transformation allows maximum power transfer in cases where the load resistance does not equal the source resistance and the load resistance cannot be changed.

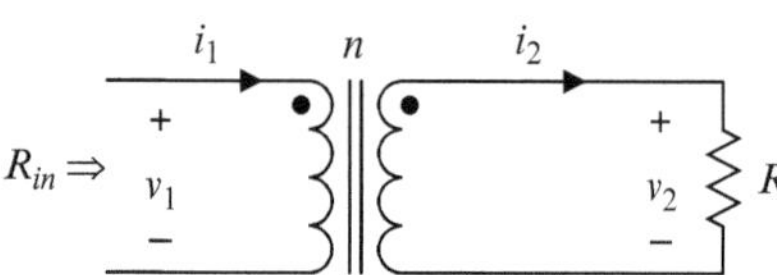

Fig. 9.42 See Example 9.13

Exercise 9.20. It is required that a fixed resistive load R_L draw maximum power from a source having fixed output resistance R_S, where $R_S \neq R_L$ (Fig. 9.43). A transformer is to be used to provide the required match. Obtain an expression for the turns ratio of the transformer (assumed ideal).

An **output transformer** usually appears as the last (output) element in a circuit; e.g., in the output stage of a multi-stage amplifier. The usual purpose of an output transformer is to match a load to a source (for maximum power transfer). The next example shows how one obtains the Thévenin equivalent for a circuit having an ideal transformer as the output element.

Example 9.14. Obtain the Thévenin equivalent at the terminals *a*–*b* for the circuit of Fig. 9.44.

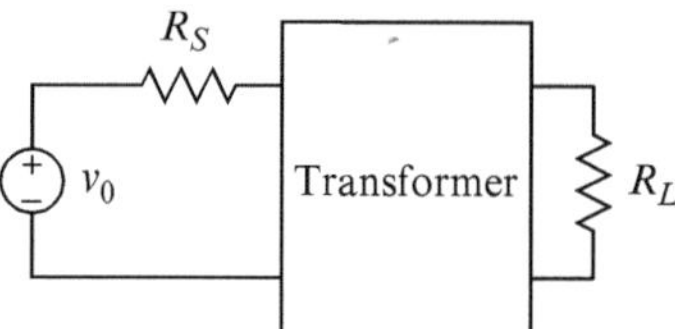

Fig. 9.43 See Exercise 9.20

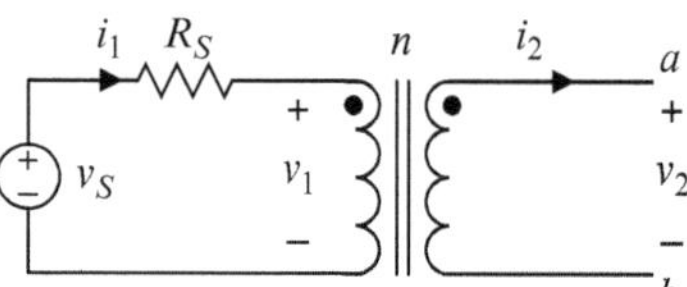

Fig. 9.44 See Example 9.14

Solution: We obtain the open-circuit voltage and the short-circuit current at the terminals a–b. With the terminals open, $i_2 = 0$ and

$$v_2 = v_{oc} = n\,v_1.$$

From (9.84), $i_2 = 0 \Rightarrow i_1 = 0$ and $v_1 = v_S$. Thus

$$v_{oc} = n\,v_S.$$

With the terminals shorted, $v_2 = 0 \Rightarrow v_1 = 0$ and $i_1 = v_S/R_S$. Thus

$$i_{sc} = \frac{v_S}{n\,R_S}.$$

The Thévenin voltage and Thévenin resistance are given by

$$v_T = v_{oc} = n\,v_S; \qquad R = \frac{v_{oc}}{i_{sc}} = n^2 R_S.$$

From the viewpoint of the load, the transformer transforms the source resistance from R_S to $n^2 R_S$. Here again, a transformer can be used to achieve maximum power transfer when the source resistance and load resistance are different and cannot be changed.

Examples 9.13 and 9.14 illustrate two useful properties of transformers: A load resistance can be made to appear smaller ($n > 1$) or larger ($n < 1$) by interposing a transformer between a source and the load. Similarly, a source and source resistance can be made to appear smaller ($n < 1$) or larger ($n > 1$) by interposing a transformer between a load and the source. In other words, we can use a transformer to change the load seen by a source or the source seen by a load, as shown in Fig. 9.45. Transforming a load as shown in Fig. 9.45(a) is called *reflecting the load to the primary* of the transformer and transforming a source as shown in Fig. 9.45(b) is called *reflecting the source to the secondary* of the transformer. In either case, the transformer disappears and the circuit is simplified for further analysis.

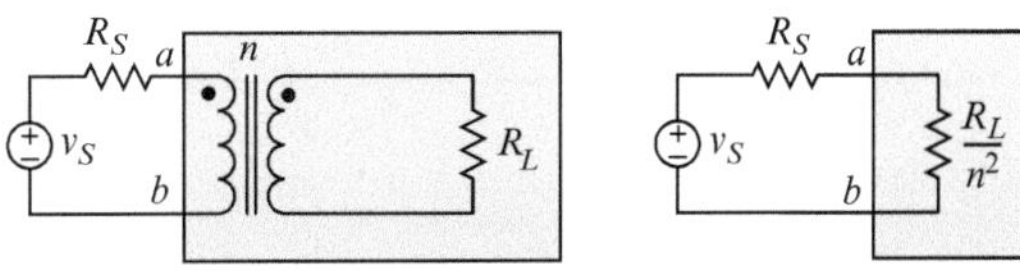

(a) load transformation

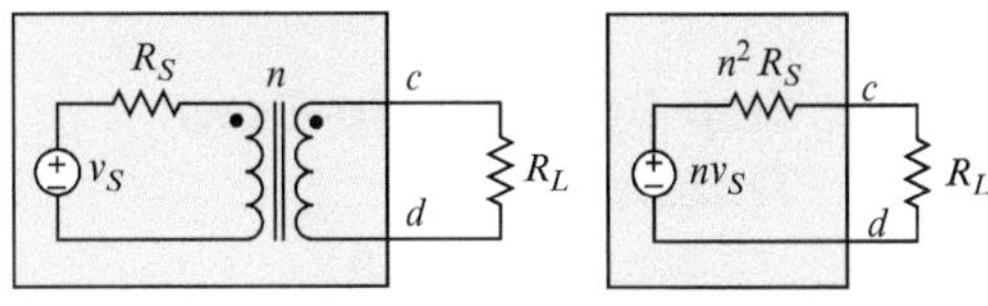

(b) source transformation

Fig. 9.45 Source and load transformation using a transformer

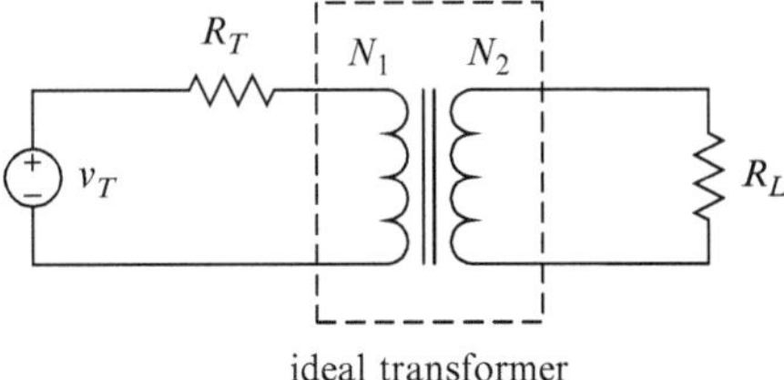

Fig. 9.46 See Example 9.15

Example 9.15. In Fig. 9.46, the source resistance is $R_T = 100\,\Omega$ and the load resistance is $R_L = 1\,\text{k}\Omega$. Specify the transformer turns ratio $n = N_2/N_1$ such that maximum power is delivered to the load.

Solution: See Fig. 9.45. From the viewpoint of the source, the load resistance is

$$R_L' = \frac{R_L}{n^2}.$$

For maximum power transfer, the apparent load resistance must equal the source resistance. Therefore,

$$\frac{R_L}{n^2} = R_S \Rightarrow n = \sqrt{\frac{R_L}{R_S}} = \sqrt{10} = 3.16.$$

Exercise 9.21. A load is connected to a source through two ideal transformers having turns ratios n_1 and n_2, as shown in Fig. 9.47. Under what condition is maximum power delivered to the load?

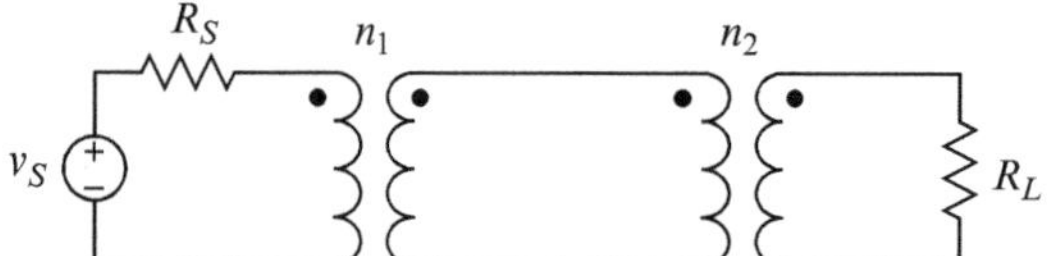

Fig. 9.47 See Exercise 9.21

9.16.2 Step-Up and Step-Down Transformers

A transformer for which $n > 1$ is called a **step-up transformer** because the secondary voltage is higher than the primary voltage. A transformer for which $n<1$ is called a **step-down transformer**. Electric utilities use step-up transformers at the origins of long-distance transmission lines and step-down transformers at the destination. This is done primarily to reduce line currents and therefore the sizes of the conductors required to transmit power from a source to a distant load, as illustrated by the following example.

Example 9.16. Obtain expressions for power loss in the transmission-line resistance in the circuits of Fig. 9.48. Assume the transformers are ideal.

Solution: (a) We transform all resistances to the primary of the first transformer (see Example 9.13), as illustrated in Fig. 9.49, where

$$i = \frac{v}{R_L + 2R/n^2}$$

and the power lost in the transmission lines is given by

$$P_a = I_{rms}{}^2 \frac{2R}{n^2} = \left(\frac{V_{rms}}{R_L + 2R/n^2}\right)^2 \frac{2R}{n^2}. \quad (9.86)$$

(b) In the circuit of Fig. 9.48b,

$$i = \frac{v}{R_L + 2R}$$

and the power loss in the transmission lines is given by

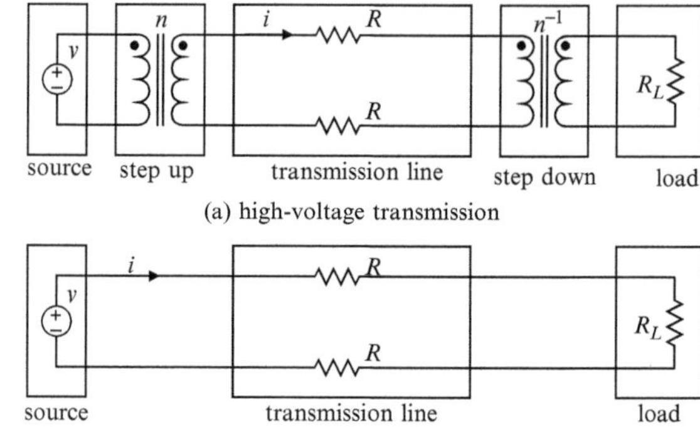

(a) high-voltage transmission

(b) low-voltage (transformerless) transmission

Fig. 9.48 See Example 9.16

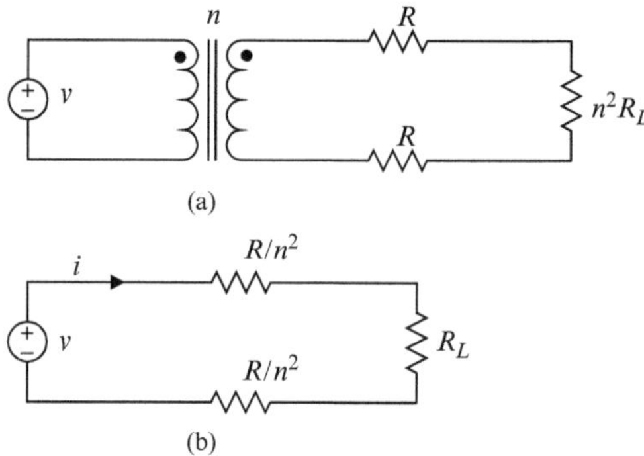

Fig. 9.49 See Example 9.16

$$P_b = I_{rms}{}^2 (2R) = \left(\frac{V_{rms}}{R_L + 2R}\right)^2 (2R). \quad (9.87)$$

The point of this example is the following: Comparing (9.86) with (9.87) shows that the effective line resistance is reduced (approximately by the factor n^{-2}) for the high-voltage transmission system. As a result, much smaller wires can be used to transmit power over long distances, allowing considerable cost savings.

Exercise 9.22. Obtain an expression for the line current i in the circuit of Fig. 9.48(a)

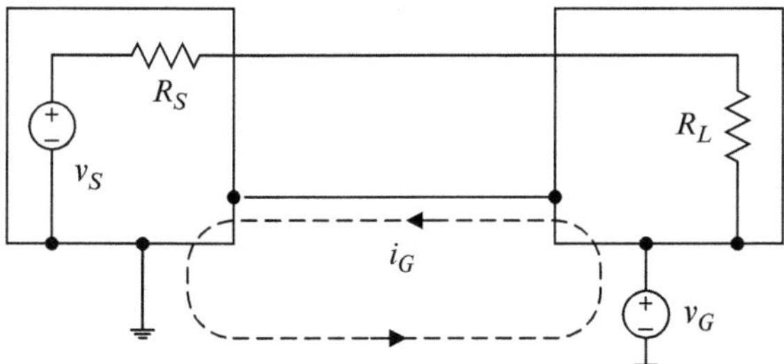

Fig. 9.50 The boxes represent the metal chassis of two circuits. The two chassis are at different potentials. Connecting the two chassis directly gives rise to a ground-loop current i_G which is almost always undesirable and can be destructive or dangerous

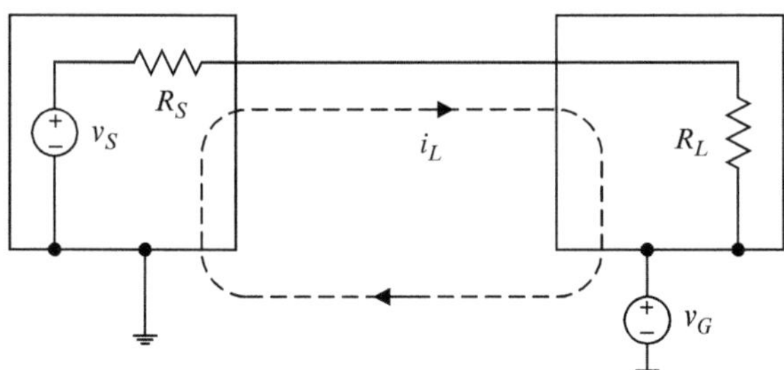

Fig. 9.51 Connecting the circuits without connecting the two chassis[17] leads to unpredictable results, because the ground-potential difference v_G might be unknown, or to undesirable results, because we do not want the load current to depend upon the ground potential difference

9.16.3 Isolation Transformers

Often, it is necessary to couple two circuits that have different ground potentials; for example, two circuits that are some distance apart, such that their earth grounds are at different potentials. Equally often, we might need to connect one circuit whose chassis (or ground plane) is connected to an earth ground to a circuit whose chassis is at a potential different from earth ground. Figure 9.50 shows two such circuits connected in the "obvious" way, where the chassis of the first is connected to the chassis of the other and the source v_G represents the potential difference. The resulting current i_G is called a **ground-loop current** and can be large enough to do damage or at least degrade performance of the circuits.

[17]Yes, the plural of chassis is chassis: Spelled the same, but pronounced differently.

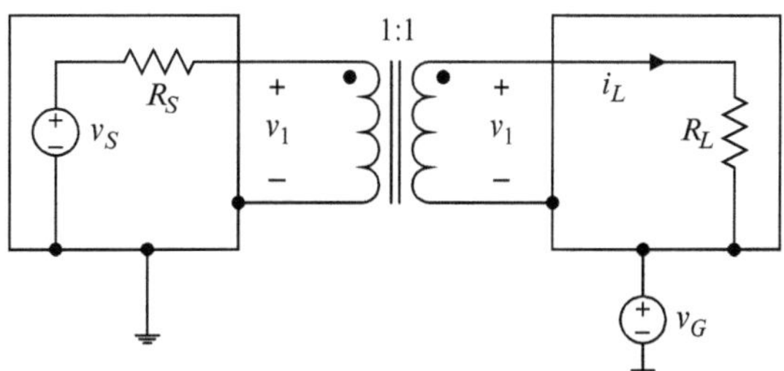

Fig. 9.52 Using a transformer allows connecting the circuits without providing a current path for the ground potential difference

The ground loop in Fig. 9.50 can be avoided by connecting the circuits as shown in Fig. 9.51, but with this connection, the load current i_L is dependent upon both the source v_S and the ground potential difference v_G. The current produced can also be destructive or dangerous if the ground potential difference is large.

If ground potential difference between two circuits is a problem, as illustrated and discussed above, a transformer can be used to isolate the ground potential difference so it has no bearing on the current transferred from one circuit to another or on a measurement of the voltage across a component in a circuit whose ground potential is different from that of the measuring instrument. Figure 9.52 shows how a transformer can isolate the non-zero ground potential from the rest of the (complete) circuit. The transformer turns ratio is one-to-one, so the secondary voltage equals the primary voltage v_1. There is no electrical path from the positive terminal to the negative terminal of the source v_G. Thus the ground potential difference v_G produces no current and in particular has no effect on the current i_L. The ground potential difference v_G is effectively cut off or *isolated* from the rest of the circuit, as illustrated by the equivalent circuit shown in Fig. 9.53. A transformer used to achieve such isolation is called an **isolation transformer**. A drawback in some applications is inability of a transformer to couple dc; i.e., in Fig. 9.52, the dc component of the source v_S (if any) is not passed to the load R_L.

Lightning can cause brief but huge ground potential differences, as illustrated by Fig. 9.54, which shows two of perhaps several buildings making up an industrial facility. The facility has shared power lines and distributed computing resources connected by various data lines. If lightning strikes nearer one building than the other, the temporary difference in ground potentials can be several thousand volts or more. Some form of protective isolation (and surge protection) is mandatory in such cases.

Isolation transformers also are used to attenuate electrical noise; for example, to prevent noise riding on power lines from reaching sensitive equipment. Transformers intended for such use are designed and constructed specifically for that purpose, and incorporate metallic shields around the primary and secondary windings to reduce capacitive coupling between the windings. The primary shield usually is connected to the transformer case, which is connected to earth ground (e.g., via the ground lead on a three-wire power cord) and the secondary shield usually is connected to the chassis ground on the equipment being isolated. Design and application of isolation transformers is a subject worthy of an entire book, and cannot be treated to depth here. Interested readers can find numerous references on the web and in libraries.

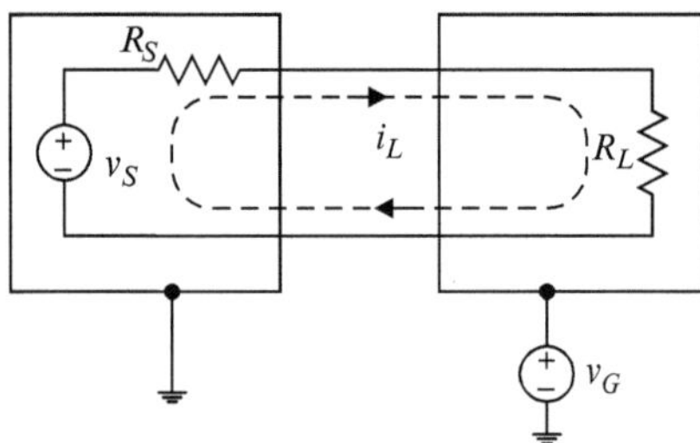

Fig. 9.53 Equivalent circuit for the circuit shown in Fig. 9.52 (assuming the transformer is ideal and the source has no dc component)

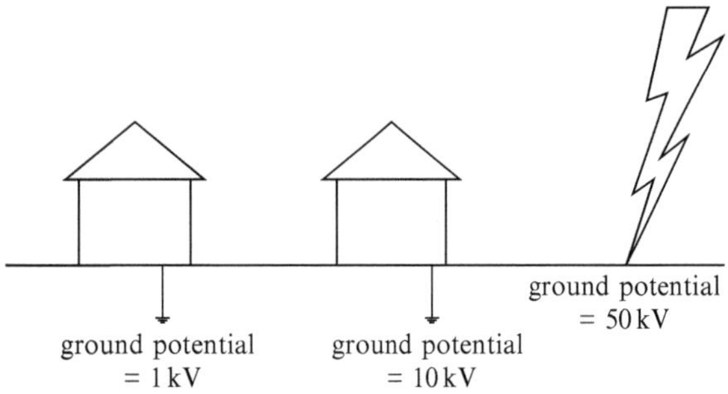

Fig. 9.54 Ground potentials at geographically separated points can be quite different, especially if lightning strikes nearer one point than the other

9.16.4 Center-Tapped Transformers and Balanced Power

Refer to Fig. 9.55(a) and focus your attention first on the transformer and the load R_L. The transformer secondary and the load share a common ground, and that ground is the reference point for all voltages in the circuit. That is, the voltage v_G always equals zero. We assume the transformer secondary voltage is sinusoidal, given by

$$v_S = V_0 \cos(2\pi f t) \tag{9.88}$$

The bottom of the secondary is pinned at zero volts and the upper (dotted) end is alternately positive and negative, which means there is alternately an excess and deficiency of charge at the upper end of the load. Because the voltage at the upper end of the load varies and the voltage at the lower end is fixed (at zero, here), the circuit (or the load) is said to be **unbalanced**.

Regard Fig. 9.55(a) as a snapshot of the circuit at an instant when the secondary voltage and the excess positive charge at the upper end of the load have their maximum values. The vertical line to the right of the load represents a grounded conductor (or metallic surface, such as a section of a chassis) near the load. The load and the conductor comprise a stray capacitance. The positively charged (by v_S) upper end of the load induces an equal and opposite negative charge on the upper end of the nearby conductor, as indicated in the figure. When the secondary voltage reaches its minimum value, as shown in Fig. 9.55(b), the induced charge on the nearby conductor is reversed. Thus there is a non-zero current i_G to ground

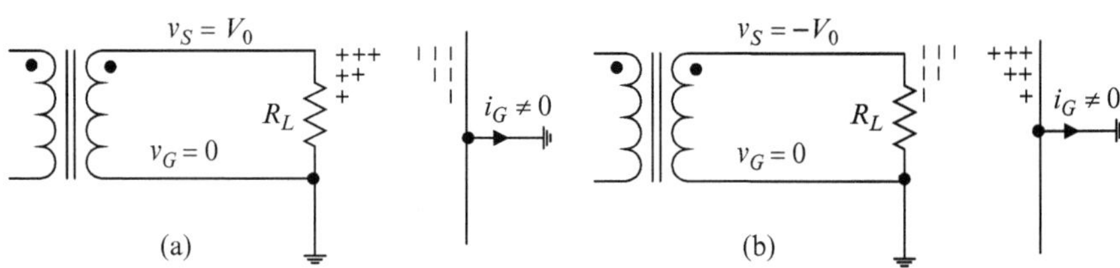

Fig. 9.55 Unbalanced load

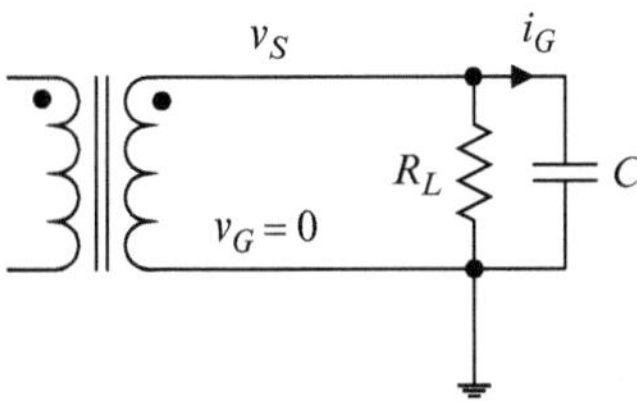

Fig. 9.56 Equivalent circuit for the circuit in Fig. 9.55

from the load, which is manifested as a shunt capacitance to ground, as shown in Fig. 9.56.

Now refer to Fig. 9.57. In Fig. 9.57, the transformer secondary has a grounded center tap. As a result, the voltages at both ends of the secondary (and the load) are alternately positive and negative, and the charge distributions on the load and the nearby conductor are always equal in magnitude and opposite in sign (assuming the center tap on the secondary is perfect). The charges (electrons) in the nearby conductor simply move back and forth along the conductor as the load voltage alternates. No charges are brought up from (or sent down to) the shared ground, so the ground current equals zero. The load (or circuit) in Fig. 9.57 is said to be **balanced**.

In practice, ground currents are common and problematic. They can cause hum and hiss in home and studio audio systems and sound bars or other disruptive effects in video. They can cause erroneous measurements in instrumentation and control systems. If the potential difference between the chassis of two circuits is large and the chassis are connected, the result can be a destructive and even dangerous current. One purpose of this and the previous section is to make you aware of such phenomena and of methods used to counter them.

There are available balanced power systems for use in audio and video studio environments. These systems are essentially equivalent to ideal center-tapped isolation transformers and provide balanced power to loads, as illustrated by Fig. 9.57. Using balanced power sources can eliminate or greatly reduce effects of ground loops and parasitic coupling. The downside of a balanced power system is that both ends of a load are hot. Consequently, a balanced power system is inherently more dangerous than a single-sided system (Fig. 9.55), where one end of the load is at ground potential.

It often is necessary to convert a balanced signal to an unbalanced one, or vice-versa. A device (or circuit) used to accomplish such a conversion is called a **balun**.[18] In most such applications, the balanced and unbalanced sides have different apparent resistances, and the balun must also provide source-load matching.

Figure 9.58 shows a balun of a kind you probably used if you ever connected a rooftop TV antenna to a coaxial cable that carries the received signal inside to a TV set. The wires would connect to the antenna and the coaxial cable would connect to the fitting on the lower right. A typical rooftop TV antenna provides a balanced output. One side (the shield) of the coaxial cable is connected to ground at the TV set, so the signal provided to the TV set is unbalanced. The apparent source resistance for the antenna is 300 Ω and the apparent load presented by the coaxial cable and TV input is 75 Ω. Baluns for this application convert the balanced output from an antenna to an unbalanced input for a TV set and also provide source-load matching.

A balun can consist of a transformer with one center-tapped winding, as illustrated by Fig. 9.59. If the apparent output resistance of the balanced source is R_S, the effective source resistance seen by the unbalanced load is given by

$$R'_S = \left(\frac{N_2}{N_1}\right)^2 R_S,$$

where N_2/N_1 is the turns ratio. For example, a balun used between a rooftop antenna and a coaxial connection to a TV would have a turns ratio of $1/2$, which would match the 300 Ω antenna to the 75 Ω coaxial cable.

9.17 Concluding Remarks

Let us distinguish between *inductance*, which is a physical phenomenon, and expressions of the form

$$v(t) = L\frac{di(t)}{dt}, \qquad i(t) = \frac{1}{L}\int_{-\infty}^{t} v(t')dt', \qquad (9.89)$$

where L is a constant having the unit of $\mathrm{V\,s\,A^{-1}}$. Equations (9.89) are *mathematical relations* between *mathematical models* for physical signals. Any device or circuit whose terminal characteristics are of the forms given by (9.89) is equivalent to an inductor, whether or not magnetic fields are involved. Various circuits that simulate

[18]Balun is a contraction of balanced-to-unbalanced.

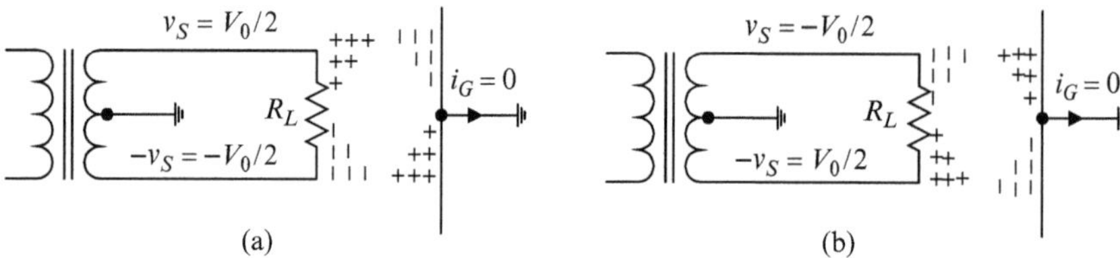

Fig. 9.57 Balanced load

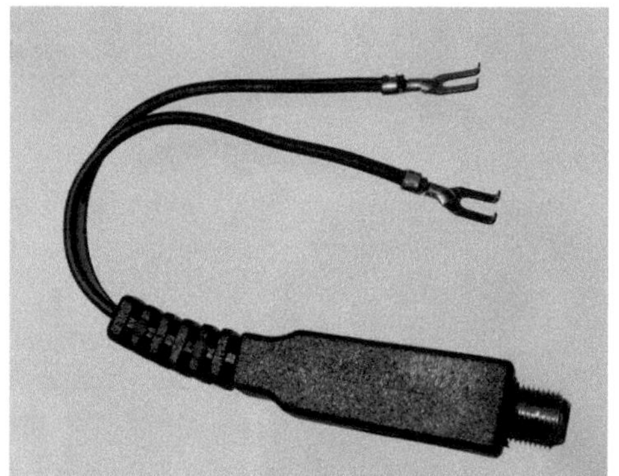

Fig. 9.58 A simple balun

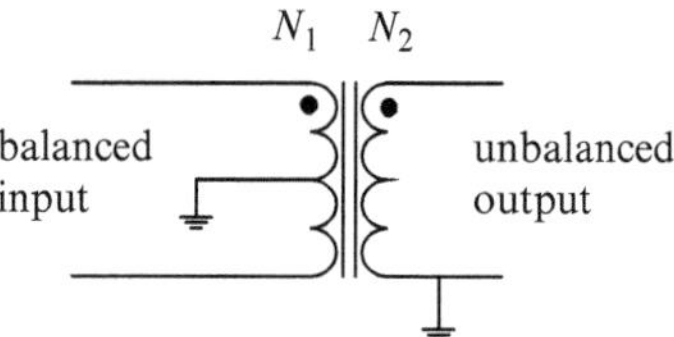

Fig. 9.59 Circuit diagram for a balun

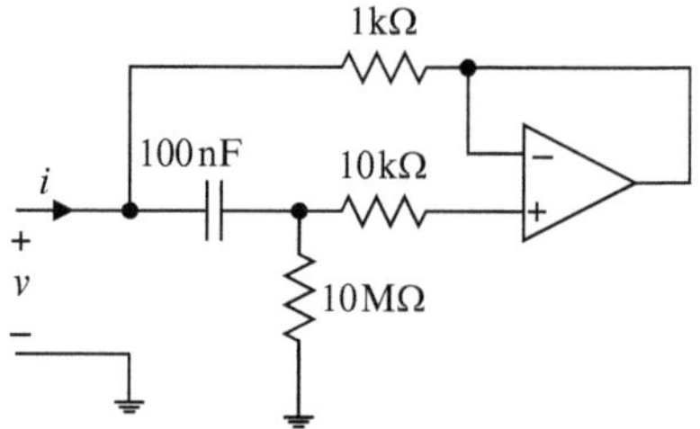

Fig. 9.60 A circuit (gyrator) whose terminal characteristic is that of an inductor

inductance have been devised. For example, for sinusoidal inputs having frequencies between about 3 and 800 Hz, the terminal characteristic of the circuit shown in Fig. 9.60 is approximately that of a 1 kH inductor.[19,20] A coil having that inductance would be huge.

The point of the remarks above is that inductive effects can be achieved without using inductors. Indeed, inductive effects are usually achieved that way in active linear signal-processing and transmission systems at frequencies below 100 MHz or so. Believe it or not, circuits such as the one shown in Fig. 9.60 can be cheaper and much smaller than coils.

Regardless of how they are achieved, inductive effects are important in virtually all linear electrical and electronic circuits. They are important in electric power systems and in machinery, where they arise naturally in the windings of transformers and motors. They are important in radio and television transmitters and receivers and in almost all high frequency – high power applications, where inductive effects cannot be achieved by electronic mimicry. They are important wherever unintentional inductive coupling can be significant; for example, in long conductors near power lines or in electronic systems where a power supply is on the same board or chassis as other circuitry.

If you intend to engage in circuit design, or to work in the power generation and distribution field, or to work as a plant engineer in a plant using electrical machinery, or to work in any field involving radio transmission and reception, or in a host of other disciplines, you need to know about inductance and its effects.

9.18 Problems

Section 9.4 is prerequisite for the following problems.

P 9.1 The inductance of a long, tightly wound air-core coil having N turns, length l, and diameter d is given by

[19] Paul Horowitz and Winfield Hill, *The Art of Electronics*, Cambridge University Press, New York, 1989, p 304.

[20] You will be able to show this after studying Chapter 12.

$$L = \frac{\pi \mu_0 N^2 d^2}{4l}.$$

The meter and Henry are inconvenient (too-large) units for specification and design of air-core inductors. Modify the relation above such that the diameter and length of the coil are expressed in centimeters and the inductance is given in microhenries.

P 9.2 *Wheeler's formula*[21] for the inductance of a single-layer air-core coil is

$$L = \frac{N^2 d^2}{18d + 40l},$$

where the coil diameter d and length l are expressed in inches and the inductance is expressed in microhenries. Rewrite this formula such that the quantities involved are expressed in SI units.

P 9.3 Modify the relation (9.8) in the text for the inductance of an air-core coil such that the diameter and length of the coil are expressed in millimeters.

P 9.4 A tightly wound, single-layer coil is stretched (like a spring). Does the inductance of the coil increase or decrease? Why?

P 9.5 If you need a specific inductance from an air-core coil and you want the coil to occupy the least volume possible, would you make a short, fat coil or a long, thin one? Or does it matter? Assume the relation (9.8) in the text is valid.

P 9.6 If M turns are removed from a long, tightly wound, single-layer air-core inductor having N turns, by what percentage is the inductance reduced?

Section 9.5 is prerequisite for the following problems.

P 9.7 The current through an inductor having inductance $L = 5\,\text{mH}$ is given by

$$i(t) = \begin{cases} 0, & t \le 0, \\ I_0\left(1 - e^{-t/\tau}\right), & t > 0, \end{cases}$$

with $I_0 = 20\,\text{mA}$ and $\tau = 100\,\mu\text{s}$. Obtain an expression for the voltage across the inductor and compute the value of the voltage for $t = \tau$. What is the maximum magnitude of the voltage and when does it occur?

P 9.8 Repeat Problem P 9.7 assuming that the winding (equivalent series) resistance of the inductor is $1\,\Omega$.

P 9.9 The current through a 50-mH inductor is given graphically in Fig. P 9.1. Draw a graph of the voltage across the inductor for $0 \le t \le 15\,\text{ms}$.

P 9.10 Repeat Problem P 9.9 assuming that the winding (equivalent series) resistance of the inductor is $1\,\Omega$.

P 9.11 The current through an inductor having inductance $L = 500\,\mu\text{H}$ is given graphically in Fig. P 9.2. Draw a graph of the voltage across the inductor for $t > 0$.

P 9.12 Repeat Problem P 9.11 assuming that the winding (equivalent series) resistance of the inductor is $1\,\Omega$.

P 9.13 In Fig. P 9.3, $R = 10\,\text{k}\Omega$, $L = 1\,\text{mH}$. The voltage across the resistor is given by

$$v_R = V_0\,[1 + 0.5\,\cos(2\pi f\,t) + 0.05\cos(20\pi f\,t)];$$
$$V_0 = 5\,\text{V}, \quad f = 5\,\text{MHz}.$$

(a) Obtain an expression for the voltage v_L across the inductor. (b) Assuming the source voltage v_S remains the same, obtain an expression for the voltage across the resistor if the inductor is replaced by a short circuit.

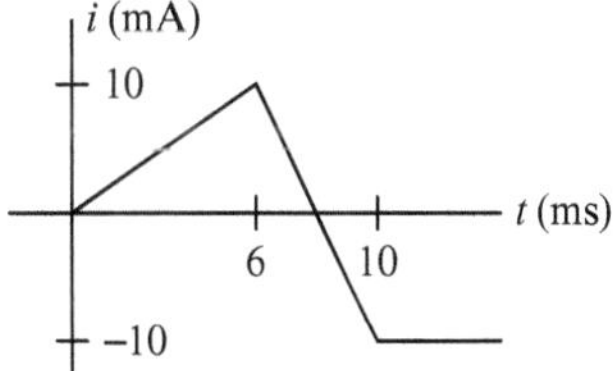

Fig. P 9.1 See Problem P 9.9

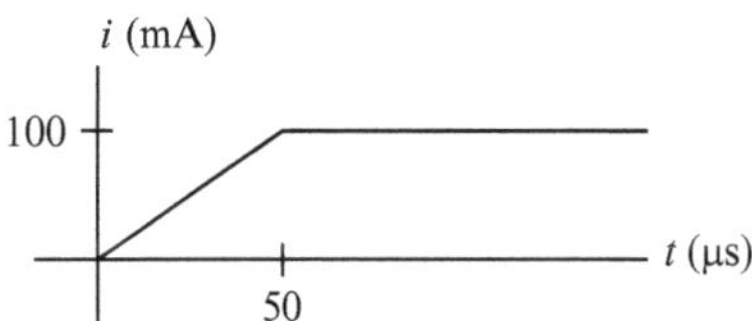

Fig. P 9.2 See Problem P 9.11

[21] Wheeler, H.A., *Simple Inductance Formulas for Radio Coils*, **Proc. IRE, vol 16**, p 1398 (Oct. 1928).

(b) P 9.14 Refer to Fig. P 9.4. The current through the inductor is given by

$$i_L(t) = I_L \cos(2\pi f t).$$

Obtain an expression for the current through the capacitor, the current through the resistor, and the total current *i*(*t*). Use trigonometric identities to express the total current as a single sinusoid of the form

$$i(t) = I \cos(2\pi f t + \theta).$$

P 9.15 In Fig. P 9.5, the fuse will melt if the current through the fuse exceeds 25 A. The switch is closed at $t = 0$. At what time does the fuse begin to melt?

P 9.16 In Fig. P 9.6, the source voltage is given by $v_S(t) = V_S \cos(2\pi f t)$, with $V_S = 25\,\text{V}$ and $f = 10\,\text{kHz}$. It is known that the voltage across the resistor has the form $v_R(t) = V_R \cos(2\pi f t + \theta)$. Find V_R and θ.

P 9.17 Refer to Fig. P 9.7. Under what condition is the output approximately proportional to the integral of the input?

P 9.18 Refer to Fig. P 9.8. Obtain an expression for the load voltage $v_L(t)$.

P 9.19 The turns in an inductor are conductors separated by a dielectric (air or some insulating material). Thus an inductor exhibits stray shunt capacitance. In Fig. P 9.9, C represents stray capacitance and $v_S = V_S \cos(2\pi f t)$. Assume that the current through the inductor equals zero for $t = 0$. Obtain an expression for the frequency f for which the rms amplitude of the current through the capacitor equals the rms amplitude of the current through the inductor.

Section 9.6 is prerequisite for the following problems.

P 9.20 Refer to Fig. P 9.10. Obtain an expression for the indicated current *i*(*t*) (i) assuming the switch has been open for a very long time and is closed at $t = 0$ and (ii) assuming the switch has been closed for a very long time and is opened at $t = 0$.

P 9.21 In Fig. P 9.11, the diode is ideal. The switch is opened at $t = 0$, having been closed for a very long

Fig. P 9.3 See Problem P 9.13

Fig. P 9.4 See Problem P 9.14

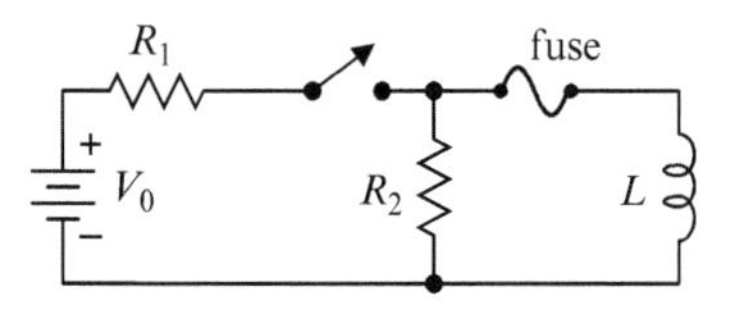

Fig. P 9.5 See Problem P 9.15

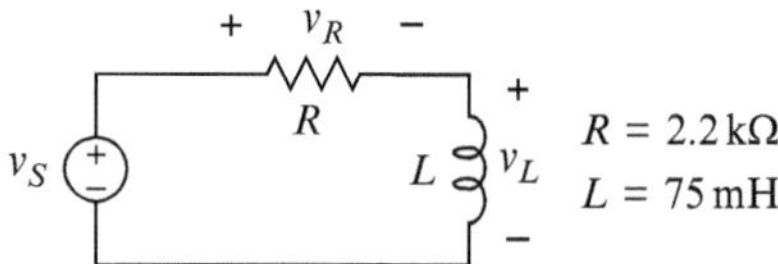

Fig. P 9.6 See Problem P 9.16

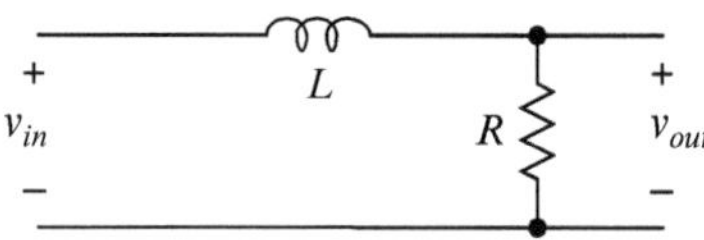

Fig. P 9.7 See Problem P 9.17

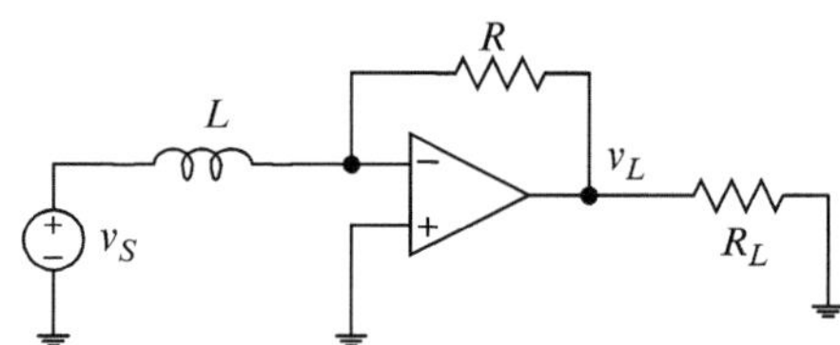

Fig. P 9.8 See Problem P 9.18

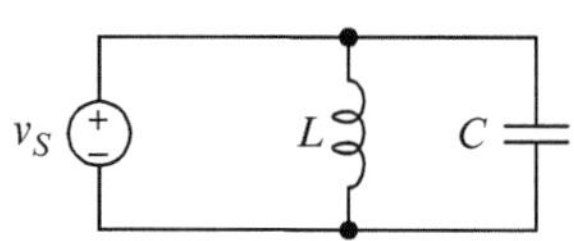

Fig. P 9.9 See Problem P 9.19

time. Obtain an expression for the current $i(t)$ though the diode for $t \geq 0$.

P 9.22 In Fig. P 9.12, the switch is closed at $t = 0$, having been open for a very long time before. Obtain an expression for the voltage $v_L(t)$.

Section 9.7 is prerequisite for the following problems. Problems in this section assume that coils connected in series or in parallel are connected in such a way that there are no magnetic flux linkages between the coils.

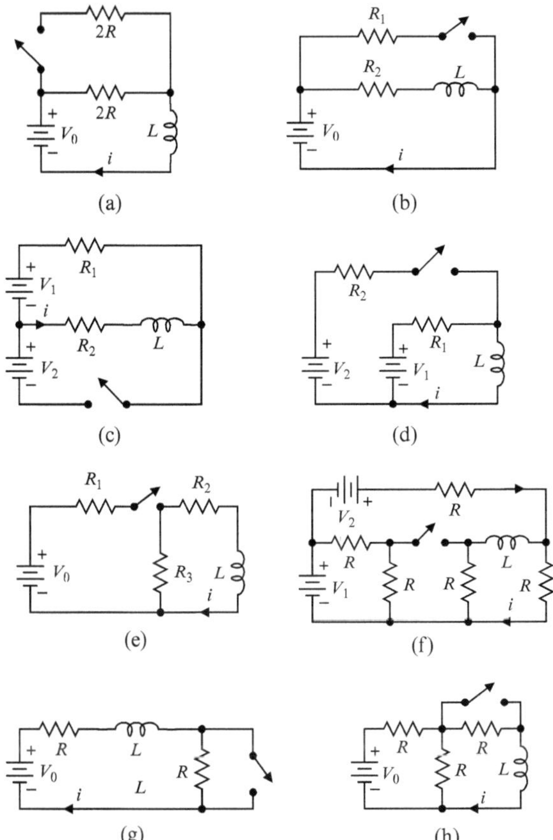

Fig. P 9.10 See Problem P 9.20

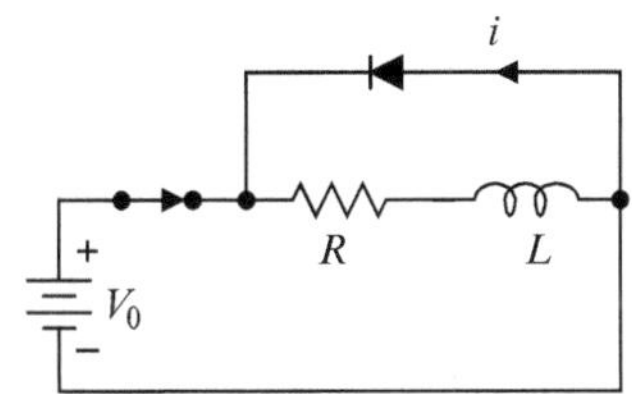

Fig. P 9.11 See Problem P 9.21

P 9.23 Show that the equivalent inductance of a parallel connection of N identical ideal inductors, each having inductance L, is given by $L_{eq} = L/N$.

P 9.24 Two identical inductors, each having inductance L and winding resistance R, are connected in parallel. Find an equivalent single inductor (equivalent inductance and equivalent winding resistance).

P 9.25 Suppose you are building a prototype circuit that requires a precision 10 μH air-core inductor. You have on hand a selection of precision 5 μH air-core inductors, having various lengths and diameters. You intend to obtain a 10 μH inductor by connecting two 5 μH inductors in series. Should you use the shortest or the longest of the 5 μH inductors you have on hand? Justify your answer.

P 9.26 In Fig. P 9.13, the switch is open for $t < 0$ and closed at $t = 0$. Obtain expressions for the voltage $v(t)$ and the currents $i_1(t)$, $i_2(t)$ for $t > 0$.

P 9.27 Two inductors having equal winding resistances and time constants τ_1 and $\tau_2 = 2\tau_1$ are connected in series. What is the time constant of the series connection? Assume the inductances add.

P 9.28 Two inductors having equal inductances and time constants τ_1 and $\tau_2 = 2\tau_1$ are connected in series. What is the time constant of the series connection? Assume the inductances add.

P 9.29 An inductor having time constant τ_1 is connected in series with one having time constant τ_2. Express the time constant of the series connection in terms of (a) the winding resistances and (b) the inductances of the individual inductors. Assume the inductances add.

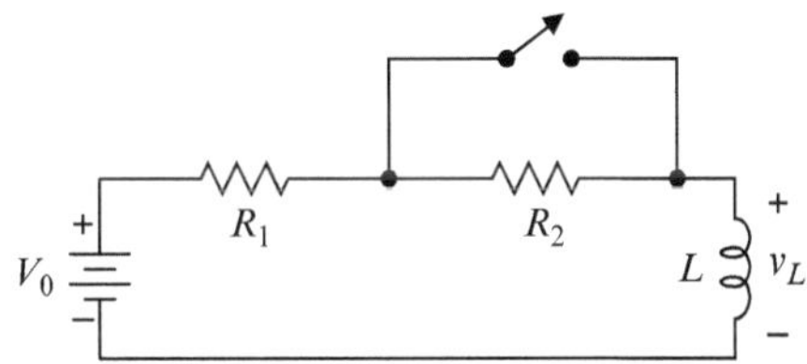

Fig. P 9.12 See Problem P 9.22

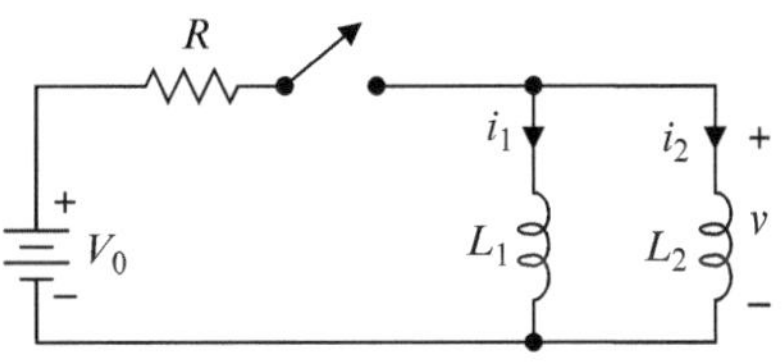

Fig. P 9.13 See Problem P 9.26

P 9.30 A long, tightly wound, single-layer, N-turn, air-core inductor having length l, diameter d, and inductance L given by

$$L = \frac{\pi \mu_0 N^2 d^2}{4l}$$

is cut into two pieces of equal length. Assume each piece is also a long coil. What is the inductance of each piece? Is the sum of the inductances equal to the inductance of the original coil?

P 9.31 A short, tightly wound, single-layer, N-turn, air-core inductor having length l, diameter d, and inductance L given by

$$L = \frac{\mu_0 N^2 d^2}{0.579\, d + 1.273\, l}$$

is cut into two pieces of equal length. (a) If the same formula applies, what is the inductance of each of the two coils? (b) Is the sum of the inductances equal to the inductance of the original coil? If not, why not? (c) Are inductors in series necessarily additive?

Section 9.8 is prerequisite for the following problems.

P 9.32 In Fig. P 9.14, the switch is closed for $t < 0$, opened at $t = 0$, and closed at $t = 5$ ms. Sketch neatly and label fully graphs of the current $i(t)$, the voltage $v_L(t)$, and the energy $w(t)$ stored in the inductor for the times during which they are changing. Use the same time scale for all three graphs and identify the times during which the inductor is absorbing energy and the times during which it is delivering energy to the circuit.

P 9.33 Refer to Fig. P 9.15, where $v_S = V_s \cos(\omega t)$. Assume the current through the inductor is given by

$$i_L(t) = \frac{1}{L} \int_0^t v_S(t')dt'$$

and obtain an expression for the total stored energy as a function of time. Show that the total stored energy is independent of time if

$$C = \frac{1}{\omega^2 L}.$$

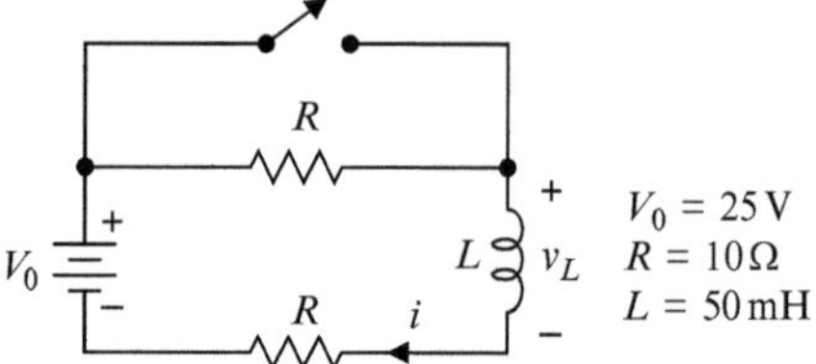

Fig. P 9.14 See Problem P 9.32

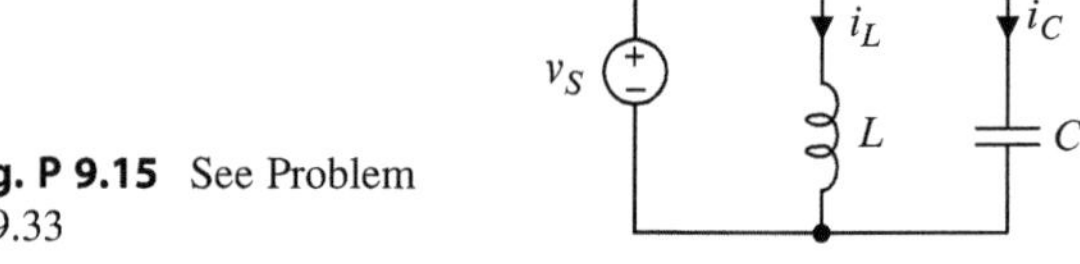

Fig. P 9.15 See Problem P 9.33

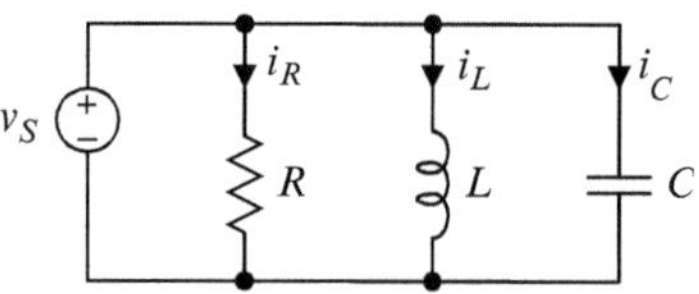

Fig. P 9.16 See Problem P 9.34

P 9.34 Refer to Fig. P 9.16, where $v_S = V_s \cos(\omega t)$ and $C = (\omega^2 L)^{-1}$. (a) Obtain an expression for the instantaneous total energy stored in the electric and magnetic fields associated with the capacitor and inductor. (b) Obtain an expression for the energy dissipated per cycle (period) of the source v_S. (c) The *quality factor* Q for a circuit in sinusoidal steady state is defined as

$$Q = 2\pi \times \frac{\text{maximum of instantaneous stored energy}}{\text{energy dissipated per period of the source}}.$$

Show that the quality factor of the circuit in Fig. P 9.16 is given by $Q = R/(\omega L)$.

P 9.35 In Fig. P 9.17, $i_S = I_s \cos(\omega_0 t)$, with $\omega_0 = 1/\sqrt{LC}$. Find the instantaneous voltage across the source. (This is an example of what is called *series resonance*. You will learn more about resonance in Chapter 12).

P 9.36 In Fig. P 9.18, $v_S = V_s \cos(\omega_0 t)$, with $\omega_0 = 1/\sqrt{LC}$. Find the instantaneous current i_S. (This is an example of what is called *parallel resonance*. You will learn more about resonance in Chapter 12).

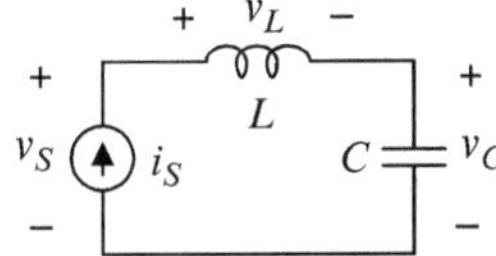

Fig. P 9.17 See Problem P 9.35

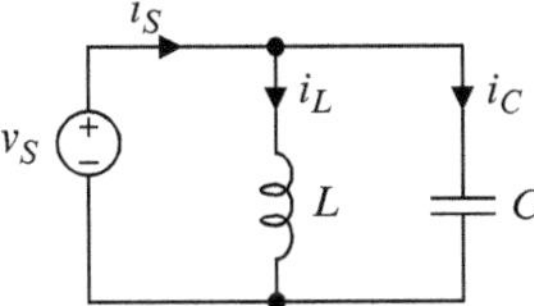

Fig. P 9.18 See Problem P 9.36

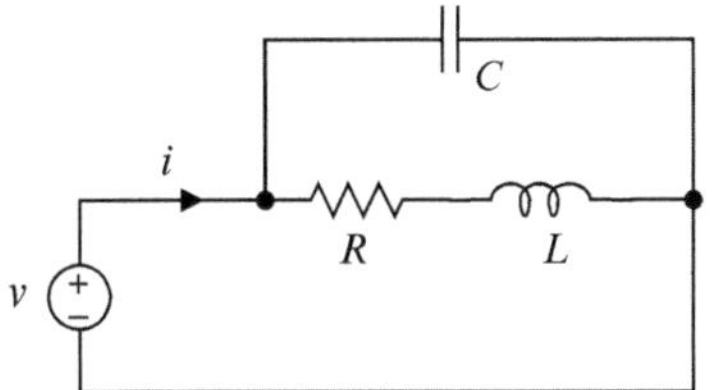

Fig. P 9.19 See Problem P 9.39

Section 9.9 is prerequisite for the following problems.

P 9.37 A sinusoidal current having frequency f and peak amplitude I passes through an axial-lead capacitor having capacitance C and series self-inductance L. Assuming a lumped-constant model is applicable, obtain an expression for the frequency at which the rms voltage drop across the self-inductance equals 1% of the rms voltage drop across the capacitance. Assume the voltage across the capacitance equals zero for $t = 0$.

P 9.38 For some purposes, a coil can be modeled as an inductance in parallel with a parasitic capacitance, the latter arising from the fact that the turns of the coil are charge-carrying conductors separated by an insulating medium. A sinusoidal voltage having frequency f and peak amplitude V_0 is impressed on a coil having inductance L and parallel parasitic capacitance C. Obtain an expression for the frequency at which the rms amplitude of the current through the capacitance equals 1% of that through the inductance. Assume the current through the inductance equals zero for $t = 0$.

P 9.39 The circuit in Fig. P 9.19 is a simplified high-frequency model for an axial-lead film resistor, where R is the nominal resistance, L is equivalent series inductance arising mainly from the internal (spiral) structure of the resistor, and C is the end-to-end capacitance. Let $R = 1\,\text{k}\Omega$, $L = 800\,\text{nH}$, $C = 0.4\,\text{pF}$, and $v = V_S \cos(2\pi f t)$, with $V_S = \sqrt{2}\,\text{V}$. Simulate the circuit and plot the ratio V_{rms}/I_{rms} versus the frequency f of the source, for $10\,\text{MHz} \le f \le 1\,\text{GHz}$. For what range of frequencies does the resistor behave like an ideal resistor?

Section 9.10 is prerequisite for the following problems.

P 9.40 The dc output of a certain power supply is 5 V. For a 0.5 Ω resistive load, the ripple factor of the load voltage is $\gamma = 0.002$. (a) What is the rms amplitude of the ac component of the load voltage? (b) Assuming the ripple is a 120 Hz sinusoid, what is the ripple factor for the load voltage if a 500 mH inductor is connected in series with the load? (c) : Simulate the circuit to check your answer.

P 9.41 The voltage across a 1Ω resistive load is given by

$$v = V_{dc} + V_{ac} \cos(2\pi f t),$$

with $f = 5\,\text{kHz}$, $V_{dc} = 25\,\text{V}$, and $V_{ac} = 25\,\text{mV}$. (a) What is the ripple factor? (b) Specify a series choke such that the rms ripple is reduced by the factor 2×10^{-4}. (c) Use a simulation to check your answer.

P 9.42 In Fig. P 9.20, $R = 0.5\,\Omega$ and $v_S = V_{dc} + V_{ac} \cos(\omega_0 t + \theta)$ with $V_{dc} = 50\,\text{V}$, $V_{ac} = 5\sqrt{2}\,\text{mV}$, and $f_0 = 120\,\text{Hz}$. The required ripple factor is 10^{-6}. Specify the inductance L.

Section 9.11 is prerequisite for the following problems.

P 9.43 The 20 A current through a 2 H inductor is reduced to zero in 20 ms by a circuit breaker. Estimate the magnitude of the average voltage across the inductor during that time.

P 9.44 In Fig. P 9.21, the fuse will melt if the current through the fuse exceeds 2.5 A. The current i_S equals zero for $t \le 0$ and is given by

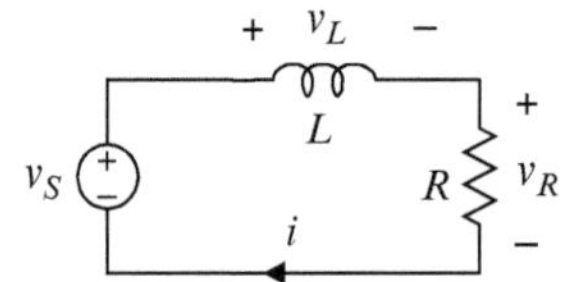

Fig. P 9.20 See Problem P 9.42

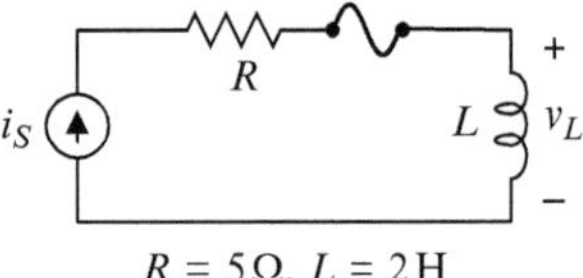

Fig. P 9.21 See Problem P 9.44

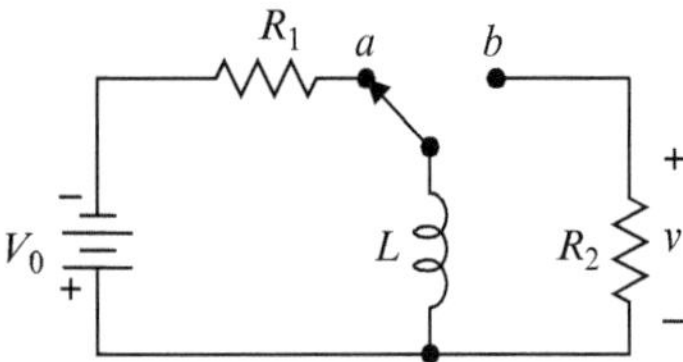

Fig. P 9.22 See Problem P 9.45

$$i_S = I_0[1 - \exp(-t/\tau)];\ I_0 = 5\,\text{A},\ \tau = 100\,\text{ms}$$

for $t > 0$. At what time will the fuse begin to melt? Assume the fuse opens completely in 15 ms and estimate the magnitude of the average voltage across the fuse body during that time.

P 9.45 Refer to Fig. P 9.22, where the switch is moved from a to b at $t = 0$, having been at a for $t < 0$. Assume a make-before-break switch, so there is no arcing. (a) Obtain an expression for the voltage $v(t)$ across the resistor R_2. (b) Let $V_0 = 3\,\text{V}$, $R_1 = 100\,\Omega$, $R_2 = 100\,\text{k}\Omega$, and $L = 500\,\text{mH}$. Construct a graph of the voltage $v(t)$ versus t for $0 \le t \le 5\tau$, where τ is the time constant for $t > 0$. (c) For the parameter values given in part (b), how much energy is eventually dissipated by the resistor R_2?

P 9.46 Refer to Fig. P 9.23, where the switch is opened at $t = 0$, having been closed for $t < 0$. (a) Obtain an expression for the voltage $v(t)$ across the resistor R_2. (b) Let $V_0 = 9\,\text{V}$, $R_1 = 10\,\Omega$, $R_2 = 10\,\text{k}\Omega$, $L = 1\,\text{H}$. Construct a graph of the voltage $v(t)$ versus t for $0 \le t \le 5\tau$, where τ is the time constant for $t > 0$.

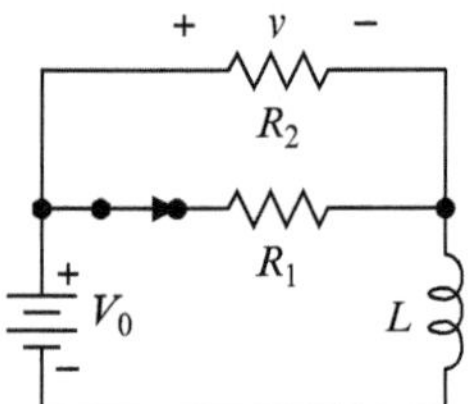

Fig. P 9.23 See Problem P 9.46

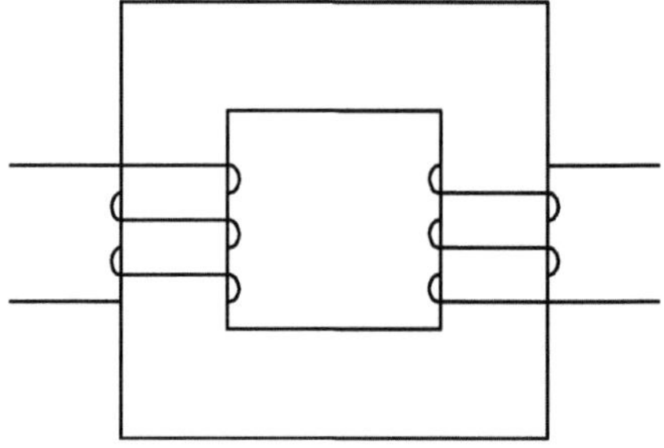

Fig. P 9.24 See Problem P 9.47

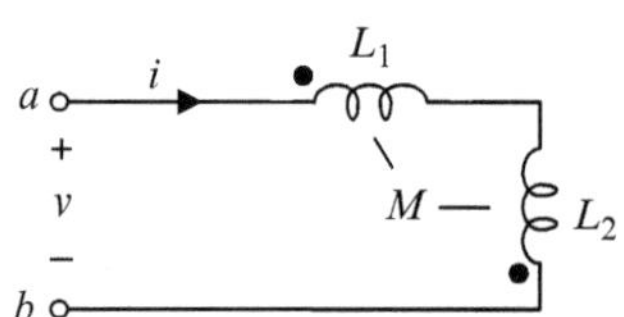

Fig. P 9.25 See Problem P 9.48

Section 9.12 is prerequisite for the following problems.

P 9.47 Figure P 9.24 depicts coils wound on a shared magnetic core. Assign dots to the terminals of the coupled coils.

P 9.48 Obtain the equivalent inductance at the terminals a–b of the circuit shown in Fig. P 9.25.

P 9.49 Two air-core coils having inductances 25 μH and 50 μH are arranged such that the mutual inductance between them is 7.1 μH. What is the coupling coefficient for the arrangement?

P 9.50 Two air-core coils having inductances 25 μH and 50 μH are connected in series, whereupon it is found that the total inductance is 90 μH. (a) What is the mutual inductance between the coils? (b) What would the total inductance become if the leads of one of the coils were reversed but the relative positions of the coils were unchanged?

P 9.51 You are given two air-core coils having inductances L_1 and L_2. If you intend to obtain an inductance $L = L_1 + L_2$ by connecting the coils in series, should you place them end to end, with the windings in the same direction, or side by side, or far apart, or how, otherwise?

P 9.52 You are given two air-core coils, each having equal inductance L. If you intend to obtain an inductance $L/2$ by connecting the coils in parallel, should you place them end to end, with the windings in the same direction, or side by side, or far apart, or how, otherwise?

P 9.53 In Fig. P 9.26,

$$v_S = \begin{cases} 0, & t \le 0, \\ V_S, & t > 0. \end{cases}$$

(a) Obtain an expression for the voltage $v(t)$ for $t > 0$. (b) Move the dot on one inductor to the other end of the inductor (reverse the leads on one inductor) and repeat part (a).

Section 9.13 is prerequisite for the following problems.

P 9.54 A pessimistic estimate for the inductance per unit length of a long straight wire distant from other conductors is 50 nHm^{-1}. Each of the two leads on a certain 10 kΩ axial-lead resistor is approximately 2.5 cm long and the current through the resistor is sinusoidal. Find the frequency for which the rms voltage due to lead inductance equals 1% of the total rms voltage across the resistor.

P 9.55 Axial and disc capacitors have relatively long leads and exhibit stray (parasitic) inductance. Inductors exhibit stray (parasitic) capacitance because the windings are neighboring current-carrying conductors. Thus, at sufficiently high frequencies, it might be necessary to model capacitors and inductors as shown in Fig. P 9.27. (1) For the model of a capacitor in Fig. P 9.27(a), assume $i(t) = I_0 \cos(2\pi f t)$ and $v_C(0) = 0$ and obtain an expression for the frequency at which the voltage across the parasitic inductance L_P equals 1% of the voltage across the capacitance C. (2) For the model of an inductor in Fig. P 9.27(b), assume $v(t) = V_0 \cos(2\pi f t)$ and $i_L(0) = 0$ and obtain an expression for the frequency at which the current through the parasitic capacitance C_p equals 1% of the current through the inductance L.

P 9.56 Refer to Fig. P 9.28. Let k denote the coupling coefficient for two coils in different branches of a certain circuit. The currents through the coils are sinusoidal, given by

$$i_1(t) = I_1 \cos(2\pi f_1 t); \quad i_2(t) = I_2 \cos(2\pi f_2 t).$$

(a) Obtain expressions for the voltages $v_1(t)$ and $v_2(t)$ across the coils, assuming both currents enter dotted terminals.
(b) Let $I_1 = I_2 = I$, $L_1 = L_2 = L$, $f_2 = a f_1$. Obtain a relation between k and a for which the rms amplitude of the voltage induced in coil 1 by the current in coil 2 equals the rms amplitude of the voltage induced in coil 1 by the current in coil 1.
(c) Let $I_2 = b I_1$, $L_1 = L_2$, $f_2 = f_1$. Obtain a relation between k and b for which the rms amplitude of the voltage induced in coil 1 by the current in coil 2 equals the rms amplitude of the voltage induced in coil 1 by the current in coil 1.
(d) Comment on the implications of the solutions to parts (b) and (c).

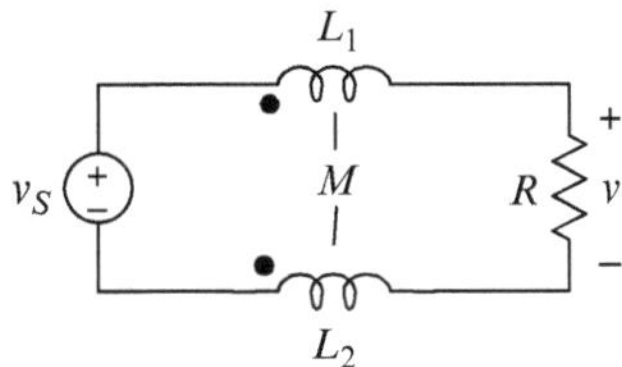

Fig. P 9.26 See Problem P 9.53

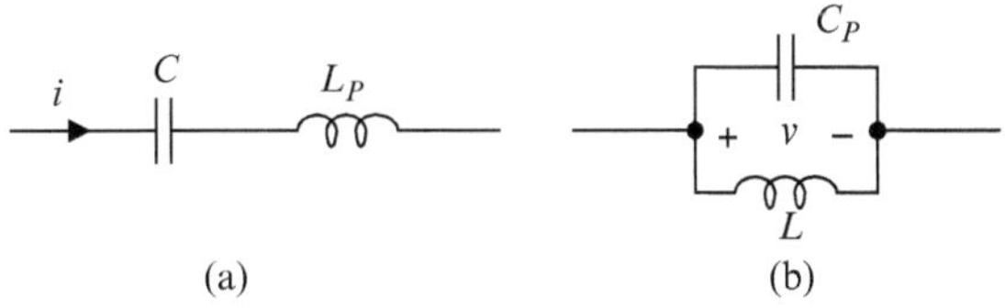

Fig. P 9.27 See Problem P 9.55

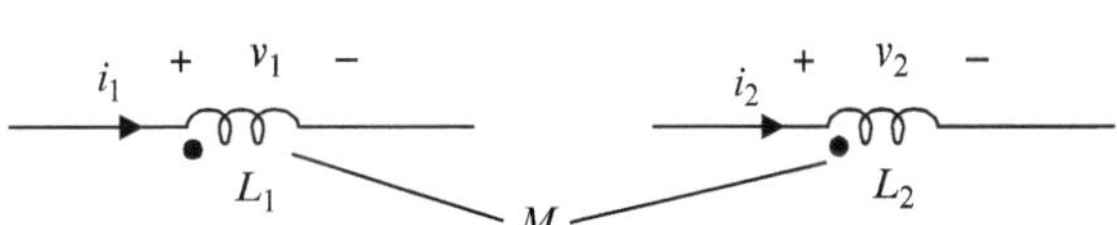

Fig. P 9.28 See Problem P 9.56

Section 9.14 is prerequisite for the following problems.

P 9.57 Obtain an expression for the equivalent inductance seen at the primary winding of a transformer whose secondary is open. Ignore winding resistance.

P 9.58 Figure P 9.29 shows a transformer whose primary and secondary are connected in parallel. Obtain an expression for the equivalent inductance at the terminals *a–b*.

P 9.59 Refer to Fig. P 9.30, where $L_1 = 25\,\text{mH}$, $L_2 = 400\,\text{mH}$, the coupling coefficient between the primary and secondary coils is 0.9, and $v(t) = V_0 \cos(2\pi f t)$, with $V_0 = 10\,\text{V}$ and $f = 100\,\text{Hz}$. Find the rms amplitude of the current $i(t)$, given that $i(0) = 0$.

P 9.60 In Fig. P 9.31, the switch is opened at $t = 0$, having been closed for a very long time. Assume the transformer coupling coefficient is nearly unity, such that $v_2 \cong n v_1$, where n is the transformer turns ratio. Obtain an expression for the voltage v_2 for $t > 0$. Calculate the maximum magnitude of v_2 if $V_0 = 12\,\text{V}$, $R_2 = 40R_1$, and $n = 50$.

Section 9.15 is prerequisite for the following problems.

P 9.61 What is the equivalent resistance at the primary of an ideal transformer whose secondary is open?

P 9.62 What is the equivalent resistance at the primary of an ideal transformer whose secondary is shorted?

P 9.63 Obtain the Thévenin and Norton equivalent circuits at the terminals *a–b* for the circuit shown in Fig. P 9.32. The transformers are ideal.

P 9.64 The output resistance for a source that is to drive a $2\,\text{k}\Omega$ resistive load is $R_S = 50\,\Omega$. Specify the turns ratio for an ideal transformer which, if inserted between the source and the load, will ensure maximum power transfer.

P 9.65 In Fig. P 9.33, the transformer is ideal and the voltage across the resistor is given by $v_R = V_0 \cos(2\pi f t)$. Obtain an expression for the source voltage v.

P 9.66 An inductor having inductance L is connected to the secondary of an ideal transformer having turns ratio $n = N_2/N_1$. Obtain an expression for the inductance seen at the primary terminals.

P 9.67 A capacitor having capacitance C is connected to the secondary of an ideal transformer having

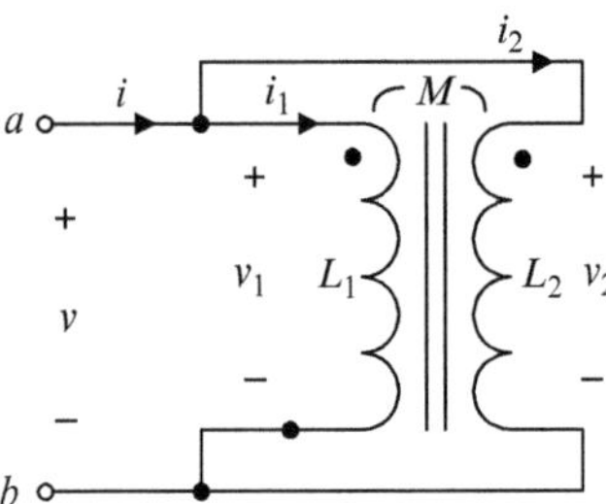

Fig. P 9.29 See Problem P 9.58

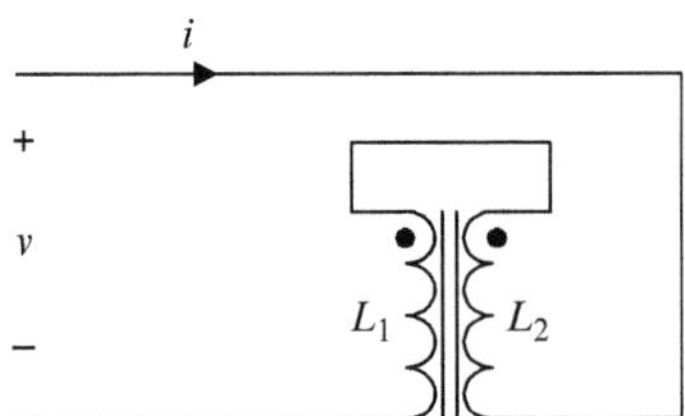

Fig. P 9.30 See Problem P 9.59

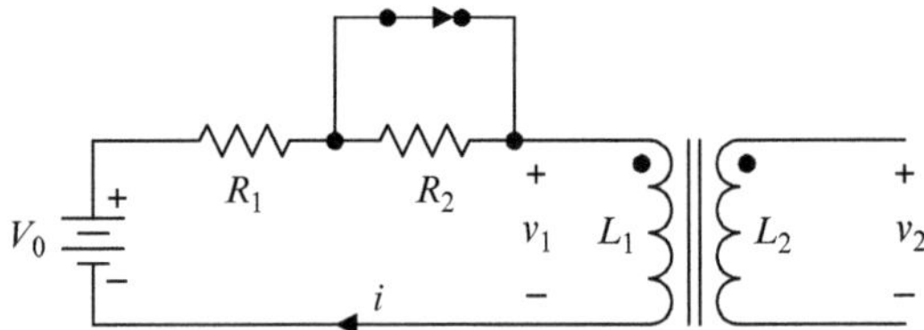

Fig. P 9.31 See Problem P 9.60

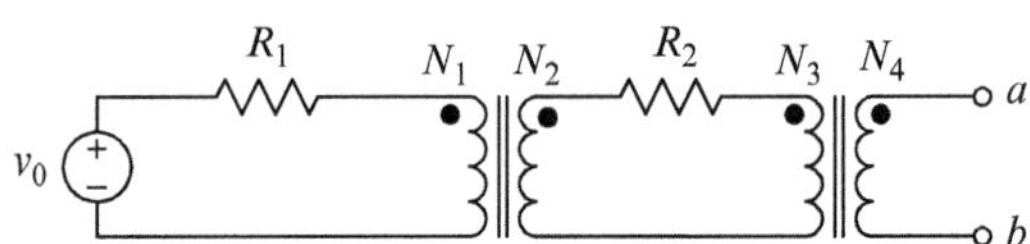

Fig. P 9.32 See Problem P 9.63

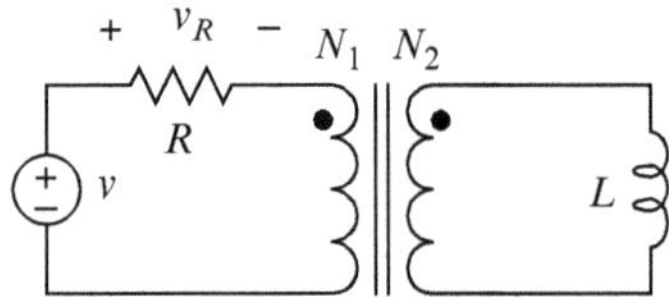

Fig. P 9.33 See Problem P 9.65

turns ratio $n = N_2/N_1$. Obtain an expression for the capacitance seen at the primary terminals.

P 9.68 Suppose you wish to use an ideal model for a certain very efficient physical transformer whose coupling coefficient is near unity. You know from measurements that the effective turns ratio of the transformer for a 1 kHz sinusoidal primary voltage is $n = 10$, what would you use for the turns ratio in the model if the voltage is a 10 kHz sinusoid?

Section 9.16 is prerequisite for the following problems.

P 9.69 A source whose output resistance is $50\,\Omega$ must be matched for maximum power transfer to a $200\,\Omega$ load. What turns ratio is required of a matching transformer?

P 9.70 The output resistance of a certain audio amplifier is $150\,\Omega$ and voltage is the output of interest. The load resistance is $3.2\,\Omega$. Specify the turns ratio for a transformer that will match the amplifier to the load (for maximum power transfer).

P 9.71 The output resistance of a certain microphone element is $2\,\text{k}\Omega$. The input resistance of an amplifier is $20\,\text{k}\Omega$. Should a transformer be used to match the microphone to the amplifier to maximize power transfer? Justify your answer.

P 9.72 A woodworker friend of yours just bought an industrial power tool having a 20 hp, 480 V (rms), 60 Hz motor. He has a 240 V outlet in his basement workshop. The outlet is protected by a 30 A circuit breaker. He intends to use a transformer to boost the 240 V to the 480 V required by the tool. Is that a good idea? Explain why or why not.

P 9.73 You are asked to specify the wire size for winding the secondary of a high-efficiency transformer whose coupling coefficient is close to unity. The numbers of primary and secondary turns in are $N_1 = 500$ and $N_2 = 50$, respectively. AWG 28 wire will be used for the primary. Copper wire is required for both windings. Is it a good idea to use the wire size that makes the primary and secondary dc resistances equal? If so, why? If not, why not, and what might be a better objective?

P 9.74 Outputs of power-plant generators are stepped up in stages to the very high voltages used for long-distance transmission to avoid excessive voltage differences and possible arcing between the primary and secondary windings of any one transformer. The output $V_{G\,rms} = 24\,\text{kV}$ of a certain power-plant generator must be stepped up to $V_{1\,rms} = 648\,\text{kV}$ for transmission. If the maximum allowable rms voltage between the primary and secondary windings of any one transformer is 220 kV, how many transformers are required and what are their turns ratios?

P 9.75 In Fig. P 9.34, $V_{G\,rms} = 24\,\text{kV}$, $V_{1rms} = 648\,\text{kV}$, $V_{L\,rms} = 440\,\text{V}$, and the load is $R_L = 500\,\text{m}\Omega$. The numbers of turns (N_1, N_2) are the same for each transformer except the last. The overall efficiency of the system is 94%, which means that 94% of the power produced by the generator is delivered to the load. Assume that each transformer is ideal and calculate the turns ratio n of the final step-down transformer and the resistance of the transmission line.

P 9.76 You want to put a speaker by your swimming pool, 50 m from your amplifier. You plan to use step-up and step-down audio transformers to allow use of small wire from the amplifier to the speaker. The effective resistance of the speaker is $8\,\Omega$ and the output resistance of the amplifier is $1\,\Omega$. The amplifier can deliver 200 W to an $8\,\Omega$ load. You have decided that the maximum loss in the wire should be no more than 5% of the power that would be delivered to the speaker if the wire resistance were zero. You have also decided to use AWG 24 copper wire to run from the amplifier to each speaker because you have that wire on hand. Specify the turns ratios for the step-up and step-down transformers. (Assume the transformers are ideal).

P 9.77 Refer to Fig. P 9.35, where the ground symbol represents earth ground. The source v_S is sinusoidal with frequency greater than zero. Suppose you wish to observe the source voltages $v_S(t)$ and $v_L(t)$ simultaneously, using a two-channel oscilloscope.

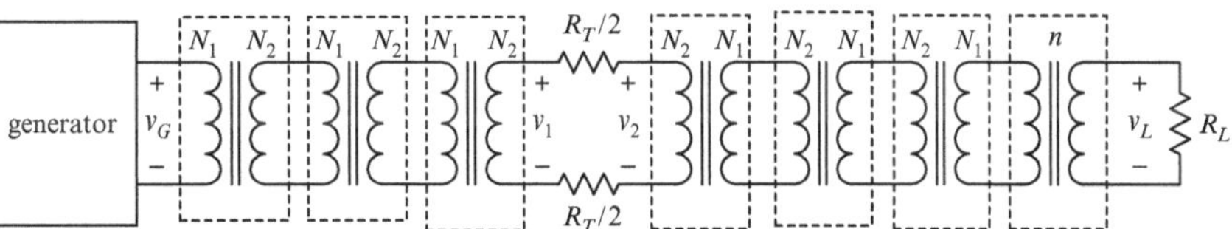

Fig. P 9.34 See Problem P 9.75

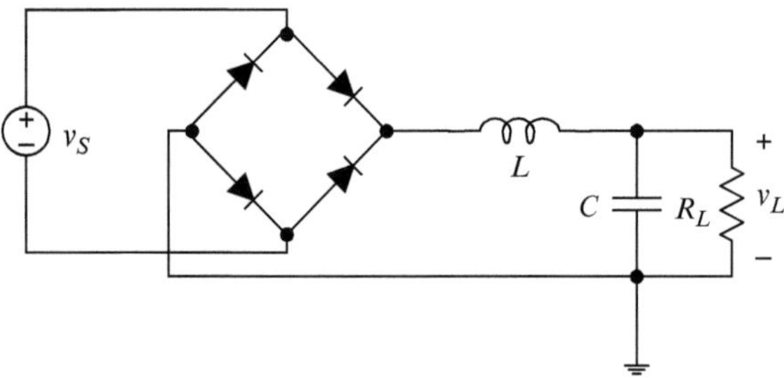

Fig. P 9.35 See Problem P 9.77

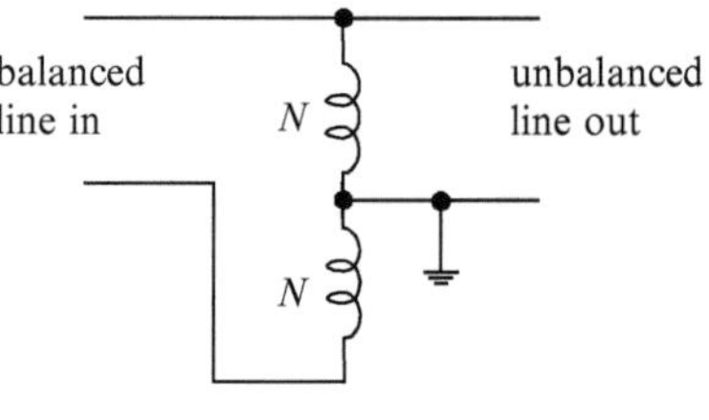

Fig. P 9.36 See Problem P 9.79

The oscilloscope is properly grounded through the three-conductor power cord and each probe has a center (positive) point and a ground clip. Describe why and how you would use an isolation transformer to perform this measurement.

P 9.78 You and a co-worker wish to test circuits you have designed for transmitting and receiving data over a long coaxial cable. You put the transmitter in one outbuilding on your industrial campus and the receiver in another, about 1 km distant, and run a coaxial cable between them. The transmitter and receiver chassis are connected to earth ground at their locations. If you connect the coaxial cable directly to both circuits, the shield on the cable will connect the chassis. Why might that be a problem? How could you use an isolation transformer to avoid a ground loop current?

P 9.79 Figure P 9.36 shows a balun consisting of a center-tapped autotransformer. The balanced input in Fig. P 9.36 is from a source having output resistance $R_S = 300\,\Omega$. Assume the balun can be modeled as an ideal transformer. What is the effective source resistance seen at the (unbalanced) output terminals?

Chapter 10
Complex Arithmetic and Algebra

In the remainder of the book, we make extensive use of complex arithmetic and algebra. This short chapter gives a brief review that subject and establishes a consistent notation. Working knowledge of the basic definitions, procedures, and relations reviewed here is essential to mastering material presented in subsequent chapters. Also, the terminology and notation associated with engineering applications of complex mathematics have not been standardized, and the terminology and notation introduced here are those used in this book.

Even if your mastery of complex arithmetic and algebra is such that you need only lightly skim this chapter, *give special attention to the definition of the four-quadrant inverse tangent in* Section 10.6 and *the definitions of the symbols* $\angle$ *and* $\measuredangle$ *in* Section 10.8.

10.1 Complex Numbers

A **complex number** is a number of the form $a+jb$, where a and b are real numbers and j is the **imaginary unit**[1] defined by

$$j=\sqrt{-1}. \tag{10.1}$$

The number a is the **real part** of the complex number and the number b is the **imaginary part**. The real part of a complex number z is denoted by $\mathrm{Re}\{z\}$ and the imaginary part by $\mathrm{Im}\{z\}$; i.e.,

$$\begin{aligned}\mathrm{Re}\{z\} &= \text{real part of } z,\\ \mathrm{Im}\{z\} &= \text{imaginary part of } z.\end{aligned} \tag{10.2}$$

Conversely,

$$z=\mathrm{Re}\{z\}+j\,\mathrm{Im}\{z\} \tag{10.3}$$

Both the real part and the imaginary part of a complex number are real. For example, if $z=3-j4$ then $\mathrm{Re}\{z\}=3$ and $\mathrm{Im}\{z\}=-4$. Do not make the common mistake of including j in the imaginary part of a complex number.

In applications, complex numbers and expressions often are dimensioned, in which case *the real part and the imaginary part must have the same dimension.*

10.2 Complex Arithmetic

The rules and results of complex arithmetic are generally consistent with those of real arithmetic; for example, the sum, difference, product, and quotient of two complex numbers (defined below) reduce to the sum, difference, product, and quotient, respectively, of the real parts of the numbers if the numbers are real (if their imaginary parts are equal to zero). But a few generalizations are necessary:

- Two complex numbers are equal if and only if the real parts are equal *and the imaginary parts* are equal.
- A complex number equals zero if and only if the real part *and the imaginary part* of the number are both equal to zero.

[1] Mathematicians, physicists, and others use i to denote the imaginary unit. Electrical engineers use j because they use i to denote electric current.

T.H. Glisson, *Introduction to Circuit Analysis and Design*,
DOI 10.1007/978-90-481-9443-8_10, © Springer Science+Business Media B.V. 2011

- The *negative* of $a + jb$ is $-a - jb$.
- The *magnitude* of $a + jb$ is $\sqrt{a^2 + b^2}$ (see Section 10.4).
- The *sum* of two (or more) complex numbers is obtained by adding the real parts and (separately) the imaginary parts; for example, $(3 + j4) + (2 - j6) = 5 - j2$. In general,

$$(a_1 + jb_1) \pm (a_2 + jb_2) = (a_1 \pm a_2) + j(b_1 \pm b_2). \quad (10.4)$$

- The *product* of two complex numbers is obtained by treating each number as a binomial; for example, $(3 + j4)(2 - j6) = 6 - j18 + j8 - j^2 24$, which, because $j^2 = -1$, reduces to $30 - j10$. In general,

$$(a_1 + jb_1)(a_2 + jb_2) = (a_1 a_2 - b_1 b_2) + j(a_1 b_2 + a_2 b_1). \quad (10.5)$$

- The *reciprocal* of a complex number $z = a + jb$ is the number $z^{-1} = x + jy$ such that $zz^{-1} = 1$. To determine x and y in terms of a and b, we must solve the equation $(a + jb)(x + jy) = 1$. This gives

$$(a + jb)(x + jy) = (ax - by) + j(bx + ay) = 1 + j0.$$

Equating the real parts and imaginary parts (separately) and solving the resulting simultaneous equations yields

$$x = \frac{a}{a^2 + b^2}, \quad y = -\frac{b}{a^2 + b^2}.$$

Thus the reciprocal of a complex number $a + jb$ is given by

$$(a + jb)^{-1} = \frac{1}{a + jb} = \frac{a - jb}{a^2 + b^2}. \quad (10.6)$$

- Using (10.6), the *quotient* of two complex numbers can be expressed as

$$\frac{z_1}{z_2} = \frac{a_1 + jb_1}{a_2 + jb_2} = (a_2 + jb_2)^{-1}(a_1 + jb_1) = \frac{(a_1 + jb_1)(a_2 - jb_2)}{a_2^2 + b_2^2} \quad (10.7)$$

Example 10.1.

$$\frac{3 + j4}{2 - j6} = \frac{(3 + j4)(2 + j6)}{2^2 + 6^2} = \frac{-18 + j26}{40} = -\frac{18}{40} + j\frac{26}{40} = -0.45 + j0.65.$$

Complex arithmetic differs from real arithmetic in another significant respect: Whereas a set of distinct real numbers can be ordered (e.g., listed from smallest to largest), *a set of complex numbers cannot be ordered.* As a result, one cannot say whether one complex number is larger or smaller than another; in other words, inequalities of the form $z_1 < z_2$ are undefined if z_1 or z_2 is complex. Otherwise, the rules of basic arithmetic for complex numbers are the same as those for real numbers. If z_1, z_2, and z_3 denote complex numbers, then

- Addition and multiplication are *commutative*, such that $z_1 + z_2 = z_2 + z_1$ and $z_1 z_2 = z_2 z_1$
- Addition and multiplication are *associative*, such that $z_1 + (z_2 + z_3) = (z_1 + z_2) + z_3$ and $z_1(z_2 z_3) = (z_1 z_2) z_3$; and
- Multiplication is *distributive* over addition, such that $z_1(z_2 + z_3) = z_1 z_2 + z_1 z_3$.

Exercise 10.1. Given $z_1 = 3 + j4$, $z_2 = 1 - j2$, $z_3 = j5$, $z_4 = -5 + j$, verify each of the following:

(*a*) $z_1 + z_2 - z_3 + z_4 = -1 - j2$, (*b*) $z_1(1 - z_2) = -8 + j6$, (*c*) $6z_1 + 4(z_2 - z_4) = 42 + j12$, (*d*) $z_1 z_2 = 11 - j2$, (*e*) $z_3 z_4 = -5 - j25$, (*f*) $z_1 z_3 = -20 + j15$, (*g*) $\frac{z_2}{z_3} = -0.4 - j0.2$, (*h*) $z_1 z_2^{-1} = -1 + j2$, (*i*) $\frac{z_1 z_3}{z_4} = 4.42 - j2.12$.

10.3 Conjugate of a Complex Number

The **conjugate** of a complex number z is obtained by changing the sign of the imaginary part of the number. For example, the conjugate of $3 + j4$ is $3 - j4$. The conjugate of a complex number z is denoted by z^*:

$$\text{conjugate of } z = z^* = \text{Re}\{z\} - j\,\text{Im}\{z\}. \quad (10.8)$$

The conjugate of a real number is the number itself.

Conjugation is distributive over addition, multiplication, and division:

$$(z_1 + z_2)^* = z_1{}^* + z_2{}^*,\ (z_1 z_2)^* = z_1{}^* z_2{}^*,\ \left(\frac{z_1}{z_2}\right)^* = \frac{z_1{}^*}{z_2{}^*}. \quad (10.9)$$

As a result, it is unnecessary to reduce an expression to the form $a + jb$ in order to obtain the conjugate. All that is necessary is to replace j by $-j$ wherever it appears in the expression. For example, the conjugate of $(j)^3(3+j4)/(2-j6)^2$ is $(-j)^3(3-j4)/(2+j6)^2$. Note that this approach requires that all of the imaginary units (js) in the expression be visible; for example, the conjugate of jz is $-jz^*$, not simply $-jz$, if z is complex.

The sum and difference of a complex number and its conjugate are of special importance because

$$z + z^* = 2\,\text{Re}\{z\},\ z - z^* = 2j\,\text{Im}\{z\}. \quad (10.10)$$

10.4 Magnitude of a Complex Number

The **magnitude** of a complex number z is denoted by $|z|$ and is defined by

$$|z| = \sqrt{\text{Re}^2\{z\} + \text{Im}^2\{z\}}; \quad (10.11)$$

that is, if $z = a + jb$, then $|z| = \sqrt{a^2 + b^2}$.

Exercise 10.2. Show that if z is real, the right side of (10.11) reduces to the absolute value of z. (Recall that $\sqrt{\ }$ denotes *positive* square root.)

The magnitude of a complex number is a nonnegative real number. Therefore, the magnitudes of complex numbers can be ordered and expressions of the form $|z_1| < |z_2|$ are meaningful. An important example is the (generalized) *triangle inequality*

$$|z_1 + z_2 + \cdots + z_n| \leq |z_1| + |z_2| + \cdots + |z_n|. \quad (10.12)$$

The magnitude of the product or quotient of two complex numbers is the product or quotient of the magnitudes:

$$|z_1 z_2| = |z_1||z_2|,\quad \left|\frac{z_1}{z_2}\right| = \frac{|z_1|}{|z_2|}. \quad (10.13)$$

The product of a complex number and its conjugate is of special importance because it equals the square of the magnitude of the number. From (10.11),

$$z z^* = (a + jb)(a - jb) = a^2 + b^2 = |z|^2. \quad (10.14)$$

This result can be used to rationalize the quotient of two complex numbers or expressions prior to performing the indicated division. From (10.14),

$$\frac{z_1}{z_2} = \frac{z_1 z_2{}^*}{z_2 z_2{}^*} = \frac{z_1 z_2{}^*}{|z_2|^2}. \quad (10.15)$$

It is easier to remember and apply the *process* expressed by (10.15) than it is to remember and use the formula (10.6).

Exercise 10.3. Let $z_1 = 3 + j4$, $z_2 = 1 - j2$, $z_3 = j5$, $z_4 = -5 + j$. Verify each of the following:

$(a)\ z_1 z_2{}^* = -5 + j10$, $(b)\ |z_1 - z_3| = 3.16$,

$(c)\ \dfrac{z_1}{z_1{}^*} = -0.28 + j0.96$, $(d)\ \left|\dfrac{z_1}{z_2}\right| = 2.24$,

$(e)\ \dfrac{|z_2|^2}{z_1 z_4{}^*} = -0.085 + j0.177$, $(f)\ \left|\dfrac{z_1 z_2}{(z_1 z_2)^*}\right| = 1.$

10.5 Arithmetic in a Complex Plane

Complex numbers can be placed in correspondence with points in a Cartesian plane called a **complex plane**. The coordinates (abscissa and ordinate) of a point in the plane are the real and imaginary parts, respectively, of the number represented by the point. The horizontal axis is called the **real axis**, often labeled $\text{Re}\{z\}$ and the vertical axis is called the **imaginary axis**, often labeled $\text{Im}\{z\}$. Figure 10.1. shows a complex plane in which several complex numbers

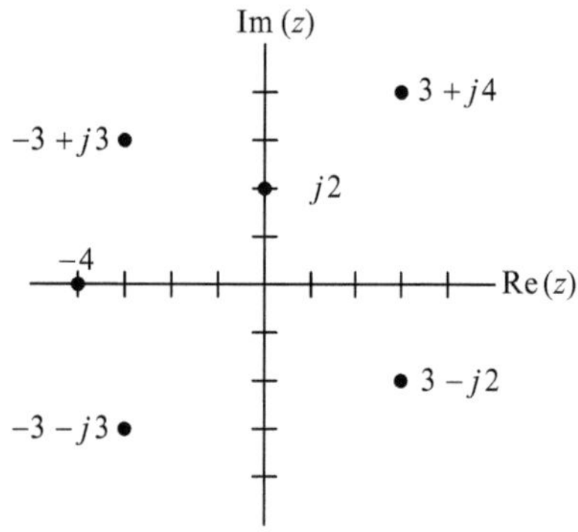

Fig. 10.1 Complex numbers represented by points in a plane

(including one real number and one imaginary number) are represented.

The complex number represented by a point is often written next to the point, as in Fig. 10.1; however, a point in a complex plane is a *representation* of the number, not the number itself. Again, *both the abscissa and ordinate of a point representing a complex number are real numbers.*

Often, it is helpful to refer to graphical descriptions of arithmetic operations using a complex plane, as illustrated in Fig. 10.2. Figure 10.2(a) shows addition of two complex numbers, illustrating that complex numbers add like vectors (because the real and imaginary parts add separately, as do the components of a vector). Figure 10.2(b) shows negation and conjugation of a single complex number. Negation reflects a number through the origin and conjugation reflects a number in the horizontal (real) axis. Figure 10.2(c) shows multiplication of a complex number by various powers of the imaginary unit j, illustrating that multiplication by j rotates the number 90° counter-clockwise about the origin.

10.6 Polar Form of a Complex Number

A complex number, when written as $z = a + jb$, is said to be expressed in **rectangular form** because the components of the number (real and imaginary parts) are the coordinates of the corresponding point (a, b) expressed in rectangular coordinates.[2] A complex number also may be expressed in **polar form** as $r\angle\theta$ as illustrated in Fig. 10.3, where

$$r = \sqrt{a^2 + b^2}, \quad \theta = \mathrm{Tan}^{-1}(a,b). \tag{10.16}$$

The function $\mathrm{Tan}^{-1}(\ )$ in (10.16) is the **four-quadrant inverse tangent**, defined by

$$\mathrm{Tan}^{-1}(a,b) = \begin{cases} \tan^{-1}(b/a), & \text{if } a>0, \\ (\pi/2)\,\mathrm{sign}(b), & \text{if } a=0 \text{ and } b\neq 0, \\ 0, & \text{if } b=0 \text{ and } a=0, \\ \pi + \tan^{-1}(b/a), & \text{if } a<0. \end{cases} \tag{10.17}$$

where $\tan^{-1}(\ )$ is the *two-quadrant inverse tangent*, defined as follows: If $\alpha = \tan(\theta)$ and $-\pi/2 < \theta \leq \pi/2$, then $\tan^{-1}(\alpha) = \theta$. The two-quadrant inverse tangent takes a single argument and returns a value in either quadrant I or quadrant IV, depending upon the sign of the argument. The four-quadrant inverse tangent takes two arguments and returns an angle in the correct one of the four quadrants.

In computer languages and mathematical software, the two-quadrant inverse tangent is usually denoted by atan and the four-quadrant inverse tangent by atan2. Also, different software packages order the arguments $x = \mathrm{Re}(z), y = \mathrm{Im}(z)$ of atan2 differently. For example, Matlab uses atan2(y, x) and Mathcad uses atan2(x, y). The definition (10.17) is consistent with Mathcad's atan2 function. Fortunately, Matlab and Mathcad both provide functions that return the angle of a complex argument: In Matlab, angle$(z) =$ atan 2[Im(z), Re(z)] and in Mathcad, arg$(z) =$ atan 2 [Re(z), Im(z)]. If you use the angle( ) function in Matlab or the arg( ) function in Mathcad when calculating angles of complex quantities, you need not remember the order of the arguments in the atan2 function.

For pocket calculators, the key labeled atan or $\tan^{-1}$ computes the two-quadrant inverse tangent, whereas the angle computed by the rectangular-to-polar conversion function is the four-quadrant inverse tangent. The two-quadrant inverse tangent function returns the correct angle of a complex number only if the number lies in the first or fourth quadrants. For an example, refer to Fig. 10.4. For $z = -1 + j$, the four-quadrant inverse tangent gives

[2] What we call rectangular coordinates also are called *cartesian coordinates* and the rectangular form of a complex number also is called the *cartesian form* of the number.

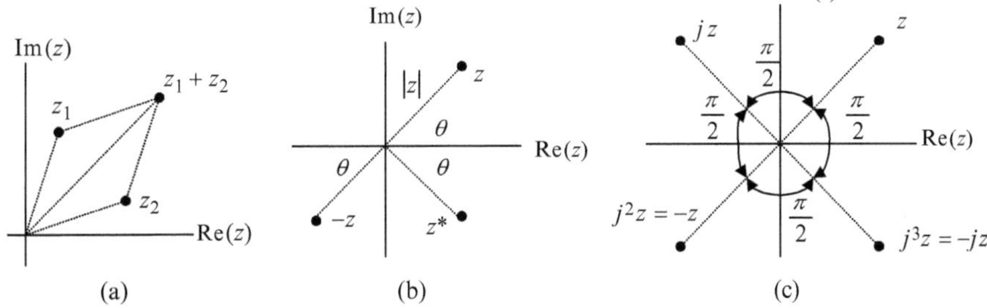

Fig. 10.2 Depiction of various operations in a complex plane

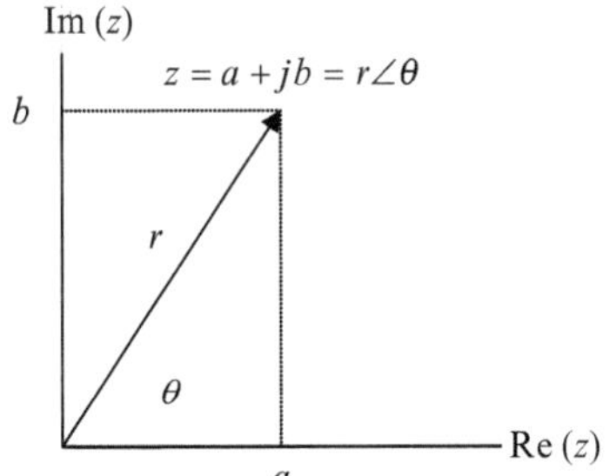

Fig. 10.3 Polar representation of a complex number in a complex plane

$$\theta = \mathrm{Tan}^{-1}(-1,1) = \pi + \tan^{-1}(-1) = \pi + \left(-\frac{\pi}{4}\right) = \frac{3\pi}{4},$$

which is correct, whereas the ordinary (two-quadrant) inverse tangent gives

$$\tan^{-1}\left(\frac{1}{-1}\right) = \tan^{-1}(-1) = -\frac{\pi}{4},$$

which is incorrect.

From (10.16), the *radial coordinate* r of a complex number is just the magnitude of the number (defined above). The *angular coordinate* θ is called the *angle* of the number; e.g., the magnitude and angle of $3 + j4$ are 5 and 0.927, respectively.[3,4] The symbol $r\angle\theta$ is read as "r angle θ." The complex-plane representation of a complex number can be shown either as a point, as in Fig. 10.2, or as a vector (an arrow from the origin to the point), as in Fig. 10.3. Usually points are used to display numbers expressed in rectangular form and vectors are used to display numbers expressed in polar form.

The relationships between the rectangular form $a + jb$ and the polar form $r\angle\theta$ of a complex number are summarized below:

$$a + jb = \sqrt{a^2 + b^2}\angle\mathrm{Tan}^{-1}(a,b) = r\angle\theta \qquad (10.18)$$

and

$$r\angle\theta = r\cos(\theta) + jr\sin(\theta). \qquad (10.19)$$

Example 10.2. Express the number $-3 + j4$ in polar form.
Solution: From (10.19),

$$\sqrt{3^2 + 4^2} = 5,\ \mathrm{Tan}^{-1}(-3,4) = 2.214$$
$$\Rightarrow -3 + j4 = 5\angle 2.214.$$

Exercise 10.4. Given $z_1 = 3 + j4, z_2 = 1 - j2, z_3 = j5, z_4 = -5 + j$, verify each of the following:
(*a*) $z_1 = 5.00\angle 0.93$, (*b*) $z_2 = 2.24\angle -1.11$,
(*c*) $z_3 = 5.00\angle 1.57$, (*d*) $z_4 = 5.10\angle 2.94$,
(*e*) $z_1^* = 5.00\angle -0.93$,
(*f*) $(z_2^*)^{-1} = 0.45\angle -1.11$.

10.7 Eulers Identity and Polar Arithmetic

Replacing $\cos(\theta)$ and $\sin(\theta)$ by their series expansions on the right side of (10.20) and comparing the result with the series expansion for $e^{j\theta}$ leads to **Eulers identity**

$$r\,e^{j\theta} \equiv r\cos(\theta) + jr\sin(\theta). \qquad (10.20)$$

[3]Because we use radian measures exclusively, we do not attach the dimensionless unit *rad* to numerical values for angles.

[4]Mathematicians refer to the magnitude as the *modulus* and to the angle as the *argument*.

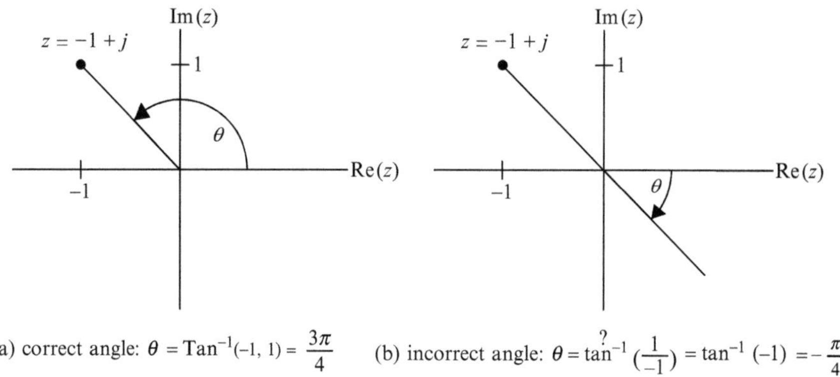

(a) correct angle: $\theta = \mathrm{Tan}^{-1}(-1, 1) = \dfrac{3\pi}{4}$ (b) incorrect angle: $\theta \overset{?}{=} \tan^{-1}\left(\dfrac{1}{-1}\right) = \tan^{-1}(-1) = -\dfrac{\pi}{4}$

Fig. 10.4 In (**a**), the angle θ is given correctly by the four-quadrant inverse tangent, in (**b**) the angle θ is given incorrectly by the two-quadrant inverse tangent

Thus,

$$r\angle\theta \equiv r\,e^{j\theta}. \tag{10.21}$$

Equation (10.22) tells us how a polar representation $r\angle\theta$ can be manipulated mathematically (e.g., in products and quotients) using the rules of ordinary arithmetic and algebra. It follows from (10.20), (10.21), and the law of exponents that

$$\begin{aligned}(r_1\angle\theta_1)(r_2\angle\theta_2) &= r_1 e^{j\theta_1} r_2 e^{j\theta_2} = r_1 r_2 e^{j(\theta_1+\theta_2)}\\ &= r_1 r_2 \angle(\theta_1+\theta_2).\end{aligned} \tag{10.22}$$

In words, the magnitude of the product is the product of the magnitudes and the angle of the product is the sum of the angles.

It also follows from (10.20), (10.21), and the law of exponents that

$$\begin{aligned}\frac{r_1\angle\theta_1}{r_2\angle\theta_2} &= \frac{r_1 e^{j\theta_1}}{r_2 e^{j\theta_2}} = \frac{r_1}{r_2}\,e^{j(\theta_1-\theta_2)}\\ &= \frac{r_1}{r_2}\angle(\theta_1-\theta_2).\end{aligned} \tag{10.23}$$

In words, the magnitude of the quotient is the quotient of the magnitudes and the angle of the quotient is the angle of the numerator minus the angle of the denominator.

Example 10.3. Obtain the product $z_1 z_2$ and the quotient z_1/z_2, where $z_1 = 2\angle\pi/6$ and $z_2 = 4\angle\pi/4$.
Solution: From (10.23) and (10.24)

$$\begin{aligned}z_1 z_2 &= (2\angle\pi/6)(4\angle\pi/4) = (2)(4)\angle\left(\frac{\pi}{6}+\frac{\pi}{4}\right)\\ &= 8\angle\frac{5\pi}{12},\\ \frac{z_1}{z_2} &= \frac{2\angle\pi/6}{4\angle\pi/4} = \frac{2}{4}\angle\left(\frac{\pi}{6}-\frac{\pi}{4}\right) = 0.5\angle-\frac{\pi}{12}.\end{aligned}$$

If the angle of a complex number is known exactly as the product of π and a rational number, the angle is often expressed that way, as in Example 10.3, above. Otherwise, angles are expressed using decimal numbers, as in Example 10.4, below.

Example 10.4. Perform the indicated operations and express the result (z) in both polar and rectangular form.

$$z = \frac{(3+j4)+(6+j8)}{5+j10}.$$

Solution:

$$\begin{aligned}z &= \frac{(3+j4)+(6+j8)}{5+j10} = \frac{9+j12}{5+j10}\\ &\cong \frac{15\angle 0.927}{11.2\angle 1.11}\\ &\cong 1.34\angle-0.180\\ &\cong 1.34\cos(-0.180)+j1.34\sin(-0.180)\\ &\cong 1.32-j0.24.\end{aligned}$$

Exercise 10.5. Let $z_1 = 3 + j4$, $z_2 = 1 - j2$, $z_3 = j5$, $z_4 = -5 + j$. Verify each of the following:

$(a)\, z_1 z_2 = 11.2\angle -0.180$, $(b)\, \dfrac{z_2}{z_3} = 0.447\angle -2.68$,

$(c)\, z_1 z_3{}^* z_4{}^{-1} = 4.90\angle 2.70$,

$(d)\, \dfrac{z_2 z_4}{z_4{}^* z_2{}^*} = 1\angle -2.61$, $(e)\, \dfrac{|z_3|^2}{z_1 z_4{}^*} = 0.981\angle 2.02$.

Nowadays, all but the simplest computations usually are done using a calculator (or computer) capable of doing complex arithmetic. When using such a calculator or computer for computations, it makes little difference whether we express complex numbers in rectangular or polar form. The choice often is dictated by ease of data entry or by the form in which the data are available. But when performing pencil-and-paper computations or *symbolic manipulations* of complex expressions, the rectangular form is generally most convenient for addition and subtraction and the polar form is best for multiplication and division. Furthermore, in a number of applications, the polar form is preferred for interpretative reasons.

10.8 The Symbols $\angle$ and $\measuredangle$

The symbol $\angle$ denotes that the number (or expression) following is an angle; e.g., when expressing a number in polar form. In that context, the symbol $\angle$ serves as a *separator* between the magnitude and angle of a number denoted by $r\angle\theta$, just as the comma serves as a separator between the x- and y-coordinates of a point denoted by (x, y).

The symbol $\measuredangle$ is an *operator* that means *the angle of the number or expression that follows*. For example, $\angle\pi/2$ denotes an angle equal to $\pi/2$ radians (90°) whereas $\measuredangle\pi/2$ means the angle of the real number $\pi/2$, which equals zero. For $z = x + jy$,

$$\measuredangle z = \measuredangle(x + jy) = \mathrm{Tan}^{-1}(x, y). \qquad (10.24)$$

A value or expression following the *symbol* $\angle$ is a real angle expressed in radians. A value or expression following the *operator* $\measuredangle$ is in general complex and can be dimensionless or dimensioned.

Example 10.5.
(a) $\measuredangle(1 + j) = \pi/4$ (b) $\measuredangle(4\angle 0.6) = 0.6$
(c) $3\angle\measuredangle j = 3\angle\pi/2$ (d) $\measuredangle 10e^{-j0.8} = -0.8$,

$$\text{(e)} \left|\frac{3 - j4}{2 + j}\right| \measuredangle\left(\frac{3 - j4}{2 + j}\right) = \frac{|3 - j4|}{|2 + j|} \times [\measuredangle(3 - j4) - \measuredangle(2 + j)] = \sqrt{5}\angle -1.39.$$

Exercise 10.6. Let $z_1 = 3 + j4$, $z_2 = 1 - j2$, $z_3 = j5$, $z_4 = -5 + j$. Verify each of the following:

$(a)\, \measuredangle\left(\dfrac{z_1 z_2}{z_3{}^*}\right) = 1.39$, $(b)\, \measuredangle(z_2{}^*)^{-2} = -2.21$,

$(c)\, \measuredangle(z_1)^4 = 4\measuredangle z_1$ (draw a picture),

$(d)\, \measuredangle(z_1 z_2 z_3 z_4) = -1.95$,

$(e)\, \measuredangle z_1 + \measuredangle z_1{}^* = 0$, $(f)\, \measuredangle\left(\dfrac{z_2}{z_2{}^*}\right) = 2\measuredangle z_2$,

$(g)\, |z_1 z_2|\angle(\measuredangle z_3) = 11.2\angle 1.57$.

10.9 Problems

Section 10.5 is prerequisite to the following problems.

P 10.1 Find the real and imaginary parts, the magnitude, the conjugate, and the reciprocal of each of the following numbers.

(a) $-j$, (b) $3 - j2$, (c) $\sqrt{-4}$, (d) $5j^2$, (e) j^3,
(f) $(2j)^{-1}$, (g) $1 - j3$, (h) $3j + 12 - \sqrt{-16}$,
(i) $j(2 - j5)$,
(j) $1 + 2j - 3j^2 + 4j^3 - 6j^4$, (k) $\dfrac{6}{2j} - 8$.

P 10.2 Reduce each of the following to the form $a + jb$. Represent each result as a point in a complex plane.

(a) $(3+j2)(5-j6)$, (b) $(2-j3)^3$, (c) $\dfrac{120-j240}{3+j4}$,

(d) $(1-j)\left[\dfrac{4+j2}{1+j}+3+j4\right]$.

P 10.3 Let $z_1 = 1-j,\ z_2 = 2+j3,\ z_3 = 6j$. Reduce each of the following to the form $a+jb$. Represent each result as a point in a complex plane.

(a) $\dfrac{z_1}{|z_2|}$, (b) $\dfrac{z_1 z_2^*}{10}$, (c) $\dfrac{(z_1 z_2 z_3)^*}{10}$, (d) $\left|\dfrac{z_3 z_1}{z_2^*}\right|$,

(e) $\dfrac{z_1 z_2 + z_1^* z_2^*}{10}$, (f) $\dfrac{z_1 z_2 - z_1^* z_2^*}{z_3}$, (g) z_2^{-1},

(h) $\dfrac{(z_1+z_3^*)z_2}{10}$, (i) $\dfrac{z_1+z_3^*}{z_2-z_1}$, (j) $\dfrac{z_3}{z_1+z_2^*}$,

(k) $\dfrac{(z_1-z_3^*)(z_1+z_3)}{3z_2-z_1^*}$, (l) $\dfrac{z_1+z_2+z_3}{(z_1+z_2+z_3)^*}$.

P 10.4 Let $z_1 = 1-j,\ z_2 = 1+j,\ z_3 = 2j$. Depict each of the following operations in a complex plane.

(a) z_1+z_2, (b) z_1-z_2, (c) z_1+z_3, (d) z_2+z_3,

(e) $z_1+z_2+z_3$, (f) $z_1 z_3$, (g) $z_2 z_3$, (h) z_3^3,

(i) $(z_1+z_2)z_3$, (j) $\dfrac{z_2-z_1}{z_3}$.

Section 10.7 is prerequisite to the following problems.

P 10.5 Refer to Problem P 10.1. Express each number given there in polar form.

P 10.6 Under what conditions on a and b do the four-quadrant inverse tangent and the two-quadrant inverse tangent give the same number for the angle of $a + jb$?

P 10.7 Use both the four-quadrant inverse tangent and the two-quadrant inverse tangent to obtain the angle of each of the numbers given in Problem P 10.1 If the angles differ, explain why.

P 10.8 Let $z_1 = 1-j,\ z_2 = 2+j3,\ z_3 = 6j$. Convert each to polar form. Then perform each of the following operations using polar arithmetic and convert the result back to rectangular form.

(a) $\dfrac{z_1}{|z_2|}$ (b) $z_1 z_2^*$ (c) $(z_1 z_2 z_3)^*$ (d) $\left|\dfrac{z_3 z_1}{z_2^*}\right|$ (e) z_2^{-1} (f) z_1^5

(g) $\left(\dfrac{z_2}{z_1}\right)^3 z_3^*$ (h) $\dfrac{z_1 z_3^2}{(z_1^2 z_2 z_3)^*}$

P 10.9 Refer to Problem P 10.1. Express each number given there in the form $r\,e^{j\theta}$.

Section 10.8 is prerequisite to the following problems.

P 10.10 Let $z_1 = 1-j,\ z_2 = 2+j3,\ z_3 = 6j$. Find each of the following, doing as little arithmetic as possible.

(a) $\measuredangle\left(\dfrac{z_1}{z_2}\right)$, (b) $\measuredangle(z_1 z_3)$, (c) $\measuredangle(z_2 z_3)$, (d) $\measuredangle(z_3^3)$,

(e) $\measuredangle(z_1 z_2 z_3)$, (f) $\measuredangle\left[\left(\dfrac{z_2 z_1^*}{z_3}\right)^*\right]$, (g) $(\measuredangle z_2)+(\measuredangle z_1)$,

(h) $(\measuredangle z_3)^2$.

P 10.11 Let $z_1 = 1-j,\ z_2 = 2+j3,\ z_3 = 6j$. Find each of the following, doing as little arithmetic as possible.

(a) $\left|\dfrac{z_1}{z_2}\right|$, (b) $|z_1 z_3|$, (c) $|z_2 z_3|$, (d) $|z_3^3|$, (e) $|z_1 z_2 z_3|$,

(f) $\left|\dfrac{z_2 z_1^*}{z_3}\right|$, (g) $|z_2|+|z_3|$, (h) $|z_2^2|$.

Chapter 11
Transient Analysis

Transient analysis examines the *transition* of a current or voltage in a circuit from one steady state to another after an *abrupt* change in excitation or circuit structure. Questions addressed by transient analysis include:

- How long does the transition take? Is the response quick or sluggish?
- What is the new steady state after the transition is complete?
- Is the transition smooth, or are there overshoots and undershoots before steady state is achieved?
- If there are overshoots or undershoots, how large are they? Might they damage some sensitive component?
- What circuit parameters influence the transition and new steady state, and in what ways?

In circuits terminology, the *order of a differential equation* relating a response (output) to an excitation (input) is the order of the highest derivative of the response, and the **order of a circuit** is the order of the differential equation relating the excitation and response of interest. In this chapter, we limit our discussion to circuits for which excitations and responses are related by first-order or second-order differential equations.

As noted above, transient analysis deals in part with responses to excitations that change abruptly from one value to another. The unit step function defined in the next section facilitates describing such excitations.

11.1 Unit Step Function

Sudden changes (jump discontinuities) in a current or voltage can be described compactly using the **unit step function**, defined by

$$u(t) = \begin{cases} 0, \ t \le 0, \\ 1, \ t > 0. \end{cases} \tag{11.1}$$

Example 11.1. A voltage given by

$$v(t) = \begin{cases} 0, & t \le 0, \\ V_0, & t > 0, \end{cases}$$

can be expressed compactly as $v(t) = V_0\, u(t)$.

Example 11.2. The voltage

$$v(t) = \begin{cases} V_1, \ t \le 0, \\ V_2, \ t > 0, \end{cases}$$

can be expressed compactly as

$$v(t) = V_1 + (V_2 - V_1)\, u(t).$$

Exercise 11.1. Use unit step functions to express

$$i(t) = \begin{cases} I_0, \ t \le t_0, \\ I_1, \ t_0 < t \le t_1, \\ I_2, \ t > t_1 \end{cases}$$

T.H. Glisson, *Introduction to Circuit Analysis and Design*,
DOI 10.1007/978-90-481-9443-8_11, © Springer Science+Business Media B.V. 2011

11.2 Notation

In transient analysis, it is often necessary to specify times that differ only infinitesimally from each other; e.g., at instants just before and just after an abrupt change in excitation. As a reminder, we denote the instants infinitesimally before and after a time $t = t_0$ by $t = t_0^-$ and $t = t_0^+$, respectively. For example, if a voltage v is defined by $v(t) = V_1\, u(t)$, then

$$v(0^-) = 0, \quad v(0^+) = V_1.$$

For any time t_0, the instants $t = t_0^-$, $t = t_0$, and $t = t_0^+$ are only infinitesimally different, so any *continuous* function has the same value at all three times. For example,

$$\exp\left(-\frac{t_0^-}{\tau}\right) = \exp\left(-\frac{t_0}{\tau}\right) = \exp\left(-\frac{t_0^+}{\tau}\right)$$

and

$$\sin\left(\omega\, t_0^-\right) = \sin(\omega\, t_0) = \sin\left(\omega\, t_0^+\right).$$

We attach units only to non-zero, finite times, e.g., $t = 5\,\text{ms}$, but $t = 0$ (no unit) and $t \to \infty$ (no unit). If $t = 0$ or $t \to \infty$, then t is zero or infinite whether expressed in seconds, microseconds, minutes, or years, so in those cases units are irrelevant. For the same reason, there is no need to associate a unit with $t = 0^-$ or $t = 0^+$. Similarly, we do not attach units to any current, voltage, power, energy, or charge that equals zero, is infinitesimal, or approaches infinity. For example, if a voltage $v(t)$ equals zero at time t_0, we write $v(t_0) = 0$, not $v(t_0) = 0\,\text{V}$.[1]

We often use a prime to denote the first derivative with respect to time of a current or voltage $y(t)$, as in

$$y' = y'(t) = \frac{dy}{dt}.$$

[1]Software packages that provide for dimensioned variables require consistent use of units (if used at all), even for zero values. Also, note that we require a unit for zero temperature, because zero Celsius, zero Fahrenheit, and zero Kelvin are all different.

Thus $y'(t_0)$ denotes the first derivative with respect to time of a current or voltage $y(t)$, evaluated at $t = t_0$:

$$y'(t_0) = \left.\frac{dy}{dt}\right|_{t=t_0}. \tag{11.2}$$

The notation in (11.2) means *differentiate first, then replace t with* t_0. If you do the reverse, the result will always be zero.

We also need a concise notation for the value of a current or voltage in the distant future; for example, a long time after an abrupt change in excitation. We use the notation $v(\infty)$ and $i(\infty)$ as shorthand for the associated limits; i.e.,

$$v(\infty) = \lim_{t\to\infty} v(t), \quad i(\infty) = \lim_{t\to\infty} i(t);$$
$$v'(\infty) = \lim_{t\to\infty} v'(t), \quad i'(\infty) = \lim_{t\to\infty} i'(t). \tag{11.3}$$

Mathematicians might object to this notation because infinity is not a specific value that can be substituted for a variable in a function. Nonetheless, the notation is compact and serves us well.

11.3 Initial Conditions

For the present, we limit our attention to obtaining and interpreting responses of stable first-order and second-order linear circuits to step excitations – excitations that change abruptly at t=0. Solving the differential equations governing such responses requires knowing the response and (for second-order circuits) the first derivative of the response at t=0^+. Such *initial conditions* are obtained using Kirchhoff's laws and the fact that currents through inductors and voltages across capacitors are continuous.

A step excitation, such as $v_S(t) = V_S\, u(t)$, is zero for *all* $t < 0$ and equals a non-zero constant V_S for *all* $t > 0$. A circuit is in **dc steady state** if all currents and voltages in the circuit are constant. In particular, voltages across capacitors and currents through inductors are constant in dc steady state. Generally, a circuit excited by a step current or voltage is in dc steady state before $t = 0$.

It follows from the terminal characteristics

$$i = C\frac{dv}{dt} = Cv', \quad v = L\frac{di}{dt} = Li',$$

that if a circuit is in dc steady state, the current through each capacitor and the voltage across each inductor equals zero. Before finding currents and voltages in a circuit in dc steady state, we may replace each capacitor by an open circuit and each inductor by a short circuit. A circuit in dc steady state is essentially resistive.

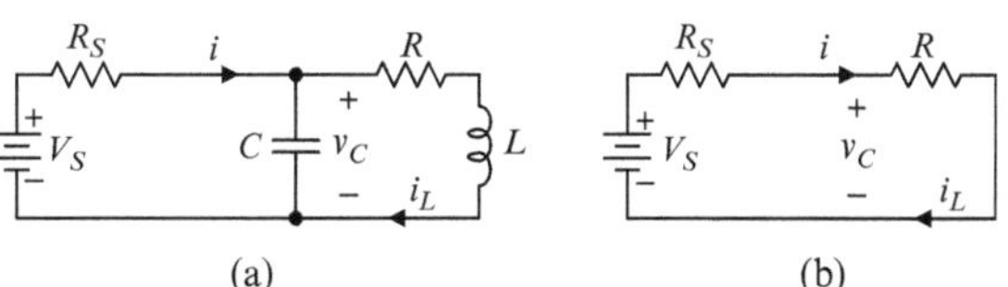

Fig. 11.1 See Example 11.3

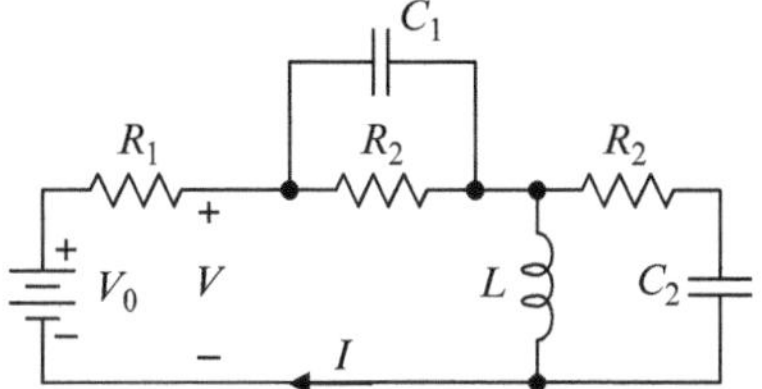

Fig. 11.2 See Exercise 11.2

Example 11.3. The circuit in Fig. 11.1(a) is in dc steady state. Obtain expressions for the current i and the voltage v_C.

Solution: Because the circuit is in dc steady state, we may replace the capacitor by an open circuit and the inductor by a short circuit. This gives the circuit in Fig. 11.1(b), where

$$i_L = i = \frac{V_S}{R_S + R}, \quad v_C = R\, i_L = \frac{RV_S}{R_S + R}.$$

Exercise 11.2. The circuit in Fig. 11.2 is in dc steady state. (a) Obtain expressions for the current I and the voltage V. (b) Obtain expressions for the voltages across the capacitors and the current through the inductor. Indicate assumed current directions and voltage polarities on the circuit diagram.

Response to a step excitation, called a **step response**, often is used to describe or specify performance, especially in applications where excitations exhibit repeated abrupt changes in amplitude. The response of a linear stable circuit to a step excitation consists of a **transient component**, which eventually vanishes, and a **steady-state component**, which remains after the transient has vanished. A step response might be described in terms of attributes such as rise time, overshoot, settling time, and steady-state amplitude. In short, step response is often used to address questions such as those posed at the beginning of this chapter. The next two sections describe step responses of first- and second-order circuits.

11.4 First-Order Circuits

A **first-order circuit** is one for which excitation and response are related by a first-order differential equation expressed in **standard form** as

$$\tau\frac{dy}{dt} + y = b_0 x + b_1\frac{dx}{dt}, \tag{11.4}$$

where $x(t)$ is an excitation and $y(t)$ is the corresponding response. For a step excitation $x(t) = X_0\, u(t)$ and for $t > 0$, the differential equation (11.4) becomes

$$\tau\frac{dy}{dt} + y = b_0 X_0, \quad t > 0. \tag{11.5}$$

The coefficient of the response y in (11.5) is unity, so every term in the standard-form differential equation has the dimension of the response. It follows that

$$\mathrm{SI}(\tau) = \mathrm{s}, \quad \mathrm{SI}(b_0) = \frac{\mathrm{SI}(y)}{\mathrm{SI}(X_0)}. \tag{11.6}$$

We treated such circuits in Chapters 8 and 9, where we showed that the response is given (for $t > 0$) by

$$y(t) = y(\infty) - [y(\infty) - y(0^+)] \exp\left(-\frac{t}{\tau}\right). \quad (11.7)$$

Example 11.4. Refer to Fig. 11.3, where $v_S = V_0\, u(t)$.

(a) Obtain a standard-form differential equation relating the voltage v_C (the response) to the excitation for $t > 0$. (b) Obtain expressions for $v_C(0^-)$ and $v_C(0^+)$. (c) Obtain an expression for $v_C(t)$.

Solution: (a) Applying Kirchhoff's current law to the circuit gives

$$\frac{v_C - V_0}{R} + C\frac{dv_C}{dt} = 0, \quad t > 0. \quad (11.8)$$

Multiplying by the resistance R (to make the coefficient of v_C equal to unity) and rearranging terms gives the standard-form differential equation

$$\tau\frac{dv_C}{dt} + v_C = V_0, \quad t > 0, \quad \tau = RC. \quad (11.9)$$

(b) The source v_S is essentially a short circuit for $t < 0$, so $v_C(0^-) = 0$. The voltage across a capacitor cannot change instantaneously, so $v_C(0^+) = v_C(0^-) = 0$. For $t > 5\tau$, the circuit is once again in dc steady state, with $v_C(\infty) = V_0$.

(c) Thus, $v_C(t) = 0$ for $t < 0$ and

$$v_C(t) = v_C(\infty) - [v_C(\infty) - v_C(0^+)] \exp\left(-\frac{t}{\tau}\right)$$
$$= V_0\left[1 - \exp\left(-\frac{t}{\tau}\right)\right], \quad t > 0, \quad \tau = RC.$$

Although we know the form of the response of a first-order circuit to a step excitation, we wish to derive that solution again, but by a different method; one that is applicable to second- and higher-order differential equations. If you have had or are taking a course in differential equations, you might be familiar with this method, called the *method of undetermined coefficients*, which proceeds as follows:

The complete solution to (11.5) is called the **complete response** or simply the **response** of the associated circuit. The response is the sum of the forced response y_f and the unforced response y_u:[2]

$$y = y_f + y_u. \quad (11.10)$$

The **forced response** *of a stable linear circuit has the form of the excitation.* For step excitation, the forced response is a constant because the excitation is a constant (for $t > 0$). Thus $y_f = Y_0$, where Y_0 is a constant. Substituting this solution for y in (11.5) gives

$$Y_0 = b_0 X_0, \quad (11.11)$$

so the forced response is given by

$$y_f = b_0 X_0 \quad (11.12)$$

The **unforced response** is the general solution to the *homogeneous equation*

$$\tau\frac{dy_u}{dt} + y_u = 0. \quad (11.13)$$

The unforced response must contain one arbitrary constant that cannot be determined from (11.13) alone. As we show below, the arbitrary constant can be determined from the initial value of the response, denoted by $y(0^+)$.

According to (11.13), a linear combination of the unforced response and its derivative (with respect to time) have the same time dependence. Otherwise the left side could not be zero for all time. The exponential function has this property, so we are led to try $y_u = Ye^{st}$ in (11.13), where Y and s are constants, as yet unknown.[3] We obtain

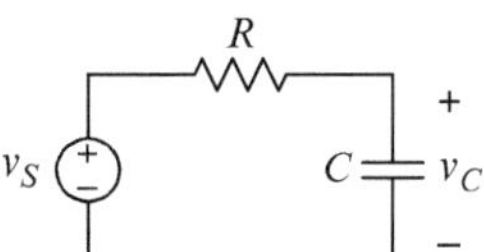

Fig. 11.3 See Example 11.4

[2]Mathematicians call the forced response the *particular integral* and the unforced response the *complementary solution* (or function). The names used here are more appropriate (and suggestive) for physical problems.

[3]Note that *s* (italic) is a variable and s (roman) is a unit (second). In written work, you might want to distinguish the unit s from the variable *s* by underlining the former.

$$\tau \frac{d(Ye^{st})}{dt} + Ye^{st} = \tau Yse^{st} + Ye^{st}$$
$$= Ye^{st}(\tau s + 1) = 0. \qquad (11.14)$$

The exponential e^{st} cannot be zero for any finite value of s. We require $Y \neq 0$ because $Y = 0$ gives the *trivial solution* $y_u = 0$, which contains no arbitrary constants and can be ignored. Thus the function $Y e^{st}$ satisfies the differential equation (11.31) if the constant s satisfies the **characteristic equation**

$$\tau s + 1 = 0, \qquad (11.15)$$

whose solution is the **characteristic root**

$$s = -\frac{1}{\tau}. \qquad (11.16)$$

The parameter τ is the **time constant** for the response.

The unforced response is given by

$$y_u = Ye^{st} = Ye^{-t/\tau}, \qquad (11.17)$$

where Y is an arbitrary constant having the dimension of the response and τ is the time constant. From (11.10), (11.12), and (11.17), the complete response is given by

$$y(t) = b_0 X_0 + Ye^{st}, \quad t > 0, \qquad (11.18)$$

where

$$y(0^+) = b_0 X_0 + Y \Rightarrow Y = y(0^+) - b_0 X_0. \quad (11.19)$$

Thus (compare with (11.7))

$$y(t) = b_0 X_0 + [y(0^+) - b_0 X_0]e^{-t/\tau}, \quad t > 0. \quad (11.20)$$

A response given by (11.20) consists of an unforced component $[y(0^+) - b_0 X_0]e^{-t/\tau}$ and a forced component $b_0 X_0$. In a stable circuit, the unforced component eventually vanishes, and only the forced component remains. Thus the forced component is the **steady-state response** and the unforced component is the **transient response**, called the **transient**, for short.

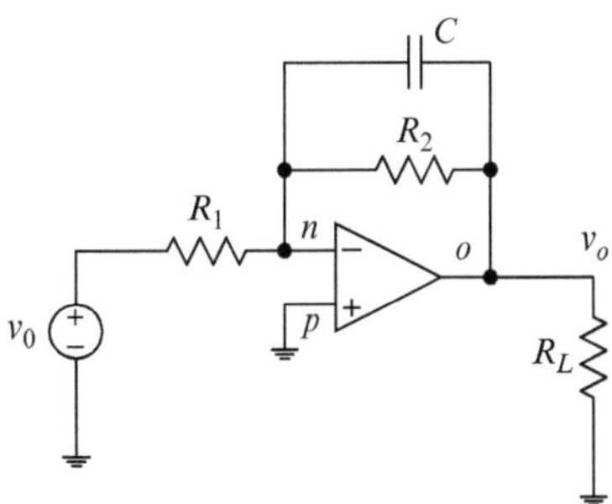

Fig. 11.4 See Example 11.5

Example 11.5. Refer to Fig. 11.4, where the op amp is ideal and $v_0(t) = V_0\, u(t)$. Obtain an expression for the voltage $v_o(t)$. Identify the transient and steady-state components of the response.

Solution: No current enters the n terminal. Kirchhoff's current law gives (for $t > 0$)

$$\frac{v_n - V_0}{R_1} + \frac{v_n - v_o}{R_2} + C\frac{d(v_n - v_o)}{dt} = 0. \qquad (11.21)$$

From the circuit diagram $v_p = 0$. The op amp is ideal, so $v_n = v_p \Rightarrow v_n = 0$. Equation (11.21) reduces to

$$R_2 C \frac{dv_o}{dt} + v_o = -\frac{R_2 V_0}{R_1}, \quad t > 0, \qquad (11.22)$$

which is (11.5), with $y = v_o$, $X_0 = V_0$, $b_0 = -R_2/R_1$, and $\tau = R_2 C$. The excitation equals zero for all $t \leq 0$, and the capacitor is shunted by a resistor, so $v_o(0^-) = 0$. Because $v_n = 0$, the voltage v_o is the voltage across the capacitor, so v_o must be continuous. It follows that

$$v_o(0^+) = v_o(0^-) = 0. \qquad (11.23)$$

From (11.20)

$$v = \begin{Bmatrix} 0, & t<0, \\ -\frac{R_2}{R_1}V_0(1-e^{-t/\tau}), & t>0 \end{Bmatrix},$$
$$\tau = R_2C. \tag{11.24}$$

The steady-state component is the forced response v_f and the transient component is the unforced response v_u:

$$v_f = -\frac{R_2}{R_1}V_0, \; v_u = \frac{R_2}{R_1}V_0e^{-t/\tau}; \; t>0.$$

The time constant of a first-order circuit is discussed at length in Chapters 8 and 9. As a reminder, the time constant τ is a measure of the time required for the transient component to vanish (for steady state to be achieved). For $t = 5\tau$, $e^{-5} \cong 0.00674$ and the magnitude of the transient component is less than 1% of its initial value. Often, 5τ is taken as the **duration of the transient** (as the time required for steady state to be achieved). For $t > 0$, a step excitation $x = X_0\,u(t)$ is a constant (dc) current or voltage given by $x = X_0$. After the transient vanishes (for practical purposes), the response of a linear stable circuit to such a step excitation is also a constant (dc) current or voltage, given by $y = b_0X_0$. Thus the magnitude of the parameter b_0 is called the **dc gain** of the circuit. The time constant τ and dc gain $|b_0|$ are constant functions of circuit parameters.

Two distinct procedures are treated in the discussion above: One addresses how to obtain an expression for the response of a *first-order circuit* to a step excitation. If that is the objective (11.7) usually provides the most direct approach. The other addresses how one can solve a differential equation using the assumed solution $Y\exp(st)$. That procedure is generalized to second-order differential equations in the sequel.

Equation (11.7) can be applied to any of the responses labeled in either of the circuits in Fig. 11.5.

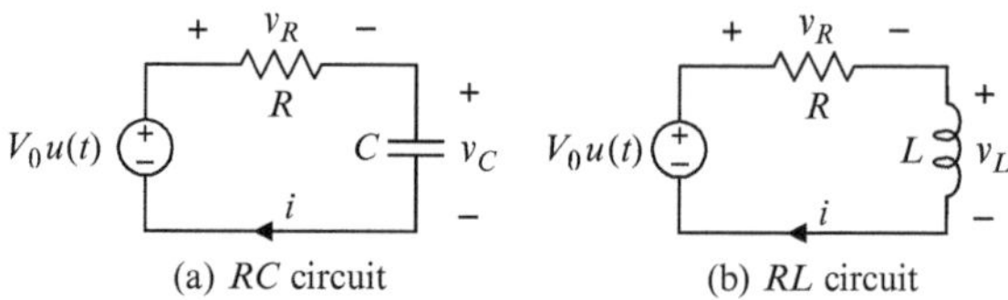

Fig. 11.5 First-order RC and RL series circuits

Example 11.6. Obtain an expression for the current $i(t)$ in Fig. 11.5(b).

Solution: The excitation (the independent source) is zero for $t < 0$, so $i(0^-) = 0$. The current though an inductor must be continuous, so $i(0^+) = i(0^-) = 0$. The time constant is $\tau = L/R$. For $t \gg \tau$, the circuit is in dc steady state, the inductor is effectively a short circuit, so $i(\infty) = V_0/R$. Thus, from (11.7), $i(t) = 0$ for $t < 0$ and

$$i(t) = i(\infty) - [i(\infty) - i(0^+)]\exp\left(-\frac{t}{\tau}\right)$$
$$= \frac{V_0}{R}\left[1 - \exp\left(-\frac{t}{\tau}\right)\right], \; t > 0.$$

The complete response would normally be written using a unit step function as

$$i(t) = \frac{V_0}{R}\left[1 - \exp\left(-\frac{t}{\tau}\right)\right]u(t), \quad \tau = \frac{L}{R}.$$

Example 11.7. Obtain an expression for the voltage $v_R(t)$ in Fig. 11.5(a).

Solution: The voltage $v_R(t)$ appears across a resistor (not a capacitor), and need not be continuous. But the voltage $v_C(t)$ must be continuous, so $v_C(0^+) = 0$. Kirchhoff's current law gives, for $t = 0^+$,

$$v_R(0^+) + v_C(0^+) = V_0 \Rightarrow v_R(0^+) = V_0.$$

The time constant is $\tau = RC$. For $t \gg \tau$, the circuit is in dc steady state, the capacitor is effectively an open circuit, so $i(\infty) = 0$, which implies (Ohm's law) $v_R(\infty) = 0$. From (11.7)

$$v_R(t) = v_R(\infty) - [v_R(\infty) - v_R(0^+)]\exp\left(-\frac{t}{\tau}\right)$$
$$= V_0\exp\left(-\frac{t}{\tau}\right), \; t>0.$$

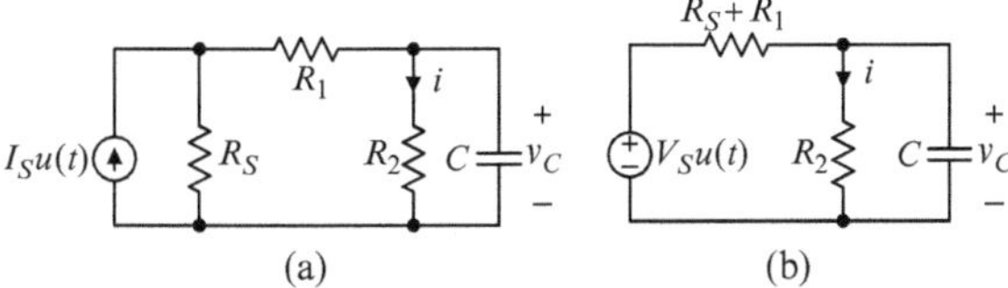

Fig. 11.6 See Example 11.8

Thus

$$i(t) = V_0 \exp\left(-\frac{t}{\tau}\right) u(t), \quad \tau = RC.$$

Any (non-pathological) circuit comprising only resistors and one capacitor or one inductor can be reduced using Thévenin's theorem to one of the circuits in Fig. 11.5. A circuit equivalent to the circuit in Fig. 11.5(a) at the terminals of the capacitor is called a **first-order *RC* circuit**. One equivalent to the circuit in Fig. 11.5(b) at the terminals of the inductor is called a **first-order *RL* circuit**.

Example 11.8. Refer to Fig. 11.6(a). Obtain an expression for the current $i(t)$.

Solution: We first obtain an expression for the voltage $v_C(t)$ and subsequently use Ohm's law to obtain an expression for the current $i(t)$. We apply a source transformation to obtain the circuit in Fig. 11.6(b), where $V_S = I_S R_S$. The Thévenin equivalent for the rest of the circuit at the terminals of the capacitor has the form of the circuit in Fig. 11.5(a), where $R = (R_S + R_1) \| R_2$. The time constant is given by

$$\tau = RC = [(R_S + R_1) \| R_2] \, C.$$

By inspection, $v_C(0^+) = 0$ and

$$v_C(\infty) = \frac{R_2 V_S}{R_1 + R_2 + R_S}$$

From (11.7)

$$v_C(t) = v_C(\infty) - [v_C(\infty) - v_C(0^+)] \exp\left(-\frac{t}{\tau}\right)$$
$$= \frac{R_2 V_S}{R_1 + R_2 + R_S}\left[1 - \exp\left(-\frac{t}{\tau}\right)\right], \; t > 0$$

or

$$v_C(t) = \frac{R_2 R_S I_S}{R_1 + R_2 + R_S}\left[1 - \exp\left(-\frac{t}{\tau}\right)\right] u(t),$$

so

$$i(t) = \frac{v_C(t)}{R_2}$$
$$= \frac{R_S I_S}{R_1 + R_2 + R_S}\left[1 - \exp\left(-\frac{t}{\tau}\right)\right] u(t).$$

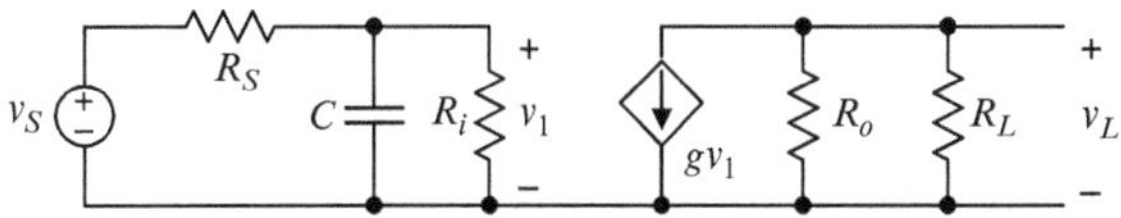

Fig. 11.7 See Example 11.9

Example 11.9. See Fig. 11.7. (a) Obtain an expression for the voltage $v_L(t)$, where $v_S = V_0\,u(t)$. (b) Let $R_S = 50\ \Omega$, $C = 22\text{nF}$, $R_i = 1\text{M}\Omega$, $g = 1\text{mS}$, $R_o = 10\text{k}\Omega$, $R_L = 5\text{k}\Omega$, and $v_S = V_0\,u(t)$, with $V_0 = 5\text{V}$. Construct a graph of the voltage v_L versus time for $-\tau \le t \le 5\tau$.

Solution: (a) The output subcircuit is resistive and has no effect on currents and voltages in the input subcircuit, so the time constant is determined by the elements in the input subcircuit. The Thévenin equivalent resistance at the terminals of the capacitor is $R_T = R_S \| R_i$, so the time constant is given by

$$\tau = (R_S \| R_i) C.$$

By inspection,

$$v_L(t) = -g\, v_1(t)\, (R_o \| R_L). \tag{11.25}$$

The voltage v_S equals zero for $t \le 0$ (the source is a short circuit for $t \le 0$), so $v_1(t) = 0$ for $t < 0$. The voltage $v_1(t)$ appears across a capacitor and must be continuous. Thus $v_1(0^+) = 0$. For $t \gg \tau$, the circuit is in dc steady state, the capacitor is effectively an open circuit, and

$$v_1(\infty) = \frac{R_i V_0}{R_i + R_S}.$$

From (11.7)

$$v_1(t) = v_1(\infty) - [v_1(\infty) - v_1(0^+)]\exp\left(-\frac{t}{\tau}\right)$$
$$= \frac{R_i V_0}{R_i + R_S}\left[1 - \exp\left(-\frac{t}{\tau}\right)\right],\ t > 0.$$

Thus, from (11.25), and incorporating a unit step to express the voltage for all time, we have

$$v_L(t) = -\frac{g R_i V_0}{R_i + R_S}\frac{R_o R_L}{R_o + R_L} \times \left[1 - \exp\left(-\frac{t}{\tau}\right)\right] u(t).$$

(b) For the given parameter values, we find

$$\frac{g R_i V_0}{R_i + R_S}\frac{R_o R_L}{R_o + R_L} \cong 16.7\,\text{V},$$
$$\tau = \frac{R_S R_i}{R_S + R_i} C \cong 1.10\,\mu\text{s}.$$

Figure 11.8 shows a graph of the response.

Exercise 11.3. In Fig. 11.9, $v_S = V_0\, u(t)$ and the response of interest is the voltage v_L.

(a) Obtain an expression for the voltage $v_L(t)$. (b) Let $R_S = 50\,\Omega$, $C = 15\,\text{nF}$, $R_1 = R_3 = 1\,\text{M}\Omega$, $\mu_1 = 15$, $\mu_2 = 25$, $R_L = 2.2\,\text{k}\Omega$, $R_2 = R_4 = 10\,\text{k}\Omega$, and $V_0 = 5\,\text{V}$. Construct a graph of the voltage v_L versus time for $-\tau \leq t \leq 5\tau$, where τ is the circuit time constant.

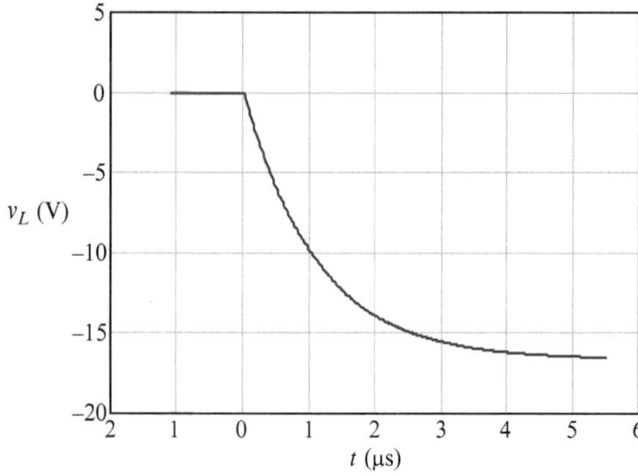

Fig. 11.8 See Example 11.9

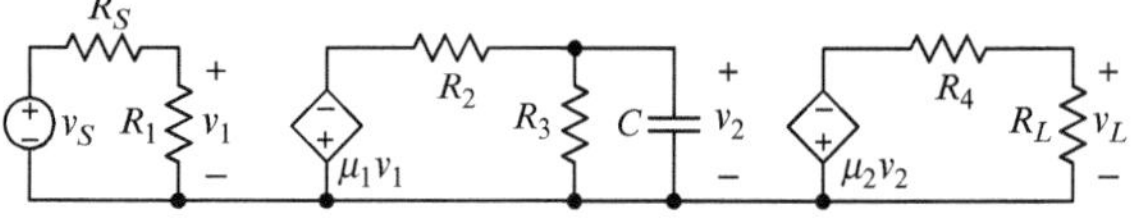

Fig. 11.9 See Exercise 11.3

11.5 Second-Order Circuits

A **second-order circuit** is one for which excitation and response are related by a second-order differential equation expressed in **standard form** as

$$a_2\frac{d^2y}{dt^2} + a_1\frac{dy}{dt} + y = b_0 x + b_1\frac{dx}{dt} + b_2\frac{d^2x}{dt^2}, \quad (11.26)$$

where $x(t)$ is an excitation and $y(t)$ is the corresponding response. For a step excitation $x(t) = X_0\, u(t)$ and for $t > 0$, (11.26) becomes

$$a_2\frac{d^2y}{dt^2} + a_1\frac{dy}{dt} + y = b_0 X_0, \quad t > 0. \quad (11.27)$$

The coefficients a_1, a_2, b_0 are constant functions of circuit parameters. The coefficient of y is unity, so every term in (11.27) has the dimension of the response. It follows that

$$\text{SI}(a_2) = \text{s}^2, \quad \text{SI}(a_1) = \text{s}, \quad \text{SI}(b_0) = \frac{\text{SI}(y)}{\text{SI}(X_0)}. \quad (11.28)$$

Example 11.10. In Fig. 11.10, $v_S(t) = V_0\, u(t)$. Obtain the standard-form differential

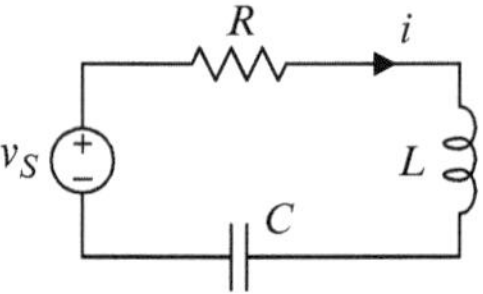

Fig. 11.10 See Example 11.10

equation relating the current i to the voltage v_S for $t > 0$.

Solution: By Kirchhoff's voltage law

$$Ri + L\frac{di}{dt} + \frac{1}{C}\int_{-\infty}^{t} i(t')dt' = V_0, \ t > 0.$$

Differentiating once with respect to time and rearranging terms gives

$$LC\frac{d^2i}{dt^2} + RC\frac{di}{dt} + i = 0, \ t > 0,$$

which has the form (11.26) with $y = i$, $a_2 = LC, a_1 = RC$, and $b_0 = 0$. As an exercise, show that the SI units of the coefficients a_1, a_2 are as specified by (11.28).

Exercise 11.4. In Fig. 11.11, $v_S(t) = V_0\, u(t)$. Obtain the standard-form differential equation relating the voltage v_C to the voltage v_S for $t > 0$. Show that the SI units of the coefficients a_1, a_2, b_0 are as specified by (11.28).

We use the method of undetermined coefficients to obtain the solution to (11.27). The **complete solution** $y(t)$ to (11.27) is the **complete response**, and is the sum of the **forced response** $y_f(t)$ and the **unforced response** $y_u(t)$. Thus

$$y(t) = y_f(t) + y_u(t). \tag{11.29}$$

The forced response has the form of the excitation, so the forced response is constant. If we let $y(t) = Y_0$ (a constant) in (11.27), we obtain $Y_0 = b_0X_0$. Therefore the forced response is given by

$$y_f(t) = b_0X_0\, u(t). \tag{11.30}$$

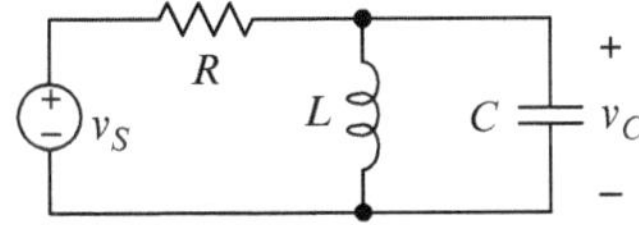

Fig. 11.11 See Exercise 11.4

The unforced response $y_u(t)$ is the solution to the *homogeneous equation*

$$a_2\frac{d^2y_u}{dt^2} + a_1\frac{dy_u}{dt} + y_u = 0 \tag{11.31}$$

obtained from (11.27) by setting the excitation X_0 to zero. Any solution to (11.31) is also a solution to (11.27) because $b_0X_0 + 0 = b_0X_0$.

If y_{u1} and y_{u2} are any two solutions to (11.31), then $y_u = Y_1 y_{u_1} + Y_2 y_{u_2}$, where Y_1 and Y_2 are arbitrary constants, is also a solution, as you can show by substitution. The general solution to (11.31) must reduce (11.31) to an identity and must contain exactly two independent arbitrary constants. (The trivial solution $y_u = 0$ meets the first requirement, but not the second.) In this context, *arbitrary* means that $y_u = Y_1 y_{u_1} + Y_2 y_{u_2}$ satisfies (11.31) for any values whatever of the constants Y_1, Y_2 and *independent* means that $y_u = Y_1 y_{u_1} + Y_2 y_{u_2}$ cannot be reduced to the form $y_u = Y y_u$, where the two arbitrary constants are reduced to a single constant.

According to (11.31), a linear combination of the unforced response $y_u(t)$ and its first two time derivatives is identically zero (for all time). That can be true only if $y_u(t)$ and its first two time derivatives have exactly the same time dependence.[4] A function having that property is the exponential function $Y e^{st}$, where Y and s are constants. Substituting $Y e^{st}$ for y_u in (11.31) yields

$$\begin{aligned} & a_2s^2Y e^{st} + a_1 s Y e^{st} + Y e^{st} \\ & = \left(a_2s^2 + a_1 s + 1\right)Y e^{st} = 0. \end{aligned} \tag{11.32}$$

The exponential e^{st} cannot be zero for any finite value of s. We assume $Y \neq 0$ because $Y = 0$ yields the *trivial solution*, which contains no arbitrary constants and can be ignored. Thus the function $Y e^{st}$ satisfies the differential equation (11.31) if the constant s satisfies the **characteristic equation**

$$a_2s^2 + a_1 s + 1 = 0. \tag{11.33}$$

[4]The sum of functions having different time dependences cannot be identically zero (cannot equal zero for all time); e.g., a sinusoidal function of time cannot cancel a real exponential function of time.

The **characteristic roots** are given by

$$s_1 = \frac{-a_1 + \sqrt{a_1^2 - 4a_2}}{2a_2},$$
$$s_2 = \frac{-a_1 - \sqrt{a_1^2 - 4a_2}}{2a_2}. \quad (11.34)$$

The SI units of the differential equation coefficients are $\mathrm{SI}(a_2) = \mathrm{s}^2$ and $\mathrm{SI}(a_1) = \mathrm{s}$, so $\mathrm{SI}(s_1) = \mathrm{SI}(s_2) = \mathrm{s}^{-1}$.[5]

Because (11.33) is satisfied by either characteristic root, the differential equation (11.31) is satisfied by $Y_1 e^{s_1 t}$ and by $Y_2 e^{s_2 t}$, where the constants Y_1, Y_2 are arbitrary. Thus

$$y_u(t) = Y_1 e^{s_1 t} + Y_2 e^{s_2 t} \quad (11.35)$$

also satisfies (11.31).

If the arbitrary constants Y_1 and Y_2 are independent, then (11.35) is the general solution to (11.31). But if $a_1^2 - 4a_2 = 0$, then from (11.34),

$$s_1 = s_2 = \frac{-a_1}{2a_2}, \quad (11.36)$$

and the right side of (11.35) reduces to

$$Y_1 e^{s_1 t} + Y_2 e^{s_2 t} = (Y_1 + Y_2) e^{s_1 t} = Y e^{s_1 t},$$

which contains only one arbitrary constant (Y) and cannot be the general solution. In that case, the general solution has the form $y_u(t) = (Y_1 + Y_2 t) e^{s_1 t}$ as you can show by substitution.

The *general solution* to (11.31) (the unforced response) is given by

$$y_u(t) = u(t) \begin{cases} Y_1 e^{s_1 t} + Y_2 e^{s_2 t}, \; a_1^2 - 4a_2 \neq 0, \\ (Y_1 + Y_2 t) e^{s_1 t}, \; a_1^2 - 4a_2 = 0, \end{cases} \quad (11.37)$$

where s_1 and s_2 are given by (11.34) and Y_1 and Y_2 are arbitrary constants. If $a_1^2 - 4a_2 \neq 0$, the SI unit of both Y_1 and Y_2 is that of the response (current or voltage). If $a_1^2 - 4a_2 = 0$, the SI unit of Y_1 is that of the response (current or voltage) and the SI unit of Y_2 is that of the response divided by time; e.g. Vs^{-1} or As^{-1}.

The **complete response** is the sum of the forced response given by (11.30) and the unforced response given by (11.35):

$$y(t) = u(t) \begin{cases} b_0 X_0 + Y_1 e^{s_1 t} + Y_2 e^{s_2 t}, \; a_1^2 - 4a_2 \neq 0, \\ b_0 X_0 + (Y_1 + Y_2 t) e^{s_1 t}, \; a_1^2 - 4a_2 = 0. \end{cases} \quad (11.38)$$

A response given by (11.38) consists of an unforced component and a forced component $b_0 X_0 u(t)$. In a stable circuit, the unforced component eventually vanishes, and only the forced component remains. Thus the forced component is the **steady-state response** and the unforced component is the **transient response**, called the **transient**, for short.

After $t = 0$, a step excitation $x = X_0 u(t)$ is a constant (dc) current or voltage given by $x = X_0$. After the transient vanishes (for practical purposes), the associated circuit is in dc steady state, with $y = b_0 X_0$. The magnitude of the parameter b_0 is called the **dc gain** of the circuit.

The function y given by (11.38) satisfies the differential equation (11.26), regardless of the values of the constants Y_1, Y_2. In any particular physical problem, the values of the constants Y_1, Y_2 are determined by the initial values denoted by y_0 and y_0', given by

$$y_0 = y(0^+), \quad y_0' = y'(0^+) = \left. \frac{dy(t)}{dt} \right|_{t=0^+}. \quad (11.39)$$

The initial values are obtained from the initial state of the circuit as follows. From (11.38),

$$y_0 = y(0^+) = \begin{cases} b_0 X_0 + Y_1 + Y_2, \; a_1^2 - 4a_2 \neq 0, \\ b_0 X_0 + Y_1, \; a_1^2 - 4a_2 = 0, \end{cases} \quad (11.40)$$

and

$$y_0' = y'(0^+) = \begin{cases} s_1 Y_1 + s_2 Y_2, \; a_1^2 - 4a_2 \neq 0, \\ s_1 Y_1 + Y_2, \; a_1^2 - 4a_2 = 0. \end{cases} \quad (11.41)$$

[5] Be aware that s (italic) is a variable and s (roman) is a unit. This usage, which is almost universal, is doubly unfortunate because the unit of s is s. In written work, you should underline the unit s to avoid confusion.

Equations (11.40) and (11.41) yield

$$Y_1 = \begin{cases} \dfrac{s_2(y_0 - b_0X_0) - y_0'}{s_2 - s_1}, & a_1^2 - 4a_2 \neq 0\ (s_1 \neq s_2), \\ y_0 - b_0X_0, & a_1^2 - 4a_2 = 0\ (s_1 = s_2), \end{cases}$$

$$Y_2 = \begin{cases} -\dfrac{s_1(y_0 - b_0X_0) - y_0'}{s2 - s_1,}, & a_1^2 - 4a_2 \neq 0\ (s_1 \neq s_2), \\ y_0' - s_1(y_0 - b_0X_0), & a_1^2 - 4a_2 = 0\ (s_1 = s_2). \end{cases} \quad (11.42)$$

Example 11.11. Figure 11.12 shows a series *RLC* circuit, where $v_S(t) = V_0\,u(t)$.

(c) Obtain an expression for the current $i(t)$.
(b) Let $R = 600\,\Omega, L = 100\,\text{mH}, C = 2\mu\text{F}$, and $V_0 = 1\text{V}$. Draw a graph of the current versus time.

Solution: (a) By Kirchhoff's voltage law, we have for $t > 0$ that

$$R\,i + L\frac{di}{dt} + \frac{1}{C}\int_{-\infty}^{t} i(t')\,dt' = V_0. \quad (11.43)$$

Differentiating (11.43) once with respect to time gives

$$R\frac{di}{dt} + L\frac{d^2i}{dt^2} + \frac{1}{C}i = 0 \quad (11.44)$$

or

$$LC\frac{d^2i}{dt^2} + RC\frac{di}{dt} + i = 0, \quad (11.45)$$

which is in standard form, with $a_2 = LC, a_1 = RC,\ b_0 = 0$. From (11.30) the forced response equals zero.

From (11.34), the characteristic roots are given by

$$s_1,\, s_2 = \frac{-a_1 \pm \sqrt{a_1^2 - 4a_2}}{2a_2} = \frac{-RC \pm \sqrt{(RC)^2 - 4LC}}{2LC}. \quad (11.46)$$

The voltage source is equivalent to a short circuit for $t \leq 0$; i.e., no energy is provided to the circuit until $t > 0$. It follows that

$$i(0^+) = 0. \quad (11.47)$$

The relation $i(0^+) = 0$ is one initial value:

$$i(0^+) = i_0 = 0. \quad (11.48)$$

We obtain the other from (11.48) and (11.43). For $t = 0^+$, (11.43) becomes

$$R\,i(0^+) + L\,i'(0^+) + \frac{1}{C}\int_{-\infty}^{0^+} i(t')\,dt' = V_0.$$

The first term vanishes because $i(0^+) = 0$. The integral vanishes because $i(t) = 0$ for $t < 0$. Thus we have the second initial value

$$i'(0^+) = i_0' = \frac{V_0}{L}. \quad (11.49)$$

From (11.38) with $b_0 = 0$, the complete response is given by

$$i(t) = (I_1e^{s_1t} + I_2e^{s_2t})u(t), \quad (11.50)$$

where, from (11.42)

$$I_1 = \frac{s_2(i_0 - b_0X_0) - i_0'}{s_2 - s_1} = -\frac{i_0'}{s_2 - s_1} = -\frac{1}{s_2 - s_1}\frac{V_0}{L} = \frac{V_0C}{\sqrt{(RC)^2 - 4LC}},$$

$$I_2 = -\frac{s_1(i_0 - b_0X_0) - i_0'}{s_2 - s_1} = \frac{i_0'}{s_2 - s_1} = \frac{1}{s_2 - s_1}\frac{V_0}{L} = -\frac{V_0C}{\sqrt{(RC)^2 - 4LC}}. \quad (11.51)$$

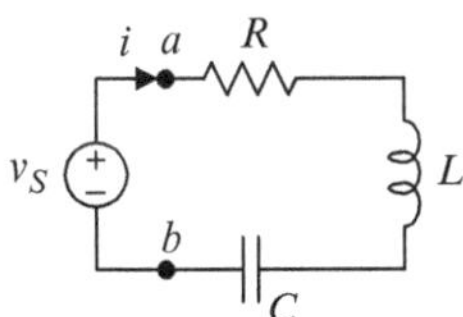

Fig. 11.12 See Example 11.11

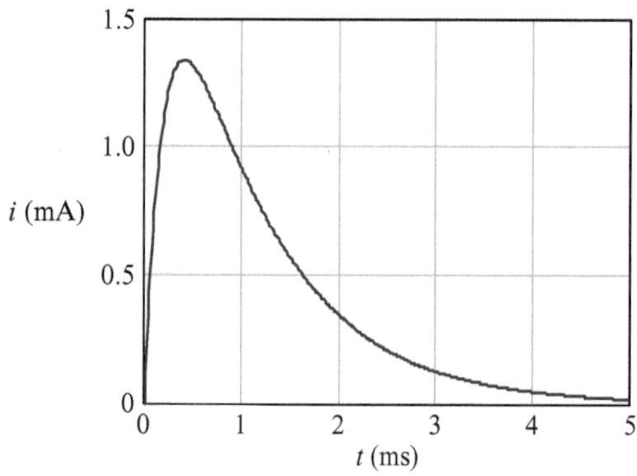

Fig. 11.13 See Example 11.11

(b) We are given $R = 600\,\Omega$, $L = 100\,\text{mH}$, $C = 2\,\mu F$, $V_0 = 1\text{V}$. For these values, the characteristic roots are

$$s_1 = -1 \times 10^3 \text{s}^{-1}, \quad s_2 = -5 \times 10^3 \text{s}^{-1}.$$

The initial values are

$$i_0 = 0, \; i_0' = \frac{V_0}{L} = 10\,\text{As}^{-1}.$$

From (11.51)

$$I_1 = 2.5\,\text{mA}, I_2 = -2.5\,\text{mA}.$$

The response is given by

$$i(t) = 2.5(e^{s_1 t} - e^{s_2 t})u(t)\,\text{mA}. \quad (11.52)$$

Figure 11.13 shows a graph of the response.

We can express the *unforced* responses given by (11.37) in more convenient forms.

If the characteristic roots are equal, it is convenient to express the unforced response as

$$y_u(t) = (Y_1 + Y_2 t)e^{-t/\tau} u(t), \quad \tau = -\frac{1}{s_1}; \; a_1^2 - 4a_2 = 0, \quad (11.53)$$

where τ is the **time constant** for the transient response.

If the characteristic roots are real and distinct, it is convenient to define

$$\tau_1 = -\frac{1}{s_1}, \quad \tau_2 = -\frac{1}{s_2}, \quad (11.54)$$

and write the unforced response as

$$y_u(t) = (Y_1 e^{-t/\tau_1} + Y_2 e^{-t/\tau_2})u(t), \quad a_1^2 - 4a_2 > 0, \quad (11.55)$$

where τ_1, τ_2 are the **time constants** for the transient response.

if the characteristic roots are complex conjugates (if $a_1^2 - 4a_2 < 0$), the expression

$$y_u(t) = (Y_1 e^{s_1 t} + Y_2 e^{s_2 t})u(t), \quad a_1^2 - 4a_2 \neq 0 \quad (11.56)$$

is correct but inconvenient because the exponential functions are complex. We can obtain a more convenient (and more easily interpreted) expression by writing the characteristic roots as

$$s, s^* = \frac{-a_1 \pm j\sqrt{4a_2 - a_1^2}}{2a_2} = -\frac{1}{\tau} \pm j\omega_0, \quad (11.57)$$

where the **time constant** τ and the **ring frequency** f_0 are given by

$$\tau = -\frac{1}{\text{Re}(s)} = \frac{2a_2}{a_1}, \; f_0 = \frac{\omega_0}{2\pi}, \quad \omega_0 = \text{Im}(s) = \frac{1}{2a_2}\sqrt{4a_2 - a_1^2}. \quad (11.58)$$

From (11.42), the coefficients in (11.56) are given by

$$Y_1 = \frac{s^*(y_0 - b_0 X_0) - y_0'}{s^* - s}, \quad Y_2 = \frac{s(y_0 - b_0 X_0) - y_0'}{s - s^*} = Y_1^*. \quad (11.59)$$

The coefficients Y_1, Y_2 are a conjugate pair, so we can simplify notation further by defining

$$Y_1 = Y; \quad Y_2 = Y^* \tag{11.60}$$

and expressing Y in polar form as

$$Y = |Y|\exp\,(j\theta_Y), \theta_Y = \measuredangle Y \tag{11.61}$$

With the definitions above, (11.56) can be written

$$\begin{aligned} y_u(t) &= [Y\exp(st) + Y^*\exp(s^*t)]u(t) \\ &= [Y\exp(st) + (Y^*\exp(st)^*]u(t) \\ &= 2\mathrm{Re}[Y\exp(st)]u(t) \\ &= 2\mathrm{Re}[|Y|\exp\,(j\theta)\exp[(-t/\tau + j\omega_0)t]]u(t) \\ &= 2|Y|\exp\,(-t/\tau\,)\,\mathrm{Re}[\exp\,(j\,\omega_0 t + j\theta_Y]u(t), \end{aligned} \tag{11.62}$$

and finally, as

$$\begin{aligned} y_u(t) &= 2|Y|\exp(-t/\tau)\cos(\omega_0 t + \theta_Y)u(t), \\ a_1^2 &- 4a_2 < 0, \end{aligned} \tag{11.63}$$

where

$$Y = \frac{s^*(y_0 - b_0X_0) - y_0'}{s^* - s}, \tag{11.64}$$

$$s = \frac{-a_1 + j\sqrt{4a_2 - a_1^2}}{2a_2}, \tag{11.65}$$

and

$$\begin{aligned} \tau &= -\frac{1}{\mathrm{Re}(s)} = \frac{2a_2}{a_1}, \\ \omega_0 &= \mathrm{Im}(s) = \frac{1}{2a_2}\sqrt{4a_2 - a_1^2}. \end{aligned} \tag{11.66}$$

As always, you should express the quantities above in terms of symbols for circuit parameters before using any numerical values. Otherwise, if you use numerical values from the outset, you will be unable (without many repeated calculations) to determine how any particular circuit parameter affects the response, which is a principal reason for doing such analyses.

11.5.1 Summary: Second-Order Circuits

The *standard form* of the differential equation governing a current or voltage in a second-order linear circuit subjected to a step excitation is

$$a_2\frac{d^2y}{dt^2} + a_1\frac{dy}{dt} + y = b_0X_0. \tag{11.67}$$

The complete response is given by

$$y = b_0X_0 + y_u, \quad t > 0, \tag{11.68}$$

where

$$\begin{aligned} X_0 &= \text{the(constant) excitation for } t > 0, \\ b_0X_0 &= \text{forced response} = \text{steady-state response}, \\ |b_0| &= \text{the dc gain of the circuit}, \\ y_u &= \text{unforced response} = \text{transient response}. \end{aligned} \tag{11.69}$$

To obtain an expression for the response, proceed as follows:

Determine the state of the circuit at $t = 0^-$. Then use Kirchhoff's laws and continuity of state to determine the initial values

$$y_0 = y(0^+), \quad y_0' = y'(0^+). \tag{11.70}$$

The form of the transient response depends upon the nature of the characteristic roots. Calculate

$$a_1^2 - 4a_2. \tag{11.71}$$

If $a_1^2 - 4a_2 > 0$, the circuit is **overdamped**, the characteristic roots are real and distinct, and the complete response is given by

$$y(t) = \left[b_0X_0 + Y_1e^{-t/\tau_1} + Y_2e^{-t/\tau_2}\right]u(t), \tag{11.72}$$

with

$$\begin{aligned} s_1 &= \frac{-a_1 + \sqrt{a_1^2 - 4a_2}}{2a_2}, \\ s_2 &= \frac{-a_1 - \sqrt{a_1^2 - 4a_2}}{2a_2}, \end{aligned} \tag{11.73}$$

$$\tau_1 = -\frac{1}{s_1}, \quad \tau_2 = -\frac{1}{s_2}, \tag{11.74}$$

and

$$Y_1 = \frac{s_2(y_0 - b_0X_0) - y_0'}{s_2 - s_1},$$
$$Y_2 = -\frac{s_1(y_0 - b_0X_0) - y_0'}{s_2 - s_1}. \tag{11.75}$$

If $a_1^2 - 4a_2 = 0$, the circuit is **critically damped**, the characteristic roots are real and equal, and the complete response is given by

$$y(t) = \left[b_0X_0 + (Y_1 + Y_2t)e^{-t/\tau}\right]u(t), \tag{11.76}$$

with

$$\tau = \frac{2a_2}{a_1}, \tag{11.77}$$

and

$$Y_1 = y_0 - b_0X_0,$$
$$Y_2 = y_0' + \frac{a_1}{2a_2}(y_0 - b_0X_0). \tag{11.78}$$

If $a_1^2 - 4a_2 < 0$, the circuit is **underdamped**, the characteristic roots are complex conjugates, and the complete response is given by

$$y(t) = \left[b_0X_0 + 2|Y|e^{-t/\tau}\cos(\omega_0 t + \theta_Y)\right]u(t), \tag{11.79}$$

where $\theta_Y = \measuredangle Y$ and

$$s = \frac{-a_1 + \sqrt{a_1^2 - 4a_2}}{2a_2}, \tag{11.80}$$

$$Y = \frac{s^*(y_0 - b_0X_0) - y_0'}{s^* - s}, \tag{11.81}$$

and

$$\tau = -\frac{1}{\mathrm{Re}(s)} = \frac{2a_2}{a_1}, \quad \omega_0 = \mathrm{Im}(s)$$
$$= \frac{1}{2a_2}\sqrt{4a_2 - a_1^2}. \tag{11.82}$$

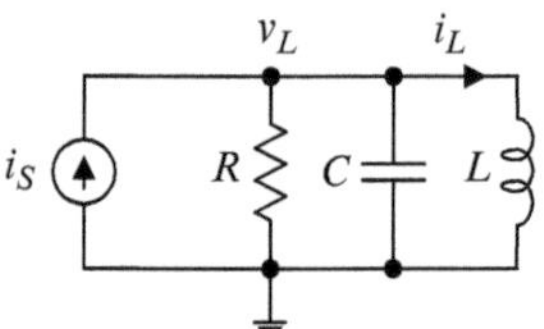

Fig. 11.14 See Example 11.12

Example 11.12. In Fig. 11.14, $C = 50\,\mathrm{pF}$, $L = 1\,\mathrm{mH}$, and $i_S(t) = I_0[1 - u(t)]$, with $I_0 = 5\,\mathrm{mA}$. The response of interest is the voltage $v_L(t)$. Give values of R and plot the corresponding responses versus time for which the response is (a) underdamped, (b) critically damped, and (c) overdamped.

Solution: For $t<0$, the circuit is in dc steady state. Thus

$$v_L(0^+) = v_L(0^-) = 0,$$
$$i_L(0^+) = i_L(0^-) = I_0.$$

Applying Kirchhoff's current law gives

$$\frac{v_L}{R} + C\frac{dv_L}{dt} + \frac{1}{L}\int_{-\infty}^{t} v_L(t')dt'$$
$$= 0, \quad t > 0, \tag{11.83}$$

which yields

$$\frac{v_L(0^+)}{R} + Cv_L'(0^+) + i_L(0^+) = 0$$
$$\Rightarrow v_L'(0^+) = -\frac{1}{C}\left[i_L(0^+) - \frac{v_L(0^+)}{R}\right]$$
$$= -\frac{I_0}{C} \cong -10^8\,\mathrm{V\,s^{-1}}.$$

Differentiating (11.83) once with respect to time and rearranging terms yields the standard-form differential equation

$$LC\frac{d^2v_L}{dt^2} + \frac{L}{R}\frac{dv_L}{dt} + v_L = 0.$$

The characteristic roots are given by

$$s_1, s_2 = \frac{1}{2LC}\left(-\frac{L}{R} \pm \sqrt{\left(\frac{L}{R}\right)^2 - 4LC}\right).$$

The characteristic roots are real and equal (the response is critically damped) if $R = R_0$, given by

$$\left(\frac{L}{R_0}\right)^2 - 4LC = 0 \Rightarrow R_0 = \frac{1}{2}\sqrt{\frac{L}{C}} \cong 2.24\,\text{k}\Omega.$$

The response is underdamped if $R > R_0$ and overdamped if $R < R_0$. We choose (arbitrarily) (a) $R = 5R_0$, (b) $R = R_0$, and (c) $R = R_0/5$.

(a) $R = 5R_0$ *(underdamped):*

$$s \cong (-8.94 + j43.82) \times 10^5\,\text{s}^{-1},$$
$$\tau = -\frac{1}{\text{Re}(s)} \cong 1.12\,\mu s,$$
$$\omega = \text{Im}(s) \cong 43.82 \times 10^6\,\text{s}^{-1},$$
$$V = \frac{s^*(v_0 - b_0 X_0) - v_0'}{s^* - s} = -\frac{v_0'}{s^* - s}$$
$$\cong j11.41\,\text{V}, 2|V| = 22.82\,\text{V}, \theta_V = \frac{\pi}{2},$$
$$v_L(t) = 2|V_1|\exp\left(-\frac{t}{\tau}\right)\cos(\omega t + \theta_V)u(t).$$

Figure 11.15(a) shows a graph of the response.

(b) $R = R_0$ *(critically damped):*

$$s \cong -4.47 \times 10^6\,\text{s}^{-1},$$
$$\tau = -\frac{1}{s} \cong 223.61\,\text{ns},$$
$$V_1 = v_L(0^+) = 0,$$
$$V_2 = v_L'(0^+) = -10^8\,\text{V}\,\text{s}^{-1},$$
$$v_L(t) = V_2 t \exp\left(-\frac{t}{\tau}\right)u(t).$$

Figure 11.15(b) shows a graph of the response.

(c) $R = R_0/5$ (overdamped):

$$s_1 \cong -4.52 \times 10^5\,\text{s}^{-1},$$
$$\tau_1 = -\frac{1}{s_1} \cong 2.21\,\mu\text{s},$$
$$s_2 \cong -4.43 \times 10^7\,\text{s}^{-1},$$
$$\tau_2 = -\frac{1}{s_2} \cong 22.59\,\text{ns},$$
$$V_1 = -\frac{v_L'(0^+)}{s_2 - s_1} \cong -2.28\,\text{V},$$
$$V_2 = -V_1 \cong 2.28\,\text{V},$$
$$v_L(t) = V_1\left[\exp\left(-\frac{t}{\tau_2}\right) - \exp\left(-\frac{t}{\tau_1}\right)\right]u(t).$$

Figure 11.15(c) shows a graph of the response.

Exercise 11.5. In Fig. 11.16, $L = 25$ mH, $C = 1$pF, and $v_S(t) = V_0\,u(t)$ with $V_0 = 5$ V. (a) Find the value R_0 of the resistance R for which the response $v_C(t)$ is critically damped. If $R < R_0$, is the response underdamped or overdamped? (b) Obtain expressions for the response $v_C(t)$ for $R = 10R_0$, $R = R_0$, and $R = 0.1R_0$.

A critically damped response is the boundary between overdamped and underdamped responses. Like any other mathematically specified response, an exactly critically damped response cannot be achieved. However, a physical circuit can be *almost* critically damped, meaning that the characteristic roots are *almost* equal. In such a case, (11.76) can be superior to either the overdamped or underdamped expressions for either computation or design. For example, if the characteristic roots of a circuit model are real and very nearly equal, using either the overdamped or underdamped models can require computing differences of nearly equal numbers, which can be troublesome for a computer (or pocket calculator). Moreover, if a circuit model believed to be accurate enough for design turns out to be critically damped, there is no reason to complicate the model just to obtain a more realistic slightly underdamped or overdamped response, when the critically damped response is close enough.

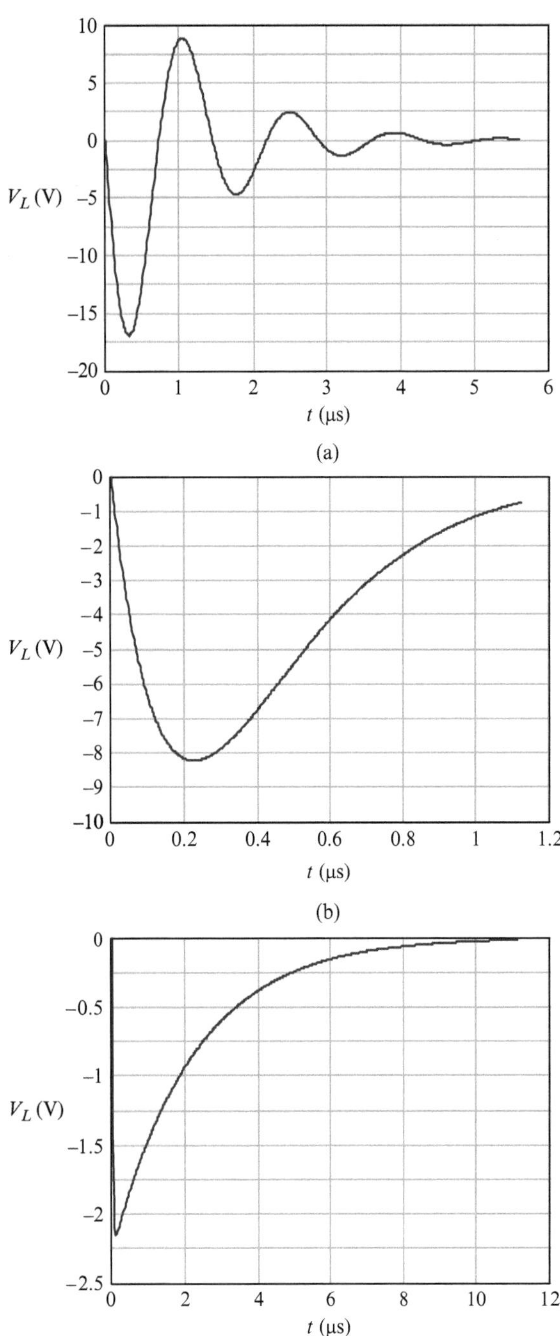

Fig. 11.15 See Example 11.12

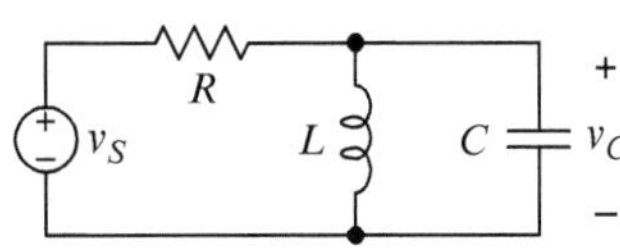

Fig. 11.16 See Exercise 11.5

The characteristic roots for a *passive* circuit consisting entirely of resistors and capacitors (no sources) must all be real, which means that such a circuit cannot be underdamped. To see why, first realize that damping increases with losses, or, conversely, the unforced response of an underdamped circuit becomes more oscillatory and lasts longer as losses diminish, because less energy is lost during each energy exchange. Losses approach zero if resistances in parallel with energy-storage elements approach infinity and resistors in series with energy-storage elements approach zero. Now imagine a passive circuit containing only resistors and capacitors. Replace every resistor that is in parallel with a capacitor with an open circuit (infinite resistance) and every resistor that is in series with a capacitor with a short circuit. If that is done, what remains is a lossless circuit that is equivalent to a single capacitor. Such a circuit is first order and cannot possibly be underdamped. Similarly, the characteristic roots for a *passive* circuit consisting entirely of resistors and inductors (no sources) must all be real, which means that such a circuit cannot be underdamped. If you obtain an underdamped response (or complex characteristic roots) for a *passive* circuit consisting of only resistors and capacitors or only of resistors and inductors, you have made an error.[6]

Often, it is unnecessary to determine a complete response, being sufficient to determine only the initial and final states and the characteristic roots. For the overdamped case, the characteristic roots are real and distinct, and the smallest characteristic root (the largest time constant) largely determines the duration of the transient response. For the critically damped case, the characteristic roots are equal, and either root determines the duration of the transient. For the underdamped case, the characteristic roots are a conjugate pair. The real part determines the duration of the transient and the imaginary part determines the frequency (or period) of the oscillations. If the characteristic roots of an underdamped circuit are expressed as $\sigma \pm j\omega$, then the number of oscillations undergone before the transient dies out is approximately three to five times the ratio of the time constant $\tau = -1/\sigma$ to the period $T = 2\pi/\omega$; i.e.,

$$\text{Number of oscillations} \cong \frac{5\tau}{T} = -\frac{5\omega}{2\pi\sigma}.$$

[6]A circuit containing only resistors, capacitors, *and dependent sources* (active devices) can be underdamped, as can a circuit containing only resistors, inductors, and dependent sources.

11.5.2 Dominant Time Constant (or Characteristic Root)

From Section 11.5.1, the unforced response of an overdamped second-order circuit to a step excitation $X_0\,u(t)$ is given by

$$y(t) = \left(Y_1 e^{-t/\tau_1} + Y_2 e^{-t/\tau_2}\right)u(t), \tag{11.84}$$

where

$$Y_1 = \frac{s_2(y_0 - b_0X_0) - y_0'}{s_2 - s_1}, \qquad Y_2 = -\frac{s_1(y_0 - b_0X_0) - y_0'}{s_2 - s_1}, \tag{11.85}$$

$$s_1 = \frac{-a_1 + \sqrt{a_1^2 - 4a_2}}{2a_2}, \qquad s_2 = \frac{-a_1 - \sqrt{a_1^2 - 4a_2}}{2a_2}, \tag{11.86}$$

and

$$\tau_1 = -\frac{1}{s_1}, \quad \tau_2 = -\frac{1}{s_2},$$

We wish to show that the duration of the transient is determined mainly by the larger of the two time constants. From (11.85) and (11.84), the transient is given for $t>0$ by

$$\begin{aligned}
y_u(t) &= \frac{s_2(y_0 - b_0X_0) - y_0'}{s_2 - s_1}e^{-t/\tau_1} - \frac{s_1(y_0 - b_0X_0) - y_0'}{s_2 - s_1}e^{-t/\tau_2} \\
&= \frac{y_0 - b_0X_0}{s_2 - s_1}\left[\left(s_2e^{-t/\tau_1} - s_1e^{-t/\tau_2}\right) - \frac{y_0'\left(e^{-t/\tau_1} - e^{-t/\tau_2}\right)}{y_0 - b_0X_0}\right] \\
&= \frac{y_0 - b_0X_0}{s_2 - s_1}\left[\left(-\frac{1}{\tau_2}e^{-t/\tau_1} + \frac{1}{\tau_1}e^{-t/\tau_2}\right) - \frac{y_0'\left(e^{-t/\tau_1} - e^{-t/\tau_2}\right)}{y_0 - b_0X_0}\right].
\end{aligned} \tag{11.87}$$

The initial amplitude of the second pair of exponentials in the brackets is zero because $e^0 - e^0 = 0$. We may assume without loss of generality that $\tau_2>\tau_1$, in which case both the initial magnitude and the duration of the term $\tau_1^{-1}e^{-t/\tau_2}$ exceed those of the term $\tau_2^{-1}e^{-t/\tau_1}$. It follows that the duration of the transient component of the step response of an overdamped second-order circuit is approximately five times the larger time constant. If one time constant is much larger than the other, we call the larger one the **dominant time constant**. The step response of an overdamped second-order circuit having a dominant time constant approximates that of a first-order circuit for all but a brief time at the beginning of the response.

Example 11.13. The response v_L of a certain circuit to an excitation v_S is governed by the differential equation

$$a_2\frac{d^2v_L}{dt^2} + a_1\frac{dv_L}{dt} + v_L = v_S,$$

where $a_2 = 7\,\mathrm{ms}^2$ and $a_1 = 11\,\mathrm{ms}$. It is known that for a step excitation $v_S = V_0\,u(t)$ with $V_0 = 1\,\mathrm{V}$, the initial conditions are

$$v_L(0^+) = 0, \quad v_L'(0^+) = 10\,\mathrm{V\,s^{-1}}.$$

We wish to show that the response is characterized by a dominant time constant.

From (11.86), the characteristic roots are

$$s_1 = \frac{-a_1 + \sqrt{a_1^2 - 4a_2}}{2a_2} \cong -96.88\,\mathrm{s}^{-1}, \qquad s_2 = \frac{-a_1 - \sqrt{a_1^2 - 4a_2}}{2a_2} \cong -1.475\times10^3\,\mathrm{s}^{-1},$$

and the time constants are

$$\tau_1 = -\frac{1}{s_1} \cong 10.3\,\mathrm{ms}, \quad \tau_2 = -\frac{1}{s_2} \cong 678\,\mu\mathrm{s}.$$

Because $\tau_1>10\tau_2$, we expect the exponential having time constant τ_1 to be dominant. From (11.85), with $b_0 = 1$,

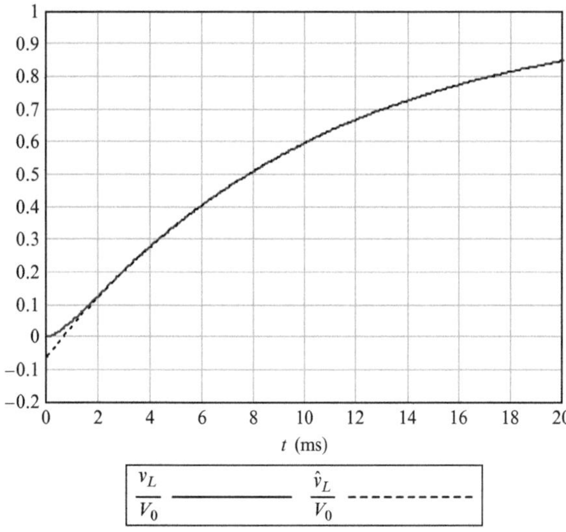

Fig. 11.17 See Example 11.13

$$Y_1 = \frac{s_2(y_0 - V_0) - y_0'}{s_2 - s_1} = -1.063\,\text{V},$$
$$Y_2 = -\frac{s_1(y_0 - V_0) - y_0'}{s_2 - s_1} = 63\,\text{mV}.$$

Thus the response is given by

$$v_L(t) = \left(b_0V_0 + Y_1e^{-t/\tau_1} + Y_2e^{-t/\tau_2}\right)u(t).$$

Figure 11.17 shows (solid line) a graph of the step response and (dotted line) a graph of only the steady-state component and first exponential, given by

$$\hat{v}_L(t) = \left(b_0V_0 + Y_1e^{-t/\tau_1}\right)u(t).$$

The graphs are indistinguishable for times greater than about 2 ms.

11.5.3 Damping Factor

The response of an *underdamped* second-order circuit has the form

$$y(t) = \left[b_0X_0 + 2|Y|e^{-t/\tau}\cos(\omega_0 t + \theta_Y)\right]u(t), \quad (11.88)$$

where

$$Y = \frac{s^*(y_0 - b_0X_0) - y_0'}{s^* - s};$$
$$s = \frac{-a_1 + \sqrt{a_1^2 - 4a_2}}{2a_2}, \quad (11.89)$$

and

$$\tau = -\frac{1}{\text{Re}(s)} = \frac{2a_2}{a_1}, \quad \omega_0 = \text{Im}(s)$$
$$= \frac{1}{2a_2}\sqrt{4a_2 - a_1^2}. \quad (11.90)$$

The steady-state (forced) component y_f and transient (unforced) component y_u of the response are given by

$$y_f(t) = b_0X_0\,u(t),$$
$$y_u(t) = 2|Y|e^{-t/\tau}\cos(\omega_0 t + \theta_Y)u(t). \quad (11.91)$$

The duration of the transient component is determined by the time constant τ. Also of interest are the maximum amplitude and the number of oscillations the transient undergoes (if any) before vanishing (before becoming insignificant). A useful measure in that regard is the dimensionless **damping factor**

$$\delta = \frac{1}{\omega_0\tau} = \frac{1}{2\pi f_0\,\tau} = \frac{T}{2\pi\tau}, \quad (11.92)$$

which, except for the factor $(2\pi)^{-1}$, is the ratio of the period $T = 2\pi/\omega_0$ of the sinusoidal factor $\cos(\omega_0 t + \theta_Y)$ to the time constant τ of the exponential factor $e^{-t/\tau}$. From (11.92), the number of cycles executed by the sinusoidal factor in (11.91) before the transient vanishes (during $0 \le t < 5\tau$) is given approximately by

$$\frac{5\tau}{T} = \frac{5}{2\pi\delta} \cong \frac{0.8}{\delta}. \quad (11.93)$$

If δ is large, the period of the sinusoidal factor is larger than the time constant of the exponential factor, there are few or no oscillations before the transient vanishes, and the circuit is relatively heavily damped. If δ is small, the period of the sinusoidal factor is smaller than the time constant of the exponential factor, there are a few or perhaps many oscillations before the transient vanishes, and the circuit is relatively lightly damped.

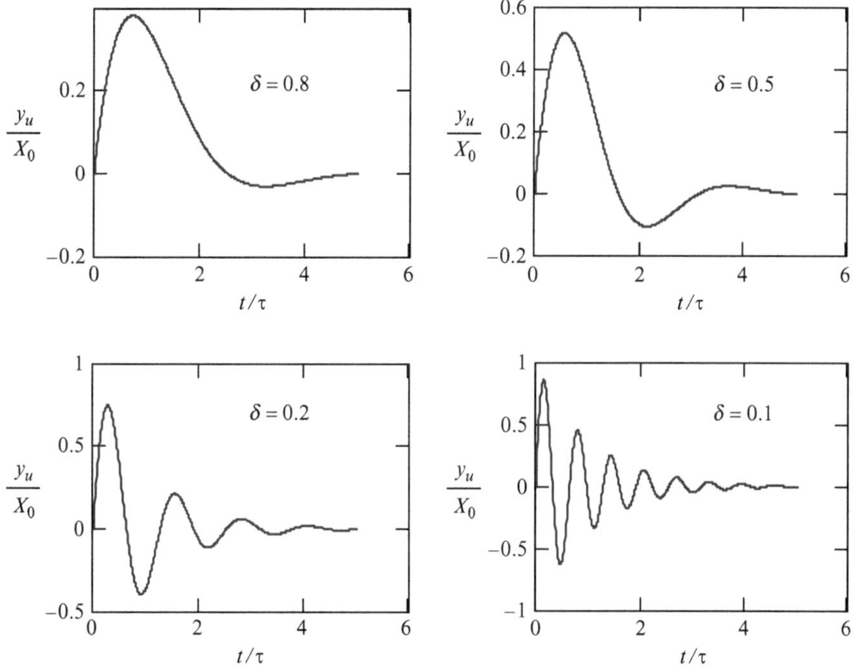

Fig. 11.18 Unforced (transient) response of a second-order circuit for various damping factors

Figure 11.18 shows graphs of an unforced response for four values of the damping factor δ. As the damping factor decreases, the number of oscillations and the maximum amplitude of the transient both increase.

Exercise 11.6. Use (11.93) to estimate the number of cycles executed before the transient vanishes for $\delta = 0.8, 0.5, 0.2, 0.1$. Are your answers in approximate agreement with Fig. 11.18?

Exercise 11.7. Show that the damping factor δ for an underdamped response is the magnitude of the ratio of the real part to the imaginary part of either characteristic root.

Exercise 11.8. What is the damping factor for a critically damped circuit?

11.5.4 Extrema of the Unforced Component of an Underdamped Response

The unforced component y_u of an underdamped response is given by:

$$y_u(t) = 2|Y|e^{-t/\tau}\cos(\omega_0 t + \theta_Y), \quad t > 0. \qquad (11.94)$$

The extrema of the unforced response are of interest. Differentiating the right side of (11.94) once with respect to time and setting the result to zero gives

$$\left[\frac{1}{\tau}\cos(\omega_0 t + \theta_Y) + \omega_0 \sin(\omega_0 t + \theta_Y)\right] = 0$$

where we have used the fact that

$$2|Y|\exp\left(-\frac{t}{\tau}\right) \neq 0$$

It follows that

$$\begin{aligned}
&\cos(\omega_0 t + \theta_Y) + \omega_0 \sin(\omega_0 t + \theta_Y) = 0 \\
&\Rightarrow \sin\left[\omega_0 t + \theta_Y + \tan^{-1}\left(\frac{1}{\omega_0 \tau}\right)\right] = 0 \\
&\Rightarrow \sin\left[\omega_0 t + \theta_Y + \tan^{-1}(\delta)\right] = 0 \\
&\Rightarrow \omega_0 t + \theta_Y + \tan^{-1}(\delta) = k\pi, \ k = 0, \pm 1, \pm 2, \cdots
\end{aligned}$$

The solutions are the times t_k given by

$$t_k = \frac{k\pi - \theta_Y - \tan^{-1}(\delta)}{\omega_0}, \qquad k = k_0,\ k_0+1,\ k_0+2, \cdots, \tag{11.95}$$

where k_0 is the smallest *nonnegative* integer for which $t_{k_0} \geq 0^+$. The corresponding extrema are given by

$$y_u(t_k) = 2|Y|e^{-t_k/\tau}\cos(\omega_0 t_k + \theta_Y), \qquad k = k_0,\ k_0+1,\ k_0+2, \cdots. \tag{11.96}$$

If the steady-state response is non-negative, extrema given by (11.96) are called **undershoots** if they are below the steady-state response and are called **overshoots** if they are above the steady-state response. The terminology is reversed if the steady-state response is negative.

Equation (11.95) is derived by solving for times at which the time derivative of the associated current or voltage equals zero. If the maximum magnitude of a current or voltage occurs at a jump discontinuity, (11.95) can give an incorrect result. For example, if $\theta_Y = 0$ in (11.94), the maximum magnitude of $y_u(t)$ occurs at $t = 0^+$, and not at any of the values given by (11.95). Thus, always evaluate (11.94) for $t = 0^+$ to see if $|y_u(0^+)|$ is larger than the largest of the values given by (11.95). If not, the maximum magnitude of the transient is given by (11.96) for $k = k_0$.

Exercise 11.9. Show that

$$y_u(t_k) = 2|Y|e^{-t_k/\tau}\cos(\omega_0 t_k + \theta_Y), \qquad k = k_0,\ k_0+1,\ k_0+2, \cdots$$

can be expressed

$$y_u(t_k) = \frac{2(-1)^k|Y|}{\sqrt{1+\delta^2}}\exp\left(-\frac{t_k}{\tau}\right), \qquad k = k_0,\ k_0+1,\ k_0+2, \cdots$$

Example 11.14. In Fig. 11.19, $R = 1\,\text{k}\Omega$, $C = 100\,\text{nF}$, $L = 25\,\text{mH}$, and $v_S = V_0\,u(t)$, with $V_0 = 5\,\text{V}$.

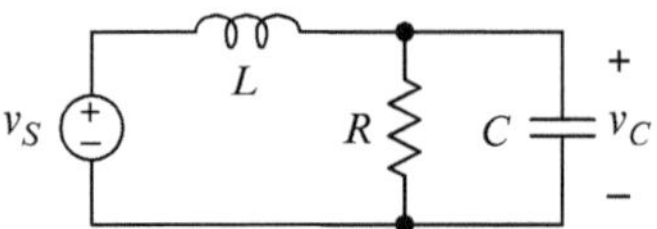

Fig. 11.19 See Example 11.14

(a) Obtain an expression for the response $v_C(t)$, assuming the response is underdamped.
(b) Find the times at which the first two extrema of the unforced response occur and the amplitudes of those extrema.
(c) Plot the response versus time for the circuit parameters given in part (b).

Solution: (a) Kirchhoff's current law gives

$$C\frac{dv_C}{dt} + \frac{1}{R}v_C + \frac{1}{L}\int_{-\infty}^{t}[v_C(t') - v_S(t')]dt' = 0. \tag{11.97}$$

Differentiating once with respect to time and rearranging terms gives, for $t > 0$

$$LC\frac{d^2v_C}{dt^2} + \frac{L}{R}\frac{dv_C}{dt} + v_C = v_S = V_0; \quad t > 0.$$

The dc gain is unity. The steady-state response is

$$v_{C\,ss} = V_0 = 5\,\text{V}.$$

The characteristic roots are

$$s,\ s^* = -\frac{1}{2RC} \pm \frac{1}{2LC}\sqrt{\left(\frac{L}{R}\right)^2 - 4LC} \cong (-5.00 + j19.36) \times 10^3\,\text{s}^{-1}.$$

The source appears as a short circuit for $t < 0$, so no energy is stored in the inductor or capacitor at $t = 0^+$. Thus, $v_C(0^+) = 0$ and $i_L(0^+) = 0$. From (11.97) we find, for $t = 0^+$

$$C\frac{dv_C}{dt}\bigg|_{t=0^+} + \frac{1}{R}v_C(0^+) + \frac{1}{L}\int_{-\infty}^{0^+}[v_C(t') - v_S(t')]dt' = C\frac{dv_C}{dt}\bigg|_{t=0^+} + 0 + 0 = 0$$

$$\Rightarrow \frac{dv_C}{dt}\bigg|_{t=0^+} = v_C'(0^+) = 0.$$

From (11.81)

$$V = \frac{s^*[v_C(0^+) - V_0] - v_C'(0^+)}{(s^* - s)} = -\frac{s^* V_0}{(s^* - s)},$$

which gives

$$2|V_1| \cong 5.16\,\text{V},\ \theta_V \cong 2.89. \tag{11.98}$$

The complete response is given by

$$v_C(t) = \left[V_0 + 2|V|\exp\left(-\frac{t}{\tau}\right) \times \cos(\omega_0 t + \theta_V)\right]u(t). \tag{11.99}$$

where

$$\tau = -\frac{1}{\text{Re}(s)} \cong 200\,\mu s, \quad \omega_0 = \text{Im}(s) \cong 19.36 \times 10^3\,\text{s}^{-1}. \tag{11.100}$$

(b) *The damping factor is*

$$\delta = \frac{1}{\omega_0 \tau} \cong 0.258,$$

From (11.95), the extrema of the unforced response occur at the times t_k given by

$$t_k = \frac{k\pi - RV - tan^{-1}(\delta)}{\omega_0}; \quad k = k_0, k_0 + 1, k_0 + 2, \cdots,$$

which are

$$t_2 = 162.2\,\mu\text{s},\ t_3 = 324.5\,\mu\text{s},$$

at which times

$$v_C(t_2) = 7.22\,\text{V},\ v_C(t_3) = 4.013\,\text{V}.$$

As a check, we compute the initial value of the unforced response, given by

$$2|V|\cos(\theta_V) = -5\,\text{V}.$$

Thus, $v_C(t_2) = 7.22\,\text{V}$ is the maximum magnitude of the unforced response. The steady-state response is $v_C(\infty) = V_0 = 5\,\text{V}$, so the first (largest) overshoot is 2.22 V and the first (largest) undershoot is 987 mV.

(c) Figure 11.20 shows a graph of the step response.

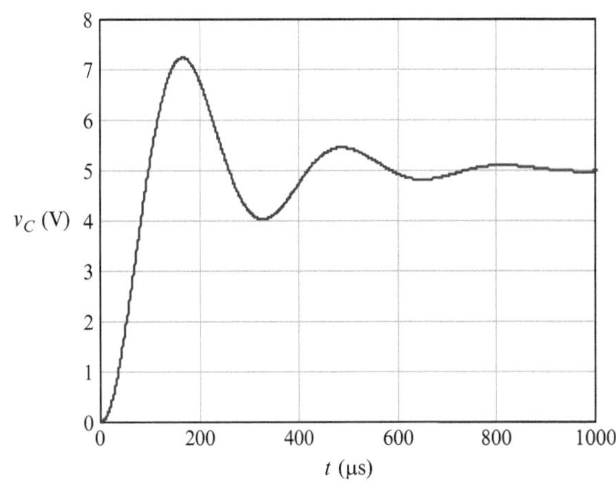

Fig. 11.20 See Example 11.14

11.6 Time Invariance, Superposition, and Pulse Response

A circuit for which the response $y(t)$ to an excitation $x(t)$ is governed by a differential equation of the form

$$a_n \frac{d^n y}{dt^n} + a_{n-1} \frac{d^{n-1} y}{dt^{n-1}} + \cdots + a_1 \frac{dy}{dt} + y$$
$$= b_n \frac{d^n x}{dt^n} + b_{n-1} \frac{d^{n-1} x}{dt^{n-1}} + \cdots + b_1 \frac{dx}{dt} + b_0 x \quad (11.101)$$

obeys the following principle of **time invariance**: If $y(t)$ is the response to an excitation $x(t)$, then the response to $x(t - t_0)$, where t_0 is a fixed delay, is $y(t - t_0)$. What this means is that *the mathematical form of the response does not depend upon the time at which the excitation is applied.* In other words, delaying the excitation simply delays the response by the same amount, but does not change its form.

Figure 11.21 illustrates time-invariance. In Fig. 11.21a a step excitation $v_S = V_0\, u(t)$ is applied as shown to a series RC circuit. The voltage across the capacitor (the response) is given by $v_C = V_0 \left(1 - e^{-t/\tau}\right) u(t)$, where $\tau = RC$. In Fig. 11.21b, a step excitation $v_S = V_0\, u(t - t_0)$ is applied to an identical circuit and the response is given by $v_C = V_0 \left[1 - e^{-(t-t_0)/\tau}\right] u(t - t_0)$.

Were it not for time-invariance, life would be a mess. A CD you played on your audio system would sound different each time you played it. Circuits laboratories would be frustrating, because identical experiments performed at different times on identical circuits would yield different results.

A two-port circuit whose response $y(t)$ to an excitation $x(t)$ is governed by an equation of the form (11.101), being linear, also obeys the following principle of **amplitude scaling**: Let $y(t)$ denote the response to an excitation $x(t)$. Then the response to $Kx(t)$, where K is a constant, is given by $Ky(t)$. The proof is trivial, as it amounts to multiplying (11.101) by K.

Example 11.15. Let $y(t)$ denote the response of a certain circuit to an excitation $x(t) = X_0\, u(t)$. Then the response to $x_1(t) = 3X_0\, u(t)$ is $y_1(t) = 3y(t)$.

Example 11.16. Let $y(t)$ denote the response of a certain circuit to excitation $x(t)$. The response of the circuit to excitation $x_1(t) = 5x(t) - 2x(t - t_0) + 10x(t - t_1)$ is given by $y_1(t) = 5y(t) - 2y(t - t_0) + 10y(t - t_1)$.

We may use amplitude scaling, time invariance, and superposition to obtain the response of a linear circuit to any piecewise constant excitation from the response of the circuit to a step excitation. A particular piecewise-constant excitation of interest in many applications is a *rectangular pulse*, illustrated by Fig. 11.22 and described by

$$v_S = V_0\, u(t) - V_0\, u(t - t_0). \quad (11.102)$$

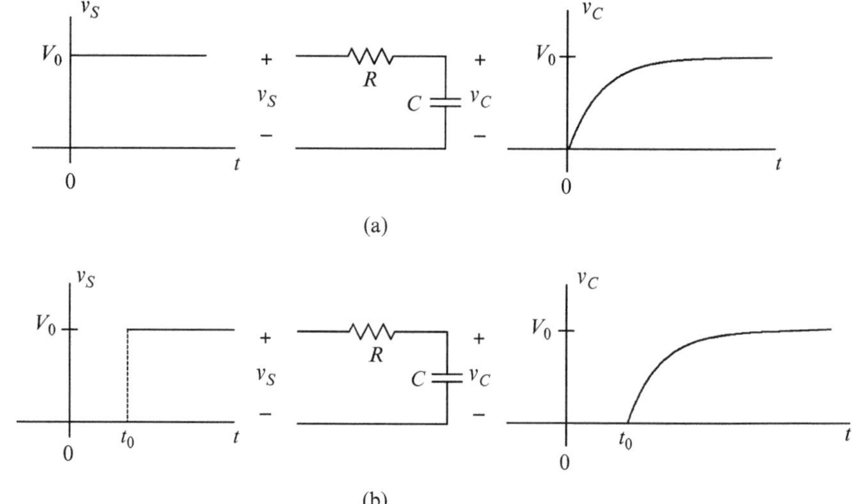

Fig. 11.21 Illustrating time-invariance

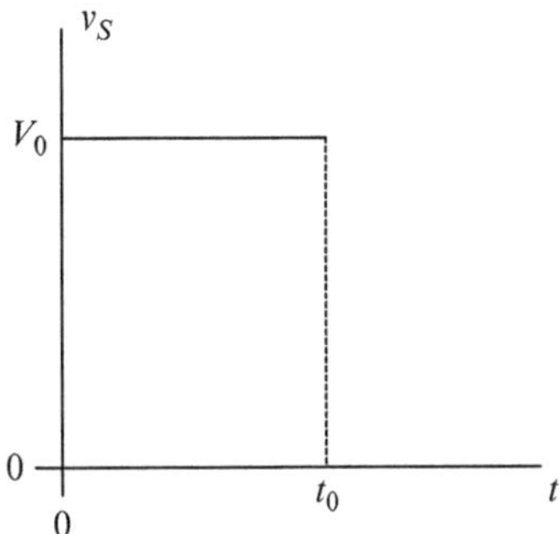

Fig. 11.22 Rectangular pulse expressed by (11.102)

To obtain the response of a linear time-invariant circuit to such an excitation, we need only obtain the response to the step $V_0\,u(t)$ and subsequently apply time-invariance and superposition. Thus if the response to $V_0\,u(t)$ is given by $v_L(t)$, then by time-invariance, the response to $V_0\,u(t-t_0)$ is given by $v_L(t-t_0)$ and by superposition, the response to $v_S = V_0\,u(t) - V_0\,u(t-t_0)$ is given by $v_L = v_L(t) - v_L(t-t_0)$.

Example 11.17. Refer to Example 11.11. Find the current $i_2(t)$ if the applied voltage is given by $v_2(t) = V_0\,u(t) - V_0\,u(t-t_0)$.

Solution: From Example 11.11, the response to $v_S(t) = V_0\,u(t)$ is given by $i(t) = I(e^{s_1t} - e^{s_2t})u(t)$. By time invariance and superposition, the response to $v_2(t) = V_0\,u(t) - V_0\,u(t-t_0) = v_S(t) - v_S(t-t_0)$ is given by

$$i_2(t) = i(t) - i(t-t_0) = I(e^{s_1t} - e^{s_2t})u(t) - I\left(e^{s_1(t-t_0)} - e^{s_2(t-t_0)}\right)u(t-t_0).$$

Figure 11.23 shows a graph of the response (compare with Fig. 11.13).

Exercise 11.10. Refer to Fig. 11.24, where $R = 100\,\text{k}\Omega$ and $C = 100\,\text{pF}$. Let $V_0 = 5\text{V}$ and $t_0 = 50\ \mu\text{s}$. Sketch neatly and label fully a graph of the voltage v_R across the resistor if

(a) $v_S(t) = V_0\,u(t)$, with $V_0 = 5\ \text{V}$;

(b) $v_S(t) = V_0\,u(t) - V_0\,u(t-t_0)$, with $V_0 = 5\ \text{V}$ and $t_0 = 50\ \mu\text{s}$; and

(c) $v_S(t) = 2V_0\,u(t) - 3V_0\,u(t-t_0) + V_0\,u(t-2t_0)$, with $V_0 = 5\ \text{V}$ and $t_0 = 50\ \mu\text{s}$.

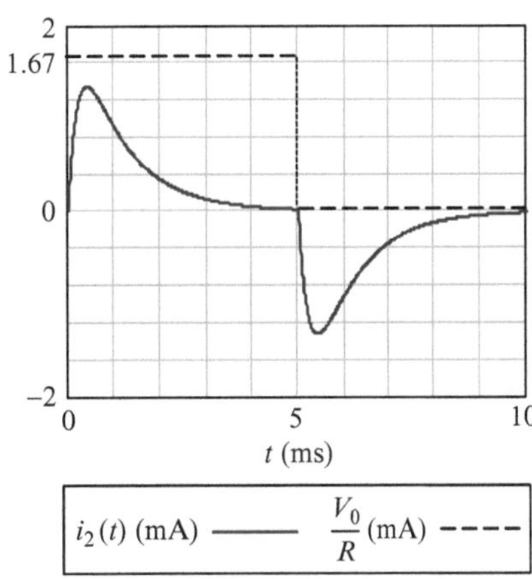

Fig. 11.23 See Example 11.17

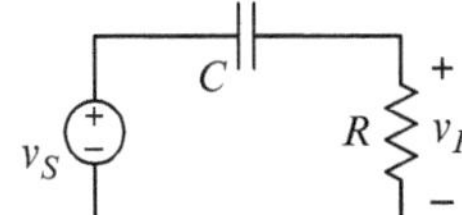

Fig. 11.24 See Exercise 11.10

11.7 Operator Notation

In **operator notation**, differentiation with respect to time is indicated by multiplication by the **operator** s: Thus

$$\frac{dy}{dt} \leftrightarrow s\,y,\quad \frac{d^2y}{dt^2} \leftrightarrow s^2y,\quad \frac{d^3y}{dt^3} \leftrightarrow s^3y,\ \cdots, \tag{11.103}$$

where y can be any function of time. The SI unit of the operator s is s^{-1}, so confusion is possible, especially where both the operator s and the unit s occur in an expression. To avoid confusion, underline the unit s in written work involving the operator s.

We may also use operator notation as shorthand for integration. Thus

$$\int_{-\infty}^{t} v(t')dt' \leftrightarrow \frac{1}{s}v(t). \tag{11.104}$$

Equation (11.104) makes sense: if s denotes differentiation, then $1/s$ should denote integration, because $(s)(1/s) = 1$ (which denotes those operations in succession). When using operator notation in this way, *be sure the limits of integration extend from* $-\infty$ to t; for example.

$$\frac{1}{L}\int_{-\infty}^{t} v_L(t')dt' \leftrightarrow \frac{1}{sL}v_L.$$

Operator notation allows us to manipulate coupled differential equations algebraically, eliminate unwanted variables, and obtain a single differential equation relating excitation and response. There is no magic here. In this context, operator notation is simply a shorthand that makes it easier to see what operations must be performed to obtain a desired (single) differential equation.

Example 11.18. In Fig. 11.25, $v_S = V_0\,u(t)$. (a) Obtain a differential equation governing the voltage v_C for $t > 0$. (b) Obtain expressions for the initial values $v_C(0^+)$ and $v_C'(0^+)$.

Solution: (a) We use Kirchhoff's current law to obtain two node equations for $t > 0$:

$$\frac{v_C - V_0}{R_1} + C\frac{dv_C}{dt} + \frac{v_C - v_L}{R_2} = 0, \quad (11.105)$$

$$\frac{v_L - v_C}{R_2} + \frac{1}{L}\int_{-\infty}^{t} v_L(t')dt' = 0. \quad (11.106)$$

Equation (11.105) is written using operator notation as

$$\left(\frac{1}{R_1} + \frac{1}{R_2} + sC\right)v_C - \frac{v_L}{R_2} = \frac{V_0}{R_1}, \quad (11.107)$$

which yields

$$(1 + s\tau_1)v_C - \frac{R_1}{R_1 + R_2}v_L = \frac{R_2}{R_1 + R_2}V_0, \quad \tau_1 = (R_1\|R_2)C. \quad (11.108)$$

Equation (11.106) is written using operator notation as

$$-\frac{L}{R_2}v_C + \left(\frac{1}{s} + \frac{L}{R_2}\right)v_L = 0,$$

which yields

$$-s\tau_2 v_C + (1 + s\tau_2)v_L = 0$$
$$\tau_2 = \frac{L}{R_2}. \quad (11.109)$$

We are interested only in the voltage v_C, so we eliminate v_L. From (11.109)

$$v_L = \frac{s\tau_2 v_C}{1 + s\tau_2}. \quad (11.110)$$

Using the right side of (11.110) for v_L in (11.108) gives

$$(1 + s\tau_1)v_C - \frac{R_1}{R_1 + R_2}\frac{s\tau_2 v_C}{1 + s\tau_2} = \frac{R_2}{R_1 + R_2}V_0, \quad (11.111)$$

which yields

$$v_C = \frac{k_2(1 + s\tau_2)V_0}{\tau_1\tau_2 s^2 + (\tau_1 + \tau_2 - k_1\tau_2)s + 1},$$

or

$$\tau_1\tau_2 s^2 v_C + (\tau_1 + \tau_2 - k_1\tau_2)s v_C + v_C = k_2(1 + s\tau_2)V_0, \quad (11.112)$$

corresponding to the standard-form differential equation

$$\tau_1\tau_2\frac{d^2v_C}{dt^2} + (\tau_1 + \tau_2 - k_1\tau_2)\frac{dv_C}{dt} + v_C = k_2\tau_2\frac{dV_0}{dt} + k_2V_0 = k_2V_0,$$

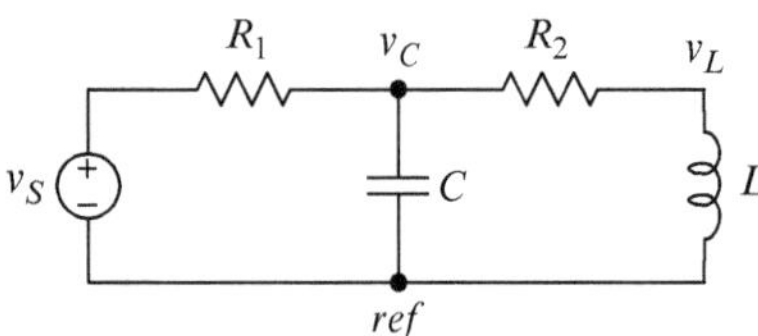

Fig. 11.25 See Example 11.18

where

$$\tau_1 = (R_1 \| R_2)C, \quad \tau_2 = \frac{L}{R_2},$$

$$k_1 = \frac{R_1}{R_1 + R_2}, \quad k_2 = \frac{R_2}{R_1 + R_2},$$

and $dV_0/dt = 0$ because V_0 is constant.

(b) No energy is supplied to the circuit before $t = 0$. Thus there can be no energy stored in the circuit for $t = 0^+$, which means $v_C(0^+) = v_C(0^-) = 0$ and $i_L(0^+) = i_L(0^-) = 0$. For $t = 0^+$, (11.105) gives

$$\frac{v_C(0^+) - V_0}{R_1} + Cv'_C(0^+) + i_L(0^+) = 0$$

$$\Rightarrow Cv'_C(0^+) = \frac{V_0}{R_1},$$

because $v_C(0^+) = 0$ and $i_L(0^+) = 0$. The initial values are

$$v_C(0^+) = 0, \quad v'_C(0^+) = \frac{V_0}{R_1 C}.$$

Exercise 11.11. The circuit in Fig. 11.26 is called a twin-T. Note that there is no connection at the apparent intersection labeled x. (a) Obtain a differential equation in standard form that relates the output v_o to the excitation v_S. (b) Let $RC = 10\,\mu\text{s}$. Find the characteristic roots and show that the circuit is overdamped. Why must this circuit be overdamped? (Why can it not be underdamped?) (c) Find the approximate duration of the transient response to an excitation of the form $v_S = V_0\, u(t)$.

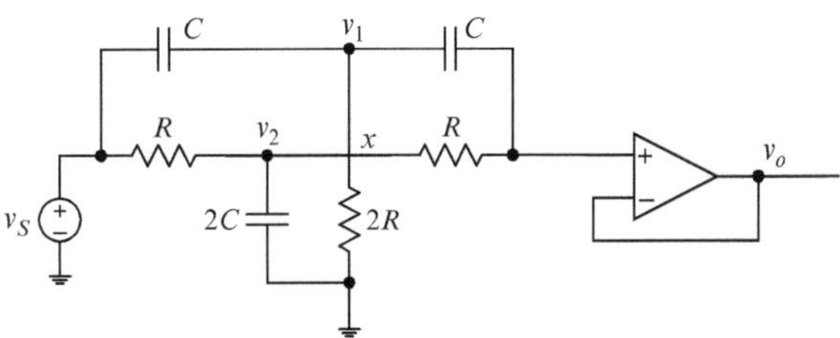

Fig. 11.26 See Exercise 11.11

11.8 Problems

Section 11.2 is prerequisite for the following problems.

P 11.1 Use unit step functions to express each of the functions below. Sketch neatly and label fully a graph of each.

$$\text{(a)}\, v = \begin{cases} -V_0, \, t \le 0, \\ 2V_0, \, t > 0; \end{cases} \quad \text{(b)}\, v = \begin{cases} 0, \, t \le 0, \\ V_0, \, 0 < t \le t_1, \\ 2V_0, \, t_1 < t \le t_2, \\ 3V_0, t > t_2; \end{cases}$$

$$\text{(c)}\, i = \begin{cases} 0, t \le 0, \\ -I_0, 0 < t \le t_1, \\ I_0, t_1 < t \le 2t_1, \\ 0, t > 2t_1; \end{cases}$$

$$\text{(d)}\, v = \begin{cases} 0, \quad t \le 0, \\ \dfrac{V_0 t}{\tau}, \, 0 < t \le \tau, \\ V_0, \quad t > \tau; \end{cases} \quad \text{(e)}\, i = \begin{cases} 0, \quad t \le -t_1, \\ I_0, \quad -t_1 < t \le t_1, \\ 0, \quad t > t_1. \end{cases}$$

P 11.2 The unit step function can be defined in various ways. Three possible definitions are

$$u_a(t) = \begin{cases} 0, & t \le 0, \\ 1, & t > 0;. \end{cases}$$

$$u_b(t) = \begin{cases} 0, & t < 0, \\ 1, & t \ge 0; \end{cases}$$

$$u_c(t) = \begin{cases} 0, & t < 0, \\ 1/2, & t = 0, \\ 1, & t > 0. \end{cases}$$

Are there any *significant* differences among these three definitions insofar as they are used to describe (approximately) physical quantities such as current and voltage? If so, what are they?

Section 11.4 is prerequisite for the following problems.

P 11.3 Refer to Fig. P 11.1, where $v_S(t) = V_0\, u(t)$. Obtain an expression for the current i for $t > 0$.

P 11.4 Refer to Fig. P 11.2, where $v_S = V_0\, u(t)$. Obtain an expression for the voltage $v(t)$.

P 11.5 In Fig. P 11.3, $i_0(t) = 2I_0 + 3I_0\, u(t)$. Obtain an expression for the voltage v_L across the inductor two ways: (1) directly from the given circuit, using Kirchhoff's current law, and (2) using a source transformation and Kirchhoff's voltage law. Identify the transient and steady-state components. If $I_0 = 1\,\text{mA}$, $L = 10\text{mH}$, and $R = 100\text{k}\Omega$, what are the duration of the transient and the maximum voltage (magnitude) across the inductor?

P 11.6 Refer to Fig. P 11.4, where the op amp is ideal and $i_S(t) = I_0\, u(t)$. Obtain an expression for the voltage $v_o(t)$.

P 11.7 In Fig. P 11.5, the op amp is ideal. Describe the response v_L to a step excitation $v_S = V_0\, u(t)$ as completely as possible, without writing a differential equation.

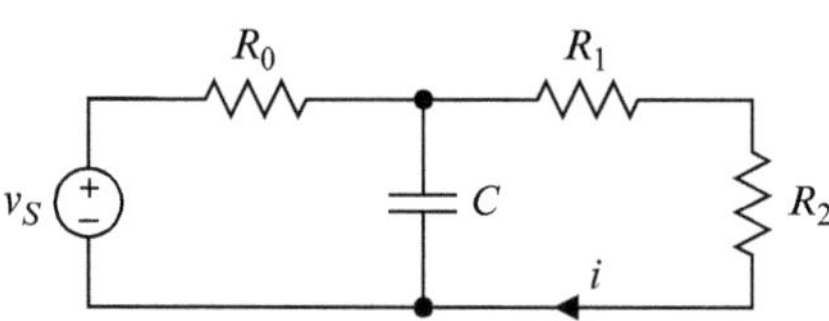

Fig. P 11.1 See Problem P 11.3

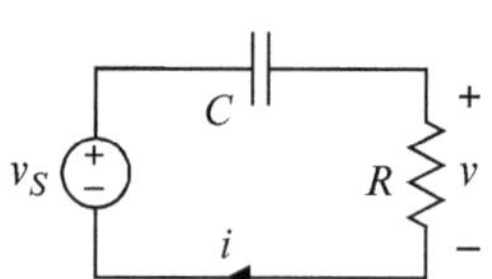

Fig. P 11.2 See Problem P 11.4

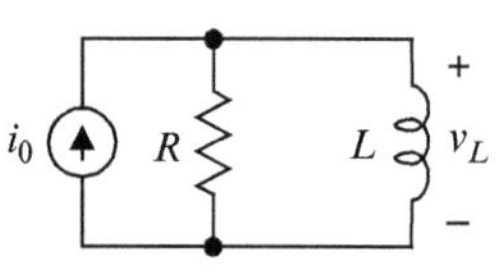

Fig. P 11.3 See Problem P 11.5

Fig. P 11.4 See Problem P 11.6

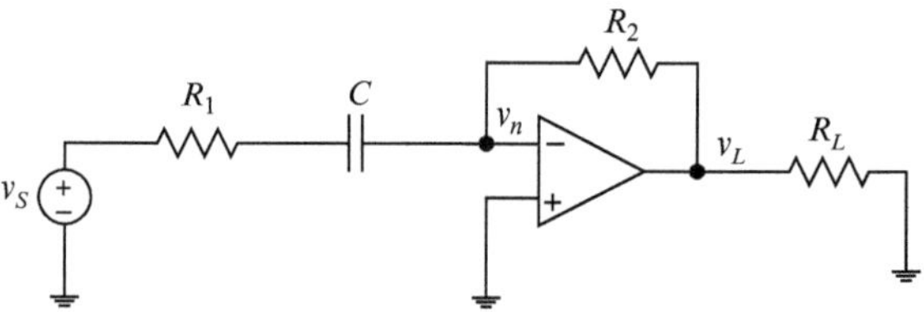

Fig. P 11.5 See Problem P 11.7

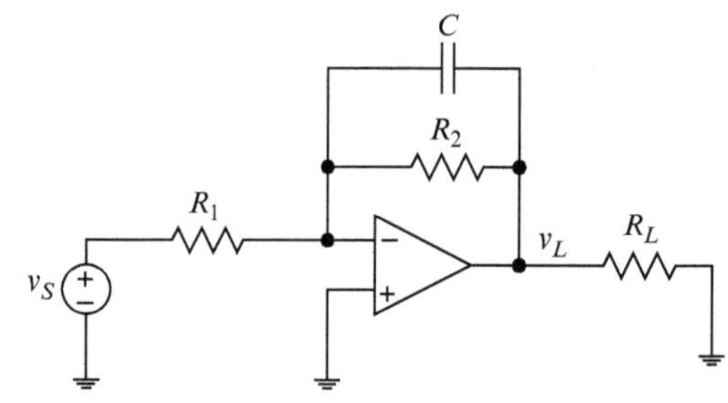

Fig. P 11.6 See Problem P 11.8

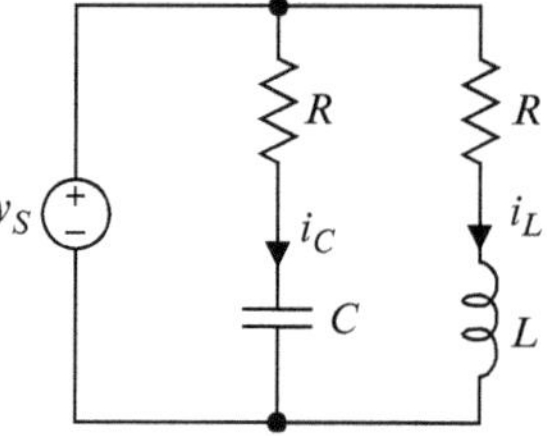

Fig. P 11.7 See Problem P 11.9

P 11.8 In Fig. P 11.6, the op amp is ideal. Describe the response v_L to a step excitation $v_S = V_0\, u(t)$ as completely as possible, without writing a differential equation.

P 11.9 In Fig. P 11.7, $v_S = V_0\,u(t)$. Without writing a differential equation, give expressions for the currents i_C, i_L.

P 11.10 In Fig. P 11.8, $i_S = I_0\,u(t)$. Without writing a differential equation, give expressions for the voltages v_C, v_L.

Section 11.5 is prerequisite for the following problems.

P 11.11 In Fig. P 11.9, $R = 10\,\Omega$, $L = 1\,\text{mH}$, $C = 100\,\text{nF}$, and $i(t) = I_0\,u(t)$, with $I_0 = 1\,\text{A}$. Obtain an expression for the voltage $v(t)$.

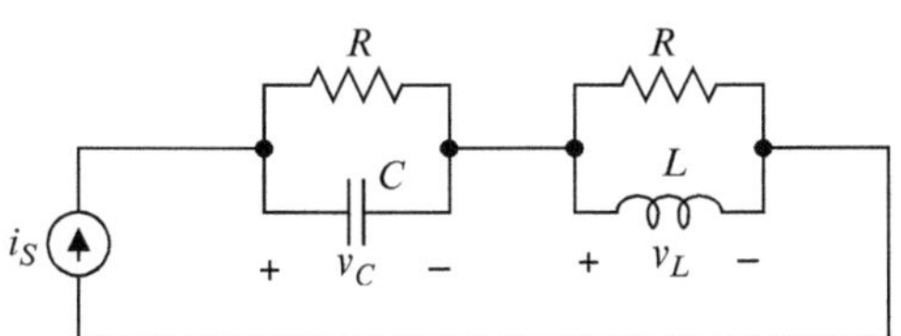

Fig. P 11.8 See Problem P 11.10

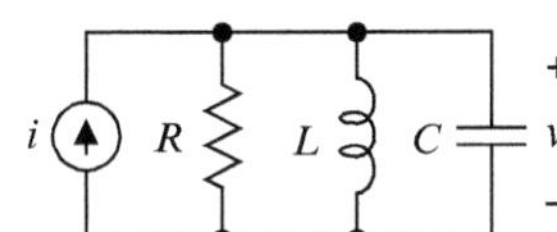

Fig. P 11.9 See Problem P 11.11

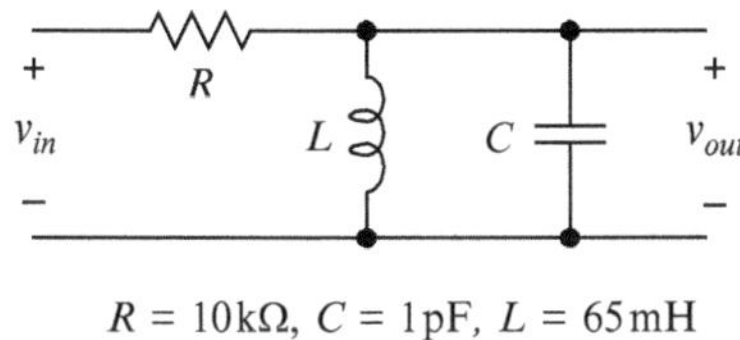

Fig. P 11.10 See Problem P 11.12

P 11.12 In Fig. P 11.10, the input is a given by $v_{in}(t) = V_0\,u(t)$, with $V_0 = 1\,\text{V}$. Sketch neatly and label fully a graph of the output $v_{out}(t)$ versus time.

P 11.13 Refer to Fig. P 11.11, where $i_S(t) = I_0\,u(t)$, $I_0 = 1\,\text{mA}$, $R_1 = 10\,\text{k}\Omega$, $R_2 = 100\,\text{k}\Omega$, $\mu_0 = 100$, $C_1 = 10\,\text{nF}$, and $C_2 = 1\,\text{nF}$. Obtain an expression for the voltage $v_C(t)$ for $t > 0$.

P 11.14 Refer to Fig. P 11.12. Show that the circuit is overdamped for $R = 7.5\,\text{k}\Omega$ and underdamped for $R = 22\,\text{k}\Omega$. For each value of R, express the voltages v_R and v_C in terms of the amplitude of the step excitation for $t > 0$.

P 11.15 In Fig. P 11.13, $v_S = V_0\,u(t)$, with $V_0 = 1\,\text{V}$.

(a) Obtain a standard-form differential equation governing the voltage v_L.
(b) Let $R = 10\,\text{k}\Omega$, $C = 1\text{nF}$, $R_F = 500\,\Omega$, $R_Q = 2.5\,\text{k}\Omega$, and $R_G = 30\,\text{k}\Omega$. Obtain an expression for the voltage $v_L(t)$.

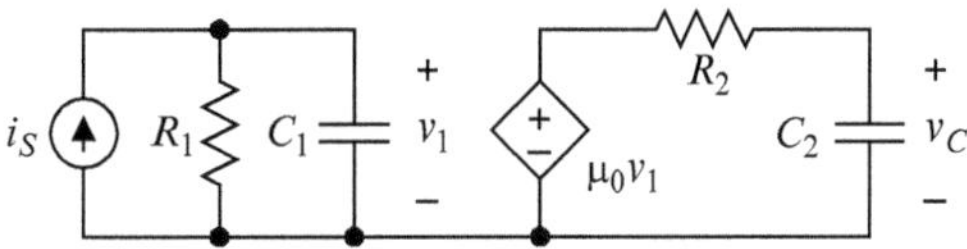

Fig. P 11.11 See Problem P 11.13

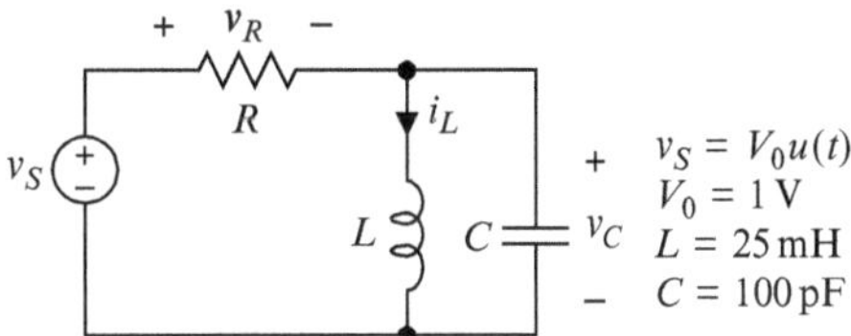

Fig. P 11.12 See Problem P 11.14

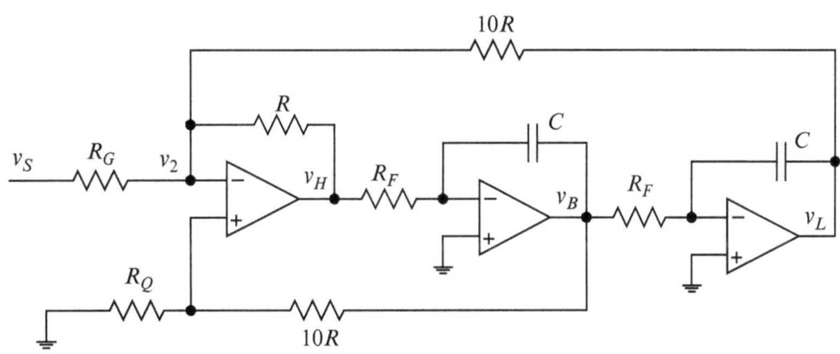

Fig. P 11.13 See Problem P 11.15

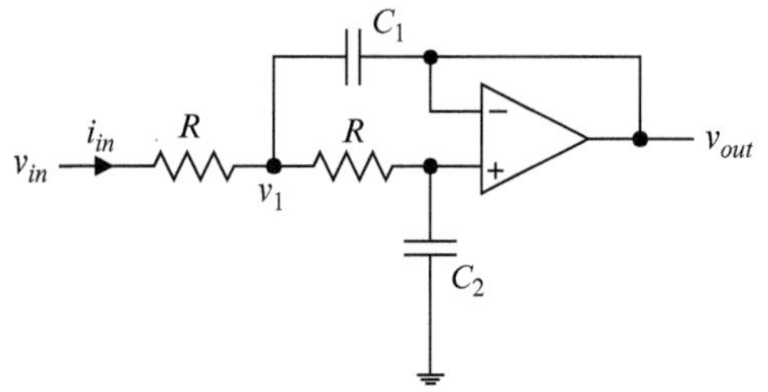

Fig. P 11.14 See Problem P 11.16

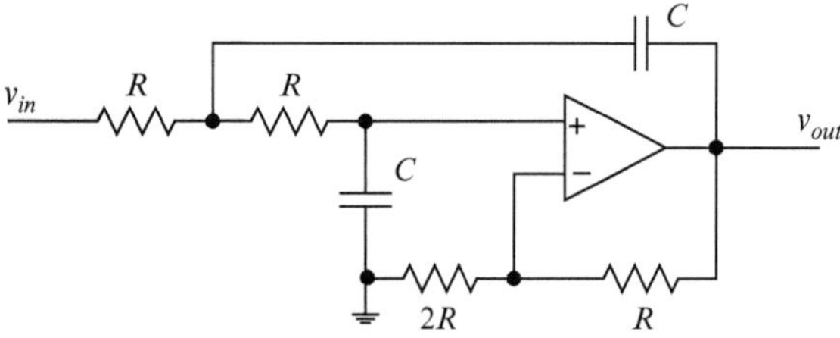

Fig. P 11.15 See Problem P 11.17

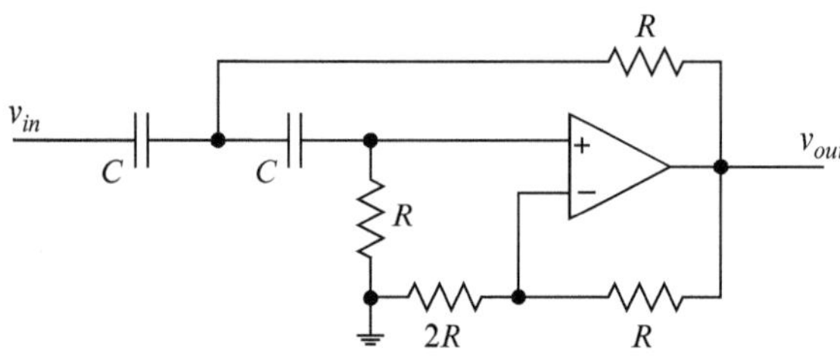

Fig. P 11.16 See Problem P 11.18

P 11.16 In Fig. P 11.14, $v_{in} = V_0\, u(t)$.

(a) Obtain a standard-form differential equation governing the voltage v_{out}.
(b) Show that the response is underdamped if $C_1 > C_2$.

P 11.17 In Fig. P 11.15, $v_{in} = V_0\, u(t)$.

(a) Obtain a standard-form differential equation governing the voltage v_{out}.
(b) Obtain an expression for the voltage $v_{out}(t)$.
(c) Let $V_0 = 1$ V $RC = 1$ μs. Construct a graph of the voltage v_{out} versus time.

P 11.18 In Fig. P 11.16, $v_{in} = V_0\, u(t)$.

(a) Obtain a standard-form differential equation governing the voltage v_{out}.

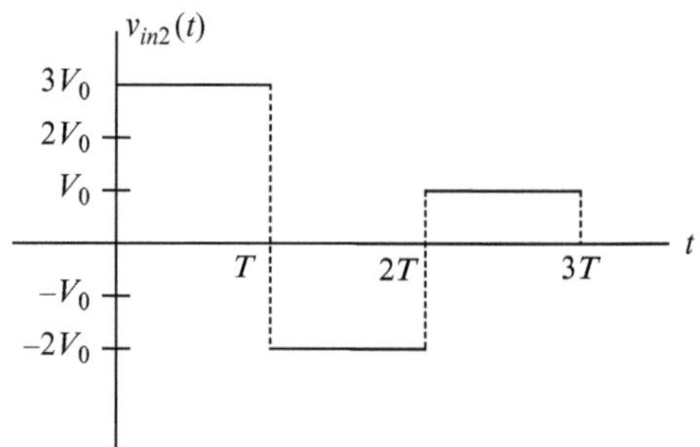

Fig. P 11.17 See Problem P 11.19

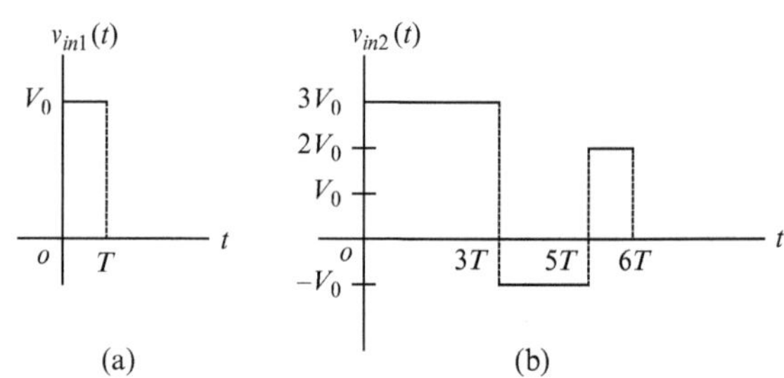

Fig. P 11.18 See Problem P 11.21

(b) Obtain an expression for the voltage $v_{out}(t)$.
(c) Let $V_0 = 1$ V $RC = 1$ μs. Construct a graph of the voltage v_{out} versus time.

Section 11.6 is prerequisite for the following problems.

P 11.19 The response of a certain circuit to a step excitation $v_{in1}(t) = V_0\, u(t)$ is given by $v_{out1}(t) = 5V_0\, e^{-t/\tau} u(t)$. Obtain an expression for the response to the excitation shown in Fig. P 11.17. The excitation is zero for times not shown.

P 11.20 The response of a certain circuit to a sinusoidal excitation $v_{in1}(t) = V_0 \cos(\omega_0 t)$ is given by $v_{out1}(t) = 100V_0 \cos(\omega_0 t - 0.5)$. Find the response of the circuit to the excitation

$$v_{in2}(t) = 2\,V_0 \cos(\omega_0 t + 0.2) + V_0 \sin(\omega_0 t).$$

P 11.21 Let $v_{out1}(t)$ denote the response of a certain circuit to the excitation $v_{in1}(t)$ shown in Fig. P 11.18 (a).

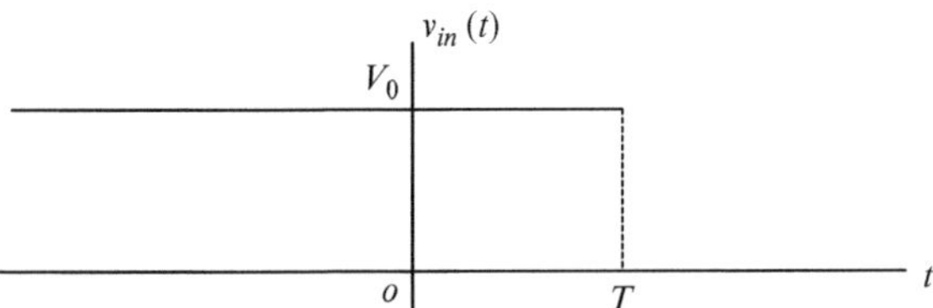

Fig. P 11.19 See Problem P 11.22

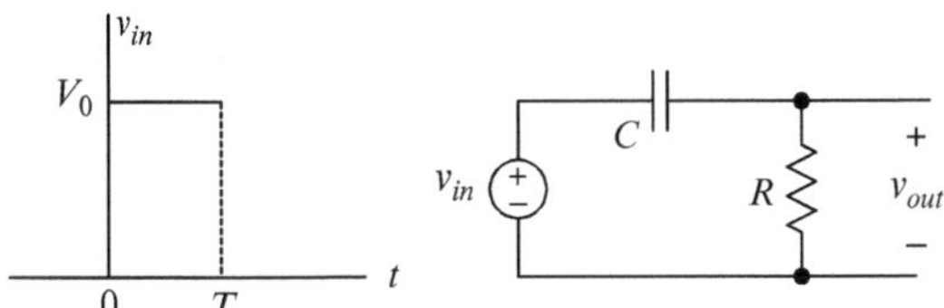

Fig. P 11.20 See Problem P 11.23

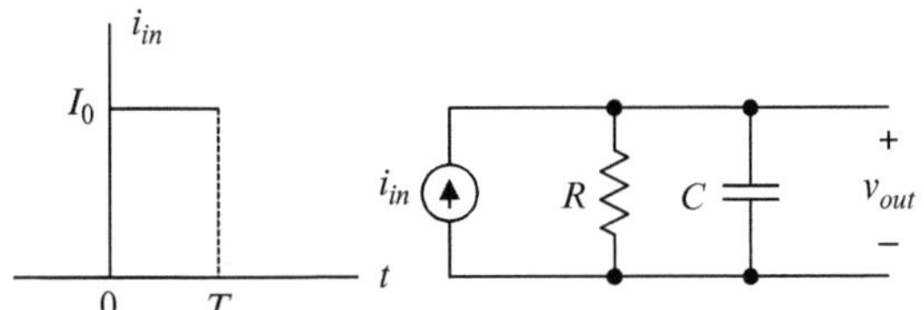

Fig. P 11.21 See Problem P 11.24, 25

Give an expression for the response $v_{out2}(t)$ of the circuit to the excitation shown in Fig. P 11.18(b).

P 11.22 The response of a certain circuit to a step excitation $v_0(t) = V_0\,u(t)$ is given by $v_1(t) = V_0\left(1 - e^{-t/\tau}\right)u(t)$. Give an expression for the response of the circuit to the excitation shown in Fig. P 11.19, where $v_{in}(t) = V_0$ for all $t < T$.

P 11.23 Refer to Fig. P 11.20. A rectangular pulse of voltage is applied to an RC circuit, as shown. Specify the time constant RC in terms of the duration T of the pulse such that $v_{out}(T^-) = 0.95 v_{out}(0^+)$.

P 11.24 Refer to Fig. P 11.21. A rectangular pulse of current is applied to an RC circuit, as shown. Specify the time constant RC in terms of the duration T of the pulse such $v_{out}(T^-) = 0.95\,RI_0$.

P 11.25 Refer to Fig. P 11.21. A rectangular pulse of current is applied to an RC circuit, as shown. Specify the time constant RC in terms of the duration T of the pulse such that no more than 5% of the total energy dissipated in the resistor is dissipated for $t \geq T$. *Hint*: Obtain expressions for the energy ε_1 dissipated for $t < T$ and the energy ε_2 dissipated for $t > T$. Then form the ratio $\varepsilon_2/(\varepsilon_1 + \varepsilon_2)$. Simplify the resulting expression by assuming $T \gg \tau$, as seems reasonable in view of the specification, and solve $\varepsilon_2/(\varepsilon_1 + \varepsilon_2) = 0.05$ for the time constant.

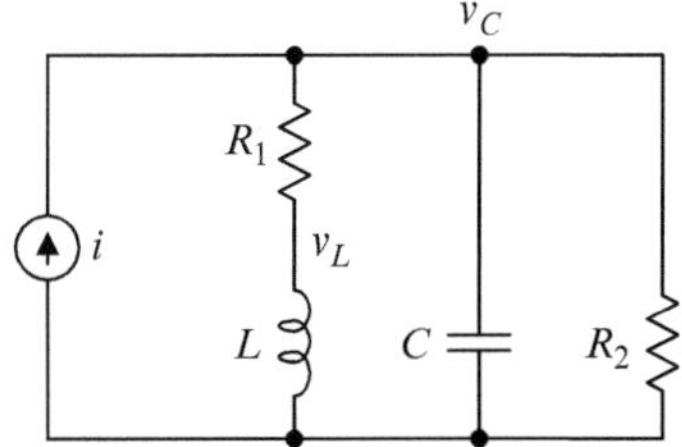

Fig. P 11.22 See Problem P 11.26

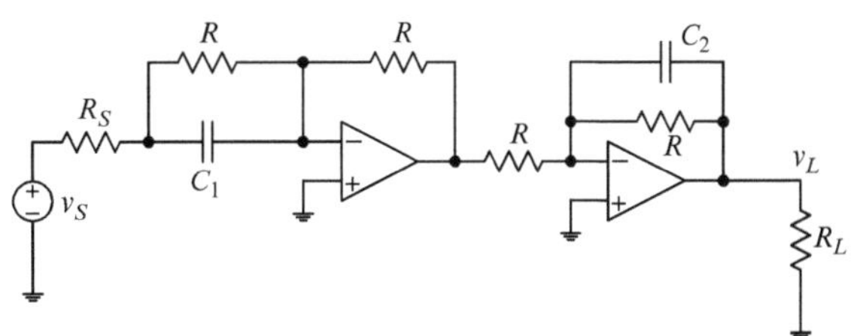

Fig. P 11.23 See Problem P 11.27

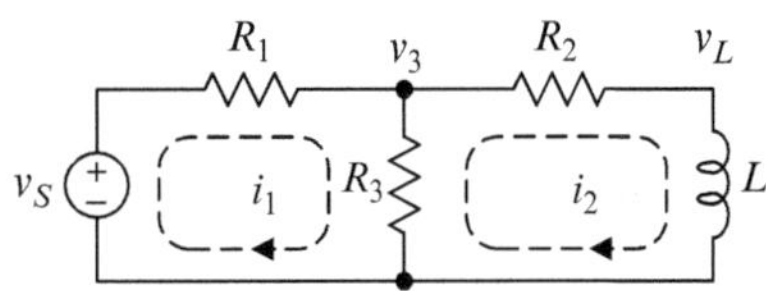

Fig. P 11.24 See Problem P 11.28, 29

Section 11.7 is prerequisite for the following problems.

P 11.26 Refer to Fig. P 11.22, where $i_S(t) = I_0\,u(t)$, $I_0 = 1\,\text{mA}$, $R_1 = 100\,\Omega$, $R = 100\,\text{k}\Omega$, $L = 100\,\text{mH}$, and $C = 100\,\text{nF}$. Obtain an expression for the voltage $v_C(t)$ for $t > 0$. Use any reasonable approximations.

P 11.27 Refer to Fig. P 11.23, where $v_S = V_0\,u(t)$. Obtain the standard-form differential equation that governs the response for $t > 0$.

P 11.28 In Fig. P 11.24, $v_S = V_0\,u(t)$. Obtain standard-form differential equations governing the indicated mesh currents and show that the time constants for the two mesh currents are the same.

P 11.29 In Fig. P 11.24, $v_S = V_0\,u(t)$. Obtain standard-form differential equations governing the indicated

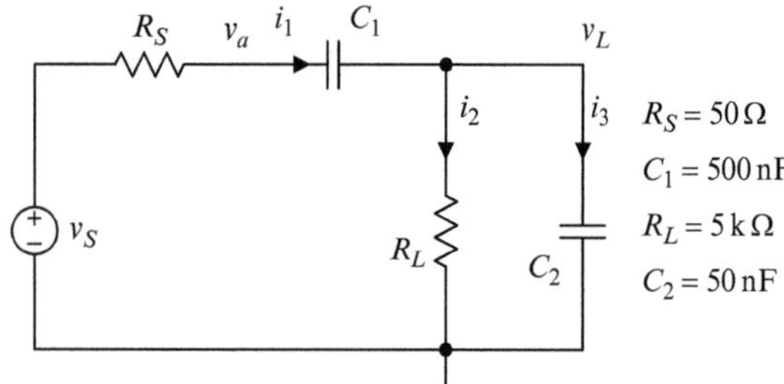

Fig. P 11.25 See Problem P 11.30

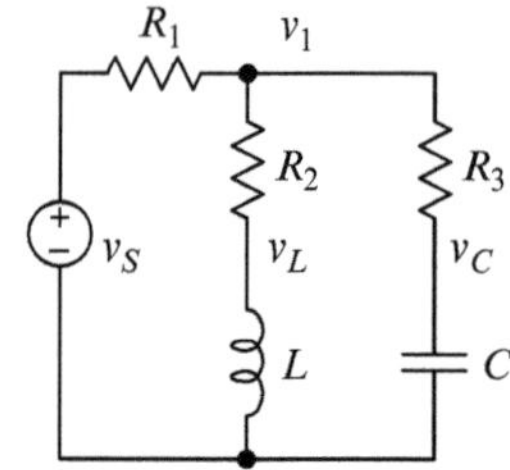

Fig. P 11.26 See Problem P 11.31

node voltages and show that the time constants for the two node voltages are the same.

P 11.30 Refer to Fig. P 11.25, where $v_S(t) = V_0\,u(t)$, with $V_0 = 5\,\text{V}$. Obtain an expression for the voltage v_L.

P 11.31 Refer to Fig. P 11.26. Obtain a standard-form differential equation relating the voltage v_C to the voltage v_S. Express the coefficients of the differential equation as functions of the circuit parameters.

P 11.32 Refer to Fig. P 11.27. (a) Obtain a standard-form differential equation relating the voltage v_C to the source voltage v_S. (b) Let $v_S = V_0\,u(t)$ and show that the equation for $t > 0$ may be written as

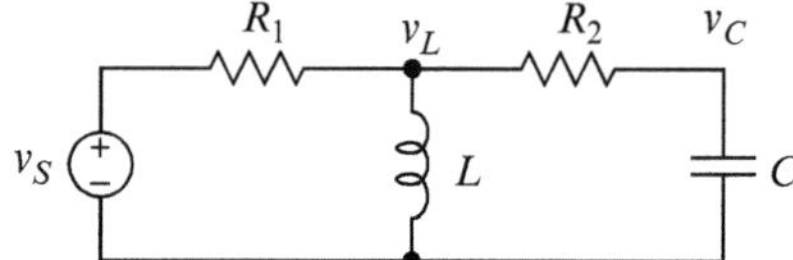

Fig. P 11.27 See Problem P 11.32

$$\omega^{-2}\frac{d^2 v_C}{dt^2} + \tau\frac{dv_C}{dt} + v_C = 0.$$

Express the parameters ω and τ in terms of the circuit parameters. (c) Show that the response is underdamped if $\omega\tau < 2$.

Chapter 12
Sinusoids, Phasors, and Impedance

This chapter describes how real sinusoids are represented using complex quantities called phasors; how circuits containing resistors, capacitors, and inductors are represented using complex quantities called impedances and admittances; and how phasors, impedances, and admittances are used for analyzing sinusoidally excited circuits. This is a long chapter because it extends methods treated in several previous chapters to sinusoidally excited circuits.

Ability to obtain the output of a circuit for a sinusoidal input is important partly because currents and voltages of interest in many systems are sinusoidal or very nearly so, at least over some time interval. Examples of such systems are electric power systems and many communication and control systems. But more important, any current or voltage (or any other physical quantity) can be expressed as a sum of sinusoidal currents or voltages using Fourier analysis.[1] Consequently, the response of a linear circuit to sinusoidal excitation can tell us a great deal about the response of the circuit to *any* excitation. For this reason, tools for analysis of sinusoidally excited circuits are far more potent and useful than they might appear to be. They are tools you must master if you wish to design circuits and systems for communication, control, signal processing, and power generation and distribution.

We begin by establishing certain conventions regarding sinusoidal voltages and currents. For economy, we use **signal** to mean *current or voltage*.

12.1 Sinusoidal Voltages and Currents

The **standard form** of a sinusoidal signal (a sinusoidal current or voltage) is

$$x(t) = X\cos(2\pi f\, t + \theta) = X\cos(\omega t + \theta), \quad (12.1)$$

where

$x(t)$ is the **instantaneous amplitude** of the sinusoid (V or A),

X is the **peak amplitude** of the sinusoid (V or A),

$2\pi ft + \theta$ is the **instantaneous phase** of the sinusoid (rad),

f is the **frequency** of the sinusoid (Hz),

$\omega = 2\pi f$ is the **angular frequency** of the sinusoid[2] (rad s^{-1}),

t is **time** (s), and

θ is the **initial phase** or **relative phase** of the sinusoid (rad).

Because $\cos(0) = 1$ and $\sin(0) = 0$, using the cosine (rather than the sine) as the standard form allows us to represent a constant (dc) current or voltage as a zero-frequency sinusoid, which lends a desirable consistency to certain developments.

By convention, *peak amplitude and frequency are non-negative* and *initial phase is confined to the interval* $-\pi \leq \theta < \pi$. These conventions cause no loss of generality and are adopted to eliminate ambiguity. For reference, the three parameters of the standard form (12.1) for a sinusoidal signal are constrained as follows:

$$\text{peak amplitude } X \geq 0, \text{ frequency } f \geq 0,$$
$$-\pi \leq \text{initial phase } \theta < \pi. \quad (12.2)$$

[1] Chapter 16 introduces Fourier analysis.

[2] Angular frequency is called *radian frequency* in some books.

DOI 10.1007/978-90-481-9443-8_12,

In practice, initial phase is expressed in both radians and degrees and preference for one or the other depends upon the field of application. In this book, except in Chapter 14, we express angles in radians, in which case the two terms of an instantaneous phase $2\pi ft+\theta$ are expressed in the same unit and the chance for computational error is reduced.[3] When expressing angles numerically, we omit the dimensionless unit rad unless confusion is otherwise possible.

In this book, as in virtually all of engineering practice, specific values for frequency are consistently expressed in Hertz. We regard angular frequency $\omega = 2\pi f$ as convenient shorthand that allows many quantities of interest to be expressed compactly. Although for economy we may write a sinusoidal signal as

$$x(t) = X\cos(2\pi ft),\ x>0, f>0. \tag{12.3}$$

We normally specify frequency, not angular frequency. When it is necessary or convenient to give the value of an angular frequency, we give the associated unit as s^{-1}, not rad/s; e.g., if we say the frequency of a sinusoid is 60 Hz, we mean f. If we say the frequency is 377 s^{-1}, we mean ω. But we rarely specify angular frequency.

Example 12.1. A branch current in a particular circuit is $i(t) = -25\sin(\omega t - \pi/6)$ mA, with $\omega > 0$. Express the current in standard form.

Solution: We use the identity $\sin(\alpha) \equiv \cos(\alpha - \pi/2)$ to express the current as

$$\begin{aligned} i(t) &= -25\cos(\omega t - \pi/6 - \pi/2) \\ &= -25\cos(\omega t - 2\pi/3). \end{aligned}$$

We use the identity $\cos(\alpha+\pi) \equiv -\cos(\alpha)$ to obtain the standard form

$$\begin{aligned} i(t) &= 25\cos(\omega t - 2\pi/3 + \pi) \\ &= 25\cos(\omega t + \pi/3)\ \text{mA}. \end{aligned}$$

The peak amplitude (25 mA) is positive and the initial phase $(\pi/3)$ is between $-\pi$ and π.

Exercise 12.1. Find the initial phase of $x(t) = X\sin(\omega t - \pi/6)$ where $X > 0$ and $\omega > 0$.

A sinusoidal signal given by (12.3) can be expressed in various other ways; for example, as

$$x(t) = X\cos[2\pi f(t - t_\theta)], \tag{12.4}$$

where

$$t_\theta = -\frac{\theta}{2\pi f} = -\frac{\theta T}{2\pi}, \tag{12.5}$$

is the **phase delay**, or as

$$x(t) = X\cos\left[\frac{2\pi(t-t_\theta)}{T}\right], \tag{12.6}$$

where

$$T = \frac{1}{f} \tag{12.7}$$

is the **period** of the sinusoid. In (12.5), initial phase θ must be expressed in radians.

Because initial phase is confined to the interval $-\pi \le \theta < \pi$, the phase delay of a sinusoid, defined by (12.5), is confined to the interval

$$-\frac{T}{2} \le t_\theta < \frac{T}{2}, \tag{12.8}$$

where T is the period of the sinusoid. A phase delay of $-T/2$ is indistinguishable from a phase delay of $T/2$. Positive phase delay (negative initial phase) corresponds to translating a sinusoid to the right along the time axis, whereas negative phase delay corresponds to translating the sinusoid to the left (which is actually an advance, rather than a delay).

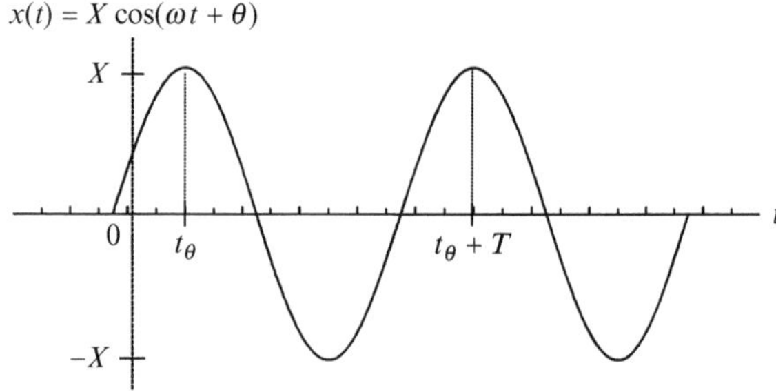

Fig. 12.1 Graph of a sinusoidal signal $x(t) = X\cos(\omega t + \theta)$

[3] In Chapter 14, we express angles in degrees, as is conventional in the electric power industry.

Figure 12.1 shows a graph of a sinusoidal signal. Of the various expressions given above, (12.6) is most directly related to a graph of instantaneous amplitude versus time because peak amplitude X, phase delay t_θ, and period T can all be read directly from such a graph.[4] Nonetheless, the expressions most often used are those given by (12.1) and (12.3).

12.2 Time Origin, Phase Reference, and Initial Phase

Generally, describing currents and voltages as functions of time implies a *time origin*; i.e., a time we designate $t = 0$. Choice of a time origin usually is suggested by some particular event or some unique feature of an input; for example, a time at which the input suddenly changes character. Events of interest are then clocked with reference to the time origin. For example, if we define the time origin as the instant we connect a resistor across the terminals of a charged capacitor, we can say that the voltage across the capacitor has dropped to less than 1% of its value at $t = 5RC$, which means five time constants after the connection was made.

A sinusoidal current or voltage (mathematical model) is periodic, having neither beginning nor end. Moreover, as we show in the sequel, *all currents and voltages in a sinusoidally excited stable linear circuit are sinusoids, and all have the frequency of the input.* A sinusoidally excited stable linear circuit is said to be in **sinusoidal steady state**. There is no such thing as a time origin in a circuit that is in sinusoidal steady state. We need some other way to clock currents and voltages in such a circuit.

In a sinusoidally excited circuit, we essentially use *relative phase* instead of time as the basis for clocking currents and voltages. We choose a particular sinusoidal current or voltage (usually, the input) as the **phase reference**, we assign zero initial phase to that sinusoid, and we specify or determine the initial phase of all other currents and voltages relative to that of the reference. For this to work properly (to avoid ambiguity), we must express all sinusoidal currents and voltages in standard form. The standard form for the sinusoid selected as the phase reference is

$$x(t) = X\cos(2\pi ft), \quad X>0, \quad f>0. \qquad (12.9)$$

By definition, the phase reference in any particular problem has zero initial phase. When we express a current or voltage as $x_1(t) = X_1\cos(\omega t + \theta_1)$, we mean that the initial phase of $x_1(t)$ exceeds that of the reference by the value θ_1.

The initial phase of a sinusoid is more properly called the **relative phase** of the sinusoid. It is the phase *relative* to that of a known reference and the known reference is assumed to have the form (12.9). Because $\cos(\alpha \pm 2\pi) \equiv \cos(\alpha)$, relative phase can be determined (by measurement) only to within a whole multiple of 2π.

Example 12.2. The output of a certain circuit is a delayed version of the input and both are voltages; that is, the output for input $v_{in}(t)$ is given by $v_{out}(t) = v_{in}(t - t_0)$. If $t_0 = 20$ ns is the delay imposed by the circuit on a 100-kHz sinusoidal voltage applied to the input, and if the input is the phase reference, (a) What is the initial phase of the output? (b) What is the largest delay that could be measured by comparing the output to the input? (c) Is this a reliable means of measuring delay?

Solution: (a) If we write the input voltage (the phase reference) as $v_{in}(t) = V_0\cos(\omega t)$, then the output voltage is given by

$$\begin{aligned} v_{out}(t) &= v_{in}(t - t_0) = V_0\cos[\omega(t - t_0)] \\ &= V_0\cos(\omega t + \theta), \end{aligned}$$

where the initial phase is

$$\begin{aligned} \theta &= -\omega t_0 = -2\pi f t_0 \\ &= -(2\pi)(100\text{kHz})(20\text{ns}) \cong -0.013. \end{aligned}$$

(b) Using a sinusoidal test input, we cannot measure a delay any larger than that corresponding to $\theta = -2\pi$. Thus the largest delay we could measure is

$$t_0 = -\frac{\theta}{2\pi f} = -\frac{(-2\pi)}{2\pi \times 100\text{ kHz}} = 10\ \mu\text{s}.$$

(c) No, because the actual delay can be the measured delay plus any whole multiple of $T = 1/f$.

[4] Determining frequency and initial phase from such a graph requires calculations, albeit simple ones.

Two sinusoids having the *same frequency* are said to be **in phase** if they have the same initial phase and **out of phase** otherwise. Two particular out-of-phase conditions have special importance: Two sinusoids having the same frequency are said to be **in phase quadrature** if their initial phases differ by $\pm\pi/2$, and **in phase opposition** if the difference equals $\pm\pi$. Thus, for example,

$V_0\cos(\omega t)$ and $I_0\cos(\omega t)$ are in phase,

$V_0\cos(\omega t-\pi/6)$ and $I_0\cos(\omega t)$ are out of phase,

$V_0\cos(\omega t)$ and $I_0\cos(\omega t\pm\pi/2)=\mp I_0\sin(\omega t)$ are in phase quadrature,

$V_0\cos(\omega t)$ and $I_0\cos(\omega t\pm\pi)=-I_0\cos(\omega t)$ are in phase opposition.

If the initial phase of one sinusoid is less than that of a second sinusoid, we say the first **lags** the second or the second **leads** the first; e.g., $V_0\cos(\omega t-\pi/6)$ lags the reference and $I_1\cos(\omega t+\pi/3)$ leads the reference. If a particular sinusoid is established as the phase reference, then any other sinusoid is said to be **lagging** or **leading** according to whether the initial phase relative to that of the reference is negative or positive, respectively.

Within any one period,[5] a positive peak of a lagging sinusoid occurs later than (is to the right of) the *nearest* positive peak of the reference. Similarly, *within any one period*, a positive peak of a leading sinusoid occurs earlier than (is to the left of) the *nearest* positive peak of the reference. From another perspective, the phase delay $t_0=-\theta/(2\pi)$ of a lagging sinusoid $(\theta<0)$ is positive, and the phase delay of a leading sinusoid $(\theta>0)$ is negative. To make the terms *lagging* and *leading* unambiguous, initial phase must be confined to the range $-\pi\le\theta<\pi$; otherwise, for example, a phase lag θ could just as well be called a phase lead $2\pi-\theta$. The terms lagging and leading are used primarily by those engaged in power system analysis and design, and there primarily in connection with certain power calculations.

Figure 12.2 illustrates definitions given above. The two sinusoidal voltages shown have the same period $T=t_2-t_0$ and thus have the same frequency $f=1/T$. When two *nearest-neighbor* peaks are compared, the peak of $v_2(t)$ occurs later than the peak of $v_1(t)$ so $v_2(t)$ lags $v_1(t)$. If $v_1(t)$ is the reference, then the phase delay $t_\theta=t_2-t_1$ is positive, $v_2(t)$ lags $v_1(t)$, and the initial phase of $v_2(t)$ is negative. If $v_2(t)$ is the reference, then the phase delay $t_\theta=t_1-t_2$ is negative, $v_1(t)$ leads $v_2(t)$, and the initial phase of $v_1(t)$ is positive.

The terminology defined above is usually applied only to sinusoids having the same frequency or very nearly the same frequency over a time interval of interest.[6]

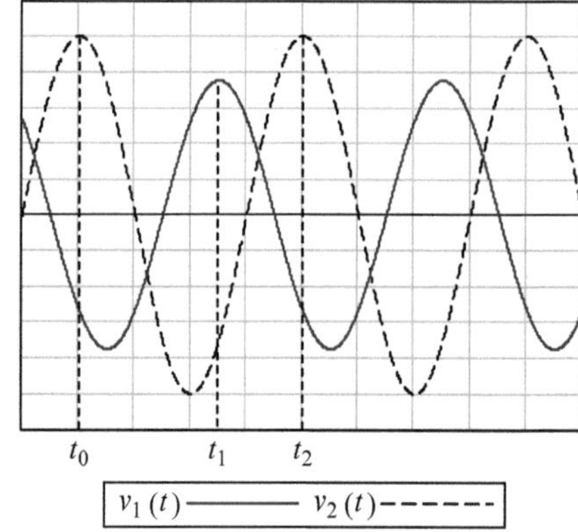

Fig. 12.2 Illustrating phase delay and relative phase for two same-frequency sinusoidal voltages. A positive peak of $v_1(t)$ occurs at time t_1 and the nearest-neighbor peak of $v_2(t)$ occurs later, at time t_2, so $v_2(t)$ lags $v_1(t)$

Example 12.3. The terminal voltage of a capacitor is sinusoidal. Does the resulting current lead or lag the terminal voltage? By what angle?

Solution: We choose the terminal voltage as the phase reference and write

$$v(t)=V_0\cos(\omega t).$$

The current through the capacitor is given by

$$\begin{aligned}i&=C\frac{dv}{dt}=C\frac{d[V_0\cos(\omega t)]}{dt}=-\omega CV_0\sin(\omega t)\\&=\omega CV_0\sin(\omega t+\pi)=\omega CV_0\cos\left(\omega t+\pi-\frac{\pi}{2}\right)\\&=\omega CV_0\cos\left(\omega t+\frac{\pi}{2}\right).\end{aligned}$$

[5]More generally, within any whole number of periods.

[6]If two sinusoids have different frequencies, then one alternately leads and lags the other. The duration of an interval during which one leads (or lags) the other depends upon the frequencies.

The current *leads* the voltage by $\pi/2$. We may also write

$$i = \omega C V_0 \cos[\omega(t - t_0)], \; t_0 = -\frac{\theta}{\omega} = -\frac{\pi}{2\omega},$$

where t_0 is negative (is an advance), meaning that each positive peak of the current occurs earlier than the nearest-neighbor positive peak of the voltage (the reference).

Exercise 12.2. The current through a certain inductor is sinusoidal. Does the current lead or lag the voltage across the inductor?

If two sinusoids have the same reference and different initial phases, we may determine which one leads the other by comparing their initial phases. Thus, if

$$x_1(t) = X_1 \cos(\omega t + \theta_1), \; x_2(t) = X_2 \cos(\omega t + \theta_2),$$

where *both are in standard form*, then x_1 leads x_2 if $\theta_1 > \theta_2$.

Example 12.4. The voltage across an element and the current into the positive terminal of the element are given by

$$v(t) = V_0 \cos\left(\omega t - \frac{\pi}{3}\right), \; i(t) = I_0 \cos[\omega(t - t_0)],$$

where $f = 15$ kHz and $t_0 = 430$ μs. Does the current lag or lead the voltage? By what angle?
Solution: The initial phase of the voltage is $\theta_v = -\pi/3 = -1.047$. The initial phase of the current is

$$\theta_i = -\omega t_0 = -2\pi(15 \text{ kHz})(430 \text{ μs}) = -2\pi \times 6.45.$$

After casting out multiples of 2π, we have

$$\theta_i = -0.45 \times 2\pi = -2.83.$$

Because $\theta_v > \theta_i$, the voltage leads the current by $\theta = \theta_v - \theta_i = 1.78$.

Example 12.5. Figure 12.3 shows graphs of voltage across and current through a two-terminal element. Express both as sinusoidal functions of time, using voltage as the phase reference. State whether the current leads or lags the voltage and by what angle.
Solution: From the graph, the peak amplitudes and period of the voltage and current are

$$V_0 = 1 \text{ V}, I_0 = 800 \text{ μA}, T = 1 \text{ ms}.$$

It follows that the frequency is

$$f = \frac{1}{T} = 1 \text{ kHz}.$$

A positive peak of the current leads (occurs earlier than) the nearest positive peak of the voltage (the reference) by approximately 2.4 ms. Thus

$$t_\theta \cong -0.24 \text{ ms}.$$

The relative phase of the current is

$$\theta = -2\pi f t_\theta \cong (-2\pi f)(-0.24 \text{ ms}) \cong 1.51$$

and the current leads the voltage. We may write

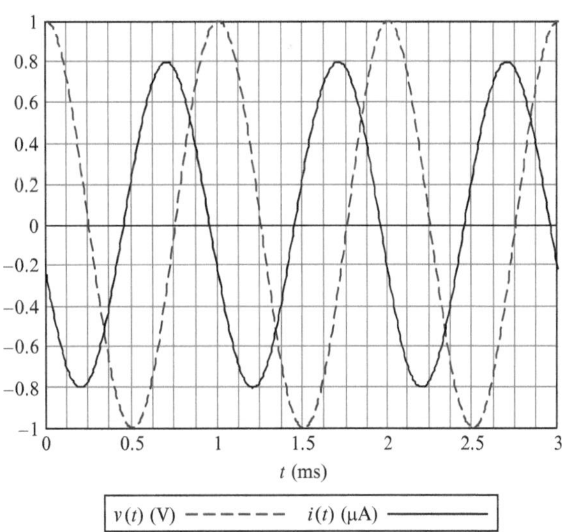

Fig. 12.3 Sinusoidal voltage and current considered in Example 12.5

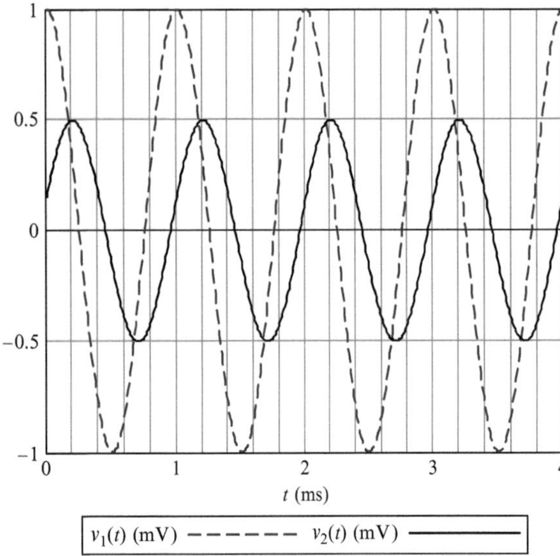

Fig. 12.4 See Exercise 12.3

$v(t) = V_0 \cos(\omega t)$; $V_0 = 1\,\text{V}$, $f = 1$ kHz,
$i(t) \cong I_0 \cos(\omega t + 1.51)$, $I_0 = 800\,\mu\text{A}$;
$f = 1\,\text{kHz}$.

Exercise 12.3. Figure 12.4 shows graphs of two voltages in a certain circuit. Choose $v_1(t)$ as the phase reference and give the peak amplitude, frequency (Hz) and relative phase of $v_2(t)$.

12.3 Phasors

By definition (Section 12.1), the standard form for a real sinusoidal signal (current or voltage) is

$$x(t) = X\cos(\omega t + \theta), \tag{12.10}$$

where $X \geq 0$, $\omega \geq 0$, and $-\pi \leq \theta < \pi$. The associated **complex representation** of the sinusoidal signal $x(t)$ above is denoted by $\tilde{x}$ and is defined by[7]

$$\tilde{x}(t) = X\,e^{j(\omega t+\theta)}. \tag{12.11}$$

Using Euler's identity, we may write (12.11) as

$$\tilde{x}(t) = X\,e^{j(\omega t+\theta)} = X\cos(\omega t + \theta) + jX\sin(\omega t + \theta),$$

which shows that *a real sinusoid is the real part of the complex representation of the sinusoid*; i.e.,

$$\begin{aligned} x(t) &= \text{Re}[\tilde{x}(t)] \\ &= \text{Re}[X\cos(\omega t + \theta) + jX\sin(\omega t + \theta)] \\ &= X\cos(\omega t + \theta). \end{aligned} \tag{12.12}$$

The complex representation of a signal has the same dimension as the signal and is expressed in the same unit as the signal.

We may use the law of exponents to write (12.11) as

$$\tilde{x}(t) = X\,e^{j(\omega t+\theta)} = X\,e^{j\theta}\,e^{j\omega t} = \tilde{X}\,e^{j\omega t}. \tag{12.13}$$

The quantity

$$\tilde{X} = X\,e^{j\,\theta} = X\angle\theta \tag{12.14}$$

is the **phasor** for the sinusoid whose complex representation is $\tilde{x}(t)$. The phasor for a sinusoidal current i or a sinusoidal voltage v is denoted $\tilde{I}$ or $\tilde{V}$, respectively. The unit of a phasor is the unit of the associated current or voltage (A or V). The phasor for a sinusoid is (in general) complex. From (12.10) and (12.14), *the magnitude and angle of a phasor are the peak amplitude and initial phase, respectively, of the associated standard-form sinusoid.*

Obtaining the phasor for a sinusoidal current or voltage is a two-step process:

- *Express the current or voltage in standard form*; i.e., as a cosine having non-negative peak amplitude, non-negative frequency, and initial phase between $-\pi$ and π.
- *The magnitude of the phasor is the peak amplitude of the cosine and the angle of the phasor is the initial phase of the cosine*:

$$x(t) = X\cos(\omega t + \theta) \Leftrightarrow \tilde{X} = X\angle\theta = X\,e^{j\theta}. \tag{12.15}$$

[7] In this book, a tilde (~) denotes a complex representation of a real quantity; thus, $\tilde{x}(t)$ denotes the complex representation of a real signal $x(t)$.

Example 12.6. Obtain the complex representation and the phasor for $x(t) = X\cos(\omega t - \pi/6)$, where $X > 0$ and $\omega > 0$.
Solution: From (12.11) and (12.14), the complex representation of the specified signal is $\tilde{x}(t) = \tilde{X}\, e^{j\omega t}$ where $\tilde{X} = X\, e^{-j\pi/6} = X\angle{-\pi/6}$ is the phasor.

Exercise 12.4. A current is given by

$$i(t) = 50\cos\left(\omega t - \frac{\pi}{4}\right) \text{ mA}.$$

Find the associated phasor.

Exercise 12.5. Obtain the phasor for a sinusoidal voltage $v(t) = V_0\, \sin(\omega t + \phi)$.
Hint: Use the identity $\sin(\alpha) \equiv \cos(\alpha - \pi/2)$:

A phasor can be expressed in rectangular form using Euler's identity. Thus

$$\begin{aligned}\tilde{V} &= V_0\, \angle\theta = V_0\, e^{j\,\theta} \\ &= V_0\cos(\theta) + j\, V_0\, \sin(\theta).\end{aligned} \tag{12.16}$$

Example 12.7. Express $\tilde{V} = 10\, e^{j\pi/3}$ V in rectangular form.
Solution: From (12.16)

$$\begin{aligned}\tilde{V} &= 10\, e^{j\,\pi/3} = 10\cos(\pi/3) + j\, 10\, \sin(\pi/3) \\ &= 5 + j5\sqrt{3}\ \text{V}.\end{aligned}$$

Exercise 12.6. Express $\tilde{I} = 50\,\angle(\pi/3)$ mA in both exponential and rectangular form. Does this sinusoid lead or lag the (implied) reference?

The phasor representation of a sinusoid does not specify the frequency of the sinusoid. In applications, the frequency is known from other considerations; e.g., in a stable linear circuit subjected to sinusoidal excitation, all currents and voltages in the circuit are sinusoidal and all have the frequency of the input.

Example 12.8. The phasor for a certain 200 kHz sinusoidal voltage is

$$\tilde{V} = 40 - j\,10 \text{ mV}.$$

Express the voltage as a real function of time.
Solution: The amplitude V and initial phase θ of the voltage are the magnitude and angle, respectively, of the phasor for the voltage:

$$\begin{aligned}V &= |\tilde{V}| = \sqrt{(40)^2 + (10)^2}\ \text{mV} \cong 41.2\ \text{mV}, \\ \theta &= \measuredangle\tilde{V} = \tan^{-1}\left(\frac{-10}{40}\right) \cong -0.245.\end{aligned}$$

It follows that

$$\begin{aligned}v(t) &= 41.2\cos(\omega t - 0.245)\ \text{mV}, \\ f &= 200\ \text{kHz}.\end{aligned}$$

Exercise 12.7. The period of the current defined in Exercise 12.6 is 250 ns. Express the current as a real function of time.

The phasor for a current or a voltage is called the **phasor current** or the **phasor voltage**, respectively. For example, if a voltage is given by $v(t) = V_0\cos(\omega t + \theta)$, the associated phasor voltage is given by $\tilde{V} = V_0\angle\theta$. A phasor is complex (in general) and is therefore non-physical; i.e., a phasor current is not a physical current and a phasor voltage is not a physical voltage. Nonetheless, for brevity, textbook authors and practicing engineers usually refer to phasor currents and phasor voltages simply as currents and voltages, and we do the same unless there is need to distinguish between a physical current or voltage and its phasor representation. As shown below, *phasor currents and voltages obey Kirchhoff's laws*, just as do physical currents or voltages. Consequently, calling (e.g.) a *phasor current* a *current* is both economical and harmless.

12.4 Phasor Diagrams

A phasor can be represented graphically as a point in a complex plane or (more often) as a vector extending from the origin to the point having the polar coordinate $X\angle\theta$, as illustrated by Fig. 12.5. Such a representation showing one or more phasors is called a **phasor diagram**. Among other things, a phasor diagram is useful for displaying or visualizing the relative phases of two or more sinusoids.

Example 12.9. Draw a phasor diagram for the quantities

$$v(t) = V_0 \cos(\omega t),\ i_1(t) = I_1 \cos(\omega t - \pi/4),$$
$$i_2(t) = I_2 \cos(\omega t + \pi/6).$$

Solution: See Fig. 12.6. The phasors for the quantities above are

$$\tilde{V} = V_0\angle 0,\ \tilde{I}_1 = I_1\angle(-\pi/4),\ \tilde{I}_2 = I_2\angle(\pi/6).$$

The voltage $v(t)$ has zero initial phase and is by implication the phase reference. The current $i_1(t)$ lags the reference by $\pi/4$ and the current $i_2(t)$ leads the reference by $\pi/6$.

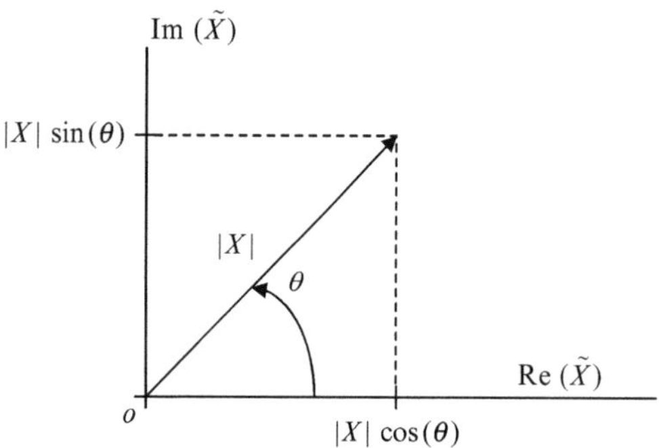

Fig. 12.5 Phasor diagram for a sinusoid $x(t) = X\cos(\omega t + \theta)$

Exercise 12.8. Draw a phasor diagram for the quantities

$$i(t) = I_0 \cos(\omega t + \pi/3),$$
$$v_1(t) = V_1 \cos(\omega t - \pi/4),$$
$$v_2(t) = V_2 \cos(\omega t).$$

As illustrated by Example 12.9, currents and voltages often are displayed in a single phasor diagram. In such mixed-dimension diagrams, only initial phase (not magnitude) can be shown to scale unless different scales are defined for current and voltage. Mixed-dimension phasor diagrams are useful primarily for displaying phase relationships among currents and voltages. In diagrams drawn for this purpose, the phasor for the phase reference, which by definition has zero initial phase, is coincident with the positive real axis. For example, Fig. 12.6 makes clear that the current i_2 leads the reference voltage v by $\pi/6$ and that the current i_1 lags the reference voltage v by $\pi/4$.[8]

A single-dimension phasor diagram (current or voltage, but not both) can provide a quick check on the correctness of a sum or difference of phasors. Phasors add like vectors, and the vector representing the sum of two phasors should look like the vector sum of the two phasors; that is, the sum should be the diagonal of the parallelogram whose sides are the components of the sum.

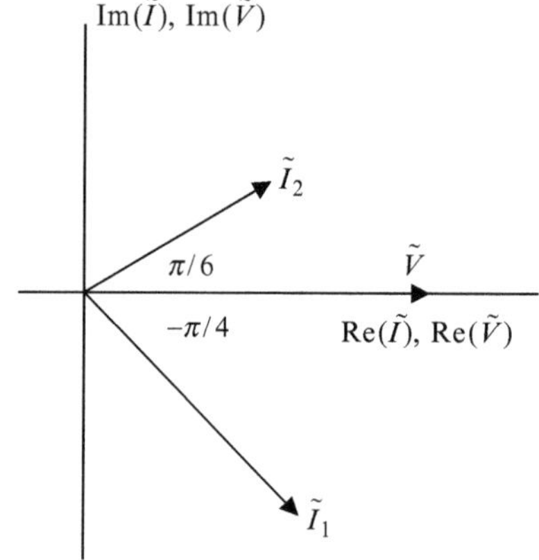

Fig. 12.6 See Example 12.9

[8] Recall that the positive direction for an angle is counter-clockwise.

Example 12.10. Draw a phasor diagram illustrating the sum of the phasors $\tilde{I}_1 = 10\angle 0.18$ mA, $\tilde{I}_2 = 5\angle -0.57$ mA.
Solution:

$$\begin{aligned}\tilde{I}_1 &= 10\cos(0.18) + j\,10\sin(0.18)\\ &= 9.84 + j\,1.79\,\text{mA},\\ \tilde{I}_2 &= 5\cos(-0.57) + j5\sin(-0.57)\\ &= 4.21 - j2.70\,\text{mA}.\end{aligned}$$

See Fig. 12.7.

Another way to display a sum of phasors using a phasor diagram is to draw the terms of the sum in head-to-tail fashion. For example, Fig. 12.8 depicts a sum

$$\tilde{V} = \tilde{V}_1 + \tilde{V}_2 + \tilde{V}_3.$$

The phasor $\tilde{V}_1$ originates at the origin, $\tilde{V}_2$ originates at the head of $\tilde{V}_1$, and $\tilde{V}_3$ originates at the head of $\tilde{V}_2$.

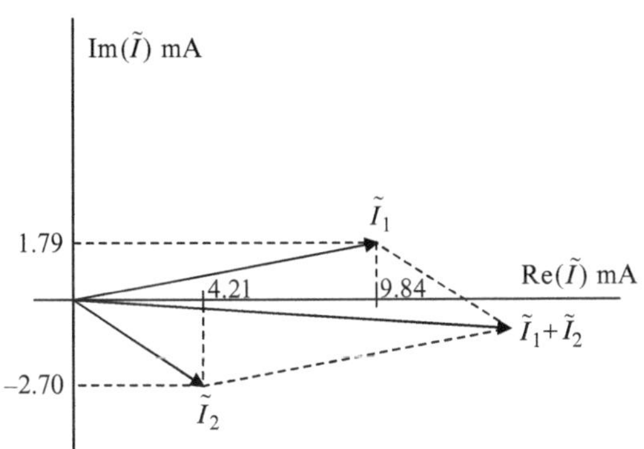

Fig. 12.7 See Example 12.10

The angle of each phasor is measured from the horizontal, as if the phasor originated at the origin. The sum $\tilde{V}$ originates at the origin and ends at the head of the last term in the head-to-tail sequence.

Example 12.11. Use a phasor diagram to illustrate that

$$V_0\angle 0 + V_0\angle 2\pi/3 + V_0\angle -2\pi/3 = 0.$$

Solution: Figure 12.9 shows the head-to-tail diagram. The last term in the sum terminates at the origin, so the sum equals zero.

12.5 Impedance and Generalized Ohm's Law

The **impedance** of a resistor, a capacitor, an inductor, or at a terminal pair (port) of a linear circuit is denoted by Z and is defined by

$$Z = \frac{\tilde{v}(t)}{\tilde{i}(t)}, \tag{12.17}$$

where $\tilde{v}$ is the complex representation of the voltage across the terminals and $\tilde{i}$ is the complex representation of the current into the positive terminal. The dimension of impedance is that of resistance and the SI unit of impedance is the ohm (Ω).

It follows from the relations

$$v = L\frac{di}{dt}, \quad i = C\frac{dv}{dt}$$

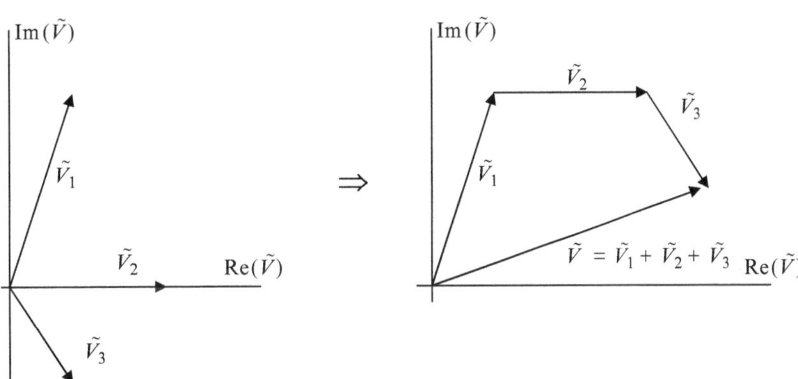

Fig. 12.8 Head-to-tail phasor diagram for a sum $\tilde{V} = \tilde{V}_1 + \tilde{V}_2 + \tilde{V}_3$

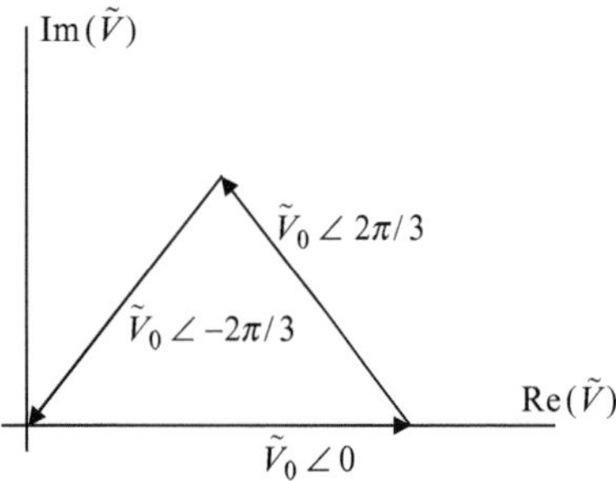

Fig. 12.9 See Example 12.11

that *if either the current through or voltage across an inductor or a capacitor is sinusoidal, then both the current and the voltage are sinusoidal and both have the same frequency*. If the complex representation for the voltage across a capacitor (or an inductor) has the form $\tilde{V} e^{j\omega t}$, then the complex representation for the current through the capacitor (or the inductor) has the form $\tilde{I} e^{j\omega t}$ (the frequencies are the same). Using $\tilde{v} = \tilde{V} e^{j\omega t}$ and $\tilde{i} = \tilde{I} e^{j\omega t}$ in (12.17) gives

$$Z = \frac{\tilde{V}}{\tilde{I}}, \tag{12.18}$$

where $\tilde{V}$ is the phasor voltage across the terminals and $\tilde{I}$ is the phasor current into the positive terminal. Because phasors are independent of time, impedance, like resistance, is independent of time. Equation (12.18) is a generalization of Ohm's law, according to which the resistance of a resistor is given by

$$R = \frac{V}{I},$$

where V and I are the dc voltage across and resulting current through the resistor, respectively. Equivalently, impedance is a generalization of resistance.

Example 12.12. Obtain an expression for the impedance of a capacitor.
First Solution: We use (12.18). The instantaneous current through a capacitor is given by

$$i = C\frac{dv}{dt},$$

where v is the instantaneous voltage across the capacitor and C is the capacitance of the capacitor. We take the voltage v as the phase reference. Thus $v(t) = V_0 \cos(\omega t)$, whence

$$i = C\frac{d}{dt}[V_0 \cos(\omega t)] = -\omega C V_0 \sin(\omega t).$$

The phasor voltage is $\tilde{V} = V_0 \angle 0$. Expressing the current given above in standard form gives

$$i = -\omega C V_0 \sin(\omega t) = \omega C V_0 \cos(\omega t + \pi/2)$$
$$\Rightarrow \tilde{I} = \omega C V_0 \angle \pi/2.$$

It follows from the definition (12.18) that the impedance of a capacitor is given by

$$Z_C = \frac{V_0 \angle 0}{\omega C V_0 \angle \pi/2} = \frac{1}{\omega C} \angle -\pi/2, \tag{12.19}$$

where C is the capacitance of the capacitor and $\omega = 2\pi f$ is the angular frequency of the applied voltage or current.
Second Solution: We use (12.17), where we again take the voltage across the capacitor as the phase reference, for

$$\tilde{v} = \tilde{V} e^{j\omega t}$$

we obtain

$$\tilde{i} = C\frac{dv}{dt} = C\frac{d}{dt}(\tilde{V} e^{j\omega t}) = j\omega C \tilde{V} e^{j\omega t}.$$

From (12.17),

$$Z = \frac{\tilde{v}}{\tilde{i}} = \frac{\tilde{V} e^{j\omega t}}{j\omega C \tilde{V} e^{j\omega t}} = \frac{1}{j\omega C} = \frac{1}{\omega C \angle \pi/2}$$
$$= \frac{1}{\omega C} \angle -\pi/2,$$

as before.

Exercise 12.9. Show that the impedance of an inductor having inductance L is

$$Z_L = \omega L \angle \pi/2 = j\omega L. \quad (12.20)$$

Exercise 12.10. Show that the impedance of a resistor having resistance R is

$$Z_R = R = R\angle 0. \quad (12.21)$$

Results obtained above are collected in Table 12.1. You should commit the entries in Table 12.1 to memory.

Example 12.13. The current through an inductor having inductance $L = 5$ mH is given by $i(t) = I_0 \cos(\omega t)$ with $I_0 = 5$ mA and $f = \omega/2\pi = 500$ kHz. Use phasors and impedance to find the voltage $v(t)$ across the inductor. Verify your answer using $v = L\,di/dt$.
Solution: From (12.15), the current phasor is

$$i(t) = I_0 \cos(\omega t) \Rightarrow \tilde{I} = I_0 \angle 0 = 5\angle 0 \text{ mA}.$$

From Table 12.1, the impedance of the inductor is

$$Z = j\omega L = j\,15.71 \text{ k}\Omega = 15.71\angle(\pi/2) \text{ k}\Omega.$$

From (12.18), the voltage phasor is

$$\tilde{V} = Z\tilde{I} = (5 \text{ mA}\angle 0)(15.71 \text{ k}\Omega\angle(\pi/2)) = 78.54\angle(\pi/2) \text{ V}.$$

It follows from (12.15) that

$$v(t) = 78.54 \cos(\omega t + \pi/2) \text{ V}.$$

Check: The voltage across the inductor is given by

$$v = L\frac{di}{dt} = L\frac{d}{dt}[I_0 \cos(\omega t)] = -\omega L I_0 \sin(\omega t).$$

Using the identity $\sin(\omega t) \equiv -\cos(\omega t + \pi/2)$ to express the voltage in standard form gives

$$v = \omega L I_0 \cos(\omega t + \pi/2) = 78.54 \cos(\omega t + \pi/2) \text{ V},$$

as above.

Table 12.1 Impedances of circuit elements

Element	*i–v* characteristic	Impedance (polar)	Impedance (rectangular)
Resistor	$v = Ri$	$Z_R = R\angle 0$	$Z_R = R$
Capacitor	$i = C\frac{dv}{dt}$	$Z_C = \frac{1}{\omega C}\angle -\pi/2$	$Z_C = \frac{1}{j\omega C}$
Inductor	$v = L\frac{di}{dt}$	$Z_L = \omega L\angle \pi/2$	$Z_L = j\omega L$

Exercise 12.11. The voltage across a capacitor having capacitance $C = 220$ nF is given by $v(t) = V_0 \cos(\omega t)$ with $V_0 = 150$ mV and $f = 100$ kHz. Use phasors and impedance to find the current through the capacitor. Verify your answer using $i = C\,dv/dt$.

From (12.18),

$$\measuredangle Z = \measuredangle\tilde{V} - \measuredangle\tilde{I}. \quad (12.22)$$

Thus the angle of an impedance is the angle by which a voltage across the impedance leads the current entering the positive terminal of the impedance. If the angle is negative, the voltage lags the current.

Example 12.14. For an impedance $Z = (300 + j400)\ \Omega$, state whether a voltage across the impedance would lead or lag the current entering the positive terminal of the impedance, and by what angle. Then assume the frequency of the applied voltage equals 1 kHz, and give the lead or lag in seconds.
Solution: The impedance Z is in the first quadrant, so $\tilde{V}$ leads $\tilde{I}$ by the angle

$$\measuredangle Z = \tan^{-1}\left(\frac{400}{300}\right) = 0.927 > 0.$$

For $f_0 = 1$ kHz,

$$\cos(\omega_0 t + 0.927) = \cos\left[\omega_0\left(t + \frac{0.927}{\omega_0}\right)\right],$$

so $v(t)$ leads $i(t)$ by

$$\frac{0.927}{2\pi \times 10^3 \text{ Hz}} \cong 148 \text{ µs}.$$

Example 12.15. Repeat Example 12.14 for $Z = (400 - j300)\,\Omega$.
Solution: The impedance Z is in the fourth quadrant, so $\tilde{V}$ leads $\tilde{I}$ by the angle

$$\measuredangle Z = \tan^{-1}\left(-\frac{300}{400}\right) = -0.644 < 0$$

or $\tilde{V}$ lags $\tilde{I}$ by 0.644. For $f_0 = 1$ kHz,

$$\cos(\omega_0 t - 0.644) = \cos\left[\omega_0\left(t - \frac{0.644}{\omega_0}\right)\right]$$

so $v(t)$ lags $i(t)$ by

$$\frac{0.644}{2\pi \times 10^3 \text{ Hz}} \cong 103 \text{ µs}.$$

Impedance can be expressed as a function of frequency f (Hz), as a function of angular frequency ω (s^{-1}), or as a function of a variable s (s^{-1}) called *complex frequency* introduced in Chapter 18. For example, the impedance of an inductor can be written as $Z = j2\pi f L$, $Z = j\omega L$, or (as shown in Chapter 18) $Z = sL$. In practice, frequency is almost always expressed in Hz. But to achieve a consistent notation, it is conventional to express impedance as a function of the variable $j\omega$, where ω is angular frequency, expressed in s^{-1}. Thus, for an inductor, $Z(j\omega) = j\omega L$.

An impedance given as $Z(j\omega)$ is expressed as a function of frequency in Hz by replacing ω with $2\pi f$ wherever it appears. For example,

$$Z(j\omega) = K\frac{j\omega/\omega_0}{1 + j\omega/\omega_0} \Rightarrow Z(j2\pi f)$$
$$= K\frac{j2\pi f/(2\pi f_0)}{1 + j2\pi f/(2\pi f_0)} = K\frac{jf/f_0}{1 + jf/f_0}.$$

For economy, we usually omit explicit indication of frequency dependence unless such is necessary for clarity. For example, we write $Z = j\omega L$ or $Z = j2\pi f L$ rather than writing $Z(j\omega) = j\omega L$ or $Z(j2\pi f) = j2\pi f L$. But where we need to express the impedance of a particular element or circuit at two or more different frequencies, we use either subscripts or explicit functional notation. For example, $Z(j\omega_1) = j\omega_1 L$ and $Z(j\omega_2) = j\omega_2 L$ or $Z_1 = j\omega_1 L$ and $Z_2 = j\omega_2 L$.

Complex numbers cannot be ordered. We cannot say whether one complex number is larger or smaller than another, so inequalities such as $Z_1 > Z_2$ are meaningless. But the magnitudes of complex numbers are real and can be ordered. For example, although we cannot say $2 + j2 > 1 + j$, we can say that $|2 + j2| > |1 + j|$. In describing circuits, we must often compare the magnitudes of various impedances. In writing or speaking, it is tedious to repeatedly write or say that (e.g.) *the magnitude of* one impedance is larger than *the magnitude of* another impedance, so we and most practicing engineers simply say that one impedance is larger or smaller than another. Because complex numbers cannot be ordered, we know that *magnitude of* is implied in such comparisons, so there is little chance for misinterpreting such statements. But in writing mathematical expressions, we must use precise language; for example, $|Z_1| > |Z_2|$, not $Z_1 > Z_2$.

When first learning how to use phasors and impedances, it is easy to forget that *phasors are complex (non-physical) representations* of currents or voltages and *impedances are complex (non-physical) representations* of real elements (R, L, or C) or of a circuit at a terminal pair. Physical quantities, including current, voltage, resistance, capacitance, and inductance are *real* (mathematical sense). *An expression for a physical current or voltage must be real* (*must not contain any j's*). In addition, it is important to keep the following limitations on phasor methods in mind:

- A phasor is a complex (non-physical) representation of a real **sinusoidal** current or voltage. *A non-sinusoidal current or voltage* (*other than a constant*) *cannot be represented by a phasor.*

- Impedance is the ratio of phasor voltage across to phasor current through a **linear** circuit element (R, L, or C) or the terminals of a linear circuit (a circuit containing only resistance, capacitance, inductance, and dependent sources). In general, *a non-linear element such as a diode cannot be fully characterized by an impedance.*
- *Impedance is a function of frequency.* Two components (or circuits) having the same impedance at one frequency do not necessarily have the same impedance at any other frequency.

12.6 Admittance

The **admittance** of an element, denoted by Y, is the reciprocal of the impedance of the element:

$$Y = \frac{1}{Z}. \tag{12.23}$$

The dimension of admittance is that of conductance and the SI unit of admittance is siemens (S). Just as impedance is a generalization of resistance, admittance is a generalization of conductance. From (12.23) and (12.18),

$$\tilde{I} = Y\tilde{V}, \tag{12.24}$$

which is a generalized form of Ohm's law. The admittances of a resistor, capacitor, and inductor can be obtained from the corresponding impedances in Table 12.1 using (12.23). The results are collected in Table 12.2. You should memorize the entries in Table 12.2, which is easy to do if you have memorized Table 12.1 and the definition (12.23).

Example 12.16. The voltage across a capacitor having capacitance $C = 100$ pF is given by $v(t) = V_0\cos(2\pi f t)$ with $V_0 = 5$ V and $f = 500$ kHz. Find the current $i(t)$ through the capacitor and draw a phasor diagram depicting the current and voltage, with the voltage as the phase reference.

Solution: The phasor $\tilde{V}$ for the voltage and the admittance Y of the capacitor are

$$\tilde{V} = V_0\angle 0 = 5\angle 0 \text{ V},$$
$$Y = j\omega C = j314\ \mu\text{S} = 314\ \mu\text{S}\angle(\pi/2).$$

From (12.24), the phasor for the current is

$$\tilde{I} = Y\tilde{V} = (5\text{ V}\angle 0)[314\ \mu\text{S}\angle(\pi/2)]$$
$$= 1.57\text{ mA}\angle(\pi/2).$$

It follows that

$$i(t) = 1.57\cos(2\pi f t + \pi/2) \text{ mA}.$$

Figure 12.10 shows a phasor diagram for the current and voltage, where the voltage is taken as the reference. The current leads the voltage by $\pi/2$. Equivalently, the voltage lags the current by $\pi/2$.

Table 12.2 Admittances of circuit elements

Element	i–v characteristic	Admittance (polar)	Admittance (rectangular)
Resistor	$v = Ri$	$Y_R = \frac{1}{R}\angle 0$	$Y_R = \frac{1}{R}$
Capacitor	$i = C\frac{dv}{dt}$	$Y_C = \omega C\angle\pi/2$	$Y_C = j\omega C$
Inductor	$v = L\frac{di}{dt}$	$Y_L = \frac{1}{\omega L}\angle -\pi/2$	$Y_L = \frac{1}{j\omega L}$

Exercise 12.12. The voltage across a 25 mH inductor is given by $v(t) = V_0\cos(\omega t)$, with $V_0 = 5$ V and $f = 5$ kHz. Use admittance to find the current through the inductor and draw a phasor diagram depicting the current and voltage, with the voltage as the phase reference.

For various reasons, it often is necessary to determine the frequencies or the values of certain circuit parameters for which an impedance or admittance

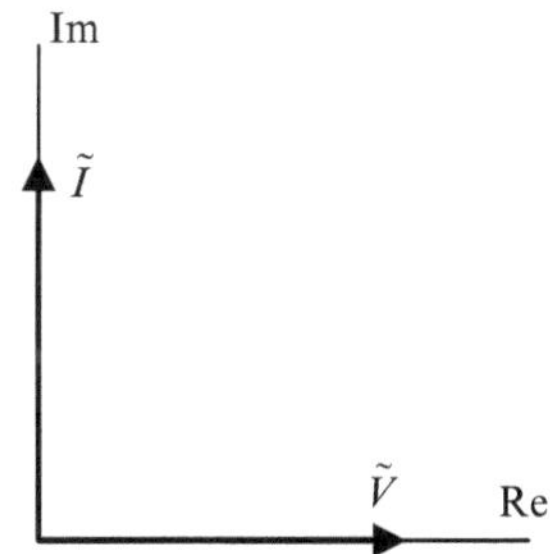

Fig. 12.10 See Example 12.16

is real. If an admittance is real (or imaginary), then so is the corresponding impedance, and vice versa.

Admittances, like impedances, are functions of frequency and are expressed as functions of angular frequency; e.g. for a capacitor, $Y(j\omega) = j\omega C$. An admittance given as a function of $j\omega$ is expressed as a function of frequency f by replacing ω by $2\pi f$ wherever it appears. For economy, we usually omit explicit frequency dependence unless such is necessary for clarity. In other words, we write

$$Y = j\omega C = j2\pi f C,$$

rather than writing either $Y(j\omega) = j\omega C$ or $Y(j2\pi f) = j2\pi f C$.

Like impedances, admittances are complex and cannot be ordered. So when we say that one admittance is larger than another, we mean that the *magnitude* of the first is larger than that of the second. But in writing mathematical expressions, we must use precise language; for example, $|Y_1| > |Y_2|$, *not* $Y_1 > Y_2$.

Also like impedances, admittances are **non-physical** representations of real elements (R, L, or C) or of a circuit at a terminal pair. You should read again the last paragraph in the preceding section, replacing *impedance* with *admittance* as you read.

12.7 Impedance and Admittance Ratios in dB

Magnitudes of impedances and admittances can range over several orders of magnitude as frequency ranges through values of interest. Consequently, it often is convenient to express an impedance in dB as

$$Z_{dB} = 20 \ \log\left|\frac{Z}{Z_0}\right|, \qquad (12.25)$$

where Z_0 is the (0 dB) reference impedance. Similarly, an admittance is expressed in dB as

$$Y_{dB} = 20 \ \log\left|\frac{Y}{Y_0}\right|, \qquad (12.26)$$

where Y_0 is the (0 dB) reference admittance. Impedances and admittances expressed in dB are dimensionless and are called **normalized** impedances and admittances; e.g., the quantity Z_{dB} defined by (12.25) is the impedance Z normalized to (or relative to) Z_0 and expressed in dB.

Example 12.17. Figure 12.11 shows a circuit and a graph of the magnitude of the impedance of the circuit normalized to the resistance R, where

$$\left|\frac{Z}{R}\right| (\text{dB}) = 20 \ \log\left|\frac{Z}{R}\right|.$$

The impedance is given by

$$Z = \frac{R + j2\pi f L}{1 - (2\pi f)^2 LC + j2\pi f RC} \Rightarrow \left|\frac{Z}{R}\right| = \frac{\sqrt{1 + \left(\frac{2\pi f L}{R}\right)^2}}{(1 - 4\pi^2 f^2 LC)^2 + (2\pi f RC)^2}.$$

The magnitude of the impedance has a range of almost 70 dB for the range of frequencies used for the graph. A 70 dB range corresponds

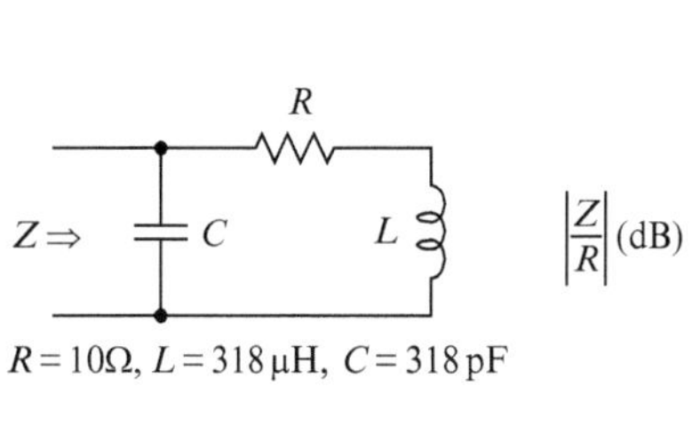

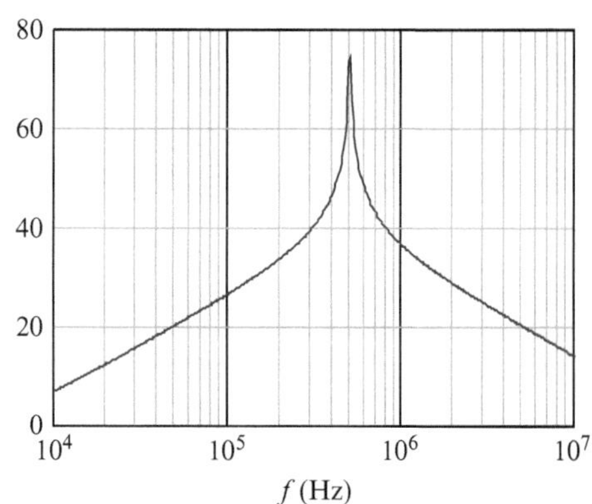

Fig. 12.11 See Example 12.17

to a range of more than three orders of magnitude. It would be impossible to usefully display such a range using a linear scale.

Exercise 12.13. The impedance of a certain two-terminal circuit is normalized to 1 kΩ and expressed in dB. The normalized impedance of the circuit equals 45 dB at 10 kHz. What is the magnitude of the actual impedance (Ω) at 10 kHz?

The dB is a convenient unit in which to express impedance and admittance ratios, which are related to current and voltage ratios. It is useful to keep in mind that the actual values of two normalized impedances *having the same normalizing impedance* that differ by 20 dB differ by a factor of ten. For example, if the normalized impedance of a certain two-terminal circuit is given as 10 dB at 100 Hz and 30 dB at 1 kHz, we know that the impedance (not dB) at 1 kHz is ten times that at 100 Hz.

Example 12.18. A variable-frequency sinusoidal voltage source is connected to the circuit in Fig. 12.11. The rms amplitude of the source is held at 10 V and the rms amplitude of the current entering the circuit is observed as the frequency of the source is varied from 1 to 500 kHz. What is the ratio of the rms current at 50 kHz to that at 300 kHz?
Solution: The impedance equals 40 dB at 300 kHz and 20 dB at 50 kHz, so the impedance at 300 kHz is 20 dB above or ten times that at 50 kHz. Thus the rms current at 50 kHz is ten times that at 300 kHz.

Exercise 12.14. The *normalized* impedance of a certain two-terminal circuit equals 15 dB at 5 kHz and 25 dB at 10 kHz. What is the ratio of the magnitude of the impedance (not dB) at 10 kHz to that at 5 kHz?

12.8 A Fundamental Relation

The following relation is fundamental in applications of phasors to circuit analysis:

$$\sum_{n=0}^{N} X_n \cos(\omega t + \theta_n) \equiv 0 \Leftrightarrow \sum_{n=0}^{N} X_n \angle \theta_n \equiv 0. \tag{12.27}$$

Recall that the double arrow in (12.27) means *implies and is implied by*. In words, if a sum of same-frequency sinusoids is identically zero, then so is the sum of the associated phasors and, conversely, if a sum of phasors is identically zero, then so is the sum of the associated same-frequency sinusoids. Equation (12.27) shows that *if Kirchhoff's current law (or Kirchhoff's voltage law) applies to a sum of sinusoidal currents (or voltages), then the law applies to the phasor representations of those currents (or voltages), which are complex constants.* As you will come to understand, (12.27) is fundamental because it implies that *analysis of sinusoidally excited linear circuits is exactly like analysis of dc resistive circuits, except that the resistances, currents, and voltages are complex.*

A proof of (12.27) follows: Assume that

$$\sum_{n} X_n \cos(\omega t + \theta_n) \equiv 0, \tag{12.28}$$

where the limits of summation are omitted and the symbol "≡" means *is identically equal to* or *equals for all time*, as is the case if (12.28) represents the sum of all currents leaving a node or the sum of all voltages around a closed path. If (12.28) is true for any time t then it must also be true for time $t - \pi/(2\omega)$. Thus

$$\begin{aligned}\sum_{n} X_n \cos\left[\omega\left(t - \frac{\pi}{2\omega}\right) + \theta_n\right] &\equiv \sum_{n} X_n \cos\left(\omega t + \theta_n - \frac{\pi}{2}\right) \\ &\equiv \sum_{n} X_n \sin(\omega t + \theta_n) \equiv 0.\end{aligned} \tag{12.29}$$

Adding (12.28) to j times the rightmost sum in (12.29) and using Euler's identity gives

$$\begin{aligned}&\sum_{n} X_n \cos(\omega t + \theta_n) + j \sum_{n} X_n \sin(\omega t + \theta_n) \\ &\equiv \sum_{n} X_n e^{j(\omega t + \theta_n)} \equiv e^{j\omega t} \sum_{n} X_n e^{j\theta_n} \equiv 0.\end{aligned}$$

Because $e^{j\omega t} \neq 0$, it follows that

$$\sum_n X_n e^{j\theta_n} \equiv \sum_n X_n \angle \theta_n \equiv 0 \qquad (12.30)$$

which proves the right implication in (12.27). To show the converse, assume that the right half of (12.27) holds and multiply that identity by $e^{j\omega t}$. This gives

$$\sum_n X_n \angle \theta_n e^{j\omega t} \equiv 0 \Rightarrow \mathrm{Re}\left\{\sum_n X_n e^{j(\omega t+\theta_n)}\right\} \equiv 0$$
$$\equiv \sum_n X_n \cos(\omega t + \theta_n).$$

The left implication in (12.27) follows.

A corollary of (12.27) is that *a sum of same-frequency sinusoids is a sinusoid having that frequency*. To show that is the case, it is necessary only to write the left half of (12.27) as

$$\sum_{n=0}^{N-1} X_n \cos(\omega t + \theta_n) = -X_N \cos(\omega t + \theta_N). \qquad (12.31)$$

Another corollary is that *the phasor for a sum of same-frequency sinusoids is the sum of the individual phasors.*[9] To show that is the case, it is necessary only to write the right half of (12.27) as

$$\sum_{n=0}^{N-1} X_n \angle \theta_n = -X_N \angle \theta_N. \qquad (12.32)$$

Example 12.19. Express $i(t) = I_1 \cos(\omega t) + I_2 \sin(\omega t)$, where $I_1 = 4$ mA and $I_2 = 3$ mA, in standard form.

Solution: From (12.32), the phasor for the sum is the sum of the phasors for the individual terms, which are

$$\tilde{I}_1 = I_1 \angle 0, \quad \tilde{I}_2 = I_2 \angle -\pi/2.$$

Because we intend to add the phasors, we express the phasors in rectangular form using Euler's identity:

$$\tilde{I}_1 = I_1 \angle 0 = I_1[\cos(0) + j\sin(0)] = I_1,$$
$$\tilde{I}_2 = I_2 \angle -\pi/2 = I_2\left[\cos\left(\frac{\pi}{2}\right) - j\sin\left(\frac{\pi}{2}\right)\right]$$
$$= -jI_2.$$

Thus, the phasor for the sum is

$$\tilde{I} = \tilde{I}_1 + \tilde{I}_2 = I_1 - jI_2 = 4 - j3\,\text{mA}.$$

To obtain the amplitude and phase of the sum, we express the phasor for the sum in polar form $I\angle\theta$:

$$I = |\tilde{I}| = \sqrt{4^2 + 3^2}\ \text{mA} = 5\ \text{mA},$$
$$\theta = \measuredangle\tilde{I} = \tan^{-1}\left(\frac{-3}{4}\right) = -0.644.$$

It follows that

$$i(t) = 5\cos(\omega t - 0.644)\ \text{mA}.$$

Again, such calculations are easy to perform using a pocket calculator.

Exercise 12.15. Express $v(t) = 50\cos(\omega t) + 100\cos(\omega t - \pi/6)$ mV in standard form.

Exercise 12.16. Find the sum $\tilde{I}$ of the phasors $\tilde{I}_1 = 10\angle 0.18$ mA, $\tilde{I}_2 = 5\angle -0.57$ mA representing 100-kHz sinusoidal currents. Express the current $\tilde{I} = \tilde{I}_1 + \tilde{I}_2$ as a real sinusoidal function of time.

[9] It is meaningless to add the phasors for sinusoids having different frequencies.

12.9 Circuit Reduction: Elements in Series and Parallel

Among other things, (12.27) provides a basis for reducing series and parallel connections of elements represented by impedances to single, *equivalent* impedances. Figure 12.12(a) shows a series connection of a resistor, a capacitor, and an inductor driven by a sinusoidal current source. Figure 12.12(b) shows the same circuit, but with the elements represented by impedances and all currents and voltages represented by phasors. By Kirchhoff's voltage law,

$$v = v_R + v_C + v_L.$$

Because the current through each element is the same and is sinusoidal, each of v_R, v_C, v_L is sinusoidal and the frequency of each is that of the current i. By (12.31) the sum of same-frequency sinusoids is sinusoidal, so the sum $v = v_R + v_C + v_L$ is sinusoidal, having the same frequency as the components of the sum. It follows from (12.27) that

$$\tilde{V} = \tilde{V}_R + \tilde{V}_C + \tilde{V}_L,$$

where

$$\tilde{V}_R = Z_R\tilde{I} = R\tilde{I}, \quad \tilde{V}_C = Z_C\tilde{I} = \frac{\tilde{I}}{j\omega C},$$
$$\tilde{V}_L = Z_L\tilde{I} = j\omega L\tilde{I}.$$

So

$$\tilde{V} = \tilde{V}_R + \tilde{V}_C + \tilde{V}_L = (R + Z_C + Z_L)\tilde{I}.$$

Thus the equivalent impedance of the series connection is given by

$$Z = \frac{\tilde{V}}{\tilde{I}} = R + Z_C + Z_L, \qquad (12.33)$$

as shown in Fig. 12.12(c).

The equivalent impedance of a series connection of resistors, capacitors, and inductors is the sum of the individual impedances. In other words, a series connection of elements can be replaced (at the terminals) by an impedance equal to the sum of the impedances of the elements.

Similarly, we can show that *the equivalent admittance of a parallel connection of resistors, capacitors, and inductors is the sum of the individual admittances.* In other words, a parallel connection of elements can be replaced (at the terminals) by a single admittance equal to the sum of the admittances of the elements in the parallel connection. With reference to Figs. 12.13 and 12.27,

$$i = i_R + i_C + i_L \Rightarrow \tilde{I} = \tilde{I}_R + \tilde{I}_C + \tilde{I}_L,$$

where

$$\tilde{I}_R = Y_R\tilde{V} = R^{-1}\tilde{V}, \; \tilde{I}_C = Y_C\tilde{V} = j\omega C\tilde{V},$$
$$\tilde{I}_L = Y_L\tilde{V} = (j\omega L)^{-1}\tilde{V}.$$

Thus the equivalent admittance of the parallel connection is given by

$$Y = \frac{\tilde{I}}{\tilde{V}} = Y_R + Y_C + Y_L. \qquad (12.34)$$

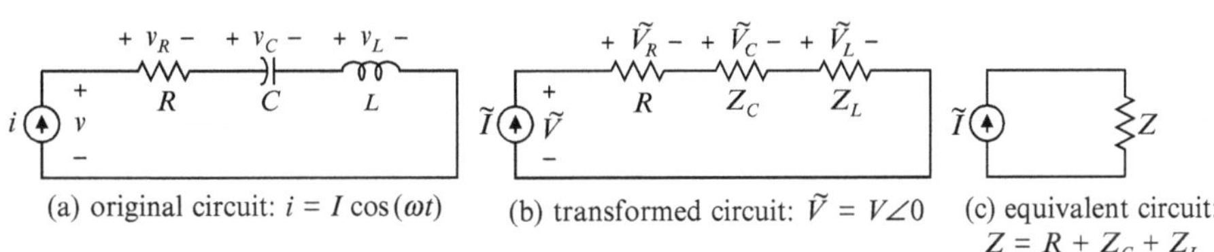

Fig. 12.12 Equivalent impedance of a series connection

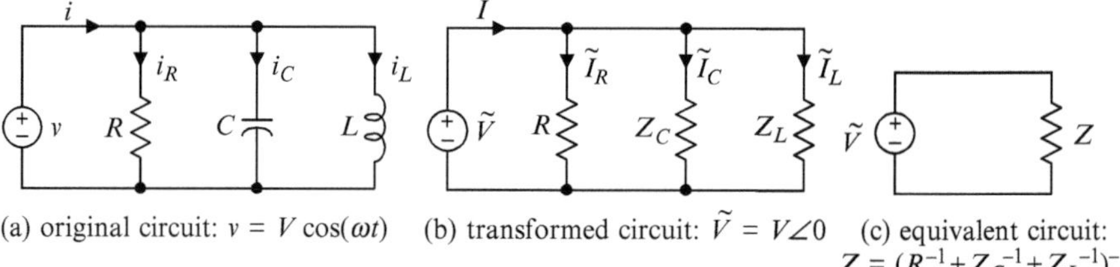

Fig. 12.13 Equivalent impedance of a parallel connection

The equivalent impedance of the parallel connection is given by

$$Z = \frac{\tilde{V}}{\tilde{I}} = Y^{-1} = (Y_R + Y_C + Y_L)^{-1} = \left(\frac{1}{R} + \frac{1}{Z_C} + \frac{1}{Z_L}\right)^{-1}. \qquad (12.35)$$

Equations (12.33) and (12.34) are easily applied to any series and parallel connections of arbitrary impedances, as illustrated by Fig. 12.14. An important special case is the parallel connection of two impedances, for which the equivalent impedance is given by

$$Z = \left(\frac{1}{Z_1} + \frac{1}{Z_2}\right)^{-1} = \frac{Z_1 Z_2}{Z_1 + Z_2}. \qquad (12.36)$$

As for resistances, the double-bar notation defined by

$$Z_1 \| Z_2 \| \cdots \| Z_N = \left(\frac{1}{Z_1} + \frac{1}{Z_2} + \cdots + \frac{1}{Z_N}\right)^{-1} \qquad (12.37)$$

often is convenient, particularly for two impedances in parallel, where

$$Z_1 \| Z_2 = \frac{Z_1 Z_2}{Z_1 + Z_2}. \qquad (12.38)$$

The two examples below illustrate reduction of series and parallel connections to an equivalent impedance. The solutions are given largely in graphical form, which is the best way to visualize the procedure.

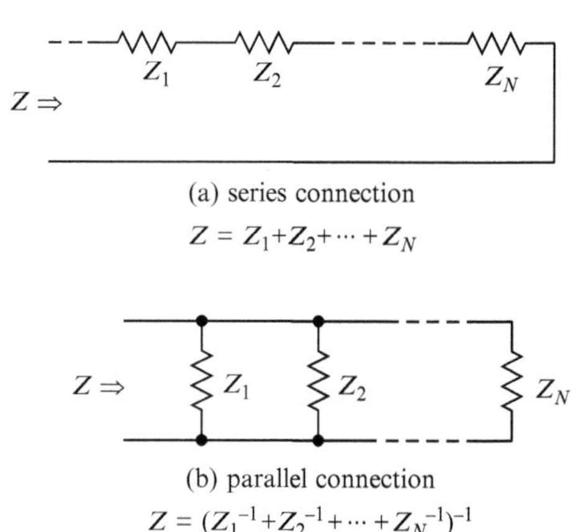

Fig. 12.14 Equivalent impedances for series and parallel connections

Example 12.20. Obtain an expression for the equivalent impedance at the terminals *a–b* of the circuit in Fig. 12.15(a). Express the impedance in rectangular form; i.e., as $a + jb$, where a and b are functions of frequency and the circuit parameters.

Solution: See Fig. 12.15(b)–(d). The equivalent impedance is

$$Z = R_1 \| Z_C + R_2 + Z_L = \frac{R_1}{1 + j\omega R_1 C} + R_2 + j\omega L = \frac{R_1(1 - j\omega R_1 C)}{1 + (\omega R_1 C)^2} + R_2 + j\omega L = \left[\frac{R_1}{1 + (\omega R_1 C)^2} + R_2\right] + j\left[\omega L - \frac{\omega R_1^2 C}{1 + (\omega R_1 C)^2}\right].$$

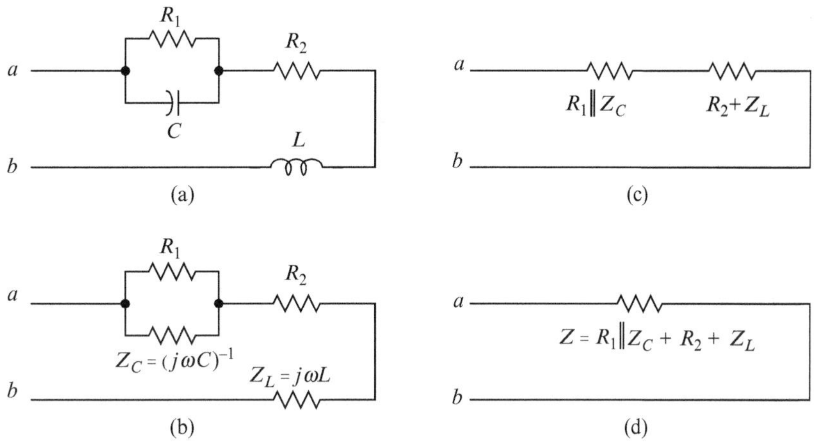

Fig. 12.15 See Example 12.20

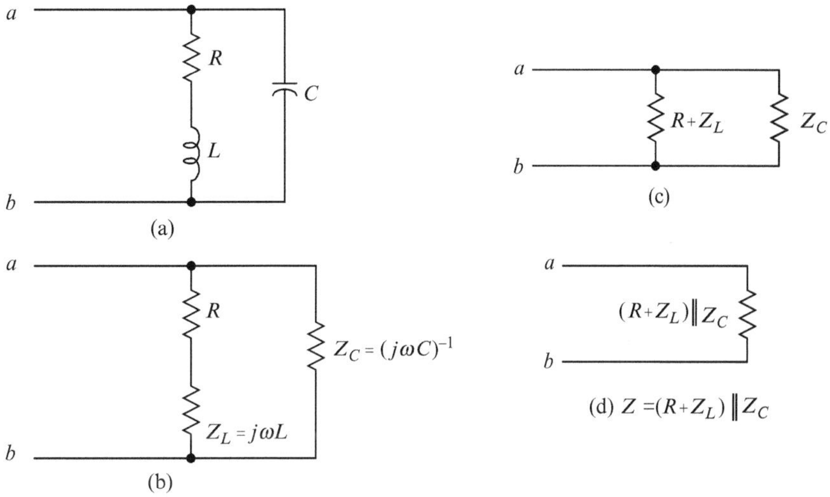

Fig. 12.16 See Example 12.21

> **Example 12.21.** Obtain an expression for the equivalent impedance at the terminals *a–b* of the circuit in Fig. 12.16(a). Express the impedance in rectangular form; i.e., as $a + jb$, where a and b are functions of frequency and the circuit parameters.
> *Solution*: See Fig. 12.16(b)–(d). The equivalent impedance is given by
>
> $$Z=(R+Z_L)\|Z_C=\frac{(R+Z_L)Z_C}{R+Z_L+Z_C}=\frac{(R+j\omega L)/(j\omega C)}{R+j\omega L+1/(j\omega C)}$$
> $$=\frac{R+j\omega L}{(1-\omega^2LC)+j\omega RC}=\frac{(R+j\omega L)\left[(1-\omega^2LC)-j\omega RC\right]}{(1-\omega^2LC)^2+(\omega RC)^2}$$
> $$=\left[\frac{R(1-\omega^2LC)+\omega^2RLC}{(1-\omega^2LC)^2+(\omega RC)^2}\right]+j\left[\frac{\omega L(1-\omega^2LC)-\omega R^2C}{(1-\omega^2LC)^2+(\omega RC)^2}\right].$$

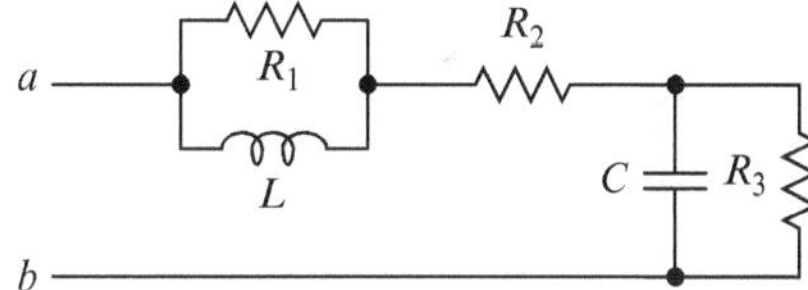

Fig. 12.17 See Exercise 12.17

> **Exercise 12.17.** Obtain an expression for the equivalent impedance at the terminals *a–b* of the circuit in Fig. 12.17. Express the impedance in rectangular form; i.e., as $a + jb$, where a and b are functions of frequency and the circuit parameters.

Recall (see (12.27) and the related discussion) that Kirchhoff's laws apply to phasor currents and

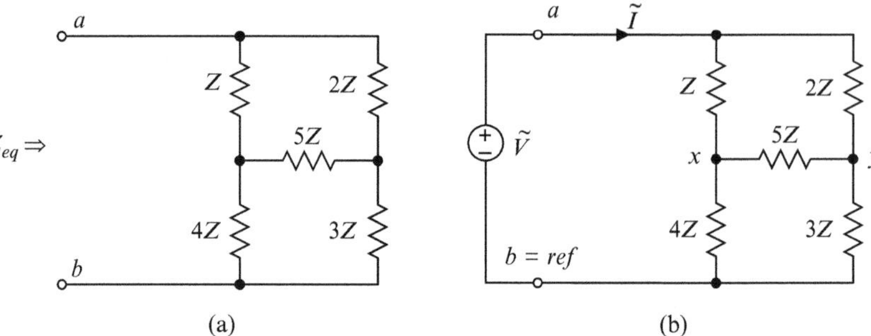

Fig. 12.18 See Example 12.22

voltages. If a circuit cannot be reduced to a single, equivalent impedance at a particular terminal pair by combining components in series and parallel, we can apply a known source to the terminals and use the definition $Z = \tilde{V}/\tilde{I}$, where $\tilde{V}$ is the terminal voltage and $\tilde{I}$ is the current entering the positive terminal.

Example 12.22. Obtain an expression for the impedance Z_{eq} at the terminals *a–b* of the circuit shown in Fig. 12.18(a).

Solution: We attach a known source to the terminals *a–b*, as shown in Fig. 12.18(b). Applying Kirchhoff's current law to nodes *x* and *y* give

$$\frac{\tilde{V}_x - \tilde{V}}{Z} + \frac{\tilde{V}_x - \tilde{V}_y}{5Z} + \frac{\tilde{V}_x}{4Z} = 0,$$
$$\frac{\tilde{V}_y - \tilde{V}}{2Z} + \frac{\tilde{V}_y - \tilde{V}_x}{5Z} + \frac{\tilde{V}_y}{3Z} = 0,$$

which yield

$$\tilde{V}_y = \frac{111}{175}\tilde{V}, \quad \tilde{V}_x = \frac{136}{175}\tilde{V}.$$

We apply Kirchhoff's current law to node *a* and obtain

$$\tilde{I} = \frac{\tilde{V} - \tilde{V}_x}{Z} + \frac{\tilde{V} - \tilde{V}_y}{2Z} = \frac{3\tilde{V}}{2Z} - \frac{136}{175Z}\tilde{V} - \frac{111}{175(2Z)}\tilde{V} = \frac{71\tilde{V}}{175Z},$$

which yields

$$\frac{\tilde{V}}{\tilde{I}} = Z_{eq} = \frac{175Z}{71} = 2.45Z.$$

Fig. 12.19 *Same-frequency* sinusoidal voltage sources in series or current sources in parallel are additive

It follows from (12.27) and Kirchhoff's laws that the phasor representing a series connection of *same-frequency* sinusoidal voltage sources or a parallel connection of *same-frequency* sinusoidal current sources is the sum of the phasors representing the individual sources, and consequently that same-frequency voltage sources in series or same-frequency current sources in parallel can be collapsed to a single source, as illustrated by Fig. 12.19; however, such connections rarely arise in practical problems, so this fact is of little use other than as a basis for textbook problems.

To summarize: Impedances in series and parallel reduce to single, equivalent impedances exactly as if they were complex resistances, and sinusoidal (phasor) voltage sources in series and current sources in parallel reduce to single, equivalent sources (additively), exactly as if they were complex dc sources.

12.10 Time Domain and Frequency Domain

A sinusoidal current or voltage can be described by a function of time or (if the frequency is recorded separately) by the corresponding phasor. We may indicate the equivalence of these descriptions as follows:

$$A\cos(\omega_0 t + \theta) \Leftrightarrow A\angle\theta; \quad \omega = \omega_0. \tag{12.39}$$

The terminal characteristic of a linear element can be expressed using either time or frequency as the independent variable. For example, the current through a capacitor is given by

$$i(t) = C\frac{dv(t)}{dt}, \quad (12.40)$$

where $v(t)$ is the voltage across the capacitor and time is the independent variable. If the voltage across the capacitor is sinusoidal, then the phasor representation of the current is given by

$$\tilde{I} = j\omega C\tilde{V} = j2\pi f C\tilde{V}, \quad (12.41)$$

where $\tilde{V}$ is the phasor representation of the voltage and frequency is the independent variable. Equation (12.40) is a **time-domain** description of a capacitor and (12.41) is the corresponding **frequency-domain** description. Similarly, the left side of (12.39) is the time-domain description of a sinusoid and the right side (phasor) is the corresponding frequency-domain description. Currents, voltages, linear elements, and entire linear circuits can be described in either the time domain or the frequency domain. Replacing the elements of a sinusoidally excited linear circuit by their impedances and representing currents and voltages in the circuit by phasors *transforms* the circuit from the time domain to the frequency domain.

Analysis using time-domain descriptions of currents, voltages, and circuit elements is called **time-domain analysis**. Analysis using frequency-domain descriptions of current, voltage, and circuit elements (e.g., phasors and impedances) is called **frequency-domain analysis**. This chapter describes how sinusoidally excited circuits are represented and analyzed in the frequency domain.

The utility of frequency-domain analysis stems from three fundamental principles:

- Using a mathematical tool called Fourier analysis, any physical current or voltage can be expressed over any finite interval as a sum of sinusoids.[10]
- Linear circuits obey superposition,[11] so we can obtain the response of a linear circuit to an input expressed as a sum of sinusoids by treating each sinusoidal component separately and adding the results.
- For every important time-domain operation on a sinusoidal waveform, there is a corresponding frequency-domain operation on the associated phasor. For example, differentiation with respect to time corresponds to multiplication by frequency ($j\omega$), as noted above. Effects of many important operations are easier to visualize and interpret in the frequency domain than in the time domain.

The time domain is the real-world domain. The frequency domain is a very useful mathematical creation. From one point of view, frequency-domain analysis is simply a computationally advantageous procedure for solving circuit problems: Whereas time-domain analysis requires solving differential equations, frequency-domain analysis requires solving only algebraic equations. But a far more important reason for using frequency-domain analysis is the *insight* fostered by frequency-domain relations among currents and voltages. In many applications, a frequency-domain description of a circuit is much more informative than the corresponding time-domain description. For example, the frequency response of an audio amplifier is much easier to interpret than would be a set of time-domain differential equations describing the same amplifier. Moreover, many real-world signals, such as currents or voltages representing speech, music, video, and radar and sonar echoes cannot be usefully described as functions of time but can be usefully described as functions of frequency. We cannot treat such descriptions in this book, so we must simply ask you to accept on faith that mastering the frequency-domain methods presented here will prepare you well for study of more complex and more general methods treated in more advanced books (and courses).

As a reminder, when we compare impedances by writing that one is larger or smaller than another, *magnitudes* are implied (because complex numbers cannot be ordered).

[10]Fourier analysis is introduced in Chapter 16.

[11]Section 12.16 treats superposition in the context of sinusoidal excitation.

12.11 Sinusoidal and DC Steady State

By definition, a linear circuit is in **sinusoidal steady state** if all currents and voltages in the circuit are sinusoidal and of the same frequency. A linear circuit in which all currents and voltages are constant is in **dc steady state**, which is a special case of sinusoidal steady state, because we may regard a constant current or voltage as a sinusoid having frequency zero.[12] The impedance of an inductor, given by $j\omega L$, equals zero for $\omega = 0$, so an inductor appears as a conductor to dc. The admittance of a capacitor, given by $j\omega C$, equals zero for $\omega = 0$, so a capacitor appears as an open circuit to dc. Thus a linear circuit in dc steady state reduces to a resistive circuit, all inductors being replaced by conductors and all capacitors being replaced by open circuits. **Sinusoidal steady-state analysis** consists of describing circuit elements by their impedances or admittances and sinusoidal currents and voltages by their phasor representations, and applying Kirchhoff's laws to those quantities, the latter being justified by the fundamental relation (12.27).

Like all mathematical descriptions of physical entities, sinusoidal currents and voltages and linear circuits themselves are abstractions having limited validity. Mathematically, a sinusoid is periodic, having neither beginning nor end. Practically, every current or voltage has a beginning and (presumably) an end. Nonetheless, sinusoidal models for currents and voltages often are realistic and lead to results in agreement with experiment. For example, a *practical sinusoidal voltage* is produced in the laboratory by a function generator. When energized, the output of such a function generator undergoes a transition from zero to sinusoidal. When such a source is connected to a stable linear circuit, all currents and voltages in the circuit also undergo transitions from zero to sinusoidal. After all such transients have vanished, the circuit is for practical purposes in sinusoidal steady state, and modeling the circuit using phasors and impedances or admittances leads to correct results. However, there are a couple of subtle points that should be addressed. The following discussion serves this purpose.

Consider Fig. 12.20, where a *lossless* (ideal) inductor is suddenly connected to a sinusoidal source, such as a function generator on a laboratory bench. Absent special circuitry, the switching will occur at a random point on the sinusoidal waveform produced by the source; e.g., at a peak, a zero-crossing, or halfway between. Therefore, if the time origin is the time at which the switch is closed, we must assign an arbitrary initial phase θ to the source voltage to account for our ignorance of the instantaneous amplitude of the sinusoid at that time. Thus

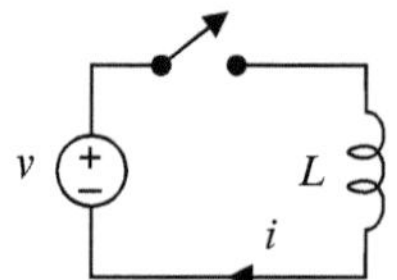

Fig. 12.20 Pertaining to (12.43) and (12.44). See text

$$v(t) = V_0 \cos(\omega_0 t + \theta),$$

where θ is unknown. The current through the inductor at any time t is given by

$$\begin{aligned} i(t) &= i(0) + \frac{1}{L}\int_0^t v(t')dt' \\ &= i(0) + \frac{V_0}{L}\int_0^t \cos(\omega_0 t' + \theta)dt', \end{aligned} \tag{12.42}$$

where $i(0) = 0$ because the inductor terminals were open before the switch was closed. It follows that

$$i(t) = \frac{V_0}{\omega_0 L}\sin(\omega_0 t + \theta) - \frac{V_0}{\omega_0 L}\sin(\theta); \quad t \geq 0. \tag{12.43}$$

We wish to discuss this result from two points of view:

1. The current given by (12.43) contains a persistent (not transient) dc component

$$I_{dc} = -\frac{V_0}{\omega_0 L}\sin(\theta), \tag{12.43}$$

which is in general non-zero and depends upon the point on the source waveform at which the switch was closed. The sinusoidal component of the current given by (12.43) is the *forced response*. The dc component is the *unforced response*. Thus

$$\begin{aligned} \textit{forced response} &= \frac{V_0}{\omega_0 L}\sin(\omega_0 t + \theta); \\ \textit{unforced response} &= -\frac{V_0}{\omega_0 L}\sin(\theta). \end{aligned}$$

[12] Indeed, this is why the cosine, rather than the sine, is chosen as the standard form for a sinusoidal current or voltage.

The forced response is maintained by the source. The unforced response arises because the current through the inductor must be continuous and persists because the inductor is lossless. But in a physical inductor, winding resistance (if nothing else) would dissipate the unforced response, there being no dc source to make up the energy lost in the winding. An unforced response cannot persist in a sinusoidally excited passive *physical* circuit. The persistent unforced response above is essentially a mathematical artifice that arose because we included no loss mechanism in the model.

2. The result (12.43) appears to be inconsistent with that obtained using phasors, which is only the forced (steady-state) component of the current obtained above. That is, using sinusoidal steady-state analysis (phasors and impedances) leads to

$$\begin{aligned}\tilde{I} &= \frac{\tilde{V}}{Z} = \frac{V_0\angle\theta}{j\omega_0 L} = \frac{V_0}{\omega_0 L}\angle\left(\theta - \frac{\pi}{2}\right)\\ &\Rightarrow \frac{V_0}{\omega_0 L}\cos\left(\omega_0 t + \theta - \frac{\pi}{2}\right)\\ &= \frac{V_0}{\omega_0 L}\sin(\omega_0 t + \theta),\end{aligned} \tag{12.44}$$

which contains no unforced component. Models such as that shown in Fig. 12.21, where a linear circuit is driven by a sinusoidal source,[13] implicitly assume that the excitation has been applied in that form for such a long time that any transient components of the currents and voltages in the circuit have vanished, *even if the circuit model is lossless.* Under that condition the circuit is in sinusoidal steady state, all currents and voltages are sinusoidal and all have the frequency of the excitation, and phasor methods can be applied to the circuit. Every current or voltage differs from every other only in peak amplitude and initial phase, which are the very things that phasors keep track of.

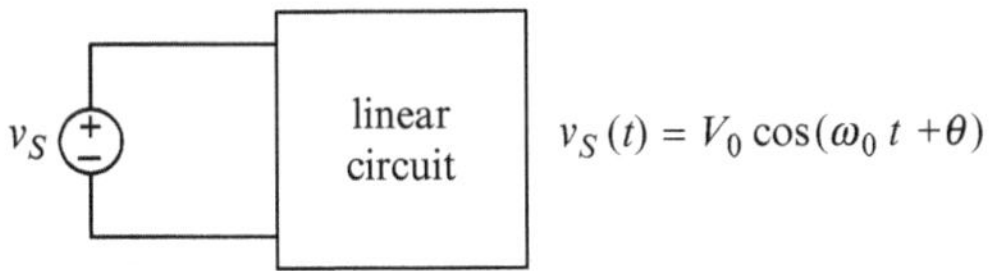

Fig. 12.21 Pertaining to the discussion of sinusoidal steady state. See text

[13] If there is more than one source, we may use superposition and consider one at a time.

Unless there is an explicit statement to the contrary, we assume henceforth that *a sinusoidally excited stable linear circuit is in sinusoidal steady state, which means that that all currents and voltages in the circuit are sinusoidal and have the frequency of the excitation.* Keep in mind that a source such as

$$v(t) = \begin{cases} 0, & t<0 \\ V_0\cos(\omega_0 t), & t \geq 0 \end{cases}$$

is not sinusoidal (mathematically). Nor are the currents and voltages in a circuit driven by such a source, because they contain both unforced (transient) and forced components having definite beginnings. Applying steady-state analysis to such a circuit will yield only the forced components of the currents and voltages in the circuit (which is ok, if those components are all you care about)

If a sinusoidally excited circuit contains no explicit losses, assuming the circuit is in sinusoidal steady state is equivalent to assuming that losses are present and are sufficient to ensure that any initial transients have vanished, but are insignificant for purposes of the analysis at hand. Such assumptions can simplify analysis and lead to meaningful results if losses are sufficiently small. Nonetheless, you should be suspicious of any lossless circuit model until you are convinced that the model is adequate for a purpose at hand or unless (in textbook problems) it is implied that the model is adequate.

12.12 Frequency-Domain Circuit Analysis

Analysis of a sinusoidally excited linear stable circuit using phasors and impedances proceeds as follows:

- *Transform all currents and voltages of interest from the time domain to the frequency domain.* Replace the excitation by its phasor representation, and denote all other currents and voltages in the circuit by their phasor representations. Usually, the source serves as the phase reference.
- *Transform the circuit elements from the time domain to the frequency domain.* Replace each capacitor and inductor in the circuit by a symbol for the impedance of the element; e.g., Z_{L3} for the impedance of inductor L_3. Because impedance is

essentially complex resistance, the circuit-diagram symbol for impedance is the same as that for resistance. Because the impedance of a resistor equals the resistance of the resistor, resistors remain unchanged; i.e., in frequency-domain representations, we denote resistance by R, not Z_R.[14]

- *Use Kirchhoff's laws to obtain equations whose solutions are phasor currents and voltages of interest*. Treat phasors (*symbolically*) as if they were dc currents and voltages and impedances as if they were resistances.
- *Solve the equations*. The results are frequency-domain descriptions of currents and voltages or of relations among currents and voltages.
- *Interpret the results*. More often than not, the results are more easily interpreted in the frequency domain than in the time domain, but if necessary, we can transform results of analysis back to the time domain. In any case, the ultimate purpose of an analysis is usually to extract some information relating to design or performance. We cannot delve too deeply into such interpretations in a first course. But keep in mind that in order to extract some answer worth interpreting, you must first be able to carry out such analyses correctly.

Example 12.23. Refer to Fig. 12.22(a), where the excitation is the phase reference and is expressed (in standard form) as $v_0(t) = V_0 \cos(\omega t)$.

(a) Obtain an expression for the phasor response $\tilde{V}_C$.

(b) Find the peak amplitude and the relative phase of the response for $V_0 = 5$ V, $f = 1$ kHz, $R = 1$ kΩ, $C = 200$ nF.

(c) Express the response in the time domain.

Solution: (a) We replace the voltages v_0, v_C by their phasor representations and we replace the capacitor by its impedance as shown in Fig. 12.22(b), where

$$\tilde{V}_0 = V_0\angle 0, \; Z_C = \frac{1}{j\omega C}.$$

The circuit is a voltage divider. Thus

$$\tilde{V}_C = \frac{Z_C}{R+Z_C}\tilde{V}_0 = \frac{1/(j\omega C)}{R+1/(j\omega C)}\tilde{V}_0$$
$$= \frac{\tilde{V}_0}{1+j\omega RC}, \; \omega RC \cong 1.257.$$

(b) The peak amplitude (magnitude) V_C and relative phase θ_C of the response are

$$V_C = |\tilde{V}_C| = \left|\frac{\tilde{V}_0}{1+j\omega RC}\right| = \frac{V_0}{\sqrt{1+(j\omega RC)^2}}$$
$$\cong \frac{5}{\sqrt{1+(1.257)^2}} \cong 3.11 \text{ V},$$
$$\theta_C = \measuredangle\tilde{V}_C = \measuredangle\tilde{V}_0 - \measuredangle(1+j\omega RC)$$
$$= -\tan^{-1}(\omega RC) \cong -\tan^{-1}(1.257)$$
$$\cong -0.899.$$

(c) It follows from the calculations above that

$$v_C(t) = 3.11\cos(\omega t - 0.899) \text{ V},$$
$$f = 1 \text{ kHz}.$$

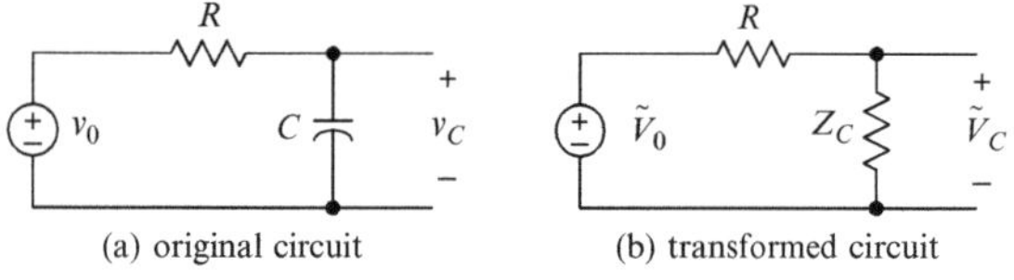

Fig. 12.22 See Example 12.23

Example 12.24. Refer to Fig. 12.23(a), where $i(t)$ is the phase reference, expressed in standard form as $i(t) = I_0 \cos(\omega t)$.

[14] Of course, we can also use admittances. But conventionally, elements are represented by their impedances in circuit diagrams, even if circuit equations are subsequently written in terms of admittances.

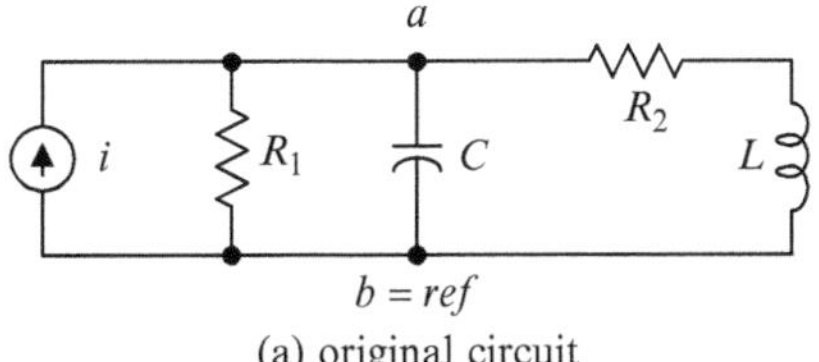

(a) original circuit

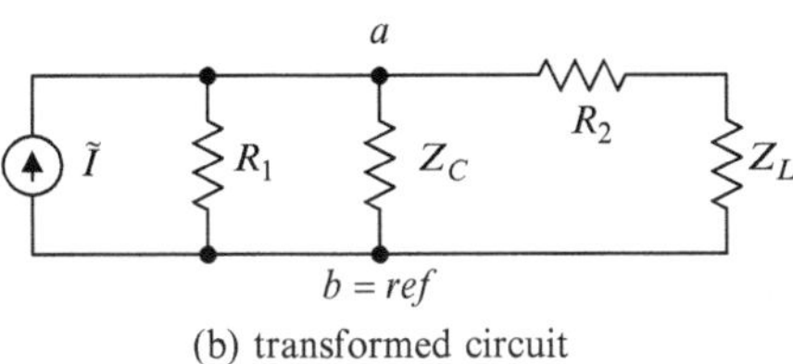

(b) transformed circuit

Fig. 12.23 See Example 12.24

(a) Obtain an *expression* for the phasor representation of the voltage v_a in terms of the phasor for the current i and the impedances of the circuit elements.

(b) Let $I_0 = 5$ mA, $f = 10$ kHz, $R_1 = 1$ kΩ, $C = 10$ nF, $R_2 = 5$ kΩ, and $L = 100$ mH. Express the voltage v_a as a real sinusoid and find the peak amplitude and initial phase of the voltage v_a.

Solution: (a) We transform the circuit as shown in Fig. 12.23(b), where

$$\tilde{I} = I_0, Z_C = \frac{1}{j\omega C}, Z_L = j\omega L$$

The equivalent impedance seen by the current source is

$$Z = \left(\frac{1}{R_1} + \frac{1}{Z_C} + \frac{1}{R_2 + Z_L}\right)^{-1}$$

and

$$\tilde{V} = I_0 Z$$

(b) For the parameter values given, we find

$$Z_C = \frac{1}{j\omega C} \cong 1.59\angle(-\pi/2)\text{k}\Omega$$
$$R_2 + Z_L = R_2 + j\omega L \cong 8.03\angle(0.90)\text{k}\Omega$$
$$Z \cong 832\angle(-0.458)\Omega.$$

Thus

$$\tilde{V}_a = I_0 Z \cong (5\,\text{mA})[832\angle(-0.458)\Omega]$$
$$\cong (4.165\angle - 0.458)\text{V}.$$

It follows that

$$v_a \cong 4.165\cos(2\pi f t - 0.458)\ \text{V},$$
$$f = 10\ \text{kHz}.$$

In general, even if a numerical result is called for, it is best to carry symbolic analysis as far as possible and reasonable, as is done in the examples above. Symbolic expressions exhibit effects of individual parameters, allow determination of critical parameters, and permit dimension and limit checking. Numerical results allow none of these things.

Exercise 12.18. Refer to Fig. 12.24.

(a) Obtain an *expression* for the phasor representation of the current i in terms of the phasor for the voltage v_S and the impedances of the circuit elements.

(b) Express the current as a real sinusoid and find the peak amplitude and initial phase of the current.

12.13 Reactance and Effective Resistance

An impedance can be expressed in rectangular form as

$$Z = R_Z + jX_Z. \tag{12.45}$$

The quantities R_Z and X_Z are *real* and in general are *functions of frequency*. The real part R_Z of impedance is called **effective resistance** and the imaginary part X_Z is called **reactance**.

$$\textit{impedance} = \textit{effective resistance} + j \times \textit{reactance}. \tag{12.46}$$

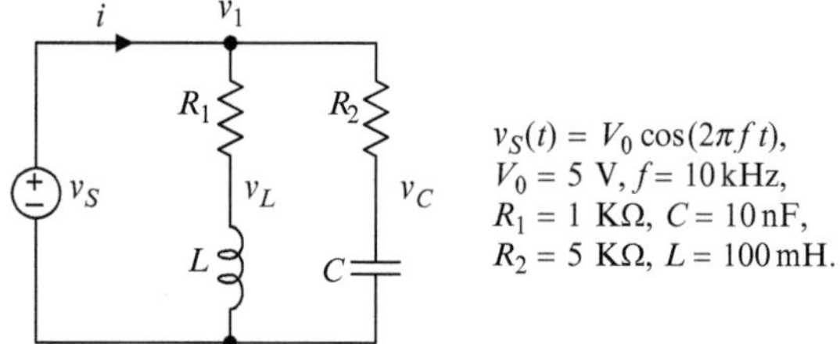

Fig. 12.24 See Exercise 12.18

Fig. 12.25 Illustrating the definitions of effective resistance and reactance. See (12.45) and (12.46)

From (12.45), and because impedances in series are additive, *the effective resistance and the reactance of a load are defined with reference to a series equivalent circuit* for *the load*, as illustrated by Fig. 12.25.

In general, **neither effective resistance nor reactance is physical**, in the sense that neither necessarily corresponds to any particular element or component in a circuit at hand.

From Table 12.1, the impedances of an inductor and a capacitor, expressed in the form (12.45), are given by

$$
\begin{aligned}
Z_L &= 0 + j(\omega L), \\
Z_C &= \frac{1}{j\omega C} = 0 - j\left(\frac{1}{\omega C}\right).
\end{aligned}
\tag{12.47}
$$

Comparing (12.47) to (12.45) shows that the effective resistances of an *ideal* inductor and an *ideal* capacitor equal zero[15] and that the reactances of an inductor and a capacitor are given by

$$
X_L = \omega L, \quad X_C = -\frac{1}{\omega C}. \tag{12.48}
$$

[15]The effective resistances of physical inductors and capacitors are not zero and their reactances are not exactly those given in (12.48). These and other departures of physical components from ideal behavior are discussed in Section 12.22.

Because frequency, inductance, and capacitance are positive quantities, *the reactance of an inductor is positive and the reactance of a capacitor is negative*. A load having a positive reactance at a particular frequency is said to be **inductive** at that frequency and a load having a negative reactance at a particular frequency is said to be **capacitive** at that frequency. A load whose impedance is real and non-zero at a particular frequency is said to be **resistive** at that frequency. A load having non-zero reactance is called a **reactive load**. A load having non-zero reactance and zero effective resistance is called a **purely reactive load**. For example, an inductor in series with a resistor is reactive and an (ideal) inductor alone is purely reactive.

Using methods described above (or Thévenin's theorem described in the sequel), we can reduce a load (a one-port circuit) containing any number of resistors, capacitors, and inductors to a single resistance in series with a single reactance, both of which are in general functions of frequency and of the various resistances, capacitances, and inductances appearing in the circuit. If the reduced (equivalent) impedance is expressed numerically *for a particular frequency*, it will be equivalent to a single resistor in series with either a single inductor (if the reactance is positive) or a single capacitor (if the reactance is negative). In any case, *an impedance expressed numerically is valid only at the frequency for which it was determined.* If the reduced impedance or admittance is expressed as a function of frequency, it is valid at all frequencies within the range of frequencies for which the original circuit model is valid. An impedance expressed as a function of frequency (and circuit parameters) can be a valuable (or essential) guide for design. An impedance expressed numerically is rarely useful in that regard, except as a specification.

Example 12.25. Refer to Fig. 12.26. (a) Obtain expressions for the effective resistance and the reactance of the parallel *RL* load as functions of the frequency of the source. (b) Calculate the effective resistance and the reactance for $f = 1$ kHz. (c) Find the equivalent *series RL* circuit for $f = 1$ kHz.

Solution: (a) The impedance of the load is given by

$$Z = \frac{Z_R Z_L}{Z_R + Z_L} = \frac{j\omega LR}{R + j\omega L}$$
$$= \frac{j\omega LR(R - j\omega L)}{R^2 + (\omega L)^2}$$
$$= \frac{(\omega L)^2 R}{R^2 + (\omega L)^2} + j\frac{\omega LR^2}{R^2 + (\omega L)^2}.$$

It follows that the effective resistance and the reactance of the load are given by

$$R_Z = \mathrm{Re}(Z) = \frac{(\omega L)^2 R}{R^2 + (\omega L)^2},$$
$$X_Z = \mathrm{Im}(Z) = \frac{\omega LR^2}{R^2 + (\omega L)^2}.$$

The effective resistance and the reactance are functions of frequency.

(b) For frequency $f = 1$ kHz, we obtain

$$Z = \frac{(2\mathrm{k}\Omega)(j2\pi)(1\mathrm{kHz})(2\,\mathrm{H})}{2\mathrm{k}\Omega + (j2\pi)(1\mathrm{kHz})(2\mathrm{H})}$$
$$= 1951 + j310\ \Omega,$$

and so

$$R_Z = 1{,}951\ \Omega,\ X_Z = 310\ \Omega.$$

(c) An equivalent series RL circuit at frequency $f = 1$ kHz has resistance $R_Z = 1{,}951\ \Omega$ and inductance

$$L_{eq} = \frac{X_Z}{\omega} = \frac{310}{2\pi 10^3} = 49\ \mathrm{mH}.$$

This particular series connection is *not* equivalent to the original load at any other frequency. Figure 12.27 shows a graph of the magnitudes of the original impedance $Z = R||(j\omega L)$ and the series impedance given by $Z_s = R_z + j\omega L_{eq}$ versus frequency. The graphs touch at 1 kHz, but are different at other frequencies.

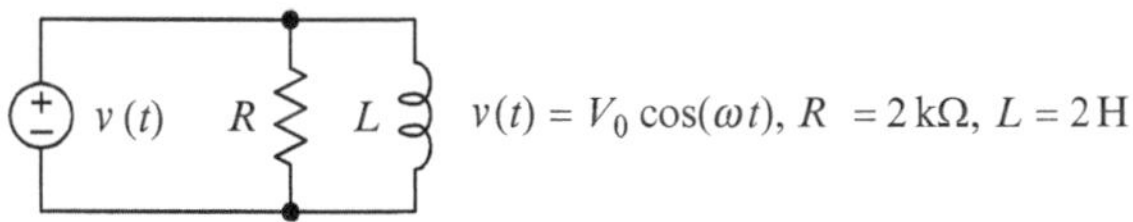

Fig. 12.26 See Example 12.25

Example 12.26. Refer to Fig. 12.28. (a) Obtain an expression for the impedance at the terminals *a*–b. (b) Calculate the value of the impedance for frequency $f = 10$ kHz, and find a series combination of resistance and capacitance or resistance and inductance having that impedance at the specified frequency. (c) Repeat (b) for $f = 50$ kHz.

Solution: Figure 12.29 illustrates steps in the solution.

(a) We replace the circuit elements by the corresponding impedances, as shown in Fig. 12.29, where

$$Z_1 = \frac{1}{j\omega C},\ Z_2 = R_2 + j\omega L.$$

We obtain the equivalent impedance of the parallel connection of Z_1, Z_2. This gives

$$Z_3 = \frac{Z_1 Z_2}{Z_1 + Z_2} = \frac{R_2 + j\omega L}{1 - \omega^2 LC + j\omega R_2 C}.$$

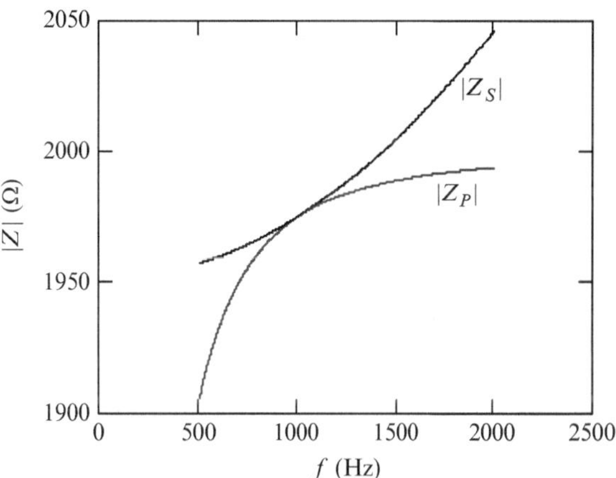

Fig. 12.27 See Example 12.25

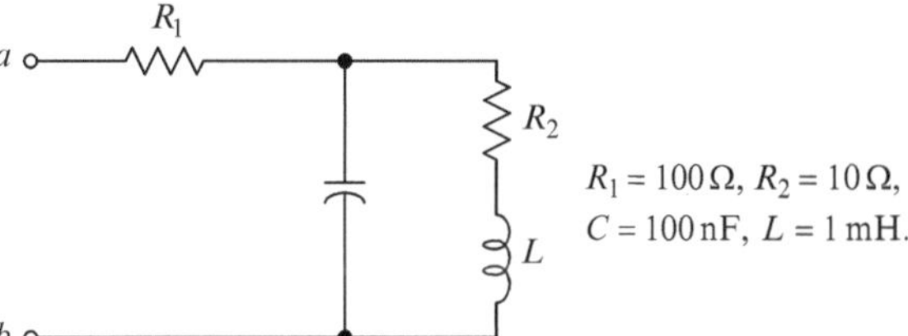

Fig. 12.28 See Example 12.26

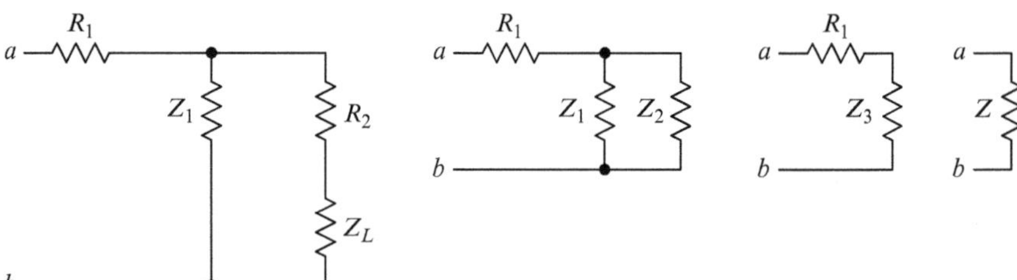

Fig. 12.29 See Example 12.26

We add the series resistance to the impedance Z_3 to obtain the equivalent impedance

$$Z = R_1 + Z_3 = R_1 + \frac{R_2 + j\omega L}{1 - \omega^2 LC + j\omega R_2 C}.$$

(b) For $f = 10$ kHz $\Rightarrow \omega = 20 \times 10^3\ \pi\ \mathrm{s}^{-1}$, we obtain

$$\begin{aligned} Z &= R_1 + \frac{R_2 + j\omega L}{1 - \omega^2 LC + j\omega R_2 C} = 127 + j101\ \Omega \\ &= R_Z + jX_Z. \end{aligned}$$

For $f = 10$ kHz, the effective resistance is $R_Z = 127\ \Omega$ and the reactance is $X_Z = 101\ \Omega$. The reactance is positive, so the circuit is equivalent (at the terminals *a–b*) to the series connection of a resistor having resistance $R' = R_Z = 127\ \Omega$ and an inductor having inductance

$$L' = \frac{X_Z}{\omega} = \frac{101\Omega}{(2\pi)(10\mathrm{kHz})} = 1.61\ \mathrm{mH}.$$

(c) Similarly, for $f = 50$ kHz $\Rightarrow \omega = 100 \times 10^3 \pi\ \mathrm{s}^{-1}$, we obtain $Z = 100 - j35.4\ \Omega$. The reactance is negative, so the circuit is equivalent (at the terminals *a–b*) to the series connection of a resistor having resistance $R' = R_Z = 100\ \Omega$ and a capacitor having capacitance

$$\begin{aligned} C &= -\frac{1}{\omega X_Z} = -\frac{1}{(2\pi)(50\,\mathrm{kHz})(-35.4\,\Omega)} \\ &= 89.9\ \mathrm{nF}. \end{aligned}$$

In general, a load can be inductive at some frequencies and capacitive at others, as illustrated by Example 12.26. *But a load containing only resistors and inductors is inductive at all frequencies and a load containing only resistors and capacitors is capacitive at all frequencies.* This follows from the fact that in a *passive* circuit, inductance arises only from energy stored in a magnetic field and capacitance arises only from energy stored in an electric field. A load containing only resistors and inductors can store energy only in magnetic fields, and must be inductive. A load containing only resistors and capacitors can store energy only in electric fields, and must be capacitive. A load containing both capacitors and inductors can be capacitive or inductive, depending upon the frequency of the source and upon which energy-storage mechanism is dominant at that frequency. Finally, *the effective resistance of a load containing only resistors, capacitors, and inductors is non-negative.* This follows from the fact that the average power dissipated at the terminals of a passive load must be non-negative.

Above, we are careful to confine remarks to passive circuits – circuits containing no dependent sources. In truth, whether a two-terminal circuit or load appears to be inductive or capacitive (whether the reactance is positive or negative) depends on the terminal characteristic. If the terminal characteristic has the form

$$i = K_1 v + K_2 \frac{dv}{dt} \leftrightarrow \tilde{I} = (K_1 + j\omega K_2)\,\tilde{V},$$

the circuit appears to be capacitive. If the terminal characteristic has the form

$$v = K_1 i + K_2 \frac{di}{dt} \leftrightarrow \tilde{V} = (K_1 + j\omega K)\,\tilde{I},$$

the circuit appears to be inductive. Certain active *RC* circuits can exhibit positive reactance and certain active *RL* circuits can exhibit negative reactance. Also, certain active circuits can exhibit negative effective resistance. We treat one such active circuit in Section 12.21.

Example 12.27. (a) Find two frequencies for which the circuit of Example 12.26 is equivalent to a resistor at the terminals *a–b* and find the effective resistance in both cases. (b) Find the range of frequencies for which the impedance is capacitive and the range of frequencies for which the impedance is inductive.

Solution: (a) From Example 12.26, the equivalent impedance at the terminals *a–b* is given by

$$Z = R_1 + \frac{R_2 + j\omega L}{1 - \omega^2 LC + j\omega R_2 C}. \quad (12.49)$$

For the equivalent impedance to be resistive (real), the imaginary part of the impedance must be zero. Because R_1 is real, this requires

$$\mathrm{Im}\left[\frac{R_2 + j\omega L}{1 - \omega^2 LC + j\omega R_2 C}\right] = 0.$$

Multiplying numerator and denominator by the conjugate of the denominator gives

$$\mathrm{Im}\left[\left(\frac{R_2 + j\omega L}{1 - \omega^2 LC + j\omega R_2 C}\right) \times \left(\frac{1 - \omega^2 LC - j\omega R_2 C}{1 - \omega^2 LC - j\omega R_2 C}\right)\right] = \frac{-\omega R_2{}^2 C + \omega L(1 - \omega^2 LC)}{|1 - \omega^2 LC + j\omega R_2 C|^2} = 0.$$

The denominator in the expression above is positive, so

$$-\omega R_2{}^2 C + \omega L\left(1 - \omega^2 LC\right) = 0.$$

This last equation is satisfied if either $\omega = 0$ or

$$R_2{}^2 C = L\left(1 - \omega^2 LC\right) \Rightarrow \omega = \sqrt{\frac{L - R_2{}^2 C}{L^2 C}} = 9.95 \times 10^4\ \mathrm{s}^{-1} \Rightarrow f = 15.8\ \mathrm{kHz}.$$

For $\omega = 0 \Rightarrow f = 0$, the effective resistance is

$$R = R_1 + R_2 = 110\ \Omega,$$

as is evident by inspection of the circuit because, for $f = 0$, the capacitor is an open circuit ($|Z_C| \to \infty$) and the inductor is a short circuit ($Z_L = 0$).

For $\omega = 9.95 \times 10^4\ \mathrm{s}^{-1}$, the effective resistance is, from (12.49),

$$R_1 + \frac{R_2 + j\omega L}{1 - \omega^2 LC + j\omega R_2 C} = 1.1\ \mathrm{k}\Omega.$$

(b) The impedance is capacitive if the reactance is negative. The frequencies for which the circuit is capacitive are given by

$$\mathrm{Im}\left[\frac{R_2 + j\omega L}{1 - \omega^2 LC + j\omega R_2 C}\right] < 0$$

$$\Rightarrow \frac{-\omega R_2{}^2 C + \omega L(1 - \omega^2 LC)}{|1 - \omega^2 LC + j\omega R_2 C|^2} < 0.$$

The denominator is the magnitude of a quantity and is positive. Angular frequency ω is non-negative. The impedance is capacitive for frequencies such that

$$-\omega R_2{}^2 C + \omega L\left(1 - \omega^2 LC\right) < 0,$$

which implies

$$\omega > \sqrt{\frac{L - R_2{}^2 C}{L^2 C}} = 9.95 \times 10^4\ \mathrm{s}^{-1} \Rightarrow f > 15.8\ \mathrm{kHz}.$$

Because the impedance is resistive for $f = 0$ and $f = 15.8$ kHz and capacitive for $f > 15.8$ kHz, and because the impedance must be one of resistive, capacitive, or inductive at each frequency, the impedance is inductive for

$$0 < f < 15.8\ \mathrm{kHz}.$$

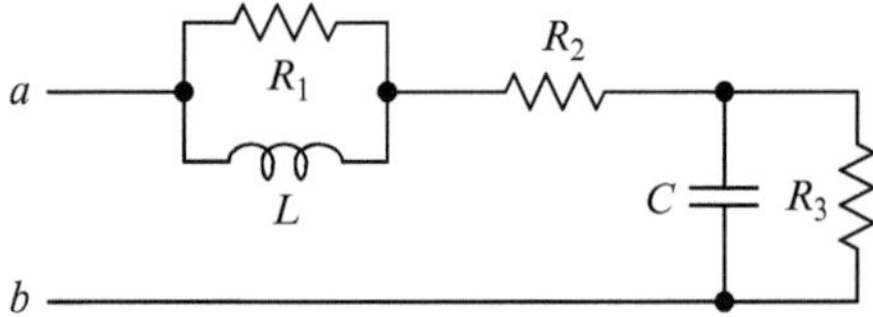

$R_1 = 1\,\text{k}\Omega$, $R_2 = 220\,\Omega$, $R_3 = 330\,\Omega$, $L = 10\,\text{mH}$, $C = 100\,\text{nF}$

Fig. 12.30 See Exercise 12.19

Exercise 12.19. Refer to Fig. 12.30. Use the result obtained in Exercise 12.17 and calculate the value of the impedance for frequency $f = 5$ kHz. Then find a series combination of resistance and capacitance or resistance and inductance having that impedance at the specified frequency. Repeat for $f = 1$ kHz.

12.14 Susceptance and Effective Conductance

An admittance can be expressed in rectangular form as

$$Y = \frac{1}{Z} = G_Y + jB_Y. \tag{12.50}$$

The real part G_Y of admittance is called **effective conductance** and the imaginary part B_Y is called **susceptance**.

$$\begin{aligned} \textit{admittance} &= \textit{effective conductance} \\ &\quad + j \times \textit{susceptance}. \end{aligned} \tag{12.51}$$

The dimension of both effective conductance and susceptance is that of conductance and the SI unit of both is siemens (S).

From (12.50), and because admittances in parallel are additive, the susceptance and effective conductance of a load are defined with reference to an equivalent *parallel* circuit for the load, as illustrated by Fig. 12.31, whereas effective resistance and reactance are defined with reference to an equivalent *series* circuit for the load. Note that in Fig. 12.31, elements are represented by their impedances, as is conventional in circuit diagrams.

Admittance is the reciprocal of impedance, but *in general effective conductance is not the reciprocal of effective resistance and effective susceptance is not the reciprocal of effective reactance.* That is, with reference to Figs. 12.25 and 12.31, $Z = Y^{-1}$, but $R_Z \neq G_Y{}^{-1}$ and $X_Z \neq B_Y{}^{-1}$. This is because

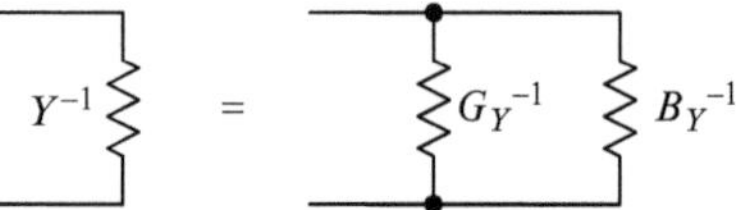

Fig. 12.31 Illustrating the definitions of effective conductance and susceptance. See (12.50) and (12.51)

$$\frac{1}{R_Z + jX_Z} \neq \frac{1}{R_Z} + \frac{1}{jX_Z}.$$

From Table 12.2, the admittances of an inductor and a capacitor are given by

$$Y_L = \frac{1}{j\omega L} = 0 - j\frac{1}{\omega L}, \quad Y_C = 0 + j\omega C. \tag{12.52}$$

It follows from (12.50) that the susceptances of an inductor and a capacitor are given by

$$B_L = -\frac{1}{\omega L}, \quad B_C = \omega C. \tag{12.53}$$

Because frequency, inductance, and capacitance are positive quantities, *the susceptance of an inductor is negative and the susceptance of a capacitor is positive.*[16] A load having a negative susceptance at a particular frequency is **inductive** at that frequency and a load having a positive susceptance at a particular frequency is **capacitive** at that frequency. A load whose admittance is real and non-zero at a particular frequency is **resistive** at that frequency. At any particular frequency, a load containing any number of resistors, capacitors, and inductors is equivalent to a single resistor in parallel with either a single inductor (if $B_Y < 0$) or a single capacitor (if $B_Y > 0$). It follows from the last paragraph in the previous section that *the susceptance of a circuit containing only resistors and inductors is negative for all frequencies and the susceptance of a circuit containing only resistors and capacitors is positive for all frequencies. Also, the*

[16] Note that for any particular load at any particular frequency, reactance and susceptance have opposite signs (if non-zero).

effective conductance of a load containing only resistors, capacitors, and inductors is non-negative.[17]

Example 12.28. Obtain expressions for the effective conductance and susceptance of the series connection of a resistor having resistance R and an inductor having inductance L.
Solution: The impedance of the series connection is given by $Z = R + j\omega L$. Hence the admittance is given by

$$Y = \frac{1}{Z} = \frac{1}{R + j\omega L} = \left(\frac{1}{R + j\omega L}\right)\left(\frac{R - j\omega L}{R - j\omega L}\right) = \frac{R}{R^2 + (\omega L)^2} - j\frac{\omega L}{R^2 + (\omega L)^2}.$$

Comparing the right side of this result with (12.50) shows that the conductance and susceptance are given by

$$G_Y = \frac{R}{R^2 + (\omega L)^2}, \quad B_Y = -\frac{\omega L}{R^2 + (\omega L)^2}.$$

Note that the susceptance is negative, as expected for an inductive load.

The next example illustrates the fact that a circuit containing both inductors and capacitors can be inductive, capacitive, or resistive, depending upon the frequency of the applied current or voltage.

Example 12.29. Obtain an expression for the admittance of the load shown in Fig. 12.32 and specify the frequencies for which the load is (i) resistive, (ii) capacitive, and (iii) inductive in terms of the circuit parameters.

Solution: The admittance of the load is given by

$$Y = \frac{1}{R} + \frac{1}{j\omega L} + j\omega C = \frac{1}{R} + j\left(\omega C - \frac{1}{\omega L}\right).$$

The susceptance is given by

$$B = \mathrm{Im}(Y) = \left(\omega C - \frac{1}{\omega L}\right).$$

The load is resistive if the admittance is real (if the susceptance equals zero), or for the frequency given by

$$B = 0 \Rightarrow \left(\omega C - \frac{1}{\omega L}\right) = 0 \Rightarrow \omega C = \frac{1}{\omega L} \Rightarrow \omega = \sqrt{\frac{1}{LC}} \Rightarrow f = \frac{1}{2\pi}\sqrt{\frac{1}{LC}}.$$

The load is capacitive if the susceptance is positive, or for frequencies given by

$$B > 0 \Rightarrow \omega > \sqrt{\frac{1}{LC}}.$$

The load is inductive if the susceptance is negative, or for frequencies given by

$$B < 0 \Rightarrow \omega < \sqrt{\frac{1}{LC}} \Rightarrow f < \frac{1}{2\pi}\sqrt{\frac{1}{LC}}.$$

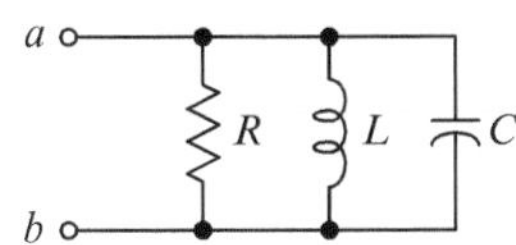

Fig. 12.32 See Example 12.29

Exercise 12.20. Refer to Fig. 12.33. Use results obtained in Exercise 12.19 and calculate the value of the impedance for frequency $f = 5$ kHz. Then find a *parallel* combination of resistance and capacitance or resistance and

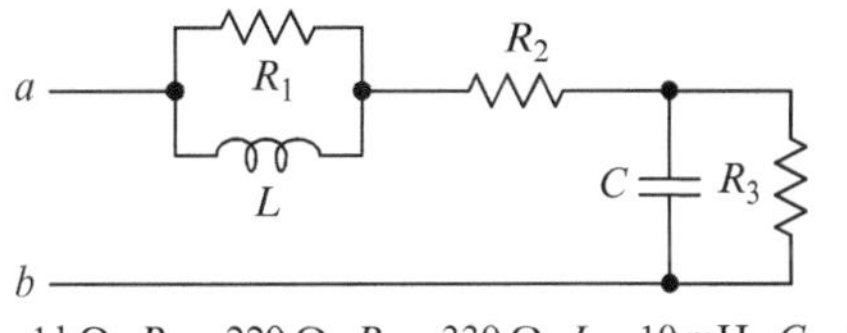

$R_1 = 1\,\mathrm{k\Omega}$, $R_2 = 220\,\Omega$, $R_3 = 330\,\Omega$, $L = 10\,\mathrm{mH}$, $C = 100\,\mathrm{nF}$

Fig. 12.33 See Exercise 12.20

[17]These remarks pertain to *passive* circuits.

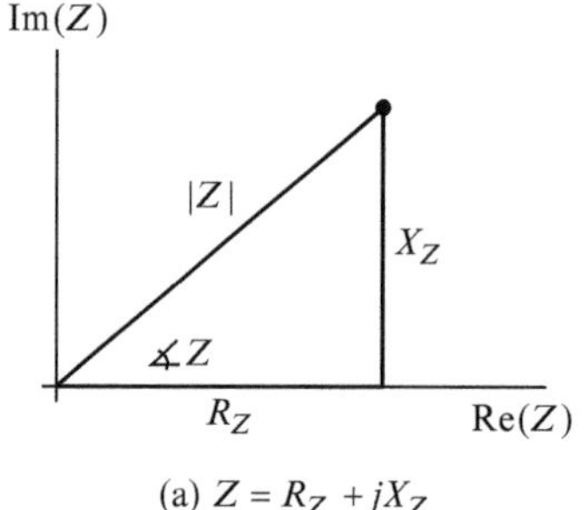

(a) $Z = R_Z + jX_Z$

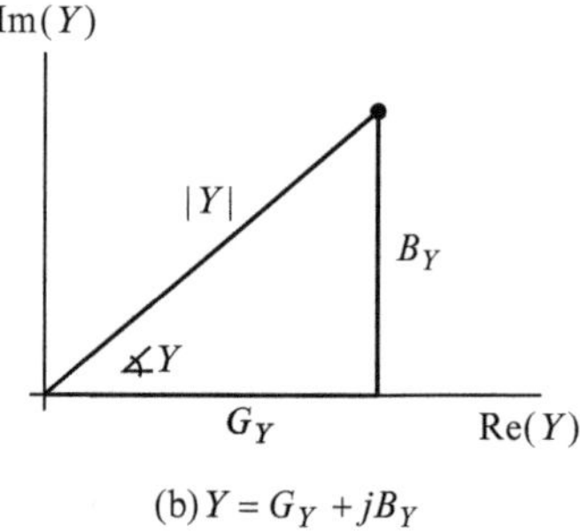

(b) $Y = G_Y + jB_Y$

Fig. 12.34 Impedance and admittance triangles

inductance having that admittance at the specified frequency. Repeat for $f = 1$ kHz.

12.15 Impedance and Admittance Triangles

From the generalized form of Ohm's law, the phasor voltage across a load having impedance Z is given by

$$\tilde{V} = Z\tilde{I},$$

where $\tilde{I}$ is the phasor current entering the positive terminal of the load. The angle of the product of complex quantities equals the sum of the angles of the factors. It follows that

$$\measuredangle\tilde{V} = \measuredangle Z + \measuredangle\tilde{I}.$$

The angle of an inductive impedance is positive because inductive reactance is positive. If the impedance is inductive, then $\measuredangle Z > 0$ and $\measuredangle\tilde{V} > \measuredangle\tilde{I}$. *The current through an inductive load lags the voltage across the load.* The angle of a capacitive impedance is negative because capacitive reactance is negative. If the impedance is capacitive, then $\measuredangle Z < 0$ and $\measuredangle\tilde{V} < \measuredangle\tilde{I}$. *The current through a capacitive load leads the voltage across the load.* If the impedance is resistive, then $\measuredangle Z = 0$ and $\measuredangle\tilde{V} = \measuredangle\tilde{I}$. *The current through a resistive load is in phase with the voltage across the load.*

An impedance $Z=R+jX$ can be represented graphically by a point (R, X) in a complex plane, as illustrated by Fig. 12.34(a). The abscissa is the effective resistance and the ordinate is the reactance. Lines drawn as shown in Fig. 12.34(a). form a right triangle called the **impedance triangle**. The length of the hypotenuse of the triangle is the magnitude of the impedance and the angle formed by the hypotenuse and the real axis is the angle of the impedance. If the angle of the impedance of a load is positive, the load is inductive and current through the load lags the voltage across the load.

Similarly, an admittance $Y = G + jB$ can be represented graphically by a point (G, B) in a complex plane, as illustrated by Fig. 12.34(b). The abscissa is the effective conductance and the ordinate is the susceptance. Lines drawn as shown in Fig. 12.34(b) form a right triangle called the **admittance triangle**.[18] The length of the hypotenuse of the triangle is the magnitude of the admittance, and the angle formed by the hypotenuse and the real axis is the angle of the admittance. If the angle of the admittance of a load is positive, the load is capacitive and current through the load leads the voltage across the load.

Exercise 12.21. Refer to Exercise 12.19 and Exercise 12.20, where you calculated the impedance and the admittance at the terminals *a–b* for $f = 1$ kHz and $f = 5$ kHz. Draw impedance and admittance triangles for both cases.

12.16 Linearity and Superposition

The **principle of superposition** (as applied to sinusoidally excited circuits) can be stated as follows:

[18]We define the admittance triangle for sake of completeness; however, admittance triangles are rarely seen in practice.

Suppose an *RLC* circuit contains two or more independent sinusoidal current or voltage sources. Then any particular current or voltage in the circuit can be obtained as the sum of the corresponding currents or voltages found by applying the independent sources one at a time, all other *independent* sources being set to zero. Do not set *dependent* sources to zero. In applications of superposition, the various independent sources may be applied in any order.

To set a voltage source to zero, replace the source with a short circuit. To set a current source to zero, replace the source with an open circuit.

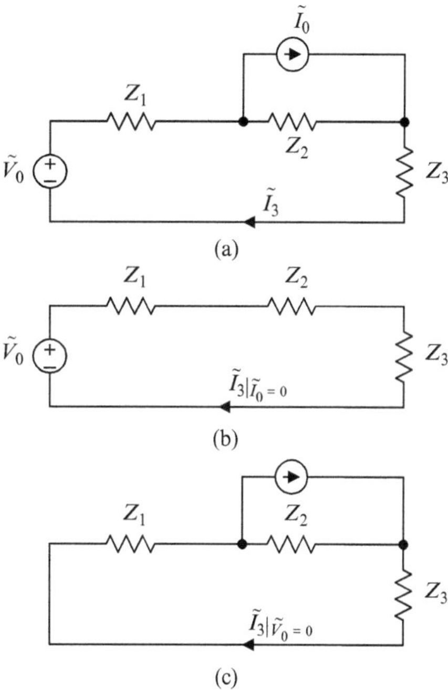

Fig. 12.35 See Example 12.30

Example 12.30. Using superposition, obtain an expression for the phasor current $\tilde{I}_3$ in the circuit shown in Fig. 12.35(a), where the independent sources have the same frequency.
Solution: We first find the current $\tilde{I}_3$ due to the voltage source $\tilde{V}_0$ acting alone; that is, with the source current $\tilde{I}_0$ set to zero (with the current source replaced by an open circuit), as illustrated by Fig. 12.35(b). This gives

$$\tilde{I}_3\big|_{\tilde{I}_0=0} = \frac{\tilde{V}_0}{Z_1+Z_2+Z_3}.$$

Next we find the current $\tilde{I}_3$ due to the current source $\tilde{I}_0$ acting alone; that is, with the source voltage $\tilde{V}_0$ set to zero (with the voltage source replaced by a short circuit), as illustrated by Fig. 12.35(c). We obtain

$$\tilde{I}_3\big|_{\tilde{V}_0=0} = \frac{\tilde{I}_0 Z_2}{Z_1+Z_2+Z_3}.$$

The phasor for the (total) current $\tilde{I}_3$ is given by

$$\begin{aligned}\tilde{I}_3 &= \tilde{I}_3\big|_{\tilde{I}_0=0} + \tilde{I}_3\big|_{\tilde{V}_0=0}\\ &= \frac{\tilde{V}_0}{Z_1+Z_2+Z_3} + \frac{\tilde{I}_0 Z_2}{Z_1+Z_2+Z_3}\end{aligned}$$

and the current is given in the time domain by

$$i_3(t) = |\tilde{I}_3|\cos(\omega_0 t + \measuredangle\tilde{I}_3),$$

where f_0 is the frequency of either source.

Example 12.31. Refer to Fig. 12.36, where

$$\begin{aligned}&R_1 = 100\ \Omega,\ C = 2\ \mu\text{F},\ R_F = 500\ \text{k}\Omega\\ &R_o = 100\ \text{k}\Omega,\ R_L = 2\ \text{k}\Omega,\ L = 400\ \text{mH},\\ &g = 0.1\ \text{S},\ v_a(t) = V_a\cos(\omega_a t),\\ &V_a = 750\ \text{mV},\ f_a = 500\ \text{Hz};\\ &v_b(t) = V_b\cos(\omega_b t + \theta),\ V_b = 20\ \text{mV},\\ &f_b = 1\ \text{kHz},\ \theta = 0.5.\end{aligned}$$

Using reasonable approximations to simplify the calculations, find the output $v_L(t)$.
Solution: Set $v_b(t)$ to zero. Do not set the dependent source to zero. Use phasors to find output due to $v_a(t)$. Kirchhoff's current law gives

$$\begin{aligned}&\left(\frac{2}{R_1}+\frac{1}{R_F}+Y_C\right)\tilde{V}_1 - \frac{1}{R_F}\tilde{V}_L = \frac{1}{R_1}\tilde{V}_a,\\ &\left(g-\frac{1}{R_F}\right)\tilde{V}_1 + \left(\frac{1}{R_o}+\frac{1}{R_F}+\frac{1}{R_L}+\frac{1}{Z_L}\right)\tilde{V}_L = 0,\\ &Y_C = j\omega C,\ Z_L = j\omega L.\end{aligned}$$

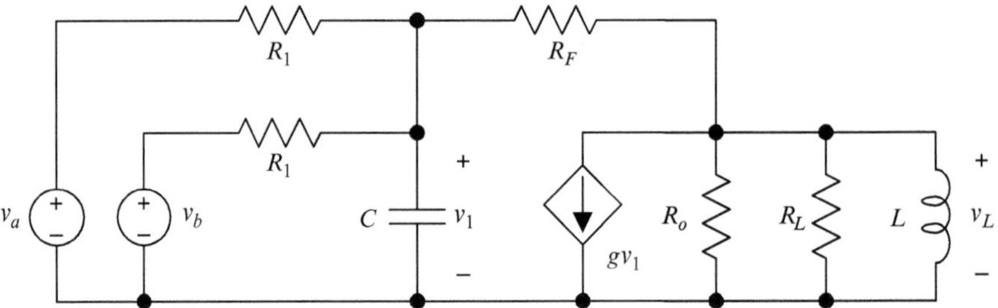

Fig. 12.36 See Example 12.31

From the given parameter values, we have

$$\frac{1}{R_1} \gg \frac{1}{R_F},\ g \gg \frac{1}{R_F},\ \frac{1}{R_L} \gg \frac{1}{R_o} > \frac{1}{R_F}.$$

so the equations above become

$$\left(\frac{2}{R_1}+Y_C\right)\tilde{V}_1 - \frac{1}{R_F}\tilde{V}_L \cong \frac{1}{R_1}\tilde{V}_a,$$
$$g\tilde{V}_1 + \left(\frac{1}{R_L}+\frac{1}{Z_L}\right)\tilde{V}_L \cong 0.$$

Eliminate $\tilde{V}_1$ to obtain

$$\tilde{V}_{La} \cong \frac{-gR_FR_LZ_L}{(R_FR_L+R_FZ_L)(2+Y_CR_1)+gR_1R_LZ_L}\tilde{V}_a$$
$$\Rightarrow \left|\tilde{V}_{La}\right| \cong 37.774\,\text{V}, \measuredangle\tilde{V}_{La} \cong -2.443.$$

Set $v_a(t)$ to zero. Do not set the dependent source to zero. Use phasors to find output due to $v_b(t)$. The equations are unchanged, except $\tilde{V}_b$ replaces $\tilde{V}_a$:

$$\tilde{V}_{Lb} \cong \frac{-gR_FR_LZ_L\tilde{V}_b}{R_F(R_L+Z_L)(2+Y_CR_1)+gR_1R_LZ_L}$$
$$\Rightarrow \left|\tilde{V}_{Lb}\right| \cong 1.308\,\text{V}, \measuredangle\tilde{V}_{Lb} \cong -2.532.$$

By superposition, the output is given by

$$\begin{aligned} v_L(t) &= \left|\tilde{V}_{La}\right|\cos\left(\omega_a t + \measuredangle\tilde{V}_{La}\right) \\ &\quad + \left|\tilde{V}_{Lb}\right|\cos\left(\omega_b t + \theta + \measuredangle\tilde{V}_{Lb}\right) \\ &\cong [37.774\cos(\omega_0 t - 2.443) \\ &\quad +1.308\cos(\omega_1 t - 2.532)]\,\text{V}. \end{aligned}$$

The principle of superposition is a consequence of Kirchhoff's laws and the fact that the terminal characteristics of resistors, capacitors, inductors, and dependent sources are linear. We can show this as follows: Applying Kirchhoff's laws to an *RLC* circuit driven by sinusoidal sources $x_1(t), x_2(t), \ldots, x_M(t)$ leads to a set of equations of the form

$$\begin{aligned} &a_{11}\tilde{Y}_1 + a_{12}\tilde{Y}_2 + \cdots + a_{1N}\tilde{Y}_N \\ &\quad = b_{11}\tilde{X}_1 + b_{12}\tilde{X}_2 + \cdots + b_{1M}\tilde{X}_M, \\ &a_{21}\tilde{Y}_1 + a_{22}\tilde{Y}_2 + \cdots + a_{2N}\tilde{Y}_N \\ &\quad = b_{21}\tilde{X}_1 + b_{22}\tilde{X}_2 + \cdots + b_{2M}\tilde{X}_M, \\ &\vdots \\ &a_{N1}\tilde{Y}_1 + a_{N2}\tilde{Y}_2 + \cdots + a_{NN}\tilde{Y}_N \\ &\quad = b_{N1}\tilde{X}_1 + b_{N2}\tilde{X}_2 + \cdots + b_{NM}\tilde{X}_M, \end{aligned} \tag{12.54}$$

where the $\tilde{X}$'s are the phasors for the source currents voltages, the $\tilde{Y}$'s are unknown phasor currents or voltages, and the a's and b's are (in general) functions of frequency and circuit parameters (impedances or admittances), but are independent of the known and unknown currents and voltages. We have temporarily ignored the fact that the various sources and responses might have different frequencies, in which case we cannot actually compute the indicated sums. (A sum of phasors representing sinusoids having different frequencies is meaningless.) However, we remedy this temporary oversight in what follows.

Formally, the solution of (12.54) has the form

$$\begin{aligned} &\tilde{Y}_1 = c_{11}\tilde{X}_1 + c_{12}\tilde{X}_2 + \cdots + c_{1M}\tilde{X}_M, \\ &\tilde{Y}_2 = c_{21}\tilde{X}_1 + c_{22}\tilde{X}_2 + \cdots + c_{2M}\tilde{X}_M, \\ &\vdots \\ &\tilde{Y}_N = c_{N1}\tilde{X}_1 + c_{N2}\tilde{X}_2 + \cdots + c_{NM}\tilde{X}_M, \end{aligned} \tag{12.55}$$

where the c's are functions of frequency and the circuit parameters, but do not depend upon the source currents and voltages (the $\tilde{X}$'s). Each term on the right side of each equation in (12.55) is that part of the associated response due to a particular source. For example, $c_{12}\tilde{X}_2$ is that part of the response $\tilde{Y}_1$ that is due to the source $\tilde{X}_2$.

If all sources have the same frequency f_0, then we can form the sums on the right side of (12.55) and obtain

$$\begin{aligned} y_1(t) &= |\tilde{Y}_1| \cos(\omega_0 t + \measuredangle\tilde{Y}_1), \\ y_2(t) &= |\tilde{Y}_2| \cos(\omega_0 t + \measuredangle\tilde{Y}_2), \\ &\vdots \\ y_N(t) &= |\tilde{Y}_N| \cos(\omega_0 t + \measuredangle\tilde{Y}_N). \end{aligned} \tag{12.56}$$

If the sources do not all have the same frequency, we must convert each phasor on the right side of (12.55) to the time domain before forming the indicated sums. In general, if all have different frequencies, we would define

$$\begin{aligned} &\tilde{Y}_{11} = c_{11}\tilde{X}_1, \tilde{Y}_{12} = c_{12}\tilde{X}_2, \ldots, \tilde{Y}_{1M} = c_{1M}\tilde{X}_M; \\ &\tilde{Y}_{21} = c_{21}\tilde{X}_1, \tilde{Y}_{22} = c_{22}\tilde{X}_2, \ldots, \tilde{Y}_{2M} = c_{2M}\tilde{X}_M; \\ &\vdots \\ &\tilde{Y}_{N1} = c_{N1}\tilde{X}_1, \tilde{Y}_{N2} = c_{N2}\tilde{X}_2, \ldots, \tilde{Y}_{NM} = c_{NM}\tilde{X}_M. \end{aligned} \tag{12.57}$$

where $\tilde{Y}_{nm}$ is the phasor for the sinusoidal component of $y_n(t)$ due to the sinusoidal source $x_m(t)$. The two phasors in each of the relations in (12.57) (e.g., $\tilde{Y}_{11}$ and $\tilde{X}_1$) represent sinusoids having the same frequency. We can convert each phasor to the equivalent time-domain sinusoidal function to obtain

$$\begin{aligned} y_{11}(t) &= |\tilde{Y}_{11}| \cos(\omega_1 t + \measuredangle\tilde{Y}_{11}), \\ y_{12}(t) &= |\tilde{Y}_{12}| \cos(\omega_2 t + \measuredangle\tilde{Y}_{12}), \ldots, \\ y_{1M}(t) &= |\tilde{Y}_{1M}| \cos(\omega_M t + \measuredangle\tilde{Y}_{1M}), \\ y_{21}(t) &= |\tilde{Y}_{21}| \cos(\omega_1 t + \measuredangle\tilde{Y}_{21}), \\ y_{22}(t) &= |\tilde{Y}_{22}| \cos(\omega_2 t + \measuredangle\tilde{Y}_{22}), \ldots, \\ y_{2M}(t) &= |\tilde{Y}_{2M}| \cos(\omega_M t + \measuredangle\tilde{Y}_{2M}), \\ &\vdots \\ y_{N1}(t) &= |\tilde{Y}_{N1}| \cos(\omega_1 t + \measuredangle\tilde{Y}_{N1}), \\ y_{N2}(t) &= |\tilde{Y}_{N2}| \cos(\omega_2 t + \measuredangle\tilde{Y}_{N2}), \ldots, \\ y_{NM}(t) &= |\tilde{Y}_{NM}| \cos(\omega_M t + \measuredangle\tilde{Y}_{NM}). \end{aligned}$$

The following example illustrates the development above.

Example 12.32. Refer to Fig. 12.37. We seek expressions for the currents $\tilde{I}_1$, $\tilde{I}_2$.
Solution: We label two paths and associated loop currents, as shown in Fig. 12.37. We write Kirchhoff's voltage law around each path to obtain

$$\begin{aligned} &Z_1\tilde{I}_1 + Z_3(\tilde{I}_1 - \tilde{I}_2) = \tilde{V}_a \\ &\Rightarrow (Z_1 + Z_3)\tilde{I}_1 - Z_3\tilde{I}_2 = \tilde{V}_a, \\ &Z_2\tilde{I}_2 - Z_3(\tilde{I}_1 - \tilde{I}_2) = -\tilde{V}_b \\ &\Rightarrow -Z_3\tilde{I}_1 + (Z_2 + Z_3)\tilde{I}_2 = -\tilde{V}_b. \end{aligned} \tag{12.58}$$

The right sides of (12.58) have the form given in (12.54), with

$$\begin{aligned} &\tilde{Y}_1 = \tilde{I}_1,\ \tilde{Y}_2 = \tilde{I}_2,\ \tilde{X}_1 = \tilde{V}_a,\ \tilde{X}_2 = \tilde{V}_b, \\ &a_{11} = a_{22} = Z_1 + Z_3,\ a_{12} = a_{21} = -Z_3, \\ &b_{11} = 1,\ b_{22} = -1. \end{aligned}$$

The solution of (12.58) is

$$\begin{aligned} \tilde{I}_1 &= \left[\frac{(Z_2 + Z_3)}{Z_1 Z_2 + Z_1 Z_3 + Z_2 Z_3}\right]\tilde{V}_a \\ &\quad + \left[\frac{-Z_3}{Z_1 Z_2 + Z_1 Z_3 + Z_2 Z_3}\right]\tilde{V}_b, \\ \tilde{I}_2 &= \left[\frac{Z_3}{Z_1 Z_2 + Z_1 Z_3 + Z_2 Z_3}\right]\tilde{V}_a \\ &\quad + \left[\frac{-(Z_1 + Z_3)}{Z_1 Z_2 + Z_1 Z_3 + Z_2 Z_3}\right]\tilde{V}_b, \end{aligned} \tag{12.59}$$

which has the form given in (12.55), with

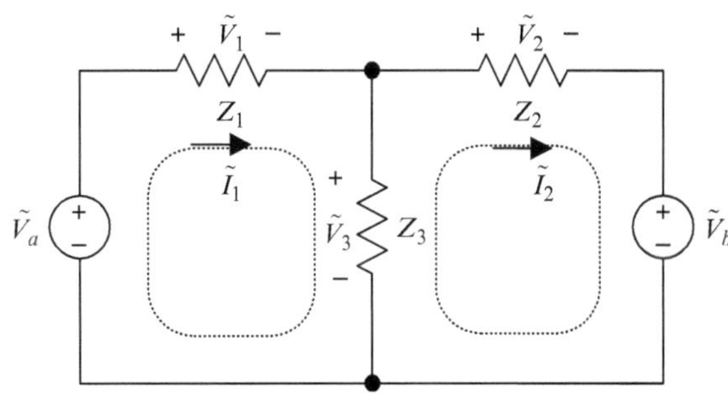

Fig. 12.37 See Example 12.32

$$c_{11} = \left[\frac{(Z_2+Z_3)}{Z_1 Z_2 + Z_1 Z_3 + Z_2 Z_3}\right],$$
$$c_{12} = \left[\frac{-Z_3}{Z_1 Z_2 + Z_1 Z_3 + Z_2 Z_3}\right],$$
$$c_{21} = \left[\frac{Z_3}{Z_1 Z_2 + Z_1 Z_3 + Z_2 Z_3}\right],$$
$$c_{22} = \left[\frac{-(Z_1+Z_3)}{Z_1 Z_2 + Z_1 Z_3 + Z_2 Z_3}\right].$$

If $\tilde{V}_a$ and $\tilde{V}_b$ have the same frequency, then both $\tilde{I}_1$ and $\tilde{I}_2$ also have that frequency, and we may use (12.59) as it stands. But if $\tilde{V}_a$ and $\tilde{V}_b$ have different frequencies, then we must rewrite (12.59) as

$$\tilde{I}_{1a} = \left[\frac{(Z_2+Z_3)}{Z_1 Z_2 + Z_1 Z_3 + Z_2 Z_3}\right]\tilde{V}_a,$$
$$\tilde{I}_{1b} = \left[\frac{-Z_3}{Z_1 Z_2 + Z_1 Z_3 + Z_2 Z_3}\right]\tilde{V}_b,$$
$$\tilde{I}_{2a} = \left[\frac{Z_3}{Z_1 Z_2 + Z_1 Z_3 + Z_2 Z_3}\right]\tilde{V}_a,$$
$$\tilde{I}_{2b} = \left[\frac{-(Z_1+Z_3)}{Z_1 Z_2 + Z_1 Z_3 + Z_2 Z_3}\right]\tilde{V}_b. \quad (12.60)$$

The next two examples further illustrate these results.

Example 12.33. In the circuit treated in Example 12.32, let $Z_1 = j\omega L$, with $L = 50$ mH, $Z_2 = (j\omega C)^{-1}$, with $C = 400$ nF, and $Z_3 = 300\ \Omega$. Find $i_a(t)$ and $i_b(t)$ if: (a), the excitations have the same frequency, and are given by

$$v_a(t) = V_a \cos(\omega_0 t),\ v_b(t) = V_b \cos(\omega_0 t + \theta),$$

with $f_0 = 1$ kHz, $V_a = 10$ V, $V_b = 5$ V, $\theta = 0.5$; and (b) the excitations have different frequencies and are given by

$$v_a(t) = V_a \cos(\omega_0 t),\ v_b(t) = V_b \cos(2\omega_0 t + \theta),$$

with $f_0 = 1$ kHz, $V_a = 10$ V, $V_b = 5$ V, $\theta = 0.5$.
Solution: From Example 12.32, we have

$$\tilde{I}_{1a} = c_{11}\tilde{V}_a, \quad \tilde{I}_{1b} = c_{12}\tilde{V}_b,$$
$$\tilde{I}_{2a} = c_{21}\tilde{V}_a, \quad \tilde{I}_{2b} = c_{22}\tilde{V}_b,$$

where

$$c_{11} = \left[\frac{Z_2+Z_3}{Z_1 Z_2 + Z_1 Z_3 + Z_2 Z_3}\right],$$
$$c_{12} = \left[\frac{-Z_3}{Z_1 Z_2 + Z_1 Z_3 + Z_2 Z_3}\right],$$
$$c_{21} = \left[\frac{Z_3}{Z_1 Z_2 + Z_1 Z_3 + Z_2 Z_3}\right],$$
$$c_{22} = \left[\frac{-(Z_1+Z_3)}{Z_1 Z_2 + Z_1 Z_3 + Z_2 Z_3}\right],$$

with

$$Z_1 = j2\pi f L,\quad Z_2 = (j2\pi f C)^{-1},\quad Z_3 = 300\,\Omega.$$

For Part (a), the excitations have the same frequency $f_0 = 1$ kHz, so

$$\tilde{I}_1 = c_{11}(f_0)\tilde{V}_a + c_{12}(f_0)\tilde{V}_b,$$
$$\tilde{I}_2 = c_{21}(f_0)\tilde{V}_a + c_{22}(f_0)\tilde{V}_b,$$

with

$$\tilde{V}_a = 10\ \text{V},\quad \tilde{V}_b = (5\angle 0.5)\ \text{V}.$$

We find

$$c_{11}(f_0) = (2.922 - j2.596)\ \text{mS},$$
$$c_{12}(f_0) = (-2.307 - j0.464)\ \text{mS},$$
$$c_{21}(f_0) = (2.307 + j0.464)\ \text{mS},$$
$$c_{22}(f_0) = (-1.821 - j2.879)\ \text{mS},$$

and so

$$\tilde{I}_1 = (39.142\angle -1.028)\ \text{mA}$$
$$\Rightarrow i_a(t) = 39.142\cos(\omega_0 t - 1.028)\ \text{mA},$$
$$\tilde{I}_2 = (25.218\angle -0.512)\ \text{mA}$$
$$\Rightarrow i_b(t) = 25.218\cos(\omega_0 t - 0.512)\ \text{mA},$$

where $v_a(t) = V_a\cos(\omega_0 t)$ is the phase reference.

In Part (b), the excitations have different frequencies: $f_0 = 1$ kHz for $v_a(t)$ and $2f_0 = 2$ kHz for $v_b(t)$. Thus we have

$$\tilde{I}_{1a} = c_{11}(f_0)\,\tilde{V}_a,\ \tilde{I}_{1b} = c_{12}(2f_0)\,\tilde{V}_b,$$
$$\tilde{I}_{2a} = c_{21}(f_0)\,\tilde{V}_a,\ \tilde{I}_{2b} = c_{22}(2f_0)\,\tilde{V}_b,$$

where $c_{11}(f_0)$ and $c_{21}(f_0)$ have the values given above, but

$$c_{12}(2f_0) = (-1.164 + j1.199)\ \text{mS},$$
$$c_{22}(2f_0) = (-3.676 - j1.238)\ \text{mS},$$

and so

$$\tilde{I}_{1a} = c_{11}(f_0)\,\tilde{V}_a = (39.084\angle -0.726)\ \text{mA}$$
$$\Rightarrow i_{1a}(t) = 39.084\cos(\omega_0 t - 0.726)\ \text{mA},$$
$$\tilde{I}_{1b} = c_{12}(2f_0)\,\tilde{V}_b = (8.357\angle 2.841)\ \text{mA}$$
$$\Rightarrow i_{1b}(t) = 8.357\cos(2\omega_0 t + 2.841)\ \text{mA},$$
$$\tilde{I}_{2a} = c_{21}(f_0)\,\tilde{V}_a = (25.530\angle 0.198)\ \text{mA}$$
$$\Rightarrow i_{2a}(t) = 25.530\cos(\omega_0 t + 0.198)\ \text{mA},$$
$$\tilde{I}_{2b} = c_{22}(2f_0)\,\tilde{V}_b = (19.395\angle -2.317)\ \text{mA}$$
$$\Rightarrow i_{2b}(t) = 19.395\cos(2\omega_0 t - 2.317)\ \text{mA}.$$

Finally,

$$\begin{aligned} i_1(t) &= i_{1a}(t) + i_{1b}(t) \\ &= 39.084\cos(\omega_0 t - 0.726)\ \text{mA} \\ &\quad + 8.357\cos(2\omega_0 t + 2.841)\ \text{mA}, \\ i_2(t) &= i_{2a}(t) + i_{2b}(t) \\ &= 25.530\cos(\omega_0 t + 0.198)\ \text{mA} \\ &\quad + 19.395\cos(2\omega_0 t - 2.317)\ \text{mA}. \end{aligned}$$

Equations obtained by applying Kirchhoff's laws to *any sinusoidally excited RLC circuit* can always be put in the form (12.54) and the solution for such a set can *always* be written in the form (12.55), provided the solution is interpreted as indicated by (12.57). Equation (12.55) [or, e.g., (12.59)] expresses each current and voltage in a sinusoidally excited *RLC* circuit as a *linear combination* of source currents and voltages. To find a particular current or voltage due to a particular source current or voltage, we set all source currents and voltages except the one of interest to zero. For example, with reference to (12.59) in the example above, the phasor for that part of the current $i_2(t)$ that can be attributed to the source $v_b(t)$ is given by

$$\tilde{I}_2\big|_{\tilde{V}_a=0} = \left[\frac{-(Z_1 + Z_3)}{Z_1 Z_2 + Z_1 Z_3 + Z_2 Z_3}\right]\tilde{V}_b. \qquad (12.61)$$

Similarly, the phasor for that part of the current $i_2(t)$ that can be attributed to the source $v_a(t)$ is given by

$$\tilde{I}_2\big|_{\tilde{V}_b=0} = \left[\frac{Z_3}{Z_1 Z_2 + Z_1 Z_3 + Z_2 Z_3}\right]\tilde{V}_a. \qquad (12.62)$$

If $v_a(t)$ and $v_b(t)$ have the same frequency, the phasor for the (total) current can be obtained by adding the right sides of (12.61) and (12.62), in agreement with the principle of superposition. If $v_a(t)$ and $v_b(t)$ have different frequencies, an expression for the total current can be obtained by converting the individual phasors to the time domain and adding the sinusoidal functions in the time-domain, again in agreement with the principle of superposition.

Equations (12.54) are linear equations, *RLC* circuits are linear circuits, and superposition is a consequence of linearity. Superposition greatly facilitates analysis of circuits excited by several sinusoidal sources having *different frequencies*, and *mainly in cases where all sources are applied to a single terminal pair (an input port)*. In such cases, it is essential to express impedances and admittances as functions of frequency (not numerically), so they can be computed individually for each sinusoidal component of the excitation.

Example 12.34. In Fig. 12.38, $i(t) = I_1\cos(\omega_1 t) + I_2\cos(\omega_2 t + \theta)$, where $\omega_1 \neq \omega_2$. Use superposition to obtain an expression for the voltage v.
Solution: Let $\tilde{I}_1 = I_1\angle 0$, $\tilde{I}_2 = I_2\angle\theta$ and obtain expressions for the responses to these excitations

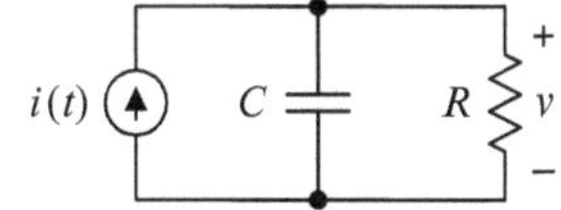

Fig. 12.38 See Example 12.34

individually, using the appropriate frequency in each case. Kirchhoff's current law yields

$$\frac{\tilde{V}}{R} + j\omega C\tilde{V} = \tilde{I}$$
$$\Rightarrow \tilde{V} = \frac{R}{1 + j\omega CR}\tilde{I}. \qquad (12.63)$$

The individual phasors for the individual responses are

$$\tilde{V}_1 = \frac{R}{1 + j\omega_1 CR}\tilde{I}_1$$
$$\Rightarrow v_1(t) = |\tilde{V}_1|\cos(\omega_1 t + \measuredangle\tilde{V}_1),$$
$$\tilde{V}_2 = \frac{R}{1 + j\omega_2 CR}\tilde{I}_2$$
$$\Rightarrow v_2(t) = |\tilde{V}_2|\cos(\omega_2 t + \theta + \measuredangle\tilde{V}_2). \qquad (12.64)$$

By superposition,

$$v(t) = v_1(t) + v_2(t)$$
$$= |\tilde{V}_1|\cos(\omega_1 t + \measuredangle\tilde{V}_1)$$
$$+ |\tilde{V}_2|\cos(\omega_2 t + \theta + \measuredangle\tilde{V}_2). \qquad (12.65)$$

Example 12.35. Refer to Fig. 12.39(a), where $R = 10\ \text{k}\Omega$, $C = 15.9$ nF and

$$v_{in}(t) = V_0 + V_1\cos(\omega_0 t)$$
$$+ V_2\cos(2\omega_0 t - \pi/6);$$
$$f_0 = 1\ \text{kHz}.$$

(a) Obtain an expression for the output $v_{out}(t)$. (b) Calculate the amplitudes and phases of the individual components of the output for $V_0 = V_1 = V_2 = 5$ V.

Solution: The best approach to a problem of this kind, where a number of sinusoidal inputs are applied at a single input port, is to first express the output phasor for a single sinusoidal input as a function of the frequency of the input, use that relation to compute the output phasor for each component of the input, and add the results (superposition). From Fig. 12.39(b), by voltage division, the phasor output $\tilde{V}_{out}$ for phasor input $\tilde{V}$ is given by

$$\tilde{V}_{out} = \frac{Z_C}{Z_C + R}\tilde{V} = \frac{\tilde{V}}{1 + j\omega RC}. \qquad (12.66)$$

Next, we use (12.66) to find the phasor output for each component of the input (individually). It is helpful to think of the necessary calculations in tabular form, as below:[19]

Frequency of input (and output)	Input phasor	Output phasor
$\omega = 0$	$\tilde{V} = \tilde{V}_0 = 5$ V	$\tilde{V}_{out} = \frac{\tilde{V}_0}{1 + j0} = 5$ V
$\omega = \omega_0$ $= 2\pi f_0$	$\tilde{V} = \tilde{V}_1 = V_0 = 5$ V	$\tilde{V}_{out} = \frac{\tilde{V}_1}{1 + j\omega_0 RC}$ $= \frac{5\ \text{V}}{1 + j}$ $= \frac{5\ \text{V}}{\sqrt{2}\angle(\pi/4)}$ $= \frac{5}{\sqrt{2}}\angle(-\pi/4)$ V
$\omega = 2\omega_0$ $= 4\pi f_0$	$\tilde{V} = \tilde{V}_2$ $= 5\angle(-\pi/6)$ V	$\tilde{V}_{out} = \frac{\tilde{V}_2}{1 + j2\omega_0 RC}$ $= \frac{5\angle(-\pi/6)}{1 + j2}$ V $= \frac{5\angle(-\pi/6)}{\sqrt{5}\angle 1.11}$ V $= \sqrt{5}\angle(-1.63)$ V

Finally, we convert each phasor component of the output to a sinusoid having the corresponding frequency and add the results. Thus

$$v_{out}(t) = \left[5 + \frac{5}{\sqrt{2}}\cos\left(\omega_0 t - \frac{\pi}{4}\right)\right.$$
$$\left. + \sqrt{5}\cos(2\omega_0 t - 1.63)\right]\ \text{V};$$
$$f_0 = 1\ \text{kHz}.$$

[19] Because the standard form for a sinusoid is a cosine, a dc component can be treated as a sinusoid whose frequency is zero.

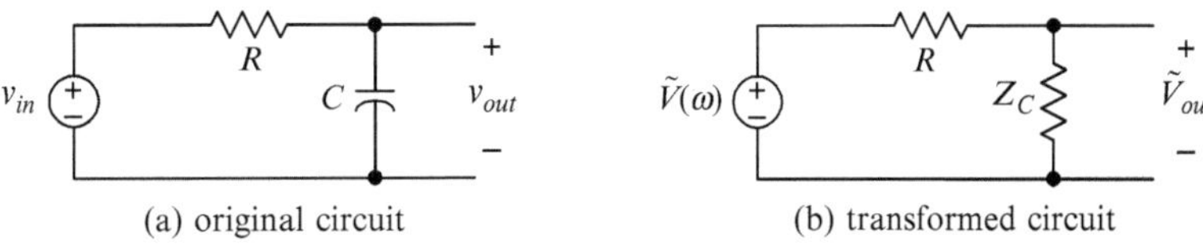

Fig. 12.39 See Example 12.35

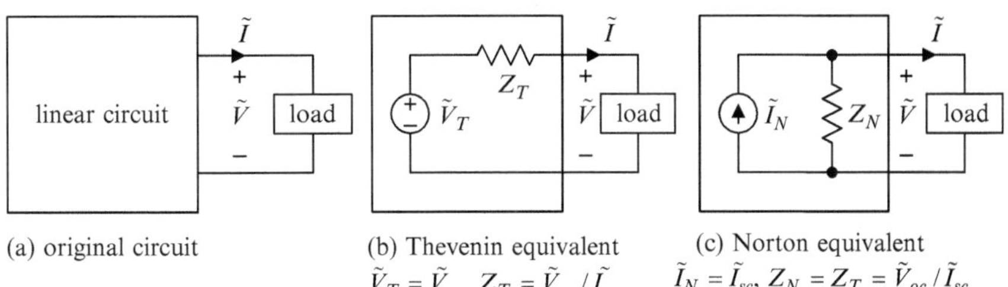

Fig. 12.40 Thévenin and Norton equivalents for a one-port linear circuit

12.17 Thévenin and Norton Equivalent Circuits: Source Transformations

The statement of Thévenin's (or Norton's) theorem for a circuit in sinusoidal steady state is essentially identical to that for a resistive circuit if the terminal characteristic is expressed in terms of phasors. Refer to Fig. 12.40(a). The terminal characteristic at a terminal pair (a port) of a linear circuit in sinusoidal steady state can be expressed as

$$\frac{\tilde{I}}{\tilde{I}_{sc}} = 1 - \frac{\tilde{V}}{\tilde{V}_{oc}}, \qquad (12.67)$$

where $\tilde{I}_{sc}$, $\tilde{V}_{oc}$ are the phasor representations of the short-circuit current and the open-circuit voltage, respectively.

With the definitions

$$\tilde{V}_T = \tilde{V}_{oc},\ Z_T = \frac{\tilde{V}_{oc}}{\tilde{I}_{sc}},\ \tilde{I}_N = \tilde{I}_{sc},\ Z_N = Z_T = \frac{\tilde{V}_{oc}}{\tilde{I}_{sc}}, \qquad (12.68)$$

Equation (12.67) can be written as

$$\tilde{V} = \tilde{V}_T - \tilde{I} Z_T, \qquad (12.69)$$

which has the form of a Kirchhoff's voltage law equation and suggests the circuit model shown in Fig. 12.40(b), or as

$$\tilde{I} = \tilde{I}_N - \frac{\tilde{V}}{Z_N}, \qquad (12.70)$$

which has the form of a Kirchhoff's current law equation and suggests the circuit model shown in Fig. 12.40 (c). The (non-physical) impedance $Z_T = Z_N$ is called (variously) the **Thévenin equivalent impedance**, the **Norton equivalent impedance**, the **output impedance** of the circuit, the **source impedance**, or the **internal impedance** of the circuit (regarded as a source), depending upon context. Proof of Thévenin's theorem expressed in terms of phasors for terminal current and voltage is virtually identical to that given for resistive circuits in Chapter 4. Note that *the same phase reference must be used when finding the open-circuit voltage and short-current current.* Otherwise, the angle of the Thévenin (Norton) impedance can be incorrect. In other words, if a circuit is driven by more than one sinusoidal source (all having the same frequency), use the same one as the phase reference for both the open-circuit and the short-circuit calculations.

Thévenin's and Norton's theorems justify source transformations for circuits in sinusoidal steady state. Such transformations are carried out (symbolically) exactly as they are for resistive circuits, but using phasor currents and voltages and impedances instead of actual (time-domain) currents and voltages and resistances.

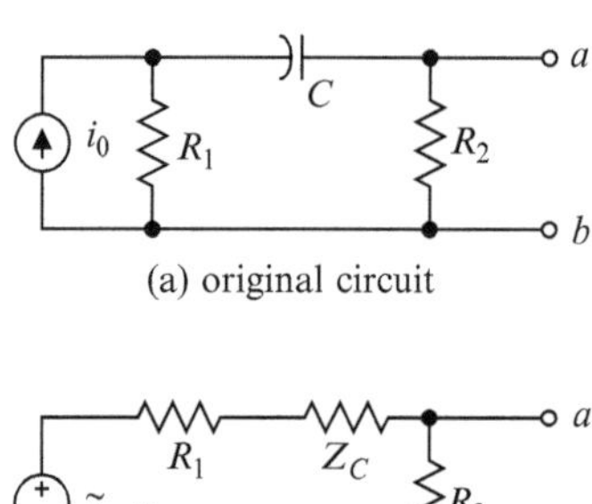

(a) original circuit

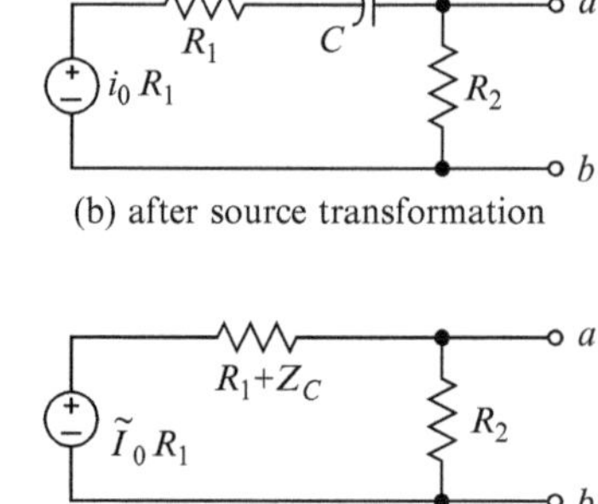

(b) after source transformation

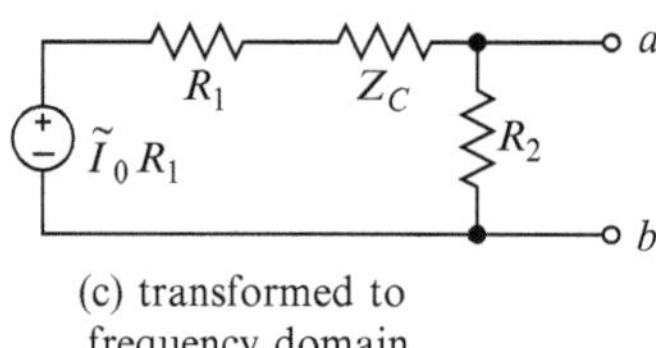

(c) transformed to frequency domain

(d) reduced to voltage divider

Fig. 12.41 See Example 12.36

Usually, it is best to express Thévenin source voltages, Norton source currents, and Thévenin or Norton equivalent impedances symbolically, as functions of frequency and circuit parameters. If the source voltage or current and equivalent impedance are expressed numerically, the model is valid for only the particular frequency for which those quantities are calculated.

Example 12.36. Refer to Fig. 12.41(a), where $i_0 = I_0 \cos(\omega t)$. Obtain the Thévenin equivalent for the circuit at the terminals *a–b*. Express the Thévenin source voltage and the Thévenin impedance in terms of frequency and circuit parameters.

Solution: Figure 12.41(b)–(d) shows steps in the solution, where

$$\tilde{I}_0 = I_0\angle 0,\ Z_C = \frac{1}{j\omega C}.$$

From Fig. 12.41(d), we obtain

$$\tilde{V}_{oc} = \frac{R_2 \tilde{I}_0 R_1}{R_1 + Z_C + R_2},\ \tilde{I}_{sc} = \frac{\tilde{I}_0 R_1}{R_1 + Z_C}.$$

It follows that

$$\tilde{V}_T = \tilde{V}_{oc} = \frac{R_2 \tilde{I}_0 R_1}{R_1 + Z_C + R_2} = \frac{j\omega C R_1 R_2 I_0}{1 + j\omega C(R_1 + R_2)},$$

$$Z_T = \frac{\tilde{V}_{oc}}{\tilde{I}_{sc}} = \frac{(R_1 + Z_C)R_2}{R_1 + Z_C + R_2} = \frac{R_2(1 + j\omega C R_1)}{1 + j\omega C(R_1 + R_2)}.$$

As one check on this result, we note that if the capacitance is set to zero, then from the expression above

$$Z_T = \frac{R_2(1+0)}{1+0} = R_2.$$

If $C = 0$ then $Z_C = 1/(j\omega C) \to \infty$ (the capacitor is replaced by an open circuit), the impedance seen at the terminals *a–b* equals R_2, and $Z_T = R_2$, in agreement with the value given under the same condition by the expression above.

Example 12.37. Refer to Fig. 12.42. (a) Obtain expressions for the Thévenin equivalent source voltage and the Thévenin equivalent impedance at the terminals *a–b*. (b) Calculate the *values* of the Thévenin source voltage and Thévenin equivalent impedance. Express the Thévenin source voltage in polar form and the Thévenin equivalent impedance in rectangular form. (c) Represent the Thévenin equivalent impedance as a resistor in series with an inductor or capacitor, as appropriate, and give the values of the effective resistance and inductance or capacitance.

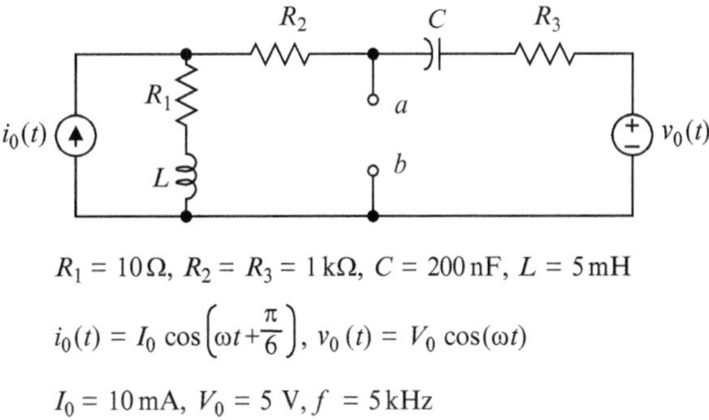

Fig. 12.42 See Example 12.37

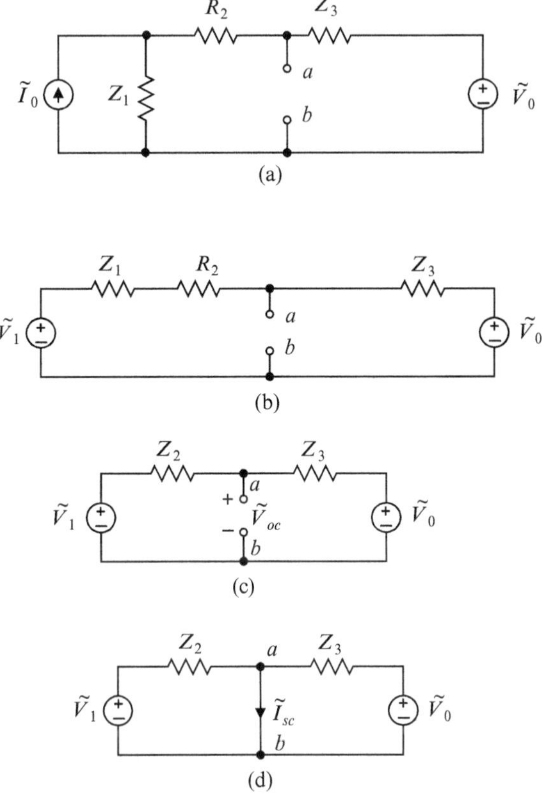

Fig. 12.43 See Example 12.37

Solution: Figure 12.43 illustrates steps in the solution. We first transform the circuit, replacing sources by their phasor representations and elements by their impedances, to obtain the circuit shown in Fig. 12.43(a), where

$$Z_1 = R_1 + j\omega L = 10.0 + j157\ \Omega,\ Z_3 = R_3 + \frac{1}{j\omega C} = 1.00 \times 10^3 - j159\ \Omega,$$
$$\tilde{I}_0 = I_0\angle(\pi/6) = 10\angle(\pi/6)\ \text{mA},\ \tilde{V}_0 = V_0\angle 0 = 5\angle 0\ \text{V}.$$

We use a source transformation to obtain the circuit shown in Fig. 12.43(b), where

$$\tilde{V}_1 = \tilde{I}_0 Z_1 = 1.57\angle 2.03 \text{ V}.$$

We combine the impedance Z_1 and the resistor R_2 to obtain the circuit shown in Fig. 12.43(c), where

$$Z_2 = Z_1 + R_2.$$

To find the open-circuit voltage, we apply Kirchhoff's current law to node a. This gives

$$\frac{\tilde{V}_{oc} - \tilde{V}_1}{Z_2} + \frac{\tilde{V}_{oc} - \tilde{V}_0}{Z_3} = 0 \Rightarrow \tilde{V}_{oc} = \frac{\tilde{V}_1 Z_3 + \tilde{V}_0 Z_2}{Z_3 + Z_2} = 2.55\angle 0.468 \text{ V}.$$

To find the short-circuit current, we short the terminals a–b, as shown in Fig. 12.43(d), and again apply Kirchhoff's current law to node a. This gives

$$-\frac{\tilde{V}_1}{Z_2} - \frac{\tilde{V}_0}{Z_3} + \tilde{I}_{sc} = 0 \Rightarrow \tilde{I}_{sc} = \frac{\tilde{V}_1 Z_3 + \tilde{V}_0 Z_2}{Z_3 Z_2} = 4.95\angle 0.471 \text{ mA}.$$

The Thévenin source voltage is

$$\tilde{V}_T = V_{oc} = 2.55\angle 0.468 \text{ V}.$$

The Thévenin equivalent impedance is

$$Z_T = \frac{\tilde{V}_{oc}}{\tilde{I}_{sc}} = \frac{Z_2 Z_3}{Z_2 + Z_3} = 515\angle -0.003 \ \Omega.$$

Figure 12.44 Shows the Thévenin equivalent circuit, where (summary)

$$Z_1 = R_1 + j\omega L, \ Z_2 = Z_1 + R_2, \ Z_3 = R_3 + \frac{1}{j\omega C},$$

$$\tilde{V}_T = \frac{\tilde{I}_0 Z_1 Z_3 + \tilde{V}_0 Z_2}{Z_2 + Z_3}, \ Z_T = \frac{Z_2 Z_3}{Z_2 + Z_3}.$$

(b) From part (a)

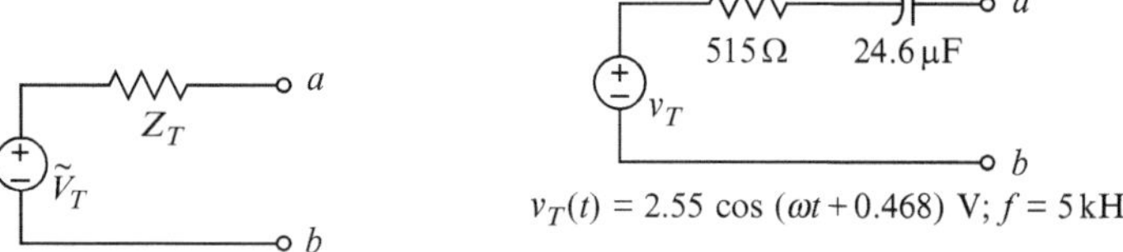

Fig. 12.44 See Example 12.37

Fig. 12.45 See Example 12.37

$$Z_1 = R_1 + j\omega L = 10.0 + j\,157\ \Omega = 157\angle 1.51\ \Omega,$$
$$Z_2 = Z_1 + R_2 \cong 1010 + j\,157\ \Omega = 1.02\angle 0.154\ \text{k}\Omega,$$
$$Z_3 = R_3 + \frac{1}{j\omega C} \cong 1.00 \times 10^3 - j\,159\ \Omega = 1.01\angle -0.158\ \text{k}\Omega,$$
$$\tilde{I}_0 = I_0\angle\frac{\pi}{6} = 10\angle\frac{\pi}{6}\ \text{mA}, \tilde{V}_0 = V_0\angle 0 = 5\angle 0\ \text{V},$$
$$\tilde{V}_T = \frac{\tilde{I}_0 Z_1 Z_3 + \tilde{V}_0 Z_2}{Z_2 + Z_3}$$
$$\cong \frac{(10\ \text{mA}\angle\pi/6)(157\ \Omega\angle 1.51)(1.01\ \text{k}\Omega\angle -0.156) + (5\ \text{V}\angle 0)(1.02\ \text{k}\Omega\angle 0.154)}{(1010 + j\,157)\ \Omega + (1.00 \times 10^3 - j159)\ \Omega}$$
$$\cong 2.55\ \text{V}\angle 0.468\ \text{V},$$
$$Z_T = \frac{Z_2 Z_3}{Z_2 + Z_3} \cong \frac{(1.02\ \text{k}\Omega\angle 0.154)(1.01\ \text{k}\Omega\angle -1.58)}{(1010 + j\,157)\ \Omega + (1.00 \times 10^3 - j159)\ \Omega} \cong 515\angle -0.0025\ \Omega$$
$$\cong 515 - j1.29\ \Omega$$

(c) The Thévenin equivalent reactance is negative, so the equivalent impedance (at the specified frequency) is a series connection of a resistor and capacitor:

$$Z_T = R_T + jX_T; \quad X_T = -\frac{1}{j\omega C_T} \cong -1.29\ \Omega,$$
$$R_T \cong 515\ \Omega, \qquad C_T = -\frac{1}{\omega X_T} \cong 24.6\ \mu\text{F}$$

Figure 12.45 shows the equivalent circuit (for $f = 5$ kHz).

Exercise 12.22. In Fig. 12.46, $v_S = V_S \cos(\omega_0 t)$, with $V_S = 5$ V and $f = 20$ kHz. Find the values of the Thévenin equivalent (open circuit) voltage and impedance at the terminals *a–b*.

The Thévenin or Norton equivalent for a circuit can be obtained *from a circuit diagram* by finding the open-circuit voltage and the short-circuit current at the terminals of the circuit. Equation (12.67) is a linear relation between terminal current $\tilde{I}$ and terminal voltage $\tilde{V}$. Given any two pairs of values of current $\tilde{I}$ and voltage $\tilde{V}$ satisfying (12.67), we can determine the short-circuit phasor current $\tilde{I}_{sc}$ and the open-circuit phasor voltage $\tilde{V}_{oc}$. Let $(\tilde{V}_1, \tilde{I}_1)$ and $(\tilde{V}_2, \tilde{I}_2)$ denote two such pairs. Then

$$\frac{\tilde{I}_1}{\tilde{I}_{sc}} + \frac{\tilde{V}_1}{\tilde{V}_{oc}} = 1, \quad \frac{\tilde{I}_2}{\tilde{I}_{sc}} + \frac{\tilde{V}_2}{\tilde{V}_{oc}} = 1, \tag{12.71}$$

which can be written

$$\begin{pmatrix} \tilde{I}_1 & \tilde{V}_1 \\ \tilde{I}_2 & \tilde{V}_2 \end{pmatrix} \begin{pmatrix} \tilde{I}_{sc}^{-1} \\ \tilde{V}_{oc}^{-1} \end{pmatrix} = \begin{pmatrix} 1 \\ 1 \end{pmatrix}$$
$$\Rightarrow \begin{pmatrix} \tilde{I}_{sc}^{-1} \\ \tilde{V}_{oc}^{-1} \end{pmatrix} = \begin{pmatrix} \tilde{I}_1 & \tilde{V}_1 \\ \tilde{I}_2 & \tilde{V}_2 \end{pmatrix}^{-1} \begin{pmatrix} 1 \\ 1 \end{pmatrix}. \tag{12.72}$$

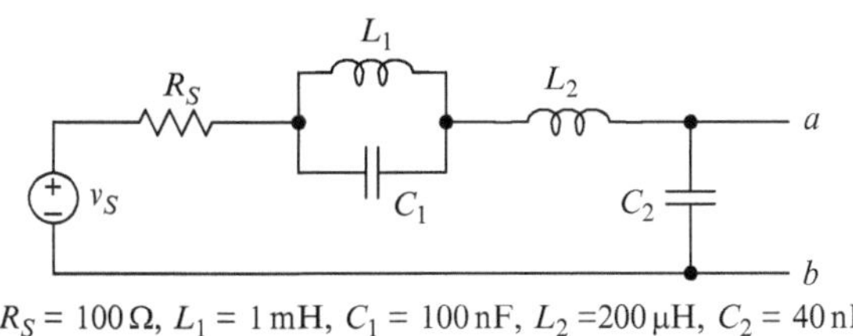

Fig. 12.46 See Exercise 12.22

The solution is

$$\tilde{V}_{oc} = \frac{\tilde{V}_2\tilde{I}_1 - \tilde{V}_1\tilde{I}_2}{\tilde{I}_1 - \tilde{I}_2}, \quad \tilde{I}_{sc} = \frac{\tilde{V}_2\tilde{I}_1 - \tilde{V}_1\tilde{I}_2}{\tilde{V}_2 - \tilde{V}_1}. \quad (12.73)$$

The analysis above culminating with (12.73) seems straightforward, but there is a practical difficulty with (12.73), in that the measurements $(\tilde{V}_1, \tilde{I}_1)$ and $(\tilde{V}_2, \tilde{I}_2)$ require a common phase reference. We cannot simply attach one load to the terminals and observe the load voltage $\tilde{V}_1$, then attach a second and again observe the load voltage $\tilde{V}_2$ because we have no means of assigning initial phases (or a relative phase) to the observed waveforms. The next example illustrates this difficulty.

Example 12.38. Refer to Fig. 12.47, where we seek to find the Thévenin parameters $\tilde{V}_T$, Z_T from measurements made at the terminals *a–b*. Unknown to us, $\tilde{V}_T = 20\angle 0$ V, $Z_T = (2 + j2)$ kΩ, and the frequency of the source is $f_0 = 5$ kHz.

We use an oscilloscope to observe the voltage $\tilde{V}$ for two different values of the load resistances R to obtain

$$R = 1\text{ k}\Omega \Rightarrow \tilde{V} = \tilde{V}_1 = (5.547\angle - 0.588)\text{ V},$$

$$\tilde{I} = \tilde{I}_1 = \frac{\tilde{V}_1}{R_1} = (5.547\angle - 0.588)\text{ mA},$$

$$R = 2\text{ k}\Omega \Rightarrow \tilde{V} = \tilde{V}_2 = (8.944\angle - 0.464)\text{ V},$$

$$\tilde{I} = \tilde{I}_1 = \frac{\tilde{V}_1}{R_1} = (4.472\angle - 0.464)\text{ mA}.$$

From (12.73), we obtain

$$\tilde{V}_{oc} = \frac{\tilde{V}_2\tilde{I}_1 - \tilde{V}_1\tilde{I}_2}{\tilde{I}_1 - \tilde{I}_2} = 20\text{ V},$$

$$\tilde{I}_{sc} = \frac{\tilde{V}_2\tilde{I}_1 - \tilde{V}_1\tilde{I}_2}{\tilde{V}_2 - \tilde{V}_1} = (7.071\angle - 0.785)\text{ mA}.$$

whence

$$\tilde{V}_T = \tilde{V}_{OC} = 20\text{ V}, Z_T = \frac{\tilde{V}_{OC}}{\tilde{I}_{SC}} = (2 + j2)\text{ k}\Omega.$$

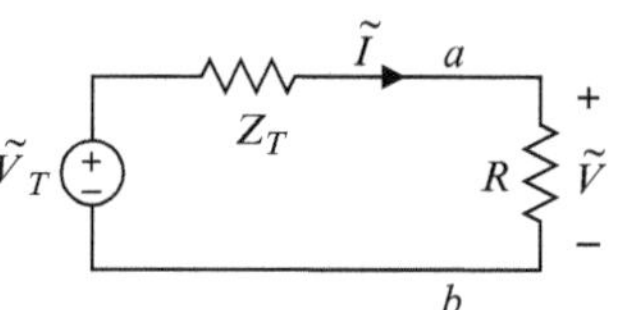

Fig. 12.47 See Example 12.38

All seems well and good. But examine the measurements of $\tilde{V}_1$, $\tilde{V}_2$ above. How did we obtain the relative phases for the two voltages? What and where is the phase reference? If we observe a *single* sinusoidal voltage (e.g., on an oscilloscope), we can obtain the amplitude and frequency, but we cannot determine a unique initial (or relative) phase. So how did we know that $\measuredangle\tilde{V}_1 = -0.588$ and $\measuredangle\tilde{V}_2 = -0.464$? We could not have known that, because we had no phase reference. The "measured" voltages used in this example were in fact secretly calculated using the resistive loads and unknown Thévenin parameters $\tilde{V}_T = 20\angle 0$ V and $Z_T = (2 + j2)$ kΩ.

The point of this example is that in applying (12.73), we must use the same phase reference for $(\tilde{V}_1, \tilde{I}_1)$ and $(\tilde{V}_2, \tilde{I}_2)$. This is easy to do mathematically, given a complete diagram for a circuit at hand, but is not as easy to do experimentally, given a circuit in a black box. For example, simply finding the current through and voltage across two different passive loads doesn't work, as illustrated above, because there is no way to ensure that the source driving the circuit has the same initial phase for both measurements.

The next example illustrates another method for finding a Thévenin equivalent experimentally, where an external source provides the required phase reference.

Example 12.39. Refer to Fig. 12.48, where a voltage source $v_0 = V_0 \cos(\omega t)$ is connected to a two-terminal linear circuit through a variable resistor having total resistance $R = R_1 + R_2 = 1$ kΩ. The circuit's open-circuit voltage is sinusoidal and the frequency of the source v_0 is made to equal that of

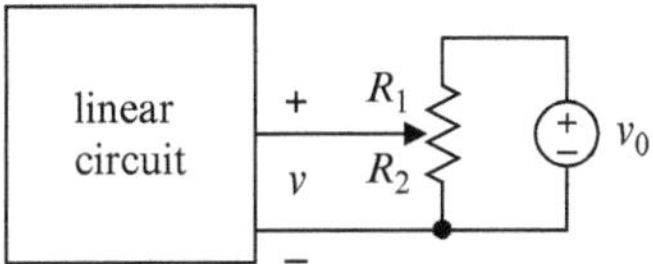

Fig. 12.48 See Example 12.39

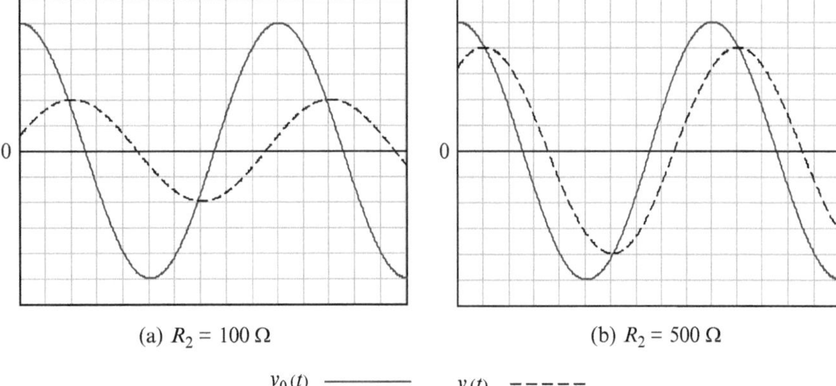

Fig. 12.49 See Example 12.39

the circuit's open-circuit voltage. The voltages v and v_0 are observed on a two-channel oscilloscope, using v_0 as the phase reference. Figure 12.49 shows the resulting oscilloscope traces for two different values of the resistance R_2. Find the Thévenin equivalent for the circuit at the terminals in question.

Solution: Refer to the graphs in Fig. 12.49. The amplitude of the source v_0 is $V_0 = 5$ V and the period of both the source and of the voltage v is 1 ms; thus the frequency of the source v_0 and of the Thévenin equivalent source is 1 kHz. For $R_2 = 100\ \Omega$, the peak amplitude V_1 and phase delay t_1 of the voltage v are $V_1 = 2\ \text{V}, t_1 = 200\ \mu\text{s}$. For $R_2 = 500\ \Omega$, the peak amplitude V_2 and phase delay t_2 of the voltage v are $V_2 = 4\ \text{V}, t_2 = 100\ \mu\text{s}$. It follows that the relative (to v_0) phases are

$$\theta_1 = -2\pi f\, t_1 = -1.257, \theta_2 = -2\pi f\, t_2 = -0.628$$

and the phasors for the measured voltages are

$$\tilde{V}_1 = 2\angle - 1.257\ \text{V}, \tilde{V}_2 = 4\angle - 0.628\ \text{V}.$$

From Kirchhoff's current law, the current exiting the positive terminal of the circuit is given by

$$\tilde{I} = \frac{\tilde{V}}{R_2} + \frac{\tilde{V} - \tilde{V}_0}{R_1},$$

and so

$$\tilde{I}_1 = \frac{\tilde{V}_1}{R_2} + \frac{\tilde{V}_1 - \tilde{V}_0}{R_1} \cong \frac{2\angle - 1.257\ \text{V}}{100\Omega} + \frac{2\angle - 1.257\ \text{V} - 5\angle 0\ \text{V}}{900\Omega} \cong 21.2\angle - 1.51\ \text{mA},$$

$$\tilde{I}_2 = \frac{\tilde{V}_2}{R_2} + \frac{\tilde{V}_2 - \tilde{V}_0}{R_1} \cong \frac{4\angle - 0.628\ \text{V}}{500\Omega} + \frac{4\angle - 0.628\ \text{V} - 5\angle 0\ \text{V}}{500\Omega} \cong 9.85\angle - 1.27\ \text{mA}.$$

From (12.73)

$$\tilde{V}_T = \frac{\tilde{V}_2\tilde{I}_1 - \tilde{V}_1\tilde{I}_2}{\tilde{I}_1 - \tilde{I}_2} = \frac{[(4\angle-0.628)(21.0\angle-1.51) - (2\angle-1.257)(9.86\angle-1.27)](10^{-3})}{(21.0\angle-1.51 - 9.86\angle-1.27)(10^{-3})}$$
$$= 5.65\angle-0.318 \text{ V},$$
$$Z_T = \frac{\tilde{V}_1 - \tilde{V}_2}{\tilde{I}_2 - \tilde{I}_1} = \frac{2\angle-1.257 - 4\angle-0.628}{(9.86\angle-1.27 - 21.0\angle-1.51)\times 10^{-3}} = 224\angle1.54 \cong j224\ \Omega.$$

The initial phase (−0.317) obtained for the Thévenin source is artificial, being relative to the initial phase of the source used for the measurements. In absence of any pre-selected external phase reference, an equally valid Thévenin equivalent source voltage is

$$\tilde{V}_T = 5.65\angle0 \text{ V}.$$

Example 12.40. Figure 12.50 illustrates another method for finding (experimentally) the Thévenin equivalent for a two-terminal linear circuit, when only the terminals of the circuit are accessible and it is known that the open-circuit terminal voltage is sinusoidal. We assume that a variable resistance R, a variable reactance X, and a voltmeter are available. We may make as many measurements as we like at the terminals *a–b*, but none inside the box.

We proceed as follows:

(1) We measure the peak amplitude of the open-circuit voltage, which we denote by V_{OC} We define the open-circuit voltage as the phase reference, so $\tilde{V}_T = V_{OC}\angle0$.
(2) We set the resistance R to any convenient non-zero value and adjust the variable reactance X until the magnitude of the voltage $\tilde{V}$ across the external resistor has its maximum value. At that point, the magnitude of the current $\tilde{I}$ also has its maximum value. Denote that reactance by X_0. Because

$$|\tilde{I}| = \frac{|\tilde{V}_T|}{\sqrt{(R_T+R)^2+(X_T+X)^2}}$$

is maximum for $X_T = -X$, we have determined that $X_T = -X_0$.
(3) Denote the maximum magnitude of the current obtained in part (2) by I_0. For $X = X_0 = -X_T$, the total series impedance equals $R_T + R$. Thus

$$I_0 = \frac{|\tilde{V}_T|}{R_T+R} \Rightarrow R_T = \frac{|\tilde{V}_T|}{I_0} - R = \frac{V_{OC}}{I_0} - R$$

and all parameters of the Thévenin equivalent have been determined.

This approach is problematic if the Thévenin equivalent resistance is large relative to the magnitude of the Thévenin equivalent reactance. In such a case, the magnitude of the voltage across the resistor in step (2) will have a broad maximum, and it can be difficult or even impossible to accurately determine the maximum value (experimentally). To illustrate

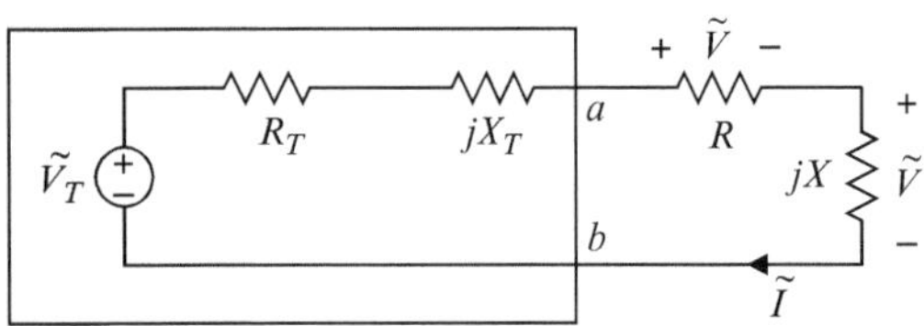

Fig. 12.50 See Example 12.40

this difficulty, we let $\tilde{V}_T = (5\angle 0)$ V, $R = 1\ \Omega$, $X_T = 100\ \Omega$ and plot

$$|\tilde{V}|_{dB} = 20\ \log\left(\left|\frac{\tilde{V}}{\tilde{V}_T}\right|\right)$$

versus $|X|$ for $1\ \Omega \leq |X| \leq 1\ \text{k}\Omega$ and for $R_T = 10\ \Omega,\ 100\ \Omega,\ 1\ \text{k}\Omega$. Figure 12.51 shows the resulting graph. The voltage $|\tilde{V}|_{dB}$ has a sharp maximum for $R_T = 10\ \Omega$ and a broader but discernible maximum for $R_T = 100\ \Omega$, but the maximum for $R_T = 1\ \text{k}\Omega$ is impossible to see. We might work around this difficulty by adding reactance that effectively increases X_T until varying X reveals a sharp peak. But this is troublesome, and there are easier ways to find the Thévenin parameters, such as that illustrated by Example 12.39, above.

Whether a Thévenin (or Norton) equivalent circuit is obtained mathematically from a circuit diagram or experimentally in a laboratory, the initial phase of the Thévenin equivalent voltage will in general be non-zero (e.g., see Example 12.39). Because the Thévenin equivalent source is the only source in the Thévenin equivalent circuit and because the Thévenin source is non-physical to begin with, the non-zero initial phase is essentially an artifice of the computational or experimental method used to obtain the Thévenin equivalent. In using the Thévenin equivalent for subsequent analysis (e.g., to examine signal transfer to a load), we may take the initial phase of the Thévenin source to be zero, which simply makes the Thévenin source the phase reference for the subsequent analysis. For example, suppose we obtain, either experimentally or from a circuit diagram, a Thévenin equivalent circuit for which

$$\tilde{V}_T = 5.8\angle -0.24\ \text{V}, Z_T = 102 + j48\ \Omega.$$

Subsequently, we might wish to compute either the current or power transferred from the circuit to a load $Z_L = R_L + jX_L$ for various values of the effective resistance R_L and the reactance X_L. In either case, we may replace the computed (or measured) Thévenin source with $\tilde{V}_T' = 5.8\angle 0$ V without ill effect. Indeed, we could assign any initial phase whatever to the Thévenin source because the initial phases found for the load current and load voltage are only relative to that of the source.

Figure 12.52 illustrates an important application of a Thévenin (or Norton) equivalent, where a linear two-port circuit containing no independent sources and driven by a sinusoidal source is represented by the Thévenin equivalent. From the viewpoint of the load Z_L, the two-port circuit and the source $\tilde{V}_0$ comprise only a single (Thévenin equivalent) impedance Z_T in series with the Thévenin equivalent source $\tilde{V}_T$. The entire circuit is effectively reduced to a voltage divider. Such a representation greatly simplifies investigating the response of a two-port to various loads.

Example 12.41. Refer to Fig. 12.53. Express the average power dissipated in the load R_L as

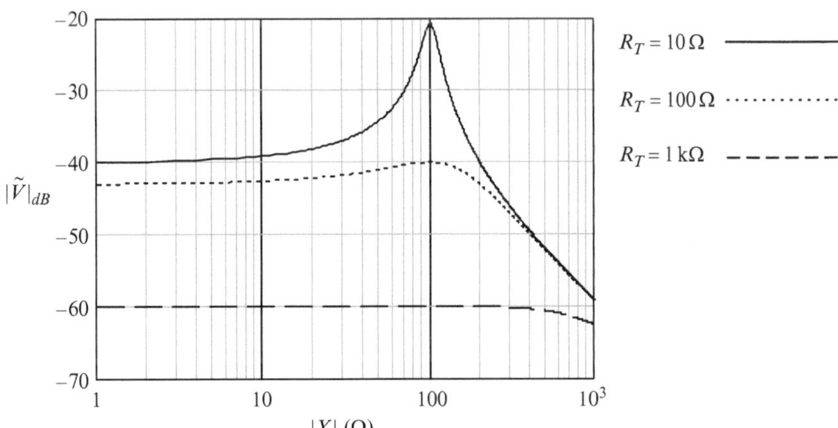

Fig. 12.51 See Example 12.40

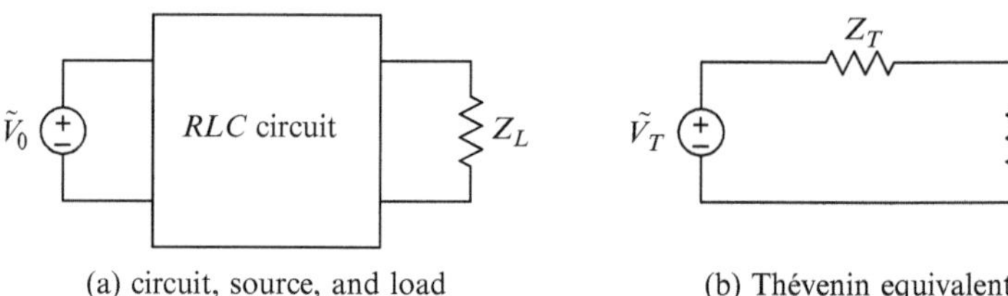

Fig. 12.52 Sinusoidally excited circuit and Thévenin equivalent

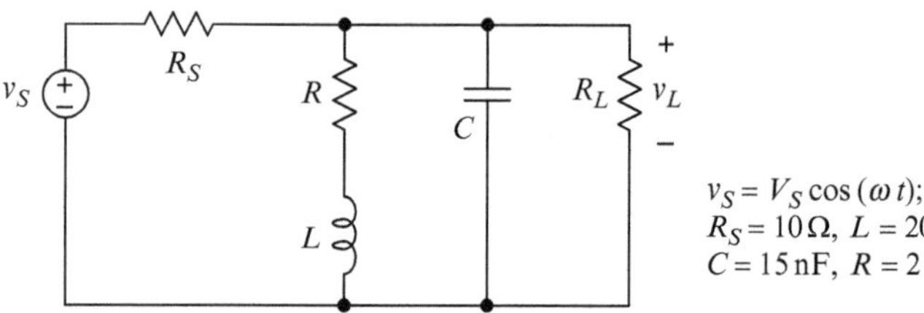

Fig. 12.53 See Example 12.41

a function of R_L and the other circuit parameters. Let the frequency of the source be $f = 30$ kHz and draw a graph of the power dissipated in the load versus the load resistance R_L, for $1\ \Omega \le R_L \le 10\ \mathrm{k}\Omega$. Use a logarithmic scale for the resistance. From the graph, determined (or estimate) the load resistance for which the power dissipated has its maximum value. Then let the load have that resistance and draw a graph of the power dissipated in the load versus the frequency f of the source, for $100\ \mathrm{Hz} \le f \le 100\ \mathrm{MHz}$ and estimate the frequency for which the power dissipated in the load is maximum.

Solution: We first obtain the Thévenin equivalent for the circuit to the left of the load R_L. Kirchhoff's current law gives

$$\tilde{V}_T\left(\frac{1}{R_S} + \frac{1}{R + j\omega L} + j\omega C\right) = \frac{V_S}{R_S}$$
$$\Rightarrow \tilde{V}_T = \frac{(R + j\omega L)V_S}{R_S + R - R_S LC\omega^2 + j\omega(L + R_S RC)},$$

where v_S is the phase reference and $\tilde{V}_T$ is the open-circuit (Thévenin equivalent) voltage. The short-circuit (Norton equivalent) current is given by

$$\tilde{I}_N = \frac{V_S}{R_S},$$

so the Thévenin equivalent impedance is given by

$$Z_T = \frac{\tilde{V}_T}{\tilde{I}_N} = \frac{(R + j\omega L)R_S}{R_S + R - R_S LC\omega^2 + j\omega(L + R_S RC)},$$

where f is the frequency of the source and $\omega = 2\pi f$. The rms *amplitude* of the voltage across a load R_L is given by

$$V_{L\,\mathrm{rms}} = \left|\frac{V_S R_L}{\sqrt{2}(Z_T + R_L)}\right|$$

and the power dissipated by the load is given by

$$P_L = \frac{{V_{L\,\mathrm{rms}}}^2}{R_L}.$$

Figure 12.54 shows a (computer-generated) graph of the power dissipated versus the load resistance. The load resistance that draws maximum power is approximately $R_L = 10\ \Omega$. To see if this result is reasonable, we calculate the Thévenin impedance for $f = 30$ kHz. We obtain $Z_T = (9.35 + j2.18)\ \Omega$, so we might expect maximum power transfer for a load in the neighborhood of $10\ \Omega$. We also expect the power dissipated to approach zero for $R_L \to 0$, in which case the load voltage approaches zero, and for $R_L \to \infty$, in which case the load current approaches zero. The graph in Fig. 12.54 is consistent with these expectations.

Figure 12.55 shows a computer-generated graph of the power dissipated in the load $R_L = 10\ \Omega$ versus frequency. Maximum power is dissipated for frequencies near 100 kHz. This is not surprising, because the magnitude of the impedance of the parallel *RLC* branch at that frequency is 678 Ω, which is much larger than both the source resistance and the load resistance. Thus the parallel *RLC* branch draws very little current for frequencies near 100 kHz. For frequencies approaching zero, the impedance of the parallel *RLC* branch approaches $R = 2\ \Omega$, so the load voltage approaches

$$V_{L\,\mathrm{rms}}(f \to 0) = \frac{R\|R_L}{R_S + R\|R_L} \frac{V_S}{\sqrt{2}} \cong 1.52\ \mathrm{V},$$

and the power dissipated approaches

$$P_L(f \to 0) = \frac{{V_{L\,\mathrm{rms}}}^2}{R_L} \cong 230\ \mathrm{mW}.$$

For frequencies much larger than 100 kHz, the impedance of the parallel *RLC* branch approaches the impedance of the capacitor, which in turn approaches zero, so the parallel *RLC* branch shunts virtually all of the current to ground and the power dissipated by the load approaches zero. The graph in Fig. 12.55 is consistent with these observations.

This example illustrates using a Thévenin equivalent to simplify studying load power (or current or voltage) as a function of the load impedance and source frequency. Using the Thévenin equivalent means we must solve equations obtained from (e.g.) Kirchhoff's current law only once. Thereafter, the circuit is simply a voltage divider, and subsequent analysis is relatively simple.

The discussion and examples above deal with obtaining the Thévenin (or Norton) equivalent for any linear circuit in which *all sources are sinusoidal and have the same frequency*. It is possible to obtain a Thévenin (or Norton) equivalent for a circuit containing sources having different frequencies, but we have no need for that generalization at this point.

12.18 Checking Your Work

Obtaining incorrect answers is a useless pursuit. Your grades while in college, your career afterward, and (in some cases) the safety of others all depend upon your ability to pose and solve engineering problems

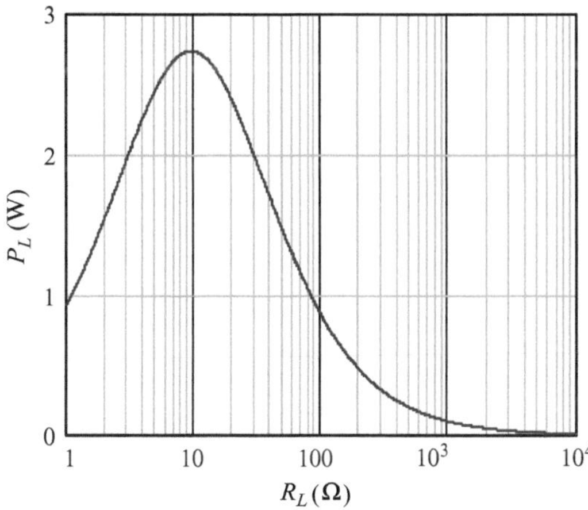

Fig. 12.54 See Example 12.41

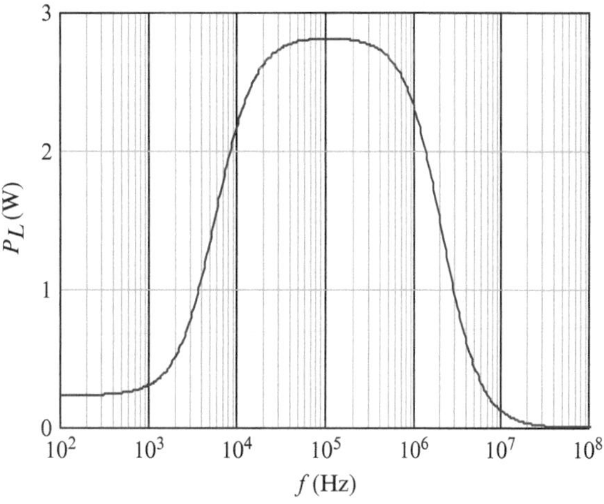

Fig. 12.55 See Example 12.41

correctly. You should cultivate the habit of checking your work. There are many ways to check a result, including:

- *Solving a problem more than one way*; e.g., once using Kirchhoff's current law and again using Kirchhoff's voltage law.
- *Kirchhoff s-law checks*: If your analysis yields all of the currents leaving a node or all of the voltages around a closed path, you can check to see if your results satisfy the applicable Kirchhoff law.
- *Checking dimensions (units)* of terms in equations, of relations obtained along the way, and of the final expression.
- *Order-of-magnitude checks* (for numerical answers). Is the magnitude of a current or voltage reasonable for a circuit at hand?
- *Frequency-limit checks*. Does an expression give correct results if the frequency of the excitation is made zero or infinite?
- *Parameter limit checks*. Does an expression give correct results if the value of a circuit parameter (e.g., resistance, capacitance, or inductance) is made zero or infinite?
- *Power-conservation check*. If your analysis yields expressions for all currents and voltages in a circuit, you can check to see if power is conserved.
- *Simulating the circuit* at hand and comparing your results with those of the simulation.
- *Building a prototype* and making measurements.

Limit checks are especially useful for relatively simple circuits because they often are easy and quick (and therefore useful on exams).

In a frequency limit check, we evaluate an expression for $f \to 0$ and $f \to \infty$ and compare results obtained with those we would expect, based upon inspection or analysis of the circuit under the same conditions. Frequency limit checks are based upon the behavior of capacitors and inductors for $f \to 0$ and $f \to \infty$

$$\lim_{f\to 0}(\omega L) = 0, \ \lim_{f\to\infty}(\omega L) = \infty,$$
$$\lim_{f\to 0}\left(\frac{1}{\omega C}\right) = \infty, \ \lim_{f\to\infty}\left(\frac{1}{\omega C}\right) = 0. \qquad (12.74)$$

In words, for $f \to 0$, a capacitor becomes an open circuit and an inductor becomes a short circuit (a conductor), whereas for $f \to \infty$, a capacitor becomes a short circuit and an inductor becomes an open circuit. At either limit, an *RLC* circuit becomes a resistive circuit.

To perform a zero-frequency limit check, re-draw the circuit for $f \to 0$, replacing capacitors with open circuits and inductors with conductors (short circuits), thereby obtaining a resistive circuit. Obtain an expression for the quantity of interest from the resistive circuit, and compare it with the limit as $f \to 0$ of the corresponding expression for the original circuit. The two expressions should be the same. To perform an infinite-frequency limit check, re-draw the circuit for $f \to \infty$, replacing inductors with open circuits and capacitors with conductors. Again, there remains a resistive circuit, for which it is relatively easy to obtain an expression for the quantity being checked. Some examples follow.

Example 12.42. Refer to Fig. 12.56(a), where $i = I\cos(\omega t)$. Obtain an expression for the phasor voltage across the inductor. Check the result by obtaining expressions for the voltage for $f = 0$ and $f \to \infty$.

Solution: We transform the current source and parallel resistance R_1 to a voltage source and series resistance R_1, and use voltage division to obtain

$$\tilde{V} = \left(\frac{R_3\|Z_L}{R_1 + R_2\|Z_C + R_3\|Z_L}\right)\tilde{I}R_1, \qquad (12.75)$$

where

$$Z_L = j\omega L, \ Z_C = \frac{1}{j\omega C}.$$

To check this result, we first examine it for $f = 0$ and $f \to \infty$. For $f = 0$, $R_3\|Z_L = 0$ and $R_2\|Z_C \to R_2$, so $\tilde{V} = 0$. For $f \to \infty$, $R_3\|Z_L \to R_3$ and $R_2\|Z_C \to 0$, so

$$\tilde{V} = \frac{R_1R_3}{R_1 + R_3}\tilde{I}.$$

Next we draw the circuit diagram for $f = 0$ and for $f \to \infty$ and obtain expressions for the voltage $\tilde{V}$ in each case. For $f = 0$, the capacitor is an open circuit and the inductor is a short

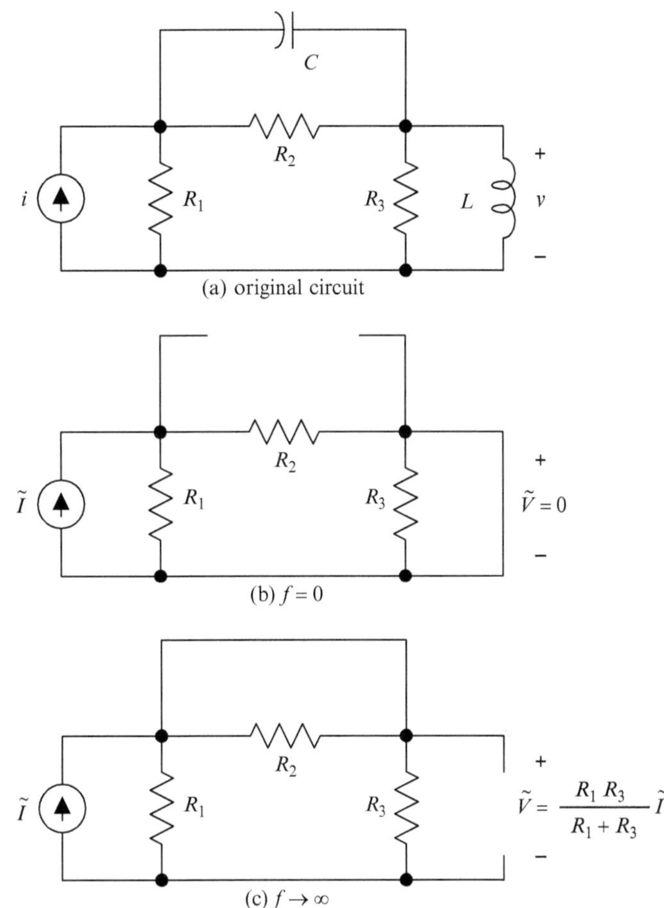

Fig. 12.56 See Example 12.42

circuit (conductor). Because the voltage drop across a conductor equals zero, $\tilde{V} = 0$.

For $f \to \infty$, the inductor is an open circuit and the capacitor is a short circuit (conductor). The voltage v is the voltage across the parallel connection of R_1 and R_3, given by

$$\tilde{V} = \frac{R_1 R_3}{R_1 + R_3} \tilde{I}.$$

These checks give us confidence in the expression (12.75).

Example 12.43. Refer to the circuit shown in Fig. 12.57(a). We find (after a little work) that the equivalent impedance at the terminals *a*–*b* is given by

$$Z = \frac{j\omega LR}{R + j\omega L - \omega^2 RLC}. \tag{12.76}$$

We may check this result several ways. First, we check dimensions (units). The SI unit of the numerator is Ω^2 and the SI unit of the denominator is Ω, so the unit of the expression (12.76) is Ω, which is correct for impedance. In checking units, it often is helpful to remember that ωRC, $\omega L/R$, and $(\omega RC)(\omega L/R) = \omega^2 LC$ are dimensionless.

Next, we perform a frequency limit check. We draw the circuit for $f \to 0$ and $f \to \infty$, as shown in Fig. 12.57(b) and (c), respectively.

For $f \to 0$ (Fig. 12.57(b)), the impedance of the capacitor approaches infinity (the capacitor becomes an open circuit) and the impedance of the inductor approaches zero (the inductor becomes a short circuit). Thus the equivalent impedance at the terminals a–b should approach zero as $f \to 0$. The expression above for the equivalent impedance gives

$$\lim_{f\to 0} Z = \lim_{f\to 0} \frac{j\omega LR}{R + j\omega L - \omega^2 RLC} = \lim_{f\to 0} \frac{j\omega LR}{R} = 0,$$

in agreement with what was deduced from Fig. 12.57(b).

For $f \to \infty$ (Fig. 12.57(c)), the impedance of the capacitor approaches zero (the capacitor becomes a short circuit) and the impedance of the inductor approaches infinity (the inductor becomes an open circuit). Thus the equivalent impedance at the terminals a–b should approach zero as $f \to \infty$. The expression above for the equivalent impedance gives

$$\lim_{f\to\infty} Z = \lim_{f\to\infty} \frac{j\omega LR}{R + j\omega L - \omega^2 LC} = \lim_{f\to\infty} \frac{j\omega LR}{-\omega^2 LC} = \lim_{f\to\infty} \frac{jR}{-\omega C} = 0,$$

in agreement with what was deduced from Fig. 12.57(c).

Finally, we determine whether the expression (12.76) reduces to the correct one if various parameters are set to zero. If the resistance is set to zero, the expression (12.76) reduces to

$$\frac{j\omega LR}{R + j\omega L - \omega^2 RLC} \to \frac{0}{0 + j\omega L - 0} = 0,$$

which is correct because a resistor having zero resistance is a conductor (is a short circuit), and a conductor in parallel with any number of other elements is equivalent to a conductor.

If the capacitance is set to zero, the expression (12.76) reduces to

$$\frac{j\omega LR}{R + j\omega L - \omega^2 RLC} \to \frac{j\omega LR}{R + j\omega L},$$

which is the impedance of the parallel connection of the resistor and inductor, and is correct because a capacitor having capacitance zero is equivalent to an open circuit.

If the inductance is set to zero, the expression (12.76) reduces to

$$\frac{j\omega LR}{R + j\omega L - \omega^2 RLC} \to \frac{0}{R} = 0,$$

which is correct because an inductor having inductance zero is equivalent to a short circuit.

The checks above give us confidence in the expression for the equivalent impedance.

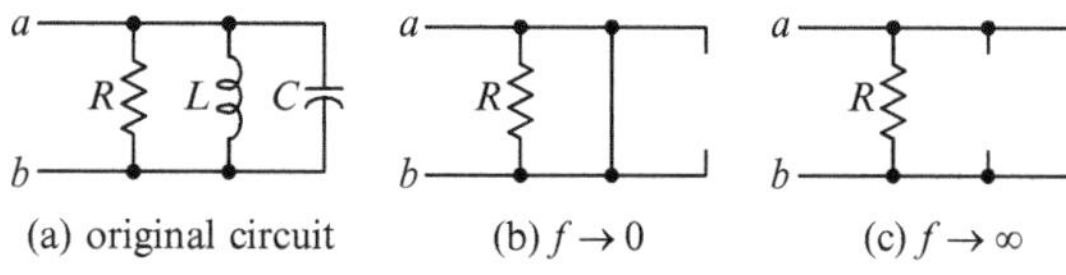

Fig. 12.57 See Example 12.43

If an expression you obtain on a homework assignment or an exam problem is dimensionally consistent, exhibits the correct behavior for $f = 0$ and $f \to \infty$, and passes at least one parameter limit check, the probability that the expression is incorrect is quite small.

Avoid applying parameter limit checks and frequency limit checks simultaneously, as such practice can lead to inconsistencies. For example, consider the circuit in Fig. 12.58, where the source is sinusoidal. Suppose, as a check on an expression we obtain for the current $\tilde{I}$, we let $L \to 0$, in which case the impedance of the inductor approaches zero. Then in the resulting degenerate circuit, we let $f \to \infty$. We have an inconsistency because

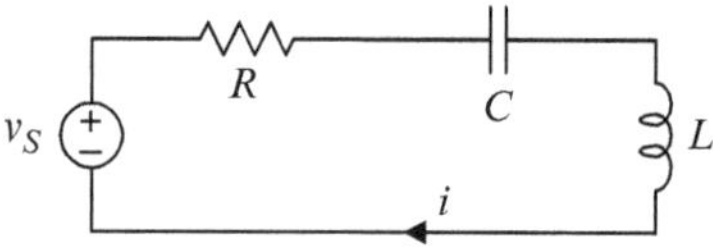

Fig. 12.58 Series *RLC* circuit used to illustrate limit checking

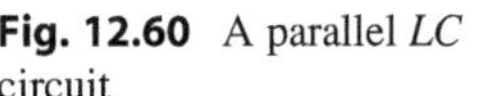

Fig. 12.59 A series LC circuit

$$\lim_{f\to\infty}\lim_{L\to 0}(\tilde{I}) = \lim_{f\to\infty}\lim_{L\to 0}\frac{\tilde{V}_S}{R+(j\omega C)^{-1}+j\omega L}$$
$$= \lim_{f\to\infty}\frac{\tilde{V}_S}{R+(j\omega C)^{-1}} = \frac{\tilde{V}_S}{R},$$

whereas (reverse the limiting operations)

$$\lim_{L\to 0}\lim_{f\to\infty}(\tilde{I}) = \lim_{L\to 0}\lim_{f\to\infty}\frac{\tilde{V}_S}{R+(j\omega C)^{-1}+j\omega L}$$
$$= \lim_{L\to 0}(0) = 0.$$

12.19 Resonance

The impedance of the series LC circuit in Fig. 12.59 is given by

$$Z = j\omega L + \frac{1}{j\omega C} = j\left(\omega L - \frac{1}{\omega C}\right) \tag{12.77}$$

and equals zero for

$$\omega = \omega_r = \frac{1}{\sqrt{LC}} \Rightarrow f_r = \frac{1}{2\pi\sqrt{LC}}. \tag{12.78}$$

The frequency f_r given by (12.78) for which the impedance of the series LC circuit equals zero is called the **resonant frequency** of the circuit and the circuit exhibits **series resonance** at that frequency.

The admittance of the parallel LC circuit in Fig. 12.60 is given by

$$Y(j\omega) = \frac{1}{j\omega L} + j\omega C = \frac{1-\omega^2 LC}{j\omega L} \tag{12.79}$$

and equals zero for

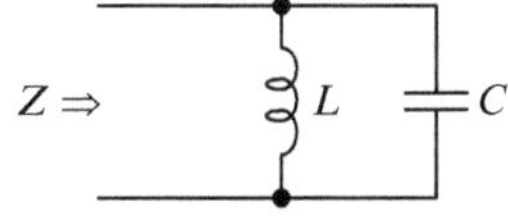

Fig. 12.60 A parallel LC circuit

$$\omega = \omega_r = \frac{1}{\sqrt{LC}} \Rightarrow f_r = \frac{1}{2\pi\sqrt{LC}}. \tag{12.80}$$

The frequency f_r given by (12.80) for which the admittance of the parallel LC circuit equals zero is called the **resonant frequency** of the circuit and the circuit exhibits **parallel resonance** at that frequency.

At resonance, the impedance of a series LC circuit and the admittance of a parallel LC circuit equal zero. Of course, these are idealizations. Physical inductors and capacitors exhibit losses and parasitic effects[20] that make the impedance of a series LC circuit and the admittance of a parallel LC circuit greater than zero. But in many applications, the losses can be made small enough that the impedance or admittance at resonance is much smaller than for frequencies that differ only slightly from the resonant frequency. We discuss practical implications of resonance in more detail in subsequent sections. In this section, we focus primarily on what resonance is.

Example 12.44. Refer to Fig. 12.61. Draw a qualitative phasor diagram for the voltages $\tilde{V}_R$, $\tilde{V}_L$, $\tilde{V}_C$ when the frequency of the sinusoidal source equals the resonant frequency of the series LC portion of the circuit. Use the source voltage $\tilde{V}$ for the phase reference.
Solution: At the resonant frequency $\omega_r = 1/\sqrt{LC}$, the impedance of the LC section equals zero, so the total voltage across the LC section equals zero. It follows that at resonance, the voltage $\tilde{V}_R$ across the resistor equals the source voltage $\tilde{V}$ and the current $\tilde{I}$ equals $\tilde{V}/R$. Because the voltage $\tilde{V}$ is the phase reference, we may write $\tilde{V} = V\angle 0$. We also use $j = 1\angle(\pi/2)$ and express the voltages across the inductor and the capacitor at resonance as

[20] For example, at sufficiently high frequencies, an inductor exhibits shunt capacitance. Section 12.22 is an introductory treatment of such parasitic effects.

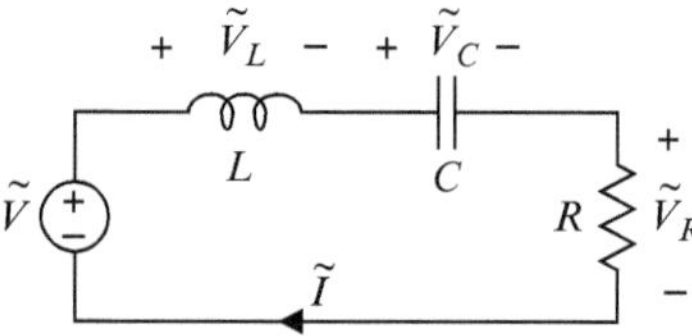

Fig. 12.61 See Example 12.44

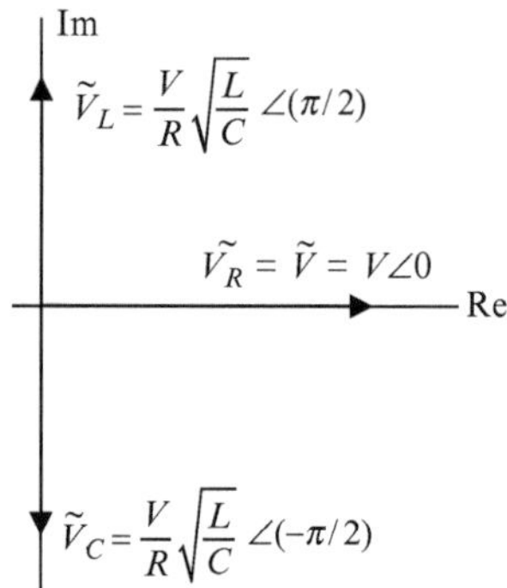

Fig. 12.62 See Example 12.44

$$\tilde{V}_L = j\omega_r L\tilde{I} = j\omega_r L\frac{\tilde{V}}{R} = \frac{1\angle(\pi/2)LV\angle 0}{R\sqrt{LC}}$$
$$= \frac{V}{R}\sqrt{\frac{L}{C}}\angle(\pi/2),$$
$$\tilde{V}_C = \frac{\tilde{I}}{j\omega_r C} = \frac{\tilde{V}}{j\omega_r RC} = \frac{(V\angle 0)\sqrt{LC}}{[1\angle(\pi/2)]RC}$$
$$= \frac{V}{R}\sqrt{\frac{L}{C}}\angle(-\pi/2).$$

At resonance, the voltages across the inductor and the capacitor have equal magnitudes and are in phase opposition. Figure 12.62 shows a phasor diagram for the voltages in the circuit at resonance.

Example 12.45. Refer to Fig. 12.63. Draw a qualitative phasor diagram for the currents $\tilde{I}_R$, $\tilde{I}_L$, $\tilde{I}_C$ when the frequency of the sinusoidal source equals the resonant frequency of the parallel LC portion of the circuit. Use the source current $\tilde{I}$ for the phase reference.

Solution: Because $\tilde{I}$ is the phase reference, we may write $\tilde{I} = I\angle 0$. At resonance, the admit-

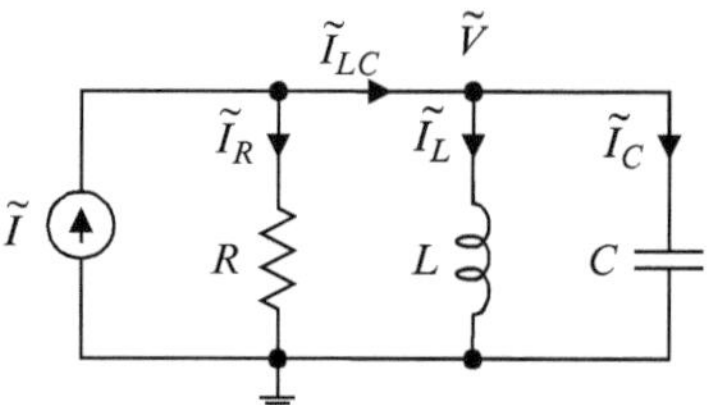

Fig. 12.63 See Example 12.45

tance of the LC section equals zero (the impedance is infinite), so the current $\tilde{I}_{LC} = \tilde{I}_L + \tilde{I}_C$ must be zero. It follows from Kirchhoff's current law that $\tilde{I}_R = \tilde{I}$ and then from Ohm's law that $\tilde{V} = R\tilde{I}_R = RI\angle 0$. Because R is real and positive and $\tilde{I}_R = \tilde{I}$ has zero initial phase, we have

$$\tilde{I}_L = \frac{RI\angle 0}{j\omega_r L} = \frac{RI\angle 0}{\omega_r L\angle(\pi/2)} = \frac{RI}{\omega_r L}\angle\left(-\frac{\pi}{2}\right),$$
$$\tilde{I}_C = j\omega_r CRI\angle 0 = [\omega_r C\angle(\pi/2)][RI\angle 0]$$
$$= \omega_r CRI\angle\left(\frac{\pi}{2}\right).$$

The resonant frequency is $\omega_r = 1/\sqrt{LC}$, so the expressions above reduce to

$$\tilde{I}_L = \frac{RI}{L/\sqrt{LC}}\angle\left(-\frac{\pi}{2}\right) = RI\sqrt{\frac{C}{L}}\angle\left(-\frac{\pi}{2}\right),$$
$$\tilde{I}_C = \frac{CRI}{\sqrt{LC}}\angle\left(\frac{\pi}{2}\right) = RI\sqrt{\frac{C}{L}}\angle\left(\frac{\pi}{2}\right).$$

At resonance, the currents $\tilde{I}_L$, $\tilde{I}_C$ have equal magnitudes and are in phase opposition. Figure 12.64 shows a phasor diagram for the currents in the circuit at resonance.

Resonance is a result of a low-loss exchange of energy between two energy-storage mechanisms. The pendulum in a grandfather clock is resonant, and the energy exchange is between the kinetic energy of the pendulum and the potential energy stored in the gravitational field. Little is lost in the bearing supporting the pendulum. In a resonant RLC circuit, the exchange often is between the electric field associated with a capacitance and the magnetic field associated with an inductance, with little energy being lost in resistance.

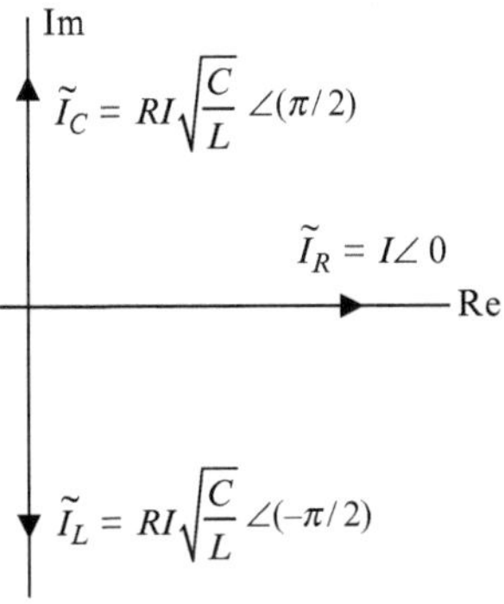

Fig. 12.64 See Example 12.45

A *passive* circuit must contain at least one capacitor and at least one inductor to exhibit resonance.[21]

The definitions given above for resonance of series and parallel *LC* circuits are specific examples of more general definitions, of which there are several, all being equivalent. For circuits containing at least one capacitor and one inductor, two useful (and equivalent) working definitions of resonance are:

A two-terminal (one-port) *RLC* circuit is in resonance when its impedance (or admittance) is real.

A two-terminal (one-port) *RLC* circuit is in resonance when the angle of its impedance (or admittance) equals zero.

For example, at resonance the impedance of the series *RLC* circuit treated in Example 12.44 equals the resistance *R* and the admittance of the parallel *RLC* circuit treated in Example 12.45 equals the conductance R^{-1}.

Example 12.46. Obtain an expression for the resonant frequency of the circuit shown in Fig. 12.65.

Solution: We may ignore the resistor R_1 because if *Z* is real, then so is $Z - R_1$. The impedance *Z* is real if

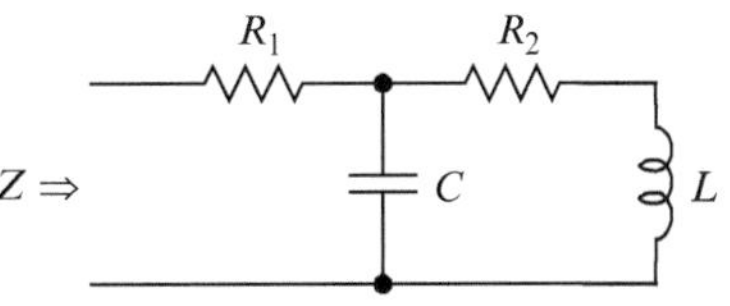

Fig. 12.65 See Example 12.46

$$\mathrm{Im}\left[\left(\frac{1}{j\omega C}\right)\middle\|(R_2 + j\omega L)\right] = 0. \quad (12.81)$$

Equation (12.81) leads to

$$\mathrm{Im}\left[\frac{R_2 + j\omega L}{1 + j\omega C(R_2 + j\omega L)}\right] = \mathrm{Im}\left[\frac{R_2 + j\omega L}{1 - \omega^2 LC + j\omega CR_2}\right] = 0,$$

which yields

$$\mathrm{Im}\left[\frac{(R_2 + j\omega L)(1 - \omega^2 LC - j\omega CR_2)}{(1 - \omega^2 LC)^2 + (\omega CR_2)^2}\right] = 0.$$

Because the denominator is real, the impedance is real if

$$\mathrm{Im}\left[(R_2 + j\omega L)\left(1 - \omega^2 LC - j\omega CR_2\right)\right] = 0$$
$$\Rightarrow \omega L\left(1 - \omega^2 LC\right) - \omega C{R_2}^2 = 0,$$

which reduces to

$$\omega^2 LC = 1 - \frac{C{R_2}^2}{L}.$$

Thus the resonant frequency is given by

$$\omega_r = \sqrt{\frac{1 - {R_2}^2(C/L)}{LC}}.$$

Exercise 12.23. Obtain an expression for the resonant frequency of the circuit shown in Fig. 12.66.

[21] As we show in a subsequent chapter, it is possible to emulate inductance using an active device and capacitance. Thus it is possible for an active circuit containing only active devices, resistors, and capacitors to exhibit resonance. Also, as discussed below, a physical component (such as a capacitor) is not ideal and can exhibit all three of resistance, capacitance and inductance and can be self-resonant at a sufficiently high frequency.

Exercise 12.24. Refer to Fig. 12.67, where $v_S = V_S \cos(\omega t)$, with $V_S = 5$ V. (a) Find the current $i(t)$ for $\omega_1 = 1/\sqrt{L_1 C_1}$ and for $\omega_2 = 1/\sqrt{L_2 C_2}$. (b) Does the circuit as a whole have a resonant frequency? If so, what is it?

12.20 Quality Factors and Common Resonant Configurations

At any particular frequency, a reactive two-terminal circuit or element can be modeled as an effective resistance R in series with either an inductor or a capacitor, depending upon whether the reactance X is positive or negative. The **quality factor** of the circuit or element at the frequency in question is defined by

$$Q = \frac{|X|}{R}, \tag{12.82}$$

where X is the (capacitive or inductive) effective series reactance and R is the effective series resistance. In particular, the quality factors of an inductor (a coil) and a capacitor at a frequency ω are given by

$$Q_L = \frac{\omega L}{R_L}, \; Q_C = \frac{1/(\omega C)}{R_C} = \frac{1}{\omega R_C C}, \tag{12.83}$$

where R_L and R_C are the effective resistances of the coil and the capacitor, respectively, at the frequency ω. Keep in mind that *the quality factor of an inductor or a capacitor (or any two-terminal reactive circuit) is a function of frequency.*

A large quality factor implies an inductive or capacitive reactance much larger than the effective series resistance. Thus the quality factors defined by (12.83) indicate how nearly a physical inductor and a physical capacitor approximate an ideal inductor and ideal capacitor, respectively.

Many practical uses of passive resonant circuits involve one or more of the three circuit configurations shown in Fig. 12.68. From (12.78), the resonant frequency of the series RLC circuit in Fig. 12.68a is given by

$$\omega_0 = \frac{1}{\sqrt{LC}}. \tag{12.84}$$

In Example 12.46 above, we show that the resonant frequency of the circuit in Fig. 12.68b is given by

$$\omega_r = \sqrt{\frac{1 - R^2(C/L)}{LC}}. \tag{12.85}$$

In an end-of-chapter problem, you are asked to show that (12.85) can be written

$$\omega_r = \omega_0 \sqrt{1 - \frac{1}{Q_0^2}}. \tag{12.86}$$

In (12.86), ω_0 is the resonant frequency of the series RLC loop in Fig. 12.68(b) and Q_0 is the quality factor of the LR coil *at the frequency* ω_0 (not at the resonant frequency ω_r of the parallel RLC circuit). However, if the quality factor of the coil is much larger than unity, then

$$\omega_r = \omega_0 \sqrt{1 - \frac{1}{Q_0^2}} \cong \omega_0 = \frac{1}{\sqrt{LC}}. \tag{12.87}$$

In other words, if the quality factor of the coil in Fig. 12.68(b) is large at the frequency $1/\sqrt{LC}$, then the

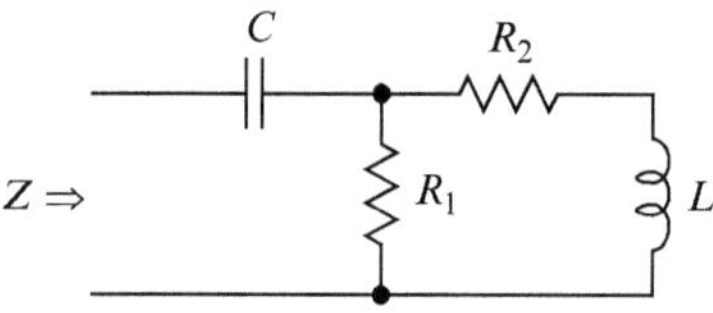

Fig. 12.66 Exercise 12.23

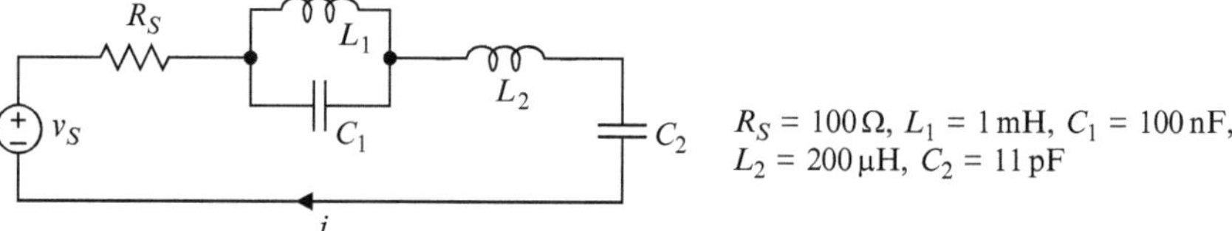

Fig. 12.67 See Exercise 12.24

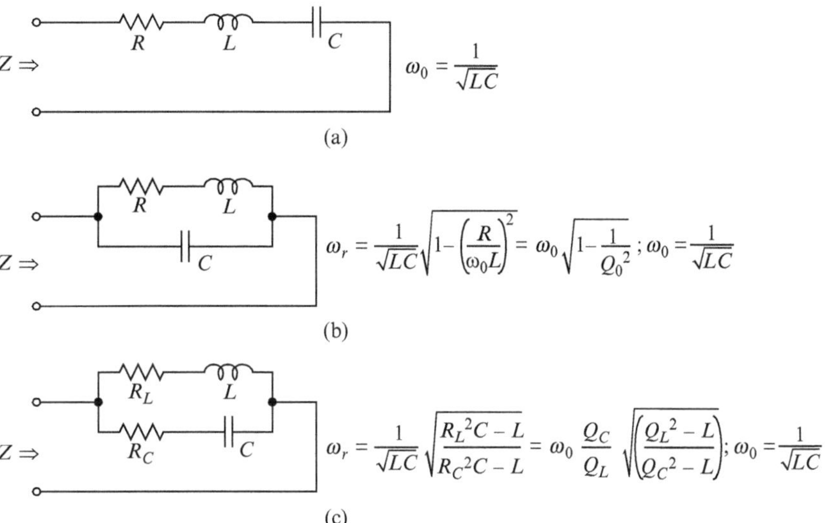

Fig. 12.68 Common resonant configurations

resonant frequency of the parallel circuit equals (to a good approximation) the resonant frequency of the *RLC* loop (considered as a series circuit). This approximation can greatly simplify some analyses and calculations.

In an end-of-chapter problem, you are asked to show that the resonant frequency of the circuit in Fig. 12.68(c) is given by

$$\omega_r = \frac{1}{\sqrt{LC}}\sqrt{\frac{R_L^2C-L}{R_C^2C-L}}. \qquad (12.88)$$

The relation (12.88) can be put in a more convenient form as follows: From (12.83),

$$R_L = \frac{\omega_0 L}{Q_L} \Rightarrow R_L^2 C = \frac{\omega_0^2 L^2 C}{Q_L^2} = \frac{L}{Q_L^2},$$
$$R_C = \frac{1}{\omega_0 C Q_C} \Rightarrow R_C^2 C = \frac{LC}{\omega_0^2 C^2 Q_C^2} = \frac{L}{Q_C^2},$$

where we have used $\omega_0{}^2 LC = 1$. Thus

$$\frac{R_L^2C-L}{R_C^2C-L} = \frac{Q_L^{-2}L-L}{Q_C^{-2}L-L} = \frac{Q_L^{-2}-1}{Q_C^{-2}-1}$$

so (12.88) can be written

$$\omega_r = \frac{1}{\sqrt{LC}}\sqrt{\frac{Q_L^{-2}-1}{Q_C^{-2}-1}}$$
$$= \frac{1}{\sqrt{LC}}\frac{Q_C}{Q_L}\sqrt{\frac{Q_L^2-1}{Q_C^2-1}}. \qquad (12.89)$$

In (12.89), $\omega_0 = 1/\sqrt{LC}$ is the resonant frequency of the series $R_L LCR_C$ loop. The quantities Q_L and Q_C are the quality factors of the coil and the capacitor, respectively, at the loop resonant frequency ω_0, and are given by

$$Q_L = \frac{X_L}{R_L} = \frac{\omega_0 L}{R_L}, \quad Q_C = \frac{|X_C|}{R_C} = \frac{1}{\omega_0 R_C C}. \qquad (12.90)$$

If both quality factors are large, the right side of (12.89) reduces to $\omega_r = \omega_0$. In other words, if the quality factors of the coil and the capacitor in Fig. 12.68(c) are both large at the frequency $1/\sqrt{LC}$, then the resonant frequency of the parallel circuit equals (to a good approximation) the resonant frequency of the series $R_L LCR_C$ loop. This approximation can greatly simplify some analyses and calculations.

Example 12.47. In Fig. 12.69, $R_L = 2.2$ kΩ and $v_S = V_S[\cos(\omega_1 t) + \cos(\omega_2 t)]$ with $f_1 =$ 100 kHz and $f_2 = 200$ kHz. The circuit is intended to block the 200 kHz component and pass the 100 kHz component of the source voltage. You have available a selection of 200 and 100 kHz coils with inductances ranging from 10 to 200 mH in steps of 10 mH. All have quality factors of about 25 at their intended operating frequencies. You also have available capacitors ranging from 10 to 91 pF and a wide selection of trimmer capacitors in the same range, both having quality factors in

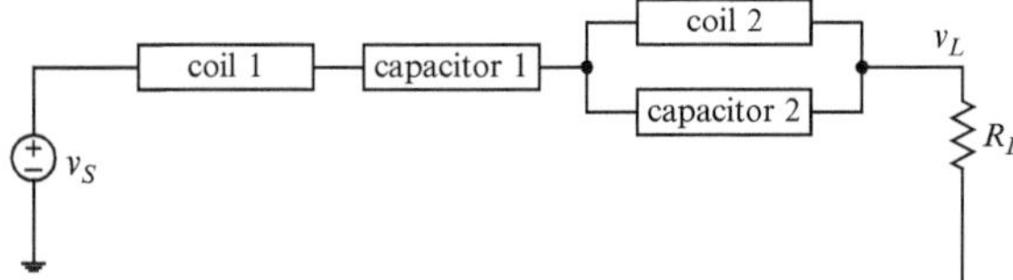

Fig. 12.69 See Example 12.47

excess of 100. Specify the circuit components such that the 100 kHz component of the load voltage is as large as possible and the 200 kHz component is as small as possible. Evaluate your design (with the help of a computer).

Solution: The impedance of a series *RLC* circuit is minimum at resonance and the impedance of a parallel *RLC* circuit is maximum at resonance. The idea here is to make the series *RLC* section resonant at 100 kHz and the parallel *RLC* section resonant at 200 kHz. For the series circuit, we must choose a 100 kHz coil and a capacitor such that

$$\frac{1}{\sqrt{L_1C_1}} = \omega_1 = 200\pi \times 10^3\ \text{s}^{-1}.$$

Given the large assortment of components and in absence of other specifications, there are a great many solutions. Choosing $C_1 = 15$ pF gives

$$L_1 = \frac{1}{\omega_1^2 C_1} = 169\ \text{mH},$$

For the parallel section, we require

$$\frac{1}{\sqrt{L_2C_2}} = \omega_2 = 400\pi \times 10^3\ \text{s}^{-1}$$

We choose $C_2 = 62$ pF and calculate

$$L_2 = \frac{1}{\omega_2^2 C_2} = 10.2\ \text{mH}.$$

For frequencies near resonance, we can express the impedance of an inductor as

$$Z_L = R_L + j\omega L = \frac{\omega L}{Q_L} + j\omega L$$
$$= j\omega L\left(1 - jQ_L{}^{-1}\right).$$

Similarly, for a capacitor,

$$Z_C = R_C + \frac{1}{j\omega C} = \frac{1}{Q_C\omega C} + \frac{1}{j\omega C}$$
$$= \frac{1}{j\omega C}\left(1 + jQ_C{}^{-1}\right).$$

The impedance of the series section in Fig. 12.69 is given by

$$Z_1 = Z_{L1} + Z_{C1}$$
$$\cong j\omega L_1\left(1 - \frac{j}{Q_L}\right) + \frac{1}{j\omega C_1}\left(1 + \frac{j}{Q_C}\right).$$

Similarly, the impedance of the parallel section is given by

$$Z_2 = \frac{Z_{L2}Z_{C2}}{Z_{L2} + Z_{C2}}$$
$$= \frac{\left[j\omega L_2\left(1 - jQ_L{}^{-1}\right)\right]\left[\left(1 + jQ_C{}^{-1}\right)\right]}{\left(1 + jQ_C{}^{-1}\right) - \omega^2 L_2 C_2\left(1 - jQ_L{}^{-1}\right)}.$$

The load voltage is given by

$$\tilde{V}_L(\omega) = \frac{R_L\tilde{V}_S}{R_L + Z_1(\omega) + Z_2(\omega)}$$

Figure 12.70 shows a graph of the normalized load voltage (or voltage gain) in dB, given by

$$A_v = 20\log\left|\frac{\tilde{V}_L}{\tilde{V}_S}\right|,$$

where $\tilde{V}_S = 1$ V. The resonant peak at 100 kHz and the resonant valley at 200 kHz are evident. Direct calculation gives $A_v(f_1) \cong -14.5$ dB, $A_v(f_2) \cong -42.9$ dB, so the 200 kHz component of the load voltage is 28.4 dB below the 100 kHz component, if the components have their specified values.

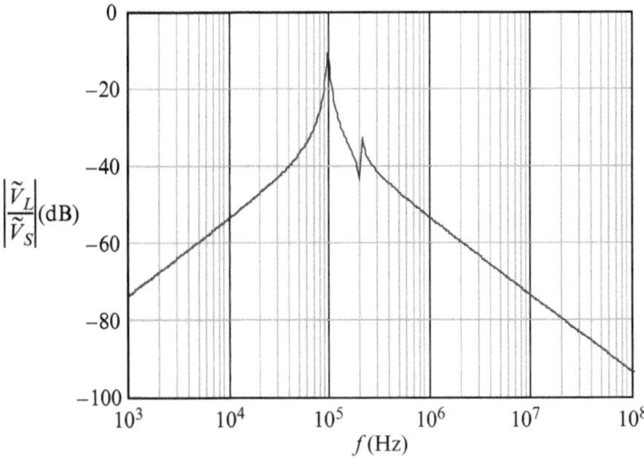

Fig. 12.70 See Example 12.47

In some important applications and as illustrated by Example 12.47, *LC* parallel resonant circuits are used to select (pass or block) sinusoidal inputs within a relatively narrow band of frequencies. Resonant circuits used in such applications are called **tuned circuits**, the implication being that the magnitude of the impedance or admittance falls off sharply for frequencies below and above the frequency to which the circuit is *tuned* (the resonant frequency). Tuned circuits are especially important in many high-frequency filtering applications and are practical applications of resonance. Quality factors of capacitors and (especially) inductors are important parameters in analysis and design of tuned *RLC* circuits.

Tuned circuits are built using coils specially designed for such applications. Coils intended for use in high-frequency tuned circuits are called a *radio-frequency coils* or **rf coils.** Any particular rf coil is designed to operate at or near a particular frequency and to exhibit a specific quality factor at that frequency. The operating frequencies of readily available (off-the-shelf) coils are mostly determined by standards for radio-frequency equipment; for example, frequencies assigned to cell-phone providers, military radar and other equipment, and commercial broadcast and cable television.

An rf coil possesses not only inductance, but also resistance and capacitance, and for many purposes can be modeled at any particular frequency as shown in Fig. 12.71. The resistance R_s is the effective resistance of the coil at the operating frequency. The capacitance C_c accounts for the end-to-end and turn-to-turn stray capacitance of the coil. Keep in mind that the model parameters R_s, L_s, C_c are functions of frequency, so

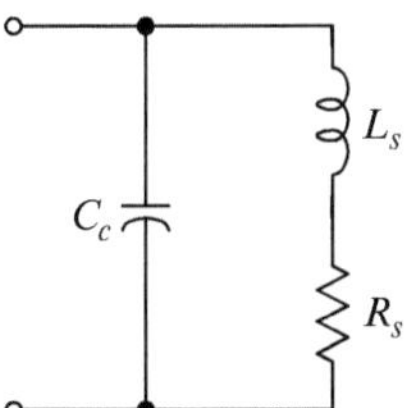

Fig. 12.71 Circuit model for an rf coil

the model is valid only for frequencies near the specified operating frequency. For example, skin and proximity effects cause the effective resistance to increase with frequency. But the dependence of the parameters on frequency usually is weak enough that the model is valid for the relatively narrow range of frequencies over which most such circuits are intended to operate.

The **quality factor** for an rf coil is denoted by Q and given by

$$Q = \frac{\omega_0 L_s}{R_s} = \frac{X_s}{R_s}, \tag{12.91}$$

where ω_0 is the operating frequency for the coil, L_s and R_s are the inductance and effective resistance of the coil, respectively, at that frequency. Because the parameters L_s, R_s are functions of frequency, the quality factor also is a function of frequency, but again, the dependence on frequency usually is weak enough that the quality factor can be assumed to be independent of frequency for frequencies near the intended operating frequency. The quality factor for a coil largely determines the degree of frequency selectivity that can be achieved using the coil in a tuned circuit. In practice, quality factors exceeding 20 are common for rf coils.

Large quality factors can be achieved using magnetic cores, many turns, and special geometries. Magnetic cores in rf coils usually are ferrite. Although a magnetic core can greatly enhance inductance, such cores also introduce losses that increase with increasing frequency, and magnetic-core coils are used mostly at frequencies below about 10–50 MHz. Air-core coils are used at higher frequencies.

In order to use results obtained previously for parallel *LC* circuits, we obtain below a parallel R_pL_p circuit that is equivalent (at the operating frequency) to the series R_sL_s branch in Fig. 12.71. Because the series model is valid only for frequencies near the operating frequency, the equivalent parallel model obtained below also is valid only for frequencies near that frequency.

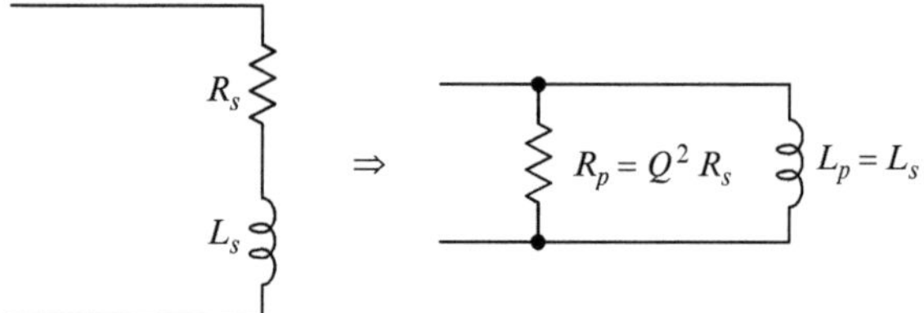

Fig. 12.72 Parallel and series RL models for a high-Q coil

Refer to Fig. 12.72. Equivalence of the circuits *at a pre-selected frequency* f_0 requires that the admittances of the circuits at that frequency be equal. Equating expressions for the admittances of the parallel and series models gives

$$\frac{1}{R_p} + \frac{1}{j\omega_0 L_p} = \frac{1}{R_s + j\omega_0 L_s}$$
$$= \frac{1}{R_S + jQR_s} = \frac{1}{R_s}\frac{1 - jQ}{1 + Q^2}. \qquad (12.92)$$

Two complex quantities are equal if and only if the real parts are equal and the imaginary parts are equal. Thus (12.92) leads to

$$\frac{1}{R_p} = \frac{1}{R_S}\frac{1}{1+Q^2};\ \frac{1}{\omega_0 L_p} = \frac{1}{R_s}\frac{Q}{1+Q^2}. \qquad (12.93)$$

For cases of interest $Q \gg 1$, and (12.93) reduces to

$$\frac{1}{R_p} \cong \frac{1}{Q^2 R_s};\ \frac{1}{\omega_0 L_p} \cong \frac{1}{QR_s} = \frac{1}{\omega_0 L_s}. \qquad (12.94)$$

It follows that

$$R_p \cong Q^2 R_s,\ L_p \cong L_s;\ Q \gg 1,\ \omega \cong \omega_0. \qquad (12.95)$$

Keep the limitations $Q \gg 1,\ \omega \cong \omega_0$ in mind. The equivalence relations expressed by (12.95) are valid only for Q large and only for frequencies near f_0. In applications, (12.95) usually is assumed to hold for $Q \geq 10$, which condition is rather easily met, and a coil is used in circuits that operate near the operating frequency for the coil (near the frequency for which the specified Q and series model are valid).

For frequencies near the intended operating frequency, we may model a high-Q coil as shown in Fig. 12.73. For ω near ω_0, the admittance of the coil is given by

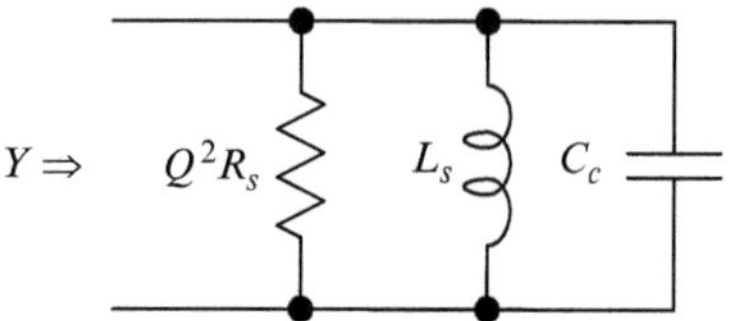

Fig. 12.73 Circuit model for a high-Q rf coil

$$Y = \frac{1}{Q^2 R_s} + \frac{1}{j\omega L_s} + j\omega C_c$$
$$= \frac{Q^2 R_s - \omega^2 L_s C_c Q^2 R_s + j\omega L_s}{j\omega L_s Q^2 R_s}. \qquad (12.96)$$

Because of the stray shunt capacitance C_c, the coil alone can exhibit resonance and is said to be **self-resonant**. The self-resonant frequency of the coil is given by

$$f_c = \frac{\omega_c}{2\pi} = \frac{1}{2\pi\sqrt{L_s C_c}}. \qquad (12.97)$$

Manufacturers specify the self-resonant frequencies of their rf offerings, partly as a way of specifying the apparent shunt capacitance. Usually, the self-resonant frequency given by (12.97) is much higher than the frequency f_0 at which the coil is intended to operate, and additional capacitance must be placed in parallel with the coil to achieve the desired resonant frequency. Thus, if C denotes the *total* parallel capacitance, the resonant frequency of the resulting tuned circuit is given by

$$f_0 = \frac{1}{2\pi\sqrt{L_s C}};\ \omega_0 = \frac{1}{\sqrt{L_s C}}. \qquad (12.98)$$

The admittance of the tuned circuit is given by

$$Y = \frac{1}{Q^2 R_s} + \frac{1}{j\omega L_s} + j\omega C$$
$$= \frac{Q^2 R_s - \omega^2 L_s C Q^2 R_s + j\omega L_s}{j\omega L_s Q^2 R_s}$$
$$= \frac{1}{Q^2 R_s}\frac{Q^2 - \omega^2 L_s C Q^2 + j(\omega L_s/R_s)}{j(\omega L_s/R_s)} \qquad (12.99)$$
$$= \frac{1}{Q^2 R_s}\frac{Q^2(1 - \omega^2/\omega_0^2) + j(\omega/\omega_0)Q}{j(\omega/\omega_0)Q}$$
$$= \frac{1}{Q^2 R_s}\frac{Q\left[1 - (\omega/\omega_0)^2\right] + j(\omega/\omega_0)}{j(\omega/\omega_0)}.$$

The impedance can be expressed as a function of frequency (Hz), as

$$Z = \frac{1}{Y} \cong \frac{j(\omega/\omega_0)Q^2R_s}{Q\left[1-(\omega/\omega_0)^2\right]+j(\omega/\omega_0)} = \frac{j(f/f_0)Q^2R_s}{Q\left[1-(f/f_0)^2\right]+j(f/f_0)}. \quad (12.100)$$

From (12.100), the magnitude of the impedance is given by

$$|Z| = \frac{Q^2R_s}{\sqrt{Q^2\left[\frac{1-(f/f_0)^2}{(f/f_0)}\right]^2+1}}.$$

The maximum impedance (magnitude) is obtained for $f = f_0$ and equals the equivalent parallel resistance

$$R = Q^2R_s. \quad (12.101)$$

The magnitude of the impedance normalized to the maximum value R is given by

$$\left|\frac{Z}{R}\right| = \frac{1}{\sqrt{Q^2\left[\frac{1-(f/f_0)^2}{(f/f_0)}\right]^2+1}}$$

We may regard this impedance as a normalized transimpedance, expressed in dB as

$$\left(\frac{|Z|}{R}\right)_{dB} \cong -10\log\left|Q^2\left[\frac{1-(f/f_0)^2}{(f/f_0)}\right]^2+1\right|. \quad (12.102)$$

Figure 12.74 shows a graph of the normalized impedance in dB versus the normalized frequency f/f_0 for three values of the quality factor Q, illustrating that the frequency selectivity of the circuit increases (the width of the peak in Fig. 12.74 decreases) with increasing Q. In the next section, we obtain a relation between the quality factor of the coil and the frequency selectivity of the circuit.

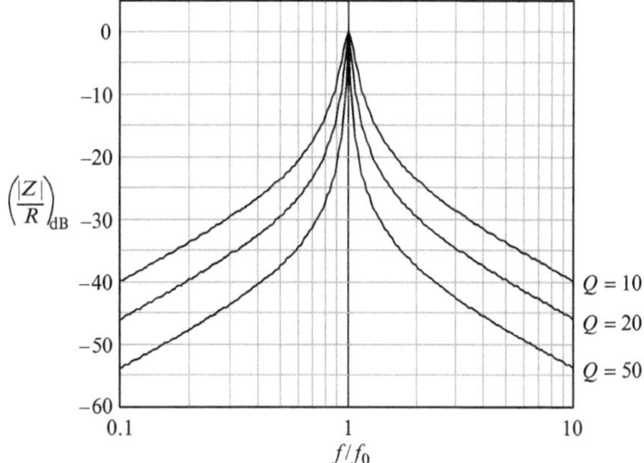

Fig. 12.74 Normalized impedance (dB) versus normalized frequency f/f_r for the parallel RLC (tuned) circuit

There are many kinds of rf coils; for example, ferrite-core and air-core "chip" inductors and ferrite-core and air-core axial-lead inductors. These have inductances ranging from a few nH to a few mH, quality factors from about 5 to about 100, operating frequencies (f_T) from about 50 kHz to almost 1 GHz, dc resistances (DCR) from a few mΩ to a few hundred Ω and self-resonant frequencies (SRF) from about 50 kHz to a few GHz.

Example 12.48. A manufacturer's specifications for one of his offerings are $L = 22$ μH, $Q = 50$, $f_T = 2.52$ MHz, and $SRF = 13$ MHz. (a) What is the ac resistance of the coil at the operating frequency? (b) What (approximately) is the shunt (stray) capacitance of the coil?

Solution:

(a)

$$Q = \frac{\omega_T L}{R_{ac}} \Rightarrow R_{ac} = \frac{\omega_T L}{Q} = \frac{2\pi(2.52\ \text{MHz})(22\ \mu\text{H})}{50} = 6.97\ \Omega.$$

(b)

$$SRF = \frac{1}{2\pi\sqrt{LC}} \Rightarrow C = \frac{1}{(2\pi SRF)^2L} = \frac{1}{4\pi^2(13\ \text{MHz})^2(22\ \mu\text{H})} = 6.81\ \text{pF}.$$

12.21 Simulating Inductance Using Active *RC* Circuits

Passive tuned circuits such as the singly tuned circuit described above are useful primarily at high (video and above) frequencies, where the inductances required are small.[22] But at audio frequencies, achieving sufficient inductance can require large, heavy, and expensive inductors. Fortunately, it is possible to build active *RC* circuits that simulate inductance and are quite small. In this section, we describe one such circuit that finds favor in the audio community.

In the vocabulary of audio engineers, a **gyrator** is an active capacitive circuit that simulates a resistance in series with an inductance. Figure 12.75 shows a model commonly presented for a gyrator, where the op amp is assumed to be ideal.[23] Thus

$$\tilde{V}_{pn} = 0,\ \tilde{I}_n = 0,\ \tilde{I}_p = 0.$$

This model is incomplete because $\tilde{I}_{out} \neq \tilde{I}_S$. To see this, note that Kirchhoff's voltage law around the loop *a-p-n-a* gives

$$\tilde{V}_{ap} + \tilde{V}_{pn} + \tilde{V}_{na} = 0 \Rightarrow \tilde{V}_{ap} = -\tilde{V}_{na} = \tilde{V}_{an},$$

which implies that the voltage across the resistor R_1 equals the voltage across the capacitor, so not all of the current $\tilde{I}_S$ passes through the capacitor and the resistor R_2. Some is diverted through R_1. But the model does not provide a closed path for the diverted current because $\tilde{I}_n = 0$ and $\tilde{I}_p = 0$.

The situation is clarified if we replace the op amp by a less-than-ideal model, as shown in Fig. 12.76, where $\mu \gg 1$. We assume the op-amp input resistance is exceedingly large and the op-amp output resistance is negligibly small. The dependent source provides a return path for the current through R_1 to the negative terminal of the independent source, so Kirchhoff can stop spinning in his grave.

Using the model in Fig. 12.76, we can show that under certain conditions the gyrator circuit in Fig. 12.75 simulates an inductor in series with a resistor. Taking *b* as the reference node and Applying Kirchhoff's current law at node *a* gives

$$\tilde{I}_S = (\tilde{V}_S - \tilde{V}_p)j\omega C + \frac{\tilde{V}_S - \tilde{V}_p}{R_1} = \left(\frac{1 + j\omega R_1 C}{R_1}\right)(\tilde{V}_S - \tilde{V}_p), \qquad (12.103)$$

where we have used $\tilde{V}_n = \tilde{V}_p$ (because the op amp is ideal). By voltage division,

$$\tilde{V}_p = \frac{R_2 \tilde{V}_S}{R_2 + \frac{1}{j\omega C}} = \frac{j\omega R_2 C}{1 + j\omega R_2 C}\tilde{V}_S. \qquad (12.104)$$

Using (12.104) to eliminate $\tilde{V}_p$ in (12.103) gives

$$\tilde{I}_S = \left(\frac{1 + j\omega R_1 C}{R_1}\right)\left(\tilde{V}_S - \frac{j\omega R_2 C}{1 + j\omega R_2 C}\tilde{V}_S\right) = \frac{1 + j\omega R_1 C}{R_1(1 + j\omega R_2 C)}\tilde{V}_S. \qquad (12.105)$$

It follows that the impedance at the terminals *a–b* of the gyrator circuit is given by

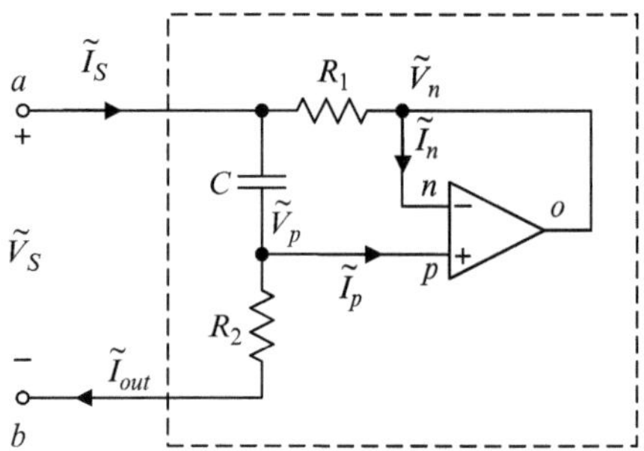

Fig. 12.75 An approximate implementation of a gyrator

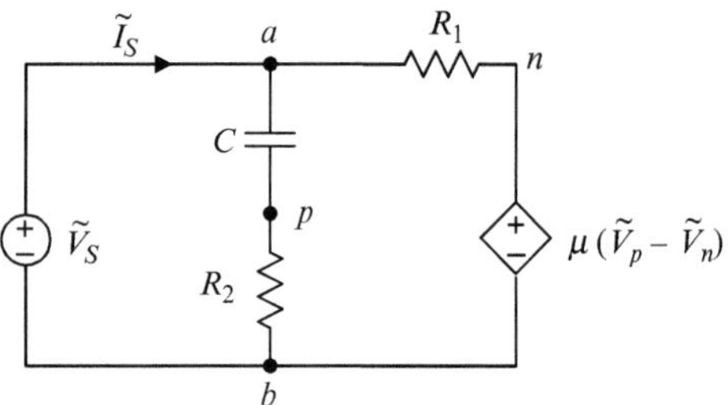

Fig. 12.76 Equivalent circuit for the gyrator in Fig. 12.75

[22]Remember that the voltage across an inductor is proportional to both the inductance *and the rate of change of the voltage*.

[23]Actually, the circuit considered here only *approximates* a true gyrator, but provides a useful approximation at minimal cost.

$$Z = \frac{\tilde{V}_S}{\tilde{I}_S} = \frac{R_1(1 + j\omega R_2 C)}{1 + j\omega R_1 C}, \tag{12.106}$$

from which we obtain

$$Z \cong R_1(1 + j\omega R_2 C); \; \omega R_1 C \ll 1. \tag{12.107}$$

We can write (12.107) as

$$Z \cong R_1 + j\omega L; \; L = R_1 R_2 C, \; \omega \ll \frac{1}{R_1 C}. \tag{12.108}$$

For frequencies well below $(2\pi R_1 C)^{-1}$, the impedance of the gyrator is that of a resistor having resistance R_1 in series with an inductor having inductance $L = R_1 R_2 C$. At any particular frequency $f_0 \ll (2\pi R_1 C)^{-1}$, the gyrator acts like a coil having quality factor

$$Q = \frac{\omega_0 R_1 R_2 C}{R_1} = \omega_0 R_2 C. \tag{12.109}$$

The development above suggests a simple approach to designing a gyrator-based singly-tuned circuit, illustrated by Fig. 12.77 and described below.

Specifications on a singly-tuned circuit normally include the resonant frequency f_0, the quality factor Q, and the impedance $Z(\omega_0)$ at resonance.[24] Thus we have three specifications on the four free parameters R_1, R_2, C, C_1. From (12.106) and (12.107), we must also require $R_2 \gg R_1$, if the gyrator is to behave like a coil for frequencies below $(2\pi R_1 C)^{-1}$.

The resonant frequency of the circuit in Fig. 12.77(b) is given by

$$\omega_0 = \frac{1}{\sqrt{LC_1}}. \tag{12.110}$$

To validate the approximations illustrated by Fig. 12.77, we require

$$\omega_0 R_1 C \ll 1. \tag{12.111}$$

The equivalent inductance of the gyrator is given by

$$L \cong R_1 R_2 C; \; \omega_0 R_1 C \ll 1. \tag{12.112}$$

The impedance of the gyrator in Fig. 12.77(a) at resonance is given by

$$Z_G = R_1 + j\omega_0 R_1 R_2 C, \tag{12.113}$$

so the quality factor of the gyrator (at resonance) is given by

$$Q = \frac{\mathrm{Im}[Z_G(\omega_0)]}{\mathrm{Re}[Z_G(\omega_0)]} = \omega_0 R_2 C. \tag{12.114}$$

The impedance of the tuned circuit in Fig. 12.77(b) at resonance is given by

$$Z_0 = R_1 + j\omega_0 R_1 R_2 C + \frac{1}{j\omega_0 C_1} = R_1, \tag{12.115}$$

which implies

$$j\omega_0 R_1 R_2 C + \frac{1}{j\omega_0 C_1} = 0 \Rightarrow R_1 R_2 C C_1 = \frac{1}{\omega_0^2}. \tag{12.116}$$

Example 12.49. Design a gyrator-based tuned circuit subject to the following specifications:

$$\text{resonant frequency} = f_0 = 400 \text{ Hz},$$
$$\text{quality factor} = Q = 10,$$
$$\text{impedance at resonance} = Z_0 = 100\ \Omega.$$

Solution: We assume a suitable op amp is available (which is almost always the case for frequencies at which the gyrator described above is useful). From (12.115) and the specification on the impedance at resonance, we have

$$R_1 = Z_0 = 100\ \Omega. \tag{12.117}$$

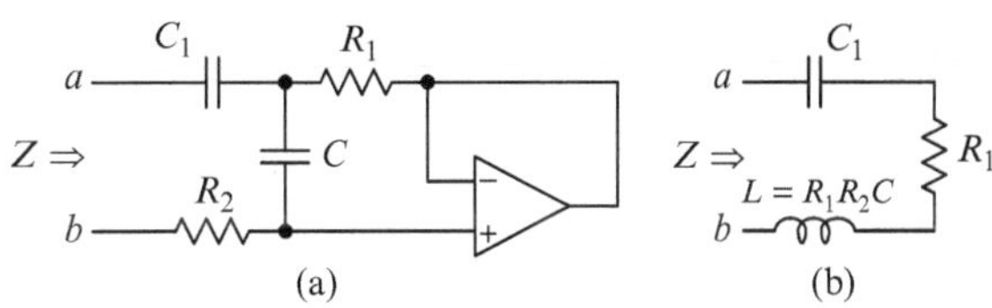

Fig. 12.77 The circuits (a) and (b) are equivalent if $\omega R_1 C \ll 1$

[24]The impedance at resonance is of interest because it determines the maximum current required of a circuit driving the impedance.

Based upon (12.111), we require

$$\frac{1}{R_1 C} = 100\omega_0,$$

which yields

$$C = \frac{1}{100\omega_0 R_1} = 39.79 \text{ nF}.$$

From (12.114)

$$R_2 = \frac{Q}{\omega_0 C} = 100 \text{ k}\Omega.$$

Finally, from (12.110)

$$C_1 = \frac{1}{\omega_0{}^2 L} = 397.89 \text{ nF}.$$

The impedance of the gyrator-based circuit is given by

$$Z_G = \frac{R_1(1 + j\omega R_2 C)}{1 + j\omega R_1 C} + \frac{1}{j\omega C_1}.$$

The impedance of an equivalent passive *RLC* circuit is given by

$$Z_{RLC} = Z_0\left[1 + jQ\left(\frac{\omega}{\omega_0} - \frac{\omega_0}{\omega}\right)\right]$$

We may express the magnitudes of the normalized impedances in dB as

$$|Z_G|_{\text{dB}} = 20 \log\left(\left|\frac{Z_G}{R_1}\right|\right),$$

$$|Z_{RLC}|_{\text{dB}} = 20 \log\left(\left|\frac{Z_{RLC}}{R_1}\right|\right).$$

Figure 12.78 shows graphs of the magnitudes of the impedances versus frequency, on the same axes. The graphs are almost indistinguishable for $f < 20$ kHz, and indicate that the design is successful. The resonant peak appears to be at 400 Hz, in good agreement with the specification.

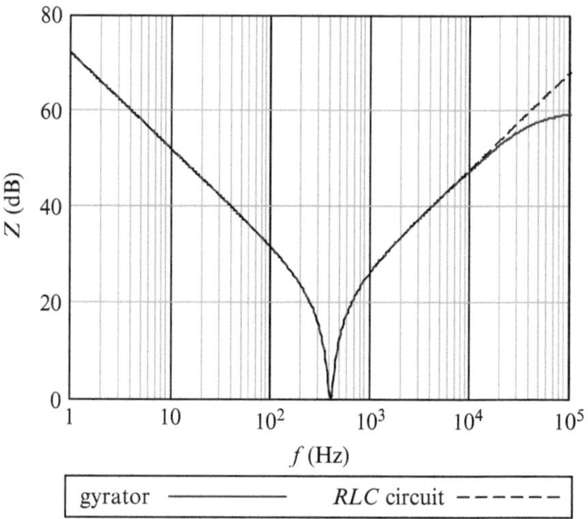

Fig. 12.78 See Example 12.49

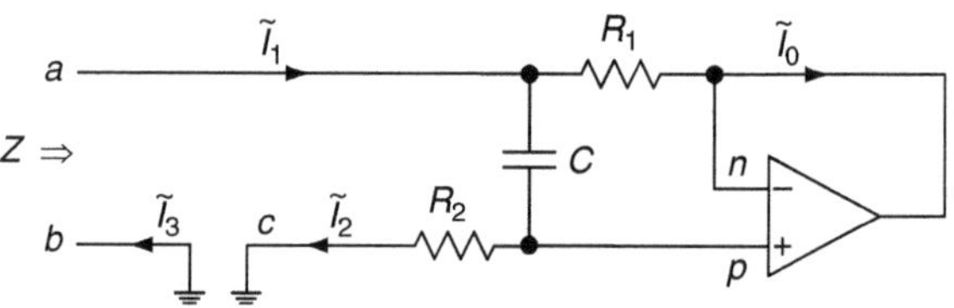

Fig. 12.79 Current paths in a gyrator

In the development and example above, we have glossed over a somewhat subtle point that can cause confusion in simulations or physical implementations of the circuit. To address this point, we again consider the gyrator in Fig. 12.75, which we have re-drawn in Fig. 12.79. In order to define an impedance at the terminals *a*–*b*, the currents $\tilde{I}_1$ and $\tilde{I}_3$ must be equal; that is, the terminal pair *a*–*b* must be a port. Recall from above that the current $\tilde{I}_1$ entering terminal *a* and the current $\tilde{I}_2$ through R_2 are not equal, because the current $\tilde{I}_0$ through R_1 is not zero, and no current enters the *n* and *p* terminals of the op amp. Thus we cannot define an impedance at the terminals *a*–*c*. Recall also that an op amp has no ground or common terminal, and that the reference or ground for the voltages in an op amp is established by the manner in which the supply ($\pm V_{CC}$) and the load are connected to the op amp. In this circuit, the current $\tilde{I}_3$ coming up from ground and out through terminal *b* consists of the current $\tilde{I}_2$ plus the current driven to ground (through the power supply) by the op amp. Showing the latter current in the circuit diagram would require using a much more complex model for the op amp. You may use a simulation program of your choosing to explore this issue.

12.22 Circuit Elements and Physical Circuit Components

To conclude this chapter, we offer a brief discussion of differences between real and ideal circuit elements. Some knowledge of such differences can help you understand and perhaps avoid many otherwise perplexing problems.

Ideal elements do not exist. Physical resistors, capacitors, and inductors are constructed such that resistance, capacitance, and inductance are their *dominant* properties under certain operating conditions, but each exhibits all three phenomena to greater or lesser degree, depending upon the frequency of the voltage across or current through the component and other factors. Secondary phenomena, such as inductance and capacitance exhibited by a physical resistor often are called *parasitic* or *residual* properties. In extreme cases, a residual property can even become dominant; for example, at sufficiently high frequencies, a physical resistor might behave more like a capacitor than a resistor.

At sufficiently high frequencies, a physical resistor, capacitor, or inductor must be modeled as a two-terminal circuit composed of three elements (in general). Such models are called *lumped-parameter models*. At even higher frequencies, even those models fail. Every lumped-parameter model is limited to wavelengths much larger than the physical dimensions of the associated component. Thus conventional discrete components having dimensions on the order of 1 cm and larger can be represented by lumped parameters only for wavelengths longer than about $\lambda_{\min} = 10$ cm; i.e., for frequencies lower than about

$$f_{\max} = \frac{c}{\lambda_{\min}} = \frac{3 \times 10^8 \text{ ms}^{-1}}{0.1 \text{ m}} = 3 \text{ GHz}.$$

The overall dimensions of and the lengths of conductors in some discrete-component circuits are much larger than those of the individual components. Applicability of lumped-parameter models for such circuits is limited to even lower frequencies – typically a few hundred MHz.

Lumped-parameter models can be used for integrated components and circuits up to much higher frequencies than for discrete components and circuits because integrated components and circuits are so small. In modern integrated circuits, dimensions of subcircuits are on the order of 1–10 μm. The corresponding limiting frequency for lumped-parameter models for subcircuits is thus on the order of

$$f_{\max} = \frac{c}{\lambda_{\min}} = \frac{3 \times 10^8 \text{ ms}^{-1}}{10^{-5} \text{ m}} = 30 \text{ THz}.$$

Stray capacitance and inductance also become more problematic at higher frequencies because of the time-derivative dependence of the resulting currents and voltages. For example, if a stray capacitance is in parallel with an inductor, the current drawn away from the inductor, given by $i_C = C\, dv_L/dt$, is proportional to the frequency of the voltage across the inductor. At a sufficiently high frequency, if the lumped-constant model holds up, the inductor will begin to behave more like a capacitor.

Other phenomena that can come into play at higher frequencies include skin effect and proximity effect, described in Chapter 2. These effects cause a current to occupy less than the whole cross-sections of a conductor, so the effective resistance of the conductor is increased. Skin effect and proximity effect have essentially the same root cause, that being eddy currents introduced in conductors by a time-varying magnetic field. Skin effect arises from self-induced eddy currents and proximity effect from mutually induced eddy currents. In both cases, the eddy currents increase with frequency, so the effective (ac) resistance of a conductor increases with frequency. The ac resistance of a conductor or coil can be ten or more times the dc resistance.

Although a principal concern, frequency is not the only concern. Circuit components also are sensitive in varying degrees to temperature, moisture, and electromagnetic interference (e.g., from other, nearby electrical equipment). A circuit designer must consider not only how to achieve the desired function (e.g., amplification), but also the particular kinds of resistors, capacitors, and other elements to be used, how they should be placed and interconnected on a circuit board, how much power each will dissipate, whether they might require forced cooling, and whether and what kind of shielding against electromagnetic interference might be necessary. You will learn more about such things as you progress through your curriculum and later in your professional life. Here, we can give only a brief and mainly descriptive introduction to a few such topics. We begin by describing how physical elements might be modeled by more complex two-terminal circuits.

12.22.1 Resistors

Resistors are made of various materials in various configurations. Ordinary wirewound resistors consist of wire wrapped around a ceramic core, which is also how many inductors are made. In so-called non-inductive wirewound resistors, half the turns are clockwise and the other, overlapping half of the turns are counterclockwise, which theoretically eliminates inductance. Such elimination is imperfect, and some inductance remains. Several other kinds of discrete resistors are made by depositing a layer of carbon or metal film on a cylindrical core and then cutting away some of the material to leave a spiral ribbon of resistive material on the core, as illustrated by Fig. 12.80. These also have a coil-like structure and become inductive at sufficiently high frequencies.

Composition resistors consist of a mix of carbon or metal granules in an insulating material such as phenol resin (phenolic resistors) or ceramic (cermet resistors). Inductance exhibited by such resistors arises primarily from the leads (recall that even straight wires exhibit inductance). Such inductance can be minimized by keeping leads short.

Some discrete surface-mount resistors are essentially miniature versions of the spiral structures described above, but encased in a small flat case and having relatively short leads. Such structures virtually eliminate lead inductance, but inductance arising from the spiral structure remains. Other surface-mount resistors (chip resistors) consist of a flat film of resistive material deposited on a flat substrate. For such resistors, capacitance typically becomes problematic at frequencies below those at which inductance would become an issue.

Chip resistors and resistors in integrated circuits are planar, consisting of a film of resistive material deposited on a flat substrate, either in a straight line or in zigzag fashion, as illustrated by Fig. 12.81. Planar resistors are typically more capacitive than inductive. Those that zigzag can be less inductive than straight-line structures because the magnetic effects that give rise to inductance tend to cancel in neighboring conductors when the currents are in opposite directions. However, a zigzag structure might exhibit more capacitance than an otherwise equivalent straight-line structure because of the proximity of the neighboring conductors. Proximity effect might also be significant.

For reasons given above and others, a resistor is not just a resistor, at least not at sufficiently high frequencies. Figure 12.82 shows a circuit model for a resistor. The resistance R is the intended resistance. The inductance L is attributed to spiral structure. In conventional discrete and surface-mount chip resistors, the capacitance C arises because (with alternating current) the ends of a resistor (and possibly neighboring turns on spiral and wirewound structures) are alternately and oppositely charged. In integrated resistors, the capacitance C arises from that effect and from the proximity of the planar film to other conducting structures on adjacent layers (stray capacitance). In conventional discrete resistors, inductance usually appears before (at a lower frequency than) capacitance, and arises

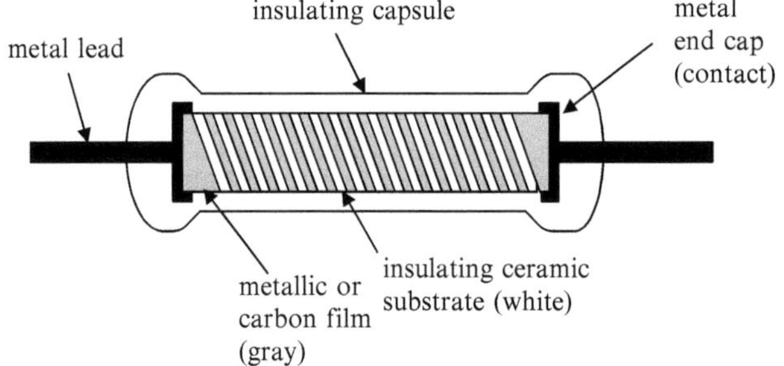

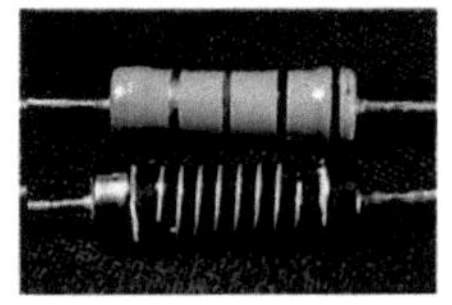

Fig. 12.80 Spiral film resistor. The inset photograph shows a carbon-film resistor with and without its insulating capsule (Photograph courtesy of Betty Fuller, on behalf of the late Jim Fuller.)

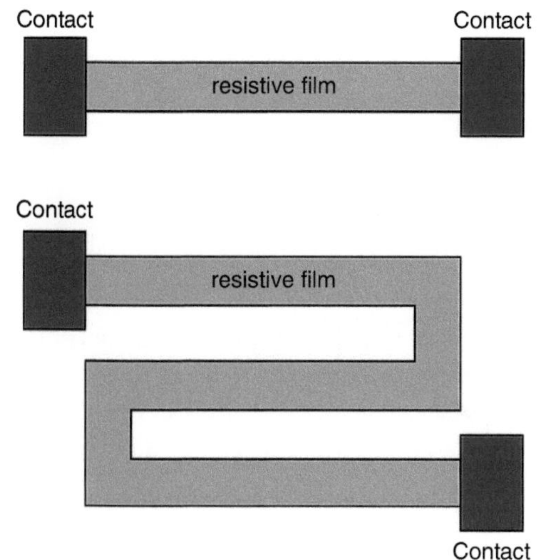

Fig. 12.81 Planar metallic-film resistor

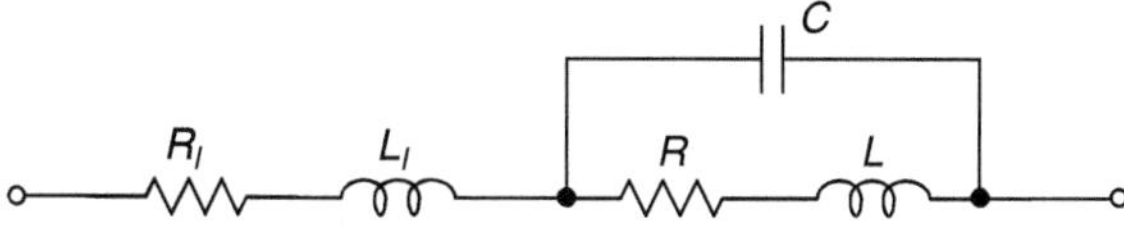

Fig. 12.82 Circuit model for a resistor

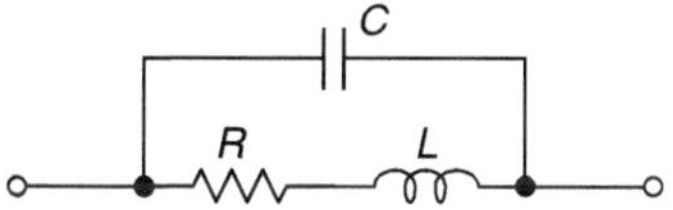

Fig. 12.83 Simplified circuit model for a resistor

from the spiral structure and (in axial resistors) from the leads. The resistance R_l and inductance L_l are attributed to the leads, depend strongly on lead length, and are virtually non-existent in surface-mount and integrated resistors. However, there can be worrisome contact effects, which are nonlinear. In the discussion below, we ignore lead resistance and inductance and contact effects, and use the model shown in Fig. 12.83.

Composition and film resistors may exhibit capacitances on the order of 1 pF, whereas the capacitance of a wirewound resistor can be on the order of 10 pF. Spiral-structure resistors, such as carbon-film and metal-film resistors can exhibit inductances on the order of 10 nH and wirewound resistors can exhibit inductances on the order of 10 μH. The inductance of composition resistors comes mainly from the leads and varies greatly with lead length, but typically is

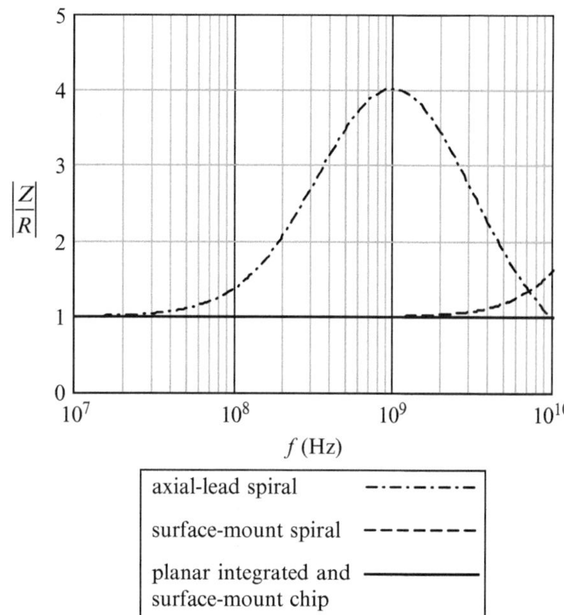

Fig. 12.84 Normalized impedance for three kinds of resistors

no more than a few nH and can be as small as a few tens of pH.

Inductance arises at lower frequencies than capacitance in discrete, axial- and orthogonal-lead structures. Within that group, inductance usually arises at lower frequencies in resistors having coil-like structures, such as wirewound resistors and spiral carbon-film resistors, and at higher frequencies in surface-mount resistors having short leads. Capacitance can arise first in planar resistors, such as those in integrated circuits.

In axial-lead spiral resistors, capacitance typically comes into play at frequencies on the order of 1 GHz, where the impedance is already far from ideal. Capacitance arises at much higher frequencies in surface mount and integrated resistors, in some cases at frequencies near those where lumped-parameter models begin to fall apart for other reasons.

Figure 12.84 shows graphs of typical impedances (magnitude, normalized to the dc resistance) versus frequency for axial spiral, surface-mount spiral, and surface-mount chip and planar integrated resistors.[25] Actual (measured) impedances will differ from the graph to varying degrees, but as a rule, we may treat all resistors except some wirewound resistors as ideal

[25]Adapted from Bogatin, Eric. *Signal and Power Integrity – Simplified*. Prentice-Hall, Englewood Cliffs, NJ. 2010.

for frequencies up to at least 10 MHz and perhaps up to 100 MHz. The excluded wirewound types are mainly those used at high power and at frequencies well below those at which inductance would come into play.

Do not assign too much significance to the models discussed above. The model components and their associations with geometry and materials are useful guides in resistor design and selection, but such association is loose, especially at very high frequencies, where the wavelengths of currents and voltages approach the physical dimensions of components. Ultimately, what matters is the actual (measured) impedance over the range of frequencies of interest. The purpose of this treatment is only to point out some of the main reasons why resistors are not always just resistors and that some kinds of resistors are much better than others for high-frequency applications. A circuit designer's selection of components is guided in part by such knowledge.

Physical resistors are sensitive to temperature, some more than others. Temperature coefficients of resistivity are useful guides to selection where high temperatures or wide-ranging temperatures are anticipated. Carbon composition and carbon-film resistors have relatively large temperature coefficients, on the order of 5000 ppm. Metal-film and wirewound resistors are better, having temperature coefficients of 100 ppm or smaller, and are preferred for that reason (and others) in applications requiring stable resistance values. Variation of resistance with temperature is one of the more important departures of real resistors from the ideal model.

Example 12.50. The temperature coefficient for a certain 10 kΩ resistor is 500 ppm at 20°C. Construct a graph showing resistance versus temperature T for $20°\text{C} \leq T \leq 200°\text{C}$, expressed as a percent of the nominal resistance.

Solution: The resistance as a function of temperature is given by

$$R = R_0[1 + \alpha(T - T_0)],$$

where

$$R_0 = 10\ \text{k}\Omega; \quad \alpha = 500 \times 10^{-6}\ (°\text{C})^{-1}; \quad T_0 = 20°\text{C}.$$

The percent variation of resistance with temperature is given by

$$\frac{100(R - R_0)}{R_0} = 100\alpha(T - T_0).$$

Figure 12.85 shows a graph of the expression above. For $T = 100\,°\text{C}$, a temperature reached or exceeded under normal operating conditions in some microelectronic circuits, the resistance is increased by 4%. Even higher temperatures are attained in a few such circuits.

Physical resistors also are sources of electrical noise. Carbon and carbon-film resistors are generally noisier than metal-film and wirewound resistors. As a rule, carbon composite and carbon-film resistors should be used only in undemanding applications, where noise and precision are relatively unimportant. Most circuit designers nowadays use metal-film resistors of either conventional or (better) surface-mount construction.

Skin effect rarely is a problem in film resistors because the films are generally thinner than their skin depths at normal operating frequencies. Exceptions arise in some microwave circuits. For example, the skin depth of nichrome at 100 GHz is in the neighborhood of 100 nm, which is on the order of the thickness of films in integrated resistors and conductors (circa 2006). Skin effect might be a problem in some

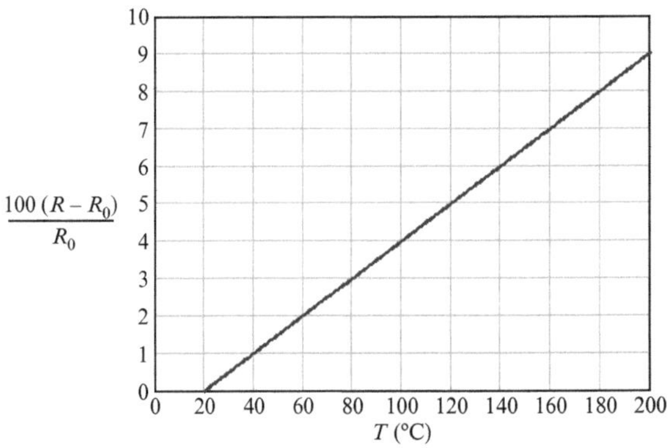

Fig. 12.85 See Example 12.50

wirewound resistors, but generally inductance limits use of such resistors to frequencies below those at which skin effect becomes important. But there are some interesting exceptions. For example, the effective cross-section area of conductors used in some inductors for high-power radio and television transmitters must be quite large to carry the currents involved. For AM radio, the inductors operate in the neighborhood of 1 MHz, where the skin depth of copper is about 50 μm, so the inductors might be made of large-diameter, thin-walled copper tube and might even be silver-plated to reduce resistivity and the associated losses. Most of a solid conductor having the same diameter would be wasted.

Finally, the tolerances of carbon composition and carbon-film resistors typically range from $\pm 20\%$ to $\pm 5\%$, whereas metal-film and wirewound resistors are typically $\pm 1\%$ and better. Tolerances of integrated film resistors might be somewhat poorer because it is difficult to achieve precise control of the composition, thicknesses, and widths of films in IC manufacturing. These problems affect every component on a particular chip in very nearly the same way, so are less problematic in circuits where only ratios of resistances and not the resistances themselves must be precise. Where individual resistances are important, one often uses resistors that can be laser trimmed to precise values.

12.22.2 Inductors

Circuit models for inductors are essentially identical to those for resistors (Fig. 12.83), except that inductance is the desired property and is dominant at low frequencies. The measured (ac) resistance of an inductor generally increases with frequency. Some of the increase is due to skin effect. At any frequency, the skin depth of the good conductors usually used in inductors is much smaller than that of the resistive materials used in resistors, so skin effect comes into play in inductors at lower frequencies than in resistors. The effective resistance of an inductor also increases with frequency because of the proximity effect and the fact that conductors in inductors can be quite close together. Adjacent turns in inductors for high-frequency operation usually are kept relatively far apart to minimize proximity effect. Finally, for iron-core and other magnetic-core inductors, the resistive component of the impedance arises partly from core losses due to eddy currents. Eddy current losses can be reduced by using laminated or granular core material.

Capacitance in the model for an inductor arises, as in a resistor, from the opposite charges at the ends of the inductor as the voltage across the inductor alternates in sign. At higher frequencies, capacitance also arises between neighboring turns. This distributed capacitance can become so large that the current in the coil varies significantly along the length of the coil, invalidating the lumped-parameter model in Fig. 12.83.

Most inductors are designed to operate at or near certain frequencies, for example, those frequencies assigned by the FCC for various uses. An important property of an inductor is its *quality factor* Q, defined as the ratio of inductive reactance to resistance at the specified operating frequency. Thus

$$Q = \frac{\omega_0 L}{R_{ac}}, \tag{12.118}$$

where R_{ac} is the effective resistance at the frequency f_0. The quality factor for an inductor is a measure of how closely the inductor approximates an ideal inductor at the intended operating frequency. The quality factor for an ideal inductor would be infinite at any frequency.

Inductors also are characterized by *self-resonance*, which occurs at a frequency f_r determined (theoretically) by the inductance, resistance, and capacitance in the model shown in Fig. 12.83:

$$\omega_r = \sqrt{\frac{1}{LC} - \left(\frac{R}{L}\right)^2} = \frac{1}{\sqrt{LC}}\sqrt{1 - \frac{1}{Q^2}}. \tag{12.119}$$

The self-resonant frequency is the highest frequency for which the coil is inductive, and thus establishes an upper limit on the frequency for which the coil can provide inductive reactance.

Manufacturers of inductors (for high-frequency applications) usually specify inductance, quality factor, dc resistance, permissible power dissipation (or maximum rms current), and (measured) self-resonant frequency. A typical inductor intended for use at or near 100 MHz might have inductance in the range from 1 to 100 nH, a quality factor in the neighborhood of 20, dc resistance of 1 mΩ to 1 Ω, a self-resonant frequency in the range 500 MHz to 5 GHz, and maximum rms current in the neighborhood of 500 mA.

The self-resonant frequency of an inductor usually is well above the intended operating frequency, for reasons given above. Specifying the self-resonant frequency is a way of specifying the highest possible operating frequency, and also is a backhanded way of specifying (approximately) the effective shunt capacitance C. From specified values for operating frequency, inductance, and operating frequency, we have from (12.118) that

$$R_{ac} = \frac{\omega_0 L}{Q}.$$

From (12.119), we obtain

$$C = \frac{Q^2 - 1}{Q^2 {\omega_r}^2 L} \cong \frac{1}{{\omega_r}^2 L}, \; Q^2 \gg 1. \tag{12.120}$$

The relation is a coarse approximation because the circuit model parameters Q, L at the self-resonant frequency are in general not the same as those at the intended operating frequency. Nonetheless, the capacitance computed using (12.120) provides a starting point for determining the capacitance to be placed in parallel with the inductor if the inductor is to be used in a tuned circuit (which is the usual case).

The model for an inductor shown in Fig. 12.83 often provides a poor fit to measurements of impedance as a function of frequency, Q as a function of frequency, and self-resonant frequency, especially for inductors designed for high-frequency operation. The model shown in Fig. 12.86 can provide a much better fit.[26] Unfortunately, few (if any) manufacturers provide parameter values for this model in their data sheets.

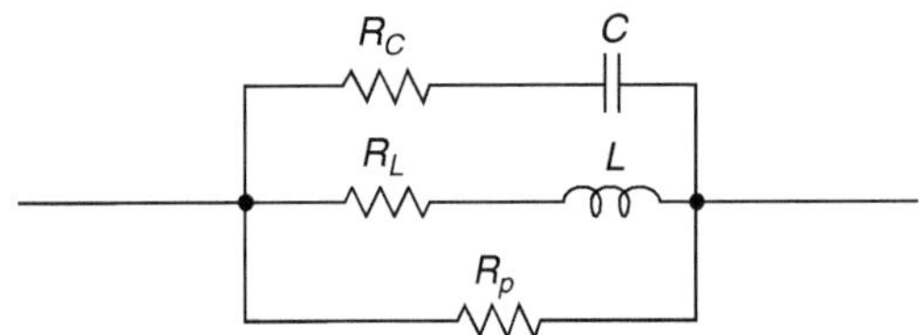

Fig. 12.86 Improved circuit model for an inductor

[26]Leslie Green, *RF Inductor modeling for the 21st Century*, ***EDN***, September 27, 2001.

Example 12.51. Figure 12.87 shows a model used by one manufacturer for air-core rf coils.[27] The resistance r_L accounts for skin effect and is given by $r_L = k\sqrt{f}$, where k is an experimentally determined parameter.

In Fig. 12.87, let $R_C = 6\ \Omega$, $R_S = 1\ \text{m}\Omega$, $C = 68\ \text{fF} = 68 \times 10^{-15}\ \text{F}$, $L = 1\ \text{nH}$, and $k = 3.4 \times 10^{-6}\ \Omega\text{Hz}^{-1/2}$. Construct a graph of the magnitude in dB of the equivalent impedance Z_{ab} at the terminals a–b versus frequency f, for 100 MHz $\leq f \leq$ 100 GHz. Use a logarithmic scale for frequency and normalize the impedance to the impedance at the self-resonant frequency. Show also a graph of the magnitude in dB of the impedance of an ideal inductor having inductance L, also normalized to the impedance at the self-resonant frequency.

Solution: From (12.88), the self-resonant frequency of the circuit is given by

$$f_r = \frac{1}{2\pi\sqrt{LC}}\sqrt{\frac{r_L^2(f_r)C - L}{R_C^2 C - L}} = \frac{1}{2\pi\sqrt{LC}}\sqrt{\frac{k^2 f_r C - L}{R_C^2 C - L}},$$

which yields

$$f_r = \frac{-k^2 C + \sqrt{k^4 C^2 - 16\pi^2 L^2 C\left(R_C^2 C - L\right)}}{8\pi^2 LC\left(L - R_C^2 C\right)} \cong 19.3\ \text{GHz}.$$

From Fig. 12.87, the equivalent impedance at the terminals a–b is given by

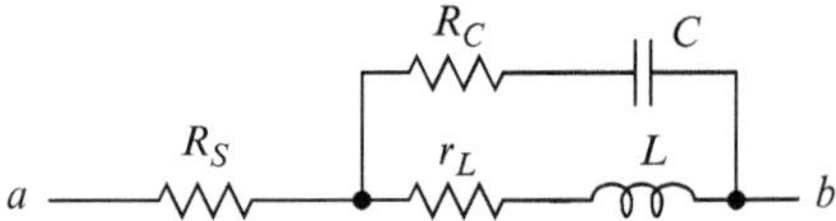

Fig. 12.87 See Example 12.51

[27]See http://www.coilcraft.com/models.cfm

$$Z = R_S + \frac{1}{\left[R_C + (j2\pi f C)^{-1}\right]^{-1} + \left[k\sqrt{f} + j2\pi f L\right]^{-1}}$$
$$= R_S + \frac{1}{\left[R_C + (j\omega C)^{-1}\right]^{-1} + \left[k\sqrt{\omega/(2\pi)} + j\omega L\right]^{-1}}. \qquad (12.121)$$

The equivalent impedance at the self-resonant frequency f_r is

$$Z_r = Z(j\omega_r) = 2.27\ \text{k}\Omega,$$

where $\omega_r = 2\pi f_r$. Figure 12.88 shows graphs of the magnitude of the normalized impedance of the model in Fig. 12.87 (solid line) and of the normalized impedance of an ideal inductor having inductance L (dashed line), where Z is given by (12.121). The normalized impedance (in dB) of the model and that of the associated ideal inductor are given by

$$|Z|(\text{dB}) = 20\log\left|\frac{Z(j\omega)}{Z(j\omega_r)}\right|,$$
$$|Z_L|(\text{dB}) = 20\log\left|\frac{j\omega L}{Z(j\omega_r)}\right|,$$

respectively. Figure 12.88 indicates that the physical inductor approximates an ideal inductor for 10 MHz $< f <$ 10 GHz. Below 10 MHz, the series resistance R_S becomes significant and above 10 GHz, the shunt capacitance is significant.

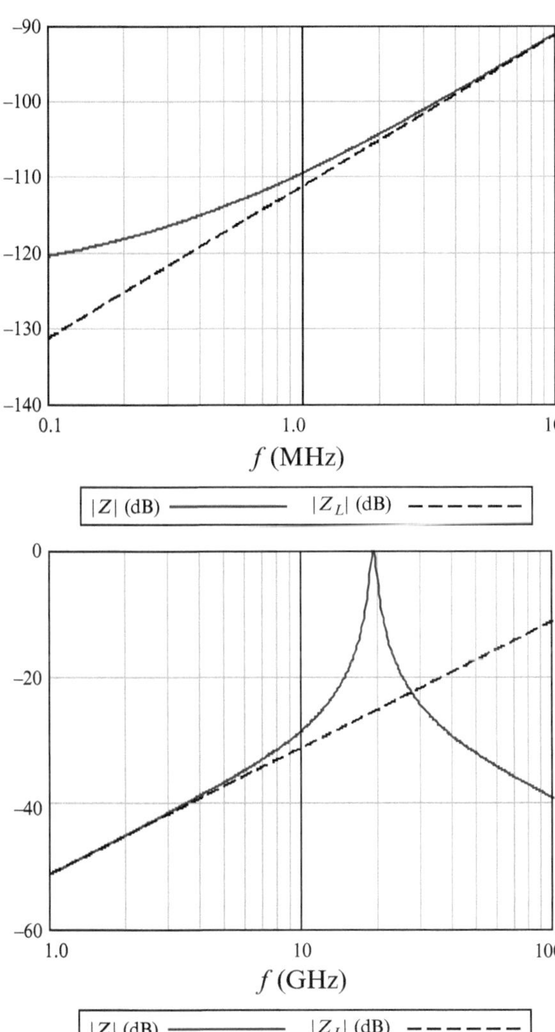

Fig. 12.88 See Example 12.51

The models discussed above are for air-core inductors. Models for magnetic-core inductors are more complex, including frequency-dependent resistances that account for core losses and (in some applications) nonlinear effects.

As pointed out above, inductors tend to be bulky, especially at low frequencies. In some cases, inductance can be simulated by active *RC* circuits, but for high-frequency and high-power applications, physical coils must be used.

12.22.3 Capacitors

A capacitor consists essentially of conducting plates separated by an insulating material, which can be air (or other gas), a liquid (usually, oil), or a solid such as paper, mica, polyester, polystyrene, or glass. The plates can consist of a roll of strips of metallic film separated by dielectric material or an alternating stack of metallic and dielectric materials. However made, capacitors (except some air-dielectric capacitors) are enclosed in some kind of air- and moisture-tight container. There are many more kinds of capacitors than

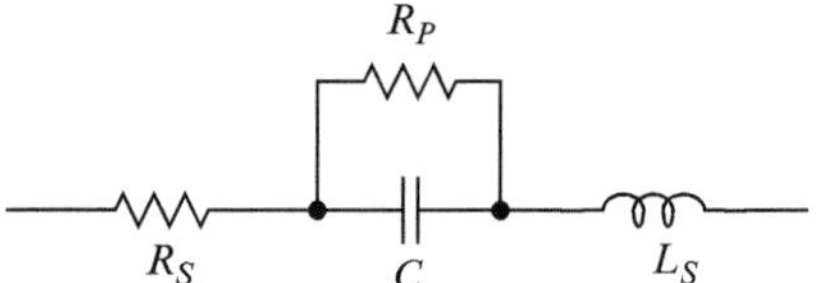

Fig. 12.89 Circuit model for a capacitor

can be described here.[28] The principal applications of capacitors are described in Chapter 8.

Figure 12.89 shows a common circuit model for a capacitor. The capacitance C is the nominal capacitance. The parallel resistance R_p is largely leakage resistance. The resistance R_S is the **equivalent series resistance (*ESR*)** and the inductance L_S is the **equivalent series inductance (*ESL*)**. *ESR* arises from lead resistance, contacts, and dielectric losses. *ESL* arises largely from leads. The parallel (leakage) resistance R_p is typically several MΩ and can often be ignored in high-frequency applications. All of the parameters in the model are frequency-dependent and temperature-dependent. They can also change with age, current, and applied voltage. We ignore those refinements, but they are important in most applications.

Power dissipation in the *ESR* generates heat. The heat can be significant, especially in power supplies where the power dissipated in the *ESR* is due to ripple in a rectified voltage. Often, *ESR* and maximum ripple current are specified for capacitors intended for use in such applications (dc current does not cause significant power dissipation in a capacitor). *ESL* causes self-resonance. Above the self-resonant frequency, a capacitor is inductive, not capacitive.

A capacitor can be characterized by a quality factor, which is the ratio of the capacitive reactance to the series resistance in the model of Fig. 12.89. Thus

$$Q = \frac{1/(\omega C_s)}{R_s} = \frac{1}{\omega R_s C_s}.$$

The reciprocal of the quality factor is the **dissipation factor**:

$$D = \frac{1}{Q} = \omega R_s C_s = \text{ dissipation factor.} \quad (12.122)$$

[28] See www.capacitorindustries.com

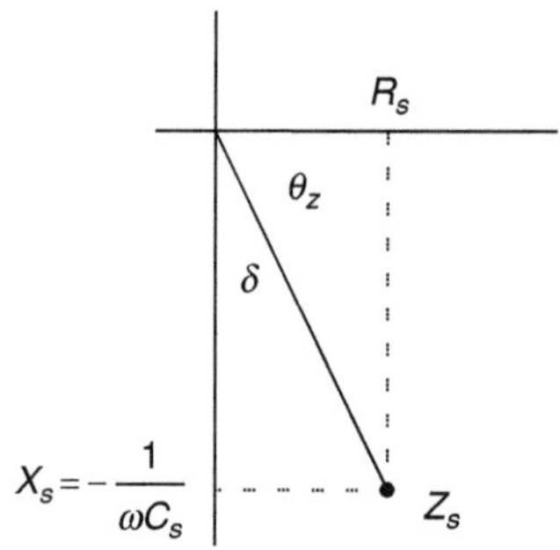

Fig. 12.90 See (12.122), (12.123), and surrounding discussion

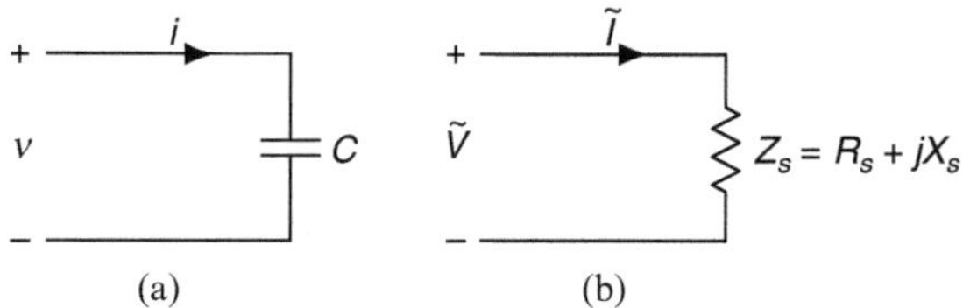

Fig. 12.91 See Example 12.52

The dissipation factor for a capacitor is (as shown below) a measure of losses in the capacitor. Figure 12.90 depicts the real and imaginary parts of the series impedance $Z_s = R_s + jZ_s$ of a capacitor at some frequency $f = \omega/(2\pi)$. *If the losses are small*, then the angle δ also is small, and the following relations exist among the dissipation factor D, the **loss angle** δ, and the angle of the equivalent impedance Z_s:

$$D = \omega R_s C_s = \frac{R_s}{-X_s} = -\tan(\delta) \cong \cos(\theta_Z). \quad (12.123)$$

Example 12.52. In Fig. 12.91(a), $C = 100\ \mu\text{F}$, $v(t) = V\cos(\omega t)$ with $f = 60$ Hz and $V = 15$ V. The dissipation factor for the capacitor is 0.03. The *ESL* and parallel resistance R_p can be neglected. Find the power dissipated in the capacitor (in the *ESR*).

Solution: We may determine the current through the model and the power dissipated in the *ESR*, as follows: we refer to the model in Fig. 12.91(b), where (see Fig. 12.90)

$$X_s = -\frac{1}{\omega C} = -\frac{1}{2\pi(60\text{ Hz})(100\ \mu\text{F})} \cong -26.5\ \Omega$$
$$R_s = -X_s D = (26.5\ \Omega)(0.03) \cong 0.80\ \Omega.$$

Thus

$$Z_s = (0.80 - j26.5)\ \Omega \cong 26.5\angle - 1.54\ \Omega$$

and

$$\tilde{I} = \frac{\tilde{V}}{Z_s} = 565\angle 1.54\ \text{mA}.$$

The power dissipated in the ESR is

$$P = I_{rms}{}^2 R_s = \frac{|\tilde{I}|^2}{2} R_s = 127\ \text{mW}.$$

Summary

You should have gleaned from the discussion above that components you might have thought simple are not so simple, after all. The behavior of physical resistors, inductors, and capacitors can be quite complex, depending upon demands placed on them in various applications.[29] Moreover, which properties of a resistor, inductor, or capacitor are important differ from one application to another. A capacitor or inductor suitable for reducing ripple in a large power supply would be completely unsuitable for use in a high-frequency tuned circuit. A high-precision, high-power wirewound resistor might be an excellent choice in an audio speaker crossover network, but a poor one in high-frequency applications, where it might behave more like an inductor than an resistor.

You should take at least three ideas away from this brief discussion:

1. There is no such thing as an ideal resistor, inductor, or capacitor. Physical resistors, inductors, and capacitors are merely components for which one of those properties is dominant under certain operating conditions.
2. There is no such thing as a universal resistor, inductor, or capacitor. For example, inductors designed for use in tuned circuits at high frequencies are quite different from those designed for smoothing currents in large power supplies. A component that is excellent in one application (e.g., for a certain range of frequencies, powers, and temperatures) might be completely unsuitable in another. Most resistors, capacitors, and inductors are designed and constructed with specific applications in mind.
3. In many applications, a designer must specify (or at least consider) not only the presumed dominant property of a component (e.g., capacitance), but also other properties relevant to the application at hand, such as size, weight, cost, precision, rated power dissipation, expected lifetime (often given as *mean time between failure* or MTBF), solderability, and resistance to moisture and mechanical and thermal shock. Some of the other properties are:

- For a resistor: Series inductance, temperature coefficient, self-heating coefficient (or a power derating chart or graph), and shunt capacitance
- For an inductor: Nominal operating frequency, quality factor, dc resistance, maximum current, and self-resonant frequency
- For a capacitor: Temperature coefficient; leakage resistance; maximum voltage; maximum current; maximum temperature; and *ESR*, *ESL*, quality factor, loss angle, or dissipation factor.

As you can see, choosing components is not a trivial task.

A circuit designer must also be aware of skin effect, proximity effect, stray capacitance and inductance, sources and effects of electrical noise, limits on lumped-parameter models, power and heat budgets, and a number of other considerations in particular applications. Those who design integrated circuits must add a large number of other items to this list. Experienced and competent circuit designers are worth their weights in gold.

Manufacturers continue to improve and add to their offerings. Components continue to get better, smaller, and less expensive. Better and less expensive are important for obvious reasons. Smaller has allowed development of personal computers, laptops, cell phones, PDA's, and a host of other gadgets now considered indispensable. Less obvious is the fact that smaller size has extended the applicability of analytical techniques treated in this book. Kirchhoff's laws, for example, can be applied to a circuit only if the smallest wavelength is much larger than the longest conductor or component in the circuit. Thus the highest frequencies for which Kirchhoff's laws can be applied to an integrated circuit are orders of magnitude

[29] http://www.eigroup.org/cmc/downloads/r2_cmc/r2_cmc_v1.0_r0.0_2005nov12.pdf

higher than those for a typical circuit having the same function in 1950. In this sense, at least, you are getting a lot more for your money than this author did when introduced to this material some years back.

12.23 Problems

Section 12.1 is prerequisite for the following problems.

P 12.1 Express each current or voltage as a single sinusoid in standard form.

(a) $-10\cos(\omega_0 t - 0.5)$ A,
(b) $5\sin(\omega_0 t + 1.57)$ mV,
(c) $[3\cos(\omega_0 t) - 4\sin(\omega_0 t)]$ V,
(d) $-12\sin(\omega_0 t - \pi/2)$ mA,
(e) $[3\sin(\omega_0 t) - 4\cos(\omega_0 t)]$ V.

P 12.2 Express each current or voltage as a single sinusoid in standard form.

(a) $v(t) = [2\cos(\omega_0 t) + 3\cos(\omega_0 t + \pi/4)]$ V,

(b) $i(t) = [\cos(\omega_0 t) - \sin(\omega_0 t + \pi/4)]$ mA,

(c) $i(t) = [10\cos(\omega_0 t) + 30\cos(\omega_0 t + \pi/4) - 20\cos(\omega_0 t - \pi/3)]$ mA,

(d) $v(t) = [\cos(\omega_0 t) + 5\cos(\omega_0 t + \pi/4) - 2\sin(\omega_0 t - \pi/3)]$ mV,

(e) $v(t) = [\cos(\omega_0 t) + 2\cos(\omega_0 t + \pi/6) + 3\cos(\omega_0 t + \pi/3)]$ V.

Section 12.2 is prerequisite for the following problems.

P 12.3 Refer to Fig. P 12.1. In each graph, the dotted line represents a sinusoidal voltage and the solid line represents a sinusoidal current. The current and the voltage have the same frequency. Does the current lead or lag the voltage? Express the voltage in standard form, using the current as the reference.

P 12.4 In a certain circuit, a 1 kHz sinusoidal branch current passes through zero with positive slope 250 μs before the 1 kHz phase reference does so. What is the relative phase of the current?

P 12.5 In a certain circuit, a positive peak of a 2 kHz sinusoid occurs 500 μs later than does the nearest positive peak of the 2 kHz phase reference. What is the relative phase of the sinusoid?

P 12.6 In a certain circuit, the relative phase of a 10 kHz sinusoid is 0.25. Does the sinusoid lead or lag the 10 kHz phase reference? By how much (time)?

P 12.7 In a certain circuit, two 5 kHz sinusoidal voltages pass through zero at the same times, but with opposite slopes. What is the initial phase of one relative to that of the other?

P 12.8 Two sinusoids having equal frequencies and equal amplitudes are in phase quadrature. Is the amplitude of their sum larger than, equal to, or smaller than the amplitude of either one alone?

P 12.9 Two sinusoids having equal frequencies and equal amplitudes are in phase opposition. Is the amplitude of their sum larger than, equal to, or smaller than the amplitude of either one alone?

Section 12.3 is prerequisite for the following problems.

P 12.10 Give the phasor representation of each sinusoid.

(a) $i(t) = -10\cos(\omega_0 t - 0.5)$ A,
(b) $v(t) = 5\sin(\omega_0 t + 1.57)$ mV,
(c) $v(t) = [3\cos(\omega_0 t) - 4\sin(\omega_0 t)]$ V,
(d) $i(t) = -12\sin(\omega_0 t - \pi/2)$ mA,
(e) $v(t) = [3\sin(\omega_0 t) - 4\cos(\omega_0 t)]$ V.

P 12.11 Express the sinusoidal currents and voltages represented by the following phasors as functions of time. Let $V_0\cos(\omega_0 t)$ denote the phase reference in each case.

(a) $\tilde{V} = (50\angle 1.57)$ V;
(b) $\tilde{V} = V_1\angle 0.5 + V_2\angle -0.3$, $V_1 = 5$ V, $V_2 = 2.5$ V;
(c) $\tilde{I} = 250\angle(\pi/4)$ mA $-$ $(120 - j200)$ mA;
(d) $\tilde{V} = (30 + j50)$ mV $+$ $45\angle(\pi/3)$ mV;
(e) $\tilde{I} = (25\angle 0.8)$ mA $-$ $(30\angle -1)$ mA $+$ $(15 + j20)$ mA.

P 12.12 Repeat Problem P 12.2, using phasors.

P 12.13 Use phasors to express each current or voltage as a single sinusoid in standard form.

(a) $i(t) = 4\mathrm{Re}[\exp(j\omega_0 t)] - 5\mathrm{Im}[\exp(j\omega_0 t)]$ A,
(b) $v(t) = \mathrm{Re}[(3 - j4)\exp(j\omega_0 t)]\mathrm{V} - \mathrm{Im}[(2 + j2)\exp(j\omega_0 t)]$ V,

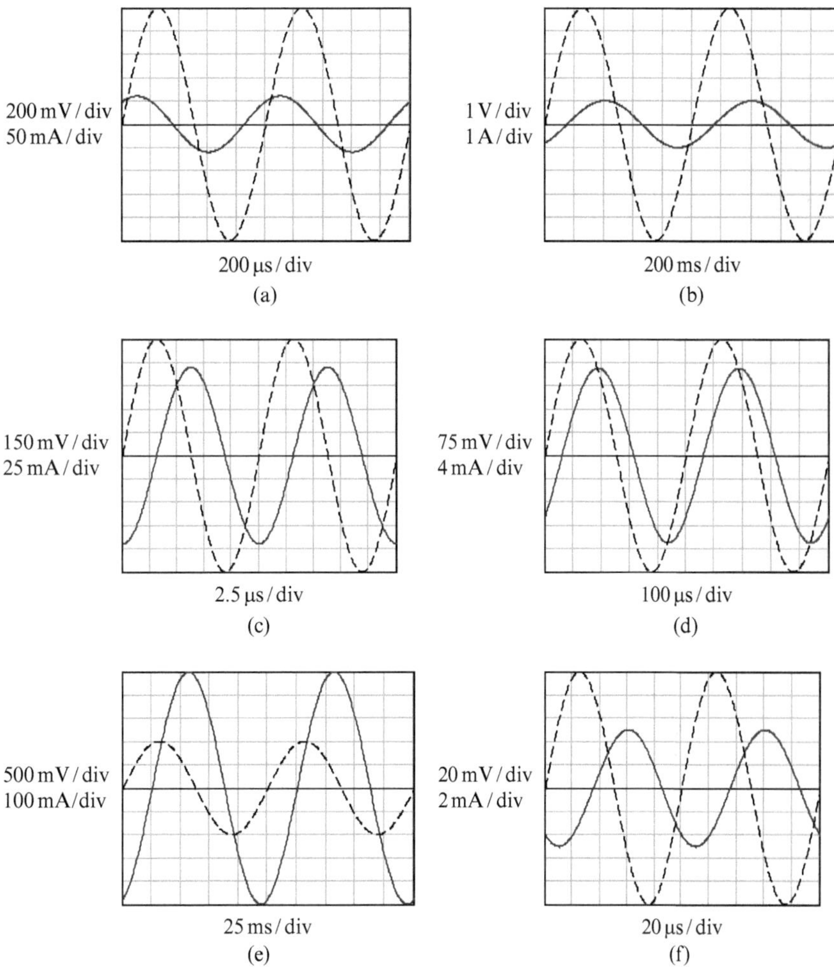

Fig. P 12.1 See Problem P 12.3, 15, 20, 27

(c) $i(t) = 4\mathrm{Re}\left\{\dfrac{j\exp(j\omega_0 t)}{4+j}\right\}$ μA $+4\mathrm{Re}\left\{\dfrac{\exp[j(\omega_0 t + \pi/2)]}{4+j}\right\}$ μA,

(d) $v(t) = [\cos(\omega_0 t) + \cos(\omega_0 t + \pi/2) + \sin(\omega_0 t + \pi/4) + \sin(\omega_0 t + \pi/3)]$ mV,

(e) $v(t) = \mathrm{Im}\left[\dfrac{5\exp(j\pi/3)(1+j)\exp(j\omega_0 t)}{2+j3}\right]$ mV,

(f) $v(t) = \mathrm{Re}\left\{\sum_{n=0}^{3}\exp\left[j\left(\omega_0 t + \dfrac{n\pi}{2}\right)\right]\right\}$ kV.

P 12.14 A certain current is given by $i(t) = I_{dc} + I_{ac}\cos(\omega_0 t)$, with $\omega_0 > 0$, $I_{dc} = 5.00$ mA, and $I_{ac} = 10.0$ mA. (a) Can the current be expressed as a phasor? (b) Can the ac and dc components of the current be represented individually by phasors? (c) What is the rms amplitude of the current? (d) Calculate the average value of the absolute value of the current.

P 12.15 Refer again to Fig. P 12.1. Assume each graph shows a sinusoidal input (dotted line) and the corresponding output (solid line) for a circuit. Use the input as the phase reference and represent the output as a phasor. Give the frequency of the output (or input).

P 12.16 The phasors for two 100 kHz sinusoidal voltages $v_1(t)$, $v_2(t)$ are $\tilde{V}_1 = (5\angle 0.2)$ V and $\tilde{V}_2 = (10\angle 0.4)$ V, respectively. Find the phasor representation $v(t) = v_1(t) + v_2(t)$. Express $v(t)$ as a function of time.

P 12.17 Can you find the product of two sinusoids by forming the product of their phasor representations? If so, show how. If not, explain why not.

P 12.18 Can you find the quotient of two sinusoids by forming the product of their phasor representations? If so, show how. If not, explain why not.

Section 12.4 is prerequisite for the following problems.

P 12.19 A fellow student tells you that the phasor representation of a sinusoid is a vector that rotates about the origin of a complex plane with angular frequency $\omega = 2\pi f$, where f is the frequency of the sinusoid. Is he correct?

P 12.20 Draw a phasor diagram for each pair of voltages in Fig. P 12.1, using the dotted voltage as the phase reference.

P 12.21 Use a phasor diagram to illustrate the fact that

$$\sum_{n=0}^{2} V_0 \cos\left(\omega_0 t + \frac{2n\pi}{3}\right) = 0$$

P 12.22 Draw a phasor diagram for each pair of sinusoids, using $v_1(t)$ as the phase reference. For scale, assume $V_1 = V_2$. Be sure to first put each sinusoid in standard form. If necessary, adjust the initial phase of each so that the reference phasor lies along the positive real axis but the phase difference remains the same.

(a) $v_1(t) = V_1 \cos(\omega_0 t),\ \ v_2(t) = V_2 \cos(\omega_0 t + \pi/3)$;
(b) $v_1(t) = V_1 \cos(\omega_0 t),\ \ v_2(t) = V_2 \cos(\omega_0 t - \pi/6)$;
(c) $v_1(t) = V_1 \cos(\omega_0 t),\ \ v_2(t) = V_2 \sin(\omega_0 t)$;
(d) $v_1(t) = V_1 \cos(\omega_0 t + \pi/6)$,
$v_2(t) = V_2 \cos(\omega_0 t - \pi/3)$;
(e) $v_1(t) = V_1 \sin(\omega_0 t),\ \ v_2(t) = V_2 \cos(\omega_0 t + \pi/4)$;
(f) $v_1(t) = V_1 \cos(\omega_0 t),\ \ f_0 = 1$ kHz,
$v_2(t) = V_2 \cos[\omega_0(t - t_0)],\ \ t_0 = 250$ μs;
(g) $v_1(t) = V_1 \cos(\omega_0 t - \pi/4),\ \ f_0 = 1$ kHz,
$v_2(t) = V_2 \sin[\omega_0(t - t_0)],\ \ t_0 = 250$ μs.

P 12.23 Fig. P 12.2 shows a phasor diagram. Express the voltages as functions of time.

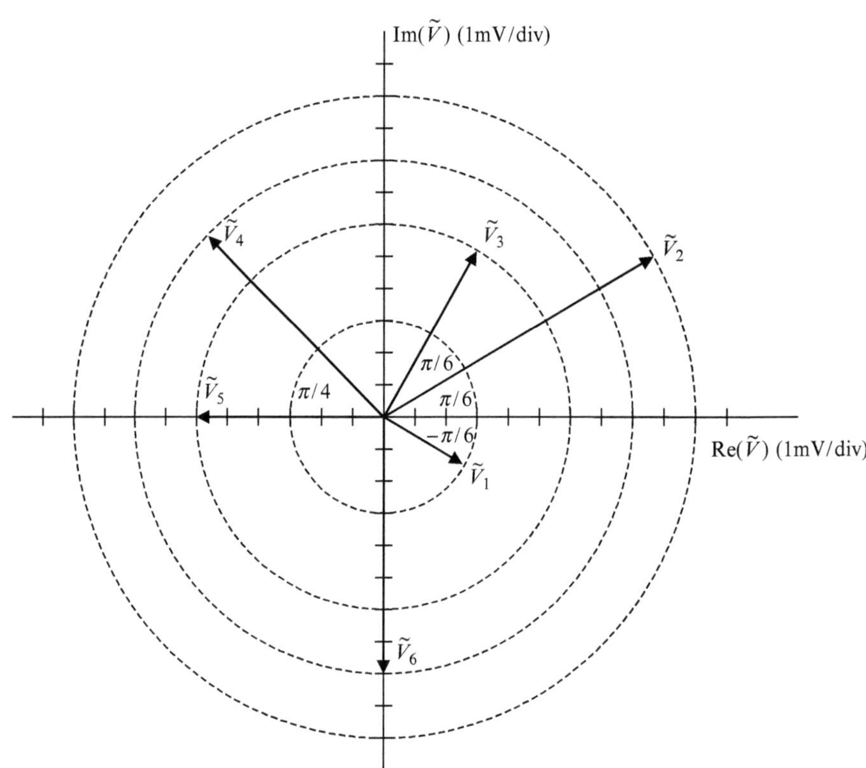

Fig. P 12.2 See Problem P 12.23

Section 12.5 is prerequisite for the following problems.

P 12.24 If we say that an impedance Z_1 is larger than another impedance Z_2, which of the following are implied? Mark all that apply.

(a) $\measuredangle Z_1 > \measuredangle Z_2$
(b) $|Z_1| > |Z_2|$
(c) $\mathrm{Re}(Z_1) > \mathrm{Re}(Z_2)$,
(d) $\mathrm{Im}(Z_1) > \mathrm{Im}(Z_2)$.

P 12.25 In general, can an impedance be described as positive or negative? Justify your answer.

P 12.26 Refer to Fig. P 12.3, which depicts a relation among three quantities. In each of (a)–(e) below, find the unspecified quantity.

(a) $\tilde{V} = (5\angle 0)$ V, $\tilde{I} = (3 + j4)$ mA,
(b) $\tilde{V} = (5\angle 0)$ V, $Z = (3 + j4)$ kΩ,
(c) $\tilde{V} = (3 + j4)$ V, $Z = 10\angle(\pi/4)$ Ω,
(d) $\tilde{V} = (3 + j4)$ V, $\tilde{I} = (4 + j3)$ mA,
(e) $Z = (3 + j4)$ Ω, $\tilde{I} = (4 + j3)$ mA.

P 12.27 Refer again to Fig. P 12.1, which shows oscilloscope traces of the current through (solid line) and voltage across (dotted line) an impedance Z. Estimate the impedance using the voltage as the phase reference.

P 12.28 Fig. P 12.4 shows a circuit and oscilloscope traces of the voltages across the 1 kΩ resistor and an unknown impedance Z. Estimate the impedance Z. Express the estimate in both rectangular and polar form.

P 12.29 A 5 V, 10 kHz sinusoidal voltage is applied to a linear element. It is observed that the current entering the positive terminal of the element has a peak amplitude of 5 mA and lags the applied voltage by 10 μs. Find the impedance of the element at 10 kHz.

Section 12.6 is prerequisite for the following problems.

P 12.30 Refer to Fig. P 12.5. In each of (a)–(e) below, find the unspecified quantity.

(a) $\tilde{V} = (5\angle 0)$ V, $\tilde{I} = (3 + j4)$ mA,
(b) $\tilde{V} = (5\angle 0)$ V, $Y = (3 + j4)$ mS,
(c) $\tilde{V} = (3 + j4)$ V, $Y = 10\angle(\pi/4)$ mS,
(d) $\tilde{V} = (3 + j4)$ V, $\tilde{I} = (4 + j3)$ mA,
(e) $Y = (3 + j4)$ mS, $\tilde{I} = (4 + j3)$ mA.

P 12.31 Each phasor diagram in Fig. P 12.6 depicts the current through and voltage across a two-terminal circuit element. The scales are 100 mV/div and 50 mA/div. Find the impedance and admittance of the element. Express both in both polar and rectangular form.

P 12.32 Let θ denote the angle by which the current entering the positive terminal of a load leads the voltage across the load. If the angle of the admittance of the load is increased, does θ increase or decrease?

P 12.33 In Fig. P 12.7, $v(t) = V_0 \cos(\omega_0 t)$ and $i(t) = I_0 \cos(\omega_0 t + \theta)$, where V_0 and ω_0 are fixed. The angle θ is to be increased by increasing or decreasing each of R, L, and C. Which ones should

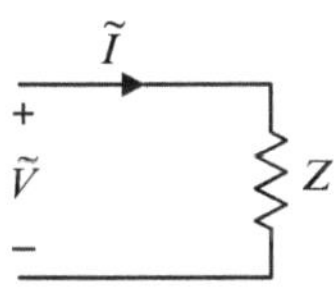

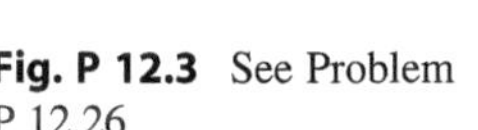

Fig. P 12.3 See Problem P 12.26

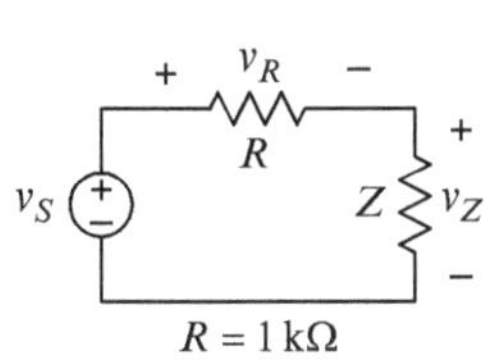

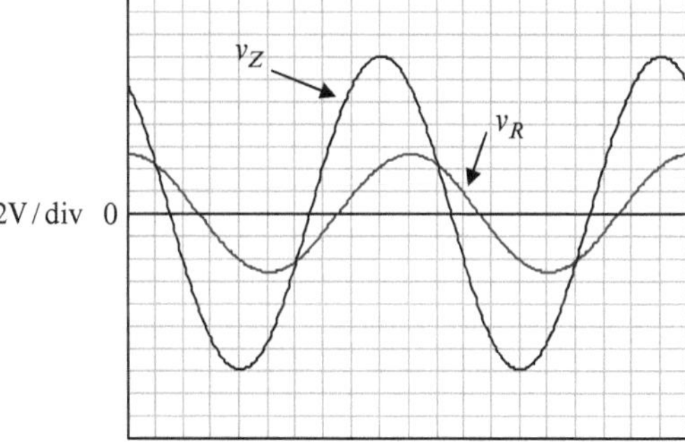

Fig. P 12.4 See Problem P 12.28

be increased and which ones should be decreased? What are the limiting values of the angle θ as $\omega_0 \to 0$ and $\omega_0 \to \infty$? What is the admittance of the circuit for $\omega_0 = 1/\sqrt{LC}$?

P 12.34 It is known that the magnitude of the impedance of a certain load equals 10 kΩ. The load is connected in series with a 1 Ω resistor and a 25 V, 1 kHz sinusoidal source $v(t)$, as illustrated by Fig. P 12.8. The voltages $v_R(t)$ across the resistor and $v_O(t)$ across the load are observed on an oscilloscope. It is found that the voltage $v_R(t)$ leads the voltage $v_O(t)$ by 125 μs. (a) Is the load capacitive or inductive? (b) Draw a phasor diagram for the voltages, with the source voltage $v(t)$ as the phase reference. (c) Find the admittance of the load. You may use any reasonable approximations.

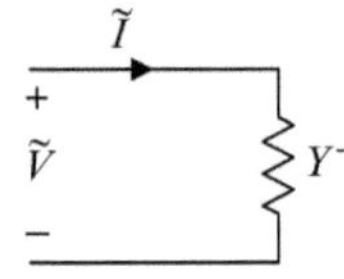

Fig. P 12.5 See Problem P 12.30

Section 12.7 is prerequisite for the following problems.

P 12.35 Express each impedance in dB at the specified frequency, using $|Z(0)|$ as the 0 dB reference.

(a) $Z = \dfrac{R}{1 + j2\pi f RC}$, $R = 10$ kΩ, $C = 50$ nF, $f = 1$ kHz;

(b) $Z = \dfrac{R + j\omega L}{1 + j\omega RC - \omega^2 LC}$, $R = 1$ kΩ, $L = 10$ mH, $C = 100$ nF, $f = 5$ kHz, $\omega = 2\pi f$;

(c) $Z = \dfrac{1}{Y}$, $Y = \dfrac{1}{R + j2\pi f L} + \dfrac{1 + j2\pi f RC}{R}$, $R = 2$ kΩ, $L = 25$ mH, $C = 500$ nF, $f = 5$ kHz;

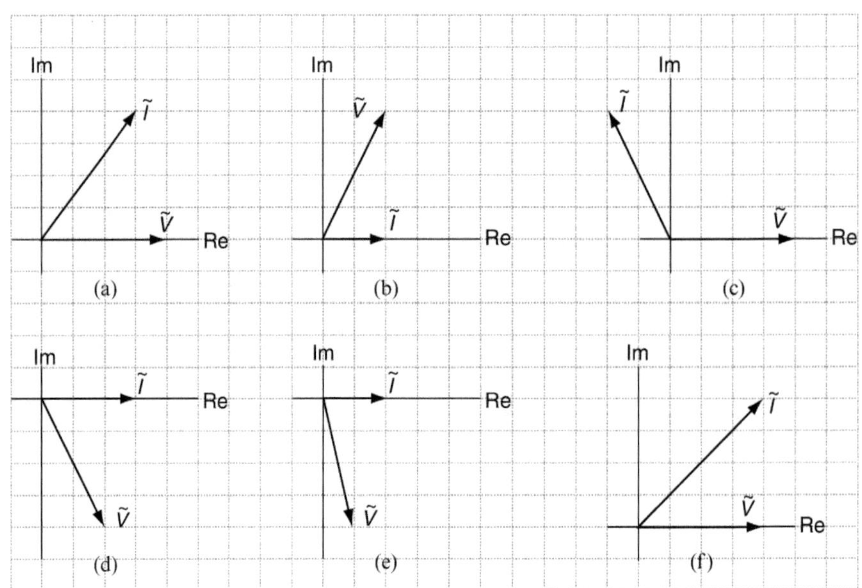

Fig. P 12.6 See Problem P 12.31

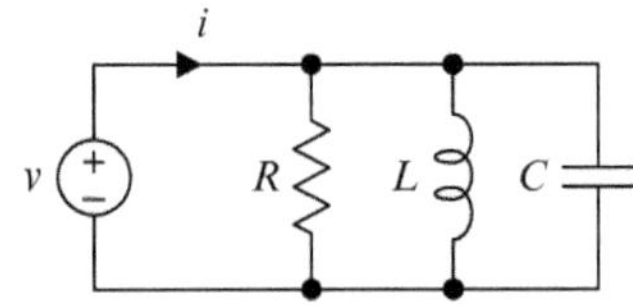

Fig. P 12.7 See Problem P 12.33

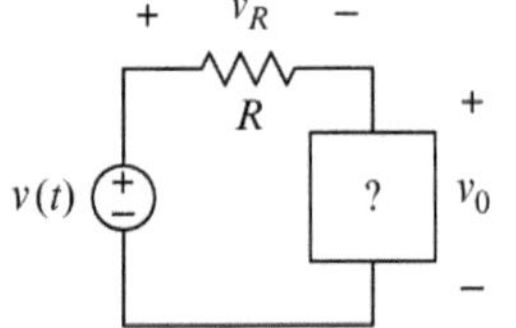

Fig. P 12.8 See Problem P 12.34

(d) $Z = \frac{1}{Y}$, $Y = \frac{1}{R_0} + \frac{1}{R_1 + j2\pi f L_1} + \frac{1}{R_2 + j2\pi f L_2}$,
$R_0 = 5\ \text{k}\Omega$, $R_1 = 100\ \Omega$, $R_2 = 50\ \Omega$,
$L_1 = 20\ \text{mH}$, $L_2 = 10\ \text{mH}$, $f = 10\ \text{kHz}$.

P 12.36 Fig. P 12.9 shows graphs representing the impedance of a certain two-terminal circuit, where $R = 5\ \text{k}\Omega$. The circuit is connected to a 10 kHz sinusoidal voltage source having peak amplitude 10 V. Using the voltage source as the phase reference, find the phasor representation of the current entering the positive terminal of the circuit. Then express the current as a function of time.

P 12.37 Construct graphs of the magnitude (dB) and angle of each impedance given in Problem P 12.35. Use $Z(0)$ as the 0 dB reference in each case.

P 12.38 The equivalent impedance for $f = f_0$ at the terminals of a certain circuit is given as

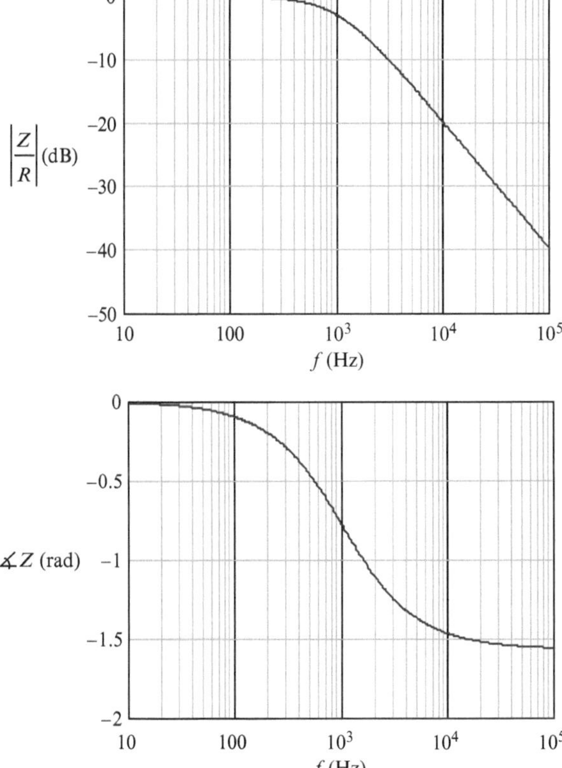

Fig. P 12.9 See Problem P 12.36, 40

$$20 \log\left|\frac{Z}{R}\right| = 20\ \text{dB},\ \measuredangle Z = -\frac{\pi}{6},$$

with $R = 1\,\text{k}\Omega$. Find the admittance for $f = f_0$ in rectangular form (not in dB).

P 12.39 The the magnitude of the impedance of a certain two-terminal load is given as 20 dB at 1 kHz and 40 dB at 10 kHz. The load is connected to the terminals of a sinusoidal voltage source, the rms amplitude of the source is set to 10 V, the frequency of the source is varied from 1 to 10 kHz, and the rms current entering the load is measured. What is the ratio of the *rms* load current at 1 kHz to that at 10 kHz?

P 12.40 Refer to Fig. P 12.9, which gives the impedance of a certain two-terminal circuit. The circuit is connected to the terminals of a sinusoidal voltage source, the rms amplitude of the source is set to 10 V, and the rms current entering the circuit is observed as the frequency of the source is increased from 10 kHz to 10 MHz. At what frequency is the rms amplitude of the current 100 times that at 100 Hz?

P 12.41 It is given that $Z_1 = (5\angle 1.2)\ \text{k}\Omega$ and that Z_2 is 35 dB greater than Z_1. What is the magnitude of Z_2?

Section 12.8 is prerequisite for the following problems.

P 12.42 Use phasors to show that for all time and for any ω_0,

$$V_0\left[\cos(\omega_0 t) + \cos\left(\omega_0 t + \frac{2\pi}{3}\right) + \cos\left(\omega_0 t - \frac{2\pi}{3}\right)\right] = 0.$$

P 12.43 Use phasors to show that

$$\cos(\omega_0 t) + \cos\left(\omega_0 t + \frac{\pi}{6}\right) + \cos\left(\omega_0 t + \frac{2\pi}{6}\right) + \cos\left(\omega_0 t + \frac{3\pi}{6}\right) + \cdots + \cos\left(\omega_0 t + \frac{11\pi}{6}\right) = 0.$$

P 12.44 Use phasors to verify the following identities:

(a) $A\cos(\omega_0 t) + B\sin(\omega_0 t)$
$\equiv \sqrt{A^2+B^2}\cos\left[\omega_0 t - \text{Tan}^{-1}(A,B)\right]$,
(b) $\cos(\omega_0 t+\theta) \equiv \cos(\theta)\cos(\omega_0 t) - \sin(\theta)\sin(\omega_0 t)$.

Section 12.9 is prerequisite for the following problems.

P 12.45 Refer to Fig. P 12.10. Obtain a symbolic expression for the equivalent impedance at the terminals a–b. Then use the parameter values given and compute the equivalent impedance at the specified frequency.

P 12.46 Is it possible for the magnitude of the impedance of a parallel connection of two impedances to be larger than that of either impedance alone? If not, show why not. If so, give an example.

P 12.47 Is it possible for the magnitude of the impedance of a series connection of two impedances to be smaller than that of either impedance alone? If not show why not. If so, give an example.

P 12.48 Refer to Fig. P 12.11, where $v(t) = V_0\cos(\omega_0 t)$ and $i_1 = i_2$. Express R_P and L_P in terms of R_S, L_S, and ω_0.

P 12.49 Show that if $\measuredangle Z_1 = \measuredangle Z_2$, then $|Z_1 + Z_2| = |Z_1| + |Z_2|$.

P 12.50 Show that if $\measuredangle Z_1 = \measuredangle Z_2 = \theta$, then the angle of $Z_1||Z_2$ also equals θ.

P 12.51 Refer to Fig. P 12.12. Show that Z is real if $Z_2 = Z_1{}^*$.

Section 12.11 is prerequisite for the following problems.

P 12.52 In Fig. P 12.13, $v_L(t) = V_0\cos(\omega_0 t)$. Can you obtain a unique expression for the current $i(t)$? If yes, do so. If not, explain why not.

P 12.53 In Fig. P 12.14, $i_S(t) = I_0 + I_1\cos(\omega_0 t)$. Obtain an expression for the voltage $v_L(t)$.

P 12.54 In Fig. P 12.15, $i(t) = I_0 + I_1\cos(\omega_0 t)$. Obtain an expression for the dc component of the voltage $v_C(t)$.

P 12.55 In Fig. P 12.16, the current through the inductor is given by $i_L(t) = I_0 + I_1\cos(\omega_0 t)$. Obtain an expression for the source current $i_S(t)$.

P 12.56 For each circuit in Fig. P 12.17, $v_S(t) = V_S\cos(\omega_0 t)$, $R = 10\ \text{k}\Omega$, $C = 100\ \text{nF}$,

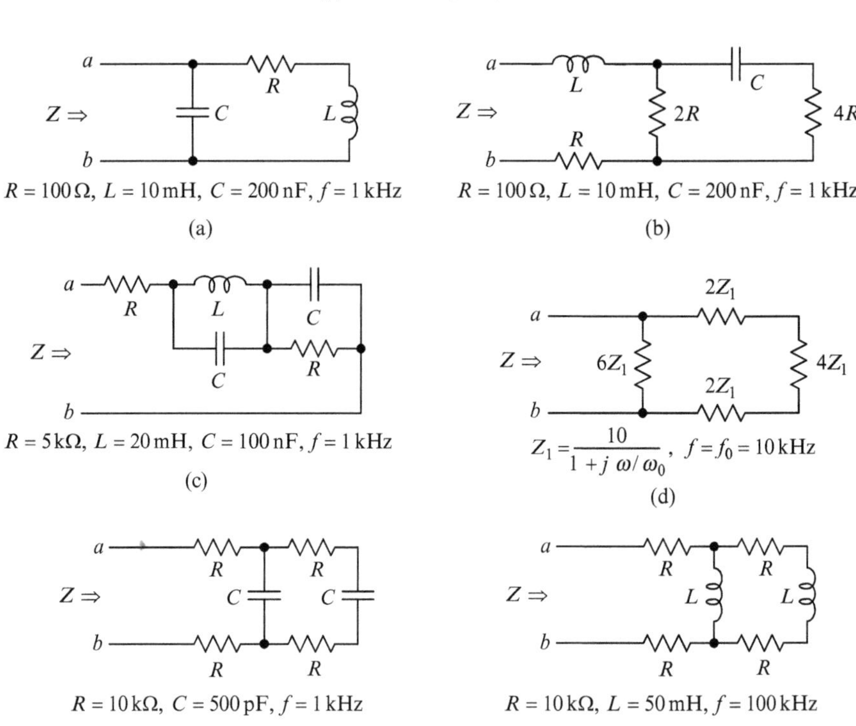

Fig. P 12.10 See Problem P 12.45

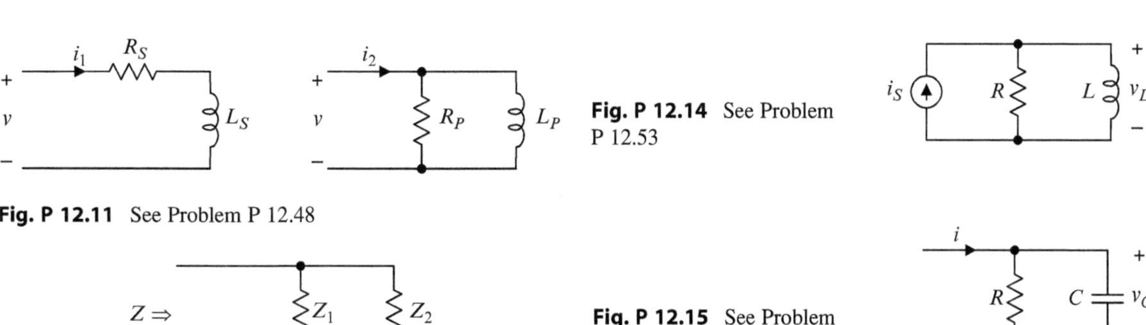

Fig. P 12.11 See Problem P 12.48

Fig. P 12.14 See Problem P 12.53

Fig. P 12.12 See Problem P 12.51

Fig. P 12.15 See Problem P 12.54

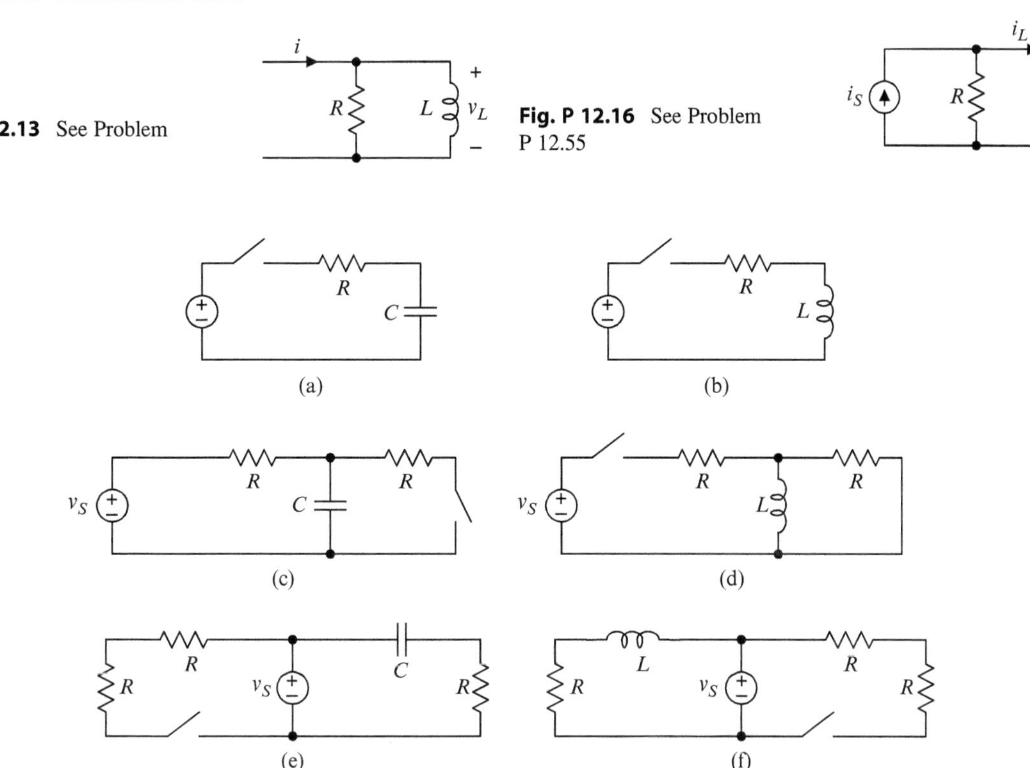

Fig. P 12.13 See Problem P 12.52

Fig. P 12.16 See Problem P 12.55

Fig. P 12.17 See Problem P 12.56

$L = 100$ mH, and the switch is closed at $t = 0$. Estimate the time at which the circuit achieves sinusoidal steady state.

Section 12.12 is prerequisite for the following problems.

P 12.57 In each of (a)–(f) below, you are given some of the six quantities $\tilde{V}_S$, $\tilde{V}_1$, $\tilde{V}_2$, $\tilde{I}$, Z_1, Z_2 defined in Fig. P 12.18. Find the remaining quantities. Express currents and voltages in polar form and impedances in rectangular form.

(a) $\tilde{V}_S = reference = 100\,\text{mV}$, $\tilde{V}_1 = 35\angle(\pi/6)\,\text{mV}$, $Z_2 = (10 + j5)\,\Omega$
(b) $\tilde{V}_S = reference = 100$ mV, $\tilde{V}_1 = 35\angle(\pi/6)$ mV, $Z_1 = (5 + j5)\ \Omega$
(c) $\tilde{V}_S = reference = 100$ mV, $\tilde{V}_1 = 35\angle(\pi/6)$ mV, $I = 5$ mA
(d) $\tilde{V}_1 = reference = 50$ mV, $\tilde{V}_2 = 100\angle(\pi/4)$ mV, $Z_1 = 4\ \Omega$

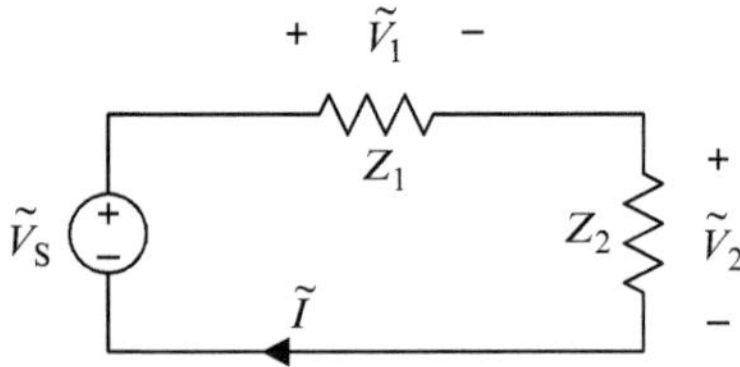

Fig. P 12.18 See Problem P 12.57

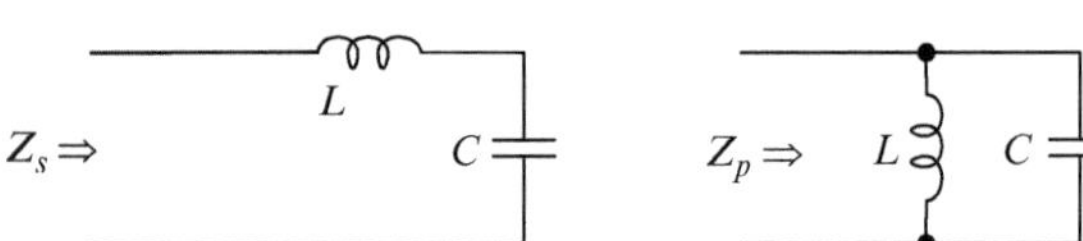

Fig. P 12.19 See Problem P 12.58

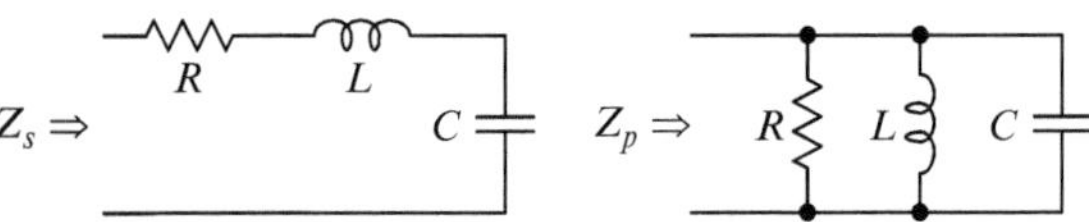

Fig. P 12.20 See Problem P 12.59

(e) $\tilde{V}_S = reference = 15$ V, $Z_1 + Z_2 = (50 + j25)\ \Omega$, $\tilde{V}_1 = 8\angle(\pi/3)$ V
(f) $\tilde{V}_S = reference = 25$ V, $\tilde{I} = 2\angle(\pi/4)$ A, $\tilde{V}_1 = 12\angle(\pi/6)$ V.

P 12.58 Find all frequencies (if any) for which the circuits shown in Fig. P 12.19 are equivalent at their terminals.

P 12.59 Find all frequencies (if any) for which the circuits shown in Fig. P 12.20 are equivalent at their terminals.

P 12.60 In Fig. P 12.21, $v_S = V_S \cos(2\pi f t)$, with $V_S = 10$ V and $f = 50$ kHz. It is required that the **rms** amplitude of the voltage v_C equal 5 V. Express the required value for the capacitance C in terms of the resistance R.

P 12.61 Refer to Fig. P 12.21, where $C = 10$ nF. The source v_S is sinusoidal, having frequency $f = 1$ kHz and rms amplitude 10 V. Specify the resistance R such that the rms amplitude of the voltage v_C equals 7.5 V.

P 12.62 Refer to Fig. P 12.21. The resistance R and the capacitance C must be such that the total impedance (magnitude) of the series connection equals 5 kΩ. The frequency of the sinusoidal source v_S is 1 kHz. Find the values of R and C such that the phase difference between the source v_S and the output v_C equals $\pi/6$. Does v_C lead or lag v_S?

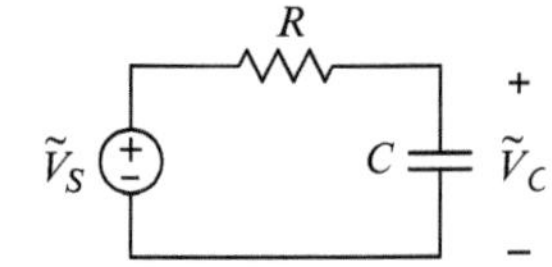

Fig. P 12.21 See Problem P 12.60, 61, 62, 63

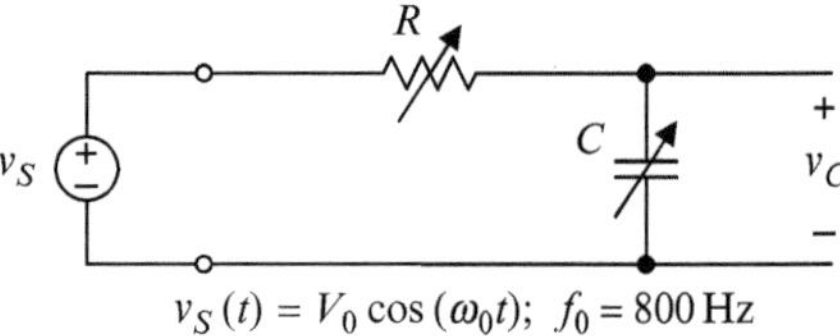

Fig. P 12.22 See Problem P 12.64

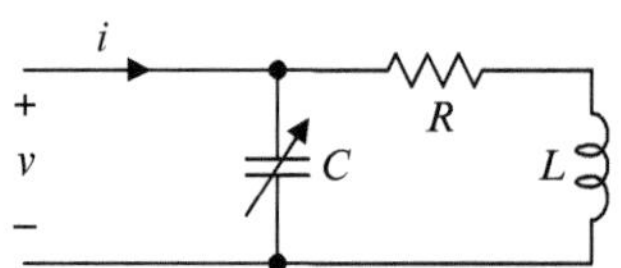

Fig. P 12.23 See Problem P 12.65

P 12.63 Refer to Fig. P 12.21. The capacitance C and the resistance R are each within 20% of their nominal (specified) values; i.e., $C = C_0(1 \pm 0.20)$ and $R = R_0(1 \pm 0.20)$, where C_0, R_0 are the nominal values. Express the ratio $\tilde{V}_C/\tilde{V}_S$ in terms of the circuit parameters, assuming that $R = R_0$, $C = C_0$. Let $v_S = V_0 \cos(2\pi f_0 t)$, with $V_0 = 1$ V and $\omega_0 = 2\pi f_0 = 1/(R_0 C_0)$ and determine the tolerances on $|\tilde{V}_C|$ and $\measuredangle \tilde{V}_C$, given the $\pm 20\%$ tolerances on R and C.

P 12.64 In Fig. P 12.22, the resistance and capacitance are to be adjusted such that the magnitude of the total series impedance is 5 kΩ and the voltage v_C lags the voltage v_S by $\pi/6$. Find the values of R and C.

P 12.65 Refer to Fig. P 12.23, where $v(t) = V_0 \cos(\omega_0 t)$. The resistance and inductance are fixed and the capacitance is variable. The capacitance is adjusted until the current $i(t)$ is in phase with the voltage $v(t)$. Express the capacitance C in terms of the resistance R, inductance L, and frequency f_0.

P 12.66 Refer to Fig. P 12.24, where $v(t) = V_0 \cos(\omega_0 t)$. The resistance and capacitance are fixed and the inductance is variable. The inductance is adjusted until the current $i(t)$ is in phase with the voltage $v(t)$. Express the inductance L in terms of the resistance R, capacitance C, and frequency f_0.

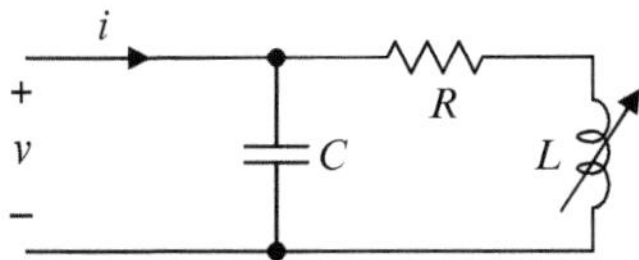

Fig. P 12.24 See Problem P 12.66

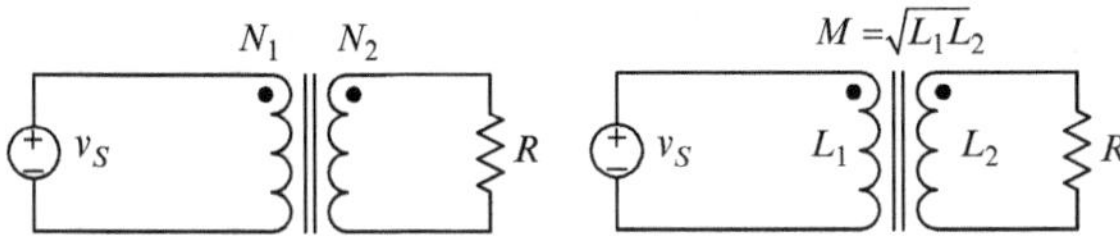

Fig. P 12.25 See Problem P 12.67

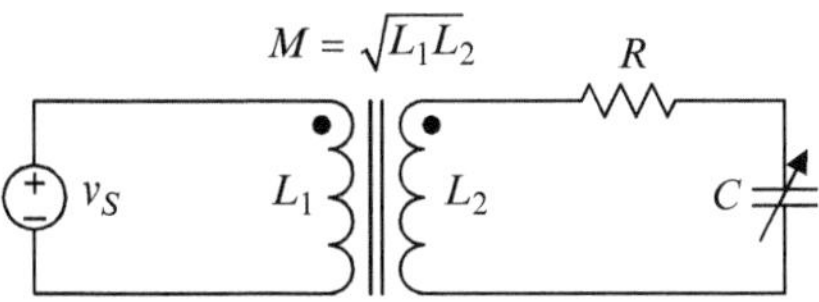

Fig. P 12.26 See Problem P 12.68

P 12.67 A transformer having unity coupling and negligible losses couples a source to a resistive load. The circuit is modeled using an ideal transformer in one case, and a unity-coupling model in another, as shown in Fig. P 12.25. Obtain expressions for the impedances seen at the primary terminals of the transformer for both circuits. Interpret the results in the context of the claim that a transformer can be modeled as ideal if the coupling coefficient equals unity and the losses are negligible.

P 12.68 A resistor and variable capacitor are connected in series and to the secondary of a transformer having negligible losses and unity coupling, as shown in Fig. P 12.26, where $R = 1$ kΩ, $L_1 = 20$ mH, $L_2 = 200$ mH, and the frequency of the source is 20 kHz. The capacitor is then adjusted such that the input impedance on the primary side of the transformer is real. Find the capacitance and the input impedance under the specified conditions.

P 12.69 In Fig. P 12.27, the transformers are ideal. (a) Express $\tilde{V}_2$, $\tilde{V}_3$, $\tilde{V}_L$ and $\tilde{I}_1$, $\tilde{I}_2$, $\tilde{I}_L$ in terms of $\tilde{V}_S$ and the circuit parameters. (b) What is the resistance seen by the source $\tilde{V}_S$? (c) What is the resistance seen at the primary *a*–*b* of the second transformer? (d) What fraction of the source voltage $\tilde{V}_S$ appears across the load? (e) If the transformers are removed, and the load is connected to the source through the transmission line, what fraction of the source voltage appears across the load?

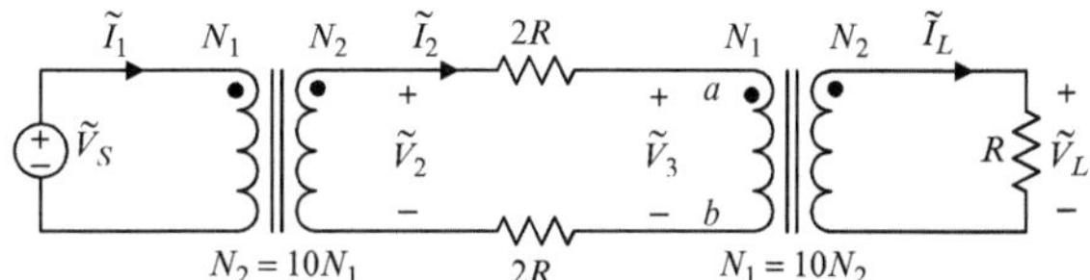

Fig. P 12.27 See Problem P 12.69

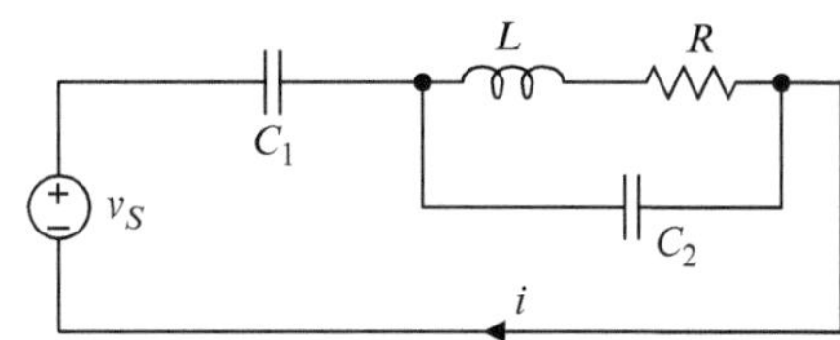

Fig. P 12.28 See Problem P 12.70

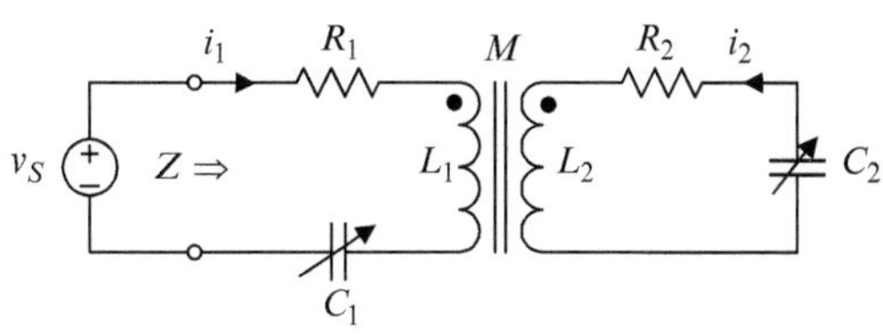

Fig. P 12.29 See Problem P 12.71

P 12.70 In Fig. P 12.28 $R = 10$ Ω, $L = 200$ μH, and $v_S = V_0[\cos(\omega_1 t) + \cos(\omega_2 t)]$, with $f_1 = 1$ MHz and $f_2 = 1.2$ MHz. It follows that $i = I_1 \cos(\omega_1 t + \theta_1) + I_2 \cos(\omega_2 t + \theta_2)$. Find the values of the capacitances $C_1\, C_2$ such that I_1/I_2 is as large as possible.

P 12.71 In Fig. P 12.29, $R_1 = R_2 = 10$ Ω, $M = 10$ μH, and $v_S = V_S \cos(\omega t)$ with $f = 1$ MHz. The capacitance C_2 is adjusted such that $(\omega C_2)^{-1} = \omega L_2$. The capacitance C_1 is then adjusted such that the impedance seen by the source is resistive. What is the impedance Z seen by the source v_S?

P 12.72 Refer to Fig. P 12.30. Write Kirchhoff's voltage law around the indicated paths for each of the four possible dot locations on the transformer.

P 12.73 Refer to Fig. P 12.31, where R_y, L_y represent a coil. The purpose of the circuit is to measure the inductance and resistance of the coil at the frequency of the source $v_S = V_S \cos(\omega_0 t)$. The capacitance C_x and the resistance R_x are adjusted until the voltage v_{ab} equals zero. Express the inductance L_y and resistance R_y in terms of the quantities R, R_x, C_x.

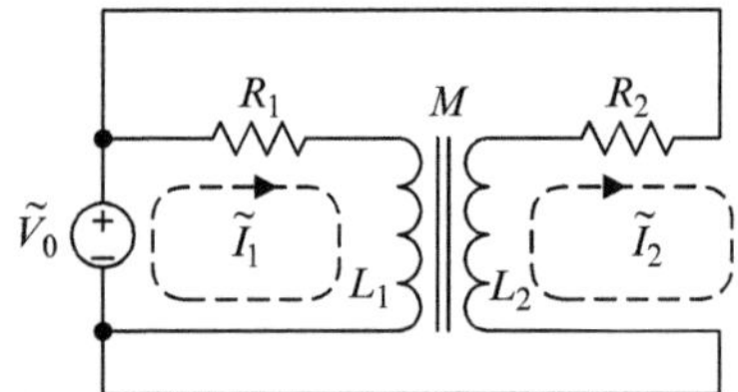

Fig. P 12.30 See Problem P 12.72

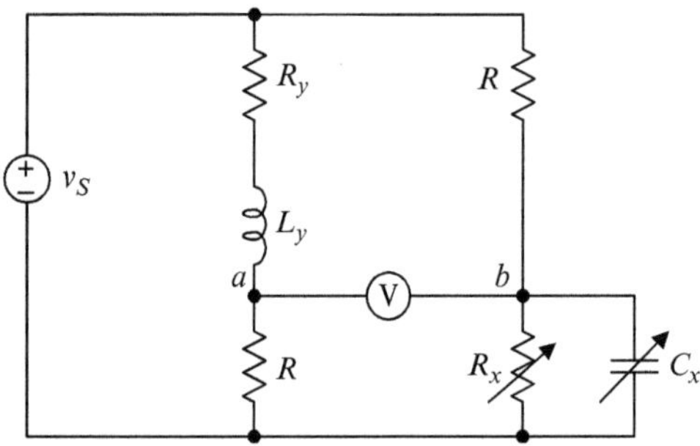

Fig. P 12.31 See Problem P 12.73

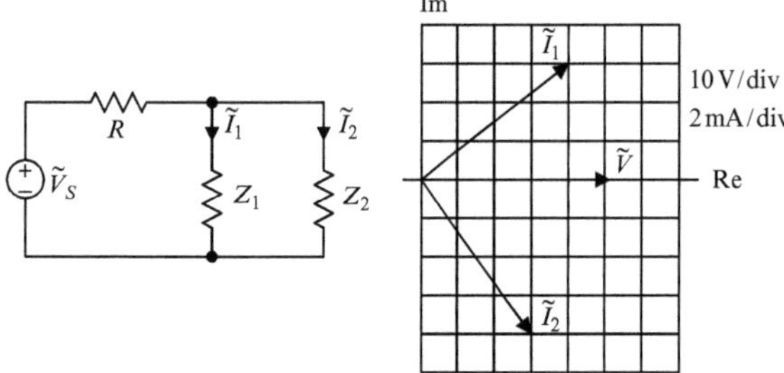

Fig. P 12.32 See Problem P 12.74

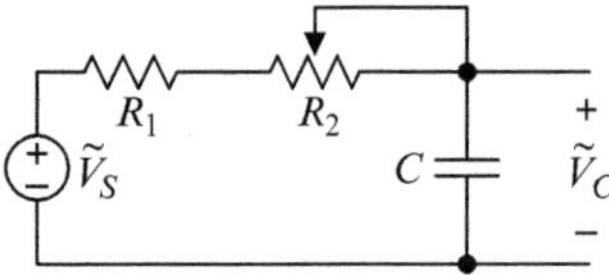

Fig. P 12.33 See Problem P 12.75

P 12.74 Fig. P 12.32 shows a circuit and associated phasor diagram. It is known that $R = 2\ \text{k}\Omega$ and that the frequency of the sinusoidal source is 5 kHz. Find the impedances Z_1, Z_2 and draw an *RLC* circuit equivalent to the one given (at 5 kHz).

P 12.75 The circuit shown in Fig. P 12.33 is intended to shift the phase of the output $\tilde{V}_C$ relative to that of the sinusoidal input $\tilde{V}_S$ by an amount that can be varied from $-\pi/6$ to $-\pi/3$. The total series resistance can be varied from R_1 to $R_1 + R_2$. Obtain expressions for the resistances R_1, R_2 in terms of the (fixed) capacitance C and the frequency of the input.

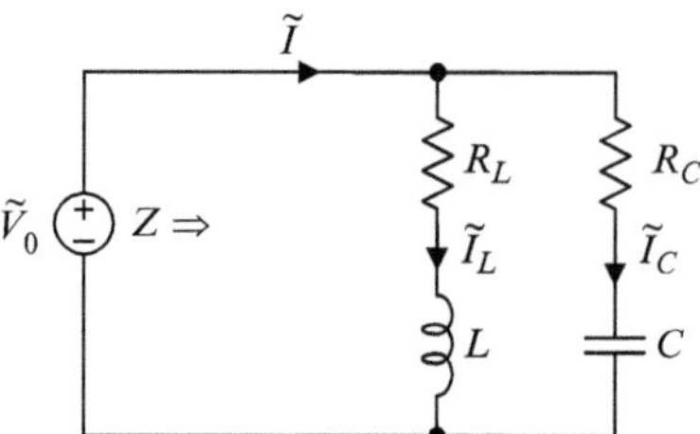

Fig. P 12.34 See Problem P 12.76

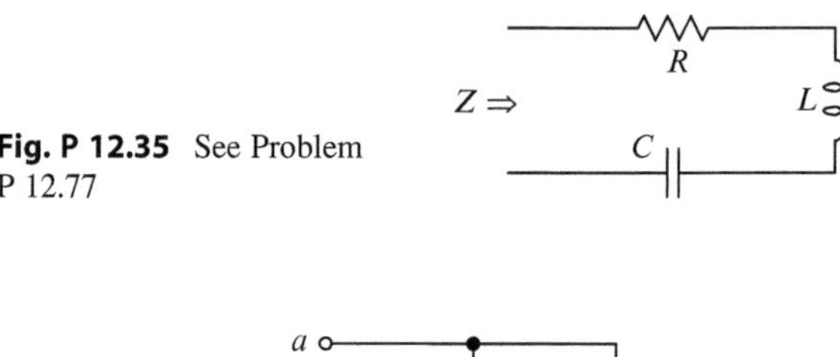

Fig. P 12.35 See Problem P 12.77

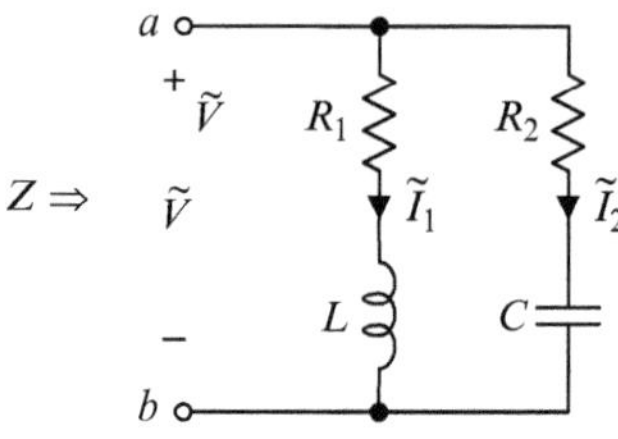

Fig. P 12.36 See Problem P 12.78

P 12.76 Refer to Fig. P 12.34. (a) Show that the impedance Z seen by the source is real for all frequencies if $R_L = R_C = \sqrt{L/C}$. (b) Assume $R_L = R_C = \sqrt{L/C}$ and give expressions for the phasor currents $\tilde{I}_L$, $\tilde{I}_C$ in terms of $\tilde{V}_0$ and the circuit parameters. (c) Let $\tilde{V}_0 = 50$ V, $C = 18$ pF, $\left(2\pi\sqrt{LC}\right)^{-1} = 800$ kHz, and $R_L = R_C = \sqrt{L/C}$. Draw a phasor diagram for the currents $\tilde{I}$, $\tilde{I}_L$, $\tilde{I}_C$, with $\tilde{V}_0$ as the phase reference.

Section 12.13 is prerequisite for the following problems.

P 12.77 Obtain expressions for the frequencies for which the impedance Z of the circuit in Fig. P 12.35 is (a) inductive, (b) resistive, and (c) capacitive.

P 12.78 For the circuit in Fig. P 12.36, the value of the inductance is given by $L = R_1 R_2 C$. Show that

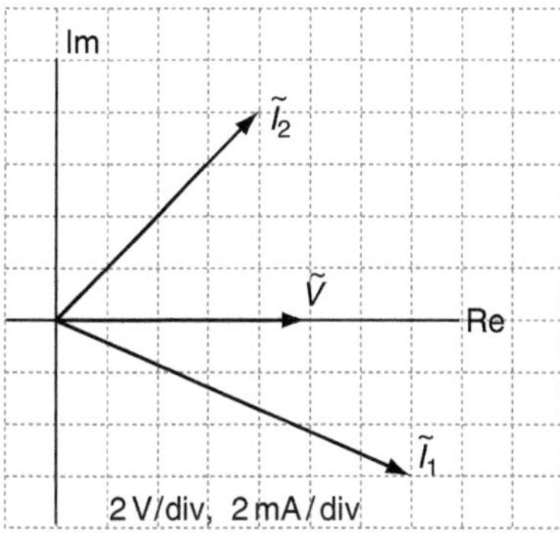

Fig. P 12.37 See Problem P 12.79

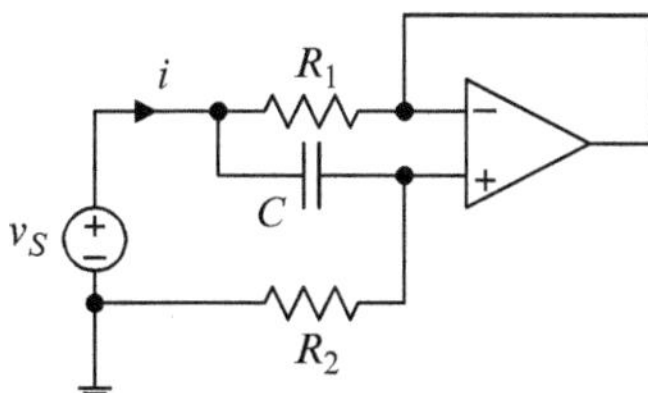

Fig. P 12.38 See Problem P 12.80

the impedance Z is inductive for all frequencies if $R_2 > R_1$ and capacitive for all frequencies if $R_2 < R_1$.

P 12.79 Fig. P 12.37 shows a phasor diagram for the currents and voltages defined in Fig. P 12.36. The frequency of the applied voltage (and phase reference) $v(t)$ is 1 kHz. (a) Is the impedance Z inductive, resistive, or capacitive? (b) What are the equivalent resistance and inductance or capacitance? (c) Find the values of the resistances R_1, R_2, the inductance L, and the capacitance C. (d) If the frequency of the applied voltage is increased but its amplitude remains the same, do the magnitudes and angles of the currents $\tilde{I}_1$, $\tilde{I}_2$ increase or decrease?

P 12.80 In Fig. P 12.38, the op amp is ideal. Obtain an expression for the impedance. $\tilde{V}_S/\tilde{I}$. Under what conditions (if any) is the impedance inductive? Capacitive?

P 12.81 The variable capacitor in Fig. P 12.39 is adjusted until the impedance Z is resistive. Express the resulting capacitance and the impedance in terms of the frequency of the applied voltage $v(t)$, the resistance R, and the inductance L.

P 12.82 Explain why the angle of the impedance Z of a *passive* two-terminal *RLC* circuit must satisfy $-\pi/2 \leq \measuredangle Z \leq \pi/2$.

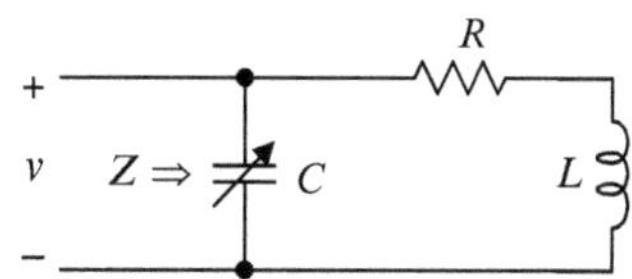

Fig. P 12.39 See Problem P 12.81

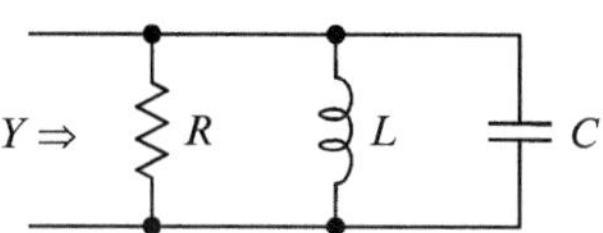

Fig. P 12.40 See Problem P 12.83

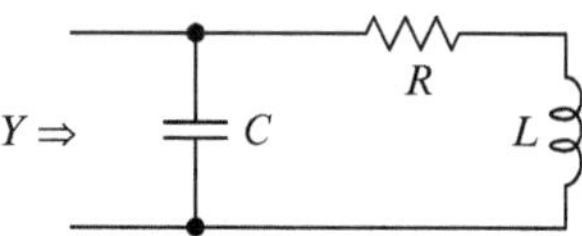

Fig. P 12.41 See Problem P 12.84

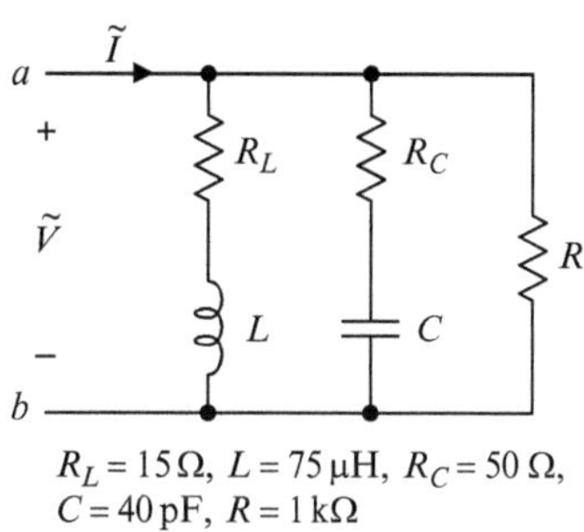

Fig. P 12.42 See Problem P 12.85

Section 12.14 is prerequisite for the following problems.

P 12.83 Obtain expressions for the frequencies for which the admittance Y of the circuit in Fig. P 12.40 is (a) inductive, (b) resistive, and (c) capacitive.

P 12.84 Obtain expressions for the frequencies for which the admittance Y of the circuit in Fig. P 12.41 is (a) inductive, (b) resistive, and (c) capacitive.

P 12.85 Refer to Fig. P 12.42, where the applied voltage $\tilde{V}$ is sinusoidal. Find the equivalent conductance, the susceptance, the equivalent resistance, and the reactance at the terminals *a–b* for frequency $f = 2.75$ MHz. Is the circuit inductive or capacitive at that frequency? Would the current $\tilde{I}$ lead or lag the voltage $\tilde{V}$?

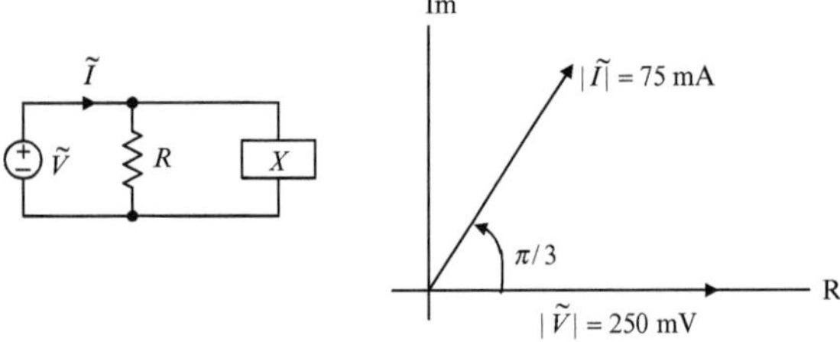

Fig. P 12.43 See Problem P 12.87

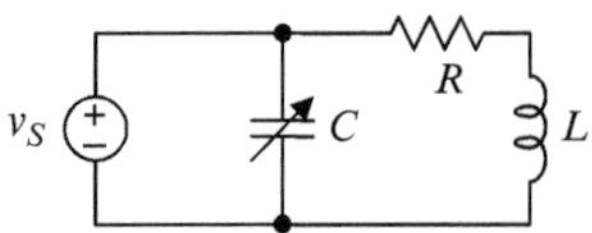

Fig. P 12.44 See Problem P 12.90

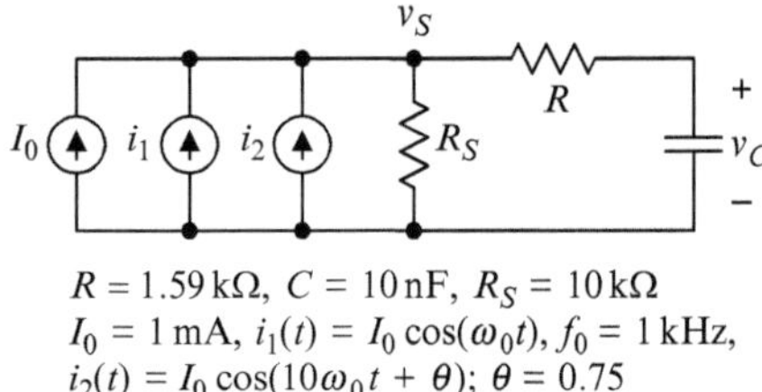

$R = 1.59\,\text{k}\Omega,\ C = 10\,\text{nF},\ R_S = 10\,\text{k}\Omega$
$I_0 = 1\,\text{mA},\ i_1(t) = I_0\cos(\omega_0 t),\ f_0 = 1\,\text{kHz},$
$i_2(t) = I_0\cos(10\omega_0 t + \theta);\ \theta = 0.75$

Fig. P 12.45 See Problem P 12.91

P 12.86 Given $Z = R + jX$, $Y = G + jB$, and $Z = Y^{-1}$, Express R and X in terms of G and B. Then express G and B in terms of R and X.

P 12.87 Fig. P 12.43 shows a sinusoidally excited two-element circuit and associated phasor diagram. (a) Is the reactive element (X) a capacitance or an inductance? (b) Find the values of R and X. (c) If the frequency of the applied voltage $\tilde{V}$ is 15 kHz, what is the capacitance or inductance represented by X? (d) If the frequency of the excitation were increased, would the angle of the current $\tilde{I}$ relative to that of the voltage $\tilde{V}$ increase or decrease?

Section 12.15 is prerequisite for the following problems.

P 12.88 For each impedance given below: (i) Draw the impedance triangle. (ii) Draw and label fully a circuit diagram for a *series* circuit having the given impedance. (iii) Assume a 10 V, 1 kHz sinusoidal voltage is applied to the circuit. Draw a phasor diagram for the resulting current and the voltages across each element of the series equivalent circuit, using the applied voltage as the phase reference.

(a) $Z = (5 + j6)\ \text{k}\Omega$,
(b) $Z = (120 - j160)\ \Omega$,
(c) $Z = (4 - j3)\ \text{k}\Omega$,
(d) $Z = (1 + j)\ \text{k}\Omega$,
(e) $Z = (10\angle\pi/3)\ \text{k}\Omega$,
(f) $Z = [(1 - j2)\ \text{mS}]^{-1}$.

P 12.89 For each admittance given below: (i) Draw the admittance triangle. (ii) Draw and label fully a circuit diagram for a *parallel* circuit having the given admittance. (iii) Assume a 10 V, 1 kHz sinusoidal voltage is applied to the circuit. Draw a phasor diagram for the resulting total current and the currents through each element of the parallel equivalent circuit, using the applied voltage as the phase reference.

(a) $Y = (500 - j600)\ \text{mS}$,
(b) $Y = (12 + j16)\ \text{S}$,
(c) $Y = (20 + j40)\ \text{mS}$,
(d) $Y = (1 + j2)\ \text{S}$,
(e) $Y = [(10 + j10)\ \text{k}\Omega]^{-1}$,
(f) $Y = (120\angle\pi/4)\ \text{mS}$.

P 12.90 In Fig. P 12.44, $R = 100\ \Omega$, $L = 250$ mH, and v_S is a 60 Hz sinusoidal voltage source. Draw the impedance and admittance triangles for the impedance Z and admittance Y seen by the source for (a) $C = 100$ nF and (b) $C = 40\ \mu$F. (c) Find the value of C for which $\measuredangle Z = 0$.

Section 12.16 is prerequisite for the following problems.

P 12.91 Refer to Fig. P 12.45. Use superposition to find the voltage $v_C(t)$.

P 12.92 Refer to Fig. P 12.46. Use superposition to find the phasor node voltages $\tilde{V}_A$, $\tilde{V}_B$.

P 12.93 Refer to Fig. P 12.47. Find the voltage $v_L(t)$. Calculate the rms amplitude of $v_L(t)$.

P 12.94 Refer to Fig. P 12.48, where $R_S = 10\ \text{k}\Omega$ and $i_S = I_0[\cos(0.5\omega_0 t) + \cos(\omega_0 t) + \cos(2\omega_0 t)]$,

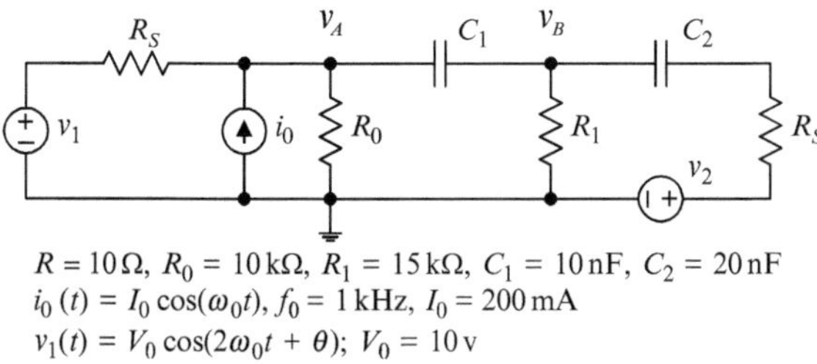

$R = 10\,\Omega$, $R_0 = 10\,\text{k}\Omega$, $R_1 = 15\,\text{k}\Omega$, $C_1 = 10\,\text{nF}$, $C_2 = 20\,\text{nF}$
$i_0(t) = I_0 \cos(\omega_0 t)$, $f_0 = 1\,\text{kHz}$, $I_0 = 200\,\text{mA}$
$v_1(t) = V_0 \cos(2\omega_0 t + \theta)$; $V_0 = 10\,\text{v}$
$v_2(t) = 2V_0 \cos(100\omega_0 t + 2\theta)$; $\theta = 0.75$

Fig. P 12.46 See Problem P 12.92

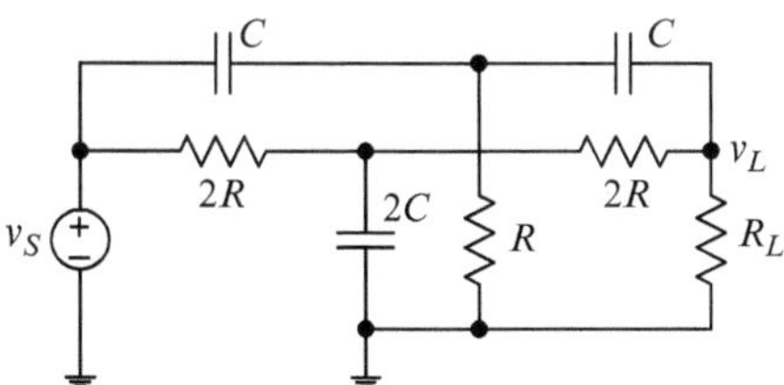

$R = 15.9\,\text{k}\Omega$, $C = 500\,\text{pF}$, $R_L = 1\,\text{M}\Omega$, $f_0 = 1\,\text{kHz}$
$v_s(t) = V_0[\cos(\omega_0 t) + \cos(10\omega_0 t) + \cos(100\omega_0 t)]$; $V_0 = 5\,\text{V}$

Fig. P 12.47 See Problem P 12.93

with $I_0 = 2.5$ mA and $f_0 = 100$ kHz. For frequencies in the range $0.5f_0 \le f \le 2f_0$, the coil has inductance $L = 200$ μH and quality factor $Q_L = \omega_0 L/R_L = 25$. The capacitor has capacitance $C = 12.67$ nF and *ESR* $R_C = 0.25\ \Omega$. Find the voltage $v(t)$.

P 12.95 Refer to Example 14.31 in the text. Use simulation to find the load voltage $v_L(t)$. Compare the results with those calculated in the example.

Section 12.17 is prerequisite for the following problems.

P 12.96 In Fig. P 12.49, $v_S = V_S \cos(\omega_0 t)$ and $i_S = I_S \cos(\omega_0 t)$. Find the Thévenin and Norton equivalent circuits at the terminals *a*–*b*.

P 12.97 Refer to Fig. P 12.50. An rms voltmeter, an rms ammeter, and a variable capacitor are connected to a 15 kHz sinusoidal source, as shown. The capacitance is varied from 100 pF to 10 μF and the voltmeter and ammeter readings are recorded, yielding the graph shown. Find the Thévenin equivalent for the source.

P 12.98 Refer to Fig. P 12.51. An rms voltmeter, an rms ammeter, and a variable inductor are connected

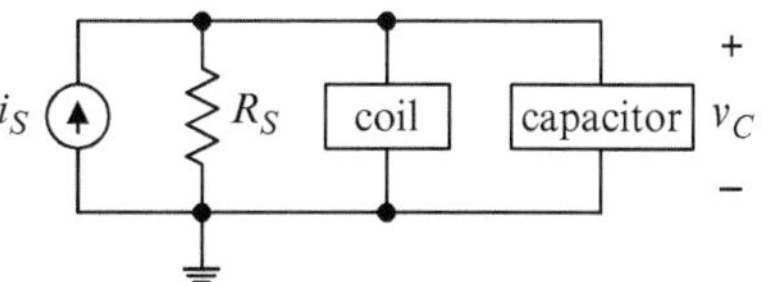

Fig. P 12.48 See Problem P 12.94

to a 15 kHz sinusoidal source, as shown. The inductance is varied from 10 nH to 100 mH and the voltmeter and ammeter readings are recorded, yielding the graph shown. Find the Thévenin equivalent for the source.

P 12.99 A co-worker suggests obtaining the Thévenin equivalent for an *RLC* circuit by measuring the rms amplitude of the voltage across a variable resistor, as a function of the resistance *R*, as illustrated by Fig. P 12.52. Will this work? Show how or explain why not.

Section 12.18 is prerequisite for the following problems.

P 12.100 If you set the inductance of an inductor to zero, does the inductor become equivalent to an open circuit or to a short circuit? If you set the capacitance of a capacitor to zero, does the capacitor become equivalent to an open circuit or to a short circuit?

P 12.101 If you let the inductance of an inductor approach infinity, does the inductor become equivalent to an open circuit or to a short circuit? If you let the capacitance of a capacitor approach infinity, does the capacitor become equivalent to an open circuit or to a short circuit?

P 12.102 Each Fig. P 12.53(a)–(f) shows a circuit diagram and a result of analysis. All sources are sinusoidal. Without repeating the analysis, state whether the result given might be correct or cannot possibly be. If the result cannot possibly be correct, state why.

P 12.103 Re-draw each circuit diagram in Fig. P 12.49 for (a) $f = 0$ and (b) $f \to \infty$.

Section 12.19 is prerequisite for the following problems.

P 12.104 Refer to Fig. P 12.54. The rms amplitude of the sinusoidal source is 25 mA. The frequency of the source equals 480 Hz, which is

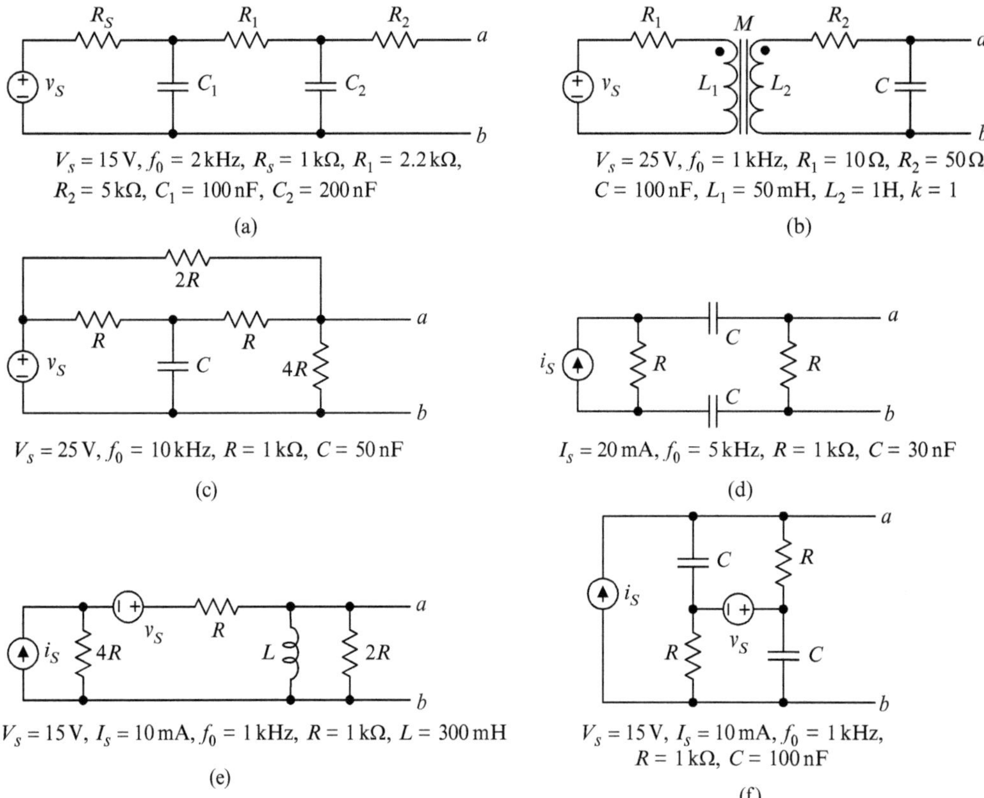

Fig. P 12.49 See Problem P 12.96,103

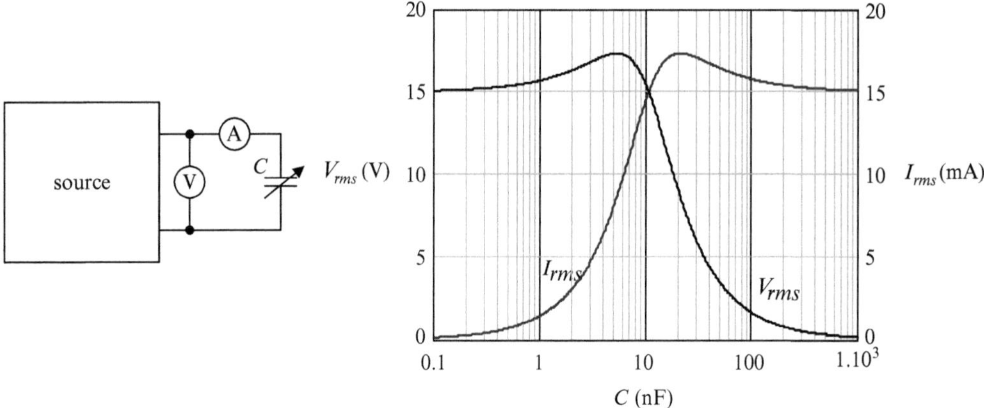

Fig. P 12.50 See Problem P 12.97

the resonant frequency of the circuit. The capacitance $C = 125$ nF and the resistance $R = 25$ kΩ. What is the rms amplitude of the current through the inductor?

P 12.105 Refer to Fig. P 12.55, where $i(t) = I_0 \cos(\omega_0 t)$, with $f_0 = 15\,\text{kHz}$, $R = 220\,\Omega$, $L = 120\,\text{mH}$. The circuit is in resonance and the rms amplitude of the voltage across the resistor is 15 V. What is the rms

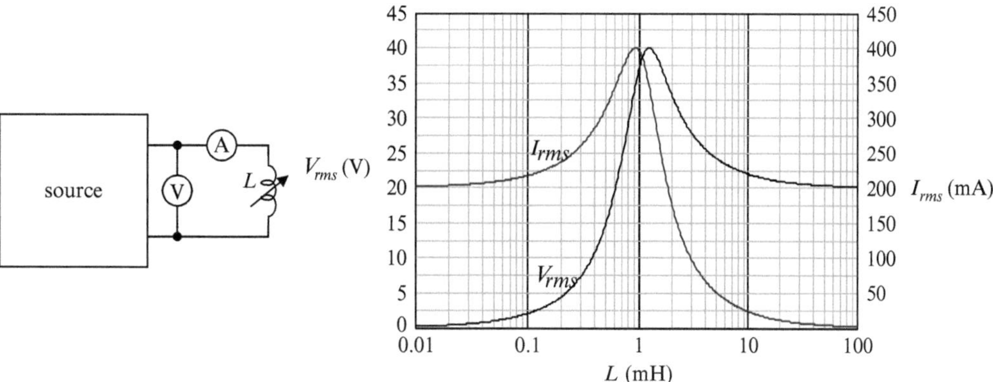

Fig. P 12.51 See Problem P 12.98

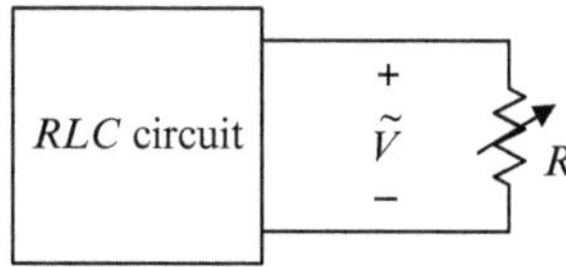

Fig. P 12.52 See Problem P 12.99

amplitude of the current through the capacitor? What is the rms amplitude of the source current? What is the phase of the current through the capacitor relative to that of the current through the resistor?

P 12.106 In Fig. P 12.56, $v = V_0 \cos(\omega_0 t)$ with $V_0 = 5\sqrt{2}$ V and $f_0 = 200$ kHz. The capacitance C is varied until the rms amplitude of the current i has its maximum value, which is found to be 5 mA. At that point the rms amplitude of the voltage across the inductor equals 10 V. Find the capacitance C, the inductance L, and the resistance R.

P 12.107 See Fig. P 12.57, where $R_S = 10$ kΩ and $i_S = I_S \cos(\omega_0 t)$, with $I_{S\,rms} = I_s/\sqrt{2} = 50$ mA and $f_0 = 1.2$ MHz. The variable capacitance C is adjusted such that the rms voltage across the source has its maximum value, at which point $C = 875.7$ pF and $V_{S\,rms} = 78.39$ V. Find the quality factor of the coil (at 1.2 MHz).

P 12.108 See Fig. P 12.58, where $R_S = 10$ kΩ and $i_S = I_S \cos(\omega_0 t)$, with $I_{S\,rms} = I_s/\sqrt{2} = 50$ mA and $f_0 = 1$ MHz. The variable capacitance C is adjusted such that the rms amplitude of the voltage v_S across the source has its minimum value, at which point $C = 1.01$ nF and $V_{S\,rms} = 400$ mV. Find the quality factor of the coil (at 1 MHz).

P 12.109 The two circuits shown in Fig. P 12.59 are independently tuned to resonance at the same frequency f_r, such that

$$\sqrt{\frac{1}{L_1 C_1} - \left(\frac{R_1}{L_1}\right)^2} = \frac{1}{\sqrt{L_2 C_2}} = 2\pi f_r.$$

The circuits are then connected at *a–b*, as indicated, to form circuit #3.

(a) Is the resonant frequency of circuit #3 larger than, smaller than, or equal to f_r?
(b) If a sinusoidal source having frequency f_r is now connected at *c–d*, as indicated in the figure, is the current i larger than, smaller than, or the same as would have been drawn by circuit #1 alone?

P 12.110 Refer to Fig. P 12.60, where the coils (in the dashed boxes) are identical and both circuits are resonant at frequency $f = f_r$. (a) Which capacitor has the larger capacitance? (b) Which circuit has the larger input impedance (magnitude) at resonance?

P 12.111 In Fig. P 12.61, $R_1 = 500$ mΩ, $L_1 = 100$ μH, $R_2 = 2.5$ Ω, $L_2 = 500$ μH and the coupling coefficient $k = 0.7$. Find the value of the capacitance C such that the circuit is resonant at $f = 100$ kHz.

P 12.112 Obtain an expression for the resonant frequency of the circuit in Fig. P 12.62.

P 12.113 Obtain an expression for the resonant frequency of the circuit in Fig. P 12.63.

P 12.114 Refer to Fig. P 12.64, where $R = 1$ kΩ and $i(t) = I_0 \cos(2\pi f_r t)$, with $I_0 = 5$ mA and $f_r = 100$ kHz is the resonant frequency of the circuit. Show that $i_L(t) = -i_C(t)$. Let $i(t)$ serve as the phase

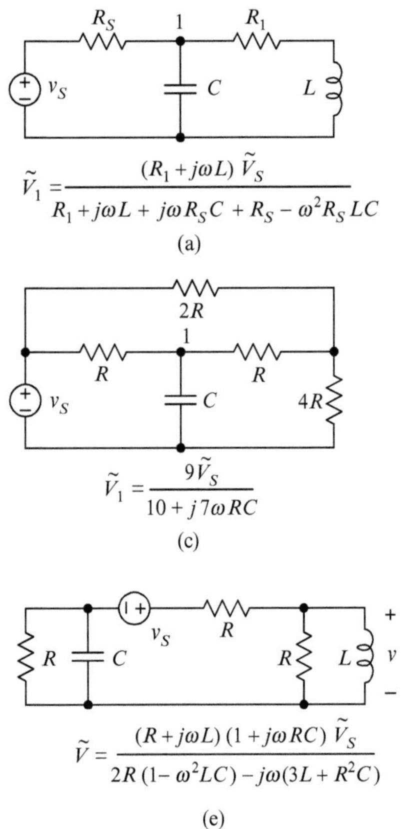

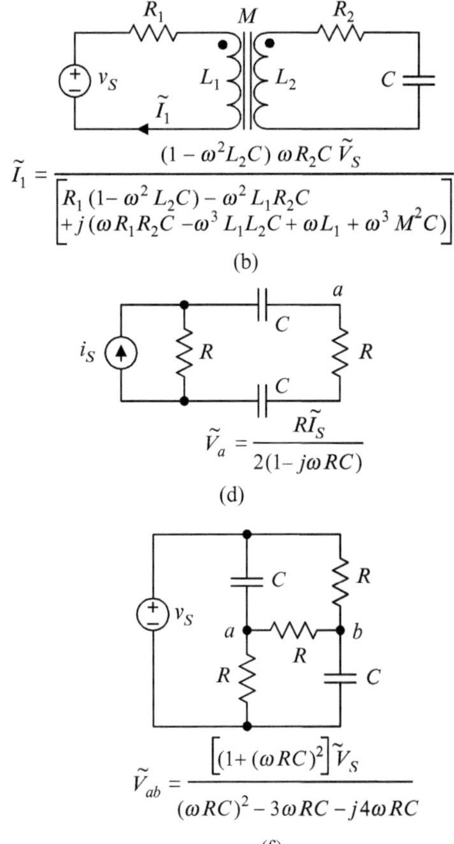

Fig. P 12.53 See Problem P 12.102

reference and draw phasor diagrams showing the relative phases of the currents i_R, i_L, i_C.

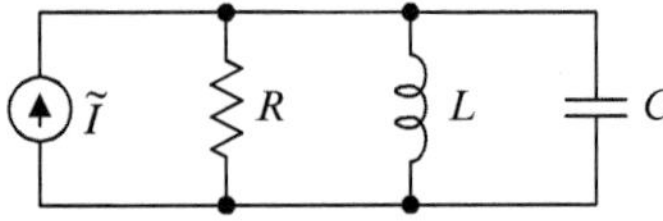

Fig. P 12.54 See Problem P 12.104

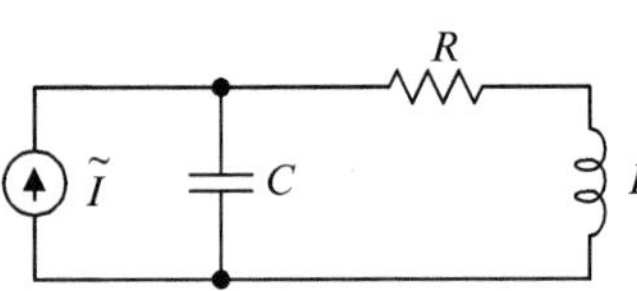

Fig. P 12.55 See Problem P 12.105

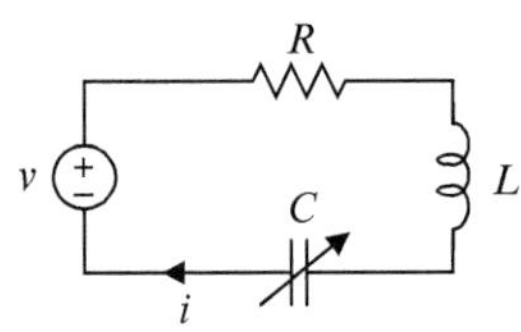

Fig. P 12.56 See Problem P 12.106

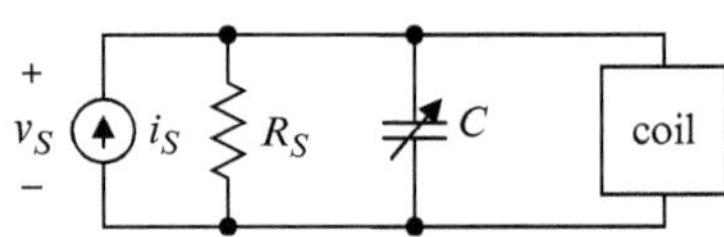

Fig. P 12.57 See Problem P 12.107

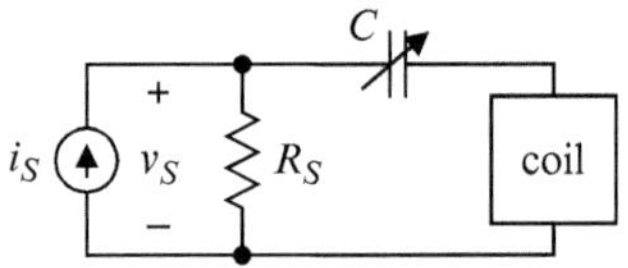

Fig. P 12.58 See Problem P 12.108

P 12.115 Refer to Fig. P 12.65, where $R = 100\ \Omega$, $L = 15$ mH, and v is a variable-frequency sinusoidal source whose rms amplitude is 5 V. It is found experimentally that the resonant frequency of the circuit is 100 kHz. Find the value of the capacitance

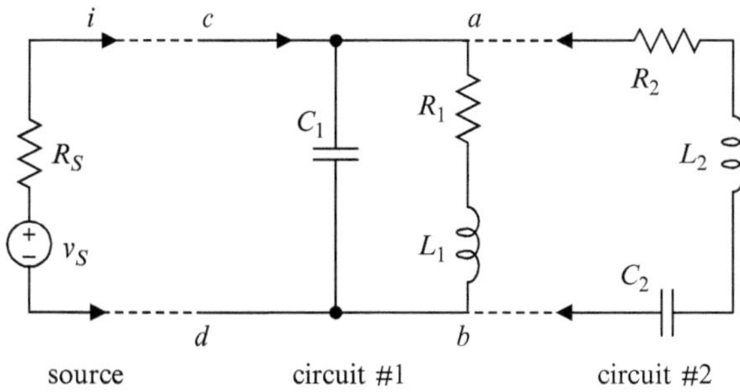

Fig. P 12.59 See Problem P 12.109

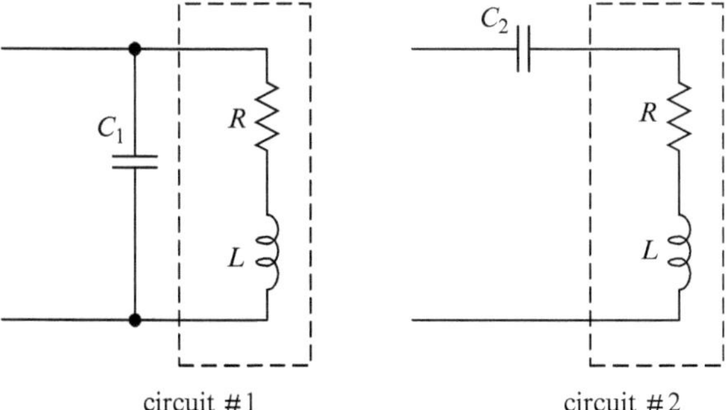

Fig. P 12.60 See Problem P 12.110

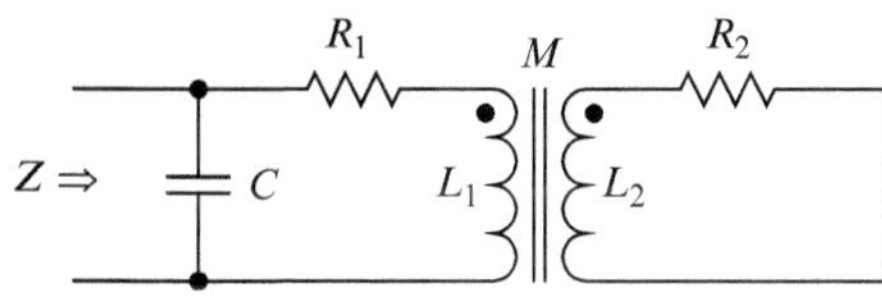

Fig. P 12.61 See Problem P 12.111

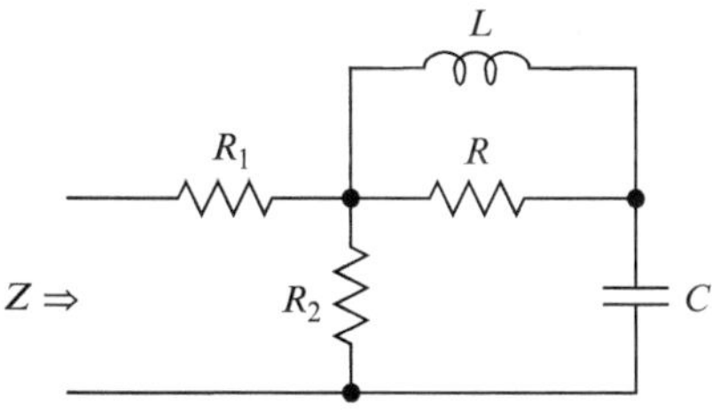

Fig. P 12.62 See Problem P 12.112

C, the impedance of the circuit at resonance, and the rms amplitude of the current i at resonance.

P 12.116 Refer to Fig. P 12.66, where $R = 200\ \Omega$, $L = 10$ mH, and i_S is a variable-frequency sinusoidal source having peak amplitude $I_S = 1$ mA. It is found experimentally that the resonant frequency of the circuit is 159 kHz. Find the value of the capacitance C, the impedance of the circuit at resonance, the voltage across the capacitor at resonance and the current through the inductor at resonance.

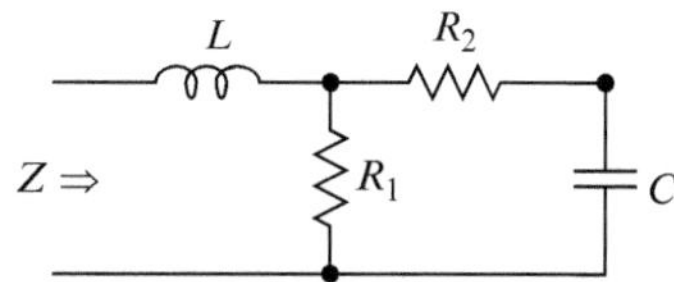

Fig. P 12.63 See Problem P 12.113

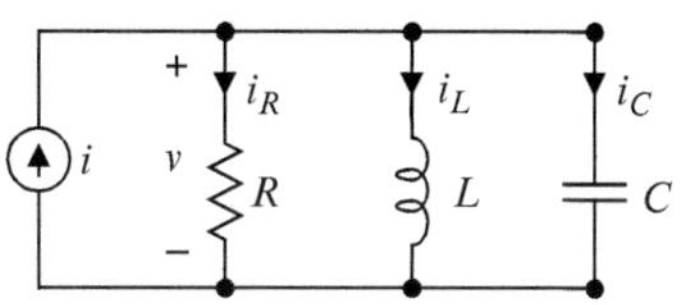

Fig. P 12.64 See Problem P 12.114

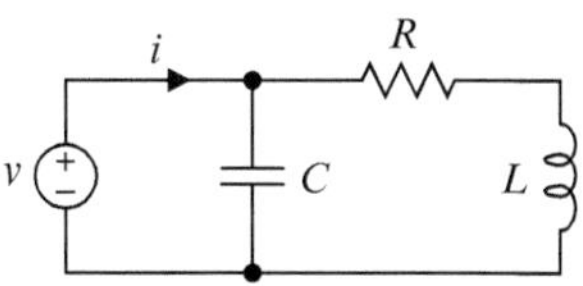

Fig. P 12.65 See Problem P 12.115

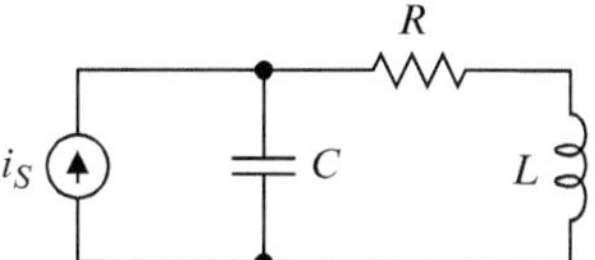

Fig. P 12.66 See Problem P 12.116

Section 12.20 is prerequisite for the following problems.

P 12.117 What is the quality factor of an ideal inductor? Of an ideal capacitor?

P 12.118 Two coils having quality factors $Q_1 = 25$, $Q_2 = 30$ and reactances $X_1 = 500\ \Omega$, $X_2 = 1\ \text{k}\Omega$ at frequency $f_0 = 10$ kHz are connected in parallel. Find the effective resistance, effective inductance, and quality factor of an equivalent (for $f = 10$ kHz) *series RL* circuit.

P 12.119 The *quality factor* for a reactive component having impedance $Z = R + jX$ is given by

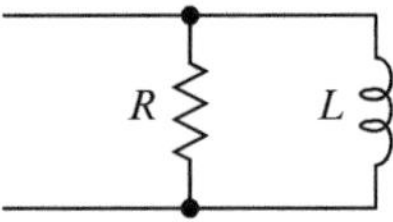

Fig. P 12.67 See Problem P 12.120

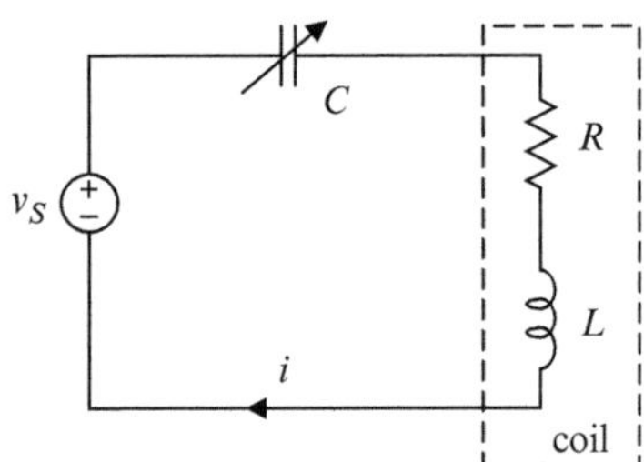

Fig. P 12.68 See Problem P 12.123

$$Q = \left|\frac{X}{R}\right|,$$

where the magnitude operation is necessary because the reactance of a capacitor is negative. Show that the quality factor is also given by

$$Q = \left|\frac{B}{G}\right|,$$

where $G + jB = Y = Z^{-1}$.

P 12.120 Show that the quality factor for the *circuit* shown in Fig. P 12.67 is given by

$$Q = \frac{R}{\omega L}.$$

P 12.121 Two identical high-Q coils are connected in parallel. Is the quality factor of the composite coil larger than, smaller than, or equal to the quality factor of either coil alone?

P 12.122 Two identical high-Q coils are connected in series. Is the quality factor of the composite coil larger than, smaller than, or equal to the quality factor of either coil alone?

P 12.123 In Fig. P 12.68, the capacitor is a laboratory-standard variable capacitor and the source is a fixed-amplitude sinusoidal source having frequency $f = 650$ kHz. The current is monitored using an accurate rms ammeter. As the capacitance is varied, the rms amplitude of the current exhibits a maximum when $C = 6.0$ pF and falls to 0.707 times its maximum value when the capacitance is decreased to 5.0 pF. Find the quality factor, the series resistance, and the inductance of the coil.

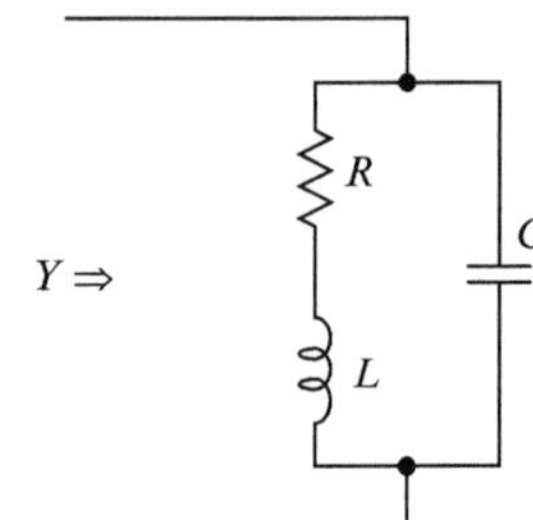

Fig. P 12.69 See Problem P 12.124

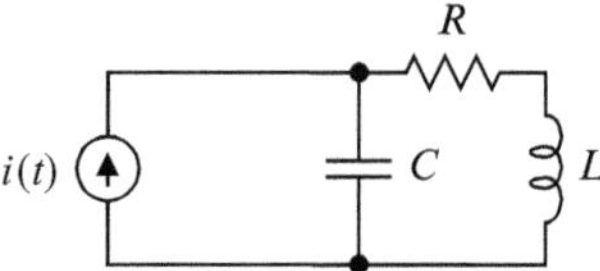

Fig. P 12.70 See Problem P 12.125

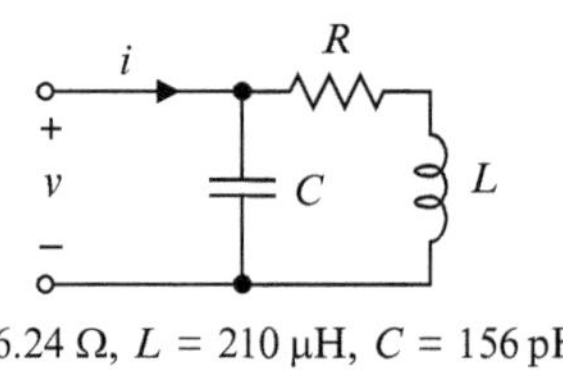

Fig. P 12.71 See Problem P 12.126

P 12.124 Refer to Fig. P 12.69. Obtain an expression for the value of C for which the magnitude of the admittance Y is maximum. Obtain an approximation to the expression for cases where $Q = \omega L/R \gg 1$ for all frequencies of interest.

P 12.125 In Fig. P 12.70, $R = 200\ \Omega$ and $L = 10.0$ mH. The current source is sinusoidal having rms amplitude $I_{rms} = 1$ mA. The circuit is resonant at $f_0 = 159$ kHz. Obtain *symbolic expressions* for and then calculate the values of: (a) the capacitance C, (b) the quality factor of the *coil* (consisting of R and L) at that frequency, (c) the impedance of the circuit at resonance, (d) the peak voltage across the capacitor at resonance, and (e) the rms current through the coil at resonance.

P 12.126 In Fig. P 12.71, the applied voltage $v(t)$ is sinusoidal. Obtain expressions for and then calculate (a) the resonant frequency, (b) the impedance of the circuit at resonance, (c) the magnitude of the impedance for $f_1 = 0.994\, f_0$, where f_0 is the resonant frequency, and (d) the phase angle of the current $i(t)$ relative to that of the applied voltage at the frequency f_1.

P 12.127 Refer to Fig. P 12.72. Obtain a symbolic expression for and then calculate the resonant

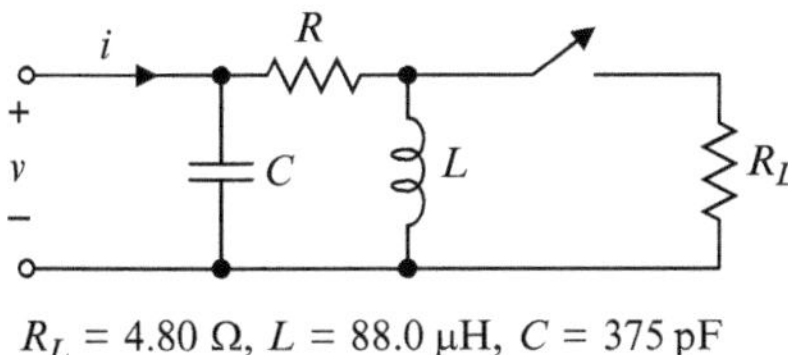

Fig. P 12.72 See Problem P 12.127

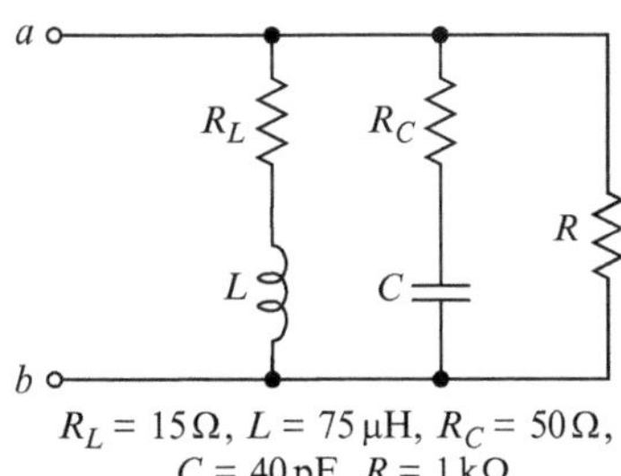

Fig. P 12.73 See Problem P 12.128

frequency with the switch open. Repeat with the switch closed.

P 12.128 Refer to Fig. P 12.73. (a) Find the frequency f_0 below which the circuit is inductive and above which the circuit is capacitive. (b) Model the circuit for $f = f_0/2$ as an inductor in series with a resistor and again as an inductor in parallel with a resistor. (c) Model the circuit for $f = 2f_0$ as a capacitor in series with a resistor and again as a capacitor in parallel with a resistor. (d) Find the equivalent resistance of the circuit for $f = f_0$. (e) Calculate the quality factor for the circuit for $f = f_0/2,\ f = f_0/4$.

P 12.129 The specifications on a particular rf coil are

$$L = 330\ \mu\text{H}, Q = 70,$$
$$SRF = 6.4\ \text{MHz}, f_T = 790\ \text{kHz},$$

where the test frequency f_T is also the intended operating frequency. (a) What is the ac resistance of the coil at the operating frequency f_T? (b) A parallel *RLC* circuit using the coil is to be tuned to 790 kHz. Estimate the capacitance required.

P 12.130 Two coils having quality factors $Q_1 = 25$, $Q_2 = 30$ and reactances $X_1 = 500\ \Omega\ X_2 = 1\ \text{k}\Omega$, at frequency $f_0 = 10$ kHz are connected in series. Find the effective resistance, effective inductance, and quality factor of an equivalent *parallel RL* circuit.

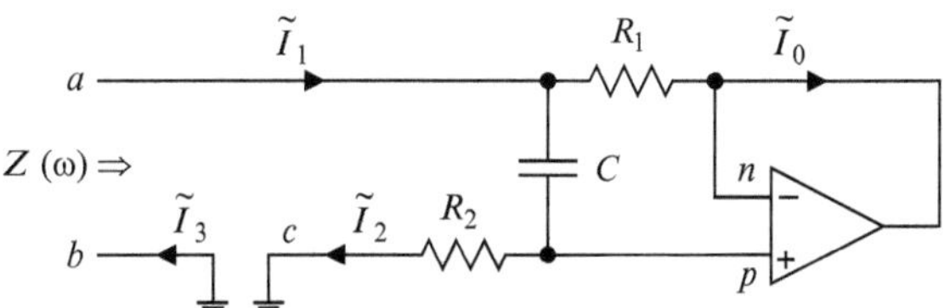

Fig. P 12.74 See Problem P 12.132

Section 12.21 is prerequisite for the following problems.

P 12.131 Using the gyrator circuit and procedure treated in the text, and with reference to Fig. 12.77, design a singly-tuned circuit meeting the following specifications:

Resonant frequency $= f_0 = 200$ Hz,

Impedance at resonance $= Z_0 = 200\ \Omega$,

Quality factor of the simulated coil $= Q = 20$.

Construct a graph of the magnitude of the normalized (to R_1) impedance in dB. On the same axes, construct a graph of a comparable *RLC* tuned circuit.

P 12.132 Refer to Fig. P 12.74. (a) Specify the circuit parameters such that the gyrator simulates an inductor having inductance $L = 4$ H and quality factor $Q = 15$ for frequencies near 100 Hz. (b) Use a simulation to verify the design. Compare the measured impedance with that implied by the specifications at 100 Hz. Show that $\tilde{I}_2$ is only a very small fraction of the current through a source attached to the terminals *a–b*.

P 12.133 Refer to Fig. P 12.75(a). In the text, we show that

$$Z(\omega) = \frac{R_1(1 + j\omega R_2 C)}{1 + j\omega R_1 C} \cong R_1 + j\omega R_1 R_2 C,\ \omega R_1 C \ll 1,$$

which implies (for $\omega R_1 C \ll 1$) that the circuit simulates an inductor having inductance $L = R_1 R_2 C$ in series with a resistor having resistance R_1, as shown in Fig. P 12.75(b). Without making any approximations (other than assuming the op amp is ideal), show that the circuit in Fig. P 12.75(a) is equivalent to the circuit in Fig. P 12.76, with

$$R_p = R_2 - R_1,\ L = R_1 C R_p.$$

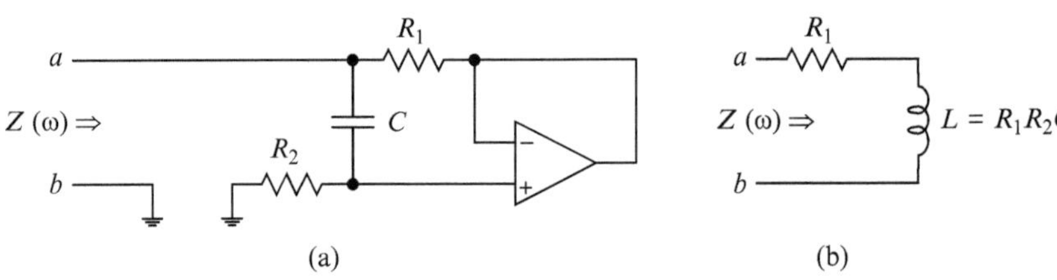

Fig. P 12.75 See Problem P 12.133

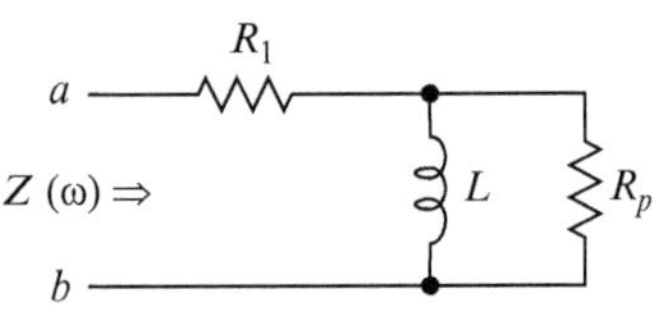

Fig. P 12.76 See Problem P 12.133

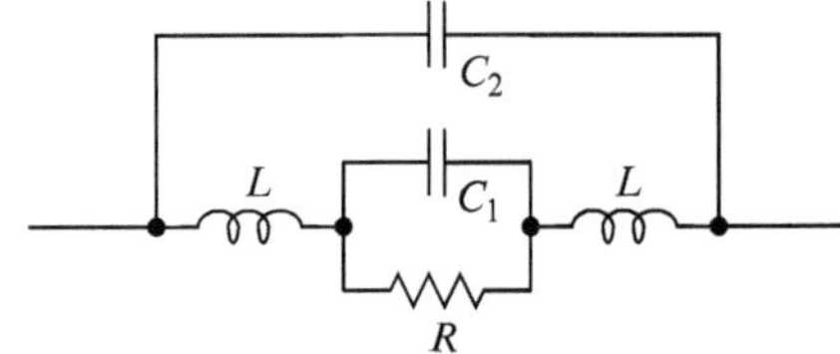

Fig. P 12.78 See Problem P 12.136

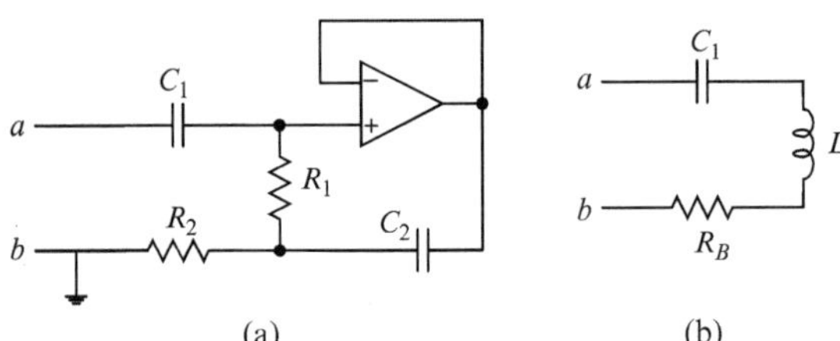

Fig. P 12.77 See Problem P 12.134

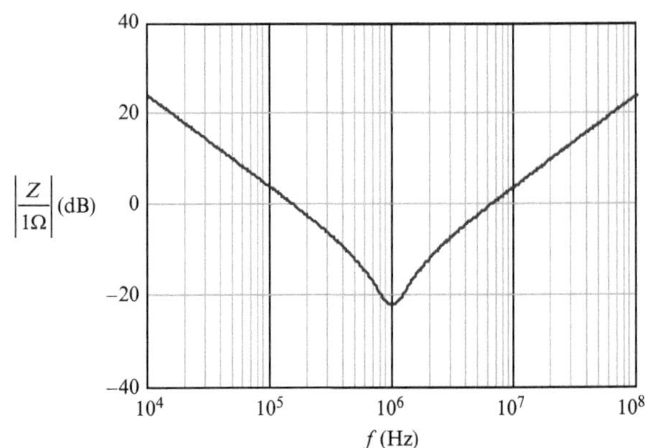

Fig. P 12.79 See Problem P 12.139

P 12.134 Show that the circuit in Fig. P 12.77(a) is equivalent to the circuit in Fig. P 12.77(b), where $R_B = R_2 + R_1$ and $L = C_2R_1R_2$.

Section 12.22 is prerequisite for the following problems.

P 12.135 A certain 10 kΩ resistor in a circuit operating at 10 MHz exhibits a series inductance of 25 pH. The series resistance and inductance appear to be shunted by a capacitance of 1.2 pF. What are the effective series resistance and reactance exhibited by the resistor?

P 12.136 For frequencies up to about 10 GHz, certain kinds of axial-lead resistors can be modeled as shown in Fig. P 12.78, where R is the nominal resistance.

(a) Obtain an expression for the impedance of the model.
(b) Let d denote the length of the resistor. Express the minimum wavelength $\lambda_{\min}$ for which the given (lumped-constant) model is valid in terms of d. Then express the maximum frequency $f_{\max}$ for which the given (lumped-constant) model is valid in terms of d. If we require $\lambda_{\min} > 10d$, and if $f_{\max} = 10$ GHz, what is the maximum value of d for which the model is valid?
(c) Let $R = 1$ kΩ, $C_1 = 1$ pF, $C_2 = 2$ pF, $L = 3$ nH. Find all frequencies for which the impedance is (i) real (resistive), (ii) inductive, and (iii) capacitive.
(d) Let d denote the length of the resistor. What is the maximum value of d for which the frequencies found in Part (c) are meaningful?
(e) Construct graphs of the magnitude and angle of the impedance for 100 kHz $\leq f \leq$ 10 GHz.

P 12.137 The self-resonant frequency of a 200 nH coil is 150 MHz. The quality factor of the coil At 80 MHz is $Q = 40$. Assume the quality factor and shunt capacitance are approximately independent of frequency for 20 MHz $\leq f \leq$ 200 MHz and estimate the shunt capacitance at 80 MHz.

P 12.138 The loss angle of a certain 4, 700 μF capacitor is $\delta = 0.106$ at 120 Hz. The power dissipated by the capacitor must not exceed 2.5 W. What is the maximum allowable 120-Hz rms ripple current?

P 12.139 Fig. P 12.79 shows a graph of the impedance of a certain capacitor. Estimate (a) the self-resonant frequency, (b) the capacitance, (c) the *ESR*, and (d) the *ESL*. (e) Using your estimates, calculate the quality factor, the dissipation factor, and the loss angle at 15 kHz.

Chapter 13
Complex Power

In this chapter, we treat power calculations for a linear circuit in sinusoidal steady state, where the input and every other current and voltage in the circuit are sinusoidal and have the same frequency. We show how to perform and interpret power calculations in terms of phasors and impedances (or admittances), without having to refer to functions of time and without making explicit use of time averages. Important applications include power-factor correction and impedance matching for maximum or efficient power transfer. We begin by defining complex power and obtaining related expressions for average power dissipated at a terminal pair of a sinusoidally excited linear circuit or device.

As a reminder, when we compare impedances by writing that one is larger or smaller than another, *magnitudes* are implied (because complex numbers cannot be ordered).

13.1 Definition of Complex Power

If the voltage across the terminals of a linear circuit or device is sinusoidal, then the current into the terminals also is sinusoidal, and vice versa, and the current and voltage have the same frequency. In Fig. 13.1, let the terminal current and voltage be expressed in standard form as

$$v(t) = V\cos(\omega t + \theta_v), \quad i(t) = I\cos(\omega t + \theta_i).$$

The average power dissipated by the circuit or device is the time average of the product of the current and voltage; i.e.,

$$\begin{aligned} P &= \overline{V\cos(\omega t + \theta_v) I\cos(\omega t + \theta_i)} \\ &= \frac{1}{2} VI\cos(\theta_v - \theta_i), \end{aligned} \tag{13.1}$$

where we have used a result obtained in Chapter 5. The average power dissipated by a sinusoidally excited linear two-terminal device or circuit equals one-half the product of the peak amplitudes of the terminal current and voltage and the cosine of the phase difference between the terminal current and voltage.

Equation (13.1) is the fundamental relation for average power dissipated in sinusoidal steady state. But in sinusoidal steady state analysis, where currents and voltages are expressed as phasors and where linear loads are represented by impedances (or admittances), it is often convenient to perform and interpret power calculations in terms of phasors and impedances (or admittances), without reference to associated functions of time.

The right side of (13.1) can be written

$$\begin{aligned} \frac{1}{2} VI\cos(\theta_v - \theta_i) &= \frac{1}{2} VI\,\mathrm{Re}\left[e^{j(\theta_v - \theta_i)}\right] \\ &= \frac{1}{2}\mathrm{Re}\left[\left(V e^{j\theta_v}\right)\left(I e^{-j\theta_i}\right)\right]. \end{aligned}$$

But $V e^{j\theta_v} = \tilde{V}$ and $I e^{-j\theta_i} = \tilde{I}^*$, where $\tilde{I}^*$ is the complex conjugate of $\tilde{I}$. Thus

$$P = \frac{1}{2}\mathrm{Re}\left[\tilde{V}\tilde{I}^*\right], \tag{13.2}$$

Equation (13.2) inspires the following definition: The **complex power** delivered to or dissipated by a linear load is denoted by S and is defined by

$$S = \frac{1}{2}\tilde{V}\tilde{I}^*, \tag{13.3}$$

where the positive direction for current is into the positive terminal of the load. Note that $\boldsymbol{S}$ (italic) denotes complex power and that S (roman) denotes

T.H. Glisson, *Introduction to Circuit Analysis and Design*,
DOI 10.1007/978-90-481-9443-8_13, © Springer Science+Business Media B.V. 2011

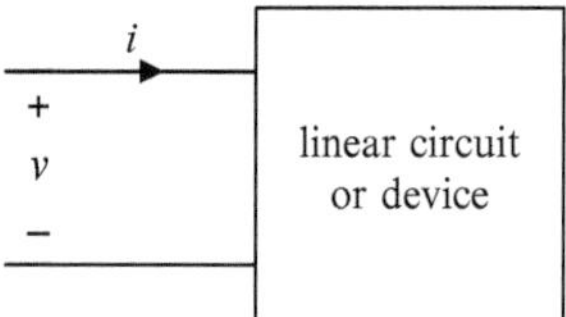

Fig. 13.1 See (13.1)

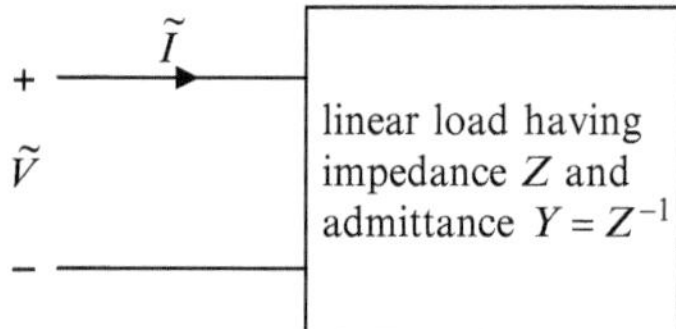

Fig. 13.2 See (13.5) and (13.6)

the unit (siemens) of conductance and admittance. It is difficult to distinguish italic and roman in written work, so where misinterpretation is otherwise possible, you should underline a handwritten letter to denote roman, indicating that the letter denotes a unit and not a quantity.

Complex power has the dimension of real power dissipated, but is non-physical and is not the same thing as real power. To highlight the distinction, the unit assigned to complex power is the **volt·ampere (VA)**, not the watt. We can use (13.3) to express (or compute) complex power from phasors for current and voltage. The real average power dissipated can then be obtained as

$$P = \mathrm{Re}(S) = \frac{1}{2}\,\mathrm{Re}\big(\tilde{V}\tilde{I}^*\big). \qquad (13.4)$$

and is expressed in watts. Useful expressions for the complex power and the real power dissipated by a linear load can be obtained using the generalized form of Ohm's law in the definition (13.3). For example, with reference to Fig. 13.2, we have

$$S = \frac{1}{2}\,\tilde{V}\tilde{I}^* = \frac{1}{2}\,\big(Z\tilde{I}\big)\,\tilde{I}^* = \frac{1}{2}\,Z\big|\tilde{I}\big|^2.$$

It follows that

$$P = \mathrm{Re}(S) = \frac{1}{2}\big|\tilde{I}\big|^2 \mathrm{Re}\,(Z). \qquad (13.5)$$

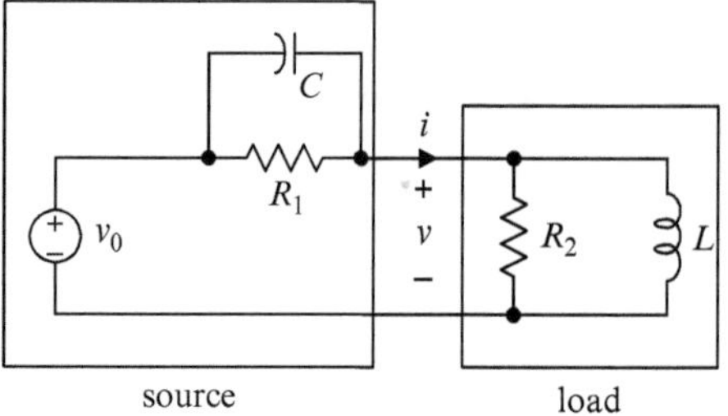

Fig. 13.3 See Example 13.1

Alternatively, from

$$S = \frac{1}{2}\,\tilde{V}\tilde{I}^* = \frac{1}{2}\tilde{V}\big(Y\tilde{V}\big)^* = \frac{1}{2}\big|\tilde{V}\big|^2\, Y^*,$$

it follows that

$$P = \mathrm{Re}(S) = \frac{1}{2}\big|\tilde{V}\big|^2 \mathrm{Re}\,(Y), \qquad (13.6)$$

where we have used the fact that $\mathrm{Re}(Y^*) = \mathrm{Re}(Y)$. Using rms amplitudes, we may write (13.5) and (13.6) as

$$P = I_{rms}{}^2\mathrm{Re}(Z) = V_{rms}{}^2\mathrm{Re}(Y), \qquad (13.7)$$

which makes clear that (13.5) and (13.6) are (for sinusoidal currents and voltages) generalizations of Joule's law

$$P = I_{rms}{}^2\, R = V_{rms}{}^2 G$$

obtained in Chapter 5.

Example 13.1. Refer to Fig. 13.3, where $v_S(t) = V_0\,\cos(\omega t)$, $V_0 = 170$ V, $f = 60$ Hz $R_1 = 100\ \Omega$, $C = 5\ \mu\mathrm{F}$, $R_2 = 1\ \mathrm{k}\Omega$, and $L = 400$ mH. Find the average power dissipated in the load.

Solution: The source and load impedances are

$$Z_S = \frac{R_1}{1 + j\omega R_1 C} \cong 96.6 - j\,18.2\ \Omega$$
$$\cong 98.3\angle -0.186\ \Omega,$$
$$Z_L = \frac{j\omega R_2 L}{R_2 + j\omega L} \cong 22.2 + j\,147\ \Omega$$
$$\cong 149\angle 1.42\ \Omega.$$

The (phasor) load current and voltage are

$$\begin{aligned}\tilde{I} &= \frac{V_0}{Z_L + Z_S} \cong 655 - j713 \text{ mA} \\ &\cong 968\angle - 0.827 \text{ mA}, \\ \tilde{V} &= Z_L\tilde{I} \cong 120 + j80.8 \text{ V} \cong 144\angle 0.594 \text{ V}.\end{aligned}$$

The complex power dissipated in the load is

$$S = \frac{1}{2}\tilde{V}\tilde{I}^* \cong 10.4 + j69.1 \text{ VA}$$

and the real power dissipated in the load is

$$P = \text{Re}(S) \cong 10.4 \text{ W}.$$

We could as well have used (13.5) or (13.6): Thus

$$\begin{aligned}S &= \frac{1}{2}|\tilde{I}|^2 Z_L \cong \frac{1}{2}(0.968\,\text{A})^2(22.2 + j147)\,\Omega \\ &\cong 10.4 + j69.1\,\text{VA}\end{aligned}$$

or

$$\begin{aligned}S &= \frac{1}{2}|\tilde{V}|^2 Y^* = \frac{1}{2}\frac{|\tilde{V}|^2}{Z_L^*} \cong \frac{1}{2}\frac{(144\,\text{V})^2}{(22.2 - j147)\,\Omega} \\ &\cong 10.4 + j69.1\,\text{VA}.\end{aligned}$$

Finally, we could have used (13.1):

$$\begin{aligned}P &= \frac{1}{2}VI\cos(\theta_v - \theta_i) \\ &= \left(\frac{1}{2}\right)(144\,\text{V})(0.968\,\text{A})\cos(0.594 + 0.827) \\ &= 10.4\,\text{W}.\end{aligned}$$

Exercise 13.1. Refer to Fig. 13.4. Find the complex and real power dissipated in the load, the complex and real power dissipated in the source impedance, and the complex and real power delivered by the voltage source v_S. Calculate each at least two ways to check your answers.

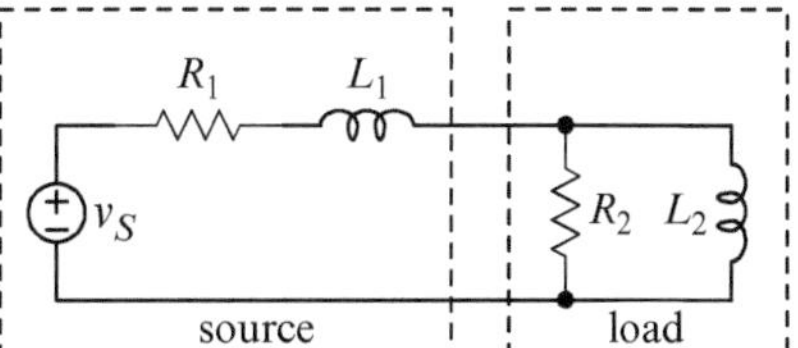

$R_1 = 50\,\Omega$, $L_1 = 100\,\text{mH}$, $R_2 = 500\,\Omega$, $L_2 = 500\,\text{mH}$, $v_S = V_0\cos(\omega_0 t)$, $V_0 = 440\,\text{V}$, $f_0 = 60\,\text{Hz}$

Fig. 13.4 See Exercise 13.1

13.2 Notation

In practice, and especially in electric power applications, amplitudes of currents and voltages often are expressed as rms quantities. Recall that the magnitude of a phasor current or voltage is indicated by removing the tilde (~) from the symbol for the phasor; i.e.,

$$|\tilde{I}| = I, \quad |\tilde{V}| = V.$$

Recall also that rms phasor currents and voltages are denoted by $\tilde{I}_{rms}$ and $\tilde{V}_{rms}$, respectively, where

$$\tilde{I}_{rms} = \frac{1}{\sqrt{2}}\tilde{I}, \quad \tilde{V}_{rms} = \frac{1}{\sqrt{2}}\tilde{V} \tag{13.8}$$

and

$$I_{rms} = |\tilde{I}_{rms}|, \quad V_{rms} = |\tilde{V}_{rms}|. \tag{13.9}$$

The relation (13.2) for real power is written in terms of rms phasor load voltage $\tilde{V}_{rms}$ and rms phasor load current $\tilde{I}_{rms}$ as

$$P = \frac{1}{2}\text{Re}(\tilde{V}\tilde{I}^*) = \text{Re}\left(\tilde{V}_{rms}\tilde{I}_{rms}{}^*\right) \tag{13.10}$$

or as

$$\begin{aligned}P &= \frac{1}{2}|\tilde{V}\tilde{I}|\cos(\theta_Z) = V_{rms}I_{rms}\cos(\theta_Z), \\ \theta_Z &= \measuredangle Z,\end{aligned} \tag{13.11}$$

where $\theta_Z = \measuredangle Z$ is the angle of the load impedance. Equations (13.5) and (13.6) are written in terms of rms quantities as

$$S = \tilde{V}_{rms}\tilde{I}_{rms}^{\ *} = \left(Z\tilde{I}_{rms}\right)\tilde{I}_{rms}^{\ *} = Z{I_{rms}}^2$$
$$\Rightarrow P = \mathrm{Re}(S) = {I_{rms}}^2\mathrm{Re}\,(Z) \qquad (13.12)$$

and

$$S = \tilde{V}_{rms}\tilde{I}_{rms}^{\ *} = \tilde{V}_{rms}\left(Y\tilde{V}_{rms}\right)^* = {V_{rms}}^2\,Y^*$$
$$\Rightarrow P = \mathrm{Re}(S) = {V_{rms}}^2\mathrm{Re}\,(Y), \qquad (13.13)$$

respectively.

When using this notation, keep in mind that I_{rms}, V_{rms} (no tildes) represent the *magnitudes* of rms phasors, and that Kirchhoff's laws apply to *phasors*, not their magnitudes. For example, Kirchhoff's voltage law for the circuit in Fig. 13.5 can be expressed as

$$R\tilde{I} + j\omega L\tilde{I} = \tilde{V}, \qquad (13.14)$$

or as (note tildes)

$$R\tilde{I}_{rms} + j\omega L\tilde{I}_{rms} = \tilde{V}_{rms}, \qquad (13.15)$$

but (no tildes)

$$RI_{rms} + j\omega LI_{rms} \neq V_{rms},$$

because V_{rms} is real.

Whether we write Kirchhoff's laws and other relations using phasors or rms phasors is a matter of convenience, and might depend upon something as minor as the number of subscripts involved. For example, if a number of node voltages or loop currents are involved, and each is subscripted, we might use phasors (not rms phasors) simply to avoid multiple subscripts and because there is no notational advantage to using rms phasors in node and mesh equations. If convenient, we can simply convert the solutions to rms phasors after solving the equations. In electric power industries and in specialized courses on power generation and distribution, it is understood that all phasors are rms phasors, and no rms subscript is appended. We do not adopt that convention because this book as a whole is not focused on that discipline.

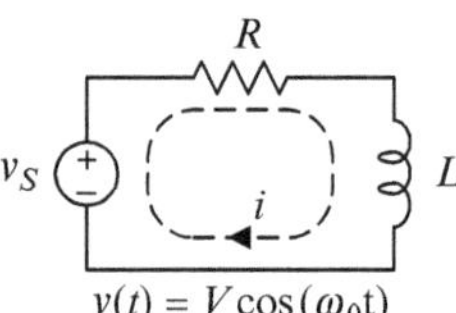

Fig. 13.5 See (13.14) and (13.15) and related discussion

13.3 Power Calculations

We can calculate the complex power dissipated in a linear load in sinusoidal steady state if we know any two of three quantities: The phasor voltage across the load, the phasor current through the load, and the load impedance (or admittance).

Equations (13.10) or (13.11) and the generalized Ohm's law $\tilde{V} = Z\tilde{I}$ are really the only relations needed for real power calculations in a sinusoidally excited *RLC* circuit. The ancillary relations (13.12) and (13.13) follow readily from those more fundamental relations.

Example 13.2. Refer to Fig. 13.6. Find the complex power dissipated in each element, including the source.

Solution: Because we are to find the complex power dissipated in each element, we do not reduce the circuit. Instead, we use Kirchhoff's current law to find the unknown node voltages, and then compute the branch currents and complex powers. Figure 13.7 shows the notation used in the solution.

We begin by writing Kirchhoff's current law at node *a*:

$$\frac{\tilde{V}_a - \tilde{V}_0}{R_1} + \frac{\tilde{V}_a}{Z_C} + \frac{\tilde{V}_a}{R_2 + Z_L} = 0.$$

Solving this equation for $\tilde{V}_a$ gives

$$\tilde{V}_a = \frac{Z_C\,(R_2 + Z_L)\,\tilde{V}_0}{(Z_L + R_2)(Z_C + R_1) + Z_C R_1}$$
$$\cong 113\angle - 0.889\ \mathrm{V},$$

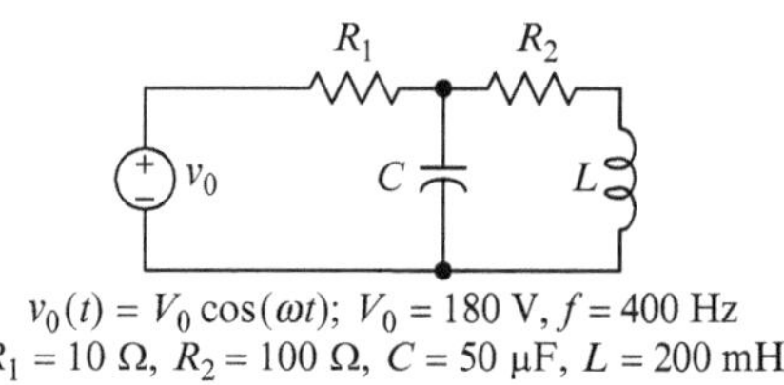

Fig. 13.6 Circuit of Example 13.2

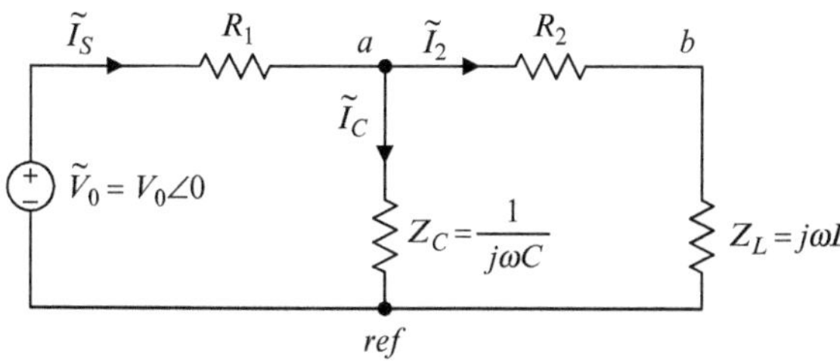

Fig. 13.7 Circuit of Example 13.2, showing notation used in the solution

from which we obtain

$$\tilde{I}_S = \frac{\tilde{V}_0 - \tilde{V}_a}{R_1} \cong 14.0\angle 0.678 \text{ A},$$

$$\tilde{I}_C = \frac{\tilde{V}_a}{Z_C} \cong 14.2\angle 0.682 \text{ A},$$

$$\tilde{I}_2 = \frac{\tilde{V}_a}{R_2 + Z_L} \cong 0.22\angle -2.26 \text{ A}.$$

The complex power dissipated in the source is

$$S_S = -\frac{1}{2}\tilde{V}_0 \tilde{I}_S^* \cong -979.5 + j789.6 \text{ VA}$$

The complex powers dissipated in the other elements are

$$R_1: \quad S_1 = \frac{1}{2}\left(\tilde{V}_0 - \tilde{V}_a\right)\tilde{I}_S^* \cong 977.1 \text{ VA},$$

$$C: \quad S_C = \frac{1}{2}\tilde{V}_a \tilde{I}_C^* \cong -j801.8 \text{ VA},$$

$$R_2: \quad S_2 = \frac{1}{2}\left|\tilde{I}_2\right|^2 R_2 \cong 2.4 \text{ VA},$$

$$L: \quad S_L = \frac{1}{2}\left|\tilde{I}_2\right|^2 Z_L \cong j12.2 \text{ VA}.$$

In this example, there is no great advantage to using rms phasors.

13.4 Reactive Power and Apparent Power

The imaginary part of complex power, denoted by Q, is called **reactive power**:

$$Q = \text{Im}(S). \qquad (13.16)$$

We may write

$$S = \tilde{V}_{rms}\tilde{I}_{rms}^{\;*} = P + jQ, \qquad (13.17)$$

where, by Euler's identity,

$$P = |S|\cos(\theta_S), \quad Q = |S|\sin(\theta_S). \qquad (13.18)$$

From (13.5), the complex power dissipated in a load Z is given by

$$S = \frac{1}{2}Z\left|\tilde{I}\right|^2 = I_{rms}^{\;2} Z = I_{rms}^{\;2}|Z|\angle\theta_Z,$$

which shows that the angle of the complex power dissipated in an impedance Z equals the angle $\theta_Z = \measuredangle Z$ of the impedance.

$$\theta_S = \measuredangle S = \measuredangle Z = \theta_Z. \qquad (13.19)$$

The sign of the angle θ_Z determines the sign of reactive power Q. The reactance of an inductance is positive and the reactance of a capacitance is negative. Thus *the reactive power dissipated in an inductive load is positive and the reactive power dissipated in a capacitive load is negative.*

Here again, because there is a finite number of symbols, and because workers in different disciplines have assigned meanings to symbols without regard to their meanings in other disciplines, the symbol Q has multiple meanings. In (13.17) and (13.16), Q denotes reactive power. In other contexts, Q denotes quality factor or electric charge. We could perhaps assign some different seldom-used symbol to reactive power or to quality factor and charge, but that would put us at odds with usual practice. We are specific where multiple interpretations are possible, but you must ultimately rely on context to discern the meaning of Q in any particular development.

The unit assigned to reactive power is the **volt·ampere-reactive** (**VAR**). The dimension of reactive power is that of real power, but reactive power and real power are different things. Real power is the rate at which electrical energy is converted to another form (often mechanical; e.g., heat or mechanical work). Reactive power refers to energy being stored in and released from magnetic and electric fields created by inductance and capacitance. This back-and-forth movement of energy is real, but does not entail conversion of any of the energy involved to heat or any other mechanical form. Nonetheless, it is conventional to refer to reactive power as being *dissipated* in (or by) a source or reactive element, although no dissipation takes place.

By way of further explanation, we offer the following: If a sinusoidal voltage is applied to a pure reactance (any combination of capacitors and inductors, but no resistors or sources), the impedance is purely imaginary, the resulting current is in phase quadrature with the applied voltage (the phase difference is $\pm\pi/2$) and the real average power dissipated equals zero, because $\cos(\theta_Z) = \cos(\pm\pi/2) = 0$ in (13.11). The difference between real power and reactive power is analogous to the difference between energy (or work) and torque. Torque, like energy, has the dimension of *force* $\times$ *length*, but torque is not the same thing as energy. Energy is a scalar and is conserved, whereas torque is a vector and is not conserved.

Exercise 13.2. Find the reactive powers dissipated in the source, the capacitor, and the inductor in Example 13.2.

The magnitude of complex power is called **apparent power**:

$$\text{apparent power} = |S| = \sqrt{P^2 + Q^2} = \left|\tilde{V}_{rms}\tilde{I}_{rms}^{\;*}\right| = V_{rms}I_{rms}. \quad (13.20)$$

The SI unit of apparent power is the volt·ampere (VA). Apparent power is significant because it determines the current required to deliver a specified real power to a specified load at a fixed source voltage. For example, refer to Fig. 13.8, where a load having impedance Z requires real power P. From (13.11), the power delivered to the load is given by

$$P = V_{rms}I_{rms}\cos(\theta_Z),\ \theta_Z = \measuredangle Z.$$

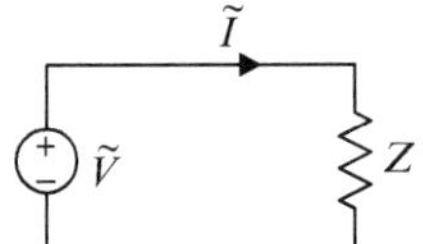

Fig. 13.8 Line current is determined by apparent power (see text)

If the rms line voltage V_{rms} is fixed, then the rms line current I_{rms} must assume the value necessary to provide the required power:

$$I_{rms} = \frac{P}{V_{rms}\cos(\theta_Z)} = \frac{|S|\cos(\theta_Z)}{V_{rms}\cos(\theta_Z)} = \frac{|S|}{V_{rms}}.$$

The rms line current I_{rms} is determined by the apparent power $|S|$, not merely by the real power P dissipated. If the real power required is fixed, the apparent power and hence the line current depend upon load reactance (see (13.34)). To reiterate: The real power delivered to a load is the real part of the complex power delivered to the load. But the current required to deliver the real power is determined by the apparent power delivered to the load. This is because reactive elements draw current but do no work. Apparent (or reactive) power is significant because the current drawn by a load causes losses in the generators, transformers, and wires needed to deliver the required power to the load.

Example 13.3. In Fig. 13.8, the load impedance is $Z = (15 + j12)\ \Omega$ at the frequency of the fixed-amplitude source voltage $\tilde{V}_{rms} = 120\ \text{V}\angle 0$. The load requires real power $P = 500$ W. What rms line current would be required if the load were $Z = 15\ \Omega$? What is the actual rms line current?

Solution: The rms current required to deliver the specified power to the load is

$$I_{rms}{}^2 R = P \Rightarrow I_{rms} = \sqrt{\frac{P}{R}} = \sqrt{\frac{500\ \text{W}}{15\Omega}} \cong 5.77\ \text{A}.$$

The actual line current is

$$I_{rms} = \frac{|S|}{V_{rms}} = \frac{V_{rms}}{|Z|} = \frac{120\ \text{V}}{\sqrt{15^2 + 12^2}\ \Omega} \cong 6.25\ \text{A}.$$

We find

$$100 \times \frac{6.25\ \text{A} - 5.77\ \text{A}}{5.77\ \text{A}} \cong 8.3\%,$$

which means that the current is about 8% larger than necessary to deliver the required power.

Also

$$100 \times \frac{(6.25)^2\ \text{A}^2 - (5.77)^2\text{A}^2}{(5.77)^2\text{A}^2} \cong 17.3\%,$$

which means that the losses in the line are about 17% greater than they would be if only the required current were delivered. We show in Section 13.8 how the load can be modified such that the reactive power required is reduced, thus reducing the line current and attendant losses.

Example 13.4. The (rms) voltmeter and ammeter readings in Fig. 13.9 are 13.67 V and 24.00 mA, respectively. Find the complex power, the apparent power, the real power, and the reactive power delivered by the independent source v_S.
Solution: We take the source as the phase reference, so

$$\tilde{V}_S = V_0\angle 0 = V_0.$$

We need the (complex) impedance of the load, which means we must find the resistance R and the capacitance C. The rms current through the capacitor (the ammeter reading) is given by

$$I_{rms} = \omega_0 C V_{rms},$$

where $\omega_0 = 2\pi f_0$ and V_{rms} is the voltmeter reading. Thus

$$C = \frac{I_{rms}}{\omega_0 V_{rms}} = \frac{24\ \text{mA}}{2\pi(5\text{kHz})(13.7\ \text{V})} \cong 56\ \text{nF}.$$

From Kirchhoff's current law,

$$\frac{\tilde{V}_a}{R} + \frac{\tilde{V}_a - V_0}{R_S} + j\omega_0 C\tilde{V}_a = 0$$
$$\Rightarrow \frac{1}{R} + \frac{1}{R_S} + j\omega_0 C = \frac{V_0}{\tilde{V}_a R_S},$$

where $\tilde{V}_a$ is the phasor voltage across the load. It follows that

$$\left(\frac{1}{R} + \frac{1}{R_S}\right)^2 + (\omega_0 C)^2 = \left(\frac{V_0}{|\tilde{V}_a| R_S}\right)^2 = \left(\frac{V_{0\,rms}}{V_{rms} R_S}\right)^2,$$

where V_{rms} is the voltmeter reading and $V_{0\,rms} = 15$ V. Thus

$$\frac{1}{R} = \sqrt{\left(\frac{V_{0\,rms}}{V_{rms} R_S}\right)^2 - (\omega_0 C)^2} - \frac{1}{R_S}$$
$$\Rightarrow R \cong 1.2\ \text{k}\Omega.$$

We now have the load impedance

$$Z_L = \frac{R}{1 + j\omega RC} \cong (221 - j465)\ \Omega$$

and the total impedance

$$Z = R_S + Z_L \cong (321 - j465)$$
$$\Omega \cong 565\angle(-0.967)\ \Omega.$$

The complex power S, apparent power $|S|$, real power P, and reactive power Q delivered by the source are

$$S = \frac{{V_{0\,rms}}^2}{Z^*} \cong \frac{(15\ \text{V})^2}{(563.8\angle 0.967)\Omega}$$
$$\cong 399\angle(-0.967)\ \text{mVA},$$
$$|S| \cong 399\ \text{mVA},$$
$$P = \text{Re}(S) \cong |S|\cos(-0.967) \cong 226\ \text{mW},$$
$$Q = \text{Im}(S) \cong |S|\sin(-0.967) \cong -329\ \text{mVAR}.$$

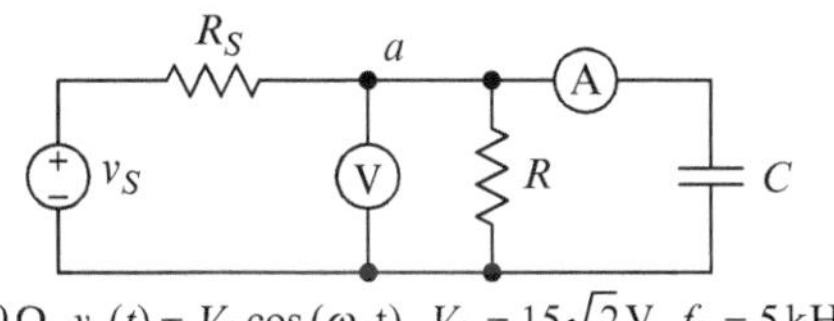

$R_S = 100\,\Omega$, $v_S(t) = V_0\cos(\omega_0 t)$, $V_0 = 15\sqrt{2}\,\text{V}$, $f_0 = 5\,\text{kHz}$

Fig. 13.9 See Example 13.4

Alternatively (or as a check),

$$\tilde{V}_{rms} = \left(\frac{1}{R}+\frac{1}{R_S}+j\omega C\right)^{-1}\frac{V_{0\,rms}}{R_S}$$
$$\cong (13.49 - j2.19)\,\mathrm{V},$$
$$\tilde{I}_{S\,rms} = \frac{V_{0\,rms} - \tilde{V}_{rms}}{R_S},$$
$$S = V_{0\,rms} I_{S\,rms}{}^{*} \cong 226 - j329\,\mathrm{mVA}$$
$$\cong 399\angle(-0.967)\ \mathrm{mVA}.$$

Apparent power is a measurable (physical) quantity, as indicated by the product $V_{rms}I_{rms}$ in (13.20), and the components of complex power delivered to a load can be computed from the apparent power and the angle of the load, denoted by $\theta_Z = \measuredangle Z$. Thus

$$\begin{aligned} P &= V_{rms}I_{rms}\cos(\theta_Z) = |S|\cos(\theta_Z), \\ Q &= V_{rms}I_{rms}\sin(\theta_Z) = |S|\sin(\theta_Z). \end{aligned} \tag{13.21}$$

When using (13.21) and similar relations involving the angle of a load or current or voltage or complex power, take care to keep track of the sign of the angle. It is easy to make mistakes in this regard because the cosine is an even function; i.e., because $\cos(-\theta) = \cos(\theta)$.

13.5 Conservation of Complex Power

Complex power is conserved; that is,

$$\sum_{n=1}^{N} S_n = 0, \tag{13.22}$$

where S_n is the complex power dissipated (according to the passive sign convention) in the nth element of an N-element circuit and the sum is over all elements in the circuit.[1]

The fact that complex power is conserved can be used to check results obtained by analysis of a circuit.

[1]Equation (13.22) is a degenerate case of *Tellegen's theorem*, which is a remarkable property of any pair of circuits having the same topology. See Artice M. Davis, *Linear Circuit Analysis*, PWS Pub. Co., (1998), pp 1052*ff*.

Example 13.5. Show that complex power is conserved in the circuit of Example 13.2.
Solution: The sum of the complex powers is

$$S_S + S_1 + S_C + S_2 + S_L \cong (-979.5 + j789.6 + 977.1 - j801.8 + 2.4 + j12.2)\mathrm{VA} \cong 0.$$

This calculation serves as a check on the accuracy of the calculated currents and voltages and individual complex powers.

Because a complex quantity equals zero if and only if both the real and imaginary parts of the quantity equal zero, conservation of complex power implies conservation of both real average power (the real part of complex power) and of average reactive power (the imaginary part of complex power). Apparent power is *not* conserved, because apparent power is non-negative.

Example 13.6. A voltage $v(t) = V_0\cos(\omega t)$ with $V_0 = 25$ V and $f = 1$ kHz is applied to a load consisting of a resistor having resistance $R = 1.5$ kΩ in parallel with an inductor having inductance $L = 50$ mH. Find the average power dissipated in the load and the average power delivered by the source.
Solution: The impedance of the load is

$$Z = \frac{j\omega LR}{R + j\omega L} = \frac{j(2\pi)(1\,\mathrm{kHz})(50\,\mathrm{mH})(1.5\,\mathrm{k\Omega})}{1.5\,\mathrm{k\Omega} + j(2\pi)(1\,\mathrm{kHz})(50\,\mathrm{mH})}$$
$$\cong 63.0 + j301\,\Omega \cong 307\angle 1.36\,\Omega.$$

The load current is

$$\tilde{I} = \frac{\tilde{V}}{Z} \cong \frac{25\ \mathrm{V}\angle 0}{307\ \Omega\angle 1.36} \cong 81.0\angle -1.36\ \mathrm{mA}.$$

The complex power is

$$S = \frac{1}{2}\tilde{V}\tilde{I}^{*} \cong \frac{1}{2}(25\ \mathrm{V}\angle 0)(81.0\ \mathrm{mA}\angle 1.36)$$
$$\cong 1.01\angle 1.36\ \mathrm{VA}.$$

The real average power dissipated in the load is

$$P = \mathrm{Re}(S) = |S| \cos(\measuredangle S) \cong 1.01 \cos(1.36) \cong 213 \text{ mW}.$$

Because complex power is conserved, the complex power S_0 delivered by the source equals the complex power dissipated in the load: $S_0 \cong 1.01\angle 1.36$ mVA.

A proof of (13.22) follows.[2] Refer to Fig. 13.10, where the box represents an element (e.g., impedance or source) connecting nodes n and m in an *arbitrary* circuit. If more than one element connects a pair of nodes, we assume the elements are lumped into one, such that each pair of nodes is connected by only one element. Thus, if there are N elements, there are N connected node pairs.

The complex power dissipated in the element joining nodes n and m is given by

$$S_n = \frac{1}{2}\tilde{V}_{nm}\tilde{I}_{nm}^{\,*} = \frac{1}{2}\tilde{V}_{mn}\tilde{I}_{mn}^{\,*}. \qquad (13.23)$$

The total complex power dissipated in the circuit is given by

$$S = \sum_{n-1}^{N} S_n = \frac{1}{2}\frac{1}{2}\sum_{m=1}^{N}\sum_{n=1}^{N}\tilde{V}_{nm}\tilde{I}_{nm}^{\,*}. \qquad (13.24)$$

In (13.24), the sums are over all nodes in the circuit. The extra factor of $1/2$ is necessary because each element is included twice, as indicated by (13.23) (e.g., the double sum on the right side of (13.24) includes both $\tilde{V}_{12}\tilde{I}_{12}^{\,*}$ and $\tilde{V}_{21}\tilde{I}_{21}^{\,*}$, each of which is the complex power dissipated in the element connecting nodes 1 and 2). By definition,

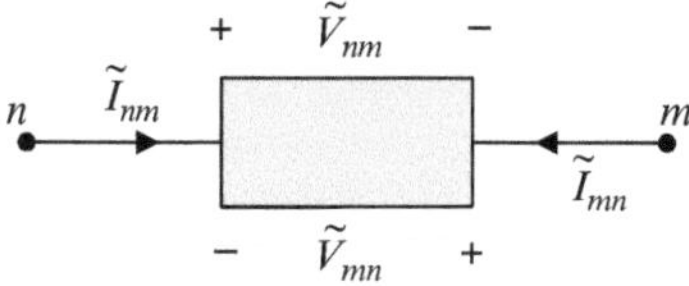

Fig. 13.10 Pertaining to the proof of (13.22)

[2]The proof can be skipped without ill effect.

$$\tilde{V}_{nm} = \tilde{V}_n - \tilde{V}_m. \qquad (13.25)$$

Using (13.25) in (13.24) gives

$$\begin{aligned} S &= \frac{1}{4}\sum_{n=1}^{N}\tilde{V}_n\sum_{m=1}^{N}\tilde{I}_{nm}^{\,*} - \frac{1}{4}\sum_{m=1}^{N}\tilde{V}_m\sum_{n=1}^{N}\tilde{I}_{nm}^{\,*} \\ &= \frac{1}{4}\sum_{n=1}^{N}\tilde{V}_n\left(\sum_{m=1}^{N}\tilde{I}_{nm}\right)^* \\ &\quad - \frac{1}{4}\sum_{m=1}^{N}\tilde{V}_m\left(\sum_{n=1}^{N}\tilde{I}_{nm}\right)^*. \end{aligned} \qquad (13.26)$$

By Kirchhoff's current law, the inner sums on the right, being the sums of currents leaving nodes n and m, respectively, equal zero. Equation (13.22) follows.

13.6 Power Relations in Resonant Circuits

Complex power, real power, and reactive power are in general functions of frequency because impedance is a function of frequency. By definition, the equivalent impedance of a circuit at resonance is real and equal to the effective resistance of the circuit at the resonant frequency. Because the equivalent impedance is real at resonance, the complex power, given by (13.12), is real at resonance. *For a circuit at resonance*,

$$S = {I_{rms}}^2 Z_r = {I_{rms}}^2 R_{eff} = \mathrm{Re}(S) = P, f = f_r, \quad (13.27)$$

where f_r is the resonant frequency, Z_r is the impedance at resonance, $\tilde{I}$ is the terminal phasor current at resonance, and I_{rms} is the rms amplitude of the terminal current. Also,

$$\mathrm{Im}(S) = Q = 0, \quad f = f_r. \qquad (13.28)$$

It follows that, *at resonance, the sum of the inductive reactive power and the capacitive reactive power equals zero.*

Example 13.7. In Fig. 13.11, $R = 2.2$ kΩ and $v = V_0\cos(\omega_0 t)$, where $V_0 = 25$ V and $f_0 = 20$ kHz equals the resonant frequency of the circuit. The reactive power dissipated in the inductor is $Q_L = 5$VAR. Find the inductance L and the capacitance C

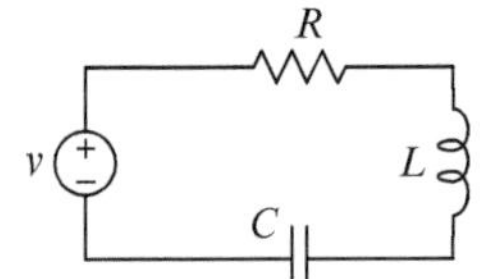

Fig. 13.11 See Example 13.7

Solution: The impedance of the circuit at resonance equals the resistance R. Thus the current through the series circuit is

$$\tilde{I} = \frac{V_0}{R} \cong 11.36 \text{ mA} \Rightarrow I_{rms} = \frac{|\tilde{I}|}{\sqrt{2}} \cong 8.04 \text{ mA}.$$

The reactive powers dissipated in the inductor and in the capacitor are equal in magnitude and opposite in sign, and are given by

$$Q_L = I_{rms}{}^2 \omega L, \quad Q_C = -\frac{I_{rms}{}^2}{\omega C}.$$

Thus $Q_C = -Q_L$ and it follows that

$$L = \frac{Q_L}{\omega_0 I_{rms}{}^2} \cong \frac{5\text{VAR}}{(2\pi)(20 \text{ kHz})(8.04 \text{ mA})^2}$$
$$\cong 616.2 \text{ mH},$$
$$C = -\frac{I_{rms}{}^2}{\omega_0 Q_C} = \frac{I_{rms}{}^2}{\omega_0 Q_L}$$
$$\cong \frac{(8.04 \text{ mA})^2}{(2\pi)(20 \text{ kHz})(5\text{VAR})} \cong 102.8 \text{ pF},$$

or

$$C = \frac{1}{\omega_0{}^2 L} \cong 102.8 \text{ pF}.$$

Example 13.8. In Fig. 13.12, the rms amplitude of the sinusoidal current source is $I_{rms} = 5$ mA and the frequency of the source is 100 kHz, which is the resonant frequency of the circuit. The coil has quality factor $Q = 34$ and effective resistance $R = 3.6\ \Omega$ at 100 kHz. Find the reactive power dissipated in the capacitor. A quality factor Q and reactive powers Q_L, Q_C appear in this example. Keep the symbols (the Q's) straight.

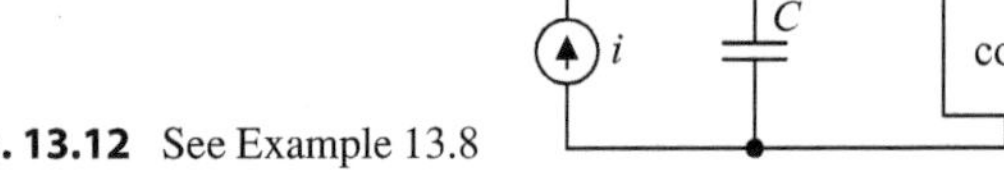

Fig. 13.12 See Example 13.8

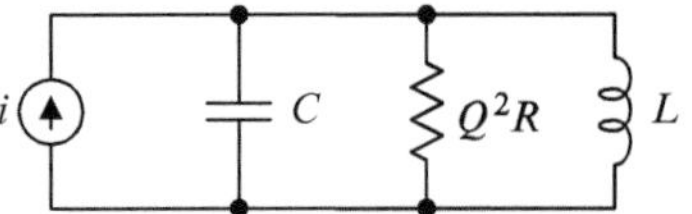

Fig. 13.13 See Example 13.8

Solution: From Chapter 12, because the quality factor of the coil is much larger than unity, the coil (and the rest of the circuit) can be represented as shown in Fig. 13.13, where

$$R = 3.6\ \Omega, \quad L = \frac{QR}{\omega} \cong \frac{(34)(3.6\ \Omega)}{2\pi(100 \text{ kHz})}$$
$$\cong 195\ \mu\text{H}.$$

The effective resistance of the circuit at resonance equals the resistance Q^2R, and the current through the resistance equals the source current. It follows that the rms voltage across the source and each element in the circuit above is given by

$$V_{rms} = \left|\tilde{V}_{rms}\right| = Q^2 R I_{rms}.$$

Thus the reactive power dissipated in the inductor is given by

$$Q_L = \frac{V_{rms}{}^2}{\omega L}.$$

At resonance, the reactive powers dissipated by the inductor and the capacitor are equal in magnitude and opposite in sign. The reactive power dissipated in the capacitor is

$$Q_C = -Q_L = -\frac{V_{rms}{}^2}{\omega L} = -\frac{Q^4 R^2 I_{rms}{}^2}{\omega L}$$
$$\cong -3.53 \text{ VAR}.$$

Alternatively, we could have obtained the capacitance from $\omega^2 = (LC)^{-1}$ and used

$$Q_C = \omega C V_{rms}{}^2$$

to obtain the same result.

13.7 Power Factor

Let $\tilde{V}$ and $\tilde{I}$ denote the voltage across and current through a load having impedance $Z = |Z|\angle\theta_Z$. From (13.5), the real power dissipated is given by

$$\begin{aligned} P &= \mathrm{Re}(S) = \mathrm{Re}\left(I_{rms}{}^2\,|Z|\,\angle\theta_Z\right) \\ &= I_{rms}{}^2\,|Z|\,\cos(\theta_Z). \end{aligned} \tag{13.29}$$

The factor $\cos(\theta_Z)$ in (13.29) is called the **power factor** for a load in question and is denoted by *pf*. *The power factor for a load is the cosine of the angle of the load impedance*:

$$\text{power factor} = pf = \cos(\measuredangle Z) = \cos(\theta_Z). \tag{13.30}$$

Equivalently, from (13.1), the power factor is the cosine of the phase difference between the voltage across and current into the terminals of a device or circuit in question. Equivalently, also, the power factor is the cosine of the angle of the complex power.

Example 13.9. A voltage $v(t) = V_0\cos(\omega t)$ is applied to the series combination of a capacitor and a resistor. Obtain expressions for the power factor and the average power dissipated in the load.

Solution: The impedance of the load is

$$\begin{aligned} Z &= R + \frac{1}{j\omega C} = R - j\frac{1}{\omega C} \\ &= \sqrt{R^2 + (\omega C)^{-2}}\,\angle\tan^{-1}\left(-\frac{1}{\omega RC}\right). \end{aligned}$$

The power factor is

$$\begin{aligned} pf &= \cos(\measuredangle Z) = \cos\left[\tan^{-1}\left(-\frac{1}{\omega RC}\right)\right] \\ &= \frac{\omega RC}{\sqrt{1 + (\omega RC)^2}}, \end{aligned}$$

as illustrated by Fig. 13.14. The load current (phasor) is given by

$$\tilde{I} = \frac{\tilde{V}}{Z} = \frac{V_0}{\sqrt{R^2 + (\omega C)^{-2}}}\angle - \tan^{-1}\left(-\frac{1}{\omega RC}\right).$$

The peak load voltage equals V_0 and the peak load current (the magnitude of the phasor $\tilde{I}$ obtained above) is given by

$$I_0 = \frac{V_0}{\sqrt{R^2 + (\omega C)^{-2}}} = \frac{\omega C V_0}{\sqrt{1 + (\omega RC)^2}}.$$

It follows from (13.11) that the average power dissipated is given by

$$\begin{aligned} P &= \frac{1}{2}(V_0 I_0)(pf) \\ &= \frac{1}{2}\frac{\omega C V_0{}^2}{\sqrt{1 + (\omega RC)^2}}\frac{\omega RC}{\sqrt{1 + (\omega RC)^2}} \\ &= \frac{1}{2}\frac{R(\omega C)^2 V_0{}^2}{\left[1 + (\omega RC)^2\right]}. \end{aligned}$$

Why is the average power dissipated equal to zero for $f = 0$? Does the expression above yield the correct average power dissipated for $f \to \infty$?

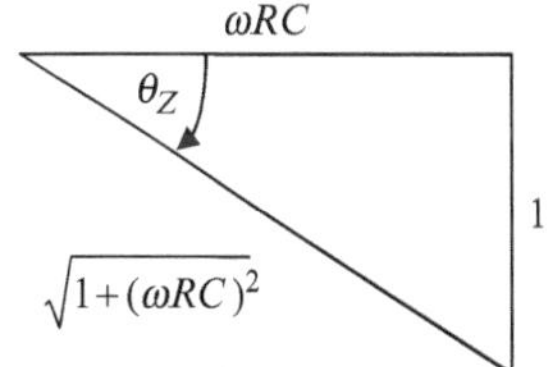

Fig. 13.14 See Example 13.9

Example 13.10. Show that the real part of the impedance of a source-free *RLC* load is non-negative. (We make use of this fact in the development following this example.)
Solution: From (13.12)

$$S = I_{rms}^{\ 2} Z = I_{rms}^{\ 2} R_Z + j I_{rms}^{\ 2} X_Z, \quad (13.31)$$

where $i(t)$ enters the positive terminal of the load. The real power dissipated by the load is given by

$$P = \mathrm{Re}(S) = I_{rms}^{\ 2} R_Z. \quad (13.32)$$

Because the load is source-free, $P \geq 0$. Because $I_{rms}^{\ 2} \geq 0$, it must be that $R_Z \geq 0$.

Because the cosine is an even function, we cannot tell from the numerical value of a power factor $\cos(\theta_Z)$ whether load current lags or leads load voltage; i.e., whether the load is inductive or capacitive. The distinction is immaterial when calculating real average power, but is important in some applications. Thus, a power factor is specified as either **lagging**, meaning that current lags voltage and $\theta_Z > 0$, or **leading**, meaning that current leads voltage and $\theta_Z < 0$. Refer to Fig. 13.15. We have

$$Z = R + jX \Rightarrow \measuredangle Z = \theta_Z = \tan^{-1}\left(\frac{X}{R}\right). \quad (13.33)$$

As shown in Example 13.10 above, the real part of the impedance of a *source-free* (passive) load is non-negative.[3] Therefore, if the reactance of a source-free load is positive, then the angle of the load impedance is positive, the angle of the load current (relative to that of the load voltage) is negative, and current lags voltage. If the reactance of a source-free load is negative, then the angle of the load impedance is negative, the angle of the load current (relative to that of the load voltage) is positive, and current leads voltage. Because the reactance of an inductor is positive and the reactance of a capacitor is negative, a load having a *lagging power factor* (positive reactance) is called **inductive** and a load having a *leading power factor* (negative reactance) is called **capacitive**. Because reactance is a function of frequency, a load containing both inductance and capacitance can be inductive (lagging power factor) at some frequencies and capacitive (leading power factor) at others. For such a load, there is at least one frequency where the impedance is real (where the power factor is unity).[4]

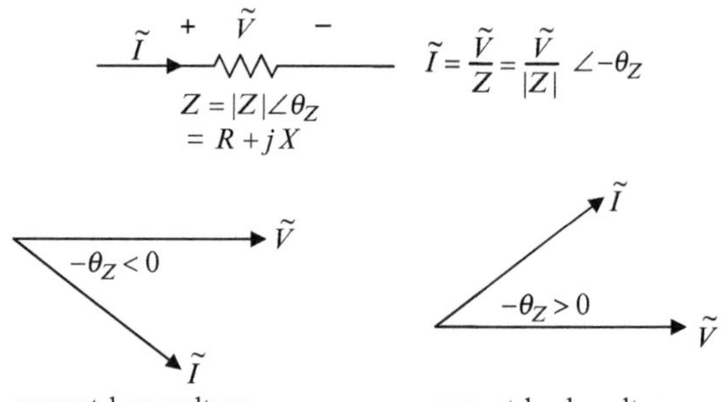

$\tilde{I}$ + $\tilde{V}$ −

$Z = |Z|\angle\theta_Z$
$= R + jX$

$\tilde{I} = \frac{\tilde{V}}{Z} = \frac{\tilde{V}}{|Z|}\angle -\theta_Z$

$-\theta_Z < 0$ $\tilde{V}$ $\tilde{I}$

current lags voltage

$\tilde{I}$ $-\theta_Z > 0$ $\tilde{V}$

current leads voltage

Fig. 13.15 Lagging and leading power factors (see text)

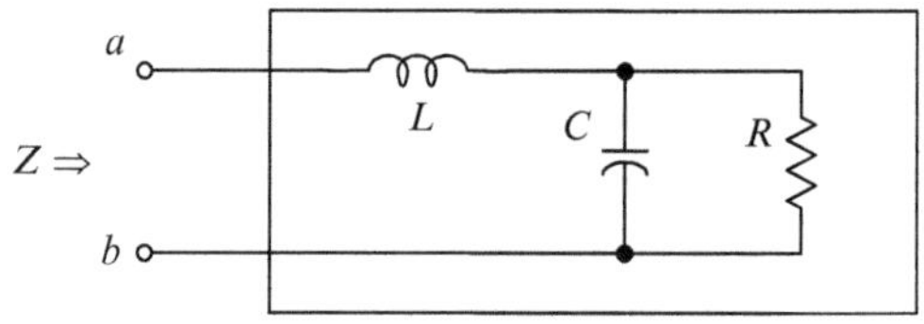

a
$Z \Rightarrow$
b
L C R

Fig. 13.16 See Example 13.11

Example 13.11. Refer to Fig. 13.16. Obtain expressions for the frequencies at which the impedance at the terminals *a–b* is real.
Solution: The impedance is given by

$$Z = j\omega L + \frac{R}{1 + j\omega RC}.$$

By inspection, the impedance is real (and equal to *R*) for $f = 0$, in which case $Z = R$.

[3] It is possible for the real part of the impedance of a load containing one or more dependent sources to be negative. The terminology introduced here is used only in references to source-free loads.

[4] Again, these remarks pertain to source-free loads. A load containing dependent sources, capacitors, and resistors (no inductors) can be made to appear inductive, as shown in a subsequent chapter.

We wish to see if there is another frequency for which the impedance is real. By inspection, the impedance Z is real if

$$\mathrm{Im}\left(\frac{R}{1+j\omega RC}\right) = -\omega L,$$

which yields

$$\frac{\omega R^2 C}{1+(\omega RC)^2} = \omega L$$
$$\Rightarrow \omega = \frac{1}{\sqrt{LC}}\sqrt{1-\left(\frac{L}{R}\right)\left(\frac{1}{RC}\right)}$$

or

$$f = \frac{1}{2\pi\sqrt{LC}}\sqrt{1-\left(\frac{\tau_L}{\tau_C}\right)};\ \tau_L = \frac{L}{R},\ \tau_C = RC.$$

If $\tau_L/\tau_C > 1$, the frequency f is imaginary, in which case there is no frequency other than zero for which the impedance is real. If $\tau_L/\tau_C = 1$, $f = 0$ is the only frequency for which the impedance is real. If $\tau_L/\tau_C < 1$, the frequency f is real and positive and the impedance at that frequency is real and given by

$$Z = R\left[1+\left(\frac{1}{\sqrt{LC}}\sqrt{1-\left(\frac{\tau_L}{\tau_C}\right)}\right)^2 (RC)^2\right]^{-1}$$
$$= \frac{R\tau_L}{\tau_C}.$$

Exercise 13.3. Obtain an expression for the power factor of an *RLC* series circuit at the resonant frequency of the circuit.

Electric utility companies are obliged to provide a certain minimum average power to customers, especially industrial customers. Because of how electric power is generated and distributed, rms line voltage is approximately constant and rms line current varies to meet power demand.[5] Line current, in turn, is responsible for losses in transmission lines and transformers, so electric utilities do not wish to provide any more current than is necessary to deliver the required power at the specified line voltage. Most industrial loads (largely motors) are inductive and exhibit lagging power factors. Electric utility companies require large industrial customers to maintain the power factor for their total loads at or above a specified minimum value to keep line losses to acceptable values. Power factors for industrial plants are corrected (made closer to unity) by adding capacitors in parallel with the feeds to the plants or in parallel with individual inductive loads. The usual method for determining the capacitance required is based on reactive power calculations, as described in the next section.

13.8 Power Triangle and Power-Factor Correction

Equation (13.20) implies that apparent power, real power, and reactive power form the sides of a right triangle, as shown in Fig. 13.17. A triangle so constructed for a load is called the **power triangle** for the load. By convention, the power triangle for a load is drawn in the first quadrant if the reactive power is positive and in the fourth quadrant if the reactive power is negative. The power triangle for a load is a useful visual aid in various applications.

The power triangle for a load is similar (geometric sense) to the *impedance triangle* for the load, because

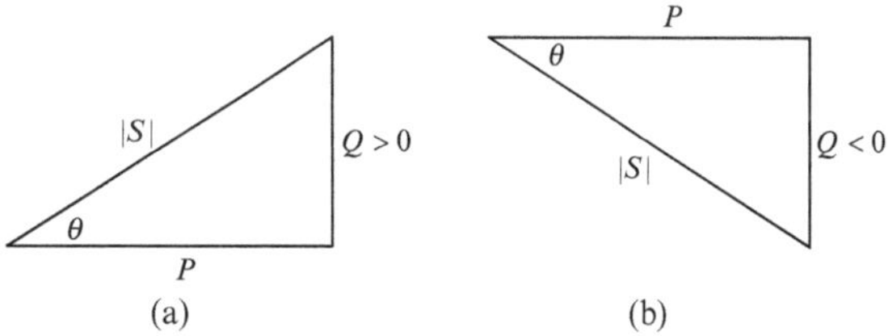

Fig. 13.17 Power Triangles: (**a**) $Q > 0$ (inductive load or lagging power factor); (**b**) $Q < 0$ (capacitive load or leading power factor)

[5]Line voltage varies somewhat with load because of losses in transmission, but not nearly to the degree that line current varies with load.

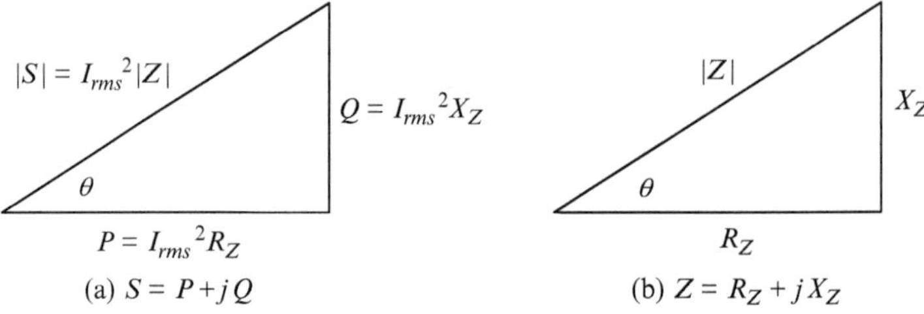

Fig. 13.18 The power triangle and impedance triangle for any particular load are geometrically similar triangles

$$S = \tilde{V}_{rms}\tilde{I}_{rms}^{\,*} = (\tilde{I}_{rms}Z)\tilde{I}_{rms}^{\,*} = |\tilde{I}_{rms}|^2 Z$$
$$= I_{rms}^{\,2}(R + jX) = P + jQ. \tag{13.34}$$

The rightmost equality shows that each side of the power triangle for a load is proportional to the corresponding side of the impedance triangle for the load (the sides of the two triangles are in the same proportion), as illustrated by Fig. 13.18.

Exercise 13.4. Show that the power triangle and the admittance triangle for any particular load are geometrically similar. Illustrate, using a picture like that in Fig. 13.18.

Again, *the real part R_Z of an impedance Z does not necessarily correspond to any particular resistor or combination of resistors in the circuit comprising the impedance*. In general, the real part of impedance is a function of not only resistance, but also of capacitance, inductance, and the frequency of an applied current or voltage. To illustrate this point, we refer to Example 13.6, where the real part of the impedance equals 63 Ω, whereas the only resistor in the circuit has resistance 1.5 kΩ.

Example 13.12. The current through a load lags the voltage across the load by $\pi/6$. The real power dissipated by the load is $P = 2$ kW. The rms amplitude of the line (load) voltage is 480 V. (a) Is the load inductive or capacitive? (b) What is the apparent power dissipated by the load? (c) What is the reactive power dissipated by the load? (d) What is the rms amplitude of the line current? (e) What is the impedance of the load?

Solution: (a) The load is inductive because the load current lags the load voltage. (b) Refer to Fig. 13.19. We are given the angle $\theta_Z = \pi/6$ and adjacent side $P = 2$ kW of the power triangle. The apparent power is

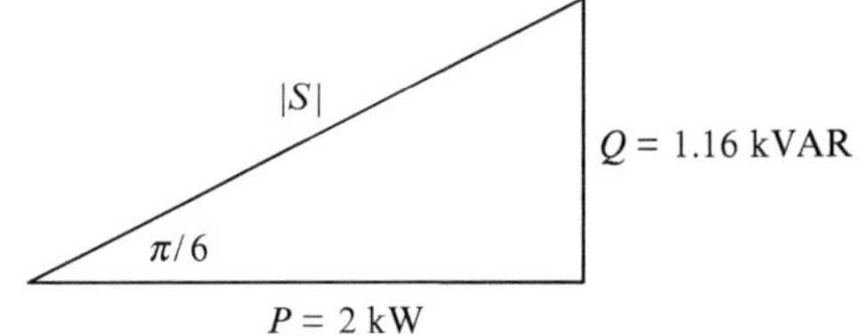

Fig. 13.19 See Example 13.12

$$|S| = \frac{P}{\cos(\theta_Z)} = \frac{2\text{ kW}}{\cos(\pi/6)} \cong 2.31\text{ kVA}.$$

(c) The reactive power is

$$Q = |S|\sin(\theta_Z) \cong (2.31\text{ kVA})\sin(\pi/6)$$
$$\cong 1.16\text{ kVAR}.$$

(d) From (13.34),

$$|S| = \frac{|VI|}{2} = V_{rms}I_{rms} \Rightarrow I_{rms} = \frac{|S|}{V_{rms}}$$
$$\cong \frac{2.31\text{ kVA}}{480\text{ V}} \cong 4.81\text{ A}.$$

(e) From (13.34),

$$P = I_{rms}^{\ 2} R \Rightarrow R = \frac{P}{I_{rms}^{\ 2}} \cong \frac{2\ \text{kW}}{(4.81\ \text{A})^2} \cong 86.4\ \Omega,$$

$$Q = I_{rms}^{\ 2} X \Rightarrow X = \frac{Q}{I_{rms}^{\ 2}} \cong \frac{1.16\ \text{kVAR}}{(4.81\ \text{A})^2} \cong 50.1\ \Omega,$$

and so the load impedance is

$$Z = R + jX \cong (86.4 + j50.1)\Omega \cong (99.9\angle 0.524)\ \Omega.$$

The reactive power dissipated by a load having equivalent series reactance X_s is given by

$$Q = I_{rms}^{\ 2} X_s = \frac{V_{rms}^{\ 2}}{X_s}, \qquad (13.35)$$

where V_{rms} and I_{rms} are the rms amplitudes of the voltage across and current through the load, respectively. It follows that if loads $Z_1, Z_2, \ldots, Z_N$ having equivalent series reactances $X_1, X_2, \ldots, X_N$ are connected in parallel across a source having rms amplitude V_{rms}, the total reactive power dissipated is given by

$$Q = \frac{V_{rms}^{\ 2}}{X_1} + \frac{V_{rms}^{\ 2}}{X_2} + \cdots + \frac{V_{rms}^{\ 2}}{X_N} = Q_1 + Q_2 + \cdots + Q_N. \qquad (13.36)$$

Equation (13.36) embodies the fundamental principle of power-factor correction

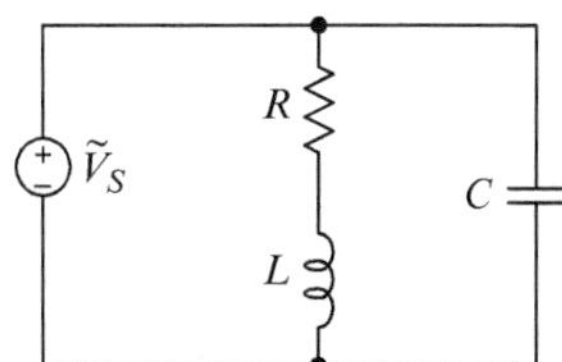

Fig. 13.20 See Example 13.13

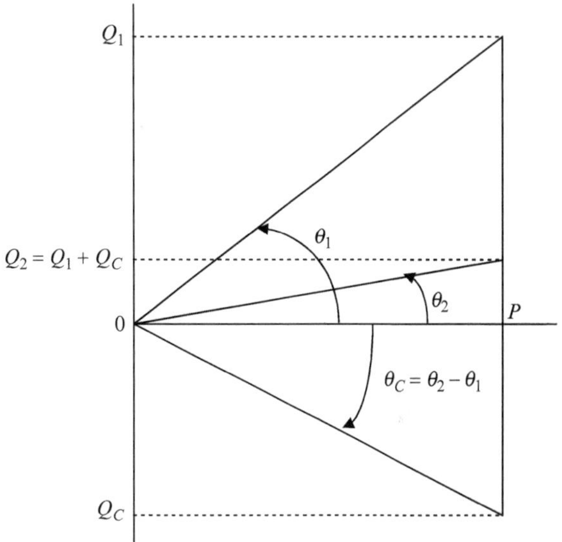

Fig. 13.21 Power triangles for Example 13.13

Example 13.13. Refer to Fig. 13.20. A certain industrial load is fed by a 60 Hz 4 kV (rms) line and requires 100 kW of (real) power. The load has a lagging power factor of 0.7. Find the shunt capacitance C that increases the power factor to 0.9. Compute the line current before and after the capacitance is added. Compute the percent reduction in line loss.

Solution: The solution is based upon (13.36), as illustrated by the power triangles shown in Fig. 13.21. The triangle having sides P and Q_1 represents the load alone. The triangle having sides P and Q_C represents the capacitor alone. The triangle having sides P, Q_2 represents the parallel connection of the load and the capacitor. The (negative) reactive power dissipated by the capacitor offsets some of the (positive) reactive power dissipated by the load. The residual (positive) reactive power Q_2 yields the desired power factor. The necessary trigonometric calculations follow:

$$\theta_1 = \cos^{-1}(pf_1) = \cos^{-1}(0.7);$$
$$Q_1 = P\tan(\theta_1) \cong 102 \text{ kVAR},$$
$$\theta_2 = \cos^{-1}(pf_2) = \cos^{-1}(0.9);$$
$$Q_2 = P\tan(\theta_2) \cong 48.4 \text{ kVAR},$$
$$Q_C = Q_2 - Q_1 \cong -53.6 \text{ kVAR} = -\omega C v_{rms}^2$$
$$\Rightarrow C = -\frac{Q_C}{\omega v_{rms}^2} \cong 8.88 \ \mu\text{F}.$$

The line currents before and after correction are

$$I_{1\,rms} = \frac{|S_1|}{V_{rms}} = \frac{\sqrt{P^2 + Q_1^2}}{V_{rms}} = 35.7 \text{ A},$$
$$I_{2\,rms} = \frac{|S_2|}{V_{rms}} = \frac{\sqrt{P^2 + Q_2^2}}{V_{rms}} = 27.8 \text{ A}.$$

The percent reduction in line loss is

$$\frac{I_{1\,rms}^2 R_{line} - I_{2\,rms}^2 R_{line}}{I_{1\,rms}^2 R_{line}} \times 100$$
$$= 100\left[1 - \left(\frac{I_{2\,rms}}{I_{1\,rms}}\right)^2\right] = 39.4\%.$$

Large industrial plants have substations on their premises, fed by a high-voltage (115 kV or more) transmission line. That voltage is stepped down at the substation to an intermediate voltage that can be several kV. Large (100 hp or larger) machines usually are powered from intermediate voltages, but smaller machines, lighting, and other loads require lower voltages, so the voltage is stepped down further, typically to one or more of 120, 260, 480, and 600 V. In some cases, so-called *bulk correction* is installed at the substation on either the intermediate-voltage or low-voltage line(s). Automated load monitoring equipment switches capacitors in or out in appropriate steps to maintain a nearly constant power factor.

In other cases, and especially in plants where not all (or any) machines run continuously and loads vary substantially, the power factor of each machine is corrected *locally* by one or more capacitors installed on the machine. In such installations, and especially for large motors, load monitoring equipment on each machine switches the capacitors in or out, automatically, to maintain a specified power factor. Such switching is necessary because the power factor of an induction motor varies with the speed and the (mechanical) load on the motor. In older plants, separate correction might be employed for fluorescent or other inductive lighting, but modern lighting systems have power factors of 0.9 or more, so correction might be unnecessary. Local correction can maintain a nearly constant overall power factor, even as individual machines are turned on and off, provided both the machine and its correction capacitor(s) are removed from the line when shut down. Generally, it is necessary to also disconnect the correction capacitor(s) from the machine to eliminate the potential for generating damaging voltages as the machine slows down.

Because the frequency of electrical power is fixed (in the US) at 60 Hz and the line voltages at which capacitors are used are standardized, power-factor correction capacitors are rated by reactive power provided at a specific line voltage. Thus specifying a capacitor requires determining the (negative) reactive power needed to offset some of the positive reactive power associated with an inductive load, the quantity being determined by the real power required and by the initial and corrected power factors.

Refer to Fig. 13.22, where $pf_1 = \cos(\theta_1)$, $pf_2 = \cos(\theta_2)$. The initial and corrected apparent powers are given by

$$|S_1| = \frac{P}{pf_1}, \quad |S_2| = \frac{P}{pf_2}.$$

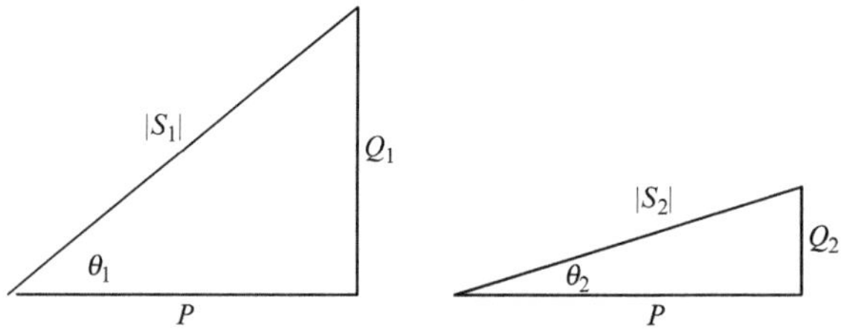

Fig. 13.22 Illustrating power-factor correction. See (13.37)

The associated reactive powers are given by

$$Q_1 = |S_1|\sqrt{1 - pf_1^{\,2}} = \frac{P}{pf_1}\sqrt{1 - pf_1^{\,2}},$$
$$Q_2 = |S_2|\sqrt{1 - pf_2^{\,2}} = \frac{P}{pf_2}\sqrt{1 - pf_2^{\,2}}.$$

The magnitude of the reactive power required of the power-factor correcting capacitor is given by

$$|Q_C| = Q_1 - Q_2 = P\left(\sqrt{\frac{1}{pf_1^{\,2}} - 1} - \sqrt{\frac{1}{pf_2^{\,2}} - 1}\right). \tag{13.37}$$

The magnitude of the reactive power Q dissipated by a capacitor having capacitance C is given by

$$|Q_C| = V_{rms}{}^2 \omega C, \tag{13.38}$$

where V_{rms} is the rms line voltage and $f = \omega/(2\pi)$ is the line frequency.

Equation (13.38) shows that the reactive power provided by a particular capacitor depends upon the capacitance and the rms amplitude and frequency of the line voltage. The frequency of the voltage (and current) provided by electric utilities in the United States is 60 Hz. In Europe it is 50 Hz. On board some aircraft and ships, higher-frequency supplies (usually 400 Hz) are used to reduce weight, primarily by allowing smaller and much lighter power supplies, transformers, and electric motors. Onboard power-factor correction capacitors are proportionally smaller and lighter, as well.

Example 13.14. The load on a 480 V line in an industrial substation is 50 kW at a power factor of 0.6. The desired power factor is 0.95. From (13.37), the additional reactive power needed is

$$|Q_C| = (50\,\text{kW})\left(\sqrt{\frac{1}{(0.6)^2} - 1} - \sqrt{\frac{1}{(0.95)^2} - 1}\right)$$
$$= 50.2\,\text{kVAR}$$

Thus we would specify a 50.2 kVAR capacitor with a rated voltage of 480 V.

Exercise 13.5. A certain capacitor provides 25 kVAR at 60 Hz and 240 V. (a) What is the reactive power provided at 50 Hz and 220 V? (b) If this capacitor is installed on a 75 hp electric motor operating at 240 V and 60 Hz, and having (uncorrected) power factor 0.7, what is the resulting power factor?

If you have mastered the material above, you have a good *basic* understanding of the objectives and means of power-factor correction. However, the subject goes far beyond the introductory treatment given here. We conclude this section by mentioning a few of the many complicating factors.

The power factor of a motor depends upon the load on the motor, being largest when the motor is fully loaded and least when the motor is idling. If the correction capacitance is sized for a fully loaded motor, the motor will be over-corrected when running at less than full load, and problems referred to above can surface. Power-factor correction usually is automated for large motors or large collections of smaller motors. Power-factor correction equipment monitors rms current, rms voltage, and real power, determines the capacitance needed on the fly, and automatically switches capacitors in or out, in small steps, as conditions warrant.

Over-correction, resulting in a leading power factor, is to be avoided, and most utilities recommend (or require) power factors in the neighborhood 0.95, to avoid over-correction. A main reason for avoiding over-correction is the following: The majority of motors used in industrial operations are induction motors. When turned off (removed from the line), a motor does not stop abruptly, but coasts to a stop in a manner determined by the inertia and friction of the mechanical load. While doing so, an induction motor can act as a generator, producing a voltage proportional to the angular velocity of the rotor. If the motor is under-corrected (lagging $pf < 1$), the resonant frequency of the circuit comprising the capacitor and motor windings is above the line frequency, but if the motor is overcorrected (leading pf), the resonant frequency of the capacitor and motor windings is below the line frequency, and the frequency of the voltage produced by the motor will pass through the resonant frequency of the circuit as the motor slows down. When that happens, voltages and currents produced can be large enough to damage the motor, the capacitor, or both. For this reason, an induction motor must be disconnected from both the line and the power-factor correction capacitor(s) when the motor is shut down. A snubbing circuit might also be employed to dissipate the energy produced by the motor as it coasts to a stop.

Finally, plants are increasingly using digital equipment to control motor speed or torque, such as variable-frequency drives (VFDs) that create variable-frequency digitally synthesized sinusoidal voltages for controlling the speed (angular velocity) of motors. There also are nonlinearities in lighting ballasts, transformers, and ac–dc converters (rectifiers) for running dc motors and appliances. Such devices introduce harmonics of the power-line frequency into the distribution system. Thus, in addition to 60 Hz, there will be 120, 180 Hz, and higher harmonics on the lines. The Cdv/dt currents produced by those voltages cause additional real power dissipation (and thus heat) in the *ESR* (equivalent series resistance) of power-factor correcting capacitors. Also, the frequency of one or more such harmonics might be near an unintentional resonance associated with motor and transformer windings and power-factor correction capacitors. Overheating will at best shorten the lives of capacitors and in severe cases (e.g., if boosted by a resonance) can be instantly destructive. When conditions warrant, plants take extra measures, such as tuned-circuit traps, to eliminate harmonics and dangerous resonances.

13.9 Superposition of Complex Power

If two or more sinusoidal currents or voltages having *different frequencies* are impressed simultaneously on a load, the *real* average power dissipated by the load is the sum of the *real* average powers that would be dissipated if the sinusoidal currents or voltages were applied individually.[6] The same is true for the reactive powers.

For example, refer to Fig. 13.23, where

$$v_1 = V_1 \cos(\omega_1 t), \quad v_2 = V_2 \cos(\omega_2 t),$$
$$V_1 = 10 \text{ V}, V_2 = 5 \text{ V}, f_1 = 100 \text{ Hz}, f_2 = 200 \text{ Hz}, \tag{13.39}$$

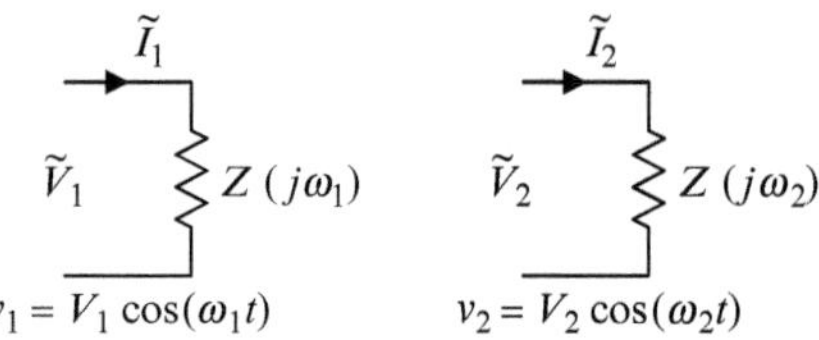

Fig. 13.23 Pertaining to superposition of complex power. See (13.39) and (13.40)

and

$$Z(j\omega) = R + j\omega L; \quad R = 100\ \Omega, \quad L = 100 \text{ mH}.$$

Thus

$$\tilde{I}_1 = \frac{\tilde{V}_1}{Z(j\omega_1)} \cong (84.67\angle -0.561) \text{ mA},$$
$$\tilde{I}_2 = \frac{\tilde{V}_2}{Z(j\omega_2)} \cong (31.13\angle -0.899) \text{ mA}.$$

The individual complex powers are given by

$$S_1 = \frac{1}{2}\tilde{V}_1\tilde{I}_1^* \cong (358 + j225) \text{ mVA},$$
$$S_2 = \frac{1}{2}\tilde{V}_2\tilde{I}_2^* \cong (48.5 + j60.9) \text{ mVA}. \tag{13.40}$$

Thus the total real and reactive powers dissipated are

$$P = \text{Re}(S_1) + \text{Re}(S_2) \cong 407 \text{ mW},$$
$$Q = \text{Im}(S_1) + \text{Im}(S_2) \cong 286 \text{ mVAR}.$$

Complex powers can be superimposed (added) as illustrated above. However, the total complex power dissipated in this example *cannot* be calculated as

$$S \neq \frac{1}{2}(\tilde{V}_1 + \tilde{V}_2)(\tilde{I}_1 + \tilde{I}_2)^*,$$

because a sum of phasors for sinusoids having different frequencies is meaningless; e.g., if $\tilde{V}_1$ and $\tilde{V}_2$ represent sinusoids having different frequencies, then $\tilde{V}_1 + \tilde{V}_2$ is meaningless.

13.10 Power Transfer

This section describes conditions for maximum power transfer efficiency and for maximum power transfer.

[6] Superposition of real power is treated in Chapter 5.

13.10.1 Power Transfer Efficiency

Refer to Fig. 13.24. We assume the elements of the model represent physical components. The current through and voltage across the load are given by

$$\tilde{I}_L = \frac{\tilde{V}_S}{Z_S + Z_L}, \; \tilde{V}_L = \frac{Z_L \tilde{V}_S}{Z_S + Z_L}. \tag{13.41}$$

From (13.5), the power delivered to the load is given by

$$P_L = \frac{1}{2} \left|\tilde{I}_L\right|^2 \mathrm{Re}(Z_L) = \frac{1}{2} \frac{\left|\tilde{V}_S\right|^2}{\left|Z_S + Z_L\right|^2} \mathrm{Re}(Z_L) \tag{13.42}$$

and the power delivered by the source $\tilde{V}_S$ is given by

$$\begin{aligned} P_S &= \frac{1}{2} \left|\tilde{I}_L\right|^2 \mathrm{Re}(Z_L + Z_S) \\ &= \frac{1}{2} \frac{\left|\tilde{V}_S\right|^2}{\left|Z_S + Z_L\right|^2} \mathrm{Re}(Z_L + Z_S). \end{aligned} \tag{13.43}$$

It follows that the power transfer efficiency is given by

$$\begin{aligned} \eta = \frac{P_L}{P_S} &= \frac{\mathrm{Re}(Z_L)}{\mathrm{Re}(Z_L + Z_S)} \\ &= \frac{\mathrm{Re}(Z_L)}{\mathrm{Re}(Z_L) + \mathrm{Re}(Z_S)}. \end{aligned} \tag{13.44}$$

Equation (13.44) shows that power transfer efficiency is near unity and nearly independent of frequency if

$$\mathrm{Re}(Z_S) \ll \mathrm{Re}(Z_L). \tag{13.45}$$

We can maximize power transfer efficiency either by minimizing $\mathrm{Re}(Z_S)$, maximizing $\mathrm{Re}(Z_L)$, or both, depending upon which of those parameters can be varied. No matter which of power transfer or power transfer efficiency is most important, minimizing $\mathrm{Re}(Z_S)$ usually is desirable because that minimizes wasted power (in the form of undesirable heat).

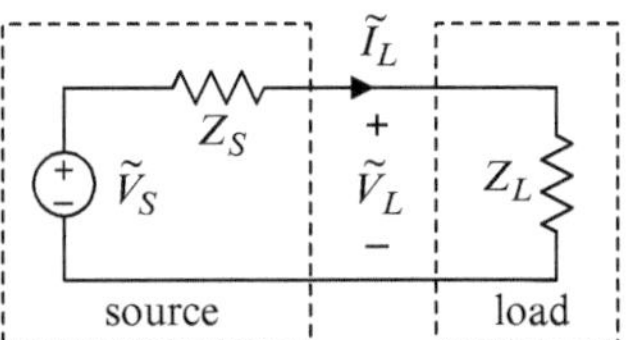

Fig. 13.24 Circuit model used to examine power transfer efficiency

On the other hand, increasing power transfer efficiency by increasing $\mathrm{Re}(Z_L)$ decreases the load current and thus decreases the power delivered to the load. Moreover, the load impedance Z_L might be beyond the control of the circuit designer, in which case minimizing $\mathrm{Re}(Z_S)$ is the only option.

Keep in mind that power dissipated by a circuit element is physically meaningful only if the element corresponds to a physical component, and that the components of a Thévenin source model do not necessarily correspond to physical components. If the source in Fig. 13.24 is a Thévenin model for a more complex physical source, such that the parameters V_S, Z_S are non-physical, then the power dissipated in the source is non-physical and not necessarily equal to the power dissipated in the associated physical source. Consequently, we cannot calculate power transfer efficiency using the (non-physical) Thévenin equivalent for a more complex physical source, as illustrated by the following example.

Example 13.15. Figure 13.25(a) shows a physical circuit, where the elements correspond to physical components. Figure 13.25(b) shows a circuit model, where the source is represented by its Thévenin equivalent. We wish to show that the power transfer efficiency calculated from the model is not equal to the true power transfer efficiency calculated from the physical circuit. For specificity, let $R_0 = R = R_L = 100\ \Omega$ and $V_0 = 100$ V. Then

$$\begin{aligned} V_T &= \frac{RV_0}{R_0 + R} = 50\ \mathrm{V}, \\ R_T &= R_0 \| R = 50\ \Omega. \end{aligned} \tag{13.46}$$

Because the physical source and the Thévenin model are equivalent at their terminals, the

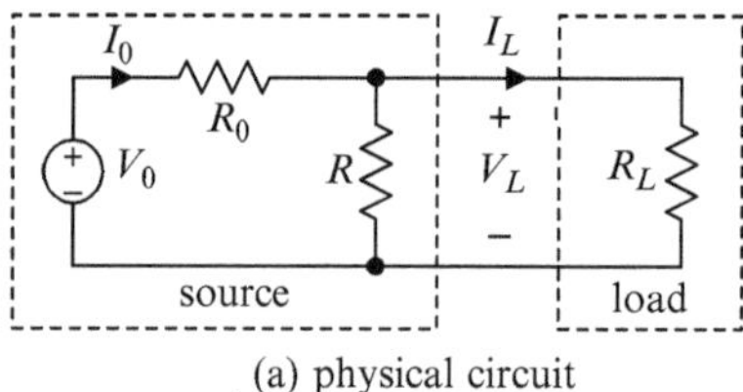

(a) physical circuit

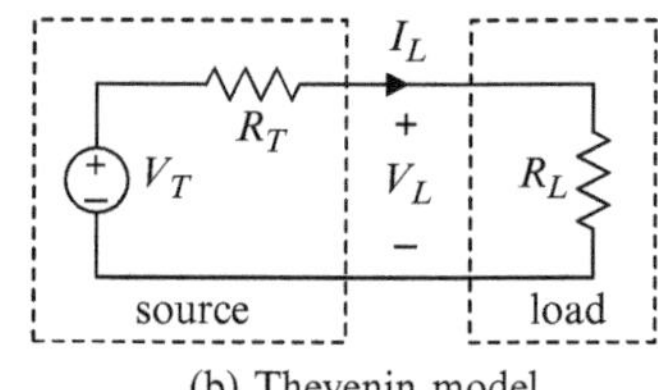

(b) Thevenin model

Fig. 13.25 See Example 13.15

load current I_L, the load voltage V_L, and the power P_L dissipated in the load are the same for both circuits. We find

$$I_L = \frac{V_T}{R_T + R_L} = 333 \text{ mA},$$
$$V_L = R_L I_L = 33.3 \text{ V},$$
$$P_L = V_L I_L = 11.1 \text{ W}.$$

The power produced by the Thévenin equivalent voltage source is

$$P_{VT} = I_L V_T = 16.7 \text{ W},$$

so the power transfer efficiency based upon the Thévenin model in Fig. 13.25(b) is

$$\eta_T = \frac{P_L}{P_{VT}} = 0.665.$$

For the model in Fig. 13.25(a), the power produced by the voltage source is

$$P_0 = I_0 V_0 = \left(\frac{V_0 - V_L}{R_0}\right) V_0 = 66.7 \text{ W},$$

so the actual power transfer efficiency is

$$\eta = \frac{P_L}{P_0} = \frac{11.1 \text{ W}}{66.7 \text{ W}} = 0.166,$$

which is far from the value obtained using the Thévenin source model.

If the components of the source model in Fig. 13.24 correspond to physical components, and if the source impedance Z_S is variable, then power transfer efficiency can be maximized by minimizing $\text{Re}(Z_S)$. But if the source is the Thévenin equivalent for a more complex physical source, minimizing $\text{Re}(Z_S)$ by varying a physical parameter will not necessarily maximize power transfer efficiency because the Thévenin equivalent voltage might depend upon the same physical parameter. For example, we can minimize the Thévenin resistance R_T in Fig. 13.25(a) by minimizing the physical resistance R, but in so doing we would minimize, not maximize, power transfer efficiency.

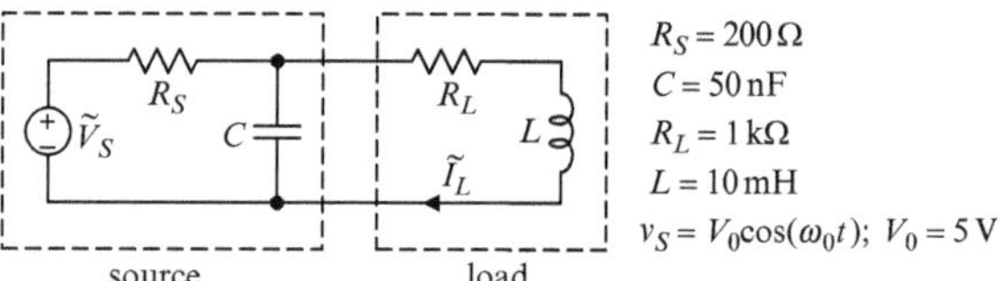

Fig. 13.26 See Example 13.16

Example 13.16. Refer to Figure 13.26.

(a) Calculate and plot the power-transfer efficiency versus the frequency f of the source for 100 Hz $\leq f \leq$ 1 MHz. Use a logarithmic scale for frequency. On the same axes, plot the (dimensionless) quantity

$$P_n(f) = \frac{P(f)}{P(f_0)},$$

where $P(f)$ is the power delivered to the load and $f_0 = 10$ kHz.

(b) Let the frequency of the source be fixed at $f = f_0 = 10$ kHz and let the capacitance C be variable. Plot the power-transfer efficiency versus the capacitance C for 5 nF $\leq C \leq$ 100 nF. On the same axes, plot the dimensionless quantity

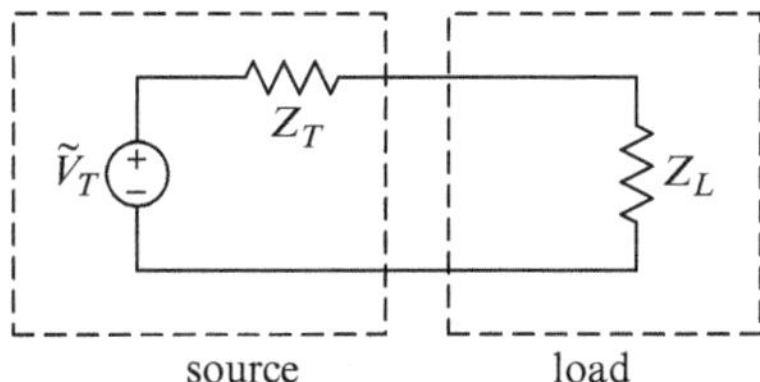

Fig. 13.27 See Example 13.16

$$P_{n0}(C) = \frac{P_0(C)}{P_0(C_0)},$$

where P_0 is the power delivered to the load as a function of the capacitance C for $f = f_0$ and $C_0 = 50$ nF.

Solution: (a) Figure 13.27 shows the circuit reduced to the Thévenin equivalent for the source and the equivalent impedance for the load, where

$$\tilde{V}_T = \frac{V_0}{1 + j2\pi f R_S C}; \; Z_T = \frac{R_S}{1 + j2\pi f R_S C};$$
$$Z_L = R_L + j2\pi f L.$$

From (13.44), the power-transfer efficiency is given by

$$\eta = \frac{\text{Re}(Z_L)}{\text{Re}(Z_L) + \text{Re}(Z_T)},$$

where, in this case,

$$\text{Re}(Z_L) = R_L, \; \text{Re}(Z_T) = \text{Re}\left(\frac{R_S}{1 + j2\pi f R_S C}\right) = \frac{R_S}{1 + (2\pi f R_S C)^2}. \qquad (13.47)$$

Thus

$$\eta = \frac{R_L}{R_L + \dfrac{R_S}{1 + (2\pi f R_S C)^2}}. \qquad (13.48)$$

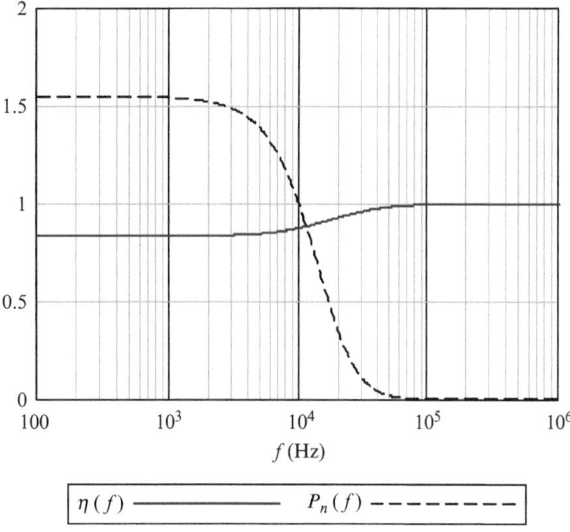

Fig. 13.28 See Example 13.16

The power dissipated in the load is given by

$$P = \frac{1}{2}\left|\tilde{I}_L\right|^2 R_L = \frac{1}{2}\left|\frac{\tilde{V}_T}{Z_T + Z_L}\right|^2 R_L.$$

Figure 13.28 shows the graph requested in part (a). As the frequency is increased from 100 Hz to 1 MHz, the power-transfer efficiency (solid line) increases from 0.833 to unity. However, the actual power transferred (dashed line) decreases from about $1.5P(f_0)$ to nearly zero, where $P(f_0) = 5.61$ mW. Depending upon the application, we might say that reasonable efficiency and power transfer are achieved for frequencies below 10 kHz. Above 10 kHz, the efficiency is approximately unity, but the power transferred is quite small, most being dissipated in the source impedance, which is an undesirable situation.

Figure 13.29 shows the graph requested in part (b). From (13.47) and (13.48), increasing the capacitance decreases the real part of the source impedance and thereby increases the power-transfer efficiency. But increasing the capacitance also decreases the magnitude of the Thévenin equivalent voltage, so we pay for increased power-transfer efficiency with a reduction in the actual power transferred to the

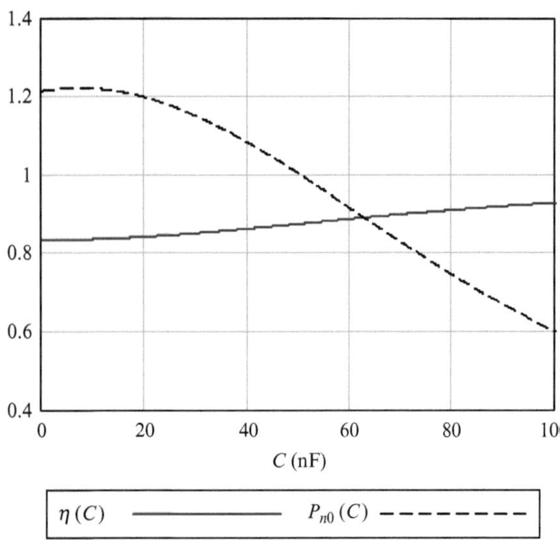

Fig. 13.29 See Example 13.16

load. Whether that is acceptable depends upon the application.

13.10.2 Power Transfer

Refer again to Fig. 13.27, where the source and load impedances are expressed in terms of their real and reactive components as

$$Z_T = R_T + jX_T, \quad Z_L = R_L + jX_L. \tag{13.49}$$

The complex power delivered to the load is given by

$$S_L = \frac{1}{2}\frac{(R_L + jX_L)\left|\tilde{V}_T\right|^2}{(R_T + R_L)^2 + (X_T + X_L)^2}. \tag{13.50}$$

Thus the real power dissipated in the load is given by

$$P_L = \text{Re}(S_L) = \frac{1}{2}\frac{R_L\left|\tilde{V}_T\right|^2}{(R_T + R_L)^2 + (X_T + X_L)^2}. \tag{13.51}$$

In what follows, we consider two main cases: (1) The source impedance Z_T is variable and the load impedance Z_L is fixed and (2) The load impedance is variable and the source impedance is fixed. In each case, we seek the condition under which the power transferred to the load is maximum.

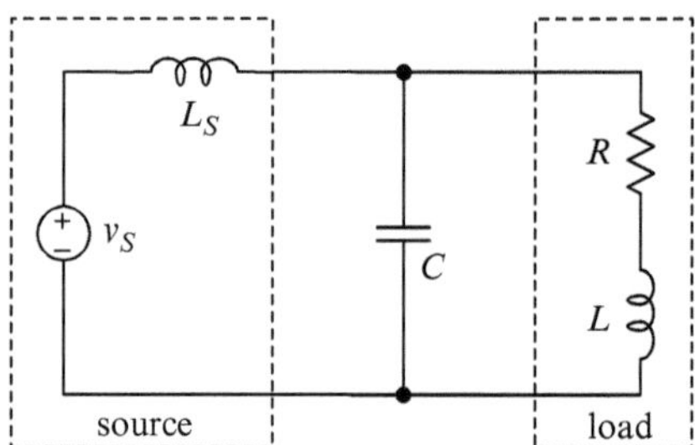

Fig. 13.30 See Example 13.17

First, it is clear from (13.51) that if the source impedance is variable and the load impedance is fixed, then the source impedance that maximizes power transfer is

$$Z_T = -jX_L, \tag{13.52}$$

because in that case, $R_T = 0$, $X_T = -X_L$, and (13.51) becomes

$$P_L = \frac{1}{2}\frac{\left|\tilde{V}_T\right|^2}{R_L},$$

which means that all of the real power delivered by the source is dissipated in the load (none in the source impedance).

Example 13.17. In Fig. 13.30, $v_S = V_0\cos(\omega_0 t)$. Obtain an expression for the capacitance that maximizes the power transferred to the load.

Solution: From the perspective of the load, the source (Thévenin equivalent) impedance is given by

$$Z_T = \frac{j\omega_0 L_S}{1 - \omega_0^2 L_S C}.$$

From (13.52), the capacitance that maximizes the power transferred to the load is that for which

$$Z_T = -jX_L = -j\omega_0 L \Rightarrow \frac{L_S}{1 - \omega_0{}^2 L_S C} = -L.$$

Solving this equation for the capacitance C gives

$$C = \frac{L_S + L}{\omega_0{}^2 L_S L}.$$

Refer again to Fig. 13.27. Suppose the source (Thévenin) impedance Z_T and source voltage $\tilde{V}_T$ are fixed, but the load is variable. We assume the reactive component of the load can be chosen independently of the real component. From (13.51), the load reactance that maximizes the power dissipated in the load for any particular load resistance is

$$X_L = -X_T, \tag{13.53}$$

in which case (13.51) becomes

$$P_L = \frac{1}{2} \frac{R_L \left|\tilde{V}_T\right|^2}{(R_T + R_L)^2}. \tag{13.54}$$

Because $\tilde{V}_T$ is fixed, the power P_L is maximized by maximizing the quantity

$$\frac{R_L}{(R_T + R_L)^2}.$$

The value of R_L for which this quantity is maximum is given by

$$\begin{aligned} &\frac{d}{dR_L}\left(\frac{R_L}{(R_T + R_L)^2}\right) \\ &= \frac{(R_T + R_L)^2 - 2R_L(R_T + R_L)}{(R_T + R_L)^4} = 0, \end{aligned} \tag{13.55}$$

which yields

$$R_L = R_T. \tag{13.56}$$

Together, (13.53) and (13.56) show that maximum power is delivered to the load if

$$Z_L = Z_T^*. \tag{13.57}$$

Keep in mind that unless the source impedance and the load impedance are both real, maximum power transfer is in general achieved at only a single frequency.

Exercise 13.6. Consider a case where the load impedance is complex and the source impedance is real. Denote the load impedance by $Z_L = R_L + jX_L$ and the source impedance by R_T. Show that the power transferred to the load is maximized by adding a series reactance $X_T = -X_L$ to the load, such that that the power transferred to the load is given by

$$\begin{aligned} P_L &= \frac{1}{2} \frac{R_L \left|\tilde{V}_T\right|^2}{(R_T + R_L)^2 + (X_T + X_L)^2} \\ &= \frac{1}{2} \frac{R_L \left|\tilde{V}_T\right|^2}{(R_T + R_L)^2}. \end{aligned} \tag{13.58}$$

Exercise 13.7. Consider a case where the load impedance is real and the source impedance is complex. Denote the source impedance by $Z_T = R_T + jX_T$ and the load impedance by R_L. Show that the power transferred to the load is maximized by adding a series reactance $X_L = -X_T$ to the load, such that the power transferred to the load is given by (13.58).

Exercise 13.8. Under what conditions on source and load impedances can power transfer be maximized at dc?

Exercise 13.9. Denote the Thévenin equivalent impedance of a source by $Z_T = R_T + jX_T$ and the impedance of the load on the source by $Z_L = R_L + jX_L$.

(a) Show that if the reactances X_T, X_L cannot be altered, then maximum power is transferred to the load if $R_L = \sqrt{R_T{}^2 + (X_T + X_L)^2}$.

(b) Consider such a case, where $R_T < R_L$. Is the power transferred to the load increased by adding a resistance $R_s = R_L - R_T$ in series with the load?
(c) If $R_T > R_L$, is the power transferred to the load increased by adding a resistance $R_s = R_T - R_L$ in series with the load?

Example 13.18. A certain amplifier having (real) output impedance $R_S = 50\ \Omega$ drives a load whose input impedance is that of a $100\ \Omega$ resistor R in parallel with a 62 pF capacitor C, as illustrated by Fig. 13.31. Neither the source impedance nor the load impedance can be modified.

(a) Use physical reasoning (not mathematics) to find the frequency f_0 for which the power delivered to the load is maximum.
(b) Assume the open-circuit voltage at the amplifier output terminals is $V_S = 20$ V, independent of frequency. Obtain an expression for the power $P_L(f)$ delivered to the load as a function of frequency. Using a logarithmic scale for frequency, plot the dimensionless quantity

$$P_n(f) = \frac{P_L(f)}{P_L(f_0)}$$

for 100 kHz $\leq f \leq$ 1 GHz, where f_0 is the frequency found in part (a).
(c) On the same axes used for the plot obtained in part (b), plot the power transfer efficiency versus frequency.

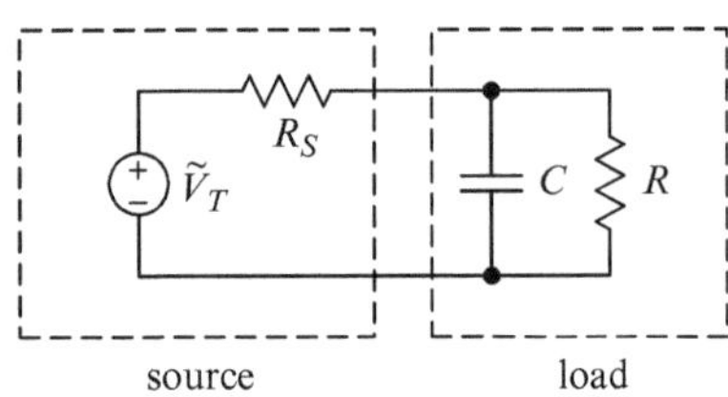

Fig. 13.31 See Example 13.18

Solution:

(a) Refer to Fig. 13.31. For any non-zero frequency, some of the current available from the source is diverted through the capacitor and contributes nothing to the power dissipated in the load (because the average power dissipated in a capacitance is zero). Thus the power dissipated in the load is maximum for $f = 0$ (at dc).
Alternatively: the average power dissipated in a capacitance is zero, so whatever power is dissipated in the load is dissipated entirely in the resistance R. The impedance of the capacitor (and thus of the load) decreases with increasing frequency, so (voltage division) the voltage across the resistance R is maximum for $f = 0$. It follows that the power dissipated in the load is maximum for $f = 0$ (at dc).
(b) Taking the source $\tilde{V}_S$ as the phase reference, the voltage across the load is given by (voltage division)

$$\tilde{V}_L = \frac{Z_L}{Z_L + R_S} V_S,$$

where $V_S = |\tilde{V}_S|$ and

$$Z_L = \frac{R}{1 + j2\pi f RC}.$$

Thus

$$\tilde{V}_L = \frac{RV_S}{R + (1 + j2\pi f RC)R_S} = \frac{RV_S}{R + R_S + j2\pi f R R_S C}.$$

The power dissipated in the load is given by

$$P_L(f) = \frac{1}{2R}|\tilde{V}_L|^2 = \frac{RV_S{}^2}{2\left[(R+R_S)^2 + (2\pi f R R_S C)^2\right]}$$

For $f = f_0 = 0$, we find

$$P_L(0) = \frac{R V_S{}^2}{2(R + R_S)^2} = 889 \text{ mW}$$

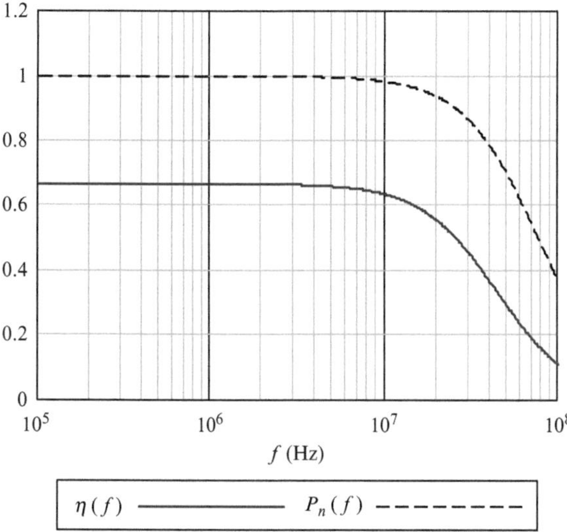

Fig. 13.32 See Example 13.18

and

$$P_n(f) = \frac{P_L(f)}{P_L(0)} = \frac{2(R+R_S)^2}{2\left[(R+R_S)^2 + (2\pi f R R_S C)^2\right]}.$$

(c) From (13.44), the power transfer efficiency is given by

$$\begin{aligned}\eta(f) &= \frac{\mathrm{Re}[Z_L(f)]}{\mathrm{Re}[Z_L(f)] + \mathrm{Re}[Z_T(f)]} \\ &= \frac{R}{R + R_S\left[1 + (2\pi f R C)^2\right]} \\ &= \left\{1 + R_S R^{-1}\left[1 + (2\pi f R C)^2\right]\right\}^{-1}.\end{aligned}$$

Figure 13.32 shows the graphs requested in parts (b) and (c).

Exercise 13.10. Repeat Example 13.18, with $R = 10\ \mathrm{k\Omega}$, $C = 6.2$ pF.

Example 13.18 illustrates that both power transfer efficiency and power transfer generally are dependent upon frequency and can be strongly so. In most electronics applications, the source contains many frequencies, spread over a band. For example, a current or voltage representing a segment of audible music contains sinusoidal components whose frequencies range from about 20 Hz to more than 15 kHz, and it is desirable that power transfer efficiency of a high-fidelity amplifier be nearly independent of frequency for frequencies in that band. In such applications, one generally tries to minimize the *magnitude* of the source impedance Z_T, not just $\mathrm{Re}(Z_T)$, so that the frequency dependence of Z_T has little effect on the frequency dependence of the overall response of the circuit.

It is worth repeating that the condition (13.57) ensures maximum power transfer from a *fixed source* to a *variable load*; that is, where the source impedance Z_T is fixed and the load impedance Z_L can be specified. It also is worth noting that the derivation above of (13.52) and the derivation of (13.58) called for in Exercise 13.6 *implicitly assume that the source voltage is independent of the source impedance*. Such is rarely the case if the model used (Fig. 13.24) is the (non-physical) Thévenin equivalent for a more complex circuit, where both the Thévenin equivalent impedance and Thévenin equivalent source voltage usually depend upon some of the same parameters of the associated physical circuit. In that case, altering the Thévenin equivalent impedance by changing one or more physical parameters almost invariably also changes the Thévenin equivalent voltage source. The correct approach to such a problem is to use a model that correctly represents the physical circuit, express the power delivered to the load as a function of actual (not Thévenin) circuit parameters and then maximize that expression subject to constraints on the parameters. Such problems require mathematical methods beyond the scope of this book. Usually, a more practical approach to modifying the output impedance of a system is to append another circuit (e.g., a voltage follower) to the output.

13.10.3 Insertion Loss

Insertion loss is the reduction (in dB) of power delivered by a source to a load when a circuit is inserted between the source and the load. Insertion loss is used

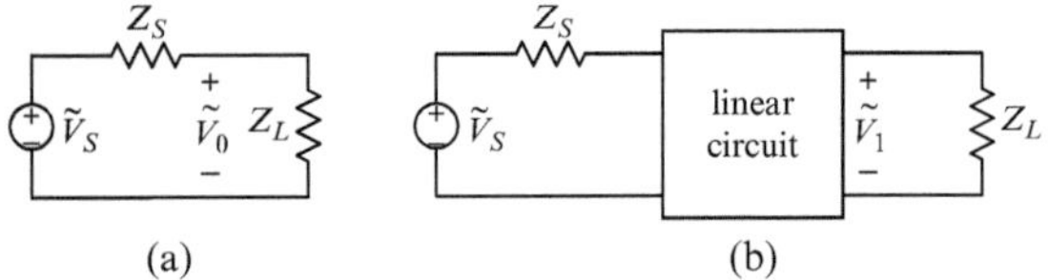

Fig. 13.33 Regarding the definition of insertion loss

by telecommunication engineers to estimate how much gain (amplification) is needed to balance losses introduced by components such as filters, splitters, and transmission lines. Generally, passive components, such as passive filters and transmission lines, introduce losses, whereas active components, such as amplifiers and active filters, can provide gains.

Refer to Fig. 13.33. The insertion loss attributed to the circuit is given by

$$\beta = 10\ \log\left(\frac{P_1}{P_0}\right).$$

where P_0 is the power delivered to the load before the circuit is inserted and P_1 is the power delivered to the load afterward. The load is the same in both cases, so

$$\beta = 10\ \log\left(\frac{P_1}{P_0}\right) = 20\ \log\left(\left|\frac{\tilde{V}_1}{\tilde{V}_0}\right|\right).$$

By voltage division and the definition of voltage gain,

$$|\tilde{V}_0| = \left|\frac{Z_L\tilde{V}_S}{Z_L + Z_S}\right|, \ |\tilde{V}_1| = A_v|\tilde{V}_S|.$$

Thus the insertion loss is given by

$$\beta\ (\text{dB}) = A_{v\,\text{dB}} - 20\log\left[\left|\frac{Z_L}{Z_S + Z_L}\right|\right]\text{dB} \quad (13.59)$$

If $|Z_S| \ll |Z_L|$, as is often the case, then

$$\beta\ (\text{dB}) \cong A_{v\,\text{dB}}, |Z_S| \ll |Z_L|. \quad (13.60)$$

A negative insertion loss (dB) indicates **insertion gain.**

Exercise 13.11. In a certain scenario, an antenna is to be connected to a receiver by a length of RG-6 coaxial cable. For frequencies of interest, the maximum signal loss in the cable is approximately 20 dB km^{-1}. The maximum tolerable loss is 6 dB. What is the maximum allowable distance from the antenna to the receiver? If the receiver must be 1.2 km from the antenna, how much gain (dB) must be provided by an auxiliary amplifier?

13.11 Impedance Matching

It is often impractical or impossible to achieve maximum power transfer or maximum power transfer efficiency by altering parameters of a given physical source or load. In such cases the desired result sometimes can be achieved by inserting a *matching circuit* between the source and load. From the perspective of the source, the purpose of the matching circuit is to change the apparent load impedance. From the perspective of the load, the purpose of the matching circuit is to change the apparent source impedance.

This section introduces impedance matching using transformers and L sections.

13.11.1 Transformers

A transformer can provide the proper conditions for maximum power transfer where the source and load impedance are both resistive, because a transformer can make a load (or the source impedance) seem larger or smaller, depending upon the turns ratio. A transformer alone cannot provide matching for a complex load or source impedance, because a transformer alone cannot introduce the necessary conjugate relationship between load and source impedance. But often, a good match can be achieved with a transformer and one or more reactive elements.

The next example uses an ideal transformer to illustrate how a transformer can provide good matching for a resistive load to a resistive source over a wide band of frequencies. But bear in mind that the ideal

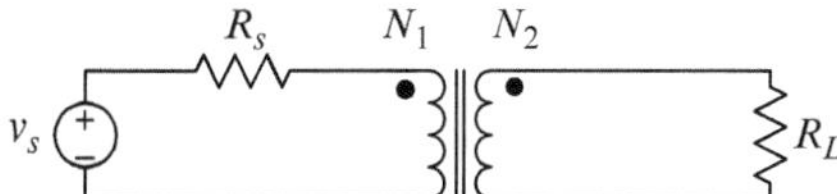

Fig. 13.34 See Example 13.19

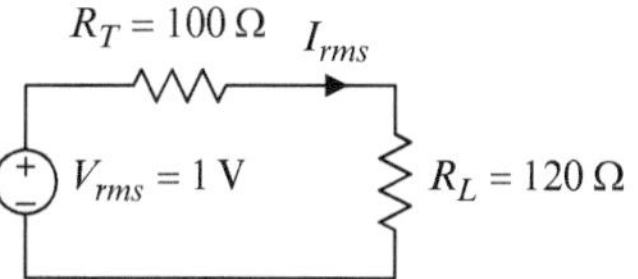

Fig. 13.35 See Example 13.20

transformer is an inadequate model for a great many matching applications. In practice, computer-aided design programs are used to design matching circuits.

Example 13.19. In the circuit shown in Fig. 13.34, the transformer is ideal. Obtain an expression for the transformer turns ratio $n = N_2/N_1$ that maximizes the power transferred to the load R_L.

Solution: The load seen by the source is that seen looking into the primary of the transformer. For maximum power transfer, that resistance must equal the source output resistance R_S. From Chapter 9, the resistance seen at the primary terminals of the transformer is given by

$$R_L' = \frac{1}{n^2} R_L. \tag{13.61}$$

The required turns ratio is given by

$$\frac{1}{n^2} R_L = R_o \Rightarrow n = \frac{N_2}{N_1} = \sqrt{\frac{R_L}{R_o}}.$$

As shown in Chapter 5, the power transferred by a *resistive source* to a matched *resistive load* has a broad maximum, in the sense that a slight mismatch between a source and a load leads to only a small decrease in power dissipated by the load. Consequently, it might not be worthwhile to correct a small mismatch, as illustrated by the following example.

Example 13.20. A 1 Vrms source having output resistance $R_S = 100\ \Omega$ is to drive a load having resistance $R_L = 120\ \Omega$. Calculate the power dissipated by the load (a) without matching and (b) if a transformer is used to achieve a match.

Solution: Refer to Fig. 13.35. The power dissipated by the load is given by

$$P_L = I_{rms}{}^2 R_L = \left(\frac{V_{rms}}{R_S + R_L}\right)^2 R_L.$$

(a) Without matching, the power dissipated by the load is

$$\begin{aligned} P_1 &= \left(\frac{V_{rms}}{R_S + R_L}\right)^2 R_L \\ &= \left(\frac{1\text{V}}{220\,\Omega}\right)^2 (120\,\Omega) \\ &\cong 2.48\ \text{mW}. \end{aligned}$$

(b) With matching, the apparent load resistance equals the source resistance. Therefore the power dissipated by the load with matching is

$$\begin{aligned} P_2 &= \left(\frac{V_{rms}}{R_S + R_S}\right)^2 R_S \\ &= \left(\frac{1\text{V}}{200\,\Omega}\right)^2 (100\,\Omega) \\ &\cong 2.50\ \text{mW}. \end{aligned}$$

The difference (20 μW) is a small fraction of the total power and would be insignificant in most applications.

An ideal transformer provides a perfect match of a resistive source to a resistive load for *all frequencies*, but such a match is unattainable using real components. The range of frequencies for which a real transformer can provide such a match is limited by the frequency response of the transformer, and in no case extends to dc. Also, it can happen that power losses in such networks approach those due to mismatch between the original source and load, in which case matching can be a waste of time and money.

Extending the procedure above to complex source and load impedances requires introducing additional elements (capacitors or inductors or both), because a transformer alone cannot introduce the required conjugate relation between source and load impedances.

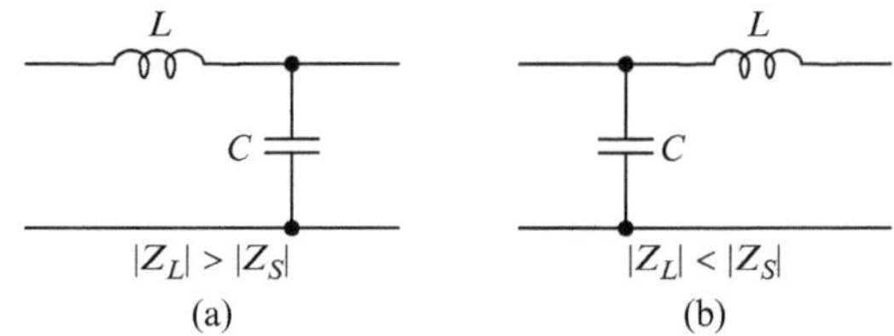

Fig. 13.36 L-section impedance matching networks

13.11.2 L Sections

The simple passive transformerless circuits shown in Fig. 13.36 can often provide good matching for frequencies at or near a particular frequency. These matching networks are called **L sections** (because of their shape). The circuit in Fig. 13.36(a) would be used if the load impedance is larger than the source impedance, because the series inductance increases the source impedance and the shunt capacitance reduces the load impedance. The circuit in Fig. 13.36(b) would be used if the load impedance is smaller than the source impedance, because the series inductance increases the load impedance and the shunt capacitance reduces the source impedance.

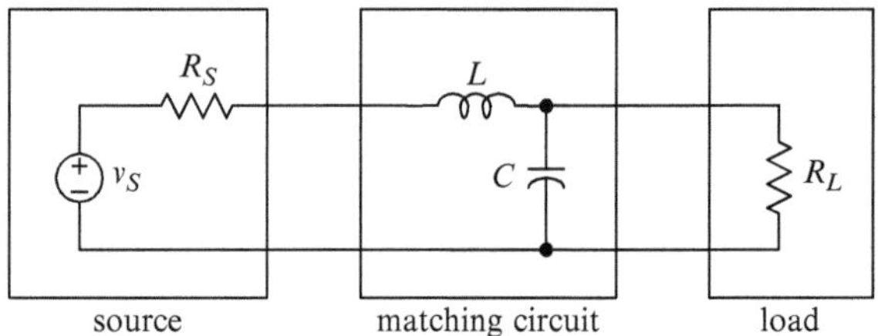

Fig. 13.37 See Example 13.21

Example 13.21. In the circuit shown in Fig. 13.37, the source output resistance is $R_S = 10\ \Omega$ and the load resistance is $R_L = 100\ \Omega$. The source voltage is sinusoidal, with frequency $f = 60$ Hz. (a) Find the values of the inductance L and capacitance C in the L-section matching circuit for which maximum power is transferred to the load. (b) Would you expect an insertion loss or insertion gain from inserting the passive matching network? Why? Calculate the insertion loss at $f = 60$ Hz.

Solution: Because the load resistance is larger than the source resistance, we associate the shunt capacitor with the load (to decrease the apparent load impedance) and the series inductor with the source (to increase the apparent source impedance). The modified source and load impedances are given by

$$Z'_S(j\omega) = R_S + j\omega L,$$
$$Z'_L(j\omega) = \frac{R_L}{1 + j\omega R_L C}.$$

For maximum power transfer at the match frequency $f_0 = 60$ Hz, we require $Z'_L(j\omega_0) = Z'^*_S(j\omega_0)$, which leads to

$$\frac{R_L(1 - j\omega_0 R_L C)}{1 + (\omega_0 R_L C)^2} = R_S - j\omega_0 L;\ \omega_0 = 2\pi f_0.$$

It follows that

$$\frac{R_L}{1 + (\omega_0 R_L C)^2} = R_S,$$
$$\frac{R_L{}^2 C}{1 + (\omega_0 R_L C)^2} = R_L R_S C = L.$$

Solving the first of these relations for the capacitance and then using the second to calculate the inductance yields

$$C = \frac{1}{\omega_0 R_L}\sqrt{\frac{R_L}{R_S} - 1} = 79.6\ \mu\text{F},$$
$$L = R_S R_L C = 79.6\ \text{mH}. \tag{13.62}$$

The power delivered to the load (as a function of frequency) is given by

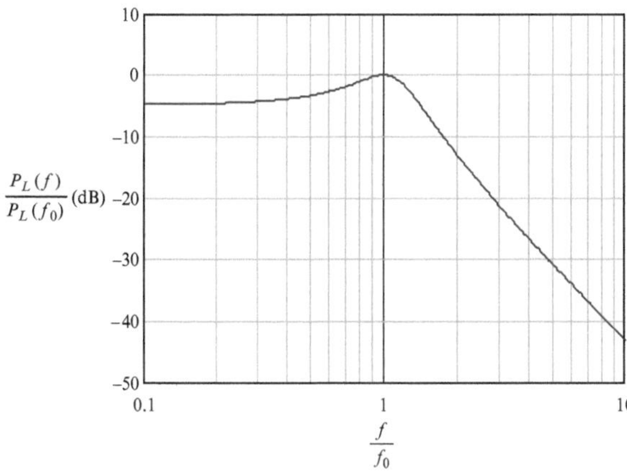

Fig. 13.38 See Example 13.21

$$P_L(f) = \frac{1}{2}|\tilde{I}|^2 \mathrm{Re}[Z_L] = \frac{1}{2}\left|\frac{\tilde{V}}{Z_o + Z_L}\right|^2 \mathrm{Re}[Z_L].$$

Figure 13.38 shows the power transferred to the load, normalized to the power transferred at the match frequency, as a function of frequency. The shape of the graph is typical of L-section matching. The match is good at the match frequency f_0 and is within 5 dB of the maximum below the match frequency, but falls off sharply above the match frequency. (b) The matching network increases the power delivered to the load, so we expect an insertion gain (negative loss), even though the matching network is passive. The voltage gain

$$A_v = 20 \log\left|\frac{Z_L'}{Z_L' + Z_S'}\right|$$

and the voltage gain at 60 Hz is

$$A_v(60 \text{ Hz}) \cong 3.98 \text{ dB}$$

The insertion loss at 60 Hz is

$$\beta(60 \text{ Hz}) = 20 \log\left(\left|\frac{R_L}{R_S + R_L}\right|\right) - A_v(60 \text{ Hz})$$
$$\cong -4.81 \text{ dB}$$

so the matching network provides an insertion *gain* of approximately 4.81 dB at the match frequency $f_0 = 60$ Hz.

When using a matching circuit to match a load to a source for maximum power transfer, you are finding the *matching-circuit parameters* for which maximum power is transferred to the load *at the specified frequency*. That is different from finding the *frequency* at which maximum power is transferred for a *specified set of matching-circuit parameters*. In other words, if you use a matching circuit that maximizes power transferred to the load at a particular frequency, it is possible for the power delivered to the load to be even larger at a different frequency, as illustrated by the next example.

Example 13.22. In the circuit shown in Fig. 13.37, the source output resistance is $R_o = 100\ \Omega$ and the load resistance is $R_L = 10\ \Omega$. The source voltage is sinusoidal, with frequency $f = 60$ Hz. Find the values of the inductance L and capacitance C in the L-section matching circuit for which maximum power is transferred to the load. For those values of L and C, find the frequency for which maximum power is transferred to the load.

Solution: Because the source resistance is larger than the load resistance, we associate the shunt capacitor with the source (to decrease the apparent source impedance) and the series inductor with the load (to increase the apparent load impedance). The modified source and load impedances are given by

$$Z_L(j\omega) = R_L + j\omega L;$$
$$Z_o(j\omega) = \frac{R_o}{1 + j\omega R_o C}, \quad \omega = 2\pi f.$$

For maximum power transfer at the match frequency $f_0 = 60$ Hz, we require $Z_L(j\omega_0) = Z_o^*(j\omega_0)$, which leads to

$$\frac{R_o(1 - j\omega_0 R_o C)}{1 + (\omega_0 R_o C)^2} = R_L - j\omega_0 L; \; \omega_0 = 2\pi f_0.$$

It follows that

$$\frac{R_o}{1+(\omega_0 R_o C)^2} = R_L,$$
$$\frac{R_o^2 C}{1+(\omega_0 R_o C)^2} = R_L R_o C = L.$$

Solving the first of these relations for the capacitance and then using the second to calculate the inductance yields

$$C = \frac{1}{\omega_0 R_o}\sqrt{\frac{R_0}{R_L} - 1} = 79.6\ \mu\text{F}, \qquad (13.63)$$
$$L = R_o R_L C = 79.6\ \text{mH}.$$

The power delivered to the load (as a function of frequency) is given by

$$P_L(f) = \frac{1}{2}\left|\tilde{I}(f)\right|^2 \text{Re}[Z_L(j2\pi f)]$$
$$= \frac{1}{2}\left|\frac{\tilde{V}(f)}{Z_o(j2\pi f) + Z_L(j2\pi f)}\right|^2 R_L.$$

Figure 13.39 shows the power transferred to the load, normalized to the power transferred at the match frequency, as a function of frequency. The power transferred to the load is maximum at a frequency slightly higher than the match frequency f_0. As an exercise, you can show that the frequency for maximum power transfer is $f_1 = 65.8$ Hz.

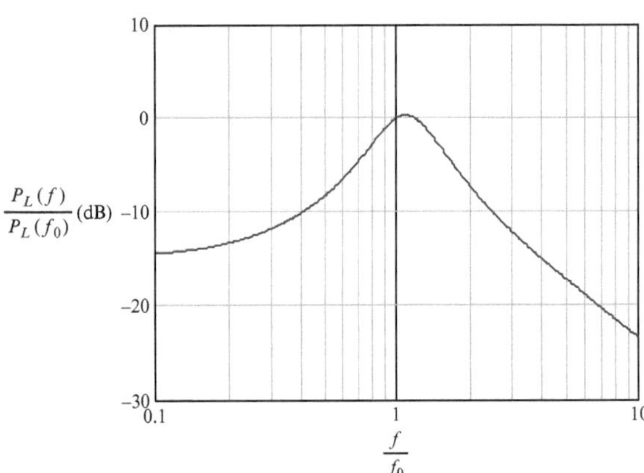

Fig. 13.39 See Example 13.22

If a load is matched to a source according to (13.57), the power dissipated by the load is exactly half of the power produced by the source *at the match frequency*, but not necessarily at any other frequency. The other half of the power produced by the source is dissipated in the source resistance (in the real part of the source impedance).

L-section matching circuits can also be used to match a reactive source to a reactive load, but the match typically is good over a narrower band of frequencies than when a resistive source is matched to a resistive load, as illustrated by the following example.

Example 13.23. Refer to Fig. 13.40, where

$$R_1 = 1\ \text{k}\Omega,\ L_1 = 10\ \text{mH},$$
$$R_2 = 100\ \Omega, C_2 = 100\ \text{nF},$$
$$v(t) = V_0\cos(\omega_0 t),\ f_0 = 10\ \text{kHz}.$$

Specify a matching circuit such that maximum power is delivered to the load.

Solution: For the parameter values given,

$$X_1 = \omega_0 L_1,\ Z_1 = Z_S = R_1 + jX_1$$
$$= R_S + jX_S \cong 1000 + j628\ \Omega,$$
$$X_2 = -\frac{1}{\omega_0 C_2},\ Z_L = \frac{jX_2 R_2}{R_2 + jX_2}$$
$$= R_L + jX_L = 71.7 - j45.0\ \Omega.$$

Because $R_S > R_L$ and because the source is inductive, we reduce the effective resistance of

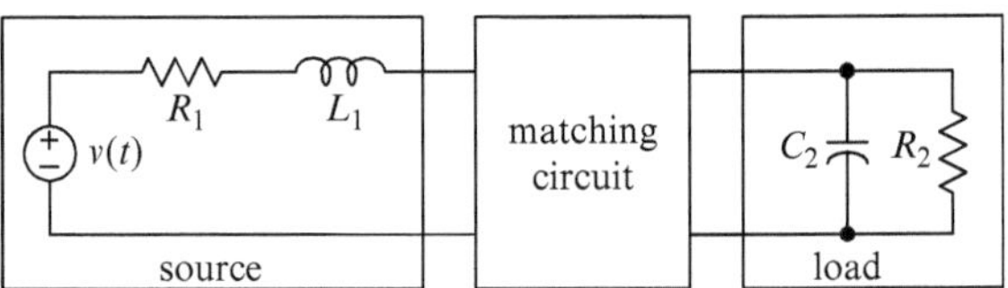

Fig. 13.40 See Example 13.23

the source by inserting a capacitor C in parallel with the source, such that the apparent source impedance is

$$Z_S' = \frac{Z_S Z_C}{Z_S + Z_C} = R_S' + jX_S'.$$

Subsequently, we complete the match by inserting an inductor in series with the load. Figure 13.41 shows the resulting matching circuit.

We determine the capacitance C such that $R_S' = R_L$. We must solve the equation

$$\text{Re}\left(Z_S'\right) = \text{Re}\left(\frac{Z_S Z_C}{Z_S + Z_C}\right) = R_L$$

for the capacitance C. We find

$$\begin{aligned}\frac{Z_S Z_C}{Z_S+Z_C} &= \frac{(R_S+jX_S)(jX_C)}{R_S+j(X_S+X_C)}\\ &= \frac{[(R_S+jX_S)(jX_C)][R_S-j(X_S+X_C)]}{{R_S}^2+(X_S+X_C)^2}\\ &= \frac{j{R_S}^2X_C - X_SX_CR_S + R_SX_C(X_S+X_C) + jX_SX_C(X_S+X_C)}{R_S^2+(X_S+X_C)^2}\\ &= \frac{R_S{X_C}^2 + j{R_S}^2X_C + jX_SX_C(X_S+X_C)}{{R_S}^2+(X_S+X_C)^2}\\ \Rightarrow \text{Re}(Z_S') &= \frac{R_S{X_C}^2}{R_S^2+(X_S+X_C)^2}.\end{aligned}$$

We require

$$\text{Re}(Z_S') = \frac{R_S{X_C}^2}{{R_S}^2 + (X_S + X_C)^2} = R_L$$

or

$$\begin{aligned} R_S{X_C}^2 &= R_L\left[{R_S}^2 + (X_S + X_C)^2\right]\\ &\Rightarrow {X_C}^2 + bX_C + c = 0;\\ b &= \frac{2R_LX_S}{R_L - R_S} = -97.05\ \Omega,\\ c &= \frac{R_L\left({R_S}^2 + {X_S}^2\right)}{R_L - R_S} = -1.077 \times 10^5\ \Omega^2,\end{aligned}$$

which yields

$$X_C = \frac{-b \pm \sqrt{b^2 - 4c}}{2} = (48.5 \pm 331.8)\ \Omega.$$

We must choose the negative solution because the reactance of a capacitor is negative. This gives

$$X_C = -283.3\,\Omega \Rightarrow C = -\frac{1}{\omega_0 X_C} = 56.2\,\text{nF}.$$

The modified source impedance is

$$Z_S' = \frac{Z_S Z_C}{Z_S + Z_C} = 71.7 - j308.0\,\Omega = R_S' + jX_S'$$

and the load impedance is

$$Z_L = \frac{jX_2R_2}{R_2 + jX_2} = R_L + jX_L = 71.7 - j45.0\,\Omega.$$

Finally, we add enough series inductance to cancel the total capacitive reactance. The required inductance is given by

$$\begin{aligned} j\omega_0 L &= j\,308 + j\,45 = j353\,\Omega\\ &\Rightarrow L = \frac{353\,\Omega}{2\pi 10^4 \text{s}^{-1}} = 5.62\ \text{mH}.\end{aligned}$$

Figure 13.41 shows the completed design.

It is instructive to calculate and plot the power delivered to the load as a function of frequency. For this purpose, we may associate all of the reactive components with the load, because they dissipate no average power. Thus we calculate the average power dissipated in the load Z in the circuit shown in Fig. 13.42, where, from Fig. 13.41,

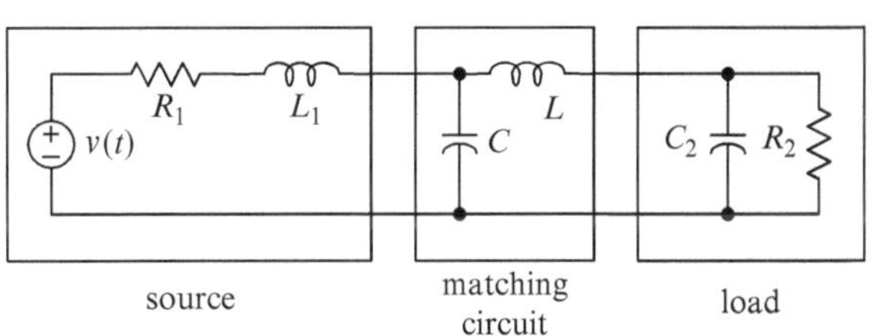

Fig. 13.41 See Example 13.23

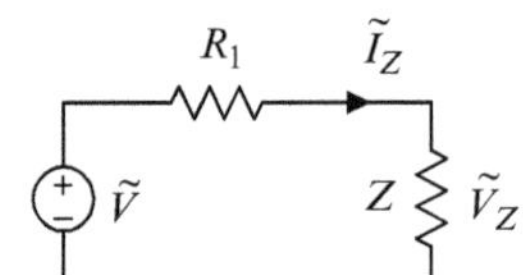

Fig. 13.42 See Example 13.23

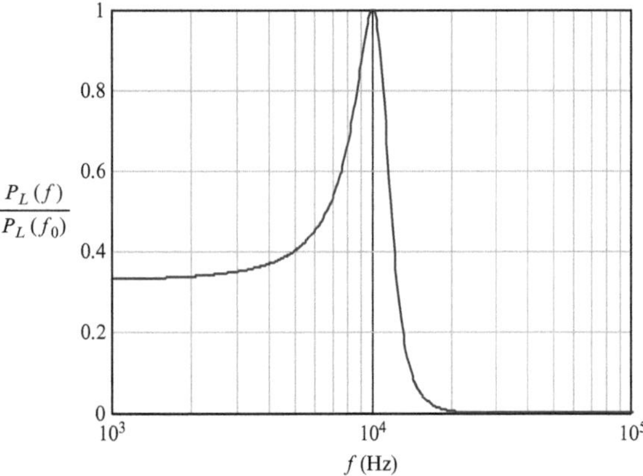

Fig. 13.43 See Example 13.23

$$Z = Z_{L1} + Z_C \| (Z_L + Z_{C2} \| Z_{R2})$$
$$= j\omega L_1 + \left(\frac{1}{j\omega C}\right) \left\| \left[j\omega L + \left(\frac{1}{j\omega C_2}\right) \| R_2 \right] \right.$$
$$= j\omega L_1 + \left(\frac{1}{j\omega C}\right) \left\| \left[j\omega L + \frac{R_2}{1 + j\omega C_2 R_2} \right] \right.$$
$$= j\omega L_1 + \left(\frac{1}{j\omega C}\right) \left\| \left[\frac{R_2 - \omega^2 L C_2 R_2 + j\omega L}{1 + j\omega C_2 R_2} \right] \right.$$
$$= j\omega L_1 + \frac{R_2 - \omega^2 L C_2 R_2 + j\omega L}{1 + j\omega C_2 R_2 + j\omega C (R_2 - \omega^2 L C_2 R_2 + j\omega L)}.$$

The power dissipated in the load is given by

$$P_L(f) = \frac{1}{2} |\tilde{I}_Z|^2 \mathrm{Re}[Z] = \frac{1}{2} \left| \frac{\tilde{V}}{R_1 + Z} \right|^2 \mathrm{Re}[Z].$$

Figure 13.43 shows a graph (computer generated) of the transfer efficiency $P_L(f)/P_L(f_0)$ versus frequency, where $P_L(f_0)$ is the maximum power transferred. The graph shows that the matching is very sensitive to frequency. The power transfer efficiency drops sharply as frequency departs from f_0.

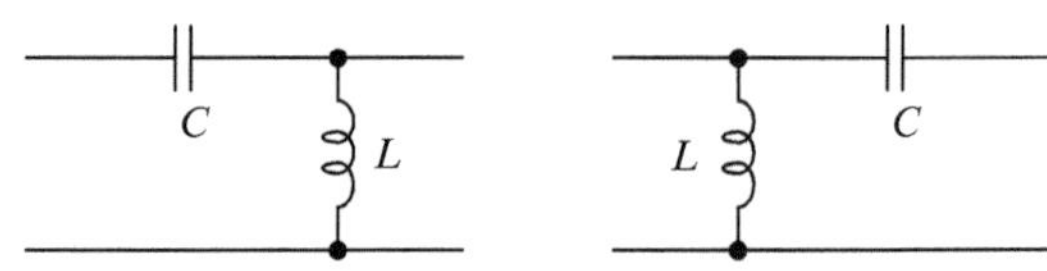

Fig. 13.44 Alternative matching L-sections

It is possible to match a source to a load using one of the L sections shown in Fig. 13.44. The principles involved are the same as those for the L sections in Fig. 13.36. The L sections in Fig. 13.44 are less-often used than those in Fig. 13.36 for low-(audio and some video) frequency matching, because the combination of series capacitance and shunt inductance tends to keep low frequency currents from reaching a load.

A circuit in which an L section matches a source impedance to a load impedance is resonant at the match frequency because (by design) the equivalent impedance of the circuit as a whole is real at that frequency. The impedance of the circuit can be sharply peaked at the resonant (match) frequency, and the bandwidth of the match provided by an L section often is inadequate.

Using more complex matching circuits, it is possible to achieve better (more wideband) matches than those obtained using only an L section. In practice, such matching circuits are designed using software for that purpose. Nonetheless, working through the examples above and a few end-of-chapter problems will help you grasp the ideas involved, which are applicable in a variety of contexts.

13.12 Problems

Section 13.3 is prerequisite for the following problems.

P 13.1 The complex power delivered to a certain load is $S = (100\angle 1.02)$ VA. (a) What is the real power delivered to the load? (b) If the load voltage is $\tilde{V} = (50\angle 0)$ V at 60 Hz, what is the load current (phasor)? (c) Construct a series *RL* or series *RC* model for the load, whichever is appropriate.

P 13.2 In Fig. P 13.1, $R_1 = R_2 = 1\ \mathrm{k\Omega}$, $C = 50$ nF, the applied voltage $\tilde{V}$ is the phase reference, and the current through the resistor R_2 is $\tilde{I}_R = (250\angle 1.2)$ mA at 1 kHz. Find the complex power dissipated by the circuit.

P 13.3 In Fig. P 13.1, let $R_1 = 1$ kΩ, $R_2 = 2.2$ kΩ, $C = 47$ nF, and $\tilde{V} = (25\angle 0)$ V at 10 kHz. Find the complex power delivered by the source and the power dissipated by each resistor.

P 13.4 Refer to Fig. P 13.2. Find the power delivered by the independent source and the power delivered to the load. Where does the additional power come from?

P 13.5 In Fig. P 13.3, $\tilde{V}_S = (25\angle 0)$ V at 2 kHz and the complex power dissipated by the circuit is $(223.18 - j96.62)$ mVA. Specify the power-dissipation rating of each resistor as either $1/8$ W or $1/4$ W.

P 13.6 Refer to Fig. P 13.4. Obtain an expression for the complex power dissipated by the circuit. Show that your expression is dimensionally consistent and that the expression behaves correctly for $f = 0$ and $f \to \infty$.

P 13.7 Refer to Fig. P 13.5. (a) Find the complex power S delivered by the source and the total power P dissipated in the three resistors. (b) Find the Thévenin equivalent for the circuit to the left of the inductor and find the complex power S_T delivered by the Thévenin equivalent source. If S_T is different from S, explain why. (c) Find the power P_T dissipated in the Thévenin equivalent resistance. If P_T is different from P, explain why.

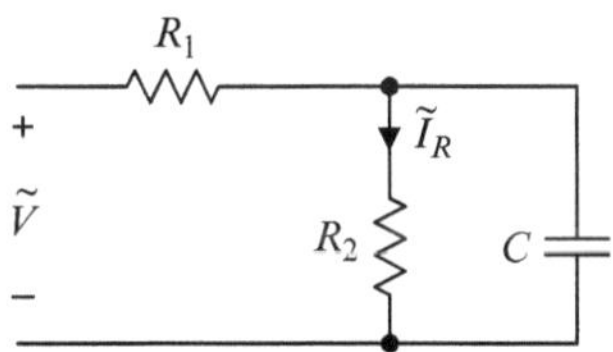

Fig. P 13.1 See Problem P 13.2, 3

P 13.8 Refer to Fig. P 13.6, where the source is sinusoidal and fixed amplitude. It is found that the power dissipated by the load R_L decreases as the frequency of the source increases, for all finite frequencies. Does the current through the load lead or lag the voltage across the load?

P 13.9 A certain 2,000 μF electrolytic capacitor has an *ESR* of 1.4 Ω. The voltage across the capacitor is given by $v_C(t) = V_{dc} + V_{ac}\cos(\omega_0 t)$, where $f_0 = 120$ Hz. The ac component is called the ripple voltage. The real power dissipated by the capacitor must not exceed 10 W. (a) If the real power dissipated has its maximum value, what is the complex power dissipated? (b) What is the maximum allowable rms amplitude of the ripple voltage?

P 13.10 A certain coil has dc resistance 1.2 Ω. At 100 kHz, the inductance and quality factor of the coil are $L = 250$ μH and $Q = 30$. The current through the coil is given by $i(t) = I_0[1 + 5\cos(\omega_0 t)]$, where $f_0 = 100$ kHz and $I_0 = 200$ mA. Find the complex and real power dissipated by the coil.

Section 13.4 is prerequisite for the following problems.

P 13.11 A 120 V (rms), 60 Hz sinusoidal voltage is applied to the terminals of a certain *RLC* circuit. The complex power delivered by the source is

$$S = (400 + j300)\ \text{VA}.$$

Model the circuit as (a) a resistor in series with an inductor and (b) a resistor in parallel with an inductor.

P 13.12 An *RL* series branch is driven by a sinusoidal source v_S. A capacitor is connected in parallel with the branch, as illustrated by Fig. P 13.7. Which of the following statements are *unconditionally* true, which

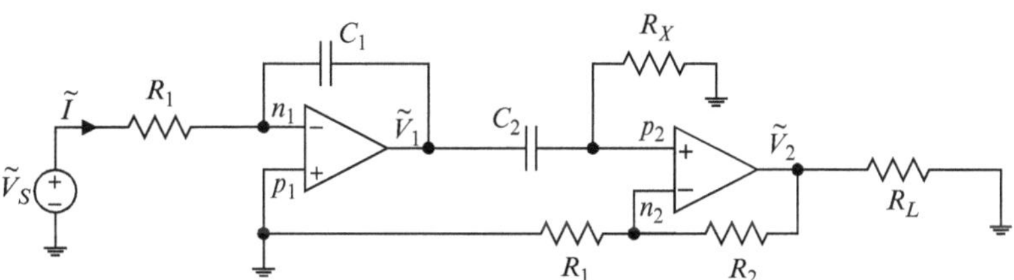

Fig. P 13.2 See Problem P 13.4

are *unconditionally* false, and which might be true or false, depending upon the capacitance?

(a) The real power delivered by the source increases.
(b) The real power delivered by the source decreases.
(c) The reactive power dissipated by the circuit increases.
(d) The reactive power dissipated by the circuit decreases.
(e) The apparent power dissipated by the circuit increases.
(f) The apparent power dissipated by the circuit decreases.
(g) The rms current drawn from the source increases.
(h) The rms current drawn from the source decreases.
(i) The angle of the impedance of the circuit increases.
(j) The angle of the impedance of the circuit decreases.
(k) The rms voltage across the resistor increases.
(l) The rms voltage across the resistor decreases.
(m) The power dissipated by the resistor increases.
(n) The power dissipated by the resistor decreases.
(o) The relative phase of the current increases.
(p) The relative phase of the current decreases.

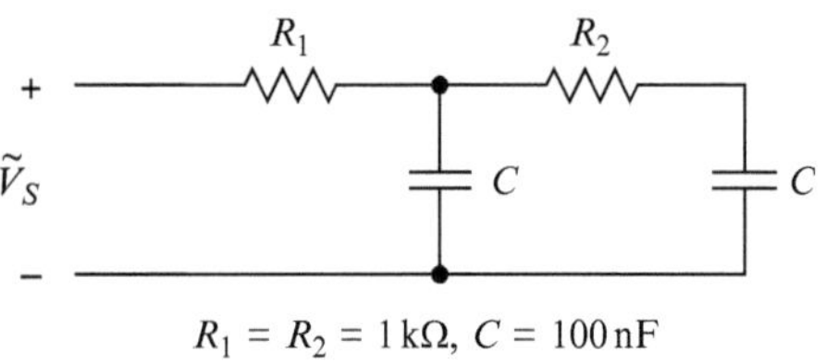

Fig. P 13.3 See Problem P 13.5

P 13.13 Refer to Fig. P 13.8. (a) Find the complex power, the real power, the reactive power and the apparent power delivered by the source. (b) Find the current through each resistor and the power dissipated in each resistor. Verify that the power dissipated in the resistors equals the real power found in part (a). (c) Find the phasor voltage across the source and use $P = V_{rms} I_{rms} \cos(\theta)$ to calculate the real power dissipated. Verify that the result agrees with those obtained in parts (a) and (b).

P 13.14 Refer to Fig. P 13.9. (a) Find the complex power, the real power, the reactive power and the apparent power delivered by the source. (b) Find the current through each resistor and the power dissipated in each resistor. Verify that the power dissipated in the resistors equals the real power found in part (a). (c) Find the phasor voltage across the source and use $P = V_{rms} I_{rms} \cos(\theta)$ to calculate the real power dissipated. Verify that the result agrees with those obtained in parts (a) and (b).

P 13.15 Refer to Fig. P 13.10, where $\tilde{V}_{S\,rms} = 240$ V. The real power dissipated by the load Z_L is $P_L = 5.62$ kW, the apparent power dissipated by the load is $AP_L = 7.02$ kVA, and the complex power delivered by the source $\tilde{V}_S$ is $S = (7.02 + j5.62)$ kVA. Find the rms phasor current $\tilde{I}_{rms}$, and the *ESR* and *ESX* of both the load and source impedance. There are two solutions. Find both.

P 13.16 Refer to Fig. P 13.10, where Z_S, Z_L are passive. Under what conditions are the current $\tilde{I}$ and voltage $\tilde{V}_S$(a) in phase, (b) in phase quadrature?

P 13.17 Refer to Fig. P 13.11, where $R_S = 2\ \Omega$ and the rms amplitude of $\tilde{V}_S$ is 240 V. The reactance X_0 is adjusted until the current $\tilde{I}$ is in phase with the voltage $\tilde{V}_S$, at which point $X_0 = -8\ \Omega$ and the real power delivered to the load Z_L is 2.5 kW. Find the load impedance, the complex power delivered by the

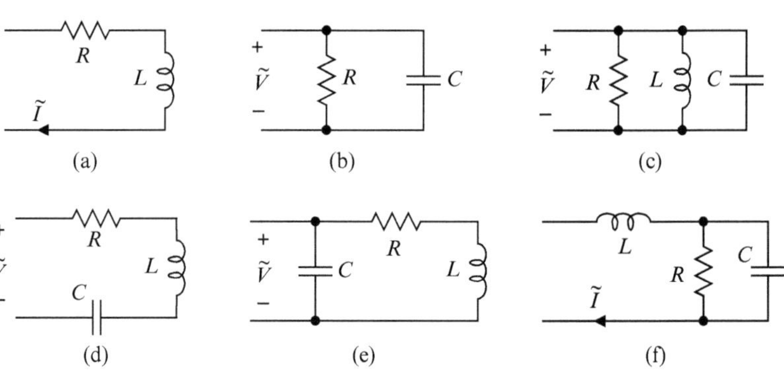

Fig. P 13.4 See Problem P 13.6

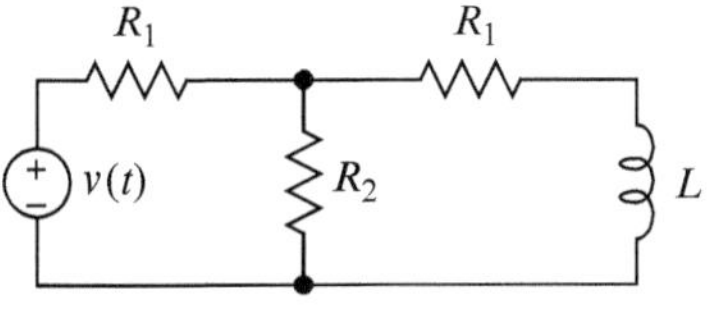

$R_1 = 150\,\Omega$, $R_2 = 220\,\text{k}\Omega$, $L = 5.5\,\text{mH}$,
$v(t) = V_0 \cos(\omega_0 t)$, $V_0 = 50\,\text{V}$, $f_0 = 10\,\text{kHz}$

Fig. P 13.5 See Problem P 13.7

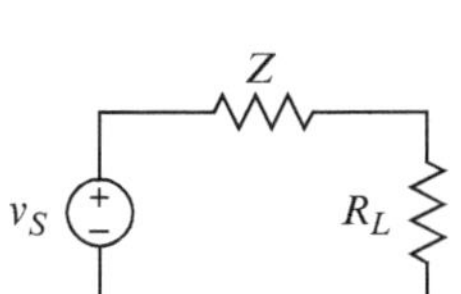

Fig. P 13.6 See Problem P 13.8

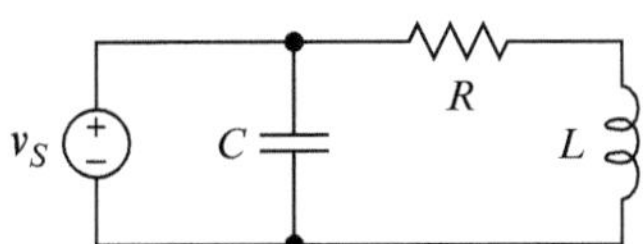

Fig. P 13.7 See Problem P 13.12

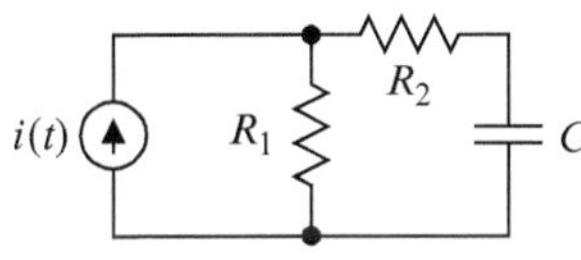

$R_1 = 22\,\text{k}\Omega$, $R_2 = 33\,\text{k}\Omega$, $C = 470\,\text{pF}$,
$i(t) = I \cos(\omega_0 t)$, $I = 25\,\text{mA}$, $f_0 = 10\,\text{kHz}$

Fig. P 13.8 See Problem P 13.13

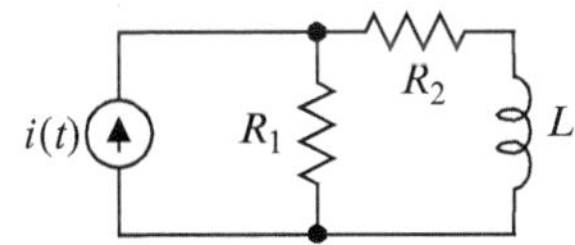

$R_1 = 33\,\text{k}\Omega$, $R_2 = 22\,\text{k}\Omega$, $L = 550\,\text{mH}$,
$i(t) = I \cos(\omega_0 \text{t})$, $I = 25\,\text{mA}$, $f_0 = 10\,\text{kHz}$

Fig. P 13.9 See Problem P 13.14

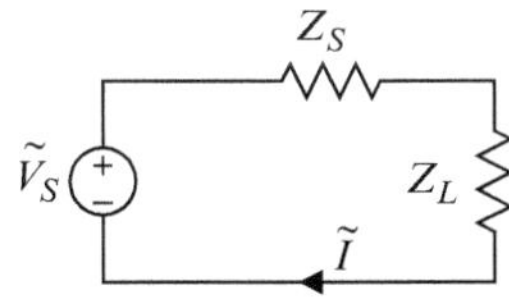

Fig. P 13.10 See Problem P 13.15, 16

Fig. P 13.11 See Problem P 13.17

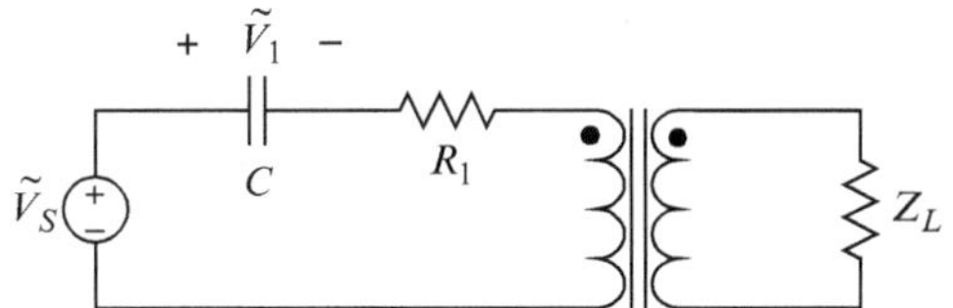

Fig. P 13.12 See Problem P 13.18

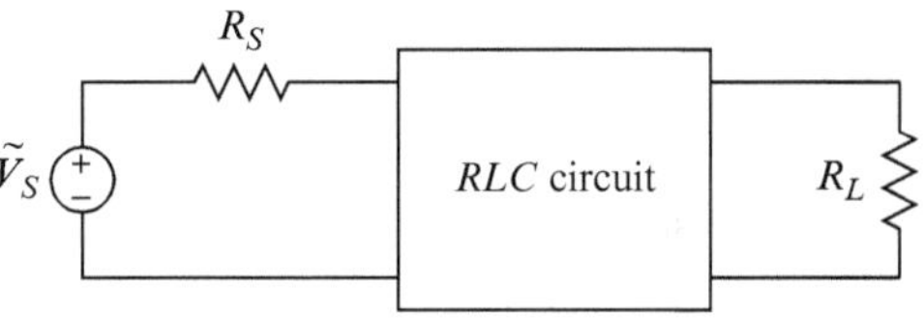

Fig. P 13.13 See Problem P 13.19

source $\tilde{V}_S$, the complex power dissipated by the load Z_L, and the reactive power dissipated by jX_0. There are two sets of solutions. Find both.

Section 13.5 is prerequisite for the following problems.

P 13.18 In Fig. P 13.12, $\tilde{V}_S = (120\angle 0)$ V at 400 Hz, $\tilde{V}_1 = (10\angle -\pi/6)$ V, $R_1 = 1\ \Omega$, $C = 500\ \mu\text{F}$, and the transformer is lossless. Find the fraction of the power delivered by the source that is transferred to the load Z_L.

P 13.19 Refer to Fig. P 13.13, where $\tilde{V}_S = (100\angle 0)$ V at 1 kHz. The input impedance of the passive *RLC* circuit (with the load attached) is $Z = (10 - j20)$ kΩ, where $|Z| \gg R_S$. What is the upper bound on the power that can be delivered to the load?

P 13.20 Refer to Fig. P 13.14. The load dissipates complex power $S_L = (6 + j4)$ kVA. If the source and

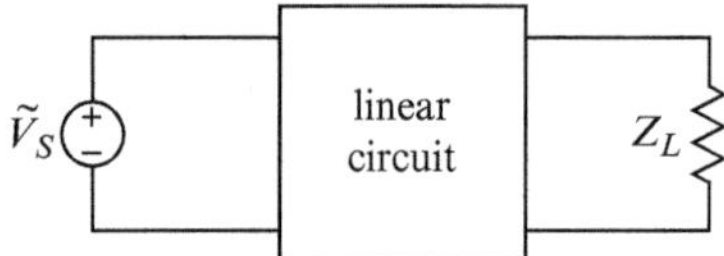

Fig. P 13.14 See Problem P 13.20

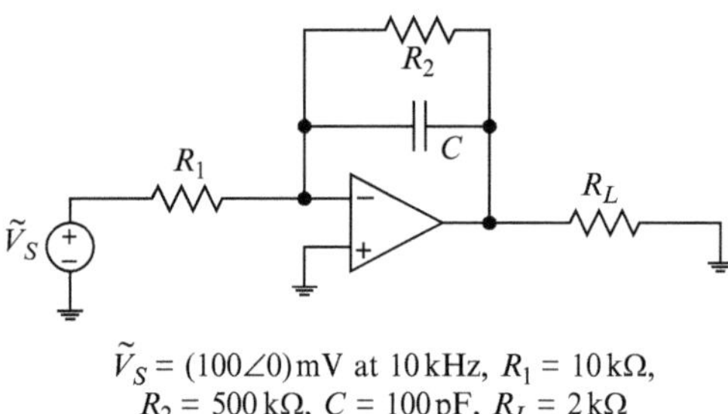

Fig. P 13.15 See Problem P 13.21

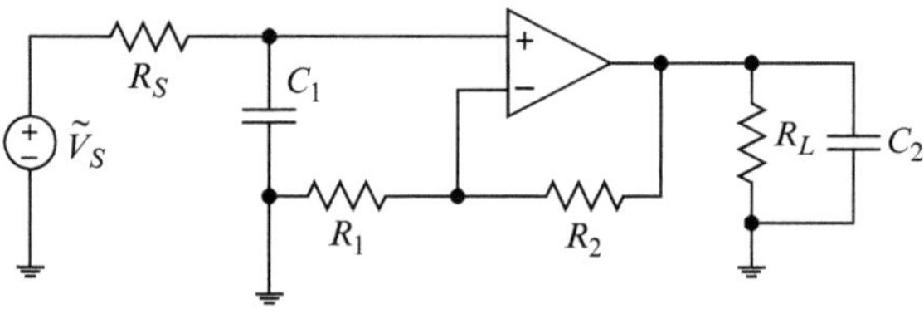

$\tilde{V}_S = 150\,\text{mV}$, $f = 1\,\text{kHz}$, $R_S = 50\,\Omega$, $R_1 = 50\,\text{k}\Omega$, $C_1 = 15\,\text{nF}$
$R_2 = 1\,\text{M}\Omega$, $C_2 = 500\,\text{nF}$, $R_L = 1\,\text{k}\Omega$

Fig. P 13.16 See Problem P 13.22, 23

circuit are replaced by a Thévenin equivalent at the terminals of the load, what is the total real power *dissipated* by the Thévenin source and the Thévenin equivalent impedance (recall the passive sign convention)?

P 13.21 Refer to Fig. P 13.15. Find the power dissipated by each resistor, the reactive power dissipated by the capacitor, and the complex power delivered by the ideal op amp. What fraction of the *real power delivered* by the op amp is dissipated in the load R_L?

P 13.22 Refer to Fig. P 13.16. Find the power dissipated by each resistor, the reactive power dissipated by each capacitor, the complex power delivered by the independent source, the complex power delivered to the load, and the complex power delivered by the ideal op amp. What fraction of the *real power delivered* by the op amp is dissipated in the resistor R_L?

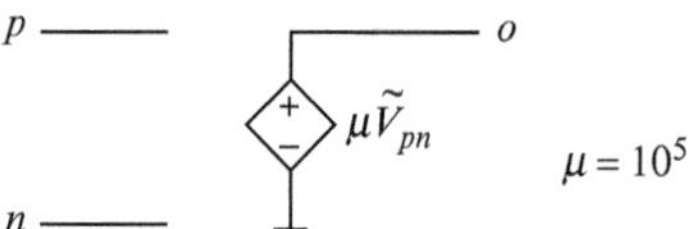

Fig. P 13.17 See Problem P 13.23

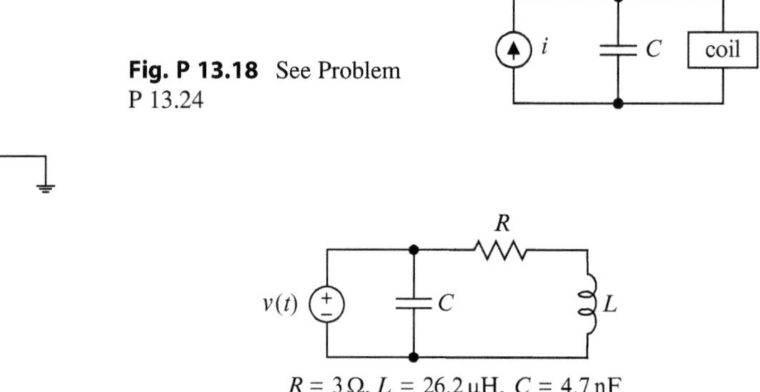

Fig. P 13.18 See Problem P 13.24

$R = 3\,\Omega$, $L = 26.2\,\mu\text{H}$, $C = 4.7\,\text{nF}$
$v(t) = V_0\cos(\omega_0 t)$, $V_0 = 50\,\text{V}$,

Fig. P 13.19 See Problem P 13.25

P 13.23 Refer to Fig. P 13.16. Replace the op amp by the linear model shown in Fig. P 13.17. Then find the power dissipated by each resistor, the reactive power dissipated by each capacitor, the complex power delivered by the independent source, the complex power delivered by the dependent source, and the complex power delivered to the load. What fraction of the *real power delivered* by the dependent source is dissipated in the resistor R_L? Verify that the total complex power dissipated equals zero (within computational accuracy).

Section 13.6 is prerequisite for the following problems.

P 13.24 Refer to Fig. P 13.18. The resonant frequency of the circuit is 10 kHz. The real power delivered to the circuit at the resonant frequency is 15 W. The quality factor of the coil at resonance is $Q = 25$ and the effective resistance of the circuit at resonance is 150 Ω. What is the rms current in the capacitor at resonance? *Hint*: Employ a reasonable approximation using the fact that $Q \gg 1$.

P 13.25 Refer to Fig. P 13.19, where the frequency of the source is variable. (a) Find the resonant

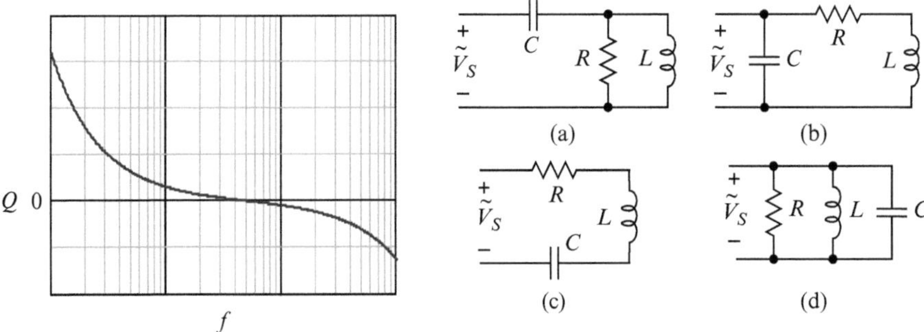

Fig. P 13.20 See Problem P 13.26

frequency f_r of the circuit. (b) Express the complex power delivered by the source as a function of the frequency of the source (use symbols). (c) Calculate the complex power delivered at the resonant frequency.

P 13.26 Fig. P 13.20 shows four circuits and a qualitative graph of reactive power versus frequency. Which of the four circuits could exhibit such behavior?

P 13.27 Fig. P 13.21 shows four qualitative graphs of real power dissipated versus frequency and five *RLC* circuits. Match each graph with one or more of the circuits.

Section 13.7 is prerequisite for the following problems.

P 13.28 A 5 hp electric motor draws 24 A from a 60 Hz 220 V source (both rms). Assume losses in the motor are negligible. Find the power factor for the motor and obtain a series *RL* model for the motor windings.

P 13.29 Two loads having lagging power factors pf_1, pf_2 are connected in parallel. When a voltage having rms amplitude V_{rms} is applied to the parallel connection, the average real powers dissipated by the loads are P_1, P_2. (a) Express the current drawn (not rms) by the parallel connection in terms of the powers dissipated, the power factors, and the applied voltage (phase reference). (b) Let $V_{rms} = 100$ V, $P_1 = 1.6$ kW, $P_2 = 2.5$ kW, $pf_1 = 0.8$, and $pf_2 = 1/\sqrt{2}$. Calculate the current. (c) What would the current be if both power factors were leading?

P 13.30 The apparent power dissipated by a certain load is 20 kVA with a lagging power factor of 0.7. Find the real power, reactive power, and complex power dissipated by the load.

P 13.31 Refer to Fig. P 13.22. Find the real, reactive, complex, and apparent power dissipated by the load and the impedance and power factor of the load (all at the frequency f_0) in each of the following cases.

(a) $v(t) = V_0 \cos(\omega_0 t)$, $i(t) = I_0 \cos(\omega_0 t)$, $V_0 = 10\text{V}, I_0 = 10$ mA;
(b) $v(t) = V_0 \cos(\omega_0 t)$, $i(t) = I_0 \cos(\omega_0 t - \pi/6)$, $V_0 = 10$ V, $I_0 = 10$ mA;
(c) $v(t) = V_0 \cos(\omega_0 t - \pi/4)$, $i(t) = I_0 \cos(\omega_0 t)$, $V_0 = 10$ V, $I_0 = 10$ mA;
(d) $v(t) = V_0 \cos(\omega_0 t - \pi/4)$, $i(t) = I_0 \cos(\omega_0 t + \pi/4)$, $V_0 = 10$ V, $I_0 = 10$ mA;
(e) $v(t) = V_0 \cos(\omega_0 t)$, $i(t) = I_0 \cos(\omega_0 t + \pi/2)$, $V_0 = 10$ V, $I_0 = 10$ mA.

P 13.32 Without making any calculations, state whether the power factor of each load in Fig. P 13.23 increases or decreases as the frequency of an applied voltage $\tilde{V}_{ab}$ is increased. Explain why.

P 13.33 In a certain industrial plant, three electric motors are connected in parallel to a 480 V (rms) source. Each motor is 95% efficient. The full-load outputs (work done under full load) of the motors are 25, 30, and 50 hp. The power factors of the motors (listed in the same order) are 0.7, 0.6, and 0.8, all lagging. Find the power factor of the three-motor load.

P 13.34 If the overall power factor for a parallel connection of several loads equals unity, does that mean that the power factor of each load equals unity? Justify your answer.

P 13.35 Two identical 25 hp, 480 V (rms), 60 Hz electric motors are connected in parallel. The power factor of the parallel connection is 0.8. What are the power factors of the individual motors? If the motors are 100% efficient, what is the current drawn by the parallel connection?

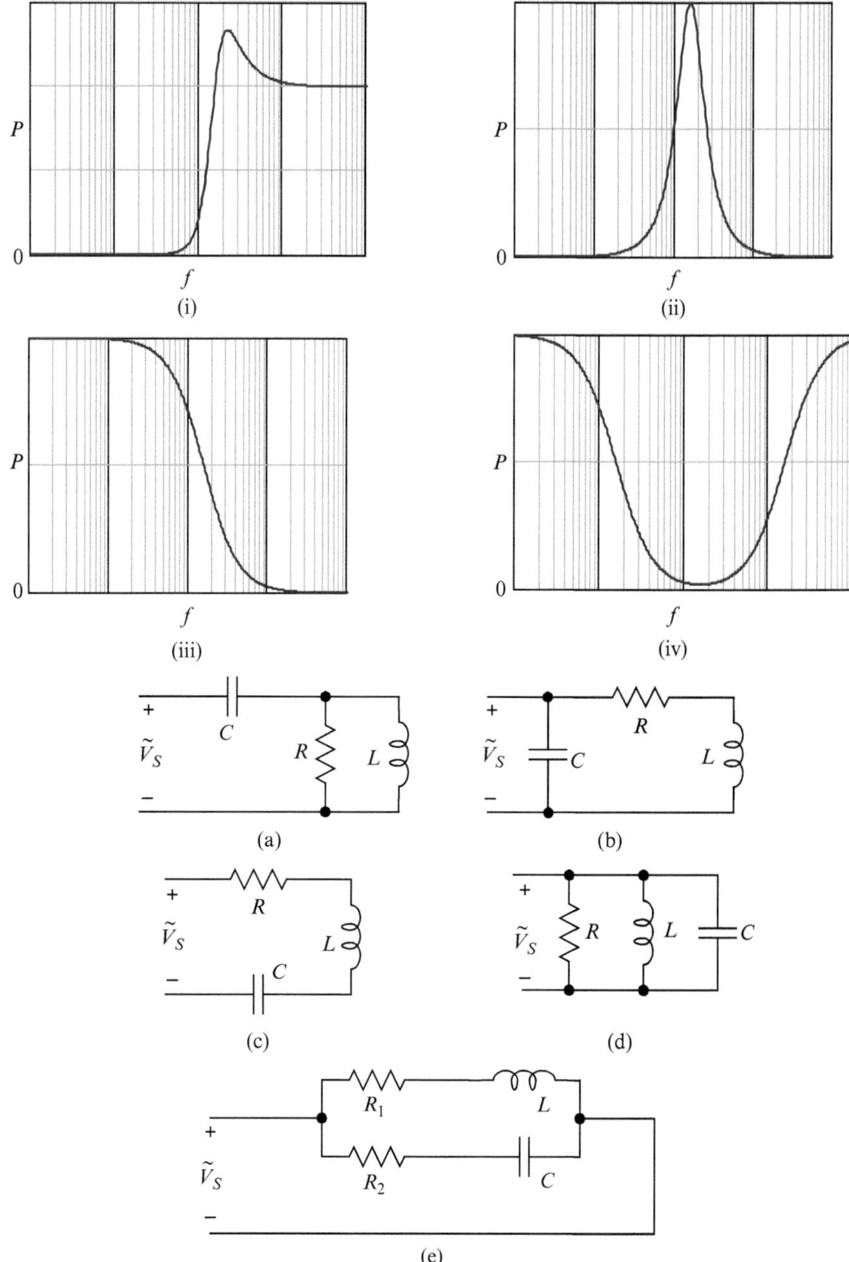

Fig. P 13.21 See Problem P 13.27

Section 13.8 is prerequisite for the following problems.

P 13.36 A certain 230 V (rms), 60 Hz, Hz, 3 hp (full-load output power) induction motor has a power factor of 0.6 and an efficiency of 0.9.

(a) Find the complex power delivered to the motor and the current drawn by the motor when the motor is operating at full load.
(b) Assume the load presented by the motor to a source can be modeled as a series *RL* circuit. Find the values of *R* and *L* for the model.
(c) Find the capacitance required to increase the power factor to 0.95.

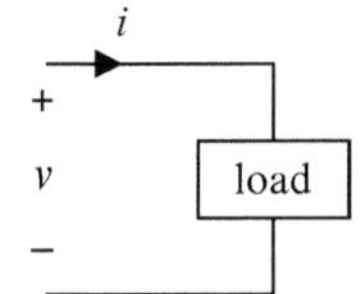

Fig. P 13.22 See Problem P 13.31

(d) Find the current drawn by the motor under full load when the power factor is corrected to 0.95.

P 13.37 For purposes of steady-state power-factor correction, an induction motor can be modeled as either a series *RL* circuit or an equivalent parallel *RL* circuit. (a) Obtain a series *RL* model for a 440 V (rms, 60 Hz) 30 hp induction motor having power factor $pf = 0.75$. (b) Find the capacitance required to obtain unity power factor. (c) Obtain a parallel *RL* model for the motor (excluding the capacitance) that is equivalent to the series *RL* model at 60 Hz. (d) Show that connecting the capacitance found in part (b) in parallel with the parallel *RL* model yields a circuit that is resonant at 60 Hz.

P 13.38 Given below are initial power factors (pf_1), load power requirements (P), line voltages (V_{rms}), line-voltage frequencies (f), and required power factors (pf_2). In each case, determine the shunt capacitance required to attain the required power factor and the resulting percent saving in line losses.

(a) $pf_1 = 0.6$, $P = 500$ kW, $V_{rms} = 10$ kV, $f = 60$Hz, $pf_2 = 1.0$;
(b) $pf_1 = 0.7$, $P = 200$ kW, $V_{rms} = 4$ kV, $f = 60$Hz, $pf_2 = 1.0$;
(c) $pf_1 = 0.5$, $P = 600$ kW, $V_{rms} = 10$ kV, $f = 60$Hz, $pf_2 = 0.9$;
(d) $pf_1 = 0.45$, $P = 100$ kW, $V_{rms} = 4$ kV, $f = 60$Hz, $pf_2 = 0.95$;
(e) $pf_1 = 0.5$, $P = 5$ kW, $V_{rms} = 440$ V, $f = 60$ Hz, $pf_2 = 0.9$;
(f) $pf_1 = 0.5$, $P = 1$ kW, $V_{rms} = 100$ V, $f = 480$ Hz, $pf_2 = 0.9$;
(g) $pf_1 = 0.6$, $P = 100$ W, $V_{rms} = 24$ V, $f = 480$Hz, $pf_2 = 1.0$;
(h) $pf_1 = 0.9$, $P = 2$ kW, $V_{rms} = 240$ V, $f = 60$Hz, $pf_2 = 1.0$.

P 13.39 Two loads having identical power factors but different power requirements are connected in parallel. Obtain an expression for the power factor of the parallel load.

P 13.40 Two loads are connected in parallel. One draws 15 kW and has power factor 0.7. The other draws 25 kW and has power factor 0.9. Find the power factor of the parallel connection.

P 13.41 All else held constant, does increasing the line voltage increase or decrease the capacitance required for power-factor correction? Does increasing the line voltage increase or decrease the actual (not percent) line losses?

P 13.42 All else held constant, does increasing the line frequency increase or decrease the capacitance required for power-factor correction?

P 13.43 The power factor of a 50 kW load is corrected to unity by a 4 μF shunt capacitor. The capacitor has a tolerance of +20%, −40%. The line voltage is 10 kV (rms) and the frequency is 60 Hz. What is the range of possible values for the actual power factor?

P 13.44 Two loads are connected in parallel across a 4 kV (rms) 60 Hz line. One load requires 2 kW and has power factor 0.6 lagging. The other requires 1.5 kW and has power factor 0.7 lagging. What capacitance is required to correct the overall power factor to 0.9?

P 13.45 Some manufacturing operations, such as electroplating, have leading power factors. How would you correct a leading power factor?

P 13.46 A small electroplating shop having a leading power factor of 0.7 is adjacent to a small machine shop having a lagging power factor of 0.5. The electroplating shop requires 8 kW and the machine shop requires 10 kW. Both are fed by the same 60 Hz, 440 V (rms) line. What capacitance is required to attain an overall power factor of 0.95?

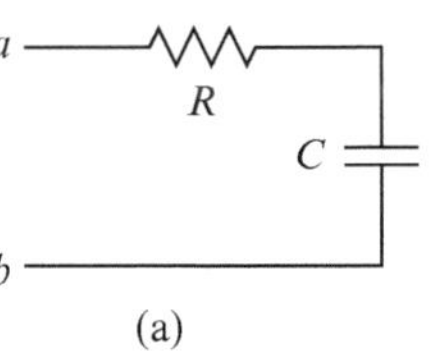

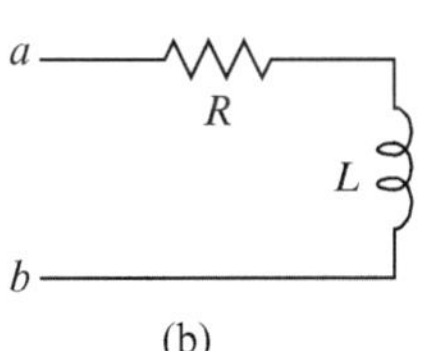

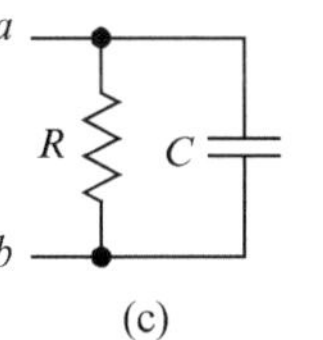

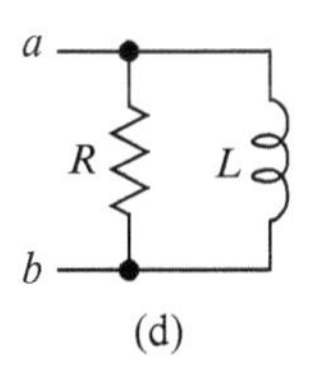

Fig. P 13.23 See Problem P 13.32

Section 13.10 is prerequisite for the following problems.

P 13.47 At the output terminals of a certain circuit, the open-circuit voltage equals 100 $V_{(rms)}$ and the output impedance is $Z_{out} = 4 - j6\ \Omega$. Find the maximum power the circuit could deliver to a load and the current through the load in that case.

P 13.48 In Fig. P 13.24, the source is sinusoidal, having frequency $f = 1.6$ kHz and rms amplitude $V_{rms} = 1$ V. (a) What load impedance connected to the terminals *a–b* would dissipate maximum power? What is the power dissipated? (b) What real (resistive) load connected to the terminals *a–b* would dissipate more power than any other real load? What is the power dissipated in that case?

P 13.49 Measurements made at 10 kHz on a certain transformer reveal that the primary and secondary resistances and inductances are $R_1 = 100\ \Omega$, $L_1 = 5$ H, $R_2 = 2\ \Omega$, and $L_2 = 100$ mH. The shunt capacitances are negligible and the coupling coefficient is approximately unity. (a) Find the impedance at the primary terminals at 10 kHz when a 100 Ω load is connected to the secondary terminals. (b) Find the turns ratio three different ways. (c) Find the power transfer efficiency of the transformer at 10 kHz when driving a 100 Ω load. (d) Obtain the Thévenin equivalent for the transformer and source at the secondary terminals.

P 13.50 Refer to Fig. P 13.25, where $v_S$is a 4 V, 1.6 kHz sinusoidal source, $R_S = 900\ \Omega$, $L = 75$ mH, and $C_1 = 50$ nF. Find the values of C_2 and R_L such that maximum power is dissipated in the load.

P 13.51 The rms voltage *available* from a certain 60 Hz source is 480 V. The output impedance of the source is $Z_{out} = 2.1 + j3.6\ \Omega$. The complex power dissipated by a certain load driven by the source is $S_L = (13.44 + j6.72)$ kVA. Find the load impedance. Why are there two solutions?

Section 13.11 is prerequisite for the following problems.

P 13.52 Refer to Fig. P 13.26, where the transformer is ideal, having turns ratio $n = 4$. It is found that the complex power delivered to the load Z_L is $S_L = (5 - j2)$ kVA when the real power delivered to the load is maximum. (a) Find the complex power dissipated by the source impedance Z_S. (b) Find the complex power delivered by the source $\tilde{V}_S$. (c) If the secondary current is $\tilde{I}_2 = (51.89\angle 0)$ A, what is the source voltage $\tilde{V}_S$?

P 13.53 A 1 kHz source having output impedance $Z_S = (2 + j)\ \Omega$ and available voltage $\tilde{V}_S = 50$ V is to drive a load having impedance $Z_L = (200 - j100)\ \Omega = 100Z_S^*$. Matching for maximum power transfer requires a turns ratio $n = 10$. A transformer having approximately the desired turns ratio $n \cong \sqrt{L_2/L_1} = 10$ is used to achieve an approximate match, as shown in Fig. P 13.27. The transformer is nearly lossless and has a coupling coefficient near unity. (a) Find the power delivered to the load. (b) What is the ratio of the power actually delivered to the load to the theoretical maximum?

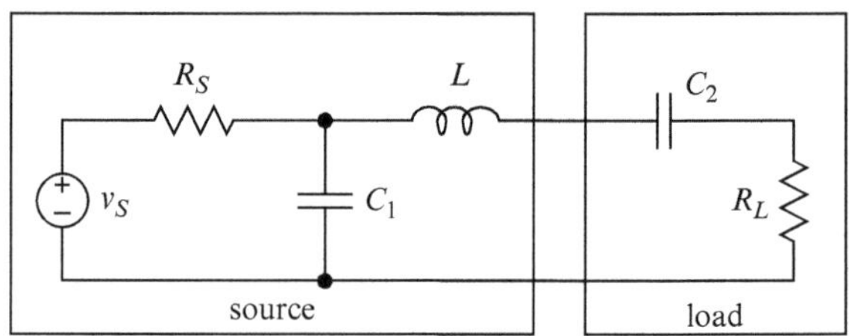

Fig. P 13.25 See Problem P 13.50

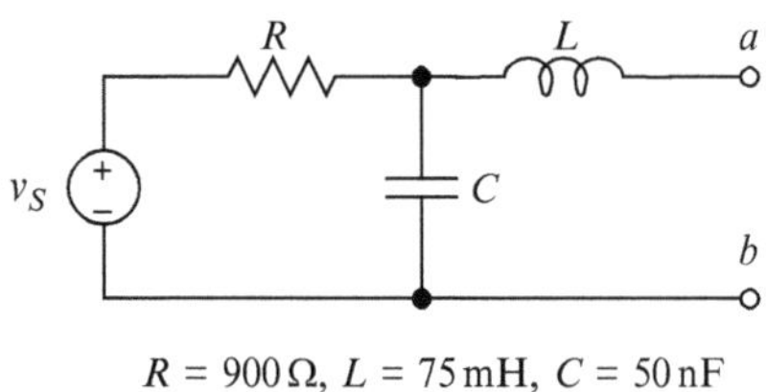

Fig. P 13.24 See Problem P 13.48

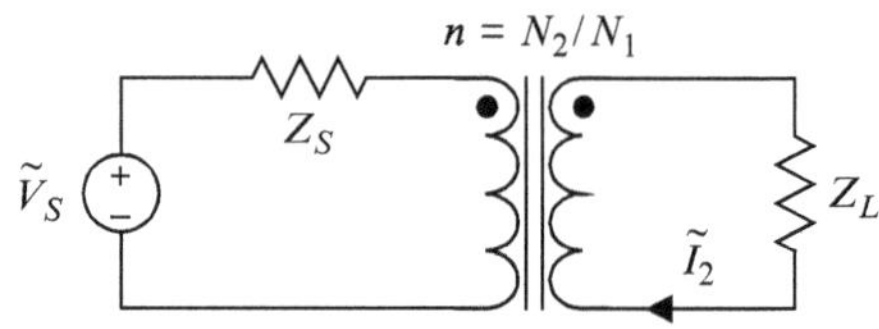

Fig. P 13.26 See Problem P 13.52

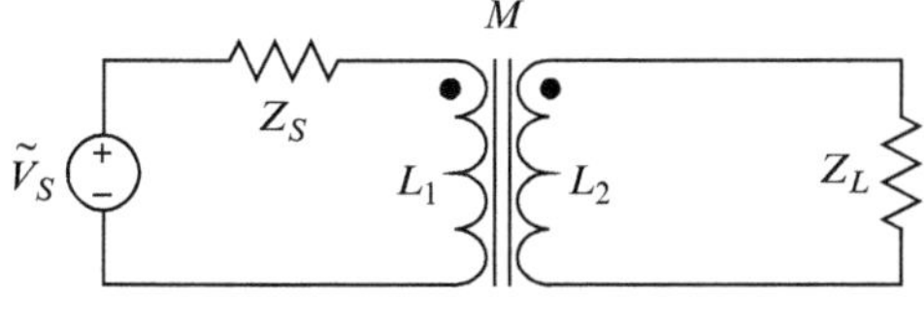

$L_1 = 5\,\text{mH}$, $L_2 = 500\,\text{mH}$, $k \cong 1$

Fig. P 13.27 See Problem P 13.53

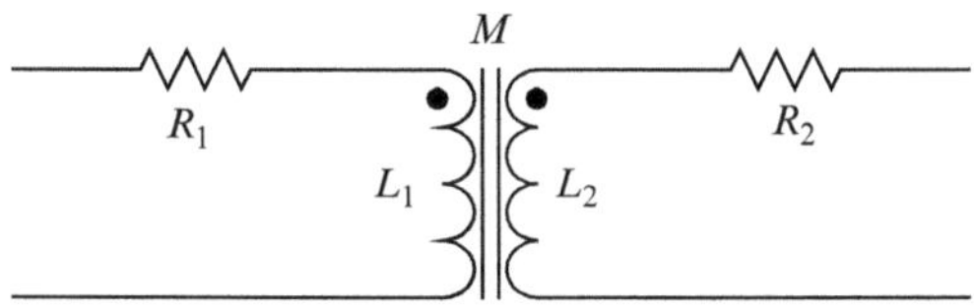

Fig. P 13.28 See Problem P 13.55

P 13.54 A certain 20 kHz source having output impedance $R_s = 20\ \Omega$ is to drive a load having input impedance $R_l = 50\ \Omega$, using a transformer to match the source to the load (for maximum power transfer). Which of the two transformers described in the table below is best suited for that application? (R_1, L_1 and R_2, L_2 denote the primary and secondary resistance and inductance at 20 kHz and k denotes the coupling coefficient.)

	R_1 (Ω)	L_1 (mH)	R_2 (Ω)	L_2 (mH)	k
Transformer 1	1.5	1.78	4.2	3.10	0.99
Transformer 2	1.0	2.60	2.5	2.80	0.98

P 13.55 A certain transformer is modeled as shown in Fig. P 13.28. The primary is driven by a source having frequency f, available voltage $\tilde{V}_S$, and output impedance Z_S. Show that the load impedance Z_L that draws the available power from the secondary is given by

$$Z_L = R_2 - j\omega L_2 + \frac{(\omega M)^2}{R_1 - j\omega L_1 + Z_S^*}$$

Hint: Find the Thévenin equivalent at the terminals of the secondary. Then $Z_L = Z_T{}^*$.

P 13.56 In Fig. P 13.29, the transformer is ideal, having turns ratio $n = N_2/N_1$. The source voltage is $\tilde{V}_S = 1\angle 0\,\text{Vrms}$, the source impedance is $Z_S = (1 + j)\ \Omega$ and the load impedance is $Z_L = (100 - j200)\ \Omega$.

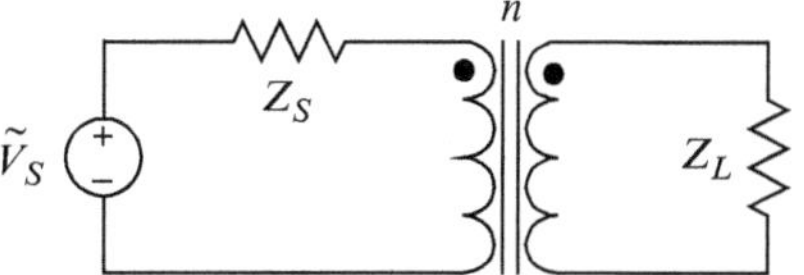

Fig. P 13.29 See Problem P 13.56

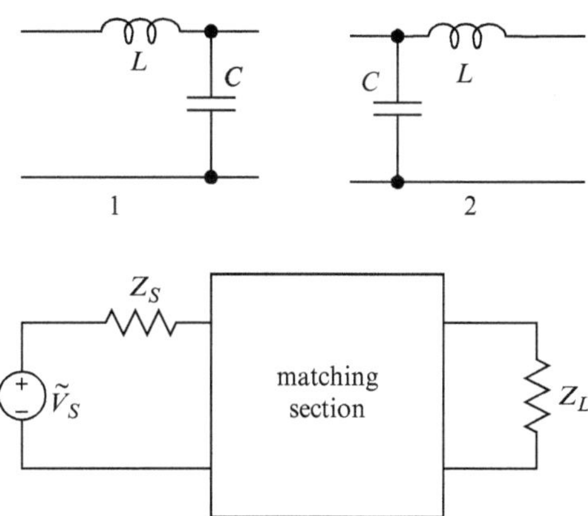

Fig. P 13.30 See Problem P 13.57

(a) Find the turns ratio for which the power delivered to the load is maximum. What fraction of the available power is delivered to the load for that turns ratio? Plot the power delivered to the load versus the turns ratio.

(b) Add positive reactance to the source impedance such that a turns ratio of 10 maximizes the power delivered to the load. Plot the power delivered to the load versus the turns ratio.

(c) Restore the original source impedance and add negative reactance to the load impedance such that a turns ratio of 10 maximizes the power delivered to the load. Plot the power delivered to the load versus the turns ratio.

P 13.57 Fig. P 13.30 shows two possible L-sections for matching a source to a load for maximum power transfer and below those, an illustration of a matching problem. For each case given in (a)–(h), state which of the two matching sections you would use, and why. Then determine the parameters of the selected matching network.

(a) $Z_S = (120 + j6)\ \Omega$, $Z_L = (50 + j25)\ \Omega$
(b) $Z_S = (20 + j50)\ \Omega$, $Z_L = (10 - j15)\ \Omega$
(c) $Z_S = (10 - j50)\ \Omega$, $Z_L = (20 - j15)\ \Omega$
(d) $Z_S = (100 + j30)\ \Omega$, $Z_L = (75 + j15)\ \Omega$

(e) $Z_S = (120 + j50)\ \Omega,\ Z_L = (200 + j150)\ \Omega$
(f) $Z_S = (10 - j30)\ \Omega,\ Z_L = (5 + j15)\ \Omega$
(g) $Z_S = (1.2 + j5.5)\ \Omega,\ Z_L = (2.0 + j15.0)\ \Omega$
(h) $Z_S = (5.2 + j3.5)\ \Omega,\ Z_L = (2.5 + j7.5)\ \Omega$

P 13.58 In Fig. P 13.31, the LC network is to match the load to the source such that maximum power is delivered to the load. The source is sinusoidal, having frequency $f = (2\pi)^{-1}$ MHz. Find the appropriate values for L and C.

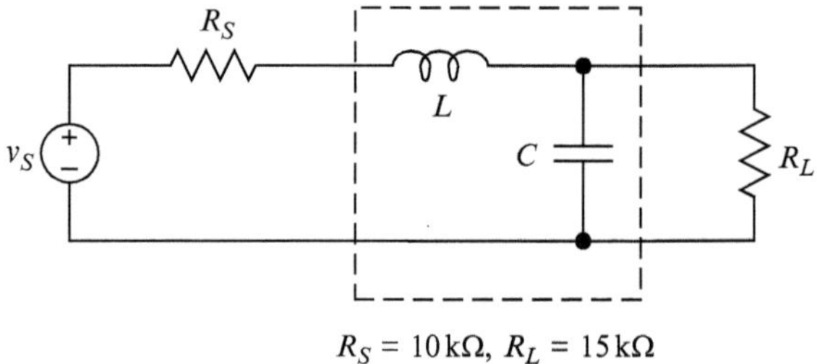

Fig. P 13.31 See Problem P 13.58

Chapter 14
Three-Phase Circuits

Three-phase circuits are important in the power industry, where electric power is distributed using three-phase voltage and current (especially over long distances) and also in almost any industrial facility using many large electric motors. The main reasons are:

(1) It is less costly to distribute three-phase power than single-phase power.
(2) Three-phase motors are generally more efficient than single-phase motors.
(3) Three-phase motors deliver constant power whereas instantaneous power delivered by a single-phase motor has an alternating component.
(4) It is easier to reverse (quickly) the direction of a three-phase motor.

On first encounter, many students find three-phase circuits confusing, primarily because they seem somehow different from ordinary circuits. That is not the case. A three-phase circuit is simply another circuit, and like other circuits succumbs to analysis using Kirchhoff's laws. This does not mean that analysis of three-phase circuits is easy, nor does it mean that there is no new terminology to be learned. But there is nothing magic about three-phase circuits. Stick with Kirchhoff's current law and Kirchhoff's voltage law, pay attention to notation and terminology, and all will work out well.

It is conventional in the electric-power industry to express voltages and currents as rms quantities. We follow that convention in this chapter. In this chapter, all phasor voltages and phasor currents are expressed as rms quantities. For example, if $\tilde{V}$ denotes a phasor voltage, then $|\tilde{V}|$ is the rms amplitude of the voltage. It also is conventional in the electric-power industry to express phase in degrees, so we follow that convention also.

14.1 Three-Phase Sources

Figure 14.1 Shows a simplified diagram of a three-phase generator and a graph of the terminal voltages. The generator windings are 120° apart but are otherwise identical.

As the magnetized rotor turns clockwise, the north pole passes the positive terminals of the three windings in the order *a*, *b*, *c*, called the *abc* sequence. If the rotor direction is reversed (or if the wires are attached differently), the voltages are generated in the sequence *a*, *c*, *b*, called the *acb* sequence. The *acb* sequence also is called the *negative sequence*. The *acb* sequence is significant because if a three-phase motor using an *abc* sequence is rewired (or switched) to the *acb* sequence, the direction of rotation is reversed. (This is what makes it easy to reverse the rotation of a three-phase motor.)

In the *abc* sequence, because the windings are 120° apart, voltage v_{bn} lags voltage v_{an} by 120° and voltage v_{cn} lags voltage v_{an} by 240°. In the *acb* sequence, voltage v_{cn} lags voltage v_{an} by 120° and voltage v_{bn} lags voltage v_{an} by 240°. The voltages and corresponding phasor representations for the *abc* sequence are given by

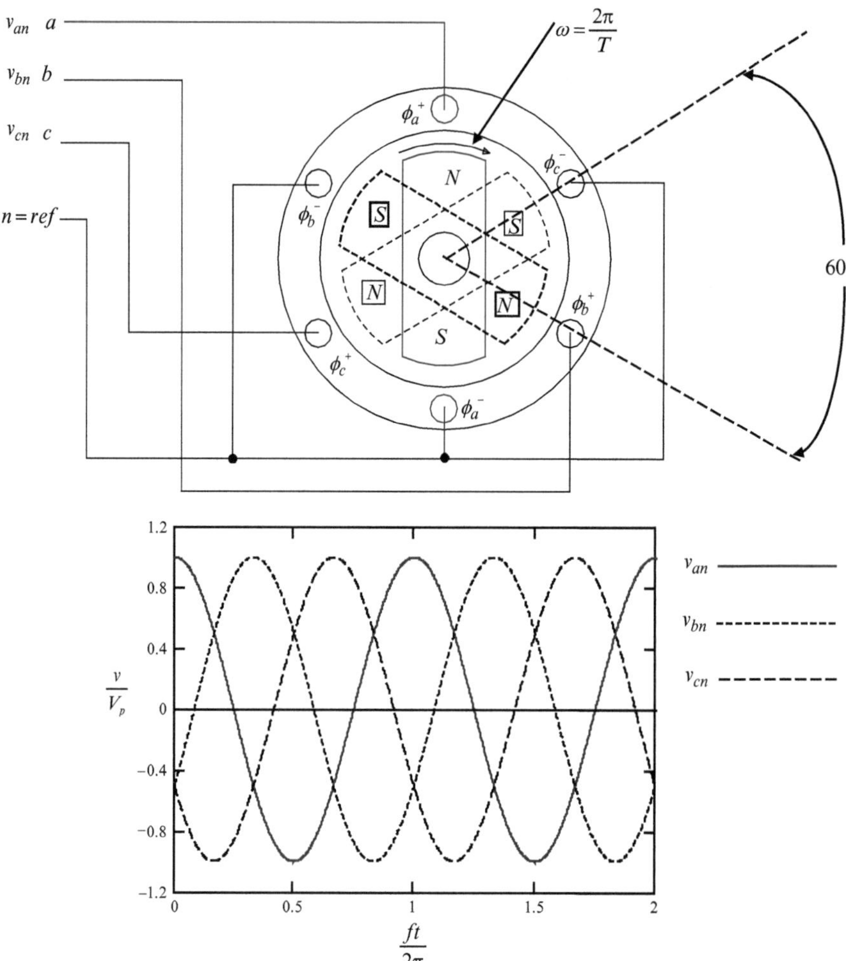

Fig. 14.1 Three-Phase generator and terminal voltages for constant rotor speed

$$
\begin{aligned}
v_{an}(t) &= V_p \cos(2\pi f t), & \tilde{V}_{an} &= V_p\angle 0, \\
v_{bn}(t) &= V_p \cos(2\pi f t - 120°), & \tilde{V}_{bn} &= V_p\angle -120°, \\
v_{cn}(t) &= V_p \cos(2\pi f t - 240°), & \tilde{V}_{cn} &= V_p\angle -240°.
\end{aligned} \tag{14.1}
$$

The voltages and corresponding phasor representations for the *acb* sequence are given by

$$
\begin{aligned}
v_{an}(t) &= V_p \cos(2\pi f t), & \tilde{V}_{an} &= V_p\angle 0, \\
v_{cn}(t) &= V_p \cos(2\pi f t - 120°), & \tilde{V}_{cn} &= V_p\angle -120° \\
v_{bn}(t) &= V_p \cos(2\pi f t - 240°), & \tilde{V}_{bn} &= V_p\angle -240°.
\end{aligned} \tag{14.2}
$$

The voltages v_{an}, v_{bn}, v_{cn} are called *phase voltages*. Figure 14.2 shows phasor diagrams for the phase voltages. The common line labeled n is called

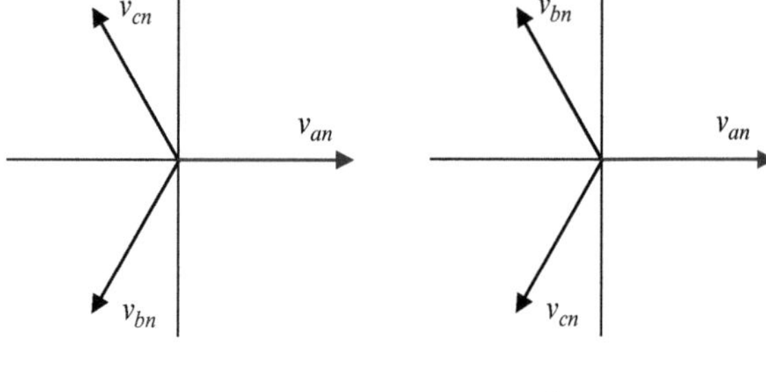

Fig. 14.2 Phasor representations of the phase voltages shown in Fig. 14.1

the neutral line or simply the *neutral*. The phase voltages all have the same magnitude but are 120° apart. A generator or other device producing such voltages is called a *balanced source*. There are few, if any, applications for unbalanced three-phase sources. *We consider only balanced sources.*

The rotor angular speed is

$$s\,\text{rpm} = \frac{2\pi s}{60}\ \text{rad s}^{-1}. \tag{14.3}$$

The north pole of the rotor passes through the 12:00 position $s/60$ times in 1 s. The frequency of the generated voltage (each phase) is

$$f = \frac{s}{60}, \tag{14.4}$$

so rotor speed $s = 3600$ rpm generates a 60-Hz voltage.

For a balanced source, the sum of the phase voltages equals zero:

$$v_{an} + v_{bn} + v_{cn} = 0 \Rightarrow \tilde{V}_{an} + \tilde{V}_{bn} + \tilde{V}_{cn} = 0. \tag{14.5}$$

This is evident from the phasor diagrams and can be shown as follows: From (14.1),

$$\begin{aligned}\tilde{V}_{an} &= V_p\angle 0 = V_p\,(1 + j0),\\ \tilde{V}_{bn} &= V_p\angle -120^\circ = V_p\left(-\frac{1}{2} - j\frac{\sqrt{3}}{2}\right),\\ \tilde{V}_{cn} &= V_p\angle -240^\circ = V_p\angle 120^\circ = V_p\left(-\frac{1}{2} + j\frac{\sqrt{3}}{2}\right).\end{aligned} \tag{14.6}$$

The sum is

$$\begin{aligned}\tilde{V}_{an} + \tilde{V}_{bn} + \tilde{V}_{cn} = V_p &\Bigg[(1 + j0)\\ &+ \left(-\frac{1}{2} - j\frac{\sqrt{3}}{2}\right) + \left(-\frac{1}{2} + j\frac{\sqrt{3}}{2}\right)\Bigg] = 0.\end{aligned} \tag{14.7}$$

A three-phase generator (or other three-phase source) usually is represented using three independent voltage sources connected in either a Y or Δ configuration, as shown in Fig. 14.3. For the Δ-connected source, Kirchhoff's voltage law requires

$$\tilde{V}_{ab} + \tilde{V}_{bc} + \tilde{V}_{ca} = 0,$$

which holds if the voltages are given by (14.1) or (14.2).

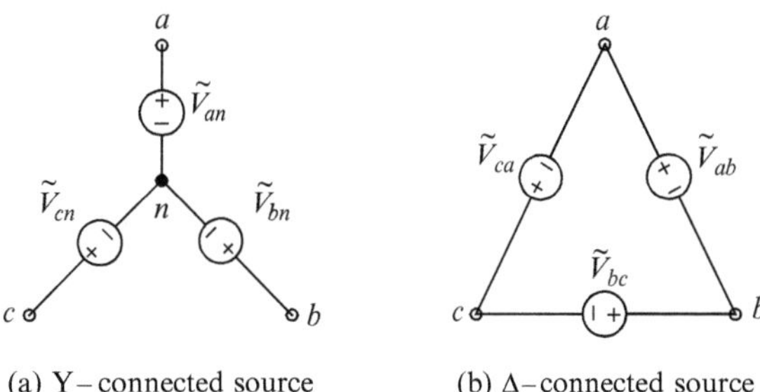

Fig. 14.3 Three-phase source models

In practice, Y-connected sources are more common than Δ-connected sources because they are less susceptible to problems arising from imbalance in the phase currents. Such imbalance can arise from driving an unbalanced load, where each source sees a different impedance. Imbalance in a Y-connected source results in a (usually small) current in the neutral (ground) conductor, whereas in a Δ-connected source, there is no neutral, and load imbalance can cause a loop current in the delta (circulating through the three sources).

14.2 Power Transmission and Distribution

Figure 14.4 shows a (greatly) simplified diagram of how three-phase power is generated, transmitted, and distributed.

Three-phase power is generated, then stepped up by a three-phase transformer to high voltage (25 kVrms or higher) for transmission. Transmitting power at high voltage and low current reduces power losses (so-called *i-squared-r losses*) in the resistance of the lines. At the destination (a substation), the voltages are stepped down to useable levels and distributed to users. Three-phase power often is supplied to industrial users at 480 Vrms or more. Power is supplied to residential users at 240 Vrms, double-ended, single-phase using a center-tapped transformer, as shown in Fig. 14.5. Electric utility companies attempt to keep loads balanced by distributing loads equally among the three phases of a transformer secondary. A reasonable balance can be achieved if there are a large number of users on each phase. A main power line from a power plant to an urban substation serves a large number of users, and the effective load on the

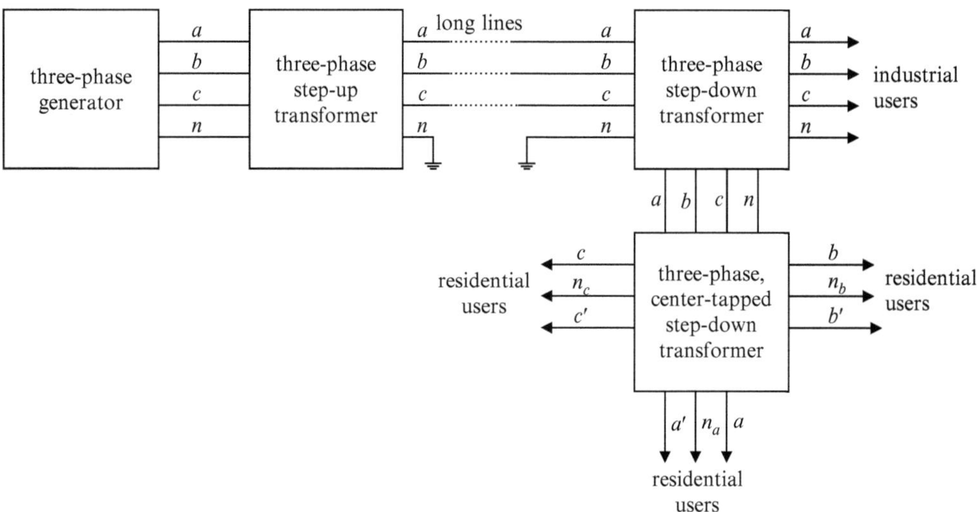

Fig. 14.4 Simplified power generation, transmission, and distribution diagram

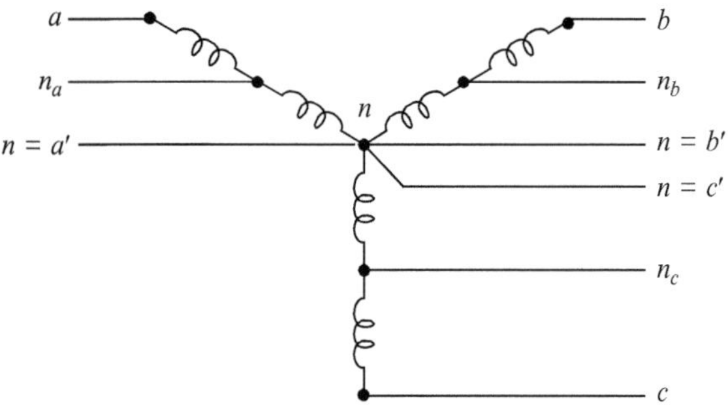

Fig. 14.5 Secondary of center-tapped three-phase transformer (Y-connected)

power plant is very nearly balanced. Balance is more difficult to maintain as the distribution lines spread out and serve fewer and fewer users.

14.3 Residential Wiring

Figure 14.6 shows how power is supplied to a residence from one leg of a center-tapped, three-phase, step-down transformer. The voltage from either end of the secondary winding to the center tap is 120 Vrms. These voltages are supplied to wall outlets, wall- and ceiling mounted lights, and other 120-Vrms loads. Electricians attempt to distribute the 120-Vrms loads equally between the two branches. The end-to-end voltage is 240 Vrms, which is supplied to loads requiring 240 Vrms, such as air conditioners, electric ranges, and 240-Vrms electric water heaters. Not shown are circuit breakers and other protective devices required by the national electrical code.

In the United States, electric services to most residences are 200–400 Arms, meaning that available full-load power (at 240 Vrms) is 48–96 kW. A load exceeding the maximum trips the main (200 or 400 Arms) circuit breaker in the distribution panel. Average residential consumption in the U.S. is approximately 1000 kWh/month.

> **Exercise 14.1.** In Fig. 14.6, assume the 120 Vrms loads are all equal. Show that the current $i(t)$ equals zero.

14.4 Three-Phase Loads

As noted above, most large industrial plants use three-phase motors which can be connected in either a Y or Δ configuration, presenting a three-phase load to a three-phase source. Plants also have single-phase loads, such as lighting, which can be fed using one or more of the three phases individually, or from an on-site center-tapped three-phase transformer, as in a residence or office. But as a whole, a large industrial plant presents a three-phase load to a three-phase source.

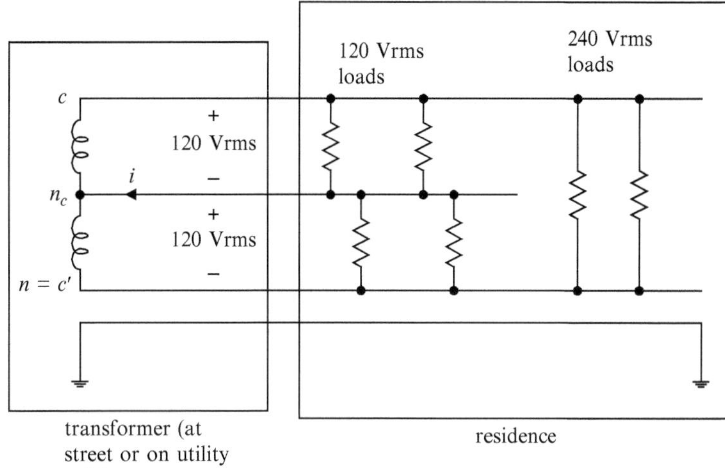

Fig. 14.6 Distribution of electric power to a residence

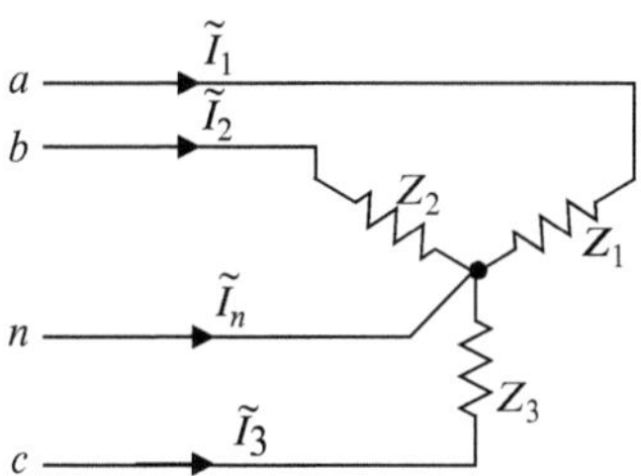

Fig. 14.7 Y-connected three-phase load

Figure 14.7 shows a Y-connected three-phase load. A **balanced** three-phase Y load is one for which $Z_1 = Z_2 = Z_3 = Z$. From (14.7), the neutral current in a balanced Y load is given by

$$\begin{aligned}\tilde{I}_n &= -(\tilde{I}_1 + \tilde{I}_2 + \tilde{I}_3)\\ &= -\left(\frac{\tilde{V}_{an} + \tilde{V}_{bn} + \tilde{V}_{cn}}{Z}\right) = 0.\end{aligned} \tag{14.8}$$

Consequently, only three wires are needed to provide 3-phase power to a balanced load. In practice, a fourth wire is used inside industrial plants, but it is smaller than the other three. Generally, power transmitted over long distances is distributed at the destination to a great many loads. Although no one load is balanced at all times, the large number of loads served by a long-distance line can be very nearly balanced, on average. Thus in long-distance, high-power transmission, only three wires are used. The earth is used as the neutral. The neutral current associated with long-distance transmission is quite small, and little would be gained by using a fourth wire.

In this chapter, we consider only balanced sources and balanced loads. In practical terms, this means we are considering effective loads on feeder lines to relatively large numbers of users or to very large manufacturing plants. It would be unusual for a small plant or a small subdivision to present a balanced load to the feeder line. If you take a subsequent course in power systems, you will learn about effects associated with unbalanced loads.

A three-phase circuit comprising a balanced source and a balanced load is called a **balanced three-phase circuit**, or in this chapter, a **balanced circuit**.

Figure 14.8 shows balanced Y and Δ loads. The currents $\tilde{I}_a$, $\tilde{I}_b$, $\tilde{I}_c$ entering the load terminals are called **line currents**. The voltages $\tilde{V}_{ab}$, $\tilde{V}_{bc}$, $\tilde{V}_{ca}$ appearing across the load terminals are called **line voltages**. Currents in the branches of the load (e.g., $\tilde{I}_1$ in the Δ load shown in Fig. 14.8) are called **branch currents**. Voltages across the branches of the load are called **branch voltages**.[1]

Phase voltages and currents are associated with a source, branch voltages and currents are associated with a load, and line voltages and currents are associated with the lines (conductors) connecting a source to a load. If the resistance of a transmission line is

[1]In some books, load branch voltages are called phase voltages and load branch currents are called phase currents. To avoid confusion, we use the terms phase voltages and phase currents only in connection with three-phase sources.

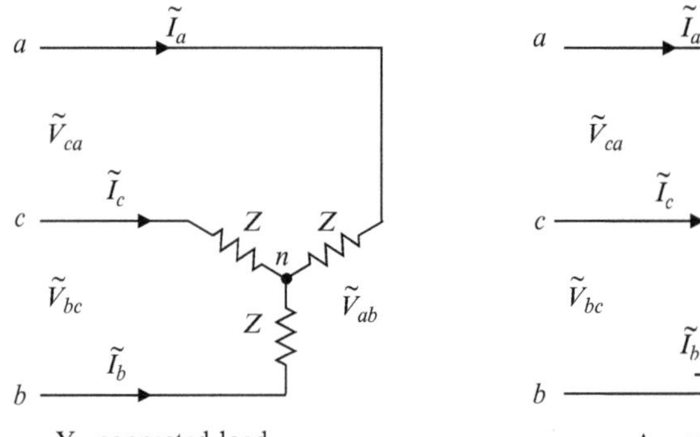

Fig. 14.8 Balanced Y- and Δ-connected loads

significant, the line voltages at the source end and at the load end are different because of voltage drop along the line.

Refer again to Fig. 14.8. In a Y load, the branch currents equal the line currents, but the branch voltages are different from the line voltages. In a Y load, the branch voltages are $\tilde{V}_{an}$, $\tilde{V}_{bn}$, and $\tilde{V}_{cn}$ and the line voltages are $\tilde{V}_{ac}$, $\tilde{V}_{ab}$, and $\tilde{V}_{bc}$. In a Y source, the phase currents equal the line currents but the phase voltages are different from the line voltages. In a balanced Y load (or source), there is no need for a neutral because the neutral current equals zero.

In a Δ load, the branch voltages equal the line voltages, but the branch currents are different from the line currents. In a Δ source, the phase voltages equal the line voltages but the phase currents are different from the line currents. There is no neutral in a Δ load (or source).

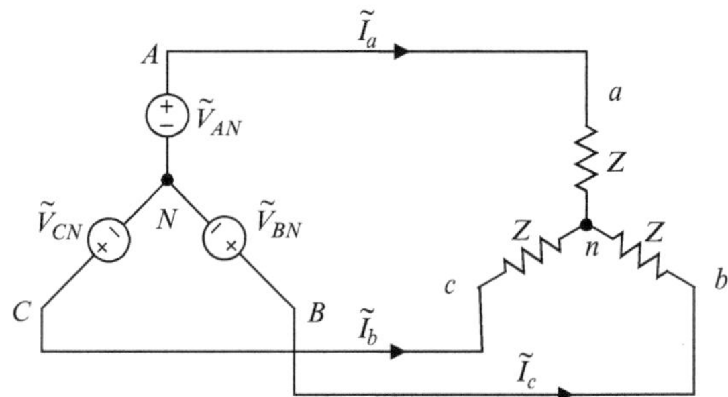

Fig. 14.9 Circuit of Example 14.1

Example 14.1. Refer to Fig. 14.9. The source and load are balanced. (a) Express each line voltage in terms of source phase voltages. Let V_p denote the magnitude of the source phase voltage and use $\tilde{V}_{AN}$ as the phase reference; i.e., $\tilde{V}_{AN} = V_p\angle 0°$. Draw a phasor diagram for the line voltages.

(b) Express the load branch voltages in terms of the source phase voltages and draw a phasor diagram for the branch voltages. Use the notation described in part (a).

Solution:

(a) The source and load are balanced, so the neutral current equals zero and there is no need for a neutral–neutral (N-n) connection. The line voltages are

$$\begin{aligned}\tilde{V}_{ca} &= \tilde{V}_{CN} - \tilde{V}_{AN} = V_p\angle 120° - V_p\angle 0 \\ &= \sqrt{3}\,V_p\,\angle 150°.\end{aligned} \tag{14.9}$$

Similarly,

$$\begin{aligned}\tilde{V}_{ab} &= \tilde{V}_{AN} - \tilde{V}_{BN} = V_p\angle 0 - V_p\angle -120° \\ &= \sqrt{3}V_p\angle 30°, \\ \tilde{V}_{bc} &= \tilde{V}_{BN} - \tilde{V}_{CN} = V_p\angle -120° \\ &\quad - V_p\angle 120° = \sqrt{3}V_p\angle -90°.\end{aligned} \tag{14.10}$$

Figure 14.10 shows the phasor diagram for the line voltages.

(b) Because the load and source are balanced, and because like terminals are connected, the source and load neutrals are at the same potential. Thus the branch voltages are

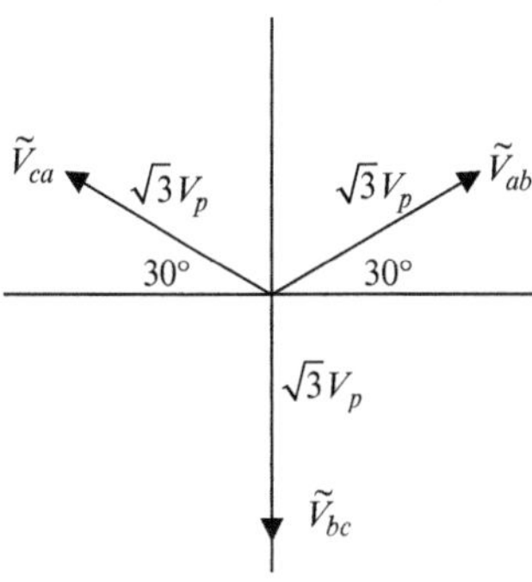

Fig. 14.10 Phasor diagram for the line voltages in Example 14.1

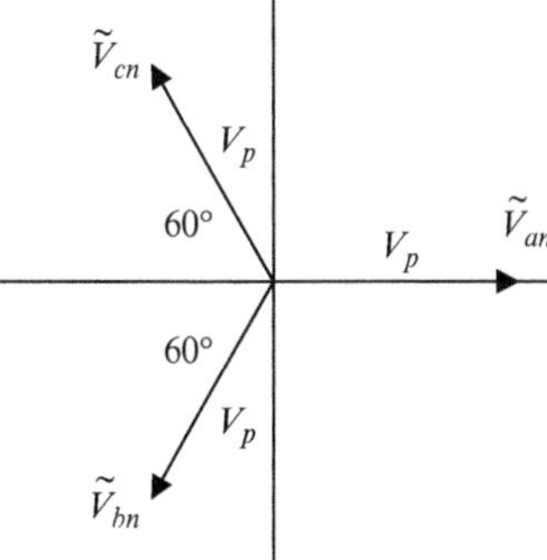

Fig. 14.11 Phasor diagram for the branch voltages in Example 14.1

$$\begin{aligned}\tilde{V}_{an} &= \tilde{V}_{AN} = V_p\,\angle 0,\\ \tilde{V}_{bn} &= \tilde{V}_{BN} = V_p\,\angle -120^\circ,\\ \tilde{V}_{cn} &= \tilde{V}_{CN} = V_p\,\angle 120^\circ.\end{aligned} \qquad (14.11)$$

Figure 14.11 shows the phasor diagram for the branch voltages.

Exercise 14.2. If, in Fig. 14.9, a conductor is connected from the source neutral N to the load neutral n, will there be a non-zero current in the conductor?

From Example 14.1, the magnitudes of the line and branch currents and voltages in a balanced Y load driven by a balanced source are related as

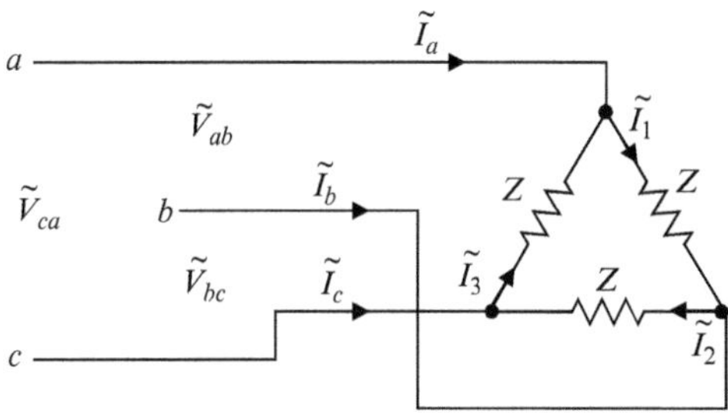

Fig. 14.12 Circuit considered in Example 14.2

$$\left|\tilde{I}_{line}\right| = \left|\tilde{I}_{branch}\right|, \quad \left|\tilde{V}_{line}\right| = \sqrt{3}\left|\tilde{V}_{branch}\right|. \qquad (14.12)$$

A typical three-phase circuit problem is that of finding branch currents, branch voltages, and branch impedances for a load, given one line voltage and one line current. In solving such problems, we make use of the facts that, in a balanced circuit,

(1) The three line voltages have the same magnitude and are 120° apart.
(2) The three line currents have the same magnitude and are 120° apart.

For example, if we take the line voltage $\tilde{V}_{ab}$ as the phase reference in an *abc* phase sequence, then $\tilde{V}_{ab} = V_0\angle 0$, $\tilde{V}_{bc} = V_0\angle -120^\circ$, $\tilde{V}_{ca} = V_0\angle 120^\circ$, where the value of the rms amplitude V_0 is determined by the source. These facts in hand, finding branch currents and voltages is a matter of applying Kirchhoff's laws.

Example 14.2. Refer to Fig. 14.12. The line voltage and line current at one terminal pair of a balanced Δ connected three-phase load are given by

$$\begin{aligned} v_{ab}(t) &= \sqrt{2}V_0\cos(\omega t), V_0 = 440\text{V},\\ i_a(t) &= \sqrt{2}I_0\cos(\omega t - 20^\circ), I_0 = 10\text{A}.\end{aligned} \qquad (14.13)$$

Assume the phase sequence is *abc*. Find (a) the branch impedance and (b) the branch currents.

Solution: We are given $\tilde{V}_{ab} = V_0\angle 0$, meaning that $\tilde{V}_{ab}$ is the phase reference for voltages and currents in the circuit. It follows that $\tilde{V}_{bc} = V_0\angle -120^\circ$ and $\tilde{V}_{ca} = V_0\angle 120^\circ$. We also are given $\tilde{I}_a = I_0\angle -20°$. It follows that $\tilde{I}_b = I_0\angle -140°$ and $\tilde{I}_c = I_0\angle 100°$.

(a) By Kirchhoff's current law,

$$-\tilde{I}_a + \frac{\tilde{V}_{ab}}{Z} + \frac{\tilde{V}_{ac}}{Z} = -\tilde{I}_a + \frac{\tilde{V}_{ab}}{Z} - \frac{\tilde{V}_{ca}}{Z} = 0, \tag{14.14}$$

which yields

$$Z = \frac{\tilde{V}_{ab} - \tilde{V}_{ca}}{\tilde{I}_a} = \frac{V_0\angle 0 - V_0\angle 120°}{\tilde{I}_0\angle -20°} = 75.1 - j\,13.2 = 76.2\angle -10°\,\Omega. \tag{14.15}$$

(b) The branch currents are (see Fig. 14.12)

$$\tilde{I}_1 = \frac{\tilde{V}_{ab}}{Z} = 5.77\angle 10°\text{ A},$$
$$\tilde{I}_2 = \frac{\tilde{V}_{bc}}{Z} = 5.77\angle -110°\text{ A},$$
$$\tilde{I}_3 = \frac{\tilde{V}_{ca}}{Z} = 5.77\angle 130°\text{ A}. \tag{14.16}$$

Exercise 14.3. In Example 14.2, what is $\tilde{I}_a + \tilde{I}_b + \tilde{I}_c$? What is $\tilde{V}_{ab} + \tilde{V}_{bc} + \tilde{V}_{ca}$?

14.5 Balanced Y–Δ and Δ–Y Load Transformations

The relations for transforming a balanced Y to an equivalent balanced Δ are obtained by requiring that impedances seen at corresponding terminal pairs are equal. Thus, with reference to Fig. 14.13, we require

$$Z_{ab} = Z_{AB} \Rightarrow 2Z_Y = \frac{Z_\Delta(2Z_\Delta)}{Z_\Delta + 2Z_\Delta} = \frac{2}{3}Z_\Delta$$

which yields

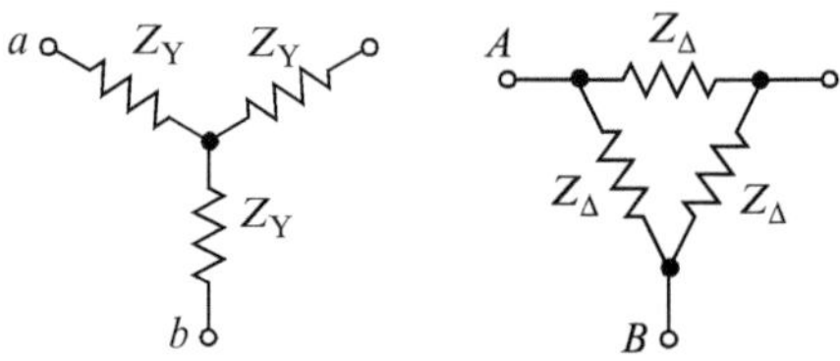

Fig. 14.13 Y and Δ circuits

$$Z_\Delta = 3\,Z_Y. \tag{14.17}$$

To transform a balanced Y to an equivalent balanced Δ, multiply each impedance by three. To transform from a balanced Δ to an equivalent balanced Y, divide each impedance by three. Balanced loads satisfying (14.17) are equivalent at their terminals (cannot be distinguished by any external measurement). *Equivalent balanced loads appear the same to a balanced source; in particular, they draw the same current and dissipate the same power.* Externally equivalent Y and Δ loads are *not* internally equivalent; e.g., a branch current in a Y load is not equal to the corresponding (or any other) branch current in the equivalent Δ load.

Example 14.3. Transform the Y load in Example 14.1 to an equivalent Δ load. Then:

(a) Express the branch voltages in the equivalent Δ load in terms of the source phase voltages. Use the assumptions and notation described in Example 14.1.
(b) Show that the branch currents (phasors) are 120° apart.
(c) Show that the magnitude of each branch current equals the magnitude of the line current, divided by $\sqrt{3}$. (The line currents in a balanced-source – balanced load configuration all have the same magnitude, as do the branch currents.)

Solution: Figure 14.14 shows the equivalent Δ load, obtained using (14.17).

(a) The branch voltages equal the line voltages. From Example 14.1, we have

$$\tilde{V}_{ca} = \tilde{V}_{CA} = \tilde{V}_{CN} - \tilde{V}_{AN} = V_p\angle 120^\circ - V_p\angle 0 = \sqrt{3}V_p\angle 150^\circ,$$
$$\tilde{V}_{ab} = \tilde{V}_{AN} - \tilde{V}_{BN} = V_p\angle 0 - V_p\angle -120^\circ = \sqrt{3}V_p\angle 30^\circ,$$
$$\tilde{V}_{bc} = \tilde{V}_{BN} - \tilde{V}_{CN} = V_p\angle -120^\circ - V_p\angle 120^\circ = \sqrt{3}V_p\angle -90^\circ. \quad (14.18)$$

(b) Each branch current equals the corresponding branch voltage divided by the branch impedance:

$$\tilde{I}_{ab} = \frac{\tilde{V}_{ab}}{3Z} = \frac{\sqrt{3}V_p\angle 30^\circ}{3|Z|\angle\theta} = \frac{\sqrt{3}V_p}{3|Z|}\angle(30^\circ - \theta),$$
$$\tilde{I}_{bc} = \frac{\tilde{V}_{bc}}{3Z} = \frac{\sqrt{3}V_p\angle -90^\circ}{3|Z|\angle\theta} = \frac{\sqrt{3}V_p}{3|Z|}\angle(-90^\circ - \theta),$$
$$\tilde{I}_{ca} = \frac{\tilde{V}_{ca}}{3Z} = \frac{\sqrt{3}V_p\angle 150^\circ}{3|Z|\angle\theta} = \frac{\sqrt{3}V_p}{3|Z|}\angle(150^\circ - \theta). \quad (14.19)$$

The branch currents are 120° apart.

(c) By Kirchhoff's current law,

$$\tilde{I}_a = \tilde{I}_{ab} - \tilde{I}_{ca} = \frac{\sqrt{3}V_p}{3|Z|}\angle(30^\circ - \theta) - \frac{\sqrt{3}V_p}{3|Z|}\angle(150^\circ - \theta)$$
$$= \left(\frac{\sqrt{3}V_p}{3|Z|}\angle -\theta\right)(1\angle 30^\circ - 1\angle 150^\circ)$$
$$= \left(\frac{\sqrt{3}V_p}{3|Z|}\angle -\theta\right)\left[\left(\frac{\sqrt{3}}{2} + j\frac{1}{2}\right) - \left(-\frac{\sqrt{3}}{2} + j\frac{1}{2}\right)\right]$$
$$\left(\frac{\sqrt{3}V_p}{3|Z|}\angle -\theta\right)\sqrt{3}$$
$$= \frac{V_p}{|Z|}\angle -\theta. \quad (14.20)$$

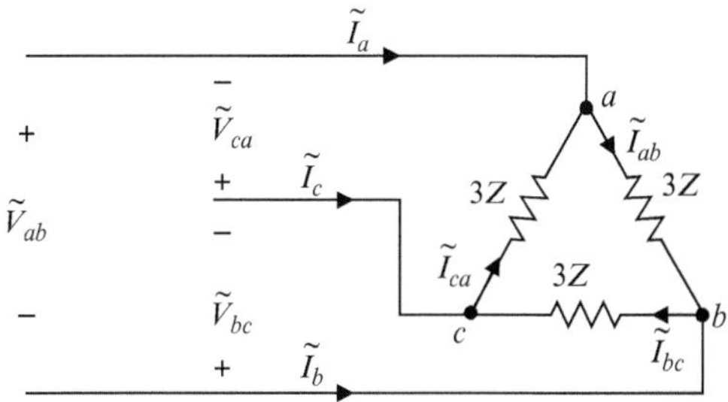

Fig. 14.14 Load considered in Example 14.3

The ratio of the magnitude of the line current to that of the branch current $\tilde{I}_a$ is

$$\frac{|\tilde{I}_a|}{|\tilde{I}_{ab}|} = \frac{|(V_p/Z)\angle -\theta|}{|(\sqrt{3}V_p)/(3Z)\angle(30^\circ - \theta)|} = \frac{|V_p/Z|}{|(\sqrt{3}V_p)/(3Z)|} = \sqrt{3}. \quad (14.21)$$

It follows that

$$|\tilde{I}_{ab}| = \frac{|\tilde{I}_a|}{\sqrt{3}}. \quad (14.22)$$

By symmetry, the same ratio holds for the other line and branch currents.

From Example 14.3, we have that for a *balanced Δ load* driven by a balanced source, the magnitudes of the line and branch currents and voltages are related as

$$|\tilde{I}_{line}| = \sqrt{3}|\tilde{I}_{branch}|, \quad |\tilde{V}_{line}| = |\tilde{V}_{branch}|, \quad (14.23)$$

whereas for a *balanced Y load* driven by a balanced source (see Example 14.1),

$$|\tilde{I}_{line}| = |\tilde{I}_{branch}|, \quad |\tilde{V}_{line}| = \sqrt{3}|\tilde{V}_{branch}|. \quad (14.24)$$

These relations are useful in power calculations.

14.6 Power Calculations for Balanced Three-Phase Loads

By symmetry, the total power delivered by a balanced source and dissipated in a balanced load is equally divided among the three phases of the source or the three branches of the load. The total complex power delivered to a balanced load by a balanced source is three times the complex power delivered by any one phase of the source. The total complex power dissipated in any one branch of a balanced load driven by a balanced source is three times the complex power dissipated in any one branch. The *magnitudes* of the currents (voltages) are the same for each phase of the source or each branch of the load. The phase angles of the currents or voltages are in general different. For a balanced load driven by a balanced source, the branch currents are 120° apart, as are the branch voltages.

We can obtain a number of expressions for complex power dissipated by using (14.23) and (14.24) in the definition of complex power:

$$S_Y = S_\Delta = 3\,\tilde{V}_{branch}\,\tilde{I}_{branch}{}^*. \tag{14.25}$$

Equation (14.25) can be written

$$S_Y = S_\Delta = 3\,\left|\tilde{I}_{branch}\right|^2 Z = 3\frac{\left|\tilde{V}_{branch}\right|^2}{Z^*}, \tag{14.26}$$

where Z is the branch impedance of the load. Using (14.23) in (14.26) yields, for a Δ load

$$S_\Delta = \left|\tilde{I}_{line}\right|^2 Z_\Delta = 3\,\frac{\left|\tilde{V}_{line}\right|^2}{Z_\Delta{}^*}, \tag{14.27}$$

where Z_Δ is the branch impedance of the load. Using (14.24) in (14.26) yields, for a Y load,

$$S_Y = 3\left|\tilde{I}_{line}\right|^2 Z_Y = \frac{\left|\tilde{V}_{line}\right|^2}{Z_Y{}^*}, \tag{14.28}$$

where Z_Y is the branch impedance. Comparing (14.27) with (14.28) shows that a balanced Δ load and a balanced Y load dissipate the same complex power if

$$Z_\Delta = 3\,Z_Y. \tag{14.29}$$

in agreement with (14.17).

From (14.25), the apparent power delivered to a balanced three-phase load is given by

$$|S_Y| = |S_\Delta| = 3\left|\tilde{V}_{branch}\right|\left|\tilde{I}_{branch}\right|. \tag{14.30}$$

Using (14.23) in (14.30) gives, for a Δ load,

$$|S_\Delta| = \sqrt{3}\left|\tilde{V}_{line}\right|\left|\tilde{I}_{line}\right|. \tag{14.31}$$

Using (14.24) in (14.30) gives, for a Y load,

$$|S_Y| = \sqrt{3}\left|\tilde{V}_{line}\right|\left|\tilde{I}_{line}\right|. \tag{14.32}$$

Thus, the *apparent power delivered to either a balanced Δ load or a balanced Y load* is given by

$$|S| = \sqrt{3}\left|\tilde{V}_{line}\right|\left|\tilde{I}_{line}\right|. \tag{14.33}$$

From (14.27),

$$\begin{aligned} S_\Delta &= \left|\tilde{I}_{line}\right|^2 Z_\Delta = \left|\tilde{I}_{line}\right|^2 (R_\Delta + jX_\Delta) \\ &= P_\Delta + jQ_\Delta. \end{aligned} \tag{14.34}$$

It follows that the real power dissipated in a balanced Δ load is given by

$$P_\Delta = \left|\tilde{I}_{line}\right|^2 R_\Delta = |S_\Delta|\,\cos(\theta_\Delta) \tag{14.35}$$

where θ_Δ is the angle of the branch impedance. From (14.28)

$$\begin{aligned} S_Y &= 3\left|\tilde{I}_{line}\right|^2 Z_Y = 3\left|\tilde{I}_{line}\right|^2 (R_Y + jX_Y) \\ &= P_Y + jQ_Y. \end{aligned} \tag{14.36}$$

It follows that the real power dissipated in a balanced Y load is given by

$$P_Y = 3\left|\tilde{I}_{line}\right|^2 R_Y = |S_Y|\,\cos(\theta_Y), \tag{14.37}$$

where θ_Y is the angle of the branch impedance. Equations (14.35) and (14.37) show that *the real power dissipated in either a Δ load or a Y load is given by*

$$P = |S|\,\cos(\theta), \tag{14.38}$$

where S is the complex power delivered to the load and θ is the angle of the branch impedance.

Example 14.4. (Continuation of Example 14.2). The line voltage and line current at one terminal pair of a balanced Δ connected three-phase load are given by

$$v_{ab}(t)=\sqrt{2}V_0\cos(\omega t),\, V_0=440\,\text{V},$$
$$i_a(t)=\sqrt{2}I_0\cos(\omega t-20^\circ),\, I_0=10\,\text{A}. \quad (14.39)$$

Assume the phase sequence is abc. Find (a) the complex power delivered to each branch, (b) the real power delivered to each branch, and (c) the apparent power delivered to the load.
Solution: (See Fig. 14.15.)

(a) From Example 14.2, the branch currents are

$$\tilde{I}_1=\frac{\tilde{V}_{ab}}{Z}=5.77\angle 10^\circ\text{ A},$$
$$\tilde{I}_2=\frac{\tilde{V}_{bc}}{Z}=5.77\angle -110^\circ\text{ A},$$
$$\tilde{I}_3=\frac{\tilde{V}_{ca}}{Z}=5.77\angle 130^\circ\text{ A}. \quad (14.40)$$

The branch voltages are the line voltages. The load is balanced, so the complex powers delivered to the branches are equal:

$$S_1=S_2=S_3=\tilde{V}_{ab}\tilde{I}_1^* = 2500-j441\text{ VA}. \quad (14.41)$$

(b) The real power dissipated in each branch is

$$P_1=P_2=P_3=2.50\,\text{kW}. \quad (14.42)$$

(c) The complex power delivered to the load is

$$S=3S_1=7500-j1323\,\text{VA}, \quad (14.43)$$

so the apparent power delivered to the load is

$$|S|=\sqrt{(7,500)^2+(1,323)^2} = 7.62\text{ kVA}. \quad (14.44)$$

An alternative is to calculate the apparent power using (14.33):

$$|S|=\sqrt{3}\,|\tilde{V}_{ab}|\,|\tilde{I}_a|=\sqrt{3}\,V_0I_0 = \sqrt{3}\,V_0I_0=\sqrt{3}\,(440)(10) = 7.62\text{ kVA}, \quad (14.45)$$

as before.

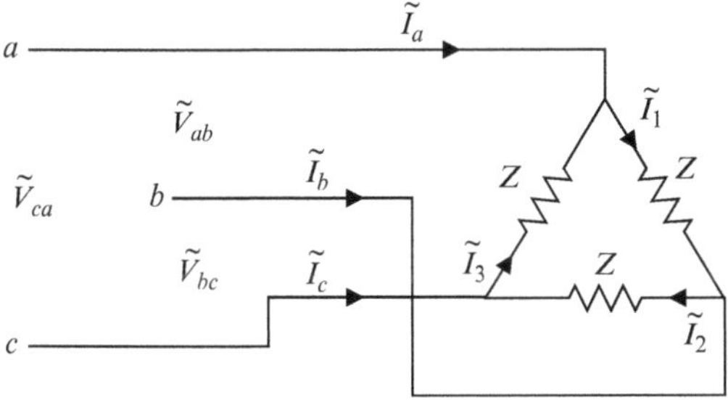

Fig. 14.15 See Example 14.4

14.7 Power-Factor Correction for Three-Phase Loads

Only certain standard line voltages are available to consumers of electric power; e.g., the line voltages available to residences and most offices (in the United States) are 120 Vrms and 240 Vrms,[2] and line voltages available to industrial users are those plus other standard values, ranging upward from 480 Vrms to about 4 kVrms or more, depending upon the size and needs of the plant. No matter what the line voltage, an industrial consumer, in particular, requires a certain real power, which may vary somewhat with time of day, day of week, and season. The supplier (the electric utility) must be able to supply that real power at the user's line voltage.

Real power, not apparent power, corresponds to work being done. But at a fixed line voltage, apparent power determines the current required to do that work and line losses are proportional to the square of line current. If the reactive component of apparent power

[2]Standard line voltages vary somewhat from one region to another and even within regions; for example, 110 Vrms, 117 Vrms, and 120 Vrms all are standard wall-outlet voltages.

equals zero (if the power factor equals unity), then real power equals apparent power and the line current and associated line losses are minimized. Most large industrial loads are primarily inductive (electric motors), so apparent power delivered to an industrial load can be much larger than real power and the line current required to deliver the apparent power can be much larger than would be necessary if the load were resistive.

Because line loss is power generated but not sold, and because power companies must be able to meet their customers' power demands, power companies wish to keep line currents as low as possible while supplying the real power requirements of their customers. To that end, power companies require large industrial companies to use power-factor correction to keep power factors within certain limits.

From Chapter 13, the magnitude of the capacitive reactive power required to increase the power factor of a single-phase load from pf_1 to pf_2 is given by

$$|Q_C| = P\left(\sqrt{\frac{1}{pf_1^{\,2}} - 1} - \sqrt{\frac{1}{pf_2^{\,2}} - 1}\right), \qquad (14.46)$$

where P is the real power required by the load. This relation applies to each branch of a three-phase load. The frequency of line voltages is fixed at 60 Hz in North America and there is a finite number of line voltages, so power-factor correction capacitors are usually specified in VARs at a particular line voltage; e.g., 50 kVAR at 480 Vrms. Thus (14.46) is usually adequate for correction calculations. The magnitude of the reactive power dissipated is given by $Q_C = \omega|\tilde{V}_L|^2 C$, so if needed, the capacitance required can be calculated using

$$C = \frac{|Q_C|}{\omega\,|\tilde{V}_L|^2}. \qquad (14.47)$$

Example 14.5. A load requiring 2 kW of real power is fed by a 60-Hz, 240-Vrms source. The power factor for the load is $pf_1 = 0.6$. (a) Specify the added reactive power that increases the power factor to $pf_2 = 0.9$. (b) Find the line current before and after correction.

Solution: (a) From (14.46)

$$\begin{aligned}|Q_C| &= P\left[\sqrt{\frac{1}{(pf_1)^2} - 1} - \sqrt{\frac{1}{(pf_2)^2} - 1}\right] \\ &= (2\,\text{kW})\left[\sqrt{\frac{1}{(0.6)^2} - 1} - \sqrt{\frac{1}{(0.9)^2} - 1}\right] \\ &\cong 1.70\ \text{kVAR}.\end{aligned}$$

We would require a capacitor that provides 1.70 kVAR at 240 Vrms. The capacitance of such a capacitor is

$$C = \frac{|Q_C|}{\omega|\tilde{V}_L|^2} = \frac{1.70\,\text{kVAR}}{2\pi(60\,\text{Hz})(240\,\text{V})^2} \cong 78.3\,\mu F.$$

(b) The apparent powers before and after correction are

$$\begin{aligned}|S_1| &= \frac{P}{pf_1} = \frac{2{,}000}{0.6} = 3.33\,\text{kVA}, \\ |S_2| &= \frac{P}{pf_2} = \frac{2{,}000}{0.9} = 2.22\,\text{kVA}.\end{aligned} \qquad (14.48)$$

The corresponding line currents are

$$\begin{aligned}|\tilde{I}_1| &= \frac{|S_1|}{|\tilde{V}_L|} = 13.9\,\text{Arms}, \\ |\tilde{I}_2| &= \frac{|S_2|}{|\tilde{V}_L|} = 9.26\,\text{Arms}.\end{aligned}$$

The line losses are reduced from $P_L = |\tilde{I}_1|^2 R$ to $P'_L = |\tilde{I}_2|^2 R$, where R is the line resistance, or by the factor

$$\left(\frac{|\tilde{I}_2|}{|\tilde{I}_1|}\right)^2 \cong \left(\frac{9.26}{13.9}\right)^2 \cong 0.44,$$

which is significant.

Refer to Fig. 14.16. For a Δ-connected load, the lines connect to the branch ends, so capacitors placed between the lines are in parallel with the branch impedances. That is not the case for a Y-connected load; but a Y-connected load can be transformed to an externally equivalent Δ-connected load having the same power factor, so line-to-line Δ-connected capacitors work in either case. Consequently, we find the values of the corrective capacitors for a three-phase load in the same manner as for a single-phase source and load. Figure 14.17 shows an assortment of three-phase power-factor correction capacitors.

Example 14.6. A balanced 60-Hz Δ source drives a balanced Y load. At the load, the line current $\tilde{I}_a$ and the terminal voltage $\tilde{V}_{ab}$ are

$$\begin{aligned}\tilde{I}_a &= 12\angle 0\,\text{Arms},\\ \tilde{V}_{ab} &= 12\angle 105^\circ\,\text{kVrms}.\end{aligned} \tag{14.49}$$

(a) Find the power factor for the load.
(b) Find the capacitance and reactive power required to increase the power factor to 0.9.
(c) Find the factor by which the line current is reduced by the correction (b).
(d) Find the factor by which the line loss is reduced by the correction (b).

Solution: (a) Refer to Fig. 14.18

For a balanced load, the line currents are 120° apart, so

$$\tilde{I}_a = I_0\angle 0,\ \tilde{I}_b = I_0\angle -120^\circ,$$
$$\tilde{I}_c = I_0\angle 120^\circ;\ I_0 = 12\,\text{Arms}.$$

By Kirchhoff's voltage law, the line voltage $\tilde{V}_{ab}$ is Given by $\tilde{V}_{ab} = \tilde{I}_a Z - \tilde{I}_b Z$, so the line-to-line impedance is

$$Z = \frac{\tilde{V}_{ab}}{\tilde{I}_a - \tilde{I}_b} = \frac{(12\angle 105^\circ)\text{kV}}{(12\angle 0)\text{A} - (12\angle -120^\circ)\text{A}}$$
$$\cong 577\angle 75^\circ\,\Omega \cong (149 + j558)\,\Omega$$

The power factor is

$$pf_1 = \cos(75^\circ) = 0.259.$$

(b) The power required by the load (one branch) is evidently

$$P = \left|\tilde{I}_a\right|^2 \text{Re}(Z) \cong 21.5\,\text{kW}.$$

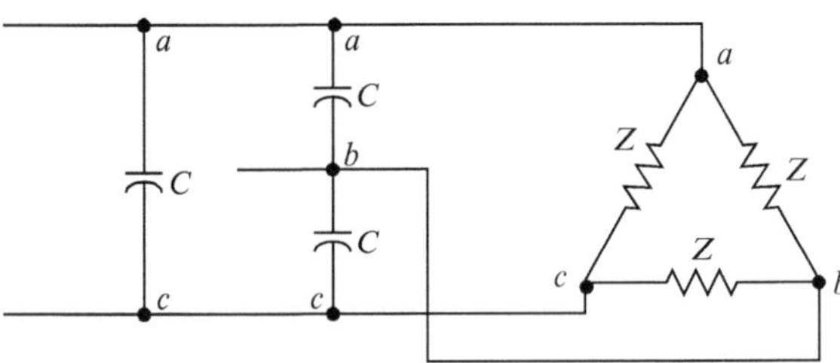

Fig. 14.16 Placement of capacitors for power-factor correction of a three-phase Δ load

Fig. 14.17 Three-phase power-factor correction capacitors. Photograph Courtesy ABB Ltd.

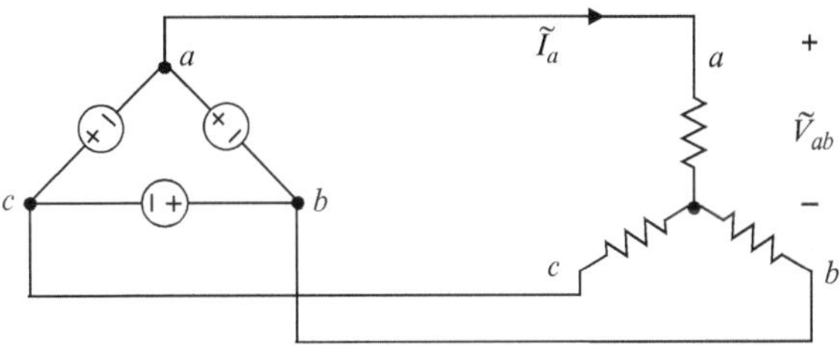

Fig. 14.18 See Example 14.6

The specified power factor is $pf_2 = 0.9$. From (14.46), the reactive power required is

$$|Q_C| = P\left(\sqrt{\frac{1}{pf_1^2} - 1} - \sqrt{\frac{1}{pf_2^2} - 1}\right)$$
$$\cong 69.8 \text{ kVAR}.$$

The frequency is $f = 60\,\text{Hz} \Rightarrow \omega \cong 377\,\text{s}^{-1}$. The capacitance required is

$$C = \frac{|Q_C|}{\omega\left|\tilde{V}_L\right|^2} \cong 1.29\,\mu\text{F}.$$

This is not a large capacitance (numerically), but a 5 μF capacitor that must operate at 12 kV is large physically.

(c)
$$|S| = \frac{P}{pf_2} \cong 23.9\,\text{kVA} = \left|\tilde{V}_{ab}\right|\left|\tilde{I}'_a\right|$$
$$\Rightarrow \left|\tilde{I}'_a\right| = \frac{|S|}{\left|\tilde{V}_{ab}\right|} \cong \frac{23.9\,\text{kVA}}{12\,\text{kV}} \cong 3.44\,\text{A}$$

The line current is reduced by the factor

$$\frac{\left|\tilde{I}'_a\right|}{\left|\tilde{I}_a\right|} \cong \frac{3.44\,\text{A}}{12\,\text{A}} \cong 0.288$$

(d) The line losses are reduced by the factor

$$\frac{\left|\tilde{I}'_a\right|^2 R}{\left|\tilde{I}_a\right|^2 R} \cong \left(\frac{3.44\,\text{A}}{12\,\text{A}}\right)^2 \cong 0.083$$

Residences in the U.S. contain many motors; e.g., air-conditioning compressor and fan motors, washing-machine and dryer motors, furnace air handling units, and ventilating fans. Thus the power factor for a residence is typically (on average) lagging. There are companies that manufacture and sell residential power-factor correction units and there are folks that buy them. But the watt-hour meter monitored by the power company records real power, not apparent power, so if you buy and install a power-factor correction unit, you are helping only the power company, not yourself. Perhaps electric utilities will eventually require residential power-factor correction, but at present it is a poor investment for a residential user.

14.8 Instantaneous Power Delivered to a Balanced Load

An advantage of a three-phase system is that the instantaneous power delivered by a balanced three-phase source to a balanced three-phase load is constant, whereas for a single-phase system instantaneous power pulsates at a rate equal to twice the line frequency, as shown below.

From Chapter 5, the instantaneous power delivered to a load by a single-phase source is given by

$$\begin{aligned} p_1(t) &= vi = VI\cos(\omega t)\cos(\omega t - \theta) \\ &= \frac{VI}{2}\left[\cos(2\,\omega t - \theta) + \cos(\theta)\right], \end{aligned} \tag{14.50}$$

where V is the peak load voltage, I is the peak load current, and θ is the angle of the load impedance. The instantaneous power has a constant (average) component

$$\frac{VI}{2}\cos(\theta)$$

and a time-varying component

$$\frac{VI}{2}\cos(2\,\omega t - \theta),$$

which pulsates at twice the frequency of the applied voltage (or current).

The instantaneous power delivered by a balanced source to a balanced three-phase load is given by

$$\begin{aligned} p_3(t) &= VI\cos(\omega t)\cos(\omega t - \theta) \\ &\quad + VI\cos(\omega t - 120)\cos(\omega t - 120 - \theta) \\ &\quad + VI\cos(\omega t - 240)\cos(\omega t - 240 - \theta) \\ &= \frac{VI}{2}\begin{bmatrix} \cos(2\,\omega t - \theta) + \cos(\theta) \\ + \cos(2\,\omega t - 240 - \theta) + \cos(\theta) \\ + \cos(2\,\omega t - 480 - \theta) + \cos(\theta) \end{bmatrix}. \end{aligned}$$

But

$$\cos(2\,\omega t - \theta) + \cos(2\,\omega t - 240 - \theta) \\ + \cos(2\,\omega t - 480 - \theta) \\ = \cos(2\,\omega t - \theta) + \cos(2\,\omega t - 240 - \theta) \\ + \cos(2\,\omega t - 120 - \theta) = 0,$$

so

$$p_3(t) = \frac{3VI}{2}\cos(\theta), \qquad (14.51)$$

which is constant (no time-varying component). The absence of a time-varying component of power translates into a smoother-running motor and makes three-phase power preferable to single-phase power in applications such as precision metal machining, where pulsating power can cause vibration and tool chatter. This is especially true for machines using variable-frequency drives, which control speed by varying the frequency of the voltage driving the motor. The power-output of a single-phase motor powered by a variable-speed drive has a time-varying component that is equal in amplitude to the constant component and which alternates in sign at twice the drive frequency. The resulting pulsating torque would be intolerable in precision machining and especially so at low frequencies (low speeds).

14.9 Problems

In the problems below, all three-phase loads and sources are balanced. In problem statements, "Y load" means "Y-connected load," "Y source" means "Y-connected source." Similar abbreviated descriptions are used for Δ-connected loads and sources.

P 14.1 Figure P 14.1 shows a balanced residential load supplied by a center-tapped secondary leg of a three-phase transformer. Show that the neutral (center-tap) current equals zero.

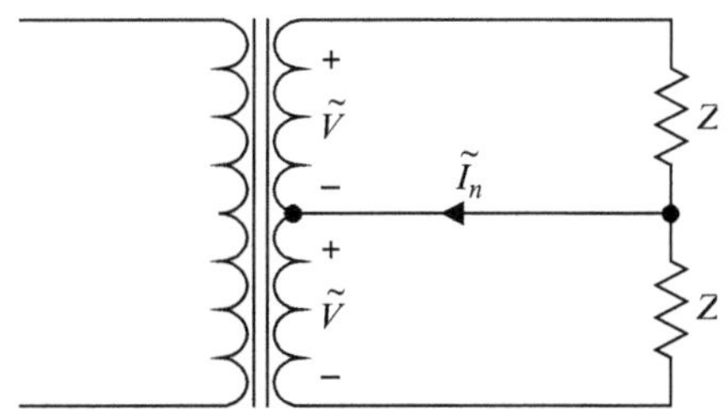

Fig. P 14.1 See Problem P 14.1

P 14.2 In Fig. P 14.2, $\tilde{V}_1 = V_0\angle 0$, $\tilde{V}_2 = V\angle -120°$, $\tilde{V}_3 = V\angle -240°$. Obtain an expression for the current $\tilde{I}$.

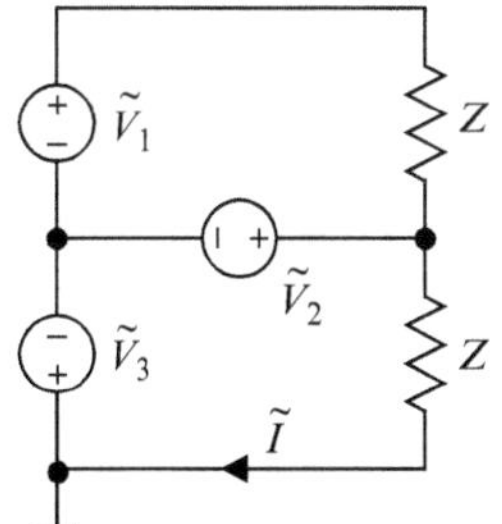

Fig. P 14.2 See Problem P 14.2

P 14.3 A three-phase Δ source drives a three-phase Δ load. The magnitude of the source phase voltage is $V_p = 4\,\text{kV}$. The branch impedance of the load is $Z = 10 + j50\,\Omega$.

(a) Find the line current (magnitude).
(b) Find the complex power and the real power delivered to the load.

P 14.4 Is there such a thing as a three-phase current source? If so, is it Y-connected or Δ-connected, or can it be either?

P 14.5 A three-phase, Δ source drives a three-phase Δ load. The magnitude of the source phase voltage is $V_p = 480\,\text{V}$. The branch impedance of the load is $Z = 5 + j25\,\Omega$. Without any other changes, the source is rewired as a Y source. Does the power delivered to the load change? If so, by what factor?

P 14.6 A three-phase, Δ source drives a three-phase Δ load. The magnitude of the source phase voltage is $V_p = 480\,\text{V}$. The branch impedance of the load is $Z = 5 + j25\,\Omega$. Without any other changes, the load is rewired as a Y load. Does the power delivered to the load change? If so, by what factor?

P 14.7 A three-phase, Y source drives a three-phase Δ load. The magnitude of the source phase voltage is $V_p = 480\,\text{V}$. The branch impedance of the load is $Z = 5 + j25\,\Omega$. Without any other changes, the load

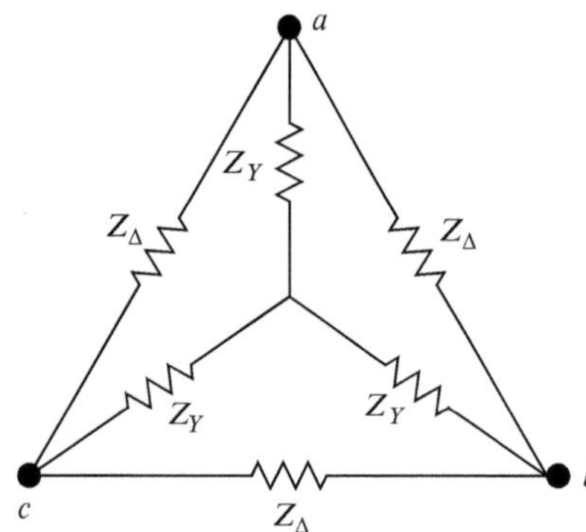

Fig. P 14.3 See Problem P 14.8

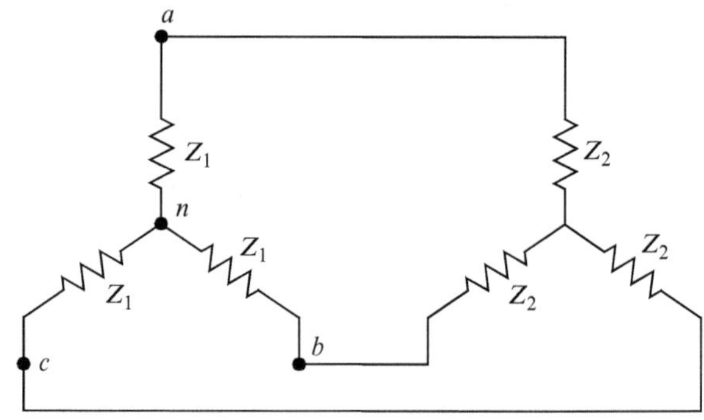

Fig. P 14.4 See Problem P 14.13

is rewired as a Y load. Does the power delivered to the load change? If so, by what factor?

P 14.8 A three-phase Y load is connected to a three-phase Δ load, as shown in Fig. P 14.3, where $Z_\Delta = 3Z_Y$. The combined load is connected to a balanced Y source having phase-a voltage $\tilde{V}_p = V_p\angle 0$. Express the line current and the branch currents in each branch in terms of the source phase voltages and the load impedances Z_Y, Z_Δ.

P 14.9 Repeat Problem P 14.8 for a Δ source.

P 14.10 If possible, obtain a Δ equivalent for the load shown in Fig. P 14.3.

P 14.11 If possible, obtain a Y equivalent for the load shown in Fig. P 14.3.

P 14.12 A Δ load is connected to a Y source. The resistance of each of the lines connecting the source to the load is 2 Ω. The source phase voltage is 10 kV and the load branch impedance is $Z = 400 + j200\,\Omega$. Find: (a) The line voltages at the load end, (b) the branch currents, (c) the real power delivered to the load, and (d) the real power lost in the lines.

P 14.13 Two Y loads are connected as shown in Fig. P 14.4. The combined load is connected at the points a, b, c to a Δ source having phase voltage (magnitude) $|\tilde{V}_p| = 480\,\text{V}$. The branch impedances are $Z_1 = 5 + j10\ \Omega$ and $Z_2 = 2 + j8\ \Omega$. Using phase voltage a as the zero-phase reference, find the line currents and line voltages and the real power delivered to each load.

P 14.14 The current through and voltage across one branch of a Y load are

$$\begin{aligned} i(t) &= 5\cos(\omega t)\ \text{A}, \\ v(t) &= 408\cos(\omega t - 30^\circ)\ \text{V}, \quad f = 60\,\text{Hz}. \end{aligned} \tag{14.52}$$

(a) Find the power factor for the load.
(b) Is the load inductive or capacitive?
(c) Find the line voltage across two of the three lines connecting the load to the source.

P 14.15 The (60-Hz) voltage across branch ab of a Δ load is $240\angle 0$ Vrms. The power factor for the load is 0.8 lagging (inductive) and the magnitude of the load impedance is $|Z| = 15\,\Omega$.

(a) Find the branch current in branch ab.
(b) Find the line current $\tilde{I}_a$.
(c) Find the total power dissipated in the load.
(d) Find the parallel capacitance that increases the power factor to 0.95.
(e) Find the line current for the new power factor.

P 14.16 The power factor for a certain Δ load is 0.7 and is increased to 0.9 by connecting a capacitor having capacitance C in parallel with each branch. The line frequency is 60 Hz and the resistive part of the load impedance is $5\,\Omega$. The line voltage is 400 Vrms.

(a) Find the capacitance C.
(b) Find the branch impedance (including the capacitor) and the branch current $\tilde{I}_a$.
(c) Find the magnitude of the line current.

P 14.17 A three-phase source feeds a 50-kW three-phase load using three transmission lines. The phase voltage for the source is 4 kVrms. The load is approximately 10 miles from the source. The lines are aluminum, having resistivity $\rho = 2.71\,\mu\Omega\,\text{cm}$. The power factor for the load is 0.7. An acceptable line loss is 1.2 kW (total). (a) Calculate the required diameter of the wires if no capacitive correction is employed. (b) Repeat, assuming the power factor is corrected to 0.9.

(c) The mass density of aluminum is approximately 2.7×10^{-3} kg cm^{-3}. What is the mass of the aluminum saved by using power-factor correction and thus the smaller wire?

P 14.18 A small plant has added some additional machinery requiring 12 kW with a lagging power factor of 0.6. Prior to adding the machinery, the plant required 50 kW and had added shunt capacitance to increase the power factor from 0.6 to 0.95. The plant is fed by a 10 kV 60 Hz line. What is the new power factor? What *additional* capacitance is required to maintain a power factor of 0.95?

P 14.19 The machinery in a small woodworking shop uses two 25 hp motors having power factors of 0.6, three 10 hp motors having power factors of 0.65, five 5 hp motors having power factors of 0.68, and eight 2 hp motors having power factors of 0.7. All power factors are lagging. All machines are connected in parallel to a 240 V line. (a) What is the power factor for the shop and what capacitance is required to correct the power factor to 0.95? (b) Because each machine is turned on and off a few times each working day, the utility suggests that the power factor for each machine be corrected individually to 0.95. Determine the capacitance required for each machine. (c) Show how a capacitor should be connected in relation to the power (on–off) switch for a machine and explain why.

Chapter 15
Transfer Functions and Frequency-Domain Analysis

This chapter describes frequency-domain representations of circuits, focusing on circuits having one input port and one output port, where the input port is driven by a source and the output port drives a load. The essential feature of any such circuit is the *relation* the circuit establishes between the available current or voltage from the source and the corresponding current or voltage impressed on the load. In the frequency domain, such relations are expressed using transfer functions, which are generalizations of the transfer ratios described in Chapter 6. The definitions and uses of *transfer functions* are among the most important things electronic circuit and system designers must know.

15.1 Transfer Functions

Refer to Fig. 15.1, where the two-port circuit (in the box) is linear and contains no independent sources, and the voltage and current sources are sinusoidal. *The Thévenin and Norton source models are equivalent*, so the models in Figs. 15.1(a) and (b) are essentially identical.

All dependent phasor currents and phasor voltages in a sinusoidally excited stable linear circuit have the frequency of the excitation, and differ only in amplitude and relative phase from the excitation. The phasor response is proportional to the phasor excitation, the proportionality being expressed as a complex function of the frequency of the excitation. It follows that the load voltage $\tilde{V}_L$ and load current $\tilde{I}_L$ in Fig. 15.1 can be expressed in terms of the *available* source voltage as

$$\tilde{V}_L = H_v \tilde{V}_S, \; \tilde{I}_L = H_y \tilde{V}_S, \tag{15.1}$$

or in terms of the available source current as

$$\tilde{V}_L = H_z \tilde{I}_S, \; \tilde{I}_L = H_i \tilde{I}_S, \tag{15.2}$$

where H_v, H_y, H_z, H_i are complex functions of the frequency of the sinusoidal source, the source and load impedances, and circuit parameters.

The relations (15.1) and (15.2) are the frequency-domain **transfer characteristics** of the two-port circuit in the box. By convention, a transfer characteristic for a circuit relates either the load current $\tilde{I}_L$ or the load voltage $\tilde{V}_L$ to either the available source current $\tilde{I}_S$ or the available source voltage $\tilde{V}_S$.[1] The quantities H_v, H_y, H_z, H_i defined implicitly by (15.1) and (15.2) are the **transfer functions** for the circuit, and are defined explicitly as follows:

Voltage Transfer Function

$$H_v = \frac{\tilde{V}_L}{\tilde{V}_S}, \tag{15.3}$$

Current Transfer Function

$$H_i = \frac{\tilde{I}_L}{\tilde{I}_S} = \frac{\tilde{V}_L/Z_L}{\tilde{V}_S/Z_S} = \frac{Z_S}{Z_L} H_v, \tag{15.4}$$

[1] Recall that the available source voltage is the open-circuit (Thévenin-equivalent) source voltage and the available source current is the short-circuit (Norton-equivalent) source current. Bear in mind that the available voltage (or current) is not necessarily the voltage across (or current entering) the input terminals of the circuit.

T.H. Glisson, *Introduction to Circuit Analysis and Design*,
DOI 10.1007/978-90-481-9443-8_15, © Springer Science+Business Media B.V. 2011

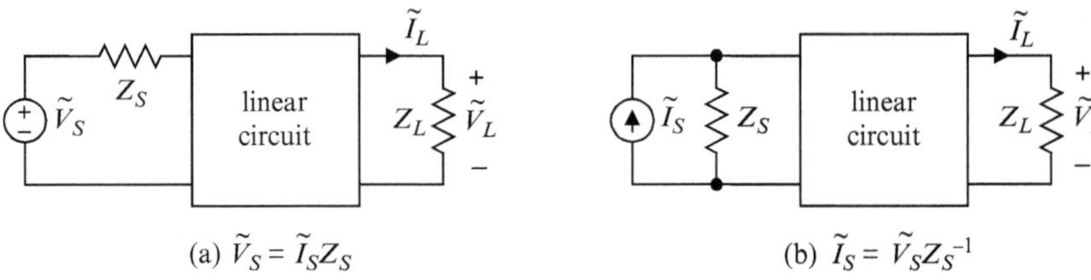

Fig. 15.1 Equivalent source models driving a loaded two-port circuit

Transimpedance

$$H_z = \frac{\tilde{V}_L}{\tilde{I}_S} = \frac{\tilde{V}_L}{\tilde{V}_S/Z_S} = Z_S H_v, \qquad (15.5)$$

Transadmittance

$$H_y = \frac{\tilde{I}_L}{\tilde{V}_S} = \frac{\tilde{V}_L/Z_L}{\tilde{V}_S} = \frac{1}{Z_L} H_v. \qquad (15.6)$$

The transfer functions defined by (15.3–15.6) are functions of the frequency of the source current i_S or source voltage v_S represented by the phasors $\tilde{I}_S$ and $\tilde{V}_S$, respectively. In this regard, we face a little notational dilemma. In practice, frequency is invariably expressed in hertz, which makes it desirable to regard a transfer function as a function of frequency f. On the other hand, as shown in Chapter 18, we achieve a desirable notational consistency if we regard a transfer function as a function of the imaginary variable $j\omega$, where $\omega = 2\pi f$. Thus, whenever it is necessary to make explicit the dependence of a transfer function on frequency, we shall write $H(j\omega)$ or $H(j2\pi f)$, rather than $H(f)$, even though the latter seems (at this point) more natural and reasonable. But unless confusion would otherwise result, we simply omit such indications. For example, we may write

$$H = \frac{K}{1 + j\omega\tau} \quad \text{or} \quad H = \frac{K}{1 + j2\pi f\tau}$$

rather than

$$H(j\omega) = \frac{K}{1 + j\omega\tau} \quad \text{or} \quad H(j2\pi f) = \frac{K}{1 + j2\pi f\tau}.$$

A transfer function is determined with a source (including the source impedance) attached to the input terminals and a load attached to the circuit output terminals. In fact, neither a current transfer function nor a transadmittance can be determined unless a load is attached to the circuit output terminals, because there must be a path for the load current that appears in the definitions of those quantities. It might turn out that a transfer function is independent of the load, but we often don't know that until an analysis is complete.

Example 15.1. Obtain expressions for the voltage transfer function and the current transfer function of the circuit in Fig. 15.2(a). Assume the op amp is ideal.

Solution: We transform the circuit to the frequency domain, as shown in Fig. 15.2(b), where

$$R_1' = R_1 + R_S$$

and

$$Z = \frac{R_2/(j\omega C)}{R_2 + 1/(j\omega C)} = \frac{R_2}{1 + j\omega R_2 C}.$$

Kirchhoff's current law gives

$$-\frac{\tilde{V}_S}{R_1'} - \frac{\tilde{V}_L}{Z} = 0 \Rightarrow \tilde{V}_L = -\frac{Z}{R_1'}\tilde{V}_S.$$

From (15.3), the voltage transfer function is

$$H_v = \frac{\tilde{V}_L}{\tilde{V}_S} = -\frac{Z}{R_1'} = -\frac{R_2/(R_1 + R_S)}{1 + j2\pi f R_2 C}.$$

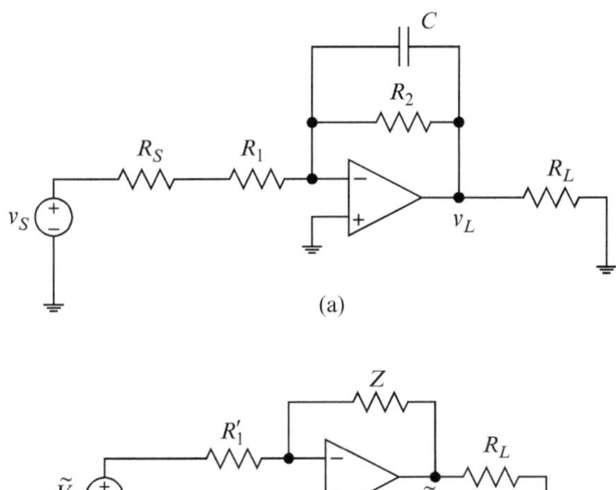

Fig. 15.2 See Example 15.1

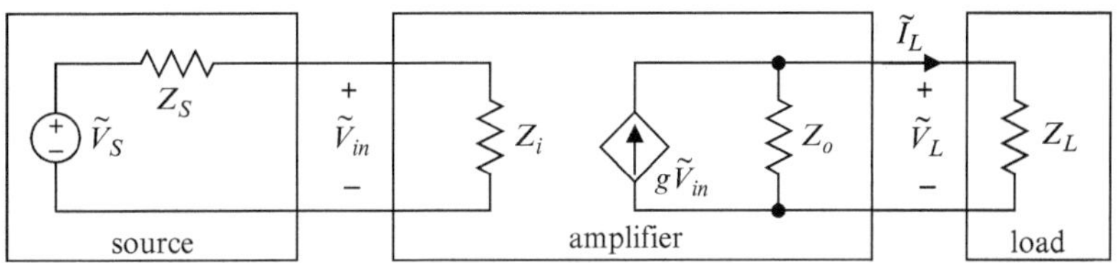

Fig. 15.3 See Example 15.2

From (15.4), the current transfer function is

$$H_i = \frac{\tilde{I}_L}{\tilde{I}_S} = \frac{\tilde{V}_L/R_L}{\tilde{V}_S/R_S} = \frac{R_S}{R_L} H_v$$

$$= -\frac{R_S}{R_L} \frac{R_2}{(R_1 + R_S)(1 + j2\pi f R_2 C)}$$

Example 15.2. (a) Obtain expressions for the voltage transfer function, the current transfer function, the transimpedance, and the transadmittance of the circuit (amplifier) in Fig. 15.3. (b) Let $Z_S = 100\ \Omega$, $Z_i = 10\ \text{k}\Omega$, $Z_o = 10\ \Omega$, $Z_L = 5\ \text{k}\Omega$, and

$$g = \frac{g_0}{1 + jf/f_0};\ g_0 = 0.2\ \text{S}, f_0 = 2\ \text{kHz}.$$

Find the load voltage for

$$v_S = V_0 + V_1 \cos(2\pi f_1 t) + V_2 \cos(2\pi f_2 t + 0.7),$$

where $V_0 = 100$ mV, $V_1 = 800$ mV, $V_2 = 400$ mV, $f_1 = 1$ kHz, $f_2 = 5$ kHz.

Solution: (a) The load and the output subcircuit comprise a current divider, for which

$$\tilde{I}_L = \frac{Z_o g \tilde{V}_{in}}{Z_o + Z_L} \Rightarrow \tilde{V}_L = \tilde{I}_L Z_L = \frac{Z_L Z_o g \tilde{V}_{in}}{Z_o + Z_L}. \tag{15.7}$$

The source and the input subcircuit comprise a voltage divider, for which

$$\tilde{V}_{in} = \frac{Z_i \tilde{V}_S}{Z_i + Z_S}. \quad (15.8)$$

Using the right side of (15.8) to replace $\tilde{V}_{in}$ in the right side of (15.7) gives

$$\tilde{V}_L = \frac{Z_i Z_o Z_L g \tilde{V}_S}{(Z_o + Z_L)(Z_i + Z_S)}.$$

From (15.3), the voltage transfer function is given by

$$H_v = \frac{Z_i Z_o Z_L g}{(Z_o + Z_L)(Z_i + Z_S)}. \quad (15.9)$$

From (15.4–15.6), we obtain the current transfer function

$$H_i = \frac{Z_S}{Z_L} H_v = \frac{Z_i Z_o Z_S g}{(Z_o + Z_L)(Z_i + Z_S)}, \quad (15.10)$$

the transimpedance

$$H_z = Z_S H_v = \frac{Z_S Z_i Z_o Z_L g}{(Z_o + Z_L)(Z_i + Z_S)},$$

and the transadmittance

$$H_y = \frac{1}{Z_L} H_v = \frac{Z_i Z_o g}{(Z_o + Z_L)(Z_i + Z_S)}.$$

(b) We use superposition. From part (a), the voltage transfer function is

$$H_v = \frac{\tilde{V}_L}{\tilde{V}_S} = \frac{Z_i Z_o Z_L g}{(Z_o + Z_L)(Z_i + Z_S)},$$

where (in this example) the impedances are all resistive. Only the intrinsic transconductance g depends upon the frequency of the excitation. For the parameter values given in the problem statement,

$$H_v = \frac{Z_i Z_o Z_L g}{(Z_o + Z_L)(Z_i + Z_S)} = \frac{(10^4\ \Omega)(10\ \Omega)(5{,}000\ \Omega)}{(5{,}010\ \Omega)(10{,}100\ \Omega)} \times \frac{0.2\ \text{S}}{1 + jf/f_0} = \frac{1.976}{1 + jf/f_0};\ f_0 = 2\ \text{kHz}.$$

For each sinusoidal component of the excitation, the associated phasor load voltage is given by

$$\tilde{V}_L = H_v \tilde{V}_S, \quad (15.11)$$

where $\tilde{V}_S$ is the phasor for the component having frequency f. We need only (1) calculate the value of the voltage transfer function H_v for each frequency represented in the excitation, (2) use (15.11) to find the corresponding component of the load voltage, and (3) add the results (superposition). Table 15.1 summarizes the required calculations. The load voltage is given by

$$v_L(t) = 0.198 + 1.414\cos(2\pi f_1 t - 0.464) + 0.294\cos(2\pi f_2 t - 0.490)\ \text{V}$$

with $f_1 = 1$ kHz, $f_2 = 5$ kHz

Exercise 15.1. (a) Obtain the voltage transfer function for the circuit in Fig. 15.4, where v_S is the input and v_L is the output. (b) Let $R_S = 50\ \Omega$, $R_i = 100\ \text{k}\Omega$, $C_1 = 50$ nF, $R_o = 100\ \Omega$, $R_L = 5\ \text{k}\Omega$, $C_2 = 10$ nF, $g = 500$ mS, and

$$v_S = V_0[1 + \cos(2\pi f_1 t) + \cos(2\pi f_2 t + 0.7) + \cos(2\pi f_3 t - 0.7)],$$

Table 15.1 See Example 15.2

f (Hz)	f/f_0	$H_v = 1.976/(1 + jf/f_0)$; $f_0 = 2$ kHz	$\tilde{V}_S$ (V)	$\tilde{V}_L$ (V) $= H_v\tilde{V}_S$
0	0	1.976	0.1	0.198
1000	0.5	$1.768\angle -0.464$	$0.8\angle 0$	$1.414\angle -0.464$
5000	2.5	$0.734\angle -1.190$	$0.4\angle 0.7$	$0.294\angle -0.490$

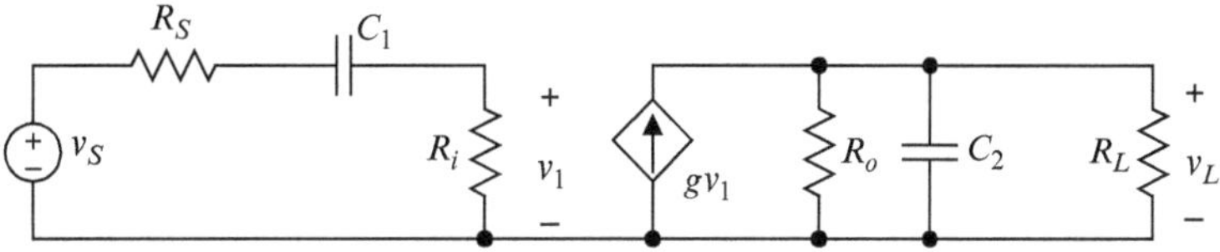

Fig. 15.4 See Exercise 15.1

with $V_0 = 250$ mV, $f_1 = 30$ Hz, $f_2 = 10$ kHz, and $f_3 = 150$ kHz. Find the load voltage $v_L(t)$.

In addition to defining current transfer, transimpedance, and transadmittance (15.3–15.6) relate those functions to voltage transfer. This does not imply that voltage transfer is any more fundamental or important than the other three. Nor does (15.3) imply that the voltage transfer function is independent of the source and load impedances. We could as well have defined current transfer first and then related voltage transfer, transimpedance, and transadmittance to current transfer, in which case we would have found (e.g.)

$$H_v = \frac{Z_L}{Z_S} H_i.$$

Exercise 15.2. Complete the table shown below, where each entry is the factor that produces the transfer function in the row heading from the transfer function in the column heading.

	H_v	H_i	H_z	H_y
H_v	1	Z_L/Z_S	Z_S^{-1}	Z_L
H_i	Z_S/Z_L	1		
H_z			1	
H_y				1

Transfer functions are defined in terms of *available voltage* or *available current* (at the input) and load voltage or current (at the output) because we want a transfer function to describe the associated circuit when loaded and driven by a realistic source. In other words, we want a transfer function to describe how a circuit behaves when actually used. We can illustrate this point with reference to Fig. 15.5. If we defined the voltage transfer function in terms of terminal voltages for an unloaded circuit (voltage divider), as shown in Fig. 15.5(a), the resulting transfer function alone does not tell us how the circuit will perform in use, because the function $R_2/(R_1 + R_2)$ contains no parameters associated with the source or load.

On the other hand, the voltage transfer function in Fig. 15.5(b), defined as load voltage (phasor) divided by available voltage (phasor) does tell us how the circuit will perform in actual use. The transfer function in Fig. 15.5(b) can be used to compute load voltage from available source voltage, whereas that in Fig. 15.5(a) cannot be so used (unless the source resistance equals zero and the load resistance approaches infinity).

In general, *the relation between available voltage (or current) and load voltage (or current) established by a circuit depends not only upon properties of the circuit alone, but also upon properties of the source and load.* Such dependence is not always desirable, and can often be minimized for any particular transfer function, as described in the next section.

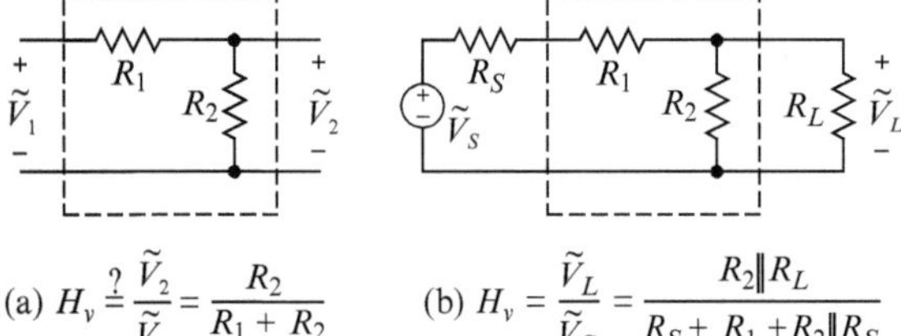

Fig. 15.5 (**a**) Unloaded and (**b**) loaded transfer functions

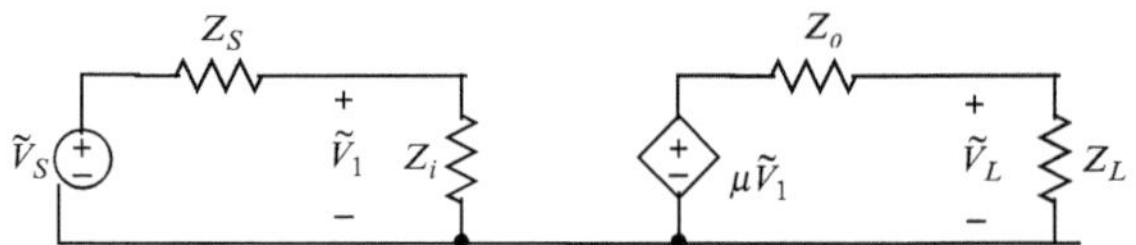

Fig. 15.6 See Example 15.3

15.2 Dependence of a Transfer Function upon Source and Load

Any *engineered*[2] two-port circuit is designed to achieve a specified relation between either the current or voltage available from a source and either the current or voltage delivered to a load. It is almost always desirable that the specified transfer function be at most only weakly dependent upon properties of the source and load, for at least two reasons:

- We would like the circuit to enforce the desired relation for *any* source and load (within reason), because in that case, the circuit can be used in a variety of applications. Also, one charged with designing the circuit can proceed without detailed knowledge of the source and load.
- The terminal characteristics of the source, the load, or both might be nonlinear, which means (loosely) that the source and load impedances depend upon the amplitudes of the input and output. But if the transfer function of interest is independent of the source and load impedances, then the desired linear relation between input and output often can be achieved even if the source and load are nonlinear.

Example 15.3. Refer to Fig. 15.6, What constraints would you impose on the input impedance Z_i and the output impedance Z_o if it is desired that the voltage transfer function be only weakly dependent upon the source impedance Z_S and the load impedance Z_L?

Solution: From Fig. 15.6, we have (by voltage division)

$$\frac{\tilde{V}_1}{\tilde{V}_S} = \frac{Z_i}{Z_i + Z_S}, \frac{\tilde{V}_L}{\tilde{V}_1} = \frac{\mu Z_L}{Z_L + Z_o} \Rightarrow H_v = \frac{\tilde{V}_L}{\tilde{V}_S} = \frac{Z_i}{Z_i + Z_S} \frac{\mu Z_L}{Z_L + Z_o}.$$

If $|Z_i| \gg |Z_S|$ and $|Z_L| \gg |Z_o|$, then

$$H_v \cong \mu,$$

independent of Z_S and Z_L

Exercise 15.3. Based upon results obtained in Example 15.3, we may say that the voltage transfer function of a two-port is independent of the source and load impedances if the input impedance approaches infinity and the output impedance equals zero. In other words, such a two port can achieve a specified voltage transfer function for wide ranges of source and load impedances. Using reasoning similar to that employed in Example 15.3, give conditions on the input impedance and the output impedance of a two port if the current transfer function is to be independent of source and load impedances. Repeat for the transimpedance and transadmittance transfer functions.

For any particular circuit, *at most one of voltage transfer, current transfer, transimpedance, and transadmittance can be independent of the source and load.* For example, from (15.3–15.6), if the voltage transfer function for a circuit is independent of both the source

[2] As opposed to arising as a model for an existing component or circuit.

impedance Z_S and the load impedance Z_L, then the other three transfer functions are not. For example, if the voltage transfer function for a two port is given by

$$H_v \cong gZ_o, \qquad (15.12)$$

as in Example 15.3, then the current transfer function is given by

$$H_i = \frac{Z_S}{Z_L} H_v \cong \frac{Z_S Z_o g}{Z_L}, \qquad (15.13)$$

and is dependent upon both the source impedance Z_S and the load impedance Z_L. Note that in order to determine conditions under which a transfer function is independent of the source and load, we must define (and determine) the transfer function in terms of *available* current or voltage and load current or voltage.

15.3 Gain and Phase Shift

A transfer function is a complex function of frequency. The magnitude of a transfer function is called **gain**, and is denoted by A.[3] The angle of a transfer function is called **phase shift** and is denoted by ϕ:

$$A = gain = |H|; \ \phi = phase\ shift = \measuredangle H. \qquad (15.14)$$

Subscripts are used to indicate which of the four transfer functions is intended; for example, $A_v = |H_v|$ and $\phi_v = \measuredangle H_v$.

It follows from (15.14) that a transfer function can be expressed in polar form as

$$H = A\angle\phi = Ae^{j\phi}. \qquad (15.15)$$

The notational consideration that leads us to regard a transfer function as a function of $j\omega$ does not exist for gain and phase shift, so we may regard gain and phase shift as functions of real frequency f(Hz). If we need to show dependence on frequency, we would write (15.15) as

$$H(j\omega) = H(j2\pi f) = A(f)\angle\phi(f) = A(f)e^{j\phi(f)}.$$

[3] Although the name seems to imply otherwise, gain is not necessarily greater than unity at any frequency.

The inspiration for the terms *gain* and *phase shift* is as follows. Suppose the voltage available from a source is given by

$$v_S(t) = V_S \cos(\omega_0 t) \Rightarrow \tilde{V}_S = V_S\angle 0, \qquad (15.16)$$

and suppose the source drives a linear circuit having voltage transfer function H_v. From (15.3) and (15.15), the output (load) voltage is given by

$$\begin{aligned}\tilde{V}_L &= H_v(j\omega_0)\tilde{V}_S = [A_v(f_0)\angle\phi_v(f_0)][V_S\angle 0]\\ &= A_v(f_0)V_S\angle\phi_v(f_0) \qquad (15.17)\\ &\Rightarrow v_L(t) = A_v(f_0)V_S\cos[2\pi f_0 t + \phi_v(f_0)].\end{aligned}$$

Comparing (15.17) and (15.16) reveals that the effect of the circuit is to scale (increase or decrease) the peak amplitude of the input by the gain $A_v(f_0) = |H_v(j\omega_0)|$ and shift the initial phase of the input by the phase shift $\phi_v(f_0) = \measuredangle H_v(j\omega_0)$, where $\omega_0 = 2\pi f_0$. Although this terminology can be applied to any of the four transfer functions defined above, gain does not have the same intuitive meaning for transimpedance and transadmittance that it has for voltage and current transfer, because current and voltage have different dimensions. We may say that one voltage is larger than another, but it makes no sense to say that a voltage is larger or smaller than a current.

A gain and associated phase shift are frequency-domain generalizations of the gains defined for resistive circuits in Chapter 6. Both gain and phase shift are real functions of frequency (and of circuit, source, and load parameters).

In general, the gains and phase shifts associated with the four transfer functions for a particular circuit exhibit different dependencies on frequency. But if the source and load impedances are resistive (real and independent of frequency), then the four phase shifts are equal and the four gains differ one from another only by a real, frequency-independent scale factor; e.g., if $Z_S = R_S$ and $Z_L = R_L$, then (15.13) becomes

$$H_i = \frac{R_S}{R_L} H_v \cong \frac{R_S Z_o g}{R_L}, \qquad (15.18)$$

so the current transfer function differs from the voltage transfer function by only the real, frequency-independent scale factor R_S/R_L. In other words, if the source and load impedances are resistive, the four transfer functions defined by (15.3–15.6) all exhibit the same frequency dependence.

Example 15.4. In Fig. 15.7, $Z_S = 100\ \Omega$ and $Z_L = R + j\omega L$, with $R = 10\ \Omega$ and $L = 80\ \mu\text{H}$. The available voltage is $v_S = V_S \cos(\omega_1 t)$, with $V_S = 750$ mV and $f_1 = 20$ kHz. The voltage transfer function of the circuit is

$$H_v = \frac{K}{1 + jf/f_0},$$

with K = 100 and $f_0 = 10$ kHz. Use the voltage gain and phase shift to find the load voltage $v_L(t)$.

Solution: The voltage gain and phase shift at the frequency of the source are

$$A_v = \frac{K}{\sqrt{1 + (f_1/f_0)^2}} = \frac{100}{\sqrt{5}} \cong 44.72,$$

$$\phi_v = -\tan^{-1}\left(\frac{f_1}{f_0}\right) = -1.107.$$

From (15.17), the load voltage is given by

$$\begin{aligned} v_L &= A_v V_S \cos[2\pi f_1 t + \phi_v] \\ &= (44.72)(750\ \text{mV})\cos(2\pi f_1 t + 1.107) \\ &= 33.54 \cos(2\pi f_1 t - 1.107)\text{V},\ f_1 = 20\ \text{kHz}. \end{aligned}$$

Exercise 15.4. Find the current transfer function for the circuit of Example 15.4. Replace the Thévenin-model source with the equivalent Norton model and use the current transfer function to find the phasor load current $\tilde{I}_L$. Then find the phasor load voltage from $\tilde{V}_L = Z_L \tilde{I}_L$. Verify that the result equals that obtained in Example 15.4.

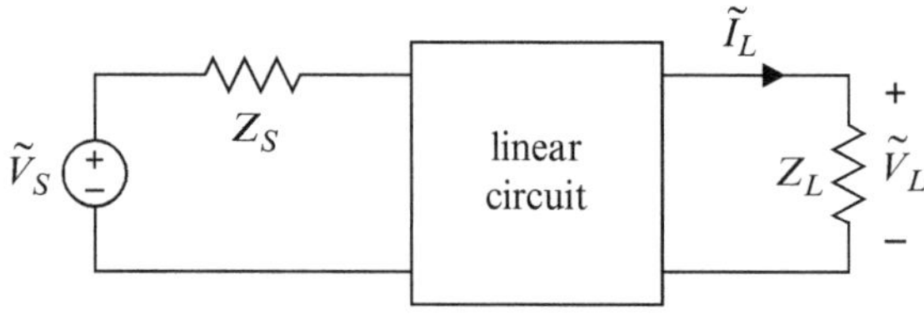

Fig. 15.7 See Example 15.4

15.4 Gain in Decibels (dB)

In applications, the ratio of one power, current, or voltage to another is often of more interest than actual power, current, or voltage. For example, suppose the total average power delivered to (dissipated in) the voice coil of a loudspeaker is given by $P = P_S + P_n$, where P_S is dissipated by a desirable *signal* (representing music) and P_n is dissipated by an undesirable *noise* signal. Whether we hear the noise depends upon the power produced by the signal *relative to that produced by the noise*. In other words, whether the noise is perceptible depends upon the ratio P_S/P_n.

As another example, suppose you are listening to your audio system at a certain loudness (output power) denoted by P_1 and you change the loudness to another loudness denoted by P_2. Whether the change in loudness is perceptible depends upon the ratio P_2/P_1.

Power ratios of interest in applications usually span several orders of magnitude. For example, for a person having normal hearing, the smallest (quietest) discernible sound corresponds to an acoustic power density at the eardrum of about 1 pW m^{-2} and the loudest sound (perhaps accompanied by pain) corresponds to an acoustic power density at the eardrum of about 1 W m^{-2}, so the range of normal human hearing spans about 12 orders of magnitude of power density; i.e. $(1\ \text{W})/(1\ \text{pW}) = 1/10^{-12} = 10^{12}$.

It is difficult to display (graphically) on a linear scale a quantity whose range exceeds more than about two orders of magnitude (e.g., from 1 to 100). Thus in almost all applications, it is impossible to display the full range of powers, currents, or voltages of interest on a linear scale. Additionally, human sensory perception is more nearly logarithmic than linear. For example, *doubling* perceived loudness requires *squaring* the power, current, or voltage delivered to a loudspeaker.

For reasons given above and others, powers, currents, and voltages often are expressed as *relative* values on a logarithmic scale and assigned a dimensionless unit called the **decibel** (dB). A power ratio in dB is defined by

$$\left(\frac{P}{P_0}\right)_{dB} = 10 \log\left(\frac{P}{P_0}\right) \text{ dB}. \tag{15.19}$$

For example, if $P = 50$ W and $P_0 = 2$ W, then

$$\left(\frac{P}{P_0}\right)_{dB} = 10\ \log\left(\frac{50\ \text{W}}{2\ \text{W}}\right) \cong 14\ \text{dB},$$

and we would say that the power P is 14 dB *above* P_0. If $P_0 = 50$ W and $P = 2$ W, then

$$\left(\frac{P}{P_0}\right)_{dB} = 10\ \log\left(\frac{2\ \text{W}}{50\ \text{W}}\right) \cong -14\ \text{dB},$$

and we would say that the power P is 14 dB *below* P_0.

The historical reason for the factor 10 (as opposed to unity) in (15.19) is that a 1 dB change in sound intensity is (approximately) the smallest change in loudness that a person having normal hearing can detect.

In some disciplines, the denominator P_0 in (15.19) is a fixed reference, understood by all who work in that discipline. In such cases, it is conventional to refer to the left side of (15.19) as *the power P expressed in decibels*. For example, in acoustics the standard reference for sound intensity is 1 pW m^{-2}, i.e., 1 pW of acoustic power incident on a 1 m^2 surface, which corresponds approximately to the threshold of normal (average) human hearing.[4] Thus sound intensity in dB at a point is given by (15.19), where $P_0 = 1$ pW and P is the actual acoustic power incident on a 1 m^2 surface at that point. For example, it has been found experimentally that the sound intensity on the stage of a performing rock band can exceed 100 dB. This means that the acoustic power incident on a 1-m^2 surface on the stage with the band is given by

$$100 = 10\ \log\left(\frac{P}{P_0}\right) = 10\ \log\left(\frac{P}{10^{-12}\ \text{W}}\right),$$

which yields

$$P = \left(10^{-12}\ \text{W}\right) \times 10^{10} = 10\ \text{mW}.$$

For perspective, the sound intensity of adult male conversational speech 1 m from a speaker is about 50–70 dB and the sound intensity 10 m from a jet airliner backing away from a terminal can exceed 110 dB. Sustained sound intensity above about 80–90 dB can cause permanent hearing loss. Sound intensities above about 120 dB are painful and can instantly rupture an eardrum.

[4]The threshold of hearing depends upon frequency as well as incident power.

Exercise 15.5. What is the sound intensity in watts per meter corresponding to 130 dB?

Current and voltage ratios also are expressed in dB, and the definitions are based on that for power ((15.19), above). Because power is proportional to the square of current or voltage, current and voltage ratios are expressed in dB as

$$\left(\frac{I_{2rms}}{I_{1rms}}\right)_{dB} = 10\ \log\left(\frac{I_{2rms}{}^2}{I_{1rms}{}^2}\right) = 20\ \log\left(\frac{I_{2rms}}{I_{1rms}}\right) \tag{15.20}$$

and

$$\left(\frac{V_{2rms}}{V_{1rms}}\right)_{dB} = 10\ \log\left(\frac{V_{2rms}{}^2}{V_{1rms}{}^2}\right) = 20\ \log\left(\frac{V_{2rms}}{V_{1rms}}\right). \tag{15.21}$$

The ratio of two currents or voltages in dB is not the same as the ratio of the corresponding powers in dB unless the loads seen by the currents or voltages are identical; that is, if i_1 and i_2 are the currents through loads R_1 and R_2, respectively, then

$$\left(\frac{P_2}{P_1}\right)_{dB} = 10\ \log\left(\frac{I_{2rms}{}^2 R_2}{I_{1rms}{}^2 R_1}\right) = 20\ \log\left(\frac{I_{2rms}}{I_{1rms}}\right) + 10\ \log\left(\frac{R_2}{R_1}\right).$$

The current (or voltage) ratio in decibels equals the power ratio in dB only if $R_1 = R_2$.

If the currents or voltages of interest are the input (i_S or v_S) and output (i_L or v_L) of a linear circuit, then (15.20) and (15.21) can be written

$$\left(\frac{I_{Lrms}}{I_{Srms}}\right)_{dB} = 20\ \log(A_i) \tag{15.22}$$

and

$$\left(\frac{V_{Lrms}}{V_{Srms}}\right)_{dB} = 20\ \log(A_v). \tag{15.23}$$

Equations (15.22) and (15.23) define **current gain in dB** and **voltage gain in dB**. Equations (15.22) and

(15.23) are general. In cases where the input and output are sinusoidal and expressed as phasors, then

$$\left(\frac{|\tilde{I}_L|}{|\tilde{I}_S|}\right)_{dB} = \left(\frac{I_L}{I_S}\right)_{dB} = 20\ \log(A_i) \qquad (15.24)$$

and

$$\left(\frac{|\tilde{V}_L|}{|\tilde{V}_S|}\right)_{dB} = \left(\frac{V_L}{V_S}\right)_{dB} = 20\ \log(A_v), \qquad (15.25)$$

where

$$V_L = |\tilde{V}_L|, I_L = |\tilde{I}_L|, V_S = |\tilde{V}_S|, I_S = |\tilde{I}_S|.$$

Instead of attaching the subscript dB to the symbol for a gain expressed in dB, we attach the unit dB to the numerical value (if given). For example, if we write $A_v = 200$, we mean actual voltage gain (not dB). If we wish to express that same gain in dB, we write $A_v = 46$ dB. *In practice, gain is almost always expressed in dB.*

Example 15.5. The voltage gain of a certain circuit at a certain frequency is 400. What is the gain in dB?

Solution: From (15.23),

$$A_v = 20\ \log(400) = 52\ \text{dB}.$$

Exercise 15.6. For $f = 20$ kHz, express the voltage gain and the current gain of the circuit considered in Example 15.4 in dB.

Example 15.6. The voltage gain of a certain circuit at a certain frequency is 55 dB. What is the actual voltage gain (not dB) at that frequency?

Solution: From (15.23) and the data given,

$$A_v = 20\ \log(A_v) = 55\ \text{dB} \Rightarrow A_v = 10^{55/20} = 562.$$

Not only gains, but also frequencies of interest usually span several orders of magnitude. For example, normal adult human hearing extends from about 20 Hz to about 15 kHz, and spans almost three orders of magnitude. As another example, frequencies of interest in an analog (not HDTV) television signal extend from dc to about 5 MHz. Consequently, it is necessary to somehow compress frequency when plotting gain versus frequency. The compression is ordinarily achieved by using a logarithmic scale on the frequency axis, as illustrated by the next example.

Example 15.7. The voltage transfer function for a certain circuit is

$$H_v = \frac{100}{1 + jf/f_0}, \ f_0 = 100\ \text{Hz}.$$

Plot the voltage gain in dB versus frequency for $1\ \text{Hz} \le f \le 10$ kHz. Use a logarithmic scale for frequency.

Solution: The voltage gain is given by

$$A_v = |H_v| = \left|\frac{100}{1 + jf/f_0}\right| = \frac{100}{\sqrt{1 + (f/f_0)^2}},$$

$$f_0 = 100\ \text{Hz}$$

and in dB by

$$\begin{aligned} A_v &= 20\ \log\left(\frac{100}{\sqrt{1 + (f/f_0)^2}}\right) \\ &= 20\ \log(100) - 20\ \log\left(\sqrt{1 + (f/f_0)^2}\right) \\ &= 40 - 20\ \log\left(\sqrt{1 + (f/f_0)^2}\right), f_0 = 100\ \text{Hz}. \end{aligned}$$

The gain in dB is a reasonably smooth function of frequency, so only a few points are needed. Calculating a few values yields the graph shown in Fig. 15.8. For future reference, note that the gain is nearly constant (and equal to 40 dB) for frequencies much smaller than f_0 and is very nearly a straight line sloping downward at 20 dB/decade of frequency for frequencies much larger than f_0.

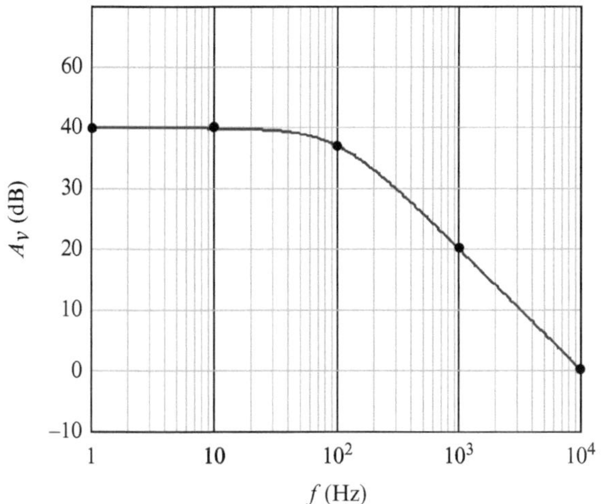

Fig. 15.8 See Example 15.7

Were we to plot the same gain function on a linear scale for 1 Hz $\leq f \leq$ 10 kHz, most of the frequency axis would be devoted to frequencies between 1 and 10 kHz, and the structure of the gain for frequencies below 1 kHz would be compressed and difficult to discern.

Neither transimpedance nor transadmittance can be expressed directly in dB because we cannot take the logarithm of a dimensioned quantity. However, transimpedance and transadmittance can be normalized to a specified impedance or admittance, after which the *normalized* transimpedance or transadmittance can be expressed in dB. For example, we can express normalized transimpedance as

$$H_z' = \frac{H_z}{Z_0}, \tag{15.26}$$

where Z_0 is a particular impedance suggested by a problem at hand. It might be the circuit input impedance, or the circuit output impedance, or the load impedance, or the value of the transimpedance H_z at a particular frequency. It might even be agreed by those involved that $Z_0 = 1\ \Omega$. In any case, the normalized transimpedance can then be expressed in dB as

$$A_z' = 20\ \log|H_z'| = 20\ \log\left|\frac{H_z}{Z_0}\right|\ \text{dB}. \tag{15.27}$$

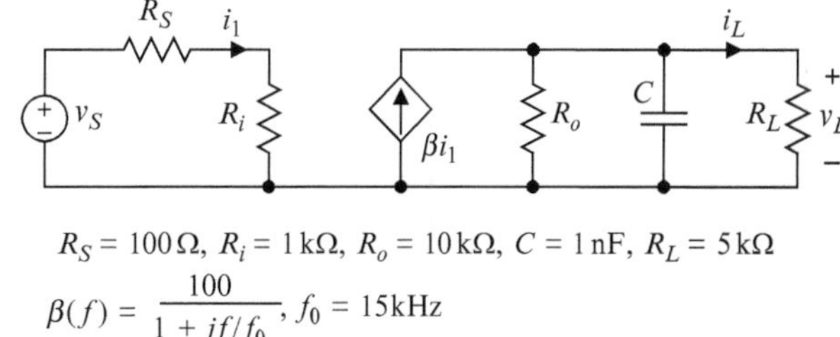

Fig. 15.9 See Example 15.8

Example 15.8. Refer to Fig. 15.9. (a) Obtain expressions for the voltage transfer function, the current transfer function, the transimpedance, and the transadmittance. (b) Express the associated gain for each in dB, where each gain is normalized to its value for $f = 0$, and construct a plot of each gain for 10 Hz $\leq f \leq$ 1 MHz.

Solution: (a) Kirchhoff's current law gives

$$\left(\frac{1}{R_o} + \frac{1}{R_L} + j\omega C\right)\tilde{V}_L = \beta(f)\tilde{I}_1 = \frac{\beta(f)\tilde{V}_S}{R_S + R_i},$$

which yields the voltage transfer function

$$\begin{aligned} H_v = \frac{\tilde{V}_L}{\tilde{V}_S} &= \frac{\beta(f)R_oR_L}{(R_S + R_i)(R_o + R_L + j\omega CR_oR_L)} \\ &= \frac{\beta(f)(R_o\|R_L)}{(R_S + R_i)[1 + j\omega C(R_o\|R_L)]} \\ &= \frac{100(R_o\|R_L)}{(1 + jf/f_0)(R_S + R_i)[1 + j2\pi fC(R_o\|R_L)]} \\ &= \frac{K_v}{(1 + jf/f_0)(1 + jf/f_1)}, \end{aligned}$$

where

$$K_v = \frac{100(R_o\|R_L)}{R_S + R_i} \cong 303,\ f_0 = 15\ \text{kHz},$$

$$f_1 = \frac{1}{2\pi C(R_o\|R_L)} \cong 47.8\ \text{kHz}.$$

The current transfer function is

$$H_i = \frac{\tilde{V}_L/R_L}{\tilde{V}_S/R_S} = \frac{R_S}{R_L}H_v = \frac{K_i}{(1 + jf/f_0)(1 + jf/f_1)},$$

$$K_i = \frac{R_S}{R_L}K_v \cong 6.06,$$

the transimpedance is

$$H_z = \frac{\tilde{V}_L}{\tilde{V}_S/R_S} = R_S H_v = \frac{K_z}{(1+jf/f_0)(1+jf/f_1)},$$
$$K_z = R_S K_v \cong 30.3\ \text{k}\Omega,$$

and the transadmittance is

$$H_y = \frac{\tilde{V}_L/R_L}{\tilde{V}_S} = \frac{1}{R_L} H_v = \frac{K_y}{(1+jf/f_0)(1+jf/f_1)},$$
$$K_y = \frac{K_v}{R_L} \cong 61\ \text{mS}.$$

(b) The values of the transfer functions for $f = 0$ are

$$H_v(0) = K_v,\ H_i(0) = K_i,$$
$$H_z(0) = K_z,\ H_y(0) = K_y.$$

Therefore the four normalized transfer functions are identical; i.e.,

$$\frac{H_v(j\omega)}{H_v(0)} = \frac{H_i(j\omega)}{H_i(0)} = \frac{H_z(j\omega)}{H_z(0)} = \frac{H_y(j\omega)}{H_y(0)}$$
$$= \frac{1}{(1+jf/f_0)(1+jf/f_1)}$$

and the four normalized gains are identical; i.e.,

$$\frac{A_v}{A_v(0)} = \frac{A_i}{A_i(0)} = \frac{A_z}{A_z(0)} = \frac{A_y}{A_y(0)}$$
$$= \frac{1}{\sqrt{1+(f/f_0)^2}\sqrt{1+(f/f_1)^2}}.$$

Thus

$$A = -10\log\left[1+\left(\frac{f}{f_0}\right)^2\right]$$
$$-10\log\left[1+\left(\frac{f}{f_1}\right)^2\right],$$
$$f_0 = 15\ \text{kHz},\ f_1 \cong 47.8\ \text{kHz}.$$

Being identical, the four normalized gains all have the same graph, shown in Fig. 15.10.

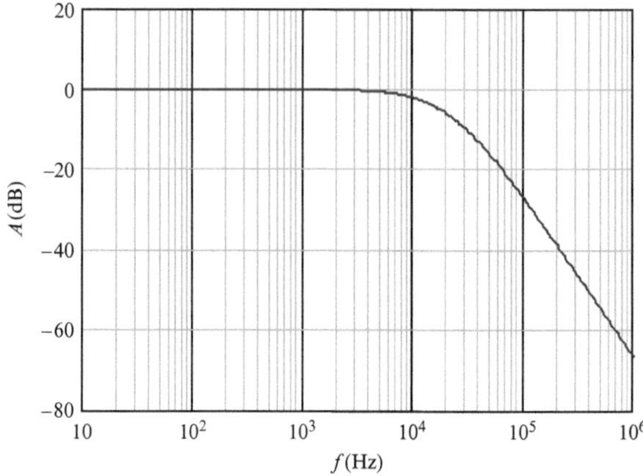

Fig. 15.10 See Example 15.8

Exercise 15.7. At a particular frequency, the transadmittance of a certain circuit, normalized to 1 mS, is 25 dB. Find the magnitude of the transadmittance in mS at that frequency.

15.5 Standard Form of a Transfer Function

The frequency dependence of a transfer function arises from inductance, capacitance or both. Because of the forms of the associated impedances ($j\omega L$ for an inductor and $1/j\omega C$ for a capacitor), a transfer function can be expressed as a ratio of polynomials in $j\omega$ having real coefficients. Thus the general form of a transfer function is

$$H = \frac{b_M(j\omega)^M + b_{M-1}(j\omega)^{M-1} + \cdots + b_1 j\omega + b_0}{a_N(j\omega)^N + a_{N-1}(j\omega)^{N-1} + \cdots + a_1 j\omega + a_0}, \tag{15.28}$$

where the coefficients $\{a_0, a_1, \cdots, a_N\}$ and $\{b_0, b_1, \cdots, b_M\}$ are real and independent of frequency. In electronic circuits, the degree of the numerator rarely exceeds that of the denominator; i.e., $M \leq N$. One reason why is that $M > N$ implies differentiation of an input; e.g., if $M = N + 1$, the right side of (15.28) can be expressed after long division as

$$H = c_M j\omega + \frac{c_{M-1}(j\omega)^{M-1} + c_{M-2}(j\omega)^{M-2} + \cdots + c_1 j\omega + c_0}{a_N(j\omega)^N + a_{N-1}(j\omega)^{N-1} + \cdots + a_1 j\omega + a_0}.$$

The term $c_M j\omega$ gives rise to a term of the form $c_M dv/dt$ or $c_M di/dt$ in the output, which means that a rapidly changing input can give rise to very large output amplitudes. This is generally undesirable for at least two reasons: First, a desirable current or voltage is always accompanied by electrical noise. The noise usually fluctuates rapidly, so differentiation enhances noise. Second, all circuits and circuit components perform as intended over only a limited range of amplitudes, and differentiation can produce amplitudes that exceed the normal operating ranges of the circuit components.

Example 15.9. Obtain the voltage transfer function for the circuit shown in Fig. 15.11. Express the transfer function in the form (15.28) and give the values of M, N, and the coefficients of the polynomials.

Solution: Kirchhoff's current law gives

$$\frac{\tilde{V}_L - \tilde{V}_S}{R_S + R} + j\omega C\tilde{V}_L + \frac{\tilde{V}_L}{R_L} = 0,$$

which yields

$$H_v = \frac{\tilde{V}_L}{\tilde{V}_S} = \frac{R_L}{R_S + R + R_L + R_L C(R + R_S)j\omega}.$$

The transfer function has the form given in (15.28), with $M = 0$, $N = 1$, $b_0 = R_L$, $a_0 = R_S + R + R_L$, and $a_1 = R_L C(R + R_S)$

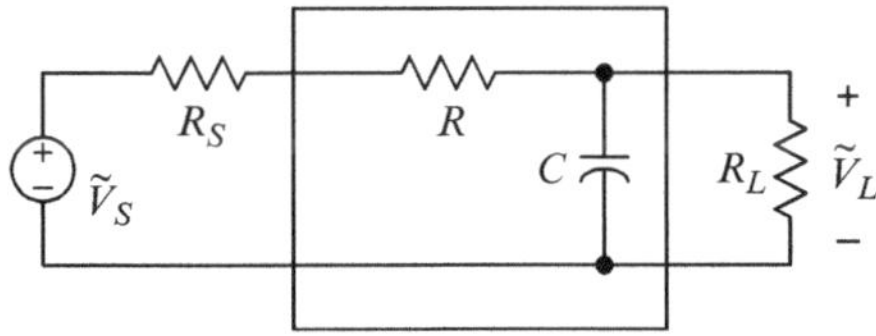

Fig. 15.11 See Example 15.9

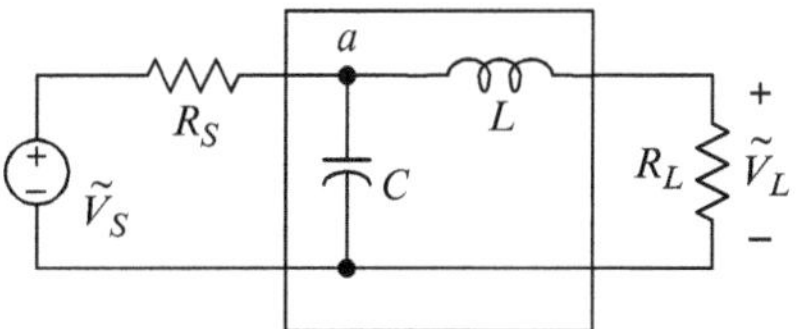

Fig. 15.12 See Example 15.10

Example 15.10. Obtain the voltage transfer function for the circuit shown in Fig. 15.12. Express the transfer function in the form (15.28) and give the values of M, N, and the coefficients of the polynomials. Find the SI unit of each coefficient and show that the transfer function is dimensionless.

Solution: Kirchhoff's current law gives

$$\frac{\tilde{V}_a - \tilde{V}_S}{R_S} + j\omega C\tilde{V}_a + \frac{\tilde{V}_a - \tilde{V}_L}{j\omega L} = 0,$$
$$\frac{\tilde{V}_L - \tilde{V}_a}{j\omega L} + \frac{\tilde{V}_L}{R_L} = 0.$$

Eliminating $\tilde{V}_a$ yields

$$\frac{\tilde{V}_L}{\tilde{V}_S} = H_v = \frac{R_L}{(j\omega)^2 R_S LC + j\omega(L + R_S R_L C) + R_S + R_L},$$

which has the form (15.28), with

$$M = 0, b_0 = R_L, N = 2, a_0 = R_S + R_L,$$
$$a_1 = L + R_S R_L C, a_2 = R_S LC.$$

The SI unit of b_0 is

$$\mathrm{SI}(b_0) = \mathrm{SI}(R_L) = \Omega.$$

The units of the denominator coefficients are

$$\mathrm{SI}(a_0) = \mathrm{SI}(R_S + R_L) = \Omega,$$
$$\mathrm{SI}(a_1) = \mathrm{SI}(L + R_S R_L C) = \mathrm{V\,s\,A^{-1}} = \Omega\,\mathrm{s}$$
$$\mathrm{SI}(a_2) = \mathrm{SI}(R_S LC) = \Omega\,\mathrm{V\,s\,A^{-1}\,A\,s\,V^{-1}} = \Omega\,\mathrm{s}^2$$

Because $\mathrm{SI}(j\omega) = \mathrm{s}^{-1}$, the unit of the denominator is Ω. Thus the transfer function is dimensionless.

It is useful to express transfer functions in a particular factored form that facilitates interpretation. Because the coefficients of the numerator and denominator polynomials in (15.28) are real, the zeros of each polynomial either are real or occur in complex-conjugate pairs. Therefore the numerator and the denominator of (15.28) can each be expressed as the product of a constant and factors of the forms

$$j\left(\frac{f}{f_0}\right), 1+j\left(\frac{f}{f_1}\right), 1+2\alpha j\left(\frac{f}{f_2}\right)-\left(\frac{f}{f_2}\right)^2, \quad (15.29)$$

where we have used $\omega = 2\pi f$ and where f_0, f_1, f_2 and α are *real and positive*. The first two factors in (15.29) are called **linear factors** and arise from the real zeros of the associated polynomial. The third is called a **quadratic factor** and arises from a pair of complex-conjugate zeros of the associated polynomial. *Each factor in* (15.29) *is dimensionless.*

The frequencies f_1 and f_2 (but not f_0) in (15.29) are called **corner frequencies** or **break frequencies**. The parameter α in a quadratic factor is called a **peaking factor**. The motivation for this terminology is explained in Sections 15.6 and 15.7, as is the reason why the frequency f_0 in (15.29) is not called a corner frequency.

A transfer function is in **standard form** if the numerator and denominator are both expressed as products of dimensionless linear and quadratic factors having the forms given in (15.29). A frequency-independent constant multiplier also usually appears in the standard form of a transfer function. The multiplier is dimensionless in current and voltage transfer functions, but is dimensioned in transimpedance and transadmittance transfer functions.

The magnitude of a transfer function for $f = 0$ is the **dc gain** of the associated circuit. For example, if the voltage transfer function for a certain circuit is

$$H_v = -\frac{100}{1+jf/f_1},$$

the dc voltage gain is $|H_v(0)| = 100$. As another example, if the current transfer function for a certain circuit is

$$H_i = \frac{10(1+jf/f_3)}{(1+jf/f_1)(1+jf/f_2)},$$

the dc current gain is $|H_i(0)| = 10$. The dc gain is undefined if a factor of the form jf/f_0 appears in the denominator of a transfer function and equals zero if a factor of the form jf/f_0 appears in the numerator.

Example 15.11. Express the transfer function obtained in Example 15.9 in standard form.

Solution: From Example 15.9,

$$H_v = \frac{R_L}{R_S+R+R_L+j2\pi f R_L C(R+R_S)}$$
$$= \frac{R_L/(R_S+R+R_L)}{1+j2\pi f R_L C(R+R_S)/(R_S+R+R_L)}$$
$$= \frac{R_L/(R_S+R+R_L)}{1+j2\pi f[R_L\|(R+R_S)]C},$$

which is written in standard form as

$$H_v = \frac{K}{1+j(f/f_0)}, \quad K = \frac{R_L}{R_S+R+R_L},$$
$$f_0 = \frac{1}{2\pi[R_L\|(R+R_S)]C}.$$

As an exercise, show that the transfer function is dimensionless.

Usually, it is best to express the frequencies $f_0, f_1, f_2, \cdots$ that appear in the linear and quadratic factors comprising a standard-form transfer functions *symbolically*, in terms of circuit parameters, so the effect of any particular circuit parameter on the transfer function can be determined. But if we do not assign numerical values to circuit parameters, we cannot determine whether a quadratic factor can be expressed as the product of two linear factors, because we cannot tell (in general) from a symbolic expression whether the zeros of the quadratic factor are real or complex. For this reason, it is best to leave quadratic factors in that form (not factored) until the associated circuit parameters are specified. Once those parameters are known, we can determine whether any quadratic forms that are present can be expressed as products of linear factors. In what follows, we show that if $\alpha \geq 1$, a quadratic factor

$$1+2\alpha j\left(\frac{f}{f_2}\right)-\left(\frac{f}{f_2}\right)^2 \quad (15.30)$$

can be expressed as the product of two linear factors. A term of the form (15.30) can be factored such that

$$(jf)^2+2\alpha f_2(jf)+f_2^2=(jf+f_a)\times(jf+f_b), \qquad (15.31)$$

where f_a, f_b must be real if the quadratic factor on the left can be expressed as the product of two linear factors. Equating like powers of (jf) on the two sides of (15.31) gives

$$f_2^2=f_af_b,\ 2\alpha f_2=f_a+f_b. \qquad (15.32)$$

From the first relation in (15.32), we find $f_b=f_2^2/f_a$, whence the second becomes

$$2\alpha f_2=\frac{f_2^2}{f_a}+f_a.$$

Solving this expression for f_a gives

$$f_a=\left(\alpha\pm\sqrt{\alpha^2-1}\right)f_2. \qquad (15.33)$$

Both values are real and positive if $\alpha\geq 1$. By symmetry, we would have obtained the same result if we had solved (15.32) for f_b. Thus

$$f_b=\left(\alpha\pm\sqrt{\alpha^2-1}\right)f_2. \qquad (15.34)$$

From (15.33) and (15.34),

$$f_a,f_b=\left(\alpha\pm\sqrt{\alpha^2-1}\right)f_2. \qquad (15.35)$$

Both f_a and f_b are real if $\alpha\geq 1$, as was to be shown. If we now divide both sides of (15.31) by $f_2^2=f_af_b$, we find

$$\frac{(jf)^2+2\alpha f_2(jf)+f_2^2}{f_2^2}=\frac{(jf+f_a)(jf+f_b)}{f_af_b},$$

which can be written

$$1+2\alpha j\left(\frac{f}{f_2}\right)-\left(\frac{f}{f_2}\right)^2=\left(1+j\frac{f}{f_a}\right)\left(1+j\frac{f}{f_b}\right). \qquad (15.36)$$

If $\alpha=1$, then from (15.35)

$$f_a=f_b=f_2 \qquad (15.37)$$

and (15.36) becomes

$$1+2\alpha j\left(\frac{f}{f_2}\right)-\left(\frac{f}{f_2}\right)^2=\left(1+\frac{jf}{f_2}\right)^2. \qquad (15.38)$$

Because $\omega/\omega_0=(2\pi f)/(2\pi f_0)=f/f_0$, we can often simplify the algebra involved in obtaining the standard form for a transfer function by using angular frequency ω, replacing ω/ω_0 by f/f_0 as the last step.

Example 15.12. Obtain the transfer function for the circuit shown in Fig. 15.13. Express the transfer function in standard form and give expressions for all corner frequencies and other parameters that appear in the standard-form expression.

Solution: The circuit is a voltage divider. The load voltage (phasor) is given by

$$\tilde{V}_L=\frac{R_L\tilde{V}_S}{R_S+R_L+Z_{LC}},\quad Z_{LC}=Z_L\|Z_C$$
$$=\frac{j\omega L/j\omega C}{j\omega L+1/j\omega C}=\frac{j\omega L}{1-\omega^2LC}.$$

Thus the transfer function is

$$H_v=\frac{\tilde{V}_L}{\tilde{V}_S}=\frac{R_L}{R_S+R_L+j\omega L/(1-\omega^2LC)}$$
$$=\frac{R_L(1-\omega^2LC)}{(R_S+R_L)(1-\omega^2LC)+j\omega L}$$
$$=\left(\frac{R_L}{R_S+R_L}\right)\frac{1-\omega^2LC}{1+j\omega L/(R_S+R_L)-\omega^2LC}$$
$$=\left(\frac{R_L}{R_S+R_L}\right)\frac{1-(\omega/\omega_0)^2}{1+2\alpha j(\omega/\omega_0)-(\omega/\omega_0)^2},$$

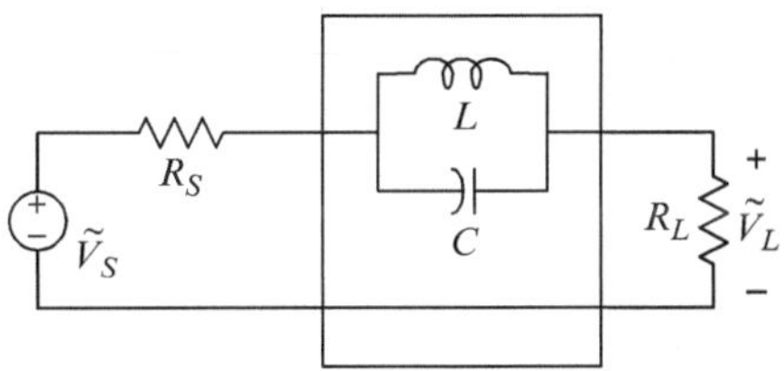

Fig. 15.13 See Example 15.12

where

$$\omega^2 LC = \left(\frac{\omega}{\omega_0}\right)^2 \Rightarrow \omega_0 = \frac{1}{\sqrt{LC}}$$

and

$$\frac{\omega L}{R_S + R_L} = 2\alpha\left(\frac{\omega}{\omega_0}\right) \Rightarrow \alpha = \frac{\omega_0 L}{2(R_S + R_L)}$$
$$= \frac{1}{\sqrt{LC}}\left(\frac{L}{2(R_S + R_L)}\right) = \frac{\sqrt{L/C}}{2(R_S + R_L)}.$$

Because $\omega/\omega_0 = f/f_0$, the transfer function is expressed in standard form as

$$H_v = K\frac{1-(f/f_0)^2}{1+2\alpha j(f/f_0)-(f/f_0)^2}, \quad (15.39)$$

where

$$K = \frac{R_L}{R_S + R_L}, \quad f_0 = \frac{\omega_0}{2\pi} = \frac{1}{2\pi\sqrt{LC}},$$
$$\alpha = \frac{\sqrt{L/C}}{2(R_S + R_L)}. \quad (15.40)$$

The numerator is a quadratic factor having corner frequency f_0 and peaking factor $\alpha = 0$. The numerator cannot be expressed as a product of two linear factors because the associated peaking factor has magnitude less than one. The denominator also is a quadratic factor having corner frequency f_0 and peaking factor given by (15.40). The denominator can be expressed as a product of linear factors if the associated peaking factor exceeds unity; i.e., if

$$\alpha \geq 1 \Rightarrow \sqrt{\frac{L}{C}} \geq 2(R_S + R_L).$$

We are given no values for the parameters R_S, L, C, R_L, so we cannot compute a value for the associated peaking factor and thus cannot tell whether the quadratic factor in the denominator can be expressed as the product of two linear factors. Thus we leave the transfer function in the form (15.39), with the denominator expressed as a quadratic factor.

Exercise 15.8. In Example 15.12, we say the quadratic factor $1-(f/f_0)^2$ cannot be expressed as a product of linear factors. Why should it not be written

$$1-\left(\frac{f}{f_0}\right)^2 = \left(1+\frac{f}{f_0}\right)\left(1-\frac{f}{f_0}\right)?$$

Example 15.13. (a) Express the transfer function obtained in Example 15.10 in standard form. (b) Let $R_L = 500\ \Omega$, $R_S = 100\ \Omega$, $L = 100\ \mu\text{H}$, $C = 50$ nF and calculate the values of all corner frequencies.

Solution: (a) From Example 15.10,

$$H_v = \frac{R_L}{(j\omega)^2 R_S LC + j\omega(L + R_S R_L C) + R_S + R_L}, \quad (15.41)$$

The transfer function is dimensionless (because it is a voltage transfer function). We first make all terms in the numerator and denominator dimensionless. We extract the factor that makes the frequency-independent term of the denominator equal to unity. Factoring $R_S + R_L$ from both the numerator and the denominator of the transfer function above gives

$$H_v = \frac{R_L/(R_S + R_L)}{(j\omega)^2 \frac{R_S LC}{R_S R_L} + j\omega\frac{(L + R_S R_L C)}{R_s + R_L} + 1}.$$

All terms in the numerator and denominator are now dimensionless. The denominator can be written

$$1 + j\omega \frac{(L/R_S + R_L C)}{R_s + R_L} + (j\omega)^2 \frac{R_S LC}{R_S + R_L}$$
$$= 1 + 2\alpha j\left(\frac{\omega}{\omega_2}\right) - \left(\frac{\omega}{\omega_2}\right)^2.$$

It follows that

$$\omega_2 = \sqrt{\frac{R_S + R_L}{R_S + LC}} \Rightarrow f_2 = \frac{1}{2\pi}\sqrt{\frac{R_S + R_L}{R_S LC}}$$
$$\frac{2\alpha}{\omega_2} = \frac{(L + R_S R_L C)}{R_S + R_L} \Rightarrow \alpha = \frac{\omega_2}{2}\frac{(L + R_S R_L C)}{R_S + R_L}$$
$$= \frac{(L + R_S R_L C)}{2\sqrt{R_S LC(R_S + R_L)}}$$

With these definitions, (15.41) can be written

$$H_v = \frac{K_v}{\left[1 + 2\alpha j(f/f_2) - (f/f_2)^2\right]},$$

(b) For the values given,

$$K_v = \frac{R_L}{R_S + R_L} = 0.833,$$
$$f_2 = \frac{1}{2\pi}\sqrt{\frac{R_S + R_L}{R_S LC}} \cong 174.3\,\text{kHz},$$
$$\alpha = \frac{(L + R_S R_L C)}{2\sqrt{R_s LC(R_S + R_L)}} \cong 2.37.$$

Because $\alpha > 1$, the quadratic factor should be expressed as the product of two linear factors. Thus,

$$1 + 2\alpha j\left(\frac{f}{f_2}\right) - \left(\frac{f}{f_2}\right)^2 = \left(1 + j\frac{f}{f_3}\right)\left(1 + j\frac{f}{f_4}\right),$$

where

$$f_3 = \left(\alpha - \sqrt{\alpha^2 - 1}\right)f_2 \cong 38.6\,\text{kHz},$$
$$f_4 = \left(\alpha + \sqrt{\alpha^2 - 1}\right)f_2 \cong 787.8\,\text{kHz} \quad (15.42)$$

For the parameter values given

$$H_v = \frac{K_v}{[1 + j(f/f_3)][1 + j(f/f_4)]},$$

where

$K_v = 0.833$, $f_3 \cong 38.6\,\text{kHz}$, $f_4 \cong 787.8\,\text{kHz}$.

15.6 Asymptotic Gain Plots: Linear Factors

Graphs of voltage gain in dB and phase shift in radians versus frequency are called **Bode plots**.[5] A Bode plot can be sketched quickly and with reasonable accuracy using asymptotic approximations. In the context of electronic circuits, the primary justification for learning how to construct such asymptotic plots is that such knowledge fosters insight and interpretation. An asymptotic plot clarifies the influence of each numerator and denominator factor on the form of a transfer function. An asymptotic plot and associated expressions for corner frequencies in terms of circuit parameters can tell us which parameters are critical and even in what direction the critical parameters must be changed to achieve a desired transfer function.

This section and the next two describe how asymptotic plots can be obtained easily from the standard form of a transfer function. This section treats asymptotic gain plots for transfer functions containing only linear factors. Section 15.7 treats asymptotic gain plots for transfer functions containing linear and quadratic factors. Section 15.8 treats asymptotic plots of phase shift.

A transfer function containing *only linear factors* can be expressed as

$$H_v = K\left(j\frac{f}{f_0}\right)^{k_0}\left(1 + j\frac{f}{f_1}\right)^{k_1}\left(1 + j\frac{f}{f_2}\right)^{k_2}\ldots, \quad (15.43)$$

where $k_0, k_1, k_2, \ldots$ are positive or negative integers, depending upon whether the associated factors appear in the numerator or denominator, respectively, of the transfer function. The magnitudes of $k_0, k_1, k_2, \ldots$ are the **multiplicities** of the associated factors. Multiplicities greater than two are rare in electronic circuits. We assume the transfer function is expressed in lowest form; that is, all factors common to both numerator and denominator have been cancelled. All critical frequencies and peaking factors associated with

[5]After Hendrik Bode (bōdee) (1905–1982), an engineer who showed how to use such plots to determine whether a system is stable and if not, how to make it so. Bode contributed much to our understanding of feedback systems.

denominator factors in a transfer function are positive if the associated circuit is stable.

The voltage gain for a transfer function of the form (15.43) is expressed in dB as

$$\begin{aligned} A_v &= 10\ \log\left(|H_v|^2\right) \\ &= 10\ \log\left\{K^2\left[\left(\frac{f}{f_0}\right)^2\right]^{k_0}\left[1+\left(\frac{f}{f_1}\right)^2\right]^{k_1}\cdots\right\} \\ &= 10\ \log(K^2) + 10k_0\ \log\left[\left(\frac{f}{f_0}\right)^2\right] \\ &\quad + 10k_1\ \log\left[1+\left(\frac{f}{f_1}\right)^2\right] + \ldots \\ &= 20\ \log|K| + 20k_0\ \log\left(\frac{f}{f_0}\right) \\ &\quad + 10k_1\ \log\left[1+\left(\frac{f}{f_1}\right)^2\right] + \ldots. \end{aligned} \tag{15.44}$$

The first term $20\ \log|K|$ is independent of frequency, represented graphically by a horizontal line at $20\ \log|K|$ dB. On a logarithmic scale for frequency, where $\log(f/f_0)$ is the independent variable, the second term, $20k_0\ \log(f/f_0)$, is a straight line having slope$20k_0$ dB/decade. Because $\log(f_0/f_0) = 0$, the line passes through the point $(f_0, 0)$, as illustrated by Fig. 15.14.

The frequency f_0 in a factor of the form $j(f/f_0)$ is arbitrary. That is, we can write the first two factors on the right side of (15.43) as

$$K\left(j\frac{f}{f_0}\right)^{k_0} = K\left(\frac{f_0'}{f_0}\right)^{k_0}\left(j\frac{f}{f_0'}\right)^{k_0} = K'\left(j\frac{f}{f_0'}\right)^{k_0},$$

where f_0' is arbitrary and

$$K' = K\left(\frac{f_0'}{f_0}\right)^{k_0}.$$

It follows that we can write the first and second terms on the right side of (15.44) as

$$\begin{aligned} &20\ \log|K| + 20k_0\ \log\left(\frac{f}{f_0}\right) \\ &\quad = 20\ \log|K'| + 20k_0\ \log\left(\frac{f}{f_0'}\right), \end{aligned} \tag{15.45}$$

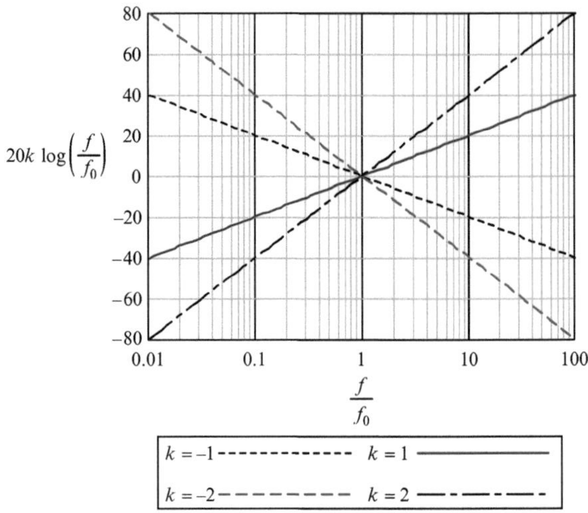

Fig. 15.14 A graph of $20\,k\ \log(f/f_0)$ versus frequency for four values of k

where f_0' is arbitrary and

$$K' = K\left(\frac{f_0'}{f_0}\right)^{k_0}. \tag{15.46}$$

The fact that

$$\begin{aligned} 20\ \log|K| + 20k_0\ \log\left(\frac{f}{f_0}\right) &= 20\ \log|K'| \\ &+ 20k_0\ \log\left(\frac{f}{f_0'}\right) \end{aligned}$$

with K' given by (15.46) means that we can assign any value we wish to the frequency f_0 by appropriately scaling the frequency-independent factor K. The parameter f_0 associated with a linear factor of the form $j(f/f_0)$ has no special significance. A linear factor of that kind is written that way primarily for sake of dimensional consistency and convenience (e.g., in drawing an asymptotic Bode plot).

Example 15.14. The transfer function for a certain circuit is

$$H_v = \frac{K(jf/f_0)^2}{(1+jf/f_1)^2(1+jf/f_2)},$$

where

$$K = 10^5, f_0 = 1\,\text{kHz}, f_1 = 100\,\text{Hz}, f_2 = 15\,\text{kHz}.$$

Find K' such that

$$H_v = \frac{K'(jf/f_0')^2}{(1+jf/f_1)^2(1+jf/f_2)},$$

where $f_0' = 10$ Hz.

Solution: From (15.46),

$$K' = K\left(\frac{f_0'}{f_0}\right)^2 = 10^5\left(\frac{10\,\text{Hz}}{1,000\,\text{Hz}}\right)^2 = 10$$

The third and every subsequent term on the right side of (15.45) has the form

$$10k\ \log\left[1 + (f/f_0)^2\right], \tag{15.47}$$

where k is a positive or negative integer and f_0 is a corner frequency. A method for quickly sketching (or visualizing) a graph of gain versus frequency is based upon the asymptotic behavior of the term (15.47) for frequencies well below and well above the corner frequency f_0. The **low-frequency asymptote** for the term (15.47) is given by

$$\begin{aligned} f \ll f_0 &\Rightarrow 10k\ \log\left[1 + (f/f_0)^2\right] \\ &\cong 10k\ \log[1] = 0\ \text{dB} \end{aligned} \tag{15.48}$$

and the **high-frequency asymptote** is given by

$$\begin{aligned} f \gg f_0 &\Rightarrow 10k\ \log\left[1 + (f/f_0)^2\right] \\ &\cong 10k\ \log\left[(f/f_0)^2\right] = 20k\ \log(f/f_0). \end{aligned} \tag{15.49}$$

The low-frequency asymptote is a horizontal line through 0 dB. On a logarithmic scale, the high-frequency asymptote is a straight line having slope $20k$ dB/decade; that is, for each tenfold increase in frequency f, the high-frequency asymptote rises (or falls) by $20k$ dB. The low-frequency asymptote and the high-frequency asymptote intersect where they are equal; i.e.,

$$20k\ \log(f/f_0) = 0 \Rightarrow f = f_0. \tag{15.50}$$

The **asymptotic approximation** to a factor of the form $10k\ \log[1 + (f/f_0)^2]$ is defined as

$$10k\ \log\left[1 + \left(\frac{f}{f_0}\right)^2\right] \cong a(f) = \begin{cases} 0\ \text{dB}, & f \le f_0, \\ 20k\ \log\left(\frac{f}{f_0}\right)\ \text{dB}, & f > f_0. \end{cases} \tag{15.51}$$

In words, the asymptotic approximation equals the low-frequency asymptote for frequencies below the corner frequency f_0 and equals the high-frequency asymptote for frequencies above the corner frequency.

Figure 15.15 shows graphs of the asymptotic approximation given by (15.51) for $k = \pm 1, \pm 2$.

Figure 15.16 shows graphs of the term $10k$ $\log[1 + (f/f_0)^2]$ and the asymptotic approximation to that term for $k = -1$ and $k = -2$. The graphs would simply be inverted (reflected in the 0 dB axis) for $k = +1$ and $k = +2$.

Figure 15.17 shows graphs of the error in the asymptotic approximation, given by

$$e(f) = \left|a(f) - 10k\ \log\left[1 + (f/f_0)^2\right]\right|. \tag{15.52}$$

The maximum error is $a(f_0) = 10k\ \log(2) \cong 3k$ dB. In virtually all applications, the error is negligible for $f < f_0/10$ and $f > 10f_0$. To obtain a reasonably accurate graph we can simply draw a smooth

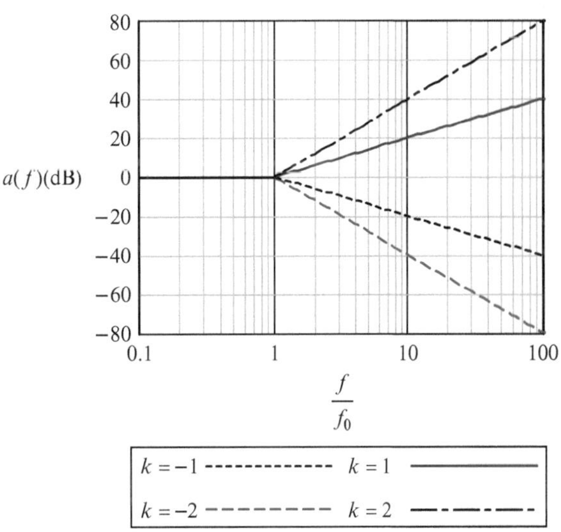

Fig. 15.15 Asymptotic approximations for a linear factor $(1 + jf/f_0)^k$ (see (15.51))

a

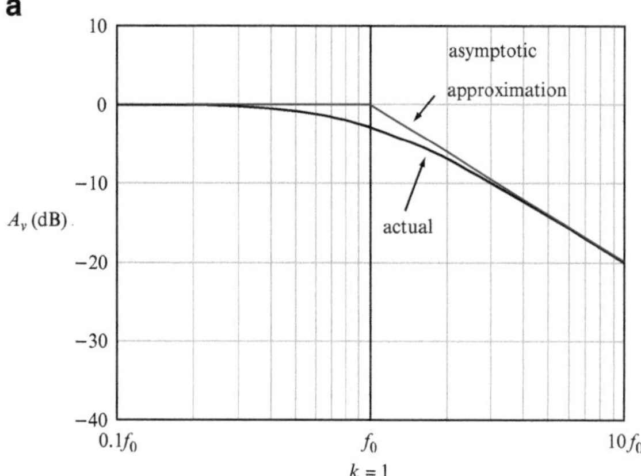

b

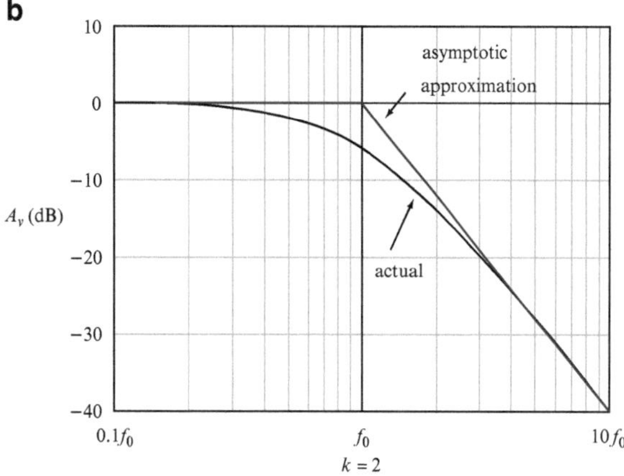

Fig. 15.16 Bode gain plot for a linear factor $H_v(f) = (1 + jf/f_0)^{-k}$

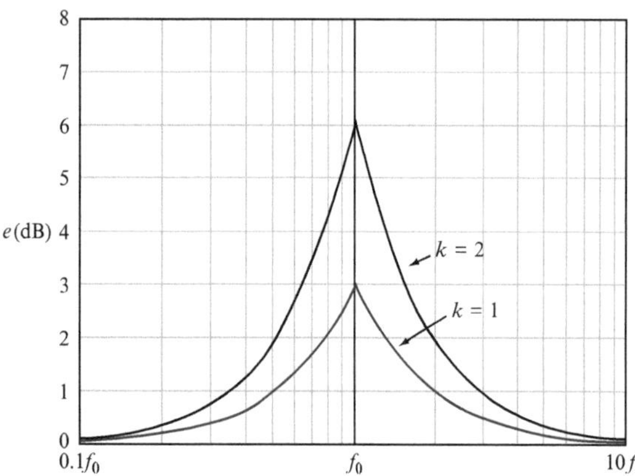

Fig. 15.17 Error in an asymptotic approximation to a plot of gain versus frequency (see (15.52))

curve tangent to the asymptotes at (approximately) $f = f_0/10$ and $f = 10f_0$ and passing through the actual value for $f = f_0$, but that step is seldom necessary. The asymptotic approximation is sufficient for most purposes. Besides, if an accurate graph is needed, it is easy to produce one with the aid of a computer (or even a modern pocket calculator).

From (15.44), an asymptotic plot of gain versus frequency is obtained by adding the asymptotic plots for each term in the transfer function. The procedure is best elaborated by examples.

Example 15.15. Draw an asymptotic gain plot for a circuit having the voltage transfer function

$$H_v = \frac{10[1 + j(f/f_0)]}{[1 + j(f/f_1)]^2},$$

where $f_0 = 100$ Hz and $f_1 = 1$ kHz.

Solution: The voltage gain is

$$\begin{aligned} A_v &= 10\log\left(|H_v|^2\right) \\ &= 10\log\left\{\frac{100\left[1+(f/f_0)^2\right]}{\left[1+(f/f_1)^2\right]^2}\right\} \\ &= 10\log(100) + 10\log\left[1+(f/f_0)^2\right] \\ &\quad - 10\log\left\{\left[1+(f/f_1)^2\right]^2\right\} \\ &= 20 + 10\log\left[1+(f/f_0)^2\right] \\ &\quad - 20\log\left[1+(f/f_1)^2\right]. \end{aligned} \tag{15.53}$$

The lowest corner frequency is $f_0 = 100$ Hz and the highest is $f_1 = 1$ kHz. As a rule of thumb, a gain plot should span a range extending from 1 decade below the lowest corner frequency to 1 decade above the highest corner frequency. Thus, we want the frequency axis to span at least the range from 10 Hz to 10 kHz (3 decades). We choose the minimum frequency to be 10 Hz, although if we were using five-cycle semi-log paper, a lower value would do.

To construct the plot, we add the asymptotic approximations for each term in (15.53). The asymptotic approximations for the individual terms are

$$a_0(f) = 10\log(100) = 20\text{ dB},$$
$$a_1(f) = \begin{cases} 0, & f \le f_0, \\ 20\log\left(\dfrac{f}{f_0}\right), & f > f_0, \end{cases} \tag{15.54}$$
$$a_2(f) = \begin{cases} 0, & f \le f_1, \\ -40\log\left(\dfrac{f}{f_1}\right), & f > f_1. \end{cases}$$

The (overall) asymptotic approximation is the sum of the individual asymptotic approximations above:

$$A_v(f) \cong a_0(f) + a_1(f) + a_2(f)$$
$$a_0(f) = 10\log(100) = 20\text{ dB},$$

$$A_v(f) \cong \begin{cases} 20\text{ dB}, & f \le f_0, \\ 20 + 20\log\left(\dfrac{f}{f_0}\right), & f_0 < f \le f_1, \\ 20 + 20\log\left(\dfrac{f}{f_0}\right) - 40\log\left(\dfrac{f}{f_1}\right), & f > f_1. \end{cases} \tag{15.55}$$

Figure 15.18 shows the individual asymptotic approximations defined by (15.54). Figure 15.19 shows the overall approximation given by (15.55) and (for comparison) the actual gain given by (15.53).

Example 15.16. Draw an asymptotic plot of gain versus frequency for

$$H_v = -\frac{\sqrt{10}j(f/f_0)}{[1 + j(f/f_1)]^2},$$

where $f_0 = 5$ kHz and $f_1 = 300$ kHz.

Solution:

$$A_v = 10\log\left(|H_v|^2\right) = 10\log\left\{\frac{10(f/f_0)^2}{\left[1+(f/f_1)^2\right]^2}\right\}$$
$$= 10 + 20\log(f/f_0) - 20\log\left[1 + (f/f_1)^2\right]. \tag{15.56}$$

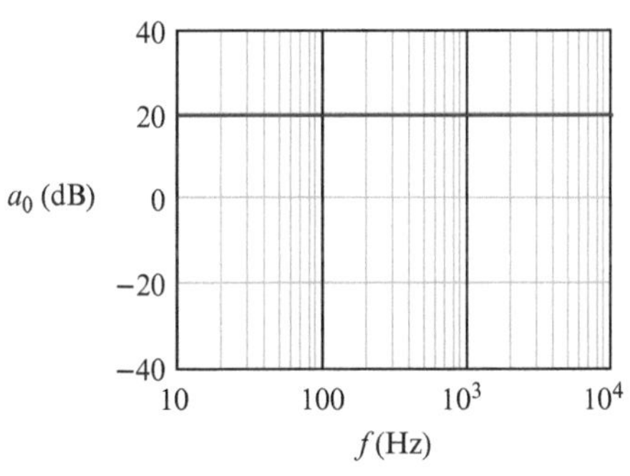

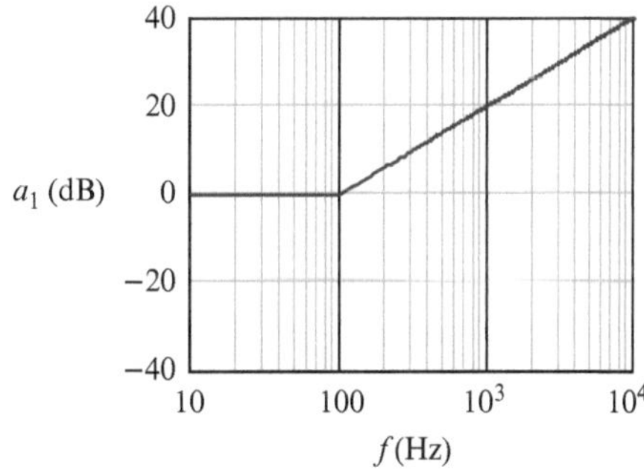

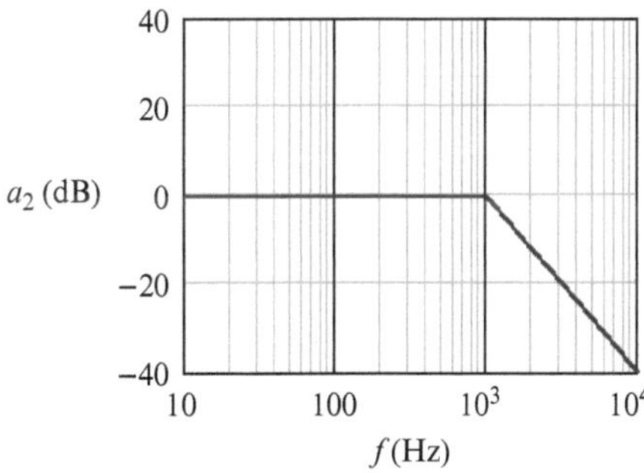

Fig. 15.18 See Example 15.15

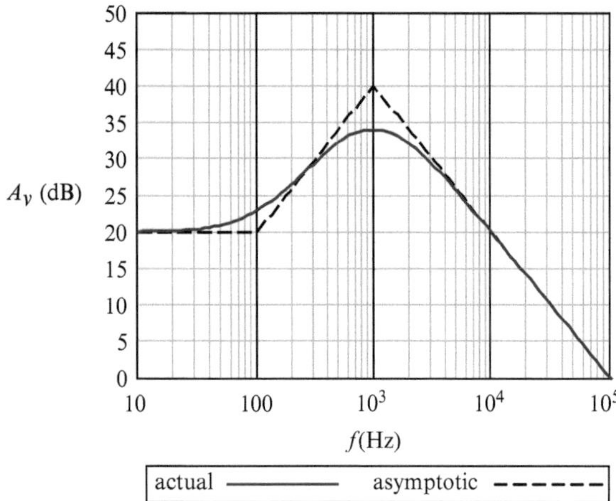

Fig. 15.19 See Example 15.15

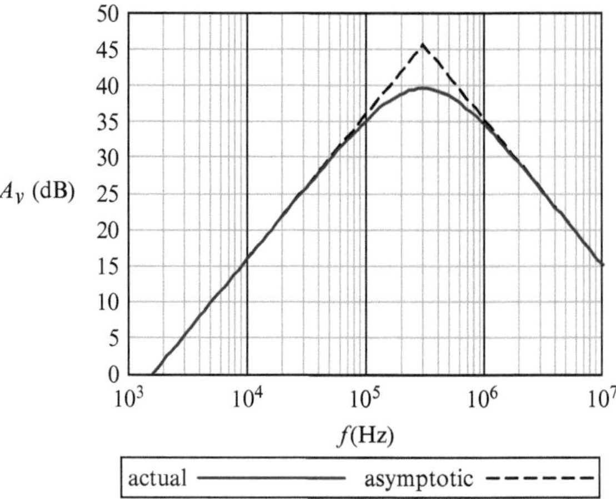

Fig. 15.20 See Example 15.16

Because the first term is independent of frequency and the second is a straight line (on a graph using a logarithmic scale for frequency), the interesting behavior of the gain is in the neighborhood of the corner frequency f_1. We choose the frequency-axis limits $f_{\min} = 1$ kHz, $f_{\max} = 10$ MHz because five-cycle semi-log paper is commonly available. The asymptotic approximation for the gain is

$$a(f) = \begin{cases} 10 + 20\log\left(\dfrac{f}{f_0}\right), & f \le f_1, \\ 10 + 20\log\left(\dfrac{f}{f_0}\right) - 40\log\left(\dfrac{f}{f_1}\right), & f > f_1. \end{cases} \tag{15.57}$$

From (15.56)

$$A_v(f_0) \cong 10 \text{ dB}, \; A_v(f_1) = 39.5 \text{ dB} \tag{15.58}$$

The first line segment begins (on our scale) at $f_{\min} = 1$ kHz passes through the point (5 kHz, 10.3 dB) with slope 20 dB/decade, and ends at $f_1 = 300$ kHz. The second line segment begins at the endpoint of the first (at $f = f_1$) and has slope -20 dB/decade (the sum of the slopes of the terms in (15.57)). Figure 15.20 shows the resulting graph.

In the two examples above, we obtained expressions for the gain in dB and for the asymptotic approximation in dB before drawing the graphs. After a little experience, those steps become unnecessary. An asymptotic gain plot can be drawn by inspection of a transfer function, as follows:

1. Express the transfer function in standard form, as

$$H = K(jf/f_0)^{k_0}(1 + jf/f_1)^{k_1}(1 + jf/f_2)^{k_2}\ldots(1 + jf/f_N)^{k_N},$$

where $f_1 < f_2 < \ldots < f_N$.

2. Set aside the magnitude $|K|$ of the frequency-independent factor K. If there is no such explicit factor, then by implication $|K| = 1$.
3. If the transfer function has a factor of the form $(jf/f_0)^{k_0}$, adjust the frequency f_0 to a frequency f_0' that is an integral power of ten and is at least 1 decade below the lowest corner frequency f_1; e.g., if $f_1 = 30$ Hz, let $f_0' = 1$ Hz. Then $(jf/f_0)^{k_0} = \left(f_0'/f_0\right)^{k_0}\left(jf/f_0'\right)^{k_0}$.
4. Absorb the factor $(f_0'/f_0)^{k_0}$ into the constant K that was set aside in 1 above; i.e., $K' = \left(f_0'/f_0\right)^{k_0}K$. Compute

$$\begin{aligned} |K'|_{dB} &= 20\ \log(|K'|) \\ &= 20\ \log|K| + 20k_0 \log\left(f_0'/f_0\right) \end{aligned}$$

and set this value aside.
5. Draw a line having slope $20\,k_0$ dB/decade from the point $(f_0', 0\text{ dB})$ at the left edge of the graph to a point whose abscissa is the lowest corner frequency f_1.

6. Draw a line having slope $20(k_0 + k_1)$dB/decade from the endpoint of the previous line to a point whose abscissa is the next lowest corner frequency f_2.
7. Draw a line having slope $20(k_0 + k_1 + k_2)$ dB/decade from the endpoint of that line to a point whose abscissa is the next lowest corner frequency f_3. Continue this process until all factors are accounted for.
8. Adjust the scale on the vertical axis to account for the factor K' by adding K'_{dB} to each label; e.g., 0 dB becomes K'_{dB}.

Example 15.17. Draw an asymptotic gain plot for the transfer function

$$H_v = \frac{10^5(jf/f_0)^2(1+jf/f_2)}{(1+jf/f_1)^2(1+jf/f_3)(1+jf/f_4)},$$

where

$$f_0 = 100\text{ Hz},\ f_1 = 20\text{ Hz},\ f_2 = 100\text{ Hz},$$
$$f_3 = 3\text{kHz},\ f_4 = 50\text{ kHz}.$$

Solution: The steps below are numbered in agreement with those in the procedure above. Figure 15.21 shows the completed plot, where the line segments and the gain adjustment are numbered in agreement with the steps below. Gain values given in steps 5–9 refer to the "original scale" shown in Fig. 15.21. Beginning with step 5, refer to Fig. 15.21 as you follow the solution.

1. The transfer function is given in standard from.
2. The constant $K = 10^5$.
3. The lowest corner frequency is $f_1 = 20$ Hz. An integral power of ten that is at least 1 decade below f_1 is $f'_0 = 1$ Hz Thus $(jf/f_0)^2 = (f'_0/f_0)^2(jf/f'_0)^2$, where $f_0 = 100$ Hz and $f'_0 = 1$ Hz.
4. $K' = K(f'_0/f_0)^2 = 10^5(1/100)^2 = 10 \Rightarrow |K'|_{\text{dB}} = 20\log(|K'|) = 20$ dB.
5. The first line segment begins at $(f'_0, 0\text{ dB}) = (1\text{ Hz}, 0\text{ dB})$, has slope 40 dB/decade, passes through the point $(10\text{ Hz}, 40\text{ dB})$, and terminates at the lowest corner frequency $f = f_1 = 20$ Hz, where $A_v \cong 52$ dB.
6. The next line segment, associated with the corner frequency f_1, begins at the end of the previous one, has slope 40 – 40 = 0 dB/decade (is horizontal), and terminates at the next corner frequency $f = f_2 = 100$ Hz.

7(a) The next line segment, associated with the corner frequency f_2, begins at the end

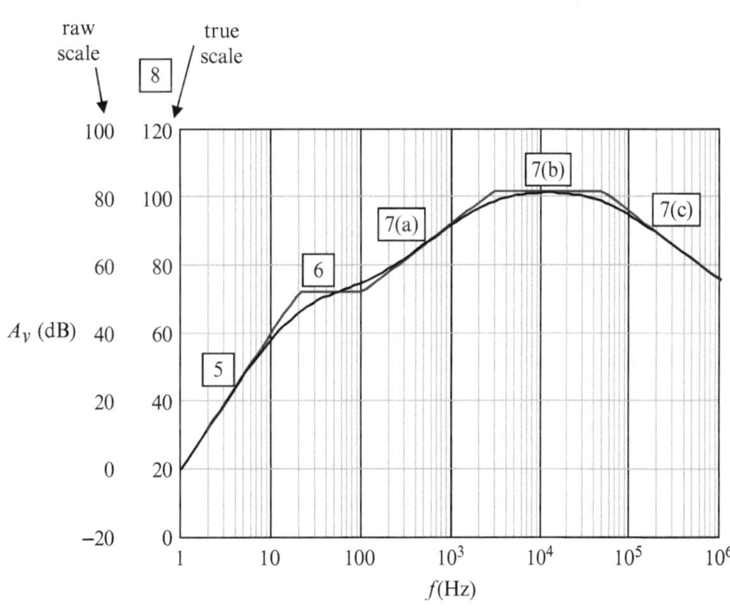

Fig. 15.21 Asymptotic gain plot (piecewise linear) obtained in Example 15.17. The true gain (the smooth curve) is shown for comparison

of the previous one, has slope $0 + 20 = 20$ dB/decade, passes through the point (1 kHz, 72 dB), and terminates at $f = f_3 = 3$ kHz, where $A_v \cong 80$ dB.

7(b) The next line segment, associated with the corner frequency f_3, begins at the end of the previous one and has slope $20 - 20 = 0$ dB/decade, and terminates at $f = f_4 = 50$ kHz, where $A_v \cong 80$ dB.

7(c) The last line segment, associated with the corner frequency f_4, begins at the end of the previous one and has slope $0 - 20 = -20$ dB/decade, which means the line passes through the point (300 kHz, 61 dB) and continues indefinitely.

8. Finally, we account for the factor K'_{dB} set aside in step 4 above by shifting the entire graph up by $K'_{dB} = 20$ dB. Actually, rather than translate the graph, we simply re-label the vertical axis by adding 20 dB to each value, as shown in Fig. 15.21.

With some practice, you will be able to sketch a graph of gain in decibels versus frequency (logarithmic scale) for a transfer function (expressed in standard form) with little effort. From symbolic expressions for corner frequencies, you will often be able to discern how various circuit parameters affect a gain of interest and even the performance of the associated circuit.

15.7 Asymptotic Gain Plots: Quadratic Factors

As noted above (review the discussion in Section 15.5), a transfer function can contain one or more *quadratic factors* that cannot be expressed as products of linear factors. The standard form for a quadratic factor having multiplicity k is

$$\left[1 + 2\alpha j\left(\frac{f}{f_0}\right) - \left(\frac{f}{f_0}\right)^2\right]^k, |\alpha| < 1. \tag{15.59}$$

We limit our discussion to quadratic factors having multiplicity one ($k = \pm 1$ in (15.59)), because quadratic factors having multiplicities greater than one rarely arise in transfer functions for electronic circuits. For a quadratic factor in the numerator of a transfer function, $k = 1$. For a quadratic factor in the denominator, $k = -1$. We show in Chapter 18 that for a stable circuit, the *peaking factor* α associated with a quadratic factor in the denominator must be positive. In this chapter, we limit our attention to stable circuits, so we assume $\alpha > 0$ for any quadratic factor appearing in the denominator of a transfer function. As we show in Section 15.5 (see (15.38)), the quadratic factor can be expressed as the product of two linear factors if $\alpha \geq 1$.

The quadratic factor (15.59) is expressed in dB as

$$a(f) = 20k \log\left|1 + 2\alpha j\left(\frac{f}{f_0}\right) - \left(\frac{f}{f_0}\right)^2\right| \text{ dB}. \tag{15.60}$$

The low- and high-frequency asymptotes for $a(f)$ are

$$a_{lo}(f) = a(f \ll f_0) = 20k \log(1) = 0 \text{ dB},$$
$$a_{hi}(f) = a(f \gg f_0) = 20k \log\left[\left(\frac{f}{f_0}\right)^2\right]$$
$$= 40k \log\left(\frac{f}{f_0}\right).$$

The low- and high-frequency asymptotes intersect at $f = f_0$. Thus the asymptotic approximation to (15.60) is given by

$$a(f) \cong \begin{cases} 20k \log(1) = 0 \text{ dB}, & f \leq f_0, \\ 40k \log\left(\dfrac{f}{f_0}\right) \text{ dB}, & f > f_0, \end{cases} \tag{15.61}$$

and is identical to the asymptotic approximation for a linear factor having multiplicity k; i.e., for

$$a(f) = 20k \log\left|\left[1 + j\left(\frac{f}{f_0}\right)\right]^2\right|$$
$$= 20k \log\left[1 + \left(\frac{f}{f_0}\right)^2\right].$$

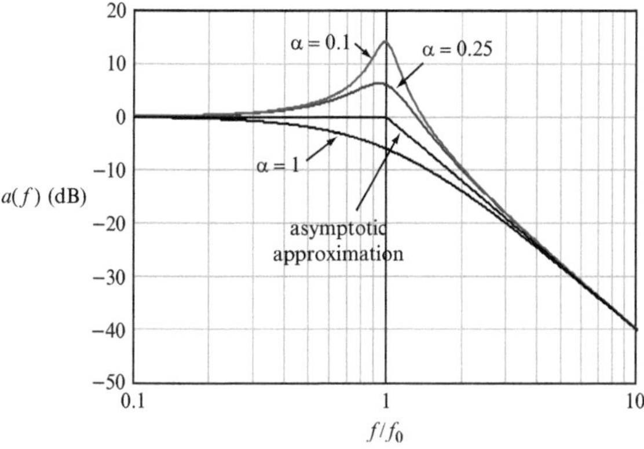

Fig. 15.22 Graphs of the quantities defined in (15.60) and (15.61) for $k = -1$

Figure 15.22 shows graphs of the quadratic factor given by (15.60) versus frequency for $k = -1$ and for selected values of α. Also shown is the asymptotic approximation (15.61), which for a quadratic factor is identical to that for a linear factor having multiplicity two. For $\alpha = 1$,

$$1 + 2\alpha j\left(\frac{f}{f_0}\right) - \left(\frac{f}{f_0}\right)^2 = 1 + 2j\left(\frac{f}{f_0}\right) - \left(\frac{f}{f_0}\right)^2 = \left[1 + j\left(\frac{f}{f_0}\right)\right]^2, \tag{15.62}$$

so $\alpha = 1$ (corresponding to a linear factor having multiplicity two) is the boundary between a (nonfactorable) quadratic factor and two linear factors. For $0.25 \leq \alpha \leq 1$, the maximum error (magnitude) in the asymptotic approximation is about 6 dB, which equals the maximum error in the asymptotic approximation for a linear factor having multiplicity two. Consequently, for an electronic circuit, if $\alpha \geq 0.25$, it is common practice to approximate a quadratic factor by a linear factor having multiplicity two, as in (15.62). If it turns out that $\alpha < 0.25$ for one or more quadratic factors and a more accurate graph is needed, we can resort to a computer-generated graph.

From Fig. 15.22, you can see that the asymptotic approximation (15.61) is quite good for $f \ll f_0$ and for $f \gg f_0$ and is within about 6 dB of the true value if $\alpha > 0.25$. But as we show at the end of this section, the magnitude of the peak and thus the error in the asymptotic approximation increases by about 20 dB/decade for $\alpha < 0.2$, from about 8 dB for $\alpha = 0.2$ to about 28 dB for $\alpha = 0.02$.

Example 15.18. Draw an asymptotic plot of voltage gain in dB versus frequency for a circuit having the transfer function

$$H_v = \frac{10j(f/f_0)}{[1 + j(f/f_1)]\left[1 + 2\alpha j(f/f_2) - (f/f_2)^2\right]}, \tag{15.63}$$

where

$$f_0 = 10\,\text{kHz}, f_1 = 200\,\text{kHz}, f_2 = 3\,\text{MHz}, \alpha = 0.3.$$

Solution: Because $\alpha = 0.3 > 0.25$, we approximate the quadratic factor as the square of a linear factor:

$$1 + 2\alpha j(f/f_2) - (f/f_2)^2 \cong [1 + j(f/f_2)]^2.$$

Thus

$$H_v = \frac{10j(f/f_0)}{[1 + j(f/f_1)][1 + j(f/f_2)]^2}.$$

We then construct the asymptotic gain plot in the usual way. Figure 15.23 shows both the asymptotic plot (dashed line) and the actual gain (solid line) computed from the transfer function (15.63). The peak in the gain at the corner frequency $f_2 = 3$ MHz is only about 6 dB above the asymptotic value.

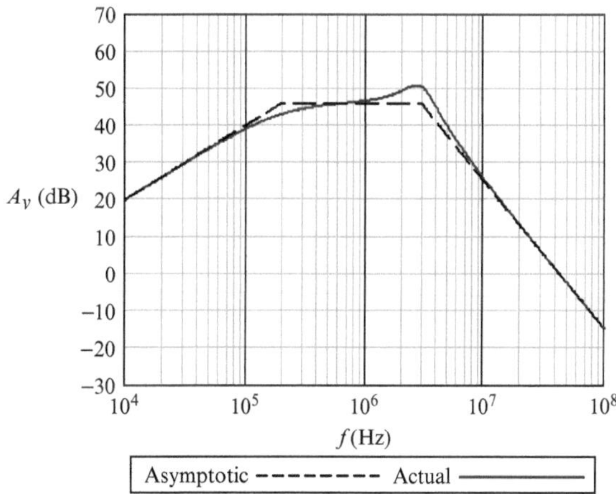

Fig. 15.23 See Example 15.18

To conclude this section, we examine the difference between the maximum magnitude of a nonrepeated quadratic factor and that of the associated asymptotic approximation.

From (15.60), the magnitude (gain) of a quadratic factor having multiplicity one and appearing in the denominator of a transfer function is given by

$$\begin{aligned} a(f) &= -20 \log \sqrt{\left[1-\left(\frac{f}{f_0}\right)^2\right]^2 + 4\alpha^2\left(\frac{f}{f_0}\right)^2} \\ &= -10 \log\left\{\left[1-\left(\frac{f}{f_0}\right)^2\right]^2 + 4\alpha^2\left(\frac{f}{f_0}\right)^2\right\}. \end{aligned} \tag{15.64}$$

The frequency at which the maximum value of $a(f)$ occurs is the solution to

$$\frac{da(f)}{df} = 0. \tag{15.65}$$

Instead of using (15.65) directly, it is easier to find the maximum of the argument of the logarithm, which is permissible because the logarithm is a monotonic function. We can simplify the problem further by letting

$$x = \left(\frac{f}{f_0}\right)^2.$$

Thus we find

$$\frac{d}{dx}\left[(1-x)^2 + 4\alpha^2 x\right] = 2(1-x)(-1) + 4\alpha^2 = 0,$$

which gives

$$x = \left(\frac{f}{f_0}\right)^2 = 1 - 2\alpha^2 \Rightarrow f = f_0\sqrt{1-2\alpha^2}. \tag{15.66}$$

If $\alpha < 1/\sqrt{2}$, the frequency f given by (15.66) is real and positive and the gain has a relative maximum (peak) at that frequency. If $\alpha \geq 1/\sqrt{2}$, the relative maximum occurs at $f = 0$ and there is no peak at a higher frequency.

For $0 < \alpha < 1/\sqrt{2}$, the magnitude of the peak at $f = f_0\sqrt{1-2\alpha^2}$ is given by

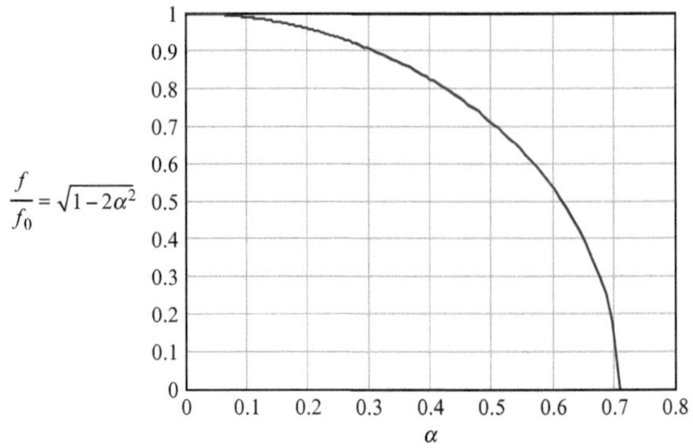

Fig. 15.24 Frequency at which the gain is maximum versus the peaking factor α

$$a\left(f_0\sqrt{1-2\alpha^2}\right) = -10 \log\left\{\left[1-\left(\frac{f_0\sqrt{1-2\alpha^2}}{f_0}\right)^2\right]^2 + 4\alpha^2\left(\frac{f_0\sqrt{1-2\alpha^2}}{f_0}\right)^2\right\},$$

which simplifies to

$$a\left(f_0\sqrt{1-2\alpha^2}\right) = -10 \log\left(4\alpha^2 - 4\alpha^4\right). \tag{15.67}$$

Figure 15.24 shows a normalized graph of the frequency given by (15.66) versus the peaking factor α. Figure 15.25 shows a graph of the maximum magnitude given by (15.67). These graphs illustrate that as α decreases from $1/\sqrt{2}$ toward zero, the maximum gain increases and the frequency of maximum gain approaches f_0.

Example 15.19. Repeat Example 15.18, but with $\alpha = 0.05$.

Solution: The asymptotic approximation is unchanged. The frequency at which the peak occurs is virtually unchanged, because

$$f_2\sqrt{1-2\alpha^2} \cong 0.997 f_2$$

Figure 15.26 shows both the asymptotic plot (dashed line) and the actual gain (solid line) computed from the transfer function given by (15.63) for $\alpha = 0.05$ and the corner frequencies given in Example 15.18. From

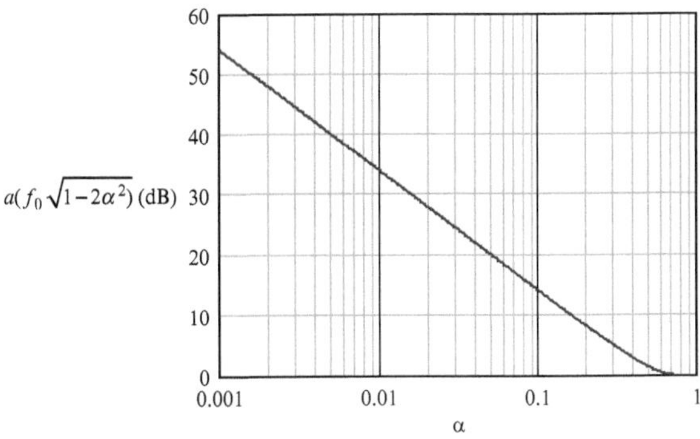

Fig. 15.25 Peak gain given by (15.67) for $0.001 \le \alpha \le 1/\sqrt{2}$

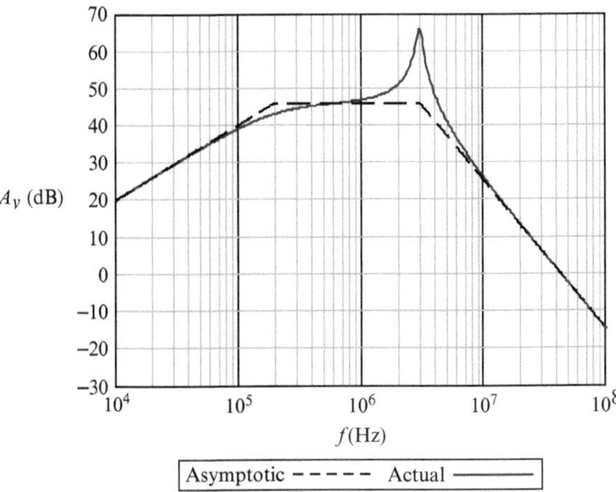

Fig. 15.26 See Example 15.19

Fig. 15.25, the maximum gain is approximately 20 dB above the asymptotic approximation, in agreement with (15.67) and Fig. 15.25.

In this example, once we have drawn the asymptotic graph, we could obtain an approximation to the actual gain by drawing a smooth curve tangent to the asymptotic graph for $f \ll f_1$, passing through a point 3 dB below the asymptotic graph at $f = f_1$, rising above the asymptotic graph at about $5f_1$, passing through a point 20 dB above the asymptotic approximation at $f_2 = 3$ MHz, and becoming tangent to the asymptotic graph at about $5f_2$

Again, the main reason for learning to draw asymptotic plots is to strengthen ability to see how the various parameters of a circuit influence the transfer function of the circuit.

Accurate graphs, if needed for other reasons, are easily obtained with the aid of a computer or powerful pocket calculator.

15.8 Asymptotic Plots of Phase Shift Versus Frequency

As a reminder, a transfer function can be expressed as

$$H_v = K\left(j\frac{f}{f_0}\right)^{k_0}\left(1+j\frac{f}{f_1}\right)^{k_1}\left(1+j\frac{f}{f_2}\right)^{k_2}\cdots \left[1+2\alpha_a\left(j\frac{f}{f_a}\right)-\left(\frac{f}{f_a}\right)^2\right]^{k_a} \left[1+2\alpha_b\left(j\frac{f}{f_b}\right)-\left(\frac{f}{f_b}\right)^2\right]^{k_b}\cdots, \tag{15.68}$$

where each of the integers $k_0, k_1, k_2, \ldots, k_a, k_b, \ldots$ is either positive or negative, depending upon whether the associated factor appears in the numerator or denominator, respectively, of the transfer function. For example, we may write

$$H_v = \frac{K(1+jf/f_1)^2\left[1+2\alpha_a jf/f_a-(f/f_a)^2\right]}{(jf/f_0)(1+jf/f_2)\left[1+2\alpha_b jf/f_b-(f/f_b)^2\right]^2}$$

as

$$H_v = K\left(j\frac{f}{f_0}\right)^{-1}\left(1+j\frac{f}{f_1}\right)^{2}\left(1+j\frac{f}{f_2}\right)^{-1} \times\left[1+2\alpha_a j\frac{f}{f_a}-\left(\frac{f}{f_a}\right)^2\right] \times\left[1+2\alpha_b j\frac{f}{f_b}-\left(\frac{f}{f_b}\right)^2\right]^{-2}.$$

The phase shift of a circuit is the angle (as a function of frequency) of the transfer function:

$$\phi_v = \measuredangle H_v. \tag{15.69}$$

The angle of a product is the sum of the angles of the individual factors. From (15.68)

$$\measuredangle H_v = \measuredangle K + \measuredangle\left[j\left(\frac{f}{f_0}\right)\right]^{k_0} + \measuredangle\left\{\left[1+\left(\frac{f}{f_1}\right)\right]^{k_1}\right\}$$
$$+ \measuredangle\left\{\left[1+\left(\frac{f}{f_2}\right)\right]^{k_2}\right\} + \dots$$
$$+ \measuredangle\left\{\left[1+2\alpha_a\left(j\frac{f}{f_a}\right)-\left(\frac{f}{f_a}\right)^2\right]^{k_a}\right\}$$
$$+ \measuredangle\left\{\left[1+2\alpha_b\left(j\frac{f}{f_b}\right)-\left(\frac{f}{f_b}\right)^2\right]^{k_b}\right\} + \dots.$$

Because $\measuredangle z^k = k\measuredangle z$ and (f/f_0) is real, the last expression can be written

$$\begin{aligned}\measuredangle H_v = \measuredangle K + \frac{k_0\pi}{2} + k_1\measuredangle\left[1+\left(\frac{f}{f_1}\right)\right] \\ + k_2\measuredangle\left[1+\left(\frac{f}{f_2}\right)\right] + \dots \\ + k_a\measuredangle\left[1+2\alpha_a\left(j\frac{f}{f_a}\right)-\left(\frac{f}{f_a}\right)^2\right] \\ + k_b\measuredangle\left\{\left[1+2\alpha_b\left(j\frac{f}{f_b}\right)-\left(\frac{f}{f_b}\right)^2\right]\right\} + \dots.\end{aligned} \tag{15.70}$$

In words, the angle of a transfer function given by (15.68) is the algebraic sum of the angles of the individual factors. Again, linear factors having multiplicities greater than two and quadratic factors having multiplicities greater than one rarely arise in transfer functions for electronic circuits.

The real part of a quadratic factor can be negative for some frequencies, so we must be careful to put the angles of the quadratic factors in the correct quadrant (by keeping track of the signs of the real and imaginary parts).

The first and second terms on the right side of (15.70) are independent of frequency. The first term is either zero (if $k \geq 0$) or π (if $K < 0$) and the second term is an positive or negative integer multiple of $\pi/2$. Thus the only effect of the first and second terms is a translation (up or down) by a multiple of $\pi/2$ radians. Frequency dependence of phase shift comes from only other linear and quadratic factors. All of the other linear factors have the same form:

$$\begin{aligned}\phi_1 &= \measuredangle(1+jf/f_1)^{k_1} = k_1\measuredangle(1+jf/f_1) \\ &= k_1\tan^{-1}\left(\frac{f}{f_1}\right).\end{aligned} \tag{15.71}$$

Figure 15.27 shows a graph of the right side of (15.71) for $k_1 = \pm 1, + \pm 2$.

Figure 15.28 shows piecewise straight-line approximations to the curves in Fig. 15.27. The actual curves also are shown (dotted) for comparison. For $k_1 = \pm 1$, the approximation consists of a horizontal line at $\phi_1 = 0$ for $f < 0.1f_1$, a line from the point $(0.1f_1, 0)$

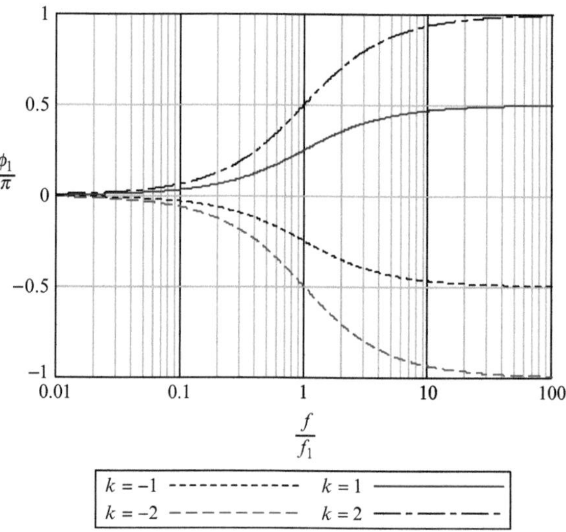

Fig. 15.27 Phase shift versus frequency for a linear factor (see (15.71))

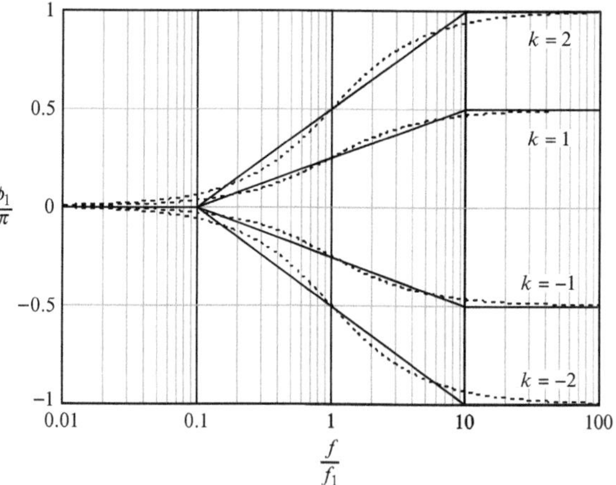

Fig. 15.28 Piecewise straight-line approximations to the phase shift introduced by a linear factor

to the point $(10f_1, \pm\pi/2)$, and a horizontal line at $\phi_1 = \pm\pi/2$ for $f > 10f_1$. For $k_1 = \pm 2$, the approximation consists of a horizontal line at $\phi_1 = 0$ for $f < 0.1f_1$, a line from the point $(0.1f_1, 0)$ to the point $(10f_1, \pm\pi)$, and a horizontal line at $\phi_1 = \pm\pi$ for $f > 10f_1$. Thus two slope changes are associated with each corner frequency, the slope changes occurring 1 decade below and 1 decade above the corner frequency. The frequencies $0.1f_0$ and $10f_0$ associated with a corner frequency f_0 are called **phase break frequencies**.

Example 15.20. Draw a piecewise straight-line graph approximating the phase shift for a circuit having the transfer function

$$H_v = \frac{-10^3(jf/f_0)(1 + jf/f_2)}{(1 + jf/f_1)(1 + jf/f_3)^2},$$

where $f_0 = 10\,\text{Hz}$, $f_1 = 300\,\text{Hz}$, $f_2 = 2\,\text{kHz}$, and $f_3 = 50\,\text{kHz}$.

Solution: The procedure is similar that illustrated by Example 15.17 for asymptotic gain plots. The principal differences lie in (1) treating factors of the form $K(jf/f_0)^k$ and (2) the piecewise straight-line approximation for the phase shift of a linear factor $(1 + jf/f_1)^{k_1}$ has two breaks: the first at $f_1/10$ and the second at $10f_1$. The solution follows.

1. The factor $-10^3(jf/f_0)$ contributes a frequency-independent phase shift $\pi + \pi/2 = 3\pi/2$. We set this aside and treat the remaining factors.
2. List the phase break frequencies, the increase in slope associated with each phase break frequency, and the net slope after each phase break frequency, being careful to account for the multiplicities of the associated factors. The slope increases associated with a factor at the first and second phase break frequencies must have equal magnitudes and opposite signs; e.g., a factor $(1 + jf/f_1)$ in the denominator decreases the slope by $\pi/4$ rad/decade at $f = f_1/10$ and increases the slope by $-\pi/4$ rad/decade at $f = 10f_1$.

Phase break frequency	Slope increase (rad/decade)	Net slope after (rad/decade)
$f_1/10 = 30$ Hz	$-\pi/4$	$-\pi/4$
$f_2/10 = 200$ Hz	$+\pi/4$	0
$10f_1 = 3$ kHz	$+\pi/4$	$+\pi/4$
$f_3/10 = 5$ kHz	$-\pi/2$	$-\pi/4$
$10f_2 = 20$ kHz	$-\pi/4$	$-\pi/2$
$10f_3 = 50 \times 10^4$ Hz	$+\pi/2$	0
$10f_3 = 500$ kHz		

3. Begin the graph at 10 Hz, which is the largest integral power of ten at least 1 decade below f_1. The initial slope is zero if you begin the graph at least 1 decade below the lowest corner frequency.

4(a) Draw a horizontal line from 10 Hz to the first break at $f_1/10 = 30$ Hz.

4(b) Draw a line having slope $-\pi/4$ rad/decade from 20 to 200 Hz.

4(c) Draw a line having slope zero (a horizontal line) from 200 Hz to 3 kHz.

4(d) Draw a line having slope $-\pi/4$ rad/decade from 3 to 5 kHz.

4(e) Draw a line having slope $-\pi/4$ rad/decade from 5 to 20 kHz.

4(f) Draw a line having slope $-\pi/2$ rad/decade from 20 to 500 kHz.

4(g) Draw a horizontal line from 500 kHz to the end of the graph.

5. Adjust the vertical scale to account for the factor set aside in step 1 above. In this case, add $3\pi/2$ to each label (e.g., 0 becomes $3\pi/2$).

Figure 15.29 shows the resulting approximate graph. The actual (computed) phase shift also is shown (dotted curve) for comparison.

The phase shift contributed by a quadratic factor (multiplicity one) is given by

$$\phi = \pm\measuredangle\left[1 + 2j\alpha\left(\frac{f}{f_0}\right) - \left(\frac{f}{f_0}\right)^2\right]. \tag{15.72}$$

where $0 < \alpha \leq 1$. We do not treat quadratic factors having multiplicity greater than one because they rarely arise in analysis of electronic circuits. In (15.72),

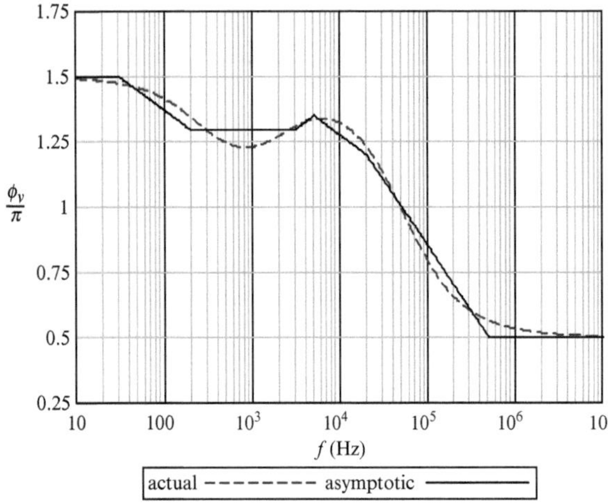

Fig. 15.29 See Example 15.20

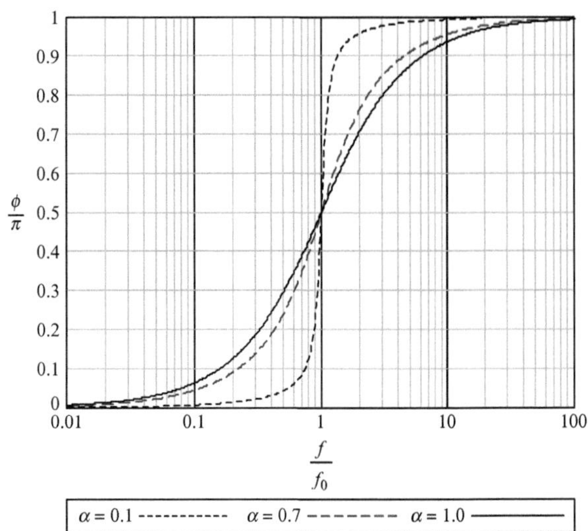

Fig. 15.30 Phase shift introduced by a quadratic factor (see (15.72))

the positive sign applies if the factor is in the numerator and the negative sign applies if the factor is in the denominator. If $\alpha = 1$, (15.72) may be written as the square of a linear factor:

$$\phi = \pm\measuredangle\left[1 + 2j\left(\frac{f}{f_0}\right) - \left(\frac{f}{f_0}\right)^2\right] = \pm\measuredangle\left[1 + j\left(\frac{f}{f_0}\right)\right]^2. \tag{15.73}$$

If $\alpha < 1$, the quadratic factor cannot be expressed as a product of two linear factors.

Figure 15.30 shows graphs of the angle given by (15.72) using the positive sign (for a numerator factor), for $\alpha = 0.1, 0.7, 1.0$. The graph indicates that a squared linear factor ($\alpha = 1.0$) is a fair approximation to a quadratic factor if α is very close to one, say $\alpha > 0.9$, and a step is a fair approximation if α is small, say $\alpha < 0.1$. For $0 < \alpha < 0.7$ About all we can do (other than resort to a computer) is to either sketch the graph based upon familiarity with Fig. 15.30 or compute a few points and draw a smooth curve through the computed points. In general, if phase shift is critical in analysis and design, it is best to use a computer or calculator to generate an accurate graph.

Example 15.21. The voltage transfer function for a certain circuit is

$$H_v = \frac{(jf/f_0)(1 + jf/f_2)}{\left[1 + 2\alpha j(f/f_1) - (f/f_1)^2\right](1 + jf/f_3)},$$

where $f_0 = 10\,\text{Hz}$, $f_1 = 200\,\text{Hz}$, $f_2 = 2\,\text{kHz}$, and $f_3 = 50\,\text{kHz}$. Construct graphs of phase shift versus frequency for $\alpha = 0.1,\ 0.7,\ 1.0$ and comment on the implications for approximating a quadratic factor as the square of linear factor.

Solution: The phase shift is given by

$$\phi = \measuredangle\frac{(jf/f_0)(1 + jf/f_2)}{\left[1 + 2\alpha j(f/f_1) - (f/f_1)^2\right](1 + jf/f_3)}$$

Figure 15.31 shows graphs of the phase shift for three values of the peaking factorα. For $\alpha = 1$, the quadratic factor is the square of a linear factor, as in (15.73). The graph suggests that approximating the quadratic factor as the square of a linear factor having the same corner frequency is not too bad if $\alpha \geq 0.7$.

In some electronics textbooks, a quadratic factor is approximated by the square of a linear factor having the same corner frequency, regardless of the value of the peaking factor α. As shown in Section 15.7, such an approximation gives reasonably good results for gain, provided the magnitude of the parameter α is larger than about 0.25. But for phase shift, such an approximation is poor if α is smaller than about 0.7.

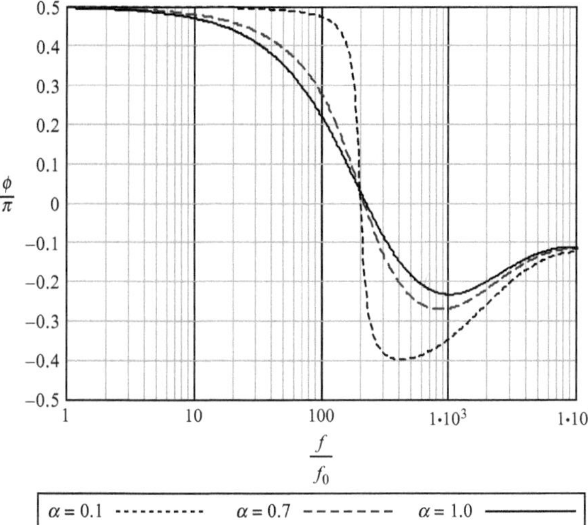

Fig. 15.31 See Example 15.21

Because phase shift is at least as important as gain in many applications, it might be necessary to use a computer or calculator to obtain a sufficiently accurate representation of phase shift versus frequency if $\alpha < 0.7$

15.9 Filters and Bandwidth

All physical circuits are *frequency selective*, meaning that any circuit attenuates or amplifies some sinusoidal inputs more than others. Some linear circuits, called filters, are designed to allow certain sinusoidal inputs to reach a load with minimal differences in attenuation or gain and to prevent others from reaching the load. Such filters are said to *pass* some sinusoidal inputs and *reject* (or *stop*) others, and are characterized by *passbands* and *stopbands*. The passband gain of a filter is approximately uniform. The stopband gain is much smaller than that in the passband. Sinusoids having frequencies in the passband all reach the load having undergone approximately the same attenuation or amplification. Sinusoids having frequencies in the stopband also reach the load, but after having undergone much more attenuation (or much less amplification) than those in the passband.

Figure 15.32 shows qualitative graphs of gain versus frequency for four important kinds of filters, named as follows:

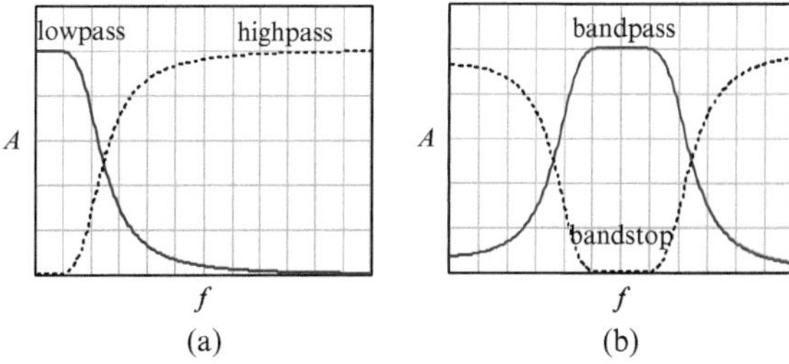

Fig. 15.32 Qualitative graphs of gains of (**a**) lowpass and bandpass circuits and (**b**) bandpass and bandstop circuits

- A **lowpass filter** has a passband that extends from $f = 0$ to some finite frequency and a stopband that extends from that frequency to infinity (ideally).
- A **highpass** filter has a stopband that extends from zero to some finite, non-zero frequency and a passband that extends from that frequency to infinity (ideally).
- A **bandpass** filter has a passband that lies between two finite, non-zero frequencies and stopbands on each side of the passband.
- A **bandstop** filter has a stopband that lies between two finite, non-zero frequencies and passbands on each side of the stopband.

The **bandwidth** of a lowpass or bandpass filter is the width (Hz) of the passband. Highpass and bandstop filters are described in terms of their stopbands, because their passbands are (ideally) infinite.

Filters are used to select a signal of interest from among many others. For example, a radio receiver uses filtering to select one of the many AM and FM broadcast stations available in any geographical area. Such selection is possible because the *significant* sinusoidal components of a signal are confined to a finite band on the frequency axis. For example, the significant components of a current or voltage representing *telephone-quality speech* lie between 300 Hz and 4 kHz. In other words, if we eliminate all of the sinusoidal components outside that band from a speech signal and listen to the signal so obtained, we will be able to understand what is said and (if the speaker is someone we know) recognize the speaker, which are the two main requirements for telephone-quality speech.

The **bandwidth** of a signal is the width of the band (Hz) occupied by the significant components of the signal. The meaning of significant depends upon both the signal and the application. For example, so far as

AM radio is concerned, the significant components of a music or speech signal lie in a band extending from zero to about 5 kHz, whereas for FM, that band extends from zero to about 15 kHz. Thus the bandwidth of a signal heard on an AM radio is about 5 kHz, whereas the bandwidth of a signal heard on an FM radio is about 15 kHz.

All filters introduce distortion, even for signals in their passbands. The quality of a filter is measured in part by the amount of distortion introduced. To define or quantify distortion, we must first describe what is meant by absence of distortion. By definition, a circuit having input v_S and output v_L provides **distortionless transmission** if the circuit time-domain transfer characteristic has the form

$$y(t) = Kx(t - t_0), \tag{15.74}$$

where $x(t)$ is an input (current or voltage) and $y(t)$ is the corresponding output (current or voltage). The parameters K and t_0 are independent of both time and frequency. In words, a circuit provides distortionless transmission for an input if the output is simply an amplitude scaled (by K) and time delayed (by t_0) version of the input. This condition can be explained in familiar terms. When you listen to recorded music using a high-quality audio system, there are two acceptable differences between what you hear and what you would have heard at the original performance. One is loudness, which you can control using the loudness control on the amplifier. The other is the time difference between recording and playback. Neither of these differences is perceived as distortion because neither alters the basic structure of the acoustic wave incident on your eardrum. In (15.74), the difference in loudness is represented by the scale factor K and the time difference by the delay t_0.

To obtain frequency-domain conditions for distortionless transmission, we must first obtain the transfer function of a distortionless circuit. To that end, we assume a transfer characteristic of the form (15.74) and a sinusoidal input represented in complex form as

$$\tilde{x} = \tilde{X}\exp(j\omega t).$$

where $\tilde{X}$ is the phasor for the input. From (15.74), the complex representation of the output is

$$\tilde{y} = K\tilde{X}\exp[j\omega(t - t_0)] = \tilde{Y}\exp(j\omega t).$$

Thus the phasor for the output is

$$Y = K\tilde{X}\exp(-j\omega t_0).$$

It follows that the transfer function is

$$H = \frac{\tilde{Y}}{\tilde{X}} = K\exp(-j\omega t). \tag{15.75}$$

The transfer function given by (15.75) might be any of the four possible kinds, depending upon the dimensions of the input $\tilde{x}(t)$ and output $\tilde{y}(t)$; e.g., if both are voltages, the transfer function is a voltage transfer function. So the condition for distortionless transmission is that the *transfer function of interest* must have the form (15.75).

In (15.75), we may assume $K > 0$, because inversion (sign change) does not introduce distortion. Thus the gain and phase shift of a distortionless circuit are given by

$$A = |H| = K, \ \phi = \measuredangle H = -2\pi f t_0, \ K > 0. \tag{15.76}$$

In view of (15.76), the condition for distortionless transmission is usually stated as follows: *A circuit provides distortionless transmission if the gain is independent of frequency and the phase shift is a linear (straight-line) function of frequency.*

No physical circuit can provide distortionless transmission for all frequencies. At best, a physical circuit can provide approximately distortionless (high-fidelity) transmission for a limited class of signals, such as signals representing music, speech, or video. Ideally, any one of the four filters described above should introduce no distortion in its passband. Distortion in a stopband is irrelevant, because signals in the stopband are greatly attenuated relative to those in the passband.

A linear filter can introduce two kinds of distortion: **amplitude distortion**, caused by frequency-dependent gain in the passband, and **phase distortion**, caused by nonlinear phase shift in the passband. The relative importance of these two kinds of distortion depends upon the signals involved. The human ear is relatively insensitive to phase distortion in speech, so phase distortion is relatively less important than amplitude distortion in some speech transmission systems. That difference is significant because it is easier (cheaper) to limit amplitude distortion than it is to limit both amplitude and phase distortion.

Even so, and fortunately, gain and phase shift of many filter circuits are related such that if gain is approximately independent of frequency over a certain band, then phase shift is approximately linear over the same or a slightly smaller band. Consequently, in many applications, gain alone is used to define circuit bandwidth, especially in cases where phase distortion is less disruptive than amplitude distortion. Below, we define circuit bandwidth in terms of gain alone. However, there are many important applications where phase distortion is at least as disruptive as amplitude distortion, so when high-fidelity transmission is required, both gain and phase shift should be considered.

A widely used quantitative definition of bandwidth is the **half-power** or **3 dB bandwidth**. For *lowpass and bandpass* circuits, the half-power bandwidth is the width (in Hz) of the band over which the squared gain is no less than half the square of a **reference gain**, usually the *maximum or average gain in the passband.* Equivalently, the half-power bandwidth is the width (in Hz) of the band over which the gain is no more than 3 dB below the reference gain. Thus **the half-power (3 dB) bandwidth of a lowpass or bandpass circuit** is the range of frequencies for which

$$\frac{A^2}{{A_0}^2} \geq \frac{1}{2} \Rightarrow A_{\mathrm{dB}} \geq A_{0\mathrm{dB}} - 3, \tag{15.77}$$

where A_0 is the reference gain and $A_{0\,\mathrm{dB}} = 20\,\log(A_0)$. The 3 dB bandwidth is convenient partly because it often corresponds to a corner frequency in an asymptotic plot of gain versus frequency.

Example 15.22. The voltage transfer function for a certain lowpass circuit is

$$H_v = \frac{1}{1 + jf/f_0}$$

Use the dc gain as the reference gain and find the half-power bandwidth of the circuit.

Solution: The squared voltage gain is given by

$${A_v}^2 = \frac{1}{1 + (f/f_0)^2}$$

The squared reference (dc) gain is ${A_0}^2 = {A_v}^2(0) = 1$. The half-power bandwidth consists of the frequencies for which

$$\frac{A^2}{{A_0}^2} \geq \frac{1}{2} \Rightarrow \frac{1}{1 + (f/f_0)^2} \geq \frac{1}{2} \Rightarrow \frac{f}{f_0} \leq 1 \Rightarrow f \leq f_0$$

The half-power bandwidth is f_0

Bandwidth is seldom used to describe highpass and bandstop filters. The width of the passband for such circuits is (theoretically) infinite. Instead, such a circuit usually is described (in part) by the width of its stopband, which is often defined as the width (in Hz) of the band over which the squared gain is no more than half the square of a *reference gain*, usually either *the maximum or average gain in the passband.* Equivalently, the width of the stopband is the width (in Hz) of the band over which the gain is at least 3 dB *below* the reference gain. Thus **the width of the stopband of a highpass or bandstop circuit** is the range of frequencies for which

$$\frac{A^2}{{A_0}^2} \leq \frac{1}{2} \Rightarrow A_{\mathrm{dB}} \leq A_{0\mathrm{dB}} - 3, \tag{15.78}$$

where A_0 is the reference gain and $A_{0\,\mathrm{dB}} = 20\,\log(A_0)$.

Example 15.23. The voltage transfer function for a certain highpass circuit is

$$H_v = \frac{jf/f_0}{1 + jf/f_0}$$

Use the maximum passband gain as the reference gain and find the half-power width of the stopband.

Solution: The squared voltage gain is given by

$${A_v}^2 = \frac{(f/f_0)^2}{1 + (f/f_0)^2}$$

The maximum gain occurs for $f \to \infty$, where $A_v(f) \to 1$, so the squared reference gain is $A_0{}^2 = A_v{}^2(\infty) = 1$. The half-power width of the stopband consists of the frequencies for which

$$\frac{A^2}{A_0{}^2} \leq \frac{1}{2} \Rightarrow \frac{(f/f_0)^2}{1+(f/f_0)^2} \leq \frac{1}{2} \Rightarrow \frac{f}{f_0} \leq 1$$
$$\Rightarrow 0 \leq f \leq f_0$$

The half-power width of the stopband is f_0

Although half-power bandwidth is a widely used definition of bandwidth, it is not the only definition in use. For example, the bandwidth of a high-quality audio amplifier might be defined as the width of the band for which the gain is within 0.5 dB of the average passband gain.

Moreover, many useful circuits are not simply lowpass, highpass, bandpass, or bandstop. For some such circuits, bandwidth is more generally defined as the *width of the band over which the circuit performs its intended function.* For example, the bandwidth of an integrating circuit refers to the band of frequencies for which the circuit performs as an integrator.

15.10 Frequency Response

Quantities (such as transfer functions) used to specify or describe circuit performance should be measurable so one can verify performance experimentally. As commonly used in textbooks, the **frequency response** of a circuit is essentially the *measured or calculated voltage transfer function* of the circuit. In practice, *frequency response* almost always refers to a measured *voltage gain in dB versus frequency.*

Various specialized and sophisticated instruments are available for measuring frequency response, but the relatively primitive scheme illustrated by Fig. 15.33 is instructive. The VF O (variable-frequency oscillator) generates a variable-frequency sinusoidal voltage at the circuit input and serves as the phase reference. The oscilloscope displays both the input and output voltages on the same time axis. The measurement is made by recording the peak

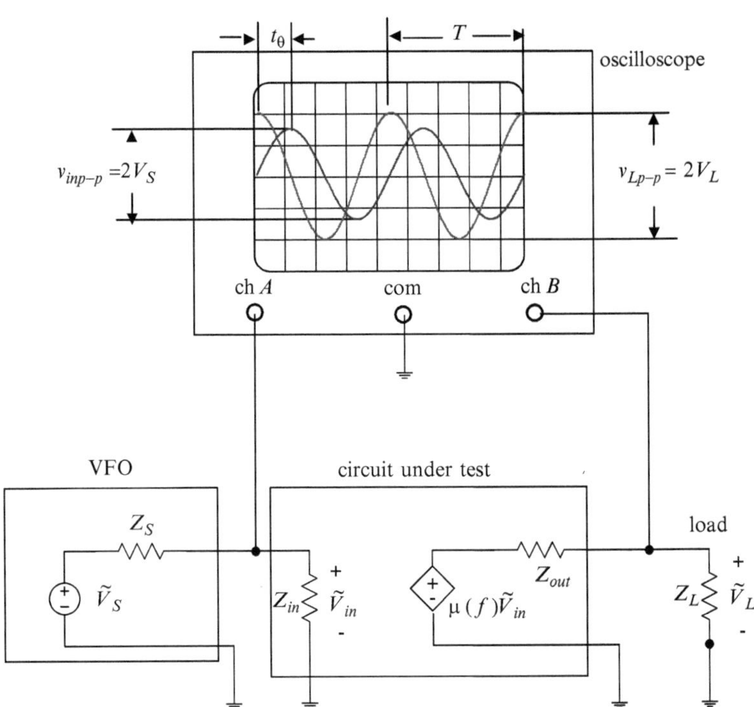

Fig. 15.33 Measurement of frequency response

amplitudes of the input and output and the phase delay of the output from the oscilloscope screen at each of several frequencies in the band of interest. The *voltage gain* at each measurement frequency *f* is then recorded as

$$|FR(f)| = \frac{V_{Lp-p}(f)}{V_{inp-p}(f)}, \tag{15.79}$$

where V_{inp-p} and V_{Lp-p} are the peak-to-peak amplitudes of the input and output, respectively.[6] The *phase shift* at each measurement frequency f is the initial phase of the output (relative to that of the input) and is recorded as

$$\measuredangle FR(f) = -2\pi f\, t_\theta(f), \tag{15.80}$$

where $t_\theta(f)$ is the measured phase delay at the frequency f. Recall that phase delay is measured from a positive peak of the input (the reference peak) to the *nearest-neighbor* positive peak of the output. (An equivalent and perhaps more accurate method is to measure the time from a positive-going or negative going zero crossing for the input to the nearest-neighbor crossing of the same kind for the output.) The phase delay is positive if the nearest peak of the output is to the right of the reference peak and negative if the nearest peak of the output is to the left of the reference peak, as it is in Fig. 15.33. The frequency can be verified by measuring the period T and using the relation $f = T^{-1}$.

A measured frequency response and the corresponding (mathematically defined) voltage transfer function are (in general) not quite the same things. The measured frequency response takes the voltage $\tilde{V}_{in}$ at the input terminals of the circuit as the input, whereas the mathematically defined frequency response takes the available (Thévenin-equivalent) voltage $\tilde{V}_S$ as the input. With reference to Fig. 15.33, the voltage $\tilde{V}_{in}$ used in the measurement is given by

$$\tilde{V}_{in} = \frac{Z_{in}}{Z_{in} + Z_S}\tilde{V}_S, \tag{15.81}$$

[6]Gain also could be expressed in terms of peak amplitudes; however, it usually is easier to obtain accurate values for peak-to-peak amplitudes from an oscilloscope display than it is to obtain accurate readings of peak amplitudes – especially if a dc component also is present.

where Z_{in} is the input impedance of the circuit, Z_S is the output impedance of the source, and $\tilde{V}_S$ is the available voltage. If the impedances Z_{in}, Z_S are known, the measured frequency response can be corrected to account for source impedance. In many cases such correction is unnecessary because $|Z_{in}| \gg |Z_S|$ and thus (from (15.81)), $\tilde{V}_{in} \cong \tilde{V}_S$.

Generally, the load impedance used in measuring a frequency response is known, so if the circuit output impedance is known, the measured frequency response can be corrected to account for another, different load impedance.

Moreover, it is usually the shape of a frequency response (selectivity) that is of interest, and not the actual magnitudes. That is, we are primarily interested in the gains at some frequencies relative to the gains at others, and not necessarily the actual gain at either. If the actual source and load impedances and the source and load impedances used for a measurement are all resistive, then the actual (in-use) voltage transfer function and the measured transfer function (frequency response) differ only by a frequency-independent scale factor.

We can in principle measure any of the four transfer functions for a circuit, but the one measured is almost invariably the voltage transfer function, primarily because it is easier to obtain a voltage source than a current source and it is easier to measure voltage than it is to measure current. Here again, if we know the impedances of our instruments and the actual source and load impedances when the circuit is in use, we often can deduce other transfer functions from the measured voltage transfer function.

15.11 Problems

Section 15.1 is prerequisite to the following problems.

P 15.1 Refer to Fig. P 15.1. Obtain expressions for the voltage transfer function, the current transfer function, the transimpedance, and the transadmittance for each circuit.

P 15.2 Refer to Fig. P 15.2. Obtain expressions for the voltage transfer function, the current transfer

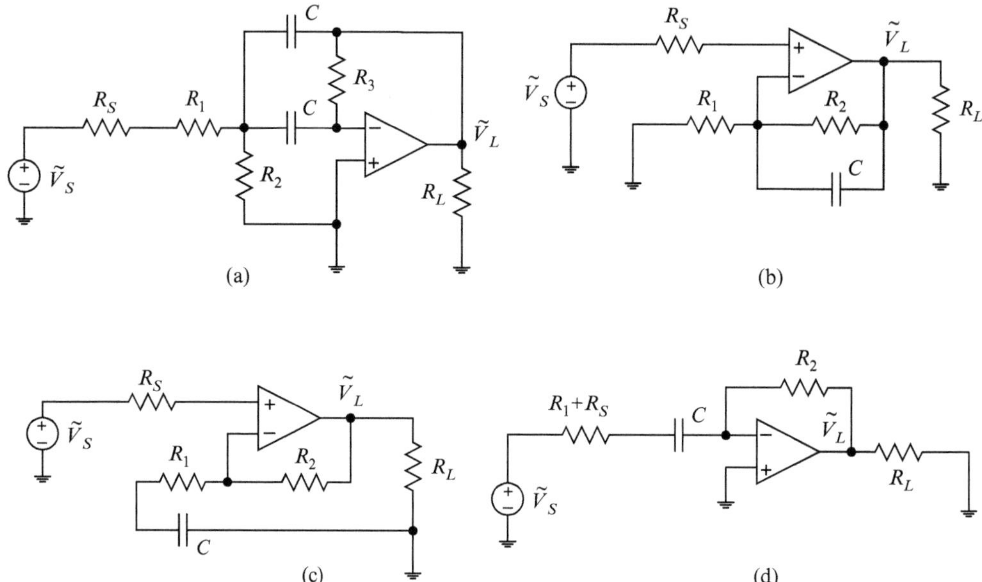

Fig. P 15.1 See Problem P 15.1

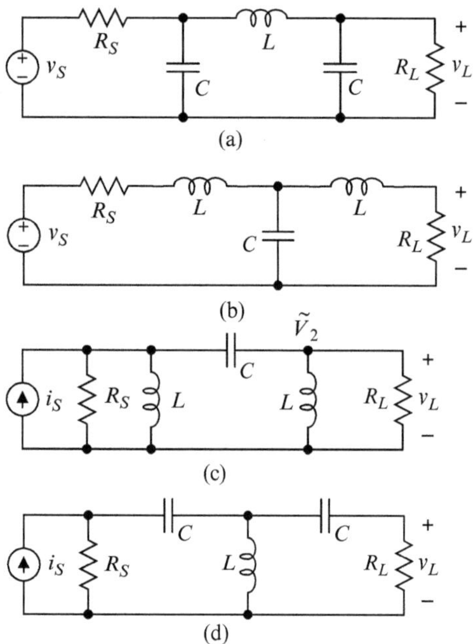

Fig. P 15.2 See Problem P 15.2

function, the transimpedance, and the transadmittance for each circuit.

P 15.3 Refer to Fig. P 15.3. For each circuit,

$$R_S = 50\ \Omega, R_i = 20\ \text{k}\Omega, C = 20\ \text{nF}, R_F = 50\ \text{k}\Omega,$$
$$R_o = 100\ \Omega, R_L = 10\ \text{k}\Omega.$$

Obtain expressions for the voltage transfer function, the current transfer function, the transimpedance, and the transadmittance. Use all reasonable approximations (10% error is tolerable). Then find the load voltage $v_L(t)$ for

$$v_S(t) = V_0[1 + \cos(\omega_0 t) + 0.5\cos(2\omega_0 t) + 0.2\cos(3\omega_0 t)];\ V_0 = 10\ \text{V}, f_0 = 100\ \text{kHz}$$

or

$$i_S(t) = I_0[1 + \cos(\omega_0 t) + 0.5\cos(2\omega_0 t) + 0.2\cos(3\omega_0 t)];\ I_0 = 200\ \text{mA}, f_0 = 100\ \text{kHz}$$

as appropriate.

Section 15.2 is prerequisite to the following problems.

P 15.4 You have asked your design team to propose various alternatives for achieving a voltage transfer function having the form

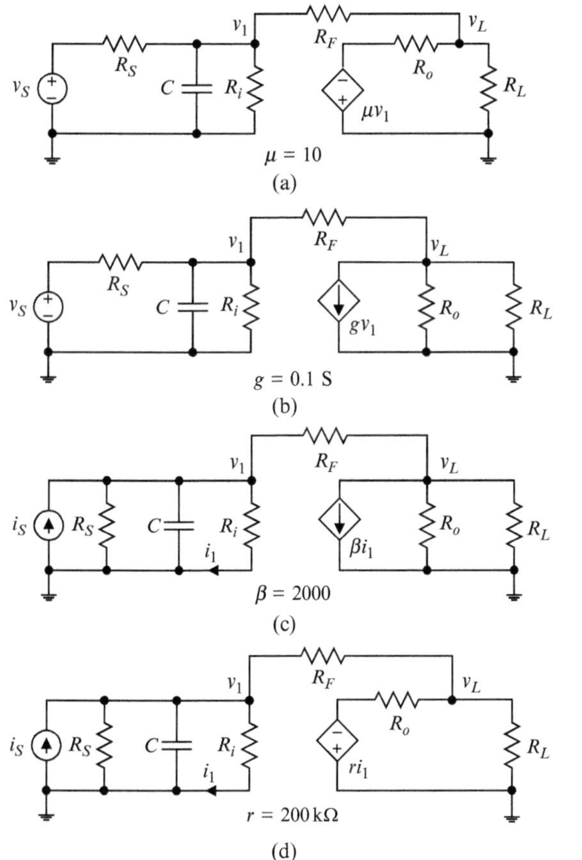

Fig. P 15.3 See Problem P 15.3

$$H_v = \frac{K}{1 + j(f/f_0)}$$

independent of the source and load impedances. The value of K is unimportant, but the parameter f_0 is specified. Your team returns with the circuits shown in Fig. P 15.4.

(a) Show that under certain conditions the transfer function of each has the form above. Express K and f_0 in terms of the circuit parameters and state conditions necessary to making the transfer function independent of the source and load.
(b) Choose the one you think is best, and explain why you made that choice.

P 15.5 The input impedance of a certain amplifier is known to be resistive and fixed for frequencies up to 10 MHz. In the laboratory, the amplifier is driven by a fixed-amplitude, variable-frequency sinusoidal voltage source having negligible output impedance. It is

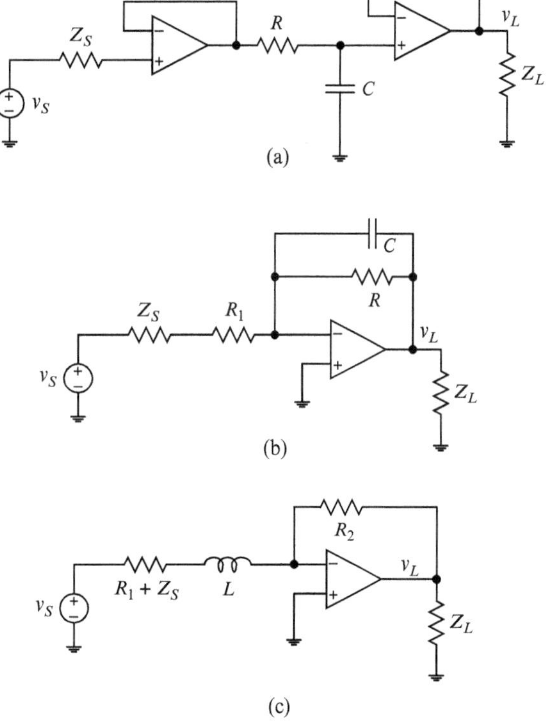

Fig. P 15.4 See Problem P 15.4

observed that the rms amplitude of the load voltage is almost independent of frequency for frequencies up to about 100 kHz, at which point the load voltage begins decreasing with increasing frequency. What, if anything, does this tell you about the amplifier, the load, or both?

P 15.6 The output impedance of a certain circuit is known to be resistive and fixed for frequencies up to 10 MHz. In the laboratory, the amplifier is driven by a fixed-amplitude, variable-frequency sinusoidal voltage source. It is observed that the rms amplitude of the load voltage is almost independent of frequency for frequencies up to about 100 kHz, at which point the load voltage begins decreasing with increasing frequency. What, if anything, does this tell you about the circuit, the load, or both?

P 15.7 For each circuit in Fig. P 15.3 (Problem P 15.3), describe how you would specify the input impedance $Z_i = R_i/(1 + j\omega R_i C)$ and the output resistance R_o relative to the source and load impedances if the circuit is to function for frequencies below 10 kHz as (a) a voltage amplifier, (b) a current amplifier, (c) a transadmittance amplifier, and (d) a transimpedance amplifier.

P 15.8 An amplifier is to drive an inductive load, where the load current is to be proportional to the

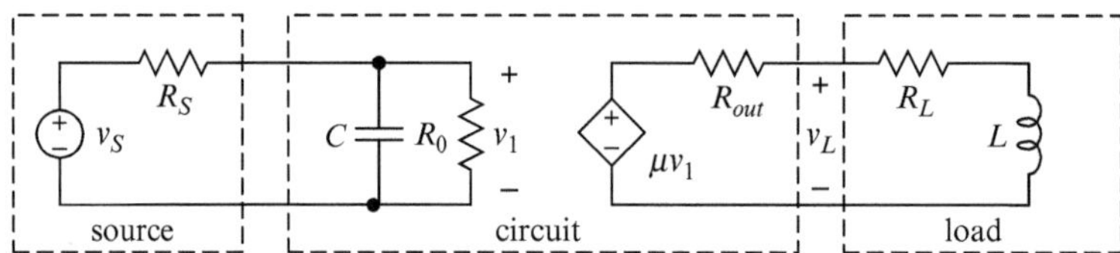

Fig. P 15.5 See Problem P 15.9

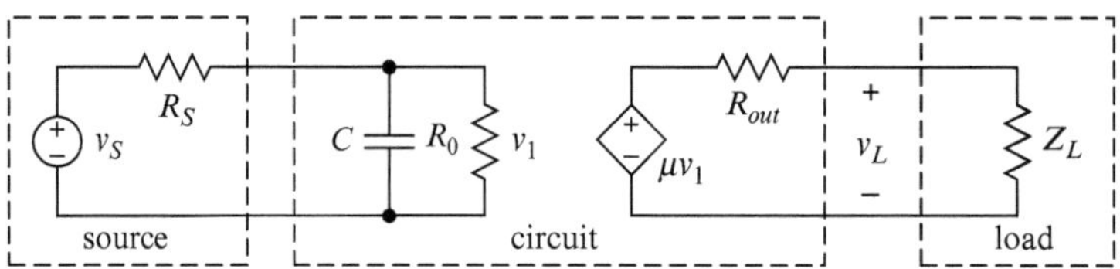

Fig. P 15.6 See Problem P 15.10

available voltage from a source. (a) Which of the four transfer functions should be independent of the source and load impedances? (b) The load impedance is given by

$$Z_L = R_L + j\omega L,$$

with $R_L = 100\ \Omega$ and $L = 500$ mH. The output impedance of the amplifier is resistive for frequencies up to 100 kHz. If the appropriate transfer function is to be independent of the load for frequencies up to 500 Hz, what is the minimum allowable value for the amplifier output resistance? Express your answer in the form $R_{out} \gg$ ____.

P 15.9 Refer to Fig. P 15.5, where

$$R_S = 100\,\Omega, C = 25\,\text{pF}, R_0 = 25\,\text{k}\Omega, \mu = 10,$$
$$R_{out} = 75\,\Omega, R_L = 200\,\Omega, L = 10\,\text{H}, v_S = V_S\cos(\omega t).$$

For what frequencies is the voltage transfer function approximately independent of frequency? Express your answer in the form ___ $\ll f \ll$ ___.

P 15.10 Refer to Fig. P 15.6, where

$$R_S = 100\ \Omega, C = 25\ \text{pF}, R_0 = 25\ \text{k}\Omega, \mu = 10,$$
$$R_{out} = 75\ \Omega, v_S = V_S\cos(\omega t)$$

and

$$Z_L = \frac{K}{1 + j(f/f_1)},\ f_1 = 10\ \text{MHz},\ K = 1\ \text{k}\Omega.$$

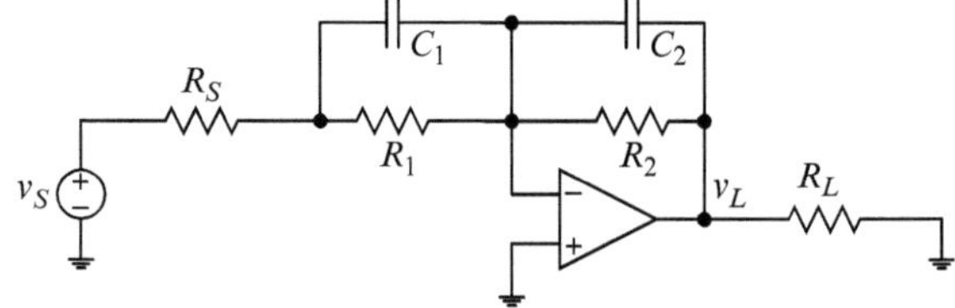

Fig. P 15.7 See Problem P 15.11

For what frequencies (if any) is the voltage transfer function approximately independent of frequency? Express your answer in the form ___ $\ll f \ll$ ___.

P 15.11 See Fig. P 15.7. (a) Obtain an expression for the voltage transfer function of the circuit. (b) Let

$$R_S = 25\ \Omega, R_1 = 10\ \text{k}\Omega, R_2 = 1\text{M}\Omega, C_1 = 12\ \text{pF},$$
$$C_2 = 30\ \text{pF}, R_L = 5\ \text{k}\Omega.$$

For what frequencies (if any) is the voltage transfer function approximately independent of frequency? Express your answer in the form ___ $\ll f \ll$ ___.

(c) Find the frequencies for which each of the current transfer function, the transadmittance, and the transimpedance are approximately independent of frequency. Express each answer in the form ___ $\ll f \ll$ ___.

Section 15.3 is prerequisite to the following problems.

P 15.12 Find the voltage gain and phase shift, the current gain and phase shift, the transadmittance and associated phase shift, and the transimpedance and associated phase shift for each circuit in Problem P 15.3 and for $f = 100, 200, 300$ H_z.

P 15.13 Refer to Fig. P 15.8, where the op amp is ideal. The specified **dc** input resistance is $R_{in} = 10$ kΩ. The rms amplitude of the output is to be no larger than 70.7% of the rms amplitude of the input for any sinusoidal input having frequency greater than $f_1 = 25$ kHz. Specify R, R_F, and C. Assume the source impedance is negligible, relative to the specified input resistance.

P 15.14 One of your associates presents you with the graph of voltage versus time shown in Fig. P 15.9, which he claims is the response of a certain low-pass filter to a 10-V step input. How do you know that he has made an error?

P 15.15 A certain amplifier is driven by a sinusoidal source at a frequency for which the output impedance of the amplifier is known to be resistive. Figure P 15.10 shows a graph of the voltage transferred by the amplifier to a resistive load versus the resistance of the load. (a) What is the output resistance of the amplifier? (b) If the output resistance were not resistive, could you deduce the magnitude of the output impedance from the graph in Fig. P 15.10?

Section 15.4 is prerequisite to the following problems. Express all gains in dB.

P 15.16 The voltage gain of a certain amplifier equals 10^4 at 10 kHz, and is down by 40 dB from that value at 100 kHz. What is the voltage gain at 100 kHz?

P 15.17 The current gain of a certain amplifier equals 10^3 at 10 kHz, which is 25 dB below the dc current gain. What is the dc current gain?

P 15.18 The transadmittance of a certain circuit is normalized to its dc value and expressed in dB (the normalized dc transadmittance equals 0 dB). The normalized transadmittance at 15 kHz is – 6 dB. What is the ratio of the transadmittance (not dB) at 15 kHz to the dc transadmittance (not dB)?

P 15.19 The transimpedance of a certain circuit is normalized to its value at 10 kHz and expressed in dB. It is found that the normalized transimpedance is 10 dB at 1 kHz and – 20 dB at 100 kHz. If the actual transimpedance at 100 kHz is 2 kΩ, what are the transimpedances (not dB) at 1 and 10 kHz?

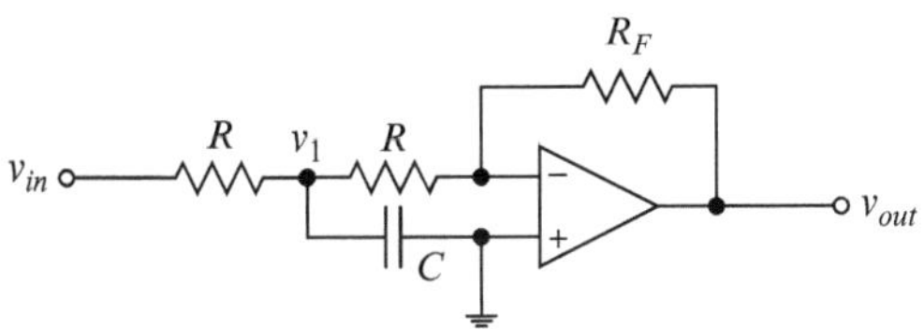

Fig. P 15.8 See Problem P 15.13

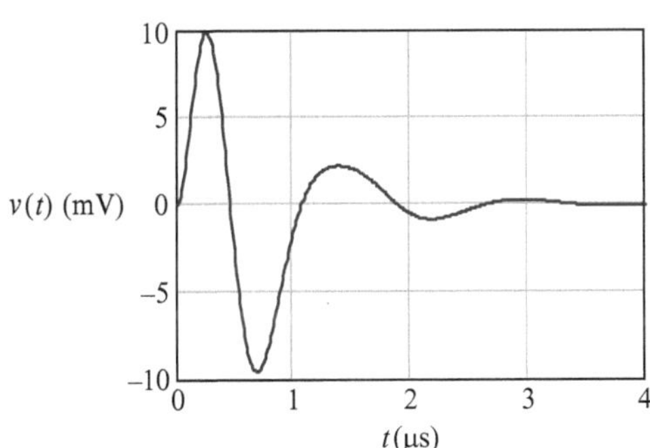

Fig. P 15.9 See Problem P 15.14

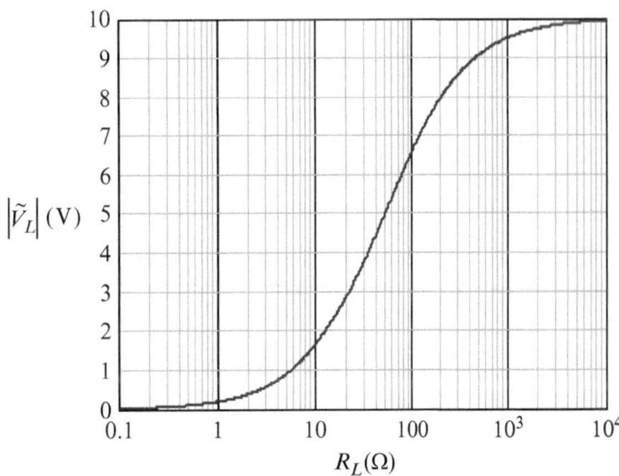

Fig. P 15.10 See Problem P 15.15

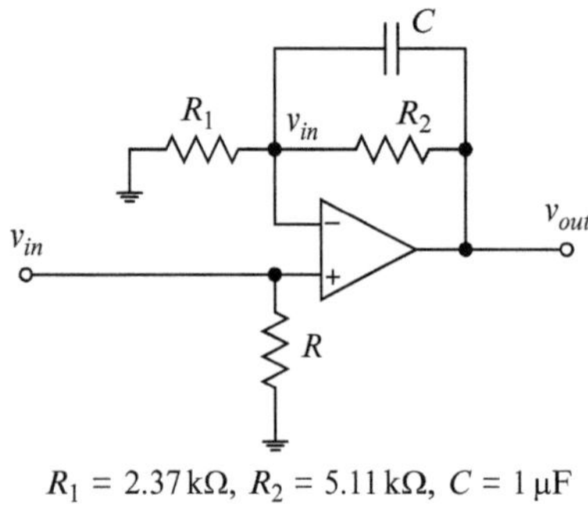

Fig. P 15.11 See Problem P 15.20

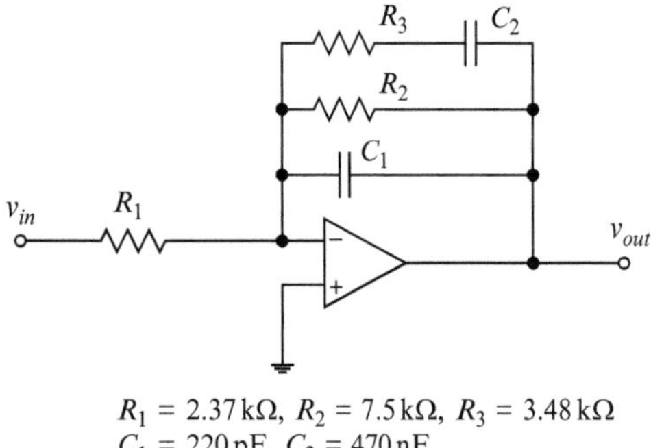

Fig. P 15.12 See Problem P 15.21

P 15.20 Figure P 15.11 shows an audio bass boost circuit. Obtain the voltage transfer function and plot voltage gain and phase shift versus frequency. Assume the op amp is ideal and that the source impedance is negligible.

P 15.21 Figure P 15.12 shows an audio bass boost circuit. Obtain the voltage transfer function and plot voltage gain and phase shift versus frequency. Assume the op amp is ideal and that the source impedance is negligible.

P 15.22 Figure P 15.13 shows an audio crossover circuit, where the output v_{hi} is to drive a small tweeter speaker and v_{low} is to drive a larger midrange + woofer speaker. You may assume the op amps are ideal. Obtain and plot the frequency response (voltage gain) for each of the two sections.

P 15.23 Three voltage amplifiers are connected in cascade and driven by a sinusoidal source having frequency f_0 and output impedance $R_{out} = 100\,\Omega$. The voltage transfer functions of the three amplifiers at the frequency of the source are

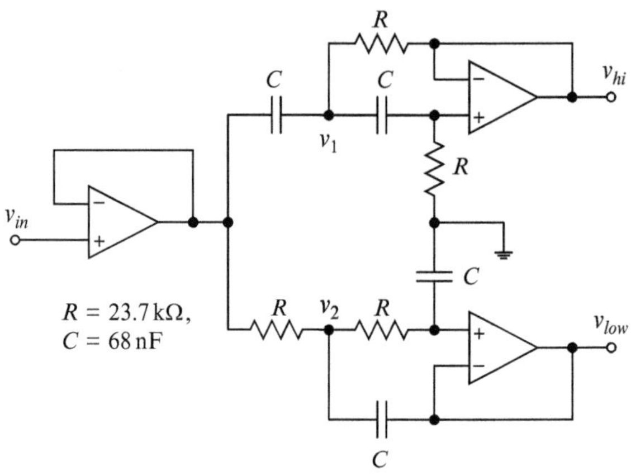

Fig. P 15.13 See Problem P 15.22

$$H_{1v} = 50\angle 0, H_{2v} = 40\angle 0.25, H_{3v} = 5\angle 0.14.$$

Each amplifier has input impedance $R_{in} = 80$ kΩ and output impedance $R_{out} = 50\Omega$. The load on the cascade amplifier is $R_L = 2$ kΩ. Find the overall voltage gain (dB) and phase shift at the frequency f_0 (a) ignoring loading and (b) including loading.

P 15.24 A sinusoidal voltage $v(t) = V_0 \cos(\omega t)$ having amplitude $V_0 = 1$ V is applied to the terminals of a loudspeaker. The resulting sound intensity is 70 dB at a point 1 m from the loudspeaker. The amplitude of the sinusoid is increased until the sound intensity at the same point reaches 100 dB. What is peak amplitude of the resulting voltage across the terminals of the loudspeaker?

P 15.25 The voltage across a resistor is tripled. What is the increase in dB of power dissipated?

P 15.26 A sinusoidal voltage is applied to the terminals of a resistor and the power dissipated by the resistor is calculated. By what factor must the voltage be increased to double the power dissipated? By how many dB?

Section 15.5 is prerequisite to the following problems.

P 15.27 Refer to Problem P 15.20, where $R_2 \gg R_1$. (a) Express the voltage transfer function in standard form as

$$H_v = K\frac{1 + j(f/f_0)}{1 + j(f/f_1)}$$

and express the parameters K, f_0, f_1 in terms of the circuit parameters. (b) Which circuit parameter(s) would you change (and in what direction) if you wished to increase f_1 without changing f_0? (c) Which circuit parameter(s) would you change (and in what direction) if you wished to increase f_0 without changing f_1? (d) Which circuit parameters (if any) affect both corner frequencies?

P 15.28 Express each transfer function below in standard form. Define all corner frequencies and other constants in terms of circuit parameters appearing in the original expression. Describe how increasing each circuit parameter affects each corner frequency.

(a) $$H_i = \frac{50}{j\omega R_0 C_0 (1 + j\omega R_1 C_1)},$$

(b) $$H_v = \frac{1}{1 + 1.8\, j\omega RC + (j\omega RC)^2},$$

(c) $$H_z = \frac{R + j\omega L}{1 + j\omega RC},$$

(d) $$H_y = \frac{1}{R} + j\omega C + \frac{1}{j\omega L},$$

(e) $$H_v = \frac{(1 + j\omega RC)(j\omega RC)^2}{(1 + j\omega RC)\left[(j\omega RC)^2 + 2 + j\omega RC\right]},$$

(f) $$H_v = \frac{R_{in} R_L \mu_0}{(R_L + R_{out} + j\omega R_L R_{out} C_L)(R_{in} + R_S + j\omega L_S)},$$

(g) $$H_v = \frac{20}{1 - \omega^2 LC + 2j\alpha\omega RC}.$$

Section 15.6 is prerequisite to the following problems.

P 15.29 Draw asymptotic plots of gain versus frequency for each of the following transfer functions:

(a) $$H_v = K\frac{1 + jf/f_0}{1 + jf/f_1},\ K = 10,\ f_0 = 200\text{ Hz},\ f_1 = 5\text{ kHz};$$

(b) $$H_v = K\left(\frac{1 + jf/f_0}{1 + jf/f_1}\right)^2,\ K = 10,\ f_0 = 200\text{ Hz},\ f_1 = 5\text{ kHz};$$

(c) $$H_v = \frac{K}{(jf/f_0)(1 + jf/f_1)},\ K = 20,\ f_0 = 1\text{ Hz},\ f_1 = 1\text{ kHz};$$

(d) $$H_v = \frac{K}{(jf/f_0)(1 + jf/f_1)^3},\ K = 20,\ f_0 = 1\text{ Hz},\ f_1 = 1\text{ kHz};$$

(e) $$H_v = \frac{120}{(1 + jf/f_0)(1 + jf/f_1)(1 + jf/f_2)},\ f_0 = 1\text{ kHz},\ f_1 = 10\text{ kHz},\ f_2 = 50\text{ kHz};$$

(f) $$H_v = \frac{K(1 + jf/f_2)}{(jf/f_0)(1 + jf/f_1)^2},\ K = 20,\ f_0 = 1\text{ Hz},\ f_1 = 1\text{ kHz},\ f_2 = 10\text{ Hz}.$$

Section 15.7 is prerequisite to the following problems.

P 15.30 Draw asymptotic plots of gain versus frequency for each of the following transfer functions:

(a) $$H_v = \frac{K}{1 + 2j\alpha f/f_0 + (jf/f_0)^2},\ \alpha = 0.9,\ K = 1,\ f_0 = 5\text{ kHz};$$

(b) $$H_v = K\frac{1 + 2j\alpha(f/f_0) - (f/f_0)^2}{j(f/f_0)},\ \alpha = 0.8,\ K = 100,\ f_0 = 20\text{ kHz};$$

(c) $$H_v = \frac{-50(1 + jf/f_2)}{(1 + jf/f_3)\left[1 + 2j\alpha(f/f_1) - (f/f_1)^2\right]},\ f_1 = 20\text{ kHz},\ f_2 = 10\text{ Hz},\ f_3 = 100\text{ Hz},\ \alpha = 0.75;$$

(d) $$H_v = -\frac{20(f/f_0)^2}{1 + 2j\alpha(f/f_0) - (f/f_0)^2},\ f_0 = 2\text{ kHz},\ \alpha = 0.7;$$

(e) $$H_v = K\frac{120\left[1 + 2j\alpha_0(f/f_0) - (f/f_0)^2\right]}{j(f/f_0)\left[1 + 2j\alpha_1(f/f_1) - (f/f_1)^2\right]},\ f_0 = 1\text{ kHz},\ f_1 = 20\text{ kHz},\ \alpha_0 = 0.8,\ \alpha_1 = 0.9;$$

(f) $$\mathrm{H_v} = \frac{-150(1 + j\,f/f_2)(j\,f/f_0)^2}{(1 + j\,f/f_3)\left[1 + 2j\alpha(f/f_1) - (f/f_1)^2\right]},$$
$\mathrm{f_0} = 10$ kHz, $\mathrm{f_1} = 20$ kHz, $\mathrm{f_2} = 10$ Hz, $\mathrm{f_3} = 100$ Hz, $\alpha = 0.7$.

Section 15.8 is prerequisite to the following problems.

P 15.31 Draw asymptotic plots of phase shift versus frequency for each of the transfer functions defined in Problem P 15.29.

P 15.32 Draw asymptotic plots of phase shift versus frequency for each of the transfer functions defined in parts (a–f) of Problem P 15.30.

Section 15.9 is prerequisite to the following problems.

P 15.33 The voltage transfer function for a certain circuit has the form

$$H_v = \frac{Kj(f/f_0)}{1 + j(f/f_0)}.$$

The circuit is to provide high-fidelity transmission for any group of sinusoids whose minimum frequency is greater than some specified frequency $f_{\min}$. Obtain a constraint on the parameter f_0.

P 15.34 The circuit in Fig. P 15.14 is an audio *bass boost* circuit. Assume the op amps are ideal and the source impedance is negligible.

(a) Obtain a standard-form expression for the voltage transfer function.
(b) Plot the voltage gain and phase shift versus frequency, using a logarithmic scale for frequency. Explain why the circuit is called a bass boost circuit. How much boost (in dB) is provided? For what frequencies? For what frequencies does the circuit provide high-fidelity transmission?
(c) Determine which circuit parameters have the greatest influence on the boost (dB) and the frequencies boosted.

P 15.35 The circuit in Fig. P 15.15 is an audio *bass boost* circuit. Assume the op amp is ideal and the source impedance is negligible.

(a) Obtain a standard-form expression for the voltage transfer function.
(b) Plot the voltage gain and phase shift versus frequency, using a logarithmic scale for frequency. Explain why the circuit is called a bass boost circuit. How much boost (in dB) is provided? For what frequencies? For what frequencies does the circuit provide high-fidelity transmission?
(c) Determine which circuit parameters have the greatest influence on the boost (dB) and the frequencies boosted.

P 15.36 The circuit shown in Fig. P 15.16 is a delay equalizer for an audio application. Obtain the voltage transfer function and plot voltage gain and phase shift versus frequency. Assume that the op

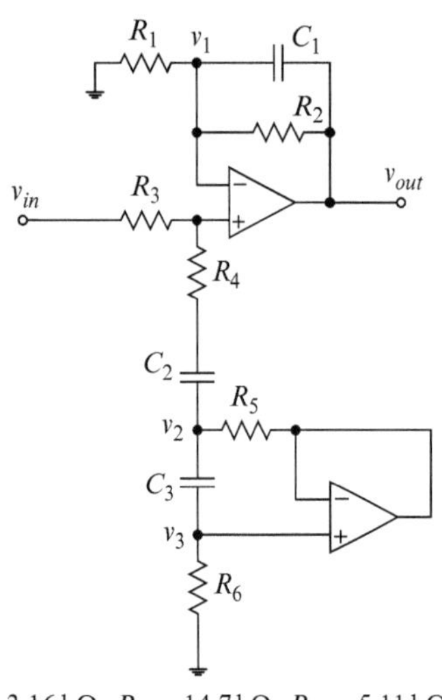

$R_6 = 3.16\,\mathrm{k\Omega}$, $R_2 = 14.7\,\mathrm{k\Omega}$, $R_3 = 5.11\,\mathrm{k\Omega}$,
$R_4 = 4.22\,\mathrm{k\Omega}$, $R_5 = 511\,\Omega$, $R_3 = 147\,\mathrm{k\Omega}$,
$C_1 = 120\,\mathrm{nF}$, $C_2 = 62\,\mathrm{nF}$, $C_3 = 33\,\mathrm{nF}$

Fig. P 15.14 See Problem P 15.34

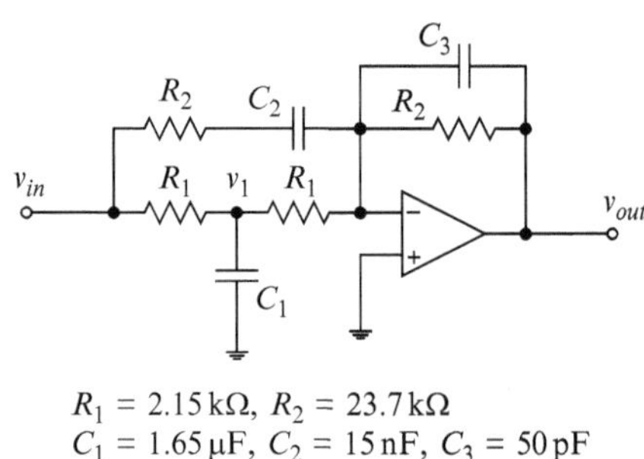

$R_1 = 2.15\,\mathrm{k\Omega}$, $R_2 = 23.7\,\mathrm{k\Omega}$
$C_1 = 1.65\,\mu\mathrm{F}$, $C_2 = 15\,\mathrm{nF}$, $C_3 = 50\,\mathrm{pF}$

Fig. P 15.15 See Problem P 15.35

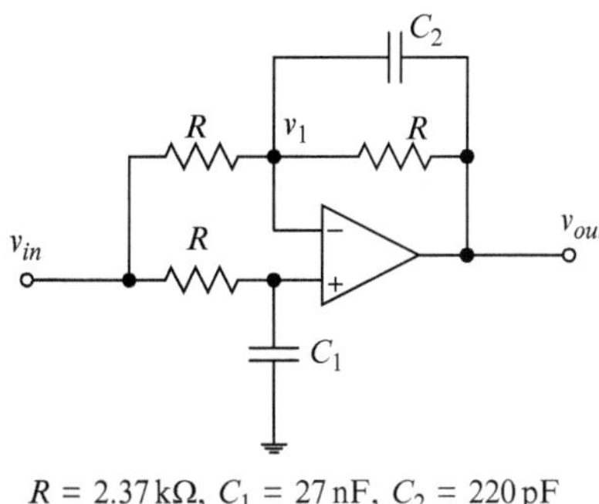

Fig. P 15.16 See Problem P 15.36

amp is ideal and that the source impedance (magnitude) is negligible.

P 15.37 The voltage transfer function for a certain filter is

$$H_v = \frac{K}{(1 + jf/f_1)(1 + jf/f_2)(1 + jf/f_3)},$$

with $f_1 = 20$ kHz, $f_2 = 500$ kHz, $f_3 = 1$ MHz, and $K = 100$. (a) Construct an asymptotic plot of the voltage gain versus frequency for the filter. What kind of filter is this? (b) Find the 3 dB bandwidth of the filter. (c) Suppose the values of the corner frequencies were unknown, but it is known that $0 < f_1 \ll f_2 \ll f_3$. In that case, what is the 3 dB bandwidth of the filter?

P 15.38 The voltage transfer function for a certain filter is

$$H_v = \frac{K(jf/f_1)}{(1 + jf/f_1)(1 + jf/f_2)},$$

with $f_1 = 20$ kHz, $f_2 = 500$ kHz, and $K = 1$. (a) Construct an asymptotic plot of the voltage gain versus frequency for the filter. What kind of filter is this? (b) Find the 3 dB bandwidth of the filter. (c) Suppose the values of the corner frequencies were unknown, but it is known that $0 < f_1 \ll f_2$. In that case, what is the 3 dB bandwidth of the filter?

P 15.39 The squared voltage gain for a certain filter has the form

$$A_v{}^2 = \frac{1}{1 + (f/f_0)^8},$$

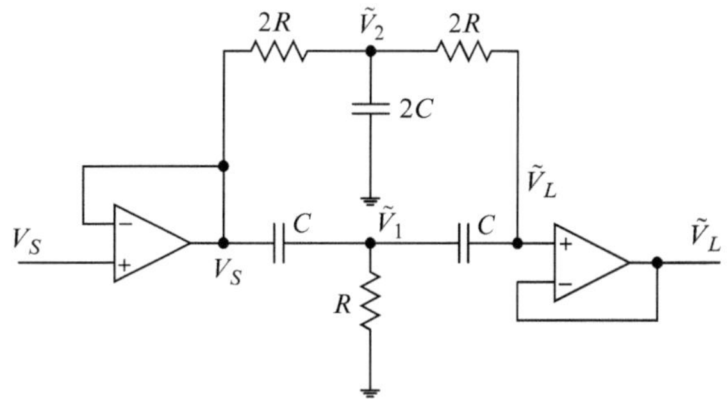

Fig. P 15.17 See Problem P 15.41

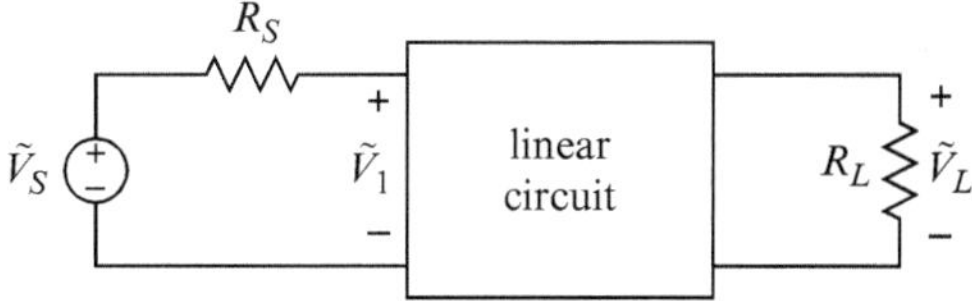

Fig. P 15.18 See Problem P 15.42

where $f_0 = 10$ kHz. Construct an asymptotic plot of the voltage gain in decibels versus frequency for the filter. What kind of filter is this? Obtain an expression for the 3 dB bandwidth of the filter.

P 15.40 The voltage transfer function for a certain circuit has the form

$$H_v = \frac{j(f/f_0)}{1 + 2j\alpha(f/f_0) - (f/f_0)^2}$$

Show that if $\alpha \ll 1$, the half-power bandwidth of the circuit is given by $2\alpha f_0$.

P 15.41 Figure P 15.17 shows a *buffered twin-T notched filter*. (a) Obtain an expression for the voltage transfer function (in standard form) and show that $H_v(f_0) = 0$, where $f_0 = (4\pi RC)^{-1}$. (b) Show that the 3 dB notch width equals $4f_0$. (c) Construct a graph of the voltage gain in dB versus f/f_0, for $0.1 \leq f/f_0 \leq 10$ and for $A_v > -100$ dB.

P 15.42 Refer to Fig. P 15.18, where the frequency of the source is known and denoted by f_0. The circuit input resistance and the source output resistance are known and denoted by R_{in} and R_S, respectively. The output resistance of the circuit is negligible. It is found by measurement that $|\tilde{V}_1| = V_{10}$ and $|\tilde{V}_L| = V_{L0}$. Express the voltage gain at the frequency f_0 in terms of the measured voltages.

P 15.43 Measurements made on a certain linear circuit as shown Fig. P 15.19 yield the values given

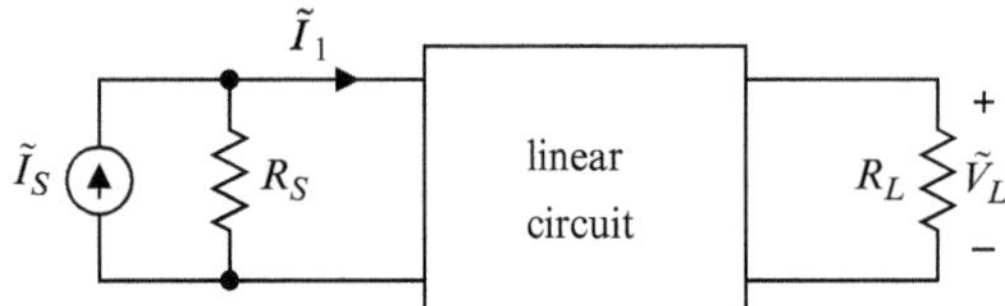

Fig. P 15.19 See Problem P 15.43

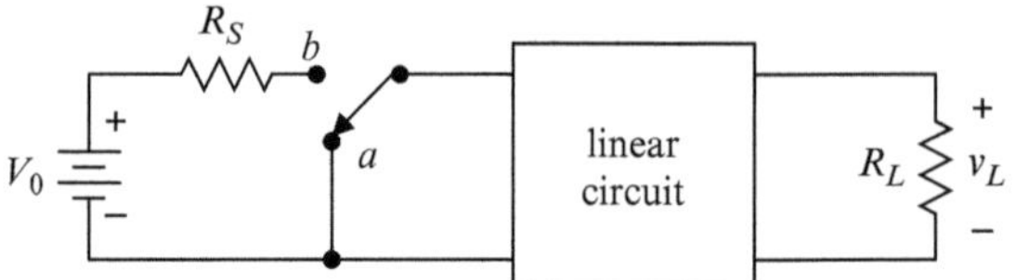

Fig. P 15.20 See Problem P 15.44

in the table below. The load resistance equals 500 Ω, the source resistance equals 50 kΩ, the circuit input resistance equals 20 kΩ, and the circuit output resistance equals 50 Ω. From these data, estimate the voltage gain, the current gain, and the magnitudes of the transimpedance and transadmittance at each of the frequencies in the table.

P 15.44 Refer to Fig. P 15.20. The switch is moved from a to b at $t = 0$, having been at a for $t < 0$. Figure P 15.21 shows a graph of the ratio v_L/V_0 versus time. The circuit is known to consist of only resistors and a single capacitor. (a) What is the dc voltage gain of the circuit? (b) The switch is returned to position a and the dc source V_0 is replaced by a sinusoidal source $\tilde{V}_S$ having the same source impedance. After a long time, the switch is moved to position b. How long does it take for the load voltage v_L to reach steady state? (c) If the load resistance were increased between the two measurements, would the time required to reach steady state increase or decrease?

f(kHz)	1	2	5	7	10	20	50	70	100
$\|\tilde{V}_L/\tilde{I}_1\|$ (MΩ)	1.81	1.78	1.63	1.49	1.29	0.81	0.36	0.25	0.20

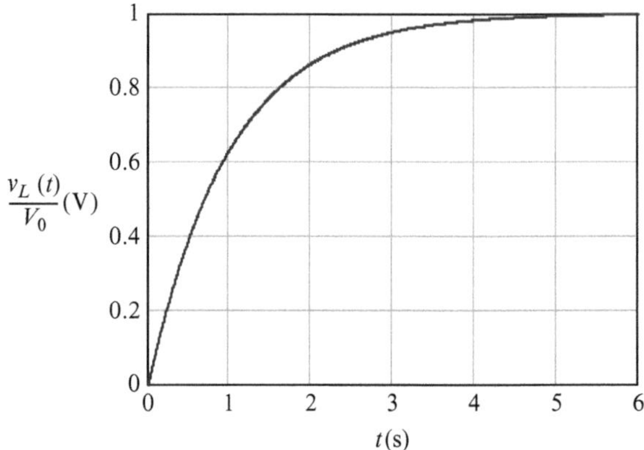

Fig. P 15.21 See Problem P 15.44

Chapter 16
Fourier Series

In this chapter we show how any finite segment of a physical signal (e.g., a current or voltage) can be expressed as a sum of sinusoids called a *Fourier series*. Such representations, in concert with superposition, allow us to apply frequency-domain analytical methods to virtually any physical signal.

16.1 Amplitude–Phase Series

A **Fourier series** is a harmonic series[1] expressed in **amplitude–phase form** as

$$x(t) = x_{dc} + \sum_{k=1}^{\infty} A_k \cos(2\pi k f_0 t + \theta_k), \qquad (16.1)$$

A Fourier series can in general be complex. We are interested only in Fourier series representations of real currents and voltages, in which case *all quantities in* (16.1) *are real.*

The frequency f_0 is the **fundamental frequency** and the frequency of every term is an integer multiple of f_0. The implied **phase reference** is

$$A_1 \cos(2\pi f_0 t)$$

but is not necessarily a term in the series (if $A_1 = 0$).

The sum on the right in (16.1) comprises the **ac component** of $x(t)$ and x_{dc} is the **dc component** (the time average) of $x(t)$.

[1]A harmonic series is a sum of sinusoids where the frequency of every component is an integer multiple of a fundamental frequency.

From (16.1),

$$x\left(t + \frac{1}{f_0}\right) = x_{dc} + \sum_{k=1}^{\infty} A_k \cos(2\pi k f_0 t + 2\pi k + \theta_k)$$
$$= x_{dc} + \sum_{k=1}^{\infty} A_k \cos(2\pi k f_0 t + \theta_k) = x(t).$$

It follows that *a Fourier series is periodic* and the **period** T equals the reciprocal of the fundamental frequency:

$$T = \frac{1}{f_0}. \qquad (16.2)$$

For any particular value of k, the term $A_k \cos(2\pi k f_0 t + \theta_k)$ is called the ***kth harmonic***.[2] The first harmonic $A_1 \cos(2\pi f_0 t + \theta_1)$ also is called the **fundamental component** or, for brevity, the **fundamental**.

We require each term of the sum to be in standard form, such that

$$A_k \geq 0,\ f_0 > 0,\ -\pi \leq \theta_k < \pi.$$

The parameters A_k and θ_k are the **peak amplitude** and **relative phase**, respectively, of the kth harmonic. The dimension and SI unit of the peak amplitudes and of the dc term are those of the current or voltage $x(t)$ represented by the series.

It is often convenient to express the dc term of a Fourier series as a sinusoid having frequency zero, peak amplitude $A_0 = |x_{dc}|$, and relative phase

[2]A tone whose frequency is an integer multiple of the frequency of another tone is called a harmonic of the latter. In music, multiples of a basic note (the fundamental frequency) are sometimes called *overtones*, where the first overtone is the second harmonic.

T.H. Glisson, *Introduction to Circuit Analysis and Design*,
DOI 10.1007/978-90-481-9443-8_16, © Springer Science+Business Media B.V. 2011

$\theta_0 = \measuredangle x_{dc}$. Thus the dc term of a Fourier series becomes the **zeroth harmonic**, expressed as

$$x_{dc} = A_0 \cos(\theta_0), \tag{16.3}$$

where

$$A_0 = |x_{dc}|, \quad \theta_0 = \measuredangle\, x_{dc} = \begin{cases} 0, & x_{dc} \geq 0, \\ -\pi, & x_{dc} < 0. \end{cases} \tag{16.4}$$

The definition (16.3) allows us to write (16.1) more compactly as

$$x(t) = \sum_{k=0}^{\infty} A_k \cos(2\pi k f_0 t + \theta_k). \tag{16.5}$$

Exercise 16.1. The fundamental frequency of a certain Fourier series is $f_0 = 2\,\text{kHz}$. The peak amplitudes and relative phases of the ac terms in the series are given by

$$A_k = \frac{500}{\sqrt{1+k^2}}\,\text{mV}, \; \theta_k = \frac{k\pi}{6}; \; k = 0, 1, 2, \cdots.$$

What is the average value of the series? What are the frequency, period, peak amplitude, and relative phase of the third harmonic? What is the rms amplitude of the fifth harmonic? What are the frequency, relative phase, and rms amplitude of the fundamental?

Exercise 16.2. The frequency of the fifth harmonic in a certain Fourier series is 1.5 kHz. What is the period of the series?

Example 16.1. The two-term series

$$x(t) = A_1 \cos(2\pi f_0 t) + A_2 \cos\left(2\pi \sqrt{2} f_0 t\right)$$

is not a Fourier series because the frequencies f_0 and $\sqrt{2} f_0$ cannot both be expressed as integer multiples of a single fundamental frequency (because $\sqrt{2}$ is an irrational number).

Example 16.2. The two-term series

$$x(t) = B_1 \cos(\pi f_1 t) + B_2 \cos(3\pi f_1 t),$$

where $B_1 > 0$ and $B_2 > 0$ is a Fourier series because the series can be expressed as

$$x(t) = A_1 \cos(2\pi f_0 t) + A_3 \cos(2\pi\, 3 f_0 t),$$

where $f_0 = f_1/2$ is the fundamental frequency, $A_1 = B_1$ is the peak amplitude of the fundamental (first harmonic), $A_3 = B_2$ is the peak amplitude of the third harmonic. All other harmonics and the dc component are zero.

Exercise 16.3. Which of the following series are Fourier series? Justify your answers.

$$x(t) = 5 + 4\cos(2\pi f_0 t) + 2\cos\left(2\pi\sqrt{3} f_0 t\right)\ \text{V},$$

$$x(t) = 10\cos(2\pi f_0 t) + 20\cos\left(2\pi^2 f_0 t\right)\mu\text{A},$$

$$x(t) = 3\cos(2\pi f_0 t) + 4\cos(2\pi\, 5 f_0 t) + 5\cos(2\pi\, 7 f_0 t)\ \text{mA}.$$

Exercise 16.4. The amplitudes and relative phases of the Fourier series for a certain periodic voltage are given by

$$A_k = \left\{ \begin{array}{ll} 0, & k = 0 \\ \dfrac{V_0}{1+k^2}, & k > 0 \end{array} \right\},$$

$$\theta_k = \frac{k\pi}{12}, \quad k = 1, 2, 3, \cdots.$$

with $V_0 = 5$ V. The voltage is applied to the terminals of a 1 kΩ resistor. (a) What is the average power dissipated in the resistor by each of the first five harmonics? (b) The total average power dissipated in the resistor is 3.835 mW. What fraction of the total average power is dissipated in the resistor by the first five harmonics?

16.2 Exponential Series and Fourier Coefficients

Using Euler's identity $2\cos(\alpha) \equiv e^{j\alpha} + e^{-j\alpha}$, we can write the amplitude–phase form of a Fourier series as

$$\begin{aligned}
x_{dc} + \sum_{k=1}^{\infty} A_k \cos(2\pi k f_0 t + \theta_k)
&= x_{dc} + \sum_{k=1}^{\infty} \left(\frac{A_k}{2} e^{j[2\pi k f_0 t + \theta_k]} + \frac{A_k}{2} e^{-j[2\pi k f_0 t + \theta_k]} \right) \\
&= x_{dc} + \sum_{k=1}^{\infty} \left(\frac{A_k}{2} e^{j\theta_k} e^{j2\pi k f_0 t} + \frac{A_k}{2} e^{-j\theta_k} e^{-j2\pi k f_0 t} \right) \\
&= x_{dc} + \sum_{k=1}^{\infty} \frac{A_k}{2} e^{j\theta_k} e^{j2\pi k f_0 t} + \sum_{k=-1}^{-\infty} \frac{A_{-k}}{2} e^{-j\theta_{-k}} e^{j2\pi k f_0 t}.
\end{aligned} \tag{16.6}$$

If $k < 0$, then $-k > 0$. Thus, for example, for $k = -1$,

$$\frac{A_{-(-1)}}{2} e^{-j\theta_{-(-1)}} = \frac{A_1}{2} e^{-j\theta_1} = \left(\frac{A_1}{2} e^{j\theta_1} \right)^*,$$

because A_1 is real. In general,

$$\frac{A_{-(-k)}}{2} e^{-j\theta_{-(-k)}} = \frac{A_k}{2} e^{-j\theta_k} = \left(\frac{A_k}{2} e^{j\theta_k} \right)^*$$

We define

$$X_k = \begin{cases} \dfrac{A_k}{2} e^{j\theta_k}, & k > 0, \\ x_{dc}, & k = 0. \\ \dfrac{A_{-k}}{2} e^{-j\theta_{-k}}, & k < 0, \end{cases} \tag{16.7}$$

and write (16.6) as

$$\begin{aligned}
x(t) &= x_{dc} + \sum_{k=1}^{\infty} X_k e^{j2\pi k f_0 t} + \sum_{k=-1}^{-\infty} X_k e^{j2\pi k f_0 t} \\
&= \sum_{k=-\infty}^{\infty} X_k e^{j2\pi k f_0 t},
\end{aligned} \tag{16.8}$$

The right side of (16.8) is the **exponential form** of a Fourier series and is the preferred form in most applications. The quantities X_k $(k = 0, \pm 1, \pm 2, \cdots)$ given by (16.7) are the **Fourier coefficients**. From (16.7), the Fourier coefficients for a real signal exhibit conjugate symmetry, such that

$$X_{-k} = X_k^*. \tag{16.9}$$

The peak amplitudes and relative phases of the ac terms of a Fourier series are expressed in terms of the Fourier coefficients by

$$A_k = 2|X_k|, \quad \theta_k = \measuredangle X_k, \quad k = 1, 2, 3, \cdots. \tag{16.10}$$

The dimension and SI unit of the Fourier coefficients are those of the signal $x(t)$ represented by the series (16.8).

Example 16.3. Obtain the Fourier coefficients for a Fourier series expressed in amplitude-phase form as

$$x(t) = 5 + 10\cos(2\pi f_0 t) + 5\cos(4\pi f_0 t + \pi/4)\ \text{V}.$$

Solution: From (16.7),

$$X_0 = 5\ \text{V}, \ X_{\pm 1} = 5\ \text{V}, \ X_{\pm 2} = 2.5\, e^{\pm j\pi/4}\ \text{V},$$

and all other coefficients are zero.

Exercise 16.5. A certain Fourier series is expressed in amplitude–phase form as

$$x(t) = -10 + 50\cos(2\pi f_0 t) + 25\cos(6\pi f_0 t - 0.785) + 12.5\cos(10\pi f_0 t - 1.57)\ \text{mV}.$$

Obtain the Fourier coefficients for the series.

Example 16.4. The Fourier coefficients for a certain Fourier series are given by

$$X_k = \frac{1}{1+jk}\ \text{V};\ k = 0, \pm 1, \pm 2, \cdots.$$

Obtain expressions for the peak amplitudes and relative phases of the amplitude–phase form of the series.

Solution: From (16.10),

$$A_0 = |X_0| = 1\,\text{V},\quad \theta_0 = \angle\{X_0\} = 0,$$
$$A_k = 2\,|X_k| = \frac{2}{\sqrt{1+k^2}}\ \text{V},\ \theta_k = \angle\{X_k\} = -\tan^{-1}(k),\quad k > 0.$$

Exercise 16.6. A periodic voltage having period $T = 2\,\text{ms}$ is expressed as a Fourier series whose Fourier coefficients are given by

$$X_k = \frac{1}{1+jk}\ \text{V};\ k = 0, \pm 1, \pm 2, \cdots.$$

What is the average value of the series? What are the frequency, period, peak amplitude, and relative phase of the third harmonic? What is the rms amplitude of the fifth harmonic? What are the frequency, relative phase, and rms amplitude of the fundamental?

Exercise 16.7. The Fourier coefficients for a certain Fourier series are given by

$$X_k = \begin{cases} 25\,\text{mA}, & k = 0, \\ \dfrac{100}{k\pi}\sin\left(\dfrac{k\pi}{4}\right)\text{mA}, & k \neq 0. \end{cases}$$

The period of the series is $T = 250\,\mu\text{s}$. Obtain the first 8 terms of the amplitude–phase form of the series.

16.3 Quadrature Series

Using the identity $\cos(\alpha+\beta) \equiv \cos(\alpha)\cos(\beta) - \sin(\alpha)\sin(\beta)$, we may write

$$\begin{aligned} x_{dc} &+ \sum_{k=1}^{\infty} A_k \cos(2\pi k f_0 t + \theta_k) \\ &= x_{dc} + \sum_{k=1}^{\infty} [A_k \cos(\theta_k)\cos(2\pi k f_0 t) - A_k \sin(\theta_k)\sin(2\pi k f_0 t)] \\ &= x_{dc} + \sum_{k=1}^{\infty} [a_k \cos(2\pi k f_0 t) + b_k \sin(2\pi k f_0 t)], \end{aligned}$$

where

$$a_k = A_k \cos(\theta_k),\quad b_k = -A_k \sin(\theta_k),\quad k = 1, 2, 3, \ldots. \tag{16.11}$$

Thus we have obtained the **quadrature form** of a Fourier series, expressed as

$$x(t) = x_{dc} + \sum_{k=1}^{\infty} [a_k \cos(2\pi k f_0 t) + b_k \sin(2\pi k f_0 t)], \tag{16.12}$$

where the **quadrature coefficients** a_k, b_k are given by (16.11). The dimensions and SI units of the quadrature coefficients are those of the signal represented by the series (16.12). Of the three forms of Fourier series, the quadrature form is probably least used.

Exercise 16.8. Show that the Fourier coefficients for a particular series can be expressed in terms of the quadrature coefficients for that series as

$$X_{\pm k} = \frac{1}{2}(a_k \mp j\, b_k);\ \ k = 1,\ 2,\ 3, \cdots \quad (16.13)$$

and conversely

$$a_k = 2\mathrm{Re}(X_k),\ \ b_k = -2\mathrm{Im}(X_k); \quad k = 1,\ 2,\ 3, \cdots. \quad (16.14)$$

Exercise 16.9. Define the quadrature coefficients a_0, b_0 in terms of the dc component x_{dc} such that (16.12) can be written

$$x(t) = \sum_{k=0}^{\infty} [a_k \cos(2\,\pi\, k f_0\, t) + b_k \sin(2\,\pi\, k f_0\, t)]$$

Exercise 16.10. The Fourier coefficients for a certain Fourier series are given by

$$X_k = \frac{2+4jk}{1+k^2}\ \mathrm{V},\ k = 0, \pm 1, \pm 2, \cdots.$$

Obtain the quadrature coefficients a_k, b_k for $k = 0,\ 1,\ 2, 3$. (See Exercise 16.8)

Exercise 16.11. The amplitude–phase form of the Fourier series for a certain signal is

$$x(t) = x_{dc} + V_0 \sum_{k=1}^{\infty} \left(\frac{1}{k}\right)^2 \cos(2\,\pi\, k f_0\, t + k\pi).$$

Give the quadrature and exponential forms of the series.

16.4 Summary: Three Forms of Fourier Series

The three forms of a Fourier series are:

(1) The **amplitude–phase form**

$$x(t) = x_{dc} + \sum_{k=1}^{\infty} A_k \cos(2\,\pi\, k f_0\, t + \theta_k), \quad (16.15)$$

(2) The **exponential form**

$$x(t) = \sum_{k=-\infty}^{\infty} X_k\, e^{j2\,\pi\, k f_0\, t}, \quad (16.16)$$

and (3) the **quadrature form**

$$x(t) = x_{dc} + \sum_{k=1}^{\infty} [a_k \cos(2\,\pi\, k f_0\, t) + b_k \sin(2\,\pi\, k f_0\, t)]. \quad (16.17)$$

The three forms of a Fourier series are identical if they have the same fundamental frequency f_0, the same dc component ($X_0 = x_{dc}$), and the parameters are related as summarized in Table 16.1. We limit our treatment to real signals, so all three series above are real.

Example 16.5. The Fourier coefficients for a certain Fourier series are given by

$$X_k = \frac{1}{1+jk}\ \mathrm{V};\ \ k = 0,\ \pm 1,\ \pm 2,\ \cdots.$$

Obtain the quadrature form of the series.
Solution: We can write the expression above for the Fourier coefficients as

$$X_k = \frac{1-jk}{(1+jk)(1-jk)} = \frac{1-jk}{1+k^2}\ \mathrm{V}; \quad k = 0, \pm 1, \pm 2, \cdots.$$

From Table 16.1,

$$a_0 = X_0 = 1\ \mathrm{V},$$
$$a_k = 2\mathrm{Re}(X_k) = \frac{2}{1+k^2}\ \mathrm{V}; \quad k = \pm 1, \pm 2, \pm 3 \cdots,$$
$$b_k = -2\mathrm{Im}(X_k) = \frac{2k}{1+k^2}\ \mathrm{V}, \quad k = \pm 1, \pm 2, \pm 3 \cdots.$$

Table 16.1 Relations among coefficients of the three forms of a Fourier series

From \ To	Exponential	Amplitude-Phase	Quadrature
Exponential		$A_k = 2\lvert X_k \rvert, k=1,2,\cdots$ $A_0 = \lvert X_0 \rvert$ $\theta_k = \measuredangle X_k$	$a_k = 2\mathrm{Re}(X_k), k=1,2,\cdots$ $b_k = -2\mathrm{Im}(X_k), k=1,2,\cdots$ $a_0 = X_0, b_0 = 0$
Amplitude-Phase	$X_{\pm k} = \frac{A_k}{2}\angle \pm \theta_k,$ $k=1,2,\cdots$ $X_0 = A_0 \cos(\theta_0) = x_{dc}$		$a_k = A_k \cos(\theta_k),$ $b_k = -A_k \sin(\theta_k),$ $k=1,2,\cdots$
Quadrature	$X_{\pm k} = \frac{1}{2}(a_k \mp jb_k),$ $k=1,2,\cdots$ $X_0 = a_0$	$A_k = \sqrt{a_k^2 + b_k^2}, k=1,2,\cdots$ $A_0 = \lvert a_0 \rvert$ $\theta_k = \measuredangle(a_k - jb_k)$	

Exercise 16.12. Write out the dc and first three ac terms of the amplitude–phase and quadrature forms of the Fourier series for a voltage having period *T*, and whose Fourier coefficients are given by

$$X_k = \frac{1000\, e^{jk\pi/6}}{\sqrt{1+k^2}}\ \mathrm{mV}.$$

16.5 Integral Formula for Fourier Coefficients

In this section, we describe how to obtain a Fourier series representing a function (e.g., a current or voltage) over an interval. We begin by deriving an important identity.

Suppose we somehow obtain (or just make up) a set of coefficients $\{X_0, X_{\pm 1}, X_{\pm 2}, \cdots\}$, and form the function $\hat{x}(t)$, given by

$$\hat{x}(t) = \sum_{k=-\infty}^{\infty} X_k\, e^{j2\pi k f_0 t}. \qquad (16.18)$$

We seek a relation between the coefficients $\{X_k;\ k = 0, \pm 1, \pm 2, \cdots\}$ of the series on the right side of (16.18) and the sum of the series (the function represented by the series) $\hat{x}(t)$ on the left side of (16.18). We multiply both sides of (16.18) by $e^{-j2\pi n f_0 t}$ and integrate over one period, from t_0 to $t_0 + T$, where $T = 1/f_0$. This gives

$$\int_{t_0}^{t_0+T} \hat{x}(t)\, e^{-j2\pi n f_0 t}\, dt = \int_{t_0}^{t_0+T} \left[\sum_{k=-\infty}^{\infty} X_k\, e^{j2\pi k f_0 t}\right] e^{-j2\pi n f_0 t}\, dt = \sum_{k=-\infty}^{\infty} X_k \int_{t_0}^{t_0+T} e^{j2\pi (k-n) f_0 t}\, dt, \qquad (16.19)$$

where k and n are integers. The last integral on the right equals T for $k = n$ because $e^0 = 1$. The integral equals zero for $k \neq n$ because

$$e^{j2\pi (k-n) f_0 t} = \cos[2\pi(k-n) f_0 t] + j \sin[2\pi(k-n) f_0 t]$$

and the integral of a sinusoid over a whole number of periods equals zero. Thus

$$\int_{t_0}^{t_0+T} e^{j2\pi (k-n) f_0 t}\, dt = \begin{cases} T, & k = n, \\ 0, & k \neq n. \end{cases} \qquad (16.20)$$

Using (16.20) in (16.19) gives

$$\int_{t_0}^{t_0+T} \hat{x}(t)\, e^{-j2\pi n f_0 t}\, dt = \sum_{k=-\infty}^{\infty} X_k \times \begin{Bmatrix} T, k=n \\ 0, k\neq n \end{Bmatrix} = TX_n,$$

or

$$X_n = \frac{1}{T}\int_{t_0}^{t_0+T} \hat{x}(t)\, e^{-j2\pi n f_0 t}\, dt;\ n = 0, \pm 1, \pm 2, \cdots. \tag{16.21}$$

Equation (16.21) is an *identity* that relates the coefficients of the exponential harmonic series (16.18) to the function $\hat{x}(t)$ represented by the series. No matter how we obtain the harmonic series (16.18), the coefficients $\{X_n;\ n = 0, \pm 1, \pm 2, \cdots\}$ of the series and the function $\hat{x}(t)$ represented by the series are related as specified by (16.21).

Now suppose we begin with a function $x(t)$ and compute a set of coefficients using (16.21); i.e.,

$$X_k = \frac{1}{T}\int_{t_0}^{t_0+T} x(t) e^{-j2\pi k f_0 t}\, dt, \quad f_0 = T^{-1}. \tag{16.22}$$

The interval $t_0 < t \le t_0 + T$ in (16.22) is called the **interval of expansion**. Having obtained a set of coefficients from a function $x(t)$ using (16.22), we can define a function $\hat{x}(t)$ as the harmonic series

$$\hat{x}(t) = \sum_{k=-\infty}^{\infty} X_k\, e^{j2\pi k f_0 t}, \quad f_0 = T^{-1}. \tag{16.23}$$

The question is whether the function $x(t)$ in (16.22) and the function $\hat{x}(t)$ represented by the series (16.23) are the same in the interval of expansion (for $t_0 < t \le t_0 + T$). Many years ago, Dirichlet[3] proved that if a function $x(t)$ has a finite number of maxima and minima and a finite number of discontinuities in the interval of expansion, and if the coefficients for a function $x(t)$ are obtained using (16.21), then the series (16.23) converges to

$$\hat{x}(t) = \frac{x(t^-) + x(t^+)}{2} \tag{16.24}$$

everywhere in the interval of expansion. This means that the series $\hat{x}(t)$ equals the function $x(t)$ wherever $x(t)$ is continuous and that the series $\hat{x}(t)$ equals the average of the left and right limits of $x(t)$ wherever $x(t)$ has a jump discontinuity. Because we assume the function $x(t)$ has a finite number of discontinuities, the function $x(t)$ and the series $\hat{x}(t)$ differ at most at a countable number of points. No physical system can distinguish two currents or voltages that differ at only a countable number of points, so we may regard an infinite Fourier series as an exact representation of any physical signal over the interval of expansion. Any current or voltage can be expressed over any finite interval (no matter how long) as a Fourier series whose coefficients are obtained using (16.22).

For all *practical* purposes, the series $\hat{x}(t)$ in (16.23) and the function $x(t)$ are indistinguishable for $t_0 < t \le t_0 + T$ if the coefficients for the series are obtained using (16.22). The series $\hat{x}(t)$ and the function $x(t)$ are not necessarily the same outside the interval $t_0 < t \le t_0 + T$, because the series $\hat{x}(t)$ is periodic with period T, whereas the function $x(t)$ is not necessarily periodic. However, at least conceptually, we may choose the interval of expansion long enough such that for many practical purposes, $\hat{x}(t) = x(t)$.

In circuit analysis, Fourier series are used in two ways: (1) to describe a periodic signal for all time and (2) to describe a non-periodic signal over some interval. In the first use, the series is an exact representation for all time. In the second use, the series is exact over the interval of expansion. We illustrate the first use, first, and begin by illustrating how the Fourier coefficients for a signal are obtained using the integral formula (16.22).

Example 16.6. Obtain the Fourier coefficients for the periodic signal (rectangular pulse train) defined in Fig. 16.1.

Solution: In (16.22), t_0 is arbitrary for a periodic signal. We usually choose a value for t_0 that makes the required integration as simple as possible. In this example, we choose $t_0 = -\tau/2$ to minimize the number of integrations required. (Any other value for which the

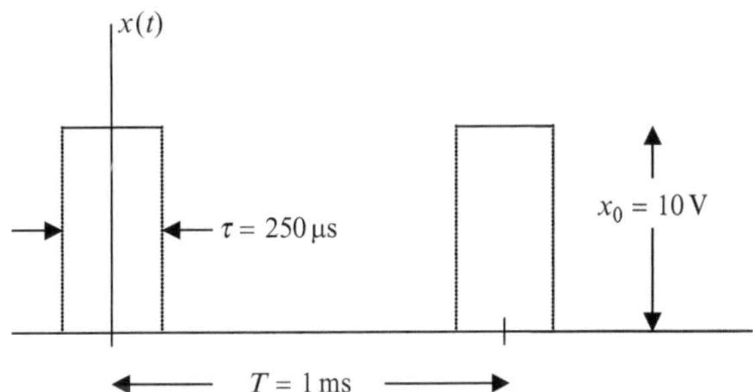

Fig. 16.1 See Example 16.6

[3] Peter Gustave Lejeune Dirichlet (1805–1859), a German mathematician.

interval $t_0 < t \le T + t_0$ encompasses one whole pulse works as well.) Thus,

$$\begin{aligned}
X_k &= \frac{1}{T}\int_{-\tau/2}^{T-\tau/2} x(t)\, e^{-j2\pi kt/T}\, dt \\
&= \frac{1}{T}\int_{-\tau/2}^{\tau/2} x_0 e^{-j2\pi kt/T}\, dt \\
&= \begin{cases} \dfrac{1}{T}\displaystyle\int_{-\tau/2}^{\tau/2} x_0\, dt = \dfrac{x_0\tau}{T}, & k=0 \\ \dfrac{x_0}{T}\dfrac{e^{-j\pi k\tau/T} - e^{j\pi k\tau/T}}{-j2\pi k/T}, & k=\pm 1, \pm 2, \cdots \end{cases} \\
&= \begin{cases} \dfrac{x_0\tau}{T}, & k=0, \\ \dfrac{x_0\tau}{T}\dfrac{\sin(k\pi\tau/T)}{k\pi\tau/T}, & k=\pm 1, \pm 2, \cdots. \end{cases}
\end{aligned}$$

Where the period of the signal is $T = 1$ ms, the pulse amplitude is $x_0 = 10$ V, and the pulse width is $\tau = 250$ μs.

Exercise 16.13. Use the integral formula (16.22) to obtain an expression for the Fourier coefficients for the square wave in Fig. 16.2

Example 16.7. Choose an appropriate interval of expansion for the signal defined graphically by Fig. 16.3.

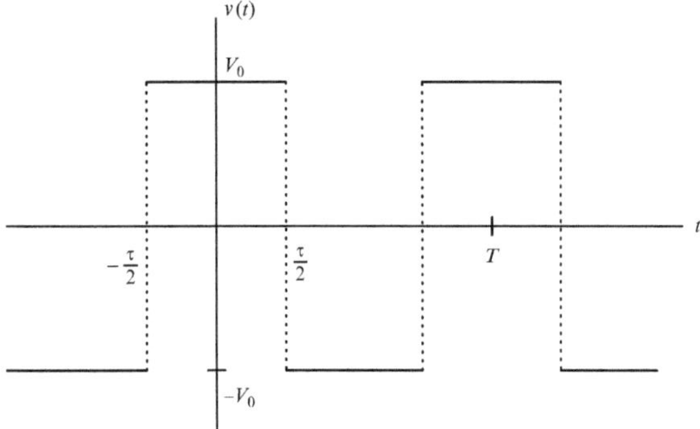

Fig. 16.2 See Exercise 16.13

Solution: Using (16.22) with $t_0 = -T/2$ to compute the coefficients requires two integrations:

$$X_k = \frac{1}{T}\int_{-T/2}^{-T/2+\tau/2} x(t)\, e^{-j2\pi k f_0 t}\, dt + \frac{1}{T} \times \int_{T/2-\tau/2}^{T/2} x(t)\, e^{-j2\pi k f_0 t}\, dt.$$

Using (16.22) with $t_0 = -T/2 - \tau/2$ requires only one integration:

$$X_k = \frac{1}{T}\int_{-T/2-\tau/2}^{-T/2+\tau/2} x(t)\, e^{-j2\pi k f_0 t}\, dt.$$

Therefore an appropriate choice for the interval of expansion is

$$-T/2 - \tau/2 \le t < T/2 - \tau/2,$$

as shown in Fig. 16.3.

Exercise 16.14. Specify another convenient interval of expansion for the signal defined in Example 16.7.

16.6 A Table of Fourier Coefficients[4]

It is rarely necessary to calculate Fourier coefficients using (16.22) and pencil and paper. Tables of Fourier coefficients for most periodic waveforms encountered

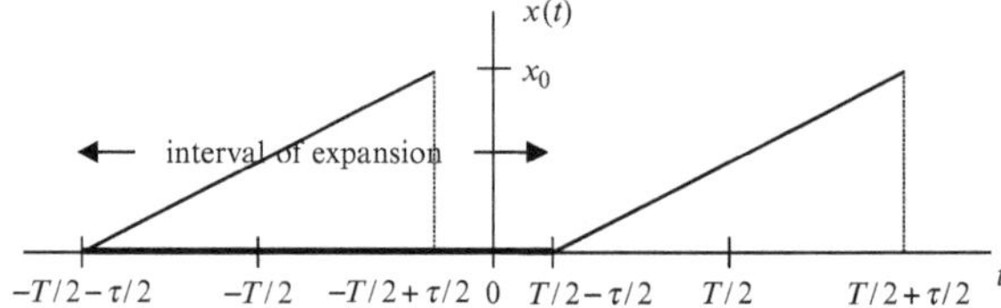

Fig. 16.3 See Example 16.7

[4]In this chapter, we use x_{rms} to denote the rms amplitude of a signal $x(t)$ because we use X to denote the Fourier coefficients of the signal.

in circuit analysis are readily available, and virtually all mathematical software packages (e.g., Maple®, Mathcad®, Matlab®) include functions or subroutines for calculating Fourier coefficients and summing (finite) Fourier series. Microsoft's Excel® also provides such tools. Table 16.2 gives the Fourier coefficients for several waveforms that arise in applications (some more than others). For example, the Fourier series for a fractional sinusoid is useful in analyzing circuits called class C amplifiers, which are found in many radio and television transmitters.

In Table 16.2, we introduce a function called the **sampling function**, defined by

$$\mathrm{sa}(x) = \begin{cases} 1, & x = 0, \\ \dfrac{\sin(x)}{x}, & x \neq 0. \end{cases} \tag{16.25}$$

In many applications of Fourier analysis, it is helpful to have in mind a picture (graph) of the sampling function. Figure 16.4 shows such a graph. The essential features of the sampling function are:

- The sampling function is an even function of its argument.
- A graph of the sampling function exhibits a **central lobe** and a number of smaller **side lobes**.
- The peak amplitude of the sampling function (the height of the central lobe) equals unity.
- The heights (magnitudes) of the side lobes decrease with increasing values of $|x|$, approximately as $1/|x|$.

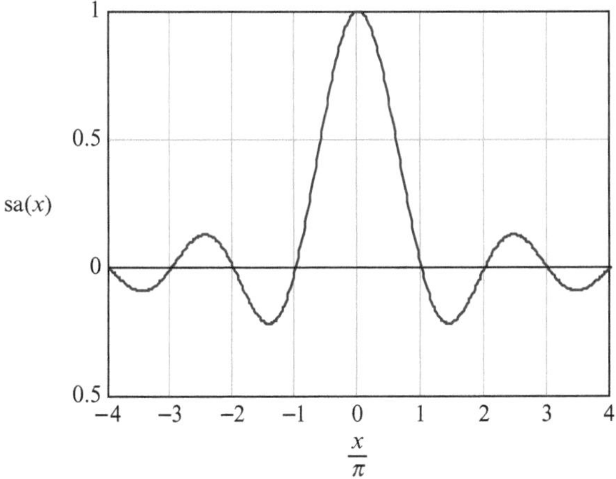

Fig. 16.4 Graph of the sampling function defined in (16.25)

- The sampling function sa(x) equals zero for $x = \pm k\pi$, $k = \pm 1, \pm 2, \cdots$. The zeros of the sampling function are also called the **nulls** of the function.

You should study the definition (16.25) and Fig. 16.4 until you have the above five features firmly in mind.

A function closely related to the sampling function is the **sine-cardinal function**, defined by

$$\mathrm{sinc}(x) = \begin{cases} 1, & x = 0, \\ \dfrac{\sin(\pi x)}{\pi x}, & x \neq 0. \end{cases} \tag{16.26}$$

Thus

$$\mathrm{sinc}(x) = \mathrm{sa}(\pi x).$$

We prefer the sampling function because in applications, π appears explicitly in the argument, which seems to lead to fewer computational errors. Others prefer the sine-cardinal function. Using one or the other is a matter of personal preference.

Exercise 16.15. Refer to Table 16.2 and express the Fourier coefficients for a triangular pulse train and those for a triangular wave in terms of the sine-cardinal function.

Example 16.8. Find the peak amplitudes of the first three non-zero terms in the amplitude-phase Fourier series for a symmetric triangular wave having peak amplitude $x_0 = 10$ V and period $T = 2$ ms.

Solution: From Table 16.2, the Fourier coefficients for a symmetric triangular wave are given by

$$X_0 = x_{dc} = 0,$$
$$X_k = x_0\,\mathrm{sa}^2\left(\frac{k\pi}{2}\right); \quad k = \pm 1, \pm 2, \pm 3, \cdots.$$

We can use a spreadsheet (e.g., Excel), a computer program (e.g., Matlab), or a pocket calculator to compute the coefficients. We obtain the values in Table 16.3:

Table 16.2 Fourier coefficients for selected waveforms[5]

Symmetric Rectangular Pulse Train

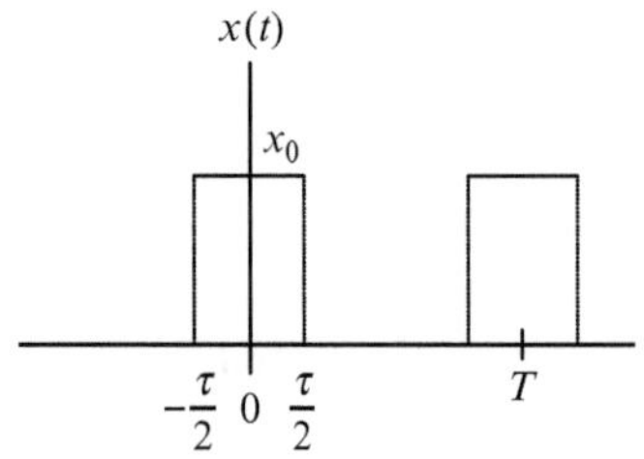

$$X_k = \frac{x_0 \tau}{T} \mathrm{sa}\left(\frac{k\pi\tau}{T}\right); \quad k = 0, \pm 1, \pm 2, \cdots$$

$$x_{rms} = \sqrt{\frac{x_0^2 \tau}{T}}$$

Symmetric Triangular Pulse Train

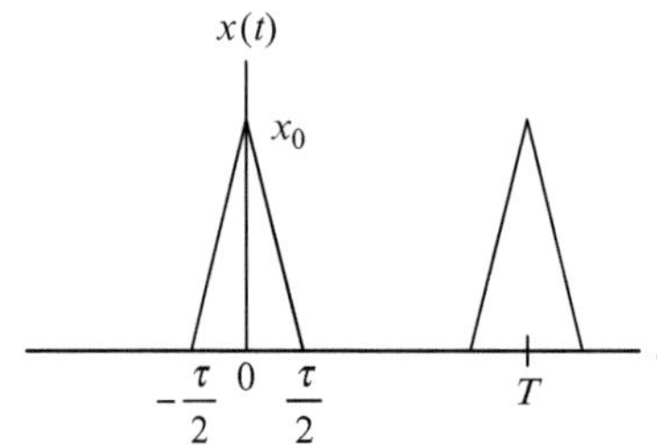

$$X_k = \frac{x_0 \tau}{2T} \mathrm{sa}^2\left(\frac{k\pi\tau}{2T}\right); \quad k = 0, \pm 1, \pm 2, \cdots$$

$$x_{rms} = \sqrt{\frac{x_0^2 \tau}{3T}}$$

Sawtooth Pulse Train

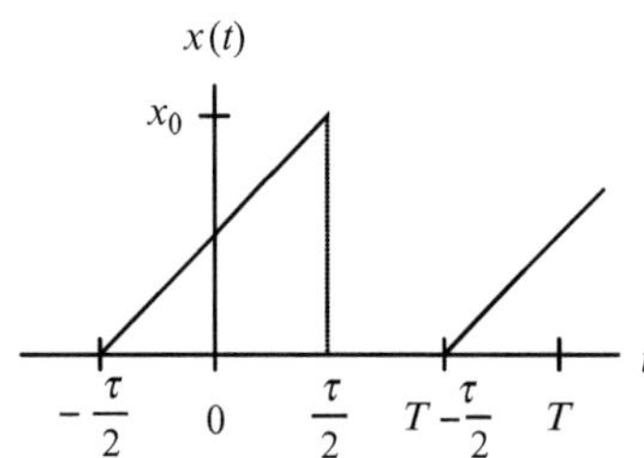

$$X_k = \frac{j x_0}{2k\pi}\left[e^{-jk\pi\tau/T} - \mathrm{sa}\left(\frac{k\pi\tau}{2T}\right)\right]; \quad k = \pm 1, \pm 2, \cdots$$

$$X_0 = \frac{x_0 \tau}{2T}, \quad x_{rms} = \sqrt{\frac{x_0^2 \tau}{3T}}$$

Symmetric Trapezoidal Pulse Train

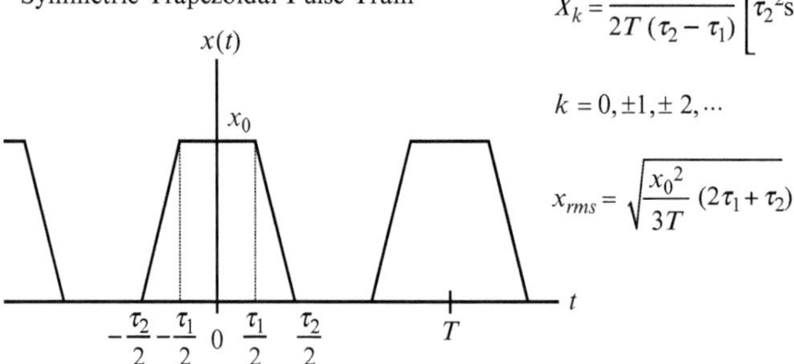

$$X_k = \frac{x_0}{2T(\tau_2 - \tau_1)}\left[\tau_2^2 \mathrm{sa}^2\left(\frac{k\pi\tau_2}{2T}\right) - \tau_1^2 \mathrm{sa}^2\left(\frac{k\pi\tau_1}{2T}\right)\right]$$

$k = 0, \pm 1, \pm 2, \cdots$

$$x_{rms} = \sqrt{\frac{x_0^2}{3T}(2\tau_1 + \tau_2)}$$

(*continued*)

[5]In the table, *symmetric* means even symmetry about the vertical axis.

Table 16.2 (continued)

Exponential Pulse Train

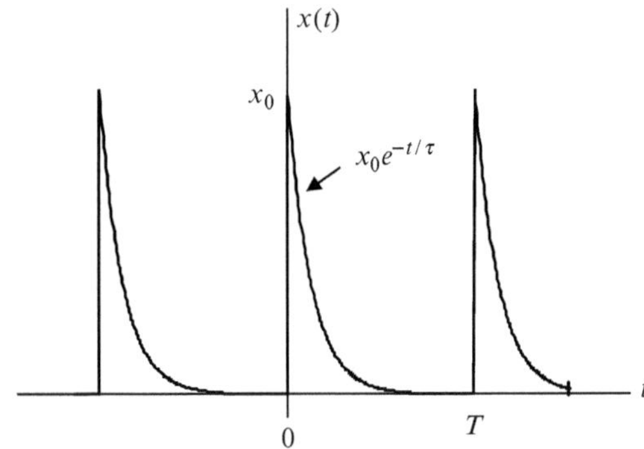

$$X_k = \frac{x_0\tau(1 - e^{-T/\tau})}{T + j2k\pi\tau}; \quad k = \pm 2, \pm 3, \pm 4 \cdots$$

$$x_{rms} = \sqrt{\frac{x_0^2\tau}{2T}(1 - e^{-2T/\tau})}$$

Symmetric Square Wave

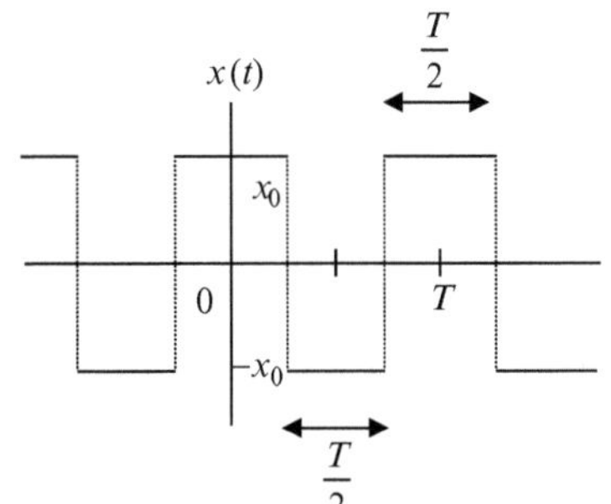

$$X_0 = 0$$

$$X_k = x_0 \,\mathrm{sa}\left(\frac{k\pi}{2}\right); \quad k = \pm 1, \pm 2, \pm 3, \ldots$$

$$x_{rms} = x_0$$

Symmetric Triangular Wave

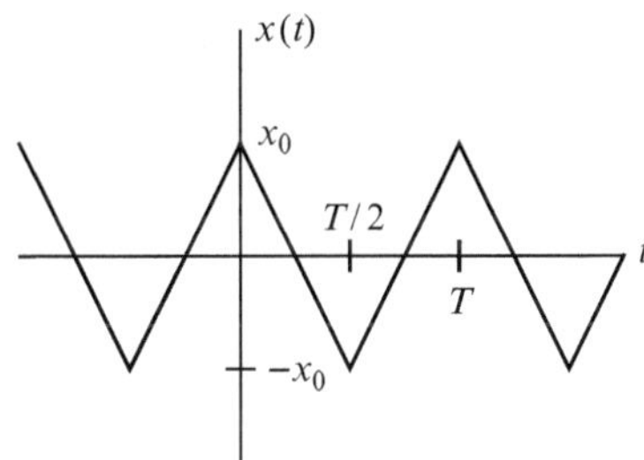

$$X_0 = 0$$

$$X_k = x_0 \,\mathrm{sa}^2\left(\frac{k\pi}{2}\right); \quad k = \pm 1, \pm 2, \pm 3, \cdots$$

$$x_{rms} = \sqrt{\frac{x_0^2}{3}}$$

Sawtooth Wave

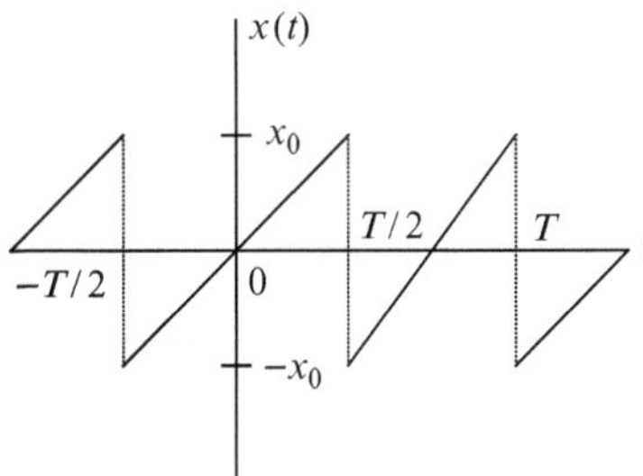

$$X_0 = 0$$

$$X_k = \frac{jx_0(-1)^k}{\pi k}; \quad k = \pm 1, \pm 2, \pm 3, \cdots$$

$$x_{rms} = \sqrt{\frac{x_0^2}{3}}$$

(continued)

Table 16.2 (continued)

Symmetric Trapezoidal Wave

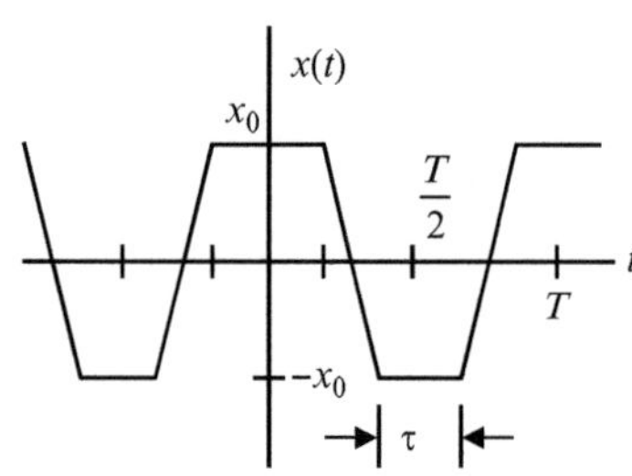

$$X_0 = 0$$

$$X_k = \frac{x_0}{T\,(T-2\tau)}\left[(T-\tau)^2\,\mathrm{sa}^2\left(\frac{k\pi\,[T-\tau]}{2T}\right) - \tau^2\mathrm{sa}^2\left(\frac{k\pi\tau}{T}\right)\right]$$

$$k = \pm 1, \pm 2, \cdots$$

$$x_{rms} = \sqrt{\frac{x_0^2}{3T}\,(T+4\tau)}$$

Symmetric Fractional Sinusoid

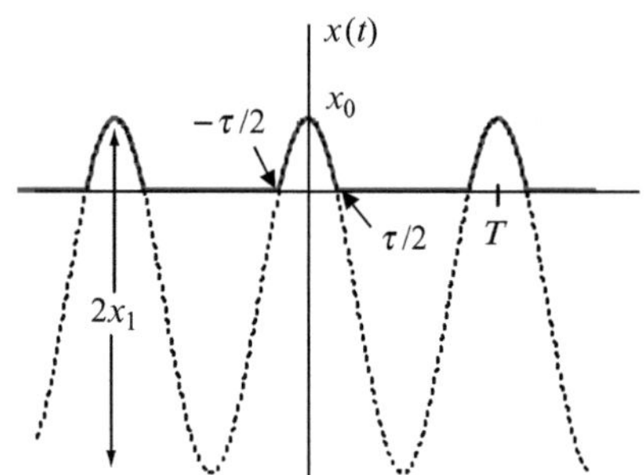

$$X_k = \frac{x_1\tau}{2T}\left\{\mathrm{sa}\left[\frac{(k+1)\pi\tau}{T}\right] + \mathrm{sa}\left[\frac{(k-1)\pi\tau}{T}\right]\right\} - \frac{(x_1-x_0)\tau}{T}\,\mathrm{sa}\left(\frac{k\pi\tau}{T}\right)$$

$$k = 0, \pm 1, \pm 2, \pm 3, \cdots$$

$$\tau = \frac{T}{\pi}\cos^{-1}\left(1 - \frac{x_0}{x_1}\right)$$

$$x_{rms} = \frac{x_1^2}{2\pi}\left\{\sin\left(\frac{\pi\tau}{T}\right)\left[\cos\left(\frac{\pi\tau}{T}\right) + 4\left(\frac{x_0}{x_1} - 1\right)\right] + \frac{\pi\tau}{T}\left[2\left(\frac{x_0}{x_1}\right)^2 - 4\frac{x_0}{x_1} + 3\right]\right\}$$

Symmetric Clipped Sinusoid

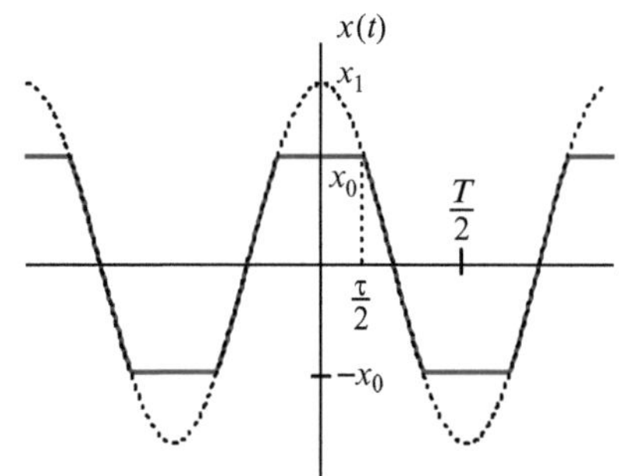

$$\tau = \frac{T}{\pi}\cos^{-1}\left(\frac{x_0}{x_1}\right)$$

$$X_0 = 0$$

$$X_{\pm 1} = \frac{x_1}{2} - \frac{x_1\tau}{T}\left[\mathrm{sa}\left(\frac{2\pi\tau}{T}\right) + 1\right] + \frac{2x_0\tau}{T}\,\mathrm{sa}\left(\frac{\pi\tau}{T}\right)$$

$$X_k = \frac{x_1\tau[(-1)^k - 1]}{2T}\left\{\mathrm{sa}\left[\frac{(k+1)\pi}{2}\right] + \mathrm{sa}\left[\frac{(k-1)\pi}{2}\right] - \frac{2x_0}{x_1}\,\mathrm{sa}\left(\frac{k\pi\tau}{T}\right)\right\}$$

$$k = \pm 2, \pm 3, \pm 4 \cdots$$

$$x_{rms} = \sqrt{\frac{2x_0^2\tau}{T} + \frac{x_1^2}{2\pi T}\left[\pi\,(T-2\tau) - T\sin\left(\frac{2\pi\tau}{T}\right)\right]}$$

Symmetric Half-Wave Rectified Sinusoid

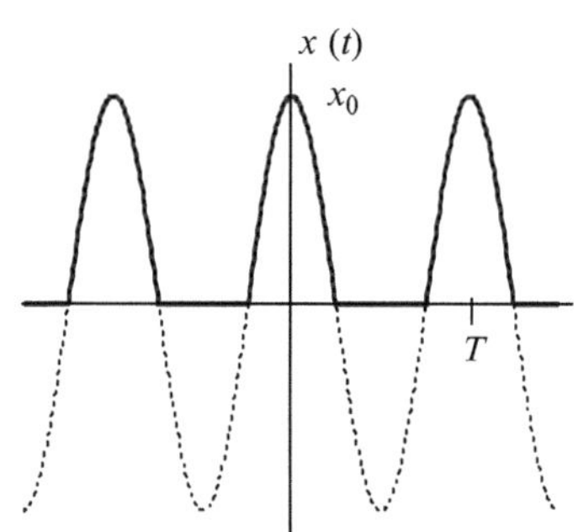

$$X_k = \frac{x_0}{4}\left\{\mathrm{sa}\left[\frac{(k+1)\pi}{2}\right] + \mathrm{sa}\left[\frac{(k-1)\pi}{2}\right]\right\};\ k = 0, \pm 1, \pm 2, \cdots$$

$$x_{rms} = \sqrt{\frac{x_0^2}{4}}$$

(*continued*)

Table 16.2 (continued)

Symmetric Full-Wave Rectified Sinusoid

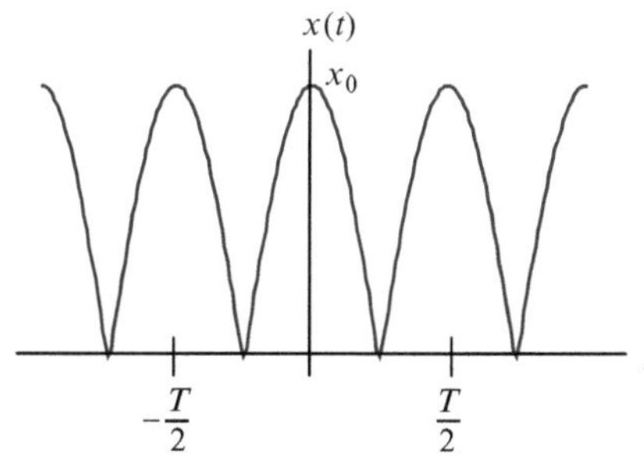

$$X_k = \frac{x_0[1+(-1)^k]}{4}\left\{\text{sa}\left[\frac{(k+1)\pi}{2}\right] + \text{sa}\left[\frac{(k-1)\pi}{2}\right]\right\}$$

$k = 0, \pm 1, \pm 2, \cdots$

$$x_{rms} = \sqrt{\frac{x_0^2}{2}}$$

Table 16.3 See Example 16.8

k	0	1	2	3	4	5
X_k (V)	0.000	4.053	0.000	0.450	0.000	0.162

The first three non-zero coefficients are X_1, X_3, X_5. The peak amplitudes of the terms in the corresponding amplitude–phase series are

$$A_1 = 2|X_1| = 8.106 \text{ V},$$
$$A_3 = 2|X_3| = 900 \text{ mV},$$
$$A_5 = 2|X_5| = 324 \text{ mV}.$$

Exercise 16.16. Find the peak amplitudes of the first three non-zero terms in the amplitude-phase Fourier series for a sawtooth wave having peak amplitude $x_0 = 10$ V and period $T = 2$ ms.

16.7 Modified Fourier Coefficients for Composite Waveforms

The Fourier coefficients given in Table 16.2 are easily modified to account for relatively minor changes to the waveforms. Figure 16.5 illustrates three such changes and the corresponding modifications of the Fourier coefficients. The figure uses a sawtooth pulse train to illustrate the changes, but the modifications to the Fourier coefficients apply to any periodic signal.

Adding a constant to a Fourier series is equivalent to translating the signal represented by the series vertically, and alters only the dc component of the series. The other harmonics are unchanged.

If

$$x(t) = \sum_{k=-\infty}^{\infty} X_k\, e^{j2\pi k f_0 t}, \tag{16.27}$$

then

$$\begin{aligned} x(t-t_1) &= \sum_{k=-\infty}^{\infty} X_k\, e^{j2\pi k f_0 (t-t_1)} \\ &= \sum_{k=-\infty}^{\infty} X_k\, e^{-j2\pi k f_0 t_1} e^{j2\pi k f_0 t} \\ &= \sum_{k=-\infty}^{\infty} X'_k\, e^{j2\pi k f_0 t}, \end{aligned}$$

where

$$X'_k = X_k\, e^{-j2\pi k f_0 t_1}. \tag{16.28}$$

In words, time translation by t_1 is equivalent to multiplying the Fourier coefficients by $\exp(-j2\pi k f_0 t_1)$, where the translation is to the right (is a delay) if $t_1 > 0$ and to the left (is an advance) if $t_1 < 0$.

Time reversal of a signal is achieved by replacing t by $-t$. Thus

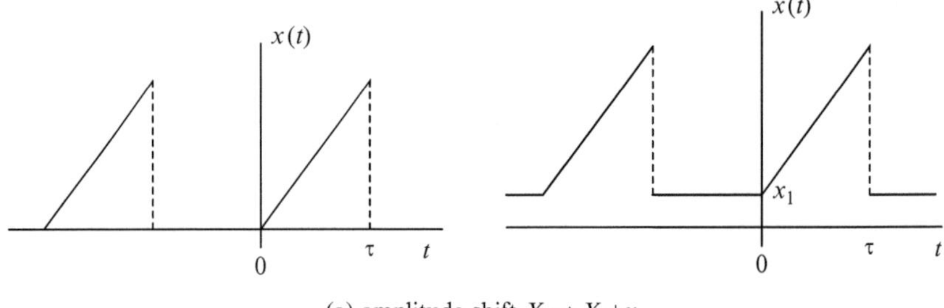

(a) amplitude shift $X_0 \rightarrow X_0 + x_1$

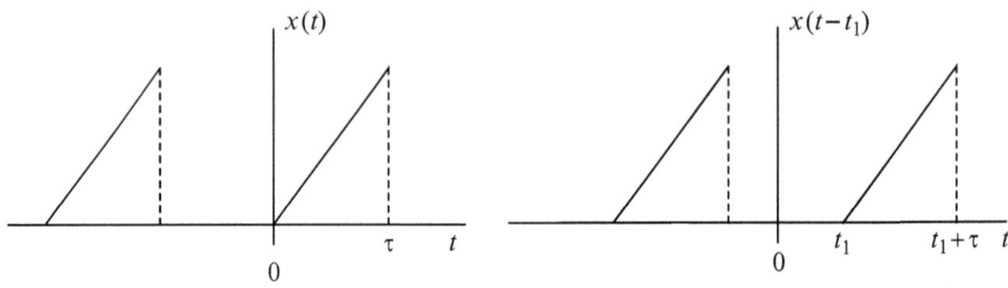

(b) time translation $(t_1 > 0)$ $X_k \rightarrow X_k \exp(-j2\pi k f_0 t_1)$

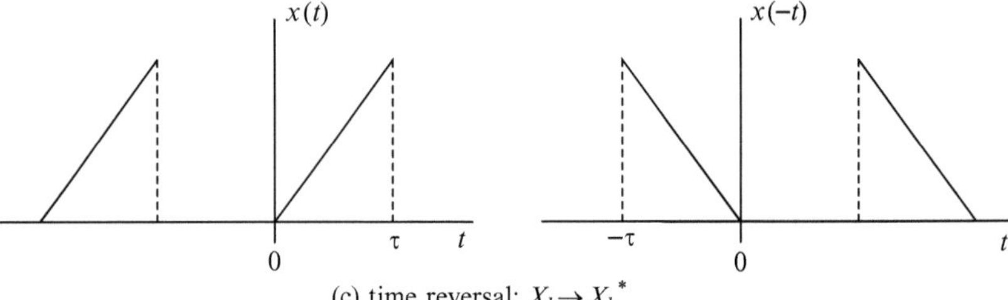

(c) time reversal: $X_k \rightarrow X_k^*$

Fig. 16.5 Illustrating how amplitude shift, time translation, and time reversal alter the Fourier coefficients of a waveform

$$x(-t) = \sum_{k=-\infty}^{\infty} X_k e^{-j2\pi k f_0 t}. \tag{16.29}$$

The signal is real, so $x(-t)^* = x(-t)$. Conjugating (16.29) gives

$$x(-t) = \sum_{k=-\infty}^{\infty} X_k^* e^{j2\pi k f_0 t},$$

which implies that the Fourier coefficients of a time-reversed signal are the conjugates of the original Fourier coefficients.

Negating a signal negates the Fourier coefficients. Multiplying a signal by a constant multiplies the Fourier coefficients by the same constant. The Fourier coefficients for the sum of two or more signals having the same period are the sums of the Fourier coefficients for the signals. These relations and those illustrated by Fig. 16.5 can be expressed as follows:

$$\text{Negation: } y(t) = -x(t) \leftrightarrow Y_k = -X_k, \quad k = 0, \pm 1, \pm 2, \cdots; \tag{16.30}$$

$$\text{Amplitude scaling: } y(t) = Ax(t) \leftrightarrow Y_k = AX_k, \quad k = 0, \pm 1, \pm 2, \cdots; \tag{16.31}$$

$$\text{Superposition: } z(t) = x(t) + y(t) \leftrightarrow Z_k = X_k + Y_k, \quad k = 0, \pm 1, \pm 2, \cdots; \tag{16.32}$$

$$\text{DC shift: } y(t) = x(t) + a \leftrightarrow Y_0 = X_0 + a, \quad Y_k = X_k, \quad k = \pm 1, \pm 2, \cdots; \tag{16.33}$$

$$\text{Time translation: } y(t) = x(t - t_1) \leftrightarrow Y_k = X_k \exp(-j2\pi k f_0 t_1), \quad k = 0, \pm 1, \pm 2, \cdots; \tag{16.34}$$

$$\text{Time reversal: } y(t) = x(-t) \leftrightarrow Y_k = X_k^*, \quad k = 0, \pm 1, \pm 2, \cdots. \tag{16.35}$$

Equations (16.30–16.35) are sometimes called *operational properties* of Fourier coefficients. We can use the properties summarized above to obtain Fourier coefficients for various waveforms not in the table but composed of those in the table.

Example 16.9. Figure 16.6(a) shows one period of an asymmetric trapezoidal pulse train, represented as the sum of a sawtooth pulse train $y(t)$ and a rectangular pulse train $z(t)$. Figure 16.6(b) shows how the sawtooth pulse train and rectangular pulse train given in Table 16.2 are modified (time-translated) to obtain the component waveforms $y(t)$ and $z(t)$ in Fig. 16.6(a). From Table 16.2 and (16.28), the Fourier coefficients of the trapezoidal pulse train are given by

$$X_0 = \frac{x_0\tau}{2T} + \frac{x_0\tau}{T} = \frac{3x_0\tau}{2T},$$

$$X_k = \frac{jx_0}{2k\pi}\left[e^{-jk\pi\tau/T} - \mathrm{sa}\left(\frac{k\pi\tau}{T}\right)\right]e^{-j\pi k f_0\tau} + \frac{x_0\tau}{T}\mathrm{sa}\left(\frac{k\pi\tau}{T}\right)e^{-j3\pi k f_0\tau},$$

$$k = \pm 1, \pm 2, \cdots$$

16.8 Convergence of Fourier Series

In any practical application of Fourier series, three questions regarding convergence are of interest:

- Whether the series converges at all
- If the series converges, whether it is equal to the signal being represented; and
- If it converges and equals the signal, whether the series converges slowly or rapidly; i.e., whether many or only a few terms are needed to represent a signal with sufficient accuracy

The first two issues are addressed by Dirichlet's theorem, described in Section 16.5. But Dirichlet's theorem says nothing about how rapidly a series converges. Put another way, Dirichlet's theorem does not tell us how many terms of a Fourier series are needed to obtain a satisfactory representation of a signal, which is an important issue in practical applications. A series might converge rapidly for some times (or in some intervals) and slowly for others. Depending upon the form of a signal (or function) and the intended use for the associated Fourier series, we might need relatively few or relatively many terms to describe the signal.

One way to specify the number of terms needed in a finite Fourier series is to compare the mean-squared amplitude (or rms amplitude) of the series with that of the signal to be represented, assuming the latter is

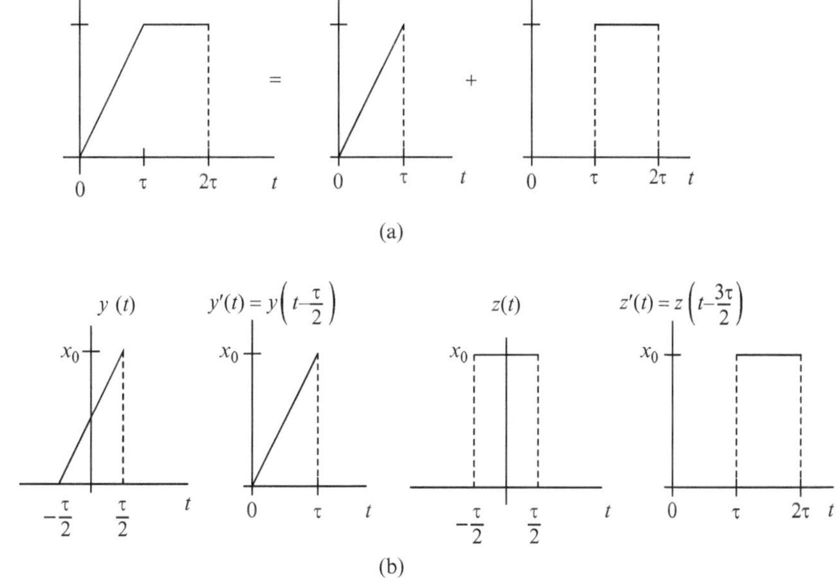

Fig. 16.6 See Example 16.9

known or can be measured. Such a comparison is equivalent to comparing the powers that the signal and the Fourier series would cause to be dissipated in a resistor. In Chapter 5, we show that superposition of power applies to a sum of sinusoids having different frequencies. Thus, the mean-squared amplitude of a signal expressed as

$$x(t) = \sum_{k=0}^{K} A_k \cos(2k\pi f_0 t + \theta_k)$$

is given by

$$x_{rms}^2 = A_0^2 + \frac{1}{2}\sum_{k=1}^{K} A_k^2 = \sum_{k=-K}^{K} |X_k|^2 \tag{16.36}$$

which holds for any integer K, including $K \to \infty$.

Equation (16.36) states that the mean-squared amplitude of a periodic current or voltage $x(t)$ equals the sum of the squared magnitudes of the Fourier coefficients for the current or voltage.

We may interpret (16.36) as follows: Suppose a periodic current or voltage $x(t)$, expressed as

$$x(t) = \sum_{k=-\infty}^{\infty} X_k e^{j2\pi k f_0 t}$$

is represented by the finite Fourier series

$$\hat{x}(t) = \sum_{k=-K}^{K} X_k e^{j2\pi k f_0 t}$$

where the Fourier coefficients are in both cases given by

$$X_k = \frac{1}{T}\int_{t_0}^{t_0+T} x(t) e^{-j2\pi k f_0 t} dt$$

The difference between the representation $\hat{x}(t)$ and the actual current or voltage $x(t)$ can be defined in terms of the average power each would cause to be dissipated in the same resistive load. If both are applied (individually) to the same resistive load, then the ratio of the powers dissipated equals the ratio of the mean-square amplitudes. By (16.36), the ratio is

$$\frac{\hat{x}_{rms}^2}{x_{rms}^2} = \frac{\sum_{k=-K}^{K} |X_k|^2}{\sum_{k=-\infty}^{\infty} |X_k|^2} = \frac{1}{x_{rms}^2}\sum_{k=-K}^{K} |X_k|^2 \tag{16.37}$$

Because $\hat{x}(t)$ approaches $x(t)$ as K approaches infinity, the ratio in (16.37) approaches unity as K approaches infinity, and can be regarded as a measure of how closely a finite number of terms from the Fourier series for $x(t)$ approximates $x(t)$. Because $X_{-k} = X_k^*$ and $|X_k| = A_k/2$, we may write (16.37) as

$$\begin{aligned} \frac{\hat{x}_{rms}^2}{x_{rms}^2} &= \frac{1}{x_{rms}^2}\left[X_0^2 + 2\sum_{k=1}^{K} |X_k|^2\right] \\ &= \frac{1}{x_{rms}^2}\left[X_0^2 + \frac{1}{2}\sum_{k=1}^{K} A_k{}^2\right] \end{aligned} \tag{16.38}$$

Example 16.10. A certain triangular pulse train has period $T = 4\,\text{ms}$ and pulse duration $\tau = 1\,\text{ms}$. How many terms are needed in a finite Fourier series to account for at least 99.9% of the power the pulse train would cause to be dissipated in a resistive load?

Solution: The Fourier coefficients of a triangular pulse train are given by

$$X_k = \frac{x_0\tau}{2T}\text{sa}^2\left(\frac{k\pi\tau}{2T}\right); \quad k = 0, \pm 1, \pm 2, \cdots$$

From Table 16.2, the mean-squared amplitude of the triangular pulse train is

$$x_{rms}^2 = \frac{x_0^2\tau}{3T} = \frac{x_0^2}{12}$$

where x_0 is the amplitude of each pulse. A finite Fourier series for the triangular pulse train is

$$\begin{aligned} \hat{x}(t) &= \frac{x_0\tau}{2T}\sum_{k=-K}^{K} \text{sa}^2\left(\frac{k\pi\tau}{2T}\right) e^{j2\pi k f_0 t} \\ &= \frac{x_0}{8}\sum_{k=-K}^{K} \text{sa}^2\left(\frac{k\pi\tau}{2T}\right) e^{j2\pi k f_0 t} \end{aligned}$$

Table 16.4 Quantities calculated in Example 16.10

k	0	1	2	3	4	5	6	7	8	9	10	11	12
$(\hat{x}_{rms}/x_{rms})^2$	0.188	0.526	0.772	0.914	0.975	0.994	0.997	0.997	0.997	0.997	0.998	0.998	0.999

From (16.37), the fraction of the mean-squared amplitude of the pulse train that is accounted for by the first K harmonics is given by

$$\frac{\hat{x}_{rms}^2}{x_{rms}^2} = \frac{12}{x_0^2}\frac{x_0^2}{64}\left[1+2\sum_{k=1}^{K}\mathrm{sa}^4\left(\frac{k\pi}{8}\right)\right]$$

$$= \frac{3}{16}\left[1+2\sum_{k=1}^{K}\mathrm{sa}^4\left(\frac{k\pi}{8}\right)\right]$$

A computer calculation yields the values given in Table 16.4, which shows that the mean-squared amplitude of a finite Fourier series including harmonics through the 12th equals 99.9% of the mean-squared amplitude of the pulse train.

Figure 16.7 shows graphs of one period of the triangular pulse train and the finite (12-harmonic) Fourier series approximation.

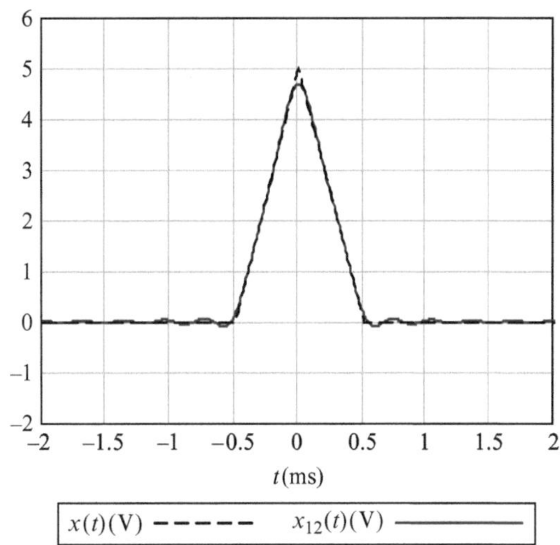

Fig. 16.7 See Example 16.10

The relation

$$x_{rms}^2 = \frac{1}{T}\int_{t_0}^{t_0+T} x^2(t)dt = A_0^2 + \frac{1}{2}\sum_{k=1}^{\infty} A_k^2 = \sum_{k=-\infty}^{\infty} |X_k|^2 \quad (16.39)$$

holds for any signal represented over $t_0 < t \leq t_0 + T$ by the Fourier series

$$x(t) = \sum_{k=-\infty}^{\infty} X_k e^{j2\pi k f_0 t}, \ X_k = \frac{1}{T}\int_{t_0}^{t_0+T} x(t)e^{-j2\pi k f_0 t}dt,$$

whether the signal is periodic or not. Therefore, the ratio

$$\frac{\sum_{k=-K}^{K} |X_k|^2}{\sum_{k=-\infty}^{\infty} |X_k|^2}$$

can be used to specify how many terms of the Fourier series for a signal are required to represent the signal over the interval of expansion, whether the signal is periodic or non-periodic.

16.9 Gibbs' Phenomenon

A Fourier series (or other series) converges *uniformly* at a point or over an interval if the magnitude of the error of the partial sums (finite-series approximations) at the point or over the interval decreases each time a term is added to the series. It can be shown that the Fourier series for a *continuous* signal converges uniformly everywhere in the interval of expansion, but the Fourier series for a signal having a jump discontinuity does not converge uniformly near the edges of the discontinuity. The series does converge to the signal (except at the discontinuity), but not until all terms (an infinite number) are included in the series. This behavior of a Fourier series is illustrated by Fig. 16.8, which shows two finite Fourier series approximations to a rectangular pulse train. The finite series exhibit oscillations on each side of each discontinuity. As the number of terms in the series is increased, the oscillations crowd closer together near

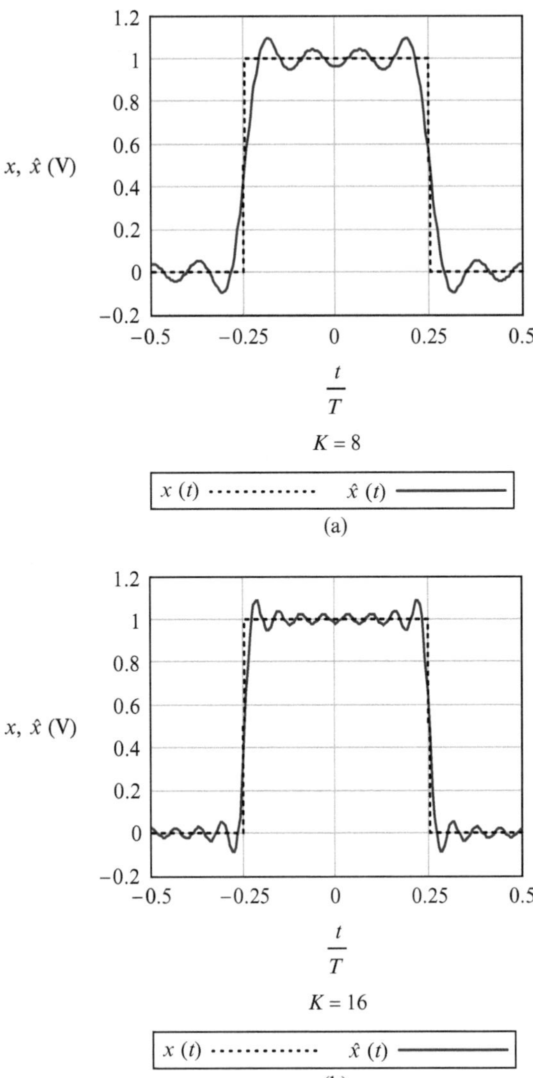

Fig. 16.8 Illustration of Gibbs' phenomenon for a rectangular pulse train (a) $K = 8$ (b) $K = 16$

the jump discontinuities, but their maximum amplitude does not decrease (uniformly). The amplitude stabilizes at about 9% of the jump below the signal on the low side of the jump and 9% of the jump above the signal on the high side. It can be shown that this behavior persists until the entire (infinite) series is summed, when the series converges to the average of the upper and lower values of the discontinuity.[6]

This phenomenon is called *Gibbs' Phenomenon*[7] and the oscillations of a finite Fourier series near a discontinuity are called *Gibbs' oscillations*.

Gibbs oscillations are troublesome in some applications, such as digital filter design, but those applications are beyond our scope. We describe Gibbs oscillations here only because you will see them if you evaluate (numerically) and plot a Fourier series for a signal having one or more jump discontinuities.

16.10 Circuit Response to Periodic Excitation

Recall from Chapter 15 that transfer functions are defined in terms of the phasor representations for a sinusoidal excitation and associated sinusoidal response. For example, the voltage transfer function for a circuit is given by

$$H_v = \frac{\tilde{V}_L}{\tilde{V}_S},$$

where $\tilde{V}_L$ is the phasor for the response (load voltage) and $\tilde{V}_S$ is the available source voltage. Multiplying both numerator and denominator by $\exp(j\omega t)$ gives

$$H_v(j\omega) = \frac{\tilde{V}_L \exp(j\omega t)}{\tilde{V}_S \exp(j\omega t)} = \frac{\tilde{v}_L(t)}{\tilde{v}_S(t)},$$

which shows that a transfer function can also be defined as the ratio of the complex representation of the response to the complex representation of the excitation. That in mind, suppose an excitation given by

$$x(t) = \sum_{k=-\infty}^{\infty} X_k \exp(2j\,k\pi f_0\,t) \qquad (16.40)$$

is applied to a circuit having transfer function $H(j2\pi f)$. Each term in the series above can be regarded as the complex representation of a sinusoid. Thus, by superposition, the corresponding response is given by

$$\begin{aligned} y(t) &= \sum_{k=-\infty}^{\infty} Y_k \exp(2j\,k\pi f_0\,t), \\ Y_k &= H(j2\pi k f_0) X_k. \end{aligned} \qquad (16.41)$$

[6]See G. Moretti, *Functions of a Complex Variable*, Prentice-Hall, 1964, pp 334–337.

[7]After Josiah Willard Gibbs (1839–1903), a mathematical physicist who first explained the phenomenon.

If an excitation is represented by the Fourier series (16.40), then the response is represented by the Fourier series (16.41).

We can use Fourier series, frequency-domain methods, and superposition to obtain the response of a linear circuit to a periodic excitation, as follows:

- Obtain the Fourier coefficients for the excitation, from Table 16.2 if possible, using (16.22) if not.
- Obtain the appropriate transfer function for the circuit.
- Obtain the Fourier coefficients for the response, using (16.41).
- If necessary, sum and plot the Fourier series for the response, using a computer.

Example 16.11. Refer to Fig. 16.9, where the op amp is ideal and the excitation v_S is a half-wave rectified 800 kHz sinusoid having peak amplitude $x_0 = 200$ mV. (a) Obtain an expression for the Fourier coefficients for the response v_L. (b) Calculate the peak amplitudes of the first four non-zero terms in the amplitude–phase series for the response.

Solution: (a) From Table 16.2, the Fourier coefficients for the excitation are given by

$$X_k = \frac{x_0}{4}\left\{\mathrm{sa}\left[\frac{(k+1)\pi}{2}\right] + \mathrm{sa}\left[\frac{(k-1)\pi}{2}\right]\right\},\quad k = 0, \pm 1, \pm 2, \cdots. \qquad (16.42)$$

Using frequency-domain (phasor) methods, we find that the voltage transfer function for the circuit is

$$H_v(j2\pi f) = \frac{\tilde{V}_L}{\tilde{V}_S} = \frac{R_2/R_1}{1 + j2\pi f R_2 C}. \qquad (16.43)$$

The Fourier coefficients for the response are given by

$$Y_k = H_v(2\pi k f_0)X_k,\quad k = 0, \pm 1, \pm 2, \cdots. \qquad (16.44)$$

Table 16.5 gives the peak amplitudes and relative phases of the first four zero terms, where

$$A_0 = |Y_0|,\ \theta_0 = \measuredangle Y_0 = 0,$$
$$A_k = 2|Y_k|,\ \theta_k = \measuredangle Y_k;\ k = 0, 1, 2, \cdots.$$

Figure 16.10 shows a graph of one period of the excitation and of the approximate response, given by

$$\hat{y}(t) = \sum_{k=-4}^{4} Y_k \exp(2j k\pi f_0 t)$$

Example 16.12. In Fig. 16.11, the op amp is ideal, $C = 220$ pF, $R_2 = 100$ kΩ, and

$$v_S = A[\cos(2\pi f_0 t) + \cos(10\pi f_0 t) + \cos(100\pi f_0 t)],$$

with $A = 10$ mV and $f_0 = 50$ kHz. (a) Express the excitation as an exponential Fourier series and obtain the Fourier coefficients for the load

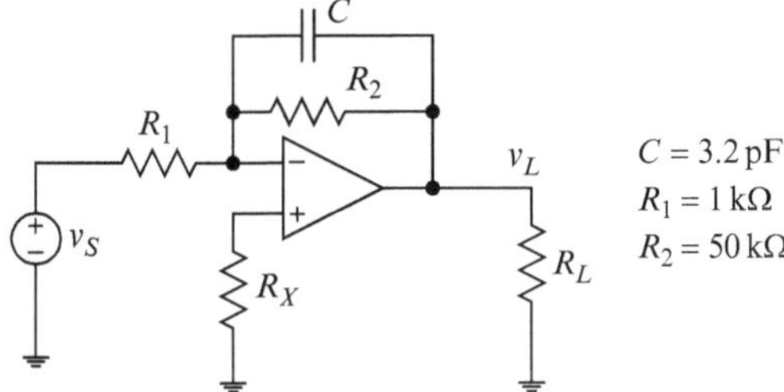

Fig. 16.9 See Example 16.11

Table 16.5 Amplitude and phase of the first five terms of the Fourier series for the response of the circuit considered in Example 16.11. The graph does not change perceptibly if the number of terms is doubled

k	0	1	2	3	4
A_k (V)	3.183	3.896	1.120	0.000	0.126
θ_k	0.000	−0.677	−1.015	0.000	1.872

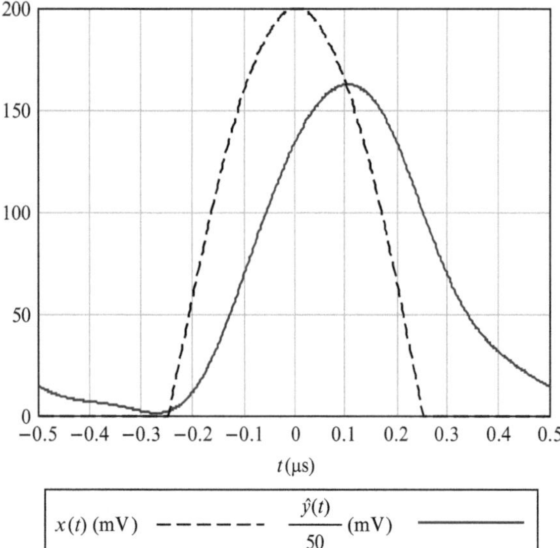

Fig. 16.10 See Example 16.11

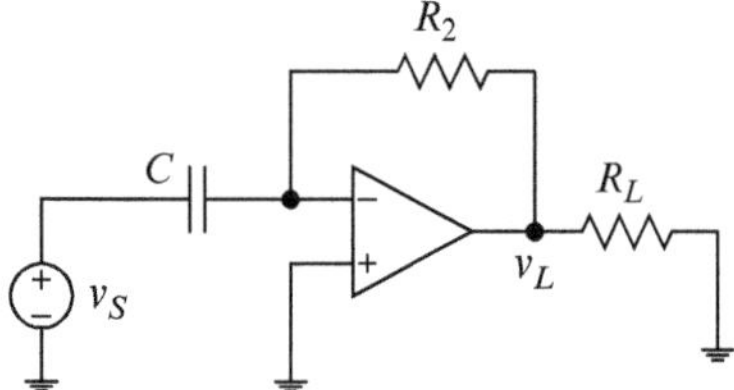

Fig. 16.11 See Example 16.12

voltage v_L. (b) Express the load voltage in amplitude–phase form.

Solution: (a) The fundamental frequency is f_0 and the harmonics present in the excitation have frequencies f_0, $5f_0$, and $50f_0$. Euler's identity (or Table 16.1) gives

$$\begin{aligned} v_S &= \frac{A}{2}\left[e^{j2\pi f_0 t} + e^{-j2\pi f_0 t} + e^{j10\pi f_0 t} + e^{-j10\pi f_0 t}\right. \\ &\quad \left. + e^{j100\pi f_0 t} + e^{-j100\pi f_0 t}\right] \\ &= \sum_{k=-\infty}^{\infty} X_k e^{j2\pi k f_0 t}, \end{aligned}$$

with

$$X_k = \begin{cases} \dfrac{A}{2}, & k = \pm 1,\ \pm 5,\ \pm 50, \\ 0, & \text{otherwise.} \end{cases}$$

We need the voltage transfer function for the circuit. Kirchhoff's current law gives

$$-j2\pi f C \tilde{V}_s = \frac{\tilde{V}_L}{R_2}$$

$$\Rightarrow H_v = \frac{\tilde{V}_L}{\tilde{V}_S} = -j2\pi f C R_2 = -j2\pi f \tau$$

with $\tau = R_2 C = 22\,\mu s$. The Fourier coefficients for the load voltage are

$$\begin{aligned} Y_k &= \left\{ \begin{matrix} H_v(kf_0)\dfrac{A}{2}, & k = \pm 1,\ \pm 5,\ \pm 50 \\ 0, & \text{all other } k \end{matrix} \right\} \\ &= \begin{cases} -j\pi k f_0 \tau A, & k = \pm 1,\ \pm 5,\ \pm 50, \\ 0, & \text{all other } k, \end{cases} \end{aligned}$$

where $f_0 = 50$ kHz, $\tau = 22\,\mu s$, and $A = 10$ mV. Thus,

$$Y_k = \begin{cases} -j35 \text{ mV}, & k = 1, \\ -j173 \text{ mV}, & k = 5, \\ -j1.728 \text{ V}, & k = 50, \\ Y_k^*, & k = -1,\ -5,\ -50, \\ 0, & \text{all other } k. \end{cases}$$

(b) From Table 16.1, the amplitude–phase parameters for the load voltage are

$$B_k = 2|Y_k| = \left\{ \begin{matrix} 69.1 \text{ mV}, & k = 1 \\ 345 \text{ mV}, & k = 5 \\ 3.456 \text{ V}, & k = 50 \\ 0, & \text{all other } k \end{matrix} \right\},$$

$$\theta_k = \measuredangle Y_k = \left\{ \begin{matrix} -\dfrac{\pi}{2}, & k = 1,\ 5,\ 50 \\ 0, & \text{all other } k \end{matrix} \right\}.$$

and the amplitude–phase series for the load voltage is

$$v_L = 0.069\cos\left(2\pi f_0 t - \frac{\pi}{2}\right) + 0.345\cos\left(2\pi 5 f_0 t - \frac{\pi}{2}\right) + 3.456\cos\left(2\pi 50 f_0 t - \frac{\pi}{2}\right) \text{ V.}$$

The components of the source voltage have equal amplitudes. The highest-frequency component of the load voltage is 25 times larger than the lowest-frequency component, illustrating that differentiation enhances the high-frequency components of a signal.

Exercise 16.17. Refer to Fig. 16.12, where $R_1 = 10$ kΩ, $C_1 = 320$ nF, $R_2 = 50$ kΩ, and $C_2 = 320$ nF. The op amp is ideal and the excitation v_S is a symmetric triangular wave having peak amplitude $V_0 = 500$ mV and period $T = 100$ ms. (a) Obtain an expression for the Fourier coefficients for the response v_L. (b) Calculate the peak amplitudes of the first four non-zero terms in the amplitude–phase series for both the excitation and response.

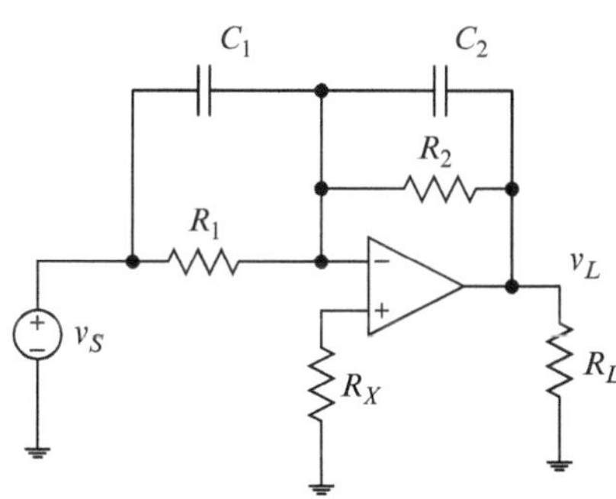

Fig. 16.12 See Exercise 16.17

16.11 Spectra and Spectral Analysis

The peak amplitudes and relative phases of the components of the *amplitude–phase series* for a signal, *regarded as functions of frequency*, comprise the **one-sided spectrum** of the signal, where *one-sided* refers to the fact that the associated frequencies are non-negative. In the sequel, we define two-sided spectra based upon the exponential Fourier series.

The peak amplitudes of an amplitude–phase series comprise the **one-sided amplitude spectrum** and the relative phases comprise the **one-sided phase spectrum**. The dimension of the one-sided amplitude spectrum for a signal is the dimension of the signal. The one-sided phase spectrum is dimensionless and (in this book) is expressed in radians.

Example 16.13. Obtain and plot the first 12 values of the amplitude spectrum and the phase spectrum for a symmetric rectangular pulse train (see Table 16.2) having pulse duration $\tau = 250\,\mu s$, period $T = 1$ ms and amplitude $V_0 = 500$ mV.

Solution: From Table 16.2, the Fourier coefficients for the specified pulse train are given by

$$X_k = \frac{V_0 \tau}{T}\text{sa}\left(\frac{k\pi\tau}{T}\right) \tag{16.45}$$

The amplitude and phase spectra are given by

$$A_k = \begin{cases} |X_0|, & k = 0 \\ 2|X_k|, & k = 1, 2, 3, \cdots \end{cases}, \qquad \theta_k = \measuredangle X_k, \quad k = 0, 1, 2, \cdots \tag{16.46}$$

Table 16.6 gives the values of the Fourier coefficients, the peak amplitude, and the relative phase of the first 12 components, where $f = kf_0 = kT^{-1}$. The values were calculated

Table 16.6 Fourier coefficients for the rectangular pulse train treated in Example 16.13

k	0	1	2	3	4	5	6	7	8	9	10	11
f (kHz)	0	1	2	3	4	5	6	7	8	9	10	11
X_k (mV)	125	112.5	79.6	37.5	0	−22.5	−26.5	−16.1	0	12.5	15.9	10.2
A_k (mV)	125	225	159	75.0	0	45.0	53.1	32.2	0	25.0	31.8	20.5
θ_k (rad)	0	0	0	0	0	3.14	3.14	3.14	3.14	0	0	0

on a computer and rounded to three significant figures.

Figures 16.13 and 16.14 show the graphs of the amplitude spectrum and the phase spectrum, respectively. Conventionally, lines are drawn from the spectral values to the horizontal axis, presumably to make such graphs resemble what one would see on the screen of a spectrum analyzer. Partly for this reason, such spectra are called **line spectra**. But only the spectral values are significant. The lines have no mathematical meaning, but are visually helpful.

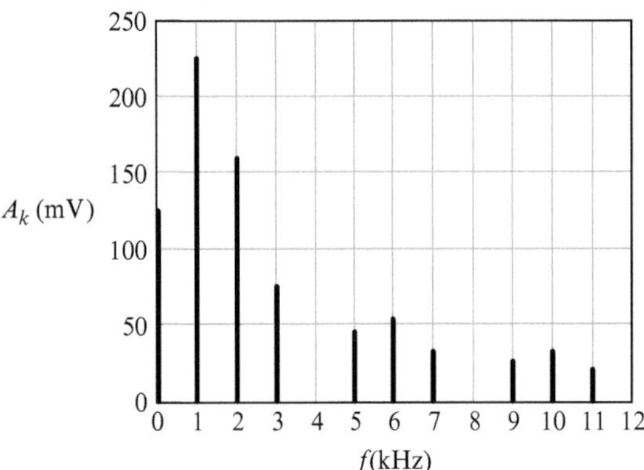

Fig. 16.13 See Example 16.13

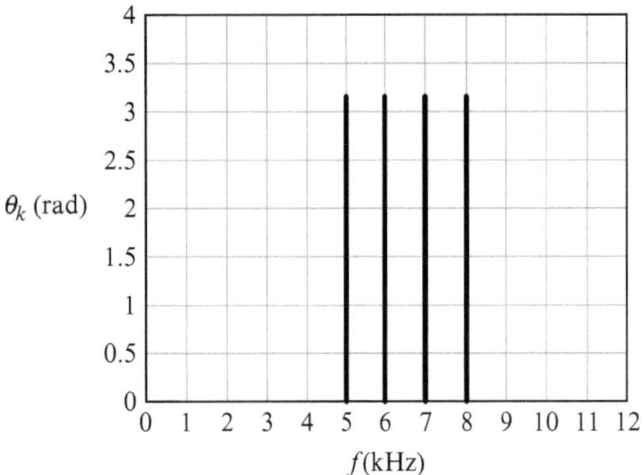

Fig. 16.14 See Example 16.13

Exercise 16.18. Plot the amplitude and phase spectra for the exponential pulse train in Table 16.2, where the pulse amplitude, time constant, and period are $v_0 = 50$ mA, $\tau = 5$ ms, and $T = 100$ ms, respectively.

The **two-sided amplitude spectrum** and the **two-sided phase spectrum** for a signal $x(t)$ are defined as the magnitudes and angles, respectively, of the Fourier coefficients for the signal, regarded as functions of frequency. Formally,

$$\begin{aligned} \alpha_x(kf_0) &= \left\{ \begin{array}{ll} |X_k|, & k = 0, \pm 1, \pm 2, \cdots \\ 0, & otherwise \end{array} \right\}, \\ \phi_x(kf_0) &= \left\{ \begin{array}{ll} \measuredangle X_k, & k = 0, \pm 1, \pm 2, \cdots \\ 0, & otherwise \end{array} \right\} \end{aligned} \tag{16.47}$$

The Fourier coefficients for a real signal have conjugate symmetry about $k = 0$ (because $X_{-k} = X_k^*$), so *the two-sided amplitude spectrum is an even function of frequency and the two-sided phase spectrum is an odd function of frequency*. It is sufficient to compute values for only $f \geq 0$, but often helpful to display both sides graphically. For most purposes, the exponential form of Fourier series and thus two-sided spectral representations are preferred.

Example 16.14. Obtain and plot the two-sided amplitude and phase spectra for the rectangular pulse train of Example 16.13 for $k = 0, \pm 1, \pm 2, \cdots, \pm 8$.

Solution: The Fourier coefficients are given in Table 16.2. The two-sided amplitude and phase spectra are given by

$$\begin{aligned} \alpha_v(f) &= \left\{ \begin{array}{ll} |X_k|, & f = kf_0 \\ 0, & otherwise \end{array} \right\}, \\ \phi_v(f) &= \left\{ \begin{array}{ll} \measuredangle X_k, & f = kf_0 \\ 0, & otherwise \end{array} \right\} \end{aligned}$$

Figure 16.15 shows the spectral plots.

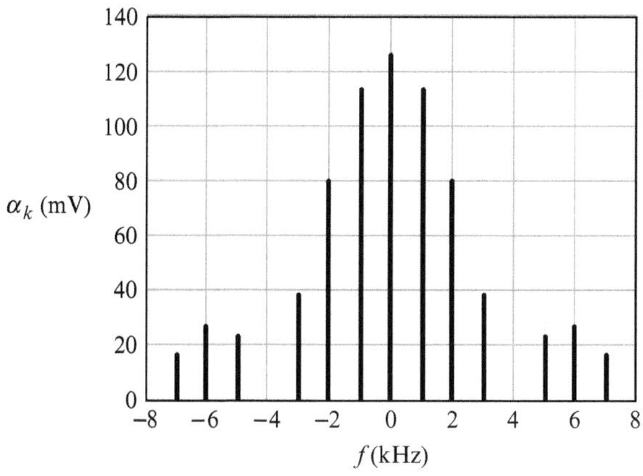

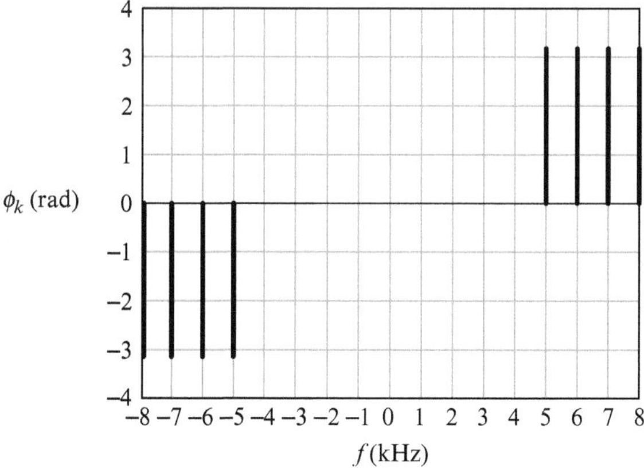

Fig. 16.15 See Example 16.14

Exercise 16.19. Plot the two-sided amplitude and phase spectra for the exponential pulse train in Table 16.2, where the pulse amplitude, time constant, and period are $v_0 = 50$ mA, $\tau = 5$ ms, and $T = 100$ ms, respectively.

The graph of a spectrum tells us at a glance the relative significance of the components.

An amplitude–phase Fourier series expresses a signal as a dc term and a sum of real sinusoids. In the associated one-sided spectrum, the independent variable f is *frequency of sinusoidal oscillation* (number of cycles per second), and is a positive quantity. The associated exponential Fourier series expresses the same signal as the dc term and a sum of complex exponentials (in conjugate pairs). Each exponential can be represented by a point orbiting the origin of a complex plane, where the frequency of the exponential is the *angular velocity* of the point (revolutions per second). This frequency can be positive (for counter-clockwise motion) or negative (for clockwise motion), and the two-sided spectrum associated with an exponential Fourier series as both positive-frequency and negative-frequency components. Neither the positive-frequency components nor the negative-frequency components are physical (real). Only the sums of conjugate pairs yield real sinusoids.

16.12 Problems

P 16.1 Provide the missing entries in the table below.

Amplitude–phase	Exponential	Quadrature
$x_{dc}=5\,\text{V}, \theta_k=0$ $A_k=\frac{10\,\text{V}}{k^2+1}, k=1,2,\cdots$		
	$X_k=\frac{100\exp(-j\pi k/3)}{1+k^2}\text{mA},$ $k=0,\pm1,\pm2,\cdots$	
		$a_0=1\,\text{V}, b_0=0$ $a_k=\frac{2\,\text{V}}{k^2+1}, k=1,2,\cdots$ $b_k=-\frac{1\,\text{V}}{k^2+1}, k=1,2,\cdots$

P 16.2 Express each composite periodic waveform shown in Fig. 16.1 in terms of waveforms given in Table 16.2, where T is the period of the composite waveform.

P 16.3 If the period of a signal $x(t)$ is T, what is the period of $x(2t)$? Of $x(t/2)$?

P 16.4 If the fundamental frequency of a Fourier series is doubled but the Fourier coefficients are not changed, how is the waveform changed? Use graphs of triangular pulse trains to illustrate your answer.

P 16.5 Let $\{X_k;\, k = 0, \pm1, \pm2, \cdots\}$ denote the Fourier coefficients for the waveform given in Fig. P 16.1. Draw the waveform represented by

$$y(t) = \sum_{k=-\infty}^{\infty} Y_k \exp(j2k\pi f_0 t)$$

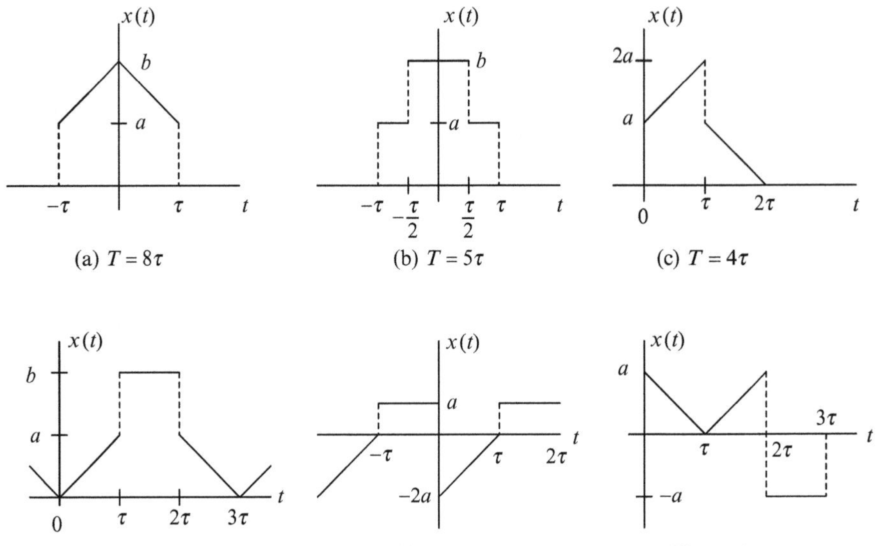

Fig. P 16.1 See Problem P 16.2

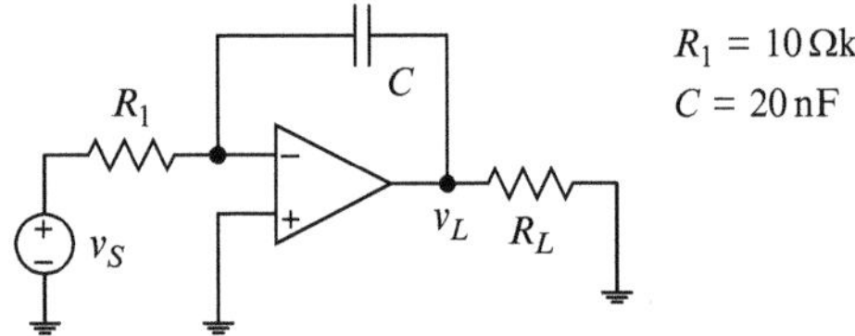

Fig. P 16.2 See Problem P 16.6

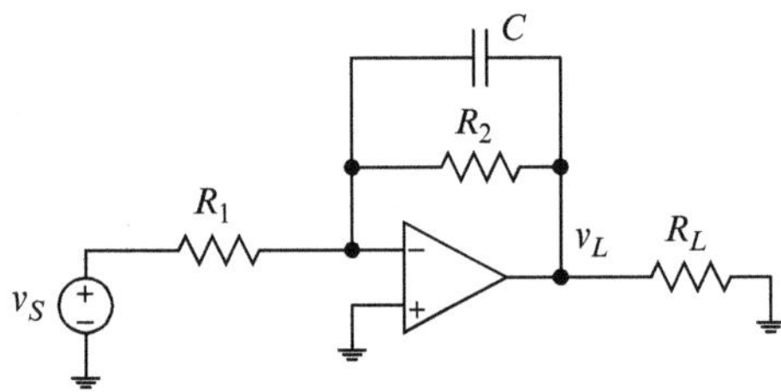

Fig. P 16.3 See Problem P 16.7

if (a) $Y_k = -X_k^*$, (b) $Y_k = X_k \exp(-j\,k\,\pi f_0\,\tau/2)$, (c) $Y_k = X_k - X_k^* \exp(-j\,k\,\pi f_0\,\tau)$

P 16.6 In Fig. P 16-2, the source voltage v_S is a symmetric triangular wave having period $T = 1$ ms and peak amplitude $x_0 = 1$ mV, with a positive peak at $t - 0$. Obtain the Fourier coefficients for the load voltage v_L. Draw graphs of the waveforms of the source and load voltages.

P 16.7 In Fig. P 16.3, the source impedance is negligible, $R_1 = 10\,\text{k}\Omega$, $R_2 = 1\,\text{M}\Omega$, $R_L = 50\,k\Omega$, and $C = 10\,\text{pF}$. The excitation $v_S(t)$ is the *ac component* of a rectangular pulse train having amplitude $V_0 = 100\,\text{mV}$, pulse duration $\tau = 10\,\mu s$, and period $T = 100\,\mu s$.

(a) Obtain an expression for the Fourier coefficients) for the excitation $v_S(t)$.
(b) The excitation is to be represented by a finite Fourier series

$$\hat{v}_S(t) = \sum_{k=-K}^{K} X_k \exp(j\,2k\,\pi f_0\,t)$$

Specify the smallest value of K for which the mean-squared value of $\hat{v}_S(t)$ is at least 99% of the mean-squared value of $v_S(t)$.

(c) Use the finite Fourier series (the value of K) found in part (b) to represent the excitation. Obtain the corresponding finite Fourier series $\hat{v}_L(t)$ for the response $v_L(t)$.
(d) Plot $R_2\hat{v}_S(t)/R_1$ and $-\hat{v}_L(t)$ on the same axes.

P 16.8 Under certain conditions, the circuit shown in Fig. P 16.4(a) functions as a differentiator. The period of the excitation shown in Fig. P 16.4(b) is

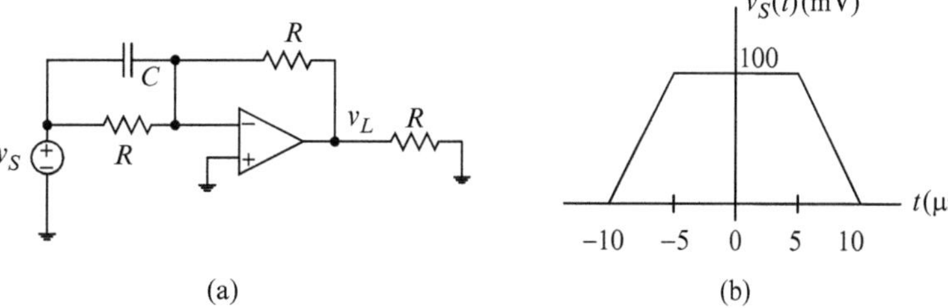

Fig. P 16.4 See Problem P 16.8

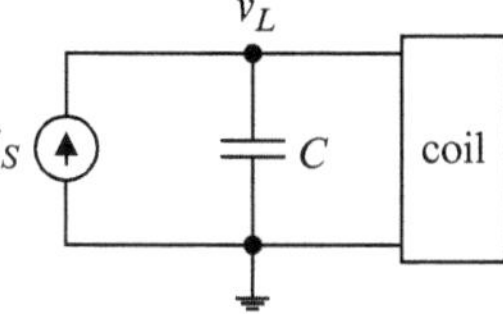

Fig. P 16.5 See Problem P 16.10

$T = 100\,\mu s$. The circuit parameters are $R = 10\,k\Omega$, $C = 100\,\text{nF}$.

(a) Obtain an expression for the voltage transfer function for the circuit.
(b) If the excitation is a sinusoid having frequency f_0, under what condition(s) is the response $v_L(t)$ approximately proportional to the derivative of the excitation? Is the response then approximately proportional to the derivative of any higher-frequency sinusoidal excitation?
(c) Assume the circuit acts as a perfect differentiator (except for gain) for the excitation shown in Fig. P 16.4(b) and draw a graph of the response.
(d) Obtain an expression for the Fourier coefficients for the excitation, express the excitation as a finite Fourier series

$$\hat{v}_S(t) = \sum_{k=-K}^{K} X_k \exp(j\,2k\,\pi f_0\,t),$$

where $K = 50$, and obtain the corresponding finite Fourier series for the response.
(e) Plot one period of the response and comment.

P 16.9 Design a circuit that will remove the dc component from a triangular pulse train having period $T = 500\,\mu s$ and pulse duration $\tau = 125\,\mu s$. Use Fourier analysis to demonstrate the effectiveness of the circuit in this task.

P 16.10 In Fig. P 16.5, the excitation $i_S(t)$ is a symmetric fractional sinusoid as described in Table 16.2, with $b = 5\,\text{mA}$, $x_0 = 1\,\text{mA}$, and $T = 2\,\mu s$. The coil has inductance $L = 10\,\text{mH}$ and quality factor $Q = 50$ at $f_0 = 500\,\text{kHz}$ and the circuit (as seen by the excitation) is resonant at 500 kHz.

(a) Use a finite Fourier series to represent the excitation as

$$\hat{i}_S(t) = \sum_{k=-K}^{K} X_k \exp(j2k\pi f_0\,t)$$

with $K = 50$. Plot $\hat{i}_S(t)$ versus t for $-T/2 \le t \le T/2$ as a check on the series representation.

(b) Approximate the voltage $v_L(t)$ as a finite Fourier series

$$\hat{v}_L(t) = \sum_{k=-K}^{K} Y_k \exp(j2k\pi f_0\,t)$$

with $K = 50$. Plot $\hat{v}_L(t)$ versus t for $-T/2 \le t \le T/2$. On the same axes, plot the fundamental component of $\hat{v}_L(t)$ and comment.

P 16.11 All amplifiers are nonlinear, to lesser or greater degree, depending upon the quality of the amplifier and the amplitude of the excitation. The response of an amplifier to sinusoidal excitation is periodic, but not perfectly sinusoidal, and consists of a fundamental, having the frequency of the excitation, and harmonics that would not be present if the amplifier were linear, because in that case the response would be sinusoidal. The standard measure of distortion for audio amplifiers is *total harmonic distortion*, or *THD*. Total harmonic distortion is measured or specified in either of two ways: One is the total rms amplitude of the second and higher harmonics, expressed as a percentage of the rms amplitude of the fundamental. The other is total power that would be dissipated in a load by the second and higher harmonics, expressed in dB relative to the power that would be dissipated in the same load by the fundamental. We denote the first by *THD%* and the second by *THD*dB. (There is a simple relation between these two measures, as you are asked to show in part (c), below.) Specifically, if the response is given by

$$v_L(t) = \sum_{k=1}^{\infty} A_k \cos(2\pi k f_0 t + \theta_k)$$

where the fundamental frequency f_0 is the frequency of the excitation, then

$$THD\% = 100\frac{\sqrt{\sum_{k=2}^{\infty} A_k^2}}{A_1}, \quad THD\text{dB} = 10 \log\left(\frac{\sum_{k=2}^{\infty} A_k^2}{A_1^2}\right)$$

In either case, the amplitude of the excitation is the specified full-scale amplitude for the amplifier under test, and THD usually is measured at three or more frequencies in the amplifier passband.

This problem considers *THD* caused by clipping (saturation), which is much more severe than would be seen in a high-quality audio amplifier subjected to a full-scale input. The purpose is to illustrate the definitions. More realistic calculations would require methods beyond our scope.

(a) Refer to the symmetric clipped sinusoid in Table 16.2. Let $T = 1$ ms, $x_1 = 1$ V, and $x_0 = \alpha x_1$. Calculate and plot *THD%* and *THD*dB versus α, for $\alpha = 0.5, 0.6, 0.7, 0.8, 0.9$.
(b) To be unnoticeable, the *THD*dB must be below about -70 dB. What are the corresponding values of α and *THD%*?
(c) Show that

$$20 \log (THD\%) = 40 + THD\text{dB}$$

Note: You do not need to calculate any Fourier coefficients other than the first (X_1) to perform the calculations called for above. Bear in mind that the mean-squared amplitudes of different-frequency sinusoids are additive.

P 16.12 Obtain and plot the two-sided amplitude and phase spectra for the sawtooth pulse train in Table 16.2 for $k = 0, 1, 2, \cdots, 8$, where the pulse amplitude is $A = 25$ mV, the pulse duration is $\tau = 250$ ns, and the period is $T = 1\,\mu$s, for $0 \le f \le 8$ MHz.

P 16.13 In Fig. P 16.6(a), $R = 100\,\text{k}\Omega$, $L = 1$ mH, and $C = 1.62$ nF. The excitation $i_S(t)$ is a symmetric triangular pulse train having period $T = 8\,\mu$s, pulse duration $\tau = 2\,\mu$s, and pulse amplitude $I_0 - 10$ mA, as shown in Fig. P 16.6(b). (a) Obtain the Fourier coefficients for the excitation $i_S(t)$ and the response $v_L(t)$. (b) Plot the one-sided amplitude spectra of the excitation and response. (c) Plot one period of the excitation and response on the same axes.

P 16.14 Refer to Fig. P 16.7, where $V_0 = 5\,V$, $R = 10\,\Omega$, and $C = 100$ nF. The switch changes position – from a to b or from b to a every $10\,\mu$s. (a) Plot one period of the current $i(t)$, assuming the switch is moved from a to b at $t = 0$. (b) Find the rms amplitude of $i(t)$. (c) Find the dc component of the current $i(t)$. (d) Obtain and tabulate the Fourier coefficients and the peak amplitudes and relative phases of the amplitude–phase series for $i(t)$, for $0 \le f \le 8/T$.

P 16.15 For each signal listed below, the period and amplitude are 1 s and 1 V, respectively. Plot the two-sided amplitude and phase spectra for each, for $-10\,\text{Hz} \le f \le 10\,\text{Hz}$.

(a) Symmetric square wave, (b) symmetric sawtooth wave, (c) symmetric triangular wave, (d) symmetric half-wave rectified sinusoid, (e) symmetric full-wave rectified sinusoid.

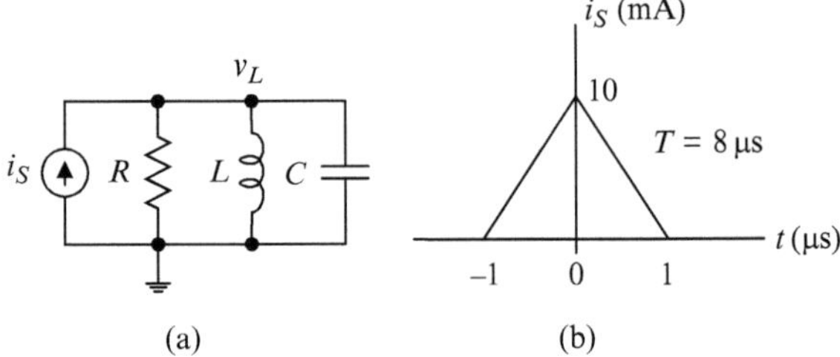

Fig. P 16.6 See Problem P 16.13

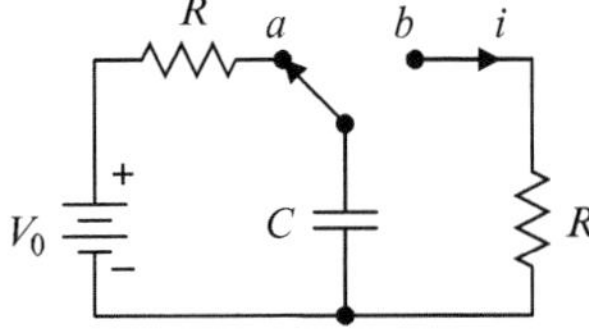

Fig. P 16.7 See Problem P 16.14

Chapter 17
Operational Amplifiers II: AC Model and Applications

Chapter 7 describes a dc model for an op amp and gives examples of dc resistive circuit design based on the model. Such a treatment is instructive but unrealistic for all but simple circuits operating at relatively low frequencies. In almost all practical applications of op amps, the frequency responses of surrounding circuitry and of op amps themselves must be considered.

In this chapter, we present a linear ac model for an op amp and describe ac analysis of op amp circuits. We present expressions for input impedance, output impedance, and the voltage transfer function for inverting, non-inverting, and voltage follower configurations. We describe slew rate, which is a limitation on an op amp's ability to track a rapidly varying input. We discuss power dissipation and power-conversion efficiency of op-amp circuits, and we discuss typical issues in linear op-amp circuit design.

As a reminder, when we compare impedances by writing that one is larger or smaller than another, *magnitudes* are implied (because complex numbers cannot be ordered).

17.1 AC Model for an Op Amp

The transfer function of the ideal op amp (model) treated in Chapter 7 is independent of frequency. The voltage gain of a real op amp decreases with frequency, the decrease beginning at about 10 Hz for typical integrated op amps. The primary cause of the decrease is a capacitor internal to most op amps, whose purpose is described in the next paragraph.

A transistor exhibits capacitance (e.g., junction capacitance) that limits the bandwidth of a transistor circuit. An op amp contains a number of transistors and a corresponding number of bandwidth-limiting capacitances. For reasons you will learn in subsequent courses, these capacitances and the gains of the transistors can cause an op amp to be unstable, even if the net external feedback is negative. Most op amps contain one (intentional) dominant capacitor, called a *compensating capacitor*, whose purpose is to eliminate this inherent instability. Op amps containing such a compensating capacitor are said to be *internally compensated*. Those that are not internally compensated must be externally compensated, and provide additional terminals for that purpose. We limit our discussion to internally compensated op amps.

The compensating capacitor in an op amp is dominant for frequencies where the op amp is useful (for frequencies where the intrinsic voltage gain exceeds unity). For purposes of analyzing linear op-amp circuits, we may assume that the **intrinsic voltage transfer function** of a compensated op amp is of the form

$$\mu(j2\pi f) = \frac{\mu_0}{1 + j(f/f_0)}, \quad f < f_T, \qquad (17.1)$$

where μ_0 is the**intrinsic dc voltage gain**, also called the **open-loop dc voltage gain**, f_0 is the **intrinsic (or open-loop) bandwidth** of the op amp, and f_T defines the upper end of the useful bandwidth of the op amp (and of the model), as described further below. For a typical general-purpose op amp, the intrinsic bandwidth f_0 is on the order of 10 Hz and the intrinsic dc gain is on the order of 10^5.

T.H. Glisson, *Introduction to Circuit Analysis and Design*,
DOI 10.1007/978-90-481-9443-8_17, © Springer Science+Business Media B.V. 2011

Exercise 17.1. Draw a Bode gain plot for the voltage gain of an op amp whose voltage transfer function is given by (17.1), with $\mu_0 = 10^5$ and $f_0 = 10\,\text{Hz}$.

Figure 17.1 shows a linear ac model for an op amp. The only differences between the dc model treated in Chapter 7 and the ac model in Fig. 17.1 are that the intrinsic voltage transfer function μ and input impedance Z_i in the ac model are functions of frequency. The output impedance of an op amp is resistive for frequencies below the unity-gain frequency. The intrinsic voltage transfer function is given by (17.1). The input impedance Z_i of an op amp is typically that of a resistor R_i in parallel with a capacitor C_i, given by

$$Z_i = \frac{R_i}{1 + jf/f_i}; \quad f_i = \frac{1}{2\pi R_i C_i}. \tag{17.2}$$

The resistance R_i is on the order of $1\,\text{M}\Omega$ for op amps having bipolar junction transistors (BJTs) in the input stage and is on the order of $1\,\text{T}\Omega$ ($10^{12}\,\Omega$) for op amps having field-effect transistors (FETs) in the input stage. The capacitance C_i is on the order of 1 pF in either case. The input impedance is usually large enough that it is inconsequential to performance of linear op-amp circuits. Keep in mind that Z_i given by (17.2) is the input impedance of an op amp, not necessarily the input impedance of a circuit in which the op amp is embedded.

The **intrinsic voltage gain** of an op amp is the magnitude of the intrinsic voltage transfer function. From (17.1),

$$\begin{aligned} |\mu(j2\pi f)| &= \left|\frac{\mu_0}{1 + j(f/f_0)}\right| \\ &= \frac{\mu_0}{\sqrt{1 + (f/f_0)^2}}, \quad f < f_T. \end{aligned} \tag{17.3}$$

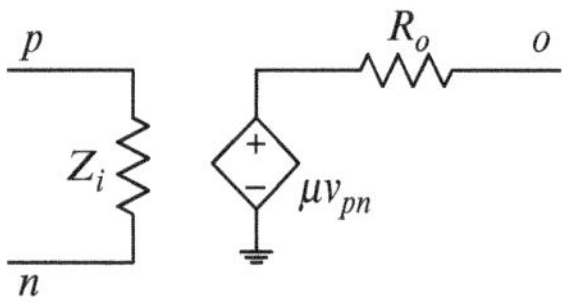

Fig. 17.1 AC linear circuit model for an op amp

The **unity-gain frequency**, denoted by f_T, is the frequency for which the intrinsic voltage gain $|\mu|$ equals unity. Thus

$$\begin{aligned} |\mu(j2\pi f_T)| &= \frac{\mu_0}{\sqrt{1 + (f_T/f_0)^2}} = 1 \\ &\Rightarrow (f_T/f_0)^2 = {\mu_0}^2 - 1. \end{aligned} \tag{17.4}$$

Because $\mu_0 \gg 1$, it follows from (17.4) that

$$\frac{f_T}{f_0} \cong \mu_0 \Rightarrow f_T \cong \mu_0 f_0. \tag{17.5}$$

The product $\mu_0 f_0$ (equal to the unity-gain frequency f_T) is called the **gain-bandwidth product** of the op amp. Gain-bandwidth products for typical general-purpose integrated op amps are on the order of 100 kHz to 1 MHz. Gain-bandwidth products for some *high-speed* op amps range from about 100 MHz to more than 1 GHz. Gain-bandwidth product is an important parameter in op-amp circuit analysis and design.

Exercise 17.2. With reference to (17.1), show that the voltage transfer function for an op amp can be expressed as

$$H_v = \frac{\mu_0}{1 + j\mu_0(f/f_T)}$$

For frequencies above the unity-gain frequency f_T, internal device capacitances cause the intrinsic voltage gain $|\mu(j2\pi f)|$ to decrease with frequency more sharply than is implied by (17.1), but that behavior is excluded from the model (17.1) because frequencies above f_T are beyond those for which the associated op amp is useful. Linear op-amp circuits are intended and designed to operate at frequencies below f_T.

We usually write μ for $\mu(j2\pi f)$ unless explicit indication of frequency dependence is necessary to avoid confusion. Also, we usually omit the condition $f < f_T$ when using the model (17.1), with the understanding that the model and relations derived from the model are valid only for $f < f_T$.

The intrinsic voltage gain defined by (17.3) is a positive, dimensionless, real function of frequency and is expressed in dB as

$$\mu_{dB} = 20\ \log|\mu|$$
$$= 20\ \log(\mu_0) - 20\ \log\left[\sqrt{1+(f/f_0)^2}\right] \quad (17.6)$$

or normalized to the intrinsic dc gain as

$$\mu_{dB0} = 20\ \log\left|\frac{\mu}{\mu_0}\right|$$
$$= -20\ \log\left[\sqrt{1+(f/f_0)^2}\right]. \quad (17.7)$$

The asymptotic approximation to the normalized intrinsic voltage gain is

$$\mu_{dB0} \cong \begin{cases} 0, & f \le f_0, \\ -20\ \log(f/f_0), & f > f_0. \end{cases} \quad (17.8)$$

Figure 17.2 shows both actual (dashed) and asymptotic Bode plots of the intrinsic gain (17.3) in dB, both normalized to the dc gain, versus frequency. For frequencies below f_0, the intrinsic gain is approximately μ_0, independent of frequency. For frequencies above f_0, the intrinsic voltage gain decreases by 20 dB/decade; e.g., as the frequency increases from $10f_0$ to $100f_0$, the voltage gain $|\mu|$ decreases from (approximately) $\mu_0/10$ (−20 dB) to $\mu_0/100$ (−40 dB).

Example 17.1. The intrinsic voltage gain of a certain op amp 2 decades above the intrinsic bandwidth (the corner frequency) is 60 dB. The gain-bandwidth product for the op amp is 1 MHz. What are the intrinsic dc voltage gain and intrinsic bandwidth?

Solution: Refer to Fig. 17.3, which displays the information given in the problem statement on an asymptotic graph of gain versus frequency. Because the gain decreases by 20 dB/decade above the intrinsic bandwidth f_0 and the gain is 60 dB 2 decades above f_0, the dc gain must be

$$\mu_{0\,dB} = 60 + 40 = 100\,\text{dB} \Rightarrow \mu_0 = 10^5.$$

Because the gain at $f = f_0$ equals 10^5 and because above f_0 the gain decreases by 20 dB/decade, the gain equals 0 dB 5 decades above f_0. Thus

$$f_T = 10^5 f_0 = 1\,\text{MHz} \Rightarrow f_0 = 10\,\text{Hz}$$

17.2 Linear Resistive-Feedback Amplifiers

A resistive-feedback amplifier is an op-amp-based inverting or non-inverting amplifier in which the elements external to the op amp are resistive (or conductors, as in a voltage follower). Figure 17.4 shows the

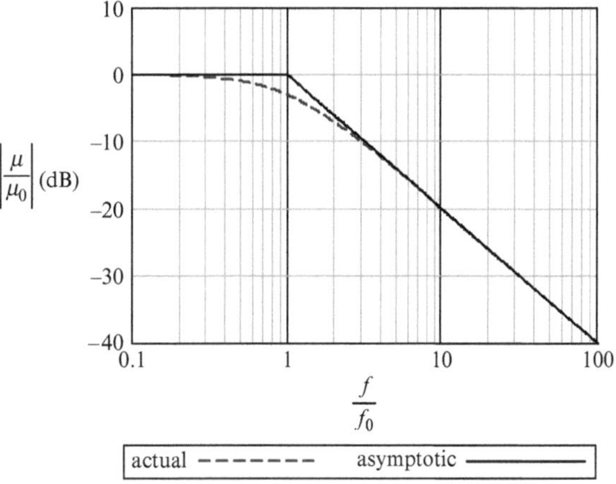

Fig. 17.2 Intrinsic voltage gain versus frequency for an op amp

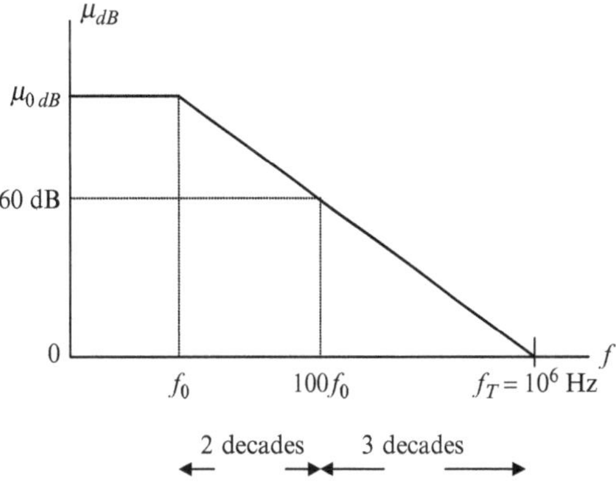

Fig. 17.3 See Example 17.1

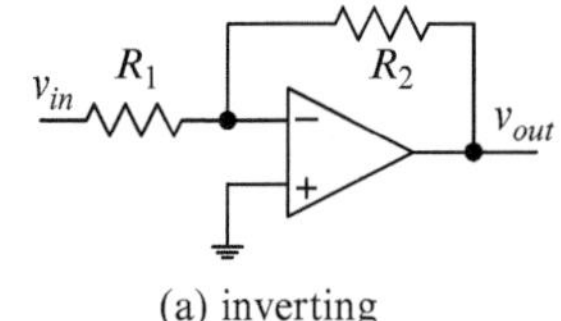

(a) inverting

(b) non-inverting

(c) follower

Fig. 17.4 Circuit diagrams for resistive-feedback amplifiers

feedback amplifiers considered in this section. To conserve space, sources and loads are not shown but assumed present. More general configurations are treated in Section 17.3.

Symbols used in references to op-amp-based resistive-feedback amplifiers are defined as follows:

R_1, R_2 are as defined in Fig. 17.4;

Z_S, Z_L are the source and load impedances;

H_v is the amplifier voltage transfer function with source and load connected;

R_i is the op-amp input resistance;

C_i is the op-amp input capacitance;

$Z_i = Z_i(j2\pi f)$ is the op-amp input impedance, given by (17.2);

R_o is the op-amp output resistance;

μ_0 is the op-amp intrinsic dc voltage gain;

f_0 is the op-amp intrinsic bandwidth;

$f_T = \mu_0 f_0$ is the op-amp gain-bandwidth product (unity-gain frequency);

$\mu = \mu(j2\pi f)$ is the op-amp intrinsic voltage transfer function, given by (17.1);

and, under conditions usually met in applications,

$A_{v0} = R_2/R_1$ is the dc voltage gain for an inverting amplifier;

$A_{v0} = 1 + R_2/R_1$ is the dc voltage gain for a non-inverting amplifier;

$W = f_T/A_{v0}$ is the approximate bandwidth of an inverting or non-inverting amplifier, provided $A_{v0} \gg 1$. See (17.9).

Under conditions met in most applications, the overall voltage transfer function H_v (including source and load resistances) for each configuration in Fig. 17.4 is to a good approximation a function of the form

$$H_v \cong \frac{H_v(0)}{1 + jf/W}; \quad W = \frac{f_T}{|H_v(0)|} = \frac{f_T}{A_{v0}}, \tag{17.9}$$

where $H_v(0)$ is the dc voltage transfer ratio, $A_{v0} = |H_v(0)|$ is the dc voltage gain, W is the half-power bandwidth, and f_T is the op-amp gain-bandwidth product (unity-gain frequency). From (17.9),

$$A_{v0}W = f_T = \mu_0 f_0 \tag{17.10}$$

Thus, under conditions met in most applications, *the gain-bandwidth product of an op-amp based inverting amplifier*, non-inverting amplifier, *or voltage follower equals that of the op amp*. If the required gain and bandwidth of an amplifier are specified, the gain-bandwidth product of the op amp is determined. Conversely, the gain-bandwidth product of an op amp determines the gain-bandwidth product of an amplifier built around the op amp.

For an *inverting amplifier*,

$$A_{v0} \cong \frac{R_2}{R_i} = -H_v(0), \tag{17.11}$$

for a *non-inverting amplifier*,

$$A_{v0} \cong \frac{R_2}{R_i} + 1 = H_v(0), \tag{17.12}$$

and for a *voltage follower*,

$$A_{v0} \cong 1 = H_v(0). \tag{17.13}$$

Table 17.1 gives the conditions under which (17.9) through (17.13) are valid for the amplifiers in Fig. 17.4.

Table 17.2 gives the approximate expressions for input impedance Z_{in} and output impedance Z_{out} and the conditions under which the expressions are valid.

Table 17.1 Conditions under which (17.9) through (17.13) are valid for basic resistive-feedback amplifiers

Amplifier	Conditions
Inverting	$R_2 \gg R_o,\ \lvert Z_L\rvert \gg R_o,\ \lvert Z_i\rvert \gg R_1 \gg \lvert Z_S\rvert,\ \mu_0 \gg A_{v0}$
Non-inverting	$R_2 \gg R_o,\ \lvert Z_L\rvert \gg R_o,\ \lvert Z_i\rvert \gg \lvert Z_S\rvert,\ \lvert Z_i\rvert \gg R_1,\ \mu_0 \gg A_{v0}$
Voltage follower	$\lvert Z_L\rvert \gg R_o,\ \lvert \mu Z_i\rvert \gg R_o,\ \lvert Z_i\rvert \gg \lvert R_o + Z_S\rvert$

Table 17.2 Input and output impedance of (direct-coupled) resistive-feedback amplifiers

Amplifier	Input impedance	Output impedance
Inverting	$Z_{in} \cong \dfrac{1+j(f/W)}{1+j(f/f_T)} R_1$	$Z_{out} \cong \dfrac{(1+jf/f_0)}{(1+jf/W)} \dfrac{A_{v0}R_o}{\mu_0}$
	$R_1 \gg R_S,\ R_2 \gg R_o,\ R_L \gg R_o,$ $\mu_0 \gg A_{v0} \gg 1,\ \mu_0\lvert Z_i\rvert \gg R_2$	$\mu_0 \gg A_{v0} \gg 1,\ A_{v0}\lvert Z_i\rvert \gg R_2 \gg R_o,$ $R_1 \gg R_S$
Non-inverting	$Z_{in} \cong \left[\dfrac{1+jf/(f_0+W)}{1+jf/f_0}\right] \dfrac{\mu_0 Z_i}{A_{v0}}$	$Z_{out} \cong \left[\dfrac{1+jf/f_0}{1+jf/(f_0+W)}\right] \dfrac{f_0 R_o}{f_0+W}$
	$R_2 \gg R_o, R_L \gg R_o, \lvert Z_i\rvert \gg R_1, \mu_0 \gg A_{v0}$	$\lvert Z_i\rvert \gg R_S + R_1,\ R_2 \gg R_o,\ \mu_0 \gg A_{v0}$
Voltage follower	$Z_{in} \cong \dfrac{1+jf/f_T}{1+jf/f_0} \mu_0 Z_i$	$Z_{out} = \left(\dfrac{1+jf/f_0}{1+jf/f_T}\right) \dfrac{R_o}{\mu_0}$
	$R_L \gg R_o, \lvert Z_i\rvert \gg R_o,\ \mu_0 \gg 1$	$R_i \gg R_o, \lvert Z_i\rvert \gg R_S,\ \mu_0 \gg 1$

Example 17.2. The parameters of a certain op amp are:
Input resistance $R_i = 3\,\text{M}\Omega$,
Input capacitance $C_i = 1.4\,\text{pF}$,
Intrinsic dc voltage gain $\mu_0 = 2 \times 10^5$,
Intrinsic bandwidth $f_0 = 10\,\text{Hz}$,
Gain-bandwidth product $f_T = \mu_0 f_0 = 2.0\,\text{MHz}$,
Output resistance $R_o = 75\,\Omega$.

The op amp is used in an inverting amplifier having the form shown in Fig. 17.4(a), with $R_1 = 10\,\text{k}\Omega$ and $R_2 = 1\,\text{M}\Omega$. The source and load are resistive. It is known that the source resistance is no larger than $100\,\Omega$ and that the load resistance is no smaller than $5\,\text{k}\Omega$.

(a) Calculate the no-load voltage gain of the amplifier at dc and at the amplifier passband edge (for $f = 0$ and $f = W$). Validate the calculations by confirming that the conditions in Table 17.1 are met.
(b) Use the expressions in Table 17.2 to calculate the amplifier input impedance and output impedance at dc and at the amplifier passband edge. Validate the calculations by confirming that the conditions in Table 17.2 are met.

Solution: (a) From (17.9) and (17.11),

$$H_{v0} = -\frac{R_2}{R_1} = -\frac{1\,\text{M}\Omega}{10\,\text{k}\Omega} = -100,$$
$$W = \frac{f_T}{|H_v(0)|} = \frac{2.0\,\text{MHz}}{100} = 20\,\text{kHz},$$
$$H_v = \frac{H_v(0)}{1+jf/W} = -\frac{100}{1+jf/W},$$

which are valid if

$$R_2 \gg R_o,\ |Z_L| \gg R_o, R_1 \gg |Z_S|,$$
$$|Z_i| \gg R_1, \mu_0 \gg A_{v0}.$$

We have $R_2 = 1\,\text{M}\Omega \gg R_o = 75\,\Omega$, $|Z_L| \geq 5\,\text{k}\Omega \gg R_o = 75\,\Omega$, $R_1 = 10\,\text{k}\Omega \gg |Z_S| \leq 100\,\Omega$. From (17.2),

$$Z_i = \frac{R_i}{1+jf/f_i};\ f_i = \frac{1}{2\pi R_i C_i} \cong 38\,\text{kHz}.$$

The minimum value of $|Z_i|$ in the amplifier passband is

$$|Z_i(j2\pi W)| = \left|\frac{R_i}{1+jW/f_i}\right| = \left|\frac{3\,\text{M}\Omega}{1+j(15\,\text{kHz}/38\,\text{kHz})}\right| \cong 2.7\,\text{M}\Omega,$$

so $|Z_i| \gg R_1$ throughout the amplifier passband (for $0 \leq f \leq W$).

(b) From Table 17.2,

$$|Z_{in}(0)| \cong R_1 = 10\,\text{k}\Omega,$$
$$|Z_{in}(j2\pi W)| \cong \left|\frac{1+j(W/W)}{1+j(W/f_T)}\right| R_1 = \left|\frac{1+j}{1+j(0.01)}\right| R_1 \cong \sqrt{2} R_1 \cong 14.1\,\text{k}\Omega.$$

The conditions that must be met for the calculation above to be valid are

$$R_1 > > R_S,\ R_2 \gg R_o, R_L \gg R_o,$$
$$\mu_0 \gg A_{v0} \gg 1,\ \mu_0|Z_i| \gg R_2.$$

As an exercise, show that the conditions are met.

The op-amp gain-bandwidth product is $f_T = \mu_0 f_0 = 2.0\,\text{MHz}$, so the op-amp intrinsic bandwidth is $f_0 = f_T/\mu_0 = 10\,\text{Hz}$. Thus, from Table 17.2,

$$|Z_{out}(0)| \cong \frac{A_{v0}R_o}{\mu_0} = 0.0375\,\Omega,$$

$$|Z_{out}(j2\pi W)| \cong \left|\frac{1+jW/f_0}{1+jW/W}\right| \frac{A_{v0}R_o}{\mu_0}$$

$$= \left|\frac{1+j2000}{1+j}\right|(0.0375\,\Omega) \cong 53\,\Omega.$$

The conditions that must be met for the calculation above to be valid are

$$\mu_0 \gg A_{v0} \gg 1,\ A_{v0}|Z_i| \gg R_2 \gg R_o,\ R_1 \gg R_S.$$

As an exercise, show that the conditions are met.

Exercise 17.3. The parameters of a certain op amp are those given in Example 17.2. The op amp is used in a non-inverting amplifier having the form shown in Fig. 17.4(b), with $R_1 = 10\,\text{k}\Omega$ and $R_2 = 1\,\text{M}\Omega$. The source and load are resistive. It is known that the source resistance is no larger than $100\,\Omega$ and that the load resistance is no smaller than $5\,\text{k}\Omega$.

(a) Calculate the no-load voltage gain of the amplifier at dc and at the amplifier passband edge (for $f = 0$ and $f = W$).
(b) Use the expressions in Table 17.2 to calculate the amplifier input impedance and output impedance at dc and at the amplifier passband edge.

Exercise 17.4. Prior to studying the following paragraphs, sketch neatly and label fully asymptotic graphs of the magnitudes of each input and output impedance in Table 17.2 versus frequency for $0.1f_0 \le f \le 10f_T$. Use a logarithmic scale for frequency. Let $R_i = 500\,\text{k}\Omega$, $R_o = 75\,\Omega$, $\mu_0 = 2\times10^5$, and $f_T = 1\,\text{MHz}$. These values are typical for a 741 BJT op amp, first introduced in 1968 and still in production. Use a source impedance of $R_S = 50\,\Omega$ and a load resistance of $R_L = 5\,\text{k}\Omega$. For inverting and non-inverting amplifiers, let $R_1 = 10\,\text{k}\Omega$ and $R_2 = 1\text{M}\Omega$, which gives each amplifier a dc voltage gain of $A_{v0} \cong 100$ and a bandwidth of $W \cong 10\,\text{kHz}$. Plot an asymptotic approximation to

$$20\log\left(\frac{|Z_{in}|}{R_1}\right)$$

versus frequency for an inverting amplifier and an asymptotic approximation to

$$20\log\left(\frac{|Z_{in}|}{R_i}\right)$$

versus frequency for a non-inverting amplifier and voltage follower. Plot an asymptotic approximation to

$$20\log\left(\frac{|Z_{out}|}{R_o}\right)$$

for both inverting and non-inverting amplifiers and for a follower. Drawing and labeling the graphs will help you follow and understand the discussion in the following paragraphs.

From Table 17.2, the magnitude of the input impedance of a direct-coupled inverting amplifier increases from R_1 at dc to

$$|Z_{in}(j2\pi W)| \cong \left|\frac{1+j(W/W)}{1+j(W/f_T)}\right| R_1 \cong \sqrt{2}R_1 \quad (17.14)$$

at the passband edge and increases with frequency thereafter (throughout the useful range of the op amp). For an inverting amplifier, the dc gain is given by $A_{v0} = R_2/R_1$. The resistance R_2 is typically on the order of $1\,\text{M}\Omega$ and the dc gain is typically on the order of 100. As a result, the minimum input resistance of an inverting amplifier (R_1) is typically on the order of $10\,\text{k}\Omega$ in the amplifier passband.

From Table 17.2, the magnitude of the input impedance of a direct-coupled non-inverting amplifier decreases from $\mu_0 R_i/A_{v0}$ at dc to

$$|Z_{in}(j2\pi W)| \cong \left|\frac{1+jW/W}{1+jW/f_0}\right|\left(\frac{\mu_0}{A_{v0}}\right)R_i$$
$$\cong \left(\frac{\mu_0 f_0}{A_{v0}W}\right)\sqrt{2}R_i \cong \sqrt{2}R_i \qquad (17.15)$$

at the passband edge, and approaches R_i at higher frequencies (assuming $W > f_0$, which is usually the case). For a typical op amp and non-inverting amplifier, μ_0 is on the order of 10^5 and A_{v0} is on the order of 100, so μ_0/A_0 is on the order of 10^3. Thus the input impedance in the passband of a direct-coupled non-inverting amplifier typically exceeds 1MΩ for BJT op amps and 1 TΩ for FET-input op amps. The input impedance of a direct-coupled non-inverting amplifier is typically much larger than that of a comparable inverting amplifier throughout the passband.[1]

From Table 17.2, the magnitude of the input impedance of a follower decreases from $\mu_0 R_i$ at dc to

$$|Z_{in}(j2\pi W)| \cong \left|\frac{1+jW/f_T}{1+jW/f_0}\right|\mu_0 R_i \cong \frac{\mu_0 f_0}{W}R_i \qquad (17.16)$$

at the passband edge W of a cascaded amplifier. If the gain-bandwidth product of the cascaded amplifier equals that of the follower, then

$$|Z_{in}(j2\pi W)| \cong \frac{\mu_0 f_0}{W}R_i = \frac{f_T}{f_T/A_{v0}}R_i = A_{v0}R_i.$$

In any case, the magnitude of the input impedance of a follower at the upper edge f_T of the follower passband is given by

$$|Z_{in}(j2\pi f_T)| \cong \left|\frac{1+jf_T/f_T}{1+jf_T/f_0}\right|\mu_0 R_i \cong \sqrt{2}R_i. \qquad (17.17)$$

Because μ_0 is typically on the order of 10^5, the dc input impedance of a follower is on the order of $10^5 R_i$, decreasing to $\sqrt{2}R_i$ for $f = f_T$, which is the bandwidth of a follower and well beyond the bandwidth of any cascaded inverting or non-inverting amplifier. This input impedance of a follower is much larger than that of a typical inverting or non-inverting amplifier throughout the passband of the amplifier.

From Table 17.2, the magnitude of the output impedance of either an inverting or non-inverting amplifier increases from $A_{v0}R_o/\mu_0$ at dc to

$$|Z_{out}(j2\pi W)| \cong \left|\frac{(1+jW/f_0)}{(1+jW/W)}\right|\frac{A_{v0}R_o}{\mu_0}$$
$$\cong \frac{R_o}{\sqrt{2}};\ W \gg f_0 \qquad (17.18)$$

at the passband edge W. For a typical general-purpose BJT op amp and amplifier, the intrinsic output resistance R_o is on the order of 100 Ω, the intrinsic dc voltage gain μ_0 is on the order of 10^5 and the amplifier gain is on the order of 100. Thus the output resistance of a typical amplifier increases from about $10^{-3}R_o$ at dc to $R_0/\sqrt{2}$ at the passband edge. To ensure efficient voltage transfer and to avoid having the load and output impedances alter the frequency response of the amplifier, the load impedance should be about $10R_o$ or larger.

From Table 17.2, the magnitude of the output impedance of a follower increases from R_o/μ_0 at dc to

$$|Z_{out}(j2\pi W)| \cong \left|\left(\frac{1+jW/f_0}{1+jW/f_T}\right)\right|\frac{R_o}{\mu_0} \cong \frac{WR_o}{f_T} \qquad (17.19)$$

at the edge of the passband of a cascaded amplifier. If the gain-bandwidth product of the cascaded amplifier equals that of the follower (f_T), then

$$|Z_{out}(j2\pi W)| \cong \frac{WR_o}{f_T} = \frac{R_o}{A_{v0}},$$

where A_{v0} is typically on the order of 100. The magnitude of the output impedance increases to

$$|Z_{out}(j2\pi f_T)| \cong \left|\left(\frac{1+jf_T/f_0}{1+jf_T/f_T}\right)\right|\frac{R_o}{\mu_0} \cong \frac{R_o}{\sqrt{2}} \qquad (17.20)$$

at the upper edge f_T of the follower passband.

Because μ_0 is typically on the order of 10^5, the dc output impedance of a follower is typically on the order of $10^{-5}R_o$ and is given to a good approximation by WR_o/f_T at the passband edge of a cascaded

[1]However, if the source is capacitance coupled to the amplifier, the required dc bias-current compensation drastically reduces the input impedance.

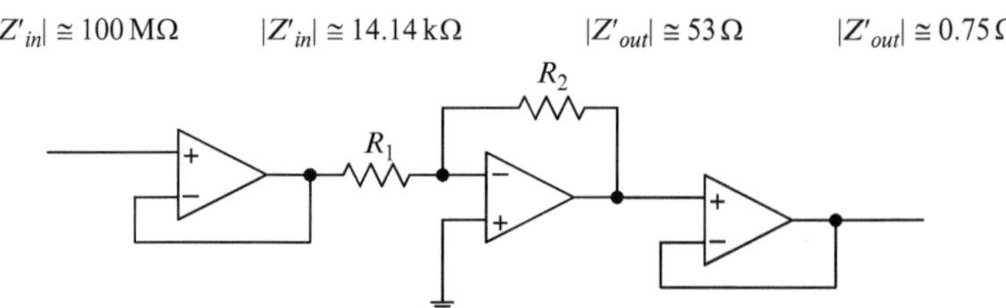

Fig. 17.5 A buffered inverting amplifier

amplifier having bandwidth W. If the follower and cascaded amplifier have equal gain-bandwidth products, then the maximum output impedance of the follower in the amplifier passband is given to a good approximation by R_o/A_{v0}, where the amplifier dc voltage gain A_{v0} is typically on the order of 100. decreasing to $R_o/\sqrt{2}$ for $f = f_T$, which is the bandwidth of a follower and well beyond the bandwidth of any cascaded inverting or non-inverting amplifier. The output impedance of a follower is much smaller than that of a typical inverting or non-inverting amplifier throughout the passband of the amplifier.

Followers can greatly increase the input impedance and greatly decrease the output impedance of an amplifier in the amplifier passband, as illustrated by Fig. 17.5, which shows input and output impedances at the passband edge (for $f = W = 10\,\text{kHz}$) of a buffered inverting amplifier for a case where $R_1 = 10\,\text{k}\Omega$, $R_2 = 1\text{M}\Omega$, and the op amps are identical, with $f_0 = 10\,\text{Hz}$, $\mu_0 = 10^5$, $R_o = 75\,\Omega$, and $R_i = 1\,\text{M}\Omega$. The follower at the input increases the input impedance from $14.14\,\text{k}\Omega$ to $100\,\text{M}\Omega$ and the follower at the output decreases the magnitude of the output impedance from 53Ω to $0.75\,\Omega$.

Input and output impedances of linear resistive op amp circuits (such as inverting and non-inverting amplifiers) are functions of frequency and depend (in general) upon parameters of the op amp and upon parameters of the feedback circuit in which the op amp is imbedded. Frequency dependence of the input impedance and output impedance of a linear op amp circuit are irrelevant in many applications. But such dependence can come into play if an amplifier either is driven by a source having a large output impedance or drives a load having a small input impedance.

Op-amp gain-bandwidth products range from about 10 kHz to more than 1 GHz. The largest gain-bandwidth products are provided by relatively expensive "high-speed" op amps. Most inexpensive general-purpose integrated op amps have gain-bandwidth products from 1 to 10 MHz. If we require a dc voltage gain on the order of 100 from a single-stage (inverting or non-inverting) amplifier using a general-purpose op amp, which is somewhat typical, the maximum bandwidth we can achieve is from 10 to 100 kHz. In other words, feedback amplifiers built around typical general-purpose op amps are limited to applications where the highest frequency of interest is on the order of 100 kHz. Such op amps are suitable for audio-frequency applications, but not for (e.g.) video-frequency applications.

The gain-bandwidth product f_T of an op amp is a critical parameter in specifying or determining the bandwidth of a resistive feedback amplifier built around the op amp.[2] If the dc voltage gain and the bandwidth of an amplifier are specified, then the required gain-bandwidth product is determined. From another point of view, if we know the required dc gain, we can easily calculate the bandwidth that can be achieved with any particular op amp. Conversely, if we know the required bandwidth, we can find the maximum dc gain that can be achieved.[3]

Op amps typically have three stages of amplification: An input stage, middle stage, and output stage. Almost all of the gain is provided by the input and middle stages, with the output stage generally providing near-unity gain and low output resistance. Op amps are built using various kinds of transistors. The primary types are constructed using (1) only bipolar junction transistors (BJTs), (2) junction field-effect transistors (JFETs) in the input stage and BJTs elsewhere, and (3) metal-oxide-semiconductor field-effect transistors (MOSFETs) throughout.

[2]Another is slew rate, discussed in Section 17.5.

[3]We assume here that the bandwidth of the feedback amplifier exceeds that of the op amp alone, which is almost always the case (except for a follower). Because the gain-bandwidth product for a feedback amplifier equals that for the op amp alone, and because the dc gain of a stable feedback amplifier is smaller than that of the op amp, the bandwidth of the amplifier must be larger than that of the op amp.

The dc voltage gain of a BJT stage is roughly an order of magnitude higher than that of a JFET or MOSFET stage. An all-BJT op amp has two BJT gain stages (input and middle). A JFET-input op amp has one JFET gain stage (input) and one BJT gain stage (middle), so the dc voltage gain of an all-BJT op amp is about an order of magnitude higher than that of a JFET-input op amp. A MOSFET op amp has two MOSFET gain stages (input and middle), so the dc voltage gain of an all-BJT op amp is roughly two orders of magnitude higher than that for a MOSFET op amp.

For a BJT op amp, R_i is on the order of 1 MΩ. For a JFET-input or MOSFET op amp, R_i is on the order of 1 TΩ. In each case, C_i is on the order of 1 pF. Thus (1) the corner frequency f_i is *on the order of* 100 kHz for a BJT op amp and *on the order of* 0.1 Hz for a JFET-input or MOSFET op amp and (2) the input impedance of a JFET-input or MOSFET op amp is many orders of magnitude larger than that for a BJT op amp throughout any passband established by external feedback circuitry.

Although the exceedingly large input impedances of JFET-input or MOSFET op amps seem attractive, input impedances of BJT op amps are adequate for most applications of general-purpose op amps. Also, because JFET-input and MOSFET op amps provide lower gains than and cost about three times as much as BJT op amps, BJT op amps are the op amps of choice for many linear applications. But where linear and MOSFET-based digital circuits are integrated in a single IC, MOSFET op amps are preferred because manufacture of such integrated circuits is less difficult if all transistors are of the same kind.

17.3 Linear Reactive-Feedback Circuits

A reactive-feedback op-amp circuit is one in which reactive elements appear in the feedback network. Examples of reactive-feedback circuits include integrators, differentiators, and virtually all op-amp based linear active filters. Circuits treated in this section have the forms shown in Fig. 17.6, where at least one of Z_1, Z_2 is complex. The source and load impedances might be real or complex.

The bandwidth of a linear reactive-feedback circuit depends upon the bandwidth of the op amp itself and also upon the bandwidth of surrounding circuitry.

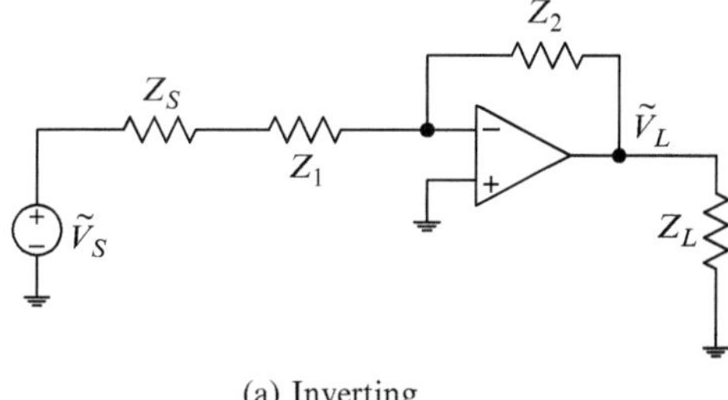

(a) Inverting

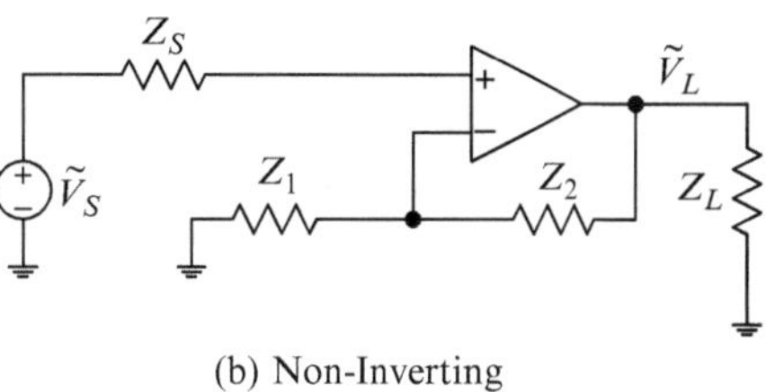

(b) Non-Inverting

Fig. 17.6 Basic inverting and non-inverting linear reactive-feedback circuits

For example, refer to the inverting configuration in Fig. 17.6(a). The voltage transfer function for the inverting reactive-feedback amplifier in Fig. 17.6(a) is

$$H_v = \frac{\tilde{V}_L}{\tilde{V}_S} \cong -\frac{Z_2}{Z_1}, \ |Z_i| \gg |Z_S| \qquad (17.21)$$

and the voltage transfer function for a non-inverting reactive feedback amplifier in Fig. 17.6(b) is

$$H_v = \frac{\tilde{V}_L}{\tilde{V}_S} = 1 + \frac{Z_2}{Z_1} \qquad (17.22)$$

provided:

1. The product of the magnitude (gain) of H_v and the bandwidth of H_v does not exceed the gain-bandwidth product of the op amp throughout the passband of H_v.
2. The op-amp input impedance Z_i is such that $|Z_i| \gg |Z_1|$ throughout the passband of H_v.
3. The op amp output resistance R_o is such that $R_o \ll |Z_L|$ and $R_o \ll |Z_2|$ throughout the passband of H_v.

In most applications, reactive-feedback amplifiers are designed such that these conditions are satisfied. Of course, (17.21) and (17.22) also are valid if the op amps are ideal (if $\mu_0 \to \infty$).

Example 17.3. In Fig. 17.7(a), the circuit and op-amp parameters are

$$R_S = 40\,\Omega,\ R_1 = 10\,\text{k}\Omega,\ R_2 = 402\,\text{k}\Omega,$$
$$C = 19.8\,\text{pF}, R_L = 2\,\text{k}\Omega$$
$$\mu_0 = 4\times 10^5, f_T = 4\,\text{MHz}, R_i = 5\,\text{M}\Omega,$$
$$C_i = 3\,\text{pF}, R_o = 75\,\Omega$$

(a) Obtain and justify an approximate expression for the voltage transfer function $H_v = \tilde{V}_o/\tilde{V}_S$. (b) Construct a linear model for the circuit and obtain an exact expression for the voltage transfer function. (c) On the same axes, draw graphs of voltage gain versus frequency as obtained from the approximate expression from part (a) and from the exact expression from part (b).

Solution: (a) For the parameters given, $R_1 \gg R_S$ and $R_L \gg R_o$. From (17.21), if the associated conditions are met, the voltage transfer function is

$$H_v \cong -\frac{Z_2}{R_1} = -\frac{R_2/R_1}{1+jf/f_2};$$
$$f_2 = \frac{1}{2\pi R_2 C} = 20\,\text{kHz} \qquad (17.23)$$

so the half-power bandwidth is 20 kHz. To validate (17.23), we must show that $|\mu| \gg 1$ and $|Z_i| \gg R_1$ for $f \le f_2$, where $R_1 = 10\,\text{k}\Omega$. Both $|\mu|$ and $|Z_i|$ decrease monotonically with increasing frequency, so the minimum values of those quantities for $f \le f_2$ are

$$|\mu| = \left|\frac{\mu_0}{1+jf_2/f_0}\right| = \left|\frac{\mu_0}{1+j2\times 10^3}\right| = 200 \gg 1,$$
$$|Z_i| = \left|\frac{R_i}{1+j2\pi f_2 R_i C_i}\right| = \left|\frac{R_i}{1+j20/10.6}\right|$$
$$= 2.34\,\text{M}\Omega \gg R_1.$$

Therefore, the ideal model should suffice.

(b) Figure 17.7(b) shows the linear model for the circuit in Fig. 17.7(a). Applying Kirchhoff's current law to nodes n and o gives

$$\frac{\tilde{V}_n - \tilde{V}_S}{R_{1S}} + \frac{\tilde{V}_n}{Z_i} + \frac{\tilde{V}_n - \tilde{V}_0}{Z_2} = 0,$$
$$\frac{\tilde{V}_o - \tilde{V}_n}{Z_2} + \frac{\tilde{V}_o + \mu\tilde{V}_n}{R_o} + \frac{\tilde{V}_o}{R_L} = 0.$$

Eliminating $\tilde{V}_n$ and solving for $H_v = \tilde{V}_o/\tilde{V}_S$ yields the voltage transfer function

$$H_v = \frac{-R_L Z_i[\mu Z_2 - R_o]}{\{Z + R_{1S}[R_L + R_o + \mu R_L]\}Z_i + R_{1S}Z}, \qquad (17.24)$$

where $R_{1S} = R_1 + R_S$ and

$$Z = R_o R_L + Z_2(R_L + R_o). \qquad (17.25)$$

(c) Figure 17.8 shows graphs of the exact voltage gain given by (17.24) and the approximate voltage gain given by (17.23). Figure 17.8(a) suggests that the approximation is quite good. But the magnified view in Fig. 17.8(b) shows that the bandwidth is somewhat less than that predicted by the approximate relation (17.23). The true half-power bandwidth is approximately 16.5 kHz, whereas that given by the approximation (17.23) is approximately 20 kHz. This difference might be significant in a few demanding applications; however, the exact gain is less than 1 dB

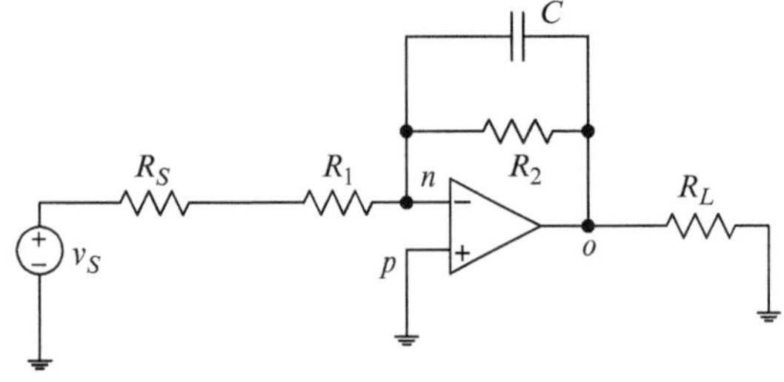

(a) circuit diagram

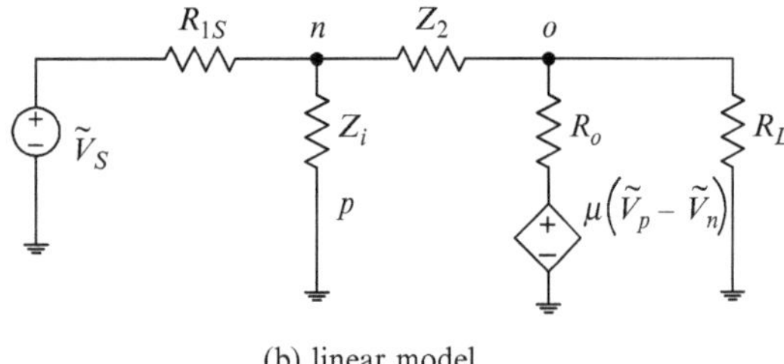

(b) linear model

Fig. 17.7 See Example 17.3

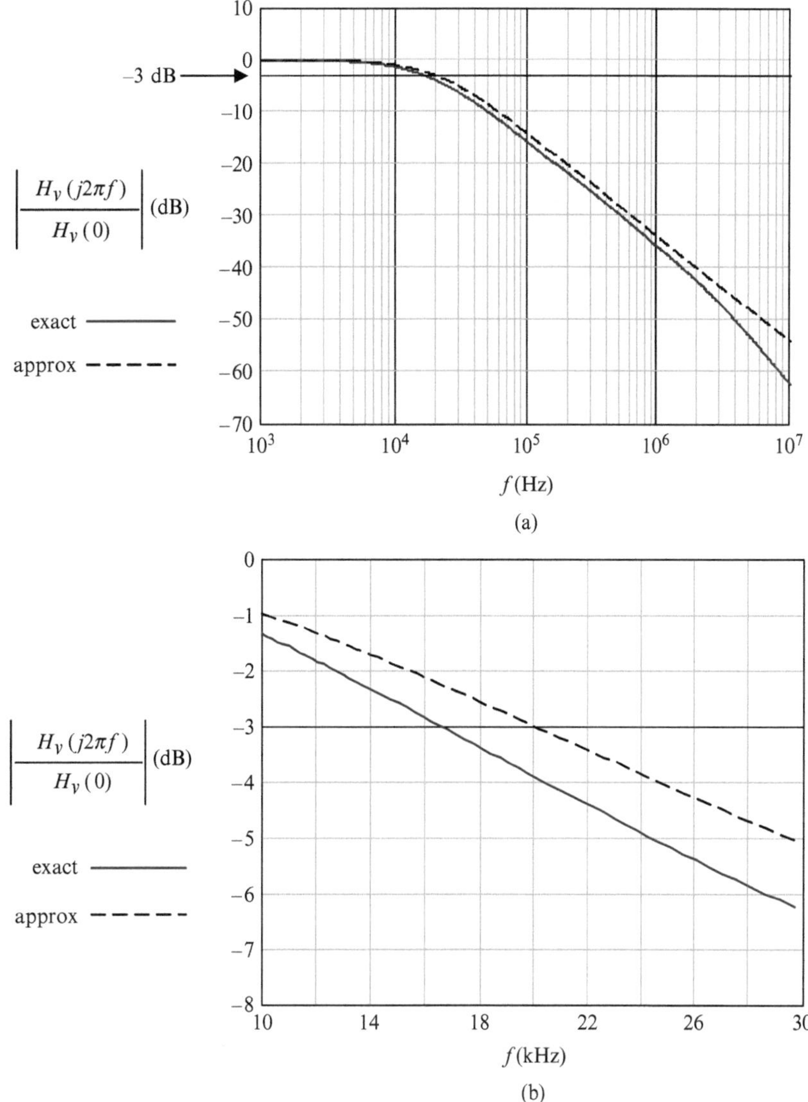

Fig. 17.8 See Example 17.3

below the approximate gain at 20 kHz, and a 1 dB error would be insignificant in many applications. Indeed, even larger errors can arise from uncertainties in circuit and op-amp parameter values.

Exercise 17.5. For the circuit shown in Fig. 17.7(a), what is the smallest value of the capacitance C for which the circuit bandwidth is determined by the impedance Z_2?

17.4 Output Swing

The **output swing** of an op amp is the largest output swing the op amp can provide, assuming an adequate symmetric supply.[4] For example, if the output swing of a rail-to-rail op amp is specified as ± 15 V, and if the supply is ± 15 V, the input and voltage gain must be such that the magnitude of the output does not attempt to exceed 15 V.

[4]The output swing of an op amp is defined as if the supply is symmetric, whether or not that is actually the case.

The required output swing for a waveform having a known peak amplitude is easy to determine. For example, If a sinusoid having peak amplitude 100 mV is applied to the input of a feedback amplifier whose voltage gain is 100 at the frequency of the sinusoid, the required output swing for that input is $\pm 10\,\text{V}$.

Example 17.4. The voltage transfer function for a certain inverting amplifier built around a rail-to-rail op amp is given by

$$H_v = \frac{A_{v0}}{1 + jf/W}; \; A_{v0} = 250, \; W = 25\,\text{kHz}$$

The input is a sinusoidal voltage having frequency $f_0 = 15\,\text{kHz}$ and rms amplitude $V_S = 50\,\text{mV}$. What output swing is required of the op amp? What supply voltage is required? (Assume a symmetric supply.)

Solution: The peak amplitude of the input is $50\sqrt{2}\,\text{mV}$. The voltage gain of the amplifier at 15 kHz is

$$|H_v(j2\pi f_0)| = \left| \frac{250}{1 + j15/25} \right| \cong 206$$

Thus the peak amplitude of the output is

$$V_{out} \cong (206)\left(50\sqrt{2}\,\text{mV}\right) \cong 14.6\,\text{V}$$

and the required output swing is $\pm 14.6\,\text{V}$. A $\pm 15\,\text{V}$ supply would be adequate.

It is impossible to describe many physical signals, such as voltages representing speech, music, or video, as simple functions of time. Instead, such signals are described probabilistically; for example, in part by a function that gives the probability that the instantaneous amplitude of the signal will exceed any specified value. When we compute (or measure) such probabilities for many common physical signals, it turns out that many **ac** signals spend most of their lives within three rms amplitudes of zero and that almost any **ac** physical signal $x(t)$ spends essentially all of its life within five rms amplitudes of zero. Consequently, when designing an amplifier for a current or voltage representing a signal such as speech, music, or video, we would *specify the gain and output swing (power supply) such that the peak-to-peak output swing is approximately six to ten times the rms amplitude of the output signal*. For a symmetric swing (symmetric supply), we might require that

$$V_{OS} \geq k\, v_{out\, rms}, \tag{17.26}$$

where $3 \leq k \leq 5$, depending on the nature of the signal to be amplified. The output swing is approximately equal to the supply voltage V_{CC} for rail-to-rail op amps and is about 1 V lower for non rail-to-rail op amps. We would design the amplifier such that the peak output does not exceed three to five times the product of the gain and the rms amplitude of the input.[5]

Example 17.5. A certain public-address amplifier whose output stage is an inverting amplifier, must deliver 25 W to an 8 Ω resistive load. If the output stage is powered by a symmetric supply, and if the op amp output can swing to within 1 V of either supply voltage, what supply voltage is required?

Solution: The power delivered to the load is given by

$$P = \frac{V_{rms}^{\;2}}{R_L}.$$

Thus the rms amplitude of the output is

$$V_{rms} = \sqrt{R_L P} = \sqrt{(8\,\Omega)(25\,\text{W})} = 14.14\,\text{V}.$$

If the output (speech or music) is almost always within 3 rms amplitudes of zero, the required output swing is $\pm (3)(14.14\,\text{V}) \cong \pm 42.4\,\text{V}$ and the required supply voltages are $\pm 43.4\,\text{V}$, so we might use a $\pm 45\,\text{V}$ supply.

[5]It also is possible to use a circuit called an *automatic gain control* (AGC), which keeps the rms amplitude of an input within specified bounds. You might learn about AGCs in a subsequent course.

17.5 Slew Rate

The **slew rate** of an op amp is the maximum rate of change (magnitude) of output voltage that can be supported by the op amp. The slew-rate limitation arises from the fact that the current available to charge or discharge the internal compensating capacitor is limited. Thus the rate of change of the capacitor voltage is limited, which imposes a limit on the rate of change of an output.

Slew rates for general-purpose integrated op amps often are expressed in units such as (e.g.) kilovolts per second ($kV\,s^{-1}$) or, equivalently, volts per millisecond ($V\,ms^{-1}$). Slew rates for integrated op amps range from about $0.1\,V\mu s^{-1}$ ($100\,kV\,s^{-1}$) to more than $1,000\,V\mu s^{-1}$. Op amps having large gain-bandwidth products and slew rates of about $100\,V\mu s^{-1}$ or more are commonly called *high-speed* op amps because they can track relatively fast changes in an input at relatively high gains.

To illustrate the limitation imposed by slew rate, we consider a sinusoidally excited op amp having slew rate SR and we let $v(t) = V_0 \cos(\omega t)$ denote the output. The rate of change of the output is given by

$$\frac{dv}{dt} = \frac{d\,V_0 \cos(\omega t)}{dt} = -2\pi f\,V_0 \sin(\omega t).$$

The maximum rate of change of the output is given by

$$\max\left(\frac{dv}{dt}\right) = 2\pi f\,V_0.$$

The maximum rate of change of the output cannot exceed the slew rate (SR) of the op amp, which requires

$$2\pi f\,V_0 < SR. \tag{17.27}$$

Equation (17.27) leads to the requirement

$$f_{\max} = \frac{SR}{2\pi V_{CC}}, \tag{17.28}$$

where V_{CC} is the maximum allowable peak amplitude of the output for linear operation, assuming a symmetric supply $\pm V_{CC}$ and rail-to-rail operation.

Figure 17.9 illustrates the effect of slew rate for a hypothetical amplifier having slew rate $SR = 4.5\,V\mu s^{-1}$. The amplifier output is unable to track a 5-V, 1 MHz sinusoid and degenerates to a triangular wave. As shown on the figure, the slope (magnitude) of each straight-line segment of the triangular wave is the slew rate.

Example 17.6. The input to a certain inverting amplifier is a sinusoid whose frequency does not exceed 10 kHz. The voltage gain of the amplifier is given by

$$H_v = \frac{K}{1 + j(f/W)};\ K = 100,\ W = 1\,\text{kHz}$$

and the maximum output swing is $\pm V_{CC}$, with $V_{CC} = 10\,V$. The amplifier must operate linearly for $0 \le f \le 10\,\text{kHz}$. Specify the maximum allowable amplitude of the input as a function of frequency and the minimum acceptable slew rate.

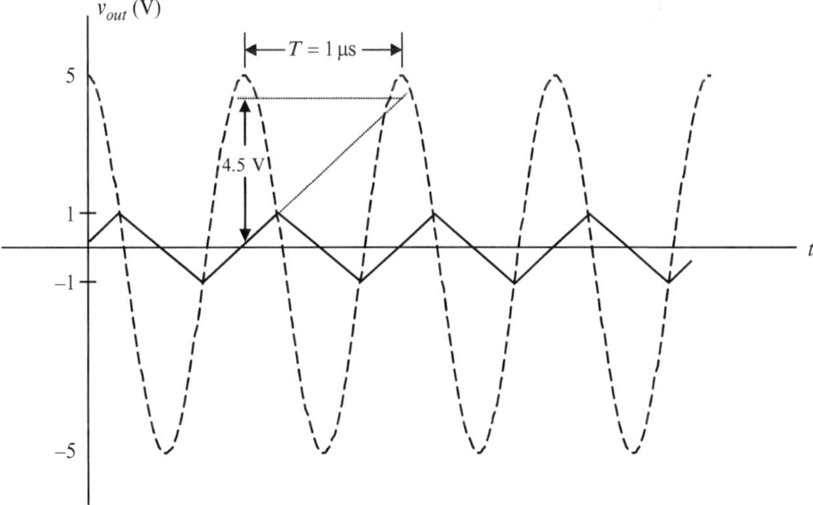

Fig. 17.9 Effect of slew rate for an amplifier having a sinusoidal input. The dashed (*sinusoidal*) waveform would be the output in absence of the slew rate limitation. The solid (*triangular*) waveform is the actual output

Solution: For any frequency, the maximum amplitude of the output may not exceed V_{CC}. Thus for frequencies below W, where the gain is approximately equal to K, the input amplitude V_{in} may not exceed $V_{CC}/K = 100\,\text{mV}$.

For frequencies above W, the voltage gain is given approximately by

$$A_v(f) = \frac{|K|}{\sqrt{1+(f/W)^2}} \cong \frac{|K|W}{f}$$

and the amplitude V_{in} of an input having frequency f must satisfy

$$\frac{|K|W}{f} V_{in} < V_{CC} \Rightarrow V_{in} < \frac{f V_{CC}}{|K|W}.$$

The amplitude of the output may not exceed V_{CC} at any frequency, so the slew rate must satisfy

$$SR > 2\pi V_{CC} f$$

for all frequencies of interest. The highest frequency of interest is 10 KHz, so we require

$$SR > 2\pi(10\,\text{V})(10\,\text{kHz}) = 628\,\text{kV}\,\text{s}^{-1}.$$

Many general-purpose op amps will meet that requirement, but many will not. In particular, both the venerable 741 op amp and the generic (virtual) simulated op amp provided by Electronic workbench have slew rates of $500\,\text{kV}\,\text{s}^{-1}$. However, if the amplitude of the input does not increase with increasing frequency and the voltage gain decreases with frequency, or if the amplifier must be linear only within the passband $f < W$, the required slew rate would be smaller – perhaps by an order of magnitude or more.

Finite bandwidth and finite slew rate both limit the ability of an op amp to respond to a rapidly changing input, but in different ways. The effect of finite bandwidth is a reduction of gain with increasing frequency; that is, as the frequency of a sinusoidal input is increased, the amplitude of the sinusoidal output decreases, but the output remains sinusoidal. But if the amplitude and frequency of a sinusoidal output are such that the slope (derivative) of the output exceeds the slew rate, the output becomes nonsinusoidal, as illustrated by Fig. 17.9. In other words, when slew rate comes into play, the amplifier becomes nonlinear. Generally, it is desirable to ensure that slew rate does not come into play at frequencies within the passband (below the bandwidth) of an amplifier. As a conservative rule of thumb, the slew rate of a rail-to-rail op amp for a linear application and sinusoidal input should satisfy the inequality

$$SR > 2\pi\, W\, V_{CC} \tag{17.29}$$

where W is the desired bandwidth. The voltage transfer functions of resistive-feedback amplifiers are of the form

$$H_v = \frac{H_v(0)}{1 + j(f/W)},$$

for which a 1-decade (factor of ten) increase in frequency of a sinusoidal output is accompanied by a 20 dB (factor of ten) decrease in the amplitude of the output. Thus, if (17.29) is satisfied, no higher-frequency input will cause the slew rate to be exceeded.

Example 17.7. Figure 17.10 shows an inverting amplifier using an LF411 op amp, where the supply voltages are $\pm 15\,\text{V}$. Typical values of the gain-bandwidth product and slew rate of an LF411 are 4 MHz and $15\,\text{V}\,\mu\text{s}^{-1}$, respectively. The maximum output swing of an LF411 is approximately $\pm(V_{CC} - 1\,\text{V})$; i.e., from $-V_{CC}+1\,\text{V}$ to $V_{CC} - 1\,\text{V}$.

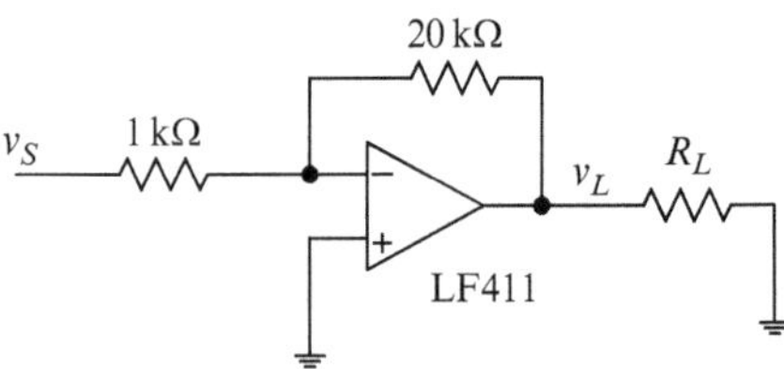

Fig. 17.10 See Example 17.7

(a) Find the bandwidth of the amplifier (b) Determine the maximum allowable amplitude of a sinusoidal output whose frequency equals the amplifier bandwidth.

Solution: (a) The dc voltage gain of the amplifier is

$$A_{v0} = \frac{R_2}{R_1} = \frac{20}{1} = 20.$$

Thus the amplifier bandwidth is

$$W \cong \frac{f_T}{A_{v0}} = \frac{4\,\text{MHz}}{20} = 200\,\text{kHz}.$$

(b) From (17.29), the maximum amplitude of a 200 kHz sinusoidal output that can be tracked by the op amp output is

$$V_{out\,\max} = \frac{SR}{2\pi W} = \frac{15\,\text{V}\,\mu\text{s}^{-1}}{(2\pi)(200\,\text{kHz})} = 11.9\,\text{V}.$$

For full-scale output (± 14 V), slew rate, not frequency response, limits the useful bandwidth. The highest frequency for which full-scale output can be obtained without slew rate distortion is

$$f_{\max} = \frac{SR}{2\pi(14\,\text{V})} = \frac{15\,\text{V}\,\mu\text{s}^{-1}}{2\pi(14\,\text{V})} = 171\,\text{kHz}. \quad (17.30)$$

The circuit is nonlinear for full-scale outputs at higher frequencies. Generally, linear op-amp circuits are designed such that gain-bandwidth product, and not slew rate, limits the bandwidth of the circuit.

Strictly, (17.29) applies only to a sinusoidal input. In most applications, inputs and outputs of interest are not sinusoidal, but often are much more complex signals such as currents or voltages representing speech, music, and similar signals. Limits on output-swing require that the rms amplitude of such a signal not exceed about $V_{OS}/3$ (see (17.26)). Let W denote the bandwidth of an amplifier for such a signal. In the worst case, the input is a single sinusoid having frequency W and rms amplitude $V_{OS}/3$. In such a case, the peak amplitude of the output is $\sqrt{2}V_{OS}/3$ and the slew rate must satisfy

$$SR > 2\pi W \frac{\sqrt{2}V_{OS}}{3} \cong 3\,WV_{OS}. \quad (17.31)$$

Equation (17.31) is a reasonable guide for preliminary design (for input voltages representing speech, music, and similar signals), with a follow-up simulation. There are more complicated ways to estimate maximum rates of change of such signals, and thus to specify required slew rates, but such calculations are beyond our scope.

Example 17.8. Refer to Example 17.7. Figure 17.11 shows an inverting amplifier using an LF411 op amp, where the supply voltages are $\pm$ 15 V. (a) Find the bandwidth of the amplifier. (b) The amplifier is to be an audio amplifier for telephone-quality speech signals having bandwidth 4 kHz. Determine the maximum allowable rms amplitude of such an input.

Solution: (a) The dc (passband) gain is

$$A_{v0} = \frac{1\,\text{M}\Omega}{1\,\text{k}\Omega} = 10^3.$$

The bandwidth of the amplifier is

$$W \cong \frac{f_T}{A_{v0}} = \frac{4\,\text{MHz}}{1000} = 4\,\text{kHz}.$$

(b) The required output swing is approximately $\pm$ three rms amplitudes of the output, which implies that the rms amplitude of the input must satisfy

$$3\,A_{v0}V_{S\,rms} = 14\,\text{V} \Rightarrow V_{S\,rms} \leq \frac{14\,\text{V}}{3000} \cong 4.7\,\text{mV}.$$

From (17.31), the required slew rate is

$$SR = 3\,W\,V_{OS} = 3(4\,\text{kHz})(14\,\text{V}) = 168\,\text{V}\,\text{ms}^{-1}.$$

Slew rates of most general-purpose op amps exceed this value. Thus output swing, not slew rate, limits the allowable rms amplitude of an input to 4.7 mV.

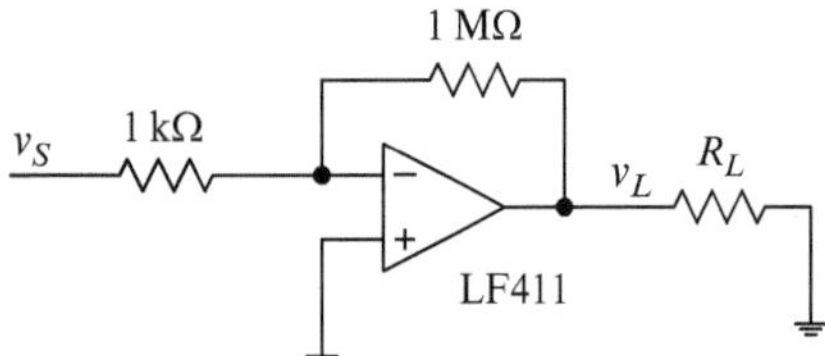

Fig. 17.11 See Example 17.7

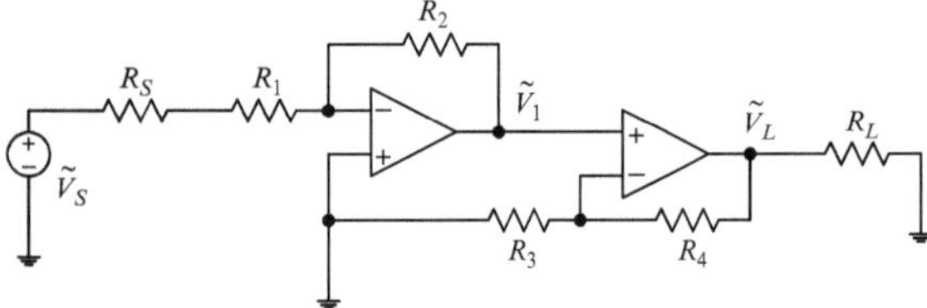

Fig. 17.12 See Example 17.9

Although generally disadvantageous, the limited slew rate of an op amp can be used to advantage in a few applications. One such application is in suppressing sharp, undesirable voltage "spikes" corrupting an input; for example, as might be seen if attempting to restore an old recording from a scratched vinyl disk. In this application, the slew rate might suppress the spikes without affecting the more slowly varying waveform representing the desirable signal (the music).

17.6 Amplifiers in Cascade

If the voltage transfer function for an inverting or non-inverting amplifier is approximately independent of the load on the amplifier, which is almost always the case (by design), the overall voltage transfer function of two or more such amplifiers in cascade equals the product of the voltage transfer functions of the individual amplifiers.

Example 17.9. Figure 17.12 shows a two-stage amplifier, where $R_1 = 5\,\text{k}\Omega$, $R_2 = 200\,\text{k}\Omega$, $R_3 = 10\,\text{k}\Omega$, and $R_4 = 400\,\text{k}\Omega$. The op amps are identical, each having gain-bandwidth product $f_T = 2\,\text{MHz}$, intrinsic dc voltage gain $\mu_0 = 10^5$, output resistance $R_o = 75\,\Omega$, and input resistance $R_i = 4\,\text{M}\Omega$. The source and load resistances are $R_S = 25\,\Omega$, $R_L = 1\,\text{k}\Omega$. (a) Using expressions given in Table 17.2, calculate the minimum input impedance and maximum output impedance (magnitudes) for each amplifier in the respective passbands. Is loading significant? (b) Obtain an expression for the voltage transfer function for the circuit. (c) Draw a Bode plot of voltage gain versus frequency for the circuit.

Solution: (a) From Table 17.2 or (equivalently) (17.14) and (17.15), the minimum input impedances of the two stages are

$$|Z_{in\,1}|_{\min} \cong R_1 = 5\,\text{k}\Omega,$$
$$|Z_{in\,2}|_{\min} \cong \sqrt{2}R_i = 5.66\,\text{M}\Omega.$$

Thus, $|Z_{in\,1}|_{\min} = 200R_S$, so the load on the source is negligible. The maximum output impedances are both equal (approximately) to

$$|Z_{out}|_{\max} \cong \frac{R_o}{\sqrt{2}} \cong 53\,\Omega.$$

Thus, $|Z_{out1}|_{\max} < 0.001\,|Z_{in\,2}|_{\min}$ and $|Z_{out}|_{\max} < 0.0002R_L$, so the loads on the first and second stages are negligible.

(b) From part (a), loading is insignificant, so the voltage transfer function for the cascade is approximately the product of the voltage transfer functions for the two stages. The dc voltage gains f the first stage (inverting amplifier) and second stage (non-inverting amplifier) are

$$A_{v01} = \frac{R_2}{R_1} = 40, \quad A_{v02} = 1 + \frac{R_4}{R_3} = 41.$$

The respective bandwidths are

$$W_1 \cong \frac{f_T}{A_{v01}} = 50\,\text{kHz}, \quad W_2 \cong \frac{f_T}{A_{v02}} = 48.8\,\text{kHz}$$

Thus the individual voltage transfer functions are

$$H_{v1} = \frac{-A_{v01}}{1 + j(f/W_1)}, \quad H_{v2} = \frac{A_{v02}}{1 + j(f/W_2)},$$

and the voltage transfer function for the cascade is

$$H_v = \frac{-A_{v01}A_{v02}}{[1+j(f/W_1)][1+j(f/W_2)]}$$

Often, a desired gain and bandwidth for an amplifier cannot be attained using a single op amp (because the desired gain-bandwidth product exceeds that for the op amp). In such cases, it might be possible to achieve the desired gain and bandwidth using a cascade of two or more stages. An alternative (usually better) is to use an op amp having an adequate gain-bandwidth product. But if for some reason cascading stages is a better approach (or the only available approach), it is useful to know how overall bandwidth of a cascade of identical stages depends upon the bandwidth of a single stage and the number of stages. An end-of-chapter problem asks you to show that the -3 dB bandwidth of a cascade of n inverting or non-inverting amplifiers (or a mix), all having the same bandwidth W, is given by

$$W_n = W\sqrt{2^{1/n}-1}. \tag{17.32}$$

For example, the bandwidth of a cascade of two identical amplifiers is about $0.64W$, where W is the bandwidth of either stage.

Example 17.10. The gain-bandwidth product for a certain op amp is 1 MHz. Using one or more of these op amps in non-inverting configuration, design an amplifier having dc gain $A_0 = 100$ and bandwidth $W \geq 40$ kHz.

Solution: The overall gain-bandwidth product required is $(100)(40\times 10^3) = 4$ MHz, so we cannot meet the specifications with a single stage using the specified op amp. Because $100 = 10^2$ and $(1\,\text{MHz})/10 = 100\,\text{kHz}$, we can meet the specification using a two-stage amplifier, each stage having gain 10 and bandwidth 100 kHz, as shown in Fig. 17.13. The overall bandwidth will be about 64 kHz. The dc voltage gain of each stage is

$$A = 1 + \frac{R_2}{R_1}.$$

If we choose $R_2 = 9\,\text{k}\Omega$, $R_1 = 1\,\text{k}\Omega$, then $A = 10$ and the overall dc gain is $A^2 = 100$.

As an aside, the number N of identical amplifiers in cascade that maximizes the overall bandwidth and gives a specified overall dc gain A is given by

$$N(A) = \frac{\ln(2)}{\ln\left[\dfrac{-2\ln(A)}{\ln(2)-2\ln(A)}\right]}. \tag{17.33}$$

For example, if we want to maximize the bandwidth of a cascade of N amplifiers, subject to the constraint that the overall gain equals 100, then the number of stages is

$$N(100) = \frac{\ln(2)}{\ln\left[\dfrac{-2\ln(100)}{\ln(2)-2\ln(100)}\right]} = 8.86,$$

so we would use either eight or nine stages, depending on which gave the largest bandwidth. From (17.32),

$$W(N) = W(1)\sqrt{2^{1/N}-1},$$

where

$$W(1) = \frac{f_T}{A^{1/N}}$$

is the gain of a single stage. Thus

$$W(8) = \frac{f_T}{100^{1/8}}\sqrt{2^{1/8}-1} = 0.169 f_T,$$
$$W(9) = \frac{f_T}{100^{1/9}}\sqrt{2^{1/9}-1} = 0.170 f_T,$$

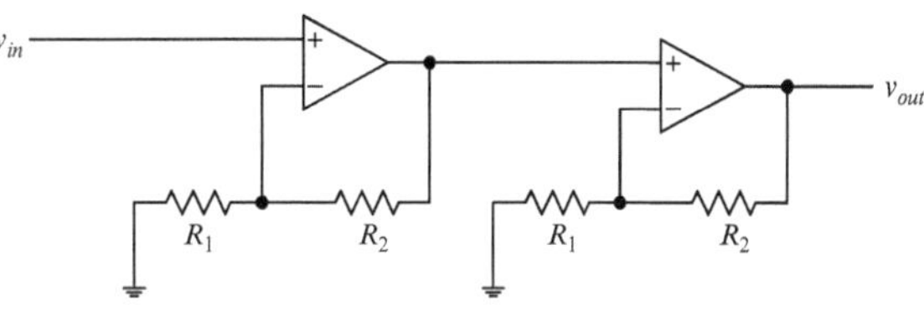

Fig. 17.13 See Example 17.10

so eight stages are about as good as nine. We would probably use eight (for simplicity and economy).

17.7 Capacitance Coupling

Where the output of one circuit is connected to the input of another, the circuits are said to be **coupled**. A source connected to an amplifier as shown in Fig. 17.14(a) is said to be **direct coupled** to the amplifier. A source connected to an amplifier through a capacitor, as shown in Fig. 17.14(b), is said to be **capacitance coupled** to the amplifier. The usual purpose of capacitance coupling is to prevent the dc component of a current or voltage from reaching the input of a subsequent amplifier (or other circuit). In this application, the capacitor is called a **coupling capacitor** or **dc blocking capacitor**.

Refer to Fig. 17.14(b). The impedance seen by the source is given by

$$Z'_{in} = Z_C + Z_{in} = \frac{1}{j2\pi f C} + Z_{in}$$

where C is the capacitance of the coupling capacitor and Z_{in} is the input impedance of the amplifier. Capacitance coupling alters the impedance seen by the source at low frequencies and thereby alters the low-frequency response of an amplifier. If $f_{\min}$ denotes the lowest frequency of interest in the input, the capacitor must be large enough that

$$\left|\frac{1}{j2\pi f_{\min} C}\right| \ll |Z_{in}(j2\pi f_{\min})| \\ \Rightarrow C \gg \frac{1}{2\pi f_{\min}|Z_{in}(j2\pi f_{\min})|} \tag{17.34}$$

For any particular value of $f_{\min}$, increasing the input impedance (magnitude) of the amplifier for $f = f_{\min}$ decreases the required size of a coupling capacitor. Consequently, if capacitance coupling is used, it is desirable that the amplifier have large input impedance (magnitude) at the lowest frequency of interest (for $f = f_{\min}$).

Example 17.11. A certain voltage source is to be capacitance coupled to an amplifier whose input impedance at dc (for $f = 0$) and throughout the passband is $10\,\text{k}\Omega$. The passband of the amplifier must extend down to $f_{\min} = 10\,\text{Hz}$. Specify the capacitance of the coupling capacitor.

Solution: We have from (17.34) that

$$C \gg \frac{1}{2\pi f_{\min}|Z_{in}(j2\pi f_{\min})|} \Rightarrow C \gg \frac{1}{2\pi(10)(10^4)} \\ = 1.59\,\mu\text{F}$$

If we interpret " $\gg$ " to mean "ten times larger than," we would require $C = 15.9\,\mu\text{F}$, which is quite a large capacitor in the context of many modern electronic circuits and especially in integrated circuits.

Capacitance coupling requires more components than direct coupling and in an integrated circuit would therefore require more real estate (a larger chip area) than an equivalent direct-coupled circuit. Consequently, direct coupling is preferred to capacitance coupling, especially in integrated circuits, and such circuits often are designed such that capacitance coupling is not required. You will learn more about this topic if you take a subsequent course in electronic circuits.

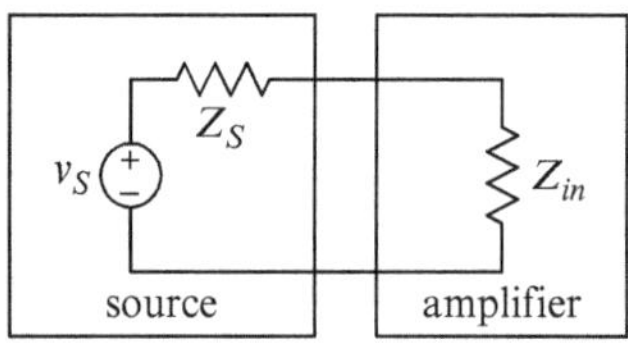

(a) direct coupling

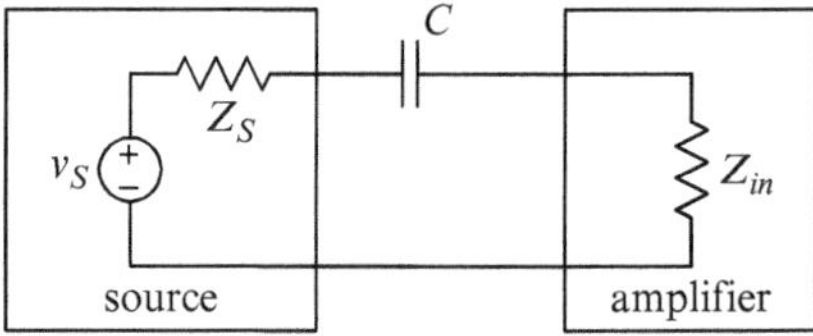

(b) capacitance coupling

Fig. 17.14 Direct and capacitance coupling

17.8 Input Bias Current Compensation in Capacitance-Coupled Amplifiers[6]

Any circuit in which an op amp is imbedded must provide dc paths to ground from both the *n* and *p* input terminals for the input dc bias currents. Also, it usually is desirable that one of the paths to ground contain a compensating resistance R_X that greatly reduces the input voltage offset that would otherwise be produced by the input bias currents. Figure 17.15 depicts usual approaches to such compensation for inverting and non-inverting amplifiers.

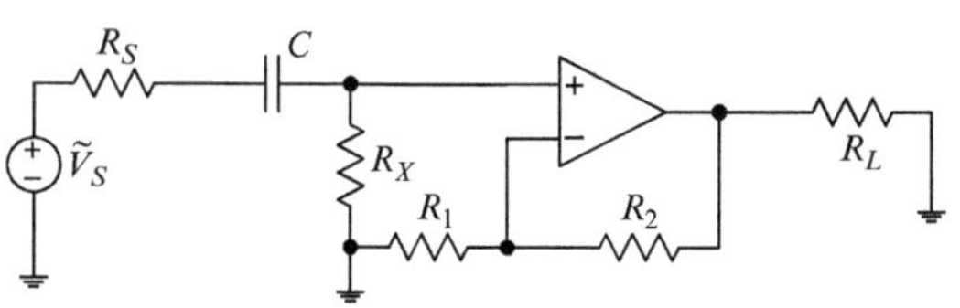

(a) capacitively coupled non-inverting amplifier: $R_X = R_1 \| R_2$

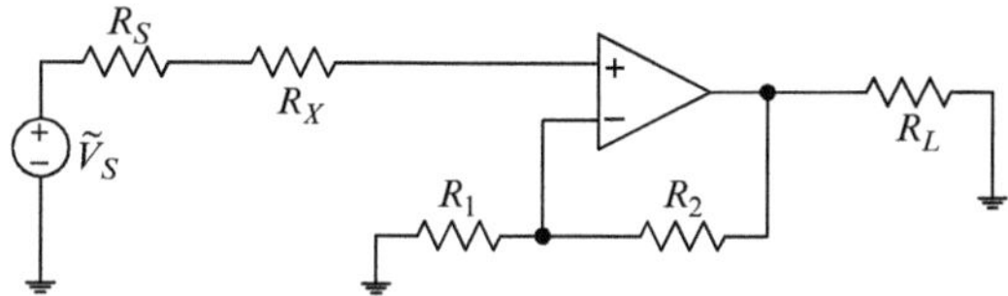

(b) direct-coupled non-inverting amplifier: $R_X = R_1 \| R_2 - R_S$

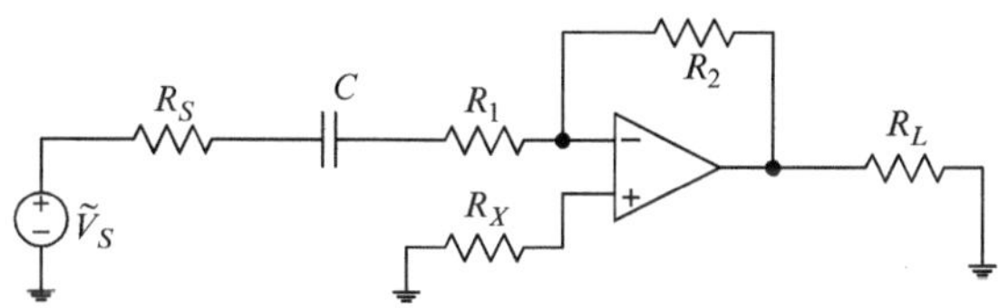

(c) capacitively coupled inverting amplifier: $R_X = R_2$

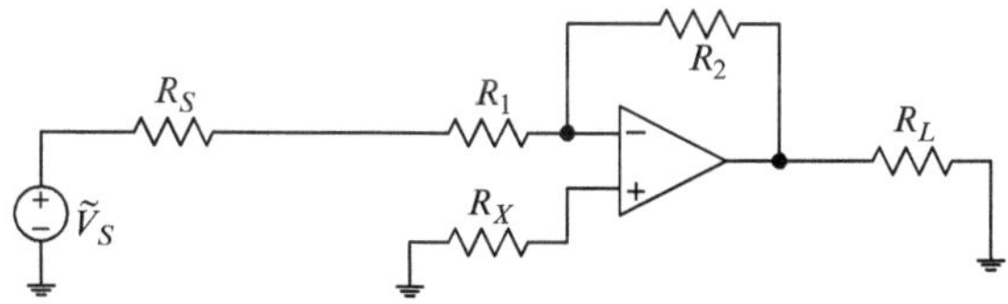

(d) direct coupled inverting amplifier: $(R_1 + R_S) \| R_2$

Fig. 17.15 Resistive compensation for input bias current in non-inverting and inverting amplifiers

[6]You might find it helpful to review Section 8.10 of Chapter 8 before proceeding.

Compensating for input bias currents might be optional in non-demanding applications of inverting and direct-coupled non-inverting amplifiers, but *compensation as shown in Fig. 17.15(a) is required for capacitance-coupled non-inverting amplifiers* because otherwise there is no dc path to ground from the positive (*p*) input of the op amp.

The compensating resistances shown in Fig. 17.15 for capacitance-coupled inverting and non-inverting amplifiers are given by[7]

$$R_X = R_1 \| R_2 \text{ (non-inverting amplifier)} \tag{17.35}$$

$$R_X = R_1 \text{ (inverting amplifier)} \tag{17.36}$$

If the op amp used has sufficiently large dc gain (if $\mu_0 \to \infty$), the input impedance in the passband (for $0 \le f < W$) of a non-inverting amplifier is approximately that of the op amp itself; i.e., $R_{in} \cong R_i$. Thus the input impedance (for $0 \le f < W$) of the capacitance coupled non-inverting amplifier in Fig. 17.15(a) is given by

$$Z_{in} \cong \frac{1}{j2\pi f C} + R_X \| R_i \cong \frac{1}{j2\pi f C} + R_X$$
$$= \frac{1}{j2\pi f C} + R_1 \| R_2$$

Similarly, if the op amp used has sufficiently large dc gain (if $\mu_0 \to \infty$), the input impedance in the passband (for $0 \le f < W$) of an inverting amplifier is approximately that of the resistor R_1. Thus the input impedance (for $0 < f < W$) of a capacitance coupled inverting amplifier is given by

$$Z_{in} \cong \frac{1}{j2\pi f C} + R_1$$

Comparing these relations shows that the input impedance of a capacitance coupled non-inverting amplifier for which $A_{v0} \gg 1 \Rightarrow R_2 \gg R_1$ is about the same as that for an inverting amplifier having the same value for R_1. If the gain is less than 10, the input impedance of a capacitance coupled non-inverting amplifier can be substantially less than that for an inverting amplifier having the same value for R_1.

[7]See Section 8.10.7 in Chapter 8.

17.9 Power Dissipation in Op Amps and Op-Amp Circuits

Removing heat from a circuit can be expensive, as can providing the wasted power that produces the heat. Circuits that operate hot generally produce more electrical noise than similar circuits that operate at lower temperatures. Also, parameter values change with increasing temperature (usually for worse), and component lifetime generally decreases with increasing operating temperature. For these and other reasons, specifying power-dissipation ratings for circuit components is an essential part of circuit design.

The power that can be dissipated (internally) by an op amp is limited. Typical small integrated op amps can dissipate from about 500 mW to a little more than 1 W. Much larger power op amps can dissipate 100 W or more, but such power op amps are typically modular, not integrated, and cost a few hundred dollars each.

The power dissipated by an op amp depends upon a number of things, including the supply voltage, the input voltage, the load (output) voltage, the load resistance, and properties of other surrounding circuitry (e.g., the resistors in a feedback network). Thus expressions for power dissipated can be complicated and difficult to interpret. An op-amp circuit designer needs a simple, approximate, and conservative relation that allows initial specification of the power dissipation rating required of an op amp. If necessary, the specification can be refined by subsequent more detailed analysis, by simulation, or by construction and testing. In this section, we derive a few simple relations that are useful in design of linear resistive op-amp circuits.

The total power delivered to an op-amp circuit consists of the power P_S delivered by the supply and the power P_{in} delivered to the circuit by an input. The total power dissipated in an op-amp circuit consists of the power P_A dissipated in the op amp and the power P_R dissipated in the load and other circuit components external to the op amp. Because power is conserved, the power delivered equals the power dissipated; that is,

$$P_S + P_{in} = P_A + P_R \tag{17.37}$$

In virtually all linear applications, the power delivered by an input is negligible ($P_{in} \ll P_S$). The power dissipated by an op amp in a linear circuit is given to a good approximation by

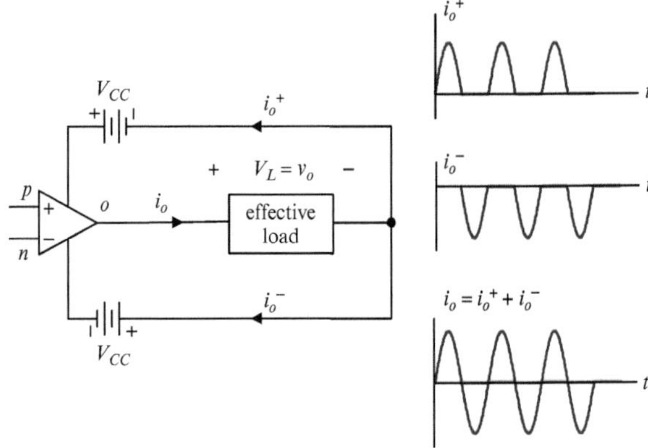

Fig. 17.16 Supply currents in a sinusoidally driven symmetrically powered op amp operating linearly

$$P_A \cong P_S - P_R \tag{17.38}$$

In what follows, we obtain an expression for the power dissipated by an op amp in a resistive circuit, assuming *sinusoidal excitation and linear operation.*

The power dissipated in circuit components external to the op amp is given to a good approximation by

$$P_R \cong \frac{V_L{}^2}{2R_L'} = \frac{V_{Lrms}{}^2}{R_L'} \tag{17.39}$$

where V_{Lrms} is the *rms amplitude* of the load voltage and R_L' is the **effective load** on the op amp, which is the resistance seen by the current exiting the output (o) terminal of the op amp. To minimize waste, we would like most of the power P_R to be dissipated in the load R_L.

Figure 17.16 shows how we obtain an expression for the power P_S delivered by the supply. Because of the internal structure of an op amp, the positive half-cycles of the output current i_o pass into the positive supply terminal and out of the op-amp output terminal.[8] The negative half-cycles pass into the negative supply terminal and out through the output terminal. The total *instantaneous* power *delivered* by the power

[8]The output stage of an op amp is typically a *push-pull* configuration of two transistors, where one transistor amplifies the positive parts of an applied voltage and the other amplifies the negative parts. There is actually small crossover voltage range near zero where there is current through both supplies. However, this current is usually quite small, relative to a typical load current.

supply (both sources in Fig. 17.16) is given approximately by[9]

$$p_S \cong V_{CC}\, i_o{}^{+} + V_{CC}\,(-i_o{}^{-}) = V_{CC}\,|i_o|$$

where i_o is the current exiting the output (o) terminal of the op amp. The effective load depends upon the actual load on the circuit and (in general) components of the feedback network. For an *inverting amplifier* and resistive load R_L, the effective load is given by

$$R_L' = R_L \| R_2 \tag{17.40}$$

For a *non-inverting amplifier* and resistive load R_L, the effective load is given by

$$R_L' = R_L \| (R_1 + R_2) \tag{17.41}$$

Usually, the specified gain is large enough that $R_2 \gg R_1$, in which case the effective load for a non-inverting amplifier is approximately $R_L' = R_L \| R_2$, or about the same as that for an inverting amplifier. To minimize the load on the op amp in an inverting or non-inverting amplifier (to maximize the effective load resistance), the feedback resistance R_2 should be as large as possible.

The average power delivered by the supply is the average of the instantaneous power, given by (the voltage V_{CC} is constant)

$$P_S = \overline{p_S} = V_{CC}\,\overline{|i_o|} \tag{17.42}$$

Let R_L' denote the effective load resistance. Then $i_o = v_L / R_L'$ and

$$P_S = \frac{V_{CC}}{R_L'}\overline{|v_L|} \tag{17.43}$$

For sinusoidal excitation and linear operation, the load voltage v_L is sinusoidal. We assume the voltage $v_L = V_L \cos(\omega t)$ is expressed in standard form, so $V_L \geq 0$. From Chapter 5, the average magnitude of the voltage v_L is given by

$$\begin{aligned}\overline{|v_L|} &= \overline{|V_L \cos(\omega t)|} = \frac{V_L}{T}\int_0^T \left|\cos\left(\frac{2\pi t}{T}\right)\right| dt \\ &= \frac{4V_L}{T}\int_0^{T/4} \cos\left(\frac{2\pi t}{T}\right) dt = \frac{2V_L}{\pi}\sin\left(\frac{2\pi t}{T}\right)\Big|_0^{T/4} \\ &= \frac{2V_L}{\pi}\end{aligned} \tag{17.44}$$

It follows from (17.43) and (17.44) that the average power delivered by the supply is given to a good approximation by

$$P_S = \frac{2V_{CC}V_L}{\pi R_L'} \tag{17.45}$$

where V_L is the amplitude of the (sinusoidal) load voltage, V_{CC} is the positive supply voltage, and R_L' is the effective load.

From (17.38), (17.39), and (17.45), the power dissipated by an op amp in a sinusoidally excited resistive circuit operating linearly is given by

$$P_A = P_S - P_R = \frac{2V_{CC}V_L}{\pi R_L'} - \frac{V_L^2}{2R_L'} \tag{17.46}$$

The power given by (17.46) is a quadratic function of the peak load voltage V_L. The power-dissipation rating for an op amp should be the maximum power dissipated. The peak load voltage for which the maximum power is dissipated by the op amp (for a fixed supply and load) is given by

$$\begin{aligned}&\frac{dP_A}{dV_L} = 0 \Rightarrow \frac{2V_{CC}}{\pi R_L'} - \frac{V_L}{R_L'} = 0 \\ &\Rightarrow V_L = \frac{2V_{CC}}{\pi} \cong 0.64\,V_{CC}\end{aligned} \tag{17.47}$$

Using the right side of (17.47) for V_L in (17.46) gives the *maximum* power dissipated by an op amp in a *sinusoidally excited* resistive circuit operating linearly:

$$P_{A\max} = \frac{4V_{CC}{}^2}{\pi^2 R_L'} - \frac{2V_{CC}{}^2}{\pi^2 R_L'} = \frac{2V_{CC}{}^2}{\pi^2 R_L'} \tag{17.48}$$

where V_{CC} is the supply voltage and R_L' is the effective load.

[9]Actually, this is only the power delivered by the supply to the output stage. A typical op amp is a three-stage amplifier, and even if no input is applied (even if $i_O = 0$), some dc power is required to keep the op amp in an active (ready) state.

A worst case is a constant (dc) input, in which case, from (17.43), the power delivered by the supply is given by

$$P_S = \frac{V_{CC}\overline{|v_L|}}{R_L'} = \frac{V_{CC}V_L}{R_L'} \tag{17.49}$$

For a dc input (and load voltage), the power delivered to resistances external to the op amp is given by

$$P_R = \frac{{V_L}^2}{R_L'} \tag{17.50}$$

and the power dissipated by the op amp for a dc input V_L is given by

$$P_A \cong P_S - P_R = \frac{V_{CC}V_L}{R_L'} - \frac{{V_L}^2}{R_L'} \tag{17.51}$$

From (17.51), the power dissipated by the op amp for a dc input is maximum for $V_L = V_{CC}/2$ and is given by

$$P_{A\max} = \frac{{V_{CC}}^2}{4R_L'} \tag{17.52}$$

Comparing (17.52) and (17.48) reveals that the maximum power dissipated in the op amp for a dc input is about 25% larger than that for a sinusoidal input. But a 25% safety margin is not unreasonable. Also, if the circuit operates linearly, (17.52) gives the maximum power that could possibly be dissipated by the op amp, regardless of the nature of the input. Consequently, (17.52) often is used for initial screening of op amps suitable for a particular application (assuming the supply voltage and the equivalent load resistance are known).

Example 17.12. (Follower) In Fig. 17.17, the supply voltage is $V_{CC} = 25\,\text{V}$ and the load resistance is $R_L = 1\,\text{k}\Omega$. Specify the power-dissipation rating for the op amp assuming (a) sinusoidal excitation and (b) dc excitation.

Solution: The current i_o equals the load current i_L because the current entering the n terminal of the op amp is negligible. Thus the effective load is the actual load on the circuit $(R_L' = R_L)$. (a) The maximum power dissipated by the op amp (the specified power-dissipation rating) for sinusoidal excitation is

$$P_{A\,ac} = \frac{2{V_{CC}}^2}{\pi^2 R_L'} = \frac{(2)(25\text{V})^2}{(\pi^2)(1\text{k}\Omega)}$$
$$= 127\,\text{mW (for sinusoidal excitation)}$$

(b) For constant (dc) excitation, the power-dissipation rating should be

$$P_{A\,dc} = \frac{{V_{CC}}^2}{4R_L'} = \frac{(25\text{V})^2}{(4)(1\text{k}\Omega)}$$
$$= 156\,\text{mW (dc excitation).}$$

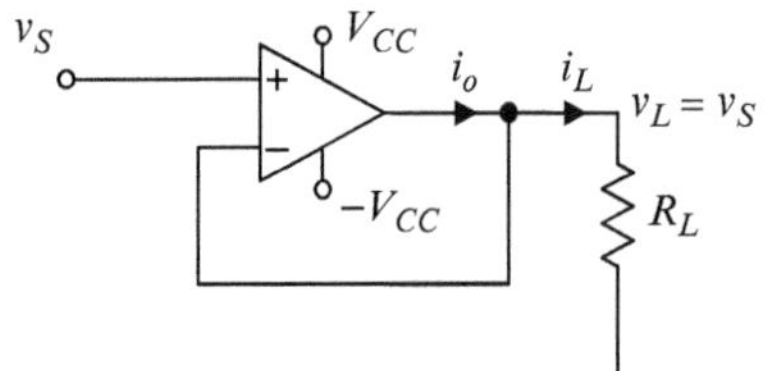

Fig. 17.17 See Example 17.12

When specifying power-dissipation ratings, keep in mind that the mathematical definition of average power is the average of instantaneous power over all time. For example, the (mathematical) average power caused to be dissipated in a resistive load R by a sinusoidal voltage having peak amplitude V is given by $P = V^2/(2R)$, regardless of the frequency of the sinusoid. *Mathematically*, the power dissipated by a resistor subjected to a sinusoidal voltage having frequency $f = 10^{-10}\,\text{Hz}$ (period $T = 10^{10}\,\text{s} > 3000$ years) is the same as would be dissipated if the frequency were $10^6\,\text{Hz}$. *Physically*, this is nonsense. So far as anyone could tell with ordinary instruments, a $10^{-10}\,\text{Hz}$ sinusoid is essentially constant, and whereas a 1 Ω, 1/2 W resistor might endure a 1 V, 10^6 Hz sinusoidal voltage for a long time, it probably would not long endure a 1 V, 10^{-10} Hz sinusoid if the voltage happened to be near a peak when applied to the resistor. In general, beware of using ac power dissipation formulas to rate components that will be subjected to very slowly-varying currents or voltages. It is safer to use peak or dc ratings in such cases.

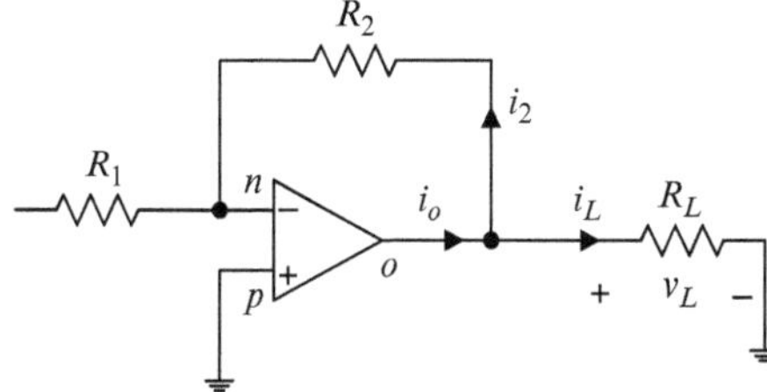

Fig. 17.18 See Example 17.13

Example 17.13. (Inverting amplifier) In Fig. 17.18, $V_{CC} = 25\,\text{V}$, $R_L = 1\,\text{k}\Omega$, $R_1 = 1\,\text{k}\Omega$, and $R_2 = 5\,\text{k}\Omega$. Specify the power-dissipation rating for the op amp assuming (a) sinusoidal excitation and (b) dc excitation.

Solution: Because $v_n = 0$, the effective resistance seen by the current i_o exiting the op amp output terminal (the effective load) is

$$R'_L = R_L \| R_2 = \frac{(5\text{k}\Omega)(1\text{k}\Omega)}{5\text{k}\Omega + 1\text{k}\Omega} = 833\,\Omega$$

and the maximum power dissipated by the op amp (the specified power-dissipation rating) is

$$P_{A\,ac} = \frac{2\,V_{CC}^2}{\pi^2 R'_L} = \frac{(2)(25\text{V})^2}{(\pi^2)(833\Omega)}$$
$$= 152\,\text{mW (for sinusoidal excitation)}$$

or

$$P_{A\,dc} = \frac{V_{CC}^2}{4R'_L} = \frac{(25\text{V})^2}{(4)(833\Omega)}$$
$$= 188\,\text{mW (dc excitation)}$$

It is safer to use the more conservative dc rating unless you are absolutely certain that the output will be sinusoidal and not too-slowly varying.

We can express the average power dissipated by an op amp as a fraction of the maximum power dissipated by the op amp. From (17.46) and (17.48), the fraction for sinusoidal excitation is given by

$$\frac{P_A}{P_{A\max}} = \left(\frac{2\,V_{CC}V_L}{\pi R'_L} - \frac{V_L^2}{2R'_L}\right)\left(\frac{2\,V_{CC}^2}{\pi^2 R'_L}\right)^{-1}$$
$$= \pi\left(\frac{V_L}{V_{CC}}\right) - \frac{\pi^2}{4}\left(\frac{V_L}{V_{CC}}\right)^2 \qquad (17.53)$$

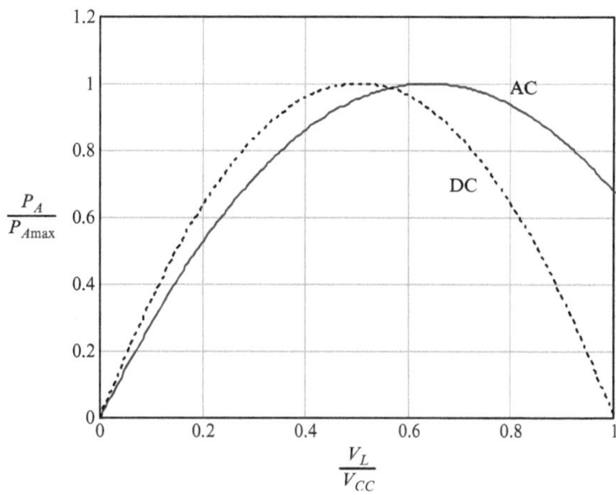

Fig. 17.19 Power-dissipation curves for an op amp in a follower circuit

From (17.51) and (17.52), the fraction is given for constant (dc) excitation by

$$\frac{P_A}{P_{A\max}} = \frac{4\,V_L}{V_{CC}}\left(1 - \frac{V_L}{V_{CC}}\right) \qquad (17.54)$$

Figure 17.19 shows graphs of the fractions $P_A/P_{A\max}$ versus V_L/V_{CC} given by (17.53) (ac) and (17.54) (dc), where V_L denotes dc voltage and the peak (not rms) amplitude of ac load voltage. For a dc input, the maximum occurs when the load voltage equals half of the supply voltage. For an ac input, the maximum occurs when the peak load voltage equals about 64% of the supply voltage (see (17.47)). The maximum power dissipated by the op amp can be calculated using (17.52) (for a dc input) or (17.48) (for an ac input). The ratio of the maximum ac power dissipated to the maximum dc power dissipated is

$$\frac{P_{A\max\,ac}}{P_{A\max\,dc}} = \left(\frac{2\,V_{CC}^2}{\pi^2 R'_L}\right)\left(\frac{4\,R'_L}{V_{CC}^2}\right) = \frac{8}{\pi^2} \cong 0.81 \qquad (17.55)$$

Again, the conservative approach is to assume dc excitation. In absence of sufficient contrary knowledge of the expected input, we should use (17.52) to specify the power dissipation rating for an op amp in a linear resistive circuit.

Example 17.14. The supply voltage to a follower is $\pm 25\,\text{V}$. The follower is to drive

a $2\,\mathrm{k}\Omega$ resistive load. Specify the power-dissipation rating for the op amp.
Solution: Because we are not told whether the excitation will be dc or ac, we assume dc. From (17.52), the power-dissipation rating for the op amp must be

$$P_{A\max} = \frac{V_{CC}{}^2}{4R_L} = \frac{(25\mathrm{V})^2}{4(2\mathrm{k}\Omega)} = 78.1\,\mathrm{mW}$$

Virtually any op amp will satisfy this requirement.

Equations (17.45) and (17.49) imply that the average power dissipated by an op amp equals zero if $v_L = 0$. Actually, an idle op amp draws a dc quiescent current, which usually is given on the manufacturer's data sheet. For a PA03 high-power op amp, for example, the quiescent supply current for a $\pm(75/2) = \pm 37.5\,\mathrm{V}$ supply is about 300 mA, which means that the average power dissipated by an idling PA03 can be as much as $(2)(75/2)(0.3) = 22.5\,\mathrm{W}$. The idle power dissipated by an op amp must be considered when specifying the capacity of the power supply and when specifying the power-dissipation rating for the op amp. In many cases, the idle power is a negligible fraction of the active power, but not always.

17.10 Power-Conversion Efficiency

A figure of merit for amplifiers is the **power conversion efficiency**, denoted by η and defined by

$$\eta = \frac{P_L}{P_S} \tag{17.56}$$

where P_L is the power delivered to the (actual, not effective) load, given by

$$P_L = \frac{V_L{}^2}{2R_L} \tag{17.57}$$

for sinusoidal excitation and by

$$P_L = \frac{V_L{}^2}{R_L} \tag{17.58}$$

for dc excitation.

The power conversion efficiency expresses the power delivered to the load as a fraction of the total power P_S provided to the circuit, and is the fraction of power supplied that is put to good use.

From (17.45) and (17.57), the power conversion efficiency for a sinusoidally-excited resistive op-amp circuit operating linearly is given by

$$\eta_{ac} = \frac{P_L}{P_S} = \frac{V_L{}^2}{2R_L}\left[\frac{2V_{CC}V_L}{\pi R'_L}\right]^{-1} = \frac{\pi V_L R'_L}{4R_L V_{CC}} \tag{17.59}$$

and for dc excitation, from (17.49) and (17.57), by

$$\eta_{dc} = \frac{V_L{}^2}{R_L}\left(\frac{V_{CC}V_L}{R'_L}\right)^{-1} = \frac{V_L R'_L}{V_{CC}R_L} \tag{17.60}$$

In (17.59) and (17.60), R_L is the actual load and R'_L is the effective load. If $R'_L \cong R_L$ and $V_L \cong V_{CC}$ in (17.59) and (17.60), then

$$\eta_{ac\max} \cong \frac{\pi}{4} = 0.785, \quad \eta_{dc\max} = 1 \tag{17.61}$$

The value above for $\eta_{ac\max}$ is the maximum possible power conversion efficiency for a resistive op-amp circuit driven by a sinusoidal input and operating linearly. For a circuit operating at maximum power conversion efficiency, 78.5% of the power supplied by the power supply is delivered to the load and 21.5% is dissipated in the op amp and surrounding circuitry, largely in the op amp and largely as heat. Note, however, that operating at maximum efficiency requires that the peak amplitude of the output equal the supply voltage. If the peak amplitude of the output is less than the supply voltage, the power dissipated by the op amp is a larger fraction of the power delivered by the supply (see Fig. 17.19) and the efficiency is less than the maximum possible value. Similar remarks hold for dc excitation.

Two useful relations are derived from (17.56) as follows: The power P_C dissipated in external circuitry other than the load is given by $P_C = P_R - P_L$. Usually, $P_C \ll P_L + P_A$ and $P_{in} \ll P_s$, in which case (17.37) gives $P_L \cong P_S - P_A$ and (17.56) gives

$$\begin{aligned}\eta &\cong \frac{P_S - P_A}{P_S} = 1 - \frac{P_A}{P_S}\\ &\Rightarrow P_S \cong \frac{P_A}{1-\eta}; P_C \ll P_L + P_A\end{aligned} \tag{17.62}$$

Similarly,

$$\eta \cong \frac{P_L}{P_L + P_A} \Rightarrow P_L \cong \frac{\eta}{1-\eta} P_A; \quad (17.63)$$
$$P_C \ll P_A + P_L.$$

17.11 Op-Amp Amplifier Circuit Design

This section describes and illustrates design of inverting and non-inverting amplifiers, including selection of an appropriate op amp from among those readily available.

Figure 17.20 defines symbols used in the following discussion.

The following rules govern design of op-amp-based feedback amplifiers:

- The amplifier *must* provide negative feedback at dc, *must* provide dc paths to ground from both the n and p input terminals of the op amp, and *must* provide more negative than positive feedback at all frequencies below the unity-gain frequency.
- The maximum possible output voltage swing must accommodate the actual output voltage, the maximum possible output current cannot be exceeded, the slew rate must be greater than the maximum slope (magnitude) of any output, and the power dissipated by any component must not exceed the dissipation rating for that component.
- The gain-bandwidth product of the amplifier cannot exceed that of the op amp.

The following considerations are relevant to designing op-amp-based feedback amplifiers:

- The input impedance of a *direct-coupled inverting amplifier, a capacitance-coupled inverting amplifier, or a capacitance coupled non-inverting amplifier* is very nearly resistive and approximately equal to R_1 throughout the amplifier passband. If necessary, a voltage follower can be used to boost the input impedance. In such a case, the input impedance of the follower in the amplifier passband is no smaller than that of the op amp (typically 1 MΩ or more), and the output impedance in the passband is no larger than that of the op amp (typically 100 Ω or less). If it is necessary to boost the input impedance of a capacitance-coupled amplifier, the follower must precede the capacitor.
- The input impedance of a *direct-coupled non-inverting amplifier* is a fairly strong complex function of frequency. The magnitude of the input impedance increases from $\mu_0|Z_i|/A_{v0}$ at dc to approximately $\sqrt{2}|Z_i|$ at the upper edge of the passband (for $f = W$). However, the magnitude throughout the passband is so large relative to that of typical source impedances that the frequency dependence of the input impedance is rarely a concern. The input impedance (magnitude) of a non-inverting direct-coupled amplifier typically exceeds 1 MΩ for BJT-input op amps and is on the order of 1 TΩ (10^{12} Ω) for FET-input op amps. For many practical purposes, the input impedance of an op amp is essentially infinite.
- For an *inverting amplifier*, the resistance R_X that compensates for dc input bias current is independent of the source resistance. Thus we do not need to know the source impedance before specifying R_X.
- For a *direct-coupled non-inverting amplifier*, the compensating resistance R_X is in series with the source, so the source resistance is in effect part of the compensating resistance.[10] If the source resistance is unknown, it might be necessary to use a variable compensating resistor and some in-situ adjustment. If the source resistance is larger than $R_1 \| R_2$, compensation is impossible or impractical.
- For some nonlinear sources, the apparent dc source resistance is effectively a function of the source voltage (amplitude), and bias-current compensation of a *direct-coupled non-inverting amplifier* driven by such a source is ineffective. In such cases, it might be necessary to use an inverting amplifier, possibly preceded by a follower if large input impedance is required.

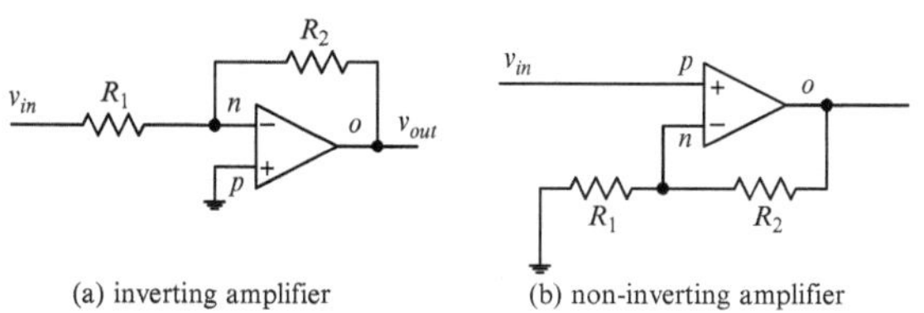

Fig. 17.20 Notation used in Section 17.11

[10] Direct coupling is used almost exclusively in integrated circuits.

- In some non-demanding applications, the offset caused by input bias currents can be tolerated and the compensating resistor R_X can be omitted. However, *the compensating resistor is required for a capacitance-coupled non-inverting amplifier*, because otherwise there is no dc path to ground from the non-inverting input and the amplifier will saturate.
- If the source output impedance is large (if the source looks more like a current source than a voltage source), an inverting amplifier (with $R_1 = 0$) can serve as a transimpedance amplifier (current in – voltage out), whereas a non-inverting amplifier is less suited for that purpose.
- The effective load on an op amp in an inverting amplifier is $R'_L = R_L \| R_2$. The effective load on an op amp in a non-inverting amplifier is $R'_L = R_L \| (R_1 + R_2)$, which, for dc gains greater than 10 (i.e., in most applications) is approximately $R'_L = R_L \| R_2$. In most applications, the effective load R'_L will be approximately the same for comparable non-inverting and inverting amplifiers.

All things considered, most practicing circuit designers would choose an inverting configuration in absence of a good reason to do otherwise. About the only cases where one would choose a non-inverting amplifier are where inversion is not allowable (rare), where a source is coupled directly to the amplifier and the input impedance must be much larger than R_1 (and using a follower is undesirable), or where the gain is less than 10 and the difference in effective load is significant (where power dissipation is critical).

Exercise 17.6. Specifications on an amplifier call for a dc voltage gain of $A_{v0} = 200$. If $R_L = 5\,\text{k}\Omega$ and it is specified that $R_2 = 1\,\text{M}\Omega$, what would be the ratio of the effective load on an inverting amplifier to that of a non-inverting amplifier in this application? What would be the ratio of the powers dissipated by the op amps in the two configurations?

As noted a few times in this book, preliminary design is easiest if the relevant governing relations are simple. For example, compare the following two expressions for the voltage transfer function of a resistive-feedback inverting amplifier:

$$H_v = \frac{-Z_i Z_L(\mu R_2 - R_o)}{\{D + Z_{1S}[(\mu+1)Z_L + R_o]\}Z_i + Z_{1S}D},$$
$$Z_{1S} = Z_1 + Z_S, \quad D = R_o Z_L + R_2 Z_L + R_2 R_o, \tag{17.64}$$

and

$$H_v \cong -\frac{R_2}{R_1}. \tag{17.65}$$

The relation (17.64) is exact for a fairly complete linear model, whereas (17.65) is an approximation valid under certain conditions. The relation (17.64) contains at least 11 parameters (μ and each complex impedance contain at least two), whereas (17.65) contains only two. It is much easier to base a preliminary design on a two-parameter model than on one containing 11 parameters, especially when (as in this case) there are also nonlinear effects to be considered. Thus it is desirable, if possible, to ensure that the conditions that validate (17.65) are met. For example, R_1 should be much larger than $|Z_S|$ (the source impedance) but no larger than about 1% of $|Z_i|$ (the op-amp input impedance), $|Z_L|$ (the load impedance) should be much larger than the op-amp output resistance R_o, and the desired dc voltage gain $A_{v0} = R_2/R_1$ should be no larger than about 1% of the op-amp voltage gain $|\mu|$ in the amplifier passband, given by $W \cong f_T/A_{v0}$, where f_T is the unity-gain frequency (or gain-bandwidth product) of the op amp. If these conditions are met, then (17.65) is a good approximation to the amplifier voltage gain in the amplifier passband. Fortunately, in many applications, these conditions are not difficult to meet.

Designing an inverting amplifier having dc voltage gain $A_{v0} < 10$ or approaching μ_0 is problematic, partly because the gain-bandwidth product $A_{v0}W$ of the amplifier is not equal to the gain-bandwidth product $f_T = \mu_0 f_0$ of the op amp unless $1 \ll A_{v0} \ll \mu_0$, where μ_0 is the intrinsic dc voltage gain of the op amp.

Although we can achieve fractional gain using an inverting amplifier (but not a non-inverting amplifier), there is little reason to do so. Fractional gain is easily

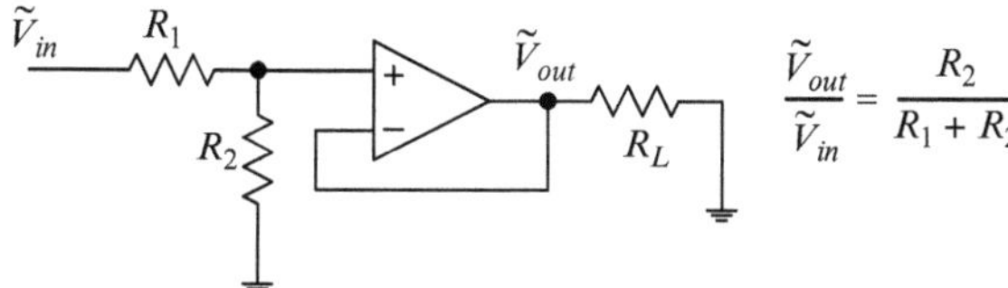

Fig. 17.21 Buffered voltage divider

obtained using a voltage divider and follower, as shown in Fig. 17.21, where we would choose R_1 much larger than the source impedance to prevent loading from affecting the desired gain.

Independent specifications on input resistance and dc gain or bandwidth might be inconsistent. For example, if particular values are specified for both input resistance and dc voltage gain of an inverting amplifier, then both R_1 and R_2 are determined, and they in turn determine bandwidth (for any particular op amp). If the resistance R_1 does not meet guidelines given above (for the specified dc gain), it might be difficult to obtain the specified bandwidth. In such cases it might be necessary to use a voltage follower as an input stage.

Generally, we would begin design of an inverting amplifier by specifying R_1, either to satisfy a specification on input resistance or to make the input resistance as large as possible. If we cannot achieve a specified minimum input resistance while keeping $R_1 \leq 0.01|Z_i|$, we might either select another op amp (e.g., one having JFET inputs), use a non-inverting amplifier (if direct coupling is possible), or precede the amplifier with a voltage follower.

After specifying R_1, we calculate the value of R_2 (based upon the specified dc gain). We then calculate the value of the input-bias-current compensating resistor R_X, either as $R_L \| R_2$ for an inverting amplifier or as $R_L \| (R_1 + R_2)$ for a non-inverting amplifier.

The input dc offset voltage (typically on the order of 1 mV) can be troublesome. Most op amps provide additional terminals for external circuitry that reduces the offset to a negligible value. The external circuitry usually consists of a voltage divider that uses a fraction of the supply voltage to null the offset. For one-of-a-kind or prototype circuits, the voltage divider can be a variable resistor (potentiometer). If the amplifier is part of a larger, high-precision integrated circuit to be produced in quantity, the voltage divider might be adjusted by laser trimming (there are automated processes for such adjustments).

Power dissipation by the various components of an amplifier are of interest for at least three reasons: One is to specify power-dissipation ratings. Another is to estimate demand on the power supply, and the third is to specify any required measures for cooling. Typically, the powers dissipated by the resistors R_1, R_2, R_X are quite small relative to that dissipated by the op amp. In an inverting amplifier, the voltage across R_1 does not exceed the input voltage, the voltage across R_2 does not exceed the output voltage, and the voltage across R_X is exceedingly small. Except in high-voltage or high-power applications, the input voltage is typically no larger than about 1 V and R_1 is typically on the order of 10 kΩ, so the average power dissipated by R_1, given by $P_1 = V^2_{in\,rms}/R_1$, is less than 100 μW. The power dissipated by R_X is even smaller. In circuits built using discrete (individually packaged) general-purpose op amps, the peak output voltage is rarely more than 25 V. If R_2 is on the order of 1 MΩ, the power dissipated by R_2 is no more than about 625 μW. Thus the total power dissipated by the three external resistors in low-power voltage amplifiers is typically less than 1 mW. The power-dissipation ratings for the resistors R_1, R_2, R_X are rarely a concern in typical low-power *discrete circuits*, where the smallest resistors one can readily obtain off the shelf have dissipation ratings of 125 mW.

The precision required of the resistances R_1 and R_2 depends upon the precision required of the dc voltage gain. For example, if the voltage gain must be within 1% of 100, then the ratio R_2/R_1 must also be within 1% of 100 for an inverting amplifier or within 1% of 99 for a non-inverting amplifier. Let α denote the maximum fraction by which either resistance departs from the specified value. For example, if we use ± 5% resistors, then $\alpha = 0.05$. Maximum departures of the dc voltage gain of an inverting amplifier from the specified value A_0 are given by

$$
\begin{aligned}
A_{0\,\max} &= \frac{R_2 + \alpha R_2}{R_1 - \alpha R_1} = \frac{R_2}{R_1}\left(\frac{1+\alpha}{1-\alpha}\right) = A_0\left(\frac{1+\alpha}{1-\alpha}\right) \\
A_{0\,\min} &= \frac{R_2 - \alpha R_2}{R_1 + \alpha R_1} = \frac{R_2}{R_1}\left(\frac{1-\alpha}{1+\alpha}\right) = A_0\left(\frac{1-\alpha}{1+\alpha}\right)
\end{aligned} \tag{17.66}
$$

where A_0 is the nominal (specified) dc voltage gain and α is the precision of the resistances R_1, R_2. The relations (17.66) can be reduced by long division to

$$
\begin{aligned}
A_{0\,\max} &= A_0\left(1+2\alpha+2\alpha^2+\cdots\right) \cong A_0(1+2\alpha) \\
A_{0\,\min} &= A_0\left(1-2\alpha+2\alpha^2+\cdots\right) \cong A_0(1-2\alpha)
\end{aligned} \tag{17.67}
$$

From (17.67), the precision (tolerance) of the resistors R_1 and R_2 must be half the desired precision of the dc voltage gain. For example, if the voltage gain must be within 2% of the specified value, then the resistances R_1 and R_2 must be within 1% of the specified values. Although derived above for an inverting amplifier, the same rule holds for a non-inverting amplifier, provided the specified dc gain is greater than about 10.

The discussion of precision above is relevant to discrete fabrication. In integrated fabrication, it is easier to achieve a specified precision on a ratio of resistances than it is to achieve a specified precision on any particular resistance. You will learn more about such things if you take a subsequent course in integrated circuit design.

Once the values of R_1 and R_2 have been determined, we must find an op amp that allows the amplifier to meet the remaining specifications. First, the gain-bandwidth product f_T of the op amp must satisfy

$$
f_T \geq A_{v0} W \tag{17.68}
$$

where A_{v0} is the specified dc (or passband) gain and W is the specified bandwidth. For a conservative design, the slew rate must satisfy

$$
SR \geq 2\pi W V_{CC} \tag{17.69}
$$

where W is the specified bandwidth and V_{CC} is supply voltage, which limits the maximum amplitude of the output. The supply voltage must be sufficient to allow the expected output swing.

For a conservative design, the power-dissipation rating for the op amp must satisfy

$$
P_{A\max} \geq \frac{V_{CC}^2}{4R_L'}, \begin{cases} R_L' = R_L \| R_2 & \text{inverting amp} \\ R_L' = R_L \| (R_1 + R_2) & \text{non-inverting amp} \end{cases} \tag{17.70}
$$

where V_{CC} is the supply voltage (assumed symmetric), and R_L' is the effective load on the op amp.

Recall from Chapter 7 that for proper operation of an op amp, the larger supply voltage V_2 must exceed the smaller V_1 by at least an amount V_B, but must not exceed V_1 by more than an amount V_A, where V_A and V_B depend upon the internal structure of the op amp, vary from one op amp to another, and are often specified on manufacturers' data sheets. If so, the supply voltages must satisfy the inequality

$$
V_A \geq V_2 - V_1 \geq V_B \tag{17.71}
$$

For a symmetric supply, $V_1 = -V_{CC}$ and $V_2 = V_{CC}$, so (17.71) implies

$$
\frac{V_A}{2} \geq V_{CC} \geq \frac{V_B}{2} \tag{17.72}
$$

which determines the maximum output swing; e.g., $\pm V_{CC} = \pm V_A/2$ for a rail-to-rail op amp, and approximately $\pm(V_{CC} - 1\,\text{V})$ otherwise.

Instead of specifying V_A and V_B, some manufacturers specify the maximum symmetric supply voltage and the output swing for a particular load resistance. For example, for their LF411 general-purpose op amp, National Semiconductor specifies a maximum supply voltage of ± 18 V and an output swing of at least ± 12 V for a 10 kΩ load, with ± 13.5 V being typical. Some manufacturers also specify a minimum supply voltage. Specifying minimum and maximum supply voltages is equivalent to specifying V_A and V_B. For a symmetric supply $\pm V_{CC}$, the maximum input voltage (for linear operation) must satisfy

$$
\max(|v_{in}|) \leq \frac{V_{CC}}{A_v}, \tag{17.73}
$$

where A_v is the passband gain. If designing an entire system from scratch, we might be able to specify the power-supply voltages, in which case we might postpone that choice until more is known about the required or desired voltage and/or power levels in the system. But often, we are called upon to design a circuit that will become part of an existing system or be powered by an existing supply, in which case there might be little or no choice regarding supply voltages. Supply voltages can be increased or decreased using additional circuitry, but at additional cost.

The current that an op amp can deliver to a load also is limited. The limit on output current is imposed intentionally by internal circuitry to protect the op amp from overload (e.g., from being destroyed by accidental shorting of the output to ground). For any particular load, the limit on output current imposes a limit on output swing. For example, if the maximum possible output current is 3 A and the supply voltage is 30 V, then the minimum permissible *effective* load resistance (for linear operation) is $30/3 = 10\,\Omega$. For a conservative design, the output current the op amp is asked to deliver must satisfy

$$\max(|i_{out}|) \geq \frac{V_{CC}}{R'_L} \tag{17.74}$$

The following example illustrates application of the design guidelines given above.

Example 17.15. Specify the parameters of the op amp and surrounding components for an inverting amplifier meeting the following specifications:

Input resistance of $10\,\text{k}\Omega$ or larger, dc voltage gain of 50 dB $\pm$ 1 dB, 3-dB bandwidth greater than 10 kHz, load resistance of $10\,\text{k}\Omega$, and output swing of $\pm 12\,\text{V}$. The available supply voltage is $\pm 15\,\text{V}$. Simulate the amplifier to verify that the 3-dB bandwidth exceeds 10 kHz.

Solution: To meet the input resistance specification, we choose $R_1 = 10\,\text{k}\Omega$. The required nominal dc voltage gain is

$$20 \log\left(\frac{R_2}{R_1}\right) = 50\,\text{dB} \Rightarrow A_{v0} = \frac{R_2}{R_1}$$
$$= 10^{50/20} = 316.2$$

Thus

$$R_2 = 316R_1 = 3.16\,\text{M}\Omega$$

The $\pm$ 1 dB tolerance on dc voltage gain in decibels corresponds to

$$A_{v0\,\min} = 10^{49/20} = 282 = A_{v0} - 34,$$
$$A_{v0\,\max} = 10^{51/20} = 355 = A_{v0} + 39$$

Because $34 < 39$, we specify $A_{v0} = 316 \pm 34$ and the gain tolerance in percent is

$$\frac{34}{316} \times 100 = 10.8\%$$

so we require

$$R_1 = 10\,\text{k}\Omega \pm 5\%, \; R_2 = 3.16\,\text{M}\Omega \pm 5\%$$

The required dc bias compensating resistance is

$$R_X = R_L \| R_2 \cong 10\,\text{k}\Omega$$

To allow adjustment, we might use a $15\,\text{k}\Omega$ potentiometer or (e.g.) a fixed $5\,\text{k}\Omega$ resistor in series with a $10\,\text{k}\Omega$ variable potentiometer.

Because the supply voltages are $\pm$ 15 V, the value of V_A for the op amp must equal or exceed 30 V and V_B must be less than 30 V (see (17.72)). In that case, the inverting amplifier easily meets the output swing specification ($\pm 12\,\text{V}$), even with a non-rail-to-rail op amp.

The required gain-bandwidth product is

$$f_T \geq A_{0\max} W = (316 + 34) \times 10\text{kHz} = 3.5\,\text{MHz}$$

The required slew rate, maximum output current, and required power-dissipation rating are

$$SR \geq 2\pi V_{CC} W = (2\pi)(15\,\text{V})(10\text{kHz}) \cong 1\,\text{V}\mu\text{s}^{-1}$$

$$I_{o\max} \geq \frac{V_{CC}}{R'_L} = \frac{V_{CC}}{R_L \| R_2} = 1.2\,\text{mA}$$

$$P_{A\max} \geq \frac{V_{CC}^2}{4R'_L} = 3.64\,\text{mW}$$

A great many relatively inexpensive op amps would meet or exceed these specifications.

Figure 17.22 shows a simulation of the circuit, where the op-amp parameters are as shown in Fig. 17.23. The voltage gain at 10 kHz relative to the dc gain is[11]

$$20 \log\left[\frac{A_v(10\,\text{kHz})}{A_v(0\,\text{Hz})}\right] = 20 \log\left(\frac{2.324\,\text{V}/10\,\text{mV}}{316}\right)$$
$$\cong -2.67\,\text{dB}$$

which is greater than $-$ 3 dB, so the 3-dB bandwidth exceeds 10 kHz, as specified.

[11] The 14.14 mV source voltage shown is peak. The meter reading is rms.

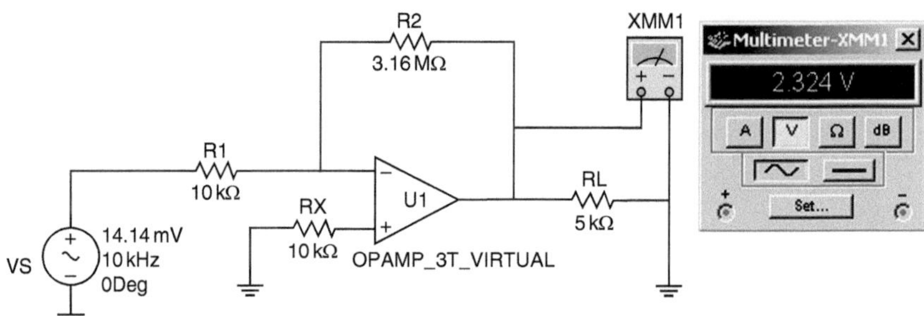

Fig. 17.22 See Example 17.15

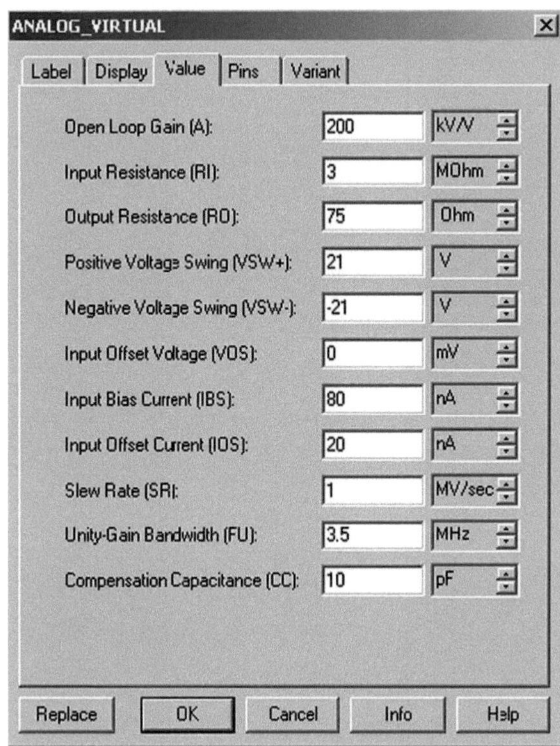

Fig. 17.23 See Example 17.15

Hundreds (if not thousands) of op amps are available from various manufacturers. To facilitate selection, manufacturers group their offerings using *somewhat* consistent descriptions. Op amps are commonly described as being *general-purpose* or *special-purpose*, the latter meaning one or more of *high-speed*, *high-precision*, *high-power*, *low-power*, *low-noise*, and *high-voltage*. General-purpose op amps have reasonably good characteristics across the board. High-speed op amps have large gain-bandwidth products and large slew rates. High-precision op amps have small dc bias currents, small input offset voltages, and other properties not treated above. High-power op amps have large power-dissipation ratings. Low-power op amps are designed to consume very little power (from a supply) when no input is applied, and have small quiescent supply currents. High-voltage op amps can accept high supply voltages, can tolerate large voltages across their input terminals, and can produce large load voltages. We do not describe low-noise op amps because a useful discussion of noise is beyond the scope of this book, but the implication is that a low-noise op amp produces relatively little electrical noise internally. In the descriptions just given for special-purpose op amps, *large* and *small* mean significantly larger or smaller than would be the case for a general-purpose op amp.

Lines of demarcation between the various types of op amps are not sharply drawn, and an op amp might belong to two or more classes; for example, an op amp might be both high-speed and low power. Although boundaries are a bit fuzzy, they are fairly consistent among various manufacturers and most manufacturers offering a wide variety of op amps group their offerings according to the categories described above. Some manufacturers' web sites also allow a prospective user to sort offerings (e.g., in order of power-dissipation rating) to further narrow a search.[12]

Table 17.3 gives selected parameters for representative members of the classes listed above. The column contents are as follows:

- *Type* is the manufacturer's designation for the op amp.
- V_B, V_A are the minimum and maximum permissible differences between the upper and lower supply voltages; e.g., for a symmetrically powered LF411,

[12]Virtually all manufacturers of op amps have extensive web sites describing their offerings.

Table 17.3 Representative op amps

Class	Type	V_B (V)	V_A (V)	$I_{dc\,bias}$ (pA)	SR ($\text{V}\mu\text{s}^{-1}$)	f_T (MHz)	$I_{o\,max}$ (mA)	$P_{A\max}$ (25°C) (mW)	Z_{in} (Ω)	Z_{out} (Ω)
General purpose	LF411	10	36	200	15	4	20	670	10^{12}	NA[a]
Precision	OPA124	10	36	1	2	2	3.5	750	10^{13}	100
High speed	AD380JH	6	20	10	330	300	10	1500	10^{11}	NA
Low power	TL061C	4	36	400	3.5	1	5	680	10^{12}	NA
High voltage	OPA445	20	90	10	15	2	25	680	10^{13}	220
High power	PA03	15	75	50	10	5	30 A	500 W	10^{11}	NA

[a]NA = not available from manufacturer's data sheet. Op amp output impedances are typically on the order of 100 Ω.

a bipolar supply must be at least $\pm 10/2 = \pm 5$ V and no larger than $\pm 36/2 = \pm 18$ V.

- $I_{dc\,bias}$ is the average of the dc bias currents entering or exiting the *n* and *p* terminals when the op amp is driven by a source.
- *SR* is the slew rate.
- f_T is the gain-bandwidth product.
- I_{omax} is the maximum current the op amp can deliver, and is generally the current delivered if the output terminals are shorted. Many op amps can endure shorted output terminals indefinitely, but some cannot.
- P_{Amax} is the maximum permissible power dissipation (at 25°C).
- Z_{in}, Z_{out} are the input and output impedances (at dc). Not all manufacturers specify a value for output impedance. Often, input impedance is capacitive at frequencies above dc, and is specified by a device data sheet as $R_{in}\|C$; e.g., as $10^{12}\,\Omega\|4\,\text{pF}$, which means that the input impedance is that of a $10^{12}\,\Omega$ resistor in parallel with a 4 pF capacitor.

In addition to intended application, op amps can be grouped by internal structure. The primary types are bipolar (BJT), JFET-input, and MOSFET (throughout), meaning (respectively) that op amps are constructed using only bipolar junction transistors (BJT's), or with junction field-effect transistors (JFET's) in the input stage and BJT's thereafter, or entirely from metal-oxide-silicon field-effect transistors (MOSFET's). All of the entries in Table 17.3 are BJT op amps having JFET inputs, as one can tell from the input impedances. BJT-input op amps have input impedances on the order of 1MΩ.

Often, linear (analog) and digital circuits are fabricated on a single chip (e.g., as in an analog-to-digital converter). Because digital circuits are almost always constructed using MOSFET's, and because fabrication is easier if all transistors on a single chip have the same basic structure, MOSFET op amps usually are preferred when op amps and digital circuits appear together in a single integrated circuit. Also, input impedances of both JFET-input and MOSFET op amps are much larger than those of BJT op amps and the input bias currents are much smaller (but more sensitive to temperature). On the downside, the gain-bandwidth products of MOSFET op amps tend to be smaller than those available from BJT op amps. As a rule of thumb, the dc gain achievable for an inverting or non-inverting amplifier using a JFET-input BJT op amp is typically about ten times the dc gain achievable with a MOSFET op amp. You will learn more about these things in subsequent courses.

Example 17.16. Determine which (if any) of the op amps in Table 17.3 could be used in an inverting amplifier subject to the following specifications:

Dc gain = 20
Bandwidth = 50 kHz
Output swing = ± 25 V

Solution: Assuming rail-to-rail operation, the specified output swing requires $V_A \geq 50$ V, so we can immediately eliminate all but the OPA445 and the PA03. None of the others can support the specified output swing. Both the OPA445 and the PA03 have the minimum required gain-bandwidth product of $20 \times 50\,\text{kHz} = 1\,\text{MHz}$.

From (17.28), the minimum acceptable slew rate is

$$SR_{\min} = 2\pi f_{\max} V_{CC} = (2\pi)(50 \times 10^3)(25)$$
$$= 7.85\,\text{V}\mu\text{s}^{-1}$$

Both the OPA445 and the PA03 can meet this condition. Based upon only the three specifications given, either the OPA445 or the PA03 would serve. Before choosing one over the other, we would need more information (e.g., cost and load and power-dissipation requirements).

Data sheets for most op amps (and other electronic devices) specify a temperature range over which the op amp will operate as described, typically from about − 50°C to about 125°C. The maximum power dissipation ratings given by Table 17.3 are the absolute maximum ratings and assume an operating temperature no greater than 25°C. In most circuits, and especially in high-speed and high-power circuits, heat sinking and forced-air cooling are needed to keep operating temperatures anywhere near 25°C(77°F). More often, operating temperatures are nearer 70°C (158°F).

Maximum power-dissipation ratings for op amps are strong functions of operating temperature and often are given by graphs of power dissipation versus operating temperature, as illustrated by Fig. 17.24 for a PA03 op amp. For a PA03, permissible internal power dissipation is 500 W at or below 25°C, but drops to 300 W at 80°C. Internal temperature-monitoring circuitry shuts the PA03 op amp down if the internal temperature rises above 80°C. Typical power-dissipation limits for small general-purpose op amps range from a few hundred mW to a little more than 1 W (at 25°C).

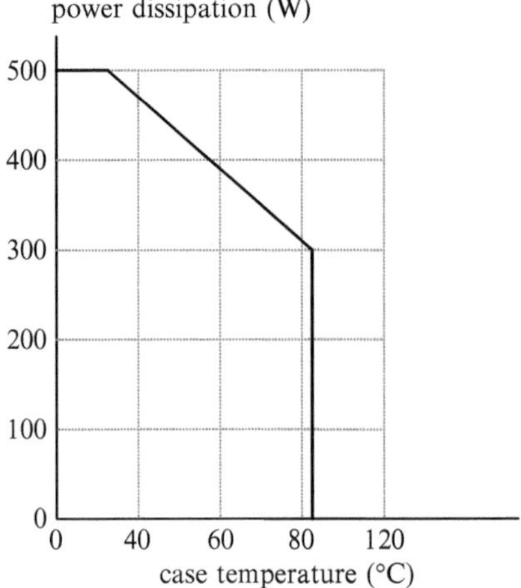

Fig. 17.24 Permissible internal power-dissipation for a PA03 op amp (Courtesy of Cirrus Logic, Inc.)

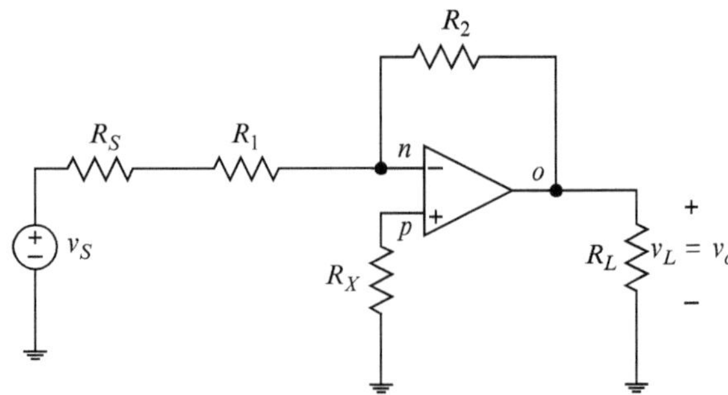

Fig. 17.25 An inverting amplifier, driven by a source and driving a load

Example 17.17.

(a) Design an inverting amplifier subject to the following specifications and constraints:

Dc voltage gain $= 100 \pm 10\%$
Input resistance $= 5\,\text{k}\Omega \pm 10\%$
Bandwidth $\geq 20\,\text{kHz}$
Available supply $= \pm 15\,\text{V}$
Load resistance $\geq 15\,\text{k}\Omega$
Source output resistance $\leq 10\,\Omega$
Input amplitude $|v_S| \leq 120\,\text{mV}$
The amplifier and source are to be directly coupled.

(b) Which, if any, of the op amps in Table 17.3 could be used in this application?

Solution: Figure 17.25 shows an inverting amplifier, driven by a source having output resistance R_S and driving a resistive load having resistance R_L.

The specifications on dc gain and input amplitude imply that the peak output can be as large as 12 V for any frequency in the passband. The ± 15 V supply should support this swing, even if the op amp is not rail-to-rail.

The input resistance equals the resistance R_1, so

$$R_1 = 5\,\text{k}\Omega$$

The source resistance is a negligible fraction of the input resistance and can be ignored. From the specification on the dc voltage gain A_0, we obtain

$$R_2 = R_1 \times A_0 = (5\,\text{k}\Omega)(100) = 500\,\text{k}\Omega$$

The dc voltage gain must satisfy $A_{v0} = 100 \pm 10\%$, which means (see (17.67)) R_1 and R_2 must satisfy $R_1 = 5\,\text{k}\Omega \pm 5\%$, $R_2 = 500\,\text{k}\Omega \pm 5\%$; i.e., if $A_{v0} = R_2/R_1$ is the nominal dc gain, then

$$A_{v0\max} = \frac{1.05R_2}{0.95R_1} = 1.105A_{v0};$$
$$A_{v0\min} = \frac{0.95R_2}{1.05R_1} = 0.905A_{v0}$$

Thus the ± 10% specification on gain, not the ± 10% specification on input resistance, determines the precision of the resistor R_1.

The compensating resistance is

$$R_X = R_L \| R_2 = 14.6\,\text{k}\Omega \cong 15\,\text{k}\Omega$$

The difference between the two (p and n) input dc bias currents can be as large as 20%. If dc-bias compensation is critical, we might decrease R_X and use a variable (trimmer) resistor in series with R_X to provide fine adjustment; for example, we could use a 10 kΩ fixed resistor in series with a 10 kΩ trimmer. Alternatively, we could eliminate the fixed resistor and use only a variable resistor whose maximum resistance is about 20 kΩ.

Together, the specified dc gain A_{v0} and bandwidth W determine the minimum acceptable gain-bandwidth product:

$$f_T \geq A_{v0} W = (100)(20\,\text{kHz}) = 2\,\text{MHz}$$

If the specified bandwidth is to be achieved for full-scale output, the slew rate must satisfy the inequality

$$SR > 2\pi W V_{CC} = (2\pi)(20\,\text{kHz})(13\text{V}) \cong 1.64\text{V}\mu\text{s}^{-1}$$

The specified supply voltage and load resistance provide a conservative estimate of the required power-dissipation rating for the op amp:

$$P_A \cong \frac{V_{CC}{}^2}{4R_L} = \frac{(13\text{V})^2}{(4)(15\,\text{k}\Omega)} \cong 2.82\,\text{mW}$$

Assuming a full-scale output, the maximum output current must satisfy

$$I_{o\max} = \frac{V_{CC}}{R_L} = \frac{13\text{V}}{15\,\text{k}\Omega} \cong 867\,\mu\text{A}$$

Thus we seek an op amp satisfying the following conditions:

$$f_T \geq 2\,\text{MHz},\ SR \geq 1.64\text{V}\mu\text{s}^{-1},$$
$$V_{CC\min} \leq 15\,\text{V}, V_{CC\max} \geq 15\,\text{V},$$
$$P_{A\max} \geq 2.82\,\text{mW},\ I_{o\max} \geq 867\,\mu\text{A}$$

Consulting Table 17.3, we find that the general-purpose LF411 op amp meets or exceeds the specifications.

For the LF 411, the dc input-bias current is 200 pA. In absence of the compensating resistor R_X, the dc offset in the output would be

$$V_{offset} = R_2 I_{dc\,bias} = (500\,\text{k}\Omega)(200\,\text{pA}) = 100\,\mu\text{V}$$

which is probably insignificant in any application specified as loosely as this one, and the compensating resistor can probably be omitted. In more demanding applications, where such a small offset might be considered significant, we would find additional specifications on things such as noise levels.

17.12 Problems

Section 17.1 is prerequisite for the following problems.

P 17.1 Figure P 17.1 shows two ac circuit models for an op amp. Show that the model in Fig. P 17.1(a) can be simplified to the ac model defined in Fig. P 17.1(b). Express the intrinsic dc gain μ_0 and the intrinsic bandwidth f_0 of the ac model in Fig. P 17.1(b) in terms of the circuit parameters g, k, R_1, C in Fig. P 17.1(a).

P 17.2 Are there values of R, C, and μ_0 for which the circuit in Fig. P 17.2 is equivalent to the model shown in Fig. P 17.1(b)? If so, prove it. If not, explain why not.

P 17.3 Are there values of R, C, and g_0 for which the circuit in Fig. P 17.3 is equivalent to the model shown in Fig. P 17.1(b)? If so, prove it. If not, explain why not.

P 17.4 Are there values of R_A, C_A, and μ_0 for which the circuit in Fig. P 17.4 is equivalent to the model shown in Fig. P 17.1(b)? If so, prove it. If not, explain why not.

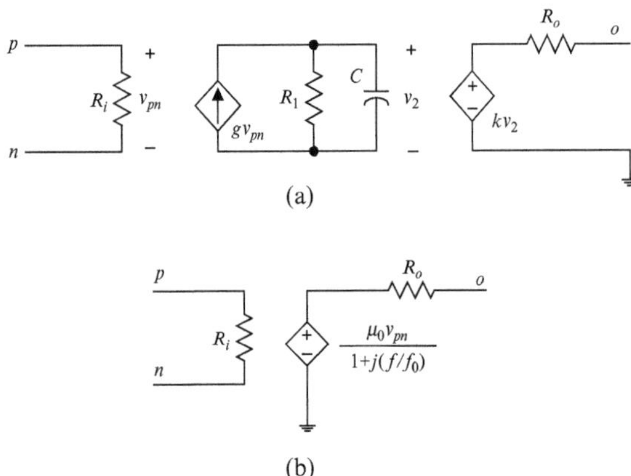

Fig. P 17.1 See Problem P 17.1

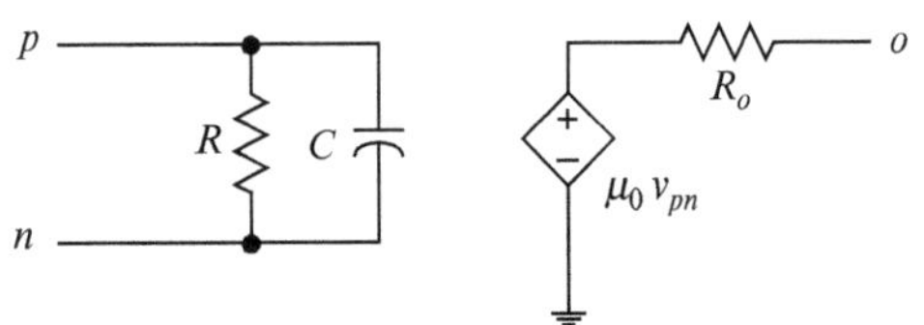

Fig. P 17.2 See Problem P 17.2

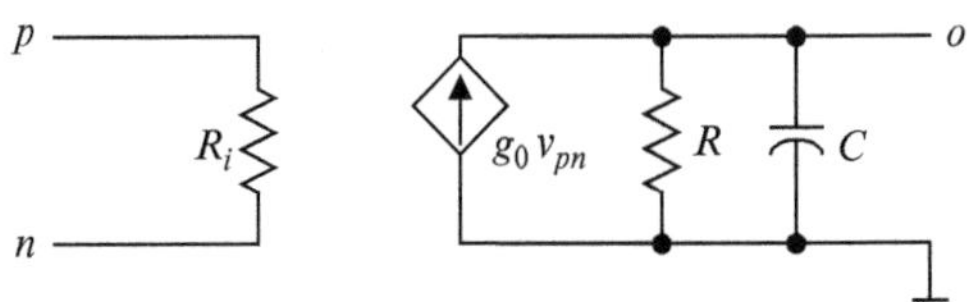

Fig. P 17.3 See Problem P 17.3

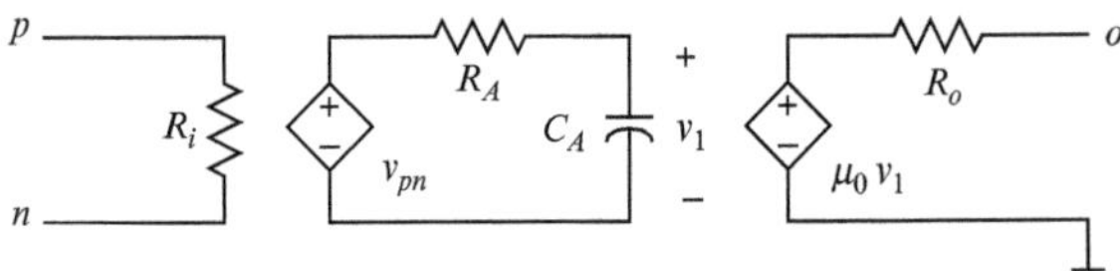

Fig. P 17.4 See Problem P 17.4

P 17.5 The intrinsic dc voltage gain and gain-bandwidth product for a certain op amp are 2×10^5 and 5 MHz, respectively. What is the intrinsic 3 dB bandwidth of the op amp?

P 17.6 The unity-gain frequency and intrinsic bandwidth for op amp A are 5 MHz and 10 Hz, respectively. The unity-gain frequency and intrinsic bandwidth for op amp B are 2.5 MHz and 50 Hz, respectively. Which op amp has the larger intrinsic dc voltage gain?

P 17.7 It is found by careful measurement that the intrinsic bandwidth of a certain op amp is 15 Hz and the intrinsic voltage gain at 500 Hz is 80 dB. What is the gain-bandwidth product for the op amp?

Section 17.2 is prerequisite for the following problems.

P 17.8 In a certain application, an inverting amplifier must have a dc voltage gain of 120 and a bandwidth of 80 kHz. What gain-bandwidth product is required of the op amp?

P 17.9 The dc voltage gain and 3-dB bandwidth of a certain non-inverting amplifier are 75 and 100 kHz, respectively. What is the voltage gain at 200 kHz?

P 17.10 For a certain project, you require an inverting amplifier having half-power bandwidth $W = 50\,\text{kHz}$. You have available an op amp having a gain-bandwidth product of 2.5 MHz. What is the maximum dc voltage gain you can achieve if you use that op amp?

P 17.11 The voltage gain of a certain op-amp-based non-inverting amplifier is 60 dB at 5 kHz and 40 dB at 50 kHz. What is the gain-bandwidth product of the op amp?

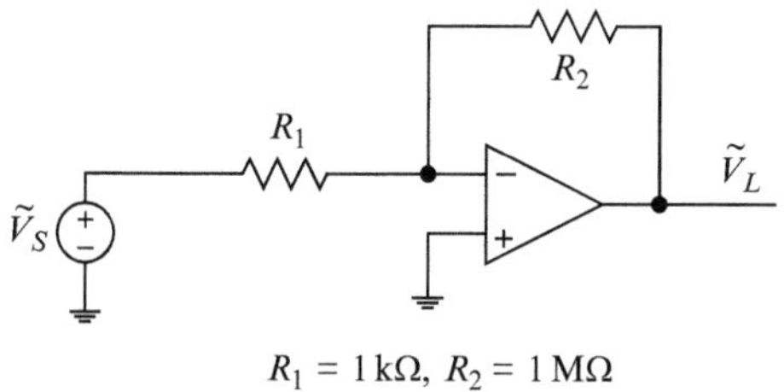

Fig. P 17.5 See Problem P 17.13

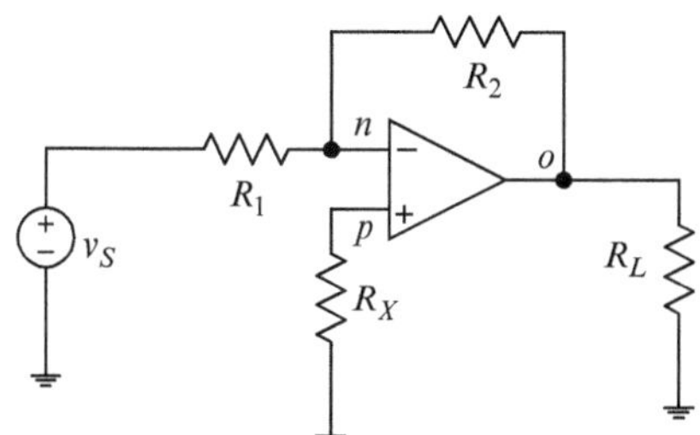

Fig. P 17.6 See Problem P 17.15

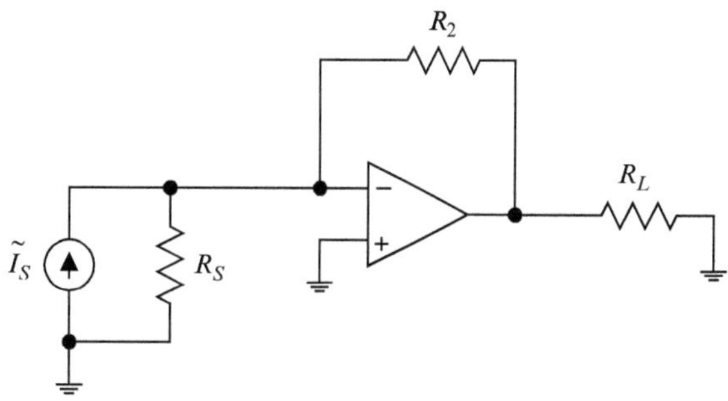

Fig. P 17.7 See Problem P 17.16

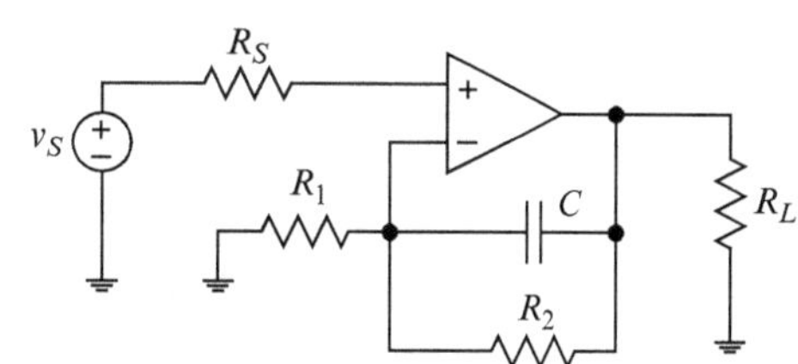

Fig. P 17.8 See Problem P 17.17

P 17.12 The gain-bandwidth product for the op amp in a certain inverting amplifier is 2 MHz. The half-power bandwidth of the amplifier is 20 kHz. What is the dc voltage gain of the amplifier?

P 17.13 Simulate the circuit shown in Fig. P 17.5, using your simulation software's *virtual*, *generic*, or *default* op amp and use trial-and-error to find the op amp's gain-bandwidth product.

P 17.14 You have available an op amp whose gain-bandwidth product exceeds 100 kHz but is otherwise unknown.

(a) Design a non-inverting amplifier having a gain of 101, such that neither resistor has a resistance exceeding 1MΩ What is the lower bound on the half-power bandwidth of the amplifier?
(b) Simulate the amplifier using your simulation software's *virtual*, *generic*, or *default* op amp and use trial-and-error to find the amplifier's half-power bandwidth.
(c) What is the maximum dc voltage gain you could obtain if the required amplifier half-power bandwidth is 1 kHz?

P 17.15 In the circuit shown in Fig. P 17.6, $R_2 = 1\,\text{M}\Omega$ and $R_L = 5\,\text{k}\Omega$. (a) Measurements reveal that for $v_S = 200\,\text{mV}$ (dc), $V_o = 10\,\text{V}$. Find the resistance R_1 and the current exiting the output (o) terminal of the op amp. (b) Further measurements reveal that for $v_S = 200\cos(\omega_1 t)\,\text{mV}$, with $f_1 = 20\,\text{kHz}$, the peak amplitude of the output v_o equals 5 V. Find the op-amp gain-bandwidth product.

P 17.16 Refer to Fig. P 17.7, which shows a current-to-voltage converter. Obtain an expression for the frequency-dependent transfer function $H_z = \tilde{V}_o/\tilde{I}_S$ based on the usual linear ac model for the op amp. Show that your expression reduces to the correct expression if the op amp is ideal. Invoke reasonable approximations to obtain a simple expression for the bandwidth of the circuit, defined as the frequency W for which

$$\left|\frac{H_z(j2\pi W)}{H_z(0)}\right|^2 = \frac{1}{2}$$

Section 17.3 is prerequisite for the following problems.

P 17.17 In Fig. P 17.8, the circuit parameters are

$$R_S = 50\,\Omega,\ R_1 = 1\,\text{k}\Omega,\ R_2 = 100\,\text{k}\Omega,\ C = 159\,\text{pF},\ R_L = 5\,\text{k}\Omega.$$

The op-amp input resistance, output resistance, and gain-bandwidth product are $R_i = 10\,\text{M}\Omega$, $R_o = 75\,\Omega$, and $f_T = 6\,\text{MHz}$.

(a) Obtain and justify an approximate expression for the voltage transfer function $H_v = \tilde{V}_o/\tilde{V}_S$. Use the expression to calculate and plot the voltage gain in dB versus frequency for $1\,\text{Hz} \leq f \leq 1\,\text{MHz}$

(b) Simulate the circuit, modifying the default (virtual) op-amp parameters as indicated by Fig. P 17.9. Set the *rms* amplitude of the sinusoidal input to 1 mV. Obtain the voltage gain for $f = 1\,\text{Hz}$, $1\,\text{kHz}$, $10\,\text{kHz}$, $100\,\text{kHz}$ and plot the values (in dB) on the graph obtained in part (a).

(c) Replace the op amp in Fig. P 17.8 with the model shown in Fig. P 17.4, using $R_A = 100\,\text{k}\Omega$ and the resistance values given above. Find the value of C_A such that the intrinsic bandwidth of the op amp (model) equals 30 Hz. Determine the value of μ_0 such that the gain-bandwidth product of the op amp (model) is 6 MHz. Then simulate the circuit and find the voltage gain in dB for $f = 1\,\text{Hz}$, $1\,\text{kHz}$, $10\,\text{kHz}$, $100\,\text{kHz}$. Compare the values with those obtained in parts (a) and (b).

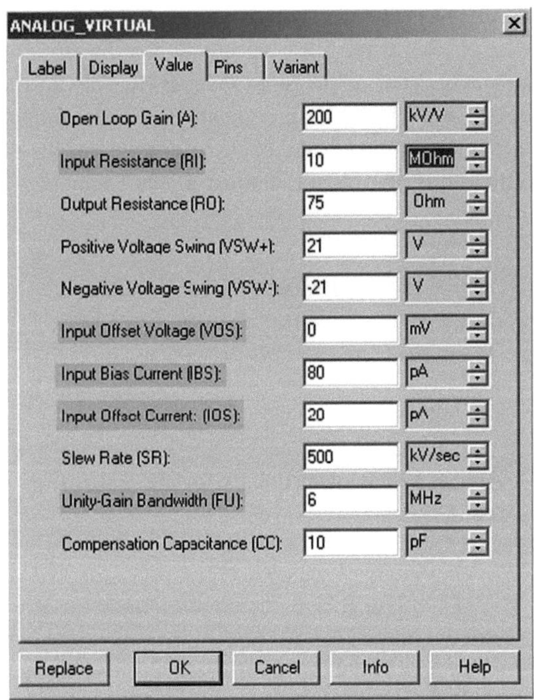

Fig. P 17.9 See Problem P 17.17

P 17.18 In Fig. P 17.10, the circuit parameters are

$$R_S = 5\,\Omega,\ R_1 = 10\,\text{k}\Omega,\ R_2 = 1\,\text{M}\Omega,$$
$$C_1 = 1.59\,\text{pF},\ C_2 = 15.9\,\text{pF},\ R_L = 5\,\text{k}\Omega.$$

(a) Assume the op amp is ideal. Obtain an expression for the voltage transfer function $H_v = \tilde{V}_o/\tilde{V}_S$. Express the transfer function in standard form and express the dc gain and corner frequencies in terms of the circuit parameters. Employ reasonable approximations to simplify the expression. Construct an asymptotic plot of voltage gain in dB versus frequency for $1\,\text{Hz} \leq f \leq 1\,\text{MHz}$.

(b) Assume the op-amp input resistance, output resistance, and gain-bandwidth product are $R_i = 10\,\text{M}\Omega$, $R_o = 75\,\Omega$, and $f_T = 6\,\text{MHz}$. Simulate the circuit, modifying the default (virtual) op-amp parameters as indicated by Fig. P 17.9. Set the *rms* amplitude of the sinusoidal input to 1 mV. Obtain from the simulation the voltage gain for $f = 10\,\text{Hz}$, $100\,\text{Hz}$, $1\,\text{kHz}$, $10\,\text{kHz}$, $100\,\text{kHz}$ and plot the values (in dB) on the graph obtained in part (a). Explain any differences between the gains obtained in part (a) and those obtained from the simulation.

P 17.19 In Fig. P 17.11, the circuit parameters are

$$R_S = 5\,\Omega,\ R_1 = 20\,\text{k}\Omega,\ R_2 = 330\,\text{k}\Omega,$$
$$C_1 = 200\,\text{nF},\ C_2 = 15.9\,\text{pF},\ R_L = 5\,\text{k}\Omega.$$

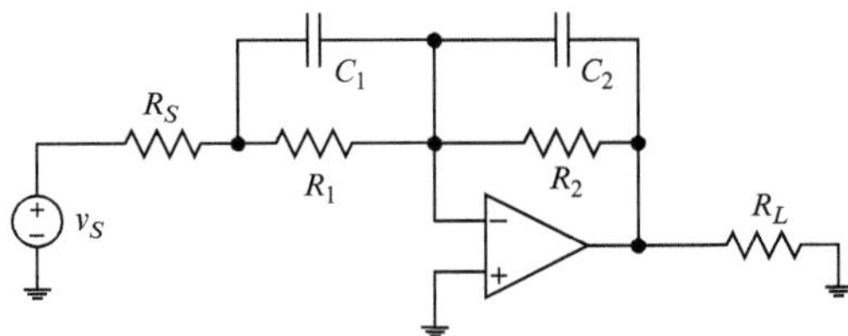

Fig. P 17.10 See Problem P 17.18

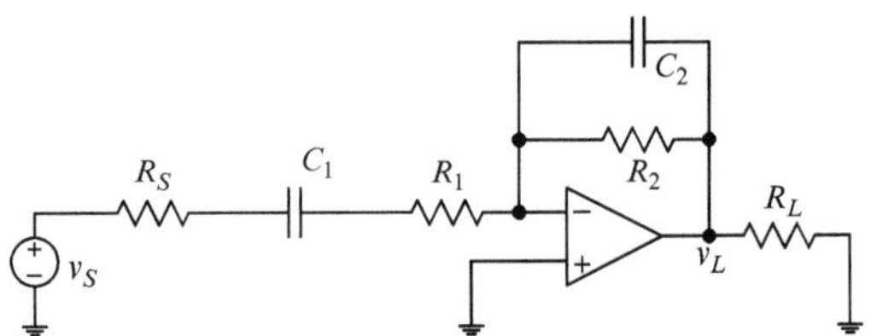

Fig. P 17.11 See Problem P 17.19

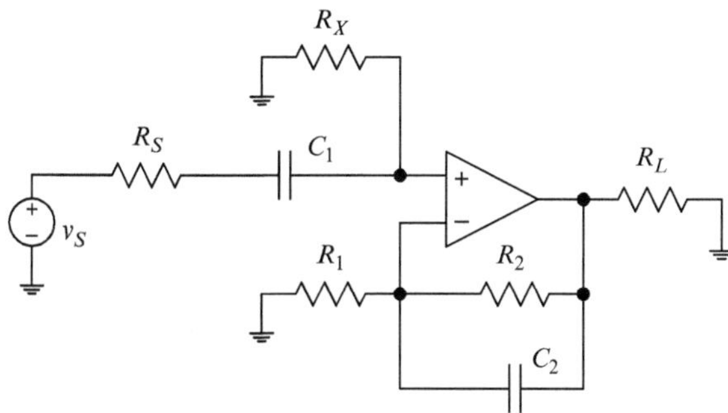

Fig. P 17.12 See Problem P 17.20

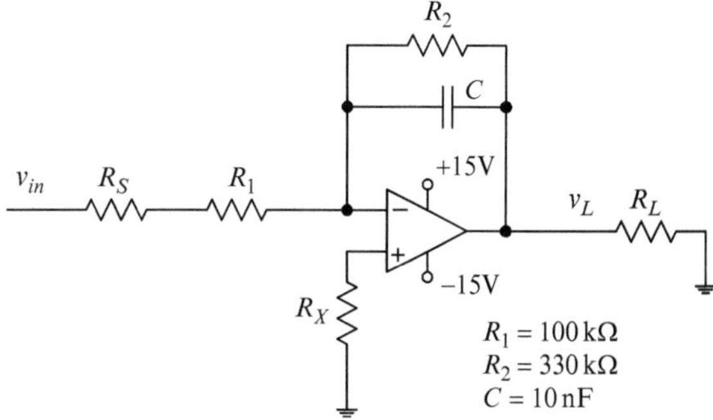

Fig. P 17.13 See Problem P 17.22

(a) Assume the op amp is ideal. Obtain an expression for the voltage transfer function $H_v = \tilde{V}_o/\tilde{V}_S$. Express the transfer function in standard form and express the dc gain and corner frequencies in terms of the circuit parameters. Employ reasonable approximations to simplify the expression. Construct an asymptotic plot of voltage gain in dB versus frequency for $1\,\text{Hz} \le f \le 1\,\text{MHz}$.

(b) 💻 Assume the op-amp input resistance, output resistance, and gain-bandwidth product are $R_i = 10\,\text{M}\Omega$, $R_o = 75\,\Omega$, and $f_T = 6\,\text{MHz}$. Simulate the circuit, modifying the default (virtual) op-amp parameters as indicated by Fig. P 17.9. Set the *rms* amplitude of the sinusoidal input to 100 mV. Obtain from the simulation the voltage gain for $f = 10\,\text{Hz}$, $100\,\text{Hz}$, $1\,\text{kHz}$, $10\,\text{kHz}$, $100\,\text{kHz}$ and plot the values (in dB) on the graph obtained in part (a). Explain any differences between the gains obtained in part (a) and those obtained from the simulation.

P 17.20 In Fig. P 17.12, the circuit parameters are

$$R_S = 5\,\Omega,\ R_1 = 10\,\text{k}\Omega,\ R_2 = 1\,\text{M}\Omega,$$
$$C_1 = 1\,\mu\text{F},\ C_2 = 15.9\,\text{pF},\ R_L = 5\,\text{k}\Omega$$

(a) Assume the op amp is ideal. Obtain an expression for the voltage transfer function $H_v = \tilde{V}_o/\tilde{V}_S$. Express the transfer function in standard form and express the dc gain and corner frequencies in terms of the circuit parameters. Employ reasonable approximations to simplify the expression. Construct an asymptotic plot of voltage gain in dB versus frequency for $1\,\text{Hz} \le f \le 1\,\text{MHz}$.

(b) 💻 Assume the op-amp input resistance, output resistance, and gain-bandwidth product are $R_i = 10\,\text{M}\Omega$, $R_o = 75\,\Omega$, and $f_T = 6\,\text{MHz}$. Simulate the circuit, modifying the default (virtual) op-amp parameters as indicated by Fig. P 17.9. Set the rms amplitude of the sinusoidal input to 100 mV. Obtain from the simulation the voltage gain for f = 10 Hz, 100 Hz, 1 kHz, 10 kHz, 100 kHz,1 MHz and plot the values (in dB) on the graph obtained in part (a). Explain any differences between the gains obtained in part (a) and those obtained from the simulation.

P 17.21 If $|Z_1| \gg |Z_S|$, $|Z_2| \gg R_o$, and $|Z_L| \gg R_o$, then the input impedance of an inverting circuit is given by

$$Z_{in} \cong Z_1 + \frac{Z_i Z_2}{(\mu + 1)Z_i + Z_2}.$$

Can this expression be simplified further if the op amp used is a JFET-input op amp? If so, show how.

Sections 17.4 and 17.5 are prerequisite for the following problems.

P 17.22 In Fig. P 17.13, v_{in} is a sinusoid. The source impedance is negligible and the load resistance is $R_L = 10\,\text{k}\Omega$. The op-amp parameters are:

Input resistance: $R_i = 1\,\text{T}\Omega$
Input capacitance: $C_i = 1.5\,\text{pF}$
Input impedance: $Z_i = R_i \| (j\omega C_i)^{-1}$
Output resistance: $R_o = 55\,\Omega$
Intrinsic dc voltage gain: $\mu_0 = 2 \times 10^5$
Unity-gain frequency: $f_T = 1.5\,\text{MHz}$
Output swing: Rail-to-rail, up to $\pm 21\,\text{V}$
Slew rate: $500\,\text{kV}\,\text{s}^{-1}$.

(a) Assuming linear operation, obtain an expression for the voltage transfer function of the circuit.
(b) Under what conditions on the peak amplitude and frequency of the input is the circuit linear?

P 17.23 If you worked Problem P 17.17, you might have noticed that the simulated voltage gains were in very good agreement with the graph of theoretical gain for $f < 10\,\text{kHz}$, but began falling below the theoretical values for higher frequencies. Was the output limited by the op amp's output swing or slew rate?

P 17.24 Refer to Problem P 17.18. Use a simulation to explore whether the limited slew rate of the op amp comes into play at 100 kHz or below.

P 17.25 Refer to Problem P 17.19. Use simulation to explore whether the limited slew rate of the op amp comes into play at 100 kHz or below.

P 17.26 Refer to Problem P 17.20. Use simulation to explore whether the limited slew rate of the op amp comes into play at 1 MHz or below.

P 17.27 In Fig. P 17.14, the circuit parameters are

$$R_S = 5\,\Omega,\ R_1 = 1\,\text{k}\Omega,\ R_2 = 100\,\text{k}\Omega,$$
$$C_1 = 159\,\text{nF},\ C_2 = 159\,\text{pF},\ R_L = 5\,\text{k}\Omega.$$

(a) Assume the op amp is ideal. Obtain an expression for the voltage transfer function $H_v = \tilde{V}_o/\tilde{V}_S$. Express the transfer function in standard form and express the dc gain and corner frequencies in terms of the circuit parameters. Employ reasonable approximations to simplify the expression. Construct an asymptotic plot of voltage gain in dB versus frequency for $1\,\text{Hz} \le f \le 1\,\text{MHz}$.
(b) Simulate the circuit after modifying the default (virtual) op-amp parameters as indicated by Fig. P 17.9. Set the rms amplitude of the sinusoidal input to 1 mV. Obtain the voltage gain for $f = 10\,\text{Hz}, 100\,\text{Hz}, 1\,\text{kHz}, 10\,\text{kHz}, 100\,\text{kHz}, 1\,\text{MHz}$ and plot the values (in dB) on the graph obtained in part (a). Explain any differences between the gains obtained in part (a) and those obtained from the simulation. In particular, explore whether the limited slew rate of the op amp comes into play at 1 MHz or below.

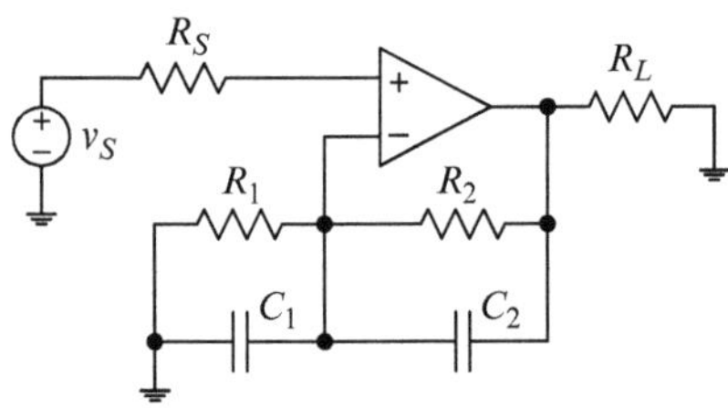

Fig. P 17.14 See Problem P 17.27

P 17.28 Is the model in Fig. P 17.4 a complete model for an op amp? Justify your answer.

P 17.29 The gain-bandwidth product and slew rate of an op amp to be used in a certain inverting amplifier are 2.5 MHz and $800\,\text{kV}\,\text{s}^{-1}$, respectively. The op amp is rail-to-rail and the power supply voltages are $\pm 15\,\text{V}$.

(a) What is the maximum voltage gain achievable by the amplifier if the bandwidth must be 50 kHz?
(b) What is the limit imposed by output swing on the peak amplitude of an input if the amplifier must remain linear?
(c) If the bandwidth is 50 kHz, what is the limit imposed by slew rate on the peak amplitude of a 25 kHz sinusoidal input if the output must be sinusoidal (if the amplifier must remain linear)?

P 17.30 The gain-bandwidth product and slew rate of an op amp to be used in a certain non-inverting amplifier are 5 MHz and $2\,\text{V}\,\mu\text{s}^{-1}$, respectively. The op amp is rail-to-rail and the power supply voltages are $\pm 20\,\text{V}$.

(a) What is the maximum gain achievable by the amplifier if the bandwidth must be 25 kHz?
(b) What is the bandwidth of the amplifier if the dc voltage gain is 100?
(c) If the dc voltage gain is100, what is the limit imposed by output swing on the peak amplitude of an input if the amplifier must remain linear?
(d) If the voltage gain and bandwidth are as given in part (b), what is the limit imposed by slew rate on the peak amplitude of a 5 kHz sinusoidal input if the output must be sinusoidal (if the amplifier must remain linear)?

P 17.31 In a certain application, a voltage follower must operate linearly for sinusoidal inputs having frequencies up to 200 kHz and peak amplitudes up to 2 V. Specify acceptable values for the op-amp gain-bandwidth product and slew rate.

P 17.32 The text gives the following expression for the input impedance of an inverting amplifier:

$$Z_{in} \cong \frac{1 + j(f/W)}{1 + j(f/f_T)} R_1;\quad \mu|Z_i| \gg R_2,$$
$$|Z_i| \gg R_o,\ R_L \gg R_o$$

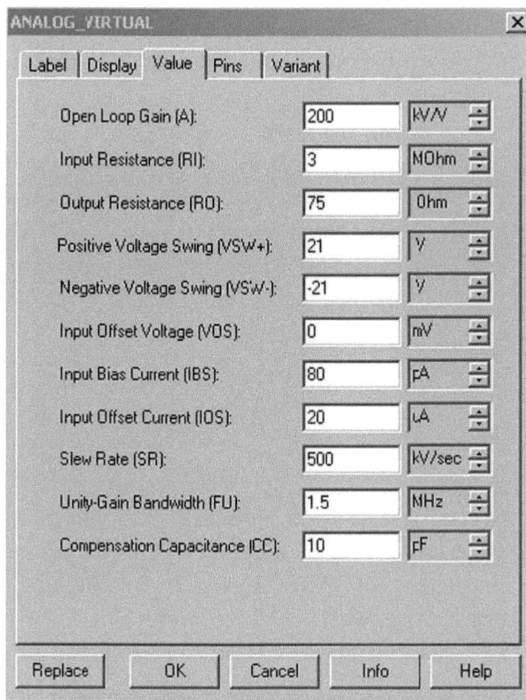

Fig. P 17.15 See Problem P 17.32

(a) Design an inverting amplifier having dc voltage gain $A_{v0} = 150$, using an op amp having gain-bandwidth product 1.5 MHz, and use the expression above to calculate the magnitude of the input impedance for $f = 0, 100\,\text{Hz}, 1\,\text{kHz}, 10\,\text{kHz}, 100$ kHz, 1 MHz.

(b) Simulate the circuit, using an op amp having the parameter values shown in Fig. P 17.15. Use the simulation to measure the magnitude of the input impedance for $f = 0, 100\,\text{Hz}, 1\,\text{kHz}, 10\,\text{kHz}$, 100 kHz, 1 MHz. Explain any significant differences between the measured values obtained from the simulation and the theoretical values obtained in part (a).

(c) Verify that the amplifier is linear for each simulation.

P 17.33 For each op amp described in Table P 17.1:

(a) Specify the parameters of an inverting amplifier whose input resistance is no smaller than $10\,\text{k}\Omega$ in the amplifier passband and whose gain is as large as possible, subject to the constraints that the feedback resistance R_2 must not exceed $2\text{M}\Omega$ and the amplifier 3 dB bandwidth must be at least 20 kHz.

(b) Find the maximum output resistance in the passband and the maximum output swing, assuming rail-to-rail operation.

(c) Find whether the slew rate is adequate, assuming the maximum amplitude allowed by the output swing limit for any input within the amplifier passband.

Table P 17.1 See Problem P 17.33

Op amp	SR ($\text{V}\mu\text{s}^{-1}$)	f_T (MHz)	R_{in} (Ω)	R_{out} (Ω)	$V_{CC\,\max}$ (V)
(a)	15	4	2×10^6	75	18
(b)	2	2	4×10^6	100	18
(c)	330	300	10^{11}	50	10
(d)	3.5	1	10^{12}	75	18
(e)	15	2	10^{13}	200	45
(f)	10	5	10^{11}	100	35

P 17.34 The input impedance of an inverting amplifier is given by

$$Z_{in} = Z_1 + Z_S + \frac{Z_i D}{[(\mu+1)Z_L + R_o]Z_i + D}$$

where

$$D = R_o Z_L + Z_2 Z_L + Z_2 R_o.$$

The expression is exact if the ac linear model for the op amp is exact. The circuit and op-amp parameters of a certain inverting amplifier are:

$$R_1 = 10\,\text{k}\Omega,\ R_2 = 1\text{M}\Omega,\ R_S = 50\,\Omega,\ R_L = 5\,\text{k}\Omega$$
$$R_i = 4\text{M}\Omega,\ R_o = 75\,\Omega,\ \mu_0 = 2 \times 10^5,\ f_T = 2\,\text{MHz}.$$

(a) Use the expression above to calculate the input impedance at the passband edge.

(b) A capacitor having capacitance $C_2 = 159\,\text{pF}$ is connected in parallel with the feedback resistor R_2. Predict whether the input impedance will increase, decrease, or remain the same. Again calculate the input impedance at the passband edge to check your prediction.

(c) Use simulations to check your calculations in parts (a) and (b). Be sure the op amp operates linearly; e.g., don't exceed the output-swing or slew-rate limits.

P 17.35 Repeat Problem P 17.34, except in part (b), connect the 159 pF capacitor in parallel with the input resistor R_1.

P 17.36 Figure P 17.16 shows a follower using an AD549L op amp. The supply voltages are $\pm$ 15 V. An AD549L has a slew rate of $3 \mathrm{V}\mu\mathrm{s}^{-1}$, a gain-bandwidth product of 1 MHz, and allows a maximum symmetric supply voltage of $\pm$ 18 V. The output can swing to within 1 V of either supply voltage. Which of slew rate and bandwidth limit the bandwidth of the follower?

Section 17.6 is prerequisite for the following problems.

P 17.37 In Fig. P 17.17, the wipers on the variable resistors are connected mechanically, such that the product of the feedback resistances is fixed and equal to ${R_2}^2$, and $0.1 \leq k < 1$. The gain-bandwidth product of each op amp is 2 MHz.

(a) Which of the two stages limits the bandwidth of the cascade?
(b) Obtain a standard-form expression for the voltage transfer function.
(c) Obtain an expression for the half-power frequency of the cascade.
(d) If $0.1 \leq k \leq 1$, what are the minimum and maximum half-power bandwidths of the cascade?

P 17.38 Show that the bandwidth of a cascade of n inverting or non-inverting amplifiers, or a cascade of a mix of such amplifiers, each having bandwidth W, is given by

$$W_n = W\sqrt{2^{1/n} - 1}$$

P 17.39 Two identical inverting amplifiers, each having dc voltage gain $A_{v0} = 100$, are connected in cascade. The op-amp parameters are $R_i = 3\,\mathrm{M\Omega}$, $R_o = 75\,\Omega$, $f_T = 1.5\,\mathrm{MHz}$, $\mu_0 = 2 \times 10^5$, and $SR = 500\,\mathrm{V\,ms^{-1}}$. The circuit parameters are $R_S = 50\,\Omega$, $R_1 = 1\,\mathrm{k\Omega}$, $R_2 = 100\,\mathrm{k\Omega}$, and $R_L = 5\,\mathrm{k\Omega}$.

(a) What are the half-power bandwidths of each amplifier and what is the half-power bandwidth W_2 of the cascade?
(b) What is the rms amplitude of the largest input having frequency W_2 for which the slew rate is adequate?
(c) What must the slew rate be if the input can be a sinusoid having rms amplitude 1 mV and any frequency up to $f_T = 1.5\,\mathrm{MHz}$?
(d) Use a simulation to check your answers to parts (a) and (b).

P 17.40 Two non-inverting amplifiers, each having dc voltage gain $A_{v0} = 100$ and bandwidth $W = 15\,\mathrm{kHz}$, are connected in cascade. What is the gain-bandwidth product of the cascade?

P 17.41 Two non-inverting amplifiers having bandwidths 10 and 15 kHz and the same dc voltage gain $A_{v0} = 100$, are connected in cascade. What is the gain-bandwidth product of the cascade?

P 17.42 Two identical inverting amplifiers, each having dc voltage gain 150 and bandwidth 20 kHz, are connected in cascade. The feedback resistance in one of the amplifiers is doubled. What is the new bandwidth of the cascade?

P 17.43 You might need a cut-and-try approach to solving parts of this problem. Refer to Fig. P 17.18, where $R_S = 100\,\Omega$. The op amps are identical, having input resistance $R_i = 1\,\mathrm{M\Omega}$, intrinsic dc voltage gain $\mu_0 = 2 \times 10^5$, unity-gain frequency $f_T = 4\,\mathrm{MHz}$, output resistance $R_o = 200\,\Omega$, slew rate $SR = 1\,\mathrm{MV\,s^{-1}}$, output voltage swing $\pm$ 20 V, and output current limit

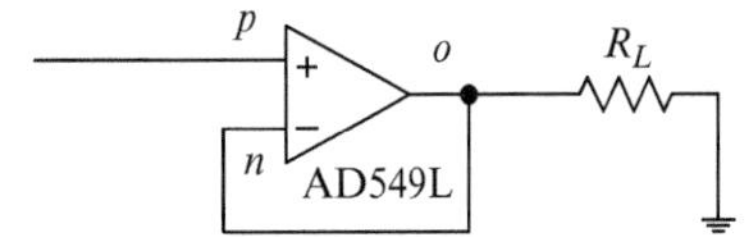

Fig. P 17.16 See Problem P 17.36

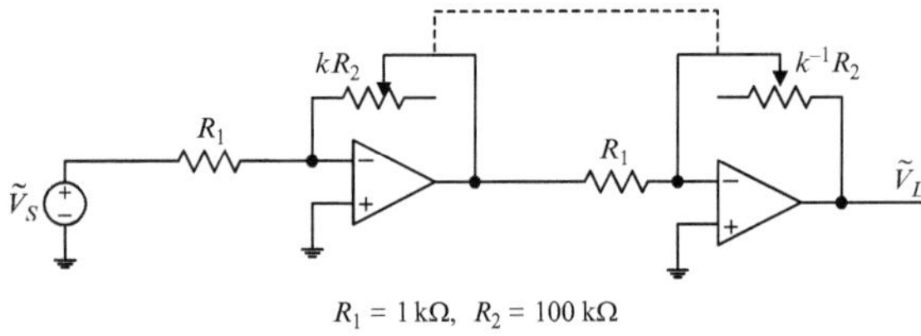

Fig. P 17.17 See Problem P 17.37

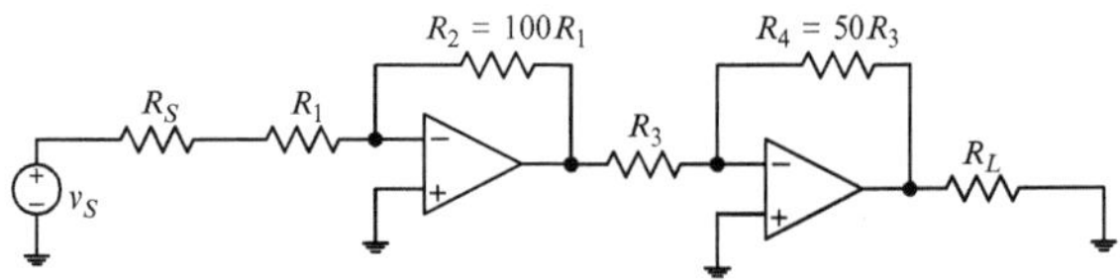

Fig. P 17.18 See Problem P 17.43

250 mA. The input bias currents and dc input voltage offset are negligible.

(a) If $R_3 = 10\,\mathrm{k}\Omega$, what is the smallest value of R_1 for which input loading reduces the dc voltage gain of the cascade by no more than 1%?
(b) If R_1 has the value found in part (a), what is the smallest value of R_3 for which interstage loading reduces the dc voltage gain of the cascade by no more than 1%?
(c) If R_1, R_3 have the values found in parts (a) and (b), what is the smallest value of R_L for which output loading reduces the dc voltage gain of the cascade by no more than 1%?
(d) If R_1, R_3, R_L have the values found in parts (a), (b) and (c), what is the largest dc input for which the amplifier is linear? What is the limiting parameter?
(e) If R_1, R_3, R_L have the values found in parts (a), (b) and (c), and if the input is a sinusoid whose frequency equals the half-power bandwidth of the cascade, what is the largest peak amplitude of the input for which the cascade is linear? You might need to refer to the solution to Problem P 17.39(a).

Sections 17.7 and 17.8 are prerequisite for the following problems.

P 17.44 A voltage to be amplified in a certain application consists of a desirable sinusoidal component whose frequency is confined to the band $30\,\mathrm{Hz} \leq f \leq 20\,\mathrm{kHz}$, plus an undesirable dc component. The rms amplitude of the sinusoidal component does not exceed 150 mV, and the source resistance is no larger than $50\,\Omega$. The voltage gain for the desirable component must be 100 or larger.

(a) Draw a circuit diagram showing the amplifier, source, and coupling capacitor.
(b) Specify the coupling capacitor and the resistances external to the op amp in an inverting amplifier for this application, including input-bias-current compensation.
(c) Specify the minimum acceptable unity-gain frequency, intrinsic dc voltage gain, output swing, and slew rate of an acceptable op amp. amplifier.
(d) Use simulation or analysis to check performance and adjust your specifications as needed.

P 17.45 Repeat Problem P 17.44, using a non-inverting amplifier.

P 17.46 Two identical cascaded inverting amplifiers are capacitively coupled, as shown in Fig. P 17.19, and driven by a sinusoidal source. The maximum passband voltage gain of the cascade is to be 10^4 and the half-power passband of the cascade is to extend from 100 Hz to 20 kHz. The source and load resistances are $R_S = 30\,\Omega$ and $R_L = 5\,\mathrm{k}\Omega$, respectively, and the rms amplitude of the available source voltage does not exceed 1.4 mV. Specify the circuit components and the relevant op-amp parameters. Use simulation to check the design.

P 17.47 Two cascaded non-inverting amplifiers are capacitively coupled, as shown in Fig. P 17.20, and driven by a sinusoidal source whose rms amplitude does not exceed 1.4 mV. The op amps are rail-to-rail, powered by a symmetric supply $\pm 25\,\mathrm{V}$, and characterized as follows:

Output swing: $\pm 22\,\mathrm{V}$
Unity-gain frequency: $f_T = 3\,\mathrm{MHz}$
Slew rate: $SR = 2\,\mathrm{V\,\mu s^{-1}}$
Intrinsic dc voltage gain: $\mu_0 = 2 \times 10^5$
Input resistance: $R_i = 2\,\mathrm{M}\Omega$
Input capacitance: $C_i = 1.5\,\mathrm{pF}$
Input impedance: $Z_i = R_i \| \left(\frac{1}{j\omega C_i}\right)$.
Output resistance: $R_o = 75\,\Omega$

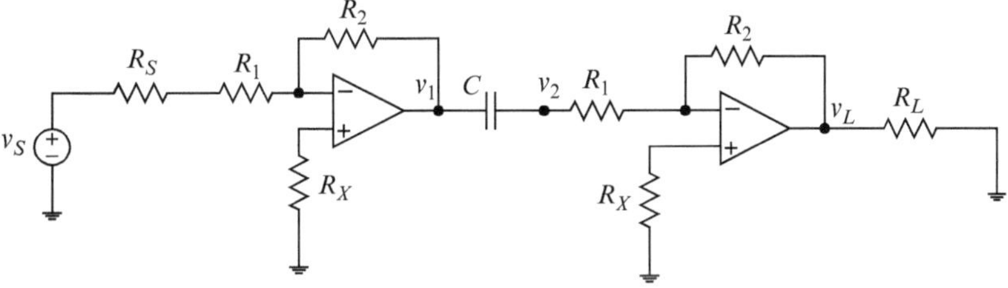

Fig. P 17.19 See Problem P 17.46

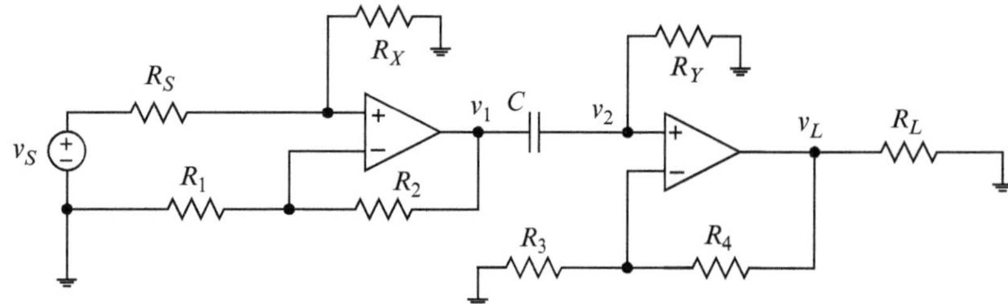

Fig. P 17.20 See Problem P 17.47

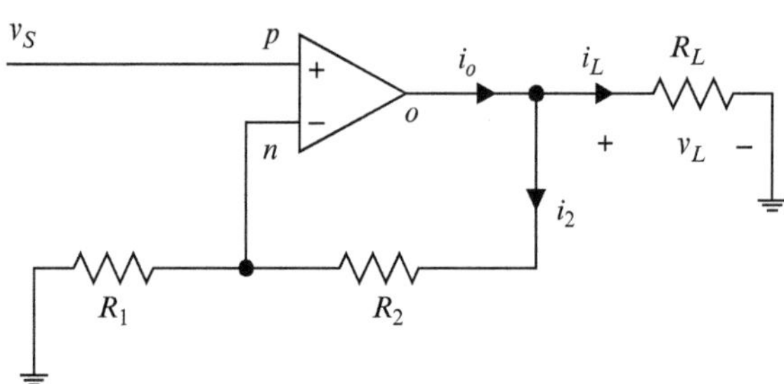

Fig. P 17.21 See Problem P 17.48

The passband of the cascade must extend down to 50 Hz. The source and load resistances are $R_S = 50\,\Omega$ and $R_L = 5\,\text{k}\Omega$, respectively.

(a) What is the maximum voltage gain that can be allowed if the passband must extend up to 25 kHz?
(b) What is the maximum bandwidth of the cascade if the voltage gain must be 10^4?
(c) What is the maximum allowable rms amplitude of the input if the cascade must have a gain of 10^4 in a passband extending from 50 Hz to 25 kHz?
(d) If $R_Y = 10\,\text{k}\Omega$, what is the highest frequency for which interstage loading reduces the voltage gain of the cascade by less than 1%?

Sections 17.9 and 17.10 are prerequisite for the following problems.

P 17.48 In Fig. P 17.21, $V_{CC} = 25\,\text{V}$, $R_L = 1\,\text{k}\Omega$, $R_1 = 1\,\text{k}\Omega$, and $R_2 = 5\,\text{k}\Omega$. Specify the power-dissipation rating for the op amp assuming (a) sinusoidal excitation and (b) dc excitation.

P 17.49 In Fig. P 17.19, the source and load resistances are $R_S = 20\,\Omega$ and $R_L = 1\,\text{k}\Omega$, respectively, and the coupling capacitor $C = 200\,\text{nF}$. The input is a 5 kHz sinusoid having peak amplitude 2 mV. The resistors in the amplifier circuits are $R_1 = R_X = 10\,\text{k}\Omega$, $R_2 = 1\,\text{M}\Omega$. The op amps are rail-to-rail, powered by a symmetric supply $\pm 25\,\text{V}$, and may be assumed to be ideal.

Calculate the total power delivered to the circuit, the power dissipated by the op amps and the power delivered to the load as functions of the rms amplitude and frequency of a sinusoidal input, assuming linear operation. Calculate the power-transfer efficiency of the circuit.

P 17.50 In Fig. P 17.20, the op-amp and circuit parameters are as given in Problem P 17.49. Calculate the total power delivered to the circuit, the power dissipated by the op amps and the power delivered to the load as functions of the rms amplitude and frequency of the input, assuming linear operation. Calculate the power-transfer efficiency of the circuit.

P 17.51 The output swing and slew rate of the op amp in a certain inverting amplifier are $\pm 16\,\text{V}$ and $500\,\text{kV s}^{-1}$, respectively. The amplifier is powered by a $\pm 20\,\text{V}$ supply. The op-amp short-circuit rms output current is 25 mA, which can be delivered continuously without damage, and the op amp can safely dissipate 100 mW in ambient air (in the open at room temperature). The op-amp unity-gain frequency is 1.5 MHz and the op-amp intrinsic dc voltage gain is 2×10^5. The dc voltage gain of the inverting amplifier is 100. The input is a 15-kHz sinusoidal voltage.

(a) The rms amplitude of the input is 18 mV. Draw a graph of the power delivered to a resistive load R_L versus the resistance the load, for $1\,\Omega \leq R_L \leq 10\,\text{k}\Omega$. Use logarithmic scales for both the load power and the load resistance.
(b) Draw a graph of the power delivered to a $500\,\Omega$ resistive load versus the rms amplitude of the input, for $100\,\mu\text{V} \leq V_{S\,rms} \leq 1\,\text{V}$. Use logarithmic scales for both the load power and the input voltage.
(c) What limits the output power in each case?

P 17.52 An inverting amplifier built around a rail-to-rail op amp is to drive a 1 kΩ resistive load. The amplifier must have an input resistance of 10 kΩ and a voltage gain of 100. The amplifier is to be powered by a ± 15 V supply.

(a) Specify the power-dissipation rating for the op amp.
(b) Find the maximum power required of the supply by the amplifier and load.
(c) Find the power-conversion efficiency of the amplifier for a 50 mV dc input.

P 17.53 Refer to Fig. P 17.22, where the op amps are powered by a symmetric supply ± 15 V,

$$R_1 = 1\,\mathrm{k\Omega},\ R_2 = 1\,\mathrm{M\Omega},\ R_S = 100\,\Omega,\ R_L = 5\,\mathrm{k\Omega}$$

and

$$v_S = V_S \cos(2\pi f_S\, t);\ V_S = 1\,\mathrm{mV},\ f_S = 5\,\mathrm{kHz}.$$

(a) Find the average power dissipated by each circuit component. Assume the op amps are ideal.
(b) Find the power-conversion efficiency of the circuit.

P 17.54 Repeat Problem P 17.53 for the circuit in Fig. P 17.23.

Section 17.11 is prerequisite for the following problems.

Some problems in this section refer to Table 17.3 in Section 17.11.

P 17.55 A non-inverting audio amplifier must deliver 50 W to an 8 Ω load. The bandwidth must be at least 20 kHz and the dc gain must be at least 60 dB. Assume a sinusoidal input and specify the power dissipation, output current, supply voltage, slew rate, and gain-bandwidth product for the op amp. If possible, select an op amp from those in Table 17.3 that would be adequate for this application.

P 17.56 An audio-frequency voltage $v(t)$ has bandwidth $W = 20\,\mathrm{kHz}$, rms amplitude $v_{rms} = 150\,\mathrm{mV}$, and zero average value (no dc component). The voltage is to be amplified using an inverting amplifier powered by a symmetric supply $\pm V_{CC} = \pm 30\mathrm{V}$. Assume the op amp used is rail-to-rail.

(a) What is the maximum allowable gain if the output swing is assumed to be ± three times the rms amplitude of the output?
(b) If that gain is achieved, what is the minimum acceptable gain-bandwidth product for the op amp?
(c) If the source resistance is $R_S = 20\,\Omega$, what values would you specify for R_1, R_2, and the bias-current compensating resistor R_X?
(d) What slew rate is required?

P 17.57 Specify the parameters of the op amp and external resistances (including tolerances on the latter) for an inverting audio amplifier meeting the following specifications:

Input resistance of 10kΩ or larger, dc voltage gain of 50 ± 1dB, 3-dB bandwidth W greater than 10 kHz, load resistance of 10 kΩ, and output swing of ± 12 V. The available supply voltage is ± 15 V. The expected source and load resistances are $R_S \cong 50\,\Omega$ and $R_L = 10\,\mathrm{k\Omega}$.

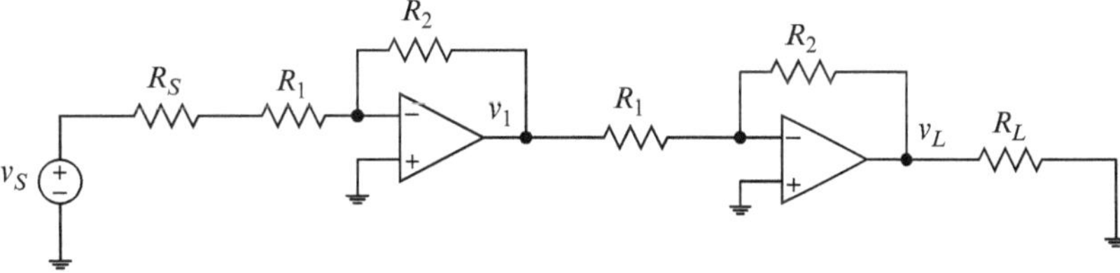

Fig. P 17.22 See Problem P 17.53

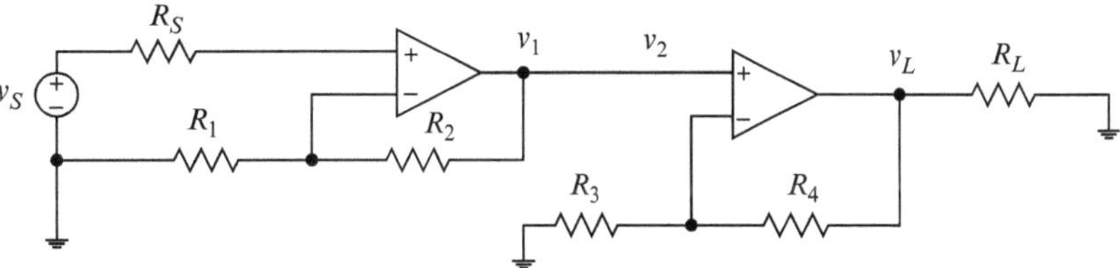

Fig. P 17.23 See Problem P 17.54

The slew rate must be adequate for a sinusoidal input having frequency W and rms amplitude 20 mV.

Simulate the amplifier to verify that the specifications are met, even under worst-case variations of the external resistors.

P 17.58 Determine which (if any) of the op amps in Table 17.3 could be used in an inverting amplifier subject to the following specifications:

Dc gain = 20
Bandwidth = 50 kHz
Output swing = ± 25 V.

P 17.59 Design a direct-coupled inverting *audio* amplifier subject to the following specifications and constraints:

Dc voltage gain $= 100 \pm 10\%$
Input resistance $> 5\,\mathrm{k\Omega}$
Bandwidth $\geq 20\,\mathrm{kHz}$
Available supply $= \pm 15\,\mathrm{V}$
Load resistance $\geq 15\,\mathrm{k\Omega}$
Source output resistance $\leq 10\,\Omega$
Input amplitude $V_{S\,rms} \leq 120\,\mathrm{mV}$.

Which, if any, of the op amps in Table 17.3 could be used in this application?

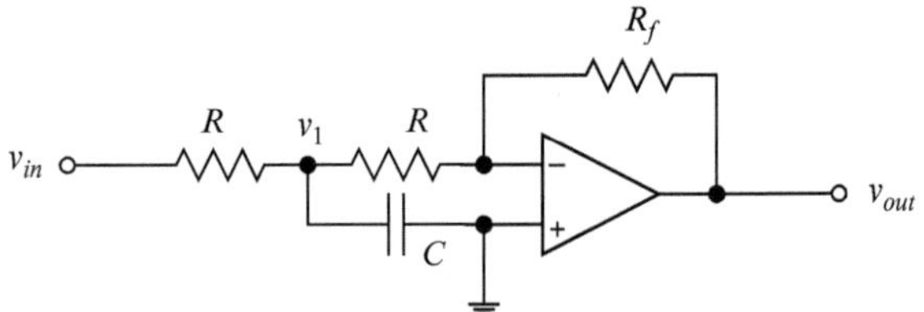

Fig. P 17.24 See Problem P 17.60

P 17.60 Refer to Fig. P 17.24. The specified dc input resistance is $R_{in} = 10\,\mathrm{k\Omega}$, the supply voltage is $\pm V_{CC} = \pm 15\,\mathrm{V}$, and the maximum amplitude of the input is 50 mV. The amplitude of the output is to be no larger than 70.7% of the amplitude of the input for any sinusoidal input having frequency greater than $f_1 = 25\,\mathrm{kHz}$. Specify R, R_f, and C.

Chapter 18
Laplace Transformation and *s*-Domain Circuit Analysis

In the present context, a **transformation** establishes a one-to-one relation between two sets of objects. Such a relation exists between sinusoidal functions of time and the corresponding phasor representations, provided a specification of the frequency of the sinusoid accompanies the phasor representation (to make the relation one-to-one). We indicate such a relation using a double-headed arrow; e.g.,

$$V\cos(\omega_0 t+\theta) \leftrightarrow V\angle\theta,\ \omega=\omega_0.$$

A transformation establishes not only a relation between two sets of items, but also a relation between operations on one set and corresponding operations on the other set. For example, transforming sinusoids to phasors also transforms differentiation with respect to time to multiplication by $j\omega$; that is,

$$\frac{d}{dt}[V\cos(\omega_0 t+\theta)] \leftrightarrow j\omega V\angle\theta,\ \omega=\omega_0.$$

We may view a transformation as a relation between two **domains**. For example, sinusoidal functions of time reside in the **time domain** and associated phasors reside in the **frequency domain**. Representing currents and voltages as functions of time, expressing Kirchhoff's laws as differential equations, and solving such equations by classical (or numerical) methods is called **time-domain analysis**. Representing sinusoidal currents and voltages as phasors, expressing Kirchhoff's laws as algebraic equations, and solving such equations is called **frequency-domain analysis**. The Laplace transformation transforms currents and voltages and operations on currents and voltages to the **complex-frequency domain**. In the time domain, the independent variable is time, denoted by t. In the frequency domain, the independent variable is frequency, denoted by f, or angular frequency, denoted by ω. In the complex-frequency domain, the independent variable is called **complex frequency**, and is denoted by s. Henceforth, for economy, we often refer to the complex-frequency domain as the ***s* domain**.

This chapter introduces circuit analysis in the s domain. We begin by defining the Laplace transformation, which connects the time domain to the s domain.

18.1 Definition of the Laplace Transformation

As used in circuit analysis, the Laplace transformation is defined by the (forward) **Laplace transformation**

$$F(s)=\int_{-\infty}^{\infty} f(t)\, e^{\ st}\, dt \tag{18.1}$$

and the **inverse Laplace transformation**[1]

$$f(t)=\frac{1}{j2\pi}\int_{c-j\infty}^{c+j\infty} F(s)\, e^{st}\, ds,\quad t>0, \tag{18.2}$$

where t is time, s is a complex variable called **complex frequency**, and c is a real constant that is independent of s. In all applications involving realistic circuit and

[1]The integral in (18.2) is an example of a *line integral*, so called because the path of integration is along a line defined by $s=c$ that is not necessarily coincident with one of the axes. You need not be alarmed by this integral because we never need to use it.

T.H. Glisson, *Introduction to Circuit Analysis and Design*,
DOI 10.1007/978-90-481-9443-8_18,

signal models,[2] (18.1) and (18.2) exist and comprise a one-to-one transformation.

From (18.1), *the dimension of a Laplace transform $F(s)$ is that of the associated function $f(t)$ multiplied by time*; for example, if $f(t)$ is a voltage the SI unit of $F(s)$ is volt-seconds (V s).

Equation (18.1) defines the transformation of a function $f(t)$ from the **time domain** to another function $F(s)$ in the **s domain**. The function $F(s)$ given by (18.1) is called the **Laplace transform** of the function $f(t)$. The function $f(t)$ given by (18.2) is called the **inverse Laplace transform** of the function $F(s)$ and the integral in (18.2) is called the *inversion integral*. Together, the functions $f(t)$ and $F(s)$ in (18.1) and (18.2) comprise a **Laplace transform pair**. This relationship is denoted by

$$f(t) \leftrightarrow F(s). \tag{18.3}$$

This notation is analogous to denoting the relationship between a sinusoidal voltage and the phasor representation of the voltage by

$$V\cos(\omega t + \theta) \leftrightarrow V\angle\theta.$$

The unit of complex frequency is s^{-1}. Here again is an opportunity for confusing a variable (italic s) and a unit (roman s), so in written work you should underline the unit s. To avoid such conflicts, some disciplines use p for complex frequency, but unfortunately, s has won out in electrical engineering.

The real and imaginary parts of complex frequency are denoted by σ and ω, respectively. Thus

$$s = \sigma + j\omega. \tag{18.4}$$

The meaning of frequency f and angular frequency $\omega = 2\pi f$ is clear. They are the frequency and angular frequency of a sinusoidal current or voltage. The meaning of complex frequency s might be made a little more clear by the following heuristic interpretation:

Recall that the complex representation of a sinusoidal voltage having angular frequency ω is $\tilde{v}(t) = \tilde{V}e^{j\omega t}$, where $\tilde{V}$ is the phasor for the voltage and ω is angular frequency. If we apply (mathematically) such a voltage to a capacitor, then the complex representation of the current entering the positive terminal of the capacitor is given by

$$\tilde{i}(t) = C\frac{d\tilde{v}(t)}{dt} = j\omega C\tilde{V}e^{j\omega t}.$$

and we are led by analogy with Ohm's law to define the impedance is of a capacitor as

$$Z_C = \frac{\tilde{v}(t)}{\tilde{i}(t)} = \frac{1}{j\omega C}.$$

We may generalize, and define (by analogy) the complex representation of an exponentially damped sinusoidal voltage as $\hat{v}(t) = \hat{V}e^{st}$, where $s = \sigma + j\omega$ is complex frequency. If we apply (mathematically) such a voltage to a capacitor, then the complex representation of the current entering the positive terminal of the capacitor is given by

$$\hat{i}(t) = C\frac{d\hat{v}(t)}{dt} = sC\hat{V}e^{st} = sC\hat{v}(t).$$

and we are led by analogy with Ohm's law to define the *generalized impedance* is of a capacitor as

$$Z_C = \frac{\hat{v}(t)}{\hat{i}(t)} = \frac{1}{sC}.$$

A real frequency ω refers to the angular frequency of the complex representation of a constant-amplitude sinusoid. A complex frequency $s = \sigma + j\omega$ refers to both the damping of the exponential factor and the angular frequency of the sinusoidal factor in the complex representation of an exponentially damped sinusoid.

The Laplace transform of a current or voltage is a **non-physical** complex function of a complex variable. The Laplace transform of a current (or voltage) is **not** a current (or voltage) and indeed does not even have the unit of a current (or voltage). However, *Laplace transforms of currents and voltages in a circuit satisfy Kirchhoff's laws* just as phasors do, and in that context it is conventional and economical to refer to Laplace transforms of currents and voltages as currents and voltages, just as it is conventional to refer to phasor

[2]In this book, *signal* means *current or voltage*.

currents and voltages as currents and voltages. For economy, we follow that convention; for example, we might say that the sum of the currents leaving a node equals zero, when we actually mean the sum of the Laplace transforms of the currents equals zero. *But in this book, where you are asked to find or obtain an expression for a particular current or voltage, you should present a real function of time, not a Laplace transform (or phasor).*

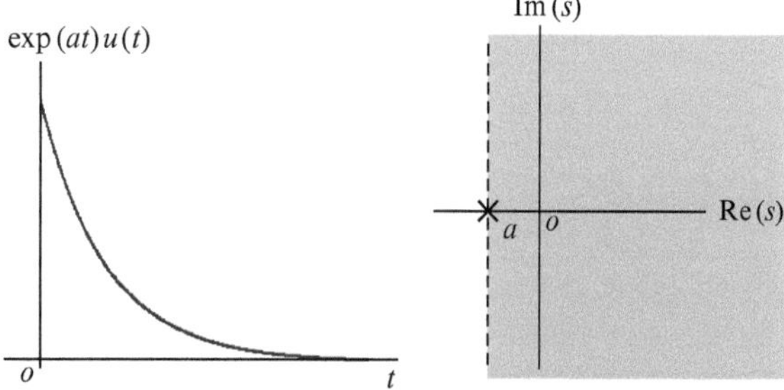

Fig. 18.1 Graphs of the function $f_1(t) = e^{at}u(t)$ and the associated region of convergence for the Laplace transform $F_1(s)$. See (18.7)

18.2 Convergence and Uniqueness

To avoid too much prestidigitation in subsequent developments, we discuss conditions under which the right side of (18.1) converges (exists) and conditions under which associated functions $f(t)$ and $F(s)$ are in one-to-one correspondence and comprise a unique transform pair.

Consider the function

$$f_1(t) = e^{at}u(t), \tag{18.5}$$

where $a < 0$. Equation (18.1) gives

$$\begin{aligned} F_1(s) &= \int_{-\infty}^{0} 0\, e^{-st}\,dt + \int_{0}^{\infty} e^{at}\,e^{-st}\,dt = \int_{0}^{\infty} e^{-(s-a)t}\,dt \\ &= -\left.\frac{e^{-(s-a)t}}{(s-a)}\right|_{t=0}^{t\to\infty} = -\lim_{t\to\infty}\left[\frac{e^{-(s-a)t}}{(s-a)}\right] + \frac{1}{(s-a)}. \end{aligned} \tag{18.6}$$

The limit exists only if $\mathrm{Re}(s-a) > 0$; i.e., only if $\mathrm{Re}(s) > a$, in which case the limit vanishes and

$$F_1(s) = \frac{1}{s-a};\ \mathrm{Re}(s) > a. \tag{18.7}$$

For $f_1(t)$ defined by (18.5), the integral on the right of (18.1) converges (exists) only for values of s whose real parts exceed a. The region defined by $\mathrm{Re}(s) > a$ is called the **region of convergence** for the Laplace transform of the function defined by (18.5). See Fig. 18.1.

Now consider the function

$$f_2(t) = -e^{at}u(-t). \tag{18.8}$$

where $a > 0$. For this function, (18.1) yields

$$\begin{aligned} F_2(s) &= -\int_{-\infty}^{0} e^{at}\,e^{-st}\,dt + \int_{0}^{\infty} 0e^{-st}\,dt \\ &= -\int_{-\infty}^{0} e^{-(s-a)t}\,dt = \left.\frac{e^{-(s-a)t}}{(s-a)}\right|_{t\to-\infty}^{t=0} \\ &= \frac{1}{(s-a)} - \lim_{t\to-\infty}\left[\frac{e^{-(s-a)t}}{(s-a)}\right] \\ &= \frac{1}{(s-a)} - \lim_{t\to\infty}\left[\frac{e^{(s-a)t}}{(s-a)}\right]. \end{aligned} \tag{18.9}$$

The limit vanishes if $\mathrm{Re}(s-a) < 0$ or $\mathrm{Re}(s) < a$. Thus the Laplace transform and the associated region of convergence for the function defined by (18.8) are

$$F_2(s) = \frac{1}{s-a},\ \mathrm{Re}(s) < a. \tag{18.10}$$

See Fig. 18.2.

For $F(s) = (s-a)^{-1}$, the inversion integral (18.2) yields the function $f_1(t) = e^{at}u(t)$ if the path of integration lies to the right of the line $s = a$ (if $c > a$) and yields the function $f_2(t) = -e^{at}u(-t)$ if the path of integration lies to the left of $s = a$ (if $c < a$).[3] In other words, the Laplace transform $F(s) = (s-a)^{-1}$ is the transform of two different functions, which one depending upon which of $\mathrm{Re}(s) > a$ and $\mathrm{Re}(s) < a$ is taken as the associated region of convergence.

For present purposes, an important point of the discussion above is that $F(s) = (s-a)^{-1}$ is the

[3] You will see this proved if you take a course in complex variables.

Laplace transform of either $f_1(t) = e^{at}u(t)$, which is non-zero only for $t > 0$ and is said to be *right-sided*, or of $f_2(t) = -e^{at}u(-t)$, which is non-zero only for $t < 0$ and is said to be *left-sided*. Keep this point in mind as you read on.

A function of time can be right-sided, left-sided, or two sided. A function of time is **right-sided** if we can choose a time origin $t = 0$ such that the function equals zero for all times less than zero. A function of time is **left-sided** if we can choose a time origin such that the function equals zero for all times greater than zero. A function of time is **two-sided** if it is neither left-sided nor right sided; that is, if the function has non-zero values both to the left and to the right of *any* particular time. A function of time that is non-zero over only some finite interval is both left-sided and right sided. We consider such functions to be right-sided. Figure 18.3 illustrates these definitions.

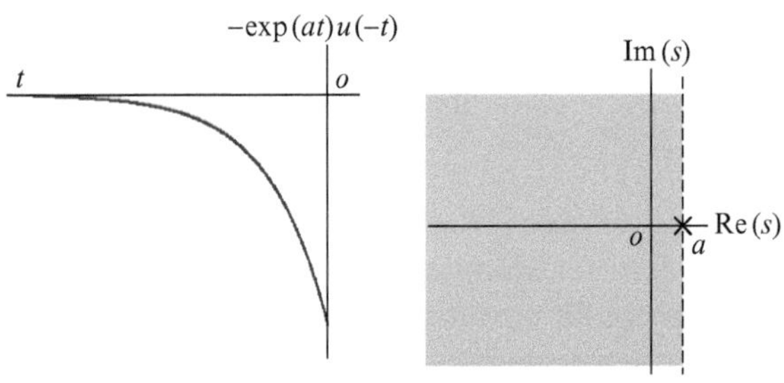

Fig. 18.2 Graphs of the function $f_2(t) = -e^{at}u(-t)$ and the associated region of convergence for the Laplace transform $F_2(s)$. See (18.10)

In general, the inversion integral (18.2) can return a right-sided function, a left-sided function, or a two-sided function, depending upon the function $F(s)$ and the path of integration (the assumed region of convergence). However, *any particular function $F(s)$ will yield at most one right-sided function of time*. If a function of time $f(t)$ is known to be right-sided, (18.1) alone establishes a one-to-one correspondence between the function $f(t)$ and the associated Laplace transform $F(s)$. If we limit application of the Laplace transformation to only right-sided functions, we are assured that a function $F(s)$ obtained using (18.1) is uniquely associated with the right-sided function $f(t)$ in the integrand. Thus, if we limit application of the Laplace transformation to only right-sided functions, we may use only (18.1) to construct a table of transforms. We do not need to use (18.2) and we need not be concerned about regions of convergence.

Exercise 18.1 Obtain the Laplace transform and associated region of convergence for the function $f(t) = \exp(-a|t|)$, where $a > 0$. Is this function right-sided, left-sided, or two-sided?

Exercise 18.2 Classify each of the following functions as right-sided, left-sided, or two-sided. (a) $\cos(2\pi f t)$, (b) $\cos(2\pi f t)u(t - t_0)$, (c) $u(t - t_1) - u(t - t_2)$, (d) $\exp(t/\tau)u(-t)$.

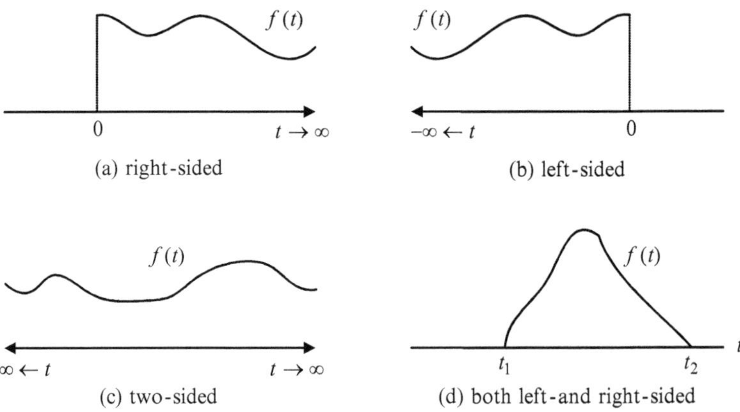

Fig. 18.3 Illustrations of the definitions of right-sided, left-sided, and two sided functions of time

18.3 One-Sided Laplace Transforms

We limit our attention to right-sided functions – functions that equal zero for $t < 0$.[4] Consequently, (18.1) becomes

$$F(s) = \int_{0^-}^{\infty} f(t)\, e^{-st}\, dt. \qquad (18.11)$$

The function $F(s)$ is the **one-sided Laplace transform** of the function $f(t)$. Recall from calculus that a lower limit of integration is not included in the range of integration. For example, in

$$\int_a^b f(t)dt$$

the range of integration $a < t \leq b$ excludes the point $t = a$. For reasons given in the sequel, we wish to include $t = 0$ in the defining integral for the one-sided Laplace transform, which is why the lower limit in (18.11) is $t = 0^-$ and not $t = 0$.

Henceforth, we use only the one-sided Laplace transform and Laplace transform means ***one-sided*** *Laplace transform.*

If a function $f(t)$ is to have a Laplace transform, the integral in (18.11) must exist for at least some values of s. Existence of the Laplace transform of a function $f(t)$ is guaranteed if there are at least some finite values of s for which[5]

$$\lim_{t \to \infty} \left[f(t)\, e^{-st}\right] = 0 \qquad (18.12)$$

such a function is said to be of **exponential order**. All functions of time that represent physical quantities (such as current and voltage) are of exponential order. For example, the function

$$f(t) = e^{at} u(t)$$

is of exponential order, because

$$\lim_{t \to \infty} \left[e^{at}\, e^{-st}\right] = \lim_{t \to \infty} \left[e^{-(s-a)\,t}\right] = 0 \text{ for } \mathrm{Re}(s-a) > 0$$

For any finite value of a, there are finite values of s for which $\mathrm{Re}(s - a) > 0$. On the other hand, the function

$$f(t) = e^{(at)^2} u(t)$$

is not of exponential order because there are no values of s for which (18.12) holds.

Exercise 18.3 Which of the following functions are of exponential order?
(a) $e^{at}\cos(\omega\, t)u(t),\ a > 0$; (b) $t^n e^{at} u(t)$, $a < 0,\ n > 0$; (c) $\cos(\omega_0 t)$.

18.4 Shorthand Notation

The operation defined by (18.11) is denoted by the symbol $\mathcal{L}$. With this shorthand, we may write (18.11) as

$$F(s) = \mathcal{L}\{f(t)\} \qquad (18.13)$$

The right side of (18.13) is an abbreviation for the right side of (18.11), just as $\ln(x)$ is an abbreviation for the infinite series that defines the natural logarithm. The right side of (18.13) means *multiply the function $f(t)$ by the exponential e^{-st} and integrate the product from $t = 0^-$ to $t \to \infty$.*

Similarly, the inverse Laplace transform of a function $F(s)$ is denoted

$$f(t) = \mathcal{L}^{-1}\{F(s)\} \qquad (18.14)$$

For our purposes, this notation is simply convenient shorthand that simplifies discussion and use of Laplace transforms.

In the next section, we digress a bit to introduce a *generalized function* that is useful in subsequent developments.

18.5 The Delta Function (Unit Impulse)

In many physical circuits, currents and voltages frequently change abruptly from one value to another. In a physical circuit, such changes cannot be instantaneous

[4]In physical problems, we are free to choose a time origin. We could generalize by treating right-sided functions that equal zero for $t < t_0$, but that would clutter the development without providing any useful increase in generality.

[5]This requirement is somewhat stronger than necessary, but is easily tested and completely satisfactory in all practical applications.

because of capacitance and inductance, which are present to some degree in every physical circuit. Nonetheless, it often is useful to use a discontinuous mathematical model such as a step function for an applied current or voltage. Because analysis sometimes entails differentiation of such models, we need a formal definition of the derivative of a function at a jump discontinuity. What follows is a heuristic justification for such a definition.[6]

Assume that the derivative of a unit step function $u(t)$ exists and let $\delta(t)$ denote the derivative. Thus

$$\delta(t) = \frac{du(t)}{dt}. \tag{18.15}$$

The entity $\delta(t)$ defined by (18.15) is called the **Dirac delta**,[7] the **delta function**, or, if it represents a current or voltage, the **unit impulse**. Below, we examine the properties of the delta function.

The derivative of $u(t)$ must be zero for $t < 0$ and for $t > 0$, because $u(t)$ is independent of time in both of those intervals; i.e.,

$$\delta(t) = 0;\ t \neq 0. \tag{18.16}$$

The converse of (18.15) is

$$\int_{-\infty}^{t} \delta(t')dt' = u(t), \tag{18.17}$$

which, in view of (18.16) implies

$$\int_{0^-}^{0^+} \delta(t)dt = 1, \tag{18.18}$$

so the area bounded by $\delta(t)$ and the t axis equals unity. Finally, the dimension of the **delta function** $\delta(t)$ is *time*$^{-1}$.[8] Thus

$$\mathrm{SI}[\delta(t)] = \mathrm{SI}\left[t^{-1}\right] = \mathrm{s}^{-1}, \tag{18.19}$$

which is implied by (18.15), (18.17), and (18.18). The delta function is unusual not only because it takes a dimensioned argument (time) but also because $\delta(t)$ itself is dimensioned. Keep (18.19) in mind when checking expressions for dimensional consistency.

Equations (18.16), (18.18), and (18.19) express the three essential properties of the delta function $\delta(t)$ defined by (18.15).

Ordinarily, where a function of time is multiplied by a constant, the constant affects the *amplitude* of the function; for example, $a\,u(t)$ is a step having amplitude a. But from (18.18),

$$\int_{0^-}^{0^+} a\delta(t)dt = a\int_{0^-}^{0^+} \delta(t)dt = a,$$

so the constant a in $a\,\delta(t)$ is the *area* bounded by the function $a\delta(t)$ and the time axis. The amplitude of $\delta(t)$ is undefined (remains infinite) at $t = 0$. To emphasize this distinction, we call a the **strength** (not the amplitude) of $a\delta(t)$.

Intuitively, you may imagine that $\delta(t)$, being the slope of a unit step function, equals zero everywhere except at $t = 0$, where $\delta(t)$ is exceedingly large, such that its area is unity. It is usually harmless and sometimes helpful to think of $\delta(t)$ that way. But *the only physically meaningful properties of a unit delta function are those given by or derived from* (18.16) *and* (18.18).

The property (18.16) implies that

$$f(t)\delta(t) = f(0)\delta(t), \tag{18.20}$$

provided $f(t)$ is defined at $t = 0$. Also from (18.16),

$$\delta(t - t_0) = 0,\ t \neq t_0. \tag{18.21}$$

It follows from (18.20) and (18.21) that

$$f(t)\delta(t - t_0) = f(t_0)\delta(t - t_0), \tag{18.22}$$

provided $f(t)$ is defined at $t = t_0$. If $f(t)$ is undefined at $t = t_0$, then the left sides of (18.22) and (18.23) are undefined.[9]

[6] Please accept on faith that the result is consistent with that of a rigorous development, such as provided by the theory of generalized functions (or distribution theory).

[7] After Paul Dirac (1902–1984), a British physicist credited with introducing the delta function.

[8] In general, the dimension of a delta function is the reciprocal of the dimension of the argument.

[9] The left sides of the referenced relations can be defined in a self-consistent manner, but we do not need those more general relations in this book.

Exercise 18.4 Show that $\cos(\omega_0 t)\delta(t) = \delta(t)$ and that $\sin(\omega_0 t)\delta(t) = 0$.

Exercise 18.5 Show that

$$\int_{-\infty}^{\infty} f(t)\delta(t-t_0)dt = f(t_0) \qquad (18.23)$$

provided $f(t)$ is defined at $t = t_0$.

From (18.23), the Laplace transform of the delta function is

$$\mathcal{L}\{\delta(t)\} = \int_{0^-}^{\infty} \delta(t)e^{-st}dt = e^0 \int_{0^-}^{\infty} \delta(t)dt = e^0 = 1.$$

Thus we have the important transform pair

$$\delta(t) \leftrightarrow 1 \qquad (18.24)$$

which we use in the next section.

18.6 Tables of Operational Properties and Transform Pairs

Operational properties of the Laplace transformation describe how certain operations on a function of time are transformed to other operations on the Laplace transform of the function. In other words, how operations in the time domain are transformed to other operations in the s domain, and vice-versa. Table 18.1 gives the operational properties of the Laplace transformation that are used in this book.

Table 18.2 gives a few useful Laplace-transform pairs which are also the only transform pairs used in this book. In examples that follow, we derive a few of the entries in each table and call for derivations of others in end-of-chapter problems. We note that all of the entries in Table 18.2 can be derived from the first entry in Table 18.2 and the operational properties of the Laplace transformation given in Table 18.1.

Example 18.1 We can derive the differentiation property in Table 18.1 as follows: By definition,

$$\mathcal{L}\left\{\frac{df(t)}{dt}\right\} = \int_{0^-}^{\infty} \frac{df(t)}{dt} e^{-st}dt = \int_{t=0^-}^{\infty} e^{-st}df(t).$$

Integration by parts yields

$$\int_{t=0^-}^{\infty} e^{-st}df(t) = e^{-st}f(t)\Big|_{t=0^-}^{t\to\infty} + s\int_{t=0^-}^{\infty} f(t)e^{-st}dt,$$

so

$$\mathcal{L}\left\{\frac{df(t)}{dt}\right\} = \lim_{t\to\infty} e^{-st}f(t) - f(0^-) + s\int_{t=0^-}^{\infty} f(t)e^{-st}d(t).$$

There are values of s for which the first term on the right vanishes. The integral in the last term is the Laplace transform of $f(t)$. Thus

$$\mathcal{L}\left\{\frac{df(t)}{dt}\right\} = sF(s) - f(0^-). \qquad (18.25)$$

Table 18.1 Selected operational properties of the Laplace transformation

Number	Name	$f(t) = \mathcal{L}^{-1}\{F(s)\}$	$F(s) = \mathcal{L}\{f(t)\}$
1	Linearity	$f_1(t) + f_2(t) + \cdots$	$F_1(s) + F_2(s) + \cdots$
2	Frequency translation	$e^{-at}f(t)$	$F(s+a)$
3	Time translation	$f(t-t_0),\ t_0 \geq 0$	$e^{-st_0}F(s)$
4	Differentiation	$\frac{df(t)}{dt}$	$sF(s) - f(0^-)$
5		$\frac{d^2f(t)}{dt^2}$	$s^2F(s) - sf(0^-) - \left.\frac{df(t)}{dt}\right\vert_{t=0^-}$
6		$\frac{d^nf(t)}{dt^n}$	$s^nF(s) - \sum_{k=1}^{n} s^{n-k} \left.\frac{d^{n-k}f(t)}{dt^{n-k}}\right\vert_{t=0^-}$
7	Integration	$\int_{0^-}^{t} f(t')\,dt'$	$\frac{1}{s}F(s)$
8	Multiplication by time	$t^nf(t)$	$(-1)^n\frac{d^nF(s)}{ds^n}$

Table 18.2 Selected Laplace-transform pairs[10]

Number	$f(t)$	$F(s)$
1	$\delta(t)$	1
2	$u(t)$	$\frac{1}{s}$
3	$e^{-at}u(t)$	$\frac{1}{s+a}$
4	$(1-e^{-at})u(t)$	$\frac{1}{s}-\frac{1}{s+a}=\frac{a}{s(s+a)}$
5	$t^n e^{-at}u(t)$	$\frac{n!}{(s+a)^{n+1}}$
6	$cos(\omega_0 t)\,u(t)$	$\frac{s}{s^2+\omega_0{}^2}$
7	$sin(\omega_0 t)\,u(t)$	$\frac{\omega_0}{s^2+\omega_0{}^2}$
8	$e^{-at}\cos(\omega_0 t)\,u(t)$	$\frac{s+a}{(s+a)^2+\omega_0{}^2}$
9	$e^{-at}\sin(\omega_0 t)\,u(t)$	$\frac{\omega_0}{(s+a)^2+\omega_0{}^2}$
10	$e^{-at}\cos(\omega_0 t+\theta)\,u(t)$	$\frac{(s+a)\cos(\theta)-\omega_0\sin(\theta)}{(s+a)^2+\omega_0{}^2}$
11[11]	$2\lvert A\rvert e^{-t/\tau}\cos(\omega_0 t+\measuredangle A)u(t)$	$\frac{A}{s-p}+\frac{A^*}{s-p^*}$, $p=-\frac{1}{\tau}+j\omega_0$
12	$Ce^{-qt/2}\cos(\omega_0 t+\theta)u(t)$ $\omega_0=\sqrt{r-(q/2)^2}$ $C=\omega_0^{-1}\sqrt{rA^2+B^2-qAB}$ $\theta=\mathrm{Tan}^{-1}\left[A\omega_0,\frac{Aq}{2}-B\right]$	$\frac{As+B}{s^2+qs+r}$
13	$\omega_0\,\beta^{-1}e^{-\alpha\omega_0 t}\sin(\beta\,\omega_0\,t)u(t)$ $\beta=\sqrt{1-\alpha^2}$	$\frac{\omega_0{}^2}{s^2+2\alpha\,\omega_0\,s+\omega_0{}^2}$

Example 18.2 We can use the transform of the unit delta and the differentiation property to obtain the transform of the unit step. By definition,

$$\delta(t)=\frac{du(t)}{dt}.$$

From the differentiation property (18.25)

$$\mathcal{L}\{\delta(t)\}=\mathcal{L}\left\{\frac{du(t)}{dt}\right\}=s\mathcal{L}\{u(t)\}-u(0^-)$$
$$=s\mathcal{L}\{u(t)\}.$$

Thus,

$$\mathcal{L}\{u(t)\}=\frac{1}{s}\mathcal{L}\{\delta(t)\}=\frac{1}{s}.$$

Exercise 18.6 Use the integration property in Table 18.1 to show that

$$\mathcal{L}\{u(t)\}=\frac{1}{s}.$$

Example 18.3 The frequency translation property in Table 18.1 is obtained as follows:

$$\begin{aligned}\mathcal{L}\{e^{-at}f(t)\}&=\int_{0^-}^{\infty}e^{-at}f(t)\,e^{-st}\,dt\\&=\int_{0^-}^{\infty}f(t)\,e^{-(s+a)\,t}\,dt\\&=\mathcal{L}\{f(t)\}|_{s\to s+a}\\&=F(s+a)\end{aligned}\qquad(18.26)$$

Multiplying a function by an exponential in the time domain is equivalent to translating the Laplace transform of the function in the s domain.

Example 18.4 Find the Laplace transform of $e^{-at}u(t)$.

Solution: From the frequency-translation property,

$$\mathcal{L}\{e^{-at}u(t)\}=\mathcal{L}\{u(t)\}|_{s\to s+a}=\frac{1}{s+a}\qquad(18.27)$$

Example 18.5 Show that

$$\mathcal{L}\{\cos(\omega t)\,u(t)\}=\frac{s}{s^2+\omega^2}.\qquad(18.28)$$

[10]Entries 11–13 are useful mainly for obtaining inverse Laplace transforms.

[11]Note that A is associated with p, which has a positive imaginary part and A^* is associated with p^*.

Solution: We first use Euler's identity to express the sinusoidal function as:

$$\cos(\omega t)\,u(t) = \frac{1}{2}\,e^{j\omega t}\,u(t) + \frac{1}{2}\,e^{-j\omega t}\,u(t).$$

From the linearity property of the Laplace transformation,

$$\begin{aligned}&\mathcal{L}\{\cos(\omega t)\,u(t)\}\\ &= \mathcal{L}\left\{\frac{1}{2}\,e^{j\omega t}\,u(t) + \frac{1}{2}\,e^{-j\omega t}\,u(t)\right\}\\ &= \frac{1}{2}\left[\mathcal{L}\{e^{j\omega t}\,u(t)\} + \mathcal{L}\{e^{-j\omega t}\,u(t)\}\right]. \quad (18.29)\end{aligned}$$

From the frequency translation property,

$$\begin{aligned}&\frac{1}{2}\left[\mathcal{L}\{e^{j\omega t}\,u(t)\} + \mathcal{L}\{e^{-j\omega t}\,u(t)\}\right]\\ &= \frac{1}{2}\left[\mathcal{L}\{u(t)\}|_{s\to s-j\omega} + \mathcal{L}\{u(t)\}|_{s\to s+j\omega}\right]\\ &= \frac{1}{2}\left(\frac{1}{s}\bigg|_{s\to s-j\omega} + \frac{1}{s}\bigg|_{s\to s+j\omega}\right)\\ &= \frac{1}{2}\left(\frac{1}{s-j\omega} + \frac{1}{s+j\omega}\right). \quad (18.30)\end{aligned}$$

Equation (18.28) follows.

Exercise 18.7 Show that

$$\mathcal{L}\{\sin(\omega t)\,u(t)\} = \frac{\omega}{s^2+\omega^2}. \quad (18.31)$$

We wish to derive the time-translation property. But a couple of subtleties are involved, so the following preamble might be helpful.

Time translation means shifting a function of time left or right on the time axis, as illustrated in Fig. 18.4. Time translation of a function $f(t)$ is accomplished (mathematically) by replacing time t with $t - t_0$, where t_0 is a constant having the dimension of time.

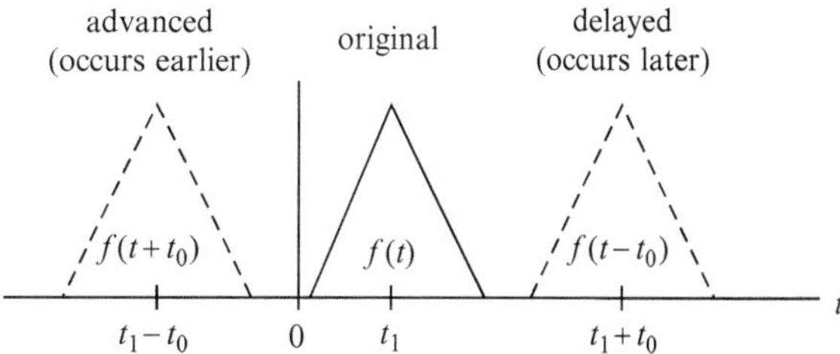

Fig. 18.4 Illustrating time translation (delay and advance), where $t_0 > 0$

If $t_0 > 0$, the function is translated to the right. If $t_0 < 0$, the function is translated to the left.

Shifting a function left makes any particular feature of the function occur at an earlier time and is called an **advance**. Shifting a function right makes any particular feature of the function occur at a later time and is called a **delay**. Conversely, **advancing** a function means translating the function to the left on the time axis and **delaying** a function means translating the function to the right on the time axis.

Two aspects of an advance are relevant to this discussion. First, advancing a *physical* quantity (e.g., a current or voltage in a physical circuit) in real time is impossible, because such an advance requires perfect knowledge of the future.[12] For example, advancing the Dow-Jones industrial average implies perfect prediction of future values of that index, and you know what that would mean. Second, if the advance is such that any nonzero part of the function is shifted to the left of the chosen time origin, the one-sided Laplace transform of the advanced function can be quite different from the one-sided Laplace transform of the original function. For example, the one-sided transform of $f(t+t_0)$ in Fig. 18.4 is identically zero, whereas the one-sided transform of $f(t)$ is not.

With that preamble, we can tackle the mathematics. The time-translation property of the one-sided Laplace transform can be derived as follows. From (18.11),

$$\mathcal{L}\{f(t-t_0)\} = \int_{0^-}^{\infty} f(t-t_0)\,e^{-st}\,dt. \quad (18.32)$$

[12]Recorded data or signals can be advanced relative to the time at which they were recorded, but that is not a true advance because it must be preceded by a delay (the recording time) sufficient to allow the advance.

Changing the variable of integration from t to $t' = t - t_0$ gives

$$\begin{aligned}\int_{0^-}^{\infty} f(t-t_0)\,e^{-st}\,dt &= \int_{-t_0}^{\infty} f(t')\,e^{-s(t'+t_0)}\,dt' \\ &= e^{-st_0}\int_{-t_0}^{\infty} f(t')\,e^{-st'}\,dt' \\ &= e^{-st_0}\int_{-t_0}^{0^-} f(t')\,e^{-st'}\,dt' \\ &\quad + e^{-st_0}\int_{0^-}^{\infty} f(t')\,e^{-st'}\,dt' \\ &= e^{-st_0}\int_{-t_0}^{0^-} f(t')\,e^{-st'}\,dt' \\ &\quad + e^{-st_0}F(s)\end{aligned} \tag{18.33}$$

We assume $f(t)$ is right-sided. If $t_0 > 0$,

$$e^{-st_0}\int_{-t_0}^{0^-} f(t')\,e^{-st'}\,dt' = 0, \tag{18.34}$$

because the integration is over times for which $f(t') = 0$. It follows (for $t_0 > 0$) that

$$\mathcal{L}\{f(t-t_0)\} = \exp\,(-st_0)\,F(s),\ t_0 > 0. \tag{18.35}$$

But if $t_0 \leq 0, f(t)$ is translated to the left. If any non-zero part of $f(t)$ is translated to the left of the time origin, then $f(t-t_0)$ is not zero for all $t<0$. In that case, the integral in (18.34) is not zero and (18.35) fails to hold

For example, let $f(t) = u(t)$ and refer to Fig. 18.5. For $t_0 > 0$ (translation to the right),

$$\begin{aligned}\mathcal{L}\{u(t-t_0)\} &= \int_{0^-}^{t_0} (0)\,e^{-st}\,dt \\ &\quad + \int_{t_0}^{\infty} (1)\,e^{-st}\,dt \\ &= \int_{t_0}^{\infty} e^{-st}\,dt = \frac{e^{-st_0}}{s},\end{aligned} \tag{18.36}$$

consistent with (18.35). But for $t_0 < 0$ (translation to the left),

$$\begin{aligned}\mathcal{L}\{u(t-t_0)\} &= \int_{0^-}^{\infty} u(t-t_0)\,e^{-st}\,dt \\ &= \int_{0^-}^{\infty} (1)\,e^{-st}\,dt = \frac{1}{s} \neq \frac{e^{-st_0}}{s},\end{aligned} \tag{18.37}$$

so (18.35) does not apply.

The discussion above shows that the time-translation property (18.35) holds (in general) only for $t_0 \geq 0$, as indicated in Table 18.1. This constraint is not much of a limitation on the practical utility of the time-translation property, because an advance is physically impossible. In some theoretical studies or where one is analyzing recorded data, it is convenient to allow $t_0 < 0$, in which case one would use the two-sided Laplace transform, given by (18.1). For the two-sided transform, (18.35) holds for both positive and negative values of t_0. But we do not need that generalization.

Example 18.6 Obtain the Laplace transform of the rectangular pulse $v(t)$ defined graphically in Fig. 18.6.

Solution: The pulse can be expressed as $v(t) = V_0\,u(t) - V_0\,u(t-\tau)$. From Table 18.2, the Laplace transform of $u(t)$ is $1/s$. By the linearity and time-translation properties in Table 18.1,

$$V(s) = \frac{V_0}{s} - \frac{V_0 e^{-s\tau}}{s} = \frac{V_0}{s}(1 - e^{-s\tau}).$$

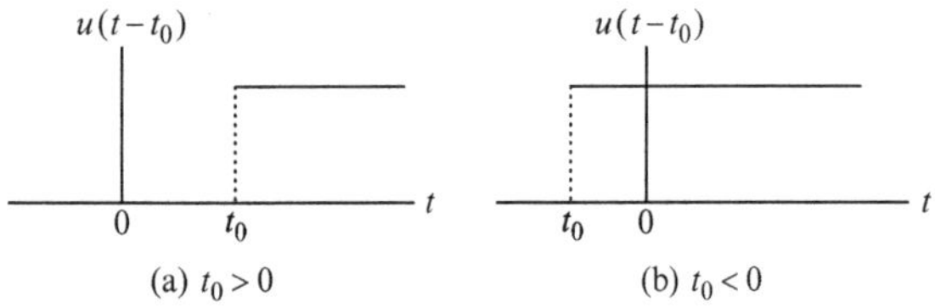

Fig. 18.5 (a) Delayed and (b) advanced unit step function

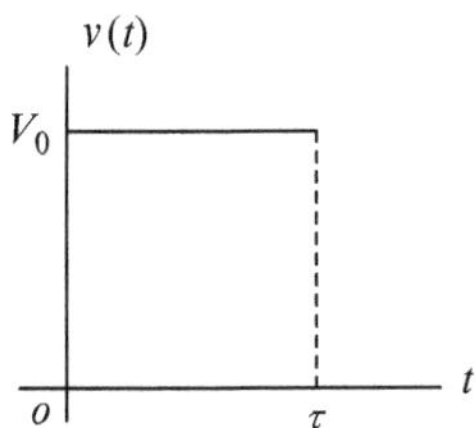

Fig. 18.6 See Example 18.6

Example 18.7 By obtaining and transforming the derivative, show that

$$\mathcal{L}\left\{\frac{d}{dt}[\cos(\omega t)u(t)]\right\} = s\mathcal{L}\{\cos(\omega t)u(t)\}$$

Solution:

$$\begin{aligned}\frac{d}{dt}[\cos(\omega t)u(t)] &= -\omega\sin(\omega t)u(t) + \cos(\omega t)\delta(t)\\ &= -\omega\sin(\omega t)u(t) + \delta(t),\end{aligned}$$

where we have used

$$\cos(\omega t)\delta(t) = \cos(0)\delta(t) = \delta(t).$$

Thus

$$\begin{aligned}\mathcal{L}\left\{\frac{d}{dt}[\cos(\omega t)u(t)]\right\} &= \mathcal{L}\{-\omega\sin(\omega t)u(t) + \delta(t)\}\\ &= -\frac{\omega^2}{s^2+\omega^2} + 1 = \frac{s^2}{s^2+\omega^2}\\ &= s\mathcal{L}\{\cos(\omega t)u(t)\},\end{aligned}$$

in agreement with the differentiation property.

Exercise 18.8 For the pulse $v(t)$ defined in Fig. 18.6, let $V_0 = 5$ V and $\tau = 50$ µs.

(a) Sketch neatly and label fully a graph of $v(t - t_0)$, where $t_0 = 75$ µs, and obtain an expression for the one-sided Laplace transform of $v(t - t_0)$.

(b) Repeat part (a) for $t_0 = -25$ µs.

18.7 Inverse Transforms Using Partial-Fraction Expansions

Using the Laplace transformation to find the response of a linear circuit to any particular excitation almost invariably requires finding the inverse Laplace transform of an expression of the form

$$F(s) = K\frac{s^m + b_{m-1}s^{m-1} + \cdots + b_1 s + b_0}{s^n + a_{n-1}s^{n-1} + \cdots + a_1 s + a_0}, \quad (18.38)$$

where the coefficients $\{a_0, a_1, \cdots, a_{n-1}\}$ and $\{b_0, b_1, \cdots, b_{m-1}\}$ are real. For subsequent reference, we note that the SI unit of each term in the numerator is s^{-m}, and the SI unit of each term in the denominator is s^{-n}. It follows that

$$\mathrm{SI}[K] = \mathrm{SI}[F(s)]\mathrm{s}^{m-n}. \quad (18.39)$$

For example, if $F(s)$ is the Laplace transform of a voltage, $m = 2$, and $n = 4$, then $\mathrm{SI}[F(s)] = \mathrm{Vs}$ and $\mathrm{SI}[K] = (\mathrm{Vs})(\mathrm{s}^{2-4}) = \mathrm{Vs}^{-1}$.

Exercise 18.9 In (18.38), let $m = 4$, $n = 3$. What is the SI unit of b_3? Of a_2?

Exercise 18.10 In (18.38), $m = 3$, $n = 5$, and $F(s)$ is the Laplace transform of a current. What is the SI unit of K?

There is an infinite number of functions of the form given in (18.38), and the number we can tabulate is finite. Fortunately, a function of the form given in (18.38) can be expressed as a linear combination of the s-domain entries in Table 18.2, after which the inverse transform can be expressed as a linear combination of the corresponding time-domain entries. This section describes how that is done.

We may write (18.38) in factored form as

$$F(s) = K\frac{(s - z_1)(s - z_2)\cdots(s - z_m)}{(s - p_1)(s - p_2)\cdots(s - p_n)}. \quad (18.40)$$

The zeros $\{z_1, z_2, \cdots, z_m\}$ of the numerator are called the **zeros** of $F(s)$ and the zeros $\{p_1, p_2, \cdots, p_n\}$ of the denominator are called the **poles** of $F(s)$. Because the coefficients $\{a_1, a_2, \cdots, a_{n-1}; b_0, b_1, \cdots, b_{m-1}\}$ are real, the zeros and poles of $F(s)$ either are real or occur in conjugate pairs. The parameter K is real, constant, and independent of s. In most applications, $m < n$. We treat that case first.

An expression of the form given in (18.38) with $n > m$ can be expanded in partial fractions. The form of the partial-fraction expansion depends upon the nature of the poles.

18.7.1 Distinct Poles

If the poles $p_1, p_2, \cdots, p_n$ are distinct (if all are different), we may write (18.38) as

$$F(s) = \frac{A_1}{s - p_1} + \frac{A_2}{s - p_2} + \cdots + \frac{A_n}{s - p_n}. \quad (18.41)$$

The **partial-fraction coefficients** $A_1, A_2, \cdots, A_n$ can be found using a method introduced by Heaviside.[13] Multiplying both sides of (18.41) by the factor $(s - p_1)$ gives

$$F(s)(s - p_1) = A_1 + \frac{A_2(s - p_1)}{s - p_2} + \cdots + \frac{A_n(s - p_1)}{s - p_n}. \quad (18.42)$$

We let $s \to p_1$ in (18.42) and obtain

$$A_1 = \lim_{s \to p_1} [F(s)(s - p_1)],$$

where the limiting operation is necessary because the denominator of $F(s)$ contains a factor $(s - p_1)$. In general, the partial-fraction coefficients for a function of the form (18.38) having distinct poles are given by

$$A_k = \lim_{s \to p_k} [F(s)(s - p_k)], \; k = 1, 2, \cdots, n. \quad (18.43)$$

In (18.40), we expressed both the numerator and the denominator in factored form in order to define the poles and zeros of a function. But it is unnecessary to factor the numerator to obtain a partial-fraction expansion.

Example 18.8 (distinct real poles) Obtain the inverse Laplace transform of

$$V(s) = \frac{p_2 p_3 V_0}{s(s - p_2)(s - p_3)},$$

where $p_2 = -10^3\ \text{s}^{-1}$, $p_3 = -2 \times 10^3\ \text{s}^{-1}$.
Solution: The poles $p_1 = 0$, $p_2 = 10^3\ \text{s}^{-1}$, $p_3 = 2 \times 10^3\ \text{s}^{-1}$ are distinct, so we can expand the function $V(s)$ in partial fractions as

$$V(s) = \frac{V_1}{s} + \frac{V_2}{s - p_2} + \frac{V_3}{s - p_3}.$$

We obtain the partial-fraction coefficients using (18.43):

$$V_1 = \lim_{s \to 0} [sV(s)] = \frac{p_2 p_3 V_0}{(0 - p_2)(0 - p_3)} = V_0,$$
$$V_2 = \lim_{s \to p_2} [(s - p_2)V(s)] = \frac{p_2 p_3 V_0}{p_2(p_2 - p_3)} = -2V_0,$$
$$V_3 = \lim_{s \to p_3} [(s - p_3)V(s)] = \frac{p_2 p_3 V_0}{p_3(p_3 - p_2)} = V_0.$$

Thus

$$v(t) = \mathcal{L}^{-1}\{V(s)\} = \mathcal{L}^{-1}\left\{\frac{V_0}{s} - \frac{2V_0}{s - p_2} + \frac{V_0}{s - p_3}\right\}$$
$$= V_0[1 - 2e^{p_2 t} + e^{p_3 t}]u(t).$$

We almost never encounter cases where $m > n$ in (18.38),[14] but occasionally, we encounter cases where $m = n$. If $m = n$, long division yields

$$F(s) = K\frac{s^n + b_{n-1}s^{n-1} + \cdots + b_1 s + b_0}{s^n + a_{n-1}s^{n-1} + \cdots + a_1 s + a_0}$$
$$= K\left[1 + \frac{(b_{n-1} - a_{n-1})s^{n-1} + (b_{n-2} - a_{n-2})s^{n-2} + \cdots + (b_1 - a_1)s + (b_0 - a_0)}{s^n + a_{n-1}s^{n-1} + \cdots + a_1 s + a_0}\right],$$

[13] After the British physicist Oliver Heaviside (1850–1925), who contributed much to electrical engineering.

[14] The ideal differentiator described in Chapter 8 is rarely (if ever) used in practice.

which can be written

$$F(s) = K + K' \frac{s^{n-1} + b'_{n-2}s^{n-2} + \cdots + b'_1 s + b'_0}{s^n + a_{n-1}s^{n-1} + \cdots + a_1 s + a_0}, \tag{18.44}$$

where

$$K' = K(b_{n-1} - a_{n-1}),\ b'_i = \frac{b_i - a_i}{b_{n-1} - a_{n-1}}. \tag{18.45}$$

Example 18.9 Obtain an expression for the inverse Laplace transform of

$$Y(s) = \frac{Ks}{s-p},$$

where $p < 0$.
Solution: By long division,

$$Y(s) = K\left(1 + \frac{p}{s-p}\right).$$

Thus

$$y(t) = K\delta(t) + Kpe^{pt}u(t).$$

Exercise 18.11 The Laplace transform of a current $i(t)$ is

$$I(s) = K\frac{s^4 + b_3 s^3 + b_2 s^2 + b_1 s + b_0}{s^4 + a_3 s^3 + a_2 s^2 + a_1 s + a_0},$$

where $b_0 = a_0 = 1$, $b_1 = 2\,\text{s}$, $b_2 = 3\,\text{s}^2$, $b_3 = 1\,\text{s}^3$, $a_1 = 4\,\text{s}$, $a_2 = 2\,\text{s}^2$, $a_3 = 6\,\text{s}^3$ and $K = 50\,\text{mAs}$. Use long division to express $I(s)$ in the form given in (18.44). Give the values and units of all of the parameters.

The fraction on the right of (18.44) has the form given in (18.38), with $m < n$. The inverse transform of $F(s)$ in (18.44) is given by

$$f(t) = K\delta(t) + K'\mathcal{L}^{-1}\left\{\frac{s^{n-1} + b'_{n-2}s^{n-2} + \cdots + b'_1 s + b'_0}{s^n + a_{n-1}s^{n-1} + \cdots a_1 s + a_0}\right\}. \tag{18.46}$$

In (18.46), $m < n$ so the indicated inverse transform can be obtained by factoring and expanding in partial fractions as described above.

Exercise 18.12 Use the differentiation property of the Laplace transformation to obtain the inverse Laplace transform of the function $Y(s)$ in Example 18.9.

18.7.2 Complex-Conjugate Poles

Usually, it is easiest to treat complex-conjugate poles in the same manner as distinct real poles. We may take advantage of the fact that the partial-fraction coefficients for a pair of complex-conjugate poles must themselves be complex conjugates, because the associated inverse transform must be real.

Example 18.10 The Laplace transform of a voltage is given by

$$V(s) = \frac{K(s - z_1)(s - z_2)}{s(s - p_1)(s^2 + \omega_0{}^2)},$$

with

$$K = 5\ \text{Vs}^{-1},\ z_1 = 1\ \text{s}^{-1},\ z_2 = 2\ \text{s}^{-1},$$
$$p_1 = -3\ \text{s}^{-1},\ \omega_0 = 1\ \text{s}^{-1}.$$

Obtain the inverse Laplace transform.
Solution: The partial-fraction expansion of $V(s)$ is

$$V(s) = \frac{K(s-z_1)(s-z_2)}{s(s-p_1)(s-j\omega_0)(s+j\omega_0)} = \frac{V_1}{s} + \frac{V_2}{s-p_1} + \frac{V_3}{s-j\omega_0} + \frac{V_3{}^*}{s+j\omega_0},$$

where

$$V_1 = \left.\frac{K(s-z_1)(s-z_2)}{(s-p_1)(s+j\omega_0)(s-j\omega_0)}\right|_{s=0} = \frac{K(-z_1)(-z_2)}{(-p_1)(j\omega_0)(-j\omega_0)} = \frac{(5\,\mathrm{V\,s^{-1}})(-1\,\mathrm{s^{-1}})(-2\,\mathrm{s^{-1}})}{(3\,\mathrm{s^{-1}})(1\,\mathrm{s^{-1}})^2} = \frac{10}{3}\ \mathrm{V},$$

$$V_2 = \left.\frac{K(s-z_1)(s-z_2)}{s(s+j\omega_0)(s-j\omega_0)}\right|_{s=p_1} = \frac{(5\,\mathrm{V\,s^{-1}})(-4\,\mathrm{s^{-1}})(-5\,\mathrm{s^{-1}})}{(-3\,\mathrm{s^{-1}})[(-3+j)\,\mathrm{s^{-1}}][(-3-j)\,\mathrm{s^{-1}}]}. = \frac{100\,\mathrm{V}}{(-3)(9+1)} = -\frac{10}{3}\mathrm{V},$$

$$V_3 = \left.\frac{K(s-z_1)(s-z_2)}{s(s-p_1)(s+j\omega_0)}\right|_{s=j\omega_0} = \frac{5\,\mathrm{V\,s^{-1}}[(j-1)\,\mathrm{s^{-1}}][(j-2)\,\mathrm{s^{-1}}]}{(j\,\mathrm{s^{-1}})[(j+3)\,\mathrm{s^{-1}}][(j+j)\,\mathrm{s^{-1}}]} = \frac{5-j15}{-6-j2}\ \mathrm{V} = j\frac{5}{2}\ \mathrm{V}.$$

Referring to Table 18.2, we find

$$\begin{aligned} v(t) &= \mathcal{L}^{-1}\left\{\frac{V_1}{s}\right\} + \mathcal{L}^{-1}\left\{\frac{V_2}{s-p_1}\right\} + \mathcal{L}^{-1}\left\{\frac{V_3}{s-j\omega_0} + \frac{V_3^*}{s+j\omega_0}\right\} \\ &= (V_1 + V_2 e^{p_1 t})u(t) + 2|V_3|\cos(\omega_0 t + \measuredangle V_3)u(t) \\ &= (V_1 + V_2 e^{p_1 t})u(t) + 2|V_3|\cos\left(\omega_0 t + \frac{\pi}{2}\right)u(t) \\ &= (V_1 + V_2 e^{p_1 t})u(t) - 2|V_3|\sin(\omega_0 t)u(t) \\ &= \left[\frac{10}{3}(1 - e^{p_1 t}) - 5\sin(\omega_0 t)\right]u(t)\,\mathrm{V}; \end{aligned}$$

$p_1 = -3\,\mathrm{s^{-1}}, \omega_0 = 1\,\mathrm{s^{-1}}$

Example 18.11 The Laplace transform of a voltage is given by

$$V(s) = \frac{K\omega_0{}^2(s-z_1)}{(s-p_1)(s^2 + 2\alpha\omega_0 s + \omega_0{}^2)},$$

with

$$K = 5\ \mathrm{Vs},\ z_1 = 1\ \mathrm{s^{-1}},\ p_1 = -2\ \mathrm{s^{-1}},\ \omega_0 = 2\ \mathrm{s^{-1}},\ \alpha = 0.4.$$

Obtain the inverse Laplace transform.
Solution: We first obtain the zeros of the quadratic factor:

$$p_0 = \frac{-2\alpha\omega_0 + \sqrt{(2\alpha\omega_0)^2 - 4\omega_0{}^2}}{2} = \omega_0\left(-\alpha + \sqrt{\alpha^2 - 1}\right) = (-0.80 + j1.83)\ \mathrm{s^{-1}}.$$

Then write

$$V(s) = \frac{K\omega_0{}^2(s-z_1)}{(s-p_1)(s-p_0)(s-p_0{}^*)} = \frac{V_1}{s-p_1} + \frac{V_0}{s-p_0} + \frac{V_0{}^*}{s-p_0{}^*}.$$

We find

$$V_1 = \frac{K\omega_0{}^2(p_1 - z_1)}{(p_1 - p_0)(p_1 - p_0^*)} = -12.5\,\mathrm{V},$$

$$V_0 = \frac{K\omega_0{}^2(p_0 - z_1)}{(p_0 - p_1)(p_0 - p_0{}^*)} = (6.25 - j1.36)\mathrm{V} = (6.40\angle - 0.22)\ \mathrm{V}.$$

Define

$$\tau_1 = -\frac{1}{\mathrm{Re}(p_1)} = 0.50\,\mathrm{s},\ \tau = -\frac{1}{\mathrm{Re}(p_0)} = 1.25\,\mathrm{s},\ \omega = \mathrm{Im}(p_0) = 1.83\,\mathrm{s^{-1}}.$$

Thus

$$v(t) = \mathcal{L}^{-1}\left\{\frac{V_1}{s-p_1}\right\} + \mathcal{L}^{-1}\left\{\frac{V_0}{s-p_0} + \frac{V_0{}^*}{s-p_0{}^*}\right\} = \left[V_1 e^{-t/\tau_1} + 2|V_0|e^{-t/\tau}\cos(\omega t + \measuredangle V_0)\right]u(t)$$

18.7.3 Repeated Poles

There are several ways of obtaining a partial-fraction expansion of a transform having repeated poles, as illustrated by the following example.

Example 18.12 (repeated real pole) Obtain the inverse Laplace transform of

$$I(s) = \frac{K\,(s - z_1)}{(s - p_1)(s - p_2)^2},$$

with

$$K = 5\text{ mA s}^{-1},\ z_1 = 1\text{ s}^{-1},$$
$$p_1 = -1\text{ s}^{-1},\ p_2 = -2\text{ s}^{-1}.$$

First *Solution:* The expansion is

$$\frac{K\,(s-z_1)}{(s-p_1)(s-p_2)^2} = \frac{I_1}{s-p_1} + \frac{I_2}{s-p_2} + \frac{I_3}{(s-p_2)^2}$$

We multiply both sides of the expansion by $(s - p_1)$ and let $s = p_1$ to obtain I_1. We multiply both sides by $(s - p_2)^2$ and let $s = p_2$ to obtain I_2. Thus

$$I_1 = \frac{K\,(p_1 - z_1)}{(p_1 - p_2)^2} = -10\,\text{mA},$$
$$I_3 = \frac{K\,(p_2 - z_1)}{(p_2 - p_1)} = 15\,\text{mA s}^{-1}$$

We again multiply both sides of the expansion by $(s - p_2)^2$ to obtain

$$\frac{K\,(s - z_1)}{(s - p_1)} = \frac{I_1(s - p_2)^2}{s - p_1} + I_2(s - p_2) + I_3$$

To obtain I_2, we differentiate this expression with respect to s. This gives

$$\frac{K(z_1 - p_1)}{(s - p_1)^2} = \frac{d}{ds}\left[\frac{I_1(s - p_2)^2}{s - p_1}\right] + I_2$$

We set $s = p_2$. The first term on the right vanishes because the derivative has a factor $(s - p_2)$. We do not need to carry out that differentiation. Thus

$$I_2 = \frac{K(z_1 - p_1)}{(p_2 - p_1)^2} = 10\,\text{mA}$$

The inverse transform is

$$i(t) = [I_1 \exp(p_1 t) + I_2 \exp(p_2 t) + I_3 t \exp(p_2 t)]u(t), \tag{18.47}$$

with

$$I_1 = -I_2 = -10\,\text{mA},\ I_3 = 15\,\text{mA s}^{-1},$$
$$p_1 = -1\,\text{s}^{-1},\ p_2 = -2\,\text{s}^{-1}.$$

Second Solution: We factor out one of the repeated factors to obtain

$$I(s) = \frac{K}{(s - p_2)}\left[\frac{(s - z_1)}{(s - p_1)(s - p_2)}\right].$$

We expand the bracketed quantity in partial fractions. This gives

$$I(s) = \frac{K}{(s - p_2)}\left[\frac{B_2}{(s - p_2)} + \frac{B_1}{(s - p_1)}\right],$$

where

$$B_1 = \frac{p_1 - z_1}{p_1 - p_2} = -2,\ B_2 = \frac{p_2 - z_1}{p_2 - p_1} = 3.$$

Thus

$$I(s) = \frac{KB_2}{(s - p_2)^2} + \frac{KB_1}{(s - p_1)(s - p_2)}.$$

We expand the last term. This gives

$$I(s) = \frac{KB_2}{(s-p_2)^2} + KB_1\left[\frac{D_1}{(s-p_1)} + \frac{D_2}{(s-p_2)}\right], \tag{18.48}$$

where

$$D_1 = \frac{1}{p_1 - p_2} = 1\,\mathrm{s}^{-1},\ D_2 = \frac{1}{p_2 - p_1} = -1\,\mathrm{s}^{-1}.$$

Thus

$$\begin{aligned} I(s) &= \frac{KB_2}{(s-p_2)^2} - \frac{KB_1D_1}{(s-p_1)} - \frac{KB_1D_2}{(s-p_2)} \\ &= \frac{I_3}{(s-p_2)^2} + \frac{I_1}{(s-p_1)} + \frac{I_2}{(s-p_2)}, \end{aligned}$$

with

$$\begin{aligned} I_1 &= KB_1D_1 = -10\,\mathrm{mA},\ I_2 = KB_1D_2 \\ &= 10\,\mathrm{mA},\ I_3 = KB_2 = 15\,\mathrm{mA\,s}^{-1} \end{aligned}$$

as obtained in the first solution above.

Third Solution: We obtain the coefficients I_1, I_3 as in the first solution above. Thus

$$I_1 = -10\,\mathrm{mA},\ I_3 = 15\,\mathrm{mA\,s}^{-1}$$

We then rationalize the expression

$$\frac{K\,(s-z_1)}{(s-p_1)(s-p_2)^2} = \frac{I_1}{s-p_1} + \frac{I_2}{s-p_2} + \frac{I_3}{(s-p_2)^2}$$

to obtain

$$\begin{aligned} K\,(s-z_1) &= I_1(s-p_2)^2 + I_2(s-p_1)(s-p_2) \\ &\quad + I_3(s-p_1) \\ &= (I_2 + I_1)s^2 \\ &\quad + (I_3 - 2I_1p_2 - I_2p_1 - I_2p_2)s \\ &\quad + I_1p_2^2 + I_2p_1p_2 - I_3p_1 \end{aligned}$$

where coefficients of like powers of s must be equal. Thus

$$(I_2 + I_1)s^2 = 0 \Rightarrow I_2 = -I_1 = 10\,\mathrm{mA}$$

There are still other ways to obtain a partial-fraction expansion. For example, if we write

$$\frac{K\,(s-z_1)}{(s-p_1)(s-p_2)^2} = \frac{I_1}{s-p_1} + \frac{I_2}{s-p_2} + \frac{I_3}{(s-p_2)^2} \tag{18.49}$$

for three different values of s (other than the pole values), we obtain three linear equations in the three unknowns I_1, I_2, I_3 which can be solved by standard methods. When solving problems, use whatever method you think easiest.

When done, *always check your work*, both dimensionally and by evaluating the original function and your expansion for a couple of values of s. In Example 18.12, we see from (18.47) that the unit of I_1 and I_2 must be A, and that the unit of I_3 must be $\mathrm{A\,s}^{-1}$. So the values obtained are dimensionally consistent. If we evaluate (18.49) for a value of s (other than the pole values), the two sides of the equation should agree. For example, for $s = 0$, we obtain

$$\frac{K\,(-z_1)}{(-p_1)(-p_2)^2} = \frac{(5\,\mathrm{mA\,s}^{-1})(-1\,\mathrm{s}^{-1})}{(1\,\mathrm{s}^{-1})(2\,\mathrm{s}^{-1})^2} = -1.25\,\mathrm{mA\,s}$$

and

$$\begin{aligned} \frac{I_1}{-p_1} + \frac{I_2}{-p_2} + \frac{I_3}{(-p_2)^2} &= \frac{-10\,\mathrm{mA}}{1\,\mathrm{s}^{-1}} + \frac{10\,\mathrm{mA}}{2\,\mathrm{s}^{-1}} \\ &\quad + \frac{15\,\mathrm{mA\,s}^{-1}}{4\,\mathrm{s}^{-2}} = -1.25\,\mathrm{mA\,s} \end{aligned}$$

so the two sides agree.

In frequency-domain analysis, we transform a circuit under consideration from the time domain to the frequency domain by replacing elements by their impedances or admittances and currents and voltages by their phasor representation. The transformed circuit is effectively a dc resistive circuit comprised of complex resistances and complex dc sources. The corresponding complex-frequency-domain procedure is essentially the same. We replace elements by their generalized impedances (defined below) and currents and voltages by their Laplace transforms. The transformed circuit is effectively a dc resistive circuit comprised of complex resistances and complex dc sources that are functions of complex frequency s.

The next section defines generalized impedance and admittance and describes how a circuit is transformed from the time domain to the s domain.

18.8 Terminal Characteristics and Equivalent Circuits

We obtain the s-**domain terminal characteristic** of a linear element by taking the Laplace transform of the time-domain terminal characteristic. For example, the *s-domain terminal characteristic of a capacitor* is obtained as follows:

$$\mathcal{L}\left\{i(t) = C\frac{dv(t)}{dt}\right\} \Rightarrow I(s)$$
$$= sCV(s) - Cv(0^-). \quad (18.50)$$

Table 18.3 gives the time-domain, frequency-domain, and complex-frequency-domain terminal characteristics for resistors, capacitors, and inductors.

The **generalized impedance** of a two-terminal linear element is denoted by Z and defined by

$$Z(s) = \frac{V(s)}{I(s)}, \quad (18.51)$$

where $V(s)$ and $I(s)$ are the Laplace transforms of the terminal current and voltage, *with initial values set to zero*; i.e., $v(0^-) = 0$ and $i(0^-) = 0$. From Table 18.3, with $v(0^-) = 0$, the *generalized impedance of a capacitor* is given by

$$\frac{V(s)}{I(s)} = \frac{1}{sC}. \quad (18.52)$$

From Table 18.3, with $i(0^-) = 0$, the *generalized impedance of an inductor* is given by

$$\frac{V(s)}{I(s)} = sL. \quad (18.53)$$

and the *generalized impedance of a resistor* is given by

$$\frac{V(s)}{I(s)} = R. \quad (18.54)$$

The **generalized admittance** of a linear two-terminal element or circuit is denoted by Y and defined as

$$Y(s) = \frac{I(s)}{V(s)} = \frac{1}{Z(s)}, \quad (18.55)$$

where $V(s)$ and $I(s)$ are the Laplace transforms of the terminal current and voltage, *with initial values set to zero* and $Z(s)$ is the generalized impedance of the element or circuit.

The generalized impedance (admittance) of a two-terminal linear element is just the frequency-domain impedance (admittance) of the element, with $j\omega$ replaced by complex frequency s.

Recall that we defined frequency-domain impedance as a function of $j\omega$, rather than f. We did that in anticipation of the definition of generalized impedance, and in order to have a consistent notation for impedance as a function of complex frequency, as a function of angular frequency, and as a function of frequency (Hz). Thus, for example, for an inductor

$$Z(s) = sL; \quad Z(j\omega) = j\omega L; \quad Z(j2\pi f) = j2\pi f L.$$

Furthermore, as we show below, everything you have learned about finding equivalent circuits in the

Table 18.3 Terminal characteristics of linear elements

Element	Time-domain	Complex-frequency-domain	Frequency-domain
Current as dependent variable			
Resistor	$i(t) = \frac{v(t)}{R}$	$I(s) = \frac{V(s)}{R}$	$\tilde{I} = \frac{\tilde{V}}{R}$
Capacitor	$i(t) = C\frac{dv}{dt}$	$I(s) = sCV(s) - Cv(0^-)$	$\tilde{I} = j\omega C\tilde{V}$
Inductor	$i(t) = i(0^-) + \frac{1}{L}\int_{0^-}^{t} v(t')\,dt'$	$I(s) = \frac{V(s)}{sL} + \frac{i(0^-)}{s}$	$\tilde{I} = \frac{\tilde{V}}{j\omega L}$
Voltage as dependent variable			
Resistor	$v(t) = R\,i(t)$	$V(s) = RI(s)$	$\tilde{V} = R\tilde{I}$
Capacitor	$v(t) = v(0^-) + \frac{1}{C}\int_{0^-}^{t} i(t')\,dt'$	$V(s) = \frac{I(s)}{sC} + \frac{v(0^-)}{s}$	$\tilde{V} = \frac{\tilde{I}}{j\omega C}$
Inductor	$v(t) = L\frac{di}{dt}$	$V(s) = sLI(s) - Li(0^-)$	$\tilde{V} = j\omega L\tilde{I}$

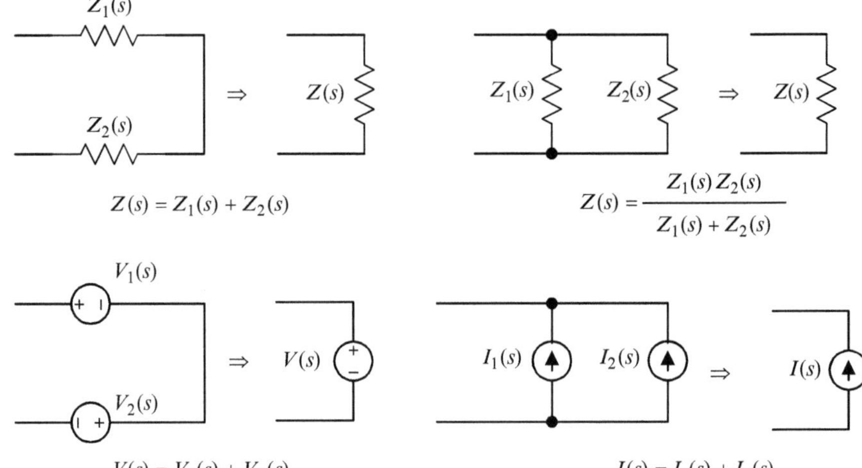

Fig. 18.7 Illustrating how impedances and voltage sources in series and parallel are reduced to equivalent single impedances or sources

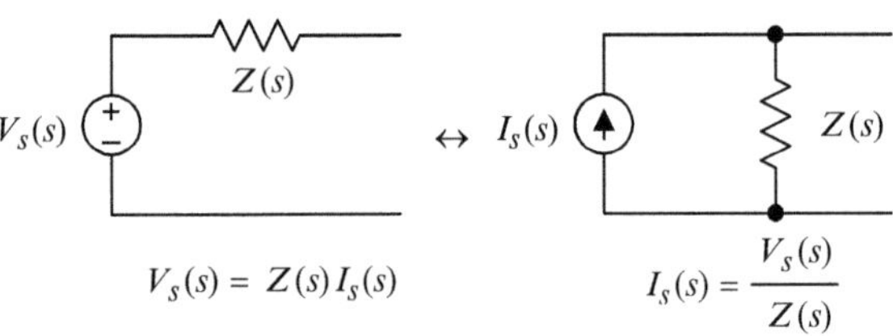

Fig. 18.8 Thévenin and Norton source models in the s domain

frequency domain you can apply in the s domain, except s replaces $j\omega$. Generalized impedances in series and parallel reduce to single, equivalent generalized impedances just like frequency-domain impedances do. All of the proofs or derivations are identical, except that s replaces $j\omega$. Thus we give without proof the results summarized in Fig. 18.7 and stated as follows:

Generalized impedances in series are additive. Generalized admittances in parallel are additive. The équivalent impedance of a parallel connection of two impedances is as given in Fig. 18.7. *Voltage sources in series are additive and current sources in parallel are additive.*

Source transformations are done in the s domain exactly as in the frequency domain, as illustrated by Fig. 18.8. Thévenin and Norton equivalent circuits are found using the same techniques one would use in the frequency domain.

A principal difference between frequency-domain (steady-state sinusoidal) analysis and complex-frequency-domain analysis is that complex-frequency-domain analysis can handle any Laplace-transformable excitations, not just sinusoidal excitations. Another is that complex-frequency-domain analysis can yield a complete response (transient and steady-state), not just steady-state response to sinusoidal or dc excitation. Prices paid for this additional generality include a slight increase in the complexity of transformed circuits (if the initial state is non-zero) and a somewhat more difficult transformation of a response from the s domain back to the time domain.[15]

18.9 Circuit Analysis in the *s* Domain

Frequency-domain circuit analysis usually begins by transforming a circuit under study to the frequency domain, where elements are represented by impedances and currents and voltages are represented by phasors. Similarly, s-domain circuit analysis usually begins by transforming a circuit under study to the s domain.

Unless initial voltages across capacitors and initial currents through inductors are all zero, transforming a circuit to the s domain can be slightly more complicated than transforming the same circuit to the frequency domain. For one thing, capacitors and inductors are represented in the s domain by generalized impedances *and associated initial-condition sources* (see Table 18.3),

[15] In truth, frequency-domain analysis can be generalized through the Fourier transformation, and resulting properties and methods allow analysis of virtually any linear circuit and transformable excitation, not just sinusoids. You will learn about the Fourier transformation if you take a subsequent course in signal processing or communication systems, where it is favored over the Laplace transformation.

and it is not always easy to determine the required initial conditions. Also, capacitors and inductors can each be represented in either of two ways, depending upon whether the dependent variable is terminal voltage, as in mesh analysis, or terminal current, as in node analysis. Table 18.4 gives the transformations one would normally use, depending upon whether nodal or mesh method are intended for subsequent analysis. Usually (but not always), a generalized impedance in series with a voltage source is most convenient for mesh or loop analysis, and a generalized impedance in parallel with a current source is most convenient for node analysis, as suggested by Table 18.4. However, the s-domain representations in each row are equivalent, so you may use whichever you prefer.

Once a circuit has been transformed to the s domain, analysis proceeds essentially as if the circuit were a resistive circuit, where the resistances are functions of complex frequency, and where currents and voltages are represented by their Laplace transforms, and are treated as if they were complex dc currents and voltages that are functions of complex frequency.

Example 18.13 In Fig. 18.9(a), $R_1 = 200\,\Omega$, $C = 200\,\text{nF}$, $R_2 = 10\,\Omega$, $L = 100\,\text{mH}$, and $v_S(t) = V_0\,u(t)$, with $V_0 = 5\,\text{V}$. Obtain an expression for the voltage $v_C(t)$.

Solution: Figure 18.9(b) shows the transformed circuit, where the circuit elements have been transformed according to Table 18.4.

Because $v_S(t) = 0$ for $t < 0$, no energy has been delivered to the circuit before $t = 0$. Thus,

$$v_C(0^-) = 0,\ i_L(0^-) = 0.$$

Applying Kirchhoff's current law to the transformed circuit gives

$$\frac{V_C(s) - V_0/s}{R_1} + sCV_C(s) + \frac{V_C(s)}{R_2 + sL} = 0,$$

which yields

Table 18.4 Transformations for circuit analysis in the complex-frequency domain

Time domain	Complex-frequency domain	
	Node transformations: current is the dependent variable	Loop or mesh transformations: voltage is the dependent variable
$i(t)$, $+$ $v(t)$ $-$, R	$I(s)$, $+$ $V(s)$ $-$, R	$I(s)$, $+$ $V(s)$ $-$, R
$i(t)$, $+$ $v(t)$ $-$, C	$I(s)$, $+$ $V(s)$ $-$, $\frac{1}{sC}$, $Cv(0^-)$	$I(s)$, $+$ $V(s)$ $-$, $\frac{1}{sC}$, $\frac{v(0^-)}{s}$
$i(t)$, $+$ $v(t)$ $-$, L	$I(s)$, $+$ $V(s)$ $-$, sL, $\frac{i(0^-)}{s}$	$I(s)$, $+$ $V(s)$ $-$, sL, $Li(0^-)$

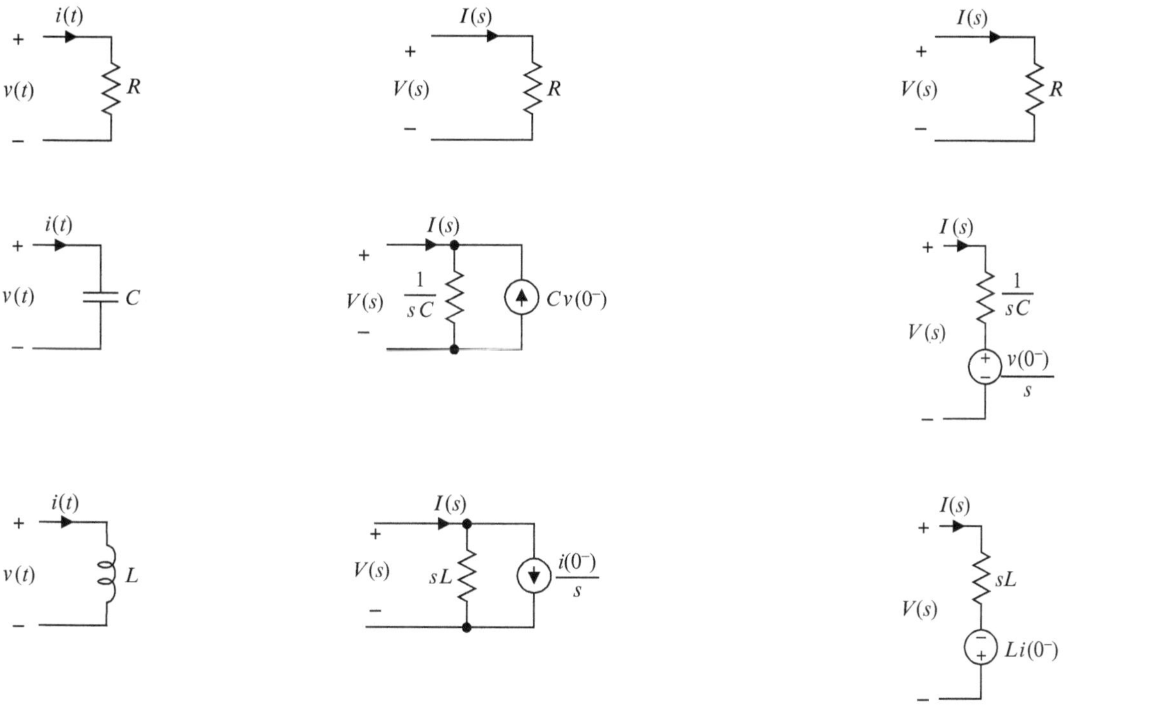

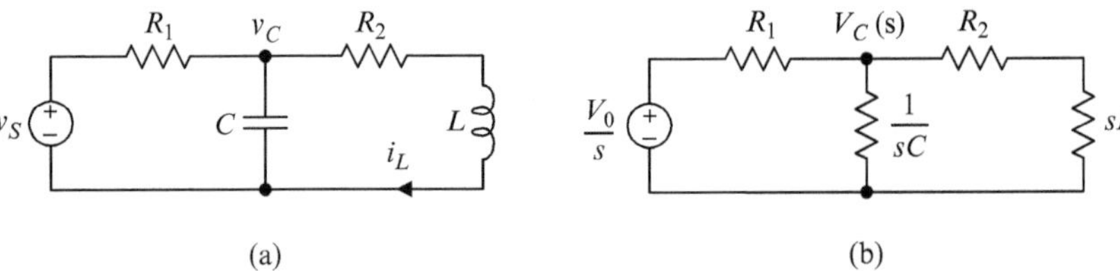

Fig. 18.9 See Example 18.13

$$V_C(s) = \frac{(R_2 + sL)V_0}{s[R_1LCs^2 + (R_1R_2C + L)s + R_1 + R_2]}. \tag{18.56}$$

The voltage $v_C(t)$ is the inverse Laplace transform of $V_C(s)$. Before finding the inverse transform, we determine whether the quadratic factor in the denominator has real or complex-conjugate roots. We write the quadratic factor as

$$R_1LCs^2 + (R_1R_2C + L)s + R_1 + R_2 = as^2 + bs + c,$$

where

$$a = R_1LC,\ b = R_1R_2C + L,\ c = R_1 + R_2.$$

We calculate

$$D = \sqrt{b^2 - 4ac} \cong 81.98\,\text{mH},$$

which shows that the roots are real and distinct. We find

$$s_1, s_2 = \frac{-b \pm D}{2a} \Rightarrow s_1 \cong -2.30 \times 10^3\,\text{s}^{-1},\quad s_2 \cong -2.28 \times 10^4\,\text{s}^{-1}.$$

Thus (18.56) becomes

$$V_C(s) = \frac{K(R_2 + sL)}{s(s - s_1)(s - s_2)};\ K = \frac{V_0}{R_1LC} \cong 1.25 \times 10^6\,\text{A}\,\text{s}^{-2},$$

which expands in partial fractions as

$$V_C(s) = \frac{V_1}{s} + \frac{V_2}{s - s_1} + \frac{V_3}{s - s_2},$$

with

$$V_1 = \frac{KR_2}{s_1s_2} \cong 0.24\,\text{V},$$
$$V_2 = \frac{K(R_2 + s_1L)}{s_1(s_1 - s_2)} \cong 5.83\,\text{V},$$
$$V_3 = \frac{K(R_2 + s_2L)}{s_2(s_2 - s_1)} \cong -6.07\,\text{V}.$$

Thus,

$$v_C(t) = \left(V_1 + V_2e^{-t/\tau_1} + V_3e^{-t/\tau_2}\right)u(t),$$
$$\tau_1 = -\frac{1}{s_1} \cong 434\,\mu\text{s},\ \tau_2 = -\frac{1}{s_2} \cong 43.9\,\mu\text{s}.$$

Example 18.14 In Fig. 18.10(a), the op amp is ideal, $R = 10\,\text{k}\Omega$, $C = 100\,\text{pF}$, and $k = 0.586$. Find the response $v_{out}(t)$ to a step input $v_{in}(t) = V_0\,u(t)$, with $V_0 = 5\,\text{V}$.
Solution: Figure 18.10(b) shows the transformed circuit, where the Laplace transforms of voltages are denoted by capital letters. Because the op amp is ideal, $V_p = V_n$. Because the excitation equals zero for $t < 0$, the voltages across the capacitors are both zero for $t = 0^-$. Replacing the capacitors with their generalized impedances and applying Kirchhoff's current law to nodes 1 and p gives

$$sC(V_1 - V_{in}) + sC(V_1 - V_p) + \frac{V_1 - V_{out}}{R} = 0,$$
$$sC(V_p - V_1) + \frac{V_p}{R} = 0.$$

We eliminate V_1 and use $V_p = V_n$ to obtain

$$\frac{s\tau(V_{in} + V_p) + V_{out}}{2s\tau + 1} = \frac{s\tau + 1}{s\tau}V_p \Rightarrow$$
$$V_p = \frac{s\tau(s\tau V_{in} + V_{out})}{s^2\tau^2 + 3s\tau + 1} = V_n.$$

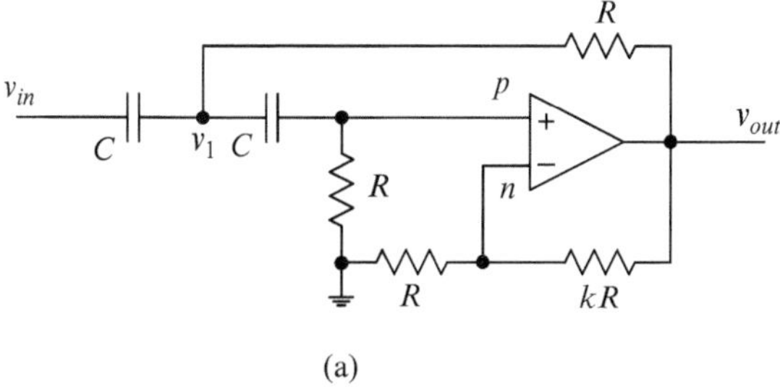

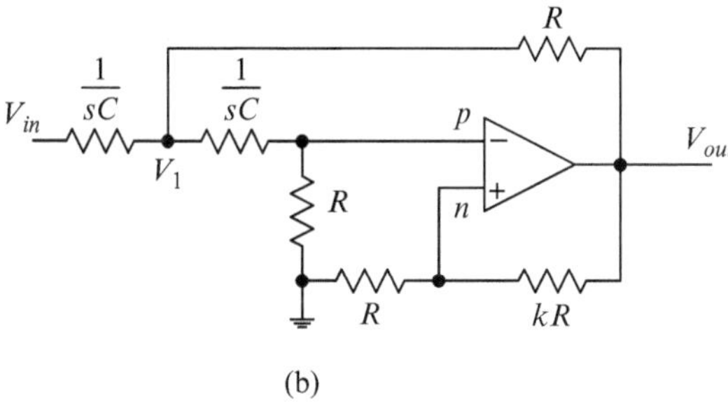

Fig. 18.10 See Example 18.14

By voltage division,

$$V_n = \frac{V_{out}}{k+1}.$$

Thus,

$$\frac{s\tau(s\tau V_{in} + V_{out})}{s^2\tau^2 + 3s\tau + 1} = \frac{V_{out}}{k+1}$$

$$\Rightarrow V_{out} = \frac{(k+1)s^2\tau^2}{s^2\tau^2 + (2-k)s\tau + 1} V_{in}.$$

The Laplace transform of the input voltage is

$$V_{in} = \mathcal{L}\{V_0 u(t)\} = \frac{V_0}{s},$$

so the Laplace transform of the output is

$$V_{out} = \frac{V_0(k+1)s^2\tau^2}{s[s^2\tau^2 + (2-k)s\tau + 1]}.$$

The zeros of the quadratic factor are

$$s_0, s_0^* = \frac{-(2-k)\tau \pm \sqrt{(2-k)^2\tau^2 - 4\tau^2}}{2\tau^2}$$
$$\cong (-0.707 \pm j0.707) \times 10^6\,\mathrm{s}^{-1},$$

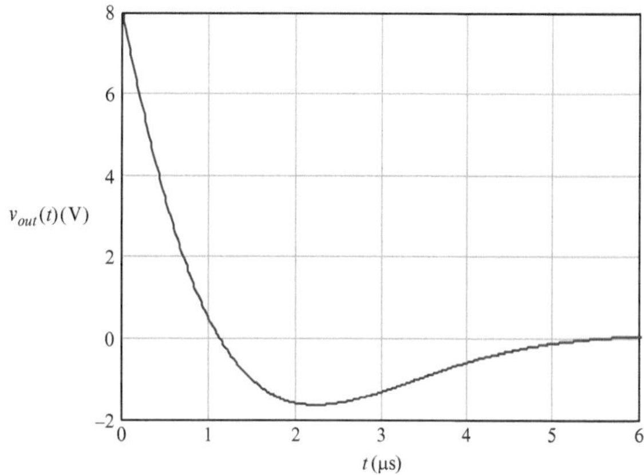

Fig. 18.11 See Example 18.14

so we expand $V_{out}(s)$ in partial fractions as

$$V_{out} = \frac{V_0(k+1)s^2\tau^2}{s\tau^2(s-s_0)(s-s_0{}^*)}$$
$$= \frac{V_0(k+1)s}{(s-s_0)(s-s_0{}^*)} = \frac{V_1}{s-s_0} + \frac{V_1^*}{s-s_0{}^*},$$

where we find that

$$V_1 = \frac{V_0(k+1)s_0}{(s_0 - s_0{}^*)} \cong 3.96(1+j)\,\mathrm{V}.$$

Thus the response is given by

$$v_{out}(t) = 2|V_1|e^{-t/\tau_0}\cos(\omega_0 t + \theta)u(t),$$

where

$$2|V_1| \cong 11.21\,\mathrm{V}, \theta = \measuredangle V_1 \cong 0.785,$$
$$\tau_0 = -\frac{1}{\mathrm{Re}(s_0)} \cong 1.414\,\mu s,\ \omega_0 = \mathrm{Im}(s_0)$$
$$\cong 7.072 \times 10^5\,\mathrm{s}^{-1}.$$

Figure 18.11 shows a graph of the output versus time. As quick checks on the reasonableness of the solution, we observe that the output approaches zero as $t \to \infty$, which is consistent with the capacitive coupling to the source (dc voltage gain = 0). Also, at $t = 0$, the output jumps instantaneously to the peak value $(k+1)V_0$, as is consistent with the fact that the voltages across the capacitors cannot change instantaneously (and $v_n = v_p$).

Example 18.15 In Fig. 18.12, $V_0 = 5\,\text{V}$, $V_1 = 10\,\text{V}$, $R = 1\,\text{k}\Omega$, $L = 150\,\text{mH}$, and $C = 200\,\text{nF}$. The switch is moved from a to b at $t = 0$, having been at a for $t < 0$. Obtain an expression for the voltage $v(t)$.

Solution: Before the switch is moved, the circuit is in dc steady state, the capacitor is an open circuit, the inductor is a short circuit, the circuit is a resistive voltage divider, and

$$v(t) = \frac{RV_0}{2R} = \frac{V_0}{2} = 2.5\,\text{V},\ t < 0.$$

Also, before the switch is moved, the current $i(t)$ is given by

$$i(t) = \frac{V_0}{2R} = 2.5\,\text{mA}.$$

Therefore, the initial conditions are

$$v_C(0^-) = V_0 - v(0^-) = 2.5\,\text{V} = v(0^-),$$
$$i(0^-) = 2.5\,\text{mA}.$$

Figure 18.13 shows the transformed circuit for t > 0 , where we have used the node formulations because we intend to use node analysis (Kirchhoff's current law).

Let $V = V(s) = \mathcal{L}\{v(t)\}$ and $V_L = V_L(s) = \mathcal{L}\{v_L(t)\}$. Kirchhoff's current law gives

$$\frac{1}{R}\left(V - \frac{V_1}{s}\right) + sC\left(V - \frac{V_1}{s}\right) + Cv(0^-) + \frac{V - V_L}{R} = 0$$

and

$$\frac{V_L - V}{R} + \frac{V_L}{sL} + \frac{i(0^-)}{s} = 0.$$

Eliminating V_L yields

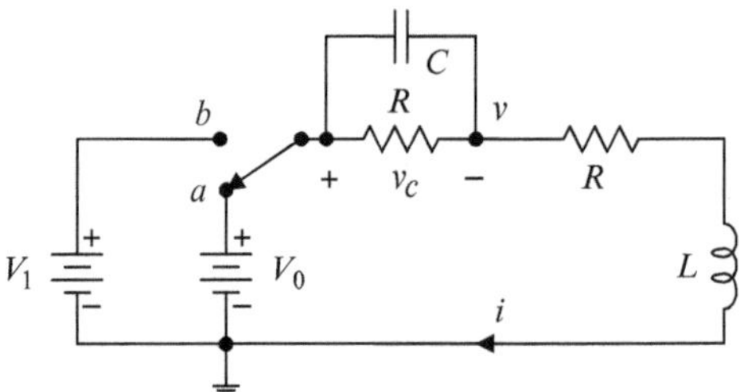

Fig. 18.12 See Example 18.15

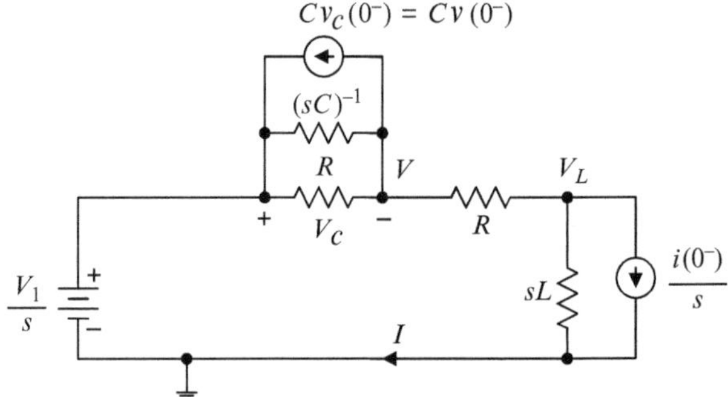

Fig. 18.13 See Example 18.15

$$V(s) = \frac{CRL[V_1 - v(0^-)]s^2 + [CR^2V_1 + LV_1 - i(0^-)RL - CR^2v(0^-)]s + RV_1}{s[CRLs^2 + (L + CR^2)s + 2R]}$$
$$= \frac{b_2s^2 + b_1s + b_0}{s(a_2s^2 + a_1s + a_0)},$$

where

$$b_0 = RV_1 = 1.00 \times 10^4\,\text{V}\,\Omega,$$
$$b_1 = CR^2V_1 + LV_1 - i(0^-)RL - CR^2v(0^-) \cong 2.63\,\text{V}\,\Omega\,\text{s},$$
$$b_2 = CRL[V_1 - v(0^-)] \cong 2.25 \times 10^{-4}\text{V}\,\Omega\,\text{s}^2,$$
$$a_0 = 2R = 2\,\text{k}\Omega,$$
$$a_1 = L + CR^2 = 0.35\,\Omega\,\text{s},$$
$$a_2 = CRL = 3.00 \times 10^{-5}\,\Omega\,\text{s}^2.$$

The zeros of the quadratic factor in the denominator are

$$p, p^* = \frac{-a_1 \pm \sqrt{a_1^2 - 4a_2a_0}}{2a_2}$$
$$\cong (5.83 \pm j5.71) \times 10^3\,\text{s}^{-1}.$$

Thus

$$V(s) = \frac{b_2s^2 + b_1s + b_0}{a_2s(s-p)(s-p^*)}$$
$$= \frac{A_1}{s} + \frac{A_2}{s-p} + \frac{A_2{}^*}{s-p^*}$$

with

$$A_1 = \frac{b_0}{a_2(-p)(-p^*)} = 5\,\text{V},$$
$$A_2 = \frac{b_2p^2 + b_1p + b_0}{a_2p(p-p^*)} \cong (1.25 - j1.28)\,\text{V}.$$

Therefore

$$v(t) = \begin{cases} V_0/2 = 2.5\,\text{V}, & t<0 \\ A_1 + 2|A_2|\exp(-t/\tau) & \\ \times\cos(\omega_0 t + \measuredangle A_2), & t>0, \end{cases}$$

with

$$V_0 = 5\,\text{V},\ A_1 = 5\,\text{V},$$
$$2|A_2| \cong 3.57\,\text{V},\ \measuredangle A_2 \cong -0.80,$$
$$\tau = -\frac{1}{\text{Re}(p)} \cong 171\,\mu\text{s},$$
$$\omega_0 = \text{Im}(p) \cong 5.71 \times 10^3\,\text{s}^{-1}.$$

18.10 Checking Your Work

A thorough way to check a solution to a problem is to work the problem a second time, using another approach; for example, if you used node analysis first, work the problem again using mesh or loop analysis. Such thorough checks can be performed on homework or later, in practice, but are impractical on exams. Quick checks that can be used on exams are checks on dimensional consistency and on agreement with initial and final values (limit checks). You should cultivate the habits of checking intermediate results for dimensional consistency as you work through a problem and of performing one or two simple limit checks where possible. Space constraints prohibit our showing such checks in every example, but below, we give one example to illustrate the process.

Example 18.16 Refer to Fig. 18.14(a), where $V_1 = 5\,\text{V}$, $V_2 = 10\,\text{V}$, $R = 10\,\text{k}\Omega$, $L = 10\,\text{mH}$, and $C = 50\,\text{nF}$. The switch is moved from V_1 to V_2 at $t = 0$. Obtain an expression for the voltage $v(t)$.

Solution: Figure 18.14(b) shows the transformed circuit for $t > 0$.

At $t = 0^-$,

$$i_L(0^-) = \frac{V_1}{R},\ v(0^-) = 0.$$

For $t > 0$, $v_S(t) = V_2$.

Kirchhoff's current law gives

$$\frac{V(s) - V_S(s)}{R} + \frac{V(s)}{sL} + \frac{i_L(0^-)}{s} + sCV(s) = 0,$$

which yields

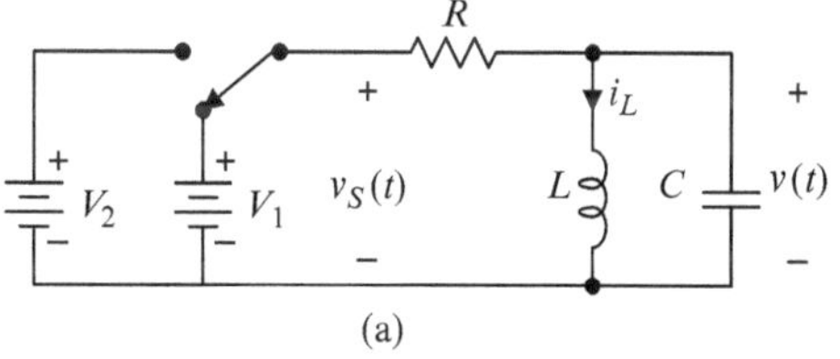

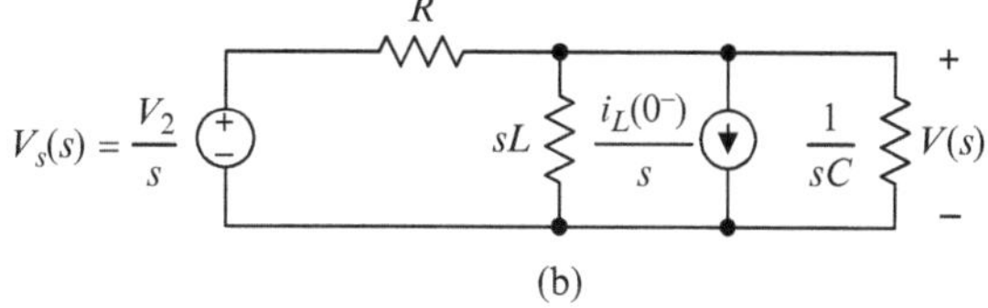

Fig. 18.14 See Example 18.16

$$V(s) = \frac{L[sV_S(s) - i_L(0^-)R]}{RLCs^2 + Ls + R} = \frac{L[V_2 - i_L(0^-)R]}{RLCs^2 + Ls + R}.$$

Because $V(s)$ is the Laplace transform of a voltage, the SI unit of $V(s)$ must be V s. We find

$$\mathrm{SI}[V(s)] = \frac{\mathrm{SI}[L[V_2 - i_L(0^-)R]]}{\mathrm{SI}[RLCs^2 + Ls + R]},$$

$$\mathrm{SI}[L[V_2 - i_L(0^-)R]] = \left(\frac{\mathrm{V\,s}}{\mathrm{A}}\right)\left(\mathrm{V} - \mathrm{A}\frac{\mathrm{V}}{\mathrm{A}}\right) = \frac{\mathrm{V}^2\,\mathrm{s}}{\mathrm{A}},$$

$$\mathrm{SI}\left[RLCs^2\right] = \left(\frac{\mathrm{V}}{\mathrm{A}}\right)\left(\frac{\mathrm{V\,s}}{\mathrm{A}}\right)\left(\frac{\mathrm{A\,s}}{\mathrm{V}}\right)\mathrm{s}^{-2} = \frac{\mathrm{V}}{\mathrm{A}},\ \mathrm{SI}[Ls] = \left(\frac{\mathrm{V\,s}}{\mathrm{A}}\right)\mathrm{s}^{-1} = \frac{\mathrm{V}}{\mathrm{A}},\ \mathrm{SI}[R] = \Omega = \frac{\mathrm{V}}{\mathrm{A}},$$

$$\mathrm{SI}\left[RLCs^2 + Ls + R\right] = \frac{\mathrm{V}}{\mathrm{A}},$$

$$\mathrm{SI}[V(s)] = \frac{\mathrm{V}^2\,\mathrm{s}}{\mathrm{A}}\left(\frac{\mathrm{V}}{\mathrm{A}}\right)^{-1} = \mathrm{V\,s}.$$

Our expression for $V(s)$ is dimensionally consistent.

The roots of the denominator are a complex-conjugate pair s_0, $s_0{}^*$, given by

$$s_0,\, s_0{}^* = \frac{-b \pm \sqrt{b^2 - 4ac}}{2a},$$

where

$$a = RLC = 5\mu\mathrm{s},\quad b = L = 10\,\mathrm{mH},\quad c = R = 10\,\mathrm{k\Omega}.$$

For dimensional consistency, we must have

$$\mathrm{SI}[s_0] = \mathrm{SI}\left[\frac{-b \pm \sqrt{b^2 - 4ac}}{2a}\right] = \mathrm{s}^{-1}.$$

We find

$$\mathrm{SI}[ac] = \mathrm{SI}\left[R^2LC\right] = \left(\frac{\mathrm{V}}{\mathrm{A}}\right)^2\left(\frac{\mathrm{V\,s}}{\mathrm{A}}\right)\left(\frac{\mathrm{A\,s}}{\mathrm{V}}\right) = \left(\frac{\mathrm{V\,s}}{\mathrm{A}}\right)^2 = \mathrm{H}^2 = \mathrm{SI}\left[b^2\right],$$

$$\Rightarrow \mathrm{SI}\left[\sqrt{b^2 - 4ac}\right] = \mathrm{SI}[b],$$

$$\mathrm{SI}\left[\frac{-b}{2a}\right] = \mathrm{SI}\left[\frac{L}{RLC}\right] = \frac{1}{\Omega\mathrm{F}} = \left(\frac{\mathrm{V}}{\mathrm{A}}\frac{\mathrm{A\,s}}{\mathrm{V}}\right)^{-1} = \mathrm{s}^{-1}.$$

So the expression for the roots is dimensionally consistent.

We calculate

$$a = RLC = 5\times10^{-6}\,\Omega\,\mathrm{s}^2,\quad b = L = 10\,\mathrm{mH} = 10^{-2}\,\Omega\,\mathrm{s},\quad c = R = 10\,\mathrm{k\Omega}$$

and

$$D = \sqrt{b^2 - 4ac} \cong j\,447\,\mathrm{mH},$$

so

$$s_0 = \frac{-b + D}{2a} \cong (-1 + j\,44.7)\times10^3\,\mathrm{s}^{-1}$$

and

$$V(s) = \frac{L[V_2 - i_L(0^-)R]}{RLCs^2 + Ls + R} = \frac{1}{RC}\frac{V_2 - i_L(0^-)R}{(s - s_0)(s - s_0{}^*)} = \frac{V_0}{s - s_0} + \frac{V_0{}^*}{s - s_0{}^*},$$

with

$$V_0 = \frac{1}{RC}\frac{V_2 - i_L(0^-)R}{(s_0 - s_0{}^*)} \cong -j112\,\mathrm{mV}.$$

We note that $\mathrm{SI}[V(s)] = \mathrm{V\,s}$ requires $\mathrm{SI}[V_0] = \mathrm{V}$. We find

$$\begin{aligned}\mathrm{SI}[V_0] &= \mathrm{SI}\left[\frac{1}{RC}\frac{V_2 - i_L(0^-)R}{(s_0 - s_0^*)}\right] \\ &= \left(\frac{\mathrm{V}}{\mathrm{A}}\frac{\mathrm{A\,s}}{\mathrm{V}}\right)^{-1}\left(\frac{\mathrm{V}}{\mathrm{s}^{-1}}\right) \\ &= \mathrm{s}^{-1}\left(\frac{\mathrm{V}}{\mathrm{s}^{-1}}\right) = \mathrm{V}.\end{aligned}$$

so our expression for the partial-fraction coefficient V_0 is dimensionally consistent. From Table 18.2,

$$v(t) = 2|V_0|e^{-t/\tau}\cos(\omega_0 t + \measuredangle V_0)u(t),$$

with

$$\begin{aligned}2|V_0| &\cong 224\,\mathrm{mV}, \\ \measuredangle V_0 &= -\frac{\pi}{2} \cong -1.57, \\ \tau &= -\frac{1}{\mathrm{Re}(s_0)} \cong 1\,\mathrm{ms}, \\ \omega_0 &= \mathrm{Im}(s_0) \cong 44.71 \times 10^3\,\mathrm{s}^{-1}.\end{aligned}$$

We note that $\mathrm{SI}[v(t)] = \mathrm{SI}[V_0] = \mathrm{V}$, as it should be. As a final check, we compute the initial and final values of the voltage. Both should be zero. We find

$$\begin{aligned}v(0^+) &= 2|V_0|\cos(\measuredangle V_0) = 2|V_0|\cos\left(\frac{\pi}{2}\right) = 0, \\ v(t \to \infty) &= 0 \text{ because } \lim_{t\to\infty}\left(e^{-t/\tau}\right) = 0\end{aligned}$$

There are two errors that are common in complex-frequency-domain circuit analysis. One is made early, when transforming a circuit to the s domain. The other is made later, when expressing a current or voltage in the time domain.

The early error is to define the transformed voltage across or current through a capacitor or inductor as only the voltage across or current through the associated generalized impedance. Refer to Table 18.4 and keep in mind that the initial-condition sources are part of the transformed element. The late error is to write the time-domain expression for a current or voltage as if the current or voltage were zero for $t < 0$ even if that is not the case. Keep in mind that an inverse transform obtained using Table 18.2 is only the right half of a function whose left half might be non-zero. In many initial-value or switched-circuit problems, a current or voltage of interest is two-sided, and an inverse transform from Table 18.2 gives only the right half of the result. We must supply the left half, either as an initial value or as the result of analysis in the previous time interval. The next example illustrates these remarks.

Example 18.17 Refer to Fig. 18.15(a), where the switch is open for $t < 0$ and closed at $t = 0$. (a) Obtain an expression for the current $i_L(t)$. (b) Use that result to obtain an expression for the voltage $v_L(t)$ across the inductor.

Solution: For $t < 0$, the circuit is in dc steady state, and

$$i_L(0^-) = \frac{2V_0}{2R} = \frac{V_0}{R}.$$

To solve for the current for $t > 0$, we first transform the circuit (with the switch closed) to the complex frequency domain. Figure 18.15(b) shows the transformed circuit. Note that *the transformed voltage $V_L(s)$ appears across both the generalized impedance of the inductor and the initial-condition source.* Applying Kirchhoff's voltage law to the transformed circuit gives

$$\begin{aligned}(R + sL)I_L(s) &= \frac{V_0}{s} + Li_L(0^-) \\ &= \frac{V_0}{s} + \frac{LV_0}{R},\end{aligned} \tag{18.57}$$

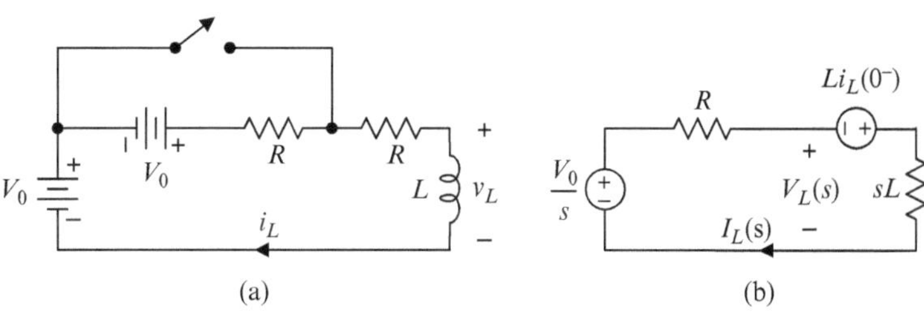

Fig. 18.15 See Example 18.17

so

$$I_L(s) = \frac{V_0}{s(R+sL)} + \frac{LV_0}{R(R+sL)},$$

which we may write as

$$I_L(s) = \frac{V_0}{Ls(s-s_1)} + \frac{V_0}{R(s-s_1)}, \ s_1 = -\frac{R}{L}.$$

Expanding the first term on the right in partial fractions gives

$$I_L(s) = -\frac{V_0}{Ls_1 s} + \frac{V_0}{Ls_1(s-s_1)} + \frac{V_0}{R(s-s_1)}.$$

From Table 18.2, the inverse transform is

$$i_L(t) = \left[-\frac{V_0}{Ls_1} + \left(\frac{V_0}{Ls_1} + \frac{V_0}{R}\right)\exp\left(-\frac{t}{\tau}\right)\right]u(t),$$
$$\tau = \frac{L}{R}. \tag{18.58}$$

According to this expression, the current $i_L(t)$ is zero for $t < 0$. But we know that

$$i_L(t) = \frac{V_0}{R}, \ t \le 0.$$

Thus the correct expression for the current is

$$i_L(t) = \begin{cases} \frac{V_0}{R}, & t \le 0, \\ \left[-\frac{V_0}{Ls_1} + \left(\frac{V_0}{Ls_1} + \frac{V_0}{R}\right)\exp\left(-\frac{t}{\tau}\right)\right], & t > 0. \end{cases} \tag{18.59}$$

(b) Although both (18.58) and (18.59) are correct for $t > 0$, the expression in (18.58) has a jump discontinuity at $t = 0$, whereas the expression in (18.59) is continuous. If we use (18.58) in

$$v_L(t) = L\frac{di_L(t)}{dt}$$

to find the voltage $v_L(t)$, we obtain an incorrect result. The discontinuity at $t = 0$ would give rise to an impulse in the derivative of (18.58). But the correct expression (18.59) is continuous, and the correct expression for the voltage across the inductor is

$$v_L(t) = L\frac{di_L}{dt}$$
$$= -\frac{1}{\tau}\left(\frac{V_0}{Ls_1} + \frac{V_0}{R}\right)\exp\left(-\frac{t}{\tau}\right)u(t).$$

It might also be tempting to write

$$V_L(s) = sLI_L(s)$$
$$= sL\left[-\frac{V_0}{Ls_1 s} + \frac{V_0}{Ls_1(s-s_1)} + \frac{V_0}{R(s-s_1)}\right].$$

But this is also incorrect, as it ignores the initial-condition source, which must be included in the s domain terminal characteristic for the inductor. The correct expression is

$$V_L(s) = sLI_L(s) - Li_L(0^-)$$
$$= sL\left[-\frac{V_0}{Ls_1 s} + \frac{V_0}{Ls_1(s-s_1)} + \frac{V_0}{R(s-s_1)}\right]$$
$$-\frac{LV_0}{R}.$$

18.11 *s*-Domain Transfer Functions

Figure 18.16 shows *s*-domain representations of a two-port circuit driven by either of *equivalent* Théve-nin and Norton source models. The circuit can be described by four ***s*-domain transfer functions**, defined as follows:

Voltage Transfer Function

$$H_v(s) = \frac{V_L(s)}{V_S(s)}. \tag{18.60}$$

Current Transfer Function

$$H_i(s) = \frac{I_L(s)}{I_S(s)} = \frac{Z_S(s)}{Z_L(s)}H_v(s). \tag{18.61}$$

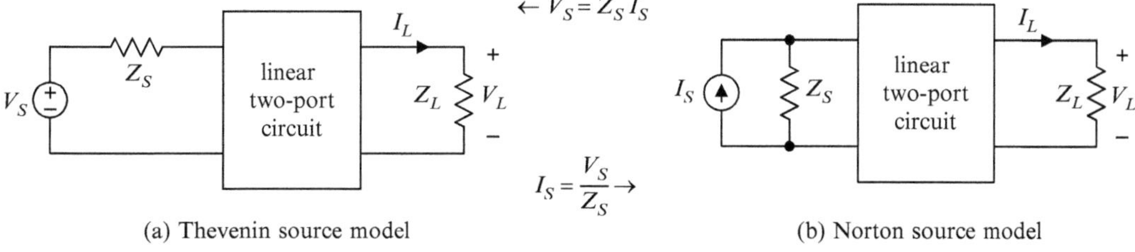

(a) Thevenin source model (b) Norton source model

Fig. 18.16 Definitions of terms in (18.60–18.63)

Transimpedance Transfer Function

$$H_z(s) = \frac{V_L(s)}{I_S(s)} = Z_S(s)H_v(s). \tag{18.62}$$

Transadmittance Transfer Function

$$H_y(s) = \frac{I_L(s)}{V_S(s)} = \frac{H_v(s)}{Z_L(s)}. \tag{18.63}$$

In each definition above, V_S and I_S are the Laplace transforms of the *available* input voltage and current, respectively. *An s-domain transfer function is determined with a source and load attached to the circuit and depends in general upon the source and load impedances.*

If all initial conditions of a circuit, including those of the associated source and load, are zero, the transform $Y(s)$ of the complete response of the circuit to an excitation (current or voltage) $X(s)$ is given by

$$Y(s) = H(s)X(s) \tag{18.64}$$

where $H(s)$ is the appropriate s-domain transfer function for the circuit; e.g., if X and Y both represent voltages, the appropriate transfer function is the voltage transfer function.

Extending transfer-function representations to cases where initial conditions are non-zero is essentially an application of superposition, where a transfer function is defined for each initial-condition source, and the complete response is then the sum of the responses to the excitation and the initial-condition sources. We do not pursue that generalization.

The definitions above of s-domain transfer functions are essentially identical to those given in Chapter 15 for frequency-domain transfer functions, except that phasor representations of the input and output are replaced by Laplace transforms of the input and output. In practice and in this book, both frequency-domain and s-domain transfer functions are referred to as transfer functions. Usually, which is meant is clear from context or from whether frequency f or complex frequency s is the independent variable. Where necessary to avoid confusion, we specify explicitly one or the other of frequency-domain or s-domain.

In general, the four transfer functions defined by (18.60–18.63) exhibit different dependencies on complex frequency s because of the various ways they depend upon the source and load impedances. But if the source and load impedances are resistive (real), the four transfer functions differ from each other only by real constant multipliers that are independent of s. For example, if $Z_S = R_S$ and $Z_L = R_L$, the current transfer function $H_i(s)$ and the voltage transfer function $H_v(s)$ are related as

$$H_i(s) = \frac{I_L(s)}{I_S(s)} = \frac{V_L(s)/R_L}{V_S(s)/R_S} = \frac{R_S}{R_L}H_v(s), \tag{18.65}$$

where the factor R_S/R_L is real, constant, and independent of s.

Under certain conditions, a transfer function can be practically independent of source and load impedances. For example, the s-domain voltage transfer function for a circuit having input impedance Z_{in} and output impedance Z_{out} is practically independent of the source impedance Z_S and the load impedance Z_L if $|Z_{in}| \gg |Z_S|$ and $|Z_{out}| \ll |Z_L|$. Such independence usually is desirable for reasons discussed in Chapter 15 in connection with f-domain transfer functions.

For economy, we often omit explicit dependence on s when writing symbols for impedances, transforms, and transfer functions. For example, we may write H_y implying $H_y(s)$, I_L implying $I_L(s)$, and so on.

The dimension of a transfer function is determined by the dimensions of the associated input and output. Voltage and current transfer functions are dimensionless, but transadmittance and transimpedance transfer functions have the dimensions of admittance (siemens) and impedance (ohms), respectively:

$$\begin{aligned} &\mathrm{SI}[H_v] = 1, \ \mathrm{SI}[H_i] = 1, \\ &\mathrm{SI}\left[H_y\right] = S, \mathrm{SI}[H_z] = \Omega. \end{aligned} \tag{18.66}$$

We omit the subscripts v, i, y, z in developments that apply equally to all of the four transfer functions defined above.

Example 18.18 Find the transadmittance transfer function for the circuit in Fig. 18.17.

Solution: The transadmittance transfer function is given by (18.63). We need to express the (transformed) load current I_L in terms of the (transformed) source voltage V_S. For $t = 0^+$, the current through the inductor and the voltage across the capacitor are both zero because no energy is provided to the circuit before $t = 0$.

Figure 18.18 shows the transformed circuit. Kirchhoff's current law gives

$$\frac{V_C - V_S}{R_S} + sCV_C + \frac{V_C}{R_L + sL} = 0,$$

which yields

$$V_C = \frac{(R_L + sL)V_S}{R_SLCs^2 + (R_SR_LC + L)s + R_L + R_S}.$$

By Ohm's law

$$\begin{aligned} I_L &= \frac{V_C}{(R_L + sL)} \\ &= \frac{V_S}{R_SLCs^2 + (R_SR_LC + L)s + R_L + R_S}. \end{aligned}$$

Thus

$$H_y = \frac{I_L}{V_S} = \frac{1}{R_SLCs^2 + (R_SR_LC + L)s + R_S + R_L}. \tag{18.67}$$

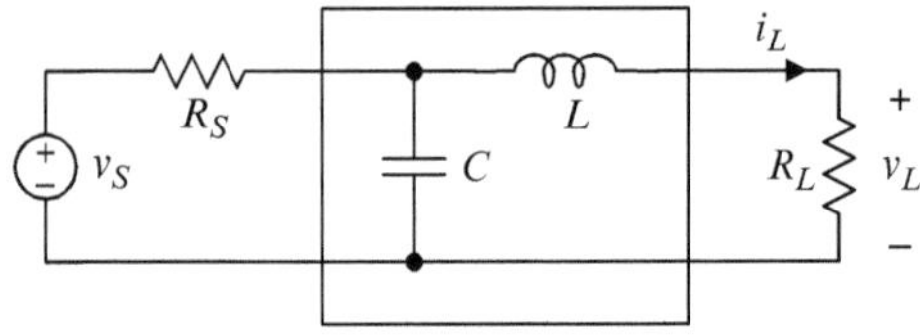

Fig. 18.17 See Example 18.18

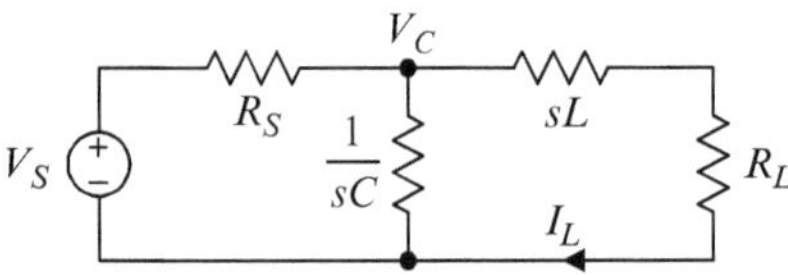

Fig. 18.18 See Example 18.18

Exercise 18.13 Use mesh analysis (Kirchhoff's voltage law) to find the transadmittance transfer function for the circuit in Fig. 18.17

Exercise 18.14 Find the current transfer function, the voltage transfer function, and the transimpedance transfer function for the circuit in Fig. 18.17

Example 18.19 Obtain all four transfer functions (see (18.60–18.63)) for the circuit shown in Fig. 18.19. Assume the op amp is ideal.

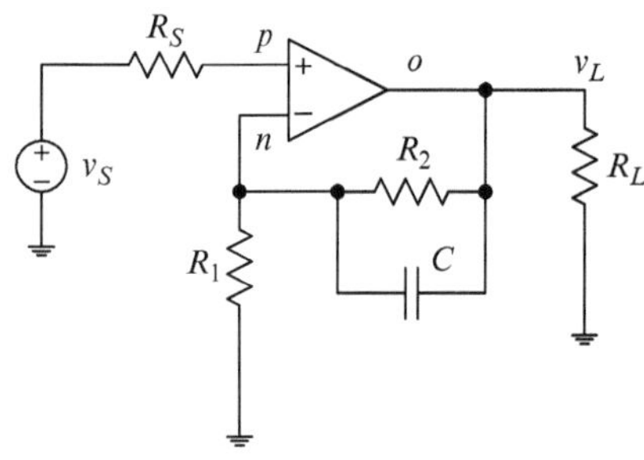

Fig. 18.19 See Example 18.19

Solution: The circuit is essentially a non-inverting amplifier having voltage transfer function

$$H_v = 1 + \frac{Z_2}{R_1},\ Z_2 = R_2 \Big\| \frac{1}{sC} = \frac{R_2}{1 + sR_2C},$$

which gives

$$\begin{aligned} H_v &= 1 + \frac{R_2}{R_1(1+sR_2C)} = \frac{R_1(1+sR_2C)+R_2}{R_1(1+sR_2C)} \\ &= \frac{R_1+R_2+sR_1R_2C}{R_1(1+sR_2C)} = \left(\frac{R_1+R_2}{R_1}\right)\frac{(1+sR'C)}{(1+sR_2C)} \\ &= \left(\frac{R_2}{R'}\right)\frac{(1+sR'C)}{(1+sR_2C)},\ R' = R_1\|R_2. \end{aligned}$$

From (18.65),

$$H_i = \frac{R_S}{R_L}H_v = \frac{R_SR_2}{R_LR'}\left(\frac{1+sR'C}{1+sR_2C}\right),\ R' = R_1\|R_2.$$

Similarly,

$$\begin{aligned} H_z &= \frac{V_L}{I_S} = \frac{V_L}{V_S/R_S} = R_SH_v \\ &= \frac{R_SR_2}{R'}\left(\frac{1+sR'C}{1+sR_2C}\right),\ R' = R_1\|R_2 \end{aligned}$$

and

$$H_y = \frac{I_L}{V_S} = \frac{V_L/R_L}{V_S} = \frac{H_v}{R_L} = \frac{R_2}{R_LR'}\left(\frac{1+sR'C}{1+sR_2C}\right),$$
$$R' = R_1\|R_2.$$

Exercise 18.15 For the circuit considered in Example 18.19, show that $H_iH_v = H_zH_y$. Is this relation true in general?

Exercise 18.16 The available voltage at the terminals of a certain source is given by $v_S(t) = V_0\,u(t)$. The source is connected to the input terminals of a certain circuit. The resulting voltage across the load on the circuit is given by $v_L(t) = V_0[1 - \exp(-t/\tau)]u(t)$. (a) Find the voltage transfer function for the circuit. (b) The source impedance is $Z_S(s) = R_S$ and the load impedance is $Z_L = R_L + sL$. Obtain expressions for the current, transadmittance, and transimpedance transfer functions.

Example 18.20 Refer to Fig. 18.20, where $R_1 = 15\,\text{k}\Omega$, $C_1 = 330\,\text{nF}$, $R_2 = 180\,\text{k}\Omega$, $C_2 = 110\,\text{nF}$, and $v_{in}(t) = V_0\exp(-a\,t)u(t)$, with $V_0 = 5\,\text{V}$ and $a = 100\ \text{s}^{-1}$. Assume the op amp is ideal and obtain an expression for the output voltage $v_{out}(t)$.

Solution: Because the input equals zero (is shorted) for all $t \leq 0$, the capacitors are initially uncharged. We transform the circuit to the complex frequency domain to obtain the circuit shown in Fig. 18.21, where

$$Z_1 = R_1 + \frac{1}{sC_1} = \frac{1 + sR_1C_1}{sC_1},\ Z_2 = \frac{R_2}{1 + sR_2C_2}.$$

Because the initial conditions are zero, we may use the voltage transfer function to obtain the transform of the output. Thus

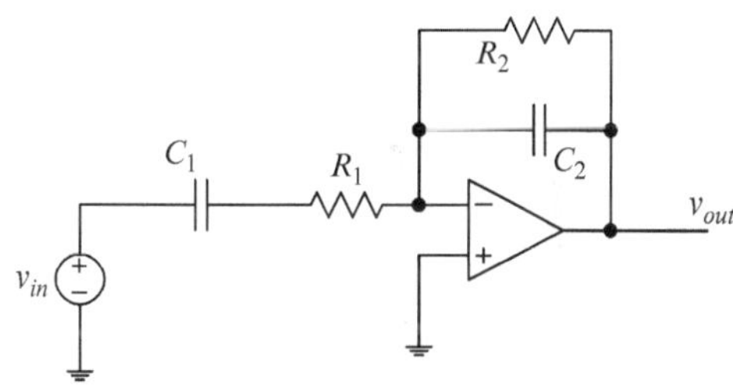

Fig. 18.20 See Example 18.20

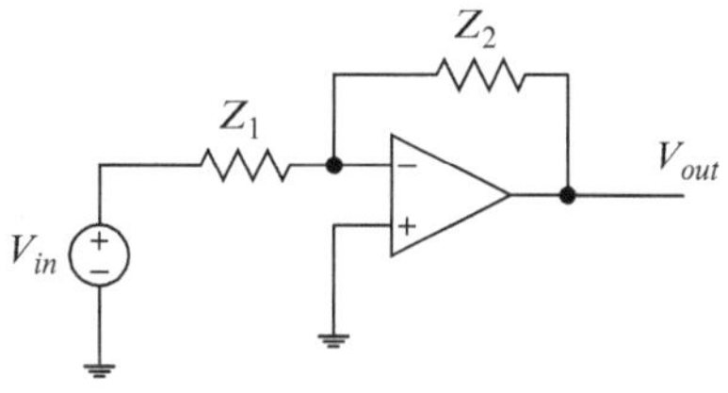

Fig. 18.21 See Example 18.20

$$H_v = \frac{V_{out}}{V_{in}} = -\frac{Z_2}{Z_1}$$
$$= -\frac{sR_2C_1}{(1+sR_2C_2)(1+sR_1C_1)}$$
$$= \frac{Ks}{(s-p_1)(s-p_2)}, \quad (18.68)$$

where

$$K = -\frac{1}{R_1C_2} \cong -606\,\text{s}^{-1},\ p_1 = -\frac{1}{R_1C_1}$$
$$\cong -202\,\text{s}^{-1},\ p_2 = -\frac{1}{R_2C_2} = -50.5\,\text{s}^{-1}.$$

The Laplace transform of the input is

$$V_{in} = \frac{V_0}{s-p_3},$$

where $V_0 = 5\,\text{V}$ and $p_3 = -a = -100\,\text{s}^{-1}$. From (18.68), the Laplace transform of the output is

$$V_{out} = H_vV_{in} = \frac{KV_0s}{(s-p_1)(s-p_2)(s-p_3)}.$$

The partial-fraction expansion is

$$V_{out} = \frac{KV_0s}{(s-p_1)(s-p_2)(s-p_3)}$$
$$= \frac{V_1}{s-p_1} + \frac{V_2}{s-p_2} + \frac{V_3}{s-p_3},$$

where

$$V_1 = \frac{KV_0p_1}{(p_1-p_2)(p_1-p_3)} \cong 39.6\,\text{V},$$
$$V_2 = \frac{KV_0p_2}{(p_2-p_1)(p_2-p_3)} \cong 20.4\,\text{V},$$

and

$$V_3 = \frac{KV_0p_3}{(p_3-p_1)(p_3-p_2)} \cong -60.0\,\text{V}.$$

The output is given by

$$v_{out}(t) = \left[V_1e^{p_1t} + V_2e^{p_2t} + V_3e^{p_3t}\right]u(t)$$
$$= \left[39.6e^{p_1t} + 20.4e^{p_2t} - 60.0e^{p_3t}\right]u(t)\,\text{V}$$

where $p_1 = -202\,\text{s}^{-1}$, $p_2 = -50.5\,\text{s}^{-1}$, and $p_3 = -100\,\text{s}^{-1}$.

Transfer-function representations are appropriate for a circuit that can be represented as a linear time-invariant circuit driven by a current or voltage source. For example, consider the circuit in Fig. 18.22(a), where the switch is moved from a to b at $t = 0$, having been at a for $t < 0$. We can deduce that all initial conditions are zero because no energy is supplied to the circuit before $t = 0$. The source can be represented by a step function and the circuit can be described by a transfer function, as shown in Fig. 18.22(b). On the other hand, the switched source driving the circuit in Fig. 18.22(c) cannot be modeled as a voltage source because the internal impedance of the switched source is infinite for $t < 0$. As a result, we might be unable to deduce the required initial conditions, in which case

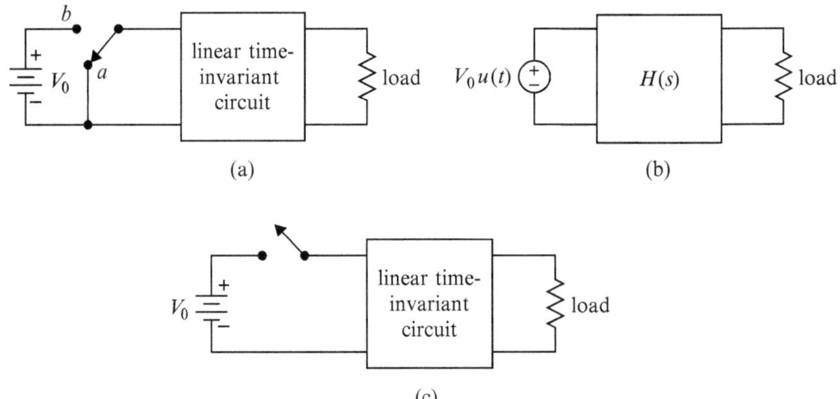

Fig. 18.22 The switched source in (**a**) can be modeled as an independent voltage source, as shown in (**b**). The switched source in (**c**) cannot be so modeled

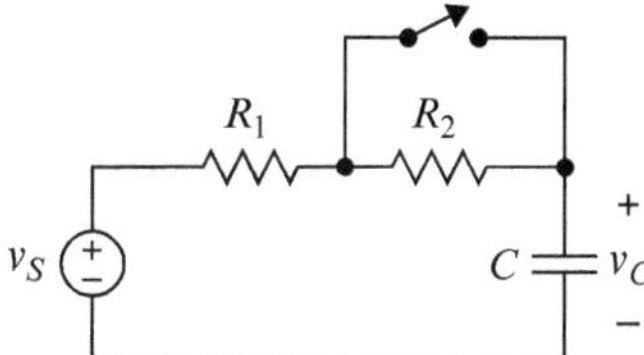

Fig. 18.23 A piecewise-linear circuit

we must be given that information by some higher authority (usually, a professor). And if any initial condition is non-zero, using methods described in this section might be awkward.

Analyses based upon transfer-function representations are most appropriate for linear, time-invariant circuits initially at rest (where all initial conditions are zero). As a rule, circuits having switched components or non-zero initial conditions are best treated using methods described in Chapter 10 or Section 18.9 in this chapter. For example, in Fig. 18.23, where the switch is periodically moved from one position to the other, the circuit is only piecewise linear. The transfer function relating v_C to v_S depends upon the switch position and the voltage v_C immediately after a switching is not necessarily zero.

18.12 Forced Response and Unforced Response

Each term in the partial-fraction expansion for a response can be attributed to either the transfer function or the excitation. Thus, if $Y(s)$ denotes thc response of a transfer having transfer function $H(s)$ to an excitation $X(s)$ and if the initial state of the associate circuit equals zero, we may write

$$Y(s) = H(s)\,X(s) = [\text{terms due to poles of } H(s)] + [\text{terms due to poles of } X(s)]. \tag{18.69}$$

Response terms that arise from the poles of the transfer function comprise the **unforced response** and terms that arise from the poles of the excitation comprise the **forced response**. Thus

$$\mathcal{L}^{-1}[\text{terms due to poles of } H(s)] = \text{Unforced response} \tag{18.70}$$

and

$$\mathcal{L}^{-1}[\text{terms due to poles of } X(s)] = \text{Forced response.} \tag{18.71}$$

Example 18.21 A circuit having a voltage transfer function of the form

$$H_v(s) = \frac{K(s - z_1)}{(s - p_1)(s - p_2)} \tag{18.72}$$

is initially at rest and driven by the excitation

$$v_{in}(t) = V_0 \cos(\omega_0 t) u(t).$$

Obtain expressions for the unforced response and the forced response.

Solution: The Laplace transform of the excitation is

$$V_{in}(s) = \mathcal{L}\{V_0 \cos(\omega_0 t) u(t)\} = \frac{V_0 s}{s^2 + \omega_0{}^2}.$$

The Laplace transform of the response is

$$V_{out}(s) = H_v(s) V_{in}(s) = \frac{K(s - z_1)}{(s - p_1)(s - p_2)} \frac{V_0 s}{(s^2 + \omega_0{}^2)}. \tag{18.73}$$

Expanding the right side in partial fractions gives

$$V_{out}(s) = \underbrace{\frac{A_1}{s - p_1} + \frac{A_2}{s - p_2}}_{\text{Unforced-response terms}} + \underbrace{\frac{A_3}{s + j\omega_0} + \frac{A_3{}^*}{s - j\omega_0}}_{\text{Forced-response terms}}, \tag{18.74}$$

where the unforced terms arise from the poles p_1, p_2 of the transfer function and the forced terms arise from the poles $\pm j\omega_0$ of the input. The complete (time-domain) response has the form

$$v_{out}(t) = \underbrace{(A_1 e^{p_1 t} + A_2 e^{p_2 t})u(t)}_{\text{Unforced response}} + \underbrace{2|A_3|\cos(\omega_0 t + \measuredangle A_3)u(t)}_{\text{Forced response}}. \quad (18.75)$$

The partial-fraction coefficients are given by

$$A_1 = \frac{K(p_1 - z_1)}{(p_1 - p_2)}\frac{V_0 p_1}{(p_1{}^2 + \omega_0{}^2)},$$
$$A_2 = \frac{K(p_2 - z_1)}{(p_2 - p_1)}\frac{V_0 p_2}{(p_2{}^2 + \omega_0{}^2)},$$
$$A_3 = \frac{K(-j\omega_0 - z_1)}{(-j\omega_0 - p_1)(-j\omega_0 - p_2)}\frac{V_0(-j\omega_0)}{(-j\omega_0 - j\omega_0)} = \frac{-K(j\omega_0 + z_1)V_0}{2(j\omega_0 + p_1)(j\omega_0 + p_2)}. \quad (18.76)$$

18.13 Impulse Response and Step Response

By definition, the **impulse response** of a linear circuit model is the response to a *unit impulse excitation* $\delta(t)$ and the **step response** is the response to a *unit step excitation* $u(t)$. Conventionally, a step response is denoted by $g(t)$ and an impulse response by $h(t)$. Figure 18.24 illustrates these definitions.

A step $u(t)$ is dimensionless and a unit impulse $\delta(t)$ has the dimension of $(time)^{-1}$. Neither the excitations $\delta(t)$, $u(t)$ nor the corresponding responses $h(t)$, $g(t)$ are physical. Nonetheless, a unit impulse or a unit step may be applied *mathematically* to a circuit *model* as if each were a current or a voltage. The response of a circuit to an impulse current or voltage is proportional to the impulse response and the response of a circuit to a step current or voltage excitation is proportional to the step response.

A physical current or voltage cannot be infinite – even for an instant. An impulse (a delta function) is an idealization of a pulse or spike whose duration is much shorter than any time constant in a circuit at hand. The impulse response of a circuit indicates how the circuit responds to such an excitation.

A physical current or voltage cannot change instantaneously, because of ubiquitous stray capacitance and inductance. A step is an idealization of a current or voltage that changes from one value to another in a time much shorter than any time constant in a circuit at hand. The step response of a circuit indicates how the circuit responds to such an excitation.

Recall that

$$\mathcal{L}\{\delta(t)\} = 1, \ \mathcal{L}\{u(t)\} = \frac{1}{s}.$$

It follows that the Laplace transforms of the impulse response and step response corresponding to a transfer function $H(s)$ are given by

$$h(t) = \mathcal{L}^{-1}\{H(s)\}, \ g(t) = \mathcal{L}^{-1}\left\{\frac{H(s)}{s}\right\} \quad (18.77)$$

A circuit can be described by any of four transfer functions, so a circuit can have any of four step responses and any of four impulse responses, depending upon which of current or voltage is taken as the input and which of current or voltage is taken as the output. Formally, we may define the four impulse responses as

$$h_v(t) = \mathcal{L}^{-1}\{H_v(s)\}, \ h_i(t) = \mathcal{L}^{-1}\{H_i(s)\},$$
$$h_z(t) = \mathcal{L}^{-1}\{H_z(s)\}, \ h_y(t) = \mathcal{L}^{-1}\{H_y(s)\}, \quad (18.78)$$

and the four step responses as

$$g_v(t) = \mathcal{L}^{-1}\left\{\frac{H_v(s)}{s}\right\}, \ g_i(t) = \mathcal{L}^{-1}\left\{\frac{H_i(s)}{s}\right\},$$
$$g_z(t) = \mathcal{L}^{-1}\left\{\frac{H_z(s)}{s}\right\}, \ g_y(t) = \mathcal{L}^{-1}\left\{\frac{H_y(s)}{s}\right\}. \quad (18.79)$$

The forms of the four impulse responses and the forms of the four step responses depend upon the

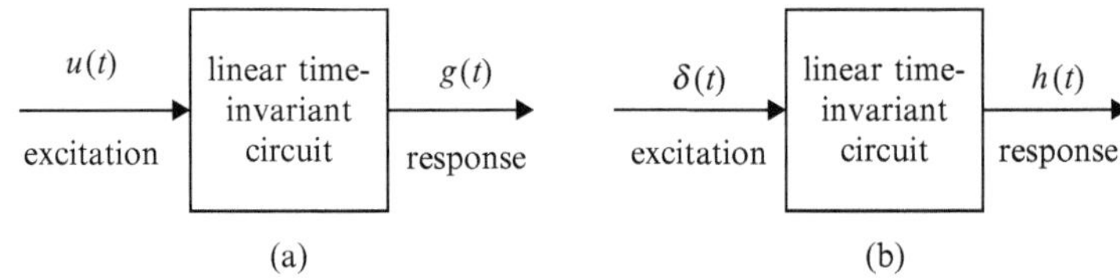

Fig. 18.24 Definitions of (**a**) step response and (**b**) impulse response

nature of the source impedance and load impedance. If the source impedance and load impedance are both resistive, then the four impulse responses differ by only a constant, frequency-independent factor, as do the four step responses.

Example 18.22 The sources in Fig. 18.25(a) are equivalent; i.e., $i_S = v_S/R_S$. Obtain expressions for each of the four impulse responses and the four step responses.
Solution: We transform the circuits to the s domain, as shown in Fig. 18.25(b). The four transfer functions are

$$\begin{aligned} H_v(s) &= \frac{R_L}{R_S(1+sR_LC)+R_L}, \\ &= \frac{1}{R_S}\frac{R_LR_S}{R_S+R_L+sR_SR_LC} \\ &= \frac{R}{R_S}\frac{1}{1+s\tau}, \quad R = R_S||R_L, \\ \tau = RC,\ H_i(s) &= \frac{R_S}{R_L}H_v(s) \\ H_z(s) = R_SH_v(s),\ H_y(s) &= \frac{1}{R_L}H_v(s). \end{aligned}$$

From (18.78), the corresponding impulse responses are

$$\begin{aligned} h_v(t) &= \frac{R}{R_S}\frac{1}{\tau}\exp(-t/\tau)u(t), \\ h_i(t) &= \frac{R_S}{R_L}h_v(t), \\ h_z(t) &= R_Sh_v(t), \\ h_y(t) &= \frac{1}{R_L}h_v(t). \end{aligned}$$

The transforms of the four step responses are

$$\begin{aligned} G_v(s) &= \frac{H_v(s)}{s} = \frac{R}{R_S}\frac{1}{s(1+s\tau)}, \\ G_i(s) &= \frac{R_S}{R_L}G_v(s), \\ G_z(s) &= R_SG_v(s), \\ G_y(s) &= \frac{1}{R_L}G_v(s). \end{aligned}$$

The corresponding step responses are given by

$$\begin{aligned} g_v(t) &= \frac{R}{R_S}[1-\exp(-t/\tau)]u(t), \\ g_i(t) &= \frac{R_S}{R_L}g_v(t), \\ g_z(t) &= R_Sg_v(t), \\ g_y(t) &= \frac{1}{R_L}g_v(t), \end{aligned}$$

In this example, the source and load impedances are resistive, so the four impulse responses differ by only constant factors that are independent of s, as do the four step responses.

In any particular application, only one of the four possible input-output relations is of primary interest. Often, the impulse response and step response of a circuit are defined assuming (implicitly or explicitly) that the input and output are both voltages. But that is not always the case. There is almost always a preferred description for a circuit, whether it be voltage transfer, current transfer, transadmittance, or transimpedance.

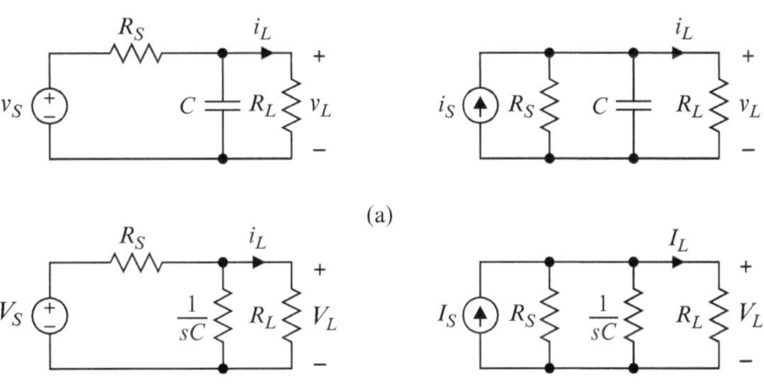

Fig. 18.25 See Example 18.22

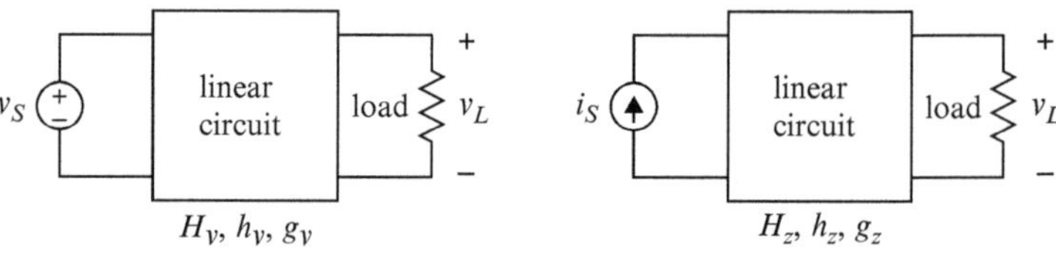

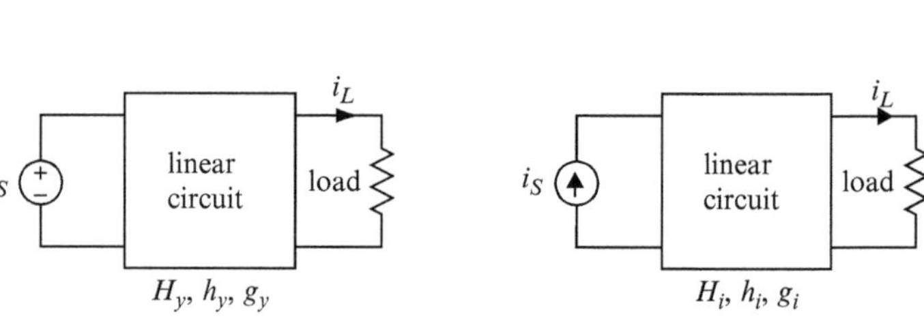

Fig. 18.26 The preferred descriptions of a circuit depend upon the dimensions of the specified input and output

In practice, the preferred description is determined by the specified function of the circuit. In textbook problems, the preferred description is implied by the manner in which a problem is presented, as illustrated by Fig. 18.26. For example, if it is given that the excitation is a current and the response is a voltage, it is implied that the preferred description is a transimpedance.

When we refer to *the* impulse response or *the* step response of a circuit, we mean the impulse response or step response corresponding to the preferred description. Thus, for example, *the* impulse response of a voltage amplifier is given by $h_v(t) = \mathcal{L}^{-1}\{H_v(s)\}$, where $H_v(s)$ is the voltage transfer function for the amplifier, and *the* impulse response of a current amplifier is $h_i(t) = \mathcal{L}^{-1}\{H_i(s)\}$, where $H_i(s)$ is the current transfer function for the amplifier.

It is helpful to know the SI units of impulse and step responses so we can check expressions for dimensional consistency as we work our way through a problem.

The SI units of the four transfer functions are

$$\begin{aligned} &\mathrm{SI}[H_v] = 1,\ \mathrm{SI}[H_i] = 1, \\ &\mathrm{SI}[H_y] = \mathrm{S}, \mathrm{SI}[\mathrm{H_z}] = \Omega. \end{aligned} \quad (18.80)$$

In general, the SI unit of the Laplace transform of a function $x(t)$ is given by

$$\mathrm{SI}[\mathcal{L}\{x(t)\}] = \mathrm{SI}[x]\mathrm{SI}[t]. \quad (18.81)$$

For example, if $x(t)$ is a voltage, then $\mathrm{SI}[X(s)] = \mathrm{V\,s}$. It follows from $H(s) = \mathcal{L}\{h(t)\}$ that

$$\mathrm{SI}[H(s)] = \mathrm{SI}[\mathcal{L}\{h(t)\}] = \mathrm{SI}[h(t)]\mathrm{SI}[t],$$

and thus that the *SI unit of an impulse response* is given by

$$\mathrm{SI}[h(t)] = \frac{\mathrm{SI}[H(s)]}{\mathrm{SI}[t]}. \quad (18.82)$$

The Laplace transform of a unit step is $1/s$, so *the SI unit of the Laplace transform of a step response* is given by

$$\mathrm{SI}[G(s)] = \mathrm{SI}\left[\frac{H(s)}{s}\right] = \mathrm{SI}[H(s)]\mathrm{SI}[t]. \quad (18.83)$$

From (18.81) and (18.83), the *SI unit of a step response* is given by

$$\mathrm{SI}[g(t)] = \mathrm{SI}[H(s)]. \quad (18.84)$$

If the excitation and response have the same dimension, the transfer function is dimensionless, so the step response is dimensionless and the dimension of the impulse response is $(time)^{-1}$.

Example 18.23 For the circuit in Fig. 18.27 (a), the excitation is defined as the available current i_S and the response of interest is the voltage v_C. Obtain expressions for the step response and impulse response. Show that your answers are dimensionally consistent.

Solution: The input–output relation is that of a transimpedance. We redefine the excitation and response as shown in Fig. 18.27(b). For

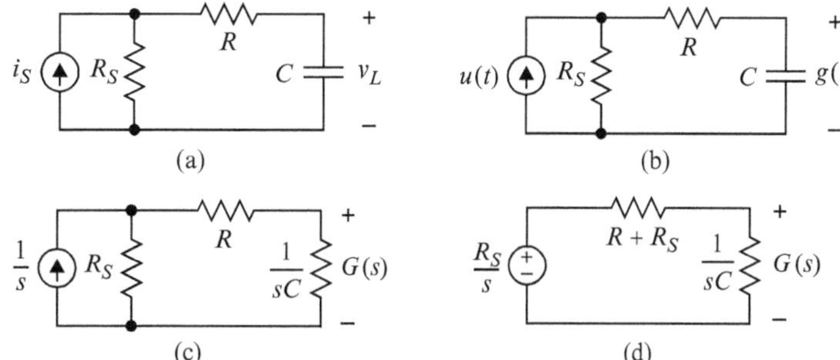

Fig. 18.27 See Example 18.23

$i_S(t) = u(t)$, the current source is effectively an open circuit for $t < 0$, so the initial charge on the capacitor must be zero; i.e., $v_C(0^-) = 0$. Thus the transformed circuit is as shown in Fig. 18.27(c). A source transformation yields the simpler circuit in Fig. 18.27(d). By voltage division,

$$G_z(s) = \frac{1/(sC)}{R + R_S + 1/(sC)} \frac{R_S}{s} = \frac{R_S}{s(1+s\tau)};$$
$$\tau = (R + R_S)C.$$

As a quick check, we find

$$\mathrm{SI}[G_z(s)] = \mathrm{SI}\left[\frac{R_S}{s(1+s\tau)}\right] = \frac{\mathrm{V\,s}}{\mathrm{A}},$$

in agreement with (18.83), which gives us confidence in this intermediate result.

We expand $G_z(s)$ in partial fractions and take the inverse Laplace transform to obtain

$$G_z(s) = \frac{R_S}{s(1+s\tau)} = \frac{R_S/\tau}{s(s+1/\tau)}$$
$$= \frac{R_S}{s} - \frac{R_S}{s+1/\tau}$$
$$\Rightarrow g_z(t) = R_S\left(1 - e^{-t/\tau}\right)u(t);$$
$$\tau = (R + R_S)C.$$

As a quick check, we find

$$\mathrm{SI}[g_z(t)] = \mathrm{SI}\left[R_S\left(1 - e^{-t/\tau}\right)u(t)\right]$$
$$= \mathrm{SI}[R_S] = \frac{\mathrm{V}}{\mathrm{A}},$$

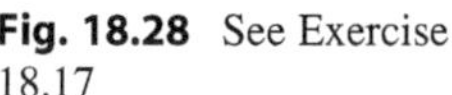

Fig. 18.28 See Exercise 18.17

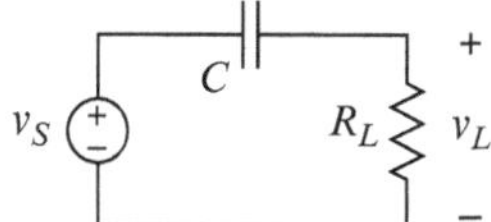

in agreement with (18.84).

To obtain the impulse response, we replace the source R_S/s in Fig. 18.27(c) with R_S. Thus

$$H_z(s) = \frac{R_S}{(1+s\tau)} = \frac{R_S}{\tau}\frac{1}{s+1/\tau} \Rightarrow h_z(t)$$
$$= \frac{R_S}{\tau}e^{-t/\tau}u(t);\ \tau = (R + R_S)C.$$

As a quick check, we find

$$\mathrm{SI}[h_z(t)] = \mathrm{SI}\left[\frac{R_S}{\tau}e^{-t/\tau}u(t)\right] = \frac{\mathrm{SI}[R_S]}{\mathrm{SI}[\tau]} = \frac{\mathrm{V}}{\mathrm{A\,s}}$$
$$= \frac{\mathrm{SI}[response]}{\mathrm{SI}[excitation]\mathrm{SI}[t]},$$

in agreement with (18.82).

Exercise 18.17 Obtain expressions for the step response and impulse response of the circuit in Fig. 18.28, where the excitation is the available voltage $v_S(t)$ and the response is the load voltage $v_L(t)$. Show that the expressions are dimensionally consistent.

Example 18.24 Obtain the impulse response and the step response corresponding to the voltage transfer function

$$H_v(s) = \frac{(k+1)s^2\tau^2}{s^2\tau^2 + (2-k)s\tau + 1}, \qquad (18.85)$$

where $\tau = 1\ \mu s$ and $k = 0.586$.

Solution: The impulse response is given by $h_v(t) = \mathcal{L}^{-1}\{H_v(s)\}$. The numerator and denominator of the transfer function have the same order, so we use long division to obtain

$$H_v(s) = (k+1)\left[1 - \frac{(2-k)s\tau + 1}{s^2\tau^2 + (2-k)s\tau + 1}\right]$$

The poles of the transfer function are

$$s_0, s_0{}^* = \frac{-(2-k)\tau \pm \sqrt{(2-k)^2\tau^2 - 4\tau^2}}{2\tau^2} \cong (-707 \pm j707) \times 10^3\ \mathrm{s}^{-1},$$

so the partial-fraction expansion is

$$H_v(s) = (k+1)\left[1 - \frac{(2-k)s\tau + 1}{\tau^2(s - s_0)(s - s_0{}^*)}\right] = (k+1)\left[1 - \left(\frac{A_1}{s - s_0} + \frac{A_1{}^*}{s - s_0{}^*}\right)\right],$$

where

$$A_1 = \frac{(2-k)s_0\tau + 1}{\tau^2(s_0 - s_0{}^*)} \cong (707 - j0.214) \times 10^3\ \mathrm{s}^{-1}.$$

Thus

$$h_v(t) = \mathcal{L}^{-1}\{H_v(s)\} = (k+1) \times \left[\delta(t) - 2|A_1|\exp\left(-\frac{t}{\tau_0}\right)\cos(\omega_0 t + \theta_1)u(t)\right],$$

where

$$k+1 = 1.586,\ |A_1| \cong 707 \times 10^3\ \mathrm{s}^{-1},$$
$$\theta_1 = \measuredangle A_1 \cong -3.02 \times 10^{-4} \cong 0,$$
$$\tau_0 = -\frac{1}{\mathrm{Re}(s_0)} \cong 1.414\ \mu s,$$
$$\omega_0 = \mathrm{Im}(s_0) \cong 7.072 \times 10^5\ \mathrm{s}^{-1}.$$

The SI unit of the impulse response is s^{-1}, as it should be (because the excitation and response have the same dimension).

The Laplace transform of the step response is given by

$$G_v(s) = \frac{H_v(s)}{s} = \frac{(k+1)s^2\tau^2}{s[s^2\tau^2 + (2-k)s\tau + 1]} = \frac{(k+1)s\tau^2}{s^2\tau^2 + (2-k)s\tau + 1} = \frac{A_2}{s - s_0} + \frac{A_2{}^*}{s - s_0{}^*}$$

where

$$A_2 = \frac{(k+1)s_0}{s_0 - s_0{}^*} = 0.793 + j0.793.$$

Thus

$$g_v(t) = \mathcal{L}^{-1}\{G_v(s)\} = 2|A_2|\exp\left(-\frac{t}{\tau_0}\right)\cos(\omega_0 t + \theta_2)u(t),$$

where

$$2|A_2| \cong 2.243, \theta_2 = \measuredangle A_2 \cong 0.785,$$

and

$$\tau_0 = -\frac{1}{\mathrm{Re}(s_0)} \cong 1.414\ \mu s,$$
$$\omega_0 = \mathrm{Im}(s_0) \cong 7.072 \times 10^5\ \mathrm{s}^{-1}.$$

The step response is dimensionless, as it should be (because the excitation and response have the same dimension).

Impulse response, in particular, has important applications in more advanced work. For example, given the impulse response of a circuit, one can calculate the response to any other excitation in the time domain by a process called *convolution*, but such topics are beyond our scope.

18.14 Relation of s-Domain to Frequency-Domain Transfer Functions

An s-domain transfer function is the corresponding frequency-domain transfer function with $j\omega = j2\pi f$ replaced by s. For example,

$$H(j\omega) = K\frac{1 + jf/f_0}{1 + jf/f_1} = K\frac{1 + j\omega/\omega_0}{1 + j\omega/\omega_1}$$
$$\Rightarrow H(s) = K\frac{1 + s/\omega_0}{1 + s/\omega_1} = K\frac{\omega_1}{\omega_0}\left(\frac{s + \omega_0}{s + \omega_1}\right).$$

Conversely, an frequency-domain transfer function for a circuit can be obtained from the corresponding s-domain transfer function by replacing s with $j\omega = j2\pi f$. For example,

$$H(s) = \frac{K}{1 + s\tau} \Rightarrow H(j2\pi f) = \frac{K}{1 + j2\pi f\tau}$$
$$= \frac{K}{1 + jf/f_\tau};\ f_\tau = \frac{1}{2\pi\tau}.$$

It is largely for this reason that we chose to define f-domain transfer functions as functions of $j\omega$ or $j2\pi f$, rather than ω or f.

Example 18.25 The f-domain transadmittance transfer function for a certain circuit is

$$H_y(j2\pi f) = \frac{K_y(1 + jf/f_0)}{(1 + jf/f_1)\left[1 + 2\alpha jf\big/f_2 - (f/f_2)^2\right]},$$

where the quadratic factor cannot be expressed as a product of linear factors. Obtain the s-domain transadmittance transfer function.

Solution: Because $\omega = 2\pi f$, we may write the f-domain transadmittance transfer function as

$$H_y(j\omega) = \frac{K_y(1 + j\omega/\omega_0)}{(1 + j\omega/\omega_1)\left[1 + 2\alpha j\omega/\omega_2 + (j\omega/\omega_2)^2\right]}.$$

We replace $j\omega$ with s to obtain the s-domain transadmittance transfer function

$$H_y(s) = \frac{K_y(1 + s/\omega_0)}{(1 + s/\omega_1)\left[1 + 2\alpha\, s\big/\omega_2 + (s/\omega_2)^2\right]}$$
$$= \frac{K_y'(s + \omega_0)}{(s + \omega_1)(s^2 + 2\alpha\omega_2 s + \omega_2{}^2)},$$

where

$$K_y' = \frac{\omega_1\omega_2{}^2}{\omega_0}K_y$$

Exercise 18.18 Obtain the f-domain transfer function corresponding to each s-domain transfer function in Example 18.19.

18.15 s-Domain Models for Op Amps and Basic Op-Amp Circuits

Using the frequency-domain description of an op amp, we can find the responses of op-amp circuits to sinusoidal excitations or (by superposition) to excitations composed of several sinusoids. Using s-domain models described in this section, we can find the response of an op-amp circuit to any Laplace transformable excitation.

From Chapter 17, the intrinsic frequency-domain voltage transfer function for an op amp is of the form

$$\mu(j2\pi f) = \frac{\mu_0}{1 + jf/f_0}$$
$$\Rightarrow \mu(j\omega) = \frac{\mu_0}{1 + j\omega/\omega_0}. \tag{18.86}$$

where f_0 is the intrinsic half-power bandwidth of the op amp. It follows that the intrinsic s-domain voltage transfer function for an op amp is of the form

$$\mu(s) = \frac{\mu_0}{1 + s/\omega_0} = \frac{\omega_0\mu_0}{s + \omega_0},\quad \omega_0 = 2\pi f_0.$$

But $\mu_0 f_0 = f_T$, so the intrinsic voltage transfer function for a non-ideal op amp is given by

$$\mu(s) = \frac{\omega_T}{s + \omega_0},\quad \omega_0 = 2\pi f_0,\ \omega_T = 2\pi f_T. \tag{18.87}$$

Also from Chapter 17, the frequency-domain voltage transfer functions for an inverting amplifier, a non-inverting amplifier, and a voltage follower are all of the form

$$H_v(j2\pi f) \cong \frac{H_v(0)}{1+jf/W}, \; W \cong \frac{f_T}{|H_v(0)|} = \frac{f_T}{A_{v0}},$$

where $H_v(0) = -R_2/R_1$ for an inverting amplifier, $H_v(0) = 1 + R_2/R_1$ for a non-inverting amplifier, and $H_v(0) = 1$ for a voltage follower. The corresponding s-domain transfer functions are all of the form

$$H_v(s) = \frac{H_v(0)}{1+s/\omega_W} = \frac{H_v(0)\omega_W}{s+\omega_W}, \; \omega_W = 2\pi W.$$

But $|H_v(0)| = A_{v0}$ and $A_{v0}\omega_W = \omega_T$, so

$$H_v(s) = \left\{ \begin{array}{l} \dfrac{-\omega_T}{s+\omega_W}, \text{ inverting amplifier} \\ \dfrac{\omega_T}{s+\omega_W}, \text{ non-inverting amplifier} \\ \dfrac{\omega_T}{s+\omega_T}, \text{ voltage follower} \end{array} \right\},$$

$$\omega_W = 2\pi W, \; \omega_T = 2\pi f_T. \tag{18.88}$$

These expressions assume no loading; i.e., that the source impedance is a negligible fraction of the amplifier input impedance and the amplifier output impedance is a negligible fraction of the load impedance. These conditions are met in virtually all practical applications.

Example 18.26 The dc voltage gain of a certain op-amp based inverting amplifier is $A_{v0} = 100$. The op amp is rail-to-rail and the supply voltages are ± 15 V. The op-amp slew rate and gain-bandwidth product are $SR = 1\,\mathrm{V\,\mu s^{-1}}$ and $f_T = 1$ MHz, Respectively. If linear operation is required, which of output swing and slew rate limits the maximum amplitude of a step input?

Solution: Denote the step input by $v_{in}(t) = V_0\,u(t)$. Then

$$V_{in}(s) = \frac{V_0}{s}.$$

The transfer function of the amplifier is given by (18.88). If the amplifier operates linearly, the Laplace transform of the response is

$$\begin{aligned} V_{out}(s) = H_v(s)V_{in}(s) &= \frac{-\omega_T V_0}{s(s+\omega_W)} \\ &= -\frac{A_{v0}V_0}{s} + \frac{A_{v0}V_0}{s+\omega_W}, \end{aligned}$$

which yields

$$\begin{aligned} v_{out} &= -A_{v0}V_0[1-\exp(-\omega_W t)]u(t) \\ &= -100V_0[1-\exp(-\omega_W t)]u(t). \end{aligned}$$

The maximum output is $\pm 100V_0$. For linear operation, the output swing must be confined to ± 15 V. Thus, for linear operation, the maximum allowable amplitude of a step input is

$$\max|V_0| = \frac{15\,\mathrm{V}}{100} = 150\,\mathrm{mV}.$$

The derivative with respect to time (the rate of change) of the response is

$$\begin{aligned} \frac{dv_{out}}{dt} &= -\omega_W A_{v0} V_0 \exp(-\omega_W t)u(t) \\ &\quad - A_{v0} V_0[1-\exp(-\omega_W t)]\delta(t) \\ &= -\omega_W A_{v0} V_0 \exp(-\omega_W t)u(t). \end{aligned}$$

The maximum magnitude of the rate of change occurs for $t = 0$ and is given by

$$\left|\frac{dv_{out}}{dt}\right|_{\max} = \omega_W A_{v0}|V_0| = \omega_T|V_0|.$$

If the slew rate is not to limit the speed of response (the rate of change of the output voltage), it is necessary that

$$SR > \omega_T|V_0| \Rightarrow |V_0| < \frac{SR}{\omega_T}.$$

For the given parameter values, the maximum allowable amplitude of a step input is

$$\max|V_O| = \frac{SR}{2\pi f_T} = \frac{1\,\text{V}\,\mu\text{s}^{-1}}{2\pi \times 10^6 s^{-1}}$$
$$= \frac{10^6\,\text{Vs}^{-1}}{2\pi \times 10^6 s^{-1}} \cong 159\,\text{mV}.$$

Thus output swing limits the amplitude of a step input (for linear operation).

18.16 Circuits in Cascade

Analysis in the s domain of two-port circuits in cascade proceeds exactly as in the frequency domain, only with frequency-domain impedances and transfer functions replaced by their s-domain counterparts.

Exercise 18.19 Show that the voltage transfer function for the circuit in Fig. 18.29 is

$$H_v = \frac{V_L}{V_S} = \left(\frac{Z_L}{Z_L + Z_4}\right)\left(\frac{Z_2}{Z_2 + Z_3}\right) \times \left(\frac{Z_1}{Z_1 + Z_S}\right)\mu_1\mu_2. \quad (18.89)$$

Then show that if both circuits are designed for efficient voltage transfer, such that $|Z_L| \gg |Z_4|$, $|Z_2| \gg |Z_3|$, and $|Z_1| \gg |Z_S|$, then the overall voltage transfer function is (to a good approximation)

$$H_v \cong \mu_1\mu_2. \quad (18.90)$$

A product of transfer functions can be performed in any order; e.g., $H_1H_2 = H_2H_1$. But this *mathematical* commutativity does not necessarily imply that the associated *physical circuits* can be cascaded in either order. For example, although the *transfer functions* for linear models of a motor and an amplifier driving the motor commute, the physical motor and the physical amplifier do not. The motor has a mechanical output and cannot drive an electronic amplifier.

Linear two ports might commute *physically*, provided they are both described by either current transfer functions or voltage transfer functions and provided loading is negligible. But a cascade might not live up to design specifications if the order of circuits is changed. For example, if the passband gain of one circuit is much different from that of the other, reversing the order of the circuits might cause one to exceed its allowable output swing or exceed its output limit or slew rate. Linear two ports of different kinds most likely do not commute; for example, a current amplifier (current in, current out) can be followed by a transimpedance amplifier (current in, voltage out), but not the other way 'round, because a current amplifier is not designed for efficient voltage transfer at its input port.

In any event, there is rarely, if ever, any reason to reverse the order of two physical circuits. If an application requires a cascade of circuits, we would simply design the circuits such that they provide the desired transfer function when cascaded in the indicated order.

Exercise 18.20 Show that the transfer function of the circuit in Fig. 18.30 equals $\mu_1\,\mu_2$, regardless of the order in which the (ideal) voltage amplifiers are cascaded.

Exercise 18.21 Draw a circuit diagram, similar to that in Fig. 18.30, for two ideal current amplifiers in cascade and give the transfer function for the cascade. Would the transfer function change if the order of the circuits were reversed?

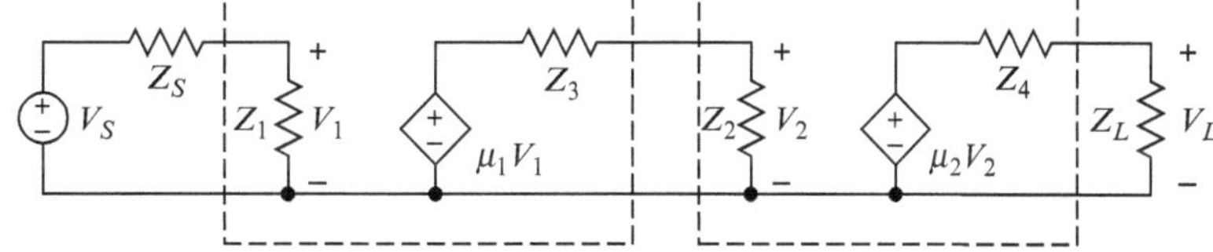

Fig. 18.29 Voltage amplifiers in cascade. See (18.89) and (18.90)

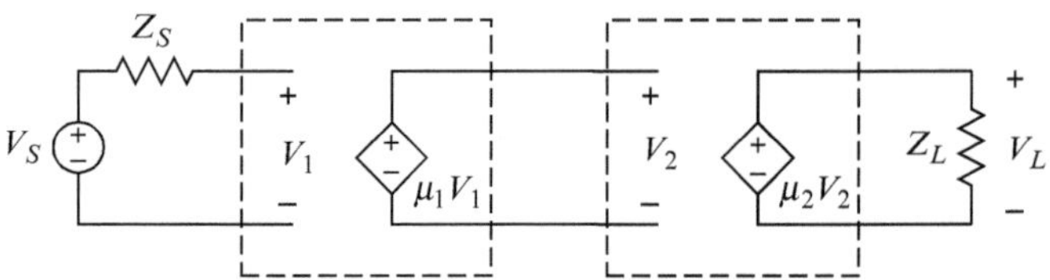

Fig. 18.30 Ideal voltage amplifiers in cascade

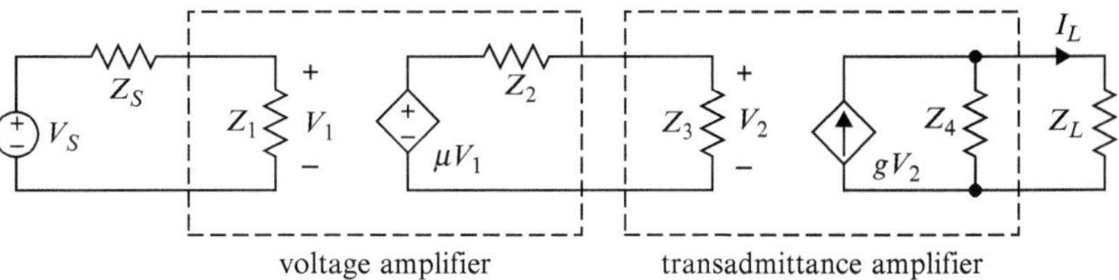

Fig. 18.31 A voltage amplifier and transadmittance amplifier in cascade

Exercise 18.22 Show that if *loading is negligible*, the transadmittance transfer function of the circuit in Fig. 18.31 is

$$H_y \cong \mu g.$$

Would the transfer function change if the order of the circuits were reversed?

18.17 Poles, Zeros, and Pole-Zero Plots

The general form of a transfer function for an *RLC* circuit is

$$H(s) = \frac{Y(s)}{X(s)} = \frac{b_m s^m + b_{m-1} s^{m-1} + \cdots + b_1 s + b_0}{a_n s^n + a_{n-1} s^{n-1} + \cdots + a_1 s + a_0}.$$

We limit our discussion to cases for which $m \leq n$. A transfer function can be expressed in factored form as

$$H(s) = K \frac{(s - z_1)(s - z_2) \cdots (s - z_m)}{(s - p_1)(s - p_2) \cdots (s - p_n)}, \quad (18.91)$$

where $p_1, p_2, \cdots, p_n$ are the finite **poles** of the transfer function and $z_1, z_2, \cdots, z_m$ are the finite **zeros**. Poles and zeros, if complex, occur in conjugate pairs.

A transfer function can also have poles or zeros at infinity. For example,

$$H(s) = K \frac{(s - z_1)}{(s - p_1)(s - p_2)}$$

has one finite zero, two finite poles and one zero at infinity, because

$$\lim_{s \to \infty} H(s) = 0.$$

It can be shown that the total number of poles equals the total number of zeros if poles or zeros at infinity are counted. Henceforth, for economy, *pole* means *finite pole* and *zero* means *finite zero*.

The poles and zeros of a transfer function can be displayed as points on a complex plane called the *s* **plane**. Such a display is called a **pole-zero plot**. The abscissa for each point is the real part of the associated pole or zero and the ordinate is the imaginary part. Conventionally, a zero is represented by a circle (○) and a pole by a cross (×). Multiplicities greater than one are written in parentheses next to the associated circle or cross.

Example 18.27 Draw a pole-zero plot for the transfer function

$$H(s) = \frac{Ks(s - z_1)}{(s - p_1)^2 (s - p_2)(s - p_2{}^*)},$$

where $z_1 = 4\,\mathrm{s}^{-1}$, $p_1 = -6\,\mathrm{s}^{-1}$, and $p_2 = -3 + j\,4\,\mathrm{s}^{-1}$.

Solution: There are two real zeros: $z_0 = 0$ and $z_1 = 4\,\mathrm{s}^{-1}$. There is one real pole $p_1 = -6\,\mathrm{s}^{-1}$ having multiplicity 2. There is a complex-conjugate pair of poles $p_2, p_2{}^* = -3 \pm j4\,\mathrm{s}^{-1}$. Figure 18.32 shows the pole-zero plot, where the scale on each axis is $1\,\mathrm{s}^{-1}$ per division. The constant K is independent of s and has no effect on the pole-zero plot.

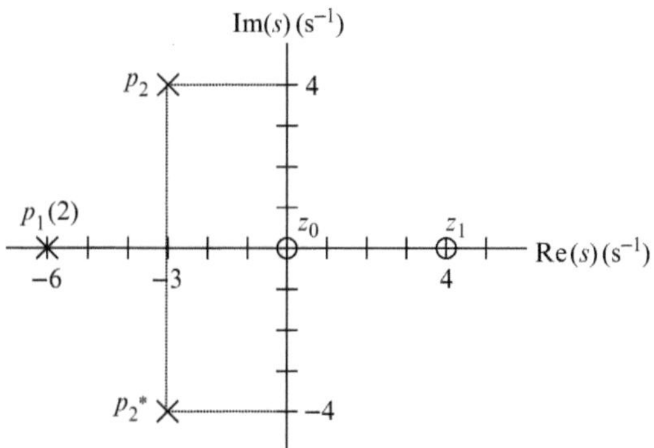

Fig. 18.32 See Example 18.27

A pole-zero plot alone does not tell us what kind of transfer function (e.g., voltage, current, transimpedance, or transadmittance) it represents. That information, if relevant, must be conveyed separately, possibly by specifying the value of the transfer function for a particular value of s or by a statement such as "Fig. 18.32 shows a pole-zero plot for the transadmittance of a certain circuit."

In discussions of pole-zero plots, we often refer to the s plane in halves or quadrants, without explicit reference to s, as illustrated by Fig. 18.33. Similarly, we refer to the $\mathrm{Re}(s)$ axis and the $\mathrm{Im}(s)$ axis as the **real axis** and the **imaginary axis**, respectively, and to the point $s = 0$ as the **origin**, again without explicit reference to s. For example, in Fig. 18.32, the pole p_2 lies in the left half-plane, the upper half-plane, and the second quadrant. The zero z_1 lies on the positive real axis. There is a zero z_0 at the origin. The double pole p_1 lies on the negative real axis. The poles are all in the left half-plane. The zero z_1 is in the right half-plane. The pole p_2 is in the upper half-plane and in the second quadrant and the pole $p_2{}^*$ is in the lower half-plane and the third quadrant.

For economy, we often use **LHP** for *left half-plane* and **RHP** for *right half-plane*.

Exercise 18.23 Sketch neatly and label fully a pole-zero plot for the transimpedance transfer function

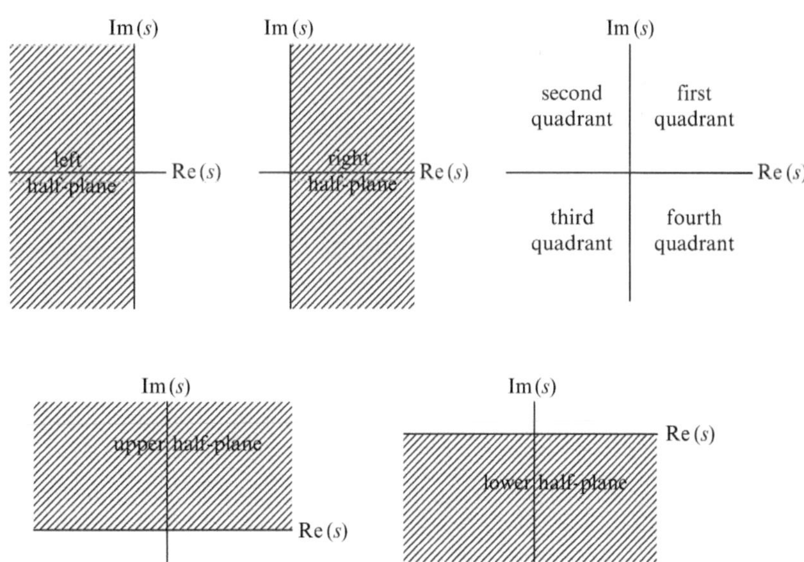

Fig. 18.33 s-plane terminology

$$H_z(s) = \frac{K(s - z_1)(s - z_2)}{(s - p_1)(s^2 + 2\alpha\omega_0 s + \omega_0{}^2)},$$

where $K = 100\Omega s^{-1}$, $z_1 = 10^3 s^{-1}$, $z_2 = -z_1$, $p_1 = -2 \times 10^3 s^{-1}$, $\alpha = 0.4$, and $\omega_0 = 2 \times 10^3 s^{-1}$

A pole-zero plot is a graphical representation of the associated s-domain transfer function, just as a Bode plot is a graphical representation of the associated f-domain transfer function. A pole-zero plot determines the associated transfer function to within a frequency-independent factor (e.g., K in Exercise 18.23), just as a normalized Bode plot determines the associated transfer function to within a frequency-independent factor. A pole-zero plot is useful as an aid in visualizing effects of parameter changes (which change pole and zero locations), for studying or describing the stability or relative stability of an associated circuit, and as a guide (in design) to specifying the poles and zeros (the transfer function) of a circuit.

It is usually difficult or impossible to draw pole-zero plots to scale because the poles and zeros of a transfer function can differ from one another by orders of magnitude. However, this is not a serious limitation on the usefulness of pole-zero plots as conceptual aids in circuit analysis and design. In subsequent sections, we discuss how various important attributes of a circuit are described in terms of or deduced from a pole-zero plot for the appropriate transfer function.

18.18 Stability

A circuit model is stable if and only if every bounded input produces a bounded output.[16]

[16]There are several definitions of stability. The one given here is called the BIBO (bounded-input-bounded output) definition and is an appropriate definition for linear circuit models. Note that no *physical* circuit can actually produce an unbounded output. Mathematical studies of stability necessarily deal with circuit *models* and thus with mathematical models of currents and voltages.

The product of a transfer function $H(s)$ and the transform $X(s)$ of an excitation can be expanded in partial fractions as

$$\begin{aligned} Y(s) &= H(s)X(s) \\ &= \underbrace{\frac{A_1}{s - p_1} + \frac{A_2}{s - p_2} + \cdots + \frac{A_n}{s - p_n}}_{\text{Unforced terms from poles of } H(s)} \\ &\quad + \underbrace{\frac{B_1}{s - q_1} + \frac{B_2}{s - q_2} + \cdots + \frac{B_m}{s - q_m}}_{\text{Forced terms from poles of } X(s)}. \end{aligned} \tag{18.92}$$

Equation (18.92) assumes all poles are distinct. Including multiple poles would clutter this discussion without adding any significant generality.

As indicated by (18.92), terms arising from the poles of the input comprise the forced response and terms arising from the poles of the transfer function comprise the unforced response. In this section, we are concerned with only the unforced response.

Each real pole of a response transform gives rise to a response term of the form

$$\mathcal{L}^{-1}\left\{\frac{A}{s - p}\right\} = A_0 \exp(pt)u(t). \tag{18.93}$$

Each complex-conjugate pair of poles of a response transform gives rise to a response term of the form

$$\begin{aligned} &\mathcal{L}^{-1}\left\{\frac{A}{s - p} + \frac{A^*}{s - p^*}\right\} \\ &\quad = 2|A|\exp(\sigma t)\cos(\omega t + \measuredangle A)u(t), \\ &p = \sigma + j\omega. \end{aligned} \tag{18.94}$$

Figure 18.34 depicts (on the left) responses of the form given by (18.93) for a *real pole* p. If $p < 0$, (if the pole lies in the LHP) the response decreases exponentially and eventually vanishes. If $p = 0$ (if the pole lies on the imaginary axis) the response is a constant (for $t > 0$). If $p > 0$ (if the pole lies in the RHP) the response grows exponentially (is unbounded), and the circuit is unstable.

Figure 18.34 depicts (on the right) responses of the form given by (18.94) for a complex-conjugate pair of poles $p, p^* = \sigma \pm j\omega$ for which $\text{Re}(p) = \sigma$. If $\sigma < 0$, the poles lie in the LHP and the response decreases exponentially and eventually vanishes. If $\sigma = 0$, the

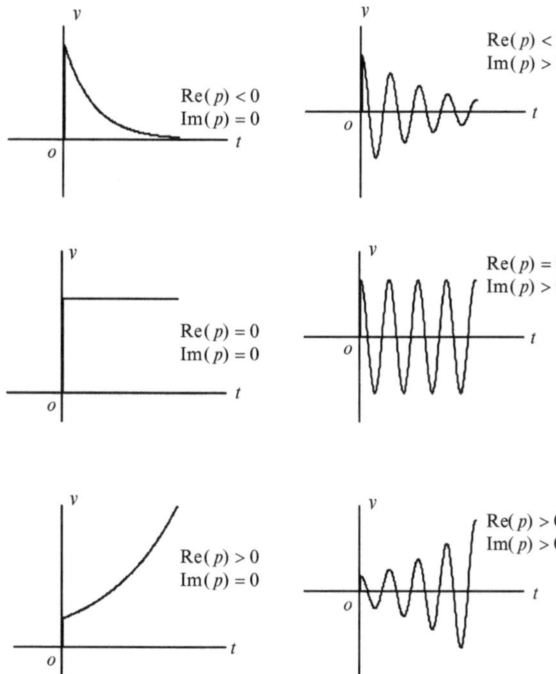

Fig. 18.34 Qualitative graphs of the right sides of (18.93) and (18.94), illustrating the condition $\mathrm{Re}(p) < 0$ for stability

poles lie on the imaginary axis and the response is a sinusoid (for $t > 0$). If $\sigma > 0$, the poles lie in the RHP, the response grows exponentially (is unbounded), and the circuit is unstable.

Figure 18.35 further illustrates effects of pole location on the form of the associated unforced term (for non-repeated poles). A pole in the LHP (one having negative real part) gives rise to a stable (decaying) response. A pole in the RHP (one having positive real part) gives rise to an unstable (growing) response. The rate of decay or growth increases as the pole is moved away from the imaginary axis (horizontally). For a complex pole, the frequency of oscillation increases as the pole is moved away from the real axis (vertically).

If any pole p of a circuit lies in the RHP (if $\mathrm{Re}(p) > 0$), the response arising from that pole will grow without bound, no matter what input is applied to the circuit, and the circuit is unstable.[17] Of course, no physical current or voltage can be unbounded. In a physical circuit, the response increases until limited by an inherent nonlinearity or until one or more components self-destruct. For example, the output of an op amp in an unstable circuit configuration might simply lock up at the supply voltage, regardless of the input.

If a non-repeated pole lies on the imaginary axis (if the real part of the pole p equals zero), the associated term in the unforced response is (for $t > 0$) either a constant (if $p = 0$) or a sinusoid (if $p = j\omega_0 \neq 0$). Physically, it is virtually impossible for the real part of a pole to become and remain exactly on the imaginary axis because a pole is a function of imprecise circuit parameters that change (however slightly) with temperature and age. As a practical matter, we cannot be assured that the pole will not drift into the RHP, in which case the circuit would become unstable. In most applications, a circuit having one or more poles on or very near the imaginary axis would be regarded as insufficiently stable.

A multiple pole on the imaginary axis having multiplicity k and imaginary part ω_0 gives rise to an unforced response term of the form $t^{k-1}\cos(\omega_0\, t)u(t)$, which increases without bound, even if the real part of the pole is exactly zero. But again, in most applications, a circuit having one or more poles in the LHP but very near the imaginary axis would be regarded as insufficiently stable.

A passive circuit cannot be unstable because a bounded input cannot produce an unbounded (mathematically) branch current or node voltage in a passive circuit. If in the course of analyzing a passive *RLC* circuit (one containing no dependent sources) you obtain a pole having a positive real part, you have made an error.

To be unstable, a circuit must have access to a source of power other than the input. In electronic circuits, such access is provided by components such as transistors and op amps, which are energized by power supplies distinct from the input. Such active components are represented using dependent sources in circuit models.

Unstable circuits can be useful, but an unstable *physical* circuit cannot be linear because the response of such a circuit is not linearly related to the excitation. If the transfer function for a linear circuit model contains a pole having a positive real part, we know immediately that the model is at best valid for only a finite time after the circuit is energized. This does not

[17] One can cook up excitations whose zeros cancel (mathematically) one or more poles of a transfer function, but in the real world, such cancellation is imperfect and all unforced terms are present, regardless of the form of the excitation.

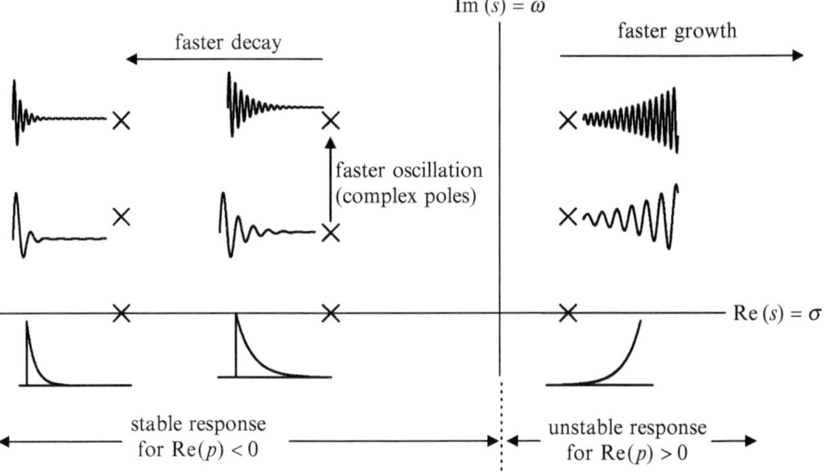

Fig. 18.35 Illustrating how the character of a first-order or second-order unforced response depends upon pole location

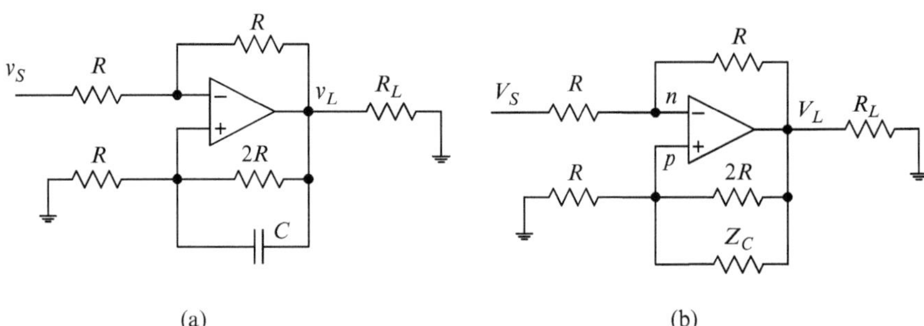

Fig. 18.36 See Example 18.28

necessarily mean the circuit is not operating correctly. Some circuits, such as oscillators, limiters, and comparators, if modeled and analyzed as linear circuits, must be unstable to operate as intended.

Example 18.28 Determine whether the circuit shown in Fig. 18.36(a) is stable. The op amp is ideal.
Solution: Figure 18.36(b) shows the circuit transformed to the s domain. Let

$$Z_C = \frac{1}{sC}, \; Z_4 = \frac{2RZ_C}{2R + Z_C} = \frac{2R}{2RCs + 1}.$$

By voltage division,

$$V_p = \frac{RV_L}{R + Z_4} = \frac{2RCs + 1}{2RCs + 3} V_L. \qquad (18.95)$$

The op amp is ideal, so $V_n = V_p$. Applying Kirchhoff's current law at node n gives

$$\frac{V_p - V_S}{R} + \frac{V_p - V_L}{R} = 0 \Rightarrow 2V_p - V_S - V_L = 0.$$

We use the right side of (18.95) for V_p in this equation and obtain

$$2\frac{2RCs + 1}{2RCs + 3} V_L - V_S - V_L = 0,$$

which yields

$$V_L = \frac{2RCs + 3}{2RCs - 1} V_S \Rightarrow H_v(s) = \frac{2RCs + 3}{2RCs - 1}.$$

The voltage transfer function has a pole $p = (2RC)^{-1}$ in the RHP, so the circuit is unstable.

18.19 Pole-Zero Cancellation

Mathematically, it appears that a circuit having a RHP pole can be stabilized by cascading the circuit with a stable circuit having a RHP zero equal to the RHP pole. Unfortunately, such cancellation is impossible because the precision required to achieve perfect cancellation cannot be achieved with physical circuit components. But even if perfect cancellation were possible, such cancellation could not stabilize an unstable *physical* circuit. For example, refer to Fig. 18.37, where

$$H_{v1}(s) = \frac{K_1}{(s-p_1)}, \; H_{v2}(s) = \frac{K_2(s-p_1)}{(s-p_2)}.$$

The pole p_2 is in the LHP but p_1 is in the RHP. The transfer function for the cascade is

$$H_v(s) = H_{v1}(s)H_{v2}(s) = \frac{K_1K_2}{s-p_2},$$

so it would appear that the cascade is stable. But such is not the case for physical circuits. If the unstable circuit is placed first in the cascade connection, as illustrated by Fig. 18.37(a), its output (the input to the second circuit) will grow exponentially, regardless of what the second circuit does. So even if the output of the second circuit (the overall output) could somehow remain bounded, the intermediate current or voltage produced by the first circuit would grow until limited by nonlinearity (or until the circuit self-destructs). If the unstable circuit is placed second in the cascade connection, as illustrated by Fig. 18.37(b), its output (the overall output) will be unbounded (mathematically), no matter what its input might be.

An unstable physical circuit cannot be made stable by cascading it with a circuit having zeros equal to the RHP poles of the unstable circuit.

The fact that an unstable circuit cannot be stabilized by cascade pole-zero cancellation does not mean that an unstable circuit cannot be stabilized by any means. Indeed, many inherently unstable circuits (op amps, for example) can be stabilized by negative feedback, as illustrated by circuits treated in Chapter 17. You will learn more about this subject if you take subsequent courses in electronic circuits and feedback control systems.

Although cascade cancellation is imperfect, and cannot stabilize an unstable circuit, such cancellation can improve the response of a stable circuit. We pursue that topic in the next section.

18.20 Dominant Poles

The *unforced* component of the step response of a circuit is regarded as the **transient response** of the circuit. This section describes how a pole-zero plot conveys important features of the transient response of the associated circuit and how the pole-zero plot can suggest ways to improve the transient response. We limit this discussion to stable circuits having at least as many poles as zeros.

The voltage transfer function represented by the pole-zero plot in Fig. 18.38 has the form

$$H_v = \frac{K}{(s-p_1)(s-p_2)}, \quad p_1 = -1\,\text{s}^{-1}, \; p_2 = -10\,\text{s}^{-1}. \tag{18.96}$$

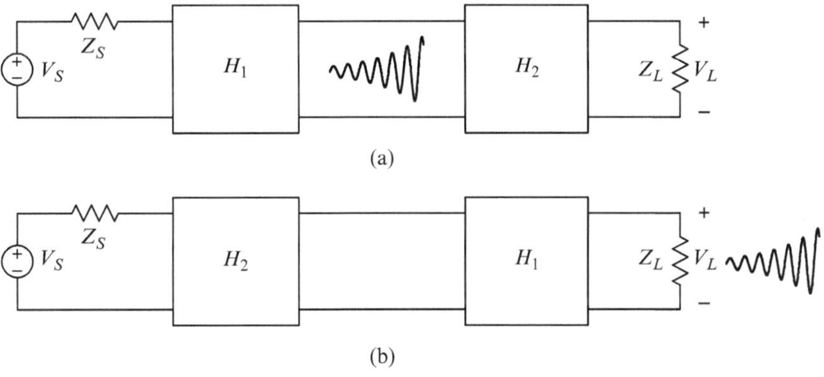

Fig. 18.37 Illustrating that an unstable circuit cannot be stabilized by another circuit in cascade

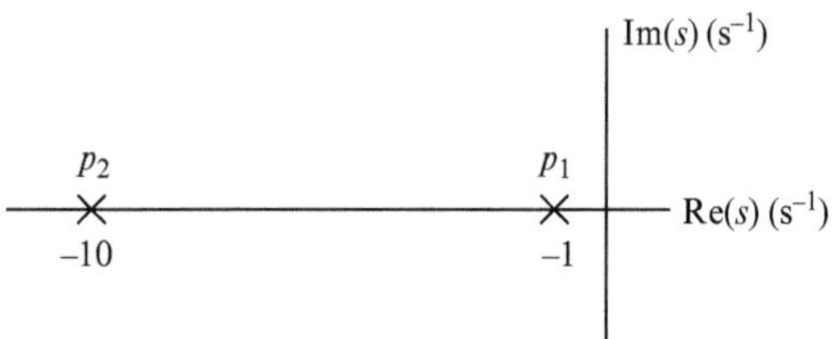

Fig. 18.38 Pole-zero plot for the voltage transfer function given by (18.96)

The transform of the step response is given by

$$G_v = \frac{K}{s(s-p_1)(s-p_2)} = \underbrace{\frac{A_0}{s}}_{\text{Forced}} + \underbrace{\frac{A_1}{s-p_1} + \frac{A_2}{s-p_2}}_{\text{Unforced}}$$

where

$$A_0 = \frac{K}{p_1p_2} = \frac{K}{10}, \quad A_1 = \frac{K}{p_1(p_1-p_2)} = -\frac{K}{9},$$
$$A_2 = \frac{K}{p_2(p_2-p_1)} = \frac{K}{90}.$$

The transient response (the unforced component of the step response) has the form

$$g_u(t) = \left[A_1 \exp\left(-\frac{t}{\tau_1}\right) + A_2 \exp\left(-\frac{t}{\tau_2}\right)\right]u(t), \tag{18.97}$$

where

$$\tau_1 = -\frac{1}{p_1} = 1\,\text{s}, \quad \tau_2 = -\frac{1}{p_2} = 0.1\,\text{s}.$$

Figure 18.39 shows graphs of the first 1 s of the transient response $g_u(t)$ (solid line) and of the term

$$g_1(t) = A_1 \exp\left(-\frac{t}{\tau_1}\right)u(t), \tag{18.98}$$

(dashed line). After a few tenths of 1 s, the transient response and the term $g_1(t)$ are virtually indistinguishable.

Focus your attention on two aspects of the transient response given by (18.97) and shown graphically in Fig. 18.39:

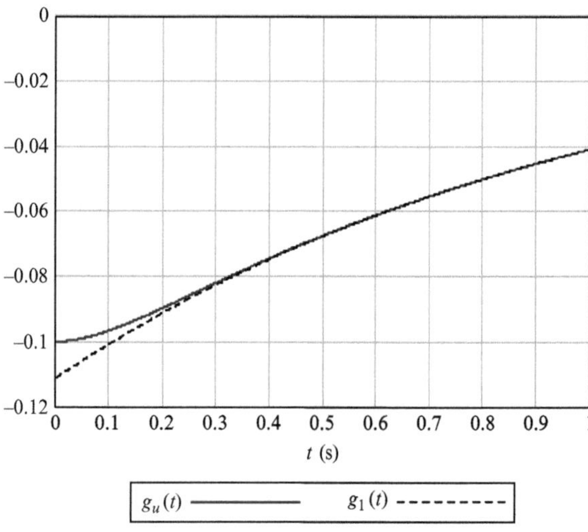

Fig. 18.39 Graphs of the transient response given by (18.97) and of the dominant term given by (18.98)

- The *larger time constant* τ_1 *is associated with the pole* p_1 *nearest the origin* and determines the duration of the transient response. In this example, $\tau_1 = 10\tau_2$, so the second term in (18.97) influences only about the first one-tenth of the transient response.
- The *larger partial-fraction coefficient* $A_1 = -K/9$ *is also associated with the pole* p_1 *nearest the origin* and is ten times larger than the other coefficient A_2. Thus the first term in (18.97) is initially ten times larger and persists ten times longer than the second.

So in brief, the first term in (18.97), associated with the pole nearest the origin, largely determines the form and duration of the transient response. The pole p_1 is the **dominant pole** of the transfer function H_v (or of the associated circuit).

In this example, the pole p_2 is ten times farther from the origin than p_1. If p_2 were moved closer to the origin but kept to the left of p_1, then p_1 would be less dominant, but would still determine the duration of the transient.

If the transient response of a circuit is too sluggish, we might be able to speed it up by cancelling (approximately) the dominant pole(s), making one or more original poles or added poles dominant, as illustrated by the next example.

Example 18.29 For the circuit in Fig. 18.40, the op amps are ideal, $R_1 = 10\,\text{k}\Omega$, $R_2 = 100\,\text{k}\Omega$, $C_1 = 1\,\text{nF}$, and $C_2 = 10\,\text{nF}$. These values cannot be changed, but we have access to all the internal connection points (nodes). The source impedance is negligible and the load impedance is effectively infinite. Construct a pole-zero plot for the circuit, identify the dominant pole (if any), and suggest a modification to the circuit (an additional component) that would speed up the transient response by canceling the dominant pole.

Solution: The excitation and response are both voltages. The circuit is a cascade of inverting amplifiers. The voltage transfer function for the cascade is

$$H_v = \left(\frac{R_2}{R_1}\right)^2 \frac{p_1 p_2}{(s-p_1)(s-p_2)},\ p_1 = -\frac{1}{R_2 C_1} = -10^4\,\text{s}^{-1},\ p_2 = -\frac{1}{R_2 C_2} = -10^3\,\text{s}^{-1}.$$

Figure 18.41 shows the pole-zero plot. The pole p_2 is dominant and determines the duration of the transient response. The duration of the transient is approximately

$$-\frac{5}{p_2} = 5\,\text{ms}.$$

We can speed up (shorten the duration of) the transient by adding a zero to cancel the pole p_2 and make p_1 dominant.

In this case, there are several ways to add a zero. One of the simplest is to add a capacitor in parallel with one of the input resistors (either R_1); e.g., as shown in Fig. 18.42. The voltage transfer function for the modified circuit is

$$H_v = -\frac{C_0}{C_1}\frac{p_1(s-z)}{(s-p_1)(s-p_2)},\ z = -\frac{1}{R_1 C_0}.$$

To cancel the pole p_2, we specify $z = p_2$, or

$$\frac{1}{R_1 C_0} = \frac{1}{R_2 C_2} \Rightarrow C_0 = \frac{R_2 C_2}{R_1} = 100\,\text{nF},$$

and the transfer function becomes

$$H_v = -\frac{C_0}{C_2}\frac{p_1}{(s-p_1)} = \frac{K}{(s-p_1)},\ p_1 = -\frac{1}{R_2 C_1} = -10^4\,\text{s}^{-1},\ K = -\frac{C_0 p_1}{C_2} = 10^5\,\text{s}^{-1}.$$

The transform of the modified step response is

$$G_v = \frac{K}{s(s-p_1)} = -\frac{K}{p_1 s} + \frac{K}{p_1(s-p_1)},$$

so the modified transient response is

$$g_v(t) = \frac{K}{p_1}\exp\left(-\frac{t}{\tau_1}\right)u(t),\ \tau_1 = -\frac{1}{p_1} = 100\,\mu\text{s}.$$

The duration of the transient is decreased by a factor of ten – from 5 ms to 500 µs.

Modifying a circuit to improve transient response or certain other aspects of performance is called **compensation**. Thus the circuit

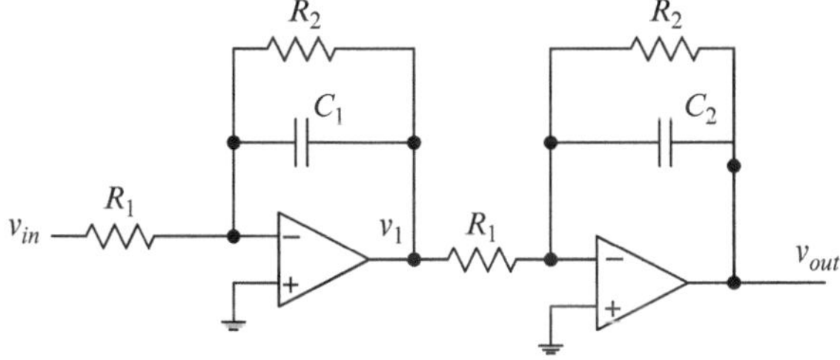

Fig. 18.40 Two-pole circuit. See Example 18.29

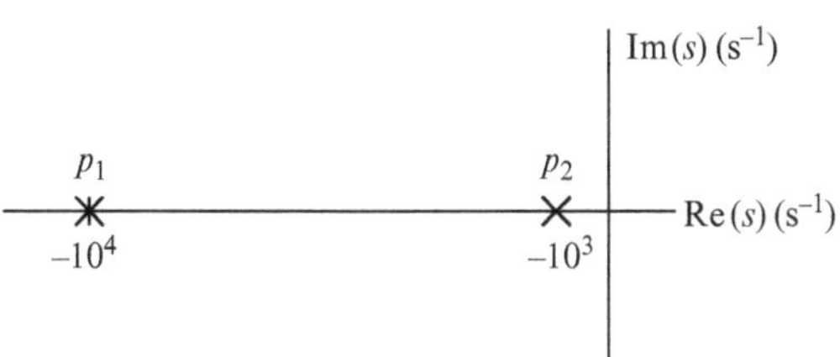

Fig. 18.41 Pole-zero plot for the circuit in Fig. 18.40

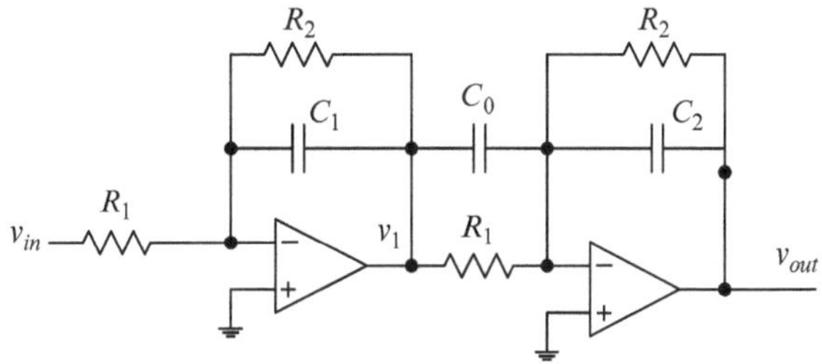

Fig. 18.42 Circuit of Fig. 18.40, compensated to speed up the transient response

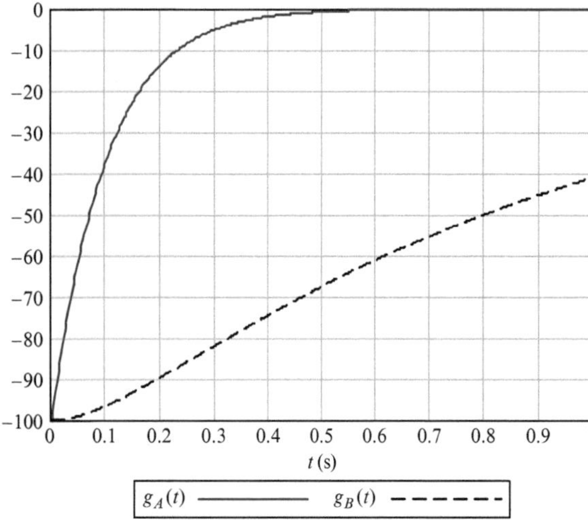

Fig. 18.43 Transient responses of the compensated $[g_A(t)]$ and uncompensated $[g_B(t)]$ circuits treated in Example 18.29

in Fig. 18.42 is a compensated version of the circuit in Fig. 18.40.

Figure 18.43 Shows the transient response before $[g_B(t)]$ and after $[g_A(t)]$ compensation. The transient response of the compensated circuit is essentially over before the transient response of the uncompensated circuit gets started.

Exercise 18.24 Repeat Example 18.29, but put the compensating capacitor in parallel with the first-stage input resistor.

Identifying the dominant pole(s) of a circuit being designed helps identify critical circuit parameters and can suggest how to compensate the associated circuit to improve the transient response. But do not get overly enthralled with pole cancellation as a means of improving transient response. For example, it might seem possible (mathematically) to use an electric circuit to cancel a sluggish pole of an electric motor and thereby improve the acceleration of the motor. However, if acceleration is limited primarily by inductance and mechanical inertia, attempting to overcome these problems electrically could lead to large currents that damage the motor.

18.21 Pole-Zero Plots and Bode Plots

If we replace s with $j\omega$ in an s-domain transfer function $H(s)$, we obtain the corresponding frequency-domain transfer function $H(j\omega)$. Thus the imaginary axis in the s plane is called the **frequency axis**, and values of $H(s)$ above the frequency axis are values of $H(j\omega) = H(j2\pi f)$. With a little practice, you will be able to visualize important features of the frequency response of a circuit by inspecting a pole-zero plot for the circuit.

Consider the transfer function

$$H(s) = \frac{K(s-z)}{s-p}, \tag{18.99}$$

having one pole p and one zero $z < 0$. The frequency-domain transfer function has the form

$$H(j2\pi f) = \frac{K(j2\pi f - z)}{j2\pi f - p} = \frac{Kz}{p}\frac{1 + jf/f_z}{1 + jf/f_p}, \tag{18.100}$$

where

$$f_z = -\frac{z}{2\pi},\ f_p = -\frac{p}{2\pi}.$$

Figure 18.44 shows pole-zero plots and Bode gain plots for two pole locations relative to the zero. In Fig. 18.44(a), $f_z = 1\,\text{Hz}$ and $f_p = 10\,\text{Hz}$. In Fig. 18.44(b), $f_p = 1\,\text{Hz}$ and $f_z = 10\,\text{Hz}$.

Because $|z| = 2\pi f_z$ and $|p| = 2\pi f_p$, the pole p and the angular frequency $\omega_p = 2\pi f_p$ lie on one circle centered on the origin and the zero z and the angular frequency $\omega_z = 2\pi f_z$ lie on another such circle, as illustrated by the pole-zero plots in Fig. 18.44. To visualize the Bode gain plot from the associated pole-zero plot, imagine calculating gain at each step of the way as you travel from the origin up the imaginary axis. Consider the case $z < p$, illustrated by Fig. 18.44(a). Initially, before you reach either circle, the (asymptotic) gain is independent of frequency. As you continue on, you encounter the *zero frequency* ω_z (f_z) first, at which point the gain begins to increase at 20 dB/decade. Then you encounter the *pole frequency* ω_p (f_p), at which point the asymptotic gain again becomes independent of frequency. If $p < z$, you encounter the pole frequency first, and the gain plot steps down instead of up, as illustrated by Fig. 18.44(b).

With a slight modification, the visualization illustrated by Fig. 18.44 extends to higher order transfer functions (circuits). Refer to Fig. 18.45, which shows a

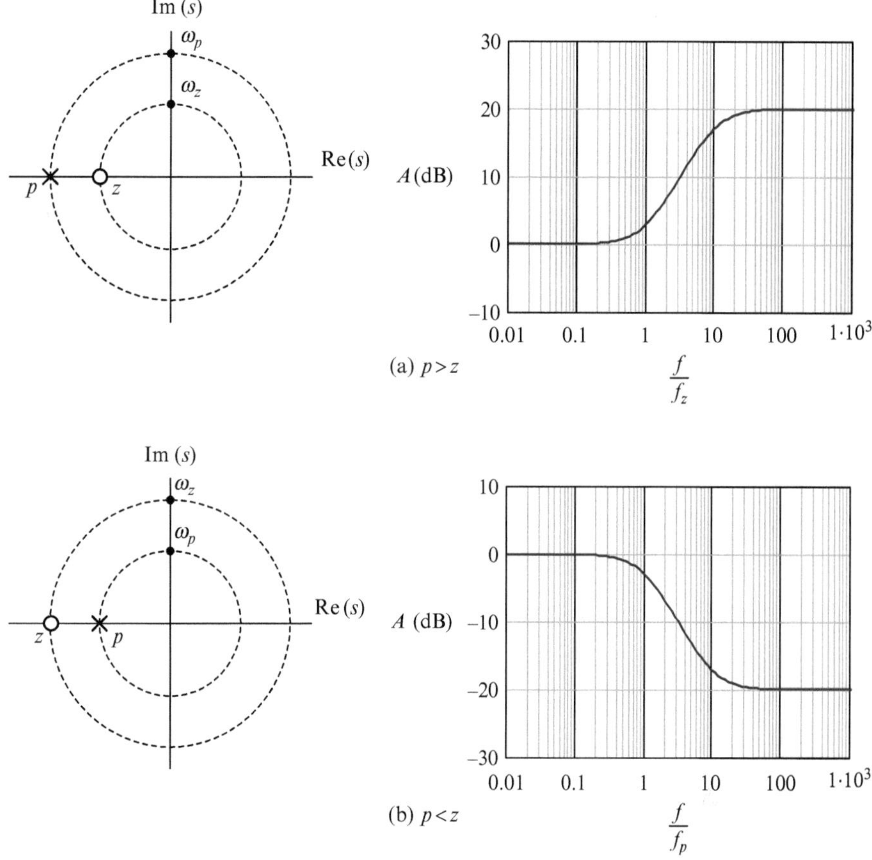

Fig. 18.44 Pole location and frequency response for a first-order circuit. See (18.100)

Fig. 18.45 Pole location and frequency response for a second-order circuit

pole-zero plot and the corresponding Bode gain plot. There is a zero on the real axis and complex-conjugate poles p, p^* on a circle having radius $\omega_p = |p| = 10\omega_z$, with $f_z = 1$ Hz. Beginning at the origin and moving up the imaginary axis, we encounter the zero frequency first, at which point the (asymptotic) gain plot begins to rise at 20 dB/decade. As we continue up the axis, we subsequently encounter the pole frequency, at which point the slope of the asymptotic plot decreases by 40 dB/decade, because there are two poles on the circle having radius ω_p. The slope of the asymptotic plot above the frequency f_p is -20 dB/decade.

Example 18.30 Figure 18.46 shows (not to scale) the pole-zero plot for a certain transfer function. Draw the corresponding asymptotic Bode gain plot.

Solution: We calculate the radii of the circles passing through the poles and zeros:

$$|p_1| = |(-10\pi + j4\pi) \times 10^3|\text{s}^{-1}$$
$$= 2\pi 10^3\sqrt{5^2 + 2^2}\ \text{s}^{-1} \cong 2\pi \times 5390\,\text{s}^{-1}$$
$$\Rightarrow f_1 = \frac{|p_1|}{2\pi} \cong 5.39\,\text{kHz}$$
$$|p_2| = |(-4\pi + j6\pi) \times 10^2|\text{s}^{-1}$$
$$= 2\pi 10^2\sqrt{2^2 + 3^2}\ \text{s}^{-1} \cong 2\pi \times 361\,\text{s}^{-1}$$
$$\Rightarrow f_2 = \frac{|p_2|}{2\pi} \cong 361\,\text{Hz}$$
$$|z| = |(-6\pi + j2\pi) \times 10|\text{s}^{-1}$$
$$= 2\pi \times 10\sqrt{3^2 + 1^2}\,\text{s}^{-1} \cong 2\pi \times 31.6\,\text{s}^{-1}$$
$$\Rightarrow f_z = \frac{|z|}{2\pi} \cong 31.6\,\text{Hz}$$

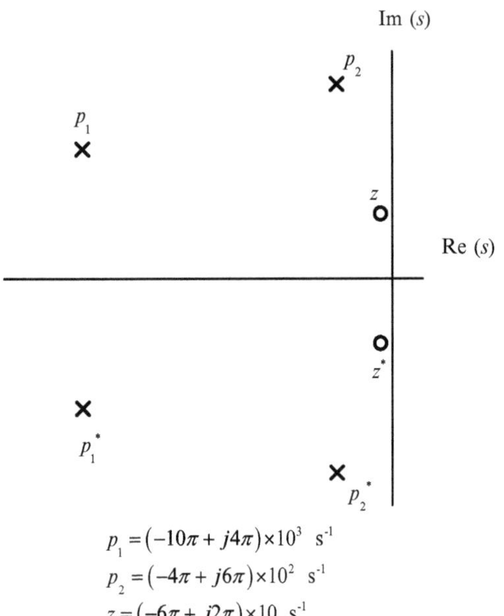

Fig. 18.46 See Example 18.30

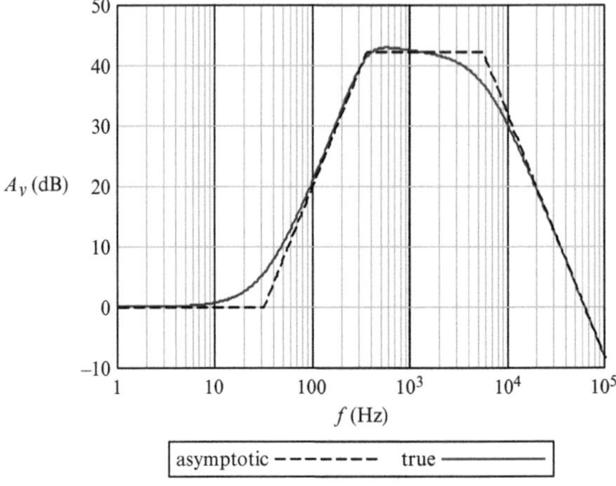

Fig. 18.47 See Example 18.30

We cannot show the pole-zero plot to scale because the poles and zero differ by two orders of magnitude. We cannot show the circles on the pole-zero plot for the same reason. But we can picture the circles and their intersections with the imaginary (frequency) axis in our mind's eye. Moving up the imaginary axis from the origin, we encounter the zero frequency $f_z = 31.6$ Hz first, at which point the gain begins to increase by 40 dB/decade (because there are two zeros on the associated circle). Next, we encounter the pole frequency $f_{p_1} = 361$ Hz, and the slope of the asymptotic plot decreases by 40 dB/decade, so the gain becomes independent of frequency. Finally, we encounter the pole frequency f_{p_2}, and from that point on, the gain decreases by 40 dB/decade. Figure 18.47 shows the asymptotic (dotted line) and actual (solid line) Bode gain plots. The problem statement did not specify the dc gain, so the Bode plots assume a dc gain of unity (zero dB) and are correct to within a frequency-independent scale factor.

The graphical method described and illustrated above is essentially a shortcut equivalent to forming $H(j2\pi f)$ in standard form and constructing the associated Bode plot in the usual way.

18.22 Problems

Section 18.4 is prerequisite for the following problems.

P 18.1 If $i(t)$ represents a current, what is the SI unit of $\mathcal{L}\{i(t)\}$?

P 18.2 Let $V(s)$ denote the Laplace transform of a voltage and use the defining integral to show that $\mathrm{SI}[\mathcal{L}^{-1}\{V(s)\}] = \mathrm{V}$.

P 18.3 If $V(s)$ is the Laplace transform of voltage and $I(s)$ is the Laplace transform of a current, what is the SI unit of $V(s)/I(s)$?

P 18.4 Which of the following functions have (one-sided) Laplace transforms? Explain your answers.

(a) $\left(\frac{t}{\tau}\right)^5 e^{-t/\tau}u(t);\ \tau > 0$ (b) $e^{\omega(t-t_0)^2}u(t);\ \omega > 0$
(c) $\cosh(\omega t)u(t);\ \omega < 0$ (d) $\sinh(\omega t)u(t);\ \omega < 0$
(e) $\cos(\omega^2 t^2)u(t)$ (f) $\sin(\omega^2 t^2)u(t)$
(g) $10^{3t/\tau}u(t);\ \tau > 0$
(h) $(e^{\omega t})^4 u(t);\ \omega > 0$ (i) $(e^{\omega^2 t^2})^{-3}u(t)$

Section 18.5 is prerequisite for the following problems.

P 18.5 Use step functions and impulses to describe each function in Fig. P 18.1. The strength of each impulse is the product of 1 μs and the length of the arrow representing the impulse.

P 18.6 Use a sketch to show that for Δt sufficiently small,

$$\frac{u(t) - u(t - \Delta t)}{\Delta t} \cong \delta(t)$$

Then use this approximation and another sketch to show that $\delta(kt) = k^{-1}\delta(t)$.

P 18.7 Simplify each function to the extent possible. Assume $t_0 > 0$

(a) $a(t) = e^{-(t-t_0)/\tau}[\cos(\omega t) - \sin(\omega t)][a\delta(t) - I_0 u(t)]$
(b) $b(t) = \mathrm{Re}\{(3 - j4)e^{j\omega t}\}[2u(t) - b\delta(t - t_0)]$
(c) $c(t) = 10\cos(\omega t)[c_1\delta(t) + c_2\delta(t - t_0) + c_3\delta(t - 2t_0) + c_4\delta(t - 3t_0)]$
$\omega = 2\pi f_0,\ f_0 = 1\,\mathrm{kHz},\ t_0 = 250\,\mu\mathrm{s}$;
(d) $d(t) = [u(t) - u(t - 5t_0)][\delta(t - t_0) + \delta(t - 4t_0) + \delta(t - 6t_0)]$
(e) $e(t) = u(3t - 6t_0)\delta(t - 3t_0)$

P 18.8 In Fig. P 18.2, the inductors are identical. The switch is opened at $t = 0$, having been closed for $t < 0$. Obtain expressions for the voltages $v_I(t),\ v_1(t),\ v_R(t),\ v_2(t)$.

P 18.9 In Fig. P 18.3, the voltages $v_1(t),\ v_2(t),\ v_3(t)$ are all zero for $t < 0$. The switch is closed at $t = 0$. Obtain expressions for the current $i(t)$ and the voltages $v_1(t),\ v_2(t),\ v_3(t)$.

P 18.10 Complete the indicated operations.

(a) $\frac{d}{dt}\{V_S[1 - \cos(\omega_0 t)]u(t)\}$
(b) $\frac{d}{dt}\{V_S\cos(\omega_0 t)\cos(\omega_1 t)u(t)\}$
(c) $\frac{d}{dt}\{V_S\exp(-t/\tau)\cos(\omega_0 t + \theta)[u(t) - u(t - t_0)]\}$
(d) $\frac{d}{dt}[u(t - t_1)u(t_2 - t)];\ t_2 > t_1 > 0$
(e) $\frac{d}{dt}\left\{V_S(t/\tau)^2\exp(-t/\tau)\cos(\omega_0 t + \theta)u(t)\right\}$
(f) $\frac{d}{dt}\left[\sum_{n=1}^{3} V_n\exp(\sigma_n t)\cos(\omega_n t)u(t)\right]$
(g) $\int_{-\infty}^{\infty} V_S\exp(-t/\tau)\cos(\omega_0 t)\delta(t)dt$
(h) $\int_{-\infty}^{\infty} V_S\exp(-t/\tau)\sin(\omega_0 t)\delta(t)dt$
(i) $\int_{-\infty}^{\infty} V_S\exp(-t/\tau)\cos(\omega_0 t + \theta)\delta(t)dt$
(j) $\int_{-\infty}^{\infty} [u(t - t_0) - u(t - t_2)]\delta(t - t_1)dt;\ t_2 > t_1 > t_0$
(k) $\int_0^t \delta(t' - t_0)dt'$
(l) $\int_{-\infty}^{\infty} V_S\exp(-t/\tau)\delta(t - t_0)dt$
(m) $\int_{-\infty}^{\infty} V_S\exp(-t/\tau)\cos(\omega t)\delta(t - t_0)dt$

Section 18.6 is prerequisite for the following problems.

P 18.11 Show that if $\mathcal{L}\{f(t)\} = F(s)$ then

$$\mathcal{L}\{tf(t)\} = -\frac{dF(s)}{ds}$$

P 18.12. Refer to Problem P 18.11. Show that

$$\mathcal{L}\{t^n f(t)\} = (-1)^n\frac{d^n F(s)}{ds^n}$$

P 18.13 Derive the integration property of the Laplace transformation.

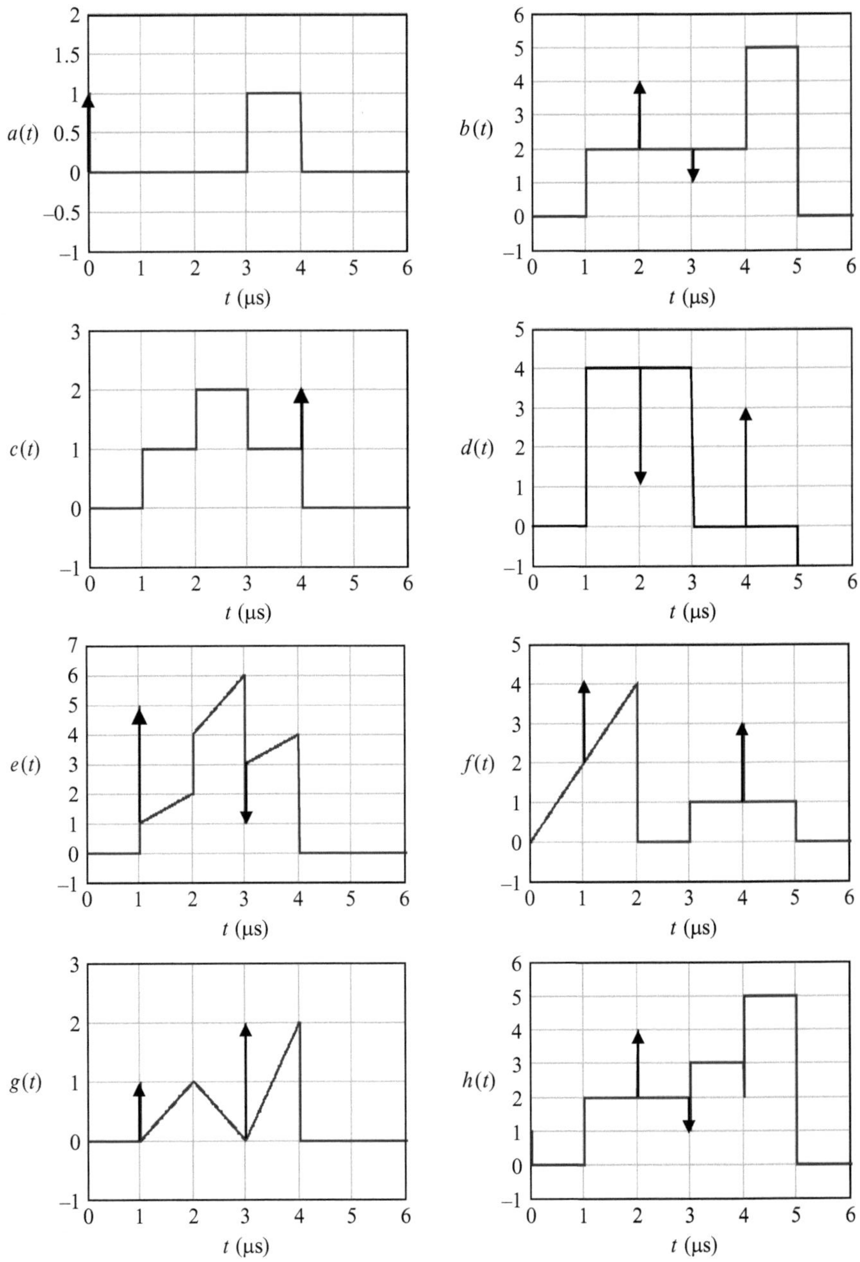

Fig. P 18.1 See Problem P 18.5

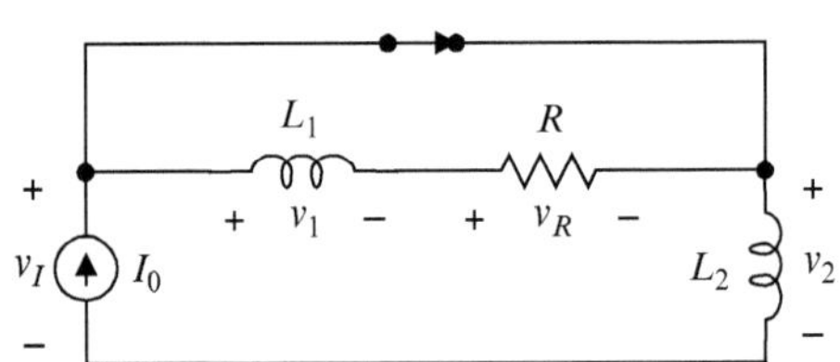

Fig. P 18.2 See Problem P 18.8

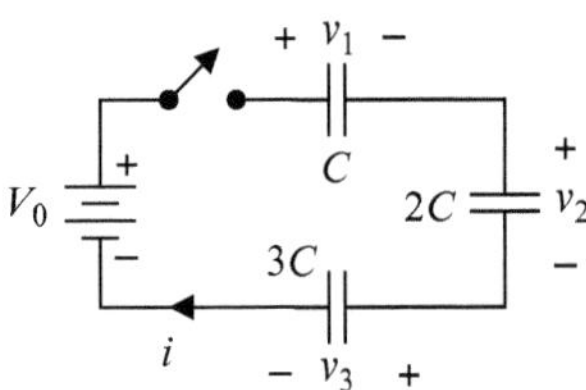

Fig. P 18.3 See Problem P 18.9

P 18.14 Use the tables in the text to obtain each of the following transforms, where $\mathcal{L}\{f(t)\} = F(s)$:

(a) $\mathcal{L}\left\{\int_{0-}^{t-t_0} f(t')dt'\right\}$ (b) $\mathcal{L}\left\{e^{-at}\int_{0-}^{t}\cos(\omega t')dt'\right\}$

(c) $\mathcal{L}\left\{\frac{d}{dt}[tf(t)]\right\}$ (d) $\mathcal{L}\left\{\int_{0-}^{t} f(t'-t_0)u(t'-t_0)dt'\right\}$

(e) $\mathcal{L}\{\cos(\omega t+\theta)f(t)\}$ (f) $\mathcal{L}\{(t/\tau)\cos(\omega t)u(t)\}$

(g) $\mathcal{L}\left\{\frac{d}{dt}[\sin(\omega t)f(t)]\right\}$ (h) $\mathcal{L}\{\cosh(t/\tau)f(t)\}$

(i) $\mathcal{L}\{(t/\tau)\exp(-t/\tau)u(t)\}$

(j) $\mathcal{L}\{[(t-t_0)/\tau]\exp[-(t-t_0)/\tau]u(t-t_0)\}$

(k) $\mathcal{L}\{(t/\tau)\cos(\omega t+\theta)u(t-t_0)\}$

(l) $\mathcal{L}\left\{(t/\tau)^2\exp(-t/\tau)\cos(\omega t+\theta)u(t)\right\}$

(m) $\mathcal{L}\{(t/\tau)a\delta(t-t_0)\}$

(n) $\mathcal{L}\left\{\frac{d}{dt}\{[(t/\tau)\exp(-t/\tau)\cos(\omega t+\theta)u(t)]\}\right\}$

P 18.15 Below, t is time; τ, t_0, t_1 are positive constants having the dimension of time; ω_0 is a constant having the dimension of $(\text{time})^{-1}$; and θ is an angle in radians. Obtain the Laplace transform of each function. Simplify each to the extent possible. Show that each transform has the dimension of time.

(a) $(t/\tau)\,u(t)$ (b) $\omega_0 t\cos(\omega_0 t+\theta)u(t)$

(c) $\omega_0 t\cos^2(\omega_0 t+\theta)u(t)$ (d) $\omega_0 t\sin(\omega_0 t+\theta)u(t)$

(e) $\sin^2(\omega_0 t+\theta)u(t)$ (f) $\sin(\omega_0 t+\theta)\tau\,\delta(t)$

(g) $\sinh(t/\tau)u(t)$ (h) $\cosh(t/\tau)u(t)$

(i) $(t/\tau)\sinh(t/\tau)u(t)$

(j) $(t/\tau)\cosh(t/\tau)u(t)$ (k) $\sinh^2(t/\tau)u(t)$

(l) $\cosh^2(t/\tau)u(t)$ (m) $\cosh\left(\frac{t-t_0}{\tau}\right)u(t-t_0)$

(n) $\frac{d}{dt}[t\sinh(t/\tau)u(t)]$

(o) $\tau\frac{d}{dt}[(t/\tau)\cosh(t/\tau)u(t)]$

(p) $[(t-t_0/\tau)]\cosh[(t-t_0/\tau)]u(t-t_0)$

(q) $\tau\frac{d}{dt}\left[\sin^2(\omega_0 t+\theta)u(t)\right]$

(r) $\tau\frac{d}{dt}[[(t-t_0/\tau)]\cosh[(t-t_0/\tau)]u(t-t_0)]$

(s) $\tau\frac{d}{dt}[\omega_0 t\cos(\omega_0 t+\theta)u(t)]$

(t) $\exp[-(t-t_0)/\tau]\sin(\omega_0 t)u(t-t_1)$

(u) $\frac{1}{\tau}\left[\int_{0-}^{t}\cosh(t'/\tau)dt'\right]u(t)$

(v) $\delta(t)\int_{0-}^{t}\sinh(t'/\tau)dt'$

(w) $\frac{d}{dt}\left[\exp(-t/\tau)u(t)\int_{0-}^{t}\cos(\omega_0 t')dt'\right]$

(x) $\frac{d}{dt}\left[\exp(-t/\tau)\cos(\omega_0 t)u(t)\int_{0-}^{t}\sin(\omega_0 t')dt'\right]$

(y) $\frac{d}{dt}\left[\tau\exp(-t/\tau)\sin(\omega_0 t)u(t)\int_{0-}^{t}\cosh(\omega_0 t')\delta(t')dt'\right]$

Section 18.6 is prerequisite for the following problems.

P 18.16 In the following transforms,

$$s_1 = -1\,\text{s}^{-1},\ s_2 = -2\,\text{s}^{-1},\ s_3 = -3\,\text{s}^{-1},$$
$$K = 1\,\text{V s},\ \omega_0 = 1\,\text{s}^{-1},\ \alpha = 0.5$$

Expand in partial fractions and obtain the inverse transform.

(a) $\frac{K\omega_0^2(s-s_1)}{s(s-s_2)(s-s_3)}$ (b) $\frac{K\omega_0^2}{s(s-s_1)}$ (c) $\frac{Ks(s-s_1)}{(s-s_2)(s-s_3)}$

(d) $\frac{K\omega_0^3}{(s-s_1)(s-s_2)(s-s_3)}$ (e) $\frac{K\omega_0^3}{s(s^2+2\alpha\omega_0 s+\omega_0^2)}$

(f) $\frac{K\omega_0(s-s_1)(s-s_2)}{(s-s_3)(s^2+2\alpha\omega_0 s+\omega_0^2)}$ (g) $\frac{K\omega_0(s-s_1)^2}{(s^2+2\alpha\omega_0 s+\omega_0^2)s}$

(h) $\frac{K\omega_0^4}{s^2(s-s_1)^2}$ (i) $\frac{K\omega_0^3(s-s_1)}{(s-s_2)^2(s-s_3)^2}$

(j) $\frac{K\omega_0^2(s-s_1)^2}{(s-s_2)^2(s^2+2\alpha\omega_0 s+\omega_0^2)}$

P 18.17 In each function below, $K = 5\,\text{V s}$, $s_n = -n\,\text{s}^{-1}$, $\omega_0 = 1\,\text{s}^{-1}$, and $\alpha = 0.5$. Obtain the inverse Laplace transform.

(a) $\frac{K\omega_0^2(s-s_1)(s-s_5)}{s(s-s_3)(s-s_4)(s-s_2)}$

(b) $\frac{K\omega_0 s^2}{(s-s_1)(s^2+2\alpha\omega_0 s+\omega_0^2)}$

(c) $\frac{K(s-s_1)^2}{(s-s_1)^2+2\alpha\omega_0(s-s_1)+\omega_0^2}$

(d) $\frac{Ks_1^2(s^2+2\alpha\omega_0 s+\omega_0^2)}{s(s-s_1)^2(s-s_2)}$

(e) $\frac{Ks(s-s_1)(s-s_2)}{(s-s_3)(s-s_4)(s-s_5)}$

(f) $\frac{Ks_1\omega_0^2(s^2+2\alpha\omega_0 s+\omega_0^2)}{s(s-s_1)^2(s-s_2)(s-s_3)}$

(g) $\frac{K\omega_0 s^2(s-s_2)}{(s-s_1)^2(s^2+2\alpha\omega_0 s+\omega_0^2)}$

(h) $\frac{K\omega_0^2(s-s_1)^2(s-s_5)^2}{s(s-s_3)^2(s-s_4)^2(s-s_2)}$

(i) $\frac{K\omega_0^2(s-s_1)^2(s-s_2)^2}{s(s-s_3)(s^2+2\alpha\omega_0 s+\omega_0^2)^2}$

(j) $\frac{K\omega_0(s-s_1)^2(s-s_2)}{(s-s_3)^3(s-s_4)}$

P 18.18 Using partial fractions, obtain expressions for the inverse Laplace transform of each of the following, where

$$\tau_1 = 1\,\mu\text{s},\ \tau_2 = 2\,\mu\text{s},\ q = 2\,\mu\text{s}^{-1},\ r = 4\,\mu\text{s}^{-2},\ K = 50\,\mu\text{V s},\ s_1 = -10^6\,\text{s}^{-1},\ s_2 = 2s_1,\ s_3 = 3s_1,\ s_4 = 4s_1,\ \omega_0 = s_1.$$

(a) $\dfrac{K\omega_0 s(1+s\tau_1)}{(s^2+qs+r)(s\tau_2+1)}$ (b) $\dfrac{K\omega_0(s^2+qs+r)}{s(s-s_1)(s-s_2)}$

(c) $\dfrac{K\omega_0(s-s_1)^2}{s(s-s_2)(s-s_3)^2}$ (d) $\dfrac{K(s-s_1)^3}{(s-s_2)^3}$

(e) $\dfrac{K(1-s/s_1)}{(1-s/s_2)(1-s/s_3)}$

(f) $\dfrac{K(1-s/s_1)}{(2-s/s_2)(3-s/s_3)(4-s/s_4)}$

(g) $\dfrac{Ks(s+q)}{(s^2+qs+r)(1+s\tau_1)}$ (h) $\dfrac{K(1+s\tau_1)}{(1+s\tau_2)^2}$

(i) $\dfrac{K(s-s_1)^2+2Ks(s-s_2)}{(s-s_3)(s-s_4)}$

Section 18.9 is prerequisite for the following problems.

P 18.19 Refer to Fig. P 18.4, where each circuit is in steady state for $t=0^-$ and the switch is closed at $t=0$. Obtain expressions for each indicated current and voltage for $t>0$.

P 18.20 In Fig. P 18.5, $L = 2.5\,\text{mH}$, $C = 100\,\text{nF}$, $V_0 = 10\,\text{V}$. The circuit is in steady state for $t<0$ and the switch is moved from a to b at $t=0$. Find the resistance $R = R_0$ for which the response $v(t)$ is critically damped for $t>0$. Obtain expressions for and graphs of the voltage $v(t)$ for $t>0$ and for each of $R = R_0$, $R = 2R_0$, $R = R_0/2$.

P 18.21 In Fig. P 18.6, $R = 10\ \text{k}\Omega$, $C = 10\,\mu\text{F}$, and $V_0 = 10\text{V}$. At $t=0$, the switch is moved from 1 to 2, having been at 1 for all previous time. Draw a neat, fully-labeled graph of the voltage v_{ab} versus time.

P 18.22 Obtain expressions for the indicated voltages in Fig. P 18.7, where $R = 100\,\Omega$, $L = 20\,\text{mH}$, $C = 50\,\text{nF}$, and $V_0 = 5V$.

P 18.23 In Fig. P 18.8, $R = 100\,\Omega$ and

$$v_S(s) = V_0\left[u(t) + \tau\delta(t)\right],\ \mu(s) = \frac{\mu_0}{1+s\tau}$$

with $V_0 = 1\,\text{mV}$, $\mu_0 = 10^4$, and $\tau = 1\,\text{ms}$. Obtain an expression for the voltage $v(t)$.

P 18.24 In Fig. P 18.9, $R = 10\,\text{k}\Omega$, $C = 100\,\text{nF}$, and $v_S(t) = V_0\,u(t)$ with $V_0 = 500\,\text{mV}$. Obtain expressions for the voltages $v_1(t)$, $v_2(t)$.

P 18.25 In Fig. P 18.10, $R_S = 100\,\Omega$, $L_1 = 1\,\text{H}$, $L_2 = 4\,\text{H}$, $R_L = 10\,\Omega$, $M = 0.9\sqrt{L_1L_1}$, and $v_S(t) = V_0\cos(\omega_0 t)u(t)$, with $V_0 = 170\,\text{V}$ and $\omega_0 = 2\pi 60\,\text{s}^{-1}$. Obtain an expression for the voltage $v_2(t)$.

P 18.26 In Fig. P 18.11, $v_S(t) = V_0\,u(t), g = -Kp/(s-p)$. Obtain an expression for the voltage $v_L(t)$.

P 18.27 In Fig. P 18.12, $R = 1\,\text{k}\Omega$, $L = 10\,\text{mH}$, $C = 200\,\text{nF}$, and $V_0 = 50\,\text{V}$. The switch is open for $t<0$ and is closed at $t=0$. Obtain expressions for the voltages $v_1(t)$, $v_2(t)$.

P 18.28 Each of 20 identical 250 nF capacitors is charged to 100 kV. The charged capacitors are connected in series, and then discharged through a circuit consisting of a $260\,\Omega$ resistor in series with a $200\,\mu\text{H}$ inductor.

(a) Obtain an expression for the current through the resistor and find the maximum value of the magnitude of the current.
(b) Set the resistance R to the value required for critical damping and find the maximum value of the magnitude of the current.

P 18.29 In Fig. P 18.13, $R_1 = 3.16\ \text{k}\Omega$, $R_2 = 17.8\ \text{k}\Omega$, $R_3 = 3.83\ \text{k}\Omega$, $R_L = 10\,\text{k}\Omega$, $C = 100\ \text{nF}$, and the op amp is ideal. Obtain an expression for and then plot the response $v_{out}(t)$ to a step input $v_S(t) = V_0\,u(t)$, with $V_0 = 1\,\text{V}$.

Section 18.11 is prerequisite for the following problems.

P 18.30 The magnitudes of the voltage, transimpedance, and transadmittance transfer functions at complex frequency s_0 for a certain circuit are

$$|H_v(s_0)| = 9.885\times 10^{-4},\ |H_z(s_0)| = 0.099\,\Omega,\ |H_y(s_0)| = 9.885\times 10^{-9}\ \text{S}$$

(a) Find the magnitudes of the source and load impedances.

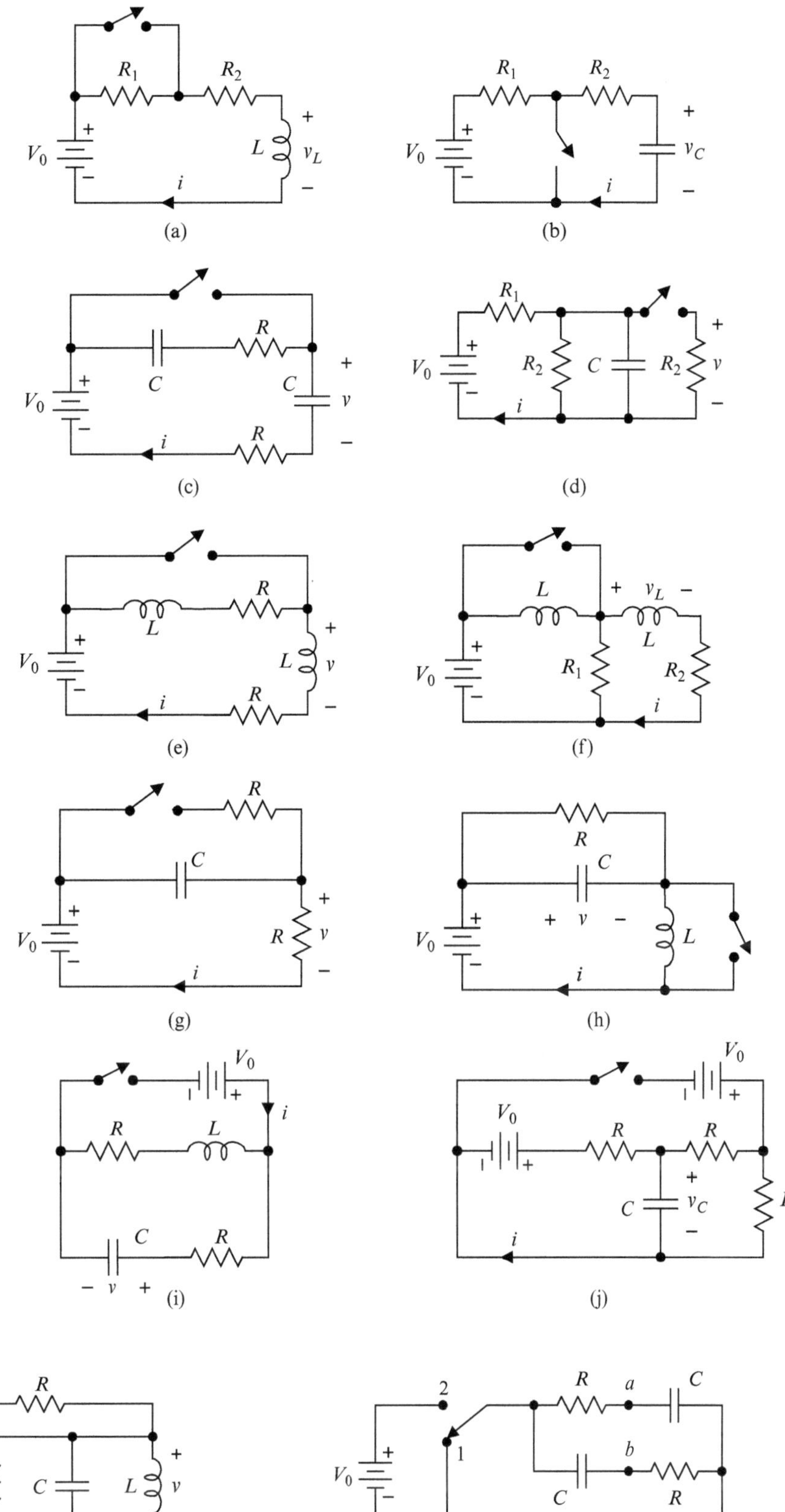

Fig. P 18.4 See Problem P 18.19

Fig. P 18.5 See Problem P 18.20

Fig. P 18.6 See Problem P 18.21

(b) What is the magnitude of the current transfer function for $s = s_0$?

P 18.31 The voltage transfer function for a certain circuit model is

$$H_v(s) = \frac{V_{out}(s)}{V_{in}(s)} = K\frac{(s-z)}{(s-p)},$$

where $z \neq 0$, $p < 0$, and $K \neq 0$. Can $v_{out}(t)$ be the voltage across a capacitor? Justify your answer.

P 18.32 In Fig. P 18.14, $R_1 = 10\,\mathrm{k\Omega}$, $R_2 = 51\,\mathrm{k\Omega}$, $C_1 = 100\,\mathrm{pF}$, $R_3 = 120\,\mathrm{k\Omega}$, $C_2 = 27\,\mathrm{pF}$, and $v_{in} = V_0\cos(\omega_0 t)u(t)$, with $V_0 = 100\,\mathrm{mV}$ and $f_0 = 40\,\mathrm{kHz}$. The source impedance is negligible. Obtain an expression for the output $v_{out}(t)$.

P 18.33 Obtain expressions for the voltage transfer function, the current transfer function, the transadmittance, and the transimpedance of each circuit in Fig. P 18.15. Unless otherwise specified, the dependent source parameters (g, μ, β, r) are independent of s.

P 18.34 In use, each circuit in Fig. P 18.16 will be driven by a source attached from terminal a to ground and will drive a load attached from terminal b to ground. The source output impedance and the load impedance are resistive, but are otherwise as yet unspecified. Obtain an expression for the voltage transfer function of each circuit. Express the transfer function in factored form, where the numerator consists of a constant, frequency-independent factor K and factors of the form $(s-z)$ and the denominator consists of factors of the form $(s-p)$. Express the poles, zeros, and the factor K in terms of the circuit parameters. Give conditions (if any) under which the voltage transfer function is approximately independent of the source impedance.

P 18.35 Two first-order circuits have voltage transfer functions

$$H_{v1}(s) = \frac{p_1}{s-p_1},\ H_{v2}(s) = \frac{p_2}{s-p_2}.$$

Second-order circuits can be constructed by connecting the two first-order circuits in cascade (series)

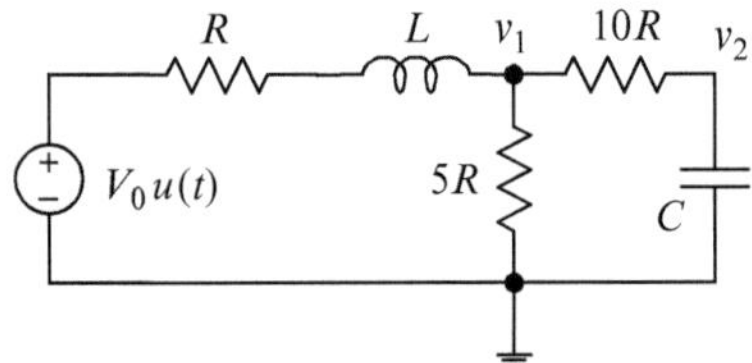

Fig. P 18.7 See Problem P 18.22

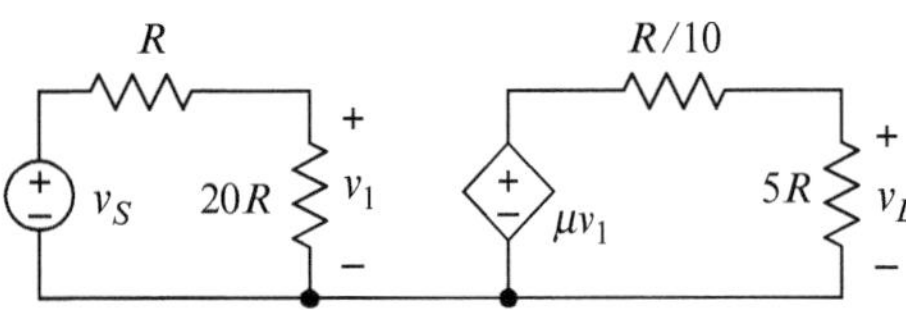

Fig. P 18.8 See Problem P 18.23

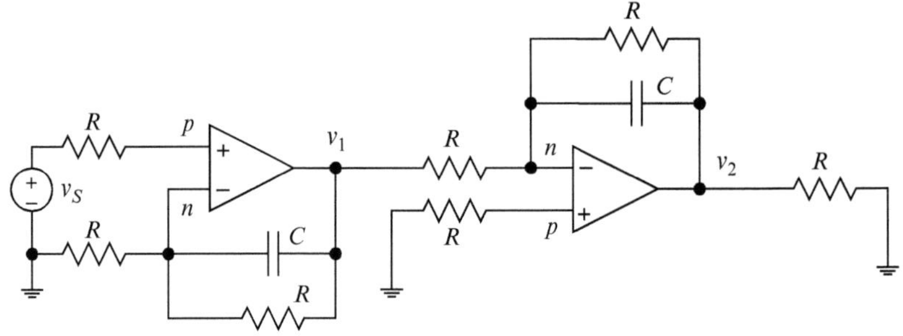

Fig. P 18.9 See Problem P 18.24

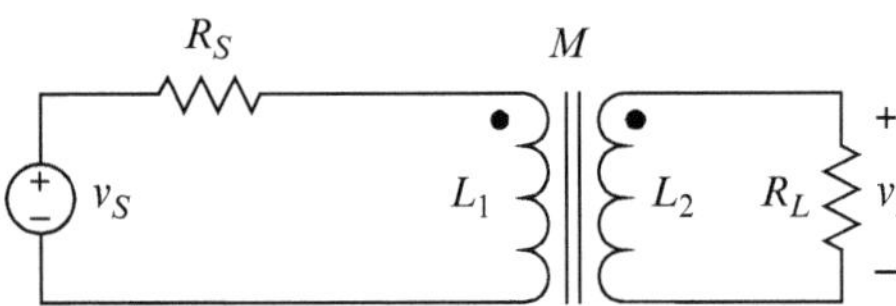

Fig. P 18.10 See Problem P 18.25

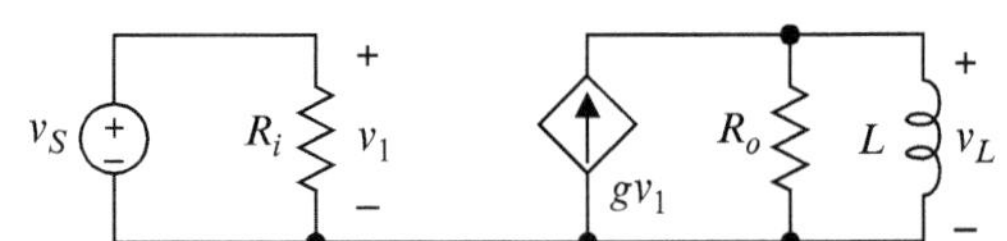

Fig. P 18.11 See Problem P 18.26

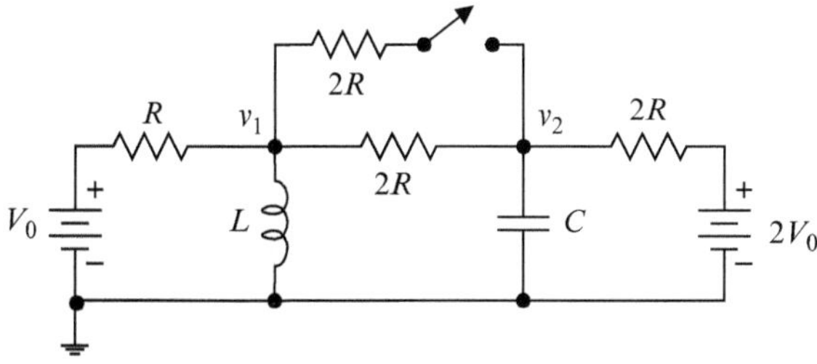

Fig. P 18.12 See Problem P 18.27

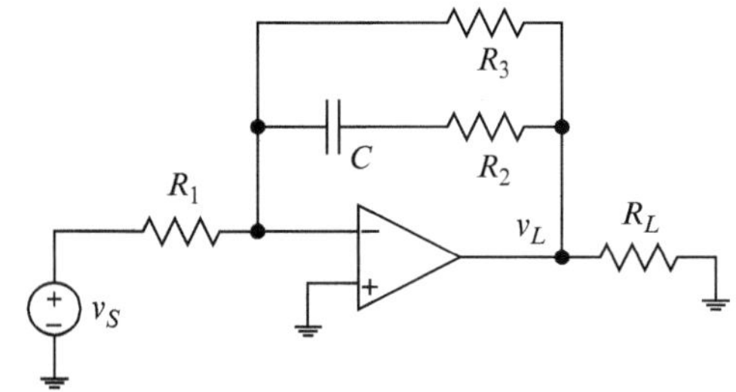

Fig. P 18.13 See Problem P 18.29

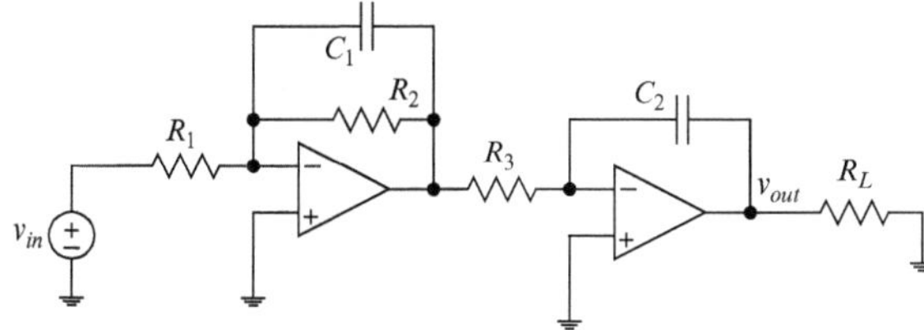

Fig. P 18.14 See Problem P 18.32

or parallel, as illustrated in Fig. P 18.17. The source resistance is $R_S = 100\,\Omega$.

(a) Use ideal op amps in inverting configurations to obtain circuits having the transfer functions H_{v1}, H_{v2}. Specify the input resistances such that the source resistance is negligible. Then redraw the diagrams in Fig. P 18.17, replacing the boxes with the circuits.

(b) Obtain the voltage transfer functions for the cascade and parallel circuits. What are the essential differences between the two?

P 18.36 Obtain expressions for the four transfer functions for the circuit in Fig. P 18.18.

P 18.37 Obtain expressions for the four transfer functions for the circuit in Fig. P 18.19.

P 18.38 Refer to Fig. P 18.20. Assume the output impedance and the source impedance are negligible. Show that the poles of the voltage transfer function are a complex conjugate pair if $C_1 > C_2$.

P 18.39 The excitation v_S and response v_L of a certain circuit are related by the differential equation

$$\tau \frac{dv_L}{dt} + v_L = v_S; \;\; \tau = 100\,\mu\text{s}$$

Obtain and plot the response of the circuit to the excitation defined in Fig. P 18.21.

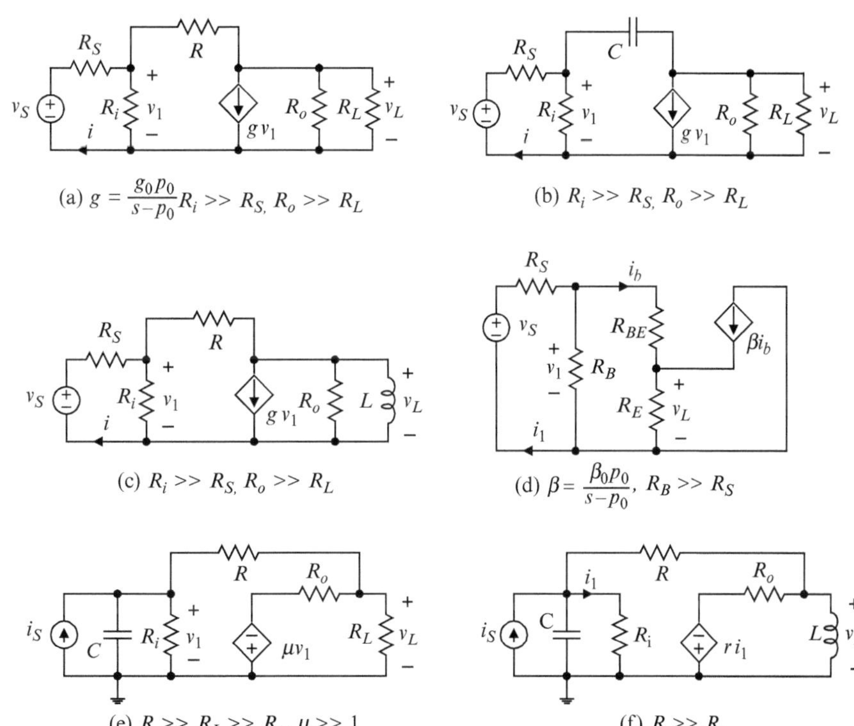

Fig. P 18.15 See Problem P 18.33

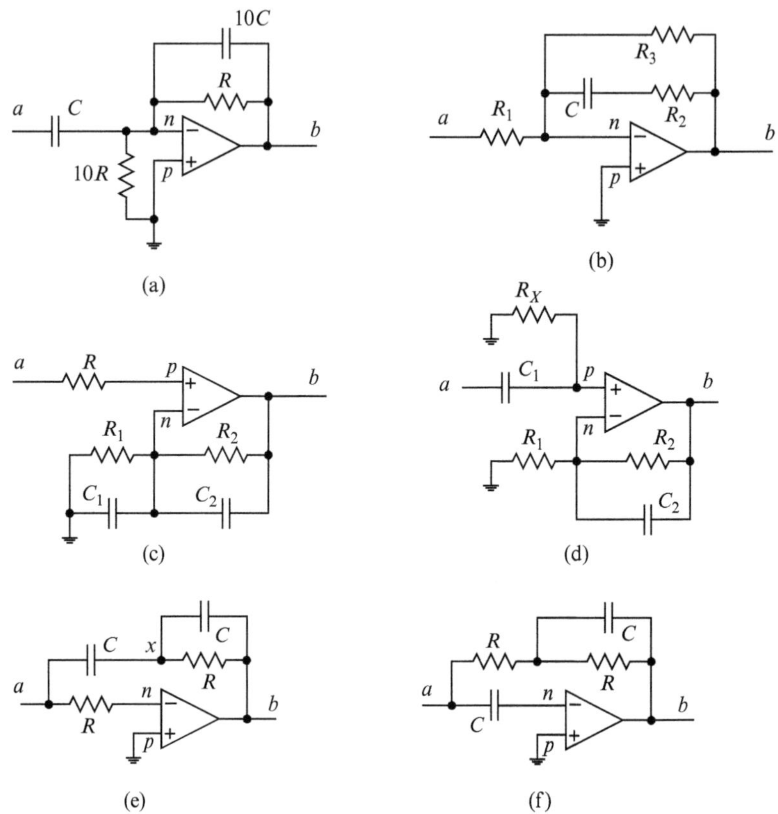

Fig. P 18.16 See Problem P 18.34

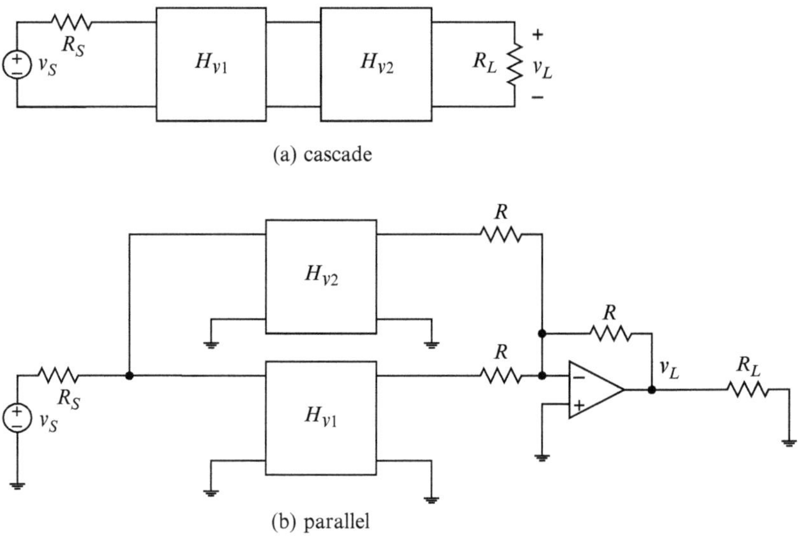

Fig. P 18.17 See Problem P 18.35

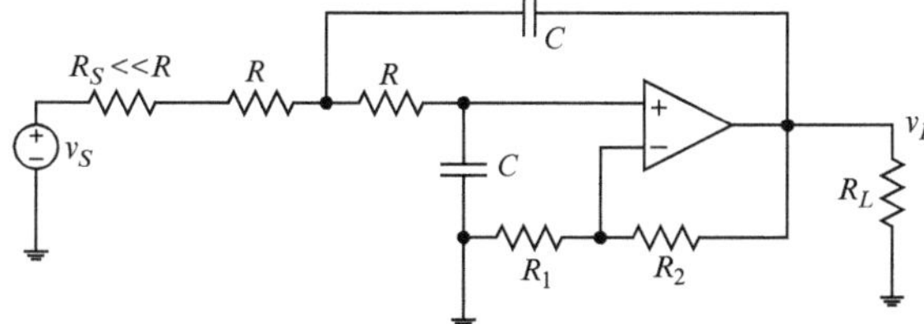

Fig. P 18.18 See Problem P 18.36

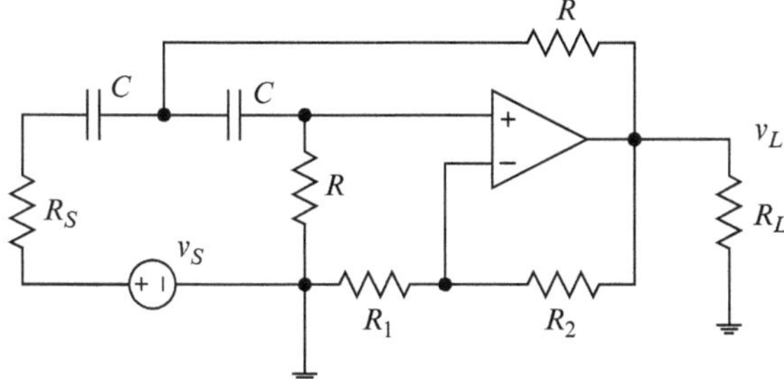

Fig. P 18.19 See Problem P 18.37

Section 18.12 is prerequisite for the following problems.

P 18.40 A circuit having transfer function

$$H(s) = \frac{Ks}{(s-p)^2}$$

is driven by an excitation having Laplace transform

$$X(s) = \frac{X_0}{s-p}.$$

Identify the forced and unforced terms in the Laplace transform of the response.

P 18.41 The (voltage) transfer function for a certain circuit is of the form

$$H_v(s) = \frac{K\omega_0 s}{s^2 + 2\alpha\omega_0 s + \omega_0{}^2} = \frac{K\omega_0 s}{(s-p)(s-p^*)},$$

where the poles are complex conjugates. What is the form of the unforced response?

P 18.42 What is the form of the unforced response corresponding to the transfer function in Problem P 18.41 if the poles are real and distinct? If the poles are real and equal?

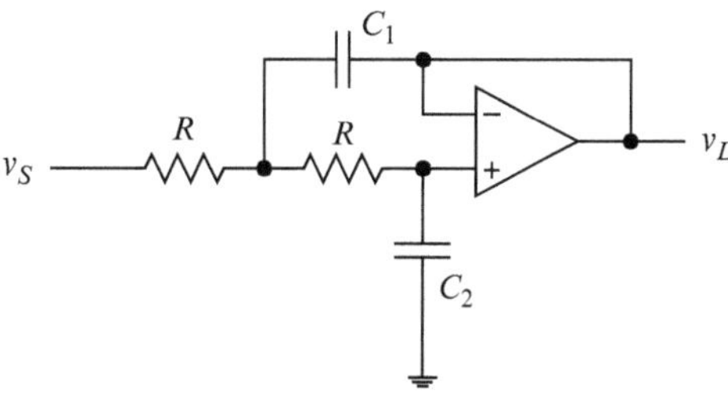

Fig. P 18.20 See Problem P 18.38

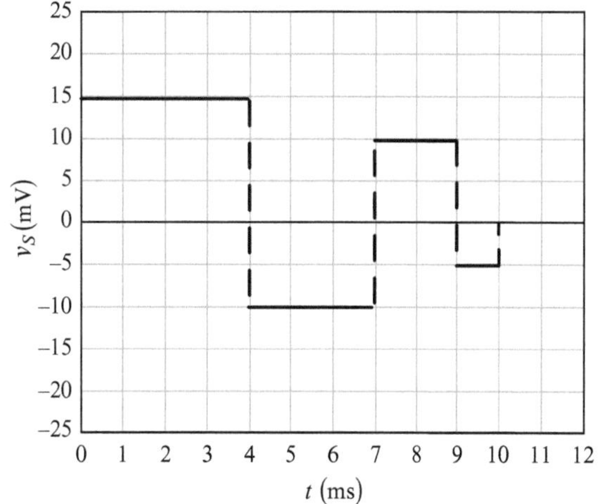

Fig. P 18.21 See Problem P 18.39

Section 18.13 is prerequisite for the following problems.

P 18.43 The step response of a certain circuit is given by

$$g(t) = e^{-t/\tau}\cos(\omega_0 t + \theta)u(t) \text{ V/V},$$
$$\tau = 500\,\mu\text{s},\ \omega_0 = 4\pi \times 10^3\ \text{s}^{-1},\ \theta = -\frac{\pi}{2}.$$

(a) Obtain an expression for the response to the excitation defined in Fig. P 18.21.
(b) Plot the response obtained in part (a).
(c) Obtain the response to the excitation

$$v_S(t) = A[\cos(2\omega_0 t) + \cos(20\omega_0 t)]u(t);\ A = 10\,\text{mV}$$

(d) Plot the response obtained in part (c).

P 18.44 The voltage impulse response of a certain circuit is given by

$$h_v(t) = (t/\tau)\exp(-t/\tau)u(t), \ \tau = 10\,\mu\text{s}$$

Find and plot the response to each of the following excitations.

(a) $v_S(t) = V_0[u(t) - u(t - t_0)], \ V_0 = 1\,\text{V}, \ t_0 = 10\,\mu\text{s}$
(b) $v_S(t) = V_0[u(t) - u(t - t_0)], \ V_0 = 1\,\text{V}, \ t_0 = 50\,\mu\text{s}$
(c) $v_S(t) = V_0[u(t) - u(t - t_0) + t_0\delta(t - 2t_0)], \ V_0 = 1\,\text{V}, \ t_0 = 30\,\mu\text{s}$
(d) $v_S(t) = V_0 t_0[\delta(t) + \delta(t - t_0) + \delta(t - 2t_0) + \cdots + \delta(t - 10t_0)], \ V_0 = 1\,\text{V}, \ t_0 = 10\,\mu\text{s}$
(e) $v_S(t) = V_0[u(t) + u(t - t_0) + u(t - 2t_0) + \cdots + u(t - 10t_0)], \ V_0 = 1\,\text{V}, \ t_0 = 10\,\mu\text{s}$

Section 18.14 is prerequisite for the following problems.

P 18.45 Obtain the corresponding f-domain transfer function from the s-domain transfer function. Give the values of the corner frequencies.

(a) $H_v(s) = \dfrac{K(s - z)}{(s - p_1)(s - p_2)}, \ K = 1\,\text{s}^{-1}, \ z = -10^2\,\text{s}^{-1}, \ p_1 = -10^3\,\text{s}^{-1}, \ p_2 = -10^4\,\text{s}^{-1}$

(b) $H_i(s) = \dfrac{K(s - z_1)(s - z_2)}{(s - p_1)(s - p_2)}, \ K = 1, \ z_1 = -10\,\text{s}^{-1}, \ z_2 = -10^2\,\text{s}^{-1}, \ p_1 = -10^3\,\text{s}^{-1}, \ p_2 = -10^4\,\text{s}^{-1}$

(c) $H_v(s) = \dfrac{K(s - z)^2}{s(s - p)^2}, \ K = 1\,\text{s}^{-1}, \ z = -10^2\,\text{s}^{-1}, \ p = -10^3\,\text{s}^{-1}$

(d) $H_z(s) = \dfrac{Rs(s - z)}{s^2 + 2\alpha\omega_0 s + \omega_0{}^2}, \ R = 1\,\text{k}\Omega, \ z = -10^2\,\text{s}^{-1}, \alpha = 0.2, \ \omega_0 = 10^4\,\text{s}^{-1}$

(e) $H_y(s) = \dfrac{g(s^2 + 2\alpha_1\omega_1 s + \omega_1^2)}{s^2 + 2\alpha_0\omega_0 s + \omega_0{}^2}, \ g = 100\,\text{mS}, \ \alpha_1 = 0.5, \omega_1 = 10^3\,\text{s}^{-1}, \alpha_0 = 0.2, \omega_0 = 10^5\,\text{s}^{-1}$.

P 18.46 Obtain the corresponding s-domain transfer function from the f-domain transfer function. Give the values of the poles and zeros.

(a) $H_v(j2\pi f) = \dfrac{K(1 + jf/f_0)}{(1 + jf/f_1)(1 + jf/f_2)}, K = 100, \ f_0 = 1\,\text{kHz}, \ f_1 = 100\,\text{Hz}, \ f_2 = 10\,\text{kHz}$

(b) $H_v(j2\pi f) = \dfrac{Kj(f/f_0)(1 + jf/f_0)}{(1 + jf/f_1)^2(1 + jf/f_2)}, \ K = 100, \ f_0 = 1\,\text{kHz}, \ f_1 = 100\,\text{Hz}, \ f_2 = 10\,\text{kHz}$

(c) $H_v(j2\pi f) = \dfrac{K(1 + jf/f_0)(1 + jf/f_1)}{j(f/f_0)(1 + jf/f_2)^2}, \ K = 100, \ f_0 = 1\,\text{kHz}, \ f_1 = 100\,\text{Hz}, \ f_2 = 10\,\text{kHz}$.

(d) $H_z(j2\pi f) = R\dfrac{(jf/f_0)(1 + jf/f_1)^2}{\left[1 - (f/f_0)^2 + 2\alpha jf/f_0\right](1 + jf/f_2)}, \ R = 10\,\text{k}\Omega, \ f_0 = 1\,\text{kHz}, \ f_1 = 100\,\text{Hz}, \ f_2 = 10\,\text{kHz}, \ \alpha = 0.2$

(e) $H_y(j2\pi f) = g\dfrac{(jf/f_0)(1 + jf/f_1)^2}{\left[1 - (f/f_0)^2 + 2\alpha_0 j(f/f_0)\right]\left[1 - (f/f_1)^2 + 2\alpha_1 j(f/f_0)\right]}, \ g = 1\,\text{S}, \ f_0 = 1\,\text{kHz}, \ f_1 = 100\,\text{Hz}, \ f_2 = 10\,\text{kHz}, \ \alpha_0 = 0.2, \ \alpha_1 = 0.3$.

Section 18.15 is prerequisite for the following problems.

P 18.47 The op amp in Fig. P 18.22 has input impedance $R_i = 10\,\text{M}\Omega$, output impedance $R_o = 100\,\Omega$, and gain-bandwidth product $f_T = 1\,\text{MHz}$. The source impedance is negligible and the load impedance is $R_L = 10\,\text{k}\Omega$.

(a) Obtain an expression for the step response of the inverting amplifier.
(b) Plot the time required for the step response to reach 90% of the steady-state value versus the amplifier dc voltage gain $A_{v0} = R_2/R_1$ for $10 \leq A_{v0} \leq 100$.
(c) Assume a 15 V symmetric supply, rail-to-rail operation, and a slew rate of $10\,\text{V}/\mu\text{s}$. Which of slew rate and output swing determines the maximum allowable amplitude of a step input (for linear operation)?

P 18.48 Assume the op amp in Fig. P 18.23 has input impedance $R_i \to \infty$, output impedance $R_o \to 0$, intrinsic bandwidth $f_0 = 10\,\text{Hz}$, and gain-bandwidth product $f_T = 1\,\text{MHz}$. The source impedance is a

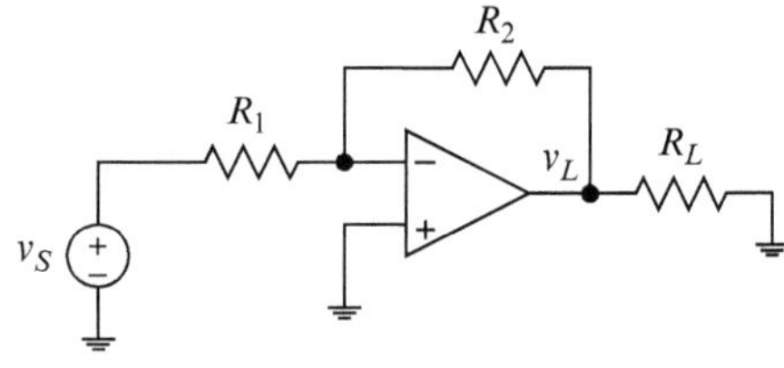

Fig. P 18.22 See Problem P 18.47

negligible fraction of R_1 and the load impedance is $R_L = 10\,\text{k}\Omega$. Show that the voltage transfer function of the circuit is

$$H_v = \frac{-\mu Z_2}{(\mu+1)R_1 + Z_2}, \; Z_2 = \frac{R_2}{1 + sR_2C}.$$

Under what conditions is

$$H_v \cong -\frac{Z_2}{R_1}$$

a good approximation?

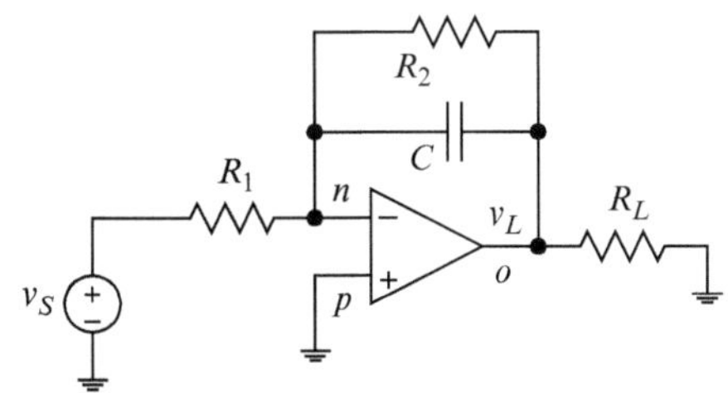

Fig. P 18.23 See Problem P 18.48

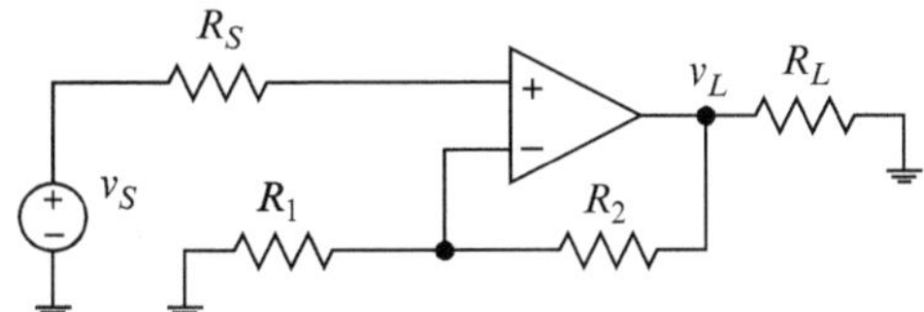

Fig. P 18.24 See Problem P 18.49

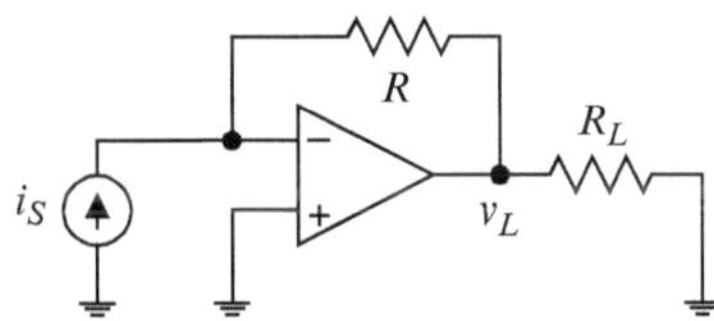

Fig. P 18.25 See Problem P 18.50

P 18.49 Assume the op amp in Fig. P 18.24 has input impedance $R_i \to \infty$, output impedance $R_o \to 0$, intrinsic bandwidth $f_0 = 10\,\text{Hz}$, and gain-bandwidth product $f_T = 1\,\text{MHz}$. The source resistance is $R_S = 200\,\Omega$ and the load resistance is $R_L = 5\,\text{k}\Omega$. (a) Obtain an expression for the step response. (b) Plot the 0–90% rise time versus the dc voltage gain A_{v0} of the circuit for $10 \le A_{v0} \le 100$. (c) If the slew rate is $5\,\text{V}/\mu\text{s}$, and the output swing is unlimited, what is the maximum amplitude of a step input $v_S = V_0\,u(t)$ for which the amplifier is linear?

P 18.50 Figure P 18.25 shows a transimpedance amplifier (current-to-voltage converter). (a) Show that the transimpedance is given by $H_z = -R$. (b) Obtain expressions for the transimpedance and step response if the op amp has input impedance $R_i \to \infty$, output impedance $R_o \to 0$, intrinsic bandwidth f_0, and gain-bandwidth product f_T.

Section 18.16 is prerequisite for the following problems.

P 18.51 Obtain an expression for the transimpedance transfer function H_z for the circuit in Fig. P 18.26. Show that the expression is dimensionally consistent and give conditions under which $H_z \cong g\,r\,Z_L$. If those conditions are met, what is the current transfer function? Which of the two transfer functions is least dependent on source and load impedance?

P 18.52 Obtain an expression for the voltage transfer function H_z for the circuit shown in Fig. P 18.27. Show that the expression is dimensionally consistent and give conditions under which $H_v \cong \beta_1\beta_0 Z_L R_S^{-1}$. If those conditions are met, is there another transfer function that is approximately independent of source and load impedance?

P 18.53 Obtain expressions for all four transfer functions for the circuit in Fig. P 18.28. Show that each expression is dimensionally consistent. Is there a

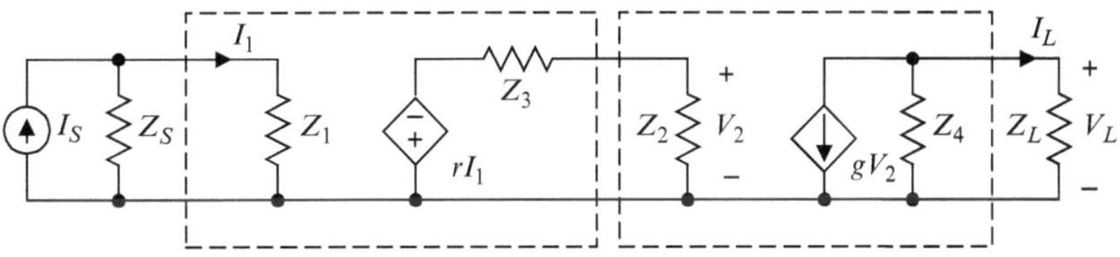

Fig. P 18.26 See Problem P 18.51

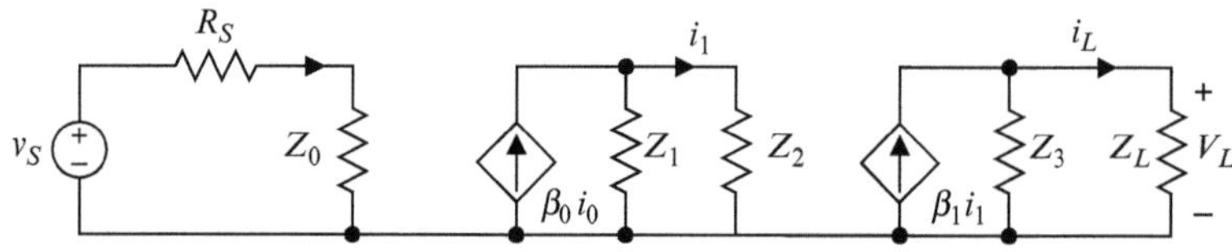

Fig. P 18.27 See Problem P 18.52

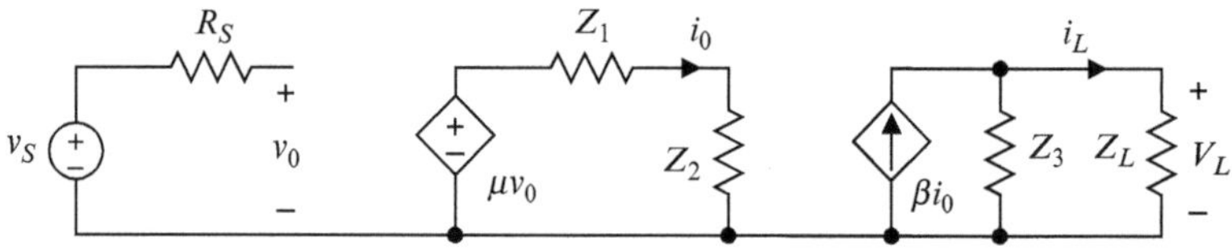

Fig. P 18.28 See Problem P 18.53

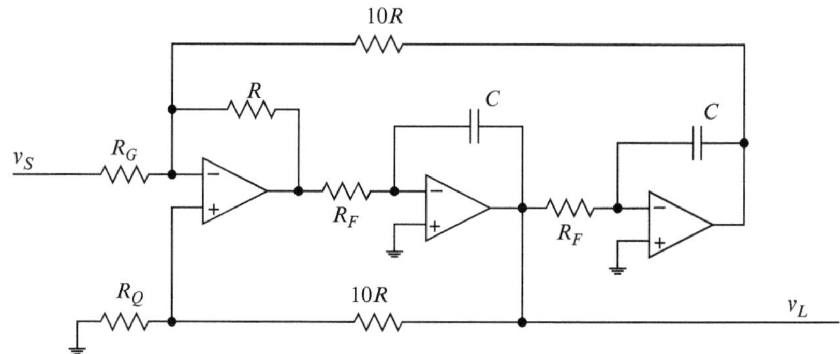

Fig. P 18.29 See Problem P 18.54

transfer function that is approximately independent of source and load impedances? Under what conditions?

Section 18.17 is prerequisite for the following problems.

P 18.54

(a) Obtain the voltage transfer function $H_v = V_L/V_S$ for the circuit in Fig. P 18.29. Express the transfer function in the form

$$H_v(s) = \frac{K\omega_0 s}{s^2 + 2\alpha\omega_0\, s + {\omega_0}^2}$$

and express the parameters K, α, and ω_0 in terms of the circuit parameters. The source and load are not shown. You may assume that the source impedance is a negligible fraction of R_G and the output impedance is a negligible fraction of the load impedance.

(b) Show that the poles are complex conjugates if $R_Q = R_G = R$.

The pole-zero plots in Fig. P 18.30 are referenced in several subsequent problems. Pay attention to the scales; e.g., in plot (a), the real pole is at $p_0 = -6\pi \times 10^4\ \text{s}^{-1}$.

P 18.55 Each pole-zero plot in Fig. P 18.30 represents the voltage transfer function of a circuit. The dc steady-state component of the response of each associated circuit to a 1 V step input is 10 V. Obtain expressions for the step responses.

P 18.56 (a) Show that if the source resistance is negligible, the form of voltage transfer function of the circuit in Fig. P 18.31 is

$$H_v(s) = K\frac{(s-z)}{(s-p)}$$

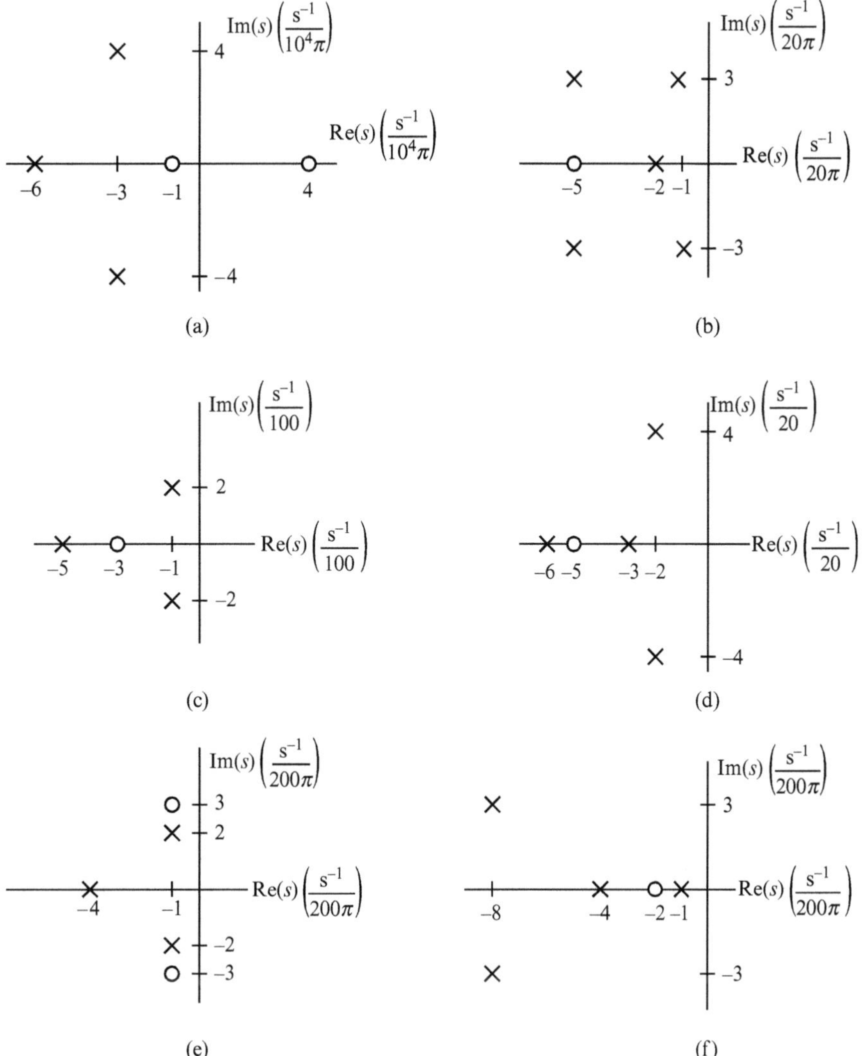

Fig. P 18.30 See Problem P 18.55

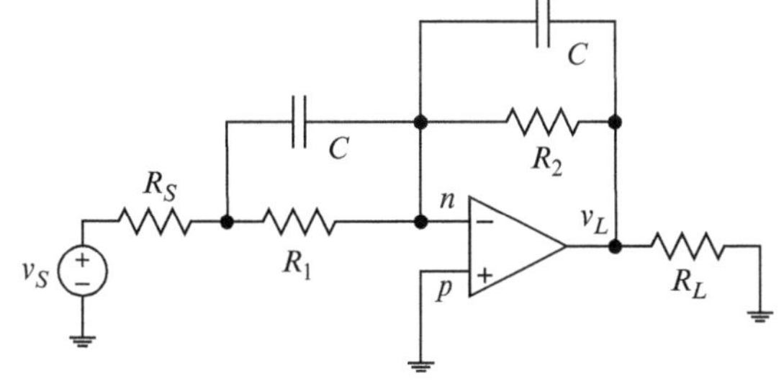

Fig. P 18.31 See Problem P 18.56

Express the pole and the zero in terms of the circuit parameters.

(b) Show how you would modify the circuit to obtain a single real pole. What is the voltage transfer function in that case?
(c) Show how you would modify the circuit to obtain a single real zero. What is the voltage transfer function in that case?

P 18.57 (a) Show that if loading is negligible, the poles of the voltage transfer function for the circuit in Fig. P 18.32 are given by

$$p_1, p_2 = \frac{1}{2\tau}\left(k - 2 \pm \sqrt{k^2 - 4k}\right), \ \tau = RC$$

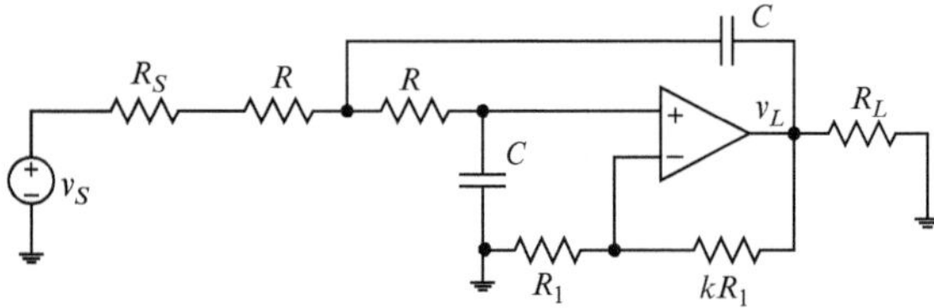

Fig. P 18.32 See Problem P 18.57

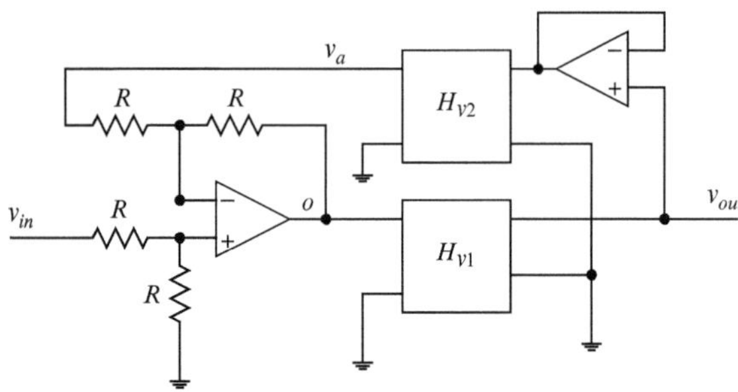

Fig. P 18.33 See Problem P 18.59

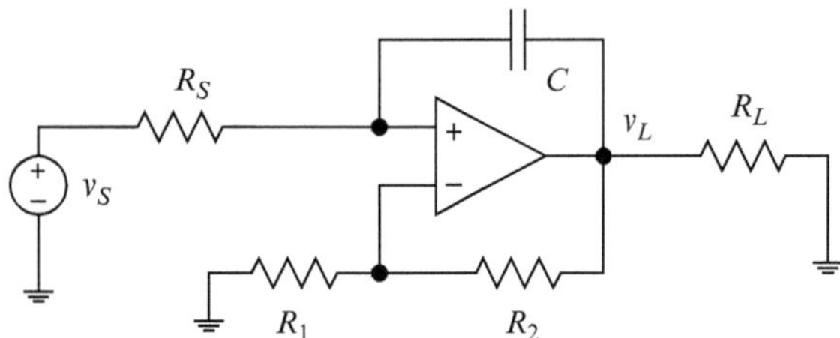

Fig. P 18.34 See Problem P 18.60

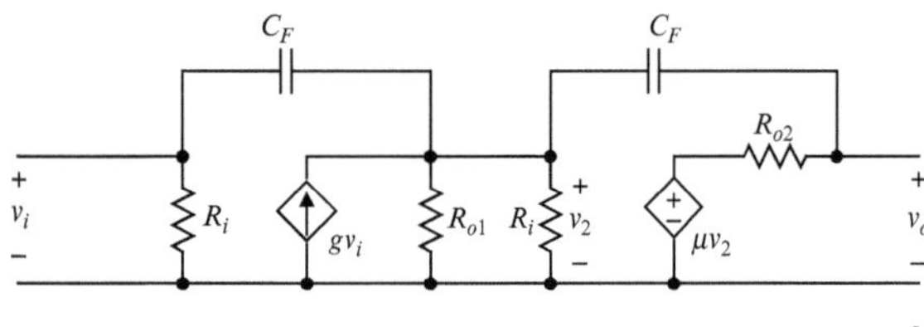

$R_i = 100\,\text{M}\Omega$, $R_{o1} = 10\,\text{k}\Omega$, $R_{o2} = 100\,\Omega$, $C_F = 50\,\text{pF}$, $g = 1\,\text{mS}$, $|\mu| = 10^5$

Fig. P 18.35 See Problem P 18.61

(b) Draw pole-zero plots for $\tau = 1/(2\pi)$ ms and for $k = 0.5, 1.0, 1.5, 2.0, 2.5$.
(c) For what values of k is the circuit stable?
(d) For what value of k is the circuit critically damped?
(e) For what values of k is the circuit underdamped?
(f) For what value of k is the damping factor equal to 1?
(g) What is the dc voltage gain of the circuit (if the circuit is stable)?
(h) Show that if the poles are complex conjugates (p, p^*), then $|p| = 1/\tau$
(i) Show that if the circuit is underdamped,

$$k = 2 - \frac{2\delta}{\sqrt{1 + \delta^2}}$$

where δ is the damping factor.
(j) Show that if the circuit is underdamped,

$$k = 2 - 2\alpha$$

where α is the peaking factor.
(k) Show that

$$k = 2 + 2\cos(\measuredangle p)$$

P 18.58 Using one or more of the circuits like that in Fig. P 18.31 (possibly with one of the capacitors omitted) and one or more of the circuits like that in Fig. P 18.32, and assuming loading is negligible, design a cascade circuit whose voltage transfer function is represented by the pole-zero plot in:

(a) Fig. P 18.30(b)
(b) Fig. P 18.30(c)
(c) Fig. P 18.30(d)
(d) Fig. P 18.30(f)

Note: We cannot use the same procedure for the pole-zero plots in Fig. P 18.30(a) and (e) because the circuits available (here) cannot provide complex or right-half plane zeros.

Section 18.18 is prerequisite for the following problems.

P 18.59 Refer to Fig. P 18.33, where circuits 1 and 2 have voltage transfer functions H_{v1}, H_{v2}, respectively, as shown. The diagram represents a larger circuit in which the output of circuit 1 is fed back through circuit 2 to the input of circuit 1.

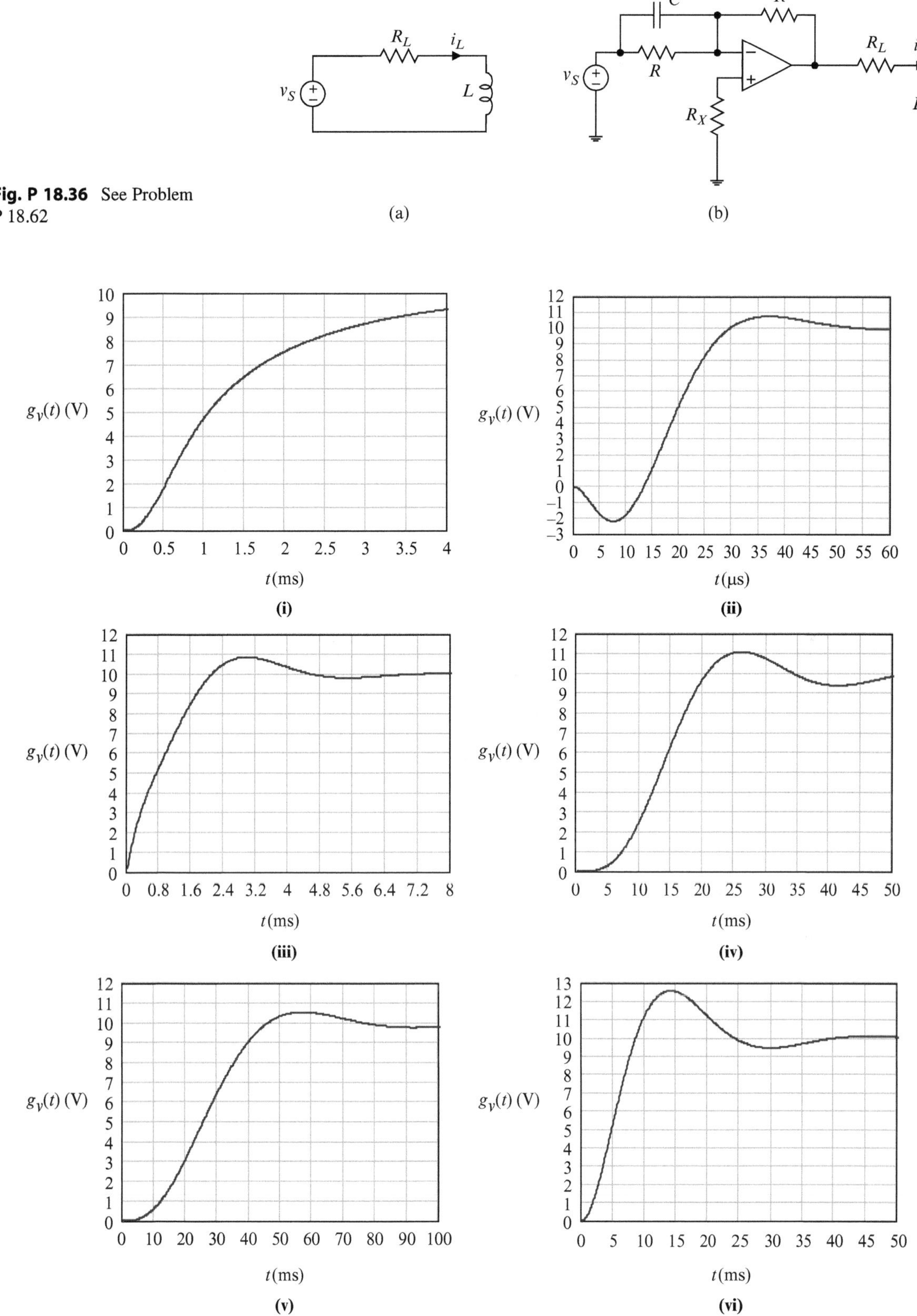

Fig. P 18.36 See Problem P 18.62

Fig. P 18.37 See Problem P 18.63

(a) Loading is negligible. Obtain an expression for the overall voltage transfer function $H_v(s) = V_{out}(s)/V_{in}(s)$.

(b) Let

$$H_{v1}(s) = \frac{K_1}{s - p_1}$$

where p_1 lies in the right-half-plane, making circuit 1 unstable. If possible, find a transfer function $H_{v2}(s)$ such that the overall system is stable.

(c) Explain how this stabilization can work (physically), in apparent contradiction of the discussion of pole-zero cancellation in the text.

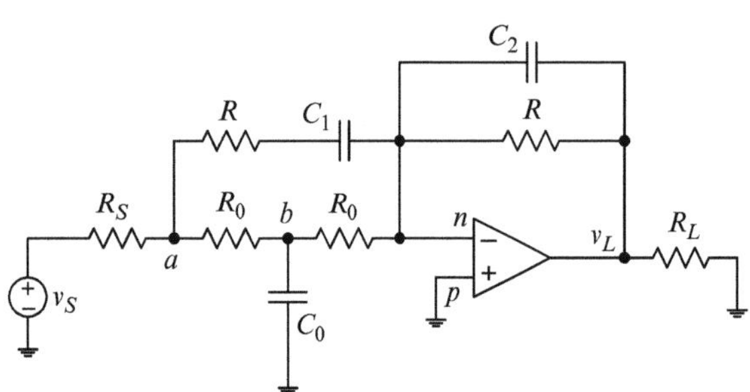

$R_0 = 2.15\,\mathrm{k\Omega}$, $R = 23.7\,\mathrm{k\Omega}$, $C_0 = 1.65\,\mu\mathrm{F}$, $C_1 = 15\,\mathrm{nF}$, $C_2 = 50\,\mathrm{pF}$

Fig. P 18.38 See Problem P 18.65

P 18.60 Obtain an expression for the voltage transfer function of the circuit shown in Fig. P 18.34. Is the circuit stable?

P 18.61 In Fig. 18.35, μ can be positive or negative. For which sign (if either) is the circuit model stable?

Section 18.20 is prerequisite for the following problems.

P 18.62 A source v_S must drive a certain inductive load, where the important output is the load current. It is found that the direct approach, illustrated by Fig. P 18.36(a), yields a step response that is too sluggish (long transient). It is proposed that the

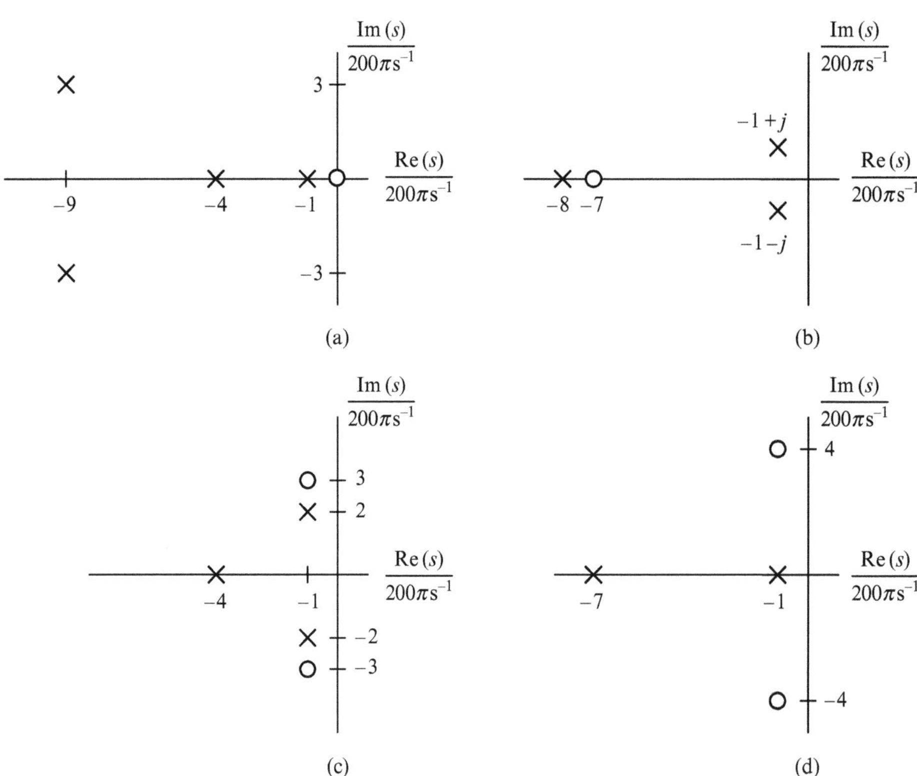

Fig. P 18.39 See Problem P 18.66

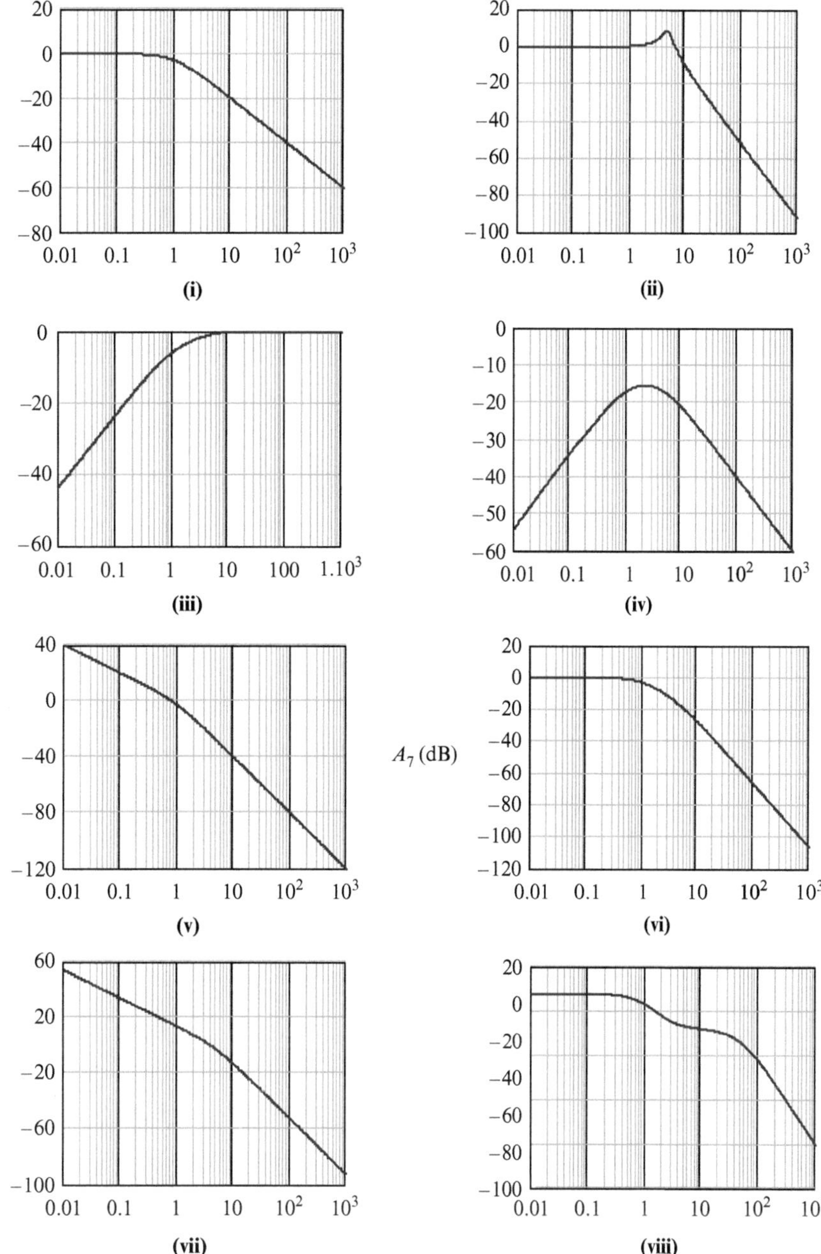

Fig. P 18.40 See Problem P 18.67

arrangement in Fig. P 18.36(b), with a suitable choice for the components R, C, will speed up the response. If this is so, show how and discuss any limitations on this approach.

P 18.63 Match each transfer-function pole-zero plot in Fig. P 18.30 to a step response in Fig. P 18.37.

P 18.64 The finite poles and zeros of the voltage transfer function for a certain circuit are

$$z_1 = 0,\ z_2 = -20\pi\,\mathrm{s}^{-1},\ p_1 = -2\pi \times 10^6\,\mathrm{s}^{-1},$$
$$p_2 = 62.8(-1 + j) \times 10^3\,\mathrm{s}^{-1},\ p_3 = {p_2}^*.$$

The voltage gain of the circuit equals 50 dB at $f = 10\,\mathrm{kHz}$.

(a) Obtain expressions for the voltage transfer function and the step response of the circuit.

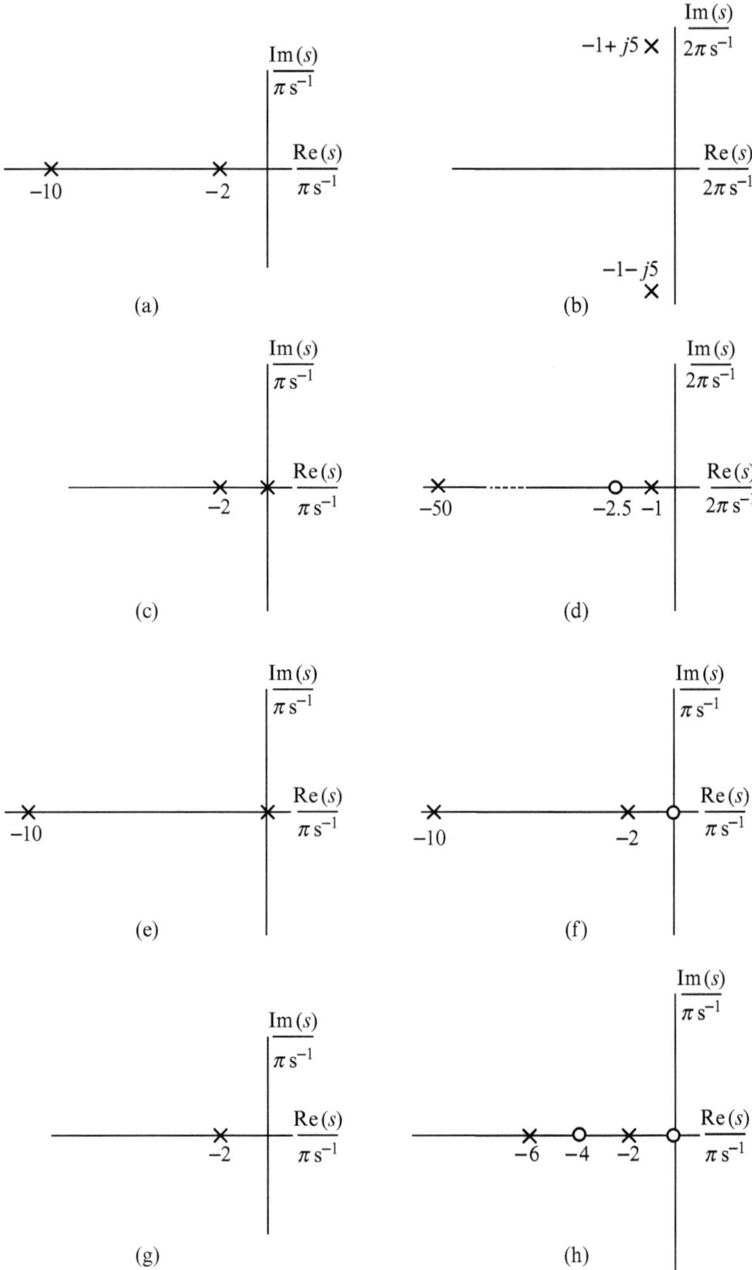

Fig. P 18.41 See Problem P 18.68

(b) Which pole(s) is (are) dominant, if any? Use graphs of the step response and an approximate step response to justify your answer.

P 18.65 Refer to Fig. P 18.38, where the source resistance R_S is negligible.

(a) Find the voltage transfer function and the poles and zeros. Is there a dominant pole?
(b) Obtain an expression for and plot the step response. If you identified a dominant pole in part (a), use graphs to justify your opinion.

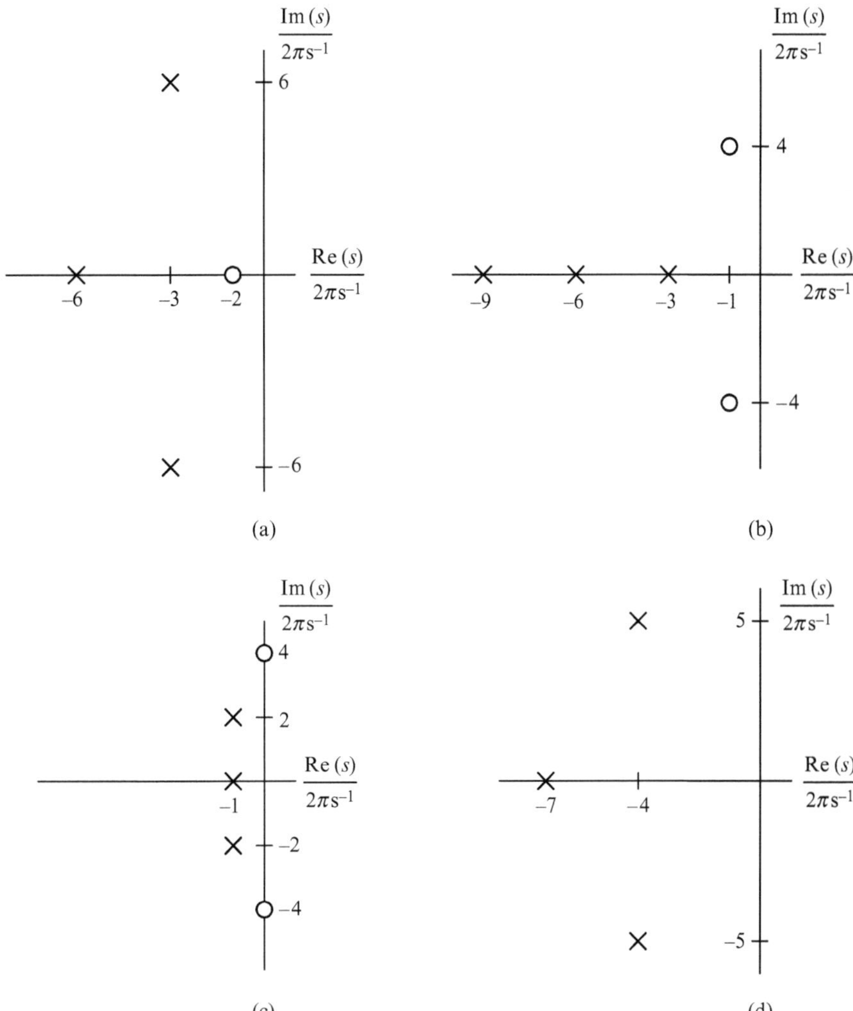

Fig. P 18.42 See Problem P 18.69

Section 18.21 is prerequisite for the following problems.

P 18.66 Figure P 18.39 shows pole-zero plots for voltage transfer functions. The dc voltage gain of each is 10. Draw a Bode voltage gain plot for the associated circuit.

P 18.67 Refer to the Bode plots of voltage gains in Fig. P 18.40, where the abscissas are frequency in Hz and the ordinates are gain in dB

(a) Which have a pole at the origin?
(b) Which have a zero at the origin?
(c) Which have one or more finite zeros not at the origin?
(d) Which of the associated step responses are underdamped and exhibit overshoot and ringing?

P 18.68 Match each Bode plot in Fig. P 18.40 with the corresponding pole-zero plot in Fig. P 18.41. Pay attention to the scales given in the labels for the pole-zero plots.

P 18.69 Figure P 18.42 shows pole-zero plots for voltage transfer functions. The value of each transfer function for $s = 0$ is 10^4 Obtain the associated *f*-domain voltage transfer function in standard form and construct the corresponding Bode gain plot.

Chapter 19
Active Filters

Fundamentally, a **filter** is a circuit whose purpose is to separate one or more signals from one or more others. Filtering allows you to select a particular radio station from the hundreds that are simultaneously present at your radio's antenna. Filtering is used to process radar and sonar echoes, to clean up old recordings, and in a host of other applications. Filters are present in virtually all electronic systems for communication, control, instrumentation, echolocation (radar, sonar, infrared), image processing, and (of course) signal processing. It is hard to think of a significant electronic system that does not involve filtering.

A filter can be implemented (built) in various ways:

- A **passive filter** contains only passive components (resistors, capacitors, and inductors).
- An **active filter** usually contains not only passive components, but also one or more active devices (e.g., op amps).
- A **digital filter** achieves filtering by operating on sample values of the input via a special-purpose digital computer (often called a **DSP**, for digital-signal-processor) and subsequently converting the output samples back to a continuous signal. In contrast, passive and active filters described above are called **analog filters**. We do not treat digital filters. However, just as active filters have displaced passive filters in many applications, digital filters are displacing active filters in a great many applications.

Passive filters require no external power supply (an advantage), but can be large and heavy at low (audio and below) frequencies if inductors are required. Also, because passive filters provide no gain (only attenuation), the number of components (which is related to performance) in a passive filter is limited. Active filters can be smaller, lighter, and more accurate than passive filters, but require an external power supply and might have shorter lives (e.g., an inductor will normally outlive an op amp). A passive filter might be preferred in applications where a power supply is not otherwise required, but active filters are especially advantageous in applications where a power supply is already present.

In this chapter, we treat a few active filter circuits that incorporate op amps as the active components. At this writing, gain-bandwidth products for the fastest off-the-shelf op amps are in the neighborhood of 1 GHz, so op-amp-based active filters are limited to relatively low-frequency applications – typically where frequencies of interest are below 100 MHz. Thus the circuits we describe are similarly limited. However, principles involved are more generally applicable and useful.

19.1 Gain

The simplest filters, and the only ones treated in this, are intended to selectively pass or reject sinusoidal inputs based upon their frequencies. Such filters are classified according to their gain characteristics.

Figure 19.1 shows idealized graphs of gain versus frequency for four common types of filters. The meanings of lowpass, bandpass, bandstop, and highpass should be evident from the graphs. On the graphs, the **minimum passband gain** is denoted by A_1 and the **maximum stopband gain** by A_2. In general, the gain $A(f)$ can be voltage gain, current gain, magnitude of transimpedance, or magnitude of transconductance,

T.H. Glisson, *Introduction to Circuit Analysis and Design*,
DOI 10.1007/978-90-481-9443-8_19, © Springer Science+Business Media B.V. 2011

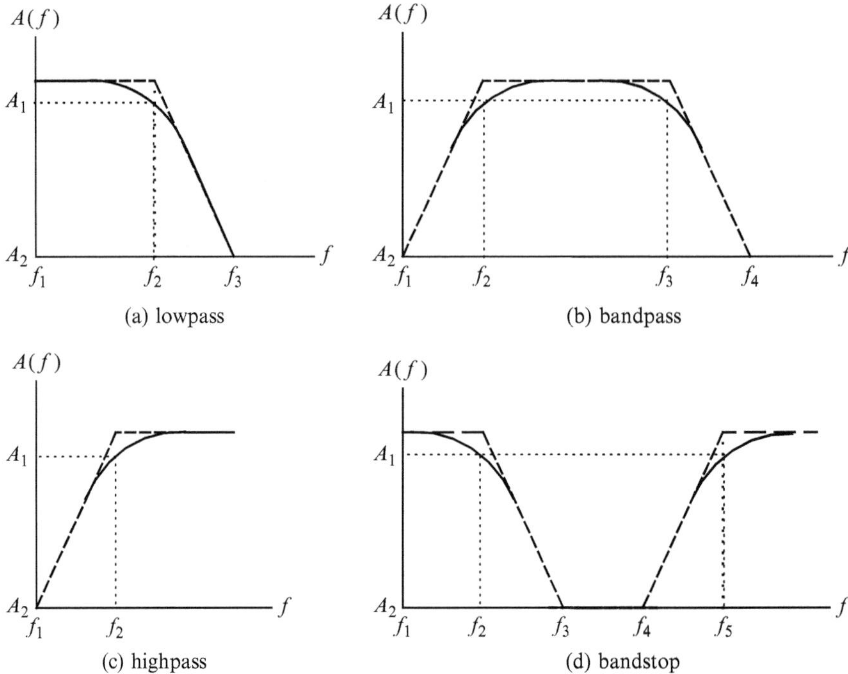

Fig. 19.1 Common types of filters

depending upon the natures of the source and load. But often, the input and output are voltages, and a filter is described by its voltage gain.

Each of the four filters in Fig. 19.1 is characterized by one or more **passbands**, where the gain equals or exceeds A_1; one or more **stopbands**, where the gain is no larger than A_2; and one or more **transition bands**, where the gain is between A_1 and A_2. For example, if the frequencies f_2, f_3 are half-power frequencies, the bandpass filter of Fig. 19.1(b) has stopbands $f < f_1$ and $f > f_4$, transition bands $f_1 < f < f_2$ and $f_3 < f < f_4$, and passband $f_2 < f < f_3$.

A lowpass or bandpass filter is described in part by its **bandwidth**, which is the width of the passband, often defined by the half-power (–3 dB) frequency or frequencies. For example, if the frequencies f_2, f_3 are half-power frequencies, the bandwidth of the bandpass filter is $f_3 - f_2$. A bandstop or highpass filter is described in part by the width of the stopband, often defined by the half-power (–3 dB) frequency or frequencies.

The transition from passband to stopband often is characterized by its slope, in decibels per decade. A filter having a steep transition (or transitions) is said to have a **sharp cutoff**. A bandpass filter whose bandwidth is much smaller than its center frequency is called a **narrowband** filter.

Filters often are described (as in Fig. 19.1) in terms of gain alone. But in many applications, a property called *group delay* is at least as important.

19.2 Group Delay

Recall (Chapter 15) that distortionless transmission requires frequency-independent gain and linear phase shift. Ideally, a filter possesses these characteristics in the passband(s), at least to a degree acceptable for the intended application. In the passband(s), the transfer function of a distortionless filter has the form

$$H = A\angle(-2\pi f t_0) = A\exp(-j2\pi f t_0) \qquad (19.1)$$

where A and t_0 are independent of frequency. The **group delay** of a filter is denoted by $D(f)$ and defined by

$$D(f) = -\frac{1}{2\pi}\frac{d\Phi(f)}{df} \qquad (19.2)$$

where $\Phi(f)$ is the phase shift of the filter. Thus, the group delay of a distortionless filter, described by (19.1), is given by

$$D = t_0 \tag{19.3}$$

which is the actual delay suffered by a signal (in the passband). Both gain and group delay are independent of frequency in the passband(s) of a distortionless filter.

A sharp-cutoff filter typically has poorer group-delay characteristics than does a less-sharp filter having the same bandwidth. It is possible to design a filter that meets demanding specifications on both gain and group delay by first designing a filter based on desired gain alone and subsequently adding (in cascade) an *all-pass delay equalizer* whose gain is frequency-independent and whose group delay complements that of the filter such that overall group delay is nearly frequency independent. A good filter (including an associated delay equalizer, if necessary) generally has sharp transitions and almost frequency-independent passband gain and group delay. The price paid for this performance is complexity (more components) and possibly a lengthy delay.

The usual application of the kinds of filters illustrated by Fig. 19.1 is separating one signal from another when the two signals occupy different bands on the frequency axis. In this chapter, we consider only applications of this kind. There are many other kinds of filters, such as integrating and differentiating filters, notch filters, comb filters, noise-weighting filters, delay equalizing filters, and matched filters. Some of these you have encountered in previous chapters, and others of which you might encounter in subsequent courses or in later practice.

Henceforth, *filter* means a circuit designed to separate one signal from another when the two signals occupy different bands on the frequency axis. Generally desirable properties of such a filter are:

1. Frequency-independent gain and group delay in the passband(s),
2. sharp transition(s) from passband(s) to stopband(s), and
3. a sufficiently large difference (dB) between nominal passband gain and maximum stopband gain, such that the desirable signal is passed to the load but other (unwanted) signals are adequately suppressed.

19.3 A Simple Two-Pole Active Filter

Refer to Fig. 19.2. We assume $R \gg R_S$, where R_S is the source resistance. By inspection of the circuit, $v_n = v_{out}$. For an ideal op amp, $v_p = v_n$ and v_{out} is independent of the load.

Applying Kirchhoff's current law to the circuit gives

$$\begin{aligned}
&\frac{V_1 - V_{in}}{R} + \frac{V_1 - V_{out}}{R} + sC_1(V_1 - V_{out}) = 0\\
&\Rightarrow (2 + s\tau_1)V_1 - (1 + s\tau_1)V_{out} = V_{in},\\
&\frac{V_{out} - V_1}{R} + sC_2V_{out} = 0 \Rightarrow V_1 = (1 + s\tau_2)V_{out},
\end{aligned} \tag{19.4}$$

where

$$\tau_1 = RC_1,\ \tau_2 = RC_2. \tag{19.5}$$

The solutions to the node Equations (19.4) are

$$\begin{aligned}
V_1 &= \frac{(1 + s\tau_2)V_{in}}{1 + 2s\tau_2 + s^2\tau_1\tau_2},\\
V_{out} &= \frac{V_{in}}{1 + 2s\tau_2 + s^2\tau_1\tau_2},
\end{aligned} \tag{19.6}$$

from which we obtain expressions for the input impedance

$$Z_{in} = \frac{V_{in}}{I_{in}} = \frac{V_{in}R}{V_{in} - V_1} = \frac{1 + 2s\tau_2 + s^2\tau_1\tau_2}{s\tau_2(1 + s\tau_1)}R \tag{19.7}$$

and the voltage transfer function

$$\begin{aligned}
H_v(s) &= \frac{V_{out}}{V_{in}} = \frac{1}{\tau_1\tau_2 s^2 + 2\tau_2 s + 1}\\
&= \frac{(\tau_1\tau_2)^{-1}}{s^2 + 2\tau_1^{-1}s + (\tau_1\tau_2)^{-1}},
\end{aligned} \tag{19.8}$$

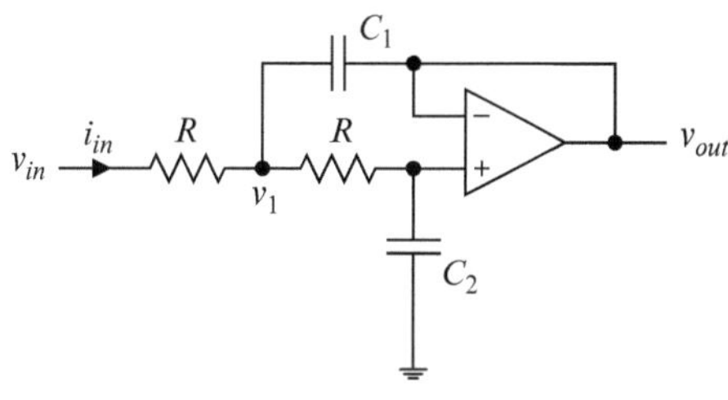

Fig. 19.2 A two-pole active filter

which is a low-pass transfer function, having dc gain $|H_v(0)| = 1$.

We consider the input impedance in the sequel, and focus first on the transfer function. The poles of the transfer function are given by

$$\begin{aligned} p_1, p_2 &= \frac{-\tau_2 \pm \sqrt{\tau_2^2 - \tau_1\tau_2}}{\tau_1\tau_2} \\ &= -\frac{1}{\tau_1} \pm \frac{1}{\tau_1}\sqrt{1 - \frac{\tau_1}{\tau_2}}. \end{aligned} \tag{19.9}$$

In filtering applications, the circuit in Fig. 19.2 is almost always used to provide complex conjugate poles. Therefore we assume $\tau_1 > \tau_2$, which from (19.9) implies that $C_1 > C_2$. The poles are given by

$$p, p^* = \frac{1}{\tau_1}\left(-1 \pm j\sqrt{\frac{\tau_1}{\tau_2} - 1}\right). \tag{19.10}$$

We may write the voltage transfer function in terms of the upper-half-plane pole p as

$$\begin{aligned} H_v(s) &= \frac{|p|^2}{(s-p)(s-p^*)} \\ &= \frac{|p|^2}{s^2 - 2\mathrm{Re}(p)s + |p|^2} \\ &= \frac{\omega_0^2}{s^2 - 2\mathrm{Re}(p)s + \omega_0^2}, \end{aligned} \tag{19.11}$$

where

$$\omega_0 = |p| = \frac{1}{\sqrt{\tau_1\tau_2}}, \quad \mathrm{Re}(p) = -\frac{1}{\tau_1}.$$

The frequency-domain transfer function is

$$H_v(j2\pi f) = \frac{1}{1 - (f/f_0)^2 + j2\sqrt{\tau_2/\tau_1}(f/f_0)}. \tag{19.12}$$

The denominator is a quadratic factor having peaking factor $\alpha = \sqrt{\tau_2/\tau_1}$, and the bandwidth is sometimes defined as the corner frequency f_0. This is the half-power bandwidth only if $\alpha = 1/\sqrt{2}$.

We may write the transfer function (19.12) as

$$H_v(j2\pi f) = \frac{Q}{Q\left[1 - (f/f_0)^2\right] + j(f/f_0)} \tag{19.13}$$

where

$$Q = \frac{1}{2}\sqrt{\frac{\tau_1}{\tau_2}} = -\frac{1}{2}\frac{|p|}{\mathrm{Re}(p)} = \frac{1}{2\cos(\theta)}, \quad \theta = \measuredangle p. \tag{19.14}$$

Comparing the right sides of (19.8) and (19.11) shows that

$$\begin{aligned} \mathrm{Re}(p) &= -\tau_1^{-1} = -\frac{1}{RC_1}, \\ |p|^2 &= (\tau_1\tau_2)^{-1} = \left(R^2C_1C_2\right)^{-1}. \end{aligned} \tag{19.15}$$

It follows that

$$\begin{aligned} C_1 &= -\frac{1}{\mathrm{Re}(p)R}, \\ C_2 &= \frac{1}{|p|^2 R^2 C_1} = -\frac{\mathrm{Re}(p)}{|p|^2 R}. \end{aligned} \tag{19.16}$$

If the resistance R and the pole p are specified, the circuit parameters are determined. We can determine R from a specification on input impedance. From (19.7),

$$Z_{in} = \frac{1 + 2s\tau_2 + s^2\tau_1\tau_2}{s\tau_2(1 + s\tau_1)}R$$

Let

$$s_0 = \frac{1}{\sqrt{\tau_1\tau_2}} \tag{19.17}$$

Then

$$Z_{in}(s) = \frac{1 + 2\tau_2 s_0(s/s_0) + (s/s_0)^2}{\tau_2 s_0(s/s_0)[1 + \tau_1 s_0(s/s_0)]}R$$

and

$$Z_{in}(j2\pi f) = \frac{\sqrt{\tau_1/\tau_2}\left[1 - (f/f_0)^2\right] + 2j(f/f_0)}{j(f/f_0)\left[1 + j\sqrt{\tau_1/\tau_2}(f/f_0)\right]}R \tag{19.18}$$

It can be shown (after some tedious mathematics) that the minimum input impedance (magnitude) in the passband occurs for $f = f_0$. From (19.18), the minimum input impedance (magnitude) is given by

$$\begin{aligned}|Z_{in}(f_0)| &= \left|\frac{2j}{j\left[1+j\sqrt{(\tau_1/\tau_2)}\right]}\right|R = \frac{2R}{\sqrt{1+(\tau_1/\tau_2)}}\\ &= \frac{2R}{\sqrt{1+(C_1/C_2)}} = \frac{2R}{\sqrt{1+[|p|/\mathrm{Re}(p)]^2}}\\ &= \frac{2R}{\sqrt{1+4Q^2}}\end{aligned} \tag{19.19}$$

Example 19.1. Design a filter having poles $p, p^* = 5\sqrt{2}\pi(1 \pm j) \times 10^3\,\mathrm{s}^{-1}$ and minimum input impedance (magnitude) $\min|Z_{in}| = 10\,\mathrm{k\Omega}$.
Solution: From (19.14),

$$Q = -\frac{1}{2}\frac{|p|}{\mathrm{Re}(p)} = \frac{\pi \times 10^4\,\mathrm{s}^{-1}}{\sqrt{2}\pi \times 10^4\,\mathrm{s}^{-1}} = \frac{1}{\sqrt{2}}$$

From (19.19),

$$\begin{aligned}R &= \frac{1}{2}\sqrt{1+4Q^2}|Z_{in}(f_0)|\\ &= \frac{1}{2}\sqrt{1+4Q^2}\min|Z_{in}| = \frac{\sqrt{3}}{2} \times 10\,\mathrm{k\Omega}\\ &\cong 8.66\,\mathrm{k\Omega}\end{aligned}$$

From (19.15),

$$\begin{aligned}\tau_1 &= -\mathrm{Re}(p)^{-1} = \left[5\sqrt{2}\pi \times 10^3\,\mathrm{s}^{-1}\right]^{-1}\\ &\cong 45.02\,\mu\mathrm{s},\\ \tau_2 &= \frac{1}{|p|^2\tau_1} = \frac{1}{(\pi \times 10^4\,\mathrm{s}^{-1})^2\tau_1} \cong 22.51\,\mu\mathrm{s}\end{aligned}$$

From (19.16)

$$C_1 = -\frac{1}{\mathrm{Re}(p)R} \cong 5.20\,\mathrm{nF},$$

$$C_2 = \frac{1}{|p|^2R^2C_1} \cong 2.60\,\mathrm{nF}$$

and the circuit external to the op amp in Fig. 19.2 is specified. Figure 19.3 shows a graph of voltage gain versus frequency for the filter.

Exercise 19.1. How many independent specifications can be met by the two-pole filter in Fig. 19.2?

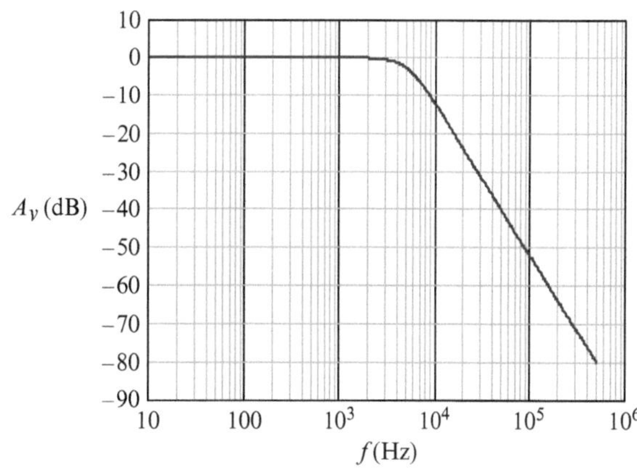

Fig. 19.3 See Example 19.1

Exercise 19.2. Design a two-pole filter for which $f_0 = 20\,\mathrm{kHz}$, peaking factor $\alpha = 0.5$, and minimum passband input resistance $\min(R_{in}) = 20\,\mathrm{k\Omega}$. Use a resistor from the E24 series.

The virtue of the two-pole circuit in Fig. 19.2 is its simplicity. But it can only provide poles (no zeros), so it can implement only lowpass filters. The slightly more complex Sallen-Key circuits treated in the next section can implement lowpass and highpass filters. In principle, lowpass and highpass Sallen-key filters can be cascaded to implement bandpass and band-stop filters, but such implementations are inefficient, requiring more components than other, comparable implementations. The state-variable biquadratic filter treated in Section 19.5 can implement a much wider variety of filters.

19.4 Sallen-Key (VCVS) Filters[1]

Figure 19.4 shows lowpass and highpass versions of second-order (two-pole) filters that are variants of what are called Sallen-Key or VCVS filters. The

[1]Invented in 1955 by R.P. Sallen and E.L. Key of MIT Lincoln Laboratory.

s-domain voltage transfer functions for the filters in Fig. 19.4 are

$$H_{v\,\mathrm{LP}}(s) = \frac{(k+1)}{(s\tau)^2 + (2-k)(s\tau) + 1}; \quad \tau = RC \quad (19.20)$$

for the lowpass filter and

$$H_{v\,\mathrm{HP}}(s) = \frac{(k+1)(s\tau)^2}{(s\tau)^2 + (2-k)(s\tau) + 1}; \quad \tau = RC \quad (19.21)$$

for the highpass filter. The poles of both transfer functions are given by

$$p, p^* = \frac{1}{2\tau}\left(-2 + k \pm j\sqrt{4k - k^2}\right), \quad \tau = RC. \quad (19.22)$$

For $k = 0$, the poles are real and equal (to $-1/\tau$). For $k > 2$, the poles are in the RHP and the circuits are unstable. We are interested primarily in stable circuits having complex-conjugate pole, so we consider values of k in the range $0 < k < 2$. The lowpass VCVS filter has no finite zeros. This property limits the kinds of lowpass filters that can be implemented as cascades of VCVS sections. A similar restriction applies to the highpass VCVS filter, which has no zeros other than a double zero at $s = 0$.

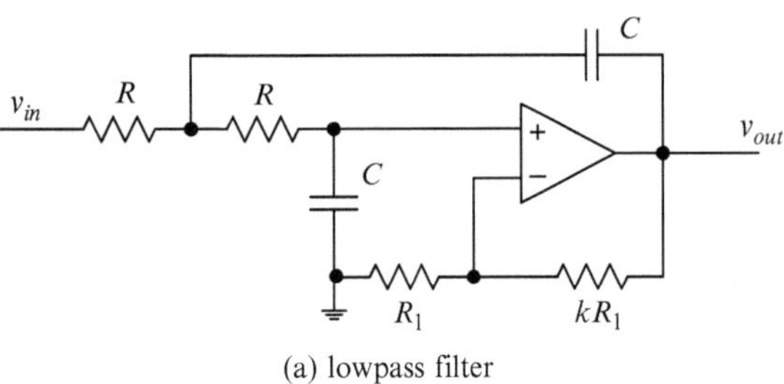

(a) lowpass filter

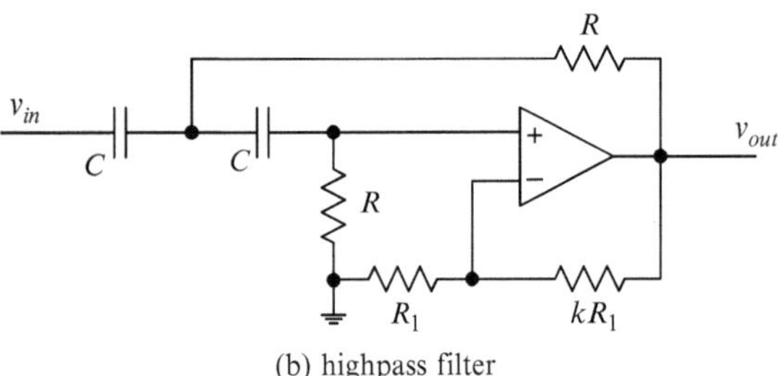

(b) highpass filter

Fig. 19.4 VCVS (Sallen-Key) filters

Exercise 19.3. Derive (19.20), (19.21), and (19.22).

Exercise 19.4. Refer to (19.22). Show that $|p| = \tau^{-1}$, independent of k.

Exercise 19.5. What are the maximum passband voltage gains of the VCVS sections described by (19.20) and (19.21)?

The frequency-domain voltage transfer functions for the VCVS filters in Fig. 19.4 are

$$H_{v\,\mathrm{LP}}(j2\pi f) = \frac{(k+1)}{1 - (f/f_\tau)^2 + j(2-k)(f/f_\tau)} = \frac{K}{Q\left[1 - (f/f_\tau)^2\right] + j(f/f_\tau)} \quad (19.23)$$

and

$$H_{v\,\mathrm{HP}}(j2\pi f) = \frac{-(k+1)(f/f_\tau)^2}{1 - (f/f_\tau)^2 + j(2-k)(f/f_\tau)} = \frac{-K(f/f_\tau)^2}{Q\left[1 - (f/f_\tau)^2\right] + j(f/f_\tau)}, \quad (19.24)$$

where

$$Q = \frac{1}{2-k}, \; K = (k+1)Q, \; f_\tau = \frac{1}{2\pi\tau} = \frac{1}{2\pi RC}. \quad (19.25)$$

The maximum passband voltage gain for both filters equals $k + 1$. The voltage gains must be accounted for to avoid overload when cascading VCVS sections; e.g., the op-amp supply voltage must be large enough to accommodate the output swing.

The phase shifts are given by

$$\Phi_{vLP}(f) = -\tan^{-1}\left\{\frac{f/f_\tau}{Q\left[1-(f/f_\tau)^2\right]}\right\} \quad (19.26)$$

and

$$\Phi_{vHP}(f) = \pi - \tan^{-1}\left\{\frac{f/f_\tau}{Q\left[1-(f/f_\tau)^2\right]}\right\}. \quad (19.27)$$

The group delays for the two filters are identical, and are given by

$$D(f) = -\frac{1}{2\pi}\frac{d\Phi_{vLP}(f)}{df} = \frac{\tau Q\left[1+(f/f_\tau)^2\right]}{Q^2\left[1-(f/f_\tau)^2\right]^2+(f/f_\tau)^2} \quad (19.28)$$

Figure 19.5 shows graphs of gain and group for a single lowpass VCVS section and for three values of the parameter k. The gain is normalized to the dc gain and the group delay is normalized to the time constant $\tau = RC$. Both are plotted versus frequency, normalized to the corner frequency $f_\tau = (2\pi\tau)^{-1}$.

Both gain and group delay appear to be almost independent of frequency for frequencies below $0.1f_\tau$, so the filters should provide high-fidelity transmission for signals confined to that band. Both gain and group delay become increasingly frequency-dependent with increasing k for frequencies above $0.1f_\tau$, and the filters would introduce both amplitude distortion and delay distortion for signals in that band. However, in many applications of low-pass filtering, the sinusoidal components of signals of interest decrease in amplitude with increasing frequency, such that components above $0.1f_\tau$ might be considerably weaker than those below $0.1f_\tau$, in which case the net distortion could be small.

The gain of the highpass VCVS filter is essentially that of the lowpass filter, flipped horizontally. The group delay of the highpass filter is identical to that of the lowpass filter. Thus, for the highpass VCVS filter, the gain and group delay are almost independent of frequency above $10f_\tau$. In many applications of high-pass filtering, the most significant components of the signals of interest lie above $10f_\tau$, in which cases net distortion could be small.

Equations (19.20) and (19.21) assume the op amp is ideal. More exact relations can be obtained using the frequency-dependent model for an op amp, but the slight improvement in accuracy is usually not worth the trouble. Figure 19.6 compares voltage gain versus frequency of a lowpass VCVS filter for ideal and non-ideal (frequency-dependent) op amps having equal dc voltage gains. The circuit parameters used are $R = 500\,\text{k}\Omega$, $C = 31.83\,\text{pF}$, $k = 0.5$, and $R_1 = 50\,\text{k}\Omega$ corresponding (approximately) to a half-power bandwidth of 10 kHz and a dc voltage gain of $1 + k = 1.5$. The ideal op amp has infinite input resistance, zero output resistance, and intrinsic voltage gain $\mu_0 = 10^5$. The non-ideal op amp has input resistance $5\,\text{M}\Omega$, output resistance $100\,\Omega$, dc voltage gain $\mu_0 = 10^5$, and bandwidth $f_0 = 10\,\text{Hz}$. You can see from the graph that the difference between the voltage gains computed using these models is slight.

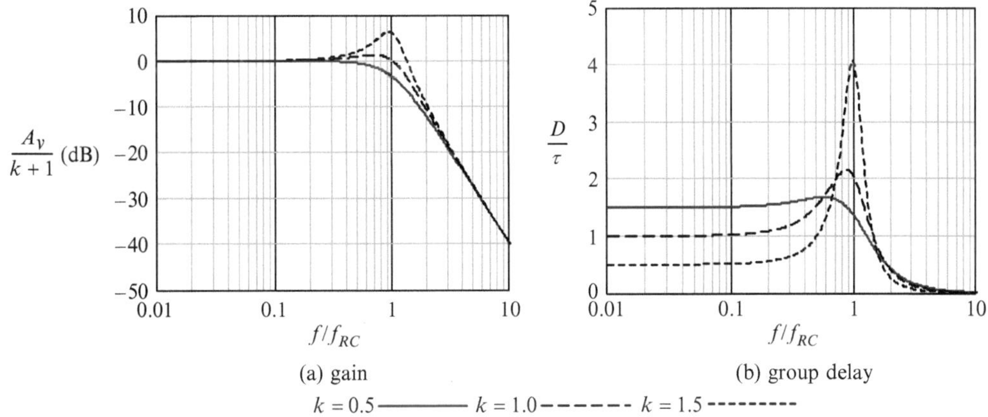

Fig. 19.5 Gain and group delay of lowpass VCVS filters. See Fig. 19.4(a)

The input impedances of the VCVS filters in Fig. 19.4 are given by

$$Z_{inLP} = \frac{1-(f/f_{RC})^2+j(2-k)(f/f_{RC})}{j(f/f_{RC})[j(f/f_{RC})+1-k]}R;\ f_{RC} = \frac{1}{2\pi RC} \tag{19.29}$$

and

$$Z_{inHP} = \frac{1-(f/f_{RC})^2+j(2-k)(f/f_{RC})}{j(f/f_{RC})[j(1-k)(f/f_{RC})+1]}R;\ f_{RC} = \frac{1}{2\pi RC} \tag{19.30}$$

Figure 19.7 shows graphs of input impedance (magnitude) versus frequency for the lowpass and highpass VCVS filters for two values of the parameter k and normalized to the resistance R. The magnitude of the input impedance of the lowpass filter decreases monotonically with increasing frequency in the passband. The magnitude of the input impedance of the highpass filter increases monotonically with increasing frequency in the passband.

The minimum magnitude of the input impedance of either filter occurs at the passband edge and is given by

$$|Z_{inHP}(j2\pi f_{RC})| = |Z_{inLP}(j2\pi f_{RC})| = \frac{(2-k)}{\sqrt{1+(1-k)^2}}R. \tag{19.31}$$

Figure 19.8 shows a graph of the input impedance (magnitude) of a single VCVS circuit at the passband edge versus the parameter k. For both the lowpass and the highpass VCVS filters, the minimum magnitude of the input impedance occurs near $f = f_{RC}$. The magnitude of the input impedance at the passband edge is a fraction of the resistance R for values of k in the neighborhood of $k = 2$. As we show in the sequel, some designs call for values of k near 2.

The output impedances of the lowpass and highpass VCVS filters in Fig. 19.4 are the same and are given by

$$Z_{out} = \frac{1-(f/f_{RC})^2+3j(f/f_{RC})}{\mu(f)R+j(f/f_{RC})R_o-(f/f_{RC})^2R_o} \times \frac{(k+1)R_oR}{1-(f/f_{RC})^2+(2-k)j(f/f_{RC})} \tag{19.32}$$

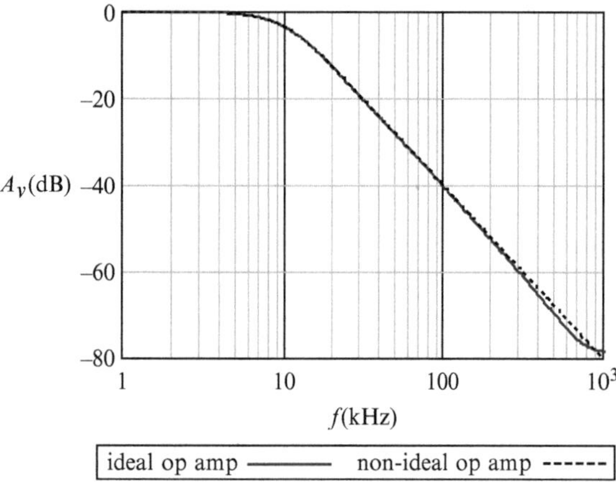

Fig. 19.6 Gain versus frequency for a typical general-purpose op amp

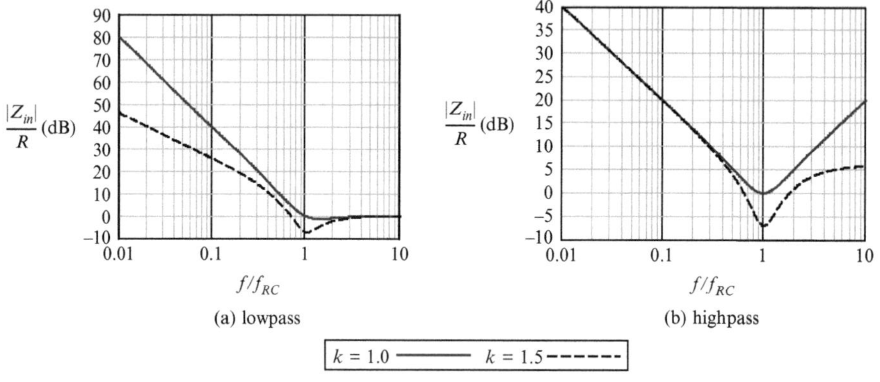

Fig. 19.7 Input impedance (magnitude) of the VCVS circuits in Fig. 19.4

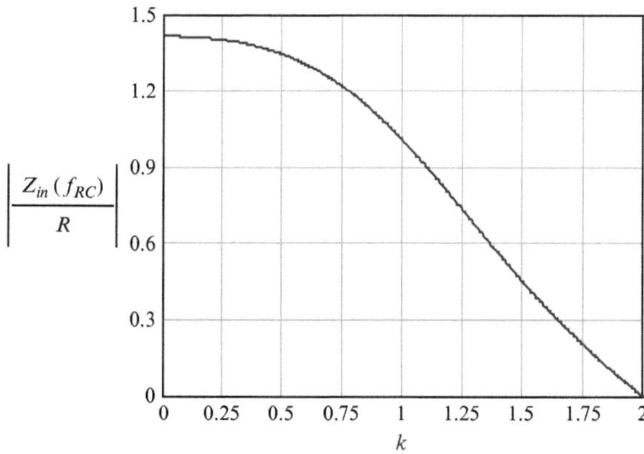

Fig. 19.8 The minimum input impedance (magnitude) of a VCVS filter decreases with increasing k

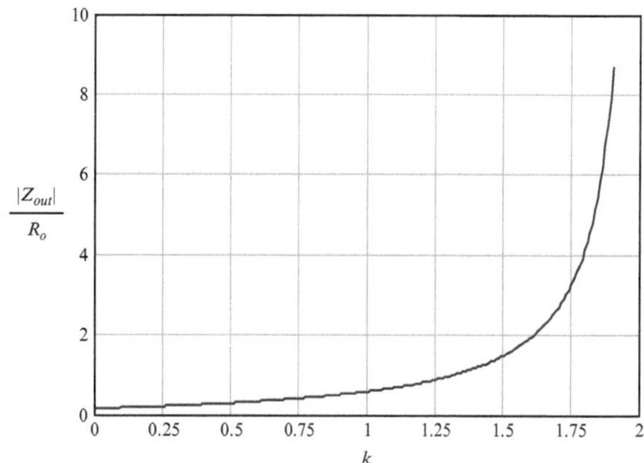

Fig. 19.9 The maximum output impedance (magnitude) of a VCVS filter increases with increasing k

where

$$f_{RC} = \frac{1}{2\pi RC} \tag{19.33}$$

and

$$\mu(f) = \frac{\mu_0}{1 + jf/f_0}, \tag{19.34}$$

and where R_o is the output resistance of the op amp, which is typically 100 Ω or less, μ_0 is the intrinsic dc voltage gain of the op amp, which is typically 10^5 or greater, and f_0 is the intrinsic bandwidth of the op amp, which is typically on the order of 10 Hz.

At dc, the output resistance is given by

$$Z_{out}(0) = \frac{(k+1)R_o}{\mu_0} \tag{19.35}$$

and is quite small.

In most applications, $f_T \gg f_{RC} \gg f_0$, $R \gg R_o$, and the frequencies of interest are those in the passband, where $f \leq f_{RC}$. Thus

$$\mu(f_{RC}) \cong \frac{\mu_0}{jf_{RC}/f_0} = \frac{\mu_0 f_0}{jf_{RC}} = \frac{f_T}{jf_{RC}}$$

and

$$\frac{f_T R}{jf_{RC}} + j\left(\frac{f}{f_{RC}}\right)R_o - \left(\frac{f}{f_{RC}}\right)^2 R_o \cong \frac{f_T R}{jf_{RC}},$$

where $f_T = \mu_0 f_0$ is the unity-gain frequency (gain-bandwidth product) of the op amp. With these approximations, (19.32) reduces to

$$Z_{out} \cong \frac{\left[1 - (f/f_{RC})^2 + 3j(f/f_{RC})\right](k+1)jf_{RC}R_o}{\left[1 - (f/f_{RC})^2 + (2-k)j(f/f_{RC})\right]f_T} \tag{19.36}$$

It can be shown that the maximum magnitude of the output impedance given by (19.36) occurs for $f = f_{RC}$; i.e., at approximately the passband edge, where

$$|Z_{out}(j2\pi f_{RC})| \cong \frac{3(k+1)}{(2-k)} \frac{f_{RC}}{f_T} R_o \tag{19.37}$$

For any value of k, the maximum magnitude of the output impedance (in the passband) occurs at the passband edge, or for $f \cong f_{RC}$, and is a function of k and of f_{RC}/f_T, where f_T is the unity-gain frequency (gain-bandwidth product) of the op amp. Figure 19.9 shows a graph of the maximum magnitude of the output impedance, normalized to R_o, versus k, for $f_T = 1$ MHz, which is a typical value for a general-purpose BJT op amp, and for $f_{RC} = 100$ kHz, which is on the order of the largest bandwidth one would attempt with such an op amp.

From Figs. 19.8 and 19.9, increasing k decreases the magnitude of the input impedance and increases the output impedance. *To minimize loading in a cascade of VCVS filters, the filters should be cascaded in the order of increasing values of k.*

19.5 State-Variable Biquadratic Filter

A function of the form

$$G(s) = \frac{b_2 s^2 + b_1 s + b_0}{a_2 s^2 + a_1 s + a_0} \tag{19.38}$$

is called a **biquadratic** function because it is the ratio of two quadratic functions of the independent variable (s, in this case).

Most major manufacturers of integrated circuits offer integrated circuits (ICs) called **biquads**, which (with a few external components) can implement a voltage transfer function having the form of (19.38). A typical biquad IC simultaneously provides outputs corresponding to the voltage transfer functions

$$\begin{aligned} H_{LP}(s) &= \frac{b_0}{a_2 s^2 + a_1 s + a_0}, \\ H_{BP}(s) &= \frac{b_1 s}{a_2 s^2 + a_1 s + a_0}, \\ H_{HP}(s) &= \frac{b_2 s^2}{a_2 s^2 + a_1 s + a_0}. \end{aligned} \tag{19.39}$$

which, as the subscripts indicate, represent lowpass, bandpass, and highpass filters. With the addition of an op amp (or summing IC) and a few resistors, the circuit also can provide a full biquadratic voltage transfer function of the form (19.38), which can be used to provide delay equalization or a more complex second-order transfer function.

In this section, we treat the *state-variable* biquad shown in Fig. 19.10. National Semiconductor's MF10 biquad IC contains two state-variable filters similar to the one shown in Fig. 19.10 in a single 20-pin DIP package about 2.6×0.6 cm. Other manufacturers offer similar circuits. The capacitors and the resistors denoted by R and $10R$ in Fig. 19.10 are internal. The resistors R_G, R_F, R_Q are provided externally to achieve a specified transfer function. The internal resistors in most such ICs are switched-capacitor resistors,[2] so an accurate external clock is required to establish the internal resistances denoted by R and $10R$. To keep things simple, we treat the internal switched-capacitor resistors as ohmic resistors, where $R = 100\,\mathrm{k}\Omega$, and we assume the capacitance is fixed at $C = 1$ nF. Below, we obtain the lowpass, bandpass, and highpass transfer functions for the filter in Fig. 19.10.

The leftmost op-amp circuit in Fig. 19.10 is a summing amplifier with two inverting inputs and one non-inverting input. The other two op-amp circuits are inverting integrators.[3] By inspection, the inputs and outputs of the integrators are related as

$$V_B = -\frac{V_H}{s\tau}, \quad V_L = -\frac{V_B}{s\tau}, \quad \tau = R_F C. \tag{19.40}$$

By voltage division,

$$V_2 = \frac{R_Q}{10R + R_Q} V_B. \tag{19.41}$$

By Kirchhoff's current law,

$$\begin{aligned} &\frac{V_2 - V_S}{R_G} + \frac{V_2 - V_H}{R} + \frac{V_2 - V_L}{10R} = 0 \\ &\Rightarrow V_2 = \frac{10RV_S + 10R_G V_H + R_G V_L}{10R + 11R_G}. \end{aligned} \tag{19.42}$$

Equating the right sides of (19.41) and (19.42) gives

$$\frac{R_Q}{10R + R_Q} V_B = \frac{10RV_S + 10R_G V_H + R_G V_L}{10R + 11R_G}. \tag{19.43}$$

We can use (19.40) to express any two of V_B, V_H, V_L in terms of the third. For example, using (19.40) to express V_L and V_H in terms of V_B leads to

$$\frac{R_Q}{10R + R_Q} V_B = \frac{[10RV_S - 10R_G V_B s\tau]s\tau - R_G V_B}{(10R + 11R_G)s\tau} \tag{19.44}$$

from which we obtain the bandpass voltage transfer function

$$H_{vBP}(s) = \frac{V_B}{V_S} = \frac{10\frac{R}{R_G}s\tau}{10(s\tau)^2 + \frac{R_Q(10R+11R_G)}{R_G(10R+R_Q)}s\tau + 1}$$

[2]Switched-capacitor resistors are discussed in Chapter 8.

[3]The resistors denoted by $10R$ provide dc feedback for the integrators.

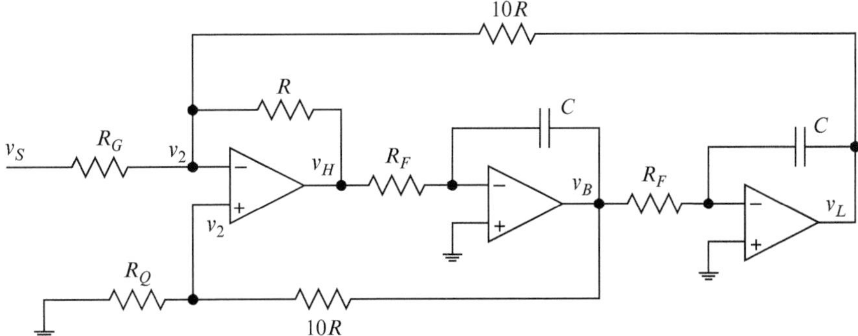

Fig. 19.10 State-variable filter

We define

$$\tau' = \tau\sqrt{10} = R_F C\sqrt{10},$$
$$Q = \frac{\sqrt{10}R_G(10R + R_Q)}{R_Q(10R + 11R_G)}, \quad K_{BP} = \sqrt{10}\frac{RQ}{R_G} \qquad (19.45)$$

to obtain

$$H_{vBP} = \frac{K_{BP}s\tau'}{Q(1 + s^2\tau'^2) + s\tau'} \qquad (19.46)$$

The bandpass voltage transfer function is expressed as a function of frequency by

$$H_{vBP}(j2\pi f) = \frac{K_{BP}(jf/f_c)}{Q\left[1 - (f/f_c)^2\right] + (jf/f_c)} \qquad (19.47)$$

where

$$f_c = \frac{1}{2\pi\tau'} = \frac{1}{2\pi\sqrt{10}R_F C} \qquad (19.48)$$

The parameter f_c is the center frequency, Q is the quality factor, and K' is the maximum passband gain (the gain for $f = f_c$). If $Q \gg 1$, the half-power bandwidth is given (approximately) by

$$W \cong \frac{f_c}{Q} \qquad (19.49)$$

The center frequency, quality factor (or bandwidth), and maximum passband gain are determined by the three external resistors R_G, R_F, R_Q. If we assume R and C are fixed, and that the center frequency, bandwidth, and maximum passband gain are specified, then from (19.45) and (19.48), the design equations are

$$R_F = \frac{1}{2\pi\sqrt{10}\,f_c\,C}, \quad Q = \frac{f_c}{W},$$
$$R_G = Q\frac{\sqrt{10}R}{K_{BP}}, \qquad (19.50)$$
$$R_Q = \frac{10\sqrt{10}R_G R}{(10R + 11R_G)Q - \sqrt{10}R_G}.$$

Example 19.2. Refer to Fig. 19.10. Assume $R = 100\,\text{k}\Omega$ and $C = 1\,\text{nF}$. Specify the external resistors R_G, R_F, R_Q to obtain a bandpass filter having center frequency $f_c = 10\,\text{kHz}$, bandwidth $W = 500\,\text{Hz}$, and maximum passband gain $K_{BP} = 100$. Construct a plot of gain versus frequency as a check on the design.
Solution: From (19.50) we find
$R_F = 5.033\,\text{k}\Omega$, $Q = 20$, $R_G = 63.246\,\text{k}\Omega$, $R_Q = 5.932\,\text{k}\Omega$. Figure 19.11 shows graphs of gain versus frequency, normalized to the maximum gain, given by $|H_{vBP}(j2\pi f_c)| = K_{BP} = 100$. Thus

$$A_v = 20\log\left|\frac{H_{vBP}(j2\pi f)}{K_{BP}}\right|\text{dB}.$$

In Fig. 19.11(a), $100\,\text{Hz} \le f \le 1\,\text{MHz}$ on a logarithmic scale. In Fig. 19.11(b), $9.75\,\text{kHz} \le f \le 10.25\,\text{kHz}$ on a linear scale, to show that the half-power (−3 dB) bandwidth is as specified.

Exercise 19.6. Show that the lowpass and highpass transfer functions for the filter in Fig. 19.10 are

$$H_{vLP}(s) = \frac{K_{LP}}{Q\left[1 + (s\tau')^2\right] + s\tau'},$$
$$\tau' = \tau\sqrt{10},$$
$$K_{LP} = -\frac{10R}{R_G}Q, \tag{19.51}$$

and

$$H_{vHP}(s) = \frac{K_{HP}(s\tau')^2}{Q\left[1 + (s\tau')^2\right] + s\tau'},$$
$$\tau' = \tau\sqrt{10},$$
$$K_{HP} = -\frac{R}{R_G}Q. \tag{19.52}$$

where Q is given by (19.45).

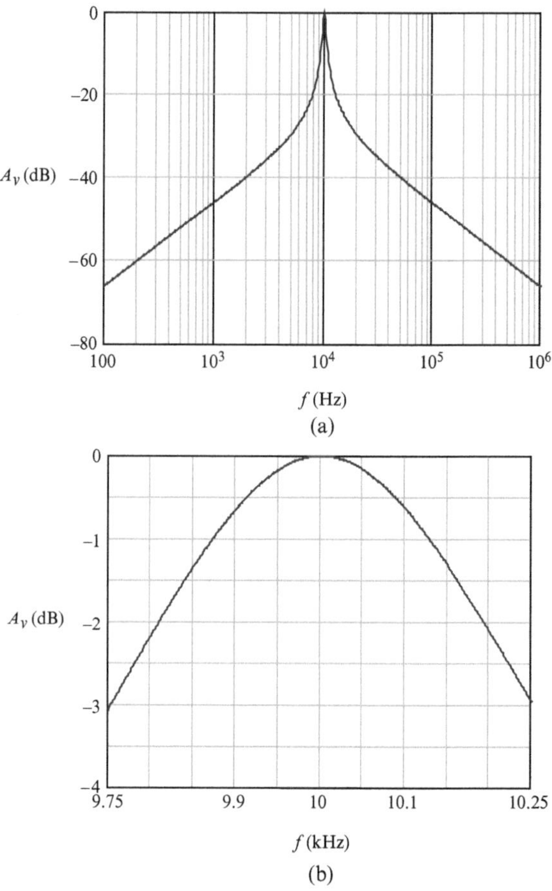

Fig. 19.11 See Example 19.2

The expressions given above for lowpass, bandpass, and highpass filters are useful for designing single-stage (one-biquad) filters. But often, the required selectivity cannot be achieved with a single stage. For applications requiring two or more stages, it is useful to express the transfer functions in terms of their poles. All three transfer functions have the same denominator and thus the same poles, given by

$$p_1, p_2 = \frac{-1 \pm \sqrt{1 - 4Q^2}}{2Q\tau'}. \tag{19.53}$$

In most applications requiring two or more poles, the poles are complex, in which case $Q > 0.5$ and

$$p, p^* = \frac{-1 \pm j\sqrt{4Q^2 - 1}}{2Q\tau'}. \tag{19.54}$$

The complex poles are determined by the two parameters Q, τ', such that

$$\mathrm{Re}(p) = \sigma = -\frac{1}{2Q\tau'}, \quad \mathrm{Im}(p) = \omega_0 = \frac{\sqrt{4Q^2 - 1}}{2Q\tau'}. \tag{19.55}$$

If the poles are specified, then we find from (19.55) that

$$Q = \frac{1}{2}\sqrt{\left(\frac{\omega_0}{\sigma}\right)^2 + 1}, \quad \tau' = -\frac{1}{2Q\sigma}. \tag{19.56}$$

Example 19.3. Using the biquad in Fig. 19.10, with $R = 100\,\mathrm{k\Omega}$ and $C = 1\,\mathrm{nF}$, design a low-pass filter having poles $p, p^* = 2\pi f_p \exp(\pm j3\pi/4)$, with $f_p = 4\,\mathrm{kHz}$, and dc voltage gain $|H_{vLP}(0)| = 10$.
Solution: From (19.51), the dc gain is

$$\left|H_{vLP}(0)\right| = \left|\frac{K_{LP}}{Q}\right| = \frac{10R}{R_G}.$$

The specified dc gain is 10. Thus

$$R_G = R = 100\,\mathrm{k\Omega}.$$

The real and imaginary parts of the pole p are

$$\sigma = \text{Re}(p) = 2\pi f_p \cos\left(\frac{3\pi}{4}\right) = -\frac{2\pi W}{\sqrt{2}},$$

$$\omega_0 = \text{Im}(p) = 2\pi f_p \sin\left(\frac{3\pi}{4}\right) = \frac{2\pi W}{\sqrt{2}}.$$

From (19.56)

$$Q = \frac{1}{2}\sqrt{\left(\frac{\omega_0}{\sigma}\right)^2 + 1} = \frac{1}{\sqrt{2}},$$

$$\tau' = -\frac{1}{2Q\sigma} = \frac{1}{2\sqrt{2}Q\pi W}.$$

From (19.50)

$$R_Q = \frac{10\sqrt{10}R_G R}{(10R + 11R_G)Q - \sqrt{10}R_G} \cong 270.58\,\text{k}\Omega,$$

$$Q = \frac{f_c}{W} \Rightarrow f_c = WQ,$$

$$R_F = \frac{1}{2\sqrt{10}\pi f_c C} = \frac{1}{2\sqrt{10}\pi WQC} \cong 17.79\,\text{k}\Omega,$$

and the filter is specified. Figure 19.12 shows a graph of the gain versus frequency. As a check, we compute

$$|H_{vLP}(j2\pi W)| = \left|\frac{K_{LP}}{Q\left[1 + (j2\pi W\tau')^2\right] + j2\pi W\tau'}\right|$$

$$\cong 7.071 \cong \frac{10}{\sqrt{2}}$$

which is 3 dB below the dc gain, as specified.

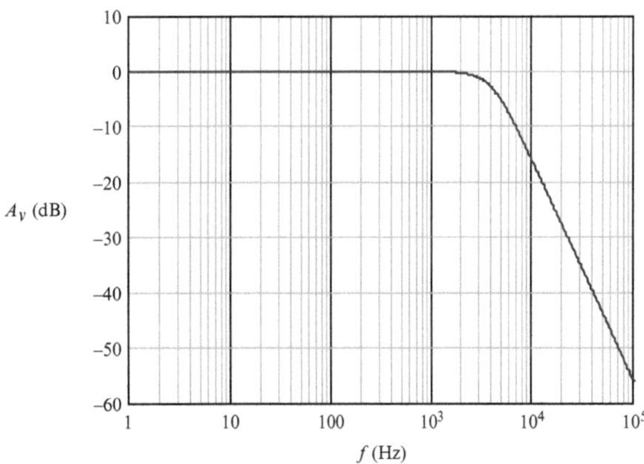

Fig. 19.12 See Example 19.3

The input impedance of the state-variable filter in Fig. 19.10 is given by

$$Z_{in}(f) = R_G\left(1 - \frac{K' j(f/f_c)}{Q\left[1 - (f/f_c)^2\right] + j(f/f_c)}\right)^{-1}, \quad (19.57)$$

where

$$K' = \frac{R_Q\sqrt{10}RQ}{R_G(10R + R_Q)}. \quad (19.58)$$

The input impedance for $f = f_c$ is given by

$$Z_{in}(f_c) = \frac{R_G(10R + R_Q)(10R + 11R_G)}{90R^2 + 110R_G R + 11R_G R_Q}. \quad (19.59)$$

Figure 19.13 shows a graph of the magnitude of the input impedance (normalized to R_G) versus frequency, in the vicinity of the center of the passband, for the filter considered in Example 19.2. The input impedance (magnitude) is maximum at the passband center, and, from (19.57), approaches R_G for $f \to 0$ and $f \to \infty$. It is good practice to calculate the input impedance to see if loading – especially frequency-selective loading – is a concern. In this case, if $R_G \gg R_S$, where R_S is the source resistance, loading might be negligible.

The output impedance for each output of the state-variable filter in Fig. 19.10 is quite small, as is evident from the fact that the associated transfer functions are independent of the load impedance. Of course, this is because the op amps are assumed to be ideal, but a tedious analysis using ac models for the op amps also shows that the output impedances are negligible if the load impedance is ten or more times the op-amp output impedance, which is typically in the neighborhood of 100 Ω.

In any case, if loading is a concern, it can be effectively eliminated by buffering with a voltage follower.

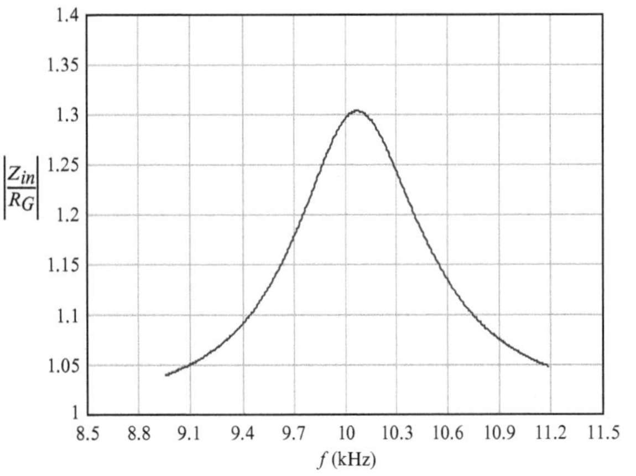

Fig. 19.13 Input impedance (normalized to R_G) versus frequency, in the vicinity of the center of the passband, for the filter considered in Example 19.2

19.6 Modern Filter Design

Modern filter design (which has been around since at least 1930) consists of various procedures for obtaining the poles of a filter from constraints on the frequency response of the filter. We limit our discussion of modern filter design to *lowpass filters having no finite zeros*.

Figure 19.14 illustrates (in part) typical specifications on gain for a lowpass filter, consisting of the upper edge of the passband (the bandwidth) W, the minimum and maximum allowable gains in the passband, the lower edge f_{stop} of the stopband, and the maximum allowable gain in the stopband.

Specifications on gain illustrated by Fig. 19.14 do not uniquely determine the transfer function of a lowpass filter. There are various ways of imposing the required additional constraints, each leading to a different filter characteristic. Three common ones are:[4]

- **Butterworth** lowpass filters, whose gains are as frequency-independent as possible in the passband and decrease monotonically with increasing frequency above the passband;
- **Bessel** lowpass filters, whose group delays are as frequency-independent as possible in the passband;
- **Type I Chebyshev** lowpass filter, whose gains ripple between specified maximum and minimum values in the passband and decrease sharply and monotonically above the passband.

Butterworth filters attempt to minimize amplitude distortion in the passband, Bessel filters attempt to minimize delay distortion in the passband, and type I Chebyshev filters attempt to achieve a sharp transition to a deep stopband. Of these three types and for any particular set of specifications on gain, imposed as illustrated by Fig. 19.14, a type I Chebyshev filter requires the fewest poles and a Bessel the most. Also for the same set of specifications, a Bessel filter has the best delay characteristic and the Chebyshev the worst. A Butterworth filter is intermediate in terms of both gain and delay, and is probably the most commonly used of the three.

Other types, not treated here, are Chebyshev Type II filters, which allow ripple in the stopband but not the passband, and elliptic (Cauer) filters, which allow ripple in both the passband and stopband. These filters have finite zeros, and cannot be implemented using the VCVS circuits described above. They can be implemented using biquads, such as the state-variable filter described in Section 19.5. You will learn more about these and other kinds of filters if you take a subsequent course in signal processing.

We note that the filter circuits treated here each provide two poles. Although it is possible to design and build first-order (single-pole) sections, there is no good reason to do so. With integrated op amps or biquads, a two-pole section is no more costly than a single-pole section and generally provides better performance. Thus, active filters (other than delay equalizers) are almost always of even order.

Horowitz and Hill[5] present a simplified design procedure for Butterworth, Chebyshev I, and Bessel filters using one or a cascade of VCVS sections, based upon the data given in Table 19.1. The procedure is described and illustrated below.

19.6.1 VCVS Butterworth Filters

For all sections of a Butterworth filter,

$$RC = \frac{1}{\omega_0} = \frac{1}{2\pi f_0} = \frac{1}{2\pi f_{RC}} \tag{19.60}$$

[4]Look up *Butterworth filter* in *Wikipedia* for more complete descriptions of these filter types.

[5]Horowitz, Paul and Winfield Hill, *The Art of Electronics*, Cambridge University Press, New York, 1989, pp 273–276.

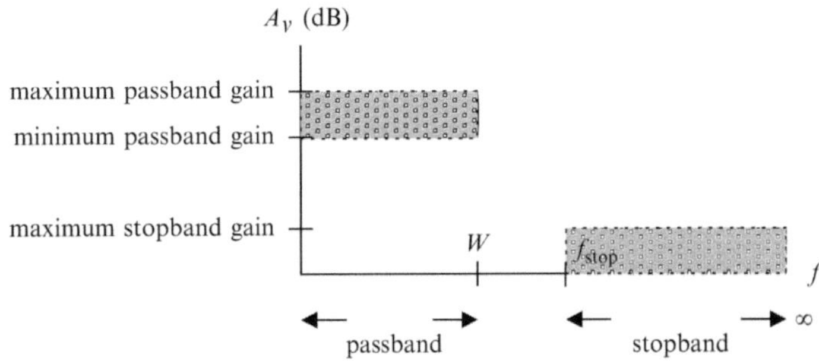

Fig. 19.14 Design specifications for lowpass filters

Table 19.1 Design parameters for VCVS filters: N is the number of poles and λ is the scaling factor for Bessel and Chebyshev sections[6]

N	Butterworth	Bessel		Chebyshev	
	k	k	λ	k	λ
2	0.586	0.268	−1.738	0.842	1.231
4	0.152	0.084	1.432	0.582	0.597
	1.235	0.759	1.606	1.660	1.031
6	0.068	0.040	1.607	0.537	0.396
	0.586	0.364	1.692	1.448	0.768
	1.483	1.023	1.908	1.846	1.011
8	0.038	0.024	1.781	0.522	0.297
	0.337	0.213	1.835	1.379	0.599
	0.889	0.593	1.956	1.711	0.861
	1.610	1.184	2.192	1.913	1.006

where $f_0 = f_{RC}$ is the specified half-power bandwidth. Refer to Table 19.1, the two rows opposite $N = 4$ in the column headed k under Butterworth. To obtain a four-pole lowpass Butterworth filter, cascade two lowpass VCVS sections. Design the first section (Fig. 19.4(a)) using $k = 0.152$ and the second using $k = 1.235$. Choose R to give an acceptable minimum input impedance and use (19.60) to determine C.

Example 19.4. Using three VCVS sections in cascade, design a six-pole lowpass Butterworth filter having half-power frequency $f_0 = 10\,\text{kHz}$ and minimum input impedance (magnitude) greater than $100\,\text{k}\Omega$.

Solution: From (19.60), for each section,

$$RC = \frac{1}{2\pi f_{RC}} = 15.92\,\mu\text{s} \qquad (19.61)$$

From Table 19.1, the values of k for the three sections are 0.068, 0.586, and 1.483. The sections should be cascaded in that order, such that the section having the largest input impedance comes first. From (19.29), the input impedance (magnitude) for $f = f_{RC}$ and $k = 0.068$ is $|Z_{in}(f_{RC})|_{k=0.068} = 1.41R$. Choosing $R = 100\,\text{k}\Omega$ will exceed the specifications. From (19.61)

$$C = \frac{1}{2\pi f_{RC} R} = \frac{1}{(2\pi)(10\,\text{kHz})(100\,\text{k}\Omega)} = 159\,\text{pF}$$

The filter is a cascade of three sections like that in Fig. 19.4(a), with

$$R = 100\,\text{k}\Omega,\ C = 159\,\text{pF}$$

for all three, and $k = 0.068$ for the first, $k = 0.586$ for the second, and $k = 1.483$ for the third. Figure 19.15 shows graphs of voltage gain, phase shift, and group delay versus frequency and step response versus time for the filter. The voltage gain and the step response are normalized to the dc gain, given by

$$H_v(0) = (1 + k_1)(1 + k_2)(1 + k_3) = 4.21 \Rightarrow A_v(0) = 12.48\,\text{dB}.$$

Note that the duration of the transient response is approximately $10.5/f_0 \cong 1\,\text{ms}$

[6]Adapted from Table 5.2 in Horowitz and Hill (ibid).

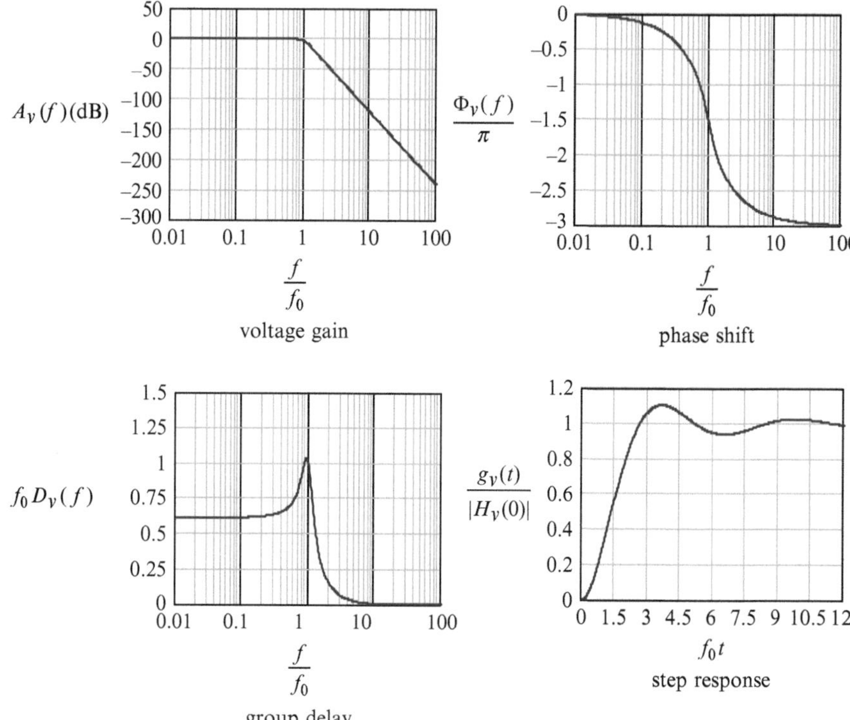

Fig. 19.15 See Example 19.4

19.6.2 VCVS Bessel Filters

The procedure for designing a Bessel filter as a cascade of VCVS sections differs from that for a Butterworth filter, in that the frequency f_{RC} for each section must be scaled by the factors λ given in Table 19.1.

Example 19.5. Using three VCVS sections in cascade, design a six-pole lowpass Bessel filter having half-power frequency $f_0 = 10\,\text{kHz}$ and minimum input impedance (magnitude) greater than $100\,\text{k}\Omega$.
Solution: From Table 19.1, the values of the scale factor λ for the three sections are 1.607, 1.692, and 1.908. This means, for example, that for the first section $f_{RC} = 1.607 f_0$, where f_0 is the specified overall half-power frequency. Thus the RC products for the three sections are:

$$(RC)_1 = \frac{1}{2\pi f_{RC1}} = \frac{1}{2\pi\lambda_1 f_0} = \frac{1}{2\pi(1.607)(10\,\text{kHz})} = 9.904\,\mu\text{s}$$

$$(RC)_2 = \frac{1}{2\pi f_{RC2}} = \frac{1}{2\pi\lambda_2 f_0} = \frac{1}{2\pi(1.692)(10\,\text{kHz})} = 9.406\,\mu\text{s}$$

$$(RC)_3 = \frac{1}{2\pi f_{RC3}} = \frac{1}{2\pi\lambda_3 f_0} = \frac{1}{2\pi(1.908)(10\,\text{kHz})} = 8.431\,\mu\text{s}$$

We choose $R = 100\,\text{k}\Omega$ for all sections, for which value the input impedance (of the first section) will meet or exceed the specification $|Z_{in}| \geq 100\,\text{k}\Omega$. The required values for the capacitances are then

$$C_1 = \frac{RC_1}{R} = \frac{9.904\,\mu\text{s}}{100\,\text{k}\Omega} = 99.04\,\text{pF}$$

$$C_2 = \frac{RC_2}{R} = \frac{9.406\,\mu\text{s}}{100\,\text{k}\Omega} = 94.06\,\text{pF}$$

$$C_3 = \frac{RC_3}{R} = \frac{8.431\,\mu\text{s}}{100\,\text{k}\Omega} = 84.31\,\text{pF}$$

The filter is a cascade of three sections like that in Fig. 19.4(a), $R = 100\,\text{k}\Omega$ and the capacitance values given above, and with $k = 0.040$ for the first, $k = 0.364$ for the second, and $k = 1.023$ for the third. Again, VCVS sections should generally be cascaded in the order implied by Table 19.1. Figure 19.16 shows graphs of voltage gain, phase shift, and group delay versus frequency and step response versus time for the filter. The voltage gain and the step response are normalized to the dc gain, given by

$$H_v(0) = (1 + k_1)(1 + k_2)(1 + k_3) = 2.87$$
$$\Rightarrow A_v(0) = 9.16\,\text{dB}$$

Exercise 19.7. The maximum amplitude of the input to an eight-pole lowpass Chebyshev filter is 500 mV. The op amps are dual-supply, rail-to-rail. What supply voltage is required?

19.6.3 VCVS Chebyshev Filters

The procedure for designing a VCVS Chebyshev filter is like that for a VCVS Bessel filter, except that the frequency f_0 is not the half-power frequency. It is the frequency at which the voltage gain begins its plunge into the stopband. Also, whereas the voltage gains of Butterworth and Bessel filters are monotonically decreasing in the passband, the voltage gain of a Chebyshev filter can ripple throughout the passband. The voltage gain of a Chebyshev filter designed using the parameters in Table 19.1 exhibits a 0.5 dB passband ripple.

Example 19.6. Design a six-pole lowpass Chebyshev filter having 0.5 dB passband ripple, bandwidth $f_0 = 10\,\text{kHz}$, and minimum input impedance (magnitude) of at least $100\,\text{k}\Omega$. Obtain the voltage transfer function for the filter and plot the gain, phase shift, and group delay versus frequency, for $0.1f_0 \leq f \leq 10f_0$.
Solution: From Table 19.1, the values of the scale factor λ for the three sections are 0.396,

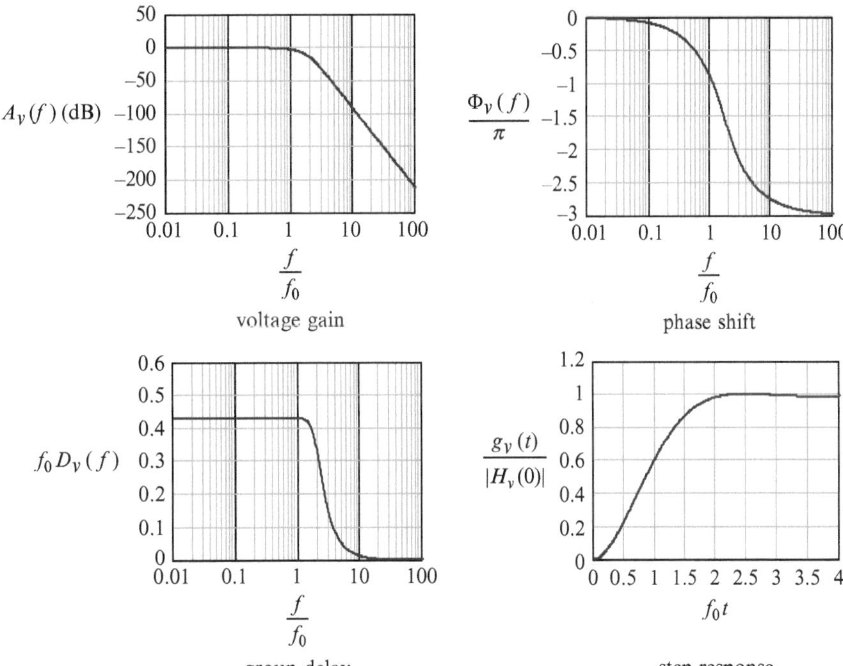

Fig. 19.16 See Example 19.5

0.768, and 1.011. This means, for example, that for the first section $f_{RC} = 0.396 f_0$, where f_0 is the specified passband edge. Thus the *RC* products for the three sections are:

$$
\begin{aligned}
(RC)_1 &= \frac{1}{2\pi f_{RC1}} = \frac{1}{2\pi\lambda_1 f_0} = \frac{1}{2\pi(0.396)(10\,\text{kHz})} \\
&= 40.19\,\mu\text{s} \\
(RC)_2 &= \frac{1}{2\pi f_{RC2}} = \frac{1}{2\pi\lambda_2 f_0} = \frac{1}{2\pi(0.768)(10\,\text{kHz})} \\
&= 20.72\,\mu\text{s} \\
(RC)_3 &= \frac{1}{2\pi f_{RC3}} = \frac{1}{2\pi\lambda_3 f_0} = \frac{1}{2\pi(1.011)(10\,\text{kHz})} \\
&= 15.74\,\mu\text{s}
\end{aligned}
$$

We choose $R = 100\,\text{k}\Omega$ for all sections, for which value the input impedance (of the first section) will meet or exceed the specification $|Z_{in}| \geq 100\,\text{k}\Omega$. The required values for the capacitances are then

$$
\begin{aligned}
C_1 &= \frac{RC_1}{R} = \frac{40.19\,\mu\text{s}}{100\,\text{k}\Omega} = 401.906\,\text{pF} \\
C_2 &= \frac{RC_2}{R} = \frac{20.72\,\mu\text{s}}{100\,\text{k}\Omega} = 207.233\,\text{pF} \\
C_3 &= \frac{RC_3}{R} = \frac{15.74\,\mu\text{s}}{100\,\text{k}\Omega} = 157.423\,\text{pF}
\end{aligned}
$$

The filter is a cascade of three sections like that in Fig. 19.4(a), with $R = 100\,\text{k}\Omega$ and the capacitance values given above, and with $k = 0.537$ for the first, $k = 1.448$ for the second, and $k = 1.846$ for the third. Again, VCVS sections should generally be cascaded in the order implied by Table 19.1. Figure 19.16 shows graphs of voltage gain, phase shift, and group delay versus frequency and step response versus time for the filter. The voltage gain and the step response are normalized to the dc gain, given by

$$
\begin{aligned}
H_v(0) &= (1 + k_1)(1 + k_2)(1 + k_3) = 10.71 \\
&\Rightarrow A_v(0) = 20.59\,\text{dB}.
\end{aligned}
$$

The 0.5 dB passband ripple is not visible on the scale used for gain in Fig. 19.16. Figure 19.18 shows the gain in the passband on a scale that makes the ripple visible.

From Figs. 19.15, 19.16, and 19.17, the voltage gain at $f = 10 f_0$ is about -120 dB for the Butterworth, about -100 dB for the Bessel, and about -140 dB for the Chebyshev.

Of the three types treated here, the Chebyshev has the steepest and the Bessel the least steep *initial* transition from passband to stopband, with the Butterworth somewhere in between. Of course, for frequencies well above the corner frequency, the slope of the gain is $-\ 20n\,\text{dB/decade}$ for all three, where n is the filter order (number of poles).

The Butterworth has the flattest passband gain and a sharp knee (downward turn) at the half-power frequency. The Bessel voltage gain exhibits a bit more droop in the passband and a gentler knee. The Chebyshev exhibits passband ripple. The others do not.

The Bessel step response exhibits the fastest rise time and no ringing, the Butterworth a slightly longer rise time and some ringing, and the Chebyshev exhibits a fast rise, but with considerable ringing and a long settling time.

The Chebyshev has (by far) the worst delay characteristic of the three. The Bessel has the best delay characteristic and the Butterworth delay characteristic is somewhere in between.

Which of the three types one chooses depends upon the relevant performance criteria. Generally, for some audio applications (such as telephone-quality speech), delay distortion is less important than gain distortion, and one might choose a Butterworth or Chebyshev. For video signals, delay distortion is at least as important as gain distortion, and Bessel might be a good choice. For digital (binary pulse) signals, delay distortion is perhaps more important than gain distortion – so much so that none of the filters described above might meet performance criteria, and one might incorporate delay equalization or use other kinds of filters not treated here.[7]

[7] Ability to achieve frequency-independent delay in the passband is one of the more important justifications for using certain kinds of digital filters.

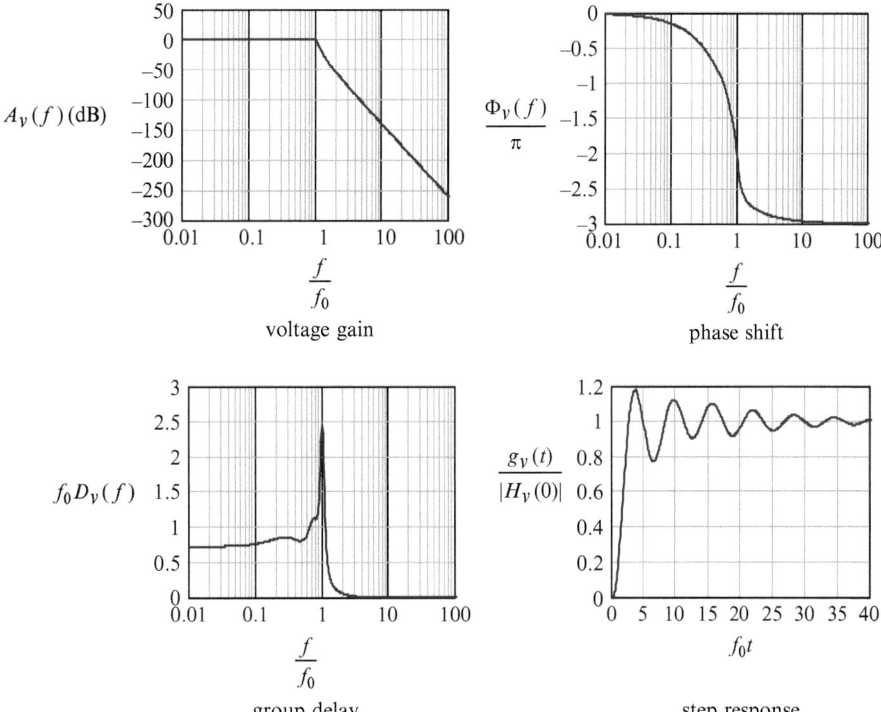

Fig. 19.17 See Example 19.6

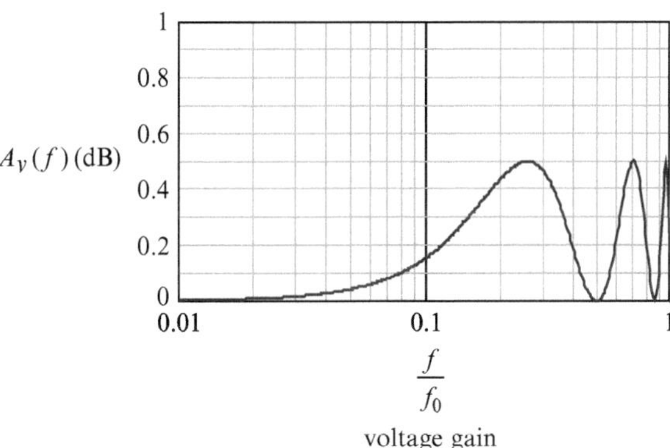

Fig. 19.18 See Example 19.6

19.7 Problems

In the problems that follow, plot all gains (normalized appropriately) in dB . Also, most problems require use of a computer and/or simulation software.

P 19.1 Refer to Fig. P 19.1. (a) By inspection, deduce the dc voltage gain and the voltage gain for $f \to \infty$. (b) Assume the source resistance is negligible and obtain the voltage transfer function for the circuit. What kind of filter is this? (c) Obtain an expression for the frequency at which the gain is 6 dB below its maximum value.

P 19.2 The four poles of a certain lowpass filter are given by

$$p_k = j2\pi W \exp\left[\frac{j(2k+1)\omega}{8}\right], \quad k = 0,\ 1,\ 2,\ 3$$

where $W = 15\,\text{kHz}$. Use two state-variable biquads in cascade to implement the filter. Construct plots of gain and group delay versus frequency, for $100\,\text{Hz} \le f \le 1.5\,\text{MHz}$.

P 19.3 See Fig. P 19.2. (a) Let $\tau = RC$ and obtain an expression for the voltage transfer function. (b) Let

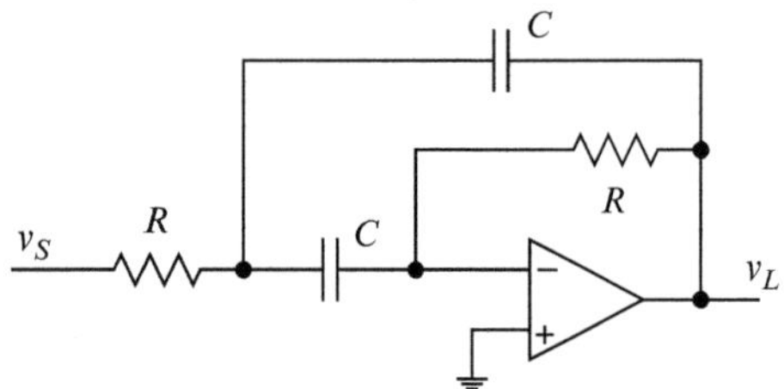

Fig. P 19.1 See Problem P 19.1

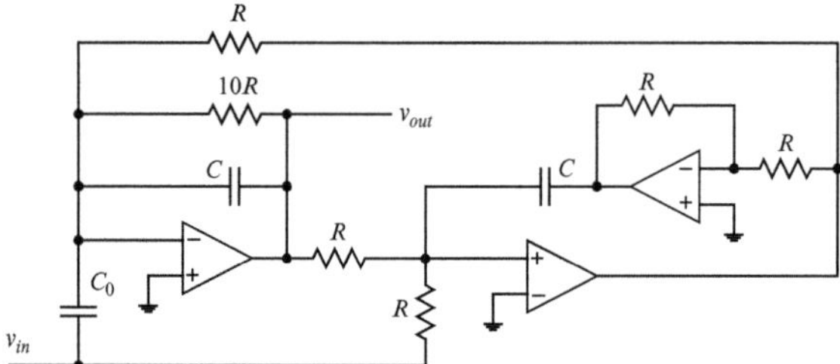

Fig. P 19.2 See Problem P 19.3

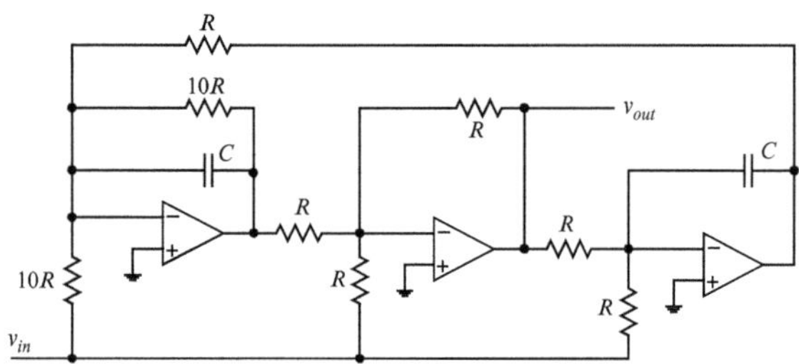

Fig. P 19.3 See Problem P 19.4

$R = 10\,\mathrm{k}\Omega$ and $C = 15.92\,\mathrm{pF}$. Plot the voltage gain in dB, normalized to the dc voltage gain, for 0.01 Hz $\leq f \leq$ 100 kHz. What kind of filter is this?

P 19.4 See Fig. P 19.3. (a) Let $\tau = RC$ and obtain an expression for the voltage transfer function. (b) Let $R = 10\,\mathrm{k}\Omega$ and $C = 15.92\,\mathrm{pF}$. Plot the voltage gain versus frequency for 1 Hz $\leq f \leq$ 100 kHz. What kind of filter is this?

P 19.5 See Fig. P 19.4. (a) Let $\tau = RC$ and obtain an expression for the voltage transfer function. (b) Let $R = 10\,\mathrm{k}\Omega$ and $C = 15.92\,\mathrm{pF}$. Plot the voltage gain versus frequency for 100 Hz $\leq f \leq$ 10 kHz. What kind of filter is this?

P 19.6 The three outputs of a state-variable biquad can be weighted and summed as illustrated by Fig. P 19.5 to provide a voltage transfer function of the form

$$H_v = K\frac{(s-z)(s-z^*)}{(s-p)(s-p^*)}.$$

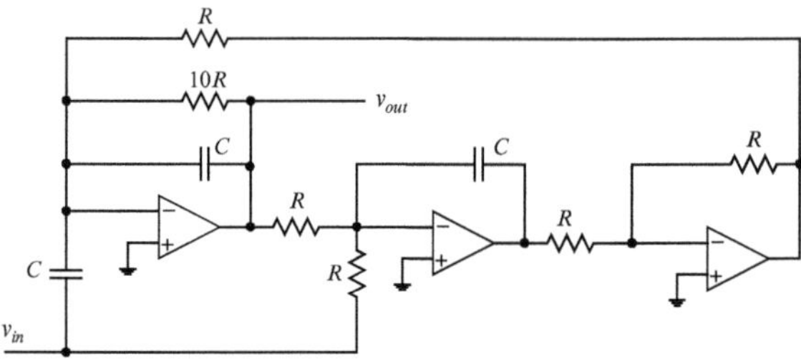

Fig. P 19.4 See Problem P 19.5

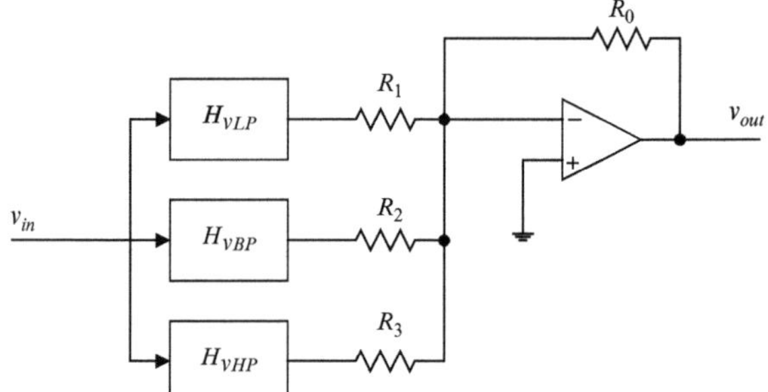

Fig. P 19.5 See Problem P 19.6

Design such a filter for which $K = 20$, $p = 2\pi W(-1 + j8)$, and $z = -p$, with $W = 5\,\mathrm{kHz}$. Construct plots of voltage gain and group delay versus frequency, for $10\,\mathrm{Hz} \leq f \leq 500\,\mathrm{kHz}$. What kind of filter is this?

P 19.7 Design a four-pole lowpass VCVS Butterworth filter having half-power frequency 5 kHz and minimum input impedance (magnitude) 200 kΩ. Construct plots of gain and group delay versus frequency for $50\,\mathrm{Hz} \leq f \leq 500\,\mathrm{kHz}$. If the supply voltage is $\pm$ 15 V, what is the maximum allowable amplitude of an input (for linear operation)?

P 19.8 Design a four-pole lowpass VCVS Bessel filter having half-power frequency 5 kHz and minimum input impedance (magnitude) 200 kΩ. Construct plots of gain and group delay versus frequency for $50\,\mathrm{Hz} \leq f \leq 500\,\mathrm{kHz}$. If the supply voltage is $\pm$ 15 V, what is the maximum allowable amplitude of an input (for linear operation)?

P 19.9 Design a four-pole lowpass VCVS Chebyshev filter having passband edge frequency 5 kHz, passband ripple 0.5 dB, and minimum input impedance (magnitude) of 200 kΩ. Construct plots of gain and group delay versus frequency for $50\,\mathrm{Hz} \leq f \leq 500\,\mathrm{kHz}$. If the supply voltage is $\pm$ 15 V, what is the maximum allowable amplitude of an input (for linear operation)?

P 19.10 Design an eight-pole lowpass VCVS Butterworth filter having half-power frequency 50 kHz and minimum input impedance (magnitude) 200 kΩ. Construct plots of gain and group delay versus frequency for $500\,\text{Hz} \le f \le 5\,\text{MHz}$. If the supply voltage is ± 15 V, what is the maximum allowable amplitude of an input (for linear operation)?

P 19.11 Design an eight-pole lowpass VCVS Bessel filter having half-power frequency 50 kHz and minimum input impedance (magnitude) 200 kΩ. Construct plots of gain and group delay versus frequency for $500\,\text{Hz} \le f \le 5\,\text{MHz}$. If the supply voltage is ± 15 V, what is the maximum allowable amplitude of an input (for linear operation)?

P 19.12 Design an eight-pole lowpass VCVS Chebyshev filter having passband edge frequency 50 kHz, passband ripple 0.5 dB, and minimum input impedance (magnitude) 200 kΩ. Construct plots of gain and group delay versus frequency for $500\,\text{Hz} \le f \le 5\,\text{MHz}$. If the supply voltage is ± 15 V, what is the maximum allowable amplitude of an input (for linear operation)?

P 19.13 Design a four-pole highpass VCVS Butterworth filter having half-power frequency 5 kHz and minimum input impedance (magnitude) 200 kΩ. Construct plots of gain and group delay versus frequency for $50\,\text{Hz} \le f \le 500\,\text{kHz}$. If the supply voltage is ± 15 V, what is the maximum allowable amplitude of an input (for linear operation)?

P 19.14 Design a four-pole highpass VCVS Bessel filter having half-power frequency 5 kHz and minimum input impedance (magnitude) 200 kΩ. Construct plots of gain and group delay versus frequency for $50\,\text{Hz} \le f \le 500\,\text{kHz}$. If the supply voltage is ± 15 V, what is the maximum allowable amplitude of an input (for linear operation)?

P 19.15 Design a four-pole highpass VCVS Chebyshev filter having passband edge frequency 5 kHz, passband ripple 0.5 dB, and minimum input impedance (magnitude) 200 kΩ. Construct plots of gain and group delay versus frequency for $50\,\text{Hz} \le f \le 500\,\text{kHz}$. If the supply voltage is ± 15 V, what is the maximum allowable amplitude of an input (for linear operation)?

P 19.16 Design an eight-pole highpass VCVS Butterworth filter having half-power frequency 50 kHz and minimum input impedance (magnitude) 200 kΩ. Construct plots of gain and group delay versus frequency for $500\,\text{Hz} \le f \le 5\,\text{MHz}$. If the supply voltage is ± 15 V, what is the maximum allowable amplitude of an input (for linear operation)?

P 19.17 Design an eight-pole highpass VCVS Bessel filter having half-power frequency 50 kHz and minimum input impedance (magnitude) 200 kΩ. Construct plots of gain and group delay versus frequency for $500\,\text{Hz} \le f \le 5\,\text{MHz}$. If the supply voltage is ± 15 V, what is the maximum allowable amplitude of an input (for linear operation)?

P 19.18 Design an eight-pole highpass VCVS Chebyshev filter having passband edge frequency 50 kHz, passband ripple 0.5 dB, and minimum input impedance (magnitude) 200 kΩ. Construct plots of gain and group delay versus frequency for $500\,\text{Hz} \le f \le 5\,\text{MHz}$. If the supply voltage is ± 15 V, what is the maximum allowable amplitude of an input (for linear operation)?

P 19.19 If two four-pole lowpass VCVS Butterworth filters are connected in cascade, is the result a Butterworth filter? Justify your answer.

P 19.20 If two two-pole lowpass VCVS Chebyshev filters are connected in cascade, is the result a Chebyshev filter? Justify your answer.

P 19.21 If two eight-pole lowpass VCVS Bessel filters are connected in cascade, is the result a Bessel filter? Justify your answer.

P 19.22 Three sinusoids having frequencies $0.01f_0$, f_0, and $100f_0$ are applied (individually) to the input terminals of a two-pole lowpass Butterworth filter having half-power bandwidth f_0 and dc gain 10 dB. The rms amplitude of each is 200 mV. What are the (approximate) peak amplitudes of the outputs?

P 19.23 Three sinusoids having frequencies $0.01f_0$, f_0, and $100f_0$ are applied (individually) to the input terminals of a four-pole lowpass Butterworth filter having half-power bandwidth f_0 and dc gain 20 dB. The rms amplitude of each is 200 mV. What are the (approximate) peak amplitudes of the outputs?

P 19.24 Three sinusoids having frequencies $0.01f_0$, f_0, and $100f_0$ are applied (individually) to the input terminals of a six-pole lowpass Butterworth filter having half-power bandwidth f_0 and dc gain 30 dB. The rms amplitude of each is 200 mV. What are the (approximate) peak amplitudes of the outputs?

P 19.25 Three sinusoids having frequencies $0.01f_0$, f_0, and $100f_0$ are applied (individually) to the input terminals of an eight-pole lowpass Butterworth filter having half-power bandwidth f_0 and dc gain 40 dB. The rms amplitude of each is 200 mV. What are the (approximate) peak amplitudes of the outputs?

Appendix: Answers to Exercises

1. Chapter 1

Exercise 1.1. Physical law.

Exercise 1.2. (a), (b), (f), (g), (i), (j) are quantities. (c), (d), (e), (n) are units. (k) is neither. (l), (m) could be quantities if we could devise suitable definitions and measurements. (h) could be a quantity if we define color in terms of frequency.

Exercise 1.3. (a) W (b) J (c) A (d) W (e) s (f) Nm = J.

Exercise 1.4. $\mathrm{SI}\left[\frac{1}{L}\int_{-\infty}^{t_0} v(t')dt'\right] = \mathrm{SI}[i] = \mathrm{A}$.

Exercise 1.5. (a) No, (b) Yes.

Exercise 1.6. (a) $10^{-4}\,\mathrm{As}^{-1}$, (b) $10^4\,\mathrm{Vs}^{-1}$, (c) $10^5\,\mathrm{V\,A}$, (d) $10^{10}\mathrm{m\,s}^{-2}$.

Exercise 1.7. (a) $5\mathrm{mVs}^{-1} = \frac{5\,\mu\mathrm{V}}{\mathrm{ms}} = \frac{5\mathrm{nV}}{\mu\mathrm{s}}$, (b) $25\mathrm{kA}\mu\mathrm{s} = 25\mathrm{Ams} = 25\mathrm{MAns}$, (c) $100\,\mathrm{mJ\,ms}^{-1} = 100\mu\mathrm{J}\,\mu\mathrm{s}^{-1} = 100\,\mathrm{Js}^{-1}$.

Exercise 1.8. (a) No space, milli (b), (e) first m is meter, second is milli (c), (d) meter

Exercise 1.9. $A_v \cong 1/b$; b determines the gain.

Exercise 1.10. $x \simeq 5.064$.

Exercise 1.11. (a) 0.495, (b) 5, (c) 1, (d) 3.02, (e) 50.

Exercise 1.12. (a) i_0, (b) 0.

Exercise 1.13. (a) 2.015×10^3, (b) 16.38×10^6, (c) 759×10^{-3}, (d) 462×10^{-6}, (e) 4.792×10^3.

2. Chapter 2

Exercise 2.1. $I = 2Nq_p/T$ where q_p is the charge of a proton. $\mathrm{SI}[2Nq_p/T] = \mathrm{Cs}^{-1} = \mathrm{A}$.

Exercise 2.2. $h = u^2/(2g)$; $\mathrm{SI}[u^2/(2g)] = \mathrm{m}$.

Exercise 2.3. $q = d\sqrt{-f/k}$; $\mathrm{SI}\left[d\sqrt{-f/k}\right] = \mathrm{C}$.

Exercise 2.4. $\Phi(x_1) = x_0\sqrt{kf}\left(\frac{1}{x_1} + \frac{1}{x_1 - x_0}\right)$; $\mathrm{SI}\left[x_0\sqrt{kf}\left(\frac{1}{x_1} - \frac{1}{x_1 - x_0}\right)\right] = \mathrm{V}$.

Exercise 2.5. $d = \frac{x_a^2 v_{ab}}{kq - x_a v_{ab}}$; $\mathrm{SI}\left\{\frac{x_a^2 v_{ab}}{kq - x_a v_{ab}}\right\} = \mathrm{m}$.

Exercise 2.6. $i_2 = -10$ mA.

Exercise 2.7. $w_1 = w_0\rho_1/\rho_0$.

Exercise 2.8. 14.4 nm.

Exercise 2.9. 1 kΩ.

Exercise 2.10. (a) 9.5 Ω (b) 127 Ω.

Exercise 2.11. 49.9 kΩ from the E96 series.

Exercise 2.12. (a) 4.3 kΩ ± 5% (b) 459 kΩ ± 0.1%.

Exercise 2.13. Various answers

Exercise 2.14. Because 298 K = 25°C, 298 K is the reference temperature.

Exercise 2.15. $\alpha_{10} = 0.0042\ \mathrm{K}^{-1}$.

Exercise 2.16. 4.

Exercise 2.17. 10.2.

Exercise 2.18. (a) (i) 0.42 Ω, (ii) 4.08 Ω; (b) (i) 0.33 Ω, (ii) 3.23 Ω

Exercise 2.19.

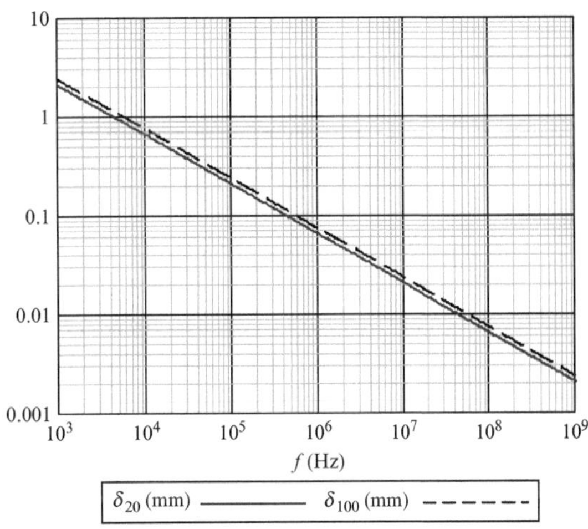

T.H. Glisson, *Introduction to Circuit Analysis and Design*, DOI 10.1007/978-90-481-9443-8, © Springer Science+Business Media B.V. 2011

3. Chapter 3

Exercise 3.1. $v_{bd} = 15$ V, $v_{cd} = 5$V, $v_{da} = -20$ V, $v_1 = -5$ V, $v_b = 15$V.

Exercise 3.2. (a) The series connection of elements 1 and 2 is in parallel with the series connection of elements 3 and 4. (b) The parallel connection of elements 1, 2, and 3 is in series with element 4.

Exercise 3.3. $v_{ab} = v_1 - v_2 - v_3$, $v_{ac} = v_1 - v_3$, $v_{bd} = v_2 + v_3$.

Exercise 3.4. No. $i_2 = i_1 + i_3$ so the sources aren't independent.

Exercise 3.5. Let $R_{nm} = \dfrac{R_n R_m}{R_n + R_m}$, $I_1 = \dfrac{R_{23} I_0}{R_1 + R_{23}}$, $I_2 = \dfrac{R_{13} I_0}{R_2 + R_{13}}$, $I_3 = \dfrac{R_{12} I_0}{R_{12} + R_3}$.

Exercise 3.6. $v_1 = \dfrac{R_1}{R_1 + R_2} v_0$.

Exercise 3.7.

$$i_1 = \frac{R_3 v_2 - (R_2 + R_3) v_1}{R_1 R_2 + R_1 R_3 + R_2 R_3}, \; i_3 = \frac{R_2}{R_3} \frac{R_3 v_1 - (R_1 + R_3) v_2}{R_1 R_2 + R_1 R_3 + R_2 R_3}.$$

Exercise 3.8. $v_{ac} \cong -0.015 v_0$, $v_{bc} \cong 0.463 v_0$.

Exercise 3.9. (a) $-i_1 + \dfrac{v_a}{R_1} + \dfrac{v_a - (v_d - v_2 - v_1)}{R_2} = 0$,

(d) $\dfrac{v_d}{R_4} + \dfrac{v_d - v_2}{R_3} + \dfrac{v_d - v_2 - v_1 - v_a}{R_2} = 0$.

Exercise 3.10. (b) $\dfrac{v_b - R_1 i_1}{R_1 + R_2} + \dfrac{v_c}{R_3} + \dfrac{v_c + v_2}{R_4} = 0$,

(c) $\dfrac{v_c}{R_3} + \dfrac{v_c + v_2}{R_4} + \dfrac{v_b - R_1 i_1}{R_1 + R_2} = 0$.

Exercise 3.11. Choose node c as reference node. $v_b = \dfrac{R_1 R_2 R_3 (i_1 + i_2) + v_1}{R_1 R_2 + R_1 R_3 + R_2 R_3}$.

Exercise 3.12. No. $v_3 = v_1 + v_2$ so the sources aren't independent.

Exercise 3.13. $i_1 = \dfrac{(R_2 + R_3) v_a - R_3 v_b}{R_1 R_2 + R_1 R_3 + R_2 R_3}$.

Exercise 3.14.

$$i_1 = \frac{R_B v_0 + (R_2 R_3 - R_2 R_4) i_0}{R_A R_B - R_2^2},$$

$$i_2 = \frac{R_2 v_0 + R_A (R_3 - 5 R_4) i_0}{R_A R_B - R_2^2},$$

$$R_A = R_1 + R_2, \; R_B = R_2 + R_3 + R_4.$$

Exercise 3.15. Answer given in problem statement.

4. Chapter 4

Exercise 4.1. $i_a = -\left(\dfrac{3}{2R_0} + \dfrac{1}{R_1}\right) v_a + \dfrac{3 v_0}{2R_0} - i_0$.

Exercise 4.2.

$$R_{eq} = \frac{R_1 (2R_0^2 + 4R_0 R_2 + R_2^2)}{2R_0^2 + 4R_0 R_2 + R_2^2 + 2R_1 (R_0 + R_2)}.$$

Exercise 4.3. $\alpha = -2$, $\beta = 19/5$, $\gamma = 6/5$, $\delta = -1$.

Exercise 4.4. 8.77 Ω.

Exercise 4.5. (a) $v_x = \dfrac{R_0 R_1 R_2}{R_1 (R_0 + R_2) + R_0 R_2} (i_0 - i_1)$

(b) $v_x \cong -\dfrac{R_1 R_2}{R_1 + R_2} i_1$ (c) 1.06%.

Exercise 4.6. (a) $R_T = 0.91R$, $v_T = 0.91 R i_0 - 0.48 v_0$ (b) $R_N = 0.91R$, $i_N = i_0 - 0.53(v_0/R)$.

Exercise 4.7. $v_T = v_0 + \dfrac{4v_1 + 12 R i_0}{7}$, $R_T = \dfrac{12}{7} R$.

Exercise 4.8. $i_N = i_0 + \dfrac{7 v_0 + 4 v_1}{12R}$, $R_N = \dfrac{12}{7} R$.

Exercise 4.9. $v_T = \dfrac{(R_2 - R_1) v_2 i_1}{R_2 i_1 - v_2}$, $R_T = \dfrac{R_2 v_2 - R_1 R_2 i_1}{R_2 i_1 - v_2}$.

Exercise 4.10. $R_N = \dfrac{R_2 (3R_1 + 2R_2)}{R_1 + R_2}$.

5. Chapter 5

Exercise 5.1. Answer given in problem statement.

Exercise 5.2.

$$p_R = (1/9R)(V_0 - 2RI_0)^2 \cos^2(\omega_0 t) \geq 0$$

$$p_{2R} = (2/9R)(V_0 + RI_0)^2 \cos^2(\omega_0 t) \geq 0$$

$$p_i = -(2/3)(V_0 I_0 + RI_0^2) \cos^2(\omega_0 t) \leq 0$$

$$p_v = -(1/3R)(V_0^2 - 2RI_0 V_0) \cos^2(\omega_0 t)$$

$p_R > 0$ and $p_{2R} > 0$ so the resistors consume energy. $p_i < 0$, so the current source produces energy. If $V_0^2 - 2RI_0V_0 > 0$, then $P_v < 0$ and the voltage source produces energy. If $V_0^2 - 2RI_0V_0 < 0$, then $p_v > 0$ and the voltage source consumes energy.

Exercise 5.3. $P = 25$ mW, $V = 5$ V, $w = 50$ mJ.

Exercise 5.4. $P \cong 547$ mW.

Exercise 5.5. Answer given in problem statement.

Exercise 5.6.

$$\hat{p} = \frac{P_0}{2}[1 + \cos(\theta)] = P_0 \cos^2\left(\frac{\theta}{2}\right)$$

Exercise 5.7. Answer given in problem statement.

Exercise 5.8.

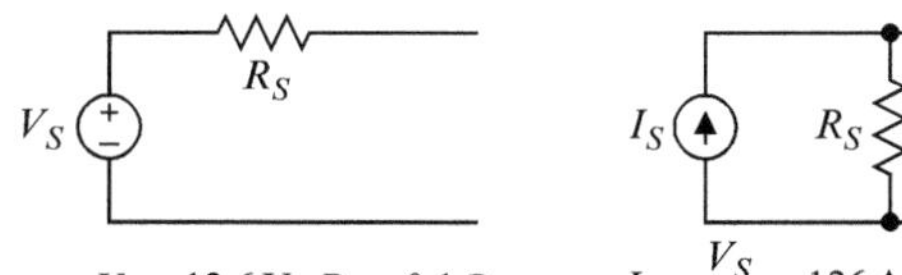

$V_S = 12.6\,\text{V},\ R_S = 0.1\,\Omega$ $\quad I_S = \dfrac{V_S}{R_S} = 126\,\text{A},\ R_S = 0.1\,\Omega$

$P_{L\,\max} = \dfrac{V_S^2}{4R_S} = \dfrac{I_S^2 R_S}{4} \cong 397\,\text{W};\quad I_{L\,\max} = I_S = 126\,\text{A}$

Exercise 5.9. There are two solutions: $R_T \cong \begin{Bmatrix} 179\Omega \\ 0.557\Omega \end{Bmatrix},\ V_T \cong \begin{Bmatrix} 189\text{V} \\ 10.6\text{V} \end{Bmatrix} \Rightarrow I_N \cong \begin{Bmatrix} 1.06\text{A} \\ 19.0\text{A} \end{Bmatrix}$.

Exercise 5.10. $1/2$.

Exercise 5.11. $1 + \cos(\theta_1 - \theta_2)$.

Exercise 5.12. Let $x_1(t) = x_2(t) = \cos(\omega t) \Rightarrow \overline{x_1 x_2} = \overline{\cos^2(\omega t)} = \frac{1}{2},\ \overline{x_1}\,\overline{x_2} = \overline{\cos(\omega t)}\ \overline{\cos(\omega t)} = 0$.

Exercise 5.13. 1.

Exercise 5.14. $p(t) = \dfrac{I_0^2 R}{2} + \dfrac{I_0^2 R}{2}\cos(4\pi f t), \hat{p} = 1.5\,\text{kW}, P = 750\,\text{W}$.

Exercise 5.15.

$$P_{R_1} = \frac{R^2}{R_1}\overline{(i_0 - i_1)^2},\ P_{R_2} = \frac{R^2}{R_2}\overline{(i_0 - i_1)^2},\ \overline{(i_0 - i_1)^2}$$
$$= \frac{1}{2}\left[I_0^2 + I_1^2 - 2I_0 I_1 \cos(\theta)\right]$$
$$P_{i_0} = -\frac{I_0^2 R}{2} + \frac{I_0 I_1 R}{2}\cos(\theta),\ P_{i_1} = -\frac{I_1^2 R}{2} + \frac{I_0 I_1 R}{2}\cos(\theta).$$

Exercise 5.16. (a) 25 mA (b) 5 V(c) $\frac{10}{\sqrt{2}}$ V (d) $\frac{10}{\sqrt{2}}$ mA.

Exercise 5.17.

$$P_R = \frac{V_0^2 + (I_0 R)^2}{4R},\ P_i = -\frac{1}{4}\left(V_0 I_0 + I_0^2 R\right),$$
$$P_v = -\frac{V_0^2 - V_0 I_0 R}{4R},\ P_R + P_i + P_v = 0.$$

Exercise 5.18. (a) 4 s, (b) 10^{-7}%.

Exercise 5.19. Yes. Measure peak and divide by $\sqrt{2}$.

Exercise 5.20. 1.037 kΩ.

Exercise 5.21. The load voltages are equal and given by $V_L = \dfrac{V_0}{2}$.

Exercise 5.22. $P_{\max} = \dfrac{V_0^2 (R + R_0)}{4RR_0},\quad P_{V_0} = -\dfrac{V_0^2}{2R_0}\left(\dfrac{R + 2R_0}{R + R_0}\right),\ P_{R_0} = \dfrac{V_0^2}{4R_0}\left(\dfrac{R + 2R_0}{R + R_0}\right)^2,\ P_R = \dfrac{V_0^2 R}{4(R + R_0)^2}$.

Exercise 5.23. 0.5.

Exercise 5.24. $P_L = \dfrac{2R_L\left(R_0^2 I_0^2 + V_0^2\right)}{(R_0 + R_1 + 2R_L)^2} = 25\,\text{mW}$.

6. Chapter 6

Exercise 6.1. $R_{in} = R_1 \| (R_2 + R_3 \| R_L) = \dfrac{R_1(R_2 R_3 + R_2 R_L + R_3 R_L)}{R_1 R_3 + R_1 R_L + R_2 R_3 + R_2 R_L + R_3 R_L}$
$R_{out} = R_3 \| (R_2 + R_1 \| R_S) = \dfrac{R_3(R_2 R_1 + R_2 R_S + R_1 R_S)}{R_1 R_3 + R_1 R_S + R_2 R_1 + R_2 R_S + R_3 R_S}$,
(b) and (c) same as (a).

Exercise 6.2. $g = \mu / R$.

Exercise 6.3. $v_{ab} = \left(\dfrac{R_o R_L}{R_o + R_L}\right)\left(\dfrac{R_i R_S}{R_i + R_S}\right) g i_s$.

Exercise 6.4. $R_{in} = \dfrac{R_1}{1 + \beta}$.

Exercise 6.5. $v_T = \left(\dfrac{R_o R_L}{R_o + R_L}\right)\left(\dfrac{R_i R_S}{R_i + R_S}\right) g i_s$, $R_T = \dfrac{R_o R_L}{R_o + R_L}$.

Exercise 6.6.

$$v_{out} = \frac{(7\mu - 1)v_1 - (1 + 4\mu)v_2 + (11\beta - 4\mu - 1)Ri_1}{\mu - 19}.$$

Exercise 6.7. $\mu_1 = 0$.

Exercise 6.8. $\mu = gR_0$.

Exercise 6.9.

$$R_{in} = \frac{R_i\left[R_f(R_L + R_o) + R_L R_o\right]}{R_f(R_L + R_o) + R_L R_o + R_i[R_o + R_L(\mu + 1)]},$$

$$R_{out} = \frac{R_o\left[R_f(R_i + R_S) + R_i R_S\right]}{(R_f + R_o)(R_i + R_S) + (\mu + 1)R_i R_S}.$$

Exercise 6.10.

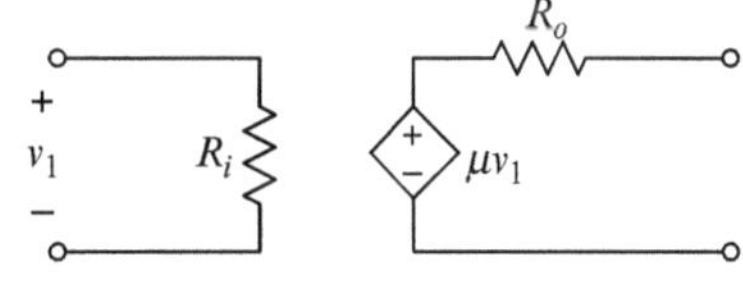

(a) $R_i = R_{i1}, R_o = R_{o2}, \mu = \dfrac{R_{o1} R_{o2}}{R_{o1} + R_{i2}} \beta_2 g_1$,

(b) VCCS: $R_i = R_{i1}, R_o = R_{o2}, g = \dfrac{R_{o1}}{R_{o1} + R_{i2}} \beta_2 g_1$,

CCVS: $R_i = R_{i1}, R_o = R_{o2}, r = \dfrac{R_{i1} R_{o1} R_{o2}}{R_{o1} + R_{i2}} \beta_2 g_1$

CCCS: $R_i = R_{i1}, R_o = R_{o2}, \beta = \dfrac{R_{i1} R_{o1}}{R_{o1} + R_{i2}} \beta_2 g_1$.

Exercise 6.11.

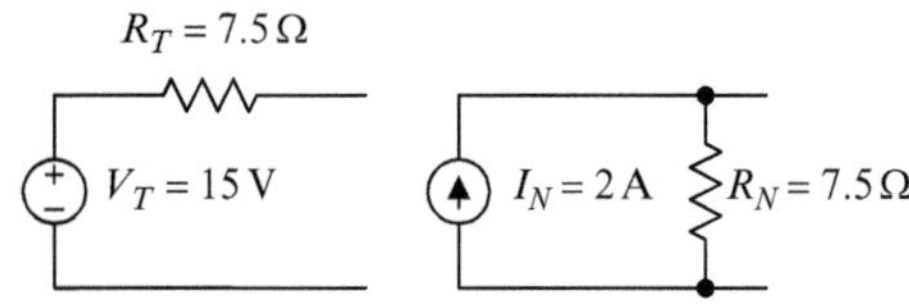

Exercise 6.12. No. Magnitudes only.
Exercise 6.13. 7.5 W.
Exercise 6.14. Two.
Exercise 6.15. $R_L/R_S \geq 19$.
Exercise 6.16. $R_L/R_S \leq 0.053$.
Exercise 6.17. $8/9 \cong 0.889$.
Exercise 6.18. $0.63 \leq R_L/R_S \leq 1.58$.
Exercise 6.19. $D = (R_i + R_S)(R_o + R_L)$

$$H_v = \frac{\mu R_i R_L}{D},\ H_r = \frac{\mu R_i R_S R_L}{D},$$
$$H_g = \frac{\mu R_i}{D},\quad H_i = \frac{\mu R_i R_S}{D}.$$

Exercise 6.20. $D = (R_S + R_i)(R_o + R_L)$,

$$A_v = \frac{r R_L}{D},\ A_i = \frac{r R_S}{D},$$
$$A_g = \frac{r}{D},\ A_r = \frac{r R_S R_L}{D}.$$

Exercise 6.21. $D = (R_S + R_{i1})(R_{o1} + R_{i2})(R_{o2} + R_L)$

$$A_v = \frac{r\beta R_{o2} R_L}{D},\ A_i = \frac{r\beta R_S R_{o2}}{D},$$
$$A_g = \frac{r\beta R_{o2}}{D},\ A_r = \frac{r\beta R_S R_{o2} R_L}{D}.$$

Exercise 6.22. $A_v = A_i R_L/R_S, A_g = A_i/R_S, A_r = R_L A_i$.
Exercise 6.23. (a) $\alpha = 1/R_L$, (b) $\alpha = 1/R_S$, (c) $\alpha = 1/R_S$, (d) $\alpha = 1$.
Exercise 6.24. Results follow immediately from relations among the gains (see Example 7.12).
Exercise 6.25. Usually no, because increasing the output resistance of the source increases the power wasted (decreases the power transferred to the load).
Exercise 6.26. No, they are consistent because $A_i = R_S A_v / R_L$.
Exercise 6.27. $A_P = 6.31 \times 10^{10}$.
Exercise 6.28. $P_A = 0.5\,\mu\text{W}$.
Exercise 6.29. Follows immediately from Exercise 6.24.
Exercise 6.30. $D = (R_o + R_L)(R_S + R_i)$

$$A_{vdB} = 20\log\left[\frac{R_i R_o R_L g}{D}\right],\ A_{idB} = 20\log\left[\frac{R_S R_i R_o g}{D}\right],$$
$$A_{PdB} = 10\log\left(\frac{4R_S}{R_L}\right) + 20\log\left[\frac{R_i R_o R_L g}{D}\right].$$

Exercise 6.31. 114 dB.

7. Chapter 7

Exercise 7.1. (a) (ii), (b) (iv), (c) (i), (d) (iii).
Exercise 7.2. (a) (iii), (b) (iv), (c) (i), (d) (ii).
Exercise 7.3. $v_L = -\frac{R_4}{R_3}\left(1 + \frac{R_2}{R_1}\right) v_S$.
Exercise 7.4. $v_L = v_1 + v_2$.
Exercise 7.5.

$$R_{in} = \frac{(\mu_0+1)R_L R_i R_1 + (R_1+R_i)[R_2(R_L+R_o)+R_o R_L] + R_i R_1 R_o}{R_2(R_o+R_L)+R_L(R_o+R_1)+R_1 R_o}.$$

Exercise 7.6. $R_{out} \cong \frac{A_v}{\mu_0 + A_v} R_o$.
Exercise 7.7. Exact: $R_{in} = 974\ \text{G}\Omega$, $R_{out} = 808\,\mu\Omega$
Approx: $R_{in} = 990\ \text{G}\Omega$, $R_{out} = 808\ \mu\Omega$.
Exercise 7.8. Answer given in problem statement.
Exercise 7.9. Inverting amp: Exact $A_v = 99; R_{in} = 10\,\text{k}\Omega$; $R_{out} = 10.01\ \text{m}\Omega$.
Approx: $A_v \cong 99.01$; $R_{in} \cong 10\,\text{k}\Omega$; $R_{out} \cong 9.9\,\text{m}\Omega$.
Non-inverting amp: $A_v = 101$; $R_{in} = 97.1\,\text{G}\Omega; R_{out} = 10.1\,\text{m}\Omega$
Approx: $A_v \cong 101$; $R_{in} \cong 99\,\text{G}\Omega$; $R_{out} \cong 10.1\,\text{m}\Omega$.
Exercise 7.10. Answer given in problem statement.
Exercise 7.11. $P_{A\max} = 78.1$ mW (virtually any op amp will satisfy this requirement).
Exercise 7.12. $R'_L = R_L \| (R_1 + R_2) \cong 998\,\Omega$,

$$P_{A\max} = \frac{V_{CC}^2}{4R'_L} \cong 157\,\text{mW}.$$

8. Chapter 8

Exercise 8.1.

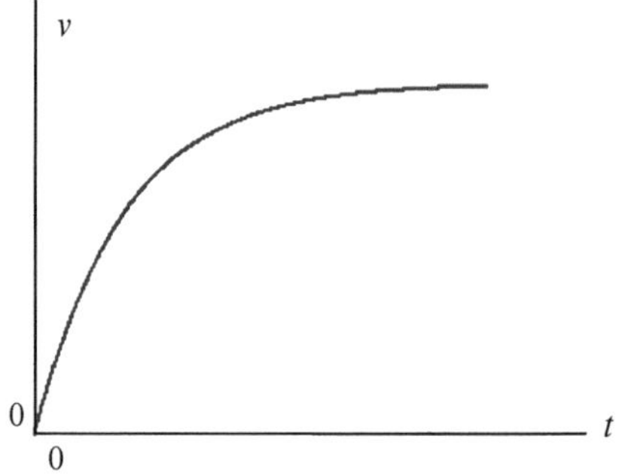

Exercise 8.2.

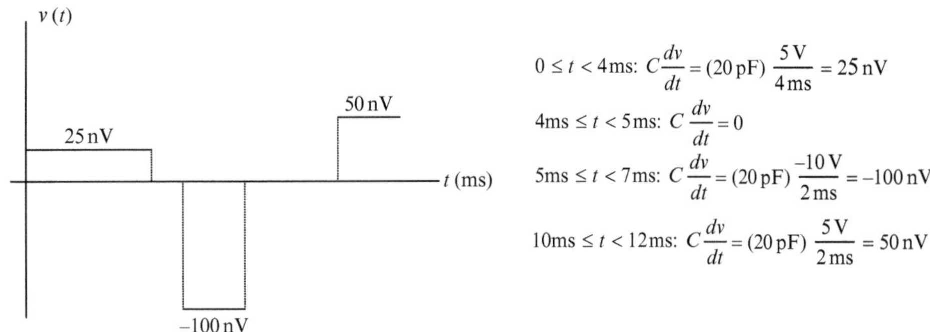

Exercise 8.3. $i(t) = 2\pi f V C \cos(2\pi f t)$, $\max|i(t)| = 2\pi f V C = 31.42\text{A}$.

Exercise 8.4.

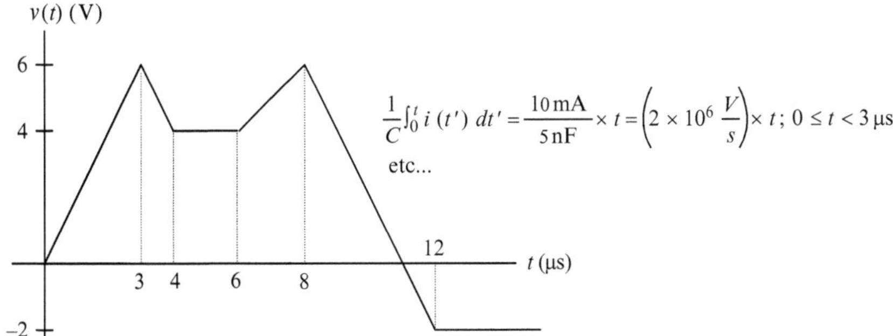

Exercise 8.5. $\lim_{t\to\infty} v_C(t) = RI_0$.

Exercise 8.6. The model is unrealistic if i_0 has a dc component, because the charge on the capacitor would grow without bound (mathematically). Also, if the voltage v_C is the quantity of interest, the resistor is irrelevant, and one must wonder why it is in the model. We cannot find the voltage $v_C(t_1)$ at any time t_1 unless we know (or are given) $v(t_0)$ for some time $t_0 \le t_1$ and an expression for the current $i_0(t)$ for $t_0 \le t \le t_1$.

Exercise 8.7. $R \cong 101\,\Omega$.

Exercise 8.8. $C \cong 39.1\,\text{pF}$.

Exercise 8.9. $\cong 50\,\mu\text{s}$ at 1, $\cong 250\,\mu\text{s}$ at 2.

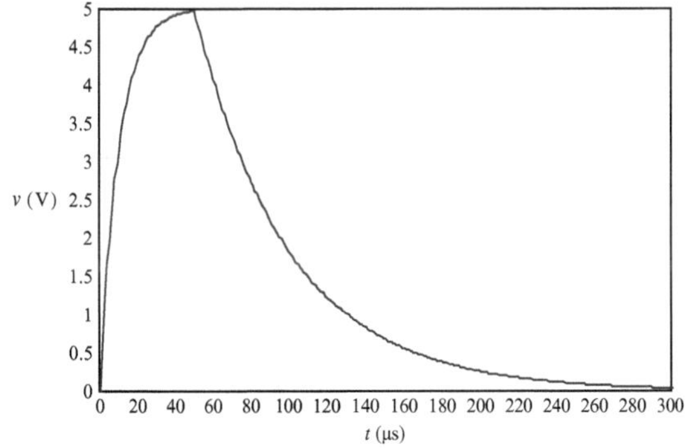

Exercise 8.10. v_C, i_2.

Exercise 8.11. $4C$.

Exercise 8.12. No, because the models would still allow capacitor voltages to change instantaneously.

Exercise 8.13.

X5R ⇒ −55°C to 85°C; ± 15%,

Y5V ⇒ −30°C to 85°C; + 22%, −82%,

Z5U ⇒ 10°C to 85°C; + 22%, −56%.

Exercise 8.14. Any combination _ _ _, where the first character is X, or Y, the second is 5 or 7, and the third is A,B,C,D,E,F,P,R,S,T, or U.

Exercise 8.15.

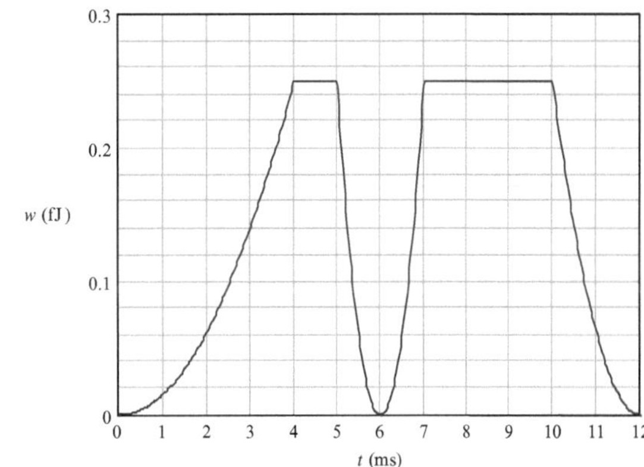

Exercise 8.16. $\tau = RC, \exp(-t/\tau) = 1/\sqrt{2} \Rightarrow t \cong 866\text{ns}$.
Exercise 8.17. $v_C(t) = 14.14 + 31.6\cos(2\pi f_0 t)$ V, $f_0 =$ 20kHz.

Exercise 8.18. (a) $\cong 10.9$ MHz, (b) $\cong 11$ V.

Exercise 8.19.

$$v_{out} = \begin{cases} 0, & t \leq 0 \\ -V_0 \sin(2\pi f t)/(2\pi f RC), & t > 0 \end{cases}$$

Exercise 8.20. (a) $V_{CC} \geq 25$ V, (b) $\cong 5$ ns.

Exercise 8.21. $C \cong 563$ μF.

Exercise 8.22. $V = IR,\ 99I = 99\frac{V}{R} = CV\omega_0 \Rightarrow C = \frac{99}{\omega_0 R} \cong 158$ μF.

Exercise 8.23. $f < 18.4$ Hz.

Exercise 8.24.

$$\tau = -\frac{Td}{\ln(1 - 2\Delta V/V_{\max})}.$$

9. Chapter 9

Exercise 9.1. H.
Exercise 9.2. 97.18 μm, 0.724 Ω.
Exercise 9.3.

$$L\frac{di}{dt} = (10\text{mH})\frac{5\text{mA}}{4\text{ms}} = 12.5\text{mV};\ 0 \leq t < 4\text{ms, etc...}$$

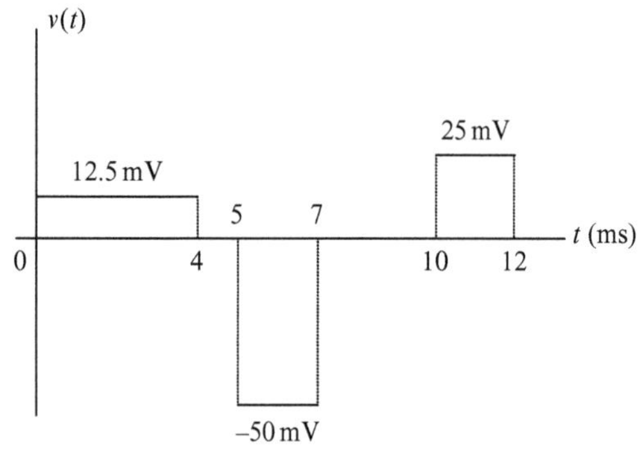

Exercise 9.4. $v(t) = 2\pi f I_1 \cos(2\pi f t + \pi/4)$.
Exercise 9.5.
$i(t) = 0,\ t < 0,\ i(t) = (V_0\tau/L)(1 - e^{-t/\tau}),\ t \geq 0.$

Exercise 9.6.

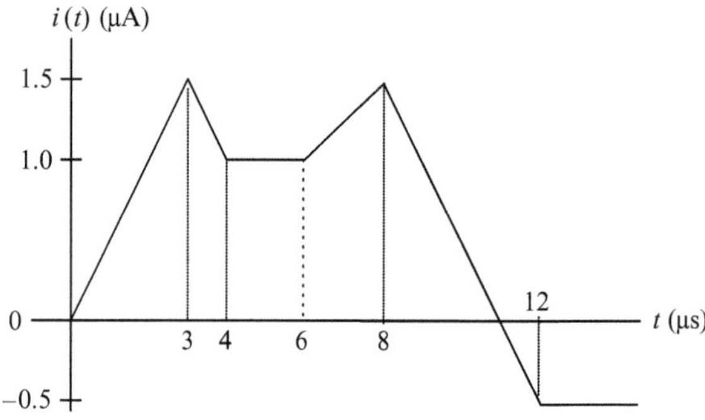

Exercise 9.7. Answer implied by problem statement.
Exercise 9.8. $v_L = \frac{R_2 V_0}{(R_1 + R_2)} \exp\left(-\frac{t}{\tau}\right)$. Other answers implied by problem statement.
Exercise 9.9.

$$i_L(t) = I_0,\ t \leq 0,\ i_L(t) = I_0 \exp\left(-\frac{t}{\tau}\right),\ t > 0,\ \tau = \frac{L}{R}$$
$$v_L(t) = 0,\ t \leq 0,\ v_L(t) = -\frac{LI_0}{\tau} \exp\left(-\frac{t}{\tau}\right),\ t > 0.$$

Exercise 9.10. L.
Exercise 9.11. $w(t) = \frac{1}{4}LI^2[1 + \cos(4\pi f t)]$.

Exercise 9.12. 0 V, V_0/R.

Exercise 9.13. 0 V, I_0.

Exercise 9.14. $f = 0.01R/(2\pi L)$.

Exercise 9.15. $V_{ac\,rms} \cong 1.2$ mV, $\gamma \cong 2.65 \times 10^{-6}$.

Exercise 9.16. $L \cong 31.8$ mH.

Exercise 9.17. Answer implied by problem statement.
Exercise 9.18. (a) Answer implied by problem statement.
(b) No. i_1 could have dc component.
Exercise 9.19. (a) Dots at tops of coils. (b) Reverse one winding.
Exercise 9.20. $n = \sqrt{R_L/R_S}$.
Exercise 9.21. $n_1 n_2 = \sqrt{R_L/R_S}$.
Exercise 9.22. $i(t) = \frac{nv(t)}{2R + n^2 R_L}$.

10. Chapter 10

Exercise 10.1. Answers given in problem statement.
Exercise 10.2. z real $\Rightarrow \sqrt{\mathrm{Re}^2(z)+\mathrm{Im}^2(z)}=\sqrt{z^2}=|z|$.

Remaining exercises: answers given in problem statement.

11. Chapter 11

Exercise 11.1. $i(t)=I_0+(I_1-I_0)u(t-t_0)+(I_2-I_1)u(t-t_1)$.

Exercise 11.2. $I_L=\dfrac{V_0}{R_1+R_2}$, $V_{C_1}=R_2 I$, $V_{C_2}=0$.

Exercise 11.3. $v_L(t)=v_L(\infty)[1-\exp(-t/\tau)]u(t)$, $\tau\cong 148.52\,\mu\mathrm{s}$, $v_L(\infty)\cong 334.75\,\mathrm{V}$.

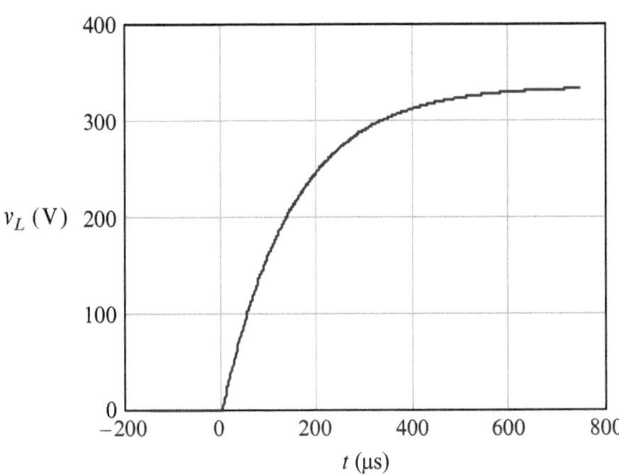

Exercise 11.4. $LC\dfrac{d^2v_C}{dt^2}+\dfrac{L}{R}\dfrac{dv_C}{dt}+v_C=0$, $t>0$; SI[LC] = s^2, SI[L/R] = s.

Exercise 11.5. (a) $R_0\cong 79.1\,\mathrm{k}\Omega$, $R<R_0\Rightarrow$ overdamped (b) $R=10R_0\Rightarrow v_C(t)=2|Y|\exp(-t/\tau)\cos(\omega_0 t+\theta)u(t), 2|Y|\cong 503\,\mathrm{mV}, \tau\cong 1.58\,\mu\mathrm{s}$, $\omega_0\cong 6.29\times 10^6\,\mathrm{s}^{-1}$, $\theta\cong -1.57$, $R=R_0\Rightarrow v_C(t)=Yt\exp(-t/\tau)u(t)$, $Y\cong 6.33\times 10^7\,\mathrm{V\,s}^{-1}$, $\tau\cong 158\,\mathrm{ns}$
$R=R_0/10\Rightarrow v_C(t)=Y[\exp(-t/\tau_1)-\exp(-t/\tau_2)]u(t)$, $Y\cong 5.03\,\mathrm{V}, \tau_1\cong 3.16\,\mu\mathrm{s}, \tau_2\cong 7.93\,\mathrm{ns}$.

Exercise 11.6. *n_cycles* ≅ 1, 1.6, 4, 8; Yes.

Exercise 11.7. Answer implied by problem statement.
Exercise 11.8. $\delta\to\infty$.
Exercise 11.9. Answer implied by problem statement.
Exercise 11.10.

a

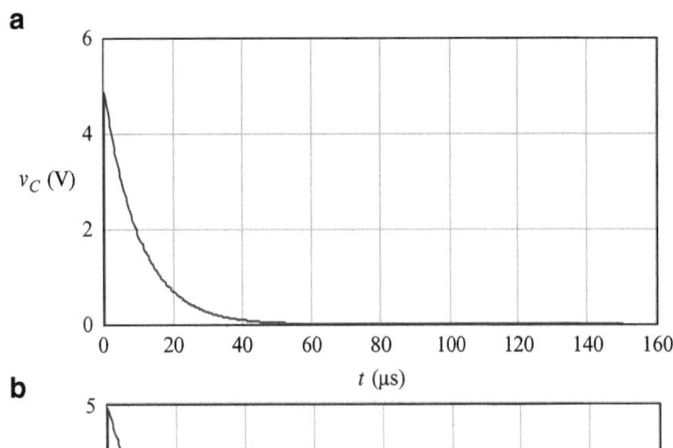

b

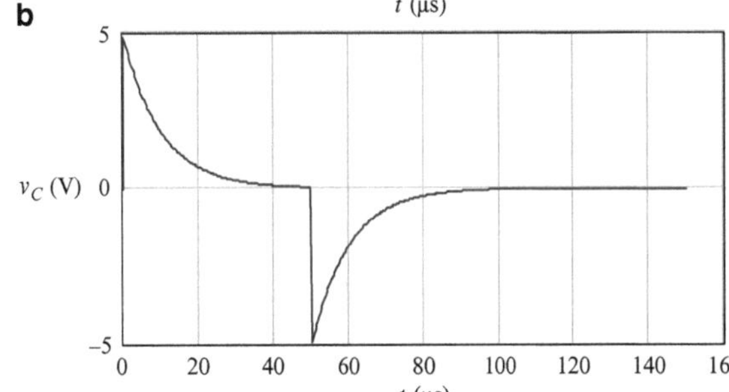

c

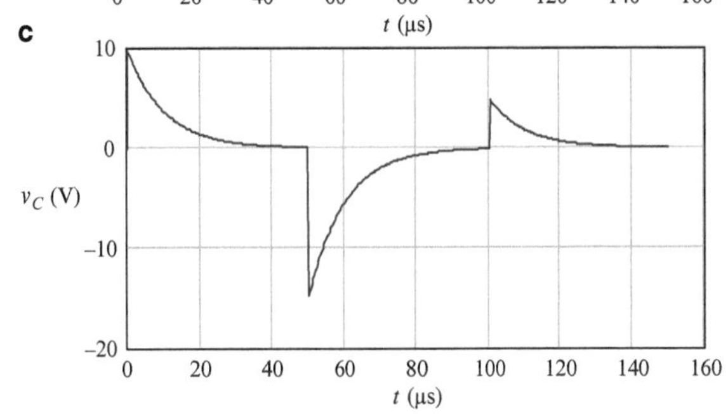

Exercise 11.11.

$$4\frac{d^3v_o}{dt^3}+14\frac{d^2v_o}{dt^2}+8\frac{dv_o}{dt}+v_o=4\frac{d^3v_S}{dt^3}+4\frac{d^2v_S}{dt^2}+4\frac{dv_S}{dt}+v_S$$

$$\mathrm{v}:=\begin{pmatrix}1\\8\\14\\4\end{pmatrix}\quad s\tau:=\mathrm{Polyroots}(\mathrm{v})\quad s\tau=\begin{pmatrix}-2.823\\-0.5\\-0.177\end{pmatrix}.$$

Characteristic roots real and negative, circuit is overdamped. The circuit cannot be underdamped because it consists of only resistors and capacitors.

Largest time constant is $\tau_3\cong\dfrac{\tau}{0.177}\cong 5.65\tau=56.5\,\mathrm{s}$.

12. Chapter 12

Exercise 12.1. $-2\pi/3$.
Exercise 12.2. Lags.
Exercise 12.3. $v_2 = V_2\cos(2\pi f t + \theta); V_2 = 500\ \mu\text{V}$, $f = 1\,\text{kHz}$, $\theta \cong 1.26$.
Exercise 12.4. $50\angle(-\pi/4)\,\text{mA}$.
Exercise 12.5. $\tilde{V} = V_0\angle(\phi - \pi/2)$.
Exercise 12.6. $\tilde{I} = 50\angle\pi/3\,\text{mA} \cong (25 + j43.3)\text{mA}$; leads.
Exercise 12.7. $i(t) = I_0\cos(2\pi f t + \theta); I_0 = 50\,\text{mA}$, $f = 4\,\text{MHz}$, $\theta = \pi/3$.

Exercise 12.8.

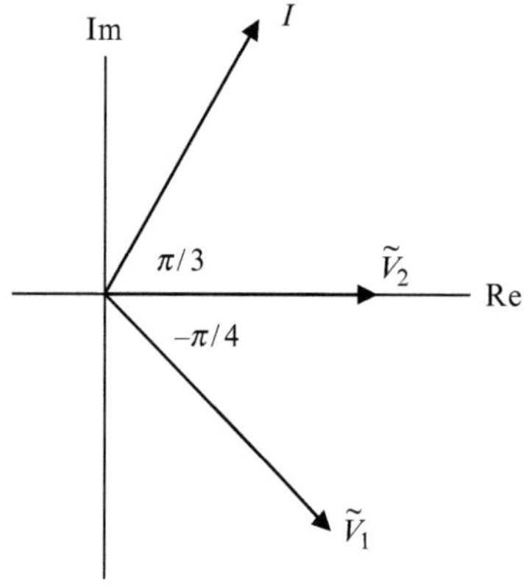

Exercise 12.9. Answer implied by problem statement.
Exercise 12.10. Answer implied by problem statement.
Exercise 12.11. $\tilde{I} \cong 20.7\angle\pi/2\,\text{mA}$
$\Rightarrow i(t) \cong -20.7\sin(2\pi f_0 t)\,\text{mA}$.

Exercise 12.12. $\tilde{I} \cong 6.37\angle -\pi/2\,\text{mA}$
$\Rightarrow i(t) \cong 6.37\cos(\omega t - \pi/2)\,\text{mA}$.

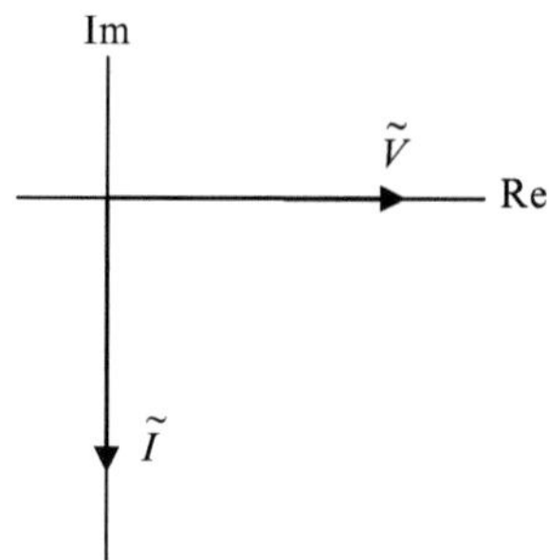

Exercise 12.13. $178\,\text{k}\Omega$.
Exercise 12.14. $\sqrt{10} \cong 3.16$.
Exercise 12.15. $v(t) \cong 146\cos(\omega t - 0.350)\,\text{mV}$.
Exercise 12.16. $\tilde{I} = \tilde{I}_1 + \tilde{I}_2 \cong 14.1\angle -0.065\,\text{mA} \Rightarrow$ $i(t) \cong 14.1\cos(\omega t - 0.065)\,\text{mA}$, $f = 100\,\text{kHz}$.

Exercise 12.17. $Z = R_2 + \dfrac{(\omega L)^2 R_1}{R_1^2 + (\omega L)^2} + \dfrac{R_3}{1 + (\omega C R_3)^2} + j\left[\dfrac{\omega L R_1^2}{R_1^2 + (\omega L)^2} - \dfrac{\omega C R_3^2}{1 + (\omega C R_3)^2}\right]$.

Exercise 12.18. $\tilde{I} = \dfrac{\tilde{V}}{Z} = \dfrac{1 - \omega^2 LC + j\omega C(R_1 + R_2)}{(R_1 + j\omega L)(1 + j\omega C R_2)} V_0$,
$i(t) = I_0\cos(\omega t + \theta); I_0 = 1.14\text{mA}, \theta = -0.44$.
Exercise 12.19. $f = 5\,\text{kHz}$, $R \cong 469\,\Omega$, $L \cong 3.85\,\text{mH}$; $f = 1\,\text{kHz}$, $R \cong 540\,\Omega$, $C \cong 52.7\,\mu\text{F}$.
Exercise 12.20. $f = 5\,\text{kHz}$, $R \cong 500\,\Omega$, $L \cong 61.7\,\text{mH}$; $f = 1\,\text{kHz}$, $R \cong 540\,\Omega$, $C \cong 1.65\,\text{nF}$.
Exercise 12.21.

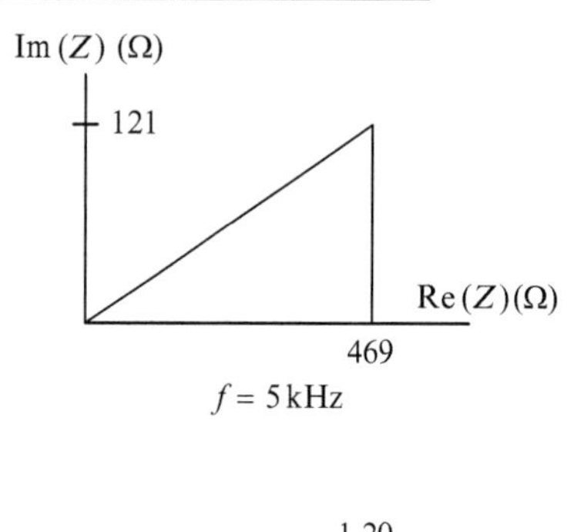

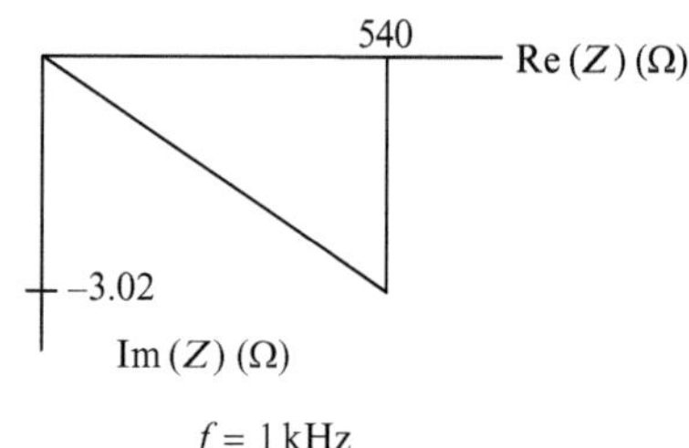

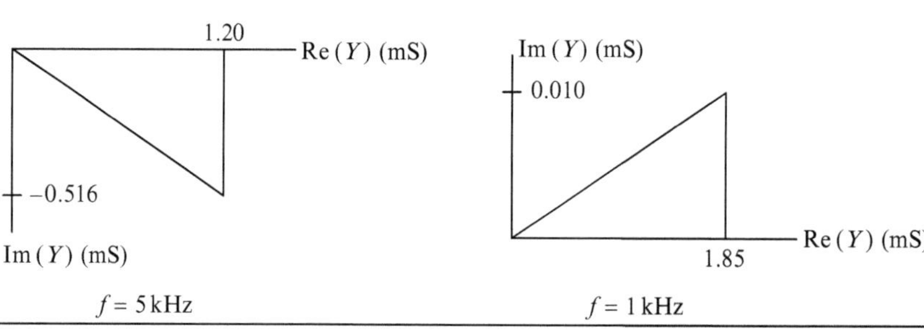

Exercise 12.22.
$\tilde{V}_T \cong (2.47\angle - 0.251)\,\text{V}$, $Z_T \cong (107\angle - 1.34)\,\Omega$.
Exercise 12.23. Resonant frequency is $\omega = \dfrac{R_1 + R_2}{\sqrt{LR_1^2C - L^2}}$ if $R_1^2C > L$. Otherwise circuit is not resonant.
Exercise 12.24. Two resonant frequencies: $f_1 \cong 67.1\,\text{kHz}$, $f_2 \cong 13.3\,\text{kHz}$.

13. Chapter 13

Exercise 13.1. Load: $S \cong (112 + j298)\,\text{VA}$, $P \cong 112\,\text{W}$
Source impedance: $S \cong (90.1 + j68.0)\,\text{VA}$, $P \cong 90.1\,\text{W}$
Voltage source: $S \cong (202 + j365)\,\text{VA}$, $P \cong 202\,\text{W}$.
Exercise 13.2. Source: $\cong 790\,\text{VAR}$, Capacitor: $\cong -802\,\text{VAR}$, Inductor: $\cong 12.2$ VAR.
Exercise 13.3. $pf = 1$.
Exercise 13.4. See figure below. Each side of the power triangle equals V_{rms}^2 times the corresponding side of the admittance triangle.
Exercise 13.5. (a) $\cong 17.5\,\text{kVAR}$ (b) $pf \cong 0.87$.

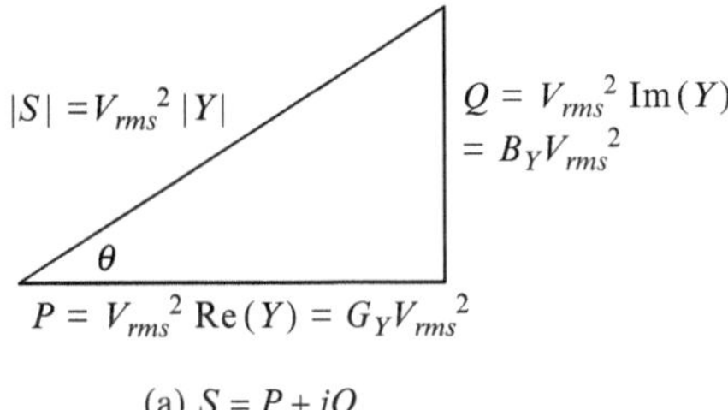

(a) $S = P + jQ$

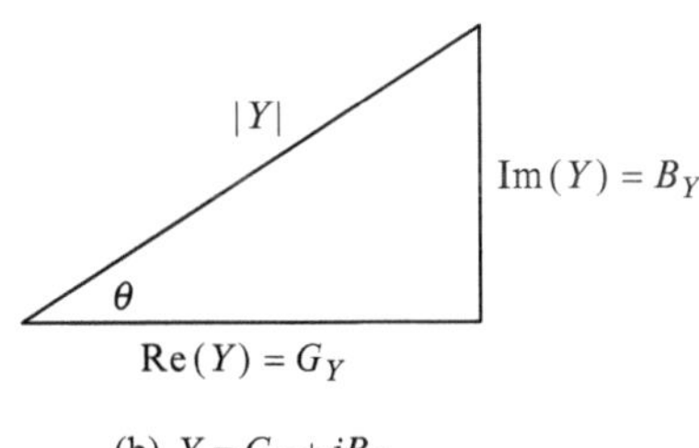

(b) $Y = G_Y + jB_Y$

Exercise 13.6. Answer implied by problem statement
Exercise 13.7. Answer implied by problem statement
Exercise 13.8. The equivalent series reactances X_L and X_T must be non-negative (capacitor doesn't pass dc) and $R_L = R_T$ if $R_T \neq 0$. If $R_T = 0$ (not realistic), then R_L should be the value that draws the maximum current from the (current-limited) source.
Exercise 13.9. (a) Answer implied by problem statement. (b) No, (c) Yes.
Exercise 13.10.
(a) $f = 0$(dc) (b).

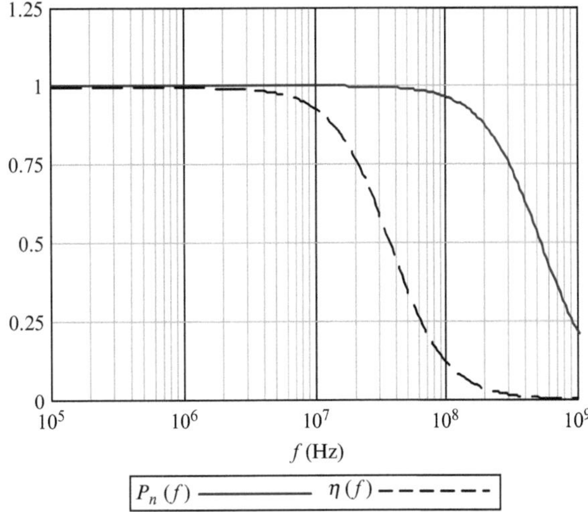

Exercise 13.11. 18 dB.

14. Chapter 14

Exercise 14.1. Answer implied by problem statement.
Exercise 14.2. No.
Exercise 14.3. Both are zero.

15. Chapter 15

Exercise 15.1. $v_L \cong 8.40\cos(\omega_1 t + 0.82) + 12.23\cos(\omega_2 t - 0.06) + 9.00\cos(\omega_3 t - 0.75)$.
Exercise 15.2.

	H_v	H_i	H_z	H_y
H_v	1	$Z_S^{-1}Z_L$	Z_S^{-1}	Z_L
H_i	$Z_L^{-1}Z_S$	1	Z_L^{-1}	Z_S
H_z	Z_S	Z_L	1	Z_LZ_S
H_y	Z_L^{-1}	Z_S^{-1}	$(Z_LZ_S)^{-1}$	1

Exercise 15.3. Current transfer: $Z_{in} \to 0$, $Z_{out} \to \infty$; Transimpedance: $Z_{in} \to 0, Z_{out} \to 0$; Transadmittance: $Z_{in} \to \infty, Z_{out} \to \infty$.
Exercise 15.4. $H_i(j\omega_1) = \dfrac{Z_S}{Z_L(j\omega_1)} \dfrac{K}{1 + j\omega_1/\omega_0}$.
Exercise 15.5. $10\,\mathrm{W\,m^{-1}}$.
Exercise 15.6. $A_v \cong 33$ dB, $A_i \cong 50$ dB.
Exercise 15.7. $|Y| \cong 17.8$ mS.
Exercise 15.8. Because $1 \pm f/f_0$ is not a linear factor (no j).

16. Chapter 16

Exercise 16.1. Average value: $x_{dc} = A_0 \angle \theta_0 = 500\,\mathrm{mV}$

Third harmonic: $f = 3f_0 = 6\,\mathrm{kHz}$, $T = \dfrac{1}{3f_0} \cong 167\,\mu\mathrm{s}$, $A_3 = \dfrac{500}{\sqrt{10}}\,\mathrm{mV}$, $\theta_3 = \dfrac{3\pi}{6} = \dfrac{\pi}{2}$.

Fifth harmonic: $A_{5\,rms} = \dfrac{A_5}{\sqrt{2}} = \dfrac{500}{\sqrt{2}\sqrt{26}}\,\mathrm{mV}$.

Fundamental: $f_0 = 2\,\mathrm{kHz}$, $\theta_1 = \dfrac{\pi}{6}$, $A_{1\,rms} = \dfrac{500}{\sqrt{2}\sqrt{2}}\,\mathrm{mV} = 250\,\mathrm{mV}$.

Exercise 16.2. $5f_0 = 1.5\,\mathrm{kHz} \Rightarrow f_0 = 300\,\mathrm{Hz} \Rightarrow T = \dfrac{1}{f_0} \cong 3.33\,\mathrm{ms}$.
Exercise 16.3. Only the third, whose frequencies are harmonics of a fundamental.
Exercise 16.4. $P_1 = 3.125\,\mathrm{mW}$, $P_2 = 500\,\mu\mathrm{W}$, $P_3 = 125\,\mu\mathrm{W}$, $P_4 = 43.25\,\mu\mathrm{W}$, $P_5 = 18.49\,\mu\mathrm{W}$.

$$P = 3.812\,\mathrm{mW}, \eta = 0.994.$$

Exercise 16.5. $X_0 = -10\,\mathrm{mV}$, $X_1 = 25\,\mathrm{mV}$, $X_3 = (12.5\angle -0.785)\,\mathrm{mV}$, $X_5 = (6.25\angle -1.57)\,\mathrm{mV}$, all others $= 0$.
Exercise 16.6. Average value = dc component = $X_0 = 1\,\mathrm{V}$

Period $T = 2\,\mathrm{ms}$ Fundamental frequency $f_0 = \dfrac{1}{T} = 500\,\mathrm{Hz}$

Third harmonic: $A_3 = 2|X_3| \cong 632\,\mathrm{mV}$, $\theta_3 = \angle\dfrac{1}{1+j3} \cong -1.25$, $\dfrac{1}{3f_0} \cong 667\,\mu s$

Fifth harmonic: $\dfrac{A_5}{\sqrt{2}} = \sqrt{2}|X_5| = 277\,\mathrm{mV}$

Fundamental: $f_0 = 500\,\mathrm{Hz}$, $\theta_1 = \angle X_1 \cong -0.785$, $A_1/\sqrt{2} = \sqrt{2}|X_1| = 1\,\mathrm{V}$.

Exercise 16.7.

f (kHz)	4	8	12	16	20	24	28	32
A_k (mA)	0.00	45.02	31.83	15.01	0.00	9.00	10.61	6.43
θ_k	0.00	0.00	0.00	0.00	0.00	3.14	3.14	3.14

Exercise 16.8. Answer implied in problem statement.
Exercise 16.9. $a_0 = x_{dc}$, $b_0 = 0$.
Exercise 16.10.

K	0	1	2	3
a_k (V)	2.0	1.0	0.4	0.2
b_k (V)	0.0	1.0	0.8	0.6

Exercise 16.11. $A_k = \dfrac{V_0}{k^2}$, $\theta_k = k\pi$, $k = 1, 2, \cdots$; $X_0 = x_{dc}$; $X_k \dfrac{V_0}{2k^2}(-1)^k$, $k = \pm1, \pm2, \cdots$; $a_0 = x_{dc}$, $b_0 = 0$; $a_k = \dfrac{V_0}{k^2}(-1)^k$, $b_k = 0$, $k = 1, 2, \cdots$.
Exercise 16.12.

k	0	1	2	3
A_k (mV)	1.00	1.41	0.89	0.63
θ_k	0.00	0.52	1.05	1.57
a_k (mV)	1.00	1.23	0.45	0.00
b_k (mV)	0.00	−0.71	−0.78	−0.63

Exercise 16.13. $X_k = V_0\,\mathrm{sa}\left(\dfrac{k\pi}{2}\right)$. $k = 0, \pm1, \pm2, \cdots$.
Exercise 16.14. $\dfrac{T}{2} - \dfrac{\tau}{2} < t \le \dfrac{T}{2} + \dfrac{\tau}{2}$.
Exercise 16.15. $X_k = \dfrac{x_0\tau}{2T}\mathrm{sa}^2\left(\dfrac{k\pi\tau}{2T}\right) = \dfrac{x_0\tau}{2T}\mathrm{sinc}^2\left(\dfrac{k\tau}{2T}\right)$, $X_k = x_0\mathrm{sa}^2\left(\dfrac{k\pi}{2}\right) = x_0\mathrm{sinc}^2\left(\dfrac{k}{2}\right)$.
Exercise 16.16.

k	1	2	3
A_k (V)	6.37	3.18	2.12
θ_k	−1.57	1.57	−1.57

Exercise 16.17.
(a) $Y_k = -\dfrac{R_2(1 + j\omega R_1 C)}{R_1(1 + j\omega R_2 C)} V_0\,\mathrm{sa}^2\left(\dfrac{k\pi}{2}\right)$, (b).

k	0	1	2	3	4	5	6	7
A_k (mV)	0.00	405.29	0.00	45.03	0.00	16.21	0.00	8.27
B_k (V)	0.00	1.46	0.00	0.83	0.00	0.02	0.00	0.01

Exercise 16.18.

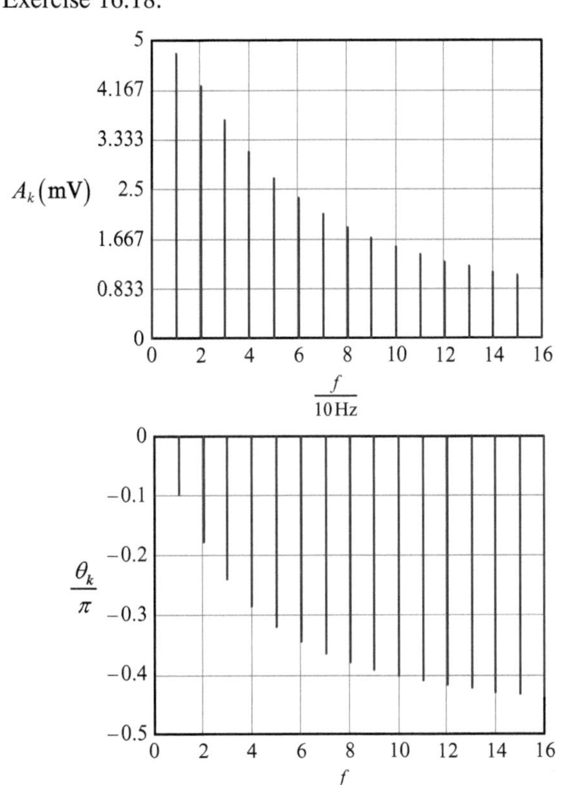

Exercise 16.19.

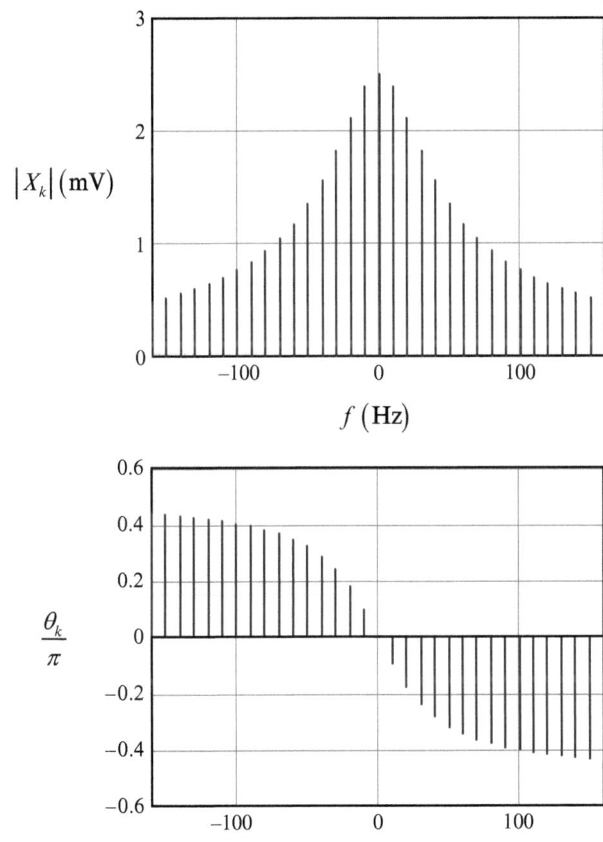

17. Chapter 17

Exercise 17.1.

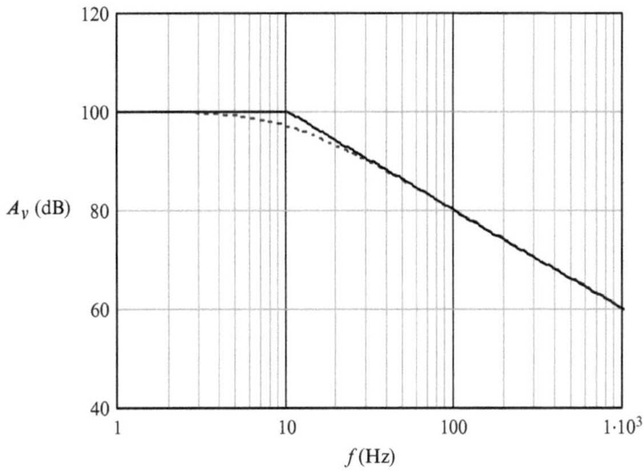

Exercise 17.2. Answer implied by problem statement.
Exercise 17.3. (a) $A_v(0) = 100$, $A_v(W) = 100/\sqrt{2}$; (b) $Z_{in}(0) \cong 10\,\text{k}\Omega$, $|Z_{in}(j2\pi W)| \cong 14.14\,\text{k}\Omega$; $Z_{out}(0) \cong 0.038\,\Omega$, $|Z_{out}(j2\pi W)| \cong 39.78\,\Omega$.

Exercise 17.4.

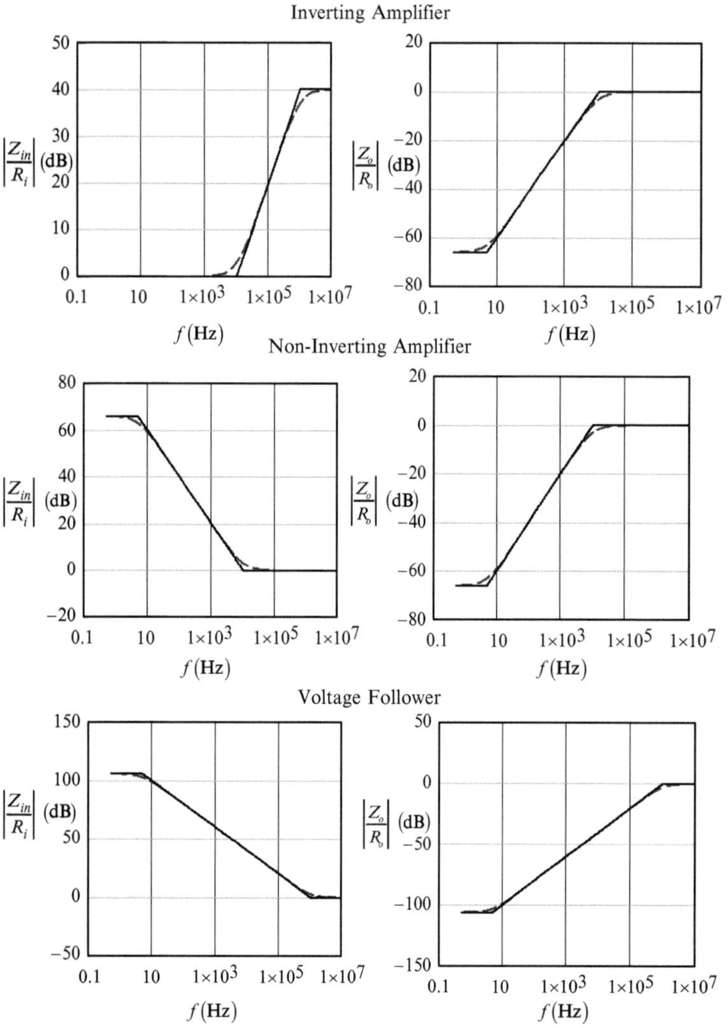

Inverting Amplifier
Non-Inverting Amplifier
Voltage Follower
$\left|\frac{Z_{in}}{R_i}\right|$ (dB)
$\left|\frac{Z_o}{R_o}\right|$ (dB)
0.1
10
1×10^3
1×10^5
1×10^7
f (Hz)

Exercise 17.5. $C \cong 3.98\,\text{pF}$.
Exercise 17.6. Both ratios $\cong 1$.

18. Chapter 18

Exercise 18.1. $F(s) = -2a/(s^2 - a^2)$, $|s| < a$, 2-sided.
Exercise 18.2. (a) Two, (b) right, (c) both, (d) left.
Exercise 18.3. All.
Exercise 18.4. Answer implied by problem statement.
Exercise 18.5. Answer implied by problem statement.
Exercise 18.6. Answer implied by problem statement.
Exercise 18.7. Answer implied by problem statement.
Exercise 18.8.

$v(t-t_0)$ (V)
5
0
75
125
t (μs)
(a)
$v(t-t_0)$ (V)
5
25
75
t (μs)
(b)

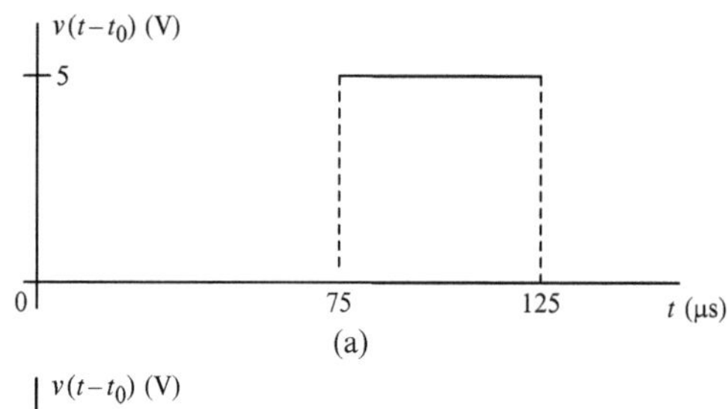

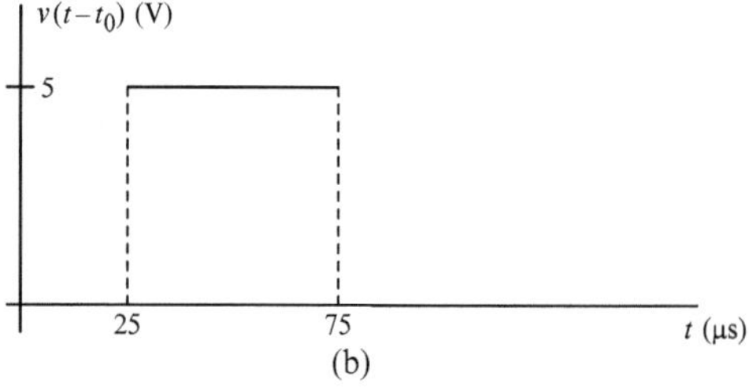

$$v(t) = V_0[u(t) - u(t-\tau)] \Rightarrow V(s) = V_0\left[\frac{1-\exp(-s\tau)}{s}\right],$$

$$\mathcal{L}\{v(t-t_0)\} = V_0\left[\frac{1-\exp(-s\tau)}{s}\right]\exp(-st_0).$$

Exercise 18.9. $\mathrm{SI}[b_3] = \mathrm{s}^{-1}$, $\mathrm{SI}[a_2] = \mathrm{s}^{-1}$.
Exercise 18.10. $\mathrm{SI}[K] = \mathrm{A\,s}^{-1}$.

Exercise 18.11. $I(s) = K\left[1 + \frac{c_3 s^3 + c_2 s^2 + c_1 s + c_0}{s^4 + a_3 s^3 + a_2 s^2 + a_1 s + a_0}\right]$,
$c_3 = b_3 - a_3 = -5\mathrm{s}^3$, $c_2 = 1\mathrm{s}^2$, $c_1 = -2\mathrm{s}$, $c_0 = 0$.
Exercise 18.12. Answer implied by problem statement.
Exercise 18.13. Answer implied by problem statement.
Exercise 18.14.

$$H_i = \frac{1}{s^2LC + (L/R_S + R_LC)s + 1 + R_L/R_S},\ H_v = \frac{R_L}{R_S}H_i,$$
$$H_z = R_L H_i.$$

Exercise 18.15. First answer implied by problem statement. Yes.

Exercise 18.16. $H_v = \frac{V_L}{V_S} = \frac{1}{(1+s\tau)},\ H_i = \frac{R_S}{R_L + jX_L}H_v,$

$$H_y = \frac{H_v}{(R_L + jX_L)(1+s\tau)},\ H_z = R_S H_v.$$

Exercise 18.17. $\tau = RC,\ p = -1/\tau$

$$g_v(t) = \exp(pt)u(t) = \exp(-t/\tau)u(t),$$
$$\mathrm{SI}[g_v(t)] = 1(\text{dimensionless})$$
$$h_v(t) = \delta(t) - (1/\tau)\exp(-t/\tau)u(t),$$
$$\mathrm{SI}[h_v(t)] = \mathrm{SI}[\delta(t)] - \mathrm{SI}[1/\tau] = \mathrm{s}^{-1}.$$

Exercise 18.18.

$$H_v = \left(\frac{R_2}{R'}\right)\frac{(1+jf/f_0)}{(1+jf/f_1)},\ f_0 = \frac{1}{2\pi R'C},\ f_1 = \frac{1}{2\pi R_2 C}$$
$$H_i = \frac{R_S R_2}{R_L R'}\frac{(1+jf/f_0)}{(1+jf/f_1)},\ H_z = \frac{R_S R_2}{R'}\frac{(1+jf/f_0)}{(1+jf/f_1)},$$
$$H_y = \frac{R_2}{R_L R'}\frac{(1+jf/f_0)}{(1+jf/f_1)}.$$

Exercise 18.19. Answer implied by problem statement.
Exercise 18.20. Answer implied by problem statement.

Exercise 18.21. $I_L = \beta_2 I_2 = \beta_1\beta_2 I_1 = \beta_1\beta_2 I_S$ No.

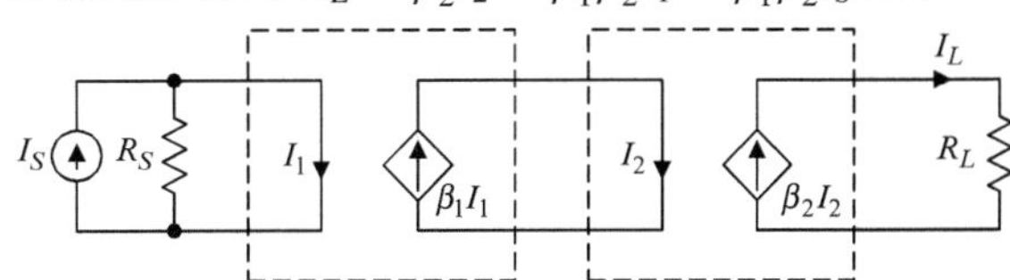

Exercise 18.22. Answer implied by problem statement; Yes.
Exercise 18.23.

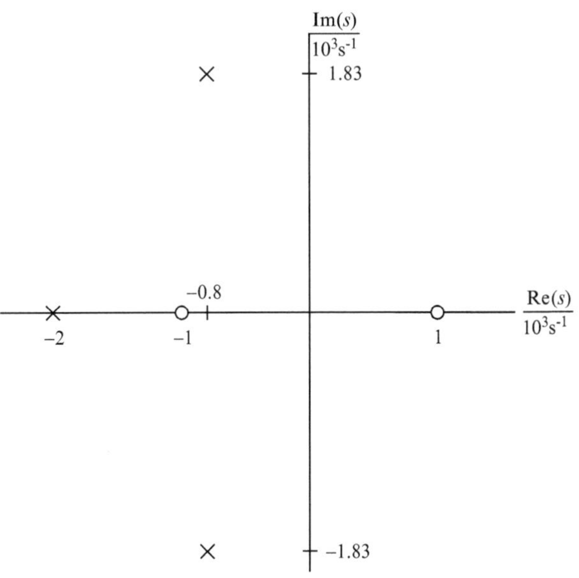

Exercise 18.24. Same as in the referenced example.

19. Chapter 19

Exercise 19.1. Three, one of which must be minimum passband input resistance. The other two must specify the poles, either directly or indirectly.
Exercise 19.2. $R = 24\,\mathrm{k\Omega}$, $C_1 \cong 663\,\mathrm{pF}$, $C_2 \cong 166\,\mathrm{pF}$.
Exercise 19.3. Answer implied by problem statement.
Exercise 19.4. Answer implied by problem statement.
Exercise 19.5. Lowpass: $H_{v\,\mathrm{LP}}(0) = k + 1$; Highpass: $H_{v\,\mathrm{HP}}(\infty) = k + 1$.
Exercise 19.6. Answer implied by problem statement.
Exercise 19.7. $V_{CC} \cong \pm 15\,\mathrm{V}$.

Index

N

O

P

MIX
Papier aus verantwortungsvollen Quellen
Paper from responsible sources
FSC® C105338

If you have any concerns about our products, you can contact us on
ProductSafety@springernature.com

In case Publisher is established outside the EU, the EU authorized representative is:
Springer Nature Customer Service Center GmbH
Europaplatz 3, 69115 Heidelberg, Germany

Printed by Libri Plureos GmbH
in Hamburg, Germany